National Cancer Institute

National Cancer Institute — Environmental Epidemiology Branch, Bethesda MD 20205

United States Environmental Protection Agency — Health Effects Research Laboratory, Research Triangle Park NC 27711 — EPA-600/1-83-015a, September 1983

Research and Development

EPA

U.S. Cancer Mortality Rates and Trends, 1950-1979

Volume I

by

Wilson B. Riggan, Ph.D., John Van Bruggen,
John F. Acquavella, Ph.D., and Jeff Beaubier, Ph.D.
Biometry Division
Health Effects Research Laboratory
U.S. Environmental Protection Agency
Research Triangle Park, NC 27711

and

Thomas J. Mason, Ph.D.
Chief, Population Studies Section
Environmental Epidemiology Section
National Cancer Institute
Bethesda, MD 20205

Published by NCI/EPA Interagency Agreement on
Environmental Carcinogenesis

For sale by the Superintendent of Documents, U.S. Government Printing Office, Washington, D.C. 20402

Disclaimer

Foreword

The many benefits of our modern, developing, industrial society are accompanied by certain hazards. Careful assessment of the relative risk of existing and new man-made environmental hazards is necessary for the establishment of sound regulatory policy. These regulations serve to enhance the quality of our environment in order to promote the public health and welfare and the productive capacity of our Nation's population.

The complexities of environmental problems originate in the deep interdependent relationships between the various physical and biological segments of man's natural and social world. Solutions to these environmental problems require an integrated program of research and development using input from a number of disciplines. The Health Effects Research Laboratory, Research Triangle Park, NC and Cincinnati, OH conducts a coordinated environmental health research program in toxicology, epidemiology and clinical studies using human volunteer subjects. Wide ranges of pollutants known or suspected to cause health problems are studied. The research focuses on air pollutants, water pollutants, toxic substances, hazardous wastes, pesticides and non-ionizing radiation. The laboratory participates in the development and revision of air and water quality criteria and health assessment documents on pollutants for which regulatory actions are being considered. Direct support to the regulatory function of the Agency is provided in the form of expert testimony and preparation of affidavits as well as expert advice to the Administrator to assure the adequacy of environmental regulatory decisions involving the protection of the health and welfare of all U.S. inhabitants.

This report contains cancer mortality rates and changes in rates for counties of the United States for the time periods of 1950-1959, 1960-1969, 1970-1979. Included are mortality data for each of 35 site-specific cancers for each sex and for white and non-white races. Further, this report ranks counties nationally by percentile (for each cancer site, race, and sex) for their percent change from 1950-1959 to 1970-1979 and for their 1970-1979 age-adjusted death rate. Knowledge of these geographic and temporal differences in cancer mortality is useful for generating hypotheses of environmental determinants of human cancer.

F. Gordon Hueter
Director
Health Effects Research Laboratory

Abstract

With the reduction in deaths from contagious disease during the past century and the reduction in deaths due to cardiovascular diseases during the past decade, the relative importance of cancer mortality has increased. Cancer is the second leading cause of death in the United States, accounting for approximately 400,000 deaths in 1979.

This report contains cancer mortality rates and changes in rates for counties of the United States for 1950-1959, 1960-1969, and 1970-1979. This report includes the number of deaths and cancer mortality rates age-adjusted to the 1970 U.S. population for each of 35 site-specific cancers for four race-sex groups. Further, counties are ranked nationally by percentile for 1970-1979 age-adjusted death rates and for percent change from 1950-1959 to 1970-1979.

Over the three periods covered by this report, there have been marked fluctuations in mortality for several site-specific cancers. Total cancer mortality rates increased 16% for white males and 39% for non-white males. Among the ten major cancer sites, lung cancer mortality increased most noticeably. White male lung cancer mortality increased 116% from 1950-1959 to 1970-1979 and white female lung cancer mortality increased 199%. Non-white lung cancer rates rose even more sharply. Non-white male and female lung cancer mortality increased 185% and 188%.

Cancer mortality data as arranged in this report have many uses. For example the geographic and temporal differences are useful in developing and examining hypotheses concerning the influence of various environmental factors. The degree of consistency of such differences across the four race-sex groups and among the age groups can furnish additional insights into possible causative factors. Elevated rates in even one race-sex group may suggest hypotheses for further investigation.

Contents

Cancer Sites ICD 9th Rev.	Site	Page Number WM	WF	NM	NF
	Volume III				
190	Cancer of the eye	001	019	035	039
191,192	Cancer of the brain and other parts of the nervous system	042	074	105	117
193	Cancer of the thyroid gland	128	148	173	178
194,164.0	Cancer of the thymus and other endocrine glands	186	203	218	222
200,202,159.1, 202.0,.1,.8,.9	Lymphosarcoma and reticulum cell sarcoma including other lymphoma	226	258	290	304
201	Hodgkin's disease	316	346	374	384
203 except 203.1	Multiple myeloma	392	422	451	464
204-208,202.4,203.1	Leukemias	476	508	540	556
152,158,159.2-.9, 164.2-.9,184,187, 195-199,202.2,.3, .5,.6	Secondary, site unspecified, and not previously listed cancers	571	604	637	656

Cancer Sites ICD 9th Rev.	Site	Page Number WM	WF	NM	NF
	Volume IV				
157	Cancer of the pancreas	001	033	065	082
188,189.3	Cancer of the bladder and other urinary organs	099	130	160	175
189 except 189.3	Cancer of the kidney and ureter	188	219	249	262

Acknowledgment

We wish to acknowledge the support and help of the National Center for Health Statistics (NCHS), especially to thank Joyce Scott, June S. Pearce, and Katheryn G. Lyndon, who checked our comparability between the seventh, eighth, and ninth revision of the International Classification of Diseases (ICD) and verified that our classifications and tables matched the format used to code cause of death from the certificates.

We wish to acknowledge the support and help of Ms. Constance Percy, Demographic Analysis Section, Biometry Branch, National Cancer Institute in facilitating the updating of our data base to include 1979 mortality, and to check our comparability between the eighth and ninth revision of the International Classification of Diseases (ICD).

We wish to acknowledge the support and help of members of the USEPA Health Effects Research Laboratory, especially Drs. Carl Hayes, Robert Chapman, and William Nelson for their encouragement, critical reviews, and suggestions, and Drs. Victor Hasselblad and John Creason for their review of the statistical procedures and suggestions.

We wish to acknowledge other scientists for their technical reviews and helpful suggestions, especially Dr. Donald F. Austin, Resource for Cancer Epidemiology Section, California Department of Health Services; Dr. John F. Finklea, Department of Public Health, University of Alabama in Birmingham; Drs. Clark Heath, Jean French, Roy Ing, Brian Plikaytis, and Mathew Zack, Centers for Disease Control; and Drs. Kenneth G. Manton and Max A. Woodbury, Center for Demographic Studies and Medical Center, Duke University.

Section 1

Introduction and Summary

Introduction

With the reduction in deaths from contagious diseases during the past century and the reduction in deaths due to cardiovascular diseases during the past decade, the relative importance of cancer mortality has increased. Cancer is the second leading cause of death in the United States, accounting for approximately 400,000 deaths in 1979 (1,2). The increase in cancer deaths over the past three decades has been accompanied by increased cigarette consumption, industrial expansion, increased use of private automobiles and other modes of transportation, and growth in chemical production, as well as changes in life style, occupation, and many advances in the knowledge, detection, and treatment of cancer (1,3,4,5). However, the lag time between exposure and clinical diagnosis suggests that the relevant exposure for current cancer deaths may have occurred 10, 15, 20 or more years in the past.

The geographic distribution of cancer has been used for various epidemiologic purposes. "U.S. Cancer Mortality by County: 1950-1969" by Mason and McKay provided 20-year average cancer mortality rates by county (6). In this publication, and also in the two companion atlases by Mason *et al.*, the importance of analyses of county mortality data is discussed (7,8). The three publications have been most useful in identifying high-risk areas, allowing correlational analyses to suggest potential risk factors, and selecting areas for further cancer studies.

We provide a similar approach for screening county mortality data; but in place of 20-year average rates, we present cancer mortality rates for counties in the United States for three decades: 1950-1959, 1960-1969, and 1970-1979. These additional data include: cancer mortality rates for each of the three decades; percent change in rates between each of the decades; percentile rank of counties for percent change between 1950-1959 and 1970-1979; percentile rank of counties by average annual cancer death rates for 1970-1979. We use an asterisk to identify counties with observed number of deaths significantly larger than the expected.

Major Results

The present report collates virtually all reported cancer deaths in the United States from 1950 through 1979, a 30-year period. More than 8 million cancer deaths occurred during this period. These results may be subject to various interpretations, yet certain features of the data are unmistakable.

Over the three decades covered by this report, there have been marked fluctuations in mortality for several site-specific cancers. Some of these changes have involved all four race/sex groups, while others have been race and/or sex specific. Total cancer mortality rates increased 16% for white males and 39% for non-white males. Female cancer mortality rates declined slightly for whites and remained virtually unchanged for non-whites.

Among the ten major cancer sites, lung cancer mortality increased most noticeabaly. White male lung cancer mortality increased 116% from 1950-1959 to 1970-1979 and white female lung cancer mortality increased 199%. Non-white lung cancer rates rose even more sharply. Non-white male and female lung cancer mortality increased 185% and 188%. Interestingly, the largest part of this increase occurred between the first two periods for white and non-white males, while this increase occurred between the second and third periods for female lung cancer mortality. We observed consistent increments in pancreatic cancer mortality for all race/sex groups. Increments were 17% and 14% for white males and females, while non-whites experienced greater increases of 37% and 47%. Mortality from cancer of the large intestine increased among white males (17%), non-white males (61%), and non-white females (24%). Only white females experienced a decline (9%) for this site. Stomach cancer mortality continued to decline. All four race/sex groups experienced a decrement of greater than 40% from 1950-59 to 1970-79. Rectal cancer also declined for all race/sex groups. For this site, the decrements for white males, white females, non-white males, and non-white females were 36%, 44%, 19%, 38%, respectively.

Marked changes over time also occurred for several less common cancer sites. Most striking among these changes was the large increment in multiple myeloma. Mortality from this cause of death increased sharply for white males (65%) and females (78%), and even more sharply for non-white males (141%) and females (140%). Melanoma mortality continued to rise in the four race-sex categories, white males (88%), white females (51%), non-white males (49%), and non-white females (15%). Laryngeal cancer mortality rose only slightly for white males (5%), but white females (53%) and especially non-white males (68%) and females (54%) exhibited larger increases. Esophageal cancer showed little change for whites, but non-whites experienced increments of 53% and 65% for males and females, respectively. Less common cancer sites characterized by decreasing mortality included cancer of the bone and, for females, cancer of the uterine corpus. Bone cancer mortality declined 41%, 43%, 32%, and 46% among the four race/sex groups. Mortality from cancer of the chorion and uterus declined 46% for whites and 57% for non-whites.

Uses of These Tables

The mortality rates and percent change in mortality rates in this report may be used for various epidemiologic studies and purposes. These data may be used to identify counties and groups of counties with high cancer mortality rates, with large increments in cancer mortality, or both. A second use of these data would be to locate counties with unique demographic, environmental, and industrial characteristics and determine if they had remarkable mortality rates and mortality trends. High mortality rates and high percent change in mortality rates occurring in both sexes suggests a possible connection with an environmental exposure or other risk factors unrelated to sex, while high mortality rates among men only suggest the existence of occupational or other sex-related risk factors. An indicated use of this report, then, along with other secondary data on ambient environmental pollution levels or emission data, demographic characteristics, etc., follows the step-wise cancer research approach as described by Blot *et al.* (9).

Researchers may use this report to identify counties and groups of counties for developing and implementing a detailed monitoring program of ambient environmental pollutants, or for performing in depth field studies, e.g., cancer case control studies. Whether the study is an environmental monitoring or an epidemiologic field study, this report provides a method of screening cancer mortality data to select areas most suitable for study.

Another important use of this report will be as a ready reference for EPA and other governmental agencies, state and local health departments, universities, and citizens' groups. When local communities are concerned about a specific cancer situation, this report provides an historic data base with which to compare the current local conditions. Thus, this data base is an important resource for responding to community questions and concerns.

Considerations in Application

The many uses and strengths of these cancer mortality data have just been discussed. There are also several characteristics that users should bear in mind in using these data. The major characteristics are related to the accuracy and completeness of the death certificate information, the effects of changes over time in classification, coding and population characteristics and the effects of migration and latent period of disease.

Several studies have compared the diagnostic classification information on the death certificate to diagnostic information from other sources such as hospital records or autopsy results (10,11,12).These studies have generally shown the overall agreement between diagnosis and underlying cause of death to be very high for the most frequent causes of mortality. Studies of completeness and consistency of the demographic information on the death certificate have also shown high agreement (11). Although these results are very encouraging, the user should be aware of potential limitations in interpretation.

Studies of temporal trends in cancer mortality require recognition of possible effects of the decennial revisions of the International Classification of Diseases (ICD) during the 30 year span of our study. We discuss our treatment of the comparability issue in Section 3,C. We generated equivalent codes for the sixth, seventh, eighth and ninth edition of the ICD.

In Section 3,A. we discuss our method to establish fixed equivalent county boundaries over the 30 years.

The rapid scientific development in medical diagnostic practice over the 30 years have increased the precision of identifying site specific cancers. At the level of aggregation used in this report, this characteristic should have little effect on the results for most cancer sites; however, users should be aware of this characteristic in using these data.

Population distribution by age, sex, and race affects cancer mortality. In Section 3,B. we discuss our use of the total U.S. population for 1970 as the standard for 1950-1979. Population proportions of the above demographic characteristics not only vary across counties but vary over time within counties due to migration, births, deaths and possibly changes in competing risks of death (13-14). The effect of relative changes in these factors is complicated by the relatively long latent period between exposure and appearance for most types of cancer. Therefore, hypotheses concerning cause and effect relationships deserve careful, and cautious, consideration.

A final reminder is that these cancer mortality data are presented in terms of geographic risk. Our major purpose is to examine the spatial and temporal differences in order to generate hypotheses about possible environmental factors. In this respect, this report differs from those studies which have been concerned with the temporal trend in cancer rates per se, such as Pollack and Horm (2).

Many of these characteristics have been discussed in more detail elsewhere in the literature (10-14). These data are the prime source in documenting geographic patterns of cancer rates and trends over the 30 year period. With these characteristics in mind, we believe these new mortality data can provide a fresh perspective to the assessment of cancer mortality differences.

Section 2

Materials and Methods

A. Source and Description of Data

This report uses data from both the National Center for Health Statistics (NCHS) and the Bureau of Census.

1. Death Records – General Characteristics:
   - Years available: We have virtually all cancer death records for 1950-1979.
   - International Classification of Diseases: We used ICD codes from the ninth revision for 1979 mortality and in our tables, and the table of contents. NCHS used the International Classification of Diseases adapted for use in the United States (ICDA) instead of the unmodified ICD codes from the eighth revision to classify and code the underlying cause of death for years 1968-1978 (15). The comparability tables follow NCHS practice in use of ICD for edition 6, 7, 9, and ICDA for edition 8. We have used ICD in place of ICDA when referring to editions other than 8, when referring to two or more editions, and in general discussion.
   - Each death record contains the county of residence, year and month of death, age in years, race (white or non-white), sex, and ICD rubric. For more detail on death records see Technical Appendix from Vital Statistics of the United States 1977 and other years (16).
2. Death Records – Special Characteristics:
   - Mortality data as coded for 1952 combined five boroughs of New York City. We used the average ratio for years 1950, 1951, 1953, and 1954 of each borough in New York City to estimate 1952 deaths in each borough.
   - New Jersey did not code race in 1962 and 1963. F. W. McKay, National Cancer Institute, gave us the estimates he used in their publications (6,7,8).
   - We include all cancer deaths during 1972 even though Vital Statistics of the United States 1972 used only a 50 percent sample in their publication (16).
   - Over the 30 years covered in this report, the International Classification of Diseases (ICD) has undergone four revisions: sixth, seventh, eighth, and ninth. We have included a table of comparative ICD codes from each revision. Also in the table, codes from the sixth and seventh editions are grouped together because for our purposes they are essentially identical. For comparability over the 30 years, we did not include Polycythemia Vera or Myelofibrosis (ICDA 208, 209) since only the eighth edition classified these as cancer.
3. Population records:
   - Census years 1950, 1960, 1970, 1980.
   - Detail: By county, race, and sex, 5 year age groups through 84, and 85+.
   - Population 1950 census: Available 1950 Bureau of Census population for each county, race, and sex were for 5 year age groups through 74, 75-85, and 85+. In order to make the 1950 county counts consistent with 1960, 1970, and 1980 data, we used the U.S. race and sex specific counts for age 75-79, and 80-84 from the U.S. Bureau of Census 20% enumeration sample to estimate the percent of age 75-84 age group who were in the 75-79 age group. Using this percentage, we estimated the county counts in the 75-79 age group and assigned the residual to the 80-84 age group.
   - Interpolation between census years: We used a simple linear interpolation to generate mid-year population estimates between census years.
4. Geographic detail:
   - "County" as used in this report refers to a county, parish in Louisiana, independent city, or a combination of counties as shown in technical notes.
   - Through the years, political area changes have taken place. In many cases, the only way to keep a consistent geographical area was to combine counties. This is especially evident in Virginia where independent cities have annexed parts of counties. Details of these changes are given in the technical notes on changes and combination of counties.

B. Calculations

1. Average annual age-adjusted mortality rate:
   - We calculated annual age-adjusted mortality rates (per 100,000) for counties listed on the death certificate as the county of usual residence by cancer site, race, and sex with the total U.S. age, specific population for 1970 used as the standard for the total study period 1950-1979. The table in Section 3B presents the standard million population used in this study. This report contains rates for each of the contiguous states for 1950-1979. We added Alaska and Hawaii rates for 1960-1979. The District of Columbia has been included as a

separate entity which is listed both as a state and a county.

- Formula for average annual age-adjusted death rate per 100,000:

$$S(i,y) = D(i,y)/R(i,y)$$

$$A(y) = \sum_{i=1}^{n} S(i,y)\,P(i)(100{,}000)/\sum_{i=1}^{n} P(i)$$

$$A(d) = \sum_{y=1}^{m} A(y)/m$$

S(i,y) = Age-specific death rates in age group i, year y.
D(i,y) = Number of deaths in age group i, in year y.
R(i,y) = Mid-year population in age group i, in year y.
A(y) = Age-adjusted death rate per 100,000 in year y.
P(i) = Number of age group i in the standard population (1970 U.S. population).
A(d) = Average annual age-adjusted death rate per decade.
y = Year index.
i = Age index.
n = Number of age groups (18 age groups: 0-4, 5-9, 10-14, 15-19, 20-24, 25-29, 29-34, 35-39, 40-44, 45-49, 50-54, 55-59, 60-64, 65-69, 70-74, 75-79, 80-84, 85+).
m = (10), number of years in period.
d = Decade (i.e., 1950-1959, 1960-1969, or 1970-1979).

2. Expected number of deaths:

- Expected number of deaths for each race, sex group by year are calculated by multiplying the U.S. age-specific death ratio for each age interval by the county population of the specific age group, adding over all age groups, and then adding over all years in each time period (1950-1959, 1960-1969, 1970-1979).
- Formula for the expected number of deaths for each race, sex, cancer site, county, and the year:

$$E(k,j,y) = \sum_{i=1}^{18} D(i,j,y)\,P(i,k,y)/P(i,y)$$

y = year index (i.e., 1950, 1951, ..., 1979)
i = age index (i.e., 0-4, 5-9, 10-14, ..., 80-84, 85+)
k = county index
j = cancer site index

E(k,j,y) = Expected number of deaths in the kth county from the jth cancer site in the yth year.
D(i,j,y) = Number of deaths in the U.S. from the jth cancer site, in the ith age group, in the yth year.
P(i,y) = The U.S. population in the ith age group in the yth year.
P(i,k,y) = The population in the ith age group, kth county, and the yth year.

$$E(k,j) = \sum_{y=1}^{m} E(k,j,y) = \lambda$$

λ = Expected number of deaths in the kth county from the jth cancer site for the decade.
m = Number of years in the decade.

- Given the expected number of deaths λ, use the function below to calculate the probability that the number of deaths is less than the observed number n.

$$Prob(X \leq n) = \sum_{X=0}^{n} (\lambda^{x}e^{-\lambda})/X!$$

n = number of observed deaths

- If the Prob(X≤n)>.975, flag the county as being statistically significant. This deviation is the upper limit of a 95% confidence interval.
- For counties with 36 or more deaths for a cancer site in a decade (1950-1959, 1960-1969, 1970-1979), we used the normal approximation for the interval of the expected value of a poisson variable (17). Formula for calculating the upper confidence limit:

$$U = \lambda + 1/2 + 3/8\,z^2 + z(\lambda + 1/2 + 1/8\,z^2)^{0.5}$$

U = Upper confidence limit
λ = expected number of deaths
z = 1.96 for the 95% confidence interval.

- Counties with an observed number of deaths larger than the upper confidence limit are identified by an asterisk in the table.

3. Percent change in mortality rates:

- In the process of generating percent change in death rates, we generated total deaths and rates for each decade, percent change from 1950-1959 to 1960-1969, from 1960-1969 to 1970-1979 and from 1950-1959 to 1970-1979. For each race and sex, counties are percentile ranked for percent change from 1950-1959 to 1970-1979 and percentile ranked for average rate per 100,000 for 1970-1979. There may be up to 31 counties in a percentile. This report is presented in three separate volumes. These volumes have cancer sites arranged by ICD number in sequential order as found in the ICD ninth edition.

4. Special cases (which were rare) are dealt with as follows:

- If any age group the population is less than the number of deaths, the population is increased to the number of deaths. This occurs in small populations if deaths occur before census count or due to anomolies of population estimates.
- Percent change must have a non-zero age-adjusted death rate for both numerator and denominator to be printed. This restriction also holds for ranking of percent change between 1950-1959 and 1970-1979.
- Death records with unknown age or other missing information were omitted from the calculations.

5. Methods for ranking counties:

- Calculation of percentile rank of rate per 100,000 population for 1970-1979 includes all counties with a rate of zero or greater listed in the tables.
- Calculation of percentile rank for percent change in rates from 1950-1959 to 1970-1979 includes all counties with percent changes listed in the table. Both of the rankings include all counties less those left blank or not printed. The formula is:

$$PC = 100 (R)/(NCR + 1)$$

where PC = percentile
R = rank given
NCR = number of counties ranked (i.e., 3065 counties for the 1970-1979 rate or less for percent change)

- The United States and individual states are not included in the ranking.

C. Report Page Layout

1. Clarification of table headings:

- The first row of the heading contains, from left to right, the race and sex, the name of site-specific cancer, and the ICD code from the ninth revision;
- Columns headed "Deaths" and "Rate" under 1950-1959, 1960-1969, and 1970-1979 contain the total number of deaths and age-adjusted average annual death rate per 100,000 occurring during each period;
- An asterisk indicates that the observed number of deaths in a specific county is significantly larger than the expected number of deaths;
- The center column lists the United States, each state, and county in alphabetical order under the state with independent cities within the state at the end of all counties;
- Columns headed 60S/50S, 70S/60S, and 70S/50S under percent change indicate the percent change in mortality rate between 1950-1959 and 1960-1969, 1960-1969 and 1970-1979, and 1950-1959 and 1970-1979; and
- Columns headed 70S/50S and 1970S under percentile contain the percentile rank of counties for the percent change between 1950-1959 and 1970-1979 and the percentile rank of counties for the average annual rate for 1970-1979. Percentiles range from 99 for the highest to 00 for the lowest.

2. Explanation of blank entries in the table:

- In case of no deaths in a decade, the period is left blank.
- In case of no deaths from a site specific cancer in the 30 year period, the line is not printed.

B. Standard Million Population

The standard population used is the total 1970 U.S. population as shown below.

United States
Standard Million Populations by Age for 1970

Age	Population
All ages	1,000,000
<5	84,416
5-9	98,204
10-14	102,304
15-19	93,845
20-24	80,561
25-29	66,320
30-34	56,249
35-39	54,656
40-44	58,958
45-49	59,622
50-54	54,643
55-59	49,077
60-64	42,403
65-69	34,406
70-74	26,789
75-79	18,871
80-84	11,241
85+	7,435

Source: U.S. Bureau of the Census, Census of Population: 1970, *General Population Characteristics,* Final Report PC(1)-B

C. Constructing Equivalent Cause of Death Categories Across Years

The rates and trends generated in the present work cover 30 years spanning four editions of the ICD. This long time span required achieving comparability between time periods covered when different rules of classification of underlying cause of death were in effect. We did not want to create artifically increased rates by attributing deaths to a site specific cancer solely because of a change in classification rules. To avoid this potential artifact, we developed comparability codes for translation between the 6th, 7th, 8th, and 9th edition (revision) of the International Classification of Diseases (see table below).

Several authors have attempted to create equivalencies between editions of the ICD (18,19,20). The National Center for Health Statistics conducted the most intensive studies to test comparability by generating comparability ratios. However, for some cancer sites no absolute equivalence can ever be achieved. At the level of aggregation we are using, this should present few problems since whatever effects remain should apply equally to all counties. However, users of these data should keep in mind potential problems relating to coding in equivalencies.

Cancer Site Comparability Table

Cancer Site	ICD 9	ICDA 8	ICD 6* & 7
Lip	140	140	140
Oral cavity including tongue	141,143-146,148,149	141,143-146,148,149	141,143-145,147,148
Salivary gland	142	142	142
Nasopharynx	147	147	146
Esophagus	150	150	150
Stomach	151	151	151
Large intestine	153,159.0	153	153
Rectum	154 except 154.3	154	154
Liver and gallbladder including bile ducts	155,156	155,156,197.8	155*
Pancreas	157	157	157
Nose, nasal cavities, middle ear and accessory sinuses	160	160	160
Larynx	161	161	161
Trachea, bronchus and lung including pleura and other respiratory sites	162,163,165	162,163.0, 163.9	162,163
Bone including jaw	170	170	196
Connective and soft tissue	164.1,171	171,192.4,192.5	197,193.3,193.4**
Malignant melanoma of skin	172	172	190
Nonmelanoma skin cancer	173,154.3	173 except 173.5	191
Breast	174,175	174	170
Cervix uteri	180	180	171

Cancer Site Comparability Table

Cancer Site	ICD 9	ICDA 8	ICD 6* & 7
Chorion, corpus uterus and uterus not otherwise specified	179,181,182	181,182	172,173,174
Ovary, fallopian tube and broad ligament	183	183	175
Prostate	185	185	177
Testis	186	186	178
Bladder and other urinary organs	188,189.3	188,189.9	181
Kidney and ureter	189 except 189.3	189.0,189.1,189.2	180
Eye	190	190	192
Brain and other parts of the nervous system	191,192	191,192 except 192.4 & 192.5	193 except 193.3 & 193.4*
Thyroid gland	193	193	194
Thymus and other endocrine glands	194,164.0	194	195
Lymphosarcoma and reticulum cell sarcoma including other lymphoma	159.1,200,202 except 202.2-.6	200,202	200,202,205
Hodgkin's disease	201	201	201
Multiple myeloma	203 except 203.1	203	203
Leukemias	204-208,202.4,203.1	204-207	204
Secondary, site unspecified and not previously listed cancers	152,158,159.2-.9,164.2-.9, 184,187,195-199, 202.2,.3,.5,.6	152,158,159,163.1,173.5 184,187,195,196,197.0-.7, 197.9,198,199	152,156,158,159,164 165,176,179,198,199
All cancers	140-208	140-207	140-205

*Code 156 ICD 6 and 7 secondary and unspecified cannot be separated. For this reason, death rates for 1950-1959 and 1970-1979 are not comparable and the comparison should not be used.
**Code 193 ICD 6 cannot be subdivided. For this reason deaths from malignant neoplasm of brain and other parts of the nervous system for years 1950-1957 contain deaths from malignant neoplasm of peripheral nerves and sympathetic nervous system.

References

1. Toxic Substances Strategy Committee. Toxic Chemicals and Public Protection – A Report to the President. Council on Environmental Quality. U.S. Govt. Print. Off., Washington, DC, May 1980.
2. Pollack, E. S., and J. W. Horm. Trends in Cancer Incidence and Mortality in the United States, 1969-1976. J. Natl. Cancer Inst., 64, 5:1091-1103, 1980.
3. Schottenfeld, D., and J. F. Haas. Carcinogens in the Work Place. Amer. Cancer Soc., 1978.
4. Shabad, L. M. Circulation of Carcinogenic Polycyclic Aromatic Hydrocarbons in the Human Environment and Cancer Prevention. J. Natl. Cancer Inst., 64, 3:405-410, 1980.
5. Cederlof, R., R. Doll, B. Fowler, L. Friberg, N. Nelson, and V. Vouk, eds. Air Pollution and Cancer: Risk Assessment Methodology and Epidemiological Evidence. Report of a Task Group, Environ. Health Perspect., 22:1-2, 1978.
6. Mason, T J., and F. W. McKay. U.S. Cancer Mortality by County: 1950-1969. DHEW Pub. No. (NIH) 74-615, U.S. Govt. Print. Off., Washington, DC, 1973.
7. Mason, T. J., F. W. McKay, R. Hoover, W. T. Blot, and J. F. Fraumeni, Jr. Atlas of Cancer Mortality for U.S. Counties: 1950-1969. DHEW Pub. No. (NIH) 75-780, U.S. Govt. Print. Off., Washington, DC, 1975.
8. Mason, T. J., F. W. McKay, R. Hoover, W. T. Blot, and J. F. Fraumeni, Jr. Atlas of Cancer Mortality Among Non-Whites: 1950-1969. DHEW Pub. No. (NIH) 76-1204, U.S. Govt. Print. Off., Washington, DC, 1976.

9. Blot, W. J., J. F. Fraumeni, Jr., T. J. Mason, and R. N. Hoover. Developing Clues to Environmental Cancer: A Stepwise Approach With the Use of Mortality Data. Environ. Health Perspect., 32:53-58, 1979.

10. Percy, C., E. Stanek, and L. Gloeckler. Accuracy of Cancer Death Certificates and Its Effect on Cancer Mortality Statistics. Amer. J. Public Health, 71, 3:242-250, 1981.

11. Marcus, S. C. Some Limitation in the Use of Cancer Mortality Data in Epidemiological Studies. Proceedings of the 71st Annual Meeting of the Air Pollution Control Association, Paper No. 78-6.2, pp. 1-12, 1978.

12. Glasser, J. H. The Quality and Utility of Death Certificate Data. Editorial, Amer. J. Public Health, 71, 3:231-233, 1981.

13. Polissar, L. The Effect of Migration on Comparison of Disease Rates in Geographic Studies in the United States. Amer. J. Epidemiol., 111:175-182, 1980.

14. Chiang, C. L. Introduction to Stochastic Processes in Biostatistics. John Wiley and Sons., Inc., New York, NY, pp. 242-268, 1968.

15. U.S. Dept. of Health, Education, and Welfare. International Classification of Diseases, adapted for use in the United States, Eighth Revision, PHS Pub. No. 1693, 1967. World Health Organization, International Classification of Diseases, sixth, seventh, and ninth revision.

16. U.S. Dept. of Health and Human Services. Vital Statistics of the United States, 1977, Volume II – Mortality, Part A. DHHS Publication No. (PHS) 81-1101. U.S. Govt. Printing Office, Washington, DC, 1981. See Technical Appendix. Also published annually.

17. Crow, E. L., and R. S. Gardner. Confidence Intervals for the Expectation of a Poisson Variable. Biometrika, 46:441-447, 1959.

18. Klebba, A. J., and J. H. Scott. Estimates of Selected Comparability Ratios Based on Dual Coding of the 1976 Death Certificates by the Eighth and Ninth Revisions of the International Classification of Diseases. Natl. Center Health Stat., Mon. Vital Stat. Report, 28, 11, Sup., Feb. 29, 1980.

19. Nectoux, J. Comparison of the 7th and 8th Revisions of the ICD and Tables of Equivalence. In: Chapter IV. Cancer Incidence in Five Continents: Vol. III. J. Waterhouse, P. Correa, C. Muir, and J. Powell, eds. Intl. Agency for Res. on Cancer, Lyon, pp. 25-38, 1976.

20. Klebba, A. J., and J. H. Scott. Comparability of Mortality Statistics, Approximate Title. Natl. Center Health Stat., in preparation.

U.S. CANCER MORTALITY RATES AND TRENDS 1950-1979 VOLUMES 1-IV

Originally published
as a three-volume set
by

The NCI/EPA Interagency Agreement on
Environmental Carcinogenesis
September, 1983

Republished by

LEGIS BOOKS
a division of
LEGIS CORP
P.O. Box 7888
The Woodlands, Texas 77380

http://www.legis.com
http://www.260press.com

Printed in the United States of America
Printing Number
1 2 3 4 5 6 7 8 9 10

Library of Congress Cataloging-in-Publication Data
Mason, Thomas et al
U.S. Cancer Mortality Rates and Trends 1950-1979 Volumes 1-IV
1. Study Aids
2. Science
3. Reference
4. Epidemiology

U.S. Cancer Mortality Rates and Trends 1950-1979
Volume I: ISBN: 1-881043-22-3
Volume II: ISBN: 1-881043-23-1
Volume III: ISBN: 1-88-1043-25-8
Volume IV: ISBN: 1-88-1043-26-6
Volumes 1-IV (Set): 1-88-1043-24-X

WHITE MALE — ALL CANCERS COMBINED — ICD 140-208

1950-1959 DEATHS	1950-1959 RATE	1960-1969 DEATHS	1960-1969 RATE	1970-1979 DEATHS	1970-1979 RATE	COUNTY AND STATE	PERCENT CHANGE 60S/50S	PERCENT CHANGE 70S/60S	PERCENT CHANGE 70S/50S	PERCENTILE %CHANGE 70S/50S	PERCENTILE RATES 1970S
1135259	176.6	1434818	190.0	1754868	204.1	UNITED STATES	8	7	16		
11958	148.6	17084	172.1	24023	199.1	ALABAMA AL	16	16	34		
69	154.5	91	169.8	144	183.9	AUTAUGA AL	10	8	19	38	42
259	172.0	369	177.7	603	203.9	BALDWIN AL	3	15	19	37	70
71	140.5	109	191.7	134	204.8	BARBOUR AL	36	7	46	78	71
61	134.4	96	189.1	134	232.5	BIBB AL	41	23	73	93	93
135	134.9	197	163.8	250	167.2	BLOUNT AL	21	2	24	46	20
20	108.1	39	215.0	36	185.0	BULLOCK AL	99	-14	71	92	43
99	145.4	126	171.2	147	187.1	BUTLER AL	18	9	29	54	46
316	140.2	559	196.6	715	198.8	CALHOUN AL	40	1	42	73	64
135	128.9	178	151.0	236	170.3	CHAMBERS AL	17	13	32	60	23
65	108.5	95	136.1	163	197.8	CHEROKEE AL	25	45	82	95	62
143	144.9	180	166.6	262	209.8	CHILTON AL	15	26	45	76	77
51	144.2	69	171.3	107	230.9	CHOCTAW AL	19	35	60	87	93
99	162.3	108	153.3	153	190.8	CLARKE AL	-6	24	18	34	52
62	104.7	76	116.8	127	183.8	CLAY AL	12	57	76	93	41
46	99.4	54	113.9	105	188.6	CLEBURNE AL	15	66	90	96	48
97	107.6	152	146.6	231	173.7	COFFEE AL	36	19	61	88	27
160	139.7	263	170.9	384	193.7	COLBERT AL	22	13	39	70	56
59	124.5	85	150.0	101	180.5	CONECUH AL	20	20	45	77	36
48	129.1	48	126.9	78	180.5	COOSA AL	-2	42	40	71	36
160	118.3	253	164.5	339	190.4	COVINGTON AL	39	16	61	88	51
65	123.6	94	170.9	132	216.0	CRENSHAW AL	38	26	75	93	83
246	138.3	305	138.6	502	178.4	CULLMAN AL	0	29	29	55	33
62	84.2	120	138.7	204	169.8	DALE AL	65	22	102	97	23
126	163.0	179	201.6	238	227.8	DALLAS AL	24	13	40	71	91
214	125.4	315	156.9	428	177.5	DE KALB AL	25	13	42	73	32
128	136.0	211	187.0	264	182.3	ELMORE AL	38	-2	34	64	39
123	146.8	182	183.6	274*	234.7	ESCAMBIA AL	25	28	60	87	94
397	142.2	594	169.0	815	195.9	ETOWAH AL	19	16	38	68	60
90	140.3	108	144.0	179	209.5	FAYETTE AL	3	45	49	81	77
135	142.6	169	159.7	250	195.2	FRANKLIN AL	12	22	37	67	59
119	144.3	143	152.9	204	187.7	GENEVA AL	6	23	30	57	47
20	156.0	24	169.2	23	158.3	GREENE AL	8	-6	2	09	12
28	95.9	57	183.4	79	231.1	HALE AL	91	26	141	98	93
53	139.6	64	154.7	88	172.7	HENRY AL	11	12	24	46	26
142	109.4	248	157.5	389	184.8	HOUSTON AL	44	17	69	91	43
185	137.7	234	151.5	338	183.8	JACKSON AL	10	21	34	63	41
2576*	186.8	3468*	200.7	4461*	227.9	JEFFERSON AL	7	14	22	43	91
70	112.7	103	154.6	133	174.2	LAMAR AL	37	13	55	84	28
210	126.8	368	167.7	553	193.6	LAUDERDALE AL	32	15	53	83	56
91	131.3	129	159.0	182	177.3	LAWRENCE AL	21	12	35	65	32
79	107.5	179	173.9	246	177.5	LEE AL	62	2	65	90	32
131	131.3	151	118.6	283	181.5	LIMESTONE AL	-10	53	38	69	38
23	159.2	30	175.7	41	237.2	LOWNDES AL	10	35	49	81	95
32	146.8	43	153.8	61	189.3	MACON AL	5	23	29	55	49
261	124.0	497	151.5	915	189.3	MADISON AL	22	25	53	83	50
56	135.0	83	177.8	114	213.0	MARENGO AL	32	20	58	86	81
88	97.1	138	123.2	205	148.2	MARION AL	27	20	53	83	06
210	129.0	313	146.4	501	177.7	MARSHALL AL	13	21	38	68	32
1084*	190.6	1607*	217.7	2178*	237.3	MOBILE AL	14	9	25	47	95
72	142.8	94	170.2	124	193.5	MONROE AL	19	14	35	65	56
469	162.2	749*	208.6	918	212.6	MONTGOMERY AL	29	2	31	59	80
294	163.2	364	159.6	549	180.9	MORGAN AL	-2	13	11	22	37
43	176.5	39	128.3	73	212.7	PERRY AL	-27	66	20	40	80
61	106.6	99	156.4	124	173.6	PICKENS AL	47	11	63	89	27
108	165.4	117	166.0	152	180.1	PIKE AL	0	9	9	19	36
74	98.4	99	120.1	118	136.8	RANDOLPH AL	22	14	39	70	03
100	135.2	191	205.0	279*	242.9	RUSSELL AL	52	18	80	94	96
119	132.8	156	151.2	313	226.5	ST CLAIR AL	14	50	70	92	91
166	168.8	185	154.9	318	189.5	SHELBY AL	-8	22	12	25	50
48	225.4	51	212.0	51	184.5	SUMTER AL	-6	-13	-18	01	42
245	170.6	336	186.3	427	193.1	TALLADEGA AL	9	4	13	26	55
142	134.5	187	156.0	244	164.5	TALLAPOOSA AL	16	5	22	44	18
297	106.8	464	134.0	695	172.8	TUSCALOOSA AL	25	29	62	89	26
310	149.6	418	172.2	581	202.7	WALKER AL	15	18	35	65	69
53	153.6	77	169.9	101	191.9	WASHINGTON AL	11	13	25	48	53
48	201.6	47	188.0	58	235.5	WILCOX AL	-7	25	17	33	95
110	165.5	108	145.4	169	182.7	WINSTON AL	-12	26	10	21	40
		653	167.5	941	184.4	ALASKA AK		10			
4992	150.9	9776	172.7	16994	182.9	ARIZONA AZ	14	6	21		
32	171.8	37	150.6	50	151.6	APACHE AZ	-12	1	-12	03	08
251	165.0	320	157.7	529	178.0	COCHISE AZ	-4	13	8	17	33
90	143.7	123	135.6	187	173.3	COCONINO AZ	-6	28	21	40	27
138	161.7	201	196.3	286	191.0	GILA AZ	21	-3	18	36	52
76	166.2	87	150.5	131	172.5	GRAHAM AZ	-9	15	4	11	26
53	166.6	38	129.0	63	182.6	GREENLEE AZ	-23	42	10	20	40
2559	150.0	5520	175.9	9814	186.4	MARICOPA AZ	17	6	24	47	45
55	146.7	115	133.7	444	175.6	MOHAVE AZ	-9	31	20	39	29
77	166.7	95	129.1	172	175.8	NAVAJO AZ	-23	36	5	14	30
1049	158.4	2094	176.8	3441	182.1	PIMA AZ	12	3	15	30	39
181	121.1	355	169.8	519	171.3	PINAL AZ	40	1	41	73	25
52	152.7	72	178.6	109	191.4	SANTA CRUZ AZ	17	7	25	49	53
217	136.2	400	179.4	725	191.1	YAVAPAI AZ	32	7	40	72	52
162	146.5	319	171.8	524	164.7	YUMA AZ	17	-4	12	25	18
9036	135.1	13136	168.1	18273	194.1	ARKANSAS AR	24	15	44		
114	140.8	166	177.5	220	210.3	ARKANSAS AR	26	18	49	81	78
84	145.3	153	214.3	195	222.9	ASHLEY AR	47	4	53	84	89
94	147.3	170	155.7	389	188.0	BAXTER AR	6	21	28	52	48
272	120.8	433	160.1	635	163.1	BENTON AR	33	2	35	65	16
80	87.8	184	164.6	302	210.1	BOONE AR	87	28	139	98	78

* SIGNIFICANTLY MORE DEATHS THAN EXPECTED

WHITE MALE — ALL CANCERS COMBINED — ICD 140-208

1950-1959 DEATHS	1950-1959 RATE	1960-1969 DEATHS	1960-1969 RATE	1970-1979 DEATHS	1970-1979 RATE	COUNTY AND STATE	PERCENT CHANGE 60S/50S	PERCENT CHANGE 70S/60S	PERCENT CHANGE 70S/50S	PERCENTILE %CHANGE 70S/50S	PERCENTILE RATES 1970S
70	147.1	107	188.5	126	195.5	BRADLEY AR	28	4	33	61	59
32	132.8	36	140.5	48	171.7	CALHOUN AR	6	22	29	56	25
89	113.1	135	153.1	186	180.9	CARROLL AR	35	18	60	87	37
77	214.4	65	153.5	86	189.9	CHICOT AR	-28	24	-11	03	50
95	122.2	152	186.2	170	186.6	CLARK AR	52	0	53	83	45
136	117.2	176	144.3	290	218.0	CLAY AR	23	51	86	95	85
56	105.4	93	143.1	147	148.4	CLEBURNE AR	36	4	41	72	06
33	102.4	50	155.8	69	184.0	CLEVELAND AR	52	18	80	94	42
86	106.7	140	150.9	225	220.8	COLUMBIA AR	42	46	107	97	87
63	101.5	136	196.8	146	175.6	CONWAY AR	94	-11	73	92	29
247	130.9	354	168.4	545*	223.4	CRAIGHEAD AR	29	33	71	92	89
126	113.4	198	153.2	298	194.9	CRAWFORD AR	35	27	72	92	58
85	159.3	153	212.7	226*	242.5	CRITTENDEN AR	33	14	52	83	96
96	156.7	109	164.4	156	212.9	CROSS AR	5	30	36	66	81
54	136.4	71	166.3	95	213.8	DALLAS AR	22	29	57	86	82
62	133.0	78	153.4	122	216.8	DESHA AR	15	41	63	89	84
72	142.1	82	149.9	120	199.2	DREW AR	5	33	40	71	64
120	125.4	188	154.2	302	186.7	FAULKNER AR	23	21	49	80	45
71	112.9	115	165.5	142	171.3	FRANKLIN AR	47	4	52	83	25
36	79.1	71	138.7	115	178.3	FULTON AR	75	29	125	98	33
455	176.0	653	195.2	949	218.4	GARLAND AR	11	12	24	47	85
45	97.0	85	167.8	131	220.3	GRANT AR	73	31	127	98	87
156	125.8	208	155.5	255	164.0	GREENE AR	24	5	30	57	17
118	147.7	171	201.3	194	198.0	HEMPSTEAD AR	36	-2	34	64	63
99	118.4	156	159.9	215	178.5	HOT SPRING AR	35	12	51	82	33
76	134.6	93	158.0	124	184.8	HOWARD AR	17	17	37	68	43
127	119.9	209	170.8	258	177.7	INDEPENDENCE AR	42	4	48	80	32
47	109.7	68	133.1	142	180.6	IZARD AR	21	36	65	90	36
148	188.1	164	178.7	220	217.4	JACKSON AR	-5	22	16	31	84
251	164.5	413*	209.6	492	206.9	JEFFERSON AR	27	-1	26	49	74
112	142.8	153	169.4	230	227.1	JOHNSON AR	19	34	59	87	91
43	122.4	57	154.0	82	207.9	LAFAYETTE AR	26	35	70	91	76
134	156.6	149	153.5	221	206.5	LAWRENCE AR	-2	35	32	60	74
54	150.7	57	154.0	91	243.2	LEE AR	2	58	61	88	97
47	147.0	49	134.0	81	205.3	LINCOLN AR	-9	53	40	71	72
59	165.8	69	177.3	90	190.3	LITTLE RIVER AR	7	7	15	30	51
127	116.9	176	150.4	211	175.2	LOGAN AR	29	16	50	81	29
111	116.3	173	162.5	223	177.8	LONOKE AR	40	9	53	84	33
62	101.9	72	113.3	106	148.0	MADISON AR	11	31	45	77	06
47	114.3	74	144.5	136	199.8	MARION AR	26	38	75	93	65
148	141.9	229	180.6	328	230.5	MILLER AR	27	28	62	89	93
235	128.8	332	183.7	397	211.8	MISSISSIPPI AR	43	15	64	90	79
61	142.7	80	170.5	120	247.1	MONROE AR	19	45	73	93	97
49	149.2	55	135.8	77	145.6	MONTGOMERY AR	-9	7	-2	06	05
76	162.1	82	159.4	108	188.8	NEVADA AR	-2	18	16	32	49
32	84.9	41	115.1	81	200.2	NEWTON AR	36	74	136	98	66
129	165.6	158	157.9	230	202.1	OUACHITA AR	-5	28	22	43	68
37	122.6	55	172.5	71	188.2	PERRY AR	41	9	53	84	48
116	163.8	155	205.1	176	222.5	PHILLIPS AR	25	9	36	66	88
54	118.6	106	199.5	101	170.5	PIKE AR	68	-15	44	75	24
131	116.2	202	174.1	264	211.5	POINSETT AR	50	21	82	95	79
97	125.2	151	165.5	199	184.3	POLK AR	32	11	47	79	42
156	149.0	200	162.8	294	186.3	POPE AR	9	14	25	48	45
60	119.0	81	163.8	101	194.9	PRAIRIE AR	38	19	64	90	58
1111	167.7	1589	192.1	2077*	212.0	PULASKI AR	15	10	26	51	80
101	137.4	99	123.9	170	179.0	RANDOLPH AR	-10	44	30	57	34
102	187.5	115	184.1	161	226.2	ST FRANCIS AR	-2	23	21	40	91
129	107.0	200	139.4	299	164.4	SALINE AR	30	18	54	84	18
50	108.2	68	137.5	92	152.8	SCOTT AR	27	11	41	73	09
43	89.7	68	138.6	97	164.9	SEARCY AR	54	19	84	95	18
394	146.6	565	179.4	776	200.5	SEBASTIAN AR	22	12	37	67	66
83	145.8	105	167.5	119	169.1	SEVIER AR	15	1	16	32	22
48	115.2	68	126.4	159	160.0	SHARP AR	10	27	39	70	13
30	84.0	46	110.0	75	138.6	STONE AR	31	26	65	90	03
211	138.6	333	189.3	390	195.3	UNION AR	37	3	41	72	59
41	85.5	72	136.2	142	178.5	VAN BUREN AR	59	31	109	97	34
315	126.6	482	157.1	669	175.7	WASHINGTON AR	24	12	39	70	30
177	106.1	327	166.9	433	185.6	WHITE AR	57	11	75	93	44
65	142.9	90	198.1	85	184.9	WOODRUFF AR	39	-7	29	56	43
87	125.6	118	140.1	176	180.8	YELL AR	12	29	44	75	37
89242	172.5	127386	188.5	164011	203.4	CALIFORNIA CA	9	8	18		
5669	179.7	7067*	200.0	7623*	212.1	ALAMEDA CA	11	6	18	36	80
4	220.4	3	246.4	5	279.7	ALPINE CA	12	14	27	51	99
77	132.0	148	226.3	161	161.0	AMADOR CA	71	-29	22	43	14
689	166.0	1022	192.2	1303	181.4	BUTTE CA	16	-6	9	20	38
101	148.9	150	177.0	218	189.7	CALAVERAS CA	19	7	27	52	50
116	182.5	155	223.5	146	219.4	COLUSA CA	22	-2	20	40	86
1765	179.9	2873	187.5	4274	202.6	CONTRA COSTA CA	4	8	13	25	69
74	137.7	110	175.5	166	204.8	DEL NORTE CA	27	17	49	80	71
180	157.3	284	162.5	557	187.4	EL DORADO CA	3	15	19	38	46
1971	167.0	2712	184.3	3391	197.9	FRESNO CA	10	7	19	37	62
150	177.0	157	170.0	230	217.1	GLENN CA	-4	28	23	44	84
616	176.0	781	191.4	889	199.4	HUMBOLDT CA	9	4	13	27	65
282	122.4	434	148.0	590	208.0	IMPERIAL CA	21	41	70	91	76
72	125.4	137	189.0	191	194.0	INYO CA	51	3	55	84	57
1271	154.4	1938	175.1	2763	206.0	KERN CA	13	18	33	62	73
282	163.1	378	183.1	440	198.3	KINGS CA	12	8	22	42	63
125	123.4	261	181.4	537	225.7	LAKE CA	47	24	83	95	90
108	167.0	135	192.2	142	177.2	LASSEN CA	15	-8	6	15	31
35353*	178.0	46525	191.5	52060*	206.9	LOS ANGELES CA	8	8	16	32	74
206	140.0	300	159.8	446	198.8	MADERA CA	14	24	42	74	64
733	177.4	1146	187.3	1606	198.6	MARIN CA	6	6	12	24	63
53	137.6	73	173.0	99	166.8	MARIPOSA CA	26	-4	21	42	20
302	126.8	421	163.1	569	193.0	MENDOCINO CA	29	18	52	83	55
426	156.9	616	183.6	933*	227.1	MERCED CA	17	24	45	76	91
53	131.8	62	146.0	70	150.6	MODOC CA	11	3	14	29	07
9	70.3	16	135.2	31	167.7	MONO CA	92	24	139	98	21
813	160.7	1281	183.7	1749	201.5	MONTEREY CA	14	10	25	49	67
500	137.7	704	153.3	980	182.6	NAPA CA	11	19	33	61	40

* SIGNIFICANTLY MORE DEATHS THAN EXPECTED

WHITE MALE — ALL CANCERS COMBINED — ICD 140-208

County and State		1950-1959 Deaths	1950-1959 Rate	1960-1969 Deaths	1960-1969 Rate	1970-1979 Deaths	1970-1979 Rate	Percent Change 60S/50S	Percent Change 70S/60S	Percent Change 70S/50S	Percentile %Change 70S/50S	Percentile Rates 1970S
NEVADA	CA	233	179.1	296	187.3	452	178.9	5	-5	0	08	34
ORANGE	CA	2241	148.2	5682	174.7	11168	205.5	18	18	39	69	72
PLACER	CA	401	151.1	587	172.5	953	206.4	14	20	37	67	74
PLUMAS	CA	83	149.7	107	183.0	153	205.1	22	12	37	68	72
RIVERSIDE	CA	1629	149.2	3148	165.9	5424	183.3	11	10	23	45	41
SACRAMENTO	CA	2575	179.7	3993*	211.4	5357*	221.6	18	5	23	45	88
SAN BENITO	CA	128	176.5	149	187.9	160	184.9	6	-2	5	13	43
SAN BERNARDINO	CA	2521	155.1	4249	175.5	5906	190.4	13	8	23	45	51
SAN DIEGO	CA	4488	163.4	7542	179.9	11941	195.8	10	9	20	39	60
SAN FRANCISCO	CA	8084*	217.4	8428*	235.9	7177*	235.7	8	0	8	18	95
SAN JOAQUIN	CA	1673	168.1	2251	189.7	2757	207.8	13	10	24	46	76
SAN LUIS OBISPO	CA	547	161.6	819	166.4	1114	166.2	3	0	3	10	19
SAN MATEO	CA	1982	171.1	3409	196.5	4571	212.0	15	8	24	46	80
SANTA BARBARA	CA	937	177.7	1437	174.3	2219	200.5	-2	15	13	26	66
SANTA CLARA	CA	2681	167.0	4630	182.7	6899	203.2	9	11	22	42	69
SANTA CRUZ	CA	837	165.1	1253	185.9	1618	197.8	13	6	20	39	62
SHASTA	CA	316	162.0	497	167.3	928	209.6	3	25	29	56	77
SIERRA	CA	21	151.9	28	204.0	33	182.7	34	-10	20	40	40
SISKIYOU	CA	268	170.3	302	170.8	391	200.9	0	18	18	36	66
SOLANO	CA	675	188.5	879	189.3	1256	210.1	0	11	11	23	78
SONOMA	CA	1216	177.2	1764	185.9	2603	198.9	5	7	12	25	64
STANISLAUS	CA	1001	148.1	1458	175.5	1937	191.0	18	9	29	55	52
SUTTER	CA	180	146.9	327	206.9	403	211.0	41	2	44	75	79
TEHAMA	CA	194	166.2	291	203.7	347	180.7	23	-11	9	18	36
TRINITY	CA	51	142.0	52	128.9	93	180.2	-9	40	27	51	36
TULARE	CA	877	139.4	1268	162.1	1766	199.6	16	23	43	75	65
TUOLUMNE	CA	94	111.6	192	180.4	308	195.6	62	8	75	93	59
VENTURA	CA	743	138.8	1605	175.2	2789	192.4	26	10	39	70	54
YOLO	CA	361	162.5	568	196.9	771	214.8	21	9	32	60	82
YUBA	CA	205	170.4	286	208.6	348	209.5	22	0	23	45	77
COLORADO	CO	9647	150.6	12158	156.4	15599	169.2	4	8	12		
ADAMS	CO	233	127.9	522	157.5	991	184.7	23	17	44	76	42
ALAMOSA	CO	50	127.3	65	151.0	65	154.7	19	3	22	42	10
ARCHULETA	CO	13	112.4	13	115.3	15	127.3	3	10	13	26	02
BACA	CO	45	138.2	55	175.3	61	177.0	27	1	28	53	31
BENT	CO	32	72.4	55	128.0	62	154.3	77	21	113	97	09
BOULDER	CO	368	145.2	487	142.0	711	150.6	-2	6	4	11	07
CHAFFEE	CO	58	131.6	74	151.3	99	181.4	15	20	38	68	38
CHEYENNE	CO	19	136.3	10	80.3	17	133.4	-41	66	-2	07	02
CLEAR CREEK	CO	23	147.5	36	186.5	41	182.0	26	-2	23	45	39
CONEJOS	CO	35	97.9	50	141.2	51	183.4	44	30	87	95	41
COSTILLA	CO	25	147.1	27	157.0	26	162.5	7	4	10	21	16
CROWLEY	CO	24	98.8	35	151.3	42	177.4	53	17	80	94	32
CUSTER	CO	12	126.5	10	111.2	17	206.6	-12	86	63	89	74
DELTA	CO	138	138.6	163	144.3	177	134.3	4	-7	-3	06	03
DENVER-ARAPAHOE#	CO	3787	174.1	4333	170.7	4999	184.6	-2	8	6	15	42
DOLORES	CO	5	73.6	10	201.9	16	202.6	174	0	175	99	69
DOUGLAS	CO	29	138.2	40	146.8	72	159.8	6	9	16	31	13
EAGLE	CO	29	159.9	24	105.8	40	169.6	-34	60	6	15	23
ELBERT	CO	35	160.6	26	105.2	29	109.9	-34	4	-32	00	00
EL PASO	CO	636	159.5	743	137.1	1247	163.6	-14	19	3	10	17
FREMONT	CO	155	121.2	228	158.2	272	155.3	31	-2	28	53	10
GARFIELD	CO	65	104.1	111	151.2	127	149.9	45	-1	44	76	07
GILPIN	CO	9	157.9	13	299.7	12	204.7	90	-32	30	56	71
GRAND	CO	20	127.0	24	139.4	32	187.7	10	35	48	80	47
GUNNISON	CO	27	127.4	30	136.9	35	176.7	7	29	39	69	31
HINSDALE	CO	1	114.7	2	97.4	7*	1075.6	-15	1005	838	99	99
HUERFANO	CO	88	163.1	74	139.4	70	154.9	-15	11	-5	05	10
JACKSON	CO	10	123.3	14	178.0	7	91.5	44	-49	-26	01	00
JEFFERSON	CO	443	152.9	894	167.5	1585	181.8	10	9	19	37	38
KIOWA	CO	6	55.4	12	112.0	18	161.4	102	44	191	99	15
KIT CARSON	CO	63	166.0	59	143.1	61	136.4	-14	-5	-18	02	03
LAKE	CO	31	162.1	37	259.6	38	204.1	60	-21	26	49	70
LA PLATA	CO	104	143.9	108	131.4	149	166.4	-9	27	16	31	20
LARIMER	CO	325	130.6	474	146.4	609	138.3	12	-6	6	14	03
LAS ANIMAS	CO	199	163.9	171	151.8	166	163.0	-7	7	-1	07	16
LINCOLN	CO	38	138.8	39	131.2	51	165.1	-5	26	19	37	18
LOGAN	CO	142	173.1	143	163.7	152	161.2	-5	-1	-7	04	15
MESA	CO	286	134.2	431	161.7	573	176.9	20	9	32	60	31
MINERAL	CO	3	117.0	3	82.0	8	364.4	-30	345	212	99	99
MOFFAT	CO	36	130.7	43	141.5	51	148.3	8	5	14	27	06
MONTEZUMA	CO	57	124.4	83	155.7	127	214.6	25	38	73	92	82
MONTROSE	CO	112	144.9	151	171.8	167	160.0	19	-7	10	21	13
MORGAN	CO	118	131.0	144	144.6	160	144.4	10	0	10	21	05
OTERO	CO	158	136.7	182	153.4	195	164.7	12	7	21	40	18
OURAY	CO	26	259.0	13	155.9	13	153.2	-40	-2	-41	00	09
PARK	CO	13	150.5	14	126.8	24	191.3	-16	51	27	51	53
PHILLIPS	CO	48	170.1	51	173.2	37	118.3	2	-32	-30	00	01
PITKIN	CO	13	163.2	19	190.1	29	228.0	17	20	40	71	92
PROWERS	CO	80	127.0	80	119.6	105	165.9	-6	39	31	58	19
PUEBLO	CO	602	129.1	799	155.2	924	172.1	20	11	33	62	25
RIO BLANCO	CO	29	168.2	26	138.9	36	182.9	-17	32	9	18	40
RIO GRANDE	CO	50	102.2	81	153.3	84	156.3	50	2	53	84	10
ROUTT	CO	52	139.1	48	145.3	44	144.5	4	-1	4	12	05
SAGUACHE	CO	28	131.1	25	132.8	21	116.1	1	-13	-11	03	00
SAN JUAN	CO	3	43.0	4	179.4	4	95.1	317	-47	121	97	00
SAN MIGUEL	CO	18	199.3	16	189.3	22	336.0	-5	77	69	91	99
SEDGWICK	CO	34	160.9	36	163.4	37	181.0	2	11	12	25	37
SUMMIT	CO	9	206.2	9	230.0	13	90.1	12	-61	-56	00	00
TELLER	CO	37	242.5	25	159.8	38	168.9	-34	6	-30	00	22
WASHINGTON	CO	43	127.8	47	122.7	54	153.1	-4	25	20	39	09
WELD	CO	394	129.6	530	154.1	609	151.3	19	-2	17	33	08
YUMA	CO	76	135.2	87	145.5	55	92.4	8	-36	-32	00	00
CONNECTICUT	CT	19880*	204.6	24582*	212.3	28491*	215.3	4	1	5		
FAIRFIELD	CT	5060*	208.4	6283*	212.1	7292*	215.4	2	2	3	11	83
HARTFORD	CT	5163*	204.1	6670*	218.7	7442*	213.7	7	-2	5	12	81
LITCHFIELD	CT	1044*	198.7	1247	197.5	1533	205.8	-1	4	4	11	73

* SIGNIFICANTLY MORE DEATHS THAN EXPECTED

\# COUNTIES COMBINED; SEE TECHNICAL NOTES

WHITE MALE

ALL CANCERS COMBINED

ICD 140-208

1950-1959 DEATHS	1950-1959 RATE	1960-1969 DEATHS	1960-1969 RATE	1970-1979 DEATHS	1970-1979 RATE	COUNTY AND STATE	PERCENT CHANGE 60S/50S	PERCENT CHANGE 70S/60S	PERCENT CHANGE 70S/50S	PERCENTILE %CHANGE RATES 70S/50S	PERCENTILE 1970S
614	161.2	855	184.7	1095	198.9	MIDDLESEX CT	15	8	23	45	64
5665*	216.2	6605*	218.3	7559*	222.8	NEW HAVEN CT	1	2	3	11	89
1333*	189.6	1655*	202.7	2003*	217.1	NEW LONDON CT	7	7	15	29	84
396	182.7	485	176.1	678	200.9	TOLLAND CT	-4	14	10	21	66
605*	195.8	782*	224.7	889*	222.0	WINDHAM CT	15	-1	13	27	88
2434	181.6	3251*	197.8	4396*	222.8	DELAWARE DE	9	13	23		
297	157.6	364	164.4	519	190.3	KENT DE	4	16	21	41	51
1722*	193.9	2369*	212.3	3070*	235.4	NEW CASTLE DE	9	11	21	42	94
415	157.9	518	171.5	807	208.1	SUSSEX DE	9	21	32	60	76
4381*	220.6	3526*	216.1	2406	209.7	DIST. OF COLUMBIA DC	-2	-3	-5		
4381*	220.6	3526*	216.1	2406	209.7	WASHINGTON DC	-2	-3	-5	05	77
24917	157.4	51900	180.1	92444	197.2	FLORIDA FL	14	10	25		
206	148.4	375	192.8	600	202.7	ALACHUA FL	30	5	37	67	69
15	93.2	49	183.8	88	210.2	BAKER FL	97	14	126	98	78
208	156.1	418	205.9	703*	230.4	BAY FL	32	12	48	79	93
57	153.0	82	177.2	124	185.2	BRADFORD FL	16	4	21	41	43
266	128.5	821	168.3	2096	201.0	BREVARD FL	31	19	56	86	67
1219	130.9	4902	171.1	11324	194.3	BROWARD FL	31	14	48	80	57
40	142.6	50	145.5	72	175.0	CALHOUN FL	2	20	23	45	28
64	101.9	403	168.5	869	171.0	CHARLOTTE FL	65	2	68	91	24
66	162.3	186	152.3	691	185.1	CITRUS FL	-6	22	14	28	43
64	96.6	128	138.8	359	213.8	CLAY FL	44	54	121	97	82
42	106.2	210	141.9	774	186.3	COLLIER FL	34	31	75	93	45
104	194.2	125	176.1	220	207.8	COLUMBIA FL	-9	18	7	16	76
5679*	186.3	10173	190.6	15425	203.4	DADE FL	2	7	9	19	69
75	148.5	91	142.1	165	192.4	DE SOTO FL	-4	35	30	56	54
28	195.5	36	199.3	51	159.7	DIXIE FL	2	-20	-18	01	13
1750*	192.4	2794*	218.9	3726*	237.1	DUVAL FL	14	8	23	45	95
514	163.9	907	192.5	1402*	219.7	ESCAMBIA FL	17	14	34	64	86
23	178.0	55*	289.4	81	167.8	FLAGLER FL	63	-42	-6	05	21
40	169.8	72	223.7	83	235.1	FRANKLIN FL	32	5	38	69	94
79	91.3	98	101.2	155	156.5	GADSDEN FL	11	55	71	92	11
13	175.8	27	176.0	52	233.4	GILCHRIST FL	0	33	33	61	94
9	127.8	14	119.8	48	171.4	GLADES FL	-6	43	34	64	25
33	200.7	60	221.4	89	240.4	GULF FL	10	9	20	39	96
30	142.5	36	148.5	60	222.0	HAMILTON FL	4	50	56	85	88
80	152.5	107	178.8	162	224.2	HARDEE FL	17	25	47	79	90
22	115.7	48	170.5	97	191.7	HENDRY FL	47	12	66	90	53
51	107.7	160	179.7	471	190.4	HERNANDO FL	67	6	77	93	51
115	140.2	261	157.1	556	181.4	HIGHLANDS FL	12	15	29	56	38
2275	177.9	3740*	204.9	5020*	213.5	HILLSBOROUGH FL	15	4	20	39	81
46	88.5	86	143.8	129	178.4	HOLMES FL	63	24	102	97	33
112	132.5	317	171.3	623	191.2	INDIAN RIVER FL	29	12	44	76	53
128	137.9	155	135.1	273	200.2	JACKSON FL	-2	48	45	77	66
27	141.2	36	172.3	60	234.1	JEFFERSON FL	22	36	66	90	94
12	94.8	23	153.0	33	179.8	LAFAYETTE FL	61	17	90	96	35
359	150.0	702	173.4	1232	181.4	LAKE FL	16	5	21	41	37
258	139.2	790	153.2	2137	179.4	LEE FL	10	17	29	55	35
172	163.5	309	186.2	497	193.3	LEON FL	14	4	18	36	56
56	157.3	92	178.0	183	209.2	LEVY FL	13	17	33	61	77
11	98.1	21	155.2	32	185.7	LIBERTY FL	58	20	89	96	44
49	139.6	69	173.1	93	195.4	MADISON FL	24	13	40	71	59
395	116.7	1116	153.3	1984	179.3	MANATEE FL	31	17	54	84	34
204	138.3	479	193.2	990	192.9	MARION FL	40	0	39	71	55
90	135.5	237	138.9	738	195.0	MARTIN FL	3	40	44	75	58
234*	232.3	359	204.9	589	197.2	MONROE FL	-12	-4	-15	02	62
46	126.2	110	222.6	189	235.2	NASSAU FL	76	6	86	95	94
111	127.0	200	131.4	495	193.1	OKALOOSA FL	3	47	52	83	55
33	209.9	50	180.8	159	219.4	OKEECHOBEE FL	-14	21	5	12	86
1169	163.1	2058	180.5	3215	204.4	ORANGE FL	11	13	25	49	71
154	116.5	299	179.2	473	189.4	OSCEOLA FL	54	6	63	89	50
1224	158.3	2843	180.3	6021	185.8	PALM BEACH FL	14	3	17	34	44
253	128.6	773	140.4	2562	177.4	PASCO FL	9	26	38	69	32
3301	152.1	7363	179.3	11608	200.1	PINELLAS FL	18	12	31	59	65
918	150.4	1645	177.4	2726	203.9	POLK FL	18	15	36	65	70
122	133.7	197	158.7	440	215.2	PUTNAM FL	19	36	61	88	83
175	175.6	249	191.4	398	200.0	ST JOHNS FL	9	4	14	28	65
174	150.4	390	177.0	736	191.8	ST LUCIE FL	18	8	27	52	53
87	131.5	147	147.7	275	181.3	SANTA ROSA FL	12	23	38	69	37
482	140.4	1468	170.1	3010	190.3	SARASOTA FL	21	12	36	66	51
134	103.6	390	168.9	848	170.8	SEMINOLE FL	63	1	65	90	24
59	142.6	99	159.3	241	224.3	SUMTER FL	12	41	57	86	90
80	167.1	94	159.2	175	214.1	SUWANNEE FL	-5	34	28	53	82
45	132.9	70	162.7	141	241.7	TAYLOR FL	22	49	82	95	96
16	76.0	27	113.5	52	208.0	UNION FL	49	83	174	99	76
867	145.8	1986	191.2	3137	195.0	VOLUSIA FL	31	2	34	63	58
25	157.3	46	201.2	87	256.8	WAKULLA FL	28	28	63	89	98
69	109.9	106	136.1	173	167.0	WALTON FL	24	23	52	83	20
53	111.4	71	133.9	127	195.1	WASHINGTON FL	20	46	75	93	59
13730	151.6	19743	172.1	28738	199.6	GEORGIA GA	14	16	32		
42	112.7	64	162.8	90	179.4	APPLING GA	44	10	59	87	34
28	157.2	33	180.7	43	202.4	ATKINSON GA	15	12	29	54	68
41	157.1	49	169.6	86	253.2	BACON GA	8	49	61	88	98
10	117.2	10	121.9	31*	302.4	BAKER GA	4	148	158	98	99
68	67.5	90	82.4	136	128.3	BALDWIN GA	22	56	90	96	02
27	97.5	49	159.8	60	161.5	BANKS GA	64	1	66	90	15
60	124.6	91	155.8	136	190.7	BARROW GA	25	22	53	84	52
109	135.3	181	171.6	303*	231.9	BARTOW GA	27	35	71	92	93
51	127.9	86	190.3	115	228.0	BEN HILL GA	49	20	78	94	92
55	139.2	78	179.0	100	195.6	BERRIEN GA	29	9	41	72	59
507	194.6	636	190.3	853*	225.4	BIBB GA	-2	18	16	31	90
28	143.9	45	169.2	69	197.7	BLECKLEY GA	18	17	37	68	62
30	162.8	39	189.4	59	187.2	BRANTLEY GA	16	-1	15	30	46

* SIGNIFICANTLY MORE DEATHS THAN EXPECTED

WHITE MALE — ALL CANCERS COMBINED — ICD 140-208

1950-1959 DEATHS	1950-1959 RATE	1960-1969 DEATHS	1960-1969 RATE	1970-1979 DEATHS	1970-1979 RATE	COUNTY AND STATE	PERCENT CHANGE 60S/50S	PERCENT CHANGE 70S/60S	PERCENT CHANGE 70S/50S	PERCENTILE %CHANGE 70S/50S	PERCENTILE RATES 1970S
54	128.3	59	135.8	84	170.3	BROOKS GA	6	25	33	61	23
27	184.8	37	224.0	49	210.1	BRYAN GA	21	-6	14	27	78
69	110.1	116	143.4	174	177.5	BULLOCH GA	30	24	61	88	32
49	161.4	67	206.3	89	228.5	BURKE GA	28	11	42	73	92
42	166.9	55	188.4	73	190.9	BUTTS GA	13	1	14	29	52
20	143.3	20	163.2	33	233.7	CALHOUN GA	14	43	63	89	94
30	204.8	37	193.5	61	217.3	CAMDEN GA	-6	12	6	15	84
33	187.4	29	122.3	57	183.7	CANDLER GA	-35	50	-2	07	41
169	143.2	216	146.8	372	205.1	CARROLL GA	3	40	43	75	71
70	132.2	146	188.6	241	208.8	CATOOSA GA	43	11	58	86	76
20	154.8	40	272.1	55*	284.5	CHARLTON GA	76	5	84	95	99
670*	200.2	846*	203.4	1122*	232.4	CHATHAM GA	2	14	16	32	93
80	126.3	117	145.2	168	178.3	CHATTOOGA GA	15	23	41	72	33
109	141.0	140	130.4	258	177.8	CHEROKEE GA	-7	36	26	50	33
149	168.7	193	181.4	280	195.7	CLARKE GA	8	8	16	32	60
15	190.9	17	204.9	16	206.5	CLAY GA	7	1	8	17	74
84	120.6	244	171.8	557	213.7	CLAYTON GA	42	24	77	94	81
15	112.0	23	125.0	49	235.5	CLINCH GA	12	88	110	97	95
332	150.5	658	178.6	1283	206.1	COBB GA	19	15	37	67	73
81	144.9	119	181.0	156	198.2	COFFEE GA	25	10	37	67	63
114	128.4	161	147.1	239	183.7	COLQUITT GA	15	25	43	74	41
41	200.8	50	146.0	153	239.2	COLUMBIA GA	-27	64	19	38	96
46	145.6	60	155.7	107	232.4	COOK GA	7	49	60	87	93
132	187.1	146	167.1	218	193.9	COWETA GA	-11	16	4	11	57
21	152.5	17	138.6	28	170.0	CRAWFORD GA	-9	23	12	23	23
74	195.5	96	214.6	125	237.3	CRISP GA	10	11	21	42	95
35	163.0	54	167.8	99	237.4	DADE GA	3	42	46	77	95
22	156.6	26	158.4	42	221.1	DAWSON GA	1	40	41	73	87
81	154.3	102	164.0	146	207.4	DECATUR GA	6	26	34	64	75
743	152.9	1451	180.9	2194	199.4	DE KALB GA	18	10	30	57	64
62	135.7	87	166.2	117	187.3	DODGE GA	22	13	38	69	46
47	156.0	52	182.0	68	213.8	DOOLY GA	17	18	37	68	82
140	164.6	251	208.8	350	207.4	DOUGHERTY GA	27	-1	26	50	75
38	86.0	93	121.7	243	205.2	DOUGLAS GA	41	69	139	98	72
42	128.5	31	92.3	94*	258.3	EARLY GA	-28	180	101	96	98
9	147.6	7	131.2	11	121.3	ECHOLS GA	-11	-8	-18	01	01
49	207.5	52	157.8	89	187.3	EFFINGHAM GA	-24	19	-10	04	46
58	118.8	106	177.6	139	206.5	ELBERT GA	49	16	74	93	74
50	105.7	67	115.8	131	190.9	EMANUEL GA	10	65	81	94	52
21	123.8	40	221.5	50	187.2	EVANS GA	79	-16	51	82	46
59	104.6	95	146.5	147	186.7	FANNIN GA	40	27	79	94	45
23	85.9	50	157.1	88	157.5	FAYETTE GA	83	0	83	95	11
315	159.5	435	178.3	572	193.1	FLOYD GA	12	8	21	41	56
58	144.8	80	150.2	162	187.1	FORSYTH GA	4	25	29	55	46
59	108.2	87	158.5	103	162.8	FRANKLIN GA	46	3	50	82	16
2367*	192.3	2895*	203.0	3015	209.8	FULTON GA	6	3	9	19	77
25	70.1	69	165.7	93	179.4	GILMER GA	136	8	156	98	34
17	182.2	18	189.0	19	177.4	GLASCOCK GA	4	-6	-3	06	32
136	177.3	203	188.7	337	216.9	GLYNN GA	6	15	22	44	84
93	125.3	127	155.4	189	178.9	GORDON GA	24	15	43	74	34
62	122.7	87	151.7	129	188.8	GRADY GA	24	24	54	84	49
55	191.9	51	169.3	62	180.0	GREENE GA	-12	6	-6	05	36
166	136.1	268	155.6	569	181.3	GWINNETT GA	14	16	33	62	37
83	142.5	107	138.6	182	189.9	HABERSHAM GA	-3	37	33	62	51
149	102.1	256	142.4	430	175.1	HALL GA	39	23	71	92	29
24	156.4	22	144.3	29	188.0	HANCOCK GA	-8	30	20	40	47
63	117.2	82	131.1	123	163.2	HARALSON GA	12	24	39	70	17
38	148.5	50	154.9	85	178.5	HARRIS GA	4	15	20	39	33
54	117.4	81	153.6	128	192.7	HART GA	31	25	64	90	55
33	143.8	26	115.4	34	124.2	HEARD GA	-20	8	-14	03	01
54	127.0	107	181.2	189	206.0	HENRY GA	43	14	62	89	73
64	141.9	141	183.8	287	187.8	HOUSTON GA	30	2	32	60	47
34	122.0	53	185.0	74	219.1	IRWIN GA	52	18	80	94	86
80	126.9	105	138.3	186	202.4	JACKSON GA	9	46	59	87	68
29	166.5	24	124.8	35	162.2	JASPER GA	-25	30	-3	06	15
20	87.5	35	121.3	83	234.3	JEFF DAVIS GA	39	93	168	99	94
52	166.2	66	176.7	92	221.0	JEFFERSON GA	6	25	33	62	87
24	120.2	45	183.4	50	188.5	JENKINS GA	53	3	57	86	48
31	129.2	37	136.2	54	174.7	JOHNSON GA	5	28	35	65	28
16	100.5	39	174.5	47	131.4	JONES GA	74	-25	31	58	02
40	151.5	52	187.2	58	170.7	LAMAR GA	24	-9	13	25	24
6	58.3	25	165.6	30	162.5	LANIER GA	184	-2	179	99	16
95	118.3	181	175.2	229	180.1	LAURENS GA	48	3	52	83	36
26*	339.0	20	233.4	41	223.2	LEE GA	-31	-4	-34	00	89
19	138.2	39	215.6	68	248.2	LIBERTY GA	56	15	80	94	97
19	135.8	32	189.1	40	181.6	LINCOLN GA	39	-4	34	63	38
17	230.8	23	190.0	34	287.4	LONG GA	-18	51	25	48	99
140	162.0	233	195.2	269	186.1	LOWNDES GA	21	-5	15	30	44
23	94.3	46	158.5	70	183.1	LUMPKIN GA	68	16	94	96	40
41	160.7	32	101.2	93	229.8	MC DUFFIE GA	-37	127	43	74	93
18	160.4	32	216.4	42	233.6	MC INTOSH GA	35	8	46	77	94
31	146.7	49	187.1	58	181.4	MACON GA	28	-3	24	46	37
52	122.6	64	132.4	121	203.9	MADISON GA	8	54	66	91	70
19	148.5	11	90.9	31	223.2	MARION GA	-39	146	50	82	89
63	163.4	98	194.5	100	174.7	MERIWETHER GA	19	-10	7	16	28
29	125.2	20	82.1	53	194.2	MILLER GA	-34	136	55	85	57
69	162.9	96	223.9	121	248.6	MITCHELL GA	37	11	53	83	97
37	144.4	49	182.7	88	252.9	MONROE GA	27	38	75	93	98
24	125.5	27	136.0	47	202.4	MONTGOMERY GA	8	49	61	88	68
39	169.0	47	163.9	63	196.4	MORGAN GA	-3	20	16	32	61
54	146.5	77	171.9	129	212.2	MURRAY GA	17	23	45	76	80
392	171.2	645	199.8	903*	232.7	MUSCOGEE#GA	17	16	36	66	94
74	140.1	130	187.5	170	177.6	NEWTON GA	34	-5	27	51	32
33	148.5	33	119.1	52	146.9	OCONEE GA	-20	23	-1	07	06
36	144.9	35	130.6	49	163.4	OGLETHORPE GA	-10	25	13	25	17
63	135.5	85	140.2	149	185.1	PAULDING GA	4	32	37	67	43
24	106.7	41	154.7	69	194.6	PEACH GA	45	26	82	95	58
54	157.5	58	145.9	79	156.7	PICKENS GA	-7	7	0	07	11
42	149.2	56	174.9	62	153.2	PIERCE GA	17	-12	3	10	09
36	164.9	39	163.9	46	157.3	PIKE GA	-1	-4	-5	05	11
126	138.8	173	164.2	240	193.5	POLK GA	18	18	39	70	56

* SIGNIFICANTLY MORE DEATHS THAN EXPECTED

COUNTIES COMBINED; SEE TECHNICAL NOTES

WHITE MALE

ALL CANCERS COMBINED

ICD 140-208

1950-1959 DEATHS	1950-1959 RATE	1960-1969 DEATHS	1960-1969 RATE	1970-1979 DEATHS	1970-1979 RATE	COUNTY AND STATE	PERCENT CHANGE 60S/50S	PERCENT CHANGE 70S/60S	PERCENT CHANGE 70S/50S	PERCENTILE %CHANGE 70S/50S	PERCENTILE RATES 1970S
25	141.4	43	213.9	66	252.4	PULASKI GA	51	18	78	94	98
37	263.8	36	219.9	57	221.2	PUTNAM GA	-17	1	-16	02	87
8	179.6	9	147.2	15	221.2	QUITMAN GA	-18	50	23	45	87
44	141.4	58	153.8	84	159.9	RABUN GA	9	4	13	26	13
26	125.2	41	195.3	50	202.6	RANDOLPH GA	56	4	62	88	69
426	176.6	645*	212.9	853*	229.1	RICHMOND GA	21	8	30	56	92
44	155.7	64	153.0	153	188.6	ROCKDALE GA	-2	23	21	42	48
10	151.5	11	142.2	16	159.5	SCHLEY GA	-6	12	5	14	13
38	125.7	60	175.4	78	192.1	SCREVEN GA	40	10	53	84	54
15	86.0	18	82.9	50	176.2	SEMINOLE GA	-4	112	105	97	30
136	164.2	196	179.1	263	185.5	SPALDING GA	9	4	13	26	44
64	123.3	102	150.2	182	206.7	STEPHENS GA	22	38	68	91	74
23	171.8	18	139.9	15	121.9	STEWART GA	-19	-13	-29	00	01
78	169.9	106	186.8	137	208.8	SUMTER GA	10	12	23	45	76
17	174.4	19	135.2	34	243.6	TALBOT GA	-22	80	40	71	97
13	157.0	8	103.1	16	216.6	TALIAFERRO GA	-34	110	38	69	84
50	97.8	73	124.7	134	221.3	TATTNALL GA	28	77	126	98	87
29	125.6	38	169.0	42	167.7	TAYLOR GA	35	-1	33	62	21
62	175.1	71	202.7	84	203.3	TELFAIR GA	16	0	16	32	69
29	122.1	50	208.1	67	255.0	TERRELL GA	70	23	109	97	98
123	127.8	159	156.8	225	203.6	THOMAS GA	23	30	59	87	70
65	117.7	127	193.5	174	198.6	TIFT GA	64	3	69	91	63
66	163.8	85	161.5	134	193.1	TOOMBS GA	-1	20	18	36	55
21	112.6	27	113.9	42	124.7	TOWNS GA	1	10	11	22	01
16	94.4	23	115.5	41	189.5	TREUTLEN GA	22	64	101	96	50
147	128.3	242	175.8	312	196.3	TROUP GA	37	12	53	84	61
35	137.4	42	157.7	54	166.3	TURNER GA	15	5	21	41	19
13	108.4	22	168.2	42	227.1	TWIGGS GA	55	35	109	97	91
25	87.1	36	99.1	79	160.6	UNION GA	14	62	84	95	14
99	169.3	112	147.3	192	203.8	UPSON GA	-13	38	20	40	70
196	143.9	314	174.7	434	194.2	WALKER GA	21	11	35	65	57
81	137.5	97	131.9	199	208.3	WALTON GA	-4	58	51	82	76
135	169.4	196	186.1	267	218.6	WARE GA	10	17	29	55	85
26	171.9	22	146.0	34	191.3	WARREN GA	-15	31	11	23	53
33	83.4	52	125.6	93	202.4	WASHINGTON GA	51	61	143	98	68
56	151.2	94	182.1	125	182.1	WAYNE GA	20	0	20	40	39
12	167.1	10	151.7	10	194.4	WEBSTER GA	-9	28	16	32	58
18	104.0	33	181.7	42	206.5	WHEELER GA	75	14	99	96	74
35	144.0	47	147.3	65	149.4	WHITE GA	2	1	4	11	07
162	144.5	270	167.7	424	201.0	WHITFIELD GA	16	20	39	70	66
24	88.8	47	169.2	54	180.4	WILCOX GA	91	7	103	97	36
52	194.5	57	199.7	61	187.5	WILKES GA	3	-6	-4	06	46
29	129.8	41	180.9	51	196.7	WILKINSON GA	39	9	52	83	61
64	181.3	54	137.9	90	176.1	WORTH GA	-24	28	-3	06	30
		1036	181.4	1593	179.4	HAWAII HI		-1			
		107	201.5	157	156.7	HAWAII HI		-22			11
		818	180.8	1263	185.5	HONOLULU HI		3			44
		44	190.5	60	173.2	KAUAI HI		-9			27
		67	162.3	113	159.8	MAUI#HI		-2			13
3714	139.8	4673	151.8	6032	166.3	IDAHO ID	9	10	19		
489	143.4	741	171.1	914	166.7	ADA ID	19	-3	16	32	20
11	65.9	27	159.6	25	151.4	ADAMS ID	142	-5	130	98	08
216	153.9	269	150.2	349	171.0	BANNOCK ID	-2	14	11	23	24
33	104.9	37	109.0	29	77.2	BEAR LAKE ID	4	-29	-26	01	00
62	161.4	53	139.3	65	165.8	BENEWAH ID	-14	19	3	10	19
109	129.2	134	132.7	167	142.4	BINGHAM ID	3	7	10	21	04
24	114.2	43	182.1	38	175.4	BLAINE ID	59	-4	54	84	29
10	96.7	12	130.8	10	78.4	BOISE ID	35	-40	-19	01	00
117	128.9	146	154.6	204	187.6	BONNER ID	20	21	46	77	47
164	142.3	210	141.0	320	166.6	BONNEVILLE ID	-1	18	17	33	20
40	130.8	47	157.0	59	180.4	BOUNDARY ID	20	15	38	69	36
13	133.1	18	161.1	29	207.3	BUTTE ID	21	29	56	85	75
9	196.6	11	322.0	9	200.3	CAMAS ID	64	-38	2	09	66
325	125.2	478	157.0	642	179.5	CANYON ID	25	14	43	75	35
27	144.3	29	134.2	39	149.0	CARIBOU ID	-7	11	3	11	06
74	132.6	89	137.4	106	141.2	CASSIA ID	4	3	7	15	04
7	214.4	8	201.4	4	97.4	CLARK ID	-6	-52	-55	00	00
67	140.0	59	125.3	65	133.4	CLEARWATER ID	-10	6	-5	05	02
27	177.7	23	147.9	26	148.0	CUSTER ID	-17	0	-17	02	06
53	164.2	87	223.1	99	207.2	ELMORE ID	36	-7	26	50	75
48	135.4	36	86.6	60	128.1	FRANKLIN ID	-36	48	-5	05	02
47	146.6	52	139.6	66	161.3	FREMONT ID	-5	16	10	21	15
80	177.5	63	121.2	115	188.3	GEM ID	-32	55	6	15	48
57	116.9	82	151.2	108	168.7	GOODING ID	29	12	44	76	21
89	151.1	106	168.1	124	171.8	IDAHO ID	11	2	14	28	25
56	156.6	55	117.0	75	134.3	JEFFERSON ID	-25	15	-14	02	03
61	141.2	67	137.6	109	169.9	JEROME ID	-3	23	20	40	23
212	139.4	272	155.0	427	184.1	KOOTENAI ID	11	19	32	60	42
135	155.8	153	160.0	157	149.9	LATAH ID	3	-6	-4	06	07
49	147.2	50	151.6	73	204.6	LEMHI ID	3	35	39	70	71
23	102.0	30	142.3	38	170.0	LEWIS ID	39	19	67	91	23
20	109.2	25	137.4	32	170.6	LINCOLN ID	26	24	56	85	24
41	131.3	38	106.6	65	155.0	MADISON ID	-19	45	18	36	10
67	148.2	83	144.4	113	163.4	MINIDOKA ID	-3	13	10	21	17
185	153.8	229	167.9	305	195.8	NEZ PERCE ID	9	17	27	52	60
19	107.0	17	83.6	40	197.0	ONEIDA ID	-22	136	84	95	61
39	136.1	47	144.2	54	143.6	OWYHEE ID	6	0	6	14	05
74	112.9	121	158.0	130	144.6	PAYETTE ID	40	-8	28	53	05
22	128.2	38	191.1	40	186.8	POWER ID	49	-2	46	77	46
171	221.6	158	196.8	176	205.8	SHOSHONE ID	-11	5	-7	04	73
9	84.3	27	301.2	21	166.1	TETON ID	257	-45	97	96	19
241	129.1	304	145.0	383	155.9	TWIN FALLS ID	12	8	21	41	10
24	187.1	21	119.3	30	137.4	VALLEY ID	-36	15	-27	01	03
68	149.1	78	159.6	92	167.6	WASHINGTON ID	7	5	12	25	21
75909*	189.6	86696*	199.6	94471*	211.3	ILLINOIS IL	5	6	11		

* SIGNIFICANTLY MORE DEATHS THAN EXPECTED

COUNTIES COMBINED; SEE TECHNICAL NOTES

WHITE MALE — ALL CANCERS COMBINED — ICD 140-208

1950-1959 DEATHS	1950-1959 RATE	1960-1969 DEATHS	1960-1969 RATE	1970-1979 DEATHS	1970-1979 RATE	COUNTY AND STATE	PERCENT CHANGE 60S/50S	PERCENT CHANGE 70S/60S	PERCENT CHANGE 70S/50S	PERCENTILE %CHANGE 70S/50S	PERCENTILE RATES 1970S
721	189.7	718	183.4	865	216.9	ADAMS IL	-3	18	14	29	84
137	205.5	123	196.6	133	237.8	ALEXANDER IL	-4	21	16	31	95
149	163.3	150	169.9	158	176.9	BOND IL	4	4	8	18	31
158	167.4	160	160.7	204	179.9	BOONE IL	-4	12	7	16	35
73	155.0	41	85.3	90	222.7	BROWN IL	-45	161	44	75	89
362	170.1	392	179.5	406	178.9	BUREAU IL	6	0	5	13	34
47	117.1	61	139.2	79	180.8	CALHOUN IL	19	30	54	84	37
166	153.1	182	158.8	214	187.2	CARROLL IL	4	18	22	44	46
141	161.6	162	187.6	163	185.1	CASS IL	16	-1	15	29	43
542	158.7	644	163.5	846	197.3	CHAMPAIGN IL	3	21	24	47	62
327	151.0	421	197.4	480*	229.0	CHRISTIAN IL	31	16	52	83	92
126	115.7	170	156.6	191	177.1	CLARK IL	35	13	53	84	31
130	136.3	140	148.2	187	192.0	CLAY IL	9	30	41	72	53
184	157.8	235	183.1	278	192.6	CLINTON IL	16	5	22	43	55
299	142.0	365	170.8	470	213.2	COLES IL	20	25	50	82	81
40230*	216.0	43840*	221.7	43080*	225.0	COOK IL	3	1	4	12	90
203	166.4	195	158.4	229	180.5	CRAWFORD IL	-5	14	8	18	36
92	152.7	98	166.3	127	203.1	CUMBERLAND IL	9	22	33	61	69
323	152.2	398	171.4	430	170.0	DE KALB IL	13	-1	12	23	23
164	170.8	176	178.6	177	181.8	DE WITT IL	5	2	6	15	38
136	149.3	151	165.5	185	191.0	DOUGLAS IL	11	15	28	53	52
1370	181.4	2233	185.6	3561	206.9	DU PAGE IL	2	11	14	28	74
224	167.5	244	188.4	247	188.7	EDGAR IL	12	0	13	25	48
66	131.6	74	149.3	88	164.1	EDWARDS IL	13	10	25	48	17
193	171.2	205	161.8	287	201.7	EFFINGHAM IL	-5	25	18	35	67
184	137.4	222	162.4	231	168.6	FAYETTE IL	18	4	23	44	21
142	161.9	187	206.4	186	203.6	FORD IL	27	-1	26	49	70
505	177.4	535	188.3	623*	229.7	FRANKLIN IL	6	22	29	56	92
339	138.2	418	168.6	477	188.2	FULTON IL	22	12	36	66	48
83	173.7	94	208.9	119*	254.1	GALLATIN IL	20	22	46	78	98
188	162.6	225	196.9	203	185.1	GREENE IL	21	-6	14	28	43
165	165.8	195	171.5	245	190.4	GRUNDY IL	3	11	15	30	51
97	138.3	118	173.8	105	154.6	HAMILTON IL	26	-11	12	24	09
220	144.3	263	182.8	276	193.4	HANCOCK IL	27	6	34	63	56
54	171.0	62	192.8	76	228.1	HARDIN IL	13	18	33	62	92
56	123.7	70	150.2	87	171.6	HENDERSON IL	21	14	39	70	25
448	170.9	492	177.8	528	183.7	HENRY IL	4	3	7	16	41
273	157.4	308	169.9	360	191.6	IROQUOIS IL	8	13	22	43	53
287	159.0	339	175.4	398	195.7	JACKSON IL	10	12	23	45	60
115	159.4	109	155.4	122	174.5	JASPER IL	-2	12	9	20	28
251	141.6	302	163.9	385	198.1	JEFFERSON IL	16	21	40	71	63
147	179.1	155	176.2	180	192.4	JERSEY IL	-2	9	7	16	54
177	146.8	246	201.5	278	228.5	JO DAVIESS IL	37	13	56	85	92
72	150.8	69	145.8	112	201.0	JOHNSON IL	-3	38	33	62	67
1341	172.2	1678	183.6	1955	202.0	KANE IL	7	10	17	34	68
524	124.3	689	161.0	822	202.0	KANKAKEE IL	30	25	63	89	68
107	160.4	130	159.4	187	170.9	KENDALL IL	-1	7	7	15	24
489	162.8	551	169.8	592	182.4	KNOX IL	4	7	12	24	39
1482	185.4	2100	194.4	2687	206.7	LAKE IL	5	6	11	23	74
945	182.8	1043	186.4	1280	215.8	LA SALLE IL	2	16	18	36	83
176	162.6	179	164.9	198	182.3	LAWRENCE IL	1	11	12	24	39
279	163.5	342	187.0	317	177.9	LEE IL	14	-5	9	18	33
296	143.6	358	166.6	399	186.4	LIVINGSTON IL	16	12	30	57	45
248	157.2	272	163.3	312	186.4	LOGAN IL	4	14	19	37	45
250	156.5	257	164.6	273	172.3	MC DONOUGH IL	5	5	10	21	25
573	187.1	773	184.5	1071	203.1	MC HENRY IL	-1	10	9	18	69
608	157.0	705	172.3	832	197.0	MC LEAN IL	10	14	25	49	61
814	177.0	1001	193.2	1137	207.6	MACON IL	9	7	17	34	75
513	176.2	554	188.4	563	193.9	MACOUPIN IL	7	3	10	21	57
1396	174.2	1876	197.3	2347*	219.8	MADISON IL	13	11	26	50	86
350	166.4	425	191.1	531*	229.1	MARION IL	15	20	38	68	92
128	168.0	126	167.1	145	181.7	MARSHALL IL	-1	9	8	17	38
155	182.4	160	169.4	215	200.5	MASON IL	-7	18	10	20	66
106	150.9	123	163.0	169	205.1	MASSAC IL	8	26	36	66	72
78	134.8	96	165.4	102	163.4	MENARD IL	23	-1	21	42	17
142	141.4	159	163.3	175	173.9	MERCER IL	16	6	23	45	27
139	183.9	171	192.3	199	199.0	MONROE IL	5	3	8	17	64
338	166.2	388	198.4	391	206.3	MONTGOMERY IL	19	4	24	47	73
271	123.6	298	142.6	350	184.5	MORGAN IL	15	29	49	81	42
98	122.3	127	158.2	172	217.4	MOULTRIE IL	29	37	78	94	84
281	159.1	327	169.6	376	178.4	OGLE IL	7	5	12	24	33
1440	180.1	1669	197.1	1965*	229.8	PEORIA IL	9	17	28	52	92
194	169.5	245	203.4	249	205.0	PERRY IL	20	1	21	41	71
133	174.2	137	168.4	161	198.4	PIATT IL	-3	18	14	28	63
208	154.8	234	177.3	228	172.6	PIKE IL	15	-3	12	23	26
41	130.4	41	134.6	57	196.3	POPE IL	3	46	51	82	61
67	151.7	81	191.3	97	237.2	PULASKI IL	26	24	56	85	95
48	162.2	62	211.7	58	202.4	PUTNAM IL	31	-4	25	48	68
249	148.7	268	153.4	329	179.1	RANDOLPH IL	3	17	20	40	34
143	164.9	169	186.4	173	180.5	RICHLAND IL	13	-3	9	20	36
1195	181.9	1363	190.4	1495	204.0	ROCK ISLAND IL	5	7	12	25	70
1688*	201.6	1977*	212.1	2088*	221.0	ST CLAIR IL	5	4	10	20	87
313	173.2	326	186.2	429*	242.9	SALINE IL	7	30	40	71	96
1201	186.6	1474*	209.8	1667*	225.2	SANGAMON IL	12	7	21	41	90
99	166.0	84	148.6	104	184.6	SCHUYLER IL	-10	24	11	23	42
74	181.3	73	184.9	67	166.2	SCOTT IL	2	-10	-8	04	19
217	143.2	238	162.9	257	177.4	SHELBY IL	14	9	24	46	32
70	144.9	68	150.7	91	212.4	STARK IL	4	41	47	78	80
372	168.1	414	181.0	507	210.3	STEPHENSON IL	8	16	25	48	78
584	185.8	720	176.5	981	198.2	TAZEWELL IL	-5	12	7	15	63
137	107.4	163	137.5	184	158.6	UNION IL	28	15	48	79	12
821	176.3	914	189.0	1077*	227.5	VERMILION IL	7	20	29	55	91
128	169.5	136	177.4	115	155.0	WABASH IL	5	-13	-9	04	10
171	147.9	177	155.1	206	176.3	WARREN IL	5	14	19	38	30
137	140.6	155	153.3	201	186.8	WASHINGTON IL	9	22	33	61	46
141	130.8	176	158.6	194	164.8	WAYNE IL	21	4	26	50	18
171	151.7	193	162.9	250	211.1	WHITE IL	7	30	39	70	79
382	157.0	489	179.1	593	199.6	WHITESIDE IL	14	11	27	52	65
1159	179.5	1384	179.0	1897	207.2	WILL IL	0	16	15	31	75
416	153.5	537	197.6	640	219.4	WILLIAMSON IL	29	11	43	74	86
1214	165.2	1627	186.3	1926	196.2	WINNEBAGO IL	13	5	19	37	60

* SIGNIFICANTLY MORE DEATHS THAN EXPECTED

WHITE MALE — ALL CANCERS COMBINED — ICD 140-208

1950-1959 DEATHS	1950-1959 RATE	1960-1969 DEATHS	1960-1969 RATE	1970-1979 DEATHS	1970-1979 RATE	COUNTY AND STATE		PERCENT CHANGE 60S/50S	PERCENT CHANGE 70S/60S	PERCENT CHANGE 70S/50S	PERCENTILE %CHANGE 70S/50S	PERCENTILE RATES 1970S
203	178.1	207	160.4	222	154.1	WOODFORD	IL	-10	-4	-13	03	09
30384	165.2	37200	184.8	44206	205.2	INDIANA	IN	12	11	24		
145	141.2	176	150.7	218	172.8	ADAMS	IN	7	15	22	44	26
1369	173.0	1773	190.9	2080	200.1	ALLEN	IN	10	5	16	31	66
250	148.1	359	183.7	420	192.0	BARTHOLOMEW	IN	24	4	30	56	53
79	137.8	112	194.5	126	218.5	BENTON	IN	41	12	59	86	85
105	154.3	135	192.8	126	170.0	BLACKFORD	IN	25	-12	10	21	23
203	148.1	228	160.7	262	169.2	BOONE	IN	9	5	14	29	22
46	135.9	54	134.1	91	168.6	BROWN	IN	-1	26	24	47	21
128	138.0	149	159.5	174	179.3	CARROLL	IN	16	12	30	57	34
310	139.6	372	175.7	396	200.1	CASS	IN	26	14	43	75	65
298	151.2	450	190.1	615	207.7	CLARK	IN	26	9	37	68	76
226	157.5	284	192.6	320	227.4	CLAY	IN	22	18	44	76	91
220	138.1	281	171.8	321	200.4	CLINTON	IN	24	17	45	77	66
65	111.6	84	153.9	96	184.8	CRAWFORD	IN	38	20	66	90	43
194	144.9	230	163.8	242	167.0	DAVIESS	IN	13	2	15	30	20
191	156.0	243	178.4	319	217.5	DEARBORN	IN	14	22	39	71	85
152	138.6	178	159.4	219	190.0	DECATUR	IN	15	19	37	68	51
203	143.7	234	164.3	280	194.1	DE KALB	IN	14	18	35	65	57
590	150.6	797	183.6	954	205.5	DELAWARE	IN	22	12	36	67	72
152	132.3	249	191.5	278	191.8	DUBOIS	IN	45	0	45	77	53
618	152.5	626	179.7	983	193.3	ELKHART	IN	18	8	27	51	56
182	178.2	208	189.9	283*	239.0	FAYETTE	IN	7	26	34	64	95
313	159.8	413	181.5	580*	232.0	FLOYD	IN	14	28	45	77	93
159	154.4	204	205.1	184	188.2	FOUNTAIN	IN	33	-8	22	43	48
116	146.6	107	130.5	163	192.6	FRANKLIN	IN	-11	48	31	59	55
156	159.4	179	181.2	190	188.0	FULTON	IN	14	4	18	36	47
269	164.5	331	195.1	368	212.2	GIBSON	IN	19	9	29	55	80
456	147.2	574	172.3	602	173.5	GRANT	IN	17	1	18	35	27
291	170.9	286	169.1	346	200.5	GREENE	IN	-1	19	17	34	66
221	139.2	333	179.2	456	187.7	HAMILTON	IN	29	5	35	65	47
169	151.5	202	158.9	254	167.9	HANCOCK	IN	5	6	11	22	21
116	117.1	138	135.2	199	177.9	HARRISON	IN	15	32	52	83	33
210	151.4	304	177.7	408	189.3	HENDRICKS	IN	17	7	25	48	50
307	143.3	365	157.2	467	190.6	HENRY	IN	10	21	33	62	52
392	161.8	581*	214.1	642	212.1	HOWARD	IN	32	-1	31	58	80
265	156.9	288	168.0	323	185.5	HUNTINGTON	IN	7	10	18	37	44
198	141.8	280	187.3	339	207.5	JACKSON	IN	32	11	46	78	75
138	162.2	131	146.7	177	181.4	JASPER	IN	-10	24	12	24	38
160	129.3	211	176.8	221	184.8	JAY	IN	37	5	43	74	43
164	128.8	180	140.7	207	157.8	JEFFERSON	IN	9	12	23	44	11
122	147.6	123	144.4	198	220.3	JENNINGS	IN	-2	53	49	81	87
241	167.3	333	184.8	503	220.8	JOHNSON	IN	10	19	32	60	87
422	187.3	464	205.6	446	198.9	KNOX	IN	10	-3	6	15	64
296	162.3	308	151.4	432	179.7	KOSCIUSKO	IN	-7	19	11	22	35
115	141.3	146	171.2	155	161.1	LAGRANGE	IN	21	-6	14	28	15
2828*	205.5	3398*	213.9	3649*	217.9	LAKE	IN	4	2	6	15	85
700	186.7	837	197.6	984*	219.2	LA PORTE	IN	6	11	17	34	86
278	163.7	332	184.5	398	201.0	LAWRENCE	IN	13	9	23	45	67
751	165.2	953	186.1	1208*	217.2	MADISON	IN	13	17	32	59	84
4010*	183.5	5187*	210.5	6036*	235.8	MARION	IN	15	12	29	54	95
230	148.5	250	151.4	346	190.5	MARSHALL	IN	2	26	28	54	51
74	147.6	80	158.4	93	175.0	MARTIN	IN	7	10	19	37	28
284	178.6	269	166.7	266	168.9	MIAMI	IN	-7	1	-5	05	22
259	136.8	395	182.8	533	202.3	MONROE	IN	34	11	48	80	68
228	141.5	293	176.1	335	197.8	MONTGOMERY	IN	25	12	40	71	62
215	169.7	264	172.1	355	192.8	MORGAN	IN	1	12	14	27	55
96	166.7	112	185.4	151	225.5	NEWTON	IN	11	22	35	65	90
226	158.3	272	189.7	309	208.8	NOBLE	IN	20	10	32	60	76
40	174.0	40	173.6	47	195.9	OHIO	IN	0	13	13	25	60
142	155.7	149	152.2	163	167.9	ORANGE	IN	-2	10	8	17	21
119	162.7	138	182.7	175	219.2	OWEN	IN	12	20	35	64	66
126	127.2	149	154.5	171	181.0	PARKE	IN	21	17	42	74	37
141	167.4	163	180.0	168	172.0	PERRY	IN	8	-4	3	10	25
135	167.2	154	189.4	174	219.0	PIKE	IN	13	16	31	58	86
308	154.7	508	205.2	653	202.4	PORTER	IN	33	-1	31	58	68
180	181.4	173	172.1	227	205.5	POSEY	IN	-5	19	13	27	72
98	146.7	115	167.6	145	204.6	PULASKI	IN	14	22	39	71	71
172	138.1	190	149.8	264	195.3	PUTNAM	IN	8	30	41	73	59
219	144.6	244	162.1	272	186.6	RANDOLPH	IN	12	15	29	55	45
163	144.5	140	121.2	215	176.8	RIPLEY	IN	-16	46	22	44	31
138	127.2	179	165.2	190	186.5	RUSH	IN	30	13	47	78	45
1690*	199.9	1934	198.7	2249*	215.8	ST JOSEPH	IN	-1	9	8	17	83
48	83.0	117	168.4	169	206.6	SCOTT	IN	103	23	149	98	74
221	143.2	293	183.7	306	180.8	SHELBY	IN	28	-2	26	50	37
146	157.6	151	160.5	161	160.2	SPENCER	IN	2	0	2	09	14
161	180.6	192	191.1	229	217.5	STARKE	IN	6	14	20	40	65
156	167.5	184	197.3	196	182.1	STEUBEN	IN	18	-8	9	18	39
236	163.3	241	178.5	249	197.6	SULLIVAN	IN	9	11	21	41	62
59	127.9	78	178.8	67	153.2	SWITZERLAND	IN	40	-14	20	39	09
489	162.5	582	171.0	816	220.4	TIPPECANOE	IN	5	29	36	66	87
119	145.0	118	142.1	156	192.6	TIPTON	IN	-2	36	33	61	55
57	163.9	49	148.2	64	198.4	UNION	IN	-10	34	21	41	63
1135	176.2	1384	192.1	1716*	222.6	VANDERBURGH	IN	9	16	26	50	88
209	179.8	217	200.9	228	217.5	VERMILLION	IN	12	8	21	41	85
952	179.1	1086*	204.1	1191*	232.1	VIGO	IN	14	14	30	56	93
258	176.6	225	148.5	298	186.0	WABASH	IN	-16	25	5	14	44
64	134.7	78	167.7	95	198.8	WARREN	IN	25	19	48	79	64
147	133.6	164	153.7	294	212.6	WARRICK	IN	15	38	59	87	80
115	122.6	140	144.7	199	181.9	WASHINGTON	IN	18	26	48	80	38
473	150.0	567	168.8	708	205.2	WAYNE	IN	13	22	37	67	72
146	133.3	157	144.8	185	162.9	WELLS	IN	9	13	22	44	16
167	167.1	183	169.3	216	175.7	WHITE	IN	1	4	5	13	30
154	156.2	153	145.4	224	193.9	WHITLEY	IN	-7	33	24	47	57
22012	162.2	24845	174.1	28440	194.2	IOWA	IA	7	12	20		
94	136.1	107	158.3	141	202.7	ADAIR	IA	16	28	49	80	69
64	136.4	68	145.0	84	193.2	ADAMS	IA	6	33	42	73	56

* SIGNIFICANTLY MORE DEATHS THAN EXPECTED

WHITE MALE — ALL CANCERS COMBINED — ICD 140-208

1950-1959 DEATHS	1950-1959 RATE	1960-1969 DEATHS	1960-1969 RATE	1970-1979 DEATHS	1970-1979 RATE	COUNTY AND STATE	PERCENT CHANGE 60S/50S	PERCENT CHANGE 70S/60S	PERCENT CHANGE 70S/50S	PERCENTILE %CHANGE 70S/50S	PERCENTILE RATES 1970S
149	159.6	160	168.5	187	184.0	ALLAMAKEE IA	6	9	15	30	42
217	180.0	191	161.3	196	185.1	APPANOOSE IA	-10	15	3	10	43
97	160.1	103	157.9	125	206.1	AUDUBON IA	-1	31	29	54	73
201	163.4	218	168.7	278	211.0	BENTON IA	3	25	29	55	79
706	160.5	872	175.8	1026	195.0	BLACK HAWK IA	10	11	22	42	58
227	138.8	258	156.5	279	175.0	BOONE IA	13	12	26	50	28
141	137.8	169	153.2	203	167.5	BREMER IA	11	9	22	42	21
153	130.6	145	127.6	236	213.1	BUCHANAN IA	-2	67	63	89	81
164	143.6	214	174.4	236	183.1	BUENA VISTA IA	21	5	27	52	40
136	143.3	137	133.5	196	179.8	BUTLER IA	-7	35	25	49	35
144	153.0	156	163.7	159	162.2	CALHOUN IA	7	-1	6	14	15
208	186.7	205	169.0	288	230.1	CARROLL IA	-9	36	23	45	93
193	177.2	179	159.2	246	208.5	CASS IA	-10	31	18	35	76
157	158.7	183	175.3	204	187.2	CEDAR IA	10	7	18	36	46
401	173.4	491	194.9	525	203.4	CERRO GORDO IA	12	4	17	34	70
155	152.9	159	156.9	190	195.8	CHEROKEE IA	3	25	28	53	60
124	151.5	122	140.7	161	187.9	CHICKASAW IA	-7	33	24	47	47
83	148.6	101	181.1	110	204.1	CLARKE IA	22	13	37	68	70
139	154.9	166	167.3	192	188.8	CLAY IA	8	13	22	43	49
195	150.4	211	158.3	235	173.4	CLAYTON IA	5	10	15	30	27
504	191.1	508	180.4	608	216.1	CLINTON IA	-6	20	13	26	83
186	184.5	205	196.1	189	176.7	CRAWFORD IA	6	-10	-4	06	31
222	160.7	252	175.5	294	198.1	DALLAS IA	9	13	23	45	63
73	121.3	84	150.9	86	150.5	DAVIS IA	24	0	24	47	07
127	164.6	107	147.2	120	171.4	DECATUR IA	-11	16	4	12	25
140	151.4	174	179.1	172	179.9	DELAWARE IA	18	0	19	37	35
380	175.4	411	175.1	478	198.5	DES MOINES IA	0	13	13	26	63
108	154.1	125	157.0	179	199.9	DICKINSON IA	2	27	30	57	65
613	189.2	743*	218.5	813*	224.2	DUBUQUE IA	16	3	19	37	90
112	158.9	122	161.6	165	216.1	EMMET IA	2	34	36	66	83
233	150.4	254	160.8	279	176.7	FAYETTE IA	7	10	17	34	31
171	147.6	202	175.4	222	196.4	FLOYD IA	19	12	33	62	61
138	169.7	109	131.3	160	183.7	FRANKLIN IA	-23	40	8	18	41
104	147.7	111	175.7	119	185.5	FREMONT IA	19	6	26	49	44
154	189.0	150	181.1	147	181.4	GREENE IA	-4	0	-4	06	38
86	124.4	126	166.3	150	183.5	GRUNDY IA	34	10	48	79	41
132	152.4	139	156.2	162	180.3	GUTHRIE IA	3	15	18	37	36
168	172.8	177	171.8	177	167.6	HAMILTON IA	-1	-2	-3	06	21
113	152.5	148	183.6	147	175.4	HANCOCK IA	20	-4	15	30	29
217	168.7	234	171.0	268	197.5	HARDIN IA	1	15	17	33	62
156	151.0	190	177.4	204	196.2	HARRISON IA	17	11	30	57	60
159	133.5	181	163.8	209	200.0	HENRY IA	23	22	50	81	65
124	164.5	129	165.3	121	152.3	HOWARD IA	0	-8	-7	04	08
119	174.4	138	188.9	121	159.6	HUMBOLDT IA	8	-16	-9	04	13
92	154.1	102	173.1	124	201.8	IDA IA	12	17	31	58	67
151	173.1	192	216.2	191	208.9	IOWA IA	25	-3	21	41	77
164	148.1	187	162.9	232	190.7	JACKSON IA	10	17	29	54	52
250	150.2	342	192.9	352	189.6	JASPER IA	28	-2	26	50	50
142	157.7	138	148.5	158	171.8	JEFFERSON IA	-6	16	9	19	25
279	154.7	325	157.9	415	189.4	JOHNSON IA	2	20	22	44	50
163	157.0	179	169.5	185	171.5	JONES IA	8	1	9	19	25
145	137.9	155	144.7	158	158.5	KEOKUK IA	5	10	15	30	12
179	150.9	215	169.6	217	168.7	KOSSUTH IA	12	-1	12	24	21
356	163.8	435	199.3	436	203.0	LEE IA	22	2	24	46	69
961	180.0	1099	183.6	1380	212.0	LINN IA	2	15	18	35	80
99	157.4	126	199.7	112	175.4	LOUISA IA	27	-12	11	23	29
112	143.0	136	179.1	129	180.7	LUCAS IA	25	1	26	50	37
103	155.1	138	186.3	153	196.9	LYON IA	20	6	27	51	61
105	133.3	128	157.5	129	155.6	MADISON IA	18	-1	17	33	10
207	150.0	215	150.2	264	187.6	MAHASKA IA	0	25	25	48	47
215	133.3	231	133.3	236	137.2	MARION IA	0	3	3	10	03
290	149.3	382	183.0	434	194.2	MARSHALL IA	23	6	30	57	57
88	127.0	77	109.6	152	227.1	MILLS IA	-14	107	79	94	91
109	133.7	145	175.7	141	171.2	MITCHELL IA	31	-3	28	53	24
114	133.3	140	159.7	140	158.9	MONONA IA	20	-1	19	38	13
123	166.8	121	168.8	114	174.3	MONROE IA	1	3	5	12	28
131	137.8	147	156.8	162	180.1	MONTGOMERY IA	14	15	31	58	36
304	165.7	344	183.6	361	191.4	MUSCATINE IA	11	4	15	31	53
183	184.3	193	177.5	197	172.7	O'BRIEN IA	-4	-3	-6	05	26
69	138.4	69	135.8	103	188.6	OSCEOLA IA	-2	39	36	67	48
187	132.3	208	152.8	250	186.8	PAGE IA	15	22	41	72	46
128	169.8	126	158.6	160	198.4	PALO ALTO IA	-7	25	17	33	63
185	163.1	230	182.8	242	180.2	PLYMOUTH IA	12	-1	10	22	36
125	167.6	165	207.5	148	182.7	POCAHONTAS IA	24	-12	9	19	40
1915*	189.3	2150	196.2	2530*	221.9	POLK IA	4	13	17	34	88
560	159.3	713	188.9	810	205.5	POTTAWATTAMIE IA	19	9	29	55	72
136	133.7	153	151.7	177	170.4	POWESHIEK IA	13	12	27	52	24
82	144.8	64	157.7	113	224.9	RINGGOLD IA	9	43	55	85	90
127	131.7	158	160.1	195	186.3	SAC IA	22	16	41	73	45
867	181.0	993	184.9	1279*	222.1	SCOTT IA	2	20	23	44	88
127	155.8	129	142.4	135	143.5	SHELBY IA	-9	1	-8	04	04
223	183.0	248	186.9	281	188.9	SIOUX IA	2	1	3	11	49
269	142.1	356	171.3	372	166.4	STORY IA	21	-3	17	34	20
209	168.1	204	158.2	271	213.0	TAMA IA	-6	35	27	51	81
110	138.9	118	158.9	133	196.2	TAYLOR IA	14	23	41	73	60
145	151.5	142	159.3	200	214.6	UNION IA	5	35	42	73	82
121	162.0	111	163.1	114	181.1	VAN BUREN IA	1	11	12	24	37
372	158.9	461	192.6	501	205.2	WAPELLO IA	21	7	29	55	72
159	165.5	179	168.5	203	165.7	WARREN IA	2	-2	0	08	19
168	152.3	191	170.5	209	178.3	WASHINGTON IA	12	5	17	34	33
109	139.4	128	173.3	103	145.7	WAYNE IA	24	-16	5	12	05
367	165.2	441	187.4	497	202.8	WEBSTER IA	13	8	23	44	69
122	171.4	127	163.1	137	168.3	WINNEBAGO IA	-5	3	-2	07	21
175	152.0	187	158.8	222	179.6	WINNESHIEK IA	4	13	18	36	35
940	183.3	972	182.7	1065	208.2	WOODBURY IA	0	14	14	27	76
93	161.2	88	143.0	112	176.8	WORTH IA	-11	24	10	20	31
170	166.4	188	178.9	219	198.7	WRIGHT IA	7	11	19	38	64
13932	144.4	16995	162.2	20658	187.1	KANSAS KS	12	15	30		
177	156.0	174	168.2	209	205.4	ALLEN KS	8	22	32	59	72

* SIGNIFICANTLY MORE DEATHS THAN EXPECTED

WHITE MALE

ALL CANCERS COMBINED

ICD 140-208

1950-1959 DEATHS	1950-1959 RATE	1960-1969 DEATHS	1960-1969 RATE	1970-1979 DEATHS	1970-1979 RATE	COUNTY AND STATE	PERCENT CHANGE 60S/50S	PERCENT CHANGE 70S/60S	PERCENT CHANGE 70S/50S	PERCENTILE %CHANGE 70S/50S	PERCENTILE RATES 1970S
89	134.8	106	163.6	97	138.9	ANDERSON KS	21	-15	3	11	03
141	129.5	150	144.8	205	215.8	ATCHISON KS	12	49	67	91	83
50	109.5	78	164.8	109	236.0	BARBER KS	51	43	116	97	95
202	184.7	246	175.4	350	225.4	BARTON KS	-5	29	22	43	90
170	149.8	175	155.3	226	210.5	BOURBON KS	4	36	40	72	78
127	138.9	121	126.8	147	162.6	BROWN KS	-9	28	17	33	16
247	147.1	306	163.9	354	170.1	BUTLER KS	11	4	16	31	23
45	156.0	37	131.1	44	158.2	CHASE KS	-16	21	1	09	12
61	122.0	69	152.4	87	212.5	CHAUTAUQUA KS	25	39	74	93	80
219	170.7	249	193.7	331*	253.6	CHEROKEE KS	13	31	49	80	98
43	161.8	35	117.5	35	125.0	CHEYENNE KS	-27	6	-23	01	01
28	137.5	45	223.2	36	164.6	CLARK KS	62	-26	20	39	18
96	132.7	103	137.0	127	169.8	CLAY KS	3	24	28	53	23
141	142.0	162	167.9	176	194.8	CLOUD KS	18	16	37	68	58
95	130.6	104	147.4	94	146.4	COFFEY KS	13	-1	12	24	05
34	174.0	36	184.6	38	193.6	COMANCHE KS	6	5	11	23	56
311	147.9	318	153.4	369	179.0	COWLEY KS	4	17	21	41	34
387	157.7	475	193.3	500	203.8	CRAWFORD KS	23	5	29	55	70
50	131.5	52	133.6	67	173.5	DECATUR KS	2	30	32	60	27
203	155.1	227	168.6	257	185.4	DICKINSON KS	9	10	20	38	43
106	168.6	93	160.9	106	177.6	DONIPHAN KS	-5	10	5	14	32
192	127.9	227	134.8	321	174.2	DOUGLAS KS	5	29	36	67	28
60	193.0	44	136.9	56	176.4	EDWARDS KS	-29	29	-9	04	30
54	124.3	60	143.6	58	158.8	ELK KS	16	11	28	53	12
98	154.1	132	163.8	180	189.0	ELLIS KS	6	15	23	44	49
69	136.3	95	198.7	73	154.7	ELLSWORTH KS	46	-22	14	27	10
82	137.0	99	150.2	150	192.2	FINNEY KS	10	28	40	72	54
150	164.2	154	148.6	186	168.9	FORD KS	-9	14	3	10	22
183	147.9	177	139.1	239	187.9	FRANKLIN KS	-6	35	27	52	47
96	139.8	127	174.0	153	203.0	GEARY KS	24	17	45	77	69
26	126.9	37	166.1	26	109.0	GOVE KS	31	-34	-14	02	00
17	68.2	54	206.0	48	176.4	GRAHAM KS	202	-14	158	98	30
19	169.2	27	153.0	28	124.2	GRANT KS	-10	-19	-27	01	01
22	98.0	26	102.6	40	157.2	GRAY KS	5	53	60	88	11
14	238.2	17	221.0	15	164.0	GREELEY KS	-7	-26	-31	00	17
130	155.1	157	195.3	141	178.5	GREENWOOD KS	24	-9	13	26	33
18	125.4	20	131.3	33	210.3	HAMILTON KS	5	60	68	91	78
79	130.5	91	155.0	100	170.9	HARPER KS	19	10	31	58	24
173	150.7	196	143.3	245	160.3	HARVEY KS	-5	12	6	15	14
12	115.7	21	170.5	30	204.8	HASKELL KS	47	20	77	94	71
14	92.0	20	118.2	20	120.4	HODGEMAN KS	28	2	31	58	01
78	108.8	116	164.6	137	191.3	JACKSON KS	51	16	76	93	53
121	158.9	120	159.7	132	162.9	JEFFERSON KS	1	2	3	10	16
92	149.6	82	150.1	71	147.8	JEWELL KS	0	-2	-1	07	06
431	139.8	872	160.3	1475	185.7	JOHNSON KS	15	16	33	61	44
20	157.9	22	158.9	23	150.1	KEARNY KS	1	-6	-5	05	07
70	130.9	86	147.5	117	196.2	KINGMAN KS	13	33	50	81	60
33	131.3	48	174.1	46	151.6	KIOWA KS	33	-13	15	31	08
219	134.5	292	186.8	329	219.3	LABETTE KS	39	17	63	89	86
21	186.9	19	129.8	28	167.8	LANE KS	-31	29	-10	03	21
282	126.3	340	151.2	403	185.7	LEAVENWORTH KS	20	23	47	79	44
63	145.7	65	159.7	78	194.2	LINCOLN KS	9	22	32	61	57
91	122.0	93	145.1	135	216.3	LINN KS	19	49	77	94	84
22	114.7	31	153.7	35	159.7	LOGAN KS	34	4	39	70	13
199	133.7	220	144.9	260	169.4	LYON KS	8	17	27	51	22
152	118.7	192	134.5	256	165.1	MC PHERSON KS	13	23	39	70	18
130	142.7	150	149.5	161	152.5	MARION KS	5	2	7	16	09
170	149.1	174	162.0	184	185.4	MARSHALL KS	9	14	24	47	44
31	127.6	54	198.5	51	170.7	MEADE KS	56	-14	34	63	24
132	109.0	175	143.7	189	157.9	MIAMI KS	32	10	45	76	12
83	141.1	94	152.1	130	227.9	MITCHELL KS	8	50	61	88	91
364	146.8	412	168.1	502	204.1	MONTGOMERY KS	15	21	39	70	70
61	118.6	83	153.8	88	157.0	MORRIS KS	30	2	32	60	11
10	95.1	23	173.8	26	164.3	MORTON KS	83	-5	73	92	17
98	123.1	118	137.7	130	165.2	NEMAHA KS	12	20	34	64	18
175	146.0	192	153.7	234	194.8	NEOSHO KS	5	27	33	63	58
53	170.6	48	143.3	53	148.5	NESS KS	-16	4	-13	03	06
73	130.5	60	110.9	93	176.3	NORTON KS	-15	59	35	65	30
116	127.7	119	130.3	177	184.7	OSAGE KS	2	42	45	76	43
73	135.1	69	131.7	82	145.5	OSBORNE KS	-2	10	8	17	05
91	179.1	75	145.5	90	178.7	OTTAWA KS	-19	23	0	08	34
52	82.1	81	134.5	89	170.2	PAWNEE KS	64	27	107	97	23
69	120.9	75	133.0	110	188.0	PHILLIPS KS	10	41	56	85	48
123	152.0	116	142.8	138	168.7	POTTAWATOMIE KS	-6	18	11	22	22
91	145.1	108	170.1	123	188.8	PRATT KS	17	11	30	57	49
34	132.9	48	155.3	55	185.9	RAWLINS KS	17	20	40	71	44
345	127.6	481	163.3	542	171.3	RENO KS	28	5	34	64	25
111	149.8	101	133.8	132	190.9	REPUBLIC KS	-11	43	27	52	52
113	148.4	130	162.1	153	187.9	RICE KS	9	16	27	51	47
147	126.7	206	163.3	222	169.3	RILEY KS	29	4	34	63	22
60	132.6	90	182.5	79	157.4	ROOKS KS	38	-14	19	37	11
57	152.1	58	154.6	76	192.6	RUSH KS	2	25	27	51	55
101	178.4	108	179.1	131	204.2	RUSSELL KS	0	14	14	29	71
220	129.9	351	187.0	402	194.2	SALINE KS	44	4	49	81	57
32	168.6	33	142.2	39	144.0	SCOTT KS	-16	1	-15	02	05
1482	161.0	2081	184.9	2611	207.1	SEDGWICK KS	15	12	29	54	75
66	168.2	76	141.0	108	187.3	SEWARD KS	-16	33	11	23	46
735	144.5	977	169.2	1168	185.8	SHAWNEE KS	17	10	29	54	44
19	102.7	32	148.0	32	140.6	SHERIDAN KS	44	-5	37	67	04
42	129.2	54	156.8	53	132.4	SHERMAN KS	21	-16	2	10	02
89	149.0	71	114.0	79	130.4	SMITH KS	-23	14	-12	03	02
68	149.4	68	146.8	55	127.8	STAFFORD KS	-2	-13	-14	02	02
9	107.5	12	144.8	17	184.6	STANTON KS	35	28	72	92	42
9	89.6	21	138.5	26	117.3	STEVENS KS	55	-15	31	58	01
222	156.1	228	156.7	299	197.3	SUMNER KS	0	26	26	50	62
48	148.5	50	139.5	93	223.8	THOMAS KS	-6	60	51	82	89
27	103.9	48	176.5	60	202.4	TREGO KS	70	15	95	96	68
67	135.2	79	147.9	93	184.7	WABAUNSEE KS	9	25	37	67	42
6	52.3	16	141.3	23	175.5	WALLACE KS	170	24	236	99	29
106	141.0	124	173.4	111	142.9	WASHINGTON KS	23	-18	1	09	04
12	108.4	14	105.2	23	158.2	WICHITA KS	-3	50	46	78	12

* SIGNIFICANTLY MORE DEATHS THAN EXPECTED

WHITE MALE — ALL CANCERS COMBINED — ICD 140-208

1950-1959 DEATHS	1950-1959 RATE	1960-1969 DEATHS	1960-1969 RATE	1970-1979 DEATHS	1970-1979 RATE	COUNTY AND STATE	PERCENT CHANGE 60S/50S	PERCENT CHANGE 70S/60S	PERCENT CHANGE 70S/50S	PERCENTILE %CHANGE 70S/50S	PERCENTILE RATES 1970S
136	147.6	134	157.3	157	183.4	WILSON KS	7	17	24	47	41
48	102.4	70	156.7	90	215.6	WOODSON KS	53	38	110	97	83
1102	183.4	1268	198.8	1448*	232.7	WYANDOTTE KS	8	17	27	51	93
16572	141.1	22212	168.6	29665	202.4	KENTUCKY KY	20	20	43		
73	100.1	118	154.1	147	178.1	ADAIR KY	54	16	78	94	33
82	110.1	89	118.6	119	149.3	ALLEN KY	8	26	36	66	07
61	140.8	81	184.2	105	194.9	ANDERSON KY	31	6	38	69	58
69	144.5	76	146.3	127	232.9	BALLARD KY	1	59	61	88	94
170	133.8	188	135.8	279	172.6	BARREN KY	1	27	29	55	26
70	154.1	61	125.7	93	177.1	BATH KY	-18	41	15	30	31
203	141.5	256	176.0	309	207.6	BELL KY	24	18	47	79	75
102	138.7	184	184.1	316*	246.4	BOONE KY	33	34	78	94	97
104	149.0	137	178.7	172	207.5	BOURBON KY	20	16	39	70	75
309	161.8	458	204.4	586*	234.1	BOYD KY	26	15	45	76	94
81	94.6	118	127.7	195	195.6	BOYLE KY	35	53	107	97	59
61	133.7	69	157.2	100	219.0	BRACKEN KY	18	39	64	90	86
62	97.0	94	138.5	116	162.0	BREATHITT KY	43	17	67	91	15
95	123.5	117	142.7	142	155.5	BRECKINRIDGE KY	15	9	26	49	10
83	154.0	125	180.1	188	187.7	BULLITT KY	17	4	22	43	47
54	92.2	91	158.9	115	184.4	BUTLER KY	72	16	100	96	42
89	131.8	114	149.1	153	198.7	CALDWELL KY	13	33	51	82	64
125	119.6	166	132.3	243	162.8	CALLOWAY KY	11	23	36	66	16
737*	207.9	880*	223.9	1012*	260.6	CAMPBELL KY	8	16	25	49	98
34	101.4	68	185.9	85	241.2	CARLISLE KY	83	30	138	98	96
48	112.7	65	145.1	72	156.5	CARROLL KY	29	8	39	70	11
118	130.1	148	155.4	213	203.0	CARTER KY	19	31	56	85	69
71	112.6	91	124.9	133	167.1	CASEY KY	11	34	48	80	20
170	109.7	216	129.5	261	149.3	CHRISTIAN KY	18	15	36	66	06
119	151.8	144	156.0	202	194.8	CLARK KY	3	25	28	54	58
59	80.5	88	113.9	162	181.2	CLAY KY	41	59	125	98	37
49	124.3	52	119.4	89	174.4	CLINTON KY	-4	46	40	72	28
72	134.3	69	128.1	123	227.4	CRITTENDEN KY	-5	77	69	91	91
43	107.4	49	125.3	66	152.2	CUMBERLAND KY	17	22	42	73	08
347	145.2	470	164.2	694	211.0	DAVIESS KY	13	28	45	77	79
51	131.8	53	121.4	87	178.6	EDMONSON KY	-8	47	36	66	34
24	85.7	42	138.2	46	536.4	ELLIOTT KY	61	288	526	99	99
73	121.1	99	147.9	118	167.1	ESTILL KY	22	13	38	69	20
531	140.3	827	175.7	1187	209.3	FAYETTE KY	25	19	49	81	77
74	113.1	92	140.8	106	156.6	FLEMING KY	25	11	38	69	11
158	104.9	240	148.8	366	193.3	FLOYD KY	42	30	84	95	56
166	150.1	198	156.0	337	222.9	FRANKLIN KY	4	43	48	80	89
98	174.3	115	212.7	101	201.6	FULTON KY	22	-5	16	31	67
38	169.6	33	158.8	50	220.4	GALLATIN KY	-6	39	30	57	87
46	103.6	57	120.2	87	162.0	GARRARD KY	16	35	56	85	15
75	130.4	111	182.8	110	172.6	GRANT KY	40	-6	32	60	26
214	132.9	256	145.3	348	181.4	GRAVES KY	9	25	36	67	38
96	110.0	128	133.4	169	152.3	GRAYSON KY	21	14	38	69	08
64	130.8	68	126.5	99	159.4	GREEN KY	-3	26	22	43	13
160	158.6	203	166.4	327	221.2	GREENUP KY	5	33	39	71	87
46	145.1	38	111.7	70	203.6	HANCOCK KY	-23	82	40	72	70
188	137.2	259	155.5	398	192.1	HARDIN KY	13	24	40	71	54
233	135.2	332	189.2	427*	244.9	HARLAN KY	40	29	81	94	97
96	124.9	143	177.8	158	183.4	HARRISON KY	42	3	47	79	41
76	105.0	90	122.4	146	188.3	HART KY	17	54	79	94	48
193	149.7	255	180.1	336	210.6	HENDERSON KY	20	17	41	72	78
84	134.7	77	118.1	100	149.6	HENRY KY	-12	27	11	22	07
52	138.7	68	169.9	79	209.0	HICKMAN KY	22	23	51	82	77
209	120.3	328	173.5	448	221.2	HOPKINS KY	44	28	84	95	87
49	98.2	63	119.4	109	182.5	JACKSON KY	22	53	86	95	39
3208	177.1	4526*	208.9	5841*	241.0	JEFFERSON KY	18	15	36	66	96
62	121.6	92	153.4	148	192.3	JESSAMINE KY	26	25	58	86	54
111	124.2	132	140.0	181	170.0	JOHNSON KY	13	21	37	67	23
919*	201.2	1200*	232.4	1410*	258.7	KENTON KY	15	11	29	54	98
64	100.5	81	122.2	137	192.4	KNOTT KY	22	57	92	96	54
127	120.6	158	136.9	247	193.7	KNOX KY	14	41	61	88	56
55	119.6	57	104.1	115	182.2	LARUE KY	-13	75	52	83	39
104	94.4	198	154.5	258	161.9	LAUREL KY	64	5	72	92	15
86	127.8	93	142.2	147	202.9	LAWRENCE KY	11	43	59	87	69
36	102.6	61	157.0	88	203.1	LEE KY	53	29	98	96	69
31	83.7	58	141.4	86	172.3	LESLIE KY	69	22	106	97	25
125	117.9	190	169.0	260	213.8	LETCHER KY	43	27	81	94	82
70	115.8	100	162.8	106	160.6	LEWIS KY	41	-1	39	69	14
98	122.1	132	153.6	149	155.8	LINCOLN KY	26	1	28	52	10
45	117.0	88	207.2	82	168.2	LIVINGSTON KY	77	-19	44	75	21
142	144.0	170	157.7	210	175.5	LOGAN KY	9	11	22	43	29
43	139.6	34	115.0	58	158.7	LYON KY	-18	38	14	27	12
302	137.8	466	185.1	567	204.7	MC CRACKEN KY	34	11	49	80	71
57	111.3	98	173.3	126	194.8	MC CREARY KY	56	12	75	93	58
61	117.5	74	136.4	98	176.9	MC LEAN KY	16	30	51	82	31
162	132.2	205	147.4	298	179.7	MADISON KY	11	22	36	66	35
43	91.4	54	108.9	67	128.5	MAGOFFIN KY	19	18	41	72	02
79	147.1	86	140.7	140	202.6	MARION KY	-4	44	38	68	68
102	139.6	143	148.8	242	181.5	MARSHALL KY	7	22	30	57	38
38	118.4	58	153.3	66	147.1	MARTIN KY	29	-4	24	47	06
120	145.0	151	179.3	155	179.1	MASON KY	24	0	23	46	34
44	105.4	83	180.4	101	181.5	MEADE KY	71	1	72	92	38
37	156.6	40	185.5	44	192.6	MENIFEE KY	18	4	23	45	55
125	176.7	131	174.4	168	196.8	MERCER KY	-1	13	11	23	61
49	110.5	61	127.2	78	147.6	METCALFE KY	15	16	34	63	06
77	128.9	94	147.9	113	164.8	MONROE KY	15	11	28	53	18
76	137.3	108	179.5	150	207.0	MONTGOMERY KY	31	15	51	82	74
60	108.3	82	143.9	91	151.3	MORGAN KY	33	5	40	71	08
208	147.3	227	153.2	312	199.8	MUHLENBERG KY	4	30	36	66	65
88	113.5	112	131.6	209	211.6	NELSON KY	16	61	86	95	79
49	122.4	69	175.3	88	216.2	NICHOLAS KY	43	23	77	93	84
152	132.4	199	167.6	213	177.1	OHIO KY	27	6	34	63	31
54	109.5	75	138.3	120	177.7	OLDHAM KY	26	28	62	89	32
52	101.4	63	126.9	84	170.4	OWEN KY	25	34	68	91	24
32	111.4	26	80.8	38	120.5	OWSLEY KY	-27	49	8	17	01

* SIGNIFICANTLY MORE DEATHS THAN EXPECTED

WHITE MALE — ALL CANCERS COMBINED — ICD 140-208

1950-1959 DEATHS	1950-1959 RATE	1960-1969 DEATHS	1960-1969 RATE	1970-1979 DEATHS	1970-1979 RATE	COUNTY AND STATE	PERCENT CHANGE 60S/50S	PERCENT CHANGE 70S/60S	PERCENT CHANGE 70S/50S	PERCENTILE %CHANGE 70S/50S	PERCENTILE RATES 1970S
65	119.1	89	157.3	118	208.6	PENDLETON KY	32	33	75	93	76
138	112.5	223	179.8	290	225.4	PERRY KY	60	25	100	96	90
267	111.7	401	156.4	571	195.5	PIKE KY	40	25	75	93	59
34	120.3	57	176.1	80	192.5	POWELL KY	46	9	60	87	54
195	123.5	256	142.3	385	182.7	PULASKI KY	15	28	48	80	40
17	103.6	19	116.4	24	147.1	ROBERTSON KY	12	26	42	74	06
58	100.7	87	138.2	151	217.0	ROCKCASTLE KY	37	57	116	97	84
58	115.6	84	147.6	126	188.4	ROWAN KY	28	28	63	89	48
55	114.4	82	152.5	143	201.7	RUSSELL KY	33	32	76	93	67
95	138.1	131	181.9	139	169.4	SCOTT KY	32	-7	23	44	23
121	140.2	113	123.8	178	174.6	SHELBY KY	-12	41	25	47	28
69	131.5	85	145.8	125	195.0	SIMPSON KY	11	34	48	80	58
33	130.2	31	131.9	48	175.4	SPENCER KY	1	33	35	64	29
69	110.6	118	163.4	142	161.7	TAYLOR KY	48	-1	46	78	15
75	143.2	96	175.6	90	161.1	TODD KY	23	-8	13	25	14
52	125.5	72	165.7	90	178.0	TRIGG KY	32	7	42	73	33
28	100.2	50	179.6	43	146.8	TRIMBLE KY	79	-18	47	78	05
117	175.4	134	195.9	137	199.5	UNION KY	12	2	14	28	65
267	148.5	325	159.9	438	187.1	WARREN KY	8	17	26	50	46
59	126.1	71	142.1	75	144.0	WASHINGTON KY	13	1	14	29	05
60	92.3	80	111.3	132	158.2	WAYNE KY	21	42	71	92	12
126	148.5	145	178.7	138	167.3	WEBSTER KY	20	-6	13	25	20
171	137.3	212	160.1	320	225.2	WHITLEY KY	17	41	64	90	90
27	89.8	36	112.7	64	179.8	WOLFE KY	26	59	100	96	35
69	158.2	83	163.0	101	165.4	WOODFORD KY	3	1	5	12	18
13753*	192.6	18888*	209.0	24310*	222.6	LOUISIANA LA	9	6	16		
287*	207.9	370*	225.5	462*	243.1	ACADIA LA	8	8	17	33	96
88	150.1	123	182.2	179	221.8	ALLEN LA	21	22	48	80	88
100	165.3	165	206.6	217	198.7	ASCENSION LA	25	-4	20	40	64
89	221.1	86	187.8	120	222.6	ASSUMPTION LA	-15	19	1	08	89
219	190.8	261	191.0	348	221.0	AVOYELLES LA	0	16	16	31	87
113	170.8	144	179.8	179	182.5	BEAUREGARD LA	5	2	7	16	39
71	128.2	114	198.1	142	234.0	BIENVILLE LA	55	18	83	95	94
126	177.7	218	202.7	372*	234.5	BOSSIER LA	14	16	32	60	94
822	178.0	1247*	211.0	1583*	231.7	CADDO LA	19	10	30	57	93
453	190.2	705	200.0	989*	217.8	CALCASIEU LA	5	9	15	29	85
55	155.7	84	205.3	79	168.9	CALDWELL LA	32	-18	8	18	22
28	122.4	47	181.1	53	153.8	CAMERON LA	48	-15	26	49	09
51	168.1	74	197.9	95	216.7	CATAHOULA LA	18	10	29	55	84
70	116.4	131	220.3	109	171.0	CLAIBORNE LA	89	-22	47	79	24
44	158.2	97*	248.2	118	219.2	CONCORDIA LA	57	-12	39	69	86
102	178.7	131	204.4	124	169.5	DE SOTO LA	14	-17	-5	05	23
544	166.8	963	196.9	1336	195.8	EAST BATON ROUGE LA	18	-1	17	34	60
42	202.9	49	227.3	36	190.1	EAST CARROLL LA	12	-16	-6	05	51
38	72.5	75	136.6	86	163.4	EAST FELICIANA LA	88	20	125	98	17
167	184.4	232	213.8	294*	238.6	EVANGELINE LA	16	12	29	56	95
99	143.4	160	198.0	217*	235.3	FRANKLIN LA	38	19	64	90	94
73	129.7	107	175.9	142	194.8	GRANT LA	36	11	50	82	58
212*	204.9	264	199.3	390*	238.5	IBERIA LA	-3	20	16	32	95
111	196.8	118	171.1	148	198.4	IBERVILLE LA	-13	16	1	08	63
82	179.0	116	197.1	153	230.8	JACKSON LA	10	17	29	55	93
678*	221.5	1316*	213.5	2418*	233.4	JEFFERSON LA	-4	9	5	14	94
142	179.0	197	218.7	259*	239.9	JEFFERSON DAVIS LA	22	10	34	64	96
282	189.1	455*	208.9	673*	222.6	LAFAYETTE LA	10	7	18	35	89
238	202.1	338	212.0	445	210.4	LAFOURCHE LA	5	-1	4	12	78
85	166.3	118	194.9	153	207.2	LA SALLE LA	17	6	25	48	75
98	153.9	119	155.4	189	207.6	LINCOLN LA	1	34	35	65	75
127	177.5	189	183.5	275	192.5	LIVINGSTON LA	3	5	8	18	55
43	198.5	43	168.1	70	234.2	MADISON LA	-15	39	18	36	94
85	137.9	142	183.9	168	171.3	MOREHOUSE LA	33	-7	24	47	25
147	165.1	172	168.1	223	186.7	NATCHITOCHES LA	2	11	13	26	45
3829*	240.8	4247*	247.4	4021*	264.3	ORLEANS LA	3	7	10	20	98
382	182.8	588*	208.6	722	204.9	OUACHITA LA	14	-2	12	24	71
69	228.0	102	223.3	139	237.9	PLAQUEMINES LA	-2	7	4	12	95
80	198.4	102	201.0	136	213.5	POINTE COUPEE LA	1	6	8	17	81
442	173.7	547	172.8	761	201.3	RAPIDES LA	-1	16	16	32	67
47	164.1	70	225.3	57	162.3	RED RIVER LA	37	-28	-1	07	15
83	150.0	134	212.5	165	212.5	RICHLAND LA	42	0	42	73	80
101	133.1	137	162.9	202	185.7	SABINE LA	22	14	40	71	44
84	192.4	213*	227.9	463*	279.6	ST BERNARD LA	18	23	45	77	99
56	166.9	103	227.0	156	204.9	ST CHARLES LA	36	-10	23	44	71
19	95.4	30	161.7	49	269.9	ST HELENA LA	69	67	183	99	98
61	208.3	60	170.5	109*	265.9	ST JAMES LA	-18	56	28	53	98
60	219.6	81	225.9	101	201.6	ST JOHN THE BAPT LA	3	-11	-8	04	67
317*	225.8	395*	224.5	478	221.8	ST LANDRY LA	-1	-1	-2	07	88
130*	227.7	182*	248.8	228*	239.1	ST MARTIN LA	9	-4	5	13	96
182*	210.7	237	217.7	321*	238.7	ST MARY LA	3	10	13	27	95
163	153.0	307	196.8	527	195.5	ST TAMMANY LA	29	-1	28	53	59
263	171.1	366	207.2	430	202.4	TANGIPAHOA LA	21	-2	18	37	68
33	179.4	38	222.0	42	195.9	TENSAS LA	24	-12	9	19	60
247*	229.7	327*	221.3	487*	247.8	TERREBONNE LA	-4	12	8	17	97
106	173.4	112	157.7	150	178.7	UNION LA	-9	13	3	11	34
268*	216.3	342*	216.3	474*	258.5	VERMILION LA	0	20	19	38	98
146	180.3	171	181.7	218	188.1	VERNON LA	1	4	4	12	48
202	201.7	241	190.8	296	207.4	WASHINGTON LA	-5	9	3	10	75
141	140.8	260	203.7	329	209.7	WEBSTER LA	45	3	49	81	77
43	200.3	58	201.0	88	252.0	WEST BATON ROUGE LA	0	25	26	49	97
91	175.9	112	201.2	128	207.1	WEST CARROLL LA	14	3	18	35	75
19	145.2	27	176.0	28	192.5	WEST FELICIANA LA	21	9	33	61	54
78	129.3	129	196.1	161	229.3	WINN LA	52	17	77	94	92
8264*	181.9	9509*	200.6	11063*	216.4	MAINE ME	10	8	19		
753*	192.4	923*	220.6	987*	218.9	ANDROSCOGGIN ME	15	-1	14	28	86
602	168.7	660	177.0	806	210.3	AROOSTOOK ME	5	19	25	48	78
1545*	188.3	1850*	212.6	2187*	229.8	CUMBERLAND ME	13	8	22	43	92
171	159.0	183	171.9	250	209.9	FRANKLIN ME	8	22	32	60	78
334	177.1	420*	212.0	453	208.6	HANCOCK ME	20	-2	18	35	76
807	189.4	835	184.3	1004	202.0	KENNEBEC ME	-3	10	7	15	68

* SIGNIFICANTLY MORE DEATHS THAN EXPECTED

WHITE MALE — ALL CANCERS COMBINED — ICD 140-208

1950-1959 DEATHS	1950-1959 RATE	1960-1969 DEATHS	1960-1969 RATE	1970-1979 DEATHS	1970-1979 RATE	COUNTY AND STATE	PERCENT CHANGE 60S/50S	PERCENT CHANGE 70S/60S	PERCENT CHANGE 70S/50S	PERCENTILE %CHANGE 70S/50S	PERCENTILE RATES 1970S
347*	201.4	373*	210.3	411	221.7	KNOX ME	4	5	10	21	88
198	164.4	262	212.8	352*	247.1	LINCOLN ME	29	16	50	82	97
392	166.1	484	206.5	532	210.5	OXFORD ME	24	2	27	51	78
868	171.3	1017	191.3	1171	213.2	PENOBSCOT ME	12	11	24	47	81
189	173.9	193	188.4	214	221.1	PISCATAQUIS ME	8	17	27	52	87
206	184.1	240	203.8	283*	233.1	SAGADAHOC ME	11	14	27	51	94
374	180.5	380	183.4	434	199.5	SOMERSET ME	2	9	10	22	65
210	166.1	258	208.6	248	195.1	WALDO ME	26	-6	17	34	59
342	173.5	364	193.4	395	206.7	WASHINGTON ME	11	7	19	38	74
926*	202.3	1067*	208.4	1336*	222.6	YORK ME	3	7	10	21	89
16329*	193.9	22795*	211.7	29796*	230.6	MARYLAND MD	9	9	19		
596	159.7	782	190.1	937	213.6	ALLEGANY MD	19	12	34	63	81
716	188.5	1351*	220.6	2195*	241.1	ANNE ARUNDEL MD	17	9	28	53	96
2151*	196.7	3459	196.3	5744*	230.5	BALTIMORE MD	0	17	17	34	93
71	186.1	112	199.6	204*	237.2	CALVERT MD	7	19	27	52	95
140	165.7	168	186.7	190	190.8	CAROLINE MD	13	2	15	30	52
335	142.0	455	162.1	647	193.1	CARROLL MD	14	19	36	66	55
256	170.3	314	179.6	414	195.2	CECIL MD	5	9	15	29	59
91	140.1	176	208.4	298*	243.0	CHARLES MD	49	17	73	93	96
197	161.9	262	205.6	315*	239.9	DORCHESTER MD	27	17	48	80	96
410	152.3	560	184.0	800*	218.9	FREDERICK MD	21	19	44	75	86
123	128.7	152	142.3	206	166.6	GARRETT MD	11	17	29	56	20
300	165.1	457	176.5	782	215.7	HARFORD MD	7	22	31	58	83
149	164.7	247	177.3	433	204.0	HOWARD MD	8	15	24	46	70
110	183.4	123	185.0	156	205.2	KENT MD	1	11	12	24	72
1194	177.9	2286	187.9	3507	195.4	MONTGOMERY MD	6	4	10	20	59
1007	173.5	2321*	219.1	3321*	247.3	PRINCE GEORGE'S MD	26	13	43	74	97
83	132.1	120	166.4	192	203.2	QUEEN ANNE'S MD	26	22	54	84	69
110	164.6	166	196.9	300*	237.7	ST MARY'S MD	20	21	44	76	95
137	177.6	168	216.1	195*	249.1	SOMERSET MD	22	15	40	72	97
163	197.7	226*	226.9	270	218.6	TALBOT MD	15	-4	11	22	85
547	153.6	763	185.1	984	200.9	WASHINGTON MD	21	8	31	58	66
218	131.5	371	195.1	500*	226.3	WICOMICO MD	48	16	72	92	91
170	191.5	177	185.4	244	211.2	WORCESTER MD	-3	14	10	21	79
7055*	236.5	7579*	262.5	6962*	279.5	BALTIMORE CITY MD	11	6	18	36	98
45446*	203.8	50149*	206.9	56966*	222.0	MASSACHUSETTS MA	2	7	9		
528	181.7	870	188.2	1698	209.5	BARNSTABLE MA	4	11	15	30	77
1264*	197.1	1442*	206.1	1588	210.6	BERKSHIRE MA	5	2	7	16	78
3909*	217.4	4106*	207.6	4931*	227.6	BRISTOL MA	-5	10	5	13	91
55	175.4	79	230.0	100	233.3	DUKES MA	31	1	33	62	94
5411*	202.2	6092*	210.2	6805*	222.9	ESSEX MA	4	6	10	21	89
468	166.3	588	194.7	697	213.0	FRANKLIN MA	17	9	28	53	81
3521*	201.6	4019*	207.5	4625*	225.2	HAMPDEN MA	3	9	12	24	90
682	162.5	800	168.9	1060	202.0	HAMPSHIRE MA	4	20	24	47	68
9916*	202.9	11230*	207.7	12676*	221.5	MIDDLESEX MA	2	7	9	19	88
31	189.9	36	206.6	84*	389.2	NANTUCKET MA	9	88	105	97	99
3615*	189.8	4585*	198.3	5948*	223.1	NORFOLK MA	4	13	18	34	89
1920	179.8	2508*	202.0	3121	211.4	PLYMOUTH MA	12	5	18	35	79
9188*	242.1	8319*	237.2	7286*	250.5	SUFFOLK MA	-2	6	3	11	97
4938*	181.4	5475	186.9	6347	206.2	WORCESTER MA	3	10	14	27	73
50929*	187.3	62313*	201.0	70243*	209.1	MICHIGAN MI	7	4	12		
61	160.1	73	164.4	139	208.7	ALCONA MI	3	27	30	57	76
107	205.2	105	200.5	100	195.2	ALGER MI	-2	-3	-5	05	59
414	162.6	469	168.6	601	187.5	ALLEGAN MI	4	11	15	30	46
214	190.5	251	205.1	319	227.3	ALPENA MI	8	11	19	38	91
114	169.2	130	171.1	197	216.5	ANTRIM MI	1	27	28	53	84
106	197.7	112	187.4	170	218.5	ARENAC MI	-5	17	11	22	85
78	198.0	87	186.3	96	183.6	BARAGA MI	-6	-1	-7	04	41
207	138.9	287	172.1	392	202.4	BARRY MI	24	18	46	78	68
790*	203.0	910*	214.4	1034*	219.2	BAY MI	6	2	8	17	86
69	154.2	100	209.8	129	224.5	BENZIE MI	36	7	46	77	90
1039*	192.9	1241	200.0	1349	202.5	BERRIEN MI	4	1	5	13	68
260	158.6	301	176.7	398	217.1	BRANCH MI	11	23	37	67	84
869	166.0	1064	190.4	1155	195.9	CALHOUN MI	15	3	18	36	60
287	192.1	296	176.0	404	210.2	CASS MI	-8	19	9	20	78
114	146.3	144	179.6	205	221.4	CHARLEVOIX MI	23	23	51	82	88
123	163.9	168	207.4	219	213.8	CHEBOYGAN MI	27	3	30	58	81
218	170.3	271	206.1	301	215.7	CHIPPEWA MI	21	5	27	51	83
79	142.4	138	173.2	263	205.2	CLARE MI	22	18	44	76	72
239	160.6	272	166.3	326	180.9	CLINTON MI	4	9	13	25	37
44	195.9	59	217.1	106*	258.1	CRAWFORD MI	11	19	32	60	98
335	193.7	345	187.0	385	196.1	DELTA MI	-3	5	1	09	60
252	187.7	262	165.8	332	198.2	DICKINSON MI	-12	20	6	14	63
301	147.1	416	184.8	474	171.3	EATON MI	26	-7	16	32	24
134	159.8	157	182.9	206	205.1	EMMET MI	14	12	28	54	72
1881	179.8	2531*	202.4	2959*	217.0	GENESEE MI	13	7	21	41	84
76	136.3	129	177.8	193	188.0	GLADWIN MI	30	6	38	69	48
287	189.8	277	180.8	305	205.1	GOGEBIC MI	-5	13	8	17	72
218	128.3	263	143.6	386	187.7	GRAND TRAVERSE MI	12	31	46	78	47
264	149.5	283	165.2	313	176.9	GRATIOT MI	11	7	18	37	31
257	146.9	306	168.6	332	166.5	HILLSDALE MI	15	-1	13	27	20
460*	194.0	398	179.6	468	213.5	HOUGHTON MI	-7	19	10	21	81
263	147.9	316	171.5	357	171.0	HURON MI	16	0	16	31	24
1173	174.5	1496	202.9	1554	200.8	INGHAM MI	16	-1	15	30	66
309	151.5	342	170.9	381	193.7	IONIA MI	13	13	28	53	56
114	161.3	159	163.9	273	199.1	IOSCO MI	2	21	23	46	64
268*	246.7	222	208.8	238	213.2	IRON MI	-15	2	-14	03	81
223	185.1	254	190.1	294	195.2	ISABELLA MI	3	3	5	14	59
927	182.8	1109	202.2	1238	215.1	JACKSON MI	11	6	18	35	83
956	165.9	1177	178.7	1405	199.9	KALAMAZOO MI	8	12	21	40	65
39	138.3	76*	252.7	99	232.4	KALKASKA MI	83	-8	68	91	93
2426	181.3	2904	191.3	3172	191.0	KENT MI	6	0	5	14	52
33	173.2	35	199.8	27	136.6	KEWEENAW MI	15	-32	-21	01	03
47	173.4	57	181.2	85	194.4	LAKE MI	4	7	12	24	57
247	141.5	327	169.9	417	196.7	LAPEER MI	20	16	39	70	61

* SIGNIFICANTLY MORE DEATHS THAN EXPECTED

WHITE MALE — ALL CANCERS COMBINED — ICD 140-208

1950-1959 DEATHS	1950-1959 RATE	1960-1969 DEATHS	1960-1969 RATE	1970-1979 DEATHS	1970-1979 RATE	COUNTY AND STATE	PERCENT CHANGE 60S/50S	PERCENT CHANGE 70S/60S	PERCENT CHANGE 70S/50S	PERCENTILE %CHANGE 70S/50S	PERCENTILE RATES 1970S
67	140.0	108	192.9	115	160.1	LEELANAU MI	38	-17	14	29	14
533	168.7	638	192.4	753	216.0	LENAWEE MI	14	12	28	53	83
253	163.1	345	176.1	558	202.3	LIVINGSTON MI	8	15	24	46	68
43	86.9	61	128.7	82	207.0	LUCE MI	48	61	138	98	75
85	183.4	84	165.4	125	221.8	MACKINAC MI	-10	34	21	41	88
1538*	185.4	2900*	212.2	4573*	221.0	MACOMB MI	14	4	19	38	87
186	184.8	194	178.2	244	193.0	MANISTEE MI	-4	8	4	12	55
437	188.5	499	203.0	503	192.5	MARQUETTE MI	8	-5	2	10	54
184	159.3	228	183.0	249	172.7	MASON MI	15	-6	8	18	26
151	145.8	183	170.7	231	188.0	MECOSTA MI	17	10	29	55	47
239	181.1	240	172.0	269	183.1	MENOMINEE MI	-5	6	1	08	40
234	182.6	322	192.2	441	205.8	MIDLAND MI	5	7	13	25	73
58	161.4	70	170.1	83	176.5	MISSAUKEE MI	5	4	9	20	31
621	186.3	758	187.0	913	198.0	MONROE MI	0	6	6	15	62
320	179.2	333	173.5	418	190.3	MONTCALM MI	-3	10	6	15	51
39	166.8	62	191.1	105	224.8	MONTMORENCY MI	15	18	35	65	90
861	176.8	1012	182.3	1308	216.0	MUSKEGON MI	3	18	22	44	83
210	174.9	235	174.4	328	199.1	NEWAYGO MI	0	14	14	28	64
2911	178.5	4814	193.7	6911	205.0	OAKLAND MI	9	6	15	30	71
148	160.1	170	179.0	190	184.4	OCEANA MI	12	3	15	30	42
89	164.4	150	223.1	212	218.3	OGEMAW MI	36	-2	33	61	85
124	194.2	96	163.8	133	211.7	ONTONAGON MI	-16	29	9	19	79
103	132.7	122	152.9	199	218.4	OSCEOLA MI	15	43	65	90	85
34	171.6	51	168.4	91	199.6	OSCODA MI	-2	19	16	32	65
54	147.7	82	176.6	128	214.7	OTSEGO MI	20	22	45	77	82
576	165.2	728	173.7	934	179.8	OTTAWA MI	5	4	9	19	35
107	178.1	118	176.3	140	171.3	PRESQUE ISLE MI	-1	-3	-4	06	24
69	166.2	111	172.6	236	226.5	ROSCOMMON MI	4	31	36	67	91
1258*	199.6	1409	197.4	1581	207.4	SAGINAW MI	-1	5	4	12	75
742	166.0	942	185.9	1247*	220.0	ST CLAIR MI	12	18	33	61	86
329	163.5	424	197.3	491	206.1	ST JOSEPH MI	21	4	26	50	73
291	156.2	292	156.9	393	187.0	SANILAC MI	0	19	20	39	46
86	180.7	91	189.6	111	213.7	SCHOOLCRAFT MI	5	13	18	36	81
385	166.0	463	197.5	523	211.4	SHIAWASSEE MI	19	7	27	52	79
277	132.9	351	165.0	403	181.1	TUSCOLA MI	24	10	36	66	37
423	177.5	454	180.0	501	181.9	VAN BUREN MI	1	1	2	10	38
717	151.7	925	167.4	1266	199.6	WASHTENAW MI	10	19	32	59	65
19064*	214.2	21756*	229.9	19237*	228.8	WAYNE MI	7	-1	7	16	92
161	165.0	173	171.2	222	186.8	WEXFORD MI	4	9	13	26	46
24798	165.3	29599	175.0	33657	182.4	MINNESOTA MN	6	4	10		
142	163.8	182	188.0	181	180.4	AITKIN MN	15	-4	10	21	36
195	113.3	424	160.4	701	189.1	ANOKA MN	42	18	67	91	49
208	158.0	228	160.6	268	170.1	BECKER MN	2	6	8	17	23
167	139.4	207	170.3	203	158.4	BELTRAMI MN	22	-7	14	27	12
114	157.8	141	160.2	188	178.9	BENTON MN	2	12	13	27	34
87	167.3	89	161.8	119	201.9	BIG STONE MN	-3	25	21	41	68
327	166.7	368	172.7	387	169.1	BLUE EARTH MN	4	-2	1	09	22
230	170.8	264	173.0	291	175.8	BROWN MN	1	2	3	10	30
195	143.1	241	168.9	264	177.1	CARLTON MN	18	5	24	46	31
134	134.9	197	168.7	196	143.0	CARVER MN	25	-15	6	14	04
148	147.6	199	159.8	240	163.8	CASS MN	8	3	11	22	17
126	145.0	163	174.6	178	176.5	CHIPPEWA MN	20	1	22	43	31
129	133.9	170	162.6	215	178.0	CHISAGO MN	21	9	33	61	33
218	159.6	258	154.1	348	176.3	CLAY MN	-3	14	10	22	30
70	120.7	84	135.2	101	153.4	CLEARWATER MN	12	13	27	52	09
21	115.6	42	242.4	34	137.8	COOK MN	110	-43	19	38	03
132	164.2	135	148.5	170	174.9	COTTONWOOD MN	-10	18	7	15	28
275	159.0	339	171.1	419	175.9	CROW WING MN	8	3	11	22	30
351	152.9	532	169.3	817	178.2	DAKOTA MN	11	5	17	33	33
121	173.3	99	135.4	129	166.0	DODGE MN	-22	23	-4	06	19
190	152.0	232	160.3	275	165.0	DOUGLAS MN	5	3	9	18	18
203	165.0	222	170.9	249	183.4	FARIBAULT MN	4	7	11	23	41
193	135.8	247	165.2	221	142.2	FILLMORE MN	22	-14	5	12	04
262	155.2	318	168.6	358	172.6	FREEBORN MN	9	2	11	23	26
289	152.9	352	171.0	384	171.1	GOODHUE MN	12	0	12	24	24
78	149.7	81	136.4	110	179.4	GRANT MN	-9	32	20	39	35
5745	180.6	6869	190.7	7574	201.1	HENNEPIN MN	6	5	11	23	67
107	127.4	151	155.4	189	178.2	HOUSTON MN	22	15	40	71	33
120	180.1	133	176.5	166	184.7	HUBBARD MN	-2	5	3	10	42
97	122.2	117	130.2	160	158.5	ISANTI MN	7	22	30	56	12
286	165.8	361	190.7	405	186.4	ITASCA MN	15	-2	12	25	45
104	133.6	151	171.6	142	156.0	JACKSON MN	28	-9	17	33	10
83	153.1	91	150.0	114	163.3	KANABEC MN	-2	9	7	15	17
204	122.5	252	137.6	307	152.1	KANDIYOHI MN	12	11	24	47	08
72	141.0	80	159.8	71	137.8	KITTSON MN	13	-14	-2	06	03
158	168.5	155	167.1	175	190.1	KOOCHICHING MN	-1	14	13	26	51
100	126.1	133	154.6	138	158.9	LAC QUI PARLE MN	23	3	26	50	13
84	178.9	103	194.2	97	160.4	LAKE MN	9	-17	-10	03	14
37	134.1	47	172.6	48	188.6	LAKE OF THE WOODS MN	29	9	41	72	48
188	167.0	188	155.0	279	205.7	LE SUEUR MN	-7	33	23	45	73
84	159.0	99	174.7	92	151.7	LINCOLN MN	10	-13	-5	05	08
158	144.2	222	179.3	227	169.0	LYON MN	24	-6	17	34	22
189	147.0	232	159.3	272	166.5	MC LEOD MN	8	5	13	27	20
43	156.3	37	117.6	57	162.0	MAHNOMEN MN	-25	38	4	11	15
116	143.7	136	154.9	130	146.5	MARSHALL MN	8	-5	2	09	05
199	163.3	205	147.0	250	162.5	MARTIN MN	-10	11	0	07	16
172	169.7	165	146.3	188	147.6	MEEKER MN	-14	1	-13	03	06
139	160.3	168	175.7	175	158.5	MILLE LACS MN	10	-10	-1	07	12
172	133.5	223	151.2	271	166.9	MORRISON MN	13	10	25	48	20
271	149.9	351	164.0	393	163.5	MOWER MN	9	0	9	19	17
104	160.7	108	150.4	126	164.2	MURRAY MN	-6	9	2	10	17
144	126.2	155	134.7	189	168.1	NICOLLET MN	7	25	33	62	21
170	166.9	191	160.7	223	172.2	NOBLES MN	-4	7	3	11	25
110	157.1	136	176.2	156	194.1	NORMAN MN	12	10	24	46	57
402	173.4	476	171.0	522	166.7	OLMSTED MN	-1	-3	-4	06	20
362	120.4	509	154.9	576	158.4	OTTER TAIL MN	29	2	32	59	12
106	147.9	125	154.8	146	172.3	PENNINGTON MN	5	11	16	32	25
173	159.1	176	156.6	239	198.3	PINE MN	-2	27	25	48	63
97	141.1	125	161.1	161	197.4	PIPESTONE MN	14	23	40	71	62

* SIGNIFICANTLY MORE DEATHS THAN EXPECTED

WHITE MALE　　ALL CANCERS COMBINED　　ICD 140-208

1950-1959 DEATHS	1950-1959 RATE	1960-1969 DEATHS	1960-1969 RATE	1970-1979 DEATHS	1970-1979 RATE	COUNTY AND STATE	PERCENT CHANGE 60S/50S	PERCENT CHANGE 70S/60S	PERCENT CHANGE 70S/50S	PERCENTILE %CHANGE 70S/50S	PERCENTILE RATES 1970S
283	147.9	338	160.7	352	164.2	POLK MN	9	2	11	22	17
105	146.4	137	178.7	135	156.8	POPE MN	22	-12	7	16	11
3063*	194.8	3672*	204.5	3792	200.2	RAMSEY MN	5	-2	3	10	66
51	161.0	66	204.3	51	143.7	RED LAKE MN	27	-30	-11	03	05
169	154.9	197	162.5	208	163.3	REDWOOD MN	5	0	5	14	17
204	158.7	222	161.1	240	169.8	RENVILLE MN	2	5	7	16	23
243	141.2	282	150.5	367	181.8	RICE MN	7	21	29	54	38
84	149.2	113	180.9	122	182.1	ROCK MN	21	1	22	43	39
104	150.4	109	146.2	135	165.1	ROSEAU MN	-3	13	10	20	18
2225*	190.8	2381*	197.8	2444	208.5	ST LOUIS MN	4	5	9	20	76
153	184.5	176	162.3	246	182.3	SCOTT MN	-12	12	-1	07	39
83	139.4	121	156.5	179	185.9	SHERBURNE MN	12	19	33	62	44
123	144.1	131	136.7	191	189.4	SIBLEY MN	-5	39	31	59	50
487	156.8	571	153.7	746	173.8	STEARNS MN	-2	13	11	22	27
191	177.3	207	163.1	246	169.3	STEELE MN	-8	4	-5	05	22
102	189.2	86	145.3	118	194.5	STEVENS MN	-23	34	3	10	58
116	148.9	161	188.9	167	177.6	SWIFT MN	27	-6	19	38	32
205	154.0	254	174.7	247	159.3	TODD MN	13	-9	3	11	13
62	153.9	62	150.2	70	160.9	TRAVERSE MN	-2	7	5	12	14
165	165.6	196	187.5	220	190.6	WABASHA MN	13	2	15	30	52
83	130.5	115	159.1	140	159.0	WADENA MN	22	0	22	43	13
124	145.3	143	159.3	173	179.5	WASECA MN	10	13	23	46	35
253	141.2	385	173.3	522	183.3	WASHINGTON MN	23	6	30	57	41
111	147.1	141	173.8	159	188.7	WATONWAN MN	18	9	28	54	49
92	192.5	71	139.1	95	177.2	WILKIN MN	-28	27	-8	04	31
329	166.9	364	170.6	434	188.3	WINONA MN	2	10	13	26	48
254	155.6	271	141.7	395	170.5	WRIGHT MN	-9	20	10	20	24
133	148.4	143	148.3	159	160.7	YELLOW MEDICINE MN	0	8	8	18	14
7396	150.1	10645	180.5	14172	202.8	MISSISSIPPI MS	20	12	35		
94	196.7	136	222.5	183	218.8	ADAMS MS	13	-2	11	23	86
123	122.2	184	162.6	229	176.8	ALCORN MS	33	9	45	76	31
54	146.9	73	173.5	78	172.7	AMITE MS	18	0	18	35	26
89	138.2	125	171.7	166	204.7	ATTALA MS	24	19	48	80	71
37	172.3	34	143.4	52	187.0	BENTON MS	-17	30	9	18	46
124	192.5	138	187.9	179	214.7	BOLIVAR MS	-2	14	12	23	82
74	126.1	113	173.8	144	204.7	CALHOUN MS	38	18	62	89	71
45	150.5	58	200.1	51	158.1	CARROLL MS	33	-21	5	13	12
70	143.6	88	159.9	116	195.4	CHICKASAW MS	11	22	36	66	59
45	122.9	53	151.3	69	176.1	CHOCTAW MS	23	16	43	75	30
19	124.1	32	222.8	36	211.9	CLAIBORNE MS	79	-5	71	92	79
89	171.5	81	150.7	112	197.0	CLARKE MS	-12	31	15	30	61
58	159.6	64	157.4	87	173.4	CLAY MS	-1	10	9	18	27
95	180.5	139	219.1	142	218.0	COAHOMA MS	21	0	21	41	85
128	182.9	181*	243.3	148	201.4	COPIAH MS	33	-17	10	21	67
53	124.4	77	174.8	120	221.9	COVINGTON MS	41	27	78	94	88
53	141.2	105	179.1	219	216.8	DE SOTO MS	27	21	54	84	84
177	141.7	310	196.1	402	215.5	FORREST MS	38	10	52	83	83
48	157.4	59	192.4	72	219.4	FRANKLIN MS	22	14	39	71	86
49	140.1	99	221.3	122	224.4	GEORGE MS	58	1	60	87	90
33	120.0	57	185.2	69	192.5	GREENE MS	54	4	60	88	54
48	140.2	83	181.9	127	217.2	GRENADA MS	30	19	55	85	84
75	152.1	118	180.4	175	194.0	HANCOCK MS	19	8	28	52	57
445	162.3	690	192.8	1024*	228.7	HARRISON MS	19	19	41	72	92
486	166.5	836*	198.9	1090	207.7	HINDS MS	19	4	25	48	76
80	190.4	88	202.3	88	199.8	HOLMES MS	6	-1	5	13	65
31	130.5	54	196.2	80*	298.9	HUMPHREYS MS	50	52	129	98	99
8	145.0	7	118.8	9	195.4	ISSAQUENA MS	-18	64	35	64	59
99	150.6	115	155.5	164	174.8	ITAWAMBA MS	3	12	16	32	28
156	142.8	282	158.4	576	210.3	JACKSON MS	11	33	47	79	78
57	132.0	76	160.5	87	172.4	JASPER MS	22	7	31	58	26
26	160.6	38	233.0	27	193.7	JEFFERSON MS	45	-17	21	40	56
47	189.1	65	208.0	82	233.4	JEFFERSON DAVIS MS	10	12	23	45	94
240	162.3	357	184.2	424	192.5	JONES MS	13	5	19	37	54
41	135.0	38	121.7	62	191.7	KEMPER MS	-10	58	42	74	53
74	136.3	109	182.5	132	179.2	LAFAYETTE MS	34	-2	31	59	34
72	158.3	108	195.4	145	209.2	LAMAR MS	23	7	32	60	77
264	146.0	324	155.6	493	212.4	LAUDERDALE MS	7	37	45	77	80
50	161.8	71	204.2	105*	259.6	LAWRENCE MS	26	27	60	88	98
77	128.6	120	183.1	146	201.7	LEAKE MS	42	10	57	86	67
205	174.3	240	165.8	368	204.8	LEE MS	-5	24	17	34	71
118	188.6	179*	252.9	208*	246.9	LEFLORE MS	34	-2	31	58	97
113	151.6	144	163.0	174	165.2	LINCOLN MS	8	1	9	19	18
112	133.0	193	186.8	253	206.6	LOWNDES MS	41	11	55	85	74
64	134.0	99	194.2	143	226.0	MADISON MS	45	16	69	91	90
82	135.2	142	195.4	180	213.6	MARION MS	45	9	58	86	81
43	133.6	68	180.3	113	213.2	MARSHALL MS	35	18	60	87	81
127	134.6	179	165.5	222	178.8	MONROE MS	23	8	33	61	34
55	137.5	66	150.0	93	197.8	MONTGOMERY MS	9	32	44	75	62
109	148.8	130	156.5	150	158.7	NESHOBA MS	5	1	7	15	12
75	115.3	100	134.6	144	172.2	NEWTON MS	17	28	49	81	25
55	246.0	54	215.9	58	217.1	NOXUBEE MS	-12	1	-12	03	84
61	141.8	88	170.8	129	205.8	OKTIBBEHA MS	20	20	45	77	73
84	133.6	125	188.4	146	181.6	PANOLA MS	41	-4	36	66	38
105	161.4	193*	225.5	211	189.4	PEARL RIVER MS	40	-16	17	34	50
37	150.3	57	180.0	61	181.8	PERRY MS	20	1	21	41	38
161	180.5	188	191.5	249	228.4	PIKE MS	6	19	27	51	92
97	129.6	122	147.5	176	190.2	PONTOTOC MS	14	29	47	79	51
89	129.6	130	153.0	184	180.1	PRENTISS MS	18	18	39	70	36
58	181.1	61	183.7	70	214.9	QUITMAN MS	1	17	19	37	82
85	114.6	156	145.3	274	173.3	RANKIN MS	27	19	51	82	27
88	154.3	95	141.0	182	239.2	SCOTT MS	-9	70	55	85	96
16	136.4	20	152.3	35	202.5	SHARKEY MS	12	33	48	80	68
97	149.5	132	178.7	196*	242.0	SIMPSON MS	20	35	62	89	96
69	123.3	94	154.7	114	174.7	SMITH MS	25	13	42	73	28
28	127.3	50	174.8	72	212.5	STONE MS	37	22	67	91	80
95	153.3	117	176.2	160	229.3	SUNFLOWER MS	15	30	50	81	92
71	171.1	92	208.3	99	220.7	TALLAHATCHIE MS	22	6	29	55	87
60	177.1	80	194.4	126	238.8	TATE MS	10	23	35	65	95
74	117.4	124	174.1	134	162.4	TIPPAH MS	48	-7	38	69	16

* SIGNIFICANTLY MORE DEATHS THAN EXPECTED

WHITE MALE — ALL CANCERS COMBINED — ICD 140-208

1950-1959 DEATHS	1950-1959 RATE	1960-1969 DEATHS	1960-1969 RATE	1970-1979 DEATHS	1970-1979 RATE	COUNTY AND STATE	PERCENT CHANGE 60S/50S	PERCENT CHANGE 70S/60S	PERCENT CHANGE 70S/50S	PERCENTILE %CHANGE 70S/50S	PERCENTILE RATES 1970S
66	105.1	118	154.7	171	189.5	TISHOMINGO MS	47	22	80	94	50
22	165.6	30	210.2	26	175.3	TUNICA MS	27	-17	6	14	29
71	101.0	142	171.4	168	172.7	UNION MS	70	1	71	92	26
53	153.1	60	149.6	90	201.3	WALTHALL MS	-2	35	32	59	67
144	178.7	212*	227.8	246	222.5	WARREN MS	28	-2	25	47	88
152	183.3	246*	235.2	290*	237.4	WASHINGTON MS	28	1	30	56	95
73	184.8	65	136.1	111	202.7	WAYNE MS	-26	49	10	20	69
70	154.9	79	159.2	95	168.9	WEBSTER MS	3	6	9	19	22
40	200.1	45	216.1	39	189.0	WILKINSON MS	8	-13	-6	05	49
75	132.4	114	184.6	125	180.3	WINSTON MS	39	-2	36	66	36
67	150.9	104	222.6	117	226.0	YALOBUSHA MS	48	2	50	81	90
97	177.0	127	219.4	138	206.6	YAZOO MS	24	-6	17	33	74
32265	166.6	38506	184.2	44671	202.2	MISSOURI MO	11	10	21		
168	148.6	166	148.6	174	155.6	ADAIR MO	0	5	5	13	10
98	134.5	127	166.0	117	143.7	ANDREW MO	23	-13	7	16	05
79	136.0	110	188.8	101	173.9	ATCHISON MO	39	-8	28	53	27
195	162.0	239	184.1	264	194.1	AUDRAIN MO	14	5	20	39	57
150	121.9	237	181.4	279	177.7	BARRY MO	49	-2	46	78	32
103	118.5	135	156.3	150	169.3	BARTON MO	32	8	43	74	22
170	139.3	202	170.3	230	189.1	BATES MO	22	11	36	66	49
97	145.2	117	147.9	180	179.7	BENTON MO	2	22	24	46	35
73	116.1	82	132.4	126	185.2	BOLLINGER MO	14	40	60	87	43
250	129.7	341	155.6	494	188.7	BOONE MO	20	21	45	77	49
838	166.4	987	203.3	1132*	242.5	BUCHANAN MO	22	19	46	78	96
254	147.6	312	166.3	416	207.8	BUTLER MO	13	25	41	72	76
103	138.8	100	142.3	140	196.3	CALDWELL MO	3	38	41	73	60
149	105.1	227	164.5	230	160.6	CALLAWAY MO	56	-2	53	83	14
76	144.9	142	178.7	233	192.3	CAMDEN MO	23	8	33	61	54
238	132.9	356	166.9	454	179.7	CAPE GIRARDEAU MO	26	8	35	65	35
147	151.2	151	164.7	163	201.3	CARROLL MO	9	22	33	62	67
51	191.8	58	205.1	51	177.7	CARTER MO	7	-13	-7	04	32
175	131.5	241	160.2	339	184.5	CASS MO	22	15	40	72	42
92	124.8	102	130.2	163	189.0	CEDAR MO	4	45	51	82	49
149	156.6	132	148.2	133	161.6	CHARITON MO	-5	9	3	11	15
101	136.4	132	164.7	174	181.9	CHRISTIAN MO	21	10	33	62	39
93	144.3	92	147.2	91	158.3	CLARK MO	2	8	10	20	12
356	145.0	586	178.4	779	175.7	CLAY MO	23	-2	21	42	29
112	143.2	114	148.2	139	169.5	CLINTON MO	3	14	18	37	23
222	135.8	296	159.1	385	183.4	COLE MO	17	15	35	65	41
105	116.1	131	150.3	167	190.5	COOPER MO	30	27	64	90	51
133	168.7	148	162.6	192	177.3	CRAWFORD MO	-4	9	5	13	32
89	145.7	90	152.9	108	170.7	DADE MO	5	12	17	34	24
89	137.8	126	162.3	140	162.7	DALLAS MO	18	0	18	36	16
132	158.9	134	177.0	132	198.0	DAVIESS MO	11	12	25	48	63
69	112.9	87	156.0	93	159.6	DE KALB MO	38	2	41	73	13
95	143.1	118	156.6	164	192.1	DENT MO	9	23	34	64	54
79	126.1	97	150.2	118	167.3	DOUGLAS MO	19	11	33	61	20
260	150.2	312	167.7	429	221.8	DUNKLIN MO	12	32	48	79	88
331	154.9	443	173.5	581	188.2	FRANKLIN MO	12	8	21	42	48
123	157.9	151	176.2	181	189.6	GASCONADE MO	12	8	20	39	50
117	159.0	125	178.7	132	202.6	GENTRY MO	12	13	27	52	69
799	146.4	1112	173.5	1430	192.4	GREENE MO	18	11	31	59	54
124	126.6	166	182.1	187	226.1	GRUNDY MO	44	24	79	94	91
165	170.6	156	173.3	151	194.9	HARRISON MO	2	12	14	29	58
209	153.3	236	166.5	255	183.7	HENRY MO	9	10	20	39	41
52	147.7	61	142.7	73	132.5	HICKORY MO	-3	-7	-10	03	02
103	153.6	96	154.1	98	170.3	HOLT MO	0	11	11	22	24
89	126.0	82	119.1	129	205.6	HOWARD MO	-5	73	63	89	73
180	142.7	199	140.5	319	193.4	HOWELL MO	-1	38	36	66	56
74	158.8	81	151.2	106	173.5	IRON MO	-5	15	9	20	27
4367*	186.6	5001*	200.9	5126*	212.0	JACKSON MO	8	6	14	27	80
725	178.5	862*	204.9	1030*	236.5	JASPER MO	15	15	32	61	95
336	155.1	528	172.5	851	199.0	JEFFERSON MO	11	15	28	54	64
180	134.9	214	159.3	202	145.6	JOHNSON MO	18	-9	8	17	05
67	131.3	75	154.8	84	195.8	KNOX MO	18	26	49	81	60
129	123.7	202	175.1	253	196.4	LACLEDE MO	42	12	59	87	61
244	166.7	275	175.2	301	179.0	LAFAYETTE MO	5	2	7	16	34
217	143.9	252	156.0	357	201.3	LAWRENCE MO	8	29	40	71	67
112	157.7	117	171.6	137	211.4	LEWIS MO	9	23	34	64	79
129	137.5	151	149.3	197	173.7	LINCOLN MO	9	16	26	50	27
178	141.9	249	201.2	236	201.4	LINN MO	42	0	42	74	67
146	143.9	158	154.4	169	171.5	LIVINGSTON MO	7	11	19	38	25
111	141.3	116	139.0	158	189.1	MC DONALD MO	-2	36	34	63	49
207	158.3	188	158.1	192	166.6	MACON MO	0	5	5	13	20
85	153.8	106	175.2	113	152.4	MADISON MO	14	-13	-1	07	08
65	154.7	49	109.3	69	141.7	MARIES MO	-29	30	-8	04	04
238	147.1	298	180.9	340	209.0	MARION MO	23	16	42	74	77
62	132.4	57	130.6	85	205.7	MERCER MO	-1	57	55	85	73
108	132.2	153	163.5	171	164.3	MILLER MO	24	1	24	47	17
105	154.3	132	179.7	151	205.3	MISSISSIPPI MO	16	14	33	62	72
85	114.9	124	164.1	122	162.8	MONITEAU MO	43	-1	42	73	16
84	112.3	98	130.1	126	194.0	MONROE MO	16	49	73	92	57
130	165.3	135	170.0	146	181.7	MONTGOMERY MO	3	7	10	20	38
88	125.6	121	157.0	213	220.0	MORGAN MO	25	40	75	93	86
158	153.7	169	163.8	235	230.4	NEW MADRID MO	7	41	50	81	93
200	131.1	268	160.5	356	181.2	NEWTON MO	22	13	38	69	37
205	142.8	217	165.0	201	159.1	NODAWAY MO	16	-4	11	23	13
76	115.2	89	130.3	126	181.1	OREGON MO	13	39	57	86	37
84	131.4	92	134.5	126	175.2	OSAGE MO	2	30	33	62	29
66	166.2	82	161.5	99	181.2	OZARK MO	-3	12	9	19	37
173	150.8	218	186.6	231	214.6	PEMISCOT MO	24	15	42	74	82
110	139.2	146	168.8	182	184.6	PERRY MO	21	9	33	61	42
285	152.2	352	177.9	392	195.3	PETTIS MO	17	10	28	54	59
152	134.3	201	154.2	321	212.9	PHELPS MO	15	38	58	86	81
161	160.1	173	172.4	187	188.6	PIKE MO	8	9	18	35	48
132	147.9	167	154.0	208	156.1	PLATTE MO	4	1	6	14	10
136	129.0	164	157.7	209	177.9	POLK MO	22	13	38	69	33
107	163.1	129	162.8	185	196.4	PULASKI MO	0	21	20	40	61
72	124.7	105	186.7	97	188.7	PUTNAM MO	50	1	51	82	49

* SIGNIFICANTLY MORE DEATHS THAN EXPECTED

WHITE MALE — ALL CANCERS COMBINED — ICD 140-208

1950-1959 DEATHS	1950-1959 RATE	1960-1969 DEATHS	1960-1969 RATE	1970-1979 DEATHS	1970-1979 RATE	COUNTY AND STATE	PERCENT CHANGE 60S/50S	PERCENT CHANGE 70S/60S	PERCENT CHANGE 70S/50S	PERCENTILE %CHANGE 70S/50S	PERCENTILE RATES 1970S
72	137.2	84	159.7	99	183.5	RALLS MO	16	15	34	63	41
216	145.0	277	183.1	323	228.0	RANDOLPH MO	26	25	57	86	92
153	147.9	147	140.3	183	173.6	RAY MO	-5	24	17	34	27
40	110.0	48	135.6	85	203.1	REYNOLDS MO	23	50	85	95	69
80	126.4	105	141.2	148	175.4	RIPLEY MO	12	24	39	70	29
268	173.7	404	186.0	714	223.0	ST CHARLES MO	7	20	28	54	89
101	139.4	114	159.1	129	190.2	ST CLAIR MO	14	20	36	67	51
79	144.1	117	179.6	136	182.3	STE GENEVIEVE MO	25	2	27	51	39
285	150.0	395	189.8	466	203.4	ST FRANCOIS MO	27	7	36	66	70
3417	185.1	5573*	198.2	7615*	211.5	ST LOUIS COUNTY MO	7	7	14	29	79
208	139.5	275	174.3	334	217.1	SALINE MO	25	25	56	85	84
50	123.5	57	146.7	66	180.5	SCHUYLER MO	19	23	46	78	36
67	129.9	85	175.7	73	160.6	SCOTLAND MO	35	-9	24	46	14
191	147.9	250	173.5	343	212.0	SCOTT MO	17	22	43	75	80
41	91.8	79	181.8	65	140.1	SHANNON MO	98	-23	52	83	04
114	152.6	101	153.2	99	166.1	SHELBY MO	0	8	9	18	19
190	135.2	253	169.2	299	182.6	STODDARD MO	25	8	35	65	40
78	159.4	86	142.3	155	157.0	STONE MO	-11	10	-1	07	11
112	154.8	114	166.0	115	178.6	SULLIVAN MO	7	8	15	31	34
73	125.0	135	151.8	271	198.3	TANEY MO	21	31	59	86	63
120	110.3	159	141.1	245	187.8	TEXAS MO	28	33	70	91	47
173	109.9	217	143.7	243	182.2	VERNON MO	31	27	66	90	39
85	149.7	88	144.0	122	160.8	WARREN MO	-4	12	7	16	14
103	146.8	121	164.1	171	214.1	WASHINGTON MO	12	30	46	78	82
82	144.2	122	181.2	167	218.1	WAYNE MO	26	20	51	82	85
118	134.5	152	156.9	162	158.2	WEBSTER MO	17	1	18	35	12
50	159.4	42	139.6	40	162.0	WORTH MO	-12	16	2	09	15
114	126.3	151	152.8	210	205.0	WRIGHT MO	21	34	62	89	71
7314*	229.2	6587*	243.7	5287*	264.8	ST LOUIS CITY MO	6	9	16	31	98
4703	155.9	5488	170.4	6254	182.7	MONTANA MT	9	7	17		
63	142.2	77	167.7	99	228.4	BEAVERHEAD MT	18	36	61	88	92
45	137.5	64	202.4	52	170.3	BIG HORN MT	47	-16	24	46	23
46	124.4	64	172.7	65	200.9	BLAINE MT	39	16	62	88	66
27	158.1	32	200.0	41	229.4	BROADWATER MT	27	15	45	77	92
95	166.6	80	138.7	105	188.0	CARBON MT	-17	36	13	26	47
17	116.7	22	167.6	18	130.5	CARTER MT	44	-22	12	24	02
429	168.2	550	184.0	655	202.9	CASCADE MT	9	10	21	40	69
51	129.0	69	179.8	57	150.6	CHOUTEAU MT	39	-16	17	33	07
109	157.9	98	137.9	133	183.7	CUSTER MT	-13	33	16	32	41
31	282.2	26	145.2	40	190.5	DANIELS MT	-49	31	-32	00	51
54	121.4	79	161.6	87	175.1	DAWSON MT	33	8	44	76	28
161	161.2	181	186.1	146	183.8	DEER LODGE MT	15	-1	14	28	42
33	202.8	37	206.0	30	169.2	FALLON MT	2	-18	-17	02	22
132	160.9	142	171.7	132	154.9	FERGUS MT	7	-10	-4	06	10
268	152.4	299	157.8	433	196.2	FLATHEAD MT	4	24	29	54	60
137	135.8	179	155.0	185	137.8	GALLATIN MT	14	-11	1	09	03
13	100.0	16	152.3	22*	484.9	GARFIELD MT	52	218	385	99	99
46	173.2	41	135.8	59	204.7	GLACIER MT	-22	51	18	36	71
9	91.2	16	215.6	12	169.1	GOLDEN VALLEY MT	136	-22	85	95	22
29	171.9	36	203.1	31	186.4	GRANITE MT	18	-8	8	18	45
106	161.8	127	178.8	122	172.5	HILL MT	11	-3	7	15	26
28	138.1	39	178.1	50	188.6	JEFFERSON MT	29	6	37	67	48
23	127.2	27	148.6	18	98.7	JUDITH BASIN MT	17	-34	-22	01	00
95	128.9	99	129.1	155	172.5	LAKE MT	0	34	34	63	26
228	183.6	270	195.8	312	197.8	LEWIS AND CLARK MT	7	1	8	17	62
9	78.4	11	130.8	14	137.2	LIBERTY MT	67	5	75	93	03
54	109.9	87	144.2	121	177.4	LINCOLN MT	31	23	61	88	32
25	133.0	19	104.6	17	118.8	MC CONE MT	-21	14	-11	03	01
47	122.7	61	168.4	75	215.3	MADISON MT	37	28	75	93	83
22	127.4	22	143.9	23	165.4	MEAGHER MT	13	15	30	57	19
23	178.2	25	189.2	27	177.4	MINERAL MT	6	-6	0	08	32
269	153.1	363	178.3	397	164.7	MISSOULA MT	16	-8	8	17	18
56	147.0	50	166.5	48	165.6	MUSSELSHELL MT	13	-1	13	25	19
107	152.3	122	167.3	132	175.4	PARK#MT	10	5	15	30	29
10	182.7	9	236.8	13*	420.7	PETROLEUM MT	30	78	130	98	99
50	142.6	61	169.7	52	147.7	PHILLIPS MT	19	-13	4	11	06
39	140.9	51	157.3	73	205.1	PONDERA MT	12	30	46	77	72
18	127.4	26	221.7	18	142.0	POWDER RIVER MT	74	-36	11	23	04
77	227.0	63	184.6	60	175.3	POWELL MT	-19	-5	-23	01	29
16	130.5	19	158.3	26	189.8	PRAIRIE MT	21	20	45	77	50
100	132.5	135	154.7	169	173.2	RAVALLI MT	17	12	31	58	27
92	183.9	69	137.5	104	205.7	RICHLAND MT	-25	50	12	24	73
59	150.9	63	143.7	58	144.4	ROOSEVELT MT	-5	0	-4	05	05
40	142.3	52	189.0	44	157.0	ROSEBUD MT	33	-17	10	21	11
57	127.0	70	147.8	96	197.7	SANDERS MT	16	34	56	85	62
39	111.7	64	164.1	73	209.5	SHERIDAN MT	47	28	87	96	77
491*	211.2	457	202.3	434	203.9	SILVER BOW MT	-4	1	-3	06	70
36	119.3	50	155.9	66	194.9	STILLWATER MT	31	25	63	89	58
21	98.0	46	196.9	42	162.7	SWEET GRASS MT	101	-17	66	91	16
56	163.1	59	154.2	55	135.6	TETON MT	-5	-12	-17	02	03
62	186.4	52	160.0	56	192.7	TOOLE MT	-14	20	3	11	55
9	125.6	10	159.0	9	191.2	TREASURE MT	27	20	52	83	52
79	140.0	86	152.2	91	176.5	VALLEY MT	9	16	26	50	31
34	162.8	38	204.4	32	186.2	WHEATLAND MT	26	-9	14	29	45
9	99.8	13	138.0	16	172.7	WIBAUX MT	38	25	73	93	26
422	162.6	565	179.2	734	195.6	YELLOWSTONE MT	10	9	20	40	59
11345	167.0	12597	171.3	14344	186.1	NEBRASKA NB	3	9	11		
243	148.5	240	139.0	336	197.1	ADAMS NB	-6	42	33	61	62
101	159.0	100	155.3	91	153.9	ANTELOPE NB	-2	-1	-3	06	09
4	199.1	7	205.0	4	135.1	ARTHUR NB	3	-34	-32	00	03
4	133.1	6	134.2	1	37.2	BANNER NB	1	-72	-72	00	00
8	190.6	7	103.4	10	147.5	BLAINE NB	-46	43	-23	01	06
91	160.9	105	182.8	97	183.6	BOONE NB	14	0	14	28	41
73	138.0	101	168.9	100	156.6	BOX BUTTE NB	22	-7	13	27	11
40	156.6	41	139.6	48	161.9	BOYD NB	-11	16	3	11	15
48	165.1	60	218.4	39	136.6	BROWN NB	32	-37	-17	02	03
203	146.9	233	153.2	303	188.3	BUFFALO NB	4	23	28	53	48

* SIGNIFICANTLY MORE DEATHS THAN EXPECTED

COUNTIES COMBINED; SEE TECHNICAL NOTES

WHITE MALE — ALL CANCERS COMBINED — ICD 140-208

1950-1959 DEATHS	1950-1959 RATE	1960-1969 DEATHS	1960-1969 RATE	1970-1979 DEATHS	1970-1979 RATE	COUNTY AND STATE	PERCENT CHANGE 60S/50S	PERCENT CHANGE 70S/60S	PERCENT CHANGE 70S/50S	PERCENTILE %CHANGE 70S/50S	PERCENTILE RATES 1970S
93	138.0	94	138.5	113	152.7	BURT NB	0	10	11	22	09
109	163.1	133	185.4	127	174.1	BUTLER NB	14	-6	7	16	27
150	153.2	172	172.6	199	192.3	CASS NB	13	11	26	49	54
102	148.8	98	132.7	135	171.5	CEDAR NB	-11	29	15	30	25
28	135.9	37	145.2	51	179.3	CHASE NB	7	23	32	60	34
57	147.4	59	146.4	65	159.8	CHERRY NB	-1	9	8	18	13
88	168.6	108	184.6	120	200.7	CHEYENNE NB	10	9	19	38	66
92	148.2	99	180.6	94	161.1	CLAY NB	22	-11	9	18	14
107	167.4	122	173.1	123	164.9	COLFAX NB	3	-5	-2	07	18
112	173.4	130	183.5	127	165.8	CUMING NB	6	-10	-4	05	19
145	135.3	172	160.0	165	156.1	CUSTER NB	18	-2	15	31	10
89	178.8	111	194.9	134	211.1	DAKOTA NB	9	8	18	36	79
91	181.9	87	162.1	84	162.6	DAWES NB	-11	0	-11	03	16
147	153.0	186	168.3	200	169.0	DAWSON NB	10	0	10	22	22
20	119.1	27	132.6	38	184.8	DEUEL NB	11	39	55	85	43
69	133.6	92	169.8	84	150.6	DIXON NB	27	-11	13	25	07
246	168.5	293	173.7	349	182.8	DODGE NB	3	5	8	18	40
2559*	204.8	2918*	209.5	3261*	223.3	DOUGLAS NB	2	7	9	19	89
24	115.9	18	80.4	33	146.1	DUNDY NB	-31	82	26	50	05
97	153.4	97	156.3	113	188.0	FILLMORE NB	2	20	23	44	47
63	157.5	73	190.1	67	177.4	FRANKLIN NB	21	-7	13	25	32
38	150.2	29	103.2	41	161.3	FRONTIER NB	-31	56	7	16	15
79	134.1	94	152.2	83	144.4	FURNAS NB	13	-5	8	17	05
252	169.5	238	150.0	310	192.8	GAGE NB	-11	28	14	27	55
23	117.1	24	115.7	39*	577.9	GARDEN NB	-1	400	394	99	99
27	160.9	35	182.7	35	197.6	GARFIELD NB	14	8	23	45	62
13	104.9	15	107.7	16	114.4	GOSPER NB	3	6	9	19	00
5	100.8	14	307.2	7	148.1	GRANT NB	205	-52	47	79	06
47	162.0	47	168.5	47	163.9	GREELEY NB	4	-3	1	09	17
323	190.3	375	186.1	497	222.0	HALL NB	-2	19	17	33	88
91	167.7	73	134.4	93	179.7	HAMILTON NB	-20	34	7	16	35
53	148.9	62	177.9	54	162.4	HARLAN NB	20	-9	9	19	16
8	78.5	10	126.6	15	622.5	HAYES NB	61	392	693	99	99
46	157.3	55	189.7	35	116.9	HITCHCOCK NB	21	-38	-26	01	01
113	146.7	127	158.1	157	194.4	HOLT NB	8	23	33	61	58
8	139.8	9	149.8	16	230.2	HOOKER NB	7	54	65	90	93
59	132.7	51	116.2	76	163.0	HOWARD NB	-12	40	23	45	16
108	140.6	117	155.0	114	150.4	JEFFERSON NB	10	-3	7	16	07
55	134.7	73	163.6	65	149.6	JOHNSON NB	21	-9	11	23	07
56	143.6	55	126.7	72	163.2	KEARNEY NB	-12	29	14	27	17
57	164.1	80	190.4	94	197.6	KEITH NB	16	4	20	40	62
20	195.0	13	134.0	14	440.7	KEYA PAHA NB	-31	229	126	98	99
45	201.5	38	151.1	63	240.0	KIMBALL NB	-25	59	19	38	96
111	134.2	134	161.4	129	147.3	KNOX NB	20	-9	10	20	06
986	167.0	1156	178.1	1332	191.4	LANCASTER NB	7	7	15	29	53
223	176.0	264	180.2	301	184.9	LINCOLN NB	2	3	5	13	43
7	102.2	8	141.5	9	398.3	LOGAN NB	38	182	290	99	99
10	159.3	13	224.6	10	387.1	LOUP NB	41	72	143	98	99
3	135.4	7	161.4	5	692.1	MC PHERSON NB	19	329	411	99	99
247	166.4	267	165.7	322	189.5	MADISON NB	0	14	14	28	50
78	150.6	82	155.9	84	157.2	MERRICK NB	4	1	4	12	11
47	125.1	74	178.3	61	146.7	MORRILL NB	42	-18	17	34	05
55	145.4	69	175.1	67	172.3	NANCE NB	20	-2	18	35	26
103	174.2	92	160.2	97	166.8	NEMAHA NB	-8	4	-4	05	20
70	135.8	85	160.2	82	157.9	NUCKOLLS NB	18	-1	16	32	12
170	168.2	177	171.6	215	203.2	OTOE NB	2	18	21	41	69
69	153.5	60	144.1	60	154.6	PAWNEE NB	-6	7	1	08	09
20	103.1	32	137.3	49	191.6	PERKINS NB	33	40	86	95	53
95	173.2	91	141.7	119	203.5	PHELPS NB	-18	44	18	35	70
68	145.6	92	177.4	90	162.6	PIERCE NB	22	-8	12	23	16
175	171.7	190	159.8	251	189.7	PLATTE NB	-7	19	11	22	50
81	163.2	69	135.0	86	178.9	POLK NB	-17	33	10	20	34
125	200.7	98	140.5	122	167.1	RED WILLOW NB	-30	19	-17	02	20
160	172.9	147	160.4	170	180.2	RICHARDSON NB	-7	12	4	12	36
20	128.5	30	204.8	32	191.6	ROCK NB	59	-6	49	81	53
173	184.9	158	161.1	194	194.6	SALINE NB	-13	21	5	13	58
96	159.6	158	175.9	239	175.6	SARPY NB	10	0	10	21	29
160	151.5	192	168.6	202	176.1	SAUNDERS NB	11	4	16	32	30
234	178.6	257	167.5	328	183.1	SCOTTS BLUFF NB	-6	9	2	10	40
130	167.3	112	137.1	128	151.8	SEWARD NB	-18	11	-9	04	08
89	195.2	72	152.0	90	181.9	SHERIDAN NB	-22	20	-7	04	38
56	167.3	43	125.8	64	179.8	SHERMAN NB	-25	43	7	16	35
20	154.7	14	118.5	10	96.1	SIOUX NB	-23	-19	-38	00	00
31	99.3	49	142.7	43	121.1	STANTON NB	44	-15	22	43	01
92	143.0	115	180.6	119	188.2	THAYER NB	26	4	32	59	48
8	118.8	4	68.7	7	129.6	THOMAS NB	-42	89	9	19	02
57	177.9	58	180.5	65	192.5	THURSTON NB	1	7	8	18	54
63	150.8	79	179.8	72	158.5	VALLEY NB	19	-12	5	13	12
97	150.4	110	152.3	115	148.2	WASHINGTON NB	1	-3	-1	07	06
72	145.6	88	171.5	97	179.5	WAYNE NB	18	5	23	45	35
79	169.7	72	150.8	88	198.7	WEBSTER NB	-11	32	17	34	64
12	183.2	11	181.3	10	171.4	WHEELER NB	-1	-5	-6	05	25
124	149.0	108	132.7	126	153.6	YORK NB	-11	16	3	11	09
1435	162.9	2608	184.1	4684	198.3	NEVADA NV	13	8	22		
50	139.1	76	171.2	120	190.7	CHURCHILL NV	23	11	37	68	52
391	144.4	1149	184.6	2477	206.6	CLARK NV	28	12	43	74	74
18	130.2	26	124.0	72	154.2	DOUGLAS NV	-5	24	18	37	09
81	149.5	98	176.4	99	164.4	ELKO NV	18	-7	10	21	18
5	95.0	6	111.5	11	259.4	ESMERALDA NV	17	133	173	99	98
9	153.9	8	143.9	10	183.1	EUREKA NV	-7	27	19	37	40
44	154.7	58	195.7	68	214.4	HUMBOLDT NV	26	10	39	69	82
14	140.6	22	217.2	21	185.0	LANDER NV	55	-15	32	59	43
32	232.0	25	191.0	26	148.0	LINCOLN NV	-18	-23	-36	00	06
40	163.2	53	155.8	88	155.8	LYON NV	-5	0	-5	05	10
47	199.7	37	150.3	55	198.3	MINERAL NV	-25	32	-1	07	63
38	161.0	31	121.0	44	127.9	NYE NV	-25	6	-21	01	02
45	245.0	51*	301.1	40	221.9	PERSHING NV	23	-26	-9	04	88
7	187.1	8	167.2	13	288.7	STOREY NV	-11	73	54	84	99
501	176.7	802	191.8	1275	204.7	WASHOE NV	9	7	16	31	71

* SIGNIFICANTLY MORE DEATHS THAN EXPECTED

WHITE MALE — ALL CANCERS COMBINED — ICD 140-208

1950-1959 DEATHS	1950-1959 RATE	1960-1969 DEATHS	1960-1969 RATE	1970-1979 DEATHS	1970-1979 RATE	COUNTY AND STATE		PERCENT CHANGE 60S/50S	PERCENT CHANGE 70S/60S	PERCENT CHANGE 70S/50S	PERCENTILE %CHANGE 70S/50S	PERCENTILE RATES 1970S
74	169.5	75	168.9	70	161.1	WHITE PINE	NV	0	-5	-5	05	15
39	176.3	83	188.2	195	197.0	ORMSBY-CARSON CITY	NV	7	5	12	24	61
5434*	194.9	6499*	209.6	8133*	222.5	NEW HAMPSHIRE	NH	8	6	14		
283	196.7	335*	215.3	411	212.6	BELKNAP	NH	9	-1	8	17	80
172	172.3	238	216.6	286	201.4	CARROLL	NH	26	-7	17	33	67
389	193.4	413	186.4	587	218.7	CHESHIRE	NH	-4	17	13	26	86
338	180.9	395	210.0	445*	234.4	COOS	NH	16	12	30	56	94
469	187.9	512	198.3	633	220.9	GRAFTON	NH	6	11	18	35	87
1624*	209.7	1984*	227.0	2423*	239.7	HILLSBOROUGH	NH	8	6	14	29	96
633	176.7	729	189.7	877	203.6	MERRIMACK	NH	7	7	15	30	70
762*	190.6	985	204.1	1372	215.6	ROCKINGHAM	NH	7	6	13	26	83
521*	217.9	633*	237.4	763*	247.9	STRAFFORD	NH	9	4	14	28	97
243	177.6	275	179.4	336	190.2	SULLIVAN	NH	1	6	7	16	51
48298*	217.8	58537*	218.3	68495*	226.1	NEW JERSEY	NJ	0	4	4		
1411*	205.8	1783*	209.1	2086*	222.7	ATLANTIC	NJ	2	6	8	17	89
5446*	212.0	7472*	218.0	8951*	222.8	BERGEN	NJ	3	2	5	13	89
1078	187.9	1620*	211.0	2198	211.3	BURLINGTON	NJ	12	0	12	25	79
2735*	208.4	3550*	224.2	4161*	237.9	CAMDEN	NJ	8	6	14	29	95
547*	210.2	720	202.8	1025	205.4	CAPE MAY	NJ	-4	1	-2	06	72
746	182.1	940*	206.1	1128*	222.8	CUMBERLAND	NJ	13	8	22	44	89
8606*	234.3	8188*	221.7	7118*	219.6	ESSEX	NJ	-5	-1	-6	05	86
838*	202.1	1088*	205.8	1417*	220.3	GLOUCESTER	NJ	2	7	9	19	86
7057*	250.8	7052*	244.9	6651*	254.2	HUDSON	NJ	-2	4	1	09	98
425	169.9	609	207.5	693	208.7	HUNTERDON	NJ	22	1	23	45	76
2143*	215.5	2485*	220.2	2804*	232.2	MERCER	NJ	2	5	8	17	93
2775*	232.2	3807*	237.1	5027*	243.5	MIDDLESEX	NJ	2	3	5	13	97
2320*	198.7	3414*	218.4	4438*	225.9	MONMOUTH	NJ	10	3	14	27	90
1655*	191.3	2218	192.0	2932	207.1	MORRIS	NJ	0	8	8	18	75
831*	196.4	1763	195.8	4246*	226.6	OCEAN	NJ	0	16	15	31	91
3716*	224.4	4264*	221.7	4483*	225.6	PASSAIC	NJ	-1	2	1	08	90
380	189.0	467	206.3	567	222.8	SALEM	NJ	9	8	18	35	89
910*	190.2	1252	198.5	1653	216.4	SOMERSET	NJ	4	9	14	28	84
406*	203.3	533	190.8	806	214.7	SUSSEX	NJ	-6	13	6	14	82
3693*	217.6	4604*	215.6	5306*	226.0	UNION	NJ	-1	5	4	12	91
580*	190.6	708*	210.6	805	213.5	WARREN	NJ	11	1	12	24	81
3068	133.7	4606	153.5	6681	175.4	NEW MEXICO	NM	15	14	31		
766	147.6	1341	166.4	2107	186.5	BERNALILLO	NM	13	12	26	50	45
18	121.2	17	129.5	23	179.8	CATRON	NM	7	39	48	80	35
157	127.1	288	175.9	454	206.3	CHAVES	NM	38	17	62	89	73
91	135.5	94	133.5	114	163.7	COLFAX	NM	-1	23	21	41	17
124	138.0	161	144.1	232	169.5	CURRY	NM	4	18	23	45	23
20	136.3	20	123.6	23	141.9	DE BACA	NM	-9	15	4	12	04
181	150.9	235	142.7	421	164.2	DONA ANA	NM	-5	29	22	43	42
188	144.5	274	174.8	360	185.7	EDDY	NM	21	6	29	54	44
118	177.3	124	162.0	163	175.6	GRANT	NM	-9	8	-1	07	29
28	115.1	28	115.1	34	154.8	GUADALUPE	NM	0	35	35	64	10
14	102.3	12	115.6	9	91.5	HARDING	NM	13	-21	-11	03	00
16	75.3	43	225.4	34	176.2	HIDALGO	NM	199	-22	134	98	30
109	119.8	249	183.2	336	193.4	LEA	NM	53	6	62	88	56
41	131.1	55	135.3	65	144.6	LINCOLN	NM	3	7	10	21	05
31	348.1	23	82.2	50	149.3	LOS ALAMOS	NM	-76	82	-57	00	07
41	117.5	77	166.2	130	169.1	LUNA	NM	41	2	44	75	22
72	206.0	69	168.8	87	187.0	MC KINLEY	NM	-18	11	-9	04	46
30	90.2	36	124.7	19	81.4	MORA	NM	38	-35	-10	04	00
67	137.1	113	140.3	202	190.0	OTERO	NM	2	35	39	69	51
75	134.4	103	175.1	96	156.4	QUAY	NM	30	-11	16	32	11
75	96.2	115	138.5	145	214.7	RIO ARRIBA	NM	44	55	123	97	82
76	113.0	107	141.8	126	168.3	ROOSEVELT	NM	25	19	49	81	21
30	95.7	46	119.0	111	150.4	SANDOVAL	NM	24	26	57	86	07
70	120.1	113	122.8	238	179.9	SAN JUAN	NM	2	47	50	81	35
119	121.0	112	108.4	150	158.9	SAN MIGUEL	NM	-10	47	31	59	13
148	115.0	281	165.8	332	153.9	SANTA FE	NM	44	-7	34	63	09
80	144.9	86	127.4	129	168.8	SIERRA	NM	-12	33	17	32	22
43	129.0	58	156.0	84	206.2	SOCORRO	NM	21	32	60	87	73
46	81.7	67	130.9	98	170.1	TAOS	NM	60	30	108	97	23
41	159.7	46	170.6	55	181.7	TORRANCE	NM	7	7	14	28	38
54	158.3	53	149.9	45	150.6	UNION	NM	-5	1	-5	05	07
99	143.1	140	138.8	209	151.3	VALENCIA	NM	-3	9	6	14	08
144674*	209.4	163121*	214.3	166829*	218.3	NEW YORK	NY	2	2	4		
2528*	211.4	2890*	219.4	3046*	224.9	ALBANY	NY	4	3	6	15	90
335	152.9	386	175.9	461	202.3	ALLEGANY	NY	15	15	32	60	68
14654*	235.5	11709	184.8	11330*	223.5	BRONX	NY	-22	21	-5	05	89
1579	178.1	1868	192.1	2156	210.1	BROOME	NY	8	9	18	36	78
699	178.6	757	188.8	870	212.7	CATTARAUGUS	NY	6	13	19	38	80
647	172.5	663	175.7	743	191.3	CAYUGA	NY	2	9	11	22	53
1252	168.4	1370	178.8	1479	190.2	CHAUTAUQUA	NY	6	6	13	26	51
760	185.1	973*	220.8	1006*	222.5	CHEMUNG	NY	19	1	20	40	88
352	166.7	397	182.0	464	200.5	CHENANGO	NY	9	10	20	40	66
396	162.3	436	174.5	562	215.7	CLINTON	NY	8	24	33	61	83
511	189.6	565	197.0	680	209.0	COLUMBIA	NY	4	6	10	21	77
295	158.1	355	184.0	415	207.0	CORTLAND	NY	16	12	31	58	75
362	156.5	371	153.6	440	168.6	DELAWARE	NY	-2	10	8	17	21
1302	169.2	1617	183.0	1933	196.2	DUTCHESS	NY	8	7	16	32	60
8746*	213.5	10188*	224.2	10715*	233.2	ERIE	NY	5	4	9	19	94
324	184.6	357	199.5	414	214.3	ESSEX	NY	8	7	16	32	82
327	152.6	407	194.7	461	221.1	FRANKLIN	NY	28	14	45	77	87
552	193.2	613*	210.7	644	213.9	FULTON	NY	9	2	11	22	82
407	165.7	458	176.6	573	210.4	GENESEE	NY	7	19	27	51	78
360*	200.3	418*	210.7	548*	241.1	GREENE	NY	5	14	20	40	96
40	156.0	60	210.5	69	208.1	HAMILTON	NY	35	-1	33	62	76
685*	194.9	677	187.3	703	194.4	HERKIMER	NY	-4	4	0	08	58
801	175.3	869	194.0	996*	228.6	JEFFERSON	NY	11	18	30	57	92
25094*	224.4	19810	177.6	20482*	212.8	KINGS(BROOKLYN)	NY	-21	20	-5	05	80

* SIGNIFICANTLY MORE DEATHS THAN EXPECTED

WHITE MALE — ALL CANCERS COMBINED — ICD 140-208

1950-1959 DEATHS	1950-1959 RATE	1960-1969 DEATHS	1960-1969 RATE	1970-1979 DEATHS	1970-1979 RATE	COUNTY AND STATE		PERCENT CHANGE 60S/50S	PERCENT CHANGE 70S/60S	PERCENT CHANGE 70S/50S	PERCENTILE %CHANGE 70S/50S	PERCENTILE RATES 1970S
190	154.4	189	162.8	212	180.1	LEWIS	NY	5	11	17	33	36
320	154.4	406	184.9	484	204.4	LIVINGSTON	NY	20	11	32	60	71
391	162.3	459	182.0	553	210.5	MADISON	NY	12	16	30	56	78
4712*	188.4	5486*	198.5	6103*	212.4	MONROE	NY	5	7	13	26	80
655*	204.3	618	181.3	703	207.4	MONTGOMERY	NY	-11	14	2	09	75
6965*	220.9	10557*	226.7	12795*	227.4	NASSAU	NY	3	0	3	11	91
20247*	252.2	29522*	396.4	13274*	225.1	NEW YORK(MANHATTAN)	NY	57	-43	-11	03	90
1693*	191.4	1937	196.8	2388*	228.5	NIAGARA	NY	3	16	19	38	92
1883	157.7	2231	175.8	2501	195.5	ONEIDA	NY	11	11	24	46	59
3259*	195.0	3796*	209.1	4398*	228.9	ONONDAGA	NY	7	9	17	34	92
507	150.6	619	173.6	731	186.1	ONTARIO	NY	15	7	24	46	45
1571*	187.9	2006*	212.8	2409*	232.4	ORANGE	NY	13	9	24	46	93
292	171.0	344	197.9	320	179.7	ORLEANS	NY	16	-9	5	13	35
671	165.4	726	171.8	945	209.8	OSWEGO	NY	4	22	27	51	77
498	169.9	506	171.2	586	189.9	OTSEGO	NY	1	11	12	24	51
259	199.7	402*	215.1	629*	236.4	PUTNAM	NY	8	10	18	37	95
15621*	219.9	16536	182.4	20445*	218.4	QUEENS	NY	-17	20	-1	07	85
1327*	202.3	1515*	216.0	1661*	233.5	RENSSELAER	NY	7	8	15	31	94
2070*	223.8	2053	190.9	3007*	239.7	RICHMOND	NY	-15	26	7	16	96
798	186.0	1161	194.9	1713	218.1	ROCKLAND	NY	5	12	17	34	85
699	145.3	901	186.3	920	190.7	ST LAWRENCE	NY	28	2	31	59	52
704	182.1	916*	209.9	1154*	217.8	SARATOGA	NY	15	4	20	38	85
1508*	210.0	1701*	220.8	1671	209.2	SCHENECTADY	NY	5	-5	0	08	77
250	189.7	250	184.8	285	191.2	SCHOHARIE	NY	-3	3	1	08	52
135	173.8	141	177.8	163	189.7	SCHUYLER	NY	2	7	9	19	50
247	133.5	248	136.2	332	183.5	SENECA	NY	2	35	37	68	41
855	170.3	934	183.0	1028	200.7	STEUBEN	NY	7	10	18	35	66
3176	162.9	6105	191.2	9646*	227.1	SUFFOLK	NY	17	19	39	71	91
494*	192.1	606*	207.5	785*	227.0	SULLIVAN	NY	8	9	18	36	91
248	152.4	324	194.4	392	211.7	TIOGA	NY	28	9	39	70	79
384	165.4	432	178.0	525	201.8	TOMPKINS	NY	8	13	22	43	67
1086*	197.0	1372*	214.6	1616*	223.8	ULSTER	NY	9	4	14	27	89
415	192.2	488*	213.9	555	221.0	WARREN	NY	11	3	15	30	87
419	170.3	485	197.6	551	216.7	WASHINGTON	NY	16	10	27	52	84
544	167.5	588	172.2	757	211.4	WAYNE	NY	3	23	26	50	79
6136*	207.3	7894*	217.5	8789*	221.4	WESTCHESTER	NY	5	2	7	16	88
288	162.2	269	149.3	372	202.9	WYOMING	NY	-8	36	25	48	69
149	141.5	184	173.7	241	216.8	YATES	NY	23	25	53	84	84
14423	132.8	22082	160.3	32974	188.4	NORTH CAROLINA	NC	21	18	42		
273	141.5	448	165.0	686	198.0	ALAMANCE	NC	17	20	40	71	63
43	64.2	92	144.7	146	178.0	ALEXANDER	NC	72	23	111	97	33
43	114.4	66	162.1	90	179.4	ALLEGHANY	NC	42	11	57	86	35
92	175.0	86	137.8	122	170.5	ANSON	NC	-21	24	-3	06	24
120	133.8	143	144.4	184	157.3	ASHE	NC	8	9	18	35	11
35	74.5	84	150.1	95	153.4	AVERY	NC	101	2	106	97	09
147	154.4	188	172.6	309*	239.4	BEAUFORT	NC	12	39	55	85	96
69	149.3	85	176.5	101	195.8	BERTIE	NC	18	11	31	59	60
64	97.4	91	126.9	141	172.1	BLADEN	NC	30	36	77	94	25
52	112.0	101	152.5	173	148.0	BRUNSWICK	NC	36	-3	32	60	06
666	144.7	939	158.8	1317	186.7	BUNCOMBE	NC	10	18	29	55	45
160	100.0	272	134.5	412	161.8	BURKE	NC	34	20	62	88	15
252	141.8	388	165.6	581	188.8	CABARRUS	NC	17	14	33	62	49
159	115.3	286	167.6	376	164.9	CALDWELL	NC	45	-2	43	74	18
22	134.3	33	181.7	47	227.3	CAMDEN	NC	35	25	69	91	91
132	153.8	252*	235.2	321	205.4	CARTERET	NC	53	-13	34	63	72
46	106.6	78	167.8	83	158.8	CASWELL	NC	57	-5	49	81	12
225	114.6	388	149.9	597	174.0	CATAWBA	NC	31	16	52	83	27
106	138.3	135	149.2	184	167.7	CHATHAM	NC	8	12	21	42	21
91	128.5	102	125.2	180	182.6	CHEROKEE	NC	-3	46	42	74	40
45	157.6	53	181.1	85	222.0	CHOWAN	NC	15	23	41	72	88
21	77.0	41	130.8	51	127.0	CLAY	NC	70	-3	65	90	01
231	139.3	353	166.1	482	180.5	CLEVELAND	NC	19	9	30	56	36
131	117.2	182	143.4	301	203.6	COLUMBUS	NC	22	42	74	93	70
150	171.8	206	178.9	326	224.0	CRAVEN	NC	4	25	30	57	89
224	121.7	412	165.7	711	206.1	CUMBERLAND	NC	36	24	69	91	73
47	194.4	53	202.2	79	217.6	CURRITUCK	NC	4	8	12	24	85
37	138.7	72	216.5	108	206.9	DARE	NC	56	-4	49	81	74
255	126.4	396	143.1	638	162.5	DAVIDSON	NC	13	14	29	54	16
50	85.9	84	120.6	158	175.9	DAVIE	NC	40	46	105	97	30
116	136.5	153	154.8	254	213.8	DUPLIN	NC	13	38	57	86	81
379	163.3	571	191.7	799*	220.9	DURHAM	NC	17	15	35	65	87
105	118.7	196	184.2	272	219.8	EDGECOMBE	NC	55	19	85	95	86
525	133.0	903	174.6	1298	191.7	FORSYTH	NC	31	10	44	76	53
91	134.2	128	175.2	157	183.2	FRANKLIN	NC	31	5	37	67	40
371	120.7	691	166.1	1000	183.8	GASTON	NC	38	11	52	83	41
17	72.9	45	190.7	50	200.1	GATES	NC	161	5	174	99	66
36	144.3	43	142.4	48	134.1	GRAHAM	NC	-1	-6	-7	04	03
79	101.0	105	123.2	144	164.7	GRANVILLE	NC	22	34	63	89	18
30	133.5	51	168.2	80	221.0	GREENE	NC	26	31	66	90	87
763	140.7	1254	169.3	1885	201.6	GUILFORD	NC	20	19	43	75	67
129	140.1	224	194.0	280	198.2	HALIFAX	NC	38	2	41	73	63
143	119.2	235	177.7	339	206.3	HARNETT	NC	49	16	73	93	73
165	115.2	272	148.1	372	164.6	HAYWOOD	NC	29	11	43	74	18
218	152.0	323	161.3	545	186.3	HENDERSON	NC	6	16	23	44	45
51	144.6	70	167.3	102	193.3	HERTFORD	NC	16	16	34	63	56
16	63.3	46	170.0	58	176.3	HOKE	NC	169	4	179	99	30
40	175.3	42	193.5	40	187.4	HYDE	NC	10	-3	7	16	46
216	121.1	327	142.2	504	171.0	IREDELL	NC	17	20	41	73	24
98	139.2	103	127.5	141	141.9	JACKSON	NC	-8	11	2	09	04
228	144.6	326	170.7	486	214.1	JOHNSTON	NC	18	25	48	80	82
31	147.9	43	169.3	58	202.7	JONES	NC	14	20	37	68	69
81	138.8	128	149.7	214	198.1	LEE	NC	8	32	43	74	63
143	159.0	203	178.5	276	200.3	LENOIR	NC	12	12	26	49	66
103	123.4	168	157.2	263	184.6	LINCOLN	NC	27	17	50	81	42
103	116.6	162	143.8	239	169.3	MC DOWELL	NC	23	18	45	77	22
59	87.5	108	128.3	181	149.3	MACON	NC	47	16	71	92	07
73	93.8	93	106.7	142	150.7	MADISON	NC	14	41	61	88	07
75	162.0	104	179.4	132	201.1	MARTIN	NC	11	12	24	47	67
878	167.6	1410	193.1	2052	208.9	MECKLENBURG	NC	15	8	25	48	77

* SIGNIFICANTLY MORE DEATHS THAN EXPECTED

WHITE MALE — ALL CANCERS COMBINED — ICD 140-208

1950-1959 DEATHS	1950-1959 RATE	1960-1969 DEATHS	1960-1969 RATE	1970-1979 DEATHS	1970-1979 RATE	COUNTY AND STATE	PERCENT CHANGE 60S/50S	PERCENT CHANGE 70S/60S	PERCENT CHANGE 70S/50S	PERCENTILE %CHANGE 70S/50S	PERCENTILE RATES 1970S
75	124.5	96	141.0	137	171.9	MITCHELL NC	13	22	38	69	25
67	130.4	113	171.6	119	151.0	MONTGOMERY NC	32	-12	16	31	08
117	117.6	199	158.6	303	163.3	MOORE NC	35	3	39	70	17
173	139.0	231	155.9	365	204.0	NASH NC	12	31	47	79	70
294	172.0	417	191.5	600	215.3	NEW HANOVER NC	11	12	25	49	83
53	114.6	98	200.2	98	180.9	NORTHAMPTON NC	75	-10	58	86	37
89	117.5	169	187.8	299	200.6	ONSLOW NC	60	7	71	92	66
115	136.9	177	157.4	247	153.8	ORANGE NC	15	-2	12	25	09
37	121.2	62	181.4	88	225.9	PAMLICO NC	50	25	86	95	90
95	163.4	127	185.2	191*	242.7	PASQUOTANK NC	13	31	49	80	96
50	130.7	95	192.2	127	204.1	PENDER NC	47	6	56	85	70
26	108.7	45	159.6	55	155.0	PERQUIMANS NC	47	-3	43	74	10
74	135.0	122	171.8	153	176.1	PERSON NC	27	3	30	58	30
173	155.9	234	166.5	380	213.8	PITT NC	7	28	37	68	82
50	119.1	63	109.9	140	182.1	POLK NC	-8	66	53	84	39
195	107.8	323	131.9	542	164.2	RANDOLPH NC	22	24	52	83	17
134	148.0	202	181.9	296	218.0	RICHMOND NC	23	20	47	79	85
165	128.9	227	158.8	351	211.2	ROBESON NC	23	33	64	90	79
248	130.7	380	162.6	608	210.5	ROCKINGHAM NC	24	29	61	88	78
308	130.3	459	146.9	664	161.8	ROWAN NC	13	10	24	47	15
203	135.7	282	152.2	388	175.0	RUTHERFORD NC	12	15	29	55	28
134	121.9	177	132.6	303	200.4	SAMPSON NC	9	51	64	90	66
67	148.7	95	186.8	126	194.3	SCOTLAND NC	26	4	31	58	57
138	117.0	215	138.6	348	183.0	STANLY NC	18	32	56	85	40
103	131.1	103	114.2	206	181.7	STOKES NC	-13	59	39	69	38
191	122.2	273	138.0	432	177.6	SURRY NC	13	29	45	77	32
45	137.5	68	190.2	78	182.9	SWAIN NC	38	-4	33	61	40
86	160.8	84	125.1	145	154.7	TRANSYLVANIA NC	-22	24	-4	06	10
10	67.1	20	145.0	32	193.9	TYRRELL NC	116	34	189	99	57
187	156.2	214	140.0	363	182.2	UNION NC	-10	30	17	33	39
87	128.1	146	190.0	228*	256.6	VANCE NC	48	35	100	96	98
475	134.3	738	157.2	1302	203.5	WAKE NC	17	29	52	82	70
58	173.4	66	183.1	93	225.6	WARREN NC	6	23	30	57	90
38	124.2	59	174.0	89	228.8	WASHINGTON NC	40	31	84	95	92
80	117.0	101	122.4	170	159.6	WATAUGA NC	5	30	36	67	13
164	126.8	282	178.3	423	210.9	WAYNE NC	41	18	66	91	78
175	115.7	226	128.9	316	143.3	WILKES NC	11	11	24	46	04
146	137.3	233	190.4	327	213.2	WILSON NC	39	12	55	85	81
79	95.5	118	117.7	222	172.7	YADKIN NC	23	47	81	94	26
69	108.0	87	132.6	125	165.0	YANCEY NC	23	24	53	84	18
4169	147.8	4859	160.6	5335	165.9	NORTH DAKOTA ND	9	3	12		
30	126.6	48	213.9	30	128.4	ADAMS ND	69	-40	1	09	02
126	144.4	149	162.0	184	192.9	BARNES ND	12	19	34	63	55
55	111.7	65	139.0	98	213.2	BENSON ND	24	53	91	96	81
9	131.9	11	209.5	7	555.5	BILLINGS ND	59	165	321	99	99
90	142.6	91	142.6	99	150.2	BOTTINEAU ND	0	5	5	14	07
30	137.3	31	147.1	35	156.5	BOWMAN ND	7	6	14	28	11
67	176.9	46	146.6	57	186.5	BURKE ND	-17	27	5	14	45
204	191.2	223	168.8	273	160.7	BURLEIGH ND	-12	-5	-16	02	14
408	150.3	528	176.8	585	179.0	CASS ND	18	1	19	38	34
56	105.1	67	131.1	88	160.1	CAVALIER ND	25	22	52	83	14
67	152.3	89	187.9	76	161.8	DICKEY ND	23	-14	6	15	15
57	150.4	37	107.7	56	160.3	DIVIDE ND	-28	49	7	15	14
30	138.3	37	165.9	36	150.8	DUNN ND	20	-9	9	19	07
37	142.4	45	162.3	43	142.8	EDDY ND	14	-12	0	08	04
52	164.0	55	164.1	72	190.4	EMMONS ND	0	16	16	32	51
46	180.3	57	203.8	44	143.4	FOSTER ND	13	-30	-20	01	04
22	125.4	38	238.1	24	161.1	GOLDEN VALLEY ND	90	-32	28	54	14
256	145.1	309	154.5	394	185.7	GRAND FORKS ND	6	20	28	53	44
43	186.0	40	149.6	50	199.0	GRANT ND	-20	33	7	16	64
47	152.5	46	135.4	44	129.0	GRIGGS ND	-11	-5	-15	02	02
39	154.0	38	151.4	43	160.9	HETTINGER ND	-2	6	4	12	14
30	128.0	42	180.0	33	141.5	KIDDER ND	41	-21	11	22	04
61	134.1	69	147.7	71	149.2	LA MOURE ND	10	1	11	23	06
36	167.3	33	156.5	37*	839.8	LOGAN ND	-6	436	402	99	99
96	166.5	98	174.9	98	177.2	MC HENRY ND	5	1	6	15	32
50	177.1	63	209.5	60	151.3	MC INTOSH ND	18	-28	-15	02	08
41	118.0	50	148.6	47	146.0	MC KENZIE ND	26	-2	24	46	05
108	159.2	104	148.0	100	135.4	MC LEAN ND	-7	-9	-15	02	03
52	169.5	60	196.3	70	186.1	MERCER ND	16	-5	10	20	45
132	158.3	158	174.4	191	183.0	MORTON ND	10	5	16	31	40
76	146.1	94	182.0	81	160.0	MOUNTRAIL ND	25	-12	10	20	14
56	120.6	72	142.0	80	157.9	NELSON ND	18	11	31	58	12
20	196.1	15	252.9	14	124.4	OLIVER ND	29	-51	-37	00	01
71	110.6	115	166.3	127	170.9	PEMBINA ND	50	3	55	84	24
56	161.6	49	145.0	60	159.1	PIERCE ND	-20	10	-12	03	13
91	133.5	98	129.0	140	173.2	RAMSEY ND	-3	34	30	57	27
69	139.2	72	133.2	88	152.0	RANSOM ND	-4	14	9	19	08
32	119.4	51	209.9	43	174.2	RENVILLE ND	76	-17	46	78	28
148	149.8	159	146.8	195	175.4	RICHLAND ND	-2	19	17	34	29
45	149.6	47	153.1	53	177.4	ROLETTE ND	2	16	19	37	32
51	136.3	66	173.5	56	157.0	SARGENT ND	27	-9	15	30	11
29	160.0	32	168.5	29	142.6	SHERIDAN ND	5	-15	-11	03	04
10	161.8	13	184.1	6	126.2	SIOUX ND	14	-31	-22	01	01
10	142.2	12	187.4	11	651.2	SLOPE ND	32	247	358	99	99
121	193.2	130	174.6	147	166.9	STARK ND	-10	-4	-14	03	20
38	157.1	38	155.5	38	161.5	STEELE ND	-1	4	3	10	15
141	111.9	185	140.7	215	163.0	STUTSMAN ND	26	16	46	78	16
40	139.5	35	123.3	44	133.8	TOWNER ND	-12	9	-4	06	02
80	137.3	107	157.4	101	138.1	TRAILL ND	15	-12	1	08	03
137	150.1	166	163.9	173	166.3	WALSH ND	9	1	11	22	19
264	160.0	332	177.0	338	162.3	WARD ND	11	-8	1	09	15
75	158.2	78	156.5	98	186.1	WELLS ND	-1	19	18	35	45
132	136.9	166	164.3	153	155.4	WILLIAMS ND	20	-5	13	27	10
65990*	180.5	80184*	199.1	90969*	212.7	OHIO OH	10	7	18		
161	130.6	193	154.3	215	168.7	ADAMS OH	18	9	29	55	22
721	174.9	826	187.5	1010*	219.0	ALLEN OH	7	17	25	48	86

* SIGNIFICANTLY MORE DEATHS THAN EXPECTED

WHITE MALE ALL CANCERS COMBINED ICD 140-208

1950-1959 DEATHS	1950-1959 RATE	1960-1969 DEATHS	1960-1969 RATE	1970-1979 DEATHS	1970-1979 RATE	COUNTY AND STATE	PERCENT CHANGE 60S/50S	PERCENT CHANGE 70S/60S	PERCENT CHANGE 70S/50S	PERCENTILE %CHANGE 70S/50S	PERCENTILE RATES 1970S
244	137.1	302	155.8	371	180.8	ASHLAND OH	14	16	32	60	37
796*	196.0	829	191.3	939	199.6	ASHTABULA OH	-2	4	2	09	65
355	152.6	381	171.6	469	224.4	ATHENS OH	12	31	47	79	90
272	162.3	279	159.0	341	179.4	AUGLAIZE OH	-2	13	11	22	34
803	175.0	946*	203.9	965	213.1	BELMONT OH	16	5	22	43	81
190	142.4	234	163.4	271	183.2	BROWN OH	15	12	29	54	40
977	159.7	1351	183.5	1831	211.3	BUTLER OH	15	15	32	60	79
160	157.7	154	143.1	190	162.5	CARROLL OH	-9	14	3	11	16
185	133.9	238	171.0	277	188.1	CHAMPAIGN OH	28	10	41	72	48
845	172.1	1083	191.8	1286	209.0	CLARK OH	11	9	21	42	77
366	156.1	585	198.4	811*	218.4	CLERMONT OH	27	10	40	71	85
177	125.4	239	166.8	309	208.2	CLINTON OH	33	25	66	90	76
825	171.5	994	195.0	1052	197.0	COLUMBIANA OH	14	1	15	30	62
234	140.9	303	179.1	354	201.9	COSHOCTON OH	27	13	43	75	67
347	165.0	405	184.4	497	220.7	CRAWFORD OH	12	20	34	63	87
12853*	218.7	14883*	232.8	14647*	230.1	CUYAHOGA OH	6	-1	5	13	93
323	140.8	370	155.1	509	199.0	DARKE OH	10	28	41	73	64
215	161.7	218	154.4	274	181.1	DEFIANCE OH	-5	17	12	24	37
212	132.4	256	153.1	400	218.5	DELAWARE OH	16	43	65	90	85
555*	193.5	635*	204.4	714	215.2	ERIE OH	6	5	11	23	83
379	136.9	573	182.6	719	200.4	FAIRFIELD OH	33	10	46	78	66
180	148.7	198	158.6	237	181.9	FAYETTE OH	7	15	22	44	38
3767*	185.5	4894*	210.6	5869*	229.0	FRANKLIN OH	14	9	23	46	92
185	134.6	197	137.8	250	162.9	FULTON OH	2	18	21	41	16
168	136.9	208	166.5	249	186.5	GALLIA OH	22	12	36	67	45
203	140.8	290	156.6	431	184.3	GEAUGA OH	11	18	31	58	42
307	135.0	476	165.9	698	191.5	GREENE OH	23	15	42	73	53
351	141.8	345	143.1	424	191.9	GUERNSEY OH	1	34	35	65	53
6160*	203.6	7479*	229.9	8128*	245.3	HAMILTON OH	13	7	20	40	97
386	162.6	465	181.1	484	177.0	HANCOCK OH	11	-2	9	19	31
251	162.3	225	151.7	324	222.0	HARDIN OH	-7	46	37	67	88
150	149.6	167	167.7	193	198.8	HARRISON OH	12	19	33	61	64
215	171.0	220	169.1	257	191.2	HENRY OH	-1	13	12	24	53
216	128.6	277	163.2	349	198.6	HIGHLAND OH	27	22	54	84	63
153	143.9	213	200.7	225	199.9	HOCKING OH	39	0	39	70	65
121	131.6	148	146.2	151	134.1	HOLMES OH	11	-8	2	09	02
323	152.7	402	189.2	430	194.9	HURON OH	24	3	28	52	58
207	140.2	261	175.3	334	224.1	JACKSON OH	25	28	60	87	89
816	186.4	949*	210.3	1008*	217.5	JEFFERSON OH	13	3	17	33	85
272	143.3	347	177.7	377	177.2	KNOX OH	24	0	24	46	31
660	183.9	1074*	207.3	1545*	221.5	LAKE OH	13	7	20	40	88
351	160.6	504*	209.2	559	207.9	LAWRENCE OH	30	-1	29	56	76
609	156.0	765	178.0	1029	214.7	LICKING OH	14	21	38	68	82
273	155.3	322	173.5	424	217.9	LOGAN OH	12	26	40	72	85
1250	184.8	1717*	215.3	1935	216.0	LORAIN OH	16	0	17	33	83
3586*	200.3	4278*	219.3	4598*	237.1	LUCAS OH	10	8	18	37	95
159	153.9	174	157.5	247	204.9	MADISON OH	2	30	33	62	71
2466*	212.5	2753*	216.7	2872*	219.4	MAHONING OH	2	1	3	11	86
380	151.5	482	185.5	566	210.7	MARION OH	22	14	39	70	78
336	151.5	448	164.8	608	188.2	MEDINA OH	9	14	24	47	48
180	138.3	225	186.2	244	208.2	MEIGS OH	35	12	51	82	76
219	150.8	261	168.4	349	210.5	MERCER OH	12	25	40	71	78
448	153.9	593	184.0	744	205.2	MIAMI OH	20	11	33	62	72
137	134.6	153	167.8	159	172.6	MONROE OH	25	3	28	54	26
2825	178.9	3608	192.9	4221	206.8	MONTGOMERY OH	8	7	16	31	74
124	156.5	110	145.4	117	154.9	MORGAN OH	-7	7	-1	07	10
132	136.4	139	140.7	155	140.8	MORROW OH	3	0	3	11	04
511	140.3	651	176.4	829*	223.7	MUSKINGUM OH	26	27	59	87	89
91	123.1	122	179.0	127	198.5	NOBLE OH	45	11	61	88	63
288	186.8	327	192.3	383	205.6	OTTAWA OH	3	7	10	21	73
128	163.7	126	158.6	151	180.3	PAULDING OH	-3	14	10	21	36
245	149.5	299	193.4	312	212.3	PERRY OH	29	10	42	74	80
218	153.7	227	149.0	312	184.0	PICKAWAY OH	-3	23	20	39	42
102	127.3	138	153.0	198	180.7	PIKE OH	20	18	42	74	36
508	171.9	638	182.6	780	186.0	PORTAGE OH	6	2	8	17	44
201	139.7	201	131.1	271	163.0	PREBLE OH	-6	24	17	33	16
205	154.2	199	149.9	247	178.1	PUTNAM OH	-3	19	16	31	33
658	160.8	837	180.8	1055	207.1	RICHLAND OH	12	15	29	55	75
347	134.4	420	144.6	555	179.4	ROSS OH	8	24	34	63	34
410	171.6	424	165.4	537	200.3	SANDUSKY OH	-4	21	17	33	66
630	171.4	797	200.8	913*	222.4	SCIOTO OH	17	11	30	56	88
444	166.5	449	166.1	535	198.3	SENECA OH	0	19	19	38	63
209	145.2	238	156.6	274	162.6	SHELBY OH	8	4	12	24	16
2374	176.7	2892	191.1	3351	206.0	STARK OH	8	8	17	33	73
3272*	187.7	4234*	206.1	4497	205.1	SUMMIT OH	10	0	9	19	71
1364*	187.4	1756*	207.0	2000	206.4	TRUMBULL OH	10	0	10	21	73
604	158.9	731	188.0	820	200.7	TUSCARAWAS OH	18	7	26	50	66
173	151.3	187	166.1	213	180.9	UNION OH	10	9	20	38	37
226	153.0	257	171.6	331	216.7	VAN WERT OH	12	26	42	73	84
67	124.6	72	129.1	114	196.9	VINTON OH	4	53	58	86	61
294	161.2	418	180.1	573	196.8	WARREN OH	12	9	22	43	61
353	143.5	426	163.5	536	192.4	WASHINGTON OH	14	18	34	64	54
418	141.0	481	143.8	667	177.9	WAYNE OH	2	24	26	50	33
220	148.9	255	165.5	289	173.2	WILLIAMS OH	11	5	16	32	27
510	176.8	558	176.5	648	181.7	WOOD OH	0	3	3	10	38
159	146.7	180	167.1	210	187.6	WYANDOT OH	14	12	28	53	47
15107	155.8	19160	175.2	24689	200.0	OKLAHOMA OK	12	14	28		
63	95.5	106	153.7	133	183.9	ADAIR OK	61	20	93	96	42
95	150.2	102	166.2	109	187.4	ALFALFA OK	11	13	25	48	46
96	166.0	103	169.6	113	168.3	ATOKA OK	2	-1	1	09	21
36	104.8	66	191.5	58	154.5	BEAVER OK	83	-19	47	79	09
171	162.1	195	171.0	221	185.8	BECKHAM OK	5	9	15	29	44
114	165.2	123	167.4	147	188.7	BLAINE OK	1	13	14	29	49
223	154.6	265	175.4	342	211.8	BRYAN OK	13	21	37	68	79
248	161.1	264	174.4	299	183.7	CADDO OK	8	5	14	28	41
166	138.8	199	147.9	313	186.8	CANADIAN OK	7	26	35	64	45
246	143.3	340	170.2	492*	232.3	CARTER OK	19	36	62	89	93
92	124.1	148	163.5	209	188.8	CHEROKEE OK	32	16	52	83	49
124	139.6	147	168.3	163	182.6	CHOCTAW OK	21	8	31	58	40

* SIGNIFICANTLY MORE DEATHS THAN EXPECTED

WHITE MALE | ALL CANCERS COMBINED | ICD 140-208

1950-1959 DEATHS	1950-1959 RATE	1960-1969 DEATHS	1960-1969 RATE	1970-1979 DEATHS	1970-1979 RATE	COUNTY AND STATE	PERCENT CHANGE 60S/50S	PERCENT CHANGE 70S/60S	PERCENT CHANGE 70S/50S	PERCENTILE %CHANGE 70S/50S	PERCENTILE RATES 1970S
28	136.3	28	126.7	31	144.5	CIMARRON OK	-7	14	6	15	05
177	102.2	295	132.9	448	150.7	CLEVELAND OK	30	13	47	79	07
75	201.6	67	179.5	96	262.9	COAL OK	-11	46	30	58	98
256	151.2	390	181.1	494	188.3	COMANCHE OK	20	4	25	47	48
95	188.4	97	198.7	91	187.9	COTTON OK	5	-5	0	08	47
134	126.3	178	185.3	181	202.2	CRAIG OK	47	9	60	87	68
314	159.1	375	181.9	501	212.4	CREEK OK	14	17	33	62	80
157	164.0	168	160.5	213	182.1	CUSTER OK	-2	13	11	22	39
104	127.8	126	123.7	253	187.6	DELAWARE OK	-3	52	47	79	47
60	145.3	57	144.1	69	171.8	DEWEY OK	-1	19	18	36	25
42	116.0	70	196.2	63	173.5	ELLIS OK	69	-12	50	81	27
355	152.4	448	171.3	541	188.8	GARFIELD OK	12	10	24	46	49
191	145.7	211	145.8	357	212.9	GARVIN OK	0	46	46	78	80
263	154.8	270	154.2	353	193.1	GRADY OK	0	25	25	48	56
92	155.8	107	195.6	95	180.0	GRANT OK	26	-8	16	31	36
109	163.4	120	189.0	153*	243.4	GREER OK	3	29	33	61	97
49	137.3	59	157.8	52	144.2	HARMON OK	15	-9	5	13	05
44	138.8	58	182.8	54	169.3	HARPER OK	32	-7	22	43	22
76	133.8	98	170.3	139	220.1	HASKELL OK	27	29	64	90	86
125	131.3	186	196.5	171	175.5	HUGHES OK	50	-11	34	63	29
144	148.4	177	160.0	220	191.7	JACKSON OK	8	20	29	55	53
81	138.9	86	150.0	96	176.4	JEFFERSON OK	8	18	27	51	31
61	121.3	105	200.9	102	184.8	JOHNSTON OK	66	-8	52	83	43
393	170.4	423	165.3	546	199.4	KAY OK	-3	21	17	33	65
73	119.2	97	150.2	142	192.7	KINGFISHER OK	26	28	62	88	55
140	154.2	143	157.9	181	199.9	KIOWA OK	2	27	30	56	65
65	149.6	81	183.2	86	166.2	LATIMER OK	22	-9	11	23	19
208	129.1	310	173.1	410	205.6	LE FLORE OK	34	19	59	87	72
201	181.2	236	194.1	263	193.6	LINCOLN OK	7	0	7	16	56
167	167.6	168	167.7	197	171.2	LOGAN OK	0	2	2	10	24
69	190.1	53	146.3	82	200.6	LOVE OK	-23	37	6	14	66
96	142.2	123	166.3	174	200.1	MC CLAIN OK	17	20	41	72	65
148	131.6	233	189.0	278	197.2	MC CURTAIN OK	44	4	50	81	62
88	138.2	105	153.2	174	204.3	MC INTOSH OK	15	29	48	80	71
81	159.0	75	148.2	84	154.0	MAJOR OK	-7	4	-3	06	09
62	148.8	78	150.1	132	191.4	MARSHALL OK	1	28	29	54	53
172	171.1	230	194.0	326	215.8	MAYES OK	13	11	26	50	83
97	141.8	118	153.1	156	195.6	MURRAY OK	8	28	38	69	59
407	166.8	530	202.3	595	215.3	MUSKOGEE OK	21	6	29	55	83
88	141.8	100	153.7	127	194.3	NOBLE OK	8	26	37	68	57
94	152.9	107	180.7	131	206.5	NOWATA OK	18	14	35	65	74
90	163.2	91	165.2	95	173.6	OKFUSKEE OK	1	5	6	15	27
2216	177.0	3082	187.2	4316*	221.4	OKLAHOMA OK	6	18	25	48	88
288	160.6	326	177.4	386	205.2	OKMULGEE OK	10	16	28	53	72
222	152.2	257	163.1	303	182.6	OSAGE OK	7	12	20	39	40
277	194.5	300	203.6	377*	247.8	OTTAWA OK	5	22	27	52	97
116	149.5	120	154.3	156	193.5	PAWNEE OK	3	25	29	56	56
272	159.5	307	163.7	402	189.4	PAYNE OK	3	16	19	37	50
294	156.7	316	160.2	481*	228.9	PITTSBURG OK	2	43	46	78	92
217	154.6	252	161.7	314	188.5	PONTOTOC OK	5	17	22	43	48
316	149.0	410	175.9	497	194.9	POTTAWATOMIE OK	18	11	31	58	58
90	151.5	111	180.2	127	185.9	PUSHMATAHA OK	19	3	23	44	44
53	166.7	43	148.6	46	153.5	ROGER MILLS OK	-11	3	-8	04	09
145	145.9	184	161.4	271	190.2	ROGERS OK	11	18	30	57	51
228	164.5	272	194.0	319	215.9	SEMINOLE OK	18	11	31	59	83
109	133.7	149	160.7	232	192.7	SEQUOYAH OK	20	20	44	76	55
236	148.5	391	208.4	478	217.8	STEPHENS OK	40	5	47	79	85
87	139.1	95	141.5	115	152.5	TEXAS OK	2	8	10	20	08
121	157.7	129	173.0	128	175.4	TILLMAN OK	10	1	11	23	29
1753	175.5	2460	194.7	3175*	213.5	TULSA OK	11	10	22	42	81
88	132.4	134	161.4	222	189.6	WAGONER OK	22	17	43	75	50
205	145.8	325	192.4	411	199.9	WASHINGTON OK	32	4	37	68	65
103	123.6	130	158.1	141	166.3	WASHITA OK	28	5	35	64	19
117	163.5	128	179.5	107	140.7	WOODS OK	10	-22	-14	02	04
109	129.4	104	118.2	134	152.4	WOODWARD OK	-9	29	18	35	08
12125	155.8	16169	173.5	21206	190.2	OREGON OR	11	10	22		
121	129.2	158	163.5	178	174.1	BAKER OR	27	6	35	65	27
153	123.9	252	166.5	318	162.5	BENTON OR	34	-2	31	59	16
716	152.7	1022	166.3	1610	192.0	CLACKAMAS OR	9	15	26	49	53
288	181.0	315	183.3	358	199.7	CLATSOP OR	1	9	10	21	65
187	154.4	242	178.4	290	178.0	COLUMBIA OR	16	0	15	30	33
351	175.0	400	163.0	597	199.7	COOS OR	-7	22	14	29	65
53	144.7	80	174.2	125	196.0	CROOK OR	20	13	35	65	60
51	117.5	87	132.7	179	176.4	CURRY OR	13	33	50	82	30
132	124.9	231	162.0	376	166.3	DESCHUTES OR	30	3	33	62	19
367	147.9	523	169.0	762	191.1	DOUGLAS OR	14	13	29	56	52
26	184.3	20	155.8	18	158.0	GILLIAM OR	-15	1	-14	02	12
54	144.2	66	165.8	74	171.7	GRANT OR	15	4	19	38	25
43	148.8	68	206.3	84	232.2	HARNEY OR	39	13	56	85	93
101	161.8	123	170.1	157	203.8	HOOD RIVER OR	5	20	26	49	70
482	140.2	764	166.9	1181	195.2	JACKSON OR	19	17	39	70	59
25	125.7	34	129.0	71	193.5	JEFFERSON OR	3	50	54	84	56
244	154.6	349	166.2	561	187.7	JOSEPHINE OR	7	13	21	42	47
246	136.4	345	169.4	454	188.6	KLAMATH OR	24	11	38	69	48
47	152.0	53	159.0	57	159.4	LAKE OR	5	0	5	13	13
823	144.0	1188	158.5	1800	181.5	LANE OR	10	15	26	50	38
166	126.6	236	147.1	351	165.6	LINCOLN OR	16	13	31	58	19
380	160.5	463	155.5	692	182.5	LINN OR	-3	17	14	27	39
133	138.2	195	175.6	230	184.6	MALHEUR OR	27	5	34	63	42
787	141.8	1067	152.9	1553	175.8	MARION OR	8	15	24	47	30
42	163.9	45	175.8	50	167.4	MORROW OR	7	-5	2	10	20
4414	176.3	5534*	203.4	5871*	217.7	MULTNOMAH OR	15	7	23	46	85
135	111.1	230	149.2	303	152.3	POLK OR	34	2	37	68	08
16	143.0	21	204.9	20	149.1	SHERMAN OR	43	-27	4	12	06
115	136.2	197	196.2	211	169.5	TILLAMOOK OR	44	-14	24	47	23
284	135.1	397	169.5	482	185.3	UMATILLA OR	26	9	37	68	43
161	163.6	163	156.2	159	139.5	UNION OR	-5	-11	-15	02	04
44	130.2	64	167.5	53	124.9	WALLOWA OR	29	-25	-4	06	01
123	139.2	172	168.5	237	204.7	WASCO OR	21	22	47	79	71

* SIGNIFICANTLY MORE DEATHS THAN EXPECTED

WHITE MALE — ALL CANCERS COMBINED — ICD 140-208

1950-1959 DEATHS	1950-1959 RATE	1960-1969 DEATHS	1960-1969 RATE	1970-1979 DEATHS	1970-1979 RATE	COUNTY AND STATE		PERCENT CHANGE 60S/50S	70S/60S	70S/50S	PERCENTILE %CHANGE 70S/50S	RATES 1970S
492	140.2	732	147.7	1284	176.6	WASHINGTON	OR	5	20	26	50	31
28	238.7	20	179.3	25	219.6	WHEELER	OR	-25	23	-8	04	86
295	165.4	313	152.7	435	175.3	YAMHILL	OR	-8	15	6	14	29
87507*	189.1	101392*	200.4	113370*	209.6	PENNSYLVANIA	PA	6	5	11		
304	143.2	365	152.1	533	184.7	ADAMS	PA	6	21	29	55	42
13331*	206.3	15557*	222.3	16706*	233.7	ALLEGHENY	PA	8	5	13	27	94
582	154.4	713	179.7	810	192.3	ARMSTRONG	PA	16	7	25	48	54
1564*	200.8	1841*	208.3	2130*	220.5	BEAVER	PA	4	6	10	20	87
261	134.0	273	129.9	388	169.7	BEDFORD	PA	-3	31	27	51	23
2168	166.8	2714	187.3	3153	196.2	BERKS	PA	12	5	18	35	60
1168	172.4	1380	195.3	1529	214.3	BLAIR	PA	13	10	24	47	82
437	160.5	459	168.7	584	202.0	BRADFORD	PA	5	20	26	49	68
1368	175.4	2142	190.0	3043	199.0	BUCKS	PA	8	5	13	27	64
802	166.3	958	178.3	1188	192.4	BUTLER	PA	7	8	16	31	54
1589	177.5	1760	187.5	1907	198.6	CAMBRIA	PA	6	6	12	24	63
57	189.2	70	235.5	64	189.5	CAMERON	PA	24	-20	0	08	50
499	184.4	550	195.2	613	202.7	CARBON	PA	6	4	10	20	69
385	156.6	453	160.9	613	187.5	CENTRE	PA	3	17	20	39	47
1166	164.5	1524	173.3	2089	191.7	CHESTER	PA	5	11	17	32	53
260	139.4	290	152.3	364	180.7	CLARION	PA	9	19	30	56	37
634	155.6	730	177.7	814	192.7	CLEARFIELD	PA	14	8	24	46	55
293	178.3	292	169.0	389	206.9	CLINTON	PA	-5	22	16	32	74
385	148.9	475	173.3	563	186.5	COLUMBIA	PA	16	8	25	49	45
661	174.5	704	178.7	845	197.4	CRAWFORD	PA	2	10	13	26	62
672	160.5	990	184.3	1339	194.5	CUMBERLAND	PA	15	6	21	42	58
1523	172.3	1974*	205.9	2135*	214.2	DAUPHIN	PA	20	4	24	47	82
3467*	202.7	4575*	208.8	5649*	227.2	DELAWARE	PA	3	9	12	25	91
282	177.8	333	196.7	376	204.2	ELK	PA	11	4	15	30	71
1985*	200.6	2239*	205.7	2485*	212.4	ERIE	PA	3	3	6	14	80
1462	177.3	1649	197.1	1759	207.4	FAYETTE	PA	11	5	17	33	75
43	167.4	44	145.0	58	156.6	FOREST	PA	-13	8	-6	05	11
500	144.2	688	170.5	924	186.7	FRANKLIN	PA	18	9	30	56	45
66	139.9	75	149.3	88	150.6	FULTON	PA	7	1	8	17	07
301	144.3	378	173.9	391	179.8	GREENE	PA	20	3	25	48	35
282	158.4	342	187.6	370	190.4	HUNTINGDON	PA	18	1	20	39	51
549	154.3	603	160.8	662	165.8	INDIANA	PA	4	3	7	16	19
373	147.0	432	167.3	454	169.9	JEFFERSON	PA	14	2	16	31	23
86	120.2	129	168.5	144	163.0	JUNIATA	PA	40	-3	36	66	16
2422*	205.6	2561*	208.9	2782*	216.1	LACKAWANNA	PA	2	3	5	13	83
1745	154.8	2151	165.3	2952	193.2	LANCASTER	PA	7	17	25	48	56
849	169.4	985	186.2	1087	198.9	LAWRENCE	PA	10	7	17	34	64
634	170.0	795	187.5	878	174.7	LEBANON	PA	10	-7	3	10	28
1737	179.5	2251	199.1	2684	205.8	LEHIGH	PA	11	3	15	29	73
3310*	197.2	3319	184.5	3792	199.9	LUZERNE	PA	-6	8	1	09	65
851	170.9	1006	188.8	1158	205.3	LYCOMING	PA	10	9	20	39	72
469	173.8	502	182.7	542	188.6	MC KEAN	PA	5	3	9	18	48
927	171.3	1074	184.7	1214	193.7	MERCER	PA	8	5	13	26	56
275	153.8	377	192.2	424	195.8	MIFFLIN	PA	25	2	27	52	60
338	186.5	427	200.6	585	200.4	MONROE	PA	8	0	7	16	66
3207	183.5	4361	189.1	5632	199.4	MONTGOMERY	PA	3	5	9	18	64
110	128.3	118	134.1	173	191.0	MONTOUR	PA	5	42	49	80	52
1686*	193.5	1998*	204.7	2230	208.1	NORTHAMPTON	PA	6	2	8	17	76
863	168.6	1030	195.7	1094	197.4	NORTHUMBERLAND	PA	16	1	17	33	62
229	196.1	219	165.1	295	196.6	PERRY	PA	-16	19	0	08	61
17693*	231.2	18018*	241.4	16116*	242.9	PHILADELPHIA	PA	4	1	5	13	96
98	180.0	152	204.7	204	188.7	PIKE	PA	14	-8	5	13	49
143	147.5	167	185.2	181	194.7	POTTER	PA	26	5	32	60	58
1628*	182.9	1740	189.7	1833	192.5	SCHUYLKILL	PA	4	1	5	13	54
132	126.0	156	128.4	222	160.1	SNYDER	PA	2	25	27	51	14
635	160.1	679	163.3	705	164.8	SOMERSET	PA	2	1	3	11	18
72	199.0	64	178.0	82	210.8	SULLIVAN	PA	-11	18	6	14	78
270	161.8	290	168.7	359	190.3	SUSQUEHANNA	PA	4	13	18	35	51
300	166.8	295	158.8	364	185.9	TIOGA	PA	-5	17	11	23	44
139	133.1	152	132.4	233	178.7	UNION	PA	-1	35	34	64	34
513	169.6	574	188.6	643	205.5	VENANGO	PA	11	9	21	42	72
326	133.8	393	157.3	424	169.2	WARREN	PA	18	8	27	51	22
1782	182.1	2135*	205.8	2332*	214.4	WASHINGTON	PA	13	4	18	35	82
275	162.1	305	173.5	378	192.0	WAYNE	PA	7	11	18	37	54
2600	177.5	3252	193.3	3836	203.9	WESTMORELAND	PA	9	5	15	30	70
144	166.7	165	182.6	208	199.5	WYOMING	PA	10	9	20	39	65
1600	167.2	1980	171.9	2569	190.5	YORK	PA	3	11	14	28	52
7399*	208.0	9025*	224.5	10527*	236.9	RHODE ISLAND	RI	8	6	14		
292*	207.3	394*	226.8	493	223.5	BRISTOL	RI	9	-1	8	17	89
798*	213.8	1182*	236.8	1591*	249.9	KENT	RI	11	6	17	33	97
476*	209.7	580*	216.1	784*	249.8	NEWPORT	RI	3	16	19	38	97
5493*	210.4	6367*	226.2	6937*	236.2	PROVIDENCE	RI	7	4	12	25	95
340	169.9	502	198.1	722	218.9	WASHINGTON	RI	17	11	29	55	86
6736	151.1	10003	173.9	15057	199.9	SOUTH CAROLINA	SC	15	15	32		
68	115.0	116	189.4	108	147.3	ABBEVILLE	SC	65	-22	28	53	06
224	160.7	361	183.3	596	208.8	AIKEN	SC	14	14	30	57	77
28	175.0	26	161.0	48	274.6	ALLENDALE	SC	-8	71	57	86	98
399	165.6	555	177.2	754	179.5	ANDERSON	SC	7	1	8	18	35
46	152.8	61	180.1	90	222.2	BAMBERG	SC	18	23	45	77	88
48	158.0	66	183.3	98	220.6	BARNWELL	SC	16	20	40	71	87
54	153.7	101	180.6	181	153.8	BEAUFORT	SC	18	-15	0	08	09
55	126.1	95	148.2	240	207.8	BERKELEY	SC	18	40	65	90	76
27	137.2	30	154.2	46	189.9	CALHOUN	SC	12	23	38	69	51
623*	209.1	964*	238.5	1197*	227.0	CHARLESTON	SC	14	-5	9	18	91
136	146.2	189	168.4	307*	237.0	CHEROKEE	SC	15	41	62	89	95
97	159.2	113	158.7	200	229.0	CHESTER	SC	0	44	44	75	92
97	129.8	131	154.6	172	174.2	CHESTERFIELD	SC	19	13	34	64	28
30	83.5	74	180.8	101	192.3	CLARENDON	SC	116	6	130	98	54
91	162.9	119	188.0	176	233.9	COLLETON	SC	15	24	44	75	94
133	148.1	158	146.0	244	179.3	DARLINGTON	SC	-1	23	21	41	34
65	123.9	95	148.8	137	193.3	DILLON	SC	20	30	56	85	56

* SIGNIFICANTLY MORE DEATHS THAN EXPECTED

WHITE MALE | ALL CANCERS COMBINED | ICD 140-208

1950-1959 DEATHS	1950-1959 RATE	1960-1969 DEATHS	1960-1969 RATE	1970-1979 DEATHS	1970-1979 RATE	COUNTY AND STATE	PERCENT CHANGE 60S/50S	PERCENT CHANGE 70S/60S	PERCENT CHANGE 70S/50S	PERCENTILE %CHANGE 70S/50S	PERCENTILE RATES 1970S
62	160.9	94	171.7	187	211.3	DORCHESTER SC	7	23	31	59	79
47	174.6	56	171.7	77	198.9	EDGEFIELD SC	-2	16	14	28	64
47	186.2	59	150.6	82	182.1	FAIRFIELD SC	-19	21	-2	06	39
206	153.5	326	186.2	501*	222.6	FLORENCE SC	21	20	45	77	89
56	136.6	86	161.8	164	197.6	GEORGETOWN SC	19	22	45	76	62
727	155.2	1200	190.0	1734	208.0	GREENVILLE SC	22	9	34	63	76
141	158.8	255	194.0	362	217.3	GREENWOOD SC	22	12	37	67	84
48	173.9	62	181.9	100	233.3	HAMPTON SC	5	28	34	64	94
134	100.0	284	156.6	517	169.8	HORRY SC	57	21	90	96	50
20	113.1	41	231.3	50	217.2	JASPER SC	105	-6	92	96	84
85	139.6	140	195.3	229*	249.1	KERSHAW SC	40	28	78	94	97
108	141.1	148	145.7	251	173.0	LANCASTER SC	3	19	23	44	26
158	143.1	216	160.3	328	203.5	LAURENS SC	12	27	42	74	70
45	142.5	59	182.2	62	173.2	LEE SC	28	-5	22	42	27
184	133.9	306	143.5	595	182.1	LEXINGTON SC	7	27	36	66	39
20	133.0	44	265.9	42	242.5	MC CORMICK SC	100	-9	82	95	96
68	148.2	92	161.9	152	233.5	MARION SC	9	44	58	86	94
68	145.4	83	159.2	122	177.5	MARLBORO SC	9	11	22	43	32
135	166.8	157	160.4	213	188.7	NEWBERRY SC	-4	18	13	26	48
164	145.4	222	164.5	343	196.6	OCONEE SC	13	20	35	65	61
148	150.2	217	180.3	306	193.0	ORANGEBURG SC	20	7	28	54	55
174	136.4	234	136.8	391	172.9	PICKENS SC	0	26	27	51	26
521	158.6	751	175.6	1077	200.2	RICHLAND SC	11	14	26	50	66
57	140.7	86	183.3	103	182.0	SALUDA SC	30	-1	29	56	39
573	145.5	769	157.1	1189	191.2	SPARTANBURG SC	8	22	31	59	52
123	157.6	178	162.5	263	192.2	SUMTER SC	3	18	22	43	54
93	119.2	147	165.4	177	189.2	UNION SC	39	14	59	87	49
67	150.4	84	160.1	145*	240.7	WILLIAMSBURG SC	6	50	60	87	96
235	157.6	353	172.1	600*	225.8	YORK SC	9	31	43	75	90
4878	156.8	5627	165.9	6179	174.4	SOUTH DAKOTA SD	6	5	11		
42	164.1	49	179.7	34	136.7	AURORA SD	10	-24	-17	02	03
165	160.9	189	172.5	252	230.3	BEADLE SD	7	34	43	74	93
23	214.4	11	94.0	16	133.8	BENNETT SD	-56	42	-38	00	02
78	148.8	90	161.9	120	201.6	BON HOMME SD	9	25	36	65	67
146	169.6	183	187.8	175	170.3	BROOKINGS SD	11	-9	0	08	23
250	174.2	292	183.6	294	170.2	BROWN SD	5	-7	-2	06	23
34	105.6	61	191.3	50	157.1	BRULE SD	81	-18	49	80	11
5	170.3	3	64.4	9*	423.0	BUFFALO SD	-62	557	148	98	99
70	174.5	64	144.2	101	212.7	BUTTE SD	-17	47	22	43	80
25	195.0	23	160.8	23	158.4	CAMPBELL SD	-18	-2	-19	01	12
94	174.8	100	181.6	90	163.9	CHARLES MIX SD	4	-10	-6	05	17
67	159.0	84	191.7	75	181.8	CLARK SD	21	-5	14	29	38
70	136.0	79	160.7	92	187.1	CLAY SD	18	16	38	68	46
149	160.7	180	173.3	179	164.5	CODINGTON SD	8	-5	2	10	18
29	174.3	41	243.7	24	175.3	CORSON SD	40	-28	1	08	29
40	145.4	45	165.6	41	141.4	CUSTER SD	14	-15	-3	06	04
131	155.7	167	184.4	195	205.6	DAVISON SD	18	11	32	60	73
89	139.4	116	175.5	120	182.3	DAY SD	26	4	31	58	39
58	146.8	74	184.1	62	154.1	DEUEL SD	25	-16	5	13	09
29	194.3	30	237.9	41*	308.8	DEWEY SD	22	30	59	87	99
32	145.9	42	166.3	58	202.6	DOUGLAS SD	14	22	39	70	69
46	165.6	56	172.3	64	167.7	EDMUNDS SD	4	-3	1	09	21
94	127.4	114	135.2	109	150.6	FALL RIVER SD	6	11	18	36	07
39	166.5	44	168.7	33	132.5	FAULK SD	1	-21	-20	01	02
74	139.8	90	158.5	86	146.8	GRANT SD	13	-7	5	13	05
75	172.9	89	200.2	80	181.4	GREGORY SD	16	-9	5	13	37
25	143.5	25	164.5	24	159.7	HAAKON SD	15	-3	11	23	13
63	152.6	64	141.2	82	183.8	HAMLIN SD	-7	30	20	40	41
43	142.7	66	200.5	50	150.7	HAND SD	40	-25	6	14	07
30	157.1	36	162.8	36	154.4	HANSON SD	4	-5	-2	07	09
14	111.1	23	212.0	20	210.0	HARDING SD	91	-1	89	96	78
52	127.5	86	184.8	103	207.2	HUGHES SD	45	12	63	89	75
82	148.6	89	128.2	105	139.3	HUTCHINSON SD	-14	9	-6	05	04
25	172.6	22	154.4	27	169.1	HYDE SD	-10	9	-2	07	22
23	196.3	23	201.6	24	199.1	JACKSON SD	3	-1	1	09	64
34	144.0	53	209.8	31	138.9	JERAULD SD	46	-34	-4	06	04
9	117.3	22	253.1	24	250.0	JONES SD	116	-1	113	97	97
80	148.3	66	107.2	102	172.2	KINGSBURY SD	-28	61	16	32	25
112	183.4	116	171.8	97	145.2	LAKE SD	-6	-15	-21	01	05
112	148.0	135	170.1	156	181.2	LAWRENCE SD	15	7	22	44	37
118	158.0	107	135.8	153	183.6	LINCOLN SD	-14	35	16	32	41
28	166.4	35	169.4	24	120.5	LYMAN SD	2	-29	-28	01	01
64	144.0	65	133.4	96	197.2	MC COOK SD	-7	48	37	67	62
52	195.3	41	133.5	43	131.7	MC PHERSON SD	-32	-1	-33	00	02
61	147.5	75	184.4	68	166.4	MARSHALL SD	25	-10	13	26	20
97	143.3	80	121.5	110	149.9	MEADE SD	-15	23	5	12	07
25	239.6	14	150.5	11	117.7	MELLETTE SD	-37	-22	-51	00	01
55	183.4	51	153.5	59*	488.8	MINER SD	-16	219	166	99	99
579	175.0	704	185.2	834	192.7	MINNEHAHA SD	6	4	10	21	55
74	158.5	70	144.2	80	172.6	MOODY SD	-9	20	9	19	26
251	172.9	270	139.1	446	195.5	PENNINGTON SD	-20	41	13	26	59
59	199.1	43	144.9	45	150.8	PERKINS SD	-27	4	-24	01	07
44	215.9	41	160.2	40	151.0	POTTER SD	-26	-6	-30	00	08
93	131.9	116	160.4	103	137.1	ROBERTS SD	22	-15	4	12	03
38	129.9	46	179.2	39	165.8	SANBORN SD	38	-8	28	52	19
6	218.3	7	155.1	4	155.4	SHANNON SD	-29	0	-29	00	10
97	158.1	113	169.4	120	192.9	SPINK SD	7	14	22	43	55
13	118.4	19	170.2	28	265.7	STANLEY SD	44	56	124	98	98
9	122.8	14	103.9	17	132.4	SULLY SD	-15	27	8	17	02
10	121.9	6	82.0	10	105.2	TODD SD	-33	28	-14	03	00
70	164.1	90	202.2	88	199.0	TRIPP SD	23	-2	21	42	64
116	171.4	146	194.3	116	143.8	TURNER SD	13	-26	-16	02	05
75	126.8	114	191.3	121	190.2	UNION SD	51	-1	50	82	51
66	204.8	63	151.2	73	154.9	WALWORTH SD	-26	2	-24	01	10
116	117.0	144	142.4	138	143.4	YANKTON SD	22	1	23	44	04
4	63.3	11	193.3	9	216.8	ZIEBACH SD	205	12	242	99	84
16215	144.7	22117	164.0	31371	193.5	TENNESSEE TN	13	18	34		

* SIGNIFICANTLY MORE DEATHS THAN EXPECTED

WHITE MALE — ALL CANCERS COMBINED — ICD 140-208

COUNTY AND STATE	1950-1959 DEATHS	1950-1959 RATE	1960-1969 DEATHS	1960-1969 RATE	1970-1979 DEATHS	1970-1979 RATE	PERCENT CHANGE 60S/50S	PERCENT CHANGE 70S/60S	PERCENT CHANGE 70S/50S	PERCENTILE %CHANGE 70S/50S	PERCENTILE RATES 1970S
ANDERSON TN	212	137.3	332	159.3	523	195.2	16	23	42	74	59
BEDFORD TN	128	132.2	157	145.5	239	190.6	10	31	44	76	52
BENTON TN	97	178.1	123	189.8	153	193.2	7	2	8	18	56
BLEDSOE TN	32	93.6	49	136.2	55	134.4	46	-1	44	75	03
BLOUNT TN	244	134.2	344	137.4	560	173.9	2	27	30	56	27
BRADLEY TN	178	156.5	219	144.0	450	211.5	-8	47	35	65	79
CAMPBELL TN	158	135.8	247	194.0	347	220.8	43	14	63	89	87
CANNON TN	54	146.0	52	118.1	76	152.3	-19	29	4	12	08
CARROLL TN	142	126.9	203	165.5	253	176.4	30	7	39	70	30
CARTER TN	176	119.1	260	146.2	404	181.1	23	24	52	83	37
CHEATHAM TN	46	110.9	74	146.0	114	162.5	32	11	47	78	16
CHESTER TN	57	137.9	54	119.8	99	191.0	-13	59	39	69	52
CLAIBORNE TN	136	156.6	169	177.3	219	199.9	13	13	28	53	65
CLAY TN	38	119.6	58	170.2	59	150.2	42	-12	26	49	07
COCKE TN	117	130.3	174	175.7	218	186.9	35	6	43	75	46
COFFEE TN	142	154.2	186	159.4	261	181.7	3	14	18	36	38
CROCKETT TN	76	127.6	123	195.4	157	218.1	53	12	71	92	85
CUMBERLAND TN	90	117.3	131	140.2	213	166.7	20	19	42	74	20
DAVIDSON TN	1707	164.2	2338	183.5	3119*	209.3	12	14	27	52	77
DECATUR TN	52	128.1	67	135.1	102	161.7	5	20	26	50	15
DE KALB TN	61	115.1	77	130.3	103	147.5	13	13	28	53	06
DICKSON TN	102	116.4	151	153.3	247	196.2	32	28	69	91	60
DYER TN	193	155.6	240	179.0	287	194.7	15	9	25	48	58
FAYETTE TN	48	130.2	50	123.9	85	153.3	-5	24	18	35	09
FENTRESS TN	68	140.8	80	140.8	115	168.3	0	20	20	38	21
FRANKLIN TN	104	113.5	177	164.3	270	206.6	45	26	82	95	74
GIBSON TN	263	144.4	332	158.7	433	189.5	10	19	31	59	50
GILES TN	113	118.1	160	155.7	176	157.5	32	1	33	62	11
GRAINGER TN	44	83.4	69	113.0	117	160.4	36	42	92	96	14
GREENE TN	187	124.8	283	155.2	390	173.6	24	12	39	70	27
GRUNDY TN	44	91.7	78	155.6	96	169.6	70	9	85	95	23
HAMBLEN TN	112	122.7	174	140.0	305	190.7	14	36	55	85	52
HAMILTON TN	1032	159.3	1452	183.1	1977	208.9	15	14	31	59	77
HANCOCK TN	35	104.1	53	153.1	57	166.4	47	9	60	87	20
HARDEMAN TN	85	115.0	113	141.8	161	187.5	23	32	63	89	47
HARDIN TN	87	119.1	133	155.4	184	179.9	31	16	51	82	35
HAWKINS TN	146	127.8	193	145.6	276	164.8	14	13	29	55	18
HAYWOOD TN	68	164.1	85	179.2	107	196.3	9	10	20	39	61
HENDERSON TN	92	128.9	119	144.1	182	187.4	12	30	45	77	46
HENRY TN	153	138.0	211	172.3	270	183.2	25	6	33	61	40
HICKMAN TN	88	160.6	100	169.8	147	224.6	6	32	40	71	90
HOUSTON TN	41	165.9	36	125.3	59	160.6	-24	28	-3	06	14
HUMPHREYS TN	77	148.9	93	155.0	160	222.8	4	44	50	81	89
JACKSON TN	61	123.8	74	146.4	93	168.7	18	15	36	66	21
JEFFERSON TN	116	149.5	152	157.4	200	161.9	5	3	8	18	15
JOHNSON TN	48	94.4	80	135.7	103	158.1	44	16	67	91	12
KNOX TN	1160	146.3	1677	170.7	2430*	210.6	17	23	44	75	78
LAKE TN	51	168.2	56	188.3	62	191.8	12	2	14	28	53
LAUDERDALE TN	104	159.2	143	198.9	172	211.5	25	6	33	61	79
LAWRENCE TN	140	120.5	205	153.5	325	208.5	27	36	73	93	76
LEWIS TN	48	191.5	47	155.7	83	217.1	-19	39	13	27	84
LINCOLN TN	143	153.2	160	158.3	224	185.0	3	17	21	41	43
LOUDON TN	109	122.5	173	166.1	235	184.7	36	11	51	82	43
MC MINN TN	189	154.2	214	144.6	305	174.8	-6	21	13	27	28
MC NAIRY TN	126	149.8	166	176.5	176	166.3	18	-6	11	22	19
MACON TN	79	130.4	87	122.2	149	187.5	-6	53	44	75	47
MADISON TN	249	131.9	384	176.9	475	193.9	34	10	47	79	57
MARION TN	94	141.7	123	153.3	188	205.4	8	34	45	76	72
MARSHALL TN	120	164.9	143	181.3	192	211.9	10	17	28	54	79
MAURY TN	196	144.4	235	149.3	366	192.9	3	29	34	63	55
MEIGS TN	25	113.9	32	140.3	51	183.3	23	31	61	88	41
MONROE TN	106	119.0	133	128.2	194	161.6	8	26	36	66	15
MONTGOMERY TN	160	137.7	218	144.9	381	193.5	5	33	41	72	56
MOORE TN	24	170.7	23	120.1	39	191.9	-30	60	12	25	53
MORGAN TN	60	99.8	91	147.6	133	187.7	48	27	88	96	47
OBION TN	161	124.2	227	152.0	343	211.6	22	39	70	92	79
OVERTON TN	106	152.1	105	134.8	132	149.1	-11	11	-2	07	06
PERRY TN	37	146.2	44	146.6	76	221.9	0	51	52	83	88
PICKETT TN	26	142.6	22	119.3	42	181.3	-16	52	27	52	37
POLK TN	62	125.6	102	196.2	119	212.3	56	8	69	91	80
PUTNAM TN	165	134.4	211	144.6	323	174.9	8	21	30	57	28
RHEA TN	90	152.5	114	157.3	170	183.7	3	17	20	40	41
ROANE TN	199	170.3	233	159.6	355	190.5	-6	19	12	24	51
ROBERTSON TN	157	148.6	150	121.5	252	168.2	-18	38	13	27	21
RUTHERFORD TN	168	114.4	268	145.5	447	180.7	27	24	58	86	36
SCOTT TN	60	98.0	91	133.9	149	192.2	37	44	96	96	54
SEQUATCHIE TN	32	149.8	38	145.3	62	192.9	-3	33	29	54	55
SEVIER TN	124	145.1	168	144.5	298	182.5	0	26	26	49	39
SHELBY TN	2182*	187.9	3030*	208.1	3865*	229.0	11	10	22	43	92
SMITH TN	88	131.0	75	103.9	115	147.8	-21	42	13	26	06
STEWART TN	55	128.7	63	135.0	88	175.9	5	30	37	67	30
SULLIVAN TN	441	142.1	686	163.6	1055	191.5	15	17	35	64	53
SUMNER TN	159	112.2	288	151.7	470	176.7	35	16	57	86	31
TIPTON TN	115	162.2	130	166.7	188	201.7	3	21	24	47	67
TROUSDALE TN	29	135.2	43	184.3	49	189.6	36	3	40	72	50
UNICOI TN	72	119.7	81	116.7	168	207.2	-3	78	73	93	75
UNION TN	45	125.3	51	125.6	64	131.0	0	4	4	12	02
VAN BUREN TN	16	103.0	21	123.4	26	125.8	20	2	22	44	01
WARREN TN	122	120.7	172	147.2	251	177.2	22	20	47	79	31
WASHINGTON TN	373	139.3	488	148.3	691	186.4	6	26	34	63	45
WAYNE TN	65	133.7	68	120.9	113	175.9	-10	46	32	59	30
WEAKLEY TN	165	112.8	224	143.9	300	176.0	28	22	56	85	30
WHITE TN	85	120.5	119	151.0	153	165.1	25	9	37	67	18
WILLIAMSON TN	140	167.0	162	142.9	265	170.5	-14	19	2	10	24
WILSON TN	143	122.3	208	151.7	301	164.1	24	8	34	64	17
TEXAS TX	43746	155.1	63759	177.8	87166	198.8	15	12	28		
ANDERSON TX	188	162.9	209	161.8	277	182.6	-1	13	12	24	40
ANDREWS TX	24	141.6	44	136.6	72	195.7	-3	43	38	69	60
ANGELINA TX	224	159.8	304	170.9	455	199.7	7	17	25	48	65

* SIGNIFICANTLY MORE DEATHS THAN EXPECTED

WHITE MALE — ALL CANCERS COMBINED — ICD 140-208

1950-1959 DEATHS	1950-1959 RATE	1960-1969 DEATHS	1960-1969 RATE	1970-1979 DEATHS	1970-1979 RATE	COUNTY AND STATE	PERCENT CHANGE 60S/50S	PERCENT CHANGE 70S/60S	PERCENT CHANGE 70S/50S	PERCENTILE %CHANGE 70S/50S	PERCENTILE RATES 1970S
38	136.8	74	163.5	137	197.6	ARANSAS TX	20	21	44	76	62
35	111.8	54	167.3	62	163.5	ARCHER TX	50	-2	46	78	17
11	91.5	26	203.9	24	163.9	ARMSTRONG TX	123	-20	79	94	17
110	147.5	139	153.7	209	192.1	ATASCOSA TX	4	25	30	57	54
102	149.9	132	162.7	170	173.1	AUSTIN TX	9	6	15	31	27
31	111.8	53	156.7	54	153.6	BAILEY TX	40	-2	37	68	09
35	128.2	52	167.9	101	228.2	BANDERA TX	31	36	78	94	92
128	168.0	148	170.6	207	196.1	BASTROP TX	2	15	17	33	60
51	148.5	50	127.5	69	183.0	BAYLOR TX	-14	44	23	45	40
109	171.7	127	168.7	172	201.4	BEE TX	-2	19	17	34	67
374	136.2	553	161.1	782	185.9	BELL TX	18	15	36	67	44
2895	161.8	4447	189.9	6033*	213.5	BEXAR TX	17	12	32	60	81
33	149.0	58	224.0	70	228.8	BLANCO TX	50	2	54	84	92
1	37.2	8	212.7	3	59.9	BORDEN TX	472	-72	61	88	00
120	155.0	147	165.1	175	165.9	BOSQUE TX	7	1	7	16	19
296	152.4	479	194.1	579	200.6	BOWIE TX	27	3	32	59	66
256	167.3	447	174.0	812	214.5	BRAZORIA TX	4	23	28	53	82
118	121.6	193	143.8	336	194.1	BRAZOS TX	18	35	60	87	57
34	142.1	56	214.5	51	166.3	BREWSTER TX	51	-22	17	33	19
25	165.3	19	135.9	26	166.2	BRISCOE TX	-18	22	1	08	19
40	157.7	31	114.4	85*	255.7	BROOKS TX	-27	124	62	89	98
234	155.9	281	173.3	354	193.9	BROWN TX	11	12	24	47	57
60	119.5	86	169.3	98	155.5	BURLESON TX	42	-8	30	57	10
78	146.8	132	176.7	223	191.3	BURNET TX	20	8	30	57	53
103	133.8	135	158.3	196	204.3	CALDWELL TX	18	29	53	83	71
57	176.6	83	169.8	143	237.6	CALHOUN TX	-4	40	35	64	95
82	149.6	99	159.6	122	172.3	CALLAHAN TX	7	8	15	30	25
529	142.4	753	152.3	1050	166.5	CAMERON TX	7	9	17	33	20
48	171.4	58	197.0	78	210.9	CAMP TX	15	7	23	45	79
37	139.6	46	143.5	57	160.0	CARSON TX	3	11	15	29	13
120	137.8	184	181.4	249	198.4	CASS TX	32	9	44	75	63
19	99.3	38	152.1	46	151.2	CASTRO TX	53	-1	52	83	08
54	202.2	66	171.6	108	207.1	CHAMBERS TX	-15	21	2	10	75
201	134.5	229	143.7	364	197.0	CHEROKEE TX	7	37	47	78	61
67	132.7	83	164.4	107	214.8	CHILDRESS TX	24	31	62	89	82
77	149.0	91	171.5	98	164.0	CLAY TX	15	-4	10	21	17
23	111.7	30	149.3	30	145.4	COCHRAN TX	34	-3	30	57	05
35	204.6	34	174.3	45	197.5	COKE TX	-15	13	-3	06	62
114	136.0	136	158.5	192	225.9	COLEMAN TX	17	42	66	91	90
300	148.2	387	164.2	533	167.4	COLLIN TX	11	2	13	26	20
59	164.1	65	193.3	45	150.2	COLLINGSWORTH TX	18	-22	-8	04	07
101	146.4	127	166.1	156	172.1	COLORADO TX	13	4	18	35	25
111	139.5	153	148.3	283	167.6	COMAL TX	6	13	20	39	21
148	154.4	134	146.1	147	157.6	COMANCHE TX	-5	8	2	09	11
36	169.3	43	186.7	33	139.5	CONCHO TX	10	-25	-18	02	04
128	121.7	223	189.6	262	190.0	COOKE TX	56	0	56	85	51
111	131.7	138	140.5	170	152.0	CORYELL TX	7	8	15	31	08
33	155.6	38	182.8	38	192.2	COTTLE TX	17	5	24	46	54
18	172.1	17	202.4	24	145.2	CRANE TX	18	-28	-16	02	05
12	132.6	29	222.2	41	259.2	CROCKETT TX	68	17	95	96	98
51	134.2	73	182.7	63	152.9	CROSBY TX	36	-16	14	28	09
7	110.5	10	114.8	12	158.3	CULBERSON TX	4	38	43	75	12
39	121.3	49	148.9	42	134.7	DALLAM TX	23	-10	11	22	03
3820	177.4	6055*	194.8	8411*	215.5	DALLAS TX	10	11	22	42	83
78	140.4	110	155.3	126	167.4	DAWSON TX	11	8	19	38	21
51	144.7	86	159.5	100	156.0	DEAF SMITH TX	10	-2	8	17	10
64	156.5	60	147.9	85	209.1	DELTA TX	-5	41	34	63	77
263	149.3	374	170.9	583	185.6	DENTON TX	15	9	24	47	44
122	126.9	185	171.4	251	218.7	DE WITT TX	35	28	72	92	86
49	166.4	49	165.7	60	215.6	DICKENS TX	0	30	30	56	83
65	198.0	43	119.6	74	176.4	DIMMIT TX	-40	48	-11	03	31
44	138.5	46	154.5	49	162.7	DONLEY TX	12	5	17	34	16
75	153.4	82	153.2	90	151.8	DUVAL TX	0	-1	-1	07	08
231	173.2	240	167.7	279	187.3	EASTLAND TX	-3	12	8	17	46
196	170.1	400	184.1	645	207.2	ECTOR TX	8	13	22	43	75
13	120.2	12	108.5	24	235.3	EDWARDS TX	-10	117	96	96	94
285	155.5	333	168.2	431	194.7	ELLIS TX	8	16	25	48	58
962	153.6	1424	169.3	2017	205.4	EL PASO TX	10	21	34	63	72
158	139.1	217	167.5	235	163.9	ERATH TX	20	-2	18	35	17
136	140.6	166	165.6	196	188.8	FALLS TX	18	14	34	64	49
182	120.3	252	156.3	281	162.7	FANNIN TX	30	4	35	65	16
164	135.1	191	142.0	255	179.3	FAYETTE TX	5	26	33	61	34
50	120.8	64	148.4	61	145.7	FISHER TX	23	-2	21	40	05
60	131.5	76	151.0	71	137.2	FLOYD TX	15	-9	4	12	03
36	186.7	26	136.4	27	154.3	FOARD TX	-27	13	-17	02	09
142	141.7	243	188.0	387	194.1	FORT BEND TX	33	3	37	68	57
33	100.0	57	152.1	70	161.2	FRANKLIN TX	52	6	61	88	15
92	159.0	102	171.1	145	202.4	FREESTONE TX	8	18	27	52	68
65	169.4	86	196.1	87	175.6	FRIO TX	16	-10	4	11	29
35	113.3	61	173.0	63	135.9	GAINES TX	53	-21	20	39	03
718*	205.9	1013*	218.7	1425*	248.5	GALVESTON TX	6	14	21	41	97
42	172.2	47	184.5	44	176.9	GARZA TX	7	-4	3	10	31
80	136.8	116	159.1	145	155.1	GILLESPIE TX	16	-3	13	27	10
4	134.9	5	116.0	14	520.0	GLASSCOCK TX	-14	348	286	99	99
28	120.7	38	146.7	63	250.0	GOLIAD TX	22	70	107	97	97
133	162.4	135	155.9	183	198.6	GONZALES TX	-4	27	22	44	63
134	151.1	215	182.0	307	222.1	GRAY TX	20	22	47	79	88
454	143.3	644	174.0	815	190.2	GRAYSON TX	21	9	33	61	51
323	178.9	468	190.3	724*	231.2	GREGG TX	6	21	29	55	93
74	146.9	99	184.9	143	248.4	GRIMES TX	26	34	69	91	97
139	139.9	232	178.8	272	159.2	GUADALUPE TX	28	-11	14	28	13
166	152.4	206	149.8	235	160.8	HALE TX	-2	7	5	14	14
67	159.5	69	188.6	69	191.8	HALL TX	18	2	20	40	53
116	184.0	129	204.6	131	180.5	HAMILTON TX	11	-12	-2	07	36
25	182.5	34	166.8	40	151.8	HANSFORD TX	-9	-9	-17	02	08
68	145.4	88	189.9	97	213.8	HARDEMAN TX	31	13	47	79	81
129	168.4	180	169.1	276	196.3	HARDIN TX	0	16	17	33	60
4617*	185.9	7662*	208.0	10631*	221.8	HARRIS TX	12	7	19	38	88
151	141.2	235	185.1	295	189.3	HARRISON TX	31	2	34	64	49
8	103.1	21	185.9	21	121.5	HARTLEY TX	80	-35	18	36	01
94	159.4	96	159.7	114	199.2	HASKELL TX	0	25	25	48	64

* SIGNIFICANTLY MORE DEATHS THAN EXPECTED

WHITE MALE — ALL CANCERS COMBINED — ICD 140-208

1950-1959 DEATHS	1950-1959 RATE	1960-1969 DEATHS	1960-1969 RATE	1970-1979 DEATHS	1970-1979 RATE	COUNTY AND STATE	PERCENT CHANGE 60S/50S	PERCENT CHANGE 70S/60S	PERCENT CHANGE 70S/50S	PERCENTILE %CHANGE 70S/50S	PERCENTILE RATES 1970S
98	138.4	119	141.3	201	196.9	HAYS TX	2	39	42	74	61
18	115.4	29	168.5	25	133.0	HEMPHILL TX	46	-21	15	30	02
149	146.0	227	176.6	357	180.3	HENDERSON TX	21	2	23	45	36
633	134.5	876	147.0	1246	152.3	HIDALGO TX	9	4	13	26	08
215	148.4	260	174.1	298	182.5	HILL TX	17	5	23	45	39
62	107.0	124	165.8	154	180.4	HOCKLEY TX	55	9	69	91	36
51	143.6	81	199.3	138	187.7	HOOD TX	39	-6	31	58	47
162	140.0	202	165.8	249	185.9	HOPKINS TX	18	12	33	61	44
85	112.1	127	159.2	174	187.5	HOUSTON TX	42	18	67	91	46
146	139.4	224	159.4	301	202.7	HOWARD TX	14	27	45	77	69
9	101.3	10	113.1	13	118.7	HUDSPETH TX	12	5	17	34	01
272	144.0	352	168.2	497	204.0	HUNT TX	17	21	42	73	70
109	145.7	156	165.4	214	175.0	HUTCHINSON TX	14	6	20	39	28
13	177.1	13	168.4	16	177.8	IRION TX	-5	6	0	08	32
66	156.5	86	184.8	98	187.6	JACK TX	18	2	20	39	47
64	140.3	90	183.0	124	227.0	JACKSON TX	30	24	62	88	91
123	165.2	150	158.1	229	185.4	JASPER TX	-4	17	12	25	44
9	109.3	6	69.4	16	165.0	JEFF DAVIS TX	-36	138	51	82	18
1075*	199.0	1497*	212.0	1896*	230.5	JEFFERSON TX	7	9	16	31	93
30	195.2	33	163.6	27	122.5	JIM HOGG TX	-16	-25	-37	00	01
156	182.6	198	173.4	228	170.1	JIM WELLS TX	-5	-2	-7	04	23
243	139.7	332	162.7	528	202.1	JOHNSON TX	16	24	45	76	68
161	160.6	173	161.3	208	189.8	JONES TX	0	18	18	36	50
91	147.5	106	158.7	133	196.7	KARNES TX	8	24	33	62	61
174	131.3	247	170.8	316	194.6	KAUFMAN TX	30	14	48	80	58
59	187.5	94	226.8	108	195.4	KENDALL TX	21	-14	4	12	59
5	194.9	3	78.4	1	27.0	KENEDY TX	-60	-66	-86	00	00
11	125.4	8	85.4	11	139.5	KENT TX	-32	63	11	23	04
129	138.1	195	141.3	360	187.8	KERR TX	2	33	36	66	47
37	168.8	48	192.9	49	166.3	KIMBLE TX	14	-14	-1	07	19
2	70.9	3	135.7	3	110.0	KING TX	91	-19	55	85	00
13	119.7	15	150.0	11	111.6	KINNEY TX	25	-26	-7	04	00
100	161.5	132	176.9	172	210.1	KLEBERG TX	10	19	30	57	78
64	159.0	77	179.5	83	205.3	KNOX TX	13	14	29	55	72
256	151.8	348	195.4	384	192.4	LAMAR TX	29	-2	27	51	54
109	151.5	130	156.3	153	168.9	LAMB TX	3	8	11	23	22
78	149.9	122	210.8	115	180.9	LAMPASAS TX	41	-14	21	41	37
58	220.1	61	235.7	48	196.6	LA SALLE TX	7	-17	-11	03	61
150	136.1	193	156.5	229	168.9	LAVACA TX	15	8	24	47	22
56	130.1	80	167.3	99	183.8	LEE TX	29	10	41	73	41
71	160.5	76	163.8	107	206.5	LEON TX	2	26	29	54	74
171	174.2	180	144.1	334	205.9	LIBERTY TX	-17	43	18	36	73
134	137.9	168	170.8	226	214.9	LIMESTONE TX	24	26	56	85	82
27	162.2	31	163.8	31	158.0	LIPSCOMB TX	1	-4	-3	06	12
46	142.0	59	163.9	74	165.8	LIVE OAK TX	15	1	17	33	19
46	138.9	102	184.8	163	175.6	LLANO TX	33	-5	26	51	29
1	84.4	4	349.9			LOVING TX	314				
462	142.2	762	167.7	1018	180.3	LUBBOCK TX	18	8	27	51	36
52	127.5	61	150.9	71	168.5	LYNN TX	18	12	32	60	21
101	173.2	119	202.2	132	195.1	MC CULLOCH TX	17	-4	13	25	59
756	151.0	1064	177.3	1365	202.1	MC LENNAN TX	17	14	34	63	68
7	107.0	16	241.1	15	257.1	MC MULLEN TX	125	7	140	98	98
56	161.5	55	161.0	102	240.5	MADISON TX	0	49	49	80	96
44	204.4	40	143.3	82	210.9	MARION TX	-30	47	3	11	78
23	135.8	22	115.6	35	161.8	MARTIN TX	-15	40	19	38	15
35	138.1	41	159.2	54	185.4	MASON TX	15	16	34	64	44
129	177.2	170	183.9	230	204.5	MATAGORDA TX	4	11	15	31	71
43	108.3	76	164.8	116	165.9	MAVERICK TX	52	1	53	84	19
125	161.7	156	167.6	211	189.6	MEDINA TX	4	13	17	34	50
42	210.1	29	136.5	39	204.8	MENARD TX	-35	50	-3	06	71
137	174.7	256	170.7	389	185.6	MIDLAND TX	-2	9	6	15	44
149	135.9	207	173.6	249	202.1	MILAM TX	28	16	49	80	68
45	111.6	62	140.0	79	166.3	MILLS TX	25	19	49	81	19
67	132.8	84	161.6	104	186.3	MITCHELL TX	22	15	40	72	45
126	128.8	180	170.8	269	222.2	MONTAGUE TX	33	30	73	92	88
136	149.3	240	165.3	546	186.9	MONTGOMERY TX	11	13	25	49	46
52	183.9	67	159.4	65	159.8	MOORE TX	-13	0	-13	03	13
48	132.0	80	169.9	116	210.3	MORRIS TX	29	24	59	87	78
26	153.7	28	158.4	33	207.1	MOTLEY TX	3	31	35	64	75
154	139.5	206	160.4	283	182.7	NACOGDOCHES TX	15	14	31	58	40
233	143.6	275	161.9	351	189.1	NAVARRO TX	13	17	32	60	49
43	118.8	66	151.1	68	129.5	NEWTON TX	27	-14	9	19	02
98	124.8	155	178.1	173	189.1	NOLAN TX	43	6	52	82	49
799	160.4	1267	195.6	1636	211.3	NUECES TX	22	8	32	60	79
26	116.0	55	176.8	64	178.2	OCHILTREE TX	52	1	54	84	33
4	60.8	13	165.9	13	230.5	OLDHAM TX	173	39	279	99	93
206	167.5	376	202.0	539*	227.8	ORANGE TX	21	13	36	66	91
171	167.5	187	162.7	248	190.8	PALO PINTO TX	-3	17	14	28	52
87	146.7	98	131.4	192	223.9	PANOLA TX	-10	70	53	83	89
154	128.5	244	168.6	328	173.2	PARKER TX	31	3	35	65	27
23	96.4	47	140.2	61	153.6	PARMER TX	45	10	59	87	09
48	163.4	52	136.8	71	163.2	PECOS TX	-16	19	0	08	17
77	149.0	94	147.6	201	188.6	POLK TX	-1	28	27	51	48
452	163.6	652	182.0	789	206.7	POTTER TX	11	14	26	50	74
28	121.1	23	93.3	25	98.0	PRESIDIO TX	-23	5	-19	01	00
26	125.1	34	145.8	54	175.9	RAINS TX	17	21	41	72	30
59	93.2	177	151.3	323	172.0	RANDALL TX	62	14	84	95	25
15	167.5	16	151.5	32*	464.4	REAGAN TX	-10	207	177	99	99
8	69.8	21	162.3	32	209.8	REAL TX	133	29	201	99	77
121	146.4	143	162.7	188	203.8	RED RIVER TX	11	25	39	70	70
58	170.5	74	154.3	72	159.3	REEVES TX	-10	3	-7	04	13
43	129.6	69	191.8	70	169.1	REFUGIO TX	48	-12	30	58	22
3	46.3	4	129.8	12	282.8	ROBERTS TX	180	118	511	99	99
81	137.3	101	155.8	135	198.0	ROBERTSON TX	13	27	44	76	63
37	143.1	47	151.7	70	171.0	ROCKWALL TX	6	13	19	38	24
109	144.0	147	176.8	142	173.1	RUNNELS TX	23	-2	20	39	27
185	136.2	279	173.8	364	186.3	RUSK TX	28	7	37	67	45
48	137.0	55	150.0	93	194.3	SABINE TX	10	30	42	73	57
31	113.7	49	143.9	69	160.7	SAN AUGUSTINE TX	27	12	41	73	14
18	91.1	45	183.1	67	157.5	SAN JACINTO TX	101	-14	73	92	11
173	145.7	282	179.6	385	206.6	SAN PATRICIO TX	23	15	42	73	74

* SIGNIFICANTLY MORE DEATHS THAN EXPECTED

WHITE MALE — ALL CANCERS COMBINED — ICD 140-208

1950-1959 DEATHS	1950-1959 RATE	1960-1969 DEATHS	1960-1969 RATE	1970-1979 DEATHS	1970-1979 RATE	COUNTY AND STATE		PERCENT CHANGE 60S/50S	PERCENT CHANGE 70S/60S	PERCENT CHANGE 70S/50S	PERCENTILE %CHANGE 70S/50S	PERCENTILE RATES 1970S
62	123.4	84	182.6	74	150.2	SAN SABA	TX	48	-18	22	42	07
17	147.9	27	229.6	31	253.1	SCHLEICHER	TX	55	10	71	92	98
101	159.8	133	173.4	154	194.1	SCURRY	TX	8	12	21	42	57
40	170.4	51	216.3	53	203.4	SHACKELFORD	TX	27	-6	19	38	70
135	135.1	157	152.2	250	212.3	SHELBY	TX	13	40	57	86	80
15	182.9	16	132.2	26	164.7	SHERMAN	TX	-28	25	-10	03	18
401	167.1	578	187.0	883	209.6	SMITH	TX	12	12	25	49	77
26	146.6	40	200.3	39	177.5	SOMERVELL	TX	37	-11	21	41	32
62	138.8	77	126.5	127	161.7	STARR	TX	-9	28	16	32	15
79	155.1	100	168.9	131	220.7	STEPHENS	TX	9	31	42	74	87
9	191.1	10	182.8	11	206.5	STERLING	TX	-4	13	8	17	74
17	106.6	28	172.1	38	226.3	STONEWALL	TX	61	31	112	97	91
24	235.9	25	162.6	33	197.6	SUTTON	TX	-31	21	-16	02	62
48	128.6	66	152.6	96	205.8	SWISHER	TX	19	35	60	87	73
2353	172.5	3657	192.3	5341*	220.4	TARRANT	TX	11	15	28	53	87
368	142.7	559	161.5	744	185.0	TAYLOR	TX	13	15	30	56	43
5	45.7	22	207.2	11	121.3	TERRELL	TX	353	-41	165	99	01
61	132.0	75	138.4	107	185.1	TERRY	TX	5	34	40	71	43
23	116.6	29	142.6	31	159.7	THROCKMORTON	TX	22	12	37	67	13
98	137.9	161	191.5	206	206.6	TITUS	TX	39	8	50	81	74
344	150.5	501	177.6	665	201.3	TOM GREEN	TX	18	13	34	63	67
784	143.5	1216	166.7	1841	196.9	TRAVIS	TX	16	18	37	68	61
29	73.8	70	183.6	88	172.6	TRINITY	TX	149	-6	134	98	26
74	144.1	109	171.2	168	194.9	TYLER	TX	19	14	35	65	58
109	133.5	176	181.6	239	198.6	UPSHUR	TX	36	9	49	80	63
27	177.4	30	184.5	33	179.7	UPTON	TX	4	-3	1	09	35
90	137.2	129	167.3	164	182.5	UVALDE	TX	22	9	33	62	39
73	125.3	111	152.2	192	205.3	VAL VERDE	TX	21	35	64	90	72
197	168.1	210	152.5	359	199.5	VAN ZANDT	TX	-9	31	19	37	65
200	186.9	281	187.7	408	214.5	VICTORIA	TX	0	14	15	29	82
55	89.2	118	181.4	153	172.8	WALKER	TX	103	-5	94	96	26
40	135.8	74	212.4	107	213.7	WALLER	TX	56	1	57	86	81
62	177.4	69	149.6	101	213.5	WARD	TX	-16	43	20	40	81
98	120.6	163	175.7	194	176.9	WASHINGTON	TX	46	1	47	79	31
276	154.2	361	159.2	462	168.1	WEBB	TX	3	6	9	19	21
178	158.4	246	182.1	288	189.6	WHARTON	TX	15	4	20	39	50
78	171.8	72	150.2	73	157.3	WHEELER	TX	-13	5	-8	04	11
498	137.0	782	181.0	921	198.8	WICHITA	TX	32	10	45	77	64
119	139.5	153	161.7	183	180.7	WILBARGER	TX	16	12	29	56	37
72	132.5	103	168.9	112	182.2	WILLACY	TX	28	8	38	68	39
248	150.0	297	154.1	431	181.6	WILLIAMSON	TX	3	18	21	41	38
102	159.4	129	186.8	143	189.1	WILSON	TX	17	1	19	37	49
42	221.7	70	226.5	90	261.6	WINKLER	TX	2	15	18	36	98
138	139.0	209	190.9	226	175.3	WISE	TX	37	-8	26	50	29
140	145.8	199	179.8	300	219.4	WOOD	TX	23	22	50	82	86
22	140.5	30	138.1	42	161.0	YOAKUM	TX	-2	17	15	29	14
118	136.7	177	177.9	226	196.6	YOUNG	TX	30	10	44	75	61
9	68.7	33	140.7	38	117.9	ZAPATA	TX	105	-16	72	92	01
53	152.4	59	138.5	79	193.8	ZAVALA	TX	-9	40	27	52	56
3613	139.3	4751	145.1	6068	146.5	UTAH	UT	4	1	5		
30	221.5	33	165.9	38	189.6	BEAVER	UT	-25	14	-14	02	50
106	143.8	110	121.1	148	129.6	BOX ELDER	UT	-16	7	-10	03	02
134	106.5	183	121.8	234	131.8	CACHE	UT	14	8	24	46	02
147	196.4	145	173.9	142	157.4	CARBON	UT	-11	-9	-20	01	11
		3	78.9			DAGGETT	UT					
137	130.5	223	131.1	363	132.3	DAVIS	UT	0	1	1	09	02
31	134.6	44	186.8	61	190.5	DUCHESNE	UT	39	2	41	73	51
33	144.0	41	158.0	48	157.8	EMERY	UT	10	0	10	20	12
13	89.6	27	188.9	21	144.9	GARFIELD	UT	111	-23	62	88	05
19	166.0	34	184.9	43	169.7	GRAND	UT	11	-8	2	10	23
44	125.8	61	148.6	66	127.0	IRON	UT	18	-15	1	08	01
41	189.1	34	149.2	35	135.6	JUAB	UT	-21	-9	-28	01	03
12	126.5	16	158.8	20	125.0	KANE	UT	26	-21	-1	07	01
48	142.6	48	126.9	49	110.0	MILLARD	UT	-11	-13	-23	01	00
16	147.9	15	128.6	21	149.3	MORGAN	UT	-13	16	1	08	07
4	73.8	8	126.3	11	151.4	PIUTE	UT	71	20	105	97	08
10	125.5	10	110.1	4	53.8	RICH	UT	-12	-51	-57	00	00
1652	149.7	2192	154.9	2804	160.6	SALT LAKE	UT	3	4	7	16	14
14	147.1	24	160.5	18	116.5	SAN JUAN	UT	9	-27	-21	01	01
93	145.4	109	163.6	107	153.2	SANPETE	UT	13	-6	5	14	09
60	137.3	63	123.3	66	94.6	SEVIER	UT	-10	-23	-31	00	00
26	113.4	32	126.0	41	142.9	SUMMIT	UT	11	13	26	49	04
82	170.3	98	166.9	126	184.8	TOOELE	UT	-2	11	8	18	43
50	179.5	52	148.8	66	129.3	UINTAH	UT	-17	-13	-28	01	02
312	112.5	452	127.7	663	135.0	UTAH	UT	13	6	20	39	03
25	119.1	29	128.8	44	147.3	WASATCH	UT	8	14	24	46	06
38	85.2	69	116.3	113	101.2	WASHINGTON	UT	37	-13	19	37	00
6	82.0	12	133.9	12	102.6	WAYNE	UT	63	-23	25	48	00
430	133.7	584	144.5	704	142.5	WEBER	UT	8	-1	7	15	04
3291	178.1	3683	191.2	4476*	211.0	VERMONT	VT	7	10	18		
147	157.4	178	184.6	244	220.3	ADDISON	VT	17	19	40	71	86
224	180.7	242	171.3	310	189.4	BENNINGTON	VT	-5	11	5	13	50
214	170.2	236	195.9	259	206.6	CALEDONIA	VT	15	5	21	42	74
479	189.4	597*	213.4	724*	226.7	CHITTENDEN	VT	13	6	20	39	91
62	187.8	48	158.0	68	206.0	ESSEX	VT	-16	30	10	20	73
268	188.9	262	181.4	324	215.7	FRANKLIN	VT	-4	19	14	29	83
28	154.7	31	168.1	34	173.9	GRAND ISLE	VT	9	3	12	25	27
90	156.4	95	164.2	137	216.4	LAMOILLE	VT	5	32	38	69	84
132	147.3	146	159.1	189	184.9	ORANGE	VT	8	16	25	49	43
202	196.0	179	178.2	239	221.1	ORLEANS	VT	-9	24	13	26	87
473*	199.9	599*	244.7	629*	232.5	RUTLAND	VT	22	-5	16	32	93
314	162.1	365	173.6	473	209.2	WASHINGTON	VT	7	21	29	55	77
287	178.5	316	193.0	359	204.2	WINDHAM	VT	8	6	14	29	71
371	174.2	389	173.7	487	195.1	WINDSOR	VT	0	12	12	24	59
15693	154.2	22509	177.2	31559	201.4	VIRGINIA	VA	15	14	31		

* SIGNIFICANTLY MORE DEATHS THAN EXPECTED

WHITE MALE — ALL CANCERS COMBINED — ICD 140-208

1950-1959 DEATHS	1950-1959 RATE	1960-1969 DEATHS	1960-1969 RATE	1970-1979 DEATHS	1970-1979 RATE	COUNTY AND STATE	PERCENT CHANGE 60S/50S	PERCENT CHANGE 70S/60S	PERCENT CHANGE 70S/50S	PERCENTILE %CHANGE 70S/50S	PERCENTILE RATES 1970S
206	167.2	299*	234.4	320*	247.0	ACCOMACK VA	40	5	48	79	97
313	177.5	356	166.2	498	188.8	ALBEMARLE#VA	-6	14	6	15	49
170	162.4	210	178.8	239	185.0	ALLEGHANY#VA	10	3	14	28	43
28	139.4	42	199.1	46	196.2	AMELIA VA	43	-1	41	72	60
89	136.5	143	180.3	174	185.0	AMHERST VA	32	3	36	65	43
42	123.8	65	170.4	87	195.1	APPOMATTOX VA	38	15	58	86	59
733	180.5	1144*	208.3	1364*	224.8	ARLINGTON#VA	15	8	25	48	90
326	121.0	459	143.5	651	172.3	AUGUSTA#VA	19	20	42	74	26
42	189.2	55	218.6	59	213.5	BATH VA	16	-2	13	26	81
495	132.8	741	164.8	1061	193.6	BEDFORD#VA	24	18	46	78	56
36	171.3	36	145.0	42	140.8	BLAND VA	-15	-3	-18	01	04
103	155.0	102	138.2	158	175.6	BOTETOURT VA	-11	27	13	27	29
55	166.7	65	196.8	66	175.4	BRUNSWICK VA	18	-11	5	13	29
91	101.6	137	128.8	204	174.5	BUCHANAN VA	27	35	72	92	28
33	90.9	53	143.9	71	174.7	BUCKINGHAM VA	58	21	92	96	28
35	111.7	54	164.2	96	242.9	CAROLINE VA	47	48	117	97	96
254	132.3	295	137.3	387	150.8	CARROLL#VA	4	10	14	28	08
11	231.3	8	165.2	9	199.1	CHARLES CITY VA	-29	20	-14	02	64
67	170.7	59	130.2	82	175.7	CHARLOTTE VA	-24	35	3	11	30
1629*	201.4	1940*	216.4	2219*	229.8	CHESTERFIELD#VA	7	6	14	28	92
46	158.5	77	231.0	96	232.2	CLARKE VA	46	1	47	78	93
18	114.0	28	151.4	40	202.5	CRAIG VA	33	34	78	94	68
71	134.3	87	135.3	160	199.9	CULPEPER VA	1	48	49	80	65
30	190.4	20	122.0	26	125.8	CUMBERLAND VA	-36	3	-34	00	01
72	108.3	119	172.8	151	193.8	DICKENSON VA	60	12	79	94	57
304	166.7	439	194.9	556	212.5	DINWIDDIE#VA	17	9	27	52	80
33	175.4	41	202.2	60	212.5	ESSEX VA	15	5	21	42	80
840	160.9	1756	183.2	3023	204.0	FAIRFAX#VA	14	11	27	51	70
122	164.6	175	200.5	216	207.9	FAUQUIER VA	22	4	26	50	76
61	118.6	76	137.2	104	164.9	FLOYD VA	16	20	39	70	18
30	115.0	44	147.7	61	180.0	FLUVANNA VA	28	22	57	86	35
100	105.3	123	114.3	201	151.3	FRANKLIN VA	9	32	44	75	08
196	143.2	309	184.1	434	206.3	FREDERICK#VA	29	12	44	76	73
75	131.1	102	146.7	177	209.0	GILES VA	12	42	59	87	77
69	172.2	74	150.3	147	224.4	GLOUCESTER VA	-13	49	30	57	90
28	116.8	30	112.7	61	190.6	GOOCHLAND VA	-3	69	63	89	52
27	143.6	23	117.9	37	160.5	GREENE VA	-18	36	12	24	14
41	176.4	55	176.2	71	178.4	GREENSVILLE#VA	0	1	1	09	33
172	185.7	174	161.1	241	194.1	HALIFAX#VA	-13	21	5	12	57
115	154.9	177	172.0	334	223.7	HANOVER VA	11	30	44	76	89
320	130.9	803	202.5	1322*	230.7	HENRICO VA	55	14	76	93	93
138	118.5	243	148.3	377	166.6	HENRY#VA	25	12	41	72	20
38	187.9	34	186.4	27	160.0	HIGHLAND VA	-1	-14	-15	02	13
50	151.1	80	204.6	101	201.0	ISLE OF WIGHT VA	35	-2	33	62	67
135	137.6	239	172.5	346	177.3	JAMES CITY#VA	25	3	29	55	32
24	146.6	25	162.2	43	205.1	KING AND QUEEN VA	11	27	40	71	72
27	130.4	45	197.1	59	213.9	KING GEORGE VA	51	9	64	90	82
50*	282.3	37	200.8	58	239.9	KING WILLIAM VA	-29	19	-15	02	96
49	162.6	58	140.5	102	204.7	LANCASTER VA	-14	46	26	49	71
166	141.9	195	156.9	271	205.6	LEE VA	11	31	45	76	73
127	146.0	176	166.2	271	198.3	LOUDOUN VA	14	19	36	66	63
44	106.5	74	152.6	130	215.9	LOUISA VA	43	42	103	97	83
56	139.5	61	158.3	75	185.1	LUNENBURG VA	13	17	33	61	43
51	157.6	53	150.7	62	142.0	MADISON VA	-4	-6	-10	03	04
61	153.6	74	170.6	104	215.4	MATHEWS VA	11	26	40	72	83
84	135.4	133	162.4	199	202.6	MECKLENBURG VA	20	25	50	81	68
40	150.9	62	215.3	75	207.2	MIDDLESEX VA	43	-4	37	68	75
163	134.8	229	149.8	360	186.2	MONTGOMERY#VA	11	24	38	69	45
71	142.4	88	174.5	99	179.9	NELSON VA	23	3	26	50	35
82	170.4	116*	248.5	125*	257.5	NORTHAMPTON VA	46	4	51	82	98
54	160.8	74	174.3	110	205.5	NORTHUMBERLAND VA	8	18	28	53	72
65	150.0	88	179.5	105	192.8	NOTTOWAY VA	20	7	29	54	55
67	145.4	92	169.1	167	234.2	ORANGE VA	16	38	61	88	94
88	133.7	106	142.0	165	187.8	PAGE VA	6	32	41	72	47
68	108.9	90	129.0	134	159.4	PATRICK VA	18	24	46	78	13
352	130.7	543	169.9	776	202.0	PITTSYLVANIA#VA	30	19	55	84	68
22	133.2	37	181.7	47	157.1	POWHATAN VA	36	-14	18	36	11
48	125.9	75	183.2	105	229.7	PRINCE EDWARD VA	45	25	82	95	92
107	150.2	228	182.7	438	218.3	PRINCE WILLIAM#VA	22	19	45	77	85
119	140.4	188	177.2	210	160.8	PULASKI VA	26	-9	15	29	14
28	128.2	36	155.1	48	172.7	RAPPAHANNOCK VA	21	11	35	64	26
32	165.8	35	161.2	49	182.4	RICHMOND VA	-3	13	10	21	39
795	156.3	1096	168.6	1562	197.5	ROANOKE#VA	8	17	26	50	62
160	163.1	197	174.9	220	168.8	ROCKBRIDGE#VA	7	-3	4	11	22
268	140.7	325	145.7	500	182.9	ROCKINGHAM#VA	4	26	30	57	40
125	128.7	159	149.7	194	157.1	RUSSELL VA	16	5	22	43	11
135	126.0	174	151.1	226	175.5	SCOTT VA	20	16	39	70	29
162	156.1	171	148.5	234	165.7	SHENANDOAH VA	-5	12	6	15	19
158	134.3	187	136.0	267	168.9	SMYTH VA	1	24	26	49	22
73	155.2	96	182.5	123	206.6	SOUTHAMPTON#VA	18	13	33	62	74
112	145.7	181	186.8	269	207.0	SPOTSYLVANIA#VA	28	11	42	74	75
61	142.0	101	165.4	153	173.2	STAFFORD VA	16	5	22	43	27
15	111.1	28	185.4	39	251.3	SURRY VA	67	36	126	98	97
28	141.0	38	177.0	67*	276.5	SUSSEX VA	25	56	96	96	98
204	137.1	280	169.4	352	184.3	TAZEWELL VA	24	9	34	64	42
84	151.3	73	119.2	195	239.9	WARREN VA	-21	101	59	86	96
274	137.6	346	149.6	483	173.0	WASHINGTON#VA	9	16	26	49	26
59	182.3	87	207.3	117	198.8	WESTMORELAND VA	14	-4	9	19	64
261	154.5	336	188.3	428*	224.5	WISE#VA	22	19	45	77	90
135	157.9	161	171.0	180	166.9	WYTHE VA	8	-2	6	14	20
1839	182.2	2823*	219.8	3720*	234.4	CHESAPEAKE CITY#VA	21	7	29	54	94
277	167.2	438	188.8	667*	243.9	HAMPTON CITY VA	13	29	46	78	97
311	180.5	432	197.4	711*	237.4	NEWPORT NEWS CITY VA	9	20	31	59	95
123	162.5	171	195.5	263*	240.1	SUFFOLK CITY VA	20	23	48	80	96
20069	168.7	24918	181.8	31175	195.8	WASHINGTON WA	8	8	16		
50	159.9	72	183.1	101	210.4	ADAMS WA	14	15	32	59	78
95	133.6	129	170.3	180	213.3	ASOTIN WA	27	25	60	87	81
256	158.5	376	173.4	558	186.8	BENTON WA	9	8	18	35	46
289	150.6	423	192.0	500	197.5	CHELAN WA	27	3	31	59	62

* SIGNIFICANTLY MORE DEATHS THAN EXPECTED

\# COUNTIES COMBINED; SEE TECHNICAL NOTES

WHITE MALE — ALL CANCERS COMBINED — ICD 140-208

1950-1959 DEATHS	1950-1959 RATE	1960-1969 DEATHS	1960-1969 RATE	1970-1979 DEATHS	1970-1979 RATE	COUNTY AND STATE	PERCENT CHANGE 60S/50S	PERCENT CHANGE 70S/60S	PERCENT CHANGE 70S/50S	PERCENTILE %CHANGE 70S/50S	PERCENTILE RATES 1970S
185	134.7	281	162.2	450	180.8	CLALLAM WA	20	11	34	64	37
605	149.2	905	184.1	1192	186.2	CLARK WA	23	1	25	48	45
51	170.5	42	143.2	47	157.8	COLUMBIA WA	-16	10	-7	04	11
374	173.9	473	173.3	631	194.7	COWLITZ WA	0	12	12	24	58
65	123.6	128	183.5	178	198.4	DOUGLAS WA	48	8	61	88	63
25	156.7	19	134.3	35	184.4	FERRY WA	-14	37	18	35	42
95	159.4	155	188.7	199	194.0	FRANKLIN WA	18	3	22	42	57
32	185.5	38	230.6	22	128.0	GARFIELD WA	24	-44	-31	00	02
129	138.2	214	150.9	330	176.3	GRANT WA	9	17	28	52	30
533	176.6	607	193.1	697	210.0	GRAYS HARBOR WA	9	9	19	37	78
111	164.1	136	148.2	269	165.7	ISLAND WA	-10	12	1	08	19
87	177.3	95	160.1	161	179.2	JEFFERSON WA	-10	12	1	08	34
6880*	191.8	8099	195.6	9459	204.1	KING WA	2	4	6	15	70
574	160.6	740	173.0	1052	197.0	KITSAP WA	8	14	23	44	62
159	143.1	189	172.1	215	190.8	KITTITAS WA	20	11	33	62	52
95	160.7	104	150.1	137	178.8	KLICKITAT WA	-7	19	11	23	34
432	172.5	491	183.8	512	178.2	LEWIS WA	7	-3	3	11	33
70	116.9	93	146.8	112	168.6	LINCOLN WA	26	15	44	76	21
124	160.9	167	167.7	256	198.0	MASON WA	4	18	23	45	62
165	133.8	231	171.0	292	195.0	OKANOGAN WA	28	14	46	78	58
162	177.9	154	149.5	238	203.2	PACIFIC WA	-16	36	14	29	69
66	156.9	65	156.1	84	188.7	PEND OREILLE WA	0	21	20	40	49
2177	167.7	2650	182.3	3491	206.6	PIERCE WA	9	13	23	45	74
32	152.1	44	155.7	61	137.9	SAN JUAN WA	2	-11	-9	04	03
349	131.5	438	153.7	586	181.7	SKAGIT WA	17	18	38	69	38
39	151.2	39	141.3	56	177.9	SKAMANIA WA	-7	26	18	35	33
1092	162.9	1580	182.1	2142	192.6	SNOHOMISH WA	12	6	18	36	55
2029	163.3	2503	184.3	2842	195.2	SPOKANE WA	13	6	20	38	59
160	154.8	151	145.0	195	169.3	STEVENS WA	-6	17	9	20	22
386	163.3	546	183.0	810	192.2	THURSTON WA	12	5	18	35	54
24	113.3	44	202.9	43	189.7	WAHKIAKUM WA	79	-7	67	91	50
287	137.2	382	169.3	422	176.3	WALLA WALLA WA	23	4	28	54	30
577	154.6	682	170.6	784	178.2	WHATCOM WA	10	4	15	30	33
214	163.6	197	145.1	258	184.2	WHITMAN WA	-11	27	13	25	42
994	162.3	1236	177.5	1548	203.3	YAKIMA WA	9	15	25	49	69
11248	147.4	14750	179.8	17838	202.9	WEST VIRGINIA WV	22	13	38		
105	117.7	139	156.3	165	184.0	BARBOUR WV	33	18	56	85	42
197	128.7	274	157.8	379	186.0	BERKELEY WV	23	18	45	76	44
146	145.5	210	175.7	269	212.9	BOONE WV	21	21	46	78	80
110	128.6	103	120.8	161	189.7	BRAXTON WV	-6	57	48	79	50
188	169.3	258	202.5	314	220.8	BROOKE WV	20	9	30	58	87
699	156.9	919	187.7	1210*	238.2	CABELL WV	20	27	52	83	95
51	112.4	61	134.2	90	184.8	CALHOUN WV	19	38	64	90	43
46	84.7	65	127.1	102	190.7	CLAY WV	50	50	125	98	52
54	112.2	60	140.9	87	195.9	DODDRIDGE WV	26	39	75	93	60
341	140.2	500	201.6	547	209.6	FAYETTE WV	44	4	49	81	77
46	101.4	76	173.2	69	148.9	GILMER WV	71	-14	47	79	06
39	105.7	74	173.8	80	166.8	GRANT WV	64	-4	58	86	20
210	144.1	246	150.2	317	174.4	GREENBRIER WV	4	16	21	41	28
77	129.3	106	167.6	159	223.1	HAMPSHIRE WV	30	33	72	92	89
268	192.2	353*	225.6	389	218.6	HANCOCK WV	17	-3	14	28	85
63	137.0	79	159.4	75	137.3	HARDY WV	16	-14	0	08	03
574	150.9	755	189.7	847	205.9	HARRISON WV	26	9	36	67	73
102	132.4	133	156.2	185	179.4	JACKSON WV	18	15	35	65	34
98	155.7	130	181.3	204	205.4	JEFFERSON WV	16	13	32	60	72
1245	157.7	1840*	202.3	2254*	226.2	KANAWHA WV	28	12	43	75	91
154	122.5	153	121.4	206	179.5	LEWIS WV	-1	48	47	78	35
122	143.8	149	161.1	204	195.1	LINCOLN WV	12	21	36	66	59
290	157.0	367	186.5	388	199.7	LOGAN WV	19	7	27	52	65
300	156.6	376	204.9	373	209.2	MC DOWELL WV	31	2	34	63	77
520	174.2	619	195.2	719	214.6	MARION WV	12	10	23	45	82
312	174.3	400*	213.3	435	223.1	MARSHALL WV	22	5	28	53	89
111	107.2	171	154.5	215	174.2	MASON WV	44	13	63	89	28
380	150.9	474	172.9	552	183.1	MERCER WV	15	6	21	42	40
131	138.0	192	184.3	247	211.1	MINERAL WV	34	15	53	84	79
198	143.0	253	176.9	289	203.1	MINGO WV	24	15	42	74	69
406	174.0	477	188.7	546	207.0	MONONGALIA WV	8	10	19	37	75
81	117.2	93	134.7	127	184.5	MONROE WV	15	37	57	86	42
58	145.8	69	170.3	116	238.2	MORGAN WV	17	40	63	89	95
105	102.3	186	168.8	221	182.0	NICHOLAS WV	65	8	78	94	39
661*	194.8	814*	238.1	808*	244.4	OHIO WV	22	3	25	49	97
46	103.9	64	142.8	64	136.4	PENDLETON WV	38	-5	31	59	03
40	117.7	53	156.6	69	193.6	PLEASANTS WV	33	24	65	90	56
62	106.5	81	140.7	99	159.6	POCAHONTAS WV	32	13	50	81	13
193	138.9	221	159.3	247	164.6	PRESTON WV	15	3	19	37	18
111	124.8	165	152.3	269	196.5	PUTNAM WV	22	29	57	86	61
386	146.7	548	182.5	661	193.8	RALEIGH WV	24	6	32	60	56
171	131.7	197	142.7	275	193.6	RANDOLPH WV	8	36	47	79	56
87	118.0	101	140.4	133	189.2	RITCHIE WV	19	35	60	88	49
109	116.7	131	140.7	150	159.9	ROANE WV	21	14	37	68	13
108	141.6	137	179.5	170	216.5	SUMMERS WV	27	21	53	84	84
138	164.2	161	194.4	182	215.5	TAYLOR WV	18	11	31	59	83
74	148.2	74	160.6	76	150.7	TUCKER WV	8	-6	2	09	07
73	114.4	107	182.4	103	177.0	TYLER WV	59	-3	55	84	31
114	121.2	131	132.2	182	171.2	UPSHUR WV	9	30	41	73	24
208	131.5	261	150.2	343	175.4	WAYNE WV	14	17	33	62	29
75	119.8	105	171.4	131	209.5	WEBSTER WV	43	22	75	93	77
147	147.2	175	172.7	223	206.8	WETZEL WV	17	20	40	72	74
37	131.3	29	102.1	53	192.3	WIRT WV	-22	88	46	78	54
470	151.1	667	185.4	819	202.3	WOOD WV	23	9	34	63	68
111	109.6	168	156.9	240	211.5	WYOMING WV	43	35	93	96	79
30392	176.8	34949	179.3	40547	189.6	WISCONSIN WI	1	6	7		
71	129.9	100	156.5	165	196.9	ADAMS WI	20	26	52	82	61
174	171.6	178	166.7	213	194.0	ASHLAND WI	-3	16	13	26	57
279	147.0	299	140.7	420	177.6	BARRON WI	-4	26	21	41	32
131	152.7	131	145.1	168	183.4	BAYFIELD WI	-5	26	20	39	41
742	173.0	955	182.9	1219	196.8	BROWN WI	6	8	14	28	61

* SIGNIFICANTLY MORE DEATHS THAN EXPECTED

WHITE MALE — ALL CANCERS COMBINED — ICD 140-208

1950-1959 DEATHS	1950-1959 RATE	1960-1969 DEATHS	1960-1969 RATE	1970-1979 DEATHS	1970-1979 RATE	COUNTY AND STATE	PERCENT CHANGE 60S/50S	PERCENT CHANGE 70S/60S	PERCENT CHANGE 70S/50S	PERCENTILE %CHANGE 70S/50S	PERCENTILE RATES 1970S
106	136.8	132	162.6	155	174.7	BUFFALO WI	19	7	28	53	28
103	162.6	108	144.6	142	164.3	BURNETT WI	-11	14	1	08	17
123	133.8	170	156.1	217	172.2	CALUMET WI	17	10	29	54	25
312	146.9	391	173.8	414	161.5	CHIPPEWA WI	18	-7	10	20	15
285	156.6	302	156.5	334	162.6	CLARK WI	0	4	4	12	16
339	164.2	367	163.0	444	183.7	COLUMBIA WI	-1	13	12	24	41
126	142.1	142	155.6	172	172.5	CRAWFORD WI	9	11	21	42	26
1267	171.9	1542	173.8	1795	175.7	DANE WI	1	1	2	10	30
516	165.8	589	173.1	675	177.6	DODGE WI	4	3	7	16	32
194	167.1	209	163.9	264	187.5	DOOR WI	-2	14	12	25	47
478	187.9	450	178.4	509	207.9	DOUGLAS WI	-5	17	11	22	76
214	152.6	215	139.0	266	160.7	DUNN WI	-9	16	5	13	14
373	149.0	486	170.4	608	189.4	EAU CLAIRE WI	14	11	27	52	50
41	205.4	43	195.1	43	168.1	FLORENCE WI	-5	-14	-18	01	21
580	166.9	607	159.6	733	178.4	FOND DU LAC WI	-4	12	7	16	33
86	174.5	89	179.0	86	161.8	FOREST WI	3	-10	-7	04	15
378	166.1	371	161.1	385	155.9	GRANT WI	-3	-3	-6	05	10
218	156.7	258	173.7	287	177.3	GREEN WI	11	2	13	26	32
139	155.1	174	171.3	229	196.7	GREEN LAKE WI	10	15	27	51	61
168	152.6	163	151.1	182	171.3	IOWA WI	-1	13	12	25	24
104*	217.7	98	195.7	99	193.9	IRON WI	-10	-1	-11	03	57
115	120.7	156	164.7	185	174.8	JACKSON WI	36	6	45	76	28
414	160.4	498	172.5	555	176.3	JEFFERSON WI	8	2	10	20	30
143	132.0	167	154.5	236	190.7	JUNEAU WI	17	23	44	76	52
753*	196.4	878	186.4	1094	207.1	KENOSHA WI	-5	11	5	14	75
130	142.3	147	151.8	198	187.3	KEWAUNEE WI	7	23	32	59	46
487	154.9	658	183.5	717	182.7	LA CROSSE WI	18	0	18	36	40
145	147.8	177	183.5	163	172.0	LAFAYETTE WI	24	-6	16	32	25
177	156.5	204	171.3	237	189.0	LANGLADE WI	9	10	21	41	49
204	164.1	260	193.4	305	201.3	LINCOLN WI	18	4	23	44	67
542	167.2	660	173.9	738	170.6	MANITOWOC WI	4	-2	2	09	24
562	147.1	744	172.0	926	193.1	MARATHON WI	17	12	31	59	55
332	178.3	375	177.4	432	183.9	MARINETTE WI	-1	4	3	11	42
86	145.1	110	164.1	131	158.5	MARQUETTE WI	13	-3	9	19	12
498	155.0	593	167.4	718	182.4	MENOMINEE#WI	8	9	18	35	39
8611*	219.1	9543*	216.2	9582*	217.9	MILWAUKEE WI	-1	1	-1	07	85
223	127.0	258	137.0	321	156.3	MONROE WI	8	14	23	45	10
208	188.6	287*	222.5	334	200.6	ONEIDA WI	18	-10	6	15	66
597	173.3	707	171.1	913	188.6	OUTAGAMIE WI	-1	10	9	18	48
221	184.0	270	156.9	380	165.5	OZAUKEE WI	-15	5	-10	03	19
62	140.8	51	109.1	72	148.7	PEPIN WI	-23	36	6	14	06
186	151.4	179	137.0	201	148.7	PIERCE WI	-9	8	-2	07	06
225	152.3	231	134.9	332	174.3	POLK WI	-11	29	14	29	28
302	181.5	312	164.2	372	170.4	PORTAGE WI	-10	4	-6	05	24
169	177.0	158	148.4	179	158.2	PRICE WI	-16	7	-11	03	12
1073*	199.3	1183	186.1	1393	201.5	RACINE WI	-7	8	1	08	67
159	154.6	147	146.0	182	176.0	RICHLAND WI	-6	21	14	28	30
797	174.4	1008	197.5	1085	193.0	ROCK WI	13	-2	11	22	55
146	167.4	151	173.5	159	164.6	RUSK WI	4	-5	-2	07	18
207	143.5	222	139.6	315	172.8	ST CROIX WI	-3	24	20	40	26
330	162.8	344	160.7	397	166.2	SAUK WI	-1	3	2	09	19
115	178.7	112	159.9	158	185.8	SAWYER WI	-10	16	4	12	44
747	180.2	782	164.5	938	182.3	SHEBOYGAN WI	-9	11	1	09	39
135	139.8	159	153.7	171	154.9	TAYLOR WI	10	1	11	22	10
205	140.1	253	161.8	297	178.2	TREMPEALEAU WI	16	10	27	52	33
248	159.4	232	142.6	308	177.1	VERNON WI	-11	24	11	23	31
86	171.1	126	172.5	213	207.5	VILAS WI	1	20	21	42	75
467	185.2	479	164.9	652	195.9	WALWORTH WI	-11	19	6	14	60
116	170.2	151	209.5	165	201.8	WASHBURN WI	23	-4	19	37	67
255	141.6	367	164.5	519	183.6	WASHINGTON WI	16	12	30	56	41
749	164.0	1138	173.7	1729	198.7	WAUKESHA WI	6	14	21	42	64
341	152.7	403	154.0	475	169.4	WAUPACA WI	1	10	11	22	22
126	130.9	151	151.1	196	170.2	WAUSHARA WI	15	13	30	57	23
779	169.5	883	167.2	1059	182.6	WINNEBAGO WI	-1	9	8	17	40
372	158.8	466	167.3	591	180.3	WOOD WI	5	8	14	27	36
1781	145.3	2169	152.5	2683	171.6	WYOMING WY	5	13	18		
100	158.2	127	162.1	156	194.0	ALBANY WY	2	20	23	44	57
75	144.0	90	153.5	102	158.1	BIG HORN WY	7	3	10	20	12
26	105.2	40	128.5	54	139.9	CAMPBELL WY	22	9	33	62	04
79	125.6	100	146.7	137	196.4	CARBON WY	17	34	56	85	61
48	162.7	45	136.6	62	167.5	CONVERSE WY	-16	23	3	10	21
23	100.1	20	102.1	34	170.8	CROOK WY	2	67	71	92	24
120	154.9	129	144.0	185	174.7	FREMONT WY	-7	21	13	26	28
68	120.8	83	133.1	117	174.2	GOSHEN WY	10	31	44	76	28
50	164.1	50	146.6	70	186.8	HOT SPRINGS WY	-11	27	14	28	46
47	164.8	53	154.4	65	172.8	JOHNSON WY	-6	12	5	13	26
231	131.6	347	163.2	464	196.1	LARAMIE WY	24	20	49	81	60
62	170.6	59	145.9	53	122.2	LINCOLN WY	-14	-16	-28	00	01
200	148.3	302	170.5	363	181.8	NATRONA WY	15	7	23	44	38
36	177.8	41	205.4	35	179.5	NIOBRARA WY	16	-13	1	08	35
97	135.1	131	154.4	164	157.9	PARK-TETON#WY	14	2	17	33	12
49	125.0	50	128.3	70	151.9	PLATTE WY	3	18	22	42	08
171	144.1	193	154.1	201	162.4	SHERIDAN WY	7	5	13	25	16
19	153.2	22	202.0	35	286.1	SUBLETTE WY	32	42	87	95	99
162	219.2	148	175.0	154	154.7	SWEETWATER WY	-20	-12	-29	00	10
39	101.3	40	101.5	46	119.6	UINTA WY	0	18	18	36	01
39	149.4	55	181.0	64	183.2	WASHAKIE WY	21	1	23	44	40

* SIGNIFICANTLY MORE DEATHS THAN EXPECTED

COUNTIES COMBINED; SEE TECHNICAL NOTES

WHITE FEMALE — ALL CANCERS COMBINED — ICD 140-208

1950-1959 DEATHS	1950-1959 RATE	1960-1969 DEATHS	1960-1969 RATE	1970-1979 DEATHS	1970-1979 RATE	COUNTY AND STATE	PERCENT CHANGE 60S/50S	PERCENT CHANGE 70S/60S	PERCENT CHANGE 70S/50S	PERCENTILE %CHANGE 70S/50S	PERCENTILE RATES 1970S
1031475	141.6	1220907	132.4	1488093	131.7	UNITED STATES	-6	-1	-7		
11403	123.1	14207	116.2	18357	115.0	ALABAMA AL	-6	-1	-7		
52	108.1	79	124.9	100	106.0	AUTAUGA AL	16	-15	-2	58	21
187	121.5	278	119.2	403	111.1	BALDWIN AL	-2	-7	-9	40	30
69	109.6	67	88.5	82	88.5	BARBOUR AL	-19	0	-19	15	05
70	140.1	58	101.1	81	112.4	BIBB AL	-28	11	-20	15	33
121	113.3	136	102.7	180	99.6	BLOUNT AL	-9	-3	-12	30	13
37	153.6	38	147.1	18	63.5	BULLOCK AL	-4	-57	-59	00	00
75	103.0	105	115.2	114	103.8	BUTLER AL	12	-10	1	66	17
382	149.9	435	117.9	609	124.5	CALHOUN AL	-21	6	-17	19	60
146	117.6	168	112.1	207	108.3	CHAMBERS AL	-5	-3	-8	42	24
74	123.7	97	128.7	130	137.6	CHEROKEE AL	4	7	11	83	85
141	130.9	132	102.2	204	128.5	CHILTON AL	-22	26	-2	59	69
40	109.6	44	97.6	37	65.0	CHOCTAW AL	-11	-33	-41	01	00
76	118.7	83	103.7	117	117.0	CLARKE AL	-13	13	-1	60	43
68	108.7	86	117.0	83	94.0	CLAY AL	8	-20	-14	27	08
60	127.4	56	104.2	60	88.5	CLEBURNE AL	-18	-15	-31	04	05
95	95.6	116	90.4	156	89.8	COFFEE AL	-6	-1	-6	47	05
155	115.0	213	116.2	274	111.8	COLBERT AL	1	-4	-3	56	31
58	109.4	72	120.6	78	107.4	CONECUH AL	10	-11	-2	59	23
60	157.3	45	106.4	55	109.5	COOSA AL	-32	3	-30	05	27
165	110.0	201	108.5	262	111.0	COVINGTON AL	-1	2	1	66	30
44	75.0	60	92.0	88	105.3	CRENSHAW AL	23	14	40	96	19
244	130.1	277	109.2	366	105.6	CULLMAN AL	-16	-3	-19	16	20
90	106.0	123	115.2	147	94.3	DALE AL	9	-18	-11	33	08
118	111.3	167	129.8	201	129.4	DALLAS AL	17	0	16	88	71
227	124.4	270	118.9	285	95.9	DE KALB AL	-4	-19	-23	11	09
143	144.6	131	102.8	177	100.6	ELMORE AL	-29	-2	-30	05	13
127	144.5	143	121.9	160	108.0	ESCAMBIA AL	-16	-11	-25	08	24
439	135.4	504	118.1	683	124.7	ETOWAH AL	-13	6	-8	42	61
109	154.8	98	117.1	109	104.7	FAYETTE AL	-24	-11	-32	03	18
124	134.6	156	129.5	185	113.6	FRANKLIN AL	-4	-12	-16	21	35
105	110.0	115	102.3	129	95.5	GENEVA AL	-7	-7	-13	27	09
20	143.7	16	86.3	24	110.6	GREENE AL	-40	28	-23	10	29
32	96.9	44	112.1	58	125.5	HALE AL	16	12	30	94	63
53	116.2	63	114.8	59	92.2	HENRY AL	-1	-20	-21	13	06
196	128.1	216	106.2	273	96.3	HOUSTON AL	-17	-9	-25	09	10
184	127.0	181	101.3	265	114.1	JACKSON AL	-20	13	-10	35	36
2214	129.8	2868	124.6	3696	129.2	JEFFERSON AL	-4	4	-1	62	70
67	105.0	81	106.8	121	125.5	LAMAR AL	2	17	20	90	63
202	102.0	296	109.0	421	111.8	LAUDERDALE AL	7	3	10	81	31
91	121.9	102	106.6	141	113.7	LAWRENCE AL	-12	7	-7	45	36
109	115.6	164	121.4	187	97.4	LEE AL	5	-20	-16	21	11
133	120.3	154	106.5	195	95.4	LIMESTONE AL	-11	-10	-21	13	09
12	64.3	20	92.6	35	147.0	LOWNDES AL	44	59	129	99	94
29	115.8	34	117.4	43	129.2	MACON AL	1	10	12	83	70
318	123.5	476	118.4	780	120.3	MADISON AL	-4	2	-3	57	51
42	90.4	59	108.5	65	92.7	MARENGO AL	20	-15	3	69	07
95	90.8	137	105.0	154	91.4	MARION AL	16	-13	1	66	06
224	124.0	315	124.7	407	116.3	MARSHALL AL	1	-7	-6	47	41
901	131.1	1243	130.4	1515	122.5	MOBILE AL	-1	-6	-7	45	56
58	107.8	75	113.3	82	93.3	MONROE AL	5	-18	-13	27	07
503	126.3	671	126.3	889	130.0	MONTGOMERY AL	0	3	3	70	72
263	127.2	292	105.2	429	111.0	MORGAN AL	-17	6	-13	29	30
34	105.1	27	69.6	35	80.8	PERRY AL	-34	16	-23	10	02
57	91.1	67	89.5	88	97.8	PICKENS AL	-2	9	7	78	11
97	122.1	106	110.3	124	108.2	PIKE AL	-10	-2	-11	32	24
79	99.9	99	101.3	107	95.5	RANDOLPH AL	1	-6	-4	52	09
144	166.2	158	130.7	175	113.7	RUSSELL AL	-21	-13	-32	04	36
108	113.9	104	88.0	182	105.7	ST CLAIR AL	-23	20	-7	44	20
113	109.6	145	107.3	217	106.9	SHELBY AL	-2	0	-3	57	22
39	136.4	36	108.7	45	109.2	SUMTER AL	-20	0	-20	14	26
210	120.2	285	127.0	355	121.0	TALLADEGA AL	6	-5	1	65	52
137	125.5	188	125.3	209	109.5	TALLAPOOSA AL	0	-13	-13	28	27
267	87.8	384	93.1	558	111.1	TUSCALOOSA AL	6	19	26	93	30
306	145.9	324	122.3	406	113.4	WALKER AL	-16	-7	-22	11	35
42	111.2	57	128.1	72	123.3	WASHINGTON AL	15	-4	11	82	58
37	132.6	33	103.8	46	119.1	WILCOX AL	-22	15	-10	35	48
86	131.8	134*	170.3	119	112.4	WINSTON AL	29	-34	-15	24	33
		391	125.5	710	130.9	ALASKA AK		4			
3986	116.3	7309	115.4	13128	118.6	ARIZONA AZ	-1	3	2		
27	151.7	19	82.6	18	58.4	APACHE AZ	-46	-29	-62	00	00
209	124.1	301	128.7	443	131.7	COCHISE AZ	4	2	6	76	74
62	110.0	87	91.0	142	107.9	COCONINO AZ	-17	19	-2	59	24
97	116.2	138	124.5	178	105.1	GILA AZ	7	-16	-10	37	19
53	118.6	65	108.0	94	112.2	GRAHAM AZ	-9	4	-5	49	33
35	96.2	51	129.7	47	119.7	GREENLEE AZ	35	-8	24	92	49
2092	117.2	4116	116.2	7763	120.2	MARICOPA AZ	-1	3	3	69	51
23	80.8	70	89.3	243	95.9	MOHAVE AZ	11	7	19	90	09
58	113.2	77	101.6	124	108.4	NAVAJO AZ	-10	7	-4	52	24
892	118.4	1651	116.8	2826	121.1	PIMA AZ	-1	4	2	69	53
113	89.4	221	111.0	366	120.7	PINAL AZ	24	9	35	95	52
47	115.3	56	95.6	126*	157.1	SANTA CRUZ AZ	-17	64	36	96	98
172	130.4	236	110.9	432	104.5	YAVAPAI AZ	-15	-6	-20	14	18
106	99.3	221	123.5	326	106.4	YUMA AZ	24	-14	7	78	21
7930	115.6	9788	111.8	13295	115.6	ARKANSAS AR	-3	3	0		
101	127.3	125	124.0	161	122.9	ARKANSAS AR	-3	-1	-3	54	57
59	92.6	69	85.1	125	114.5	ASHLEY AR	-8	35	24	92	37
69	120.8	99	92.8	245	118.7	BAXTER AR	-23	28	-2	59	47
228	99.5	314	104.6	428	95.4	BENTON AR	5	-9	-4	53	09
99	112.6	138	109.1	214	121.1	BOONE AR	-3	11	8	78	53

* SIGNIFICANTLY MORE DEATHS THAN EXPECTED

WHITE FEMALE

ALL CANCERS COMBINED

ICD 140-208

1950-1959 DEATHS	1950-1959 RATE	1960-1969 DEATHS	1960-1969 RATE	1970-1979 DEATHS	1970-1979 RATE	COUNTY AND STATE	PERCENT CHANGE 60S/50S	PERCENT CHANGE 70S/60S	PERCENT CHANGE 70S/50S	PERCENTILE %CHANGE 70S/50S	PERCENTILE RATES 1970S
53	101.5	90	137.0	92	111.4	BRADLEY AR	35	-19	10	81	30
26	118.1	24	86.0	38	112.3	CALHOUN AR	-27	31	-5	51	33
103	131.7	96	95.4	132	107.4	CARROLL AR	-28	13	-18	17	23
42	107.9	57	132.3	58	110.6	CHICOT AR	23	-16	2	69	29
94	116.5	106	112.8	112	89.2	CLARK AR	-3	-21	-23	10	05
140	123.2	153	116.4	198	123.4	CLAY AR	-6	6	0	64	58
59	114.1	56	86.8	118	120.7	CLEBURNE AR	-24	39	6	75	52
34	110.6	34	97.4	42	97.3	CLEVELAND AR	-12	0	-12	31	11
79	90.9	117	110.2	143	109.6	COLUMBIA AR	21	-1	21	90	27
73	111.0	75	97.3	107	103.8	CONWAY AR	-12	7	-6	46	17
220	115.2	263	107.4	386	120.1	CRAIGHEAD AR	-7	12	4	72	50
132	125.4	177	129.0	190	104.0	CRAWFORD AR	3	-19	-17	19	17
60	107.5	96	114.8	147	126.8	CRITTENDEN AR	7	10	18	89	66
78	127.7	83	110.1	116	129.5	CROSS AR	-14	18	1	67	71
55	134.7	49	104.0	71	133.6	DALLAS AR	-23	28	-1	62	78
57	136.4	55	98.1	73	107.1	DESHA AR	-28	9	-21	12	22
50	91.0	60	97.9	94	124.7	DREW AR	8	27	37	96	61
112	109.2	142	102.8	218	112.2	FAULKNER AR	-6	9	3	70	33
67	104.4	80	106.3	93	105.7	FRANKLIN AR	2	-1	1	66	20
34	83.1	55	99.8	72	99.6	FULTON AR	20	0	20	90	12
367	154.4	420	124.3	665	136.3	GARLAND AR	-19	10	-12	31	83
38	102.9	40	77.8	86	123.2	GRANT AR	-24	58	20	90	58
130	103.9	151	100.6	217	114.4	GREENE AR	-3	14	10	82	37
110	133.8	143	144.3	145	118.3	HEMPSTEAD AR	8	-18	-12	32	46
88	101.0	119	107.3	165	112.5	HOT SPRING AR	6	5	11	83	34
60	104.6	75	109.5	103	129.0	HOWARD AR	5	18	23	92	70
92	83.9	154	109.5	176	98.4	INDEPENDENCE AR	31	-10	17	88	12
53	121.7	40	78.8	81	102.3	IZARD AR	-35	30	-16	21	15
105	126.9	117	119.6	160	133.7	JACKSON AR	-6	12	5	74	79
227	126.7	254	106.6	356	115.5	JEFFERSON AR	-16	8	-9	39	39
99	126.3	106	117.2	150	122.7	JOHNSON AR	-7	5	-3	56	56
42	118.6	39	99.6	40	85.9	LAFAYETTE AR	-16	-14	-28	07	03
92	111.7	112	107.2	158	124.3	LAWRENCE AR	-4	16	11	83	60
25	71.5	53	133.7	67	145.5	LEE AR	87	9	103	99	93
29	89.1	29	83.9	53	114.9	LINCOLN AR	-6	37	29	93	38
45	125.1	50	122.9	72	126.3	LITTLE RIVER AR	-2	3	1	66	65
120	117.6	146	126.4	166	114.6	LOGAN AR	7	-9	-3	57	37
99	105.5	136	115.6	163	105.0	LONOKE AR	10	-9	-1	63	19
59	109.0	55	87.8	65	85.1	MADISON AR	-19	-3	-22	11	03
37	109.4	62	133.0	96	128.4	MARION AR	22	-3	17	88	69
155	132.6	188	127.6	228	125.5	MILLER AR	-4	-2	-5	49	63
195	112.9	273	128.6	286	116.7	MISSISSIPPI AR	14	-9	3	71	42
40	97.4	54	109.1	64	112.9	MONROE AR	12	4	16	87	34
33	103.2	43	109.3	49	94.6	MONTGOMERY AR	6	-13	-8	40	08
64	128.0	66	116.6	91	137.8	NEVADA AR	-9	18	8	78	85
29	96.3	31	93.6	34	80.4	NEWTON AR	-3	-14	-16	20	02
119	138.3	119	101.7	161	110.8	OUACHITA AR	-26	9	-20	14	30
28	100.2	37	105.7	53	134.8	PERRY AR	5	28	34	95	80
69	94.7	115	129.1	147	135.1	PHILLIPS AR	36	5	43	96	81
42	94.1	78	138.0	102	134.0	PIKE AR	47	-3	42	96	79
112	106.5	143	115.2	185	127.1	POINSETT AR	8	10	19	90	67
93	116.3	105	110.2	118	100.2	POLK AR	-5	-9	-14	26	13
126	120.7	159	115.4	222	116.2	POPE AR	-4	1	-4	54	41
45	98.1	69	144.5	57	96.3	PRAIRIE AR	47	-33	-2	59	10
1054	132.4	1291	119.0	1757	126.6	PULASKI AR	-10	6	-4	52	66
67	99.9	81	101.2	118	113.0	RANDOLPH AR	1	12	13	85	35
70	124.6	70	101.8	99	111.6	ST FRANCIS AR	-18	10	-10	35	31
113	89.7	138	85.4	225	101.3	SALINE AR	-5	19	13	85	14
40	88.3	51	98.4	68	109.0	SCOTT AR	11	11	23	92	26
38	82.6	56	111.4	57	91.6	SEARCY AR	35	-18	11	82	06
360	113.2	515	124.8	625	116.0	SEBASTIAN AR	10	-7	2	69	41
63	111.6	96	142.4	111	133.5	SEVIER AR	28	-6	20	90	78
38	90.3	46	86.3	93	97.5	SHARP AR	-4	13	8	79	11
28	88.7	31	81.2	58	107.1	STONE AR	-8	32	21	91	22
189	116.2	246	120.2	332	122.5	UNION AR	3	2	5	75	56
52	123.4	47	88.5	69	86.2	VAN BUREN AR	-28	-3	-30	05	04
316	123.6	356	103.9	527	112.0	WASHINGTON AR	-16	8	-9	37	32
172	103.6	220	103.4	313	109.5	WHITE AR	0	6	6	75	27
56	133.7	43	91.8	61	107.9	WOODRUFF AR	-31	18	-19	15	24
83	122.2	107	121.4	128	111.4	YELL AR	-1	-8	-9	39	31
83285	136.7	111746	131.5	148727*	139.1	CALIFORNIA CA	-4	6	2		
5605	140.3	6536*	137.8	7422*	147.5	ALAMEDA CA	-2	7	5	74	94
2	259.3	5*	269.0	4	201.7	ALPINE CA	4	-25	-22	11	99
67	136.3	79	117.1	124	107.2	AMADOR CA	-14	-8	-21	12	23
463	115.8	700	121.9	1080	132.0	BUTTE CA	5	8	14	85	75
54	101.7	85	110.0	152	126.5	CALAVERAS CA	8	15	24	92	66
78	147.2	81	126.8	103	138.9	COLUSA CA	-14	9	-6	48	86
1547	140.5	2451	131.5	3818*	137.9	CONTRA COSTA CA	-6	5	-2	59	85
60	148.1	91	139.6	113	132.2	DEL NORTE CA	-6	-5	-11	34	76
104	106.0	207	117.8	389	116.3	EL DORADO CA	11	-1	10	81	41
1557	125.3	2149	126.4	2656	123.1	FRESNO CA	1	-3	-2	59	57
86	112.4	108	109.9	163	131.8	GLENN CA	-2	20	17	88	75
495	150.2	571	134.6	751*	147.7	HUMBOLDT CA	-10	10	-2	60	94
200	122.8	295	123.8	407	140.9	IMPERIAL CA	1	14	15	86	89
59	135.4	86	127.6	167*	154.8	INYO CA	-6	21	14	86	97
1104	131.3	1553	126.6	2155	134.0	KERN CA	-4	6	2	68	79
234	138.5	296	136.1	321	119.7	KINGS CA	-2	-12	-14	27	49
98	113.3	162	118.0	307	118.6	LAKE CA	4	0	5	73	47
62	113.7	73	109.2	127	144.2	LASSEN CA	-4	32	27	93	92
36572	141.2	44638*	135.0	51111*	144.2	LOS ANGELES CA	-4	7	2	68	92
199	143.6	221	118.2	312	125.0	MADERA CA	-18	6	-13	28	62
702	143.4	1052	132.5	1692*	152.8	MARIN CA	-8	15	7	77	97
19	75.6	37	106.9	86	146.9	MARIPOSA CA	41	37	94	98	94
242	115.7	281	107.2	429	129.0	MENDOCINO CA	-7	20	12	83	70
298	119.0	472	134.2	627	132.0	MERCED CA	13	-2	11	82	75
42	133.6	32	79.0	57	118.3	MODOC CA	-41	50	-12	32	46
3	69.0	19	190.5	26	140.8	MONO CA	176	-26	104	99	89
701	126.5	1085	128.4	1569	138.3	MONTEREY CA	1	8	9	80	86
358	107.4	528	116.0	778	129.8	NAPA CA	8	12	21	91	71

* SIGNIFICANTLY MORE DEATHS THAN EXPECTED

WHITE FEMALE — ALL CANCERS COMBINED — ICD 140-208

1950-1959 DEATHS	1950-1959 RATE	1960-1969 DEATHS	1960-1969 RATE	1970-1979 DEATHS	1970-1979 RATE	COUNTY AND STATE	PERCENT CHANGE 60S/50S	PERCENT CHANGE 70S/60S	PERCENT CHANGE 70S/50S	PERCENTILE %CHANGE 70S/50S	PERCENTILE RATES 1970S
147	116.6	230	136.9	374	131.7	NEVADA CA	17	-4	13	85	74
2340	122.5	5328	122.5	10478*	136.2	ORANGE CA	0	11	11	83	83
281	112.0	453	125.7	677	126.5	PLACER CA	12	1	13	85	66
42	95.1	72	136.8	101	126.0	PLUMAS CA	44	-8	32	94	65
1388	120.9	2418	115.8	4259	121.1	RIVERSIDE CA	-4	5	0	64	53
1965	131.9	3079*	137.6	4566*	143.0	SACRAMENTO CA	4	4	8	79	91
89	130.7	101	118.1	134	130.3	SAN BENITO CA	-10	10	0	63	72
2250	127.5	3471	123.0	4872	125.5	SAN BERNARDINO CA	-4	2	-2	60	63
4450	134.4	6697	126.4	10641	134.3	SAN DIEGO CA	-6	6	0	64	80
6877*	160.0	7062*	156.2	6835*	165.8	SAN FRANCISCO CA	-2	6	4	71	98
1131	122.5	1553	124.1	2076	132.0	SAN JOAQUIN CA	1	6	8	78	75
367	110.8	566	111.5	949	121.5	SAN LUIS OBISPO CA	1	9	10	81	53
1949	142.4	3026	137.3	4442*	150.4	SAN MATEO CA	-4	10	6	75	96
852	129.4	1368	127.5	2014	132.5	SANTA BARBARA CA	-1	4	2	69	76
2528	128.4	4249	127.8	6552*	137.6	SANTA CLARA CA	0	8	7	78	85
688	124.0	1036	126.9	1480	136.4	SANTA CRUZ CA	2	8	10	81	83
242	132.3	365	119.5	658	130.0	SHASTA CA	-10	9	-2	59	72
10	98.1	12	86.6	18	132.6	SIERRA CA	-12	53	35	95	77
172	131.0	213	130.3	284	135.8	SISKIYOU CA	-1	4	4	71	82
527	141.4	702	133.3	1115*	150.1	SOLANO CA	-6	13	6	76	96
908	129.5	1310	123.5	2243*	141.3	SONOMA CA	-5	14	9	80	89
801	117.1	1041	110.7	1561	123.6	STANISLAUS CA	-6	12	6	75	59
156	137.4	238	146.3	267	118.0	SUTTER CA	6	-19	-14	25	45
125	117.3	189	130.8	287	132.6	TEHAMA CA	12	1	13	85	76
28	102.7	39	118.5	55	110.6	TRINITY CA	15	-7	8	78	29
757	121.6	1020	122.3	1364	129.4	TULARE CA	1	6	6	77	71
80	113.4	139	127.6	226	130.9	TUOLUMNE CA	13	3	15	87	73
718	117.3	1276	111.8	2420	128.0	VENTURA CA	-5	14	9	80	68
272	140.9	373	123.9	602*	143.7	YOLO CA	-12	16	2	68	91
134	132.0	167	123.1	231	125.2	YUBA CA	-7	2	-5	50	62
9324	131.4	10882	116.5	13884	115.9	COLORADO CO	-11	-1	-12		
219	123.3	479	124.8	761	115.3	ADAMS CO	1	-8	-6	46	39
47	140.8	56	126.8	50	97.6	ALAMOSA CO	-10	-23	-31	04	11
10	86.4	11	99.4	9	73.7	ARCHULETA CO	15	-26	-15	24	01
43	163.1	27	87.4	41	114.6	BACA CO	-46	31	-30	05	38
46	138.3	47	134.2	46	125.9	BENT CO	-3	-6	-9	39	64
368	122.2	525	115.2	678	103.0	BOULDER CO	-6	-11	-16	21	16
53	119.8	53	105.2	80	119.8	CHAFFEE CO	-12	14	0	64	50
19	141.1	15	98.5	12	95.5	CHEYENNE CO	-30	-3	-32	03	09
28	168.6	34	188.4	26	110.2	CLEAR CREEK CO	12	-41	-35	03	28
41	121.8	37	105.7	46	142.6	CONEJOS CO	-13	35	17	88	90
15	90.3	24	144.3	26	170.3	COSTILLA CO	60	18	89	98	99
30	133.4	25	102.8	31	119.8	CROWLEY CO	-23	17	-10	35	50
12	152.0	6	81.4	9	96.8	CUSTER CO	-46	19	-36	02	10
129	141.7	112	96.7	150	97.0	DELTA CO	-32	0	-32	04	10
3868	141.4	4235	124.8	4707	122.3	DENVER-ARAPAHOE#CO	-12	-2	-14	27	55
10	183.6	9	122.3	8	109.8	DOLORES CO	-33	-10	-40	01	27
27	136.0	35	118.5	54	97.5	DOUGLAS CO	-13	-18	-28	06	11
24	154.9	22	112.0	26	107.8	EAGLE CO	-28	-4	-30	05	24
18	95.4	21	93.7	27	104.1	ELBERT CO	-2	11	9	80	18
695	133.4	785	106.2	1281	121.4	EL PASO CO	-20	14	-9	39	53
184	146.4	157	99.0	232	111.7	FREMONT CO	-32	13	-24	10	31
62	102.8	90	120.2	89	92.7	GARFIELD CO	17	-23	-10	36	07
7	242.3	4	59.2	5	68.0	GILPIN CO	-76	15	-72	00	01
16	134.2	10	75.5	21	111.4	GRAND CO	-44	47	-17	19	31
15	95.5	17	88.2	23	89.2	GUNNISON CO	-8	1	-7	45	05
1	112.4	1	74.3	1	122.7	HINSDALE CO	-34	65	9	80	56
83*	181.1	55	113.9	47	96.5	HUERFANO CO	-37	-15	-47	00	10
9	198.4	12	151.4	12	172.7	JACKSON CO	-24	14	-13	28	99
455	138.1	783	119.5	1465	125.4	JEFFERSON CO	-13	5	-9	38	63
14	128.2	16	131.1	13	122.9	KIOWA CO	2	-6	-4	53	57
48	148.9	51	130.9	66	143.0	KIT CARSON CO	-12	9	-4	53	91
26	132.2	34	187.4	36	174.3	LAKE CO	42	-7	32	94	99
74	105.2	118	130.9	138	125.6	LA PLATA CO	24	-4	19	90	64
315	118.6	413	111.3	586	103.0	LARIMER CO	-6	-7	-13	28	16
144	135.1	141	124.3	139	120.9	LAS ANIMAS CO	-8	-3	-11	34	52
24	96.0	35	110.6	41	101.0	LINCOLN CO	15	-9	5	74	14
110	135.0	104	106.1	108	93.0	LOGAN CO	-21	-12	-31	04	07
261	122.6	304	103.7	419	107.2	MESA CO	-15	3	-13	29	23
1	47.4	7	256.1	2	48.3	MINERAL CO	440	-81	2	68	00
26	110.2	38	120.0	43	113.5	MOFFAT CO	9	-5	3	70	35
43	95.9	54	100.0	75	104.8	MONTEZUMA CO	4	5	9	80	19
102	148.1	105	118.3	136	116.1	MONTROSE CO	-20	-2	-22	12	41
94	112.0	105	97.0	150	115.0	MORGAN CO	-13	19	3	69	39
161	132.1	160	115.0	185	124.9	OTERO CO	-13	9	-5	49	62
10	107.5	11	121.9	8	82.9	OURAY CO	13	-32	-23	11	03
10	169.6	12	148.9	17	110.8	PARK CO	-12	-26	-35	03	30
32	114.8	31	99.0	42	120.0	PHILLIPS CO	-14	21	4	73	50
15	177.9	10	82.8	17	97.1	PITKIN CO	-53	17	-45	01	10
81	130.9	92	130.0	108	141.1	PROWERS CO	-1	9	8	78	89
547	110.3	666	113.7	710	107.5	PUEBLO CO	3	-5	-2	57	23
19	121.7	21	110.7	28	117.2	RIO BLANCO CO	-9	6	-4	54	43
54	118.2	60	113.8	79	143.6	RIO GRANDE CO	-4	26	21	91	91
40	134.8	37	125.8	47	136.2	ROUTT CO	-7	8	1	66	83
22	123.2	19	99.7	30	154.0	SAGUACHE CO	-19	55	25	92	97
7	180.5	4	116.5	7	274.8	SAN JUAN CO	-35	136	52	97	99
11	148.4	5	58.6	10	103.2	SAN MIGUEL CO	-60	76	-30	05	16
21	96.6	28	117.4	32	139.5	SEDGWICK CO	22	19	44	96	87
5	77.6	6	127.8	10	120.4	SUMMIT CO	65	-6	55	97	51
20	130.0	21	121.4	30	115.5	TELLER CO	-7	-5	-11	33	40
49	167.5	30	91.1	49	129.8	WASHINGTON CO	-46	43	-22	11	71
352	115.0	391	100.3	477	96.0	WELD CO	-13	-4	-16	20	09
52	99.6	66	104.1	83	124.6	YUMA CO	4	20	25	92	61
17792*	154.4	20515*	139.2	25420*	140.2	CONNECTICUT CT	-10	1	-9		
4606*	156.8	5430*	141.5	6760*	143.1	FAIRFIELD CT	-10	1	-9	39	91
4589*	153.6	5378*	138.3	6670*	139.3	HARTFORD CT	-10	1	-9	38	87
951*	156.9	1070	135.6	1327	135.3	LITCHFIELD CT	-14	0	-14	26	82

* SIGNIFICANTLY MORE DEATHS THAN EXPECTED

\# COUNTIES COMBINED; SEE TECHNICAL NOTES

WHITE FEMALE — ALL CANCERS COMBINED — ICD 140-208

1950-1959 DEATHS	1950-1959 RATE	1960-1969 DEATHS	1960-1969 RATE	1970-1979 DEATHS	1970-1979 RATE	COUNTY AND STATE	PERCENT CHANGE 60S/50S	PERCENT CHANGE 70S/60S	PERCENT CHANGE 70S/50S	PERCENTILE %CHANGE 70S/50S	PERCENTILE RATES 1970S
528	119.5	711	123.8	959	131.5	MIDDLESEX CT	4	6	10	82	74
4959*	160.6	5457*	142.7	6558*	140.6	NEW HAVEN CT	-11	-1	-12	29	88
1193	141.7	1449*	140.9	1830*	147.8	NEW LONDON CT	-1	5	4	72	94
323	132.2	419	127.2	558	125.5	TOLLAND CT	-4	-1	-5	50	63
643*	170.7	601	131.4	758	138.9	WINDHAM CT	-23	6	-19	16	87
2256	142.8	2913*	139.6	3794*	140.5	DELAWARE DE	-2	1	-2		
226	110.6	331	128.1	451	128.2	KENT DE	16	0	16	87	69
1609	148.5	2079*	143.6	2696*	147.7	NEW CASTLE DE	-3	3	-1	62	94
421	145.0	503	133.3	647	125.0	SUSSEX DE	-8	-6	-14	26	62
4492*	155.2	3623*	143.5	2835*	150.0	DIST. OF COLUMBIA DC	-8	4	-3		
4492*	155.2	3623*	143.5	2835*	150.0	WASHINGTON DC	-8	4	-3	55	96
19358	115.1	37075	115.3	69088	123.3	FLORIDA FL	0	7	7		
200	125.2	284	116.9	458	119.0	ALACHUA FL	-7	2	-5	50	48
14	68.6	30	113.9	64	131.0	BAKER FL	66	15	91	98	73
169	115.7	294	125.8	455	120.1	BAY FL	9	-4	4	72	50
35	91.7	52	100.9	86	120.1	BRADFORD FL	10	19	31	94	50
221	99.7	635	114.6	1620	132.5	BREVARD FL	15	16	33	94	76
948	96.6	3389	111.5	8393	126.0	BROWARD FL	15	13	30	94	65
25	95.8	43	132.2	48	93.9	CALHOUN FL	38	-29	-2	58	07
61	124.2	210	109.5	548	106.7	CHARLOTTE FL	-12	-3	-14	25	22
29	69.8	109	107.1	394	104.5	CITRUS FL	53	-2	50	97	18
52	95.5	89	90.0	213	105.3	CLAY FL	-6	17	10	82	19
29	84.0	121	84.7	545	122.7	COLLIER FL	1	45	46	96	56
68	126.4	86	106.4	115	93.6	COLUMBIA FL	-16	-12	-26	08	07
4250	129.7	7684	125.1	13108*	137.4	DADE FL	-4	10	6	76	84
62	116.9	69	92.1	94	92.2	DE SOTO FL	-21	0	-21	13	06
20	151.1	15	81.6	44	143.0	DIXIE FL	-46	75	-5	49	91
1463	127.9	2224	132.2	3082*	137.6	DUVAL FL	3	4	8	78	85
422	113.0	746	128.5	1015	120.5	ESCAMBIA FL	14	-6	7	77	51
11	107.9	23	112.4	42	89.8	FLAGLER FL	4	-20	-17	20	05
30	132.4	38	120.5	64	144.1	FRANKLIN FL	-9	20	9	80	92
80	77.7	94	77.5	93	69.2	GADSDEN FL	0	-11	-11	33	01
18	166.9	9	58.0	31	128.2	GILCHRIST FL	-65	121	-23	10	69
8	167.4	7	80.7	26	92.5	GLADES FL	-52	15	-45	01	07
26	120.2	36	111.8	63	145.3	GULF FL	-7	30	21	91	93
35	155.3	17	67.1	46	135.0	HAMILTON FL	-57	101	-13	28	81
54	109.7	75	116.7	92	108.2	HARDEE FL	6	-7	-1	60	24
25	129.4	45	158.7	69	138.9	HENDRY FL	23	-12	7	78	87
51	144.1	96	117.9	215	88.0	HERNANDO FL	-18	-25	-39	01	04
93	113.8	172	101.8	347	97.0	HIGHLANDS FL	-10	-5	-15	23	10
1709	126.1	2589	123.5	3747	126.3	HILLSBOROUGH FL	-2	2	0	64	65
65	112.3	67	100.1	66	78.2	HOLMES FL	-11	-22	-30	05	02
76	85.5	216	112.6	410	112.4	INDIAN RIVER FL	32	0	31	94	33
114	109.2	119	90.3	170	98.5	JACKSON FL	-17	9	-10	36	12
19	83.6	28	114.5	41	124.1	JEFFERSON FL	37	8	49	97	60
18	144.0	16	113.5	23	117.9	LAFAYETTE FL	-21	4	-18	17	45
298	119.1	482	110.0	861	119.6	LAKE FL	-8	9	0	65	49
185	101.0	502	100.3	1442	111.9	LEE FL	-1	12	11	82	32
162	113.8	245	113.4	423	116.4	LEON FL	0	3	2	69	42
41	117.8	57	110.5	104	114.6	LEVY FL	-6	4	-3	56	37
6	48.9	14	102.7	16	92.0	LIBERTY FL	110	-10	88	98	06
38	105.7	42	91.2	49	78.3	MADISON FL	-14	-14	-26	08	02
308	92.2	737	98.7	1417	105.4	MANATEE FL	7	7	14	86	20
164	106.6	299	116.6	613	109.6	MARION FL	9	-6	3	70	27
50	82.4	148	99.8	471	117.0	MARTIN FL	21	17	42	96	43
139	137.5	243	139.4	369	122.4	MONROE FL	1	-12	-11	33	56
44	121.0	61	107.1	119	125.0	NASSAU FL	-11	17	3	71	62
92	96.2	192	110.2	363	115.5	OKALOOSA FL	15	5	20	90	40
17	133.8	40	125.3	101	139.4	OKEECHOBEE FL	-6	11	4	72	87
935	113.1	1621	115.0	2748	130.9	ORANGE FL	2	14	16	87	73
100	80.0	210	108.9	368	121.2	OSCEOLA FL	36	11	52	97	53
911	109.3	2003	111.8	4608	122.3	PALM BEACH FL	2	9	12	84	55
194	107.8	464	93.7	1534	105.6	PASCO FL	-13	13	-2	58	20
2497	108.7	5119	113.0	8734	124.1	PINELLAS FL	4	10	14	86	60
683	108.0	1146	111.1	1832	114.1	POLK FL	3	3	6	75	36
106	111.1	163	118.5	275	121.8	PUTNAM FL	7	3	10	81	54
151	131.5	196	119.5	320	126.9	ST JOHNS FL	-9	6	-3	54	66
118	112.2	279	129.9	465	106.3	ST LUCIE FL	16	-18	-5	50	21
72	96.3	102	93.7	195	108.7	SANTA ROSA FL	-3	16	13	85	25
371	104.5	936	105.2	2254	125.4	SARASOTA FL	1	19	20	90	63
134	98.2	284	108.5	691	114.1	SEMINOLE FL	11	5	16	88	36
53	137.4	73	121.5	115	107.5	SUMTER FL	-12	-12	-22	12	23
69	130.6	69	100.0	95	93.7	SUWANNEE FL	-23	-6	-28	06	07
48	153.8	52	119.2	74	112.6	TAYLOR FL	-22	-6	-27	07	34
13	92.8	13	65.6	30	115.0	UNION FL	-29	75	24	92	38
729	115.2	1411	118.7	2359	122.6	VOLUSIA FL	3	3	6	77	56
18	119.8	12	55.7	46	118.2	WAKULLA FL	-54	112	-1	61	46
70	107.4	77	94.4	102	86.6	WALTON FL	-12	-8	-19	15	04
42	93.1	56	101.0	68	88.3	WASHINGTON FL	8	-13	-5	50	05
13300	120.1	17065	114.3	22435	112.5	GEORGIA GA	-5	-2	-6		
42	111.1	46	91.2	50	84.0	APPLING GA	-18	-8	-24	09	03
27	138.7	22	105.5	29	118.6	ATKINSON GA	-24	12	-14	24	47
33	127.3	38	111.7	48	108.6	BACON GA	-12	-3	-15	24	25
9	107.1	10	101.7	14	121.3	BAKER GA	-5	19	13	85	53
57	39.5	60	39.1	116	78.3	BALDWIN GA	-1	100	98	98	02
27	95.3	30	86.7	44	99.0	BANKS GA	-9	14	4	72	12
54	89.7	96	118.3	100	95.1	BARROW GA	32	-20	6	76	08
114	126.7	146	116.3	202	119.7	BARTOW GA	-8	3	-6	49	49
67	136.1	76	120.1	83	110.4	BEN HILL GA	-12	-8	-19	16	29
39	98.5	52	104.9	56	87.1	BERRIEN GA	7	-17	-12	32	04
465	127.7	531	109.4	711	120.3	BIBB GA	-14	10	-6	48	51
42	168.7	41	107.1	53	105.7	BLECKLEY GA	-37	-1	-37	02	20
23	145.1	25	114.3	44	141.5	BRANTLEY GA	-21	24	-2	57	89

* SIGNIFICANTLY MORE DEATHS THAN EXPECTED

WHITE FEMALE — ALL CANCERS COMBINED — ICD 140-208

1950-1959 DEATHS	1950-1959 RATE	1960-1969 DEATHS	1960-1969 RATE	1970-1979 DEATHS	1970-1979 RATE	COUNTY AND STATE	PERCENT CHANGE 60S/50S	PERCENT CHANGE 70S/60S	PERCENT CHANGE 70S/50S	PERCENTILE %CHANGE 70S/50S	PERCENTILE RATES 1970S
50	115.9	51	97.2	75	102.2	BROOKS GA	-16	5	-12	31	15
18	131.2	24	165.9	34	135.2	BRYAN GA	26	-18	3	70	81
68	104.9	93	95.8	126	96.7	BULLOCH GA	-9	1	-8	42	10
32	93.5	32	78.2	63	117.0	BURKE GA	-16	50	25	92	43
41	137.1	48	120.2	58	112.0	BUTTS GA	-12	-7	-18	17	32
19	130.7	20	119.9	19	108.6	CALHOUN GA	-8	-9	-17	19	25
18	142.6	29	138.8	35	102.8	CAMDEN GA	-3	-26	-28	06	16
23	116.7	32	126.4	29	77.8	CANDLER GA	8	-38	-33	03	02
188	136.0	213	115.7	246	101.2	CARROLL GA	-15	-12	-26	08	14
78	138.1	104	115.2	148	101.5	CATOOSA GA	-17	-12	-26	08	14
21	165.5	13	91.7	31	135.0	CHARLTON GA	-45	47	-18	17	81
581	133.8	775	130.5	908	125.6	CHATHAM GA	-2	-4	-6	47	63
93	120.9	108	118.9	151	126.4	CHATTOOGA GA	-2	6	5	73	65
110	121.7	149	117.8	190	106.3	CHEROKEE GA	-3	-10	-13	29	21
131	105.2	161	106.7	257	114.8	CLARKE GA	1	8	9	80	38
11	106.5	12	104.5	10	88.7	CLAY GA	-2	-15	-17	20	05
86	100.0	183	103.0	428	122.1	CLAYTON GA	3	19	22	91	55
15	139.4	26	145.3	27	115.9	CLINCH GA	4	-20	-17	19	40
313	119.9	552	112.3	987	116.0	COBB GA	-6	3	-3	55	40
70	115.2	70	92.0	100	100.5	COFFEE GA	-20	9	-13	28	13
108	108.0	157	119.6	185	110.0	COLQUITT GA	11	-8	2	68	28
21	102.3	45	113.3	88	110.5	COLUMBIA GA	11	-3	8	79	29
46	127.3	42	94.2	47	82.6	COOK GA	-26	-12	-35	02	03
114	129.7	135	119.6	172	110.0	COWETA GA	-8	-8	-15	22	28
14	108.8	14	93.6	16	73.1	CRAWFORD GA	-14	-22	-33	03	01
46	99.8	57	95.2	70	87.8	CRISP GA	-5	-8	-12	30	04
33	126.6	45	122.9	72	138.5	DADE GA	-3	13	9	81	86
13	81.1	16	88.0	29	128.1	DAWSON GA	8	45	58	97	68
72	126.1	80	107.6	90	99.5	DECATUR GA	-15	-7	-21	13	12
808	116.5	1429	123.4	2044	122.4	DE KALB GA	6	-1	5	74	56
71	141.2	59	93.1	85	103.2	DODGE GA	-34	11	-27	07	16
39	120.9	48	130.4	41	95.0	DOOLY GA	8	-27	-21	12	08
128	120.3	182	103.8	293	119.4	DOUGHERTY GA	-14	15	-1	62	49
67	136.3	92	101.7	158	103.2	DOUGLAS GA	-25	1	-24	09	16
32	90.8	51	117.2	48	91.3	EARLY GA	29	-22	1	65	06
5	86.8	7	83.0	4	37.7	ECHOLS GA	-4	-55	-57	00	00
30	106.8	28	74.4	46	80.1	EFFINGHAM GA	-30	8	-25	09	02
72	118.3	85	113.2	100	105.9	ELBERT GA	-4	-6	-10	35	20
56	100.8	60	88.9	82	91.2	EMANUEL GA	-12	3	-10	37	06
19	91.9	36	145.8	29	88.9	EVANS GA	59	-39	-3	55	05
73	123.1	81	106.3	98	99.2	FANNIN GA	-14	-7	-19	15	12
30	116.9	50	129.7	75	103.2	FAYETTE GA	11	-20	-12	31	16
333	139.5	376	117.9	475	114.7	FLOYD GA	-15	-3	-18	18	38
55	114.1	74	115.7	98	96.8	FOPSYTH GA	1	-16	-15	22	10
75	127.2	92	131.9	116	133.8	FRANKLIN GA	4	1	5	74	79
2378	138.8	2801	133.1	2851	126.5	FULTON GA	-4	-5	-9	39	66
34	83.7	56	119.2	78	126.3	GILMER GA	42	6	51	97	65
14	133.4	11	95.6	13	86.4	GLASCOCK GA	-28	-10	-35	02	04
101	115.0	150	107.5	236	112.1	GLYNN GA	-7	4	-3	57	33
88	111.5	110	108.7	175	125.1	GOPDON GA	-2	15	12	84	62
55	95.8	64	95.5	81	94.8	GPADY GA	0	-1	-1	61	08
44	129.8	46	124.6	53	116.7	GREENE GA	-4	-6	-10	36	42
134	98.3	216	102.6	461	111.9	GWINNETT GA	4	9	14	85	32
69	97.2	89	98.9	110	90.1	HABERSHAM GA	2	-9	-7	44	05
174	103.7	220	94.5	330	99.4	HALL GA	-9	5	-4	53	12
14	73.8	24	122.9	23	98.3	HANCOCK GA	66	-20	33	95	11
67	109.7	93	122.9	102	101.7	HARALSON GA	12	-17	-7	44	15
31	103.8	46	126.4	57	106.1	HARRIS GA	22	-16	2	69	21
56	111.1	67	103.5	91	103.7	HART GA	-7	0	-7	45	17
26	129.5	35	134.8	34	102.6	HEARD GA	4	-24	-21	13	16
57	117.6	78	109.7	120	103.5	HENRY GA	-7	-6	-12	31	17
61	125.9	111	89.9	225	113.4	HOUSTON GA	-29	26	-10	36	35
22	85.1	36	104.4	50	109.6	IRWIN GA	23	5	29	93	27
93	119.7	101	103.1	148	118.6	JACKSON GA	-14	15	-1	61	47
22	110.8	28	135.7	26	82.8	JASPER GA	22	-39	-25	08	03
24	94.0	35	115.3	53	121.6	JEFF DAVIS GA	23	5	29	93	54
59	140.9	55	118.3	66	106.1	JEFFERSON GA	-16	-10	-25	09	21
22	122.0	28	119.9	33	120.3	JENKINS GA	-2	0	-1	60	51
31	112.5	41	120.4	35	82.3	JOHNSON GA	7	-32	-27	07	03
16	84.5	28	99.2	31	67.8	JONES GA	17	-32	-20	14	00
42	131.4	35	91.3	56	106.1	LAMAR GA	-30	16	-19	15	21
16	104.9	14	82.8	20	96.8	LANIER GA	-21	17	-8	42	10
96	117.0	112	100.8	151	103.3	LAURENS GA	-14	2	-12	32	16
18	226.8	9	73.2	25	95.0	LEE GA	-68	30	-58	00	08
16	99.1	18	90.0	40	117.0	LIBERTY GA	-9	30	18	89	43
28	174.3	19	93.1	23	91.6	LINCOLN GA	-47	-2	-47	00	06
8	75.4	12	94.7	20	137.3	LONG GA	26	45	82	98	84
106	102.1	167	113.0	238	116.3	LOWNDES GA	11	3	14	85	41
29	109.1	29	86.2	44	95.7	LUMPKIN GA	-21	11	-12	30	09
31	107.3	43	104.9	67	107.6	MC DUFFIE GA	-2	3	0	65	23
11	106.9	13	87.7	34	174.9	MC INTOSH GA	-18	99	64	98	99
28	105.8	36	103.8	25	68.7	MACON GA	-2	-34	-35	02	01
47	98.2	78	141.2	77	108.6	MADISON GA	44	-23	11	82	25
14	102.1	21	158.2	13	81.7	MARION GA	55	-48	-20	14	02
57	105.4	70	105.4	92	110.4	MERIWETHER GA	0	5	5	73	29
23	98.6	33	110.2	30	85.1	MILLER GA	12	-23	-14	26	03
47	96.6	50	85.9	71	94.5	MITCHELL GA	-11	10	-2	58	08
39	141.2	51	142.0	45	82.2	MONROE GA	1	-42	-42	01	02
19	94.0	24	106.6	20	75.0	MONTGOMERY GA	13	-30	-20	14	01
37	122.9	39	108.9	42	100.1	MORGAN GA	-11	-8	-19	16	13
39	98.6	52	107.2	85	122.6	MURPAY GA	9	14	24	92	56
386	114.6	573	123.3	735	123.6	MUSCOGEE#GA	8	0	8	78	59
55	84.4	115	135.2	145	115.0	NEWTON GA	60	-15	36	96	38
27	106.0	36	115.2	45	105.8	OCONEE GA	9	-8	0	64	20
29	103.3	41	134.9	46	124.2	OGLETHORPE GA	31	-8	20	90	60
63	131.4	68	102.1	94	94.2	PAULDING GA	-22	-8	-28	06	08
26	101.2	31	92.6	61	135.9	PEACH GA	-9	47	34	95	82
51	138.9	59	129.1	81	125.0	PICKENS GA	-7	-3	-10	36	62
34	125.2	42	112.2	49	100.8	PIERCE GA	-10	-10	-19	15	14
25	101.1	32	114.1	34	91.1	PIKE GA	13	-20	-10	36	06
138	128.4	158	118.2	217	126.5	POLK GA	-8	7	-1	60	66

COUNTIES COMBINED; SEE TECHNICAL NOTES

WHITE FEMALE — ALL CANCERS COMBINED — ICD 140-208

1950-1959 DEATHS	1950-1959 RATE	1960-1969 DEATHS	1960-1969 RATE	1970-1979 DEATHS	1970-1979 RATE	COUNTY AND STATE	PERCENT CHANGE 60S/50S	PERCENT CHANGE 70S/60S	PERCENT CHANGE 70S/50S	PERCENTILE %CHANGE 70S/50S	PERCENTILE RATES 1970S
26	138.9	34	127.1	42	121.5	PULASKI GA	-8	-4	-12	29	54
16	91.7	21	97.7	32	105.5	PUTNAM GA	7	8	15	87	20
5	120.1	4	67.2	10	120.3	QUITMAN GA	-44	79	0	64	51
32	97.2	42	97.8	63	98.6	RABUN GA	1	1	1	67	12
38	147.4	35	109.0	44	125.6	RANDOLPH GA	-26	15	-15	23	64
403	128.4	510	122.7	717	131.4	RICHMOND GA	-4	7	2	69	74
43	151.5	52	101.5	114	113.3	ROCKDALE GA	-33	12	-25	09	35
20	218.1	8	87.9	20	163.8	SCHLEY GA	-60	86	-25	09	98
38	100.9	39	93.1	46	84.9	SCREVEN GA	-8	-9	-16	21	03
29	147.9	21	88.0	35	96.9	SEMINOLE GA	-41	10	-34	03	10
166	153.4	175	113.7	214	106.7	SPALDING GA	-26	-6	-30	05	22
67	99.9	104	116.6	118	98.1	STEPHENS GA	17	-16	-2	59	11
18	106.8	20	127.4	24	139.1	STEWART GA	19	9	30	94	87
84	138.4	90	105.2	105	98.8	SUMTER GA	-24	-6	-29	06	12
20	142.5	19	111.8	11	58.6	TALBOT GA	-22	-48	-59	00	00
13	114.6	6	49.8	11	147.9	TALIAFERRO GA	-57	197	29	93	95
41	86.3	50	88.4	73	94.1	TATTNALL GA	2	6	9	80	08
26	102.3	32	119.1	33	102.7	TAYLOR GA	16	-14	0	65	16
48	126.0	42	89.8	77	134.6	TELFAIR GA	-29	50	7	77	80
29	92.7	42	118.8	47	110.8	TERRELL GA	28	-7	19	90	30
101	101.6	129	103.8	166	109.0	THOMAS GA	2	5	7	78	26
82	128.7	79	91.2	125	106.1	TIFT GA	-29	16	-18	18	21
49	88.3	53	70.5	108	110.1	TOOMBS GA	-20	56	25	92	28
22	119.5	24	96.2	35	97.3	TOWNS GA	-20	1	-19	16	11
25	151.9	21	100.0	30	106.6	TREUTLEN GA	-34	7	-30	05	22
178	117.8	240	128.4	262	114.1	TROUP GA	9	-11	-3	55	36
22	82.0	43	123.4	36	87.0	TURNER GA	51	-30	6	76	04
12	82.8	10	57.8	21	87.7	TWIGGS GA	-30	52	6	76	04
26	85.3	41	111.8	43	76.2	UNION GA	31	-32	-11	34	01
91	118.2	97	101.8	146	111.0	UPSON GA	-14	9	-6	47	30
191	124.7	250	119.9	323	113.9	WALKER GA	-4	-5	-9	40	36
81	113.7	97	100.5	112	87.9	WALTON GA	-12	-13	-23	11	04
107	112.5	166	127.4	178	105.0	WARE GA	13	-18	-7	45	19
21	112.7	20	86.9	25	105.0	WARREN GA	-23	21	-7	45	19
50	106.8	55	95.5	66	95.4	WASHINGTON GA	-11	0	-11	34	09
47	101.9	72	118.1	89	104.5	WAYNE GA	16	-12	3	69	18
8	132.4	12	148.7	9	112.9	WEBSTER GA	12	-24	-15	24	34
21	143.3	21	103.5	31	124.8	WHEELER GA	-28	21	-13	28	61
29	106.5	24	65.2	39	74.3	WHITE GA	-39	14	-30	05	01
207	147.2	231	115.5	335	121.9	WHITFIELD GA	-22	6	-17	19	54
38	149.9	29	97.0	29	75.1	WILCOX GA	-35	-23	-50	00	01
37	113.2	37	92.0	42	90.5	WILKES GA	-19	-2	-20	14	06
26	110.1	32	111.6	39	118.7	WILKINSON GA	1	6	8	78	47
65	160.6	58	119.1	70	113.7	WORTH GA	-26	-5	-29	06	36
		805	119.7	1252	122.4	HAWAII HI		2			
		51	89.8	114	107.4	HAWAII HI		20			23
		680	123.4	1009	126.3	HONOLULU HI		2			65
		18	80.0	47	123.7	KAUAI HI		55			59
		56	135.7	82	100.5	MAUI#HI		-26			13
2998	120.0	3639	112.8	4671	111.9	IDAHO ID	-6	-1	-7		
468	125.6	554	106.0	881	125.0	ADA ID	-16	18	-1	63	62
14	98.8	21	170.1	21	131.3	ADAMS ID	72	-23	33	94	74
185	125.7	230	120.2	291	115.7	BANNOCK ID	-4	-4	-8	42	40
34	128.7	29	86.9	49	131.2	BEAR LAKE ID	-32	51	2	68	73
43	188.7	33	114.0	41	115.1	BENEWAH ID	-40	1	-39	01	39
82	101.7	107	101.0	148	111.9	BINGHAM ID	-1	11	10	82	32
20	96.1	35	143.5	39	123.8	BLAINE ID	49	-14	29	93	59
3	56.9	6	75.8	8	64.0	BOISE ID	33	-16	12	84	00
90	127.6	97	114.3	116	103.3	BONNER ID	-10	-10	-19	16	17
133	114.1	187	109.7	232	99.4	BONNEVILLE ID	-4	-9	-13	28	12
32	124.9	32	114.2	37	100.3	BOUNDARY ID	-9	-12	-20	15	13
7	93.5	11	106.3	14	92.9	BUTTE ID	14	-13	-1	62	07
4	79.2	7	194.3	5	149.8	CAMAS ID	145	-23	89	98	96
314	119.5	408	122.4	445	103.1	CANYON ID	2	-16	-14	26	16
11	53.1	30	130.4	28	92.3	CARIBOU ID	146	-29	74	98	06
73	137.1	78	110.6	96	108.5	CASSIA ID	-19	-2	-21	13	25
9*	219.4	3	82.2	3	112.3	CLARK ID	-63	37	-49	00	33
31	109.3	33	95.4	52	122.4	CLEARWATER ID	-13	28	12	84	56
11	84.2	18	127.3	19	111.5	CUSTER ID	51	-12	32	94	31
37	140.8	48	118.5	64	110.7	ELMORE ID	-16	-7	-21	12	29
36	104.3	49	119.3	44	91.9	FRANKLIN ID	14	-23	-12	31	06
33	108.0	37	100.3	37	80.8	FREMONT ID	-7	-19	-25	08	02
51	122.1	61	115.4	76	113.4	GEM ID	-5	-2	-7	44	35
51	119.4	49	94.3	79	118.0	GOODING ID	-21	25	-1	61	45
60	131.1	70	123.0	67	89.1	IDAHO ID	-6	-28	-32	04	05
41	122.3	49	110.3	57	102.3	JEFFERSON ID	-10	-7	-16	20	15
45	105.2	75	137.3	85	122.9	JEROME ID	30	-11	17	88	57
170	122.8	224	125.7	333	122.9	KOOTENAI ID	2	-2	0	64	57
99	120.2	114	113.8	150	119.6	LATAH ID	-5	5	0	63	49
30	129.0	30	105.0	47	134.9	LEMHI ID	-19	28	5	73	81
15	79.4	22	112.4	36	157.1	LEWIS ID	41	40	98	98	98
18	118.7	12	73.5	14	77.0	LINCOLN ID	-38	5	-35	02	01
32	107.1	25	60.4	39	72.7	MADISON ID	-44	21	-32	04	01
59	135.0	60	103.4	78	101.1	MINIDOKA ID	-23	-2	-25	09	14
152	130.5	192	130.8	242	135.2	NEZ PERCE ID	0	3	4	71	81
20	114.7	19	86.6	21	89.2	ONEIDA ID	-24	3	-22	11	05
21	103.1	41	154.4	32	87.4	OWYHEE ID	50	-43	-15	22	04
68	114.3	81	103.2	103	106.3	PAYETTE ID	-10	3	-7	44	21
16	99.4	37*	191.2	25	106.8	POWER ID	92	-44	8	78	22
97	152.4	95	117.7	127	139.7	SHOSHONE ID	-23	19	-8	40	88
10	127.5	10	90.4	10	78.9	TETON ID	-29	-13	-38	02	02
212	116.7	249	104.9	293	100.5	TWIN FALLS ID	-10	-4	-14	26	13
20	142.3	16	105.1	25	129.4	VALLEY ID	-26	23	-9	39	71
41	92.0	55	106.5	62	106.1	WASHINGTON ID	16	0	15	87	21
67861*	150.3	74501*	140.0	82457*	137.8	ILLINOIS IL	-7	-2	-8		

* SIGNIFICANTLY MORE DEATHS THAN EXPECTED

COUNTIES COMBINED; SEE TECHNICAL NOTES

WHITE FEMALE — ALL CANCERS COMBINED — ICD 140-208

1950-1959 DEATHS	1950-1959 RATE	1960-1969 DEATHS	1960-1969 RATE	1970-1979 DEATHS	1970-1979 RATE	COUNTY AND STATE	PERCENT CHANGE 60S/50S	PERCENT CHANGE 70S/60S	PERCENT CHANGE 70S/50S	PERCENTILE %CHANGE 70S/50S	PERCENTILE RATES 1970S
580	134.6	615	127.7	683	130.8	ADAMS IL	-5	2	-3	56	73
111	145.6	111	149.3	120	161.1	ALEXANDER IL	3	8	11	82	98
123	132.5	116	114.5	138	125.2	BOND IL	-14	9	-6	49	62
158	147.8	171	144.3	217*	155.3	BOONE IL	-2	8	5	74	97
71	141.1	69	143.5	60	121.3	BROWN IL	2	-15	-14	25	53
336	147.0	337	134.3	341	123.3	BUREAU IL	-9	-8	-16	21	58
38	111.7	36	90.7	61	140.3	CALHOUN IL	-19	55	26	93	88
165	144.8	174	133.7	191	135.1	CARROLL IL	-8	1	-7	45	81
150	153.9	133	127.4	146	124.5	CASS IL	-17	-2	-19	16	60
611	140.9	588	110.6	753	121.2	CHAMPAIGN IL	-22	10	-14	25	53
301	133.5	319	126.3	350	122.0	CHRISTIAN IL	-5	-3	-9	40	55
161	134.6	172	133.6	150	108.1	CLARK IL	-1	-19	-20	15	24
140	138.0	133	119.5	119	95.2	CLAY IL	-13	-20	-31	04	09
138	118.4	155	109.0	209	114.5	CLINTON IL	-8	5	-3	55	37
330	138.6	338	121.3	405	130.7	COLES IL	-12	8	-6	48	73
34776*	161.6	37332*	151.1	38585*	147.4	COOK IL	-6	-2	-9	39	94
194	147.3	168	111.2	187	118.7	CRAWFORD IL	-24	7	-19	15	47
82	134.8	79	111.9	106	142.6	CUMBERLAND IL	-17	27	6	75	90
341	145.9	405	145.0	422	125.7	DE KALB IL	-1	-13	-14	26	64
131	123.3	139	120.4	174	133.3	DE WITT IL	-2	11	8	79	78
152	146.3	135	114.0	142	106.3	DOUGLAS IL	-22	-7	-27	07	21
1385	151.1	2075	133.1	3316	137.6	DU PAGE IL	-12	3	-9	39	85
230	152.3	204	126.1	216	121.1	EDGAR IL	-17	-4	-20	13	53
68	129.3	82	141.7	74	111.1	EDWARDS IL	10	-22	-14	25	30
138	114.6	180	123.5	229	123.7	EFFINGHAM IL	8	0	8	79	59
204	157.4	186	129.0	206	124.9	FAYETTE IL	-18	-3	-21	13	62
129	135.0	140	130.1	144	123.0	FORD IL	-4	-6	-9	39	57
385	143.4	447	143.1	480	132.7	FRANKLIN IL	0	-7	-7	43	77
354	135.7	332	114.6	460	142.7	FULTON IL	-16	25	5	74	91
69	158.6	89*	190.3	52	100.6	GALLATIN IL	20	-47	-37	02	13
166	133.7	159	122.9	162	121.7	GREENE IL	-8	-1	-9	39	54
161	149.0	167	130.9	184	119.3	GRUNDY IL	-12	-9	-20	14	48
88	119.9	94	124.0	85	100.6	HAMILTON IL	3	-19	-16	21	14
248	141.3	209	119.8	217	111.4	HANCOCK IL	-15	-7	-21	13	31
53	161.5	53	153.5	52	132.8	HARDIN IL	-5	-14	-18	18	77
48	113.8	64	129.2	61	108.9	HENDERSON IL	13	-16	-4	52	26
412	147.4	435	134.8	476	125.0	HENRY IL	-9	-7	-15	22	62
285	154.8	253	119.4	323	134.8	IROQUOIS IL	-23	13	-13	28	80
270	141.5	310	137.7	338	129.7	JACKSON IL	-3	-6	-8	40	71
102	136.8	87	112.2	120	143.6	JASPER IL	-18	28	5	74	91
274	149.9	257	121.6	336	131.9	JEFFERSON IL	-19	8	-12	31	75
119	147.3	123	126.6	154	132.3	JERSEY IL	-14	4	-10	35	76
187	149.6	166	122.2	188	129.1	JO DAVIESS IL	-18	6	-14	26	70
55	111.6	63	124.0	89	145.0	JOHNSON IL	11	17	30	94	93
1339	137.9	1560	132.9	1780	132.1	KANE IL	-4	-1	-4	52	75
473	105.2	576	114.1	690	131.5	KANKAKEE IL	9	15	25	92	74
110	154.4	137	147.0	164	125.3	KENDALL IL	-5	-15	-19	16	63
498	143.4	501	124.3	511	116.1	KNOX IL	-13	-7	-19	16	41
1293	149.4	1676	133.0	2284	137.1	LAKE IL	-11	3	-8	41	84
892*	152.2	959	139.4	1083	137.9	LA SALLE IL	-8	-1	-9	37	85
139	116.5	159	113.6	180	121.8	LAWRENCE IL	-2	7	5	73	54
251	126.5	274	124.7	279	115.0	LEE IL	-1	-8	-9	38	39
303	134.4	317	122.8	388	132.2	LIVINGSTON IL	-9	8	-2	60	76
232	126.1	232	107.3	297	123.2	LOGAN IL	-15	15	-2	58	58
263	145.8	243	123.4	295	136.3	MC DONOUGH IL	-15	11	-7	46	83
461	140.0	675	142.1	885	134.3	MC HENRY IL	2	-5	-4	53	80
631	134.7	706	131.0	808	130.7	MC LEAN IL	-3	0	-3	56	73
819	145.4	947	140.9	973	127.8	MACON IL	-3	-9	-12	30	68
358	120.0	384	114.8	477	126.0	MACOUPIN IL	-4	10	5	74	64
1169	130.0	1519	131.1	1842	130.0	MADISON IL	1	-1	0	64	72
362	157.2	380	142.1	394	128.9	MARION IL	-10	-9	-18	17	70
93	118.7	84	95.4	114	116.5	MARSHALL IL	-20	22	-2	59	42
134	143.4	135	123.1	154	121.3	MASON IL	-14	-1	-15	22	53
134*	181.3	130	152.5	125	121.4	MASSAC IL	-16	-20	-33	03	53
111*	182.6	87	131.7	77	94.2	MENARD IL	-28	-28	-48	00	08
150	149.4	138	124.3	168	136.4	MERCER IL	-17	10	-9	39	83
122	152.6	103	103.6	162	132.0	MONROE IL	-32	27	-13	27	75
284	128.1	307	135.3	314	123.9	MONTGOMERY IL	6	-8	-3	55	59
281	111.9	283	106.4	287	109.2	MORGAN IL	-5	3	-2	57	26
93	107.8	135	141.5	107	97.2	MOULTRIE IL	31	-31	-10	36	11
262	140.3	296	133.0	335	127.2	OGLE IL	-5	-4	-9	38	67
1329	141.1	1495	137.7	1626	135.6	PEORIA IL	-2	-2	-4	54	82
151	129.7	178	128.9	194	123.1	PERRY IL	-1	-4	-5	50	57
117	135.7	107	110.5	144	141.6	PIATT IL	-19	28	4	72	90
170	119.6	184	120.3	195	121.4	PIKE IL	1	1	2	67	53
39	118.7	36	122.6	38	125.9	POPE IL	3	3	6	76	64
65	141.4	67	143.2	74	165.5	PULASKI IL	1	16	17	88	98
41	164.8	43	152.5	49	152.1	PUTNAM IL	-7	0	-8	43	96
229	139.7	215	112.0	267	122.0	RANDOLPH IL	-20	9	-13	28	54
123	124.2	158	138.3	146	119.2	RICHLAND IL	11	-14	-4	53	48
1072	149.2	1139	132.8	1261	129.0	ROCK ISLAND IL	-11	-3	-14	26	70
1386	148.3	1634*	142.8	1771*	140.2	ST CLAIR IL	-4	-2	-5	49	88
273	148.8	263	124.1	353	141.0	SALINE IL	-17	14	-5	49	89
1207*	156.5	1339*	140.8	1509	138.3	SANGAMON IL	-10	-2	-12	32	86
87	145.0	73	108.7	76	105.6	SCHUYLER IL	-25	-3	-27	07	20
66	146.8	57	121.9	52	101.9	SCOTT IL	-17	-16	-31	04	15
234	144.6	218	129.3	219	125.3	SHELBY IL	-11	-3	-13	27	63
64	123.2	63	114.1	77	129.3	STARK IL	-7	13	5	74	71
376	139.2	418	141.2	411	124.7	STEPHENSON IL	1	-12	-10	35	61
464	132.8	632	133.2	774	122.4	TAZEWELL IL	0	-8	-8	42	56
146	116.7	133	102.0	173	130.4	UNION IL	-13	28	12	83	72
726	148.3	731	132.0	866	139.6	VERMILION IL	-11	6	-6	48	87
133	158.1	142	160.3	111	117.1	WABASH IL	1	-27	-26	08	43
181	133.3	165	111.7	184	116.8	WARREN IL	-16	5	-12	29	43
119	123.1	122	107.7	165	129.1	WASHINGTON IL	-13	20	5	73	70
136	122.9	145	115.3	156	114.3	WAYNE IL	-6	-1	-7	44	37
141	117.1	182	140.6	189	122.3	WHITE IL	20	-13	4	73	55
371	137.4	405	127.2	477	128.8	WHITESIDE IL	-7	1	-6	47	70
1009	143.2	1270	138.3	1688*	141.0	WILL IL	-3	2	-2	60	89
354	126.0	432	136.1	554	144.4	WILLIAMSON IL	8	6	15	86	92
1117	134.6	1423	133.4	1702	132.2	WINNEBAGO IL	-1	-1	-2	59	76

* SIGNIFICANTLY MORE DEATHS THAN EXPECTED

WHITE FEMALE — ALL CANCERS COMBINED — ICD 140-208

1950-1959 DEATHS	1950-1959 RATE	1960-1969 DEATHS	1960-1969 RATE	1970-1979 DEATHS	1970-1979 RATE	COUNTY AND STATE	PERCENT CHANGE 60S/50S	PERCENT CHANGE 70S/60S	PERCENT CHANGE 70S/50S	PERCENTILE %CHANGE 70S/50S	PERCENTILE RATES 1970S
159	127.4	189	117.9	262	135.1	WOODFORD IL	-7	15	6	76	81
29007	140.6	33077*	134.2	37474	131.0	INDIANA IN	-5	-2	-7		
152	128.1	169	120.0	184	113.0	ADAMS IN	-6	-6	-12	31	35
1437	148.8	1672	135.9	2014*	139.6	ALLEN IN	-9	3	-6	47	87
253	131.8	339	142.6	332	113.6	BARTHOLOMEW IN	8	-20	-14	26	36
74	115.6	105	151.2	96	129.6	BENTON IN	31	-14	12	84	71
112	142.1	125	142.1	107	108.9	BLACKFORD IN	0	-23	-23	10	25
226	142.2	217	116.0	269	125.1	BOONE IN	-18	8	-12	30	62
38	119.5	46	115.8	71	126.2	BROWN IN	-3	9	6	75	65
121	122.8	125	110.1	150	117.1	CARROLL IN	-10	6	-5	51	43
312	121.3	335	121.1	343	124.3	CASS IN	0	3	2	69	60
292	135.2	348	119.8	497	128.1	CLARK IN	-11	7	-5	49	69
214	135.2	257	147.3	269	144.3	CLAY IN	9	-2	7	77	92
257	139.7	262	128.3	267	111.7	CLINTON IN	-8	-13	-20	14	31
58	108.9	70	137.8	76	120.9	CRAWFORD IN	27	-12	11	83	52
199	135.2	192	116.0	216	111.2	DAVIESS IN	-14	-4	-18	18	30
202	149.2	222	142.8	289*	159.4	DEARBORN IN	-4	12	7	77	98
127	109.1	187	142.5	180	121.0	DECATUR IN	31	-15	11	82	52
195	126.3	233	138.0	236	124.2	DE KALB IN	9	-10	-2	59	60
612	138.6	670	121.5	803	124.8	DELAWARE IN	-12	3	-10	36	61
193	160.8	199	136.4	212	117.3	DUBOIS IN	-15	-14	-27	07	44
614	130.7	747	128.4	847	122.1	ELKHART IN	-2	-5	-7	45	55
180	163.7	183	135.8	194	123.8	FAYETTE IN	-17	-9	-24	09	59
332	139.3	361	125.4	456	132.6	FLOYD IN	-10	6	-5	51	77
167	154.5	172	144.6	171	136.1	FOUNTAIN IN	-6	-6	-12	31	83
121	142.1	119	132.8	118	119.3	FRANKLIN IN	-7	-10	-16	21	48
160	150.7	159	144.6	157	125.8	FULTON IN	-4	-13	-16	20	64
249	141.1	248	124.3	288	126.5	GIBSON IN	-12	2	-10	35	66
455	137.6	531	139.0	500	113.0	GRANT IN	1	-19	-18	17	35
241	139.5	257	136.6	266	119.9	GREENE IN	-2	-12	-14	25	50
250	139.9	293	126.0	393	119.6	HAMILTON IN	-10	-5	-14	24	49
161	124.0	202	128.2	242	122.7	HANCOCK IN	3	-4	-1	61	56
118	125.9	103	95.9	159	124.2	HARRISON IN	-24	29	-1	60	60
198	133.0	249	120.8	324	116.9	HENDRICKS IN	-9	-3	-12	30	43
337	142.2	337	121.2	377	115.6	HENRY IN	-15	-5	-19	16	40
418	148.1	447	128.6	542	129.6	HOWARD IN	-13	1	-13	29	71
284	135.6	297	125.5	375*	147.6	HUNTINGTON IN	-7	18	9	80	94
182	120.8	249	141.6	250	117.7	JACKSON IN	17	-17	-3	57	44
116	136.3	132	135.2	143	121.2	JASPER IN	-1	-10	-11	33	53
166	125.0	185	125.0	198	130.2	JAY IN	0	4	4	72	72
159	114.1	153	99.8	216	127.2	JEFFERSON IN	-13	27	11	83	67
95	126.8	110	125.8	115	108.8	JENNINGS IN	-1	-14	-14	25	25
217	120.5	312	125.0	438	128.0	JOHNSON IN	4	2	6	76	68
333	135.2	368	129.4	412	130.4	KNOX IN	-4	1	-4	54	72
299	145.8	333	137.1	354	118.9	KOSCIUSKO IN	-6	-13	-18	17	47
90	109.8	111	120.0	141	125.5	LAGRANGE IN	9	5	14	86	63
2040*	152.0	2577*	146.0	2931*	142.6	LAKE IN	-4	-2	-6	47	90
582	148.1	682	138.4	758	130.8	LA PORTE IN	-7	-5	-12	32	73
257	137.9	289	133.4	343	135.1	LAWRENCE IN	-3	1	-2	58	81
712	135.5	802	124.2	1046	137.7	MADISON IN	-8	11	2	67	85
4174*	151.7	4765*	142.9	5212*	140.6	MARION IN	-6	-2	-7	43	88
252	152.2	231	119.9	317	128.9	MARSHALL IN	-21	7	-15	22	70
62	139.6	54	105.6	88	153.8	MARTIN IN	-24	46	10	82	97
251	140.7	260	132.3	263	125.3	MIAMI IN	-6	-5	-11	34	63
292	141.7	334	126.8	431	124.8	MONROE IN	-10	-2	-12	31	61
272	147.8	296	138.0	288	127.5	MONTGOMERY IN	-7	-8	-14	26	68
174	127.4	228	128.0	286	123.9	MORGAN IN	0	-3	-3	56	59
69	114.6	101	146.1	114	138.9	NEWTON IN	27	-5	21	91	86
191	120.1	211	120.1	238	120.5	NOBLE IN	0	0	0	65	51
32	130.2	37	134.7	37	121.9	OHIO IN	4	-10	-6	46	54
126	132.6	143	139.2	137	115.1	ORANGE IN	5	-17	-13	27	39
88	118.9	121	146.8	107	115.1	OWEN IN	24	-22	-3	55	39
149	150.2	145	148.4	138	125.6	PARKE IN	-1	-15	-16	20	63
117	133.9	126	124.5	150	125.4	PERRY IN	-7	1	-6	46	63
122	146.6	150*	173.2	120	130.6	PIKE IN	18	-25	-11	34	73
287	141.0	408	142.3	567	135.4	PORTER IN	1	-5	-4	53	82
160	142.0	143	117.7	185	133.8	POSEY IN	-17	14	-6	48	79
110	162.5	93	126.1	125	141.3	PULASKI IN	-22	12	-13	28	89
169	134.4	176	119.9	183	109.1	PUTNAM IN	-11	-9	-19	16	26
205	123.9	230	129.9	238	128.5	RANDOLPH IN	5	-1	4	71	69
152	131.8	154	119.1	160	111.5	RIPLEY IN	-10	-6	-15	22	31
166	146.3	164	130.0	203	150.7	RUSH IN	-11	16	3	70	96
1411	147.1	1715*	143.3	1888	136.0	ST JOSEPH IN	-3	-5	-8	43	83
67	120.4	102	141.3	137	141.6	SCOTT IN	17	0	18	89	90
199	115.8	273	135.0	265	120.7	SHELBY IN	17	-11	4	72	52
126	140.9	115	117.0	126	116.1	SPENCER IN	-17	-1	-18	18	41
125	147.9	132	122.4	190	150.9	STARKE IN	-17	23	2	68	96
154	160.6	124	113.8	186	138.2	STEUBEN IN	-29	21	-14	26	86
186	121.0	235	147.9	210	124.6	SULLIVAN IN	22	-16	3	70	61
75	158.9	62	137.0	53	108.9	SWITZERLAND IN	-14	-20	-31	04	26
463	128.1	560	129.9	661	129.8	TIPPECANOE IN	1	0	1	67	71
118	134.0	143	145.8	145	137.0	TIPTON IN	9	-6	2	69	84
40	111.4	46	123.1	49	118.6	UNION IN	10	-4	6	77	47
1161	142.8	1342	140.4	1501	133.9	VANDERBURGH IN	-2	-5	-6	47	79
155	137.2	189	157.9	177	130.0	VERMILLION IN	15	-18	-5	49	72
949*	155.3	996*	146.3	1030	140.2	VIGO IN	-6	-4	-10	37	88
254	141.1	284	132.9	305	136.3	WABASH IN	-6	3	-3	55	83
52	119.0	65	138.1	77	140.7	WARREN IN	16	2	18	69	88
163	137.6	190	139.2	194	112.6	WARRICK IN	1	-19	-18	17	34
125	126.4	116	106.9	161	126.7	WASHINGTON IN	-15	19	0	65	66
481	124.0	553	126.1	615	127.1	WAYNE IN	2	1	2	69	67
172	146.1	156	122.7	182	117.6	WELLS IN	-16	-4	-20	15	44
144	134.8	172	137.4	185	123.6	WHITE IN	2	-10	-8	40	59
160	151.9	156	125.0	185	122.6	WHITLEY IN	-18	-2	-19	15	56
20260	135.2	21816	127.2	24028	124.8	IOWA IA	-6	-2	-8		
82	114.3	108	135.9	103	116.9	ADAIR IA	19	-14	2	69	43
54	108.1	58	109.1	63	114.2	ADAMS IA	1	5	6	75	36

* SIGNIFICANTLY MORE DEATHS THAN EXPECTED

WHITE FEMALE — ALL CANCERS COMBINED — ICD 140-208

1950-1959 DEATHS	1950-1959 RATE	1960-1969 DEATHS	1960-1969 RATE	1970-1979 DEATHS	1970-1979 RATE	COUNTY AND STATE	PERCENT CHANGE 60S/50S	PERCENT CHANGE 70S/60S	PERCENT CHANGE 70S/50S	PERCENTILE %CHANGE 70S/50S	PERCENTILE RATES 1970S
133	138.2	143	131.2	123	96.0	ALLAMAKEE IA	-5	-27	-31	04	09
181	135.5	195	133.1	184	133.1	APPANOOSE IA	-2	0	-2	59	78
83	134.3	75	110.5	94	127.7	AUDUBON IA	-18	16	-5	50	68
162	124.5	193	129.6	194	116.8	BENTON IA	4	-10	-6	47	42
615	120.8	806	128.3	912	126.6	BLACK HAWK IA	6	-1	5	73	66
211	117.5	218	105.6	231	101.6	BOONE IA	-10	-4	-14	26	14
151	138.1	145	110.0	194	129.0	BREMER IA	-20	17	-7	45	70
142	118.5	150	118.0	149	111.9	BUCHANAN IA	0	-5	-6	48	32
177	147.9	184	127.4	189	113.4	BUENA VISTA IA	-14	-11	-23	10	35
114	113.1	140	123.9	132	103.8	BUTLER IA	10	-16	-8	41	17
126	127.5	165	151.8	142	115.0	CALHOUN IA	19	-24	-10	36	38
175	141.2	169	116.1	218	128.3	CARROLL IA	-18	10	-9	38	69
151	129.0	175	133.6	190	129.1	CASS IA	4	-3	0	64	70
147	145.7	135	112.1	146	110.2	CEDAR IA	-23	-2	-24	09	28
362	138.7	372	117.2	448	125.1	CERRO GORDO IA	-16	7	-10	36	62
144	127.9	144	122.0	166	134.6	CHEROKEE IA	-5	10	5	74	80
113	134.4	94	97.6	107	98.9	CHICKASAW IA	-27	1	-26	08	12
93	148.5	94	152.7	89	124.5	CLARKE IA	3	-18	-16	21	60
123	129.7	137	123.7	167	125.0	CLAY IA	-5	1	-4	54	62
199	153.4	223*	155.4	188	118.2	CLAYTON IA	1	-24	-23	11	46
425	147.9	473*	146.2	505	139.3	CLINTON IA	-1	-5	-6	48	87
163	160.3	186*	160.1	144	115.3	CRAWFORD IA	0	-28	-28	06	39
167	110.2	210	125.9	254	127.5	DALLAS IA	14	1	16	87	68
85	137.3	83	125.2	96	140.3	DAVIS IA	-9	12	2	68	88
107	127.5	101	130.5	98	121.7	DECATUR IA	2	-7	-5	51	54
112	117.6	115	107.5	154	135.2	DELAWARE IA	-9	26	15	86	82
361	137.5	382	126.8	401	119.2	DES MOINES IA	-8	-6	-13	27	48
90	132.7	112	129.8	121	110.2	DICKINSON IA	-2	-15	-17	19	28
640*	156.0	693*	145.9	812*	147.0	DUBUQUE IA	-6	1	-6	48	94
102	139.0	124	135.9	113	109.4	EMMET IA	-2	-19	-21	12	26
218	133.1	228	128.0	242	120.1	FAYETTE IA	-4	-6	-10	36	50
158	128.5	160	113.3	165	110.3	FLOYD IA	-12	-3	-14	25	29
114	130.3	112	116.0	125	119.3	FRANKLIN IA	-11	3	-8	40	48
112	154.1	83	113.6	88	113.4	FREMONT IA	-26	0	-26	08	35
135	146.6	111	109.7	136	122.0	GREENE IA	-25	11	-17	20	54
112	143.1	121	125.9	140	124.6	GRUNDY IA	-12	-1	-13	28	61
110	117.0	126	123.0	136	110.2	GUTHRIE IA	5	-10	-6	48	28
157	143.6	152	126.9	185	135.0	HAMILTON IA	-12	6	-6	47	81
95	128.1	126	145.8	131	125.1	HANCOCK IA	14	-14	-2	58	62
177	125.2	204	126.7	227	122.6	HARDIN IA	1	-3	-2	58	56
136	123.1	191*	155.4	140	102.2	HARRISON IA	26	-34	-17	19	15
140	103.3	141	107.9	159	115.9	HENRY IA	4	7	12	84	40
103	134.0	98	118.7	94	96.3	HOWARD IA	-11	-19	-28	06	10
110	157.8	107	127.3	123	130.1	HUMBOLDT IA	-19	2	-18	18	72
68	112.5	73	112.3	95	119.4	IDA IA	0	6	6	76	48
140	145.0	121	115.3	130	111.2	IOWA IA	-20	-4	-23	10	30
173	160.7	158	126.6	187	134.9	JACKSON IA	-21	7	-16	21	80
227	127.1	250	117.6	282	112.3	JASPER IA	-8	-5	-12	32	33
123	123.8	125	114.7	141	115.3	JEFFERSON IA	-7	1	-7	45	39
307	143.1	310	119.4	419	135.5	JOHNSON IA	-17	14	-5	49	82
146	137.2	148	121.3	169	126.7	JONES IA	-12	4	-8	43	66
156	136.1	139	120.4	156	130.4	KEOKUK IA	-12	8	-4	52	72
201*	166.1	153	108.7	165	109.8	KOSSUTH IA	-35	1	-34	03	27
379	153.6	393	143.5	398	136.7	LEE IA	-7	-5	-11	33	84
858	134.4	1060	134.4	1194	130.1	LINN IA	0	-3	-3	55	72
74	121.2	92	135.6	90	116.0	LOUISA IA	12	-14	-4	52	41
120	141.6	111	122.7	97	104.7	LUCAS IA	-13	-15	-26	08	18
99	146.7	89	112.9	120	136.6	LYON IA	-23	21	-7	45	84
113	132.0	122	129.5	127	127.4	MADISON IA	-2	-2	-3	54	67
219	136.2	215	126.5	244	135.4	MAHASKA IA	-7	7	-1	62	82
220	147.3	209	127.5	222	113.6	MARION IA	-13	-11	-23	11	35
262	121.0	306	124.7	349	124.6	MARSHALL IA	3	0	3	70	61
81	107.0	88	116.0	89	108.4	MILLS IA	8	-7	1	67	24
108	126.2	131	136.8	137	124.4	MITCHELL IA	8	-9	-1	60	60
103	126.2	105	113.5	111	102.8	MONONA IA	-10	-9	-18	17	16
107	145.9	107	135.8	109	128.5	MONROE IA	-7	-5	-12	31	69
115	109.5	127	111.0	176	144.9	MONTGOMERY IA	1	31	32	94	92
271	132.6	310	136.2	331	132.1	MUSCATINE IA	3	-3	0	63	75
163	152.6	179	138.9	168	113.1	O'BRIEN IA	-9	-19	-26	08	35
64	132.9	72	130.0	73	109.8	OSCEOLA IA	-2	-15	-17	18	27
182	115.2	193	113.8	227	126.2	PAGE IA	-1	11	10	81	65
102	138.6	118	134.6	115	115.9	PALO ALTO IA	-3	-14	-16	20	40
168	143.3	177	124.6	240	143.5	PLYMOUTH IA	-13	15	0	64	91
105	134.0	122	132.1	123	119.7	POCAHONTAS IA	-1	-9	-11	34	49
1867*	149.4	1891	127.5	2134	127.5	POLK IA	-15	0	-15	24	68
511	131.7	590	131.5	646	126.2	POTTAWATTAMIE IA	0	-4	-4	53	65
123	108.1	149	120.2	162	117.5	POWESHIEK IA	11	-2	9	80	44
81	141.9	50	85.2	70	115.6	RINGGOLD IA	-40	36	-18	17	40
107	106.9	133	112.5	164	128.5	SAC IA	5	14	20	90	69
825	147.9	940	139.9	1053	137.6	SCOTT IA	-5	-2	-7	44	85
117	140.3	111	114.2	126	107.7	SHELBY IA	-19	-6	-23	10	23
173	139.2	191	125.5	229	123.2	SIOUX IA	-10	-2	-12	32	57
280	125.8	326	125.8	388	123.3	STORY IA	0	-2	-2	59	58
158	122.7	203	145.3	172	109.8	TAMA IA	18	-24	-11	34	27
107	122.4	92	107.4	116	134.1	TAYLOR IA	-12	25	10	81	79
134	125.6	137	116.9	157	125.5	UNION IA	-7	7	0	64	63
99	126.2	106	145.7	94	125.3	VAN BUREN IA	15	-14	-1	62	63
360	135.3	392	129.8	465	144.7	WAPELLO IA	-4	12	7	77	92
137	131.3	147	121.6	178	110.9	WARREN IA	-7	-9	-16	22	30
160	123.3	167	116.6	176	114.2	WASHINGTON IA	-5	-2	-7	43	36
95	108.1	115	127.4	99	121.9	WAYNE IA	18	-4	13	85	54
327	132.0	427	141.4	426	123.5	WEBSTER IA	7	-13	-6	46	58
97	131.3	89	97.8	130	115.1	WINNEBAGO IA	-26	18	-12	29	39
159	132.3	128	93.9	151	101.7	WINNESHIEK IA	-29	8	-23	10	15
809	143.3	843	132.6	944	137.2	WOODBURY IA	-7	3	-4	52	84
47	79.8	70	107.0	86	113.6	WORTH IA	34	6	42	96	35
166	151.1	154	120.6	160	114.1	WRIGHT IA	-20	-5	-24	09	36
13065	122.7	15006	119.0	17043	117.9	KANSAS KS	-3	-1	-4		
141	117.9	174	130.1	145	110.5	ALLEN KS	10	-15	-6	47	29

* SIGNIFICANTLY MORE DEATHS THAN EXPECTED

WHITE FEMALE — ALL CANCERS COMBINED — ICD 140-208

1950-1959 DEATHS	1950-1959 RATE	1960-1969 DEATHS	1960-1969 RATE	1970-1979 DEATHS	1970-1979 RATE	COUNTY AND STATE		PERCENT CHANGE 60S/50S	PERCENT CHANGE 70S/60S	PERCENT CHANGE 70S/50S	PERCENTILE %CHANGE 70S/50S	PERCENTILE RATES 1970S
77	111.2	80	108.9	103	118.1	ANDERSON	KS	-2	8	6	76	46
159	132.9	160	123.3	187	151.0	ATCHISON	KS	-7	22	14	85	96
57	121.9	56	98.8	57	87.4	BARBER	KS	-19	-11	-28	06	04
177	133.4	210	130.2	235	120.6	BARTON	KS	-2	-7	-10	37	51
125	99.9	167	132.8	180	131.0	BOURBON	KS	33	-1	31	94	73
127	120.2	131	118.6	119	104.1	BROWN	KS	-1	-12	-13	27	18
200	113.8	249	115.6	315	119.0	BUTLER	KS	2	3	5	73	48
35	122.6	27	88.5	41	118.1	CHASE	KS	-28	33	-4	54	45
61	122.9	42	86.0	73	127.6	CHAUTAUQUA	KS	-30	48	4	71	68
207	140.2	210	127.2	213	116.6	CHEROKEE	KS	-9	-8	-17	19	42
25	92.7	29	85.4	26	79.0	CHEYENNE	KS	-8	-8	-15	23	02
35	168.3	26	121.9	24	91.2	CLARK	KS	-28	-25	-46	00	06
89	107.0	120	127.0	115	118.2	CLAY	KS	19	-7	10	82	46
139	122.9	151	121.6	140	106.6	CLOUD	KS	-1	-12	-13	27	22
83	117.7	90	131.8	84	114.2	COFFEY	KS	12	-13	-3	56	37
26	127.9	26	100.7	31	108.0	COMANCHE	KS	-21	7	-16	22	24
274	112.9	321	124.1	308	106.5	COWLEY	KS	10	-14	-6	48	21
317	118.3	345	115.6	401	122.3	CRAWFORD	KS	-2	6	3	71	55
46	123.3	44	106.9	55	120.1	DECATUR	KS	-13	12	-3	57	50
191	135.5	184	109.9	200	108.9	DICKINSON	KS	-19	-1	-20	15	25
66	110.4	84	134.5	74	118.9	DONIPHAN	KS	22	-12	8	78	47
234	130.6	241	113.5	262	105.5	DOUGLAS	KS	-13	-7	-19	15	20
37	109.5	44	121.2	47	107.8	EDWARDS	KS	11	-11	-2	60	24
50	110.3	56	116.3	45	83.0	ELK	KS	5	-29	-25	09	03
93	132.7	102	108.1	141	114.6	ELLIS	KS	-19	6	-14	26	38
70	130.3	79	136.0	72	114.7	ELLSWORTH	KS	4	-16	-12	31	38
91	148.7	96	129.2	110	114.5	FINNEY	KS	-13	-11	-23	11	37
133	134.5	144	118.8	158	105.0	FORD	KS	-12	-12	-22	12	19
163	119.1	133	88.5	185	110.8	FRANKLIN	KS	-26	25	-7	44	30
101	130.1	104	111.5	141	144.6	GEARY	KS	-14	30	11	83	92
15	78.6	21	98.2	23	92.3	GOVE	KS	25	-6	17	89	07
25	104.5	32	118.4	36	113.3	GRAHAM	KS	13	-4	8	79	35
10	95.3	18	76.4	33	132.9	GRANT	KS	-20	74	39	96	77
12	51.0	28	107.4	36	114.4	GRAY	KS	111	6	124	99	37
6	92.5	8	94.5	11	106.7	GREELEY	KS	2	13	15	87	22
105	122.1	101	109.9	109	123.9	GREENWOOD	KS	-10	13	1	67	59
19	152.8	10	67.9	28	146.4	HAMILTON	KS	-56	116	-4	52	94
75	118.2	72	99.7	85	108.6	HARPER	KS	-16	9	-8	41	25
172	124.4	207	117.6	223	105.1	HARVEY	KS	-5	-11	-16	22	19
8	102.3	20	140.8	22	129.4	HASKELL	KS	38	-8	27	93	71
11	72.3	19	124.7	19	103.5	HODGEMAN	KS	73	-17	43	96	17
83	113.8	94	121.3	104	130.2	JACKSON	KS	7	7	14	86	72
96	119.7	85	107.4	89	92.5	JEFFERSON	KS	-10	-14	-23	11	07
83	132.7	49	80.6	58	104.3	JEWELL	KS	-39	29	-21	12	18
441	115.7	829	118.5	1265	115.3	JOHNSON	KS	2	-3	0	63	39
9	80.1	18	126.1	14	91.9	KEARNY	KS	58	-27	15	86	06
76	128.6	67	96.8	92	125.3	KINGMAN	KS	-25	29	-3	57	63
34	130.4	43	125.5	45	122.8	KIOWA	KS	-4	-2	-6	48	57
251	134.3	237	115.7	255	123.4	LABETTE	KS	-14	7	-8	41	58
13	109.0	12	74.4	13	62.8	LANE	KS	-32	-16	-42	01	00
228	121.0	285	130.1	342	140.8	LEAVENWORTH	KS	8	8	16	88	89
52	123.8	55	116.7	59	129.3	LINCOLN	KS	-6	11	4	72	70
77	113.3	78	125.4	76	108.4	LINN	KS	11	-14	-4	52	24
27	160.2	32	150.6	29	104.2	LOGAN	KS	-6	-31	-35	03	18
215	129.5	232	123.3	225	110.8	LYON	KS	-5	-10	-14	24	30
178	125.3	202	119.5	220	109.2	MC PHERSON	KS	-5	-9	-13	28	26
124	121.2	138	125.5	141	104.3	MARION	KS	4	-17	-14	26	18
156	136.6	154	127.9	142	110.2	MARSHALL	KS	-6	-14	-19	15	28
26	98.1	29	86.6	46	117.9	MEADE	KS	-12	36	20	90	45
150	114.3	145	101.9	158	106.0	MIAMI	KS	-11	4	-7	44	21
72	106.3	79	112.6	79	106.8	MITCHELL	KS	6	-5	0	65	22
350	125.3	373	121.8	420	123.4	MONTGOMERY	KS	-3	1	-2	60	58
65	118.1	69	118.9	84	131.2	MORRIS	KS	1	10	11	83	73
9	87.6	12	80.6	25	151.9	MORTON	KS	-8	89	73	98	96
99	112.3	94	101.5	112	107.8	NEMAHA	KS	-10	6	-4	53	24
158	116.1	170	122.3	202	128.9	NEOSHO	KS	5	5	11	83	70
47	145.8	55	145.7	38	95.2	NESS	KS	0	-35	-35	03	08
71	124.2	73	120.7	80	106.3	NORTON	KS	-3	-12	-14	24	21
114	127.3	114	115.0	114	101.0	OSAGE	KS	-10	-12	-21	13	14
59	104.5	55	86.6	66	115.7	OSBORNE	KS	-17	34	11	82	40
79	145.9	62	107.3	57	88.1	OTTAWA	KS	-26	-18	-40	01	04
59	80.8	79	110.9	78	126.4	PAWNEE	KS	37	14	57	97	66
55	98.0	64	109.8	70	103.2	PHILLIPS	KS	12	-6	5	74	16
107	135.0	90	107.8	98	96.4	POTTAWATOMIE	KS	-20	-11	-29	06	10
75	115.3	89	112.7	99	107.1	PRATT	KS	-2	-5	-7	44	22
40	151.7	31	104.6	20	60.3	RAWLINS	KS	-31	-42	-60	00	00
358	119.1	400	110.5	548	125.4	RENO	KS	-7	13	5	74	63
91	121.7	110	126.0	94	114.2	REPUBLIC	KS	4	-9	-6	47	36
100	121.5	130	143.6	138	126.1	RICE	KS	18	-12	4	71	65
139	102.5	190	121.5	197	108.7	RILEY	KS	19	-11	6	76	25
47	95.9	55	99.0	79	128.6	ROOKS	KS	3	30	34	95	69
56	154.5	41	92.3	53	109.0	RUSH	KS	-40	18	-29	05	26
76	131.2	87	128.2	109	139.5	RUSSELL	KS	-2	9	6	76	87
255	129.2	286	114.7	312	108.0	SALINE	KS	-11	-6	-16	20	24
18	89.2	30	117.1	48	142.3	SCOTT	KS	31	22	60	98	90
1473	128.3	1914	126.9	2211	123.6	SEDGWICK	KS	-1	-3	-4	54	59
40	91.8	85	136.1	74	98.1	SEWARD	KS	48	-28	7	77	11
802	126.2	925	121.1	1053	121.0	SHAWNEE	KS	-4	0	-4	53	52
26	158.0	28	129.2	40	165.1	SHERIDAN	KS	-18	28	5	73	98
41	119.1	47	110.8	49	100.4	SHERMAN	KS	-7	-9	-16	21	13
64	106.8	81	123.7	56	67.3	SMITH	KS	16	-46	-37	02	00
52	105.5	67	131.9	78	127.3	STAFFORD	KS	25	-4	21	91	67
6	158.9	9	106.3	8	74.9	STANTON	KS	-33	-30	-53	00	01
12	115.3	16	82.4	28	111.5	STEVENS	KS	-29	35	-3	55	31
178	114.5	229	129.0	248	123.1	SUMNER	KS	13	-5	7	78	57
36	107.5	45	106.8	58	112.4	THOMAS	KS	-1	5	5	73	33
17	66.8	35	118.6	38	105.3	TREGO	KS	78	-11	58	97	19
46	97.2	51	99.8	57	93.8	WABAUNSEE	KS	3	-6	-3	54	07
15	162.9	15	125.9	18	136.4	WALLACE	KS	-23	8	-16	20	83
106	133.4	104	124.8	90	105.1	WASHINGTON	KS	-6	-16	-21	13	19
12	127.7	18	136.8	15	99.1	WICHITA	KS	7	-28	-22	11	12

WHITE FEMALE — ALL CANCERS COMBINED — ICD 140-208

1950-1959 DEATHS	1950-1959 RATE	1960-1969 DEATHS	1960-1969 RATE	1970-1979 DEATHS	1970-1979 RATE	COUNTY AND STATE		PERCENT CHANGE 60S/50S	PERCENT CHANGE 70S/60S	PERCENT CHANGE 70S/50S	PERCENTILE %CHANGE 70S/50S	PERCENTILE RATES 1970S
119	119.2	115	111.9	138	124.9	WILSON	KS	-6	12	5	73	62
55	129.2	59	128.4	57	129.8	WOODSON	KS	-1	1	0	65	71
960	138.3	1080	133.8	1222	140.9	WYANDOTTE	KS	-3	5	2	68	89
16781	131.1	19208	125.0	23741	128.1	KENTUCKY	KY	-5	2	-2		
61	89.2	97	122.9	95	103.3	ADAIR	KY	38	-16	16	87	16
104	143.1	94	114.2	87	92.7	ALLEN	KY	-20	-19	-35	02	07
65	141.6	54	107.2	80	121.6	ANDERSON	KY	-24	13	-14	25	54
63	130.2	69	120.1	85	122.8	BALLARD	KY	-8	2	-6	48	57
186	141.9	201	131.0	264	135.2	BARREN	KY	-8	3	-5	51	81
66	143.9	68	133.2	87	146.4	BATH	KY	-7	10	2	68	94
177	121.4	241	148.2	263	140.7	BELL	KY	22	-5	16	87	88
92	124.4	150	139.6	207	131.2	BOONE	KY	12	-6	6	75	74
91	114.9	125	135.1	156	142.1	BOURBON	KY	18	5	24	92	90
353*	159.6	379	138.8	471	140.7	BOYD	KY	-13	1	-12	31	88
112	112.5	135	118.3	169	126.1	BOYLE	KY	5	7	12	84	65
55	122.3	55	108.6	72	134.7	BRACKEN	KY	-11	24	10	82	80
49	79.5	65	95.8	75	100.2	BREATHITT	KY	21	5	26	93	13
109	140.0	114	135.0	116	122.1	BRECKINRIDGE	KY	-4	-10	-13	28	55
70	133.2	69	122.1	162	141.2	BULLITT	KY	-8	16	6	76	89
67	132.4	63	106.6	99	145.1	BUTLER	KY	-20	36	10	81	93
132*	172.6	118	131.8	131	121.3	CALDWELL	KY	-24	-8	-30	05	53
156	137.8	172	117.4	195	105.2	CALLOWAY	KY	-15	-10	-24	10	19
731*	166.6	793*	155.4	837*	153.7	CAMPBELL	KY	-7	-1	-8	42	97
41	117.3	31	86.5	45	101.8	CARLISLE	KY	-26	18	-13	27	15
70	153.9	65	133.9	80	151.9	CARROLL	KY	-13	13	-1	61	96
101	115.8	105	104.6	156	132.1	CARTER	KY	-10	26	14	86	75
91	141.7	82	111.6	80	91.7	CASEY	KY	-21	-18	-35	02	06
193	111.9	244	121.3	300	127.1	CHRISTIAN	KY	8	5	14	85	67
129	145.8	116	105.8	167	123.5	CLARK	KY	-27	17	-15	22	58
63	96.4	100	133.1	120	129.6	CLAY	KY	38	-3	34	95	71
48	128.7	62	141.6	65	117.7	CLINTON	KY	10	-17	-9	40	44
77	129.5	70	112.2	82	121.5	CRITTENDEN	KY	-13	8	-6	47	53
36	98.2	61	143.0	63	133.1	CUMBERLAND	KY	46	-7	36	95	78
382	138.6	458	127.6	573	127.0	DAVIESS	KY	-8	0	-8	40	67
36	101.4	43	97.2	54	107.9	EDMONSON	KY	-4	11	6	77	24
30	122.6	27	92.2	36*	272.2	ELLIOTT	KY	-25	195	122	99	99
79	130.0	78	107.1	95	107.5	ESTILL	KY	-18	0	-17	18	23
618	132.5	744	119.9	1104	139.0	FAYETTE	KY	-9	16	5	73	87
76	124.1	67	97.6	71	85.6	FLEMING	KY	-21	-12	-31	04	03
165	127.4	192	116.9	256	118.5	FLOYD	KY	-8	1	-7	44	46
165	127.4	226	139.6	232	114.8	FRANKLIN	KY	10	-18	-10	36	38
82	128.2	89	120.6	100	137.0	FULTON	KY	-6	14	7	77	84
31	165.8	33	147.0	34	141.5	GALLATIN	KY	-11	-4	-15	24	89
43	96.9	65	127.0	67	107.7	GARRARD	KY	31	-15	11	83	23
73	132.9	81	133.0	96	141.9	GRANT	KY	0	7	7	77	90
235	136.4	228	109.0	277	112.3	GRAVES	KY	-20	3	-18	18	33
103	122.0	98	100.3	132	112.3	GRAYSON	KY	-18	12	-8	42	33
55	107.2	63	112.4	78	111.5	GREEN	KY	5	-1	4	72	31
130	126.2	194	147.1	240	136.6	GREENUP	KY	17	-7	8	79	83
34	101.5	38	116.5	45	125.2	HANCOCK	KY	15	7	23	92	62
175	128.7	217	120.9	282	116.4	HARDIN	KY	-6	-4	-10	37	42
202	117.1	225	115.3	252	118.1	HARLAN	KY	-2	2	1	66	45
114	135.6	114	126.1	119	114.2	HARRISON	KY	-7	-9	-16	21	36
77	118.8	85	116.5	97	112.1	HART	KY	-2	-4	-6	48	33
182	126.4	207	115.7	275	128.0	HENDERSON	KY	-9	11	1	66	68
89	140.2	74	111.4	94	123.8	HENRY	KY	-21	11	-12	32	59
48	119.6	70	165.5	69	152.1	HICKMAN	KY	38	-8	27	93	96
229	123.1	261	117.0	311	113.7	HOPKINS	KY	-5	-3	-8	43	36
42	96.4	43	84.0	75	116.4	JACKSON	KY	-13	39	21	91	42
3366	143.4	3874	131.3	5035*	143.1	JEFFERSON	KY	-8	9	0	63	91
59	108.1	77	116.7	121	135.6	JESSAMINE	KY	8	16	25	93	82
108	121.9	129	128.1	138	116.0	JOHNSON	KY	5	-9	-5	51	40
896*	156.1	1083*	158.9	1257*	160.1	KENTON	KY	2	1	3	69	98
47	79.8	55	82.4	94	119.9	KNOTT	KY	3	45	50	97	50
121	109.7	134	107.7	161	107.7	KNOX	KY	-2	0	-2	59	23
55	111.9	65	113.1	79	111.2	LARUE	KY	1	-2	-1	62	30
115	110.3	157	117.2	210	115.2	LAUREL	KY	6	-2	4	72	39
76	112.4	78	109.1	96	125.8	LAWRENCE	KY	-3	15	12	84	64
39	115.0	44	110.7	55	111.6	LEE	KY	-4	1	-3	56	31
31	94.6	38	91.1	52	94.2	LESLIE	KY	-4	3	0	63	08
125	121.6	130	115.2	164	122.0	LETCHER	KY	-5	6	0	65	54
80	138.4	81	137.8	59	86.9	LEWIS	KY	0	-37	-37	02	04
100	130.7	101	115.2	120	115.2	LINCOLN	KY	-12	0	-12	31	39
56	139.5	64	134.6	71	124.0	LIVINGSTON	KY	-4	-8	-11	33	60
137	128.6	149	124.1	155	110.2	LOGAN	KY	-3	-11	-14	25	28
42	150.7	37	119.5	36	89.3	LYON	KY	-21	-25	-41	01	05
359	141.9	394	124.3	520	136.3	MC CRACKEN	KY	-12	10	-4	53	83
68	144.4	79	135.9	107	147.3	MC CREARY	KY	-6	8	2	68	94
62	121.4	58	100.1	92	138.7	MC LEAN	KY	-18	38	14	86	86
157	118.5	165	100.5	247	117.7	MADISON	KY	-15	17	-1	62	44
33	79.0	43	94.5	40	71.7	MAGOFFIN	KY	20	-24	-9	38	01
110	156.9	98	121.5	96	107.9	MARION	KY	-23	-11	-31	04	24
99	135.2	121	117.8	184	117.9	MARSHALL	KY	-13	0	-13	28	45
45	147.8	35	92.7	51	105.5	MARTIN	KY	-37	14	-29	06	20
110	122.9	124	128.1	151	133.9	MASON	KY	4	5	9	80	79
44	103.6	61	130.2	62	103.3	MEADE	KY	26	-21	0	63	16
15	70.7	30	133.3	38	170.5	MENIFEE	KY	89	28	141	99	99
98	126.9	120	137.0	129	120.1	MERCER	KY	8	-12	-5	49	51
45	121.9	59	131.9	51	82.6	METCALFE	KY	8	-37	-32	03	03
73	132.6	79	123.4	74	96.3	MONROE	KY	-7	-22	-27	07	10
100	163.0	111	147.6	126	134.9	MONTGOMERY	KY	-9	-9	-17	19	81
44	79.1	69	123.3	86	129.7	MORGAN	KY	56	5	64	98	71
162	118.3	194	119.2	235	120.0	MUHLENBERG	KY	1	1	1	67	50
101	124.0	131	130.4	142	114.3	NELSON	KY	5	-12	-8	42	37
46	111.5	46	104.3	52	117.0	NICHOLAS	KY	-6	12	5	73	43
118	105.5	174	141.1	178	125.3	OHIO	KY	34	-11	19	89	63
45	93.5	63	107.2	87	110.1	OLDHAM	KY	15	3	18	89	28
65	139.1	54	114.2	65	113.7	OWEN	KY	-18	0	-18	17	36
17	59.6	24	83.3	39	119.2	OWSLEY	KY	40	43	100	99	48

* SIGNIFICANTLY MORE DEATHS THAN EXPECTED

WHITE FEMALE — ALL CANCERS COMBINED — ICD 140-208

1950-1959 DEATHS	1950-1959 RATE	1960-1969 DEATHS	1960-1969 RATE	1970-1979 DEATHS	1970-1979 RATE	COUNTY AND STATE	PERCENT CHANGE 60S/50S	PERCENT CHANGE 70S/60S	PERCENT CHANGE 70S/50S	PERCENTILE %CHANGE 70S/50S	PERCENTILE RATES 1970S
80	142.1	68	115.8	73	114.1	PENDLETON KY	-18	-1	-20	15	36
121	107.2	134	106.1	152	103.5	PERRY KY	-1	-2	-3	55	17
221	101.4	360	135.2	396	119.7	PIKE KY	33	-11	18	89	49
39	148.5	42	128.4	56	122.1	POWELL KY	-14	-5	-18	18	55
186	117.2	213	111.2	270	107.9	PULASKI KY	-5	-3	-8	42	24
20	113.7	15	87.9	23	142.5	ROBERTSON KY	-23	62	25	93	90
66	115.4	93	147.5	100	135.1	ROCKCASTLE KY	28	-8	17	88	81
58	123.6	57	95.3	96	121.2	ROWAN KY	-23	27	-2	59	53
54	106.8	71	121.1	79	95.0	RUSSELL KY	13	-22	-11	33	08
97	138.1	92	114.8	134	133.9	SCOTT KY	-17	17	-3	55	79
83	97.9	119	119.5	132	107.7	SHELBY KY	22	-10	10	81	23
78	131.7	72	102.2	109	129.8	SIMPSON KY	-22	27	-1	60	71
40	168.3	28	111.1	31	102.9	SPENCER KY	-34	-7	-39	01	16
72	109.7	92	107.3	121	106.2	TAYLOR KY	-2	-1	-3	55	21
62	113.4	65	107.9	86	124.4	TCDD KY	-5	15	10	81	60
45	111.1	49	102.9	65	114.4	TRIGG KY	-7	11	3	70	37
23	94.6	39	143.2	32	92.8	TRIMBLE KY	51	-35	-2	59	07
99	137.4	79	99.2	118	133.7	UNION KY	-28	35	-3	57	79
268	131.7	294	117.8	379	122.4	WARREN KY	-11	4	-7	44	56
60	115.0	76	135.3	72	108.4	WASHINGTON KY	18	-20	-6	48	24
72	113.6	80	111.0	98	105.1	WAYNE KY	-2	-5	-7	43	19
115	129.9	113	114.5	151	150.2	WEBSTER KY	-12	31	16	87	96
176	139.6	181	120.3	197	106.2	WHITLEY KY	-14	-12	-24	09	21
30	103.4	41	131.1	44	119.5	WOLFE KY	27	-9	16	87	49
64	135.1	69	114.3	99	125.3	WOODFORD KY	-15	10	-7	44	63
10759	128.6	13845	122.8	18362	125.5	LOUISIANA LA	-4	2	-2		
229	153.0	280	143.0	345	139.8	ACADIA LA	-7	-2	-9	40	88
70	118.2	96	127.8	118	121.8	ALLEN LA	8	-5	3	70	54
69	102.7	93	99.3	172	125.1	ASCENSION LA	-3	26	22	91	62
39	78.3	64	111.1	68	102.6	ASSUMPTION LA	42	-8	31	94	15
147	121.4	176	109.4	225	110.1	AVOYELLES LA	-10	1	-9	38	28
89	132.3	115	131.6	124	109.4	BEAUREGARD LA	0	-17	-17	19	26
65	115.6	83	127.6	103	136.1	BIENVILLE LA	10	7	18	89	83
82	96.5	185	136.3	269	130.0	BOSSIER LA	41	-5	35	95	72
720	126.8	948	119.1	1241	123.7	CADDO LA	-6	4	-2	57	59
370	134.6	513	123.9	713	122.3	CALCASIEU LA	-8	-1	-9	38	55
50	145.0	54	120.2	52	89.2	CALDWELL LA	-17	-26	-39	01	05
19	90.8	32	104.4	35	92.0	CAMERON LA	15	-12	1	67	06
31	113.9	35	98.2	58	120.2	CATAHOULA LA	-14	22	6	75	51
65	102.5	86	123.0	98	112.2	CLAIBORNE LA	20	-9	9	81	33
31	135.4	63	136.4	81	130.4	CONCORDIA LA	1	-4	-4	54	72
87	140.2	82	111.9	109	113.9	DE SOTO LA	-20	2	-19	16	36
513	122.7	786	117.5	1157	121.8	EAST BATON ROUGE LA	-4	4	-1	62	54
27	126.9	42	193.8	39	146.7	EAST CARROLL LA	53	-24	16	87	94
41	67.7	42	59.5	57	88.1	EAST FELICIANA LA	-12	48	30	94	05
133	145.3	190	153.1	215	135.1	EVANGELINE LA	5	-12	-7	44	81
86	120.8	95	107.8	132	121.7	FRANKLIN LA	-11	13	1	66	54
73	142.0	88	137.4	108	130.1	GRANT LA	-3	-5	-8	40	72
126	107.3	206	121.1	301	142.8	IBERIA LA	13	18	33	95	91
63	108.1	83	106.0	105	115.6	IBERVILLE LA	-2	9	7	77	40
48	96.0	81	119.0	74	87.1	JACKSON LA	24	-27	-9	38	04
485	123.5	937	117.8	1815	127.0	JEFFERSON LA	-5	8	3	70	67
112	129.8	142	126.2	185	133.9	JEFFERSON DAVIS LA	-3	6	3	70	79
231	122.4	383	138.4	501	119.0	LAFAYETTE LA	13	-14	-3	56	48
151	107.0	237	121.4	327	118.2	LAFOURCHE LA	14	-3	11	82	46
65	123.7	90	138.4	101	107.3	LA SALLE LA	12	-22	-13	27	23
92	117.4	105	99.4	138	109.8	LINCOLN LA	-15	11	-6	46	27
55	81.7	107	102.0	149	85.0	LIVINGSTON LA	25	-17	4	72	03
24	113.2	25	86.3	41	106.7	MADISON LA	-24	24	-6	48	22
63	99.7	98	105.4	143	109.7	MOREHOUSE LA	6	4	10	81	27
108	118.3	129	113.1	153	110.4	NATCHITOCHES LA	-4	-2	-7	45	29
3217*	147.9	3352*	138.8	3448*	148.3	ORLEANS LA	-6	7	0	65	95
297	118.6	405	114.9	579	119.3	OUACHITA LA	-3	4	1	65	48
35	120.0	49	98.8	79	123.0	PLAQUEMINES LA	-18	25	3	69	57
53	115.4	78	135.8	97	135.4	POINTE COUPEE LA	18	0	17	88	82
366	126.4	431	108.0	589	115.9	RAPIDES LA	-15	7	-8	40	40
32	115.8	38	108.9	55	119.3	RED RIVER LA	-6	10	3	70	48
75	128.6	83	114.8	124	135.2	RICHLAND LA	-11	18	5	74	81
83	111.7	92	103.6	119	99.4	SABINE LA	-7	-4	-11	33	12
53	91.8	142	122.3	297	133.3	ST BERNARD LA	33	9	45	96	78
46	132.6	66	119.8	93	101.0	ST CHARLES LA	-10	-16	-24	10	14
16	82.9	22	93.5	23	84.1	ST HELENA LA	13	-10	1	67	03
53	137.1	60	129.2	53	94.4	ST JAMES LA	-6	-27	-31	04	08
42	129.9	51	118.5	83	129.5	ST JOHN THE BAPT LA	-9	9	0	63	71
177	105.8	272	124.2	353	124.7	ST LANDRY LA	17	0	18	89	61
81	120.2	116	133.2	142	120.9	ST MARTIN LA	11	-9	1	65	52
128	136.2	170	131.1	228	130.7	ST MARY LA	-4	0	-4	53	73
143	133.4	209	117.6	399	125.0	ST TAMMANY LA	-12	6	-6	46	62
228	141.1	253	120.8	367	132.7	TANGIPAHOA LA	-14	10	-6	47	77
17	95.7	28	133.8	31	130.2	TENSAS LA	40	-3	36	95	72
151	125.1	192	111.8	327	132.0	TERREBONNE LA	-11	18	6	75	75
81	128.4	81	98.3	123	109.4	UNION LA	-23	11	-15	23	26
193	139.7	218	118.7	287	118.2	VERMILION LA	-15	0	-15	22	46
106	139.2	116	120.2	154	114.3	VERNON LA	-14	-5	-18	17	37
125	124.4	181	119.2	237	123.1	WASHINGTON LA	-4	3	-1	61	57
122	117.1	159	107.3	263	132.4	WEBSTER LA	-8	23	13	85	76
31	144.4	36	110.4	56	134.5	WEST BATON ROUGE LA	-24	22	-7	45	80
56	109.5	71	119.5	91	124.1	WEST CARROLL LA	9	4	13	85	60
20	177.6	16	131.4	13	97.7	WEST FELICIANA LA	-26	-26	-45	01	11
73	131.4	87	119.3	100	118.1	WINN LA	-9	-1	-10	35	45
7906*	154.6	8220*	142.5	9501*	139.9	MAINE ME	-8	-2	-9		
797*	168.2	799*	147.0	870	136.5	ANDROSCOGGIN ME	-13	-7	-19	16	83
478	145.9	511	130.6	527	114.0	APOOSTOOK ME	-10	-13	-22	12	36
1548*	148.4	1729*	147.4	2012*	148.7	CUMBERLAND ME	-1	1	0	64	95
169	153.7	173	142.5	184	127.0	FRANKLIN ME	-7	-11	-17	18	67
330*	163.4	338	148.4	420*	145.7	HANCOCK ME	-9	-2	-11	34	93
768*	157.9	751	133.5	919	140.0	KENNEBEC ME	-15	5	-11	32	88

* SIGNIFICANTLY MORE DEATHS THAN EXPECTED

WHITE FEMALE — ALL CANCERS COMBINED — ICD 140-208

1950-1959 DEATHS	1950-1959 RATE	1960-1969 DEATHS	1960-1969 RATE	1970-1979 DEATHS	1970-1979 RATE	COUNTY AND STATE	PERCENT CHANGE 60S/50S	PERCENT CHANGE 70S/60S	PERCENT CHANGE 70S/50S	PERCENTILE %CHANGE 70S/50S	PERCENTILE RATES 1970S
335*	160.5	330	142.9	364	138.8	KNOX ME	-11	-3	-14	27	86
210*	168.7	225*	160.8	240	135.3	LINCOLN ME	-5	-16	-20	14	82
351	149.7	355	137.5	417	135.0	OXFORD ME	-8	-2	-10	36	81
869*	150.9	835	129.1	1000	136.7	PENOBSCOT ME	-14	6	-9	37	84
154	143.7	172*	152.5	170	141.9	PISCATAQUIS ME	6	-7	-1	61	90
193	151.8	209	154.9	232	150.7	SAGADAHOC ME	2	-3	-1	62	96
324	155.2	333	144.7	413*	153.1	SOMERSET ME	-7	6	-1	60	97
216	155.1	189	136.9	238*	149.3	WALDO ME	-12	9	-4	54	95
296	150.2	292	141.9	368*	159.0	WASHINGTON ME	-6	12	6	75	98
868*	159.4	979*	151.8	1127	137.3	YORK ME	-5	-10	-14	26	84
15523*	150.1	19828*	141.3	25737*	143.9	MARYLAND MD	-6	2	-4		
647	147.5	721	140.3	854*	143.9	ALLEGANY MD	-5	3	-2	57	92
630	151.6	1060	140.1	1865*	153.3	ANNE ARUNDEL MD	-8	9	1	66	97
1959	143.8	3104	132.4	4904*	142.6	BALTIMORE MD	-8	8	-1	62	90
55	146.2	78	127.9	160	155.2	CALVERT MD	-13	21	6	76	97
137	147.2	159	149.1	185	155.3	CAROLINE MD	1	4	5	75	97
335	123.0	380	111.7	550	126.2	CARROLL MD	-9	13	3	69	65
241	159.5	260	137.6	354	142.4	CECIL MD	-14	4	-11	34	90
68	121.0	134	146.6	206	134.4	CHARLES MD	21	-8	11	83	80
165	126.1	175	111.9	249	148.5	DORCHESTER MD	-11	33	18	89	95
418	132.6	558*	149.3	638	131.9	FREDERICK MD	13	-12	-1	62	75
91	99.2	126	112.9	171	118.2	GARRETT MD	14	5	19	90	46
276	138.8	382	125.9	598	125.9	HARFORD MD	-9	0	-9	38	64
123	118.0	240	145.0	377	128.8	HOWARD MD	23	-11	9	80	69
84	132.3	104	139.4	119	133.4	KENT MD	5	-4	1	66	78
1398*	154.7	2378	135.2	3705*	139.6	MONTGOMERY MD	-13	3	-10	37	87
1032	134.4	1966	133.9	2920*	149.1	PRINCE GEORGE'S MD	0	11	11	82	95
73	120.5	93	114.1	158	145.6	QUEEN ANNE'S MD	-5	28	21	91	93
78	122.2	119	121.2	186	123.8	ST MARY'S MD	-1	2	1	67	59
123	149.5	131	143.8	150	143.1	SOMERSET MD	-4	0	-4	52	91
151	152.9	172	140.6	210	132.5	TALBOT MD	-8	-6	-13	27	76
599	143.2	727	139.0	891	135.1	WASHINGTON MD	-3	-3	-6	48	81
267	133.5	293	117.8	436	141.1	WICOMICO MD	-12	20	6	75	89
122	124.3	136	114.7	183	122.3	WORCESTER MD	-8	7	-2	60	55
6451*	166.6	6332*	163.2	5668*	157.1	BALTIMORE CITY MD	-2	-4	-6	48	98
44071*	154.4	46259*	140.2	52942*	141.3	MASSACHUSETTS MA	-9	1	-8		
607*	171.6	776	133.7	1559*	146.6	BARNSTABLE MA	-22	10	-15	24	94
1315*	163.9	1350*	146.3	1479*	139.1	BERKSHIRE MA	-11	-5	-15	23	87
3698*	160.4	3687*	143.5	4379*	137.6	BRISTOL MA	-11	-4	-14	25	85
52	120.9	56	114.6	84	137.2	DUKES MA	-5	20	13	85	84
5212*	152.0	5515*	140.0	6395*	142.8	ESSEX MA	-8	2	-6	47	91
479	139.8	515	130.5	579	131.5	FRANKLIN MA	-7	1	-6	47	74
3323*	152.7	3676*	141.2	4143*	138.9	HAMPDEN MA	-8	-2	-9	38	87
673	126.4	815	128.8	961	132.9	HAMPSHIRE MA	2	3	5	74	77
10184*	156.1	10653*	139.6	12245*	142.1	MIDDLESEX MA	-11	2	-9	39	90
37	148.7	34	117.8	82*	230.6	NANTUCKET MA	-21	96	55	97	99
3967*	152.5	4541*	137.1	5653*	140.4	NORFOLK MA	-10	2	-8	42	88
1958*	150.7	2214	137.8	2956*	143.7	PLYMOUTH MA	-9	4	-5	51	91
7993*	166.1	7248*	152.7	6680*	152.3	SUFFOLK MA	-8	0	-8	41	97
4573	138.9	4979	130.8	5747	133.3	WORCESTER MA	-6	2	-4	53	78
43295*	149.5	49621*	137.1	57369*	133.4	MICHIGAN MI	-8	-3	-11		
38	120.5	40	93.8	79	115.2	ALCONA MI	-22	23	-4	52	39
56	142.5	52	118.3	65	131.9	ALGER MI	-17	11	-7	43	75
375	144.1	413	135.4	467	121.8	ALLEGAN MI	-6	-10	-15	22	54
153	141.8	179	133.2	232	137.6	ALPENA MI	-6	3	-3	56	85
93	154.7	85	119.3	138	137.1	ANTRIM MI	-23	15	-11	32	84
71	150.4	87	156.2	87	111.9	ARENAC MI	4	-28	-26	08	32
36	113.2	46	119.3	70	134.9	BARAGA MI	5	13	19	90	81
222	143.6	253	132.7	291	127.5	BARRY MI	-8	-4	-11	33	68
674*	157.3	799*	152.5	800	129.2	BAY MI	-3	-15	-18	17	70
67	158.1	75	144.1	96	138.3	BENZIE MI	-9	-4	-13	29	86
844	143.7	984	134.1	1095	126.0	BERRIEN MI	-7	-6	-12	30	65
254	142.1	263	129.7	288	125.0	BRANCH MI	-9	-4	-12	31	62
830	142.9	822	120.6	1048	136.2	CALHOUN MI	-16	13	-5	51	83
240	153.4	241	130.0	290	126.1	CASS MI	-15	-3	-18	18	65
110	152.1	109	126.3	159	147.6	CHARLEVOIX MI	-17	17	-3	56	94
100	139.8	95	112.3	142	122.3	CHEBOYGAN MI	-20	9	-12	29	55
172	144.2	184	138.7	227	144.6	CHIPPEWA MI	-4	4	0	65	92
85	158.1	91	114.7	172	124.7	CLARE MI	-27	9	-21	13	61
203	133.3	227	123.5	268	120.7	CLINTON MI	-7	-2	-9	37	52
28	134.3	43	136.7	71	159.0	CRAWFORD MI	2	16	18	89	98
240	148.9	258	131.5	330	141.4	DELTA MI	-12	8	-5	50	89
196	149.9	212	136.4	267	142.0	DICKINSON MI	-9	4	-5	49	90
304	137.0	363	138.0	452	130.0	EATON MI	1	-6	-5	50	72
115	130.6	133	127.6	180	138.9	EMMET MI	-2	9	6	77	87
1693	147.5	2080	138.4	2478*	137.9	GENESEE MI	-6	0	-7	46	85
56	116.6	76	111.1	136	131.0	GLADWIN MI	-5	18	12	84	73
194	149.3	197	131.6	225	131.3	GOGEBIC MI	-12	0	-12	30	74
188	101.9	247	114.5	297	112.2	GRAND TRAVERSE MI	12	-2	10	82	33
275	144.9	260	126.8	264	117.6	GRATIOT MI	-13	-7	-19	16	44
234	126.0	263	129.1	296	120.3	HILLSDALE MI	2	-7	-4	51	51
402*	181.1	341*	146.1	341	139.3	HOUGHTON MI	-19	-5	-23	10	87
249	151.6	260	132.4	309	125.6	HURON MI	-13	-5	-17	19	64
1199	150.0	1301	133.6	1422	127.5	INGHAM MI	-11	-5	-15	23	68
311	151.2	273	119.4	326	127.0	IONIA MI	-21	6	-16	21	67
92	154.8	105	112.8	199	139.8	IOSCO MI	-27	24	-10	37	88
132	148.2	134	132.1	179	144.2	IRON MI	-11	9	-3	56	92
169	135.9	204	137.1	206	111.4	ISABELLA MI	1	-19	-18	17	31
780	138.6	963*	143.7	1059	138.4	JACKSON MI	4	-4	0	63	86
905	131.6	1097	130.1	1266	127.7	KALAMAZOO MI	-1	-2	-3	56	68
31	133.3	34	121.9	69	150.9	KALKASKA MI	-9	24	13	85	96
2440*	154.0	2580	132.7	2940	128.1	KENT MI	-14	-3	-17	20	68
19	128.8	22	152.4	22	141.1	KEWEENAW MI	18	-7	10	81	89
36	148.8	31	107.1	55	137.0	LAKE MI	-28	28	-8	42	84
266	150.8	254	120.2	354	136.1	LAPEER MI	-20	13	-10	37	83

* SIGNIFICANTLY MORE DEATHS THAN EXPECTED

WHITE FEMALE — ALL CANCERS COMBINED — ICD 140-208

1950-1959 DEATHS	1950-1959 RATE	1960-1969 DEATHS	1960-1969 RATE	1970-1979 DEATHS	1970-1979 RATE	COUNTY AND STATE	PERCENT CHANGE 60S/50S	PERCENT CHANGE 70S/60S	PERCENT CHANGE 70S/50S	PERCENTILE %CHANGE 70S/50S	PERCENTILE RATES 1970S
55	126.8	69	117.8	125*	167.2	LEELANAU MI	-7	42	32	94	99
533	152.0	583	142.7	640	134.2	LENAWEE MI	-6	-6	-12	32	80
235	151.3	261	121.6	410	122.9	LIVINGSTON MI	-20	1	-19	16	57
39	91.6	38	86.3	56	124.4	LUCE MI	-6	44	36	95	60
56	132.2	66	126.7	83	136.7	MACKINAC MI	-4	8	3	71	84
1271*	155.1	2176	137.9	3767*	139.1	MACOMB MI	-11	1	-10	35	87
165	152.8	171	139.8	212	140.7	MANISTEE MI	-9	1	-8	42	88
333	143.4	383	141.1	426	134.5	MARQUETTE MI	-2	-5	-6	47	80
173	147.8	174	127.6	192	114.6	MASON MI	-14	-10	-22	11	37
133	137.9	148	128.4	186	127.9	MECOSTA MI	-7	0	-7	44	68
191	153.0	209	144.4	227	132.0	MENOMINEE MI	-6	-9	-14	26	75
163	119.6	305*	154.4	360	125.9	MIDLAND MI	29	-18	5	74	64
51	162.9	49	130.5	67	133.3	MISSAUKEE MI	-20	2	-18	17	78
519	149.2	583	129.8	723	126.6	MONROE MI	-13	-3	-15	23	66
294*	158.5	275	130.4	303	116.5	MONTCALM MI	-18	-11	-26	08	42
23	125.1	46	154.5	53	110.2	MONTMORENCY MI	24	-29	-12	31	28
773	147.8	935	141.3	1067	133.8	MUSKEGON MI	-4	-5	-9	37	79
174	155.8	182	133.4	269	144.3	NEWAYGO MI	-14	8	-7	43	92
2739	146.4	4212	135.3	6232*	137.0	OAKLAND MI	-8	1	-6	46	84
108	123.6	116	119.8	144	123.9	OCEANA MI	-3	3	0	65	59
59	122.6	80	118.5	138	143.8	OGEMAW MI	-3	21	17	88	92
64	146.1	70	135.3	82	130.7	ONTONAGON MI	-7	-3	-11	34	73
105	143.2	110	131.8	155	154.9	OSCEOLA MI	-8	17	8	79	97
19	124.4	24	90.0	58	142.7	OSCODA MI	-28	59	15	86	91
41	133.4	58	142.8	87	126.9	OTSEGO MI	7	-11	-5	51	66
569	152.7	710*	145.0	791	119.7	OTTAWA MI	-5	-17	-22	12	49
62	115.7	85	126.5	118	138.6	PRESQUE ISLE MI	9	10	20	90	86
56	161.2	71	133.9	171*	164.6	ROSCOMMON MI	-17	23	2	68	98
1114*	155.7	1182	136.7	1253	124.7	SAGINAW MI	-12	-9	-20	14	61
747*	158.6	796	137.0	929	131.7	ST CLAIR MI	-14	-4	-17	19	74
342	159.6	346	136.6	392	128.1	ST JOSEPH MI	-14	-6	-20	15	68
240	140.3	260	133.1	305	127.4	SANILAC MI	-5	-4	-9	38	67
66	159.8	73	150.3	86	149.0	SCHOOLCRAFT MI	-6	-1	-7	45	95
347	139.2	377	134.4	409	125.9	SHIAWASSEE MI	-3	-6	-10	37	64
295	143.6	284	121.7	317	117.8	TUSCOLA MI	-15	-3	-18	17	45
344	140.9	349	123.1	389	115.9	VAN BUREN MI	-13	-6	-18	18	40
760	125.9	876	119.6	1180	133.8	WASHTENAW MI	-5	12	6	76	79
14725*	156.1	15951*	145.1	15032*	138.9	WAYNE MI	-7	-4	-11	33	86
135	141.3	162	145.2	178	126.5	WEXFORD MI	3	-13	-10	34	66
22005	139.6	25182	129.7	29013	124.6	MINNESOTA MN	-7	-4	-11		
76	107.4	102	128.9	128	133.3	AITKIN MN	20	3	24	92	78
204	110.0	362	123.3	550	116.1	ANOKA MN	12	-6	6	75	41
148	135.6	175	131.4	195	116.3	BECKER MN	-3	-11	-14	25	41
127	131.1	143	122.9	193	137.3	BELTRAMI MN	-6	12	5	73	84
79	115.0	118	137.6	163	140.8	BENTON MN	20	2	23	91	89
72	150.9	79	134.9	78	104.8	BIG STONE MN	-11	-22	-31	05	19
317	140.0	359	133.6	378	125.3	BLUE EARTH MN	-5	-6	-10	34	63
183	126.2	231	132.1	268	131.3	BROWN MN	5	-1	4	72	74
145	120.6	175	121.4	251	145.5	CARLTON MN	1	20	21	91	93
135	140.8	148	119.8	172	107.5	CARVER MN	-15	-10	-24	10	23
82	96.4	136	120.6	171	117.7	CASS MN	25	-2	22	91	44
110	132.6	123	124.7	150	121.0	CHIPPEWA MN	-6	-3	-9	39	52
106	129.0	132	136.5	178	137.0	CHISAGO MN	6	0	6	76	84
183	134.4	220	118.9	298	127.1	CLAY MN	-11	7	-5	49	67
59	142.3	63	130.2	56	102.6	CLEARWATER MN	-9	-21	-28	06	16
14	128.8	18	128.5	29	137.4	COOK MN	0	7	7	77	84
116	142.9	129	125.2	125	104.9	COTTONWOOD MN	-12	-16	-27	08	19
205	131.2	287	144.5	302	115.3	CROW WING MN	10	-20	-12	30	39
300	127.0	463	124.1	718	121.9	DAKOTA MN	-2	-2	-4	53	54
90	127.5	100	124.9	103	114.6	DODGE MN	-2	-8	-10	36	37
174	150.1	194	135.6	216	116.0	DOUGLAS MN	-10	-14	-23	11	40
166	130.6	190	129.3	188	110.7	FARIBAULT MN	-1	-14	-15	22	29
189	130.7	193	120.6	208	115.9	FILLMORE MN	-8	-4	-11	32	40
207	120.9	262	122.2	250	96.4	FREEBORN MN	1	-21	-20	14	10
263	134.0	303	134.5	289	106.0	GOODHUE MN	0	-21	-21	13	21
47	101.6	79	144.6	78	124.6	GRANT MN	42	-14	23	92	61
5725*	146.6	6400	131.7	7345	132.0	HENNEPIN MN	-10	0	-10	36	75
103	130.3	107	112.5	126	104.1	HOUSTON MN	-14	-7	-20	14	18
62	120.4	85	129.4	113	139.3	HUBBARD MN	7	8	16	87	87
85	124.7	101	122.5	108	100.4	ISANTI MN	-2	-18	-19	15	13
185	137.9	242	137.9	264	116.7	ITASCA MN	0	-15	-15	22	42
111	146.2	124	133.3	164*	154.3	JACKSON MN	-9	16	6	75	97
59	140.4	57	100.5	79	111.6	KANABEC MN	-28	11	-21	13	31
162	105.3	204	111.6	234	106.8	KANDIYOHI MN	6	-4	1	67	22
64	152.1	48	106.7	68	136.3	KITTSON MN	-30	28	-10	35	83
63	108.3	74	96.8	116	125.9	KOOCHICHING MN	-11	30	16	88	64
90	126.1	104	125.2	121	135.3	LAC QUI PARLE MN	-1	8	7	78	82
67	158.7	78	145.0	98	146.0	LAKE MN	-9	1	-8	42	93
23	114.6	19	91.1	25	108.6	LAKE OF THE WOODS MN	-21	19	-5	50	25
153	141.4	174	134.9	185	121.1	LE SUEUR MN	-5	-10	-14	25	53
62	126.8	71	124.1	81	109.6	LINCOLN MN	-2	-12	-14	26	27
144	132.3	180	136.7	200	123.1	LYON MN	3	-10	-7	44	57
159	123.2	204	128.0	218	118.2	MC LEOD MN	4	-8	-4	53	46
28	120.4	41	154.6	36	122.1	MAHNOMEN MN	28	-21	1	67	55
90	144.7	104	143.5	87	101.5	MARSHALL MN	-1	-29	-30	05	14
160	128.1	197	125.4	221	114.5	MARTIN MN	-2	-9	-11	34	37
144	149.9	146	125.5	162	109.8	MEEKER MN	-16	-12	-27	07	27
127	165.7	109	121.1	161	132.4	MILLE LACS MN	-27	9	-20	14	76
172	144.6	198	136.9	235	132.8	MORRISON MN	-5	-3	-8	41	77
279	138.2	320	128.0	329	111.0	MOWER MN	-7	-13	-20	15	30
77	129.1	81	113.6	102	121.8	MURRAY MN	-12	7	-6	48	54
117	98.2	146	110.7	176	132.5	NICOLLET MN	13	20	35	95	76
129	121.2	148	113.1	180	111.8	NOBLES MN	-7	-1	-8	43	32
109	165.7	95	118.0	114	130.0	NORMAN MN	-29	10	-22	12	72
401	140.0	405	109.0	551	118.8	OLMSTED MN	-22	9	-15	23	47
340	122.4	370	116.3	469	119.2	OTTER TAIL MN	-5	2	-3	57	48
74	117.6	92	116.2	123	125.6	PENNINGTON MN	-1	8	7	77	64
103	113.4	148	147.6	135	110.0	PINE MN	30	-25	-3	56	28
93	138.9	114	136.5	119	123.4	PIPESTONE MN	-2	-10	-11	33	58

* SIGNIFICANTLY MORE DEATHS THAN EXPECTED

WHITE FEMALE — ALL CANCERS COMBINED — ICD 140-208

1950-1959 DEATHS	1950-1959 RATE	1960-1969 DEATHS	1960-1969 RATE	1970-1979 DEATHS	1970-1979 RATE	COUNTY AND STATE	PERCENT CHANGE 60S/50S	PERCENT CHANGE 70S/60S	PERCENT CHANGE 70S/50S	PERCENTILE %CHANGE 70S/50S	PERCENTILE RATES 1970S
262	146.0	255	118.3	295	118.6	POLK MN	-19	0	-19	16	47
71	111.3	84	103.1	105	113.3	POPE MN	-7	10	2	68	35
3043*	152.0	3337	135.8	3705	130.4	RAMSEY MN	-11	-4	-14	25	72
38	145.2	36	121.9	49	144.2	RED LAKE MN	-16	18	-1	62	92
134	122.0	172	133.0	161	102.3	REDWOOD MN	9	-23	-16	21	15
151	127.2	184	128.0	210	128.9	RENVILLE MN	1	1	1	67	70
249	128.9	277	122.9	336	129.0	RICE MN	-5	5	0	64	70
70	134.7	76	115.7	90	111.2	ROCK MN	-14	-4	-17	18	30
87	162.2	84	134.7	86	114.5	ROSEAU MN	-17	-15	-29	05	37
1603	147.0	1854*	142.9	1989	135.4	ST LOUIS MN	-3	-5	-8	42	82
145	166.1	157	137.2	196	122.0	SCOTT MN	-17	-11	-27	07	55
64	114.9	89	115.5	136	124.7	SHERBURNE MN	1	8	9	80	61
92	113.5	111	116.0	137	119.1	SIBLEY MN	2	3	5	73	48
458	148.0	516	133.5	561	116.1	STEARNS MN	-10	-13	-22	12	41
147	123.5	175	121.0	210	118.7	STEELE MN	-2	-2	-4	54	47
57	113.3	87	144.5	77	99.3	STEVENS MN	28	-31	-12	30	12
125	168.8	114	126.5	118	116.1	SWIFT MN	-25	-8	-31	04	41
151	123.7	190	134.1	207	131.2	TODD MN	8	-2	6	76	73
43	116.8	47	109.2	64	126.0	TRAVERSE MN	-7	15	8	79	65
151	149.8	147	123.8	154	112.2	WABASHA MN	-17	-9	-25	09	33
84	143.3	88	122.2	105	109.8	WADENA MN	-15	-10	-23	10	27
117	136.1	137	130.9	150	131.3	WASECA MN	-4	0	-4	54	74
267	149.2	324	134.2	426	122.2	WASHINGTON MN	-10	-9	-18	17	55
108	140.7	119	135.4	136	144.1	WATONWAN MN	-4	6	2	69	92
66	146.1	59	106.6	79	122.6	WILKIN MN	-27	15	-16	21	56
343	155.5	321	125.8	368	116.3	WINONA MN	-19	-8	-25	09	41
225	148.0	255	138.1	272	105.6	WRIGHT MN	-7	-24	-29	06	20
97	122.1	117	124.6	111	103.9	YELLOW MEDICINE MN	2	-17	-15	23	17
6686	121.4	8335	117.6	10320	113.3	MISSISSIPPI MS	-3	-4	-7		
73	115.0	123	142.8	151	134.9	ADAMS MS	24	-6	17	88	81
156	144.4	170	127.2	204	117.9	ALCORN MS	-12	-7	-18	17	45
42	102.0	45	102.0	54	105.8	AMITE MS	0	4	4	71	20
87	121.2	112	134.9	105	103.7	ATTALA MS	11	-23	-14	24	17
29	128.8	32	117.2	31	97.4	BENTON MS	-9	-17	-24	09	11
118	165.0	93	101.9	108	96.6	BOLIVAR MS	-38	-5	-41	01	10
83	129.2	67	90.7	106	115.3	CALHOUN MS	-30	27	-11	34	39
30	94.9	29	84.3	38	101.1	CARROLL MS	-11	20	7	77	14
65	118.2	86	130.0	111	138.3	CHICKASAW MS	10	6	17	88	86
49	132.7	47	108.1	53	120.1	CHOCTAW MS	-19	11	-10	37	50
19	121.5	16	100.1	17	82.0	CLAIBORNE MS	-18	-18	-33	03	02
67	123.5	66	109.0	69	94.6	CLARKE MS	-12	-13	-23	10	08
40	93.0	80	143.1	95	131.5	CLAY MS	54	-8	41	96	74
79	119.1	106	129.6	108	114.6	COAHOMA MS	9	-12	-4	54	37
94	120.6	107	116.3	120	114.9	COPIAH MS	-4	-1	-5	51	38
55	120.9	46	83.2	69	100.0	COVINGTON MS	-31	20	-17	19	13
52	127.8	64	102.4	158	123.7	DE SOTO MS	-20	21	-3	55	59
183	124.7	236	113.3	328	123.2	FORREST MS	-9	9	-1	61	58
38	132.5	56	168.6	42	105.5	FRANKLIN MS	27	-37	-20	14	20
52	148.5	57	132.2	77	123.2	GEORGE MS	-11	-7	-17	19	58
25	97.4	38	128.8	51	124.7	GREENE MS	32	-3	28	93	61
54	131.9	69	107.7	91	111.9	GRENADA MS	-18	4	-15	23	32
56	104.9	93	129.6	110	106.7	HANCOCK MS	23	-18	2	67	22
326	125.3	525	135.8	738	137.8	HARRISON MS	8	2	10	81	85
473	117.3	722	122.6	918	116.8	HINDS MS	5	-5	0	63	43
74	146.3	68	122.7	75	132.4	HOLMES MS	-16	8	-10	37	76
34	134.3	40	139.2	35	106.4	HUMPHREYS MS	4	-24	-21	13	21
5	102.0	7	152.9	5	104.8	ISSAQUENA MS	50	-31	3	70	19
84	119.9	95	110.3	131	119.9	ITAWAMBA MS	-8	9	0	64	50
147	122.0	199	102.2	383	116.6	JACKSON MS	-16	14	-4	52	42
48	107.6	67	128.9	70	108.7	JASPER MS	20	-16	1	66	25
15	83.6	28	140.2	18	102.1	JEFFERSON MS	68	-27	22	91	15
36	128.0	37	104.6	42	86.9	JEFFERSON DAVIS MS	-18	-17	-32	04	04
216	120.8	257	110.7	343	119.4	JONES MS	-8	8	-1	61	49
26	83.0	32	86.4	41	105.2	KEMPER MS	4	22	27	93	19
57	93.6	77	102.8	103	102.1	LAFAYETTE MS	10	-1	9	80	15
46	93.8	67	114.7	72	86.1	LAMAR MS	22	-25	-8	41	03
249	108.0	331	114.3	355	103.3	LAUDERDALE MS	6	-10	-4	52	16
52	154.3	55	140.2	73	146.4	LAWRENCE MS	-9	4	-5	50	94
82	141.3	87	120.2	91	105.6	LEAKE MS	-15	-12	-25	08	20
183	130.5	224	121.9	249	101.9	LEE MS	-7	-16	-22	12	15
90	112.7	123	118.8	154	126.4	LEFLORE MS	5	6	12	84	65
103	117.5	97	89.0	171	123.4	LINCOLN MS	-24	39	5	74	58
129	128.5	188	141.3	190	106.9	LOWNDES MS	10	-24	-17	20	22
83	173.9	63	107.7	93	111.5	MADISON MS	-38	3	-36	02	31
82	133.2	105	127.4	94	83.3	MARION MS	-4	-35	-37	02	03
47	115.9	58	114.6	87	128.5	MARSHALL MS	-1	12	11	82	69
134	126.9	144	110.9	167	101.4	MONROE MS	-13	-9	-20	14	14
67	138.6	71	135.9	69	109.7	MONTGOMERY MS	-2	-19	-21	13	27
95	121.5	119	122.3	135	115.6	NESHOBA MS	1	-5	-5	51	40
89	125.7	103	119.1	106	102.6	NEWTON MS	-5	-14	-18	17	15
31	114.9	43	136.9	41	115.5	NOXUBEE MS	19	-16	0	65	39
47	89.8	76	115.9	101	119.4	OKTIBBEHA MS	29	3	33	94	49
90	134.5	90	114.2	101	98.6	PANOLA MS	-15	-14	-27	07	12
63	87.7	101	110.2	155	119.0	PEARL RIVER MS	26	8	36	95	48
31	109.4	39	118.8	38	96.2	PERRY MS	9	-19	-12	31	10
145	136.0	114	87.6	180	118.3	PIKE MS	-36	35	-13	28	46
73	92.3	116	121.2	115	98.7	PONTOTOC MS	31	-19	7	77	12
91	114.8	128	126.6	125	95.4	PRENTISS MS	10	-25	-17	19	09
48	149.5	44	117.1	48	118.0	QUITMAN MS	-22	1	-21	13	45
90	106.1	101	85.1	190	98.8	RANKIN MS	-20	16	-7	45	12
65	114.3	79	101.4	104	106.9	SCOTT MS	-11	5	-6	46	22
11	83.0	23	135.9	22	121.6	SHARKEY MS	64	-11	46	96	54
73	113.8	100	128.1	115	114.7	SIMPSON MS	13	-10	1	66	38
49	94.5	67	112.0	76	104.4	SMITH MS	19	-7	11	82	18
30	132.2	27	94.9	45	114.4	STONE MS	-28	21	-13	27	37
88	132.7	87	109.6	107	116.3	SUNFLOWER MS	-17	6	-12	30	41
67	150.8	62	120.0	72	123.3	TALLAHATCHIE MS	-20	3	-18	17	58
48	121.3	59	119.0	67	100.5	TATE MS	-2	-16	-17	19	13
75	113.1	105	134.6	106	108.6	TIPPAH MS	19	-19	-4	54	25

* SIGNIFICANTLY MORE DEATHS THAN EXPECTED

WHITE FEMALE — ALL CANCERS COMBINED — ICD 140-208

1950-1959 DEATHS	1950-1959 RATE	1960-1969 DEATHS	1960-1969 RATE	1970-1979 DEATHS	1970-1979 RATE	COUNTY AND STATE		PERCENT CHANGE 60S/50S	PERCENT CHANGE 70S/60S	PERCENT CHANGE 70S/50S	PERCENTILE %CHANGE 70S/50S	PERCENTILE RATES 1970S
76	112.7	118	138.5	130	119.7	TISHOMINGO	MS	23	-14	6	76	49
16	106.0	29	160.9	26	126.5	TUNICA	MS	52	-21	19	90	66
96	117.5	97	97.2	123	98.3	UNION	MS	-17	1	-16	20	11
41	124.4	49	111.4	54	93.5	WALTHALL	MS	-10	-16	-25	09	07
144	137.4	170	127.8	209	125.8	WARREN	MS	-7	-2	-8	40	64
111	106.3	186	135.4	210	120.7	WASHINGTON	MS	27	-11	14	85	52
42	102.2	69	124.5	70	105.0	WAYNE	MS	22	-16	3	69	19
56	122.2	69	126.1	76	116.6	WEBSTER	MS	3	-7	-5	51	42
31	138.7	24	98.2	56*	201.1	WILKINSON	MS	-29	105	45	96	99
78	134.8	85	122.5	89	98.3	WINSTON	MS	-9	-20	-27	07	11
66	130.5	69	123.8	61	81.8	YALOBUSHA	MS	-5	-34	-37	02	02
96	151.3	82	106.3	104	109.8	YAZOO	MS	-30	3	-27	07	27
30207	136.9	32963	128.1	37168	126.9	MISSOURI	MO	-6	-1	-7		
151	119.2	192	140.2	178	122.8	ADAIR	MO	18	-12	3	70	57
95	118.2	94	106.7	112	117.1	ANDREW	MO	-10	10	-1	61	43
95	154.6	78	117.1	101	138.7	ATCHISON	MO	-24	18	-10	35	86
190	137.4	194	117.0	218	120.5	AUDRAIN	MO	-15	3	-12	30	51
153	120.9	169	114.6	215	117.9	BARRY	MO	-5	3	-2	57	45
111	129.0	107	115.8	117	101.8	BARTON	MO	-10	-12	-21	13	15
151	128.3	157	111.6	164	112.0	BATES	MO	-13	0	-13	29	32
68	113.1	87	108.7	129	126.4	BENTON	MO	-4	16	12	83	65
72	131.2	79	131.8	90	130.5	BOLLINGER	MO	0	-1	-1	62	73
276	121.7	354	125.5	428	118.9	BOONE	MO	3	-5	-2	58	47
819	132.6	891	140.1	888	135.1	BUCHANAN	MO	6	-4	2	68	81
195	121.9	248	125.5	297	125.1	BUTLER	MO	3	0	3	69	62
110	143.8	101	125.8	94	118.4	CALDWELL	MO	-13	-6	-18	18	46
138	98.9	173	116.3	198	117.6	CALLAWAY	MO	18	1	19	90	44
60	123.8	76	101.3	156	124.6	CAMDEN	MO	-18	23	1	65	61
246	121.2	318	121.2	385	119.9	CAPE GIRARDEAU	MO	0	-1	-1	61	50
156	147.2	126	121.5	146	130.4	CARROLL	MO	-17	7	-11	32	72
33	156.3	28	107.9	41	133.7	CARTER	MO	-31	24	-14	24	79
189	140.4	189	112.1	279	122.1	CASS	MO	-20	9	-13	28	55
93	127.4	92	123.4	127	127.1	CEDAR	MO	-3	3	0	63	67
117	124.2	119	132.6	103	100.5	CHARITON	MO	7	-24	-19	16	13
96	131.1	97	112.7	119	102.2	CHRISTIAN	MO	-14	-9	-22	11	15
74	119.1	85	137.4	88	137.8	CLARK	MO	15	0	16	87	85
336	122.7	457	109.1	690	118.9	CLAY	MO	-11	9	-3	55	47
139	158.4	140	147.9	128	121.7	CLINTON	MO	-7	-18	-23	10	54
202	113.6	266	115.3	324	112.0	COLE	MO	1	-3	-1	60	32
99	102.2	122	118.9	128	118.4	COOPER	MO	16	0	16	87	46
106	151.5	115	121.7	147	122.5	CRAWFORD	MO	-20	1	-19	15	56
90	145.5	91	141.8	73	86.7	DADE	MO	-3	-39	-40	01	04
65	109.7	83	107.1	113	124.0	DALLAS	MO	-2	16	13	85	59
115	146.8	100	130.6	96	120.1	DAVIESS	MO	-11	-8	-18	17	50
68	115.4	60	96.7	86	129.4	DE KALB	MO	-16	34	12	84	71
90	137.1	95	117.1	109	111.4	DENT	MO	-15	-5	-19	16	31
60	109.8	74	123.0	75	106.0	DOUGLAS	MO	12	-14	-3	54	21
224	128.4	261	122.7	294	122.4	DUNKLIN	MO	-4	0	-5	51	55
304	142.2	340	124.4	495	136.0	FRANKLIN	MO	-12	9	-4	52	83
100	125.0	118	124.1	113	105.7	GASCONADE	MO	-1	-15	-15	22	20
102	125.4	108	127.8	101	120.5	GENTRY	MO	2	-6	-4	53	51
847	130.8	986	118.7	1283	122.6	GREENE	MO	-9	3	-6	46	56
123	121.0	106	102.9	148	116.1	GRUNDY	MO	-15	13	-4	53	41
124	127.5	129	121.8	110	109.9	HARRISON	MO	-4	-10	-14	26	28
186	127.9	215	130.5	193	112.6	HENRY	MO	2	-14	-12	31	34
29	83.9	51	128.9	53	110.6	HICKORY	MO	53	-14	32	94	29
90	135.1	90	145.4	78	116.0	HOLT	MO	8	-20	-14	25	41
94	129.7	87	101.8	97	123.6	HOWARD	MO	-22	21	-5	51	59
162	132.4	177	120.8	217	116.9	HOWELL	MO	-9	-3	-12	32	43
76	138.9	77	126.3	87	114.9	IRON	MO	-9	-9	-17	19	38
4191	141.5	4488	131.5	4754	133.0	JACKSON	MO	-7	1	-6	47	77
714	140.2	731	126.8	848	137.2	JASPER	MO	-10	8	-2	58	84
297	137.6	424	126.4	620	122.1	JEFFERSON	MO	-8	-3	-11	33	55
205	149.5	188	118.0	185	108.6	JOHNSON	MO	-21	-8	-27	07	25
65	136.5	65	124.8	58	106.6	KNOX	MO	-9	-15	-22	11	22
118	117.6	125	100.0	192	126.9	LACLEDE	MO	-15	27	8	79	66
180	117.5	214	115.6	253	118.9	LAFAYETTE	MO	-2	3	1	66	48
202	127.8	210	113.2	278	127.5	LAWRENCE	MO	-11	13	0	63	68
95	127.7	84	112.8	88	119.8	LEWIS	MO	-12	6	-6	47	50
123	132.1	127	113.9	148	106.3	LINCOLN	MO	-14	-7	-20	15	21
183	133.2	212	143.8	197	123.2	LINN	MO	8	-14	-7	43	58
147	128.2	143	122.9	169	134.8	LIVINGSTON	MO	-4	10	5	74	80
96	126.0	95	112.1	124	125.8	MC DONALD	MO	-11	12	0	64	64
158	111.8	145	95.8	162	110.2	MACON	MO	-14	15	-1	60	28
78	146.8	82	121.4	118	141.6	MADISON	MO	-17	17	-4	54	90
54	143.9	41	92.9	35	71.9	MARIES	MO	-35	-23	-50	00	01
249	131.9	252	121.1	301	134.6	MARION	MO	-8	11	2	68	80
54	120.5	48	117.4	57	112.2	MERCER	MO	-3	-4	-7	45	33
93	114.2	104	102.5	160	134.6	MILLER	MO	-10	31	18	89	80
84	125.5	98	120.6	106	113.9	MISSISSIPPI	MO	-4	-6	-9	38	36
97	120.9	105	130.0	83	86.1	MONITEAU	MO	8	-34	-29	06	03
84	102.0	106	131.7	99	110.7	MONROE	MO	29	-16	8	79	29
126	147.2	127	131.2	127	134.0	MONTGOMERY	MO	-11	2	-9	39	79
74	117.0	98	129.4	114	112.3	MORGAN	MO	11	-13	-4	53	33
153	148.7	121	106.4	142	114.4	NEW MADRID	MO	-28	8	-23	10	37
197	123.4	197	102.6	285	117.6	NEWTON	MO	-17	15	-5	51	44
207	133.6	209	124.7	179	110.0	NODAWAY	MO	-7	-12	-18	18	28
67	114.5	74	106.5	101	118.8	OREGON	MO	-7	12	4	71	47
75	123.3	78	121.7	99	130.1	OSAGE	MO	-1	7	5	75	72
36	95.2	55	125.0	71	132.2	OZARK	MO	31	6	39	96	76
148	123.7	150	114.1	155	114.4	PEMISCOT	MO	-8	0	-8	43	37
82	109.4	129	141.9	136	121.2	PERRY	MO	30	-15	11	82	53
283	135.4	315	129.0	315	117.0	PETTIS	MO	-5	-9	-14	26	43
145	129.4	153	109.9	216	121.0	PHELPS	MO	-15	10	-7	46	52
144	137.5	154	135.4	175	143.2	PIKE	MO	-2	6	4	72	91
108	122.8	146	125.3	162	102.6	PLATTE	MO	2	-18	-16	20	16
106	97.1	122	105.0	137	96.0	POLK	MO	8	-9	-1	61	09
87	135.5	103	111.3	131	112.5	PULASKI	MO	-18	1	-17	19	34
88	149.7	76	121.4	96	145.9	PUTNAM	MO	-19	20	-3	57	93

* SIGNIFICANTLY MORE DEATHS THAN EXPECTED

WHITE FEMALE — ALL CANCERS COMBINED — ICD 140-208

1950-1959 DEATHS	1950-1959 RATE	1960-1969 DEATHS	1960-1969 RATE	1970-1979 DEATHS	1970-1979 RATE	COUNTY AND STATE	PERCENT CHANGE 60S/50S	PERCENT CHANGE 70S/60S	PERCENT CHANGE 70S/50S	PERCENTILE %CHANGE 70S/50S	PERCENTILE RATES 1970S
68	127.3	71	129.3	61	95.1	RALLS MO	2	-26	-25	08	08
213	126.1	241	132.8	203	100.6	RANDOLPH MO	5	-24	-20	14	13
124	118.8	140	124.4	143	109.0	RAY MO	5	-12	-8	41	26
36	121.8	35	101.2	46	109.1	REYNOLDS MO	-17	8	-10	35	26
62	113.8	52	79.2	81	93.7	RIPLEY MO	-30	18	-18	18	07
192	107.2	329	121.4	509	116.2	ST CHARLES MO	13	-4	8	79	41
77	114.5	75	105.8	93	118.4	ST CLAIR MO	-8	12	3	71	46
66	121.9	60	90.3	81	97.2	STE GENEVIEVE MO	-26	8	-20	14	11
278	131.6	331	133.5	408	139.8	ST FRANCOIS MO	1	5	6	76	88
3367	140.8	5120	136.6	7090*	140.2	ST LOUIS COUNTY MO	-3	3	0	63	88
218	128.5	254	141.5	252	124.1	SALINE MO	10	-12	-3	55	60
60	129.0	57	121.6	52	117.9	SCHUYLER MO	-6	-3	-9	40	45
77	144.1	64	118.6	56	99.6	SCOTLAND MO	-18	-16	-31	04	12
162	122.4	208	128.5	249	118.2	SCOTT MO	5	-8	-3	55	46
45	118.1	57	138.4	40	79.7	SHANNON MO	17	-42	-33	03	02
102	124.6	111	141.2	84	110.6	SHELBY MO	13	-22	-11	33	29
175	129.1	181	111.1	211	108.7	STODDARD MO	-14	-2	-16	21	25
49	109.2	52	86.4	94	94.7	STONE MO	-21	10	-13	27	08
101	133.2	124*	164.2	98	134.0	SULLIVAN MO	23	-18	1	65	79
65	120.0	108	131.9	176	125.3	TANEY MO	10	-5	4	72	63
107	108.1	125	105.7	151	103.4	TEXAS MO	-2	-2	-4	52	17
166	99.4	191	107.4	190	116.9	VERNON MO	8	9	18	89	43
66	120.4	78	124.1	98	118.7	WARREN MO	3	-4	-1	60	47
80	122.5	103	142.9	106	119.6	WASHINGTON MO	17	-16	-2	57	49
54	105.7	71	114.1	96	121.3	WAYNE MO	8	6	15	86	53
109	127.8	116	119.4	114	93.3	WEBSTER MO	-7	-22	-27	07	07
44	128.1	30	92.0	41	127.4	WORTH MO	-28	38	-1	62	67
112	123.8	119	120.0	145	123.5	WRIGHT MO	-3	3	0	63	58
6720*	160.9	5664*	145.6	4613*	141.5	ST LOUIS CITY MO	-10	-3	-12	30	89
3464	129.3	3985	121.4	4711	119.7	MONTANA MT	-6	-1	-7		
30	100.1	47	127.1	52	117.6	BEAVERHEAD MT	27	-7	17	89	44
23	87.4	35	124.8	38	121.8	BIG HORN MT	43	-2	39	96	54
36	127.8	30	94.2	27	88.0	BLAINE MT	-26	-7	-31	04	04
16	122.0	15	110.1	19	117.4	BROADWATER MT	-10	7	-4	54	44
61	118.1	70	126.6	72	109.1	CARBON MT	7	-14	-8	43	26
17	153.6	18	147.0	12	90.6	CARTER MT	-4	-38	-41	01	06
364	147.3	422	131.8	529	133.7	CASCADE MT	-10	1	-9	38	78
43	142.2	38	115.7	31	88.4	CHOUTEAU MT	-19	-24	-38	02	05
83	137.9	98	140.8	83	100.6	CUSTER MT	2	-29	-27	07	13
14	91.7	22	116.3	21	92.3	DANIELS MT	27	-21	1	65	07
65	159.0	62	124.3	62	106.3	DAWSON MT	-22	-14	-33	03	21
93	108.0	99	107.5	115	120.0	DEER LODGE MT	0	12	11	83	50
20	126.4	31	174.1	26	133.6	FALLON MT	38	-23	6	75	78
92	130.6	89	108.6	97	102.6	FERGUS MT	-17	-6	-21	12	16
202	139.3	223	120.7	308	124.7	FLATHEAD MT	-13	3	-10	35	61
131	128.5	133	107.8	202	126.7	GALLATIN MT	-16	18	-1	60	66
7	84.5	9	109.3	11*	288.0	GARFIELD MT	29	163	241	99	99
20	91.0	31	106.1	42	134.2	GLACIER MT	17	27	47	97	80
4	131.3	8	123.2	7	108.5	GOLDEN VALLEY MT	-6	-12	-17	18	25
18	142.5	20	133.3	15	96.3	GRANITE MT	-6	-28	-32	03	10
57	98.7	87	116.0	91	109.6	HILL MT	18	-6	11	83	27
23	125.6	24	103.4	35	117.9	JEFFERSON MT	-18	14	-6	47	45
11	78.1	10	66.7	16	100.3	JUDITH BASIN MT	-15	50	28	93	13
63	100.9	70	92.3	102	100.3	LAKE MT	-8	9	-1	62	13
165	127.0	231	141.3	267	131.2	LEWIS AND CLARK MT	11	-7	3	71	73
10	192.2	11	118.8	11	94.5	LIBERTY MT	-38	-20	-51	00	08
55	147.2	58	118.1	77	115.4	LINCOLN MT	-20	-2	-22	12	39
10	79.5	18	132.3	19	142.3	MC CONE MT	67	8	79	98	90
37	126.6	40	132.3	48	124.3	MADISON MT	5	-6	-2	59	60
12	139.3	14	131.1	11	100.9	MEAGHER MT	-6	-23	-28	07	14
7	93.8	10	91.1	21	139.1	MINERAL MT	-3	53	48	97	87
229	132.2	293	126.8	367	126.3	MISSOULA MT	-4	0	-4	52	65
34	123.4	36	127.0	41	135.3	MUSSELSHELL MT	3	7	10	81	82
77	121.6	92	128.0	89	103.1	PARK#MT	5	-19	-15	23	16
2	55.4	6	148.0	3	123.0	PETROLEUM MT	167	-17	122	99	57
36	134.6	38	132.4	40	122.4	PHILLIPS MT	-2	-8	-9	38	56
34	131.0	40	129.2	39	101.6	PONDERA MT	-1	-21	-22	11	15
8	95.4	17	155.4	7	54.7	POWDER RIVER MT	63	-65	-43	01	00
34	115.4	35	104.4	48	125.7	POWELL MT	-9	20	9	80	64
6	58.2	18	156.2	15	132.2	PRAIRIE MT	168	-15	127	99	76
87	126.3	82	90.7	140	111.1	RAVALLI MT	-28	23	-12	31	30
44	106.6	61	113.2	67	111.0	RICHLAND MT	6	-2	4	72	30
39	118.9	40	100.2	43	103.8	ROOSEVELT MT	-16	4	-13	29	17
26	121.4	32	131.7	33	123.8	ROSEBUD MT	9	-6	2	68	59
44	140.9	29	77.3	44	99.2	SANDERS MT	-45	28	-30	05	12
31	122.0	45	127.7	54	138.0	SHERIDAN MT	5	8	13	85	85
417*	162.8	339	127.0	434*	156.1	SILVER BOW MT	-22	23	-4	53	97
24	90.0	31	103.0	36	104.1	STILLWATER MT	14	1	16	87	17
22	125.4	29	140.6	24	116.0	SWEET GRASS MT	12	-18	-8	43	40
33	118.0	39	116.9	53	136.1	TETON MT	-1	16	15	87	83
18	68.3	36	112.5	37	114.3	TOOLE MT	65	2	67	98	37
5	120.3	6	107.9	4	71.8	TREASURE MT	-10	-33	-40	01	01
46	104.0	57	111.2	59	108.7	VALLEY MT	7	-2	4	73	25
18	120.6	26	160.9	19	108.3	WHEATLAND MT	33	-33	-10	35	24
9	135.0	12	146.2	6	67.8	WIBAUX MT	8	-54	-50	00	01
352	136.4	473	128.2	542	111.5	YELLOWSTONE MT	-6	-13	-18	17	31
9847	135.4	10874	126.7	12168	123.4	NEBRASKA NB	-6	-3	-9		
195	109.1	261	125.0	278	113.2	ADAMS NB	15	-9	4	71	35
68	105.3	78	107.1	77	111.5	ANTELOPE NB	2	4	6	76	31
2	71.3	2	58.9	2	51.3	ARTHUR NB	-17	-13	-28	06	00
3	59.1	3	73.9	5	148.7	BANNER NB	25	101	152	99	95
5	112.4	3	46.5	6	114.9	BLAINE NB	-59	147	2	68	38
71	128.8	85	142.8	93	149.7	BOONE NB	11	5	16	88	95
72	124.9	92	134.8	90	110.4	BOX BUTTE NB	8	-18	-12	32	29
47	195.1	36	125.0	39	123.9	BOYD NB	-36	-1	-37	02	59
32	113.5	57*	176.3	44	139.7	BROWN NB	55	-21	23	92	88
192	131.1	197	114.0	246	120.6	BUFFALO NB	-13	6	-8	41	51

* SIGNIFICANTLY MORE DEATHS THAN EXPECTED

\# COUNTIES COMBINED; SEE TECHNICAL NOTES

WHITE FEMALE — ALL CANCERS COMBINED — ICD 140-208

1950-1959 DEATHS	1950-1959 RATE	1960-1969 DEATHS	1960-1969 RATE	1970-1979 DEATHS	1970-1979 RATE	COUNTY AND STATE	PERCENT CHANGE 60S/50S	PERCENT CHANGE 70S/60S	PERCENT CHANGE 70S/50S	PERCENTILE %CHANGE 70S/50S	PERCENTILE RATES 1970S
85	130.6	100	128.8	101	129.1	BURT NB	-1	0	-1	61	70
110	159.9	127*	173.2	99	122.2	BUTLER NB	8	-29	-24	10	55
148	147.8	148	133.5	141	113.2	CASS NB	-10	-15	-23	10	35
77	114.9	115	148.1	97	117.3	CEDAR NB	29	-21	2	68	44
33	134.1	31	99.1	32	83.0	CHASE NB	-26	-16	-38	02	03
45	128.5	58	143.5	42	89.4	CHERRY NB	12	-38	-30	05	05
70	139.4	68	104.0	74	100.9	CHEYENNE NB	-25	-3	-28	07	14
64	96.5	80	115.3	75	111.9	CLAY NB	19	-3	16	87	32
89	140.5	121	151.4	112	124.8	COLFAX NB	8	-18	-11	33	62
91	136.7	81	100.1	129	143.8	CUMING NB	-27	44	5	74	92
144	137.0	136	120.3	159	119.8	CUSTER NB	-12	0	-13	29	49
63	135.0	84	140.8	120	157.4	DAKOTA NB	4	12	17	88	98
79	161.7	55	92.3	72	108.2	DAWES NB	-43	17	-33	03	24
125	128.8	150	119.9	167	112.3	DAWSON NB	-7	-6	-13	28	33
19	99.7	27	122.5	32	132.1	DEUEL NB	23	8	33	94	75
65	129.0	76	137.4	76	128.0	DIXON NB	7	-7	-1	62	68
203	124.2	271	130.6	331	135.8	DODGE NB	5	4	9	81	82
2244*	153.5	2438	135.9	2892*	139.4	DOUGLAS NB	-11	3	-9	38	87
21	101.9	26	105.8	30	107.5	DUNDY NB	4	2	5	75	23
88	132.8	84	113.8	90	125.0	FILLMORE NB	-14	10	-6	48	62
60	148.5	59	134.4	54	104.9	FRANKLIN NB	-10	-22	-29	05	19
30	112.7	35	125.7	34	98.5	FRONTIER NB	12	-22	-13	29	12
67	115.3	71	115.7	90	139.0	FURNAS NB	0	20	21	91	87
208	122.6	243	128.3	284	134.9	GAGE NB	5	5	10	81	80
19	101.6	19	90.1	29*	459.8	GARDEN NB	-11	410	352	99	99
17	105.3	27	155.3	32	143.0	GARFIELD NB	47	-8	36	95	91
12	116.6	15	111.4	10	54.4	GOSPER NB	-4	-51	-53	00	00
5	125.2	9	176.7	3	96.9	GRANT NB	41	-45	-23	11	10
29	124.1	27	106.7	24	87.4	GREELEY NB	-14	-18	-30	05	04
252	134.3	324	135.2	412	141.5	HALL NB	1	5	5	74	89
76	138.0	69	112.5	71	112.3	HAMILTON NB	-18	0	-19	16	33
42	115.3	39	98.5	43	100.8	HARLAN NB	-15	2	-13	29	14
18	226.6	12	134.2	10	278.1	HAYES NB	-41	107	23	92	99
35	124.9	32	106.3	41	143.7	HITCHCOCK NB	-15	35	15	86	91
85	114.1	95	112.2	111	108.7	HOLT NB	-2	-3	-5	51	25
8	127.0	9	145.6	9	120.9	HOOKER NB	15	-17	-5	51	52
54	143.0	77*	170.0	44	97.6	HOWARD NB	19	-43	-32	04	11
109	133.3	102	115.5	115	112.3	JEFFERSON NB	-13	-3	-16	21	33
60	133.7	80*	160.5	70	135.2	JOHNSON NB	20	-16	1	66	81
40	91.0	60	121.2	71	119.9	KEARNEY NB	33	-1	32	94	50
49	148.3	70	155.6	65	107.8	KEITH NB	5	-31	-27	07	24
8	92.9	10	122.0	7	342.5	KEYA PAHA NB	31	181	269	99	99
17	98.3	35	134.2	36	115.6	KIMBALL NB	37	-14	18	89	40
88	120.8	91	108.8	90	94.9	KNOX NB	-10	-13	-21	12	08
1005	135.0	1132	128.5	1190	118.7	LANCASTER NB	-5	-8	-12	31	47
159	124.2	185	115.8	226	114.6	LINCOLN NB	-7	-1	-8	42	37
6	112.8	7	129.9	5	126.9	LOGAN NB	15	-2	13	84	66
4	74.6	9	201.2	10*	980.4	LOUP NB	170	387	1215	99	99
6	143.7	2	53.7	1	490.8	MC PHERSON NB	-63	813	242	99	99
204	131.5	228	127.4	259	120.7	MADISON NB	-3	-5	-8	41	52
77	149.8	66	123.7	82	133.2	MERRICK NB	-17	8	-11	33	78
39	115.5	48	123.9	58	124.8	MORRILL NB	7	1	8	79	62
58	160.2	44	106.6	54	117.5	NANCE NB	-33	10	-27	07	44
93	144.5	75	111.7	93	117.2	NEMAHA NB	-23	5	-19	16	43
82	153.9	52	87.6	81	120.0	NUCKOLLS NB	-43	37	-22	11	50
178*	171.1	177	144.4	149	110.1	OTOE NB	-16	-24	-36	02	28
76	164.7	48	104.7	45	99.9	PAWNEE NB	-36	-5	-39	01	13
22	123.7	17	74.5	23	87.3	PERKINS NB	-40	17	-29	05	04
83	141.6	78	107.3	89	107.1	PHELPS NB	-24	0	-24	09	22
59	126.3	64	115.2	63	99.4	PIERCE NB	-9	-14	-21	12	12
136	132.5	144	106.8	190	114.6	PLATTE NB	-19	7	-13	27	38
57	113.7	65	112.0	59	97.4	POLK NB	-2	-13	-14	24	11
85	124.8	105	132.9	123	133.5	RED WILLOW NB	7	0	7	77	78
142	137.9	127	111.5	162	131.9	RICHARDSON NB	-19	18	-4	52	75
23	155.7	25	154.9	24	131.9	ROCK NB	-1	-15	-15	22	75
124	135.5	150	139.1	148	118.9	SALINE NB	3	-15	-12	30	47
65	97.9	105	91.3	213	118.1	SARPY NB	-7	29	21	90	45
149	151.0	149	134.7	173	128.2	SAUNDERS NB	-11	-5	-15	23	69
201	142.0	250	138.8	271	120.0	SCOTTS BLUFF NB	-2	-14	-16	22	50
104	131.7	103	118.9	101	95.2	SEWARD NB	-10	-20	-28	06	09
65	143.0	62	113.5	62	97.7	SHERIDAN NB	-21	-14	-32	04	11
45	146.0	47	140.8	44	125.4	SHERMAN NB	-4	-11	-14	25	63
9	76.4	9	98.5	7	67.3	SIOUX NB	29	-32	-12	31	00
34	119.1	37	115.5	48	124.8	STANTON NB	-3	8	5	73	61
83	125.7	95	135.8	91	120.1	THAYER NB	8	-12	-4	51	51
9	184.8	10	193.2	8	138.8	THOMAS NB	5	-28	-25	09	86
44	161.8	42	129.3	48	137.4	THURSTON NB	-20	6	-15	23	84
53	117.4	67	138.0	54	102.7	VALLEY NB	18	-26	-12	29	16
93	131.2	100	120.0	111	121.0	WASHINGTON NB	-9	1	-8	42	52
53	106.9	68	121.6	87	128.2	WAYNE NB	14	5	20	90	69
58	115.1	65	119.8	76	136.2	WEBSTER NB	4	14	18	89	83
7	146.5	5	82.7	9	147.2	WHEELER NB	-44	78	1	65	94
119	130.3	112	113.2	125	116.6	YORK NB	-13	3	-11	34	42
905	120.1	1746	125.3	3466	131.9	NEVADA NV	4	5	10		
34	118.7	44	115.2	81	132.7	CHURCHILL NV	-3	15	12	83	77
275	116.3	739	114.7	1798	131.9	CLARK NV	-1	15	13	85	75
15	147.9	24	114.9	56	108.3	DOUGLAS NV	-22	-6	-27	07	24
46	119.4	64	130.6	82	129.6	ELKO NV	9	-1	9	80	71
2	94.5	3	159.3	2	50.2	ESMERALDA NV	69	-68	-47	00	00
8	256.4	2	82.4	2	51.9	EUREKA NV	-68	-37	-80	00	00
19	105.3	30	129.6	34	109.3	HUMBOLDT NV	23	-16	4	72	26
6	87.5	7	82.0	9	86.9	LANDER NV	-6	6	-1	62	04
13	137.3	17	135.1	22	137.5	LINCOLN NV	-2	2	0	64	85
18	90.2	26	93.1	76	140.8	LYON NV	3	51	56	97	89
19	116.7	26	118.6	38	136.6	MINERAL NV	2	15	17	88	83
23	162.5	31	196.5	29	101.5	NYE NV	21	-48	-38	02	14
12	97.3	13	99.3	23	156.8	PERSHING NV	2	58	61	98	98
5	139.1	11*	315.5	8	132.0	STOREY NV	127	-58	-5	50	75
346	125.1	594	133.7	981	134.6	WASHOE NV	7	1	8	78	80

* SIGNIFICANTLY MORE DEATHS THAN EXPECTED

WHITE FEMALE — ALL CANCERS COMBINED — ICD 140-208

1950-1959 DEATHS	1950-1959 RATE	1960-1969 DEATHS	1960-1969 RATE	1970-1979 DEATHS	1970-1979 RATE	COUNTY AND STATE	PERCENT CHANGE 60S/50S	PERCENT CHANGE 70S/60S	PERCENT CHANGE 70S/50S	PERCENTILE %CHANGE 70S/50S	PERCENTILE RATES 1970S
36	108.1	48	134.2	72	164.0	WHITE PINE NV	24	22	52	97	98
28	124.1	67	143.1	153	132.6	ORMSBY-CARSON CITY NV	15	-7	7	77	77
5030*	152.6	5614*	143.1	6957*	139.6	NEW HAMPSHIRE NH	-6	-2	-8		
257	149.7	278	141.0	334	134.0	BELKNAP NH	-6	-5	-11	34	79
159	150.8	204*	160.6	257	143.7	CARROLL NH	7	-11	-5	51	91
351	151.2	398	144.2	488	138.1	CHESHIRE NH	-5	-4	-9	40	85
260	140.3	313	149.5	322	133.0	COOS NH	7	-11	-5	50	77
407	147.7	418	133.5	527	135.6	GRAFTON NH	-10	2	-8	41	82
1554*	159.3	1751*	149.9	2140*	146.1	HILLSBOROUGH NH	-6	-3	-8	41	94
602	133.8	681	132.3	840	138.1	MERRIMACK NH	-1	4	3	70	85
712*	156.3	814	137.1	1153	137.0	ROCKINGHAM NH	-12	0	-12	30	84
469*	161.8	515*	151.0	605	140.0	STRAFFORD NH	-7	-7	-13	27	88
259*	167.5	242	136.6	291	132.8	SULLIVAN NH	-18	-3	-21	13	77
42484*	162.1	51464*	151.9	60651*	147.9	NEW JERSEY NJ	-6	-3	-9		
1314*	160.8	1601*	151.4	1930*	152.4	ATLANTIC NJ	-6	1	-5	49	97
5175*	162.8	6809*	152.1	8319*	150.6	BERGEN NJ	-7	-1	-8	43	96
993	150.9	1419*	144.7	1998*	142.9	BURLINGTON NJ	-4	-1	-5	49	91
2568*	165.0	3107*	151.8	3503*	144.0	CAMDEN NJ	-8	-5	-13	29	92
451*	164.3	595	142.1	867	141.5	CAPE MAY NJ	-13	0	-14	26	89
725*	152.9	829*	144.3	923	134.0	CUMBERLAND NJ	-6	-7	-12	29	79
7568*	164.9	7723*	160.8	6830*	151.1	ESSEX NJ	-2	-6	-8	40	96
740*	158.8	945*	146.6	1181	138.8	GLOUCESTER NJ	-8	-5	-13	29	86
5536*	172.0	5524*	154.0	5694*	156.3	HUDSON NJ	-10	1	-9	38	97
447*	164.6	488	142.1	598	141.1	HUNTERDON NJ	-14	-1	-14	25	89
1878*	158.8	2112*	147.2	2425*	142.8	MERCER NJ	-7	-3	-10	36	91
2185*	163.6	3112*	152.8	4211*	152.3	MIDDLESEX NJ	-7	0	-7	44	96
2253*	159.1	3123*	153.8	4255*	156.0	MONMOUTH NJ	-3	1	-2	59	97
1517	144.4	2120*	140.6	2782*	141.8	MORRIS NJ	-3	1	-2	59	90
668	150.9	1383*	146.2	3145*	143.8	OCEAN NJ	-3	-2	-5	51	92
3028*	160.0	3654*	154.2	3946*	147.4	PASSAIC NJ	-4	-4	-8	42	94
340	158.3	399*	152.4	420	131.4	SALEM NJ	-4	-14	-17	19	74
772	147.6	1070	140.0	1423*	143.1	SOMERSET NJ	-5	2	-3	55	91
316	146.9	489*	150.2	698*	147.0	SUSSEX NJ	2	-2	0	64	94
3481*	166.5	4324*	155.8	4778*	148.1	UNION NJ	-6	-5	-11	33	95
329*	159.3	638*	153.6	725*	141.7	WARREN NJ	-4	-8	-11	33	90
2899	122.6	4007	118.6	5858	126.9	NEW MEXICO NM	-3	7	3		
762	129.8	1232	123.1	1985	133.6	BERNALILLO NM	-5	9	3	70	78
11	93.9	6	46.1	9	72.8	CATRON NM	-51	58	-22	11	01
154	111.6	198	102.2	318	120.4	CHAVES NM	-8	18	8	78	51
76	124.8	78	115.5	92	126.3	COLFAX NM	-7	9	1	66	65
112	116.6	134	99.3	202	111.9	CURRY NM	-15	13	-4	53	32
16	101.9	23	137.0	16	75.3	DE BACA NM	34	-45	-26	08	01
176	133.1	250	127.7	352	125.7	DONA ANA NM	-4	-2	-6	49	64
161	117.8	182	97.0	319	135.2	EDDY NM	-18	39	15	86	81
119	171.6	111	135.8	139	124.7	GRANT NM	-21	-8	-27	07	61
29	151.5	33	168.4	25	111.7	GUADALUPE NM	11	-34	-26	08	31
19	221.0	6	99.0	8	108.5	HARDING NM	-55	10	-51	00	25
20	108.5	21	111.1	31	117.8	HIDALGO NM	2	6	9	80	45
106	96.9	182	110.4	263	126.8	LEA NM	14	15	31	94	66
43	139.5	38	97.9	46	91.3	LINCOLN NM	-30	-7	-35	03	06
24	117.4	39	137.1	47	100.8	LOS ALAMOS NM	17	-26	-14	25	14
57	164.8	60	118.6	113	137.7	LUNA NM	-28	16	-16	20	85
34	90.9	53	98.9	74	123.4	MC KINLEY NM	9	25	36	95	58
34	130.9	33	139.6	33	161.6	MORA NM	7	16	24	92	98
67	120.4	114	120.8	141	106.6	OTERO NM	0	-12	-11	32	22
63	119.9	62	95.9	98	145.5	QUAY NM	-20	52	21	91	93
65	94.0	112	143.0	120	164.6	RIO ARRIBA NM	52	15	75	98	98
71	104.5	87	111.2	100	111.4	ROOSEVELT NM	6	0	7	77	31
28	113.4	36	100.1	90	110.8	SANDOVAL NM	-12	11	-2	58	29
51	84.0	102	90.5	181	108.5	SAN JUAN NM	8	20	29	93	25
117	128.4	128	125.1	147	142.1	SAN MIGUEL NM	-3	14	11	82	90
195	140.0	287	137.5	404	142.0	SANTA FE NM	-2	3	1	67	90
39	94.3	75	138.5	77	103.2	SIERRA NM	47	-25	10	81	16
40	122.0	48	131.0	66	149.4	SOCORRO NM	7	14	22	91	95
56	105.3	81	121.8	105*	164.9	TAOS NM	16	35	57	97	98
30	124.2	31	117.4	33	101.2	TORRANCE NM	-5	-14	-18	17	14
47	156.6	41	119.7	33	85.3	UNION NM	-24	-29	-46	01	03
77	123.7	124	132.7	191	126.4	VALENCIA NM	7	-5	2	68	65
130034*	161.3	143170*	150.4	154483*	149.1	NEW YORK NY	-7	-1	-8		
2285*	157.4	2483*	145.4	2803*	145.9	ALBANY NY	-8	0	-7	43	93
329	135.9	324	124.6	395	132.8	ALLEGANY NY	-8	7	-2	58	77
13344*	182.1	10465	130.9	10990*	156.3	BRONX NY	-28	19	-14	25	97
1491	142.3	1603	129.5	1949	137.5	BROOME NY	-9	6	-3	55	85
631	149.0	639	134.9	747*	143.0	CATTARAUGUS NY	-9	6	-4	53	91
586	134.8	668*	146.3	681	133.6	CAYUGA NY	9	-9	-1	61	78
1171	138.6	1232	132.8	1333	130.7	CHAUTAUQUA NY	-4	-2	-6	48	73
738	143.9	834*	143.2	932*	143.1	CHEMUNG NY	0	0	-1	62	91
357	145.0	338	124.3	411	138.1	CHENANGO NY	-14	11	-5	51	85
356	141.1	412	139.6	472	135.9	CLINTON NY	-1	-3	-4	54	82
400	136.5	483	144.5	589*	147.6	COLUMBIA NY	6	2	8	79	94
318	145.8	331	136.2	343	128.9	CORTLAND NY	-7	-5	-12	32	70
366	143.8	364	131.0	449	139.4	DELAWARE NY	-9	6	-3	55	87
1067	116.4	1429	129.4	1750	136.4	DUTCHESS NY	11	5	17	88	83
7438*	156.4	8154*	142.8	9341*	146.5	ERIE NY	-9	3	-6	46	94
301	155.3	325*	152.4	377*	153.9	ESSEX NY	-2	1	-1	61	97
309	132.9	370	150.6	415*	152.2	FRANKLIN NY	13	1	15	86	96
531*	153.9	497	135.4	603*	153.3	FULTON NY	-12	13	0	63	97
380	139.6	403	131.9	465	132.4	GENESEE NY	-5	0	-5	50	76
321*	164.6	339	144.9	417*	148.1	GREENE NY	-12	2	-10	36	95
45	195.2	36	124.9	53	144.4	HAMILTON NY	-36	16	-26	08	92
581	152.9	597	139.0	655	137.3	HERKIMER NY	-9	-1	-10	35	84
750	141.1	755	131.0	868*	143.1	JEFFERSON NY	-7	9	1	67	91
22463*	176.0	17614	129.1	19095*	150.7	KINGS(BROOKLYN) NY	-27	17	-14	24	96

* SIGNIFICANTLY MORE DEATHS THAN EXPECTED

WHITE FEMALE — ALL CANCERS COMBINED — ICD 140-208

1950-1959 DEATHS	1950-1959 RATE	1960-1969 DEATHS	1960-1969 RATE	1970-1979 DEATHS	1970-1979 RATE	COUNTY AND STATE		PERCENT CHANGE 60S/50S	PERCENT CHANGE 70S/60S	PERCENT CHANGE 70S/50S	PERCENTILE %CHANGE 70S/50S	PERCENTILE RATES 1970S
165	133.1	177	138.6	168	119.6	LEWIS	NY	4	-14	-10	35	49
298	128.2	351	134.3	398	131.8	LIVINGSTON	NY	5	-2	3	70	75
386	144.9	372	124.0	533*	157.3	MADISON	NY	-14	27	9	80	98
4423*	146.3	5012*	138.9	5728*	139.7	MONROE	NY	-5	1	-4	51	88
568	150.7	584	139.1	603	134.5	MONTGOMERY	NY	-8	-3	-11	34	80
6540*	166.5	9815*	158.7	12488*	160.8	NASSAU	NY	-5	1	-3	54	98
17840*	189.7	26288*	281.5	13700*	162.7	NEW YORK(MANHATTAN)	NY	48	-42	-14	25	98
1364	141.9	1642*	140.9	1990*	145.8	NIAGARA	NY	-1	4	3	70	93
1794	133.9	1997	128.5	2320	134.2	ONEIDA	NY	-4	4	0	64	80
2919*	151.0	3111	133.7	3865*	144.5	ONONDAGA	NY	-11	8	-4	52	92
496	137.8	527	125.5	678	136.2	ONTARIO	NY	-9	9	-1	61	83
1418	145.6	1705*	148.9	2064*	154.2	ORANGE	NY	2	4	6	75	97
250	133.6	274	132.9	279	125.4	ORLEANS	NY	-1	-6	-6	47	63
608	138.1	669	135.8	826*	146.6	OSWEGO	NY	-2	8	6	76	94
472	142.1	517*	145.2	529	136.1	OTSEGO	NY	2	-6	-4	52	83
249*	175.7	338*	155.8	478*	145.6	PUTNAM	NY	-11	-7	-17	19	93
13778*	160.2	14363	123.7	19285*	149.8	QUEENS	NY	-23	21	-6	46	95
1296*	157.8	1393*	152.4	1462*	147.6	RENSSELAER	NY	-3	-3	-6	46	94
1559*	160.8	1561	118.9	2520*	149.2	RICHMOND	NY	-26	25	-7	44	95
731	131.8	1069	131.8	1576*	139.6	ROCKLAND	NY	0	6	6	76	87
724	137.0	772	134.5	892	141.4	ST LAWRENCE	NY	-2	5	3	70	89
661	150.3	725	133.9	1002*	144.2	SARATOGA	NY	-11	8	-4	53	92
1284*	152.1	1402*	141.3	1585*	144.2	SCHENECTADY	NY	-7	2	-5	50	92
185	132.6	183	124.6	271*	156.6	SCHOHARIE	NY	-6	26	18	89	97
125	154.2	102	111.2	133	123.0	SCHUYLER	NY	-28	11	-20	14	57
214	102.7	260	120.7	256	108.6	SENECA	NY	17	-10	6	75	25
742	144.7	771	134.8	833	134.8	STEUBEN	NY	-7	0	-7	45	80
2830	126.8	5167	132.8	8727*	155.8	SUFFOLK	NY	5	17	23	92	97
359	146.5	456*	155.2	554*	145.1	SULLIVAN	NY	6	-7	-1	61	93
271	147.9	282	138.9	355*	147.0	TIOGA	NY	-6	6	-1	62	94
390	135.6	415	128.8	477	130.3	TOMPKINS	NY	-5	1	-4	53	72
994*	157.2	1130*	147.8	1369*	145.1	ULSTER	NY	-6	-2	-8	43	93
396	152.8	443*	149.4	516*	150.0	WARREN	NY	-2	0	-2	59	96
413	149.8	450*	150.5	472	139.9	WASHINGTON	NY	0	-7	-7	45	88
505	137.5	502	121.9	568	122.4	WAYNE	NY	-11	0	-11	33	56
5790*	155.3	7239*	152.2	8346*	151.5	WESTCHESTER	NY	-2	-1	-2	57	96
265	139.7	239	117.0	301	135.1	WYOMING	NY	-16	15	-3	55	81
174	147.6	172	133.4	203	138.2	YATES	NY	-10	4	-6	46	86
14698	114.9	19146	110.1	26643	112.5	NORTH CAROLINA	NC	-4	2	-2		
294	113.0	389	108.0	622	126.7	ALAMANCE	NC	-4	17	12	84	66
61	104.5	56	72.6	109	102.3	ALEXANDER	NC	-31	41	-2	58	15
48	118.7	63	128.3	70	105.4	ALLEGHANY	NC	8	-18	-11	33	19
92	140.6	98	120.3	97	94.3	ANSON	NC	-14	-22	-33	03	08
87	92.6	104	96.3	125	95.5	ASHE	NC	4	-1	3	70	09
28	52.1	61	104.0	92	124.5	AVERY	NC	100	20	139	99	60
135	121.2	172	126.3	218	124.8	BEAUFORT	NC	4	-1	3	70	61
58	108.8	59	96.1	90	129.2	BERTIE	NC	-12	34	19	90	70
64	100.2	75	96.3	108	107.4	BLADEN	NC	-4	12	7	78	23
46	106.9	68	105.5	127	108.3	BRUNSWICK	NC	-1	3	1	67	24
711	127.1	859	116.7	1153	118.9	BUNCOMBE	NC	-8	2	-6	46	47
168	100.4	236	92.7	322	96.1	BURKE	NC	-8	4	-4	52	09
253	114.1	336	109.8	451	105.6	CABARRUS	NC	-4	-4	-7	43	20
178	120.6	200	96.3	293	102.2	CALDWELL	NC	-20	6	-15	22	15
17	103.6	21	112.4	32	137.7	CAMDEN	NC	9	23	33	95	85
97	100.9	164	124.4	235	122.7	CARTERET	NC	23	-1	22	91	56
48	110.2	52	98.9	64	95.8	CASWELL	NC	-10	-3	-13	28	09
239	109.5	313	93.7	558	118.2	CATAWBA	NC	-14	26	8	79	46
106	128.8	119	112.9	183	131.2	CHATHAM	NC	-12	16	2	68	74
95	127.9	99	111.9	116	102.0	CHEROKEE	NC	-13	-9	-20	14	15
41	130.3	37	104.2	64	134.1	CHOWAN	NC	-20	29	3	70	79
23	92.2	37	114.7	30	71.8	CLAY	NC	24	-37	-22	11	01
260	121.0	325	117.9	410	111.3	CLEVELAND	NC	-3	-6	-8	41	30
111	90.4	137	93.5	198	103.7	COLUMBUS	NC	3	11	15	86	17
146	134.7	182	120.2	230	114.9	CRAVEN	NC	-11	-4	-15	24	38
217	106.5	363	112.3	538	116.1	CUMBERLAND	NC	5	3	9	80	41
32	143.5	30	107.5	51	110.1	CURRITUCK	NC	-25	2	-23	10	28
28	101.6	48	126.2	87	137.6	DARE	NC	24	9	35	95	85
250	107.6	381	109.2	509	102.7	DAVIDSON	NC	2	-6	-5	51	16
84	129.3	83	100.0	106	95.1	DAVIE	NC	-23	-5	-26	08	08
97	96.1	113	90.5	180	113.1	DUPLIN	NC	-6	25	18	89	35
384	121.6	515	124.2	647	117.8	DURHAM	NC	2	-5	-3	55	45
128	117.8	176	128.8	195	114.4	EDGECOMBE	NC	9	-11	-3	56	37
599	123.0	800	112.9	1148	115.4	FORSYTH	NC	-8	2	-6	47	39
81	113.3	95	102.9	130	104.9	FRANKLIN	NC	-9	2	-7	43	19
416	108.7	579	106.2	840	112.2	GASTON	NC	-2	6	3	71	33
30	117.9	24	80.8	37	108.5	GATES	NC	-31	34	-8	42	25
19	93.2	28	99.7	39	103.0	GRAHAM	NC	7	3	11	82	16
77	80.1	94	86.8	104	86.0	GRANVILLE	NC	8	-1	7	78	03
30	91.2	44	125.4	57	118.0	GREENE	NC	37	-6	29	93	45
847	122.0	1156	117.1	1586	118.0	GUILFORD	NC	-4	1	-3	55	45
123	104.0	192	123.8	230	115.5	HALIFAX	NC	19	-7	11	83	39
149	115.9	197	115.7	266	113.0	HARNETT	NC	0	-2	-3	57	35
167	109.7	225	109.4	271	95.2	HAYWOOD	NC	0	-13	-13	27	08
215	129.1	267	113.8	398	106.7	HENDERSON	NC	-12	-6	-17	19	22
50	124.7	48	82.7	88	122.0	HERTFORD	NC	-34	47	-2	58	54
19	72.6	28	88.7	38	87.9	HOKE	NC	22	-1	21	91	04
30	141.9	28	109.7	25	98.3	HYDE	NC	-23	-10	-31	04	11
238	110.1	314	107.6	414	104.6	IREDELL	NC	-2	-3	-5	50	18
86	121.1	92	111.3	124	108.4	JACKSON	NC	-8	-3	-10	35	25
213	117.7	228	94.0	370	117.9	JOHNSTON	NC	-20	25	0	64	45
27	134.4	20	83.1	37	109.7	JONES	NC	-38	32	-18	17	27
74	103.5	116	110.9	154	103.8	LEE	NC	7	-6	0	65	17
164	150.5	198	131.8	240	115.1	LENOIR	NC	-12	-13	-24	10	39
106	103.2	164	124.9	210	114.2	LINCOLN	NC	21	-9	11	82	37
117	123.5	129	94.9	212	117.2	MC DOWELL	NC	-23	23	-5	50	43
67	97.0	98	105.3	131	96.0	MACON	NC	9	-9	-1	61	09
84	108.4	95	103.2	95	85.4	MADISON	NC	-5	-17	-21	13	03
49	96.0	75	110.6	99	109.4	MARTIN	NC	15	-1	14	86	26
812	113.3	1212	114.2	1714	116.8	MECKLENBURG	NC	1	2	3	70	43

* SIGNIFICANTLY MORE DEATHS THAN EXPECTED

WHITE FEMALE — ALL CANCERS COMBINED — ICD 140-208

1950-1959 DEATHS	1950-1959 RATE	1960-1969 DEATHS	1960-1969 RATE	1970-1979 DEATHS	1970-1979 RATE	COUNTY AND STATE		PERCENT CHANGE 60S/50S	PERCENT CHANGE 70S/60S	PERCENT CHANGE 70S/50S	PERCENTILE %CHANGE 70S/50S	PERCENTILE RATES 1970S
81	133.4	69	92.1	120	125.9	MITCHELL	NC	-31	37	-6	49	64
69	112.4	88	111.0	129	122.8	MONTGOMERY	NC	-1	11	9	80	57
142	120.1	160	102.2	263	111.7	MOORE	NC	-15	9	-7	44	31
164	117.3	224	117.2	301	117.9	NASH	NC	0	1	1	65	45
287	134.8	351	121.3	512	127.5	NEW HANOVER	NC	-10	5	-5	49	68
55	98.3	75	106.9	97	129.5	NORTHAMPTON	NC	9	21	32	94	71
91	101.7	133	110.1	229	120.0	ONSLOW	NC	8	9	18	89	50
120	121.6	123	82.0	241	107.3	ORANGE	NC	-33	31	-12	31	23
27	87.2	29	71.5	49	103.0	PAMLICO	NC	-18	44	18	89	16
86	117.2	99	109.5	156	135.6	PASQUOTANK	NC	-7	24	16	87	82
41	96.4	52	96.2	96	130.7	PENDER	NC	0	36	36	95	73
24	90.0	30	87.4	54	133.6	PERQUIMANS	NC	-3	53	48	97	78
80	127.3	101	119.4	135	116.0	PERSON	NC	-6	-3	-9	39	40
151	110.1	232	123.2	305	118.8	PITT	NC	12	-4	8	79	47
65	115.8	91	118.5	112	109.0	POLK	NC	2	-8	-6	48	26
256	130.0	303	105.6	422	102.3	RANDOLPH	NC	-19	-3	-21	12	15
127	118.1	181	128.6	196	103.2	RICHMOND	NC	9	-20	-13	29	16
185	119.8	194	105.0	249	105.7	ROBESON	NC	-12	1	-12	32	20
238	110.8	331	114.6	440	111.9	ROCKINGHAM	NC	3	-2	1	66	32
338	123.0	424	108.6	554	104.1	ROWAN	NC	-12	-4	-15	22	18
210	121.6	263	113.6	359	115.4	RUTHERFORD	NC	-7	2	-5	50	39
144	114.8	191	126.4	185	95.4	SAMPSON	NC	10	-25	-17	19	09
68	128.2	70	101.5	126	134.7	SCOTLAND	NC	-21	33	5	74	80
134	93.9	201	107.8	321	126.8	STANLY	NC	15	18	35	95	66
89	112.3	94	91.2	126	89.2	STOKES	NC	-19	-2	-21	13	05
213	120.2	270	113.5	352	110.1	SURRY	NC	-6	-3	-8	40	28
37	111.3	39	96.7	54	111.2	SWAIN	NC	-13	15	0	64	30
86	159.5	84	114.6	128	111.5	TRANSYLVANIA	NC	-28	-3	-30	05	31
12	84.8	18	106.8	25	151.0	TYRRELL	NC	26	41	78	98	96
164	123.6	206	110.0	252	96.7	UNION	NC	-11	-12	-22	12	10
109	133.4	122	115.4	144	105.3	VANCE	NC	-14	-9	-21	13	19
524	121.3	735	113.0	1114	117.6	WAKE	NC	-7	4	-3	56	44
46	113.7	55	114.4	66	127.0	WARREN	NC	1	11	12	83	67
38	118.1	51	132.9	62	132.2	WASHINGTON	NC	13	0	12	84	76
68	96.5	84	90.1	134	103.5	WATAUGA	NC	-7	15	7	78	17
168	104.3	245	112.0	324	110.3	WAYNE	NC	7	-2	6	75	29
144	85.1	185	86.1	285	100.2	WILKES	NC	1	16	18	89	13
132	98.3	208	119.6	303	133.9	WILSON	NC	22	12	36	95	79
89	98.1	140	122.8	167	109.3	YADKIN	NC	25	-11	11	83	26
61	101.0	78	108.1	89	100.0	YANCEY	NC	7	-8	-1	61	13
3330	130.1	3749	122.3	4313	118.9	NORTH DAKOTA	ND	-6	-3	-9		
20	123.8	25	114.9	35	131.7	ADAMS	ND	-7	15	6	76	74
121	149.1	111	118.1	106	95.3	BARNES	ND	-21	-19	-36	02	09
34	84.7	53	124.5	58	131.8	BENSON	ND	47	6	56	97	75
6	131.5	8	230.4	6	617.5	BILLINGS	ND	75	168	370	99	99
71	138.9	70	120.4	84	128.9	BOTTINEAU	ND	-13	7	-7	44	70
33	190.9	19	85.9	27	111.8	BOWMAN	ND	-55	30	-41	01	32
48	162.0	45	146.6	40	136.7	BURKE	ND	-9	-7	-16	22	84
143	126.3	227	134.9	276	122.6	BURLEIGH	ND	7	-9	-3	56	56
394	142.4	435	126.3	553	129.6	CASS	ND	-11	3	-9	39	71
62	137.1	53	112.3	73	133.3	CAVALIER	ND	-18	19	-3	56	78
44	114.3	42	91.2	74	134.6	DICKEY	ND	-20	48	18	89	80
29	136.8	24	91.2	41	135.9	DIVIDE	ND	-33	49	-1	62	82
23	111.9	26	123.7	24	111.3	DUNN	ND	10	-10	-1	62	30
27	121.2	28	106.7	37	117.9	EDDY	ND	-12	10	-3	56	45
43	160.2	42	122.3	54	130.4	EMMONS	ND	-24	7	-19	16	72
20	100.9	40	141.5	40	134.9	FOSTER	ND	40	-5	34	95	81
27	177.3	22	142.8	15	76.7	GOLDEN VALLEY	ND	-19	-46	-57	00	01
225	125.0	267	117.8	303	115.3	GRAND FORKS	ND	-6	-2	-8	42	39
16	80.6	29	124.4	38	134.7	GRANT	ND	54	8	67	98	80
33	135.1	29	104.9	30	92.7	GRIGGS	ND	-22	-12	-31	04	07
28	132.2	28	110.0	26	91.6	HETTINGER	ND	-17	-17	-31	04	06
21	109.7	38	178.7	27	114.5	KIDDER	ND	63	-36	4	72	37
42	110.2	69	160.0	53	99.0	LA MOURE	ND	45	-38	-10	35	12
32	157.6	39	176.4	27*	488.3	LOGAN	ND	12	177	210	99	99
63	125.7	77	143.7	71	122.1	MC HENRY	ND	14	-15	-3	56	55
42	151.1	39	112.7	54	129.3	MC INTOSH	ND	-25	15	-14	24	71
36	143.5	28	102.7	53	159.3	MC KENZIE	ND	-28	55	11	82	98
66	118.0	78	126.5	80	110.4	MC LEAN	ND	7	-13	-6	46	29
37	138.3	36	115.4	49	119.5	MERCER	ND	-17	4	-14	26	49
91	113.8	115	121.4	133	109.0	MORTON	ND	7	-10	-4	52	26
53	133.1	61	138.2	77	156.1	MOUNTRAIL	ND	4	13	17	88	97
41	107.8	53	121.5	54	109.7	NELSON	ND	13	-10	2	68	27
14	153.4	11	170.8	8	95.8	OLIVER	ND	11	-44	-38	02	09
72	119.4	78	111.8	75	99.7	PEMBINA	ND	-6	-11	-17	20	13
49	154.3	35	94.6	51	120.7	PIERCE	ND	-39	28	-22	12	52
74	123.5	84	113.2	110	123.1	RAMSEY	ND	-8	9	0	63	57
42	88.0	75	142.1	70	135.1	RANSOM	ND	61	-5	53	97	81
32	151.6	34	170.3	36	158.5	RENVILLE	ND	12	-7	5	73	98
140	153.6	142	131.5	145	116.9	RICHLAND	ND	-14	-11	-24	09	43
29	118.5	21	76.3	46	136.4	ROLETTE	ND	-36	79	15	86	83
38	121.7	47	138.5	41	116.3	SARGENT	ND	14	-16	-4	52	41
21	124.5	21	127.2	18	97.6	SHERIDAN	ND	2	-23	-22	12	11
1	7.0	7	128.0	5	87.6	SIOUX	ND	1738	-32	1158	99	04
10	145.8	9	171.8	6	445.5	SLOPE	ND	18	159	205	99	99
94	151.8	83	100.6	136	126.8	STARK	ND	-34	26	-17	20	66
25	137.4	21	93.6	31	128.4	STEELE	ND	-32	37	-7	46	69
114	96.2	141	105.7	152	98.6	STUTSMAN	ND	10	-7	2	69	12
41	157.6	42	148.6	42	130.0	TOWNER	ND	-6	-13	-18	18	72
64	120.3	83	118.1	91	109.4	TRAILL	ND	-2	-7	-9	39	26
119	141.4	122	123.8	145	131.4	WALSH	ND	-12	6	-7	44	74
224	145.0	250	122.5	308	123.1	WARD	ND	-15	0	-15	23	57
57	131.7	62	135.5	59	105.7	WELLS	ND	3	-22	-20	15	20
99	126.1	125	128.4	120	107.5	WILLIAMS	ND	2	-16	-15	23	23
61375*	148.2	68965*	139.2	78375*	137.2	OHIO	OH	-6	-1	-7		
149	131.3	174	146.0	188	140.8	ADAMS	OH	11	-4	7	78	89
658	140.8	762	139.6	859	136.0	ALLEN	OH	-1	-3	-3	54	82

* SIGNIFICANTLY MORE DEATHS THAN EXPECTED

WHITE FEMALE — ALL CANCERS COMBINED — ICD 140-208

1950-1959 DEATHS	1950-1959 RATE	1960-1969 DEATHS	1960-1969 RATE	1970-1979 DEATHS	1970-1979 RATE	COUNTY AND STATE	PERCENT CHANGE 60S/50S	PERCENT CHANGE 70S/60S	PERCENT CHANGE 70S/50S	PERCENTILE %CHANGE 70S/50S	PERCENTILE RATES 1970S
259	128.2	286	122.5	325	120.5	ASHLAND OH	-4	-2	-6	47	51
703*	156.8	707	137.5	743	125.4	ASHTABULA OH	-12	-9	-20	14	63
334	131.2	328	125.1	383	144.2	ATHENS OH	-5	15	10	81	92
236	129.0	270	127.4	285	114.7	AUGLAIZE OH	-1	-10	-11	33	38
633	130.9	742	139.8	799	141.6	BELMONT OH	7	1	8	79	90
170	124.0	177	119.7	217	129.6	BROWN OH	-3	8	4	73	71
1020	144.6	1213	133.5	1549	134.7	BUTLER OH	-8	1	-7	45	80
147	145.3	138	120.4	164	123.6	CARROLL OH	-17	3	-15	23	59
213	144.6	223	135.4	252	132.7	CHAMPAIGN OH	-6	-2	-8	41	77
868	149.7	939	131.0	1092	132.0	CLARK OH	-12	1	-12	31	75
328	135.3	450	133.4	641	138.5	CLERMONT OH	-1	4	2	69	86
205	129.1	228	132.8	259	133.0	CLINTON OH	3	0	3	70	77
823	152.4	861	140.2	952	139.4	COLUMBIANA OH	-8	-1	-9	40	87
288*	162.4	266	135.7	281	124.3	COSHOCTON OH	-16	-8	-24	10	60
359	152.7	357	131.5	402	130.6	CRAWFORD OH	-14	-1	-14	24	73
10800*	157.6	12361*	151.7	12725*	143.3	CUYAHOGA OH	-4	-6	-9	38	91
353	140.0	328	115.8	432	127.9	DARKE OH	-17	10	-9	39	68
221	150.7	231	132.2	264	132.6	DEFIANCE OH	-12	0	-12	30	77
253	148.0	244	124.2	295	123.5	DELAWARE OH	-16	-1	-17	20	58
459	155.9	497	142.9	574	141.1	ERIE OH	-8	-1	-9	37	89
408	135.3	497	134.4	638	136.4	FAIRFIELD OH	-1	1	1	66	83
175	131.5	212	140.3	226	133.9	FAYETTE OH	7	-5	2	68	79
3992*	155.6	4471*	140.4	5361*	141.8	FRANKLIN OH	-10	1	-9	39	90
184	124.6	190	114.7	231	116.7	FULTON OH	-8	2	-6	46	42
159	123.6	153	108.8	199	126.2	GALLIA OH	-12	16	2	68	65
220	153.4	300	146.5	360	124.1	GEAUGA OH	-4	-15	-19	16	60
337	135.5	419	119.1	617	130.0	GREENE OH	-12	9	-4	53	72
335	129.0	320	121.2	349	126.2	GUERNSEY OH	-6	4	-2	58	65
6315*	162.1	6776*	150.5	7581*	155.3	HAMILTON OH	-7	3	-4	52	97
428	152.1	420	128.4	454	120.2	HANCOCK OH	-16	-6	-21	13	51
231	140.9	248	136.0	255	132.2	HARDIN OH	-3	-3	-6	47	76
135	130.5	135	125.1	143	119.0	HARRISON OH	-4	-5	-9	39	48
194	150.8	189	126.5	210	124.0	HENRY OH	-16	-2	-18	18	59
206	115.2	240	126.0	286	133.9	HIGHLAND OH	9	6	16	88	79
166	154.4	180*	155.1	200	150.0	HOCKING OH	0	-3	-3	56	96
119	127.0	147	133.4	146	115.0	HOLMES OH	5	-14	-10	37	38
309	130.6	347	134.8	366	127.2	HURON OH	3	-6	-3	57	67
225	150.0	254	152.9	236	125.6	JACKSON OH	2	-18	-16	20	64
631	151.1	720*	146.1	865*	152.2	JEFFERSON OH	-3	4	1	66	96
302	139.2	306	126.1	346	125.7	KNOX OH	-9	0	-10	37	64
619	149.9	933*	145.6	1292*	142.1	LAKE OH	-3	-2	-5	50	90
326	142.3	411*	150.4	462	139.3	LAWRENCE OH	6	-7	-2	58	87
614	141.2	734*	142.0	866	140.5	LICKING OH	1	-1	0	63	88
259	131.7	297	135.4	360	146.1	LOGAN OH	3	8	11	82	94
1053	145.0	1345*	139.8	1583	132.8	LORAIN OH	-4	-5	-8	40	77
3099*	148.8	3300	134.3	3895*	145.2	LUCAS OH	-10	8	-2	57	93
148	136.4	165	132.3	198	132.3	MADISON OH	-3	0	-3	55	76
1845*	151.1	2128*	144.1	2252	132.5	MAHONING OH	-5	-8	-12	30	76
422	146.3	423	127.0	492	135.7	MARION OH	-13	7	-7	43	82
343	145.2	437	137.3	564	133.9	MEDINA OH	-5	-3	-8	42	79
198	153.9	174	132.2	200	138.7	MEIGS OH	-14	5	-10	36	86
206	138.4	196	110.2	283	135.9	MERCER OH	-20	23	-2	59	82
507	145.8	539	132.1	644	132.7	MIAMI OH	-9	0	-9	39	77
129	131.2	138	142.7	141	132.2	MONROE OH	9	-7	1	66	76
2784*	148.5	3187	134.4	3707	133.2	MONTGOMERY OH	-9	-1	-10	35	78
85	105.0	117	137.0	84	93.9	MORGAN OH	30	-31	-11	34	07
133	135.3	142	133.0	141	113.7	MORROW OH	-2	-15	-16	21	36
601	144.7	612	136.5	717*	143.8	MUSKINGUM OH	-6	5	-1	62	92
92	126.3	95	132.3	93	122.1	NOBLE OH	5	-8	-3	55	55
230	143.9	253	132.0	344*	148.9	OTTAWA OH	-8	13	3	71	95
109	138.7	131	144.5	113	111.8	PAULDING OH	4	-23	-19	15	32
213	134.0	247	147.2	249	138.6	PERRY OH	10	-6	3	71	86
219	143.9	214	125.7	274	139.9	PICKAWAY OH	-13	11	-3	56	88
103	148.9	135	147.9	159	132.1	PIKE OH	-1	-11	-11	32	75
427	138.0	523	131.1	658	128.3	PORTAGE OH	-5	-2	-7	44	69
220	142.2	192	109.6	263	126.0	PREBLE OH	-23	15	-11	32	65
192	141.4	210	140.1	220	132.7	PUTNAM OH	-1	-5	-6	47	77
665	144.7	742	132.0	888	132.9	RICHLAND OH	-9	1	-8	41	77
387	148.9	421	144.2	471	141.3	ROSS OH	-3	-2	-5	50	89
394	144.6	403	126.0	428	122.8	SANDUSKY OH	-13	-3	-15	23	57
660*	162.9	691*	147.8	784*	148.8	SCIOTO OH	-9	1	-9	40	95
415	137.9	417	127.2	455	126.5	SENECA OH	-8	-1	-8	41	66
211	133.7	234	128.9	302	139.0	SHELBY OH	-4	8	4	72	87
2014	136.3	2361	129.9	2782	130.5	STARK OH	-5	0	-4	52	73
2927*	148.5	3552*	142.3	3934	134.0	SUMMIT OH	-4	-6	-10	37	79
1036	139.7	1320	136.6	1624	132.4	TRUMBULL OH	-2	-3	-5	50	76
570	144.5	595	133.3	656	127.1	TUSCARAWAS OH	-8	-5	-12	31	67
183	144.6	199	141.5	186	117.8	UNION OH	-2	-17	-19	16	45
229	135.4	287*	150.4	259	125.9	VAN WERT OH	11	-16	-7	44	64
76	142.2	71	131.2	66	100.7	VINTON OH	-8	-23	-29	06	14
281	136.7	349	126.2	448	118.5	WARREN OH	-8	-6	-13	27	47
365	128.8	387	122.6	480	135.8	WASHINGTON OH	-5	11	5	74	82
469	144.0	471	117.5	573	121.0	WAYNE OH	-18	3	-16	21	52
221	134.6	227	119.0	271	125.7	WILLIAMS OH	-12	6	-7	45	64
402	127.5	460	121.7	576	123.3	WOOD OH	-5	1	-3	55	58
144	117.4	194	145.6	188	120.3	WYANDOT OH	24	-17	2	69	51
13108	124.8	15744	119.8	19417	120.9	OKLAHOMA OK	-4	1	-3		
54	93.0	72	104.2	97	121.4	ADAIR OK	12	17	30	94	53
90	136.3	83	116.4	83	109.8	ALFALFA OK	-15	-6	-19	15	27
73	137.9	68	110.8	64	96.7	ATOKA OK	-20	-13	-30	05	10
41	114.8	47	125.5	44	97.0	BEAVER OK	9	-23	-16	22	10
126	111.5	147	102.3	172	110.3	BECKHAM OK	-8	8	-1	61	28
91	135.8	108	131.7	121	119.5	BLAINE OK	-3	-9	-12	30	49
167	115.3	206	120.4	260	128.6	BRYAN OK	4	7	12	83	69
202	132.2	198	116.2	228	117.4	CADDO OK	-12	1	-11	33	44
130	104.2	182	121.0	246	117.2	CANADIAN OK	16	-3	13	84	43
250	132.9	276	119.1	308	113.7	CARTER OK	-10	-4	-14	24	36
91	134.8	100	107.8	138	105.9	CHEROKEE OK	-20	-2	-21	12	20
76	90.5	104	112.0	131	125.1	CHOCTAW OK	24	12	38	96	62

* SIGNIFICANTLY MORE DEATHS THAN EXPECTED

WHITE FEMALE | ALL CANCERS COMBINED | ICD 140-208

1950-1959 DEATHS	1950-1959 RATE	1960-1969 DEATHS	1960-1969 RATE	1970-1979 DEATHS	1970-1979 RATE	COUNTY AND STATE	PERCENT CHANGE 60S/50S	PERCENT CHANGE 70S/60S	PERCENT CHANGE 70S/50S	PERCENTILE %CHANGE 70S/50S	PERCENTILE RATES 1970S
18	97.0	18	85.8	32	122.9	CIMARRON OK	-11	43	27	93	57
187	97.4	253	95.9	430	111.5	CLEVELAND OK	-2	16	15	86	31
51	152.4	60	141.8	44	100.1	COAL OK	-7	-29	-34	03	13
219	112.2	338	130.8	422	124.6	COMANCHE OK	17	-5	11	83	61
65	126.1	54	99.8	65	119.5	COTTON OK	-21	20	-5	49	49
110	103.1	124	113.9	141	132.2	CRAIG OK	10	16	28	93	76
289	140.5	320	128.4	380	124.4	CREEK OK	-9	-3	-11	32	60
131	126.8	131	105.5	126	82.3	CUSTER OK	-17	-22	-35	02	02
70	107.4	113	118.3	147	105.7	DELAWARE OK	10	-11	-2	60	20
56	128.5	47	102.3	73	149.2	DEWEY OK	-20	46	16	87	95
40	101.7	49	114.0	49	104.5	ELLIS OK	12	-8	3	70	18
355	127.8	396	115.3	504	124.8	GARFIELD OK	-10	8	-2	57	61
172	124.2	229	136.5	231	115.5	GARVIN OK	10	-15	-7	44	40
196	109.1	208	102.6	223	94.4	GRADY OK	-6	-8	-14	27	08
77	121.6	73	116.9	77	112.0	GRANT OK	-4	-4	-8	42	32
73	111.9	103	133.4	89	104.8	GREER OK	19	-21	-6	46	18
39	104.9	47	107.1	45	104.9	HARMON OK	2	-2	0	64	19
34	119.5	42	122.4	34	88.4	HARPER OK	2	-28	-26	08	05
65	115.5	73	117.5	93	122.0	HASKELL OK	2	4	6	75	55
111	109.9	144	129.5	152	126.0	HUGHES OK	18	-3	15	86	65
112	105.3	134	98.6	179	118.3	JACKSON OK	-6	20	12	84	46
82	134.6	79	112.4	83	117.7	JEFFERSON OK	-16	5	-13	29	45
61	119.4	74	140.4	65	96.0	JOHNSTON OK	18	-32	-20	15	09
332	126.8	392	124.5	500	135.8	KAY OK	-2	9	7	78	82
69	108.6	55	75.0	89	96.8	KINGFISHER OK	-31	29	-11	34	10
114	117.1	117	102.4	137	110.1	KIOWA OK	-13	7	-6	47	28
53	126.9	49	102.2	69	119.0	LATIMER OK	-19	16	-6	47	48
189	114.8	221	111.1	302	122.7	LE FLORE OK	-3	10	7	77	56
153	131.6	167	125.1	155	100.0	LINCOLN OK	-5	-20	-24	09	13
162	136.8	152	119.4	184	119.1	LOGAN OK	-13	0	-13	28	48
44	132.7	38	95.0	35	76.7	LOVE OK	-28	-19	-42	01	01
82	120.4	88	108.7	119	117.6	MC CLAIN OK	-10	8	-2	58	44
119	112.5	151	115.2	188	114.6	MC CURTAIN OK	2	-1	2	68	38
72	123.7	89	130.7	106	105.3	MC INTOSH OK	6	-19	-15	23	19
51	102.8	66	115.7	59	97.0	MAJOR OK	13	-16	-6	48	10
55	126.3	67	118.3	83	100.7	MARSHALL OK	-6	-15	-20	14	14
111	115.3	153	123.1	201	119.3	MAYES OK	7	-3	3	71	48
75	116.4	69	80.8	107	122.1	MURRAY OK	-31	51	5	73	55
356	133.9	436	136.4	503	138.2	MUSKOGEE OK	2	1	3	70	86
82	131.5	100	142.7	96	114.8	NOBLE OK	8	-20	-13	29	38
88	147.3	78	115.1	93	126.5	NOWATA OK	-22	10	-14	25	66
72	122.7	66	103.5	81	118.9	OKFUSKEE OK	-16	15	-3	55	47
1947	125.2	2714	124.7	3574	129.8	OKLAHOMA OK	0	4	4	71	71
284	151.9	278	126.6	305	127.5	OKMULGEE OK	-17	1	-16	21	68
168	113.5	178	103.3	215	103.6	OSAGE OK	-9	0	-9	39	17
238	151.8	263	143.1	285	137.1	OTTAWA OK	-6	-4	-10	37	84
110	149.4	95	114.4	117	119.5	PAWNEE OK	-23	4	-20	14	49
265	137.1	272	115.7	301	106.2	PAYNE OK	-16	-8	-23	11	21
253	133.8	275	122.5	312	117.0	PITTSBURG OK	-8	-4	-13	29	43
205	128.7	223	119.8	268	116.2	PONTOTOC OK	-7	-3	-10	37	41
289	123.1	313	108.8	396	112.5	POTTAWATOMIE OK	-12	3	-9	40	34
66	124.2	71	114.0	97	123.2	PUSHMATAHA OK	-8	8	-1	62	58
51	166.6	31	96.8	35	92.9	ROGER MILLS OK	-42	-4	-44	01	07
115	117.6	152	119.7	223	125.5	ROGERS OK	2	5	7	77	63
185	130.5	206	126.5	199	109.8	SEMINOLE OK	-3	-13	-16	21	27
101	126.4	145	138.7	153	108.1	SEQUOYAH OK	10	-22	-14	24	24
190	113.7	243	105.1	359	127.3	STEPHENS OK	-8	21	12	84	67
69	119.7	105	140.9	108	119.5	TEXAS OK	18	-15	0	63	49
106	125.9	98	106.9	119	123.5	TILLMAN OK	-15	16	-2	59	58
1650	136.2	2142	128.8	2856	133.0	TULSA OK	-5	3	-2	58	77
65	104.6	81	94.5	138	105.6	WAGONER OK	-10	12	1	66	20
196	119.6	283	130.0	345	122.2	WASHINGTON OK	9	-6	2	68	55
87	101.9	107	114.0	87	84.0	WASHITA OK	12	-26	-18	18	03
96	123.8	119	132.9	107	104.6	WOODS OK	7	-21	-16	22	18
89	104.5	96	97.3	136	119.6	WOODWARD OK	-7	23	15	86	49
10378	129.6	12756	121.5	16721	122.7	OREGON OR	-6	1	-5		
102	120.8	106	104.2	113	103.9	BAKER OR	-14	0	-14	25	17
137	100.6	214	118.0	297	121.5	BENTON OR	17	3	21	91	54
563	116.8	845	120.2	1253	119.4	CLACKAMAS OR	3	-1	2	69	49
223	146.7	245	132.7	282	137.4	CLATSOP OR	-10	4	-6	46	85
131	129.4	178	136.5	210	122.1	COLUMBIA OR	6	-11	-6	48	55
259	138.2	268	107.7	447	135.0	COOS OR	-22	25	-2	58	81
39	119.5	46	104.3	72	108.8	CROOK OR	-13	4	-9	39	25
37	103.3	64	109.0	106	107.1	CURRY OR	6	-2	4	71	22
134	138.1	154	107.8	275	116.6	DESCHUTES OR	-22	8	-16	22	42
288	129.5	331	111.2	543	127.5	DOUGLAS OR	-14	15	-2	60	68
9	73.6	18	134.5	13	98.7	GILLIAM OR	83	-27	34	95	12
41	137.1	42	121.0	41	95.7	GRANT OR	-12	-21	-30	05	09
25	120.4	33	114.5	33	91.2	HARNEY OR	-5	-20	-24	09	06
71	115.4	92	126.2	108	124.4	HOOD RIVER OR	9	-1	8	78	60
410	124.7	583	124.4	847	125.6	JACKSON OR	0	1	1	66	64
15	79.1	23	91.9	38	91.3	JEFFERSON OR	16	-1	15	87	06
165	116.8	254	124.2	411	126.0	JOSEPHINE OR	6	2	8	78	65
185	125.1	254	126.7	318	118.6	KLAMATH OR	1	-6	-5	49	47
38	162.8	35	123.3	36	101.7	LAKE OR	-24	-18	-38	02	15
705	120.2	977	114.8	1463	122.5	LANE OR	-4	7	2	68	56
132	116.1	186	118.9	280	120.3	LINCOLN OR	2	1	4	71	51
275	114.7	355	110.5	539	121.7	LINN OR	-4	10	6	76	54
102	138.2	115	112.6	179	132.6	MALHEUR OR	-19	18	-4	53	77
740	121.1	962	117.5	1288	116.2	MARION OR	-3	-1	-4	53	41
25	119.9	29	108.8	44	137.2	MORROW OR	-9	26	14	86	84
3984	137.7	4579	133.3	4883	131.1	MULTNOMAH OR	-3	-2	-5	51	73
155	130.3	177	104.3	273	110.6	POLK OR	-20	6	-15	23	29
15	132.7	13	118.1	15	122.2	SHERMAN OR	-11	3	-8	42	55
99	124.8	102	102.2	166	124.3	TILLAMOOK OR	-18	22	0	63	60
284	149.2	290	120.1	355	119.4	UMATILLA OR	-19	-1	-20	14	49
134	142.4	115	103.6	164	129.1	UNION OR	-27	25	-9	38	70
35	119.2	45	120.4	50	106.0	WALLOWA OR	1	-12	-11	33	21
116	134.7	138	122.9	179	132.2	WASCO OR	-9	8	-2	59	76

WHITE FEMALE | ALL CANCERS COMBINED | ICD 140-208

1950-1959 DEATHS	1950-1959 RATE	1960-1969 DEATHS	1960-1969 RATE	1970-1979 DEATHS	1970-1979 RATE	COUNTY AND STATE	PERCENT CHANGE 60S/50S	PERCENT CHANGE 70S/60S	PERCENT CHANGE 70S/50S	PERCENTILE %CHANGE 70S/50S	PERCENTILE RATES 1970S
433	128.1	628	110.6	1064	113.2	WASHINGTON OR	-14	2	-12	32	35
10	103.9	13	146.7	17	159.0	WHEELER OR	41	8	53	97	98
262	146.4	247	108.7	319	104.9	YAMHILL OR	-26	-3	-28	06	19
81184*	152.6	89586*	142.7	99781*	137.2	PENNSYLVANIA PA	-7	-4	-10		
331	137.0	393	136.7	450	125.2	ADAMS PA	0	-8	-9	40	62
11275*	154.3	12843*	146.9	14324*	146.0	ALLEGHENY PA	-5	-1	-5	49	93
552	146.0	560	129.3	690	134.1	ARMSTRONG PA	-11	4	-8	41	80
1171*	154.8	1369	139.6	1571	132.3	BEAVER PA	-10	-5	-14	24	76
285	148.2	270	119.7	309	112.6	BEDFORD PA	-19	-6	-24	09	34
2273*	153.0	2455*	138.1	2779	131.9	BERKS PA	-10	-4	-14	26	75
1244*	155.8	1256*	140.0	1337	135.8	BLAIR PA	-10	-3	-13	28	82
432	144.9	433	133.5	473	129.1	BRADFORD PA	-8	-3	-11	33	70
1242	140.3	1882	134.7	2865*	141.4	BUCKS PA	-4	5	1	66	89
732	144.4	805	129.4	950	121.6	BUTLER PA	-10	-6	-16	21	54
1313	141.4	1448	135.0	1555	126.1	CAMBRIA PA	-5	-7	-11	34	65
43	130.9	43	111.8	51	116.0	CAMERON PA	-15	4	-11	32	40
470*	159.2	460	134.9	558	141.3	CARBON PA	-15	5	-11	33	89
406	148.2	416	122.7	505	117.9	CENTRE PA	-17	-4	-20	14	45
1176	144.1	1428	132.5	1929	136.5	CHESTER PA	-8	3	-5	50	83
252	134.0	272	129.8	315	132.0	CLARION PA	-3	2	-2	60	75
592	144.3	646	141.8	677	130.7	CLEARFIELD PA	-2	-8	-9	37	73
247	139.6	297	147.2	361*	154.4	CLINTON PA	5	5	11	82	97
414	140.3	471	141.7	484	124.0	COLUMBIA PA	1	-12	-12	32	59
643	150.0	596	126.5	691	126.7	CRAWFORD PA	-16	0	-16	22	66
751	143.4	1036*	146.1	1230	127.9	CUMBERLAND PA	2	-12	-11	34	68
1619	148.2	1814*	142.1	1939	135.9	DAUPHIN PA	-4	-4	-8	41	82
3314*	152.5	4173*	140.6	5107*	142.7	DELAWARE PA	-8	1	-6	46	90
246	150.8	283	148.8	329	146.4	ELK PA	-1	-2	-3	56	94
1715*	151.1	1918*	143.3	2210*	142.7	ERIE PA	-5	0	-6	49	91
1201	147.9	1325*	145.2	1447	138.3	FAYETTE PA	-2	-5	-7	46	86
36	144.8	50	174.1	48	151.7	FOREST PA	20	-13	5	73	96
562	139.6	691	135.3	821	125.8	FRANKLIN PA	-3	-7	-10	36	64
62	130.8	74	137.6	79	120.8	FULTON PA	5	-12	-8	43	52
295	141.0	332	142.1	334	129.6	GREENE PA	1	-9	-8	41	71
259	134.5	299	142.6	317	129.2	HUNTINGDON PA	6	-9	-4	53	70
472	130.8	536	130.1	573	118.6	INDIANA PA	0	-9	-9	38	47
393	148.3	354	120.8	415	125.6	JEFFERSON PA	-19	4	-15	22	63
80	105.6	118	130.5	136	119.9	JUNIATA PA	24	-8	13	85	50
2391*	167.0	2474*	152.2	2598*	139.2	LACKAWANNA PA	-9	-9	-17	20	87
1877	137.6	2196	130.0	2737	131.9	LANCASTER PA	-6	1	-4	53	75
782	149.6	832	137.2	932	133.2	LAWRENCE PA	-8	-3	-11	33	78
688*	158.9	805*	152.0	853	131.1	LEBANON PA	-4	-14	-18	18	73
1663	145.6	1894	132.2	2264	127.7	LEHIGH PA	-9	-3	-12	30	68
3162*	156.1	3373*	142.9	3658	134.0	LUZERNE PA	-8	-6	-14	25	79
840	139.4	1067*	157.6	1058	139.1	LYCOMING PA	13	-12	0	63	87
397	133.3	437	131.3	490	135.4	MC KEAN PA	-2	3	2	67	82
836	143.3	893	131.9	1021	128.1	MERCER PA	-8	-3	-11	34	69
310	149.5	328	135.0	334	111.5	MIFFLIN PA	-10	-17	-25	08	31
315	156.1	447*	174.2	492	132.8	MONROE PA	12	-24	-15	23	77
3168	143.0	4236*	136.8	5576*	139.8	MONTGOMERY PA	-4	2	-2	58	88
136	121.7	110	88.9	162	118.2	MONTOUR PA	-27	33	-3	56	46
1457*	148.3	1606	134.2	1949	135.8	NORTHAMPTON PA	-9	1	-8	40	82
930*	151.2	948	139.4	1067	138.8	NORTHUMBERLAND PA	-8	0	-8	41	86
219*	179.9	218	148.2	215	121.5	PERRY PA	-18	-18	-32	03	53
16078*	169.0	15478*	158.7	14103*	149.6	PHILADELPHIA PA	-6	-6	-11	32	95
109*	179.7	105	142.0	162	132.2	PIKE PA	-21	-7	-26	08	76
124	129.1	123	120.4	136	117.7	POTTER PA	-7	-2	-9	39	44
1649*	164.0	1672*	152.2	1743	136.5	SCHUYLKILL PA	-7	-10	-17	20	83
171	146.4	180	127.7	195	112.3	SNYDER PA	-13	-12	-23	10	33
574	146.1	602	131.9	615	114.5	SOMERSET PA	-10	-13	-22	12	37
65	177.1	48	128.3	46	106.9	SULLIVAN PA	-28	-17	-40	01	22
263	147.0	276	138.7	290	125.9	SUSQUEHANNA PA	-6	-9	-14	24	64
299	154.3	279	135.0	263	110.3	TIOGA PA	-13	-18	-29	06	28
148	121.3	191	134.5	208	121.3	UNION PA	11	-10	0	64	53
514	146.4	535	144.8	547	136.5	VENANGO PA	-1	-6	-7	45	83
330	118.0	363	122.5	377	119.0	WARREN PA	4	-3	1	66	48
1346	141.3	1594*	140.1	1856	136.0	WASHINGTON PA	-1	-3	-4	54	83
264	150.0	246	120.1	352	145.6	WAYNE PA	-20	21	-3	56	93
2123	142.9	2513	132.4	2972	125.0	WESTMORELAND PA	-7	-6	-13	29	62
168*	172.8	161	149.0	181	137.1	WYOMING PA	-14	-8	-21	13	84
1690*	151.8	2010*	143.7	2470*	140.8	YORK PA	-5	-2	-7	44	89
7029*	157.8	7725*	145.1	9128*	144.9	RHODE ISLAND RI	-8	0	-8		
233	143.3	296	135.5	406	144.2	BRISTOL RI	-5	6	1	65	92
714*	163.2	956*	149.5	1257*	142.4	KENT RI	-8	-5	-13	29	90
490*	169.3	503	140.0	645*	144.2	NEWPORT RI	-17	3	-15	23	92
5252*	158.1	5522*	146.1	6186*	145.6	PROVIDENCE RI	-8	0	-8	42	93
340	145.3	448	140.3	634*	144.4	WASHINGTON RI	-3	3	-1	62	92
6540	118.0	8735	115.7	11868	113.8	SOUTH CAROLINA SC	-2	-2	-4		
72	110.8	101	126.1	118	111.6	ABBEVILLE SC	14	-11	1	66	31
216	133.4	334	129.9	445	116.3	AIKEN SC	-3	-10	-13	28	41
24	117.9	26	106.4	26	86.0	ALLENDALE SC	-10	-19	-27	07	03
371	120.8	462	111.2	684	118.4	ANDERSON SC	-8	6	-2	58	46
30	84.3	51	117.9	54	102.9	BAMBERG SC	40	-13	22	91	16
42	113.8	50	100.5	71	113.9	BARNWELL SC	-12	13	0	64	36
63	162.8	60	91.8	137	101.0	BEAUFORT SC	-44	10	-38	02	14
33	77.9	59	78.3	157	106.4	BERKELEY SC	1	36	37	96	21
16	66.2	28	107.9	35	106.5	CALHOUN SC	63	-1	61	98	21
532	130.1	773	138.7	937	124.2	CHARLESTON SC	7	-10	-4	51	60
136	129.6	195	137.9	211	116.5	CHEROKEE SC	6	-16	-10	36	42
86	103.1	132	127.8	126	97.6	CHESTER SC	24	-24	-5	49	11
107	121.4	127	121.2	160	116.8	CHESTERFIELD SC	0	-4	-4	54	42
26	63.1	52	102.3	55	85.9	CLARENDON SC	62	-16	36	95	03
68	118.8	89	116.3	110	110.1	COLLETON SC	-2	-5	-7	43	28
118	105.1	140	95.5	241	117.6	DARLINGTON SC	-9	23	12	84	44
78	123.7	87	115.7	91	91.6	DILLON SC	-6	-21	-26	08	06

* SIGNIFICANTLY MORE DEATHS THAN EXPECTED

WHITE FEMALE — ALL CANCERS COMBINED — ICD 140-208

DEATHS 1950-1959	RATE 1950-1959	DEATHS 1960-1969	RATE 1960-1969	DEATHS 1970-1979	RATE 1970-1979	COUNTY AND STATE		PERCENT CHANGE 60S/50S	70S/60S	70S/50S	PERCENTILE %CHANGE 70S/50S	PERCENTILE RATES 1970S
51	106.2	51	74.5	150	124.5	DORCHESTER	SC	-30	67	17	88	60
31	92.0	38	89.2	45	81.0	EDGEFIELD	SC	-3	-9	-12	31	02
49	130.6	56	113.9	81	133.0	FAIRFIELD	SC	-13	17	2	68	77
219	125.6	283	118.4	371	115.2	FLORENCE	SC	-6	-3	-8	41	39
50	92.6	78	110.6	107	100.4	GEORGETOWN	SC	19	-9	8	79	13
741	122.3	1091	126.4	1439	119.7	GREENVILLE	SC	3	-5	-2	58	49
169	133.7	210	119.6	277	114.8	GREENWOOD	SC	-11	-4	-14	25	38
46	132.0	49	114.6	62	118.1	HAMPTON	SC	-13	3	-10	34	46
128	82.3	216	99.0	352	103.3	HORRY	SC	20	4	25	93	16
22	117.9	25	105.0	28	87.0	JASPER	SC	-11	-17	-26	08	04
90	128.7	122	127.5	144	111.4	KERSHAW	SC	-1	-13	-13	27	30
112	109.6	140	103.8	208	107.6	LANCASTER	SC	-5	4	-2	59	23
178	125.0	175	97.4	259	111.9	LAURENS	SC	-22	15	-10	35	32
50	149.5	43	103.2	39	79.1	LEE	SC	-31	-23	-47	00	02
171	103.5	244	96.5	467	107.5	LEXINGTON	SC	-7	11	4	72	23
14	89.4	19	102.1	21	88.3	MC CORMICK	SC	14	-14	-1	61	05
64	103.6	91	124.0	85	88.5	MARION	SC	20	-29	-15	24	05
91	139.1	76	98.3	117	123.6	MARLBORO	SC	-29	26	-11	33	59
136	133.6	155	124.1	149	95.9	NEWBERRY	SC	-7	-23	-28	06	09
144	113.6	210	127.0	256	114.3	OCONEE	SC	12	-10	1	65	37
134	107.3	183	111.4	259	116.0	ORANGEBURG	SC	4	4	8	79	40
150	101.0	210	99.4	317	104.5	PICKENS	SC	-2	5	3	71	18
531	118.9	701	113.2	941	118.9	RICHLAND	SC	-5	5	0	64	48
46	99.6	63	113.7	69	98.0	SALUDA	SC	14	-14	-2	59	11
555	115.8	756	117.1	1053	121.1	SPARTANBURG	SC	1	3	5	73	53
140	146.3	188	132.4	233	120.1	SUMTER	SC	-10	-9	-18	18	50
120	137.4	142	128.2	148	103.5	UNION	SC	-7	-19	-25	09	17
49	85.7	77	118.7	102	123.2	WILLIAMSBURG	SC	39	4	44	96	58
241	121.7	277	103.7	431	114.6	YORK	SC	-15	10	-6	48	38
3975	132.5	4401	122.3	5014	119.5	SOUTH DAKOTA	SD	-8	-2	-10		
32	132.9	42	151.5	32	106.8	AURORA	SD	14	-29	-20	15	22
142	143.9	153	121.7	171	117.7	BEADLE	SD	-15	-3	-18	17	44
10	110.5	10	89.5	11	99.3	BENNETT	SD	-19	11	-10	35	12
72	136.6	75	130.2	76	119.4	BON HOMME	SD	-5	-8	-13	29	48
111	131.6	114	109.1	134	111.5	BROOKINGS	SD	-17	2	-15	22	31
213	135.5	241	126.9	263	116.8	BROWN	SD	-6	-8	-14	26	42
40	135.7	27	81.1	39	103.4	BRULE	SD	-40	27	-24	09	17
3	100.9	4	160.7	4	308.7	BUFFALO	SD	59	92	206	99	99
41	116.1	52	116.1	54	94.0	BUTTE	SD	0	-19	-19	16	08
15	96.1	22	170.1	13	81.1	CAMPBELL	SD	77	-52	-16	22	02
65	128.4	79	142.3	70	118.0	CHARLES MIX	SD	11	-17	-8	41	45
41	111.3	53	135.1	49	115.5	CLARK	SD	21	-15	4	71	39
79	150.7	76	132.4	73	122.4	CLAY	SD	-12	-8	-19	16	56
118	123.2	164	135.9	158	113.3	CODINGTON	SD	10	-17	-8	41	35
16	120.8	22	136.6	18	122.5	CORSON	SD	13	-10	1	67	56
28	119.7	41	151.0	22	65.1	CUSTER	SD	26	-57	-46	00	00
132	138.0	155	142.3	164	124.5	DAVISON	SD	3	-13	-10	36	61
78	138.3	78	127.5	88	120.7	DAY	SD	-8	-5	-13	28	52
37	108.2	43	112.4	50	116.8	DEUEL	SD	4	4	8	79	42
19	167.6	13	104.1	16	123.5	DEWEY	SD	-38	19	-26	08	58
31	124.4	35	123.5	28	82.8	DOUGLAS	SD	-1	-33	-33	03	03
35	140.6	46	145.3	43	112.0	EDMUNDS	SD	3	-23	-20	14	32
62	128.0	67	119.7	65	122.7	FALL RIVER	SD	-6	2	-4	53	56
27	137.0	23	100.2	33	129.7	FAULK	SD	-27	29	-5	49	71
62	125.7	72	119.7	95	151.0	GRANT	SD	-5	26	20	90	96
48	124.8	50	112.3	44	87.7	GREGORY	SD	-10	-22	-30	05	04
16	115.5	24	158.5	21	129.7	HAAKON	SD	37	-18	12	84	71
50	136.7	50	118.7	59	117.5	HAMLIN	SD	-13	-1	-14	25	44
33	127.8	43	133.4	41	106.0	HAND	SD	4	-21	-17	19	21
30	136.5	20	93.5	27	113.3	HANSON	SD	-32	21	-17	19	35
3	29.3	4	59.9	7	66.9	HARDING	SD	105	12	128	99	00
68	176.1	65	119.0	83	125.0	HUGHES	SD	-32	5	-29	06	62
78	131.9	85	116.9	139*	156.6	HUTCHINSON	SD	-11	34	19	90	97
14	110.0	19	132.2	11	51.9	HYDE	SD	20	-61	-53	00	00
12	102.8	18	147.3	8	68.5	JACKSON	SD	43	-53	-33	03	01
29	118.6	28	100.6	43	145.2	JERAULD	SD	-15	44	23	91	93
8	88.6	10	107.9	7	76.3	JONES	SD	22	-29	-14	26	01
70	141.9	66	122.3	78	132.7	KINGSBURY	SD	-14	8	-6	46	77
86	144.2	106	150.2	90	112.0	LAKE	SD	4	-25	-22	11	32
109	134.5	104	110.8	139	121.0	LAWRENCE	SD	-18	9	-10	36	52
103	145.6	101	124.4	126	123.7	LINCOLN	SD	-15	-1	-15	23	59
24	141.6	21	113.4	22	116.7	LYMAN	SD	-20	3	-18	18	42
66	154.5	54	105.6	73	139.8	MC COOK	SD	-32	32	-10	37	88
27	95.3	29	87.0	36	89.6	MC PHERSON	SD	-9	3	-6	47	05
43	136.5	44	117.5	54	125.2	MARSHALL	SD	-14	7	-8	41	62
72	156.4	63	108.6	72	96.8	MEADE	SD	-31	-11	-38	02	10
8	106.4	8	109.3	8	112.4	MELLETTE	SD	3	3	6	75	33
37	134.1	35	106.8	36	313.3	MINER	SD	-20	193	134	99	99
521	143.3	600	129.5	775	131.6	MINNEHAHA	SD	-10	2	-8	41	74
63	141.3	61	126.3	72	139.6	MOODY	SD	-11	11	-1	61	87
209	137.7	263	119.3	330	114.9	PENNINGTON	SD	-13	-4	-17	20	38
22	83.0	32	108.0	47	142.4	PERKINS	SD	30	32	72	98	90
29	141.8	29	114.0	37	122.2	POTTER	SD	-20	7	-14	26	55
73	121.4	80	114.1	91	109.5	ROBERTS	SD	-6	-4	-10	36	27
30	121.9	31	135.5	37	156.8	SANBORN	SD	11	16	29	93	97
3	114.8	7	225.3	5	150.4	SHANNON	SD	96	-33	31	94	96
67	109.5	82	114.0	85	106.5	SPINK	SD	4	-7	-3	56	21
12	133.3	18	197.5	21*	221.4	STANLEY	SD	48	12	66	98	99
11	144.5	17	160.5	12	114.5	SULLY	SD	11	-29	-21	13	37
6	105.4	9	133.2	13	173.1	TODD	SD	26	30	64	98	99
48	124.1	50	111.4	64	130.8	TRIPP	SD	-10	17	5	74	73
90	135.4	116	153.7	128	149.1	TURNER	SD	13	-3	10	82	95
75	135.0	70	111.4	90	128.0	UNION	SD	-17	15	-5	50	68
43	129.1	64	150.1	64	117.5	WALWORTH	SD	16	-22	-9	39	44
125	123.7	116	96.6	144	115.2	YANKTON	SD	-22	19	-7	44	39
5	142.3	2	91.7	6	173.9	ZIEBACH	SD	-36	90	22	91	99
16098	125.6	19652	118.9	24837	116.4	TENNESSEE	TN	-5	-2	-7		

* SIGNIFICANTLY MORE DEATHS THAN EXPECTED

WHITE FEMALE — ALL CANCERS COMBINED — ICD 140-208

1950-1959 DEATHS	1950-1959 RATE	1960-1969 DEATHS	1960-1969 RATE	1970-1979 DEATHS	1970-1979 RATE	COUNTY AND STATE	PERCENT CHANGE 60S/50S	PERCENT CHANGE 70S/60S	PERCENT CHANGE 70S/50S	PERCENTILE %CHANGE 70S/50S	PERCENTILE RATES 1970S
216	121.2	294	121.5	416	119.6	ANDERSON TN	0	-2	-1	60	49
124	113.4	161	121.8	199	127.3	BEDFORD TN	7	5	12	84	67
79	132.8	74	99.7	100	104.7	BENTON TN	-25	5	-21	13	18
29	91.3	44	127.0	51	114.4	BLEDSOE TN	39	-10	25	93	37
243	121.8	329	118.2	432	107.7	BLOUNT TN	-3	-9	-12	32	23
196	149.5	217	117.7	300	111.8	BRADLEY TN	-21	-5	-25	09	32
153	126.8	158	106.6	258	133.9	CAMPBELL TN	-16	26	6	75	79
47	114.4	41	84.1	61	97.6	CANNON TN	-27	16	-15	24	11
134	109.8	177	118.1	189	101.0	CARROLL TN	8	-15	-8	41	14
174	111.1	204	100.4	272	97.9	CARTER TN	-10	-3	-12	31	11
59	151.8	50	96.1	95	118.4	CHEATHAM TN	-37	23	-22	11	46
48	108.2	59	113.4	74	109.9	CHESTER TN	5	-3	2	67	28
119	139.0	122	124.4	156	128.4	CLAIBORNE TN	-10	3	-8	43	69
21	61.3	45	119.8	52	116.4	CLAY TN	95	-3	90	98	42
144	172.1	140	132.2	165	119.1	COCKE TN	-23	-10	-31	04	48
178*	176.2	173	124.3	213	110.6	COFFEE TN	-29	-11	-37	02	29
64	99.2	85	105.1	108	115.4	CROCKETT TN	6	10	16	88	39
78	103.8	107	105.8	162	109.9	CUMBERLAND TN	2	4	6	75	27
1804	132.3	2152	120.2	2628	118.2	DAVIDSON TN	-9	-2	-11	34	46
59	144.2	49	91.0	76	111.9	DECATUR TN	-37	23	-22	11	32
68	123.2	67	101.3	78	86.8	DE KALB TN	-18	-14	-30	05	04
105	117.9	123	108.2	172	109.9	DICKSON TN	-8	2	-7	45	28
163	122.2	175	109.5	219	113.8	DYER TN	-10	4	-7	45	36
39	101.0	41	87.6	59	98.4	FAYETTE TN	-13	12	-3	57	12
45	88.7	75	125.0	79	107.2	FENTRESS TN	41	-14	21	91	22
158	152.5	153	119.7	177	107.8	FRANKLIN TN	-22	-10	-29	06	24
274	132.8	308	121.0	328	105.0	GIBSON TN	-9	-13	-21	13	19
107	100.5	162	129.4	185	124.0	GILES TN	29	-4	23	92	59
60	108.7	75	114.1	87	104.7	GRAINGER TN	5	-8	-4	54	18
187	115.1	229	105.5	316	109.8	GREENE TN	-8	4	-5	51	27
64	126.3	74	130.1	81	113.8	GRUNDY TN	3	-13	-10	36	36
119	108.3	180	118.5	232	109.6	HAMBLEN TN	9	-7	1	66	27
1128	139.9	1322	125.1	1651	122.2	HAMILTON TN	-11	-2	-13	29	55
43	134.9	38	106.2	40	96.3	HANCOCK TN	-21	-9	-29	06	10
85	97.1	90	87.2	109	95.1	HARDEMAN TN	-10	9	-2	58	08
74	100.5	100	105.7	135	104.6	HARDIN TN	5	-1	4	72	18
124	103.3	189	122.0	238	116.2	HAWKINS TN	18	-5	12	84	41
60	117.4	45	77.2	80	109.3	HAYWOOD TN	-34	42	-7	45	26
59	80.2	106	117.4	136	113.5	HENDERSON TN	46	-3	41	96	35
163	135.5	189	133.5	238	133.8	HENRY TN	-1	0	-1	61	79
66	125.5	55	89.6	102	122.8	HICKMAN TN	-29	37	-2	58	57
20	83.3	27	81.4	37	93.4	HOUSTON TN	-2	15	12	84	07
68	135.6	71	109.9	92	107.6	HUMPHREYS TN	-19	-2	-21	13	23
61	126.4	52	93.1	64	103.4	JACKSON TN	-26	11	-18	17	17
112	131.1	135	125.1	140	89.3	JEFFERSON TN	-5	-29	-32	04	05
63	121.5	66	108.4	68	86.7	JOHNSON TN	-11	-20	-29	06	04
1186	121.9	1595	124.4	1982	120.6	KNOX TN	2	-3	-1	61	51
38	120.2	36	105.8	67	162.8	LAKE TN	-12	54	35	95	98
91	125.3	117	133.0	116	112.4	LAUDERDALE TN	6	-16	-10	35	33
149	123.0	191	125.0	205	106.2	LAWRENCE TN	2	-15	-14	26	21
44	171.6	41	121.9	52	112.7	LEWIS TN	-29	-7	-34	03	34
124	119.3	168	133.7	186	117.8	LINCOLN TN	12	-12	-1	61	45
129	129.5	144	113.5	207	122.7	LOUDON TN	-12	8	-5	50	57
176	129.4	223	130.3	252	112.7	MC MINN TN	1	-14	-13	28	34
95	110.2	129	123.3	154	121.2	MC NAIRY TN	12	-2	10	81	53
75	123.5	76	103.5	78	79.7	MACON TN	-16	-23	-35	02	02
251	111.9	302	105.0	424	121.3	MADISON TN	-6	16	8	79	53
97	129.6	106	117.7	125	107.7	MARION TN	-9	-9	-17	19	23
126	143.7	109	103.8	136	104.1	MARSHALL TN	-28	0	-28	07	18
203	134.1	219	116.1	299	121.5	MAURY TN	-13	5	-9	38	53
20	104.6	29	127.7	35	121.9	MEIGS TN	22	-5	17	88	54
105	113.9	139	124.8	145	100.5	MONROE TN	10	-19	-12	31	13
161	122.5	193	107.3	270	107.4	MONTGOMERY TN	-12	0	-12	30	23
22	125.6	25	124.6	27	126.7	MOORE TN	-1	2	1	66	66
52	90.9	70	108.8	105	134.8	MORGAN TN	20	24	48	97	80
143	100.1	188	103.5	215	101.5	OBION TN	3	-2	1	67	14
79	109.7	91	108.6	117	110.2	OVERTON TN	-1	2	0	65	28
28	107.3	32	99.5	51	114.9	PERRY TN	-7	15	7	78	38
24	118.7	17	80.5	22	83.3	PICKETT TN	-32	3	-30	05	03
66	130.4	76	123.8	83	114.3	POLK TN	-5	-8	-12	30	37
164	126.9	197	118.2	255	105.6	PUTNAM TN	-7	-11	-17	20	20
90	129.5	112	131.6	127	112.6	RHEA TN	2	-14	-13	28	34
164	129.8	220	126.6	314	130.0	ROANE TN	-3	3	0	64	72
155	133.4	161	116.4	212	119.3	ROBERTSON TN	-13	2	-11	34	48
189	118.9	248	117.2	349	116.2	RUTHERFORD TN	-1	-1	-2	58	41
59	104.4	85	123.3	123	138.0	SCOTT TN	18	12	32	94	85
35	159.5	20	71.9	37	92.2	SEQUATCHIE TN	-55	28	-42	01	06
118	125.9	154	122.0	177	92.8	SEVIER TN	-3	-24	-26	08	07
2067	133.8	2800	133.6	3445	135.2	SHELBY TN	0	1	1	66	81
88	131.5	65	86.2	103	112.5	SMITH TN	-34	31	-14	24	34
39	95.2	57	125.9	46	80.6	STEWART TN	32	-36	-15	22	02
502	138.4	637	118.4	875	118.5	SULLIVAN TN	-14	0	-14	24	47
178	120.8	237	116.5	344	108.9	SUMNER TN	-4	-7	-10	36	25
92	120.5	117	127.2	128	106.1	TIPTON TN	6	-17	-12	31	21
24	97.4	27	93.4	36	112.4	TROUSDALE TN	-4	20	15	87	34
73	112.7	89	109.7	105	101.8	UNICOI TN	-3	-7	-10	37	15
37	101.8	46	106.3	58	105.3	UNION TN	4	-1	3	71	19
18	125.5	25	142.8	30	133.0	VAN BUREN TN	14	-7	6	76	77
146	141.1	155	116.6	182	102.4	WARREN TN	-17	-12	-27	07	15
314	128.0	376	113.6	486	110.1	WASHINGTON TN	-11	-3	-14	25	28
55	112.0	59	101.7	83	112.0	WAYNE TN	-9	10	0	64	32
173	114.2	183	104.3	230	109.1	WEAKLEY TN	-9	5	-4	52	26
77	105.5	109	120.0	117	102.4	WHITE TN	14	-15	-3	56	15
136	139.3	146	110.1	217	106.8	WILLIAMSON TN	-21	-3	-23	10	22
137	110.0	183	111.9	255	111.8	WILSON TN	2	0	2	67	32
38885	121.3	51133	116.6	68879	120.6	TEXAS TX	-4	3	-1		
139	105.2	168	104.4	220	112.6	ANDERSON TX	-1	8	7	77	34
14	127.8	37	103.2	51	103.4	ANDREWS TX	-19	0	-19	16	17
183	127.4	196	96.9	305	109.1	ANGELINA TX	-24	13	-14	24	26

* SIGNIFICANTLY MORE DEATHS THAN EXPECTED

WHITE FEMALE — ALL CANCERS COMBINED — ICD 140-208

1950-1959 DEATHS	1950-1959 RATE	1960-1969 DEATHS	1960-1969 RATE	1970-1979 DEATHS	1970-1979 RATE	COUNTY AND STATE		PERCENT CHANGE 60S/50S	70S/60S	70S/50S	PERCENTILE %CHANGE 70S/50S	RATES 1970S
26	112.4	54	128.9	95	133.1	ARANSAS	TX	15	3	18	89	78
33	119.6	39	106.0	42	92.3	ARCHER	TX	-11	-13	-23	11	07
13	114.7	13	78.9	22	156.8	ARMSTRONG	TX	-31	99	37	96	97
84	112.0	105	111.5	151	124.0	ATASCOSA	TX	0	11	11	82	59
97	124.5	102	110.8	146	124.1	AUSTIN	TX	-11	12	0	63	60
31	113.0	43	128.1	42	98.1	BAILEY	TX	13	-23	-13	27	11
23	89.2	34	96.9	43	81.2	BANDERA	TX	9	-16	-9	39	02
90	110.3	112	119.9	164	135.2	BASTROP	TX	9	13	23	92	81
40	112.9	39	93.7	45	88.5	BAYLOR	TX	-17	-6	-22	12	05
99	133.4	94	99.3	152	136.9	BEE	TX	-26	38	3	69	84
322	112.7	408	105.8	550	107.9	BELL	TX	-6	2	-4	52	24
3056	136.4	3999	129.7	5252	134.3	BEXAR	TX	-5	4	-1	60	80
40	170.8	29	93.3	35	101.6	BLANCO	TX	-45	9	-41	01	14
2	54.8	3	93.5	2	38.8	BORDEN	TX	71	-59	-29	06	00
79	93.1	108	102.0	125	89.1	BOSQUE	TX	10	-13	-4	52	05
277	125.2	334	112.4	457	118.9	BOWIE	TX	-10	6	-5	50	48
194	121.0	325	115.8	544	120.0	BRAZORIA	TX	-4	4	-1	62	50
128	120.6	163	104.4	255	110.5	BRAZOS	TX	-13	6	-8	40	29
34	122.5	40	125.8	55	149.8	BREWSTER	TX	3	19	22	91	95
9	72.5	23	136.1	25	145.9	BRISCOE	TX	88	7	101	99	93
38	148.0	42	122.9	42	104.5	BROOKS	TX	-17	-15	-29	06	18
187	110.8	235	117.1	274	112.0	BROWN	TX	6	-4	1	66	32
56	113.3	58	102.5	79	109.6	BURLESON	TX	-10	7	-3	55	27
75	131.8	86	98.4	136	104.2	BURNET	TX	-25	6	-21	13	18
91	103.4	93	94.7	142	116.0	CALDWELL	TX	-8	22	12	84	40
37	121.2	72	149.3	99	143.2	CALHOUN	TX	23	-4	18	89	91
71	122.5	68	97.1	93	111.8	CALLAHAN	TX	-21	15	-9	40	32
476	111.0	673	119.5	852	113.6	CAMERON	TX	6	-5	2	69	35
45	142.1	44	105.7	55	120.8	CAMP	TX	-26	14	-15	23	52
26	97.0	36	105.6	46	112.9	CARSON	TX	9	7	16	88	34
119	128.7	114	92.2	145	93.7	CASS	TX	-28	2	-27	07	07
18	116.8	24	93.7	28	77.1	CASTRO	TX	-20	-18	-34	03	01
31	117.5	25	67.7	64	105.8	CHAMBERS	TX	-42	56	-10	36	20
163	101.1	188	99.5	225	94.2	CHEROKEE	TX	-2	-5	-7	45	08
57	108.8	62	109.8	67	110.5	CHILDRESS	TX	1	1	2	67	29
50	93.8	79	128.5	74	100.3	CLAY	TX	37	-22	7	77	13
21	117.3	21	92.7	34	147.9	COCHRAN	TX	-21	60	26	93	94
17	82.0	29	128.6	41	138.2	COKE	TX	57	7	69	98	86
112	124.7	122	121.4	122	125.5	COLEMAN	TX	-3	3	1	65	63
274	122.1	338	117.0	444	104.1	COLLIN	TX	-4	-11	-15	23	18
37	93.4	39	93.2	37	105.3	COLLINGSWORTH	TX	0	13	13	84	19
82	111.0	81	87.0	100	86.8	COLORADO	TX	-22	0	-22	12	04
95	107.5	127	99.1	244	119.8	COMAL	TX	-8	21	11	83	50
106	105.1	120	107.6	115	95.5	COMANCHE	TX	2	-11	-9	38	09
22	103.6	34	118.7	32	125.7	CONCHO	TX	15	6	21	91	64
144	122.1	158	108.7	192	110.7	COOKE	TX	-11	2	-9	38	29
93	98.1	98	84.3	178	118.7	CORYELL	TX	-14	41	21	91	47
22	94.8	26	111.7	31	108.5	COTTLE	TX	18	-3	14	86	25
12	165.7	14	89.7	31	161.2	CRANE	TX	-46	80	-3	57	98
12	98.4	19	121.6	25	138.8	CROCKETT	TX	24	14	41	96	86
51	133.4	41	86.2	62	118.9	CROSBY	TX	-35	38	-11	34	47
8	131.9	7	66.0	18	163.2	CULBERSON	TX	-50	147	24	92	98
35	124.3	28	84.0	29	84.4	DALLAM	TX	-32	0	-32	04	03
3703	129.9	5269	123.1	7204	127.0	DALLAS	TX	-5	3	-2	58	67
83	127.5	78	96.2	107	116.8	DAWSON	TX	-25	21	-8	40	42
52	139.9	59	100.4	98	118.4	DEAF SMITH	TX	-28	18	-15	22	46
60	129.0	54	99.5	49	88.6	DELTA	TX	-23	-11	-31	04	05
220	111.2	308	111.8	458	110.7	DENTON	TX	1	-1	0	63	29
116	102.1	122	95.2	180	117.6	DE WITT	TX	-7	24	15	87	44
32	108.0	35	109.2	41	112.4	DICKENS	TX	1	3	4	72	33
45	139.8	41	115.1	49	113.9	DIMMIT	TX	-18	-1	-19	16	36
35	115.8	33	102.2	36	90.6	DONLEY	TX	-12	-11	-22	12	06
55	110.0	78	131.2	74	111.9	DUVAL	TX	19	-15	2	68	32
168	123.5	214	124.0	228	120.5	EASTLAND	TX	0	-3	-2	57	51
140	101.4	295	113.7	465	120.7	ECTOR	TX	12	6	19	90	52
9	82.3	14	120.2	16	129.8	EDWARDS	TX	46	8	58	97	71
213	103.1	256	102.0	332	107.4	ELLIS	TX	-1	5	4	72	23
1021	133.7	1450	132.5	2011*	155.9	EL PASO	TX	-1	18	17	88	97
137	113.3	166	110.0	148	79.1	ERATH	TX	-3	-28	-30	05	02
120	113.2	113	95.2	132	100.2	FALLS	TX	-16	5	-12	32	13
199	115.8	217	111.7	217	102.1	FANNIN	TX	-4	-9	-12	31	15
120	90.2	162	107.8	179	107.8	FAYETTE	TX	20	0	20	90	23
47	105.1	52	108.9	56	110.9	FISHER	TX	4	2	6	75	30
64	129.1	51	85.2	66	109.5	FLOYD	TX	-34	29	-15	23	27
13	65.3	25	118.9	31	148.2	FOARD	TX	82	25	127	99	95
134	142.1	161	111.4	290	114.1	FORT BEND	TX	-22	2	-20	15	36
29	91.9	42	106.3	63	112.4	FRANKLIN	TX	16	6	22	91	34
55	87.9	85	110.4	85	97.3	FREESTONE	TX	26	-12	11	82	11
46	110.3	63	128.1	67	112.6	FRIO	TX	16	-12	2	68	34
29	92.2	53	126.7	56	107.9	GAINES	TX	37	-15	17	88	24
483	127.1	687	126.0	1036*	142.5	GALVESTON	TX	-1	13	12	84	90
26	104.9	24	82.3	40	128.4	GARZA	TX	-22	56	22	91	69
64	88.3	96	104.5	125	103.8	GILLESPIE	TX	18	-1	18	89	17
3	90.7	7	195.7	5	112.4	GLASSCOCK	TX	116	-43	24	92	34
31	124.2	42	142.7	36	106.6	GOLIAD	TX	15	-25	-14	25	22
103	111.9	108	107.8	136	126.3	GONZALES	TX	-4	17	13	85	65
101	105.8	181	131.1	192	110.8	GRAY	TX	24	-15	5	73	30
470	123.5	518	107.8	681	112.8	GRAYSON	TX	-13	5	-9	39	34
264	127.2	368	122.1	539	127.1	GREGG	TX	-4	4	0	64	67
65	119.4	72	115.8	77	108.6	GRIMES	TX	-3	-6	-9	39	25
135	123.8	169	110.2	264	120.4	GUADALUPE	TX	-11	9	-3	56	51
138	114.2	187	115.9	209	111.0	HALE	TX	1	-4	-3	56	30
41	99.2	51	109.6	57	114.2	HALL	TX	10	4	15	87	37
78	114.5	102	127.7	112	116.7	HAMILTON	TX	12	-9	2	68	42
23	128.1	20	86.5	31	102.4	HANSFORD	TX	-32	18	-20	14	15
68	140.5	64	105.3	72	113.0	HARDEMAN	TX	-25	7	-20	15	34
85	111.4	110	97.0	190	117.1	HARDIN	TX	-13	21	5	74	43
4042	136.4	5966	126.0	8587	130.8	HARRIS	TX	-8	4	-4	53	73
133	108.1	178	108.9	257	118.3	HARRISON	TX	1	9	9	81	46
9	95.8	10	102.1	19	95.6	HARTLEY	TX	7	-6	0	63	09
75	121.1	79	116.0	80	115.4	HASKELL	TX	-4	-1	-5	51	39

* SIGNIFICANTLY MORE DEATHS THAN EXPECTED

WHITE FEMALE — ALL CANCERS COMBINED — ICD 140-208

1950-1959 DEATHS	1950-1959 RATE	1960-1969 DEATHS	1960-1969 RATE	1970-1979 DEATHS	1970-1979 RATE	COUNTY AND STATE	PERCENT CHANGE 60S/50S	PERCENT CHANGE 70S/60S	PERCENT CHANGE 70S/50S	PERCENTILE %CHANGE 70S/50S	PERCENTILE RATES 1970S
83	105.7	114	113.3	145	111.4	HAYS TX	7	-2	5	74	31
21	121.5	23	136.0	23	103.4	HEMPHILL TX	12	-24	-15	23	17
129	122.1	153	107.3	225	96.1	HENDERSON TX	-12	-10	-21	12	09
573	113.3	756	113.0	997	105.2	HIDALGO TX	0	-7	-7	44	19
192	120.8	202	106.3	220	107.1	HILL TX	-12	1	-11	32	22
68	118.0	73	90.1	108	110.4	HOCKLEY TX	-24	22	-6	46	29
34	103.3	72	156.3	99	117.2	HOOD TX	51	-25	14	85	43
155	126.1	153	111.6	191	111.4	HOPKINS TX	-12	0	-12	32	31
82	105.3	86	99.0	126	115.3	HOUSTON TX	-6	16	10	81	39
116	97.7	169	109.8	193	105.6	HOWARD TX	12	-4	8	79	20
5	73.0	12	135.5	12	127.0	HUDSPETH TX	86	-6	74	98	67
241	108.6	300	111.3	364	109.4	HUNT TX	2	-2	1	65	26
104	115.9	126	119.8	185	125.4	HUTCHINSON TX	3	5	8	79	63
4	58.6	5	57.7	9	106.8	IRION TX	-2	85	82	98	22
37	80.1	48	90.0	69	121.3	JACK TX	12	35	52	97	53
48	111.6	65	115.0	73	96.8	JACKSON TX	3	-16	-13	27	10
91	119.9	101	93.9	159	110.3	JASPER TX	-22	17	-8	41	29
5	70.9	4	38.8	10	93.2	JEFF DAVIS TX	-45	140	31	94	07
845	128.9	1117	125.9	1469	133.2	JEFFERSON TX	-2	6	3	71	78
29	163.6	31	137.8	33	118.8	JIM HOGG TX	-16	-14	-27	07	47
110	124.0	153	118.9	192	121.8	JIM WELLS TX	-4	2	-2	59	54
230	120.2	265	106.6	411	117.2	JOHNSON TX	-11	10	-2	57	43
129	125.7	150	119.4	162	114.8	JONES TX	-5	-4	-9	40	38
91	138.1	76	96.1	101	116.2	KARNES TX	-30	21	-16	21	41
182	123.4	198	109.1	237	108.2	KAUFMAN TX	-12	-1	-12	30	24
40	110.4	46	103.1	93	140.5	KENDALL TX	-7	36	27	93	88
1	39.0	2	66.3			KENEDY TX	70				
7	69.3	12	113.2	14	145.7	KENT TX	63	29	110	99	93
100	103.9	132	78.7	227	96.8	KERR TX	-24	23	-7	45	10
29	117.4	30	105.4	36	132.7	KIMBLE TX	-10	26	13	85	77
3	193.8					KING TX					
9	92.3	14	144.3	12	106.4	KINNEY TX	56	-26	15	87	21
65	87.1	128	133.0	136	120.9	KLEBERG TX	53	-9	39	96	52
25	62.8	60	130.5	55	94.1	KNOX TX	108	-28	50	97	08
249	129.2	279	123.1	316	119.2	LAMAR TX	-5	-3	-8	43	48
80	117.1	111	115.6	114	102.1	LAMB TX	-1	-12	-13	28	15
68	120.2	92	147.9	101	129.9	LAMPASAS TX	23	-12	8	79	72
41	154.1	26	101.8	39	149.1	LA SALLE TX	-34	47	-3	55	95
153	123.8	129	86.8	155	98.9	LAVACA TX	-30	14	-20	14	12
49	109.7	47	93.2	79	129.1	LEE TX	-15	39	18	89	70
48	111.0	55	104.3	68	96.1	LEON TX	-6	-8	-13	27	10
119	129.0	142	107.4	226	119.5	LIBERTY TX	-17	11	-7	43	49
126	104.3	121	95.4	144	96.1	LIMESTONE TX	-9	1	-8	42	10
18	103.8	20	95.1	22	91.0	LIPSCOMB TX	-8	-4	-12	30	06
30	92.3	41	104.0	55	111.5	LIVE OAK TX	13	7	21	91	31
42	117.4	59	110.1	118	132.4	LLANO TX	-6	20	13	85	76
1	119.2	2	151.2			LOVING TX	27				
431	114.0	616	103.3	876	114.2	LUBBOCK TX	-9	11	0	64	36
41	102.4	41	89.0	46	92.1	LYNN TX	-13	3	-10	36	06
69	111.9	81	112.3	94	114.8	MC CULLOCH TX	0	2	3	69	38
658	115.7	861	115.7	1004	109.4	MC LENNAN TX	0	-5	-5	49	26
7	137.9	5	80.3	8	150.5	MC MULLEN TX	-42	87	9	80	96
37	109.7	51	120.7	53	92.6	MADISON TX	10	-23	-16	22	07
15	63.9	37	120.2	49	104.4	MARION TX	88	-13	63	98	18
27	165.8	27	130.6	36	152.3	MARTIN TX	-21	17	-8	41	96
25	83.3	44	135.2	51	138.4	MASON TX	62	2	66	98	86
88	117.6	123	116.6	161	116.4	MATAGORDA TX	-1	0	-1	61	42
52	128.8	63	107.5	80	92.5	MAVERICK TX	-17	-14	-28	06	07
100	126.7	122	123.4	152	117.2	MEDINA TX	-3	-5	-8	43	44
15	70.8	26	109.0	32	145.3	MENARD TX	54	33	105	99	93
117	118.6	213	114.4	308	107.2	MIDLAND TX	-4	-6	-10	37	23
138	117.5	147	110.0	166	103.7	MILAM TX	-6	-6	-12	31	17
43	109.5	48	103.6	62	135.5	MILLS TX	-5	31	24	92	82
62	112.5	58	93.9	80	110.0	MITCHELL TX	-17	17	-2	58	28
113	108.3	156	125.8	166	117.5	MONTAGUE TX	16	-7	9	80	44
89	99.2	145	99.9	376	112.6	MONTGOMERY TX	1	13	14	85	34
23	65.9	50	109.7	65	104.1	MOORE TX	67	-5	58	97	18
43	106.2	66	120.1	87	123.4	MORRIS TX	13	3	16	87	58
25	162.8	25	130.5	20	95.5	MOTLEY TX	-20	-27	-41	01	09
124	103.7	186	121.5	230	113.5	NACOGDOCHES TX	17	-7	9	81	35
238	128.1	235	107.7	284	109.0	NAVARRO TX	-16	1	-15	23	26
31	98.0	44	98.1	42	74.7	NEWTON TX	0	-24	-24	10	01
108	123.4	110	106.8	157	129.3	NOLAN TX	-14	21	5	73	70
751	136.0	996	124.5	1358	134.9	NUECES TX	-8	8	-1	62	81
28	110.0	47	136.6	48	108.5	OCHILTREE TX	24	-21	-1	60	25
10	144.2	6	80.6	16	187.2	OLDHAM TX	-44	132	30	94	99
190	138.4	248	119.3	349	116.8	ORANGE TX	-14	-2	-16	21	42
155	140.8	158	119.0	187	117.2	PALO PINTO TX	-15	-2	-17	20	43
80	122.6	94	114.4	110	97.1	PANOLA TX	-7	-15	-21	13	10
152	121.1	189	116.9	260	112.5	PARKER TX	-3	-4	-7	44	34
22	85.2	46	134.0	55	123.5	PARMER TX	57	-8	45	96	58
30	111.4	51	113.6	60	116.8	PECOS TX	2	3	5	73	43
59	115.4	80	118.2	132	117.2	POLK TX	2	-1	2	67	43
421	125.2	524	116.1	624	117.9	POTTER TX	-7	2	-6	48	45
27	103.8	30	109.2	37	130.7	PRESIDIO TX	5	20	26	93	73
21	98.6	25	110.3	42	132.5	RAINS TX	12	20	34	95	76
66	94.3	144	102.5	266	105.5	RANDALL TX	9	3	12	84	20
14	190.2	17	174.8	27*	218.8	REAGAN TX	-8	25	15	86	99
10	106.6	11	105.0	11	59.4	REAL TX	-1	-43	-44	01	00
99	112.8	101	94.6	129	101.5	RED RIVER TX	-16	7	-10	36	14
70	180.3	62	109.5	68	122.8	REEVES TX	-39	12	-32	04	57
37	101.0	55	130.3	52	105.3	REFUGIO TX	29	-19	4	72	19
1	17.7	5	74.8	10	168.1	ROBERTS TX	324	125	851	99	99
83	132.1	77	105.2	96	118.5	ROBERTSON TX	-20	13	-10	35	46
30	98.6	40	106.0	61	108.8	ROCKWALL TX	8	3	10	82	25
92	106.1	87	84.1	117	120.8	RUNNELS TX	-21	44	14	85	52
162	107.3	193	103.8	243	99.9	RUSK TX	-3	-4	-7	44	13
40	122.5	54	139.3	58	111.0	SABINE TX	14	-20	-9	38	30
34	118.5	41	111.0	45	102.2	SAN AUGUSTINE TX	-6	-8	-14	26	15
15	82.5	19	83.8	38	100.2	SAN JACINTO TX	2	20	21	91	13
168	132.9	194	118.1	269	125.6	SAN PATRICIO TX	-11	6	-5	49	64

* SIGNIFICANTLY MORE DEATHS THAN EXPECTED

WHITE FEMALE

ALL CANCERS COMBINED

ICD 140-208

1950-1959 DEATHS	1950-1959 RATE	1960-1969 DEATHS	1960-1969 RATE	1970-1979 DEATHS	1970-1979 RATE	COUNTY AND STATE	PERCENT CHANGE 60S/50S	PERCENT CHANGE 70S/60S	PERCENT CHANGE 70S/50S	PERCENTILE %CHANGE 70S/50S	PERCENTILE RATES 1970S
64	131.4	58	114.0	72	112.2	SAN SABA TX	-13	-2	-15	24	33
16	136.4	15	104.1	26	149.2	SCHLEICHER TX	-24	43	9	81	95
71	108.8	113	130.4	129	128.0	SCURRY TX	20	-2	18	89	68
41	160.2	27	83.1	38	109.0	SHACKELFORD TX	-48	31	-32	04	26
99	98.7	128	109.6	155	103.8	SHELBY TX	11	-5	5	74	17
13	215.2	19	159.8	23	144.2	SHERMAN TX	-26	-10	-33	03	92
325	116.7	478	121.7	653	117.1	SMITH TX	4	-4	0	65	43
23	122.1	38	157.2	24	81.9	SOMERVELL TX	29	-48	-33	03	02
40	80.8	64	94.9	85	93.3	STARR TX	17	-2	16	87	07
88	153.1	91	136.4	88	116.6	STEPHENS TX	-11	-15	-24	10	42
8	177.0	10	161.2	7	85.4	STERLING TX	-9	-47	-52	00	03
14	97.5	17	94.3	22	120.7	STONEWALL TX	-3	28	24	92	52
20	156.1	23	153.5	24	130.6	SUTTON TX	-2	-15	-16	20	73
35	92.9	52	105.7	72	123.1	SWISHER TX	14	17	32	94	57
2171	129.2	3049	121.5	4295	125.3	TARRANT TX	-6	3	-3	56	63
365	116.1	500	113.7	610	110.9	TAYLOR TX	-2	-2	-4	51	30
7	67.6	6	75.1	10	110.5	TERRELL TX	11	47	63	98	29
58	125.2	48	80.9	95	137.5	TERRY TX	-35	70	10	81	85
18	89.7	28	125.9	20	84.6	THROCKMORTON TX	40	-33	-6	48	03
82	107.0	116	116.7	148	120.2	TITUS TX	9	3	12	84	51
277	108.6	394	113.1	514	117.1	TOM GREEN TX	4	4	8	79	43
765	113.5	1069	111.7	1595	121.0	TRAVIS TX	-2	8	7	77	53
50	123.8	42	97.7	58	87.7	TRINITY TX	-21	-10	-29	06	04
37	78.3	54	77.1	116	124.4	TYLER TX	-2	61	59	97	60
113	135.2	113	101.3	167	113.2	UPSHUR TX	-25	12	-16	20	35
6	40.8	24	128.4	17	85.4	UPTON TX	215	-33	110	99	03
77	104.8	112	125.3	114	107.4	UVALDE TX	20	-14	2	69	23
93	132.9	116	135.2	162	142.3	VAL VERDE TX	2	5	7	78	90
134	112.3	165	118.1	225	106.7	VAN ZANDT TX	5	-10	-5	50	22
135	109.3	200	107.5	285	114.8	VICTORIA TX	-2	7	5	74	38
50	104.2	65	96.2	105	101.1	WALKER TX	-8	5	-3	56	14
42	134.8	48	116.1	67	118.6	WALLER TX	-14	2	-12	31	47
44	106.9	52	108.7	74	126.0	WARD TX	2	16	18	89	65
103	117.6	132	112.5	201	149.7	WASHINGTON TX	-4	33	27	93	95
342	151.8	377	131.2	492	130.3	WEBB TX	-14	-1	-14	25	72
115	95.6	165	107.7	192	104.5	WHARTON TX	13	-3	9	80	18
46	100.0	64	122.1	55	90.2	WHEELER TX	22	-26	-10	37	06
466	111.8	645	117.0	761	117.6	WICHITA TX	5	1	5	74	44
103	111.6	102	86.9	139	109.6	WILBARGER TX	-22	26	-2	59	27
52	105.3	71	102.2	102	145.2	WILLACY TX	-3	42	38	96	93
193	106.8	221	97.4	330	105.8	WILLIAMSON TX	-9	9	-1	61	20
75	115.9	98	129.2	110	127.4	WILSON TX	12	-1	10	81	67
33	140.3	46	131.8	67	160.5	WINKLER TX	-6	22	14	86	98
107	107.8	137	117.8	168	113.0	WISE TX	9	-4	5	73	34
120	117.7	138	107.3	186	104.6	WOOD TX	-9	-3	-11	33	18
16	153.8	27	105.9	36	126.2	YOAKUM TX	-31	19	-18	17	65
92	99.0	124	101.9	164	113.2	YOUNG TX	3	11	14	86	35
7	41.1	23	115.0	27	88.3	ZAPATA TX	180	-23	115	99	05
43	113.2	51	120.5	61	139.5	ZAVALA TX	6	16	23	92	87
3157	109.0	4071	105.2	5154	99.9	UTAH UT	-4	-5	-8		
23	120.6	25	122.3	18	71.2	BEAVER UT	1	-42	-41	01	01
78	98.2	119	116.9	157	118.3	BOX ELDER UT	19	1	20	90	46
139	97.8	191	109.7	176	79.3	CACHE UT	12	-28	-19	16	02
74	125.7	79	102.6	112	117.0	CARBON UT	-18	14	-7	45	43
2	301.1	3	108.6	1	47.2	DAGGETT UT	-64	-57	-84	00	00
141	118.3	203	99.0	305	92.0	DAVIS UT	-16	-7	-22	11	06
22	86.2	33	130.3	34	95.9	DUCHESNE UT	51	-26	11	83	09
20	85.1	30	114.2	27	76.8	EMERY UT	34	-33	-10	36	01
12	89.3	16	115.3	11	65.4	GARFIELD UT	29	-43	-27	07	00
13	131.3	23	130.7	25	95.8	GRAND UT	0	-27	-27	07	09
43	143.8	32	73.7	61	101.0	IRON UT	-49	37	-30	05	14
24	96.8	26	105.5	23	80.0	JUAB UT	9	-24	-17	18	02
11	126.4	11	104.8	9	50.0	KANE UT	-17	-52	-60	00	00
50	146.9	49	127.6	34	72.6	MILLARD UT	-13	-43	-51	00	01
13	115.6	12	85.2	13	80.7	MORGAN UT	-26	-5	-30	05	02
7	111.1	6	88.3	6	74.5	PIUTE UT	-21	-16	-33	03	01
6	101.6	4	47.5	8	88.9	RICH UT	-53	87	-12	29	05
1488	114.3	1972	110.9	2391	104.1	SALT LAKE UT	-3	-6	-9	39	18
10	78.2	12	72.8	13	66.9	SAN JUAN UT	-7	-8	-14	24	00
63	98.2	66	89.3	79	89.2	SANPETE UT	-9	0	-9	38	05
73	154.3	58	102.5	74	101.6	SEVIER UT	-34	-1	-34	03	14
39	158.0	26	97.7	39	132.5	SUMMIT UT	-38	36	-16	21	76
40	89.0	89	143.7	97	116.2	TOOELE UT	62	-19	31	94	41
29	99.2	37	102.3	67	124.2	UINTAH UT	3	21	25	93	60
299	93.5	426	101.0	578	97.4	UTAH UT	8	-4	4	72	11
20	98.6	20	85.5	26	82.0	WASATCH UT	-13	-4	-17	20	02
43	103.2	56	91.2	93	79.3	WASHINGTON UT	-12	-13	-23	10	02
5	101.7	3	39.3	8	86.6	WAYNE UT	-61	120	-15	23	04
370	102.7	444	90.5	669	105.8	WEBER UT	-12	17	3	70	20
3141	146.3	3402*	141.0	3781	133.0	VERMONT VT	-4	-6	-9		
141	143.5	144	131.1	162	118.5	ADDISON VT	-9	-10	-17	18	46
232	157.4	242	137.1	304	145.0	BENNINGTON VT	-13	6	-8	42	93
214	152.9	225*	155.1	218	130.2	CALEDONIA VT	1	-16	-15	23	72
432	135.0	553	142.7	658	138.7	CHITTENDEN VT	6	-3	3	70	86
41	137.6	48	162.5	51	149.1	ESSEX VT	18	-8	8	79	95
251	159.1	252	145.8	248	127.4	FRANKLIN VT	-8	-13	-20	14	67
28	167.4	18	118.0	34	150.6	GRAND ISLE VT	-30	28	-10	36	96
93	153.1	85	132.1	97	113.0	LAMOILLE VT	-14	-14	-26	08	34
126	133.7	143	139.8	168	133.4	ORANGE VT	5	-5	0	63	78
173	158.1	170	144.5	208	149.2	ORLEANS VT	-9	3	-6	48	95
443	151.0	454	139.0	498	132.0	RUTLAND VT	-8	-5	-13	29	75
384	153.6	385	137.6	425	131.8	WASHINGTON VT	-10	-4	-14	25	75
264	136.2	318*	147.7	315	134.0	WINDHAM VT	8	-9	-2	60	79
319	138.5	365	137.5	395	125.4	WINDSOR VT	-1	-9	-9	37	63
15327	127.6	19923	123.1	26671	124.9	VIRGINIA VA	-4	2	-2		

* SIGNIFICANTLY MORE DEATHS THAN EXPECTED

WHITE FEMALE — ALL CANCERS COMBINED — ICD 140-208

1950-1959 DEATHS	1950-1959 RATE	1960-1969 DEATHS	1960-1969 RATE	1970-1979 DEATHS	1970-1979 RATE	COUNTY AND STATE	PERCENT CHANGE 60S/50S	PERCENT CHANGE 70S/60S	PERCENT CHANGE 70S/50S	PERCENTILE %CHANGE 70S/50S	PERCENTILE RATES 1970S
149	105.4	217	139.2	267	143.0	ACCOMACK VA	32	3	36	95	91
273	125.7	363	126.7	469	124.5	ALBEMARLE#VA	1	-2	-1	61	61
161	136.0	181	125.0	202	113.8	ALLEGHANY#VA	-8	-9	-16	20	36
21	108.0	18	80.3	32	114.1	AMELIA VA	-26	42	6	75	36
84	122.0	91	98.2	112	93.4	AMHERST VA	-20	-5	-23	10	07
38	111.8	45	106.1	47	87.3	APPOMATTOX VA	-5	-18	-22	12	04
914	149.7	1237*	145.0	1440*	146.0	ARLINGTON#VA	-3	1	-2	57	93
389	117.7	431	103.3	580	112.6	AUGUSTA#VA	-12	9	-4	52	34
39	143.7	41	142.7	55	167.9	BATH VA	-1	18	17	88	99
580	127.8	653	114.3	829	112.2	BEDFORD#VA	-11	-2	-12	30	33
26	111.1	30	109.4	37	110.8	BLAND VA	-2	1	0	63	29
77	117.1	113	134.9	119	108.9	BOTETOURT VA	15	-19	-7	44	25
38	93.3	57	124.6	70	133.5	BRUNSWICK VA	34	7	43	96	78
97	121.7	108	101.2	153	120.3	BUCHANAN VA	-17	19	-1	61	51
28	83.4	31	85.3	48	107.4	BUCKINGHAM VA	2	26	29	93	23
51	146.6	45	126.4	56	119.8	CAROLINE VA	-14	-5	-18	17	50
248	119.8	280	113.6	309	98.9	CARROLL#VA	-5	-13	-18	18	12
6	101.5	7	128.9	12	157.9	CHARLES CITY VA	27	22	55	97	98
51	131.3	52	116.5	68	125.9	CHARLOTTE VA	-11	8	-4	53	64
1621	139.3	1717	124.8	1972	127.8	CHESTERFIELD#VA	-10	2	-8	41	68
41	137.4	47	128.3	68	131.7	CLARKE VA	-7	3	-4	52	74
30	167.6	28	123.5	27	109.2	CRAIG VA	-26	-12	-35	03	26
70	112.5	90	104.3	121	109.6	CULPEPER VA	-7	5	-3	57	27
15	105.3	17	88.0	32	120.9	CUMBERLAND VA	-16	37	15	86	52
64	111.7	75	108.6	102	116.6	DICKENSON VA	-3	7	4	72	42
282	124.9	343	119.4	477	131.4	DINWIDDIE#VA	-4	10	5	74	74
29	153.3	35	131.7	42	117.6	ESSEX VA	-14	-11	-23	10	44
874	127.9	1790	126.5	3027	133.1	FAIRFAX#VA	-1	5	4	72	78
106	127.7	138	135.2	171	131.9	FAUQUIER VA	6	-2	3	71	75
69	123.0	71	112.7	77	103.7	FLOYD VA	-8	-8	-16	21	17
24	91.0	21	74.4	51	140.2	FLUVANNA VA	-18	88	54	97	88
97	104.1	117	101.4	147	94.7	FRANKLIN VA	-3	-7	-9	38	08
198	121.4	255	124.4	369	135.9	FREDERICK#VA	2	9	12	84	82
76	118.9	106	134.2	129	128.4	GILES VA	13	-4	8	79	69
51	123.6	77	138.9	95	116.1	GLOUCESTER VA	12	-16	-6	47	41
24	100.7	26	90.3	43	120.4	GOOCHLAND VA	-10	33	20	90	51
29	145.7	26	122.3	27	95.0	GREENE VA	-16	-22	-35	03	08
40	130.1	39	98.4	68	129.2	GREENSVILLE#VA	-24	31	-1	62	70
186*	178.9	136	106.0	181	115.8	HALIFAX#VA	-41	9	-35	02	40
86	106.3	143	118.3	242	131.6	HANOVER VA	11	11	24	92	74
328	118.0	647	121.9	1078	128.8	HENRICO VA	3	6	9	80	69
170	131.8	224	114.3	321	110.6	HENRY#VA	-13	-3	-16	21	29
24	111.2	33	165.0	27	121.8	HIGHLAND VA	48	-26	10	81	54
52	143.8	65	133.1	82	128.0	ISLE OF WIGHT VA	-7	-4	-11	33	68
108	92.6	160	91.9	275	108.1	JAMES CITY#VA	-1	18	17	88	24
22	140.5	14	80.0	17	74.4	KING AND QUEEN VA	-43	-7	-47	00	01
19	87.0	24	88.6	35	103.0	KING GEORGE VA	2	16	18	89	16
17	96.3	47*	202.2	35	112.7	KING WILLIAM VA	110	-44	17	88	34
36	114.2	54	123.5	71	112.3	LANCASTER VA	8	-9	-2	60	33
145	120.6	163	126.9	177	111.0	LEE VA	5	-13	-8	42	30
142	147.7	159	127.9	253	147.1	LOUDOUN VA	-13	15	0	63	94
46	102.8	63	117.9	101	147.8	LOUISA VA	15	25	44	96	94
51	131.5	49	99.8	69	124.5	LUNENBURG VA	-24	25	-5	49	61
48	141.4	62	158.7	57	113.8	MADISON VA	12	-28	-20	15	36
67	146.8	65	140.5	80	119.9	MATHEWS VA	-4	-15	-18	17	50
103	140.8	100	103.1	151	122.8	MECKLENBURG VA	-27	19	-13	28	57
33	127.5	49	144.5	57	126.5	MIDDLESEX VA	13	-12	-1	62	66
169	117.5	228	118.4	302	115.0	MONTGOMERY#VA	1	-3	-2	58	39
65	128.4	67	119.5	58	87.1	NELSON VA	-7	-27	-32	04	04
76	146.0	83	136.1	95	138.8	NORTHAMPTON VA	-7	2	-5	50	86
32	82.5	56	116.1	81	125.7	NORTHUMBERLAND VA	41	8	52	97	64
47	94.4	69	111.3	68	88.4	NOTTOWAY VA	18	-21	-6	46	05
71	135.3	102	158.7	119	137.5	ORANGE VA	17	-13	2	67	85
93	128.8	128	143.8	141	124.8	PAGE VA	12	-13	-3	55	61
60	96.8	77	106.9	102	104.6	PATRICK VA	10	-2	8	79	18
368	111.5	474	113.4	557	103.8	PITTSYLVANIA#VA	2	-8	-7	45	17
21	130.7	30	145.6	38	127.6	POWHATAN VA	11	-12	-2	58	68
47	109.1	50	92.0	73	118.0	PRINCE EDWARD VA	-16	28	8	79	45
104	127.4	203	131.0	373	129.3	PRINCE WILLIAM#VA	3	-1	1	67	71
144	140.7	164	124.7	201	114.8	PULASKI VA	-11	-8	-18	17	38
42	188.1	37	155.1	34	112.0	RAPPAHANNOCK VA	-18	-28	-40	01	32
22	102.4	29	100.4	59	172.6	RICHMOND VA	-2	72	68	98	99
818	129.4	1031	121.6	1312	116.8	ROANOKE#VA	-6	-4	-10	37	42
156	137.6	164	119.5	245	144.8	ROCKBRIDGE#VA	-13	21	5	74	92
287	126.5	329	113.1	423	111.0	ROCKINGHAM#VA	-11	-2	-12	30	30
106	108.4	124	111.9	192	136.2	RUSSELL VA	3	22	26	93	83
114	102.3	162	123.4	174	110.4	SCOTT VA	21	-11	8	79	29
154	127.7	150	105.3	189	106.7	SHENANDOAH VA	-18	1	-16	20	22
137	112.5	179	113.6	192	97.4	SMYTH VA	1	-14	-13	27	11
75	140.9	82	122.6	114	137.7	SOUTHAMPTON#VA	-13	12	-2	58	85
133	140.0	148	115.9	235	131.7	SPOTSYLVANIA#VA	-17	14	-6	48	74
52	121.7	86	132.0	142	132.6	STAFFORD VA	8	0	9	80	77
20	130.7	17	102.3	29	137.2	SURRY VA	-22	34	5	74	84
30	129.7	38	143.1	44	146.0	SUSSEX VA	10	2	13	84	93
213	143.8	279	146.2	317	132.2	TAZEWELL VA	2	-10	-8	41	76
81	140.7	99	135.2	135	132.9	WARREN VA	-4	-2	-6	49	77
249	113.8	336	119.4	421	115.5	WASHINGTON#VA	5	-3	2	67	40
36	108.0	44	94.7	93	144.8	WESTMORELAND VA	-12	53	34	95	92
263	151.1	248	115.5	356	141.4	WISE#VA	-24	22	-6	46	89
130	125.8	134	112.1	151	101.2	WYTHE VA	-11	-10	-20	15	14
1619	129.4	2333	135.3	3137*	139.2	CHESAPEAKE CITY#VA	5	3	8	78	87
237	136.6	370	142.1	453	132.1	HAMPTON CITY VA	4	-7	-3	55	75
259	131.4	349	115.3	562	132.0	NEWPORT NEWS CITY VA	-12	15	0	65	75
104	114.2	149	124.7	195	127.3	SUFFOLK CITY VA	9	2	12	83	67
16527	135.3	19693	125.1	25659	128.2	WASHINGTON WA	-8	2	-5		
32	110.7	46	111.8	70	120.6	ADAMS WA	1	8	9	80	51
79	106.9	95	105.0	133	122.9	ASOTIN WA	-2	17	15	86	57
147	90.8	274	111.1	433	120.3	BENTON WA	22	8	33	94	51
243	123.8	288	119.0	366	124.5	CHELAN WA	-4	5	1	65	61

* SIGNIFICANTLY MORE DEATHS THAN EXPECTED

COUNTIES COMBINED; SEE TECHNICAL NOTES

WHITE FEMALE — ALL CANCERS COMBINED — ICD 140-208

1950-1959 DEATHS	1950-1959 RATE	1960-1969 DEATHS	1960-1969 RATE	1970-1979 DEATHS	1970-1979 RATE	COUNTY AND STATE	PERCENT CHANGE 60S/50S	PERCENT CHANGE 70S/60S	PERCENT CHANGE 70S/50S	PERCENTILE %CHANGE 70S/50S	PERCENTILE RATES 1970S
137	112.6	222	129.4	327	121.3	CLALLAM WA	15	-6	8	78	53
567	139.1	670	118.9	1032	129.1	CLARK WA	-15	9	-7	44	70
38	139.5	34	120.1	40	115.1	COLUMBIA WA	-14	-4	-18	18	39
284	127.1	344	114.0	525	135.2	COWLITZ WA	-10	19	6	76	81
41	80.3	81	113.3	90	91.4	DOUGLAS WA	41	-19	14	85	06
16	136.4	17	138.1	17	102.3	FERRY WA	1	-26	-25	09	15
69	128.6	97	112.3	155	122.3	FRANKLIN WA	-13	9	-5	50	55
23	139.0	16	84.6	24	116.0	GARFIELD WA	-39	37	-17	20	40
91	99.5	137	95.7	260	130.5	GRANT WA	-4	36	31	94	73
380	148.9	456	144.9	528	140.0	GRAYS HARBOR WA	-3	-3	-6	47	88
76	141.3	97	104.0	221	126.7	ISLAND WA	-26	22	-10	35	66
56	127.5	59	102.7	89	102.8	JEFFERSON WA	-19	0	-19	15	16
5932*	146.9	6896	131.9	8391	131.6	KING WA	-10	0	-10	35	74
491	136.1	579	122.1	825	127.3	KITSAP WA	-10	4	-6	46	67
123	127.0	139	127.1	154	122.3	KITTITAS WA	0	-4	-4	54	55
70	126.0	72	111.3	92	116.6	KLICKITAT WA	-12	5	-7	43	42
343	140.8	354	125.3	407	120.4	LEWIS WA	-11	-4	-14	24	51
69	128.6	68	112.6	98	137.1	LINCOLN WA	-12	22	7	77	84
97	138.2	112	113.2	198	131.6	MASON WA	-18	16	-5	51	74
121	113.7	159	124.3	187	120.9	OKANOGAN WA	9	-3	6	76	52
108	132.5	115	114.2	175	138.3	PACIFIC WA	-14	21	4	72	86
36	116.3	38	107.3	56	124.4	PEND OREILLE WA	-8	16	7	77	60
1784	132.9	2244	132.8	2926	136.8	PIERCE WA	0	3	3	70	84
22	110.6	28	105.9	58	125.1	SAN JUAN WA	-4	18	13	85	62
270	108.0	340	111.6	431	112.9	SKAGIT WA	3	1	5	73	34
17	75.2	25	104.8	23	71.4	SKAMANIA WA	39	-32	-5	50	01
908	136.9	1202	125.6	1761	126.8	SNOHOMISH WA	-8	1	-7	43	66
1664	130.0	1953	122.6	2429	128.9	SPOKANE WA	-6	5	-1	62	70
94	115.6	91	96.9	128	106.5	STEVENS WA	-16	10	-8	42	21
357	143.7	403	116.4	698	128.2	THURSTON WA	-19	10	-11	34	69
16	87.5	30	155.0	23	98.0	WAHKIAKUM WA	77	-37	12	84	11
264	127.5	296	116.0	394	131.3	WALLA WALLA WA	-9	13	3	70	74
507	129.8	547	120.9	644	117.7	WHATCOM WA	-7	-3	-9	38	44
165	121.4	173	112.6	208	128.8	WHITMAN WA	-7	14	6	76	69
790	130.8	896	119.9	1043	116.6	YAKIMA WA	-8	-3	-11	34	42
10434	132.7	11786	127.1	14078	128.6	WEST VIRGINIA WV	-4	1	-3		
106	127.3	85	90.5	122	119.9	BARBOUR WV	-29	32	-6	48	50
221	142.9	260	144.6	302	130.0	BERKELEY WV	1	-10	-9	38	72
131	134.2	129	106.3	161	113.0	BOONE WV	-21	6	-16	21	34
88	106.5	84	99.7	111	118.2	BRAXTON WV	-6	19	11	83	46
146	132.1	199	142.7	221	128.9	BROOKE WV	8	-10	-2	57	70
734	139.4	805	126.7	935	131.4	CABELL WV	-9	4	-6	48	74
51	121.6	56	127.8	70	143.8	CALHOUN WV	5	12	18	89	92
54	110.6	44	89.8	52	91.1	CLAY WV	-19	1	-18	18	06
49	106.1	64	149.0	73	146.0	DODDRIDGE WV	40	-2	38	96	93
326	143.7	349	130.1	422	131.7	FAYETTE WV	-10	1	-8	40	74
51	120.5	47	103.6	57	116.0	GILMER WV	-14	12	-4	54	40
47	124.8	63	141.8	60	109.5	GRANT WV	14	-23	-12	30	27
205	135.0	186	101.0	266	117.4	GREENBRIER WV	-25	16	-13	28	44
78	141.6	94	148.0	86	109.9	HAMPSHIRE WV	5	-26	-22	11	28
199	142.2	248	140.2	312	141.7	HANCOCK WV	-1	1	0	63	90
53	120.9	56	115.0	66	108.4	HARDY WV	-5	-6	-10	35	25
554	135.8	645	137.2	749	138.1	HARRISON WV	1	1	2	67	86
75	95.1	128	139.2	134	110.0	JACKSON WV	46	-21	16	87	28
91	123.0	109	122.1	162	134.3	JEFFERSON WV	-1	10	9	80	80
1198	138.1	1416	126.9	1824	137.4	KANAWHA WV	-8	8	0	63	85
160	118.7	160	112.9	185	126.8	LEWIS WV	-5	12	7	77	66
93	122.6	106	115.2	143	128.3	LINCOLN WV	-6	11	5	73	69
225	135.2	269	137.5	340*	148.8	LOGAN WV	2	8	10	81	95
223	133.5	254	138.0	269	131.6	MC DOWELL WV	3	-5	-1	60	74
444	137.8	489	129.4	595	135.2	MARION WV	-6	4	-2	59	81
258	145.0	304	148.1	289	119.8	MARSHALL WV	2	-19	-17	18	50
103	100.9	145	123.3	156	109.3	MASON WV	22	-11	8	79	26
352	133.4	404	122.6	480	117.4	MERCER WV	-8	-4	-12	30	44
155	156.2	141	118.7	182	122.7	MINERAL WV	-24	3	-21	12	56
174	132.9	224*	151.9	237	145.8	MINGO WV	14	-4	10	81	93
344	143.7	357	124.9	432	126.0	MONONGALIA WV	-13	1	-12	30	65
94	149.4	73	98.9	109	133.5	MONROE WV	-34	35	-11	34	78
62	151.6	51	107.0	77	120.5	MORGAN WV	-29	13	-20	13	51
115	118.8	123	110.7	194	141.1	NICHOLAS WV	-7	28	19	90	89
591	145.5	703*	159.1	719*	151.7	OHIO WV	9	-5	4	72	96
37	92.2	46	100.7	75	143.2	PENDLETON WV	9	42	55	97	91
41	125.4	49	125.2	52	119.0	PLEASANTS WV	0	-5	-5	50	48
69	139.1	75	132.2	71	103.8	POCAHONTAS WV	-5	-22	-25	08	17
155	119.3	161	110.7	197	113.0	PRESTON WV	-7	2	-5	49	34
85	99.2	122	106.2	185	118.4	PUTNAM WV	7	11	19	90	46
304	120.0	363	110.5	537	124.3	RALEIGH WV	-8	12	4	71	60
165	129.7	193	135.8	200	119.3	RANDOLPH WV	5	-12	-8	41	48
80	98.6	98	121.8	98	113.6	RITCHIE WV	24	-7	15	87	35
118	138.0	109	113.0	105	98.6	ROANE WV	-18	-13	-29	06	12
115	149.1	107	121.8	105	106.8	SUMMERS WV	-18	-12	-28	06	22
136	149.2	131	131.9	145	129.6	TAYLOR WV	-12	-2	-13	27	71
73	147.6	76	144.0	67	112.2	TUCKER WV	-2	-22	-24	09	33
89	133.5	74	102.6	90	123.7	TYLER WV	-23	21	-7	43	59
130	130.4	142	129.5	150	118.2	UPSHUR WV	-1	-9	-9	38	46
154	100.5	206	113.6	261	116.4	WAYNE WV	13	2	16	87	42
65	119.7	84	143.1	87	133.9	WEBSTER WV	19	-6	12	83	79
144	147.2	149	135.6	189	149.7	WETZEL WV	-8	10	2	67	95
34	134.0	41	152.1	44	145.8	WIRT WV	14	-4	9	80	93
497	136.2	554	123.4	672	123.1	WOOD WV	-9	0	-10	37	57
93	111.2	136	123.8	156	115.6	WYOMING WV	11	-7	4	72	40
26718*	145.5	30156*	134.2	34653	129.2	WISCONSIN WI	-8	-4	-11		
60	137.7	84	152.5	108	131.4	ADAMS WI	11	-14	-5	51	74
158	163.8	158	142.8	176	141.3	ASHLAND WI	-13	-1	-14	26	89
218	125.4	292	137.0	318	123.3	BARRON WI	9	-10	-2	60	58
93	132.8	107	146.8	119	133.3	BAYFIELD WI	10	-9	0	65	78
693	142.3	824	127.0	1031	125.1	BROWN WI	-11	-2	-12	30	62

* SIGNIFICANTLY MORE DEATHS THAN EXPECTED

WHITE FEMALE ALL CANCERS COMBINED ICD 140-208

1950-1959 DEATHS	1950-1959 RATE	1960-1969 DEATHS	1960-1969 RATE	1970-1979 DEATHS	1970-1979 RATE	COUNTY AND STATE	PERCENT CHANGE 60S/50S	PERCENT CHANGE 70S/60S	PERCENT CHANGE 70S/50S	PERCENTILE %CHANGE 70S/50S	PERCENTILE RATES 1970S
116	158.8	111	131.3	115	122.5	BUFFALO WI	-17	-7	-23	11	56
67	126.2	91	142.6	99	112.0	BURNETT WI	13	-21	-11	33	33
116	125.6	129	114.4	184	130.0	CALUMET WI	-9	14	3	71	72
240	116.3	287	114.5	359	123.5	CHIPPEWA WI	-1	8	6	76	58
201	121.0	253	133.9	277	129.3	CLARK WI	11	-3	7	77	71
313	145.6	352	141.9	332	114.6	COLUMBIA WI	-3	-19	-21	12	38
100	122.3	102	115.1	132	122.5	CRAWFORD WI	-6	6	0	64	56
1219	137.8	1445	126.4	1706	119.7	DANE WI	-8	-5	-13	27	49
453	142.2	495	132.1	570	124.6	DODGE WI	-7	-6	-12	29	61
162	150.3	158	119.6	194	115.7	DOOR WI	-20	-3	-23	10	40
375	157.0	419*	156.7	442*	149.2	DOUGLAS WI	0	-5	-5	50	95
185	137.8	213	133.4	217	120.9	DUNN WI	-3	-9	-12	30	52
391	136.1	479	135.0	525	123.1	EAU CLAIRE WI	-1	-9	-10	37	57
30	182.6	32	174.1	34	133.6	FLORENCE WI	-5	-23	-27	07	78
620	148.7	636	132.8	714	128.4	FOND DU LAC WI	-11	-3	-14	26	69
41	112.8	50	122.2	75	140.6	FOREST WI	8	15	25	92	88
300	124.1	384	136.7	371	116.2	GRANT WI	10	-15	-6	46	41
206	139.1	210	123.1	245	121.6	GREEN WI	-12	-1	-13	29	54
145	160.0	144	129.2	180	130.7	GREEN LAKE WI	-19	1	-18	17	73
160	144.2	141	117.9	155	119.5	IOWA WI	-18	1	-17	19	49
45	112.4	54	110.0	83	142.5	IRON WI	-2	30	27	93	90
120	150.5	135	147.8	152	145.3	JACKSON WI	-2	-2	-3	54	93
438	156.2	454	134.1	483	117.7	JEFFERSON WI	-14	-12	-25	09	45
131	132.2	158	140.2	161	113.2	JUNEAU WI	6	-19	-14	24	35
590	149.3	671	125.2	886	134.5	KENOSHA WI	-16	7	-10	36	80
124	142.1	125	123.0	167	141.7	KEWAUNEE WI	-13	15	0	63	90
569	151.5	614	136.0	754	137.8	LA CROSSE WI	-10	1	-9	39	85
119	123.1	123	115.6	133	116.2	LAFAYETTE WI	-6	1	-6	48	41
155	149.3	161	131.9	172	123.9	LANGLADE WI	-12	-6	-17	19	59
167	146.9	189	135.1	254	149.7	LINCOLN WI	-8	11	2	68	95
538	150.6	608	137.6	671	127.0	MANITOWOC WI	-9	-8	-16	21	67
499	137.1	601	127.9	709	122.4	MARATHON WI	-7	-4	-11	34	56
296	158.3	279	124.0	350	128.3	MARINETTE WI	-22	3	-19	16	69
91	159.0	97	143.3	92	112.5	MARQUETTE WI	-10	-22	-29	06	34
413	142.7	463	135.0	523	124.0	MENOMINEE#WI	-5	-8	-13	27	59
7385*	159.9	8056*	145.1	8433*	138.3	MILWAUKEE WI	-9	-5	-14	27	86
217	132.5	219	116.8	255	118.7	MONROE WI	-12	2	-10	35	47
155	154.2	187	137.0	247	132.6	ONEIDA WI	-11	-3	-14	25	77
553	139.2	656	129.2	801	128.1	OUTAGAMIE WI	-7	-1	-8	42	68
202	149.8	245	125.7	361	127.3	OZAUKEE WI	-16	1	-15	23	67
43	108.2	35	76.5	61	111.8	PEPIN WI	-29	46	3	71	32
136	115.4	163	116.0	166	105.6	PIERCE WI	1	-9	-8	40	20
153	117.9	214	132.4	239	115.7	POLK WI	12	-13	-2	59	40
224	128.5	285	138.3	362	139.7	PORTAGE WI	8	1	9	80	88
94	120.0	109	120.1	157	136.9	PRICE WI	0	14	14	86	84
921*	156.4	1014	134.1	1154	128.8	RACINE WI	-14	-4	-18	18	69
129	126.8	152	137.0	141	105.3	RICHLAND WI	8	-23	-17	19	19
752	143.8	904*	143.3	1018	134.0	ROCK WI	0	-7	-7	45	79
101	138.6	98	112.5	132	133.3	RUSK WI	-19	18	-4	54	78
157	109.0	198	119.6	252	121.0	ST CROIX WI	10	1	11	83	52
301	144.4	310	129.2	315	109.9	SAUK WI	-11	-15	-24	09	28
60	126.8	86	146.3	120	157.5	SAWYER WI	15	8	24	92	98
678	148.3	723	129.7	867	133.2	SHEBOYGAN WI	-13	3	-10	35	78
95	122.6	98	104.9	131	119.9	TAYLOR WI	-14	14	-2	58	50
167	123.0	215	144.0	228	121.2	TREMPEALEAU WI	17	-16	-2	60	53
184	124.4	210	122.8	216	108.3	VERNON WI	-1	-12	-13	28	24
80*	174.4	78	129.0	144	128.7	VILAS WI	-26	0	-26	08	69
416	154.0	462	136.4	528	124.6	WALWORTH WI	-11	-9	-19	16	61
79	136.2	92	133.5	118	132.6	WASHBURN WI	-2	-1	-3	57	76
263	138.7	331	133.0	406	119.1	WASHINGTON WI	-4	-11	-14	25	48
736	145.0	1021	131.6	1467	133.4	WAUKESHA WI	-9	1	-8	42	78
310	141.9	319	122.3	396	127.3	WAUPACA WI	-14	4	-10	35	67
108	124.3	109	111.7	162	131.3	WAUSHARA WI	-10	18	6	75	74
717	136.9	810	126.1	977	127.7	WINNEBAGO WI	-8	1	-7	45	68
345	144.9	429	137.0	531	131.8	WOOD WI	-5	-4	-9	38	74
1358	121.2	1605	110.6	2076	114.1	WYOMING WY	-9	3	-6		
87	124.2	113	129.5	119	114.6	ALBANY WY	4	-12	-8	43	38
43	83.3	54	93.9	67	93.0	BIG HORN WY	13	-1	12	83	07
31	147.3	25	83.9	42	91.2	CAMPBELL WY	-43	9	-38	02	06
59	108.4	82	128.0	90	124.2	CARBON WY	18	-3	15	86	60
38	131.3	33	95.2	58	134.8	CONVERSE WY	-27	42	3	69	80
20	109.8	27	144.6	34	149.7	CROOK WY	32	3	36	96	95
77	127.3	106	116.8	155	128.4	FREMONT WY	-8	10	1	66	69
53	100.2	61	93.8	67	83.2	GOSHEN WY	-6	-11	-17	19	03
23	92.0	38	106.8	49	123.5	HOT SPRINGS WY	16	16	34	95	58
27	110.3	44	139.9	42	112.0	JOHNSON WY	27	-20	2	67	32
190	111.5	271	115.4	351	117.9	LARAMIE WY	3	2	6	75	45
27	89.6	38	104.8	54	120.0	LINCOLN WY	17	15	34	95	50
200	145.0	212	106.9	306	121.7	NATRONA WY	-26	14	-16	21	54
19	113.9	27	138.7	29	128.4	NIOBRARA WY	22	-7	13	84	69
90	139.9	103	118.9	125	98.4	PARK-TETON#WY	-15	-17	-30	05	12
45	127.8	42	106.9	55	111.5	PLATTE WY	-16	4	-13	28	31
128	125.2	137	114.5	176	123.4	SHERIDAN WY	-9	8	-1	60	58
10	112.5	13	122.9	13	79.2	SUBLETTE WY	9	-36	-30	05	02
95	155.7	86	110.2	133	124.0	SWEETWATER WY	-29	12	-20	14	59
33	98.5	28	74.0	40	102.4	UINTA WY	-25	38	4	72	15
38	173.4	40	130.1	39	99.6	WASHAKIE WY	-25	-23	-43	01	12

* SIGNIFICANTLY MORE DEATHS THAN EXPECTED # COUNTIES COMBINED; SEE TECHNICAL NOTES

NON-WHITE MALE — ALL CANCERS COMBINED — ICD 140-208

1950-1959 DEATHS	1950-1959 RATE	1960-1969 DEATHS	1960-1969 RATE	1970-1979 DEATHS	1970-1979 RATE	COUNTY AND STATE	PERCENT CHANGE 60S/50S	PERCENT CHANGE 70S/60S	PERCENT CHANGE 70S/50S	PERCENTILE %CHANGE 70S/50S	PERCENTILE RATES 1970S
102662	166.3	160896	209.5	230120	231.6	UNITED STATES	26	11	39		
4234	124.9	5886	166.5	8388	224.1	ALABAMA AL	33	35	79		
30	98.4	44	156.8	46	180.1	AUTAUGA AL	59	15	83	71	53
46	175.1	56	169.5	95	223.6	BALDWIN AL	-3	32	28	39	68
34	68.8	71	163.8	78	183.8	BARBOUR AL	138	12	167	90	55
26	141.1	27	157.5	28	203.3	BIBB AL	12	29	44	50	61
2	56.2	5	172.9	6	212.1	BLOUNT AL	207	23	277	96	64
41	100.1	53	141.4	84	245.9	BULLOCK AL	41	74	146	88	76
48	122.3	48	124.2	67	178.7	BUTLER AL	2	44	46	51	53
57	109.3	103	176.8	142	214.3	CALHOUN AL	62	21	96	77	65
43	86.0	79	163.2	79	163.3	CHAMBERS AL	90	0	90	74	48
5	104.6	7	163.1	12	227.2	CHEROKEE AL	56	39	117	82	70
17	107.1	18	116.6	33	217.1	CHILTON AL	9	86	103	79	66
33	92.7	52	153.1	57	166.8	CHOCTAW AL	65	9	80	70	49
63	121.6	80	168.0	109	224.6	CLARKE AL	38	34	85	72	68
5	82.4	10	125.7	15	245.1	CLAY AL	52	95	197	93	75
1	70.1	5	312.7	5	177.6	CLEBURNE AL	346	-43	153	89	53
11	55.7	37	183.5	52	234.1	COFFEE AL	229	28	320	97	72
37	116.4	52	156.4	75	206.9	COLBERT AL	34	32	78	69	63
53	157.3	46	147.4	48	157.9	CONECUH AL	-6	7	0	25	47
21	175.7	22	162.4	30	188.6	COOSA AL	-8	16	7	28	56
19	91.3	42	193.3	46	234.1	COVINGTON AL	112	21	156	89	72
14	74.8	30	152.9	28	159.1	CRENSHAW AL	104	4	113	81	47
2	90.0	2	218.4	2	53.2	CULLMAN AL	143	-76	-41	11	31
16	97.5	25	137.0	39	182.6	DALE AL	40	33	87	73	54
133	107.0	182	155.5	211	185.3	DALLAS AL	45	19	73	66	55
10*	252.4	7	190.3	7	257.3	DE KALB AL	-25	35	2	26	79
38	98.9	64	177.3	62	177.0	ELMORE AL	79	0	79	69	52
59	154.6	60	133.3	90	202.4	ESCAMBIA AL	-14	52	31	41	61
64	141.3	91	179.6	146	262.4	ETOWAH AL	27	46	86	73	80
14	139.4	16	180.8	13	114.2	FAYETTE AL	30	-37	-18	18	39
7	123.8	9	239.9	6	133.6	FRANKLIN AL	94	-44	8	28	42
10	83.0	20	170.0	28	215.6	GENEVA AL	105	27	160	90	65
59	123.4	45	107.1	71	176.4	GREENE AL	-13	65	43	49	52
49	94.2	79	154.2	92	181.7	HALE AL	64	18	93	75	54
24	107.3	40	179.6	42	192.3	HENRY AL	67	7	79	69	58
41	84.5	73	151.0	98	178.0	HOUSTON AL	79	18	111	81	53
6	67.6	8	92.9	16	172.4	JACKSON AL	37	86	155	89	51
1220	159.7	1653	193.1	2534*	274.2	JEFFERSON AL	21	42	72	65	85
6	65.5	10	111.5	19	202.6	LAMAR AL	70	82	209	93	61
35	120.5	48	159.9	69	223.0	LAUDERDALE AL	33	39	85	72	68
23	118.8	24	126.1	37	184.8	LAWRENCE AL	6	47	56	56	55
65	117.9	90	157.9	122	200.5	LEE AL	34	27	70	64	61
20	73.5	36	128.4	55	205.0	LIMESTONE AL	75	60	179	91	62
53	107.9	38	87.1	43	117.9	LOWNDES AL	-19	35	9	29	40
99	95.0	115	109.7	178	172.8	MACON AL	16	57	82	71	51
64	96.5	95	127.5	193	217.1	MADISON AL	32	70	125	84	66
87	117.1	114	171.0	131	206.4	MARENGO AL	46	21	76	68	63
4	203.4	6	383.5	7	236.3	MARION AL	89	-38	16	32	72
7	171.4	2	39.5	15	370.3	MARSHALL AL	-77	839	116	82	95
391	149.9	703	214.2	980*	262.6	MOBILE AL	43	23	75	67	80
38	90.8	52	132.5	65	162.7	MONROE AL	46	23	79	70	48
266	125.9	390	181.4	583	247.9	MONTGOMERY AL	44	37	97	77	76
33	105.6	50	147.9	59	177.9	MORGAN AL	40	20	68	63	53
46	93.4	38	86.3	76	167.0	PERRY AL	-8	93	79	69	49
41	119.4	39	112.2	66	175.4	PICKENS AL	-6	56	47	52	52
41	91.3	65	168.2	82	217.4	PIKE AL	84	29	138	87	66
26	144.1	24	135.9	35	185.2	RANDOLPH AL	-6	36	29	40	55
69	114.4	128	198.9	136	192.2	RUSSELL AL	74	-3	68	63	58
23	161.9	29	182.2	21	142.6	ST CLAIR AL	13	-22	-12	20	44
39	159.2	32	125.1	55	203.4	SHELBY AL	-21	63	28	39	61
68	106.3	73	121.2	127	228.1	SUMTER AL	14	88	114	82	70
66	114.0	105	168.2	126	184.7	TALLADEGA AL	48	10	62	60	55
28	79.1	36	112.7	72	197.4	TALLAPOOSA AL	42	75	149	88	60
110	120.5	132	127.7	274	232.9	TUSCALOOSA AL	6	82	93	75	71
42	148.7	52	190.7	49	197.0	WALKER AL	28	3	32	42	60
17	99.2	26	150.7	48	248.7	WASHINGTON AL	52	65	151	89	76
69	123.0	72	141.5	72	144.2	WILCOX AL	15	2	17	33	44
		1	344.1	1	344.1	WINSTON AL		0			94
		195	150.9	315	177.4	ALASKA AK		18			
391	110.5	668	154.4	872	114.5	ARIZONA AZ	40	-26	4		
36	64.5	65	113.2	100	117.3	APACHE AZ	76	4	82	71	39
6	165.2	18*	373.3	15	117.8	COCHISE AZ	126	-68	-29	14	39
15	51.3	29	88.3	37	84.5	COCONINO AZ	72	-4	65	61	35
7	62.9	17	137.4	16	95.6	GILA AZ	118	-30	52	55	36
6	145.0	5	89.5	4	48.0	GRAHAM AZ	-38	-46	-67	04	31
1	59.0	1	94.4			GREENLEE AZ	60				
161	160.9	266	187.1	379	133.5	MARICOPA AZ	16	-29	-17	18	42
1	49.1	4	146.9	2	43.0	MOHAVE AZ	199	-71	-12	20	30
22	44.2	46	77.9	58	76.8	NAVAJO AZ	76	-1	74	66	34
79	144.0	122	185.0	156	118.3	PIMA AZ	29	-36	-18	18	40
36	121.9	57	166.4	50	97.8	PINAL AZ	36	-41	-20	18	36
4	150.5	5	242.1	3	44.4	SANTA CRUZ AZ	61	-82	-71	03	31
4	104.1	6	154.5	3	33.2	YAVAPAI AZ	48	-79	-68	03	30
13	119.0	27	173.2	49	166.2	YUMA AZ	45	-4	40	47	49
2057	116.2	2856	162.0	3511	202.2	ARKANSAS AR	39	25	74		
27	122.9	50	188.2	61	234.9	ARKANSAS AR	53	25	91	75	72
.33	91.9	58	158.9	72	199.2	ASHLEY AR	73	25	117	82	60
		3	523.9	1	28.1	BENTON AR		-95			29
1*	188.7					BOONE AR					
21	91.1	34	149.0	50	204.2	BRADLEY AR	64	37	124	84	62

* SIGNIFICANTLY MORE DEATHS THAN EXPECTED

NON-WHITE MALE

ALL CANCERS COMBINED

ICD 140-208

1950-1959 DEATHS	1950-1959 RATE	1960-1969 DEATHS	1960-1969 RATE	1970-1979 DEATHS	1970-1979 RATE	COUNTY AND STATE		PERCENT CHANGE 60S/50S	PERCENT CHANGE 70S/60S	PERCENT CHANGE 70S/50S	PERCENTILE %CHANGE 70S/50S	PERCENTILE RATES 1970S
15	136.7	12	122.2	12	116.9	CALHOUN	AR	-11	-4	-14	19	39
				1	212.0	CARROLL	AR					64
87	137.9	102	170.0	105	197.9	CHICOT	AR	23	16	44	50	60
25	94.6	40	159.1	43	173.1	CLARK	AR	68	9	83	71	51
1*	112.4					CLAY	AR					
12	149.0	8	111.4	9	92.6	CLEVELAND	AR	-25	-17	-38	12	36
43	106.0	74	170.9	89	201.2	COLUMBIA	AR	61	18	90	74	61
30	184.5	26	176.4	22	140.4	CONWAY	AR	-4	-20	-24	16	43
20	241.9	14	197.0	12	159.3	CRAIGHEAD	AR	-19	-19	-34	13	47
10	281.2	3	91.8	7	163.1	CRAWFORD	AR	-67	78	-42	11	48
89	72.1	155	133.4	195	189.4	CRITTENDEN	AR	85	42	163	90	57
39	133.7	37	144.4	43	179.4	CROSS	AR	8	24	34	44	53
36	220.4	26	139.3	50	234.2	DALLAS	AR	-37	68	6	28	72
49	97.0	77	171.5	77	178.2	DESHA	AR	77	4	84	72	53
28	103.7	23	84.7	49	199.5	DREW	AR	-18	135	92	75	60
11	104.6	15	123.1	23	167.9	FAULKNER	AR	18	36	60	59	49
		1	94.4			FRANKLIN	AR					
42	168.0	66	245.2	65	234.2	GARLAND	AR	46	-5	39	47	72
6	267.5			3	195.4	GRANT	AR			-27	15	59
		1	188.7	1	188.7	GREENE	AR		0			57
45	106.1	57	144.4	62	154.6	HEMPSTEAD	AR	36	7	46	51	46
16	149.8	21	157.9	25	172.2	HOT SPRING	AR	5	9	15	31	51
5	39.4	21	173.3	29	218.6	HOWARD	AR	340	26	455	98	67
5	164.2	6	264.5	8	283.2	INDEPENDENCE	AR	61	7	72	66	87
1	596.2					IZARD	AR					
19	119.0	29	189.4	31	181.2	JACKSON	AR	59	-4	52	55	54
220	134.1	305	192.6	367	223.0	JEFFERSON	AR	44	16	66	62	68
3	239.8	4	326.0	7*	595.3	JOHNSON	AR	36	83	148	88	99
18	78.5	37	161.1	47	205.5	LAFAYETTE	AR	105	28	162	90	62
5	379.6			1	27.0	LAWRENCE	AR			-93	00	29
55	90.3	100	167.2	109	214.9	LEE	AR	85	29	138	87	65
34	96.9	41	131.1	35	123.5	LINCOLN	AR	35	-6	27	39	40
11	66.9	15	101.5	32	178.0	LITTLE RIVER	AR	52	75	166	90	53
1	38.2	3	136.8	1	60.6	LOGAN	AR	258	-56	58	57	32
22	91.9	33	135.0	45	198.2	LONOKE	AR	47	47	116	82	60
				1	490.8	MARION	AR					98
58	156.2	61	168.3	72	195.1	MILLER	AR	8	16	25	38	59
72	82.1	106	130.0	155	215.3	MISSISSIPPI	AR	58	66	162	90	65
47	119.6	64	165.8	62	182.5	MONROE	AR	39	10	53	55	54
23	121.5	21	96.2	51	264.9	NEVADA	AR	-21	176	118	83	81
53	105.4	82	145.5	114	192.1	OUACHITA	AR	38	32	82	71	58
1	70.1					PERRY	AR					
128	104.8	187	154.1	199	188.8	PHILLIPS	AR	47	23	80	70	57
3	145.0	3	142.9	4	263.3	PIKE	AR	-1	84	82	71	80
11	86.8	17	134.8	21	172.9	POINSETT	AR	55	28	99	78	51
				1	172.0	POLK	AR					51
4	161.8	5	140.4	6	174.5	POPE	AR	-13	24	8	28	52
7	71.4	12	131.1	13	159.5	PRAIRIE	AR	84	22	123	84	47
338	161.8	466	202.5	581	233.5	PULASKI	AR	25	15	44	50	71
		2	139.6	2	130.6	RANDOLPH	AR		-6			42
84	103.9	121	158.3	141	204.4	ST FRANCIS	AR	52	29	97	77	62
5	61.8	6	50.0	7	92.9	SALINE	AR	-19	86	50	54	36
1	163.6					SCOTT	AR					
1*	188.7					SEARCY	AR					
24	114.2	51	241.4	55	240.0	SEBASTIAN	AR	111	-1	110	81	74
5	91.7	8	197.9	16	322.4	SEVIER	AR	116	63	252	96	92
				1	490.8	SHARP	AR					98
				1	424.0	STONE	AR					97
62	103.0	91	149.9	137	211.9	UNION	AR	45	41	106	80	64
2	166.6	3	148.7	9	242.8	WASHINGTON	AR	-11	63	46	51	74
10	184.7	13	214.3	21	272.1	WHITE	AR	16	27	47	52	84
31	108.9	34	131.9	44	208.6	WOODRUFF	AR	21	58	91	75	63
2	105.6	6	263.9	8	365.5	YELL	AR	150	38	246	95	95
5319	173.1	9753	187.0	16384	161.3	CALIFORNIA	CA	8	-14	-7		
544	175.6	1064	192.3	1908	204.6	ALAMEDA	CA	10	6	17	33	62
				1	5.0	ALPINE	CA					29
		2	61.2			AMADOR	CA					
15	199.5	15	127.2	29	154.0	BUTTE	CA	-36	21	-23	16	46
2	131.2			1	49.7	CALAVERAS	CA			-62	05	31
8	267.2	6	451.4	4	111.2	COLUSA	CA	69	-75	-58	06	38
78	129.5	199	180.2	399	167.9	CONTRA COSTA	CA	39	-7	30	40	49
1	70.7	2	69.8	2	63.2	DEL NORTE	CA	-1	-9	-11	21	32
1	112.4			2	24.9	EL DORADO	CA			-78	02	29
216*	192.8	284	210.0	348	138.3	FRESNO	CA	9	-34	-28	14	43
4	436.0	3	183.5	3	129.2	GLENN	CA	-58	-30	-70	03	41
11	118.3	9	67.4	19	112.5	HUMBOLDT	CA	-43	67	-5	23	39
34	126.4	47	173.6	65	105.4	IMPERIAL	CA	37	-39	-17	19	37
4	98.5	5	135.3	4	88.3	INYO	CA	37	-35	-10	21	35
140	200.8	199	202.8	272	146.6	KERN	CA	1	-28	-27	15	44
24	175.3	21	126.9	45	135.5	KINGS	CA	-28	7	-23	16	43
1	42.6	1	31.3	7	99.5	LAKE	CA	-27	218	134	86	37
1	114.7	3	91.2	1	23.6	LASSEN	CA	-20	-74	-79	01	29
2280	181.4	4572	204.1	7668	180.7	LOS ANGELES	CA	13	-12	0	25	54
19	129.8	28	158.1	41	129.1	MADERA	CA	22	-18	-1	25	41
10	162.5	14	124.4	29	143.9	MARIN	CA	-23	16	-11	21	44
1	109.3	2	280.8			MARIPOSA	CA	157				
4	58.5	11	120.5	12	131.4	MENDOCINO	CA	106	9	125	84	42
31	145.4	62	239.9	75	150.1	MERCED	CA	65	-37	3	26	45
2	125.7	3	239.6	2	145.0	MODOC	CA	91	-39	15	32	44
2	238.4	1	89.3	1	25.4	MONO	CA	-63	-72	-89	00	29
57	137.5	119	187.8	131	96.3	MONTEREY	CA	37	-49	-30	14	36
3	34.3	7	90.0	5	27.6	NAPA	CA	162	-69	-20	18	29
		1	117.9	3	117.0	NEVADA	CA		-1			39
22	146.1	55	105.5	162	60.4	ORANGE	CA	-28	-43	-59	06	32
16	122.0	12	96.0	18	103.8	PLACER	CA	-21	8	-15	19	37
				4	219.0	PLUMAS	CA					67
76	173.6	135	168.4	245	133.3	RIVERSIDE	CA	-3	-21	-23	16	42
215*	220.9	302	188.7	524	173.0	SACRAMENTO	CA	-15	-8	-22	17	51
6	179.8	3	128.1	2	16.6	SAN BENITO	CA	-29	-87	-91	00	29

* SIGNIFICANTLY MORE DEATHS THAN EXPECTED

NON-WHITE MALE — ALL CANCERS COMBINED — ICD 140-208

1950-1959 DEATHS	1950-1959 RATE	1960-1969 DEATHS	1960-1969 RATE	1970-1979 DEATHS	1970-1979 RATE	COUNTY AND STATE	PERCENT CHANGE 60S/50S	PERCENT CHANGE 70S/60S	PERCENT CHANGE 70S/50S	PERCENTILE %CHANGE 70S/50S	PERCENTILE RATES 1970S
57	130.5	113	135.2	219	103.9	SAN BERNARDINO CA	4	-23	-20	17	37
167	177.1	317	183.7	538	134.2	SAN DIEGO CA	4	-27	-24	16	43
675	172.1	1200	195.2	1980	198.0	SAN FRANCISCO CA	13	1	15	32	60
166	153.8	288	174.1	385	144.2	SAN JOAQUIN CA	13	-17	-6	23	44
15	145.2	21	112.7	27	86.2	SAN LUIS OBISPO CA	-22	-24	-41	11	35
47	137.0	114	149.5	288	132.7	SAN MATEO CA	9	-11	-3	24	42
38	181.0	40	141.6	85	112.0	SANTA BARBARA CA	-22	-21	-38	12	38
118	212.6	159	162.3	302	100.6	SANTA CLARA CA	-24	-38	-53	07	37
26	153.3	41	189.8	42	96.6	SANTA CRUZ CA	24	-49	-37	12	36
4	78.1	11	261.3	14	153.7	SHASTA CA	235	-41	97	77	46
10	162.7	7	91.6	13	172.1	SISKIYOU CA	-44	88	6	27	51
36	115.6	66	151.4	175	177.6	SOLANO CA	31	17	54	55	53
15	163.5	26	182.2	44	147.1	SONOMA CA	11	-19	-10	21	45
9	105.8	15	117.1	34	94.3	STANISLAUS CA	11	-19	-11	21	36
10	111.4	7	89.0	16	126.2	SUTTER CA	-20	42	13	31	41
3	475.1	1	68.3	2	81.2	TEHAMA CA	-86	19	-83	01	34
50	136.6	70	149.4	79	77.9	TULARE CA	9	-48	-43	10	34
2	263.1	2	129.0			TUOLUMNE CA	-51				
17	119.1	33	146.9	63	61.6	VENTURA CA	23	-58	-48	09	32
11	146.4	16	114.3	22	65.0	YOLO CA	-22	-43	-56	07	32
15	222.3	19	214.5	24	152.8	YUBA CA	-4	-29	-31	14	46
266	189.2	353	173.2	536	124.0	COLORADO CO	-8	-28	-34		
9	193.9	8	131.1	20	78.0	ADAMS CO	-32	-41	-60	05	34
1	114.7	1	267.9			ALAMOSA CO	134				
				1	136.6	BACA CO					43
3	339.4	1	106.0	3	32.6	BOULDER CO	-69	-69	-90	00	29
				1	188.7	CLEAR CREEK CO					57
		1	74.3	2	94.0	COSTILLA CO		26			36
2*	1080.3					CROWLEY CO					
2	814.3	1	62.9			DELTA CO	-92				
192*	217.1	275	197.8	414	173.5	DENVER-ARAPAHOE#CO	-9	-12	-20	17	51
1*	344.1					EAGLE CO					
17	215.6	25	155.1	41	133.3	EL PASO CO	-28	-14	-38	12	42
				1	9.1	FREMONT CO					29
2	256.5	2	305.1	1	27.0	HUERFANO CO	19	-91	-89	00	29
1	172.0	1	74.3	10	118.3	JEFFERSON CO	-57	59	-31	14	40
2	243.4	1	14.3	3	25.8	LA PLATA CO	-94	81	-89	00	29
				2	47.7	LARIMER CO					31
1	112.4	2	145.5	1	17.2	LAS ANIMAS CO	29	-88	-85	00	29
1	86.0	1	112.4			LOGAN CO	31				
2	148.9	1	57.8	2	33.0	MESA CO	-61	-43	-78	02	29
1	53.0	1	10.8	3	83.0	MONTEZUMA CO	-80	671	57	57	34
1	344.1					MORGAN CO					
3	130.4	3	146.4	2	56.0	OTERO CO	12	-62	-57	06	31
3*	869.1			1	267.9	PROWERS CO			-69	03	83
14	101.8	23	180.8	24	78.3	PUEBLO CO	78	-57	-23	16	34
		2	186.8			ROUTT CO					
2	200.9			1	85.2	SEDGWICK CO			-58	06	35
6	101.2	4	101.7	3	12.8	WELD CO	1	-87	-87	00	29
545*	247.9	894*	243.0	1476*	258.0	CONNECTICUT CT	-2	6	4		
179*	268.2	287*	253.4	437	248.7	FAIRFIELD CT	-5	-2	-7	22	76
165*	281.6	268*	247.9	453*	255.1	HARTFORD CT	-12	3	-9	21	78
8	222.3	6	91.5	16	264.8	LITCHFIELD CT	-59	189	19	34	81
5	111.5	18	245.3	21	203.9	MIDDLESEX CT	120	-17	83	71	62
162*	224.1	278*	252.5	479*	288.6	NEW HAVEN CT	13	14	29	40	88
20	157.2	32	173.2	60	215.8	NEW LONDON CT	10	25	37	46	66
1	112.4	1	84.8	2	45.2	TOLLAND CT	-25	-47	-60	05	31
5	485.4	4	219.3	8	170.6	WINDHAM CT	-55	-22	-65	04	50
419*	207.3	680*	278.0	924*	317.4	DELAWARE DE	34	14	53		
60	159.2	90	220.0	139*	285.9	KENT DE	38	30	80	70	87
282*	241.8	456*	310.3	583*	327.9	NEW CASTLE DE	28	6	36	45	92
77	163.2	134	240.0	202*	317.2	SUSSEX DE	47	32	94	76	91
2577*	252.0	4212*	295.0	6039*	349.0	DIST. OF COLUMBIA DC	17	18	39		
2577*	252.0	4212*	295.0	6039*	349.0	WASHINGTON DC	17	18	39	47	94
3840	150.2	6832*	210.7	10775*	252.3	FLORIDA FL	40	20	68		
107	157.9	186*	246.8	252*	282.8	ALACHUA FL	56	15	79	69	87
6	86.4	11	162.6	9	116.5	BAKER FL	88	-28	35	44	39
36	159.3	60	196.4	104	269.2	BAY FL	23	37	69	64	83
8	95.8	26	239.6	33	274.2	BRADFORD FL	150	14	186	92	85
47	164.3	98	224.4	162	249.3	BREVARD FL	37	11	52	54	77
150	157.2	352	209.8	667*	243.2	BROWARD FL	33	16	55	56	75
3	85.1	6	174.2	5	100.6	CALHOUN FL	105	-42	18	34	37
7	205.1	11	311.7	20	416.4	CHARLOTTE FL	52	34	103	79	97
17	240.2	11	151.7	17	190.8	CITRUS FL	-37	26	-21	17	57
8	94.8	36*	355.2	32	278.1	CLAY FL	275	-22	193	92	86
15	147.6	33	255.0	63	306.7	COLLIER FL	73	20	108	80	90
26	102.6	36	148.7	57	212.3	COLUMBIA FL	45	43	107	80	65
486*	170.3	1065*	237.3	1783	223.5	DADE FL	39	-6	31	42	68
11	114.5	24	211.3	28	221.2	DE SOTO FL	85	5	93	75	68
6	168.4	3	136.4	9	246.9	DIXIE FL	-19	81	47	52	76
607*	177.5	1003*	248.3	1364*	291.4	DUVAL FL	40	17	64	61	88
145	159.8	215	188.3	395*	269.9	ESCAMBIA FL	18	43	69	63	84
7	100.2	12	202.3	8	171.2	FLAGLER FL	102	-15	71	65	50
12	153.7	10	166.9	19	307.8	FRANKLIN FL	9	84	100	78	91
73	91.3	99	114.2	142	161.0	GADSDEN FL	25	41	76	68	47
2	183.8	2	123.2	2	313.2	GILCHRIST FL	-33	154	70	65	91
2	97.7	10	361.8	2	73.8	GLADES FL	271	-80	-24	16	33
9	113.2	16	162.4	20	183.4	GULF FL	43	13	62	60	54
19	132.9	19	137.9	30	225.7	HAMILTON FL	4	64	70	64	69
2	69.9	11	236.1	17	223.3	HARDEE FL	238	-5	219	94	68
12	168.1	15	143.3	38	243.1	HENDRY FL	-15	70	45	50	75

* SIGNIFICANTLY MORE DEATHS THAN EXPECTED

\# COUNTIES COMBINED; SEE TECHNICAL NOTES

NON-WHITE MALE — ALL CANCERS COMBINED — ICD 140-208

COUNTY AND STATE	1950-1959 DEATHS	1950-1959 RATE	1960-1969 DEATHS	1960-1969 RATE	1970-1979 DEATHS	1970-1979 RATE	PERCENT CHANGE 60S/50S	PERCENT CHANGE 70S/60S	PERCENT CHANGE 70S/50S	PERCENTILE %CHANGE 70S/50S	PERCENTILE RATES 1970S
HERNANDO FL	9	107.2	14	151.3	31	305.8	41	102	185	92	90
HIGHLANDS FL	23	142.5	41	208.5	80*	314.1	46	51	120	83	91
HILLSBOROUGH FL	325*	176.3	487	210.1	823*	284.5	19	35	61	59	87
HOLMES FL	1	42.1	3	380.5	7	381.7	804	0	806	99	95
INDIAN RIVER FL	19	112.7	46	271.7	67	281.4	141	4	150	88	86
JACKSON FL	36	86.8	70	173.6	101	227.0	100	31	161	90	69
JEFFERSON FL	27	127.6	21	102.4	38	180.1	-20	76	41	48	53
LAFAYETTE FL	1	94.4	3	241.2	3	196.8	156	-18	109	80	59
LAKE FL	64	176.7	94	212.1	135	239.1	20	13	35	44	73
LEE FL	42	157.9	66	183.3	127	274.0	16	49	74	66	85
LEON FL	87	129.3	135	173.7	207	228.1	34	31	76	68	70
LEVY FL	14	89.2	25	158.9	46	303.5	78	91	240	95	90
LIBERTY FL	4	203.4	6	350.7	1	37.2	72	-89	-82	01	30
MADISON FL	31	133.0	48	202.1	63	252.5	52	25	90	74	78
MANATEE FL	40	116.6	106*	281.3	143*	315.6	141	12	171	91	91
MARION FL	81	121.7	136	185.9	230*	269.8	53	45	122	84	84
MARTIN FL	20	162.3	42	284.0	51	276.2	75	-3	70	65	85
MONROE FL	33	231.0	41	279.2	53	300.2	21	8	30	41	89
NASSAU FL	13	96.8	44	278.3	44	251.1	187	-10	159	90	77
OKALOOSA FL	9	145.2	15	140.6	35	214.9	-3	53	48	53	65
OKEECHOBEE FL	5	126.1	9	167.1	30*	406.7	32	143	223	94	96
ORANGE FL	127	129.4	266	194.3	449	234.5	50	21	81	71	72
OSCEOLA FL	2	28.7	21	228.3	30	288.1	696	26	904	99	88
PALM BEACH FL	193	130.5	397	205.4	644*	263.5	57	28	102	78	81
PASCO FL	20	162.7	40	282.1	38	194.6	73	-31	20	35	59
PINELLAS FL	133	159.3	289	223.7	466*	271.1	40	21	70	65	84
POLK FL	164	144.8	265	189.2	424	240.1	31	27	66	62	74
PUTNAM FL	50	132.2	85	204.9	107	239.6	55	17	81	70	74
ST JOHNS FL	51	179.3	50	166.5	82	251.9	-7	51	41	48	78
ST LUCIE FL	43	143.1	71	167.3	176*	297.5	17	78	108	80	89
SANTA ROSA FL	8	115.6	14	211.3	19	267.1	83	26	131	85	82
SARASOTA FL	27	127.7	70	254.3	98*	287.8	99	13	125	85	87
SEMINOLE FL	89	186.6	132	238.3	166	237.7	28	0	27	39	73
SUMTER FL	13	96.5	24	166.4	47	266.3	72	60	176	91	82
SUWANNEE FL	29	153.7	26	142.8	47	275.2	-7	93	79	69	85
TAYLOR FL	23	149.5	29	205.8	46	289.2	38	41	93	76	88
UNION FL	8	76.4	6	53.2	12	192.2	-30	261	152	89	58
VOLUSIA FL	125	170.1	188	219.1	288*	276.3	29	26	62	60	85
WAKULLA FL	7	113.9	6	100.6	9	125.0	-12	24	10	30	41
WALTON FL	10	133.1	17	232.2	16	184.2	74	-21	38	47	55
WASHINGTON FL	11	144.7	14	162.3	18	201.0	12	24	39	47	61
GEORGIA GA	4698	134.3	6838	179.5	9630	225.5	34	26	68		
APPLING GA	9	99.0	18	210.6	16	171.5	113	-19	73	66	50
ATKINSON GA	6	103.8	2	41.4	17	265.6	-60	542	156	89	81
BACON GA	5	137.7	11	313.2	3	76.5	127	-76	-44	10	34
BAKER GA	14	140.5	9	111.5	13	178.7	-21	60	27	39	53
BALDWIN GA	47	92.8	51	93.7	81	165.1	1	76	78	69	49
BANKS GA	2	113.5			1	70.1			-38	12	33
BARROW GA	4	60.2	6	61.5	21	210.1	2	242	249	95	64
BARTOW GA	27	203.5	34	227.5	34	216.7	12	-5	6	28	66
BEN HILL GA	18	101.5	28	143.8	30	141.5	42	-2	39	47	44
BERRIEN GA	11	213.4	11	226.1	16	243.7	6	8	14	31	75
BIBB GA	207	149.4	344	216.1	363	205.9	45	-5	38	46	62
BLECKLEY GA	6	58.7	15	200.7	14	195.0	242	-3	232	94	59
BRANTLEY GA	5	132.7	5	169.6	9	308.3	28	82	132	86	91
BROOKS GA	22	79.1	45	170.4	64	246.7	115	45	212	93	76
BRYAN GA	8	110.4	19	267.2	17	252.3	142	-6	129	85	78
BULLOCH GA	28	99.2	51	152.7	71	212.1	54	39	114	82	64
BURKE GA	67	125.0	58	127.2	80	200.9	2	58	61	59	61
BUTTS GA	14	121.3	10	70.2	38	260.4	-42	271	115	82	80
CALHOUN GA	21	117.4	31	177.4	29	181.0	51	2	54	56	54
CAMDEN GA	18	146.7	28	236.1	34	206.3	61	-13	41	48	62
CANDLER GA	7	92.2	9	114.8	14	147.8	25	29	60	59	45
CARROLL GA	18	99.7	35	172.7	44	192.7	73	12	93	76	58
CATOOSA GA	1	60.7	3	217.5	6	586.2	258	170	865	99	98
CHARLTON GA	5	83.9	15	272.9	19	283.3	225	4	238	94	87
CHATHAM GA	332	154.3	535*	227.6	744*	287.3	47	26	86	73	87
CHATTOOGA GA	10	196.3	7	109.7	16	198.8	-44	81	1	26	60
CHEROKEE GA	2	87.2	6	174.1	11	344.6	100	98	295	97	94
CLARKE GA	53	150.0	66	170.9	117	255.0	14	49	70	64	78
CLAY GA	12	106.6	10	103.6	14	151.4	-3	46	42	49	46
CLAYTON GA	13	109.9	32	257.1	45	267.6	134	4	144	87	82
CLINCH GA	9	109.0	10	105.8	17	216.4	-3	105	99	77	66
COBB GA	28	145.0	61	230.5	70	235.8	59	2	63	60	72
COFFEE GA	25	138.3	30	156.1	42	202.8	13	30	47	52	61
COLQUITT GA	31	139.6	50	210.0	68	278.1	50	32	99	77	86
COLUMBIA GA	17	101.2	21	136.8	33	196.5	35	44	94	76	59
COOK GA	11	89.5	20	182.1	26	222.8	103	22	149	88	68
COWETA GA	40	113.5	46	128.3	63	176.2	13	37	55	56	52
CRAWFORD GA	16	140.9	12	123.2	17	169.1	-13	37	20	35	50
CRISP GA	33	128.1	50	205.8	48	204.6	61	-1	60	58	62
DADE GA			1	344.1	2	587.6		71			98
DECATUR GA	45	128.0	71	198.2	78	212.3	55	7	66	62	65
DE KALB GA	78	150.8	197	194.7	460	234.4	29	20	55	56	72
DODGE GA	19	118.3	23	154.4	35	219.5	31	42	86	73	67
DOOLY GA	24	120.8	22	120.4	36	204.2	0	70	69	64	62
DOUGHERTY GA	83	125.6	142	176.0	264*	268.7	40	53	114	82	83
DOUGLAS GA	2	37.5	12	157.2	17	179.7	319	14	379	98	53
EARLY GA	24	93.6	33	138.8	40	164.7	48	19	76	68	49
ECHOLS GA	3	144.4	2	95.2	6	350.1	-34	268	143	87	94
EFFINGHAM GA	9	74.9	18	160.6	19	156.9	115	-2	110	81	47
ELBERT GA	15	75.0	22	111.5	27	147.4	49	32	96	77	45
EMANUEL GA	29	128.0	32	145.5	40	187.4	14	29	46	52	56
EVANS GA	14	169.7	29*	317.6	28	291.1	87	-8	72	65	88
FANNIN GA					2	611.9					99
FAYETTE GA	3	49.0	8	118.8	9	145.8	142	23	198	93	44
FLOYD GA	57	166.7	59	165.5	103	269.8	-1	63	62	59	84
FORSYTH GA					2*	914.8					99
FRANKLIN GA	4	100.4	11	218.5	16	364.7	118	67	263	96	95

* SIGNIFICANTLY MORE DEATHS THAN EXPECTED

NON-WHITE MALE — ALL CANCERS COMBINED — ICD 140-208

1950-1959 DEATHS	1950-1959 RATE	1960-1969 DEATHS	1960-1969 RATE	1970-1979 DEATHS	1970-1979 RATE	COUNTY AND STATE	PERCENT CHANGE 60S/50S	PERCENT CHANGE 70S/60S	PERCENT CHANGE 70S/50S	PERCENTILE %CHANGE 70S/50S	PERCENTILE RATES 1970S
954*	177.0	1506*	226.5	1999*	258.9	FULTON GA	28	14	46	51	79
1*	188.7			1*	112.4	GILMER GA			-40	11	38
2	112.0	2	87.0			GLASCOCK GA	-22				
58	145.6	91	194.0	152*	279.4	GLYNN GA	33	44	92	75	86
9	270.7	10	231.7	7	138.3	GORDON GA	-14	-40	-49	09	43
29	135.7	39	171.8	41	175.3	GRADY GA	27	2	29	40	52
16	76.0	18	92.4	42	201.2	GREENE GA	22	118	165	90	61
17	188.9	21	207.5	27	205.4	GWINNETT GA	10	-1	9	29	62
4	140.1	4	197.4	3	109.9	HABERSHAM GA	41	-44	-22	17	38
18	152.7	20	128.4	44	260.4	HALL GA	-16	103	70	65	80
32	148.8	32	135.6	48	188.5	HANCOCK GA	-9	39	27	38	56
4	117.0	5	134.3	5	99.4	HARALSON GA	15	-26	-15	19	37
23	115.2	31	156.0	39	178.8	HARRIS GA	35	15	55	56	53
9	73.4	16	149.5	19	144.4	HART GA	104	-3	97	77	44
5	105.6	4	96.8	4	124.7	HEARD GA	-8	29	18	33	41
27	127.3	23	109.7	48	204.8	HENRY GA	-14	87	61	59	62
22	87.2	33	125.5	71	209.1	HOUSTON GA	44	67	140	87	63
9	65.9	24	237.2	12	132.9	IRWIN GA	260	-44	101	78	42
9	124.3	13	156.1	22	250.9	JACKSON GA	26	61	102	78	77
14	111.0	22	212.0	22	204.1	JASPER GA	91	-4	84	72	62
4	76.3	7	97.1	10	139.2	JEFF DAVIS GA	27	43	83	71	43
35	118.4	56	179.1	80	237.9	JEFFERSON GA	51	33	101	78	73
11	65.1	15	98.8	27	178.1	JENKINS GA	52	80	174	91	53
17	209.5	18	337.4	14	175.4	JOHNSON GA	61	-48	-16	19	52
28	239.5	26	190.4	32	181.3	JONES GA	-20	-5	-24	16	54
22	165.4	9	69.4	26	180.9	LAMAR GA	-58	161	9	30	54
8	156.0	13	271.2	8	165.2	LANIER GA	74	-39	6	27	49
50	115.2	68	159.7	76	169.8	LAURENS GA	39	6	47	52	50
20	136.2	16	141.4	22	201.0	LEE GA	4	42	48	53	61
19	100.7	34	171.8	51	220.0	LIBERTY GA	71	28	119	83	67
11	176.0	14	214.4	17	169.4	LINCOLN GA	22	-21	-4	24	50
9	163.4	3	47.8	8	166.3	LONG GA	-71	248	2	26	49
58	109.4	105	188.2	133	214.0	LOWNDES GA	72	14	96	77	65
1	490.8			1	56.2	LUMPKIN GA			-89	00	32
13	87.2	34	204.2	40	224.3	MC DUFFIE GA	134	10	157	89	68
14	96.8	28	204.8	32	214.8	MC INTOSH GA	112	5	122	84	65
37	137.8	49	190.7	54	212.5	MACON GA	38	11	54	56	65
8	128.1	9	155.4	14	291.7	MADISON GA	21	88	128	85	88
10	92.4	10	95.1	25	284.4	MARION GA	3	199	208	93	87
39	117.4	48	142.6	81	223.9	MERIWETHER GA	21	57	91	75	68
8	122.0	10	205.0	15	223.1	MILLER GA	68	9	83	71	68
38	116.0	45	147.1	58	185.1	MITCHELL GA	27	26	60	58	55
12	77.2	23	154.1	37	216.1	MONROE GA	100	40	180	91	66
9	98.0	10	114.8	12	177.4	MONTGOMERY GA	17	54	81	70	52
13	69.0	21	127.7	33	197.0	MORGAN GA	85	54	186	92	60
				1	424.0	MURRAY GA					97
146	157.6	244	201.3	374	242.6	MUSCOGEE#GA	28	20	54	55	74
21	99.8	42	182.6	53	206.9	NEWTON GA	83	13	107	80	63
5	83.4	8	199.5	3	79.3	OCONEE GA	139	-60	-5	23	34
15	138.0	14	143.5	12	124.9	OGLETHORPE GA	4	-13	-9	21	41
3	86.2	4	106.7	10	214.1	PAULDING GA	24	101	149	88	65
30	140.8	43	175.6	71	267.5	PEACH GA	25	52	90	74	82
2	201.7	2	180.0	2	176.0	PICKENS GA	-11	-2	-13	20	52
9	117.0	8	135.2	20	302.5	PIERCE GA	16	124	158	89	90
8	74.4	15	151.0	23	215.9	PIKE GA	103	43	190	92	66
13	89.0	19	112.0	41	247.9	POLK GA	26	121	179	91	76
13	103.5	16	127.1	28	211.7	PULASKI GA	23	67	105	79	64
22	186.5	11	87.9	19	128.9	PUTNAM GA	-53	47	-31	14	41
6	129.4	3	59.6	11	210.1	QUITMAN GA	-54	253	62	60	64
1	89.3	3	530.8	1	141.3	RABUN GA	494	-73	58	57	44
40	140.8	40	163.5	56	235.0	RANDOLPH GA	16	44	67	63	72
229	181.3	314	213.1	424	239.0	RICHMOND GA	18	12	32	42	73
13	208.3	13	179.5	27	251.4	ROCKDALE GA	-14	40	21	36	77
6	81.1	7	131.8	11	225.3	SCHLEY GA	63	71	178	91	69
16	55.4	35	135.2	48	195.0	SCREVEN GA	144	44	252	96	59
7	75.6	11	122.8	18	169.8	SEMINOLE GA	62	38	125	84	50
31	113.9	44	141.7	89	261.2	SPALDING GA	24	84	129	85	80
4	63.0	18	219.9	20	289.8	STEPHENS GA	249	32	360	97	88
23	118.1	20	119.5	30	191.7	STEWART GA	1	60	62	60	58
44	99.3	62	141.6	121	278.9	SUMTER GA	43	97	181	92	86
18	121.2	17	111.2	23	133.4	TALBOT GA	-8	20	10	30	42
12	119.3	10	135.2	16	234.1	TALIAFERRO GA	13	73	96	77	71
15	92.7	25	138.4	28	168.7	TATTNALL GA	49	22	82	71	50
12	91.2	19	141.0	21	188.0	TAYLOR GA	55	33	106	80	56
16	109.4	25	166.4	40	271.0	TELFAIR GA	52	63	148	88	84
55	189.6	36	135.8	55	214.3	TERRELL GA	-28	58	13	31	65
45	91.8	97	183.0	131	237.9	THOMAS GA	99	30	159	90	73
23	105.8	31	132.5	62	231.7	TIFT GA	25	75	119	83	71
22	151.9	35	206.6	39	216.7	TOOMBS GA	36	5	43	49	66
				1*	546.6	TOWNS GA					98
8	117.0	9	139.3	8	127.9	TREUTLEN GA	19	-8	9	29	41
59	113.6	84	154.9	102	193.1	TROUP GA	36	25	70	64	58
8	87.1	12	133.6	22	248.0	TURNER GA	53	86	185	92	76
10	59.7	28	195.0	30	192.7	TWIGGS GA	227	-1	223	94	58
19	98.2	30	129.8	40	152.0	UPSON GA	32	17	55	56	46
12	132.4	22	234.0	18	182.2	WALKER GA	77	-22	38	46	54
19	116.9	29	193.7	41	219.1	WALTON GA	66	13	87	73	67
48	152.9	63	187.2	86	257.6	WARE GA	22	38	68	63	79
18	123.3	25	176.5	28	182.9	WARREN GA	43	4	48	53	54
54	154.6	43	124.3	74	214.5	WASHINGTON GA	-20	73	39	47	65
14	137.3	23	186.2	33	227.5	WAYNE GA	36	22	66	62	70
11	161.8	3	60.2	6	140.6	WEBSTER GA	-63	134	-13	20	43
6	105.4	3	30.6	8	143.4	WHEELER GA	-71	368	36	45	44
2	122.0	1	70.7	2	286.7	WHITE GA	-42	306	135	86	87
10	209.4	12	208.0	21	261.7	WHITFIELD GA	-1	26	25	38	80
9	80.5	10	97.5	17	183.1	WILCOX GA	21	88	127	85	54
35	145.5	28	135.9	37	164.4	WILKES GA	-7	21	13	31	48
23	163.5	24	177.9	27	168.1	WILKINSON GA	9	-5	3	26	50
41	163.3	37	167.0	40	177.4	WORTH GA	2	6	9	29	52

* SIGNIFICANTLY MORE DEATHS THAN EXPECTED

COUNTIES COMBINED; SEE TECHNICAL NOTES

NON-WHITE MALE 1950-1959		1960-1969		1970-1979		ALL CANCERS COMBINED COUNTY AND STATE	PERCENT CHANGE			ICD 140-208 PERCENTILE %CHANGE RATES	
DEATHS	RATE	DEATHS	RATE	DEATHS	RATE		60S/50S	70S/60S	70S/50S	70S/50S	1970S
		2998	176.9	3888	163.5	HAWAII HI		-8			
		422	161.1	488	157.6	HAWAII HI		-2			47
		2074	184.6	2808	166.6	HONOLULU HI		-10			49
		201	158.3	217	140.6	KAUAI HI		-11			43
		301	170.5	375	164.7	MAUI#HI		-3			49
40	106.4	63	162.2	49	73.0	IDAHO ID	52	-55	-31		
2	87.1	6	275.4	9	161.9	ADA ID	216	-41	86	73	48
6	95.5	14	248.4	9	112.1	BANNOCK ID	160	-55	17	33	38
1	70.1			1	112.4	BENEWAH ID			60	58	38
5	85.7	8	117.0	9	117.1	BINGHAM ID	37	0	37	45	39
1*	490.8			1	165.8	BLAINE ID			-66	04	49
3	311.2	4	194.7	3	97.0	BONNEVILLE ID	-37	-50	-69	03	36
2	125.8	10	330.9	5	60.3	CANYON ID	163	-82	-52	08	32
		1	112.4			CASSIA ID					
1	188.7					CLEARWATER ID					
		2	189.0			ELMORE ID					
1	133.9	1	47.2			FREMONT ID	-65				
		1	188.7	2	186.8	JEROME ID		-1			55
		1	114.7			KOOTENAI ID					
2*	380.3	1	141.3			LEWIS ID	-63				
1	546.4					LINCOLN ID					
1	188.7	1	94.4			MADISON ID	-50				
1	188.7	2	301.1	1	21.3	MINIDOKA ID	60	-93	-89	00	29
4	97.8	1	47.2	4	121.8	NEZ PERCE ID	-52	158	24	37	40
1	182.1					OWYHEE ID					
2	133.9	2	152.8	1	36.4	PAYETTE ID	14	-76	-73	03	30
1	37.2	1	98.2			POWER ID	164				
1*	188.7					SHOSHONE ID					
		1*	546.6			TETON ID					
2	201.7	2	278.0	2	69.8	TWIN FALLS ID	38	-75	-65	04	33
				1	212.0	VALLEY ID					64
2	76.1	4	316.2	1	53.6	WASHINGTON ID	315	-83	-30	14	31
5778*	209.2	9230*	240.7	13471*	262.4	ILLINOIS IL	15	9	25		
17	220.3	24	310.0	33*	411.1	ADAMS IL	41	33	87	73	96
43	130.8	65	208.1	57	253.7	ALEXANDER IL	59	22	94	76	78
		2	190.9	6	315.6	BOND IL		65			91
1	112.4					BUREAU IL					
1	273.2	2	536.4	2	235.7	CARROLL IL	96	-56	-14	19	72
		1*	188.7			CASS IL					
36	214.3	48	233.6	77	269.8	CHAMPAIGN IL	9	15	26	38	83
2	225.3	1	589.6	3	263.7	CHRISTIAN IL	162	-55	17	33	81
		2*	532.8	1*	546.4	CLARK IL		3			98
				1	267.9	CLAY IL					83
1	94.4	3	213.3	3	180.1	CLINTON IL	126	-16	91	75	53
2	273.1	2	192.9	1	70.7	COLES IL	-29	-63	-74	02	33
4795*	223.4	7699*	249.8	11230*	267.6	COOK IL	12	7	20	35	82
				2*	691.9	CUMBERLAND IL					99
3	203.9	6	388.2	5	231.8	DE KALB IL	90	-40	14	31	71
3	93.7	12	250.3	18	71.2	DU PAGE IL	167	-72	-24	16	33
6*	689.0	2	208.3	1	172.0	EDGAR IL	-70	-17	-75	02	51
1	344.1					EDWARDS IL					
				1	245.4	EFFINGHAM IL					75
1	98.2					FAYETTE IL					
2*	688.1					FORD IL					
		1	74.3	1	133.9	FRANKLIN IL		80			42
		3	679.8	2	420.8	FULTON IL		-38			97
2	141.3	4*	859.1			GALLATIN IL	508				
				1	74.3	HARDIN IL					34
				1	182.2	HENDERSON IL					54
4	196.9	7	271.7	5	191.7	HENRY IL	38	-29	-3	24	58
1	44.6	1	56.2	1	112.4	IROQUOIS IL	26	100	152	89	38
21	138.1	31	185.4	54*	352.0	JACKSON IL	34	90	155	89	94
8	195.0	13	284.0	21*	368.7	JEFFERSON IL	46	30	89	74	95
1	490.8	1	112.4	1	98.2	JOHNSON IL	-77	-13	-80	01	37
22	169.1	32	152.6	60	222.5	KANE IL	-10	46	32	42	68
19	56.5	60	121.3	108	191.2	KANKAKEE IL	115	58	239	95	57
15	174.1	22	222.9	28	287.1	KNOX IL	28	29	65	61	87
41	160.8	54	134.7	144	220.7	LAKE IL	-16	64	37	46	67
4	285.6	3	213.2	9	392.3	LA SALLE IL	-25	84	37	46	96
3	226.2	2	131.7	2	544.3	LAWRENCE IL	-42	313	141	87	98
2	124.0	2	113.5	5	209.8	LEE IL	-8	85	69	64	63
1	267.9	2	286.9			LIVINGSTON IL	7				
1	188.7	3	352.7	2	175.3	LOGAN IL	87	-50	-7	22	52
				1	74.3	MC DONOUGH IL					34
				1	112.4	MC HENRY IL					38
14*	345.3	11	311.7	17	373.5	MC LEAN IL	-10	20	8	29	95
23	144.8	55	247.3	87	306.0	MACON IL	71	24	111	81	90
1	212.0			2	107.6	MACOUPIN IL			-49	08	38
72	165.0	104	209.9	132	238.2	MADISON IL	27	13	44	50	73
9	148.5	22*	341.7	16	253.1	MARION IL	130	-26	70	65	78
10	135.1	18	253.6	16	235.3	MASSAC IL	88	-7	74	67	72
		1	344.1			MENARD IL					
4*	464.3	1	141.3	1	141.3	MONTGOMERY IL	-70	0	-70	03	44
14	166.7	17	214.9	15	259.9	MORGAN IL	29	21	56	56	79
				1	84.8	OGLE IL					35
57	211.9	81	230.2	130	265.5	PEORIA IL	9	15	25	38	81
4	129.0	7	308.1	6	256.7	PERRY IL	139	-17	99	77	79
1	490.8					PIKE IL					
3	305.0					POPE IL					
30	114.1	33	148.7	40	221.8	PULASKI IL	30	49	94	76	68
15	170.7	9	149.4	13	237.2	RANDOLPH IL	-12	59	39	47	72
24	226.1	45	272.1	50	179.4	ROCK ISLAND IL	20	-34	-21	17	53
231	162.7	413	210.2	632*	281.5	ST CLAIR IL	29	34	73	66	86
15	216.4	21*	315.5	12	257.0	SALINE IL	46	-19	19	34	79
67*	270.1	84*	302.1	88	268.5	SANGAMON IL	12	-11	-1	25	83

* SIGNIFICANTLY MORE DEATHS THAN EXPECTED

COUNTIES COMBINED; SEE TECHNICAL NOTES

NON-WHITE MALE — ALL CANCERS COMBINED — ICD 140-208

1950-1959 DEATHS	1950-1959 RATE	1960-1969 DEATHS	1960-1969 RATE	1970-1979 DEATHS	1970-1979 RATE	COUNTY AND STATE	PERCENT CHANGE 60S/50S	PERCENT CHANGE 70S/60S	PERCENT CHANGE 70S/50S	PERCENTILE %CHANGE 70S/50S	PERCENTILE RATES 1970S
		1*	267.9			SCHUYLER IL					
6	166.6	8	176.1	18	270.5	STEPHENSON IL	6	54	62	60	84
				1	37.2	TAZEWELL IL					30
2	69.6	2	106.3	1	74.3	UNION IL	53	-30	7	28	34
35	167.3	41	171.9	57	231.8	VERMILION IL	3	35	39	47	71
5	259.6	7	343.2	5	272.0	WARREN IL	32	-21	5	27	84
		1*	74.3			WAYNE IL					
2	306.4	1	112.4	1	267.9	WHITE IL	-63	138	-13	20	83
		3	352.8	1	37.2	WHITESIDE IL		-89			30
33	120.1	63	228.6	111	216.7	WILL IL	90	-5	80	70	66
8	136.0	14	254.7	10	264.7	WILLIAMSON IL	87	4	95	76	81
34	258.8	57	224.3	120	283.2	WINNEBAGO IL	-13	28	11	30	88
1590*	196.8	2544*	236.3	3599*	268.1	INDIANA IN	20	13	36		
44	210.5	72	202.1	152	276.1	ALLEN IN	-4	37	31	42	85
4	708.8	1	42.1	1	23.8	BARTHOLOMEW IN	-94	-44	-97	00	29
		1*	589.6			BLACKFORD IN					
3	153.1	3	171.8	8	589.1	CASS IN	12	243	285	96	98
22	209.8	29	228.7	39	289.0	CLARK IN	9	26	38	46	88
2	161.3	3	314.3	3	177.3	CLAY IN	95	-44	10	30	52
		1	172.0	2	324.1	CLINTON IN		88			92
3	435.9	2	212.0	2	397.6	DAVIESS IN	-51	88	-9	22	96
1	114.7	1	99.4	4	277.7	DEARBORN IN	-13	179	142	87	85
36	167.6	80*	301.4	92*	310.8	DELAWARE IN	80	3	86	73	91
11	259.3	16	162.2	26	192.4	ELKHART IN	-37	19	-26	15	58
9*	312.9	8	302.6	12*	500.6	FAYETTE IN	-3	65	60	58	98
15	188.7	17	190.9	30	331.5	FLOYD IN	1	74	76	68	93
				3	270.1	FULTON IN					84
9	205.0	9	190.8	6	213.8	GIBSON IN	-7	12	4	26	65
13	118.0	31	231.6	40	239.3	GRANT IN	96	3	103	79	73
				1	267.9	GREENE IN					83
5	202.2	5	161.2	2	80.9	HAMILTON IN	-20	-50	-60	05	34
				1	74.3	HANCOCK IN					34
				3	161.2	HARRISON IN					47
1	136.6			1	59.6	HENDRICKS IN			-56	06	32
5	219.4	4	180.1	3	188.1	HENRY IN	-18	4	-14	19	56
13	176.2	21	228.7	23	211.2	HOWARD IN	30	-8	20	35	64
1	74.3	1	112.4			HUNTINGTON IN	51				
4*	668.9	1	122.7	2	382.6	JACKSON IN	-82	212	-43	10	96
1*	424.0					JASPER IN					
		1	294.8	2	276.1	JAY IN		-6			85
5	221.8	3	103.8	6	263.8	JEFFERSON IN	-53	154	19	34	81
1	133.9	3	485.3	3	291.0	JENNINGS IN	262	-40	117	83	88
3	321.9	4	200.1	3	146.0	JOHNSON IN	-38	-27	-55	07	44
3	145.4	2	188.7	4	300.0	KNOX IN	30	59	106	80	89
1	188.7	1	198.7	1	94.4	KOSCIUSKO IN	5	-53	-50	08	36
390*	193.2	664	224.4	1036*	257.6	LAKE IN	16	15	33	43	79
19	132.9	38	200.3	54	243.7	LA PORTE IN	51	22	83	71	75
1	84.8	2	536.4	2	171.4	LAWRENCE IN	533	-68	102	78	50
20	182.6	26	162.4	53	259.0	MADISON IN	-11	59	42	49	79
680*	213.0	1122*	264.4	1453*	282.2	MARION IN	24	7	32	42	86
				1	37.5	MARSHALL IN					30
2	127.9	3	177.4	4	231.5	MIAMI IN	39	31	81	70	71
9	320.4	9	326.1	14	332.3	MONROE IN	2	2	4	26	93
3	354.7	1	56.2			MONTGOMERY IN	-84				
				1	562.5	MORGAN IN					98
				1	267.9	NOBLE IN					83
		2	188.7			OHIO IN					
3	192.0	2	226.9	3	189.4	ORANGE IN	18	-17	-1	25	57
1	112.4			1	267.9	OWEN IN			138	87	83
1	344.1					PARKE IN					
				1	344.1	PERRY IN					94
		1	188.7			PIKE IN					
2*	558.7	1	344.1	1	89.3	PORTER IN	-38	-74	-84	01	35
2	143.5	6	306.9	6	400.8	POSEY IN	114	31	179	91	96
		1	188.7	2	191.3	PULASKI IN		1			57
2	115.3	2	87.6	6	413.0	PUTNAM IN	-24	371	258	96	96
2	301.1	1	112.4	2	384.0	RANDOLPH IN	-63	242	28	39	96
		4	353.9	3	306.4	RUSH IN		-13			90
79*	235.5	124	232.5	192*	277.1	ST JOSEPH IN	-1	19	18	33	85
3	135.1	3	266.0	8	413.8	SHELBY IN	97	56	206	93	97
2	108.9	1	62.9	1	62.9	SPENCER IN	-42	0	-42	11	32
2*	691.9					STEUBEN IN					
		1	74.3	1	133.9	SULLIVAN IN		80			42
2	73.3	9	305.0	9	299.4	TIPPECANOE IN	316	-2	308	97	89
88	196.9	107	225.4	151*	307.9	VANDERBURGH IN	14	37	56	57	91
41	169.7	46	186.8	68	277.4	VIGO IN	10	48	63	61	85
		2	168.6			WABASH IN					
		5	360.3	4	230.7	WARRICK IN		-36			70
20	151.4	41	250.9	45	255.9	WAYNE IN	66	2	69	64	79
				1	294.8	WELLS IN					89
1*	344.1					WHITLEY IN					
227*	213.1	288	235.3	324	217.1	IOWA IA	10	-8	2		
				1	344.1	ALLAMAKEE IA					94
1	63.9			2	306.6	APPANOOSE IA			380	98	90
29*	227.1	33	209.6	55	299.3	BLACK HAWK IA	-8	43	32	42	89
		1	344.1	1	94.4	BOONE IA		-73			36
		1	424.0	1	267.9	BUCHANAN IA		-37			83
1	344.1	1*	188.7			CASS IA	-45				
1*	344.1					CEDAR IA					
6	363.9	7	413.7	5	192.5	CERRO GORDO IA	14	-53	-47	09	58
				1*	74.3	CLARKE IA					34
2	210.3	2	100.8	5	446.3	CLINTON IA	-52	343	112	81	97
3	246.4	1	141.3	1	188.7	DALLAS IA	-43	34	-23	16	57
		1	112.4			DAVIS IA					
9*	428.8	10	354.1	5	182.0	DES MOINES IA	-17	-49	-58	06	54
1	188.7			2	147.8	DUBUQUE IA			-22	17	45
		1	245.4	1	267.9	FAYETTE IA		9			83

* SIGNIFICANTLY MORE DEATHS THAN EXPECTED

NON-WHITE MALE — ALL CANCERS COMBINED — ICD 140-208

1950-1959 DEATHS	1950-1959 RATE	1960-1969 DEATHS	1960-1969 RATE	1970-1979 DEATHS	1970-1979 RATE	COUNTY AND STATE		PERCENT CHANGE 60S/50S	PERCENT CHANGE 70S/60S	PERCENT CHANGE 70S/50S	PERCENTILE %CHANGE 70S/50S	PERCENTILE RATES 1970S
		1*	344.1			GRUNDY	IA					
				1	188.7	HAMILTON	IA					57
1	68.8			2	456.6	HENRY	IA			564	99	97
		1	112.4			JASPER	IA					
		1	45.4	3	94.6	JOHNSON	IA		109			36
12	186.1	23*	374.7	13	196.8	LEE	IA	101	-47	6	27	59
13	277.3	14	295.7	14	289.3	LINN	IA	7	-2	4	27	88
1	490.8					LUCAS	IA					
		1	133.9	2	246.4	MAHASKA	IA		84			76
2	198.4					MARION	IA					
2	187.5	2	59.5	4	162.1	MARSHALL	IA	-68	172	-14	20	48
		2	299.3			MONROE	IA					
				1	424.0	MONTGOMERY	IA					97
				3	302.2	MUSCATINE	IA					90
2	141.3			1	344.1	PAGE	IA			143	87	94
89*	225.6	134*	300.6	140	269.7	POLK	IA	33	-10	20	35	83
11	331.0	8	267.9	5	129.9	POTTAWATTAMIE	IA	-19	-51	-61	05	41
11	193.0	13	175.9	20	129.3	SCOTT	IA	-9	-27	-33	13	41
2	294.8	2	303.4	2	200.4	STORY	IA	3	-34	-32	13	61
1	44.6	3	123.7	3	238.5	TAMA	IA	177	93	434	98	73
1*	74.3					TAYLOR	IA					
10*	327.7	9	418.0	7	386.4	WAPELLO	IA	28	-8	18	33	96
4	317.2	2	111.8	5	317.9	WEBSTER	IA	-65	184	0	25	92
12	197.8	14	192.9	18	235.8	WOODBURY	IA	-2	22	19	34	72
720*	188.1	922	212.5	1158	231.9	KANSAS	KS	13	9	23		
5	269.6	9*	529.2	5	267.1	ALLEN	KS	96	-50	-1	25	82
1	267.9					ANDERSON	KS					
15	152.2	18	247.2	21	343.2	ATCHISON	KS	62	39	126	85	94
2	61.0	12*	478.3	6	258.7	BARTON	KS	685	-46	324	97	79
6	144.1	9	274.3	7	165.0	BOURBON	KS	90	-40	15	31	49
3	105.2	3	97.2	5	149.7	BROWN	KS	-8	54	42	49	45
1	89.3	3	241.4	2	334.9	BUTLER	KS	170	39	275	96	93
		2	263.1	2	183.5	CHAUTAUQUA	KS		-30			54
6	193.6	5	234.0	6	200.4	CHEROKEE	KS	21	-14	3	26	61
1*	344.1					CLAY	KS					
10	172.9	6	90.1	17	224.9	COWLEY	KS	-48	150	30	41	69
9	299.5	6	197.2	6	165.2	CRAWFORD	KS	-34	-16	-45	10	49
2	279.8	1	74.3	1	94.4	DICKINSON	KS	-73	27	-66	04	36
3	80.2	4	133.4	10*	561.3	DONIPHAN	KS	66	321	600	99	98
21	200.7	27	227.6	24	189.4	DOUGLAS	KS	13	-17	-6	23	57
1*	188.7					EDWARDS	KS					
1*	267.9					ELK	KS					
		1	74.3			ELLIS	KS					
		1	74.3			ELLSWORTH	KS					
4	413.7	5	506.7	2	111.8	FINNEY	KS	22	-78	-73	02	38
3	228.3	3	157.4	4	142.3	FORD	KS	-31	-10	-38	12	44
2	127.2	4	282.9	7	320.7	FRANKLIN	KS	122	13	152	89	92
17*	318.9	17	234.1	29	276.1	GEARY	KS	-27	18	-13	20	85
4	209.7	10*	558.5	2	181.7	GRAHAM	KS	166	-67	-13	20	54
		1*	424.0			GPAY	KS					
				1	267.9	GREELEY	KS					83
				2	834.8	GREENWOOD	KS					99
7	397.8	4	157.8	5	130.5	HARVEY	KS	-60	-17	-67	03	42
		1	327.3			HODGEMAN	KS					
6	222.3	4	142.5	5	171.6	JACKSON	KS	-36	20	-23	16	51
1	98.2	1	163.6	3	70.0	JEFFERSON	KS	67	-57	-29	14	33
		1*	344.1			JEWELL	KS					
10	307.8	5	119.9	5	97.8	JOHNSON	KS	-61	-18	-68	03	37
15	211.0	24*	290.0	18	307.7	LABETTE	KS	37	6	46	51	90
55	242.6	45	176.1	40	177.2	LEAVENWORTH	KS	-27	1	-27	15	52
3	425.8			2	324.1	LINN	KS			-24	16	92
1	106.0	1	112.4			LOGAN	KS	6				
9	266.5	10	302.0	6	244.1	LYON	KS	13	-19	-8	22	75
1	132.6	1	424.0			MC PHERSON	KS	220				
		1	112.4	1	74.3	MARION	KS		-34			34
1	344.1					MAPSHALL	KS					
9	214.4	13	344.1	8	208.8	MIAMI	KS	60	-39	-3	24	63
41	222.5	41	226.3	51	281.7	MONTGOMERY	KS	2	24	27	38	86
2	235.7	2	301.1	1	37.2	MORRIS	KS	28	-88	-84	01	30
1*	188.7					MORTON	KS					
				3	607.1	NEMAHA	KS					99
2	121.5	5	219.3	6	327.9	NEOSHO	KS	81	50	170	91	92
2*	1020.2					NORTON	KS					
1	188.7	1	344.1			OSAGE	KS	82				
				1	344.1	OTTAWA	KS					94
2	107.1	3	231.2	1	56.2	PAWNEE	KS	116	-76	-48	09	32
1	938.4					PHILLIPS	KS					
1	34.4	3	189.0	2	72.0	PRATT	KS	449	-62	109	81	33
9	209.1	8	155.8	14	226.1	RENO	KS	-26	45	8	29	69
1*	546.4					REPUBLIC	KS					
		2	163.2	2	218.5	RICE	KS		34			67
4	136.6	4	70.2	10	242.5	RILEY	KS	-49	246	78	68	74
1	546.4	2	362.2			ROOKS	KS	-34				
1	188.7	1	112.4			RUSSELL	KS	-40				
5	186.5	9	224.6	21*	409.1	SALINE	KS	20	82	119	83	96
				1	267.9	SCOTT	KS					83
72	146.2	127	182.3	207	220.2	SEDGWICK	KS	25	21	51	54	67
		4	472.5	3	110.8	SEWARD	KS		-77			38
74	186.6	106	231.9	130	239.6	SHAWNEE	KS	24	3	28	40	73
		1*	267.9			SHERMAN	KS					
				1	74.3	STAFFORD	KS					34
				4	345.8	STEVENS	KS					94
		4	249.8	2	51.5	SUMNER	KS		-79			31
4	207.7	4	539.6	1	37.2	WABAUNSEE	KS	160	-93	-82	01	30
				1	112.4	WALLACE	KS					38
1	149.1					WILSON	KS					
260	195.3	337	223.8	444*	274.2	WYANDOTTE	KS	15	23	40	48	85

* SIGNIFICANTLY MORE DEATHS THAN EXPECTED

NON-WHITE MALE — ALL CANCERS COMBINED — ICD 140-208

1950-1959 DEATHS	1950-1959 RATE	1960-1969 DEATHS	1960-1969 RATE	1970-1979 DEATHS	1970-1979 RATE	COUNTY AND STATE		PERCENT CHANGE 60S/50S	PERCENT CHANGE 70S/60S	PERCENT CHANGE 70S/50S	PERCENTILE %CHANGE 70S/50S	PERCENTILE RATES 1970S
1764*	175.6	2452*	238.8	2917*	282.4	KENTUCKY	KY	36	18	61		
7	178.0	8	261.5	2	40.1	ADAIR	KY	47	-85	-77	02	30
		3	182.6	1	37.5	ALLEN	KY		-79			30
2	68.3	5	350.4	7	303.5	ANDERSON	KY	413	-13	345	97	90
6	114.2	5	216.5	3	159.5	BALLARD	KY	90	-26	40	47	47
13	122.8	18	190.9	25	270.0	BARREN	KY	56	41	120	83	84
3	87.6	4	159.4	9*	492.7	BATH	KY	82	209	462	98	98
15	202.8	16	223.3	19	356.1	BELL	KY	10	59	76	68	94
6*	509.9	5*	552.5	3	217.5	BOONE	KY	8	-61	-57	06	66
33	210.4	41	279.4	42*	339.2	BOURBON	KY	33	21	61	59	93
9	239.8	16*	319.4	9	211.2	BOYD	KY	33	-34	-12	20	64
20	144.3	29	199.4	35	261.2	BOYLE	KY	38	31	81	70	80
2	513.3	1	112.4	1	133.9	BRACKEN	KY	-78	19	-74	02	42
1	114.7	1	62.9	1	188.7	BREATHITT	KY	-45	200	65	61	57
3	81.9	9	283.4	4	168.6	BRECKINRIDGE	KY	246	-41	106	80	50
2	116.1	1	57.3	2	173.8	BULLITT	KY	-51	203	50	53	51
1	805.6			2	219.8	BUTLER	KY			-73	03	67
12	171.5	11	208.4	14	299.8	CALDWELL	KY	22	44	75	67	89
4	110.0	9	218.3	12	375.8	CALLOWAY	KY	98	72	242	95	95
7	131.5	10	225.1	10	194.7	CAMPBELL	KY	71	-14	48	53	59
1	663.2	1	112.4	2	306.4	CARLISLE	KY	-83	173	-54	07	90
1	94.4	4	410.2	3	346.3	CARROLL	KY	335	-16	267	96	94
42	76.7	81	162.0	99	205.5	CHRISTIAN	KY	111	27	168	90	62
26	227.0	36*	315.5	27	262.6	CLARK	KY	39	-17	16	32	80
2	176.3	3	159.4	8*	520.0	CLAY	KY	-10	226	195	93	98
				1	245.4	CLINTON	KY					75
2	269.4	2	153.0	5*	545.3	CRITTENDEN	KY	-43	256	102	79	98
8*	404.9	7	376.8	3	147.0	CUMBERLAND	KY	-7	-61	-64	04	44
43*	284.6	39	245.4	34	238.1	DAVIESS	KY	-14	-3	-16	19	73
1	56.2					EDMONSON	KY					
2	439.5	1	163.6			ESTILL	KY	-63				
183*	193.5	235*	242.4	334*	327.5	FAYETTE	KY	25	35	69	64	92
4	159.0	10*	491.2	4	209.2	FLEMING	KY	209	-57	32	42	63
3	99.0	3	173.6	8*	934.2	FLOYD	KY	75	438	844	99	99
17	231.0	22	243.9	13	130.2	FRANKLIN	KY	6	-47	-44	10	41
11	123.2	16	199.0	15	203.4	FULTON	KY	62	2	65	61	62
1	133.9			1	163.6	GALLATIN	KY			22	36	48
9	199.3	12	284.3	8	230.8	GARRARD	KY	43	-19	16	32	71
1	281.2	2	456.5	1	245.4	GRANT	KY	62	-46	-13	20	75
11	126.3	16	177.3	21	213.6	GRAVES	KY	40	20	69	64	65
		1	267.9	2	301.1	GRAYSON	KY		12			90
2	63.2	7	272.1	5	211.2	GREEN	KY	330	-22	234	94	64
1	267.9	1	188.7	3	315.4	GREENUP	KY	-30	67	18	33	91
1	188.7	1	136.6	1	141.3	HANCOCK	KY	-28	3	-25	15	44
11	191.5	13	179.0	19	230.0	HARDIN	KY	-7	29	20	35	70
22	109.2	32	177.2	21	153.7	HARLAN	KY	62	-13	41	48	46
4	112.5	13*	473.8	6	288.0	HARRISON	KY	321	-39	156	89	88
6	139.6	9	162.9	12	216.9	HART	KY	17	33	55	56	66
22	125.7	42	259.0	44	314.8	HENDERSON	KY	106	22	150	88	91
11	294.9	11	272.4	11	303.4	HENRY	KY	-8	11	3	26	90
3	104.4	3	69.8	8	257.2	HICKMAN	KY	-33	269	146	88	79
44	200.2	50	260.9	53	277.6	HOPKINS	KY	30	6	39	47	85
664*	214.6	1006*	283.3	1279*	326.0	JEFFERSON	KY	32	15	52	55	92
6	85.2	2	29.6	11	186.3	JESSAMINE	KY	-65	528	119	83	55
				1	188.7	JOHNSON	KY					57
44	227.3	68*	360.3	66*	379.4	KENTON	KY	59	5	67	63	95
1	89.3	1	344.1	2	330.2	KNOTT	KY	285	-4	270	96	93
4	186.7	7	271.0	7	281.7	KNOX	KY	45	4	51	54	86
3	139.3	9*	393.2	3	150.6	LARUE	KY	182	-62	8	29	45
4*	629.6	2	143.5			LAUREL	KY	-77				
1	112.4	3*	598.7			LAWRENCE	KY	433				
1*	112.4					LESLIE	KY					
9	193.2	7	118.3	9	203.3	LETCHER	KY	-39	72	5	27	61
				1	172.0	LEWIS	KY					51
9	139.8	11	223.8	12	212.3	LINCOLN	KY	60	-5	52	54	65
2	423.3	1	112.4			LIVINGSTON	KY	-73				
12	102.1	13	106.4	27	223.0	LOGAN	KY	4	110	118	83	68
2	40.8	3	102.6	8	376.4	LYON	KY	151	267	822	99	95
53	163.1	55	189.1	91*	309.3	MC CRACKEN	KY	16	64	90	74	91
2	357.2			2	546.7	MC LEAN	KY			53	55	98
22	148.3	31	197.6	40	268.7	MADISON	KY	33	36	81	70	83
4	62.4	10	208.9	17	328.6	MARION	KY	235	57	427	98	93
10	100.7	12	158.3	19	252.2	MASON	KY	57	59	150	88	78
3	232.2	2	163.6	4	210.0	MEADE	KY	-30	28	-10	21	64
				1*	267.9	MENIFEE	KY					83
14	257.4	15	326.8	6	130.8	MERCER	KY	27	-60	-49	09	42
		4	181.7	3	161.2	METCALFE	KY		-11			48
2	126.3	3	171.3	2	123.6	MONROE	KY	36	-28	-2	24	40
19*	282.9	12	203.1	12	233.3	MONTGOMERY	KY	-28	15	-18	18	71
19	231.8	12	159.0	21	335.3	MUHLENBERG	KY	-31	111	45	50	93
15	227.6	19	282.3	20	288.0	NELSON	KY	24	2	27	38	87
5	520.6	3	290.9	3	261.1	NICHOLAS	KY	-44	-10	-50	08	80
2	61.2	6	432.6	2	128.5	OHIO	KY	607	-70	110	81	41
5	101.0	9	164.2	6	149.0	OLDHAM	KY	63	-9	47	52	45
6	328.5	3	111.3	3	182.8	OWEN	KY	-66	64	-44	10	54
1	188.7					OWSLEY	KY					
		2	450.0	1	188.7	PENDLETON	KY		-58			57
7	146.4	11	324.2	8	250.4	PERRY	KY	122	-23	71	65	77
6	138.9	7	259.2	5	154.9	PIKE	KY	87	-40	12	30	46
2	325.3	1	188.7	2	400.7	POWELL	KY	-42	112	23	37	96
2	60.3	8	305.8	8	204.4	PULASKI	KY	407	-33	239	95	62
		2	697.2	4*	537.4	RUSSELL	KY		-23			98
15	142.4	22	207.4	20	238.1	SCOTT	KY	46	15	67	63	73
20	180.9	22	214.4	20	201.9	SHELBY	KY	18	-6	12	30	61
13	206.7	11	150.5	11	159.2	SIMPSON	KY	-27	6	-23	16	47
1	49.2	4	311.1	2	194.0	SPENCER	KY	533	-38	295	97	59
2	44.6	13	285.9	13	245.3	TAYLOR	KY	540	-14	449	98	75
18	169.2	9	113.5	17	177.9	TODD	KY	-33	57	5	27	53
8	98.0	10	157.1	11	160.0	TRIGG	KY	60	2	63	60	47

* SIGNIFICANTLY MORE DEATHS THAN EXPECTED

NON-WHITE MALE — ALL CANCERS COMBINED — ICD 140-208

1950-1959 DEATHS	1950-1959 RATE	1960-1969 DEATHS	1960-1969 RATE	1970-1979 DEATHS	1970-1979 RATE	COUNTY AND STATE		PERCENT CHANGE 60S/50S	PERCENT CHANGE 70S/60S	PERCENT CHANGE 70S/50S	PERCENTILE %CHANGE 70S/50S	PERCENTILE RATES 1970S
16	195.2	23*	359.4	14	238.7	UNION	KY	84	-34	22	37	73
24	129.3	53	252.2	62	282.7	WARREN	KY	95	12	119	83	87
6	138.2	13	368.2	15	378.6	WASHINGTON	KY	166	3	174	91	95
1	74.3	3	176.9	4	163.4	WAYNE	KY	138	-8	120	83	48
13	136.0	25	289.1	28*	460.1	WEBSTER	KY	113	59	238	95	97
2	186.9	1	267.9	3	302.6	WHITLEY	KY	43	13	62	59	90
23	267.0	18	235.6	18	230.3	WOODFORD	KY	-12	-2	-14	19	70
6274*	185.0	8290*	218.1	10526*	253.3	LOUISIANA	LA	18	16	37		
64	212.3	82	235.8	100	275.2	ACADIA	LA	11	17	30	41	85
37	218.8	47	237.6	50	243.7	ALLEN	LA	9	3	11	30	75
62	179.0	69	201.1	81	234.0	ASCENSION	LA	12	16	31	41	71
45	167.7	63	247.7	72*	309.4	ASSUMPTION	LA	48	25	84	72	91
53	149.9	77	201.9	85	220.2	AVOYELLES	LA	35	9	47	52	67
23	149.7	45	277.4	37	232.9	BEAUREGARD	LA	85	-16	56	56	71
47	133.7	66	169.7	66	174.5	BIENVILLE	LA	27	3	31	41	52
81	150.6	106	180.5	111	201.4	BOSSIER	LA	20	12	34	43	61
436	162.4	627	199.4	857*	252.2	CADDO	LA	23	27	55	56	78
178*	226.2	208	216.2	309*	268.6	CALCASIEU	LA	-4	24	19	34	83
17	133.7	12	98.5	20	177.9	CALDWELL	LA	-26	81	33	43	53
4	159.6	2	130.6	8	380.2	CAMERON	LA	-18	191	138	87	95
28	164.4	43	271.4	38	265.8	CATAHOULA	LA	65	-2	62	59	81
37	79.3	66	141.2	98	195.1	CLAIBORNE	LA	78	38	146	88	59
52	137.2	67	163.5	79	196.4	CONCORDIA	LA	19	20	43	50	59
87	148.6	125	193.6	140	219.3	DE SOTO	LA	30	13	48	53	67
340	180.5	449	186.8	693	237.4	EAST BATON ROUGE	LA	4	27	32	42	73
66	159.0	69	178.6	73	218.6	EAST CARROLL	LA	12	22	38	46	67
48	102.5	49	104.3	74	181.6	EAST FELICIANA	LA	2	74	77	68	54
22	106.3	49	227.4	64	251.2	EVANGELINE	LA	114	10	136	86	77
56	154.3	60	164.8	64	175.9	FRANKLIN	LA	7	7	14	31	52
23	165.1	27	195.9	29	189.6	GRANT	LA	19	-3	15	31	57
113*	234.9	134*	262.3	192*	336.4	IBERIA	LA	12	28	43	50	93
103	177.2	124	220.6	145	259.1	IBERVILLE	LA	24	17	46	51	79
21	125.1	40	195.3	45	188.1	JACKSON	LA	56	-4	50	54	56
155*	220.0	246	235.4	427*	287.9	JEFFERSON	LA	7	22	31	41	87
38	186.5	50	246.6	68*	310.9	JEFFERSON DAVIS	LA	32	26	67	62	91
108*	240.8	170*	293.3	199*	284.6	LAFAYETTE	LA	22	-3	18	34	87
52*	262.1	66*	312.1	67	284.3	LAFOURCHE	LA	19	-9	8	29	87
10	142.8	14	180.2	16	184.8	LA SALLE	LA	26	3	29	40	55
44	132.0	79	189.8	92	177.5	LINCOLN	LA	44	-6	34	44	52
15	146.8	25	217.8	23	163.1	LIVINGSTON	LA	48	-25	11	30	48
72	145.2	116	243.3	109	251.2	MADISON	LA	68	3	73	66	77
73	128.2	101	168.5	110	181.9	MOREHOUSE	LA	31	8	42	49	54
84	125.2	99	160.1	113	183.0	NATCHITOCHES	LA	28	14	46	51	54
1774*	256.6	2293*	272.4	2910*	306.1	ORLEANS	LA	6	12	19	34	90
182	168.6	226	181.0	297	228.3	OUACHITA	LA	7	26	35	44	70
31	161.9	41	226.4	60*	320.8	PLAQUEMINES	LA	40	42	98	77	92
77	162.3	74	167.1	106	251.4	POINTE COUPEE	LA	3	50	55	56	77
201	152.5	273	199.0	330	234.2	RAPIDES	LA	31	18	54	55	72
25	109.9	26	126.3	30	145.9	RED RIVER	LA	15	16	33	43	44
65	157.4	65	157.1	80	192.1	RICHLAND	LA	0	22	22	36	58
26	135.5	37	200.3	44	211.0	SABINE	LA	48	5	56	56	64
17*	267.5	22	315.0	23	260.0	ST BERNARD	LA	18	-17	-3	24	80
35	192.3	32	147.2	75	291.9	ST CHARLES	LA	-23	98	52	54	88
15	96.2	20	101.1	28	136.4	ST HELENA	LA	5	35	42	49	43
58	187.3	60	202.1	81	268.5	ST JAMES	LA	8	33	43	50	83
46	152.2	95*	309.0	101	281.9	ST JOHN THE BAPT	LA	103	-9	85	73	86
174	178.1	252	237.2	307	262.5	ST LANDRY	LA	33	11	47	52	80
57	198.6	86*	276.6	97	272.2	ST MARTIN	LA	39	-2	37	45	84
93	168.9	134	231.3	189*	315.6	ST MARY	LA	37	36	87	73	91
57	176.7	82	212.3	108	235.9	ST TAMMANY	LA	20	11	33	43	72
123	188.0	144	196.7	158	201.6	TANGIPAHOA	LA	5	2	7	28	61
47	121.3	59	167.7	64	210.6	TENSAS	LA	38	26	74	66	64
74	204.0	100*	277.0	124*	300.4	TERREBONNE	LA	36	8	47	52	89
31	134.1	41	152.7	58	213.5	UNION	LA	14	40	59	57	65
36	237.5	47	256.7	57	272.5	VERMILION	LA	8	6	15	31	84
17	150.1	24	222.6	22	199.4	VERNON	LA	48	-10	33	43	60
106*	227.6	98	182.7	108	192.6	WASHINGTON	LA	-20	5	-15	19	58
71	144.4	124	211.5	140	215.6	WEBSTER	LA	46	2	49	53	65
39	139.3	54	200.0	75	281.5	WEST BATON ROUGE	LA	44	41	102	78	86
26	197.0	27	211.7	25	235.7	WEST CARROLL	LA	7	11	20	35	72
40	152.5	47	189.2	32	151.2	WEST FELICIANA	LA	24	-20	-1	25	46
37	168.7	35	140.9	43	166.5	WINN	LA	-16	18	-1	25	49
20	153.0	27	177.3	34	124.5	MAINE	ME	16	-30	-19		
1	172.0			3	518.0	ANDROSCOGGIN	ME			201	93	98
2	138.7	3	144.7	3	51.4	AROOSTOOK	ME	4	-64	-63	05	31
4	122.0	8	251.3	8	143.1	CUMBERLAND	ME	106	-43	17	33	44
				2	204.0	HANCOCK	ME					62
		3	285.4	1	112.4	KENNEBEC	ME		-61			38
1	344.1					OXFORD	ME					
2	58.2	2	68.9	9	242.7	PENOBSCOT	ME	18	252	317	97	74
				1	24.8	PISCATAQUIS	ME					29
				1	133.9	SOMERSET	ME					42
1*	344.1					WALDO	ME					
7*	381.4	10*	797.0	4	249.9	WASHINGTON	ME	109	-69	-34	13	77
2	206.8	1	74.3	2	76.4	YORK	ME	-64	3	-63	05	34
3049*	201.6	5047*	257.3	7654*	298.4	MARYLAND	MD	28	16	48		
10	169.2	13	264.7	20	374.1	ALLEGANY	MD	56	41	121	84	95
141	151.8	243	216.1	333	251.4	ANNE ARUNDEL	MD	42	16	66	62	77
119	204.6	169	239.2	273	246.7	BALTIMORE	MD	17	3	21	36	76
27	154.6	36	183.7	46	184.4	CALVERT	MD	19	0	19	34	55
38	219.3	43	250.8	50	267.8	CAROLINE	MD	14	7	22	36	82
14	134.7	17	141.3	32	241.7	CARROLL	MD	5	71	79	70	74
23	188.8	33	234.1	33	225.1	CECIL	MD	24	-4	19	34	69
39	145.5	66	208.5	105	255.1	CHARLES	MD	43	22	75	68	79
61	166.8	77	179.2	129*	279.0	DORCHESTER	MD	7	56	67	63	86

* SIGNIFICANTLY MORE DEATHS THAN EXPECTED

NON-WHITE MALE — ALL CANCERS COMBINED — ICD 140-208

1950-1959 DEATHS	1950-1959 RATE	1960-1969 DEATHS	1960-1969 RATE	1970-1979 DEATHS	1970-1979 RATE	COUNTY AND STATE		PERCENT CHANGE 60S/50S	PERCENT CHANGE 70S/60S	PERCENT CHANGE 70S/50S	PERCENTILE %CHANGE 70S/50S	PERCENTILE RATES 1970S
34	181.7	55	259.8	80*	333.4	FREDERICK	MD	43	28	84	72	93
36	165.2	74	275.2	100*	290.3	HARFORD	MD	67	5	76	68	88
25	164.6	30	189.4	48	210.4	HOWARD	MD	15	11	28	39	64
29	159.0	36	179.1	46	221.1	KENT	MD	13	23	39	47	67
71	162.8	137	256.5	265	225.1	MONTGOMERY	MD	58	-12	38	46	69
107	138.4	244	188.0	674	232.1	PRINCE GEORGE'S	MD	36	23	68	63	71
31	166.5	44	199.4	46	188.4	QUEEN ANNE'S	MD	20	-6	13	31	56
29	146.4	59	254.6	82	256.5	ST MARY'S	MD	74	1	75	67	79
50	152.3	61	188.1	108*	343.2	SOMERSET	MD	24	82	125	84	93
46	185.1	69	226.3	84	281.7	TALBOT	MD	22	24	52	55	86
20	215.2	33*	351.6	26	248.7	WASHINGTON	MD	63	-29	16	32	76
66	183.8	83	178.2	168*	317.8	WICOMICO	MD	-3	78	73	66	91
51	169.5	72	227.1	98	290.2	WORCESTER	MD	34	28	71	65	88
1982*	228.4	3353*	291.3	4808*	340.8	BALTIMORE CITY	MD	28	17	49	53	93
875*	239.4	1086	222.2	1600	222.7	MASSACHUSETTS	MA	-7	0	-7		
13	134.2	24	243.0	32	190.5	BARNSTABLE	MA	81	-22	42	49	57
13	283.1	17	247.8	29	326.3	BERKSHIRE	MA	-12	32	15	32	92
73*	311.1	47	191.6	72	172.7	BRISTOL	MA	-38	-10	-44	10	51
6*	354.1	2	50.4	6	174.1	DUKES	MA	-86	246	-51	08	52
29*	273.3	34	238.5	34	123.1	ESSEX	MA	-13	-48	-55	07	40
1	172.0	6*	802.2	2	127.7	FRANKLIN	MA	366	-84	-26	15	41
63	230.0	93	209.1	182	248.5	HAMPDEN	MA	-9	19	8	28	76
4	184.6	3	100.9	2	69.4	HAMPSHIRE	MA	-45	-31	-62	05	33
125*	256.2	126	192.5	142	141.3	MIDDLESEX	MA	-25	-27	-45	10	43
4*	405.4	3	469.6	4	791.3	NANTUCKET	MA	16	69	95	76	99
17	221.2	21	148.4	39	145.4	NORFOLK	MA	-33	-2	-34	13	44
32	148.6	47	190.6	56	165.6	PLYMOUTH	MA	28	-13	11	30	49
476*	248.8	646*	254.4	947*	270.4	SUFFOLK	MA	2	6	9	29	84
19	155.9	17	96.4	53	218.3	WORCESTER	MA	-38	127	40	48	66
3480*	197.6	6312*	231.5	9757*	259.6	MICHIGAN	MI	17	12	31		
		1	490.8	1	172.0	ALGER	MI		-65			51
4	75.4	16	248.4	17	175.6	ALLEGAN	MI	230	-29	133	86	52
				1	112.4	ALPENA	MI					38
		1	188.7			ANTRIM	MI					
1	57.3	4	312.3	2	78.8	BARAGA	MI	445	-75	37	46	34
				1	84.8	BARRY	MI					35
2	90.9	4	113.9	8	301.4	BAY	MI	25	165	232	94	90
		1	114.7			BENZIE	MI					
69	213.5	132*	265.5	194*	283.3	BERRIEN	MI	24	7	33	42	87
1	42.7					BRANCH	MI					
44	198.5	81	255.3	113	238.2	CALHOUN	MI	29	-7	20	35	73
31	174.9	41	187.3	80*	313.1	CASS	MI	7	67	79	69	91
2	173.0			1	188.7	CHARLEVOIX	MI			9	29	57
1	141.3			1	62.9	CHEBOYGAN	MI			-55	07	32
1	44.6	2	58.8	7	122.8	CHIPPEWA	MI	32	109	175	91	40
1	490.8					CLINTON	MI					
1	344.1					CRAWFORD	MI					
3*	606.1	1	212.0	1	133.9	DELTA	MI	-65	-37	-78	02	42
				1	74.3	DICKINSON	MI					34
		2	157.1	7	202.2	EATON	MI		29			61
4	127.2	1	67.0	4	226.1	EMMET	MI	-47	238	78	69	69
131*	217.1	260*	263.7	436*	274.1	GENESEE	MI	21	4	26	38	85
1	344.1					GOGEBIC	MI					
1	54.2	3	114.9	2	168.7	GRAND TRAVERSE	MI	112	47	211	93	50
				4*	753.0	HILLSDALE	MI					99
				2	532.8	HOUGHTON	MI					98
29	216.7	51	207.4	105	240.8	INGHAM	MI	-4	16	11	30	74
2	53.0	2	36.4	1	112.4	IONIA	MI	-31	209	112	81	38
1	344.1			1	245.4	IOSCO	MI			-29	14	75
1*	344.1					IRON	MI					
3	121.7	5	234.7	6	208.4	ISABELLA	MI	93	-11	71	65	63
29	156.4	46	175.0	73	241.6	JACKSON	MI	12	38	54	56	74
31	217.7	43	213.1	79	254.4	KALAMAZOO	MI	-2	19	17	33	78
68	217.1	119	235.3	208*	280.3	KENT	MI	8	19	29	40	86
21	172.8	31	260.2	41	282.3	LAKE	MI	51	8	63	61	87
2	219.8	2	171.1	2	149.7	LAPEER	MI	-22	-13	-32	14	45
3	293.9	2	109.1	3	242.8	LEELANAU	MI	-63	123	-17	18	74
2	185.4	2	205.2	3	109.8	LENAWEE	MI	11	-46	-41	11	38
1	74.5	1	67.0	2	186.8	LIVINGSTON	MI	-10	179	151	88	55
1	344.1					LUCE	MI					
3	193.1	1	89.3	1	38.3	MACKINAC	MI	-54	-57	-80	01	30
33	196.0	42	178.9	58	168.6	MACOMB	MI	-9	-6	-14	19	50
				1	68.8	MANISTEE	MI					33
1	22.8	1	86.0	4	147.6	MARQUETTE	MI	278	72	548	99	45
1	81.8	2	224.8	2	159.6	MASON	MI	175	-29	95	76	47
4	321.2	2	100.9	3	184.4	MECOSTA	MI	-69	83	-43	10	55
				2	163.2	MENOMINEE	MI					48
2	688.1	1	245.4	2	124.3	MIDLAND	MI	-64	-49	-82	01	40
1	188.7					MISSAUKEE	MI					
7	121.6	15	167.1	23	207.9	MONROE	MI	37	24	71	65	63
2	342.2			4	325.2	MONTCALM	MI			-5	23	92
46	161.5	95	235.8	153	257.4	MUSKEGON	MI	46	9	59	57	79
10	312.4	9	217.4	18	298.0	NEWAYGO	MI	-30	37	-5	23	89
77	149.5	172	255.5	258	211.7	OAKLAND	MI	71	-17	42	48	64
1	141.3			5	410.5	OCEANA	MI			190	92	96
1*	74.3					OGEMAW	MI					
		1	424.0	1	188.7	OSCEOLA	MI		-55			57
				1*	267.9	OTSEGO	MI					83
3	334.1	1	188.7	3	31.0	OTTAWA	MI	-44	-84	-91	00	29
		1*	74.3			PRESQUE ISLE	MI					
56	147.4	113	212.1	204	227.3	SAGINAW	MI	44	7	54	56	70
11	142.5	33	315.1	40	300.2	ST CLAIR	MI	121	-5	111	81	89
3	70.4	24*	391.6	19	260.4	ST JOSEPH	MI	456	-33	270	96	80
		1	424.0	1	133.9	SANILAC	MI		-68			42
				1	141.3	SCHOOLCRAFT	MI					44
		1	141.3	1	112.4	SHIAWASSEE	MI		-20			38
1	267.9	1	267.9	3	118.0	TUSCOLA	MI	0	-56	-56	06	40

* SIGNIFICANTLY MORE DEATHS THAN EXPECTED

NON-WHITE MALE — ALL CANCERS COMBINED — ICD 140-208

1950-1959 DEATHS	1950-1959 RATE	1960-1969 DEATHS	1960-1969 RATE	1970-1979 DEATHS	1970-1979 RATE	COUNTY AND STATE	PERCENT CHANGE 60S/50S	PERCENT CHANGE 70S/60S	PERCENT CHANGE 70S/50S	PERCENTILE %CHANGE 70S/50S	PERCENTILE RATES 1970S
24	164.0	45	180.6	69	220.3	VAN BUREN MI	10	22	34	44	67
58	177.7	97	220.5	139	230.3	WASHTENAW MI	24	4	30	40	70
2643*	206.0	4798*	235.1	7334*	268.1	WAYNE MI	14	14	30	41	83
		2*	456.6			WEXFORD MI					
249*	203.9	288	180.2	396	165.9	MINNESOTA MN	-12	-8	-19		
1	490.8					AITKIN MN					
4*	581.4			3	71.9	ANOKA MN			-88	00	33
6	166.4	7	175.0	5	87.5	BECKER MN	5	-50	-47	09	35
5	71.2	11	161.3	14	147.8	BELTRAMI MN	127	-8	108	80	45
				2	377.0	BLUE EARTH MN					95
3	204.2	3	196.7			CARLTON MN	-4				
7	126.3	5	79.1	5	78.1	CASS MN	-37	-1	-38	12	34
1*	267.9					CLAY MN					
1	114.7	3	193.4	2	107.2	CLEARWATER MN	69	-45	-7	22	38
1	172.0	1	267.9			CROW WING MN	56				
1	589.6	1	84.8	4	86.1	DAKOTA MN	-86	1	-85	00	35
				1	344.1	DOUGLAS MN					94
		1	212.0			GOODHUE MN					
98*	221.0	136	223.1	194	193.3	HENNEPIN MN	1	-13	-13	20	59
		1	424.0			HOUSTON MN					
4	136.9	2	65.3	5	152.2	ITASCA MN	-52	133	11	30	46
1	141.3	1	119.2			KOOCHICHING MN	-16				
2*	456.6					LE SUEUR MN					
		2*	380.3			LYON MN					
2	70.5	6	179.0	5	121.3	MAHNOMEN MN	154	-32	72	65	40
		1*	422.1			MARTIN MN					
1	57.3	2	127.5			MILLE LACS MN	122				
				1	267.9	NOBLES MN					83
1*	663.2					NORMAN MN					
3*	446.6			4	328.5	OLMSTED MN			-26	15	92
1	133.9			1	119.2	OTTER TAIL MN			-11	21	40
1	112.4			1	43.0	PINE MN			-62	05	30
1	188.7					POLK MN					
91*	280.9	93	212.3	132	224.2	RAMSEY MN	-24	6	-20	17	68
2	142.2	1	47.2	1	62.9	REDWOOD MN	-67	33	-56	06	32
10	161.1	10	177.2	12	169.4	ST LOUIS MN	10	-4	5	27	50
1*	341.0					SCOTT MN					
				1	188.7	STEELE MN					57
				1*	546.4	SWIFT MN					98
				1	112.4	TRAVERSE MN					38
		1	182.2	1	267.9	YELLOW MEDICINE MN		47			83
4332	128.2	5503	162.9	6879	201.1	MISSISSIPPI MS	27	23	57		
84	129.5	152	219.3	226*	299.1	ADAMS MS	69	36	131	85	89
18	116.8	21	133.9	32	206.6	ALCORN MS	15	54	77	68	63
37	118.2	39	121.0	50	156.8	AMITE MS	2	30	33	42	47
34	97.0	49	137.5	52	139.8	ATTALA MS	42	2	44	50	43
10	78.9	15	116.6	18	150.8	BENTON MS	48	29	91	75	45
208	137.4	237	170.3	240	201.2	BOLIVAR MS	24	18	46	52	61
11	92.5	17	140.5	26	188.2	CALHOUN MS	52	34	103	79	56
27	95.8	29	128.2	30	130.6	CARROLL MS	34	2	36	45	42
28	99.7	45	167.2	41	161.9	CHICKASAW MS	68	-3	62	60	48
17	176.7	11	105.6	22	182.1	CHOCTAW MS	-40	72	3	26	54
42	134.4	45	138.0	48	144.2	CLAIBORNE MS	3	5	7	28	44
26	97.1	32	110.6	60	218.0	CLARKE MS	14	97	124	84	66
34	97.7	63	172.0	62	169.0	CLAY MS	76	-2	73	66	50
189	141.7	211	169.8	256	238.7	COAHOMA MS	20	41	68	63	73
61	120.0	87	171.7	120	227.6	COPIAH MS	43	33	90	74	70
20	102.6	27	143.5	27	137.4	COVINGTON MS	40	-4	34	44	43
48	83.9	48	93.4	49	100.3	DE SOTO MS	11	7	20	35	37
74	145.7	124	216.9	167*	282.7	FORREST MS	49	30	94	76	87
17	106.3	32	206.2	48*	341.0	FRANKLIN MS	94	65	221	94	93
8	193.6	12	244.0	12	193.6	GEORGE MS	26	-21	0	25	59
13	179.3	6	76.6	9	88.4	GREENE MS	-57	15	-51	08	35
43	118.0	57	147.4	76	207.9	GRENADA MS	25	41	76	68	63
14	148.4	19	186.6	34	319.8	HANCOCK MS	26	71	116	82	92
84	166.6	138	223.9	183	218.3	HARRISON MS	34	-2	31	41	66
309	146.1	461	182.0	754	246.8	HINDS MS	25	36	69	64	76
116	140.3	118	149.2	140	193.8	HOLMES MS	6	30	38	46	59
68	134.5	75	181.1	93	228.1	HUMPHREYS MS	35	26	70	64	70
19	133.7	19	217.1	16	243.6	ISSAQUENA MS	62	12	82	71	75
2	62.3	7	146.3	14	250.2	ITAWAMBA MS	135	71	302	97	77
48	209.5	73	228.6	107	240.3	JACKSON MS	9	5	15	31	74
32	106.3	52	172.1	47	136.8	JASPER MS	62	-21	29	40	43
47	137.7	57	172.8	77	233.0	JEFFERSON MS	25	35	69	64	71
43	184.3	35	135.1	48	175.1	JEFFERSON DAVIS MS	-27	30	-5	23	52
59	114.5	84	149.3	105	170.5	JONES MS	30	14	49	53	50
34	122.0	24	93.5	32	122.2	KEMPER MS	-23	31	0	25	40
32	123.9	42	159.4	43	156.2	LAFAYETTE MS	29	-2	26	38	47
15	169.8	22	271.3	16	184.0	LAMAR MS	60	-32	8	29	55
108	133.6	141	168.0	188	218.3	LAUDERDALE MS	26	30	63	61	66
26	172.9	28	173.8	37	236.6	LAWRENCE MS	1	36	37	45	72
32	127.4	36	126.3	67	205.5	LEAKE MS	-1	63	61	59	62
53	136.5	68	191.4	83	213.8	LEE MS	40	12	57	57	65
125	107.9	190	182.4	203	204.4	LEFLORE MS	69	12	90	74	62
40	126.9	49	153.0	62	175.6	LINCOLN MS	21	15	38	46	52
63	110.4	81	136.5	108	176.0	LOWNDES MS	24	29	59	58	52
87	102.2	99	126.9	150	189.5	MADISON MS	24	49	85	73	57
36	134.8	47	162.5	92*	305.0	MARION MS	21	88	126	85	90
52	92.4	86	150.6	85	145.5	MARSHALL MS	63	-3	57	57	44
43	95.2	48	102.8	71	166.0	MONROE MS	8	61	74	67	49
28	128.7	34	138.9	31	123.7	MONTGOMERY MS	8	-11	-4	24	40
30	167.1	25	155.6	38	178.2	NESHOBA MS	-7	15	7	28	53
39	149.3	36	134.9	50	173.2	NEWTON MS	-10	28	16	32	51
57	121.9	52	124.0	69	177.8	NOXUBEE MS	2	43	46	51	53
44	123.2	69	193.1	80	225.4	OKTIBBEHA MS	57	17	83	71	69
95	150.9	102	156.7	131	203.3	PANOLA MS	4	30	35	44	61
20	117.8	23	117.8	48	265.5	PEARL RIVER MS	0	126	125	85	81

* SIGNIFICANTLY MORE DEATHS THAN EXPECTED

NON-WHITE MALE — ALL CANCERS COMBINED — ICD 140-208

1950-1959 DEATHS	1950-1959 RATE	1960-1969 DEATHS	1960-1969 RATE	1970-1979 DEATHS	1970-1979 RATE	COUNTY AND STATE	PERCENT CHANGE 60S/50S	PERCENT CHANGE 70S/60S	PERCENT CHANGE 70S/50S	PERCENTILE %CHANGE 70S/50S	PERCENTILE RATES 1970S
12	130.1	9	80.3	21	216.9	PERRY MS	-38	170	67	62	66
73	138.7	97	173.8	123	198.9	PIKE MS	25	14	43	50	60
18	130.7	18	140.7	24	185.9	PONTOTOC MS	8	32	42	49	55
7	94.3	13	168.8	11	123.3	PRENTISS MS	79	-27	31	41	40
68	129.5	71	141.6	78	191.5	QUITMAN MS	9	35	48	53	58
55	109.8	68	128.8	80	138.0	RANKIN MS	17	7	26	38	43
27	91.3	40	124.1	56	157.9	SCOTT MS	36	27	73	66	47
51	167.1	47	180.8	38	168.3	SHARKEY MS	8	-7	1	25	50
43	181.5	48	176.2	54	190.3	SIMPSON MS	-3	8	5	27	57
13	144.6	23	255.8	11	108.8	SMITH MS	77	-57	-25	16	38
10	138.3	10	167.0	29*	406.5	STONE MS	21	143	194	92	96
164	144.9	181	174.6	164	191.7	SUNFLOWER MS	21	10	32	42	58
61	103.8	83	142.9	83	168.3	TALLAHATCHIE MS	38	18	62	60	50
27	83.4	45	130.5	66	204.1	TATE MS	56	56	145	87	62
3	28.3	16	131.5	17	123.2	TIPPAH MS	364	-6	335	97	40
5	127.0	6	133.5	6	152.0	TISHOMINGO MS	5	14	20	35	46
81	128.8	80	156.5	62	179.3	TUNICA MS	22	15	39	47	53
13	98.2	15	101.0	27	184.0	UNION MS	3	82	87	73	55
30	146.9	39	197.1	44	216.6	WALTHALL MS	34	10	47	53	66
156	169.0	185	202.9	222	241.0	WARREN MS	20	19	43	49	74
262	153.2	329	202.3	346	216.8	WASHINGTON MS	32	7	42	48	66
33	157.9	37	180.9	44	192.7	WAYNE MS	15	7	22	36	58
8	88.2	15	165.4	23	237.5	WEBSTER MS	88	44	169	90	73
26	74.2	48	135.1	77	206.4	WILKINSON MS	82	53	178	91	62
24	90.3	35	111.1	66	191.8	WINSTON MS	23	73	112	81	58
23	94.5	18	79.3	35	160.0	YALOBUSHA MS	-16	102	69	64	47
90	114.7	128	173.6	99	147.4	YAZOO MS	51	-15	29	40	45
2964*	203.4	4126*	238.4	5179*	271.6	MISSOURI MO	17	14	34		
		1	172.0			ADAIR MO					
1*	112.4					ANDREW MO					
14	164.0	24	294.4	25	302.5	AUDRAIN MO	79	3	84	72	90
1*	188.7	1	844.2			BARRY MO	347				
4*	473.5	1	141.3	1	267.9	BATES MO	-70	90	-43	10	83
29	173.0	44	265.0	46	254.7	BOONE MO	53	-4	47	52	78
28	154.0	24	176.9	37	245.2	BUCHANAN MO	15	39	59	57	75
19	123.3	23	161.0	36	285.7	BUTLER MO	31	77	132	86	87
13	83.4	17	120.3	33	287.6	CALLAWAY MO	44	139	245	95	87
7	116.5	13	244.7	22	386.7	CAPE GIRARDEAU MO	110	58	232	94	96
6	310.5	1	65.5	5	395.8	CARROLL MO	-79	504	27	39	96
3	419.3	2	197.0	4	464.3	CASS MO	-53	136	11	30	97
16	258.6	14	281.0	9	140.8	CHARITON MO	9	-50	-46	09	43
		1	546.4			CHRISTIAN MO					
2*	1050.9					CLARK MO					
19*	392.1	12	246.8	14	271.3	CLAY MO	-37	10	-31	14	84
7*	461.9	6	486.9	5	217.2	CLINTON MO	5	-55	-53	07	66
16	189.6	24	263.1	18	195.7	COLE MO	39	-26	3	26	59
14	165.0	10	130.5	15	269.9	COOPER MO	-21	107	64	61	84
1*	490.8	1	344.1			CRAWFORD MO	-30				
2	603.2	1	114.7	1	267.9	DADE MO	-81	134	-56	07	83
1*	424.0					DALLAS MO					
3*	254.7					DAVIESS MO					
4	58.1	17	246.3	11	147.5	DUNKLIN MO	324	-40	154	89	45
5	198.4	2	45.5	13	258.8	FRANKLIN MO	-77	468	30	41	79
14	108.9	26	199.2	38	261.4	GREENE MO	83	31	140	87	80
1*	546.4					HARRISON MO					
2	69.2	5	288.0	2	106.3	HENRY MO	223	-63	19	34	37
		1*	490.8			HICKORY MO					
		1*	112.4			HOLT MO					
23*	311.6	16	249.1	21	372.4	HOWARD MO	-20	49	20	35	95
		3	797.4			HOWELL MO					
				1	62.9	IRON MO					32
638*	212.2	875*	230.3	1157*	262.8	JACKSON MO	9	14	24	37	80
13	230.0	14	222.1	10	151.1	JASPER MO	-3	-32	-34	13	46
8	212.4	19*	453.6	10	201.7	JEFFERSON MO	114	-56	-5	23	61
12*	298.9	10	315.2	8	239.5	JOHNSON MO	5	-24	-20	18	73
1	74.3			1	267.9	KNOX MO			260	96	83
		1	56.2	2	183.7	LACLEDE MO		227			55
14	179.4	19	279.3	12	232.8	LAFAYETTE MO	56	-17	30	41	71
		1	98.2			LAWRENCE MO					
5	181.6	5	209.9	2	92.2	LEWIS MO	16	-56	-49	09	36
8	174.1	7	152.8	11	255.6	LINCOLN MO	-12	67	47	52	79
9*	595.3	5	244.8	3	174.1	LINN MO	-59	-29	-71	03	51
3	87.1	7	360.6	4	262.4	LIVINGSTON MO	314	-27	201	93	80
3	94.1	5	296.9	5	331.3	MACON MO	216	12	252	96	93
2	1087.0			2	384.0	MADISON MO			-65	04	96
19	184.9	32*	351.8	21	263.5	MARION MO	90	-25	43	49	81
		1	112.4			MILLER MO					
38	168.0	29	157.1	29	238.5	MISSISSIPPI MO	-7	52	42	49	73
2	232.6	4	440.9	4	341.9	MONITEAU MO	90	-22	47	52	93
3	106.6	6	241.5	6	322.0	MONROE MO	126	33	202	93	92
13*	412.4	10	314.6	12	477.8	MONTGOMERY MO	-24	52	16	32	98
1	74.5	3	384.5	6*	754.9	MORGAN MO	416	96	913	99	99
36	155.6	26	127.6	39	208.6	NEW MADRID MO	-18	64	34	44	63
1	57.3	2	131.7	4	241.7	NEWTON MO	130	83	321	97	74
		1*	424.0			OSAGE MO					
46	112.0	72	180.4	62	199.1	PEMISCOT MO	61	10	78	69	60
1	344.1			1	267.9	PERRY MO			-22	17	83
24	214.7	29*	288.6	24	262.8	PETTIS MO	34	-9	22	37	80
2	382.6					PHELPS MO					
18	188.5	24	220.3	20	216.9	PIKE MO	17	-2	15	32	66
2	121.9	4	228.6	3	253.7	PLATTE MO	87	11	108	80	78
1	54.6	1	15.3	4	250.9	PULASKI MO	-72	1541	359	97	77
2	201.7	2	81.2	3	366.2	RALLS MO	-60	351	82	71	95
19	223.8	16	213.5	27*	384.3	RANDOLPH MO	-5	80	72	65	96
8	256.1	9	245.4	8	324.8	RAY MO	-4	32	27	39	92
1*	589.6					RIPLEY MO					
8	144.0	2	39.9	17	250.6	ST CHARLES MO	-72	527	74	67	77
1	86.0			2	791.9	ST CLAIR MO			821	99	99
2	314.2	2	228.2	1	546.4	STE GENEVIEVE MO	-27	139	74	67	98

* SIGNIFICANTLY MORE DEATHS THAN EXPECTED

NON-WHITE MALE — ALL CANCERS COMBINED — ICD 140-208

COUNTY AND STATE	1950-1959 DEATHS	1950-1959 RATE	1960-1969 DEATHS	1960-1969 RATE	1970-1979 DEATHS	1970-1979 RATE	PERCENT CHANGE 60S/50S	PERCENT CHANGE 70S/60S	PERCENT CHANGE 70S/50S	PERCENTILE %CHANGE 70S/50S	PERCENTILE RATES 1970S
ST FRANCOIS MO	3	458.9			3	264.9			-42	11	81
ST LOUIS COUNTY MO	142	189.0	226	214.2	476	228.6	13	7	21	36	70
SALINE MO	29*	251.3	22	225.8	39*	424.4	-10	88	69	63	97
SCOTT MO	10	89.6	18	138.8	33	276.1	55	99	208	93	85
SHELBY MO	2	278.3	3	625.4	6*	1009.6	125	61	263	96	99
STODDARD MO	3	55.2	7	138.7	12	315.9	151	128	472	98	91
STONE MO	1*	344.1									
SULLIVAN MO	1*	424.0									
TANEY MO	2*	424.0									
VERNON MO					1	60.7					32
WARREN MO	3	184.1	4	314.5	7	476.1	71	51	159	90	97
WASHINGTON MO	4*	652.8	1	424.0	5*	861.7	-35	103	32	42	99
WEBSTER MO	1*	490.8									
WRIGHT MO	1	212.0	3	155.6	3	203.4	-27	31	-4	24	62
ST LOUIS CITY MO	1543*	220.6	2308*	262.9	2717*	299.0	19	14	36	45	89
MONTANA MT	113	161.1	113	141.9	143	156.5	-12	10	-3		
BIG HORN MT	21*	238.2	17	168.2	19	209.0	-29	24	-12	20	63
BLAINE MT	8	151.1	9	164.6	5	104.3	9	-37	-31	14	37
CASCADE MT	6	199.3	5	218.1	14	231.7	9	6	16	32	71
CHOUTEAU MT	1	344.1			1	424.0			23	37	97
CUSTER MT	2	478.0	1	112.4			-76				
DEER LODGE MT	2	128.0	3	222.4	2	119.2	74	-46	-7	22	40
FERGUS MT	1	267.9									
FLATHEAD MT					1	56.2					32
GALLATIN MT	1	490.8	2	168.6	1	267.9	-66	59	-45	09	83
GLACIER MT	13	108.6	16	112.8	17	124.1	4	10	14	31	40
GRANITE MT	1*	344.1			1*	546.4			59	57	98
HILL MT	4	137.5	6	169.1	5	104.1	23	-38	-24	16	37
LAKE MT	11	217.2	5	115.9	16	209.0	-47	80	-4	24	63
LEWIS AND CLARK MT	4	217.2	3	193.2	4	247.8	-11	28	14	31	76
MADISON MT					1	267.9					83
MEAGHER MT	1*	267.9									
MINERAL MT	1*	188.7									
MISSOULA MT			2	114.7							
PARK#MT			1	245.4	1	424.0		73			97
PHILLIPS MT	4	299.8	5	270.6	3	249.2	-10	-8	-17	18	77
PONDERA MT	2	94.0	1	62.9	1	112.4	-33	79	20	35	38
POWELL MT	1	74.3	1	74.5	1	163.6	0	120	120	83	48
RAVALLI MT					1	267.9					83
RICHLAND MT					1	94.4					36
ROOSEVELT MT	7	102.6	11	126.2	26	304.0	23	141	196	93	90
ROSEBUD MT	3	58.9	4	105.2	6	301.0	79	186	411	98	89
SANDERS MT	1	60.7	4	427.7	1	44.6	604	-90	-27	15	31
SHERIDAN MT			1	106.0	2	224.8		112			69
SILVER BOW MT	2	198.4	4	205.2	3	132.2	3	-36	-33	13	42
TETON MT	1*	74.3			1	172.0			131	86	51
TOOLE MT	1	546.4									
VALLEY MT	6*	314.4	3	70.9	1	112.4	-77	58	-64	04	38
WHEATLAND MT	1*	267.9									
YELLOWSTONE MT	7	291.1	9	263.7	8	141.1	-9	-46	-52	08	43
NEBRASKA NB	240*	217.0	336*	247.9	390	229.5	14	-7	6		
ADAMS NB	2	173.0	6*	307.3	4	445.5	78	45	158	89	97
BOONE NB	1*	546.4									
BOX BUTTE NB	2	360.7	4	352.2	2	126.8	-2	-64	-65	04	41
BOYD NB	1*	112.4									
BURT NB					1	112.4					38
BUTLER NB	1*	267.9									
CHERRY NB			2	394.2							
CHEYENNE NB			2*	439.5							
DAKOTA NB	1	112.4									
DAWES NB	1	109.3	2	224.8			106				
DAWSON NB	1	344.1									
DODGE NB			1	267.9							
DOUGLAS NB	184*	224.5	269*	267.2	327*	271.3	19	2	21	36	84
GAGE NB	1	188.7	1	74.3			-61				
GOSPER NB			1*	267.9							
HALL NB	2	297.5	2	228.3	2	106.7	-23	-53	-64	04	38
KEITH NB	1*	112.4									
KNOX NB	1	112.4	4	243.9	5	338.9	117	39	201	93	93
LANCASTER NB	22*	290.1	15	171.3	21	190.2	-41	11	-34	13	57
LINCOLN NB	2	249.1	1	109.3	2	83.1	-56	-24	-67	04	34
MADISON NB			2	179.4							
OTOE NB					1*	344.1					94
POLK NB	1*	490.8									
RICHARDSON NB			1	112.4	2	258.1		130			79
SALINE NB			1*	112.4							
SARPY NB	1	39.3	1	212.0	3	79.2	439	-63	102	78	34
SCOTTS BLUFF NB	3	87.7	5	172.0	6	104.2	96	-39	19	34	37
SHERIDAN NB			2	237.9	3	252.3		6			78
SIOUX NB	1*	344.1									
THURSTON NB	10	162.8	12	208.8	11	168.2	28	-19	3	26	50
WASHINGTON NB			1	112.4							
YORK NB	1	424.0	1	344.1			-19				
NEVADA NV	59	135.1	120	152.9	298	179.6	13	17	33		
CHURCHILL NV	3	135.3	1	81.8	3	150.7	-40	84	11	30	45
CLARK NV	28	152.3	75	182.5	206	195.1	20	7	28	39	59
DOUGLAS NV			1	112.4	1	172.0		53			51
ELKO NV	4	90.9	8	145.2	9	129.1	60	-11	42	49	41
ESMERALDA NV			1	424.0							
EUREKA NV					1	112.4					38
HUMBOLDT NV	1	62.9	2	120.6	2	43.4	92	-64	-31	14	30
LANDER NV			2	311.2							
LYON NV			1	89.3	2	78.6		-12			34
MINERAL NV	3	114.8	4	103.9	6	162.1	-9	56	41	48	48
NYE NV	1	62.9									

* SIGNIFICANTLY MORE DEATHS THAN EXPECTED

COUNTIES COMBINED; SEE TECHNICAL NOTES

NON-WHITE MALE — ALL CANCERS COMBINED — ICD 140-208

1950-1959 DEATHS	1950-1959 RATE	1960-1969 DEATHS	1960-1969 RATE	1970-1979 DEATHS	1970-1979 RATE	COUNTY AND STATE	PERCENT CHANGE 60S/50S	PERCENT CHANGE 70S/60S	PERCENT CHANGE 70S/50S	PERCENTILE %CHANGE 70S/50S	PERCENTILE RATES 1970S
2	366.2	1	136.6	1	424.0	PERSHING NV	-63	210	16	32	97
15	217.8	23	174.1	60	201.5	WASHOE NV	-20	16	-7	22	61
1	61.3			1	81.8	WHITE PINE NV			33	43	34
1	89.3	1	112.4	6	196.6	ORMSBY-CARSON CITY NV	26	75	120	83	59
8	152.9	9	146.6	13	77.1	NEW HAMPSHIRE NH	-4	-47	-50		
1	196.5					BELKNAP NH					
				1	133.9	CARROLL NH					42
1	212.0	1	212.0			CHESHIRE NH	0				
		1	188.7			GRAFTON NH					
		1	188.7	1	24.4	HILLSBOROUGH NH		-87			29
1	267.9	1	94.4	2	165.2	MERRIMACK NH	-65	75	-38	12	49
4	348.6	5	88.9	9	152.2	ROCKINGHAM NH	-74	71	-56	06	46
1	490.8					STRAFFORD NH					
2904*	219.0	4939*	261.6	7142*	262.8	NEW JERSEY NJ	19	0	20		
281*	232.7	332*	237.9	400*	276.4	ATLANTIC NJ	2	16	19	34	85
107*	220.3	208*	283.2	291	223.8	BERGEN NJ	29	-21	2	26	68
72	203.7	113	226.6	202	253.1	BURLINGTON NJ	11	12	24	37	78
235*	228.3	384*	272.5	541*	282.8	CAMDEN NJ	19	4	24	37	87
34	206.3	35	171.0	57	232.2	CAPE MAY NJ	-17	36	13	31	71
61	147.3	84	165.7	156	240.4	CUMBERLAND NJ	12	45	63	60	74
939*	230.6	1632*	263.7	2348*	276.2	ESSEX NJ	14	5	20	35	85
60	139.4	131	247.1	178	272.7	GLOUCESTER NJ	77	10	96	77	84
217*	258.5	364*	297.5	493	224.1	HUDSON NJ	15	-25	-13	20	68
4	161.6	3	95.5	13	263.8	HUNTERDON NJ	-41	176	63	60	81
156	209.7	276	232.7	463*	269.2	MERCER NJ	11	16	28	40	83
74	190.8	178*	306.0	230	220.7	MIDDLESEX NJ	60	-28	16	32	67
206*	215.2	326*	266.7	440*	286.1	MONMOUTH NJ	24	7	33	43	87
43	259.5	43	195.5	77	232.6	MORRIS NJ	-25	19	-10	21	71
13	138.6	34	225.0	81*	307.7	OCEAN NJ	62	37	122	84	91
95*	210.0	257*	327.7	363	249.2	PASSAIC NJ	56	-24	19	34	77
58	214.6	89	269.9	128*	318.9	SALEM NJ	26	18	49	53	92
23	197.5	38	211.0	58	197.2	SOMERSET NJ	7	-7	0	25	60
1	188.7	1	70.1	18*	1094.6	SUSSEX NJ	-63	1461	480	99	99
219*	226.5	409*	300.2	593*	268.6	UNION NJ	33	-11	19	34	83
6	321.9	2	49.1	12	319.6	WARREN NJ	-85	551	-1	25	92
164	93.2	233	107.8	383	79.3	NEW MEXICO NM	16	-26	-15		
21	145.5	28	117.1	71	72.2	BERNALILLO NM	-19	-38	-50	08	33
7	183.8	11	234.3	13	113.2	CHAVES NM	27	-52	-38	12	39
1	74.3					COLFAX NM					
8	371.9	9	269.4	14	249.2	CURRY NM	-28	-8	-33	13	77
				1	267.9	DE BACA NM					83
6	150.9	9	239.3	6	30.7	DONA ANA NM	59	-87	-80	01	29
3	117.3	5	109.3	7	51.1	EDDY NM	-7	-53	-56	06	31
		1	147.4	1	37.8	GRANT NM		-74			30
				1	188.7	HIDALGO NM					57
11	314.5	14	135.5	34	243.6	LEA NM	-57	80	-23	17	75
				2	86.3	LINCOLN NM					35
				1	163.6	LOS ALAMOS NM					48
1	54.6	1	70.7	1	33.5	LUNA NM	29	-53	-39	12	30
39	78.0	60	116.0	80	100.2	MC KINLEY NM	49	-14	28	40	37
7	144.1	6	65.4	7	83.8	OTERO NM	-55	28	-42	11	35
		1	114.7			QUAY NM					
5	84.6	8	108.2	14	49.4	RIO ARRIBA NM	28	-54	-42	11	31
				1	16.7	ROOSEVELT NM					29
19	96.1	18	90.6	40	129.8	SANDOVAL NM	-6	43	35	44	41
22	76.7	35	88.6	39	67.6	SAN JUAN NM	15	-24	-12	21	33
		1	163.6			SAN MIGUEL NM					
4	85.2	2	41.1	7	44.0	SANTA FE NM	-52	7	-48	09	31
1	490.8					SIERRA NM					
		1	59.6	6	158.9	SOCORRO NM		166			47
2	53.0	3	50.5	11	69.5	TAOS NM	-5	38	31	41	33
1*	663.2					TORRANCE NM					
6	35.9	20	103.2	26	75.2	VALENCIA NM	187	-27	110	81	34
8151*	228.3	13411*	247.4	19391*	242.2	NEW YORK NY	8	-2	6		
63*	233.0	83	208.7	162*	291.3	ALBANY NY	-10	40	25	38	88
1	117.9			4	823.9	ALLEGANY NY			599	99	99
842*	258.0	1354	205.4	2473	219.5	BRONX NY	-20	7	-15	19	67
28*	472.5	19	242.1	19	211.4	BROOME NY	-49	-13	-55	07	64
7	127.3	13	214.7	23	291.2	CATTARAUGUS NY	69	36	129	85	88
4	245.4	4	108.5	10	213.3	CAYUGA NY	-56	97	-13	20	65
11	214.3	11	244.0	25	355.2	CHAUTAUQUA NY	14	46	66	62	94
16	279.4	22	398.9	36	372.1	CHEMUNG NY	43	-7	33	43	95
2	194.2	2	246.4	7	626.0	CHENANGO NY	27	154	222	94	99
		1	26.5	5	180.9	CLINTON NY		583			54
11	180.0	12	167.6	18	250.2	COLUMBIA NY	-7	49	39	47	77
		3	413.5	1	67.0	CORTLAND NY		-84			33
				1	49.1	DELAWARE NY					31
40	136.2	73	165.4	100	192.0	DUTCHESS NY	21	16	41	48	58
381*	210.3	704*	265.2	1078*	304.8	ERIE NY	26	15	45	50	90
1	85.2	2	154.8	1	112.4	ESSEX NY	82	-27	32	42	38
3	52.8	14	217.2	11	149.6	FRANKLIN NY	311	-31	183	92	45
3	281.8	4	364.8	4	165.1	FULTON NY	29	-55	-41	11	49
6	139.0	13	253.5	18	276.4	GENESEE NY	82	9	99	77	85
5	216.8	8	228.9	7	161.3	GREENE NY	6	-30	-26	15	48
				1*	267.9	HAMILTON NY					83
				1	39.0	HERKIMER NY					30
		2	170.8	1	133.9	JEFFERSON NY		-22			42
1530*	217.1	2304	186.0	4299	235.7	KINGS(BROOKLYN) NY	-14	27	9	29	72
7	358.2	6	167.4	3	59.6	LIVINGSTON NY	-53	-64	-83	01	32
3	285.4	1	54.7	3	127.7	MADISON NY	-81	134	-55	07	41
69	215.2	160	217.2	348	263.3	MONROE NY	1	21	22	37	80
2	270.5	2	181.2	3	269.3	MONTGOMERY NY	-33	49	0	25	83
136	215.2	315*	268.3	529	229.5	NASSAU NY	25	-14	7	28	70

* SIGNIFICANTLY MORE DEATHS THAN EXPECTED

NON-WHITE MALE — ALL CANCERS COMBINED — ICD 140-208

1950-1959 DEATHS	1950-1959 RATE	1960-1969 DEATHS	1960-1969 RATE	1970-1979 DEATHS	1970-1979 RATE	COUNTY AND STATE	PERCENT CHANGE 60S/50S	PERCENT CHANGE 70S/60S	PERCENT CHANGE 70S/50S	PERCENTILE %CHANGE 70S/50S	PERCENTILE RATES 1970S
3661*	250.3	5809*	332.1	5539*	270.6	NEW YORK(MANHATTAN) NY	33	-19	8	28	84
25	135.8	50	222.8	89	249.7	NIAGARA NY	64	12	84	72	77
18	168.4	31	169.6	39	172.3	ONEIDA NY	1	2	2	26	51
32	134.4	98	255.1	178*	286.4	ONONDAGA NY	90	12	113	82	87
5	274.9	8	242.2	14	269.3	ONTARIO NY	-12	11	-2	24	83
51	250.3	77	202.8	120	220.6	ORANGE NY	-19	9	-12	21	67
3	102.8	12	352.2	18	300.2	ORLEANS NY	243	-15	192	92	89
				4	213.0	OSWEGO NY					65
2	370.8	2	180.4	3	245.8	OTSEGO NY	-51	36	-34	13	76
1	136.6	2	206.7	3	220.6	PUTNAM NY	51	7	61	59	67
542*	200.9	1060	171.0	2374	207.3	QUEENS NY	-15	21	3	26	63
20*	358.0	21	246.3	26	206.5	RENSSELAER NY	-31	-16	-42	10	63
58*	231.5	47	146.9	132	222.1	RICHMOND NY	-37	51	-4	24	68
26	166.6	42	138.9	94	186.0	ROCKLAND NY	-17	34	12	30	55
1	98.2	1	49.1	3	217.8	ST LAWRENCE NY	-50	343	122	84	66
12	330.1	11	192.1	15	197.0	SARATOGA NY	-42	3	-40	11	60
17	263.1	25	265.6	40	264.3	SCHENECTADY NY	1	0	0	25	81
2	558.0	2	384.0			SCHOHAFIE NY	-31				
2	408.5			2	228.2	SCHUYLER NY			-44	10	70
2	98.8	2	64.2	1	47.2	SENECA NY	-35	-27	-52	08	31
7	174.0	11	206.0	9	154.9	STEUBEN NY	18	-25	-11	21	46
91	93.9	253	159.7	491	227.2	SUFFOLK NY	70	42	142	87	70
4	120.9	18	258.2	34	267.7	SULLIVAN NY	114	4	121	84	82
4	339.4	3	311.6	2	123.5	TIOGA NY	-8	-60	-64	04	40
12	257.6	15	264.4	12	155.3	TOMPKINS NY	3	-41	-40	11	46
27	229.6	27	149.1	44	194.2	ULSTER NY	-35	30	-15	19	59
		3	383.8	3	232.5	WARREN NY		-39			71
1	29.5	3	349.8			WASHINGTON NY	1087				
4	114.1	19*	351.8	22	220.1	WAYNE NY	208	-37	93	75	67
346*	250.8	626*	276.9	884*	266.1	WESTCHESTER NY	10	-4	6	28	82
1	5.8	1	9.5	3	102.4	WYOMING NY	63	982	1660	99	37
1	89.3			2	329.4	YATES NY			269	96	93
4060	126.3	6476	175.4	10375*	239.1	NORTH CAROLINA NC	39	36	89		
58	157.5	108	219.0	160	245.1	ALAMANCE NC	39	12	56	56	75
1	27.3	11	357.4	16	363.6	ALEXANDER NC	1208	2	1231	99	95
3	222.4	1	56.2	2	265.3	ALLEGHANY NC	-75	372	19	34	81
47	114.4	59	152.7	70	169.9	ANSON NC	33	11	49	53	50
4	388.4	1	38.5	2	340.2	ASHE NC	-90	783	-12	20	93
59	123.8	78	162.2	138	262.2	BEAUFORT NC	31	62	112	81	80
46	111.7	76	169.8	106	218.4	BERTIE NC	52	29	95	76	67
46	125.1	54	138.5	78	180.0	BLADEN NC	11	30	44	50	53
17	74.9	21	87.0	56	195.3	BRUNSWICK NC	16	125	161	90	59
115	186.6	161	246.0	207*	321.3	BUNCOMBE NC	32	31	72	66	92
12	109.2	34	237.9	43	227.4	BURKE NC	118	-4	108	80	70
30	116.3	65	204.1	103	250.3	CABARRUS NC	76	23	115	82	77
19	191.7	29	270.7	23	170.9	CALDWELL NC	41	-37	-11	21	50
10	149.1	12	143.9	14	192.7	CAMDEN NC	-3	34	29	40	58
16	166.5	27	268.4	32	264.0	CARTERET NC	61	-2	59	57	81
34	124.7	50	175.1	72	223.4	CASWELL NC	40	28	79	70	68
18	99.4	42	212.5	66	234.0	CATAWBA NC	114	10	135	86	71
26	111.9	43	152.0	66	187.6	CHATHAM NC	36	23	68	63	56
2	202.1	8*	554.0	4	220.3	CHEROKEE NC	174	-60	9	29	67
24	156.4	21	121.5	56	280.6	CHOWAN NC	-22	131	79	70	86
				3*	532.8	CLAY NC					98
39	107.1	66	147.8	107	225.8	CLEVELAND NC	38	53	111	81	69
39	86.4	82	157.4	132	212.4	COLUMBUS NC	82	35	146	88	65
56	107.1	115	207.0	154	247.7	CRAVEN NC	93	20	131	85	76
90	126.8	167	161.3	322	232.6	CUMBERLAND NC	27	44	83	72	71
17	198.2	18	204.4	23	263.4	CURRITUCK NC	3	29	33	43	81
1	133.9	5	362.1	5	296.3	DARE NC	170	-18	121	84	89
28	127.3	46	156.4	70	208.0	DAVIDSON NC	23	33	63	61	63
14	218.2	8	102.7	22	232.6	DAVIE NC	-53	127	7	28	71
51	119.7	93	202.4	110	207.0	DUPLIN NC	69	2	73	66	63
162	162.3	269	201.8	485*	281.0	DURHAM NC	24	39	73	66	86
88	127.4	133	175.2	265*	297.9	EDGECOMBE NC	38	70	134	86	89
223	165.6	323	196.6	478*	254.9	FORSYTH NC	19	30	54	55	78
36	93.6	64	157.7	84	182.2	FRANKLIN NC	68	16	95	76	54
62	139.3	81	160.5	152	236.0	GASTON NC	15	47	69	64	72
16	99.1	16	86.0	34	176.5	GATES NC	-13	105	78	69	52
		2	200.5			GRAHAM NC					
28	72.2	69	140.5	113	196.5	GRANVILLE NC	95	40	172	91	59
21	126.0	24	129.4	35	153.1	GREENE NC	3	18	21	36	46
230	193.3	388	219.0	537	247.8	GUILFORD NC	13	13	28	39	76
116	132.6	115	127.7	212	210.8	HALIFAX NC	-4	65	59	57	64
41	119.9	63	161.3	132*	299.2	HARNETT NC	35	85	149	88	89
2	87.4	3	78.3	13	339.2	HAYWOOD NC	-10	333	288	97	93
16	203.9	29*	335.8	25	265.3	HENDERSON NC	65	-21	30	41	81
52	129.6	76	181.0	117	225.4	HERTFORD NC	40	25	74	67	69
25	107.0	22	89.3	52	183.8	HOKE NC	-17	106	72	65	55
4	37.2	10	115.6	17	187.9	HYDE NC	211	63	406	98	56
33	101.6	70	191.8	96	220.0	IREDELL NC	89	15	116	82	67
6	133.8	5	80.9	14	172.4	JACKSON NC	-40	113	29	40	51
38	99.6	74	172.7	113	233.6	JOHNSTON NC	73	35	134	86	71
7	42.9	19	118.9	41	246.4	JONES NC	177	107	474	98	76
23	140.2	42	226.0	67	263.1	LEE NC	61	16	88	74	80
72	113.4	93	130.3	191	237.4	LENOIR NC	15	82	109	81	73
13	141.0	13	137.4	28	248.4	LINCOLN NC	-3	81	76	68	76
7	128.5	16	264.8	17	263.8	MC DOWELL NC	106	0	105	80	81
2	123.4	1	196.5	3	282.3	MACON NC	59	44	129	85	87
		3	338.4	2	227.1	MADISON NC		-33			70
33	103.7	71	196.0	76	171.6	MARTIN NC	89	-12	66	62	51
264	152.6	397	185.9	675*	256.8	MECKLENBURG NC	22	38	68	63	79
1	261.2					MITCHELL NC					
21	173.9	22	142.9	48	262.2	MONTGOMERY NC	-18	84	51	54	80
38	131.0	66	191.5	117	275.8	MOORE NC	46	44	111	81	85
70	103.2	118	165.4	191	253.6	NASH NC	60	53	146	88	78
105	155.5	177*	252.1	213*	266.8	NEW HANOVER NC	62	6	72	65	82
47	102.9	79	151.4	126	225.0	NORTHAMPTON NC	47	49	119	83	69
15	77.3	25	108.9	84	280.5	ONSLOW NC	41	158	263	96	86

* SIGNIFICANTLY MORE DEATHS THAN EXPECTED

NON-WHITE MALE — ALL CANCERS COMBINED — ICD 140-208

1950-1959 DEATHS	1950-1959 RATE	1960-1969 DEATHS	1960-1969 RATE	1970-1979 DEATHS	1970-1979 RATE	COUNTY AND STATE		PERCENT CHANGE 60S/50S	PERCENT CHANGE 70S/60S	PERCENT CHANGE 70S/50S	PERCENTILE %CHANGE 70S/50S	PERCENTILE RATES 1970S
38	130.6	62	185.1	90	209.1	ORANGE	NC	42	13	60	58	63
8	73.7	13	121.4	21	149.1	PAMLICO	NC	65	23	102	78	45
46	146.2	81	227.8	134*	342.6	PASQUOTANK	NC	56	50	134	86	93
22	71.2	51	173.4	75	225.4	PENDER	NC	144	30	217	93	69
13	87.6	24	149.3	42	230.2	PERQUIMANS	NC	70	54	163	90	70
31	124.5	49	183.4	62	191.6	PERSON	NC	47	4	54	55	58
90	111.4	146	169.6	277*	286.4	PITT	NC	52	69	157	89	87
9	141.0	8	125.9	19	327.5	POLK	NC	-11	160	132	86	92
16	117.9	38	214.2	43	192.6	RANDOLPH	NC	82	-10	63	61	58
40	113.4	57	145.9	99	211.9	RICHMOND	NC	29	45	87	73	64
124	102.2	198	152.6	348	211.6	ROBESON	NC	49	39	107	80	64
57	135.0	99	203.9	147	261.0	ROCKINGHAM	NC	51	28	93	76	80
55	113.9	109	214.0	150	250.4	ROWAN	NC	88	17	120	83	77
25	147.4	28	158.7	45	197.5	RUTHERFORD	NC	8	24	34	44	60
54	101.5	75	130.5	125	191.3	SAMPSON	NC	29	47	88	74	57
35	112.0	44	135.5	95	246.9	SCOTLAND	NC	21	82	121	84	76
19	131.6	30	193.0	47	268.1	STANLY	NC	47	39	104	79	83
9	155.9	5	71.7	26	338.6	STOKES	NC	-54	372	117	82	93
12	132.2	16	160.3	32	291.4	SURRY	NC	21	82	120	83	88
6	121.8	9	141.1	15	182.7	SWAIN	NC	16	29	50	54	54
4	214.5	9	259.8	7	192.1	TRANSYLVANIA	NC	21	-26	-10	21	58
5	85.5	8	110.8	11	164.6	TYRRELL	NC	30	49	93	75	48
25	91.7	39	140.7	67	186.6	UNION	NC	53	33	104	79	55
62	145.4	83	179.0	139	254.8	VANCE	NC	23	42	75	67	78
160	128.9	292	187.8	501*	263.7	WAKE	NC	46	40	105	79	81
47	100.1	76	171.1	140*	304.9	WARREN	NC	71	78	205	93	90
19	129.5	36	202.7	39	187.3	WASHINGTON	NC	57	-8	45	50	56
1	163.6	1	273.2	2	278.0	WATAUGA	NC	67	2	70	64	86
101	114.0	140	140.9	234	228.4	WAYNE	NC	24	62	100	78	70
14	127.0	21	227.8	28	245.2	WILKES	NC	79	8	93	75	75
69	101.0	142	188.4	185	239.6	WILSON	NC	86	27	137	86	73
4	107.3	7	160.4	13	261.1	YADKIN	NC	49	63	143	87	80
1	74.3	1	141.3	2	146.4	YANCEY	NC	90	4	97	77	44
46	159.8	50	156.2	68	150.9	NORTH DAKOTA	ND	-2	-3	-6		
6	161.5	6	136.8	4	142.5	BENSON	ND	-15	4	-12	21	44
		1	133.9	1	23.8	BURLEIGH	ND		-82			29
1	267.9	2	248.8	4	103.6	CASS	ND	-7	-58	-61	05	37
2	456.6	1	28.9			DUNN	ND	-94				
1	267.9					GRAND FORKS	ND					
1	267.9	1	86.0	1	112.4	MC KENZIE	ND	-68	31	-58	06	38
1	188.7	1	112.4	4	381.5	MC LEAN	ND	-40	239	102	78	95
				1	212.0	MERCER	ND					64
1	112.4					MORTON	ND					
2	434.1	2	157.1	5	223.9	MOUNTRAIL	ND	-64	43	-48	09	68
1*	188.7					OLIVER	ND					
		2	764.0	1	267.9	RAMSEY	ND		-65			83
		1	281.2			RICHLAND	ND					
21	175.0	22	172.5	32	195.7	ROLETTE	ND	-1	13	12	31	59
4	70.9	9	334.0	10	194.3	SIOUX	ND	371	-42	174	91	59
1*	344.1					STARK	ND					
4	372.5	1	133.9	1	114.7	WARD	ND	-64	-14	-69	03	39
		1	112.4	4	272.6	WILLIAMS	ND		142			84
4736*	211.2	7781*	254.4	10865*	291.8	OHIO	OH	20	15	38		
2	235.1	1	133.9	1	344.1	ADAMS	OH	-43	157	46	51	94
25	133.8	54	200.5	88	292.0	ALLEN	OH	50	46	118	83	88
1	136.6	1	73.7	3	374.3	ASHLAND	OH	-46	408	174	91	95
9	166.7	10	135.1	23	197.0	ASHTABULA	OH	-19	46	18	34	60
8	141.2	9	201.0	7	164.2	ATHENS	OH	42	-18	16	32	48
26	213.3	19	171.9	27	259.5	BELMONT	OH	-19	51	22	36	79
10	215.0	5	167.3	3	102.6	BROWN	OH	-22	-39	-52	08	37
70*	223.4	96	267.1	121*	300.5	BUTLER	OH	20	13	35	44	89
2	144.8	3	266.5	4	684.2	CARROLL	OH	84	157	372	98	99
8	124.1	14	277.6	10	162.3	CHAMPAIGN	OH	124	-42	31	41	48
97	209.3	127	236.1	182*	317.8	CLARK	OH	13	35	52	54	91
7	167.1	12	292.6	15	316.1	CLERMONT	OH	75	8	89	74	91
5	153.6	4	233.8	13*	553.5	CLINTON	OH	52	137	260	96	98
14	211.0	14	186.1	25	321.5	COLUMBIANA	OH	-12	73	52	55	92
1	68.3	3	272.9	6	322.5	COSHOCTON	OH	299	18	372	98	92
7*	621.5	5	317.6	4	319.8	CRAWFORD	OH	-49	1	-49	09	92
1315*	198.4	2459*	261.6	3576*	297.3	CUYAHOGA	OH	32	14	50	54	89
2	135.8	1	91.1	1	62.9	DARKE	OH	-33	-31	-54	07	32
				1	188.7	DEFIANCE	OH					57
6	157.2	11	277.9	7	147.7	DELAWARE	OH	77	-47	-6	23	45
20	206.3	37	242.8	34	174.5	ERIE	OH	18	-28	-15	19	52
2	232.6	1	212.0	3	225.1	FAIRFIELD	OH	-9	6	-3	24	69
9	197.7	10	256.0	7	187.8	FAYETTE	OH	30	-27	-5	23	56
522*	214.7	793*	248.5	1108*	289.4	FRANKLIN	OH	16	16	35	44	88
2*	612.7			2	151.9	FULTON	OH			-75	02	46
14	205.3	10	142.6	16	236.8	GALLIA	OH	-31	66	15	32	72
5	132.1	5	92.9	13	249.6	GEAUGA	OH	-30	169	89	74	77
47	219.3	62	250.7	70	263.8	GREENE	OH	14	5	20	35	81
4	119.8	4	101.6	11	384.2	GUERNSEY	OH	-15	278	221	94	96
919*	236.2	1544*	303.7	1894*	321.5	HAMILTON	OH	29	6	36	45	92
2	192.7	1	89.3	6	330.0	HANCOCK	OH	-54	269	71	65	93
		5	475.2	1	133.9	HARDIN	OH		-72			42
		2	65.9	4	191.0	HARRISON	OH		190			57
1*	344.1					HENRY	OH					
14*	319.6	8	166.4	8	221.1	HIGHLAND	OH	-48	33	-31	14	67
1	344.1	2	424.4	3	412.7	HOCKING	OH	23	-3	20	35	96
				1	74.3	HOLMES	OH					34
7	370.5	4	143.3	8	242.3	HURON	OH	-61	69	-35	13	74
1	27.0	2	135.8	6*	418.1	JACKSON	OH	404	208	1451	99	97
49	257.4	61	252.2	72	293.1	JEFFERSON	OH	-2	16	14	31	89
8	372.3	9	416.2	6	291.5	KNOX	OH	12	-30	-22	17	88
10	186.4	24	392.7	28	239.0	LAKE	OH	111	-39	28	39	73
19	225.9	19	199.3	16	184.0	LAWRENCE	OH	-12	-8	-19	18	55
9	196.4	15	277.6	19	259.1	LICKING	OH	41	-7	32	42	79

* SIGNIFICANTLY MORE DEATHS THAN EXPECTED

NON-WHITE MALE — ALL CANCERS COMBINED — ICD 140-208

1950-1959 DEATHS	1950-1959 RATE	1960-1969 DEATHS	1960-1969 RATE	1970-1979 DEATHS	1970-1979 RATE	COUNTY AND STATE	PERCENT CHANGE 60S/50S	PERCENT CHANGE 70S/60S	PERCENT CHANGE 70S/50S	PERCENTILE %CHANGE 70S/50S	PERCENTILE RATES 1970S
4	95.1	3	85.1	8	239.6	LOGAN OH	-10	181	152	89	73
60	191.8	82	184.2	161	264.5	LORAIN OH	-4	44	38	46	81
260*	222.9	433*	262.0	593*	286.4	LUCAS OH	18	9	28	40	87
8	159.6	5	65.3	3	64.4	MADISON OH	-59	-1	-60	05	32
216	197.8	359*	245.2	507*	301.3	MAHONING OH	24	23	52	55	90
10	276.3	12	350.6	14	262.2	MARION OH	27	-25	-5	23	80
4	296.2	7	264.0	4	123.4	MEDINA OH	-11	-53	-58	06	40
5	131.1	5	279.4	2	265.6	MEIGS OH	113	-5	103	79	81
13	205.9	20	324.4	21	271.1	MIAMI OH	58	-16	32	42	84
345*	222.7	563*	239.3	876*	289.0	MONTGOMERY OH	7	21	30	41	88
2	116.1	4	230.3	9	312.4	MORGAN OH	98	36	169	90	91
		1	74.3			MORROW OH					
26	189.7	33	223.3	57*	389.3	MUSKINGUM OH	18	74	105	79	96
		1*	424.0			NOBLE OH					
6*	360.6	1	66.2	3	170.9	OTTAWA OH	-82	158	-53	07	50
3	222.5	5	290.3	2	91.2	PAULDING OH	30	-69	-59	06	36
		4	264.6	2	175.1	PERRY OH		-34			52
7	327.2	6	361.7	2	91.8	PICKAWAY OH	11	-75	-72	03	36
3	179.5	4	215.7	5	360.4	PIKE OH	20	67	101	78	95
9	135.8	25	291.3	18	180.0	PORTAGE OH	114	-38	33	42	53
		2	281.3			PREBLE OH					
		1*	267.9			PUTNAM OH					
35*	240.9	41	224.8	69	260.3	RICHLAND OH	-7	16	8	28	80
18	142.8	18	103.8	32	168.1	ROSS OH	-27	62	18	33	49
10*	377.7	8	223.4	8	151.7	SANDUSKY OH	-41	-32	-60	05	46
18	212.3	11	134.2	36*	453.7	SCIOTO OH	-37	238	114	82	97
4	222.4	9	355.2	10	205.9	SENECA OH	60	-42	-7	22	62
3	255.6	4	241.7	6	352.6	SHELBY OH	-5	46	38	46	94
75	131.4	142	207.0	234*	280.7	STARK OH	58	36	114	82	86
214*	214.9	373*	262.2	528*	287.3	SUMMIT OH	22	10	34	43	87
58	198.6	84	187.1	151*	291.6	TRUMBULL OH	-6	56	47	52	88
7	202.6	17*	425.9	11	250.7	TUSCARAWAS OH	110	-41	24	37	77
1	62.9	2	131.5	1	47.2	UNION OH	109	-64	-25	15	31
4*	473.3	1	273.2			VAN WERT OH	-42				
9	234.0	10	286.6	6	139.4	WARREN OH	22	-51	-40	11	43
10	224.6	8	246.3	12	240.6	WASHINGTON OH	10	-2	7	28	74
4	124.1	8	199.5	15	350.2	WAYNE OH	61	75	182	92	94
				1	424.0	WILLIAMS OH					97
5	662.4	2	147.9	5	215.8	WOOD OH	-78	46	-67	03	66
		1*	424.0			WYANDOT OH					
1211	137.9	1714	166.1	2371	185.2	OKLAHOMA OK	20	12	34		
24	227.5	20	141.0	23	113.8	ADAIR OK	-38	-19	-50	08	39
7	120.7	8	128.2	12	173.6	ATOKA OK	6	35	44	50	51
2	121.1	3	228.9	2	102.5	BECKHAM OK	89	-55	-15	19	37
13	195.5	15	176.4	14	211.4	BLAINE OK	-10	20	8	29	64
11	165.4	7	96.3	15	128.9	BRYAN OK	-42	34	-22	17	41
16	123.9	30	197.5	35	197.5	CADDO OK	59	0	59	58	60
8	128.2	8	131.2	16	257.1	CANADIAN OK	2	96	101	78	79
29	153.6	36	167.2	46	187.8	CARTER OK	9	12	22	37	56
24	178.9	16	102.2	35	139.6	CHEROKEE OK	-43	37	-22	17	43
23	110.6	21	98.6	30	134.7	CHOCTAW OK	-11	37	22	36	43
		2*	848.1			CIMARRON OK					
3	143.4	2	94.1	8	116.8	CLEVELAND OK	-34	24	-19	18	39
2	97.8	4	89.1	7	200.0	COAL OK	-9	125	105	79	61
27	145.1	47	195.4	74	206.3	COMANCHE OK	35	6	42	49	62
1	89.3	2	82.7	2	176.3	COTTON OK	-7	113	97	77	52
4	70.1	9	81.4	12	88.7	CRAIG OK	16	9	27	38	35
22	103.8	43	179.6	44	173.0	CREEK OK	73	-4	67	62	51
5	88.2	7	121.5	9	161.7	CUSTER OK	38	33	83	71	48
3	28.9	10	75.3	20	99.0	DELAWARE OK	160	31	242	95	37
1	267.9	2	535.8	3	326.7	DEWEY OK	100	-39	22	36	92
16*	279.7	15	201.2	20	216.2	GARFIELD OK	-28	7	-23	17	66
9	171.0	15	222.3	12	127.3	GARVIN OK	30	-43	-26	15	41
14	167.8	15	184.0	21	224.9	GRADY OK	10	22	34	44	69
2	148.0			4	212.7	GREER OK			44	50	65
1	267.9	2	138.4	7	428.0	HARMON OK	-48	209	60	58	97
1	37.7	2	63.4	3	63.6	HASKELL OK	68	0	68	63	32
14	161.2	14	143.8	24	214.6	HUGHES OK	-11	49	33	43	65
13	219.1	15	206.7	28	334.0	JACKSON OK	-6	62	52	55	93
3	302.4	2	245.4	3	376.5	JEFFERSON OK	-19	53	24	37	95
3	54.9	3	97.2	10	186.0	JOHNSTON OK	77	91	239	95	55
14	183.8	18	167.3	32	246.8	KAY OK	-9	48	34	44	76
8	130.9	9	199.6	12	207.6	KINGFISHER OK	52	4	59	57	63
7	118.5	14	251.2	7	98.7	KIOWA OK	112	-61	-17	18	37
2	38.7	1	39.3	6	129.7	LATIMER OK	1	230	235	94	41
14	152.0	14	119.6	18	112.0	LE FLORE OK	-21	-6	-26	15	38
18	217.9	9	153.0	11	162.9	LINCOLN OK	-30	6	-25	15	48
31	133.2	40	212.2	37	202.2	LOGAN OK	59	-5	52	54	61
3	150.6	2	73.9	6	176.7	LOVE OK	-51	139	17	33	52
1	22.3	3	147.5	9	245.1	MC CLAIN OK	561	66	998	99	75
42	122.6	39	117.3	60	166.1	MC CURTAIN OK	-4	42	35	45	49
14	85.5	19	117.0	19	118.0	MC INTOSH OK	37	1	38	46	40
2	102.7	1	43.0	5	154.5	MARSHALL OK	-58	259	50	54	46
7	96.9	11	116.0	18	124.4	MAYES OK	20	7	28	40	40
1	70.7	4	142.7	2	37.0	MURRAY OK	102	-74	-48	09	30
113	129.9	155	190.1	174	226.8	MUSKOGEE OK	46	19	75	67	69
6	145.7	5	142.4	13	362.8	NOBLE OK	-2	155	149	88	95
11	165.0	20	281.6	8	76.9	NOWATA OK	71	-73	-53	07	34
20	89.8	33	169.9	30	146.3	OKFUSKEE OK	89	-14	63	60	44
213	188.6	348	203.1	533	225.4	OKLAHOMA OK	8	11	20	34	69
63	160.3	71	179.8	76	186.2	OKMULGEE OK	12	4	16	32	55
12	130.7	27	209.2	28	138.5	OSAGE OK	60	-34	6	27	43
4	87.4	3	36.8	16	127.5	OTTAWA OK	-58	246	46	51	41
10	271.5	7	160.2	13	223.6	PAWNEE OK	-41	40	-18	18	68
4	63.1	9	140.7	19	249.6	PAYNE OK	123	77	296	97	77
28	147.9	30	145.0	42	177.2	PITTSBURG OK	-2	22	20	35	52
12	163.9	12	164.7	15	128.1	PONTOTOC OK	1	-22	-22	17	41
14	122.4	22	168.4	32	214.3	POTTAWATOMIE OK	38	27	75	67	65
8	218.1	5	128.5	7	129.1	PUSHMATAHA OK	-41	0	-41	11	41

* SIGNIFICANTLY MORE DEATHS THAN EXPECTED

NON-WHITE MALE — ALL CANCERS COMBINED — ICD 140-208

1950-1959 DEATHS	1950-1959 RATE	1960-1969 DEATHS	1960-1969 RATE	1970-1979 DEATHS	1970-1979 RATE	COUNTY AND STATE	PERCENT CHANGE 60S/50S	PERCENT CHANGE 70S/60S	PERCENT CHANGE 70S/50S	PERCENTILE %CHANGE 70S/50S	PERCENTILE RATES 1970S
		2	135.1	3	365.3	ROGER MILLS OK		170			95
7	108.0	16	221.7	10	73.4	ROGERS OK	105	-67	-32	13	33
29	145.9	39	164.3	58	232.7	SEMINOLE OK	13	42	59	58	71
9	89.1	16	132.9	23	125.3	SEQUOYAH OK	49	-6	41	48	41
3	93.6	8	180.6	16	217.8	STEPHENS OK	93	21	133	86	66
		1	546.4	1	60.3	TEXAS OK		-89			32
14	162.2	13	172.8	16	169.2	TILLMAN OK	7	-2	4	27	50
135	124.9	262	179.3	403	211.5	TULSA OK	44	18	69	64	64
17	83.0	30	138.6	36	186.5	WAGONER OK	67	34	125	84	55
13	151.1	19	182.3	16	112.2	WASHINGTON OK	21	-38	-26	15	38
1	188.7	1	424.0	1	49.1	WASHITA OK	125	-88	-74	02	31
		1*	490.8			WOODS OK					
				1	273.2	WOODWARD OK					84
198	168.8	272	170.2	445	174.0	OREGON OR	1	2	3		
				1	163.6	BAKER OR					48
1	245.4	1	37.2	1	94.4	BENTON OR	-85	154	-62	05	36
1	70.1	3	62.1	11	100.0	CLACKAMAS OR	-11	61	43	49	37
2	202.8	4	179.8	2	77.5	CLATSOP OR	-11	-57	-62	05	34
		1	212.0			COLUMBIA OR					
2	113.6			5	129.5	COOS OR			14	31	41
				1	64.8	CROOK OR					35
1	112.4	1	89.3	1	172.0	DESCHUTES OR	-21	93	53	55	51
		1	49.2	8	404.1	DOUGLAS OR		722			96
				2	253.8	HARNEY OR					78
2	46.8			2	63.2	HOOD PIVER OR			35	44	32
2	199.5	1	112.4	3	40.6	JACKSON OR	-44	-64	-80	01	30
3	123.1	2	120.0	3	128.7	JEFFEPSON OR	-3	7	4	27	41
		1	106.0	1	56.2	JOSEPHINE OR		-47			32
2	26.7	5	100.9	6	90.0	KLAMATH OR	278	-11	238	94	36
5	285.2	1	84.8	9	85.0	LANE OR	-70	0	-70	03	35
		1	89.3	3	155.6	LINCOLN OP		74			46
				1	188.7	LINN OR					57
10	168.5	11	140.1	13	139.6	MALHEUP OR	-17	0	-17	18	43
2	72.2	4	93.5	11	91.4	MARION OR	30	-2	27	38	36
				1	163.6	MOPROW OR					48
154*	218.5	220	218.5	338	228.8	MULTNOMAH OR	0	5	5	27	70
				1	34.4	POLK OR					30
6	143.5	9	162.0	9	199.8	UMATILLA OR	13	23	39	47	60
2	251.9	1	172.0	1	112.4	UNION OR	-32	-35	-55	07	38
				1	267.9	WALLOWA OR					83
2	53.6	2	116.8	5	246.7	WASCO OR	118	111	361	97	76
1	67.0	1	49.1	5	80.5	WASHINGTON OR	-27	64	20	35	34
		2	164.1			YAMHILL OR					
5878*	222.7	9199*	266.0	11962*	295.9	PENNSYLVANIA PA	19	11	33		
2	210.6	3	198.3	4	205.9	ADAMS PA	-6	4	-2	24	62
1065*	206.3	1693*	263.9	2084*	308.9	ALLEGHENY PA	28	17	50	53	91
9	235.2	13	274.8	13	254.4	ARMSTRONG PA	17	-7	8	29	78
75*	256.0	110*	263.6	145*	299.6	BEAVER PA	3	14	17	33	89
		3	282.3	3	401.8	BEDFORD PA		42			96
32	238.6	56*	277.8	69	247.6	BERKS PA	16	-11	4	26	76
12	282.5	15	338.8	21*	422.0	BLAIR PA	20	25	49	53	97
				3	579.8	BRADFORD PA					98
30	253.7	41	229.3	75	252.6	BUCKS PA	-10	10	0	25	78
4	103.3	2	131.5	11	332.2	BUTLER PA	27	153	221	94	93
25	241.1	32	232.2	34	226.2	CAMBRIA PA	-4	-3	-6	23	69
2*	814.3			1	188.7	CARBON PA			-77	02	57
1	133.9	2	116.4	3	224.7	CENTRE PA	-13	93	68	63	68
117	200.0	187*	271.3	246*	293.5	CHESTER PA	36	8	47	52	89
				5*	812.6	CLARION PA					99
5	332.3	3	258.2	2	188.7	CLEARFIELD PA	-22	-27	-43	10	57
		1	267.9			CLINTON PA					
		1	589.6	6*	757.0	COLUMBIA PA		28			99
5	97.4	9	177.7	17	332.0	CRAWFORD PA	83	87	241	95	93
19*	373.3	9	186.3	18	279.2	CUMBERLAND PA	-50	50	-25	15	86
163*	243.7	216*	256.6	276*	269.5	DAUPHIN PA	5	5	11	30	83
270*	227.0	389*	248.7	567*	309.7	DELAWARE PA	10	25	36	45	91
				2	814.3	ELK PA					99
39*	254.3	58	293.4	98*	334.5	ERIE PA	15	14	32	42	93
81	194.4	116*	285.0	103*	291.3	FAYETTE PA	47	2	50	54	88
12	176.6	23	275.0	26	294.4	FRANKLIN PA	56	7	67	62	89
				3*	636.0	FULTON PA					99
5	180.5	6	249.8	1	38.3	GREENE PA	38	-85	-79	01	30
2	226.9	6	293.1	6	197.3	HUNTINGDON PA	29	-33	-13	20	60
5	200.3	4	164.9	5	143.2	INDIANA PA	-18	-13	-28	14	44
		1	196.5			JEFFERSON PA					
9	226.9	12	321.6	15	271.3	LACKAWANNA PA	42	-16	20	35	84
25	194.5	43	302.9	69	310.5	LANCASTER PA	56	3	60	58	91
22	201.5	29	235.7	33	277.2	LAWRENCE PA	17	18	38	46	85
4	415.4	7	319.5	12	280.1	LEBANON PA	-23	-12	-33	13	86
6	220.1	12	176.5	23	222.0	LEHIGH PA	-20	26	1	25	68
15*	311.5	15	221.1	25	306.3	LUZERNE PA	-29	39	-2	24	90
12	235.3	16	320.4	15	211.1	LYCOMING PA	36	-34	-10	21	64
3*	468.5	1	163.6	2	235.7	MC KEAN PA	-65	44	-50	08	72
28	176.3	44	226.0	55	245.4	MERCER PA	28	9	39	47	75
4*	477.1	2	204.0	1	89.3	MIFFLIN PA	-57	-56	-81	01	35
6	232.6	3	126.1	14	448.7	MONROE PA	-46	256	93	75	97
146*	238.5	191	239.9	334*	312.3	MONTGOMERY PA	1	30	31	41	91
1	589.6			8*	1802.9	MONTCUR PA			206	93	99
11	151.4	32*	329.0	45	307.1	NOPTHAMPTON PA	117	-7	103	79	90
4*	399.2	1	84.4	14*	1392.3	NORTHUMBERLAND PA	-79	1549	249	95	99
1*	490.8			2*	456.6	PERRY PA			-7	22	97
3407*	230.0	5515*	272.0	7085*	292.6	PHILADELPHIA PA	18	8	27	39	88
		1	273.3	2	436.8	POTTER PA		60			97
3	263.1	9*	381.4	13*	871.4	SCHUYLKILL PA	45	128	231	94	99
				1	133.9	SNYDER PA					42
3	319.7	2	162.2	3	116.5	SOMERSET PA	-49	-28	-64	04	39
		3*	775.5	2	187.8	SUSQUEHANNA PA		-76			56

* SIGNIFICANTLY MORE DEATHS THAN EXPECTED

NON-WHITE MALE — ALL CANCERS COMBINED — ICD 140-208

1950-1959 Deaths	1950-1959 Rate	1960-1969 Deaths	1960-1969 Rate	1970-1979 Deaths	1970-1979 Rate	County and State	Percent Change 60S/50S	Percent Change 70S/60S	Percent Change 70S/50S	Percentile %Change 70S/50S	Percentile Rates 1970S
1	140.6	1	182.1	2	244.9	TIOGA PA	30	34	74	67	75
6	288.9	5	225.3	8	415.5	VENANGO PA	-22	84	44	50	97
				3	254.9	WARREN PA					78
87	188.2	117	252.7	144*	327.2	WASHINGTON PA	34	30	74	66	92
2	58.1	3	76.4	3	113.0	WAYNE PA	32	48	95	76	39
60	209.4	81	244.7	82	247.9	WESTMORELAND PA	17	1	18	34	76
				3*	1259.0	WYOMING PA					99
32	216.8	55	301.0	97*	370.1	YORK PA	39	23	71	65	95
151*	246.5	191*	257.4	242	220.2	RHODE ISLAND RI	4	-14	-11		
2	556.0					BRISTOL RI					
4	283.0	9*	421.4	6	183.6	KENT RI	49	-56	-35	12	54
23	264.0	21	240.5	24	219.2	NEWPORT RI	-9	-9	-17	18	67
114*	252.0	146*	260.1	199	232.2	PROVIDENCE RI	3	-11	-8	22	71
8	169.6	15	272.1	13	187.8	WASHINGTON RI	60	-31	11	30	56
3102	133.4	4076	166.0	6237	223.1	SOUTH CAROLINA SC	24	34	67		
26	106.0	31	140.8	36	149.1	ABBEVILLE SC	33	6	41	48	45
103	161.3	119	166.2	188	233.6	AIKEN SC	3	40	45	50	71
35	157.2	54	242.3	57	251.3	ALLENDALE SC	54	4	60	58	77
66	104.6	95	142.1	161	223.0	ANDERSON SC	36	57	113	82	68
35	141.3	35	131.4	73	242.0	BAMBERG SC	-7	84	71	65	74
28	98.5	28	103.2	38	132.0	BARNWELL SC	5	28	34	44	42
74	152.0	90	188.5	160*	282.3	BEAUFORT SC	24	50	86	73	86
71	139.7	87	161.2	125	187.8	BERKELEY SC	15	16	34	44	56
33	130.5	26	113.6	54	232.4	CALHOUN SC	-13	105	78	69	71
329*	164.2	398	179.1	715*	266.7	CHARLESTON SC	9	49	62	60	82
26	127.1	36	157.8	45	185.9	CHEROKEE SC	24	18	46	51	55
50	108.7	55	139.3	89	234.8	CHESTER SC	28	68	116	82	72
54	153.2	64	180.4	102	264.5	CHESTERFIELD SC	18	47	73	66	81
50	93.4	63	124.9	81	150.2	CLARENDON SC	34	20	61	59	45
47	113.7	58	125.3	102	189.5	COLLETON SC	10	51	67	62	57
67	122.6	104	167.5	144	202.8	DARLINGTON SC	37	21	65	62	61
48	138.9	46	131.5	83	228.8	DILLON SC	-5	74	65	61	70
40	119.9	62	178.9	102	221.4	DORCHESTER SC	49	24	85	72	68
36	119.3	45	179.5	60	220.7	EDGEFIELD SC	50	23	85	72	67
57	147.2	63	162.1	76	186.0	FAIRFIELD SC	10	15	26	38	55
118	130.1	163	165.9	283	241.8	FLORENCE SC	27	46	86	73	74
63	148.5	80	170.0	129	216.9	GEORGETOWN SC	14	28	46	51	66
144	140.1	249	213.9	374*	272.4	GREENVILLE SC	53	27	94	76	84
72	177.2	81	185.3	110	206.5	GREENWOOD SC	5	11	17	32	63
38	135.6	42	145.1	68	213.2	HAMPTON SC	7	47	57	57	65
34	93.4	77	173.9	87	165.2	HORRY SC	86	-5	77	68	49
20	84.2	41	149.6	70	229.3	JASPER SC	78	53	172	91	70
49	109.5	79	192.9	107	238.0	KERSHAW SC	76	23	117	83	73
64	204.3	46	147.3	71	194.5	LANCASTER SC	-28	32	-5	23	59
74	162.9	82	177.3	112	225.5	LAURENS SC	9	27	38	46	69
28	87.8	41	117.6	69	204.2	LEE SC	34	74	133	86	62
39	129.4	55	161.3	89	221.5	LEXINGTON SC	25	37	71	65	68
26	151.8	19	116.9	37	229.6	MC CORMICK SC	-23	96	51	54	70
66	128.5	93	186.3	127	225.4	MARION SC	45	21	75	68	69
43	113.7	53	151.2	114	256.8	MARLBORO SC	33	70	126	85	79
50	134.9	77	206.4	84	217.7	NEWBERRY SC	53	5	61	59	66
19	123.3	32	199.4	38	220.9	OCONEE SC	62	11	79	70	67
168	153.8	206	184.2	319	245.8	ORANGEBURG SC	20	33	60	58	75
30	196.3	31	169.6	54	243.0	PICKENS SC	-14	43	24	37	74
213	124.1	323	156.0	522	215.9	RICHLAND SC	26	38	74	67	66
34	169.6	32	184.3	26	142.0	SALUDA SC	9	-23	-16	19	44
137	132.4	184	152.2	289	221.1	SPARTANBURG SC	15	45	67	63	67
100	103.8	153	155.2	219	191.2	SUMTER SC	50	23	84	72	57
52	177.7	56	190.6	61	206.7	UNION SC	7	8	16	32	63
71	100.8	105	158.7	118	167.2	WILLIAMSBURG SC	57	5	66	62	49
75	120.2	117	182.6	169	243.7	YORK SC	52	33	103	79	75
118	143.5	132	150.2	148	141.9	SOUTH DAKOTA SD	5	-6	-1		
		1	490.8			BEADLE SD					
6	189.7	5	286.3	2	97.1	BENNETT SD	51	-66	-49	09	36
1	188.7					BON HOMME SD					
		1	56.2			BRULE SD					
1	47.1	2	71.0	4	114.9	BUFFALO SD	51	62	144	87	39
1	344.1					BUTTE SD					
9	250.8	4	129.1	11	245.4	CHARLES MIX SD	-49	90	-2	24	75
4	60.2	6	135.6	9	205.4	CORSON SD	125	51	241	95	62
				1	188.7	CUSTER SD					57
2	198.4	2	400.7	3	505.3	DAY SD	102	26	155	89	98
5	90.3	15	275.4	6	85.1	DEWEY SD	205	-69	-6	23	35
2	197.2			3	114.3	FALL RIVER SD			-42	11	39
2	226.2			2	130.6	GREGORY SD			-42	11	42
2	172.0	4	304.2	5	391.2	HUGHES SD	77	29	127	85	96
		1*	267.9			HYDE SD					
4	254.2	4	216.5	5	165.2	JACKSON SD	-15	-24	-35	12	49
				1*	596.2	JONES SD					99
				1	112.4	LAWRENCE SD					38
1*	546.4	1*	112.4			LINCOLN SD	-79				
1	112.4	2	93.7	2	98.8	LYMAN SD	-17	5	-12	20	37
		1	62.9	3	435.1	MARSHALL SD		592	.		97
2	284.4			1	106.0	MEADE SD			-63	05	37
7	170.4	4	99.9	6	345.2	MELLETTE SD	-41	246	103	79	94
2	159.6	1	62.9	2	132.9	MINNEHAHA SD	-61	111	-17	18	42
2	203.6	2	532.8	1	49.7	MOODY SD	162	-91	-76	02	31
5	194.1	2	48.3	10	153.2	PENNINGTON SD	-75	217	-21	17	46
				2*	566.0	POTTER SD					98
12	263.1	3	59.9	6	88.1	ROBERTS SD	-77	47	-66	04	35
23	124.6	43	209.0	37	155.0	SHANNON SD	68	-26	24	37	46
1	424.0			1	67.0	STANLEY SD			-84	01	33
16	194.0	12	114.3	16	117.1	TODD SD	-41	2	-40	12	39
1	112.4	7*	631.2	1	267.9	TRIPP SD	462	-58	138	87	83
1	267.9	2	301.1	2	136.2	WALWORTH SD	12	-55	-49	09	43

* SIGNIFICANTLY MORE DEATHS THAN EXPECTED

NON-WHITE MALE — ALL CANCERS COMBINED — ICD 140-208

1950-1959		1960-1969		1970-1979		COUNTY AND STATE	PERCENT CHANGE			PERCENTILE	
DEATHS	RATE	DEATHS	RATE	DEATHS	RATE		60S/50S	70S/60S	70S/50S	%CHANGE 70S/50S	RATES 1970S
2	99.7	1	61.3	1	86.0	YANKTON SD	-38	40	-14	19	35
3	73.9	6	215.4	4	118.1	ZIEBACH SD	191	-45	60	58	40
3214	146.2	4657	192.2	6840*	261.0	TENNESSEE TN	31	36	79		
4	113.0	18	403.4	29*	320.3	ANDERSON TN	257	-21	183	92	92
24	187.2	24	180.3	32	260.0	BEDFORD TN	-4	44	39	47	80
2	255.7	2	148.9	8*	411.2	BENTON TN	-42	176	61	59	96
2	204.0	1	133.9	3	355.5	BLEDSOE TN	-34	165	74	67	94
19	172.5	26	208.2	37	279.6	BLOUNT TN	21	34	62	60	86
8	126.1	14	197.5	21	235.8	BRADLEY TN	57	19	87	73	72
5	276.3	6*	509.8	2	180.4	CAMPBELL TN	85	-65	-35	12	54
2	133.0	2	151.8	1	89.3	CANNON TN	14	-41	-33	13	35
16	97.4	27	159.4	31	205.4	CARROLL TN	64	29	111	81	62
		2	110.4	6	292.6	CARTER TN		165			88
		7	199.9	9	278.0	CHEATHAM TN		39			86
4	62.9	10	170.4	14	223.8	CHESTER TN	171	31	256	96	68
2	152.1	3	398.5	3	244.1	CLAIBORNE TN	162	-39	60	59	75
1	149.1	3	519.6	2	112.4	CLAY TN	249	-78	-25	16	38
4	223.7	4	154.0	7	243.7	COCKE TN	-31	58	9	29	75
6	196.9	6	146.4	12	262.4	COFFEE TN	-26	79	33	43	80
15	110.8	24	171.7	27	205.9	CROCKETT TN	55	20	86	73	62
474	168.5	682	214.4	1070*	300.5	DAVIDSON TN	27	40	78	69	89
6	340.0	4	246.8	5	193.4	DECATUR TN	-27	-22	-43	10	59
3	160.0	5	293.1	4	322.0	DE KALB TN	83	10	101	78	92
8	131.7	11	189.8	14	264.0	DICKSON TN	44	39	100	78	81
38	176.1	39	176.7	39	186.9	DYER TN	0	6	6	28	56
66	101.1	56	96.7	85	147.7	FAYETTE TN	-4	53	46	51	45
				1*	344.1	FENTRESS TN					94
17	180.5	16	198.7	20	227.3	FRANKLIN TN	10	14	26	38	70
54	134.9	70	168.0	109	255.5	GIBSON TN	25	52	89	74	79
28	134.2	29	158.3	44	238.6	GILES TN	18	51	78	69	73
2	227.7			1	163.6	GRAINGER TN			-28	14	48
7	255.4	8	168.6	6	169.7	GREENE TN	-34	1	-34	13	50
6	87.2	9	116.5	26	304.5	HAMBLEN TN	34	161	249	95	90
261	153.2	405	221.6	587*	288.0	HAMILTON TN	45	30	88	74	88
		2	360.7	1	196.4	HANCOCK TN		-46			59
34	108.2	24	71.7	58	152.3	HARDEMAN TN	-34	112	41	48	46
9	131.8	9	129.7	17	300.3	HARDIN TN	-2	132	128	85	89
9	199.4	11	220.3	15	327.0	HAWKINS TN	10	48	64	61	92
50	98.0	49	95.5	82	181.4	HAYWOOD TN	-3	90	85	72	54
10	161.7	11	145.4	21	263.8	HENDERSON TN	-10	81	63	60	81
25	137.4	31	172.8	47	261.0	HENRY TN	26	51	90	74	80
6	161.1	4	108.1	7	249.7	HICKMAN TN	-33	131	55	56	77
1	38.3	8*	543.1	4	186.2	HOUSTON TN	1319	-66	387	98	55
4	138.7	2	58.4	7	400.3	HUMPHREYS TN	-58	585	189	92	96
1	89.3					JACKSON TN					
6	133.1	11	263.6	8	232.7	JEFFERSON TN	98	-12	75	67	71
2	153.0	2	362.2	1	94.4	JOHNSON TN	137	-74	-38	12	36
140	155.9	228	230.5	296*	279.5	KNOX TN	48	21	79	70	86
16	193.4	23	269.1	24	344.6	LAKE TN	39	28	78	69	94
40	107.9	46	122.3	68	193.0	LAUDERDALE TN	13	58	79	69	58
1	39.3	4	171.0	4	171.7	LAWRENCE TN	335	0	337	97	51
1	133.9	1	267.9			LEWIS TN	100				
15	117.7	17	131.6	21	161.6	LINCOLN TN	12	23	37	45	48
5	207.2	9	356.7	7	335.4	LOUDON TN	72	-6	62	59	93
8	114.9	8	107.9	32*	407.7	MC MINN TN	-6	278	255	96	96
7	128.4	10	162.6	15	285.3	MC NAIRY TN	27	75	122	84	87
		1	490.8	1	56.2	MACON TN		-89			32
127	162.2	158	176.6	210	224.6	MADISON TN	9	27	38	47	68
12	197.4	13	216.4	20	360.8	MARION TN	10	67	83	71	95
10	103.2	17	203.7	18	198.0	MARSHALL TN	97	-3	92	75	60
51	127.4	74	190.9	101	250.2	MAURY TN	50	31	96	77	77
		3	246.7	1	114.7	MEIGS TN		-54			39
3	111.1	9	228.9	14	380.9	MONROE TN	106	66	243	95	95
47	115.6	47	124.2	115	283.0	MONTGOMERY TN	7	128	145	88	87
1	168.7	5	568.0	1	62.9	MOORE TN	201	-89	-67	04	32
17	107.3	22	154.8	45	314.6	OBION TN	44	103	193	92	91
		1	56.2	1	589.6	OVERTON TN		949			99
		3	312.6			PERRY TN					
		1*	344.1			POLK TN					
5	218.7	9	387.2	7	324.7	PUTNAM TN	77	-16	48	53	92
3	118.5	7	257.6	9	388.7	RHEA TN	117	51	228	94	96
8	123.3	14	188.1	18	241.6	ROANE TN	53	28	96	77	74
15	66.5	28	132.1	45	192.8	ROBERTSON TN	99	46	190	92	58
52	180.5	57	184.5	74	225.6	RUTHERFORD TN	2	22	25	38	69
		2*	456.6			SCOTT TN					
1	84.8	1	344.1	2	153.2	SEVIER TN	306	-55	81	70	46
1155	149.9	1910	208.5	2786*	270.0	SHELBY TN	39	29	80	70	84
5	192.8	4	124.2	8	274.7	SMITH TN	-36	121	42	49	85
1	67.0	1	70.7	3	407.4	STEWART TN	6	476	508	99	96
24*	270.6	29	292.9	36*	359.7	SULLIVAN TN	8	23	33	43	95
39	219.6	37	187.7	57	267.1	SUMNER TN	-15	42	22	36	82
49	115.5	41	95.9	91	218.0	TIPTON TN	-17	127	89	74	66
2	44.8	6	129.1	9	251.7	TROUSDALE TN	188	95	462	98	77
9	156.0	5	87.1	7	148.3	WARREN TN	-44	70	-5	23	45
30	176.6	41	222.2	47*	345.3	WASHINGTON TN	26	55	96	76	94
3	273.7	4	257.1	1	68.8	WAYNE TN	-6	-73	-75	02	33
15	144.9	19	227.0	24	345.7	WEAKLEY TN	57	52	139	87	94
2	110.8	6	257.7	3	167.6	WHITE TN	133	-35	51	54	49
33	155.5	32	155.7	48	213.2	WILLIAMSON TN	0	37	37	45	65
21	144.6	25	145.2	48	241.2	WILSON TN	0	66	67	62	74
6215	148.5	9590	193.9	13647	197.8	TEXAS TX	31	2	33		
54	127.6	67	154.8	100	234.6	ANDERSON TX	21	52	84	72	72
		2	424.0	4	283.8	ANDREWS TX		-33			87
35	163.8	50	148.9	62	161.3	ANGELINA TX	-9	8	-2	24	48
		2	173.1	4	160.6	ARANSAS TX		-7			47
1	114.7	1	86.0			ATASCOSA TX	-25				
28	182.7	32	210.6	38	248.8	AUSTIN TX	15	18	36	45	76

* SIGNIFICANTLY MORE DEATHS THAN EXPECTED

NON-WHITE MALE — ALL CANCERS COMBINED — ICD 140-208

1950-1959 DEATHS	1950-1959 RATE	1960-1969 DEATHS	1960-1969 RATE	1970-1979 DEATHS	1970-1979 RATE	COUNTY AND STATE	PERCENT CHANGE 60S/50S	PERCENT CHANGE 70S/60S	PERCENT CHANGE 70S/50S	PERCENTILE %CHANGE 70S/50S	PERCENTILE RATES 1970S
		2	166.7	3	137.9	BAILEY TX		-17			43
34	111.8	48	165.5	65	226.1	BASTROP TX	48	37	102	79	69
		3	585.9	2	186.8	BAYLOR TX		-68			55
4	215.9	5	189.8	4	81.1	BEE TX	-12	-57	-62	05	34
41	140.5	39	109.4	89	165.8	BELL TX	-22	52	18	33	49
258	188.9	470*	255.8	614	171.6	BEXAR TX	35	-33	-9	22	50
		1	56.2	1	133.9	BLANCO TX		138			42
4	187.7	1	53.0	5	143.0	BOSQUE TX	-72	170	-24	16	44
99	158.7	120	177.4	143	198.3	BOWIE TX	12	12	25	37	60
57	178.0	82	213.9	85	156.1	BRAZORIA TX	20	-27	-12	20	46
44	111.3	58	133.8	105	211.8	BRAZOS TX	20	58	90	75	64
1	112.4			1	42.4	BREWSTER TX			-62	05	30
		1	344.1			BRISCOE TX					
7	232.5	9	214.3	6	101.5	BROWN TX	-8	-53	-56	06	37
17	83.3	36	184.8	35	192.0	BURLESON TX	122	4	130	85	58
1	89.3	1	147.4			BURNET TX	65				
25	169.9	30	186.2	45	253.6	CALDWELL TX	10	36	49	53	78
5	198.4	7	316.7	7	199.8	CALHOUN TX	60	-37	1	25	60
10	195.8	12	253.7	13	45.8	CAMERON TX	30	-82	-77	02	31
24	153.0	25	153.3	43	252.6	CAMP TX	0	65	65	61	78
39	117.0	44	122.8	77	210.4	CASS TX	5	71	80	70	64
1	106.0	2	84.2	2	33.8	CASTRO TX	-21	-60	-68	03	30
14	206.0	9	98.3	22	218.0	CHAMBERS TX	-52	122	6	27	66
37	89.9	52	120.1	84	182.1	CHEROKEE TX	34	52	103	79	54
8	279.0	8	293.4	6	373.3	CHILDRESS TX	5	27	34	43	95
1	74.3	1	424.0	2	394.2	CLAY TX	470	-7	430	98	96
1	212.0	1	94.4	1	35.3	COCHRAN TX	-55	-63	-83	01	30
3	139.8	2	362.2	4	191.7	COLEMAN TX	159	-47	37	45	58
21	110.7	24	128.7	39	197.6	COLLIN TX	16	54	79	69	60
2	72.2	4	121.6	2	68.9	COLLINGSWORTH TX	69	-43	-5	23	33
37	179.3	34	149.8	56	250.3	COLORADO TX	-16	67	40	47	77
2	325.7	5	494.1	4	109.6	COMAL TX	52	-78	-66	04	38
8	192.2	8	234.5	14	306.5	COOKE TX	22	31	59	58	90
1	38.3	2	102.4	7	212.7	CORYELL TX	168	108	456	98	65
2	176.7	4	367.8	3	261.0	COTTLE TX	108	-29	48	53	80
		1	188.7	2	240.0	CRANE TX		27			74
		2	329.2			CROCKETT TX					
		7	226.0	4	92.9	CROSBY TX		-59			36
		1	281.2			DALLAM TX					
568	146.8	1166*	222.0	1888*	249.8	DALLAS TX	51	13	70	64	77
4	142.7	9	208.1	9	157.0	DAWSON TX	46	-25	10	30	47
4	129.3	6	155.5	7	192.4	DELTA TX	20	24	49	53	58
18	172.8	18	166.6	40	271.4	DENTON TX	-4	63	57	57	84
28	171.2	42	243.7	46	250.1	DE WITT TX	42	3	46	51	77
3	163.5	4	199.9	3	92.1	DICKENS TX	22	-54	-44	10	36
3	365.1	1	37.2	1	89.3	DONLEY TX	-90	140	-76	02	35
7*	390.2	4	243.9	7	258.4	EASTLAND TX	-37	6	-34	13	79
8	90.2	22	163.4	47	220.6	ECTOR TX	81	35	145	88	67
55	119.4	72	161.8	95	219.0	ELLIS TX	36	35	83	71	67
26	137.5	35	152.0	73	39.0	EL PASO TX	11	-74	-72	03	30
3	583.4	2	131.8	2	227.1	ERATH TX	-77	72	-61	05	70
42	116.1	54	173.1	62	189.1	FALLS TX	49	9	63	60	57
28	185.9	17	118.0	27	202.0	FANNIN TX	-37	71	9	29	61
36	173.8	33	189.3	35	235.5	FAYETTE TX	9	24	36	45	72
4	160.9	5	234.3	1	34.4	FISHER TX	46	-85	-79	02	30
2	85.7	5	255.0	4	121.6	FLOYD TX	198	-52	42	49	40
1	74.5	1	37.7	2	232.4	FOARD TX	-49	516	212	93	71
68	192.7	60	168.3	81	159.8	FORT BEND TX	-13	-5	-17	18	47
1	37.5	3	183.3	2	84.9	FRANKLIN TX	389	-54	127	85	35
24	92.2	43	175.0	47	169.6	FREESTONE TX	90	-3	84	72	50
				4	209.8	GAINES TX					64
172	197.9	288*	238.8	372	244.9	GALVESTON TX	21	3	24	37	75
3	438.7	1	57.3	4	180.7	GARZA TX	-87	215	-59	06	54
				1	267.9	GILLESPIE TX					83
5	126.2	7	284.7	5	121.0	GOLIAD TX	125	-57	-4	23	40
20	113.8	19	109.6	44	278.6	GONZALES TX	-4	154	145	88	86
7	211.5	8	250.9	7	154.0	GRAY TX	19	-39	-27	15	46
58	176.8	62	197.1	61	189.5	GRAYSON TX	11	-4	7	28	57
74	130.2	140	212.8	143	200.7	GREGG TX	63	-6	54	55	61
35	124.7	54	182.9	51	202.3	GRIMES TX	47	11	62	60	61
17	94.7	48	272.5	50	230.9	GUADALUPE TX	188	-15	144	87	71
6	167.5	9	185.2	8	73.7	HALE TX	11	-60	-56	06	33
1	22.8	3	89.7	7	190.6	HALL TX	294	112	737	99	57
6	155.5	6	148.5	6	159.3	HARDEMAN TX	-4	7	2	26	47
22	139.6	22	115.5	49	268.8	HARDIN TX	-17	133	93	75	83
1202*	191.0	1979*	220.3	3093	229.0	HARRIS TX	15	4	20	35	70
103	117.0	137	150.3	187	208.6	HARRISON TX	28	39	78	69	63
1	21.1	9	432.8	5	150.8	HASKELL TX	1955	-65	616	99	45
13	225.3	14	265.2	19	169.1	HAYS TX	18	-36	-25	16	50
22	109.0	41	198.8	51	224.3	HENDERSON TX	82	13	106	80	68
8	234.6	1	29.8	7	11.9	HIDALGO TX	-87	-60	-95	00	29
18	94.1	23	136.6	47	299.9	HILL TX	45	120	219	94	89
5	171.6	7	138.8	10	141.7	HOCKLEY TX	-19	2	-17	18	44
2*	433.4	1	188.7			HOOD TX	-56				
11	114.0	16	163.9	24	235.5	HOPKINS TX	44	44	107	80	72
35	92.7	35	95.4	59	148.6	HOUSTON TX	3	56	60	58	45
2	58.7	3	63.8	14	136.1	HOWARD TX	9	113	132	86	43
32	142.3	35	142.2	56	207.5	HUNT TX	0	46	46	51	63
2	103.9	5	133.9	5	109.3	HUTCHINSON TX	29	-18	5	27	38
		1*	188.7			IRION TX					
3*	512.2	1	424.0			JACK TX	-17				
14	190.9	17	213.5	24	236.4	JACKSON TX	12	11	24	37	72
30	129.5	42	160.8	43	157.8	JASPER TX	24	-2	22	36	47
320*	187.3	503*	234.0	700*	274.7	JEFFERSON TX	25	17	47	52	85
5	386.7	3	185.6	5	62.1	JIM WELLS TX	-52	-67	-84	01	32
7	81.9	16	210.5	19	206.9	JOHNSON TX	157	-2	153	89	63
5	87.8	12	228.2	9	127.9	JONES TX	160	-44	46	51	41
4	188.0	6	321.7	5	87.9	KARNES TX	71	-73	-53	07	35
32	81.4	59	141.3	64	178.0	KAUFMAN TX	74	26	119	83	53
		2	778.3	1	112.4	KENDALL TX		-86			38
1	163.6					KENT TX					

* SIGNIFICANTLY MORE DEATHS THAN EXPECTED

NON-WHITE MALE — ALL CANCERS COMBINED — ICD 140-208

1950-1959 DEATHS	1950-1959 RATE	1960-1969 DEATHS	1960-1969 RATE	1970-1979 DEATHS	1970-1979 RATE	COUNTY AND STATE	PERCENT CHANGE 60S/50S	PERCENT CHANGE 70S/60S	PERCENT CHANGE 70S/50S	PERCENTILE %CHANGE 70S/50S	PERCENTILE RATES 1970S
8	342.3	8	234.5	14	223.5	KERR TX	-32	-5	-35	12	68
1*	490.8					KING TX					
1	89.3	2	257.5	2	188.6	KINNEY TX	188	-27	111	81	56
4	110.5	3	72.1	11	150.6	KLEBERG TX	-35	109	36	45	45
2	86.3	3	164.3	2	41.7	KNOX TX	90	-75	-52	08	30
30	91.0	56	163.7	82	242.1	LAMAR TX	80	48	166	90	74
5	125.5	13	310.7	11	133.0	LAMB TX	148	-57	6	27	42
2	149.1	1	84.8	2	85.0	LAMPASAS TX	-43	0	-43	10	35
13	105.4	26	211.2	22	195.2	LAVACA TX	100	-8	85	72	59
16	151.0	18	129.4	18	147.0	LEE TX	-14	14	-3	24	45
27	123.0	31	153.9	32	179.7	LEON TX	25	17	46	51	53
50	177.9	50	168.5	76	240.7	LIBERTY TX	-5	43	35	44	74
32	101.8	43	139.8	48	177.8	LIMESTONE TX	37	27	75	67	53
		1	198.7	1	74.3	LLANO TX		-63			34
39	137.1	60	175.2	88	136.5	LUBBOCK TX	28	-22	0	25	43
2	71.0	3	140.6	8	247.3	LYNN TX	98	76	248	95	76
3	276.0	4	301.5	3	82.4	MC CULLOCH TX	9	-73	-70	03	34
145	151.2	207	196.5	296	243.0	MC LENNAN TX	30	24	61	59	74
12	108.2	21	190.5	21	197.4	MADISON TX	76	4	82	71	60
24	102.8	46	197.3	44	168.6	MARION TX	92	-15	64	61	50
		2	390.6	1	86.0	MARTIN TX		-78			35
				2	141.0	MASON TX					43
32	142.8	58	222.3	77	233.2	MATAGORDA TX	56	5	63	61	71
2	167.0	1	38.2	3	56.2	MEDINA TX	-77	47	-66	04	32
16	148.2	29	158.3	61	222.1	MIDLAND TX	7	40	50	54	68
40	180.5	35	171.5	41	225.7	MILAM TX	-5	32	25	38	69
5	172.3	11	433.6	4	95.6	MITCHELL TX	152	-78	-45	10	36
47	165.4	60	211.4	68	212.7	MONTGOMERY TX	28	1	29	40	65
		1*	188.7			MOORE TX					
12	114.2	21	150.7	27	157.7	MORRIS TX	32	5	38	46	47
1	49.7			2	192.0	MOTLEY TX			287	97	58
46	154.5	49	146.2	74	195.9	NACOGDOCHES TX	-5	34	27	39	59
55	123.5	78	189.5	91	237.7	NAVARRO TX	53	25	92	75	73
19	111.6	31	192.3	22	144.5	NEWTON TX	72	-25	30	40	44
6	277.4	6	207.4	11	208.2	NOLAN TX	-25	0	-25	15	63
32	119.3	75	175.7	147	155.5	NUECES TX	47	-11	30	41	46
28	208.6	37	197.3	50	226.1	ORANGE TX	-5	15	8	29	69
8	213.1	12	350.2	8	193.4	PALO PINTO TX	64	-45	-9	22	59
21	96.8	35	154.9	46	197.5	PANOLA TX	60	28	104	79	60
1	57.3	2	156.1	4	173.9	PARKER TX	172	11	203	93	51
				1	44.6	PARMER TX					31
		1	212.0	1	28.9	PECOS TX		-86			29
33	156.4	27	134.4	45	208.5	POLK TX	-14	55	33	43	63
22	140.9	49	217.7	63	205.1	POTTER TX	55	-6	46	51	62
2*	589.4	1	188.7	1	44.6	PRESIDIO TX	-68	-76	-92	00	31
1	70.1	7	471.7	1	33.5	RAINS TX	573	-93	-52	08	30
2	229.8	2	270.5	2	106.6	REAGAN TX	18	-61	-54	07	38
22	109.6	30	155.2	36	178.7	RED RIVER TX	42	15	63	60	53
2	84.2	3	117.6	9	142.0	REEVES TX	40	21	69	63	44
9	236.7	8	171.7	4	64.6	REFUGIO TX	-27	-62	-73	03	32
41	112.9	55	185.0	48	150.9	ROBERTSON TX	64	-18	34	43	45
6	92.2	5	120.6	9	261.3	ROCKWALL TX	31	117	184	92	80
5	222.7	7	279.6	2	91.5	RUNNELS TX	26	-67	-59	06	36
65	139.7	85	167.2	103	201.5	RUSK TX	20	21	44	50	61
14	185.7	10	116.3	15	166.8	SABINE TX	-37	43	-10	21	49
11	96.1	20	152.2	31	255.6	SAN AUGUSTINE TX	58	68	166	90	79
18	99.4	22	133.3	30	173.9	SAN JACINTO TX	34	30	75	67	51
1	34.4	9	317.7	9	118.1	SAN PATRICIO TX	823	-63	243	95	40
1	546.4	2	612.7	4	276.5	SAN SABA TX	12	-55	-49	08	85
		1	424.0			SCHLEICHER TX					
2	112.6	2	209.6	7	251.9	SCURRY TX	86	20	124	84	78
1	163.6	1	212.0	3	285.3	SHACKELFORD TX	30	35	74	67	87
20	89.9	42	170.0	42	172.7	SHELBY TX	89	2	92	75	51
115	133.4	144	143.5	246	210.4	SMITH TX	8	47	58	57	64
2	72.8	3	108.3	4	194.8	STEPHENS TX	49	80	168	90	59
3*	682.8	1	344.1			STONEWALL TX	-50				
		1*	344.1			SUTTON TX					
1	119.2	2	103.2	2	71.2	SWISHER TX	-13	-31	-40	11	33
290	154.4	537	223.2	774	237.9	TARRANT TX	45	7	54	55	73
26	239.3	31	236.2	36	168.9	TAYLOR TX	-1	-28	-29	14	50
		6	262.8	3	71.2	TERRY TX		-75			33
14	104.9	20	154.5	37	273.2	TITUS TX	47	77	160	90	84
20	174.4	19	132.8	28	99.5	TOM GREEN TX	-24	-25	-43	10	37
143	155.7	209	192.8	329	200.0	TRAVIS TX	24	4	28	40	60
18	136.9	26	225.9	33	253.5	TRINITY TX	65	12	85	72	78
13	137.1	19	178.3	24	243.1	TYLER TX	30	36	77	68	74
24	106.1	44	186.2	45	176.2	UPSHUR TX	76	-5	66	62	52
1	100.7	4	384.0	2	106.1	UPTON TX	281	-72	5	27	37
2	516.1	4	332.4			UVALDE TX	-36				
2	216.8	1	94.4	3	110.0	VAL VERDE TX	-56	17	-49	09	38
9	130.5	8	123.7	12	153.7	VAN ZANDT TX	-5	24	18	33	46
29	165.1	52	288.8	50	177.9	VICTORIA TX	75	-38	8	28	53
37	105.3	52	131.1	74	161.2	WALKER TX	25	23	53	55	48
26	115.7	39	160.9	52	187.1	WALLER TX	39	16	62	59	56
1	70.3	2	103.4	3	65.7	WARD TX	47	-36	-7	22	33
48	134.5	58	166.8	75	262.2	WASHINGTON TX	24	57	95	76	80
1	114.7	2	541.1			WEBB TX	372				
46	130.7	87	234.9	68	159.8	WHARTON TX	80	-32	22	36	47
2	161.6	1	49.2	1	53.6	WHEELER TX	-70	9	-67	04	31
48	168.0	57	175.5	98	239.6	WICHITA TX	-7	37	27	39	73
9	161.4	12	169.0	17	226.1	WILBARGER TX	5	34	40	48	69
1	112.4	1	163.6	2	40.2	WILLACY TX	46	-75	-64	04	30
35	162.8	43	195.0	58	207.6	WILLIAMSON TX	20	6	28	39	63
4	172.5	1	86.0	1	21.0	WILSON TX	-50	-76	-88	00	29
3*	688.1	1	267.9	3	117.1	WINKLER TX	-61	-56	-83	01	39
1	84.2	1	133.9	3	197.3	WISE TX	59	47	134	86	60
11	79.8	19	121.6	33	220.3	WOOD TX	52	81	176	91	67
				1	188.7	YOAKUM TX					57
3	604.0	2	107.5	2	94.1	YOUNG TX	-82	-13	-84	01	36
		1	91.1	1	14.1	ZAVALA TX		-85			29

* SIGNIFICANTLY MORE DEATHS THAN EXPECTED

NON-WHITE MALE — ALL CANCERS COMBINED — ICD 140-208

1950-1959 DEATHS	1950-1959 RATE	1960-1969 DEATHS	1960-1969 RATE	1970-1979 DEATHS	1970-1979 RATE	COUNTY AND STATE	PERCENT CHANGE 60S/50S	PERCENT CHANGE 70S/60S	PERCENT CHANGE 70S/50S	PERCENTILE %CHANGE 70S/50S	PERCENTILE RATES 1970S
96	174.4	102	159.3	103	101.4	UTAH UT	-9	-36	-42		
5	141.5	3	82.7	4	73.2	BOX ELDER UT	-42	-11	-48	09	33
				1	163.6	CACHE UT					48
2	88.3	5	289.9	2	61.0	CARBON UT	228	-79	-31	14	32
6	179.6	3	77.7	7	130.4	DAVIS UT	-57	68	-27	15	42
		1	424.0			DUCHESNE UT					
				1	188.7	IRON UT					57
				1	94.4	MILLARD UT					36
54*	218.9	56	203.6	52	114.4	SALT LAKE UT	-7	-44	-48	09	39
2	32.4	6	46.0	8	56.0	SAN JUAN UT	42	22	73	66	31
1*	74.3			1	245.4	SEVIER UT			230	94	75
1	62.6	1	267.9			TOOELE UT	328				
4	113.7			2	47.0	UINTAH UT			-59	06	31
1	94.4	2	306.4	1	2.3	UTAH UT	225	-99	-98	00	29
		2	611.9			WASHINGTON UT					
20	192.7	23	215.1	23	120.2	WEBER UT	12	-44	-38	12	40
9	208.6	9	210.0	6	96.6	VERMONT VT	1	-54	-54		
1	112.4	1	344.1			ADDISON VT	206				
2*	848.1	1	172.0	2	435.2	BENNINGTON VT	-80	153	-49	09	97
2	126.9	3	192.9	2	224.8	CHITTENDEN VT	52	17	77	68	69
		1	188.7			FRANKLIN VT					
1*	267.9					LAMOILLE VT					
1	133.9					RUTLAND VT					
1	133.9			1	188.7	WASHINGTON VT			41	48	57
		2	150.6			WINDHAM VT					
1	89.3	1	172.0	1	163.6	WINDSOR VT	93	-5	83	71	48
4583	166.6	6896*	223.6	9683*	270.9	VIRGINIA VA	34	21	63		
66	139.7	104	216.3	147*	290.0	ACCOMACK VA	55	34	108	80	88
84	209.2	123*	268.0	158*	298.4	ALBEMARLE#VA	28	11	43	49	89
32*	245.5	42*	326.3	33	230.5	ALLEGHANY#VA	33	-29	-6	23	70
28	150.3	25	143.5	50	277.9	AMELIA VA	-5	94	85	72	85
27	127.3	35	161.8	46	182.1	AMHERST VA	27	13	43	49	54
16	171.6	14	136.7	24	211.5	APPOMATTOX VA	-20	55	23	37	64
53	226.0	75	271.2	112	264.3	ARLINGTON#VA	20	-3	17	33	81
40	201.9	52	223.4	59	198.0	AUGUSTA#VA	11	-11	-2	24	60
10	298.7	6	202.8	9	252.3	BATH VA	-32	24	-16	19	78
137	155.3	233	234.7	322*	292.0	BEDFORD#VA	51	24	88	74	88
2	283.1	1	267.9	1	172.0	BLAND VA	-5	-36	-39	12	51
8	110.4	16	232.9	19	256.5	BOTETOURT VA	111	10	132	86	79
55	152.7	72	195.5	89	226.0	BRUNSWICK VA	28	16	48	53	69
				1*	596.2	BUCHANAN VA					99
36	179.3	29	149.8	34	161.1	BUCKINGHAM VA	-16	8	-10	21	47
41	172.5	51	196.5	75	238.3	CAROLINE VA	14	21	38	46	73
4	100.1	7	164.3	9	211.8	CARROLL#VA	64	29	112	81	64
18	141.1	28	182.4	36	197.7	CHARLES CITY VA	29	8	40	48	60
28	149.6	40	187.8	55	248.7	CHARLOTTE VA	26	32	66	62	76
628*	195.5	1003*	258.0	1435*	332.1	CHESTERFIELD#VA	32	29	70	64	93
12	215.4	13	223.2	13	192.5	CLARKE VA	4	-14	-11	21	58
29	188.4	36	220.5	47	264.8	CULPEPER VA	17	20	41	48	81
27	158.9	28	157.6	45	254.3	CUMBERLAND VA	-1	61	60	58	78
		1	112.4			DICKENSON VA					
185	127.6	273	179.8	425*	270.2	DINWIDDIE#VA	41	50	112	81	84
20	159.9	18	122.3	41	264.3	ESSEX VA	-23	116	65	62	81
111	205.0	215*	288.5	346	266.9	FAIRFAX#VA	41	-7	30	41	82
42	171.6	57	242.7	82*	304.8	FAUQUIER VA	41	26	78	69	90
2	145.5	3	147.3	2	89.6	FLOYD VA	1	-39	-38	12	36
16	200.6	21	223.9	34	271.7	FLUVANNA VA	12	21	35	45	84
13	114.9	32	276.4	28	188.7	FRANKLIN VA	141	-32	64	61	56
15	280.6	11	146.9	24	292.7	FREDERICK#VA	-48	99	4	27	89
2	86.0	2	85.4	2	119.0	GILES VA	-1	39	38	46	40
25	150.8	28	156.2	58	301.8	GLOUCESTER VA	4	93	100	78	90
21	123.0	32	216.3	37	204.8	GOOCHLAND VA	76	-5	66	62	62
6	203.5	6	245.7	7	249.1	GREENE VA	21	1	22	37	77
33	119.1	50	182.0	74	236.0	GREENSVILLE#VA	53	30	98	77	72
102	177.0	109	187.2	129	208.1	HALIFAX#VA	6	11	18	33	63
41	166.5	52	218.8	74	267.0	HANOVER VA	31	22	60	58	82
40	176.4	74	253.8	120	258.7	HENRICO VA	44	2	47	52	79
51	145.1	77	175.4	133	226.5	HENRY#VA	21	29	56	56	69
1	172.0					HIGHLAND VA					
52	223.0	58	221.0	83	266.4	ISLE OF WIGHT VA	-1	21	19	34	82
71	192.3	107	218.0	145	242.2	JAMES CITY#VA	13	11	26	38	74
21	135.1	25	176.8	31	218.3	KING AND QUEEN VA	31	23	62	59	66
13	216.1	17	216.4	16	177.7	KING GEORGE VA	0	-18	-18	18	53
24	163.7	31	216.3	46	311.3	KING WILLIAM VA	32	44	90	74	91
21	130.5	33	198.1	65*	361.5	LANCASTER VA	52	82	177	91	95
2	697.2			4	510.4	LEE VA			-27	15	98
33	213.4	32	184.3	51	275.6	LOUDOUN VA	-14	50	29	40	85
35	156.4	45	202.0	48	184.1	LOUISA VA	29	-9	18	33	55
27	131.7	23	107.5	52	245.0	LUNENBURG VA	-18	128	86	73	75
3	38.9	13	144.5	18	214.2	MADISON VA	271	48	451	98	65
18	232.3	24	266.4	29	341.0	MATHEWS VA	15	28	47	52	93
71	141.1	94	177.7	133	237.8	MECKLENBURG VA	26	34	69	63	73
19	131.7	37	285.2	37	264.0	MIDDLESEX VA	117	-7	101	78	81
13	137.0	26	268.1	26	247.7	MONTGOMERY#VA	96	-8	81	70	76
16	102.5	23	136.5	31	199.7	NELSON VA	33	46	95	76	60
80*	212.2	89	241.6	98	255.6	NORTHAMPTON VA	14	6	20	36	79
36	215.1	38	235.8	44	245.3	NORTHUMBERLAND VA	10	4	14	31	75
30	114.4	62	219.4	86	288.4	NOTTOWAY VA	92	31	152	89	88
22	160.3	29	225.3	29	195.4	ORANGE VA	41	-13	22	36	59
6	284.0	7	237.7	6	187.1	PAGE VA	-16	-21	-34	13	56
3	71.7	8	172.1	3	48.9	PATRICK VA	140	-72	-32	14	31
126	145.2	193	198.0	258	228.6	PITTSYLVANIA#VA	36	15	57	57	70
20	157.0	21	156.1	25	200.4	POWHATAN VA	-1	28	28	39	61
33	125.4	44	174.8	66	253.9	PRINCE EDWARD VA	39	45	102	79	78
23	193.2	31	236.7	62	296.6	PRINCE WILLIAM#VA	23	25	53	55	89

* SIGNIFICANTLY MORE DEATHS THAN EXPECTED

COUNTIES COMBINED; SEE TECHNICAL NOTES

NON-WHITE MALE — ALL CANCERS COMBINED — ICD 140-208

1950-1959 DEATHS	1950-1959 RATE	1960-1969 DEATHS	1960-1969 RATE	1970-1979 DEATHS	1970-1979 RATE	COUNTY AND STATE	PERCENT CHANGE 60S/50S	PERCENT CHANGE 70S/60S	PERCENT CHANGE 70S/50S	PERCENTILE %CHANGE 70S/50S	PERCENTILE RATES 1970S
19	228.8	11	145.6	22	304.1	PULASKI VA	-36	109	33	43	90
6	101.3	6	115.4	16	350.4	RAPPAHANNOCK VA	14	204	246	95	94
7	98.7	14	142.7	31	279.1	RICHMOND VA	45	96	183	92	86
144	187.8	199	222.4	260	255.0	ROANOKE#VA	18	15	36	45	78
17	177.0	26	278.3	33*	382.4	ROCKBRIDGE#VA	57	37	116	82	96
7	122.0	13	211.0	14	245.0	ROCKINGHAM#VA	73	16	101	78	75
3	141.9	4	202.5	3	264.4	RUSSELL VA	43	31	86	73	81
2	281.3	3	327.3	6*	670.6	SCOTT VA	16	105	138	87	99
2	135.7	4	205.5	5	240.1	SHENANDOAH VA	51	17	77	68	74
10*	575.8	6	438.7	3	124.3	SMYTH VA	-24	-72	-78	02	40
68	140.8	101	205.8	149	266.4	SOUTHAMPTON#VA	46	29	89	74	82
32	167.6	42	194.9	72	270.8	SPOTSYLVANIA#VA	16	39	62	59	84
9	157.3	20*	338.6	15	184.4	STAFFORD VA	115	-46	17	33	55
17	120.0	20	152.6	35	239.7	SURRY VA	27	57	100	78	74
40	159.9	57	206.5	71	255.1	SUSSEX VA	29	24	60	58	78
22	195.6	20	240.2	21	276.0	TAZEWELL VA	23	15	41	48	85
6	122.5	9	219.4	13	254.8	WARREN VA	79	16	108	80	78
17	217.9	23	274.5	27	308.2	WASHINGTON#VA	26	12	41	48	91
33	188.2	43	214.0	59	272.7	WESTMORELAND VA	14	27	45	50	84
14	146.1	27*	389.7	16	327.6	WISE#VA	167	-16	124	84	92
7	207.4	8	217.1	10	312.6	WYTHE VA	5	44	51	54	91
794*	172.5	1342*	252.4	1887*	298.1	CHESAPEAKE CITY#VA	46	18	73	66	89
93	128.8	196	222.5	320*	305.1	HAMPTON CITY VA	73	37	137	86	90
171	164.6	315*	251.5	403*	274.6	NEWPORT NEWS CITY VA	53	9	67	62	85
118	161.3	183	220.3	265*	280.8	SUFFOLK CITY VA	37	27	74	67	86
505	177.7	736	186.6	1072	166.4	WASHINGTON WA	5	-11	-6		
		1	74.3	1	45.5	ADAMS WA		-39			31
1*	267.9					ASOTIN WA					
1	54.6	2	110.8	3	63.0	BENTON WA	103	-43	15	32	32
		2	162.7	1	67.0	CHELAN WA		-59			33
4	158.9	1	10.8	4	100.6	CLALLAM WA	-93	832	-37	12	37
1	53.6	1	20.4	14	167.8	CLARK WA	-62	721	213	93	49
1*	596.2			1	344.1	COLUMBIA WA			-42	10	94
2	156.7	1	109.3	2	90.6	COWLITZ WA	-30	-17	-42	11	36
		1	94.4	1	133.9	DOUGLAS WA		42			42
2	59.6	3	94.5	3	97.8	FERRY WA	59	3	64	61	37
4	138.1	11	187.2	16	174.2	FRANKLIN WA	36	-7	26	38	52
4	226.5	13	257.3	11	166.2	GRANT WA	14	-35	-27	15	49
7	312.1	4	89.2	9	169.5	GRAYS HARBOR WA	-71	90	-46	09	50
		2	106.9			ISLAND WA					
2	166.6	1	133.9			JEFFERSON WA	-20				
325*	205.3	456	193.2	696	192.0	KING WA	-6	-1	-6	22	58
12	178.3	15	114.1	20	116.5	KITSAP WA	-36	2	-35	13	39
1	245.4					KITTITAS WA					
1	56.2	1	74.3	1	60.6	KLICKITAT WA	32	-19	8	28	32
1	67.0			3	127.7	LEWIS WA			91	75	41
		1	74.3	1	74.3	LINCOLN WA		0			34
2	131.8	3	189.6	3	119.8	MASON WA	44	-37	-9	22	40
3	34.1	10	211.7	10	184.8	OKANOGAN WA	520	-13	442	98	55
		1	114.7	2	113.7	PACIFIC WA		-1			39
				1	344.1	PEND OREILLE WA					94
47	184.4	76	218.2	125	171.2	PIERCE WA	18	-22	-7	22	50
1	22.3	5	150.7	4	80.0	SKAGIT WA	575	-47	259	96	34
		1	112.4	1	273.3	SKAMANIA WA		143			84
5	201.4	15	266.3	12	101.2	SNOHOMISH WA	32	-62	-50	08	37
39	225.5	44	208.5	50	178.7	SPOKANE WA	-8	-14	-21	17	53
2	71.4	4	164.2	9	280.0	STEVENS WA	130	71	292	97	86
2	63.9	5	263.2	5	84.2	THURSTON WA	312	-68	32	42	35
1	34.4	9	353.5	3	61.0	WALLA WALLA WA	928	-83	77	68	32
2	58.3	6	126.7	8	98.0	WHATCOM WA	117	-23	68	63	37
		2	532.8			WHITMAN WA					
32	175.7	39	166.5	52	119.4	YAKIMA WA	-5	-28	-32	13	40
807	168.0	957	214.3	981*	245.6	WEST VIRGINIA WV	28	15	46		
2	121.6	3	412.2	1	490.8	BARBOUR WV	239	19	304	97	98
11	103.5	18	181.8	16	164.4	BERKELEY WV	76	-10	59	57	48
2	66.4	3	138.1	2	129.8	BOONE WV	108	-6	95	76	41
1	188.7	1	273.2			BRAXTON WV	45				
4	143.6	5	219.6	1	149.1	BROOKE WV	53	-32	4	26	45
56*	233.2	70*	276.7	80*	341.0	CABELL WV	19	23	46	51	93
73	158.7	89	251.9	77	246.4	FAYETTE WV	59	-2	55	56	76
1*	188.7			1*	424.0	GILMER WV			125	84	97
		5	565.7	1	91.1	GRANT WV		-84			36
13	138.5	15	174.8	13	154.5	GREENBRIER WV	26	-12	12	30	46
		1	172.0	3	718.0	HAMPSHIRE WV		317			99
13	234.7	17	361.6	18	355.7	HANCOCK WV	54	-2	52	54	94
1	201.4	4	359.4	7	413.5	HARDY WV	78	15	105	80	96
13	145.6	19	273.8	22	354.1	HARRISON WV	88	29	143	87	94
26	215.2	23	189.7	33	273.5	JEFFERSON WV	-12	44	27	39	85
134*	209.8	167	238.2	190*	279.0	KANAWHA WV	14	17	33	43	86
1	424.0			2	166.1	LEWIS WV			-61	05	49
42	179.2	36	168.5	31	151.0	LOGAN WV	-6	-10	-16	19	45
115	142.7	140	219.5	125	235.1	MC DOWELL WV	54	7	65	61	72
33	187.3	39	188.5	37	231.6	MARION WV	1	23	24	37	71
1	49.2	3	155.0	1	133.9	MARSHALL WV	215	-14	173	91	42
2	44.1	1	47.2	2	159.5	MASON WV	7	238	262	96	47
54	155.6	56	148.1	67	199.5	MERCER WV	-5	35	28	39	60
3	134.1	2	70.1	10	305.6	MINERAL WV	-48	336	128	85	90
25	199.1	17	136.1	21	209.2	MINGO WV	-32	54	5	27	63
11	129.8	28*	412.7	16	229.3	MONONGALIA WV	218	-44	77	68	70
6	260.6	3	91.4	3	73.1	MONROE WV	-65	-20	-72	03	33
1	172.0	3	512.2			MORGAN WV	198				
15	127.5	34*	314.4	37*	375.8	OHIO WV	147	20	195	92	95
8	262.9	1	54.6	5	419.6	POCAHONTAS WV	-79	668	60	58	97
2	434.1					PRESTON WV					
100	185.6	110	214.7	120	256.3	RALEIGH WV	16	19	38	46	79
5	406.7			3	182.6	RANDOLPH WV			-55	07	54
8	154.8	12	298.9	11	331.5	SUMMERS WV	93	11	114	82	93

* SIGNIFICANTLY MORE DEATHS THAN EXPECTED

COUNTIES COMBINED; SEE TECHNICAL NOTES

1950-1959 DEATHS	1950-1959 RATE	1960-1969 DEATHS	1960-1969 RATE	1970-1979 DEATHS	1970-1979 RATE	COUNTY AND STATE	PERCENT CHANGE 60S/50S	PERCENT CHANGE 70S/60S	PERCENT CHANGE 70S/50S	PERCENTILE %CHANGE 70S/50S	PERCENTILE RATES 1970S
		2	680.4	2	364.3	TAYLOR WV		-46			95
1	112.4					TUCKER WV					
				1*	112.4	TYLER WV					38
1	24.8					UPSHUR WV					
		1	490.8	1	74.3	WAYNE WV		-85			34
1*	344.1	1*	490.8			WETZEL WV	43				
				1*	546.4	WIRT WV					98
7	195.6	12	349.9	11	237.9	WOOD WV	79	-32	22	36	73
15	171.3	16	380.7	9	296.5	WYOMING WV	122	-22	73	66	89
351*	203.4	525	182.8	995	218.4	WISCONSIN WI	-10	19	7		
		2*	532.8	1	267.9	ADAMS WI		-50			83
5	177.2	1	49.7	7	240.2	ASHLAND WI	-72	383	36	45	74
6	321.4	3	193.3	4	158.4	BAYFIELD WI	-40	-18	-51	08	47
8	210.3	6	100.3	10	174.1	BROWN WI	-52	74	-17	18	52
1	53.6	2	120.6	3	188.3	BURNETT WI	125	56	251	95	56
1	188.7			1	89.3	CLARK WI			-53	07	35
8	302.7	10	165.8	32	234.3	DANE WI	-45	41	-23	17	72
1	267.9	1	62.9			DODGE WI	-77				
1	132.6	1	140.6	1	133.9	DOOR WI	6	-5	1	26	42
2	351.4	4	421.6	6	337.8	DOUGLAS WI	20	-20	-4	24	93
2*	556.1			1	267.9	EAU CLAIRE WI			-52	08	83
				1	86.0	FOND DU LAC WI					35
1	188.7	3	274.9			FOREST WI	46				
1	596.2	1	188.7			GRANT WI	-68				
1	67.0	1	70.1	3	226.2	JACKSON WI	5	223	238	94	69
				2	89.3	JEFFERSON WI					35
		1	106.0	1	46.8	JUNEAU WI		-56			31
1	89.3	8	204.4	14	164.0	KENOSHA WI	129	-20	84	72	48
1*	344.1					LAFAYETTE WI					
1	74.3					LANGLADE WI					
1	163.6	1	112.4			MANITOWOC WI	-31				
1	344.1	1	188.7			MAPATHON WI	-45				
1	424.0			1	188.7	MARINETTE WI			-55	07	57
		1*	344.1			MARQUETTE WI					
20	201.1	10	99.4	21	151.6	MENOMINEE#WI	-51	52	-25	16	46
230*	216.9	386	198.3	760*	248.7	MILWAUKEE WI	-9	25	15	31	76
		3	222.5	1	172.0	MONROE WI		-23			51
				3	477.2	ONEIDA WI					98
3	72.9	4	133.9	4	67.1	OUTAGAMIE WI	84	-50	-8	22	33
				1	141.3	OZAUKEE WI					44
2	305.4	1	94.4	2	150.6	POLK WI	-69	60	-51	08	45
16	170.3	36	313.0	65	227.1	RACINE WI	84	-27	33	43	70
16	268.7	17	218.4	28	243.8	ROCK WI	-19	12	-9	22	75
				1	424.0	RUSK WI					97
2	298.0					SAUK WI					
7	161.2	9	232.2	4	153.1	SAWYER WI	44	-34	-5	23	46
		1	141.3	1	39.3	SHEBOYGAN WI		-72			30
1*	344.1					TAYLOR WI					
4	166.0	2	45.2	5	180.0	VILAS WI	-73	299	8	29	53
		4	223.5	3	69.5	WALWORTH WI		-69			33
				1	161.1	WASHBURN WI					47
1*	188.7	2	788.9			WASHINGTON WI	318				
2	79.5	1	94.4	3	125.3	WAUKESHA WI	19	33	58	57	41
1	188.7			2	112.0	WINNEBAGO WI			-41	11	38
2	198.4	2	250.1	2	156.7	WOOD WI	26	-37	-21	17	47
35	148.1	29	117.4	39	115.0	WYOMING WY	-21	-2	-22		
		2	192.0	1	44.6	ALBANY WY		-77			31
1	424.0			1	56.2	BIG HORN WY			-87	00	32
1	133.9	2	190.4	1	35.3	CARBON WY	42	-81	-74	02	30
14	179.4	11	107.4	15	157.1	FPEMONT WY	-40	46	-12	20	47
1	589.6	1	133.9			HOT SPPINGS WY	-77				
1*	267.9					JOHNSON WY					
5	141.3	6	123.4	9	132.5	LARAMIE WY	-13	7	-6	22	42
1	267.9					LINCOLN WY					
6	359.6	3	112.7	4	148.4	NATPONA WY	-69	32	-59	06	45
		2	286.4			PARK-TETON#WY					
		1	74.3	2	278.0	SHEPIDAN WY		274			86
2	71.4	1	56.2	3	86.0	SWEETWATER WY	-21	53	20	36	35
3*	814.3			2	215.7	WASHAKIE WY			-74	02	65

NON-WHITE FEMALE — ALL CANCERS COMBINED — ICD 140-208

1950-1959 DEATHS	1950-1959 RATE	1960-1969 DEATHS	1960-1969 RATE	1970-1979 DEATHS	1970-1979 RATE	COUNTY AND STATE	PERCENT CHANGE 60S/50S	PERCENT CHANGE 70S/60S	PERCENT CHANGE 70S/50S	PERCENTILE %CHANGE 70S/50S	PERCENTILE RATES 1970S
98414	142.9	129161	142.3	169793	133.1	UNITED STATES	0	-6	-7		
5122	126.9	5754	131.2	6753	134.4	ALABAMA AL	3	2	6		
35	101.5	42	122.5	41	112.5	AUTAUGA AL	21	-8	11	68	55
39	113.4	37	92.1	69	130.8	BALDWIN AL	-19	42	15	71	67
52	89.5	91	154.3	82	126.6	BARBOUR AL	72	-18	42	83	64
29	136.5	20	107.4	27	152.4	BIBB AL	-21	42	12	68	78
3	82.4	6	223.9	2	52.9	BLOUNT AL	172	-76	-36	22	34
52	103.3	71	137.6	55	115.9	BULLOCK AL	33	-16	12	69	57
38	80.2	55	120.4	48	104.3	BUTLER AL	50	-13	30	79	50
104	166.4	110	144.3	119	131.3	CALHOUN AL	-13	-9	-21	33	67
77	137.8	72	117.7	71	101.0	CHAMBERS AL	-15	-14	-27	28	48
8	158.6	7	111.7	8	134.9	CHEROKEE AL	-30	21	-15	40	70
24	153.3	18	103.1	20	116.3	CHILTON AL	-33	13	-24	30	57
30	85.2	45	121.7	41	103.0	CHOCTAW AL	43	-15	21	74	49
64	125.5	65	122.5	77	127.8	CLARKE AL	-2	4	2	58	65
10	117.7	13	194.1	10	113.0	CLAY AL	65	-42	-4	52	55
1	53.6	4	149.4	2	83.7	CLEBURNE AL	179	-44	56	87	42
20	77.0	36	131.5	45	138.2	COFFEE AL	71	5	80	91	72
47	130.0	50	119.2	61	124.1	COLBERT AL	-8	4	-5	51	62
36	93.1	43	104.4	50	116.3	CONECUH AL	12	11	25	76	57
23	179.2	25	184.2	16	109.4	COOSA AL	3	-41	-39	20	53
22	81.5	43	159.4	46	154.8	COVINGTON AL	96	-3	90	91	79
22	105.7	23	111.3	25	105.3	CRENSHAW AL	5	-5	0	55	51
		3	151.5	3	125.7	CULLMAN AL		-17			63
22	115.7	21	91.3	33	115.9	DALE AL	-21	27	0	56	57
203	129.3	170	111.0	203	122.9	DALLAS AL	-14	11	-5	50	62
5	204.3	3	78.3	11	207.7	DE KALB AL	-62	165	2	58	91
60	134.5	58	133.9	68	153.1	ELMORE AL	0	14	14	69	79
55	128.3	64	128.3	71	118.7	ESCAMBIA AL	0	-7	-8	48	59
70	125.6	99	158.7	104	141.1	ETOWAH AL	26	-11	12	69	73
13	125.3	12	108.3	18	131.2	FAYETTE AL	-14	21	5	62	67
6	101.3	6	91.0	10	143.6	FRANKLIN AL	-10	58	42	83	74
18	119.3	13	86.8	16	110.3	GENEVA AL	-27	27	-8	48	54
68	124.7	54	110.2	57	110.4	GREENE AL	-12	0	-11	44	54
77	121.4	72	126.5	77	128.2	HALE AL	4	1	6	63	65
46	158.5	29	110.2	24	80.7	HENRY AL	-31	-27	-49	14	41
63	107.7	78	120.8	73	97.2	HOUSTON AL	12	-20	-10	46	47
8	89.2	17	164.4	19	166.5	JACKSON AL	84	1	87	91	84
1313	139.5	1513	141.4	1911*	153.4	JEFFERSON AL	1	9	10	67	79
9	117.0	17	173.5	15	146.9	LAMAR AL	48	-15	26	77	76
43	127.9	48	125.5	60	132.4	LAUDERDALE AL	-2	6	4	60	68
31	127.0	40	159.7	36	141.5	LAWRENCE AL	26	-11	11	68	73
114	161.3	99	128.2	109	125.1	LEE AL	-21	-2	-22	32	63
34	115.3	46	140.6	47	145.5	LIMESTONE AL	22	3	26	77	75
51	95.8	53	113.3	32	61.6	LOWNDES AL	18	-46	-36	22	36
116	127.2	115	117.2	160	136.9	MACON AL	-8	17	8	64	71
108	135.8	129	142.5	170	141.5	MADISON AL	5	-1	4	61	73
103	124.7	71	90.6	115	134.8	MARENGO AL	-27	49	8	65	70
5	194.5	6	205.3	7	185.8	MARION AL	6	-9	-4	51	88
6	182.3	4	84.1	6	109.9	MARSHALL AL	-54	31	-40	20	53
405	118.2	620	145.5	785*	152.8	MOBILE AL	23	5	29	79	79
46	98.3	64	133.6	67	133.1	MONROE AL	36	0	35	82	68
394	142.6	384	126.4	450	129.3	MONTGOMERY AL	-11	2	-9	46	66
57	157.0	47	122.6	72	168.3	MORGAN AL	-22	37	7	64	84
47	83.2	47	87.6	60	106.0	PERRY AL	5	21	27	78	51
47	109.9	54	120.0	47	90.7	PICKENS AL	9	-24	-17	38	45
69	131.3	74	152.2	73	131.2	PIKE AL	16	-14	0	56	67
17	85.0	31	151.4	39	161.5	RANDOLPH AL	78	7	90	91	82
91	110.1	134	147.2	121	122.0	RUSSELL AL	34	-17	11	67	61
15	89.6	24	139.3	24	134.2	ST CLAIR AL	56	-4	50	85	69
32	130.8	32	112.2	42	126.8	SHELBY AL	-14	13	-3	53	64
86	119.3	83	118.1	96	129.4	SUMTER AL	-1	10	8	65	66
106	152.6	123	158.4	111	121.2	TALLADEGA AL	4	-23	-21	34	61
45	109.0	44	99.0	53	105.7	TALLAPOOSA AL	-9	7	-3	53	51
155	144.9	157	116.5	217	133.0	TUSCALOOSA AL	-20	14	-8	48	68
40	133.1	41	137.3	43	134.7	WALKER AL	3	-2	1	57	70
13	64.8	13	62.0	27	113.9	WASHINGTON AL	-4	84	76	90	56
72	106.3	66	109.9	56	92.3	WILCOX AL	3	-16	-13	42	45
2*	691.9					WINSTON AL					
		155	132.4	248	153.4	ALASKA AK		16			
307	102.4	501	127.1	673	77.9	ARIZONA AZ	24	-39	-24		
38	75.1	82	141.4	98	106.4	APACHE AZ	88	-25	42	83	52
8	282.6	9	174.7	7	31.6	COCHISE AZ	-38	-82	-89	00	31
13	53.9	24	98.1	34	80.1	COCONINO AZ	82	-18	49	85	41
9	117.1	10	86.5	12	73.3	GILA AZ	-26	-15	-37	21	39
5	410.7	5	84.7	3	28.2	GRAHAM AZ	-79	-67	-93	00	31
123	141.3	177	128.9	263	78.5	MARICOPA AZ	-9	-39	-44	17	41
1	11.6			4	44.6	MOHAVE AZ			286	97	33
26	59.7	43	83.2	71	87.6	NAVAJO AZ	39	5	47	85	44
56	113.4	89	139.2	119	74.7	PIMA AZ	23	-46	-34	23	40
18	92.9	44	203.0	45	89.4	PINAL AZ	118	-56	-4	52	45
				1	8.6	SANTA CRUZ AZ					30
1	29.8	2	77.5	7	145.6	YAVAPAI AZ	160	88	389	98	75
9	114.4	16	159.6	9	30.3	YUMA AZ	40	-81	-74	04	31
2161	119.0	2496	130.7	2671	126.2	ARKANSAS AR	10	-3	6		
25	118.8	30	114.1	36	128.5	ARKANSAS AR	-4	13	8	65	65
28	70.9	42	106.5	43	94.3	ASHLEY AR	50	-11	33	80	46
		1*	344.1			BOONE AR					
26	122.6	34	158.5	36	143.1	BRADLEY AR	29	-10	17	72	74
13	120.1	8	78.0	13	116.8	CALHOUN AR	-35	50	-3	53	58
68	106.9	87	129.1	90	129.8	CHICOT AR	21	0	21	74	66

* SIGNIFICANTLY MORE DEATHS THAN EXPECTED

NON-WHITE FEMALE — ALL CANCERS COMBINED — ICD 140-208

1950-1959 DEATHS	1950-1959 RATE	1960-1969 DEATHS	1960-1969 RATE	1970-1979 DEATHS	1970-1979 RATE	COUNTY AND STATE	PERCENT CHANGE 60S/50S	PERCENT CHANGE 70S/60S	PERCENT CHANGE 70S/50S	PERCENTILE %CHANGE 70S/50S	PERCENTILE RATES 1970S
28	116.8	34	129.0	30	99.4	CLARK AR	10	-23	-15	40	48
9	92.3	12	159.8	8	101.6	CLEVELAND AR	73	-36	10	67	49
48	111.7	48	95.5	52	91.8	COLUMBIA AR	-14	-4	-18	37	45
28	156.2	17	90.0	31	152.1	CONWAY AR	-42	69	-3	53	78
9	110.1	9	82.9	19	168.3	CRAIGHEAD AR	-25	103	53	86	84
4	89.3	6	132.5	4	52.2	CRAWFORD AR	48	-61	-42	18	34
163	142.2	113	95.1	138	116.8	CRITTENDEN AR	-33	23	-18	37	58
30	100.9	29	107.9	30	108.0	CROSS AR	7	0	7	64	52
21	103.1	35	148.9	40	173.3	DALLAS AR	44	16	68	89	86
54	103.8	67	143.7	71	135.8	DESHA AR	38	-6	31	80	70
39	154.4	28	114.6	16	51.0	DREW AR	-26	-55	-67	06	34
19	168.6	15	118.6	20	136.6	FAULKNER AR	-30	15	-19	36	71
1	196.5	1	114.7			FRANKLIN AR	-42				
41	133.9	77*	236.0	54	140.3	GARLAND AR	76	-41	5	62	73
6	346.1	2	111.4	2	105.3	GRANT AR	-68	-5	-70	05	51
57	136.2	44	112.2	45	104.6	HEMPSTEAD AR	-18	-7	-23	31	50
15	127.3	18	135.2	25	147.8	HOT SPRING AR	6	9	16	71	76
17	152.6	16	119.2	18	120.3	HOWARD AR	-22	1	-21	33	60
4	86.2	2	25.3	8	116.7	INDEPENDENCE AR	-71	362	35	82	58
17	103.6	33	207.7	22	107.9	JACKSON AR	100	-48	4	61	52
174	100.7	239	131.7	266	127.2	JEFFERSON AR	31	-3	26	77	64
1	86.0	2	140.2	3	183.1	JOHNSON AR	63	31	113	93	88
36	152.3	34	138.3	35	123.5	LAFAYETTE AR	-9	-11	-19	36	62
2	159.9	2	131.7			LAWRENCE AR	-18				
56	91.4	84	137.8	65	100.7	LEE AR	51	-27	10	67	48
35	115.8	34	123.1	29	103.8	LINCOLN AR	6	-16	-10	45	50
19	103.1	26	156.3	23	130.2	LITTLE RIVER AR	52	-17	26	77	66
3	140.4	1	22.5	3	121.5	LOGAN AR	-84	440	-14	42	61
31	125.0	24	88.3	39	138.0	LONOKE AR	-29	56	10	67	72
57	143.6	58	135.3	73	150.9	MILLER AR	-6	11	5	62	78
65	79.4	110	130.0	132	158.3	MISSISSIPPI AR	64	22	99	92	81
53	129.7	50	131.8	68	166.5	MONROE AR	2	26	28	79	84
				1	74.3	MONTGOMERY AR					39
19	106.9	31	162.7	30	144.3	NEVADA AR	52	-11	35	81	74
1*	344.1					NEWTON AR					
68	122.9	75	119.3	79	110.8	OUACHITA AR	-3	-7	-10	46	54
156	135.1	159	132.3	147	121.8	PHILLIPS AR	-2	-8	-10	46	61
3	125.5	2	105.9	7*	454.2	PIKE AR	-16	329	262	97	98
11	83.8	21	151.4	19	117.1	POINSETT AR	81	-23	40	83	58
7	172.5	3	67.5	7	108.1	POPE AR	-61	60	-37	21	52
9	96.4	17	193.9	12	128.6	PRAIRIE AR	101	-34	33	81	65
342	143.5	421	154.0	466	141.7	PULASKI AR	7	-8	-1	55	73
1	117.9			2	173.0	RANDOLPH AR			47	85	86
99	129.3	99	126.0	96	128.6	ST FRANCIS AR	-3	2	-1	55	65
1	12.8	3	31.8	9	160.1	SALINE AR	147	404	1146	99	81
1	424.0					SCOTT AR					
		2*	708.6			SEARCY AR					
29	121.5	35	131.4	41	129.6	SEBASTIAN AR	8	-1	7	64	66
6	126.9	7	141.8	7	142.8	SEVIER AR	12	1	13	69	74
1*	188.7	1*	424.0			SHARP AR	125				
73	114.7	96	139.7	103	132.9	UNION AR	22	-5	16	71	68
				2	438.4	VAN BUREN AR					98
4	177.8	3	93.3	7	107.9	WASHINGTON AR	-48	16	-39	20	52
7	115.8	10	173.3	10	105.1	WHITE AR	50	-39	-9	47	50
20	66.3	33	119.3	36	140.1	WOODRUFF AR	80	17	111	93	72
1	20.2	6	227.4	3	88.0	YELL AR	1023	-61	335	98	44
3917	136.4	6861	126.0	11857	98.7	CALIFORNIA CA	-8	-22	-28		
485	139.3	783	125.5	1349	122.4	ALAMEDA CA	-10	-2	-12	43	61
1	424.0					ALPINE CA					
1	112.4			2	190.4	AMADOR CA			69	89	89
11	242.0	9	106.9	16	95.7	BUTTE CA	-56	-11	-60	09	47
		1	188.7			CALAVERAS CA					
		2	128.5	1	43.0	COLUSA CA		-67			33
75	136.5	150	127.7	293	107.3	CONTRA COSTA CA	-6	-16	-21	33	52
		1	112.4	3	66.5	DEL NORTE CA		-41			37
1	112.4	2	136.2			EL DORADO CA	21				
102	128.4	145	129.3	215	86.1	FRESNO CA	1	-33	-33	24	43
1	182.1	2	162.3	2	69.3	GLENN CA	-11	-57	-62	09	38
8	77.7	18	132.5	9	49.9	HUMBOLDT CA	71	-62	-36	22	34
20	135.2	18	113.0	24	46.4	IMPERIAL CA	-16	-59	-66	07	33
10*	349.6	2	34.9	7	110.4	INYO CA	-90	216	-68	06	54
87	174.0	125	163.8	127	73.1	KERN CA	-6	-55	-58	10	39
15	164.4	24	207.3	23	74.6	KINGS CA	26	-64	-55	12	40
4	558.9	2	98.8	5	117.5	LAKE CA	-82	19	-79	02	58
		2	128.3	2	55.9	LASSEN CA		-56			35
2036	142.2	3709	134.0	6143	109.2	LOS ANGELES CA	-6	-18	-23	31	53
9	69.5	15	93.7	23	75.2	MADERA CA	35	-20	8	65	40
11	146.7	19	153.7	20	73.3	MARIN CA	5	-52	-50	14	39
		1	424.0	2	230.7	MARIPOSA CA		-46			93
8	137.7	7	106.0	9	73.6	MENDOCINO CA	-23	-31	-47	16	39
22	157.3	25	121.8	30	59.6	MERCED CA	-23	-51	-62	08	36
2	245.4	1	89.3	3	195.6	MODOC CA	-64	119	-20	34	90
1	212.0	1	114.7	1	86.0	MONO CA	-46	-25	-59	09	43
16	79.1	47	131.1	80	58.3	MONTEREY CA	66	-56	-26	28	36
1	12.0	3	57.6	3	20.8	NAPA CA	378	-64	73	90	30
		2	176.2			NEVADA CA					
9	89.4	32	64.6	137	49.0	ORANGE CA	-28	-24	-45	17	34
2	20.3	11	159.1	21	115.5	PLACER CA	685	-27	470	99	57
3	209.0	3	230.0	2	72.9	PLUMAS CA	10	-68	-65	07	39
59	166.2	90	118.6	159	73.3	RIVERSIDE CA	-29	-38	-56	11	39
104	148.5	161	118.0	314	94.3	SACRAMENTO CA	-21	-20	-36	21	46
2	223.1	1	47.2	6	68.3	SAN BENITO CA	-79	45	-69	05	38
47	106.7	102	120.1	170	73.7	SAN BERNARDINO CA	13	-39	-31	25	39
118	126.0	223	119.6	456	84.9	SAN DIEGO CA	-5	-29	-33	24	43
368	134.0	668	117.5	1252	115.2	SAN FRANCISCO CA	-12	-2	-14	41	57
75	117.1	95	106.8	140	75.2	SAN JOAQUIN CA	-9	-30	-36	22	40
4	138.5	5	67.6	8	33.7	SAN LUIS OBISPO CA	-51	-50	-76	03	32
28	91.4	73	88.5	202	84.7	SAN MATEO CA	-3	-4	-7	48	43
18	123.2	21	77.1	50	68.6	SANTA BARBARA CA	-37	-11	-44	17	38

* SIGNIFICANTLY MORE DEATHS THAN EXPECTED

NON-WHITE FEMALE — ALL CANCERS COMBINED — ICD 140-208

1950-1959 DEATHS	1950-1959 RATE	1960-1969 DEATHS	1960-1969 RATE	1970-1979 DEATHS	1970-1979 RATE	COUNTY AND STATE	PERCENT CHANGE 60S/50S	PERCENT CHANGE 70S/60S	PERCENT CHANGE 70S/50S	PERCENTILE %CHANGE 70S/50S	PERCENTILE RATES 1970S
40	89.1	90	89.0	222	56.1	SANTA CLAPA CA	0	-37	-37	21	35
10	183.8	12	108.4	11	25.3	SANTA CRUZ CA	-41	-77	-86	01	31
6	131.7	6	93.4	7	64.8	SHASTA CA	-29	-31	-51	13	37
7	136.6	5	78.0	7	130.0	SISKIYOU CA	-43	67	-5	51	66
30	112.5	57	152.6	119	109.6	SOLANO CA	36	-28	-3	53	53
12	201.6	13	107.2	26	68.5	SONOMA CA	-47	-36	-66	07	38
5	104.2	2	17.4	19	38.9	STANISLAUS CA	-83	123	-63	08	33
		3	113.0	8	60.6	SUTTER CA		-46			36
1	78.1	1	53.0	3	74.4	TEHAMA CA	-32	40	-5	51	40
				1	84.8	TPINITY CA					43
19	121.3	32	149.3	43	55.1	TULARE CA	23	-63	-55	12	35
		1	112.4			TUOLUMNE CA					
9	70.2	18	91.8	52	37.5	VENTUPA CA	31	-59	-47	16	32
8	158.9	8	91.9	13	42.8	YOLO CA	-42	-53	-73	04	33
6	185.9	12	213.8	17	139.7	YUBA CA	15	-35	-25	30	72
192	134.0	270	118.0	413	78.4	COLORADO CO	-12	-34	-42		
5	143.9	6	148.6	13	59.6	ADAMS CO	3	-60	-59	10	36
		1	114.7			ALAMOSA CO					
				1	74.3	ARCHULETA CO					39
		1*	344.1			BENT CO					
2	183.1	1	21.8	1	5.5	BOULDER CO	-88	-75	-97	00	30
		1	596.2			DELTA CO					
125	125.8	209	126.6	300	95.4	DENVER-ARAPAHOE#CO	1	-25	-24	30	47
21	178.8	17	97.1	48	111.3	EL PASO CO	-46	15	-38	21	54
1*	490.8	1	546.4			FREMONT CO	11				
		1*	267.9			GARFIELD CO					
		1	24.8	6	50.3	JEFFERSON CO		103			34
4	344.8	2	87.4	3	46.4	LA PLATA CO	-75	-47	-87	01	33
		1	59.0	1	7.7	LARIMER CO		-87			30
3	427.0	2	189.0			LAS ANIMAS CO	-56				
1	424.0			1	54.2	LOGAN CO			-87	01	35
1	74.3	1	41.4	2	38.4	MESA CO	-44	-7	-48	15	33
1	74.3	4	347.3	7	171.0	MONTEZUMA CO	367	-51	130	94	85
4	278.0	1	44.6	4	45.1	OTEPO CO	-84	1	-84	01	33
		1	267.9			PROWEPS CO					
19	170.0	13	94.2	20	68.3	PUEBLO CO	-45	-28	-60	09	38
		1	546.6	1	424.0	ROUTT CO		-22			98
1	344.1			2	254.9	SEDGWICK CO			-26	29	94
4	105.8	5	141.8	3	25.7	WELD CO	34	-82	-76	03	31
424	165.6	621	135.1	1062	135.7	CONNECTICUT CT	-18	0	-18		
113	135.0	211	139.2	332	130.8	FAIPFIELD CT	3	-6	-3	52	67
140*	205.0	175	133.7	287	119.9	HARTFORD CT	-35	-10	-42	18	60
4	118.2	7	146.5	11	124.5	LITCHFIELD CT	24	-15	5	62	63
5	113.9	9	114.9	18	125.1	MIDDLESEX CT	1	9	10	67	63
138	170.6	185	134.7	360*	158.4	NEW HAVEN CT	-21	18	-7	49	81
22	182.1	30	152.6	49	156.4	NEW LONDON CT	-16	2	-14	41	80
		1	29.5	2	34.2	TOLLAND CT		16			32
2	119.2	3	177.6	3	52.1	WINDHAM CT	49	-71	-56	11	34
325*	160.6	448*	171.4	623*	171.5	DELAWARE DE	7	0	7		
53	161.2	61	136.8	95	151.1	KENT DE	-15	10	-6	49	78
196	157.4	289*	181.0	376*	167.2	NEW CASTLE DE	15	-8	6	63	84
76	168.1	98	174.2	152*	208.0	SUSSEX DE	4	19	24	76	91
2284*	178.1	3158*	168.1	4170*	171.8	DIST. OF COLUMBIA DC	-6	2	-4		
2284*	178.1	3158*	168.1	4170*	171.8	WASHINGTON DC	-6	2	-4	52	85
3713	129.9	5237	137.4	7032	129.8	FLORIDA FL	6	-6	0		
117	154.1	146	160.9	146	126.3	ALACHUA FL	4	-21	-18	37	64
4	62.8	7	115.9	4	42.5	BAKER FL	85	-63	-32	25	33
32	123.2	47	118.9	72	134.5	BAY FL	-3	13	9	66	70
19	190.5	15	131.2	16	128.7	BRADFOPD FL	-31	-2	-32	24	65
32	106.1	79	151.9	91	115.2	BREVARD FL	43	-24	9	65	57
137	140.3	253	134.7	400	114.4	BROWAPD FL	-4	-15	-18	37	56
4	77.1	7	123.8	7	110.6	CALHOUN FL	61	-11	43	84	54
3	113.9	7	190.5	3	31.0	CHAPLOTTE FL	67	-84	-73	04	31
11	158.6	7	93.4	9	74.8	CITRUS FL	-41	-20	-53	12	40
11	136.9	18	176.4	12	90.5	CLAY FL	29	-49	-34	23	45
9	114.5	14	184.4	23	118.3	COLLIER FL	61	-36	3	60	59
28	102.5	28	95.1	44	131.8	COLUMBIA FL	-7	39	29	79	67
434	130.8	770	138.3	1152	110.8	DADE FL	6	-20	-15	40	54
17	169.9	15	175.8	15	104.4	DE SOTO FL	4	-41	-39	20	50
3	126.7	5	180.5	3	90.8	DIXIE FL	43	-50	-28	27	45
585	141.2	798*	155.1	969*	150.5	DUVAL FL	10	-3	7	63	78
170	147.0	219	143.1	317*	153.1	ESCAMBIA FL	-3	7	4	61	79
7	108.1	6	127.3	7	124.4	FLAGLER FL	18	-2	15	71	63
10	123.6	9	108.8	13	161.8	FRANKLIN FL	-12	49	31	80	82
90	96.0	108	98.1	128	105.4	GADSDEN FL	2	7	10	67	51
1	60.7	2	165.2	1	37.5	GILCHRIST FL	172	-77	-38	20	32
3	142.8	6	188.6	2	51.4	GLADES FL	32	-73	-64	08	34
9	128.2	7	72.6	12	111.3	GULF FL	-43	53	-13	42	54
17	108.7	16	114.2	20	120.7	HAMILTON FL	5	6	11	68	60
3	110.7	6	147.7	4	86.6	HAPDEE FL	33	-41	-22	32	43
8	135.8	10	99.1	23	171.8	HENDRY FL	-27	73	27	78	85
6	108.2	13	148.6	10	80.1	HERNANDO FL	37	-46	-26	29	41
15	86.6	25	122.7	40	148.9	HIGHLANDS FL	42	21	72	89	77
279	131.1	375	139.1	483	132.8	HILLSBOROUGH FL	6	-5	1	57	68
3	79.9	7	191.8	4	116.8	HOLMES FL	140	-39	46	85	58
10	79.6	22	111.9	38	129.3	INDIAN RIVER FL	41	16	62	88	66
78	141.9	49	90.5	52	85.4	JACKSON FL	-36	-6	-40	19	43
26	93.9	41	157.9	24	91.5	JEFFERSON FL	68	-42	-3	53	45
1	59.0			2	348.5	LAFAYETTE FL			491	99	97
48	136.2	66	152.0	84	138.2	LAKE FL	12	-9	1	57	72
21	95.3	55	148.8	82	166.2	LEE FL	56	12	74	90	83

* SIGNIFICANTLY MORE DEATHS THAN EXPECTED

\# COUNTIES COMBINED; SEE TECHNICAL NOTES

NON-WHITE FEMALE — ALL CANCERS COMBINED — ICD 140-208

1950-1959 DEATHS	1950-1959 RATE	1960-1969 DEATHS	1960-1969 RATE	1970-1979 DEATHS	1970-1979 RATE	COUNTY AND STATE	PERCENT CHANGE 60S/50S	PERCENT CHANGE 70S/60S	PERCENT CHANGE 70S/50S	PERCENTILE %CHANGE 70S/50S	PERCENTILE RATES 1970S
87	101.9	138	138.6	182	143.5	LEON FL	36	4	41	83	74
25	160.9	18	121.7	27	168.8	LEVY FL	-24	39	5	62	84
4	280.4	2	80.6	2	98.9	LIBERTY FL	-71	23	-65	07	48
44	161.7	39	122.2	34	92.7	MADISON FL	-24	-24	-43	18	46
27	72.6	62	142.5	67	119.3	MANATEE FL	96	-16	64	88	59
91	132.4	114	138.0	143	137.6	MARION FL	4	0	4	60	71
19	163.4	21	132.4	22	114.8	MARTIN FL	-19	-13	-30	26	56
31	194.4	28	155.1	36	175.7	MONROE FL	-20	13	-10	46	86
33*	199.1	33	172.1	41	176.2	NASSAU FL	-14	2	-11	44	86
5	59.3	21	139.3	28	125.3	OKALOOSA FL	135	-10	111	93	63
3	83.7	10	259.5	9	182.9	OKEECHOBEE FL	210	-30	119	94	88
137	131.9	196	123.9	315	132.2	ORANGE FL	-6	7	0	56	68
4	70.9	15	178.3	27*	216.5	OSCEOLA FL	151	21	205	96	92
178	116.9	283	136.9	392	131.5	PALM BEACH FL	17	-4	13	69	67
18	146.1	16	109.9	22	107.6	PASCO FL	-25	-2	-26	28	52
133	124.0	210	135.4	314	137.5	PINELLAS FL	9	2	11	68	71
154	138.2	201	141.5	270	144.0	POLK FL	2	2	4	61	74
49	116.4	62	133.3	79	150.0	PUTNAM FL	15	12	29	79	77
51	142.3	53	137.8	67	161.2	ST JOHNS FL	-3	17	13	69	82
46	139.1	68	155.6	112*	177.3	ST LUCIE FL	12	14	27	78	87
13	181.9	10	151.3	13	131.5	SANTA ROSA FL	-17	-13	-28	28	67
23	95.4	50	165.1	69	152.6	SARASOTA FL	73	-8	60	88	78
73	137.4	82	126.1	105	119.3	SEMINOLE FL	-8	-5	-13	42	59
13	105.1	12	83.9	20	118.4	SUMTER FL	-20	41	13	69	59
26	121.9	21	97.8	31	139.1	SUWANNEE FL	-20	42	14	70	72
17	110.9	25	143.2	29	158.5	TAYLOR FL	29	11	43	84	81
6	139.6	4	89.3	4	72.7	UNION FL	-36	-19	-48	15	39
123	136.7	151	139.3	234*	168.1	VOLUSIA FL	2	21	23	76	84
9	126.0	4	68.0	8	91.5	WAKULLA FL	-46	35	-27	28	45
12	126.4	15	149.3	10	77.4	WALTON FL	18	-48	-39	20	41
8	88.1	15	153.1	8	68.8	WASHINGTON FL	74	-55	-22	32	38
5875	131.7	6766	133.0	7668	126.5	GEORGIA GA	1	-5	-4		
11	122.9	21	183.8	14	125.2	APPLING GA	49	-32	2	58	63
1	21.0	5	92.3	14	226.1	ATKINSON GA	339	145	976	99	92
4	113.1	4	121.5	3	63.7	BACON GA	7	-48	-44	18	37
12	115.4	7	70.6	9	103.6	BAKER GA	-39	47	-10	45	50
51	67.9	50	63.2	68	104.2	BALDWIN GA	-7	65	53	86	50
2	120.8	1	49.7	1	60.6	BANKS GA	-59	22	-50	14	36
6	69.1	10	89.2	24	183.3	BARROW GA	29	105	165	96	88
14	82.5	27	136.5	29	119.8	BARTOW GA	65	-12	45	84	60
20	92.4	38	159.8	25	97.8	BEN HILL GA	73	-39	6	63	48
8	117.7	9	121.7	13	141.9	BERRIEN GA	3	17	21	74	73
285	141.6	302	131.4	324	118.2	BIBB GA	-7	-10	-17	38	59
21	165.3	14	134.9	12	115.0	BLECKLEY GA	-27	-15	-38	20	56
4	126.9	4	153.5			BPANTLEY GA	21				
38	119.3	50	160.8	47	137.5	BROOKS GA	35	-14	15	71	71
11	120.5	13	215.2	15	190.4	BPYAN GA	79	-12	58	87	89
30	92.8	38	98.0	38	82.5	BULLOCH GA	6	-16	-11	44	42
88	140.9	60	106.6	68	121.6	BURKE GA	-24	14	-14	42	61
15	97.2	29	183.7	23	109.8	BUTTS GA	89	-40	13	69	53
35	158.4	37	174.8	38	154.8	CALHOUN GA	10	-11	-2	53	80
19	130.2	26	174.0	22	110.9	CAMDEN GA	34	-36	-15	41	54
7	64.8	16	153.7	12	97.3	CANDLER GA	137	-37	50	86	48
16	68.5	26	109.4	41	125.5	CAPROLL GA	60	15	83	91	63
2	247.4	1	91.1	1	212.0	CATCOSA GA	-63	133	-14	41	92
10	203.0	12	189.0	6	82.4	CHARLTON GA	-7	-56	-59	09	42
412*	149.4	468	147.6	577*	156.3	CHATHAM GA	-1	6	5	61	80
5	68.0	7	78.8	20*	232.2	CHATTOOGA GA	16	195	242	97	93
1	26.0	2	42.5	5	76.5	CHEROKEE GA	63	80	194	96	40
68	135.3	71	120.9	102	144.4	CLARKE GA	-11	19	7	63	74
17	123.0	18	142.0	15	101.9	CLAY GA	15	-28	-17	38	49
11	69.9	17	93.3	31	138.0	CLAYTON GA	33	48	97	92	72
8	139.0	7	130.5	4	56.4	CLINCH GA	-6	-57	-59	09	35
48	159.3	59	166.1	61	134.7	COBB GA	4	-19	-15	40	70
20	103.6	25	127.6	32	126.0	COFFEE GA	23	-1	22	75	64
35	121.4	37	116.9	45	127.1	COLQUITT GA	-4	9	5	62	64
12	76.5	13	75.9	18	78.8	COLUMBIA GA	-1	4	3	59	41
12	97.9	15	106.2	13	76.6	COOK GA	8	-28	-22	32	40
54	129.6	53	121.7	66	133.8	COWETA GA	-6	10	3	59	69
14	116.9	11	104.3	12	101.2	CRAWFOPD GA	-11	-3	-13	42	49
39	114.1	52	148.2	41	107.0	CRISP GA	30	-28	-6	49	52
1	298.1			1	119.2	DADE GA			-60	09	59
1*	546.4					DAWSON GA					
65	137.6	53	111.0	48	94.3	DECATUR GA	-19	-15	-31	25	46
106	148.4	217	160.9	379	133.9	DE KALB GA	8	-17	-10	46	69
20	102.3	15	84.1	23	107.3	DODGE GA	-18	28	5	62	52
33	124.2	33	125.4	37	141.6	DOOLY GA	1	13	14	70	73
130	139.4	167	141.2	203	132.7	DOUGHERTY GA	1	-6	-5	50	68
8	112.6	12	121.2	11	87.7	DOUGLAS GA	8	-28	-22	32	44
33	114.3	34	123.3	34	113.0	EARLY GA	8	-8	-1	55	55
2	76.7	1	67.0	2	146.9	ECHOLS GA	-13	119	91	92	76
15	124.2	7	54.4	18	127.6	EFFINGHAM GA	-56	134	3	59	64
25	110.6	27	110.6	23	88.3	ELBERT GA	0	-20	-20	34	44
31	124.2	32	120.8	32	100.3	EMANUEL GA	-3	-17	-19	35	48
17	171.7	11	102.0	7	50.4	EVANS GA	-41	-51	-71	05	34
		1	546.6	1	133.9	FANNIN GA		-75			69
14	184.6	8	100.2	8	109.4	FAYETTE GA	-46	9	-41	19	53
62	156.7	64	134.5	85	163.1	FLOYD GA	-14	21	4	61	82
7	135.4	15*	265.8	9	145.3	FRANKLIN GA	96	-45	7	64	75
1204*	161.4	1455*	154.4	1637*	140.1	FULTON GA	-4	-9	-13	42	72
2	54.1	2	129.9			GLASCOCK GA	140				
56	123.1	92	156.5	86	125.0	GLYNN GA	27	-20	1	57	63
9	169.0	6	108.1	9	150.4	GORDON GA	-36	39	-11	45	77
34	133.3	31	113.4	35	113.9	GRADY GA	-15	0	-15	41	56
29	106.1	33	120.0	29	104.1	GREENE GA	13	-13	-2	54	50
20	166.0	25	183.7	20	116.3	GWINNETT GA	11	-37	-30	26	57
4	111.0	2	60.6	8	165.0	HABERSHAM GA	-45	172	49	85	83
24	137.3	36	171.3	37	131.8	HALL GA	25	-23	-4	52	67
26	99.0	26	92.4	35	113.4	HANCOCK GA	-7	23	15	70	56

* SIGNIFICANTLY MORE DEATHS THAN EXPECTED

NON-WHITE FEMALE — ALL CANCERS COMBINED — ICD 140-208

1950-1959 DEATHS	1950-1959 RATE	1960-1969 DEATHS	1960-1969 RATE	1970-1979 DEATHS	1970-1979 RATE	COUNTY AND STATE	PERCENT CHANGE 60S/50S	PERCENT CHANGE 70S/60S	PERCENT CHANGE 70S/50S	PERCENTILE %CHANGE 70S/50S	PERCENTILE RATES 1970S
7	134.9	5	93.3	7	113.2	HARALSON GA	-31	21	-16	39	56
32	141.5	24	100.5	38	132.0	HARRIS GA	-29	31	-7	49	68
13	116.0	11	88.0	29	176.2	HART GA	-24	100	52	86	86
8	124.6	5	115.1	6	110.6	HEARD GA	-8	-4	-11	44	54
28	130.9	21	79.2	30	94.4	HENRY GA	-40	19	-28	28	46
31	130.0	34	102.9	51	115.0	HOUSTON GA	-21	12	-12	44	56
10	79.7	19	152.9	15	128.3	IRWIN GA	92	-16	61	88	65
8	82.0	15	160.8	15	121.8	JACKSON GA	96	-24	49	85	61
18	117.1	11	83.4	22	157.2	JASPER GA	-29	89	34	81	80
8	136.6	8	96.8	11	103.2	JEFF DAVIS GA	-29	7	-24	30	49
45	115.4	48	118.9	46	95.2	JEFFERSON GA	3	-20	-18	38	47
26	128.2	21	103.6	17	90.1	JENKINS GA	-19	-13	-30	26	45
7	69.4	15	143.6	11	101.2	JOHNSON GA	107	-30	46	85	49
9	58.0	20	105.9	24	102.6	JONES GA	82	-3	77	90	49
19	124.1	25	142.8	24	117.6	LAMAR GA	15	-18	-5	50	58
5	91.8	5	89.7	9	163.8	LANIER GA	-2	83	78	91	83
39	74.3	63	117.3	54	89.4	LAURENS GA	58	-24	20	74	45
17	106.4	18	122.9	18	133.2	LEE GA	16	8	25	77	68
31	144.6	26	104.9	36	126.6	LIBERTY GA	-27	21	-12	43	64
11	122.9	17	212.3	12	120.5	LINCOLN GA	73	-43	-2	54	60
6	160.2	8	136.0	10	181.1	LONG GA	-15	33	13	69	87
73	116.1	105	145.7	111	127.0	LOWNDES GA	26	-13	9	66	64
2	282.7	2	273.2	1	106.0	LUMPKIN GA	-3	-61	-63	08	51
25	109.2	42	180.1	23	77.0	MC DUFFIE GA	65	-57	-29	27	40
23	135.7	11	67.8	23	117.0	MC INTOSH GA	-50	73	-14	41	58
38	105.4	44	121.2	33	78.8	MACON GA	15	-35	-25	29	41
11	158.6	11	188.5	9	107.0	MADISON GA	19	-43	-33	24	52
14	117.4	10	76.0	18	134.7	MARION GA	-35	77	15	70	70
57	141.4	44	102.1	56	126.3	MERIWETHER GA	-28	24	-11	45	64
10	113.1	9	126.6	9	111.4	MILLER GA	12	-12	-1	54	54
60	149.6	47	108.9	60	136.8	MITCHELL GA	-27	26	-9	47	71
22	115.1	36	181.2	27	105.1	MONROE GA	57	-42	-9	47	50
12	122.3	12	128.0	12	125.7	MONTGOMERY GA	5	-2	3	59	63
38	170.4	34	150.3	30	141.8	MORGAN GA	-12	-6	-17	38	73
197	135.0	227	121.0	326	132.5	MUSCOGEE#GA	-10	9	-2	54	68
33	135.6	42	140.8	42	117.2	NEWTON GA	4	-17	-14	42	58
4	75.3	7	113.0	9	171.9	OCONEE GA	50	52	128	94	85
15	99.8	16	130.1	14	118.5	OGLETHORPE GA	30	-9	19	73	59
5	152.3	6	152.3	3	56.6	PAULDING GA	0	-63	-63	08	36
36	127.6	39	118.7	42	106.1	PEACH GA	-7	-11	-17	38	51
3	147.8	3	188.7	3	99.6	PICKENS GA	28	-47	-33	24	48
9	109.1	12	125.9	9	86.4	PIERCE GA	15	-31	-21	33	43
12	97.3	18	144.7	12	92.9	PIKE GA	49	-36	-5	51	46
21	125.9	25	112.8	36	142.1	POLK GA	-10	26	13	69	73
27	172.4	16	99.8	25	139.1	PULASKI GA	-42	39	-19	35	72
23	150.9	20	112.7	17	92.4	PUTNAM GA	-25	-18	-39	20	45
15	235.2	7	101.8	9	133.3	QUITMAN GA	-57	31	-43	18	68
2	184.6	1	94.4	1	298.1	RABUN GA	-49	216	62	88	96
45	132.4	33	93.1	31	94.2	RANDOLPH GA	-30	1	-29	27	46
251	145.1	312	150.9	374	144.2	RICHMOND GA	4	-4	-1	55	74
13	134.6	15	139.7	16	106.1	ROCKDALE GA	4	-24	-21	33	51
3	43.5	9	127.1	7	85.1	SCHLEY GA	192	-33	96	92	43
33	97.9	40	130.5	37	115.0	SCREVEN GA	33	-12	17	72	57
13	123.0	10	88.1	10	74.5	SEMINOLE GA	-28	-15	-39	20	40
50	130.7	58	134.7	72	132.7	SPALDING GA	3	-1	2	57	68
19	203.5	23	225.6	13	104.8	STEPHENS GA	11	-54	-48	15	50
14	59.9	21	94.6	21	95.3	STEWART GA	58	1	59	88	47
72	122.4	84	134.8	82	122.7	SUMTER GA	10	-9	0	56	62
30	170.2	22	122.0	14	69.2	TALBOT GA	-28	-43	-59	10	38
14	125.4	12	121.7	9	103.1	TALIAFERRO GA	-3	-15	-18	37	49
14	125.6	24	182.8	22	128.5	TATTNALL GA	46	-30	2	58	65
15	103.2	13	75.7	13	76.4	TAYLOR GA	-27	1	-26	29	40
12	72.1	28	154.7	25	135.3	TELFAIR GA	115	-13	88	91	70
57	154.0	43	124.7	51	151.6	TERRELL GA	-19	22	-2	54	78
79	132.8	88	133.9	100	136.2	THOMAS GA	1	2	3	59	71
45	167.9	33	108.9	36	103.6	TIFT GA	-35	-5	-38	20	50
24	129.8	31	138.9	31	117.4	TOOMBS GA	7	-16	-10	46	58
5	55.1	6	89.9	9	102.3	TREUTLEN GA	63	14	86	91	49
89	136.4	106	148.0	100	126.8	TROUP GA	9	-14	-7	49	64
19	163.2	15	124.6	7	50.5	TURNER GA	-24	-60	-69	06	34
16	108.2	12	72.4	15	80.5	TWIGGS GA	-33	11	-26	29	41
24	98.6	43	138.5	42	127.6	UPSON GA	40	-8	29	79	64
8	79.2	13	120.5	20	163.8	WALKER GA	52	36	107	93	83
28	128.6	32	140.8	37	147.0	WALTON GA	9	4	14	70	76
43	102.8	67	151.6	62	128.1	WARE GA	47	-15	25	76	65
26	156.8	19	103.0	17	68.1	WARREN GA	-34	-14	-44	17	44
57	125.0	69	153.8	77	159.5	WASHINGTON GA	23	4	28	78	81
16	160.0	13	83.6	18	111.1	WAYNE GA	-48	33	-31	26	54
8	108.2	7	107.5	12	198.6	WEBSTER GA	-1	85	84	91	90
8	120.5	5	76.5	7	96.3	WHEELER GA	-37	26	-20	34	47
7*	657.7			2	164.1	WHITE GA			-75	04	83
18*	267.2	13	134.7	18	158.2	WHITFIELD GA	-50	17	-41	19	81
13	112.6	7	68.3	17	134.6	WILCOX GA	-39	97	20	73	70
31	113.1	41	151.2	47	156.8	WILKES GA	34	4	39	83	80
19	119.7	29	171.6	25	127.7	WILKINSON GA	43	-26	7	64	64
28	101.5	45	170.5	21	76.9	WORTH GA	68	-55	-24	30	40
		1801	119.9	2545	109.0	HAWAII HI		-9			
		222	117.9	285	113.2	HAWAII HI		-4			55
		1328	121.3	1931	108.9	HONOLULU HI		-10			53
		101	118.6	129	108.4	KAUAI HI		-9			53
		150	114.4	200	101.9	MAUI#HI		-11			49
32	117.1	40	121.1	45	71.0	IDAHO ID	3	-41	-39		
		1	114.7	1	8.9	ADA ID		-92			30
8	193.7	9	202.9	9	106.1	BANNOCK ID	5	-48	-45	17	51
2	163.6	3	281.3	1	141.3	BENEWAH ID	72	-50	-14	42	73
8	182.0	7	104.7	8	86.2	BINGHAM ID	-42	-18	-53	13	43
4	316.6	1	29.5	5	177.4	BONNEVILLE ID	-91	502	-44	17	87

* SIGNIFICANTLY MORE DEATHS THAN EXPECTED

COUNTIES COMBINED; SEE TECHNICAL NOTES

NON-WHITE FEMALE — ALL CANCERS COMBINED — ICD 140-208

1950-1959 DEATHS	1950-1959 RATE	1960-1969 DEATHS	1960-1969 RATE	1970-1979 DEATHS	1970-1979 RATE	COUNTY AND STATE	PERCENT CHANGE 60S/50S	PERCENT CHANGE 70S/60S	PERCENT CHANGE 70S/50S	PERCENTILE %CHANGE 70S/50S	PERCENTILE RATES 1970S
		1	112.4			BOUNDARY ID					
3	130.5	2	59.9	5	67.7	CANYON ID	-54	13	-48	15	37
		1	11.0	1	60.6	ELMORE ID		449			36
				1	74.3	GEM ID					39
		1*	424.0			GOODING ID					
1	172.0	3	252.9	1	32.8	KOOTENAI ID	47	-87	-81	02	32
		1	490.8			LATAH ID					
		1	182.2			LEMHI ID					
				2	273.1	LEWIS ID					95
				1	106.0	LINCOLN ID					51
1	344.1					MADISON ID					
5	139.2	6	187.3	7	170.1	NEZ PERCE ID	34	-9	22	75	85
		1	267.9			OWYHEE ID					
		1	344.1			PAYETTE ID					
				1	70.7	TWIN FALLS ID					38
		1	114.7	2	84.5	WASHINGTON ID		-26			42
5496*	170.3	7546*	162.8	10038*	150.0	ILLINOIS IL	-4	-8	-12		
18*	257.6	12	162.1	22*	266.7	ADAMS IL	-37	64	4	60	95
50	145.6	39	127.2	48	196.8	ALEXANDER IL	-13	55	35	81	90
		2	98.2	3	105.1	BOND IL		7			50
				1	49.7	BOONE IL					34
		1	188.7			BUREAU IL					
1	74.3			2	180.4	CARROLL IL			143	95	87
		1*	267.9			CASS IL					
37	253.4	42	160.6	51	129.7	CHAMPAIGN IL	-37	-19	-49	14	66
1	188.7	1	196.5	2	66.4	CHRISTIAN IL	4	-66	-65	07	37
				3	400.9	CLINTON IL					98
3	315.0			2	115.3	COLES IL			-63	08	57
4574*	174.8	6347*	167.3	8406*	152.9	COOK IL	-4	-9	-13	43	79
		1	112.4	5	302.4	DE KALB IL		169			96
1	546.6			2	180.4	DE WITT IL			-67	06	87
4	189.6	5	94.7	22	77.8	DU PAGE IL	-50	-18	-59	10	41
3	344.8			2	249.0	EDGAR IL			-28	28	94
		1*	546.4			EDWARDS IL					
		1*	344.1			EFFINGHAM IL					
				1	141.3	FORD IL					73
1	182.1			1	188.7	FRANKLIN IL			4	60	89
		3	342.1	2	125.8	FULTON IL		-63			64
				1	114.7	GALLATIN IL					56
		1	172.0			GREENE IL					
1*	424.0					HANCOCK IL					
1	212.0					HARDIN IL					
2	78.6	3	148.0	3	66.1	HENRY IL	88	-55	-16	39	37
1	133.9	1	89.3	2	123.0	IROQUOIS IL	-33	38	-8	48	62
27	185.8	24	126.2	24	108.6	JACKSON IL	-32	-14	-42	18	53
7	131.8	17*	295.3	14	169.5	JEFFERSON IL	124	-43	29	79	84
				1	273.2	JERSEY IL					95
1	109.3	1	546.4			JOHNSON IL	400				
20	149.4	21	95.1	45	113.7	KANE IL	-36	20	-24	31	56
17	42.6	52	99.3	78	127.3	KANKAKEE IL	133	28	199	96	64
13	158.1	12	121.4	17	134.0	KNOX IL	-23	10	-15	40	69
47	167.3	74	151.8	110	132.0	LAKE IL	-9	-13	-21	33	68
6	295.6	3	173.9	6	199.4	LA SALLE IL	-41	15	-33	24	90
		2	133.6	3	183.6	LAWRENCE IL		37			88
1	172.0	3	169.2	3	90.9	LEE IL	-2	-46	-47	16	45
1	54.6	1	49.7			LIVINGSTON IL	-9				
1	56.2	4	293.6	2	34.1	LOGAN IL	422	-88	-39	20	32
1	133.9	1	196.5			MC DONOUGH IL	47				
1	141.3			3	246.5	MC HENRY IL			74	90	94
10	204.8	8	127.9	14	157.7	MC LEAN IL	-38	23	-23	31	80
43*	248.2	33	136.2	65	154.4	MACON IL	-45	13	-38	21	79
1	344.1	1	68.8	1	53.0	MACOUPIN IL	-80	-23	-85	01	35
71	164.0	65	118.5	82	115.9	MADISON IL	-28	-2	-29	27	57
17*	243.8	16	221.0	13	133.4	MARION IL	-9	-40	-45	16	68
13	132.7	12	152.4	17	213.3	MASSAC IL	15	40	61	88	92
1	188.7	1*	74.3	1	589.6	MENARD IL	-61	693	212	97	99
3	422.9	2	212.0	3	148.8	MONTGOMERY IL	-50	-30	-65	07	77
12	205.9	9	133.5	11	155.6	MORGAN IL	-35	17	-24	30	80
				1	78.1	OGLE IL					41
38	155.2	72	174.5	95	159.9	PEORIA IL	12	-8	3	59	81
10*	271.5	4	111.2	2	44.6	PERRY IL	-59	-60	-84	01	33
				1	546.6	PIKE IL					99
				1	112.4	POPE IL					55
35	134.4	30	129.9	37	163.3	PULASKI IL	-3	26	21	74	82
		1*	546.4			PUTNAM IL					
6	164.9	9	276.0	6	162.3	RANDOLPH IL	67	-41	-2	54	82
15	158.1	27	146.9	33	101.4	ROCK ISLAND IL	-7	-31	-36	22	49
211	129.3	324	140.1	439*	150.8	ST CLAIR IL	8	8	17	72	78
5	90.9	13	241.5	7	153.7	SALINE IL	166	-36	69	89	79
58*	212.8	64	183.4	90*	197.7	SANGAMON IL	-14	8	-7	49	90
		1*	188.7			SHELBY IL					
5	140.4	17*	280.2	9	93.6	STEPHENSON IL	100	-67	-33	23	46
1	196.5			1	26.8	TAZEWELL IL			-86	01	31
				2	302.5	UNION IL					96
34	177.5	47*	210.2	43	151.0	VERMILION IL	18	-28	-15	41	78
				1*	273.2	WABASH IL					95
6	339.9	2	80.7	4	174.8	WARREN IL	-76	117	-49	15	86
				1	163.6	WASHINGTON IL					83
				1	136.6	WAYNE IL					71
1	273.2	3*	791.6			WHITE IL	190				
		1	78.1	2	76.3	WHITESIDE IL		-2			40
26	157.8	56	181.4	85	120.9	WILL IL	15	-33	-23	31	60
5	78.4	12	207.7	6	95.4	WILLIAMSON IL	165	-54	22	75	47
28	155.3	40	114.2	77	133.6	WINNEBAGO IL	-26	17	-14	41	69
1484*	174.3	1924*	156.8	2764*	164.4	INDIANA IN	-10	5	-6		
				1	267.9	ADAMS IN					95
50*	201.0	58	131.4	120*	164.1	ALLEN IN	-35	25	-18	37	83

* SIGNIFICANTLY MORE DEATHS THAN EXPECTED

NON-WHITE FEMALE — ALL CANCERS COMBINED — ICD 140-208

1950-1959		1960-1969		1970-1979		COUNTY AND STATE		PERCENT CHANGE			PERCENTILE	
DEATHS	RATE	DEATHS	RATE	DEATHS	RATE			60S/50S	70S/60S	70S/50S	%CHANGE 70S/50S	RATES 1970S
1	141.3	4	192.2	4	116.5	BARTHOLOMEW	IN	36	-39	-18	38	57
1	94.4			1	62.9	BOONE	IN			-33	23	37
		1*	281.2			CARROLL	IN					
1	89.3	2	90.5	4	171.0	CASS	IN	1	89	92	92	85
18	172.3	20	151.8	27	143.0	CLARK	IN	-12	-6	-17	38	74
2	187.8	4	279.0	1	47.2	CLAY	IN	49	-83	-75	04	33
1	1023.0	1	37.2	1	109.3	CLINTON	IN	-96	194	-89	00	53
2	152.2	1	94.4			DAVIESS	IN	-38				
		4*	463.5	2	200.9	DEARBORN	IN		-57			91
				1	112.4	DECATUR	IN					55
1	89.3			1	74.3	DE KALB	IN			-17	38	39
39	198.9	39	136.1	45	129.3	DELAWARE	IN	-32	-5	-35	23	66
6	108.7	8	72.1	15	94.7	ELKHART	IN	-34	31	-13	42	46
9*	300.4	6	200.8	8	240.1	FAYETTE	IN	-33	20	-20	34	93
16	189.1	17	164.4	26*	232.8	FLOYD	IN	-13	42	23	76	93
				1	21.0	FULTON	IN					30
10*	285.6	6	162.2	6	121.3	GIBSON	IN	-43	-25	-58	11	61
18	192.2	24	178.8	21	107.2	GRANT	IN	-7	-40	-44	17	52
3	135.4	2	145.7	3	77.4	HAMILTON	IN	8	-47	-43	18	40
1	188.7	1	74.3			HANCOCK	IN	-61				
1	267.9	1	469.2			HARRISON	IN	75				
		3*	643.0	1	344.1	HENDRICKS	IN		-46			97
3	148.7	5	181.3	6	187.2	HENRY	IN	22	3	26	77	89
11	117.4	21	173.7	20	122.8	HOWARD	IN	48	-29	5	61	62
		1	589.6			HUNTINGTON	IN					
		3	373.0	1	37.2	JACKSON	IN		-90			32
		1*	424.0			JASPER	IN					
		1	94.4	1	37.2	JAY	IN		-61			32
4	210.4			4	119.0	JEFFERSON	IN			-43	18	59
2	255.7	1	68.3	1	56.2	JENNINGS	IN	-73	-18	-78	03	35
3	157.4	5	221.7	5	158.1	JOHNSON	IN	41	-29	0	56	81
4	378.9			6*	388.4	KNOX	IN			3	58	97
1	112.4	1	56.2	4	403.0	KOSCIUSKO	IN	-50	617	258	97	98
				1*	112.4	LAGRANGE	IN					55
349	160.2	538*	157.7	811*	167.2	LAKE	IN	-2	6	4	61	84
12	152.5	19	107.3	38	161.6	LA PORTE	IN	-30	51	6	63	82
2	348.7			2	558.0	LAWRENCE	IN			60	88	99
16	163.5	20	104.0	42	154.7	MADISON	IN	-36	49	-5	50	79
624*	172.0	791*	156.9	1157*	173.9	MARION	IN	-9	11	1	57	86
		3*	1214.3	1	38.3	MARSHALL	IN		-97			32
2	110.1	2	69.4	5	179.3	MIAMI	IN	-37	158	63	88	87
2	90.4	2	54.9	9	144.8	MONROE	IN	-39	164	60	88	75
1	106.0	4	263.9	2	275.3	MONTGOMERY	IN	149	4	160	95	95
		1	424.0			MORGAN	IN					
1*	267.9					NOBLE	IN					
2*	354.2					OHIO	IN					
2	127.8	1	68.8	1	62.9	ORANGE	IN	-46	-9	-51	13	37
2*	456.5					OWEN	IN					
1	112.4			2*	253.8	PARKE	IN			126	94	94
		1	182.1	1	23.7	PORTER	IN		-87			31
7*	419.4	2	267.9	6*	333.7	POSEY	IN	-36	25	-20	34	97
1*	188.7					PULASKI	IN					
1	99.4	1	86.0	3	157.7	PUTNAM	IN	-13	83	59	88	80
		1	94.4	1	112.4	RANDOLPH	IN		19			55
4	269.9	3	166.7	1	67.0	RUSH	IN	-38	-60	-75	03	37
76*	217.5	86	149.6	135	155.0	ST JOSEPH	IN	-31	4	-29	27	80
				1	74.3	SCOTT	IN					39
6	272.3	6	315.9	4	169.3	SHELBY	IN	16	-46	-38	20	84
2	126.9	1	38.3	2	185.4	SPENCER	IN	-70	384	46	85	88
		1	112.4			SULLIVAN	IN					
2*	611.9					SWITZERLAND	IN					
9*	281.2	7	196.9	5	86.7	TIPPECANOE	IN	-30	-56	-69	05	44
				1	663.2	UNION	IN					99
86*	187.2	90	164.5	108*	166.7	VANDERBURGH	IN	-12	1	-11	44	84
38	153.8	63*	231.3	43	129.8	VIGO	IN	50	-44	-16	40	66
2*	603.2					WABASH	IN					
2	205.3	3	193.8	2	163.6	WARRICK	IN	-6	-16	-20	34	83
24	162.0	37	209.6	42*	194.6	WAYNE	IN	29	-7	20	74	90
				1	112.4	WELLS	IN					55
186*	181.0	192	144.8	245	131.8	IOWA	IA	-20	-9	-27		
1*	267.9	1*	74.3			ADAIR	IA	-72				
		1*	112.4			ALLAMAKEE	IA					
2	129.9	1	140.3			APPANOOSE	IA	8				
14	183.0	23	148.4	31	121.1	BLACK HAWK	IA	-19	-18	-34	23	60
1*	344.1	1*	267.9			BUCHANAN	IA	-22				
				1	188.7	CALHOUN	IA					89
5	379.0	4	234.4			CERRO GORDO	IA	-38				
2	125.5	5	313.7	2	73.1	CLINTON	IA	150	-77	-42	18	39
				1	163.6	DALLAS	IA					83
1*	424.0			1	112.4	DAVIS	IA			-73	04	55
4	145.5	7	267.3	4	101.4	DES MOINES	IA	84	-62	-30	26	49
5*	562.2	5*	836.8	2	234.3	DUBUQUE	IA	49	-72	-58	10	93
				1*	490.8	FLOYD	IA					98
		1*	112.4	1*	490.8	HARRISON	IA		337			98
1	47.2	2	669.4	2	278.0	HENRY	IA	1319	-58	489	99	95
1*	424.0					HUMBOLDT	IA					
1	490.8			1	53.0	JASPER	IA			-89	00	35
				1	56.2	JEFFERSON	IA					35
1	344.1			3	211.9	JOHNSON	IA			-38	20	91
11	187.6	8	128.1	9	119.4	LEE	IA	-32	-7	-36	22	59
8	159.7	7	110.3	9	84.1	LINN	IA	-31	-24	-47	15	42
1	70.7			2	381.2	MAHASKA	IA			439	98	97
1	81.8					MARION	IA					
2	141.3			3	114.4	MARSHALL	IA			-19	36	56
				1*	562.5	MONONA	IA					99
1	94.4	2	206.8	2*	201.7	MONROE	IA	119	-2	114	93	91
		1	589.6			MONTGOMERY	IA					
2	410.5	1	86.0			MUSCATINE	IA	-79				
1	136.6	1	114.7	1	112.4	PAGE	IA	-16	-2	-18	38	55

* SIGNIFICANTLY MORE DEATHS THAN EXPECTED

NON-WHITE FEMALE — ALL CANCERS COMBINED — ICD 140-208

1950-1959 DEATHS	1950-1959 RATE	1960-1969 DEATHS	1960-1969 RATE	1970-1979 DEATHS	1970-1979 RATE	COUNTY AND STATE	PERCENT CHANGE 60S/50S	PERCENT CHANGE 70S/60S	PERCENT CHANGE 70S/50S	PERCENTILE %CHANGE 70S/50S	PERCENTILE RATES 1970S
		1*	546.6			PLYMOUTH IA					
90*	211.1	80	155.9	113*	166.1	POLK IA	-26	7	-21	33	83
6	176.1	7	161.2	6	113.6	POTTAWATTAMIE IA	-8	-30	-35	22	56
4	98.9	20	196.0	20	136.7	SCOTT IA	98	-30	38	82	71
		1	98.2	3	140.8	STORY IA		43			73
3	181.1			5	243.1	TAMA IA			34	81	93
2	82.7	4	133.7	5	179.8	WAPELLO IA	62	34	118	94	87
2	255.0	1	61.3			WEBSTER IA	-76				
13	229.5	7	86.2	15	177.8	WOODBURY IA	-62	106	-23	32	87
582	147.6	701	146.3	814	130.3	KANSAS KS	-1	-11	-12		
4	222.0	1	60.6	2	82.4	ALLEN KS	-73	36	-63	08	42
		1	37.2	1	112.4	ANDERSON KS		202			55
7	86.7	13	164.2	11	138.3	ATCHISON KS	89	-16	59	88	72
6	245.1	6	200.4	4	129.0	BARTON KS	-18	-36	-47	15	65
7	174.1	10	188.3	3	65.9	BOURBON KS	8	-65	-62	08	37
6	186.8	2	69.0	2	49.3	BROWN KS	-63	-29	-74	04	34
		2	114.2	2	71.6	BUTLER KS		-37			38
		1	188.7			CHASE KS					
1	182.1			1	56.2	CHAUTAUQUA KS			-69	05	35
1	68.3	8	231.9	4	119.4	CHEROKEE KS	239	-49	75	90	59
1*	424.0	1*	546.4			CLAY KS	29				
5	93.8	12	187.8	8	83.6	COWLEY KS	100	-55	-11	45	42
7	240.7	4	132.3	6	129.6	CRAWFORD KS	-45	-2	-46	16	66
1	188.7	1	62.9	1	53.0	DICKINSON KS	-67	-16	-72	05	35
2	71.0	3	145.7	2	50.4	DONIPHAN KS	105	-65	-29	27	34
19	155.8	23	191.3	19	126.3	DOUGLAS KS	23	-34	-19	36	64
				1	74.3	EDWARDS KS					39
1	172.0					ELLSWORTH KS					
1	37.2	3	361.6	4	260.8	FINNEY KS	873	-28	602	99	95
4*	517.7	3	206.4	1	31.5	FORD KS	-60	-85	-94	00	31
4	157.0	6	267.5	6	232.8	FRANKLIN KS	70	-13	48	85	93
13	210.5	14	149.0	19	159.8	GEARY KS	-29	7	-24	30	81
1	49.7			2	71.1	GRAHAM KS			43	84	38
				1*	212.0	GRAY KS					92
1*	596.2					HARPER KS					
7	255.9	5	159.7	7	171.2	HARVEY KS	-38	7	-33	23	85
3	167.9	5	161.3	2	49.6	JACKSON KS	-4	-69	-70	05	34
1	99.4	2	362.5			JEFFERSON KS	265				
9	196.4	11	170.7	6	44.7	JOHNSON KS	-13	-74	-77	03	33
		1*	344.1			KIOWA KS					
8	81.4	12	150.6	11	89.0	LABETTE KS	85	-41	9	66	44
28	177.4	22	134.2	44*	242.3	LEAVENWORTH KS	-24	80	37	82	93
		2	294.6			LINN KS					
1	188.7					LOGAN KS					
5	144.0	7	287.3	3	85.1	LYON KS	100	-70	-41	19	43
				1	188.7	MC PHERSON KS					89
1	188.7					MARSHALL KS					
6	122.1	2	56.3	4	79.8	MIAMI KS	-54	42	-35	23	41
34	188.9	37	183.2	29	118.1	MONTGOMERY KS	-3	-36	-37	21	58
1	94.4					MORRIS KS					
		1	141.3			NEMAHA KS					
3	142.2	4	208.8	4	145.6	NEOSHO KS	47	-30	2	58	75
		1*	140.6			NORTON KS					
		1	546.4			OSAGE KS					
		1	74.3	2	178.6	PAWNEE KS		140			87
		1	112.4	1	196.5	POTTAWATOMIE KS		75			90
3	375.5	2	81.3	2	63.9	PRATT KS	-78	-21	-83	02	37
4	63.6	11	217.9	17	232.7	RENO KS	243	7	266	97	93
3	94.2	3	69.1	7	137.8	RILEY KS	-27	99	46	85	72
1	424.0					ROOKS KS					
1	344.1	1	188.7			RUSSELL KS	-45				
3	65.9	13	215.8	7	76.8	SALINE KS	228	-64	17	71	40
68	138.2	98	115.8	158	128.1	SEDGWICK KS	-16	11	-7	48	65
1	267.9	1	42.6	1	40.9	SEWARD KS	-84	-4	-85	01	33
52	115.3	73	138.0	101	146.4	SHAWNEE KS	20	6	27	78	75
				1	74.3	SHERMAN KS					39
2*	624.7	1	212.0			STEVENS KS	-66				
1	109.3	1	53.6			SUMNER KS	-51				
1	81.8			2	246.1	WABAUNSEE KS			201	96	94
1	122.7					WILSON KS					
242	163.7	268	153.7	304	144.5	WYANDOTTE KS	-6	-6	-12	44	74
1674*	156.6	1895*	161.0	2280*	173.4	KENTUCKY KY	3	8	11		
4	83.5	3	85.2	4	115.5	ADAIR KY	2	35	38	82	57
3	142.9	2	191.2	5*	355.2	ALLEN KY	34	86	149	95	97
4	180.5	3	171.4	5	228.4	ANDERSON KY	-5	33	27	78	93
4	174.0	3	98.5	5	172.8	BALLARD KY	-43	75	-1	55	86
19	171.3	17	154.1	17	139.5	BARREN KY	-10	-10	-19	36	72
4	213.6	2	114.7	4	238.9	BATH KY	-46	108	12	68	93
13	161.5	16	224.2	17*	233.0	BELL KY	39	4	44	84	93
		1	344.1	1	84.2	BOONE KY		-76			42
29	182.9	29	180.8	33*	179.2	BOURBON KY	-1	-1	-2	53	87
8	141.9	8	132.8	15*	239.4	BOYD KY	-6	80	69	89	93
14	96.2	25	155.6	26	150.4	BOYLE KY	62	-3	56	87	77
2	141.0	2	230.6	1	188.7	BRACKEN KY	64	-18	34	81	89
		1	546.4			BREATHITT KY					
4	153.8	9*	310.6	3	106.5	BRECKINRIDGE KY	102	-66	-31	25	52
3	382.4					BULLITT KY					
				1	133.9	BUTLER KY					69
11	166.0	10	148.1	11	117.0	CALDWELL KY	-11	-21	-30	27	58
7	232.9	6	131.6	10	258.1	CALLOWAY KY	-43	96	11	67	95
8	158.8	5	102.8	10	249.5	CAMPBELL KY	-35	143	57	87	94
2	928.3	2	353.9	1	596.2	CARLISLE KY	-62	68	-36	22	99
		2	115.2	1	54.5	CARROLL KY		-53			35
				1	149.1	CARTER KY					77
1*	188.7	1	182.1			CASEY KY	-3				
56	99.2	73	132.6	106*	187.2	CHRISTIAN KY	34	41	89	91	89
18	159.3	20	147.3	20	143.6	CLARK KY	-8	-3	-10	46	74

* SIGNIFICANTLY MORE DEATHS THAN EXPECTED

NON-WHITE FEMALE — ALL CANCERS COMBINED — ICD 140-208

1950-1959 DEATHS	1950-1959 RATE	1960-1969 DEATHS	1960-1969 RATE	1970-1979 DEATHS	1970-1979 RATE	COUNTY AND STATE		PERCENT CHANGE 60S/50S	70S/60S	70S/50S	PERCENTILE %CHANGE 70S/50S	PERCENTILE RATES 1970S
2	226.9			2	130.7	CLAY	KY			-42	18	67
1	74.3					CLINTON	KY					
3	262.9			2	63.2	CRITTENDEN	KY			-76	03	37
1	37.8	2	86.0	6	185.4	CUMBERLAND	KY	128	116	391	98	88
34	192.9	29	153.7	27	130.8	DAVIESS	KY	-20	-15	-32	25	67
2*	423.1					EDMONSON	KY					
1	198.7					ESTILL	KY					
157	157.4	193*	173.0	219*	170.4	FAYETTE	KY	10	-2	8	65	85
2	84.3	5	206.4	2	113.3	FLEMING	KY	145	-45	34	81	56
2	76.3	2	166.0	4	163.9	FLOYD	KY	117	-1	115	93	83
18	157.6	18	138.2	28*	200.9	FRANKLIN	KY	-12	45	27	78	91
13	139.8	10	110.1	16	176.9	FULTON	KY	-21	61	27	78	87
				1	74.3	GALLATIN	KY					39
5	108.4	13*	204.7	10	286.5	GARRARD	KY	89	40	164	96	96
1	70.7			2	324.1	GRANT	KY			359	98	97
14	153.3	20	199.4	19	187.1	GRAVES	KY	30	-6	22	75	89
				1	245.4	GRAYSON	KY					94
6	257.8	1	49.2	1	42.0	GREEN	KY	-81	-14	-84	01	33
		1	37.2	1	74.3	GREENUP	KY		100			39
1	172.0	1	56.2			HANCOCK	KY	-67				
8	102.0	19*	233.0	14	134.1	HARDIN	KY	128	-42	32	80	69
23	130.2	26	150.9	21	172.1	HARLAN	KY	16	14	32	80	85
6	154.8	2	41.6	5	155.0	HARRISON	KY	-73	273	0	56	80
7	129.4	11	218.2	12	275.4	HART	KY	69	26	113	93	95
19	94.2	34	179.1	36	184.3	HENDERSON	KY	90	3	96	92	88
7	149.5	6	158.2	10	215.0	HENRY	KY	6	36	44	84	92
4	134.5	6	165.6	2	68.6	HICKMAN	KY	23	-59	-49	14	38
49*	221.8	33	147.2	44	171.1	HOPKINS	KY	-34	16	-23	32	85
616*	171.2	778*	179.2	975*	182.8	JEFFERSON	KY	5	2	7	64	88
11	167.1	10	149.8	15	222.8	JESSAMINE	KY	-10	49	33	80	92
43	187.4	41	181.1	45	188.0	KENTON	KY	-3	4	0	56	89
				1	344.1	KNOTT	KY					97
4	151.9	3	56.9	5	179.9	KNOX	KY	-63	216	18	73	87
4	133.8	4	176.8	5	170.9	LARUE	KY	32	-3	28	78	85
1	62.9			2	131.7	LAUREL	KY			109	93	67
				2	246.4	LAWRENCE	KY					94
		1*	596.2			LESLIE	KY					
5	85.6	2	34.5	8	180.5	LETCHER	KY	-60	423	111	93	87
1*	490.8					LEWIS	KY					
10	176.5	2	24.0	11	182.5	LINCOLN	KY	-86	661	3	60	88
1	267.9			2*	255.7	LIVINGSTON	KY			-5	51	94
12	90.4	20	151.4	16	136.3	LOGAN	KY	68	-10	51	86	71
3	174.9	1	112.5	2	153.2	LYON	KY	-36	36	-12	43	79
54	147.7	60	160.1	59	157.3	MC CRACKEN	KY	8	-2	6	63	80
		1*	490.8			MC CREARY	KY					
3*	1391.2	1	188.7			MC LEAN	KY	-86				
28	191.5	29	188.4	29	167.8	MADISON	KY	-2	-11	-12	43	84
3	49.8	6	102.5	8	117.5	MARION	KY	106	15	136	94	58
12	131.5	11	133.1	16	172.2	MASON	KY	1	29	31	80	85
1	141.3	2	122.5	3	140.3	MEADE	KY	-13	14	-1	55	72
5	97.5	10	177.4	14*	228.6	MERCER	KY	82	29	135	94	93
2	160.7	2	202.2	1	49.7	METCALFE	KY	26	-75	-69	06	34
2	120.6	1	24.8	4	240.2	MONROE	KY	-79	869	99	92	93
22*	257.0	9	116.9	12	142.9	MONTGOMERY	KY	-55	22	-44	17	74
19	199.7	10	120.9	13	173.5	MUHLENBERG	KY	-39	44	-13	42	86
8	102.8	14	156.8	14	162.8	NELSON	KY	53	4	58	87	82
3	177.6	7*	508.9	2	198.4	NICHOLAS	KY	187	-61	12	68	90
3	84.7	2	61.0	8*	481.2	OHIO	KY	-28	689	468	99	98
6	133.6	4	116.8	3	79.9	OLDHAM	KY	-13	-32	-40	19	41
3	287.3			2	89.4	OWEN	KY			-69	06	45
1	490.8			1	212.0	PENDLETON	KY			-57	11	92
13*	349.2	3	74.1	2	56.2	PERRY	KY	-79	-24	-84	01	35
6	296.2	11*	440.6	4	149.0	PIKE	KY	49	-66	-50	14	77
1	112.4					POWELL	KY					
3	73.9	16*	403.4	10	202.8	PULASKI	KY	446	-50	174	96	91
				1	37.2	ROWAN	KY					32
23	202.2	13	110.1	19	211.3	SCOTT	KY	-46	92	4	61	91
8	76.8	18	162.5	19	163.5	SHELBY	KY	112	1	113	93	82
7	94.1	11	133.4	13	147.5	SIMPSON	KY	42	11	57	87	76
2	101.8	2	144.1	1	122.7	SPENCER	KY	41	-15	20	74	62
3	120.5	8	180.5	9	161.6	TAYLOR	KY	50	-10	34	81	82
4	37.7	11	111.3	9	81.1	TODD	KY	196	-27	115	93	42
10	177.0	10	153.5	10	131.9	TRIGG	KY	-13	-14	-25	29	68
12	179.3	10	177.5	20*	255.7	UNION	KY	-1	44	43	84	94
39	169.6	41	169.4	56*	212.8	WARREN	KY	0	26	25	77	92
8	197.5	6	137.6	13*	281.7	WASHINGTON	KY	-30	105	43	84	96
1	60.7	1	44.6	2	69.5	WAYNE	KY	-26	56	15	70	38
17	162.7	14	162.1	12	131.1	WEBSTER	KY	0	-19	-19	35	67
2	217.5	2	208.7	2	175.3	WHITLEY	KY	-4	-16	-19	35	86
29*	288.4	15	145.9	18	216.6	WOODFORD	KY	-49	48	-25	30	92
5634	146.1	6830*	149.0	7836*	145.7	LOUISIANA	LA	2	-2	0		
52	144.7	56	132.9	74	156.3	ACADIA	LA	-8	18	8	65	80
25	130.4	45	204.8	39	152.6	ALLEN	LA	57	-25	17	72	78
50	136.6	66	164.2	74	167.9	ASCENSION	LA	20	2	23	75	84
50	176.9	44	149.2	54*	188.8	ASSUMPTION	LA	-16	27	7	64	89
46	113.9	66	151.6	76	150.8	AVOYELLES	LA	33	-1	32	80	78
15	98.5	27	147.1	28	135.3	BEAUREGARD	LA	49	-8	37	82	70
55	150.5	53	136.2	54	119.8	BIENVILLE	LA	-9	-12	-20	34	60
70	122.3	77	118.5	77	107.2	BOSSIER	LA	-3	-10	-12	43	52
468	143.0	517	129.3	674	142.9	CADDO	LA	-10	11	0	56	74
119	149.4	155	140.2	209	150.2	CALCASIEU	LA	-6	7	1	56	77
18	158.0	18	145.5	12	94.3	CALDWELL	LA	-8	-35	-40	19	46
1	141.3	1	94.4	3	101.5	CAMERON	LA	-33	8	-28	27	49
28	142.8	26	122.2	23	106.3	CATAHOULA	LA	-14	-13	-26	29	51
59	118.5	70	136.8	66	112.7	CLAIBORNE	LA	15	-18	-5	50	55
48	108.4	72	155.1	66	131.3	CONCORDIA	LA	43	-15	21	74	67
76	121.0	100	134.9	100	131.5	DE SOTO	LA	11	-3	9	65	67
306	128.5	421	136.6	535	135.4	EAST BATON ROUGE	LA	6	-1	5	63	70
62	133.5	70	140.3	72	147.5	EAST CARROLL	LA	5	5	10	67	76

* SIGNIFICANTLY MORE DEATHS THAN EXPECTED

1950-1959 DEATHS	1950-1959 RATE	1960-1969 DEATHS	1960-1969 RATE	1970-1979 DEATHS	1970-1979 RATE	COUNTY AND STATE	PERCENT CHANGE 60S/50S	PERCENT CHANGE 70S/60S	PERCENT CHANGE 70S/50S	PERCENTILE %CHANGE 70S/50S	PERCENTILE RATES 1970S
41	78.3	50	94.9	45	95.3	EAST FELICIANA LA	21	0	22	75	47
23	118.8	35	150.9	45	154.0	EVANGELINE LA	27	2	30	79	79
54	138.6	46	112.3	41	88.1	FRANKLIN LA	-19	-22	-36	22	44
20	122.3	19	117.0	26	145.7	GRANT LA	-4	24	19	73	75
78	156.7	118*	185.2	142*	194.7	IBERIA LA	18	5	24	76	90
82	124.2	74	110.0	116	161.4	IBERVILLE LA	-11	47	30	79	82
17	100.0	28	121.4	23	77.2	JACKSON LA	21	-36	-23	32	40
106	140.2	185	154.2	233	134.0	JEFFERSON LA	10	-13	-4	51	69
28	152.4	35	153.6	47	185.3	JEFFERSON DAVIS LA	1	21	22	75	88
85	158.9	105	142.3	160*	167.4	LAFAYETTE LA	-10	18	5	62	84
38	190.7	48	201.1	47	164.7	LAFOURCHE LA	5	-18	-14	41	83
8	107.6	11	136.9	17	184.3	LA SALLE LA	27	35	71	89	88
56	134.1	55	113.5	84	131.5	LINCOLN LA	-15	16	-2	54	67
11	106.3	20	144.8	22	148.6	LIVINGSTON LA	36	3	40	83	77
52	103.5	66	124.6	72	119.1	MADISON LA	20	-4	15	70	59
74	117.1	105	144.6	89	112.5	MOREHOUSE LA	23	-22	-4	52	55
78	109.9	84	115.5	88	105.8	NATCHITOCHES LA	5	-8	-4	52	51
1676*	188.2	1910*	173.6	2253*	172.1	ORLEANS LA	-8	-1	-9	47	85
172	140.4	239	154.6	210	115.4	OUACHITA LA	10	-25	-18	37	57
25	157.7	23	135.2	33	171.3	PLAQUEMINES LA	-14	27	9	65	85
75	157.5	74	147.5	68	128.7	POINTE COUPEE LA	-6	-13	-18	37	65
185	130.4	249	156.3	236	130.7	RAPIDES LA	20	-16	0	56	67
28	128.6	24	114.5	18	77.3	RED RIVER LA	-11	-33	-40	19	40
58	142.5	58	125.1	67	129.8	RICHLAND LA	-12	4	-9	47	66
22	121.6	33	169.2	28	106.2	SABINE LA	39	-37	-13	43	51
10	138.1	15	203.0	14	115.9	ST BERNARD LA	47	-43	-16	39	57
28	128.6	38	156.6	31	102.5	ST CHARLES LA	22	-35	-20	34	49
10	52.4	12	56.7	19	74.8	ST HELENA LA	8	32	43	84	40
41	116.6	78*	215.2	57	143.6	ST JAMES LA	85	-33	23	76	74
47	153.3	58	157.2	74*	175.2	ST JOHN THE BAPT LA	3	11	14	70	86
162	148.3	190	150.9	210	147.4	ST LANDRY LA	2	-2	-1	55	76
42	134.8	53	144.6	65	145.6	ST MARTIN LA	7	1	8	65	75
90	148.1	114	172.3	159*	217.9	ST MARY LA	16	26	47	85	92
44	127.5	63	148.8	80	150.5	ST TAMMANY LA	17	1	18	72	77
94	141.5	98	120.5	119	118.8	TANGIPAHOA LA	-15	-1	-16	39	59
54	130.8	58	150.0	50	122.5	TENSAS LA	15	-18	-6	49	61
57	147.5	64	143.9	77	145.8	TERREBONNE LA	-2	1	-1	55	75
26	106.3	39	138.3	45	122.9	UNION LA	30	-11	16	71	62
25	150.2	26	121.2	43	163.6	VERMILION LA	-19	35	9	66	82
16	156.0	19	153.4	18	126.8	VERNON LA	-2	-17	-19	36	64
67	133.8	73	117.2	76	108.6	WASHINGTON LA	-12	-7	-19	36	53
78	155.3	111	168.7	110	136.6	WEBSTER LA	9	-19	-12	43	71
36	125.0	53	175.7	56	167.5	WEST BATON ROUGE LA	41	-5	34	81	84
13	93.8	21	148.8	19	141.6	WEST CARROLL LA	59	-5	51	86	73
23	104.9	26	118.7	29	164.1	WEST FELICIANA LA	13	38	57	87	83
31	153.0	43	170.3	39	139.6	WINN LA	11	-18	-9	47	72
30*	262.9	18	98.6	33	109.8	MAINE ME	-62	11	-58		
				2	71.4	ANDROSCOGGIN ME					38
1	11.2	3	20.1	3	30.1	AROOSTOOK ME	78	50	167	96	31
8*	393.2	6	185.9	7	113.6	CUMBERLAND ME	-53	-39	-71	05	56
4*	932.7			2	108.9	KENNEBEC ME			-88	00	53
1*	112.4					LINCOLN ME					
				1	490.8	OXFORD ME					98
8	277.5	4	92.5	9	199.5	PENOBSCOT ME	-67	116	-28	28	90
1*	273.2	1*	112.4			PISCATAQUIS ME	-59				
				1	273.2	SOMERSET ME					95
				1	546.4	WALDO ME					98
6*	342.4	1	267.9	6	328.9	WASHINGTON ME	-22	23	-4	52	97
1	172.0	3	245.7	1	94.4	YORK ME	43	-62	-45	17	46
2582*	160.0	3492*	156.8	5227*	160.1	MARYLAND MD	-2	2	0		
12	234.0	8	135.7	14	195.8	ALLEGANY MD	-42	44	-16	39	90
120	132.9	158	135.8	207	139.7	ANNE ARUNDEL MD	2	3	5	62	72
96	157.7	139*	184.4	213*	161.3	BALTIMORE MD	17	-13	2	58	82
16	95.2	35	166.1	45	173.1	CALVERT MD	75	4	82	91	86
25	155.7	33	185.1	35	169.4	CAROLINE MD	19	-8	9	65	84
13	181.8	11	110.6	12	78.5	CARROLL MD	-39	-29	-57	11	41
11	102.8	15	147.0	23	173.3	CECIL MD	43	18	69	89	86
27	113.0	38	123.5	62	133.2	CHARLES MD	9	8	18	72	68
33	84.8	48	111.2	90*	186.9	DORCHESTER MD	31	68	120	94	88
33	171.9	34	150.7	49	184.2	FREDERICK MD	-12	22	7	64	88
33	160.0	53	179.8	53	129.0	HARFORD MD	12	-28	-19	35	65
20	146.7	12	93.8	45	144.3	HOWARD MD	-36	54	-2	54	74
25	153.4	28	160.3	28	135.2	KENT MD	4	-16	-12	43	70
66	163.2	87	128.0	211	136.3	MONTGOMERY MD	-22	6	-16	39	71
109	141.0	171	121.4	489	129.2	PRINCE GEORGE'S MD	-14	6	-8	48	66
25	152.1	31	151.1	30	120.6	QUEEN ANNE'S MD	-1	-20	-21	33	60
30	155.5	29	127.9	43	136.6	ST MARY'S MD	-18	7	-12	43	71
41	132.6	55	157.0	70*	189.4	SOMERSET MD	18	21	43	84	89
42	168.8	38	117.1	54	145.7	TALBOT MD	-31	24	-14	41	75
24*	237.0	20	185.3	24	191.2	WASHINGTON MD	-22	3	-19	35	89
56	163.8	68	142.3	119*	181.8	WICOMICO MD	-13	28	11	68	87
39	153.6	40	130.4	62	160.0	WORCESTER MD	-15	23	4	61	81
1686*	169.3	2341*	168.2	3249*	173.0	BALTIMORE CITY MD	-1	3	2	58	86
642*	164.6	833	138.1	1136	114.6	MASSACHUSETTS MA	-16	-17	-30		
13	196.8	16	165.8	15	80.7	BARNSTABLE MA	-16	-51	-59	10	41
10	153.3	13	146.2	21	181.9	BERKSHIRE MA	-5	24	19	73	87
46*	248.3	40	156.6	39	80.1	BRISTOL MA	-37	-49	-68	06	41
3	368.1	7	286.9	7	122.9	DUKES MA	-22	-57	-67	07	62
15	131.7	14	71.1	32	84.9	ESSEX MA	-46	19	-36	22	43
		1	188.7			FRANKLIN MA					
54	155.2	91	154.5	131	123.2	HAMPDEN MA	0	-20	-21	34	62
2	141.3	2	68.1	4	54.1	HAMPSHIRE MA	-52	-21	-62	09	35
92	158.0	103	121.6	149	106.2	MIDDLESEX MA	-23	-13	-33	24	51
2	274.7	1	196.5	4*	459.7	NANTUCKET MA	-28	134	67	89	98
15	231.1	17	107.8	30	82.7	NORFOLK MA	-53	-23	-64	08	42

* SIGNIFICANTLY MORE DEATHS THAN EXPECTED

NON-WHITE FEMALE — ALL CANCERS COMBINED — ICD 140-208

1950-1959 DEATHS	1950-1959 RATE	1960-1969 DEATHS	1960-1969 RATE	1970-1979 DEATHS	1970-1979 RATE	COUNTY AND STATE		PERCENT CHANGE 60S/50S	70S/60S	70S/50S	PERCENTILE %CHANGE 70S/50S	RATES 1970S
18	116.0	31	169.4	42	107.7	PLYMOUTH	MA	46	-36	-7	49	52
354	167.9	477	143.1	626	124.7	SUFFOLK	MA	-15	-13	-26	29	63
18	139.2	20	97.5	36	106.0	WORCESTER	MA	-30	9	-24	31	51
3174*	166.3	4657*	149.7	6641*	143.7	MICHIGAN	MI	-10	-4	-14		
1*	188.7					ALCONA	MI					
12*	346.2	9	182.4	20*	233.9	ALLEGAN	MI	-47	28	-32	24	93
3*	439.9					ANTRIM	MI					
		1	172.0			ARENAC	MI					
2	171.2	1	59.6	4	179.8	BARAGA	MI	-65	202	5	62	87
1	89.3	4	151.1	6	116.9	BAY	MI	69	-23	31	80	58
2	384.0	1	114.7	1	89.3	BENZIE	MI	-70	-22	-77	03	44
54	154.2	78	149.3	118	153.8	BERRIEN	MI	-3	3	0	56	79
50	194.9	51	137.7	71	125.0	CALHOUN	MI	-29	-9	-36	22	63
31	188.7	23	107.3	39	142.8	CASS	MI	-43	33	-24	30	74
2	458.7	3	339.5	2	114.2	CHARLEVOIX	MI	-26	-66	-75	03	56
				1	122.7	CHEBOYGAN	MI					62
5	299.4	4	151.2	7	176.3	CHIPPEWA	MI	-49	17	-41	19	86
1*	596.2	1*	267.9			CLARE	MI	-55				
1	182.2					CLINTON	MI					
1	73.7	1	114.7	4	217.8	DELTA	MI	56	90	196	96	92
				1	245.4	DICKINSON	MI					94
		1	84.8	5	116.8	EATON	MI		38			58
8	267.6	4	117.4	5	186.6	EMMET	MI	-56	59	-30	26	88
103	155.1	168	135.2	263	128.6	GENESEE	MI	-13	-5	-17	38	65
1	267.9	1	212.0			GOGEBIC	MI	-21				
		1	86.0	2	155.7	GRAND TRAVERSE	MI		81			80
		1	188.7			GRATIOT	MI					
24	208.7	35	131.1	55	94.4	INGHAM	MI	-37	-28	-55	12	46
		1	62.9			IONIA	MI					
1	188.7					IOSCO	MI					
				1	596.2	IRON	MI					99
3	144.9			4	103.4	ISABELLA	MI			-29	27	49
23	162.9	27	116.5	56	160.2	JACKSON	MI	-28	38	-2	54	81
25	171.2	34	138.3	42	102.3	KALAMAZOO	MI	-19	-26	-40	19	49
				2*	894.3	KALKASKA	MI					99
57	178.7	89	141.4	139	144.8	KENT	MI	-21	2	-19	36	75
20	164.1	26	151.3	25	167.1	LAKE	MI	-8	10	2	58	84
2	310.2			2	156.1	LAPEER	MI			-50	14	80
3	315.0	3*	442.9	2	308.0	LEELANAU	MI	41	-30	-2	53	96
2	274.7	6*	551.8	1	19.6	LENAWEE	MI	101	-96	-93	00	30
2	177.2	1	59.6	2	75.9	LIVINGSTON	MI	-66	27	-57	11	40
				1	163.6	LUCE	MI					83
1	112.4			3	234.8	MACKINAC	MI			109	93	93
22	134.3	37	141.5	69	166.0	MACOMB	MI	5	17	24	76	83
1	65.5	1	84.2	2	139.4	MANISTEE	MI	29	66	113	93	72
1	11.3			1	35.1	MARQUETTE	MI			212	97	32
1	188.7	1	37.2	2	81.0	MASON	MI	-80	118	-57	11	42
1	62.9	2	173.7	1	86.0	MECOSTA	MI	176	-50	37	82	43
				1	344.1	MENOMINEE	MI					97
				1	15.7	MIDLAND	MI					30
1*	546.4					MISSAUKEE	MI					
5	112.6	9	139.2	8	74.1	MONROE	MI	24	-47	-34	23	39
				2	162.7	MONTCALM	MI					82
41	150.8	60	139.2	107	155.5	MUSKEGON	MI	-8	12	3	59	80
3	113.7	7	197.2	12	160.6	NEWAYGO	MI	74	-19	41	83	81
76	126.5	100	111.3	203	129.2	OAKLAND	MI	-12	16	2	58	65
2	325.3	3	318.3	1	112.4	OCEANA	MI	-2	-65	-65	07	55
1	188.7			1	18.2	OTTAWA	MI			-90	00	30
				1*	424.0	PRESQUE ISLE	MI					98
				1	596.2	ROSCOMMON	MI					99
70*	196.2	99	174.9	139	134.5	SAGINAW	MI	-11	-23	-31	25	69
11	128.7	16	121.6	16	94.7	ST CLAIR	MI	-5	-22	-26	28	46
8	278.0	7	121.2	16	155.9	ST JOSEPH	MI	-56	29	-44	17	80
				1	245.4	SCHOOLCRAFT	MI					94
1	273.2	1	212.0	1	66.2	TUSCOLA	MI	-22	-69	-76	03	37
20	144.3	29	123.7	55	152.4	VAN BUREN	MI	-14	23	6	63	78
44	132.8	69	135.0	120	157.2	WASHTENAW	MI	2	16	18	73	80
2423*	166.9	3641*	155.0	4996*	147.7	WAYNE	MI	-7	-5	-11	44	76
1*	490.8					WEXFORD	MI					
175	152.4	204	123.1	321	112.8	MINNESOTA	MN	-19	-8	-26		
2*	456.6	1	112.4	1	114.7	AITKIN	MN	-75	2	-75	04	56
1	196.5			4	115.2	ANOKA	MN			-41	19	57
12*	276.4	5	121.9	5	97.3	BECKER	MN	-56	-20	-65	07	47
9	160.7	8	101.5	14	153.0	BELTRAMI	MN	-37	51	-5	51	79
				2	362.2	BLUE EARTH	MN					97
4	274.7			2	84.6	CARLTON	MN			-69	05	42
14*	240.2	7	161.8	13	218.1	CASS	MN	-33	35	-9	47	92
		1	73.7	1	245.4	CLAY	MN		233			94
3	272.3			1	86.0	CLEARWATER	MN			-68	06	43
1	212.0			2	167.8	COOK	MN			-21	33	84
				1	106.0	CROW WING	MN					51
				3	55.8	DAKOTA	MN					35
1	589.6					FREEBORN	MN					
62	154.8	96	137.0	146	116.5	HENNEPIN	MN	-12	-15	-25	30	57
		1	163.6	1	188.7	HUBBARD	MN		15			89
4	141.8	1	13.0	2	101.1	ITASCA	MN	-91	679	-29	27	48
1	172.0			2	286.9	KOOCHICHING	MN			67	89	96
		1*	424.0			LAC QUI PARLE	MN					
2	76.9	6	218.3	2	67.5	MAHNOMEN	MN	184	-69	-12	43	37
		2	203.5	1	188.7	MILLE LACS	MN		-7			89
				2*	540.6	NICOLLET	MN					98
		1	424.0	1	188.7	NOBLES	MN		-55			89
				1*	273.3	NORMAN	MN					95
				3	225.9	OLMSTED	MN					92
		1	298.1	1	98.3	OTTER TAIL	MN		-67			48
1	172.0	1	267.9			PINE	MN	56				
				1	546.4	PIPESTONE	MN					98

* SIGNIFICANTLY MORE DEATHS THAN EXPECTED

NON-WHITE FEMALE — ALL CANCERS COMBINED — ICD 140-208

COUNTY AND STATE		1950-1959 DEATHS	1950-1959 RATE	1960-1969 DEATHS	1960-1969 RATE	1970-1979 DEATHS	1970-1979 RATE	PERCENT CHANGE 60S/50S	PERCENT CHANGE 70S/60S	PERCENT CHANGE 70S/50S	PERCENTILE %CHANGE 70S/50S	PERCENTILE RATES 1970S
POLK	MN	1*	94.7	1	344.1			263				
POPE	MN					1*	94.4					46
RAMSEY	MN	50	160.0	55	120.5	91	124.0	-25	3	-22	32	62
REDWOOD	MN	1	80.4	1	112.4	1	112.4	40	0	40	83	55
RICE	MN			1	74.3							
ROCK	MN			1*	490.8							
ROSEAU	MN			1	172.0							
ST LOUIS	MN	5	123.9	10	281.7	9	73.8	127	-74	-40	19	39
SHERBURNE	MN					3*	516.5					98
STEARNS	MN					1	17.1					30
STEELE	MN					1	147.4					76
WASHINGTON	MN			1	86.0	1	7.9		-91			30
WINONA	MN	1*	74.3									
WRIGHT	MN					1	74.3					39
YELLOW MEDICINE	MN			1	94.4							
MISSISSIPPI	MS	4919	133.2	5212	134.6	5297	123.8	1	-8	-7		
ADAMS	MS	141*	169.4	161*	180.0	168*	160.3	6	-11	-5	50	81
ALCORN	MS	32	173.4	21	109.8	20	97.7	-37	-11	-44	18	48
AMITE	MS	32	108.3	44	145.7	28	91.1	34	-37	-16	39	45
ATTALA	MS	44	115.8	56	141.1	54	118.0	22	-16	2	58	58
BENTON	MS	14	141.2	11	105.1	9	68.1	-26	-35	-52	13	37
BOLIVAR	MS	213	128.4	188	118.8	187	122.4	-8	3	-5	51	61
CALHOUN	MS	18	133.8	18	137.3	17	116.8	3	-15	-13	43	58
CARROLL	MS	31	108.6	30	128.8	22	107.7	19	-16	-1	55	52
CHICKASAW	MS	18	60.1	32	114.0	32	106.3	90	-7	77	90	51
CHOCTAW	MS	14	128.8	9	84.2	12	96.0	-35	14	-25	29	47
CLAIBORNE	MS	50	148.7	40	112.2	53	132.5	-25	18	-11	45	68
CLARKE	MS	40	141.3	51	172.0	35	115.3	22	-33	-18	37	57
CLAY	MS	58	148.4	57	128.8	63	127.8	-13	-1	-14	41	65
COAHOMA	MS	193	137.1	176	125.5	178	127.1	-8	1	-7	48	64
COPIAH	MS	75	122.9	86	141.4	81	115.7	15	-18	-6	50	57
COVINGTON	MS	24	132.3	23	119.1	27	114.4	-10	-4	-14	42	56
DE SOTO	MS	69	125.1	73	138.8	47	88.7	11	-36	-29	27	44
FORREST	MS	90	148.2	116	163.4	112	134.4	10	-18	-9	46	69
FRANKLIN	MS	28	159.2	29	158.9	34*	182.2	0	15	14	70	88
GEORGE	MS	5	96.6	7	164.1	8	136.3	70	-17	41	83	71
GREENE	MS	6	94.0	13	168.1	9	92.0	79	-45	-2	53	45
GRENADA	MS	66	168.6	58	133.9	54	116.3	-21	-13	-31	25	57
HANCOCK	MS	13	139.2	8	70.3	21	152.8	-49	117	10	67	79
HARRISON	MS	84	132.9	105	138.6	159	145.7	4	5	10	66	75
HINDS	MS	389	154.3	390	126.1	596*	144.9	-18	15	-6	49	75
HOLMES	MS	148	162.8	107	118.7	117	125.9	-27	6	-23	32	64
HUMPHREYS	MS	58	114.8	60	127.3	55	106.9	11	-16	-7	49	52
ISSAQUENA	MS	11	85.8	14	142.5	14	176.4	66	24	106	92	86
ITAWAMBA	MS	2	98.9	4	90.8	9	127.9	-8	41	29	79	65
JACKSON	MS	48	170.9	81*	214.9	78	139.3	26	-35	-18	37	72
JASPER	MS	27	82.6	51	151.0	26	62.4	83	-59	-25	30	36
JEFFERSON	MS	35	98.5	45	121.6	53	134.2	23	10	36	82	69
JEFFERSON DAVIS	MS	38	137.0	39	127.1	39	110.1	-7	-13	-20	35	54
JONES	MS	74	132.8	92	141.2	95	122.5	6	-13	-8	48	61
KEMPER	MS	33	109.7	27	93.0	36	112.8	-15	21	3	59	55
LAFAYETTE	MS	35	121.4	43	140.9	46	127.3	16	-10	5	62	64
LAMAR	MS	8	87.3	15	147.1	15	146.4	69	-1	68	89	75
LAUDERDALE	MS	137	137.2	170	150.1	150	124.9	9	-17	-9	47	63
LAWRENCE	MS	24	144.9	21	113.0	24	118.7	-22	5	-18	37	59
LEAKE	MS	30	120.2	41	133.1	39	111.9	11	-16	-7	49	54
LEE	MS	64	154.4	80	173.0	69	133.9	12	-23	-13	42	69
LEFLORE	MS	151	125.6	159	131.4	147	118.0	5	-10	-6	49	58
LINCOLN	MS	44	128.5	55	151.8	48	108.6	18	-28	-16	40	53
LOWNDES	MS	74	101.6	113	148.4	95	107.4	46	-28	6	63	52
MADISON	MS	106	118.2	120	132.1	116	118.5	12	-10	0	56	59
MARION	MS	49	184.3	47	139.6	40	102.0	-24	-27	-45	17	49
MARSHALL	MS	62	105.3	79	129.2	69	95.9	23	-26	-9	47	47
MONROE	MS	71	130.9	57	107.1	59	104.6	-18	-2	-20	34	50
MONTGOMERY	MS	33	149.6	20	80.3	44	137.2	-46	71	-8	48	71
NESHOBA	MS	24	113.9	33	154.9	30	120.1	36	-22	5	63	60
NEWTON	MS	46	172.7	31	106.9	37	118.1	-38	10	-32	25	58
NOXUBEE	MS	59	113.7	61	125.7	53	112.1	11	-11	-1	54	54
OKTIBBEHA	MS	62	143.0	65	140.8	63	120.4	-2	-14	-16	39	60
PANOLA	MS	99	153.2	81	126.0	79	120.0	-18	-5	-22	33	60
PEARL RIVER	MS	19	101.8	26	134.5	21	89.2	32	-34	-12	43	44
PERRY	MS	9	84.6	12	103.3	7	60.4	22	-42	-29	27	36
PIKE	MS	73	121.0	99	149.9	99	129.1	24	-14	7	64	65
PONTOTOC	MS	26	217.3	11	79.2	20	135.8	-64	71	-38	21	70
PRENTISS	MS	2	24.0	17	160.8	20	163.0	569	14	661	99	88
QUITMAN	MS	82	163.3	66	139.5	54	119.3	-15	-14	-27	28	59
RANKIN	MS	46	81.9	51	80.5	44	67.8	-2	-16	-17	38	37
SCOTT	MS	38	134.4	58	178.7	48	124.5	33	-30	-7	48	63
SHARKEY	MS	39	124.6	35	125.1	31	131.8	0	5	6	63	67
SIMPSON	MS	33	123.9	48	173.4	36	104.8	40	-40	-15	40	50
SMITH	MS	11	155.2	15	150.9	8	69.3	-3	-54	-55	12	38
STONE	MS	8	113.0	9	118.1	14	170.7	4	45	51	86	85
SUNFLOWER	MS	152	134.6	153	138.6	139	138.9	3	0	3	59	72
TALLAHATCHIE	MS	99	163.7	65	114.7	77	132.0	-30	15	-19	35	68
TATE	MS	52	143.8	56	156.1	43	110.2	9	-29	-23	31	54
TIPPAH	MS	14	133.0	11	100.6	21	158.6	-24	58	19	73	81
TISHOMINGO	MS	6	175.4	5	224.8	3	79.6	28	-65	-55	12	41
TUNICA	MS	70	118.4	89*	184.2	56	133.1	56	-28	12	69	68
UNION	MS	21	131.7	28	174.3	18	107.1	32	-39	-19	36	52
WALTHALL	MS	21	106.5	23	112.5	17	69.1	6	-39	-35	22	38
WARREN	MS	149	135.4	168	149.4	187*	154.4	10	3	14	70	79
WASHINGTON	MS	293	147.7	300	149.9	316*	148.6	2	-1	1	56	77
WAYNE	MS	34	154.5	17	67.5	34	136.5	-56	102	-12	44	71
WEBSTER	MS	9	105.8	11	105.4	18	170.9	0	62	62	88	85
WILKINSON	MS	41	122.9	38	101.2	41	104.5	-18	3	-15	40	50
WINSTON	MS	24	89.0	33	107.8	39	104.7	21	-3	18	72	50
YALOBUSHA	MS	33	133.3	34	149.9	30	108.1	12	-28	-19	36	52
YAZOO	MS	116	136.4	121	144.2	85	97.8	6	-32	-28	27	48

* SIGNIFICANTLY MORE DEATHS THAN EXPECTED

NON-WHITE FEMALE

ALL CANCERS COMBINED

ICD 140-208

1950-1959 DEATHS	1950-1959 RATE	1960-1969 DEATHS	1960-1969 RATE	1970-1979 DEATHS	1970-1979 RATE	COUNTY AND STATE	PERCENT CHANGE 60S/50S	PERCENT CHANGE 70S/60S	PERCENT CHANGE 70S/50S	PERCENTILE %CHANGE 70S/50S	PERCENTILE RATES 1970S
2649*	167.0	3230*	159.8	3851*	155.4	MISSOURI MO	-4	-3	-7		
				1	245.4	ADAIR MO					94
		1*	267.9			ANDREW MO					
8	99.8	20*	222.4	17	171.7	AUDRAIN MO	123	-23	72	90	85
2	283.0	1	163.6	2	200.4	BATES MO	-42	22	-29	27	90
		1	196.5			BENTON MO					
31	178.3	40*	214.8	52*	224.9	BOONE MO	20	5	26	77	92
32	168.6	19	112.7	32	172.1	BUCHANAN MO	-33	53	2	58	85
19	143.3	20	140.2	23	161.4	BUTLER MO	-2	15	13	69	82
10	74.7	19	157.1	17	151.0	CALLAWAY MO	110	-4	102	92	78
				1	114.7	CAMDEN MO					56
11	149.2	12	176.8	8	113.5	CAPE GIRARDEAU MO	18	-36	-24	31	56
3	135.5	2	134.8	1	17.9	CARROLL MO	-1	-87	-87	01	30
				1*	562.5	CARTER MO					99
1	273.2	2	195.3			CASS MO	-29				
8	139.2	7	151.7	6	152.7	CHARITON MO	9	1	10	66	79
8	205.6	6	147.6	8	112.1	CLAY MO	-28	-24	-45	16	54
9*	718.9	3	174.0	4	179.1	CLINTON MO	-76	3	-75	04	87
15	216.4	13	174.5	15	134.1	COLE MO	-19	-23	-38	20	69
11	142.5	7	101.3	10	163.6	COOPER MO	-29	61	15	70	82
1	344.1	2	308.9			DADE MO	-10				
		2*	301.1			DAVIESS MO					
		6	91.6	13	135.1	DUNKLIN MO		48			70
3	86.8	5	175.2	4	77.6	FRANKLIN MO	102	-56	-11	45	41
14	106.9	28	203.9	22	123.5	GREENE MO	91	-39	16	71	62
1	298.1	1*	112.4			GRUNDY MO	-62				
6*	452.4	2	121.6	2	129.4	HENRY MO	-73	6	-71	05	66
		1*	546.6			HOLT MO					
21*	261.2	20*	266.5	11	113.3	HOWARD MO	2	-57	-57	11	56
2*	446.6	1	74.3	1	85.2	HOWELL MO	-83	15	-81	02	43
1	114.7	1	1023.0			IRON MO	792				
546*	169.2	684*	154.0	878*	153.9	JACKSON MO	-9	0	-9	47	79
11	154.2	11	128.3	14	129.0	JASPER MO	-17	1	-16	39	65
9	196.9	6	152.9	11	235.1	JEFFERSON MO	-22	54	19	73	93
7	149.7	2	53.9	3	74.1	JOHNSON MO	-64	38	-50	13	39
		1*	344.1			KNOX MO					
		1	70.3	2	139.1	LACLEDE MO		98			72
19*	216.6	17*	223.5	12	172.0	LAFAYETTE MO	3	-23	-21	33	85
		1	596.2	1	84.8	LAWRENCE MO		-86			43
3	156.1	1	24.8	2	85.5	LEWIS MO	-84	245	-45	17	43
4	134.6	3	76.2	1	22.3	LINCOLN MO	-43	-71	-83	01	30
6*	428.5	2	84.6	2	105.8	LINN MO	-80	25	-75	03	51
5	271.1	3	85.7	3	85.6	LIVINGSTON MO	-68	0	-68	06	43
				1	57.3	MC DONALD MO					36
4	308.2	2	136.5	2	191.1	MACON MO	-56	40	-38	20	89
1	562.5	3*	612.7			MADISON MO	9				
22	203.2	17	173.5	25*	237.5	MARION MO	-15	37	17	72	93
24	129.6	20	102.1	20	106.6	MISSISSIPPI MO	-21	4	-18	37	52
2	147.7	3	187.8	4	231.6	MONITEAU MO	27	23	57	87	93
5	243.5	11*	522.6	3	168.9	MONROE MO	115	-68	-31	25	84
6	146.5	7	201.8	7	185.2	MONTGOMERY MO	38	-8	26	77	88
1	68.8	5*	268.3	4	210.2	MORGAN MO	290	-22	205	96	91
23	113.7	21	95.3	26	129.5	NEW MADRID MO	-16	36	14	69	66
2	200.4	1	38.2	2	64.2	NEWTON MO	-81	68	-68	06	37
1*	546.4					OSAGE MO					
54	147.3	62	167.4	41	99.1	PEMISCOT MO	14	-41	-33	24	48
				1*	136.6	PERRY MO					71
26*	218.9	16	124.9	22	204.9	PETTIS MO	-43	64	-6	49	91
		3	266.9	2	279.8	PHELPS MO		5			96
12	152.7	15	172.5	13	154.4	PIKE MO	13	-10	1	57	79
1	112.4	3	276.8	1	56.2	PLATTE MO	146	-80	-50	14	35
2*	933.6					POLK MO					
1	109.3	2	28.0	3	41.2	PULASKI MO	-74	47	-62	08	33
2	82.1	1	117.9	5*	319.1	RALLS MO	44	171	288	97	97
17	193.8	15	194.9	22*	282.9	RANDOLPH MO	1	45	46	85	96
5	151.2	8	326.7	9*	393.4	RAY MO	116	20	160	95	98
8	184.3	7	133.2	20*	253.8	ST CHARLES MO	-28	91	38	82	94
1	281.2	2	456.5	2*	687.8	ST CLAIR MO	62	51	145	95	99
1	74.3	1	267.9			STE GENEVIEVE MO	260				
3	206.1	2	139.1	1	112.4	ST FRANCOIS MO	-32	-19	-45	16	55
133	147.9	146	115.5	388*	149.9	ST LOUIS COUNTY MO	-22	30	1	57	77
20	189.5	25*	243.1	31*	287.1	SALINE MO	28	18	51	86	96
13	124.1	26	198.5	21	141.4	SCOTT MO	60	-29	14	70	73
		1	70.7	1	141.3	SHELBY MO		100			73
4	85.5	4	104.3	2	85.7	STODDARD MO	22	-18	0	56	43
1*	112.4					SULLIVAN MO					
2*	342.2			1	188.7	TANEY MO			-45	17	89
1*	344.1			1	596.2	TEXAS MO			73	90	99
1*	589.6					VERNON MO					
2	126.4	2	245.9	5	212.7	WARREN MO	94	-13	68	89	92
1	57.3					WASHINGTON MO					
		1	106.0	2	273.1	WRIGHT MO		158			95
1420*	171.3	1824*	168.3	1983*	160.4	ST LOUIS CITY MO	-2	-5	-6	49	81
89	156.7	118	165.4	161	161.0	MONTANA MT	6	-3	3		
				1	424.0	BEAVERHEAD MT					98
10	174.5	16	210.1	15	117.1	BIG HORN MT	20	-44	-33	24	58
4	89.5	16*	347.0	8	149.0	BLAINE MT	288	-57	66	89	77
				1*	589.6	BROADWATER MT					99
3	129.6	8	228.9	12	171.6	CASCADE MT	77	-25	32	80	85
2*	1037.2					CUSTER MT					
1	424.0	2	266.8			DEER LODGE MT	-37				
				1	112.4	FERGUS MT					55
1	212.0			1	70.7	FLATHEAD MT			-67	06	38
2*	685.2	1	546.4	3*	355.3	GALLATIN MT	-20	-35	-48	15	97
				1	67.0	GARFIELD MT					37
17	146.3	20	152.7	29	184.2	GLACIER MT	4	21	26	77	88
2	95.6	10*	279.6	10	206.0	HILL MT	192	-26	115	93	91

* SIGNIFICANTLY MORE DEATHS THAN EXPECTED

NON-WHITE FEMALE

ALL CANCERS COMBINED

ICD 140-208

1950-1959 DEATHS	1950-1959 RATE	1960-1969 DEATHS	1960-1969 RATE	1970-1979 DEATHS	1970-1979 RATE	COUNTY AND STATE	PERCENT CHANGE 60S/50S	PERCENT CHANGE 70S/60S	PERCENT CHANGE 70S/50S	PERCENTILE %CHANGE 70S/50S	PERCENTILE RATES 1970S
8	144.2	8	136.8	18*	209.4	LAKE MT	-5	53	45	84	91
2	146.9	6*	310.9	1	42.4	LEWIS AND CLARK MT	112	-86	-71	05	33
1*	267.9	1	490.8	1	188.7	LINCOLN MT	83	-62	-30	26	89
3	286.4	3	101.1	3	73.7	MISSOULA MT	-65	-27	-74	04	39
2	190.5	2	112.4	3	200.8	PHILLIPS MT	-41	79	5	63	90
1	85.2	1	84.8	2	184.2	PONDERA MT	0	117	116	93	88
				1	85.2	RAVALLI MT					43
18*	303.5	11	139.5	22*	231.4	ROOSEVELT MT	-54	66	-24	31	93
4	162.1	5	185.0	11	192.3	ROSEBUD MT	14	4	19	73	90
1	37.5	3	202.8	2	168.3	SANDERS MT	441	-17	349	98	84
				1	267.9	SHERIDAN MT					95
1	122.7			1	47.1	SILVER BOW MT			-62	09	33
1	424.0	1	281.2	1	182.1	TETON MT	-34	-35	-57	11	87
				2*	565.1	TOOLE MT					99
1	133.9	2	401.8	3	200.1	VALLEY MT	200	-50	49	85	90
4	207.4	2	61.6	7	88.2	YELLOWSTONE MT	-70	43	-57	11	44
189	167.4	293*	192.5	274	130.1	NEBRASKA NB	15	-32	-22		
3	229.9	2	149.5	6*	569.5	ADAMS NB	-35	281	148	95	99
2	186.1	3	190.7	1	344.1	BOX BUTTE NB	2	80	85	91	97
		1*	596.2			CASS NB					
1	546.4			1	182.1	CHERRY NB			-67	07	87
		1	188.7			CHEYENNE NB					
1	99.4	1	49.2	1	62.9	DAKOTA NB	-51	28	-37	21	37
1	136.6	2	365.1	1	298.1	DAWES NB	167	-18	118	94	96
1	267.9					DODGE NB					
141	163.2	233*	201.2	228	147.4	DOUGLAS NB	23	-27	-10	46	76
				1	94.4	GAGE NB					46
5*	351.5	6*	445.5	2	118.7	KNOX NB	27	-73	-66	07	59
12	186.0	14	142.8	11	83.1	LANCASTER NB	-23	-42	-55	12	42
1	117.9	2	183.4			LINCOLN NB	56				
1	267.9					OTOE NB					
				1*	562.5	PHELPS NB					99
4*	1036.0					POLK NB					
1	344.1	1	84.8			RICHARDSON NB	-75				
		3	64.7	4	70.9	SARPY NB		10			38
4	217.0	5	215.5	5	68.3	SCOTTS BLUFF NB	-1	-68	-69	06	38
2	400.7			1	122.7	SHERIDAN NB			-69	05	62
9	168.5	18*	352.6	9	118.7	THURSTON NB	109	-66	-30	27	59
		1	56.2	1	56.2	WASHINGTON NB		0			35
				1	196.5	YORK NB					90
62	146.1	118	169.4	167	97.5	NEVADA NV	16	-42	-33		
2	80.5	1	43.0	2	67.5	CHURCHILL NV	-47	57	-16	39	37
33	181.1	76	193.1	128	115.9	CLARK NV	7	-40	-36	22	57
2	361.0	1	245.4			DOUGLAS NV	-32				
5	101.6	6	118.9	4	53.1	ELKO NV	17	-55	-48	15	35
1*	344.1					ESMERALDA NV					
1	99.4	3	414.6	2	109.7	HUMBOLDT NV	317	-74	10	67	53
1	212.0	1	74.3			LANDER NV	-65				
		1*	74.3			LINCOLN NV					
2	342.4	1	109.3	4	227.9	LYON NV	-68	109	-33	23	92
3	92.4	4	134.6	4	163.9	MINERAL NV	46	22	77	91	83
		2	173.8	2	166.6	NYE NV		-4			84
		1	117.3	1	65.5	PERSHING NV		-44			37
11	144.9	17	140.1	17	51.7	WASHOE NV	-3	-63	-64	07	34
		1	112.4	1	43.0	WHITE PINE NV		-62			33
1	62.9	3	254.5	2	66.6	ORMSBY-CARSON CITY NV	305	-74	6	63	37
6	234.8	10	113.5	13	57.7	NEW HAMPSHIRE NH	-52	-49	-75		
				1	37.2	CARROLL NH					32
				1	188.7	COOS NH					89
		1	490.8	1	49.1	GRAFTON NH		-90			34
2	298.0	3	102.7	2	27.8	HILLSBOROUGH NH	-66	-73	-91	00	31
		2	178.6	1	33.1	MERRIMACK NH		-81			32
3	332.1	4	160.0	6	104.7	ROCKINGHAM NH	-52	-35	-68	06	50
1	344.1			1	91.1	STRAFFORD NH			-74	04	45
2624*	171.1	3468	142.7	5472*	147.6	NEW JERSEY NJ	-17	3	-14		
199	144.4	266	156.3	368*	175.8	ATLANTIC NJ	8	12	22	75	86
89	144.2	157	145.5	264	142.1	BERGEN NJ	1	-2	-1	54	73
55	150.3	64	118.5	150	142.2	BURLINGTON NJ	-21	20	-5	50	73
197*	181.3	289*	161.8	451*	174.0	CAMDEN NJ	-11	8	-4	52	86
35	196.4	34	143.8	48	151.4	CAPE MAY NJ	-27	5	-23	31	78
66	170.9	71	130.8	107	131.1	CUMBERLAND NJ	-23	0	-23	31	67
914*	178.9	1203	141.2	1783*	146.9	ESSEX NJ	-21	4	-18	37	76
59	156.4	72	130.7	116	148.1	GLOUCESTER NJ	-16	13	-5	50	76
223*	218.0	287*	170.4	384	121.1	HUDSON NJ	-22	-29	-44	17	60
4	167.8	6	159.8	4	83.6	HUNTERDON NJ	-5	-48	-50	14	42
134	162.3	197	142.6	304	145.0	MERCER NJ	-12	2	-11	45	75
69	190.3	103	153.2	158	121.0	MIDDLESEX NJ	-20	-21	-36	22	60
149	138.0	175	119.2	335*	163.1	MONMOUTH NJ	-14	37	18	73	82
29	133.5	36	104.4	75	153.1	MORRIS NJ	-22	47	15	70	79
19	215.6	21	121.5	60*	179.5	OCEAN NJ	-44	48	-17	38	87
95	190.3	151	134.6	272	130.4	PASSAIC NJ	-29	-3	-31	25	66
44	176.9	45	139.7	88*	194.4	SALEM NJ	-21	39	10	67	90
20	224.4	24	131.9	44	124.4	SOMERSET NJ	-41	-6	-45	17	63
1	119.2	2	96.2	9*	295.0	SUSSEX NJ	-19	207	147	95	96
220*	173.3	264	144.9	436	146.1	UNION NJ	-16	1	-16	40	75
3	128.5	1	38.3	16*	298.0	WARREN NJ	-70	679	132	94	96
144	97.9	207	102.0	329	61.2	NEW MEXICO NM	4	-40	-38		
19	124.4	31	130.1	51	50.3	BERNALILLO NM	5	-61	-60	09	34
8	239.2	9	170.0	11	89.0	CHAVES NM	-29	-48	-63	08	44
6	244.8	9	171.8	5	47.5	CURRY NM	-30	-72	-81	02	34
3	126.3	6	193.3	2	6.1	DONA ANA NM	53	-97	-95	00	30

* SIGNIFICANTLY MORE DEATHS THAN EXPECTED

NON-WHITE FEMALE — ALL CANCERS COMBINED — ICD 140-208

1950-1959 DEATHS	1950-1959 RATE	1960-1969 DEATHS	1960-1969 RATE	1970-1979 DEATHS	1970-1979 RATE	COUNTY AND STATE		PERCENT CHANGE 60S/50S	PERCENT CHANGE 70S/60S	PERCENT CHANGE 70S/50S	PERCENTILE %CHANGE 70S/50S	PERCENTILE RATES 1970S
1	24.6	6	127.7	8	61.7	EDDY	NM	420	-52	151	95	36
		2	160.2			GRANT	NM					
				1	89.3	GUADALUPE	NM					44
1*	112.4					HIDALGO	NM					
7	77.9	20	191.9	20	119.8	LEA	NM	146	-38	54	87	60
		1	98.3	2	103.3	LINCOLN	NM		5			49
3*	490.7	2	122.7	5	104.5	LUNA	NM	-75	-15	-79	02	50
48	122.2	51	102.9	92	109.0	MC KINLEY	NM	-16	6	-11	45	53
8	147.3	7	121.4	11	75.3	OTERO	NM	-18	-38	-49	14	40
				3	74.6	QUAY	NM					40
8	141.1	6	78.7	11	32.9	RIO ARRIBA	NM	-44	-58	-77	03	32
3	25.7	9	51.9	25	82.2	SANDOVAL	NM	102	58	220	97	42
16	61.7	33	98.6	48	84.7	SAN JUAN	NM	60	-14	37	82	43
				1	5.5	SAN MIGUEL	NM					30
2	74.3			1	7.6	SANTA FE	NM			-90	00	30
				2	33.3	SOCORRO	NM					32
5	118.7	3	51.0	5	25.0	TAOS	NM	-57	-51	-79	02	31
6	42.0	12	65.7	25	64.0	VALENCIA	NM	57	-3	52	86	37
7780*	171.0	11130*	151.4	15986	135.7	NEW YORK	NY	-11	-10	-21		
58*	205.5	58	122.9	90	128.0	ALBANY	NY	-40	4	-38	21	65
1	133.9			1	68.8	ALLEGANY	NY			-49	15	38
828*	177.3	1169	115.9	2281	124.1	BRONX	NY	-35	7	-30	26	62
8	178.7	15	176.5	19	151.2	BROOME	NY	-1	-14	-15	40	78
5	98.0	12	161.8	8	79.9	CATTARAUGUS	NY	65	-51	-19	36	41
3	145.4	8*	406.9	7	179.6	CAYUGA	NY	180	-56	24	76	87
5	119.7	3	54.4	2	22.9	CHAUTAUQUA	NY	-55	-58	-81	02	31
7	111.6	10	107.6	21	154.4	CHEMUNG	NY	-4	44	38	83	79
3	424.2	1	188.7	1	188.7	CHENANGO	NY	-56	0	-56	12	89
		1	267.9	2	94.1	CLINTON	NY		-65			46
6	128.2	11	200.7	7	89.5	COLUMBIA	NY	57	-55	-30	26	45
1	196.5	1	94.4	2	321.3	CORTLAND	NY	-52	241	63	88	97
2	672.9	1	98.3	1	112.4	DELAWARE	NY	-85	14	-83	01	55
30	102.7	44	95.3	76	118.7	DUTCHESS	NY	-7	25	16	71	59
325*	178.5	472	155.8	693*	159.3	ERIE	NY	-13	2	-11	45	81
3*	274.7			2	404.5	ESSEX	NY			47	85	98
13	219.9	9	110.1	12	143.1	FRANKLIN	NY	-50	30	-35	23	74
7*	457.7	4	185.0	6	191.6	FULTON	NY	-60	4	-58	10	89
7	182.8	6	114.3	6	74.3	GENESEE	NY	-37	-35	-59	09	39
8	305.0	6	172.0	11*	288.5	GREENE	NY	-44	68	-5	50	96
1	267.9	1	106.0	4	231.4	JEFFERSON	NY	-60	118	-14	42	93
1615*	166.0	2247	121.5	3880	133.6	KINGS(BROOKLYN)	NY	-27	10	-20	35	69
1	844.2					LEWIS	NY					
2	157.1	8	220.0	7	148.0	LIVINGSTON	NY	40	-33	-6	50	76
1	114.7	2	109.3	1	32.7	MADISON	NY	-5	-70	-71	05	32
57	128.8	117	134.3	227	131.1	MONROE	NY	4	-2	2	58	67
1	68.8	3	241.9	4	168.2	MONTGOMERY	NY	252	-30	144	95	84
151	142.8	248	118.0	507	145.1	NASSAU	NY	-17	23	2	57	75
3398*	187.8	4572*	205.4	4236*	148.6	NEW YORK(MANHATTAN)	NY	9	-28	-21	33	77
27	244.6	30	120.2	57	133.8	NIAGARA	NY	-51	11	-45	16	69
14	143.4	24	125.4	35	130.3	ONEIDA	NY	-13	4	-9	47	66
45	183.6	58	118.4	124	135.6	ONONDAGA	NY	-36	14	-26	28	70
2	79.3	4	135.6	4	61.5	ONTARIO	NY	71	-55	-23	32	36
27	113.3	42	111.6	76	133.4	ORANGE	NY	-2	19	18	72	68
3	160.9	8	176.7	7	175.3	ORLEANS	NY	10	-1	9	66	86
2	556.1	2	172.3	1	53.6	OSWEGO	NY	-69	-69	-90	00	35
		1	93.7	4	206.4	OTSEGO	NY		120			91
				1	44.6	PUTNAM	NY					33
565	144.1	1001	110.6	2068	121.8	QUEENS	NY	-23	10	-15	40	61
9	147.5	13	137.5	21	124.7	RENSSELAER	NY	-7	-9	-15	40	63
41	181.3	47	118.8	102	122.5	RICHMOND	NY	-34	3	-32	24	61
24	113.9	41	96.9	80	112.9	ROCKLAND	NY	-15	17	-1	55	55
		2	93.1	4	102.9	ST LAWRENCE	NY		11			49
14*	270.3	12	178.4	15	131.7	SARATOGA	NY	-34	-26	-51	13	67
11	145.6	19	153.8	24	128.7	SCHENECTADY	NY	6	-16	-12	44	65
				3	299.0	SCHOHARIE	NY					96
1	188.7			1	114.7	SCHUYLER	NY			-39	20	56
		1	24.8	1	23.6	SENECA	NY		-5			31
3	149.0	3	114.4	8	132.9	STEUBEN	NY	-23	16	-11	45	68
90	87.7	188	99.2	345	124.9	SUFFOLK	NY	13	26	42	84	63
2	35.0	9	100.3	24	194.0	SULLIVAN	NY	187	93	454	99	90
1	112.4			1	19.5	TIOGA	NY			-83	02	30
5	103.0	10	130.1	22*	227.6	TOMPKINS	NY	26	75	121	94	92
9	85.7	21	142.3	36	144.9	ULSTER	NY	66	2	69	89	75
				5	397.8	WARREN	NY					98
1	424.0	2	279.0	1	94.4	WASHINGTON	NY	-34	-66	-78	03	46
1	188.7	6	201.9	13	130.0	WAYNE	NY	7	-36	-31	25	66
334	158.8	557	162.2	788*	151.9	WESTCHESTER	NY	2	-6	-4	51	78
1	298.1					WYOMING	NY					
1	56.2			1	109.3	YATES	NY			94	92	53
4743	123.4	5944	129.6	7682	132.4	NORTH CAROLINA	NC	5	2	7		
62	134.4	89	154.5	118	139.6	ALAMANCE	NC	15	-10	4	60	72
7	234.3	6	157.3	9	140.3	ALEXANDER	NC	-33	-11	-40	19	73
5*	351.9	2	115.5	1	80.4	ALLEGHANY	NC	-67	-30	-77	03	41
52	110.1	59	117.3	67	112.5	ANSON	NC	7	-4	2	58	55
1	67.0	1	112.4	3	296.3	ASHE	NC	68	164	342	98	96
		2	150.6	1	89.3	AVERY	NC		-41			44
80	152.2	68	117.0	84	124.9	BEAUFORT	NC	-23	7	-18	37	63
62	128.3	55	103.5	65	107.1	BERTIE	NC	-19	3	-16	39	52
27	74.3	51	118.9	49	89.7	BLADEN	NC	60	-25	21	74	45
26	119.7	28	100.7	43	116.8	BRUNSWICK	NC	-16	16	-2	53	58
111	152.2	128	156.9	134	139.8	BUNCOMBE	NC	3	-11	-8	48	72
15	129.1	22	117.0	22	86.5	BURKE	NC	-9	-26	-33	23	43
45	123.2	68	141.3	82	140.7	CABARRUS	NC	15	0	14	70	73
14	115.1	18	131.8	23	135.6	CALDWELL	NC	14	3	18	72	70
9	121.3	7	86.1	14	150.3	CAMDEN	NC	-29	75	24	76	77
21	175.2	22	172.0	30	176.4	CARTERET	NC	-2	3	1	56	86
31	127.0	30	92.6	47	122.4	CASWELL	NC	-27	32	-4	52	61

* SIGNIFICANTLY MORE DEATHS THAN EXPECTED

NON-WHITE FEMALE — ALL CANCERS COMBINED — ICD 140-208

1950-1959 DEATHS	1950-1959 RATE	1960-1969 DEATHS	1960-1969 RATE	1970-1979 DEATHS	1970-1979 RATE	COUNTY AND STATE	PERCENT CHANGE 60S/50S	PERCENT CHANGE 70S/60S	PERCENT CHANGE 70S/50S	PERCENTILE %CHANGE 70S/50S	PERCENTILE RATES 1970S
34	187.1	47	171.8	53	146.3	CATAWBA NC	-8	-15	-22	32	75
38	148.1	32	107.7	52	124.3	CHATHAM NC	-27	15	-16	39	63
2	123.5	2	113.7	2	57.0	CHEROKEE NC	-8	-50	-54	12	36
22	112.2	32	147.2	31	123.8	CHOWAN NC	31	-16	10	67	62
				1	112.4	CLAY NC					55
42	94.1	71	136.0	92	136.1	CLEVELAND NC	45	0	45	84	70
50	88.5	71	116.9	89	117.3	COLUMBUS NC	32	0	33	80	58
74	114.2	97	138.4	107	131.2	CRAVEN NC	21	-5	15	70	67
98	98.4	138	111.6	248	123.7	CUMBERLAND NC	13	11	26	77	62
9	123.5	16	195.8	11	118.4	CURRITUCK NC	59	-40	-4	52	59
2	234.6	1	29.8	4	110.8	DARE NC	-87	272	-53	13	54
21	88.8	45	137.1	68	149.7	DAVIDSON NC	54	9	69	89	77
9	113.1	13	140.0	7	62.7	DAVIE NC	24	-55	-45	17	36
62	121.7	68	122.8	85	127.4	DUPLIN NC	1	4	5	61	64
190	140.8	250	135.3	367*	152.5	DURHAM NC	-4	13	8	65	78
114	126.0	130	125.5	155	120.0	EDGECOMBE NC	0	-4	-5	51	60
269	147.0	351*	154.9	443*	160.0	FORSYTH NC	5	3	9	66	81
54	128.0	61	127.4	64	109.1	FRANKLIN NC	0	-14	-15	41	53
61	99.4	100	130.9	116	121.3	GASTON NC	32	-7	22	75	61
18	93.2	20	110.6	32	146.5	GATES NC	19	32	57	87	75
		1	74.5	2	123.7	GRAHAM NC		66			62
60	132.6	66	123.8	81	119.1	GRANVILLE NC	-7	-4	-10	45	59
33	140.1	33	166.4	30	126.3	GREENE NC	19	-24	-10	46	64
230	137.8	315	141.6	439*	150.6	GUILFORD NC	3	6	9	66	78
127	131.6	117	109.4	156	122.1	HALIFAX NC	-17	12	-7	48	61
45	117.0	56	110.1	79	136.7	HARNETT NC	-6	24	17	72	71
5	164.9	3	58.0	8	139.5	HAYWOOD NC	-65	140	-15	40	72
14	152.3	12	107.6	25	203.0	HENDERSON NC	-29	89	33	81	91
57	143.2	55	107.7	74	114.6	HERTFORD NC	-25	6	-20	35	56
22	84.2	28	88.2	35	95.5	HOKE NC	5	8	13	69	47
4	37.6	8	81.8	7	50.3	HYDE NC	118	-38	34	81	34
53	137.9	55	121.8	92	158.0	IREDELL NC	-12	30	15	70	81
5	81.1	14	204.9	10	125.7	JACKSON NC	153	-39	55	87	63
48	105.1	68	128.6	69	108.5	JOHNSTON NC	22	-16	3	59	53
11	63.1	16	88.8	31	152.3	JONES NC	41	72	141	95	78
31	150.0	28	100.1	43	121.5	LEE NC	-33	21	-19	36	61
87	127.4	106	129.5	147	136.2	LENOIR NC	2	5	7	64	70
12	106.3	15	114.7	17	116.4	LINCOLN NC	8	2	10	66	57
16*	263.6	7	117.7	4	45.2	MC DOWELL NC	-55	-62	-83	02	33
1	54.6	2	119.8	2	60.4	MACON NC	119	-50	11	67	36
1	37.5	2	237.9	1	37.2	MADISON NC	535	-84	-1	55	32
51	126.8	55	124.8	69	128.4	MARTIN NC	-2	3	1	57	65
295	134.8	366	125.1	577*	150.8	MECKLENBURG NC	-7	21	12	68	78
				1	212.0	MITCHELL NC					92
10	72.0	29	156.3	27	106.4	MONTGOMERY NC	117	-32	48	85	51
45	124.3	59	134.1	74	130.2	MOORE NC	8	-3	5	62	66
90	118.1	100	120.8	140	141.9	NASH NC	2	17	20	74	73
124	143.0	140	144.6	170	142.6	NEW HANOVER NC	1	-1	0	55	74
57	98.0	82	143.8	96	142.5	NORTHAMPTON NC	47	-1	45	84	73
23	133.6	31	106.9	50	118.6	ONSLOW NC	-20	11	-11	44	59
45	148.9	51	131.3	68	123.7	ORANGE NC	-12	-6	-17	38	62
8	69.6	14	113.2	19	121.9	PAMLICO NC	63	8	75	90	61
36	104.8	60	148.8	68	140.7	PASQUOTANK NC	42	-5	34	81	73
25	73.8	51	143.1	49	107.3	PENDER NC	94	-25	45	84	52
26	153.7	29	160.7	28	130.6	PERQUIMANS NC	5	-19	-15	40	66
36	133.7	40	124.3	41	98.8	PERSON NC	-7	-20	-26	29	48
118	111.4	163	143.1	162	117.1	PITT NC	28	-18	5	62	58
11	176.4	4	52.9	17	185.0	POLK NC	-70	250	5	62	88
25	162.0	31	146.3	39	135.0	RANDOLPH NC	-10	-8	-17	38	70
55	124.1	65	123.0	85	137.3	RICHMOND NC	-1	12	11	67	71
164	110.0	183	112.1	250	114.9	ROBESON NC	2	3	4	61	56
47	87.4	103	164.5	118	154.8	ROCKINGHAM NC	88	-6	77	91	80
76	159.2	66	107.5	119	153.9	ROWAN NC	-32	43	-3	52	79
23	111.5	31	138.0	44	145.3	RUTHERFORD NC	24	5	30	80	75
57	98.2	95	147.6	104	124.5	SAMPSON NC	50	-16	27	78	63
62	168.7	61	147.3	72	138.6	SCOTLAND NC	-13	-6	-18	37	72
25	179.0	17	103.5	35	132.2	STANLY NC	-42	28	-26	29	68
4	57.5	10	127.8	7	82.3	STOKES NC	122	-36	43	84	42
10	120.0	12	102.8	24	150.1	SURRY NC	-14	46	25	77	77
4	93.8	6	137.2	8	116.1	SWAIN NC	46	-15	24	76	57
6	170.8	3	84.0	7	164.9	TRANSYLVANIA NC	-51	96	-3	52	83
4	56.6	13	186.3	11	118.8	TYPRELL NC	229	-36	110	93	59
42	130.7	48	130.0	51	110.6	UNION NC	0	-15	-15	40	54
56	107.0	83	142.6	101	142.8	VANCE NC	33	0	33	81	74
197	129.8	251	127.9	352	134.2	WAKE NC	-1	5	3	60	69
43	89.3	47	96.5	65	114.6	WARREN NC	8	19	28	79	56
24	134.3	32	150.5	36	156.3	WASHINGTON NC	12	4	16	71	80
1	122.7	1	56.2	3	157.4	WATAUGA NC	-54	180	28	79	80
119	109.5	160	130.1	151	107.8	WAYNE NC	19	-17	-2	54	52
7	68.4	19	161.7	11	80.9	WILKES NC	136	-50	18	73	42
89	115.2	117	127.4	157	152.8	WILSON NC	11	20	33	80	79
4	145.7	8	163.4	4	66.5	YADKIN NC	12	-59	-54	12	37
				1	267.9	YANCEY NC					95
39	141.0	71*	207.4	83	163.8	NORTH DAKOTA ND	47	-21	16		
6	168.6	9	217.2	11	205.0	BENSON ND	29	-6	22	75	91
				1	32.2	BURLEIGH ND					31
1	490.8			2	122.6	CASS ND			-75	04	61
1	81.8	2	147.4	1	267.9	DUNN ND	80	82	228	97	95
1	81.8			3	65.1	GRAND FORKS ND			-20	34	37
		1*	424.0	1*	267.9	GRANT ND		-37			95
1*	188.7					MC HENRY ND					
		2	92.1	7*	550.8	MC KENZIE ND		498			99
		2	224.8	6*	425.3	MC LEAN ND		89			98
		1	267.9	1	245.4	MORTON ND		-8			94
1	68.3	6*	383.7	3	110.4	MOUNTRAIL ND	462	-71	62	88	54
		2*	814.3			PEMBINA ND					
1	490.8	1	70.7	1	65.5	RAMSEY ND	-86	-7	-87	01	37
		1	117.9			RICHLAND ND					
17	144.1	25	199.8	33	190.2	ROLETTE ND	39	-5	32	80	89

* SIGNIFICANTLY MORE DEATHS THAN EXPECTED

NON-WHITE FEMALE — ALL CANCERS COMBINED — ICD 140-208

1950-1959 DEATHS	1950-1959 RATE	1960-1969 DEATHS	1960-1969 RATE	1970-1979 DEATHS	1970-1979 RATE	COUNTY AND STATE		PERCENT CHANGE 60S/50S	PERCENT CHANGE 70S/60S	PERCENT CHANGE 70S/50S	PERCENTILE %CHANGE 70S/50S	PERCENTILE RATES 1970S
				1*	490.8	SARGENT	ND					98
5	138.7	14*	290.6	6	104.0	SIOUX	ND	109	-64	-25	30	50
2*	263.1					STARK	ND					
1*	490.8					TOWNER	ND					
1	133.9	4	625.8	4	121.6	WARD	ND	367	-81	-9	47	61
1	188.7	1	114.7	2	96.0	WILLIAMS	ND	-39	-16	-49	14	47
4116*	170.5	5643*	160.9	7440*	157.3	OHIO	OH	-6	-2	-8		
1	133.9	2*	201.7	1	188.7	ADAMS	OH	51	-6	41	83	89
15	81.0	45	173.8	53	159.0	ALLEN	OH	115	-8	96	92	81
		3*	550.4	1	68.8	ASHLAND	OH		-87			38
6	131.8	15	162.1	17	135.8	ASHTABULA	OH	23	-16	3	59	70
4	193.5	5	126.1	7	152.5	ATHENS	OH	-35	21	-21	33	78
26*	240.3	19	161.0	21	176.8	BELMONT	OH	-33	10	-26	28	86
6	198.4	3	110.5	3	94.3	BROWN	OH	-44	-15	-52	13	46
60	174.3	71	163.5	93	163.7	BUTLER	OH	-6	0	-6	49	83
		1	68.8	1	94.4	CARROLL	OH		37			46
13	244.5	15*	247.0	7	115.9	CHAMPAIGN	OH	1	-53	-53	13	57
76	148.5	121*	183.7	135*	172.5	CLARK	OH	24	-6	16	71	86
8	224.6	5	126.5	5	76.2	CLERMONT	OH	-44	-40	-66	07	40
5	156.6	6	158.8	6	171.8	CLINTON	OH	1	8	10	66	85
18*	251.5	18	195.6	12	107.6	COLUMBIANA	OH	-22	-45	-57	11	52
3	152.8	5	199.4	8	239.6	COSHOCTON	OH	31	20	57	87	93
2	127.9	4	203.0	3	129.6	CRAWFORD	OH	59	-36	1	57	66
1190*	157.0	1802*	159.4	2428*	156.7	CUYAHOGA	OH	2	-2	0	56	80
4	221.1	2	294.7	2	133.9	DARKE	OH	33	-55	-39	20	69
6	148.0	9	227.8	8	144.1	DELAWARE	OH	54	-37	-3	53	74
12	110.9	18	132.5	34	174.2	ERIE	OH	19	31	57	87	86
1	56.2	2	171.4	1	133.9	FAIRFIELD	OH	205	-22	138	95	69
8	154.9	8	224.8	9	201.3	FAYETTE	OH	45	-10	30	79	91
459*	177.9	630*	170.2	787*	157.1	FRANKLIN	OH	-4	-8	-12	44	80
		1*	546.4			FULTON	OH					
3	48.8	3	48.2	10	194.3	GALLIA	OH	-1	303	298	98	90
1	74.3	2	129.6	10	165.1	GEAUGA	OH	74	27	122	94	83
50*	216.6	48	169.0	49	147.0	GREENE	OH	-22	-13	-32	25	76
1	39.0	6	96.7	5	120.7	GUERNSEY	OH	148	25	209	96	60
835*	190.5	1090*	179.5	1422*	182.4	HAMILTON	OH	-6	2	-4	51	88
3	211.8	4	229.5	3	95.4	HANCOCK	OH	8	-58	-55	12	47
3	165.5	2	100.4	4	257.2	HARDIN	OH	-39	156	55	87	95
4	166.2	5	296.8	4	172.4	HARRISON	OH	79	-42	4	60	85
				1	65.5	HENRY	OH					37
8	154.2	9	246.4	2	31.6	HIGHLAND	OH	60	-87	-79	02	31
		1	114.7	2	479.9	HOCKING	OH		318			98
3	165.3	1	49.7	7	193.9	HURON	OH	-70	290	17	72	90
4	245.9			4	324.2	JACKSON	OH			32	80	97
30	154.2	44	168.3	42	143.4	JEFFERSON	OH	9	-15	-7	49	74
3	81.0	1	38.5	2	61.0	KNOX	OH	-52	58	-25	30	36
6	89.2	9	96.7	16	115.7	LAKE	OH	8	20	30	79	57
9	120.5	16	186.5	15	147.4	LAWRENCE	OH	55	-21	22	75	76
11	233.1	17	198.5	25*	271.5	LICKING	OH	-15	37	17	71	95
4	135.1	7	209.3	8	208.4	LOGAN	OH	55	0	54	87	91
49	164.2	71	153.0	87	124.6	LORAIN	OH	-7	-19	-24	30	63
221*	169.0	301	158.5	382	145.4	LUCAS	OH	-6	-8	-14	41	75
8	202.1	10	232.7	11*	238.4	MADISON	OH	15	2	18	72	93
191*	176.1	214	143.8	319*	162.9	MAHONING	OH	-18	13	-8	48	82
3	239.2	8	241.6	8	167.5	MARION	OH	1	-31	-30	26	84
		1	53.6	6	147.1	MEDINA	OH		175			76
2	96.7	2	96.4			MEIGS	OH	0				
				1	298.1	MERCER	OH					96
11	160.9	14	168.6	16	179.9	MIAMI	OH	5	7	12	68	87
				1*	546.6	MONROE	OH					99
288*	175.4	393	143.0	572*	148.0	MONTGOMERY	OH	-18	3	-16	40	76
2	123.2	4	233.5	3	115.9	MORGAN	OH	90	-50	-6	50	57
		1*	424.0			MORROW	OH					
15	129.5	24	158.0	30	162.1	MUSKINGUM	OH	22	3	25	77	82
3	367.3	4	252.3	6*	327.4	OTTAWA	OH	-31	30	-11	45	97
1	245.4	1	60.7	5	303.8	PAULDING	OH	-75	400	24	76	96
3	375.9	2	183.7	1	68.8	PERRY	OH	-51	-63	-82	02	38
2	75.0	1	74.3	5	193.5	PICKAWAY	OH	-1	160	158	95	90
1	43.0	3	214.6	4	257.6	PIKE	OH	399	20	499	99	95
17*	249.5	17	223.2	15	131.8	PORTAGE	OH	-11	-41	-47	16	67
				2	339.9	PREBLE	OH					97
1*	267.9					PUTNAM	OH					
27	205.2	22	110.3	56	173.7	RICHLAND	OH	-46	58	-15	40	86
19	205.1	11	100.8	32*	247.1	ROSS	OH	-51	145	20	74	94
2	101.4	1	28.1	6	87.7	SANDUSKY	OH	-72	212	-14	42	44
12	144.0	15	163.6	14	152.7	SCIOTO	OH	14	-7	6	63	79
1	114.7	4	142.7	7	146.8	SENECA	OH	24	3	28	78	75
4*	385.9	5*	301.2	2	104.5	SHELBY	OH	-22	-65	-73	04	50
81	168.8	114	149.5	139	131.8	STARK	OH	-11	-12	-22	32	67
178	157.2	217	133.5	324	137.3	SUMMIT	OH	-15	3	-13	43	71
47	170.6	69	156.2	82	135.3	TRUMBULL	OH	-8	-13	-21	33	70
6	173.1	6	166.8	6	133.7	TUSCARAWAS	OH	-4	-20	-23	32	69
3	266.0	3	408.7	3	108.3	UNION	OH	54	-73	-59	10	53
1	86.0			1	149.1	VAN WERT	OH			73	90	77
3	91.0	12*	349.2	3	57.8	WARREN	OH	284	-83	-36	22	36
7	138.5	6	131.7	6	108.8	WASHINGTON	OH	-5	-17	-21	33	53
3	172.8	6	98.0	8	130.5	WAYNE	OH	-43	33	-24	30	66
3	123.6	6*	370.7	2	50.5	WOOD	OH	200	-86	-59	10	34
1371	147.4	1528	129.3	1911	117.2	OKLAHOMA	OK	-12	-9	-20		
33*	369.1	21	165.7	26	119.5	ADAIR	OK	-55	-28	-68	06	60
7	131.9	8	153.1	9	111.3	ATOKA	OK	16	-27	-16	40	54
1	44.6	5	192.0	7*	360.3	BECKHAM	OK	330	88	708	99	97
16	201.8	20*	259.6	11	143.5	BLAINE	OK	29	-45	-29	27	74
5	55.8	9	96.8	8	56.4	BRYAN	OK	74	-42	1	57	35
40*	281.5	34	191.9	45	189.0	CADDO	OK	-32	-2	-33	24	89
12	213.9	9	112.1	16	157.5	CANADIAN	OK	-48	41	-26	28	80
25	120.1	42	163.4	39	121.1	CARTER	OK	36	-26	1	57	60

* SIGNIFICANTLY MORE DEATHS THAN EXPECTED

NON-WHITE FEMALE — ALL CANCERS COMBINED — ICD 140-208

1950-1959 DEATHS	1950-1959 RATE	1960-1969 DEATHS	1960-1969 RATE	1970-1979 DEATHS	1970-1979 RATE	COUNTY AND STATE	PERCENT CHANGE 60S/50S	PERCENT CHANGE 70S/60S	PERCENT CHANGE 70S/50S	PERCENTILE %CHANGE 70S/50S	PERCENTILE RATES 1970S
24	175.5	25	153.8	33	109.9	CHEROKEE OK	-12	-29	-37	21	53
21	99.5	23	97.0	26	96.7	CHOCTAW OK	-3	0	-3	53	47
2	118.0	1	26.5	6	50.9	CLEVELAND OK	-78	92	-57	11	34
6	197.0	3	95.8	2	28.3	COAL OK	-51	-70	-86	01	31
35	163.6	50	167.6	55	114.5	COMANCHE OK	2	-32	-30	26	56
1	49.1	3	147.4	1	36.4	COTTON OK	200	-75	-26	29	32
7	111.7	10	82.7	13	74.5	CRAIG OK	-26	-10	-33	23	40
31	133.1	33	137.5	32	103.2	CREEK OK	3	-25	-22	32	49
8	145.5	6	102.6	13	212.8	CUSTER OK	-30	107	46	85	92
16	222.8	11	102.2	17	93.8	DELAWARE OK	-54	-8	-58	10	46
		1	86.0	1	78.1	DEWEY OK		-9			41
1*	546.4					ELLIS OK					
14	193.7	20	205.4	25	196.9	GARFIELD OK	6	-4	2	57	90
12	211.0	8	121.6	9	75.9	GARVIN OK	-42	-38	-64	08	40
8	73.0	7	69.1	25	205.5	GRADY OK	-5	198	181	96	91
3	187.3	2	89.7	7	252.4	GREER OK	-52	181	35	81	94
1	66.2	1	89.3	2	98.2	HARMON OK	35	10	48	85	48
7*	336.1	3	114.0	9	163.2	HASKELL OK	-66	43	-51	13	82
20	210.1	16	125.1	16	125.7	HUGHES OK	-40	0	-40	19	63
3	47.8	9	123.0	13	128.9	JACKSON OK	157	5	169	96	65
1	106.0	1	62.9	3	218.9	JEFFERSON OK	-41	248	107	92	92
5	131.4	9	254.0	5	88.1	JOHNSTON OK	93	-65	-33	23	44
14	177.0	18	147.1	23	156.3	KAY OK	-17	6	-12	44	80
13*	259.4	7	135.1	4	96.5	KINGFISHER OK	-48	-29	-63	08	47
8	137.7	9	147.9	10	163.6	KIOWA OK	7	11	19	73	83
6	123.6	4	97.6	6	113.2	LATIMER OK	-21	16	-8	47	55
10	107.1	11	93.0	23	116.4	LE FLORE OK	-13	25	9	65	57
7	87.6	12	150.5	9	91.2	LINCOLN OK	72	-39	4	60	45
36	170.4	25	102.9	23	90.6	LOGAN OK	-40	-12	-47	16	45
3	190.5	2	166.9			LOVE OK	-12				
1	26.5	3	74.3	6	137.4	MC CLAIN OK	181	85	418	98	71
51	161.0	36	103.5	42	97.1	MC CURTAIN OK	-36	-6	-40	19	47
31	172.2	21	113.9	24	91.0	MC INTOSH OK	-34	-20	-47	16	45
2	98.0	2	85.0	4	80.5	MARSHALL OK	-13	-5	-18	37	41
5	79.2	13	110.2	15	81.1	MAYES OK	39	-26	2	58	42
6	163.4	4	156.7	6	119.7	MURRAY OK	-4	-24	-27	28	60
116	132.0	125	129.9	131	125.4	MUSKOGEE OK	-2	-3	-5	50	63
3	66.3	4	84.2	11*	228.1	NOBLE OK	27	171	244	97	93
11	175.7	11	140.5	10	92.4	NOWATA OK	-20	-34	-47	15	46
28	136.2	41	209.6	39	172.4	OKFUSKEE OK	54	-18	27	78	85
213	158.9	273	132.4	380	121.5	OKLAHOMA OK	-17	-8	-24	31	61
67	166.5	60	125.9	74	147.9	OKMULGEE OK	-24	17	-11	44	76
21	209.9	18	126.9	31	106.2	OSAGE OK	-40	-16	-49	14	51
3	49.3	5	59.9	8	39.6	OTTAWA OK	22	-34	-20	35	33
10	282.3	10	186.7	6	101.5	PAWNEE OK	-34	-46	-64	08	49
8	201.0	12	246.3	12	146.9	PAYNE OK	23	-40	-27	28	76
20	116.6	29	142.5	33	109.6	PITTSBURG OK	22	-23	-6	50	53
8	111.4	11	110.7	13	95.3	PONTOTOC OK	-1	-14	-14	41	47
20	168.1	17	121.4	15	75.7	POTTAWATOMIE OK	-28	-38	-55	12	40
6	147.5	5	89.7	7	93.3	PUSHMATAHA OK	-39	4	-37	21	46
2	308.0	1	267.9	2	162.2	ROGER MILLS OK	-13	-39	-47	15	82
8	109.8	7	80.6	17	88.1	ROGERS OK	-27	9	-20	35	44
31	127.8	49	191.5	47	159.3	SEMINOLE OK	50	-17	25	76	81
21*	232.7	15	113.8	19	93.4	SEQUOYAH OK	-51	-18	-60	09	46
3	102.7	6	120.7	8	91.1	STEPHENS OK	17	-24	-11	44	45
7	76.5	12	162.4	18	200.3	TILLMAN OK	112	23	162	95	90
156	128.9	225	126.2	325	124.1	TULSA OK	-2	-2	-4	52	62
36	186.2	16	89.5	12	52.0	WAGONER OK	-52	-42	-72	04	34
14	176.3	19	167.2	23	104.7	WASHINGTON OK	-5	-37	-41	19	50
1	117.9					WASHITA OK					
				1	106.0	WOODWARD OK					51
135	151.4	170	119.2	277	101.2	OREGON OR	-21	-15	-33		
		1	344.1	1	267.9	BAKER OR		-22			95
				1	67.0	BENTON OR					37
1	86.0	2	57.1	14	131.0	CLACKAMAS OR	-34	130	52	86	67
		1	78.1			CLATSOP OR					
2*	617.3	1	78.1	2	208.5	COLUMBIA OR	-87	167	-66	07	91
2	316.1	3	163.9	2	64.3	COOS OR	-48	-61	-80	02	37
				1	106.0	DESCHUTES OR					51
3*	417.9	2	283.1	1	7.3	DOUGLAS OR	-32	-97	-98	00	30
2	658.8			1	94.4	HARNEY OR			-86	01	46
		2	224.8	3	111.5	HOOD RIVER OR		-50			54
2	175.4			4	59.3	JACKSON OR			-66	07	36
4	249.2	2	71.9	5	113.1	JEFFERSON OR	-71	57	-55	12	55
		1	66.2	1	32.1	JOSEPHINE OR		-51			31
12	254.7	7	165.3	11	148.3	KLAMATH OR	-35	-10	-42	18	77
1	25.6	1	27.1	8	83.2	LANE OR	6	207	225	97	42
1	117.9	3	314.3	1	49.2	LINCOLN OR	167	-84	-58	10	34
		1	109.3	1	62.9	LINN OR		-42			37
6	102.3			10	101.7	MALHEUR OR			-1	55	49
1	49.1	1	12.8	6	37.6	MARION OR	-74	195	-23	31	32
91	178.8	133	142.8	189	120.7	MULTNOMAH OR	-20	-15	-32	24	60
1	73.1			2	87.7	POLK OR			20	74	44
				1	94.4	TILLAMOOK OR					46
4	101.5	6	128.0	8	102.6	UMATILLA OR	26	-20	1	57	49
		1	66.2			WASCO OR					
1	42.0			4	66.7	WASHINGTON OR			59	88	37
1	37.2	2	163.6			YAMHILL OR	340				
4998*	166.9	6527*	158.1	8868*	164.5	PENNSYLVANIA PA	-5	4	-1		
		3	145.7	3	120.2	ADAMS PA		-17			60
842*	161.7	1052*	156.4	1342*	162.8	ALLEGHENY PA	-3	4	1	57	82
5	97.6	5	99.6	7	130.7	ARMSTRONG PA	2	31	34	81	66
49	156.7	73	167.9	103*	178.7	BEAVER PA	7	6	14	70	87
3	203.5			5	310.2	BEDFORD PA			52	86	97
21	136.0	38	171.9	54	151.0	BERKS PA	26	-12	11	68	78
8	159.7	7	97.1	13	179.9	BLAIR PA	-39	85	13	69	87
2	485.4	1	267.9	3	275.5	BRADFORD PA	-45	3	-43	18	95

* SIGNIFICANTLY MORE DEATHS THAN EXPECTED

NON-WHITE FEMALE — ALL CANCERS COMBINED — ICD 140-208

1950-1959 DEATHS	1950-1959 RATE	1960-1969 DEATHS	1960-1969 RATE	1970-1979 DEATHS	1970-1979 RATE	COUNTY AND STATE		PERCENT CHANGE 60S/50S	PERCENT CHANGE 70S/60S	PERCENT CHANGE 70S/50S	PERCENTILE %CHANGE 70S/50S	PERCENTILE RATES 1970S
13	137.1	27	129.0	78*	188.5	BUCKS	PA	-6	46	37	82	89
1	68.8	6	223.0	6	139.2	BUTLER	PA	224	-38	102	92	72
20	229.4	22	142.5	21	121.0	CAMBRIA	PA	-38	-15	-47	15	60
				1	163.6	CARBON	PA					83
1	78.1			6	304.3	CENTRE	PA			290	97	96
110*	195.6	112	155.3	162*	168.0	CHESTER	PA	-21	8	-14	41	84
				3*	981.1	CLARION	PA					99
1	34.4	4	306.9	5	203.9	CLEAPFIELD	PA	792	-34	493	99	91
2*	734.5			2	848.1	CLINTON	PA			15	71	99
2	412.8	1	47.2	4*	358.4	COLUMBIA	PA	-89	660	-13	42	97
9	189.9	5	91.2	9	135.0	CRAWFORD	PA	-52	48	-29	27	70
15	194.2	12	146.6	12	99.7	CUMBERLAND	PA	-24	-32	-49	14	48
118	187.2	161	163.3	249*	184.3	DAUPHIN	PA	-13	13	-2	54	88
205	157.9	258	140.2	407*	166.4	DELAWAPE	PA	-11	19	5	62	84
1*	168.7	1	344.1	7*	1909.8	ELK	PA	82	455	912	99	99
31*	192.4	36	141.1	60	154.5	ERIE	PA	-27	10	-20	35	79
64	172.6	66	178.3	80*	203.0	FAYETTE	PA	3	14	18	72	91
				1*	596.2	FOREST	PA					99
11	181.6	12	142.4	22*	205.7	FRANKLIN	PA	-22	44	13	69	91
1	267.9	1	596.2			FULTON	PA	123				
1	57.3	6	273.3	2	71.2	GREENE	PA	377	-74	24	76	38
2	88.1	4	166.8	8	300.3	HUNTINGDON	PA	89	80	241	97	96
1	38.5	3	152.2	5	209.0	INDIANA	PA	295	37	442	98	91
1	188.7			3*	773.8	JEFFERSON	PA			310	98	99
9	186.2	9	216.1	16*	226.7	LACKAWANNA	PA	16	5	22	75	92
29*	220.5	30	199.2	48	179.6	LANCASTER	PA	-10	-10	-19	36	87
19	157.0	22	170.7	31*	212.1	LAWRENCE	PA	9	24	35	81	92
2	146.6	2	95.9	5	161.4	LEBANON	PA	-35	68	10	67	82
4	101.3	6	134.2	16	124.6	LEHIGH	PA	32	-7	23	75	63
5	112.4	6	88.9	23*	250.3	LUZERNE	PA	-21	182	123	94	94
12	200.8	10	117.2	14	154.4	LYCOMING	PA	-42	32	-23	31	79
2	271.4	4*	556.2	4*	712.0	MC KEAN	PA	105	28	162	96	99
16	111.7	29	160.9	35	147.3	MERCER	PA	44	-8	32	80	76
1	81.8			3	232.1	MIFFLIN	PA			184	96	93
3	149.9	3	100.9	3	114.4	MONPOE	PA	-33	13	-24	31	56
95	135.9	132	128.8	239*	166.3	MONTGOMERY	PA	-5	29	22	75	83
1	163.6	2	360.7	6*	1300.9	MONTOUR	PA	121	261	695	99	99
8	113.5	18	155.0	20	111.0	NORTHAMPTON	PA	37	-28	-2	53	54
2	161.4	2	168.3	16*	946.9	NORTHUMBERLAND	PA	-7	463	422	98	99
				2*	556.1	PERRY	PA					99
3103*	169.5	4153*	161.3	5452*	161.5	PHILADELPHIA	PA	-5	0	-5	51	82
2*	814.3			1	344.1	PIKE	PA			-58	10	97
1	32.7	2	81.0	4	82.7	SCHUYLKILL	PA	148	2	153	95	42
		1	424.0	2*	282.0	SNYDER	PA		-33			96
1	37.5	1	37.2	3	176.8	SOMERSET	PA	-1	376	372	98	86
				1*	74.3	SULLIVAN	PA					39
1	344.1	2*	189.0	6*	747.0	SUSQUEHANNA	PA	-45	295	117	94	99
		2	168.6	1	294.8	TIOGA	PA		75			96
				1	136.6	UNION	PA					71
6	191.2	3	153.3	5	185.7	VENANGO	PA	-20	21	-3	53	88
1	141.3			2	339.7	WARREN	PA			140	95	97
66	168.9	68	155.6	112*	237.4	WASHINGTON	PA	-8	53	41	83	93
		1	490.8	1	112.4	WAYNE	PA		-77			55
44	153.0	54	168.6	60	159.5	WESTMORELAND	PA	10	-5	4	61	81
25	162.1	49*	228.7	53	165.4	YORK	PA	41	-28	2	58	83
100	166.1	111	124.6	184	126.3	RHODE ISLAND	RI	-25	1	-24		
1	245.4			1	54.2	BRISTOL	RI			-78	03	35
3	176.5	4	185.5	5	119.5	KENT	RI	5	-36	-32	25	59
21	210.8	19	148.9	20	131.2	NEWPORT	PI	-29	-12	-38	21	67
66	150.4	85	128.1	149	129.8	PROVIDENCE	RI	-15	1	-14	41	66
9	190.7	3	59.3	9	99.4	WASHINGTON	RI	-69	68	-48	15	48
3609	124.5	4024	125.8	4881	126.6	SOUTH CAROLINA	SC	1	1	2		
41	137.0	41	135.8	38	112.8	ABBEVILLE	SC	-1	-17	-18	38	55
100	122.5	131	144.2	141	127.4	AIKEN	SC	18	-12	4	61	64
35	114.0	33	114.8	39	124.3	ALLENDALE	SC	1	8	9	66	63
89	119.0	117	138.9	106	105.3	ANDERSON	SC	17	-24	-12	44	51
43	136.4	33	102.8	40	102.2	BAMBERG	SC	-25	-1	-25	29	49
31	83.2	34	101.4	36	105.3	BARNWELL	SC	22	4	26	77	50
86	133.5	97	147.8	107	136.9	BEAUFORT	SC	11	-7	3	59	71
86	140.5	60	92.6	82	98.2	BERKELEY	SC	-34	6	-30	26	48
39	117.7	37	124.9	37	115.3	CALHOUN	SC	6	-8	-2	53	57
365	133.5	398	132.2	549*	143.1	CHARLESTON	SC	-1	8	7	64	74
34	126.6	39	130.0	50	147.6	CHEROKEE	SC	3	14	17	72	76
68	125.5	76	147.0	90	167.6	CHESTER	SC	17	14	33	81	84
58	145.5	63	141.7	68	127.9	CHESTERFIELD	SC	-3	-10	-12	43	65
78	125.7	71	106.1	95	129.9	CLARENDON	SC	-16	22	3	60	66
61	119.1	52	93.1	81	126.8	COLLETON	SC	-22	36	6	63	64
104	144.0	100	119.9	120	117.8	DARLINGTON	SC	-17	-2	-18	37	58
52	117.2	33	73.2	66	121.9	DILLON	SC	-38	66	4	60	61
45	111.9	54	119.4	64	103.5	DORCHESTER	SC	7	-13	-8	48	50
38	125.7	34	117.2	55	166.8	EDGEFIELD	SC	-7	42	33	80	84
59	128.2	71	152.2	69	132.2	FAIPFIELD	SC	19	-13	3	59	68
144	124.2	192	143.0	190	114.3	FLORENCE	SC	15	-20	-8	48	56
75	133.4	77	120.2	96	123.5	GEOPGETOWN	SC	-10	3	-7	48	62
184	141.6	216	137.4	293	146.8	GREENVILLE	SC	-3	7	4	60	75
85	168.6	85	139.1	99	136.8	GREENWOOD	SC	-17	-2	-19	36	71
39	111.5	49	133.7	51	123.7	HAMPTON	SC	20	-7	11	68	62
37	81.9	59	106.1	68	89.2	HORRY	SC	30	-16	9	66	44
28	109.4	47	154.8	54	141.4	JASPER	SC	42	-9	29	79	73
68	128.0	73	134.3	63	103.4	KERSHAW	SC	5	-23	-19	36	50
34	87.7	46	109.3	64	121.2	LANCASTER	SC	25	11	38	82	61
62	111.8	84	141.2	84	122.3	LAURENS	SC	26	-13	9	66	61
43	100.9	42	95.3	59	129.7	LEE	SC	-6	36	29	79	66
58	156.0	59	139.3	66	125.9	LEXINGTON	SC	-11	-10	-19	35	64
24	121.7	28	151.6	28	145.8	MC CORMICK	SC	25	-4	20	74	75
80	121.7	80	115.3	91	110.4	MARION	SC	-5	-4	-9	46	54
65	120.5	75	143.2	76	118.3	MARLBORO	SC	19	-17	-2	54	59

* SIGNIFICANTLY MORE DEATHS THAN EXPECTED

NON-WHITE FEMALE | ALL CANCERS COMBINED | ICD 140-208

1950-1959 DEATHS	1950-1959 RATE	1960-1969 DEATHS	1960-1969 RATE	1970-1979 DEATHS	1970-1979 RATE	COUNTY AND STATE		PERCENT CHANGE 60S/50S	PERCENT CHANGE 70S/60S	PERCENT CHANGE 70S/50S	PERCENTILE %CHANGE 70S/50S	PERCENTILE RATES 1970S
56	124.7	64	138.8	62	122.3	NEWBERRY	SC	11	-12	-2	53	61
27	146.7	22	117.7	34	150.4	OCONEE	SC	-20	28	2	58	77
169	125.2	188	126.9	236	128.5	ORANGEBURG	SC	1	1	3	59	65
31	181.6	40	174.3	40	138.1	PICKENS	SC	-4	-21	-24	31	72
248	118.3	303	114.8	445	136.1	RICHLAND	SC	-3	19	15	71	70
25	132.5	27	142.1	28	132.0	SALUDA	SC	7	-7	0	55	68
149	115.1	208	138.2	265	148.1	SPARTANBURG	SC	20	7	29	79	76
157	129.8	132	96.8	173	106.8	SUMTER	SC	-25	10	-18	38	52
37	92.2	50	129.2	52	123.6	UNION	SC	40	-4	34	81	62
72	91.0	92	108.8	104	105.7	WILLIAMSBURG	SC	20	-3	16	71	51
100	124.3	112	128.7	127	123.7	YORK	SC	4	-4	-1	55	62
127	176.2	175*	206.0	179	153.6	SOUTH DAKOTA	SD	17	-25	-13		
		1	546.4	1	56.2	BEADLE	SD		-90			35
8*	326.0	9*	320.6	10*	323.8	BENNETT	SD	-2	1	-1	55	97
		2*	536.4			BON HOMME	SD					
1*	344.1					BRULE	SD					
1	54.5	2	71.4	4	77.9	BUFFALO	SD	31	9	43	84	41
1*	562.5					BUTTE	SD					
6	138.4	7	154.0	7	144.7	CHARLES MIX	SD	11	-6	5	61	75
		1*	546.4			CODINGTON	SD					
6	201.1	15*	465.9	9	158.8	CORSON	SD	142	-67	-21	33	81
		1	112.4	1	84.4	CUSTER	SD		-25			42
				1	490.8	DAVISON	SD					98
1	149.1	1	106.0	2	143.8	DAY	SD	-29	36	-4	52	74
10	201.2	15	230.9	14	163.3	DEWEY	SD	15	-29	-19	36	82
				1*	546.6	DOUGLAS	SD					99
		1*	546.6			EDMUNDS	SD					
1	20.1	1	45.4	2	98.2	FALL RIVER	SD	126	117	389	98	48
		1*	663.2			GRANT	SD					
1	68.8	4*	368.5	2	200.8	GREGORY	SD	436	-46	192	96	90
3	633.6	1	74.3	5	333.4	HUGHES	SD	-88	348	-47	15	97
				1*	172.0	HYDE	SD					85
4	266.7	5	254.5	4	106.4	JACKSON	SD	-5	-58	-60	09	51
1*	546.4					LAKE	SD					
				1	188.7	LAWRENCE	SD					89
		3	166.5			LYMAN	SD					
1	490.8	1	74.3	1	109.3	MARSHALL	SD	-85	47	-78	03	53
1	490.8			1	267.9	MEADE	SD			-45	16	95
3	135.7	7	283.6	1	29.5	MELLETTE	SD	109	-90	-78	02	31
1	141.3	2	175.3	3	175.2	MINNEHAHA	SD	24	0	24	76	86
1	84.8	2	435.1	3	401.6	MOODY	SD	413	-8	373	98	98
10	266.1	5	135.3	18	185.3	PENNINGTON	SD	-49	37	-30	26	88
				1*	74.3	POTTER	SD					39
15*	345.3	11	220.3	10	184.5	ROBERTS	SD	-36	-16	-47	16	88
26	183.2	32	167.9	44	180.2	SHANNON	SD	-8	7	-2	54	87
		3*	675.0	1	212.0	STANLEY	SD		-69			92
19*	223.7	22	215.3	18	118.9	TODD	SD	-4	-45	-47	16	59
2	151.9	8*	627.5	1	84.2	TRIPP	SD	313	-87	-45	17	42
		1*	267.9	1*	267.9	TURNER	SD		0			95
				1	255.8	UNION	SD					94
2*	814.3	4*	872.9			WALWORTH	SD	7				
				3	189.0	YANKTON	SD					89
2	71.6	7	215.9	7	299.7	ZIEBACH	SD	202	39	319	98	96
3412	136.6	4378*	152.6	4984*	145.9	TENNESSEE	TN	12	-4	7		
11	267.6	11	195.0	14	158.6	ANDERSON	TN	-27	-19	-41	19	81
16	105.5	18	127.8	18	106.0	BEDFORD	TN	21	-17	0	56	51
		1	42.4	5	288.5	BENTON	TN		580			96
2	311.4	1	68.8	4	210.7	BLEDSOE	TN	-78	206	-32	24	91
9	72.0	21	147.3	26	165.7	BLOUNT	TN	105	12	130	94	83
10	107.3	17	169.2	19	151.2	BRADLEY	TN	58	-11	41	83	78
2	137.4	4	242.5	2	245.9	CAMPBELL	TN	76	1	79	91	94
3	302.6	1	49.7			CANNON	TN	-84				
20	123.0	14	80.4	24	112.7	CARROLL	TN	-35	40	-8	47	55
2	123.7	5	214.8	3	96.1	CARTER	TN	74	-55	-22	32	47
3	90.0	4	123.7	3	86.1	CHEATHAM	TN	38	-30	-4	51	43
8	132.4	11	169.1	6	86.0	CHESTER	TN	28	-49	-35	23	43
2	205.8	1	67.0	2	213.6	CLAIBORNE	TN	-67	219	4	60	92
1	74.3	1	94.4	2	259.4	CLAY	TN	27	175	249	97	95
2	53.8	4	85.7	10*	297.4	COCKE	TN	59	247	453	98	96
7	150.6	4	104.2	11	174.4	COFFEE	TN	-31	67	16	71	86
16	128.2	30*	211.7	11	88.3	CROCKETT	TN	65	-58	-31	25	44
501	146.8	628*	157.6	760*	154.7	DAVIDSON	TN	7	-2	5	62	79
		5	165.2	4	117.5	DECATUR	TN		-29			58
4	214.6	5	189.6	3	165.5	DE KALB	TN	-12	-13	-23	32	83
10	163.6	11	191.2	9	132.1	DICKSON	TN	17	-31	-19	35	68
35	140.4	36	142.8	40	142.6	DYER	TN	2	0	2	57	74
61	97.8	69	115.3	47	74.0	FAYETTE	TN	18	-36	-24	30	39
20	206.1	14	169.1	11	97.8	FRANKLIN	TN	-18	-42	-53	13	48
65	144.2	64	131.4	86	157.9	GIBSON	TN	-9	20	10	66	80
22	96.1	28	124.9	30	133.7	GILES	TN	30	7	39	83	69
		3	302.6	2	248.6	GRAINGER	TN		-18			94
5	101.6	4	63.5	8	103.1	GREENE	TN	-37	62	2	57	49
8	93.3	8	85.8	12	105.1	HAMBLEN	TN	-8	22	13	69	50
282	130.3	350	140.7	439*	151.6	HAMILTON	TN	8	8	16	71	78
49	134.4	49	123.5	54	128.3	HARDEMAN	TN	-8	4	-5	51	65
4	65.5	12	213.3	8	126.8	HARDIN	TN	226	-41	94	92	64
10	183.8	10	177.9	10	150.1	HAWKINS	TN	-3	-16	-18	37	77
57	104.5	70	129.5	64	117.9	HAYWOOD	TN	24	-9	13	69	58
7	88.0	13	155.1	15	150.5	HENDERSON	TN	76	-3	71	89	77
36	199.9	21	100.1	22	94.3	HENRY	TN	-50	-6	-53	12	46
4	105.0	3	63.8	5	134.2	HICKMAN	TN	-39	110	28	78	69
3	164.2	3	189.1	6*	461.9	HOUSTON	TN	15	144	181	96	98
3	95.7	5	176.2	1	39.0	HUMPHREYS	TN	84	-78	-59	10	33
2	611.9					JACKSON	TN					
4	88.1	12	246.6	9	172.9	JEFFERSON	TN	180	-30	96	92	86
		1	94.4			JOHNSON	TN					
140	127.6	209*	165.3	223	147.4	KNOX	TN	29	-11	15	71	76

* SIGNIFICANTLY MORE DEATHS THAN EXPECTED

NON-WHITE FEMALE — ALL CANCERS COMBINED — ICD 140-208

1950-1959 DEATHS	1950-1959 RATE	1960-1969 DEATHS	1960-1969 RATE	1970-1979 DEATHS	1970-1979 RATE	COUNTY AND STATE		PERCENT CHANGE 60S/50S	70S/60S	70S/50S	PERCENTILE %CHANGE 70S/50S	PERCENTILE RATES 1970S
12	136.4	12	164.1	13	151.0	LAKE	TN	20	-8	11	67	78
40	118.4	61	166.6	33	76.8	LAUDERDALE	TN	41	-54	-35	22	40
5	218.7	5	164.9	2	38.7	LAWRENCE	TN	-25	-77	-82	02	33
		2	306.0	1	147.4	LEWIS	TN		-52			76
15	109.6	16	119.1	30	177.5	LINCOLN	TN	9	49	62	88	87
4	174.8	5	188.6	8*	367.7	LOUDON	TN	8	95	110	93	97
10	186.9	12	133.4	10	95.0	MC MINN	TN	-29	-29	-49	14	47
6	133.9	6	105.6	6	96.1	MC NAIRY	TN	-21	-9	-28	27	47
136	154.1	153	148.7	178	151.0	MADISON	TN	-3	2	-2	53	78
3	45.4	9	135.0	13	184.4	MARION	TN	197	37	306	98	88
18	198.8	6	61.7	16	151.0	MARSHALL	TN	-69	145	-24	30	78
62	143.6	75	172.6	70	136.4	MAURY	TN	20	-21	-5	50	71
5	159.0	6	140.8	8	138.5	MONROE	TN	-11	-2	-13	42	72
49	115.3	60	128.7	90	168.2	MONTGOMERY	TN	12	31	46	84	84
1	122.7	2	230.6	1	61.3	MOORE	TN	88	-73	-50	14	36
1	424.0	1	546.4			MORGAN	TN	29				
20	120.4	43*	265.7	36*	212.6	OBION	TN	121	-20	77	90	92
1	74.3	1	94.4	1	74.3	OVERTON	TN	27	-21	0	56	39
		1	106.0	1	133.9	PERRY	TN		26			69
		1*	424.0			POLK	TN					
2	57.1	8	286.7	3	92.9	PUTNAM	TN	402	-68	63	88	46
4	141.6	9	240.1	5	136.8	RHEA	TN	70	-43	-3	52	71
16	233.2	10	119.0	8	85.0	ROANE	TN	-49	-29	-64	08	43
29	128.0	22	90.8	23	80.8	ROBERTSON	TN	-29	-11	-37	21	41
43	138.7	48	148.9	45	110.5	RUTHERFORD	TN	7	-26	-20	34	54
1	141.3			3	259.8	SEVIER	TN			84	91	95
1223	135.3	1806*	165.2	2155*	160.1	SHELBY	TN	22	-3	18	73	81
9*	253.6	2	59.5	5	96.7	SMITH	TN	-77	62	-63	08	47
		2	565.4			STEWART	TN					
23	208.4	24	194.4	23	144.4	SULLIVAN	TN	-7	-26	-31	25	74
41*	203.7	30	134.9	21	83.3	SUMNER	TN	-34	-38	-59	10	42
53	128.6	52	116.9	43	88.6	TIPTON	TN	-9	-24	-31	25	44
8	245.2	5	111.9	8	240.6	TROUSDALE	TN	-54	115	-2	54	93
1*	546.4					UNICOI	TN					
1*	490.8					VAN BUREN	TN					
13	218.4	6	103.7	7	96.9	WARREN	TN	-53	-7	-56	12	47
24	187.2	16	113.1	24	130.4	WASHINGTON	TN	-40	15	-30	26	66
3	207.0	2	146.6	1	182.1	WAYNE	TN	-29	24	-12	43	87
12	117.8	15	147.0	15	140.8	WEAKLEY	TN	25	-4	20	73	73
6	188.9	6	278.2	3	78.0	WHITE	TN	47	-72	-59	10	41
33	168.2	24	105.8	27	104.4	WILLIAMSON	TN	-37	-1	-38	20	50
33	180.9	33	164.5	20	73.8	WILSON	TN	-9	-55	-59	10	39
5823	127.2	7624	133.4	9754	114.4	TEXAS	TX	5	-14	-10		
50	113.6	61	122.8	69	142.4	ANDERSON	TX	8	16	25	77	73
				1	47.1	ANDREWS	TX					33
30	109.9	52	147.7	52	118.0	ANGELINA	TX	34	-20	7	64	58
		2	189.0	3	74.1	ARANSAS	TX		-61			39
		1	147.4	1	10.9	ATASCOSA	TX		-93			30
17	108.6	12	61.6	20	96.2	AUSTIN	TX	-43	56	-11	44	47
2	161.4	1	109.3	1	74.3	BAILEY	TX	-32	-32	-54	12	39
49	161.2	49	156.3	46	135.9	BASTROP	TX	-3	-13	-16	39	70
5	194.7	3	84.4	7	75.7	BEE	TX	-57	-10	-61	09	40
39	132.9	46	118.0	71	115.0	BELL	TX	-11	-3	-13	42	56
234	136.7	373	150.5	533	108.3	BEXAR	TX	10	-28	-21	33	53
3*	464.7	1	89.3	1	37.2	BLANCO	TX	-81	-58	-92	00	32
		3	124.0	2	87.5	BOSQUE	TX		-29			44
105	149.8	96	124.9	108	123.4	BOWIE	TX	-17	-1	-18	38	62
32	113.3	45	126.2	73	134.4	BRAZORIA	TX	11	7	19	73	69
52	114.2	55	109.0	63	98.6	BRAZOS	TX	-5	-10	-14	42	48
1	147.4					BRISCOE	TX					
6	120.8	10	179.1	4	58.5	BROWN	TX	48	-67	-52	13	36
26	113.7	24	110.4	28	123.5	BURLESON	TX	-3	12	9	65	62
1	109.3	1	112.4	1	39.0	BURNET	TX	3	-65	-64	07	33
24	141.9	23	118.2	27	125.2	CALDWELL	TX	-17	6	-12	43	63
1	28.3	2	72.3	4	51.7	CALHOUN	TX	156	-29	83	91	34
4	65.1	3	72.8	12	32.6	CAMERON	TX	12	-55	-50	14	32
25	167.5	26	152.2	35*	198.9	CAMP	TX	-9	31	19	73	90
46	134.3	39	102.5	43	107.9	CASS	TX	-24	5	-20	35	52
		2	404.4	1	35.1	CASTRO	TX		-91			32
7	96.4	6	64.0	14	132.7	CHAMBERS	TX	-34	107	38	82	68
49	98.2	53	114.0	68	134.8	CHEROKEE	TX	16	18	37	82	70
1	17.6	3	116.2	4	175.2	CHILDRESS	TX	559	51	894	99	86
2	185.4			3*	427.3	CLAY	TX			131	94	98
1	56.3	2	201.0	1	47.1	COCHRAN	TX	257	-77	-16	39	33
2	91.7	5	445.3	8	246.3	COLEMAN	TX	386	-45	169	96	94
20	118.9	24	133.8	27	108.5	COLLIN	TX	13	-19	-9	47	53
2	111.2	2	124.8	2	88.3	COLLINGSWORTH	TX	12	-29	-21	34	44
38	151.8	20	73.0	32	119.1	COLORADO	TX	-52	63	-22	33	59
1	59.0	2	57.8	2	51.6	COMAL	TX	-2	-11	-12	43	34
11	203.7	12	232.4	10	162.0	COOKE	TX	14	-30	-20	34	82
1	54.6	3	58.9	5	47.8	CORYELL	TX	8	-19	-13	43	34
		2	151.8	1	122.7	COTTLE	TX		-19			62
		1	114.7	2	301.1	CRANE	TX		163			96
2*	174.1	2	227.1	1	62.9	CROCKETT	TX	30	-72	-64	08	37
1	31.4	5	221.3	2	54.4	CROSBY	TX	605	-75	73	90	35
		1	112.4	1	114.7	DALLAM	TX		2			56
510	121.4	899	140.5	1246	125.5	DALLAS	TX	16	-11	3	60	63
4	317.8	2	40.1	9	167.3	DAWSON	TX	-87	317	-47	15	84
1	490.8			1	34.4	DEAF SMITH	TX			-93	00	32
1	23.6	4	79.1	6	171.0	DELTA	TX	236	116	626	99	85
9	81.9	14	105.0	19	90.9	DENTON	TX	28	-13	11	68	45
22	102.5	22	109.2	23	99.0	DE WITT	TX	7	-9	-3	52	48
2	195.3	1	74.3	1	84.8	DICKENS	TX	-62	14	-57	11	43
		1*	344.1			DIMMIT	TX					
				3	249.4	DONLEY	TX					94
				1	17.2	DUVAL	TX					30
7*	268.9	2	45.8	2	45.5	EASTLAND	TX	-83	0	-83	02	33
15	212.0	21	136.6	25	105.7	ECTOR	TX	-36	-23	-50	14	51
67	142.6	70	151.4	59	109.3	ELLIS	TX	6	-28	-23	31	53

* SIGNIFICANTLY MORE DEATHS THAN EXPECTED

NON-WHITE FEMALE

ALL CANCERS COMBINED

ICD 140-208

COUNTY AND STATE	1950-1959 DEATHS	1950-1959 RATE	1960-1969 DEATHS	1960-1969 RATE	1970-1979 DEATHS	1970-1979 RATE	PERCENT CHANGE 60S/50S	PERCENT CHANGE 70S/60S	PERCENT CHANGE 70S/50S	PERCENTILE %CHANGE 70S/50S	PERCENTILE RATES 1970S
EL PASO TX	23	137.0	39	153.8	54	25.9	12	-83	-81	02	31
ERATH TX	1	133.9	1	187.5	1	49.2	40	-74	-63	08	34
FALLS TX	47	111.4	53	123.4	52	120.4	11	-2	8	65	60
FANNIN TX	26	160.8	13	94.2	22	159.5	-41	69	-1	55	81
FAYETTE TX	27	135.7	22	120.5	24	145.6	-11	21	7	64	75
FISHER TX	1	49.1	2	112.1	1	54.6	128	-51	11	68	35
FLOYD TX			6	293.4	2	36.4		-88			32
FOARD TX	3	697.3	1	89.3			-87				
FORT BEND TX	46	127.3	34	85.4	58	94.2	-33	10	-26	29	46
FRANKLIN TX	5	334.3	3	180.2	4	161.7	-46	-10	-52	13	82
FREESTONE TX	37	140.1	35	109.3	29	82.0	-22	-25	-41	18	42
GAINES TX			1	45.4	1	28.8		-37			31
GALVESTON TX	130	129.2	190	138.8	243	132.6	7	-4	3	59	68
GARZA TX					3	79.1					41
GOLIAD TX	5	131.8	8	228.8	9	187.9	74	-18	43	84	89
GONZALES TX	17	88.8	14	75.9	22	96.8	-14	28	9	66	47
GRAY TX	3	153.2	2	132.3	6	97.6	-14	-26	-36	22	48
GRAYSON TX	42	113.3	44	109.2	52	117.0	-4	7	3	59	58
GPEGG TX	84	115.9	128	162.0	144	157.7	40	-3	36	82	80
GRIMES TX	32	105.0	62*	213.9	46	133.9	104	-37	28	78	69
GUADALUPE TX	26	140.7	33	171.1	23	80.6	22	-53	-43	18	41
HALE TX	3	90.9	5	104.2	3	24.7	15	-76	-73	04	31
HALL TX	1	9.7	8*	346.0	6	293.8	3466	-15	2928	99	96
HAMILTON TX					1	149.1					77
HARDEMAN TX	6	201.5	2	39.6	8	250.1	-80	532	24	76	94
HARDIN TX	15	105.1	27	145.6	22	111.3	38	-24	6	63	54
HARRIS TX	1038	144.6	1515	139.2	2182	129.3	-4	-7	-11	45	66
HARRISON TX	136	136.6	114	110.6	172	148.3	-19	34	9	65	77
HASKELL TX	3	331.9	2	71.4	3	110.8	-78	55	-67	07	54
HAYS TX	10	131.2	10	118.0	6	35.0	-10	-70	-73	04	32
HENDERSON TX	24	106.8	27	114.2	38	128.3	7	12	20	74	65
HIDALGO TX	1	26.5	6	168.7	4	6.8	537	-96	-74	04	30
HILL TX	26	148.9	23	124.5	27	126.9	-16	2	-15	41	64
HOCKLEY TX	2	43.8	3	70.2	4	62.1	60	-12	42	83	36
HOOD TX	1	589.6	1	281.2			-52				
HOPKINS TX	10	105.6	9	79.9	15	112.5	-24	41	7	63	55
HOUSTON TX	36	92.9	35	82.1	63	140.4	-12	71	51	86	73
HOWARD TX	7	164.3	7	212.5	8	53.4	29	-75	-68	06	35
HUNT TX	24	96.3	37	135.3	53	169.7	40	25	76	90	85
HUTCHINSON TX	3	156.9	3	72.4	4	103.8	-54	43	-34	23	50
JACK TX					2	540.2					98
JACKSON TX	12	138.7	15	156.9	13	103.0	13	-34	-26	29	49
JASPER TX	34	171.2	31	117.0	39	119.9	-32	3	-30	26	60
JEFFERSON TX	268	141.1	360	149.8	485*	157.6	6	5	12	68	80
JIM WELLS TX	2	118.1	1	54.6	2	32.0	-54	-41	-73	04	31
JOHNSON TX	9	95.4	11	145.0	13	116.3	52	-20	22	75	57
JONES TX	7	138.0	8	187.4	5	82.9	36	-56	-40	19	42
KAPNES TX	2	57.8	4	187.0	4	70.4	223	-62	22	75	38
KAUFMAN TX	36	90.6	52	120.4	38	84.3	33	-30	-7	49	42
KENDALL TX	1	106.0			1	86.0			-19	36	43
KENEDY TX	1*	344.1									
KENT TX			1	490.8							
KEPR TX	2	86.9	5	99.9	7	106.4	15	7	22	75	52
KING TX	2*	483.5									
KINNEY TX					3	233.0					93
KLEBERG TX	6	155.1	7	105.2	4	26.8	-32	-75	-83	02	31
KNOX TX	2	80.4	4	181.0	1	31.3	125	-83	-61	09	31
LAMAR TX	39	97.9	52	130.5	62	147.3	33	13	50	86	76
LAMB TX	3	50.1	4	79.3	11	129.7	58	64	159	95	66
LAMPASAS TX	1	54.5	3	253.8	8*	329.7	366	30	505	99	97
LA SALLE TX					1	23.6					31
LAVACA TX	23	188.1	19	168.8	14	100.8	-10	-40	-46	16	48
LEE TX	22	165.5	10	69.5	17	86.2	-58	24	-48	15	43
LEON TX	27	119.3	22	94.3	28	161.2	-21	71	35	81	82
LIBERTY TX	28	108.1	40	127.6	58	165.4	18	30	53	86	83
LIMESTONE TX	46	140.5	62	175.9	56	147.7	25	-16	5	62	76
LLANO TX					1	273.2					95
LUBBOCK TX	35	127.6	43	97.0	75	102.6	-24	6	-20	35	49
LYNN TX	1	89.3	4	400.5	3	70.2	348	-82	-21	33	38
MC CULLOCH TX	3	204.6	2	106.3	5	105.6	-48	-1	-48	15	51
MC LENNAN TX	141	134.8	160	128.3	200	126.5	-5	-1	-6	49	64
MADISON TX	14	120.2	26*	218.0	17	126.4	81	-42	5	62	64
MARION TX	27	106.5	23	89.2	29	104.6	-16	17	-2	54	50
MARTIN TX					1	67.0					37
MASON TX	1*	490.8									
MATAGORDA TX	29	128.2	34	126.3	31	81.1	-2	-36	-37	21	42
MAVERICK TX	1*	281.4									
MEDINA TX	1	106.0	3	180.3	2	32.6	70	-82	-69	05	31
MENARD TX	2*	281.3									
MIDLAND TX	17	148.5	21	99.6	27	77.6	-33	-22	-48	15	41
MILAM TX	18	74.0	42	187.4	38	125.8	153	-33	70	89	63
MITCHELL TX	3	137.2	1	24.8	3	50.7	-82	104	-63	08	34
MONTAGUE TX	1*	344.1									
MONTGOMERY TX	40	135.9	29	89.1	45	108.5	-34	22	-20	34	53
MOORE TX	1*	281.4									
MORRIS TX	18	148.7	24	157.3	29	159.3	6	1	7	64	81
MOTLEY TX			2	228.3	1	114.7		-50			56
NACOGDOCHES TX	37	114.5	43	115.8	54	108.7	1	-6	-5	50	53
NAVAPRO TX	55	109.7	69	142.5	67	136.7	30	-4	25	76	71
NEWTON TX	15	105.6	15	88.9	12	60.6	-16	-32	-43	18	36
NOLAN TX	4	152.7	8	275.2	3	46.9	80	-83	-69	05	33
NUECES TX	31	95.0	74	154.2	100	89.9	62	-42	-5	50	45
ORANGE TX	24	123.7	33	140.9	29	103.5	14	-27	-16	39	50
PALO PINTO TX	8	181.8	8	233.9	5	70.0	29	-70	-62	09	38
PANOLA TX	31	132.5	25	100.1	32	106.3	-24	6	-20	35	51
PARKER TX	4	188.8	1	65.5	6	225.2	-65	244	19	73	92
PARMER TX	1	344.1			2	62.5			-82	02	36
PECOS TX			1	344.1	1	11.9		-97			30
POLK TX	32	158.6	35	173.5	24	107.1	9	-38	-32	24	52
POTTER TX	24	153.4	36	126.0	52	132.1	-18	5	-14	41	68
PRESIDIO TX	1	74.3									

* SIGNIFICANTLY MORE DEATHS THAN EXPECTED

NON-WHITE FEMALE — ALL CANCERS COMBINED — ICD 140-208

1950-1959 DEATHS	1950-1959 RATE	1960-1969 DEATHS	1960-1969 RATE	1970-1979 DEATHS	1970-1979 RATE	COUNTY AND STATE	PERCENT CHANGE 60S/50S	PERCENT CHANGE 70S/60S	PERCENT CHANGE 70S/50S	PERCENTILE %CHANGE 70S/50S	PERCENTILE RATES 1970S
		3	146.6			RAINS TX					
				2	143.8	REAGAN TX					74
20	86.5	27	140.6	27	119.6	RED RIVER TX	63	-15	38	82	60
2	123.3	3	210.1	6	85.2	REEVES TX	70	-59	-31	25	43
7	182.2	8	158.1	14	207.1	REFUGIO TX	-13	31	14	69	91
1*	596.2					ROBERTS TX					
55	141.6	41	105.8	46	115.4	ROBERTSON TX	-25	9	-19	37	57
2	39.5	3	60.0	3	123.4	ROCKWALL TX	52	106	213	97	62
3	77.6	1	31.5	3	70.3	RUNNELS TX	-59	124	-9	46	38
57	112.9	78	145.2	61	100.9	RUSK TX	29	-31	-11	45	48
7	72.7	18	210.9	12	157.8	SABINE TX	190	-25	117	94	80
16	151.9	14	101.0	20	123.7	SAN AUGUSTINE TX	-34	22	-19	36	62
11	60.6	22	117.3	11	53.7	SAN JACINTO TX	94	-54	-11	44	35
3	83.4	4	334.4	4	35.8	SAN PATRICIO TX	301	-89	-57	11	32
1	182.1	1	133.9	1	53.0	SAN SABA TX	-26	-60	-71	05	35
				1	212.0	SCHLEICHER TX					92
		7*	542.6	4	79.6	SCURRY TX		-85			41
1	74.3	1	74.3	1	172.0	SHACKELFORD TX	0	131	131	94	85
16	64.3	39	156.9	25	88.5	SHELBY TX	144	-44	38	82	44
144	148.3	133	117.7	183	133.2	SMITH TX	-21	13	-10	45	68
3	100.7	4	168.8	9*	250.2	STEPHENS TX	68	48	149	95	94
1	424.0	3*	653.9	3	245.2	STONEWALL TX	54	-62	-42	18	94
		1*	344.1			SUTTON TX					
1	112.5					SWISHER TX					
301	142.7	412	143.1	591	140.2	TARRANT TX	0	-2	-2	54	72
10	76.3	20	102.8	33	131.4	TAYLOR TX	35	28	72	90	67
1	45.5	2	109.0	5	169.7	TERRY TX	139	56	273	97	85
13	90.8	29	196.7	15	87.8	TITUS TX	117	-55	-3	52	44
25	163.1	22	132.1	22	76.0	TOM GREEN TX	-19	-43	-53	12	40
128	105.0	198	134.1	250	112.9	TRAVIS TX	28	-16	8	64	55
14	106.9	12	99.7	21	164.1	TRINITY TX	-7	65	53	86	83
9	115.6	12	123.0	12	129.5	TYLER TX	6	5	12	68	66
25	105.9	32	129.3	26	95.8	UPSHUR TX	22	-26	-10	46	47
1	106.0	1	122.7	2	128.7	UPTON TX	16	5	21	74	65
1	68.3					UVALDE TX					
		4	448.6	3	19.4	VAL VERDE TX		-96			30
9	156.7	10	137.4	6	68.2	VAN ZANDT TX	-12	-50	-57	11	38
30	150.6	46*	208.4	35	103.5	VICTORIA TX	38	-50	-31	25	50
30	91.4	40	115.2	41	99.5	WALKER TX	26	-14	9	66	48
36	154.4	30	119.6	26	83.1	WALLER TX	-23	-30	-46	16	42
2	109.7	1	45.5	5	169.2	WARD TX	-58	272	54	87	84
53	139.3	53	137.8	56	166.3	WASHINGTON TX	-1	21	19	73	83
1	490.8			1	2.7	WEBB TX			-99	00	30
56	145.0	45	103.8	51	97.5	WHARTON TX	-28	-6	-33	24	48
1	60.7	2	159.7	1	62.9	WHEELER TX	163	-61	4	60	37
38	132.7	62	160.7	76	133.5	WICHITA TX	21	-17	1	56	68
8	110.9	7	90.1	9	104.6	WILBARGER TX	-19	16	-6	50	50
1	546.4					WILLACY TX					
31	117.3	38	137.5	28	79.2	WILLIAMSON TX	17	-42	-32	24	41
2	141.3	3	219.6	2	61.9	WILSON TX	55	-72	-56	11	36
1	212.0	3	323.4	2	62.7	WINKLER TX	53	-81	-70	05	36
2	969.2			3	299.4	WISE TX			-69	05	96
20	149.1	18	114.7	27	155.4	WOOD TX	-23	35	4	61	80
		3*	672.4			YOAKUM TX					
1	49.7	1	89.3	4	169.6	YOUNG TX	80	90	241	97	85
53	153.2	48	99.6	82	72.0	UTAH UT	-35	-28	-53		
2	58.1	1	31.5	4	91.3	BOX ELDER UT	-46	190	57	87	45
4	156.9	2	31.1	8	131.8	DAVIS UT	-80	325	-16	39	67
5*	921.1			1	109.3	DUCHESNE UT			-88	01	53
1*	344.1					EMERY UT					
		2	392.8	1	133.9	IRON UT		-66			69
21	164.8	21	112.5	42	80.0	SALT LAKE UT	-32	-29	-51	13	41
		5	93.1	4	31.2	SAN JUAN UT		-66			31
1	140.3					SEVIER UT					
1	114.7			2	36.4	TOOELE UT			-68	06	32
10*	335.3	3	228.4	4	96.0	UINTAH UT	-32	-58	-71	05	47
		1	188.7			WASHINGTON UT					
8	135.1	13	116.4	16	75.5	WEBER UT	-14	-35	-44	17	40
3	120.1	8	161.9	7	57.8	VERMONT VT	35	-64	-52		
2*	778.3			1	546.4	ADDISON VT			-30	26	98
1	331.6	1	344.1			BENNINGTON VT	4				
		1*	267.9			CALEDONIA VT					
		1	141.3	3	133.9	CHITTENDEN VT		-5			69
		1	424.0			FRANKLIN VT					
		1	56.2	1	53.0	RUTLAND VT		-6			35
		1	94.4			WASHINGTON VT					
		2	109.8			WINDHAM VT					
				2	131.9	WINDSOR VT					67
4093	137.2	5147	145.5	6522*	146.0	VIRGINIA VA	6	0	6		
40	84.9	69	139.1	84	140.2	ACCOMACK VA	64	1	65	89	72
83	177.8	74	131.4	111	161.5	ALBEMARLE#VA	-26	23	-9	47	82
33*	236.4	27	176.7	23	149.3	ALLEGHANY#VA	-25	-16	-37	21	77
23	154.7	10	61.6	20	111.9	AMELIA VA	-60	82	-28	28	54
28	144.5	28	128.4	30	105.9	AMHERST VA	-11	-18	-27	28	51
7	85.5	11	110.9	13	103.1	APPOMATTOX VA	30	-7	21	74	49
61*	245.7	68	173.7	100	154.0	ARLINGTON#VA	-29	-11	-37	21	79
36	149.3	42	141.0	47	136.8	AUGUSTA#VA	-6	-3	-8	47	71
6	267.1	4	163.2	9*	272.0	BATH VA	-39	67	2	58	95
135	129.5	186	157.7	211	150.3	BEDFORD#VA	22	-5	16	71	77
6	95.9	16	211.9	12	134.2	BOTETOURT VA	121	-37	40	83	69
57	162.0	47	131.1	52	124.2	BRUNSWICK VA	-19	-5	-23	31	62
20	99.5	22	112.5	36	155.0	BUCKINGHAM VA	13	38	56	87	80
30	137.7	34	133.1	50	151.7	CAROLINE VA	-3	14	10	67	78
8	163.7	2	86.1	7	121.5	CARROLL#VA	-47	41	-26	29	61
16	144.6	20	133.9	17	97.6	CHARLES CITY VA	-7	-27	-33	24	48

* SIGNIFICANTLY MORE DEATHS THAN EXPECTED

COUNTIES COMBINED; SEE TECHNICAL NOTES

NON-WHITE FEMALE — ALL CANCERS COMBINED — ICD 140-208

1950-1959 DEATHS	1950-1959 RATE	1960-1969 DEATHS	1960-1969 RATE	1970-1979 DEATHS	1970-1979 RATE	COUNTY AND STATE	PERCENT CHANGE 60S/50S	PERCENT CHANGE 70S/60S	PERCENT CHANGE 70S/50S	PERCENTILE %CHANGE 70S/50S	PERCENTILE RATES 1970S
28	136.1	28	130.6	19	86.5	CHARLOTTE VA	-4	-34	-36	22	43
579	144.6	763*	154.5	970*	158.6	CHESTERFIELD#VA	7	3	10	66	81
8	159.5	10	156.3	4	56.5	CLARKE VA	-2	-64	-65	07	36
		1*	344.1			CRAIG VA					
15	99.1	30	178.3	26	125.3	CULPEPER VA	80	-30	27	77	63
19	124.9	21	137.4	22	121.0	CUMBERLAND VA	10	-12	-3	53	60
2	385.6	2	192.3			DICKENSON VA	-50				
184	110.4	229	126.6	275	140.2	DINWIDDIE#VA	15	11	27	78	72
14	109.7	26	178.9	21	122.5	ESSEX VA	63	-32	12	68	61
77	138.9	143	163.3	247	148.5	FAIRFAX#VA	18	-9	7	64	77
39	177.5	47	192.6	39	141.4	FAUQUIER VA	8	-27	-20	34	73
1	37.5	1	42.0	2	90.2	FLOYD VA	12	115	140	95	45
19	169.6	10	86.8	27*	199.5	FLUVANNA VA	-49	130	18	72	90
28*	210.5	20	147.5	24	147.0	FRANKLIN VA	-30	0	-30	26	76
17*	228.3	13	159.8	16	153.3	FREDERICK#VA	-30	-4	-33	24	79
2	135.1	3	121.5	2	147.5	GILES VA	-10	21	9	66	76
19	121.2	13	77.9	25	109.5	GLOUCESTER VA	-36	41	-10	46	53
16	112.1	16	103.4	26	134.6	GOOCHLAND VA	-8	30	20	74	70
4	98.0	5	137.6	7	208.3	GREENE VA	40	51	113	93	91
30	90.8	35	105.2	47	119.3	GREENSVILLE#VA	16	13	31	80	59
91	147.8	82	126.6	100	135.3	HALIFAX#VA	-14	7	-8	47	70
37	162.7	34	132.6	38	120.3	HANOVER VA	-18	-9	-26	29	60
32	146.6	43	147.3	84	123.6	HENRICO VA	0	-16	-16	39	62
50	112.8	77	142.3	108	144.6	HENRY#VA	26	2	28	79	75
27	111.7	54	172.5	47	119.0	ISLE OF WIGHT VA	54	-31	6	63	59
48	136.1	68	139.0	76	108.9	JAMES CITY#VA	2	-22	-20	34	53
19	153.2	25	199.9	27	186.6	KING AND QUEEN VA	31	-7	22	75	88
7	104.5	5	58.1	20	205.5	KING GEORGE VA	-44	254	97	92	91
17	126.7	15	104.3	21	122.6	KING WILLIAM VA	-18	18	-3	52	61
12	85.2	20	117.6	33	164.5	LANCASTER VA	38	40	93	92	83
4	250.0	2	107.2	1	37.5	LEE VA	-57	-65	-85	01	32
29	185.0	28	174.5	42	189.8	LOUDOUN VA	-6	9	3	59	89
24	106.1	20	86.9	46	186.0	LOUISA VA	-18	114	75	90	88
25	127.2	19	91.9	30	120.4	LUNENBURG VA	-28	31	-5	50	60
7	78.9	12	162.4	6	57.8	MADISON VA	106	-64	-27	28	36
8	84.4	17	165.6	15	128.5	MATHEWS VA	96	-22	52	86	65
67	137.7	65	121.4	73	115.1	MECKLENBURG VA	-12	-5	-16	39	57
14	112.1	24	175.3	21	120.5	MIDDLESEX VA	56	-31	7	64	60
22	211.6	22	188.8	17	118.8	MONTGOMERY#VA	-11	-37	-44	17	59
11	71.2	16	106.6	16	99.2	NELSON VA	50	-7	39	83	48
48	130.8	62	160.4	71	166.3	NORTHAMPTON VA	23	4	27	78	83
11	66.3	14	82.0	26	124.8	NORTHUMBERLAND VA	24	52	88	91	63
24	94.9	34	119.1	52	143.7	NOTTOWAY VA	25	21	51	86	74
22	152.2	28*	198.9	21	137.7	ORANGE VA	31	-31	-9	46	71
4	147.9	2	116.4	4	102.6	PAGE VA	-21	-12	-31	26	49
5	123.6	6	122.7	12	180.9	PATRICK VA	-1	47	46	85	87
171	153.9	146	119.4	203	136.8	PITTSYLVANIA#VA	-22	15	-11	44	71
16	153.6	10	88.2	24*	196.7	POWHATAN VA	-43	123	28	78	90
32	117.1	38	143.1	49	159.3	PRINCE EDWARD VA	22	11	36	82	81
11	96.2	28	189.0	28	87.4	PRINCE WILLIAM#VA	97	-54	-9	47	44
14	152.2	17	172.6	14	135.7	PULASKI VA	13	-21	-11	45	70
9	207.8	10	211.0	4	88.9	RAPPAHANNOCK VA	2	-58	-57	11	44
11	124.9	13	134.4	16	145.7	RICHMOND VA	8	8	17	72	75
126	141.9	162	152.2	223*	174.5	ROANOKE#VA	7	15	23	76	86
15	135.9	15	118.8	21	160.0	ROCKBRIDGE#VA	-13	35	18	72	81
17*	312.9	10	163.9	14	180.3	ROCKINGHAM#VA	-48	10	-42	18	87
1	44.6	1	24.8	1	28.1	RUSSELL VA	-44	13	-37	21	31
1	133.9	1	294.8	4	291.8	SCOTT VA	120	-1	118	94	96
2	106.4	5	297.2	1	37.7	SHENANDOAH VA	179	-87	-65	07	32
3	106.4	2	107.9	3	69.4	SMYTH VA	1	-36	-35	23	38
62	129.7	68	122.0	76	117.2	SOUTHAMPTON#VA	-6	-4	-10	46	58
26	131.7	40	164.2	54	170.9	SPOTSYLVANIA#VA	25	4	30	79	85
6	101.1	14	211.7	10	128.8	STAFFORD VA	109	-39	27	78	65
20	143.4	23	164.0	15	88.7	SURRY VA	14	-46	-38	20	44
27	103.4	39	134.7	35	111.7	SUSSEX VA	30	-17	8	65	54
20	200.4	14	152.8	11	95.1	TAZEWELL VA	-24	-38	-53	13	47
6	128.4	6	122.3	7	98.1	WARREN VA	-5	-20	-24	31	48
20	185.0	12	98.0	23	169.5	WASHINGTON#VA	-47	73	-8	48	84
21	126.1	25	123.5	35	143.8	WESTMORELAND VA	-2	16	14	70	74
9	106.1	13	172.7	12	191.9	WISE#VA	63	11	81	91	90
16*	285.3	11	174.0	9	135.0	WYTHE VA	-39	-22	-53	13	70
718	135.5	1025*	158.1	1324*	158.8	CHESAPEAKE CITY#VA	17	0	17	72	81
92	160.1	132	157.4	205	152.7	HAMPTON CITY VA	-2	-3	-5	51	79
151	138.9	243*	169.0	308*	166.3	NEWPORT NEWS CITY VA	22	-2	20	74	83
115	135.0	154	145.0	165	139.1	SUFFOLK CITY VA	7	-4	3	59	72
336	154.6	436	122.6	749	107.1	WASHINGTON WA	-21	-13	-31		
				3	28.1	BENTON WA					31
		1	60.7	3	110.3	CHELAN WA		82			54
5	161.5	3	90.5	4	92.3	CLALLAM WA	-44	2	-43	18	45
		2	45.4	5	48.3	CLARK WA		6			34
2	125.1	3	144.4	4	94.9	COWLITZ WA	15	-34	-24	30	47
5	217.5	7	220.0	3	90.8	FERRY WA	1	-59	-58	10	45
6	249.6	3	98.1	10	127.1	FRANKLIN WA	-61	30	-49	14	64
2	80.9	4	119.8	10	133.5	GRANT WA	48	11	65	89	68
3	115.4	4	143.7	6	121.2	GRAYS HARBOR WA	25	-16	5	62	61
		3	222.9	1	38.5	ISLAND WA		-83			33
1	163.6	1	70.1	3	146.1	JEFFERSON WA	-57	108	-11	45	75
179	157.7	244	122.7	430	112.0	KING WA	-22	-9	-29	27	54
12	197.4	11	191.6	16	83.9	KITSAP WA	-3	-56	-57	11	42
				4*	394.3	KITTITAS WA					98
6*	442.6	4	278.8	3	238.5	KLICKITAT WA	-37	-14	-46	16	93
3*	565.7					LEWIS WA					
1*	188.7					LINCOLN WA					
		2	125.8	4	147.1	MASON WA		17			76
7	147.1	10	157.2	7	88.1	OKANOGAN WA	7	-44	-40	19	44
2	183.6					PACIFIC WA					
		1	182.1	2	322.7	PEND OREILLE WA		77			97
41	205.6	56	139.5	115	129.6	PIERCE WA	-32	-7	-37	21	66
2	119.5	3	133.9	5	107.7	SKAGIT WA	12	-20	-10	46	52

* SIGNIFICANTLY MORE DEATHS THAN EXPECTED

\# COUNTIES COMBINED; SEE TECHNICAL NOTES

NON-WHITE FEMALE — ALL CANCERS COMBINED — ICD 140-208

1950-1959 DEATHS	1950-1959 RATE	1960-1969 DEATHS	1960-1969 RATE	1970-1979 DEATHS	1970-1979 RATE	COUNTY AND STATE	PERCENT CHANGE 60S/50S	PERCENT CHANGE 70S/60S	PERCENT CHANGE 70S/50S	PERCENTILE %CHANGE 70S/50S	PERCENTILE RATES 1970S
		1	74.3			SKAMANIA WA					
7	240.5	14	206.8	16	74.3	SNOHOMISH WA	-14	-64	-69	06	39
14	94.7	19	94.3	31	93.3	SPOKANE WA	0	-1	-1	55	46
3	153.3	4	95.9	8	159.7	STEVENS WA	-37	67	4	61	81
1	53.8	3	184.1	11	223.3	THURSTON WA	242	21	315	98	92
1	98.2	3	258.2	5	134.0	WALLA WALLA WA	163	-48	36	82	69
4	99.0	3	101.7	6	103.1	WHATCOM WA	3	1	4	61	49
29	174.8	27	128.8	34	71.8	YAKIMA WA	-26	-44	-59	10	38
668	155.5	697	152.2	742*	154.8	WEST VIRGINIA WV	-2	2	0		
6	335.4	2	220.0	1	133.9	BARBOUR WV	-34	-39	-60	09	69
11	154.7	11	160.4	11	148.2	BERKELEY WV	4	-8	-4	51	77
3	166.3			6	249.4	BOONE WV			50	86	94
		1	114.7	1	114.7	BRAXTON WV		0			56
6	308.3	2	67.0	3	151.5	BROOKE WV	-78	126	-51	13	78
42	170.4	38	119.1	68*	202.5	CABELL WV	-30	70	19	73	91
55	155.7	56	154.9	45	120.7	FAYETTE WV	0	-22	-22	32	60
4*	510.0	3	323.4	1	68.8	GRANT WV	-37	-79	-87	01	38
20	230.7	14	131.1	20	193.6	GREENBRIER WV	-43	48	-16	39	90
1	81.8			1	136.6	HAMPSHIRE WV			67	89	71
6	113.7	6	96.1	15*	218.6	HANCOCK WV	-16	127	92	92	92
2	148.9	2	133.9	3	244.1	HARDY WV	-10	82	64	88	94
16	160.0	10	109.9	18	177.9	HARRISON WV	-31	62	11	68	87
16	126.5	26*	200.1	20	153.8	JEFFERSON WV	58	-23	22	75	79
103	148.3	134	165.8	165*	186.8	KANAWHA WV	12	13	26	77	88
2	189.0					LEWIS WV					
		1*	596.2			LINCOLN WV					
40	198.2	44*	210.1	18	94.7	LOGAN WV	6	-55	-52	13	46
103	159.5	97	178.0	85	148.2	MC DOWELL WV	12	-17	-7	49	77
28	171.2	25	147.9	35*	176.2	MARION WV	-14	19	3	59	86
1	147.4	3	223.1	1	117.9	MARSHALL WV	51	-47	-20	34	58
		1	26.5	1	32.6	MASON WV		23			31
65	171.4	57	137.4	56	128.7	MERCER WV	-20	-6	-25	30	65
7	290.7	5	205.6	5	145.2	MINERAL WV	-29	-29	-50	14	75
14	132.6	14	115.5	14	143.9	MINGO WV	-13	25	9	65	74
9	146.3	15	208.9	10	140.2	MONONGALIA WV	43	-33	-4	51	72
4	186.7	1	44.6	4	152.0	MONROE WV	-76	241	-19	36	78
1	172.0					MORGAN WV					
				1	546.4	NICHOLAS WV					98
13	95.4	18	145.1	28*	226.7	OHIO WV	52	56	138	95	92
		3*	729.7	1	188.7	PENDLETON WV		-74			89
1	54.5	3	133.6			POCAHONTAS WV	145				
56	126.5	77	145.6	84	141.8	RALEIGH WV	15	-3	12	68	73
2	187.5	3	191.3	2	91.8	RANDOLPH WV	2	-52	-51	13	45
4	69.1	6	144.8	6	119.5	SUMMERS WV	110	-17	73	90	59
2	438.4	3	320.2			TAYLOR WV	-27				
1	188.7	1*	424.0			TUCKER WV	125				
1	267.9					TYLER WV					
3	715.6	1	141.3			UPSHUR WV	-80				
				1	298.1	WAYNE WV					96
				1*	267.9	WIRT WV					95
10	199.4	7	137.0	3	82.0	WOOD WV	-31	-40	-59	10	42
10	187.9	7	156.4	8	195.0	WYOMING WV	-17	25	4	60	90
287*	173.3	420	134.7	764	137.5	WISCONSIN WI	-22	2	-21		
		2*	606.2	1	344.1	ADAMS WI		-43			97
7	260.3	3	111.5	9*	262.6	ASHLAND WI	-57	136	1	57	95
1	273.3			1	546.4	BARRON WI			100	92	98
4	221.3	2	179.5	6	239.4	BAYFIELD WI	-19	33	8	65	93
7	183.4	7	157.8	13	135.9	BROWN WI	-14	-14	-26	29	70
1*	344.1					BUFFALO WI					
2	253.2					BURNETT WI					
3*	1052.7					CALUMET WI					
1	74.3					CHIPPEWA WI					
				1	99.4	COLUMBIA WI					48
7	205.9	6	83.6	15	81.4	DANE WI	-59	-3	-60	09	42
		1	212.0	1	47.2	DODGE WI		-78			33
1	196.5					DOOR WI					
1	212.0	1	344.1	5	263.6	DOUGLAS WI	62	-23	24	76	95
				1	14.7	EAU CLAIRE WI					30
1	91.1	4*	470.0	1	94.4	FOREST WI	416	-80	4	60	46
1*	344.1					GRANT WI					
				1	344.1	GREEN WI					97
1*	298.1					IRON WI					
2	401.8	1	68.8			JACKSON WI	-83				
				1	84.8	JEFFERSON WI					43
2	513.3	1	114.7	1	141.3	JUNEAU WI	-78	23	-72	04	73
5	321.1	2	31.4	14	148.3	KENOSHA WI	-90	373	-54	12	77
1*	982.0					KEWAUNEE WI					
1	267.9	3*	361.4			LA CROSSE WI	35				
1	589.6					LANGLADE WI					
				2*	438.8	LINCOLN WI					98
1	74.3	1	85.2	2	94.9	MANITOWOC WI	15	11	28	78	47
1	424.0					MARATHON WI					
1*	267.9	1	294.8	1	141.3	MARINETTE WI	10	-52	-47	16	73
17	204.1	18	180.6	20	143.3	MENOMINEE#WI	-12	-21	-30	26	74
175	165.2	306	140.5	575*	149.3	MILWAUKEE WI	-15	6	-10	46	77
				1	74.3	MONROE WI					39
2*	496.8			2	156.1	ONEIDA WI			-69	06	80
3	69.6	6	144.6	7	106.0	OUTAGAMIE WI	108	-27	52	86	51
1*	74.3					PIERCE WI					
				1	122.7	POLK WI					62
1*	424.0					PORTAGE WI					
1*	344.1					PRICE WI					
12	181.7	19	115.3	39	131.7	RACINE WI	-37	14	-27	28	67
9	135.7	17	162.1	14	88.8	ROCK WI	19	-45	-35	23	44
		1	109.3			ST CROIX WI					
		1	94.4	2	219.3	SAUK WI		132			92
4	125.5	2	107.7	9	222.1	SAWYER WI	-14	106	77	90	92

* SIGNIFICANTLY MORE DEATHS THAN EXPECTED

\# COUNTIES COMBINED; SEE TECHNICAL NOTES

NON-WHITE FEMALE — ALL CANCERS COMBINED — ICD 140-208

1950-1959 DEATHS	1950-1959 RATE	1960-1969 DEATHS	1960-1969 RATE	1970-1979 DEATHS	1970-1979 RATE	COUNTY AND STATE	PERCENT CHANGE 60S/50S	PERCENT CHANGE 70S/60S	PERCENT CHANGE 70S/50S	PERCENTILE %CHANGE 70S/50S	PERCENTILE RATES 1970S
6	231.7	8*	284.8	6	151.1	VILAS WI	23	-47	-35	23	78
2	210.2	4	207.9	2	47.1	WALWORTH WI	-1	-77	-78	03	33
		2	367.6	3	55.7	WAUKESHA WI		-85			35
				1	133.9	WAUSHARA WI					69
1	85.2			4	280.4	WINNEBAGO WI			229	97	96
		1	172.0	2	106.5	WOOD WI		-38			52
28	167.0	24	116.0	39	94.3	WYOMING WY	-31	-19	-44		
2	596.1			1	32.7	ALBANY WY			-95	00	32
1*	245.4					BIG HORN WY					
		3	381.0	3	250.8	CARBON WY		-34			94
8	130.5	17	248.2	18	163.4	FREMONT WY	90	-34	25	77	82
		1	182.1			HOT SPRINGS WY					
8	257.7	2	39.4	6	56.2	LARAMIE WY	-85	43	-78	03	35
				1	198.7	LINCOLN WY					90
4	213.1	1	23.8	5	92.1	NATRONA WY	-89	288	-57	11	45
1	546.4					PARK-TETON#WY					
3	141.4			4	147.0	SWEETWATER WY			4	60	76
1	245.4			1	56.2	WASHAKIE WY			-77	03	35

* SIGNIFICANTLY MORE DEATHS THAN EXPECTED

COUNTIES COMBINED; SEE TECHNICAL NOTES

WHITE MALE — CANCER OF THE LIP — ICD 140

1950-1959 DEATHS	1950-1959 RATE	1960-1969 DEATHS	1960-1969 RATE	1970-1979 DEATHS	1970-1979 RATE	COUNTY AND STATE	PERCENT CHANGE 60S/50S	PERCENT CHANGE 70S/60S	PERCENT CHANGE 70S/50S	PERCENTILE %CHANGE 70S/50S	PERCENTILE RATES 1970S
3126	.5	1624	.2	1429	.2	UNITED STATES	-57	-26	-68		
32	.5	29	.3	14	.1	ALABAMA AL	-37	-65	-78		
1	.7			1	.3	BALDWIN AL			-59	38	81
		1*	1.2			BUTLER AL					
		2*	1.1	2*	.6	CALHOUN AL		-51			86
2*	3.9					CHEROKEE AL					
1	.9			1*	.7	CHILTON AL			-16	72	89
1	1.8	1*	1.2			CLARKE AL	-32				
		1*	1.2			CLAY AL					
1*	2.5	1*	2.2			CLEBURNE AL	-13				
		1*	1.1			COFFEE AL					
2*	1.3	1	1.1			COLBERT AL	-13				
1	1.9					CONECUH AL					
1	.6	1	.6			COVINGTON AL	-6				
				1*	1.5	CRENSHAW AL					95
		1	.4			CULLMAN AL					
				1*	.7	DALE AL					89
1	.5	2*	1.0			DE KALB AL	108				
		1*	1.0			ESCAMBIA AL					
		1	.3	1	.2	ETOWAH AL		-36			78
1	2.1					FRANKLIN AL					
		1*	1.8			GENEVA AL					
1	1.3					HOUSTON AL					
				1	.5	JACKSON AL					85
5	.6	2	.1	3	.1	JEFFERSON AL	-81	36	-74	19	76
		2*	.9			LAUDERDALE AL					
1*	18.6					MACON AL					
2	2.1					MADISON AL					
1*	2.3					MARENGO AL					
		1	.9			MARION AL					
		1	.4			MARSHALL AL					
2	.3	2	.4	1	.1	MOBILE AL	20	-71	-65	31	75
		2	.4	1	.3	MONTGOMERY AL		-18			81
4*	3.0	1	.3			MORGAN AL	-89				
1	1.3					RANDOLPH AL					
1	1.1					ST CLAIR AL					
		1	.9			SHELBY AL					
		1*	3.7			SUMTER AL					
1	.4	1	.2			TUSCALOOSA AL	-32				
1	.7					WALKER AL					
				1*	4.1	WILCOX AL					99
		1	.3	1	.5	ALASKA AK		53			
11	.4	10	.2	20	.2	ARIZONA AZ	-58	28	-47		
		2*	1.1	1	.3	COCHISE AZ		-73			81
2*	4.6					COCONINO AZ					
				1*	2.4	GREENLEE AZ					98
6	.5	6	.2	10	.2	MARICOPA AZ	-64	20	-57	39	78
1*	2.4					MOHAVE AZ					
				1*	1.1	NAVAJO AZ					93
				4	.2	PIMA AZ					78
1	.5	1	.3	2*	.5	PINAL AZ	-45	64	-10	78	85
1	.6					YAVAPAI AZ					
		1	.6	1	.4	YUMA AZ		-34			83
40	.6	18	.2	23	.2	ARKANSAS AR	-65	5	-63		
1	1.3					ARKANSAS AR					
1	1.5					ASHLEY AR					
2	.9			1	.2	BENTON AR			-78	13	78
		1	1.2	1	.6	BOONE AR		-51			87
		1*	3.5			CALHOUN AR					
1	1.3	2*	2.2			CARROLL AR	74				
		1*	1.2			CLEBURNE AR					
1	1.7	2*	2.5			CONWAY AR	46				
1	.6					CRAIGHEAD AR					
2*	1.8					CRAWFORD AR					
1*	3.4					CRITTENDEN AR					
				1*	2.5	DALLAS AR					98
1	3.1					DREW AR					
		1	.7			FAULKNER AR					
1	.3	1	.2	2	.4	GARLAND AR	-38	93	20	87	83
				1*	1.8	GRANT AR					96
1	.9	1	.7	3*	1.9	GREENE AR	-15	154	115	96	97
				1	.6	INDEPENDENCE AR					87
				1*	1.1	JACKSON AR					93
1	.6					JEFFERSON AR					
1	1.7					JOHNSON AR					
1*	3.2					LINCOLN AR					
2*	2.3					LOGAN AR					
1	1.1	1	.9	1*	.7	LONOKE AR	-19	-15	-31	60	89
				1*	1.3	MADISON AR					94
				1*	1.5	MARION AR					95
4*	2.0			1	.5	MISSISSIPPI AR			-77	14	84
1*	4.6					MONROE AR					
1	2.2					NEVADA AR					
				1*	2.3	NEWTON AR					97
1	1.1					OUACHITA AR					
1	2.2					PHILLIPS AR					
1	1.0			1*	.9	POINSETT AR			-2	81	91
1	1.9			1*	.8	POLK AR			-57	39	90
		1	.9			POPE AR					
1*	1.4					PRAIRIE AR					
3	.4	2	.3	1	.1	PULASKI AR	-27	-56	-68	27	76

* SIGNIFICANTLY MORE DEATHS THAN EXPECTED

WHITE MALE

CANCER OF THE LIP

ICD 140

1950-1959 DEATHS	1950-1959 RATE	1960-1969 DEATHS	1960-1969 RATE	1970-1979 DEATHS	1970-1979 RATE	COUNTY AND STATE	PERCENT CHANGE 60S/50S	PERCENT CHANGE 70S/60S	PERCENT CHANGE 70S/50S	PERCENTILE %CHANGE 70S/50S	PERCENTILE RATES 1970S
				1*	.8	RANDOLPH AR					89
3*	3.4					SALINE AR					
1	2.1					SCOTT AR					
1	2.1					SEARCY AR					
1	.4	1	.3			SEBASTIAN AR	-30				
1	1.5					SEVIER AR					
		1*	1.8			STONE AR					
		1	.7			UNION AR					
		1	.3	2	.5	WASHINGTON AR		77			85
				1*	.9	YELL AR					90
194	.4	117	.2	118	.2	CALIFORNIA CA	-55	-17	-63		
8	.2	7	.2	5	.2	ALAMEDA CA	-8	-32	-37	56	76
		1	.2	1	.1	BUTTE CA		-31			75
1	1.3	1*	1.3			CALAVERAS CA	1				
6	.7	1	.1	1	.1	CONTRA COSTA CA	-88	-25	-91	02	74
9	.8	5	.3	2	.1	FRESNO CA	-60	-64	-86	05	75
		2	.5			HUMBOLDT CA					
				2*	.9	IMPERIAL CA					91
6	.7	2	.2	4	.3	KERN CA	-73	79	-52	43	81
2	1.4	1	.5			KINGS CA	-62				
		1	.5			LAKE CA					
		1*	1.3	1*	1.5	LASSEN CA		8			95
68	.4	43	.2	41	.2	LOS ANGELES CA	-47	-12	-54	42	77
1	.6					MADERA CA					
2	.5	1	.2	1	.2	MARIN CA	-66	-1	-66	29	77
1*	3.1	1*	2.2			MARIPOSA CA	-29				
2	.8					MENDOCINO CA					
1	.3					MEPCED CA					
1*	3.0					MODOC CA					
2	.8					MONTEREY CA					
1	.3			1	.2	NAPA CA			-35	57	77
5	.4	3	.1	5	.1	ORANGE CA	-70	2	-69	25	75
2	.7	1	.3	2	.4	PLACER CA	-54	19	-45	51	83
1	.1	2	.2	3	.1	RIVERSIDE CA	91	-35	23	88	75
12*	1.2	2	.1	4	.2	SACRAMENTO CA	-91	116	-81	10	79
				1*	1.0	SAN BENITO CA					92
8	.5	3	.1	7	.2	SAN BERNARDINO CA	-72	66	-53	43	80
8	.4	5	.1	6	.1	SAN DIEGO CA	-60	-42	-77	15	74
18	.5	11	.3	6	.2	SAN FRANCISCO CA	-33	-44	-62	35	78
2	.2	2	.2	2	.1	SAN JOAQUIN CA	-34	-13	-43	53	76
1	.3					SAN LUIS OBISPO CA					
2	.2	2	.1	3	.1	SAN MATEO CA	-48	43	-25	65	76
1	.2			3	.3	SANTA BARBARA CA			52	92	80
8	.6	1	.0	1	.0	SANTA CLARA CA	-96	68	-93	00	73
2	.4	1	.1			SANTA CRUZ CA	-65				
2	1.8			1	.2	SHASTA CA			-89	03	78
				1	.5	SISKIYOU CA					85
2	.7			1	.1	SOLANO CA			-80	11	76
3	.5	3	.4	2	.2	SONOMA CA	-28	-56	-69	26	76
		2	.2	3	.3	STANISLAUS CA		28			81
1	1.2					SUTTER CA					
		1	.7			TEHAMA CA					
4	.7	7*	.9	4*	.5	TULARE CA	35	-50	-32	60	84
				1	.8	TUOLUMNE CA					90
		4	.6	3	.2	VENTURA CA		-57			80
1	.6					YOLO CA					
32	.6	17	.2	10	.1	COLORADO CO	-59	-51	-80		
		1	.1	1	.2	ADAMS CO		136			79
2	.9					BOULDER CO					
1*	2.3					CHAFFEE CO					
		1*	2.6			CONEJOS CO					
11	.5	3	.1	4	.1	DENVER-ARAPAHOE#CO	-78	23	-73	21	76
1*	4.5	1*	4.5			ELBERT CO	0				
5*	1.4	1	.2			EL PASO CO	-85				
		1*	2.0			HUERFANO CO					
1	.4	2	.4	2	.2	JEFFERSON CO	-4	-35	-38	55	79
1	1.6			1*	1.0	LA PLATA CO			-34	58	92
1	.4	1	.3			LAPIMER CO	-26				
		1	1.0			LAS ANIMAS CO					
1*	10.6					LINCOLN CO					
		1*	1.0			LOGAN CO					
2	.9					MESA CO					
				1*	1.7	MONTEZUMA CO					96
				1*	1.0	MONTROSE CO					91
1	1.4					MORGAN CO					
2*	1.9					OTERO CO					
		1*	2.7			PHILLIPS CO					
		1*	14.1			PITKIN CO					
1	1.5					PROWERS CO					
1	.2	2	.4			PUEBLO CO	106				
1	.4					WELD CO					
47	.5	18	.2	15	.1	CONNECTICUT CT	-67	-32	-77		
9	.4	6	.2	3	.1	FAIRFIELD CT	-43	-52	-73	21	75
13	.6	1	.1	3	.1	HARTFORD CT	-90	47	-85	05	74
3	.6	3	.5	2	.3	LITCHFIELD CT	-22	-33	-47	49	81
1	.3	1	.2			MIDDLESEX CT	-30				
13	.6	7	.3	4	.1	NEW HAVEN CT	-51	-59	-80	12	75
4	.6			3	.4	NEW LONDON CT			-44	52	82
4*	1.4					WINDHAM CT					
11	.9	1	.1	7*	.4	DELAWARE DE	-90	343	-55		
3*	1.8	1	.5	1	.4	KENT DE	-72	-11	-75	17	84
8	.9			3	.3	NEW CASTLE DE			-73	20	80
				3*	.7	SUSSEX DE					89

* SIGNIFICANTLY MORE DEATHS THAN EXPECTED

\# COUNTIES COMBINED; SEE TECHNICAL NOTES

WHITE MALE — CANCER OF THE LIP — ICD 140

1950-1959 DEATHS	1950-1959 RATE	1960-1969 DEATHS	1960-1969 RATE	1970-1979 DEATHS	1970-1979 RATE	COUNTY AND STATE	PERCENT CHANGE 60S/50S	PERCENT CHANGE 70S/60S	PERCENT CHANGE 70S/50S	PERCENTILE %CHANGE 70S/50S	PERCENTILE RATES 1970S
5	.2	1	.1	2	.2	DIST. OF COLUMBIA DC	-56	88	-18		
5	.2	1	.1	2	.2	WASHINGTON DC	-56	88	-18	70	77
51	.3	45	.2	36	.1	FLORIDA FL	-52	-55	-78		
1	.2	2	.6	1	.1	BREVARD FL	137	-76	-44	52	75
		2	.1	3	.1	BROWARD FL		-24			74
1	1.0					CHARLOTTE FL					
1	1.1					CLAY FL					
				1*	1.2	COLUMBIA FL					94
9	.3	6	.1	3	.0	DADE FL	-63	-73	-90	02	73
				2*	1.9	DE SOTO FL					97
8*	1.0	2	.2			DUVAL FL	-84				
2	.7	2	.4	1	.2	ESCAMBIA FL	-47	-39	-68	28	79
		1*	.9			GADSDEN FL					
				1*	2.5	GULF FL					98
		2*	.9			HIGHLANDS FL					
6	.5	4	.2	3	.1	HILLSBOROUGH FL	-60	-36	-74	19	76
1	.6					INDIAN RIVER FL					
		1*	.8			JACKSON FL					
1	.3	1	.2	1	.1	LEE FL	-46	-48	-72	22	74
1	.7			1	.3	LEON FL			-51	45	82
1*	2.7					LEVY FL					
1	.6					MANATEE FL					
		2*	.8			MARION FL					
1*	3.3					NASSAU FL					
				1*	1.0	OKEECHOBEE FL					91
2	.3	5	.5	3	.2	ORANGE FL	77	-63	-35	58	77
1	.5	1	.4			OSCEOLA FL	-13				
2	.3	3	.2	3	.1	PALM BEACH FL	-17	-58	-65	30	74
		1	.1	1	.0	PASCO FL		-67			73
7	.4	5	.1	3	.1	PINELLAS FL	-73	-49	-86	05	74
2	.3			5	.4	POLK FL			32	89	83
				1	.5	PUTNAM FL					84
		2	.4	1	.0	SARASOTA FL		-89			74
1	.7			1	.2	SEMINOLE FL			-66	30	80
		1*	3.4			TAYLOR FL					
2	.5	2	.3			VOLUSIA FL	-43				
33	.3	23	.2	22	.2	GEORGIA GA	-36	-23	-51		
				1*	5.6	BACON GA					99
1	.7					BARTOW GA					
2	.8	2	.8			BIBB GA	-1				
				1*	2.2	BROOKS GA					97
		2*	9.4			BRYAN GA					
1*	14.9					CAMDEN GA					
1	.5	1	.3			CHATHAM GA	-44				
1	1.1					CHEROKEE GA					
		1*	1.1			CLARKE GA					
1	1.2					CLAYTON GA					
				1*	.7	COLQUITT GA					88
1*	2.5					CRISP GA					
1*	7.0			1*	5.1	DAWSON GA			-27	64	99
		1*	1.9			DECATUR GA					
3	.5	3	.4	1	.1	DE KALB GA	-16	-77	-80	11	75
				1*	1.6	DODGE GA					95
1*	3.7	1*	4.6			DOOLY GA	27				
1	1.3	1*	.4			DOUGHERTY GA	-70				
1*	2.5					EARLY GA					
2	1.0	1	.6	1	.3	FLOYD GA	-45	-37	-65	30	82
				1*	1.0	FORSYTH GA					92
3	.2	3	.2	2	.2	FULTON GA	6	-22	-17	71	77
1	.9					GLYNN GA					
1	1.4					HABERSHAM GA					
				1	.5	HALL GA					85
		1*	1.6	1*	1.3	HARALSON GA		-19			94
		1*	4.6			HEARD GA					
				1*	3.0	IRWIN GA					98
				1*	1.4	JACKSON GA					95
1*	4.2					JENKINS GA					
1*	3.4					MC DUFFIE GA					
				1*	6.1	MC INTOSH GA					99
		1*	1.9			MERIWETHER GA					
2	1.2			1	.5	MUSCOGEE#GA			-61	36	84
1*	3.8					OGLETHORPE GA					
1	1.0	1*	.9			POLK GA	-11				
1*	5.4					RANDOLPH GA					
1	.5	1	.3	1	.3	RICHMOND GA	-33	-7	-38	55	81
				1*	3.9	SCREVEN GA					99
				1*	1.3	STEPHENS GA					94
		1	.6	2*	1.4	TROUP GA		117			95
2*	1.8					WARE GA					
				1	.5	WHITFIELD GA					85
1*	4.3	1*	3.8			WILKES GA	-12				
				1	.1	HAWAII HI					
				1*	1.2	HAWAII HI					94
20	.8	14*	.5	9	.3	IDAHO ID	-36	-47	-66		
		1	.3			ADA ID					
1	.6	1	.6	1	.5	BANNOCK ID	-3	-17	-19	70	85
		1*	2.7			BENEWAH ID					
2*	2.3					BINGHAM ID					
		1*	3.1			BLAINE ID					
2*	1.8			1	.6	BONNEVILLE ID			-66	30	87
1*	6.0					BUTTE ID					

* SIGNIFICANTLY MORE DEATHS THAN EXPECTED

\# COUNTIES COMBINED; SEE TECHNICAL NOTES

WHITE MALE — CANCER OF THE LIP — ICD 140

1950-1959 DEATHS	1950-1959 RATE	1960-1969 DEATHS	1960-1969 RATE	1970-1979 DEATHS	1970-1979 RATE	COUNTY AND STATE	PERCENT CHANGE 60S/50S	PERCENT CHANGE 70S/60S	PERCENT CHANGE 70S/50S	PERCENTILE %CHANGE 70S/50S	PERCENTILE RATES 1970S
2	.9	2	.7	2	.5	CANYON ID	-25	-18	-38	55	86
		1*	1.8			CLEARWATER ID					
		1*	2.0			GEM ID					
2*	3.5					IDAHO ID					
1*	2.2	1*	1.8			JEFFERSON ID	-19				
2	1.6	2*	1.5	1	.4	KOOTENAI ID	-11	-73	-76	17	83
1	1.4	1*	1.0			LATAH ID	-28				
				1*	2.5	MADISON ID					98
				1*	1.2	MINIDOKA ID					93
1	.9	1	.7	1	.7	NEZ PERCE ID	-27	-1	-28	63	88
1	1.5					PAYETTE ID					
2*	2.9	1*	1.4	1*	1.7	SHOSHONE ID	-53	20	-44	52	96
2	1.1					TWIN FALLS ID					
168	.5	121*	.3	82	.2	ILLINOIS IL	-34	-39	-60		
3	.8	4*	1.1			ADAMS IL	26				
1	1.2					ALEXANDER IL					
2*	2.3					BOND IL					
1	.9					BOONE IL					
		1	.4			BUREAU IL					
1	1.0			1*	.9	CARROLL IL			-13	75	90
				1*	1.0	CASS IL					92
1	.3	2	.6			CHAMPAIGN IL	89				
1	.4	2*	.9	1	.4	CHRISTIAN IL	138	-50	18	87	84
		2*	2.1	1*	1.0	CLAY IL		-52			92
1	1.3	1	.7			CLINTON IL	-48				
2	.9	2*	.9			COLES IL	1				
52	.3	43	.2	27	.1	COOK IL	-29	-41	-58	38	76
		1	.7	1*	.7	CRAWFORD IL		-2			88
1	.5					DE KALB IL					
		3*	3.0			DE WITT IL					
1	2.0					DOUGLAS IL					
2	.5	2	.2	1	.1	DU PAGE IL	-52	-50	-76	16	75
3*	2.4	1	.9			EDGAR IL	-63				
2*	1.7					EFFINGHAM IL					
				1*	.9	FORD IL					90
1	.4			3*	1.2	FRANKLIN IL			227	98	94
3	1.3			3*	1.1	FULTON IL			-15	74	93
1	.8			2*	1.6	GREENE IL			99	95	96
1	1.4					HAMILTON IL					
		1	.6	1	.8	HANCOCK IL		30			89
				1*	1.8	HENDERSON IL					97
1	.3					HENRY IL					
2	1.2	1	.5	1	.5	JACKSON IL	-56	-16	-63	34	84
				1*	1.6	JASPER IL					96
		1*	1.0			JERSEY IL					
				1*	.8	JO DAVIESS IL					90
1	3.0					JOHNSON IL					
2	.3	4	.5	1	.1	KANE IL	109	-76	-50	47	75
3	.7					KANKAKEE IL					
2	.6					KNOX IL					
3	.4	2	.3	1	.1	LAKE IL	-17	-82	-85	05	74
3	.6					LA SALLE IL					
		1	.8			LAWRENCE IL					
3*	2.2	1	.5			LEE IL	-75				
2	.9	1	.4			LIVINGSTON IL	-50				
2	1.3	2*	1.4			LOGAN IL	4				
				1	.2	MC HENRY IL					77
2	.5	2	.5	1	.3	MC LEAN IL	5	-49	-46	50	80
1	.2	4*	.8	2	.4	MACON IL	249	-55	56	92	82
1	.3					MACOUPIN IL					
1	.1	2	.4	4	.4	MADISON IL	195	9	223	98	83
1	.6	3*	1.4			MARION IL	145				
1	1.3	1*	1.3	1*	1.1	MARSHALL IL	0	-14	-14	74	93
		1*	1.1			MASON IL					
2*	3.3					MASSAC IL					
2	.8					MONTGOMERY IL					
		1	.6	1	.6	MORGAN IL		6			87
1	1.0					MOULTRIE IL					
7	1.0	3	.5			PEORIA IL	-53				
		1	.8			PERRY IL					
		2*	2.4			PIATT IL					
1*	2.4					POPE IL					
1	.6					RANDOLPH IL					
1	1.2					RICHLAND IL					
5	.8					ROCK ISLAND IL					
6	.9	3	.3	3	.3	ST CLAIR IL	-62	-10	-66	29	81
1	.5	1	.5			SALINE IL	1				
5	.9	1	.1	7*	1.0	SANGAMON IL	-84	589	8	84	92
				1*	1.7	SCHUYLER IL					96
1*	2.8					SCOTT IL					
2	1.2					SHELBY IL					
		4*	1.9	2*	.8	STEPHENSON IL		-56			90
2	.6	1	.3			TAZEWELL IL	-53				
1	.7	1	.8			UNION IL	17				
3	.7	1	.2	1	.2	VERMILION IL	-67	-8	-70	24	78
1	1.3	1*	1.2			WABASH IL	-8				
2*	1.8	2*	1.8	1*	.8	WAYNE IL	0	-53	-54	42	90
		2*	1.8	1*	1.0	WHITE IL		-46			91
1	.4	1	.4			WHITESIDE IL	5				
5	.7	2	.3	1	.1	WILL IL	-54	-62	-83	08	75
1	.3			4*	1.4	WILLIAMSON IL			331	99	94
5	.6	3	.4	2	.2	WINNEBAGO IL	-31	-43	-61	37	79
1	.9					WOODFORD IL					
98	.6	55	.3	39	.2	INDIANA IN	-49	-34	-66		
4	.6	3	.3			ALLEN IN	-45				
2	1.5	2*	1.1	1	.4	BARTHOLOMEW IN	-27	-60	-71	24	84
		1	.6			BOONE IN					

* SIGNIFICANTLY MORE DEATHS THAN EXPECTED

WHITE MALE | CANCER OF THE LIP | ICD 140

1950-1959 DEATHS	1950-1959 RATE	1960-1969 DEATHS	1960-1969 RATE	1970-1979 DEATHS	1970-1979 RATE	COUNTY AND STATE		PERCENT CHANGE 60S/50S	PERCENT CHANGE 70S/60S	PERCENT CHANGE 70S/50S	PERCENTILE %CHANGE 70S/50S	PERCENTILE RATES 1970S
		2*	5.4			BROWN	IN					
2	1.0					CASS	IN					
1	.6					CLAY	IN					
		1*	3.2			CRAWFORD	IN					
1	.6	2*	1.4			DAVIESS	IN	129				
1	.8					DEARBORN	IN					
		1	.6			DE KALB	IN					
2	.5					DELAWARE	IN					
				1*	.6	DUBOIS	IN					87
1	.3			1	.2	ELKHART	IN			-26	64	79
				1	.4	FLOYD	IN					84
		1*	1.0			FOUNTAIN	IN					
2	1.3					GIBSON	IN					
3	1.1	2	.6			GRANT	IN	-39				
3*	1.9	1	.5	1	.7	GREENE	IN	-71	27	-64	33	88
		1	.6			HAMILTON	IN					
				1*	.7	HANCOCK	IN					89
1	1.1					HARRISON	IN					
1	.9	1	.7			HENDRICKS	IN	-24				
1	.5					HENRY	IN					
1	.4	1	.4			HOWARD	IN	-2				
				1	.6	JACKSON	IN					87
1	1.2	1*	1.0			JASPER	IN	-17				
		2*	1.5			JAY	IN					
1	.8			1*	.8	JEFFERSON	IN			0	82	90
1	1.5					JENNINGS	IN					
2	1.4	1	.6	1	.5	JOHNSON	IN	-61	-5	-63	35	86
1	.5			1	.4	KNOX	IN			-8	79	83
13*	1.4	3	.3	4	.3	LAKE	IN	-81	8	-80	12	80
2	.6	1	.3	1	.3	LA PORTE	IN	-44	-20	-55	41	80
2	1.2			2*	1.0	LAWRENCE	IN			-16	72	92
		1	.3	2	.5	MADISON	IN		59			84
8	.4	6	.4	1	.1	MARION	IN	-20	-85	-88	03	74
1	.7					MARSHALL	IN					
3*	6.3					MARTIN	IN					
2	1.5					MIAMI	IN					
1	.5	1	.5	1	.4	MONROE	IN	-14	-13	-25	64	83
2	1.1	1	.6			MONTGOMERY	IN	-51				
2	1.4	1	.7			MORGAN	IN	-50				
		1*	1.9			NEWTON	IN					
4*	2.8	1	.7			NOBLE	IN	-76				
2*	2.0					ORANGE	IN					
				1*	1.0	PARKE	IN					92
				2*	2.0	PERRY	IN					97
		1	.4			PORTER	IN					
		1*	1.0			POSEY	IN					
				1*	1.4	PULASKI	IN					95
		1	.8			PUTNAM	IN					
				1	.6	RANDOLPH	IN					88
				1*	.8	RIPLEY	IN					90
1	.9	1	.7	2*	1.7	RUSH	IN	-14	135	101	95	96
1	.2	2	.2	3	.4	ST JOSEPH	IN	48	50	121	96	83
3*	2.0					SHELBY	IN					
4*	4.2	1*	.9	3*	2.6	SPENCER	IN	-77	179	-37	56	98
1	1.9					SWITZERLAND	IN					
2	.8	3*	1.0	1	.3	TIPPECANOE	IN	23	-71	-64	32	80
3	.5	2	.4	1	.1	VANDERBURGH	IN	-27	-68	-77	15	75
6*	1.2	2	.4	1	.2	VIGO	IN	-69	-48	-84	07	78
1	.6					WABASH	IN					
		1*	2.5			WARREN	IN					
				1*	.8	WARRICK	IN					90
1	.3	1	.3			WAYNE	IN	-13				
1	.9					WHITLEY	IN					
100*	.8	52*	.4	38*	.3	IOWA	IA	-53	-30	-67		
		1*	2.4			ADAMS	IA					
1	.9					ALLAMAKEE	IA					
3*	2.5	3*	2.3	1*	.8	APPANOOSE	IA	-8	-67	-70	25	89
1	1.6			1*	1.4	AUDUBON	IA			-16	72	95
1	.8					BENTON	IA					
3	.6	2	.4	2	.4	BLACK HAWK	IA	-35	-5	-38	55	83
2*	2.1			1*	.7	BREMER	IA			-66	28	89
				1*	.9	BUCHANAN	IA					91
2*	2.0					BUTLER	IA					
1	1.0			1*	.6	CARROLL	IA			-38	54	87
1	1.1			2*	1.5	CASS	IA			45	90	95
1	1.3					CEDAR	IA					
2	1.0	1	.4	2*	.8	CERRO GORDO	IA	-64	118	-21	68	90
1	1.1	1*	1.4			CHEROKEE	IA	28				
1	2.0	2*	3.2			CLARKE	IA	61				
2	1.5					CLAYTON	IA					
3	1.6	1	.3			CLINTON	IA	-80				
1	1.1	1*	1.4			CRAWFORD	IA	25				
1	.8			1	.7	DALLAS	IA			-15	73	88
3*	4.5			2*	3.5	DAVIS	IA			-23	66	99
1	1.9	1*	1.1			DECATUR	IA	-40				
3*	3.3	1*	1.0			DELAWARE	IA	-68				
2	.9			1	.4	DES MOINES	IA			-56	40	83
2	.8	2	.6	1	.3	DUBUQUE	IA	-17	-57	-65	32	80
1	1.4					EMMET	IA					
4*	2.5	1	.6	1	.8	FAYETTE	IA	-78	37	-69	25	89
1	1.0					FLOYD	IA					
3*	4.4					FRANKLIN	IA					
1	1.3	1*	1.4	1*	1.4	FREMONT	IA	13	-1	12	85	95
1	1.2					GREENE	IA					
1	1.0			1*	1.0	GUTHRIE	IA			-2	81	91
1	.7					HARDIN	IA					
		2*	1.7	1*	.9	HENRY	IA		-48			91
1	1.3					HUMBOLDT	IA					
		1*	1.4			IDA	IA					

* SIGNIFICANTLY MORE DEATHS THAN EXPECTED

WHITE MALE CANCER OF THE LIP ICD 140

1950-1959 DEATHS	1950-1959 RATE	1960-1969 DEATHS	1960-1969 RATE	1970-1979 DEATHS	1970-1979 RATE	COUNTY AND STATE		PERCENT CHANGE 60S/50S	PERCENT CHANGE 70S/60S	PERCENT CHANGE 70S/50S	PERCENTILE %CHANGE 70S/50S	PERCENTILE RATES 1970S
		1*	1.0			IOWA	IA					
1	1.0	1	1.0			JACKSON	IA	8				
1	.9					JASPER	IA					
		1*	1.0	1*	1.4	JEFFERSON	IA		42			95
1	.5			1	.5	JOHNSON	IA			-5	80	84
		1	.5			LEE	IA					
6*	1.1	1	.2	1	.2	LINN	IA	-85	-2	-85	05	77
		1*	1.4			LOUISA	IA					
1	1.2					LUCAS	IA					
		1*	1.2			LYON	IA					
1	1.1			1*	1.0	MADISON	IA			-9	78	92
1	.7			1	.7	MAHASKA	IA			-7	79	88
1	.5	1	.5	1	.4	MARSHALL	IA	3	-18	-16	72	83
1	1.1					MONONA	IA					
1	1.2					MONROE	IA					
4*	2.1	2*	1.0			MUSCATINE	IA	-55				
		1	.8			O'BRIEN	IA					
1*	1.7	1*	1.8			OSCEOLA	IA	4				
				1*	.7	PAGE	IA					89
		1*	1.2			PALO ALTO	IA					
6	.8	4	.5	2	.2	POLK	IA	-39	-61	-76	15	78
2	.6	1	.3	1	.3	POTTAWATTAMIE	IA	-56	16	-49	48	81
		1*	1.0			POWESHIEK	IA					
		1*	1.8			RINGGOLD	IA					
2	.6	1	.2	1	.2	SCOTT	IA	-69	-11	-72	22	77
1	1.2					SHELBY	IA					
1	.9			1	.9	SIOUX	IA			-4	80	90
2	1.1					STORY	IA					
		1	.7	1*	.8	TAMA	IA		20			90
		1*	1.0			TAYLOR	IA					
2*	2.3	1*	1.0			UNION	IA	-54				
1	1.3					VAN BUREN	IA					
2	.9	1	.4	1	.4	WAPELLO	IA	-56	-3	-58	39	83
1	1.1					WASHINGTON	IA					
3	1.3	3*	1.3			WEBSTER	IA	1				
4	.8	1	.2	2	.4	WOODBURY	IA	-78	113	-52	44	83
		1*	1.5	2*	2.4	WORTH	IA		56			98
1	1.1	1	1.0			WRIGHT	IA	-10				
81*	.9	34*	.3	24	.2	KANSAS	KS	-63	-33	-75		
2	1.7					ALLEN	KS					
2*	3.3	1*	1.3			ANDERSON	KS	-60				
		2*	1.8			ATCHISON	KS					
3*	3.6	2*	1.9	1	.7	BARTON	KS	-47	-64	-81	10	88
4*	2.8					BUTLER	KS					
				1*	1.8	CHAUTAUQUA	KS					96
				2*	1.5	CHEROKEE	KS					95
2*	7.3					CHEYENNE	KS					
2*	2.6					CLAY	KS					
2*	1.9					CLOUD	KS					
		1*	4.3			COMANCHE	KS					
2	.9	1	.5			COWLEY	KS	-48				
2	1.0	1	.4			CRAWFORD	KS	-60				
1*	2.7					DECATUR	KS					
		1	.8			DICKINSON	KS					
1	1.4	1*	1.3			DONIPHAN	KS	-7				
1	.7			1	.5	DOUGLAS	KS			-25	65	86
1	1.4	1*	1.0			ELLIS	KS	-32				
				1*	4.2	ELLSWORTH	KS					99
2*	1.9					FORD	KS					
1	1.5					GEARY	KS					
1	1.6	1*	1.3			HARPER	KS	-18				
1	.8	1	.7			HARVEY	KS	-5				
				1*	6.3	HASKELL	KS					99
3*	4.0					JACKSON	KS					
2*	2.5	1*	2.3			JEWELL	KS	-11				
1	.3					JOHNSON	KS					
				2*	2.9	KINGMAN	KS					98
1*	3.9	1*	4.0			KIOWA	KS	2				
1	.5	1	.6			LABETTE	KS	19				
		1*	2.0			LINCOLN	KS					
1	1.3					LINN	KS					
1	.7					LYON	KS					
2	1.7			2*	1.4	MC PHERSON	KS			-19	70	95
1	1.1					MARION	KS					
1	.8			2*	2.6	MARSHALL	KS			247	99	98
1	.7			1*	1.1	MIAMI	KS			52	91	93
1	1.7					MITCHELL	KS					
1	.4	2*	.8	1	.4	MONTGOMERY	KS	129	-55	2	83	82
1	2.5					MORRIS	KS					
		1*	1.1	1*	1.3	NEMAHA	KS		18			94
2	1.7					NEOSHO	KS					
3*	3.8					OSAGE	KS					
1	2.0			1*	1.8	OTTAWA	KS			-6	80	97
				1*	1.6	PHILLIPS	KS					95
1*	3.7					RAWLINS	KS					
1	.3	5*	1.7			RENO	KS	414				
1	1.4					RICE	KS					
2	1.8					RILEY	KS					
3*	2.0					SALINE	KS					
1*	5.4					SCOTT	KS					
8	1.1	4	.4	1	.1	SEDGWICK	KS	-62	-84	-94	00	74
1*	2.7					SEWARD	KS					
2	.4	3	.5	1	.2	SHAWNEE	KS	33	-69	-59	38	77
		1*	4.0			STEVENS	KS					
4*	3.2	1	.8	2*	1.3	SUMNER	KS	-75	68	-58	38	94
1	2.2					WABAUNSEE	KS					
				1*	1.4	WASHINGTON	KS					94
1	2.1					WOODSON	KS					
1	.2			1	.1	WYANDOTTE	KS			-10	78	76

* SIGNIFICANTLY MORE DEATHS THAN EXPECTED

WHITE MALE — CANCER OF THE LIP — ICD 140

COUNTY AND STATE	1950-1959 DEATHS	1950-1959 RATE	1960-1969 DEATHS	1960-1969 RATE	1970-1979 DEATHS	1970-1979 RATE	PERCENT CHANGE 60S/50S	PERCENT CHANGE 70S/60S	PERCENT CHANGE 70S/50S	PERCENTILE %CHANGE 70S/50S	PERCENTILE RATES 1970S
KENTUCKY KY	60	.6	38	.3	37*	.3	-47	-11	-52		
ADAIR KY	1	1.2									
ANDERSON KY					1*	2.1					97
BARREN KY	1	.8	3*	2.5			223				
BATH KY	1	1.7									
BOONE KY					1*	1.1					93
BOYD KY	2	1.0	1	.5	1	.5	-55	0	-55	41	84
BOYLE KY					1*	1.0					92
BRACKEN KY					1*	2.1					97
CALDWELL KY	1	1.5	1*	1.2			-21				
CAMPBELL KY	4	1.4									
CARLISLE KY	1*	9.3									
CASEY KY			2*	3.0							
CHRISTIAN KY			1	.5							
CLARK KY	1	1.4	1*	1.1	1*	1.1	-17	-6	-22	67	93
CRITTENDEN KY	1	1.6			2*	4.1			152	97	99
DAVIESS KY	1	.4	1	.4	1	.3	-17	-9	-25	65	81
FAYETTE KY	2	.6	2	.4	2	.3	-38	-22	-52	44	81
FLEMING KY			1*	1.8							
FRANKLIN KY			1	.8	1*	.7		-16			89
GARRARD KY					2*	3.3					98
GRAVES KY	3*	1.9									
GRAYSON KY	1	1.4									
GREEN KY	1*	2.5			1*	2.2			-12	75	97
GREENUP KY					1*	.6					87
HANCOCK KY	1*	4.0									
HARDIN KY	1	.8	2*	1.4			89				
HARRISON KY					3*	3.1					98
HART KY			2*	2.8	1*	1.4		-52			95
HENDERSON KY	1	.9									
HENRY KY	1	2.0									
HOPKINS KY	2	1.4	1	.5	1	.5	-67	-2	-67	28	84
JEFFERSON KY	14	.7	6	.3	3	.2	-60	-45	-78	13	77
JOHNSON KY					1*	1.2					94
KENTON KY	2	.6									
KNOX KY	1	.8									
LARUE KY	1	2.4			1*	1.8			-23	66	97
LAUREL KY	1	1.3	2*	1.6			19				
LESLIE KY					2*	4.7					99
LEWIS KY	1	1.6	1*	1.5			-6				
LINCOLN KY	1	1.3									
LOGAN KY	1	.9			1*	.9			1	82	91
MC CRACKEN KY			1	.6							
MC LEAN KY			2*	4.0							
MADISON KY	2*	1.7			4*	2.6			48	91	98
MARTIN KY	1*	3.5									
MASON KY			1*	1.6							
MEADE KY			1*	2.6							
MENIFEE KY	1*	3.7			1*	3.5			-5	80	99
MONTGOMERY KY	1	1.8									
MUHLENBERG KY					1	.6					87
NELSON KY	1	1.7									
NICHOLAS KY	1*	4.4									
OHIO KY	1	.9									
PENDLETON KY	1	2.1									
PERRY KY	1	1.0									
TAYLOR KY			1*	1.1							
TRIMBLE KY					1*	3.0					98
UNION KY			1*	1.3							
WARREN KY	1	1.0	2*	1.0	1	.5	3	-51	-49	48	85
WEBSTER KY			1*	1.0							
LOUISIANA LA	40	.6	25	.3	21	.2	-49	-34	-66		
ACADIA LA			1	.5							
ASCENSION LA	2*	3.3									
ASSUMPTION LA	1*	2.9									
AVOYELLES LA			1	.7							
BEAUREGARD LA	1	2.6									
BOSSIER LA			1*	1.5							
CADDO LA	3	.8	2	.3	1	.1	-61	-59	-84	06	75
CALCASIEU LA	1	.4	1	.4	1	.3	6	-38	-34	58	80
CATAHOULA LA			1*	2.2							
CLAIBORNE LA			1*	1.5							
EAST BATON ROUGE LA			1	.1							
EAST CARROLL LA	1*	5.2									
EAST FELICIANA LA					2*	4.0					99
FRANKLIN LA	1	3.2									
IBERIA LA			1	.5							
JEFFERSON LA	3	1.0	2	.2	2	.2	-76	-1	-76	15	79
JEFFERSON DAVIS LA	1	1.0	1*	1.7	2*	2.4	66	44	139	97	98
LAFAYETTE LA	1	.3	1	.3			-12				
LAFOURCHE LA	4*	4.0			1	.7			-83	07	88
LA SALLE LA					1*	1.6					96
LINCOLN LA			1*	1.3							
ORLEANS LA	9	.6	3	.2	2	.1	-68	-29	-77	14	76
OUACHITA LA	2	1.3									
PLAQUEMINES LA	1*	2.8									
POINTE COUPEE LA	1*	1.8	1*	1.7			-6				
RAPIDES LA	2	.7	1	.4	1	.3	-43	-17	-53	43	81
RICHLAND LA	1	1.6									
ST BERNARD LA	2*	2.8									
ST LANDRY LA			2*	1.0	1	.4		-64			82
ST MARY LA	2*	2.4									
TANGIPAHOA LA			2*	1.5	2*	.9		-36			91
TENSAS LA	1*	4.2									
TERREBONNE LA					2*	.7					88
UNION LA					1*	1.2					94
VERNON LA			1*	1.3							

* SIGNIFICANTLY MORE DEATHS THAN EXPECTED

WHITE MALE

CANCER OF THE LIP — ICD 140

1950-1959 DEATHS	1950-1959 RATE	1960-1969 DEATHS	1960-1969 RATE	1970-1979 DEATHS	1970-1979 RATE	COUNTY AND STATE	PERCENT CHANGE 60S/50S	PERCENT CHANGE 70S/60S	PERCENT CHANGE 70S/50S	PERCENTILE %CHANGE 70S/50S	PERCENTILE RATES 1970S
				2*	1.3	WEBSTER LA					94
43*	1.0	21*	.5	16*	.3	MAINE ME	-54	-29	-67		
3	.8	1	.3	1	.2	ANDROSCOGGIN ME	-70	-10	-73	21	79
3	1.0	3*	.8	1	.3	AROOSTOOK ME	-17	-62	-69	27	81
4	.5	4	.5	2	.2	CUMBERLAND ME	3	-57	-56	40	78
2*	1.7			1*	.8	FRANKLIN ME			-51	46	90
6*	3.4			2*	.8	HANCOCK ME			-76	16	90
5	1.4	5*	1.2	1	.2	KENNEBEC ME	-13	-82	-84	06	79
		2*	1.0			KNOX ME					
				1*	.6	LINCOLN ME					87
4*	1.7	1	.4			OXFORD ME	-74				
7*	1.4					PENOBSCOT ME					
		1	.9			SAGADAHOC ME					
2	1.0			2*	.9	SOMERSET ME			-9	78	91
2	1.4	2*	1.7	1*	.8	WALDO ME	24	-52	-40	53	90
1	.4			2*	1.0	WASHINGTON ME			132	97	92
4	.9	2	.4	2	.3	YORK ME	-55	-14	-61	36	82
40	.5	16	.2	18	.2	MARYLAND MD	-69	-6	-71		
3	1.2					ALLEGANY MD					
3	1.1	1	.1	1	.1	ANNE ARUNDEL MD	-89	3	-88	04	75
7	.8			3	.1	BALTIMORE MD			-82	09	76
1*	2.5					CALVERT MD					
1	1.2			1*	.9	CAROLINE MD			-18	71	91
1	.4	1	.3			CAPROLL MD	-13				
		1	.5	1	.6	CECIL MD		9			87
2	1.5	1	.7	1*	.7	DORCHESTER MD	-53	-4	-55	41	88
1	.4			2*	.7	FREDERICK MD			80	94	88
		2*	.9			HAPFORD MD					
2*	2.9	2*	1.6			HOWARD MD	-46				
1	1.5	1*	1.9	1*	1.3	KENT MD	25	-33	-17	72	94
3	.3	3	.5			PRINCE GEORGE'S MD	63				
				1*	.8	ST MARY'S MD					89
				1*	.7	TALBOT MD					89
1	.3			1	.2	WASHINGTON MD			-28	63	78
2	1.2					WICOMICO MD					
1	1.1			1*	1.1	WORCESTEP MD			1	82	93
11	.5	4	.2	4	.2	BALTIMORE CITY MD	-68	16	-62	35	77
103	.5	53	.2	30	.1	MASSACHUSETTS MA	-56	-50	-78		
				2	.3	BARNSTABLE MA					80
1	.1	2	.3			BERKSHIRE MA	133				
15*	1.1	4	.2	4	.2	BPISTOL MA	-80	-16	-83	07	77
17	.7	10	.4	2	.1	ESSEX MA	-49	-83	-91	02	74
2	.7					FRANKLIN MA					
9	.7	8	.4	1	.0	HAMPDEN MA	-34	-89	-93	01	74
		1	.3	1	.2	HAMPSHIRE MA		-25			78
12	.3	13	.3	8	.1	MIDDLESEX MA	-14	-43	-51	45	76
1*	7.0					NANTUCKET MA					
16*	1.0	2	.1	2	.1	NORFOLK MA	-89	-30	-93	01	74
2	.2	3	.3	2	.1	PLYMOUTH MA	40	-46	-24	66	76
15	.5	5	.1	5	.2	SUFFOLK MA	-68	5	-66	29	76
13	.6	5	.2	3	.1	WORCESTER MA	-66	-47	-82	09	75
145	.7	60	.2	55	.2	MICHIGAN MI	-68	-21	-75		
1*	2.7					ALCONA MI					
		1	.4	2*	.6	ALLEGAN MI		77			88
3*	3.1					ALPENA MI					
1	2.0			1*	1.4	ARENAC MI			-33	59	94
2	1.6	1	.6			BARRY MI	-60				
5*	1.5					BAY MI					
1*	2.0					BENZIE MI					
4	.8	2	.3	1	.1	BERRIEN MI	-59	-58	-83	08	76
3*	1.7					BPANCH MI					
1	.2	2	.3			CALHOUN MI	79				
				1	.5	CASS MI					84
2	1.5					CHIPPEWA MI					
		1*	1.1			CLARE MI					
3*	1.8	1	.7			CLINTON MI	-63				
		1	.8			DELTA MI					
1	.7			1	.5	DICKINSON MI			-31	60	85
				2*	.8	EATON MI					90
		3*	3.7			EMMET MI					
8	1.1	2	.2	3	.2	GENESEE MI	-83	11	-81	10	78
1	.6	3*	1.6	1	.5	GRAND TRAVERSE MI	167	-70	-20	69	85
		1	.6			GRATIOT MI					
1	.6					HILLSDALE MI					
4*	1.5					HOUGHTON MI					
1	.8	2*	1.2			HURON MI	43				
4	.7			3	.4	INGHAM MI			-40	54	83
				2*	1.0	IONIA MI					92
2	2.0	2*	1.5	1*	.6	ISABELLA MI	-25	-58	-68	27	87
4	.9					JACKSON MI					
1	.2	5*	.9	1	.1	KALAMAZOO MI	379	-85	-27	64	75
2*	7.3					KALKASKA MI					
9	.7	4	.3	1	.1	KENT MI	-64	-77	-92	01	74
1	.6			1	.5	LAPEER MI			-12	76	85
1	.4					LENAWEE MI					
2*	6.1					MACKINAC MI					
1	.1	2	.2	2	.1	MACOMB MI	96	-38	21	88	76
1	.9	1*	.9			MANISTEE MI	-7				
1	.5	1	.4			MARQUETTE MI	-27				
				1	.6	MASON MI					87
		1	.9			MECOSTA MI					
1	.9					MENOMINEE MI					
1	.8					MIDLAND MI					

* SIGNIFICANTLY MORE DEATHS THAN EXPECTED

WHITE MALE

CANCER OF THE LIP

ICD 140

1950-1959 DEATHS	1950-1959 RATE	1960-1969 DEATHS	1960-1969 RATE	1970-1979 DEATHS	1970-1979 RATE	COUNTY AND STATE	PERCENT CHANGE 60S/50S	PERCENT CHANGE 70S/60S	PERCENT CHANGE 70S/50S	PERCENTILE %CHANGE 70S/50S	PERCENTILE RATES 1970S
1*	4.1	1*	3.2	1*	2.5	MISSAUKEE MI	-22	-24	-40	53	98
1	.4	1	.3	1	.2	MONROE MI	-27	-30	-49	48	78
4*	2.1	2*	1.1			MONTCALM MI	-49				
1*	4.7					MONTMORENCY MI					
1	.3			2	.4	MUSKEGON MI			13	86	83
6	.4	3	.1	4	.1	OAKLAND MI	-67	-11	-70	24	75
1	1.6					OGEMAW MI					
2*	2.6					OSCEOLA MI					
1*	3.2			1*	1.7	OTSEGO MI			-48	49	96
2	.7	1	.3			OTTAWA MI	-61				
4	.7	3	.5	2	.3	SAGINAW MI	-36	-43	-64	33	80
1	.2			1	.2	ST CLAIR MI			-24	65	77
2	1.1			1	.4	ST JOSEPH MI			-65	31	83
1	.5			1	.5	SANILAC MI			-8	79	84
2	1.0	3*	1.4	1	.5	SHIAWASSEE MI	35	-66	-54	42	85
1	.6					TUSCOLA MI					
1	.4			1	.4	VAN BUREN MI			8	84	83
1	.2	1	.1	2	.4	WASHTENAW MI	-39	156	57	92	82
37	.6	8	.1	13	.2	WAYNE MI	-85	71	-75	18	76
2*	2.2	1*	.9			WEXFORD MI	-59				
86	.6	33	.2	45	.2	MINNESOTA MN	-69	24	-62		
		1*	1.3			AITKIN MN					
				1	.2	ANOKA MN					76
				1	.6	BECKER MN					86
1	.8	1	.7	1*	.8	BELTRAMI MN	-3	4	1	83	89
5*	3.9					BROWN MN					
1	.7					CARLTON MN					
1	1.7					CARVER MN					
		1	.6	1	.5	CLAY MN		-9			86
1	1.2			1*	.9	COTTONWOOD MN			-31	61	90
4*	3.0			2*	.8	CROW WING MN			-72	23	90
				1	.2	DAKOTA MN					78
3*	4.2					DODGE MN					
		1	.7			DOUGLAS MN					
		1	1.2			FARIBAULT MN					
3*	1.9			1	.5	FILLMORE MN			-72	22	86
				1	.4	FREEBORN MN					83
3*	1.6					GOODHUE MN					
11	.4	4	.1	8	.2	HENNEPIN MN	-76	104	-52	45	79
1	1.4					HUBBARD MN					
1	.5	1	.7	1	.5	ITASCA MN	47	-33	-1	82	85
1	2.1			1*	1.6	KANABEC MN			-22	67	96
2	1.4	1	.6	2*	.9	KANDIYOHI MN	-60	60	-35	57	91
				1*	1.9	KITTSON MN					97
1	2.2					KOOCHICHING MN					
2*	2.3					LAC QUI PARLE MN					
1	1.0			1*	.7	LE SUEUR MN			-32	60	88
1	2.7					LINCOLN MN					
1	1.5	1	.8			LYON MN	-46				
1	.8					MARTIN MN					
1	1.0			2*	1.2	MORRISON MN			26	88	94
2	1.2			1	.4	MOWER MN			-63	33	84
		2*	.7			OLMSTED MN					
3	1.0	2	.5			OTTER TAIL MN	-44				
1	1.7					PENNINGTON MN					
2*	1.7					PINE MN					
		1*	1.6			PIPESTONE MN					
3*	1.4					POLK MN					
2*	2.8	1*	1.1			POPE MN	-61				
8	.6	1	.1	4	.2	RAMSEY MN	-91	309	-65	31	79
1	.9	2*	1.6	2*	1.5	REDWOOD MN	74	-6	64	93	95
1	.7	1	.7	1	.7	RENVILLE MN	-7	5	-2	81	89
		1	.5			RICE MN					
		1*	1.3			ROSEAU MN					
6	.6	4	.4	3	.3	ST LOUIS MN	-31	-34	-54	42	80
		1*	1.0	1*	.5	SCOTT MN		-50			85
1	1.5					SIBLEY MN					
1	.9					STEELE MN					
1	1.2					SWIFT MN					
2	1.6					TODD MN					
		2*	1.6	1*	.9	WABASHA MN		-41			91
2*	3.0			1*	1.1	WASECA MN			-64	33	93
1	.5			1	.4	WASHINGTON MN			-23	67	83
				1*	.9	WATONWAN MN					91
		1*	2.2			WILKIN MN					
2	1.0			1	.4	WINONA MN			-58	39	83
1	.6	1	.6			WRIGHT MN	14				
				2*	1.6	YELLOW MEDICINE MN					96
32	.7	20*	.4	15	.2	MISSISSIPPI MS	-48	-37	-67		
				1*	1.3	ADAMS MS					94
2*	1.9					ALCORN MS					
1*	2.3	1*	2.1			AMITE MS	-7				
		1*	1.2	1*	1.1	ATTALA MS		-9			93
		1*	1.1			BOLIVAR MS					
2*	2.9					CALHOUN MS					
1	2.6					CHICKASAW MS					
		1*	1.7			CLARKE MS					
		1*	1.3			COAHOMA MS					
1	.6			1	.5	FORREST MS			-15	74	86
1*	3.2					FRANKLIN MS					
				1*	1.7	GEORGE MS					96
2	.8	2	.7	2	.5	HARRISON MS	-13	-27	-36	57	85
1	1.2			1*	1.0	ITAWAMBA MS			-11	76	92
2*	1.7	2*	1.8			JACKSON MS	4				
		2*	1.4			JONES MS					
1	2.0					LAFAYETTE MS					
1	.6					LAUDERDALE MS					

* SIGNIFICANTLY MORE DEATHS THAN EXPECTED

WHITE MALE | CANCER OF THE LIP | ICD 140

COUNTY AND STATE	1950-1959 DEATHS	1950-1959 RATE	1960-1969 DEATHS	1960-1969 RATE	1970-1979 DEATHS	1970-1979 RATE	PERCENT CHANGE 60S/50S	PERCENT CHANGE 70S/60S	PERCENT CHANGE 70S/50S	PERCENTILE %CHANGE 70S/50S	PERCENTILE RATES 1970S
LEAKE MS	1	1.3									
LEE MS					1	.5					85
LEFLORE MS			2*	2.5							
LOWNDES MS			2*	1.9							
NESHOBA MS	1	1.4									
NOXUBEE MS					1*	4.2					99
PEARL RIVER MS	1	1.4			1*	1.0			-30	62	92
PERRY MS	2*	10.7									
PIKE MS	1	1.4			1*	.9			-40	54	90
PONTOTOC MS			1*	1.2	1*	1.0		-14			92
PRENTISS MS	1	1.2	1*	1.1			-13				
SCOTT MS			1*	1.4							
SIMPSON MS	1	2.2			1*	1.7			-24	65	96
SUNFLOWER MS	2*	3.6									
TALLAHATCHIE MS	1*	2.0									
TATE MS	1*	3.2	1*	2.6			-20				
TIPPAH MS					1*	1.3					94
WALTHALL MS	1*	3.7									
WARREN MS	1	1.1									
WASHINGTON MS	2*	3.9									
WEBSTER MS					1*	2.6					98
YALOBUSHA MS			1*	2.3							
YAZOO MS	1	1.9									
MISSOURI MO	114	.6	59	.3	47	.2	-54	-28	-67		
ADAIR MO			2*	1.8	1*	.9		-51			90
AUDRAIN MO	1	.7									
BARRY MO			1	.8							
BARTON MO	2*	2.7									
BATES MO	1	.7									
BENTON MO			1*	1.2							
BOONE MO	1	.5									
BUCHANAN MO	2	.4									
CALLAWAY MO	1	.9	1	.7	2*	1.4	-18	93	58	92	95
CAMDEN MO	1	1.3			1*	.7			-46	50	88
CAPE GIRARDEAU MO	2	1.2			1	.4			-70	24	82
CARTER MO	1*	4.6									
CASS MO	1	.8									
CEDAR MO	1	1.7									
CHRISTIAN MO			2*	2.4							
CLARK MO			1*	1.2							
CLAY MO	3	1.3	1	.4			-67				
CLINTON MO	1	1.2									
COLE MO	1	.9			1	.5			-45	51	86
COOPER MO					2*	2.3					97
CRAWFORD MO					1*	.8					89
DADE MO			1*	1.7							
DE KALB MO			1*	1.4							
DENT MO	1	1.5	1*	1.3			-18				
DOUGLAS MO			1*	1.4							
DUNKLIN MO	3*	1.8									
FRANKLIN MO	1	.7									
GENTRY MO					1*	1.2					94
GREENE MO	3	.7			1	.1			-81	10	75
GRUNDY MO	1	.9									
HARRISON MO			1*	1.0							
HENRY MO	1	.9	2*	1.4			65				
HICKORY MO			1*	2.8							
HOWARD MO			1*	1.3							
IRON MO			1*	1.7	1*	1.5		-11			95
JACKSON MO	12	.5	6	.3	5	.2	-47	-29	-62	35	78
JASPER MO	7*	2.0	1	.3			-86				
JEFFERSON MO	1	.7									
JOHNSON MO			1	.8							
KNOX MO	1	1.6	1*	1.9			20				
LACLEDE MO					1*	.8					89
LAFAYETTE MO	3*	1.8									
LAWRENCE MO	1	.7	2*	1.5	2*	1.0	106	-31	42	90	92
LEWIS MO	1	1.2									
LINCOLN MO					1*	.8					89
LINN MO	1	.7			1*	.9			30	89	90
MC DONALD MO	2*	2.5									
MACON MO	1	.6	1	1.0			73				
MADISON MO	1	1.6									
MARIES MO	1*	2.3									
MERCER MO	1	2.0			2*	4.5			123	96	99
MILLER MO			1*	1.0							
MISSISSIPPI MO			1*	1.4	1*	1.6		13			95
MONITEAU MO	1	1.3			1*	1.1			-13	75	93
MONTGOMERY MO			1*	1.1							
NEW MADRID MO	1	1.8	2*	2.0			13				
NEWTON MO			1	.5							
NODAWAY MO			1	.7							
OREGON MO	1	1.5									
OSAGE MO	1	1.6									
PEMISCOT MO	1	2.3	2*	2.0			-15				
PERRY MO	1	1.3			2*	1.8			36	89	96
PETTIS MO	4*	2.4			1	.5			-80	11	85
PIKE MO	2*	1.9			1*	1.0			-48	49	91
PLATTE MO	1	1.7									
POLK MO	2*	1.7									
PUTNAM MO	1	2.2									
RAY MO			2*	1.7	1*	.9		-47			91
RIPLEY MO	1	1.2									
ST CHARLES MO	1	.6	1	.4	2*	.6	-22	47	15	86	88
ST CLAIR MO	2*	2.9	1*	1.1			-62				
STE GENEVIEVE MO	1	1.6			1*	1.3			-20	69	94
ST LOUIS COUNTY MO	3	.3	5	.3	3	.1	-16	-68	-73	20	74
SALINE MO			2*	1.2							
SCOTT MO	2	1.6									

* SIGNIFICANTLY MORE DEATHS THAN EXPECTED

WHITE MALE — CANCER OF THE LIP — ICD 140

1950-1959 DEATHS	1950-1959 RATE	1960-1969 DEATHS	1960-1969 RATE	1970-1979 DEATHS	1970-1979 RATE	COUNTY AND STATE	PERCENT CHANGE 60S/50S	PERCENT CHANGE 70S/60S	PERCENT CHANGE 70S/50S	PERCENTILE %CHANGE 70S/50S	PERCENTILE RATES 1970S
		1*	2.9			SHELBY MO					
2	1.3			2*	1.4	STODDARD MO			7	84	95
1	2.0					STONE MO					
1	1.7	1*	1.6			SULLIVAN MO	-2				
1	.7					TEXAS MO					
		1	.7	1	.6	VERNON MO		-13			86
1	1.8					WARREN MO					
1	1.4					WASHINGTON MO					
1	3.2					WAYNE MO					
				1*	.9	WEBSTER MO					91
1*	2.7					WORTH MO					
				1*	.9	WRIGHT MO					90
17	.6	5	.2	5	.2	ST LOUIS CITY MO	-67	11	-63	34	79
17	.6	14*	.4	7	.2	MONTANA MT	-29	-55	-68		
1*	2.1					BLAINE MT					
2*	12.9					BROADWATER MT					
		1*	1.6	2*	3.7	CARBON MT		137			99
				1	.3	CASCADE MT					81
		1*	5.5			DANIELS MT					
		1*	1.8			DAWSON MT					
1	1.0					DEER LODGE MT					
		1*	5.5			FALLON MT					
		2*	1.2			FLATHEAD MT					
		1	.8			GALLATIN MT					
				1*	1.7	LAKE MT					96
3*	2.8	1	.7			LEWIS AND CLARK MT	-75				
				1*	7.8	MEAGHER MT					99
		1*	4.5			MUSSELSHELL MT					
		1*	1.6			PARK#MT					
1*	2.1	1*	3.0			PHILLIPS MT	44				
1*	2.8					PONDERA MT					
		1*	2.1			ROOSEVELT MT					
1*	4.2					ROSEBUD MT					
2	1.6	1	.4			SILVER BOW MT	-76				
3*	5.5					VALLEY MT					
2	1.0	1	.3	2	.5	YELLOWSTONE MT	-68	71	-45	51	86
57*	.9	17	.2	26*	.3	NEBRASKA NB	-74	44	-62		
		1	.5	1	.6	ADAMS NB		7			86
				1*	1.6	BOONE NB					96
1*	5.7					BOYD NB					
1	1.0					BUFFALO NB					
				1*	1.1	BURT NB					93
		1*	.8	1*	.9	CASS NB		12			91
		1*	1.3			CEDAR NB					
1*	6.6			1*	4.0	CHASE NB			-40	53	99
1*	2.1					CHERRY NB					
1	1.6					CHEYENNE NB					
1	1.4					COLFAX NB					
2*	4.3					CUMING NB					
3*	2.9					CUSTER NB					
1	2.3					DAKOTA NB					
1	1.4					DAWSON NB					
				1*	1.8	DIXON NB					96
				1	.6	DODGE NB					87
10	.8	5	.4	2	.1	DOUGLAS NB	-54	-59	-81	10	76
1*	5.0	1*	3.8			DUNDY NB	-24				
1	1.6					FURNAS NB					
1	.8					GAGE NB					
				1*	3.4	GREELEY NB					98
				1	.5	HALL NB					84
1*	2.9					HITCHCOCK NB					
2*	4.0					HOWARD NB					
2*	2.4			2*	2.7	JEFFERSON NB			9	84	98
1*	2.6					KEITH NB					
1*	4.0					KIMBALL NB					
3*	3.4			1*	1.1	KNOX NB			-67	28	93
3	.5	2	.3	4*	.6	LANCASTER NB	-36	74	11	85	87
1	.7	1	.6	1	.6	LINCOLN NB	-12	-3	-15	73	87
2	1.2			1	.6	MADISON NB			-51	46	87
		1*	2.1			MERRICK NB					
		1*	2.5			MORRILL NB					
				1*	2.1	NANCE NB					97
1	1.6					NEMAHA NB					
3*	3.1	1*	1.0	1*	1.1	OTOE NB	-69	15	-64	32	93
1*	2.6					PAWNEE NB					
				1*	1.7	PIERCE NB					96
1	1.0			2*	1.6	PLATTE NB			61	93	96
2*	3.5					RED WILLOW NB					
1	.9					SALINE NB					
		1	.9			SAUNDERS NB					
1	1.3			1*	1.5	SEWARD NB			10	85	95
		1*	1.6			THAYER NB					
1*	2.9					VALLEY NB					
4*	7.7					WEBSTER NB					
5	.7	3	.2	8*	.4	NEVADA NV	-64	51	-46		
1	.5			3	.2	CLARK NV			-56	40	79
		1*	2.2	1*	1.9	ELKO NV		-13			97
1*	12.3					EUREKA NV					
				2*	4.6	HUMBOLDT NV					99
1*	4.1					LYON NV					
				1*	6.2	PERSHING NV					99
2	.8	2	.5	1	.2	WASHOE NV	-43	-54	-74	20	79
10	.4	5	.2	11*	.3	NEW HAMPSHIRE NH	-58	78	-24		

* SIGNIFICANTLY MORE DEATHS THAN EXPECTED

COUNTIES COMBINED; SEE TECHNICAL NOTES

WHITE MALE

CANCER OF THE LIP — ICD 140

1950-1959 DEATHS	1950-1959 RATE	1960-1969 DEATHS	1960-1969 RATE	1970-1979 DEATHS	1970-1979 RATE	COUNTY AND STATE	PERCENT CHANGE 60S/50S	PERCENT CHANGE 70S/60S	PERCENT CHANGE 70S/50S	PERCENTILE %CHANGE 70S/50S	PERCENTILE RATES 1970S
1	.5			1	.5	CHESHIRE NH			-2	81	85
3*	1.8	1	.7	2*	1.1	COOS NH	-63	61	-41	53	93
2	.4	2	.2	3	.3	HILLSBOROUGH NH	-48	37	-28	63	81
1	.3	1	.3	2	.5	MERRIMACK NH	5	71	79	94	84
2	.6	1	.2			ROCKINGHAM NH	-60				
1	.7			2*	1.0	STRAFFORD NH			48	91	92
				1	.5	SULLIVAN NH					86
89	.5	45	.2	45	.2	NEW JERSEY NJ	-59	-10	-64		
1	.2	2	.2			ATLANTIC NJ	8				
7	.3	5	.2	3	.1	BERGEN NJ	-48	-47	-73	21	74
1	.2	5*	.9			BURLINGTON NJ	300				
8	.7	5	.4	3	.2	CAMDEN NJ	-46	-35	-65	31	79
				4*	.8	CAPE MAY NJ					89
1	.3	3*	.6			CUMBERLAND NJ	149				
18	.5	2	.1	3	.1	ESSEX NJ	-90	74	-82	09	74
1	.2			1	.2	GLOUCESTER NJ			1	83	78
13	.6	6	.2	4	.2	HUDSON NJ	-67	-6	-69	26	78
		2	.7	2*	.6	HUNTERDON NJ		-10			87
2	.2	2	.2	3	.3	MERCER NJ	-2	48	44	90	80
5	.6	2	.1	7*	.4	MIDDLESEX NJ	-76	190	-30	62	83
6	.5			2	.1	MONMOUTH NJ			-78	14	75
1	.1			3	.3	MORRIS NJ			142	97	81
3	.7	2	.5	3	.1	OCEAN NJ	-36	-69	-80	11	76
8	.5	6	.3	3	.2	PASSAIC NJ	-33	-54	-69	25	77
1	.9					SALEM NJ					
1	.4					SOMERSET NJ					
3*	1.8	1	.5			SUSSEX NJ	-71				
8	.5	2	.1	3	.2	UNION NJ	-85	109	-69	25	77
1	.4			1	.3	WARREN NJ			-31	61	80
7	.3	12*	.5	5	.1	NEW MEXICO NM	38	-69	-58		
2	.4	3	.5	1	.1	BERNALILLO NM	6	-76	-74	19	75
		1	.6			CHAVES NM					
1	1.2					COLFAX NM					
1	.9	2*	1.5	1	.5	DONA ANA NM	72	-67	-44	52	85
1	.7	1	.8			EDDY NM	19				
1	2.5	1*	1.0			GRANT NM	-59				
				1	.7	LEA NM					88
		1*	5.3			LUNA NM					
				1*	.9	OTERO NM					91
		1*	1.3			ROOSEVELT NM					
		1	.6			SANTA FE NM					
				1*	2.4	SIERRA NM					98
1*	3.2	1*	4.0			UNION NM	25				
253	.4	108	.2	76	.1	NEW YORK NY	-64	-34	-77		
9	1.0	3	.3	1	.1	ALBANY NY	-66	-77	-92	01	74
1	.5					ALLEGANY NY					
14	.3	4	.1	2	.0	BRONX NY	-83	-39	-90	03	73
3	.4	3	.3	2	.2	BROOME NY	-19	-39	-51	46	78
2	.5	2	.5			CATTARAUGUS NY	-1				
1	.2	3*	.8			CAYUGA NY	232				
3	.5	1	.1	1	.1	CHAUTAUQUA NY	-73	8	-71	23	76
3	.7			1	.2	CHEMUNG NY			-67	28	79
2	1.0					CHENANGO NY					
3	1.3	4*	2.1			CLINTON NY	57				
1	.4					COLUMBIA NY					
		1	.5			CORTLAND NY					
2	.8	1	.4			DELAWARE NY	-50				
5	.7	1	.1			DUTCHESS NY	-84				
23	.7	4	.1	3	.1	ERIE NY	-87	-28	-91	02	74
2	1.5					ESSEX NY					
1	.6	1	.5			FRANKLIN NY	-17				
5*	2.0	1	.3	1	.3	FULTON NY	-83	-6	-84	06	81
		1	.5			GENESEE NY					
1	.5	1	.7			GREENE NY	43				
2	.5					HERKIMER NY					
3	.7	1	.2			JEFFERSON NY	-68				
32	.3	14	.1	12	.1	KINGS(BROOKLYN) NY	-58	-12	-63	34	75
		1	.8			LEWIS NY					
		1	.5			LIVINGSTON NY					
3	1.5					MADISON NY					
5	.2	2	.1	3	.1	MONROE NY	-63	42	-47	49	75
3	1.2	2	.6			MONTGOMERY NY	-53				
11	.4	1	.0	3	.1	NASSAU NY	-92	119	-83	08	74
27	.4	13	.2	8	.1	NEW YORK(MANHATTAN) NY	-51	-31	-66	29	75
3	.4	1	.2	2	.2	NIAGARA NY	-56	13	-50	47	78
7	.7	2	.2	1	.1	ONEIDA NY	-76	-54	-89	03	74
5	.3	4	.2	3	.2	ONONDAGA NY	-30	-28	-50	47	77
5	.6	3	.3	1	.1	ORANGE NY	-52	-67	-84	06	74
				1	.6	ORLEANS NY					87
1	.2			1	.2	OSWEGO NY			-12	75	79
1	.3	2	.7			OTSEGO NY	127				
1	.8					PUTNAM NY					
12	.2	8	.1	9	.1	QUEENS NY	-48	28	-33	59	75
2	.4					RENSSELAER NY					
2	.2			2	.2	RICHMOND NY			-11	77	78
2	.6	1	.2	1	.2	ROCKLAND NY	-66	-20	-73	21	77
3	.6	2	.4			ST LAWRENCE NY	-29				
4	1.1	1	.2	1	.2	SARATOGA NY	-80	-34	-86	04	76
7	1.0					SCHENECTADY NY					
2	.4	1	.2	1	.2	STEUBEN NY	-54	-4	-55	41	78
9	.5	5	.2	8	.2	SUFFOLK NY	-64	10	-60	37	79
2	.8					SULLIVAN NY					
				1	.5	TIOGA NY					85
2	.9			1	.3	TOMPKINS NY			-63	34	82
1	.2	2	.4	4*	.6	ULSTER NY	77	72	205	98	87

* SIGNIFICANTLY MORE DEATHS THAN EXPECTED

WHITE MALE — CANCER OF THE LIP — COUNTY AND STATE — ICD 140

1950-1959 DEATHS	1950-1959 RATE	1960-1969 DEATHS	1960-1969 RATE	1970-1979 DEATHS	1970-1979 RATE	COUNTY AND STATE	PERCENT CHANGE 60S/50S	PERCENT CHANGE 70S/60S	PERCENT CHANGE 70S/50S	PERCENTILE %CHANGE 70S/50S	PERCENTILE RATES 1970S
		1	.4	1	.4	WARREN NY		-15			83
3	1.2			1	.4	WASHINGTON NY			-66	30	83
1	.3	1	.3			WAYNE NY	-5				
10	.5	8	.3			WESTCHESTER NY	-39				
1	.5					WYOMING NY					
47	.5	33	.2	34	.2	NORTH CAROLINA NC	-51	-13	-57		
		1	.3			ALAMANCE NC					
1	2.6					ALEXANDER NC					
2*	2.7			1*	.8	ASHE NC			-70	24	90
				1*	1.3	AVERY NC					94
1	1.0			1*	1.1	BEAUFORT NC			1	83	93
1*	2.6					BERTIE NC					
1	1.6					BLADEN NC					
4	.9	2	.3	2	.3	BUNCOMBE NC	-62	-8	-65	31	81
2	1.3	1	.4	1	.3	CABARRUS NC	-70	-27	-78	12	80
1	.6					CALDWELL NC					
				2*	1.4	CARTERET NC					95
		1	.4	1	.3	CATAWBA NC		-8			82
		1*	1.5			CHEROKEE NC					
				1*	2.8	CHOWAN NC					98
		1	.9			COLUMBUS NC					
2*	3.5					CRAVEN NC					
		1*	2.6			DARE NC					
1	.5	2*	.8			DAVIDSON NC	56				
				1*	1.0	DUPLIN NC					92
3*	1.2	1	.3			DURHAM NC	-75				
1	.2	2	.6	2	.2	FORSYTH NC	237	-56	47	91	80
2*	3.1			1*	1.7	FRANKLIN NC			-44	51	96
1	.5					GASTON NC					
				1*	6.2	GATES NC					99
1*	2.2			1*	2.1	GREENE NC			-7	80	97
2	.3	1	.1	2	.2	GUILFORD NC	-56	79	-21	67	78
1	1.9			2*	1.3	HARNETT NC			-30	61	94
				1	.4	HAYWOOD NC					84
1*	4.6					HYDE NC					
2	1.4					IREDELL NC					
		2*	.9			JOHNSTON NC					
2*	2.8					LENOIR NC					
		1*	1.1			LINCOLN NC					
1	1.2	1*	1.1			MADISON NC	-9				
		1	.1	3	.4	MECKLENBURG NC		266			82
2*	3.6					MITCHELL NC					
				1	.6	MOORE NC					86
		1	.5			NASH NC					
				1	.3	NEW HANOVER NC					81
1	1.6					ORANGE NC					
1	.9			2*	1.1	PITT NC			17	87	93
1	.4					RANDOLPH NC					
				1	.5	ROBESON NC					86
1	.4			2	.6	ROWAN NC			59	93	87
1	.6	2*	1.0			RUTHERFORD NC	61				
		1*	1.8			SCOTLAND NC					
1	1.0	2*	1.4	1	.4	STANLY NC	37	-68	-56	40	84
1	1.0					STOKES NC					
		1	.5			SURRY NC					
				1*	1.2	VANCE NC					94
2	.5	4*	.7			WAKE NC	51				
				1	.5	WILKES NC					86
2*	2.9	1*	.9			WILSON NC	-70				
		1*	1.2			YADKIN NC					
1	2.7	1*	1.5			YANCEY NC	-42				
18	.6	11	.4	7	.2	NORTH DAKOTA ND	-36	-46	-65		
		2*	2.4			BARNES ND					
1	.4	4*	1.6	1	.3	CASS ND	277	-81	-30	62	81
1	3.0					CAVALIER ND					
1*	2.4			1*	2.7	DICKEY ND			9	84	98
1*	2.3					DIVIDE ND					
		1*	2.9			FOSTER ND					
1	.5	1	.7	1	.4	GRAND FORKS ND	47	-46	-21	68	83
2*	5.4					GRIGGS ND					
				1*	5.7	HETTINGER ND					99
		1*	2.8			MC KENZIE ND					
1	1.6					MC LEAN ND					
4*	4.4					MORTON ND					
				1*	1.3	PEMBINA ND					94
		1*	3.9			SHERIDAN ND					
				1*	5.3	STEELE ND					99
2*	2.8	1*	1.6			TRAILL ND	-44				
2*	2.7					WALSH ND					
				1	.5	WARD ND					84
1*	1.6					WELLS ND					
1	.9					WILLIAMS ND					
171	.5	85	.2	49	.1	OHIO OH	-57	-48	-77		
1	.3			1	.2	ALLEN OH			-18	70	78
2	.6	1	.2	1	.2	ASHTABULA OH	-67	8	-64	32	79
1	.4					ATHENS OH					
3*	2.4			1	.6	AUGLAIZE OH			-74	19	87
1	.2	3*	.6			BELMONT OH	199				
		1	.6			BROWN OH					
3	.7					BUTLER OH					
1	1.1					CARROLL OH					
1	.8	1	.7			CHAMPAIGN OH	-10				
2	.4	3	.6			CLARK OH	39				
1	.4			1	.3	CLERMONT OH			-31	61	81
2	1.3			1	.7	CLINTON OH			-48	49	88

* SIGNIFICANTLY MORE DEATHS THAN EXPECTED

WHITE MALE

CANCER OF THE LIP — ICD 140

1950-1959 DEATHS	1950-1959 RATE	1960-1969 DEATHS	1960-1969 RATE	1970-1979 DEATHS	1970-1979 RATE	COUNTY AND STATE	PERCENT CHANGE 60S/50S	PERCENT CHANGE 70S/60S	PERCENT CHANGE 70S/50S	PERCENTILE %CHANGE 70S/50S	PERCENTILE RATES 1970S
3	.6	3	.6	2	.4	COLUMBIANA OH	0	-36	-37	56	82
				1	.6	COSHOCTON OH					86
1	.5	1	.5			CRAWFORD OH	-9				
28	.6	7	.1	6	.1	CUYAHOGA OH	-77	-28	-84	07	74
		1	.4	1	.4	DARKE OH		-12			82
				1	.6	DEFIANCE OH					88
2	.8					ERIE OH					
3	1.1	1	.3	1	.3	FAIRFIELD OH	-70	-17	-75	18	80
				2*	1.5	FAYETTE OH					95
10	.6	7	.3			FRANKLIN OH	-39				
		2*	1.4			FULTON OH					
		3*	2.4			GALLIA OH					
1	1.0	1	.8			GEAUGA OH	-13				
		1	.1			GREENE OH					
1	.3	1	.3			GUERNSEY OH	-8				
10	.4	5	.2	5	.2	HAMILTON OH	-55	-13	-61	36	76
3	1.2					HANCOCK OH					
1	.5	1	.6			HARDIN OH	3				
1	.5	1	.5			HIGHLAND OH	-1				
1	.8	1	.7	1*	1.1	HOCKING OH	-6	45	36	90	93
1	1.0					HOLMES OH					
1	.4					HURON OH					
				2*	1.3	JACKSON OH					94
1	.2	3*	.7	2	.5	JEFFERSON OH	190	-22	125	96	85
1	.2			2	.2	LAKE OH			-11	77	79
2	.8	4*	1.7			LAWRENCE OH	100				
1	.2	1	.2			LICKING OH	-5				
				1	.5	LOGAN OH					85
3	.4	2	.3	2	.3	LORAIN OH	-32	6	-27	63	81
9	.7	5	.4	1	.0	LUCAS OH	-46	-86	-93	01	74
		1*	.8			MADISON OH					
6	.6	2	.2			MAHONING OH	-73				
1	.4					MARION OH					
3	1.6	1	.5			MEDINA OH	-70				
1	.9					MERCER OH					
1	.3	1	.3			MIAMI OH	-7				
1	.8					MONROE OH					
2	.1	2	.1	3	.2	MONTGOMERY OH	30	58	106	96	77
1	1.0					MORGAN OH					
2*	2.0					MORROW OH					
1	.3	1	.3			MUSKINGUM OH	6				
2	1.1			1	.7	PERRY OH			-37	56	88
				1	.6	PICKAWAY OH					87
1	1.1					PIKE OH					
1	.3					PORTAGE OH					
2	1.6					PREBLE OH					
2	1.3					PUTNAM OH					
1	.2	2	.5	1	.2	RICHLAND OH	87	-62	-30	62	77
				2*	.7	ROSS OH					88
3	1.3					SANDUSKY OH					
3	1.1	2	.5	1	.3	SCIOTO OH	-57	-40	-75	18	81
1	.7					SHELBY OH					
4	.3	2	.2			STARK OH	-54				
9	.6	2	.1			SUMMIT OH	-81				
4	.8	4	.5			TRUMBULL OH	-30				
4	1.1			1	.2	TUSCARAWAS OH			-78	13	79
1	.8					UNION OH					
				1*	1.8	VINTON OH					96
4*	2.5	1	.3			WARREN OH	-88				
3	1.1	1	.4	1	.4	WASHINGTON OH	-68	-2	-68	27	82
1	.4	3*	.9	2	.5	WAYNE OH	153	-45	39	90	85
1	.6					WILLIAMS OH					
2	.9					WOOD OH					
1	.8					WYANDOT OH					
76*	.8	33	.3	35*	.3	OKLAHOMA OK	-63	-7	-65		
		1*	2.0			ADAIR OK					
1	1.8					ALFALFA OK					
		1*	1.4			ATOKA OK					
1*	3.1					BEAVER OK					
1	.8					BECKHAM OK					
2*	2.9					BLAINE OK					
2	1.3	1	.6	2*	1.1	BRYAN OK	-54	77	-18	70	93
		1	.6			CADDO OK					
1	.8					CANADIAN OK					
1	1.3	1*	1.1	1*	1.0	CHEROKEE OK	-17	-12	-26	64	92
		1	.4	1	.3	CLEVELAND OK		-32			81
1*	2.6	1*	2.5	1*	3.1	COAL OK	-1	23	22	88	98
1	.7	1	.5			COMANCHE OK	-23				
1	2.2					COTTON OK					
		1*	.9			CRAIG OK					
2	1.0					CREEK OK					
1	1.4					DELAWARE OK					
1*	2.3					DEWEY OK					
		1*	2.8	1*	2.4	ELLIS OK		-16			98
3	1.3			1	.3	GARFIELD OK			-73	20	82
3*	2.8			1	.5	GARVIN OK			-81	09	86
4*	2.3	1	.5			GRADY OK	-78				
				1*	1.8	HASKELL OK					96
				1*	.9	HUGHES OK					91
1	1.6					JEFFERSON OK					
		5*	2.1	3*	1.0	KAY OK		-50			92
1	1.8					KINGFISHER OK					
2*	2.3	1*	1.1			KIOWA OK	-53				
				1*	1.8	LATIMER OK					97
2*	2.2	1	.7			LINCOLN OK	-68				
1	1.0	1	.9	1*	.9	LOGAN OK	-10	2	-9	79	91
1	1.5					MC CLAIN OK					
1	.9	1	.7	1*	.9	MC CURTAIN OK	-21	26	-1	82	91
1*	2.2					MARSHALL OK					

* SIGNIFICANTLY MORE DEATHS THAN EXPECTED

WHITE MALE

CANCER OF THE LIP

ICD 140

1950-1959 DEATHS	1950-1959 RATE	1960-1969 DEATHS	1960-1969 RATE	1970-1979 DEATHS	1970-1979 RATE	COUNTY AND STATE	PERCENT CHANGE 60S/50S	PERCENT CHANGE 70S/60S	PERCENT CHANGE 70S/50S	PERCENTILE %CHANGE 70S/50S	PERCENTILE RATES 1970S
1	.9					MAYES OK					
1	1.2					MURRAY OK					
2	.9			1	.4	MUSKOGEE OK			-61	36	82
1	2.1					NOBLE OK					
2*	2.7					NOWATA OK					
10	.8	5	.3	4	.2	OKLAHOMA OK	-63	-16	-69	26	79
1	.5	1	.7	2*	1.0	OKMULGEE OK	35	49	101	95	92
1	1.0					OSAGE OK					
1	.9			1	.7	OTTAWA OK			-21	68	88
1	1.2	1*	1.1	1*	1.1	PAWNEE OK	-12	-1	-13	75	93
				2*	.9	PAYNE OK					91
2	1.0	1	.6			PITTSBURG OK	-43				
				1	.5	PONTOTOC OK					86
4*	1.8			3*	1.0	POTTAWATOMIE OK			-45	51	92
		3*	2.3			SEMINOLE OK					
1	1.1					SEQUOYAH OK					
2	1.5					STEPHENS OK					
7	.7	2	.2	4	.3	TULSA OK	-79	92	-59	38	81
2*	2.9					WOODS OK					
1	1.2					WOODWARD OK					
43	.6	20	.2	13	.1	OREGON OR	-63	-45	-80		
2*	2.5	1*	.8			BAKER OR	-67				
2*	1.9					BENTON OR					
3	.7	1	.1	2	.3	CLACKAMAS OR	-79	88	-60	37	80
		1	.7			COLUMBIA OR					
2	.9	2*	1.0			COOS OR	9				
1	.7					DESCHUTES OR					
2	1.1	1	.3	2	.6	DOUGLAS OR	-73	102	-46	50	87
2	.7			1	.2	JACKSON OR			-78	12	76
2	1.7					JOSEPHINE OR					
1	.6			2*	1.0	KLAMATH OR			74	94	92
3	.7	1	.2	1	.1	LANE OR	-75	-30	-83	08	75
5*	2.2			1	.3	LINN OR			-88	04	80
1	.9					MALHEUR OR					
5	1.0	1	.1	1	.1	MARION OR	-88	-15	-90	02	74
		2*	7.4			MORROW OR					
10	.4	8	.3	3	.1	MULTNOMAH OR	-23	-66	-74	20	75
1	.7					POLK OR					
1	.4	1	.4			UMATILLA OR	-4				
		1*	.9			WASCO OR					
193	.5	63	.1	67	.1	PENNSYLVANIA PA	-71	-3	-72		
1	.4			1	.4	ADAMS PA			-15	73	83
26	.5	8	.1	7	.1	ALLEGHENY PA	-70	-19	-76	16	75
1	.2	1	.2	1	.3	ARMSTRONG PA	-5	18	12	86	80
3	.4					BEAVER PA					
1	.5					BEDFORD PA					
2	.2	1	.1			BERKS PA	-74				
2	.3			2	.3	BLAIR PA			-10	78	80
1	.3	1	.4			BRADFORD PA	7				
6	.9	1	.1	2	.2	BUCKS PA	-85	37	-79	12	78
2	.4			1	.2	BUTLER PA			-55	41	77
3	.4	3	.4	2	.2	CAMBRIA PA	5	-50	-47	50	78
		1*	3.1			CAMERON PA					
2	.6	2*	.7			CARBON PA	17				
1	.4					CENTRE PA					
4	.6	1	.1	2	.2	CHESTER PA	-78	16	-74	19	77
2	1.1					CLARION PA					
1	.3			2	.5	CLEARFIELD PA			84	94	86
2	.8					COLUMBIA PA					
2	.6					CRAWFORD PA					
2	.2	1	.1			DAUPHIN PA	-37				
11	.8	2	.1			DELAWARE PA	-87				
2	1.3					ELK PA					
12*	1.1	2	.2	2	.2	ERIE PA	-81	-18	-84	06	77
3	.3	1	.1	2	.2	FAYETTE PA	-54	64	-24	66	80
		1	.2			FRANKLIN PA					
1*	2.0					FULTON PA					
1	.3			1	.3	INDIANA PA			-7	79	80
				1	.4	JEFFERSON PA					82
				1*	1.7	JUNIATA PA					96
8	.9	2	.2	2	.2	LACKAWANNA PA	-82	1	-82	09	77
4	.4	2	.2	4	.3	LANCASTER PA	-60	81	-29	63	80
6*	1.6					LAWRENCE PA					
2	.2			2	.2	LEHIGH PA			-12	76	77
9	.7	1	.1	2	.1	LUZERNE PA	-91	45	-87	04	74
1	.3	1	.2			LYCOMING PA	-40				
1	.3			1	.3	MC KEAN PA			1	83	81
1	.2	3	.5			MERCER PA	144				
1	.5					MIFFLIN PA					
7	.6	2	.1	5	.2	MONTGOMERY PA	-84	116	-66	29	78
4	.6	3	.3	2	.2	NORTHAMPTON PA	-39	-39	-63	35	79
1	.2	2	.4	1	.3	NORTHUMBERLAND PA	144	-20	96	95	82
23	.3	8	.1	10	.2	PHILADELPHIA PA	-62	29	-51	45	76
1	1.0					POTTER PA					
9*	.9	4	.6	2	.2	SCHUYLKILL PA	-37	-64	-77	14	78
				1	.6	SUSQUEHANNA PA					86
2	1.3					TIOGA PA					
2*	2.1					UNION PA					
2	1.2	1	.3	1	.3	VENANGO PA	-74	-5	-75	18	81
2	.9	2*	.8			WARREN PA	-13				
8	1.0	1	.1	2	.2	WASHINGTON PA	-91	79	-84	07	77
		1	.6			WAYNE PA					
1	.1	4	.3	3	.2	WESTMORELAND PA	253	-25	165	98	78
1	1.2					WYOMING PA					
3	.3			2	.1	YORK PA			-49	48	76

* SIGNIFICANTLY MORE DEATHS THAN EXPECTED

WHITE MALE CANCER OF THE LIP ICD 140

1950-1959 DEATHS	1950-1959 RATE	1960-1969 DEATHS	1960-1969 RATE	1970-1979 DEATHS	1970-1979 RATE	COUNTY AND STATE	PERCENT CHANGE 60S/50S	PERCENT CHANGE 70S/60S	PERCENT CHANGE 70S/50S	PERCENTILE %CHANGE 70S/50S	PERCENTILE RATES 1970S
27*	.9	11	.3	10	.3	RHODE ISLAND RI	-62	-22	-70		
2	1.8					BRISTOL RI					
3	1.0	2	.7	2	.4	KENT RI	-31	-37	-56	40	84
1	.4			2*	.5	NEWPORT RI			31	89	86
21*	.9	8	.3	6	.2	PROVIDENCE RI	-66	-28	-75	17	79
		1	.4			WASHINGTON RI					
23	.6	22*	.4	24*	.3	SOUTH CAROLINA SC	-25	-24	-43		
2*	2.0	1	.7			AIKEN SC	-63				
		1	.2	3*	.6	ANDERSON SC		161			87
1*	4.3	1*	2.5			BAMBERG SC	-41				
1*	3.7					BARNWELL SC					
1	.4			2	.3	CHARLESTON SC			-13	74	82
		1*	.8	2*	1.5	CHEROKEE SC		89			95
		2*	2.9			CHESTER SC					
2*	2.5	1*	1.4			CHESTERFIELD SC	-45				
		1*	5.0			CLARENDON SC					
1*	3.3	1*	1.7			DILLON SC	-48				
3	.5			4*	.5	GREENVILLE SC			-14	74	84
				3*	2.3	GREENWOOD SC					97
		3*	2.2			HORRY SC					
		1*	1.7	1*	.9	LANCASTER SC		-46			91
		1	1.2			LAURENS SC					
		1	.3			LEXINGTON SC					
				1*	1.2	MARION SC					93
1*	1.9			1*	1.7	MARLBORO SC			-11	76	96
1	1.2			1*	1.0	NEWBERRY SC			-20	69	92
3*	3.0					OCONEE SC					
1	.6	1	.4	2*	1.0	PICKENS SC	-33	178	87	95	93
2	1.1	3*	.8			RICHLAND SC	-25				
4	1.1	1	.2	3*	.5	SPARTANBURG SC	-78	122	-52	44	85
		1*	1.0			SUMTER SC					
				1*	1.7	WILLIAMSBURG SC					96
		1	.3			YORK SC					
22	.7	5	.1	8	.2	SOUTH DAKOTA SD	-81	63	-69		
1	.9			1*	1.0	BEADLE SD			10	85	92
1	1.3					BROOKINGS SD					
3*	2.7					BROWN SD					
1	1.7					CHARLES MIX SD					
1	1.9			2*	4.5	CLAY SD			131	97	99
1	1.1					CODINGTON SD					
1	1.1					DAVISON SD					
1*	5.9					FAULK SD					
		1*	1.5			GRANT SD					
1*	2.5					GREGORY SD					
1*	8.3					HYDE SD					
		1*	1.8			MC COOK SD					
4*	1.1	2	.6	3*	.7	MINNEHAHA SD	-47	20	-36	56	88
1	.7	1	.6			PENNINGTON SD	-7				
				1*	4.1	POTTER SD					99
1	1.4					ROBERTS SD					
2*	2.7					UNION SD					
				1*	2.2	WALWORTH SD					97
2*	2.1					YANKTON SD					
60	.6	38	.3	45*	.3	TENNESSEE TN	-53	5	-51		
2*	2.5					BEDFORD TN					
		1	.3			BLOUNT TN					
		1	.7			BRADLEY TN					
		1*	2.5			CANNON TN					
2	1.3			2*	.9	CARTER TN			-33	59	90
1*	2.3					CHEATHAM TN					
		4*	4.2			CLAIBORNE TN					
1	1.1					COCKE TN					
		1*	.9			COFFEE TN					
				1*	1.4	CROCKETT TN					95
				1*	1.0	CUMBERLAND TN					92
4	.4	2	.1	3	.2	DAVIDSON TN	-63	77	-34	59	79
1	1.2					DICKSON TN					
				1*	.8	DYER TN					89
1*	2.3	1*	2.5			FAYETTE TN	6				
1*	1.9					FENTRESS TN					
				1*	.6	FRANKLIN TN					86
				1	.5	GIBSON TN					86
		1	.6	2*	.9	GREENE TN		55			91
		1	.8			HAMBLEN TN					
4	.7			1	.2	HAMILTON TN			-76	17	77
		1*	2.7			HANCOCK TN					
				2*	2.2	HARDEMAN TN					97
				1*	1.1	HARDIN TN					93
				1	.7	HAWKINS TN					88
				1*	1.7	HAYWOOD TN					96
1	1.6					HENDERSON TN					
1	1.0			2*	1.4	HENRY TN			47	91	95
		1*	1.6			HICKMAN TN					
1	2.3					HUMPHREYS TN					
1	1.5					JACKSON TN					
1	1.9	1*	1.4	1*	1.7	JOHNSON TN	-26	20	-11	77	96
4	1.0	1	.1	2	.2	KNOX TN	-90	76	-83	08	77
1*	3.3					LAKE TN					
1	.8					LAWRENCE TN					
				1*	2.7	LEWIS TN					98
2*	1.9					LINCOLN TN					
		1	.7			MC MINN TN					
1	1.2	1*	1.2	1*	1.2	MC NAIRY TN	-1	-1	-2	81	94

* SIGNIFICANTLY MORE DEATHS THAN EXPECTED

WHITE MALE — CANCER OF THE LIP — ICD 140

1950-1959 DEATHS	1950-1959 RATE	1960-1969 DEATHS	1960-1969 RATE	1970-1979 DEATHS	1970-1979 RATE	COUNTY AND STATE		PERCENT CHANGE 60S/50S	PERCENT CHANGE 70S/60S	PERCENT CHANGE 70S/50S	PERCENTILE %CHANGE 70S/50S	PERCENTILE RATES 1970S
1	1.6					MACON	TN					
1	.5	2*	.9	2*	.8	MADISON	TN	102	-8	86	94	90
1	1.4	1*	1.1			MARION	TN	-24				
				1	.6	MAURY	TN					87
1*	4.3					MEIGS	TN					
				1*	.8	MONROE	TN					89
1	.7					MONTGOMERY	TN					
3*	2.3					OBION	TN					
1*	4.4			2*	4.9	PERRY	TN			12	86	99
1*	5.7					PICKETT	TN					
1	.8					PUTNAM	TN					
1	1.7					RHEA	TN					
1	.9			1	.5	ROANE	TN			-40	54	86
2*	2.5	2*	1.6	1	.7	ROBERTSON	TN	-36	-55	-71	23	89
1	.6	2*	.8	1	.4	RUTHERFORD	TN	39	-53	-34	58	83
				1*	1.3	SCOTT	TN					94
4	.5	4	.3	6*	.4	SHELBY	TN	-48	58	-18	71	83
				1*	1.6	SMITH	TN					96
1*	3.0					STEWART	TN					
		5*	1.2	1	.1	SULLIVAN	TN		-88			76
1	.9					SUMNER	TN					
1	1.5					TIPTON	TN					
1*	5.9					VAN BUREN	TN					
2	.7	2	.6	1	.4	WASHINGTON	TN	-24	-36	-51	46	82
1	.8			1	.6	WEAKLEY	TN			-21	68	87
1	1.3					WHITE	TN					
1	.8	1	1.0			WILSON	TN	23				
153	.6	91	.3	103*	.2	TEXAS	TX	-51	-9	-56		
1	.8			1	.9	ANDERSON	TX			15	86	91
2	1.3	2*	1.1			ANGELINA	TX	-20				
1*	3.4					ARANSAS	TX					
				1*	2.3	ARCHER	TX					98
1	1.4					AUSTIN	TX					
				1*	3.7	BANDERA	TX					99
1	1.1					BASTROP	TX					
		1*	2.3			BAYLOR	TX					
		1*	.8			BEE	TX					
1	.3	1	.3	4*	1.0	BELL	TX	7	180	199	98	92
2	.2	7	.3	6	.3	BEXAR	TX	80	-14	55	92	80
				1*	.7	BOSQUE	TX					89
1	.7	1	.5	1	.3	BOWIE	TX	-30	-32	-52	44	82
		2*	.7	3*	1.1	BRAZORIA	TX		51			93
1	.7			1	.5	BROWN	TX			-31	60	85
				1*	.7	BURNET	TX					88
1	1.2					CALDWELL	TX					
1	2.0	1*	1.7			CALLAHAN	TX	-14				
1	.7	1	.2			CAMERON	TX	-75				
		1*	3.2	2*	5.6	CARSON	TX		74			99
2*	1.9					CASS	TX					
1	.8	1	.5			CHEROKEE	TX	-38				
				2*	11.6	COCHRAN	TX					99
1*	5.9					COKE	TX					
3*	1.4			2*	.6	COLLIN	TX			-57	39	87
		1*	2.0			COLORADO	TX					
1	1.2					COMAL	TX					
1	1.2					COMANCHE	TX					
1*	3.7					CONCHO	TX					
		1	1.1	1*	.7	COOKE	TX		-33			89
1	1.4					CORYELL	TX					
				1*	5.2	COTTLE	TX					99
1*	2.5			1*	2.1	CROSBY	TX			-15	73	97
1*	3.0					DALLAM	TX					
8	.4	6	.2	11*	.2	DALLAS	TX	-38	3	-36	57	79
1	1.7					DAWSON	TX					
		2*	4.5			DELTA	TX					
		1	.5	1	.3	DENTON	TX		-45			80
		1*	1.6			DE WITT	TX					
1*	3.0					DICKENS	TX					
1	3.2	1	.3			ECTOR	TX	-92				
3*	1.6			1	.4	ELLIS	TX			-72	22	84
4	.7	1	.1	3	.3	EL PASO	TX	-82	163	-51	45	82
1	.9			1*	.8	FALLS	TX			-12	76	89
1	.6					FANNIN	TX					
1	.7			1	.8	FAYETTE	TX			26	89	90
		2*	1.1	1	.5	FORT BEND	TX		-52			86
2*	3.9					FREESTONE	TX					
1*	4.3	1*	1.8			GAINES	TX	-58				
2	.8	1	.2			GALVESTON	TX	-82				
1	1.9					GILLESPIE	TX					
1	1.1	1*	1.1			GONZALES	TX	0				
		1*	1.3			GRAY	TX					
3	.9	2	.5			GRAYSON	TX	-44				
				1	.4	GREGG	TX					82
2*	1.7	1	.7			GUADALUPE	TX	-57				
1	.9					HALE	TX					
1	1.2	1*	.7			HARDIN	TX	-40				
11	.4	3	.1	4	.1	HARRIS	TX	-83	70	-72	23	75
				2*	1.3	HARRISON	TX					94
1	1.4					HASKELL	TX					
2*	2.8			1	.4	HENDERSON	TX			-87	04	82
3*	1.9					HILL	TX					
1*	2.5					HOOD	TX					
		2*	1.6	1*	1.1	HOPKINS	TX		-28			93
				1*	.7	HOWARD	TX					88
2	1.1	1	.5	1	.6	HUNT	TX	-55	19	-47	50	86
2*	4.4			1*	2.2	JACKSON	TX			-50	47	97
7*	1.9					JEFFERSON	TX					
1	1.4					JIM WELLS	TX					
				1	.4	JOHNSON	TX					83

* SIGNIFICANTLY MORE DEATHS THAN EXPECTED

WHITE MALE

CANCER OF THE LIP — ICD 140

1950-1959 DEATHS	1950-1959 RATE	1960-1969 DEATHS	1960-1969 RATE	1970-1979 DEATHS	1970-1979 RATE	COUNTY AND STATE	PERCENT CHANGE 60S/50S	PERCENT CHANGE 70S/60S	PERCENT CHANGE 70S/50S	PERCENTILE %CHANGE 70S/50S	PERCENTILE RATES 1970S
1	1.0			2*	1.6	JONES TX			61	93	96
		2*	3.6			KARNES TX					
1	.8	2*	1.7	1	.5	KAUFMAN TX	114	-70	-36	57	86
1	.8	1	.6	1	.5	KERR TX	-19	-19	-35	58	85
1*	2.0					KNOX TX					
1	.6	1	.5	4*	2.0	LAMAR TX	-11	304	259	99	97
				2*	3.5	LAMPASAS TX					99
1	.8					LAVACA TX					
1*	2.6			1*	1.8	LEE TX			-29	62	96
1	.8					LIBERTY TX					
		1*	1.0			LIMESTONE TX					
5*	.9			2	.4	LUBBOCK TX			-60	37	82
				1*	2.5	MC CULLOCH TX					98
6*	1.1	3	.5	3	.4	MC LENNAN TX	-56	-7	-60	37	84
		1*	5.3			MASON TX					
1	2.2					MATAGORDA TX					
2*	2.8					MIDLAND TX					
2*	1.8					MILAM TX					
		1*	2.1			MITCHELL TX					
				2*	.7	MONTGOMERY TX					88
1*	5.9					MOTLEY TX					
		2*	1.4	2*	1.3	NACOGDOCHES TX		-8			94
1	.6	1	.5	1	.5	NAVARRO TX	-10	-9	-18	71	85
				1*	1.4	NOLAN TX					95
2	.3			1	.1	NUECES TX			-76	16	74
		1	1.4			ORANGE TX					
2	1.6					PARKER TX					
1*	3.6					PARMER TX					
		1*	3.0			PECOS TX					
2	.7	1	.2			POTTER TX	-66				
1	1.5					RED RIVER TX					
1*	14.1					ROBERTS TX					
1	1.7					ROBERTSON TX					
				1*	1.1	RUNNELS TX					93
		1	.6			RUSK TX					
1*	3.1			1*	1.5	SABINE TX			-52	43	95
		1	.6			SAN PATRICIO TX					
		2*	2.4	2*	2.6	SCURRY TX		8			98
1	.3	1	.5			SMITH TX	40				
5	.4	6	.4	6	.2	TARRANT TX	3	-33	-31	61	80
1	.5					TAYLOR TX					
				1*	1.1	TITUS TX					93
2	.7	2*	.6	1	.2	TOM GREEN TX	-11	-65	-69	26	79
1	.2	2	.2	1	.2	TRAVIS TX	-6	-28	-33	59	77
		1*	3.7			TRINITY TX					
		2*	2.9			TYLER TX					
2*	2.2	1*	1.1			UPSHUR TX	-51				
				1	.5	VAN ZANDT TX					85
2*	1.6					VICTORIA TX					
1	1.0	3*	3.5			WASHINGTON TX	256				
1	.5					WEBB TX					
		1	.7	1	.8	WHARTON TX		7			89
1*	2.0					WHEELER TX					
2	.5	1	.3			WICHITA TX	-27				
2*	2.3					WILBARGER TX					
				2*	1.0	WILLIAMSON TX					92
1	1.1					WISE TX					
1	1.0					WOOD TX					
				1*	.7	YOUNG TX					89
1*	2.2					ZAVALA TX					
15	.8	6	.2	3	.1	UTAH UT	-73	-65	-91		
		1*	1.2			BOX ELDER UT					
1	2.0			1	.5	DAVIS UT			-77	15	84
1*	3.2					IRON UT					
7	1.0	2	.2	2	.1	SALT LAKE UT	-83	-34	-89	03	75
2*	3.4					TOOELE UT					
1*	4.2					UINTAH UT					
2	.7	1	.3			UTAH UT	-54				
1	.3	2	.5			WEBER UT	68				
3	.2	4	.2	4	.2	VERMONT VT	37	-10	24		
1	.7	1	.8			CALEDONIA VT	5				
		1	.4			CHITTENDEN VT					
				1*	2.8	ESSEX VT					98
1	.7					FRANKLIN VT					
				1*	2.2	LAMOILLE VT					97
				1*	.9	ORLEANS VT					91
		2*	.9	1	.4	RUTLAND VT		-58			82
1	.4					WINDSOR VT					
49	.5	32	.3	22	.1	VIRGINIA VA	-44	-52	-73		
1	.6	1	.5			ALBEMARLE#VA	-20				
		1*	1.1			AMHERST VA					
		1	.3			ARLINGTON#VA					
1	.4					AUGUSTA#VA					
3	.8	1	.3			BEDFORD#VA	-68				
1	1.6					BOTETOURT VA					
2	1.1					CARROLL#VA					
				1*	2.1	CHARLOTTE VA					97
8*	1.2	3	.3	1	.1	CHESTERFIELD#VA	-72	-74	-93	00	74
				1*	2.7	CLARKE VA					98
2	1.9					DINWIDDIE#VA					
4	.9	1	.2	1	.1	FAIRFAX#VA	-81	-66	-93	00	74
				1*	.9	FAUQUIER VA					91
		1	.8	1	.5	FREDERICK#VA		-31			86
		1*	2.0			GLOUCESTER VA					
				1*	3.3	GREENSVILLE#VA					98

* SIGNIFICANTLY MORE DEATHS THAN EXPECTED

COUNTIES COMBINED; SEE TECHNICAL NOTES

WHITE MALE — CANCER OF THE LIP — ICD 140

1950-1959 DEATHS	1950-1959 RATE	1960-1969 DEATHS	1960-1969 RATE	1970-1979 DEATHS	1970-1979 RATE	COUNTY AND STATE	PERCENT CHANGE 60S/50S	PERCENT CHANGE 70S/60S	PERCENT CHANGE 70S/50S	PERCENTILE %CHANGE 70S/50S	PERCENTILE RATES 1970S
1	1.0					HALIFAX#VA					
2*	2.3					HANOVER VA					
1	.7	2	.8	1	.1	HENRICO VA	20	-82	-78	13	76
		3*	1.9			HENRY#VA					
1*	4.2					ISLE OF WIGHT VA					
1	1.1	2*	1.4			JAMES CITY#VA	36				
		1*	10.6			KING GEORGE VA					
1*	4.5	1*	4.5			KING WILLIAM VA	0				
				1*	.7	LEE VA					88
		1*	1.0	1*	.6	LOUDOUN VA		-40			87
		1	.6			MONTGOMERY#VA					
				1*	2.0	NELSON VA					97
		1*	2.7			NORTHAMPTON VA					
				1*	3.4	NORTHUMBERLAND VA					98
				1*	2.0	NOTTOWAY VA					97
1	1.5					PAGE VA					
		1*	1.3			PATRICK VA					
				2*	.5	PITTSYLVANIA#VA					85
				1	.9	PRINCE WILLIAM#VA					90
1	.7					PULASKI VA					
1*	3.8					RICHMOND VA					
3	.8	2	.4			ROANOKE#VA	-50				
				2*	1.4	ROCKBRIDGE#VA					95
1	1.2	1*	.9			RUSSELL VA	-22				
2*	1.8					SHENANDOAH VA					
1	.8					SMYTH VA					
1*	6.5					SUSSEX VA					
				1	.5	TAZEWELL VA					85
1	1.9					WARREN VA					
1	.6	1	.5			WASHINGTON#VA	-16				
1	.5					WISE#VA					
1	1.1	1*	1.4			WYTHE VA	32				
5	.4	4	.3	2	.1	CHESAPEAKE CITY#VA	-17	-70	-75	17	74
				1*	.7	SUFFOLK CITY VA					89
81*	.8	22	.2	24	.2	WASHINGTON WA	-78	-5	-79		
				2*	2.2	ASOTIN WA					97
		1	.3			BENTON WA					
1	.6					CHELAN WA					
1	.6			1	.4	CLALLAM WA			-43	52	82
5*	1.5	3*	.7			CLARK WA	-53				
3*	1.3	1	.4			COWLITZ WA	-71				
				1*	1.1	DOUGLAS WA					93
1*	5.1					FERRY WA					
1	.3	1	.3			GRAYS HARBOR WA	-2				
1	1.3			2*	1.5	ISLAND WA			21	87	95
30*	1.0	3	.1	9	.2	KING WA	-92	187	-78	13	78
3	.9	2	.5	2	.4	KITSAP WA	-41	-16	-51	46	84
2*	1.8					KITTITAS WA					
3	1.1					LEWIS WA					
1*	2.2					PEND OREILLE WA					
6	.5			3	.2	PIERCE WA			-64	32	78
1	.7	1	.4			SKAGIT WA	-34				
4	.7	2	.3			SNOHOMISH WA	-58				
6	.5	4	.3	2	.1	SPOKANE WA	-36	-55	-71	23	75
1	1.1	1*	1.2			STEVENS WA	8				
2	.8			1	.3	THURSTON WA			-63	34	80
1	.7	1	.4			WALLA WALLA WA	-41				
2	.6	2	.5	1	.2	WHATCOM WA	-21	-52	-62	36	79
1	.9					WHITMAN WA					
5	1.1					YAKIMA WA					
32	.4	19	.2	10	.1	WEST VIRGINIA WV	-46	-53	-74		
		1*	1.0	2*	1.9	BARBOUR WV		90			97
2	1.7	1	.5			BERKELEY WV	-71				
3	.9	1	.3			CABELL WV	-68				
		1*	2.1			DODDRIDGE WV					
1	.4	1	.6			FAYETTE WV	52				
		1	.8			HANCOCK WV					
2	.5			1	.2	HARRISON WV			-53	43	79
1	1.3			1*	1.0	JACKSON WV			-18	71	92
1	1.8					JEFFERSON WV					
2	.3	3	.3			KANAWHA WV	14				
1	.9			1*	.7	LEWIS WV			-20	69	89
1	.9	1	.5			LOGAN WV	-47				
1	.3	2*	1.4	2*	1.2	MC DOWELL WV	332	-14	272	99	93
1	.3			1	.3	MARION WV			-11	77	81
		2*	1.0	1	.5	MARSHALL WV		-54			84
1	1.0					MASON WV					
		1	.3			MERCER WV					
1	1.4					MINERAL WV					
1	.7					MINGO WV					
2	.9	2*	.8			MONONGALIA WV	-13				
1	1.2					MONROE WV					
2*	2.1					NICHOLAS WV					
1	.3	1	.3			OHIO WV	-2				
1	.7					PRESTON WV					
1	1.2					PUTNAM WV					
1	.9					RANDOLPH WV					
1	1.1					TAYLOR WV					
1	2.0					TYLER WV					
1	1.0					UPSHUR WV					
1	.6			1	.5	WAYNE WV			-15	73	86
		1	.3			WOOD WV					
84	.5	39	.2	45	.2	WISCONSIN WI	-58	-4	-59		
		1*	1.2			ADAMS WI					
4*	3.8					ASHLAND WI					

* SIGNIFICANTLY MORE DEATHS THAN EXPECTED

\# COUNTIES COMBINED; SEE TECHNICAL NOTES

WHITE MALE CANCER OF THE LIP ICD 140

1950-1959 DEATHS	1950-1959 RATE	1960-1969 DEATHS	1960-1969 RATE	1970-1979 DEATHS	1970-1979 RATE	COUNTY AND STATE	PERCENT CHANGE 60S/50S	PERCENT CHANGE 70S/60S	PERCENT CHANGE 70S/50S	PERCENTILE %CHANGE 70S/50S	PERCENTILE RATES 1970S
2	1.2	2*	.9			BARRON WI	-23				
1	.3	1	.3	2	.4	BROWN WI	29	34	72	93	83
				1*	1.0	BURNETT WI					93
		2*	1.9	1*	.8	CALUMET WI		-56			90
3*	1.4			1	.4	CHIPPEWA WI			-69	26	84
1	.5					CLARK WI					
1	1.4					CRAWFORD WI					
8*	1.1	1	.2			DANE WI	-85				
		2	.6			DODGE WI					
3*	3.1					DOOR WI					
		2*	.7	2*	.8	DOUGLAS WI		13			90
1	.6	1	.9			DUNN WI	51				
3	1.4			1	.3	EAU CLAIRE WI			-76	16	81
2	.6	3*	.9			FOND DU LAC WI	57				
				3*	1.2	GRANT WI					94
		1	.7	1	.6	GREEN WI		-16			87
		1*	1.2			GREEN LAKE WI					
1	.8	1	.8			IOWA WI	-4				
1	1.0					JACKSON WI					
				1	.3	JEFFERSON WI					81
1	1.0					JUNEAU WI					
1	.2	2	.5	1	.2	KENOSHA WI	123	-64	-20	69	77
				1*	.9	KEWAUNEE WI					91
2	.6	1	.3			LA CROSSE WI	-58				
				1*	1.0	LAFAYETTE WI					92
1	.9			1*	.7	LANGLADE WI			-22	67	88
1	.3					MANITOWOC WI					
2	.5	1	.2	3*	.6	MARATHON WI	-58	178	18	87	87
3*	1.8					MARINETTE WI					
3	.9			2	.5	MENOMINEE#WI			-49	48	84
13	.3	8	.2	7	.2	MILWAUKEE WI	-42	-14	-50	47	77
		2*	1.3			MONROE WI					
				1	.3	OUTAGAMIE WI					80
		1	.6			POLK WI					
1	.6					PORTAGE WI					
5	1.1	2	.4	2	.3	RACINE WI	-67	-23	-75	18	81
1	.9			1*	1.0	RICHLAND WI			11	85	92
4	.8	1	.2	2	.4	ROCK WI	-77	104	-54	42	82
1	1.8					RUSK WI					
1	.5			1	.4	SAUK WI			-21	68	82
		1	.2	3*	.6	SHEBOYGAN WI		194			86
1	1.1					TAYLOR WI					
				1	.4	TREMPEALEAU WI					83
1	.4					WALWORTH WI					
1	.7	1	.6			WASHINGTON WI	-12				
4	1.0			2	.2	WAUKESHA WI			-78	14	79
3	1.6			2*	.5	WAUPACA WI			-68	27	85
				1*	.7	WAUSHARA WI					89
1	.2	1	.2			WINNEBAGO WI	-6				
2	1.1					WOOD WI					
5	.5	5	.4	3	.2	WYOMING WY	-20	-47	-57		
				1*	1.5	BIG HORN WY					95
1*	2.8					CONVERSE WY					
1	.8					FREMONT WY					
		2*	1.1			LARAMIE WY					
				1*	2.3	PLATTE WY					97
1	1.1	2*	1.9	1*	.6	SHERIDAN WY	77	-65	-39	54	88
2*	5.6					SWEETWATER WY					
		1*	2.1			WASHAKIE WY					

* SIGNIFICANTLY MORE DEATHS THAN EXPECTED

\# COUNTIES COMBINED; SEE TECHNICAL NOTES

WHITE FEMALE

CANCER OF THE LIP — COUNTY AND STATE

ICD 140

1950-1959 DEATHS	1950-1959 RATE	1960-1969 DEATHS	1960-1969 RATE	1970-1979 DEATHS	1970-1979 RATE	COUNTY AND STATE	PERCENT CHANGE 60S/50S	PERCENT CHANGE 70S/60S	PERCENT CHANGE 70S/50S	PERCENTILE %CHANGE 70S/50S	PERCENTILE RATES 1970S
328	.0	198	.0	200	.0	UNITED STATES	-58	-25	-69		
17*	.2	5	.0	9*	.1	ALABAMA AL	-79	36	-72		
				1*	2.4	BULLOCK AL					99
1*	.8					CHAMBERS AL					
1*	.7			1*	.3	DE KALB AL			-55	47	98
1*	1.0					ESCAMBIA AL					
2*	.7					ETOWAH AL					
				2*	.7	HOUSTON AL					99
		1*	.6			JACKSON AL					
4*	.3			1	.0	JEFFERSON AL			-85	06	95
		1*	.4			LAUDERDALE AL					
1*	.9	1*	.8			LEE AL	-18				
				1*	.2	MADISON AL					97
		1*	.7			MARION AL					
1*	.6			1*	.3	MARSHALL AL			-46	58	98
				1*	.7	MONROE AL					98
1*	.3					MONTGOMERY AL					
1*	.5					MORGAN AL					
1*	1.8			1*	.7	PIKE AL			-61	41	98
1*	1.2					RANDOLPH AL					
1*	1.3					RUSSELL AL					
1*	.4					WALKER AL					
		1*	1.5			WINSTON AL					
3	.1			3	.0	ARIZONA AZ			-82		
1*	.6					COCHISE AZ					
1	.1			1	.0	MARICOPA AZ			-77	15	94
				2*	.1	PIMA AZ					96
1*	2.1					YUMA AZ					
4	.1	7*	.1	3	.0	ARKANSAS AR	33	-70	-60		
1*	1.8					BRADLEY AR					
				1*	.5	CRAWFORD AR					98
		1*	1.2			FULTON AR					
		1*	.7			GREENE AR					
1*	1.5					HEMPSTEAD AR					
		2*	1.4	1*	.4	MISSISSIPPI AR		-73			98
		1*	1.4			NEVADA AR					
		2*	1.8			POLK AR					
				1*	.1	PULASKI AR					95
1*	1.5					RANDOLPH AR					
1*	.5					WHITE AR					
21	.0	15	.0	24	.0	CALIFORNIA CA	-56	17	-49		
1	.0	1	.0	1	.0	ALAMEDA CA	-12	-26	-35	71	94
1	.1					CONTRA COSTA CA					
1	.2			1	.0	FRESNO CA			-68	23	95
10	.0	10	.0	9	.0	LOS ANGELES CA	-32	-25	-49	56	95
		1*	.1	1*	.1	MARIN CA		-44			96
				2	.0	ORANGE CA					95
1	.1					RIVERSIDE CA					
2*	.1	1	.1	3*	.1	SACRAMENTO CA	-59	69	-31	73	96
2	.1	1	.0	2	.0	SAN DIEGO CA	-73	12	-69	21	95
1	.0	1	.0	1	.0	SAN FRANCISCO CA	-6	-2	-8	89	94
				1	.1	SANTA BARBARA CA					96
				1	.0	SANTA CLARA CA					95
				2*	.1	SANTA CRUZ CA					96
1	.2					TULARE CA					
1*	.7					YOLO CA					
2	.0	1	.0			COLORADO CO	-61				
2	.1	1	.0			DENVER-ARAPAHOE#CO	-60				
3	.0	2	.0	4	.0	CONNECTICUT CT	-49	35	-32		
		1	.0	1	.0	FAIRFIELD CT		-42			94
1	.0			1	.0	HARTFORD CT			-55	50	94
1	.1			1*	.1	LITCHFIELD CT			-16	84	96
1	.0	1	.0			NEW HAVEN CT	-20				
				1*	.1	NEW LONDON CT					96
1	.1					DELAWARE DE					
1*	.5					SUSSEX DE					
1	.0	1	.0	3*	.1	DIST. OF COLUMBIA DC	-15	302	243		
1	.0	1	.0	3*	.1	WASHINGTON DC	-15	302	243	97	96
5	.0	4	.0	10	.0	FLORIDA FL	-56	9	-52		
				1*	.3	BAY FL					97
1	.1			1	.0	BROWARD FL			-78	10	94
1	.0			2	.0	DADE FL			-65	34	94
		1	.1	1	.0	DUVAL FL		-48			95
2*	.7					ESCAMBIA FL					
				1*	3.2	GULF FL					99
				1*	5.0	LAFAYETTE FL					99
		1	.1	1	.0	PALM BEACH FL		-69			94
1	.0	1	.0			PINELLAS FL	-11				
		1*	.1			POLK FL					
				1*	.9	SANTA ROSA FL					99
				1	.0	VOLUSIA FL					95

* SIGNIFICANTLY MORE DEATHS THAN EXPECTED

\# COUNTIES COMBINED; SEE TECHNICAL NOTES

WHITE FEMALE

CANCER OF THE LIP

ICD 140

1950-1959 DEATHS	1950-1959 RATE	1960-1969 DEATHS	1960-1969 RATE	1970-1979 DEATHS	1970-1979 RATE	COUNTY AND STATE	PERCENT CHANGE 60S/50S	PERCENT CHANGE 70S/60S	PERCENT CHANGE 70S/50S	PERCENTILE %CHANGE 70S/50S	PERCENTILE RATES 1970S
10*	.1	6	.0	9*	.0	GEORGIA GA	-60	10	-56		
				1*	.2	BIBB GA					97
				1*	.8	BROOKS GA					99
		1*	.2			CHATHAM GA					
		1*	.6			CLARKE GA					
2*	43.5					CLINCH GA					
				1*	.1	COBB GA					96
1*	1.9					COWETA GA					
1	.1					DE KALB GA					
				1*	1.2	EMANUEL GA					99
				1*	.2	FLOYD GA					97
1	.1					FULTON GA					
		1*	.5			GWINNETT GA					
2*	3.1					HARALSON GA					
				1*	2.1	JOHNSON GA					99
				1*	5.2	MC INTOSH GA					99
				1*	1.8	MADISON GA					99
		1*	1.8			MERIWETHER GA					
1*	3.0					MORGAN GA					
1*	1.4					POLK GA					
				1*	.2	RICHMOND GA					97
		1*	2.1			ROCKDALE GA					
1*	5.2					TOWNS GA					
		1*	.6			WALKER GA					
1	.1					IDAHO ID					
1*	5.9					LEWIS ID					
14	.0	8	.0	4	.0	ILLINOIS IL	-59	-56	-82		
1*	1.2					CASS IL					
1*	.2					CHAMPAIGN IL					
4	.0	6	.0	2	.0	COOK IL	11	-75	-72	19	94
				1	.1	DU PAGE IL					95
1*	.8					FRANKLIN IL					
1*	.2					KNOX IL					
1*	.5					LEE IL					
1*	1.3					MONROE IL					
				1*	.4	OGLE IL					98
		1	.1			PEORIA IL					
		1*	.6			PERRY IL					
1*	3.1					POPE IL					
1*	.5					SALINE IL					
1*	.4					WHITESIDE IL					
1*	.4					WILLIAMSON IL					
9	.0	7	.0	1	.0	INDIANA IN	-41	-88	-93		
				1	.1	ALLEN IN					95
1*	.5					GREENE IN					
		1*	.4			HENRY IN					
		1*	.3			HOWARD IN					
		1*	.3			KOSCIUSKO IN					
1	.0					MARION IN					
1*	.6					MONROE IN					
1*	.5					MONTGOMERY IN					
1*	1.1					PIKE IN					
		1*	.5			RANDOLPH IN					
1*	.8					RUSH IN					
		1*	.8			SPENCER IN					
		1*	.7			TIPTON IN					
1	.2	1*	.1			VIGO IN	-45				
1*	.7					WARRICK IN					
1*	.9					WASHINGTON IN					
4	.0	4	.0	1	.0	IOWA IA	-30	-81	-87		
		1*	.7			BUTLER IA					
		1*	1.4			DAVIS IA					
		1*	.1			DUBUQUE IA					
1*	1.2					PLYMOUTH IA					
1	.1	1	.1			POLK IA	-32				
				1*	.1	SCOTT IA					96
1*	1.0					WINNESHIEK IA					
1	.2					WOODBURY IA					
7	.1			8*	.0	KANSAS KS			-32		
1*	3.5					DECATUR KS					
				1*	1.0	ELLIS KS					99
				1*	1.2	HARPER KS					99
				1*	.1	JOHNSON KS					96
1*	1.3					MARION KS					
1*	.3			1*	.2	RENO KS			-30	78	97
				1*	.3	SALINE KS					98
2*	.2			1	.1	SEDGWICK KS			-68	26	95
1	.2			1*	.1	SHAWNEE KS			-35	69	96
1*	4.8					SHERMAN KS					
				1*	.9	WABAUNSEE KS					99
17*	.1	8*	.0	3	.0	KENTUCKY KY	-67	-74	-91		
1*	1.5	1*	1.1			ALLEN KY	-30				
1*	.7					BARREN KY					
		1*	1.8			BUTLER KY					
1*	1.2					CALDWELL KY					
				1*	.6	ESTILL KY					98
1*	.2					FAYETTE KY					

* SIGNIFICANTLY MORE DEATHS THAN EXPECTED

WHITE FEMALE — CANCER OF THE LIP — COUNTY AND STATE — ICD 140

1950-1959 DEATHS	1950-1959 RATE	1960-1969 DEATHS	1960-1969 RATE	1970-1979 DEATHS	1970-1979 RATE	COUNTY AND STATE	PERCENT CHANGE 60S/50S	PERCENT CHANGE 70S/60S	PERCENT CHANGE 70S/50S	PERCENTILE %CHANGE 70S/50S	PERCENTILE RATES 1970S
1*	.7					GRAVES KY					
1*	.4					HARLAN KY					
		1*	.4			HOPKINS KY					
2*	5.8					JACKSON KY					
3*	.1	1	.0			JEFFERSON KY	-78				
1*	.9					LOGAN KY					
1*	1.8					MC LEAN KY					
1*	1.2			1*	.5	MARSHALL KY			-55	45	98
1*	1.9	2*	3.0			MORGAN KY	60				
		1*	.6			PULASKI KY					
		1*	1.3			SIMPSON KY					
1*	1.7			1*	.8	TAYLOR KY			-49	54	99
1*	.9					WHITLEY KY					
4	.1	7*	.1	3	.0	LOUISIANA LA	-5	-69	-71		
1*	1.4					ACADIA LA					
		1*	1.4			ASCENSION LA					
		1*	.1			CADDO LA					
1*	.5					CALCASIEU LA					
		1*	3.1			CATAHOULA LA					
		1*	1.0			EVANGELINE LA					
		1*	.5			IBERIA LA					
				1*	.3	LAFAYETTE LA					97
				1*	.5	LIVINGSTON LA					98
				1*	1.0	NATCHITOCHES LA					99
1	.1	2*	.1			ORLEANS LA	40				
1*	2.3					RICHLAND LA					
2	.0	2	.0	1	.0	MAINE ME	-24	-57	-68		
1*	.6					AROOSTOOK ME					
		1	.1			CUMBERLAND ME					
				1*	.1	KENNEBEC ME					96
		1*	.1			PENOBSCOT ME					
1*	.7					WALDO ME					
2	.0	3	.0	4	.0	MARYLAND MD	-6	4	-2		
		1*	.1			ANNE ARUNDEL MD					
		1*	.6			CECIL MD					
		1*	1.5			CHARLES MD					
				1*	.5	DORCHESTER MD					98
				1*	.5	HOWARD MD					98
				1*	.9	ST MARY'S MD					99
				1*	.1	WASHINGTON MD					96
2	.1					BALTIMORE CITY MD					
10	.0	5	.0	5	.0	MASSACHUSETTS MA	-61	-25	-71		
		2*	.2			BERKSHIRE MA					
2	.1					BRISTOL MA					
2	.1	2	.0	2	.0	ESSEX MA	-17	-9	-24	82	95
1	.0					HAMPDEN MA					
1	.2					HAMPSHIRE MA					
1	.0	1	.0			MIDDLESEX MA	-16				
1	.0			2	.0	NORFOLK MA			-9	86	95
1	.1			1	.0	PLYMOUTH MA			-51	52	95
1	.0					WORCESTER MA					
7	.0	10	.0	5	.0	MICHIGAN MI	-12	-66	-70		
		2*	.3			CALHOUN MI					
				1*	1.7	CRAWFORD MI					99
				1*	.2	HOUGHTON MI					97
1	.2					INGHAM MI					
		1	.0			KENT MI					
1*	.3					LENAWEE MI					
		1*	.4			LIVINGSTON MI					
		1	.1			MACOMB MI					
1*	.7					MARQUETTE MI					
1*	.2					MUSKEGON MI					
1	.1	4*	.2	1	.0	OAKLAND MI	99	-89	-78	13	94
				1*	.2	OTTAWA MI					97
				1*	.1	SAGINAW MI					96
2*	.5					ST CLAIR MI					
		1*	.1			WASHTENAW MI					
5	.0	3	.0	5	.0	MINNESOTA MN	-62	28	-51		
1*	.9					BECKER MN					
				1*	1.2	BIG STONE MN					99
				1*	.3	BLUE EARTH MN					98
				1*	.4	CHIPPEWA MN					98
1*	.9					LE SUEUR MN					
2	.1	1	.0	1	.0	RAMSEY MN	-72	-33	-81	08	95
1	.1	1	.1	1	.1	ST LOUIS MN	-21	-5	-25	80	96
		1*	1.0			WATONWAN MN					
5	.1	3	.0	4*	.0	MISSISSIPPI MS	-65	-1	-65		
1*	3.4					CARROLL MS					
		1*	.3	1*	.2	HARRISON MS		-19			97
1*	3.1			1*	3.2	HUMPHREYS MS			6	93	99
1*	5.1					JEFFERSON MS					
		2*	.6	1*	.2	LAUDERDALE MS		-73			97
1*	1.5					LEAKE MS					
				1*	.8	LEFLORE MS					99
1*	1.0					WARREN MS					

* SIGNIFICANTLY MORE DEATHS THAN EXPECTED

WHITE FEMALE — CANCER OF THE LIP — ICD 140

1950-1959 DEATHS	1950-1959 RATE	1960-1969 DEATHS	1960-1969 RATE	1970-1979 DEATHS	1970-1979 RATE	COUNTY AND STATE	PERCENT CHANGE 60S/50S	PERCENT CHANGE 70S/60S	PERCENT CHANGE 70S/50S	PERCENTILE %CHANGE 70S/50S	PERCENTILE RATES 1970S
10	.0	5	.0	7	.0	MISSOURI MO	-63	15	-57		
		1*	1.2			BENTON MO					
1*	.5					BOONE MO					
				1*	.7	CAMDEN MO					98
1*	.9					CARROLL MO					
1*	.7					CASS MO					
				1*	.7	CHRISTIAN MO					98
1*	.6					DUNKLIN MO					
				1*	.8	GASCONADE MO					99
				1*	.1	GREENE MO					96
1	.0	1	.0			JACKSON MO	-19				
				1*	.2	JEFFERSON MO					97
1*	.6					LAWRENCE MO					
		1*	.8			MILLER MO					
		1*	1.1			MISSISSIPPI MO					
1*	1.3					MORGAN MO					
1*	1.6					RIPLEY MO					
1*	1.5					ST CLAIR MO					
1*	3.4					STONE MO					
		1*	.7			TEXAS MO					
				1*	1.1	WASHINGTON MO					99
				1	.0	ST LOUIS CITY MO					94
2	.1					MONTANA MT					
1*	.6					CASCADE MT					
1*	1.4					LAKE MT					
3	.0	2	.0	1	.0	NEBRASKA NB	-48	-57	-78		
				1	.0	DOUGLAS NB					95
1*	1.4					HOLT NB					
1*	2.5					KEARNEY NB					
		1*	.1			LANCASTER NB					
		1*	1.9			NUCKOLLS NB					
1*	2.1					WAYNE NB					
				1	.0	NEVADA NV					
				1*	.1	CLARK NV					96
1	.0	1	.0			NEW HAMPSHIRE NH	-13				
		1*	.2			ROCKINGHAM NH					
1*	.3					STRAFFORD NH					
7	.0	5	.0	4	.0	NEW JERSEY NJ	-40	-48	-69		
1	.1					ATLANTIC NJ					
1	.0	1	.0			BERGEN NJ	-21				
		1*	.2			CUMBERLAND NJ					
		1	.0	1	.0	ESSEX NJ		-20			94
1*	.2					GLOUCESTER NJ					
		1	.0	1	.0	HUDSON NJ		-44			95
		1	.0			MONMOUTH NJ					
1	.1					MORRIS NJ					
1*	.3			1	.0	OCEAN NJ			-92	02	95
1	.1			1	.0	PASSAIC NJ			-40	67	95
1	.0					UNION NJ					
3*	.2	2*	.1			NEW MEXICO NM	-63				
		1*	.1			BERNALILLO NM					
1*	1.0					DONA ANA NM					
1*	1.1					EDDY NM					
1*	2.6	1*	2.0			VALENCIA NM	-21				
23	.0	16	.0	8	.0	NEW YORK NY	-47	-61	-79		
1	.0	2	.0			BRONX NY	56				
1	.1	1	.1			BROOME NY	-3				
1	.1					DUTCHESS NY					
3	.1	1	.0			ERIE NY	-76				
1*	.3					HERKIMER NY					
2	.0	2	.0	3	.0	KINGS(BROOKLYN) NY	-5	30	24	95	94
1*	.3					MADISON NY					
		1	.0			MONROE NY					
1*	.2					MONTGOMERY NY					
4*	.1					NASSAU NY					
2	.0					NEW YORK(MANHATTAN) NY					
		1	.1	2*	.1	ONEIDA NY		92			96
				1	.0	ONONDAGA NY					95
		2	.0			QUEENS NY					
		1*	.1			RENSSELAER NY					
1	.1	1	.1			RICHMOND NY	-12				
		1*	.1	1*	.1	ROCKLAND NY		-46			96
1	.1	1	.0	1	.0	SUFFOLK NY	-58	-20	-67	32	94
1*	.5					SULLIVAN NY					
		2*	.2			ULSTER NY					
2	.1					WESTCHESTER NY					
1*	.5					WYOMING NY					
12*	.1	6	.0	6	.0	NORTH CAROLINA NC	-65	-35	-77		
1*	1.7					AVERY NC					
1*	.2	1*	.1	1*	.1	BUNCOMBE NC	-32	-12	-40	65	96
1*	.6					BURKE NC					
1*	1.8					CHEROKEE NC					
1*	.7					CUMBERLAND NC					

* SIGNIFICANTLY MORE DEATHS THAN EXPECTED

1950-1959 DEATHS	1950-1959 RATE	1960-1969 DEATHS	1960-1969 RATE	1970-1979 DEATHS	1970-1979 RATE	COUNTY AND STATE	PERCENT CHANGE 60S/50S	PERCENT CHANGE 70S/60S	PERCENT CHANGE 70S/50S	PERCENTILE %CHANGE 70S/50S	PERCENTILE RATES 1970S
				1*	.2	DAVIDSON NC					97
				1*	.9	DAVIE NC					99
1*	.9					GRANVILLE NC					
				1*	.1	GUILFORD NC					96
1*	.5					HENDERSON NC					
1*	4.0					HOKE NC					
1	.2					MECKLENBURG NC					
1*	2.3					MONTGOMERY NC					
		1*	.3			NEW HANOVER NC					
		1*	.6			PITT NC					
				1*	.3	ROCKINGHAM NC					97
				1*	.3	RUTHERFORD NC					97
		1*	1.1			STOKES NC					
1*	2.7					WARREN NC					
		1*	.5			WILKES NC					
1*	1.6	1*	1.4			YANCEY NC	-12				
2	.1					NORTH DAKOTA ND					
1*	8.3					DIVIDE ND					
1*	3.9					WELLS ND					
17	.0	9	.0	7	.0	OHIO OH	-58	-36	-73		
				1*	.2	ASHTABULA OH					97
1*	.3					ATHENS OH					
1*	.2	1*	.2			BELMONT OH	13				
				1*	.1	BUTLER OH					96
1	.1					CLARK OH					
2	.0	3	.0	1	.0	CUYAHOGA OH	39	-77	-68	28	94
1*	.6					DELAWARE OH					
1*	.3					FAIRFIELD OH					
		1	.0	1	.0	FRANKLIN OH		-23			95
1*	.3					GUERNSEY OH					
2	.1	2	.0	1	.0	HAMILTON OH	-13	-62	-67	30	94
1*	.4					HANCOCK OH					
1*	.4					HURON OH					
				1*	.2	JEFFERSON OH					97
		1*	.1			LORAIN OH					
1*	.3			1*	.3	MARION OH			-1	91	97
1*	.2					MUSKINGUM OH					
		1*	1.2			NOBLE OH					
1*	.6					PUTNAM OH					
1	.1					STARK OH					
1	.1					SUMMIT OH					
5	.0	5	.0	5	.0	OKLAHOMA OK	-30	-11	-37		
1*	.6	1*	.6			CADDO OK	-1				
		1*	.4			CARTER OK					
				1*	.4	GARVIN OK					98
1*	.5					GRADY OK					
		1*	.6			MC CURTAIN OK					
1*	.4					MUSKOGEE OK					
				1*	2.0	OKFUSKEE OK					99
1	.1	1	.0	1	.0	OKLAHOMA OK	-47	4	-44	63	95
				1*	.3	PONTOTOC OK					97
1*	1.8					PUSHMATAHA OK					
				1*	.3	STEPHENS OK					97
		1*	1.0			WAGONER OK					
3	.0	1	.0	3	.0	OREGON OR	-72	55	-56		
1*	.2					CLACKAMAS OR					
				1*	.3	COOS OR					97
1*	1.4					DESCHUTES OR					
1*	.7					DOUGLAS OR					
		1*	.1	1*	.1	MARION OR		-61			95
				1	.0	MULTNOMAH OR					95
14	.0	8	.0	7	.0	PENNSYLVANIA PA	-58	-29	-70		
		2	.0	2	.0	ALLEGHENY PA		-16			94
1	.1					BLAIR PA					
				1	.1	BUCKS PA					95
1*	.2					BUTLER PA					
				1*	.1	CAMBRIA PA					96
1	.1					DELAWARE PA					
				1*	.4	ELK PA					98
1*	.3					INDIANA PA					
		1	.1			LACKAWANNA PA					
1	.1					LANCASTER PA					
2*	.4					LAWRENCE PA					
1*	.2					LEBANON PA					
		1	.1			LEHIGH PA					
1	.1	1	.0			LUZERNE PA	-20				
1	.2					NORTHUMBERLAND PA					
2	.0			1	.0	PHILADELPHIA PA			-61	39	94
2*	.3					SCHUYLKILL PA					
		1*	.2			VENANGO PA					
		1*	.1			WASHINGTON PA					
				1*	.3	WAYNE PA					97
		1	.1			WESTMORELAND PA					
3	.1	4*	.1	1	.0	RHODE ISLAND RI	9	-83	-81		
		1*	.2	1*	.1	KENT RI		-39			96
3	.1	3*	.1			PROVIDENCE RI	-14				
1	.0	3	.0	1	.0	SOUTH CAROLINA SC	160	-77	-40		

* SIGNIFICANTLY MORE DEATHS THAN EXPECTED

1950-1959 DEATHS	1950-1959 RATE	1960-1969 DEATHS	1960-1969 RATE	1970-1979 DEATHS	1970-1979 RATE	COUNTY AND STATE	PERCENT CHANGE 60S/50S	PERCENT CHANGE 70S/60S	PERCENT CHANGE 70S/50S	PERCENTILE %CHANGE 70S/50S	PERCENTILE RATES 1970S
				1*	.1	CHARLESTON SC					96
		1*	1.4			KERSHAW SC					
		1*	.5			LAURENS SC					
1*	1.1	1*	1.0			UNION SC	-10				
		2	.1	1	.0	SOUTH DAKOTA SD		-77			
				1*	.7	BON HOMME SD					98
		1*	.5			BROWN SD					
		1*	.2			MINNEHAHA SD					
20*	.2	4	.0	7	.0	TENNESSEE TN	-86	26	-82		
1*	.9					BRADLEY TN					
3*	2.6					CAMPBELL TN					
1*	1.7					CLAIBORNE TN					
1*	2.7					FENTRESS TN					
1*	1.0					FRANKLIN TN					
1*	.5	2*	.8			GIBSON TN	50				
1*	2.0					GRAINGER TN					
1	.1					HAMILTON TN					
				1*	1.0	HAYWOOD TN					99
		1	.1	1	.0	KNOX TN		-41			95
1*	.9					LOUDON TN					
1*	1.2					MC MINN TN					
				1*	.7	MARION TN					99
1*	.6					PUTNAM TN					
1*	.6					RUTHERFORD TN					
				1*	.6	SEVIER TN					98
1	.1			1	.0	SHELBY TN			-62	36	95
1*	1.6					SMITH TN					
		1*	.2			SULLIVAN TN					
1*	1.5					UNICOI TN					
				1*	4.3	VAN BUREN TN					99
				1*	.5	WARREN TN					98
1*	.6					WASHINGTON TN					
1*	.5					WEAKLEY TN					
1*	.8					WILSON TN					
23*	.1	14	.0	12	.0	TEXAS TX	-61	-41	-77		
		1*	3.6			BAILEY TX					
1	.0	2*	.1			BEXAR TX	71				
		1*	.8			BOSQUE TX					
1*	6.2					CALHOUN TX					
4*	.1	4*	.1			DALLAS TX	-39				
1*	2.7					DELTA TX					
1	.1					EL PASO TX					
1*	2.5					FRIO TX					
				1*	.5	GILLESPIE TX					98
2*	.6					GRAYSON TX					
1*	3.0					HARDEMAN TX					
3	.1	1	.0	1	.0	HARRIS TX	-83	-44	-90	04	94
				1*	.1	HIDALGO TX					96
		1*	.2			HUNT TX					
1*	2.1					JACK TX					
1*	.2					JEFFERSON TX					
2*	1.0					LAMAR TX					
		1*	.2			LUBBOCK TX					
		1*	.8	1*	.4	MIDLAND TX		-53			98
		1*	.2	1*	.1	NUECES TX		-43			96
				1*	.4	PARKER TX					98
				1*	.2	POTTER TX					97
				1*	1.0	ROCKWALL TX					99
				1*	.5	RUNNELS TX					98
				1*	.1	SMITH TX					97
		1	.0			TARRANT TX					
				1*	.2	TAYLOR TX					97
2*	.3					TRAVIS TX					
1*	1.0					UVALDE TX					
1*	.8					VAN ZANDT TX					
				1*	.1	WICHITA TX					96
5	.0	5	.0	4	.0	VIRGINIA VA	-25	-50	-63		
1*	.3					AUGUSTA#VA					
		1*	.4			CARROLL#VA					
1	.1			1	.0	CHESTERFIELD#VA			-57	43	95
		1*	.8	1*	.6	HALIFAX#VA		-25			98
				1*	.5	HANOVER VA					98
		1*	2.4			MIDDLESEX VA					
1*	.4	1*	.3			PITTSYLVANIA#VA	-38				
1*	.6					PULASKI VA					
1	.2					ROANOKE#VA					
		1	.1	1	.0	CHESAPEAKE CITY#VA		-40			95
7	.1	3	.0	6	.0	WASHINGTON WA	-73	75	-52		
		1*	.5	1*	.4	BENTON WA		-24			98
1*	.7					CHELAN WA					
				1*	1.0	DOUGLAS WA					99
1*	.4			1*	.2	GRAYS HARBOR WA			-45	60	97
3	.1	1	.0	1	.0	KING WA	-77	0	-77	17	94
1	.1			1	.0	PIERCE WA			-30	76	95
		1*	.3			THURSTON WA					
				1*	.3	WALLA WALLA WA					97
1*	.2					WHATCOM WA					
4	.1			5*	.0	WEST VIRGINIA WV			-41		
1*	.3					CABELL WV					

* SIGNIFICANTLY MORE DEATHS THAN EXPECTED

COUNTIES COMBINED; SEE TECHNICAL NOTES

WHITE FEMALE 1950-1959 DEATHS	RATE	1960-1969 DEATHS	RATE	1970-1979 DEATHS	RATE	CANCER OF THE LIP COUNTY AND STATE	PERCENT CHANGE 60S/50S	70S/60S	70S/50S	ICD 140 PERCENTILE %CHANGE 70S/50S	RATES 1970S
1*	2.0					HAMPSHIRE WV					
				1*	.2	HARRISON WV					97
				2*	.1	KANAWHA WV					96
1*	1.1					MINERAL WV					
				1*	.6	MINGO WV					98
				1*	.7	MONROE WV					98
1*	.3					OHIO WV					
9	.1	2	.0	4	.0	WISCONSIN WI	-85	33	-80		
				1*	.3	DOUGLAS WI					97
1*	.4					EAU CLAIRE WI					
1*	.2					FOND DU LAC WI					
1*	1.1					JUNEAU WI					
1*	.5					MANITOWOC WI					
				1	.0	MILWAUKEE WI					94
1*	1.0					POLK WI					
1*	.6					PORTAGE WI					
1	.3	1*	.1			RACINE WI	-50				
				1*	.3	ST CROIX WI					97
				1*	.1	WAUKESHA WI					96
1*	.4	1*	.3			WAUPACA WI	-23				
1	.2					WINNEBAGO WI					
				1	.1	WYOMING WY					
				1*	4.2	CROOK WY					99

* SIGNIFICANTLY MORE DEATHS THAN EXPECTED

1950-1959 DEATHS	1950-1959 RATE	1960-1969 DEATHS	1960-1969 RATE	1970-1979 DEATHS	1970-1979 RATE	COUNTY AND STATE	PERCENT CHANGE 60S/50S	PERCENT CHANGE 70S/60S	PERCENT CHANGE 70S/50S	PERCENTILE %CHANGE RATES 70S/50S	PERCENTILE 1970S
46	.1	34	.1	26	.0	UNITED STATES	-40	-42	-65		
		1	.0	4*	.1	ALABAMA AL		269			
				1	.1	JEFFERSON AL					99
				1*	2.3	LOWNDES AL					99
		1*	.4	1*	.4	MONTGOMERY AL		-4			99
				1*	1.6	SUMTER AL					99
2*	.6			1*	.1	ARIZONA AZ			-82		
1*	26.8					GRAHAM AZ					
1*	1.8					PIMA AZ					
				1*	2.5	YUMA AZ					99
2	.1	2	.1			ARKANSAS AR	3				
1*	4.1					CLARK AR					
		1*	15.1			CRAIGHEAD AR					
1*	11.2					HOT SPRING AR					
		1*	6.6			SEBASTIAN AR					
		1	.0	3	.0	CALIFORNIA CA		24			
		1*	.7			FRESNO CA					
				1	.0	LOS ANGELES CA					99
				1*	2.5	MERCED CA					99
				1*	.1	SAN FRANCISCO CA					99
1*	.7					CONNECTICUT CT					
1*	2.0					NEW HAVEN CT					
		2*	.2			DIST. OF COLUMBIA DC					
		2*	.2			WASHINGTON DC					
4	.1	1	.0			FLORIDA FL	-82				
1*	1.4					ALACHUA FL					
		1*	.2			DUVAL FL					
1*	.4					HILLSBOROUGH FL					
1*	1.1					ORANGE FL					
1*	1.5					VOLUSIA FL					
2	.1			1	.0	GEORGIA GA			-72		
1*	7.9					EFFINGHAM GA					
				1*	4.6	TOOMBS GA					99
1*	9.3					WALTON GA					
2	.1	2	.0			ILLINOIS IL	-45				
1	.0	2	.1			COOK IL	51				
1*	1.1					ST CLAIR IL					
4*	.7	1	.1			INDIANA IN	-89				
1*	9.9					CLARK IN					
1*	12.5					GRANT IN					
2*	1.4	1*	.3			LAKE IN	-81				
2	.2					KENTUCKY KY					
1*	.3					JEFFERSON KY					
1*	267.9					MC LEAN KY					
4	.1	1	.0	2	.1	LOUISIANA LA	-77	82	-57		
1*	2.4					ASCENSION LA					
				1*	2.4	AVOYELLES LA					99
		1*	.3			CADDO LA					
1*	2.1					EAST FELICIANA LA					
2*	.3					ORLEANS LA					
				1*	.7	RAPIDES LA					99
				1*	5.9	MAINE ME					
				1*	38.2	WASHINGTON ME					99
2	.2	3*	.2			MARYLAND MD	-9				
		1*	3.5			HARFORD MD					
		1*	3.3			TALBOT MD					
1*	3.9					WORCESTER MD					
1	.1	1	.1			BALTIMORE CITY MD	-22				
1	.2			1	.1	MICHIGAN MI			-72		
				1*	1.7	GENESEE MI					99
1	.4					WAYNE MI					
1	.0	3	.1	1	.0	MISSISSIPPI MS	81	-58	-25		
1*	.8					HINDS MS					
		1*	1.1			HOLMES MS					
		1*	1.7			HUMPHREYS MS					
		1*	.7			LEFLORE MS					
				1*	1.0	WARREN MS					99

* SIGNIFICANTLY MORE DEATHS THAN EXPECTED

NON-WHITE MALE — CANCER OF THE LIP — COUNTY AND STATE — ICD 140

1950-1959 DEATHS	1950-1959 RATE	1960-1969 DEATHS	1960-1969 RATE	1970-1979 DEATHS	1970-1979 RATE	COUNTY AND STATE	PERCENT CHANGE 60S/50S	PERCENT CHANGE 70S/60S	PERCENT CHANGE 70S/50S	PERCENTILE %CHANGE 70S/50S	PERCENTILE RATES 1970S
1	.1					MISSOURI MO					
1	.1					ST LOUIS CITY MO					
		1*	1.4			MONTANA MT					
		1*	60.6			CASCADE MT					
1*	.8					NEBRASKA NB					
1*	109.3					DAWES NB					
3*	.2					NEW JERSEY NJ					
2*	2.4					CAMDEN NJ					
1*	4.4					MORRIS NJ					
3	.1	5*	.1	4	.0	NEW YORK NY	124	-69	-29		
		3*	.7			BRONX NY					
		1*	14.9			BROOME NY					
1*	.2					ERIE NY					
				1	.0	KINGS(BROOKLYN) NY					99
1*	1.2			1*	.6	NASSAU NY			-50	33	99
1	.1			1	.0	NEW YORK(MANHATTAN) NY			-31	66	99
		1*	.2			QUEENS NY					
				1*	.3	WESTCHESTER NY					99
1	.0	3	.1	2	.0	NORTH CAROLINA NC	182	-44	57		
				1*	6.0	BURKE NC					99
		1*	1.7			EDGECOMBE NC					
		1*	15.6			MC DOWELL NC					
1*	1.2					PITT NC					
		1*	1.9			VANCE NC					
				1*	1.1	WAYNE NC					99
		1*	3.2			NORTH DAKOTA ND					
		1*	12.4			ROLETTE ND					
1	.0	4*	.2	1	.0	OHIO OH	226	-72	-10		
		1	.2			CUYAHOGA OH					
		1*	12.5			FAYETTE OH					
				1*	5.4	GREENE OH					99
		1*	.2			HAMILTON OH					
		1*	.6			MAHONING OH					
1*	.7					MONTGOMERY OH					
		1	.0	2	.1	PENNSYLVANIA PA		137			
		1*	4.4			BERKS PA					
				1*	.9	DELAWARE PA					99
				1	.0	PHILADELPHIA PA					99
1	.0	1	.1			SOUTH CAROLINA SC	45				
1*	2.3					LAURENS SC					
		1*	4.2			MARION SC					
2	.1					TENNESSEE TN					
1	.1					SHELBY TN					
1*	7.8					WEAKLEY TN					
2	.0	1	.0	2	.0	TEXAS TX	-46	18	-36		
		1*	2.3			CASS TX					
				1*	.6	GALVESTON TX					99
				1*	6.1	HARDIN TX					99
1	.1					HARRIS TX					
1*	22.9					HOCKLEY TX					
3	.1			1	.0	VIRGINIA VA			-71		
1*	12.1					APPOMATTOX VA					
				1*	2.7	HENRY#VA					99
1*	12.5					WASHINGTON#VA					
1*	4.8					WESTMORELAND VA					
1*	1.0					WISCONSIN WI					
1*	267.9					POLK WI					

* SIGNIFICANTLY MORE DEATHS THAN EXPECTED

COUNTIES COMBINED; SEE TECHNICAL NOTES

NON-WHITE FEMALE — CANCER OF THE LIP — ICD 140

1950-1959 DEATHS	1950-1959 RATE	1960-1969 DEATHS	1960-1969 RATE	1970-1979 DEATHS	1970-1979 RATE	COUNTY AND STATE	PERCENT CHANGE 60S/50S	PERCENT CHANGE 70S/60S	PERCENT CHANGE 70S/50S	PERCENTILE %CHANGE 70S/50S	PERCENTILE RATES 1970S
32	.1	24	.0	22	.0	UNITED STATES	-45	-38	-66		
3	.1	5*	.1	2	.0	ALABAMA AL	72	-71	-50		
		1*	3.0			AUTAUGA AL					
		1*	1.6			BARBOUR AL					
				1*	1.1	BULLOCK AL					99
		1*	2.8			COFFEE AL					
		1*	1.5			ETOWAH AL					
				1*	.1	JEFFERSON AL					99
1*	.9	1*	1.0			MACON AL	5				
1*	.2					MOBILE AL					
1*	4.7					SHELBY AL					
1	.1					ARKANSAS AR					
1*	18.6					CRAIGHEAD AR					
1	.0	1	.0			CALIFORNIA CA	-33				
1*	1.7					FRESNO CA					
		1*	4.9			MONTEREY CA					
				2	.1	FLORIDA FL					
				1*	.8	GADSDEN FL					99
				1*	1.2	MARION FL					99
1	.0	2	.0	3	.0	GEORGIA GA	4	6	11		
				1*	6.3	APPLING GA					99
		1*	2.8			BULLOCH GA					
				1*	1.9	DECATUR GA					99
				1*	.1	FULTON GA					99
1*	3.3					JEFFERSON GA					
		1*	.5			RICHMOND GA					
1	.1					ILLINOIS IL					
1	.1					COOK IL					
1	.1	1	.1			INDIANA IN	16				
		1*	.7			LAKE IN					
1*	.3					MARION IN					
		1	.1			KENTUCKY KY					
		1*	9.8			BELL KY					
3	.1	3	.1	2	.0	LOUISIANA LA	45	-45	-20		
		1*	3.7			ASSUMPTION LA					
				1*	.2	CADDO LA					99
1*	1.2					EVANGELINE LA					
		1*	2.8			FRANKLIN LA					
1*	1.4					NATCHITOCHES LA					
		1	.2	1*	.1	ORLEANS LA		-50			99
1*	.6					ST LANDRY LA					
1	.1	2*	.1	1	.0	MARYLAND MD	74	-63	-36		
		1*	2.9			CHARLES MD					
1	.1	1	.1	1	.1	BALTIMORE CITY MD	9	-40	-35	66	99
4	.1	1	.0	1	.0	MISSISSIPPI MS	-82	0	-82		
		1*	2.7			JEFFERSON DAVIS MS					
1*	1.3					MADISON MS					
1*	1.4					MARSHALL MS					
1*	27.0					SMITH MS					
				1*	.9	SUNFLOWER MS					99
1*	2.6					WILKINSON MS					
				2*	.1	MISSOURI MO					
				1*	3.7	NEW MADRID MO					99
				1*	.1	ST LOUIS CITY MO					99
1	.1	2*	.1			NEW JERSEY NJ	-1				
		1*	.8			BERGEN NJ					
		1*	.7			CAMDEN NJ					
1*	17.7					OCEAN NJ					
2	.0	1	.0	1	.0	NEW YORK NY	-23	-64	-73		
1*	.4					ERIE NY					
				1	.0	KINGS(BROOKLYN) NY					99
1	.1	1	.1			NEW YORK(MANHATTAN) NY	59				
2	.1	1	.0	3*	.1	NORTH CAROLINA NC	-60	111	-15		
		1*	2.5			CABARRUS NC					
				1*	.6	DURHAM NC					99
				1*	4.0	JONES NC					99
1*	9.2					RANDOLPH NC					
1*	.8					WAYNE NC					
				1*	.8	WILSON NC					99

* SIGNIFICANTLY MORE DEATHS THAN EXPECTED

1950-1959 DEATHS	1950-1959 RATE	1960-1969 DEATHS	1960-1969 RATE	1970-1979 DEATHS	1970-1979 RATE	COUNTY AND STATE	PERCENT CHANGE 60S/50S	PERCENT CHANGE 70S/60S	PERCENT CHANGE 70S/50S	PERCENTILE %CHANGE 70S/50S	PERCENTILE RATES 1970S
4*	.2	1	.0			OHIO OH	-79				
1	.2	1	.1			CUYAHOGA OH	-36				
1*	.2					HAMILTON OH					
1*	.8					MONTGOMERY OH					
1*	1.1					SUMMIT OH					
				1	.0	OKLAHOMA OK					
				1*	.3	OKLAHOMA OK					99
2	.1	1	.0	1	.0	PENNSYLVANIA PA	-57	-38	-73		
		1*	.2	1*	.1	ALLEGHENY PA		-28			99
2	.1					PHILADELPHIA PA					
				2	.1	SOUTH CAROLINA SC					
				1*	.5	ORANGEBURG SC					99
				1*	1.3	WILLIAMSBURG SC					99
1	.1					TENNESSEE TN					
1*	14.1					ROANE TN					
3	.1	1	.0	1	.0	TEXAS TX	-82	-38	-89		
1*	.7			1*	.1	DALLAS TX			-87	33	99
		1*	1.2			HARRISON TX					
1*	1.2					SMITH TX					
1*	4.6					WILLIAMSON TX					
1	.0	1	.0			VIRGINIA VA	-8				
		1*	.7			DINWIDDIE#VA					
1*	1.1					ROANOKE#VA					

* SIGNIFICANTLY MORE DEATHS THAN EXPECTED

COUNTIES COMBINED; SEE TECHNICAL NOTES

WHITE MALE — CANCER OF THE ORAL CAVITY INCLUDING TONGUE — ICD 141,143-146,148,149

1950-1959		1960-1969		1970-1979		COUNTY AND STATE	PERCENT CHANGE			PERCENTILE	
DEATHS	RATE	DEATHS	RATE	DEATHS	RATE		60S/50S	70S/60S	70S/50S	%CHANGE 70S/50S	RATES 1970S
30701	4.8	36622	4.8	40461	4.6	UNITED STATES	0	-3	-3		
320	4.1	387	3.9	473	3.9	ALABAMA AL	-4	0	-4		
1	2.1	1	1.5	6	7.7	AUTAUGA AL	-29	408	262	95	93
2	1.1	6	2.9	13	4.4	BALDWIN AL	173	51	311	96	67
1	1.9			2	2.7	BARBOUR AL			39	66	38
		2	3.8	1	1.6	BIBB AL		-59			22
1	.8	4	3.2	2	1.7	BLOUNT AL	287	-47	104	82	23
2	9.2	1	4.3	3*	16.0	BULLOCK AL	-54	277	75	76	99
1	1.4	2	2.8	1	1.2	BUTLER AL	94	-57	-16	35	18
7	3.7	10	3.3	17	4.7	CALHOUN AL	-10	41	28	61	71
6	6.7	4	3.3	5	3.3	CHAMBERS AL	-52	1	-51	12	48
		1	1.4	1	1.1	CHEROKEE AL		-20			17
3	2.9	3	2.7	4	3.1	CHILTON AL	-6	15	8	53	45
4	13.8	2	4.7	6*	11.9	CHOCTAW AL	-66	152	-14	37	98
4	7.3	3	4.0			CLARKE AL	-45				
1	1.8			1	1.2	CLAY AL			-32	23	18
1	2.1			1	2.3	CLEBURNE AL			10	54	31
3	3.1	2	2.3	9	6.3	COFFEE AL	-25	169	102	81	87
2	1.5	5	3.7	10	5.1	COLBERT AL	146	38	239	94	77
1	1.9	1	1.6	5	8.6	CONECUH AL	-15	442	358	97	96
2	4.3	4	9.9			COOSA AL	133				
3	2.0	4	2.6	4	2.2	COVINGTON AL	30	-15	11	54	30
1	1.7	1	1.7	7*	11.6	CRENSHAW AL	-1	600	592	99	98
2	1.1	4	1.8	7	2.4	CULLMAN AL	63	35	120	84	34
4	4.7	1	1.5	3	2.4	DALE AL	-68	57	-50	13	33
2	2.5	9*	10.2	6	5.5	DALLAS AL	301	-46	116	83	81
3	1.6	11	5.5	6	2.7	DE KALB AL	248	-51	70	75	38
1	1.0	5	4.6	5	3.4	ELMORE AL	379	-26	255	94	50
4	4.6	2	2.3	7	6.2	ESCAMBIA AL	-50	173	35	64	86
7	2.8	9	2.3	13	2.9	ETOWAH AL	-17	26	5	51	42
1	1.4	1	1.3	2	2.6	FAYETTE AL	-7	94	81	77	36
2	2.0	1	.9	4	3.4	FRANKLIN AL	-55	281	72	75	50
5	5.3			1	1.1	GENEVA AL			-79	01	17
		1	5.9	1	5.8	GREENE AL		-2			83
		1	3.2	4*	12.5	HALE AL		286			99
1	2.4	5*	12.0	2	3.5	HENRY AL	405	-71	47	68	52
5	3.7	6	3.4	8	3.4	HOUSTON AL	-10	2	-8	40	50
3	2.3	5	3.0	4	2.8	JACKSON AL	31	-8	21	59	39
94*	7.3	95	5.5	87	4.4	JEFFERSON AL	-25	-20	-40	19	67
1	1.7	3	4.3			LAMAR AL	158				
6	4.6	10	5.9	9	3.3	LAUDERDALE AL	27	-44	-29	25	48
2	2.7	1	1.2	1	1.0	LAWRENCE AL	-57	-17	-64	06	16
2	2.8	6	6.2	6	4.1	LEE AL	125	-33	51	70	63
		1	.7	5	3.2	LIMESTONE AL		373			47
		1	6.1	3*	16.6	LOWNDES AL		170			99
		2	10.0	1	2.6	MACON AL		-74			37
8	3.7	8	2.8	13	2.3	MADISON AL	-24	-20	-39	19	31
3	6.5	1	2.1	4	7.4	MARENGO AL	-68	254	13	55	92
3	3.7	2	2.0	1	.8	MARION AL	-45	-63	-80	01	15
5	3.0	4	1.8	6	2.0	MARSHALL AL	-39	11	-33	23	27
35	6.9	49	6.3	71*	7.9	MOBILE AL	-8	25	15	56	94
4	12.5	3	5.5	3	5.0	MONROE AL	-56	-8	-60	08	76
18	5.7	29*	8.1	20	4.7	MONTGOMERY AL	42	-42	-18	33	71
8	4.3	7	3.6	3	1.0	MORGAN AL	-15	-73	-77	02	16
				2	5.7	PERRY AL					82
1	1.6	3	4.8	3	4.2	PICKENS AL	195	-14	153	88	63
7*	10.4	6	7.8	2	2.3	PIKE AL	-25	-70	-77	02	32
		1	1.2	3	3.4	RANDOLPH AL		195			50
3	3.4	6	5.7	8	6.0	RUSSELL AL	68	5	77	76	85
		3	3.0	2	1.9	ST CLAIR AL		-35			27
5	5.3	4	3.3	4	2.5	SHELBY AL	-38	-25	-54	11	34
		1	4.0	1	3.5	SUMTER AL		-12			52
5	3.7	5	2.4	8	4.0	TALLADEGA AL	-34	66	9	53	60
		5	4.3	4	2.4	TALLAPOOSA AL		-43			34
9	3.6	8	2.1	19	4.7	TUSCALOOSA AL	-41	125	32	63	71
9	4.1	5	2.1	6	2.1	WALKER AL	-48	-3	-50	13	28
2	7.9			3	5.5	WASHINGTON AL			-31	24	81
1	3.6			2	8.9	WILCOX AL			151	87	96
3	3.7	1	1.3	2	2.5	WINSTON AL	-65	94	-31	23	35
		17	4.4	17	3.3	ALASKA AK		-24			
88	2.7	194	3.4	337	3.6	ARIZONA AZ	24	8	34		
1	1.7					APACHE AZ					
3	2.0	8	4.6	12	4.0	COCHISE AZ	126	-13	96	80	61
		5	5.1	4	3.7	COCONINO AZ		-29			55
1	1.1	6	5.7	4	2.2	GILA AZ	404	-61	98	81	30
		2	3.4	1	1.3	GRAHAM AZ		-62			19
1	9.3					GREENLEE AZ					
54	3.3	114	3.5	201	3.8	MARICOPA AZ	5	9	15	56	57
3	10.8	2	1.7	11	5.0	MOHAVE AZ	-84	169	-54	11	75
2	4.6	3	4.3	2	1.7	NAVAJO AZ	-5	-61	-63	07	23
11	1.4	31	2.4	65	3.5	PIMA AZ	72	42	144	87	51
1	.6	5	2.6	9	2.9	PINAL AZ	326	9	362	97	41
3	7.9	1	1.8	3	4.8	SANTA CRUZ AZ	-77	166	-39	19	73
4	2.2	5	2.4	14	4.1	YAVAPAI AZ	6	72	82	78	62
4	3.3	12	8.6	11	3.3	YUMA AZ	162	-61	1	48	48
221	3.4	266	3.4	290	3.0	ARKANSAS AR	1	-10	-10		
		8	8.2	4	3.9	ARKANSAS AR		-53			59
3	6.9	2	3.2	5	5.7	ASHLEY AR	-54	78	-18	33	82
3	3.7	2	1.9	7	3.1	BAXTER AR	-50	67	-16	35	46
4	1.9	6	2.2	4	1.0	BENTON AR	19	-53	-44	16	16
2	2.2	5	4.3	3	2.1	BOONE AR	91	-51	-8	41	28

* SIGNIFICANTLY MORE DEATHS THAN EXPECTED

WHITE MALE — CANCER OF THE ORAL CAVITY INCLUDING TONGUE — ICD 141,143-146,148,149

1950-1959 DEATHS	1950-1959 RATE	1960-1969 DEATHS	1960-1969 RATE	1970-1979 DEATHS	1970-1979 RATE	COUNTY AND STATE	PERCENT CHANGE 60S/50S	PERCENT CHANGE 70S/60S	PERCENT CHANGE 70S/50S	PERCENTILE %CHANGE 70S/50S	PERCENTILE RATES 1970S
2	4.7	2	3.6	2	3.2	BRADLEY AR	-24	-10	-31	23	47
				1	3.0	CALHOUN AR					43
2	2.4	3	2.8	1	1.0	CARROLL AR	17	-64	-58	09	16
1	2.8	4	9.3	1	2.1	CHICOT AR	234	-78	-26	27	28
2	2.5	4	4.8	4	4.5	CLARK AR	95	-6	83	78	68
3	3.7	3	2.3	4	3.1	CLAY AR	-37	34	-16	35	46
1	1.9			1	1.1	CLEBURNE AR			-40	18	17
2	2.3			6	5.5	COLUMBIA AR			135	86	81
3	6.1	4	5.8	2	2.2	CONWAY AR	-5	-61	-63	07	30
5	3.1	7	3.4	8	3.3	CRAIGHEAD AR	9	-2	6	52	48
4	3.5	2	1.6	2	1.3	CRAWFORD AR	-53	-22	-63	07	19
3	5.5	4	5.5	4	4.5	CRITTENDEN AR	0	-18	-18	33	68
1	1.4	4	5.7	3	3.8	CROSS AR	303	-33	171	89	57
		2	4.7			DALLAS AR					
2	4.0			3	5.9	DESHA AR			48	69	84
2	4.3	2	4.0	1	1.6	DREW AR	-7	-61	-64	07	21
2	2.0	6	4.8	1	.7	FAULKNER AR	141	-85	-64	07	15
2	3.4	2	3.1	3	3.4	FRANKLIN AR	-11	11	-1	47	50
		1	1.9	1	1.2	FULTON AR		-35			18
17	6.4	9	2.7	14	3.5	GARLAND AR	-58	31	-45	16	52
		1	2.1	2	3.6	GRANT AR		75			54
3	2.2	5	3.7	4	2.4	GREENE AR	71	-36	10	53	33
3	3.6	1	1.0	2	1.6	HEMPSTEAD AR	-71	55	-56	10	22
1	1.7	5	5.5	4	3.1	HOT SPRING AR	230	-43	87	79	45
1	2.0			2	2.3	HOWARD AR			19	58	32
3	2.9	2	1.7	4	2.6	INDEPENDENCE AR	-40	49	-11	39	36
		2	4.2	4	5.0	IZARD AR		19			75
4	6.1			2	1.9	JACKSON AR			-69	05	26
11	7.9	13	7.0	12	4.8	JEFFERSON AR	-11	-32	-39	19	72
3	3.5	1	1.2	5	6.4	JOHNSON AR	-64	411	82	78	88
2	6.6			2	5.0	LAFAYETTE AR			-25	27	74
3	3.5	4	4.3	2	2.2	LAWRENCE AR	23	-49	-37	20	30
3	7.4			1	2.3	LEE AR			-69	05	31
		2	7.7			LINCOLN AR					
2	8.2			1	1.9	LITTLE RIVER AR			-77	02	26
4	3.2	1	1.1	5	4.8	LOGAN AR	-64	320	51	70	73
2	1.9	1	1.0	5	3.8	LONOKE AR	-49	279	94	80	56
3	5.4	4	6.7	1	1.2	MADISON AR	24	-82	-78	02	18
2	4.7	1	1.8			MARION AR	-63				
5	6.0	7	6.0	5	3.5	MILLER AR	0	-41	-41	18	53
11	6.8	8	4.4	7	3.7	MISSISSIPPI AR	-35	-16	-45	15	55
2	5.1	1	2.0	4	9.2	MONROE AR	-60	354	81	77	97
1	3.3			2	2.8	MONTGOMERY AR			-13	37	40
		3	5.7	1	1.9	NEVADA AR		-66			26
				2	4.8	NEWTON AR					73
6	7.3	3	3.2	2	1.7	OUACHITA AR	-56	-48	-77	02	23
				3	8.0	PERRY AR					94
5	6.5	2	2.5	2	2.4	PHILLIPS AR	-62	-3	-63	07	33
		2	3.3			PIKE AR					
2	1.6	7	5.8	4	3.1	POINSETT AR	262	-47	94	80	45
		2	1.8	4	3.4	POLK AR		90			51
5	4.7	3	2.6	3	2.1	POPE AR	-46	-17	-55	11	29
		2	3.8	2	3.7	PRAIRIE AR		-2			55
28	4.9	50	5.9	47	4.7	PULASKI AR	20	-21	-5	43	71
1	1.4	1	1.2	1	1.5	RANDOLPH AR	-17	28	7	52	21
3	4.9	2	3.2	2	2.8	ST FRANCIS AR	-35	-14	-44	16	39
5	4.9	7	4.7	4	2.1	SALINE AR	-4	-56	-57	09	28
1	2.1			2	2.7	SCOTT AR			28	62	38
8	2.8	15	4.7	16	4.1	SEBASTIAN AR	72	-13	50	69	62
				4	5.7	SEVIER AR					82
1	2.4			3	4.4	STONE AR			82	78	67
8	6.1	7	3.8	8	3.7	UNION AR	-38	-3	-40	19	55
		3	5.2	4	5.1	VAN BUREN AR		-1			77
6	2.6	8	2.6	7	1.9	WASHINGTON AR	0	-27	-27	26	26
6	3.9	4	2.1	4	1.6	WHITE AR	-46	-24	-59	08	21
		2	4.2			WOODRUFF AR					
1	1.3	2	2.4	4	4.2	YELL AR	88	73	225	93	63
2575*	4.9	3391	4.9	3919	4.8	CALIFORNIA CA	-1	-3	-3		
168	5.2	199	5.5	188	5.1	ALAMEDA CA	6	-7	-1	47	77
2	3.4	3	4.1	2	2.2	AMADOR CA	20	-47	-36	21	29
18	4.0	33	6.4	24	3.5	BUTTE CA	60	-45	-11	39	53
4	5.7	3	3.2	2	1.9	CALAVERAS CA	-43	-42	-67	05	26
4	6.1	3	4.1			COLUSA CA	-33				
56	5.5	61	4.1	86	4.0	CONTRA COSTA CA	-26	-2	-27	26	61
3	5.9	4	7.7	1	1.8	DEL NORTE CA	30	-76	-69	04	25
4	3.0	7	3.3	9	3.1	EL DORADO CA	9	-6	3	49	45
53	4.5	80	5.4	81	4.7	FRESNO CA	19	-12	4	50	71
7	7.8	10*	10.6	3	2.9	GLENN CA	35	-73	-63	07	41
17	5.7	18	4.0	18	3.8	HUMBOLDT CA	-30	-6	-34	22	56
8	5.0	12	4.2	15	5.1	IMPERIAL CA	-16	21	2	48	77
4	7.0	6	7.5	5	5.4	INYO CA	8	-29	-23	28	80
49	6.4	43	3.8	61	4.5	KERN CA	-41	20	-29	25	68
4	2.5	8	3.9	7	3.1	KINGS CA	57	-21	25	60	44
		7	5.1	13	6.1	LAKE CA		21			86
3	4.3	1	1.5	3	3.3	LASSEN CA	-64	112	-23	29	48
986	4.9	1291*	5.1	1404*	5.4	LOS ANGELES CA	4	6	11	54	80
4	2.8	11	5.9	6	2.5	MADERA CA	115	-58	-9	40	35
23	5.0	36	5.2	41	4.7	MARIN CA	5	-10	-5	43	71
1	2.0			2	3.2	MARIPOSA CA			61	72	48
13	5.9	6	2.2	19	6.5	MENDOCINO CA	-62	191	11	54	88
6	2.2	15	4.8	27	6.5	MERCED CA	125	35	204	91	89
2	4.5			1	2.6	MODOC CA			-42	17	36
25	4.6	36	5.2	32	3.7	MONTEREY CA	13	-30	-20	31	55
18	5.0	22	4.9	30	5.7	NAPA CA	-1	17	15	56	83
4	3.1	4	2.3	6	2.6	NEVADA CA	-25	15	-14	37	37
48	3.1	98	3.0	191	3.5	ORANGE CA	-3	15	11	54	51
12	4.4	14	4.2	20	4.5	PLACER CA	-5	7	2	48	67
2	3.6	3	4.9	4	4.8	PLUMAS CA	35	-2	33	63	72
37	3.6	72	4.0	87	3.1	RIVERSIDE CA	12	-23	-15	36	45

* SIGNIFICANTLY MORE DEATHS THAN EXPECTED

WHITE MALE — CANCER OF THE ORAL CAVITY INCLUDING TONGUE — ICD 141,143-146,148,149

1950-1959 DEATHS	1950-1959 RATE	1960-1969 DEATHS	1960-1969 RATE	1970-1979 DEATHS	1970-1979 RATE	COUNTY AND STATE	PERCENT CHANGE 60S/50S	PERCENT CHANGE 70S/60S	PERCENT CHANGE 70S/50S	PERCENTILE %CHANGE 70S/50S	PERCENTILE RATES 1970S
57	3.9	117*	6.0	132	5.2	SACRAMENTO CA	53	-14	32	63	78
3	4.2	4	4.8	1	1.1	SAN BENITO CA	15	-78	-74	03	17
63	3.7	106	4.4	113	3.6	SAN BERNARDINO CA	17	-18	-4	45	54
121	4.6	165	3.9	276	4.4	SAN DIEGO CA	-15	13	-4	44	67
352*	9.3	331*	9.0	260*	8.7	SAN FRANCISCO CA	-3	-3	-6	42	96
66*	6.4	57	4.8	79	5.9	SAN JOAQUIN CA	-25	24	-7	41	85
16	4.9	21	4.0	20	3.1	SAN LUIS OBISPO CA	-18	-22	-37	21	46
52	4.6	101	5.5	114	5.1	SAN MATEO CA	19	-8	10	53	76
19	3.3	48	6.0	48	4.4	SANTA BARBARA CA	78	-26	31	63	66
77	5.0	84	3.5	163	4.7	SANTA CLARA CA	-30	34	-6	42	70
17	3.8	39	6.5	33	4.5	SANTA CRUZ CA	73	-31	20	59	68
6	2.6	10	3.2	18	3.7	SHASTA CA	24	16	44	67	56
		1	5.2			SIERRA CA					
7	4.4	11	5.9	7	3.8	SISKIYOU CA	33	-36	-14	37	57
18	4.8	16	3.1	27	4.3	SOLANO CA	-35	37	-10	39	65
31	4.9	44	4.7	55	4.2	SONOMA CA	-4	-11	-15	36	63
26	3.8	40	4.7	45	4.5	STANISLAUS CA	23	-5	16	57	67
3	2.2	4	2.4	5	2.4	SUTTER CA	7	2	10	53	34
3	2.7	5	3.5	10	5.0	TEHAMA CA	30	44	87	79	75
1	2.6	2	4.0	2	2.9	TRINITY CA	56	-29	11	54	41
13	2.1	28	3.6	43	4.9	TULARE CA	76	36	140	86	74
2	2.2	3	2.7	11	7.2	TUOLUMNE CA	27	162	232	93	92
17	3.2	29	3.1	46	3.2	VENTURA CA	-3	4	1	48	47
14	6.8	17	6.6	16	4.6	YOLO CA	-4	-30	-33	22	69
6	4.8	2	1.7	7	4.1	YUBA CA	-64	135	-14	36	61
162	2.5	209	2.7	243	2.6	COLORADO CO	4	-2	2		
1	.9	11	3.0	9	1.5	ADAMS CO	245	-51	69	75	20
				2	5.1	ALAMOSA CO					77
		2	6.1			BACA CO					
1	3.0	2	5.4	1	2.2	BENT CO	77	-60	-29	25	30
4	1.9	7	2.0	13	2.8	BOULDER CO	7	36	46	68	39
1	1.8			2	3.1	CHAFFEE CO			70	75	45
				1	7.5	CHEYENNE CO					93
1	5.5					CLEAR CREEK CO					
2	3.6			2	8.3	CONEJOS CO			128	85	95
		1	6.7			COSTILLA CO					
		2	1.6	3	2.5	DELTA CO		56			34
81	3.7	96	3.7	102	3.7	DENVER-ARAPAHOE#CO	0	-1	-1	46	55
				1	24.8	DOLORES CO					99
				1	1.8	DOUGLAS CO					25
				1	5.2	ELBERT CO					78
13	3.3	11	2.0	15	1.8	EL PASO CO	-39	-9	-45	16	25
2	1.4	5	3.5	3	1.4	FREMONT CO	148	-58	3	49	20
		2	2.6	2	2.5	GARFIELD CO		-5			34
1	1.6					HUERFANO CO					
		1	10.0			JACKSON CO					
6	1.4	18	3.4	24	3.0	JEFFERSON CO	135	-11	109	82	43
				3	7.9	KIT CARSON CO					94
		2	2.6	2	2.2	LA PLATA CO		-16			30
5	2.0	5	1.5	7	1.6	LARIMER CO	-25	5	-21	30	22
2	1.6	5	3.8	2	3.0	LAS ANIMAS CO	130	-20	84	78	44
1	2.9	1	4.1			LINCOLN CO	45				
7	8.3	2	2.1	2	1.9	LOGAN CO	-74	-11	-77	02	26
3	1.7	4	1.6	3	1.0	MESA CO	-9	-34	-40	19	16
1	3.0	1	2.8	2	5.2	MOFFAT CO	-6	83	71	75	78
1	2.1	1	1.8	2	3.2	MONTEZUMA CO	-15	81	54	71	47
		2	2.2	2	2.1	MONTROSE CO		-5			28
		1	1.0			MORGAN CO					
2	1.8	2	1.9	7	5.8	OTERO CO	7	208	229	93	83
1	14.9					OURAY CO					
2	7.4	1	4.7	2	7.8	PHILLIPS CO	-36	66	6	51	94
				1	7.0	PITKIN CO					91
		2	2.9			PROWERS CO					
15	3.2	12	2.3	12	2.2	PUEBLO CO	-31	-2	-32	23	30
1	6.2					RIO BLANCO CO					
2	3.9	1	2.1	1	1.8	RIO GRANDE CO	-47	-14	-54	11	24
1	3.9			1	3.8	SEDGWICK CO			-1	46	57
2	11.0	1	5.0			TELLER CO	-55				
3	1.0	7	2.0	10	2.5	WELD CO	111	24	163	88	35
		1	1.4	2	3.5	YUMA CO		148			51
733*	7.5	879*	7.4	872*	6.4	CONNECTICUT CT	-2	-13	-15		
206*	8.4	227*	7.5	234*	6.7	FAIRFIELD CT	-11	-10	-20	31	89
193*	7.8	249*	7.9	238*	6.6	HARTFORD CT	1	-16	-15	36	89
25	4.9	43	6.5	46	6.0	LITCHFIELD CT	32	-8	22	60	85
19	4.8	25	5.3	29	5.1	MIDDLESEX CT	10	-5	5	51	76
218*	8.2	238*	7.7	235*	6.8	NEW HAVEN CT	-6	-12	-17	34	90
44	6.6	50	6.4	49	5.2	NEW LONDON CT	-4	-19	-22	30	77
10	4.4	21*	7.3	19	5.5	TOLLAND CT	65	-25	24	60	81
18	5.9	26*	7.3	22	5.5	WINDHAM CT	23	-24	-6	42	81
72	5.7	84	5.3	110	5.3	DELAWARE DE	-7	1	-6		
10	5.5	9	4.2	13	4.6	KENT DE	-25	10	-17	34	69
53	6.5	61	5.9	75	5.3	NEW CASTLE DE	-10	-9	-18	33	79
9	3.3	14	4.6	22	5.6	SUSSEX DE	39	22	70	75	82
172*	8.4	188*	11.1	104*	9.2	DIST. OF COLUMBIA DC	32	-17	9		
172*	8.4	188*	11.1	104*	9.2	WASHINGTON DC	32	-17	9	53	96
740	4.8	1418	5.1	2145	4.8	FLORIDA FL	8	-7	0		
11	8.2	9	4.5	16	5.4	ALACHUA FL	-45	21	-33	22	80
1	5.7	1	2.9	3	7.6	BAKER FL	-49	160	33	63	93
7	4.4	13	6.2	9	3.0	BAY FL	39	-52	-33	23	43
2	5.1	3	6.1	3	4.0	BRADFORD FL	21	-35	-21	30	60
6	2.8	16	3.8	57	4.9	BREVARD FL	37	28	75	76	74

* SIGNIFICANTLY MORE DEATHS THAN EXPECTED

\# COUNTIES COMBINED; SEE TECHNICAL NOTES

WHITE MALE — CANCER OF THE ORAL CAVITY INCLUDING TONGUE — ICD 141,143-146,148,149

1950-1959 DEATHS	1950-1959 RATE	1960-1969 DEATHS	1960-1969 RATE	1970-1979 DEATHS	1970-1979 RATE	COUNTY AND STATE		PERCENT CHANGE 60S/50S	PERCENT CHANGE 70S/60S	PERCENT CHANGE 70S/50S	PERCENTILE %CHANGE 70S/50S	PERCENTILE RATES 1970S
29	3.5	109	4.2	220	4.1	BROWARD	FL	22	-3	18	58	62
2	7.8	1	3.2	1	2.0	CALHOUN	FL	-59	-37	-74	03	28
3	5.2	5	2.2	17	3.4	CHARLOTTE	FL	-58	55	-34	22	50
1	3.3	4	3.1	16	3.5	CITRUS	FL	-5	10	5	51	51
1	2.1	2	2.6	8	5.3	CLAY	FL	22	105	151	87	79
		8	5.7	14	3.3	COLLIER	FL		-42			49
2	3.4	4	5.1	10*	9.4	COLUMBIA	FL	52	83	178	90	97
161	5.3	272	5.2	458*	6.3	DADE	FL	-3	22	19	58	87
4	7.2	3	5.9	6	6.4	DE SOTO	FL	-17	8	-11	39	88
		1	8.0	2	5.8	DIXIE	FL		-28			83
72*	8.0	118*	9.0	109*	7.0	DUVAL	FL	12	-23	-13	37	91
16	5.4	26	5.3	40	6.6	ESCAMBIA	FL	-1	23	22	59	89
				3	7.9	FLAGLER	FL					94
2	8.5	1	3.3			FRANKLIN	FL	-61				
1	1.4	2	2.0	6	5.9	GADSDEN	FL	45	189	318	96	84
				1	4.9	GILCHRIST	FL					73
				1	3.9	GLADES	FL					59
		3	8.4	3	6.8	GULF	FL		-19			90
2	8.5	1	3.7	3	11.6	HAMILTON	FL	-56	212	37	65	98
3	5.8	5	8.1	4	4.9	HARDEE	FL	39	-39	-15	36	74
1	6.1			3	4.7	HENDRY	FL			-23	29	71
2	5.3	7	8.1	10	3.4	HERNANDO	FL	55	-58	-35	22	51
5	6.6	5	3.2	13	5.0	HIGHLANDS	FL	-51	55	-24	28	75
75	6.1	111*	6.2	120	5.1	HILLSBOROUGH	FL	2	-18	-16	35	77
2	4.2	2	3.2	1	1.2	HOLMES	FL	-23	-62	-71	04	18
2	2.0	8	4.6	17	5.2	INDIAN RIVER	FL	131	13	161	88	78
1	1.0	3	2.5	8	5.7	JACKSON	FL	141	129	453	98	82
1	6.5	4*	20.2			JEFFERSON	FL	211				
		1	6.5	2	10.6	LAFAYETTE	FL		63			98
6	3.1	15	4.0	27	4.6	LAKE	FL	32	14	50	69	69
10	5.6	27	6.0	50	4.3	LEE	FL	7	-27	-22	29	66
6	4.8	11	5.7	15	5.5	LEON	FL	19	-4	14	56	81
1	2.1	4	9.0	3	4.2	LEVY	FL	326	-54	97	80	63
		1	8.6	2	11.2	LIBERTY	FL		30			98
6*	17.3			2	3.8	MADISON	FL			-78	02	57
6	1.3	17	1.9	31	2.6	MANATEE	FL	53	37	110	82	37
11	7.8	17	7.5	24	4.9	MARION	FL	-4	-35	-38	20	74
4	5.7	11	7.5	17	5.2	MARTIN	FL	30	-30	-9	40	78
12*	16.1	14	7.4	16	5.0	MONROE	FL	-54	-32	-69	05	76
2	3.9	7*	20.9	5	6.4	NASSAU	FL	433	-69	64	73	88
1	1.1	6	4.9	16	6.7	OKALOOSA	FL	335	36	493	98	90
		2	8.1	6	10.1	OKEECHOBEE	FL		24			97
29	4.0	72*	6.5	83	5.1	ORANGE	FL	60	-21	27	61	77
2	.9	8	4.3	10	3.7	OSCEOLA	FL	350	-13	292	96	56
39	5.6	87	5.8	128	4.5	PALM BEACH	FL	4	-23	-20	32	68
7	3.0	11	2.1	47	3.5	PASCO	FL	-30	67	17	58	52
82	4.4	179	5.0	248	4.8	PINELLAS	FL	13	-3	10	54	73
34	5.6	36	3.9	57	4.4	POLK	FL	-30	12	-22	30	66
2	2.0	6	4.9	6	3.1	PUTNAM	FL	150	-37	57	71	45
8	7.8	7	5.4	11	5.5	ST JOHNS	FL	-31	2	-29	25	81
4	2.9	14	5.5	13	2.9	ST LUCIE	FL	87	-46	1	48	42
2	4.1	4	4.3	5	2.9	SANTA ROSA	FL	5	-33	-29	25	41
19	5.8	28	3.7	37	2.9	SARASOTA	FL	-37	-21	-50	13	41
5	3.2	13	5.4	14	2.8	SEMINOLE	FL	70	-48	-12	38	39
1	1.5	4	5.4	3	3.6	SUMTER	FL	250	-33	135	86	54
1	1.8	3	4.9	2	3.0	SUWANNEE	FL	179	-38	74	76	44
2	6.9	4	9.8	3	5.0	TAYLOR	FL	41	-48	-27	26	76
		1	3.8	1	2.6	UNION	FL		-31			37
21	3.5	52	4.9	70	4.4	VOLUSIA	FL	40	-12	24	60	66
		2	8.7	2	5.3	WAKULLA	FL		-39			79
1	2.2	4	4.8	4	3.9	WALTON	FL	117	-19	75	76	58
1	1.9	3	6.5	3	4.0	WASHINGTON	FL	243	-39	109	82	60
452	5.1	564	5.0	705	4.8	GEORGIA	GA	-2	-4	-6		
4	11.6	3	10.3			APPLING	GA	-11				
		1	4.8			ATKINSON	GA					
		2	7.0	1	2.6	BACON	GA		-63			36
2	2.8	1	.8	3	2.7	BALDWIN	GA	-73	258	-5	43	38
		1	3.1	1	2.5	BANKS	GA		-20			35
1	2.0	5	7.9			BARROW	GA	302				
4	4.9	4	3.2	9	6.3	BARTOW	GA	-36	98	27	61	87
2	4.7	1	2.0	1	1.6	BEN HILL	GA	-57	-22	-67	06	21
2	6.2	3	8.6	4	8.1	BERRIEN	GA	40	-6	32	63	95
21*	7.7	21	6.5	31*	7.7	BIBB	GA	-15	19	1	48	94
1	4.1	3	8.9	2	5.3	BLECKLEY	GA	118	-41	29	62	79
		1	5.1	2	6.2	BRANTLEY	GA		22			86
		3	6.9	2	3.8	BROOKS	GA		-45			57
				1	3.3	BRYAN	GA					49
3	4.8	3	4.5	6	6.3	BULLOCH	GA	-7	39	30	62	87
2	6.8	5*	17.0	1	2.3	BURKE	GA	151	-87	-67	06	31
2	8.8	2	8.0	3	6.9	BUTTS	GA	-9	-13	-21	30	91
		2	18.1			CALHOUN	GA					
		2	11.6	4*	15.0	CAMDEN	GA		29			99
2	7.5	2	9.8	1	3.1	CANDLER	GA	30	-68	-59	08	46
3	2.4	4	2.6	8	4.5	CARROLL	GA	9	74	90	79	68
1	2.7	1	1.3	1	.7	CATOOSA	GA	-52	-47	-75	03	15
2*	15.2	4*	26.4	3*	13.5	CHARLTON	GA	73	-49	-11	39	99
23	7.4	35*	8.2	28	5.2	CHATHAM	GA	11	-36	-29	25	79
1	1.5	2	2.1	2	1.9	CHATTOOGA	GA	38	-10	24	60	26
4	4.7	5	4.5	2	1.1	CHEROKEE	GA	-3	-75	-76	02	17
9*	11.4	3	2.9	3	2.0	CLARKE	GA	-75	-31	-83	00	27
				1	7.9	CLAY	GA					94
2	2.0	9	8.8	10	3.7	CLAYTON	GA	339	-58	83	78	54
		2	8.8	1	5.8	CLINCH	GA		-34			84
5	2.2	18	4.8	26	3.9	COBB	GA	117	-20	74	76	58
3	5.2	4	7.0	2	2.9	COFFEE	GA	33	-59	-45	15	41
2	2.0	5	4.6	2	1.3	COLQUITT	GA	128	-72	-36	21	19
3*	13.3	2	4.9	4	4.1	COLUMBIA	GA	-63	-16	-69	04	62
1	3.1	1	2.9	4	7.9	COOK	GA	-5	170	156	88	94
6	8.7	3	3.8	2	1.7	COWETA	GA	-57	-54	-80	01	24

* SIGNIFICANTLY MORE DEATHS THAN EXPECTED

WHITE MALE — CANCER OF THE ORAL CAVITY INCLUDING TONGUE — ICD 141,143-146,148,149

1950-1959 DEATHS	1950-1959 RATE	1960-1969 DEATHS	1960-1969 RATE	1970-1979 DEATHS	1970-1979 RATE	COUNTY AND STATE	PERCENT CHANGE 60S/50S	PERCENT CHANGE 70S/60S	PERCENT CHANGE 70S/50S	PERCENTILE %CHANGE 70S/50S	PERCENTILE RATES 1970S
		1	10.5			CRAWFORD GA					
3	7.4			5	11.1	CRISP GA			51	70	98
1	8.0	1	3.3			DADE GA	-59				
2	14.9					DAWSON GA					
		6	10.8	6	8.6	DECATUR GA		-20			96
23	4.9	37	4.9	46	4.0	DE KALB GA	-1	-18	-19	32	60
		2	4.2	6	9.2	DODGE GA		119			96
2	6.2	1	2.9			DOOLY GA	-54				
2	2.0	8	6.8	11	5.3	DOUGHERTY GA	235	-22	160	88	79
1	2.2	1	1.1	5	5.6	DOUGLAS GA	-50	413	154	88	81
		1	2.6	3	7.3	EARLY GA		185			92
1	14.0			1	11.2	ECHOLS GA			-20	31	98
4*	19.0			4	11.3	EFFINGHAM GA			-41	18	98
3	5.5	2	3.2	2	2.9	ELBERT GA	-41	-10	-47	15	42
6*	10.4	1	1.5	2	2.8	EMANUEL GA	-86	86	-73	03	39
1	5.6	3*	14.8	2	6.6	EVANS GA	163	-55	17	58	89
2	3.1	3	4.4	2	2.5	FANNIN GA	44	-44	-19	32	34
		1	3.6	3	3.6	FAYETTE GA		0			53
10	5.1	13	5.2	14	4.7	FLOYD GA	3	-10	-7	41	71
1	3.1	1	1.9	5	5.2	FORSYTH GA	-39	177	68	74	79
1	1.9	1	1.7	2	2.6	FRANKLIN GA	-9	54	40	66	37
81*	6.6	109*	7.6	93*	6.3	FULTON GA	15	-17	-4	44	87
		1	2.4	2	3.8	GILMER GA		60			57
1	1.1	5	5.0	9	5.2	GLYNN GA	350	3	365	97	78
6	10.4	2	2.4	5	4.7	GORDON GA	-77	96	-55	11	71
1	1.7	2	4.1	5	7.2	GRADY GA	141	74	319	97	92
1	3.7					GREENE GA					
7	5.3	6	3.3	13	4.1	GWINNETT GA	-38	25	-22	29	63
1	2.0	2	2.5	2	2.0	HABERSHAM GA	24	-21	-3	45	27
7	4.7	6	3.0	8	2.8	HALL GA	-37	-5	-40	19	40
1	6.5					HANCOCK GA					
				2	2.6	HARALSON GA					36
2	6.9	1	2.7	2	4.0	HARRIS GA	-60	46	-42	17	60
3	6.1	1	2.0	2	3.6	HART GA	-67	74	-42	17	53
				1	2.8	HEARD GA					39
1	2.1	4	7.1	6	6.7	HENRY GA	243	-6	223	92	90
3	5.9	2	1.7	12	7.0	HOUSTON GA	-71	310	19	58	91
		1	3.0	3	8.8	IRWIN GA		192			96
3	4.3	4	4.5	2	2.0	JACKSON GA	4	-56	-55	11	27
1	5.2	1	5.4			JASPER GA	3				
2	11.1	1	3.0	5*	13.6	JEFF DAVIS GA	-73	353	22	60	99
1	2.6	1	2.4			JEFFERSON GA	-8				
		2	8.2	1	3.5	JENKINS GA		-58			51
1	3.7	2	6.3			JOHNSON GA	69				
				1	2.3	JONES GA					31
1	3.6	1	2.8	2	5.0	LAMAR GA	-23	78	37	65	74
		1	7.2			LANIER GA					
5	6.3	4	4.4	9	6.2	LAURENS GA	-30	40	-2	46	86
		1	6.1			LEE GA					
2	13.9			2	6.4	LIBERTY GA			-54	11	87
				1	4.4	LINCOLN GA					66
				3*	21.7	LONG GA					99
6	8.0	8	7.5	6	4.2	LOWNDES GA	-6	-45	-48	14	63
1	2.9	1	3.0	1	2.1	LUMPKIN GA	3	-31	-29	25	28
		1	3.2	3	9.1	MC DUFFIE GA		188			96
				1	4.4	MC INTOSH GA					66
		1	5.0	2	6.7	MACON GA		35			90
1	2.5	2	4.2	3	5.2	MADISON GA	73	23	114	83	78
1	7.5					MARION GA					
2	8.1	6*	11.6	2	4.0	MERIWETHER GA	44	-65	-50	13	61
1	4.4					MILLER GA					
1	2.3	1	2.6	2	4.2	MITCHELL GA	15	60	84	78	64
				1	3.4	MONROE GA					50
2	11.9			2	8.9	MONTGOMERY GA			-25	27	96
1	3.9			1	3.8	MORGAN GA			0	47	58
1	1.3	1	2.2	2	2.6	MURRAY GA	65	21	100	81	37
23*	10.5	16	4.7	26	6.6	MUSCOGEE#GA	-56	42	-38	20	89
2	3.7	1	1.2	4	3.6	NEWTON GA	-67	187	-5	43	53
2	11.0			1	2.8	OCONEE GA			-75	03	39
				2	8.5	OGLETHORPE GA					95
1	1.7	2	3.1	4	4.4	PAULDING GA	76	43	152	87	66
1	4.6	1	4.2	1	7.4	PEACH GA	-9	77	62	73	92
2	6.0	1	2.5	1	2.1	PICKENS GA	-59	-14	-64	06	29
		1	3.5	3	6.5	PIERCE GA		87			89
1	6.2					PIKE GA					
3	2.8	6	5.1	7	5.7	POLK GA	80	12	101	81	83
3*	13.8	1	3.6	1	4.6	PULASKI GA	-74	28	-66	06	70
1	7.6	2	8.8	1	3.9	PUTNAM GA	17	-55	-48	14	59
1	3.9	2	5.5	4	7.6	RABUN GA	42	38	96	80	93
1	4.8			2	8.6	RANDOLPH GA			77	76	95
20*	10.5	30*	11.1	28*	7.5	RICHMOND GA	5	-32	-29	25	93
1	3.4	2	3.9	4	4.1	ROCKDALE GA	14	5	20	59	61
		1	11.1			SCHLEY GA					
1	2.7	3	8.0	3	7.3	SCREVEN GA	202	-9	175	90	92
3*	16.7			3	13.9	SEMINOLE GA			-16	35	99
2	2.0	2	1.6	4	2.8	SPALDING GA	-22	78	39	66	40
1	2.0	1	1.1	4	3.9	STEPHENS GA	-44	247	93	80	58
2	16.2			1	11.2	STEWART GA			-31	24	98
		8*	15.0	6	9.3	SUMTER GA		-38			97
2*	43.4	2	13.7			TALBOT GA	-68				
1	11.6			2*	22.4	TALIAFERRO GA			92	80	99
1	2.3	2	3.3	7*	14.3	TATTNALL GA	43	339	530	98	99
				2	7.8	TAYLOR GA					94
2	6.3	1	3.4	2	4.6	TELFAIR GA	-46	35	-27	26	69
2	7.9	2	8.2			TERRELL GA	3				
7	7.4	3	2.8	4	3.5	THOMAS GA	-63	24	-53	11	51
1	2.4	3	4.7	5	4.9	TIFT GA	92	5	101	81	73
		1	1.7	4	5.9	TOOMBS GA		253			84
				1	2.5	TOWNS GA					34
1	5.2	1	4.1	1	3.5	TREUTLEN GA	-21	-15	-33	22	51
3	3.1	2	1.3	1	.6	TROUP GA	-58	-57	-82	01	14

* SIGNIFICANTLY MORE DEATHS THAN EXPECTED

COUNTIES COMBINED; SEE TECHNICAL NOTES

WHITE MALE — CANCER OF THE ORAL CAVITY INCLUDING TONGUE — ICD 141,143-146,148,149

1950-1959 DEATHS	1950-1959 RATE	1960-1969 DEATHS	1960-1969 RATE	1970-1979 DEATHS	1970-1979 RATE	COUNTY AND STATE	PERCENT CHANGE 60S/50S	PERCENT CHANGE 70S/60S	PERCENT CHANGE 70S/50S	PERCENTILE %CHANGE 70S/50S	PERCENTILE RATES 1970S
1	3.2	1	5.3	3	9.0	TURNER GA	66	70	181	90	96
				2	13.5	TWIGGS GA					99
3	12.0	1	2.7			UNION GA	-78				
4	4.7	1	.9	5	4.4	UPSON GA	-81	390	-6	42	67
7	4.1	14	7.5	7	2.9	WALKER GA	84	-62	-29	25	41
		1	1.2	8	8.9	WALTON GA		676			96
4	4.3	3	2.5	10	7.8	WARE GA	-42	212	82	78	94
2	12.5	1	7.9	2	11.7	WARREN GA	-37	48	-7	42	98
1	2.6	1	2.1	2	3.8	WASHINGTON GA	-18	79	46	68	57
1	6.2			5	7.9	WAYNE GA			28	61	94
1	5.4	1	5.2			WHEELER GA	-4				
1	4.4			1	2.6	WHITE GA			-40	19	37
10*	9.1	6	3.8	6	2.8	WHITFIELD GA	-58	-26	-69	05	40
		3	11.3	1	2.7	WILCOX GA		-76			38
1	3.2	1	3.5			WILKES GA	9				
2	7.2	2	11.2	2	7.0	WILKINSON GA	55	-38	-4	45	91
1	2.4	4	11.4	5	8.7	WORTH GA	373	-24	261	94	96
		43*	7.1	60*	6.3	HAWAII HI		-11			
		5	10.8	4	3.2	HAWAII HI		-70			47
		36*	7.3	51*	7.1	HONOLULU HI		-3			91
				2	5.0	KAUAI HI					75
		2	4.1	3	4.8	MAUI#HI		17			73
56	2.0	70	2.3	94	2.6	IDAHO ID	11	13	25		
6	1.8	21	4.8	26	4.5	ADA ID	174	-7	156	88	68
1	5.4					ADAMS ID					
2	1.7	2	1.1	4	1.8	BANNOCK ID	-34	65	9	53	25
2	5.9	1	3.0	1	3.1	BEAR LAKE ID	-49	5	-47	15	45
1	2.5	2	5.1	1	2.2	BENEWAH ID	103	-58	-14	37	29
		1	1.1	4	3.1	BINGHAM ID		198			46
		1	3.7			BLAINE ID					
1	11.2					BOISE ID					
3	3.5	2	2.2	2	1.9	BONNER ID	-37	-17	-48	14	26
2	1.7	6	3.6	5	2.5	BONNEVILLE ID	115	-31	50	69	35
6	2.3	6	2.0	9	2.4	CANYON ID	-11	18	5	51	33
4	7.1	2	2.7	3	4.0	CASSIA ID	-62	51	-43	17	61
3	6.4			3	5.4	CLEARWATER ID			-16	35	80
1	5.6	1	6.7			CUSTER ID	19				
1	3.8	2	5.4	1	1.7	ELMORE ID	43	-69	-56	10	23
				3	6.9	FRANKLIN ID					91
				1	2.3	FREMONT ID					32
		1	1.6	4	7.6	GEM ID		380			93
		1	2.0	1	1.4	GOODING ID		-31			20
		3	4.3	4	5.7	IDAHO ID		32			82
1	1.9	1	3.1	2	3.3	JEFFERSON ID	69	7	80	77	49
1	2.0			1	1.4	JEROME ID			-28	25	20
3	1.9	3	1.7	1	.3	KOOTENAI ID	-12	-80	-83	00	14
1	1.0	1	1.2	3	2.9	LATAH ID	18	138	180	90	41
		1	3.1	1	3.1	LEMHI ID		2			46
		1	5.2			LEWIS ID					
				2	11.6	LINCOLN ID					98
1	2.0					MADISON ID					
4	8.0			1	1.6	MINIDOKA ID			-80	01	22
1	.7	3	2.1	2	1.2	NEZ PERCE ID	187	-44	60	72	18
		1	2.9	1	2.8	OWYHEE ID		-4			39
1	1.5	1	1.7	2	2.3	PAYETTE ID	12	31	46	68	31
4	4.3	3	4.3	1	.9	SHOSHONE ID	0	-79	-79	01	15
4	2.1	2	1.0	4	1.7	TWIN FALLS ID	-51	73	-16	35	24
2	5.5	1	1.9	1	1.5	WASHINGTON ID	-66	-20	-73	03	21
2197*	5.5	2391*	5.4	2308*	5.0	ILLINOIS IL	-2	-7	-8		
22	5.5	17	4.3	19	4.7	ADAMS IL	-23	11	-15	36	71
4	7.5	5	8.1	4	6.7	ALEXANDER IL	9	-18	-11	39	89
4	4.2			1	.9	BOND IL			-78	01	15
2	2.0	1	1.0	2	1.6	BOONE IL	-49	62	-17	34	23
3	5.8	1	2.9			BROWN IL	-50				
11	5.0	8	3.5	13	5.8	BUREAU IL	-29	64	16	56	83
1	3.1	1	1.9	2	4.3	CALHOUN IL	-39	130	40	66	65
4	3.3	2	1.7	3	3.1	CARROLL IL	-51	86	-8	41	45
3	3.0	3	3.7	4	5.2	CASS IL	22	40	71	75	78
18	5.4	13	3.4	20	4.6	CHAMPAIGN IL	-37	36	-14	37	70
4	1.8	9	4.5	13	6.1	CHRISTIAN IL	147	36	236	94	86
1	1.1	3	2.6	3	2.9	CLARK IL	125	11	148	87	41
3	3.2	4	4.1	6	6.2	CLAY IL	31	50	96	80	86
6	5.1	5	3.7	5	3.5	CLINTON IL	-28	-5	-31	23	52
14	6.7	12	5.7	7	3.2	COLES IL	-15	-45	-53	12	47
1314*	6.9	1410*	6.8	1240*	6.2	COOK IL	-1	-10	-11	39	86
10	8.1	2	1.5	1	.9	CRAWFORD IL	-81	-41	-89	00	15
1	1.6	2	3.7	2	3.5	CUMBERLAND IL	136	-5	124	84	52
5	2.3	9	3.9	9	3.5	DE KALB IL	67	-12	48	69	51
2	2.0	5	4.9	1	1.0	DE WITT IL	153	-79	-47	15	16
2	1.9	3	3.2	1	1.0	DOUGLAS IL	66	-68	-47	15	16
25	3.1	41	3.2	76	4.3	DU PAGE IL	3	35	39	66	65
6	4.6	4	3.0	6	4.5	EDGAR IL	-34	50	-1	46	68
		2	5.1			EDWARDS IL					
6	5.5	6	4.7	4	2.9	EFFINGHAM IL	-15	-39	-48	14	41
5	4.1	5	3.6	5	3.4	FAYETTE IL	-12	-7	-18	33	50
1	1.0	3	3.3	3	3.0	FORD IL	218	-10	186	91	43
14	5.3	10	3.1	17	6.9	FRANKLIN IL	-42	123	30	62	90
6	2.4	5	2.1	7	2.8	FULTON IL	-12	36	20	59	40
1	1.9	2	5.0	1	2.8	GALLATIN IL	159	-44	44	68	39
6	5.0	5	4.3	1	1.0	GREENE IL	-15	-77	-80	01	16
3	2.7	6	5.3	5	3.7	GRUNDY IL	94	-31	34	64	54
2	3.1	2	2.6	4	5.9	HAMILTON IL	-16	130	93	80	84
5	3.2	5	3.5	5	3.3	HANCOCK IL	9	-5	3	49	48
1	2.8			1	3.4	HARDIN IL			24	60	50
1	2.5	1	2.2			HENDERSON IL	-12				

* SIGNIFICANTLY MORE DEATHS THAN EXPECTED

COUNTIES COMBINED; SEE TECHNICAL NOTES

WHITE MALE — CANCER OF THE ORAL CAVITY INCLUDING TONGUE — ICD 141,143-146,148,149

1950-1959 DEATHS	1950-1959 RATE	1960-1969 DEATHS	1960-1969 RATE	1970-1979 DEATHS	1970-1979 RATE	COUNTY AND STATE		PERCENT CHANGE 60S/50S	PERCENT CHANGE 70S/60S	PERCENT CHANGE 70S/50S	PERCENTILE %CHANGE 70S/50S	PERCENTILE RATES 1970S
9	3.4	10	3.5	8	2.8	HENRY	IL	5	-20	-16	35	40
6	3.7	9	5.0	8	4.2	IROQUOIS	IL	35	-16	14	56	64
6	4.0	5	2.2	7	3.3	JACKSON	IL	-44	49	-17	34	49
3	5.5	2	3.1			JASPER	IL	-43				
10	5.4	9	4.8	7	3.7	JEFFERSON	IL	-10	-23	-31	24	56
3	4.0	5	5.9	3	3.2	JERSEY	IL	50	-46	-19	33	47
4	3.2	1	.7	5	4.3	JO DAVIESS	IL	-78	513	37	65	65
1	1.7	1	1.7	1	1.8	JOHNSON	IL	2	4	6	52	24
27	3.3	35	3.9	39	4.1	KANE	IL	19	4	25	60	61
11	2.7	19	4.4	15	3.7	KANKAKEE	IL	64	-17	36	65	54
		2	2.9	2	2.0	KENDALL	IL		-32			27
16	5.6	11	3.4	10	3.1	KNOX	IL	-39	-10	-45	15	45
37	4.8	52	4.9	72	5.1	LAKE	IL	2	5	7	52	77
23	4.5	35	6.2	29	4.8	LA SALLE	IL	36	-22	6	52	73
3	3.0	2	1.8	4	4.0	LAWRENCE	IL	-38	121	36	65	61
4	2.1	8	4.4	10	5.5	LEE	IL	114	27	170	89	81
5	2.3	5	2.5	11	5.4	LIVINGSTON	IL	7	118	133	85	80
7	4.3	6	3.6	8	4.9	LOGAN	IL	-18	37	13	55	74
5	3.0	4	2.7	6	3.8	MC DONOUGH	IL	-10	43	28	62	58
5	1.9	18	4.1	20	3.8	MC HENRY	IL	118	-8	101	81	56
15	4.0	12	3.0	15	3.5	MC LEAN	IL	-24	15	-12	38	52
29	6.3	23	4.4	25	4.5	MACON	IL	-30	1	-29	25	67
14	4.8	9	3.0	13	4.7	MACOUPIN	IL	-37	56	-2	46	71
48	6.3	40	4.3	47	4.4	MADISON	IL	-32	2	-30	24	66
12	5.6	11	5.0	11	5.3	MARION	IL	-11	5	-6	42	79
3	3.8	3	4.0	3	4.2	MARSHALL	IL	4	6	10	54	64
3	4.2	3	2.9	1	1.2	MASON	IL	-32	-58	-71	04	18
9*	13.2			4	4.7	MASSAC	IL			-64	06	71
3	5.9	4	7.5			MENARD	IL	27				
4	3.8	1	.9			MERCER	IL	-75				
2	2.4	5	5.6	3	2.9	MONROE	IL	138	-48	24	60	42
5	2.4	5	2.8	4	2.0	MONTGOMERY	IL	16	-28	-17	34	27
10	4.8	3	1.4	8	4.3	MORGAN	IL	-71	213	-10	40	65
3	3.7	2	2.5	2	2.6	MOULTRIE	IL	-31	5	-28	26	38
2	1.1	6	3.1	10	4.8	OGLE	IL	177	57	336	97	73
48	6.0	55	6.4	41	4.7	PEORIA	IL	7	-27	-23	29	70
3	3.5	4	3.1	7	5.7	PERRY	IL	-13	83	60	72	82
2	2.9	3	3.8	3	4.0	PIATT	IL	30	6	38	65	60
3	2.4	5	3.7	3	2.5	PIKE	IL	52	-33	1	48	34
3	10.1			1	3.6	POPE	IL			-65	06	53
2	4.7	3	7.4	3	7.4	PULASKI	IL	56	1	58	72	93
2	6.1					PUTNAM	IL					
8	4.7	6	3.6	9	4.6	RANDOLPH	IL	-24	30	-1	47	70
3	3.2			4	4.2	RICHLAND	IL			29	62	63
19	2.8	41	5.5	34	4.5	ROCK ISLAND	IL	98	-18	63	73	68
67*	7.8	62*	6.6	62*	6.4	ST CLAIR	IL	-15	-4	-18	33	88
3	1.4	5	2.4	5	3.4	SALINE	IL	72	42	144	87	51
28	4.5	47	6.9	36	4.7	SANGAMON	IL	53	-31	5	51	72
1	1.8			3	4.8	SCHUYLER	IL			169	89	72
2	5.7	1	2.2	2	7.2	SCOTT	IL	-62	231	25	60	92
2	1.3	6	4.0	3	2.0	SHELBY	IL	204	-49	55	71	28
1	3.0	1	4.1			STARK	IL	38				
2	.9	13	5.6	5	2.0	STEPHENSON	IL	546	-64	134	85	28
17	5.2	17	4.0	20	4.0	TAZEWELL	IL	-24	1	-23	29	61
5	3.8	3	2.2	4	3.9	UNION	IL	-42	76	3	49	59
20	4.4	20	4.0	30	6.2	VERMILION	IL	-9	54	41	67	87
2	2.4			1	1.4	WABASH	IL			-41	17	20
4	3.4	4	3.7	3	2.6	WARREN	IL	8	-30	-25	27	36
8	7.3	2	1.7	4	4.1	WASHINGTON	IL	-77	145	-44	16	62
5	4.3	2	1.7	2	1.5	WAYNE	IL	-61	-8	-65	06	21
5	4.4	5	4.5	8	6.5	WHITE	IL	1	45	46	68	88
5	2.1	6	2.1	10	3.3	WHITESIDE	IL	0	59	58	72	49
19	3.3	34	4.5	43	4.9	WILL	IL	36	9	48	69	73
13	5.0	11	3.9	10	3.5	WILLIAMSON	IL	-21	-11	-30	24	52
30	4.4	52	5.7	47	4.6	WINNEBAGO	IL	29	-19	5	50	70
2	1.8	5	3.9	3	2.1	WOODFORD	IL	119	-45	20	59	29
715	3.9	806	4.0	874	4.0	INDIANA	IN	3	0	3		
1	.9	2	1.7	8	6.4	ADAMS	IN	81	277	583	99	87
30	3.6	51	5.5	40	3.9	ALLEN	IN	51	-29	7	52	59
4	2.7	5	2.7	5	2.2	BARTHOLOMEW	IN	0	-18	-18	33	30
3	5.0	3	5.6	2	3.7	BENTON	IN	12	-34	-26	27	55
2	2.8	1	1.6	2	2.8	BLACKFORD	IN	-45	77	-2	46	39
2	1.6	5	3.5	1	.6	BOONE	IN	122	-82	-60	08	14
3	3.0	2	2.4	4	4.3	CARROLL	IN	-18	78	45	68	65
4	1.7	8	3.6	9	4.4	CASS	IN	116	22	164	89	67
9	4.7	13	5.8	21	7.1	CLARK	IN	25	21	52	70	91
2	1.4	4	2.6	3	2.5	CLAY	IN	79	-3	73	75	34
5	3.3	6	3.6	4	2.5	CLINTON	IN	9	-29	-23	29	35
2	2.7	1	1.6	2	3.9	CRAWFORD	IN	-41	143	43	67	58
5	3.8	4	2.7	4	2.7	DAVIESS	IN	-29	-1	-29	25	38
4	3.1	7	5.1	6	4.0	DEARBORN	IN	61	-21	27	61	60
4	3.5	4	3.5	4	3.6	DECATUR	IN	0	3	3	49	54
2	1.4	4	2.7	6	4.1	DE KALB	IN	93	54	197	91	63
18	4.5	13	2.9	15	3.3	DELAWARE	IN	-34	11	-27	26	48
4	3.7	6	4.6	5	3.5	DUBOIS	IN	25	-24	-5	43	53
6	1.4	20	4.2	17	3.1	ELKHART	IN	190	-25	118	84	46
1	.9	3	2.5	9	7.6	FAYETTE	IN	172	202	720	99	93
11	5.5	16	6.8	18	7.3	FLOYD	IN	25	7	33	63	92
3	3.1	5	4.9	3	3.1	FOUNTAIN	IN	59	-36	2	48	46
2	2.9	2	2.5	10*	12.7	FRANKLIN	IN	-13	404	337	97	99
2	1.9	8	8.2	3	3.0	FULTON	IN	339	-63	62	72	44
8	4.7	9	5.4	7	4.2	GIBSON	IN	14	-22	-11	39	64
3	.9	9	2.8	11	3.1	GRANT	IN	195	11	229	93	46
5	3.0	4	2.6	5	3.0	GREENE	IN	-11	13	1	48	43
1	.6	3	1.6	5	2.1	HAMILTON	IN	157	32	239	94	29
6	5.5	4	3.3	1	.7	HANCOCK	IN	-40	-78	-87	00	15
2	2.2	4	4.3	6	5.3	HARRISON	IN	91	23	135	86	79
3	2.1	6	3.6	2	1.0	HENDRICKS	IN	73	-73	-54	11	16
8	3.6	10	4.2	2	.8	HENRY	IN	15	-80	-77	02	15

* SIGNIFICANTLY MORE DEATHS THAN EXPECTED

WHITE MALE — CANCER OF THE ORAL CAVITY INCLUDING TONGUE — ICD 141,143-146,148,149

1950-1959 DEATHS	1950-1959 RATE	1960-1969 DEATHS	1960-1969 RATE	1970-1979 DEATHS	1970-1979 RATE	COUNTY AND STATE		PERCENT CHANGE 60S/50S	PERCENT CHANGE 70S/60S	PERCENT CHANGE 70S/50S	PERCENTILE %CHANGE 70S/50S	PERCENTILE RATES 1970S
3	1.1	9	3.4	10	3.2	HOWARD	IN	198	-5	183	90	48
4	2.7	6	3.5	2	1.1	HUNTINGTON	IN	33	-68	-57	09	17
3	2.1	7	4.6	11	7.0	JACKSON	IN	117	52	230	93	91
1	1.1	1	1.0	3	2.9	JASPER	IN	-7	186	165	89	42
2	1.7	2	1.7	2	1.7	JAY	IN	-3	3	0	47	23
		4	3.2	6	4.5	JEFFERSON	IN		43			68
1	1.1	1	1.1	9*	9.7	JENNINGS	IN	-2	766	751	99	97
7	4.8	5	3.0	3	1.5	JOHNSON	IN	-38	-48	-68	05	21
11	4.8	8	3.4	6	2.5	KNOX	IN	-29	-26	-47	14	35
9	4.7	7	3.5	5	2.0	KOSCIUSKO	IN	-25	-42	-57	10	28
3	3.9	2	2.4	2	2.2	LAGRANGE	IN	-40	-8	-44	16	30
81	6.4	71	4.4	71	3.9	LAKE	IN	-31	-11	-38	20	59
15	4.0	29	6.6	24	5.2	LA PORTE	IN	65	-22	28	62	77
7	4.1	10	5.7	4	2.1	LAWRENCE	IN	40	-63	-49	14	29
16	3.7	20	4.0	21	3.6	MADISON	IN	8	-9	-2	46	54
119	5.4	138	5.5	148*	5.7	MARION	IN	2	3	5	51	82
7	4.6	1	.6	6	3.3	MARSHALL	IN	-86	426	-28	26	48
2	3.7	1	2.1	3	5.6	MARTIN	IN	-43	169	52	70	81
4	2.7	1	.6	3	2.0	MIAMI	IN	-77	213	-28	26	27
7	3.6	8	3.7	13	5.3	MONROE	IN	3	45	49	69	79
5	3.3	3	1.7	5	2.9	MONTGOMERY	IN	-48	69	-11	39	42
6	4.7	6	4.0	7	3.7	MORGAN	IN	-14	-7	-20	31	56
2	3.1	1	1.7	2	2.9	NEWTON	IN	-44	72	-5	44	42
3	2.0			5	3.5	NOBLE	IN			75	76	52
1	4.2	1	5.6			OHIO	IN	35				
1	1.5	5	5.3	5	5.1	ORANGE	IN	250	-3	238	94	77
		4	4.7	1	1.1	OWEN	IN		-76			17
2	2.2	7	7.1	3	2.9	PARKE	IN	216	-59	30	62	42
1	1.0	5	5.5	5	5.3	PERRY	IN	426	-3	409	98	79
4	5.7	3	3.9	6	8.1	PIKE	IN	-30	105	43	67	95
4	2.5	5	2.0	8	2.5	PORTER	IN	-18	21	-1	47	34
5	5.0	1	1.0	7	6.5	POSEY	IN	-79	517	30	62	88
1	1.3	1	1.8	3	3.9	PULASKI	IN	41	119	209	91	58
4	3.1	5	3.8	4	2.9	PUTNAM	IN	23	-25	-8	41	41
1	.6	1	.7	1	.8	RANDOLPH	IN	19	6	26	61	15
7	6.4	2	1.8	2	1.5	RIPLEY	IN	-71	-19	-77	02	21
1	1.2	4	3.7	3	2.8	RUSH	IN	220	-26	138	86	39
45	5.3	53	5.3	53	4.9	ST JOSEPH	IN	1	-8	-7	41	74
2	3.6			2	2.4	SCOTT	IN			-33	23	34
8	5.5	5	3.1	3	1.7	SHELBY	IN	-43	-46	-69	05	23
3	3.4	2	2.4	3	3.0	SPENCER	IN	-29	24	-12	38	44
3	3.5	2	1.7	3	2.5	STARKE	IN	-51	48	-28	26	35
1	1.0	2	2.2	7	7.0	STEUBEN	IN	121	226	622	99	91
4	2.8	4	2.5	3	2.3	SULLIVAN	IN	-11	-10	-19	32	31
		1	2.1	3	6.4	SWITZERLAND	IN		203			88
10	3.4	11	3.0	17	4.3	TIPPECANOE	IN	-11	44	28	61	65
		1	1.2	3	3.7	TIPTON	IN		215			55
1	2.8	1	3.7	1	2.9	UNION	IN	30	-21	3	49	42
62*	9.4	44	6.0	56*	7.3	VANDERBURGH	IN	-36	21	-22	29	92
8	6.6	2	2.0	6	5.8	VERMILLION	IN	-69	185	-12	38	84
19	3.6	22	4.1	27	5.3	VIGO	IN	12	29	45	68	79
2	1.4	3	2.0	2	1.2	WABASH	IN	40	-40	-17	34	18
		7	5.8	3	2.1	WARRICK	IN		-63			29
3	3.0	2	1.7			WASHINGTON	IN	-42				
16	5.1	11	3.3	11	3.3	WAYNE	IN	-35	-2	-36	21	48
5	4.8	4	3.6	3	2.6	WELLS	IN	-25	-27	-45	16	37
5	5.0			8	6.5	WHITE	IN			31	63	88
3	2.9	2	1.9	1	.8	WHITLEY	IN	-36	-55	-71	04	15
427	3.1	475	3.4	486	3.4	IOWA	IA	7	1	8		
1	2.0	1	1.7	3	4.1	ADAIR	IA	-15	147	111	83	63
2	3.6	2	4.0	1	3.1	ADAMS	IA	9	-21	-13	37	46
2	2.4	4	5.3	7	8.2	ALLAMAKEE	IA	122	54	241	94	95
9	7.0	4	3.1	2	1.9	APPANOOSE	IA	-55	-39	-73	03	26
3	5.1	6	9.0	1	1.6	AUDUBON	IA	76	-82	-69	05	22
9	7.2	5	4.1	5	4.2	BENTON	IA	-43	2	-42	17	63
16	3.8	15	3.0	17	3.1	BLACK HAWK	IA	-20	2	-19	32	44
2	1.2	6	3.8	6	4.1	BOONE	IA	222	6	242	94	62
1	1.0	2	2.0	3	2.8	BREMER	IA	91	41	170	89	39
3	2.7			7	6.4	BUCHANAN	IA			139	86	88
4	3.6	3	2.7	3	2.6	BUENA VISTA	IA	-24	-7	-30	24	36
		3	3.0	4	3.7	BUTLER	IA		21			55
3	3.1	2	2.5	3	4.0	CALHOUN	IA	-21	61	28	61	60
9	7.9	6	5.3	4	3.7	CARROLL	IA	-33	-30	-53	12	56
5	5.2	4	3.5	3	3.0	CASS	IA	-32	-15	-43	17	43
1	1.0	1	1.0	2	1.8	CEDAR	IA	-5	90	81	77	25
7	2.9	2	.8	10	3.9	CERRO GORDO	IA	-72	386	35	64	59
2	1.7	3	3.2	4	4.1	CHEROKEE	IA	84	29	137	86	62
1	1.3	4	4.7	4	5.2	CHICKASAW	IA	263	11	302	96	78
2	3.5					CLARKE	IA					
2	1.9	2	2.1	7	7.2	CLAY	IA	11	243	281	95	92
4	3.3	4	2.7	1	.7	CLAYTON	IA	-17	-75	-79	01	14
10	3.3	9	3.1	10	3.5	CLINTON	IA	-6	13	6	51	53
1	1.0	5	5.5	1	1.1	CRAWFORD	IA	474	-81	11	54	17
5	3.4	4	3.0	1	.7	DALLAS	IA	-10	-77	-79	01	15
3	4.5					DAVIS	IA					
1	1.4	2	3.2	2	3.3	DECATUR	IA	130	2	136	86	48
3	3.2	2	1.9	7	7.6	DELAWARE	IA	-40	289	133	85	93
13	6.2	3	1.3	12	5.0	DES MOINES	IA	-79	290	-20	32	75
3	4.0	2	2.7	4	4.1	DICKINSON	IA	-33	54	3	49	62
23*	7.0	26*	7.7	15	4.0	DUBUQUE	IA	10	-48	-43	17	60
		1	1.5	1	1.3	EMMET	IA		-9			19
3	1.9	6	4.0	3	2.0	FAYETTE	IA	112	-51	4	50	27
3	2.6	4	3.4	3	2.6	FLOYD	IA	30	-22	2	48	38
1	1.3	1	.9	1	1.2	FRANKLIN	IA	-31	37	-5	44	18
4	5.8	4	6.0	1	1.6	FREMONT	IA	2	-73	-72	04	22
3	3.5	1	1.1	2	2.4	GREENE	IA	-70	132	-30	24	34
1	1.4	3	3.8	1	1.3	GRUNDY	IA	181	-66	-4	44	19
2	2.2	1	1.1			GUTHRIE	IA	-50				
2	2.1	3	3.4	3	2.9	HAMILTON	IA	63	-14	40	66	42

* SIGNIFICANTLY MORE DEATHS THAN EXPECTED

WHITE MALE — CANCER OF THE ORAL CAVITY INCLUDING TONGUE — ICD 141,143-146,148,149

1950-1959 DEATHS	1950-1959 RATE	1960-1969 DEATHS	1960-1969 RATE	1970-1979 DEATHS	1970-1979 RATE	COUNTY AND STATE	PERCENT CHANGE 60S/50S	PERCENT CHANGE 70S/60S	PERCENT CHANGE 70S/50S	PERCENTILE %CHANGE 70S/50S	PERCENTILE RATES 1970S
1	1.3	1	1.1	3	3.5	HANCOCK IA	-15	223	175	90	53
5	4.2	2	1.7	4	3.3	HARDIN IA	-59	88	-22	29	48
3	2.8	1	1.1	3	2.7	HARRISON IA	-61	146	-3	45	39
4	3.5	4	3.6	2	1.8	HENRY IA	2	-50	-49	13	24
1	1.3	6	7.5	3	4.3	HOWARD IA	495	-42	246	94	66
1	1.2	5	7.0			HUMBOLDT IA	467				
		1	1.7	3	4.3	IDA IA		159			65
3	3.6	6	6.6	5	5.6	IOWA IA	84	-15	56	71	82
6	5.6	5	4.8	3	2.4	JACKSON IA	-15	-51	-58	09	32
		2	1.1	13	6.9	JASPER IA		528			91
7	7.3	2	2.1	3	3.7	JEFFERSON IA	-71	79	-49	14	56
5	2.6	5	2.6	6	2.9	JOHNSON IA	1	11	12	55	41
2	1.9	4	3.9	2	1.9	JONES IA	110	-53	-1	46	26
		3	2.7	1	1.2	KEOKUK IA		-56			18
2	1.6	2	1.8	2	1.6	KOSSUTH IA	15	-13	0	47	21
8	3.6	6	2.7	8	3.8	LEE IA	-27	42	4	50	57
15	2.8	29	4.7	32	4.9	LINN IA	70	3	76	76	74
1	1.5			1	1.5	LOUISA IA			0	47	21
		1	1.2	1	1.5	LUCAS IA		27			21
1	1.4	4	5.6			LYON IA	303				
		1	1.3	3	3.5	MADISON IA		172			52
6	4.3	5	3.5	3	2.5	MAHASKA IA	-19	-27	-41	18	35
2	1.1	3	1.6	7	3.9	MARION IA	54	140	268	95	59
10	5.7	11	5.4	11	4.9	MARSHALL IA	-6	-8	-14	37	74
2	2.9	2	2.9			MILLS IA	-1				
1	1.3	3	3.8	1	1.2	MITCHELL IA	199	-67	-1	46	18
1	1.3	1	1.0	1	1.0	MONONA IA	-19	-3	-21	31	16
1	1.5	2	3.2	2	3.5	MONROE IA	122	9	141	87	53
1	1.0	2	2.1	2	2.0	MONTGOMERY IA	106	-5	95	80	27
3	1.7	6	3.2	7	3.7	MUSCATINE IA	95	16	125	84	56
1	.8	5	5.1	3	2.6	O'BRIEN IA	520	-49	215	92	37
2	4.0	1	1.8			OSCEOLA IA	-56				
1	.6	2	1.5	4	3.0	PAGE IA	151	105	414	98	44
1	1.5	5	6.6	2	2.8	PALO ALTO IA	343	-58	86	78	39
1	.7	4	3.2	2	1.5	PLYMOUTH IA	329	-54	98	81	21
2	2.6	4	4.8	2	3.0	POCAHONTAS IA	89	-38	17	57	44
46	4.8	44	4.0	44	3.7	POLK IA	-15	-8	-22	30	56
14	4.1	13	3.4	13	3.4	POTTAWATTAMIE IA	-16	-2	-17	34	50
3	3.0	5	5.0	1	1.0	POWESHIEK IA	66	-79	-66	06	16
				2	4.4	RINGGOLD IA					66
3	3.5	5	5.0	4	4.3	SAC IA	43	-13	24	60	65
26	5.3	19	3.5	26	4.5	SCOTT IA	-35	31	-15	36	69
3	3.6	3	3.6	5	6.0	SHELBY IA	-2	67	65	73	85
3	2.5	7	5.2	1	.5	SIOUX IA	104	-90	-80	01	14
1	.5	8	3.8	7	3.2	STORY IA	596	-15	492	98	48
6	4.8	3	2.4	6	5.2	TAMA IA	-49	114	10	54	78
1	1.0	3	4.0	4	7.0	TAYLOR IA	302	74	602	99	91
2	1.8	4	3.7	4	5.2	UNION IA	106	42	193	91	78
		2	2.3	6	9.3	VAN BUREN IA		300			97
7	2.8	12	4.9	8	3.1	WAPELLO IA	74	-38	7	52	44
4	4.0	2	1.9	3	2.4	WARREN IA	-53	28	-40	19	34
1	.8	4	3.6	6	4.9	WASHINGTON IA	330	37	488	98	74
1	1.1	2	3.0	1	1.4	WAYNE IA	175	-54	26	61	19
7	3.0	4	1.9	8	3.3	WEBSTER IA	-38	75	9	53	49
1	1.3	2	2.3			WINNEBAGO IA	79				
5	4.3	2	1.6	7	6.4	WINNESHIEK IA	-62	298	50	69	88
20	3.9	26	4.9	15	2.9	WOODBURY IA	28	-41	-25	28	42
1	1.7	1	1.4	2	2.9	WORTH IA	-16	100	68	74	41
3	3.0	2	2.0	4	4.2	WRIGHT IA	-34	110	39	66	64
296	3.1	354	3.4	395	3.6	KANSAS KS	10	6	16		
6	5.1	2	1.9	4	3.7	ALLEN KS	-63	95	-29	25	55
1	1.4	2	2.8	1	1.9	ANDERSON KS	103	-32	37	65	26
6	5.5	6	5.5	8	8.7	ATCHISON KS	1	57	58	72	96
		1	2.5	2	3.7	BARBER KS		51			55
3	2.7	5	3.6	12	7.6	BARTON KS	37	108	185	91	93
4	3.2	4	3.2	4	3.6	BOURBON KS	-1	13	12	55	53
3	2.9			3	3.2	BROWN KS			12	55	48
2	1.2	4	2.1	7	3.2	BUTLER KS	77	56	176	90	48
1	3.4					CHASE KS					
1	2.3	1	1.7	1	1.9	CHAUTAUQUA KS	-28	12	-19	32	26
5	3.7	5	3.8	5	4.1	CHEROKEE KS	3	8	11	54	62
1	5.4			2	6.0	CHEYENNE KS			12	55	85
2	3.0			1	1.4	CLAY KS			-54	11	20
5	4.7	1	1.5	1	1.3	CLOUD KS	-69	-12	-73	03	19
		4	5.5			COFFEY KS					
1	5.4			1	5.6	COMANCHE KS			5	51	82
4	1.8	7	3.5	9	4.4	COWLEY KS	93	26	142	87	66
10	4.0	20*	7.9	11	4.8	CRAWFORD KS	94	-39	19	58	73
				1	4.8	DECATUR KS					72
3	2.2	2	1.5	2	1.3	DICKINSON KS	-34	-10	-40	18	19
6	10.1			1	2.1	DONIPHAN KS			-79	01	29
4	2.6	5	3.0	7	3.8	DOUGLAS KS	16	28	48	69	58
1	3.0	4*	12.5	3	9.5	EDWARDS KS	316	-24	217	92	97
1	1.7	1	1.2	2	7.0	ELK KS	-28	459	304	96	91
4	8.1	3	4.1	7	7.2	ELLIS KS	-49	75	-11	39	92
		2	3.9	1	2.3	ELLSWORTH KS		-42			31
1	1.7	2	2.9	5	6.3	FINNEY KS	72	116	272	95	87
4	3.8	1	1.0	6	5.7	FORD KS	-72	448	52	70	83
3	2.2	3	2.2	1	1.0	FRANKLIN KS	-3	-54	-55	10	16
3	5.3	2	2.3	3	4.2	GEARY KS	-56	80	-21	30	64
				1	4.0	GOVE KS					61
		1	3.9			GRAHAM KS					
				3	10.2	GRAY KS					98
				1	9.2	GREELEY KS					97
1	1.1	6	7.3	2	2.4	GREENWOOD KS	566	-67	117	83	33
		1	5.6	2	12.3	HAMILTON KS		118			99
		2	3.0	1	1.4	HARPER KS		-52			20
		5	3.7			HARVEY KS					
2*	29.9					HASKELL KS					

* SIGNIFICANTLY MORE DEATHS THAN EXPECTED

WHITE MALE

CANCER OF THE ORAL CAVITY INCLUDING TONGUE — ICD 141,143-146,148,149

1950-1959 DEATHS	1950-1959 RATE	1960-1969 DEATHS	1960-1969 RATE	1970-1979 DEATHS	1970-1979 RATE	COUNTY AND STATE		PERCENT CHANGE 60S/50S	PERCENT CHANGE 70S/60S	PERCENT CHANGE 70S/50S	PERCENTILE %CHANGE 70S/50S	PERCENTILE RATES 1970S
1	7.3	3*	20.5	2	10.7	HODGEMAN	KS	182	-48	47	68	98
1	1.1	3	4.0	2	2.5	JACKSON	KS	275	-39	130	85	34
4	5.9	2	3.0			JEFFERSON	KS	-49				
		3	4.2			JEWELL	KS					
12	4.0	15	3.2	22	2.8	JOHNSON	KS	-20	-13	-30	24	40
1	9.4	3*	20.0	1	5.8	KEARNY	KS	114	-71	-38	20	84
1	1.8	1	1.5	3	5.0	KINGMAN	KS	-19	244	180	90	75
1	4.0	3	11.8	2	8.2	KIOWA	KS	192	-31	102	82	95
7	4.1	9	5.7	5	3.2	LABETTE	KS	38	-45	-24	28	47
		1	6.1	2	12.6	LANE	KS		107			99
13	6.2	11	4.8	11	5.1	LEAVENWORTH	KS	-23	7	-18	33	77
1	1.9	1	1.7	3	7.4	LINCOLN	KS	-9	330	290	96	92
2	2.3			1	1.3	LINN	KS			-41	18	19
5	3.1	4	2.8	4	2.7	LYON	KS	-8	-5	-13	37	38
1	.8	2	1.5	4	2.9	MC PHERSON	KS	91	89	262	95	42
2	2.0			1	.8	MARION	KS			-59	08	15
3	2.2	3	3.0	6	5.8	MARSHALL	KS	34	93	159	88	83
		1	3.5	1	3.3	MEADE	KS		-6			48
4	3.1	6	4.8	1	.9	MIAMI	KS	53	-81	-71	04	15
1	1.4	4	5.6	2	4.3	MITCHELL	KS	314	-23	217	92	65
11	4.5	8	3.5	10	4.2	MONTGOMERY	KS	-21	17	-7	41	63
4	8.4			2	3.4	MORRIS	KS			-60	08	50
3	3.7	3	3.1	3	3.8	NEMAHA	KS	-15	22	3	49	57
1	1.0	2	1.8	4	3.5	NEOSHO	KS	86	95	262	95	53
1	3.3			2	4.7	NESS	KS			44	68	71
		2	3.4	2	4.6	NORTON	KS		36			70
		1	.9	2	1.9	OSAGE	KS		109			26
4	7.9	5	10.4			OSBORNE	KS	31				
1	1.8	1	1.6	2	4.0	OTTAWA	KS	-10	148	124	84	60
		2	3.3			PAWNEE	KS					
				4	10.0	PHILLIPS	KS					97
2	2.6	2	3.1	3	3.7	POTTAWATOMIE	KS	18	20	42	67	55
3	4.8	1	1.7	1	1.8	PRATT	KS	-64	3	-63	07	24
9	3.7	5	1.7	12	3.7	RENO	KS	-55	120	0	47	56
6	8.3	1	1.1			REPUBLIC	KS	-87				
2	2.7	2	2.1	5	6.8	RICE	KS	-25	228	146	87	90
5	4.6	3	2.4	6	4.8	RILEY	KS	-48	100	4	50	72
1	2.0	1	2.0			ROOKS	KS	-3				
				2	5.2	RUSH	KS					78
3	5.2	1	1.7	1	2.3	RUSSELL	KS	-67	34	-56	10	31
1	.5	9	4.8	8	4.0	SALINE	KS	771	-17	626	99	60
				1	2.9	SCOTT	KS					42
33	4.2	49	4.5	44	3.3	SEDGWICK	KS	8	-25	-19	32	49
		2	2.8	1	1.5	SEWARD	KS		-47			21
19	3.7	20	3.4	19	3.0	SHAWNEE	KS	-7	-12	-18	33	44
				1	3.4	SHERIDAN	KS					51
		1	3.7			SHERMAN	KS					
		2	3.3	2	5.8	SMITH	KS		77			84
		3	6.5	3	7.0	STAFFORD	KS		8			91
2	20.9					STEVENS	KS					
3	1.9	4	3.1	6	4.2	SUMNER	KS	65	36	125	84	63
1	3.3	2	4.8	1	2.1	THOMAS	KS	48	-57	-36	21	29
1	4.5	1	6.2	2	8.5	TREGO	KS	39	36	89	79	95
		2	3.7			WABAUNSEE	KS					
1	1.4	1	1.1	1	1.1	WASHINGTON	KS	-21	-6	-26	27	16
				1	7.0	WICHITA	KS					91
1	1.0	2	2.1	3	3.6	WILSON	KS	107	72	256	94	53
2	4.1	1	2.8	4	9.4	WOODSON	KS	-32	234	129	85	97
28	4.7	38	6.3	43*	6.8	WYANDOTTE	KS	35	7	44	68	90
474	4.0	655	5.0	695	4.8	KENTUCKY	KY	23	-4	18		
1	1.4	4	5.6	3	3.7	ADAIR	KY	311	-34	172	90	55
7	9.3	2	2.5	2	2.5	ALLEN	KY	-73	-1	-73	03	35
1	2.2	1	2.1	3	5.9	ANDERSON	KY	-6	180	163	89	84
		3	6.1	2	3.2	BALLARD	KY		-48			47
2	1.4	5	3.5	2	1.2	BARREN	KY	141	-66	-17	34	18
		2	4.3			BATH	KY					
5	3.5	4	2.7	5	3.4	BELL	KY	-25	29	-3	45	50
2	3.4	6	4.8	12*	9.2	BOONE	KY	44	89	172	89	96
4	5.8	6	7.7	3	3.7	BOURBON	KY	33	-52	-36	21	56
5	2.5	8	3.4	9	3.8	BOYD	KY	36	12	52	70	57
2	2.2	1	1.0	5	5.1	BOYLE	KY	-54	412	136	86	77
1	2.2	2	4.4	1	2.6	BRACKEN	KY	102	-40	20	59	37
3	4.0	2	3.1	3	4.2	BREATHITT	KY	-22	36	6	52	64
2	2.7	1	1.4			BRECKINRIDGE	KY	-47				
2	4.9	3	4.5	2	2.1	BULLITT	KY	-9	-53	-57	09	29
				3	5.0	BUTLER	KY					75
2	2.8	2	2.7	4	5.1	CALDWELL	KY	-5	91	81	77	76
6	5.3	1	.7	3	2.0	CALLOWAY	KY	-86	172	-62	07	28
29*	7.9	47*	11.8	29*	7.4	CAMPBELL	KY	50	-38	-7	42	92
		3	8.1			CARLISLE	KY					
1	2.1	2	4.1			CARROLL	KY	96				
1	1.2	7	7.3	2	2.4	CARTER	KY	524	-68	101	81	33
				2	2.5	CASEY	KY					35
5	3.2	6	3.8	10	5.7	CHRISTIAN	KY	18	52	80	77	83
3	3.7	4	4.5	2	1.8	CLARK	KY	19	-59	-52	12	25
1	1.5			3	3.4	CLAY	KY			129	85	50
				1	1.8	CLINTON	KY					24
		1	1.6	1	1.7	CRITTENDEN	KY		11			24
4	8.8					CUMBERLAND	KY					
9	3.9	24*	8.6	16	5.1	DAVIESS	KY	123	-41	31	63	76
1	2.5	2	4.6	3	6.2	EDMONSON	KY	86	34	150	87	87
1	4.1					ELLIOTT	KY					
2	3.1	3	4.8	1	1.8	ESTILL	KY	56	-63	-43	17	24
10	2.5	22	4.6	39*	6.7	FAYETTE	KY	88	45	172	90	90
		1	1.9	1	2.0	FLEMING	KY		6			28
1	.8	2	1.3	3	1.6	FLOYD	KY	69	27	114	83	22
7	6.2	6	4.8	3	1.8	FRANKLIN	KY	-23	-62	-70	04	25
1	1.6	3	5.5	3	6.0	FULTON	KY	242	9	271	95	85
		1	5.4	2	8.8	GALLATIN	KY		62			96

* SIGNIFICANTLY MORE DEATHS THAN EXPECTED

WHITE MALE — CANCER OF THE ORAL CAVITY INCLUDING TONGUE — ICD 141,143-146,148,149

1950-1959 DEATHS	1950-1959 RATE	1960-1969 DEATHS	1960-1969 RATE	1970-1979 DEATHS	1970-1979 RATE	COUNTY AND STATE		PERCENT CHANGE 60S/50S	PERCENT CHANGE 70S/60S	PERCENT CHANGE 70S/50S	PERCENTILE %CHANGE 70S/50S	PERCENTILE RATES 1970S
1	2.4	1	2.2	2	4.3	GARRARD	KY	-11	99	78	77	65
3	5.2	1	1.7	1	1.7	GRANT	KY	-67	-3	-68	05	23
10	6.3	3	1.6	9	4.6	GRAVES	KY	-75	192	-27	26	69
4	5.2	3	3.4	3	2.6	GRAYSON	KY	-35	-23	-50	13	37
1	2.2			3	4.8	GREEN	KY			117	83	73
3	2.8	3	2.3	7	4.7	GREENUP	KY	-18	104	67	74	71
		1	2.5			HANCOCK	KY					
6	5.0	8	4.8	12	5.8	HARDIN	KY	-5	22	16	57	84
3	2.6	8	4.7	10	5.4	HARLAN	KY	78	14	104	82	80
3	4.8	5	6.8	3	3.5	HARRISON	KY	43	-49	-27	26	51
		1	1.6	2	2.4	HART	KY		44			33
5	4.7	6	4.1	3	1.8	HENDERSON	KY	-13	-55	-61	08	25
2	2.8	1	1.5			HENRY	KY	-48				
2	4.9			1	2.2	HICKMAN	KY			-55	11	30
7	3.8	10	5.3	11	5.3	HOPKINS	KY	39	0	38	65	79
				1	1.8	JACKSON	KY					25
163*	8.8	214*	9.4	212*	8.7	JEFFERSON	KY	8	-8	-1	47	96
1	1.7	2	3.2	4	5.4	JESSAMINE	KY	83	69	210	92	80
1	2.0	1	1.1	1	1.3	JOHNSON	KY	-43	14	-34	22	19
44*	8.6	65*	12.1	52*	9.3	KENTON	KY	40	-22	9	53	97
				2	2.9	KNOTT	KY					42
		3	2.4	4	3.4	KNOX	KY		41			51
1	2.1	2	3.3	3	4.6	LARUE	KY	57	37	116	83	69
3	3.1	6	5.0	2	1.3	LAUREL	KY	63	-74	-58	09	19
		1	1.4	2	2.6	LAWRENCE	KY		89			36
		1	2.6	1	2.4	LEE	KY		-8			33
		1	3.0	2	4.0	LESLIE	KY		35			60
4	3.5	2	2.0	6	4.9	LETCHER	KY	-42	141	41	66	74
		3	4.9	2	3.0	LEWIS	KY		-39			43
2	3.0	2	2.2	1	.9	LINCOLN	KY	-26	-59	-70	04	16
1	3.4	3	7.5	4	8.5	LIVINGSTON	KY	121	14	152	87	95
2	1.8	1	.9	2	1.9	LOGAN	KY	-52	112	2	49	26
2	7.0					LYON	KY					
12	6.1	13	5.1	14	5.0	MC CRACKEN	KY	-17	-1	-18	33	76
2	5.5	2	3.7	4	5.4	MC CREARY	KY	-34	48	-1	46	80
1	1.9	1	1.6			MC LEAN	KY	-17				
2	1.6	5	3.7	4	2.3	MADISON	KY	128	-36	46	68	32
2	3.9	1	2.4	1	2.2	MAGOFFIN	KY	-40	-5	-43	16	31
4	6.6	7*	11.3	3	4.5	MARION	KY	72	-60	-32	23	67
1	1.2			6	4.9	MARSHALL	KY			297	96	74
1	1.3	6	7.1	4	4.5	MASON	KY	444	-37	245	94	67
		1	2.2	5	9.2	MEADE	KY		327			97
1	5.0			2	11.0	MENIFEE	KY			121	84	98
1	1.5	4	5.1	1	1.1	MERCER	KY	231	-78	-26	27	17
1	1.9	3	6.5	1	1.7	METCALFE	KY	245	-75	-12	38	23
		1	1.5	3	4.4	MONROE	KY		197			66
3	5.3	2	3.0	3	4.0	MONTGOMERY	KY	-43	32	-25	28	60
				2	3.2	MORGAN	KY					47
6	3.9	1	.6	6	3.9	MUHLENBERG	KY	-84	533	-2	46	58
		3	3.5	7	7.5	NELSON	KY		114			93
1	2.3	2	4.8	1	2.9	NICHOLAS	KY	112	-40	27	61	42
2	1.7	2	1.5	2	1.8	OHIO	KY	-9	18	8	52	25
		3	5.9			OLDHAM	KY					
				2	3.7	OWEN	KY					55
				2	6.0	OWSLEY	KY					85
				1	1.9	PENDLETON	KY					26
3	2.5	4	3.1	7	5.9	PERRY	KY	23	92	136	86	84
6	2.9	4	1.5	10	3.4	PIKE	KY	-47	121	17	57	50
1	3.8	1	3.4	1	2.8	POWELL	KY	-10	-17	-25	27	40
6	3.6	1	.5	5	2.4	PULASKI	KY	-87	398	-33	22	33
		3	4.3	2	2.4	ROCKCASTLE	KY		-45			33
1	2.2	4	7.5	2	3.3	ROWAN	KY	239	-56	51	70	49
1	3.9	1	1.9	4	5.6	RUSSELL	KY	-50	187	43	67	82
2	3.2	4	5.2	3	3.8	SCOTT	KY	63	-27	19	58	57
1	1.1	2	2.5	1	.9	SHELBY	KY	133	-64	-16	35	15
1	1.7	2	3.5	5	8.5	SIMPSON	KY	109	141	403	98	95
		1	3.6			SPENCER	KY					
		1	1.4			TAYLOR	KY					
1	1.7	1	1.7	1	1.6	TODD	KY	0	-8	-8	40	22
				2	3.7	TRIGG	KY					55
		1	3.8			TRIMBLE	KY					
4	6.0	3	4.4	1	1.4	UNION	KY	-25	-67	-76	02	20
3	1.7	8	4.1	9	3.8	WARREN	KY	145	-7	128	85	57
		1	2.0	1	1.8	WASHINGTON	KY		-9			24
2	2.8	1	1.3	2	2.7	WAYNE	KY	-55	115	-3	45	38
1	.9	5	6.0	1	1.0	WEBSTER	KY	530	-83	7	52	16
6	4.8	2	1.6	10	7.0	WHITLEY	KY	-66	333	46	68	91
		3	5.7	3	4.6	WOODFORD	KY		-19			69
426*	6.2	499*	5.5	637*	5.7	LOUISIANA	LA	-12	5	-7		
9	7.3	7	4.3	9	4.7	ACADIA	LA	-41	8	-36	21	70
		2	2.8	2	2.5	ALLEN	LA		-10			35
3	5.7	7	9.3	7	6.5	ASCENSION	LA	64	-31	14	56	88
		4	9.1	2	4.0	ASSUMPTION	LA		-56			60
4	3.1	5	4.1	10	6.1	AVOYELLES	LA	34	49	98	81	86
2	2.6	4	5.0	3	3.6	BEAUREGARD	LA	94	-28	40	66	53
		2	3.5	1	1.7	BIENVILLE	LA		-51			23
5	8.2	3	1.8	9	7.1	BOSSIER	LA	-78	295	-14	37	91
27	6.5	31	5.3	37	5.4	CADDO	LA	-18	1	-17	34	80
11	5.0	20	5.1	35*	7.0	CALCASIEU	LA	2	39	41	67	91
4	11.8	4	10.5	2	5.0	CALDWELL	LA	-12	-52	-58	09	75
		1	2.7			CAMERON	LA					
		3	7.7			CATAHOULA	LA					
2	2.9	1	1.7	5	7.9	CLAIBORNE	LA	-43	377	172	89	94
		1	1.8	2	3.0	CONCORDIA	LA		65			43
3	5.1	3	4.7	1	1.3	DE SOTO	LA	-8	-72	-75	03	19
17	5.4	25	4.9	38	5.6	EAST BATON ROUGE	LA	-10	15	4	50	82
2	9.6	1	4.2	1	5.9	EAST CARROLL	LA	-57	41	-39	19	84
1	1.7	3	4.6	2	3.6	EAST FELICIANA	LA	172	-22	114	83	54
3	4.8	6	4.7	5	3.7	EVANGELINE	LA	-2	-21	-23	29	56

* SIGNIFICANTLY MORE DEATHS THAN EXPECTED

WHITE MALE — CANCER OF THE ORAL CAVITY INCLUDING TONGUE — ICD 141,143-146,148,149

1950-1959 DEATHS	1950-1959 RATE	1960-1969 DEATHS	1960-1969 RATE	1970-1979 DEATHS	1970-1979 RATE	COUNTY AND STATE	PERCENT CHANGE 60S/50S	PERCENT CHANGE 70S/60S	PERCENT CHANGE 70S/50S	PERCENTILE %CHANGE 70S/50S	PERCENTILE RATES 1970S
2	2.5	1	1.4	2	2.5	FRANKLIN LA	-45	84	1	48	35
1	1.8	3	4.4	5	6.5	GRANT LA	150	47	266	95	88
4	3.4	9	6.8	10	6.5	IBERIA LA	100	-5	89	79	88
3	4.7	5	6.8	4	5.9	IBERVILLE LA	44	-13	25	60	84
2	4.8	2	3.7	5	7.8	JACKSON LA	-24	112	62	73	94
12	5.0	35	6.2	56	5.1	JEFFERSON LA	24	-17	3	50	77
2	2.1	3	3.2	4	4.0	JEFFERSON DAVIS LA	51	26	90	79	60
7	5.0	10	3.9	17	5.6	LAFAYETTE LA	-21	44	13	55	82
8	7.3	8	4.8	15	6.8	LAFOURCHE LA	-35	41	-8	41	90
		1	1.5	5	6.4	LA SALLE LA		321			88
2	3.6			4	4.7	LINCOLN LA			29	62	71
2	2.4	4	3.4	2	1.2	LIVINGSTON LA	44	-65	-49	14	18
1	2.7	1	4.1			MADISON LA	50				
6	10.2	4	5.6	5	5.1	MOREHOUSE LA	-45	-10	-50	13	76
2	2.5	5	4.8	8	6.5	NATCHITOCHES LA	95	35	163	89	89
182*	11.7	175*	10.0	182*	11.9	ORLEANS LA	-15	20	2	49	99
8	3.8	12	4.6	13	3.4	OUACHITA LA	22	-26	-10	40	50
1	3.9	2	2.7	4	5.0	PLAQUEMINES LA	-32	87	27	61	75
		2	4.0	2	3.8	POINTE COUPEE LA		-5			57
19*	7.4	13	4.0	20	5.0	RAPIDES LA	-45	25	-32	23	76
2	7.5	1	4.4			RED RIVER LA	-42				
				4	5.0	RICHLAND LA					75
1	1.5	3	3.2	3	2.9	SABINE LA	112	-10	91	79	42
1	1.0	5	5.5	11	6.5	ST BERNARD LA	464	18	567	99	89
1	2.4	2	2.4	2	2.3	ST CHARLES LA	0	-5	-5	43	32
				1	4.7	ST HELENA LA					71
2	8.0	1	2.5	4	9.4	ST JAMES LA	-69	274	18	58	97
2	7.9	4	14.4	3	5.0	ST JOHN THE BAPT LA	82	-65	-37	20	75
15*	11.6	12	6.6	9	5.0	ST LANDRY LA	-43	-24	-56	10	76
3	6.6	1	1.1	6	5.6	ST MARTIN LA	-83	408	-15	35	82
5	5.9	4	4.5	11	9.2	ST MARY LA	-24	104	55	71	96
1	1.3	10	6.5	10	3.8	ST TAMMANY LA	391	-41	191	91	58
6	4.4	5	3.4	3	1.4	TANGIPAHOA LA	-24	-58	-68	05	20
2	10.8					TENSAS LA					
6	5.4	4	3.2	8	4.5	TERREBONNE LA	-41	41	-17	34	67
3	6.5	1	1.4	1	1.1	UNION LA	-78	-20	-83	01	17
2	1.1	7	4.7	9	4.5	VERMILION LA	325	-5	303	96	67
5	6.0	3	3.4	6	4.8	VERNON LA	-44	44	-19	32	73
4	4.1	5	4.2	5	3.9	WASHINGTON LA	3	-9	-6	43	58
5	4.8	3	2.2	2	1.4	WEBSTER LA	-54	-34	-70	04	20
				4	12.6	WEST BATON ROUGE LA					99
1	2.0			1	1.5	WEST CARROLL LA			-26	27	21
2	3.3					WINN LA					
204	4.5	217	4.6	297*	5.8	MAINE ME	2	27	29		
36*	9.1	23	5.5	24	5.3	ANDROSCOGGIN ME	-40	-3	-42	17	79
5	1.4	20	5.7	13	3.3	AROOSTOOK ME	306	-42	137	86	49
49	5.9	47	5.4	70*	7.4	CUMBERLAND ME	-9	37	24	60	92
6	5.3	4	3.6	2	1.7	FRANKLIN ME	-32	-53	-68	05	23
6	3.2	5	2.6	10	5.2	HANCOCK ME	-19	98	61	72	78
24	5.9	25	5.5	44*	6.9	KENNEBEC ME	-6	62	52	70	96
6	3.3	7	4.1	11	6.1	KNOX ME	27	48	88	79	86
		3	2.8	5	3.6	LINCOLN ME		31			54
6	2.7	11	4.5	10	3.8	OXFORD ME	66	-15	41	67	57
11	2.1	23	4.3	32	5.8	PENOBSCOT ME	101	34	170	89	83
2	1.9	6	6.4	6	6.2	PISCATAQUIS ME	234	-3	223	92	87
3	2.5	6	5.4	4	3.4	SAGADAHOC ME	120	-37	38	65	50
15	7.5	10	4.7	8	3.6	SOMERSET ME	-37	-23	-52	12	54
3	2.6	6	5.2	12*	9.6	WALDO ME	98	86	268	95	97
6	3.0	5	2.6	12	6.6	WASHINGTON ME	-11	150	123	84	89
26	5.6	16	3.0	34	5.8	YORK ME	-46	91	4	50	83
499*	6.0	643*	5.8	834*	6.2	MARYLAND MD	-3	7	4		
24	6.8	20	4.9	19	4.2	ALLEGANY MD	-28	-14	-38	20	64
29*	8.2	37	6.0	52	5.5	ANNE ARUNDEL MD	-27	-8	-33	23	81
62	6.1	70	3.7	130	5.1	BALTIMORE MD	-39	36	-17	34	76
1	3.0	2	3.2	9*	11.0	CALVERT MD	5	248	265	95	98
		2	2.3	9*	9.1	CAROLINE MD		294			96
10	4.4	13	4.6	13	3.7	CARROLL MD	4	-20	-17	34	55
6	4.2	5	2.6	7	3.3	CECIL MD	-38	26	-21	30	49
7*	9.2	5	5.0	10	9.5	CHARLES MD	-46	90	3	49	97
5	3.9	5	3.9	7	5.0	DORCHESTER MD	-1	28	26	61	75
13	4.7	13	4.1	22	5.9	FREDERICK MD	-12	45	28	61	85
3	3.2	1	.9	2	1.6	GARRETT MD	-72	85	-49	14	22
4	2.6	7	2.8	12	2.8	HARFORD MD	6	2	9	53	40
6	9.7	11	7.9	7	2.7	HOWARD MD	-18	-66	-72	04	38
5	9.2	1	1.8	2	2.5	KENT MD	-80	40	-73	03	35
33	4.3	66	5.2	104	5.7	MONTGOMERY MD	22	9	33	63	82
44*	6.8	87*	7.8	116*	7.8	PRINCE GEORGE'S MD	15	0	15	56	94
1	1.6	1	1.1	4	4.1	QUEEN ANNE'S MD	-29	264	160	88	62
2	3.1	8	9.3	7	6.5	ST MARY'S MD	197	-30	107	82	88
2	2.5	3	4.0	2	2.5	SOMERSET MD	62	-36	3	50	35
3	3.5	8	8.1	7	5.1	TALBOT MD	132	-38	45	68	76
7	2.0	16	3.7	22	4.4	WASHINGTON MD	88	19	123	84	66
2	1.2	10	4.9	6	2.7	WICOMICO MD	303	-46	119	84	38
2	2.5	3	3.0	9	7.5	WORCESTER MD	22	147	202	91	93
228*	7.6	249*	8.5	256*	10.1	BALTIMORE CITY MD	11	19	33	63	97
1650*	7.5	1663*	6.8	1762*	6.8	MASSACHUSETTS MA	-9	0	-9		
8	2.8	21	5.0	33	4.3	BARNSTABLE MA	81	-13	58	72	65
41	6.5	40	5.7	45	5.9	BERKSHIRE MA	-12	4	-9	40	84
142*	8.4	96	4.8	112	5.0	BRISTOL MA	-43	6	-40	19	76
		3	9.2	3	8.6	DUKES MA		-7			95
197*	7.5	225*	7.7	203*	6.5	ESSEX MA	2	-15	-13	37	89
9	3.4	28*	9.5	13	4.1	FRANKLIN MA	181	-57	21	59	62
112*	6.6	117*	5.9	167*	8.0	HAMPDEN MA	-10	34	21	59	94
16	4.3	23	4.9	25	4.5	HAMPSHIRE MA	14	-8	5	51	68
358*	7.4	364*	6.7	409*	7.1	MIDDLESEX MA	-10	6	-5	44	91

* SIGNIFICANTLY MORE DEATHS THAN EXPECTED

WHITE MALE — CANCER OF THE ORAL CAVITY INCLUDING TONGUE — ICD 141,143-146,148,149

1950-1959 DEATHS	1950-1959 RATE	1960-1969 DEATHS	1960-1969 RATE	1970-1979 DEATHS	1970-1979 RATE	COUNTY AND STATE	PERCENT CHANGE 60S/50S	PERCENT CHANGE 70S/60S	PERCENT CHANGE 70S/50S	PERCENTILE %CHANGE 70S/50S	PERCENTILE RATES 1970S
1	5.2	4*	28.4	4*	17.1	NANTUCKET MA	443	-40	226	93	99
110	5.7	115	4.9	173*	6.4	NORFOLK MA	-15	32	12	55	88
60	5.7	84*	6.7	93*	6.3	PLYMOUTH MA	18	-5	12	55	87
441*	11.5	363*	10.3	308*	10.8	SUFFOLK MA	-10	5	-5	43	98
155*	5.8	180*	6.1	174*	5.6	WORCESTER MA	5	-8	-3	45	82
1348	5.0	1517	4.8	1608	4.7	MICHIGAN MI	-4	-3	-7		
				1	1.4	ALCONA MI					20
2	3.7	1	2.0	3	6.1	ALGER MI	-45	201	65	74	86
7	2.9	2	.7	8	2.4	ALLEGAN MI	-75	246	-15	36	34
4	3.6	3	2.4	9	6.1	ALPENA MI	-32	148	68	74	86
3	3.9	3	3.8	3	3.1	ANTRIM MI	-4	-17	-20	31	46
		1	2.1	2	3.0	ARENAC MI		43			44
2	6.1			6*	11.7	BARAGA MI			94	80	98
2	1.4	5	3.0	5	2.6	BARRY MI	114	-14	83	78	37
24	6.5	20	4.5	27	5.4	BAY MI	-31	20	-17	34	80
2	4.6			4	5.9	BENZIE MI			29	62	84
29	5.5	28	4.5	33	4.7	BERRIEN MI	-19	6	-14	36	71
8	5.3	6	3.5	6	3.2	BRANCH MI	-33	-10	-39	19	47
18	3.7	30	5.3	23	3.8	CALHOUN MI	42	-28	2	49	57
3	1.9	5	3.2	8	4.3	CASS MI	69	34	126	84	65
4	5.7			2	2.0	CHARLEVOIX MI			-65	06	27
2	2.8	1	1.3	6	5.8	CHEBOYGAN MI	-54	350	108	82	83
5	4.5	9	7.0	9	6.5	CHIPPEWA MI	54	-7	43	67	88
3	5.1	1	1.6	2	1.5	CLARE MI	-69	-8	-72	04	20
6	4.2	3	1.8	8	4.4	CLINTON MI	-57	145	4	50	67
2	11.3	1	2.5	1	1.5	CRAWFORD MI	-78	-38	-86	00	21
8	4.8	7	3.8	12	6.4	DELTA MI	-22	72	35	64	88
6	3.9	5	2.7	7	4.7	DICKINSON MI	-30	70	19	58	70
7	3.3	8	3.6	12	4.5	EATON MI	7	27	35	64	68
4	4.8	4	4.9			EMMET MI	2				
50	5.4	47	3.7	48	3.3	GENESEE MI	-32	-10	-39	20	49
1	2.2	1	1.1	2	2.6	GLADWIN MI	-49	135	20	59	37
7	6.4	12	8.3	5	3.6	GOGEBIC MI	30	-57	-44	16	54
5	3.0	5	2.8	4	1.9	GRAND TRAVERSE MI	-7	-33	-38	20	26
6	3.4	4	2.2	4	2.3	GRATIOT MI	-33	4	-30	24	32
4	2.1	11	6.2	7	3.4	HILLSDALE MI	193	-44	64	73	51
12	5.0	6	2.4	7	3.0	HOUGHTON MI	-52	25	-40	19	44
8	4.6	5	2.9	4	2.1	HURON MI	-37	-29	-55	10	28
20	3.2	23	2.9	23	3.1	INGHAM MI	-10	7	-4	44	45
6	2.9	9	4.5	6	3.2	IONIA MI	57	-30	10	53	47
4	5.6	3	2.7	10	8.0	IOSCO MI	-52	198	44	67	95
4	3.7	6	5.5	8	7.6	IRON MI	50	38	107	82	93
7	6.1	4	3.1	4	2.7	ISABELLA MI	-50	-13	-56	10	38
26	5.0	24	4.2	29	4.9	JACKSON MI	-16	18	-1	46	74
20	3.4	31	4.9	39	5.5	KALAMAZOO MI	45	12	62	73	81
		1	2.7	2	3.2	KALKASKA MI		20			47
50	3.7	77	4.9	93	5.5	KENT MI	34	11	49	69	81
		1	5.4			KEWEENAW MI					
2	8.2			2	3.9	LAKE MI			-52	12	59
1	.8	2	1.1	6	2.8	LAPEER MI	51	142	264	95	39
		1	1.5	3	4.6	LEELANAU MI		195			69
10	3.2	7	2.1	17	4.7	LENAWEE MI	-34	125	47	69	71
5	3.5	6	2.9	14	4.9	LIVINGSTON MI	-17	71	42	67	74
2	3.6	2	4.0			LUCE MI	10				
1	1.9	3	5.1	3	5.3	MACKINAC MI	168	5	181	90	79
26	3.2	60	4.0	102	4.6	MACOMB MI	23	14	41	66	69
5	4.3	4	3.5	2	1.5	MANISTEE MI	-20	-56	-64	06	21
11	5.0	5	2.0	8	3.1	MARQUETTE MI	-60	54	-38	20	45
6	5.4	8	6.3	2	1.6	MASON MI	17	-74	-70	04	23
5	4.9	5	4.9			MECOSTA MI	0				
6	4.4	4	2.8	6	4.1	MENOMINEE MI	-35	44	-7	42	61
5	4.4	1	.8	5	2.4	MIDLAND MI	-81	182	-47	15	32
		1	2.9			MISSAUKEE MI					
15	4.5	23	5.7	28	5.6	MONROE MI	26	-2	23	60	82
8	4.5	8	3.9	7	3.1	MONTCALM MI	-13	-21	-32	23	45
		1	3.9			MONTMORENCY MI					
16	3.2	26	4.8	24	3.9	MUSKEGON MI	48	-19	19	59	58
4	3.4	2	1.4	4	2.3	NEWAYGO MI	-59	64	-32	23	32
71	4.9	103	4.1	158	4.5	OAKLAND MI	-15	9	-7	41	68
3	3.2	3	3.1	1	1.2	OCEANA MI	-3	-61	-62	07	18
3	5.4	1	1.1	3	3.1	OGEMAW MI	-79	172	-43	16	45
4	6.9	2	3.9	1	1.2	ONTONAGON MI	-44	-69	-83	00	18
2	2.3	1	1.5	3	3.7	OSCEOLA MI	-32	139	62	73	55
1	5.7			2	5.8	OSCODA MI			1	48	63
5*	13.2	1	2.6	2	3.1	OTSEGO MI	-80	22	-76	02	46
9	2.6	10	2.4	4	.8	OTTAWA MI	-7	-67	-70	04	15
6	9.9	2	3.3	3	4.1	PRESQUE ISLE MI	-66	24	-58	09	62
		1	1.7	3	2.0	ROSCOMMON MI		17			27
30	5.1	20	2.7	34	4.4	SAGINAW MI	-47	63	-13	37	67
18	4.2	17	3.2	29	5.0	ST CLAIR MI	-24	56	18	58	75
		6	2.8	6	2.6	ST JOSEPH MI		-9			36
5	2.8	7	3.5	6	3.0	SANILAC MI	27	-16	7	52	43
1	2.6	2	3.6	1	2.2	SCHOOLCRAFT MI	39	-39	-15	36	30
5	2.3	3	1.4	2	1.0	SHIAWASSEE MI	-42	-28	-58	09	16
7	3.4	7	3.4	9	4.4	TUSCOLA MI	1	27	28	62	66
4	1.7	8	3.2	7	2.5	VAN BUREN MI	91	-23	46	68	34
24	5.5	21	3.7	34	5.3	WASHTENAW MI	-34	44	-4	44	79
639*	7.1	718*	7.4	573*	6.6	WAYNE MI	5	-11	-6	42	89
2	2.3	2	1.8	2	1.8	WEXFORD MI	-22	-1	-24	28	24
453	3.0	544	3.2	653	3.6	MINNESOTA MN	8	12	21		
1	1.0			3	3.5	AITKIN MN			236	94	52
4	2.7	9	3.0	17	4.7	ANOKA MN	13	57	78	76	71
5	3.7	5	3.5	4	2.2	BECKER MN	-3	-38	-40	19	30
		1	.9	1	.7	BELTRAMI MN		-22			14
3	4.1	2	2.1	5	5.3	BENTON MN	-48	154	31	63	79
1	1.8	1	2.6	3	5.9	BIG STONE MN	43	130	230	93	84
4	1.9	3	1.4	11	4.7	BLUE EARTH MN	-24	229	149	87	71

* SIGNIFICANTLY MORE DEATHS THAN EXPECTED

WHITE MALE — CANCER OF THE ORAL CAVITY INCLUDING TONGUE — ICD 141,143-146,148,149

1950-1959 DEATHS	RATE	1960-1969 DEATHS	RATE	1970-1979 DEATHS	RATE	COUNTY AND STATE		PERCENT CHANGE 60S/50S	70S/60S	70S/50S	PERCENTILE %CHANGE 70S/50S	RATES 1970S
8	5.5	1	.8	7	4.7	BROWN	MN	-86	507	-16	35	70
2	1.5	6	4.2	4	2.7	CARLTON	MN	186	-37	80	77	38
2	2.1	1	.8	4	3.1	CARVER	MN	-62	287	48	69	45
3	3.7	2	1.4	2	1.4	CASS	MN	-64	1	-63	07	19
2	2.0			5	4.9	CHIPPEWA	MN			143	87	73
1	.9	3	3.2	5	4.6	CHISAGO	MN	273	42	430	98	69
4	2.9	3	1.9	3	1.6	CLAY	MN	-35	-16	-46	15	21
				1	3.2	CLEARWATER	MN					48
		1	4.2	2	9.0	COOK	MN		114			96
2	2.4	2	2.4	5	4.7	COTTONWOOD	MN	4	91	98	81	70
10	6.2	9	5.2	5	2.3	CROW WING	MN	-15	-55	-62	07	32
5	2.3	9	3.1	17	3.5	DAKOTA	MN	33	13	50	69	52
1	1.4	3	4.1	2	2.6	DODGE	MN	187	-37	81	77	36
1	.6	3	2.0	4	2.6	DOUGLAS	MN	209	27	294	96	36
2	1.6	1	.7	2	1.4	FARIBAULT	MN	-55	98	-11	38	20
4	2.8	4	2.6	2	1.6	FILLMORE	MN	-8	-38	-43	16	22
2	1.2	4	2.0	2	.9	FREEBORN	MN	75	-54	-20	31	16
5	2.6	5	2.4	6	2.5	GOODHUE	MN	-7	4	-4	44	35
		3	5.6	1	2.3	GRANT	MN		-58			32
133	4.0	165	4.5	195	5.1	HENNEPIN	MN	13	14	29	62	77
1	1.2	2	2.4	4	4.4	HOUSTON	MN	99	81	261	94	66
2	3.2	1	1.1	3	3.0	HUBBARD	MN	-66	181	-5	43	43
3	3.8	1	1.0	2	2.2	ISANTI	MN	-74	124	-43	17	29
5	2.8	7	3.5	7	3.2	ITASCA	MN	23	-7	15	56	48
2	2.9	2	2.5	2	2.2	JACKSON	MN	-16	-12	-26	27	30
2	3.1	2	3.7	2	3.1	KANABEC	MN	20	-17	0	47	45
2	1.1	1	.6	6	2.9	KANDIYOHI	MN	-50	410	157	88	41
4	4.2	6	6.7	4	4.4	KOOCHICHING	MN	60	-35	4	50	66
1	1.2	5	5.7	3	3.8	LAC QUI PARLE	MN	379	-32	225	93	57
2	3.8	2	3.7	1	2.1	LAKE	MN	-4	-42	-45	16	29
1	2.8			1	2.8	LAKE OF THE WOODS	MN			-1	46	40
2	1.7	4	3.3	4	2.7	LE SUEUR	MN	90	-18	56	71	38
		1	2.1	1	1.2	LINCOLN	MN		-41			18
1	1.0	3	2.5			LYON	MN	149				
		4	2.8	3	1.8	MC LEOD	MN		-37			24
				2	6.6	MAHNOMEN	MN					89
1	1.2	2	2.5	3	3.7	MARSHALL	MN	115	48	219	92	55
3	2.3	5	3.4	2	1.3	MARTIN	MN	49	-61	-41	17	19
1	.9	4	3.5	3	3.1	MEEKER	MN	275	-12	228	93	45
3	3.5	3	3.0	2	1.8	MILLE LACS	MN	-13	-41	-48	14	24
2	1.4	1	1.0	2	1.2	MORRISON	MN	-32	20	-18	33	18
4	2.3	1	.5	8	3.3	MOWER	MN	-79	601	46	68	49
1	1.4	2	2.5			MURRAY	MN	75				
2	1.7	1	.8	2	1.8	NICOLLET	MN	-51	111	2	49	24
3	3.0	2	1.7	2	1.7	NOBLES	MN	-44	2	-43	17	23
1	1.6			4	5.4	NORMAN	MN			234	93	80
6	2.8	7	2.5	9	2.9	OLMSTED	MN	-11	15	2	49	41
1	.3	7	2.2	7	1.9	OTTER TAIL	MN	533	-15	441	98	26
3	4.1	2	2.5	2	1.8	PENNINGTON	MN	-39	-26	-55	11	25
3	2.8	3	2.9	2	1.8	PINE	MN	3	-38	-36	21	25
1	1.2			2	3.1	PIPESTONE	MN			156	88	46
5	2.7	5	2.5	5	2.5	POLK	MN	-6	0	-6	42	35
2	2.6	1	1.7	2	2.4	POPE	MN	-32	35	-8	40	32
83	5.2	89	4.9	82	4.3	RAMSEY	MN	-5	-12	-16	35	65
1	3.9			1	4.1	RED LAKE	MN			6	52	62
2	1.8	3	2.6	3	2.3	REDWOOD	MN	40	-9	28	61	32
6	5.0	5	3.7	2	1.6	RENVILLE	MN	-26	-57	-68	05	22
3	1.8	7	3.7	6	3.4	RICE	MN	107	-8	91	79	50
		4	6.4	4	6.1	ROCK	MN		-5			86
1	1.4					ROSEAU	MN					
29	2.3	41	3.5	69	6.0	ST LOUIS	MN	52	73	162	88	85
2	2.4	5	4.6	6	4.4	SCOTT	MN	91	-5	82	78	67
2	3.7			1	1.2	SHERBURNE	MN			-68	05	18
9	2.9	11	3.2	8	1.9	STEARNS	MN	9	-40	-34	22	26
2	2.0	4	3.1	3	2.1	STEELE	MN	54	-31	7	52	29
2	4.2			1	1.7	STEVENS	MN			-60	08	23
2	2.7	3	3.6	2	2.0	SWIFT	MN	35	-43	-23	28	28
6	4.4	4	2.8	2	1.7	TODD	MN	-35	-39	-60	08	24
				1	2.6	TRAVERSE	MN					36
3	3.1	2	1.8	4	3.6	WABASHA	MN	-41	98	16	57	54
		1	1.4			WADENA	MN					
1	1.2	2	2.3	4	4.2	WASECA	MN	96	86	264	95	64
6	3.5	8	3.5	16	6.0	WASHINGTON	MN	1	69	70	75	85
2	2.4	1	1.1	5	6.7	WATONWAN	MN	-54	510	179	90	89
3	7.0	3	5.8	1	1.7	WILKIN	MN	-17	-71	-76	02	23
5	2.4	13	6.0	6	2.6	WINONA	MN	148	-57	5	51	36
5	3.2	3	1.7	3	1.2	WRIGHT	MN	-47	-32	-64	06	17
3	3.2	2	2.0	3	3.3	YELLOW MEDICINE	MN	-36	62	4	50	49
170	3.5	227	3.8	268	3.8	MISSISSIPPI	MS	9	-2	7		
2	4.0			1	1.2	ADAMS	MS			-71	04	17
1	1.1	2	1.7	3	2.2	ALCORN	MS	64	29	113	83	30
		2	4.0	1	2.3	AMITE	MS		-44			31
3	4.3	3	3.9	1	1.3	ATTALA	MS	-10	-68	-71	04	18
1	4.0			1	4.1	BENTON	MS			2	48	62
3	7.0	2	3.3	6	7.8	BOLIVAR	MS	-53	140	12	55	94
1	1.5	1	1.3	2	3.0	CALHOUN	MS	-12	123	95	80	43
1	2.8			2	7.8	CARROLL	MS			184	91	94
3	5.5	2	3.7	3	5.9	CHICKASAW	MS	-31	57	8	52	84
1	2.1	3	8.7	1	2.3	CHOCTAW	MS	318	-73	12	55	32
1	12.4	3*	20.4	1	5.2	CLAIBORNE	MS	65	-74	-58	09	78
2	3.5	2	3.8	2	4.1	CLARKE	MS	8	8	17	57	62
1	2.8	1	3.0	1	1.8	CLAY	MS	7	-40	-36	21	25
6*	13.0	2	3.2	4	6.1	COAHOMA	MS	-75	90	-53	11	86
2	2.7	3	3.8	2	3.1	COPIAH	MS	40	-17	16	57	45
1	2.3	2	4.1	1	1.5	COVINGTON	MS	73	-64	-38	20	20
1	2.5	3	7.3	9*	10.5	DE SOTO	MS	194	45	324	97	98
3	2.5	10	6.3	7	3.9	FORREST	MS	145	-38	53	70	59
1	3.0	1	3.2	2	6.7	FRANKLIN	MS	7	108	122	84	89
1	3.1	1	2.0	3	6.0	GEORGE	MS	-36	198	91	79	85

* SIGNIFICANTLY MORE DEATHS THAN EXPECTED

WHITE MALE — CANCER OF THE ORAL CAVITY INCLUDING TONGUE — ICD 141,143-146,148,149

1950-1959 DEATHS	1950-1959 RATE	1960-1969 DEATHS	1960-1969 RATE	1970-1979 DEATHS	1970-1979 RATE	COUNTY AND STATE	PERCENT CHANGE 60S/50S	PERCENT CHANGE 70S/60S	PERCENT CHANGE 70S/50S	PERCENTILE %CHANGE 70S/50S	PERCENTILE RATES 1970S
				1	2.8	GREENE MS					41
2	5.9	1	1.7	3	5.6	GRENADA MS	-71	232	-4	44	82
3	5.8	4	6.1	4	4.5	HANCOCK MS	6	-26	-22	30	68
18	7.7	18	5.1	28	5.8	HARRISON MS	-33	13	-25	27	83
12	4.3	20	4.9	23	4.0	HINDS MS	15	-18	-5	43	61
3	8.6	2	5.0	1	2.0	HOLMES MS	-42	-60	-76	02	28
				1*	18.1	ISSAQUENA MS					99
2	3.4	2	2.5	3	2.9	ITAWAMBA MS	-24	13	-15	36	41
6	4.5	2	1.3	12	4.1	JACKSON MS	-72	224	-10	40	62
1	2.1	2	4.7	1	2.7	JASPER MS	130	-43	32	63	38
		1	6.1			JEFFERSON MS					
1	3.6			1	3.4	JEFFERSON DAVIS MS			-5	43	50
3	2.4	9	4.5	4	1.9	JONES MS	88	-58	-21	30	26
				5*	16.5	KEMPER MS					99
		1	1.7	2	2.4	LAFAYETTE MS		37			33
1	2.2			2	3.2	LAMAR MS			47	68	47
5	2.7	12	5.8	7	2.9	LAUDERDALE MS	113	-49	8	52	42
3	12.5	1	2.8			LAWRENCE MS	-78				
3	5.2	2	3.0	1	1.3	LEAKE MS	-42	-57	-75	03	19
7	6.3	3	2.2	12	6.6	LEE MS	-65	201	5	51	89
2	2.7	2	2.9	6	7.3	LEFLORE MS	7	153	171	89	92
2	3.0	7	8.1	3	2.5	LINCOLN MS	173	-69	-15	35	35
6	6.6			10	7.7	LOWNDES MS			16	57	93
3	6.2	4	7.4	2	2.6	MADISON MS	19	-65	-58	09	37
3	4.9	4	5.1	5	5.1	MARION MS	4	0	4	50	76
		1	2.6			MARSHALL MS					
2	2.0	4	3.7	6	4.8	MONROE MS	81	31	137	86	72
1	2.6	3	6.2			MONTGOMERY MS	140				
1	1.3			4	4.1	NESHOBA MS			216	92	62
3	5.2	1	1.4	3	3.2	NEWTON MS	-73	130	-38	20	47
1	3.5	1	3.6	1	3.7	NOXUBEE MS	4	2	5	51	55
1	2.2	3	6.1	1	1.5	OKTIBBEHA MS	175	-75	-31	23	21
1	1.4	2	3.0	4	4.6	PANOLA MS	112	55	229	93	70
1	1.5	8*	9.5	6	4.8	PEARL RIVER MS	552	-50	226	93	72
		2	5.6	1	2.8	PERRY MS		-50			40
5	4.9	5	4.5	5	4.4	PIKE MS	-9	-2	-10	40	66
1	1.2	3	3.6	1	1.2	PONTOTOC MS	191	-67	-3	45	18
1	1.3	2	2.5	2	2.0	PRENTISS MS	89	-22	47	68	27
		3	8.0			QUITMAN MS					
		9	7.4	6	3.4	RANKIN MS		-54			50
				1	1.1	SCOTT MS					17
		1	18.6			SHARKEY MS					
1	1.6	3	4.1	2	2.5	SIMPSON MS	163	-39	59	72	35
1	1.9	1	1.5	2	3.2	SMITH MS	-20	111	68	74	46
		3	9.6	1	3.4	STONE MS		-64			50
5	7.8	3	3.9	2	2.4	SUNFLOWER MS	-50	-38	-69	05	33
1	2.4	2	4.4	2	4.6	TALLAHATCHIE MS	83	4	91	79	69
		1	2.7			TATE MS					
		4	5.6	2	2.7	TIPPAH MS		-52			38
2	3.5	1	1.4	2	1.9	TISHOMINGO MS	-59	34	-46	15	26
1	6.5			1	6.5	TUNICA MS			0	47	89
3	5.4	1	1.2	2	2.4	UNION MS	-79	109	-55	10	33
				2	4.9	WALTHALL MS					73
2	2.7	7	6.4	2	1.6	WARREN MS	138	-75	-40	19	22
		5	3.4	6	4.2	WASHINGTON MS		22			63
		1	1.7	4	7.3	WAYNE MS		328			92
3	6.7			3	4.9	WEBSTER MS			-27	26	74
3*	17.2	2	9.2			WILKINSON MS	-46				
4	6.5	4	6.4			WINSTON MS	-2				
1	1.9			2	3.7	YALOBUSHA MS			91	79	55
4	6.8	1	1.5	2	2.6	YAZOO MS	-77	69	-62	07	37
1057*	5.5	994	4.8	933	4.2	MISSOURI MO	-13	-11	-23		
3	3.2	6	5.4	1	.8	ADAIR MO	70	-86	-76	02	15
5	6.6	3	3.6	5	6.4	ANDREW MO	-46	78	-3	45	88
1	1.6	2	2.9			ATCHISON MO	81				
4	3.5	7	5.5	2	1.5	AUDRAIN MO	58	-73	-57	09	21
3	2.6	6	4.4	5	3.1	BARRY MO	69	-28	21	59	46
4	4.6	1	2.1	1	1.6	BARTON MO	-56	-25	-66	06	21
4	3.0	4	3.0	2	1.8	BATES MO	0	-41	-41	18	24
		3	3.7	1	1.1	BENTON MO		-71			17
1	1.3	5	9.0	1	1.3	BOLLINGER MO	594	-86	0	47	19
9	4.8	10	4.6	5	2.0	BOONE MO	-5	-56	-58	09	28
26	5.2	25	5.1	31*	6.7	BUCHANAN MO	-3	32	28	62	89
12	6.4	8	4.3	5	2.3	BUTLER MO	-33	-46	-64	07	32
1	1.2			3	4.5	CALDWELL MO			270	95	68
4	2.8	7	5.0	3	2.1	CALLAWAY MO	75	-58	-26	26	28
3	7.2			3	2.6	CAMDEN MO			-64	06	36
9	5.0	12	5.6	11	4.2	CAPE GIRARDEAU MO	11	-24	-15	36	64
3	3.2	3	3.4	5	5.9	CARROLL MO	7	73	86	78	84
		3	9.5	1	3.0	CARTER MO		-69			43
3	2.2	1	.5	10	5.4	CASS MO	-79	1076	149	87	80
2	2.5	3	4.0	3	3.4	CEDAR MO	59	-16	34	64	50
2	2.0	4	5.9	3	3.1	CHARITON MO	195	-47	56	71	46
6	8.3	3	3.9	3	3.3	CHRISTIAN MO	-53	-15	-60	08	49
2	2.8	1	1.4	1	2.2	CLARK MO	-50	53	-23	28	30
8	3.5	14	4.6	10	2.2	CLAY MO	32	-53	-38	20	30
2	1.9	2	2.4	3	3.6	CLINTON MO	25	47	84	78	53
6	3.6	11	5.7	11	5.0	COLE MO	59	-12	40	66	76
3	3.3	6	6.8	2	1.8	COOPER MO	107	-73	-44	16	25
3	3.5	3	3.4	2	1.4	CRAWFORD MO	-4	-57	-59	08	20
3	3.8	5	7.2	1	1.9	DADE MO	89	-73	-49	13	27
1	1.7	6	7.9	1	.9	DALLAS MO	375	-89	-46	15	15
1	1.0	1	1.1	3	4.1	DAVIESS MO	5	288	306	96	62
3	4.2	1	1.9	1	1.8	DE KALB MO	-55	-6	-57	09	25
1	1.3	1	1.1	4	3.9	DENT MO	-17	259	198	91	58
3	4.6	3	4.5			DOUGLAS MO	0				
7	4.0	11	5.9	9	4.8	DUNKLIN MO	50	-19	21	59	72
20*	9.2	19	7.5	13	4.3	FRANKLIN MO	-18	-43	-53	11	65
5	5.6	4	4.3	9*	10.0	GASCONADE MO	-23	132	79	77	97

* SIGNIFICANTLY MORE DEATHS THAN EXPECTED

WHITE MALE — CANCER OF THE ORAL CAVITY INCLUDING TONGUE — ICD 141,143-146,148,149

1950-1959 DEATHS	1950-1959 RATE	1960-1969 DEATHS	1960-1969 RATE	1970-1979 DEATHS	1970-1979 RATE	COUNTY AND STATE	PERCENT CHANGE 60S/50S	PERCENT CHANGE 70S/60S	PERCENT CHANGE 70S/50S	PERCENTILE %CHANGE 70S/50S	PERCENTILE RATES 1970S
1	1.2			1	1.5	GENTRY MO			30	62	21
18	3.3	26	4.2	28	3.8	GREENE MO	25	-9	14	56	57
				3	3.9	GRUNDY MO					59
9*	8.6	3	3.0	1	1.1	HARRISON MO	-65	-65	-88	00	17
4	3.2	7	4.0	1	.7	HENRY MO	25	-82	-77	02	15
1	1.9			3	6.7	HICKORY MO			254	94	90
2	2.6	2	3.1			HOLT MO	20				
1	1.4					HOWARD MO					
4	2.9	2	1.7	6	3.7	HOWELL MO	-42	115	26	61	55
4	8.6			1	2.3	IRON MO			-73	03	31
158*	6.9	129	5.1	113	4.6	JACKSON MO	-26	-9	-33	22	69
23	5.8	19	4.5	26	6.0	JASPER MO	-22	33	4	50	85
5	2.9	13	4.6	11	2.7	JEFFERSON MO	58	-40	-5	43	39
4	2.9	4	2.7	1	.6	JOHNSON MO	-9	-76	-78	02	14
1	2.0	3	6.0	2	4.6	KNOX MO	200	-24	127	85	69
1	1.5	2	1.6	4	3.0	LACLEDE MO	12	87	109	82	44
7	4.6	4	2.8	7	4.0	LAFAYETTE MO	-39	42	-13	38	60
3	1.8	3	2.1	3	2.0	LAWRENCE MO	13	-4	8	53	27
6	8.4	5	7.2	2	3.9	LEWIS MO	-14	-46	-54	11	59
4	4.3	1	1.3	1	1.0	LINCOLN MO	-70	-21	-76	02	16
		2	1.5	4	2.9	LINN MO		95			43
1	.9	2	2.0	6	5.7	LIVINGSTON MO	128	188	556	99	82
3	4.6	3	3.8	1	1.0	MC DONALD MO	-17	-74	-79	01	16
3	2.5	4	3.5	1	.7	MACON MO	39	-80	-72	04	14
4	7.3	3	5.5			MADISON MO	-25				
2	4.5	1	2.3			MARIES MO	-50				
4	2.6	5	3.3	7	4.7	MARION MO	29	42	83	78	72
				1	2.3	MERCER MO					32
4	5.8	1	1.0	4	4.4	MILLER MO	-83	349	-24	28	66
5	8.2	1	1.4	4	4.8	MISSISSIPPI MO	-83	247	-42	17	72
2	2.6					MONITEAU MO					
2	2.5	1	1.2	1	1.5	MONROE MO	-53	27	-41	18	21
2	2.1	3	3.9	3	3.5	MONTGOMERY MO	83	-10	64	73	51
3	4.4	1	1.1	3	4.0	MORGAN MO	-76	274	-10	40	60
5	5.4	1	.9	3	3.1	NEW MADRID MO	-84	259	-42	17	46
3	1.9	3	1.9	5	2.4	NEWTON MO	-1	29	27	61	33
4	2.7	3	2.3			NODAWAY MO	-17				
1	1.1	1	1.3	2	2.7	OREGON MO	25	102	153	88	38
5	7.4			3	3.9	OSAGE MO			-48	14	58
		1	1.8	1	1.3	OZARK MO		-26			19
7	7.3	6	4.7	5	5.0	PEMISCOT MO	-35	6	-31	24	75
7	9.0	4	4.5	3	4.0	PERRY MO	-50	-11	-56	10	60
5	2.6	6	3.1	17*	8.2	PETTIS MO	20	167	220	92	95
3	2.9	4	3.3	7	4.4	PHELPS MO	13	34	51	70	66
5	5.0	2	1.8	3	3.0	PIKE MO	-64	68	-40	18	44
3	3.0	4	3.4	3	2.4	PLATTE MO	15	-31	-21	31	33
2	1.7	3	2.6			POLK MO	58				
3	4.4	3	3.8	2	1.9	PULASKI MO	-14	-49	-57	10	26
2	3.1	1	1.3	3	5.0	PUTNAM MO	-58	280	60	72	75
3	6.3	1	2.0	3	5.2	RALLS MO	-68	154	-18	33	78
9	5.5	7	5.4	14*	9.8	RANDOLPH MO	-2	81	78	77	97
1	1.2	3	2.5	2	2.0	RAY MO	109	-22	62	73	27
1	3.2			1	2.8	REYNOLDS MO			-15	36	39
3	4.5	3	3.7	3	3.2	RIPLEY MO	-16	-15	-29	25	47
7	4.4	16	7.5	22	6.7	ST CHARLES MO	72	-10	54	71	90
5	6.4	2	3.3	1	1.1	ST CLAIR MO	-49	-66	-83	00	17
3	5.0	1	1.3	2	2.6	STE GENEVIEVE MO	-73	97	-47	14	37
5	2.7	13	6.4	7	3.1	ST FRANCOIS MO	141	-52	15	56	44
100	5.6	133	4.9	168	4.6	ST LOUIS COUNTY MO	-13	-6	-18	33	69
4	2.6	10	6.0	8	5.8	SALINE MO	129	-4	119	84	83
2	4.9			1	2.2	SCHUYLER MO			-55	10	30
1	1.5	1	1.9			SCOTLAND MO	28				
7	5.2	8	5.4	4	2.6	SCOTT MO	3	-51	-50	13	37
1	2.0	1	2.1	3	6.1	SHANNON MO	5	193	207	91	86
1	1.3	1	1.2	4	5.7	SHELBY MO	-5	363	342	97	83
5	3.5	4	2.7	7	4.5	STODDARD MO	-22	63	27	61	67
4	7.3	1	2.3	1	.6	STONE MO	-68	-74	-92	00	14
3	4.0	1	1.0			SULLIVAN MO	-75				
5	8.4	4	4.4	7	4.9	TANEY MO	-47	10	-42	17	74
2	1.8	2	1.5	6	4.1	TEXAS MO	-18	175	125	84	62
4	2.3	2	1.6	4	3.1	VERNON MO	-32	96	33	64	44
4	7.6	1	1.7	2	3.3	WARREN MO	-78	98	-57	10	48
3	4.2	5	6.7	2	3.0	WASHINGTON MO	61	-55	-27	26	44
2	2.9	5	7.8	4	6.7	WAYNE MO	166	-14	129	85	90
1	1.1			1	.9	WEBSTER MO			-15	36	15
1	3.7	1	2.5			WORTH MO	-32				
4	4.3	1	1.2	3	2.8	WRIGHT MO	-73	140	-34	22	40
347*	10.9	271*	9.9	173*	8.5	ST LOUIS CITY MO	-9	-14	-22	30	95
79	2.6	85	2.7	107	3.1	MONTANA MT	3	15	19		
1	2.1	4	7.7	2	3.8	BEAVERHEAD MT	259	-50	80	77	57
		1	3.5	2	7.6	BIG HORN MT		116			93
2	5.4			1	4.3	BLAINE MT			-21	30	65
2	13.3					BROADWATER MT					
		2	3.0	1	1.7	CARBON MT		-41			24
8	2.8	9	3.1	15	4.6	CASCADE MT	8	50	62	73	69
				3	7.4	CHOUTEAU MT					92
3	4.2	1	1.4	3	4.4	CUSTER MT	-68	222	4	50	66
1	2.3	3	6.3	2	4.1	DAWSON MT	174	-36	77	76	62
3	3.0	4	3.8	2	2.1	DEER LODGE MT	27	-46	-31	23	28
5	6.1	3	3.3	1	1.3	FERGUS MT	-46	-62	-79	01	19
3	1.5	3	1.6	4	1.7	FLATHEAD MT	7	9	16	57	23
3	2.8	1	.9	2	1.6	GALLATIN MT	-68	80	-43	16	22
		2	8.3	2	7.1	GLACIER MT		-15			91
3	4.2	2	2.6	2	2.5	HILL MT	-39	-3	-41	18	34
				1	6.4	JUDITH BASIN MT					87
1	1.2			5	5.5	LAKE MT			362	97	81
3	2.6	5	3.4	4	2.6	LEWIS AND CLARK MT	32	-22	3	49	37
		2	3.2	1	1.1	LINCOLN MT		-65			17
		1	3.0	1	2.9	MADISON MT		-1			43

* SIGNIFICANTLY MORE DEATHS THAN EXPECTED

WHITE MALE — CANCER OF THE ORAL CAVITY INCLUDING TONGUE — ICD 141,143-146,148,149

1950-1959 DEATHS	1950-1959 RATE	1960-1969 DEATHS	1960-1969 RATE	1970-1979 DEATHS	1970-1979 RATE	COUNTY AND STATE	PERCENT CHANGE 60S/50S	PERCENT CHANGE 70S/60S	PERCENT CHANGE 70S/50S	PERCENTILE %CHANGE 70S/50S	PERCENTILE RATES 1970S
		1	8.3			MINERAL MT					
5	3.1	4	2.1	9	3.6	MISSOULA MT	-33	72	16	56	54
				1	2.3	MUSSELSHELL MT					31
3	4.4	1	1.4	5	6.8	PARK#MT	-69	397	56	71	90
2	5.4	2	6.6	3	12.0	PHILLIPS MT	22	82	123	84	99
		1	2.8	3	8.8	PONDERA MT		209			96
1	6.7	1	8.0			POWDER RIVER MT	20				
2	5.9	1	2.5			POWELL MT	-57				
1	1.4	2	2.9	2	1.9	RAVALLI MT	112	-34	40	66	26
1	3.0	1	2.3	2	4.0	RICHLAND MT	-21	71	36	65	60
		2	4.5			ROOSEVELT MT					
		2	7.7	1	2.8	ROSEBUD MT		-64			40
2	5.0			1	1.8	SANDERS MT			-64	06	25
1	3.0					SHERIDAN MT					
13	5.2	9	3.8	9	4.2	SILVER BOW MT	-27	9	-20	31	63
		1	3.2	3	8.4	STILLWATER MT		161			95
1	5.4	1	4.8			SWEET GRASS MT	-11				
1	2.6	2	5.4	1	2.4	TETON MT	107	-55	-7	41	34
				1	2.8	TOOLE MT					41
2	3.2	1	1.6	1	1.6	VALLEY MT	-49	0	-49	13	22
6	2.1	10	3.3	11	2.8	YELLOWSTONE MT	56	-16	31	63	39
248	3.6	257	3.5	308	4.1	NEBRASKA NB	-2	17	14		
6	3.6	6	4.0	7	4.3	ADAMS NB	10	9	20	59	65
4	6.4			1	1.4	ANTELOPE NB			-78	02	20
1	1.8			4	6.5	BOX BUTTE NB			263	95	88
		1	3.3	1	3.0	BOYD NB		-10			43
		2	7.2			BROWN NB					
6	4.5	2	1.5	6	3.9	BUFFALO NB	-68	166	-14	36	58
2	3.4	2	2.7			BURT NB	-22				
1	1.5	3	4.8			BUTLER NB	212				
1	.9	3	3.5	3	3.1	CASS NB	290	-11	248	94	46
3	4.4	1	1.4			CEDAR NB	-68				
		1	2.8	1	3.8	CHASE NB		32			56
1	2.5	1	2.3			CHERRY NB	-6				
1	1.7	3	5.2	2	3.8	CHEYENNE NB	211	-26	129	85	57
1	1.5			1	2.2	CLAY NB			46	68	30
4	5.5	3	5.4	3	4.7	COLFAX NB	-2	-13	-15	36	70
3	4.0	2	2.5	7*	9.9	CUMING NB	-39	302	145	87	97
5	5.3	1	1.2	1	.9	CUSTER NB	-77	-30	-84	00	15
1	2.6	1	2.1	4	6.0	DAKOTA NB	-20	193	135	86	85
		1	2.3			DAWES NB					
5	5.1	2	1.6	4	3.8	DAWSON NB	-69	144	-25	28	57
				1	3.2	DEUEL NB					47
		1	1.7	1	1.6	DIXON NB		-3			22
7	4.7	4	2.4	10	5.4	DODGE NB	-49	121	13	55	80
87*	6.7	106*	7.4	97*	6.5	DOUGLAS NB	10	-12	-3	45	88
				1	6.3	DUNDY NB					87
6	8.5	1	1.3	2	3.5	FILLMORE NB	-85	178	-58	09	53
3	6.9	1	2.7			FRANKLIN NB	-62				
		1	3.3			FRONTIER NB					
1	2.0	2	3.4	2	2.7	FURNAS NB	73	-19	39	66	38
4	2.6	6	4.0	6	3.9	GAGE NB	53	-3	48	69	58
				1	7.1	GARFIELD NB					91
				1*	21.5	GRANT NB					99
1	3.2			1	3.0	GREELEY NB			-5	43	44
6	3.5	6	3.0	13	6.1	HALL NB	-13	101	74	76	86
				1	1.4	HAMILTON NB					20
1	2.5			1	2.6	HARLAN NB			4	50	36
				1*	14.0	HAYES NB					99
1	3.4	1	4.1			HITCHCOCK NB	20				
5	6.5	2	2.7			HOLT NB	-59				
				1	1.8	HOWARD NB					24
3	4.1	3	3.6	2	3.2	JEFFERSON NB	-13	-10	-21	30	48
1	2.5	2	4.9	1	2.9	JOHNSON NB	96	-40	17	58	43
1	3.0			1	2.6	KEARNEY NB			-13	37	37
2	5.5	1	2.6	2	4.5	KEITH NB	-53	75	-19	32	68
1	7.5	1	4.2	1	3.7	KIMBALL NB	-44	-11	-50	13	56
1	1.1	2	2.0			KNOX NB	85				
19	3.2	22	3.5	33	4.7	LANCASTER NB	11	32	46	68	70
1	.7	12	8.6	8	5.0	LINCOLN NB	1077	-42	580	99	75
1	16.4					LOUP NB					
4	2.6	7	4.7	7	4.0	MADISON NB	83	-14	57	71	61
		2	3.4	1	1.7	MERRICK NB		-51			23
		1	2.3	2	6.4	MORRILL NB		176			88
1	2.4					NANCE NB					
1	1.5			1	1.3	NEMAHA NB			-18	33	18
		2	4.3			NUCKOLLS NB					
6	6.0	8	7.8	4	3.5	OTOE NB	29	-55	-41	18	53
2	4.1					PAWNEE NB					
				1	3.5	PERKINS NB					52
1	1.8			6*	10.7	PHELPS NB			500	98	98
2	4.1	1	1.7	1	1.5	PIERCE NB	-58	-11	-63	07	21
2	1.8	1	.8	6	4.5	PLATTE NB	-53	426	147	87	67
				1	3.3	POLK NB					49
2	2.9	1	1.3	7*	10.3	RED WILLOW NB	-57	716	252	94	98
2	2.1	2	2.4	4	3.8	RICHARDSON NB	17	59	86	78	58
1	5.5					ROCK NB					
4	4.2	3	2.7	3	3.2	SALINE NB	-35	17	-24	28	47
3	6.9	4	4.5	7	5.0	SARPY NB	-35	12	-27	26	76
1	1.1	5	4.3	6	5.7	SAUNDERS NB	284	35	418	98	83
5	3.2	3	1.9	3	1.6	SCOTTS BLUFF NB	-43	-12	-50	13	22
3	4.3	2	2.8	3	3.9	SEWARD NB	-34	40	-8	41	59
1	2.0			2	5.1	SHERIDAN NB			156	88	77
				1	2.6	SHERMAN NB					36
				1	2.8	STANTON NB					41
5	8.6			1	1.8	THAYER NB			-79	01	24
1	2.6			2	6.9	THURSTON NB			162	88	90
		2	3.7			VALLEY NB					
1	1.6			1	1.3	WASHINGTON NB			-18	34	19

* SIGNIFICANTLY MORE DEATHS THAN EXPECTED

COUNTIES COMBINED; SEE TECHNICAL NOTES

WHITE MALE — CANCER OF THE ORAL CAVITY INCLUDING TONGUE — ICD 141,143-146,148,149

1950-1959 DEATHS	1950-1959 RATE	1960-1969 DEATHS	1960-1969 RATE	1970-1979 DEATHS	1970-1979 RATE	COUNTY AND STATE	PERCENT CHANGE 60S/50S	PERCENT CHANGE 70S/60S	PERCENT CHANGE 70S/50S	PERCENTILE %CHANGE 70S/50S	PERCENTILE RATES 1970S
1	2.5	1	1.5	2	3.5	WAYNE NB	-40	137	41	67	52
		2	3.8			WEBSTER NB					
2	2.8	1	1.2	2	2.8	YORK NB	-59	144	0	47	40
36	3.9	61	3.9	111	4.4	NEVADA NV	1	12	13		
		2	5.0	3	4.8	CHURCHILL NV		-3			72
12	5.8	20	2.8	48	3.8	CLARK NV	-51	35	-35	22	57
				4	8.4	DOUGLAS NV					95
2	2.9	1	1.4	2	3.0	ELKO NV	-54	120	2	48	43
1	12.6					EUREKA NV					
		1	3.1	1	3.9	HUMBOLDT NV		25			59
		1	13.6			LANDER NV					
		3	6.9	2	4.2	LYON NV		-40			63
2	7.0	2	5.9	1	3.0	MINERAL NV	-16	-50	-58	09	43
2	7.8			1	3.0	NYE NV			-62	07	43
		1	5.0	1	5.3	PERSHING NV		7			79
1*	38.5			2*	40.2	STOREY NV			4	50	99
14	4.4	26	5.7	40	6.0	WASHOE NV	30	5	37	65	85
2	5.2	4	9.0	1	2.1	WHITE PINE NV	71	-76	-59	08	29
				5	5.9	ORMSBY-CARSON CITY NV					85
146	5.2	182*	5.8	231*	6.3	NEW HAMPSHIRE NH	11	8	20		
6	4.3	5	3.1	8	4.5	BELKNAP NH	-28	44	4	50	67
3	3.1	4	3.5	7	4.2	CARROLL NH	13	20	35	64	64
6	3.0	11	4.6	18	6.1	CHESHIRE NH	50	35	102	81	86
7	3.6	9	4.7	10	5.2	COOS NH	30	10	42	67	77
9	3.7	14	5.4	21*	7.6	GRAFTON NH	45	42	107	82	93
64*	8.3	78*	9.1	80*	7.7	HILLSBOROUGH NH	9	-15	-7	41	94
18	5.0	22	5.8	21	4.8	MERRIMACK NH	16	-17	-3	45	73
12	2.9	15	3.0	37	5.8	ROCKINGHAM NH	3	93	99	81	83
16	6.4	19	7.2	19	6.4	STRAFFORD NH	12	-11	0	47	88
5	3.2	5	3.1	10	5.8	SULLIVAN NH	-3	88	81	77	84
1407*	6.4	1541*	5.7	1672*	5.3	NEW JERSEY NJ	-10	-7	-17		
37	5.7	44	5.1	42	4.6	ATLANTIC NJ	-10	-11	-21	31	69
159*	6.5	188	5.8	189	4.5	BERGEN NJ	-11	-22	-30	24	68
32	5.7	47	6.1	43	3.7	BURLINGTON NJ	8	-39	-34	22	56
79	6.2	100*	6.3	101	5.6	CAMDEN NJ	1	-11	-11	39	81
12	5.3	9	2.8	26	5.7	CAPE MAY NJ	-47	104	8	53	82
24	5.8	16	3.5	16	3.2	CUMBERLAND NJ	-39	-10	-45	16	47
267*	7.2	232*	6.2	216*	6.5	ESSEX NJ	-13	4	-10	40	88
23	5.5	17	3.3	30	4.9	GLOUCESTER NJ	-41	48	-12	38	73
256*	8.9	240*	8.1	222*	8.0	HUDSON NJ	-10	-1	-10	39	94
10	3.9	20	6.7	12	3.4	HUNTERDON NJ	71	-49	-13	38	51
77*	7.7	73*	6.7	91*	7.4	MERCER NJ	-13	11	-3	45	93
78*	6.4	92	5.8	132*	6.2	MIDDLESEX NJ	-10	8	-3	45	87
43	3.7	86	5.5	101	5.2	MONMOUTH NJ	50	-7	40	66	77
33	4.1	44	3.7	60	4.0	MORRIS NJ	-10	8	-3	45	60
20	5.0	22	2.6	82	5.1	OCEAN NJ	-48	95	2	49	77
115*	6.9	129*	6.7	109	5.2	PASSAIC NJ	-1	-22	-23	28	78
4	1.9	8	3.5	6	2.3	SALEM NJ	81	-35	18	58	31
16	3.4	29	4.5	34	4.5	SOMERSET NJ	31	1	32	63	68
6	3.4	10	3.5	9	2.2	SUSSEX NJ	4	-36	-33	23	31
105*	6.4	115	5.4	134	5.3	UNION NJ	-15	-1	-16	35	79
11	3.8	20	5.8	17	4.3	WARREN NJ	53	-26	14	56	65
43	2.1	78	2.5	119	3.0	NEW MEXICO NM	20	19	43		
5	1.0	18	2.2	30	2.5	BERNALILLO NM	111	12	137	86	34
2	14.3			1	7.8	CATRON NM			-45	15	94
6	4.9	6	3.3	9	4.1	CHAVES NM	-33	26	-16	35	63
1	1.3	1	1.7			COLFAX NM	31				
3	4.9	1	1.1	5	3.8	CURRY NM	-78	255	-23	29	57
				1	4.6	DE BACA NM					70
1	1.2	4	2.4	8	3.0	DONA ANA NM	105	22	150	87	43
2	1.3	6	3.2	4	2.0	EDDY NM	152	-38	57	71	27
2	3.7	2	2.8	4	4.4	GRANT NM	-25	56	17	57	66
		1	4.4	1	4.1	GUADALUPE NM		-5			63
		1	4.8	1	4.2	HIDALGO NM		-14			63
2	3.1	9	6.8	10	5.4	LEA NM	117	-21	72	75	80
				2	6.0	LINCOLN NM					85
1	2.7			1	1.6	LOS ALAMOS NM			-42	17	22
2	5.4	1	2.1	4	4.3	LUNA NM	-62	111	-20	31	65
		1	1.8	1	1.6	MC KINLEY NM		-11			22
		1	3.7			MORA NM					
		2	2.7	3	2.7	OTERO NM		2			38
2	3.5	2	3.3	4	6.4	QUAY NM	-4	92	84	78	87
1	1.3			4	5.2	RIO ARRIBA NM			314	96	78
1	1.4	2	2.6	4	5.4	ROOSEVELT NM	87	106	287	95	80
1	3.7	1	2.1	2	3.6	SANDOVAL NM	-42	71	-1	46	54
		1	1.7	4	2.4	SAN JUAN NM		43			33
4	4.3	4	4.0	1	1.4	SAN MIGUEL NM	-7	-64	-66	06	20
3	2.3	10	5.9	5	2.1	SANTA FE NM	162	-64	-6	42	29
1	1.2	1	1.7	2	1.8	SIERRA NM	37	9	50	69	25
1	8.3			3	8.0	SOCORRO NM			-4	45	94
1	3.3			1	2.9	UNION NM			-11	39	42
1	1.1	3	3.3	4	2.5	VALENCIA NM	204	-23	135	86	35
4039*	5.9	4244*	5.5	4083*	5.3	NEW YORK NY	-6	-4	-10		
91*	8.0	110*	8.4	69	5.0	ALBANY NY	5	-40	-37	20	75
3	1.3	9	4.3	13	5.7	ALLEGANY NY	231	33	341	97	83
413*	6.7	302	4.8	287*	5.9	BRONX NY	-28	23	-11	39	84
41	4.5	44	4.4	54	5.1	BROOME NY	-4	18	13	55	77
11	2.9	23	5.8	18	4.4	CATTARAUGUS NY	99	-24	52	70	67
19	5.1	20	5.2	13	3.3	CAYUGA NY	2	-37	-35	21	49
33	4.4	28	3.7	29	3.8	CHAUTAUQUA NY	-16	2	-14	37	56
23	5.8	28	6.3	14	3.1	CHEMUNG NY	8	-51	-47	15	45

* SIGNIFICANTLY MORE DEATHS THAN EXPECTED

WHITE MALE — CANCER OF THE ORAL CAVITY INCLUDING TONGUE — ICD 141,143-146,148,149

1950-1959 DEATHS	1950-1959 RATE	1960-1969 DEATHS	1960-1969 RATE	1970-1979 DEATHS	1970-1979 RATE	COUNTY AND STATE		PERCENT CHANGE 60S/50S	PERCENT CHANGE 70S/60S	PERCENT CHANGE 70S/50S	PERCENTILE %CHANGE 70S/50S	PERCENTILE RATES 1970S
7	3.3	7	3.1	6	2.6	CHENANGO	NY	-6	-18	-23	29	36
7	2.7	8	3.2	12	4.6	CLINTON	NY	18	45	71	75	70
15	5.5	17	5.8	15	4.4	COLUMBIA	NY	5	-25	-21	31	66
9	5.2	9	4.8	9	4.5	CORTLAND	NY	-7	-7	-13	37	68
6	2.5	6	2.5	8	3.1	DELAWARE	NY	-1	24	23	60	45
41	5.2	33	3.6	49	4.8	DUTCHESS	NY	-30	33	-7	42	72
313*	7.6	318*	6.8	310*	6.6	ERIE	NY	-11	-3	-13	37	89
7	3.9	7	3.7	6	3.1	ESSEX	NY	-4	-17	-21	30	44
9	4.4	6	2.8	11	5.2	FRANKLIN	NY	-36	84	19	58	78
14	4.9	11	3.7	13	4.2	FULTON	NY	-24	16	-12	38	64
13	5.1	16	6.1	9	3.5	GENESEE	NY	20	-43	-32	23	51
7	4.0	8	4.6	12	5.6	GREENE	NY	15	22	40	66	82
2	6.7	1	2.9	6*	21.7	HAMILTON	NY	-56	635	224	92	99
8	2.1	14	4.1	19	5.1	HERKIMER	NY	90	26	140	86	77
22	4.9	20	4.5	18	4.1	JEFFERSON	NY	-8	-10	-17	34	61
568	5.2	462	4.1	440	4.6	KINGS(BROOKLYN)	NY	-21	11	-12	38	69
2	1.8	4	3.6	1	.9	LEWIS	NY	101	-76	-52	12	15
11	5.1	5	2.3	14	5.9	LIVINGSTON	NY	-54	156	17	57	85
2	1.0	15	5.9	13	5.1	MADISON	NY	509	-14	422	98	76
144*	5.8	141	5.1	140	4.9	MONROE	NY	-12	-6	-17	34	73
12	4.1	7	2.1	11	3.2	MONTGOMERY	NY	-50	54	-22	29	47
165	5.4	221	4.6	277	4.6	NASSAU	NY	-15	-1	-15	36	69
725*	8.9	933*	12.5	488*	8.5	NEW YORK(MANHATTAN)	NY	41	-32	-4	44	95
55	6.6	45	4.6	54	5.0	NIAGARA	NY	-30	8	-24	28	75
57	4.7	58	4.6	67	5.2	ONEIDA	NY	-3	12	9	53	77
85	5.0	115*	6.4	110	5.7	ONONDAGA	NY	29	-12	14	56	82
9	2.7	15	4.2	18	4.7	ONTARIO	NY	54	14	76	76	72
40	4.8	49	5.1	71*	6.8	ORANGE	NY	7	32	41	67	90
9	5.3	8	4.6	7	4.1	ORLEANS	NY	-13	-12	-23	28	61
19	4.7	18	4.4	16	3.6	OSWEGO	NY	-7	-18	-24	28	53
10	3.3	8	2.7	9	2.9	OTSEGO	NY	-17	6	-12	38	42
12*	10.1	12	6.4	5	1.7	PUTNAM	NY	-37	-73	-83	00	24
368	5.4	347	3.6	471	5.0	QUEENS	NY	-33	40	-7	41	75
35	5.3	55*	7.8	55*	7.8	RENSSELAER	NY	49	0	49	69	94
72*	7.5	47	4.3	69	5.4	RICHMOND	NY	-43	26	-28	26	80
15	3.8	20	3.7	36	4.3	ROCKLAND	NY	-1	16	14	56	65
19	4.1	17	3.5	15	3.0	ST LAWRENCE	NY	-14	-13	-25	27	44
12	3.2	28	6.4	38*	7.5	SARATOGA	NY	99	18	134	85	93
44	6.5	51	6.6	36	4.5	SCHENECTADY	NY	1	-32	-31	23	67
3	2.2	2	1.4	6	4.3	SCHOHARIE	NY	-36	196	91	79	65
5	6.5	5	6.2	5	5.8	SCHUYLER	NY	-5	-6	-10	40	84
3	1.6	5	2.6	10	5.5	SENECA	NY	66	113	254	94	81
18	3.6	19	3.8	25	4.9	STEUBEN	NY	5	29	36	65	74
91	4.8	135	4.3	240*	5.5	SUFFOLK	NY	-10	29	16	57	81
12	4.4	16	5.4	17	5.0	SULLIVAN	NY	22	-7	14	56	75
2	1.2	8	4.7	9	4.8	TIOGA	NY	293	4	307	96	73
11	4.8	8	3.3	9	3.7	TOMPKINS	NY	-31	13	-22	29	55
35	6.4	38	6.1	40	5.5	ULSTER	NY	-5	-10	-14	37	61
16	7.3	12	5.1	17	6.7	WARREN	NY	-29	30	-8	41	89
9	3.8	10	4.1	6	2.3	WASHINGTON	NY	7	-43	-39	20	32
18	5.8	15	4.4	14	3.9	WAYNE	NY	-24	-12	-33	22	58
199*	6.7	235*	6.4	213	5.2	WESTCHESTER	NY	-4	-18	-21	30	78
10	5.4	6	3.3	5	2.8	WYOMING	NY	-38	-17	-48	14	39
3	3.2	4	3.6	6	5.4	YATES	NY	11	52	68	74	80
374	3.4	553	4.0	748	4.2	NORTH CAROLINA	NC	17	5	23		
10	5.6	7	2.4	15	3.7	ALAMANCE	NC	-58	57	-34	22	55
				1	1.3	ALEXANDER	NC					19
3	8.2	2	5.8	2	3.9	ALLEGHANY	NC	-29	-33	-53	12	58
1	1.7			3	3.8	ANSON	NC			127	85	58
2	2.1	1	1.0	2	1.6	ASHE	NC	-53	65	-22	29	22
1	2.0	2	4.2	2	3.6	AVERY	NC	110	-13	82	78	54
7	7.2	7	6.0	8	6.1	BEAUFORT	NC	-16	1	-15	36	86
		3	6.1	3	6.0	BERTIE	NC		0			85
1	1.5	4	5.8	2	2.1	BLADEN	NC	281	-63	41	66	29
1	1.6	2	3.5	4	3.7	BRUNSWICK	NC	117	3	125	84	54
15	3.1	18	2.9	32	4.4	BUNCOMBE	NC	-7	53	43	67	67
4	2.2	5	2.9	11	4.2	BURKE	NC	37	44	97	80	64
8	4.0	5	1.8	6	2.4	CABARRUS	NC	-54	28	-40	18	32
2	1.3	2	1.1	7	3.5	CALDWELL	NC	-19	229	167	89	51
2	14.2	2	10.9			CAMDEN	NC	-23				
1	2.1	3	3.3	11	6.6	CARTERET	NC	54	103	212	92	89
		1	1.8	4	7.1	CASWELL	NC		293			91
6	3.6	11	3.9	9	2.6	CATAWBA	NC	7	-34	-30	24	36
1	.9	3	3.5	3	2.6	CHATHAM	NC	307	-27	195	91	36
		3	4.4	3	3.5	CHEROKEE	NC		-20			52
3	10.6	3	8.8	1	2.1	CHOWAN	NC	-17	-76	-80	01	29
				1	2.4	CLAY	NC					33
2	1.0	5	2.4	8	3.2	CLEVELAND	NC	141	30	215	92	47
6	7.6	3	2.3	7	5.0	COLUMBUS	NC	-70	120	-34	22	76
5	8.1	8	10.0	12*	8.2	CRAVEN	NC	23	-18	1	48	95
10	5.6	7	3.5	16	4.3	CUMBERLAND	NC	-37	22	-23	29	65
2	7.1	1	3.2			CURRITUCK	NC	-54				
1	3.4	2	6.5	3	7.2	DARE	NC	91	10	110	82	92
4	1.7	8	2.5	9	2.0	DAVIDSON	NC	45	-20	16	57	27
		2	2.7	2	2.1	DAVIE	NC		-23			29
3	2.9	11*	10.3	3	2.3	DUPLIN	NC	251	-77	-21	30	32
14	5.6	20	6.7	18	4.9	DURHAM	NC	18	-26	-13	37	74
3	4.3	8	7.3	10	8.8	EDGECOMBE	NC	71	20	105	82	96
15	4.0	22	4.4	23	3.5	FORSYTH	NC	12	-21	-11	39	52
2	2.6	3	3.8	6	7.1	FRANKLIN	NC	46	90	178	90	92
8	2.7	16	4.3	21	3.9	GASTON	NC	59	-10	44	67	58
		1	4.3			GATES	NC					
1	3.7			1	2.7	GRAHAM	NC			-26	27	39
1	1.1	2	2.2	4	4.7	GRANVILLE	NC	94	112	312	96	71
				1	2.2	GREENE	NC					29
10	1.5	35	4.4	48	5.1	GUILFORD	NC	185	17	233	93	77
4	3.2	4	3.2	9	6.1	HALIFAX	NC	-2	92	88	79	86
5	3.8	8	5.0	13	7.6	HARNETT	NC	33	52	102	81	93
4	2.8	6	3.2	7	2.9	HAYWOOD	NC	14	-9	4	50	43

* SIGNIFICANTLY MORE DEATHS THAN EXPECTED

WHITE MALE — CANCER OF THE ORAL CAVITY INCLUDING TONGUE — ICD 141,143-146,148,149

1950-1959 DEATHS	1950-1959 RATE	1960-1969 DEATHS	1960-1969 RATE	1970-1979 DEATHS	1970-1979 RATE	COUNTY AND STATE		PERCENT CHANGE 60S/50S	PERCENT CHANGE 70S/60S	PERCENT CHANGE 70S/50S	PERCENTILE %CHANGE 70S/50S	PERCENTILE RATES 1970S
2	1.5	10	5.2	6	2.0	HENDERSON	NC	251	-60	39	66	28
		2	4.7	1	1.6	HERTFORD	NC		-65			22
1	3.3	1	3.0			HOKE	NC	-8				
1	3.8	1	6.0			HYDE	NC	57				
7	3.8	8	3.4	10	3.1	IREDELL	NC	-9	-10	-19	32	44
1	1.4			4	3.9	JACKSON	NC			189	91	59
6	3.2	10	5.4	19*	8.5	JOHNSTON	NC	70	58	169	89	95
				3	10.2	JONES	NC					97
		3	4.4	6	4.8	LEE	NC		10			72
6	6.7	5	4.6	9	6.2	LENOIR	NC	-31	35	-7	42	87
3	2.6	3	2.6	4	3.2	LINCOLN	NC	-1	23	21	59	47
3	3.4	3	2.6	1	.6	MC DOWELL	NC	-24	-76	-82	01	14
		1	1.1	3	2.5	MACON	NC		135			34
1	1.2	2	2.2	3	3.3	MADISON	NC	75	50	163	89	48
3	6.0	4	7.5	2	2.7	MARTIN	NC	25	-64	-54	11	39
22	4.5	42	5.5	56	5.2	MECKLENBURG	NC	22	-5	16	57	78
2	3.6	1	1.4	1	1.1	MITCHELL	NC	-61	-21	-69	05	17
1	3.1	1	1.4	1	1.3	MONTGOMERY	NC	-55	-4	-57	09	19
5	5.0	5	3.8	3	1.8	MOORE	NC	-23	-54	-65	06	24
3	2.4	6	4.0	12	6.8	NASH	NC	65	70	181	90	90
17*	9.6	28*	13.2	15	5.0	NEW HANOVER	NC	37	-62	-48	14	75
2	4.7	2	4.7	5	9.7	NORTHAMPTON	NC	1	105	108	82	97
5	8.0	8	9.7	14*	10.6	ONSLOW	NC	22	10	34	64	98
3	3.1	5	4.6	4	2.6	ORANGE	NC	50	-44	-16	35	36
2	6.3	3	8.2	1	2.7	PAMLICO	NC	30	-67	-57	09	38
2	3.0	3	4.4	1	1.1	PASQUOTANK	NC	47	-75	-62	07	17
2	5.3	4	7.9	5	7.4	PENDER	NC	48	-7	38	65	92
4*	15.6			1	2.2	PERQUIMANS	NC			-86	00	30
1	1.5			4	5.3	PERSON	NC			246	94	79
11*	8.8	10	7.6	12	6.0	PITT	NC	-13	-21	-31	23	85
2	6.1	1	1.3	5	5.8	POLK	NC	-78	341	-4	44	84
4	1.9	6	2.6	12	4.0	RANDOLPH	NC	40	55	117	83	61
1	.9	5	4.0	6	5.0	RICHMOND	NC	341	26	457	98	75
7	4.9	9	6.1	7	4.2	ROBESON	NC	24	-31	-15	36	64
6	3.3	7	2.8	11	3.5	ROCKINGHAM	NC	-16	25	5	51	53
5	2.2	17	4.7	12	2.8	ROWAN	NC	111	-40	27	61	41
3	1.9	8	4.5	10	4.7	RUTHERFORD	NC	130	5	141	86	70
2	1.6	5	4.1	8	5.4	SAMPSON	NC	154	34	240	94	80
3	4.7	7*	12.1	7*	11.0	SCOTLAND	NC	157	-9	135	86	98
1	.6	3	1.9	8	4.2	STANLY	NC	209	124	590	99	63
		2	2.0	6	5.1	STOKES	NC		154			77
2	1.4	1	.5	6	2.3	SURRY	NC	-60	318	68	74	31
		2	5.6	1	1.8	SWAIN	NC		-68			25
2	3.1	3	3.9	5	4.9	TRANSYLVANIA	NC	25	25	56	71	74
		2	12.0	1	5.4	TYRRELL	NC		-55			80
6	4.8	3	2.1	5	2.4	UNION	NC	-55	11	-50	13	33
3	4.7	4	4.7	7	7.0	VANCE	NC	0	48	48	69	91
13	4.3	12	2.5	38	5.8	WAKE	NC	-42	131	34	64	83
3	8.0	3	10.1	2	4.6	WARREN	NC	27	-54	-42	17	70
2	7.9	2	5.0			WASHINGTON	NC	-37				
		1	1.2	1	1.1	WATAUGA	NC		-14			17
7	4.5	10	5.5	9	4.3	WAYNE	NC	22	-22	-5	43	65
2	1.2	2	1.1	11	5.0	WILKES	NC	-8	340	303	96	75
13*	13.0	7	5.3	9	5.2	WILSON	NC	-59	-1	-60	08	79
2	3.4	3	2.9	4	3.3	YADKIN	NC	-15	13	-4	44	48
1	1.8	1	1.7			YANCEY	NC	-10				
55	2.0	65	2.1	80	2.6	NORTH DAKOTA	ND	9	20	31		
		2	9.3			ADAMS	ND					
1	1.1	1	1.5	4	5.4	BARNES	ND	31	270	385	98	80
				1	2.1	BENSON	ND					29
		1	13.7			BILLINGS	ND					
3	5.0	3	4.9			BOTTINEAU	ND	-3				
1	4.9	1	4.4	2	9.8	BOWMAN	ND	-10	122	101	81	97
		3	9.5			BURKE	ND					
3	2.8	3	2.4	4	2.2	BURLEIGH	ND	-12	-10	-21	30	30
6	2.3	12	4.1	8	2.4	CASS	ND	79	-41	6	51	34
1	1.7	1	1.8	1	1.6	CAVALIER	ND	7	-11	-4	44	23
2	3.9			1	2.1	DICKEY	ND			-45	15	29
		1	1.7	1	2.9	DIVIDE	ND		68			41
				1	3.0	EDDY	ND					44
1	2.4					EMMONS	ND					
3	2.1	4	2.0	8	3.7	GRAND FORKS	ND	-5	90	80	77	56
1	4.0			2	7.7	GRANT	ND			92	79	93
1	4.0	2	7.2	1	3.2	HETTINGER	ND	81	-55	-19	32	48
1	4.1			1	2.1	LA MOURE	ND			-50	13	28
1	3.7			1	49.7	LOGAN	ND			1258	99	99
3	4.7					MC HENRY	ND					
				1	2.8	MC INTOSH	ND					41
		3	11.1			MC KENZIE	ND					
1	1.7	2	2.8	2	2.8	MC LEAN	ND	71	-3	66	74	39
1	1.1	2	2.2	2	1.8	MORTON	ND	92	-18	58	72	25
		2	4.1	5	10.3	MOUNTRAIL	ND		151			98
		2	3.1	3	7.0	NELSON	ND		128			91
1	1.9	2	2.7			PEMBINA	ND	38				
1	3.4					PIERCE	ND					
		1	1.5	3	4.1	RAMSEY	ND		165			62
1	2.1	2	3.1	3	5.9	RANSOM	ND	46	89	176	90	84
1	4.3					RENVILLE	ND					
2	2.2	3	2.6	1	1.1	RICHLAND	ND	16	-58	-51	12	17
		1	2.6	2	7.2	SARGENT	ND		179			92
				1	5.7	SHERIDAN	ND					83
4	5.5			7	8.4	STARK	ND			51	70	95
1	10.6					STEELE	ND					
3	2.3	3	2.2	4	3.5	STUTSMAN	ND	-1	57	55	71	52
1	4.2	2	6.7			TOWNER	ND	62				
1	1.5	1	1.3	3	4.2	TRAILL	ND	-15	230	179	90	64
1	.8			2	2.2	WALSH	ND			160	88	30
6	3.6	3	1.6	3	1.5	WARD	ND	-56	-7	-59	08	21
1	1.7					WELLS	ND					

* SIGNIFICANTLY MORE DEATHS THAN EXPECTED

WHITE MALE CANCER OF THE ORAL CAVITY INCLUDING TONGUE ICD 141,143-146,148,149

1950-1959 DEATHS	1950-1959 RATE	1960-1969 DEATHS	1960-1969 RATE	1970-1979 DEATHS	1970-1979 RATE	COUNTY AND STATE	PERCENT CHANGE 60S/50S	PERCENT CHANGE 70S/60S	PERCENT CHANGE 70S/50S	PERCENTILE %CHANGE 70S/50S	PERCENTILE RATES 1970S
1	.9	2	1.7	2	2.0	WILLIAMS ND	89	15	117	83	27
1744	4.7	2115*	5.2	2139*	4.9	OHIO OH	10	-6	4		
		2	1.8	4	3.5	ADAMS OH		96			51
18	4.3	20	4.6	18	3.8	ALLEN OH	7	-17	-11	39	57
6	3.4	7	3.6	7	3.4	ASHLAND OH	8	-5	2	49	51
11	2.6	15	3.5	14	2.9	ASHTABULA OH	34	-17	12	55	42
10	4.1	6	2.8	8	4.0	ATHENS OH	-33	43	-4	44	59
6	3.9	6	3.5	5	2.6	AUGLAIZE OH	-10	-26	-33	22	36
24	4.9	16	3.6	27	5.9	BELMONT OH	-27	64	20	59	84
2	1.4	4	2.9	7	4.8	BROWN OH	102	65	233	93	73
19	3.2	27	3.7	40	4.5	BUTLER OH	16	22	41	67	68
2	1.9	1	.9	2	1.8	CARROLL OH	-53	102	-6	42	24
6	4.3	5	3.5	9	6.3	CHAMPAIGN OH	-19	79	46	68	87
25	5.0	25	4.4	38	6.2	CLARK OH	-13	42	23	60	86
6	2.8	13	4.8	15	4.2	CLERMONT OH	70	-12	49	69	64
3	2.1	5	3.4	7	4.5	CLINTON OH	65	31	118	83	68
16	3.3	29	5.7	24	4.5	COLUMBIANA OH	73	-21	36	65	67
3	1.6	15*	9.0	9	5.0	COSHOCTON OH	470	-44	219	92	75
6	3.1	8	3.7	14	6.2	CRAWFORD OH	19	65	96	80	86
430*	7.1	462*	7.0	443*	6.7	CUYAHOGA OH	-2	-4	-6	42	89
2	.8	2	.8	6	2.3	DARKE OH	-8	200	175	90	31
7	5.2	1	.7	3	2.0	DEFIANCE OH	-86	172	-61	08	27
2	1.3	2	1.2	7	3.9	DELAWARE OH	-8	231	205	91	58
34*	11.5	20	6.2	30*	9.0	ERIE OH	-46	45	-22	29	96
7	2.6	14	4.5	9	2.5	FAIRFIELD OH	69	-43	-4	44	35
4	3.4	6	4.9	4	3.1	FAYETTE OH	43	-37	-10	40	45
116	5.7	134	5.9	145	5.5	FRANKLIN OH	4	-6	-3	45	81
3	2.2	5	3.5	1	.6	FULTON OH	58	-82	-71	04	14
4	3.8	3	2.3	5	3.9	GALLIA OH	-39	72	5	51	59
6	3.9	8	4.3	10	3.7	GEAUGA OH	10	-13	-4	44	56
4	2.1	10	3.0	10	2.6	GREENE OH	41	-13	22	60	36
12	4.7	5	2.4	4	1.8	GUERNSEY OH	-48	-26	-62	07	25
223*	7.1	269*	8.1	194*	5.8	HAMILTON OH	13	-28	-19	32	83
10	4.4	6	2.3	7	2.6	HANCOCK OH	-48	12	-41	18	36
3	2.0	3	2.0	3	2.0	HARDIN OH	2	-1	1	48	28
6	5.8	3	3.5	8	7.9	HARRISON OH	-41	130	37	65	94
3	2.7	5	4.0	4	3.2	HENRY OH	50	-22	17	58	47
6	3.3	4	2.3	6	3.4	HIGHLAND OH	-31	51	5	51	50
2	2.0	2	2.1	3	2.8	HOCKING OH	3	34	38	65	40
1	.9	4	4.3			HOLMES OH	362				
8	3.9	8	3.8	8	3.6	HURON OH	-2	-6	-8	41	53
3	2.1	1	.7	6	4.0	JACKSON OH	-65	451	92	80	61
21	4.9	25	5.2	21	4.8	JEFFERSON OH	7	-8	-1	46	72
3	1.6	8	4.1	7	3.3	KNOX OH	159	-18	112	83	49
14	3.8	34	6.4	39	5.4	LAKE OH	70	-16	42	67	80
10	4.5	13	5.4	13	4.8	LAWRENCE OH	22	-11	8	52	72
13	3.3	15	3.5	23	4.7	LICKING OH	4	34	39	66	70
7	4.1	9	5.0	3	1.5	LOGAN OH	22	-69	-62	07	21
31	4.6	48	6.2	46	4.6	LORAIN OH	33	-25	-1	47	69
103	5.8	134*	6.7	142*	7.1	LUCAS OH	16	7	24	60	91
4	3.6	3	2.7	4	2.9	MADISON OH	-24	5	-20	31	41
51	4.4	77	5.9	79	6.0	MAHONING OH	34	0	34	64	85
8	3.1	6	2.4	2	.8	MARION OH	-22	-67	-74	03	15
8	3.8	10	3.7	9	2.7	MEDINA OH	-3	-28	-30	24	38
1	.8	3	2.6	5	4.7	MEIGS OH	226	82	495	98	71
4	2.9	6	4.1	8	4.7	MERCER OH	44	14	64	73	71
10	3.3	10	3.0	13	3.6	MIAMI OH	-10	21	9	53	54
3	2.6	3	3.3	3	3.5	MONROE OH	25	6	33	63	52
71	4.3	115*	6.2	102	4.8	MONTGOMERY OH	43	-23	11	54	72
3	4.1	1	1.1	4	6.2	MORGAN OH	-74	481	50	69	87
1	.9	2	2.1	3	2.7	MORROW OH	122	27	182	90	38
10	2.8	18	4.9	11	2.9	MUSKINGUM OH	77	-41	5	51	42
2	3.1	2	3.4	1	1.4	NOBLE OH	11	-59	-54	11	20
4	2.5	6	3.6	13	6.6	OTTAWA OH	45	83	166	89	89
2	2.6	4	5.2	2	2.5	PAULDING OH	100	-52	-5	43	34
4	2.5	7	4.9	7	5.0	PERRY OH	99	2	102	82	75
4	3.0	3	1.8	3	1.8	PICKAWAY OH	-39	-4	-41	18	24
1	1.1	3	3.1	4	3.7	PIKE OH	175	18	226	93	55
6	2.1	18	5.2	11	2.4	PORTAGE OH	155	-54	17	57	33
3	2.2	2	1.3	2	1.2	PREBLE OH	-38	-10	-44	16	18
1	.8	4	3.2	4	3.0	PUTNAM OH	292	-6	270	95	43
8	2.2	11	2.5	21	4.2	RICHLAND OH	18	64	93	80	63
11	4.0	13	4.4	10	3.4	ROSS OH	10	-23	-16	35	49
14	5.6	9	3.4	11	4.1	SANDUSKY OH	-39	21	-26	27	63
19	5.3	28*	6.9	18	4.4	SCIOTO OH	31	-36	-16	35	67
10	3.6	9	3.4	11	4.1	SENECA OH	-4	20	15	56	62
3	2.1	5	3.3	6	3.6	SHELBY OH	58	10	74	76	54
69	5.0	70	4.6	78	4.7	STARK OH	-9	3	-6	42	71
78	4.2	95	4.6	106	4.7	SUMMIT OH	8	3	12	55	71
25	3.8	48	5.7	59	5.9	TRUMBULL OH	50	5	57	71	85
13	3.4	14	3.6	21	5.1	TUSCARAWAS OH	5	43	50	69	76
3	2.8	3	2.7	5	4.3	UNION OH	-4	60	54	71	65
5	3.4	1	.7	4	2.7	VAN WERT OH	-80	286	-21	30	38
				1	1.9	VINTON OH					26
		6	2.4	6	2.1	WARREN OH		-12			29
6	2.6	7	2.8	7	2.5	WASHINGTON OH	6	-10	-4	44	35
4	1.5	10	2.9	6	1.6	WAYNE OH	97	-44	9	53	22
		1	.6	4	2.3	WILLIAMS OH		287			32
10	3.5	14	4.4	12	3.2	WOOD OH	27	-27	-7	42	48
4	3.7	1	1.0	5	4.7	WYANDOT OH	-72	352	27	61	70
336	3.6	383	3.5	477	3.9	OKLAHOMA OK	-2	10	8		
1	1.3	3	4.2	1	1.4	ADAIR OK	226	-68	5	51	19
		5	8.2			ALFALFA OK					
2	3.1			3	5.0	ATOKA OK			59	72	74
		1	3.4	1	2.3	BEAVER OK		-33			31
4	3.8	4	3.3	4	3.1	BECKHAM OK	-13	-6	-18	33	45
		2	3.0	1	1.1	BLAINE OK		-62			17

* SIGNIFICANTLY MORE DEATHS THAN EXPECTED

WHITE MALE

CANCER OF THE ORAL CAVITY INCLUDING TONGUE

ICD 141,143-146,148,149

1950-1959 DEATHS	1950-1959 RATE	1960-1969 DEATHS	1960-1969 RATE	1970-1979 DEATHS	1970-1979 RATE	COUNTY AND STATE	PERCENT CHANGE 60S/50S	PERCENT CHANGE 70S/60S	PERCENT CHANGE 70S/50S	PERCENTILE %CHANGE 70S/50S	PERCENTILE RATES 1970S
4	2.8	5	3.0	6	3.5	BRYAN OK	5	19	25	60	52
4	2.7	4	2.4	5	3.5	CADDO OK	-11	44	29	62	51
4	3.4	2	1.6	5	2.9	CANADIAN OK	-54	84	-15	36	41
1	.6	7	3.7	8	3.8	CARTER OK	552	2	568	99	56
3	4.5			3	2.3	CHEROKEE OK			-48	14	32
6	7.1	2	2.0	4	4.4	CHOCTAW OK	-72	123	-37	20	67
		1	4.8			CIMARRON OK					
4	2.4	9	4.4	10	3.3	CLEVELAND OK	81	-25	35	64	48
4	10.1			2	7.4	COAL OK			-26	27	92
7	4.1	11	5.9	10	3.9	COMANCHE OK	45	-35	-6	43	58
6*	11.2	2	4.2	1	2.0	COTTON OK	-62	-52	-82	01	28
7	6.7	3	2.8	6	6.5	CRAIG OK	-58	130	-4	44	88
7	4.2	5	2.4	11	4.8	CREEK OK	-41	98	16	57	73
3	3.1	1	.9	5	4.7	CUSTER OK	-71	425	51	70	71
3	3.4	2	1.4	3	2.1	DELAWARE OK	-60	52	-39	19	28
		1	2.2	1	2.0	DEWEY OK		-8			27
1	2.6			1	4.9	ELLIS OK			93	80	74
7	3.0	8	3.1	10	3.5	GARFIELD OK	4	12	16	57	52
3	2.3	1	.7	4	2.0	GARVIN OK	-71	202	-12	38	28
5	3.0	5	3.0	8	4.8	GRADY OK	-1	59	58	72	72
3	5.7	2	3.2			GRANT OK	-43				
3	5.0	3	4.5	2	2.8	GREER OK	-11	-37	-44	16	40
		1	2.7			HARMON OK					
1	1.7	1	1.6	6	9.5	HASKELL OK	-4	491	467	98	97
4	3.8	3	3.0	1	1.8	HUGHES OK	-22	-41	-54	11	24
1	1.0	2	1.8	3	2.6	JACKSON OK	93	39	168	89	36
1	1.3	1	1.5	2	3.6	JEFFERSON OK	14	137	170	89	54
2	3.5	2	3.7	1	2.8	JOHNSTON OK	4	-23	-19	32	40
5	2.3	10	4.0	6	2.2	KAY OK	73	-43	-2	45	30
3	5.0	3	4.9	2	2.9	KINGFISHER OK	-2	-40	-42	17	42
3	3.4	1	1.1	3	2.7	KIOWA OK	-66	139	-19	32	38
2	7.7	2	5.9	4	8.5	LATIMER OK	-23	43	11	54	95
3	1.6	8	4.5	9	4.6	LE FLORE OK	177	3	187	91	70
2	1.9	4	3.6	7	5.4	LINCOLN OK	90	50	185	91	80
8	7.6	1	.9	4	3.5	LOGAN OK	-88	282	-54	11	51
1	2.9	1	2.6	3	9.1	LOVE OK	-11	253	215	92	96
2	3.0	4	5.2			MC CLAIN OK	75				
4	3.5	6	5.0	5	3.7	MC CURTAIN OK	42	-27	4	50	54
1	2.5	6	8.3	7	7.9	MC INTOSH OK	233	-4	219	92	94
2	4.9	1	1.8	3	4.6	MARSHALL OK	-62	147	-7	42	69
3	3.0	2	1.8	8	5.5	MAYES OK	-41	210	83	78	81
2	3.2	6	7.3	4	4.3	MURRAY OK	130	-42	34	64	64
12	4.9	12	4.6	10	3.7	MUSKOGEE OK	-8	-19	-26	27	55
3	4.5	1	1.6	2	2.9	NOBLE OK	-65	83	-35	21	42
		3	6.0	2	2.8	NOWATA OK		-54			39
3	5.5			1	1.4	OKFUSKEE OK			-75	03	19
49	4.1	71	4.2	98	5.0	OKLAHOMA OK	3	18	22	59	75
6	3.3	6	3.3	11	6.1	OKMULGEE OK	1	83	85	78	86
9	6.6	4	2.5	6	3.5	OSAGE OK	-63	43	-47	15	52
4	2.8	2	1.3	3	2.0	OTTAWA OK	-54	53	-30	24	27
4	4.9	3	3.3	5	6.9	PAWNEE OK	-32	105	40	66	90
4	1.9	4	2.1	8	3.8	PAYNE OK	16	76	104	82	57
7	4.5	7	3.5	5	2.2	PITTSBURG OK	-23	-36	-51	12	30
7	5.2	4	2.4	5	3.0	PONTOTOC OK	-54	25	-43	16	43
2	.8	5	2.2	9	3.5	POTTAWATOMIE OK	158	61	314	96	52
2	2.9			1	2.3	PUSHMATAHA OK			-22	30	31
2	7.6					ROGER MILLS OK					
1	.9	6	5.1	10	7.2	ROGERS OK	495	41	741	99	92
4	3.3	7	5.1	7	4.3	SEMINOLE OK	57	-17	30	63	64
1	1.1	4	4.8	6	4.7	SEQUOYAH OK	335	-2	328	97	71
4	2.9	12	6.4	7	3.3	STEPHENS OK	123	-48	16	57	49
1	1.8	1	1.4	1	1.3	TEXAS OK	-23	-12	-31	23	18
2	2.6	3	3.8	6	8.7	TILLMAN OK	45	128	229	93	96
49	5.2	64	5.2	63	4.2	TULSA OK	-1	-18	-19	32	64
5	7.3	4	5.2	5	4.8	WAGONER OK	-29	-8	-35	22	72
9	7.1	4	2.2	8	3.6	WASHINGTON OK	-69	64	-49	14	54
				2	2.9	WASHITA OK					43
4	5.2	2	2.6			WOODS OK	-51				
				3	3.3	WOODWARD OK					48
216	2.8	346	3.6	427	3.8	OREGON OR	31	5	38		
1	1.0	1	1.1	5	5.5	BAKER OR	8	384	423	98	81
1	.6	4	2.3	7	4.0	BENTON OR	288	76	582	99	61
11	2.4	19	3.0	30	3.5	CLACKAMAS OR	26	19	50	69	53
6	5.0	9	5.3	8	4.0	CLATSOP OR	7	-25	-20	31	59
7	5.3	7	5.4	8	4.9	COLUMBIA OR	1	-9	-8	40	74
1	.5	12	4.9	14	4.8	COOS OR	799	-1	791	99	73
		1	1.3	2	2.8	CURRY OR		121			39
		4	2.6			DESCHUTES OR					
3	1.5	6	1.7	12	2.9	DOUGLAS OR	8	73	88	79	41
1	7.2			1	9.0	GILLIAM OR			24	60	96
		1	2.6			GRANT OR					
4*	11.5	2	5.3			HARNEY OR	-54				
1	1.7	4	5.5	6	7.2	HOOD RIVER OR	228	30	327	97	92
8	2.4	18	3.8	23	3.8	JACKSON OR	61	-1	60	72	57
3	1.8	6	3.5	10	3.6	JOSEPHINE OR	97	2	100	81	53
2	1.5	9	4.2	6	2.2	KLAMATH OR	169	-47	42	67	30
1	2.4			1	2.3	LAKE OR			-5	43	31
9	1.8	20	2.7	34	3.4	LANE OR	47	27	87	79	50
2	1.4	6	3.5	6	3.1	LINCOLN OR	146	-13	114	83	44
1	.4	5	1.9	12	3.1	LINN OR	367	68	682	99	46
				4	3.2	MALHEUR OR					47
8	1.6	20	2.9	19	2.2	MARION OR	82	-23	40	66	30
		1	3.4	1	2.9	MORROW OR		-16			41
112	4.4	151	5.5	183*	6.8	MULTNOMAH OR	25	24	55	71	90
				2	1.0	POLK OR					16
1	14.9			1	7.0	SHERMAN OR			-53	12	91
2	2.1	4	4.0	3	2.5	TILLAMOOK OR	92	-38	19	58	34
10	4.4	8	3.7	8	3.0	UMATILLA OR	-15	-19	-31	23	44
4	4.0	3	2.9	2	1.9	UNION OR	-28	-32	-51	12	27

* SIGNIFICANTLY MORE DEATHS THAN EXPECTED

WHITE MALE — CANCER OF THE ORAL CAVITY INCLUDING TONGUE — ICD 141,143-146,148,149

1950-1959 DEATHS	1950-1959 RATE	1960-1969 DEATHS	1960-1969 RATE	1970-1979 DEATHS	1970-1979 RATE	COUNTY AND STATE	PERCENT CHANGE 60S/50S	PERCENT CHANGE 70S/60S	PERCENT CHANGE 70S/50S	PERCENTILE %CHANGE 70S/50S	PERCENTILE RATES 1970S
1	2.4	2	5.2	1	2.3	WALLOWA OR	117	-56	-5	44	31
1	1.0	3	2.8	3	2.4	WASCO OR	171	-14	131	85	33
9	2.4	17	3.3	13	1.9	WASHINGTON OR	35	-42	-22	29	26
1	6.1			1	6.1	WHEELER OR			0	47	86
5	3.1	3	1.4	1	.5	YAMHILL OR	-56	-67	-86	00	14
2373*	5.2	2665*	5.2	2643	4.8	PENNSYLVANIA PA	0	-8	-8		
5	2.6	9	3.9	1	.4	ADAMS PA	51	-90	-85	00	14
483*	7.6	520*	7.2	488*	6.6	ALLEGHENY PA	-5	-9	-14	37	89
12	3.2	12	3.0	13	3.0	ARMSTRONG PA	-8	0	-8	41	43
28	4.0	40	4.3	53	5.4	BEAVER PA	8	26	35	64	80
10	5.1	6	2.8	4	1.6	BEDFORD PA	-46	-41	-68	05	23
50	3.8	50	3.3	57	3.4	BERKS PA	-12	2	-10	40	50
24	3.3	30	4.2	25	3.5	BLAIR PA	25	-16	4	50	52
4	1.5	1	.4	10	3.4	BRADFORD PA	-75	839	133	85	51
29	3.8	39	3.5	52	3.1	BUCKS PA	-8	-12	-19	32	45
15	3.0	19	3.4	31	5.0	BUTLER PA	14	45	66	74	75
42	4.8	46	4.7	43	4.2	CAMBRIA PA	-3	-11	-13	38	64
1	3.0	1	3.9			CAMERON PA	30				
9	3.0	15	5.4	14	4.5	CARBON PA	82	-16	53	70	68
8	3.5	10	3.7	14	3.9	CENTRE PA	8	4	12	55	59
31	4.5	40	4.6	34	2.9	CHESTER PA	1	-36	-35	21	42
4	2.1	5	2.8	5	2.5	CLARION PA	36	-13	19	58	34
15	3.7	14	3.3	9	2.2	CLEARFIELD PA	-10	-35	-42	17	30
6	4.1	4	2.2	4	2.0	CLINTON PA	-47	-9	-52	12	27
1	.4	7	2.5	16	5.0	COLUMBIA PA	556	103	1232	99	76
14	4.0	14	3.5	19	4.5	CRAWFORD PA	-11	26	12	55	67
15	3.6	23	4.3	28	4.0	CUMBERLAND PA	20	-7	11	54	61
36	3.9	46	4.7	54	5.2	DAUPHIN PA	21	11	34	64	78
91	5.5	114	5.1	123	4.9	DELAWARE PA	-8	-4	-12	38	74
10	6.7	6	3.6	10	5.4	ELK PA	-47	51	-20	31	80
57	5.7	65	6.0	75*	6.1	ERIE PA	5	2	7	52	86
30	3.8	42	5.2	38	4.5	FAYETTE PA	39	-13	20	59	68
1	3.5			1	2.3	FOREST PA			-36	21	31
6	1.7	17	4.2	12	2.4	FRANKLIN PA	142	-43	39	65	33
4	2.0	9	3.9	2	.9	GREENE PA	102	-78	-56	10	15
4	2.3	9	5.0	6	3.1	HUNTINGDON PA	115	-38	34	64	45
7	2.3	9	2.3	10	2.5	INDIANA PA	1	8	10	53	35
4	1.7	2	.7	3	1.0	JEFFERSON PA	-58	41	-40	19	16
		1	1.2	3	3.4	JUNIATA PA		188			50
71	6.2	93*	7.4	96*	7.4	LACKAWANNA PA	19	1	20	59	92
33	3.0	38	2.9	47	3.0	LANCASTER PA	-3	2	-1	47	44
15	2.9	19	3.6	25	4.5	LAWRENCE PA	21	26	53	71	68
10	2.7	10	2.2	7	1.4	LEBANON PA	-18	-38	-49	13	19
43	4.5	60	5.3	65	4.8	LEHIGH PA	17	-9	7	52	73
118*	6.9	97	5.1	121*	6.3	LUZERNE PA	-27	24	-9	40	87
21	4.3	25	4.6	26	4.5	LYCOMING PA	9	-2	7	52	69
9	3.3	8	2.9	8	2.7	MC KEAN PA	-13	-7	-18	33	38
19	3.6	22	3.8	32	5.1	MERCER PA	6	33	41	67	76
5	3.0	4	1.9	3	1.6	MIFFLIN PA	-38	-15	-47	15	22
9	4.7	9	4.3	7	2.3	MONROE PA	-9	-47	-52	12	31
82	4.9	104	4.4	114	4.0	MONTGOMERY PA	-10	-9	-18	33	60
3	3.7	3	3.1	2	2.4	MONTOUR PA	-16	-23	-36	21	33
49	5.6	40	4.0	45	4.1	NORTHAMPTON PA	-28	2	-27	26	62
19	3.5	26	4.7	27	4.8	NORTHUMBERLAND PA	35	1	37	65	72
6	4.4	2	1.5	6	3.9	PERRY PA	-66	159	-12	38	59
591*	8.1	584*	7.8	434*	6.5	PHILADELPHIA PA	-4	-16	-20	31	88
2	3.4	2	2.0			PIKE PA	-40				
1	1.3	5	5.1	4	3.8	POTTER PA	298	-25	199	91	58
49	5.6	54	5.5	42	4.5	SCHUYLKILL PA	-2	-18	-20	31	68
4	4.6	4	3.0	2	1.4	SNYDER PA	-34	-53	-69	05	20
10	2.3	14	3.3	14	3.1	SOMERSET PA	41	-6	33	64	45
				3	8.1	SULLIVAN PA					95
7	4.3	8	4.5	8	4.1	SUSQUEHANNA PA	6	-9	-3	45	63
3	1.7	4	2.3	5	2.6	TIOGA PA	37	13	55	71	36
1	1.0	2	1.7	5	3.7	UNION PA	79	113	281	95	56
11	3.5	5	1.6	10	3.2	VENANGO PA	-54	99	-8	40	47
8	3.2	8	3.2	13	5.1	WARREN PA	0	59	59	72	77
37	3.5	49	4.8	68	6.1	WASHINGTON PA	37	25	72	75	86
2	1.1	11	6.1	9	4.8	WAYNE PA	443	-21	329	97	73
58	3.8	88	5.2	86	4.5	WESTMORELAND PA	35	-14	17	57	67
2	2.2	5	5.4	8	7.5	WYOMING PA	141	39	235	93	93
19	2.0	37	3.0	44	3.1	YORK PA	55	3	59	72	45
250*	7.3	314*	7.8	306*	6.8	RHODE ISLAND RI	7	-12	-6		
9	8.5	8	3.9	10	4.4	BRISTOL RI	-54	12	-48	14	66
26	6.7	32	6.5	42	6.5	KENT RI	-3	0	-2	45	89
16	7.3	16	5.9	14	4.3	NEWPORT RI	-19	-27	-40	18	65
191*	7.5	252*	8.9	227*	7.7	PROVIDENCE RI	19	-14	2	49	93
8	4.1	6	2.2	13	4.1	WASHINGTON RI	-48	89	-1	46	61
219	5.0	249	4.3	365	4.8	SOUTH CAROLINA SC	-14	11	-5		
2	3.1	1	1.8	3	3.5	ABBEVILLE SC	-43	96	11	54	52
7	7.3	8	4.6	16	5.2	AIKEN SC	-38	13	-29	25	78
2	11.1	1	4.3	2	10.9	ALLENDALE SC	-61	155	-2	46	98
8	3.8	11	3.5	30*	6.9	ANDERSON SC	-7	95	81	77	90
1	2.8	2	9.8	2	5.6	BAMBERG SC	248	-43	98	81	82
2	6.9	1	2.7	4	8.6	BARNWELL SC	-61	221	26	61	96
1	2.4	1	1.9	3	3.0	BEAUFORT SC	-18	57	28	62	44
1	1.9	2	5.3	1	1.5	BERKELEY SC	174	-72	-23	29	21
				1	3.2	CALHOUN SC					47
30*	10.8	28	7.1	28	4.8	CHARLESTON SC	-34	-32	-56	10	72
3	3.1	4	3.4	13*	9.6	CHEROKEE SC	11	183	213	92	97
3	3.5	5	7.0	6	5.7	CHESTER SC	102	-19	64	73	82
3	4.5	3	4.1	9	9.5	CHESTERFIELD SC	-10	135	110	83	97
2	5.6	1	2.2	1	1.8	CLARENDON SC	-61	-17	-67	05	25
3	4.6	3	4.2	3	3.7	COLLETON SC	-9	-11	-19	32	56
3	3.3	5	4.2	4	2.8	DARLINGTON SC	27	-33	-14	37	40

* SIGNIFICANTLY MORE DEATHS THAN EXPECTED

WHITE MALE

CANCER OF THE ORAL CAVITY INCLUDING TONGUE ICD 141,143-146,148,149

1950-1959 DEATHS	1950-1959 RATE	1960-1969 DEATHS	1960-1969 RATE	1970-1979 DEATHS	1970-1979 RATE	COUNTY AND STATE	PERCENT CHANGE 60S/50S	PERCENT CHANGE 70S/60S	PERCENT CHANGE 70S/50S	PERCENTILE %CHANGE 70S/50S	PERCENTILE RATES 1970S
1	1.7	3	4.1	2	2.8	DILLON SC	149	-32	70	75	40
1	2.2	3	7.2	6	5.8	DORCHESTER SC	220	-20	157	88	83
1	2.8	3	8.0	6*	14.6	EDGEFIELD SC	182	84	419	98	99
3	9.0	2	4.4	3	5.3	FAIRFIELD SC	-52	22	-41	18	79
9	6.3	6	3.5	7	3.0	FLORENCE SC	-44	-16	-53	12	43
5*	11.3	1	1.9	3	5.2	GEORGETOWN SC	-83	170	-54	11	78
19	4.3	34	5.3	33	4.0	GREENVILLE SC	23	-25	-8	40	59
5	5.9	4	2.5	2	1.2	GREENWOOD SC	-57	-51	-79	01	18
				3	5.8	HAMPTON SC					83
6	4.0	8	4.1	12	3.7	HORRY SC	2	-8	-5	43	56
		1	5.0	2	7.4	JASPER SC		48			92
2	2.5	6	8.9	3	2.4	KERSHAW SC	250	-73	-5	43	33
		2	1.3	4	3.9	LANCASTER SC		203			59
3	3.3	4	3.0	11	7.6	LAURENS SC	-8	150	129	85	93
1	3.0					LEE SC					
7	4.7	6	2.8	6	1.8	LEXINGTON SC	-40	-35	-61	08	25
1	1.9	1	1.8	5	7.9	MARION SC	-2	332	323	97	94
3	6.1	3	4.3	7*	10.3	MARLBORO SC	-29	138	70	75	98
3	3.4	3	2.7	7	6.4	NEWBERRY SC	-21	138	88	79	88
9	7.5	4	3.4	8	4.5	OCONEE SC	-54	32	-39	19	69
7	7.7	8	5.7	6	3.4	ORANGEBURG SC	-27	-40	-56	10	50
		1	.8	13	6.1	PICKENS SC		649			85
31*	9.6	31*	6.4	37	6.9	RICHLAND SC	-33	7	-28	25	90
2	6.3					SALUDA SC					
13	3.4	10	2.0	22	3.1	SPARTANBURG SC	-40	51	-10	40	44
1	1.1	7	7.9	7	4.6	SUMTER SC	598	-41	310	96	70
3	3.4	4	4.2	3	3.1	UNION SC	23	-26	-9	40	46
3	5.9	3	6.1	3	4.9	WILLIAMSBURG SC	3	-20	-17	34	73
9	7.1	15	7.3	18	7.3	YORK SC	3	-1	3	49	92
53	1.7	78	2.3	90	2.6	SOUTH DAKOTA SD	38	15	58		
1	3.8			2	8.1	AURORA SD			113	83	95
2	1.7	4	3.6			BEADLE SD	107				
1	2.1			2	3.8	BON HOMME SD			82	78	57
1	1.2	2	1.9	4	4.0	BROOKINGS SD	53	112	223	92	61
1	.8	4	2.3	5	2.9	BROWN SD	198	26	276	95	43
		1	2.8	1	3.0	BRULE SD		9			43
				1	2.1	BUTTE SD					29
1	1.8	1	1.8	2	4.1	CHARLES MIX SD	-4	133	123	84	63
		1	2.0	1	1.8	CLARK SD		-7			25
4	7.9	2	4.1	2	4.5	CLAY SD	-48	9	-43	16	67
1	.9	2	1.8	1	1.0	CODINGTON SD	96	-43	11	54	16
1	3.1			2	6.3	CUSTER SD			106	82	87
				2	2.5	DAVISON SD					34
3	4.3	1	1.8	1	1.5	DAY SD	-59	-12	-64	06	21
1	2.5	2	4.9			DEUEL SD	93				
1	5.2	1	8.6	1	6.5	DEWEY SD	66	-24	26	60	89
2	8.3	1	3.6			DOUGLAS SD	-57				
1	3.2					EDMUNDS SD					
1	1.0	4	4.7	3	4.0	FALL RIVER SD	363	-14	296	96	60
				1	1.7	GRANT SD					23
		1	2.4			GREGORY SD					
1	5.4					HAAKON SD					
1	2.8	2	3.3	2	4.9	HAMLIN SD	18	47	73	75	74
		2	5.5			HAND SD					
1	4.7			1	5.1	HANSON SD			9	53	77
				2	4.2	HUGHES SD					64
				3	4.7	HUTCHINSON SD					71
		1	4.7	1	5.1	HYDE SD		7			76
				2	3.6	KINGSBURY SD					53
1	2.0			3	4.5	LAKE SD			132	85	69
		4	5.1	1	1.0	LAWRENCE SD		-81			16
2	2.7	1	1.5			LINCOLN SD	-44				
				1	4.2	LYMAN SD					63
2	4.9	3	6.0			MC COOK SD	22				
				1	3.1	MC PHERSON SD					45
		1	2.2	1	2.3	MARSHALL SD		5			32
2	3.0	1	1.5	2	2.6	MEADE SD	-49	74	-11	39	37
1	6.2			1	3.7	MINER SD			-40	19	56
12	3.6	16	4.2	12	2.9	MINNEHAHA SD	16	-31	-20	31	41
		1	2.1			MOODY SD					
2	1.2	5	2.4	12	5.1	PENNINGTON SD	100	115	329	97	76
		2	7.0			PERKINS SD					
1	1.3	1	1.4	1	1.3	ROBERTS SD	7	-8	-1	46	18
1	3.6					SANBORN SD					
1	1.6	2	3.2	3	5.4	SPINK SD	103	66	238	94	80
		1	6.5	1	7.4	STANLEY SD		14			92
				1	10.9	SULLY SD					98
		1	9.4			TODD SD					
				1	2.9	TRIPP SD					43
1	1.3	4	5.6	2	2.6	TURNER SD	333	-54	101	81	37
		1	1.8	2	3.8	UNION SD		109			56
2	1.9	2	1.8	3	3.1	YANKTON SD	-4	73	67	74	46
449	4.0	554	4.1	653	4.0	TENNESSEE TN	1	-1	0		
5	3.4	6	3.2	9	3.3	ANDERSON TN	-6	3	-3	45	49
3	2.9	1	.9	7	5.1	BEDFORD TN	-70	492	76	76	76
4	7.0	1	1.5	1	1.2	BENTON TN	-79	-23	-84	00	17
		1	2.1	3	7.6	BLEDSOE TN		258			93
6	3.8	9	3.1	8	2.3	BLOUNT TN	-17	-28	-40	18	31
2	1.7	5	3.3	12	5.3	BRADLEY TN	99	60	219	92	79
4	3.7	8	6.7	12	7.7	CAMPBELL TN	80	15	107	82	94
1	2.1	2	4.1	2	4.7	CANNON TN	94	14	122	84	70
1	.9	2	1.6	3	2.0	CARROLL TN	80	26	128	85	27
3	2.7	4	2.5	8	3.5	CARTER TN	-9	40	28	62	51
1	2.2	2	3.7	4	5.9	CHEATHAM TN	69	59	170	89	84
1	2.6	1	2.4	2	3.6	CHESTER TN	-9	51	36	65	53
4	4.8	2	1.9	4	4.0	CLAIBORNE TN	-60	110	-16	35	60
2	5.7			1	2.3	CLAY TN			-59	08	32

* SIGNIFICANTLY MORE DEATHS THAN EXPECTED

WHITE MALE — CANCER OF THE ORAL CAVITY INCLUDING TONGUE — ICD 141,143-146,148,149

1950-1959 DEATHS	1950-1959 RATE	1960-1969 DEATHS	1960-1969 RATE	1970-1979 DEATHS	1970-1979 RATE	COUNTY AND STATE		PERCENT CHANGE 60S/50S	PERCENT CHANGE 70S/60S	PERCENT CHANGE 70S/50S	PERCENTILE %CHANGE 70S/50S	PERCENTILE RATES 1970S
6	6.1	2	2.0	4	3.3	COCKE	TN	-68	67	-46	15	49
4	4.0	3	2.7	4	2.8	COFFEE	TN	-32	1	-31	24	39
		1	1.3			CROCKETT	TN					
5	6.6	4	4.3	3	1.9	CUMBERLAND	TN	-35	-57	-72	04	26
61	6.0	76	6.0	88	6.0	DAVIDSON	TN	0	0	0	47	65
6*	15.3	1	1.8	3	4.5	DECATUR	TN	-88	153	-70	04	68
		2	3.7			DE KALB	TN					
1	1.0	5	4.8	8	5.9	DICKSON	TN	395	23	509	98	84
4	3.6	6	4.3	6	4.0	DYER	TN	17	-5	11	54	61
2	5.1	1	2.5	1	1.3	FAYETTE	TN	-51	-49	-75	03	19
1	1.9	2	3.4			FENTRESS	TN	73				
6	5.9	5	4.9	7	5.2	FRANKLIN	TN	-16	6	-11	39	78
6	3.4	13	6.1	7	3.1	GIBSON	TN	78	-49	-9	40	46
3	3.6	1	1.0	4	3.7	GILES	TN	-74	285	2	48	55
2	3.4	2	3.0	2	2.6	GRAINGER	TN	-11	-13	-23	29	37
4	2.4	5	2.4	3	1.3	GREENE	TN	3	-48	-47	15	18
1	2.0	2	3.3	4	6.6	GRUNDY	TN	68	99	236	93	89
1	1.4	3	2.4	3	1.7	HAMBLEN	TN	64	-29	17	57	23
40	6.3	47	6.2	48	4.8	HAMILTON	TN	-2	-22	-23	28	73
2	6.3					HANCOCK	TN					
1	1.2	1	1.3	4	4.8	HARDEMAN	TN	5	275	294	96	72
3	4.6	2	2.4	2	1.8	HARDIN	TN	-49	-23	-61	08	25
5	5.2	3	2.3	3	2.0	HAWKINS	TN	-56	-11	-61	08	28
				1	1.8	HAYWOOD	TN					25
3	4.2	1	1.2	2	2.1	HENDERSON	TN	-70	70	-50	13	28
6	6.0	1	.8	4	3.1	HENRY	TN	-87	298	-48	14	46
2	3.2	1	1.8	3	4.9	HICKMAN	TN	-43	167	51	70	74
3	12.1	1	3.3	1	3.2	HOUSTON	TN	-73	-1	-73	03	48
5	9.2	2	3.4	1	1.1	HUMPHREYS	TN	-63	-66	-88	00	17
5	10.1	2	3.7			JACKSON	TN	-63				
2	2.4	3	2.9	5	4.2	JEFFERSON	TN	21	44	74	75	63
1	2.0	1	1.8	1	1.4	JOHNSON	TN	-10	-21	-30	24	20
31	3.4	43	4.2	62	5.4	KNOX	TN	23	30	61	72	80
		2	5.7	1	2.8	LAKE	TN		-51			39
2	3.7	4	5.4	2	2.4	LAUDERDALE	TN	46	-55	-35	21	33
5	4.5	6	4.4	5	3.3	LAWRENCE	TN	-1	-25	-26	27	49
1	3.0	1	3.1	2	5.2	LEWIS	TN	6	65	75	76	78
6	6.4	4	3.6	5	4.0	LINCOLN	TN	-44	11	-38	20	60
6	7.0	1	1.2	4	3.4	LOUDON	TN	-83	190	-52	12	50
3	2.8	11	7.2	6	3.7	MC MINN	TN	158	-49	32	63	54
1	1.4	2	2.1	2	2.1	MC NAIRY	TN	52	1	53	70	28
		1	1.6	3	3.8	MACON	TN		140			57
3	1.6	6	2.8	7	2.9	MADISON	TN	72	4	78	77	41
3	4.4	4	4.6	3	2.8	MARION	TN	5	-39	-35	21	40
1	1.4	2	2.6	1	1.0	MARSHALL	TN	89	-60	-25	27	16
9	6.4	6	3.6	9	5.0	MAURY	TN	-43	39	-21	30	76
5	5.3	4	4.1	2	1.8	MONROE	TN	-24	-57	-67	06	24
4	3.1	9	6.9	9	5.3	MONTGOMERY	TN	120	-23	68	74	79
1	5.7	1	6.7	1	4.3	MOORE	TN	18	-37	-26	27	64
3	5.2	1	1.6	3	4.1	MORGAN	TN	-69	157	-21	31	62
4	3.4	6	3.8	11	7.5	OBION	TN	13	96	121	84	93
4	5.2	1	1.1	5	5.5	OVERTON	TN	-78	390	6	52	81
		1	2.6	2	6.3	PERRY	TN		148			87
		2	13.2			PICKETT	TN					
3	6.1	1	2.0	4	6.1	POLK	TN	-67	205	1	48	86
2	1.5	5	3.4	8	4.3	PUTNAM	TN	127	29	193	91	65
2	3.9	4	5.7	4	4.8	RHEA	TN	45	-15	23	60	72
7	5.9	10	7.1	13	7.6	ROANE	TN	22	7	30	63	93
4	3.7	4	3.1	5	3.4	ROBERTSON	TN	-16	10	-8	41	49
3	2.4	7	3.8	13	5.5	RUTHERFORD	TN	61	45	132	85	81
1	1.7	1	1.5	2	2.9	SCOTT	TN	-11	92	71	75	42
1	4.5	2	8.6			SEQUATCHIE	TN	91				
		3	2.6	8	4.8	SEVIER	TN		82			72
75*	6.4	101*	7.0	77	4.6	SHELBY	TN	10	-34	-28	26	69
		2	2.6	3	3.9	SMITH	TN		52			58
1	2.1			1	2.2	STEWART	TN			1	48	30
4	1.6	9	2.8	17	3.1	SULLIVAN	TN	79	13	102	81	46
3	2.0	4	2.0	5	1.8	SUMNER	TN	-1	-12	-12	38	24
1	1.3	1	1.2	5	5.3	TIPTON	TN	-8	351	315	96	79
		2	7.9			TROUSDALE	TN					
		1	1.4	5	6.4	UNICOI	TN		365			87
		1	2.4	1	2.4	UNION	TN		3			34
1	1.1	2	1.6	2	1.4	WARREN	TN	50	-11	33	63	20
11	3.8	18	5.4	15	4.0	WASHINGTON	TN	43	-27	5	51	60
3	6.9			2	3.0	WAYNE	TN			-56	10	44
3	2.2	4	2.6	5	2.5	WEAKLEY	TN	18	-5	12	55	35
		1	1.4	3	3.3	WHITE	TN		135			49
1	1.0	5	4.5	9	6.3	WILLIAMSON	TN	355	39	532	99	87
3	2.7	5	3.6	3	1.7	WILSON	TN	31	-54	-39	19	23
1126	4.1	1644	4.6	1964	4.5	TEXAS	TX	13	-3	10		
5	4.3	2	1.4	1	.6	ANDERSON	TX	-67	-56	-85	00	14
		1	1.8			ANDREWS	TX					
6	4.2	12	6.6	5	2.3	ANGELINA	TX	60	-66	-45	16	31
		2	4.7	3	4.1	ARANSAS	TX		-13			62
1	3.8			2	5.1	ARCHER	TX			37	65	77
		1	7.5			ARMSTRONG	TX					
3	3.6	2	2.2	8	7.8	ATASCOSA	TX	-37	251	119	84	94
2	3.1	1	1.4	6	6.5	AUSTIN	TX	-54	362	111	83	89
		1	2.1	1	2.7	BAILEY	TX		32			38
2	7.2	1	3.6	2	3.9	BANDERA	TX	-50	10	-45	16	59
1	1.2	5	5.3	2	1.6	BASTROP	TX	358	-70	39	66	22
3	8.8	2	5.3			BAYLOR	TX	-40				
3	5.7	3	4.6	5	6.3	BEE	TX	-19	37	10	54	87
6	2.0	15	4.1	21	5.2	BELL	TX	100	28	156	88	78
117*	6.6	153*	6.8	158	5.5	BEXAR	TX	2	-18	-17	34	81
		2	7.4	1	3.8	BLANCO	TX		-48			57
1	1.2	6	6.5	4	2.8	BOSQUE	TX	441	-57	134	85	40
11	5.8	8	3.4	7	2.5	BOWIE	TX	-42	-25	-56	10	36
5	2.3	13	4.9	19	4.4	BRAZORIA	TX	110	-9	92	79	67

* SIGNIFICANTLY MORE DEATHS THAN EXPECTED

WHITE MALE — CANCER OF THE ORAL CAVITY INCLUDING TONGUE — ICD 141,143-146,148,149

1950-1959 DEATHS	1950-1959 RATE	1960-1969 DEATHS	1960-1969 RATE	1970-1979 DEATHS	1970-1979 RATE	COUNTY AND STATE	PERCENT CHANGE 60S/50S	PERCENT CHANGE 70S/60S	PERCENT CHANGE 70S/50S	PERCENTILE %CHANGE 70S/50S	PERCENTILE RATES 1970S
3	3.7	6	4.9	11	6.1	BRAZOS TX	30	25	62	73	86
		2	7.8	3	9.8	BREWSTER TX		27			97
		1	5.4			BRISCOE TX					
				1	2.8	BROOKS TX					40
5	3.5	7	4.3	8	4.2	BROWN TX	21	-2	19	58	64
		6*	11.7	3	4.9	BURLESON TX		-58			74
2	3.2	2	2.4	3	1.7	BURNET TX	-25	-28	-46	15	23
		4	5.0	3	3.1	CALDWELL TX		-37			46
		1	1.8	2	4.0	CALHOUN TX		122			59
4	7.2	1	1.4	1	1.2	CALLAHAN TX	-81	-11	-83	00	18
20	5.6	16	3.2	23	3.6	CAMERON TX	-42	11	-36	21	53
1	3.6	2	5.8	2	6.0	CAMP TX	59	3	65	73	85
		2	5.1			CARSON TX					
2	1.9	4	3.8	5	4.2	CASS TX	98	11	121	84	64
1	4.4	1	4.9			CASTRO TX	10				
2	8.6			2	4.4	CHAMBERS TX			-49	13	66
3	2.2	3	1.7	6	3.1	CHEROKEE TX	-24	82	39	66	46
		1	2.1			CHILDRESS TX					
1	1.9	1	1.6	1	1.5	CLAY TX	-14	-5	-19	33	21
		1	5.6			COCHRAN TX					
		1	4.1	1	3.9	COKE TX		-4			59
4	4.7	2	3.3	2	2.2	COLEMAN TX	-31	-31	-53	12	30
3	1.6	12	5.3	6	2.0	COLLIN TX	235	-63	24	60	27
1	2.3			1	2.5	COLLINGSWORTH TX			11	54	35
1	1.4	1	1.3	2	2.4	COLORADO TX	-6	82	71	75	34
2	2.3	4	4.0	5	2.8	COMAL TX	72	-31	18	58	39
3	3.0	3	3.2	3	2.4	COMANCHE TX	7	-24	-19	32	33
2	11.3			1	4.4	CONCHO TX			-61	08	66
1	.8	2	1.7	4	2.7	COOKE TX	107	59	228	93	38
2	2.2	4	4.2			CORYELL TX	92				
				2	10.2	COTTLE TX					98
		3	7.2	2	4.3	CROSBY TX		-40			65
1	2.6			1	2.8	DALLAM TX			6	52	39
127*	6.0	202*	6.6	221*	5.7	DALLAS TX	10	-13	-5	43	83
2	3.0	3	4.1	1	1.3	DAWSON TX	38	-68	-57	10	19
		3	5.7	2	2.7	DEAF SMITH TX		-53			38
				2	4.1	DELTA TX					62
8	4.4	5	2.7	10	3.3	DENTON TX	-39	25	-24	28	49
5	4.8	5	4.4	11*	9.5	DE WITT TX	-9	117	98	81	97
		1	2.8	2	8.2	DICKENS TX		199			95
2	5.6	3	8.4	2	5.3	DIMMIT TX	50	-37	-6	43	79
1	2.7			1	5.7	DONLEY TX			111	83	83
2	4.5	1	1.5	5	9.2	DUVAL TX	-68	525	102	81	96
3	2.1	6	3.5	3	1.6	EASTLAND TX	65	-54	-24	28	22
1	.2	8	2.4	12	3.8	ECTOR TX	902	59	1490	99	56
				1	12.4	EDWARDS TX					99
6	3.7	7	3.7	11	4.9	ELLIS TX	-2	34	32	63	74
15	2.3	42	5.2	45	4.7	EL PASO TX	123	-9	103	82	71
2	1.9	2	1.5	5	3.5	ERATH TX	-24	138	81	78	52
2	1.9	5	4.6	3	3.2	FALLS TX	141	-30	69	74	47
2	1.9	6	3.4	3	1.5	FANNIN TX	75	-57	-25	28	20
2	1.6	5	4.1	7	5.1	FAYETTE TX	151	26	216	92	77
1	3.0	1	2.4	1	2.2	FISHER TX	-22	-8	-29	25	30
		2	3.6	1	1.8	FLOYD TX		-50			25
				1	5.0	FOARD TX					75
1	1.0	10	7.3	6	2.4	FORT BEND TX	647	-67	147	87	33
		1	2.9	1	1.7	FRANKLIN TX		-41			23
2	2.9	1	1.2	5	7.7	FREESTONE TX	-57	517	164	89	94
		2	5.0	1	2.1	FRIO TX		-58			29
				2	3.4	GAINES TX					51
36*	10.0	33*	6.9	43*	7.7	GALVESTON TX	-31	12	-23	28	93
2	6.7					GARZA TX					
		5	7.0	5	7.3	GILLESPIE TX		5			92
		3	10.7	1	6.3	GOLIAD TX		-41			87
1	1.4	2	2.8	2	2.5	GONZALES TX	101	-10	81	77	35
2	2.2	5	6.6	6	4.2	GRAY TX	203	-36	94	80	64
14	4.2	11	2.9	9	2.2	GRAYSON TX	-30	-25	-48	14	30
9	5.9	3	1.1	14	4.3	GREGG TX	-81	275	-28	25	64
		3	5.2	5	8.9	GRIMES TX		71			96
3	3.5	5	3.9	10	5.9	GUADALUPE TX	12	51	69	74	84
1	.6	2	1.3	3	2.2	HALE TX	123	71	283	95	30
1	2.9	1	2.6	2	5.2	HALL TX	-11	101	80	77	78
1	1.3	1	1.3	2	3.5	HAMILTON TX	8	161	180	90	52
1	5.5					HANSFORD TX					
1	2.2	3	6.0			HARDEMAN TX	176				
4	5.7	6	5.7	5	4.0	HARDIN TX	-1	-29	-29	25	61
133	5.4	243*	6.3	294*	6.0	HARRIS TX	16	-5	10	54	85
2	1.6	4	3.3	8	4.9	HARRISON TX	108	50	212	92	74
				1	6.8	HARTLEY TX					90
3	5.0	1	1.5	3	5.1	HASKELL TX	-69	231	1	48	76
1	1.4	2	2.2			HAYS TX	55				
				1	4.6	HEMPHILL TX					70
3	2.8			8	4.7	HENDERSON TX			70	75	72
21	4.6	19	3.2	34	4.3	HIDALGO TX	-32	35	-8	41	65
5	3.0	7	4.3	5	3.0	HILL TX	40	-30	-2	46	43
		3	4.8	1	1.1	HOCKLEY TX		-76			17
				7	7.7	HOOD TX					94
4	3.2	8	6.8	5	3.1	HOPKINS TX	113	-54	-2	45	46
1	1.1	1	1.1	5	5.3	HOUSTON TX	-1	371	366	97	79
6	5.5	5	3.0	6	3.7	HOWARD TX	-45	21	-33	22	55
		1	8.2			HUDSPETH TX					
10	5.8	8	4.1	7	2.8	HUNT TX	-30	-31	-51	12	40
5	6.8	1	1.0	4	4.1	HUTCHINSON TX	-86	313	-40	18	62
1	16.1	1	14.9			IRION TX	-7				
4	11.6	2	5.0			JACK TX	-56				
1	4.1			2	3.1	JACKSON TX			-25	27	45
2	2.4	1	1.0	7	5.6	JASPER TX	-60	467	130	85	81
				1	7.5	JEFF DAVIS TX					93
21	4.3	42	5.5	53	6.3	JEFFERSON TX	29	15	48	69	87
1	11.2	2	8.3	1	3.5	JIM HOGG TX	-27	-58	-69	05	51
7	8.1	2	1.6	8	5.5	JIM WELLS TX	-81	250	-32	23	81

* SIGNIFICANTLY MORE DEATHS THAN EXPECTED

WHITE MALE — CANCER OF THE ORAL CAVITY INCLUDING TONGUE — ICD 141,143-146,148,149

1950-1959 DEATHS	1950-1959 RATE	1960-1969 DEATHS	1960-1969 RATE	1970-1979 DEATHS	1970-1979 RATE	COUNTY AND STATE	PERCENT CHANGE 60S/50S	PERCENT CHANGE 70S/60S	PERCENT CHANGE 70S/50S	PERCENTILE %CHANGE 70S/50S	PERCENTILE RATES 1970S
5	2.9	6	3.0	9	3.5	JOHNSON TX	3	18	22	59	53
2	1.7	3	2.6	4	3.4	JONES TX	49	32	97	80	50
2	3.1	4	7.1	2	2.6	KARNES TX	127	-63	-17	34	37
5	4.0	5	3.5	4	2.5	KAUFMAN TX	-11	-29	-37	20	35
1	3.5	2	6.0	2	3.5	KENDALL TX	70	-43	-2	45	51
		1*	23.4			KENEDY TX					
1	12.6					KENT TX					
2	1.8	6	4.7	8	4.1	KERR TX	170	-14	132	85	62
1	7.8					KINNEY TX					
		4	4.8	5	8.3	KLEBERG TX		74			95
		2	4.9	1	2.3	KNOX TX		-54			31
7	4.1	7	3.7	4	1.9	LAMAR TX	-9	-48	-53	12	27
2	3.0	1	1.2			LAMB TX	-61				
1	2.1			2	3.8	LAMPASAS TX			79	77	58
2	10.6	2	7.8			LA SALLE TX	-26				
1	.9	6	4.9	3	2.3	LAVACA TX	479	-53	170	89	31
2	4.6	1	2.0	2	3.8	LEE TX	-56	90	-17	34	57
3	6.1	3	6.7	2	4.1	LEON TX	10	-40	-34	22	61
3	2.7	5	4.1	5	3.1	LIBERTY TX	50	-25	13	55	45
2	2.5	2	2.0	8	8.9	LIMESTONE TX	-17	335	261	94	96
1	6.5			1	5.7	LIPSCOMB TX			-12	38	83
1	3.1	1	2.5			LIVE OAK TX	-21				
2	5.5	2	5.0	1	1.4	LLANO TX	-10	-73	-75	03	20
5	1.3	13	3.1	16	2.9	LUBBOCK TX	129	-8	112	83	41
1	1.9	2	3.3	3	5.3	MC CULLOCH TX	77	60	184	91	79
23	4.9	25	4.2	27	3.9	MC LENNAN TX	-14	-7	-20	31	59
3	10.9	1	3.0	2	8.3	MADISON TX	-73	179	-23	28	95
1	4.6	3	12.6	1	2.5	MARION TX	173	-80	-47	15	34
2	7.5			4*	11.9	MASON TX			57	72	98
3	4.2	5	5.0	9	7.6	MATAGORDA TX	18	53	80	77	93
		3	5.0	4	5.3	MAVERICK TX		6			79
1	1.2	6	6.3	6	5.3	MEDINA TX	426	-16	342	97	79
1	5.0					MENARD TX					
1	2.2	4	2.6	7	2.6	MIDLAND TX	15	0	15	56	36
3	2.9	2	1.5	2	1.7	MILAM TX	-48	10	-43	17	23
		1	3.1			MILLS TX					
3	5.9	2	3.8	2	3.5	MITCHELL TX	-36	-7	-41	18	52
3	3.0	3	2.5	3	2.6	MONTAGUE TX	-15	4	-11	39	37
3	3.0	16*	10.5	7	2.3	MONTGOMERY TX	244	-78	-25	28	31
1	12.5	1	2.3	2	2.8	MOORE TX	-82	22	-78	02	39
		2	4.1	2	3.7	MORRIS TX		-10			55
		1	4.5			MOTLEY TX					
3	2.6	1	.7	1	.7	NACOGDOCHES TX	-72	-8	-74	03	14
8	5.0	3	1.6	2	1.1	NAVARRO TX	-68	-34	-79	01	16
		1	2.2	2	3.3	NEWTON TX		52			48
		2	2.0	5	5.5	NOLAN TX		167			81
27	6.3	35	5.7	45	5.9	NUECES TX	-10	4	-6	42	84
2	16.0			1	2.1	OCHILTREE TX			-87	00	29
7	6.4	8	4.2	11	5.6	ORANGE TX	-35	33	-13	38	81
5	4.7	8	6.8	9	6.9	PALO PINTO TX	45	1	46	68	90
2	2.8	3	3.6	1	1.4	PANOLA TX	32	-61	-49	13	20
4	3.6	2	1.3	4	2.1	PARKER TX	-63	60	-41	18	29
1	6.2					PARMER TX					
1	3.5			1	2.5	PECOS TX			-29	25	35
3	6.5	5	8.3	7	6.3	POLK TX	28	-23	-2	46	87
9	3.9	11	2.8	17	4.4	POTTER TX	-28	57	13	55	66
1	4.1					RAINS TX					
1	1.4	5	5.6	2	1.0	RANDALL TX	290	-82	-31	24	16
1	11.5					REAGAN TX					
				1	6.9	REAL TX					90
		2	2.6	2	2.3	RED RIVER TX		-12			31
2	8.1	1	1.6	2	4.8	REEVES TX	-80	201	-40	18	73
2	7.0	1	2.8	2	4.5	REFUGIO TX	-60	61	-36	21	68
2	3.6	4	5.3	3	5.1	ROBERTSON TX	48	-5	41	67	76
1	3.6	2	6.1	5*	10.6	ROCKWALL TX	72	73	198	91	98
2	2.5	2	2.4	1	1.0	RUNNELS TX	-5	-60	-62	07	16
2	1.1	5	3.2	6	3.4	RUSK TX	185	7	205	91	51
		1	2.5	1	2.4	SABINE TX		-4			32
		2	5.6	1	2.8	SAN AUGUSTINE TX		-50			39
		1	2.7	1	2.8	SAN JACINTO TX		5			40
7	6.8	6	4.0	15*	8.2	SAN PATRICIO TX	-41	105	21	59	95
		1	1.7	2	3.5	SAN SABA TX		107			52
		1	11.8			SCHLEICHER TX					
1	.9	3	3.8	6	7.4	SCURRY TX	308	96	701	99	92
1	4.6	1	4.3			SHACKELFORD TX	-7				
1	.9	2	2.1	2	1.8	SHELBY TX	130	-16	93	80	24
9	3.7	8	2.5	18	4.1	SMITH TX	-33	67	11	55	63
3*	16.4					SOMERVELL TX					
1	2.6	1	1.5	4	5.4	STARR TX	-44	271	107	82	80
2	3.9	2	4.9	2	3.7	STEPHENS TX	27	-24	-4	44	56
1	7.3	3*	20.2	3*	18.9	SUTTON TX	176	-6	158	88	99
1	2.6			2	4.5	SWISHER TX			76	76	68
79	6.0	103	5.5	142*	5.7	TARRANT TX	-9	3	-6	43	82
8	3.0	11	3.2	10	2.3	TAYLOR TX	7	-26	-21	31	32
1	2.4	1	1.4			TERRY TX	-39				
		1	4.5	1	3.9	THROCKMORTON TX		-14			58
4	5.3	4	4.6	5	4.2	TITUS TX	-13	-9	-21	30	64
2	.8	11	3.9	12	3.7	TOM GREEN TX	373	-5	348	97	54
24	4.2	32	4.6	38	4.1	TRAVIS TX	10	-12	-4	44	61
1	2.4	3	7.5	1	2.2	TRINITY TX	209	-70	-7	41	31
3	6.1			3	3.3	TYLER TX			-47	15	48
		6	6.1	6	5.7	UPSHUR TX		-7			82
2	16.9	3*	20.1			UPTON TX	19				
3	4.1	5	6.4	3	3.6	UVALDE TX	56	-43	-10	39	54
		3	4.3	4	3.9	VAL VERDE TX		-9			59
3	2.7			5	3.1	VAN ZANDT TX			18	58	46
10*	9.2	10	7.1	5	2.6	VICTORIA TX	-23	-63	-72	04	37
2	3.2	1	1.2	9*	9.8	WALKER TX	-63	718	204	91	97
		1	2.7	5	10.1	WALLER TX		270			97
1	1.4	1	1.5	1	1.6	WARD TX	6	6	13	55	22
2	2.4	7	7.7	3	3.1	WASHINGTON TX	222	-60	30	62	45

* SIGNIFICANTLY MORE DEATHS THAN EXPECTED

WHITE MALE — CANCER OF THE ORAL CAVITY INCLUDING TONGUE — ICD 141,143-146,148,149

1950-1959 DEATHS	1950-1959 RATE	1960-1969 DEATHS	1960-1969 RATE	1970-1979 DEATHS	1970-1979 RATE	COUNTY AND STATE	PERCENT CHANGE 60S/50S	PERCENT CHANGE 70S/60S	PERCENT CHANGE 70S/50S	PERCENTILE %CHANGE 70S/50S	PERCENTILE RATES 1970S
6	3.6	10	4.6	15	5.5	WEBB TX	30	20	55	71	81
5	4.3	9	6.7	10	6.8	WHARTON TX	58	1	59	72	90
2	4.2	2	4.3			WHEELER TX	1				
13	4.1	15	3.5	24	5.2	WICHITA TX	-14	47	27	61	78
1	1.0	3	2.8	2	1.8	WILBARGER TX	197	-36	89	79	25
1	2.5	4	5.6	1	1.3	WILLACY TX	126	-77	-47	15	19
6	3.6	10	5.3	11	4.8	WILLIAMSON TX	49	-10	34	64	72
3	4.2	5	7.6	4	5.4	WILSON TX	82	-29	30	62	80
2	6.0	3	9.0	2	4.1	WINKLER TX	50	-54	-31	24	63
2	2.0	8	6.4	2	1.4	WISE TX	218	-78	-29	25	20
5	6.6	1	.8	6	4.0	WOOD TX	-87	388	-38	20	61
		3	3.0	3	2.6	YOUNG TX		-13			37
				1	2.6	ZAPATA TX					37
1	2.5	2	4.2	1	2.0	ZAVALA TX	67	-52	-19	32	28
74	2.9	76	2.3	82	2.0	UTAH UT	-21	-14	-32		
				1	4.0	BEAVER UT					60
2	2.9	1	1.2	5	4.8	BOX ELDER UT	-59	311	67	74	73
1	.8	3	2.1	3	1.7	CACHE UT	159	-18	112	83	23
1	.9	3	3.4	1	1.2	CARBON UT	265	-65	34	64	18
		4	1.8	4	1.4	DAVIS UT		-26			19
		1	4.0			DUCHESNE UT					
1	11.8			1	3.9	GRAND UT			-67	05	58
		1	2.2	2	3.7	IRON UT		67			55
2	10.0	1	4.5			JUAB UT	-55				
1	2.7					MILLARD UT					
1	15.7					RICH UT					
43	4.0	34	2.4	45	2.6	SALT LAKE UT	-40	9	-35	21	37
1	1.4	1	1.4	1	1.2	SANPETE UT	-6	-10	-15	36	18
		1	2.2	1	1.3	SEVIER UT		-42			18
2	3.2	1	1.1	2	2.1	TOOELE UT	-65	88	-33	22	29
		1	5.7	1	1.9	UINTAH UT		-67			26
3	1.2	6	1.5	7	1.4	UTAH UT	24	-8	13	56	20
2	3.7			1	.8	WASHINGTON UT			-78	02	15
14	4.1	18	4.5	7	1.4	WEBER UT	11	-69	-65	06	20
65	3.5	80	4.2	100	4.7	VERMONT VT	19	13	34		
4	4.5	7	7.3	4	3.7	ADDISON VT	63	-48	-16	35	56
4	3.4	4	2.7	5	3.1	BENNINGTON VT	-22	18	-7	41	46
3	2.4	7	5.7	5	3.9	CALEDONIA VT	141	-32	65	73	59
5	2.2	13	4.7	26*	7.4	CHITTENDEN VT	109	60	234	93	93
		1	2.8	2	5.6	ESSEX VT		99			82
6	4.3	6	4.1	5	3.3	FRANKLIN VT	-5	-18	-23	29	49
1	5.1	2	10.0			GRAND ISLE VT	97				
1	1.5			2	3.2	LAMOILLE VT			113	83	47
2	2.3	2	2.0	4	4.1	ORANGE VT	-11	98	77	76	61
1	.9	2	2.2	4	4.0	ORLEANS VT	155	84	369	97	61
12	5.0	13	5.4	16	5.8	RUTLAND VT	8	8	17	58	84
8	4.2	7	3.1	7	3.1	WASHINGTON VT	-26	0	-26	26	45
6	3.8	7	4.4	8	4.6	WINDHAM VT	17	4	22	59	70
12	5.6	9	4.0	12	4.9	WINDSOR VT	-28	22	-13	38	73
447	4.5	599	4.5	765	4.7	VIRGINIA VA	1	4	4		
7	6.0	2	1.4	6	4.9	ACCOMACK VA	-76	240	-18	33	74
11	5.9	9	4.1	13	4.7	ALBEMARLE#VA	-30	13	-21	30	70
5	4.1	3	2.2	6	4.7	ALLEGHANY#VA	-46	112	14	56	71
1	5.4					AMELIA VA					
2	3.0	2	2.7	7	6.8	AMHERST VA	-10	155	129	85	90
		2	4.7	2	4.4	APPOMATTOX VA		-6			66
22	6.0	50*	8.3	34	5.6	ARLINGTON#VA	38	-33	-7	42	82
6	2.3	10	3.2	23	6.0	AUGUSTA#VA	36	91	159	88	85
2	7.3	1	3.3	1	3.2	BATH VA	-55	-4	-57	09	47
6	1.6	19	4.2	24	4.2	BEDFORD#VA	159	1	162	88	64
1	6.8					BLAND VA					
3	4.5			2	1.9	BOTETOURT VA			-57	09	27
1	2.4			2	4.6	BRUNSWICK VA			90	79	69
				2	1.5	BUCHANAN VA					21
				1	2.6	BUCKINGHAM VA					36
1	3.3	1	2.7	2	5.6	CAROLINE VA	-17	106	70	75	82
4	2.0	2	.9	8	3.1	CARROLL#VA	-56	266	59	72	46
1	17.5					CHARLES CITY VA					
2	4.1	1	2.1	3	6.3	CHARLOTTE VA	-50	206	52	70	87
53*	6.7	74*	8.0	68*	6.9	CHESTERFIELD#VA	20	-15	3	49	90
		1	2.8	1	2.5	CLARKE VA		-11			35
1	5.7					CRAIG VA					
2	3.4	2	2.8	2	2.3	CULPEPER VA	-15	-19	-31	24	32
2	13.4	1	6.1			CUMBERLAND VA	-55				
		1	1.2	1	1.1	DICKENSON VA		-10			17
6	3.1	8	3.7	15	6.1	DINWIDDIE#VA	19	66	97	80	86
1	5.0			1	2.9	ESSEX VA			-42	17	42
33	7.4	54	5.4	82	5.2	FAIRFAX#VA	-28	-4	-31	24	77
7	11.3	4	4.7	6	5.7	FAUQUIER VA	-58	20	-50	13	82
2	3.6	2	3.4	1	1.6	FLOYD VA	-6	-53	-56	10	22
				1	2.8	FLUVANNA VA					39
4	4.2	3	2.9	6	4.7	FRANKLIN VA	-30	64	14	56	72
6	4.4	2	1.1	17*	7.6	FREDERICK#VA	-75	600	73	75	93
		3	6.1	6	6.9	GILES VA		13			90
1	3.0			3	4.5	GLOUCESTER VA			53	70	68
2	8.6	1	3.8	1	3.6	GOOCHLAND VA	-56	-4	-58	09	54
1	6.1					GREENE VA					
2	8.4	2	6.2	2	5.0	GREENSVILLE#VA	-26	-19	-40	18	76
7	7.2	5	4.9	7	5.5	HALIFAX#VA	-32	13	-23	29	81
3	5.6	8	7.6	4	2.9	HANOVER VA	37	-62	-47	15	42
15	6.5	17	3.7	38	6.0	HENRICO VA	-44	64	-8	41	85
2	2.3	7	4.4	19*	8.1	HENRY#VA	90	84	249	94	95
1	5.8					HIGHLAND VA					
1	2.7			4	7.3	ISLE OF WIGHT VA			172	90	92
7	6.7	6	4.1	8	3.5	JAMES CITY#VA	-39	-15	-48	14	51

* SIGNIFICANTLY MORE DEATHS THAN EXPECTED

\# COUNTIES COMBINED; SEE TECHNICAL NOTES

WHITE MALE — CANCER OF THE ORAL CAVITY INCLUDING TONGUE — ICD 141,143-146,148,149

1950-1959 DEATHS	1950-1959 RATE	1960-1969 DEATHS	1960-1969 RATE	1970-1979 DEATHS	1970-1979 RATE	COUNTY AND STATE	PERCENT CHANGE 60S/50S	PERCENT CHANGE 70S/60S	PERCENT CHANGE 70S/50S	PERCENTILE %CHANGE 70S/50S	PERCENTILE RATES 1970S
				4*	17.3	KING GEORGE VA					99
2	15.0			3	10.9	KING WILLIAM VA			-27	26	98
2	6.2	2	5.5	4	10.5	LANCASTER VA	-11	91	71	75	98
5	4.3	2	1.6	1	.7	LEE VA	-62	-57	-84	00	15
3	3.4	6	5.4	8	6.0	LOUDOUN VA	59	11	77	76	85
2	4.3	2	6.5	5	8.9	LOUISA VA	51	38	108	82	96
2	4.7			1	1.9	LUNENBURG VA			-60	08	26
1	3.4			1	2.4	MADISON VA			-30	24	33
2	6.2	3	7.2	1	3.7	MATHEWS VA	16	-48	-40	19	56
2	2.3	5	5.3	4	4.2	MECKLENBURG VA	133	-22	82	78	63
1	4.0	1	4.7	2	6.7	MIDDLESEX VA	18	42	67	74	89
2	1.5	2	1.4	10	4.9	MONTGOMERY#VA	-5	253	235	93	73
1	2.1	2	4.1	1	1.7	NELSON VA	96	-58	-18	33	23
1	2.1	2	4.3	2	4.4	NORTHAMPTON VA	109	1	111	83	66
4	14.4	1	1.9			NORTHUMBERLAND VA	-87				
		3	6.3	2	4.0	NOTTOWAY VA		-36			60
2	4.2	1	1.7	5	6.9	ORANGE VA	-60	306	64	73	90
5	7.2	3	3.9	5	5.2	PAGE VA	-46	33	-28	26	78
2	3.7	3	4.3	3	3.3	PATRICK VA	17	-23	-10	40	49
9	3.1	13	4.0	19	4.7	PITTSYLVANIA#VA	30	18	53	70	70
				1	5.7	POWHATAN VA					83
1	2.3	3	7.2	3	6.1	PRINCE EDWARD VA	211	-15	163	89	86
1	1.1	7	5.8	10	4.8	PRINCE WILLIAM#VA	420	-18	328	97	72
1	1.6	5	4.5	3	2.8	PULASKI VA	176	-37	74	76	40
1	4.3	2	9.4	2	6.9	RAPPAHANNOCK VA	121	-26	63	73	91
1	4.7	1	6.2	3	9.2	RICHMOND VA	31	49	96	80	97
18	3.5	29	4.2	37	4.5	ROANOKE#VA	20	8	29	62	68
3	3.1	5	4.4	6	4.2	ROCKBRIDGE#VA	44	-6	37	65	64
9	4.9	11	4.8	6	2.0	ROCKINGHAM#VA	-2	-58	-59	08	28
2	2.1	3	2.7	1	.8	RUSSELL VA	28	-71	-63	07	15
1	.9	3	2.4	1	.7	SCOTT VA	177	-71	-19	32	15
2	1.9	4	3.6	7	5.2	SHENANDOAH VA	92	44	177	90	78
		2	1.3	8	5.0	SMYTH VA		299			76
4	10.2	1	1.7	5	8.2	SOUTHAMPTON#VA	-83	375	-19	32	95
4	4.1	4	3.5	12*	8.4	SPOTSYLVANIA#VA	-15	138	103	82	95
3	6.9	2	2.5	2	1.6	STAFFORD VA	-64	-36	-77	02	22
1	4.4			2	6.8	SUSSEX VA			54	71	90
4	4.0	4	2.2	1	.5	TAZEWELL VA	-46	-77	-88	00	14
				1	1.0	WARREN VA					16
5	2.9	9	4.0	8	2.9	WASHINGTON#VA	39	-28	0	48	42
1	2.9	3	6.8	2	3.0	WESTMORELAND VA	135	-55	4	50	44
4	2.2	5	2.6	7	3.3	WISE#VA	22	26	53	71	49
		3	3.0	4	3.8	WYTHE VA		27			57
77*	8.0	96*	7.5	85	5.1	CHESAPEAKE CITY#VA	-7	-31	-36	21	77
11	6.5	20*	8.2	18	5.8	HAMPTON CITY VA	26	-30	-12	38	83
9	4.7	17	7.2	12	3.9	NEWPORT NEWS CITY VA	52	-46	-18	33	58
2	3.2	6	5.3	8	7.9	SUFFOLK CITY VA	68	49	150	87	94
373	3.2	564	4.1	604	3.7	WASHINGTON WA	29	-8	18		
		2	7.9	1	1.6	ADAMS WA		-80			22
		1	1.7	2	2.0	ASOTIN WA		15			27
2	1.6	11	5.2	18	6.0	BENTON WA	226	16	278	95	85
8	4.1	7	3.1	10	4.0	CHELAN WA	-25	30	-2	45	60
2	1.3	3	1.7	8	3.1	CLALLAM WA	32	78	135	86	46
12	3.2	17	3.4	16	2.3	CLARK WA	6	-31	-27	26	32
				1	4.6	COLUMBIA WA					69
5	2.7	8	3.5	4	1.3	COWLITZ WA	30	-63	-52	12	19
1	1.9	1	1.5	2	1.9	DOUGLAS WA	-19	23	-1	47	26
1	6.6					FERRY WA					
1	2.0	3	3.8	7	6.4	FRANKLIN WA	92	68	223	92	88
		1	.8	3	1.3	GRANT WA		59			19
10	3.6	19	6.2	11	3.4	GRAYS HARBOR WA	71	-45	-6	42	50
1	1.7	3	3.6	10	5.7	ISLAND WA	111	59	234	93	83
		2	3.4	3	4.1	JEFFERSON WA		22			62
166	4.6	233*	5.4	200	4.2	KING WA	17	-23	-10	39	63
9	2.3	20	4.7	26	4.4	KITSAP WA	104	-6	93	80	67
2	1.6	6	5.2	4	3.5	KITTITAS WA	228	-32	123	84	53
		1	1.3			KLICKITAT WA					
7	2.7	4	1.5	5	1.7	LEWIS WA	-45	14	-37	20	23
2	3.7	2	3.3	2	2.8	LINCOLN WA	-11	-16	-25	27	40
2	3.0	1	1.0	3	1.8	MASON WA	-66	77	-40	19	24
1	.8	4	3.0	4	2.7	OKANOGAN WA	295	-8	263	95	38
3	3.0	2	2.3	4	3.5	PACIFIC WA	-23	51	16	57	53
1	2.0	2	4.5			PEND OREILLE WA	125				
37	2.9	60	4.1	83	4.8	PIERCE WA	42	18	67	74	72
		2	8.4	2	4.3	SAN JUAN WA		-49			65
5	1.9	9	3.0	8	2.6	SKAGIT WA	58	-14	35	64	36
		2	5.1	1	2.4	SKAMANIA WA		-53			33
19	2.9	32	3.6	42	4.0	SNOHOMISH WA	24	10	37	65	60
36	2.9	55	4.1	52	3.6	SPOKANE WA	43	-12	27	61	54
4	3.8	4	3.5	1	.8	STEVENS WA	-8	-77	-79	01	15
2	.8	7	2.2	15	3.5	THURSTON WA	166	56	314	96	52
		2	10.5			WAHKIAKUM WA					
3	1.5	6	2.6	7	2.9	WALLA WALLA WA	78	11	97	81	42
4	1.1	10	2.6	13	3.0	WHATCOM WA	136	15	173	90	43
3	2.2	5	3.6	4	2.9	WHITMAN WA	60	-20	29	62	41
24	4.0	17	2.4	32	4.1	YAKIMA WA	-40	72	3	49	62
241	3.2	285	3.4	299	3.4	WEST VIRGINIA WV	9	-1	8		
2	2.1	2	2.9	1	1.1	BARBOUR WV	37	-63	-48	14	17
2	1.3	6	3.3	14	6.7	BERKELEY WV	158	107	434	98	90
1	1.6	2	1.6	7	5.9	BOONE WV	2	270	277	95	84
1	1.2	2	3.0	1	1.0	BRAXTON WV	147	-66	-16	35	16
7	5.9	7	5.2	3	1.9	BROOKE WV	-12	-63	-67	05	26
22	4.6	26	5.1	23	4.6	CABELL WV	11	-10	-1	47	69
				3	6.3	CALHOUN WV					87
				2	3.6	CLAY WV					54
1	1.7			1	2.6	DODDRIDGE WV			52	70	36
9	3.5	9	3.8	12	4.8	FAYETTE WV	7	26	36	64	72

* SIGNIFICANTLY MORE DEATHS THAN EXPECTED

\# COUNTIES COMBINED; SEE TECHNICAL NOTES

WHITE MALE — CANCER OF THE ORAL CAVITY INCLUDING TONGUE — ICD 141,143-146,148,149

1950-1959 DEATHS	1950-1959 RATE	1960-1969 DEATHS	1960-1969 RATE	1970-1979 DEATHS	1970-1979 RATE	COUNTY AND STATE	PERCENT CHANGE 60S/50S	PERCENT CHANGE 70S/60S	PERCENT CHANGE 70S/50S	PERCENTILE %CHANGE 70S/50S	PERCENTILE RATES 1970S
		2	5.0			GILMER WV					
1	2.8	1	2.5	1	1.8	GRANT WV	-9	-27	-34	22	25
6	4.4	1	.6	2	1.1	GREENBRIER WV	-85	74	-74	03	17
1	2.0	2	3.5			HAMPSHIRE WV	80				
4	3.3	8	5.4	10	5.5	HANCOCK WV	65	2	69	74	81
		2	3.9	1	1.6	HARDY WV		-58			23
14	3.7	17	4.3	5	1.2	HARRISON WV	15	-73	-69	05	17
3	4.0	3	3.7	6	6.1	JACKSON WV	-8	66	52	70	86
1	2.5	4	4.7	6	5.9	JEFFERSON WV	88	26	136	86	84
41	5.8	38	4.3	41	4.1	KANAWHA WV	-26	-5	-29	25	61
5	4.2	1	.8	2	1.8	LEWIS WV	-80	118	-56	10	25
1	1.0	5	5.1	2	1.8	LINCOLN WV	415	-66	77	76	24
7	4.4	7	3.7	3	1.6	LOGAN WV	-16	-58	-64	06	22
5	2.1	7	4.3	9	5.2	MC DOWELL WV	103	23	149	87	78
8	2.6	9	2.7	15	4.6	MARION WV	7	69	80	77	69
13	6.9	4	2.3	12	6.1	MARSHALL WV	-66	163	-11	39	86
		2	1.8	1	.8	MASON WV		-58			15
5	1.8	10	3.6	13	4.4	MERCER WV	104	23	151	87	66
1	1.1	6	5.7	5	4.3	MINERAL WV	422	-24	296	96	65
3	2.3	7	4.8			MINGO WV	110				
9	4.2	14	5.6	12	4.6	MONONGALIA WV	33	-18	9	53	70
				2	2.6	MONROE WV					37
		1	2.4	1	1.8	MORGAN WV		-24			25
2	1.9	1	.9	1	.8	NICHOLAS WV	-51	-14	-58	09	15
18	5.2	24	6.7	19	5.7	OHIO WV	29	-15	10	53	83
		2	3.9	1	2.0	PENDLETON WV		-49			27
2	3.3					POCAHONTAS WV					
5	3.5	1	.6	4	2.8	PRESTON WV	-82	337	-20	31	40
2	2.1	1	.9	4	2.9	PUTNAM WV	-57	223	38	65	42
5	2.4	4	1.3	8	2.4	RALEIGH WV	-45	80	-1	46	33
1	.8	1	.8	8	5.5	RANDOLPH WV	-6	585	544	99	80
1	1.3	1	1.8	1	1.2	RITCHIE WV	35	-35	-12	38	17
		2	2.4	1	.9	ROANE WV		-62			16
3	4.4	3	3.8	6	8.4	SUMMERS WV	-13	120	91	79	95
2	2.1	3	3.7	1	1.5	TAYLOR WV	75	-60	-31	24	21
2	3.5	1	2.5			TUCKER WV	-27				
		1	2.0			TYLER WV					
1	1.1	1	.9	3	2.8	UPSHUR WV	-15	203	158	88	40
7	4.5	7	4.0	9	4.7	WAYNE WV	-11	17	5	51	71
3	4.5	1	1.5			WEBSTER WV	-66				
3	2.8	4	4.1	5	4.6	WETZEL WV	45	12	62	73	69
1	2.6	1	3.1			WIRT WV	20				
9	2.9	18	4.9	10	2.5	WOOD WV	74	-50	-13	37	34
1	.8	3	3.0	2	2.4	WYOMING WV	291	-18	221	92	34
772	4.4	874	4.5	930	4.4	WISCONSIN WI	1	-2	-1		
1	2.0	1	1.3	3	3.6	ADAMS WI	-34	174	80	77	53
4	4.5	2	1.9	2	1.6	ASHLAND WI	-57	-17	-64	06	22
3	1.6	3	1.4	1	.4	BARRON WI	-10	-72	-75	03	14
2	2.5	1	.9	3	3.3	BAYFIELD WI	-63	252	31	63	49
19	4.2	27	5.1	27	4.5	BROWN WI	20	-11	7	52	68
1	1.3	1	1.5	4	5.4	BUFFALO WI	16	260	316	96	80
2	3.3	3	4.1			BURNETT WI	23				
6	6.1	5	4.8	4	3.0	CALUMET WI	-22	-37	-51	12	44
2	.9	6	2.9	6	2.3	CHIPPEWA WI	209	-19	151	87	32
8	4.3	3	1.6	1	.5	CLARK WI	-62	-68	-88	00	14
11	5.5	8	3.6	3	1.6	COLUMBIA WI	-33	-56	-70	04	22
4	4.7	1	1.0	4	3.9	CRAWFORD WI	-79	303	-16	35	59
29	3.9	29	3.3	44	4.3	DANE WI	-15	31	11	54	65
9	3.0	12	3.5	13	3.5	DODGE WI	15	-1	14	56	51
4	3.8	3	2.5	4	2.7	DOOR WI	-35	10	-29	25	39
8	3.3	18	7.0	13	5.4	DOUGLAS WI	114	-23	66	74	80
5	3.7	4	2.8	2	1.2	DUNN WI	-23	-57	-67	05	18
6	2.5	13	4.6	13	4.1	EAU CLAIRE WI	88	-13	64	73	61
1	5.4					FLORENCE WI					
14	4.0	18	4.7	15	3.7	FOND DU LAC WI	18	-22	-8	41	55
1	1.5	4	8.0	2	3.6	FOREST WI	428	-55	140	86	54
5	2.0	7	3.2	8	3.6	GRANT WI	55	12	74	75	53
9	6.0	4	2.7	8	5.2	GREEN WI	-54	88	-14	36	77
1	1.3	5	5.3	3	3.6	GREEN LAKE WI	320	-32	183	90	53
1	1.0	5	5.1	4	4.1	IOWA WI	431	-19	331	97	63
4	7.4	2	3.5	1	3.9	IRON WI	-52	11	-47	14	59
		2	2.5	2	2.0	JACKSON WI		-19			28
11	4.3	7	2.5	15	4.9	JEFFERSON WI	-43	96	12	55	74
2	2.3	2	2.0	3	2.6	JUNEAU WI	-15	30	10	54	36
17	4.2	15	3.2	38*	7.1	KENOSHA WI	-25	123	68	74	91
		3	3.3	1	.9	KEWAUNEE WI		-72			16
13	4.1	16	4.6	16	4.1	LA CROSSE WI	11	-11	-1	46	62
1	1.0	4	4.3	6	6.2	LAFAYETTE WI	329	44	519	98	87
3	2.4	4	4.2	5	4.2	LANGLADE WI	74	0	74	76	64
4	3.1	6	4.6	6	3.6	LINCOLN WI	47	-20	17	57	54
18	5.2	5	1.3	10	2.3	MANITOWOC WI	-75	72	-56	10	31
11	3.1	12	2.8	18	3.7	MARATHON WI	-11	34	19	58	56
6	2.9	8	4.1	8	3.2	MARINETTE WI	41	-23	9	53	47
4	6.2	1	1.2	3	4.0	MARQUETTE WI	-81	242	-36	21	60
5	1.5	12	3.2	8	2.1	MENOMINEE#WI	109	-36	35	64	28
312*	7.6	341*	7.6	319*	7.1	MILWAUKEE WI	0	-6	-7	42	91
2	1.3	6	3.3	4	2.1	MONROE WI	148	-35	60	72	29
3	3.0	9	7.3	7	4.7	ONEIDA WI	141	-36	55	71	70
15	4.5	21	4.9	22	4.6	OUTAGAMIE WI	10	-6	3	49	69
6	4.7	9	5.3	6	2.3	OZAUKEE WI	14	-56	-50	13	32
				3	6.0	PEPIN WI					85
		1	.7	1	.8	PIERCE WI		4			15
1	.8	3	2.0	4	1.8	POLK WI	166	-11	135	86	25
8	4.7	6	3.1	6	2.9	PORTAGE WI	-34	-7	-38	20	42
3	3.4	2	1.9	2	1.7	PRICE WI	-43	-10	-49	13	24
22	4.0	30	4.6	42	6.1	RACINE WI	14	31	50	70	86
4	3.6	2	2.0	3	2.9	RICHLAND WI	-45	42	-21	30	41
11	2.3	22	4.3	22	3.9	ROCK WI	82	-10	64	73	58
4	4.8	2	2.2	2	2.4	RUSK WI	-54	7	-51	13	33

* SIGNIFICANTLY MORE DEATHS THAN EXPECTED

COUNTIES COMBINED; SEE TECHNICAL NOTES

1950-1959 DEATHS	1950-1959 RATE	1960-1969 DEATHS	1960-1969 RATE	1970-1979 DEATHS	1970-1979 RATE	COUNTY AND STATE	PERCENT CHANGE 60S/50S	PERCENT CHANGE 70S/60S	PERCENT CHANGE 70S/50S	PERCENTILE %CHANGE 70S/50S	PERCENTILE RATES 1970S
6	4.4	1	.7	2	1.2	ST CROIX WI	-84	69	-72	03	18
6	3.1	7	3.5	3	1.4	SAUK WI	14	-59	-53	11	20
5	7.3	2	2.1	2	2.6	SAWYER WI	-71	24	-64	07	37
24	5.7	23	5.0	18	3.6	SHEBOYGAN WI	-14	-27	-37	21	54
		1	1.1	2	1.8	TAYLOR WI		64			24
8	5.3	7	4.8	4	2.4	TREMPEALEAU WI	-9	-50	-54	11	33
5	3.0	5	3.1	4	2.5	VERNON WI	3	-20	-17	34	35
1	1.6	2	2.1	3	2.6	VILAS WI	30	28	67	74	37
8	3.1	10	3.6	16	5.0	WALWORTH WI	17	40	64	73	75
3	4.1	1	1.7	2	3.4	WASHBURN WI	-58	100	-16	35	50
3	1.5	7	3.4	18	6.5	WASHINGTON WI	129	90	336	97	88
19	4.5	33	5.4	30	3.4	WAUKESHA WI	18	-37	-26	27	50
8	3.9	7	2.7	11	4.0	WAUPACA WI	-30	46	2	48	60
2	1.8	2	1.6	7	7.0	WAUSHARA WI	-12	331	280	95	91
22	4.8	19	3.6	23	4.0	WINNEBAGO WI	-26	12	-17	34	59
6	2.8	11	3.9	15	4.7	WOOD WI	38	20	65	74	70
29	2.4	35	2.4	39	2.4	WYOMING WY	3	1	4		
1	1.6	2	3.0	2	2.3	ALBANY WY	82	-22	41	67	31
1	2.3	2	3.1	1	1.5	BIG HORN WY	32	-53	-38	20	20
				1	2.0	CAMPBELL WY					27
		5	8.0			CARBON WY					
1	3.3	3	9.3	2	6.6	CONVERSE WY	187	-30	102	81	89
2	2.7	2	2.3	2	1.7	FREMONT WY	-16	-27	-38	20	23
2	3.1	1	1.7	2	3.3	GOSHEN WY	-45	92	6	51	48
2	6.0			1	2.9	HOT SPRINGS WY			-52	12	41
		1	3.8			JOHNSON WY					
7	3.4	4	2.0	9	3.5	LARAMIE WY	-41	75	4	50	52
		2	4.4	2	4.4	LINCOLN WY		0			66
3	3.3	4	2.0	6	3.3	NATRONA WY	-39	64	0	47	49
				1	4.4	NIOBRARA WY					66
3	4.3	4	4.7	2	1.9	PARK-TETON#WY	8	-60	-57	10	26
				1	2.1	PLATTE WY					29
3	2.8	2	1.5	2	1.5	SHERIDAN WY	-47	4	-45	16	21
3	4.5	1	1.0	5	5.0	SWEETWATER WY	-79	418	10	53	75
		2	5.3			UINTA WY					

COUNTIES COMBINED; SEE TECHNICAL NOTES

WHITE FEMALE — CANCER OF THE ORAL CAVITY INCLUDING TONGUE — ICD 141,143-146,148,149

1950-1959 DEATHS	RATE	1960-1969 DEATHS	RATE	1970-1979 DEATHS	RATE	COUNTY AND STATE	PERCENT CHANGE 60S/50S	70S/60S	70S/50S	PERCENTILE %CHANGE 70S/50S	RATES 1970S
7739	1.1	11429	1.2	16928	1.5	UNITED STATES	15	21	39		
205*	2.4	200*	1.7	265	1.6	ALABAMA AL	-30	-4	-33		
1	2.4	1	1.9	1	1.0	AUTAUGA AL	-23	-47	-59	13	49
3	1.9	1	.5	3	.6	BALDWIN AL	-76	40	-67	09	37
2	3.2			1	.9	BARBOUR AL			-71	06	47
2*	4.1	1	1.8	4*	5.6	BIBB AL	-57	216	34	65	98
		1	.6			BLOUNT AL					
		2	1.9			BUTLER AL					
5	1.9	4	1.2	8	1.6	CALHOUN AL	-38	42	-12	43	71
4*	3.7	3	2.0	2	.9	CHAMBERS AL	-46	-54	-75	04	47
2	3.1	4*	5.0	4*	3.9	CHEROKEE AL	62	-21	28	62	96
3*	2.8	5*	3.7	3	1.7	CHILTON AL	32	-53	-38	28	73
2*	5.5	1	2.2			CHOCTAW AL	-60				
2	3.3	1	1.1	4*	4.1	CLARKE AL	-65	257	25	61	96
1	1.6	2	2.6	2	2.6	CLAY AL	64	1	65	76	89
2*	4.8	1	2.1			CLEBURNE AL	-56				
6*	6.0	3	2.5	1	.6	COFFEE AL	-58	-75	-90	00	37
2	1.6	4	2.3	1	.4	COLBERT AL	44	-82	-74	05	33
1	1.5			3*	4.6	CONECUH AL			219	95	97
1	2.9	1	1.9	1	1.6	COOSA AL	-33	-19	-45	22	68
6*	4.2	1	.5	6	2.3	COVINGTON AL	-88	364	-45	22	85
1	1.6	1	1.6	1	.9	CRENSHAW AL	0	-45	-45	23	45
6*	3.5	7*	2.7	5	1.3	CULLMAN AL	-23	-52	-63	11	59
3*	3.8	4*	4.2	3	2.0	DALE AL	9	-52	-48	21	80
3*	2.7	4*	2.9			DALLAS AL	6				
5*	3.0	1	.5	3	1.0	DE KALB AL	-85	110	-68	08	48
5*	5.1	1	.9	2	.8	ELMORE AL	-82	-7	-84	01	44
2	3.3	3	2.3	4	2.4	ESCAMBIA AL	-32	5	-29	35	86
8*	2.9	7	1.7	9	1.6	ETOWAH AL	-43	-2	-44	24	71
1	1.5	2	2.5	1	.8	FAYETTE AL	61	-67	-47	21	43
3*	3.6	5*	4.0	3	1.4	FRANKLIN AL	10	-66	-62	11	62
2	1.9	1	.9	2	1.9	GENEVA AL	-56	120	-3	48	77
1*	6.2					GREENE AL					
2*	4.5	3*	5.8	2	2.9	HENRY AL	29	-50	-36	29	91
5*	3.1	1	.5	3	1.0	HOUSTON AL	-85	104	-68	08	49
3	2.1	5*	2.9	1	.4	JACKSON AL	40	-86	-80	02	33
37*	2.4	39	1.7	62*	2.1	JEFFERSON AL	-31	28	-12	44	83
		1	1.7	1	1.0	LAMAR AL		-38			51
		8*	3.4	10	2.7	LAUDERDALE AL		-21			89
1	1.4	1	1.0	1	.7	LAWRENCE AL	-24	-32	-48	20	39
1	1.3	1	.8	2	1.0	LEE AL	-40	30	-22	38	50
1	1.0	2	1.4	3	1.5	LIMESTONE AL	42	13	60	74	68
2*	11.2	1	3.9	1	3.1	MACON AL	-65	-22	-73	05	92
1	.4	4	1.3	11	1.6	MADISON AL	235	17	292	97	69
1	1.9	2	3.5			MARENGO AL	80				
1	.8			1	.5	MARION AL			-43	24	34
7*	4.5	2	.8	8	2.2	MARSHALL AL	-83	185	-50	19	84
12	1.8	9	1.0	28*	2.2	MOBILE AL	-44	125	26	61	84
3*	5.8	3*	5.4	1	.9	MONROE AL	-7	-83	-84	01	47
6	1.8	7	1.3	16	2.4	MONTGOMERY AL	-28	86	34	65	87
5*	2.7	3	1.3	4	1.0	MORGAN AL	-51	-24	-63	11	50
2*	6.5	1	2.3			PERRY AL	-65				
1	1.3	1	1.0			PICKENS AL	-25				
3*	5.0	4*	4.4	4	3.5	PIKE AL	-12	-21	-30	33	94
3*	5.3	4*	4.1	3	2.2	RANDOLPH AL	-22	-46	-58	14	84
4*	5.0	3	2.7	4	2.5	RUSSELL AL	-45	-9	-50	19	87
3*	3.2	3	2.5	4	2.5	ST CLAIR AL	-20	-1	-20	39	88
		3	2.8	2	1.1	SHELBY AL		-59			54
				1	3.2	SUMTER AL					93
2	1.3	1	.5	2	.9	TALLADEGA AL	-60	76	-30	33	46
5*	5.1	2	1.4	3	1.5	TALLAPOOSA AL	-73	13	-70	07	68
4	1.5	3	.7	7	1.3	TUSCALOOSA AL	-51	84	-9	45	61
2	1.0	6	2.3	2	.5	WALKER AL	142	-79	-50	19	34
		2*	4.0			WASHINGTON AL					
				1	2.8	WILCOX AL					90
3*	4.3	3*	3.4			WINSTON AL	-22				
		2	1.0	6	1.1	ALASKA AK		14			
33	1.0	83	1.3	167	1.5	ARIZONA AZ	37	12	53		
				1	2.1	APACHE AZ					83
		3	1.3	6	1.9	COCHISE AZ		46			77
1	1.4	2	1.7	2	1.3	GILA AZ	25	-26	-8	45	59
		2	2.9	2	2.5	GRAHAM AZ		-13			87
		1	3.5			GREENLEE AZ					
24	1.4	43	1.1	90	1.3	MARICOPA AZ	-18	18	-3	48	61
1	3.9			5	2.2	MOHAVE AZ			-44	24	84
4	.4	22	1.7	39	1.6	PIMA AZ	346	-10	303	97	69
2	1.2	2	1.4	8	2.8	PINAL AZ	16	105	138	89	90
		2	3.0	4*	5.0	SANTA CRUZ AZ		67			98
1	.6	4	1.8	6	1.3	YAVAPAI AZ	178	-29	96	83	59
		2	.8	4	1.4	YUMA AZ		77			65
107*	1.6	116	1.3	178	1.4	ARKANSAS AR	-19	12	-9		
3*	3.7					ARKANSAS AR					
2	4.0	1	1.1	1	.9	ASHLEY AR	-74	-13	-77	03	46
		2	2.1	4	2.1	BAXTER AR		1			83
2	.9	4	1.6	3	.7	BENTON AR	74	-57	-25	37	38
1	.8	2	1.1			BOONE AR	35				
1	1.5	3*	4.9			BRADLEY AR	224				
				3*	8.4	CALHOUN AR					99
		1	1.0	1	1.0	CARROLL AR		-4			50
2*	4.7					CHICOT AR					
1	1.0	3	2.8	1	.5	CLARK AR	192	-83	-49	20	34

* SIGNIFICANTLY MORE DEATHS THAN EXPECTED

WHITE FEMALE — CANCER OF THE ORAL CAVITY INCLUDING TONGUE — ICD 141,143-146,148,149

1950-1959 DEATHS	1950-1959 RATE	1960-1969 DEATHS	1960-1969 RATE	1970-1979 DEATHS	1970-1979 RATE	COUNTY AND STATE		PERCENT CHANGE 60S/50S	PERCENT CHANGE 70S/60S	PERCENT CHANGE 70S/50S	PERCENTILE %CHANGE 70S/50S	PERCENTILE RATES 1970S
3	3.3	2	1.3	2	1.0	CLAY	AR	-62	-21	-70	07	49
				2	1.5	CLEBURNE	AR					66
		1	2.7	1	1.5	CLEVELAND	AR		-43			67
				4	3.3	COLUMBIA	AR					93
				2	1.4	CONWAY	AR					64
5*	2.6	3	1.2	10*	3.0	CRAIGHEAD	AR	-54	151	16	57	92
2	2.0			4	2.2	CRAWFORD	AR			10	54	84
2*	4.3	1	2.2	2	2.0	CRITTENDEN	AR	-50	-9	-54	17	79
2	5.3					CROSS	AR					
		1	1.7	2	3.6	DALLAS	AR		111			95
2*	5.5					DESHA	AR					
1	1.9	1	1.4	3	3.5	DREW	AR	-28	152	82	80	94
2	2.0	5*	3.1	1	.6	FAULKNER	AR	56	-80	-68	08	37
1	1.6	1	.9	1	.9	FRANKLIN	AR	-45	8	-41	26	48
				1	.8	FULTON	AR					44
3	1.3	8	2.4	8	1.4	GARLAND	AR	84	-42	8	53	64
				3*	4.1	GRANT	AR					96
1	.8	2	1.3	1	.4	GREENE	AR	65	-73	-56	16	32
2	2.4	4*	3.2	3	1.8	HOT SPRING	AR	31	-45	-28	35	74
		1	1.3			HOWARD	AR					
1	.9			4	2.1	INDEPENDENCE	AR			126	87	81
		1	1.6	1	1.0	IZARD	AR		-39			48
1	1.2	2	1.9			JACKSON	AR	62				
3	1.7	3	1.2	8	2.5	JEFFERSON	AR	-31	118	50	71	88
1	3.0	1	2.0	1	2.2	LAFAYETTE	AR	-32	8	-27	35	83
1	.8			2	1.3	LAWRENCE	AR			51	71	59
				2	4.6	LEE	AR					97
1	2.9			1	2.1	LINCOLN	AR			-28	35	83
3	3.2	5*	4.1	2	1.1	LOGAN	AR	29	-72	-64	10	54
				2	1.0	LONOKE	AR					51
1	1.8			1	1.1	MADISON	AR			-42	25	52
				2	1.8	MARION	AR					76
2	1.5	1	.6	3	1.7	MILLER	AR	-63	205	14	56	74
3	2.3	2	1.0	6	2.4	MISSISSIPPI	AR	-55	129	2	51	86
		1	1.6	2	3.9	MONROE	AR		142			96
		1	1.9	2	4.9	MONTGOMERY	AR		153			98
1	2.0	1	2.9	1	1.1	NEVADA	AR	46	-63	-46	22	52
		2*	8.3			NEWTON	AR					
		1	.8	1	.5	OUACHITA	AR		-35			35
2*	6.4	1	2.5	1	2.3	PERRY	AR	-61	-7	-64	11	86
		1	1.2	5*	4.2	PHILLIPS	AR		253			97
7*	6.9	3	2.7	2	1.1	POINSETT	AR	-62	-57	-83	01	55
		2	2.6	2	1.3	POLK	AR		-50			61
1	.9	1	.7	3	1.1	POPE	AR	-12	50	31	64	54
1	2.3			1	1.7	PRAIRIE	AR			-24	37	73
17*	2.3	18	1.7	19	1.4	PULASKI	AR	-27	-18	-40	27	62
				1	.7	RANDOLPH	AR					40
				1	1.3	ST FRANCIS	AR					59
		1	.5	2	1.0	SALINE	AR		83			49
1	2.2			2	3.1	SCOTT	AR			40	67	92
1	3.2			1	1.9	SEARCY	AR			-41	26	78
7*	2.4	4	1.0	11	1.9	SEBASTIAN	AR	-58	92	-19	40	78
4*	6.8	2	3.2	3	3.4	SEVIER	AR	-54	9	-50	20	94
1	2.1	1	1.4	2	1.6	SHARP	AR	-35	20	-22	38	70
		1	2.4			STONE	AR					
3	2.1	3	1.3	4	1.3	UNION	AR	-36	-5	-39	27	58
		2	3.3	3	3.6	VAN BUREN	AR		8			95
1	.4	5	1.3	5	.9	WASHINGTON	AR	248	-32	135	88	46
4	2.4	2	1.0	3	1.0	WHITE	AR	-58	-1	-59	14	49
		1	2.1	2	3.5	WOODRUFF	AR		63			94
1	1.3	1	1.0	1	.8	YELL	AR	-18	-23	-37	28	42
695	1.1	1302*	1.5	2209*	2.1	CALIFORNIA	CA	37	37	86		
44	1.1	87*	1.8	117*	2.5	ALAMEDA	CA	64	37	124	87	87
		1*	32.7			ALPINE	CA					
		2	3.0	2	1.4	AMADOR	CA		-52			65
1	.3	7	1.3	16	1.8	BUTTE	CA	403	40	602	99	76
3*	5.5	1	1.1	2	2.0	CALAVERAS	CA	-80	83	-63	11	81
		3*	5.1	1	1.3	COLUSA	CA		-75			58
14	1.1	29	1.5	56	2.0	CONTRA COSTA	CA	33	32	75	79	80
1	1.7			3	3.6	DEL NORTE	CA			109	85	95
2	2.1	2	1.1			EL DORADO	CA	-48				
13	1.1	22	1.3	35	1.6	FRESNO	CA	23	24	52	71	71
		3	3.0			GLENN	CA					
8*	2.5	3	.8	8	1.6	HUMBOLDT	CA	-67	96	-36	29	69
		3	1.1	3	1.0	IMPERIAL	CA		-14			49
		2	2.7	6*	5.7	INYO	CA		113			98
12	1.4	19	1.7	15	.9	KERN	CA	17	-43	-33	31	48
2	1.2	5	2.2	2	.7	KINGS	CA	79	-66	-40	27	40
1	1.2	2	1.6	8*	3.0	LAKE	CA	33	89	151	90	92
1	1.5			1	1.1	LASSEN	CA			-29	34	51
313*	1.2	457*	1.4	791*	2.3	LOS ANGELES	CA	17	63	90	82	85
3	2.1			6	2.1	MADERA	CA			1	50	82
6	1.4	16	2.0	25*	2.3	MARIN	CA	51	11	68	77	85
		2*	5.6			MARIPOSA	CA					
2	.9	5	1.9	10*	3.1	MENDOCINO	CA	114	67	257	96	92
4	2.1	5	1.5	5	1.1	MERCED	CA	-28	-29	-49	20	52
		1	2.6	3*	6.2	MODOC	CA		143			99
9	1.9	9	1.1	26*	2.4	MONTEREY	CA	-43	124	27	62	86
5	1.5	6	1.4	12	2.0	NAPA	CA	-4	42	37	66	80
2	1.6	1	.5	6	2.1	NEVADA	CA	-69	312	27	62	81
23	1.2	40	.9	127	1.7	ORANGE	CA	-23	84	41	68	72
2	.9	5	1.3	9	1.8	PLACER	CA	33	42	88	82	75
1	1.3			2	2.1	PLUMAS	CA			56	72	82
2	.2	28	1.4	58	1.7	RIVERSIDE	CA	723	24	923	99	73
15	1.0	47*	2.2	81*	2.6	SACRAMENTO	CA	115	20	158	90	88
		3*	3.3	2	2.4	SAN BENITO	CA		-26			87
15	.8	46	1.7	59	1.5	SAN BERNARDINO	CA	113	-10	92	82	66
30	.8	88*	1.7	157*	2.1	SAN DIEGO	CA	102	23	149	89	81
71*	1.7	132*	3.0	136*	3.7	SAN FRANCISCO	CA	80	24	122	87	95

* SIGNIFICANTLY MORE DEATHS THAN EXPECTED

WHITE FEMALE — CANCER OF THE ORAL CAVITY INCLUDING TONGUE — ICD 141,143-146,148,149

1950-1959 DEATHS	1950-1959 RATE	1960-1969 DEATHS	1960-1969 RATE	1970-1979 DEATHS	1970-1979 RATE	COUNTY AND STATE	PERCENT CHANGE 60S/50S	PERCENT CHANGE 70S/60S	PERCENT CHANGE 70S/50S	PERCENTILE %CHANGE 70S/50S	PERCENTILE RATES 1970S
7	.7	20	1.6	22	1.4	SAN JOAQUIN CA	138	-14	104	85	64
		4	.8	17	2.1	SAN LUIS OBISPO CA		149			81
18	1.6	52*	2.4	79*	2.7	SAN MATEO CA	51	14	73	78	89
6	.9	10	1.0	24	1.7	SANTA BARBARA CA	11	68	88	82	73
18	.9	46	1.3	92*	2.0	SANTA CLARA CA	46	47	114	86	79
2	.5	11	1.5	25*	2.8	SANTA CRUZ CA	237	82	515	99	91
4	1.7	4	1.2	8	1.6	SHASTA CA	-31	34	-8	45	70
		2	1.1	5	2.3	SISKIYOU CA		103			85
4	.9	5	.9	10	1.4	SOLANO CA	4	44	49	71	62
4	.6	16	1.6	50*	3.1	SONOMA CA	177	101	458	99	93
9	1.3	14	1.4	20	1.5	STANISLAUS CA	6	8	14	56	67
		2	1.1	3	1.3	SUTTER CA		15			59
3	2.7			6	2.6	TEHAMA CA			-1	49	89
		1	2.3	1	1.9	TRINITY CA		-19			77
6	.9	12	1.4	26*	2.5	TULARE CA	51	72	159	91	87
1	1.4	2	1.9	2	.8	TUOLUMNE CA	42	-56	-38	28	44
4	.6	14	1.2	23	1.2	VENTURA CA	107	-2	103	84	55
2	1.2	3	1.0	6	1.4	YOLO CA	-14	42	22	60	64
2	1.6	2	1.5			YUBA CA	-10				
59	.9	78	.8	138	1.2	COLORADO CO	-3	41	37		
2	1.0	3	.5	8	1.1	ADAMS CO	-48	103	5	53	51
				1	2.3	ALAMOSA CO					86
		1	3.5			BACA CO					
1	3.2	1	2.2	1	1.3	BENT CO	-30	-41	-59	14	60
5	1.8	2	.4	5	.7	BOULDER CO	-77	83	-58	15	40
2*	4.3	1	1.8	1	1.7	CHAFFEE CO	-59	-5	-61	12	72
		1*	5.0			CLEAR CREEK CO					
				1*	5.9	COSTILLA CO					98
		1	3.5			CROWLEY CO					
1	1.0	2	1.7	1	.5	DELTA CO	77	-68	-43	24	35
27	1.0	31	.9	64	1.7	DENVER-ARAPAHOE#CO	-8	84	70	77	73
				1	1.6	DOUGLAS CO					69
		1	3.6			ELBERT CO					
4	.8	2	.3	12	1.2	EL PASO CO	-59	274	54	72	57
1	1.1	1	.8	1	.3	FREMONT CO	-27	-68	-77	04	31
		2	2.6			GARFIELD CO					
				1	3.8	GRAND CO					95
1	2.6					HUERFANO CO					
		1*	14.9			JACKSON CO					
		9	1.4	16	1.4	JEFFERSON CO		3			64
1	2.8					KIT CARSON CO					
		1	1.0	2	2.0	LA PLATA CO		97			79
1	.7	3	.7	2	.2	LARIMER CO	5	-75	-73	05	30
		1	.7			LAS ANIMAS CO					
				1	3.0	LINCOLN CO					92
2	2.2					LOGAN CO					
3	1.4	1	.4	4	1.1	MESA CO	-73	191	-21	39	52
		1	1.9	1	1.5	MONTEZUMA CO		-21			68
		1	1.1	1	1.1	MONTROSE CO		-5			52
1	1.4	1	.9	3	2.1	MORGAN CO	-37	133	47	70	82
1	.8	1	.8	1	.5	OTERO CO	-2	-38	-39	27	34
				1	1.3	PROWERS CO					60
3	.6	6	1.0	4	.6	PUEBLO CO	57	-38	-3	48	37
				1	4.3	RIO BLANCO CO					97
1	1.8					RIO GRANDE CO					
1*	18.7					SAN MIGUEL CO					
		1	4.3	1	4.4	SEDGWICK CO		2			97
1*	7.0					TELLER CO					
				1	3.2	WASHINGTON CO					93
		2	.5	1	.2	WELD CO		-56			31
				1	1.5	YUMA CO					66
116	1.0	201	1.4	283	1.6	CONNECTICUT CT	32	16	54		
33	1.2	50	1.3	70	1.5	FAIRFIELD CT	10	19	31	63	67
19	.6	51	1.4	85	1.8	HARTFORD CT	117	30	182	92	75
9	1.5	7	1.0	20	2.0	LITCHFIELD CT	-38	114	33	64	81
6	1.3	8	1.4	7	1.1	MIDDLESEX CT	8	-24	-18	40	53
41	1.4	60	1.5	65	1.4	NEW HAVEN CT	11	-8	3	51	64
8	1.0	13	1.2	25	2.1	NEW LONDON CT	25	70	112	86	83
		6	2.0	2	.4	TOLLAND CT		-82			32
		6	1.3	9	1.7	WINDHAM CT		32			72
12	.8	24	1.1	40	1.4	DELAWARE DE	52	25	90		
1	.5			2	.4	KENT DE			-19	40	32
9	.8	19	1.3	31	1.7	NEW CASTLE DE	57	27	99	84	71
2	.6	5	1.5	7	1.3	SUSSEX DE	150	-14	115	86	60
31	1.1	51*	2.1	62*	3.3	DIST. OF COLUMBIA DC	90	59	202		
31	1.1	51*	2.1	62*	3.3	WASHINGTON DC	90	59	202	94	93
183	1.1	498*	1.6	1071*	1.9	FLORIDA FL	43	22	75		
1	.8	7*	3.2	5	1.3	ALACHUA FL	320	-59	72	78	61
1*	4.7	1	4.5			BAKER FL	-4				
2	1.9	13*	6.2	7	1.8	BAY FL	223	-71	-7	46	75
				5*	6.9	BRADFORD FL					99
3	1.1	7	1.5	36*	2.8	BREVARD FL	43	89	169	92	91
4	.2	39	1.3	128*	1.9	BROWARD FL	408	53	677	99	78
1	4.7	2*	5.5	1	1.5	CALHOUN FL	17	-73	-68	08	66
1	1.5	5	3.5	5	1.2	CHARLOTTE FL	134	-66	-20	39	56
				10	2.3	CITRUS FL					85
1	2.5			1	.8	CLAY FL			-66	09	44
		3	1.2	1	.1	COLLIER FL		-91			30
		2	2.6	2	1.5	COLUMBIA FL		-41			67
43	1.3	116*	1.9	186*	2.0	DADE FL	42	7	51	71	81
				2	2.3	DE SOTO FL					85

* SIGNIFICANTLY MORE DEATHS THAN EXPECTED

\# COUNTIES COMBINED; SEE TECHNICAL NOTES

1950-1959 DEATHS	1950-1959 RATE	1960-1969 DEATHS	1960-1969 RATE	1970-1979 DEATHS	1970-1979 RATE	COUNTY AND STATE	PERCENT CHANGE 60S/50S	PERCENT CHANGE 70S/60S	PERCENT CHANGE 70S/50S	PERCENTILE %CHANGE 70S/50S	PERCENTILE RATES 1970S
15	1.4	38*	2.3	66*	3.0	DUVAL FL	65	31	116	86	92
4	1.3	10	2.1	13	1.5	ESCAMBIA FL	58	-24	19	58	68
				2	4.1	FLAGLER FL					96
				1	1.8	FRANKLIN FL					75
		2	1.8	4	2.9	GADSDEN FL		62			91
				1	4.6	GILCHRIST FL					97
		1*	12.4	1	2.8	GLADES FL		-77			90
				3*	7.5	GULF FL					99
1*	3.9					HAMILTON FL					
2*	3.4					HARDEE FL					
		1	4.7	1	1.9	HENDRY FL		-60			77
1	5.7			3	1.4	HERNANDO FL			-76	04	62
		1	.8	2	.4	HIGHLANDS FL		-54			32
18	1.3	42*	2.1	57	2.0	HILLSBOROUGH FL	58	-6	48	70	79
3*	5.8	1	1.5	1	.9	HOLMES FL	-74	-42	-85	01	46
		4	2.2	8	1.7	INDIAN RIVER FL		-20			73
3*	3.1	1	.7	1	.6	JACKSON FL	-77	-22	-82	02	36
1*	5.3	1	4.6			JEFFERSON FL	-13				
1*	7.0			1	4.7	LAFAYETTE FL			-33	32	97
2	.6	9	2.3	11	1.4	LAKE FL	307	-42	136	88	62
1	.3	6	1.0	20	1.5	LEE FL	219	41	349	97	65
2	1.8	4	1.8	6	1.7	LEON FL	0	-2	-2	49	73
		1	1.7	3	3.2	LEVY FL		86			93
				1	3.7	LIBERTY FL					95
1	2.3					MADISON FL					
3	1.1	7	.9	27	2.2	MANATEE FL	-22	153	98	84	84
3	2.9	6	2.0	8	1.2	MARION FL	-30	-41	-59	13	55
		1	.5	7	1.6	MARTIN FL		260			71
3*	4.2	3	2.2	2	.7	MONROE FL	-49	-66	-83	01	40
				3	3.0	NASSAU FL					92
2	2.4	4	2.9	4	1.5	OKALOOSA FL	21	-48	-37	29	67
1*	21.0			3	3.1	OKEECHOBEE FL			-85	00	93
10	1.0	17	1.3	47*	2.3	ORANGE FL	25	84	129	88	85
		2	1.1	2	1.1	OSCEOLA FL		-1			54
8	.8	19	1.1	82*	2.2	PALM BEACH FL	28	105	162	91	84
1	.6	4	.7	14	.8	PASCO FL	22	13	39	66	44
11	.5	62	1.4	136*	2.2	PINELLAS FL	200	54	362	98	83
8	1.5	8	.8	32	2.1	POLK FL	-42	142	40	67	81
		1	.8			PUTNAM FL					
3	2.6	3	1.7	11*	4.8	ST JOHNS FL	-33	175	86	81	97
		3	1.7	4	.8	ST LUCIE FL		-56			41
2	3.6	2	1.6	5	3.5	SANTA ROSA FL	-54	113	-3	48	94
6	1.2	11	1.4	35	2.0	SARASOTA FL	13	49	69	77	81
1	.7	1	.3	5	.9	SEMINOLE FL	-59	197	21	59	46
1	1.9			2	1.4	SUMTER FL			-28	35	62
1	2.3			1	1.1	SUWANNEE FL			-51	19	54
2*	6.9	1	2.6			TAYLOR FL	-63				
3	.4	21	2.1	42	2.3	VOLUSIA FL	416	10	467	99	86
		1	5.2	1	2.9	WAKULLA FL		-45			91
2	3.5	1	1.3	2	1.6	WALTON FL	-62	24	-53	17	71
		3*	4.6	1	.9	WASHINGTON FL		-80			46
209*	2.0	298*	2.0	368*	1.8	GEORGIA GA	3	-10	-8		
2*	5.2	5*	10.0			APPLING GA	91				
1	4.3	1	3.1	1	1.9	BACON GA	-28	-40	-57	15	77
1*	12.5			2*	20.0	BAKER GA			60	74	99
1	.5			4	2.5	BALDWIN GA			391	98	87
1	4.6			2	3.8	BANKS GA			-19	40	95
1	1.6	4*	4.8			BARROW GA	190				
1	3.1	3	2.5	3	1.7	BARTOW GA	-21	-31	-46	22	72
		1	1.6	2	2.4	BEN HILL GA		54			86
1	2.2	1	2.2	3*	4.7	BERRIEN GA	-1	113	111	86	97
9*	2.4	8	1.7	17*	2.9	BIBB GA	-31	72	19	58	91
				1	2.1	BLECKLEY GA					82
		1	4.0	1	3.2	BRANTLEY GA		-20			93
		1	2.3			BROOKS GA					
		1	.9	2	1.7	BULLOCH GA		80			72
2*	6.2	1	2.5	1	1.6	BURKE GA	-59	-37	-74	05	70
1	3.7			1	1.6	BUTTS GA			-57	15	70
1*	5.5					CALHOUN GA					
		3*	14.5			CAMDEN GA					
1*	5.0			1	3.1	CANDLER GA			-37	28	93
3	2.4	1	.6	2	.8	CARROLL GA	-74	30	-67	08	42
2*	3.6	1	1.4	2	1.5	CATOOSA GA	-60	2	-59	13	65
		1*	10.2			CHARLTON GA					
7	1.6	13*	2.2	14	2.0	CHATHAM GA	39	-12	21	59	79
				4	3.5	CHATTOOGA GA					94
2	2.3	2	1.5			CHEROKEE GA	-34				
1	.6	4	2.2	3	1.5	CLARKE GA	246	-31	137	88	67
1*	9.4					CLAY GA					
5*	7.3			6	2.0	CLAYTON GA			-73	05	79
		1*	5.8			CLINCH GA					
8*	3.0	5	1.2	18	2.2	COBB GA	-60	88	-25	37	84
		1	1.4	3	2.9	COFFEE GA		112			91
		1	.7	8*	4.6	COLQUITT GA		551			97
		1	4.3	1	1.3	COLUMBIA GA		-70			59
		2*	4.4			COOK GA					
5*	5.5			2	1.2	COWETA GA			-77	03	58
2*	11.6					CRAWFORD GA					
1	3.8			1	1.9	DADE GA			-49	20	78
2*	3.3	5*	7.0	3	3.0	DECATUR GA	113	-56	-7	46	92
10	1.6	17	1.5	28	1.7	DE KALB GA	-3	10	7	53	72
3*	5.6	1	1.6	3	4.0	DODGE GA	-72	158	-28	35	96
3*	3.5	2	1.5	8*	3.4	DOUGHERTY GA	-58	128	-3	48	94
		3*	3.5	4	2.5	DOUGLAS GA		-30			87
		1	2.3	2	4.3	EARLY GA		85			97
				1*	6.2	ECHOLS GA					99
		1	3.6			EFFINGHAM GA					
1	1.4	2	2.8	1	.7	ELBERT GA	99	-74	-49	20	40
1	2.6			2	1.7	EMANUEL GA			-35	30	72

* SIGNIFICANTLY MORE DEATHS THAN EXPECTED

WHITE FEMALE — CANCER OF THE ORAL CAVITY INCLUDING TONGUE — ICD 141,143-146,148,149

1950-1959 DEATHS	1950-1959 RATE	1960-1969 DEATHS	1960-1969 RATE	1970-1979 DEATHS	1970-1979 RATE	COUNTY AND STATE	PERCENT CHANGE 60S/50S	PERCENT CHANGE 70S/60S	PERCENT CHANGE 70S/50S	PERCENTILE %CHANGE 70S/50S	PERCENTILE RATES 1970S
2	3.1	3*	4.7	2	1.6	FANNIN GA	52	-65	-47	21	70
		1	2.4	1	1.1	FAYETTE GA		-55			52
1	.5	12*	3.9	5	1.2	FLOYD GA	720	-69	155	90	57
2*	4.2	5*	8.2			FORSYTH GA	96				
4*	8.4	2	2.4	3	3.7	FRANKLIN GA	-71	54	-56	16	95
30*	1.9	36	1.7	43	1.9	FULTON GA	-12	12	-1	49	77
		2*	5.0	1	1.6	GILMER GA		-67			71
				1	6.5	GLASCOCK GA					99
2	2.2	4	3.0	5	2.7	GLYNN GA	32	-9	20	59	89
4*	4.8	1	1.2	5*	3.4	GORDON GA	-75	187	-29	35	94
		2	3.3	1	1.5	GRADY GA		-55			66
1	2.3					GREENE GA					
2	1.4	4	2.2	9	2.6	GWINNETT GA	51	21	82	80	88
2	2.7	1	1.3			HABERSHAM GA	-51				
2	1.2	3	1.3	7	2.2	HALL GA	15	63	87	81	83
		1	3.4			HANCOCK GA					
3*	4.6	2	2.8	1	1.1	HARALSON GA	-40	-61	-77	04	53
1	3.2			1	2.0	HARRIS GA			-38	27	80
		1	1.3			HART GA					
1*	6.2	1	3.6	1	2.2	HEARD GA	-42	-39	-65	10	83
		1	1.2	1	.9	HENRY GA		-19			47
		1	.9	2	.8	HOUSTON GA		-13			42
				1	1.7	IRWIN GA					71
2	2.5	2	2.0	4	3.0	JACKSON GA	-22	50	17	57	92
				1	2.3	JASPER GA					86
1*	2.9	1	2.8			JEFF DAVIS GA	-6				
		3*	6.6			JEFFERSON GA					
1	2.8	3*	7.2	1	1.9	JOHNSON GA	160	-74	-33	32	77
		1	2.4			JONES GA					
1	2.8					LAMAR GA					
1	2.0	2	1.6	1	.9	LAURENS GA	-21	-44	-56	15	45
		1*	7.9	1	4.7	LEE GA		-41			97
		1	6.6			LIBERTY GA					
2	2.6	5*	3.4	8*	4.2	LOWNDES GA	32	24	64	75	97
				1	2.8	LUMPKIN GA					91
1	3.6	1	3.7			MC DUFFIE GA	4				
				1	4.6	MC INTOSH GA					97
1	3.8	3*	6.8			MACON GA	79				
		2	4.4	2	3.0	MADISON GA		-31			92
		1*	6.0	1	6.2	MARION GA		3			99
				1	1.2	MERIWETHER GA					57
1	2.0	2	3.5	2	2.9	MITCHELL GA	74	-17	44	69	91
1	3.3					MONROE GA					
1*	6.0					MONTGOMERY GA					
1	2.7					MORGAN GA					
1	2.8	1	2.1	2	2.8	MURRAY GA	-26	37	1	50	91
5	1.7	14*	3.5	24*	4.0	MUSCOGEE#GA	99	16	130	88	96
		4*	5.3			NEWTON GA					
1	3.9					OCONEE GA					
4*	8.6	1	1.5	1	.8	PAULDING GA	-83	-47	-91	00	42
				1	2.2	PEACH GA					83
				1	1.0	PICKENS GA					48
				2	3.6	PIERCE GA					95
		1	3.1	1	2.8	PIKE GA		-10			90
2	1.9	1	.7	5	2.8	POLK GA	-66	321	44	69	90
1*	4.1			2*	5.4	PULASKI GA			32	64	98
		1	2.2	1	1.3	RABUN GA		-39			61
		1	2.6	1	4.8	RANDOLPH GA		82			97
8*	2.5	14*	3.4	11	2.1	RICHMOND GA	34	-39	-18	40	82
		2*	3.7	2	1.6	ROCKDALE GA		-56			71
2*	5.0					SCREVEN GA					
				1	2.1	SEMINOLE GA					83
3	2.7	3	2.1	3	1.4	SPALDING GA	-21	-32	-46	22	65
3*	4.4	1	1.0	1	.5	STEPHENS GA	-77	-48	-88	00	35
1*	6.2					STEWART GA					
1	1.6	1	1.1			SUMTER GA	-31				
1*	8.3					TALBOT GA					
1*	9.0					TALIAFERRO GA					
1	1.8	1	1.7	2	2.4	TATTNALL GA	-7	40	30	63	87
		1	2.9			TAYLOR GA					
2*	6.4			2	2.9	TELFAIR GA			-55	16	91
1	2.6	1	2.4			TERRELL GA	-8				
2	2.2	6*	4.7	2	1.3	THOMAS GA	112	-73	-42	25	60
2	2.8	2	2.1	3	2.2	TIFT GA	-25	5	-21	39	84
1	2.4			3	3.5	TOOMBS GA			48	70	94
2*	11.5					TREUTLEN GA					
3	2.2	6*	3.3	6	2.5	TROUP GA	50	-24	14	56	88
		1	2.6	1	2.0	TURNER GA		-22			80
		2*	5.6			UNION GA					
1	1.5	2	2.3	2	1.3	UPSON GA	54	-43	-12	43	60
1	.4	6*	2.9	4	1.5	WALKER GA	544	-49	227	95	65
		3	2.8	1	.8	WALTON GA		-71			42
2	2.2	1	.8	2	1.1	WARE GA	-62	36	-48	21	55
		1	1.6			WASHINGTON GA					
		3*	4.9	1	1.4	WAYNE GA		-72			63
				1*	9.0	WEBSTER GA					99
		1	5.6	1	3.4	WHEELER GA		-39			94
		2*	5.0			WHITE GA					
4*	3.5	5	2.7	4	1.5	WHITFIELD GA	-22	-44	-56	16	68
		1	4.0	1	2.0	WILCOX GA		-50			80
				1	1.4	WILKES GA					62
1	3.9					WILKINSON GA					
1	2.1	2*	4.3	1	2.1	WORTH GA	111	-52	2	50	82
		20*	2.8	37*	3.6	HAWAII HI		27			
				2	2.2	HAWAII HI					84
		15*	2.7	33*	4.1	HONOLULU HI		51			96
		5*	10.8	2	2.9	MAUI#HI		-73			92

* SIGNIFICANTLY MORE DEATHS THAN EXPECTED

COUNTIES COMBINED; SEE TECHNICAL NOTES

WHITE FEMALE — CANCER OF THE ORAL CAVITY INCLUDING TONGUE — ICD 141,143-146,148,149

1950-1959 DEATHS	1950-1959 RATE	1960-1969 DEATHS	1960-1969 RATE	1970-1979 DEATHS	1970-1979 RATE	COUNTY AND STATE		PERCENT CHANGE 60S/50S	PERCENT CHANGE 70S/60S	PERCENT CHANGE 70S/50S	PERCENTILE %CHANGE 70S/50S	PERCENTILE RATES 1970S
19	.8	33	1.0	52	1.3	IDAHO	ID	36	22	65		
4	1.1	8	1.6	13	1.8	ADA	ID	37	18	62	75	76
2	1.4			4	1.6	BANNOCK	ID			14	56	70
		1	2.8	1	2.7	BENEWAH	ID		-3			89
		1	1.2			BINGHAM	ID					
				1	.8	BONNER	ID					41
		3	1.5	3	1.3	BONNEVILLE	ID		-15			59
				1	2.8	BOUNDARY	ID					90
3	1.0	3	.8	4	.8	CANYON	ID	-22	6	-17	41	44
		1	4.5			CARIBOU	ID					
		1	1.1	2	2.3	CASSIA	ID		100			85
		1	3.3	1	2.8	CLEARWATER	ID		-15			90
1*	10.5					CUSTER	ID					
1	2.7			1	2.4	ELMORE	ID			-12	44	86
1	4.2					FRANKLIN	ID					
		1	2.5	1	2.2	FREMONT	ID		-13			84
		1	1.6	1	1.8	GOODING	ID		8			74
				1	1.7	IDAHO	ID					74
1	3.3					JEFFERSON	ID					
				1	1.5	JEROME	ID					68
		1	.7	6	2.2	KOOTENAI	ID		200			83
1	1.1			2	1.6	LATAH	ID			40	67	68
		1	3.3			LEMHI	ID					
2*	4.5	1	2.1			MINIDOKA	ID	-53				
		3	2.2	2	1.3	NEZ PERCE	ID		-40			61
				1	1.2	PAYETTE	ID					57
				2*	7.2	POWER	ID					99
1	.8	3*	3.8			SHOSHONE	ID	402				
1*	10.3					TETON	ID					
1	.6	2	.7	3	.9	TWIN FALLS	ID	15	26	46	69	45
		1*	8.9			VALLEY	ID					
				1	1.7	WASHINGTON	ID					72
431	1.0	571	1.1	815	1.4	ILLINOIS	IL	10	29	42		
3	.7	5	1.0	5	.8	ADAMS	IL	54	-19	24	61	43
		1	1.1	2	2.9	ALEXANDER	IL		156			91
1	1.1			1	1.3	BOND	IL			13	56	60
		1	.7	1	.9	BOONE	IL		23			45
2	1.0	2	.7	3	1.2	BUREAU	IL	-27	72	26	61	58
				1	3.5	CALHOUN	IL					94
		2	1.9			CARROLL	IL					
1	.9					CASS	IL					
3	.7	6	1.1	6	.9	CHAMPAIGN	IL	62	-19	31	64	46
3	1.5			4	1.6	CHRISTIAN	IL			6	53	69
1	.7	1	.6	2	1.1	CLARK	IL	-13	86	61	74	52
2	2.1	1	.7			CLAY	IL	-65				
1	.8			2	.8	CLINTON	IL			-4	47	41
1	.4			4	1.3	COLES	IL			253	96	59
221	1.1	303	1.2	447*	1.7	COOK	IL	17	40	64	75	73
				4	2.4	CRAWFORD	IL					86
1	1.6					CUMBERLAND	IL					
3	1.3			5	1.4	DE KALB	IL			7	53	62
		1	.6	1	.9	DE WITT	IL		52			46
1	1.1	1	.8			DOUGLAS	IL	-29				
2	.2	17	1.2	26	1.0	DU PAGE	IL	462	-17	368	98	48
1	.5	1	.5			EDGAR	IL	-14				
1	1.5					EDWARDS	IL					
2	1.5			3	1.5	EFFINGHAM	IL			0	50	66
2	1.8	1	.6	2	1.3	FAYETTE	IL	-69	143	-25	37	61
1	1.0			1	.7	FORD	IL			-29	34	39
2	.8	3	1.1	5	1.2	FRANKLIN	IL	38	16	60	73	57
2	.7	1	.3	1	.2	FULTON	IL	-59	-28	-70	06	31
		1	1.7	1	1.6	GALLATIN	IL		-7			69
3	2.0	1	.5			GREENE	IL	-75				
				1	.4	GRUNDY	IL					33
1	1.3	1	1.2	2	2.0	HAMILTON	IL	-9	75	59	73	81
5*	2.9			2	1.2	HANCOCK	IL			-59	13	57
		1	2.1			HENDERSON	IL					
1	.4	2	.7	2	.4	HENRY	IL	71	-38	5	53	32
4	2.2	1	.5	1	.5	IROQUOIS	IL	-76	-4	-77	03	35
1	.6	3	1.4	4	1.7	JACKSON	IL	148	25	210	94	74
2	1.1	1	.4	2	.8	JEFFERSON	IL	-59	89	-23	38	43
2	2.2			3	2.1	JERSEY	IL			-3	48	82
1	.7	1	.5			JO DAVIESS	IL	-18				
		2	3.1	1	1.4	JOHNSON	IL		-55			63
6	.6	14	1.2	13	1.0	KANE	IL	94	-18	58	73	48
4	.8	3	.6	7	1.3	KANKAKEE	IL	-23	110	62	74	60
				1	.8	KENDALL	IL					42
2	.5	4	1.1	4	.9	KNOX	IL	100	-13	73	78	47
9	.9	8	.6	36	2.2	LAKE	IL	-33	249	134	88	84
7	1.2	5	.7	9	1.1	LA SALLE	IL	-38	55	-4	47	53
		4	2.6	3	2.0	LAWRENCE	IL		-24			80
		2	.9	1	.3	LEE	IL		-61			31
		1	.5	2	.8	LIVINGSTON	IL		63			42
1	.5	1	.3	1	.3	LOGAN	IL	-38	-14	-47	21	31
1	.5	3	1.4	3	1.4	MC DONOUGH	IL	191	2	197	93	64
2	.7	5	1.1	7	1.0	MC HENRY	IL	65	-3	60	74	51
3	.7	4	.8	9	1.6	MC LEAN	IL	14	111	141	89	70
8	1.4	9	1.2	11	1.6	MACON	IL	-12	27	12	55	69
		1	.3	2	.5	MACOUPIN	IL		99			35
8	.9	7	.6	16	1.1	MADISON	IL	-34	89	24	60	54
2	.7	3	.9	2	.8	MARION	IL	20	-14	3	51	41
1	1.1					MARSHALL	IL					
2	1.9			1	.6	MASON	IL			-68	08	36
4*	6.0	1	1.1	2	1.4	MASSAC	IL	-81	23	-77	03	62
2	1.7	5*	5.1			MERCER	IL	198				
				1	1.0	MONROE	IL					50
1	.4	2	.9	2	.8	MONTGOMERY	IL	120	-14	89	82	41

* SIGNIFICANTLY MORE DEATHS THAN EXPECTED

WHITE FEMALE — CANCER OF THE ORAL CAVITY INCLUDING TONGUE — ICD 141,143-146,148,149

1950-1959 DEATHS	1950-1959 RATE	1960-1969 DEATHS	1960-1969 RATE	1970-1979 DEATHS	1970-1979 RATE	COUNTY AND STATE		PERCENT CHANGE 60S/50S	PERCENT CHANGE 70S/60S	PERCENT CHANGE 70S/50S	PERCENTILE %CHANGE 70S/50S	PERCENTILE RATES 1970S
		2	.7	3	.8	MORGAN	IL		13			43
		2	1.0	1	.2	OGLE	IL		-80			30
8	.9	12	1.1	12	1.1	PEORIA	IL	24	-4	18	57	51
2	1.6	2	1.3	2	1.0	PERRY	IL	-23	-17	-37	28	51
2	2.4			2	1.6	PIATT	IL			-34	30	69
1	.6	5*	2.8			PIKE	IL	369				
1	2.0	1	2.2	1	1.9	PULASKI	IL	8	-14	-7	46	77
				1	3.9	PUTNAM	IL					96
2	1.3					RANDOLPH	IL					
		1	.8			RICHLAND	IL					
5	.7	11	1.3	12	1.3	ROCK ISLAND	IL	85	-3	80	80	58
7	.9	14	1.2	11	.8	ST CLAIR	IL	43	-30	-1	49	44
3	2.1	2	1.0	2	.7	SALINE	IL	-51	-32	-66	09	39
9	1.2	14	1.4	19	1.8	SANGAMON	IL	23	28	57	73	76
		2	2.2			SCHUYLER	IL					
1	1.7					SCOTT	IL					
		1	.4	3	1.1	SHELBY	IL		167			54
1	.4	6	1.9	1	.2	STEPHENSON	IL	334	-89	-50	19	31
4	1.0	2	.4	5	.7	TAZEWELL	IL	-60	76	-29	34	40
3	3.0	2	1.2	1	1.0	UNION	IL	-60	-20	-68	08	48
9	1.8	7	1.3	3	.6	VERMILION	IL	-29	-57	-69	07	35
		1	1.4			WABASH	IL					
3	2.1			2	.7	WARREN	IL			-64	10	40
		1	.8	1	.7	WASHINGTON	IL		-21			38
1	1.0	2	1.7	2	1.2	WAYNE	IL	69	-27	23	60	57
1	.7	2	1.4			WHITE	IL	110				
3	1.0	2	.6	7	2.0	WHITESIDE	IL	-38	220	97	83	81
9	1.4	14	1.5	20	1.7	WILL	IL	9	12	23	60	73
4	1.3	4	1.2	7	1.7	WILLIAMSON	IL	-8	41	30	63	74
11	1.3	16	1.6	17	1.3	WINNEBAGO	IL	19	-19	-4	47	59
1	.8			2	1.0	WOODFORD	IL			32	64	49
161	.8	254	1.0	297	1.0	INDIANA	IN	31	-1	30		
1	.8	1	.7	1	.5	ADAMS	IN	-20	-28	-42	25	34
8	.8	10	.8	16	1.1	ALLEN	IN	-3	37	34	65	54
2	1.0	4	1.8	1	.3	BARTHOLOMEW	IN	79	-81	-67	09	31
				1	1.1	BENTON	IN					54
				1	1.2	BLACKFORD	IN					56
		4	2.1	1	.6	BOONE	IN		-71			37
1	.9	1	1.1			CARROLL	IN	12				
1	.4			4	1.3	CASS	IN			193	93	58
		5	1.8	6	1.5	CLARK	IN		-14			68
2	1.2	1	.7			CLAY	IN	-40				
4	2.1	2	.8	1	.6	CLINTON	IN	-60	-28	-71	06	37
		1	1.7	1	1.3	CRAWFORD	IN		-24			59
1	.7			1	.5	DAVIESS	IN			-35	30	34
1	.7	3	2.0	1	.5	DEARBORN	IN	203	-77	-31	32	34
1	.8	2	1.3	1	.7	DECATUR	IN	69	-46	-9	44	39
1	.6	1	.4	1	.5	DE KALB	IN	-37	37	-14	43	35
7	1.8	8	1.5	8	1.1	DELAWARE	IN	-16	-23	-35	29	54
				1	.7	DUBOIS	IN					38
1	.2	3	.5	6	.9	ELKHART	IN	123	82	306	97	45
		3	2.3			FAYETTE	IN					
1	.4	5	1.8	2	.6	FLOYD	IN	309	-67	36	65	36
2	1.7	1	.6	1	.8	FOUNTAIN	IN	-64	31	-53	18	43
1	.9					FULTON	IN					
2	1.2			3	1.5	GIBSON	IN			24	60	65
1	.3	2	.5	3	.8	GRANT	IN	61	57	153	90	41
1	.7	1	.6	3	1.1	GREENE	IN	-18	87	53	72	53
		1	.4	2	.5	HAMILTON	IN		41			35
		4	2.8	2	1.0	HANCOCK	IN		-63			50
				1	.3	HENDRICKS	IN					31
1	.4			3	.8	HENRY	IN			104	84	43
2	.8	2	.6	6	1.5	HOWARD	IN	-29	164	88	82	68
3	1.2	2	.7			HUNTINGTON	IN	-44				
		2	1.2	4	1.9	JACKSON	IN		58			78
				1	.8	JASPER	IN					42
1	.6	2	1.2			JAY	IN	96				
1	.6	3	2.0			JEFFERSON	IN	211				
2	2.8	1	1.2			JENNINGS	IN	-59				
2	1.2	2	.5	4	1.2	JOHNSON	IN	-58	148	4	52	57
		5	1.5	4	1.3	KNOX	IN		-17			58
2	.9	5	2.1	3	1.0	KOSCIUSKO	IN	118	-49	10	54	50
1	1.1	2	2.3			LAGRANGE	IN	105				
3	.2	12	.7	26	1.2	LAKE	IN	259	65	493	99	57
6	1.6	1	.2	7	1.1	LA PORTE	IN	-87	437	-30	34	54
2	1.0	3	1.4	3	1.1	LAWRENCE	IN	40	-26	4	52	51
5	1.0	9	1.4	3	.3	MADISON	IN	37	-77	-69	08	31
29	1.0	45	1.4	44	1.2	MARION	IN	33	-13	16	57	55
1	.6			2	.8	MARSHALL	IN			43	68	42
1	2.4					MARTIN	IN					
1	.6	3	1.4	1	.4	MIAMI	IN	149	-75	-38	27	32
2	1.0	4	1.8	3	1.1	MONROE	IN	70	-40	2	51	52
4	2.2			1	.3	MONTGOMERY	IN			-83	00	31
2	1.4	1	.5	2	.9	MORGAN	IN	-62	67	-37	28	45
1	1.9	1	1.6	1	1.0	NEWTON	IN	-13	-41	-49	20	48
		1	.5	1	.3	NOBLE	IN		-47			31
2	2.1	3	3.3	1	.8	ORANGE	IN	58	-77	-64	11	41
				1	1.2	OWEN	IN					57
2	1.8					PARKE	IN					
		1	.8	2	2.0	PERRY	IN		161			81
2	2.2			2	1.8	PIKE	IN			-15	42	76
3	1.8	6	2.1	4	1.0	PORTER	IN	16	-51	-44	23	50
1	.8	2	1.7	1	.6	POSEY	IN	129	-64	-19	40	37
1	1.8	1	1.8	1	1.3	PULASKI	IN	-3	-28	-30	34	59
				2	.9	PUTNAM	IN					45
1	.6			3	1.6	RANDOLPH	IN			160	91	69
				2	1.4	RIPLEY	IN					65
				1	.4	RUSH	IN					33
8	.8	14	1.2	23	1.6	ST JOSEPH	IN	47	41	108	85	71

* SIGNIFICANTLY MORE DEATHS THAN EXPECTED

WHITE FEMALE — CANCER OF THE ORAL CAVITY INCLUDING TONGUE — ICD 141,143-146,148,149

1950-1959 DEATHS	1950-1959 RATE	1960-1969 DEATHS	1960-1969 RATE	1970-1979 DEATHS	1970-1979 RATE	COUNTY AND STATE	PERCENT CHANGE 60S/50S	PERCENT CHANGE 70S/60S	PERCENT CHANGE 70S/50S	PERCENTILE %CHANGE 70S/50S	PERCENTILE RATES 1970S
		1	1.3	2	1.5	SCOTT IN		15			68
2	1.2			2	1.1	SHELBY IN			-8	45	54
1	1.0	2	2.5			SPENCER IN	137				
		2	1.9	4	3.3	STARKE IN		74			94
1	1.0	2	1.8	2	1.5	STEUBEN IN	90	-18	55	72	66
		2	1.4	3	1.7	SULLIVAN IN		28			73
2	.6	2	.4	6	1.3	TIPPECANOE IN	-30	187	100	84	59
		1	1.2	1	.7	TIPTON IN		-37			40
8	1.0	21*	2.2	14	1.2	VANDERBURGH IN	115	-46	17	57	57
		2	1.7			VERMILLION IN					
6	1.0	12	1.7	15	2.3	VIGO IN	62	35	120	87	85
2	1.1	1	.4			WABASH IN	-63				
2	1.6	2	1.6	1	.6	WARRICK IN	-4	-65	-66	09	36
1	.9	2	1.6	1	.8	WASHINGTON IN	66	-49	-16	42	42
4	1.0	5	1.3	4	.8	WAYNE IN	21	-35	-21	39	43
				3	2.1	WELLS IN					82
1	.9			5	2.8	WHITE IN			216	94	90
		1	.8	3	1.5	WHITLEY IN		93			66
119	.8	160	.9	222	1.2	IOWA IA	15	28	48		
1	1.1					ADAIR IA					
1	1.9					ADAMS IA					
1	.9	1	.7	1	.8	ALLAMAKEE IA	-24	20	-9	45	42
1	.9	2	1.1	1	.8	APPANOOSE IA	19	-28	-15	42	41
				2	2.7	AUDUBON IA					89
2	1.5					BENTON IA					
2	.5	5	.8	9	1.4	BLACK HAWK IA	60	78	186	93	62
1	.5	3	1.0	1	.4	BOONE IA	93	-57	-17	41	33
		1	.7	2	1.4	BREMER IA		110			64
2	1.7			2	1.8	BUCHANAN IA			6	53	74
1	.8			1	.2	BUENA VISTA IA			-69	07	31
1	1.1	1	.7	2	1.4	CALHOUN IA	-38	107	28	62	63
		1	.9	1	.5	CARROLL IA		-43			34
				2	2.1	CASS IA					82
2	2.2	1	1.0	1	.5	CEDAR IA	-53	-54	-78	02	34
		1	.4	6	1.6	CERRO GORDO IA		304			69
		1	.7			CHEROKEE IA					
				1	.7	CHICKASAW IA					38
		1	1.2	1	1.5	CLARKE IA		17			65
				2	1.3	CLAY IA					61
		3	1.9	2	1.6	CLAYTON IA		-16			71
2	.7	5	1.7	7	2.0	CLINTON IA	141	20	190	93	80
		2	1.7	1	1.2	CRAWFORD IA		-32			55
1	.6	2	1.2	2	.9	DALLAS IA	109	-20	66	76	47
1	1.6			1	1.2	DAVIS IA			-26	36	55
		1	.8	1	1.3	DECATUR IA		57			60
1	1.0	1	1.0	2	1.3	DELAWARE IA	1	27	28	63	61
5	2.0	4	1.6	3	.9	DES MOINES IA	-19	-43	-54	17	46
				3	2.6	DICKINSON IA					88
6	1.4	5	1.0	10	1.7	DUBUQUE IA	-32	72	17	57	72
1	.6	2	1.0	1	.3	FAYETTE IA	69	-75	-58	15	31
2	1.6	2	1.7	4	2.4	FLOYD IA	3	43	46	70	86
1	1.3	2	2.1			FRANKLIN IA	68				
2	2.5					FREMONT IA					
		2	1.6	1	.6	GREENE IA		-64			36
1	1.3			1	1.1	GRUNDY IA			-18	41	52
		1	.7			GUTHRIE IA					
1	.8	1	.7			HAMILTON IA	-10				
		2	2.8			HANCOCK IA					
		1	1.5	3	2.1	HARDIN IA		42			82
1	.8	1	1.0	2	2.0	HARRISON IA	27	112	169	92	80
		2	1.2	3	2.1	HENRY IA		70			82
		1	.9			HOWARD IA					
2	3.1			3	3.7	HUMBOLDT IA			19	58	95
				2	3.4	IDA IA					94
2	2.1					IOWA IA					
2	1.7			1	.9	JACKSON IA			-45	23	47
1	.6	2	.8	2	.7	JASPER IA	45	-8	34	65	40
2	1.6			2	2.1	JEFFERSON IA			35	65	83
		2	.9	6	1.8	JOHNSON IA		108			75
				1	.4	JONES IA					32
1	.7	1	.5			KEOKUK IA	-28				
2	1.6	1	.8			KOSSUTH IA	-53				
		2	.8	1	.3	LEE IA		-66			31
4	.6	9	1.1	13	1.5	LINN IA	92	35	159	91	67
				2	1.8	LOUISA IA					75
1	1.1					LUCAS IA					
1	1.4					LYON IA					
		1	.9	2	1.0	MADISON IA		13			48
1	.7	3	1.3	1	.4	MAHASKA IA	80	-67	-41	26	33
				2	.8	MARION IA					43
1	.4	1	.4	5	1.9	MARSHALL IA	-13	402	338	97	77
1	1.2			1	1.0	MITCHELL IA			-14	42	50
		1	1.3	1	.4	MONONA IA		-67			33
				1	1.3	MONROE IA					58
				2	1.1	MONTGOMERY IA					53
1	.6	2	.7	2	.8	MUSCATINE IA	23	13	39	67	42
		1	.8	1	.5	O'BRIEN IA		-39			34
		1	1.5			OSCEOLA IA					
2	1.1	4	2.1	2	1.0	PAGE IA	93	-53	-10	44	49
		2	2.0	1	.7	PALO ALTO IA		-65			38
4*	3.3	1	.8	2	1.5	PLYMOUTH IA	-76	84	-55	16	66
2	2.6	2	2.1	2	1.4	POCAHONTAS IA	-19	-34	-46	22	63
12	1.0	16	1.0	28	1.7	POLK IA	6	63	72	78	72
7	1.9	3	.6	9	1.7	POTTAWATTAMIE IA	-69	194	-10	44	72
2	1.6	1	.8			POWESHIEK IA	-51				
		2	2.9			RINGGOLD IA					
1	1.0	2	1.3	1	1.2	SAC IA	32	-8	21	59	56
10	1.8	10	1.5	10	1.4	SCOTT IA	-18	-8	-24	37	63
1	1.2	1	1.3	1	.9	SHELBY IA	8	-28	-22	38	47

* SIGNIFICANTLY MORE DEATHS THAN EXPECTED

WHITE FEMALE — CANCER OF THE ORAL CAVITY INCLUDING TONGUE — ICD 141,143-146,148,149

1950-1959 DEATHS	1950-1959 RATE	1960-1969 DEATHS	1960-1969 RATE	1970-1979 DEATHS	1970-1979 RATE	COUNTY AND STATE		PERCENT CHANGE 60S/50S	PERCENT CHANGE 70S/60S	PERCENT CHANGE 70S/50S	PERCENTILE %CHANGE 70S/50S	PERCENTILE RATES 1970S
2	1.7	1	.8	1	.6	SIOUX	IA	-54	-27	-67	08	36
2	.8	4	1.4	4	1.6	STORY	IA	71	13	93	83	71
		1	.5			TAMA	IA					
		1	.7	1	.7	TAYLOR	IA		-5			38
1	1.2			2	1.2	UNION	IA			-4	47	56
1	.9					VAN BUREN	IA					
		3	1.0	4	1.4	WAPELLO	IA		45			65
1	.9	1	.6	3	1.0	WARREN	IA	-30	54	8	53	48
1	.8	1	1.0	3	1.7	WASHINGTON	IA	34	69	127	87	73
		1	.8			WAYNE	IA					
		2	.8	1	.4	WEBSTER	IA		-50			33
		1	.8	2	1.5	WINNEBAGO	IA		86			68
1	.7	1	.8	1	.7	WINNESHIEK	IA	14	-10	3	51	40
6	1.1	12	1.8	9	1.3	WOODBURY	IA	68	-28	21	59	61
1	2.0			1	.9	WORTH	IA			-52	18	48
		2	1.6	1	1.1	WRIGHT	IA		-34			52
85	.8	121	1.0	131	.9	KANSAS	KS	20	-4	15		
		2	1.6	3	2.0	ALLEN	KS		22			80
2	3.4	3*	3.8			ANDERSON	KS	13				
2	1.8	2	1.7	1	.9	ATCHISON	KS	-5	-44	-47	21	48
1	2.0					BARBER	KS					
1	.7	4	2.4	4	2.3	BARTON	KS	226	-2	218	94	85
3	2.5	1	.5	1	.4	BOURBON	KS	-78	-27	-84	01	32
1	.9	1	.8	2	1.2	BROWN	KS	-11	52	36	65	56
		5	2.4	2	.8	BUTLER	KS		-69			41
1	2.8			1	1.7	CHAUTAUQUA	KS			-38	27	73
2	1.1	1	.5	1	.7	CHEROKEE	KS	-60	56	-37	28	39
		1	.7			CLAY	KS					
		2	1.1	2	.6	CLOUD	KS		-47			36
3	1.2	1	.3			COWLEY	KS	-72				
3	1.0	3	1.2	3	1.1	CRAWFORD	KS	24	-13	8	53	52
2	1.2					DICKINSON	KS					
1	1.8			1	1.8	DONIPHAN	KS			-1	49	74
		4	1.7	2	.8	DOUGLAS	KS		-49			44
		1	2.4	1	2.7	EDWARDS	KS		10			89
1	1.5					ELLIS	KS					
		1	.7	1	.9	FORD	KS		23			45
2	1.3			1	.4	FRANKLIN	KS			-73	05	32
2	3.0			5*	5.6	GEARY	KS			88	82	98
		1	4.5			GOVE	KS					
1	4.1	1	3.4			GRAHAM	KS	-16				
1	1.0	2	1.7			GREENWOOD	KS	60				
				1	5.1	HAMILTON	KS					98
2	4.0			1	1.2	HARPER	KS			-69	07	58
2	1.3	2	1.0	1	.2	HARVEY	KS	-25	-83	-88	00	30
		2*	12.5			HODGEMAN	KS					
				2	2.0	JEFFERSON	KS					81
1	1.2					JEWELL	KS					
2	.4	7	1.2	12	1.1	JOHNSON	KS	180	-12	148	89	51
				1	.9	KINGMAN	KS					46
1	.4	1	.5	2	.9	LABETTE	KS	7	98	112	86	48
2	1.0	3	1.4	3	1.1	LEAVENWORTH	KS	42	-21	12	55	53
		1	1.6			LINCOLN	KS					
1	1.0	1	1.0	1	.9	LINN	KS	-7	-7	-13	43	46
				2*	7.8	LOGAN	KS					99
		1	.3			LYON	KS					
		3	1.7	3	1.7	MC PHERSON	KS		0			73
1	.9	1	.7	1	.4	MARION	KS	-19	-50	-59	13	32
2	1.7					MARSHALL	KS					
1	.7	2	1.5	1	.8	MIAMI	KS	97	-48	2	50	41
4	1.5	1	.2	1	.2	MONTGOMERY	KS	-85	-31	-89	00	30
1	1.6					MORRIS	KS					
				2	2.8	NEMAHA	KS					90
1	.7	1	.7	4	2.7	NEOSHO	KS	-6	304	280	96	90
				1	1.0	NORTON	KS					50
1	1.9			1	1.9	OSBORNE	KS			-1	49	78
1	2.5					OTTAWA	KS					
		1	1.1			PAWNEE	KS					
		1	1.4	1	1.0	PHILLIPS	KS		-28			49
2	2.2			1	1.8	POTTAWATOMIE	KS			-19	40	75
				2	2.3	PRATT	KS					85
1	3.9					RAWLINS	KS					
2	.8	4	1.2	5	1.1	RENO	KS	46	-3	42	68	54
1	1.4					REPUBLIC	KS					
1	1.3	2	1.7	1	1.1	RICE	KS	30	-36	-17	41	53
3	2.2			2	1.4	RILEY	KS			-37	29	64
		1	1.4	1	1.2	RUSSELL	KS		-18			56
4	1.9	1	.4	2	.7	SALINE	KS	-79	66	-65	10	38
8	.7	24	1.6	26	1.5	SEDGWICK	KS	135	-5	122	87	67
3	.4	9	1.3	5	.5	SHAWNEE	KS	181	-63	3	51	34
		1	1.2			SMITH	KS					
		1	2.2			STAFFORD	KS					
				1*	7.5	STANTON	KS					99
		2	1.3	6	2.4	SUMNER	KS		88			87
1	2.7	1	2.2			THOMAS	KS	-16				
				1	1.4	WASHINGTON	KS					64
1	.8			2	1.7	WILSON	KS			109	85	72
1	2.1					WOODSON	KS					
5	.8	11	1.4	4	.4	WYANDOTTE	KS	73	-68	-44	23	33
154	1.2	220*	1.4	287	1.5	KENTUCKY	KY	16	7	24		
2	2.7	2	2.4	1	.8	ADAIR	KY	-12	-67	-71	06	42
1	1.2			3	3.3	ALLEN	KY			180	92	93
1	1.8			2	3.5	ANDERSON	KY			92	83	94
		1	1.4			BALLARD	KY					
1	1.1	3	1.9	6	2.8	BARREN	KY	70	45	145	89	90
3*	6.2					BATH	KY					
1	.7	4	2.8	4	1.9	BELL	KY	322	-31	190	93	78

* SIGNIFICANTLY MORE DEATHS THAN EXPECTED

WHITE FEMALE — CANCER OF THE ORAL CAVITY INCLUDING TONGUE — ICD 141,143-146,148,149

1950-1959 DEATHS	1950-1959 RATE	1960-1969 DEATHS	1960-1969 RATE	1970-1979 DEATHS	1970-1979 RATE	COUNTY AND STATE		PERCENT CHANGE 60S/50S	PERCENT CHANGE 70S/60S	PERCENT CHANGE 70S/50S	PERCENTILE %CHANGE 70S/50S	PERCENTILE RATES 1970S
		1	1.0	3	2.1	BOONE	KY		108			83
				1	1.0	BOURBON	KY					48
3	1.4	3	1.0	4	1.1	BOYD	KY	-26	4	-23	38	52
		2	1.6	2	1.4	BOYLE	KY		-8			64
				2	2.6	BREATHITT	KY					88
		1	1.0			BRECKINRIDGE	KY					
				1	.7	BULLITT	KY					40
4*	7.1			3*	4.2	BUTLER	KY			-41	25	96
		4*	4.0	1	.6	CALDWELL	KY		-86			36
2	2.0	2	1.2			CALLOWAY	KY	-41				
12*	2.7	11	2.1	5	1.0	CAMPBELL	KY	-22	-51	-62	12	50
2*	5.4					CARLISLE	KY					
		2*	3.8	1	1.7	CARROLL	KY		-55			72
		1	1.1	2	1.5	CARTER	KY		33			66
2	3.3	1	1.5	2	2.1	CASEY	KY	-55	42	-36	29	82
2	1.2	3	1.4	3	1.3	CHRISTIAN	KY	22	-11	9	54	59
		2	1.8	1	.9	CLARK	KY		-50			46
				1	1.1	CLAY	KY					51
1	1.4					CRITTENDEN	KY					
1	2.0	2*	4.4			CUMBERLAND	KY	119				
2	.7	3	.8	6	1.3	DAVIESS	KY	27	54	95	83	60
2	2.9					ESTILL	KY					
5	1.1	11	1.7	17	2.2	FAYETTE	KY	60	29	106	85	84
1	1.3	1	1.4	2	2.0	FLEMING	KY	4	44	49	70	80
		2	1.4	7*	3.3	FLOYD	KY		136			93
1	.8	3	2.0	4	2.1	FRANKLIN	KY	150	5	162	91	81
1	1.7					FULTON	KY					
		1	4.4			GALLATIN	KY					
				1	1.3	GARRARD	KY					61
1	1.5	2	3.4	1	1.9	GRANT	KY	123	-44	24	60	77
6*	3.5	4	1.6	4	1.3	GRAVES	KY	-53	-21	-63	11	59
2	2.1	2	2.2	1	.9	GRAYSON	KY	8	-60	-56	15	46
1	2.1	2	3.8	1	1.2	GREEN	KY	86	-69	-42	25	56
		1	.9			GREENUP	KY					
				1	2.1	HANCOCK	KY					82
2	1.6	2	1.0	3	1.3	HARDIN	KY	-34	22	-19	40	58
3	1.5	3	1.8	3	1.3	HARLAN	KY	22	-26	-10	44	61
		1	1.4			HART	KY					
1	.7	2	1.2	6	2.5	HENDERSON	KY	80	102	265	96	87
		1	1.7			HICKMAN	KY					
4	2.2	2	.8	2	.5	HOPKINS	KY	-65	-36	-78	03	35
1	2.4	1	2.0	3*	4.5	JACKSON	KY	-18	129	87	81	97
31	1.3	45	1.6	78*	2.3	JEFFERSON	KY	20	47	76	79	85
1	1.8	2	3.0	1	1.1	JESSAMINE	KY	70	-65	-41	26	51
				3	2.6	JOHNSON	KY					89
3	.5	13	1.9	15	1.9	KENTON	KY	275	-2	269	96	78
				5	3.1	KNOX	KY					92
		2	3.8			LARUE	KY					
				1	.3	LAUREL	KY					31
1	1.6			1	1.8	LAWRENCE	KY			15	57	76
		2*	5.3			LEE	KY					
1	3.6					LESLIE	KY					
		1	1.0	4	2.7	LETCHER	KY		170			90
				1	1.5	LEWIS	KY					66
1	1.2	3*	3.5	1	.9	LINCOLN	KY	184	-76	-31	32	44
		2*	4.4	2	2.6	LIVINGSTON	KY		-41			88
1	.8	1	.9			LOGAN	KY	15				
2	.9	2	.6	7	1.8	MC CRACKEN	KY	-31	186	97	83	76
4*	12.6	2	3.7	7*	9.7	MC CREARY	KY	-71	164	-23	37	99
		1	1.9			MC LEAN	KY					
		1	.7	2	.8	MADISON	KY		21			43
		2	2.4			MARION	KY					
3*	4.5	1	.8	4	2.1	MARSHALL	KY	-82	163	-53	18	82
2*	6.7					MARTIN	KY					
		2	2.1	2	1.3	MASON	KY		-36			61
1	2.1			1	1.4	MEADE	KY			-30	34	65
				1	4.2	MENIFEE	KY					97
1	1.2	2	2.0			MERCER	KY	64				
3*	11.0	1	2.0			METCALFE	KY	-82				
2*	3.8			2	2.5	MONROE	KY			-34	31	88
		2	2.5	1	.7	MONTGOMERY	KY		-71			40
1	1.5	3*	4.7			MORGAN	KY	212				
1	.7	2	1.4	4	1.9	MUHLENBERG	KY	94	33	157	90	77
3*	3.8	3	3.1	2	1.4	NELSON	KY	-19	-55	-64	10	63
1	2.8					NICHOLAS	KY					
1	.8	1	1.0	1	1.2	OHIO	KY	34	14	53	71	56
		1	1.8	1	1.1	PENDLETON	KY		-38			53
3*	3.8	3	2.5	2	1.5	PERRY	KY	-36	-41	-62	12	65
3	1.8	7*	2.8	3	1.0	PIKE	KY	58	-63	-41	25	50
		1	2.9	1	1.7	POWELL	KY		-42			72
		2	1.0	2	.9	PULASKI	KY		-8			46
		1*	3.9			ROBERTSON	KY					
1	1.9			1	1.3	ROCKCASTLE	KY			-29	34	60
		2	3.7	2	2.0	RUSSELL	KY		-45			80
		1	1.2	5*	4.9	SCOTT	KY		294			98
1	1.4			1	.5	SHELBY	KY			-67	08	34
				2	2.4	SIMPSON	KY					87
		1	4.6			SPENCER	KY					
				2	2.0	TAYLOR	KY					81
2*	3.3					TODD	KY					
				1	1.6	TRIGG	KY					70
		1	1.0	2	1.9	UNION	KY		88			77
6*	2.9	3	1.2	2	.5	WARREN	KY	-58	-55	-81	02	35
2	3.1	2	2.9	1	1.2	WAYNE	KY	-6	-60	-62	12	55
		2	1.8			WEBSTER	KY					
1	.7	7*	4.4	2	1.0	WHITLEY	KY	552	-78	45	69	49
		1	3.1			WOLFE	KY					
				1	.7	WOODFORD	KY					40
111*	1.4	160	1.4	234	1.6	LOUISIANA	LA	3	12	15		

* SIGNIFICANTLY MORE DEATHS THAN EXPECTED

WHITE FEMALE — CANCER OF THE ORAL CAVITY INCLUDING TONGUE — ICD 141,143-146,148,149

1950-1959 DEATHS	1950-1959 RATE	1960-1969 DEATHS	1960-1969 RATE	1970-1979 DEATHS	1970-1979 RATE	COUNTY AND STATE		PERCENT CHANGE 60S/50S	PERCENT CHANGE 70S/60S	PERCENT CHANGE 70S/50S	PERCENTILE %CHANGE 70S/50S	PERCENTILE RATES 1970S
2	1.5	2	1.0	2	.8	ACADIA	LA	-31	-24	-47	21	42
1	1.9	1	1.1	3	2.8	ALLEN	LA	-44	155	43	68	90
				1	.8	ASCENSION	LA					41
1	.7	2	1.2	4	2.1	AVOYELLES	LA	57	81	185	92	82
2	2.9	5*	5.8	3	2.3	BEAUREGARD	LA	100	-59	-18	40	86
2	3.1			1	1.1	BIENVILLE	LA			-66	09	52
		3	2.4	3	1.3	BOSSIER	LA		-44			61
12*	2.2	11	1.3	16	1.5	CADDO	LA	-39	12	-32	32	67
1	.1	4	1.0	7	1.2	CALCASIEU	LA	559	24	718	99	57
		1	2.2			CALDWELL	LA					
1*	5.2			2*	5.8	CAMERON	LA			12	55	98
2*	8.0					CATAHOULA	LA					
1	1.7	1	1.1			CLAIBORNE	LA	-37				
		2*	5.4	1	1.7	CONCORDIA	LA		-69			72
1	2.2	1	1.7			DE SOTO	LA	-20				
6	1.5	10	1.4	26*	2.7	EAST BATON ROUGE	LA	-5	93	83	81	89
1	1.6	1	1.7	1	1.7	EAST FELICIANA	LA	4	0	3	51	72
1	.8	4*	3.6	1	.6	EVANGELINE	LA	342	-82	-20	39	37
				1	1.9	GRANT	LA					78
1	1.1	4	2.5	3	1.5	IBERIA	LA	134	-40	40	67	67
		1	1.4	2	2.2	IBERVILLE	LA		55			84
				2	2.7	JACKSON	LA					89
7	1.9	8	1.0	17	1.2	JEFFERSON	LA	-48	18	-39	27	56
3*	3.2	3	2.6	1	.7	JEFFERSON DAVIS	LA	-19	-73	-78	03	38
2	1.3	5	2.0	7	1.6	LAFAYETTE	LA	62	-20	30	63	71
3	2.4	2	1.2	4	1.5	LAFOURCHE	LA	-50	21	-39	27	66
1	2.5			2	2.5	LA SALLE	LA			0	50	88
1	1.2	1	.8	1	.9	LINCOLN	LA	-33	8	-27	35	45
1	1.5	1	1.0	1	.7	LIVINGSTON	LA	-31	-30	-52	18	39
2	2.4	2	2.2	1	.7	MOREHOUSE	LA	-6	-71	-73	05	38
		1	.8	2	1.3	NATCHITOCHES	LA		55			60
29	1.4	41	1.7	52*	2.4	ORLEANS	LA	25	39	75	79	87
3	1.2	5	1.5	9	1.9	OUACHITA	LA	25	30	62	75	79
				2	2.3	PLAQUEMINES	LA					85
		1	2.1			POINTE COUPEE	LA					
6	2.4	8	2.0	7	1.4	RAPIDES	LA	-17	-31	-43	24	63
1	1.8	1	1.2	2	2.3	RICHLAND	LA	-33	86	26	61	85
1	1.4			2	1.3	SABINE	LA			-9	45	60
				5	2.9	ST BERNARD	LA					91
				1	1.1	ST CHARLES	LA					51
				1	1.3	ST JAMES	LA					59
		1	2.1	1	1.4	ST JOHN THE BAPT	LA		-31			65
1	.5	1	.4	4	1.4	ST LANDRY	LA	-18	213	158	90	63
2	3.6	1	1.3	3	2.5	ST MARTIN	LA	-64	93	-30	33	88
		2	1.3	3	1.6	ST MARY	LA		23			69
2	1.8	3	1.7	5	1.4	ST TAMMANY	LA	-3	-19	-22	38	63
1	.4	4	1.9	2	.9	TANGIPAHOA	LA	358	-53	115	86	46
		1	4.8	1	5.2	TENSAS	LA		7			98
1	.5	3	2.0	2	.8	TERREBONNE	LA	273	-61	44	69	42
1	1.7			1	.8	UNION	LA			-54	17	41
2	1.5	3	1.7	1	.4	VERMILION	LA	8	-78	-76	04	32
1	1.2	2	2.0	2	1.4	VERNON	LA	70	-30	19	58	64
2	2.0	1	.6	7*	3.4	WASHINGTON	LA	-69	425	64	76	94
1	.9	3	2.0	3	1.4	WEBSTER	LA	124	-33	51	71	62
		2*	6.1	1	2.4	WEST BATON ROUGE	LA		-60			87
				1	1.4	WEST CARROLL	LA					62
2*	4.0	1	1.5	1	.9	WINN	LA	-64	-36	-77	03	47
56	1.1	61	1.1	88	1.3	MAINE	ME	-3	20	17		
4	.9	8	1.4	7	1.0	ANDROSCOGGIN	ME	64	-27	19	58	50
3	1.0	5	1.5	8	1.6	AROOSTOOK	ME	54	10	70	77	71
13	1.3	12	1.1	24	1.8	CUMBERLAND	ME	-17	74	45	69	76
2	1.7	1	.7	1	.8	FRANKLIN	ME	-57	9	-53	17	42
3	1.4			1	.4	HANCOCK	ME			-73	05	32
6	1.2	5	.9	5	.7	KENNEBEC	ME	-25	-26	-44	23	38
3	1.2	3	1.7	3	1.0	KNOX	ME	42	-39	-13	43	50
3	2.0	1	1.0	3	1.7	LINCOLN	ME	-52	81	-14	43	74
1	.4			2	.7	OXFORD	ME			80	80	40
4	.8	5	.8	4	.7	PENOBSCOT	ME	1	-13	-12	43	38
		2	1.6	4	2.6	PISCATAQUIS	ME		66			88
		2	1.5	4	2.1	SAGADAHOC	ME		43			82
1	.4	1	.5	4	1.5	SOMERSET	ME	13	192	230	95	66
2	1.3	3	1.8	7*	4.2	WALDO	ME	44	129	229	95	97
6*	3.0	3	1.3	1	.3	WASHINGTON	ME	-56	-74	-89	00	31
5	.8	10	1.5	10	1.1	YORK	ME	80	-28	31	63	53
115	1.1	186	1.3	348*	2.0	MARYLAND	MD	16	49	73		
4	.8	7	1.3	8	1.2	ALLEGANY	MD	58	-8	44	69	56
6	1.8	14	1.8	31*	2.6	ANNE ARUNDEL	MD	-2	46	43	68	88
12	.8	22	1.0	49	1.4	BALTIMORE	MD	14	46	66	76	63
		2	2.9	1	1.0	CALVERT	MD		-64			50
		1	1.2			CAROLINE	MD					
2	.7	5	1.3	5	1.2	CARROLL	MD	83	-10	64	75	56
2	1.3	3	1.5	4	1.6	CECIL	MD	17	3	20	58	69
		2	1.7	4	2.8	CHARLES	MD		67			90
		2	1.1	2	1.7	DORCHESTER	MD		47			72
1	.3	3	.8	5	1.1	FREDERICK	MD	158	38	256	96	54
		1	.6	3	1.9	GARRETT	MD		221			78
2	1.1	4	1.3	4	.8	HARFORD	MD	26	-39	-24	37	43
1	.9			4	1.2	HOWARD	MD			34	65	58
				3	2.9	KENT	MD					91
11	1.3	26	1.5	54	2.0	MONTGOMERY	MD	12	39	56	72	81
12	1.9	19	1.3	47*	2.4	PRINCE GEORGE'S	MD	-29	83	29	63	87
		2	2.5	1	.6	QUEEN ANNE'S	MD		-76			37
		3	3.7			ST MARY'S	MD					
				1	.8	SOMERSET	MD					43
2	2.4	3	2.7	6*	3.6	TALBOT	MD	13	36	53	71	95
3	.7	9	1.7	4	.6	WASHINGTON	MD	150	-63	-8	45	37
1	.5	3	1.2	5	1.6	WICOMICO	MD	156	39	255	96	71

* SIGNIFICANTLY MORE DEATHS THAN EXPECTED

WHITE FEMALE — CANCER OF THE ORAL CAVITY INCLUDING TONGUE — ICD 141,143-146,148,149

1950-1959 DEATHS	1950-1959 RATE	1960-1969 DEATHS	1960-1969 RATE	1970-1979 DEATHS	1970-1979 RATE	COUNTY AND STATE	PERCENT CHANGE 60S/50S	PERCENT CHANGE 70S/60S	PERCENT CHANGE 70S/50S	PERCENTILE %CHANGE 70S/50S	PERCENTILE RATES 1970S
		1	.8	2	1.8	WORCESTER MD		134			74
56*	1.5	54	1.4	105*	3.0	BALTIMORE CITY MD	-2	111	108	85	92
313	1.1	443	1.4	565	1.6	MASSACHUSETTS MA	24	14	42		
7	1.8	7	1.5	31*	3.0	BARNSTABLE MA	-16	98	67	77	92
5	.6	11	1.2	14	1.4	BERKSHIRE MA	89	14	115	86	62
30	1.2	16	.6	35	1.1	BRISTOL MA	-53	85	-12	43	53
		1	2.5	4*	8.8	DUKES MA		257			99
38	1.1	43	1.1	70	1.6	ESSEX MA	1	40	42	68	68
2	.5	8	2.0	6	1.3	FRANKLIN MA	270	-33	149	89	61
19	.9	42	1.7	49	1.7	HAMPDEN MA	90	3	96	83	73
1	.2	5	.8	10	1.4	HAMPSHIRE MA	377	67	697	99	62
71	1.1	106	1.4	126	1.5	MIDDLESEX MA	30	6	38	66	67
		1	4.4	2*	6.6	NANTUCKET MA		50			99
29	1.1	45	1.3	56	1.4	NORFOLK MA	20	9	30	63	64
11	.9	29*	1.8	26	1.4	PLYMOUTH MA	110	-26	56	72	62
75*	1.6	86*	1.9	92*	2.2	SUFFOLK MA	19	19	42	68	84
25	.8	43	1.1	44	1.1	WORCESTER MA	47	-5	40	67	52
289	1.0	426	1.2	541	1.3	MICHIGAN MI	14	8	23		
		1	1.8			ALCONA MI					
2	.8	3	1.0	4	1.1	ALLEGAN MI	32	8	42	68	53
		1	.7	3	1.8	ALPENA MI		161			75
1	2.0					ANTRIM MI					
1	1.7	1	1.6	2	2.8	ARENAC MI	-6	75	64	75	90
		1	2.6	2	4.6	BARAGA MI		76			97
1	.8			3	1.3	BARRY MI			60	74	59
		4	.7	9	1.5	BAY MI		103			66
		1	2.4			BENZIE MI					
7	1.2	11	1.5	7	.8	BERRIEN MI	31	-47	-30	33	43
1	.5	1	.5	3	1.2	BRANCH MI	14	125	157	90	56
4	.7	9	1.3	10	1.3	CALHOUN MI	79	2	82	80	61
5*	3.1	1	.5	4	1.6	CASS MI	-83	215	-48	21	70
1	1.2	1	1.3	2	1.6	CHARLEVOIX MI	12	26	42	68	71
1	1.4			3	3.0	CHEBOYGAN MI			109	65	92
3	2.5	1	.7	3	1.9	CHIPPEWA MI	-71	167	-23	38	78
1	1.7	2	2.4	1	.7	CLAPE MI	41	-73	-62	11	38
		1	.5	2	.8	CLINTON MI		41			41
				2	3.6	CRAWFORD MI					95
		2	1.0	2	.8	DELTA MI		-25			41
		3	2.0	1	.4	DICKINSON MI		-82			32
3	1.4	3	1.2	3	.8	EATON MI	-11	-32	-39	27	44
7	.7	13	.9	15	.8	GENESEE MI	29	-7	19	58	44
1	2.0			1	1.0	GLADWIN MI			-52	18	48
3	2.8	1	.6	1	1.3	GOGEBIC MI	-80	120	-56	16	58
2	1.1	4	1.7	4	1.8	GRAND TRAVERSE MI	55	4	61	74	75
3	1.6			1	.5	GRATIOT MI			-66	09	35
1	.4	2	.9	3	1.4	HILLSDALE MI	107	58	226	95	65
4	1.7	3	1.4	2	.7	HOUGHTON MI	-19	-50	-59	13	38
		3	1.4	2	.9	HURON MI		-40			45
7	1.0	14	1.5	14	1.3	INGHAM MI	49	-9	36	66	61
		2	.9	2	.9	IONIA MI		7			47
1	1.5	1	1.4	1	.7	IOSCO MI	-7	-50	-53	17	38
3*	4.1	1	1.2	1	.7	IRON MI	-70	-42	-82	02	39
		1	.6	5	2.9	ISABELLA MI		357			91
3	.6	12	1.8	11	1.5	JACKSON MI	211	-16	162	91	66
7	1.0	9	1.1	14	1.4	KALAMAZOO MI	8	28	39	66	62
		1	2.9			KALKASKA MI					
12	.7	24	1.2	28	1.2	KENT MI	67	1	69	77	58
				1	10.1	KEWEENAW MI					99
		1	3.1			LAKE MI					
1	.6			2	.8	LAPEER MI			44	68	44
1	2.1			2	2.4	LEELANAU MI			19	58	87
3	.7	4	1.0	4	.9	LENAWEE MI	35	-5	28	62	47
1	.7	2	.8	3	1.0	LIVINGSTON MI	20	25	50	71	49
				1	2.0	LUCE MI					79
				2	3.2	MACKINAC MI					93
10	1.5	12	.7	38	1.4	MACOMB MI	-53	95	-9	44	63
3	3.1	1	.9	5	3.1	MANISTEE MI	-71	238	-1	49	92
4	1.6	5	1.9	5	1.5	MARQUETTE MI	18	-23	-9	45	66
1	1.0	2	1.5	4	2.7	MASON MI	57	79	181	92	89
1	1.1	1	.8	3	1.8	MECOSTA MI	-22	121	72	78	75
		1	.6	1	.7	MENOMINEE MI		12			39
1	.7	2	1.0	3	1.1	MIDLAND MI	39	4	44	69	52
2*	6.9	2*	4.4			MISSAUKEE MI	-36				
3	.9	6	1.3	6	1.0	MONROE MI	41	-26	4	52	48
2	1.2	1	.6	2	.7	MONTCALM MI	-46	21	-35	30	41
				2	4.5	MONTMORENCY MI					97
7	1.4	5	.7	8	1.0	MUSKEGON MI	-48	37	-29	34	49
				6*	3.3	NEWAYGO MI					94
21	1.2	40	1.3	66	1.4	OAKLAND MI	12	11	24	61	65
1	2.7			2	2.1	OGEMAW MI			-22	33	82
2*	5.0	1	2.3	1	1.5	ONTONAGON MI	-54	-36	-70	07	66
1	1.3					OSCEOLA MI					
		3*	8.4	1	1.6	OSCODA MI		-81			69
8*	2.2	4	.8	5	.6	OTTAWA MI	-61	-26	-71	06	37
				1	1.0	PRESQUE ISLE MI					50
1	2.4			3	4.6	ROSCOMMON MI			89	82	97
6	.8	11	1.3	6	.6	SAGINAW MI	60	-56	-30	34	36
7	1.5	3	.5	8	1.2	ST CLAIR MI	-65	119	-23	37	55
2	.9	2	.8	2	.7	ST JOSEPH MI	-17	-14	-29	35	38
1	.8	1	.5	1	.4	SANILAC MI	-37	-17	-48	20	32
1	2.2	2*	4.1	1	1.6	SCHOOLCRAFT MI	87	-61	-27	35	70
3	1.0	2	.8	6	1.8	SHIAWASSEE MI	-22	132	80	80	76
2	.9	1	.4	1	.4	TUSCOLA MI	-53	-13	-59	13	32
4	1.6	1	.3	2	.7	VAN BUREN MI	-80	116	-57	15	38
4	.7	5	.7	12	1.4	WASHTENAW MI	6	97	108	85	64
99	1.1	165*	1.5	154	1.4	WAYNE MI	31	-3	27	62	64
1	1.0	2	1.8			WEXFORD MI	82				

* SIGNIFICANTLY MORE DEATHS THAN EXPECTED

WHITE FEMALE — CANCER OF THE ORAL CAVITY INCLUDING TONGUE — ICD 141,143-146,148,149

1950-1959 DEATHS	1950-1959 RATE	1960-1969 DEATHS	1960-1969 RATE	1970-1979 DEATHS	1970-1979 RATE	COUNTY AND STATE	PERCENT CHANGE 60S/50S	PERCENT CHANGE 70S/60S	PERCENT CHANGE 70S/50S	PERCENTILE %CHANGE 70S/50S	PERCENTILE RATES 1970S
116	.8	163	.8	242	1.0	MINNESOTA MN	10	25	37		
				1	.9	AITKIN MN					46
1	.7	2	.6	6	1.3	ANOKA MN	-10	104	83	81	58
1	.9	3	2.0			BECKER MN	113				
		3	2.5	1	.9	BELTRAMI MN		-65			45
1	2.2					BIG STONE MN					
1	.4	2	.8	3	1.1	BLUE EARTH MN	98	40	178	92	53
2	1.4			1	.3	BROWN MN			-74	04	31
3	2.4	2	1.4	3	1.9	CARLTON MN	-40	35	-19	40	78
		1	.6	4	2.5	CARVER MN		284			87
		1	.8			CASS MN					
				1	.4	CHIPPEWA MN					33
				3	1.3	CLAY MN					60
				1	6.2	COOK MN					99
1	1.2			1	.5	COTTONWOOD MN			-61	12	34
2	1.9	2	1.1	4	1.2	CROW WING MN	-43	16	-34	31	57
1	.4	3	.8	8	1.4	DAKOTA MN	95	78	248	95	62
1	.8	2	1.4	2	1.2	DOUGLAS MN	67	-15	41	67	55
		1	.7	1	.7	FARIBAULT MN		0			39
1	.6			1	.4	FILLMORE MN			-32	32	33
1	.6	1	.4	1	.2	FREEBORN MN	-24	-52	-63	11	31
1	.5	3	1.1	1	.5	GOODHUE MN	135	-52	13	56	35
1	3.5			1	1.5	GRANT MN			-59	14	66
42	1.1	43	.9	76	1.4	HENNEPIN MN	-21	64	29	63	65
		1	1.5			HUBBARD MN					
				2	1.3	ISANTI MN					61
1	1.2	3	1.7	2	.8	ITASCA MN	37	-50	-31	32	44
1	1.5	1	.9	2	1.5	JACKSON MN	-39	72	4	52	68
				1	1.0	KANABEC MN					48
		1	1.8			KITTSON MN					
		1	.8	1	1.4	LAC QUI PARLE MN		64			63
		1	2.1			LAKE MN					
		1	9.3			LAKE OF THE WOODS MN					
				1	2.1	LINCOLN MN					83
1	.8	2	1.5	1	.5	LYON MN	79	-64	-35	30	35
1	.8	1	.5	2	1.2	MC LEOD MN	-36	134	49	70	55
1	2.1					MARSHALL MN					
1	.6	4	2.2	1	.6	MARTIN MN	245	-72	-4	48	37
1	1.1	1	1.0	1	1.1	MEEKER MN	-4	4	0	50	51
		1	.8			MILLE LACS MN					
1	.9	2	1.8	3	1.4	MORRISON MN	94	-21	54	72	63
		4	1.6	2	.7	MOWER MN		-59			38
1	1.1			2	1.4	NICOLLET MN			35	65	64
		1	.7	1	.5	NOBLES MN		-27			35
1	1.6	1	1.0			NORMAN MN	-37				
2	.7			3	.7	OLMSTED MN			1	50	39
		2	.5	2	.4	OTTER TAIL MN		-24			32
1	1.6			1	1.8	PENNINGTON MN			12	55	75
		1	1.0			PINE MN					
1	2.6					PIPESTONE MN					
		4	1.8			POLK MN					
		1	.9	2	1.6	POPE MN		73			70
19	1.0	30	1.2	45	1.5	RAMSEY MN	28	23	58	73	67
		1	2.9			RED LAKE MN					
1	1.0	1	1.0	2	1.0	REDWOOD MN	-4	9	5	52	50
1	.8	1	.6	5	3.6	RENVILLE MN	-32	531	330	97	95
		2	.9	3	1.2	RICE MN		33			57
		1	1.5	2	2.8	ROCK MN		92			91
11	1.0	16	1.2	15	1.1	ST LOUIS MN	26	-12	11	55	53
				1	.8	SCOTT MN					43
				1	.5	SHERBURNE MN					34
		1	1.1	4*	3.3	SIBLEY MN		187			93
3	1.0	1	.3	1	.1	STEARNS MN	-70	-56	-87	00	30
		1	.8			STEELE MN					
				1	.6	STEVENS MN					36
				2	2.6	SWIFT MN					89
1	.8					TODD MN					
				3	2.0	WABASHA MN					80
				1	.5	WASECA MN					35
				2	.5	WASHINGTON MN					34
		2	2.0			WATONWAN MN					
2	.9			4	1.0	WINONA MN			11	55	49
3	1.8	2	1.0			WRIGHT MN	-42				
1	1.1	1	.8			YELLOW MEDICINE MN	-27				
73	1.4	106	1.5	143	1.5	MISSISSIPPI MS	8	1	9		
1	2.4	3*	3.2	3	2.8	ADAMS MS	32	-12	16	57	90
2	1.7	4	2.8	3	1.9	ALCORN MS	58	-29	12	55	79
1	2.1	1	1.9			AMITE MS	-12				
				2	2.2	ATTALA MS					84
1*	4.0					BENTON MS					
1	2.0	1	1.4	1	.9	BOLIVAR MS	-33	-32	-54	17	47
1	1.2	1	1.3			CALHOUN MS	4				
1	3.4					CARROLL MS					
		1	1.1			CHICKASAW MS					
1	2.9			1	1.6	CHOCTAW MS			-45	23	69
1*	6.6			1	5.5	CLAIBORNE MS			-17	41	98
				1	1.2	CLARKE MS					57
2	3.1	2	2.6			COAHOMA MS	-17				
1	1.2			1	.8	COPIAH MS			-34	31	42
1	2.4	1	2.1	1	1.1	COVINGTON MS	-14	-48	-56	16	52
		1	2.0	2	1.6	DE SOTO MS		-22			69
4*	2.8	4	2.1	4	1.6	FORREST MS	-24	-24	-43	24	69
		1	2.3	1	1.2	GEORGE MS		-45			58
1	4.5	1	6.2	1	2.0	GREENE MS	38	-68	-56	15	80
1	2.4			3	3.9	GRENADA MS			64	75	96
1	1.4	2	2.6			HANCOCK MS	82				
1	.4	8	2.2	12	2.3	HARRISON MS	446	4	470	99	86

* SIGNIFICANTLY MORE DEATHS THAN EXPECTED

WHITE FEMALE — CANCER OF THE ORAL CAVITY INCLUDING TONGUE — ICD 141,143-146,148,149

1950-1959 DEATHS	1950-1959 RATE	1960-1969 DEATHS	1960-1969 RATE	1970-1979 DEATHS	1970-1979 RATE	COUNTY AND STATE	PERCENT CHANGE 60S/50S	PERCENT CHANGE 70S/60S	PERCENT CHANGE 70S/50S	PERCENTILE %CHANGE 70S/50S	PERCENTILE RATES 1970S
1	.3	13*	2.2	14	1.8	HINDS MS	543	-18	427	98	75
1	2.2	2	4.6	2	3.1	HOLMES MS	115	-32	45	69	93
		1	3.4			HUMPHREYS MS					
		2	2.1	4	3.4	ITAWAMBA MS		61			94
		2	1.1	7	2.4	JACKSON MS		119			86
1	2.4	2	3.7	1	1.2	JASPER MS	51	-68	-51	19	56
		5	2.3	7	2.2	JONES MS		-3			83
				1	.6	LAFAYETTE MS					36
1	2.2			1	1.5	LAMAR MS			-30	33	67
4	1.7	4	1.2	6	1.6	LAUDERDALE MS	-27	32	-5	47	70
				2	2.1	LEAKE MS					83
5*	4.0	2	1.2	4	1.6	LEE MS	-69	32	-59	13	70
		1	.8	3	1.8	LEFLORE MS		124			76
		2	1.9	1	1.0	LINCOLN MS		-45			51
1	1.1			2	.8	LOWNDES MS			-25	37	43
		1	1.6	3	3.4	MADISON MS		116			94
1	1.8	1	1.3	1	.7	MARION MS	-25	-48	-61	12	39
2*	4.8			3*	3.6	MARSHALL MS			-24	37	95
3	3.2	3	2.2	4	2.4	MONROE MS	-30	9	-24	37	86
1	2.0	1	1.6	2	2.7	MONTGOMERY MS	-19	62	31	63	89
2	2.0	2	2.0	4	2.7	NESHOBA MS	0	33	34	64	89
		2	2.2	1	.8	NEWTON MS		-65			41
1	3.0	2*	7.5			NOXUBEE MS	146				
		1	1.4	2	2.4	OKTIBBEHA MS		75			87
				2	1.8	PANOLA MS					75
1	1.2			1	.8	PEARL RIVER MS			-35	29	42
1	3.6	1	3.3	2*	5.6	PERRY MS	-9	72	57	73	98
5*	4.5	3	2.2	2	1.1	PIKE MS	-51	-50	-75	04	53
1	1.3	5*	5.1			PONTOTOC MS	304				
		1	1.0	1	.7	PRENTISS MS		-23			40
1	1.0			4	2.1	RANKIN MS			100	84	82
2	3.6	1	1.3			SCOTT MS	-65				
2	2.9	1	1.2			SIMPSON MS	-60				
				1	1.1	SMITH MS					51
				1	2.5	STONE MS					87
1	1.4	1	1.2	1	.8	SUNFLOWER MS	-13	-37	-45	22	42
		1	2.1	2	3.3	TALLAHATCHIE MS		57			94
3*	8.4					TATE MS					
2	3.0			1	.8	TISHOMINGO MS			-74	05	42
1	1.1	2	1.8	3	2.0	UNION MS	60	9	74	79	79
		2*	4.4			WALTHALL MS					
1	.8	2	1.5	2	1.2	WARREN MS	92	-18	58	73	57
2	2.8	1	.8	3	1.6	WASHINGTON MS	-73	106	-44	24	69
		2	3.5			WAYNE MS					
3*	4.8	2	3.0	3	2.9	WINSTON MS	-38	-2	-39	27	91
				1	1.2	YALOBUSHA MS					57
1	1.7	1	1.1	1	1.0	YAZOO MS	-34	-13	-43	24	48
184	.8	254	1.0	350	1.2	MISSOURI MO	19	21	43		
				3	2.0	ADAIR MO					79
1	1.2			1	1.4	ANDREW MO			17	57	64
1	1.6			2	2.8	ATCHISON MO			79	80	91
		1	.4	3	2.1	AUDRAIN MO		416			81
3	2.1			2	.9	BARRY MO			-56	16	47
1	1.0			1	.5	BARTON MO			-52	18	35
1	1.2	1	.4	2	1.3	BATES MO	-62	187	10	54	59
		1	1.2	1	.9	BENTON MO		-29			45
		2	2.9	1	1.3	BOLLINGER MO		-55			60
		2	.7	4	.8	BOONE MO		9			41
4	.6	5	.8	12	1.8	BUCHANAN MO	24	141	200	94	76
1	1.2			6	2.7	BUTLER MO			124	87	90
				1	.6	CALDWELL MO					37
1	.6			1	.4	CALLAWAY MO			-31	33	33
1	1.9			2	1.8	CAMDEN MO			-4	47	75
		3	1.3	3	.8	CAPE GIRARDEAU MO		-41			41
2	1.5					CASS MO					
1	1.9			2	2.0	CEDAR MO			8	54	80
		2	2.0	3	4.1	CHARITON MO		108			96
2	2.3	1	1.3	2	2.2	CHRISTIAN MO	-46	78	-3	48	84
1	1.3	1	1.4			CLARK MO	11				
4	1.3	2	.5	4	.7	CLAY MO	-61	35	-47	21	38
		1	.9	2	1.9	CLINTON MO		115			77
2	1.2	1	.5	2	.7	COLE MO	-62	54	-42	25	39
				2	1.8	COOPER MO					75
				1	.6	CRAWFORD MO					37
		1	1.2			DADE MO					
				2	1.7	DALLAS MO					74
1	1.1	1	1.2	1	.8	DAVIESS MO	6	-35	-31	33	41
		3*	4.9			DE KALB MO					
1	1.7					DOUGLAS MO					
2	1.0	4	1.6	2	.9	DUNKLIN MO	57	-46	-15	42	45
3	1.3	2	.7	3	.6	FRANKLIN MO	-46	-9	-51	19	37
1	1.3	2	2.0			GASCONADE MO	51				
		1	2.0	1	.7	GENTRY MO		-65			39
12	2.0	9	1.1	11	.9	GREENE MO	-44	-18	-55	16	45
1	.8	1	.7			GRUNDY MO	-13				
		1	.7			HARRISON MO					
2	1.1	1	.5	4	2.2	HENRY MO	-55	369	112	86	84
				1	1.8	HICKORY MO					76
1	1.2					HOLT MO					
2	1.6	1	.8	4	1.9	HOWELL MO	-50	142	21	59	79
30	1.0	29	.8	61	1.8	JACKSON MO	-17	116	79	79	75
4	.7	7	1.2	3	.4	JASPER MO	71	-65	-41	26	33
1	.5	3	.7	8	1.6	JEFFERSON MO	37	112	190	93	69
3	1.8			2	.9	JOHNSON MO			-53	17	45
1	1.5	1	2.9	2	2.6	KNOX MO	96	-11	74	79	88
1	.8	1	.8			LACLEDE MO	9				
1	.5	1	.8	1	.3	LAFAYETTE MO	40	-65	-51	18	31
		1	.9	2	2.4	LEWIS MO		158			86
		1	1.1			LINCOLN MO					

* SIGNIFICANTLY MORE DEATHS THAN EXPECTED

WHITE FEMALE — CANCER OF THE ORAL CAVITY INCLUDING TONGUE — ICD 141,143-146,148,149

1950-1959 DEATHS	1950-1959 RATE	1960-1969 DEATHS	1960-1969 RATE	1970-1979 DEATHS	1970-1979 RATE	COUNTY AND STATE	PERCENT CHANGE 60S/50S	PERCENT CHANGE 70S/60S	PERCENT CHANGE 70S/50S	PERCENTILE %CHANGE 70S/50S	PERCENTILE RATES 1970S
1	.9	1	1.4			LINN MO	47				
1	.8	2	1.7	2	1.1	LIVINGSTON MO	118	-32	49	70	54
				1	.9	MC DONALD MO					45
2	1.1					MACON MO					
		2	3.0			MADISON MO					
		1	.5	6	2.8	MARION MO		405			90
		2	1.9	2	1.3	MILLER MO		-30			60
		3*	3.3	2	1.8	MISSISSIPPI MO		-45			75
		1	1.5			MONITEAU MO					
1	.9					MONROE MO					
1	1.0			1	1.7	MONTGOMERY MO			75	79	73
1	1.5	2	2.4			MORGAN MO	55				
		2	2.1	2	1.5	NEW MADRID MO		-28			67
				3	1.2	NEWTON MO					56
		1	.6			NODAWAY MO					
1	2.0	1	1.7			OREGON MO	-18				
1	2.6	1	1.9	1	1.7	OZARK MO	-25	-10	-32	32	73
2	1.6	3	2.1	1	.8	PEMISCOT MO	27	-60	-50	20	43
1	.5	2	.7	4	1.1	PETTIS MO	30	70	120	87	54
2	2.0	1	.7	2	1.0	PHELPS MO	-64	38	-51	19	49
3	2.8	1	1.0	2	1.6	PIKE MO	-62	57	-41	26	71
2	2.4			3	2.0	PLATTE MO			-17	41	80
1	1.0			2	1.1	POLK MO			12	55	54
		1	1.3	2	2.1	PULASKI MO		55			81
		1	.5	2	1.4	RANDOLPH MO		168			65
		2	1.9	1	.4	RAY MO		-79			33
1	1.3					RIPLEY MO					
3	1.6	1	.5	4	.8	ST CHARLES MO	-70	67	-49	20	44
				1	1.3	STE GENEVIEVE MO					59
1	.4	2	.8	2	.8	ST FRANCOIS MO	80	11	99	84	44
19	.8	44	1.2	64	1.3	ST LOUIS COUNTY MO	36	11	52	71	59
		4	1.9	5	2.3	SALINE MO		22			85
				3*	4.8	SCHUYLER MO					98
		2	1.1			SCOTT MO					
1	2.5					SHANNON MO					
		1	2.7	1	.8	SHELBY MO		-71			42
1	.6	3	1.6	3	1.4	STODDARD MO	172	-16	129	88	62
		2	3.1	3	3.3	STONE MO		7			93
				1	1.7	SULLIVAN MO					72
		2	2.0	1	.5	TANEY MO		-75			35
		4	2.2			VERNON MO					
				1	1.7	WARREN MO					73
1	1.8					WAYNE MO					
				2*	3.5	WORTH MO					94
1	1.1	3	3.7			WRIGHT MO	245				
43	1.1	62	1.7	49	1.6	ST LOUIS CITY MO	54	-3	50	71	70
28	1.1	31	.9	51	1.3	MONTANA MT	-12	39	22		
		2*	5.9	1	1.5	BEAVERHEAD MT		-75			66
		1	1.5			CARBON MT					
1*	8.0					CARTER MT					
5	2.1	4	1.1	8	2.2	CASCADE MT	-48	100	3	52	83
1	3.0	1	3.3	2*	5.5	CHOUTEAU MT	9	66	81	80	98
2	3.2					CUSTER MT					
1*	4.7	1*	3.5			DANIELS MT	-26				
		1	2.0	1	2.0	DAWSON MT		-1			80
		1	1.1	1	.9	DEER LODGE MT		-10			47
1	.7	1	.5	8*	3.1	FLATHEAD MT	-29	527	343	97	93
		1	3.3	1	2.9	GLACIER MT		-14			91
		2	2.5			HILL MT					
		1*	7.0			JUDITH BASIN MT					
		1	1.6			LAKE MT					
				2	2.8	LINCOLN MT					90
				1*	9.0	MC CONE MT					99
				1*	6.0	MINERAL MT					98
2	1.1			2	.8	MISSOULA MT			-34	30	41
		1	3.2			MUSSELSHELL MT					
		1	1.5			PARK# MT					
3*	12.0			1	2.7	PHILLIPS MT			-78	03	89
1	4.0					PONDERA MT					
1	3.0					POWELL MT					
1	1.5			1	1.0	RAVALLI MT			-33	32	49
		1	1.9	2	3.8	RICHLAND MT		98			95
				1	3.8	ROSEBUD MT					95
1	2.9	1	2.6	1	1.8	SANDERS MT	-11	-31	-38	27	75
				1	3.1	SHERIDAN MT					93
5	1.8	4	1.5	5	2.1	SILVER BOW MT	-19	40	14	56	82
				1	5.0	SWEET GRASS MT					98
		1	3.1	1	2.7	TETON MT		-12			90
				1	2.2	VALLEY MT					83
1*	7.2					WHEATLAND MT					
2	1.1	5	1.3	8	1.6	YELLOWSTONE MT	16	21	40	67	68
59	.8	71	.8	97	1.0	NEBRASKA NB	-2	26	24		
2	1.1	1	.3			ADAMS NB	-71				
		1	2.0	1	1.0	ANTELOPE NB		-50			50
1	1.9	1	1.1	1	1.7	BOONE NB	-41	60	-6	46	74
		1	1.2			BOX BUTTE NB					
				1	2.3	BOYD NB					86
		1	2.9	1	2.7	BROWN NB		-6			90
1	.7	2	1.0	1	.8	BUFFALO NB	39	-16	17	57	43
		1	1.2	1	1.5	BURT NB		27			66
		1	1.3	1	.6	BUTLER NB		-55			36
2	2.0	3	2.9	1	1.1	CASS NB	43	-61	-44	23	54
1	1.6	1	1.5			CEDAR NB	-3				
				2*	6.6	CHASE NB					99
		2*	5.5	1	1.1	CHERRY NB		-80			52
1	2.8	1	1.3			CHEYENNE NB	-52				
1	1.4	1	1.1	2	3.8	CLAY NB	-24	254	169	92	95

* SIGNIFICANTLY MORE DEATHS THAN EXPECTED

\# COUNTIES COMBINED; SEE TECHNICAL NOTES

WHITE FEMALE — CANCER OF THE ORAL CAVITY INCLUDING TONGUE — ICD 141,143-146,148,149

1950-1959 DEATHS	1950-1959 RATE	1960-1969 DEATHS	1960-1969 RATE	1970-1979 DEATHS	1970-1979 RATE	COUNTY AND STATE		PERCENT CHANGE 60S/50S	PERCENT CHANGE 70S/60S	PERCENT CHANGE 70S/50S	PERCENTILE %CHANGE 70S/50S	PERCENTILE RATES 1970S
		1	1.0	2	2.9	COLFAX	NB		200			92
				1	.6	CUMING	NB					36
				1	.6	CUSTER	NB					37
		1	1.6			DAKOTA	NB					
1	.8			1	.6	DAWSON	NB			-24	37	37
		1	1.5	1	2.3	DIXON	NB		49			85
3	1.6	2	1.2	1	.5	DODGE	NB	-24	-57	-67	08	35
16	1.1	15	.8	30	1.6	DOUGLAS	NB	-27	91	38	66	69
1*	4.6					DUNDY	NB					
1	2.1	1	1.1			FILLMORE	NB	-48				
1	2.6					FRANKLIN	NB					
		1	2.7			FRONTIER	NB					
		1	1.2	2	2.4	FURNAS	NB		104			86
1	.5	1	.4	1	.4	GAGE	NB	-29	4	-26	36	33
				2*	9.2	GARDEN	NB					99
1	4.5					GREELEY	NB					
2	1.1	3	1.2	3	.9	HALL	NB	12	-25	-16	42	46
1	1.4			1	1.3	HAMILTON	NB			-9	45	60
				1	1.3	HARLAN	NB					61
2	2.8	1	.9			JEFFERSON	NB	-66				
		1	2.7			JOHNSON	NB					
				2	4.4	KEARNEY	NB					97
		1	2.9			KEITH	NB					
1	1.6			1	.8	KNOX	NB			-46	22	44
4	.5	11	1.2	13	1.2	LANCASTER	NB	122	4	132	88	58
		4	2.6	1	.6	LINCOLN	NB		-77			36
				1*	42.0	LOUP	NB					99
1	.9			1	.4	MADISON	NB			-59	14	32
1	1.8	1	1.9			MERRICK	NB	6				
				1	1.8	MORRILL	NB					76
		1	2.0	1	1.4	NEMAHA	NB		-32			63
		2	1.3	1	1.2	OTOE	NB		-4			56
1*	7.9					PERKINS	NB					
				1	1.9	PHELPS	NB					77
1	1.0			3	1.5	PLATTE	NB			44	69	66
1	1.7					POLK	NB					
1	.8					RICHARDSON	NB					
				1	3.7	ROCK	NB					95
1	.9	1	.7	1	.7	SALINE	NB	-20	-2	-21	39	40
2	3.3	1	1.4	1	.8	SARPY	NB	-57	-44	-76	04	43
		1	.8			SAUNDERS	NB					
1	.7	1	.5	2	.7	SCOTTS BLUFF	NB	-22	35	5	52	40
				2*	5.2	STANTON	NB					98
1	1.4					THAYER	NB					
1	1.9					VALLEY	NB					
1	1.5			2	2.8	WASHINGTON	NB			87	81	90
1	1.7					WEBSTER	NB					
1	1.1	1	.8	2	2.5	YORK	NB	-22	208	139	89	88
8	1.0	26*	1.6	55*	2.1	NEVADA	NV	57	30	105		
		1	2.6	2	3.4	CHURCHILL	NV		32			94
3	1.4	9	1.2	25	1.7	CLARK	NV	-17	49	24	60	73
1	2.2	3*	5.7			ELKO	NV	155				
				1	9.1	LINCOLN	NV					99
				3*	5.3	LYON	NV					98
		1	3.0	1	3.2	MINERAL	NV		9			93
2	.6	11*	2.2	22*	3.2	WASHOE	NV	288	45	463	99	93
1	2.0					WHITE PINE	NV					
1	5.9	1	2.0	1	1.5	ORMSBY-CARSON CITY	NV	-66	-25	-74	05	68
30	.9	50	1.2	77	1.7	NEW HAMPSHIRE	NH	40	34	86		
2	1.1			3	1.1	BELKNAP	NH			-4	47	53
				3	1.5	CARROLL	NH					67
1	.4	4	1.4	2	.5	CHESHIRE	NH	261	-62	37	66	35
2	1.2	3	1.5	4	1.8	COOS	NH	29	17	52	71	74
3	1.0	5	1.4	7	1.8	GRAFTON	NH	37	32	81	80	76
8	.9	19	1.6	26	2.0	HILLSBOROUGH	NH	80	29	132	88	80
3	.6	7	1.2	9	1.8	MERRIMACK	NH	103	47	200	94	76
5	1.0	6	.9	15	1.8	ROCKINGHAM	NH	-9	96	78	79	76
2	.7	6	1.8	5	1.2	STRAFFORD	NH	160	-34	70	78	55
4	2.7			3	1.3	SULLIVAN	NH			-52	18	59
245	.9	380	1.1	657	1.6	NEW JERSEY	NJ	18	43	70		
12	1.4	20	2.0	15	1.1	ATLANTIC	NJ	40	-46	-25	37	52
28	.9	45	1.0	83	1.5	BERGEN	NJ	3	54	59	73	66
2	.2	8	.8	17	1.2	BURLINGTON	NJ	245	53	428	98	58
9	.6	21	1.1	40	1.6	CAMDEN	NJ	71	47	151	90	70
5	1.6	4	1.1	10	1.6	CAPE MAY	NJ	-32	50	3	51	71
6	1.3	3	.5	6	.8	CUMBERLAND	NJ	-61	53	-40	26	41
34	.7	61	1.3	82	1.8	ESSEX	NJ	72	46	151	90	76
3	.6	4	.6	9	1.0	GLOUCESTER	NJ	-1	73	72	78	50
35	1.2	43	1.2	57	1.6	HUDSON	NJ	0	32	32	64	68
2	.7	2	.6	3	.8	HUNTERDON	NJ	-17	36	13	56	41
13	1.1	17	1.2	30	1.8	MERCER	NJ	7	54	65	76	76
16	1.2	15	.8	44	1.6	MIDDLESEX	NJ	-39	110	29	63	69
10	.7	29	1.4	60*	2.2	MONMOUTH	NJ	107	55	221	95	84
11	1.0	16	1.1	37	1.9	MORRIS	NJ	8	78	93	83	78
7	1.7	11	.9	41	1.7	OCEAN	NJ	-45	88	3	52	74
26	1.3	30	1.3	42	1.6	PASSAIC	NJ	-5	31	24	60	71
3	1.5	1	.4	7	2.3	SALEM	NJ	-74	497	58	73	86
4	.7	7	.9	11	1.2	SOMERSET	NJ	24	30	61	74	55
1	.4	3	.9	8	1.9	SUSSEX	NJ	157	98	409	98	77
16	.8	39	1.5	49	1.5	UNION	NJ	84	4	91	82	67
2	.6	1	.3	6	1.0	WARREN	NJ	-49	241	75	79	50
28	1.3	25	.8	54	1.2	NEW MEXICO	NM	-42	53	-12		
7	1.2	7	.7	28	2.0	BERNALILLO	NM	-43	192	66	76	79

* SIGNIFICANTLY MORE DEATHS THAN EXPECTED

WHITE FEMALE — CANCER OF THE ORAL CAVITY INCLUDING TONGUE — ICD 141,143-146,148,149

1950-1959 DEATHS	1950-1959 RATE	1960-1969 DEATHS	1960-1969 RATE	1970-1979 DEATHS	1970-1979 RATE	COUNTY AND STATE		PERCENT CHANGE 60S/50S	PERCENT CHANGE 70S/60S	PERCENT CHANGE 70S/50S	PERCENTILE %CHANGE 70S/50S	PERCENTILE RATES 1970S
		3	1.7	3	1.1	CHAVES	NM		-38			51
2	1.9	1	.7	1	.5	CURRY	NM	-65	-30	-75	04	34
2	2.0	2	.8	2	.7	DONA ANA	NM	-58	-17	-65	10	38
2	1.7	1	.5	2	.7	EDDY	NM	-72	53	-57	15	39
1	1.6					GRANT	NM					
		1	9.3			GUADALUPE	NM					
1*	13.2	1*	14.1			HARDING	NM	6				
				1	3.5	HIDALGO	NM					95
		1	.5	3	1.4	LEA	NM		194			63
				2	3.9	LINCOLN	NM					96
1	3.1			1	1.5	LUNA	NM			-52	18	67
		1	1.8			MC KINLEY	NM					
1	1.7	1	1.4			OTERO	NM	-19				
2*	4.0	1	1.6			QUAY	NM	-59				
2	2.7	2	2.6	1	.8	ROOSEVELT	NM	-2	-70	-71	06	42
				2	2.6	SANDOVAL	NM					88
1	3.0	1	.4	1	.5	SAN JUAN	NM	-88	40	-83	01	35
2	2.4	1	1.0	1	1.1	SAN MIGUEL	NM	-58	14	-52	18	55
1	.6			3	1.1	SANTA FE	NM			89	82	53
1	2.2	1	1.2	1	2.2	SIERRA	NM	-45	84	2	50	84
1	2.7			1	1.5	TAOS	NM			-43	24	67
1	3.4					UNION	NM					
				1	.8	VALENCIA	NM					41
855	1.1	1241	1.3	1609	1.6	NEW YORK	NY	21	20	46		
20	1.4	25	1.5	38	2.0	ALBANY	NY	4	33	39	67	79
2	.9	1	.4	1	.4	ALLEGANY	NY	-57	-3	-58	14	32
84	1.2	86	1.1	117	1.8	BRONX	NY	-8	64	51	71	74
6	.6	12	1.0	22	1.6	BROOME	NY	72	66	185	93	70
3	.7	4	.9	11	2.0	CATTARAUGUS	NY	23	136	190	93	81
3	.7	6	1.2	9	1.6	CAYUGA	NY	76	36	140	89	71
9	1.0	9	.9	13	1.3	CHAUTAUQUA	NY	-10	39	26	61	59
4	.8	6	1.0	8	1.2	CHEMUNG	NY	28	27	63	75	57
		2	.5	3	.9	CHENANGO	NY		71			47
1	.4	4	1.5	4	1.3	CLINTON	NY	297	-14	239	95	58
2	.7	3	.9	9	2.1	COLUMBIA	NY	25	148	210	94	83
2	.8	1	.5	5	2.0	CORTLAND	NY	-35	301	159	91	81
3	1.2	2	.8	4	1.1	DELAWARE	NY	-31	32	-9	45	53
10	1.1	13	1.2	21	1.6	DUTCHESS	NY	14	36	55	72	71
45	.9	72	1.3	91	1.4	ERIE	NY	36	13	54	72	64
3	1.5	1	.4	5	2.2	ESSEX	NY	-70	399	47	70	84
3	1.4	1	.4	1	.4	FRANKLIN	NY	-70	-6	-72	06	32
4	1.2	3	.9	5	1.2	FULTON	NY	-22	33	3	51	58
3	1.1	1	.4	4	1.0	GENESEE	NY	-66	173	-7	46	50
3	1.4	3	1.1	5	2.4	GREENE	NY	-20	112	71	78	87
1*	5.0					HAMILTON	NY					
6	1.5	6	1.2	12	2.4	HERKIMER	NY	-19	99	62	75	87
9	1.6	4	.6	12	2.2	JEFFERSON	NY	-60	240	36	65	84
120	1.0	149	1.1	135	1.0	KINGS(BROOKLYN)	NY	3	-2	1	50	51
1	.7			1	.5	LEWIS	NY			-35	30	34
2	.8	3	1.0	3	1.0	LIVINGSTON	NY	34	-5	27	62	48
1	.4	3	1.0	4	1.3	MADISON	NY	144	28	212	94	60
23	.8	34	1.0	65	1.6	MONROE	NY	27	67	112	86	70
5	1.3	1	.2	3	.5	MONTGOMERY	NY	-84	125	-63	11	34
36	.9	101*	1.6	149*	1.9	NASSAU	NY	83	18	116	86	78
161*	1.6	262*	2.9	168*	2.1	NEW YORK(MANHATTAN)	NY	74	-29	24	61	81
8	1.0	11	.9	10	.7	NIAGARA	NY	-3	-25	-27	36	39
14	1.1	9	.6	20	1.2	ONEIDA	NY	-49	114	9	54	56
16	.9	24	1.1	36	1.4	ONONDAGA	NY	23	30	60	74	63
2	.5	6	1.4	4	.9	ONTARIO	NY	185	-33	90	82	46
11	1.1	17	1.5	21	1.5	ORANGE	NY	31	2	34	65	68
2	1.0	4	1.9	3	1.3	ORLEANS	NY	93	-34	27	62	59
6	1.3	4	.7	8	1.5	OSWEGO	NY	-45	99	10	54	66
4	1.1	1	.3	4	.9	OTSEGO	NY	-70	164	-20	39	47
1	.7	1	.5	3	.9	PUTNAM	NY	-22	57	22	59	45
75	.9	101	.8	190	1.5	QUEENS	NY	-8	76	62	75	66
11	1.3	15	1.5	12	1.2	RENSSELAER	NY	15	-22	-11	44	56
11	1.0	16	1.2	44*	2.6	RICHMOND	NY	17	116	153	90	89
3	.5	6	.8	13	1.2	ROCKLAND	NY	51	62	144	89	57
8	1.5	12	2.1	16*	2.5	ST LAWRENCE	NY	38	19	63	75	87
3	.7	6	1.1	9	1.4	SARATOGA	NY	57	22	92	83	62
11	1.3	19	1.9	20	1.9	SCHENECTADY	NY	45	1	46	69	78
				2	1.3	SCHOHARIE	NY					60
1	1.2	2	2.1	1	1.1	SCHUYLER	NY	72	-48	-11	44	53
2	.8	1	.3	2	.9	SENECA	NY	-61	159	2	51	45
6	1.2	3	.5	7	1.1	STEUBEN	NY	-54	107	-4	47	54
12	.5	42	1.1	107*	2.0	SUFFOLK	NY	102	76	256	96	79
2	.6	3	1.2	2	.5	SULLIVAN	NY	42	-59	-42	25	34
1	.5	1	.5	1	.4	TIOGA	NY	4	-21	-18	41	32
		6	1.9	8	2.4	TOMPKINS	NY		29			87
10	1.6	13	1.8	10	1.1	ULSTER	NY	12	-40	-33	31	51
6	2.6	5	1.7	11*	3.6	WARREN	NY	-34	107	38	66	95
4	1.4	2	.6	1	.2	WASHINGTON	NY	-54	-62	-83	01	31
4	1.0	4	1.0	8	1.6	WAYNE	NY	-1	65	63	75	70
43	1.1	87*	1.8	101	1.8	WESTCHESTER	NY	62	3	67	76	76
2	1.0	1	.6	6	2.6	WYOMING	NY	-44	360	158	90	89
1	1.0	1	.6	1	.8	YATES	NY	-41	37	-18	40	43
214*	1.8	279*	1.7	425*	1.8	NORTH CAROLINA	NC	-9	7	-3		
3	1.4	3	.8	12	2.3	ALAMANCE	NC	-38	176	72	78	86
2*	3.0			2	1.7	ALEXANDER	NC			-45	22	72
3*	7.1			4*	6.1	ALLEGHANY	NC			-15	42	98
1	1.5	1	1.2	1	.7	ANSON	NC	-18	-39	-50	19	40
2	2.1	3	2.7	3	2.0	ASHE	NC	30	-24	-1	49	81
1	2.5	1	2.0	2	2.3	AVERY	NC	-21	17	-8	45	85
4*	3.8	3	2.3	8*	4.3	BEAUFORT	NC	-40	87	13	56	97
				1	1.4	BERTIE	NC					63
3*	5.2	4*	4.9	1	1.0	BLADEN	NC	-5	-79	-80	02	51
1	1.8			4	3.3	BRUNSWICK	NC			82	81	94

* SIGNIFICANTLY MORE DEATHS THAN EXPECTED

WHITE FEMALE — CANCER OF THE ORAL CAVITY INCLUDING TONGUE — ICD 141,143-146,148,149

1950-1959 DEATHS	1950-1959 RATE	1960-1969 DEATHS	1960-1969 RATE	1970-1979 DEATHS	1970-1979 RATE	COUNTY AND STATE	PERCENT CHANGE 60S/50S	PERCENT CHANGE 70S/60S	PERCENT CHANGE 70S/50S	PERCENTILE %CHANGE 70S/50S	PERCENTILE RATES 1970S
6	1.1	9	1.3	20	2.0	BUNCOMBE NC	15	55	79	79	79
2	1.2	3	1.2	3	1.0	BURKE NC	1	-18	-17	41	49
3	1.6	9*	3.0	10	2.3	CABARRUS NC	90	-24	44	68	85
2	1.5	2	1.1	2	.8	CALDWELL NC	-29	-31	-51	19	41
1*	4.8			1	3.3	CAMDEN NC			-31	32	93
		1	1.1	4	2.1	CARTERET NC		92			82
1	2.1			3*	4.5	CASWELL NC			110	85	97
3	1.4	3	.8	7	1.5	CATAWBA NC	-43	83	4	52	66
3*	3.6	1	.9	1	.7	CHATHAM NC	-74	-18	-79	02	41
2	2.7	2	2.1	3	2.6	CHEROKEE NC	-22	23	-4	48	88
				2	4.8	CHOWAN NC					98
		1	3.2			CLAY NC					
4	2.2	4	1.5	6	1.4	CLEVELAND NC	-30	-9	-37	29	63
3	3.5	1	.8	2	.9	COLUMBUS NC	-79	23	-74	05	47
3	2.9	2	1.1			CRAVEN NC	-63				
5*	2.7	7	2.9	8	1.9	CUMBERLAND NC	6	-34	-30	33	77
2*	8.5			1	2.5	CURRITUCK NC			-70	07	88
		2*	5.2	2	3.3	DARE NC		-36			93
1	.5	9*	2.6	9	1.9	DAVIDSON NC	384	-25	263	96	78
1	1.7	2	2.5	3	2.8	DAVIE NC	52	10	67	76	90
4*	4.6	2	1.5	3	1.9	DUPLIN NC	-67	21	-60	12	77
8*	2.7	7	1.6	15*	2.8	DURHAM NC	-40	74	4	52	91
2	2.1	3	2.0	1	.7	EDGECOMBE NC	-3	-65	-67	09	39
7	1.8	13	1.9	15	1.4	FORSYTH NC	8	-26	-20	39	64
1	1.3			1	.6	FRANKLIN NC			-55	16	36
4	.9	9	1.6	15	2.0	GASTON NC	75	24	118	87	81
		1	3.0	2*	5.9	GRAHAM NC		101			98
2	2.2	2	1.8	2	1.9	GRANVILLE NC	-16	2	-14	42	76
6	.9	21*	2.2	25	1.7	GUILFORD NC	154	-20	102	84	74
3	2.7	6*	4.1	3	1.4	HALIFAX NC	54	-66	-48	20	63
3	2.1	4	2.2	4	2.0	HARNETT NC	3	-8	-5	47	80
3	2.1	4	1.9	6	1.9	HAYWOOD NC	-8	-2	-10	44	78
1	.6	4	1.5	2	.4	HENDERSON NC	136	-71	-31	32	33
1	2.1	1	1.4	1	1.1	HERTFORD NC	-37	-18	-49	20	53
1	3.6					HOKE NC					
		2*	5.9			HYDE NC					
		5	1.8	6	1.5	IREDELL NC		-13			67
		2	2.6	2	1.6	JACKSON NC		-37			71
2	2.3	4	1.5	10*	3.1	JOHNSTON NC	-36	113	37	66	92
1*	4.1					JONES NC					
2	3.0	2	2.3	6*	4.1	LEE NC	-25	80	36	65	96
2	2.0	4	2.8	7*	3.7	LENOIR NC	36	33	81	80	95
2	2.1	1	.8	2	1.0	LINCOLN NC	-61	29	-49	20	51
1	.9			1	.5	MC DOWELL NC			-47	21	34
2	3.0	2	2.0	1	.6	MACON NC	-34	-71	-81	02	36
2	2.8	1	1.1			MADISON NC	-62				
1	1.8	1	1.5	1	1.1	MARTIN NC	-19	-24	-38	27	54
8	1.3	13	1.3	24	1.6	MECKLENBURG NC	-1	27	26	61	70
1	2.1	1	1.4	2	2.4	MITCHELL NC	-32	67	14	56	87
		1	1.3	2	1.9	MONTGOMERY NC		41			77
2	1.7	3	1.8	2	.9	MOORE NC	7	-51	-48	21	46
2	1.4	2	1.2	4	1.5	NASH NC	-13	27	11	54	67
9*	5.3	6	2.2	15*	3.7	NEW HANOVER NC	-58	67	-30	33	95
3*	4.9	2	2.5	4*	5.6	NORTHAMPTON NC	-49	128	16	57	98
1	1.0	6*	6.2	6*	3.1	ONSLOW NC	514	-50	204	94	92
2	2.1	2	1.5			ORANGE NC	-29				
1	3.6	1	2.4	1	2.2	PAMLICO NC	-34	-7	-39	27	84
		1	1.0	2	1.7	PASQUOTANK NC		64			72
1	3.4	2	3.3	4*	4.6	PENDER NC	-3	41	36	65	97
2*	8.1			2	3.9	PERQUIMANS NC			-52	18	96
1	3.0	2	2.7			PERSON NC	-8				
4*	4.2	2	1.1	4	1.7	PITT NC	-74	52	-61	12	72
		2	2.6	1	.7	POLK NC		-73			39
2	1.3	1	.4	4	1.0	RANDOLPH NC	-72	177	-21	39	49
3*	3.3	5*	3.9	2	1.1	RICHMOND NC	17	-71	-66	09	54
3	2.1	3	1.7	3	1.2	ROBESON NC	-20	-28	-42	24	56
1	.8	2	.7	7	1.6	ROCKINGHAM NC	-3	117	111	86	70
5	1.8	4	.9	5	1.1	ROWAN NC	-48	13	-42	25	51
4	2.5	3	1.2	2	.7	RUTHERFORD NC	-53	-39	-72	06	39
6*	4.5	1	.6	5	2.8	SAMPSON NC	-87	372	-37	28	91
4*	8.8	1	1.4	3	3.2	SCOTLAND NC	-85	135	-64	10	93
1	1.1	4	2.6	4	1.5	STANLY NC	144	-42	41	67	66
				1	.7	STOKES NC					39
3	1.6	4	1.8	3	.9	SURRY NC	11	-50	-45	23	46
2*	5.5			1	1.6	SWAIN NC			-71	06	70
1	1.8	1	1.3	1	.8	TRANSYLVANIA NC	-27	-40	-57	15	42
2	1.9	4	2.0	1	.5	UNION NC	7	-78	-76	04	34
2	2.4	4*	4.1	1	.7	VANCE NC	72	-82	-69	07	40
3	.6	8	1.4	18	2.0	WAKE NC	147	47	264	96	81
1	3.1			1	1.5	WARREN NC			-50	19	68
2*	7.1			2	4.3	WASHINGTON NC			-40	26	97
1	1.2			5*	3.5	WATAUGA NC			194	93	94
2	1.4	5	2.5	7	2.6	WAYNE NC	75	6	86	81	89
3	1.9	4	2.0	5	1.8	WILKES NC	9	-13	-5	47	75
3	2.3	2	1.3	10*	4.4	WILSON NC	-44	241	91	82	97
		2	1.8	5	3.1	YADKIN NC		67			92
		1	1.3			YANCEY NC					
17	.7	16	.5	33	.9	NORTH DAKOTA ND	-30	87	31		
1	1.1					BARNES ND					
				1	3.0	BENSON ND					92
				2	2.1	BOTTINEAU ND					82
1	3.6					BURKE ND					
3	2.5	3	2.0	2	.9	BURLEIGH ND	-19	-57	-65	10	45
1	.4	4	1.1	4	1.2	CASS ND	171	10	197	94	56
		1	4.0			DIVIDE ND					
				2*	5.4	EMMONS ND					98
1*	9.4					GOLDEN VALLEY ND					
1	.5			2	.8	GRAND FORKS ND			60	74	44
				1	2.8	GRANT ND					90

* SIGNIFICANTLY MORE DEATHS THAN EXPECTED

WHITE FEMALE — CANCER OF THE ORAL CAVITY INCLUDING TONGUE — ICD 141,143-146,148,149

1950-1959 DEATHS	1950-1959 RATE	1960-1969 DEATHS	1960-1969 RATE	1970-1979 DEATHS	1970-1979 RATE	COUNTY AND STATE	PERCENT CHANGE 60S/50S	PERCENT CHANGE 70S/60S	PERCENT CHANGE 70S/50S	PERCENTILE %CHANGE 70S/50S	PERCENTILE RATES 1970S
				1	3.7	KIDDER ND					95
1	2.5	1	2.4	1	1.7	LA MOURE ND	-7	-28	-33	32	73
1	3.8					MC INTOSH ND					
1	3.9					MC KENZIE ND					
		2	3.0			MC LEAN ND					
				1	2.5	MERCER ND					88
1	1.1					MORTON ND					
		1	1.1	1	1.9	PEMBINA ND		72			78
				1	2.9	RANSOM ND					91
1	1.1			1	.6	RICHLAND ND			-40	26	37
1	1.4	1	1.3	1	.7	STARK ND	-8	-42	-47	21	40
				4	2.8	STUTSMAN ND					91
		1	1.5			TRAILL ND					
1	1.6			2	1.5	WALSH ND			-6	47	67
1	.6	2	1.0	3	1.1	WARD ND	70	12	92	82	52
1	2.7					WELLS ND					
				3	2.8	WILLIAMS ND					91
389	.9	534	1.1	782	1.4	OHIO OH	13	29	46		
		1	1.0	4	2.8	ADAMS OH		190			90
3	.7	8	1.4	9	1.4	ALLEN OH	107	0	106	85	63
1	.4	2	.8	6	2.0	ASHLAND OH	76	154	349	97	80
3	.6	4	.8	6	1.0	ASHTABULA OH	29	25	61	74	50
4	1.4	3	1.1	3	1.3	ATHENS OH	-23	27	-2	49	61
1	.5					AUGLAIZE OH					
2	.4	7	1.2	4	.7	BELMONT OH	203	-47	62	74	38
				2	1.3	BROWN OH					61
6	.9	8	.9	15	1.4	BUTLER OH	2	65	67	76	64
1	.9	1	.9	3	2.2	CARROLL OH	-1	138	136	88	83
1	.6	1	.7	3	1.6	CHAMPAIGN OH	20	136	182	92	71
4	.7	5	.7	18	2.2	CLARK OH	10	206	235	95	84
2	.7	3	.8	6	1.3	CLERMONT OH	21	62	97	83	60
1	.6	3	2.1	1	.4	CLINTON OH	247	-79	-27	35	33
8	1.5	10	1.7	9	1.4	COLUMBIANA OH	12	-18	-8	45	63
2	1.2	2	1.0	3	1.3	COSHOCTON OH	-20	30	4	52	60
1	.5					CRAWFORD OH					
71	1.0	120	1.5	141	1.6	CUYAHOGA OH	40	10	53	72	70
2	.7			1	.4	DARKE OH			-44	23	32
2	1.5	2	1.2	1	.5	DEFIANCE OH	-19	-57	-65	10	35
2	1.1			5	1.8	DELAWARE OH			59	73	74
		4	1.2	10	2.7	ERIE OH		118			89
2	.7	5	1.4	1	.2	FAIRFIELD OH	114	-84	-65	10	31
3	2.1	3	1.8	1	.4	FAYETTE OH	-17	-75	-80	02	33
32	1.3	30	.9	62	1.7	FRANKLIN OH	-25	78	34	65	72
3	2.0			1	.3	FULTON OH			-86	00	31
		1	.7	2	1.1	GALLIA OH		59			54
1	.4	2	1.1	3	1.1	GEAUGA OH	226	-4	213	94	53
2	1.0	4	1.3	8	1.7	GREENE OH	25	29	61	74	72
2	.7	2	.8	2	.8	GUERNSEY OH	8	4	13	56	43
51	1.3	53	1.2	85	1.8	HAMILTON OH	-12	49	32	64	74
1	.4	1	.3	3	.8	HANCOCK OH	-20	163	111	86	43
4	2.3	1	.6	1	.3	HARDIN OH	-72	-46	-85	00	31
1	.7	1	.8	3	2.3	HENRY OH	12	184	219	95	86
		4	2.1	5	2.1	HIGHLAND OH		3			82
1	1.0	1	.9			HOCKING OH	-11				
3	1.5			7	2.5	HURON OH			73	78	88
1	.6	5*	3.0	1	.3	JACKSON OH	380	-90	-52	18	31
2	.7	3	.6	5	.8	JEFFERSON OH	-9	36	24	60	44
3	1.3	3	1.1	2	.7	KNOX OH	-16	-36	-46	22	39
2	.5	10	1.5	15	1.6	LAKE OH	235	6	254	96	70
1	.5	4	1.4	4	1.3	LAWRENCE OH	212	-13	170	92	58
6	1.4	7	1.3	11	1.9	LICKING OH	-8	46	35	65	78
		1	.3	3	1.2	LOGAN OH		234			55
8	1.2	8	.8	15	1.3	LORAIN OH	-29	57	11	55	61
20	1.0	32	1.3	49	1.9	LUCAS OH	37	44	98	84	77
		1	.8			MADISON OH					
12	1.0	13	.9	17	1.0	MAHONING OH	-10	16	4	52	49
		5	1.6	5	1.5	MARION OH		-3			68
		2	.7	4	1.0	MEDINA OH		48			48
2	1.5			2	.8	MEIGS OH			-43	24	44
2	1.3	2	1.2	4	1.9	MERCER OH	-11	64	46	69	78
3	.8	3	.7	3	.6	MIAMI OH	-17	-12	-26	36	36
1	1.2			2	2.0	MONROE OH			69	77	81
24	1.3	25	1.1	38	1.4	MONTGOMERY OH	-17	30	8	54	63
		1	.9	1	1.6	MORGAN OH		78			70
1	1.1	2	2.1			MORROW OH	89				
4	1.0	6	1.3	7	1.4	MUSKINGUM OH	33	5	39	67	63
1	1.4			1	1.7	NOBLE OH			25	61	73
		1	.5	2	.9	OTTAWA OH		79			46
		1	.9	2	1.9	PAULDING OH		122			79
		2	1.0			PERRY OH					
1	.6			2	1.0	PICKAWAY OH			80	80	50
2	3.2	2	2.0			PIKE OH	-38				
2	.6	4	1.0	5	.9	PORTAGE OH	70	-14	46	70	46
1	.6			3	1.7	PREBLE OH			167	91	72
2	1.6	3	2.1			PUTNAM OH	26				
5	1.0	6	1.0	4	.6	RICHLAND OH	-2	-37	-38	28	37
2	.8	5	1.8	4	1.2	ROSS OH	131	-34	54	72	55
1	.3	4	1.2	6	1.8	SANDUSKY OH	243	54	428	98	75
3	.8	3	.7	4	.7	SCIOTO OH	-13	7	-7	46	39
1	.3	2	.7	3	.5	SENECA OH	94	-19	56	73	35
2	1.1	4	2.2	1	.3	SHELBY OH	107	-85	-68	08	31
10	.7	10	.6	19	.9	STARK OH	-17	66	38	66	47
23	1.2	26	1.0	44	1.5	SUMMIT OH	-16	52	28	62	68
		11	1.1	21	1.6	TRUMBULL OH		47			71
3	.7	4	.8	10	2.1	TUSCARAWAS OH	9	165	190	93	82
		3	1.8	1	.8	UNION OH		-57			42
1	.6	1	.5	1	.5	VAN WERT OH	-12	-16	-26	36	34
				1	2.0	VINTON OH					79
3	1.8	1	.4	4	1.2	WARREN OH	-75	168	-34	31	55

* SIGNIFICANTLY MORE DEATHS THAN EXPECTED

WHITE FEMALE — CANCER OF THE ORAL CAVITY INCLUDING TONGUE — ICD 141,143-146,148,149

1950-1959 DEATHS	1950-1959 RATE	1960-1969 DEATHS	1960-1969 RATE	1970-1979 DEATHS	1970-1979 RATE	COUNTY AND STATE	PERCENT CHANGE 60S/50S	PERCENT CHANGE 70S/60S	PERCENT CHANGE 70S/50S	PERCENTILE %CHANGE 70S/50S	PERCENTILE RATES 1970S
5	1.8	6	1.8	4	1.0	WASHINGTON OH	-1	-44	-44	23	50
2	.6	5	1.2	7	1.6	WAYNE OH	96	35	166	91	70
				2	1.1	WILLIAMS OH					54
4	1.2	3	.7	8	1.5	WOOD OH	-45	121	23	60	68
		2	1.7	1	.8	WYANDOT OH		-51			43
105	1.0	146	1.1	183	1.1	OKLAHOMA OK	9	5	14		
2	3.2	2	2.5	1	.9	ADAIR OK	-23	-62	-70	06	48
		1	1.0			ALFALFA OK					
		2	2.8	2	3.1	ATOKA OK		9			92
3	2.0	1	.4	2	1.1	BECKHAM OK	-80	172	-45	22	53
		2	2.4			BLAINE OK					
6*	3.9	2	1.0	2	1.2	BRYAN OK	-73	15	-69	07	56
3	2.0	1	.5	2	.8	CADDO OK	-73	52	-58	14	43
		1	.6			CANADIAN OK					
4	2.3	4	1.6	3	1.0	CARTER OK	-29	-41	-58	14	48
		1	1.0			CHEROKEE OK					
		1	.9	3	2.3	CHOCTAW OK		168			86
2	1.1	2	.9	6	1.5	CLEVELAND OK	-19	68	36	66	67
1	3.0					COAL OK					
1	.6			4	1.2	COMANCHE OK			100	84	58
				3*	3.9	COTTON OK					96
1	.9	1	.9	1	1.4	CRAIG OK	6	53	62	74	63
1	.4	4	1.9	2	.7	CREEK OK	367	-67	56	73	38
3	2.9	1	.7	1	.4	CUSTER OK	-76	-38	-85	00	33
		1	.8	1	.5	DELAWARE OK		-37			35
		1	1.4	1	2.2	DEWEY OK		57			84
1	.4	1	.3	3	.7	GARFIELD OK	-26	163	94	83	39
1	1.0	2	1.1	3	1.0	GARVIN OK	13	-12	0	50	48
4	2.3	5	2.1	2	.9	GRADY OK	-10	-55	-59	13	47
				2	1.6	GREER OK					71
1	1.7	3*	4.2			HASKELL OK	146				
1	.9			3	2.0	HUGHES OK			134	88	80
1	1.1	2	1.4	2	1.3	JACKSON OK	35	-6	26	61	61
		2	2.3	1	.9	JEFFERSON OK		-58			47
1	1.7	1	1.6	1	2.0	JOHNSTON OK	-3	25	21	59	80
1	.4	3	1.0	2	.4	KAY OK	140	-59	-2	49	33
		2	1.7	1	.6	KIOWA OK		-62			37
1	2.3	1	1.9	1	1.5	LATIMER OK	-17	-21	-35	30	68
1	.6	4	2.1	3	1.1	LE FLORE OK	223	-46	75	79	54
1	.8					LINCOLN OK					
1	.6	2	2.6	2	1.8	LOGAN OK	304	-31	180	92	75
2	2.5					MC CLAIN OK					
1	.9	4	3.0	3	1.9	MC CURTAIN OK	243	-37	115	86	77
1	1.8	2	2.9	1	.8	MC INTOSH OK	58	-70	-53	17	44
1	2.3	2	2.8			MARSHALL OK	20				
		2	1.5	3	1.4	MAYES OK		-5			65
		1	1.2			MURRAY OK					
2	.7	3	.7	7	1.7	MUSKOGEE OK	5	142	154	90	74
1	1.5	1	1.0			NOBLE OK	-32				
1	1.7			1	1.2	NOWATA OK			-28	35	57
1	1.3			1	1.2	OKFUSKEE OK			-12	44	55
12	.8	20	.9	41	1.5	OKLAHOMA OK	13	63	84	81	67
3	1.8	2	.7	4	1.4	OKMULGEE OK	-61	99	-22	38	65
1	.7	2	1.2	4	2.0	OSAGE OK	68	67	180	92	80
		3	1.3	2	1.1	OTTAWA OK		-20			51
1	1.2					PAWNEE OK					
				3	1.1	PAYNE OK					53
3	1.3	4	1.6	4	1.6	PITTSBURG OK	21	1	22	59	70
2	1.1	2	.8	1	.4	PONTOTOC OK	-27	-56	-68	08	32
		6	1.8	3	1.0	POTTAWATOMIE OK		-44			50
		2	2.9	2	2.4	PUSHMATAHA OK		-16			87
		1	3.3			ROGER MILLS OK					
2	2.0			1	.6	ROGERS OK			-72	06	36
1	.7	5*	2.9	4	1.8	SEMINOLE OK	302	-38	149	89	76
1	1.1	1	.9	1	.5	SEQUOYAH OK	-20	-48	-58	14	34
3	1.7	2	.9	2	.8	STEPHENS OK	-49	-8	-53	17	42
2*	3.4	1	1.7	1	1.3	TEXAS OK	-50	-20	-60	12	62
		1	1.1			TILLMAN OK					
19	1.6	18	1.1	30	1.4	TULSA OK	-33	31	-13	43	64
		2	2.2	1	.5	WAGONER OK		-76			35
2	1.2	2	.9	3	1.1	WASHINGTON OK	-26	32	-3	48	54
1	1.2	1	.9			WASHITA OK	-27				
83	1.0	139	1.3	196	1.4	OREGON OR	28	8	37		
2	2.6	1	1.2	1	.9	BAKER OR	-55	-20	-64	10	47
1	1.0	3	1.8	2	.7	BENTON OR	77	-60	-30	34	39
6	1.3	7	.9	13	1.2	CLACKAMAS OR	-27	31	-4	47	56
3	1.9	5	2.9	3	1.5	CLATSOP OR	57	-50	-22	38	65
		2	1.5	2	1.1	COLUMBIA OR		-26			54
5*	2.7	4	1.5	9	2.8	COOS OR	-43	84	5	53	90
				2	2.6	CROOK OR					88
		1	1.7	3	2.4	CURRY OR		38			86
		2	1.1	4	1.6	DESCHUTES OR		37			69
3	1.3	5	1.7	8	1.9	DOUGLAS OR	31	12	47	70	78
2*	7.1	1	3.7			GRANT OR	-48				
1	1.5					HOOD RIVER OR					
5	1.6	5	1.2	12	1.7	JACKSON OR	-26	51	12	55	74
				1	2.3	JEFFERSON OR					85
1	1.0	1	.6	2	.8	JOSEPHINE OR	-39	36	-18	41	43
1	.4	7*	3.4	5	1.9	KLAMATH OR	728	-43	370	98	78
4	.8	8	.9	17	1.4	LANE OR	14	49	70	78	64
2	1.7	5*	3.2	5	2.3	LINCOLN OR	88	-28	36	66	85
1	.4	4	1.3	4	.8	LINN OR	213	-38	95	83	43
		1	.9	1	.7	MALHEUR OR		-19			40
2	.3	6	.8	7	.6	MARION OR	147	-26	82	81	36
		1	3.0	1	2.7	MORROW OR		-12			89
37	1.2	55	1.7	63	1.8	MULTNOMAH OR	34	8	44	69	75
1	.7			2	.7	POLK OR			-2	48	39

* SIGNIFICANTLY MORE DEATHS THAN EXPECTED

WHITE FEMALE — CANCER OF THE ORAL CAVITY INCLUDING TONGUE — ICD 141,143-146,148,149

1950-1959 DEATHS	1950-1959 RATE	1960-1969 DEATHS	1960-1969 RATE	1970-1979 DEATHS	1970-1979 RATE	COUNTY AND STATE		PERCENT CHANGE 60S/50S	PERCENT CHANGE 70S/60S	PERCENT CHANGE 70S/50S	PERCENTILE %CHANGE 70S/50S	PERCENTILE RATES 1970S
				1*	9.0	SHERMAN	OR					99
		1	.9	3	1.7	TILLAMOOK	OR		83			72
1	.5	2	.9	4	1.4	UMATILLA	OR	70	54	162	91	64
1	1.2	4*	3.8	2	1.4	UNION	OR	223	-62	24	60	65
1	1.4	4*	3.4	5*	3.5	WASCO	OR	146	3	153	90	94
2	.5	3	.5	9	.9	WASHINGTON	OR	-6	81	70	77	45
1	.5	1	.4	5	1.9	YAMHILL	OR	-10	332	291	96	78
442	.9	648	1.0	845	1.2	PENNSYLVANIA	PA	20	12	35		
2	.8	2	.7			ADAMS	PA	-16				
66	.9	92	1.0	125	1.3	ALLEGHENY	PA	15	21	40	67	59
3	.9	1	.2	10	1.8	ARMSTRONG	PA	-75	648	86	81	74
8	1.1	12	1.3	18	1.5	BEAVER	PA	17	14	33	64	67
1	.6			3	1.3	BEDFORD	PA			128	87	60
10	.8	18	1.0	15	.7	BERKS	PA	25	-25	-6	46	40
8	1.0	8	.9	12	1.2	BLAIR	PA	-14	40	20	58	56
2	.6	1	.3	1	.2	BRADFORD	PA	-54	-33	-69	07	30
3	.4	14	1.0	22	1.2	BUCKS	PA	140	15	175	92	56
2	.4	4	.6	10	1.3	BUTLER	PA	53	106	215	94	60
1	.1	2	.2	11	.9	CAMBRIA	PA	108	344	826	99	46
		3	.8	7	1.8	CARBON	PA		122			76
1	.4	2	.6			CENTRE	PA	45				
8	1.0	9	.8	16	1.2	CHESTER	PA	-17	49	23	60	56
1	.6	1	.5			CLARION	PA	-8				
2	.5	3	.7	2	.4	CLEARFIELD	PA	44	-47	-23	38	32
1	.7	1	.5	3	1.2	CLINTON	PA	-26	149	85	81	57
3	1.2	3	.9	3	.7	COLUMBIA	PA	-26	-21	-42	25	38
3	.7	5	1.1	2	.4	CRAWFORD	PA	49	-66	-50	19	32
3	.6	10	1.3	9	1.0	CUMBERLAND	PA	123	-25	68	77	49
14	1.3	18	1.4	19	1.3	DAUPHIN	PA	9	-4	5	53	61
17	.7	38	1.3	53	1.5	DELAWARE	PA	72	15	98	83	66
1	.6	3	1.6	2	.9	ELK	PA	156	-42	49	70	47
13	1.2	19	1.5	22	1.5	ERIE	PA	24	2	27	61	66
5	.7	10	1.2	12	1.1	FAYETTE	PA	74	-5	67	76	53
1	3.4	1	3.4			FOREST	PA	2				
4	1.1	5	1.0	4	.6	FRANKLIN	PA	-13	-35	-43	24	37
		2	.9	1	.2	GREENE	PA		-72			31
		1	.5	1	.4	HUNTINGDON	PA		-22			32
		3	.7	4	.8	INDIANA	PA		12			44
4	1.4			1	.3	JEFFERSON	PA			-78	03	31
13	1.0	10	.6	23	1.2	LACKAWANNA	PA	-40	96	16	57	57
7	.5	9	.6	23	1.1	LANCASTER	PA	10	89	108	85	52
5	1.0	9	1.5	6	.9	LAWRENCE	PA	42	-42	-18	41	45
3	.7	5	.9	6	1.0	LEBANON	PA	31	8	41	68	49
10	.9	13	.9	19	1.1	LEHIGH	PA	2	18	21	59	51
12	.7	17	.7	29	1.1	LUZERNE	PA	8	52	64	75	52
2	.4	9	1.4	7	1.0	LYCOMING	PA	289	-27	186	93	50
5	1.7	1	.3	5	1.4	MC KEAN	PA	-81	327	-17	41	64
4	.7	9	1.2	6	.7	MERCER	PA	86	-40	12	55	40
1	.6	5	2.0	2	.7	MIFFLIN	PA	259	-65	26	61	39
2	1.0	7*	2.7	3	.7	MONROE	PA	164	-73	-28	35	40
20	.9	32	1.0	62	1.6	MONTGOMERY	PA	13	60	80	80	70
		1	.6	3	2.6	MONTOUR	PA		311			88
12	1.3	10	.8	13	.9	NORTHAMPTON	PA	-37	10	-31	33	46
4	.7	8	1.1	7	.8	NORTHUMBERLAND	PA	72	-29	23	60	43
		1	.7	1	.5	PERRY	PA		-30			35
108	1.2	138	1.4	142	1.5	PHILADELPHIA	PA	22	6	29	63	67
				1	.7	PIKE	PA					39
1	1.3			3	2.3	POTTER	PA			82	81	85
8	.9	11	1.0	17	1.3	SCHUYLKILL	PA	10	34	48	70	59
2	.6	5	1.1	4	.6	SOMERSET	PA	100	-43	14	56	37
		1	2.9	1	1.7	SULLIVAN	PA		-41			73
4	2.3	3	1.4			SUSQUEHANNA	PA	-37				
1	.5	3	1.5	1	.4	TIOGA	PA	207	-71	-11	44	33
		1	.5			UNION	PA					
4	1.1	3	.9	6	1.4	VENANGO	PA	-20	66	33	64	65
		2	.7	2	.7	WARREN	PA		1			40
7	.8	9	.8	19	1.4	WASHINGTON	PA	1	71	73	78	62
1	.5			1	.5	WAYNE	PA			-14	42	34
11	.8	19	1.0	27	1.2	WESTMORELAND	PA	34	18	58	73	56
1	.9	2	2.3	2	1.6	WYOMING	PA	148	-32	69	77	68
7	.7	14	1.0	16	.9	YORK	PA	46	-9	32	64	47
49	1.1	70	1.3	86	1.4	RHODE ISLAND	RI	18	5	24		
		3	1.4	5	1.9	BRISTOL	RI		40			78
6	1.4	9	1.4	12	1.4	KENT	RI	2	-2	-1	49	63
5	1.7	2	.6	8	1.6	NEWPORT	RI	-67	183	-6	46	70
37	1.2	51	1.4	56	1.4	PROVIDENCE	RI	20	-2	17	57	62
1	.4	5	1.7	5	1.2	WASHINGTON	RI	282	-29	171	92	55
92*	1.8	109	1.5	193*	1.9	SOUTH CAROLINA	SC	-16	24	4		
1	1.4			1	1.0	ABBEVILLE	SC			-30	33	50
2	1.1	12*	4.4	5	1.3	AIKEN	SC	304	-70	20	58	61
4	2.1	4	.9	18*	3.3	ANDERSON	SC	-55	252	59	73	93
		1	2.0	2	3.1	BAMBERG	SC		53			92
				2	1.4	BEAUFORT	SC					62
1	2.4			2	1.6	BERKELEY	SC			-33	31	69
4	1.1	4	.7	14	1.8	CHARLESTON	SC	-34	155	69	77	76
3	2.7	1	.7	5	2.5	CHEROKEE	SC	-74	257	-7	46	88
2	2.4	1	1.0	3	2.3	CHESTER	SC	-61	146	-4	47	86
6*	6.5	1	.8	5*	3.8	CHESTERFIELD	SC	-87	355	-42	25	96
1	3.4	2	4.4			CLARENDON	SC	30				
1	6.2			2	2.1	COLLETON	SC			-65	10	83
		3	2.2	4	1.9	DARLINGTON	SC		-16			77
2	4.2	3*	4.3	1	.8	DILLON	SC	2	-80	-80	02	44
		1	1.2	4	2.9	DORCHESTER	SC		142			91
3*	7.0			1	1.6	EDGEFIELD	SC			-77	04	71
1	2.5	1	2.2	1	1.4	FAIRFIELD	SC	-10	-38	-44	24	64

* SIGNIFICANTLY MORE DEATHS THAN EXPECTED

WHITE FEMALE — CANCER OF THE ORAL CAVITY INCLUDING TONGUE — ICD 141,143-146,148,149

1950-1959 DEATHS	1950-1959 RATE	1960-1969 DEATHS	1960-1969 RATE	1970-1979 DEATHS	1970-1979 RATE	COUNTY AND STATE		PERCENT CHANGE 60S/50S	PERCENT CHANGE 70S/60S	PERCENT CHANGE 70S/50S	PERCENTILE %CHANGE 70S/50S	PERCENTILE RATES 1970S
4	2.6	4	2.2	4	1.4	FLORENCE	SC	-16	-34	-44	23	64
1	4.1	1	1.4	3	2.7	GEORGETOWN	SC	-66	93	-35	30	89
11*	1.9	12	1.5	15	1.3	GREENVILLE	SC	-20	-16	-33	31	59
2	1.4	8*	5.0	1	.4	GREENWOOD	SC	255	-91	-69	07	33
2	1.3	3	1.4	8	2.3	HORRY	SC	5	65	74	79	85
1	1.5	1	1.1	3	2.5	KERSHAW	SC	-28	131	65	76	87
2	2.1			3	1.3	LANCASTER	SC			-38	28	60
1	.9	2	1.0	7	3.1	LAURENS	SC	18	205	261	96	93
		1	2.1			LEE	SC					
3	1.8	10*	3.6	6	1.2	LEXINGTON	SC	96	-66	-33	31	57
		1	1.3	1	1.1	MARION	SC		-14			53
		2	2.5	3	3.3	MARLBORO	SC		34			94
		1	.7	1	.6	NEWBERRY	SC		-20			36
3	2.4	4	2.8	5	2.3	OCONEE	SC	16	-18	-5	47	85
3	2.2	3	1.7	2	1.0	ORANGEBURG	SC	-22	-43	-56	16	49
4*	2.8	1	.5	6	1.8	PICKENS	SC	-81	247	-34	30	75
10*	2.1	5	.8	21*	2.7	RICHLAND	SC	-60	217	28	62	89
				2	2.3	SALUDA	SC					85
5	1.1	6	.8	16	1.9	SPARTANBURG	SC	-23	124	72	78	77
2	1.8	2	1.6	3	1.7	SUMTER	SC	-10	3	-7	46	72
2	2.4	2	1.8	3	1.7	UNION	SC	-22	-6	-27	36	73
1	1.7					WILLIAMSBURG	SC					
4	2.8	6	2.5	10	2.7	YORK	SC	-12	10	-3	48	89
18	.6	22	.6	33	.8	SOUTH DAKOTA	SD	4	24	28		
		1	3.5			AURORA	SD					
1	.7	2	2.0	1	.9	BEADLE	SD	188	-56	27	62	46
2	2.1			2	1.4	BROOKINGS	SD			-36	29	62
1*	6.6					CAMPBELL	SD					
		1	3.0			CLARK	SD					
1	1.7			1	2.1	CLAY	SD			24	60	82
1	.8			1	.5	CODINGTON	SD			-42	25	34
				1*	4.1	CORSON	SD					96
1	1.2	1	1.1	3	2.4	DAVISON	SD	-8	122	105	85	87
1*	7.8					HAAKON	SD					
				1	3.1	HAMLIN	SD					92
1	2.9					HAND	SD					
1*	3.9	1	3.2			HANSON	SD	-18				
				1	.8	HUGHES	SD					44
1	1.4					HUTCHINSON	SD					
		2	3.0	1	1.1	LAKE	SD		-62			54
				1	.9	LAWRENCE	SD					45
		2	3.5			LINCOLN	SD					
				2	1.7	MC COOK	SD					73
2	.6	7	1.4	11	1.8	MINNEHAHA	SD	156	29	230	95	76
				1	1.9	MOODY	SD					77
1	.6	2	1.0	1	.4	PENNINGTON	SD	72	-57	-26	36	33
				1	1.4	SPINK	SD					63
				1*	10.3	STANLEY	SD					99
		1*	11.6			TODD	SD					
		1	1.3	1	.7	TURNER	SD		-45			40
2*	3.6					UNION	SD					
2	2.1			2	1.6	YANKTON	SD			-21	39	71
		1*	74.3			ZIEBACH	SD					
227*	1.9	280*	1.7	362*	1.7	TENNESSEE	TN	-11	-3	-13		
2	1.6	5	2.3	5	1.6	ANDERSON	TN	44	-29	3	51	70
2	2.0	2	1.4			BEDFORD	TN	-29				
2	3.6			1	.6	BENTON	TN			-85	01	36
		1	2.4			BLEDSOE	TN					
8*	5.9	6	2.2	10	2.4	BLOUNT	TN	-63	8	-60	12	86
3	2.6	5	2.8	2	.8	BRADLEY	TN	7	-70	-68	08	44
1	.7	2	1.3	6	3.8	CAMPBELL	TN	84	190	434	98	96
2*	4.5			1	1.6	CANNON	TN			-65	10	68
3	2.4	3	2.1	3	1.4	CARROLL	TN	-12	-33	-41	25	64
1	.7	3	1.5	4	1.4	CARTER	TN	115	-5	105	85	63
2*	5.1			2	2.3	CHEATHAM	TN			-56	16	85
				1	1.3	CHESTER	TN					58
2	2.6	2	2.0	8*	6.7	CLAIBORNE	TN	-22	226	154	90	99
				2	3.7	CLAY	TN					95
3*	3.3	3	4.2	1	.8	COCKE	TN	28	-82	-77	03	41
2	2.4	2	1.5	1	.6	COFFEE	TN	-37	-61	-75	04	36
1	1.5			1	1.0	CROCKETT	TN			-37	28	48
		3	2.7	2	1.5	CUMBERLAND	TN		-43			67
19	1.5	26	1.5	35	1.5	DAVIDSON	TN	-6	5	-2	49	67
3*	7.5			1	1.5	DECATUR	TN			-81	02	65
3*	5.1					DE KALB	TN					
1	1.0	2	1.8	3	2.1	DICKSON	TN	79	17	109	85	82
		1	.5	3	2.2	DYER	TN		313			84
1	2.6			4*	7.3	FAYETTE	TN			180	92	99
1	2.0	2	3.5	1	1.2	FENTRESS	TN	78	-66	-39	27	56
5*	4.7	4*	3.3	3	1.8	FRANKLIN	TN	-30	-44	-61	12	76
4	2.0	6	2.3	3	1.0	GIBSON	TN	12	-58	-53	17	48
2	1.7	1	.7	1	.6	GILES	TN	-60	-17	-67	09	36
1	2.1	1	1.3	1	1.0	GRAINGER	TN	-40	-22	-53	17	49
2	1.3	3	1.4	4	1.3	GREENE	TN	7	-9	-3	48	59
4*	7.4			1	1.1	GRUNDY	TN			-85	01	54
1	.8	2	1.4	4	1.8	HAMBLEN	TN	72	30	123	87	74
22*	3.0	18	1.7	31*	2.2	HAMILTON	TN	-43	31	-26	36	84
2	2.3	3	2.6	1	.6	HARDEMAN	TN	11	-77	-75	04	36
		1	1.0	3	2.5	HARDIN	TN		160			87
1	.9	3	1.8	3	1.5	HAWKINS	TN	97	-18	62	75	65
1	2.2			3*	5.0	HAYWOOD	TN			126	87	98
1	1.2	3	3.4	5*	3.7	HENDERSON	TN	187	10	217	94	95
		2	1.3	4	1.8	HENRY	TN		41			75
		2	3.1	2	2.6	HICKMAN	TN		-17			88
2*	7.0					HOUSTON	TN					
				1	1.2	HUMPHREYS	TN					55
3*	5.6	1	2.5	2	3.9	JACKSON	TN	-56	58	-30	34	96

* SIGNIFICANTLY MORE DEATHS THAN EXPECTED

WHITE FEMALE — CANCER OF THE ORAL CAVITY INCLUDING TONGUE — ICD 141,143-146,148,149

1950-1959 DEATHS	1950-1959 RATE	1960-1969 DEATHS	1960-1969 RATE	1970-1979 DEATHS	1970-1979 RATE	COUNTY AND STATE	PERCENT CHANGE 60S/50S	PERCENT CHANGE 70S/60S	PERCENT CHANGE 70S/50S	PERCENTILE %CHANGE 70S/50S	PERCENTILE RATES 1970S
1	1.3	3	3.0	1	.6	JEFFERSON TN	120	-81	-58	15	36
1	1.9			1	1.2	JOHNSON TN			-36	29	57
19*	2.1	31*	2.4	34	2.0	KNOX TN	15	-19	-7	46	79
				1	1.3	LAKE TN					59
2	2.6	2	2.3	1	.8	LAUDERDALE TN	-12	-65	-69	07	43
3	3.0	2	1.3	4	2.0	LAWRENCE TN	-58	59	-33	31	80
4*	3.7	5*	3.9			LINCOLN TN	7				
		4*	3.2	5	2.9	LOUDON TN		-9			91
2	1.5	6*	3.6	8*	3.8	MC MINN TN	133	5	145	89	95
5*	5.7	6*	5.6	1	.9	MC NAIRY TN	-3	-83	-84	01	47
2	4.6			1	1.1	MACON TN			-77	03	51
4	2.1	5	1.8	3	.7	MADISON TN	-16	-60	-66	09	39
		3	3.6			MARION TN					
		4*	3.5	5*	3.8	MARSHALL TN		7			96
1	.6			6	2.4	MAURY TN			285	96	86
		2*	9.9			MEIGS TN					
2	2.0	5*	4.9	2	1.5	MONROE TN	148	-70	-26	36	65
		1	.7	1	.4	MONTGOMERY TN		-40			33
		2	3.5	4*	5.0	MORGAN TN		44			98
2	1.2	3	1.4	2	1.1	OBION TN	9	-18	-10	44	54
3*	4.0	1	1.4	1	1.2	OVERTON TN	-66	-8	-69	07	58
				1	5.2	PICKETT TN					98
1	1.8			1	1.2	POLK TN			-34	31	56
6*	4.6	3	1.6	5	2.1	PUTNAM TN	-66	34	-54	17	82
		3*	3.5	4	3.6	RHEA TN		3			95
2	1.7	1	.6	13*	5.7	ROANE TN	-64	835	234	95	98
6*	5.5			2	1.1	ROBERTSON TN			-81	02	52
6*	3.8	1	.5	6	1.7	RUTHERFORD TN	-86	238	-54	17	73
		3*	4.8	1	1.4	SCOTT TN		-71			62
4*	22.1	1	3.6			SEQUATCHIE TN	-84				
		1	.8	4	2.1	SEVIER TN		158			83
21	1.5	38*	1.8	34	1.4	SHELBY TN	23	-26	-9	45	62
		1	1.3			SMITH TN					
7	2.1	5	1.0	14	1.9	SULLIVAN TN	-51	88	-8	46	79
		3	1.6	1	.4	SUMNER TN		-77			32
1	2.2					TIPTON TN					
		1	1.3	2	1.5	UNICOI TN		18			68
		1	2.1	1	2.1	UNION TN		1			82
		1*	8.3			VAN BUREN TN					
1	1.1	2	1.5	3	1.7	WARREN TN	34	15	54	72	72
		3	.9	9	2.0	WASHINGTON TN		111			79
2*	6.6	1	1.9	1	1.1	WAYNE TN	-72	-39	-83	01	55
2	1.2	3	1.6	6	2.1	WEAKLEY TN	25	36	70	77	82
1	1.2	1	1.0			WHITE TN	-15				
2	2.0	3	2.0	3	1.5	WILLIAMSON TN	3	-28	-26	36	65
1	.6	1	.7	1	.3	WILSON TN	7	-50	-46	22	31
395*	1.3	518	1.2	830	1.5	TEXAS TX	-9	22	12		
4*	3.0	2	1.1	2	.9	ANDERSON TX	-64	-18	-70	07	45
3	2.1	2	.8	4	1.3	ANGELINA TX	-61	51	-41	26	58
		1	2.9	2	1.9	ARANSAS TX		-34			77
1	1.5			1	1.0	ATASCOSA TX			-35	30	48
				3	2.9	AUSTIN TX					91
				2	3.8	BANDERA TX					96
				2	1.8	BASTROP TX					75
		2*	5.1			BAYLOR TX					
				1	1.1	BEE TX					51
2	.7	3	.9	4	.9	BELL TX	17	1	18	57	45
35*	1.5	52*	1.7	70	1.8	BEXAR TX	10	7	19	58	76
2	2.3	1	.7	2	1.0	BOSQUE TX	-70	37	-58	14	48
3	1.6	4	1.2	7	1.7	BOWIE TX	-21	39	9	54	73
2	1.2	1	.3	10	2.3	BRAZORIA TX	-71	548	88	82	85
1	1.7			3	1.5	BRAZOS TX			-12	43	65
		2*	6.7			BREWSTER TX					
				1	2.3	BROOKS TX					86
2	1.2			4	1.3	BROWN TX			12	55	61
1	1.6	2	3.2	1	.8	BURLESON TX	93	-76	-53	17	42
1	1.4	1	1.2	1	.6	BURNET TX	-18	-50	-59	13	36
1	1.4	1	.7	2	1.6	CALDWELL TX	-48	113	10	54	69
				1	1.4	CALHOUN TX					63
2	4.1					CALLAHAN TX					
2	.4	9	1.7	7	1.0	CAMERON TX	298	-43	126	87	48
1	3.7	1	2.5	1	1.6	CAMP TX	-34	-37	-58	14	68
		1	2.2	1	2.6	CARSON TX		17			88
		1	.9	3	1.6	CASS TX		83			69
		1	.7	5	2.2	CHEROKEE TX		207			83
		1	1.5			CHILDRESS TX					
3*	5.9	2	3.0	1	1.3	CLAY TX	-50	-56	-78	03	60
				1	2.7	COKE TX					89
2	2.1			3	2.0	COLEMAN TX			-7	46	79
2	.8	1	.3	6	1.6	COLLIN TX	-56	355	98	84	69
1	2.5			1	2.2	COLLINGSWORTH TX			-10	44	84
				1	1.1	COLORADO TX					53
		1	1.0	1	.5	COMAL TX		-49			35
1	1.0	1	.7			COMANCHE TX	-23				
1	.8	4	2.8	2	1.0	COOKE TX	267	-64	32	64	50
		1*	8.6			CRANE TX					
2*	5.8					CROSBY TX					
45*	1.7	55	1.3	98	1.7	DALLAS TX	-22	34	5	52	74
1	3.4			1	1.3	DEAF SMITH TX			-63	11	58
		1	1.5			DELTA TX					
1	.6	1	.3	6	1.5	DENTON TX	-48	341	131	88	65
1	.8	1	.5	2	1.2	DE WITT TX	-37	134	48	70	58
1	3.7	1	2.6			DIMMIT TX	-30				
3*	7.8			1	1.5	DUVAL TX			-80	02	68
2	1.1	1	.4	2	.8	EASTLAND TX	-63	86	-31	32	41
1	2.4	4	1.7	6	2.0	ECTOR TX	-31	16	-20	39	79
3	2.1	3	1.1	3	1.1	ELLIS TX	-47	0	-47	21	53
4	.5	14	1.3	24	1.9	EL PASO TX	193	45	325	97	78
2	1.9	3	1.8	1	.3	ERATH TX	-7	-85	-87	00	31

* SIGNIFICANTLY MORE DEATHS THAN EXPECTED

WHITE FEMALE — CANCER OF THE ORAL CAVITY INCLUDING TONGUE — ICD 141,143-146,148,149

1950-1959 DEATHS	1950-1959 RATE	1960-1969 DEATHS	1960-1969 RATE	1970-1979 DEATHS	1970-1979 RATE	COUNTY AND STATE	PERCENT CHANGE 60S/50S	PERCENT CHANGE 70S/60S	PERCENT CHANGE 70S/50S	PERCENTILE %CHANGE 70S/50S	PERCENTILE RATES 1970S
1	.9	3	2.0			FALLS TX	114				
2	1.5	4	1.7	3	1.0	FANNIN TX	17	-40	-30	34	50
1	1.5					FISHER TX					
		1	1.9	1	1.8	FLOYD TX		-4			75
1	.9			2	.7	FORT BEND TX			-23	38	38
		2*	5.7	3*	4.3	FRANKLIN TX		-25			97
2	2.5			1	2.0	FREESTONE TX			-21	39	79
2	.5	11	2.1	19*	2.6	GALVESTON TX	294	24	390	98	88
1	1.1	1	.6	1	.9	GILLESPIE TX	-50	61	-20	40	46
				2*	7.1	GOLIAD TX					99
				1	1.0	GONZALES TX					51
		1	.7	2	1.1	GRAY TX		57			52
1	.3	5	.9	5	.8	GRAYSON TX	206	-4	194	93	44
1	.7	7	2.4	4	.9	GREGG TX	258	-62	37	66	46
1	1.4	1	1.4			GRIMES TX	-2				
		1	.8	1	.4	GUADALUPE TX		-45			33
2	1.7	3	2.0	3	1.2	HALE TX	15	-40	-31	32	55
				2	2.9	HALL TX					91
				1	.7	HAMILTON TX					38
				1	3.4	HANSFORD TX					94
1	2.0			2	2.4	HARDEMAN TX			20	59	86
1	.9	1	1.0			HARDIN TX	11				
27	.9	71	1.6	107	1.7	HARRIS TX	67	5	75	79	72
1	.8			4	1.9	HARRISON TX			149	89	78
		2	2.6			HASKELL TX					
2	2.7			4	3.2	HAYS TX			20	58	93
		2	1.5	5	2.0	HENDERSON TX		30			80
12*	2.7	4	.6	9	.9	HIDALGO TX	-76	46	-65	10	47
3	1.7	1	.3	3	1.2	HILL TX	-80	246	-30	34	55
		1	1.1			HOCKLEY TX					
1	2.6			1	1.0	HOOD TX			-59	13	51
2	1.5	2	1.3	1	.6	HOPKINS TX	-18	-51	-60	12	37
1	1.1			1	.5	HOUSTON TX			-51	19	35
1	1.0	1	.9			HOWARD TX	-11				
3	1.2	6	2.0	5	1.2	HUNT TX	64	-40	-2	48	57
1	1.9	1	.8			HUTCHINSON TX	-60				
		1	1.4	2	2.3	JACK TX		67			85
3*	3.9	2	1.9	2	1.1	JASPER TX	-53	-42	-73	05	52
7	1.3	7	.8	19	1.8	JEFFERSON TX	-38	126	40	67	74
		2*	10.0	1	4.0	JIM HOGG TX		-60			96
2	2.0	4*	3.7	1	.6	JIM WELLS TX	87	-83	-69	07	37
		1	.4	3	.9	JOHNSON TX		122			45
2	1.9	1	.8	2	1.3	JONES TX	-59	68	-32	32	60
1	2.3			1	.9	KARNES TX			-59	13	47
3	1.9	2	.9	5	2.2	KAUFMAN TX	-53	144	15	56	84
1	2.6	1	3.2			KENDALL TX	26				
1	1.5			6	2.8	KERR TX			88	82	90
1	4.2			1	2.8	KIMBLE TX			-34	30	90
1*	11.2					KINNEY TX					
		3	3.5	3	2.6	KLEBERG TX		-24			89
2*	4.8	2*	3.9			KNOX TX	-18				
3	1.5	3	1.7	3	.7	LAMAR TX	15	-57	-51	19	39
				1	.9	LAMB TX					45
				3	4.0	LAMPASAS TX					96
1	.7			1	.8	LAVACA TX			16	57	43
1	2.1					LEE TX					
1	2.2					LEON TX					
2	1.8	2	1.4	3	1.6	LIBERTY TX	-20	10	-13	43	69
1	.7	3	1.8			LIMESTONE TX	167				
		1	4.9			LIPSCOMB TX					
				1	2.1	LIVE OAK TX					82
		1	1.2	5*	12.1	LLANO TX		892			99
7	2.1	6	.9	12	1.6	LUBBOCK TX	-56	67	-27	36	68
				1	2.0	LYNN TX					80
		1	2.1			MC CULLOCH TX					
7	1.2	7	1.0	10	1.0	MC LENNAN TX	-18	4	-15	42	50
1	3.2			1	1.9	MADISON TX			-40	26	78
1	3.7	1	3.3			MARION TX	-11				
1*	6.5					MARTIN TX					
				1	3.5	MASON TX					94
				2	1.5	MATAGORDA TX					68
1	2.5			1	1.2	MAVERICK TX			-52	18	57
3*	4.0	2	2.4	3	2.1	MEDINA TX	-41	-13	-48	20	81
2	2.1	2	.9	4	1.2	MIDLAND TX	-59	40	-42	24	58
2	1.6	1	1.1	1	.4	MILAM TX	-32	-66	-77	04	32
		2*	7.8			MILLS TX					
3*	5.8	1	1.5	2	2.1	MITCHELL TX	-74	39	-64	11	82
1	1.0	4*	2.5	2	1.3	MONTAGUE TX	160	-47	37	66	61
1	1.0	3	1.5	5	1.3	MONTGOMERY TX	52	-13	33	64	61
1	1.3			1	2.4	MOORE TX			93	83	87
1	2.2			1	1.8	MORRIS TX			-22	38	74
1*	14.9					MOTLEY TX					
1	.7	1	.7	4	1.9	NACOGDOCHES TX	-4	169	159	90	77
4	2.0	1	.4	6	2.5	NAVARRO TX	-79	508	28	62	88
		1	2.1			NEWTON TX					
1	1.1	1	.7	5*	4.3	NOLAN TX	-30	487	308	97	97
13*	2.8	7	1.0	19	1.8	NUECES TX	-64	84	-34	30	76
				1	2.3	OCHILTREE TX					85
		1*	12.7			OLDHAM TX					
2	1.3	2	1.0	8	2.7	ORANGE TX	-27	177	101	84	89
2	1.9	2	1.6			PALO PINTO TX	-14				
1	1.6	3*	3.5	1	.9	PANOLA TX	122	-74	-42	25	46
1	.8	2	1.1	3	1.4	PARKER TX	45	25	81	80	63
				2	3.9	PARMER TX					96
				1	1.7	PECOS TX					73
3	1.2	7	1.6	7	1.2	POTTER TX	33	-26	-1	49	56
				1	4.0	PRESIDIO TX					96
1	2.1	1	.8	3	1.3	RANDALL TX	-60	59	-37	28	60
		1*	5.4			REAGAN TX					
		1	.7	1	.6	RED RIVER TX		-5			38
2*	5.4	2	3.6	2	3.6	REEVES TX	-33	-1	-33	31	95

* SIGNIFICANTLY MORE DEATHS THAN EXPECTED

WHITE FEMALE — CANCER OF THE ORAL CAVITY INCLUDING TONGUE — ICD 141,143-146,148,149

1950-1959 DEATHS	1950-1959 RATE	1960-1969 DEATHS	1960-1969 RATE	1970-1979 DEATHS	1970-1979 RATE	COUNTY AND STATE	PERCENT CHANGE 60S/50S	PERCENT CHANGE 70S/60S	PERCENT CHANGE 70S/50S	PERCENTILE %CHANGE 70S/50S	PERCENTILE RATES 1970S
1	3.1	1	2.0			REFUGIO TX	-34				
		1	1.5			ROCKWALL TX					
		1	.8			RUNNELS TX					
4*	2.9	2	1.0	3	1.2	RUSK TX	-66	22	-59	14	56
1	2.5	1	2.4			SAN AUGUSTINE TX	-4				
		1	3.2			SAN JACINTO TX					
5*	5.0			4	1.9	SAN PATRICIO TX			-63	11	77
		1	1.6	1	1.5	SAN SABA TX		-5			68
				1	5.4	SCHLEICHER TX					98
2	3.9	2	2.2			SCURRY TX	-44				
2*	7.3	1	4.5	1	2.2	SHACKELFORD TX	-38	-52	-70	06	83
1	1.1	1	.8	2	1.1	SHELBY TX	-25	38	3	51	53
		1*	7.9			SHERMAN TX					
5	2.0	4	1.0	4	.8	SMITH TX	-50	-24	-62	12	42
1*	5.6			1	2.3	SOMERVELL TX			-59	13	85
1	2.0	1	1.5			STARR TX	-27				
1	2.2	2	4.2			STEPHENS TX	96				
18	1.0	42	1.7	73*	2.1	TARRANT TX	60	29	106	85	83
2	.7	6	1.4	6	1.2	TAYLOR TX	83	-15	55	72	55
		1*	9.3			TERRELL TX					
2*	4.5					TERRY TX					
1*	5.2					THROCKMORTON TX					
1	1.4	1	.9	2	1.6	TITUS TX	-34	71	13	56	69
4	1.9	2	.6	6	1.3	TOM GREEN TX	-69	133	-29	35	61
12	1.8	5	.5	16	1.3	TRAVIS TX	-72	152	-29	35	59
				1	1.3	TRINITY TX					59
1	1.8			1	.7	TYLER TX			-59	14	40
1	1.0	2	1.5	1	.5	UPSHUR TX	50	-64	-46	22	35
		2	1.9	1	.9	UVALDE TX		-54			45
1	1.4			2	1.8	VAL VERDE TX			24	60	74
2	1.5	2	1.9	2	.8	VAN ZANDT TX	26	-57	-45	22	44
		2	1.1	2	.9	VICTORIA TX		-24			44
				1	1.2	WALKER TX					55
				1	1.9	WALLER TX					77
1	3.4			1	2.0	WARD TX			-41	26	80
		1	.8	2	1.1	WASHINGTON TX		34			52
3	1.4	3	1.0	3	.8	WEBB TX	-30	-16	-41	26	43
		1	.6	2	1.2	WHARTON TX		93			56
2*	4.6					WHEELER TX					
4	.9	8	1.5	7	1.3	WICHITA TX	62	-13	40	67	58
1	.8	1	1.3	3	2.0	WILBARGER TX	50	57	135	88	80
3	1.8			3	1.2	WILLIAMSON TX			-35	30	55
		1	1.6			WILSON TX					
1	7.9					WINKLER TX					
2	1.8	4*	3.3	5	2.7	WISE TX	81	-19	47	70	89
2	2.5	1	.9	2	.9	WOOD TX	-65	-3	-66	09	45
				1	.4	YOUNG TX					33
				1	2.3	ZAVALA TX					85
19	.7	26	.7	47	.9	UTAH UT	-1	35	34		
1	1.3			2	1.6	BOX ELDER UT			22	59	69
				1	.4	CACHE UT					33
1	.7	1	1.4			CARBON UT	112				
1	.9	2	1.1	3	1.0	DAVIS UT	19	-7	11	54	49
1*	8.3					GRAND UT					
		1	2.6			MILLARD UT					
8	.7	12	.7	26	1.2	SALT LAKE UT	-3	74	68	77	55
1	1.3	1	1.2	2	2.0	SANPETE UT	-5	63	56	72	79
		2	2.9	2	2.4	TOOELE UT		-18			87
1	3.6					UINTAH UT					
1	.5	1	.2	3	.5	UTAH UT	-52	122	7	53	35
				1	.8	WASHINGTON UT					44
4	1.0	6	1.3	7	1.0	WEBER UT	34	-23	3	51	49
22	1.0	31	1.2	46	1.7	VERMONT VT	15	47	70		
1	.9	1	.9	5*	4.0	ADDISON VT	2	348	357	98	96
1	.8	1	.6	7*	3.2	BENNINGTON VT	-26	474	324	97	93
1	.7	1	.5	2	1.2	CALEDONIA VT	-30	151	77	79	57
2	.6	8	2.1	5	1.1	CHITTENDEN VT	260	-49	85	81	52
1	4.8			1	3.0	ESSEX VT			-36	29	92
3	2.1	2	1.0	2	1.1	FRANKLIN VT	-51	12	-45	23	55
		1	1.3	3	3.5	LAMOILLE VT		169			94
1	1.3			1	.5	ORANGE VT			-66	09	34
		4*	2.8	3	2.4	ORLEANS VT		-14			86
2	.5	6	1.4	5	1.4	RUTLAND VT	163	1	167	91	64
4	1.5	1	.4	4	1.4	WASHINGTON VT	-75	304	-1	49	65
2	.9	3	1.3	2	1.0	WINDHAM VT	53	-27	11	55	49
4	1.7	3	1.0	6	2.0	WINDSOR VT	-44	114	20	58	81
137	1.2	204	1.3	320	1.5	VIRGINIA VA	8	18	27		
3	2.4	1	.6	3	2.0	ACCOMACK VA	-77	254	-17	41	80
2	.9	2	.7	3	.7	ALBEMARLE#VA	-19	4	-16	42	40
		1	.7	1	.6	ALLEGHANY#VA		-7			37
		1	2.5			APPOMATTOX VA					
4	.7	23*	2.7	27*	2.8	ARLINGTON#VA	311	4	326	97	90
3	1.1	6	1.4	5	.9	AUGUSTA#VA	31	-35	-16	42	47
1	3.6					BATH VA					
9	1.9	5	.9	12	1.6	BEDFORD#VA	-54	85	-16	42	71
1	1.7					BOTETOURT VA					
		1	2.9			BRUNSWICK VA					
		1	.9	2	1.6	BUCHANAN VA		76			71
3	1.4	2	.8	4	1.1	CARROLL#VA	-40	30	-22	38	53
1	2.6			1	1.9	CHARLOTTE VA			-25	37	78
15	1.3	14	1.0	34*	2.1	CHESTERFIELD#VA	-19	104	65	76	83
1	1.4			1	1.1	CULPEPER VA			-17	41	54
1	2.5	1	1.7	1	1.1	DICKENSON VA	-33	-34	-56	15	53
3	1.4	3	1.1	4	1.1	DINWIDDIE#VA	-26	1	-25	36	52
2	.3	25	1.6	44	1.9	FAIRFAX#VA	441	22	560	99	79

* SIGNIFICANTLY MORE DEATHS THAN EXPECTED

COUNTIES COMBINED; SEE TECHNICAL NOTES

WHITE FEMALE

CANCER OF THE ORAL CAVITY INCLUDING TONGUE — ICD 141,143-146,148,149

1950-1959 DEATHS	1950-1959 RATE	1960-1969 DEATHS	1960-1969 RATE	1970-1979 DEATHS	1970-1979 RATE	COUNTY AND STATE	PERCENT CHANGE 60S/50S	PERCENT CHANGE 70S/60S	PERCENT CHANGE 70S/50S	PERCENTILE %CHANGE 70S/50S	PERCENTILE RATES 1970S
3*	3.6	1	1.1			FAUQUIER VA	-70				
1	2.0			1	2.0	FLOYD VA			3	51	81
		1	.8			FRANKLIN VA					
		3	1.6	7	2.4	FREDERICK#VA		50			87
		1	1.6			GILES VA					
2*	4.9	1	1.9	3	3.7	GLOUCESTER VA	-60	88	-25	37	95
				1	3.5	GREENE VA					95
1	3.5	1	2.3			GREENSVILLE#VA	-34				
6*	5.6	1	.8			HALIFAX#VA	-85				
1	1.2	1	.8	2	1.0	HANOVER VA	-35	33	-13	43	51
3	1.8	4	.7	17	1.9	HENRICO VA	-58	158	8	54	77
3	2.7	3	1.4	2	.7	HENRY#VA	-48	-48	-73	05	40
1	3.4			2	3.8	ISLE OF WIGHT VA			11	55	96
1	.7	1	.5	4	1.6	JAMES CITY#VA	-27	211	128	88	70
1*	4.7			1	4.8	KING AND QUEEN VA			1	50	98
				1	3.0	KING WILLIAM VA					92
1	3.2	1	1.7			LANCASTER VA	-47				
		4	2.9	1	.6	LEE VA		-80			36
1	1.1	1	.7	1	.6	LOUDOUN VA	-39	-10	-45	23	36
				3*	4.3	LOUISA VA					97
		1	2.4	1	1.7	MADISON VA		-28			73
1	1.9	2	3.9	1	1.5	MATHEWS VA	103	-60	-19	40	68
		1	1.1	4	2.7	MECKLENBURG VA		140			89
		1	2.4	1	2.2	MIDDLESEX VA		-9			83
		2	1.2	3	1.0	MONTGOMERY#VA		-13			50
1	2.2	1	1.6			NELSON VA	-27				
				1	1.5	NORTHAMPTON VA					66
2*	4.9					NORTHUMBERLAND VA					
		1	2.1			NOTTOWAY VA					
		2	2.8	1	1.0	ORANGE VA		-65			48
		3	3.0	2	1.7	PAGE VA		-45			72
1	1.7			2	1.5	PATRICK VA			-13	43	67
7*	2.3	2	.6	6	1.0	PITTSYLVANIA#VA	-76	80	-57	15	49
				1	2.8	POWHATAN VA					90
		1	1.4			PRINCE EDWARD VA					
		2	1.8	4	1.8	PRINCE WILLIAM#VA		-3			74
		3	2.2	2	1.1	PULASKI VA		-51			52
1	5.4					RAPPAHANNOCK VA					
1*	4.3			1	3.0	RICHMOND VA			-30	34	92
10	1.6	9	1.1	12	1.2	ROANOKE#VA	-31	7	-26	36	56
2	2.0			1	.6	ROCKBRIDGE#VA			-71	06	36
		3	1.0	3	.6	ROCKINGHAM#VA		-33			37
2	2.5			1	.7	RUSSELL VA			-73	05	38
3	2.8			1	.7	SCOTT VA			-77	03	38
1	.7			2	1.3	SHENANDOAH VA			80	80	60
1	1.0	1	.7	4	2.0	SMYTH VA	-26	175	103	84	79
2*	3.6			1	1.4	SOUTHAMPTON#VA			-62	12	62
1	.9	2	1.4	2	.9	SPOTSYLVANIA#VA	54	-35	1	50	47
		1	1.5	1	1.0	STAFFORD VA		-35			49
		1	3.5	1	3.0	SUSSEX VA		-15			92
2	1.4	1	.7	4	1.7	TAZEWELL VA	-50	143	22	59	72
				1	.8	WARREN VA					43
		5	1.8	2	.5	WASHINGTON#VA		-74			34
		2*	4.2			WESTMORELAND VA					
4	2.4	4	1.7	4	1.7	WISE#VA	-28	-4	-31	33	72
2	1.9	1	.8	2	1.6	WYTHE VA	-61	108	-18	41	69
11	.9	31*	1.9	48*	2.1	CHESAPEAKE CITY#VA	97	15	127	87	83
1	.7	2	.6	7	2.0	HAMPTON CITY VA	-17	224	169	91	79
6*	3.5	8*	2.7	5	1.2	NEWPORT NEWS CITY VA	-22	-56	-65	10	57
1	1.1	1	.9			SUFFOLK CITY VA	-22				
120	1.0	226*	1.4	320	1.6	WASHINGTON WA	47	12	64		
				1	1.8	ADAMS WA					75
1	1.6					ASOTIN WA					
		6	2.5	5	1.4	BENTON WA		-45			62
2	.9			3	1.0	CHELAN WA			11	55	48
1	.7	2	1.2	1	.5	CLALLAM WA	63	-59	-33	31	34
5	1.1	9	1.5	7	.9	CLARK WA	40	-43	-20	39	45
		5	1.6	6	1.5	COWLITZ WA		-8			66
				3	3.2	DOUGLAS WA					93
				1	5.4	FERRY WA					98
				3	2.6	FRANKLIN WA					88
				1	3.3	GARFIELD WA					93
		1	.4	4	1.9	GRANT WA		354			77
6*	2.2	9*	2.8	8	2.2	GRAYS HARBOR WA	29	-21	1	50	84
		1	1.1	4	1.9	ISLAND WA		81			78
				3	3.2	JEFFERSON WA					93
47	1.2	86*	1.7	123*	2.0	KING WA	43	17	67	76	79
3	.8	6	1.3	6	.8	KITSAP WA	58	-34	5	52	44
1	1.0	2	1.7	2	1.3	KITTITAS WA	70	-24	29	63	60
				1	1.1	KLICKITAT WA					53
1	.4	4	1.4	6	2.0	LEWIS WA	278	43	440	98	80
2	2.4	3	2.8	2	1.4	MASON WA	17	-50	-41	25	64
1	1.2	5*	4.8	2	1.5	PACIFIC WA	310	-70	24	61	65
		4*	11.8	3*	7.5	PEND OREILLE WA		-36			99
10	.8	24	1.5	40	1.9	PIERCE WA	89	30	144	89	78
		1	3.8	1	1.7	SAN JUAN WA		-55			73
		1	.3	6	1.8	SKAGIT WA		451			74
				1	2.7	SKAMANIA WA					89
7	1.1	17	1.7	20	1.4	SNOHOMISH WA	64	-20	32	64	63
15	1.2	19	1.2	22	1.2	SPOKANE WA	5	2	7	53	58
2	2.4					STEVENS WA					
1	.3			18*	3.3	THURSTON WA			846	99	93
		1	5.5			WAHKIAKUM WA					
1	.4	4	1.6	5	1.8	WALLA WALLA WA	252	15	305	97	75
6	1.5	3	.6	3	.7	WHATCOM WA	-58	1	-57	15	38
1	.8	1	.8	1	.4	WHITMAN WA	-6	-41	-45	23	33
7	1.1	12	1.6	8	.9	YAKIMA WA	52	-43	-13	43	47

* SIGNIFICANTLY MORE DEATHS THAN EXPECTED

COUNTIES COMBINED; SEE TECHNICAL NOTES

WHITE FEMALE | CANCER OF THE ORAL CAVITY INCLUDING TONGUE | ICD 141,143-146,148,149

1950-1959 DEATHS	1950-1959 RATE	1960-1969 DEATHS	1960-1969 RATE	1970-1979 DEATHS	1970-1979 RATE	COUNTY AND STATE	PERCENT CHANGE 60S/50S	PERCENT CHANGE 70S/60S	PERCENT CHANGE 70S/50S	PERCENTILE %CHANGE 70S/50S	PERCENTILE RATES 1970S
96	1.3	130	1.4	131	1.2	WEST VIRGINIA WV	12	-17	-7		
3*	3.6	1	.9	1	.7	BARBOUR WV	-74	-20	-79	02	40
4*	2.8	1	.5	1	.4	BERKELEY WV	-80	-20	-84	01	33
		2	2.1			BOONE WV					
3*	3.6	2	2.2	1	.8	BRAXTON WV	-39	-65	-78	03	42
1	.8	4	2.7	2	1.2	BROOKE WV	242	-57	49	70	55
12*	2.3	10	1.6	3	.4	CABELL WV	-32	-75	-83	01	32
1	3.4			1	2.1	CALHOUN WV			-37	28	83
1	1.5			1	1.5	CLAY WV			1	50	66
1	2.3					DODDRIDGE WV					
1	.4	4	1.6	4	1.3	FAYETTE WV	283	-18	215	94	60
2*	4.7					GILMER WV					
4*	2.7	1	.6	4	1.8	GREENBRIER WV	-78	200	-34	30	75
		1	1.8			HAMPSHIRE WV					
1	.7	1	.9	1	.5	HANCOCK WV	30	-52	-38	28	34
1	2.3					HARDY WV					
9*	2.5	11*	2.3	7	1.1	HARRISON WV	-7	-53	-56	15	52
		1	1.0			JACKSON WV					
1	1.2	2	2.3	1	.8	JEFFERSON WV	90	-67	-36	29	41
5	.5	13	1.2	22	1.6	KANAWHA WV	133	37	220	95	71
2	1.1	3	2.1	2	1.1	LEWIS WV	86	-44	4	52	55
2	2.5	1	1.1	2	1.7	LINCOLN WV	-58	65	-30	33	74
1	.3	4	2.1	2	.8	LOGAN WV	513	-61	138	89	43
2	.9	1	1.1	5	3.0	MC DOWELL WV	22	179	242	95	92
4	1.2	4	1.0	8	1.8	MARION WV	-14	79	54	72	76
3	1.8	3	1.4	4	1.4	MARSHALL WV	-22	-6	-26	36	62
		2	1.8	1	.8	MASON WV		-57			41
1	.3	5	1.5	5	1.3	MERCER WV	370	-13	309	97	59
1	1.1	1	.9	3	2.0	MINERAL WV	-19	114	72	78	80
		2	1.4	1	.6	MINGO WV		-53			37
4	1.7	3	1.1	3	.8	MONONGALIA WV	-38	-29	-56	16	41
		1	1.1	3	3.4	MONROE WV		225			94
		1	.9			NICHOLAS WV					
3	.7	9	2.0	9	2.2	OHIO WV	165	10	192	93	83
2*	4.9			1	2.1	PENDLETON WV			-58	14	81
2*	6.1			1	2.3	PLEASANTS WV			-62	12	86
		1	2.1	1	1.5	POCAHONTAS WV		-31			65
		1	.8	1	.3	PRESTON WV		-63			31
		1	1.0	1	.7	PUTNAM WV		-26			39
5	1.8	3	1.0	4	.8	RALEIGH WV	-47	-17	-56	16	43
		6*	4.4	6*	3.7	RANDOLPH WV		-15			95
		1	.9			RITCHIE WV					
1	1.0	1	1.2			ROANE WV	22				
1	1.2	2	1.9			SUMMERS WV	55				
2	1.8	2	1.6	4	2.9	TAYLOR WV	-12	83	62	75	91
1	2.1	2	3.4	1	1.3	TUCKER WV	63	-61	-37	29	60
1	.9	3	2.6	1	.8	UPSHUR WV	179	-67	-9	45	44
1	.8	4	2.2	2	.9	WAYNE WV	158	-59	5	53	45
				1	2.1	WEBSTER WV					81
1	.9	1	.9	5*	4.0	WETZEL WV	-7	358	325	97	96
		2*	7.3	2*	6.1	WIRT WV		-17			99
6	1.8	6	1.3	2	.5	WOOD WV	-27	-65	-74	05	34
				1	.9	WYOMING WV					45
175	1.0	179	.8	337	1.3	WISCONSIN WI	-17	59	32		
		1	.9	1	.6	ASHLAND WI		-28			37
2	1.1	2	.9	2	.9	BARRON WI	-22	3	-19	40	46
2	3.2	1	1.2	2	1.8	BAYFIELD WI	-64	51	-46	22	74
7	1.3	1	.2	5	.6	BROWN WI	-87	268	-52	18	37
		1	1.5	2	2.8	BURNETT WI		78			90
1	1.4	1	.9	3	1.8	CALUMET WI	-37	100	27	62	76
				3	.9	CHIPPEWA WI					47
1	.7	1	.7	2	.7	CLARK WI	3	5	8	54	39
		1	.4	1	.3	COLUMBIA WI		-37			31
				3	3.0	CRAWFORD WI					92
11	1.3	8	.7	17	1.2	DANE WI	-45	68	-8	45	57
1	.3	3	.9	1	.2	DODGE WI	203	-72	-14	42	31
3	2.9					DOOR WI					
1	.4	1	.4	2	.6	DOUGLAS WI	7	40	50	71	36
		2	1.1	3	2.0	DUNN WI		80			79
6	2.4	3	.9	1	.2	EAU CLAIRE WI	-65	-76	-91	00	30
				1	3.4	FLORENCE WI					94
2	.4	2	.4	12	2.2	FOND DU LAC WI	-18	527	414	98	84
				1	1.4	FOREST WI					64
2	.7	3	1.1	3	1.0	GRANT WI	47	-5	40	67	50
		1	.4	2	1.0	GREEN WI		147			49
				5*	4.1	GREEN LAKE WI					96
		2	1.6			IOWA WI					
1	2.7			1	1.9	IRON WI			-31	33	78
1	1.3			1	1.0	JACKSON WI			-20	40	50
4	1.5	3	.9	3	.8	JEFFERSON WI	-44	-8	-48	21	42
		1	1.0	2	1.8	JUNEAU WI		79			74
3	.9	9	1.6	10	1.5	KENOSHA WI	83	-7	69	77	67
1	1.0			1	.5	KEWAUNEE WI			-49	20	35
3	.8	2	.5	9	1.8	LA CROSSE WI	-44	289	119	87	75
1	.9	1	.8	1	.6	LAFAYETTE WI	-16	-20	-33	31	37
		2	1.8	1	.5	LANGLADE WI		-72			34
1	.8			2	1.2	LINCOLN WI			45	69	57
2	.6	4	1.0	9	1.8	MANITOWOC WI	57	85	191	93	76
2	.5	3	.6	5	.9	MARATHON WI	11	49	66	76	46
				3	1.1	MARINETTE WI					51
2	.7	1	.3	2	.6	MENOMINEE#WI	-62	124	-16	42	36
64	1.4	63	1.1	102	1.7	MILWAUKEE WI	-20	52	21	59	73
2	1.2	1	.4			MONROE WI	-71				
1	1.0	2	1.5	1	.4	ONEIDA WI	41	-73	-63	11	32
1	.2	4	.8	8	1.3	OUTAGAMIE WI	252	56	448	98	59
		4	1.9	8	2.9	OZAUKEE WI		50			91
1	2.8					PEPIN WI					

* SIGNIFICANTLY MORE DEATHS THAN EXPECTED

COUNTIES COMBINED; SEE TECHNICAL NOTES

WHITE FEMALE — CANCER OF THE ORAL CAVITY INCLUDING TONGUE — ICD 141,143-146,148,149

1950-1959 DEATHS	1950-1959 RATE	1960-1969 DEATHS	1960-1969 RATE	1970-1979 DEATHS	1970-1979 RATE	COUNTY AND STATE	PERCENT CHANGE 60S/50S	PERCENT CHANGE 70S/60S	PERCENT CHANGE 70S/50S	PERCENTILE %CHANGE 70S/50S	PERCENTILE RATES 1970S
2	1.5					PIERCE WI					
		1	.5	1	.4	POLK WI		-22			33
1	.5	1	.5	4	1.7	PORTAGE WI	-12	275	229	95	73
				2	1.9	PRICE WI					77
7	1.1	4	.6	12	1.4	RACINE WI	-48	129	20	58	62
3	2.7	3	2.2	4	2.6	RICHLAND WI	-17	16	-4	47	88
6	1.2	5	.8	12	1.6	ROCK WI	-29	100	41	68	71
1	1.2	1	.9	1	1.1	RUSK WI	-22	19	-8	46	54
				4	2.0	ST CROIX WI					81
		3	1.4	2	.7	SAUK WI		-49			39
2*	4.4	1	1.5	1	.9	SAWYER WI	-65	-41	-79	02	46
4	.9			8	1.2	SHEBOYGAN WI			43	68	58
				1	.7	TAYLOR WI					38
		1	.4	3	1.9	TREMPEALEAU WI		369			77
1	.6	1	.7	2	1.0	VERNON WI	10	54	70	77	51
				1	.7	VILAS WI					40
2	.8	3	.8	6	1.4	WALWORTH WI	4	73	80	80	63
1	1.6	2	3.2	3	3.2	WASHBURN WI	105	-1	102	84	93
		2	.9	4	1.2	WASHINGTON WI		39			57
4	.8	6	.8	12	1.1	WAUKESHA WI	0	28	27	62	51
3	1.4	4	1.4	5	1.8	WAUPACA WI	0	30	30	63	75
1	1.3	1	.8	1	.8	WAUSHARA WI	-35	-8	-40	26	42
5	.9	3	.5	5	.6	WINNEBAGO WI	-49	26	-35	29	37
3	1.3	3	1.0	2	.4	WOOD WI	-26	-55	-66	09	33
10	.9	17	1.2	20	1.1	WYOMING WY	33	-8	22		
3*	3.9	2	2.5	1	1.1	ALBANY WY	-36	-55	-71	06	53
		2*	6.9	1	2.4	CAMPBELL WY		-65			66
		2	3.2	1	1.4	CARBON WY		-58			62
		1	3.1			CONVERSE WY					
				1	.8	FREMONT WY					41
2*	3.8					GOSHEN WY					
1	3.7			1	1.5	HOT SPRINGS WY			-59	13	67
1	.4	4	2.0	3	1.1	LARAMIE WY	385	-45	165	91	52
				2	4.1	LINCOLN WY					96
1	.6	2	1.1	4	1.6	NATRONA WY	76	52	167	91	70
		1	1.0	1	.7	PARK-TETON#WY		-31			38
1	1.0			2	1.3	SHERIDAN WY			28	62	61
		3*	3.9	1	1.0	SWEETWATER WY		-74			49
				1	2.5	UINTA WY					88
				1	2.6	WASHAKIE WY					89

* SIGNIFICANTLY MORE DEATHS THAN EXPECTED

COUNTIES COMBINED; SEE TECHNICAL NOTES

NON-WHITE MALE CANCER OF THE ORAL CAVITY INCLUDING TONGUE ICD 141,143-146,148,149

1950-1959		1960-1969		1970-1979		COUNTY AND STATE	PERCENT CHANGE			PERCENTILE	
										%CHANGE	RATES
DEATHS	RATE	DEATHS	RATE	DEATHS	RATE		60S/50S	70S/60S	70S/50S	70S/50S	1970S
2623	4.1	4372	5.4	6787	6.5	UNITED STATES	30	20	56		
75	2.3	86	2.4	138	3.8	ALABAMA AL	9	56	69		
4*	12.9			1	3.2	BALDWIN AL			-75	01	75
		3	6.6	2	3.9	BARBOUR AL		-41			77
		2	10.4			BIBB AL					
				1*	21.5	BLOUNT AL					97
				2	8.6	BULLOCK AL					88
3	7.7					BUTLER AL					
2	3.6	2	3.0	1	1.4	CALHOUN AL	-17	-54	-62	05	71
1	2.1	4	8.6	1	2.4	CHAMBERS AL	305	-72	13	40	73
				1	3.5	CHOCTAW AL					75
1*	70.1					CLEBURNE AL					
		2	10.3			COFFEE AL					
3	9.4	1	3.1	1	2.5	COLBEPT AL	-67	-18	-73	02	73
		1	3.4			CONECUH AL					
1	4.0					COVINGTON AL					
		2	9.6			CRENSHAW AL					
1	.6	2	2.1	3	2.8	DALLAS AL	249	33	365	94	74
1	2.9	2	5.2	1	2.7	ELMORE AL	79	-47	-5	30	74
		1	2.7	1	2.4	ESCAMBIA AL		-10			73
2	5.6			4	7.1	ETOWAH AL			27	49	85
		3*	26.8	1	7.3	GENEVA AL		-73			85
2	4.5					GREENE AL					
1	1.6			1	1.6	HALE AL			-2	32	71
		2	8.7	2	10.0	HENRY AL		15			90
1	1.5	3	6.5	2	3.2	HOUSTON AL	320	-51	107	74	75
25	3.6	18	1.9	41	4.5	JEFFERSON AL	-47	137	26	47	79
				1	11.2	LAMAR AL					92
1	3.4					LAUDERDALE AL					
		2	12.4			LAWRENCE AL					
		2	3.3	6	12.0	LEE AL		268			93
1	3.4			1	4.1	LIMESTONE AL			21	44	77
				1	3.9	LOWNDES AL					76
2	1.8					MACON AL					
2	2.8	2	2.5	8	9.1	MADISON AL	-11	267	227	87	88
				1	2.4	MARENGO AL					73
1*	15.9			1	18.4	MARSHALL AL			16	42	96
7	3.5	16	4.6	26	6.5	MOBILE AL	31	42	87	69	83
2	4.8	1	2.5			MONROE AL	-48				
1	.6	4	1.7	13	5.4	MONTGOMERY AL	185	214	798	99	81
1	2.1	2	6.2	2	6.5	MORGAN AL	189	6	206	85	84
		1	2.6	1	1.6	PERRY AL		-38			71
		1	3.0	1	4.0	PICKENS AL		35			77
		1	3.1	1	2.4	PIKE AL		-24			73
1	5.5	1	4.9	1	7.6	RANDOLPH AL	-12	57	38	55	86
		3	5.1			RUSSELL AL					
1	1.8			1	3.1	SUMTER AL			73	64	75
2	2.9					TALLADEGA AL					
2	2.3			6	4.9	TUSCALOOSA AL			113	75	79
		1	6.3	1	6.1	WASHINGTON AL		-4			83
3	5.4	1	1.8			WILCOX AL	-67				
		1	.7	3	1.4	ALASKA AK		89			
6	2.0	10	2.3	15	2.0	ARIZONA AZ	15	-13	1		
				1	4.1	COCHISE AZ					77
1	.9	2	1.2	7	2.7	MARICOPA AZ	43	121	216	87	73
2	5.6	8*	11.8	3	2.2	PIMA AZ	109	-81	-60	05	72
1	3.8					PINAL AZ					
1*	34.4					SANTA CRUZ AZ					
1	7.9			4	11.8	YUMA AZ			50	58	93
42	2.4	32	2.0	50	3.3	ARKANSAS AR	-15	66	41		
		2	8.6			ARKANSAS AR					
1	2.1			3	7.7	ASHLEY AR			267	90	86
1	4.0	1	3.7	1	3.3	BRADLEY AR	-6	-9	-15	26	75
1	6.7					CALHOUN AR					
3	5.8			1	2.5	CHICOT AR			-57	07	73
				1	7.4	CLARK AR					85
		1	14.9			CLEVELAND AR					
1	1.9			1	1.7	COLUMBIA AR			-7	29	72
1	5.8	1	8.1	1	11.9	CONWAY AR	38	48	104	74	93
3	2.4			4	4.4	CRITTENDEN AR			81	67	78
		1	3.6			CROSS AR					
2	3.7					DESHA AR					
		1	3.2	2	7.6	GARLAND AR		138			86
1	1.9			2	6.8	HEMPSTEAD AR			259	89	84
8	4.9	1	.5	3	2.6	JEFFERSON AR	-89	386	-47	12	73
				2*	151.5	JOHNSON AR					99
1	4.7					LAFAYETTE AR					
2	3.9					LEE AR					
				1	6.9	LONOKE AR					85
1	2.7	2	5.6	2	5.7	MILLER AR	107	1	109	75	81
3	3.3	4	4.7			MISSISSIPPI AR	44				
2	6.4			2	6.7	MONROE AR			5	35	84
		2	3.3	1	1.4	OUACHITA AR		-57			71
1	.7	1	1.0	2	2.5	PHILLIPS AR	60	138	281	91	73
6	2.9	9	4.1	15	6.3	PULASKI AR	42	51	115	76	83
2	2.2			2	3.8	ST FRANCIS AR			74	64	76
1	4.3	2	10.2	2	8.2	SEBASTIAN AR	138	-20	91	70	87
		3	5.1	1	2.0	UNION AR		-60			72
		1*	35.1			WASHINGTON AR					
				1	11.1	WHITE AR					92
1	3.9					WOODRUFF AR					

* SIGNIFICANTLY MORE DEATHS THAN EXPECTED

NON-WHITE MALE — CANCER OF THE ORAL CAVITY INCLUDING TONGUE — ICD 141,143-146,148,149

1950-1959 DEATHS	1950-1959 RATE	1960-1969 DEATHS	1960-1969 RATE	1970-1979 DEATHS	1970-1979 RATE	COUNTY AND STATE	PERCENT CHANGE 60S/50S	PERCENT CHANGE 70S/60S	PERCENT CHANGE 70S/50S	PERCENTILE %CHANGE 70S/50S	PERCENTILE RATES 1970S
128	3.8	261	4.5	462	4.3	CALIFORNIA CA	18	-5	12		
13	3.8	34	5.2	52	4.9	ALAMEDA CA	37	-5	30	50	79
1	13.5			1	8.9	BUTTE CA			-34	18	88
2	8.2	2	1.8	11	3.8	CONTRA COSTA CA	-78	106	-54	09	76
1	.9	4	3.0	14	6.2	FRESNO CA	222	106	566	98	83
				1	5.3	HUMBOLDT CA					80
		1	3.3			IMPERIAL CA					
5	6.1	2	1.7	5	2.9	KERN CA	-73	73	-53	09	74
		1	5.8	1	2.4	KINGS CA		-58			73
63	4.9	122	4.9	241	5.2	LOS ANGELES CA	0	5	5	35	80
1	5.8	2	10.6	1	4.5	MADERA CA	82	-58	-23	23	79
				1	1.3	MARIN CA					71
1	3.1	2	8.9	2	5.1	MERCED CA	185	-43	62	62	80
4	4.9	3	4.4	2	1.8	MONTEREY CA	-10	-60	-64	04	72
				1*	38.3	NEVADA CA					98
		1	.9	4	1.7	ORANGE CA		83			71
				1	3.4	PLACER CA					75
1	2.8	5	5.6	3	1.9	RIVERSIDE CA	103	-65	-30	21	72
		7	4.2	11	3.8	SACRAMENTO CA		-10			76
1*	14.9			1	11.5	SAN BENITO CA			-23	22	92
1	1.3	3	4.0	3	1.6	SAN BERNARDINO CA	211	-59	26	48	71
2	2.0	11	5.5	14	3.7	SAN DIEGO CA	171	-32	84	67	76
18	3.9	37	6.7	52	5.2	SAN FRANCISCO CA	72	-22	33	53	80
4	2.1	8	4.9	12	4.7	SAN JOAQUIN CA	131	-4	120	76	79
		1	3.2	1	2.9	SAN LUIS OBISPO CA		-11			74
		5	5.6	5	2.9	SAN MATEO CA		-49			74
1	3.8			3	3.7	SANTA BARBARA CA			-3	31	76
5	8.4			5	1.5	SANTA CLARA CA			-83	00	71
1*	19.1					SHASTA CA					
1	10.6			1	11.1	SISKIYOU CA			4	35	92
1	11.1	4	5.9	5	6.0	SOLANO CA	-47	2	-46	12	82
		2	11.7			SONOMA CA					
		1	6.9	3	9.6	STANISLAUS CA		38			89
1	2.3	3	5.3	3	2.8	TULARE CA	133	-47	22	45	74
				2	2.1	VENTURA CA					72
3	2.1	9	4.2	15	3.1	COLORADO CO	101	-27	46		
				1	3.2	ADAMS CO					75
				1	4.2	BOULDER CO					77
3	3.3	9	6.2	10	3.6	DENVER-ARAPAHOE#CO	88	-42	10	39	76
				3	13.7	PUEBLO CO					94
10	3.9	44*	10.7	58	9.3	CONNECTICUT CT	172	-13	138		
4	4.5	15*	11.8	17	8.2	FAIRFIELD CT	158	-30	80	66	87
3	3.4	9	6.3	19	8.8	HARTFORD CT	83	40	157	80	88
		1	16.6			LITCHFIELD CT					
		2*	28.1			MIDDLESEX CT					
3	4.5	17*	14.8	20*	13.2	NEW HAVEN CT	230	-11	195	84	94
				1	2.1	NEW LONDON CT					72
				1	27.1	WINDHAM CT					97
13	5.9	14	5.3	37*	11.5	DELAWARE DE	-10	118	95		
		2	5.3	4	8.5	KENT DE		62			88
12*	9.6	10	5.9	22*	10.8	NEW CASTLE DE	-39	84	13	40	91
1	1.4	2	3.7	11*	15.5	SUSSEX DE	176	313	1039	99	95
110*	9.9	202*	12.5	315*	16.0	DIST. OF COLUMBIA DC	26	28	62		
110*	9.9	202*	12.5	315*	16.0	WASHINGTON DC	26	28	62	62	95
106	4.1	216*	6.3	394*	8.8	FLORIDA FL	55	40	118		
4	5.4	1	1.4	10	11.9	ALACHUA FL	-75	775	123	77	93
1	14.3			1	14.5	BAKER FL			1	34	95
		1	2.8	1	2.6	BAY FL		-7			73
1	9.0					BRADFORD FL					
		2	6.0	4	7.0	BREVARD FL		16			85
4	6.4	7	3.9	27	9.8	BROWARD FL	-39	149	52	59	90
1*	32.6					CHARLOTTE FL					
				1	13.3	CITRUS FL					94
		1	10.5	1	7.4	CLAY FL		-29			86
1	10.1	1	6.6	3	11.0	COLLIER FL	-34	66	9	37	91
				1	4.3	COLUMBIA FL					78
15	3.4	45*	8.7	87*	9.8	DADE FL	156	12	187	83	90
		1	8.6			DE SOTO FL					
23*	6.3	35*	8.3	53*	11.1	DUVAL FL	32	35	77	65	92
7	8.1	6	4.6	17*	11.2	ESCAMBIA FL	-43	143	39	56	92
		1	16.9			FRANKLIN FL					
		2	2.4	2	2.2	GADSDEN FL		-9			72
		1*	53.6			GLADES FL					
				1	5.8	HENDRY FL					82
13*	7.2	16	6.6	29*	9.9	HILLSBOROUGH FL	-9	51	38	54	90
1	2.0	3	7.1	5	12.7	JACKSON FL	264	78	548	98	94
1	3.7			1	4.8	JEFFERSON FL			28	49	79
3	7.9	1	1.7	2	4.1	LAKE FL	-79	143	-48	11	77
1	1.9	1	3.4	3	5.9	LEE FL	78	72	206	85	82
4	7.2	5	5.8	4	4.0	LEON FL	-20	-31	-45	12	77
1	6.8	1	7.8	1	10.5	LEVY FL	15	34	55	60	91
1*	36.8					LIBERTY FL					
				2	7.5	MADISON FL					86
2	4.8	1	2.7	3	6.3	MANATEE FL	-44	139	33	52	83
1	1.7	6	8.1	7	8.3	MARION FL	386	2	395	95	87
		1	8.5	1	5.2	MARTIN FL		-40			80
		2	12.7	2	10.4	MONROE FL		-18			91
		1	7.7	2	11.6	NASSAU FL		51			92
				3	17.4	OKALOOSA FL					96

* SIGNIFICANTLY MORE DEATHS THAN EXPECTED

COUNTIES COMBINED; SEE TECHNICAL NOTES

NON-WHITE MALE — CANCER OF THE ORAL CAVITY INCLUDING TONGUE — ICD 141,143-146,148,149

1950-1959 DEATHS	1950-1959 RATE	1960-1969 DEATHS	1960-1969 RATE	1970-1979 DEATHS	1970-1979 RATE	COUNTY AND STATE	PERCENT CHANGE 60S/50S	PERCENT CHANGE 70S/60S	PERCENT CHANGE 70S/50S	PERCENTILE %CHANGE 70S/50S	PERCENTILE RATES 1970S
				3*	36.5	OKEECHOBEE FL					98
2	1.9	9	6.0	21*	10.4	ORANGE FL	212	74	441	96	91
		1	9.4	1	10.3	OSCEOLA FL		9			90
4	1.9	26*	12.8	25*	9.5	PALM BEACH FL	578	-26	402	96	89
1	5.8	2	11.4	2	10.9	PASCO FL	97	-4	88	69	91
2	2.5	11	7.9	17	9.6	PINELLAS FL	220	21	287	91	89
1	.8	3	2.1	10	5.0	POLK FL	174	143	566	98	80
		3	6.6	7*	18.3	PUTNAM FL		178			96
2	5.0	4*	11.6	4	12.7	ST JOHNS FL	131	9	152	80	94
		3	5.3	6	9.2	ST LUCIE FL		73			89
3*	9.0	4*	14.9	5	15.4	SARASOTA FL	65	3	70	63	95
2	3.6	1	1.6	5	7.1	SEMINOLE FL	-55	334	97	72	85
				1	5.6	SUMTER FL					81
		1	6.1			SUWANNEE FL					
1	4.6			2	13.4	TAYLOR FL			188	83	94
2	3.2	6	6.8	11	10.7	VOLUSIA FL	109	58	231	88	91
1	8.4					WASHINGTON FL					
96	2.8	162	4.2	268	6.2	GEORGIA GA	47	49	118		
		1	8.5			APPLING GA					
				1	14.3	ATKINSON GA					95
		1*	26.5			BACON GA					
1	7.8					BAKER GA					
5	10.8			3	5.6	BALDWIN GA			-48	11	81
4	2.5	10	5.8	12	6.7	BIBB GA	132	15	168	81	84
		1*	34.4			BRANTLEY GA					
1	3.8	1	3.5	2	11.7	BROOKS GA	-8	236	210	86	93
1	1.7	2	4.6	1	3.1	BURKE GA	167	-33	80	66	75
				1	6.7	BUTTS GA					84
		1	6.4			CAMDEN GA					
		1*	119.2			CATOOSA GA					
6	3.2	16	6.2	32*	12.1	CHATHAM GA	95	95	280	90	93
1	2.5	2	4.5	2	4.2	CLARKE GA	81	-5	73	64	78
1	6.7			2	8.1	CLAYTON GA			20	44	87
1	14.1	1	8.8	1	16.6	CLINCH GA	-38	88	17	43	96
2	5.2			4	9.9	COBB GA			90	70	90
		1	4.0	1	5.8	COFFEE GA		47			82
1	5.1	2	7.9	1	4.0	COLQUITT GA	56	-49	-21	24	77
		1	5.6			COLUMBIA GA					
				2	5.6	COWETA GA					81
1	4.7	1	4.6	1	4.4	CRISP GA	-2	-4	-5	30	78
4*	10.5			4	12.6	DECATUR GA			19	43	93
1	1.5	2	1.3	13	5.7	DE KALB GA	-16	356	283	91	81
1	4.7			1	5.1	DOOLY GA			9	38	80
1	1.7	5	5.7	2	2.3	DOUGHERTY GA	233	-60	32	52	72
				2*	19.9	DOUGLAS GA					96
		1	9.5			EFFINGHAM GA					
		3	13.4			EMANUEL GA					
				1	12.0	EVANS GA					93
		1	15.7	1	16.1	FAYETTE GA		2			95
1	2.2	1	2.7	1	2.5	FLOYD GA	25	-6	17	43	73
17	2.9	40	5.9	64	7.6	FULTON GA	104	27	160	81	86
		8*	15.0	3	5.9	GLYNN GA		-61			82
		2*	39.4			GORDON GA					
				1	4.2	GRADY GA					78
		2*	26.0	1	6.9	GWINNETT GA		-73			84
				5*	27.4	HALL GA					97
		1	4.5			HANCOCK GA					
				1	4.1	HARRIS GA					77
1	12.9	1	12.5			HART GA	-3				
				1	3.6	HOUSTON GA					75
		1	8.3			JASPER GA					
				1	12.9	JEFF DAVIS GA					94
1	3.8	1	2.9	1	3.4	JEFFERSON GA	-25	19	-11	29	75
2	16.8			1	10.0	JENKINS GA			-40	15	90
		1	6.6	3	17.6	JONES GA		166			96
1*	15.0					LANIER GA					
1	2.6	2	4.2	4	7.8	LAURENS GA	61	88	202	85	86
		1	7.5	3*	30.4	LEE GA		306			98
1	5.0			1	4.8	LIBERTY GA			-4	30	79
1	12.3			1	13.5	LINCOLN GA			10	39	94
1*	17.9					LONG GA					
2	3.3	2	3.7	4	6.4	LOWNDES GA	10	74	92	70	83
				1	5.2	MC DUFFIE GA					80
1	5.4	2	15.0	1	8.2	MC INTOSH GA	176	-46	50	59	87
3	10.6	2	9.1	1	4.9	MACON GA	-14	-46	-54	08	79
1	2.8			2	4.8	MERIWETHER GA			69	63	79
				2	10.3	MONROE GA					90
				1	6.8	MORGAN GA					84
3	4.9	4	2.5	15	9.4	MUSCOGEE# GA	-49	277	94	71	89
		2	6.3	1	3.3	NEWTON GA		-47			75
		1*	22.5	1	24.5	PAULDING GA		9			97
1	4.0			2	7.4	PEACH GA			84	68	85
2*	24.0					PIERCE GA					
				2	9.3	POLK GA					89
5	3.4	12	7.7	19*	10.0	RICHMOND GA	129	29	196	84	90
		1	10.3			ROCKDALE GA					
				2	9.3	SCREVEN GA					89
				1	8.0	SEMINOLE GA					87
		1	2.9	3	7.4	SPALDING GA		152			85
1	5.3	1	6.1			STEWART GA	17				
1	2.3			1	2.2	SUMTER GA			-4	30	72
		1	6.7			TALBOT GA					
		1	12.3			TALIAFERRO GA					
		1	5.8			TATTNALL GA					
				1	11.2	TAYLOR GA					92
				3*	22.3	TELFAIR GA					97
1	3.9	1	4.1			TERRELL GA	3				
3	6.5	2	4.2	2	3.4	THOMAS GA	-36	-19	-49	11	75
				1	3.9	TIFT GA					77

* SIGNIFICANTLY MORE DEATHS THAN EXPECTED

\# COUNTIES COMBINED; SEE TECHNICAL NOTES

NON-WHITE MALE — CANCER OF THE ORAL CAVITY INCLUDING TONGUE — ICD 141,143-146,148,149

1950-1959 DEATHS	1950-1959 RATE	1960-1969 DEATHS	1960-1969 RATE	1970-1979 DEATHS	1970-1979 RATE	COUNTY AND STATE	PERCENT CHANGE 60S/50S	PERCENT CHANGE 70S/60S	PERCENT CHANGE 70S/50S	PERCENTILE %CHANGE 70S/50S	PERCENTILE RATES 1970S
1	8.1			2	9.9	TOOMBS GA			22	45	90
4	9.7	4	7.7	1	1.9	TROUP GA	-20	-75	-80	00	72
		1	3.9	1	4.3	UPSON GA		11			78
1	11.5	1	10.3			WALKER GA	-10				
		1	9.4	2	10.3	WALTON GA		10			91
1	2.8	2	5.9	7*	20.4	WARE GA	110	244	623	98	96
		1	6.5			WARREN GA					
2	5.9	2	7.4	1	2.4	WASHINGTON GA	27	-67	-59	06	73
				1	6.1	WAYNE GA					83
1*	20.2			1	9.9	WHITFIELD GA			-51	10	90
				1	10.1	WILCOX GA					90
2	9.1			1	4.0	WILKES GA			-56	07	77
		1	6.7	1	6.5	WILKINSON GA		-2			83
1	3.7			1	4.7	WORTH GA			26	47	79
		80	4.7	92	3.8	HAWAII HI		-18			
		11	4.2	9	2.6	HAWAII HI		-38			73
		56	4.9	67	4.0	HONOLULU HI		-20			77
		8	6.0	3	1.7	KAUAI HI		-73			71
		5	3.0	13	5.7	MAUI#HI		91			81
148	5.2	279*	6.7	413*	7.1	ILLINOIS IL	29	7	38		
2*	22.1			2*	24.8	ADAMS IL			12	40	97
1	2.8	2	16.9			ALEXANDER IL	497				
		2	11.2	1	3.8	CHAMPAIGN IL		-66			76
125*	5.5	228*	6.7	363*	7.6	COOK IL	22	14	39	55	86
				1	1.5	DU PAGE IL					71
				1*	122.7	FULTON IL					99
		1*	172.0			GALLATIN IL					
				1*	29.8	HENRY IL					98
		1	6.0	1	6.6	JACKSON IL		11			84
1*	21.5	1	15.8			JEFFERSON IL	-26				
		2	9.7	1	1.2	KANE IL		-88			71
				2	3.8	KANKAKEE IL					76
		1	11.9			KNOX IL					
		1	3.7	4	6.2	LAKE IL		68			83
		1	17.0			MC LEAN IL					
1	6.0			3	10.3	MACON IL			70	63	90
1*	212.0			1*	44.6	MACOUPIN IL			-79	01	98
1	4.0	4	7.5	4	7.5	MADISON IL	86	0	86	69	86
		3*	41.8			MARION IL					
1	17.5					MASSAC IL					
		1*	344.1			MENARD IL					
		2*	19.8			MORGAN IL					
2	7.8	2	5.0			PEORIA IL	-36				
1	3.4			1	2.9	PULASKI IL			-14	26	74
		1	5.7	1	4.0	ROCK ISLAND IL		-29			77
4	3.3	11	5.8	16	7.1	ST CLAIR IL	74	23	115	75	85
				1	37.3	SALINE IL					98
3*	11.9	6*	22.7	3	9.4	SANGAMON IL	91	-59	-21	24	89
1*	33.5	1	15.2			STEPHENSON IL	-55				
1*	37.5					UNION IL					
1	5.0	2	7.4	3	13.6	VERMILION IL	49	83	172	82	94
		3	8.4	2	1.7	WILL IL		-80			71
1*	24.8	1*	20.6			WILLIAMSON IL	-17				
1	6.9	2	6.6	1	1.4	WINNEBAGO IL	-3	-79	-80	01	71
44	5.1	61	5.2	94	6.5	INDIANA IN	2	25	27		
1	4.4	2	7.0	2	2.8	ALLEN IN	58	-61	-37	16	74
1	5.9			4*	35.3	CLARK IN			499	97	98
1	3.7			2	6.9	DELAWARE IN			87	69	84
		2*	20.9	1	13.0	FLOYD IN		-38			94
				1	7.3	GRANT IN					85
				1*	74.3	HANCOCK IN					99
				1*	89.3	HENRY IN					99
				1	7.8	HOWARD IN					87
1*	37.7					JEFFERSON IN					
8	3.7	11	2.8	19	3.8	LAKE IN	-22	34	4	35	76
		1	6.0	3	13.5	LA PORTE IN		124			94
1	9.4	2	10.4	3	10.6	MADISON IN	11	2	13	40	91
25*	7.4	33	7.4	41	7.4	MARION IN	0	0	0	32	86
				1*	47.2	ORANGE IN					99
		5	7.5	4	4.3	ST JOSEPH IN		-42			78
4	8.2			6	12.8	VANDERBURGH IN			55	60	94
2	7.7	2	6.8	2	10.2	VIGO IN	-12	51	33	52	90
		3*	17.0	2	11.1	WAYNE IN		-35			92
8	8.0	13*	10.9	8	5.4	IOWA IA	37	-51	-32		
		2	10.5	2	8.2	BLACK HAWK IA		-22			87
		1*	344.1			BOONE IA					
				1*	37.8	CERRO GORDO IA					98
		1*	38.3			CLINTON IA					
2*	83.7					DES MOINES IA					
		1	13.7			LEE IA					
		1	22.5	1	14.6	LINN IA		-35			95
1*	86.0					MARION IA					
3	9.4	4	8.5	3	6.1	POLK IA	-9	-29	-36	17	82
1*	24.4					POTTAWATTAMIE IA					
		2*	49.9	1	8.5	SCOTT IA		-83			88
1*	29.8					WAPELLO IA					
		1	14.5			WOODBURY IA					
17	4.1	15	3.5	27	5.4	KANSAS KS	-15	56	32		
		1*	81.8			ALLEN KS					
1	10.1					ATCHISON KS					
		1*	45.5			BARTON KS					

* SIGNIFICANTLY MORE DEATHS THAN EXPECTED

COUNTIES COMBINED; SEE TECHNICAL NOTES

NON-WHITE MALE — CANCER OF THE ORAL CAVITY INCLUDING TONGUE — ICD 141,143-146,148,149

1950-1959 DEATHS	1950-1959 RATE	1960-1969 DEATHS	1960-1969 RATE	1970-1979 DEATHS	1970-1979 RATE	COUNTY AND STATE		PERCENT CHANGE 60S/50S	70S/60S	70S/50S	PERCENTILE %CHANGE 70S/50S	PERCENTILE RATES 1970S
1*	28.9					BOURBON	KS					
2*	20.7	2*	21.2			DOUGLAS	KS	3				
		1*	81.8			FINNEY	KS					
1*	117.9			1*	49.2	FORD	KS			-58	07	99
				1	17.2	GEARY	KS					96
1	28.1	1	15.7			LABETTE	KS	-44				
2	8.5	1	2.5	1	4.7	LEAVENWORTH	KS	-71	90	-45	13	79
				2	13.6	MONTGOMERY	KS					94
				1*	68.3	NEOSHO	KS					99
1*	16.4					RENO	KS					
				2*	30.5	SALINE	KS					98
3	4.9	2	2.9	2	2.4	SEDGWICK	KS	-41	-17	-51	10	73
1	2.5	1	2.2			SHAWNEE	KS	-11				
4	2.7	5	3.4	17*	11.2	WYANDOTTE	KS	26	233	321	93	92
71*	7.0	98*	9.6	111*	11.1	KENTUCKY	KY	36	16	59		
1*	49.7	1*	86.0			ANDERSON	KY	73				
		1	9.9			BARREN	KY					
1	13.0					BELL	KY					
1	6.5	2	16.1	3*	21.7	BOURBON	KY	147	35	234	88	97
				1	22.3	BOYD	KY					97
		1	6.3			BOYLE	KY					
		1*	45.9			BRECKINRIDGE	KY					
		1	44.2			CALDWELL	KY					
				1*	45.5	CALLOWAY	KY					98
1*	22.3	1*	32.1			CAMPBELL	KY	44				
		7*	14.8	6	12.6	CHRISTIAN	KY		-15			94
3*	22.7	5*	40.1	1	8.0	CLARK	KY	77	-80	-65	03	87
4*	25.1	1	6.8	2	13.7	DAVIESS	KY	-73	101	-45	12	94
1*	298.1					ESTILL	KY					
5	5.0	6	6.4	10	9.8	FAYETTE	KY	27	54	97	72	90
1*	37.7					FLEMING	KY					
				1*	141.3	FLOYD	KY					99
2*	20.5			1	8.0	FRANKLIN	KY			-61	05	87
				2*	62.6	GARRARD	KY					99
1	16.4	2*	17.6	1	10.9	GRAVES	KY	8	-38	-34	19	91
		2*	67.1			HARRISON	KY					
				2	18.1	HENDERSON	KY					96
3*	15.9	1	8.5			HOPKINS	KY	-46				
29*	8.7	50*	13.6	61*	15.3	JEFFERSON	KY	56	12	75	65	95
6*	31.3	5*	23.5	3	14.3	KENTON	KY	-25	-39	-54	08	95
				1*	23.6	LETCHER	KY					97
		1	3.9	3	13.3	MC CRACKEN	KY		238			94
1	6.4	2	16.9	3*	19.7	MADISON	KY	164	16	207	85	96
		1	22.8			MARION	KY					
1	13.3	1	14.4			MASON	KY	9				
1	9.0					MONTGOMERY	KY					
3*	49.4	1	14.1	1	11.4	NELSON	KY	-71	-19	-77	01	92
				1*	45.5	PULASKI	KY					98
2*	21.4	1	8.4	1	12.7	SCOTT	KY	-61	52	-40	15	94
				1	9.6	SHELBY	KY					89
		1	7.7	1	7.9	TODD	KY		3			87
1	3.4			2	10.8	WARREN	KY			216	87	91
				1	24.4	WASHINGTON	KY					97
1	7.5	1	6.5			WEBSTER	KY	-14				
		1*	267.9			WHITLEY	KY					
2*	34.6	1	9.9	1	11.6	WOODFORD	KY	-71	17	-66	02	92
139	4.0	208	5.5	309*	7.7	LOUISIANA	LA	35	41	90		
2	5.5	2	6.2	5	15.0	ACADIA	LA	11	144	171	82	95
3	9.1	1	2.9	4	12.4	ASCENSION	LA	-69	335	36	54	93
		1	7.5	5*	20.4	ASSUMPTION	LA		172			96
		3	7.6	2	5.9	AVOYELLES	LA		-23			82
				1	6.9	BEAUREGARD	LA					84
1	2.8			2	5.5	BIENVILLE	LA			100	73	81
1	1.9	1	3.2	1	2.4	BOSSIER	LA	71	-26	26	47	73
5	2.0	14	4.5	25	7.8	CADDO	LA	123	74	288	92	87
5	6.3	4	3.6	14*	11.6	CALCASIEU	LA	-42	219	84	67	92
				1*	36.8	CAMERON	LA					98
1	5.0	1	7.3	1	7.3	CATAHOULA	LA	47	1	48	57	85
		1	1.8	2	4.7	CLAIBORNE	LA		158			79
1	1.9	1	2.0	3	10.0	CONCORDIA	LA	5	396	419	96	90
6*	11.1	1	1.6			DE SOTO	LA	-85				
8	4.0	12	4.7	28*	9.2	EAST BATON ROUGE	LA	17	95	129	77	89
				1	4.3	EAST CARROLL	LA					78
		1	1.8	1	2.1	EAST FELICIANA	LA		18			72
1	5.8			2	7.1	EVANGELINE	LA			22	45	85
2	5.9					FRANKLIN	LA					
				1	6.3	GRANT	LA					83
2	3.6	3	5.5	8*	14.5	IBERIA	LA	54	164	308	92	95
4	6.5	2	3.7	3	5.4	IBERVILLE	LA	-43	48	-16	25	81
				2	5.9	JACKSON	LA					82
7*	8.2	11*	8.8	8	4.7	JEFFERSON	LA	8	-47	-43	13	79
		3	13.5			JEFFERSON DAVIS	LA					
2	6.1	6	10.3	4	6.1	LAFAYETTE	LA	70	-41	1	33	83
1	4.4	2	9.5	2	7.9	LAFOURCHE	LA	117	-16	82	67	87
		1	9.8			LA SALLE	LA					
1	3.3	1	2.6			LINCOLN	LA	-22				
				1	5.7	LIVINGSTON	LA					81
1	2.6	3	6.9	2	5.2	MADISON	LA	171	-26	102	74	80
				1	2.9	MOREHOUSE	LA					74
3	4.5	1	2.3	2	2.9	NATCHITOCHES	LA	-49	24	-36	17	74
50*	7.2	79*	9.4	122*	12.8	ORLEANS	LA	30	37	78	66	94
6	4.8	5	4.1	10	8.5	OUACHITA	LA	-15	108	76	65	88
		4*	24.0	1	4.2	PLAQUEMINES	LA		-82			78
1	2.8	1	2.3	2	5.7	POINTE COUPEE	LA	-18	146	103	74	81
2	1.4	10	7.4	10	7.7	RAPIDES	LA	419	4	442	97	86
		1	8.0			RED RIVER	LA					
		1	1.9			RICHLAND	LA					

* SIGNIFICANTLY MORE DEATHS THAN EXPECTED

NON-WHITE MALE — CANCER OF THE ORAL CAVITY INCLUDING TONGUE — ICD 141,143-146,148,149

1950-1959 DEATHS	1950-1959 RATE	1960-1969 DEATHS	1960-1969 RATE	1970-1979 DEATHS	1970-1979 RATE	COUNTY AND STATE		PERCENT CHANGE 60S/50S	PERCENT CHANGE 70S/60S	PERCENT CHANGE 70S/50S	PERCENTILE %CHANGE 70S/50S	PERCENTILE RATES 1970S
				1	6.9	SABINE	LA					85
		1	10.3			ST BERNAPD	LA					
2	10.8	4*	17.9	1	4.1	ST CHARLES	LA	66	-77	-62	04	77
				1	3.9	ST JAMES	LA					77
		1	2.9	1	2.7	ST JOHN THE BAPT	LA		-7			74
1	1.1	3	2.7	7	6.0	ST LANDRY	LA	139	120	426	96	82
2	5.8	3	7.8	3	7.4	ST MARTIN	LA	35	-6	27	49	86
4	7.1	5	9.3	4	7.1	ST MARY	LA	31	-24	0	33	85
		3	7.3	1	2.1	ST TAMMANY	LA		-71			72
3	3.7	2	2.6	3	3.7	TANGIPAHOA	LA	-29	39	-1	32	76
1	2.0					TENSAS	LA					
3	9.2	1	2.8	2	4.5	TERREBONNE	LA	-70	59	-51	09	79
1	4.9			1	3.1	UNION	LA			-37	17	74
		1	4.4	2	8.7	VERMILION	LA		100			88
1	9.4	1	10.3	1	4.5	VERNON	LA	9	-57	-53	09	78
1	1.8	1	2.0	2	4.1	WASHINGTON	LA	10	110	132	78	77
		3	6.0	2	4.7	WEBSTER	LA		-22			79
2*	13.9					WEST CAPROLL	LA					
1	4.9	1	3.7			WEST FELICIANA	LA	-23				
1	2.9	1	3.9			WINN	LA	37				
		1	5.8	3	10.1	MAINE	ME		73			
				1*	89.3	HANCOCK	ME					99
				1	37.5	PENOBSCOT	ME					98
				1*	24.8	PISCATAQUIS	ME					97
		1*	42.0			WASHINGTON	ME					
90*	5.5	151*	6.8	277*	9.5	MARYLAND	MD	23	40	72		
				2*	37.0	ALLEGANY	MD					98
5	5.5	7	5.9	16*	10.9	ANNE ARUNDEL	MD	7	86	99	72	91
3	5.6	6	8.5	8	6.1	BALTIMORE	MD	52	-29	8	37	82
		1	6.5	2	10.6	CAROLINE	MD		63			91
		2	15.2	1	7.7	CARROLL	MD		-49			86
		1	6.4			CECIL	MD					
4*	13.5	2	5.7	3	6.2	CHARLES	MD	-58	8	-54	08	83
		2	4.8	5	11.3	DORCHESTER	MD		138			92
3*	15.6	1	6.0	3	12.5	FREDERICK	MD	-62	110	-20	25	93
2	6.9	2	6.9	3	8.9	HARFORD	MD	0	29	29	50	88
				2	5.7	HOWARD	MD					81
1	10.6			2	9.4	KENT	MD			-12	28	89
1	1.9	5	6.8	4	2.3	MONTGOMERY	MD	251	-66	21	44	72
6	5.6	13	8.2	24	6.6	PRINCE GEORGE'S	MD	45	-19	18	43	84
1	6.2	2	9.3	1	4.3	QUEEN ANNE'S	MD	50	-53	-30	20	78
1	5.0	3	12.9	2	5.4	ST MARY'S	MD	155	-58	6	36	80
1	3.4			2	6.1	SOMERSET	MD			81	67	82
2	7.4	1	2.8	3	10.7	TALBOT	MD	-62	284	44	56	91
		1	10.4	3*	33.2	WASHINGTON	MD		219			98
1	2.5	2	4.4	3	6.0	WICOMICO	MD	74	36	136	78	82
		1	3.9	4	11.8	WORCESTER	MD		202			93
59*	6.4	99*	7.3	184*	11.9	BALTIMORE CITY	MD	15	61	85	68	93
27*	6.9	38	7.3	78*	10.6	MASSACHUSETTS	MA	6	45	53		
				1	6.6	BARNSTABLE	MA					84
		1	10.3	2	18.7	BERKSHIRE	MA		82			96
2	7.2			1	3.0	BRISTOL	MA			-59	07	74
		3*	22.5	1	2.5	ESSEX	MA		-89			73
		2	3.4	15*	20.2	HAMPDEN	MA		501			96
3	6.1	3	4.3	5	5.4	MIDDLESEX	MA	-29	24	-12	28	81
1*	122.7			1*	141.3	NANTUCKET	MA			15	41	99
				2	6.1	NORFOLK	MA					82
1	3.6	1	5.1	1	1.9	PLYMOUTH	MA	41	-62	-47	12	72
19*	9.3	27*	9.3	46*	12.9	SUFFOLK	MA	0	39	39	56	94
1	6.6	1	7.8	3	9.1	WORCESTER	MA	19	16	38	54	88
92	4.7	125	4.2	251	6.1	MICHIGAN	MI	-11	45	29		
2	6.7	2	4.5	2	2.7	BERRIEN	MI	-33	-39	-60	06	74
3	16.8	4	13.2	5	9.6	CALHOUN	MI	-22	-27	-43	14	89
1	4.5			5*	28.0	CASS	MI			527	97	97
2	3.6	1	.9	6	2.9	GENESEE	MI	-75	223	-20	25	74
1	4.6	2	6.4	3	6.1	INGHAM	MI	41	-6	33	53	83
				3	7.4	KALAMAZOO	MI					86
3	8.0	1	1.8	8	10.6	KENT	MI	-77	481	33	52	91
1	7.6	2*	14.5			LAKE	MI	91				
				3	6.5	MACOMB	MI					83
				1*	22.9	MARQUETTE	MI					97
		2	3.3	4	6.0	MUSKEGON	MI		84			82
				1	12.2	NEWAYGO	MI					93
2	2.8	4	4.1	7	5.8	OAKLAND	MI	46	41	106	74	81
		3	3.6	5	4.9	SAGINAW	MI		36			80
				1	7.7	ST CLAIR	MI					86
		1	26.8			ST JOSEPH	MI					
				2	7.4	VAN BUREN	MI					86
2	3.0	3	4.9	4	5.1	WASHTENAW	MI	67	3	72	64	80
75	5.1	100	4.6	191	6.4	WAYNE	MI	-11	41	26	48	83
10*	8.9	7	4.4	7	2.6	MINNESOTA	MN	-50	-41	-71		
1	18.9					BELTRAMI	MN					
				1	67.0	DAKOTA	MN					99
3	6.8	5	8.4	4	3.3	HENNEPIN	MN	25	-61	-51	10	75
6*	18.7	1	2.6	2	3.0	RAMSEY	MN	-86	17	-84	00	74
		1	16.7			ST LOUIS	MN					
60	1.9	61	1.9	96	3.1	MISSISSIPPI	MS	0	67	67		
4	6.1	2	2.4	5	7.4	ADAMS	MS	-60	201	20	44	85
				1	2.5	ATTALA	MS					73

* SIGNIFICANTLY MORE DEATHS THAN EXPECTED

NON-WHITE MALE — CANCER OF THE ORAL CAVITY INCLUDING TONGUE — ICD 141,143-146,148,149

1950-1959 DEATHS	1950-1959 RATE	1960-1969 DEATHS	1960-1969 RATE	1970-1979 DEATHS	1970-1979 RATE	COUNTY AND STATE	PERCENT CHANGE 60S/50S	PERCENT CHANGE 70S/60S	PERCENT CHANGE 70S/50S	PERCENTILE %CHANGE 70S/50S	PERCENTILE RATES 1970S
4	2.4	2	1.4	4	3.9	BOLIVAR MS	-42	179	61	61	77
		2	9.0	1	4.1	CARROLL MS		-55			77
		1	3.2	1	3.2	CLAIBORNE MS		-1			75
				1	2.9	CLARKE MS					74
6	4.5	3	2.5	3	4.3	COAHOMA MS	-44	73	-4	31	78
				1	1.4	COPIAH MS					71
				1	6.6	COVINGTON MS					84
2	3.3					DE SOTO MS					
1	3.1	1	1.8	7	12.1	FORREST MS	-43	580	286	91	93
1*	28.1			1	16.7	GEORGE MS			-40	15	96
		1	3.0	4	11.2	GRENADA MS		272			92
1	14.1			2	18.8	HANCOCK MS			33	53	96
2	3.1	4	5.3	4	5.4	HARRISON MS	73	1	74	64	80
4	2.1	4	1.4	12	4.2	HINDS MS	-32	195	102	73	78
4	5.2	1	1.3	1	2.0	HOLMES MS	-74	51	-61	05	72
		3	6.6			HUMPHREYS MS					
1	7.9	2*	17.2	1	15.7	ISSAQUENA MS	118	-9	99	73	95
1	6.8	1	2.0	4	7.4	JACKSON MS	-71	275	9	38	85
		1	3.1			JASPER MS					
				1	2.7	JEFFERSON MS					73
1	5.4					JEFFERSON DAVIS MS					
2	4.9	1	2.1	1	1.5	JONES MS	-57	-32	-71	02	71
				1	3.2	KEMPER MS					75
				1	5.4	LAFAYETTE MS					81
				2*	25.1	LAMAR MS					97
3	3.5	1	1.1	2	2.5	LAUDERDALE MS	-69	132	-28	21	73
				1	10.9	LAWRENCE MS					91
1	2.8					LEE MS					
		1	.8			LEFLORE MS					
		1	2.9			LINCOLN MS					
2	3.9	2	3.3	2	2.7	LOWNDES MS	-15	-19	-31	20	73
1	1.1	1	1.3			MADISON MS	24				
				1	4.4	MARION MS					78
1	2.0	1	1.9	1	2.2	MARSHALL MS	-1	14	13	40	72
		1	3.8			MONTGOMERY MS					
1	3.6					NEWTON MS					
				1	2.8	NOXUBEE MS					74
				1	1.3	PANOLA MS					71
		2	4.9	1	1.5	PIKE MS		-69			71
1	6.3	1	6.7			PONTOTOC MS	7				
3	7.5			2	4.5	QUITMAN MS			-40	16	79
1	1.5					RANKIN MS					
				1	2.6	SCOTT MS					73
				1	6.8	SHARKEY MS					84
2	9.1					SIMPSON MS					
				1	18.8	STONE MS					96
2	1.7	4	3.6	1	1.7	SUNFLOWER MS	112	-51	3	34	72
		1	1.3	3	9.2	TALLAHATCHIE MS		597			89
				1	3.2	TATE MS					75
3	4.8	3	6.1	1	3.5	TUNICA MS	27	-42	-26	22	75
1	7.4					WALTHALL MS					
		6	6.9	8	9.1	WARREN MS		32			88
2	1.2	4	2.2	5	3.6	WASHINGTON MS	84	64	202	84	75
		1	6.2			WAYNE MS					
1	13.3	1	12.1			WEBSTER MS	-8				
1	3.7					WINSTON MS					
		1	1.2	2	3.4	YAZOO MS		177			75
97*	6.7	141*	8.1	171*	9.1	MISSOURI MO	21	12	36		
1	9.2	2*	28.5			AUDRAIN MO	208				
2	10.6	3*	16.3	3	13.6	BOONE MO	54	-16	29	50	94
2	9.9			1	5.9	BUCHANAN MO			-40	15	82
		1	5.5	2	25.0	BUTLER MO		350			97
		1	4.9	2	20.4	CALLAWAY MO		317			96
		1	33.1	1	11.5	CAPE GIRARDEAU MO		-65			92
1*	49.2					CARROLL MO					
1*	163.6					CASS MO					
		1	18.4			CHARITON MO					
1*	17.2					CLAY MO					
1*	133.9			1*	53.6	CLINTON MO			-60	06	99
1	6.0	1	7.3	1	7.6	COLE MO	22	4	27	49	86
1	15.6	1	14.5			COOPER MO	-7				
				1	9.9	DUNKLIN MO					90
				1	23.4	FRANKLIN MO					97
1	8.3	3*	22.1			GREENE MO	165				
				1	32.1	HOWARD MO					98
		1*	109.3			HOWELL MO					
23*	8.3	28	7.1	49*	10.8	JACKSON MO	-14	52	30	51	91
1	23.8	2*	30.7			JASPER MO	29				
		1*	37.8			LINCOLN MO					
2*	103.3					LINN MO					
		1*	74.5			LIVINGSTON MO					
				1*	35.1	MACON MO					98
				1*	212.0	MADISON MO					99
				1	12.3	MARION MO					93
		1	7.9			MISSISSIPPI MO					
1*	172.0					MONITEAU MO					
1*	60.7	1*	81.8			MONROE MO	35				
1*	62.5			1*	99.4	MONTGOMERY MO			59	61	99
				1	4.8	NEW MADRID MO					79
1	1.9	1	2.7	2	7.2	PEMISCOT MO	39	173	281	91	85
1	14.2	1	9.0	3*	31.3	PETTIS MO	-37	248	120	77	98
1*	267.9					PHELPS MO					
1	10.1			1	9.4	PIKE MO			-7	29	89
1	13.7	1	21.1	1	7.9	RANDOLPH MO	54	-63	-43	14	87
		1*	27.0			RAY MO					
				1	12.3	ST CHARLES MO					93
1*	68.8					STE GENEVIEVE MO					
5	6.5	7	6.5	9	4.1	ST LOUIS COUNTY MO	-1	-37	-38	16	77
		1	13.0			SALINE MO					

* SIGNIFICANTLY MORE DEATHS THAN EXPECTED

NON-WHITE MALE — CANCER OF THE ORAL CAVITY INCLUDING TONGUE — ICD 141,143-146,148,149

1950-1959 DEATHS	1950-1959 RATE	1960-1969 DEATHS	1960-1969 RATE	1970-1979 DEATHS	1970-1979 RATE	COUNTY AND STATE	PERCENT CHANGE 60S/50S	PERCENT CHANGE 70S/60S	PERCENT CHANGE 70S/50S	PERCENTILE %CHANGE 70S/50S	PERCENTILE RATES 1970S
				2	17.4	SCOTT MO					96
45*	6.5	80*	9.0	84*	9.6	ST LOUIS CITY MO	38	6	46	57	89
1	2.0	1	1.3	2	2.1	MONTANA MT	-34	61	7		
		1	11.7			BIG HORN MT					
				1*	94.4	RICHLAND MT					99
				1	11.9	ROOSEVELT MT					93
1*	37.2					VALLEY MT					
14*	13.0	10	7.6	13	7.6	NEBRASKA NB	-41	0	-41		
13*	16.5	9	9.4	13	10.7	DOUGLAS NB	-43	13	-35	18	91
		1*	163.6			SHERIDAN NB					
1*	15.7					THURSTON NB					
		3	2.9	5	2.9	NEVADA NV		-3			
		1*	81.8			CHURCHILL NV					
		2	3.2	4	3.8	CLARK NV		19			76
				1	2.4	WASHOE NV					73
				1	9.6	NEW HAMPSHIRE NH					
				1	24.4	HILLSBOROUGH NH					97
95*	7.2	138	6.4	262*	8.5	NEW JERSEY NJ	-10	33	19		
15*	12.5	10	7.8	12	8.3	ATLANTIC NJ	-38	7	-34	19	87
5	8.8	8	8.6	10	5.8	BERGEN NJ	-2	-33	-35	18	81
		6	13.3	4	5.7	BURLINGTON NJ		-57			81
3	3.0	9	6.2	21	9.7	CAMDEN NJ	110	56	227	87	90
1	5.7	1	5.8	2	7.1	CAPE MAY NJ	2	24	26	48	85
4	10.3	1	1.4	4	7.7	CUMBERLAND NJ	-87	461	-25	22	86
31*	7.6	41	5.7	84*	8.4	ESSEX NJ	-25	46	10	39	87
1	1.8	1	1.4	6	8.3	GLOUCESTER NJ	-22	503	371	94	87
9*	11.6	9	5.9	26*	10.3	HUDSON NJ	-49	73	-11	28	90
				2*	36.5	HUNTERDON NJ					98
2	2.1	9	6.4	19	8.9	MERCER NJ	197	40	316	92	88
4	13.3	7	9.9	8	6.7	MIDDLESEX NJ	-26	-33	-50	11	84
7	6.9	10	7.3	14	7.5	MONMOUTH NJ	6	3	9	38	86
1	4.9			2	5.6	MORRIS NJ			15	41	81
				3	9.4	OCEAN NJ					89
1	1.8	11*	10.6	20*	12.0	PASSAIC NJ	489	14	571	98	93
1	3.9	2	5.5			SALEM NJ	40				
1	7.3			1	2.1	SOMERSET NJ			-71	02	72
				1*	27.3	SUSSEX NJ					97
9*	10.2	13	6.9	21	10.0	UNION NJ	-32	44	-2	31	90
				2*	49.4	WARREN NJ					99
		5	2.1	5	.9	NEW MEXICO NM		-58			
		2	6.5	2	1.6	BERNALILLO NM		-75			71
				2	9.8	LEA NM					90
		1	13.2			OTERO NM					
		1*	163.6			SAN MIGUEL NM					
				1	5.8	SANTA FE NM					82
		1	18.6			TAOS NM					
267*	7.0	546*	8.4	775*	8.5	NEW YORK NY	19	1	21		
1	4.3	5	12.3	8*	13.9	ALBANY NY	186	13	224	87	94
				1*	490.8	ALLEGANY NY					99
43*	12.4	57*	7.8	121*	8.9	BRONX NY	-37	15	-28	21	88
		1	11.2	1	6.9	BROOME NY		-38			84
		1	11.5	1	11.6	CATTARAUGUS NY		1			92
				2	18.7	CHEMUNG NY					96
		1	16.4			COLUMBIA NY					
2	5.6	7*	15.2	7	10.4	DUTCHESS NY	169	-31	84	68	91
14*	5.7	18	5.8	26	6.6	ERIE NY	2	14	16	42	84
				1	16.1	GREENE NY					95
39	4.6	97*	6.3	156	7.3	KINGS(BROOKLYN) NY	37	16	59	61	85
2	4.0	4	4.0	13	6.8	MONROE NY	0	71	72	63	84
				1*	119.2	MONTGOMERY NY					99
5	5.8	11	9.2	23	9.7	NASSAU NY	59	5	68	63	90
119*	7.5	259*	12.6	247*	11.2	NEW YORK(MANHATTAN) NY	67	-11	49	58	92
				2	3.6	NIAGARA NY					76
1	9.6	1	6.7	2	10.4	ONEIDA NY	-31	56	8	37	91
2	10.6	2	3.1	8	13.2	ONONDAGA NY	-71	326	24	45	94
				1	37.7	ONTARIO NY					98
2	12.8	1	1.7	7	13.6	ORANGE NY	-87	688	6	36	94
		1*	8.8			ORLEANS NY					
15	4.9	38	4.8	81	5.8	QUEENS NY	-1	20	19	44	82
1	12.4					RENSSELAER NY					
2	8.8	1	3.0	3	4.4	RICHMOND NY	-66	48	-50	10	78
1	4.2	1	2.2	2	4.5	ROCKLAND NY	-47	105	8	36	79
				1	12.9	SARATOGA NY					94
		1	28.1	1	6.0	SCHENECTADY NY		-79			82
		1*	212.0			SCHOHARIE NY					
		1	16.1	1	15.7	STEUBEN NY		-3			95
4	5.1	7	4.2	19	7.6	SUFFOLK NY	-18	80	48	58	86
				1	3.7	SULLIVAN NY					76
		1*	53.6			TIOGA NY					
2*	49.9					TOMPKINS NY					
				1	3.8	ULSTER NY					76
				1	12.4	WAYNE NY					93
12*	9.3	29*	10.8	36	9.6	WESTCHESTER NY	16	-11	3	34	89
71	2.2	160	4.1	258	5.7	NORTH CAROLINA NC	88	41	165		
1	1.7	3	5.3	4	5.4	ALAMANCE NC	214	0	215	86	80

* SIGNIFICANTLY MORE DEATHS THAN EXPECTED

NON-WHITE MALE — CANCER OF THE ORAL CAVITY INCLUDING TONGUE — ICD 141,143-146,148,149

1950-1959 DEATHS	1950-1959 RATE	1960-1969 DEATHS	1960-1969 RATE	1970-1979 DEATHS	1970-1979 RATE	COUNTY AND STATE	PERCENT CHANGE 60S/50S	PERCENT CHANGE 70S/60S	PERCENT CHANGE 70S/50S	PERCENTILE %CHANGE RATES 70S/50S	PERCENTILE 1970S
		1	2.6			ANSON NC					
		3	6.3	4	8.7	BEAUFORT NC		38			88
		3	7.3			BERTIE NC					
		1	2.1	2	4.5	BLADEN NC		113			79
5	11.2	2	2.8	5	8.7	BUNCOMBE NC	-75	214	-22	23	88
		3*	17.8	2	10.6	BURKE NC		-41			91
2	7.4			2	4.4	CABARRUS NC			-41	15	78
				2	15.7	CALDWELL NC					95
				4*	27.5	CARTERET NC					97
		2	7.4	2	6.9	CATAWBA NC		-7			84
1	2.3			3	7.9	CHATHAM NC			248	89	87
1	9.4	1	4.9	1	6.1	CHOWAN NC	-48	24	-36	17	83
1	2.9	3	7.6	2	3.8	CLEVELAND NC	159	-50	29	50	76
		3	5.2	4	6.5	COLUMBUS NC		26			83
		2	3.3	5	7.1	CRAVEN NC		115			85
2	2.2	3	2.3	11	6.9	CUMBERLAND NC	7	195	216	86	84
1	13.3					CURRITUCK NC					
1	3.7	1	3.0			DAVIDSON NC	-18				
				3*	28.0	DAVIE NC					97
1	2.5	1	1.9	3	5.8	DUPLIN NC	-24	203	130	78	81
1	.9	7	5.9	8	4.5	DURHAM NC	541	-24	390	95	79
3	4.4	2	2.5	10	11.9	EDGECOMBE NC	-43	376	170	82	93
3	1.6	12	6.6	9	4.5	FORSYTH NC	311	-32	178	82	78
				2	4.2	FRANKLIN NC					78
2	4.6	1	1.8	4	5.6	GASTON NC	-62	220	23	45	81
		1	4.4			GATES NC					
		2	4.2			GRANVILLE NC					
				1	4.2	GREENE NC					78
3	4.7	8	3.5	13	6.1	GUILFORD NC	-26	77	30	51	83
3	2.9	3	2.9	4	4.0	HALIFAX NC	2	35	38	55	77
1	2.4	4	8.0	2	4.3	HARNETT NC	227	-46	76	65	78
		1	9.3	1	10.9	HENDERSON NC		18			91
		2	5.8	1	2.6	HERTFORD NC		-55			73
		1	3.8	1	4.1	HOKE NC		8			77
1	2.0	1	2.4	3	6.2	IREDELL NC	18	160	208	86	83
		2	3.8	3	5.7	JOHNSTON NC		52			81
				1	3.7	LEE NC					76
1	1.3	1	1.2	3	3.7	LENOIR NC	-4	200	189	83	76
				2*	30.1	MC DOWELL NC					98
		1	3.0	1	2.6	MARTIN NC		-13			73
5	3.1	9	3.1	19	6.6	MECKLENBURG NC	-1	113	110	75	84
		1	5.0			MONTGOMERY NC					
2	6.1			8*	21.1	MOORE NC			244	89	97
		3	4.2	8	10.0	NASH NC		139			90
10*	14.0	9*	12.5	10*	12.2	NEW HANOVER NC	-11	-2	-13	27	93
		1	1.6	5	8.8	NORTHAMPTON NC		459			88
				4	10.7	ONSLOW NC					91
		2	5.1	2	4.3	ORANGE NC		-16			78
		1	7.6			PAMLICO NC					
1	3.3	6*	17.9	6*	16.3	PASQUOTANK NC	443	-9	392	95	95
				3	9.4	PENDER NC					89
2	4.3	1	6.2	1	3.9	PERSON NC	45	-38	-10	29	77
1	.8	1	1.0	8	8.2	PITT NC	25	705	902	99	87
				2*	40.9	POLK NC					98
				2	9.7	RANDOLPH NC					90
2	4.9	4	9.5	2	3.9	RICHMOND NC	91	-59	-21	24	77
1	1.0	3	1.9	5	2.8	ROBESON NC	97	47	189	84	74
2	5.2	2	3.8	4	6.5	ROCKINGHAM NC	-26	69	24	46	83
		4	8.4	2	3.6	ROWAN NC		-57			76
		2	3.2	4	5.2	SAMPSON NC		61			80
		1	2.8	3	7.7	SCOTLAND NC		174			86
				2	10.6	STANLY NC					91
		1	10.4			SURRY NC					
		1	2.9	2	5.4	UNION NC		85			81
1	2.0	2	3.9	4	7.1	VANCE NC	94	81	251	89	85
4	3.1	11	6.9	11	5.8	WAKE NC	125	-17	87	69	81
		4	9.1	2	4.8	WARREN NC		-48			79
1	6.2					WASHINGTON NC					
4	4.2	6	6.1	7	5.8	WAYNE NC	44	-4	38	54	82
1	1.5	4	4.4	4	4.9	WILSON NC	190	13	228	88	80
		1	2.7	2	4.4	NORTH DAKOTA ND		65			
				2	10.7	ROLETTE ND					91
		1	16.4			SIOUX ND					
114	4.9	220*	6.7	315*	7.8	OHIO OH	37	17	59		
		1	3.3	1	3.0	ALLEN OH		-7			74
				1	6.9	ASHTABULA OH					84
		2*	16.7			BELMONT OH					
1*	18.9					BROWN OH					
4*	11.1	1	2.5	5	11.1	BUTLER OH	-77	337	0	33	92
		1	18.4			CHAMPAIGN OH					
2	4.9	1	2.0	5	8.0	CLARK OH	-60	308	63	62	87
				1	11.2	COLUMBIANA OH					92
				1*	70.1	COSHOCTON OH					99
37	5.3	73	7.3	109*	8.2	CUYAHOGA OH	37	13	54	60	87
		1*	19.6			DELAWARE OH					
1	8.7	2	12.7	1	7.5	ERIE OH	47	-41	-13	27	86
				2*	57.7	FAYETTE OH					99
6	2.4	18	5.3	32	8.0	FRANKLIN OH	125	49	236	88	87
1	15.3					GALLIA OH					
		1	21.9			GEAUGA OH					
2	9.4	2	7.9	1	3.2	GREENE OH	-16	-60	-66	03	75
		1*	28.1			GUERNSEY OH					
29*	6.8	50*	9.0	63*	10.3	HAMILTON OH	34	13	52	59	90
		1*	122.7			HARDIN OH					
1*	49.7					HIGHLAND OH					
1*	74.3					HURON OH					
				2	8.4	JEFFERSON OH					88

* SIGNIFICANTLY MORE DEATHS THAN EXPECTED

NON-WHITE MALE — CANCER OF THE ORAL CAVITY INCLUDING TONGUE — ICD 141,143-146,148,149

1950-1959 DEATHS	1950-1959 RATE	1960-1969 DEATHS	1960-1969 RATE	1970-1979 DEATHS	1970-1979 RATE	COUNTY AND STATE	PERCENT CHANGE 60S/50S	PERCENT CHANGE 70S/60S	PERCENT CHANGE 70S/50S	PERCENTILE %CHANGE 70S/50S	PERCENTILE RATES 1970S
		1	9.6			LAWRENCE OH					
				1	9.3	LICKING OH					89
		3	8.2	2	2.8	LORAIN OH		-66			74
6	7.4	8	4.7	21	9.5	LUCAS OH	-36	102	30	51	89
3	2.0	9	6.1	16	9.2	MAHONING OH	196	51	349	94	89
				1	17.6	MEDINA OH					96
1*	15.8	1	12.7	2*	24.3	MIAMI OH	-19	91	54	59	97
7	4.6	15	6.0	16	4.5	MONTGOMERY OH	29	-24	-2	31	79
		3*	18.1			MUSKINGUM OH					
1*	47.2	1*	53.6			PAULDING OH	14				
		1*	109.3			PREBLE OH					
				1	2.8	RICHLAND OH					74
2*	15.2	2	10.6	4*	17.0	ROSS OH	-30	60	12	39	96
				1	13.2	SANDUSKY OH					94
				2*	26.9	SCIOTO OH					97
				1	22.9	SENECA OH					97
		1*	49.1			SHELBY OH					
1	1.3	2	2.6	5	5.8	STARK OH	107	123	360	94	82
6	4.1	17*	10.0	13	7.2	SUMMIT OH	147	-28	78	66	85
2	4.7	1	2.0	4	7.6	TRUMBULL OH	-58	287	63	62	86
				1	19.3	WAYNE OH					96
31	3.4	30	3.0	71	5.8	OKLAHOMA OK	-13	93	67		
1	14.1	1	15.0			BRYAN OK	6				
2*	17.1	1	7.4	1	6.1	CADDO OK	-57	-19	-65	03	82
				1	11.6	CANADIAN OK					92
		3*	15.6			CARTER OK					
1	10.1	1	8.3			CHEROKEE OK	-18				
1	3.6			1	6.6	CHOCTAW OK			85	68	84
1*	37.7					CLEVELAND OK					
				2	5.6	COMANCHE OK					81
1	10.0					CRAIG OK					
				1	3.9	CREEK OK					77
				1	4.3	DELAWARE OK					78
				1	12.2	GARFIELD OK					93
		1	10.7	2*	20.3	GARVIN OK		90			96
				2	16.9	HUGHES OK					96
		1	14.8			JACKSON OK					
		1	9.0	2	16.5	KAY OK		82			95
		1*	19.1			KINGFISHER OK					
1*	15.7					KIOWA OK					
1	10.4			1	8.0	LE FLORE OK			-24	22	87
1	3.9			1	4.2	LOGAN OK			9	38	78
				2	4.9	MC CURTAIN OK					80
				1	5.1	MC INTOSH OK					80
2	1.9	2	2.5	7	11.7	MUSKOGEE OK	34	358	514	97	93
1	7.2	1	4.5			OKFUSKEE OK	-38				
4	2.8	5	2.7	20	8.5	OKLAHOMA OK	-6	217	199	84	88
4*	11.6	3	7.2	4	9.6	OKMULGEE OK	-38	33	-17	25	89
		1	8.0	1	5.1	OSAGE OK		-36			80
				1	15.8	PAYNE OK					95
3*	12.7			1	4.8	PITTSBURG OK			-62	04	79
				2	12.5	POTTAWATOMIE OK					93
				1	11.6	ROGERS OK					92
				1	3.6	SEMINOLE OK					76
1	11.2			4*	20.7	SEQUOYAH OK			85	68	96
				1	13.0	TILLMAN OK					94
5	4.2	6	4.0	8	4.2	TULSA OK	-5	5	-1	32	78
1	7.2	1	6.5	1	9.0	WAGONER OK	-9	38	26	46	88
		1	14.0			WASHINGTON OK					
1	.6	5	3.3	7	2.3	OREGON OR	435	-30	272		
				1	11.5	JACKSON OR					92
1	.9	5	5.4	6	3.5	MULTNOMAH OR	484	-35	280	90	75
180*	6.1	265*	7.3	373*	8.8	PENNSYLVANIA PA	19	21	43		
34*	5.7	32	5.1	58	8.6	ALLEGHENY PA	-11	70	50	58	88
2	6.6	4	7.4	2	3.7	BEAVER PA	12	-50	-44	13	76
1	4.5	1	4.8			BERKS PA	6				
1*	16.6					BLAIR PA					
		2	10.1	2	5.4	BUCKS PA		-46			81
				1	43.0	BUTLER PA					98
		1	5.7			CAMBRIA PA					
4	7.6	5	6.3	4	4.0	CHESTER PA	-18	-37	-48	11	77
				1	18.7	CRAWFORD PA					96
				1	10.6	CUMBERLAND PA					91
6*	9.4	6	7.0	11	10.0	DAUPHIN PA	-25	42	7	36	90
5	3.2	18*	10.7	15	7.3	DELAWARE PA	232	-32	127	77	85
2	9.2			4	11.7	ERIE PA			26	48	93
		3	6.8	3	8.2	FAYETTE PA		21			87
				1*	38.2	HUNTINGDON PA					98
		1*	196.5			JEFFERSON PA					
				1	16.4	LACKAWANNA PA					95
		4*	32.2	4	16.1	LANCASTER PA		-50			95
		4*	35.9	1	9.2	LAWRENCE PA		-74			89
		1*	74.3			LEBANON PA					
		1	20.2			LEHIGH PA					
		2*	24.0			LUZERNE PA					
		1*	163.6			MC KEAN PA					
3*	16.4	1	4.2	3	12.4	MERCER PA	-74	195	-24	22	93
2	2.8	3	3.3	7	6.1	MONTGOMERY PA	16	86	116	76	82
		1	11.6			NORTHAMPTON PA					
117*	7.2	172*	7.9	244*	9.4	PHILADELPHIA PA	10	20	31	51	89
1	1.9	2	4.2	5	11.4	WASHINGTON PA	125	169	507	97	92
1	3.2			1	3.7	WESTMORELAND PA			16	42	76
1	5.3			4	15.1	YORK PA			184	83	95

* SIGNIFICANTLY MORE DEATHS THAN EXPECTED

NON-WHITE MALE — CANCER OF THE ORAL CAVITY INCLUDING TONGUE — ICD 141,143-146,148,149

1950-1959 DEATHS	1950-1959 RATE	1960-1969 DEATHS	1960-1969 RATE	1970-1979 DEATHS	1970-1979 RATE	COUNTY AND STATE		PERCENT CHANGE 60S/50S	PERCENT CHANGE 70S/60S	PERCENT CHANGE 70S/50S	PERCENTILE %CHANGE 70S/50S	PERCENTILE RATES 1970S
5	8.5	6	7.5	15*	11.8	RHODE ISLAND	RI	-11	56	39		
1	8.3					NEWPORT	RI					
3	6.1	4	7.0	15*	14.9	PROVIDENCE	RI	16	112	147	79	95
1*	22.9	2*	29.8			WASHINGTON	RI	30				
57	2.5	86	3.5	135	4.7	SOUTH CAROLINA	SC	39	35	87		
1	3.3	1	4.4	1	3.5	ABBEVILLE	SC	35	-21	6	36	75
1	1.3	3	4.5	5	5.8	AIKEN	SC	240	30	341	93	81
2	11.3	1	4.9			ALLENDALE	SC	-56				
1	1.5	3	3.9	6	7.5	ANDERSON	SC	153	94	392	95	86
1	5.0			1	3.4	BAMBERG	SC			-31	20	75
1	3.7	1	3.3	2	9.0	BARNWELL	SC	-11	171	141	79	88
3	6.6	3	6.3	3	5.2	BEAUFORT	SC	-5	-17	-21	24	80
3	6.2	2	3.3	1	1.1	BERKELEY	SC	-46	-66	-82	00	71
		2	8.1	2	11.5	CALHOUN	SC		41			92
6	3.2	8	3.3	15	5.2	CHARLESTON	SC	6	55	64	62	80
1	3.8	2	8.6			CHEROKEE	SC	126				
2	5.4			2	5.3	CHESTER	SC			-1	32	80
		1	2.6	5	16.0	CHESTERFIELD	SC		521			95
				1	2.0	CLARENDON	SC					72
				2	4.6	COLLETON	SC					79
2	3.2	2	3.1	2	2.8	DARLINGTON	SC	-3	-9	-11	28	74
				2	5.0	DILLON	SC					80
1	2.5			2	3.9	DORCHESTER	SC			58	60	77
		1	4.0			EDGEFIELD	SC					
1	2.0	1	2.3			FAIRFIELD	SC	11				
2	2.5	1	.9	8	6.7	FLORENCE	SC	-65	671	168	81	84
2	6.7	2	3.6	2	3.2	GEORGETOWN	SC	-47	-9	-51	10	75
4	3.0	5	3.7	6	3.9	GREENVILLE	SC	23	4	28	50	77
1	3.0	2	4.8	3	5.8	GREENWOOD	SC	60	21	94	71	81
1	1.5	1	1.5	4	6.1	HORRY	SC	2	292	300	92	82
				2	6.6	JASPER	SC					84
		2	5.2	3	7.4	KERSHAW	SC		43			85
		1	2.5	4	8.2	LAURENS	SC		234			87
1	2.5			2	5.9	LEE	SC			140	79	82
1	4.0			1	1.8	LEXINGTON	SC			-57	07	72
1	9.3	1	6.5	2	11.7	MC CORMICK	SC	-30	79	26	46	93
4	5.8	3	5.5	6	11.4	MARION	SC	-5	105	96	72	92
1	2.3	1	2.3	2	4.4	MARLBORO	SC	3	88	94	71	78
		2	5.2	2	6.3	NEWBERRY	SC		22			83
1	8.6	1	6.5			OCONEE	SC	-25				
2	2.1	6	5.5	8	5.8	ORANGEBURG	SC	164	6	179	83	82
		2	11.0	3	14.1	PICKENS	SC		28			95
3	1.8	18*	8.1	13	4.8	RICHLAND	SC	342	-41	159	81	79
		1	5.7			SALUDA	SC					
2	2.4	1	.7	5	3.9	SPARTANBURG	SC	-70	426	58	61	76
3	2.8	1	1.1	3	2.2	SUMTER	SC	-61	100	-21	24	72
1	3.0	1	3.7	1	3.3	UNION	SC	22	-9	12	39	75
1	1.2	2	3.0	3	5.0	WILLIAMSBURG	SC	145	69	315	92	80
		1	1.4			YORK	SC					
1	1.8	1	1.1	1	1.2	SOUTH DAKOTA	SD	-41	13	-33		
1*	188.7					GREGORY	SD					
				1	4.9	SHANNON	SD					79
		1*	43.0			TRIPP	SD					
90	4.0	116	4.8	183	7.3	TENNESSEE	TN	20	52	82		
1*	24.6			1	9.4	ANDERSON	TN			-62	05	89
2*	17.7	1	6.7			BEDFORD	TN	-62				
				2*	19.8	BRADLEY	TN					96
1	5.8			1	8.4	CARROLL	TN			46	56	88
				1*	29.5	CHEATHAM	TN					98
1	15.7					CHESTER	TN					
				1*	45.5	COCKE	TN					98
		1	8.5			CROCKETT	TN					
13	4.7	37*	11.1	53*	14.7	DAVIDSON	TN	133	33	210	86	95
				1	21.1	DICKSON	TN					97
1	3.6			1	7.8	DYER	TN			116	76	86
		1	11.5	2*	22.6	FRANKLIN	TN		97			97
1	2.5	1	3.2	1	2.8	GIBSON	TN	29	-12	14	41	74
		1	10.9	1	9.9	HAMBLEN	TN		-9			90
7	3.4	10	5.4	31*	15.7	HAMILTON	TN	61	189	364	94	95
2	4.6					HARDEMAN	TN					
1	14.5	1	12.5			HARDIN	TN	-14				
2	3.8			1	1.7	HAYWOOD	TN			-55	08	71
1	5.3					HENRY	TN					
		1*	86.0			HOUSTON	TN					
9*	10.2	7	7.2	9	8.8	KNOX	TN	-30	23	-14	27	88
				1	10.3	LAKE	TN					91
				1	3.7	LAUDERDALE	TN					76
				1	7.3	LINCOLN	TN					85
4	4.7	1	.9	5	6.3	MADISON	TN	-80	573	34	54	83
		1	12.4			MARSHALL	TN					
		3	7.6	3	8.4	MAURY	TN		11			87
		2	6.0	1	2.1	MONTGOMERY	TN		-65			72
		1	6.0	1	7.3	OBION	TN		23			85
		1*	47.1			PUTNAM	TN					
		1	11.1			ROANE	TN					
				2	7.1	ROBERTSON	TN					85
2	7.6	1	3.2	1	2.8	RUTHERFORD	TN	-58	-13	-64	04	74
35	4.5	41	4.5	53	5.4	SHELBY	TN	0	18	18	43	80
				1*	24.8	SMITH	TN					97
1*	67.0					STEWART	TN					
1	8.6	1	8.9	3*	31.6	SULLIVAN	TN	3	255	266	90	98
1	4.3			2	8.5	SUMNER	TN			99	72	88
1	2.4	1	2.6	2	5.9	TIPTON	TN	6	131	146	79	82

* SIGNIFICANTLY MORE DEATHS THAN EXPECTED

NON-WHITE MALE — CANCER OF THE ORAL CAVITY INCLUDING TONGUE — ICD 141,143-146,148,149

1950-1959 DEATHS	1950-1959 RATE	1960-1969 DEATHS	1960-1969 RATE	1970-1979 DEATHS	1970-1979 RATE	COUNTY AND STATE	PERCENT CHANGE 60S/50S	PERCENT CHANGE 70S/60S	PERCENT CHANGE 70S/50S	PERCENTILE %CHANGE 70S/50S	PERCENTILE RATES 1970S
		1	8.0			WASHINGTON TN					
1	9.2					WEAKLEY TN					
2	9.8					WILLIAMSON TN					
142	3.4	217	4.3	312	4.3	TEXAS TX	26	0	26		
				3	5.8	ANDERSON TX					82
1	3.1			2	4.8	ANGELINA TX			56	60	79
1	7.6	1	7.2	1	11.3	AUSTIN TX	-5	58	50	58	92
1	4.0			2	9.4	BASTROP TX			131	78	89
1*	49.2					BEE TX					
2	7.0			4	5.4	BELL TX			-22	23	81
8	5.6	12	7.6	21	5.7	BEXAR TX	35	-24	3	34	81
2	3.0	3	4.3	1	1.1	BOWIE TX	41	-75	-64	03	71
2	6.4			2	4.4	BRAZORIA TX			-31	20	78
				2	4.7	BRAZOS TX					79
				1	20.2	BROWN TX					96
		1	4.2			BURLESON TX					
		1	6.5			CALDWELL TX					
1*	43.0					CALHOUN TX					
1	3.3					CASS TX					
1	11.2					CHAMBERS TX					
1	2.0			1	2.6	CHEROKEE TX			27	49	73
				2	10.2	COLLIN TX					90
1	5.3	1	4.4	1	6.2	COLORADO TX	-18	42	17	42	83
		1*	49.7			COMAL TX					
				2*	33.8	COOKE TX					98
13	3.2	23	3.9	52	6.1	DALLAS TX	24	54	91	70	83
3*	16.2			1	4.1	DE WITT TX			-75	01	77
				1*	66.2	EASTLAND TX					99
		2	25.7			ECTOR TX					
2	4.6			1	3.1	ELLIS TX			-33	19	74
				2	.5	EL PASO TX					71
1	3.1	4	14.0	2	7.7	FALLS TX	357	-45	150	80	86
1	5.1					FANNIN TX					
2	10.8	1	4.7	1	4.3	FAYETTE TX	-56	-8	-60	06	78
		1*	80.4			FISHER TX					
		5*	14.6	4	8.4	FORT BEND TX		-42			88
1	3.8	2	8.2	3	16.4	FREESTONE TX	114	101	330	93	95
				1*	44.6	GAINES TX					98
2	6.5	4	2.9	11	6.6	GALVESTON TX	-56	131	2	34	84
				2	16.2	GONZALES TX					95
1	2.6			2	6.8	GRAYSON TX			161	81	84
3	5.3	4	6.5	2	3.3	GREGG TX	22	-49	-38	16	75
3	10.2	2	6.8	1	3.0	GRIMES TX	-33	-56	-70	02	74
				2	11.2	GUADALUPE TX					92
1	6.4	1	5.2			HARDIN TX	-18				
33	5.9	53	5.1	78	5.1	HARRIS TX	-14	0	-14	26	80
1	1.7	2	2.6	2	2.2	HARRISON TX	48	-15	26	47	72
1*	17.2			1	6.9	HAYS TX			-60	06	84
		1	4.5	1	4.4	HENDERSON TX		-2			78
1*	44.6					HIDALGO TX					
		1	5.5			HILL TX					
				1	8.3	HOPKINS TX					87
1	2.4	1	2.3	1	2.4	HOUSTON TX	-2	2	0	33	73
1*	14.0	1	26.8			HOWARD TX	91				
				3	11.3	HUNT TX					92
1*	35.1					HUTCHINSON TX					
1	15.7					JACKSON TX					
		1	3.0	2	7.6	JASPER TX		149			86
10	5.9	20*	8.4	16	6.2	JEFFERSON TX	43	-26	6	35	83
		1	11.2			JOHNSON TX					
		1	2.0	1	2.6	KAUFMAN TX		27			73
1	3.2	2	5.1			LAMAR TX	57				
1	8.8					LAVACA TX					
1	6.6	2	6.5	1	9.4	LEON TX	-1	44	42	56	89
1	4.2	3	9.9	2	6.1	LIBERTY TX	135	-38	46	57	83
		1	3.0			LIMESTONE TX					
		1	1.7	2	2.5	LUBBOCK TX		45			73
3	2.9	2	2.1	4	3.6	MC LENNAN TX	-25	67	25	46	75
		1	4.9	1	3.1	MARION TX		-35			75
1	5.1	1	3.9			MILAM TX	-23				
1	3.2	2	7.4	3	11.0	MONTGOMERY TX	131	48	243	88	91
				1	2.1	NACOGDOCHES TX					72
1	2.2	1	1.7	1	2.1	NAVARRO TX	-23	22	-6	29	72
1	5.0	4	9.4	4	4.0	NUECES TX	88	-58	-20	25	77
		1	3.4	1	3.7	ORANGE TX		9			76
1	4.6	1	8.6	1	6.4	POLK TX	87	-26	38	55	83
		1	3.8	2	4.9	POTTER TX		29			80
		1*	188.7			REAGAN TX					
2	9.9					RED RIVER TX					
		1*	25.8			REEVES TX					
1	3.3					ROBERTSON TX					
3	6.4	1	2.1	2	3.6	RUSK TX	-67	68	-44	13	75
1	6.9			1	7.4	SAN AUGUSTINE TX			8	37	86
				1*	74.3	SHACKELFORD TX					99
		1	.8			SMITH TX					
8	4.5	21*	8.6	22	6.0	TARRANT TX	90	-30	33	53	82
				1	8.2	TITUS TX					87
		1	4.9			TOM GREEN TX					
3	2.9	4	3.5	8	4.3	TRAVIS TX	21	22	48	57	78
				2	7.9	UPSHUR TX					87
		1	4.2	3	9.3	VICTORIA TX		119			89
2	4.9	1	2.7	3	6.2	WALKER TX	-45	130	26	48	83
		3	13.5	2	8.5	WALLER TX		-37			88
1	3.0	1	2.5	1	3.0	WASHINGTON TX	-19	21	-2	31	74
1	2.6	3	8.9	1	3.0	WHARTON TX	242	-66	15	41	74
1	3.0	2	5.6	1	2.3	WICHITA TX	86	-58	-23	23	72
		1	4.3	2	9.4	WILLIAMSON TX		118			89
1	9.3	1	6.7			WOOD TX	-27				

* SIGNIFICANTLY MORE DEATHS THAN EXPECTED

NON-WHITE MALE — CANCER OF THE ORAL CAVITY INCLUDING TONGUE — ICD 141,143-146,148,149

1950-1959 DEATHS	1950-1959 RATE	1960-1969 DEATHS	1960-1969 RATE	1970-1979 DEATHS	1970-1979 RATE	COUNTY AND STATE	PERCENT CHANGE 60S/50S	PERCENT CHANGE 70S/60S	PERCENT CHANGE 70S/50S	PERCENTILE %CHANGE 70S/50S	PERCENTILE RATES 1970S
		2	3.1	1	.8	UTAH UT		-75			
		1	3.3	1	1.8	SALT LAKE UT		-45			72
		1	10.1			WEBER UT					
126	4.4	201	6.1	276*	7.3	VIRGINIA VA	41	18	67		
3	6.6	1	1.9			ACCOMACK VA	-72				
2	4.2	6*	12.0	10*	18.7	ALBEMARLE#VA	183	55	340	93	96
1	5.8	4*	30.1	1	6.6	ALLEGHANY#VA	423	-78	15	41	84
1	5.8			2	11.7	AMELIA VA			101	73	93
		1	4.5	3	13.4	AMHERST VA		199			94
				2	16.5	APPOMATTOX VA					95
		3	9.9	1	1.2	ARLINGTON#VA		-87			71
1	4.8			1	4.1	AUGUSTA#VA			-15	26	77
				1*	28.8	BATH VA					98
2	2.0	13*	12.4	12	10.5	BEDFORD#VA	514	-16	419	96	91
		2*	25.4			BOTETOURT VA					
1	2.7	2	4.9	3	7.4	BRUNSWICK VA	82	52	177	82	86
1	5.2			1	3.5	BUCKINGHAM VA			-33	20	75
2	8.7	2	7.2	2	5.3	CAROLINE VA	-17	-27	-39	16	80
1	7.9					CHARLES CITY VA					
		2	9.0	1	6.1	CHARLOTTE VA		-32			83
21*	6.7	34*	8.5	46*	10.1	CHESTERFIELD#VA	28	19	52	59	90
		3*	16.7	1	6.0	CULPEPER VA		-64			82
1	8.0	1	9.4			CUMBERLAND VA	17				
4	2.7	9	5.6	11	6.4	DINWIDDIE#VA	106	15	137	78	83
		1	5.7	1	6.6	ESSEX VA		16			84
6	10.5	5	6.5	10	6.6	FAIRFAX#VA	-38	2	-37	17	84
1	4.3	2	8.0	2	7.7	FAUQUIER VA	85	-4	78	65	86
		3*	19.9			FRANKLIN VA					
		1	14.9	1	11.9	FREDERICK#VA		-20			93
				1	5.2	GLOUCESTER VA					80
		1	5.3			GOOCHLAND VA					
1	2.9			1	3.7	GREENSVILLE#VA			31	51	76
1	1.2	3	4.9	5	8.8	HALIFAX#VA	311	81	645	99	88
1	2.9	2	7.4			HANOVER VA	153				
		1	3.2	1	2.7	HENRICO VA		-14			74
2	4.7	3	6.8	3	4.4	HENRY#VA	45	-34	-5	30	78
2	8.9	1	10.6	5	17.0	ISLE OF WIGHT VA	20	60	92	71	96
3	8.0	5	9.2	2	2.8	JAMES CITY#VA	14	-69	-65	03	74
				1	11.2	KING GEORGE VA					92
				2	14.1	KING WILLIAM VA					94
				1	5.8	LANCASTER VA					82
				1	4.5	LOUDOUN VA					79
		2	8.5	1	3.7	LOUISA VA		-57			76
				1	4.2	LUNENBURG VA					78
2*	20.5	1	10.6	1	12.1	MATHEWS VA	-48	14	-41	14	93
2	3.0	2	3.8	1	1.8	MECKLENBURG VA	26	-54	-42	14	72
				1	5.0	MIDDLESEX VA					80
		1	12.6			MONTGOMERY#VA					
				1	4.7	NELSON VA					79
2	5.2			1	2.4	NORTHAMPTON VA			-54	08	73
				2	13.2	NORTHUMBERLAND VA					94
2	7.6	2	6.6	3	9.5	NOTTOWAY VA	-13	44	25	46	89
		2	15.2			ORANGE VA					
6	6.7	6	5.8	7	5.8	PITTSYLVANIA#VA	-14	0	-14	27	81
1	5.4	1	5.0	1	3.6	PRINCE EDWARD VA	-9	-27	-33	19	76
				3	17.8	PRINCE WILLIAM#VA					96
				1	12.5	PULASKI VA					93
				1	10.3	RICHMOND VA					91
7*	7.6	6	5.9	16*	15.2	ROANOKE#VA	-21	155	101	73	95
2*	24.2	1	7.3			ROCKBRIDGE#VA	-70				
		1	17.1			ROCKINGHAM#VA					
		1*	133.9			SCOTT VA					
				1*	54.2	SHENANDOAH VA					99
		1*	36.4			SMYTH VA					
2	3.3	1	1.8	7	11.4	SOUTHAMPTON#VA	-47	546	244	89	92
1	5.7	1	3.9	1	4.0	SPOTSYLVANIA#VA	-33	4	-30	21	77
				1	24.4	STAFFORD VA					97
1	10.6			2	6.9	SUSSEX VA			-35	18	65
1	6.4	1	13.0			TAZEWELL VA	104				
		4*	43.3			WASHINGTON#VA					
				2	9.9	WESTMORELAND VA					90
				2*	66.3	WISE#VA					99
		1*	35.1	2*	71.3	WYTHE VA		103			99
30*	5.8	41	7.4	56	8.1	CHESAPEAKE CITY#VA	27	9	38	55	67
3	3.7	7	7.3	10	9.0	HAMPTON CITY VA	98	24	145	79	88
4	2.9	4	3.2	12	7.2	NEWPORT NEWS CITY VA	9	127	148	80	85
2	2.9	3	3.5	4	3.8	SUFFOLK CITY VA	21	11	34	53	76
9	2.7	13	2.9	22	3.2	WASHINGTON WA	10	8	19		
2*	69.0					CLALLAM WA					
		1*	109.3			COWLITZ WA					
				1	7.2	FRANKLIN WA					85
				1	9.3	GRANT WA					89
6	3.0	11	4.4	12	3.2	KING WA	49	-27	8	37	75
		1	2.8	3	2.1	PIERCE WA		-26			72
				3	8.8	SPOKANE WA					88
1	9.0			2	5.0	YAKIMA WA			-45	13	80
19	3.6	16	4.1	29	8.4	WEST VIRGINIA WV	14	105	133		
1*	54.6					BARBOUR WV					
				2	23.8	BERKELEY WV					97
2	7.0	1	3.5	3	13.3	CABELL WV	-50	284	91	70	94
		4	13.9	4	25.5	FAYETTE WV		84			97
				1*	424.0	GILMER WV					99
				1*	70.7	HARDY WV					99

* SIGNIFICANTLY MORE DEATHS THAN EXPECTED

COUNTIES COMBINED; SEE TECHNICAL NOTES

NON-WHITE MALE — CANCER OF THE ORAL CAVITY INCLUDING TONGUE — ICD 141,143-146,148,149

1950-1959 DEATHS	1950-1959 RATE	1960-1969 DEATHS	1960-1969 RATE	1970-1979 DEATHS	1970-1979 RATE	COUNTY AND STATE	PERCENT CHANGE 60S/50S	PERCENT CHANGE 70S/60S	PERCENT CHANGE 70S/50S	PERCENTILE %CHANGE 70S/50S	PERCENTILE RATES 1970S
		1	16.6	3*	56.8	HARRISON WV		243			99
2*	17.8					JEFFERSON WV					
6	8.2	6	10.0	4	5.8	KANAWHA WV	22	-42	-29	21	81
				2	3.6	MC DOWELL WV					76
1	4.0			1	16.6	MARION WV			318	93	96
		1	2.3	4	10.4	MERCER WV		347			91
1*	44.6					MINERAL WV					
1	7.4					MINGO WV					
		1	26.5			MONONGALIA WV					
1*	35.3					MONROE WV					
		1	7.3			OHIO WV					
1*	42.4			1*	37.5	POCAHONTAS WV			-12	27	98
1	2.1			1	1.8	RALEIGH WV			-14	26	72
1*	28.9			2*	64.4	SUMMERS WV			123	77	99
		1*	133.9			TAYLOR WV					
1*	34.7					WOOD WV					
8	4.4	15	5.0	31	6.0	WISCONSIN WI	14	18	35		
				1	8.8	DANE WI					88
1*	344.1					EAU CLAIRE WI					
1	11.2					MENOMINEE#WI					
5	3.7	12	6.5	27	8.0	MILWAUKEE WI	74	23	114	75	87
1	9.3	1	6.6	3	6.1	RACINE WI	-28	-8	-34	18	83
		1*	28.9			SAWYER WI					
		1*	172.0			WOOD WI					
				1	2.0	WYOMING WY					
				1	7.4	FREMONT WY					86

* SIGNIFICANTLY MORE DEATHS THAN EXPECTED

COUNTIES COMBINED; SEE TECHNICAL NOTES

NON-WHITE FEMALE — CANCER OF THE ORAL CAVITY INCLUDING TONGUE — ICD 141,143-146,148,149

1950-1959 DEATHS	1950-1959 RATE	1960-1969 DEATHS	1960-1969 RATE	1970-1979 DEATHS	1970-1979 RATE	COUNTY AND STATE	PERCENT CHANGE 60S/50S	PERCENT CHANGE 70S/60S	PERCENT CHANGE 70S/50S	PERCENTILE %CHANGE 70S/50S	PERCENTILE RATES 1970S
863	1.3	1340	1.4	2143	1.7	UNITED STATES	12	16	30		
39	1.0	49	1.1	57	1.2	ALABAMA AL	11	8	19		
				1	2.5	AUTAUGA AL					90
		2	3.2	1	1.2	BARBOUR AL		-61			86
1	2.0	1	2.5			BULLOCK AL	24				
		1	1.3	1	1.2	CALHOUN AL		-9			86
1	1.7			2	3.6	CHAMBERS AL			111	82	94
1*	18.6					CHEROKEE AL					
		1	1.8			CLARKE AL					
				1	3.5	COFFEE AL					93
				2	4.6	COLBERT AL					95
		1	2.0			CONECUH AL					
		1*	7.6			COOSA AL					
1	3.7			2*	7.8	COVINGTON AL			112	83	97
1	3.2					CRENSHAW AL					
1	.6	1	.8	2	1.3	DALLAS AL	38	66	130	86	86
				1	2.1	ELMORE AL					89
2	3.0					ETOWAH AL					
		2*	16.6	1*	8.0	FAYETTE AL		-52			97
1*	8.0					GENEVA AL					
1	1.7	1	1.5	1	1.5	GREENE AL	-8	0	-8	50	87
1	4.7					HENRY AL					
3*	5.3	1	1.4	1	1.5	HOUSTON AL	-73	6	-71	08	87
5	.6	6	.6	22	1.8	JEFFERSON AL	4	193	204	93	88
1	1.8	2	2.6	1	.9	LEE AL	48	-64	-47	19	84
1	1.9					LOWNDES AL					
1	1.1	1	1.1			MACON AL	-8				
1	1.3	1	1.3	2	1.5	MADISON AL	0	15	15	58	87
1	1.2					MARENGO AL					
4	1.3	10	2.1	6	1.1	MOBILE AL	60	-46	-14	45	85
				1	2.8	MONROE AL					92
5	1.7	2	.6	3	1.0	MONTGOMERY AL	-67	79	-41	24	84
		1	2.5			MORGAN AL					
		1	1.7			PICKENS AL					
		2	4.0	2	3.3	PIKE AL		-18			93
1	1.6	4*	4.6			RUSSELL AL	185				
				1	5.8	ST CLAIR AL					96
1	4.3	1	2.8			SHELBY AL	-35				
2	2.4	2	2.3	1	1.2	SUMTER AL	-1	-48	-48	18	85
		1	1.2	2	2.5	TALLADEGA AL		109			91
		1	2.4			TALLAPOOSA AL					
		2	1.4			TUSCALOOSA AL					
1	3.3					WALKER AL					
1	1.4					WILCOX AL					
		1	1.5	1	.9	ALASKA AK		-44			
2	.6	5	1.5	5	.7	ARIZONA AZ	141	-54	11		
		1	2.1			APACHE AZ					
1	.9	2	1.7	5	1.9	MARICOPA AZ	87	9	104	81	89
1	2.1	2	3.6			PIMA AZ	70				
22	1.2	22	1.1	18	.8	ARKANSAS AR	-4	-28	-30		
1	2.7					ASHLEY AR					
				1*	7.0	CALHOUN AR					97
1	1.5	1	1.1	1	2.1	CHICOT AR	-25	87	40	69	89
1	5.7					CLARK AR					
2	1.7					CRITTENDEN AR					
				1	1.3	DESHA AR					86
1	4.2					DREW AR					
		1*	8.2			FAULKNER AR					
1	3.5	1	3.7			GARLAND AR	5				
1	2.0	1	2.0	1	3.4	HEMPSTEAD AR	3	67	72	77	93
				1	3.7	HOWARD AR					94
2	1.0	2	1.1	2	.8	JEFFERSON AR	8	-27	-21	38	84
		1	4.2			LAFAYETTE AR					
1	1.4	2	3.1			LEE AR	124				
1	2.5					MILLER AR					
2	3.4					MISSISSIPPI AR					
				1	1.6	MONROE AR					88
				1	4.4	NEVADA AR					95
		3*	4.9			OUACHITA AR					
1	.9	4	3.4	2	2.1	PHILLIPS AR	267	-40	120	83	89
4	1.6	5	1.8	4	1.1	PULASKI AR	17	-38	-28	34	85
				2	2.6	ST FRANCIS AR					91
1	3.5	1	3.7			SEBASTIAN AR	8				
				1*	25.2	SEVIER AR					99
2	4.2					UNION AR					
33	1.2	67	1.1	143	1.1	CALIFORNIA CA	-8	1	-7		
5	1.6	11	1.7	21	1.8	ALAMEDA CA	5	8	13	57	88
1	.8	2	1.2	3	.9	CONTRA COSTA CA	53	-23	18	59	84
1	1.3	3	2.2	2	1.1	FRESNO CA	69	-50	-15	44	85
				1	1.0	KERN CA					84
12	1.0	35	1.2	69	1.1	LOS ANGELES CA	22	-2	20	61	85
1	10.3			1	1.1	MONTEREY CA			-89	01	85
				2	.3	ORANGE CA					83
		2	2.7	3	1.5	RIVERSIDE CA		-43			87
		3	2.0	4	1.0	SACRAMENTO CA		-47			84
		1	1.6	1	.6	SAN BERNARDINO CA		-64			83
3	2.6	2	.8	5	.7	SAN DIEGO CA	-68	-14	-72	07	84
7	2.2	6	.9	13	1.3	SAN FRANCISCO CA	-59	43	-42	23	86
		1	.9	4	2.9	SAN JOAQUIN CA		217			92

* SIGNIFICANTLY MORE DEATHS THAN EXPECTED

NON-WHITE FEMALE — CANCER OF THE ORAL CAVITY INCLUDING TONGUE — ICD 141,143-146,148,149

1950-1959 DEATHS	1950-1959 RATE	1960-1969 DEATHS	1960-1969 RATE	1970-1979 DEATHS	1970-1979 RATE	COUNTY AND STATE	PERCENT CHANGE 60S/50S	PERCENT CHANGE 70S/60S	PERCENT CHANGE 70S/50S	PERCENTILE %CHANGE 70S/50S	PERCENTILE RATES 1970S
1	7.0			5	1.3	SAN MATEO CA			-81	04	86
1*	6.2					SANTA BARBARA CA					
				3	.5	SANTA CLARA CA					83
		1	.7	3	2.2	SOLANO CA		202			90
				1	3.1	SUTTER CA					92
				2	3.6	TULARE CA					94
1*	12.5					VENTURA CA					
1	.6	1	.5	3	.6	COLORADO CO	-20	7	-14		
1	.9	1	.7	2	.6	DENVER-ARAPAHOE#CO	-16	-21	-34	27	83
				1	3.3	PUEBLO CO					93
3	1.3	13*	2.1	22*	2.5	CONNECTICUT CT	60	21	93		
1	1.8	4	1.6	10*	3.6	FAIRFIELD CT	-14	127	96	80	94
1	1.7	5*	2.6	3	1.2	HARTFORD CT	54	-53	-27	36	86
1	.8	4	2.8	9*	3.3	NEW HAVEN CT	229	17	285	96	93
		8*	2.7	13*	3.6	DELAWARE DE		29			
		1	2.5	2	3.7	KENT DE		49			94
		5	2.8	6	2.6	NEW CASTLE DE		-7			91
		2	2.9	5*	6.6	SUSSEX DE		130			97
15	1.2	43*	1.9	73*	2.9	DIST. OF COLUMBIA DC	60	47	135		
15	1.2	43*	1.9	73*	2.9	WASHINGTON DC	60	47	135	89	92
37	1.3	60	1.5	113*	2.1	FLORIDA FL	18	39	64		
				1	.9	ALACHUA FL					84
		1	3.2	1	1.6	BAY FL		-49			88
		1*	12.4	1*	8.0	BRADFORD FL		-35			97
				1	1.2	BREVARD FL					86
1	.6	1	.4	3	.9	BROWARD FL	-28	109	51	72	84
		1*	11.9			CITRUS FL					
1*	13.4					CLAY FL					
				1	3.2	COLLIER FL					93
6	1.8	10	1.2	27*	2.6	DADE FL	-30	113	49	72	91
8	2.1	8	1.4	14	2.2	DUVAL FL	-33	49	0	54	89
1	.7			3	1.3	ESCAMBIA FL			79	78	86
3	2.8	3	2.6	4	3.4	GADSDEN FL	-4	27	22	62	93
		1*	6.3			HAMILTON FL					
		1*	24.5			HARDEE FL					
		1	3.9			HIGHLANDS FL					
		6	2.0	6	1.6	HILLSBOROUGH FL		-18			88
1	2.2	1	1.9			JACKSON FL	-13				
				1	5.3	JEFFERSON FL					96
1	2.7			4*	7.3	LEE FL			169	90	97
3*	2.4	1	1.0			LEON FL	-61				
				2*	13.3	LEVY FL					98
				1	3.0	MADISON FL					92
1	7.5	2	5.3			MANATEE FL	-30				
				1	1.1	MARION FL					85
1*	9.1			1	4.3	MARTIN FL			-53	17	95
1*	7.0	1	5.0			MONROE FL	-28				
1	.9	1	.8	5	2.2	ORANGE FL	-19	187	132	87	90
2	.9	6	3.0	10*	3.4	PALM BEACH FL	219	14	264	95	93
				1	5.8	PASCO FL					96
3	2.0	3	1.7	10*	4.4	PINELLAS FL	-14	163	128	85	95
1	2.0	4	3.0	4	2.1	POLK FL	54	-31	5	55	89
		2	5.9			PUTNAM FL					
		1	2.9			ST JOHNS FL					
				1	1.4	ST LUCIE FL					87
				2	4.4	SARASOTA FL					95
		1	1.5	2	2.5	SEMINOLE FL		67			90
1*	7.9					TAYLOR FL					
1	1.5	2	1.8	6*	5.0	VOLUSIA FL	17	182	229	94	96
		1*	11.8			WALTON FL					
64	1.4	81	1.6	99	1.7	GEORGIA GA	12	5	18		
		1*	10.1			APPLING GA					
				2	3.9	BALDWIN GA					94
		1	4.6	1	4.6	BARTOW GA		-1			95
		1	4.6			BEN HILL GA					
3	1.3	3	1.2	3	1.2	BIBB GA	-3	-6	-8	49	85
		1	3.4	1	3.5	BROOKS GA		4			93
1	5.0	1	3.5			BULLOCH GA	-32				
1	1.1			1	1.4	BURKE GA			24	64	86
				2*	6.9	CALHOUN GA					97
		1	3.8			CARROLL GA					
5	1.7	6	2.1	17*	4.9	CHATHAM GA	24	137	195	92	96
				1*	11.5	CHEROKEE GA					98
1	1.6	1	1.7			CLARKE GA	3				
1	3.5	1	3.1			COBB GA	-12				
		2*	6.7	2*	5.3	COLQUITT GA		-21			96
				1	4.7	COLUMBIA GA					95
1	1.8					COWETA GA					
1	2.0					CRISP GA					
1	3.1	2	5.6	1	1.9	DECATUR GA	82	-66	-38	25	89
1	.8			1	.3	DE KALB GA			-57	14	83
3	2.8	2	1.6	2	1.3	DOUGHERTY GA	-41	-20	-53	16	86
2*	5.7			1	4.8	EARLY GA			-17	42	95
1	3.0			1	3.0	ELBERT GA			0	52	92
				1	2.6	EMANUEL GA					91
2*	3.8	3*	6.4	2	4.3	FLOYD GA	71	-34	14	57	95
13	1.8	22*	2.4	21	1.8	FULTON GA	34	-23	3	55	88
2*	4.4	2	3.2	1	1.4	GLYNN GA	-28	-55	-68	09	87
1	4.1			1	2.3	GRADY GA			-43	22	90

* SIGNIFICANTLY MORE DEATHS THAN EXPECTED

\# COUNTIES COMBINED; SEE TECHNICAL NOTES

NON-WHITE FEMALE — CANCER OF THE ORAL CAVITY INCLUDING TONGUE — ICD 141,143-146,148,149

1950-1959 DEATHS	1950-1959 RATE	1960-1969 DEATHS	1960-1969 RATE	1970-1979 DEATHS	1970-1979 RATE	COUNTY AND STATE	PERCENT CHANGE 60S/50S	PERCENT CHANGE 70S/60S	PERCENT CHANGE 70S/50S	PERCENTILE %CHANGE 70S/50S	PERCENTILE RATES 1970S
				1	6.2	GWINNETT GA					97
				1	3.7	HALL GA					94
		1	2.9	1	3.6	HANCOCK GA		22			93
				2	5.2	HOUSTON GA					96
				1*	9.0	IRWIN GA					98
1	3.1	2*	5.2			JEFFERSON GA	68				
1	5.5					JENKINS GA					
		2	3.5			LAURENS GA					
		1*	9.4			LEE GA					
1	5.1			1	4.1	LIBERTY GA			-19	41	94
1*	20.2					LONG GA					
				1	1.3	LOWNDES GA					86
1	2.9	2*	5.1			MACON GA	78				
1*	19.5					MADISON GA					
				2*	13.1	MARION GA					98
1	1.9	2*	4.5			MERIWETHER GA	135				
1	2.3	1	2.3			MITCHELL GA	0				
		1*	10.6			MONTGOMERY GA					
1	4.8			1	5.8	MORGAN GA			23	63	96
1	.8	2	1.1	2	.7	MUSCOGEE#GA	32	-40	-21	38	84
1	3.3					NEWTON GA					
1*	47.1					PAULDING GA					
				1	3.6	POLK GA					93
		1*	16.9			QUITMAN GA					
		1	3.8	1	2.3	RANDOLPH GA		-38			90
6*	3.1	3	1.4	3	1.2	RICHMOND GA	-53	-15	-61	13	85
1*	8.3			1*	5.9	SEMINOLE GA			-28	33	96
		2	4.8			SPALDING GA					
1*	8.3					STEPHENS GA					
		1	5.4	1	6.0	STEWART GA		11			97
		1	1.8	3*	5.0	SUMTER GA		174			96
				1*	8.6	TALIAFERRO GA					98
1*	14.9	1*	7.6			TATTNALL GA	-49				
				1	7.4	TAYLOR GA					97
1	2.4	1	1.4			THOMAS GA	-43				
				1	3.1	TIFT GA					92
		1	4.2			TOOMBS GA					
1*	12.4					TREUTLEN GA					
1	1.3	3*	4.2	1	1.3	TROUP GA	225	-69	0	53	86
				2*	6.3	UPSON GA					97
				1	4.0	WALTON GA					94
		1	2.0	1	1.8	WARE GA		-12			88
				2	4.0	WASHINGTON GA					94
				2*	8.0	WILKES GA					97
		1	3.2	1	2.7	WORTH GA		-17			91
		15	1.0	19	.9	HAWAII HI		-16			
		1	.6	1	.3	HAWAII HI		-46			83
		13	1.2	16	1.0	HONOLULU HI		-15			84
		1	.9	2	1.1	MAUI#HI		17			85
				1	1.3	IDAHO ID					
				1*	27.3	BONNEVILLE ID					99
50	1.6	62	1.2	148*	2.2	ILLINOIS IL	-23	73	34		
1	3.5					ALEXANDER IL					
38	1.4	50	1.2	131*	2.3	COOK IL	-14	86	60	73	90
		1*	19.1			DU PAGE IL					
		1*	546.4			EDWARDS IL					
		1	3.6			KANE IL					
		2	3.9			KANKAKEE IL					
1*	14.6					KNOX IL					
		1	2.4			LAKE IL					
1*	62.9					LA SALLE IL					
				1*	13.0	MC LEAN IL					98
2*	18.7					MACON IL					
		2	3.3	3	4.5	MADISON IL		34			95
				1*	10.6	MARION IL					98
1*	19.2					MORGAN IL					
				2	3.4	PEORIA IL					93
1*	32.6					PERRY IL					
		1*	45.4			RANDOLPH IL					
1	.4			5	1.9	ST CLAIR IL			354	97	89
				1*	23.4	SALINE IL					99
1	3.8	2*	6.5	2	5.1	SANGAMON IL	70	-21	34	68	96
				1	3.6	VERMILION IL					94
2*	14.0	1	1.5	1	1.2	WILL IL	-89	-24	-92	00	85
1*	3.0					WINNEBAGO IL					
12	1.3	25	2.0	30	1.8	INDIANA IN	57	-12	38		
1	4.1			1	1.3	ALLEN IN			-69	09	86
				1	3.5	DELAWARE IN					93
		1*	37.5			FAYETTE IN					
		1*	9.0			FLOYD IN					
				1	4.8	GRANT IN					95
		1*	117.9			JACKSON IN					
3	1.0	8	2.9	13	2.7	LAKE IN	200	-8	176	90	91
				1	4.6	LA PORTE IN					95
1*	7.0					MADISON IN					
5	1.1	11	2.1	7	1.1	MARION IN	87	-50	-6	51	85
				1*	141.3	PARKE IN					99
1*	56.2					POSEY IN					
				3	3.6	ST JOSEPH IN					93
1	1.9	2	3.5	2	3.6	VANDERBURGH IN	82	4	89	79	94
		1	5.8			WAYNE IN					

* SIGNIFICANTLY MORE DEATHS THAN EXPECTED

\# COUNTIES COMBINED; SEE TECHNICAL NOTES

NON-WHITE FEMALE

CANCER OF THE ORAL CAVITY INCLUDING TONGUE — ICD 141,143-146,148,149

1950-1959 DEATHS	1950-1959 RATE	1960-1969 DEATHS	1960-1969 RATE	1970-1979 DEATHS	1970-1979 RATE	COUNTY AND STATE	PERCENT CHANGE 60S/50S	PERCENT CHANGE 70S/60S	PERCENT CHANGE 70S/50S	PERCENTILE %CHANGE 70S/50S	PERCENTILE RATES 1970S
		4	3.1	3	1.9	IOWA IA		-39			
		2*	10.3			BLACK HAWK IA					
				1*	106.0	HENRY IA					99
		1*	16.6			LINN IA					
		1	2.0	1	1.4	POLK IA		-30			87
				1*	17.0	WOODBURY IA					99
1	.2	10	2.0	5	.9	KANSAS KS	795	-57	284		
		1*	8.3			ATCHISON KS					
		1*	14.9			GEARY KS					
		2*	11.9			LEAVENWORTH KS					
				1*	22.3	MIAMI KS					99
1*	15.6					PENO KS					
		1*	17.2	1*	11.9	SALINE KS		-31			98
				1	.9	SEDGWICK KS					84
		2	3.8			SHAWNEE KS					
		3	1.7	2	.9	WYANDOTTE KS		-47			84
19	1.8	28*	2.4	30*	2.5	KENTUCKY KY	39	2	41		
1*	17.2					BALLARD KY					
		1*	9.4			BARREN KY					
		1*	18.2			BELL KY					
		1*	18.2			BOYD KY					
1*	7.7					BOYLE KY					
2	3.6	1	1.8	2	4.3	CHRISTIAN KY	-51	142	19	61	95
		1	5.1			DAVIESS KY					
1	1.0	2	1.8	2	1.6	FAYETTE KY	89	-12	66	74	87
				1*	54.2	HARRISON KY					99
		1*	15.0	1*	14.9	HART KY		0			98
1	4.9	1	2.7			HOPKINS KY	-45				
8	2.2	13*	2.8	17*	3.2	JEFFERSON KY	28	16	47	71	93
1	4.2	2*	9.7			KENTON KY	127				
		1*	10.7			LOGAN KY					
		1*	112.5			LYON KY					
				1	3.4	MC CRACKEN KY					93
				1	4.6	MADISON KY					95
1*	12.3			1*	8.6	MONTGOMERY KY			-30	32	97
1*	9.6			1*	16.1	SCOTT KY			67	75	99
		1	3.4	2*	9.6	WARREN KY		186			98
1*	28.4			1*	23.6	WASHINGTON KY			-17	42	99
1*	8.6	1*	19.5			WEBSTER KY	126				
63	1.6	83	1.8	115*	2.2	LOUISIANA LA	11	18	31		
		1	3.3			ACADIA LA					
				1	2.4	ASCENSION LA					90
		1	3.0			ASSUMPTION LA					
1	2.4			1	1.6	AVOYELLES LA			-34	27	87
1*	6.6			1	5.6	BEAUREGAPD LA			-16	43	96
1	2.8	1	2.0			BIENVILLE LA	-29				
				3*	4.0	BOSSIER LA					94
4	1.5	3	.8	10	2.0	CADDO LA	-44	151	39	69	89
1	1.4	2	1.7	2	1.3	CALCASIEU LA	22	-25	-8	48	86
		2	3.3	2	4.6	CONCORDIA LA		38			95
				1	1.4	DE SOTO LA					86
5	2.3	12*	4.3	11	2.8	EAST BATON ROUGE LA	86	-34	22	63	92
2*	4.6	1	1.8	1	1.1	EAST CARPOLL LA	-60	-40	-76	05	85
		1	3.7	1	3.7	EVANGELINE LA		1			94
1	1.8			1	2.2	FRANKLIN LA			20	62	90
		3*	4.7	1	1.6	IBERIA LA		-66			87
				4*	5.6	IBERVILLE LA					96
1*	4.1					JACKSON LA					
2	2.7	3	3.5	4	2.7	JEFFERSON LA	29	-22	0	53	91
2	3.4	1	1.5	6*	6.2	LAFAYETTE LA	-56	318	83	78	97
				2*	7.6	LAFOURCHE LA					97
1	2.0			1	2.2	LINCOLN LA			6	56	89
		1*	7.1			LIVINGSTON LA					
1	1.9	2	4.5	1	1.4	MADISON LA	137	-70	-28	33	86
		1	1.6			MOREHOUSE LA					
24*	2.7	25*	2.3	33*	2.5	ORLEANS LA	-14	6	-8	49	91
4*	3.4	2	1.2	5	3.0	OUACHITA LA	-65	154	-12	46	92
				1	4.5	PLAQUEMINES LA					95
2	3.6	1	1.8	1	2.5	POINTE COUPEE LA	-49	37	-30	30	91
1	.9	3	1.9	10*	5.6	RAPIDES LA	122	197	559	99	96
1	5.1					RED RIVER LA					
1	3.5			1	1.6	RICHLAND LA			-54	16	88
				1*	7.4	ST BERNARD LA					97
		1	3.8			ST CHARLES LA					
		2*	4.8			ST JAMES LA					
		3*	8.2	1	3.1	ST JOHN THE BAPT LA		-62			93
3	3.3	1	.9	3	2.3	ST LANDRY LA	-74	172	-30	31	90
		2*	5.3			ST MARTIN LA					
		3*	4.3	2	2.7	ST MARY LA		-38			91
				1	1.4	ST TAMMANY LA					87
2*	4.0	3*	6.3			TENSAS LA	58				
				1	2.4	TERREBONNE LA					90
		1	3.5			UNION LA					
1*	8.0					VERMILION LA					
				1	1.8	WASHINGTON LA					88
1	3.2					WEST BATON ROUGE LA					
		1	3.4			WEST FELICIANA LA					
		1*	6.2	2*	6.1	MAINE ME		-2			
		1*	18.6	1*	24.8	PENOBSCOT ME		33			99
				1*	23.3	WASHINGTON ME					99

* SIGNIFICANTLY MORE DEATHS THAN EXPECTED

NON-WHITE FEMALE — CANCER OF THE ORAL CAVITY INCLUDING TONGUE — ICD 141,143-146,148,149

1950-1959 DEATHS	1950-1959 RATE	1960-1969 DEATHS	1960-1969 RATE	1970-1979 DEATHS	1970-1979 RATE	COUNTY AND STATE	PERCENT CHANGE 60S/50S	PERCENT CHANGE 70S/60S	PERCENT CHANGE 70S/50S	PERCENTILE %CHANGE 70S/50S	PERCENTILE RATES 1970S
26	1.7	42	1.7	75*	2.2	MARYLAND MD	4	26	31		
				1*	14.8	ALLEGANY MD					98
				3	2.0	ANNE ARUNDEL MD					89
		1	1.2	3	2.4	BALTIMORE MD		88			90
1*	5.8			1	5.5	CAROLINE MD			-6	51	96
				1*	8.9	CECIL MD					98
1	8.3			1	1.9	CHARLES MD			-77	05	88
		1	2.5	1	1.9	DORCHESTER MD		-23			89
1	5.2			1	3.0	HARFORD MD			-44	21	92
		1	5.4			KENT MD					
		1	.7	5	3.8	MONTGOMERY MD		430			94
		2	1.5	8	2.6	PRINCE GEORGE'S MD		76			91
1*	8.6					QUEEN ANNE'S MD					
1	4.5			2*	5.7	TALBOT MD			26	65	96
				1*	9.3	WASHINGTON MD					98
1	3.2					WICOMICO MD					
1	6.6	2*	5.3			WORCESTER MD	-20				
19	1.6	34*	2.3	47*	2.3	BALTIMORE CITY MD	43	1	44	70	90
6	1.5	11	1.8	23	2.2	MASSACHUSETTS MA	16	26	46		
1*	17.0					BERKSHIRE MA					
		1	4.5			BRISTOL MA					
1*	122.7					DUKES MA					
1	1.7	1	1.8	4	4.1	HAMPDEN MA	6	130	144	90	94
1	3.0			2	1.6	MIDDLESEX MA			-45	20	87
		1	5.0			NORFOLK MA					
2	.6	8	2.2	17*	3.0	SUFFOLK MA	260	37	392	98	92
24	1.4	44	1.3	102*	2.2	MICHIGAN MI	-6	68	58		
				2	2.5	BERRIEN MI					91
		1	6.2	1	2.9	CASS MI		-54			92
3*	4.7	2	2.2	7	3.2	GENESEE MI	-52	44	-31	29	93
1*	13.5					INGHAM MI					
1*	7.4			1	3.6	JACKSON MI			-52	18	94
1*	2.3			1	2.8	KALAMAZOO MI			23	63	92
		1	1.1	2	2.0	KENT MI		80			89
				1*	53.6	MACKINAC MI					99
1*	8.3	1	4.3	1	1.7	MACOMB MI	-47	-62	-80	04	88
		1	1.1	1	1.2	MUSKEGON MI		9			86
1*	26.8					NEWAYGO MI					
		1	1.0	4	2.5	OAKLAND MI		164			91
1	3.0	1	1.7	1	.8	SAGINAW MI	-43	-51	-72	08	84
		1*	7.0			ST CLAIR MI					
		1	1.3	3	4.7	WASHTENAW MI		253			95
15	1.1	34	1.2	77*	2.2	WAYNE MI	14	80	105	82	90
		3	1.7	3	1.4	MINNESOTA MN		-18			
				1*	24.9	ANOKA MN					99
		2	2.4	1	1.0	HENNEPIN MN		-57			84
		1	2.1	1	1.8	RAMSEY MN		-15			88
47	1.3	40	1.0	42	1.0	MISSISSIPPI MS	-20	-4	-24		
3*	4.0	2	2.3	5*	5.2	ADAMS MS	-43	126	29	67	96
		1	2.8			ATTALA MS					
5*	2.6	2	1.7	2	1.1	BOLIVAR MS	-35	-38	-60	13	85
1*	4.6					CALHOUN MS					
		2*	7.0			CLARKE MS					
3	2.4	5*	3.3	1	.6	COAHOMA MS	37	-81	-74	06	84
				1	1.1	COPIAH MS					85
2	4.3					DE SOTO MS					
1	1.8	1	1.2			FORREST MS	-31				
1*	6.6					FRANKLIN MS					
		2*	45.8			GEORGE MS					
		1*	8.8	1*	12.2	GREENE MS		38			98
		1	1.7	1	1.0	HARRISON MS		-44			84
4	1.8	1	.3	6	1.4	HINDS MS	-83	356	-21	39	87
1	1.3					HOLMES MS					
1	1.6	2	4.0			HUMPHREYS MS	144				
1	4.4			1	1.8	JACKSON MS			-59	14	88
		1	2.8			JEFFERSON DAVIS MS					
				3	3.6	JONES MS					94
1	4.2					LAFAYETTE MS					
		2	1.9	1	1.0	LAUDERDALE MS		-45			84
1	2.2	3*	6.1	1	2.2	LEE MS	181	-64	1	54	90
		2	1.7			LEFLORE MS					
1	1.1			1	1.4	LOWNDES MS			29	67	87
				1	1.3	MADISON MS					86
1	3.5					MARION MS					
				1	1.2	MARSHALL MS					86
1	1.4					MONROE MS					
		1	4.7	2*	10.8	NESHOBA MS		131			98
1	3.2					NEWTON MS					
				1	1.8	PANOLA MS					88
1*	4.6	1	4.7	1	3.6	PEARL RIVER MS	3	-24	-22	37	93
		1	1.6			PIKE MS					
1	2.1					QUITMAN MS					
		1	1.8			RANKIN MS					
				2*	5.4	SCOTT MS					96
1	3.9					SHARKEY MS					
				2*	21.7	STONE MS					99
1	.6	3	2.4			SUNFLOWER MS	302				
				2	4.3	TALLAHATCHIE MS					95
1	1.4	2	4.4			TUNICA MS	223				
2*	12.9	1	4.0			WALTHALL MS	-69				
3	2.9			3	2.5	WARREN MS			-12	45	91

* SIGNIFICANTLY MORE DEATHS THAN EXPECTED

NON-WHITE FEMALE — CANCER OF THE ORAL CAVITY INCLUDING TONGUE — ICD 141,143-146,148,149

1950-1959 DEATHS	RATE	1960-1969 DEATHS	RATE	1970-1979 DEATHS	RATE	COUNTY AND STATE	PERCENT CHANGE 60S/50S	70S/60S	70S/50S	PERCENTILE %CHANGE 70S/50S	RATES 1970S
6*	2.8	1	.5	1	.4	WASHINGTON MS	-80	-30	-86	02	83
2*	6.0	1	2.6	1	2.4	WILKINSON MS	-57	-8	-61	12	90
				1	1.1	YAZOO MS					85
27	1.6	45*	2.2	44	1.8	MISSOURI MO	35	-16	13		
		1	6.5			BOONE MO					
				1	7.7	BUCHANAN MO					97
1*	273.2					CASS MO					
1*	29.8					CLAY MO					
		1*	14.9			COOPER MO					
2	.5	15*	3.1	7	1.2	JACKSON MO	498	-61	132	87	85
				1*	7.8	JASPER MO					97
				1*	22.3	JEFFERSON MO					99
				1*	6.8	JOHNSON MO					97
		1*	596.2			LAWRENCE MO					
				2*	17.0	MARION MO					99
1*	4.4			1	3.7	MISSISSIPPI MO			-14	45	94
		1*	24.8			MONTGOMERY MO					
1	1.8	1	2.7			PEMISCOT MO	54				
		1*	37.2			RAY MO					
		1*	112.4			ST CLAIR MO					
		1	.9	3	1.2	ST LOUIS COUNTY MO		35			85
1*	74.3					TANEY MO					
20*	2.3	22	1.9	27	2.3	ST LOUIS CITY MO	-17	22	0	54	90
				3	2.9	MONTANA MT					
				2*	10.8	GLACIER MT					98
				1*	21.2	MISSOULA MT					99
2	1.6			8*	3.7	NEBRASKA NB			133		
2	2.1			8*	5.0	DOUGLAS NB			143	89	96
		1*	5.3			NEW HAMPSHIRE NH					
		1*	60.7			MERRIMACK NH					
16	1.1	32	1.2	92*	2.4	NEW JERSEY NJ	13	94	119		
3	2.1	2	1.3	9*	4.6	ATLANTIC NJ	-39	266	122	84	95
2	5.4	1	.8	2	1.4	BERGEN NJ	-86	83	-74	06	87
		2	3.5	2	1.6	BURLINGTON NJ		-55			87
		5	2.8	15*	5.1	CAMDEN NJ		82			96
		1	4.2			CAPE MAY NJ					
1	3.0					CUMBERLAND NJ					
6	1.0	8	.9	19	1.6	ESSEX NJ	-10	81	63	73	87
				1	1.1	GLOUCESTER NJ					85
		1	.4	9	2.6	HUDSON NJ		498			91
2	2.3	2	1.6	4	1.8	MERCER NJ	-32	16	-21	37	88
		1	1.5	3	2.9	MIDDLESEX NJ		95			92
1	1.1			7	3.3	MONMOUTH NJ			198	92	93
		1	2.9			MORRIS NJ					
		6*	4.7	7	3.6	PASSAIC NJ		-25			93
				2	4.2	SALEM NJ					95
		1	9.6			SOMERSET NJ					
1	.6	1	.4	11*	3.3	UNION NJ	-26	627	440	99	93
				1*	10.5	WARREN NJ					98
		1	.4	3	.4	NEW MEXICO NM		-9			
				3	1.6	BERNALILLO NM					87
		1*	7.1			LEA NM					
63	1.5	134	1.7	246*	1.9	NEW YORK NY	16	13	32		
2*	9.8	3*	5.1	1	1.2	ALBANY NY	-48	-76	-88	01	86
6	1.0	14	1.4	30	1.5	BRONX NY	36	12	51	72	87
1*	19.6					BROOME NY					
				1*	6.5	CHEMUNG NY					97
4	3.0	7	2.1	10	2.1	ERIE NY	-30	0	-31	30	89
				1*	22.9	GREENE NY					99
7	.5	22	1.1	59	1.9	KINGS(BROOKLYN) NY	109	71	257	94	89
		3	2.1	5	2.3	MONROE NY		9			90
4*	3.4	4	1.6	2	.5	NASSAU NY	-54	-66	-84	03	83
27	1.7	62*	2.7	82*	2.8	NEW YORK(MANHATTAN) NY	58	4	65	74	92
				1	3.6	ONEIDA NY					94
		1	1.0	1	1.2	ONONDAGA NY		20			85
1	3.3			1	1.5	ORANGE NY			-54	15	87
8	1.8	8	.9	24	1.2	QUEENS NY	-48	30	-33	28	85
		1	1.7	4*	3.3	RICHMOND NY		101			93
				4*	6.9	ROCKLAND NY					97
				1*	12.4	SARATOGA NY					98
				1	5.7	SCHENECTADY NY					96
		1*	27.1			STEUBEN NY					
2	2.0			3	1.1	SUFFOLK NY			-45	20	85
				1	3.1	ULSTER NY					92
		1*	30.3			WAYNE NY					
1	1.5	7	2.0	14	2.7	WESTCHESTER NY	27	38	76	77	91
34	1.0	41	.9	82	1.4	NORTH CAROLINA NC	-10	58	42		
		1	2.1	2	2.2	ALAMANCE NC		0			89
		2	3.3	1	1.5	ANSON NC		-54			87
				1	1.5	BERTIE NC					87
				1	2.6	BRUNSWICK NC					91
2	4.0	3	3.3			BUNCOMBE NC	-18				
				2*	8.5	BURKE NC					97
1	5.1	1	2.5			CABARRUS NC	-51				
				1	2.3	CASWELL NC					90

* SIGNIFICANTLY MORE DEATHS THAN EXPECTED

NON-WHITE FEMALE — CANCER OF THE ORAL CAVITY INCLUDING TONGUE — ICD 141,143-146,148,149

1950-1959 DEATHS	1950-1959 RATE	1960-1969 DEATHS	1960-1969 RATE	1970-1979 DEATHS	1970-1979 RATE	COUNTY AND STATE	PERCENT CHANGE 60S/50S	PERCENT CHANGE 70S/60S	PERCENT CHANGE 70S/50S	PERCENTILE %CHANGE 70S/50S	PERCENTILE RATES 1970S
				1	2.7	CATAWBA NC					91
1	6.1					CHOWAN NC					
		1	2.1	2	3.0	CLEVELAND NC		47			92
1	2.1	1	1.9			COLUMBUS NC	-6				
1	1.9					CRAVEN NC					
2	1.6	3	2.6	4	2.1	CUMBERLAND NC	61	-20	29	66	89
1	4.0			1	1.9	DAVIDSON NC			-54	15	88
		1	1.7			DUPLIN NC					
		1	.4	4	1.7	DURHAM NC		283			88
1	1.1			1	.8	EDGECOMBE NC			-28	34	84
1	.4	3	1.3	4	1.7	FORSYTH NC	253	32	366	98	88
		2	4.3	1	1.8	FRANKLIN NC		-59			88
		1	2.0			GASTON NC					
				1	1.4	GRANVILLE NC					86
3	1.9	3	1.1	2	.7	GUILFORD NC	-42	-39	-65	10	84
1	1.1			6*	4.8	HALIFAX NC			320	96	95
				1	1.8	HARNETT NC					88
		1	2.7			HERTFORD NC					
				1*	11.5	JACKSON NC					98
		1	2.1			JOHNSTON NC					
				3*	8.9	LEE NC					98
				2	1.8	LENOIR NC					88
		1	2.2			MARTIN NC					
		2	.6	5	1.3	MECKLENBURG NC		104			86
		2	4.4	1	2.3	MOORE NC		-47			90
				1	1.0	NASH NC					84
2	2.3	1	.9	3	2.9	NEW HANOVER NC	-62	229	25	64	92
		2	3.6	1	1.6	NORTHAMPTON NC		-57			87
		1	3.4	2	3.8	ORANGE NC		10			94
1	3.1			1	1.8	PASQUOTANK NC			-41	23	88
1	3.1			1	2.1	PENDER NC			-33	28	89
3	4.8			5	3.5	PITT NC			-27	35	93
1*	13.3					POLK NC					
				1	4.2	RANDOLPH NC					94
1	2.7	1	1.9			RICHMOND NC	-28				
4	3.0	1	.6	3	1.1	ROBESON NC	-79	78	-63	11	85
1	4.2					RUTHERFORD NC					
		1	1.4	2	2.3	SAMPSON NC		64			90
1	3.7			1	1.8	SCOTLAND NC			-52	17	88
				1	3.7	SURRY NC					94
		1*	22.3			SWAIN NC					
		1	2.4	2	4.4	UNION NC		86			95
		1	.5	6	2.3	WAKE NC		397			90
1*	4.8					WASHINGTON NC					
1	1.0	1	.6	2	1.6	WAYNE NC	-40	188	72	76	88
2	2.8			2	2.0	WILSON NC			-29	32	89
		1	1.4	1	1.8	NORTH DAKOTA ND		29			
		1*	5.4			ROLETTE ND					
				1*	53.0	WILLIAMS ND					99
23	1.0	63	1.7	103*	2.1	OHIO OH	62	25	102		
				1*	11.7	BELMONT OH					98
		1	3.8	3*	5.3	BUTLER OH		38			96
		1	1.5			CLARK OH					
1*	33.5					CLINTON OH					
		1*	49.7			CRAWFORD OH					
7	1.1	26*	2.4	35	2.1	CUYAHOGA OH	120	-12	94	79	89
				1	3.5	ERIE OH					93
		7	1.6	11	2.2	FRANKLIN OH		33			90
9	2.6	6	.9	24*	3.1	HAMILTON OH	-66	246	18	59	92
1*	24.6					HIGHLAND OH					
		2*	7.5			JEFFERSON OH					
				1*	8.3	LICKING OH					97
				2	2.2	LORAIN OH					89
2	1.0	3	1.3	9*	3.7	LUCAS OH	23	192	259	95	94
		1	.7	3	1.4	MAHONING OH		100			87
				1*	9.8	MIAMI OH					98
		8	2.9	8	1.8	MONTGOMERY OH		-37			88
		1*	6.7	1	7.0	MUSKINGUM OH		4			97
		1	4.1			RICHLAND OH					
		1*	7.7	1*	8.9	ROSS OH		16			98
3	2.7	1	.4	1	.4	SUMMIT OH	-85	-6	-86	02	83
		1	2.2	1	2.1	TRUMBULL OH		-5			89
		2*	45.9			TUSCARAWAS OH					
14	1.8	6	.5	21	1.3	OKLAHOMA OK	-69	148	-24		
				1*	17.1	ATOKA OK					99
				1	2.6	CARTER OK					91
		1	7.3	1	4.0	CHEROKEE OK		-45			94
1	4.3					CHOCTAW OK					
		2*	9.0			CREEK OK					
				1	4.7	DELAWARE OK					95
1*	15.7					GARVIN OK					
1*	23.6					JOHNSTON OK					
1	3.5			1	2.8	MC CURTAIN OK			-21	39	92
1*	5.4					MC INTOSH OK					
				1*	29.8	MURRAY OK					99
				1	.5	MUSKOGEE OK					83
1	4.1			2*	6.9	OKFUSKEE OK			67	75	97
1	1.1	1	.7	3	1.0	OKLAHOMA OK	-36	39	-11	47	84
				1	4.3	OSAGE OK					95
1*	18.9					PAYNE OK					
				1	5.7	PITTSBURG OK					96
1*	14.8					PONTOTOC OK					
1*	11.2			1	3.0	SEQUOYAH OK			-74	07	92
		2*	22.2			TILLMAN OK					
3	3.3			4	1.6	TULSA OK			-51	18	87

* SIGNIFICANTLY MORE DEATHS THAN EXPECTED

NON-WHITE FEMALE — CANCER OF THE ORAL CAVITY INCLUDING TONGUE — ICD 141,143-146,148,149

1950-1959 DEATHS	1950-1959 RATE	1960-1969 DEATHS	1960-1969 RATE	1970-1979 DEATHS	1970-1979 RATE	COUNTY AND STATE	PERCENT CHANGE 60S/50S	PERCENT CHANGE 70S/60S	PERCENT CHANGE 70S/50S	PERCENTILE %CHANGE 70S/50S	PERCENTILE RATES 1970S
1*	6.2			1	7.0	WAGONER OK			13	56	97
				1	4.2	WASHINGTON OK					95
1	1.3	4	3.5	4	1.4	OREGON OR	182	-59	15		
				1*	18.9	KLAMATH OR					99
1	2.1	3	4.3	3	1.8	MULTNOMAH OR	102	-58	-15	43	88
		1*	34.4			UMATILLA OR					
30	1.0	69	1.5	119*	2.2	PENNSYLVANIA PA	61	42	128		
7	1.1	8	1.1	8	1.0	ALLEGHENY PA	-3	-12	-15	44	84
		2	4.3	1	2.1	BEAVER PA		-52			89
1*	6.1					BERKS PA					
		1*	12.1			BLAIR PA					
		1	2.0			BUCKS PA					
		1*	53.6			BUTLER PA					
				1	5.6	CAMBRIA PA					96
		1	1.2	5*	4.8	CHESTER PA		298			96
1	1.3	1	.9	4	2.7	DAUPHIN PA	-33	204	105	81	91
				6	2.3	DELAWARE PA					90
				1	3.1	ERIE PA					92
		1*	18.8			FRANKLIN PA					
				1*	20.6	LACKAWANNA PA					99
1*	10.1					LAWRENCE PA					
		1*	27.3			LEHIGH PA					
				1	3.7	MERCER PA					94
				1*	141.3	MIFFLIN PA					99
1	1.0			2	1.2	MONTGOMERY PA			19	60	85
				1	5.3	NORTHAMPTON PA					96
				1*	78.1	NORTHUMBERLAND PA					99
19	1.1	51	1.8	84*	2.5	PHILADELPHIA PA	69	35	128	86	91
				1*	37.2	SOMERSET PA					99
		1	2.1			WASHINGTON PA					
				1	3.4	YORK PA					93
1	1.6	1	1.4	4	3.1	RHODE ISLAND RI	-12	122	94		
1*	10.1			1	7.6	NEWPORT RI			-24	36	97
		1	1.8	3	3.0	PROVIDENCE RI		64			92
49*	1.8	53	1.7	58	1.6	SOUTH CAROLINA SC	-6	-6	-12		
1	4.0			1	2.4	ABBEVILLE SC			-39	24	90
1	1.3	2	2.1	3	2.9	AIKEN SC	62	37	123	84	92
1	2.9					ALLENDALE SC					
		2	2.3	1	1.2	ANDERSON SC		-49			85
1	2.4	1	2.9	2	5.7	BAMBERG SC	20	96	134	88	96
1	2.0			1	2.9	BARNWELL SC			43	70	92
4*	6.7	4*	5.9			BEAUFORT SC	-11				
1	1.5	2	3.0	1	1.2	BERKELEY SC	96	-59	-19	40	86
6	2.5	8	2.9	6	1.8	CHARLESTON SC	16	-40	-30	31	88
				1	3.2	CHEROKEE SC					93
1	3.0			1	2.1	CHESTER SC			-31	29	89
		2	4.5			CHESTERFIELD SC					
1	1.3					CLARENDON SC					
				3*	4.4	COLLETON SC					95
1	2.2	2	2.8			DARLINGTON SC	30				
		1	1.9	1	1.9	DILLON SC		-2			88
2*	6.7	1	2.3			DORCHESTER SC	-66				
				1	3.4	EDGEFIELD SC					93
				1	1.6	FAIRFIELD SC					87
3	2.4	1	.9	2	1.3	FLORENCE SC	-62	45	-45	21	86
2	4.2			1	1.3	GEORGETOWN SC			-69	09	86
2	1.6	1	.6	5	2.6	GREENVILLE SC	-62	346	69	76	91
1	1.2	3*	4.7			GREENWOOD SC	303				
1	3.2	2*	5.4	1	2.1	HAMPTON SC	70	-62	-35	27	89
		2	3.4			HORRY SC					
		2*	6.5			JASPER SC					
1	1.6	1	1.5			KERSHAW SC	-10				
		1	2.2			LANCASTER SC					
1	1.3	1	1.4	2	3.0	LAURENS SC	9	115	135	88	92
		1	2.0	1	2.5	LEE SC		27			91
1	2.9	1	2.3			LEXINGTON SC	-20				
1	3.9					MC CORMICK SC					
				2	2.3	MARION SC					90
		1	1.5			MARLBORO SC					
6*	4.5	3	2.0	7*	4.0	ORANGEBURG SC	-55	96	-11	46	94
1*	5.7					PICKENS SC					
2	1.6	7	3.0	10*	3.1	RICHLAND SC	89	4	97	80	92
1	.7			1	.6	SPARTANBURG SC			-17	41	83
1	1.3	1	.8			SUMTER SC	-40				
4*	5.7			1	.9	WILLIAMSBURG SC			-84	03	84
				2	2.0	YORK SC					89
2	3.0					SOUTH DAKOTA SD					
1*	35.3					MELLETTE SD					
1*	5.6					SHANNON SD					
37	1.6	38	1.3	42	1.3	TENNESSEE TN	-15	-1	-16		
		1*	8.4			BRADLEY TN					
		2*	29.6			CHESTER TN					
				1*	42.0	COCKE TN					99
		1*	28.9			COFFEE TN					
2	.5	6	1.4	7	1.6	DAVIDSON TN	200	7	222	93	87
				1	4.8	DYER TN					95
				1	1.7	FAYETTE TN					88
1*	17.5					GREENE TN					
		1*	9.5			HAMBLEN TN					

* SIGNIFICANTLY MORE DEATHS THAN EXPECTED

NON-WHITE FEMALE — CANCER OF THE ORAL CAVITY INCLUDING TONGUE — ICD 141,143-146,148,149

1950-1959 DEATHS	1950-1959 RATE	1960-1969 DEATHS	1960-1969 RATE	1970-1979 DEATHS	1970-1979 RATE	COUNTY AND STATE	PERCENT CHANGE 60S/50S	PERCENT CHANGE 70S/60S	PERCENT CHANGE 70S/50S	PERCENTILE %CHANGE 70S/50S	PERCENTILE RATES 1970S
1	.5	3	1.3	6	2.1	HAMILTON TN	172	63	344	97	89
		1	2.5			HARDEMAN TN					
1*	15.7					HARDIN TN					
1	2.1			1	2.7	HAYWOOD TN			32	68	92
4*	20.4					HENRY TN					
		4	3.1	2	1.4	KNOX TN		-55			87
1*	23.6					MC NAIRY TN					
1	1.5	3	3.1	1	1.1	MADISON TN	112	-65	-26	36	85
1*	6.9					MARSHALL TN					
		1	2.1	1	2.9	MAURY TN		41			92
1	2.1	1	2.5	1	1.9	MONTGOMERY TN	19	-24	-9	48	89
		2*	12.2			OBION TN					
		1*	12.3			ROANE TN					
2*	6.6					RUTHERFORD TN					
17	2.1	10	.9	18	1.4	SHELBY TN	-56	44	-37	25	86
1*	21.0					SMITH TN					
				1	4.5	SUMNER TN					95
				1	1.9	TIPTON TN					89
2*	10.4	1	4.2			WILLIAMSON TN	-59				
1*	6.5					WILSON TN					
53	1.1	67	1.2	78	.9	TEXAS TX	4	-20	-16		
1	3.7	1	1.5			ANDERSON TX	-59				
				1	2.4	ANGELINA TX					90
		1	2.4	1	2.7	BASTROP TX		14			91
3	1.5	1	.4	7	1.4	BEXAR TX	-72	232	-6	52	87
1	1.4	2	2.9	1	1.2	BOWIE TX	103	-60	-20	40	85
1	3.3	1	3.1	2	3.1	BRAZORIA TX	-7	0	-6	50	92
		1	1.9			BRAZOS TX					
				1	2.7	BURLESON TX					92
				1	5.6	CALDWELL TX					96
1*	6.1	1*	6.2	1	4.5	CAMP TX	1	-28	-27	35	95
		1	2.3			CASS TX					
		1	2.1			CHEROKEE TX					
		1*	19.5			COOKE TX					
4	.8	4	.5	9	.9	DALLAS TX	-37	85	17	58	84
				1	2.2	ECTOR TX					89
		1	2.1			ELLIS TX					
		2*	4.6			EL PASO TX					
		1	11.6			FAYETTE TX					
1	2.7					FORT BEND TX					
		1	2.6			FREESTONE TX					
2	1.8	2	1.2	2	1.1	GALVESTON TX	-30	-9	-36	26	85
				1	8.9	GONZALES TX					98
		1	2.9			GRAYSON TX					
		1	1.1	3	3.5	GREGG TX		204			93
1	3.7					GRIMES TX					
1*	8.5	1	5.9			HARDIN TX	-30				
16*	2.0	18	1.5	19	1.1	HARRIS TX	-27	-25	-45	19	85
2	2.0					HARRISON TX					
1*	29.8					HASKELL TX					
1*	5.2					HILL TX					
				1*	12.6	HOCKLEY TX					98
		1	3.1			HOUSTON TX					
				1*	37.2	HUTCHINSON TX					99
2	.9	6	2.4	8	2.7	JEFFERSON TX	156	12	188	91	91
1	2.4	1	2.5			KAUFMAN TX	3				
1	2.0	1	2.1			LAMAR TX	6				
		1	3.3			LEON TX					
		1	1.9			LUBBOCK TX					
		1	.7	1	.7	MC LENNAN TX		-9			84
				1	2.2	MIDLAND TX					90
		1	4.1			MILAM TX					
1	4.1					MONTGOMERY TX					
1	3.1					NACOGDOCHES TX					
				1	2.7	NAVARRO TX					91
				3	2.0	NUECES TX					89
1	4.0					PANOLA TX					
		1	3.1			ROBERTSON TX					
				1	1.4	RUSK TX					87
1*	28.8					SAN PATRICIO TX					
		1	2.9			SHELBY TX					
1	.8	1	1.0			SMITH TX	25				
5	2.2	6	2.1	5	1.3	TARRANT TX	-2	-41	-42	22	86
				1	3.8	TAYLOR TX					94
1*	12.4	1*	6.1			TITUS TX	-51				
				3	1.3	TRAVIS TX					86
				1*	10.0	TRINITY TX					98
		1	4.3			UPSHUR TX					
				1	1.7	VICTORIA TX					88
2*	4.7					WHARTON TX					
		1	1.8	1	.9	UTAH UT		-48			
		1	3.4			SALT LAKE UT					
				1	5.1	WEBER UT					96
30	.9	53	1.5	63	1.4	VIRGINIA VA	63	-1	60		
		1	2.0			ACCOMACK VA					
		1	2.1	3*	4.7	ALBEMARLE#VA		128			95
				1	5.4	AMELIA VA					96
		1*	11.5			APPOMATTOX VA					
		1	1.8			ARLINGTON#VA					
		3	2.7	1	.7	BEDFORD#VA		-73			84
		1*	5.7			CHARLES CITY VA					
6	1.1	7	1.4	10	1.6	CHESTERFIELD#VA	30	12	46	71	87
		1*	7.4			CUMBERLAND VA					
1	.4	4	2.0	2	1.0	DINWIDDIE#VA	370	-52	126	85	84
2	2.7	2	2.0	5	2.4	FAIRFAX#VA	-23	18	-9	47	90

* SIGNIFICANTLY MORE DEATHS THAN EXPECTED

\# COUNTIES COMBINED; SEE TECHNICAL NOTES

NON-WHITE FEMALE

CANCER OF THE ORAL CAVITY INCLUDING TONGUE

ICD 141,143-146,148,149

1950-1959 DEATHS	1950-1959 RATE	1960-1969 DEATHS	1960-1969 RATE	1970-1979 DEATHS	1970-1979 RATE	COUNTY AND STATE	PERCENT CHANGE 60S/50S	PERCENT CHANGE 70S/60S	PERCENT CHANGE 70S/50S	PERCENTILE %CHANGE 70S/50S	PERCENTILE RATES 1970S
				2*	7.5	FAUQUIER VA					97
				1*	12.1	FREDERICK#VA					98
				1	2.3	GREENSVILLE#VA					90
1	1.9					HALIFAX#VA					
		1	2.8			HENRICO VA					
1	3.5	1	2.1	1	1.3	HENRY#VA	-42	-35	-62	11	86
1	3.3					ISLE OF WIGHT VA					
				1*	7.2	KING AND QUEEN VA					97
				1*	12.3	KING GEORGE VA					98
		1*	5.5			LANCASTER VA					
1*	54.5	1*	53.6			LEE VA	-2				
1*	4.4					LOUDOUN VA					
				1*	10.9	MATHEWS VA					98
1	2.1					MECKLENBURG VA					
1*	8.3					MIDDLESEX VA					
				2*	9.9	NORTHUMBERLAND VA					98
2	2.0	3	2.6	1	.7	PITTSYLVANIA#VA	30	-72	-63	10	84
1	.9	3	2.7	3	2.6	ROANOKE#VA	198	-6	180	91	91
				1*	7.0	ROCKBRIDGE#VA					97
		1*	16.6			ROCKINGHAM#VA					
1	5.5					SPOTSYLVANIA#VA					
		1*	8.6			SURRY VA					
		2*	7.4	1	3.3	SUSSEX VA		-56			93
7	1.3	16*	2.2	22*	2.6	CHESAPEAKE CITY#VA	71	19	104	81	91
				1	.6	HAMPTON CITY VA					83
1	.8	1	.9	2	1.0	NEWPORT NEWS CITY VA	16	10	27	66	84
2	1.6					SUFFOLK CITY VA					
4	2.2	2	.4	7	.9	WASHINGTON WA	-80	102	-60		
3	2.8	2	.8	4	1.1	KING WA	-73	44	-61	12	85
				1*	14.0	OKANOGAN WA					98
1	9.9			1	.6	PIERCE WA			-94	00	84
				1	3.3	SNOHOMISH WA					93
11*	2.6	5	1.1	9	2.2	WEST VIRGINIA WV	-57	100	-14		
				1	4.4	CABELL WV					95
				1	2.5	FAYETTE WV					91
1*	11.7					HARRISON WV					
		1*	8.2			JEFFERSON WV					
4*	6.2			3	3.9	KANAWHA WV			-37	26	94
		1	4.3			LOGAN WV					
3*	3.4	1	1.6	2	4.0	MC DOWELL WV	-52	147	19	60	94
		1*	6.1			MARION WV					
		1	2.5	1	4.9	MERCER WV		96			96
1*	29.8					MONONGALIA WV					
1*	7.8					OHIO WV					
				1	1.3	RALEIGH WV					86
1*	29.8					WYOMING WV					
2	1.1	4	1.5	10	1.8	WISCONSIN WI	37	22	68		
2	1.8	3	1.7	9	2.3	MILWAUKEE WI	-4	31	27	65	90
				1*	122.7	POLK WI					99
		1*	172.0			WOOD WI					
		1	8.6			WYOMING WY					
		1*	18.7			FREMONT WY					

* SIGNIFICANTLY MORE DEATHS THAN EXPECTED

COUNTIES COMBINED: SEE TECHNICAL NOTES

WHITE MALE

CANCER OF THE SALIVARY GLAND

ICD 142

1950-1959 DEATHS	1950-1959 RATE	1960-1969 DEATHS	1960-1969 RATE	1970-1979 DEATHS	1970-1979 RATE	COUNTY AND STATE	PERCENT CHANGE 60S/50S	PERCENT CHANGE 70S/60S	PERCENT CHANGE 70S/50S	PERCENTILE %CHANGE 70S/50S	PERCENTILE RATES 1970S
2967	.5	3483	.5	3447	.4	UNITED STATES	0	-14	-14		
51*	.7	53	.5	58	.5	ALABAMA AL	-16	-10	-24		
		1	.4	1	.4	BALDWIN AL		-16			66
		2*	3.8			BARBOUR AL					
1	.9	2*	1.5	1	.6	BLOUNT AL	66	-59	-32	39	76
2*	2.9			1	1.0	BUTLER AL			-66	10	85
5*	2.5	4*	1.7	2	.5	CALHOUN AL	-31	-68	-78	04	74
				2*	1.6	CHAMBERS AL					93
		1	3.5			CHEROKEE AL					
				2*	1.6	CHILTON AL					93
2*	3.0					CLARKE AL					
2*	3.2	2*	3.1	3*	3.7	CLAY AL	-3	17	14	71	98
2*	1.2	3*	1.7			COLBERT AL	35				
1*	1.6	1	1.6			CONECUH AL	2				
1	.8			1	.5	COVINGTON AL			-44	27	70
		1	.4	2	.7	CULLMAN AL		59			77
		1	1.2			DALE AL					
		1	1.3			DALLAS AL					
1	.5			1	.4	DE KALB AL			-29	40	67
1	1.0			1	.9	ELMORE AL			-13	54	83
1	1.8					ESCAMBIA AL					
2	.8	2	.5	6*	1.4	ETOWAH AL	-46	219	71	85	92
		1	.9			FRANKLIN AL					
1	1.1	1	.9			GENEVA AL	-15				
1*	7.3					GREENE AL					
		2*	5.0	2*	4.4	HENRY AL		-13			99
1	.8	2	1.4	1	.4	HOUSTON AL	77	-72	-50	20	67
14*	1.1	4	.2	6	.3	JEFFERSON AL	-81	38	-74	05	64
		1	1.4			LAMAR AL					
		1	.4	3	1.5	LAUDERDALE AL		285			92
		1	.8	2*	1.3	LIMESTONE AL		68			90
1*	14.9					LOWNDES AL					
1*	4.6					MACON AL					
		3	1.2	1	.1	MADISON AL		-89			60
				1	.7	MARION AL					78
1	.4	1	.4	2	.8	MARSHALL AL	10	80	98	88	80
1	.2	4	.5	6	.6	MOBILE AL	223	6	243	98	74
4*	1.2	4*	1.1	3	.9	MONTGOMERY AL	-3	-17	-19	49	84
1	.5	3*	1.7	2	.8	MORGAN AL	244	-55	54	83	80
2*	3.7					PIKE AL					
		1	1.0			ST CLAIR AL					
		2	1.2			TALLADEGA AL					
		1	.8	1	.6	TALLAPOOSA AL		-16			77
1	.3			2	.5	TUSCALOOSA AL			75	86	73
				3	1.1	WALKER AL					87
1	1.4					WINSTON AL					
				2	.3	ALASKA AK					
10	.2	21	.4	33	.4	ARIZONA AZ	48	5	56		
		1	.6	1	.3	COCHISE AZ		-46			64
		1	.8	3*	2.3	GILA AZ		175			96
1*	2.3					GREENLEE AZ					
6	.3	14	.5	18	.4	MARICOPA AZ	54	-22	20	74	66
				1	.8	NAVAJO AZ					82
		1	.1	7	.4	PIMA AZ		371			68
		2	.8	2	.8	PINAL AZ		1			81
		1*	1.9	1*	1.8	SANTA CRUZ AZ		-6			95
3*	1.8	1	.4			YAVAPAI AZ	-79				
29	.4	32	.4	39	.4	ARKANSAS AR	-1	-1	-2		
1	1.7			1	.5	BAXTER AR			-70	08	72
1	1.5	1	1.0	3*	4.5	CARROLL AR	-32	353	209	96	99
		2*	2.5			CLARK AR					
		1	.8	4*	2.7	CLAY AR		243			97
		1	1.5			CLEBURNE AR					
1*	2.7					CLEVELAND AR					
		1	1.0			COLUMBIA AR					
				1	1.2	CONWAY AR					89
2	1.4	4*	2.1	1	.4	CRAIGHEAD AR	48	-81	-71	07	69
				1	.7	CRAWFORD AR					78
1*	1.1			1	.9	CRITTENDEN AR			-18	50	84
				1*	2.8	DALLAS AR					97
1*	1.7			1*	1.3	DREW AR			-21	48	91
		1	.8	1	.5	FAULKNER AR		-35			72
1	.3			1	.2	GARLAND AR			-40	30	61
1	1.0			1	.5	GREENE AR			-47	23	72
1	1.2					HEMPSTEAD AR					
1	.9	1	.9	1	.8	INDEPENDENCE AR	-1	-12	-12	55	82
1	1.0					JACKSON AR					
1	.7	2	1.3	2	.8	JEFFERSON AR	79	-41	6	67	80
		1*	2.6			LAFAYETTE AR					
		2*	1.8			LAWRENCE AR					
				1	.8	MILLER AR					80
		2	1.5			MISSISSIPPI AR					
		1*	2.1			MONTGOMERY AR					
1*	1.8					NEVADA AR					
				1	1.3	PHILLIPS AR					90
		1	2.2			PIKE AR					
1	.8			1	.8	POINSETT AR			0	65	81
1	1.3					POLK AR					
2*	2.1	1	.9	2	1.3	POPE AR	-56	37	-40	30	90
7*	1.2	3	.5	6	.7	PULASKI AR	-58	40	-42	28	78
		1	1.3			SALINE AR					

* SIGNIFICANTLY MORE DEATHS THAN EXPECTED

WHITE MALE — CANCER OF THE SALIVARY GLAND — ICD 142

1950-1959 DEATHS	1950-1959 RATE	1960-1969 DEATHS	1960-1969 RATE	1970-1979 DEATHS	1970-1979 RATE	COUNTY AND STATE	PERCENT CHANGE 60S/50S	PERCENT CHANGE 70S/60S	PERCENT CHANGE 70S/50S	PERCENTILE %CHANGE 70S/50S	PERCENTILE RATES 1970S
				1*	2.6	SEARCY AR					97
1	.3	1	.4	2	.5	SEBASTIAN AR	27	40	77	86	73
1	2.0	1	1.3	1	1.3	SEVIER AR	-35	-4	-37	32	90
		2	1.1			UNION AR					
2	.7	1	.3	2	.4	WASHINGTON AR	-58	35	-44	27	68
		1	.5			WHITE AR					
				2*	1.9	YELL AR					95
231	.4	274	.4	321	.4	CALIFORNIA CA	-5	-5	-9		
16	.5	8	.2	12	.3	ALAMEDA CA	-54	40	-35	36	65
				1	1.4	AMADOR CA					92
1	.2	1	.2			BUTTE CA	4				
3	.3	6	.4	8	.4	CONTRA COSTA CA	63	-14	41	79	66
1	.7					EL DORADO CA					
5	.4	8	.6	7	.4	FRESNO CA	39	-31	-3	63	68
		3	.6	2	.4	HUMBOLDT CA		-42			67
4*	1.5	1	.3	2	.6	IMPERIAL CA	-80	111	-59	15	76
5	.6	8	.8	9	.6	KERN CA	37	-16	15	72	77
		1	.5	1	.4	KINGS CA		-12			69
95	.5	89	.4	105	.4	LOS ANGELES CA	-19	7	-13	55	69
1	.6			3*	1.2	MADERA CA			98	88	89
3	.7	4	.6	2	.3	MARIN CA	-14	-58	-64	12	63
2	.8	1	.4			MENDOCINO CA	-52				
		3	.9	2	.6	MERCED CA		-38			75
2	.3	3	.4	2	.2	MONTEREY CA	38	-56	-39	31	61
2	.5	1	.2	3	.5	NAPA CA	-59	123	-8	60	72
1	1.4	2	1.1			NEVADA CA	-25				
4	.3	20	.6	35*	.6	ORANGE CA	133	0	134	92	76
		4*	1.2	1	.3	PLACER CA		-78			63
1	1.3					PLUMAS CA					
7	.6	7	.4	10	.3	RIVERSIDE CA	-36	-5	-39	31	66
12*	.9	8	.4	12	.6	SACRAMENTO CA	-55	40	-37	33	74
				1	1.0	SAN BENITO CA					86
8	.5	9	.4	12	.4	SAN BERNARDINO CA	-12	-3	-15	53	68
4	.2	15	.4	29	.5	SAN DIEGO CA	113	35	188	95	72
18	.4	20	.6	12	.4	SAN FRANCISCO CA	34	-28	-4	62	69
10*	1.0	6	.5	11*	.8	SAN JOAQUIN CA	-47	51	-19	50	82
1	.4	4	.8	4	.8	SAN LUIS OBISPO CA	127	-6	115	90	81
3	.2	7	.3	4	.2	SAN MATEO CA	52	-42	-12	56	61
2	.3	3	.3	1	.1	SANTA BARBARA CA	7	-75	-73	06	60
4	.2	10	.4	8	.2	SANTA CLARA CA	79	-36	15	72	62
3	.5	1	.1	4	.4	SANTA CRUZ CA	-76	252	-16	52	69
1	.4	1	.3	1	.3	SHASTA CA	-35	11	-27	42	65
1	.7	1	.5	1	.6	SISKIYOU CA	-21	8	-15	52	75
		1	.2	2	.3	SOLANO CA		36			64
5	.8	6	.7	2	.2	SONOMA CA	-10	-72	-75	05	61
1	.1	5	.6	3	.3	STANISLAUS CA	358	-51	126	91	64
		2	1.2			SUTTER CA					
				1	.5	TEHAMA CA					71
2	.3	2	.3	3	.4	TULARE CA	12	14	27	77	66
2	.4	3	.3	5	.4	VENTURA CA	-15	23	4	66	67
1	.4					YOLO CA					
31	.5	29	.4	30	.3	COLORADO CO	-22	-11	-31		
		2	.5	2	.4	ADAMS CO		-21			67
1	.5			1	.2	BOULDER CO			-50	21	63
1*	1.8	1*	1.9			CHAFFEE CO	6				
1*	10.6					CHEYENNE CO					
				1	.7	DELTA CO					79
12	.5	7	.3	6	.2	DENVER-ARAPAHOE#CO	-45	-16	-54	17	62
1	.3	1	.2	5	.6	EL PASO CO	-19	184	130	92	76
1	.7			1	.5	FREMONT CO			-28	41	72
		1	1.4			GARFIELD CO					
		1*	1.5			HUERFANO CO					
1	.4	3	.6	3	.3	JEFFERSON CO	54	-45	-15	53	65
		1*	24.8			LAKE CO					
		1	.8	2*	2.5	LA PLATA CO		204			97
		2	.7	2	.5	LARIMER CO		-31			71
2*	1.5	1	1.0			LAS ANIMAS CO	-33				
1*	3.3	1*	2.4			LINCOLN CO	-26				
				1	.3	MESA CO					66
				1*	2.6	MOFFAT CO					97
		1	1.9			MONTEZUMA CO					
3*	2.6					OTERO CO					
		1*	3.5			PHILLIPS CO					
1	1.6	1	1.7			PROWERS CO	6				
2	.4			1	.2	PUEBLO CO			-51	20	61
		1	2.1			RIO GRANDE CO					
1*	3.4					ROUTT CO					
1*	4.0					SAGUACHE CO					
				1*	4.4	SEDGWICK CO					99
2	.6	3	.9	3	.7	WELD CO	44	-16	22	75	80
43	.5	53	.5	67	.5	CONNECTICUT CT	3	9	12		
10	.5	11	.4	16	.5	FAIRFIELD CT	-21	23	-3	63	72
7	.3	14	.5	13	.4	HARTFORD CT	89	-22	46	80	67
5	.9	6	1.1	8*	1.1	LITCHFIELD CT	27	-3	23	75	88
1	.3	1	.2	1	.2	MIDDLESEX CT	-16	-5	-20	48	62
11	.4	11	.3	24*	.7	NEW HAVEN CT	-17	114	78	86	79
4	.6	5	.6	4	.5	NEW LONDON CT	8	-21	-15	52	71
4*	2.0	1	.4			TOLLAND CT	-81				
1	.5	4*	1.1	1	.3	WINDHAM CT	129	-77	-47	23	63
12*	.9	7	.4	9	.4	DELAWARE DE	-60	19	-52		
1	.5			3*	1.2	KENT DE			114	90	89
8*	1.1	6	.5	4	.3	NEW CASTLE DE	-55	-32	-69	08	65
3	1.1	1	.3	2	.4	SUSSEX DE	-68	29	-59	15	70

* SIGNIFICANTLY MORE DEATHS THAN EXPECTED

COUNTIES COMBINED; SEE TECHNICAL NOTES

WHITE MALE — CANCER OF THE SALIVARY GLAND — COUNTY AND STATE — ICD 142

1950-1959 DEATHS	1950-1959 RATE	1960-1969 DEATHS	1960-1969 RATE	1970-1979 DEATHS	1970-1979 RATE	COUNTY AND STATE	PERCENT CHANGE 60S/50S	PERCENT CHANGE 70S/60S	PERCENT CHANGE 70S/50S	PERCENTILE %CHANGE 70S/50S	PERCENTILE RATES 1970S
13	.7	16*	1.0	5	.4	DIST. OF COLUMBIA DC	47	-57	-37		
13	.7	16*	1.0	5	.4	WASHINGTON DC	47	-57	-37	33	69
63	.4	126	.5	188	.4	FLORIDA FL	20	-10	7		
1	.5	2	1.1	1	.4	ALACHUA FL	109	-67	-32	38	66
1*	6.3					BAKER FL					
1	.7					BAY FL					
1	.1	10	.4	16	.3	BROWARD FL	306	-16	241	97	64
1*	3.9	1*	2.6			CALHOUN FL	-32				
1	1.8			3	.8	CHARLOTTE FL			-56	16	81
				3	.9	CITRUS FL					83
				4	.9	COLLIER FL					85
1	1.7	1	1.4			COLUMBIA FL	-17				
7	.2	16	.3	36	.5	DADE FL	78	63	189	95	72
				1	1.1	DE SOTO FL					87
5	.7	10	.9	8	.6	DUVAL FL	36	-36	-13	54	75
6*	2.2	3	.6	3	.4	ESCAMBIA FL	-71	-38	-82	02	68
		1*	6.0	1*	1.1	FLAGLER FL		-81			88
2*	2.1			2*	2.1	GADSDEN FL			-1	64	96
1*	2.9					GULF FL					
1*	5.2					HAMILTON FL					
1*	2.0			1	1.4	HARDEE FL			-32	38	91
		1	.6			HERNANDO FL					
				3	1.2	HIGHLANDS FL					88
6	.6	13	.8	10	.4	HILLSBOROUGH FL	33	-48	-30	40	69
1	1.4	1	.5	2	.6	INDIAN RIVER FL	-66	24	-58	15	75
		3	1.0	3	.4	LAKE FL		-61			67
1	.5	3	.4	4	.5	LEE FL	-22	23	-4	62	72
1	1.2	1	.8	3*	1.1	LEON FL	-31	36	-6	61	88
		2*	4.9			MADISON FL					
2	.6	4	1.0	3	.2	MANATEE FL	57	-77	-65	11	62
		1	.4	1	.1	MARION FL		-65			60
				1	.8	MARTIN FL					81
				1	.2	MONROE FL					62
		1*	1.6	1	1.7	NASSAU FL		6			94
5	.7	3	.3	9	.6	ORANGE FL	-62	101	-23	45	74
		4*	1.6	1	.5	OSCEOLA FL		-70			71
2	.3	6	.4	5	.1	PALM BEACH FL	62	-69	-49	22	60
		3	1.0	7	.6	PASCO FL		-42			74
5	.2	14	.4	33	.6	PINELLAS FL	98	55	207	96	75
3	.4	7	.8	4	.3	POLK FL	75	-66	-40	30	63
		1	.7	2	1.0	PUTNAM FL		48			86
2*	2.4					ST JOHNS FL					
				2	.5	ST LUCIE FL					72
				1	.7	SANTA ROSA FL					80
		2	.5	4	.2	SARASOTA FL		-61			61
		2	.9	2	.3	SEMINOLE FL		-70			64
		1	1.4	1	.7	SUMTER FL		-50			79
1*	1.7	1	1.7			SUWANNEE FL	-1				
4	.5	7	.6	6	.3	VOLUSIA FL	5	-44	-42	29	64
		1	2.1			WASHINGTON FL					
46	.5	58	.5	63	.5	GEORGIA GA	1	-17	-15		
		2*	3.8	1*	1.7	APPLING GA		-57			94
				1*	4.9	ATKINSON GA					99
2*	1.7	1	1.1			BALDWIN GA	-39				
		1*	3.3			BANKS GA					
		1*	2.4			BEN HILL GA					
1	.3	1	.4	1	.2	BIBB GA	15	-42	-33	38	62
		1*	4.6			BRANTLEY GA					
		2*	4.7			BROOKS GA					
				1*	8.6	BRYAN GA					99
1*	5.8					CAMDEN GA					
1*	8.6					CANDLER GA					
		1	.7			CARROLL GA					
2*	3.5					CATOOSA GA					
2	.6	3	.8	2	.4	CHATHAM GA	39	-53	-35	36	67
				1	1.1	CHATTOOGA GA					87
		1	.8			CHEROKEE GA					
1	1.6			1	1.0	CLARKE GA			-38	32	86
				3*	1.5	CLAYTON GA					93
2	.9	1	.5	5	.8	COBB GA	-44	63	-9	58	82
		1	2.5			COFFEE GA					
				1	2.1	COLUMBIA GA					96
1*	3.3					COOK GA					
1	1.2					COWETA GA					
1*	4.0					CRISP GA					
1*	3.5					DADE GA					
1*	7.4					DAWSON GA					
		4	.9	8	.8	DE KALB GA		-13			81
		1*	1.4			DODGE GA					
1*	3.6					DOOLY GA					
		1	.8			DOUGHERTY GA					
		2*	5.2			EFFINGHAM GA					
				1*	2.0	FAYETTE GA					95
2	1.0	2	.7	1	.3	FLOYD GA	-35	-50	-68	09	65
				1	1.2	FORSYTH GA					90
8	.7	8	.6	1	.1	FULTON GA	-11	-87	-88	00	60
				2*	3.9	GILMER GA					99
1	1.2					GLYNN GA					
1*	4.5					GREENE GA					
		1	.3			GWINNETT GA					
				4*	4.2	HABERSHAM GA					99
		3*	1.8	1	.3	HALL GA		-82			65
				1	1.0	HARALSON GA					86
				1*	2.7	HARRIS GA					97
1*	3.8					HEARD GA					

* SIGNIFICANTLY MORE DEATHS THAN EXPECTED

WHITE MALE CANCER OF THE SALIVARY GLAND ICD 142

1950-1959 DEATHS	1950-1959 RATE	1960-1969 DEATHS	1960-1969 RATE	1970-1979 DEATHS	1970-1979 RATE	COUNTY AND STATE	PERCENT CHANGE 60S/50S	PERCENT CHANGE 70S/60S	PERCENT CHANGE 70S/50S	PERCENTILE %CHANGE 70S/50S	PERCENTILE RATES 1970S
		1	2.0			HENRY GA					
		1	.7			HOUSTON GA					
				2*	2.2	JACKSON GA					96
1*	2.9	1*	2.2			JEFFERSON GA	-24				
				1*	2.2	LAMAR GA					96
1	1.7	1	1.3			LAURENS GA	-23				
				1*	1.5	MADISON GA					93
		1*	3.4	1*	1.9	MITCHELL GA		-43			95
1*	3.6					MONROE GA					
1*	3.7					MORGAN GA					
				2	.5	MUSCOGEE#GA					70
		1	1.6			PAULDING GA					
1*	12.4	1*	6.5	1*	5.7	PULASKI GA	-47	-12	-54	18	99
				1*	2.6	RABUN GA					97
1	.8			1	.2	RICHMOND GA			-73	06	61
				2*	1.7	ROCKDALE GA					94
				2*	5.4	SCREVEN GA					99
		3*	2.9			SPALDING GA					
		2*	3.1	1	.9	STEPHENS GA		-70			85
				1	1.7	SUMTER GA					94
		1	2.0			TATTNALL GA					
				1*	5.6	TAYLOR GA					99
1*	2.6					TELFAIR GA					
				1*	3.4	TERRELL GA					98
2*	2.6			3*	2.8	THOMAS GA			9	69	97
1	1.4					TROUP GA					
				1*	3.5	TURNER GA					98
		1*	2.9			UNION GA					
1	.8	2	.9			WALKER GA	19				
1	2.0					WALTON GA					
1	1.1			1	.8	WARE GA			-27	42	81
1*	2.5					WASHINGTON GA					
1*	1.5	1*	1.5			WAYNE GA	3				
		1*	13.2			WEBSTER GA					
				1*	6.8	WHEELER GA					99
		2	1.1			WHITFIELD GA					
				1*	1.9	WORTH GA					95
		3	.6	10*	1.0	HAWAII HI		78			
		2	.4	10*	1.3	HONOLULU HI		225			90
		1*	5.4			KAUAI HI					
10	.4	10	.3	8	.2	IDAHO ID	-11	-36	-43		
2	.5			1	.2	ADA ID			-66	10	60
				1*	3.2	BEAR LAKE ID					98
		1	1.1	1	.8	BONNER ID		-23			82
2	.8	2	.7	1	.4	CANYON ID	-19	-44	-54	17	66
		1	1.6	1	1.5	CASSIA ID		-7			93
1*	3.7			1*	3.0	ELMORE ID			-19	49	98
1*	1.6					GOODING ID					
		1	1.7			JEROME ID					
1	.9					KOOTENAI ID					
1	1.0	3*	3.2			LATAH ID	236				
1*	2.6					MADISON ID					
				1*	7.8	ONEIDA ID					99
1	1.6					PAYETTE ID					
				1	.4	TWIN FALLS ID					67
		2*	4.3			WASHINGTON ID					
210	.5	218	.5	199	.5	ILLINOIS IL	-7	-10	-16		
2	.5	2	.6	2	.5	ADAMS IL	8	-14	-7	60	71
		2*	1.7			BOND IL					
		1	1.0			BOONE IL					
				2*	3.8	BROWN IL					99
2	.9	2	.9	1	.4	BUREAU IL	2	-56	-55	17	68
		1*	2.4			CALHOUN IL					
				1	.8	CARROLL IL					81
1	.9			1	1.2	CASS IL			28	77	89
2	.6	5*	1.3	1	.2	CHAMPAIGN IL	142	-85	-63	12	61
		1	.5			CHRISTIAN IL					
1	1.0					CLAY IL					
112*	.6	118*	.6	75	.4	COOK IL	-6	-35	-39	31	67
		1	.8	1	.7	CRAWFORD IL		-19			78
				1	1.6	CUMBERLAND IL					93
1	.5	1	.4			DE KALB IL	-16				
2*	2.3					DE WITT IL					
2*	2.1	1	1.1			DOUGLAS IL	-48				
5	.7	5	.5	6	.4	DU PAGE IL	-38	-19	-50	21	67
		1	.7			EDGAR IL					
1	1.1					FAYETTE IL					
2	.7	2	.7	4*	1.6	FRANKLIN IL	5	124	134	92	93
		3	1.3			FULTON IL					
1	.7					GREENE IL					
		1	.9			GRUNDY IL					
				1	.7	HANCOCK IL					78
		1*	3.2			HARDIN IL					
1	.4					HENRY IL					
3*	1.7					IROQUOIS IL					
		3*	1.6	1	.4	JACKSON IL		-72			70
1	1.2					JASPER IL					
1	.6	1	.6	3*	1.5	JEFFERSON IL	-5	168	154	94	93
		1	.8			JO DAVIESS IL					
1*	2.0					JOHNSON IL					
2	.3	4	.5	5	.5	KANE IL	96	8	111	90	74
		2	.5	2	.5	KANKAKEE IL		14			74
				1	.9	KENDALL IL					83
1	.5	4*	1.3	1	.3	KNOX IL	150	-79	-46	24	64
1	.3	5	.4	10*	.8	LAKE IL	22	108	154	94	82

* SIGNIFICANTLY MORE DEATHS THAN EXPECTED

COUNTIES COMBINED; SEE TECHNICAL NOTES

WHITE MALE — CANCER OF THE SALIVARY GLAND — ICD 142

1950-1959 DEATHS	1950-1959 RATE	1960-1969 DEATHS	1960-1969 RATE	1970-1979 DEATHS	1970-1979 RATE	COUNTY AND STATE	PERCENT CHANGE 60S/50S	PERCENT CHANGE 70S/60S	PERCENT CHANGE 70S/50S	PERCENTILE %CHANGE 70S/50S	PERCENTILE RATES 1970S
7*	1.5	3	.5	4	.6	LA SALLE IL	-66	24	-57	16	77
		2*	1.9			LAWRENCE IL					
1	.5					LIVINGSTON IL					
1	.8					LOGAN IL					
3*	1.8	1	.6			MC DONOUGH IL	-68				
3	1.2	4	1.0	3	.6	MC HENRY IL	-16	-42	-52	19	75
4	.9	4	1.0	3	.7	MC LEAN IL	4	-27	-24	45	79
1	.2	1	.2	5	.9	MACON IL	-16	377	301	98	84
				2	.6	MACOUPIN IL					75
1	.1	4	.5	3	.3	MADISON IL	406	-37	217	97	64
		1	.6			MARION IL					
2*	2.5					MARSHALL IL					
2*	2.3					MASON IL					
				1	1.7	MASSAC IL					94
				2*	1.9	MONROE IL					95
		3*	1.4			MONTGOMERY IL					
1	.4	1	.5	1	.6	MORGAN IL	25	17	46	80	74
1	1.7	2*	2.8	1	1.4	MOULTRIE IL	65	-50	-17	51	91
1	.5			1	.5	OGLE IL			-7	60	71
12*	1.6	2	.3	2	.3	PEORIA IL	-81	-6	-83	02	63
1	.9	1	.7			PERRY IL	-15				
1	1.1					PIATT IL					
2	1.3	1	.6	1	1.2	PIKE IL	-51	86	-9	59	89
2	1.1			1	.5	RANDOLPH IL			-55	17	72
1	1.2					RICHLAND IL					
4	.6	3	.5	7*	1.0	ROCK ISLAND IL	-15	113	82	87	87
2	.2	4	.4	3	.4	ST CLAIR IL	78	-10	61	84	66
		1	.5	1	.7	SALINE IL		47			78
2	.5	2	.3	6	.9	SANGAMON IL	-34	173	80	87	83
1	1.6	1	2.1			SCHUYLER IL	26				
1	.5			2*	1.2	SHELBY IL			129	91	90
				1*	1.9	STARK IL					95
1	.4	1	.4	1	.4	STEPHENSON IL	6	-9	-3	63	68
1	.2			4	.7	TAZEWELL IL			357	99	79
				1	1.2	UNION IL					90
		1	.2	3	.7	VERMILION IL		226			77
				1	1.5	WABASH IL					92
2*	2.0					WARREN IL					
				1	.9	WASHINGTON IL					83
				1	.7	WHITE IL					80
				3	1.0	WHITESIDE IL					87
2	.3	2	.2	5	.6	WILL IL	-7	153	135	93	76
		1	.4	4*	1.4	WILLIAMSON IL		258			92
2	.3	4	.4	5	.5	WINNEBAGO IL	33	21	60	84	72
				1	.9	WOODFORD IL					83
86	.5	115*	.6	83	.4	INDIANA IN	24	-34	-18		
9*	1.3	4	.4	3	.3	ALLEN IN	-70	-26	-78	03	63
1	.5					BARTHOLOMEW IN					
		1	1.6			BENTON IN					
3*	4.5	1	1.4	2*	2.8	BLACKFORD IN	-70	109	-37	33	97
1	.7			1	.6	BOONE IN			-12	56	75
1*	3.5			1*	1.8	BROWN IN			-48	23	95
1	.5	2	1.0			CASS IN	95				
1	.5	3*	1.1			CLARK IN	110				
		2	1.7			CLAY IN					
1	.7	1	.7	1	.7	CLINTON IN	-2	-6	-8	59	78
		1	2.6			CRAWFORD IN					
1	.8	1	.7	1	.6	DAVIESS IN	-6	-14	-19	49	77
				1	.7	DEARBORN IN					79
1	.8					DECATUR IN					
2	1.3			1	.6	DE KALB IN			-52	19	77
		4	.9	2	.5	DELAWARE IN		-42			73
		1	.7			DUBOIS IN					
1	.3	5*	1.1	3	.5	ELKHART IN	326	-52	106	89	73
2*	2.0	2*	2.1	1	.9	FAYETTE IN	7	-59	-56	16	83
				2	.8	FLOYD IN					80
2*	2.0					FOUNTAIN IN					
1	.9	1	.8			FULTON IN	-8				
		3*	1.8			GIBSON IN					
		3	.9			GRANT IN					
1	.7	1	.5	1	.7	GREENE IN	-31	36	-6	61	77
		1	.5			HAMILTON IN					
1	.9	1	.8	1	.6	HANCOCK IN	-7	-23	-29	41	77
1	.4	1	.5			HENRY IN	19				
		1	.4			HOWARD IN					
		2	1.1	1	.6	HUNTINGTON IN		-47			75
1	.7					JACKSON IN					
1	1.8	1	1.2			JASPER IN	-35				
3*	2.4					JAY IN					
		4*	3.1			JEFFERSON IN					
				1	.4	JOHNSON IN					70
		1	.4			KOSCIUSKO IN					
2	.2	7	.5	6	.3	LAKE IN	156	-39	57	83	64
2	.5	2	.5	2	.4	LA PORTE IN	-7	-4	-10	57	70
		2	1.1			LAWRENCE IN					
2	.4	2	.3	3	.5	MADISON IN	-18	44	18	73	72
14	.6	20*	.8	10	.4	MARION IN	37	-54	-36	34	67
		2	1.3			MARSHALL IN					
1*	2.3	1*	1.9	1*	1.9	MARTIN IN	-18	-2	-19	50	95
		4*	2.7	1	.6	MIAMI IN		-77			76
1	.5	2	.9	1	.4	MONROE IN	68	-55	-25	44	68
2	1.3			1	.6	MONTGOMERY IN			-50	20	77
1	.8			3*	1.7	MORGAN IN			104	89	94
1	1.9					NEWTON IN					
		2*	1.9	1	1.0	ORANGE IN		-45			87
1	1.1					PERRY IN					
		1	.4	4*	1.3	PORTER IN		204			90
		2*	2.0	1	.9	POSEY IN		-54			84
1	2.1					PULASKI IN					

* SIGNIFICANTLY MORE DEATHS THAN EXPECTED

1950-1959 DEATHS	1950-1959 RATE	1960-1969 DEATHS	1960-1969 RATE	1970-1979 DEATHS	1970-1979 RATE	COUNTY AND STATE	PERCENT CHANGE 60S/50S	PERCENT CHANGE 70S/60S	PERCENT CHANGE 70S/50S	PERCENTILE %CHANGE 70S/50S	PERCENTILE RATES 1970S
1	.8			1	.8	PUTNAM IN			-4	63	81
4*	2.5	2	1.4	1	.6	RANDOLPH IN	-45	-52	-74	05	77
6	.7	3	.3	5	.5	ST JOSEPH IN	-53	44	-33	38	71
		1	.7			SHELBY IN					
1	1.2	1	.9			SPENCER IN	-24				
1	1.3					STEUBEN IN					
3	1.0			3	.8	TIPPECANOE IN			-20	49	82
				1	1.1	TIPTON IN					87
1	.1	2	.3	3	.4	VANDERBURGH IN	109	49	212	97	68
				2*	1.8	VERMILLION IN					94
4	.7	4	.7	3	.6	VIGO IN	-7	-15	-21	47	75
1	.6	2	1.4			WABASH IN	122				
		1*	2.1			WARREN IN					
		2*	1.7	2*	1.3	WARRICK IN		-24			90
		1	.3	2	.6	WAYNE IN		61			74
		1	.8	1	.8	WELLS IN		2			83
				1	.8	WHITE IN					82
				1	.8	WHITLEY IN					82
64	.5	53	.4	65	.4	IOWA IA	-22	18	-8		
		1	1.7			ADAIR IA					
		1*	2.3			ADAMS IA					
		2*	1.8			ALLAMAKEE IA					
		1	2.2			APPANOOSE IA					
		1	1.9			AUDUBON IA					
5*	1.4	4	.8	3	.6	BLACK HAWK IA	-43	-28	-59	15	75
				1	.5	BOONE IA					72
				3*	2.2	BUENA VISTA IA					96
		1	.9			CALHOUN IA					
2*	1.7			1	1.0	CARROLL IA			-43	28	85
		1	.7			CASS IA					
				1	1.0	CEDAR IA					85
2	1.3					CERRO GORDO IA					
1	.8					CHEROKEE IA					
		1	1.3			CHICKASAW IA					
		1	.9			CLAY IA					
1	.7					CLAYTON IA					
		2	.7			CLINTON IA					
		1	1.2	1	1.4	DECATUR IA		14			92
3*	1.3					DES MOINES IA					
3	.9	1	.3	4*	1.1	DUBUQUE IA	-66	265	26	76	88
		1	1.6			EMMET IA					
		1	.7	2	1.2	FAYETTE IA		64			89
				1	.9	FLOYD IA					84
				3*	4.8	FREMONT IA					99
3*	3.8	2*	2.3	1	1.2	GREENE IA	-39	-47	-67	10	90
				1	1.2	GRUNDY IA					89
1	.9			1	1.0	GUTHRIE IA			10	70	86
				1	.9	HAMILTON IA					84
				2*	1.6	HARDIN IA					93
2*	1.8					HARRISON IA					
1	1.3					HOWARD IA					
				1	1.2	IOWA IA					89
		1	.6			JASPER IA					
1	1.4	1	1.0			JEFFERSON IA	-29				
		1	.5			JOHNSON IA					
1	.9	1	1.0			JONES IA	13				
1	1.0					KEOKUK IA					
2*	1.7					KOSSUTH IA					
2	.9					LEE IA					
3	.5			4	.6	LINN IA			10	70	76
1	1.3			1	1.2	LUCAS IA			-7	61	89
		2*	2.6	2*	2.5	LYON IA		-5			97
2	1.3	1	.9			MARION IA	-34				
1	.6			1	.4	MARSHALL IA			-26	43	69
		1	1.4	1	1.4	MILLS IA		1			91
				1	1.0	MITCHELL IA					85
2*	2.3					MONONA IA					
1	1.3					MONTGOMERY IA					
				1*	1.9	OSCEOLA IA					95
				1	.7	PAGE IA					80
				1	1.1	PALO ALTO IA					87
1	1.1	1	.7	1	.9	PLYMOUTH IA	-39	29	-21	47	83
		1	1.2			POCAHONTAS IA					
7	.7	7	.6	9*	.8	POLK IA	-17	27	5	67	81
				1	.2	POTTAWATTAMIE IA					63
2*	2.2			1	.8	POWESHIEK IA			-61	13	82
				1	.9	SAC IA					83
2	.4	3	.6	2	.3	SCOTT IA	54	-46	-17	51	65
				2*	1.5	SIOUX IA					92
				3*	1.3	STORY IA					91
		2*	1.6			TAMA IA					
				1	.9	UNION IA					84
1	1.2	1	1.6			VAN BUREN IA	30				
1	.4			1	.4	WAPELLO IA			-14	54	66
1	.8					WASHINGTON IA					
1	.4			1	.4	WEBSTER IA			8	69	69
1	1.4	1	1.2			WINNEBAGO IA	-14				
		2*	1.7			WINNESHIEK IA					
4	.7	5	1.0	2	.4	WOODBURY IA	45	-63	-46	24	67
1	1.5					WORTH IA					
1	.8					WRIGHT IA					
44	.5	50	.5	47	.4	KANSAS KS	2	-13	-11		
3*	2.2			3*	3.4	ALLEN KS			54	83	98
				1	1.2	ANDERSON KS					89
		2*	1.8	1	.9	ATCHISON KS		-49			84
2*	1.4					BARTON KS					
				1	.7	BOURBON KS					79

* SIGNIFICANTLY MORE DEATHS THAN EXPECTED

WHITE MALE — CANCER OF THE SALIVARY GLAND — ICD 142

1950-1959 DEATHS	1950-1959 RATE	1960-1969 DEATHS	1960-1969 RATE	1970-1979 DEATHS	1970-1979 RATE	COUNTY AND STATE	PERCENT CHANGE 60S/50S	PERCENT CHANGE 70S/60S	PERCENT CHANGE 70S/50S	PERCENTILE %CHANGE 70S/50S	PERCENTILE RATES 1970S
1	1.0					BROWN KS					
2	1.2					BUTLER KS					
1*	1.7	1*	1.6			CHAUTAUQUA KS	-7				
				1	.8	CHEROKEE KS					82
		1	.9			CLOUD KS					
1	1.2	1	1.0			COFFEY KS	-18				
2	.9	3*	1.3	2	1.1	COWLEY KS	58	-19	27	76	88
2	.8	1	.5	1	.5	CRAWFORD KS	-40	7	-36	35	73
1	.7			1	.5	DOUGLAS KS			-20	48	73
				1*	2.1	ELK KS					96
				1	1.0	ELLIS KS					87
		1*	2.0			ELLSWORTH KS					
1	.9					FRANKLIN KS					
		1*	4.3			GOVE KS					
1*	4.2					GRANT KS					
1	1.1					GREENWOOD KS					
		1*	5.6			HAMILTON KS					
				1	.6	HARVEY KS					77
		1*	8.9			HASKELL KS					
1*	6.1					HODGEMAN KS					
		1	1.1			JACKSON KS					
1	1.3	2*	2.3			JEFFERSON KS	85				
		4	.5	4	.5	JOHNSON KS		-6			71
		1	.9	2*	1.2	LABETTE KS		30			88
2	1.1	2	1.0	1	.5	LEAVENWORTH KS	-14	-53	-60	14	71
		1	.6			LYON KS					
				1	.6	MARION KS					76
1	1.1					MARSHALL KS					
		1*	3.6			MEADE KS					
		2*	1.4			MIAMI KS					
		1	1.7			MITCHELL KS					
1	.6	2	.8	1	.4	MONTGOMERY KS	44	-54	-34	37	66
		1	.7			NEOSHO KS					
				2*	3.3	NORTON KS					98
		1	1.6			PAWNEE KS					
		1	2.4			PHILLIPS KS					
				1	1.1	POTTAWATOMIE KS					87
2*	7.0					RAWLINS KS					
2	.8	3	1.1	1	.3	RENO KS	41	-71	-59	15	65
2*	2.4			1	1.3	REPUBLIC KS			-48	23	90
		1	.8	1	.7	RILEY KS		-15			79
		1*	1.9			ROOKS KS					
				1*	2.0	RUSH KS					96
1	1.5					RUSSELL KS					
		4*	2.2	2	.9	SALINE KS		-56			85
				1*	4.2	SCOTT KS					99
7	.8	5	.5	7	.5	SEDGWICK KS	-38	-9	-44	26	71
1	.2	1	.2	1	.2	SHAWNEE KS	-27	-3	-29	41	60
2*	4.2					STAFFORD KS					
2	1.4			1	.5	SUMNER KS			-61	13	73
				1*	3.1	TREGO KS					98
				1	1.0	WASHINGTON KS					86
		1	1.1			WILSON KS					
1	.2	1	.1	3	.5	WYANDOTTE KS	-13	293	242	97	74
38	.3	54	.4	75	.5	KENTUCKY KY	25	24	55		
2*	2.3			1	1.2	ADAIR KY			-47	24	90
1	1.3					ALLEN KY					
		1	2.5			BALLARD KY					
		2	1.4	1	.7	BARREN KY		-51			78
1	.6					BELL KY					
		1	.5	1	1.1	BOONE KY		96			87
		1	.5	1	.4	BOYD KY		-24			66
1	2.0					BOYLE KY					
				1*	2.0	BRACKEN KY					96
				1	1.0	BRECKINRIDGE KY					85
		1	1.5	1	1.5	BULLITT KY		-2			93
		2*	2.7	1	1.1	CALDWELL KY		-58			88
				1	.3	CAMPBELL KY					64
1	.7	1	.6	1	.6	CHRISTIAN KY	-16	4	-12	55	76
				1	1.0	CLARK KY					85
				1	1.2	CLAY KY					89
		1	1.6			CRITTENDEN KY					
1*	2.3					CUMBERLAND KY					
1	.5	1	.4	1	.3	DAVIESS KY	-26	-4	-29	40	66
		3	.5	6*	1.0	FAYETTE KY		91			86
		1	1.1			FLEMING KY					
2	2.1					FLOYD KY					
2*	1.9			2*	1.4	FRANKLIN KY			-28	41	91
1	2.2					FULTON KY					
		1*	2.0			GARRARD KY					
1	1.4					GRANT KY					
		1	.5	3*	1.7	GRAVES KY		213			94
		1	1.2			GRAYSON KY					
				1	1.7	GREEN KY					94
2*	2.4	1	1.4	1	.5	GREENUP KY	-45	-60	-78	04	74
				1	.6	HARLAN KY					77
				3*	3.7	HARRISON KY					98
				1	1.4	HART KY					92
1	.7	1	.7			HENDERSON KY	-4				
				1	1.4	HENRY KY					92
		2	1.0			HOPKINS KY					
4	.3	9	.5	8	.3	JEFFERSON KY	76	-24	33	78	66
1	1.1			1	.9	JOHNSON KY			-21	46	83
		4	.8	6*	1.2	KENTON KY		62			90
		1	.8			KNOX KY					
1	.8	1	1.0	1	.6	LAUREL KY	25	-45	-31	39	74
		1	1.6			LAWRENCE KY					
1	1.4			1	.8	LETCHER KY			-40	30	82
		1	1.4			LINCOLN KY					

* SIGNIFICANTLY MORE DEATHS THAN EXPECTED

WHITE MALE — CANCER OF THE SALIVARY GLAND — ICD 142

1950-1959 DEATHS	1950-1959 RATE	1960-1969 DEATHS	1960-1969 RATE	1970-1979 DEATHS	1970-1979 RATE	COUNTY AND STATE	PERCENT CHANGE 60S/50S	PERCENT CHANGE 70S/60S	PERCENT CHANGE 70S/50S	PERCENTILE %CHANGE 70S/50S	PERCENTILE RATES 1970S
		1*	2.1			LIVINGSTON KY					
				2*	1.6	LOGAN KY					94
				1*	2.6	LYON KY					97
1	.3			1	.4	MC CRACKEN KY			4	67	66
1	.8	2	1.4			MADISON KY	71				
1	1.5			1	1.3	MARION KY			-15	53	90
1	1.2	1	1.1			MARSHALL KY	-9				
1	1.1			1	1.0	MASON KY			-3	63	87
				2*	3.9	MEADE KY					99
				1	1.1	MERCER KY					87
				1	1.3	MONROE KY					91
		2*	3.1			MONTGOMERY KY					
1	1.9					MORGAN KY					
		1	.7	1	.6	MUHLENBERG KY		-10			77
				1*	1.7	OWEN KY					94
				1*	1.5	PENDLETON KY					92
		1	.8			PERRY KY					
1	.4			2	.7	PIKE KY			95	88	79
2	1.8			2	1.0	PULASKI KY			-44	26	86
		1*	5.6			ROBERTSON KY					
				1	1.4	RUSSELL KY					92
				2*	2.5	SCOTT KY					97
1	1.1			3*	2.6	SHELBY KY			136	93	97
		1	1.5	1	1.7	SIMPSON KY		12			94
1*	2.2					TRIGG KY					
		1	1.3			UNION KY					
		2	1.0	1	.5	WARREN KY		-53			71
1*	1.9					WASHINGTON KY					
1	1.3					WAYNE KY					
1	1.3	2	1.4	1	.8	WHITLEY KY	9	-48	-43	28	80
32	.5	59*	.7	49	.5	LOUISIANA LA	40	-31	-4		
		1	.5	1	.8	ACADIA LA		64			81
				1	1.1	ALLEN LA					88
1	2.6			1	.8	ASCENSION LA			-69	08	81
1	1.1					AVOYELLES LA					
				1	1.4	BEAUREGARD LA					91
				1*	1.5	BIENVILLE LA					93
		2*	2.9	1	.7	BOSSIER LA		-78			77
		4	.7	4	.7	CADDO LA		-2			78
1	.7	1	.3	1	.2	CALCASIEU LA	-57	-42	-75	05	61
		1*	2.2			CALDWELL LA					
		1	1.3	1	1.3	DE SOTO LA		-3			90
1	.3	4	.8	1	.3	EAST BATON ROUGE LA	225	-64	16	72	64
1	1.8					EAST FELICIANA LA					
1	.8	1	1.5	1	.9	EVANGELINE LA	75	-38	10	70	84
		1	2.2			GRANT LA					
				1	1.0	IBERIA LA					86
1*	2.0					JACKSON LA					
		6*	1.0	5	.6	JEFFERSON LA		-46			75
1	1.2	1	1.3			JEFFERSON DAVIS LA	5				
1	.6	1	.3			LAFAYETTE LA	-48				
1	1.0	1	1.2	1	.5	LAFOURCHE LA	23	-60	-51	20	72
		1	1.8			LA SALLE LA					
		1	1.9	1	1.0	LINCOLN LA		-44			86
		3*	2.9	1	.5	LIVINGSTON LA		-81			74
		1*	4.2			MADISON LA					
1	1.2	1	.9	1	1.6	MOREHOUSE LA	-23	78	37	78	94
9	.6	17*	1.1	10	.7	ORLEANS LA	82	-37	14	71	79
1	.5	1	.3			OUACHITA LA	-40				
		1*	3.9			PLAQUEMINES LA					
1*	3.7					POINTE COUPEE LA					
		1	.3	2	.6	RAPIDES LA		85			74
2*	3.4					RICHLAND LA					
		1	1.1	1	.7	SABINE LA		-36			78
				1	.6	ST CHARLES LA					75
1*	6.8					ST JOHN THE BAPT LA					
1	.6					ST LANDRY LA					
2*	2.0			2	.6	ST TAMMANY LA			-69	09	76
		1	.4	1	.5	TANGIPAHOA LA		18			71
		1*	12.4			TENSAS LA					
		2	1.3			TERREBONNE LA					
		1	1.2	1	1.3	UNION LA		2			90
				3*	1.5	VERMILION LA					93
				1	1.1	VERNON LA					88
2*	3.0	1	.7	2*	1.4	WASHINGTON LA	-78	117	-52	19	92
2*	3.7			1	1.2	WINN LA			-68	09	89
23	.5	16	.3	12	.2	MAINE ME	-33	-34	-56		
3	.8	2	.5	2	.4	ANDROSCOGGIN ME	-40	-7	-44	27	70
		1	.3	1	.2	AROOSTOOK ME		-4			63
6	.8	6	.8	2	.2	CUMBERLAND ME	-1	-74	-74	06	61
1	1.6	1	.9			FRANKLIN ME	-43				
		1	.5	1	.4	HANCOCK ME		-8			69
5*	1.2			1	.2	KENNEBEC ME			-83	01	62
1	.8			3*	2.0	LINCOLN ME			147	93	96
1	.4					OXFORD ME					
3	.6	4	.8			PENOBSCOT ME	33				
				1	.8	SAGADAHOC ME					83
2	.9	1	.5			SOMERSET ME	-47				
				1	.6	WALDO ME					77
1	.5					WASHINGTON ME					
53	.7	55	.5	54	.4	MARYLAND MD	-17	-27	-39		
2	.7	1	.2			ALLEGANY MD	-67				
1	.4			3	.3	ANNE ARUNDEL MD			-21	46	64
7	.4	8	.4	14	.5	BALTIMORE MD	2	13	15	72	72
1	.4	1	.3	3	.8	CARROLL MD	-22	174	112	90	82

* SIGNIFICANTLY MORE DEATHS THAN EXPECTED

WHITE MALE — CANCER OF THE SALIVARY GLAND — ICD 142

1950-1959 DEATHS	1950-1959 RATE	1960-1969 DEATHS	1960-1969 RATE	1970-1979 DEATHS	1970-1979 RATE	COUNTY AND STATE	PERCENT CHANGE 60S/50S	PERCENT CHANGE 70S/60S	PERCENT CHANGE 70S/50S	PERCENTILE %CHANGE 70S/50S	PERCENTILE RATES 1970S
		1	.9	2	.7	CECIL MD		-23			78
1	1.5			1	.6	CHARLES MD			-62	13	75
				1	.7	DORCHESTER MD					80
		2	.7	3	.8	FREDERICK MD		25			82
		5*	2.0	1	.2	HARFORD MD		-90			61
				2	.6	HOWARD MD					76
1	1.3					KENT MD					
8*	1.8	5	.3	7	.4	MONTGOMERY MD	-81	22	-77	04	69
		6	.6	4	.3	PRINCE GEORGE'S MD		-58			63
		1	1.7			QUEEN ANNE'S MD					
1	1.4	1	1.3			ST MARY'S MD	-6				
2*	2.7	1	1.3			SOMERSET MD	-50				
		1	.9			TALBOT MD					
5*	1.4	2	.5	1	.2	WASHINGTON MD	-62	-65	-87	00	61
1	.6	2	1.2			WICOMICO MD	113				
		1	1.2	1	1.1	WORCESTER MD		-11			87
23*	.8	17	.6	11	.5	BALTIMORE CITY MD	-28	-22	-44	26	70
98	.5	121	.5	113	.4	MASSACHUSETTS MA	10	-12	-4		
2	.6	1	.1	4	.7	BARNSTABLE MA	-79	412	7	69	78
4	.8	5	.7	7*	.9	BERKSHIRE MA	-7	31	22	75	85
6	.3	17*	.9	10	.5	BRISTOL MA	183	-49	43	80	71
				1*	2.4	DUKES MA					97
10	.4	13	.4	8	.3	ESSEX MA	9	-42	-37	33	63
1	.4	3	1.0	2	.6	FRANKLIN MA	154	-41	49	82	74
12	.7	7	.4	7	.3	HAMPDEN MA	-50	-9	-54	18	65
3	.7	2	.4	2	.4	HAMPSHIRE MA	-45	1	-44	26	68
22	.5	23	.4	22	.4	MIDDLESEX MA	-7	-7	-14	53	68
6	.3	13	.6	13	.5	NORFOLK MA	107	-11	84	87	72
1	.2	7	.5	12*	.8	PLYMOUTH MA	221	49	378	99	81
20	.6	13	.4	6	.2	SUFFOLK MA	-35	-44	-64	12	61
11	.4	17	.6	19	.6	WORCESTER MA	54	3	58	83	76
102	.4	136	.4	109	.3	MICHIGAN MI	15	-26	-14		
		1*	3.2			ALCONA MI					
		2	.8	1	.4	ALLEGAN MI		-53			66
				1	.8	ALPENA MI					82
		1	1.1			ANTRIM MI					
1*	2.3	2*	4.4			BARAGA MI	89				
1	.6	1	.5			BARRY MI	-15				
1	.3	2	.5	4	.9	BAY MI	58	85	191	95	84
				1*	1.6	BENZIE MI					93
2	.3	4	.6	3	.4	BERRIEN MI	61	-25	20	74	69
1	.6	1	.7	1	.5	BRANCH MI	1	-20	-19	49	73
2	.4	2	.4	3	.5	CALHOUN MI	4	30	35	78	73
		1	.5			CASS MI					
				1	1.3	CHEBOYGAN MI					90
		1	.9	1	.6	CHIPPEWA MI		-29			76
1	1.9					CLARE MI					
				3*	1.4	CLINTON MI					92
				1*	3.1	CRAWFORD MI					98
		2	1.5	1	.8	DICKINSON MI		-50			80
1	1.3			1	1.2	EMMET MI			-11	57	89
5	.7	5	.5	5	.4	GENESEE MI	-27	-26	-46	24	66
		1	.5	1	.6	GOGEBIC MI		18			76
1	.6	1	.5			GRAND TRAVERSE MI	-6				
1	.5			1	.6	GRATIOT MI			16	72	75
		1	.5			HILLSDALE MI					
		1	.5	1	.4	HOUGHTON MI		-23			67
1	.7	2	1.1	2	.8	HURON MI	55	-23	19	73	82
2	.2	3	.5	3	.4	INGHAM MI	139	-18	95	88	70
1	1.4					IOSCO MI					
				1	.8	IRON MI					83
2*	1.8					ISABELLA MI					
1	.2	4	.8	1	.2	JACKSON MI	278	-76	-9	58	61
3	.6	2	.3	2	.3	KALAMAZOO MI	-42	-11	-48	22	64
4	.3	10	.7	4	.2	KENT MI	129	-66	-21	46	62
2	1.1					LAPEER MI					
1	.1	4	.3	9	.4	MACOMB MI	177	45	302	98	69
1	1.0	1	1.0	1	.9	MANISTEE MI	-2	-6	-7	60	84
2	.8			2	.7	MARQUETTE MI			-13	55	79
				1	.8	MASON MI					80
1	1.2					MECOSTA MI					
1	.7	1	.7	1	.7	MENOMINEE MI	-2	-5	-7	60	79
		1	.8			MIDLAND MI					
1	.3	1	.2			MONROE MI	-32				
1	.7			1	.5	MONTCALM MI			-36	34	71
				1*	2.4	MONTMORENCY MI					97
1	.2	2	.4	2	.4	MUSKEGON MI	87	-5	78	86	67
		1	.7	1	.6	NEWAYGO MI		-15			75
7	.3	12	.4	5	.2	OAKLAND MI	39	-64	-49	21	60
1	1.0					OCEANA MI					
1	1.4					ONTONAGON MI					
2	.5	3	.7			OTTAWA MI	39				
2	.4	3	.5	4	.5	SAGINAW MI	23	13	40	79	73
1	.2	2	.4	3	.5	ST CLAIR MI	49	44	115	90	73
2	.9			2	.7	ST JOSEPH MI			-23	46	80
1	.4	2	1.2	1	.4	SANILAC MI	170	-62	3	66	70
1	.4			1	.4	SHIAWASSEE MI			-5	62	68
3*	1.3					TUSCOLA MI					
		1	.4	2	.8	VAN BUREN MI		138			82
1	.2			1	.1	WASHTENAW MI			-40	30	60
37	.5	52	.5	28	.3	WAYNE MI	14	-36	-27	43	66
1	.9					WEXFORD MI					
50	.3	58	.4	67	.4	MINNESOTA MN	6	5	12		
				1	.8	AITKIN MN					82
				3	.9	ANOKA MN					84

* SIGNIFICANTLY MORE DEATHS THAN EXPECTED

WHITE MALE | CANCER OF THE SALIVARY GLAND | ICD 142

1950-1959 DEATHS	1950-1959 RATE	1960-1969 DEATHS	1960-1969 RATE	1970-1979 DEATHS	1970-1979 RATE	COUNTY AND STATE	PERCENT CHANGE 60S/50S	PERCENT CHANGE 70S/60S	PERCENT CHANGE 70S/50S	PERCENTILE %CHANGE 70S/50S	PERCENTILE RATES 1970S
				1	.9	BECKER MN					85
		1	.7			BELTRAMI MN					
		1	1.5			BENTON MN					
2	1.3	1	.7			BLUE EARTH MN	-49				
1	.7					CARLTON MN					
1	1.0					CASS MN					
1	1.0	1	.9			CHIPPEWA MN	-11				
				2	1.0	CLAY MN					86
		1	.9			COTTONWOOD MN					
1	.6			1	.6	CROW WING MN			3	66	77
				1	.3	DAKOTA MN					63
1	1.0			1	.6	DOUGLAS MN			-42	29	74
		2*	1.5			FARIBAULT MN					
		2	1.4			FILLMORE MN					
				2	1.1	FREEBORN MN					87
				1	.4	GOODHUE MN					69
7	.2	10	.3	19	.5	HENNEPIN MN	46	72	151	93	72
				1	.9	HOUSTON MN					84
		1	1.3	1	.9	ISANTI MN		-31			83
				1	.5	ITASCA MN					71
				1	1.1	JACKSON MN					87
		1	1.8			KANABEC MN					
		1	.6	2	1.1	KANDIYOHI MN		78			87
		1	1.0	1	1.1	LAC QUI PARLE MN		9			88
		1	1.0	1	.7	LE SUEUR MN		-27			79
1	1.6					LINCOLN MN					
1	.7					MC LEOD MN					
1	.7					MARTIN MN					
1	.9	1	.7			MEEKER MN	-19				
1	.8	1	.7			MORRISON MN	-14				
1	.5			1	.6	MOWER MN			14	71	75
1	.8					NICOLLET MN					
		1	.3	1	.3	OLMSTED MN		0			65
1	.4	2	.7	2	1.0	OTTER TAIL MN	90	55	194	96	86
1	1.3	1	1.1			PENNINGTON MN	-16				
1	1.4					PIPESTONE MN					
				1	.4	POLK MN					67
10	.6	7	.4	4	.2	RAMSEY MN	-42	-41	-66	11	62
		1*	3.3			RED LAKE MN					
		2*	1.7			REDWOOD MN					
				1	.9	RENVILLE MN					83
				4*	2.3	RICE MN					96
				1	1.5	ROCK MN					93
				2*	2.3	ROSEAU MN					96
6	.5	6	.5	6	.5	ST LOUIS MN	-1	7	6	67	72
1	1.4					SCOTT MN					
		1	.9			SIBLEY MN					
1	.3	3	.8	1	.3	STEARNS MN	160	-65	-8	59	63
		2*	3.3	1*	1.5	STEVENS MN		-54			93
		2	1.5	1	.7	TODD MN		-52			79
2	.8	1	.4			WASHINGTON MN	-46				
1*	1.9					WILKIN MN					
4*	2.1					WINONA MN					
				1	.5	WRIGHT MN					71
1	1.0	3*	3.2			YELLOW MEDICINE MN	215				
37*	.9	28	.5	28	.4	MISSISSIPPI MS	-48	-12	-54		
1*	1.5					ADAMS MS					
				1	.8	ALCORN MS					81
1	2.5					ATTALA MS					
1*	12.4					BENTON MS					
1	1.4			1	1.3	BOLIVAR MS			-10	57	90
				1	1.2	CALHOUN MS					89
				1*	2.2	CHOCTAW MS					96
		1	1.8			CLARKE MS					
2*	3.2	1	1.6			COPIAH MS	-51				
		1*	2.0			COVINGTON MS					
1*	3.3			2*	2.8	DE SOTO MS			-14	53	97
2*	1.5					FORREST MS					
				1*	2.2	GEORGE MS					96
		1*	2.2			GRENADA MS					
1*	3.4	1	1.6			HANCOCK MS	-53				
2	.7	5*	1.1	3	.7	HARRISON MS	62	-39	0	65	78
1	.5			3	.6	HINDS MS			40	79	77
		1*	3.4			HUMPHREYS MS					
2*	3.7	2	.9			JACKSON MS	-75				
2	1.5	4*	2.1	1	.5	JONES MS	42	-78	-69	08	70
1	2.1					LAFAYETTE MS					
		2*	3.3	1	1.8	LAMAR MS		-46			95
1	.6			1	.4	LAUDERDALE MS			-24	45	69
1	1.6	1	1.4			LEAKE MS	-16				
1	2.9			1	.5	LEE MS			-82	02	73
2*	2.4					LINCOLN MS					
				1	1.2	MADISON MS					89
		2*	5.5			MARSHALL MS					
1	1.0					MONROE MS					
		1	1.4			NEWTON MS					
1	1.5					PANOLA MS					
2*	2.9	1	1.3			PEARL RIVER MS	-55				
		1*	3.7			PERRY MS					
				2*	1.6	PIKE MS					94
				1	1.3	SCOTT MS					91
1	1.3					SIMPSON MS					
1	1.3			1	1.5	SUNFLOWER MS			17	73	92
		1	1.5			TIPPAH MS					
1	1.5					UNION MS					
1*	2.5					WALTHALL MS					
		1	1.2	3*	2.9	WARREN MS		133			98
2*	2.8					WASHINGTON MS					
1*	3.1			1*	2.3	WAYNE MS			-25	44	96

* SIGNIFICANTLY MORE DEATHS THAN EXPECTED

WHITE MALE

CANCER OF THE SALIVARY GLAND

ICD 142

1950-1959 DEATHS	1950-1959 RATE	1960-1969 DEATHS	1960-1969 RATE	1970-1979 DEATHS	1970-1979 RATE	COUNTY AND STATE	PERCENT CHANGE 60S/50S	PERCENT CHANGE 70S/60S	PERCENT CHANGE 70S/50S	PERCENTILE %CHANGE 70S/50S	PERCENTILE RATES 1970S
1*	2.2					WEBSTER MS					
1*	8.3					WILKINSON MS					
				1	1.3	WINSTON MS					90
1*	2.0	1*	1.8			YALOBUSHA MS	-7				
				1	1.5	YAZOO MS					93
89	.5	95	.5	79	.4	MISSOURI MO	-2	-20	-22		
		1	.7			ADAIR MO					
		1	1.4			ANDREW MO					
		2*	3.8			ATCHISON MO					
1	.8	2*	1.6			AUDRAIN MO	107				
				1	1.2	BARRY MO					89
1	.9					BARTON MO					
		1	.6			BATES MO					
1	1.4			2*	1.7	BENTON MO			24	76	94
		1	.5	1	.3	BOONE MO		-36			65
1	.3	1	.2	2	.4	BUCHANAN MO	-31	74	21	74	69
1	.5	1	.5	1	.5	BUTLER MO	-11	15	2	66	73
1	.7	1	.7	1	.6	CALLAWAY MO	1	-11	-10	58	76
				1	.8	CAMDEN MO					81
2	1.1	1	.5			CAPE GIRARDEAU MO	-57				
		1*	3.0			CARTER MO					
		1	.7			CASS MO					
1	1.5	1	1.1			CHRISTIAN MO	-30				
1	2.0					CLARK MO					
1	.4	1	.4	1	.2	CLAY MO	-2	-40	-42	29	62
				2*	2.9	CLINTON MO					98
		1	.5			COLE MO					
		1	1.2			DALLAS MO					
1	2.3			1*	1.3	DE KALB MO			-43	28	90
				3*	4.3	DENT MO					99
				1	1.3	DOUGLAS MO					90
				2	1.0	DUNKLIN MO					86
2	1.1	1	.4	1	.3	FRANKLIN MO	-65	-21	-73	06	64
1	1.3	1	1.2	1	1.1	GENTRY MO	-12	-8	-19	50	87
3	.5	5	.8	5	.7	GREENE MO	58	-19	28	77	77
				1	.9	GRUNDY MO					84
2*	1.9			1	1.1	HARRISON MO			-45	26	87
		1	.6			HENRY MO					
				1	1.6	IRON MO					93
8	.3	8	.3	5	.2	JACKSON MO	-1	-35	-36	35	62
5*	1.2	4	1.0	3	.7	JASPER MO	-21	-31	-45	25	78
1	.5	3	.9	3	.7	JEFFERSON MO	67	-18	37	78	80
		1	.8			JOHNSON MO					
		1	.9			LACLEDE MO					
1	.8	1	.6	2	1.2	LAFAYETTE MO	-27	94	41	79	88
2	1.4	1	.7	1	.5	LAWRENCE MO	-49	-30	-65	11	72
				1	1.4	LEWIS MO					91
1	.6					LINN MO					
		2*	1.8			LIVINGSTON MO					
1	.7					MACON MO					
2	1.3					MARION MO					
1*	2.1					MERCER MO					
1	1.4	2*	1.8			MILLER MO	31				
				1	1.7	MISSISSIPPI MO					94
1	2.4			1	1.1	MONITEAU MO			-53	19	88
		1	1.1			MONROE MO					
1	.8	2*	1.9	1	1.0	NEW MADRID MO	157	-48	34	78	86
2	1.3	2	1.4	2*	1.4	NODAWAY MO	7	2	9	69	92
		1	2.2			OSAGE MO					
1*	2.7					OZARK MO					
		1	.7	1	1.0	PEMISCOT MO		29			85
2	1.1					PETTIS MO					
2*	2.3					PHELPS MO					
1	.8					PIKE MO					
				1	.7	PLATTE MO					78
				1	1.0	POLK MO					85
		1	1.2			PULASKI MO					
				1	.8	RANDOLPH MO					82
				1	1.2	RIPLEY MO					89
1	.7	2	.9			ST CHARLES MO	42				
				1	1.3	STE GENEVIEVE MO					91
8	.4	9	.3	11	.3	ST LOUIS COUNTY MO	-19	1	-18	51	66
				1	.5	SALINE MO					71
		1*	1.9			SCHUYLER MO					
				1	.7	SCOTT MO					78
				1*	2.0	SHANNON MO					96
1	1.1	1	1.2			SHELBY MO	1				
		1	1.4			STONE MO					
		1	1.8			TANEY MO					
1	1.0					TEXAS MO					
2	1.2	2	1.1			VERNON MO	-12				
		1	1.2			WASHINGTON MO					
1	1.9					WAYNE MO					
		1	1.1	1	1.0	WEBSTER MO		-5			86
1	1.4	1	1.2			WRIGHT MO	-13				
21	.6	18	.7	9	.5	ST LOUIS CITY MO	3	-27	-25	44	71
8	.3	15	.5	8	.2	MONTANA MT	59	-51	-22		
				1*	2.9	BLAINE MT					98
		1	1.3	1	1.5	CUSTER MT		10			92
		1	.8	2*	3.0	DEER LODGE MT		267			98
		2*	1.7			FERGUS MT					
1	.7	2	1.3	1	.4	FLATHEAD MT	84	-70	-44	26	68
2*	2.1	1	.9			GALLATIN MT	-58				
1*	5.7			1*	5.2	GRANITE MT			-9	59	99
		1*	7.4			JEFFERSON MT					
		1	.6			LEWIS AND CLARK MT					
1*	2.9					MADISON MT					

* SIGNIFICANTLY MORE DEATHS THAN EXPECTED

WHITE MALE — CANCER OF THE SALIVARY GLAND — ICD 142

1950-1959 DEATHS	1950-1959 RATE	1960-1969 DEATHS	1960-1969 RATE	1970-1979 DEATHS	1970-1979 RATE	COUNTY AND STATE		PERCENT CHANGE 60S/50S	PERCENT CHANGE 70S/60S	PERCENT CHANGE 70S/50S	PERCENTILE %CHANGE 70S/50S	PERCENTILE RATES 1970S
		1	.4			MISSOULA	MT					
		1*	2.0			ROOSEVELT	MT					
1*	5.7					SHERIDAN	MT					
2	.8	2	.8			SILVER BOW	MT	0				
		1*	3.6	1*	2.9	STILLWATER	MT		-18			98
		1	1.8			VALLEY	MT					
				1	.2	YELLOWSTONE	MT					62
37	.6	33	.5	35	.4	NEBRASKA	NB	-18	-1	-19		
2	1.2			1	.8	ADAMS	NB			-38	32	80
				1*	3.5	BROWN	NB					98
1	.8	2	1.2	1	.6	BUFFALO	NB	45	-48	-25	43	76
		4*	6.0	1	1.4	BURT	NB		-77			91
				1	1.5	BUTLER	NB					93
		1	1.4			CEDAR	NB					
		1*	2.5	1*	2.5	CHERRY	NB		0			97
2*	3.0	1	1.5			CLAY	NB	-49				
1	1.7					CUMING	NB					
				1*	1.6	DAWES	NB					93
				1	1.1	DAWSON	NB					87
		1	2.7			DIXON	NB					
1	.7	1	.6	4*	2.0	DODGE	NB	-9	229	200	96	96
8	.6	4	.3	5	.4	DOUGLAS	NB	-51	24	-40	30	66
		1	1.4	1	2.1	FURNAS	NB		52			96
1	.7			1	.7	GAGE	NB			3	66	79
				1*	3.7	GARFIELD	NB					99
1	.6	1	.5	2	.9	HALL	NB	-15	81	53	82	84
1*	3.1					HITCHCOCK	NB					
2*	2.3					HOLT	NB					
				1	1.2	JEFFERSON	NB					89
		2*	4.2			JOHNSON	NB					
				1*	1.8	KEARNEY	NB					95
1*	4.6					KEITH	NB					
2	.4	1	.2	3	.4	LANCASTER	NB	-49	136	20	74	70
		1	.6			LINCOLN	NB					
2	1.3	2	1.1			MADISON	NB	-19				
1*	2.7					MORRILL	NB					
				1*	2.5	NANCE	NB					97
		1	1.7			NUCKOLLS	NB					
		2*	2.0			OTOE	NB					
				1*	2.1	PAWNEE	NB					96
		1	1.3			PHELPS	NB					
				1*	1.6	PIERCE	NB					94
		1	1.0			PLATTE	NB					
1*	2.7					POLK	NB					
3*	4.9					RED WILLOW	NB					
1	1.1	1	.9			SALINE	NB	-20				
1	1.7	2*	2.5	1	1.0	SARPY	NB	48	-61	-42	28	86
2*	1.5	1	.6	1	.6	SCOTTS BLUFF	NB	-60	-8	-63	13	74
2*	3.0			1	1.6	THAYER	NB			-46	25	94
		1	1.4			WASHINGTON	NB					
1*	2.0			1*	1.8	WAYNE	NB			-13	54	95
				1	1.2	YORK	NB					89
7	.8	4	.3	7	.3	NEVADA	NV	-66	-11	-69		
1*	2.4	1*	2.1			CHURCHILL	NV	-12				
1	.2	2	.3	2	.1	CLARK	NV	26	-47	-33	37	60
2*	3.0			1	1.1	ELKO	NV			-64	12	88
		1*	11.2			LINCOLN	NV					
1	.3			3	.4	WASHOE	NV			71	85	70
1*	1.9					WHITE PINE	NV					
1*	8.3			1	1.1	ORMSBY-CARSON CITY	NV			-87	00	87
26*	.9	11	.4	15	.4	NEW HAMPSHIRE	NH	-61	13	-56		
1	.7	3*	2.2	2	1.0	BELKNAP	NH	231	-56	47	81	85
2	1.1			4*	1.7	CHESHIRE	NH			54	83	94
2	1.2					COOS	NH					
1	.4	1	.4			GRAFTON	NH	8				
9*	1.2	4	.5	3	.3	HILLSBOROUGH	NH	-59	-36	-74	06	65
4	1.1	1	.3	1	.2	MERRIMACK	NH	-76	-22	-81	03	62
4	1.0	2	.4	5	.7	ROCKINGHAM	NH	-62	93	-27	42	79
2	.8					STRAFFORD	NH					
1	.6					SULLIVAN	NH					
125*	.6	116	.4	116	.4	NEW JERSEY	NJ	-23	-11	-32		
6	.8	7	.8	5	.5	ATLANTIC	NJ	-3	-35	-37	33	72
12	.5	13	.4	13	.4	BERGEN	NJ	-18	-5	-22	46	66
2	.4	4	.5	8	.8	BURLINGTON	NJ	40	51	111	90	80
13*	.9	5	.3	10	.6	CAMDEN	NJ	-65	93	-33	37	75
				4	.9	CAPE MAY	NJ					84
		1	.2	2	.4	CUMBERLAND	NJ		99			68
27*	.7	17	.4	10	.3	ESSEX	NJ	-34	-29	-53	18	65
2	.5	1	.4	1	.1	GLOUCESTER	NJ	-26	-61	-71	07	60
25*	1.0	6	.2	16	.6	HUDSON	NJ	-76	170	-35	36	76
		1	.4	1	.2	HUNTERDON	NJ		-34			63
3	.4	9	.7	5	.4	MERCER	NJ	94	-44	8	69	68
4	.3	7	.3	5	.3	MIDDLESEX	NJ	4	-22	-19	50	63
5	.4	6	.4	6	.3	MONMOUTH	NJ	11	-31	-23	45	64
4	.6	5	.5	8	.6	MORRIS	NJ	-22	25	-2	64	75
1	.3	4	.4	5	.3	OCEAN	NJ	62	-36	5	67	64
4	.2	14	.7	8	.4	PASSAIC	NJ	190	-44	63	84	68
		1	.5	2	.7	SALEM	NJ		42			79
4	1.1	1	.2	1	.2	SOMERSET	NJ	-83	-6	-84	01	60
1	.4			1	.3	SUSSEX	NJ			-34	36	63
10	.7	12	.6	3	.1	UNION	NJ	-16	-76	-80	03	60
2	.6	2	.5	2	.6	WARREN	NJ	-11	10	-3	64	75

* SIGNIFICANTLY MORE DEATHS THAN EXPECTED

WHITE MALE — CANCER OF THE SALIVARY GLAND — ICD 142

1950-1959 DEATHS	1950-1959 RATE	1960-1969 DEATHS	1960-1969 RATE	1970-1979 DEATHS	1970-1979 RATE	COUNTY AND STATE	PERCENT CHANGE 60S/50S	PERCENT CHANGE 70S/60S	PERCENT CHANGE 70S/50S	PERCENTILE %CHANGE 70S/50S	PERCENTILE RATES 1970S
11	.5	12	.4	14	.4	NEW MEXICO NM	-15	0	-15		
4	.5	5	.7	7	.7	BERNALILLO NM	33	-2	30	77	79
1	.6	1	.6	1	.3	CHAVES NM	-4	-42	-45	25	65
1	1.1			2	1.0	DONA ANA NM			-12	56	86
				1	.7	EDDY NM					77
1	1.4					GPANT NM					
1	1.3	2*	1.8			LEA NM	37				
		1*	1.9			LUNA NM					
3*	9.6					MC KINLEY NM					
		1	1.7			OTERO NM					
		1	.9			RIO ARRIBA NM					
				1	1.2	ROOSEVELT NM					89
				1	.4	SAN JUAN NM					69
				1	.5	SANTA FE NM					71
		1	.8			VALENCIA NM					
316	.5	356	.5	291	.4	NEW YORK NY	5	-22	-18		
5	.5	5	.4	3	.2	ALBANY NY	-24	-35	-51	20	62
1	.4	3*	1.3	1	.5	ALLEGANY NY	210	-61	21	74	73
28	.4	25	.4	24	.5	BPONX NY	6	6	13	71	70
5	.6	8	.9	1	.1	BROOME NY	61	-89	-83	01	60
2	.5	1	.2	4	1.0	CATTARAUGUS NY	-55	336	97	88	85
3	.8	3	.9	2	.5	CAYUGA NY	16	-45	-36	35	72
3	.4	6	.8			CHAUTAUQUA NY	96				
3	.7	1	.2	2	.4	CHEMUNG NY	-69	96	-39	31	69
		1	.4	3*	1.3	CHENANGO NY		196			91
		2	1.0			CLINTON NY					
		1	.3	2	.6	COLUMBIA NY		66			74
2	1.0	2	1.1	3*	1.5	COPTLAND NY	9	36	47	81	93
		1	.4	1	.4	DELAWARE NY		-7			68
2	.3	3	.3	3	.3	DUTCHESS NY	11	-9	1	65	64
19	.5	26	.6	21	.5	ERIE NY	28	-22	0	65	71
		1	.7	1	.5	ESSEX NY		-23			72
1	.4	2	1.1	2	.9	FRANKLIN NY	145	-15	108	89	84
2	.8	4*	1.5	1	.3	FULTON NY	93	-80	-62	13	64
2	.7			2	.8	GENESEE NY			5	67	80
1	.7	1	.5			GREENE NY	-21				
2	.5	1	.3	2	.6	HERKIMER NY	-50	136	18	73	77
1	.2	1	.2	2	.5	JEFFERSON NY	21	82	120	91	70
47	.4	27	.3	32	.3	KINGS(BROOKLYN) NY	-39	18	-28	42	65
		1	1.0			LEWIS NY					
		1	.4	1	.4	LIVINGSTON NY		-11			68
		1	.4	2	.8	MADISON NY		99			81
13	.6	10	.3	7	.2	MONROE NY	-42	-27	-58	16	63
2	.6	4*	1.1	1	.3	MONTGOMERY NY	103	-75	-50	21	64
12	.3	12	.3	21	.4	NASSAU NY	-3	14	10	70	66
39	.5	68*	.9	24	.4	NEW YORK(MANHATTAN) NY	77	-56	-21	47	68
3	.3	5	.6	4	.4	NIAGARA NY	101	-35	31	77	67
7	.6	3	.3	3	.2	ONEIDA NY	-57	-6	-59	14	63
9	.5	7	.4	5	.3	ONONDAGA NY	-20	-37	-50	21	63
1	.3	1	.3	5*	1.2	ONTARIO NY	-14	341	280	98	90
4	.5	7	.8	3	.3	OPANGE NY	62	-61	-36	35	64
1	.5			1	.6	ORLEANS NY			24	76	77
3	.7			1	.2	OSWEGO NY			-72	07	62
2	.7					OTSEGO NY					
1	1.7			1	.4	PUTNAM NY			-77	04	67
27	.4	34	.4	39	.4	QUEENS NY	-4	11	6	68	69
5	.7	5	.7	1	.1	RENSSELAER NY	2	-81	-80	03	60
3	.4	5	.4	5	.4	RICHMOND NY	25	-21	-1	64	66
		1	.2	2	.2	ROCKLAND NY		28			62
4	.9	2	.4	2	.4	ST LAWRENCE NY	-52	3	-50	20	69
1	.3			2	.3	SARATOGA NY			10	70	65
2	.3	5	.6	4	.5	SCHENECTADY NY	134	-17	95	88	72
2	1.5	2	1.4			SCHOHARIE NY	-11				
2	1.1	2	1.0			SENECA NY	-8				
1	.3	2	.4	4	.8	STEUBEN NY	23	103	149	93	82
8	.3	10	.3	18	.4	SUFFOLK NY	-6	29	22	74	69
1	.3	2	.8	1	.2	SULLIVAN NY	169	-74	-30	40	62
1	.7	5*	3.0	1	.5	TIOGA NY	348	-82	-20	48	73
1	.5	1	.4	2	.7	TOMPKINS NY	-16	76	47	81	79
		8*	1.5	1	.1	ULSTER NY		-91			60
1	.4	1	.4	2	.8	WARREN NY	0	90	90	87	81
2	.9	4*	1.6			WASHINGTON NY	78				
3	1.0	1	.3	6*	1.7	WAYNE NY	-72	519	71	86	94
24*	.7	21	.6	8	.2	WESTCHESTER NY	-15	-67	-72	07	62
1	.5			1	.5	WYOMING NY			2	65	74
1	.9			1	.9	YATES NY			-3	63	83
60	.6	62	.5	74	.4	NORTH CAROLINA NC	-17	-7	-23		
		2	.7	2	.7	ALAMANCE NC		3			79
				1	1.1	ALEXANDER NC					87
1	1.0					ASHE NC					
				1	.8	BEAUFORT NC					81
2*	3.9			1*	2.0	BERTIE NC			-50	21	95
		1	1.3	2*	2.6	BLADEN NC		99			97
		1	1.5			BRUNSWICK NC					
6*	1.3	2	.4	2	.3	BUNCOMBE NC	-69	-22	-76	04	64
				1	.4	BURKE NC					68
2	.6			1	.7	CABARRUS NC			10	70	78
1	1.0			1	.4	CALDWELL NC			-60	14	69
		2*	1.8	1	.7	CARTERET NC		-63			77
1	.7	1	.4	6*	1.6	CATAWBA NC	-37	271	135	93	94
1	1.1	1	.4			CLEVELAND NC	-61				
1	.9	2*	1.6	3*	2.4	COLUMBUS NC	89	48	180	95	97
		1	.3	1	.2	CUMBERLAND NC		-29			61
				2*	5.7	CURRITUCK NC					99
		1*	2.5			DARE NC					

* SIGNIFICANTLY MORE DEATHS THAN EXPECTED

WHITE MALE — CANCER OF THE SALIVARY GLAND — ICD 142

1950-1959 DEATHS	1950-1959 RATE	1960-1969 DEATHS	1960-1969 RATE	1970-1979 DEATHS	1970-1979 RATE	COUNTY AND STATE		PERCENT CHANGE 60S/50S	PERCENT CHANGE 70S/60S	PERCENT CHANGE 70S/50S	PERCENTILE %CHANGE 70S/50S	PERCENTILE RATES 1970S
1	.4	1	.3			DAVIDSON	NC	-26				
				2*	1.9	DUPLIN	NC					95
1	.5	1	.7	1	.3	DURHAM	NC	41	-58	-41	29	64
1	1.1	1	1.6			EDGECOMBE	NC	45				
6*	1.8	2	.5	1	.1	FORSYTH	NC	-73	-77	-94	00	60
1	1.5			1	1.2	FRANKLIN	NC			-17	52	89
		1	.4			GASTON	NC					
				1*	2.7	GRAHAM	NC					97
1	.2	1	.1	9*	1.0	GUILFORD	NC	-30	716	475	99	65
		2*	1.7			HALIFAX	NC					
				1	.5	HARNETT	NC					73
1	.6			1	.6	HAYWOOD	NC			-8	59	74
2	1.3	1	.5	1	.3	HENDERSON	NC	-62	-36	-76	04	65
3*	1.5	1	.3			IREDELL	NC	-77				
		2	.9	2	1.1	JOHNSTON	NC		22			87
				1*	2.7	JONES	NC					97
				2*	1.4	LENOIR	NC					92
1	1.1	1	.8			LINCOLN	NC	-24				
				1	.6	MC DOWELL	NC					76
1	1.9	1	1.8	1	1.3	MACON	NC	-5	-24	-28	42	91
1	1.7					MADISON	NC					
3	.4	6	.9	4	.4	MECKLENBURG	NC	162	-57	13	71	68
				1	1.2	MITCHELL	NC					90
1	2.1					MONTGOMERY	NC					
		1	.9			MOORE	NC					
		3*	2.9	1	.6	NASH	NC		-78			77
1	.9	1	.5	3	1.2	NEW HANOVER	NC	-46	162	41	80	90
		2*	1.3	2*	1.6	ONSLOW	NC		26			93
		2*	1.7			ORANGE	NC					
1*	3.2					PAMLICO	NC					
1	2.3					PASQUOTANK	NC					
1*	2.2	1*	2.0			PENDER	NC	-10				
		1*	4.2			PERQUIMANS	NC					
1	1.3					PITT	NC					
1*	1.9					POLK	NC					
2	1.0					RANDOLPH	NC					
		1	.8			RICHMOND	NC					
1	.7					ROBESON	NC					
		2	.7	3	1.0	ROCKINGHAM	NC		36			86
1	.5	1	.4	1	.2	ROWAN	NC	-16	-48	-56	16	61
1	.6	1	1.0	1	.9	SAMPSON	NC	57	-6	49	82	84
				1	1.1	SCOTLAND	NC					88
				1	.6	STANLY	NC					77
		1	1.0			STOKES	NC					
2	1.4	3*	1.3			SURRY	NC	-9				
3*	6.5			1	.8	TRANSYLVANIA	NC			-88	00	81
1	1.0	2	1.2	1	.6	UNION	NC	15	-53	-46	25	74
2*	2.6					VANCE	NC					
1	.3	2	.5	2	.2	WAKE	NC	79	-61	-31	39	61
				1	.9	WATAUGA	NC					85
1	.5	1	1.0			WAYNE	NC	77				
		1	.6			WILKES	NC					
				1	.7	WILSON	NC					80
		1	1.6			YANCEY	NC					
9	.3	15	.5	11	.4	NORTH DAKOTA	ND	45	-28	4		
		2*	4.2	1*	3.4	BENSON	ND		-18			98
2*	3.8					BOTTINEAU	ND					
				1*	6.5	BOWMAN	ND					99
1	.3	1	.4			CASS	ND	20				
		1*	2.2			DICKEY	ND					
		1*	2.3			DIVIDE	ND					
		2	1.0	2	1.0	GRAND FORKS	ND		-1			85
1*	2.2					LA MOURE	ND					
				1	1.4	MC LEAN	ND					91
		1	.8			MORTON	ND					
		1	2.1			MOUNTRAIL	ND					
				1	1.0	PEMBINA	ND					86
2*	2.7			1	1.4	RAMSEY	ND			-49	22	91
				1*	2.3	RANSOM	ND					96
		1*	4.4			RENVILLE	ND					
				1	1.2	RICHLAND	ND					89
		1*	18.9			SLOPE	ND					
		2*	2.1	1	1.0	STARK	ND		-51			86
2*	1.6					STUTSMAN	ND					
		1	1.5			TRAILL	ND					
1	.6	1	.5			WARD	ND	-22				
				1	1.2	WILLIAMS	ND					89
191	.5	209	.5	172	.4	OHIO	OH	0	-25	-25		
2*	2.0					ADAMS	OH					
3	.7	5*	1.1	1	.2	ALLEN	OH	62	-79	-66	10	62
1	.6			3*	1.6	ASHLAND	OH			173	94	93
4	1.1	3	.7			ASHTABULA	OH	-36				
1	.4	1	.5			ATHENS	OH	20				
1	.6					AUGLAIZE	OH					
8*	1.8	2	.4	2	.5	BELMONT	OH	-78	21	-74	06	71
1	.6	1	.7	1	.8	BROWN	OH	1	16	18	73	80
4	.7	1	.1	3	.3	BUTLER	OH	-84	180	-54	18	65
		1	.9	1	.8	CHAMPAIGN	OH		-14			80
6*	1.2	3	.5	1	.2	CLARK	OH	-59	-66	-86	00	61
		3	1.1			CLERMONT	OH					
2	1.3					CLINTON	OH					
3	.6	6*	1.2	2	.4	COLUMBIANA	OH	114	-70	-36	34	66
1	.5					COSHOCTON	OH					
36	.6	29	.5	28	.4	CUYAHOGA	OH	-27	-8	-32	38	70
2	.9			2	.7	DARKE	OH			-17	51	79
1	.8	1	.7			DEFIANCE	OH	-12				
				1	.5	DELAWARE	OH					74

* SIGNIFICANTLY MORE DEATHS THAN EXPECTED

WHITE MALE

CANCER OF THE SALIVARY GLAND — ICD 142

1950-1959 DEATHS	1950-1959 RATE	1960-1969 DEATHS	1960-1969 RATE	1970-1979 DEATHS	1970-1979 RATE	COUNTY AND STATE	PERCENT CHANGE 60S/50S	PERCENT CHANGE 70S/60S	PERCENT CHANGE 70S/50S	PERCENTILE %CHANGE 70S/50S	PERCENTILE RATES 1970S
3	1.0	3	1.0	2	.5	ERIE OH	-5	-43	-46	24	73
2	.8	2	.7	4*	1.0	FAIRFIELD OH	-13	58	37	79	86
		2*	1.5			FAYETTE OH					
13	.6	13	.6	10	.4	FRANKLIN OH	-1	-37	-38	32	68
1	.7			1	.6	FULTON OH			-21	47	74
1	.6	1	.3	2	.4	GREENE OH	-49	35	-32	38	69
		2	.8			GUERNSEY OH					
16	.5	17	.5	13	.4	HAMILTON OH	-2	-27	-29	41	67
1	.5	2	.7	3	1.1	HANCOCK OH	58	46	130	92	88
				2*	1.5	HARDIN OH					92
		1	.9			HENRY OH					
		2	1.3	1	.5	HIGHLAND OH		-61			72
		1	1.0			HOLMES OH					
		1	.5	1	.4	HURON OH		-7			69
2	.6	5*	1.1	1	.2	JEFFERSON OH	75	-81	-67	10	61
		2	1.0	3*	1.3	KNOX OH		30			90
2	.6			1	.2	LAKE OH			-74	05	60
1	.4	1	.4	1	.3	LAWRENCE OH	-3	-23	-26	43	65
2	.5	3	.7	2	.3	LICKING OH	29	-50	-35	36	66
				1	.5	LOGAN OH					73
2	.3	5	.8	9*	.9	LORAIN OH	157	16	197	96	84
11	.7	12	.6	11	.6	LUCAS OH	-13	-1	-14	53	75
		1	1.0	1	.9	MADISON OH		-9			84
5	.5	5	.4	8	.6	MAHONING OH	-23	51	16	73	75
3	1.1	1	.4			MARION OH	-66				
3*	1.4					MEDINA OH					
1	.7			1	.7	MEIGS OH			6	68	80
				2	1.2	MERCER OH					89
		3	.9	1	.2	MIAMI OH		-74			62
				1	1.2	MONROE OH					89
12	.7	6	.3	7	.3	MONTGOMERY OH	-51	7	-48	23	66
		2*	2.1			MORROW OH					
2	.6	4	1.1	2	.5	MUSKINGUM OH	86	-51	-8	59	73
				1	1.4	NOBLE OH					91
2	1.6					OTTAWA OH					
1	.7			2	1.0	PICKAWAY OH			51	82	86
		1	.3	2	.6	PORTAGE OH		85			75
		1	.6			PREBLE OH					
		1	.3	1	.3	ROSS OH		4			65
		2	.8	1	.4	SANDUSKY OH		-49			67
3	.8	4	1.1	1	.3	SCIOTO OH	37	-77	-68	09	63
1	.3	6*	2.2	4*	1.5	SENECA OH	551	-33	334	99	92
5	.4	16*	1.1	5	.3	STARK OH	167	-72	-25	44	65
9	.5	12	.6	5	.2	SUMMIT OH	27	-66	-57	16	62
3	.4	1	.1	3	.3	TRUMBULL OH	-76	233	-19	50	64
2	.5	3	.7			TUSCARAWAS OH	47				
		1	.8	1	1.0	UNION OH		21			85
1	.7	3*	2.0			VAN WERT OH	165				
1	.6			1	.4	WARREN OH			-25	44	70
		2	.7	3	1.1	WASHINGTON OH		52			88
				3	.8	WAYNE OH					80
2	1.3					WILLIAMS OH					
1	.4	3	.9	3	.8	WOOD OH	146	-4	135	92	82
1	.8					WYANDOT OH					
46	.5	57	.5	61	.5	OKLAHOMA OK	8	-3	5		
		1	2.0	1	1.5	ATOKA OK		-22			93
1*	2.2	1*	2.4			BEAVER OK	12				
1	1.0					BECKHAM OK					
1	1.5	2*	2.4			BLAINE OK	64				
				1	.6	CADDO OK					76
2*	1.8			1	.3	CANADIAN OK			-82	02	65
4*	2.2			3*	1.5	CARTER OK			-33	37	93
1	1.6	1	.9	1	.8	CHEROKEE OK	-43	-12	-49	22	81
1*	2.5	1*	2.6			COAL OK	1				
1	.8			1	.4	COMANCHE OK			-43	28	70
		1*	2.4			COTTON OK					
				2	.8	CREEK OK					82
				2*	1.7	CUSTER OK					94
				1	1.5	DELAWARE OK					92
1*	2.3					DEWEY OK					
1	.4	2	.7			GARFIELD OK	69				
2*	1.4			3*	1.7	GARVIN OK			25	76	94
				2	1.1	GRADY OK					87
		1	1.7			GRANT OK					
		1	1.3			GREER OK					
1	1.0			1	.8	JACKSON OK			-21	47	81
1	1.5	2*	3.3			JEFFERSON OK	122				
1	1.7					JOHNSTON OK					
		2	.8			KAY OK					
				1	.9	KIOWA OK					83
				1*	1.8	LATIMER OK					95
		1	.5	1	.7	LE FLORE OK		41			78
		1	1.2			LOGAN OK					
		1	1.3	1	1.4	MC CLAIN OK		9			92
1	.9					MC CURTAIN OK					
				1	1.1	MC INTOSH OK					88
		1*	2.1			MARSHALL OK					
				4*	2.9	MAYES OK					97
1	1.2			1	1.1	MURRAY OK			-12	56	88
		2	.8			MUSKOGEE OK					
		1	1.5			NOBLE OK					
		1	2.5	1*	1.7	OKFUSKEE OK		-30			94
8	.7	9	.6	10	.6	OKLAHOMA OK	-11	-3	-14	54	76
1	.5			2	1.1	OKMULGEE OK			137	93	88
2	1.6	1	.6			OSAGE OK	-63				
2	1.5					OTTAWA OK					
		1	1.1	1	1.1	PAWNEE OK		0			87
1	.6	1	.6	2	.9	PAYNE OK	-5	56	47	81	84
		2	1.1			PITTSBURG OK					

* SIGNIFICANTLY MORE DEATHS THAN EXPECTED

WHITE MALE — CANCER OF THE SALIVARY GLAND — ICD 142

1950-1959 DEATHS	1950-1959 RATE	1960-1969 DEATHS	1960-1969 RATE	1970-1979 DEATHS	1970-1979 RATE	COUNTY AND STATE	PERCENT CHANGE 60S/50S	PERCENT CHANGE 70S/60S	PERCENT CHANGE 70S/50S	PERCENTILE %CHANGE 70S/50S	PERCENTILE RATES 1970S
				2	1.0	PONTOTOC OK					86
1	.4	1	.6	1	.5	POTTAWATOMIE OK	26	-15	7	68	71
		1	1.7			PUSHMATAHA OK					
				1*	5.3	ROGER MILLS OK					99
1	1.0	1	.9	2*	1.6	ROGERS OK	-17	83	51	82	93
		1	.7	1	.7	SEMINOLE OK		4			79
		1	1.2			SEQUOYAH OK					
		1	1.3			TEXAS OK					
1	1.2	1	1.2			TILLMAN OK	6				
6	.7	10	.8	8	.5	TULSA OK	17	-32	-20	48	73
		1	1.1			WAGONER OK					
		1	.7	1	.4	WASHINGTON OK		-41			68
2*	2.6	1	1.4			WOODS OK	-46				
27	.3	33	.4	33	.3	OREGON OR	9	-20	-13		
1	.8					BENTON OR					
2	.4	2	.4	1	.1	CLACKAMAS OR	0	-73	-73	06	60
		1	.6			CLATSOP OR					
1	.8			2	1.3	COLUMBIA OR			68	85	91
		2	1.0	1	.3	COOS OR		-67			65
				2	1.2	DESCHUTES OR					89
1	.4			1	.3	DOUGLAS OR			-23	45	63
				1*	2.4	GRANT OR					97
1	1.6			1	1.3	HOOD RIVER OR			-20	49	91
3	.8	3	.7	4	.7	JACKSON OR	-21	0	-21	46	78
				3	1.2	JOSEPHINE OR					89
2	.3	2	.3	2	.2	LANE OR	-14	-26	-36	34	61
		2	1.7			LINCOLN OR					
		1	.5			LINN OR					
1	.9	1	.9	1	.9	MALHEUR OR	2	-5	-4	63	83
2	.4	1	.1	2	.2	MARION OR	-71	82	-47	23	62
9	.3	10	.4	5	.2	MULTNOMAH OR	10	-51	-46	24	61
		1	1.0			TILLAMOOK OR					
		1	.4			UMATILLA OR					
				3*	2.9	UNION OR					98
2*	2.0			1	.9	WASCO OR			-55	17	84
2	.7	4	.9	1	.1	WASHINGTON OR	30	-89	-85	01	60
		2	1.0	2	.7	YAMHILL OR		-26			80
224	.5	273*	.6	204	.4	PENNSYLVANIA PA	10	-32	-25		
1	.5	2	.9	1	.4	ADAMS PA	77	-54	-18	50	68
32	.5	34	.5	15	.2	ALLEGHENY PA	6	-58	-55	17	62
1	.3	1	.2	2	.5	ARMSTRONG PA	-15	100	70	85	70
10*	1.5	4	.4	4	.4	BEAVER PA	-72	-7	-74	05	68
		2	.9	2	.9	BEDFORD PA		3			84
4	.3	4	.3	4	.2	BERKS PA	5	-17	-12	55	63
6	.9	5	.7	6	.8	BLAIR PA	-12	8	-5	62	82
1	.4					BRADFORD PA					
4	.7	7	.5	7	.5	BUCKS PA	-27	-6	-31	39	71
1	.2			4	.6	BUTLER PA			196	96	76
		1	.1	4	.4	CAMBRIA PA		303			70
1	.3	3	1.3	3	.9	CARBON PA	271	-25	179	95	85
1	.5	1	.3	3	1.0	CENTRE PA	-41	234	98	89	86
2	.3	7	.8	6	.5	CHESTER PA	158	-41	52	82	72
				1	.6	CLARION PA					74
1	.3	2	.5	1	.2	CLEARFIELD PA	98	-54	-10	58	62
1	.6			3*	1.6	CLINTON PA			158	94	94
1	.3	2	.8	3	.9	COLUMBIA PA	126	23	177	94	85
		3	.9	2	.5	CRAWFORD PA		-44			72
4	1.4	1	.2	3	.4	CUMBERLAND PA	-85	114	-69	08	70
7	1.0	6	.6	6	.6	DAUPHIN PA	-35	-9	-41	29	75
8	.5	15	.7	8	.3	DELAWARE PA	35	-58	-44	27	64
		2	1.3	1	.5	ELK PA		-63			71
9*	.9	5	.5	4	.4	ERIE PA	-48	-23	-60	14	66
7	.8	4	.5	2	.2	FAYETTE PA	-39	-53	-71	07	62
5*	1.5	4	.9	3	.6	FRANKLIN PA	-39	-41	-64	11	74
		1	.4	2	.9	GREENE PA		125			83
2	1.1	1	.8			HUNTINGDON PA	-28				
2	.6	2	.5	1	.3	INDIANA PA	-11	-46	-52	19	63
		1	.3	3	1.1	JEFFERSON PA		229			88
2	.1	7	.6	7	.5	LACKAWANNA PA	321	-11	276	98	73
4	.4	5	.4	2	.1	LANCASTER PA	5	-65	-63	12	60
2	.5	3	.6	1	.2	LAWRENCE PA	19	-68	-62	13	61
4	1.3	1	.2			LEBANON PA	-81				
4	.4	6	.6	3	.2	LEHIGH PA	59	-63	-41	29	62
6	.3	6	.3	5	.3	LUZERNE PA	-11	-16	-25	44	63
		4	.8	4	.7	LYCOMING PA		-12			78
2	.7	3	1.0	1	.3	MC KEAN PA	47	-68	-53	18	65
2	.4	7*	1.2			MERCER PA	225				
3*	2.0			1	.9	MIFFLIN PA			-57	16	83
2	1.4	1	.4			MONROE PA	-73				
7	.5	19*	.8	12	.4	MONTGOMERY PA	69	-47	-11	57	69
		1	1.1			MONTOUR PA					
1	.1	2	.2			NORTHAMPTON PA	203				
3	.6	3	.7	2	.3	NORTHUMBERLAND PA	19	-53	-44	26	66
45	.6	48*	.6	32	.5	PHILADELPHIA PA	8	-26	-20	48	71
				1	.8	PIKE PA					81
3	.3	6	.8	4	.5	SCHUYLKILL PA	168	-40	61	84	71
		1	.7	2*	1.7	SNYDER PA		148			94
3	.7	2	.4	1	.2	SOMERSET PA	-40	-44	-66	11	63
				1	.5	SUSQUEHANNA PA					71
1	.3	3	.9	2	.6	VENANGO PA	209	-34	104	89	76
		1	.4	2	.8	WARREN PA		101			80
5	.5	6	.6	4	.4	WASHINGTON PA	30	-36	-17	52	68
1	.5	1	.7	1	.6	WAYNE PA	26	-16	6	68	74
7	.5	9	.5	4	.2	WESTMORELAND PA	-7	-62	-65	11	61
1	1.2	1	1.1	1	1.0	WYOMING PA	-6	-9	-15	53	85
5	.5	7	.6	7	.5	YORK PA	9	-9	-1	64	73

* SIGNIFICANTLY MORE DEATHS THAN EXPECTED

WHITE MALE — CANCER OF THE SALIVARY GLAND — ICD 142

1950-1959 DEATHS	1950-1959 RATE	1960-1969 DEATHS	1960-1969 RATE	1970-1979 DEATHS	1970-1979 RATE	COUNTY AND STATE	PERCENT CHANGE 60S/50S	PERCENT CHANGE 70S/60S	PERCENT CHANGE 70S/50S	PERCENTILE %CHANGE 70S/50S	PERCENTILE RATES 1970S
10	.3	17	.4	21	.5	RHODE ISLAND RI	54	9	68		
		1	.4			BRISTOL RI					
		4	.7	2	.3	KENT RI		-55			65
1	.4	1	.4	4*	1.3	NEWPORT RI	-10	237	204	96	91
9	.3	11	.4	13	.4	PROVIDENCE RI	20	5	27	76	69
				2	.7	WASHINGTON RI					78
17	.5	29	.5	43*	.6	SOUTH CAROLINA SC	8	18	28		
				1	1.9	ABBEVILLE SC					95
		1	.4	1	.4	AIKEN SC		2			68
				1*	4.6	ALLENDALE SC					99
		2	.5			ANDERSON SC					
		1*	3.9			BARNWELL SC					
				2*	1.6	BEAUFORT SC					93
				2*	3.6	BERKELEY SC					98
2	.6	5*	1.0	2	.3	CHARLESTON SC	76	-68	-44	27	65
				1	.8	CHEROKEE SC					80
				2*	2.0	CHESTER SC					96
				1	1.3	COLLETON SC					90
1	1.0			1	.7	DARLINGTON SC			-34	37	78
		1	1.3			DILLON SC					
		2*	6.8	1*	1.9	EDGEFIELD SC		-71			95
				1*	1.9	FAIRFIELD SC					95
1	1.0	2	.8	3*	1.4	FLORENCE SC	-11	67	48	81	92
2	1.0			5	.7	GREENVILLE SC			-31	39	78
		1	.7			GREENWOOD SC					
1*	2.6					HAMPTON SC					
				2	.5	HORRY SC					74
1	1.9			1	1.1	KERSHAW SC			-38	32	88
2*	3.4	1	1.5			LANCASTER SC	-56				
				3*	2.1	LAURENS SC					96
				1*	2.3	LEE SC					97
				2	.8	LEXINGTON SC					83
1*	2.4	1	1.2			MARION SC	-51				
				1	.7	NEWBERRY SC					77
		2*	2.0	1	.5	OCONEE SC		-73			73
		3*	2.6			ORANGEBURG SC					
				1	.4	PICKENS SC					66
3	1.2	1	.2	4	.8	RICHLAND SC	-84	338	-31	39	81
1	.4	2	.5	2	.3	SPARTANBURG SC	16	-33	-22	46	65
		1	.7			SUMTER SC					
		2*	1.9			UNION SC					
2	1.2	1	.3	1	.4	YORK SC	-74	41	-63	13	70
16	.5	7	.2	18	.5	SOUTH DAKOTA SD	-62	146	-7		
1	1.4			2*	1.8	BEADLE SD			24	76	95
1*	1.5					BON HOMME SD					
				1	.9	BROOKINGS SD					85
				2	1.1	BROWN SD					88
		1*	2.5			BUTTE SD					
1*	2.2					CLAY SD					
1	1.0	1	1.2			CODINGTON SD	24				
				1	1.1	DAVISON SD					87
				1	1.0	DAY SD					86
1*	7.0					DEWEY SD					
		1*	2.4	1*	2.5	EDMUNDS SD		3			97
1	1.8					FALL RIVER SD					
1*	4.2					FAULK SD					
1*	4.3			1*	2.9	HAND SD			-34	37	98
		1	1.7			HUTCHINSON SD					
1*	9.3					JERAULD SD					
				1*	2.4	KINGSBURY SD					97
		2*	2.4			LAWRENCE SD					
1*	1.8					MC COOK SD					
2	.6	1	.2	4	1.0	MINNEHAHA SD	-68	410	64	85	85
				3*	1.3	PENNINGTON SD					91
1	1.3					ROBERTS SD					
1	1.6					SPINK SD					
1	1.4					TURNER SD					
1	1.8					UNION SD					
				1*	2.0	WALWORTH SD					96
49	.4	68	.5	75	.5	TENNESSEE TN	23	-10	10		
		1	.7			ANDERSON TN					
				1	.9	BEDFORD TN					83
		1	1.4			BENTON TN					
				4*	1.1	BLOUNT TN					88
				2	1.1	BRADLEY TN					88
		2*	1.5			CAMPBELL TN					
		2	1.3	1	.5	CARTER TN		-63			71
1*	2.2	1*	2.4			CHESTER TN	11				
		1	1.3			CLAIBORNE TN					
				1*	3.1	CLAY TN					98
				1	.8	COFFEE TN					81
		1	1.1	1	.9	CUMBERLAND TN		-18			84
9*	.8	13*	1.0	10	.7	DAVIDSON TN	23	-33	-18	51	78
				1	1.3	DECATUR TN					91
1*	1.8					DE KALB TN					
1*	1.7					FENTRESS TN					
				1	.9	FRANKLIN TN					85
1	.6	3*	1.4			GIBSON TN	138				
1	1.2	2*	1.8			GILES TN	50				
				1	2.0	GRAINGER TN					95
2	1.2			2	.8	GREENE TN			-27	42	82
				2*	3.5	GRUNDY TN					98
1	1.3	1	1.1	1	.8	HAMBLEN TN	-13	-26	-36	35	83

* SIGNIFICANTLY MORE DEATHS THAN EXPECTED

WHITE MALE — CANCER OF THE SALIVARY GLAND — ICD 142

1950-1959 DEATHS	1950-1959 RATE	1960-1969 DEATHS	1960-1969 RATE	1970-1979 DEATHS	1970-1979 RATE	COUNTY AND STATE	PERCENT CHANGE 60S/50S	PERCENT CHANGE 70S/60S	PERCENT CHANGE 70S/50S	PERCENTILE %CHANGE 70S/50S	PERCENTILE RATES 1970S
2	.2	3	.3	4	.4	HAMILTON TN	48	57	131	92	70
				1	1.1	HARDEMAN TN					88
				1	1.1	HARDIN TN					87
				1	.5	HAWKINS TN					72
3*	4.2					HENDERSON TN					
2*	1.7	1	.8	1	.7	HENRY TN	-55	-9	-59	15	79
1	1.8	1	1.5			HICKMAN TN	-18				
1*	4.5					HOUSTON TN					
				1*	1.5	JACKSON TN					92
		1	1.2	2*	1.7	JEFFERSON TN		48			94
		1	1.8	1	1.5	JOHNSON TN		-16			93
2	.2	1	.2	4	.3	KNOX TN	-21	104	61	84	65
1	1.4					LAUDERDALE TN					
1	.9					LAWRENCE TN					
2*	2.0					LINCOLN TN					
		1	1.1			LOUDON TN					
		1	.7	1	.9	MC MINN TN		20			83
1	1.4					MC NAIRY TN					
1	1.7					MACON TN					
		2	.9	1	.4	MADISON TN		-53			69
		1	1.1			MARION TN					
1	1.3					MARSHALL TN					
1	1.1	1	1.0			MONROE TN	-8				
1*	6.4					MOORE TN					
		3*	2.0			OBION TN					
		1	1.3			OVERTON TN					
		1*	4.0			PERRY TN					
				1	.6	PUTNAM TN					76
				1	1.0	RHEA TN					86
1	1.0			1	.5	ROANE TN			-49	22	72
1	1.0					ROBERTSON TN					
		1	.7	2	.8	RUTHERFORD TN		19			82
		1*	3.9			SEQUATCHIE TN					
1	1.2			1	.7	SEVIER TN			-45	25	77
5	.5	8	.6	11	.7	SHELBY TN	18	13	33	77	79
		3	.8	1	.2	SULLIVAN TN		-80			60
		2	1.1	1	.4	SUMNER TN		-62			68
				1	1.1	UNICOI TN					88
		1*	2.9	1*	2.0	UNION TN		-29			96
1	.5	2	.6	5*	1.4	WASHINGTON TN	17	146	188	95	92
1*	1.4	1	1.5			WAYNE TN	11				
1	1.2	1	1.3	1	1.0	WHITE TN	9	-25	-18	51	85
		1	.9	1	.6	WILLIAMSON TN		-32			76
1	.9					WILSON TN					
131	.5	177	.5	184	.4	TEXAS TX	8	-13	-6		
2*	1.5	1	.7	1	.6	ANDERSON TX	-57	-15	-63	12	74
1*	5.0					ANDREWS TX					
2	1.3			3*	1.5	ANGELINA TX			15	72	92
1	1.3					ATASCOSA TX					
				3*	2.7	AUSTIN TX					97
				1*	1.5	BANDERA TX					93
				1	1.7	BEE TX					94
		2	.6	1	.3	BELL TX		-46			64
7	.4	8	.4	10	.4	BEXAR TX	-5	-7	-11	56	67
				1	.9	BOSQUE TX					84
1	.5	3	1.3	1	.4	BOWIE TX	159	-70	-23	45	67
1	.4					BRAZORIA TX					
				1*	3.1	BREWSTER TX					98
1	.7	1	.5			BROWN TX	-22				
1*	1.8					BURLESON TX					
1	1.1					CALDWELL TX					
1	1.6					CALLAHAN TX					
2	.7	7*	1.4	2	.3	CAMERON TX	108	-77	-53	19	65
				1	.7	CASS TX					77
				1*	4.6	CASTRO TX					99
1	.8	1	.6	2	1.2	CHEROKEE TX	-29	100	41	80	89
		1*	2.6			CHILDRESS TX					
				1*	4.3	COKE TX					99
				2*	2.9	COLEMAN TX					98
1	.6	1	.4			COLLIN TX	-32				
		1*	2.5			COLLINGSWORTH TX					
3*	4.8					COLORADO TX					
		1	.9	2	1.4	COMAL TX		50			92
		1	1.0			COMANCHE TX					
		1*	4.1			CONCHO TX					
1	.9					COOKE TX					
1	1.6			1	.9	CORYELL TX			-44	27	84
				1*	7.2	CRANE TX					99
11	.6	13	.5	16	.5	DALLAS TX	-29	0	-29	40	70
				1	1.3	DAWSON TX					91
1	.6			1	.4	DENTON TX			-38	32	66
1	1.0					DE WITT TX					
		1*	2.9			DIMMIT TX					
		1	.6	2*	2.1	EASTLAND TX		269			96
1	.3	1	.2	1	.3	ECTOR TX	-43	67	-5	62	64
2	1.0					ELLIS TX					
2	.3	2	.3	5	.5	EL PASO TX	8	85	100	89	73
		1	.8	2*	1.2	ERATH TX		55			89
2	1.2	1	.6			FANNIN TX	-50				
1	1.0			1	.8	FAYETTE TX			-11	57	82
				1*	1.9	FLOYD TX					95
				2	1.0	FORT BEND TX					85
				1*	2.2	FRANKLIN TX					96
				1	1.3	FREESTONE TX					90
				1*	2.3	FRIO TX					96
		1*	2.5			GAINES TX					
1	.6	2	.4	2	.4	GALVESTON TX	-41	11	-35	36	67
1*	4.9					GARZA TX					
				1	1.1	GILLESPIE TX					87

* SIGNIFICANTLY MORE DEATHS THAN EXPECTED

WHITE MALE

CANCER OF THE SALIVARY GLAND — ICD 142

1950-1959 DEATHS	1950-1959 RATE	1960-1969 DEATHS	1960-1969 RATE	1970-1979 DEATHS	1970-1979 RATE	COUNTY AND STATE	PERCENT CHANGE 60S/50S	PERCENT CHANGE 70S/60S	PERCENT CHANGE 70S/50S	PERCENTILE %CHANGE 70S/50S	PERCENTILE RATES 1970S
		1*	15.7			GLASSCOCK TX					
2*	2.4	1	1.0	1	.9	GONZALES TX	-56	-9	-61	14	84
2*	2.5					GRAY TX					
		1	.3	2	.5	GRAYSON TX		59			71
		1	.3	2	.7	GREGG TX		125			79
				2*	3.4	GRIMES TX					98
		2*	1.4			GUADALUPE TX					
1	.9					HALE TX					
		1*	6.9			HANSFORD TX					
1*	2.0			1*	3.2	HARDEMAN TX			59	83	98
14	.5	14	.4	16	.5	HARRIS TX	-24	10	-17	52	70
		2*	1.6	1	.7	HARRISON TX		-56			79
1	1.6	1	1.5			HAYS TX	-8				
				2*	11.4	HEMPHILL TX					99
		2*	1.4			HENDERSON TX					
3	.6	3	.4	3	.4	HIDALGO TX	-27	-14	-37	33	67
				1	.5	HILL TX					73
		1	1.2			HOCKLEY TX					
				1	1.1	HOOD TX					88
		2*	2.4	2*	2.0	HOUSTON TX		-16			96
1	1.1	1	1.1			HOWARD TX	-4				
1	.4	3*	1.4	1	.4	HUNT TX	224	-71	-7	61	67
		1*	1.6			JACK TX					
		1	1.0			JASPER TX					
1	.2	2	.3	5	.7	JEFFERSON TX	41	132	228	97	78
1	1.3					JIM WELLS TX					
1	.5	4*	2.0	1	.5	JOHNSON TX	299	-74	6	68	73
1	1.1			1	1.0	JONES TX			-9	58	85
1	1.4					KAUFMAN TX					
1	1.2	1	.5			KERR TX	-56				
		2*	2.6			KLEBERG TX					
		1*	2.5			KNOX TX					
1	.6	1	.6			LAMAR TX	0				
		1	.9			LAMB TX					
				1	.6	LAVACA TX					76
		1*	1.9	1*	1.4	LEON TX		-27			91
2*	2.0					LIBERTY TX					
1	.9	2*	2.5	1	.8	LIMESTONE TX	173	-67	-11	57	82
		1*	2.7			LIVE OAK TX					
				1	1.2	LLANO TX					89
2	.9	3	.7	4	.8	LUBBOCK TX	-21	15	-9	58	81
1*	3.1			1*	2.3	LYNN TX			-26	43	96
2*	3.4	1	1.4			MC CULLOCH TX	-60				
2	.3	5	.8	6*	.9	MC LENNAN TX	141	9	164	94	83
1	1.3			1	.8	MATAGORDA TX			-39	31	81
				1	.8	MEDINA TX					82
				2	1.1	MIDLAND TX					87
		2*	1.4			MILAM TX					
				1*	1.9	MILLS TX					95
		1	2.9			MITCHELL TX					
		1	.8	4*	3.5	MONTAGUE TX		330			98
1*	5.9					MOTLEY TX					
		1	.7			NACOGDOCHES TX					
		3*	1.5	1	.5	NAVARRO TX		-65			74
3	.5	3	.6	2	.3	NUECES TX	22	-47	-36	35	64
		1*	4.4			OCHILTREE TX					
2*	1.8					ORANGE TX					
1	1.2	2*	1.7			PALO PINTO TX	42				
		1	1.3	1	1.0	PANOLA TX		-24			86
		1	.8	1	.6	PARKER TX		-20			76
1*	4.2					PARMER TX					
				1	1.3	POLK TX					91
		3	1.2			POTTER TX					
		1*	3.2			RAINS TX					
		1	.5			RANDALL TX					
1	1.0					RED RIVER TX					
		2*	3.8			REFUGIO TX					
				2*	2.3	RUNNELS TX					96
1	.6	1	.6	1	.5	RUSK TX	9	-19	-12	56	72
		1*	2.4			SABINE TX					
		1	.6	1	.5	SAN PATRICIO TX		-15			73
		2*	2.7			SCURRY TX					
		2*	1.6	1	.7	SHELBY TX		-55			79
1	.3	3	.9			SMITH TX	182				
		3*	14.7			SOMERVELL TX					
7	.5	8	.4	11	.5	TARRANT TX	-9	22	12	70	73
1	.4	3	.9			TAYLOR TX	134				
				1*	1.9	TERRY TX					95
				1	1.0	TITUS TX					85
		1	.4	2	.6	TOM GREEN TX		60			75
2	.4	2	.3	2	.3	TRAVIS TX	-23	-14	-34	36	63
				2*	1.7	UPSHUR TX					94
				1	1.0	UVALDE TX					86
1	.8			1	.5	VAN ZANDT TX			-37	33	73
2*	2.0	1	.8	1	.4	VICTORIA TX	-61	-49	-80	03	68
		1	1.2			WALKER TX					
				1*	1.7	WALLER TX					94
1	.4	4*	1.8	4*	1.4	WEBB TX	309	-23	213	97	91
		1	.7	1	.6	WHARTON TX		-16			76
1	.2			1	.2	WICHITA TX			-12	56	62
2*	2.0	2*	2.1	1	.9	WILBARGER TX	6	-56	-53	18	84
3*	1.6	1	.5			WILLIAMSON TX	-67				
1	1.7					WILSON TX					
1	1.0					WISE TX					
1	.8	1	1.3	2*	1.9	WOOD TX	61	40	124	91	95
2*	2.6	1	1.1	1	1.0	YOUNG TX	-59	-4	-61	14	86
10	.4	16	.5	15	.4	UTAH UT	22	-24	-8		
1	1.4					BOX ELDER UT					
		2	1.0	1	.6	CACHE UT		-42			75

* SIGNIFICANTLY MORE DEATHS THAN EXPECTED

WHITE MALE — CANCER OF THE SALIVARY GLAND — ICD 142

1950-1959 DEATHS	1950-1959 RATE	1960-1969 DEATHS	1960-1969 RATE	1970-1979 DEATHS	1970-1979 RATE	COUNTY AND STATE	PERCENT CHANGE 60S/50S	PERCENT CHANGE 70S/60S	PERCENT CHANGE 70S/50S	PERCENTILE %CHANGE 70S/50S	PERCENTILE RATES 1970S
				1	1.3	CARBON UT					91
2*	2.6	2	2.0			DAVIS UT	-24				
				1*	6.1	KANE UT					99
		1*	2.6			MILLARD UT					
		1*	8.1			MORGAN UT					
3	.2	4	.3	7	.4	SALT LAKE UT	32	37	81	87	68
1	1.4					SANPETE UT					
1*	2.0					SEVIER UT					
		3*	5.5			TOOELE UT					
				1*	2.8	UINTAH UT					97
2	.8	1	.3	2	.5	UTAH UT	-56	37	-40	30	71
				1	.8	WASHINGTON UT					82
		2	.4	1	.2	WEBER UT		-58			61
7	.4	10	.5	3	.1	VERMONT VT	29	-73	-65		
		1	.7	1	.7	BENNINGTON VT		-12			77
2*	1.6	1	.9			CALEDONIA VT	-44				
2	1.0	2	.7	1	.2	CHITTENDEN VT	-28	-70	-78	03	62
		2*	3.5			LAMOILLE VT					
		1	.9			ORLEANS VT					
1	.4	2	.9			RUTLAND VT	104				
1	.6					WASHINGTON VT					
		1	.7			WINDHAM VT					
1	.5			1	.4	WINDSOR VT			-29	41	67
45	.5	47	.4	58	.4	VIRGINIA VA	-14	-3	-16		
		1	.7			ACCOMACK VA					
				2	.9	ALBEMARLE#VA					83
1*	6.8					APPOMATTOX VA					
4	1.0	1	.1	1	.2	ARLINGTON#VA	-89	34	-85	01	60
1	.4	2	.7			AUGUSTA#VA	84				
				1*	3.3	BATH VA					98
2	.5	2	.4	3	.6	BEDFORD#VA	-19	31	6	68	75
		1	1.1			BOTETOURT VA					
				3*	2.6	BUCHANAN VA					97
1	.5	1	.4	1	.4	CARROLL#VA	-23	-9	-30	40	67
3	.3	5	.6	4	.5	CHESTERFIELD#VA	90	-14	63	84	71
2*	2.8					DICKENSON VA					
		1	.4	1	.3	DINWIDDIE#VA		-30			64
				1*	3.1	ESSEX VA					98
1	.2	4	.5	6	.4	FAIRFAX#VA	106	-8	89	87	69
1*	1.4			1	1.6	FLOYD VA			8	69	93
		1	1.0			FRANKLIN VA					
		1	.6			FREDERICK#VA					
1	1.4	1	1.1	1	.9	GILES VA	-22	-19	-37	34	84
1	1.0	1	.9			HALIFAX#VA	-11				
1	2.1			2*	1.0	HANOVER VA			-51	20	87
1	.7	3	.5	4	.7	HENRICO VA	-35	42	-7	60	78
1	.8			2	.9	HENRY#VA			14	71	83
				1*	2.0	LANCASTER VA					96
1	.8	1	.7	1	.9	LEE VA	-16	39	16	73	84
1	1.0					LOUDOUN VA					
		1*	2.5			MATHEWS VA					
		1	1.1			MECKLENBURG VA					
3*	2.1					MONTGOMERY#VA					
1*	2.1					NORTHAMPTON VA					
1*	2.1	1*	2.0			NOTTOWAY VA	-1				
				1	1.0	PAGE VA					87
1	.3	1	.3	1	.3	PITTSYLVANIA#VA	2	-8	-6	61	63
1*	3.6					PRINCE EDWARD VA					
1	1.3					PRINCE WILLIAM#VA					
1	1.6					PULASKI VA					
		1	.1	4	.5	ROANOKE#VA		313			72
1	.8	1	1.0			ROCKBRIDGE#VA	22				
1	.5	1	.4	2	.7	ROCKINGHAM#VA	-15	72	46	80	79
				1	1.1	SCOTT VA					88
1	.9					SHENANDOAH VA					
				1	.7	SMYTH VA					79
		1*	2.1			SOUTHAMPTON#VA					
		2*	2.1			SPOTSYLVANIA#VA					
1*	2.4			1	.4	STAFFORD VA			-81	02	70
				1*	4.6	SUSSEX VA					99
		2	1.3	2	1.0	TAZEWELL VA		-22			86
1	.7	2	.9			WASHINGTON#VA	31				
				3*	2.0	WISE#VA					95
		2*	2.0	1	1.0	WYTHE VA		-49			86
7	.6	3	.3	3	.2	CHESAPEAKE CITY#VA	-47	-44	-70	08	61
				1	.3	HAMPTON CITY VA					63
1	1.1	1	.5	1	.2	NEWPORT NEWS CITY VA	-49	-66	-83	01	61
		1	1.6			SUFFOLK CITY VA					
40	.3	65	.5	62	.4	WASHINGTON WA	39	-16	17		
1*	3.2					ADAMS WA					
		1	.5	1	.3	BENTON WA		-45			63
				2	.8	CHELAN WA					81
		2	.4	2	.3	CLARK WA		-15			65
				1*	3.6	COLUMBIA WA					98
		1	.6	4*	1.3	COWLITZ WA		125			91
				1*	6.7	FERRY WA					99
1*	5.9					GARFIELD WA					
2*	3.6	1	1.2	2	.8	ISLAND WA	-68	-33	-79	03	80
18	.5	20	.5	17	.4	KING WA	-3	-23	-25	43	66
1	.3	1	.2	2	.4	KITSAP WA	-33	83	22	75	68
		1	.9			KITTITAS WA					
		2*	2.6			KLICKITAT WA					
				2	.7	LEWIS WA					78
				1	1.5	LINCOLN WA					92
		1	.7			MASON WA					

* SIGNIFICANTLY MORE DEATHS THAN EXPECTED

\# COUNTIES COMBINED; SEE TECHNICAL NOTES

WHITE MALE — CANCER OF THE SALIVARY GLAND — ICD 142

1950-1959 DEATHS	1950-1959 RATE	1960-1969 DEATHS	1960-1969 RATE	1970-1979 DEATHS	1970-1979 RATE	COUNTY AND STATE	PERCENT CHANGE 60S/50S	PERCENT CHANGE 70S/60S	PERCENT CHANGE 70S/50S	PERCENTILE %CHANGE 70S/50S	PERCENTILE RATES 1970S
1	1.2			1	.8	PACIFIC WA			-37	34	81
		1*	2.2			PEND OREILLE WA					
4	.3	10	.7	11	.7	PIERCE WA	133	-2	129	91	78
1	.4			2	.7	SKAGIT WA			56	83	77
1	.1	3	.4	2	.2	SNOHOMISH WA	175	-51	36	78	61
3	.2	6	.4	3	.2	SPOKANE WA	102	-54	-6	61	61
				1	.9	STEVENS WA					83
3*	1.5	2	.9			WALLA WALLA WA	-38				
1	.3	2	.5	2	.4	WHATCOM WA	71	-16	44	80	70
1	.7	1	.7	1	.7	WHITMAN WA	3	-4	-1	64	79
2	.3	10*	1.4	4	.5	YAKIMA WA	366	-62	75	86	73
20	.3	28	.3	26	.3	WEST VIRGINIA WV	28	-13	11		
1	.6	1	.6			BERKELEY WV	-6				
				1	2.0	BRAXTON WV					95
1	.2	3	.7	1	.2	CABELL WV	269	-72	3	66	61
				1*	1.8	CALHOUN WV					95
2*	1.2			2	1.2	HANCOCK WV			1	65	90
1*	2.3					HARDY WV					
		2	.5	1	.3	HARRISON WV		-49			63
1	1.4					JEFFERSON WV					
3	.4	4	.4	2	.2	KANAWHA WV	4	-50	-48	23	62
		1	.9	1	.4	MC DOWELL WV		-52			69
		3	.9	2	.6	MARION WV		-37			75
		1	.5	1	.6	MARSHALL WV		13			74
				1	.8	MASON WV					80
		1	.4			MERCER WV					
1	1.0					MINERAL WV					
				1	.7	MINGO WV					78
2	1.2					MONONGALIA WV					
				1	1.5	MONROE WV					93
				1	1.0	NICHOLAS WV					85
		2	.6	2	.6	OHIO WV		-2			75
		1*	2.8	1*	3.4	PLEASANTS WV		23			98
				1	1.4	POCAHONTAS WV					92
		1	.8			PRESTON WV					
1	.3	4*	1.5	1	.3	RALEIGH WV	349	-81	-13	55	64
1	.8			2*	1.4	RANDOLPH WV			77	86	91
		2*	4.1			RITCHIE WV					
1	1.2			1	1.5	SUMMERS WV			23	75	92
		1	1.0			TAYLOR WV					
1*	2.1					TUCKER WV					
1	2.4					TYLER WV					
		1	.9	1	.8	WETZEL WV		-7			82
1*	3.4					WIRT WV					
2	.7			1	.2	WOOD WV			-67	10	62
67	.4	79	.4	87	.4	WISCONSIN WI	-1	1	1		
		1	2.1			ADAMS WI					
				1	.8	ASHLAND WI					80
				1	.4	BARRON WI					67
		1	1.6	1	1.8	BAYFIELD WI		9			95
2	.5	3	.6	4	.6	BROWN WI	10	11	22	75	77
		1	.9	1	.8	CALUMET WI		-14			81
1	.5			1	.4	CHIPPEWA WI			-8	60	69
		1	.5			COLUMBIA WI					
		1	1.2	1	.8	CRAWFORD WI		-28			83
5	.6	4	.5	2	.2	DANE WI	-28	-54	-67	10	62
				1	.4	DOUGLAS WI					69
3*	3.1			1	.6	DUNN WI			-82	02	75
1	.4	1	.3	1	.3	EAU CLAIRE WI	-4	-9	-13	54	65
2	.6	6*	1.6			FOND DU LAC WI	183				
3*	1.5	1	.4	2	.8	GRANT WI	-74	95	-49	22	80
				1	.7	GREEN WI					78
				1	1.4	GREEN LAKE WI					91
				1	.8	IOWA WI					81
				1	1.5	JACKSON WI					93
		3	1.1			JEFFERSON WI					
		1	1.2	1	.8	JUNEAU WI		-39			80
3	1.0					KENOSHA WI					
2*	2.6			1	.8	KEWAUNEE WI			-68	09	82
1	.4	3	.8	3	.8	LA CROSSE WI	116	-3	110	90	81
1	1.1	2*	2.1			LAFAYETTE WI	89				
				1	.7	LINCOLN WI					78
1	.3	1	.2	5*	1.2	MANITOWOC WI	-9	401	353	99	89
1	.3	2	.5	3	.6	MARATHON WI	97	14	123	91	76
2	1.0			1	.5	MARINETTE WI			-45	25	74
				1	1.1	MARQUETTE WI					87
1	.3	2	.5	5*	1.5	MENOMINEE#WI	101	176	455	99	92
13	.3	13	.3	19	.4	MILWAUKEE WI	-3	53	48	81	70
		1	.6			MONROE WI					
1	.9			1	.6	ONEIDA WI			-30	40	76
3	.6	3	.7	2	.5	OUTAGAMIE WI	13	-34	-25	43	70
1	.7	2	1.0			OZAUKEE WI	41				
				1*	1.5	PEPIN WI					92
1	.7					PIERCE WI					
		2	1.0			PORTAGE WI					
3	.5	6	1.0	1	.1	RACINE WI	87	-85	-73	07	60
				1	.9	RICHLAND WI					83
3	.6	3	.5	2	.4	ROCK WI	-15	-28	-39	31	67
1	1.3					RUSK WI					
1	.5			1	.4	SAUK WI			-5	62	70
2	.5	2	.4	2	.4	SHEBOYGAN WI	-18	-3	-21	47	67
				2*	1.9	TAYLOR WI					95
1	.9			2	1.3	TREMPEALEAU WI			38	79	90
		1	.5	1	.9	VERNON WI		76			84
				1	1.9	VILAS WI					95
1	.4			4*	1.2	WALWORTH WI			248	98	90
		1	1.1			WASHBURN WI					

* SIGNIFICANTLY MORE DEATHS THAN EXPECTED

COUNTIES COMBINED; SEE TECHNICAL NOTES

WHITE MALE — CANCER OF THE SALIVARY GLAND — ICD 142

1950-1959 DEATHS	1950-1959 RATE	1960-1969 DEATHS	1960-1969 RATE	1970-1979 DEATHS	1970-1979 RATE	COUNTY AND STATE	PERCENT CHANGE 60S/50S	PERCENT CHANGE 70S/60S	PERCENT CHANGE 70S/50S	PERCENTILE %CHANGE 70S/50S	PERCENTILE RATES 1970S
1	.5	2	1.1			WASHINGTON WI	131				
1	.2	4	.6	2	.3	WAUKESHA WI	293	-57	70	85	63
		2	.8	2	.6	WAUPACA WI		-25			75
1	1.2	1	.9			WAUSHARA WI	-23				
3	.6	2	.4	1	.2	WINNEBAGO WI	-39	-48	-68	09	61
1	.5					WOOD WI					
3	.2	4	.3	1	.1	WYOMING WY	41	-76	-67		
		1*	2.5			CONVERSE WY					
1*	3.3					CROOK WY					
		1	.4			LARAMIE WY					
1	.6			1	.6	NATRONA WY			3	66	76
		1	1.5			PARK-TETON#WY					
		1	1.0			SWEETWATER WY					
1*	2.7					WASHAKIE WY					

WHITE FEMALE | CANCER OF THE SALIVARY GLAND | ICD 142

COUNTY AND STATE

1950-1959 DEATHS	1950-1959 RATE	1960-1969 DEATHS	1960-1969 RATE	1970-1979 DEATHS	1970-1979 RATE	COUNTY AND STATE	PERCENT CHANGE 60S/50S	PERCENT CHANGE 70S/60S	PERCENT CHANGE 70S/50S	PERCENTILE %CHANGE 70S/50S	PERCENTILE RATES 1970S
1780	.2	2166	.2	2272	.2	UNITED STATES	-5	-17	-21		
21	.2	29	.2	27	.2	ALABAMA AL	6	-34	-30		
				1	.3	BALDWIN AL					80
		1*	1.7			BIBB AL					
				1	.5	BLOUNT AL					88
		1*	3.2			BULLOCK AL					
		1	.3	1	.2	CALHOUN AL		-31			76
		1*	1.2			CLAY AL					
1	.9					COLBERT AL					
				1*	1.1	CONECUH AL					95
1*	2.1					COOSA AL					
		1	.5	1	.3	COVINGTON AL		-44			81
		1*	1.4			CRENSHAW AL					
1	.9			1	.7	DALLAS AL			-27	53	91
1*	1.4					ELMORE AL					
2	.7	2	.5	1	.2	ETOWAH AL	-24	-58	-68	12	76
				1	.4	FRANKLIN AL					85
				1*	1.1	HALE AL					95
		1*	1.8			HENRY AL					
1	.7	1	.4			HOUSTON AL	-42				
				2*	.9	JACKSON AL					93
3	.1	4	.2	3	.1	JEFFERSON AL	42	-40	-15	62	71
1	.6	1	.5			LAUDERDALE AL	-17				
		1	.7			LEE AL					
		1*	4.3			MACON AL					
1	.5			1	.2	MADISON AL			-71	11	73
1*	.9					MARION AL					
1	.5					MARSHALL AL					
3	.5	3	.3	1	.1	MOBILE AL	-40	-75	-85	02	69
1*	1.8	2*	2.8			MONROE AL	51				
1	.1	2	.4	3	.4	MONTGOMERY AL	180	9	204	97	86
		1	.3	1	.2	MORGAN AL		-38			76
				1*	1.8	PERRY AL					98
				1*	.6	PICKENS AL					89
		1*	.8			PIKE AL					
		1	.8	1	.4	ST CLAIR AL		-48			85
		1	.7			SHELBY AL					
				1	.3	TALLADEGA AL					80
1	.3	1	.3			TUSCALOOSA AL	-20				
				1	.3	WALKER AL					82
1*	1.7			2*	2.1	WINSTON AL			25	78	98
6	.2	19	.3	20	.2	ARIZONA AZ	46	-35	-5		
				1	.8	GILA AZ					92
6	.4	7	.2	14	.2	MARICOPA AZ	-48	10	-43	37	76
				1	.4	MOHAVE AZ					85
		2*	2.7			NAVAJO AZ					
		8*	.5	4	.2	PIMA AZ		-62			75
		1*	1.8			SANTA CRUZ AZ					
		1	.4			YAVAPAI AZ					
21	.3	26	.3	35*	.3	ARKANSAS AR	3	-6	-4		
		1*	1.0			ARKANSAS AR					
1*	1.8					BAXTER AR					
1	.4			1	.3	BENTON AR			-31	51	79
		1*	1.5	1*	.6	BRADLEY AR		-59			89
				2*	1.2	CARROLL AR					96
		1	.7			CLAY AR					
1*	1.3	1*	2.3			CONWAY AR	76				
1	.6	1	.4	1	.3	CRAIGHEAD AR	-41	-17	-51	29	81
		1	.5			CRAWFORD AR					
1*	.9	1*	1.2			CRITTENDEN AR	30				
1*	2.4					DALLAS AR					
				1*	1.3	DESHA AR					96
		1*	.9			FULTON AR					
		1	.3	1	.2	GARLAND AR		-48			74
		1*	.9	1*	.7	HEMPSTEAD AR		-19			91
				1*	.6	HOWARD AR					89
		1	.7			INDEPENDENCE AR					
				1*	.9	IZARD AR					94
		1*	1.0	1*	.8	JACKSON AR		-19			93
1	.6	1	.5	3*	.9	JEFFERSON AR	-24	106	57	85	94
				1*	1.5	JOHNSON AR					97
				1	.6	LAWRENCE AR					90
2*	2.0					LONOKE AR					
				1*	1.1	MADISON AR					95
		1*	1.5			MARION AR					
1*	1.4	1*	1.0	2*	1.9	PHILLIPS AR	-31	99	37	82	98
2*	4.7	1*	2.0			PIKE AR	-57				
		1	1.0	1	.5	POPE AR		-50			87
4	.5	3	.3	4	.3	PULASKI AR	-46	17	-37	43	81
1*	1.6					RANDOLPH AR					
				1*	1.0	ST FRANCIS AR					95
		2*	1.3	1	.4	SALINE AR		-71			83
		2	.5	3	.6	SEBASTIAN AR		25			89
				1*	1.4	SEVIER AR					97
1	.5					UNION AR					
1*	2.1	1*	1.6			VAN BUREN AR	-22				
1	.4	1	.4	3*	.6	WASHINGTON AR	-11	68	49	84	89
				2*	.6	WHITE AR					90
1*	2.7					WOODRUFF AR					
173	.3	186	.2	212	.2	CALIFORNIA CA	-25	-12	-34		
14	.3	10	.2	9	.2	ALAMEDA CA	-37	-20	-50	31	73

* SIGNIFICANTLY MORE DEATHS THAN EXPECTED

WHITE FEMALE — CANCER OF THE SALIVARY GLAND — ICD 142

1950-1959 DEATHS	1950-1959 RATE	1960-1969 DEATHS	1960-1969 RATE	1970-1979 DEATHS	1970-1979 RATE	COUNTY AND STATE	PERCENT CHANGE 60S/50S	PERCENT CHANGE 70S/60S	PERCENT CHANGE 70S/50S	PERCENTILE %CHANGE 70S/50S	PERCENTILE RATES 1970S
1	.3	2	.4	3	.2	BUTTE CA	60	-54	-26	54	75
1	.1	2	.1	6	.2	CONTRA COSTA CA	65	148	310	99	77
		2*	3.7	1*	1.1	DEL NORTE CA		-71			95
1*	.9			4*	1.5	EL DORADO CA			66	87	97
3	.3	1	.1	3	.2	FRESNO CA	-81	177	-48	32	72
2*	2.2					GLENN CA					
		1	.2	2	.4	HUMBOLDT CA		49			83
1*	3.5	1*	1.3			INYO CA	-64				
2	.3	5	.4	1	.1	KERN CA	58	-83	-73	09	70
83*	.3	78	.2	79	.2	LOS ANGELES CA	-29	-8	-35	45	77
				1	.4	MADERA CA					84
1	.1	2	.2	4	.3	MARIN CA	77	21	115	94	81
		1	.4	2	.7	MENDOCINO CA		78			91
				2	.4	MERCED CA					84
1*	4.4					MODOC CA					
1	.2	4	.5	2	.2	MONTEREY CA	169	-72	-26	55	72
				2	.3	NAPA CA					81
5	.3	11	.3	17	.2	ORANGE CA	2	-26	-25	56	77
1	.4	2	.6			PLACER CA	60				
		1*	2.8			PLUMAS CA					
4	.3	3	.1	11	.3	RIVERSIDE CA	-52	107	-1	69	81
1	.0	7	.3	6	.2	SACRAMENTO CA	612	-42	313	99	74
				3*	3.1	SAN BENITO CA					99
8	.4	4	.1	4	.1	SAN BERNARDINO CA	-68	-30	-77	06	70
8	.3	12	.2	6	.1	SAN DIEGO CA	-14	-69	-73	09	69
10	.2	13	.3	8	.2	SAN FRANCISCO CA	28	-34	-15	61	75
6*	.7	1	.1	3	.2	SAN JOAQUIN CA	-90	154	-73	08	74
				2	.3	SAN LUIS OBISPO CA					79
4	.3			5	.2	SAN MATEO CA			-51	30	72
		3	.2	2	.1	SANTA BARBARA CA		-43			72
3	.1	5	.2	9	.2	SANTA CLARA CA	15	17	35	81	75
3	.5	1	.1	1	.1	SANTA CRUZ CA	-82	-40	-89	01	69
2*	1.1	1	.5	1	.2	SHASTA CA	-59	-53	-80	04	76
		3	.6			SOLANO CA					
1	.1	2	.2	3	.2	SONOMA CA	44	27	84	91	76
3	.5	2	.2	3	.3	STANISLAUS CA	-59	40	-42	38	81
		2*	1.3			SUTTER CA					
		2	.2	1	.1	TULARE CA		-66			70
3	.5	1	.1	1	.1	VENTURA CA	-82	-47	-90	00	69
				3*	.8	YOLO CA					92
		1	.9	2*	1.1	YUBA CA		25			95
17	.2	10	.1	12	.1	COLORADO CO	-55	-7	-58		
1*	1.2			1	.5	DELTA CO			-56	23	88
9	.3	5	.1	5	.1	DENVER-ARAPAHOE#CO	-53	-10	-58	22	71
1	.2			2	.2	EL PASO CO			0	71	73
				1	.1	JEFFERSON CO					69
				1*	.9	LA PLATA CO					93
1	.3					LARIMER CO					
				1*	.7	LAS ANIMAS CO					90
1*	1.2					LOGAN CO					
1	.5	1	.3			MESA CO	-33				
1*	1.5					MONTROSE CO					
		1	.7			MORGAN CO					
		1	.6			OTERO CO					
1*	12.8					PARK CO					
1	.2			1	.2	PUEBLO CO			-24	56	73
		2	.6			WELD CO					
22	.2	29	.2	31	.2	CONNECTICUT CT	2	-14	-12		
4	.1	8	.2	4	.1	FAIRFIELD CT	52	-61	-41	38	70
9	.3	6	.1	8	.2	HARTFORD CT	-48	8	-44	35	72
1	.2			3	.3	LITCHFIELD CT			79	90	82
1	.2	1	.2	2	.3	MIDDLESEX CT	-28	54	12	74	80
4	.1	10	.3	10	.2	NEW HAVEN CT	101	-16	68	87	77
2	.2	1	.1	3	.2	NEW LONDON CT	-58	121	-8	64	76
1	.3	3	.6	1	.2	WINDHAM CT	134	-73	-36	45	73
2	.2	2	.1	8	.3	DELAWARE DE	-43	226	86		
				2	.5	KENT DE					87
1	.1	1	.1	5	.3	NEW CASTLE DE	-35	308	163	96	79
1	.5	1	.3	1	.2	SUSSEX DE	-43	-36	-63	18	74
12	.4	8	.3	4	.2	DIST. OF COLUMBIA DC	-33	-20	-46		
12	.4	8	.3	4	.2	WASHINGTON DC	-33	-20	-46	34	78
31	.2	57	.2	106	.2	FLORIDA FL	4	8	12		
1	.6			2	.5	ALACHUA FL			-18	60	87
2*	1.7			1	.3	BAY FL			-85	03	79
				5	.4	BREVARD FL					84
2	.1	3	.1	12	.2	BROWARD FL	-12	54	35	81	74
				2	.7	CITRUS FL					91
		1*	1.1			CLAY FL					
		1*	1.0			COLUMBIA FL					
5	.1	5	.1	23	.2	DADE FL	-31	206	111	93	78
1	.1	4	.3	6	.3	DUVAL FL	140	2	146	96	80
1	.3	3	.5	2	.3	ESCAMBIA FL	91	-49	-3	69	80
1	.9			1	.8	GADSDEN FL			-14	62	92
				1*	1.2	HARDEE FL					96
		1*	4.4			HENDRY FL					
2	.2	7	.4	6	.2	HILLSBOROUGH FL	97	-47	5	72	75
				1	.2	INDIAN RIVER FL					75
		1*	5.8			LAFAYETTE FL					
				2	.2	LAKE FL					76
		1	.2	1	.0	LEE FL		-78			69
		1	.3	2	.2	MANATEE FL		-40			74

* SIGNIFICANTLY MORE DEATHS THAN EXPECTED

COUNTIES COMBINED; SEE TECHNICAL NOTES

1950-1959 DEATHS	1950-1959 RATE	1960-1969 DEATHS	1960-1969 RATE	1970-1979 DEATHS	1970-1979 RATE	COUNTY AND STATE	PERCENT CHANGE 60S/50S	PERCENT CHANGE 70S/60S	PERCENT CHANGE 70S/50S	PERCENTILE %CHANGE 70S/50S	PERCENTILE RATES 1970S
		1	.3			MARION FL					
		2*	1.0	1	.2	MARTIN FL		-83			73
		1	.5	1	.2	MONROE FL		-54			78
3	.3	2	.2	2	.1	ORANGE FL	-52	-45	-74	08	70
1	.5	1	.6	2	.6	OSCEOLA FL	12	11	24	78	89
1	.1	3	.2	3	.2	PALM BEACH FL	91	-6	79	90	73
		2	.2	2	.1	PASCO FL		-66			70
5	.2	10	.2	15	.2	PINELLAS FL	0	30	30	80	78
3	.4			3	.1	POLK FL			-64	17	72
1*	1.1					PUTNAM FL					
		2*	1.0	1	.4	ST JOHNS FL		-58			85
		1	.7			ST LUCIE FL					
		1	.1	4	.2	SARASOTA FL		303			76
				2	.3	SEMINOLE FL					82
1*	5.8					UNION FL					
		3	.3	3	.1	VOLUSIA FL		-53			71
1*	1.6					WALTON FL					
30	.3	37	.3	35	.2	GEORGIA GA	-6	-36	-40		
		1	.6			BALDWIN GA					
		1*	3.2			BANKS GA					
1*	1.6					BARROW GA					
1*	2.2			2*	2.5	BEN HILL GA			14	75	98
				1*	1.3	BERRIEN GA					97
1	.4			1	.2	BIBB GA			-61	20	73
2	.5	1	.2			CHATHAM GA	-67				
1*	1.4			1	.3	CLAYTON GA			-77	06	82
1	.2			2	.3	COBB GA			11	74	79
		1*	2.3			COLUMBIA GA					
3	.5	3	.3	2	.1	DE KALB GA	-44	-61	-78	05	70
		1*	1.3			DODGE GA					
		1	.7			DOUGHERTY GA					
				1	.6	DOUGLAS GA					89
1*	3.3					EFFINGHAM GA					
		1*	1.5			ELBERT GA					
		1*	1.7			EMANUEL GA					
		1*	1.2			FANNIN GA					
		1	.3	2	.5	FLOYD GA		51			86
		1*	1.4			FORSYTH GA					
		1*	1.5	1*	1.1	FRANKLIN GA		-26			95
7	.4	4	.2	4	.2	FULTON GA	-55	-18	-63	19	73
				1*	1.8	GILMER GA					98
		1	.7			GLYNN GA					
1*	.9					GORDON GA					
1*	1.9					GRADY GA					
				1	.2	GWINNETT GA					77
				1	.2	HALL GA					77
		1	1.0			HOUSTON GA					
				2*	7.5	JENKINS GA					99
1*	4.2			1*	1.5	JONES GA			-64	17	97
		1*	2.1			LAMAR GA					
		1	.8			LAURENS GA					
		1*	5.7			LIBERTY GA					
		1*	3.5			LINCOLN GA					
		1	.7			LOWNDES GA					
				1*	2.2	LUMPKIN GA					98
2	.5	1	.2	1	.2	MUSCOGEE#GA	-47	-36	-66	15	73
1*	2.4					PAULDING GA					
				1*	1.3	PICKENS GA					96
1	.9	1	.8	3*	1.5	POLK GA	-7	80	67	87	97
1*	6.1					PULASKI GA					
1*	6.5					PUTNAM GA					
1*	3.6					RANDOLPH GA					
1	.3			1	.2	RICHMOND GA			-43	36	74
				1	.4	SPALDING GA					85
		1*	.9			STEPHENS GA					
				1*	5.3	STEWART GA					99
				1*	1.7	TELFAIR GA					98
		1	.6			THOMAS GA					
		1*	1.1			TIFT GA					
				1*	.6	TOOMBS GA					89
		1	.6			TROUP GA					
1	.6					WALKER GA					
		1*	.8			WALTON GA					
		2*	1.8			WARE GA					
		1	.5			WHITFIELD GA					
		1*	4.7			WILKINSON GA					
				1*	1.2	WORTH GA					96
		1	.1	1	.1	HAWAII HI		-12			
		1	.1	1	.1	HONOLULU HI		-9			72
8	.3	8	.3	6	.1	IDAHO ID	-24	-46	-59		
3*	.8	1	.2	1	.2	ADA ID	-75	-2	-75	07	75
		1	.6			BANNOCK ID					
				1*	2.1	BEAR LAKE ID					98
2	.7			1	.2	CANYON ID			-67	14	77
1*	1.2	1*	1.7			CASSIA ID	37				
				1*	2.1	ELMORE ID					98
		1*	1.6			IDAHO ID					
		2*	3.9			JEROME ID					
		1*	1.6			MINIDOKA ID					
		1	.6	1	.5	NEZ PERCE ID		-6			88
1*	1.9			1*	.5	PAYETTE ID			-72	11	88
1	1.0					TWIN FALLS ID					
121	.3	144	.3	139	.2	ILLINOIS IL	-2	-14	-15		

* SIGNIFICANTLY MORE DEATHS THAN EXPECTED

COUNTIES COMBINED; SEE TECHNICAL NOTES

WHITE FEMALE — CANCER OF THE SALIVARY GLAND — ICD 142

1950-1959 DEATHS	1950-1959 RATE	1960-1969 DEATHS	1960-1969 RATE	1970-1979 DEATHS	1970-1979 RATE	COUNTY AND STATE	PERCENT CHANGE 60S/50S	PERCENT CHANGE 70S/60S	PERCENT CHANGE 70S/50S	PERCENTILE %CHANGE 70S/50S	PERCENTILE RATES 1970S
2	.5	1	.3	1	.2	ADAMS IL	-43	-34	-63	19	74
1*	.9					BOND IL					
1	.8					BOONE IL					
1*	1.8					BROWN IL					
		1	.4			BUREAU IL					
		1	.7	1	.7	CARROLL IL		5			91
		2	.4			CHAMPAIGN IL					
1	.5	1	.3	1	.5	CHRISTIAN IL	-47	69	-10	63	86
1	.9					CLAY IL					
1	.8					CLINTON IL					
1	.5	1	.3			COLES IL	-38				
57	.3	72	.3	63	.2	COOK IL	5	-19	-15	61	78
				1	.6	CRAWFORD IL					89
		1	.4			DE KALB IL					
				1	.8	DE WITT IL					93
1	.8					DOUGLAS IL					
3	.3	5	.3	7	.3	DU PAGE IL	-3	-3	-6	67	81
				1	.4	EFFINGHAM IL					85
		1	.6			FAYETTE IL					
		1	.2	3*	.9	FRANKLIN IL		304			94
		1	.3			FULTON IL					
				1*	1.5	GALLATIN IL					97
2*	1.6					GREENE IL					
				1	.6	GRUNDY IL					90
		1	.9			HANCOCK IL					
				1*	1.5	HENDERSON IL					97
1	.6	1	.4	1	.3	IROQUOIS IL	-28	-30	-49	31	80
		1	.5	2*	1.1	JACKSON IL		125			95
		1*	1.0			JASPER IL					
				1*	.8	JERSEY IL					93
5	.5	3	.2	2	.1	KANE IL	-57	-39	-74	08	72
		2	.4	3*	.6	KANKAKEE IL		79			90
1*	1.5	1*	1.1			KENDALL IL	-27				
2	.6	1	.3	2	.3	KNOX IL	-50	13	-43	37	83
1	.2	1	.1	4	.2	LAKE IL	-62	224	22	77	78
2	.4	3	.5	1	.1	LA SALLE IL	33	-84	-79	05	70
1	.5	1	.5			LEE IL	13				
1	.5			1	.5	LOGAN IL			-3	68	87
		1	.4			MC DONOUGH IL					
1	.3	2	.5	1	.1	MC HENRY IL	70	-69	-47	33	72
1	.2			2	.3	MC LEAN IL			91	92	82
2	.3	2	.3	5*	.6	MACON IL	-12	95	72	89	88
1	.3	3*	.8	2	.5	MACOUPIN IL	169	-39	63	86	86
4	.4	1	.1	4	.3	MADISON IL	-79	201	-36	45	80
				1*	1.1	MARSHALL IL					95
		1*	1.2			MASSAC IL					
				1	1.1	MONROE IL					95
2*	1.0	1	.5			MONTGOMERY IL	-49				
1	.3					MORGAN IL					
1	.5	3*	1.4			OGLE IL	187				
3	.3	1	.1	2	.2	PEORIA IL	-69	54	-53	27	73
1	.8			1	.8	PERRY IL			-4	68	92
				1*	.8	PIATT IL					92
1	1.1					PIKE IL					
		1*	3.2	2*	10.2	POPE IL		217			99
		1*	4.2			PUTNAM IL					
1	.6					RANDOLPH IL					
1	.1	4	.5	1	.1	ROCK ISLAND IL	222	-75	-21	58	71
1	.1	2	.2	3	.2	ST CLAIR IL	62	35	118	94	77
		1	.4	2*	.8	SALINE IL		94			92
3	.4	4	.4	3	.3	SANGAMON IL	22	-44	-31	50	79
				1	.4	SHELBY IL					83
1	.3			1	.2	TAZEWELL IL			-50	30	73
2	.4	3	.5			VERMILION IL	16				
		2*	1.9	1*	1.1	WABASH IL		-41			96
1	.9					WAYNE IL					
		2*	1.4			WHITE IL					
1	.4			1	.3	WHITESIDE IL			-25	55	81
4	.5	3	.3	1	.1	WILL IL	-43	-74	-85	02	70
		1	.3	1	.2	WILLIAMSON IL		-30			77
1	.1	1	.1	3	.3	WINNEBAGO IL	-38	239	111	93	79
1	.9					WOODFORD IL					
48	.2	61	.2	56	.2	INDIANA IN	9	-22	-15		
1	.8	1	.6			ADAMS IN	-22				
				3	.2	ALLEN IN					76
		2*	.7	1	.3	BARTHOLOMEW IN		-50			82
				1*	1.2	BLACKFORD IN					96
1*	3.9					BROWN IN					
		1	.9			CARROLL IN					
		1	.3			CASS IN					
1	.5			1	.2	CLARK IN			-48	33	78
		1	.8			CLAY IN					
1	.5			2*	.5	CLINTON IN			-7	65	86
				1*	1.6	CRAWFORD IN					97
		1	.5	1	.6	DE KALB IN		14			69
1	.2			1	.2	DELAWARE IN			-39	40	72
		1	.6			DUBOIS IN					
1	.2	1	.2			ELKHART IN	3				
				1	.3	FLOYD IN					79
				1	.8	FOUNTAIN IN					93
1	.8					FULTON IN					
2	.6	2	.5	1	.2	GRANT IN	-11	-70	-73	09	73
1	.6	2*	1.3	1	.4	GREENE IN	140	-74	-37	44	83
1*	1.0					HARRISON IN					
				1	.3	HENDRICKS IN					81
		1	.4	1	.2	HENRY IN		-47			75
		1	.3	2	.5	HOWARD IN		40			86
				1	.3	HUNTINGTON IN					79
1	.8					JACKSON IN					

* SIGNIFICANTLY MORE DEATHS THAN EXPECTED

WHITE FEMALE — CANCER OF THE SALIVARY GLAND — ICD 142

1950-1959 DEATHS	1950-1959 RATE	1960-1969 DEATHS	1960-1969 RATE	1970-1979 DEATHS	1970-1979 RATE	COUNTY AND STATE	PERCENT CHANGE 60S/50S	PERCENT CHANGE 70S/60S	PERCENT CHANGE 70S/50S	PERCENTILE %CHANGE 70S/50S	PERCENTILE RATES 1970S
1*	1.3					JASPER IN					
1	.7					JAY IN					
1	.5					JEFFERSON IN					
		1	.2			JOHNSON IN					
1	.3			2	.8	KNOX IN			139	95	93
1	.6	3*	1.2			KOSCIUSKO IN	88				
				1*	.8	LAGRANGE IN					93
2	.1	2	.1	5	.2	LAKE IN	7	99	112	93	78
5*	1.2	3	.6			LA PORTE IN	-49				
		3	.5	3	.4	MADISON IN		-15			84
8	.3	11	.3	6	.2	MARION IN	8	-45	-40	39	74
1	.6	1	.5	1	.5	MARSHALL IN	-24	13	-14	62	88
		1	.4			MIAMI IN					
		1	.4	1	.2	MONROE IN		-33			79
1	.6			1	.5	MONTGOMERY IN			-20	59	86
1	.6	1	.7			MORGAN IN	7				
1	.7					NOBLE IN					
1*	1.0					ORANGE IN					
		2*	3.0			PIKE IN					
				1	.2	PORTER IN					77
		1	.6			PUTNAM IN					
		1	.7	1	.5	RANDOLPH IN		-33			86
1	.8					RIPLEY IN					
1	.7	1	.6			RUSH IN	-13				
4	.4	1	.1	3	.2	ST JOSEPH IN	-74	135	-39	41	77
				1*	1.1	SCOTT IN					96
		1	.5	1	.3	SHELBY IN		-32			82
		1*	.8	1*	.5	SPENCER IN		-39			87
		1	.8			STARKE IN					
1	.3			2	.4	TIPPECANOE IN			65	86	85
1	.1	4	.4	2	.2	VANDERBURGH IN	234	-60	35	81	73
2	.3	3	.5	1	.1	VIGO IN	50	-79	-68	13	70
		1*	3.0			WARREN IN					
		1	.9	1	.9	WASHINGTON IN		2			94
				1	.2	WAYNE IN					76
		1	.7			WELLS IN					
1	.8			1	.7	WHITE IN			-18	59	91
27	.2	40	.2	31	.2	IOWA IA	26	-31	-14		
		1*	1.0			ADAIR IA					
		1	.6	1	.5	APPANOOSE IA		-20			87
				1*	.5	AUDUBON IA					86
				1	.2	BLACK HAWK IA					73
				1	.2	BOONE IA					75
		1	.9			BREMER IA					
		1	.7	1	.8	BUCHANAN IA		5			92
		1	.6			BUENA VISTA IA					
				1	.8	BUTLER IA					92
		1*	1.0			CHICKASAW IA					
2*	4.8			1*	2.2	CLARKE IA			-54	26	98
		1	1.0			CLAYTON IA					
		1	.3	1	.3	CLINTON IA		33			82
				1*	.7	DECATUR IA					91
				1*	.8	DELAWARE IA					92
				2	.5	DES MOINES IA					87
		1	.2			DUBUQUE IA					
2*	1.2					FAYETTE IA					
1*	1.2					FREMONT IA					
2*	2.1					GREENE IA					
1*	.8			2*	1.1	GUTHRIE IA			28	79	95
		1	1.0			HENRY IA					
2*	3.1					HUMBOLDT IA					
1*	1.0					IOWA IA					
1	.9			1	.6	JEFFERSON IA			-38	43	88
1	.5					JOHNSON IA					
1	1.0					JONES IA					
		3*	1.1	2*	.6	LEE IA		-47			89
3	.4	3	.4	1	.1	LINN IA	-7	-83	-84	03	70
		1	.6			MAHASKA IA					
1	.8					MARION IA					
		2*	.7			MARSHALL IA					
				1*	.7	MITCHELL IA					91
1	.6	1	.3			MUSCATINE IA	-39				
				1	.7	PLYMOUTH IA					91
1*	1.0			1*	.6	POCAHONTAS IA			-40	39	90
3	.3	6	.4	4	.3	POLK IA	46	-25	9	73	80
		4*	.9	1	.2	POTTAWATTAMIE IA		-80			74
1	.9			1	.7	POWESHIEK IA			-25	55	91
		4*	.6	2	.2	SCOTT IA		-60			78
		1*	1.0			SHELBY IA					
1	.8					SIOUX IA					
				1	.2	WAPELLO IA					75
1	.8					WARREN IA					
		2	.6			WEBSTER IA					
		1*	1.1			WINNEBAGO IA					
		1	.7			WINNESHIEK IA					
1	.2	1	.1			WOODBURY IA	-32				
				1	1.1	WRIGHT IA					96
32	.3	30	.2	28	.2	KANSAS KS	-21	-26	-42		
				1	.5	ALLEN KS					87
				1*	2.4	BARBER KS					98
1	.8	1	.5			BARTON KS	-35				
1	1.0	1	.7			BOURBON KS	-28				
		1	.5			BROWN KS					
		2*	.9	1	.2	BUTLER KS		-71			79
		1	.4			CHEROKEE KS					
				1*	.8	COFFEY KS					92
2*	.8	2	.7			COWLEY KS	-17				

* SIGNIFICANTLY MORE DEATHS THAN EXPECTED

WHITE FEMALE — CANCER OF THE SALIVARY GLAND — ICD 142

1950-1959 DEATHS	1950-1959 RATE	1960-1969 DEATHS	1960-1969 RATE	1970-1979 DEATHS	1970-1979 RATE	COUNTY AND STATE	PERCENT CHANGE 60S/50S	PERCENT CHANGE 70S/60S	PERCENT CHANGE 70S/50S	PERCENTILE %CHANGE 70S/50S	PERCENTILE RATES 1970S
1	.3			1	.3	CRAWFORD KS			-8	64	82
1	.5			1	.5	DOUGLAS KS			0	70	87
		1	.9	2*	1.2	FORD KS		32			96
		1*	1.0			GEARY KS					
				1*	5.2	GOVE KS					99
1*	1.4	1*	1.7			JEFFERSON KS	23				
		2	.3	1	.1	JOHNSON KS		-81			69
		1*	1.1			KINGMAN KS					
1*	3.6					KIOWA KS					
1	.5			1	.3	LABETTE KS			-38	42	80
2*	1.0					LEAVENWORTH KS					
				1	.9	LYON KS					94
				1	.5	MC PHERSON KS					86
				1	.6	MARION KS					89
				1*	1.1	MITCHELL KS					95
2	.6	2	.6			MONTGOMERY KS	-5				
				1*	.7	MORRIS KS					91
				1*	.6	NEMAHA KS					89
1*	.8	1	.7			OSAGE KS	-9				
		1*	1.2			OSBORNE KS					
1*	1.4					POTTAWATOMIE KS					
		1*	1.0			PRATT KS					
				2	.5	RENO KS					87
		1*	1.6	1*	2.9	REPUBLIC KS		80			99
		1*	2.1			ROOKS KS					
1*	3.4					RUSH KS					
				1*	.7	RUSSELL KS					92
				1	.3	SALINE KS					80
				1*	2.7	SCOTT KS					99
6*	.6	3	.2	2	.1	SEDGWICK KS	-68	-44	-82	03	71
1*	2.6					SEWARD KS					
4*	.6	3	.5			SHAWNEE KS	-25				
		1*	1.1			SMITH KS					
				1*	.9	STAFFORD KS					94
1	.6					SUMNER KS					
				1*	1.1	WILSON KS					95
4	.6	2	.2	1	.1	WYANDOTTE KS	-57	-65	-85	02	70
35	.3	42	.3	44	.2	KENTUCKY KY	-4	-15	-19		
		1*	1.1			ALLEN KY					
		1	.6			BARREN KY					
				1*	1.6	BATH KY					97
				1	.7	BOONE KY					91
		1	.3	1	.3	BOYD KY		-5			81
				1	.7	BOYLE KY					91
		2*	1.8	1*	1.6	BRECKINRIDGE KY		-16			97
		1*	1.7			BUTLER KY					
		1*	1.1			CALDWELL KY					
2	.5			2	.4	CAMPBELL KY			-9	63	85
4*	2.5					CHRISTIAN KY					
				2*	1.4	CLARK KY					97
1*	4.6					CLINTON KY					
				1*	3.2	CUMBERLAND KY					99
		1	.2			DAVIESS KY					
1	.2	2	.3	1	.1	FAYETTE KY	72	-79	-63	18	70
				1*	.9	FLEMING KY					94
				1	.5	FLOYD KY					86
1	.8	1	.7			FRANKLIN KY	-18				
1*	2.7					GARRARD KY					
		1*	1.3			GRANT KY					
1	.5	1	.4			GRAVES KY	-14				
		1*	1.0	2*	1.5	GRAYSON KY		55			97
				1*	1.1	GREEN KY					95
		1	.7			GREENUP KY					
		1	.6			HARLAN KY					
		3*	1.6			HENDERSON KY					
		1*	1.9			HICKMAN KY					
				1	.4	HOPKINS KY					84
8	.3	5	.2	13*	.3	JEFFERSON KY	-51	111	5	71	83
1*	1.1	1*	.9	1*	.7	JOHNSON KY	-15	-27	-38	42	90
3	.6	2	.3			KENTON KY	-50				
				1*	1.5	KNOTT KY					97
1	.9					KNOX KY					
				2*	1.3	LAUREL KY					96
				1*	1.7	LESLIE KY					98
				1	.6	LETCHER KY					90
				1*	1.2	LINCOLN KY					96
1*	1.9					LIVINGSTON KY					
				1	.6	LOGAN KY					90
1*	3.7					LYON KY					
		1	.3	1	.2	MC CRACKEN KY		-34			75
		1*	1.7			MC CREARY KY					
		1	.6	1	.6	MADISON KY		2			90
		1*	1.0			MARION KY					
		1*	.9	1	.6	MARSHALL KY		-37			88
1*	1.2					MASON KY					
		1*	1.4			MONTGOMERY KY					
1	.8					MUHLENBERG KY					
		1*	1.0			NELSON KY					
				1*	2.0	NICHOLAS KY					98
		1	.8			OHIO KY					
2*	3.7					PENDLETON KY					
		2*	1.4			PERRY KY					
		1*	3.3			POWELL KY					
		1	.5			PULASKI KY					
1*	2.3					RUSSELL KY					
1*	1.2					SCOTT KY					
2*	2.2			1	.8	SHELBY KY			-66	15	92
				1*	3.0	TRIMBLE KY					99
1	.5	2*	.7			WARREN KY	47				

* SIGNIFICANTLY MORE DEATHS THAN EXPECTED

WHITE FEMALE — CANCER OF THE SALIVARY GLAND — ICD 142

1950-1959 DEATHS	1950-1959 RATE	1960-1969 DEATHS	1960-1969 RATE	1970-1979 DEATHS	1970-1979 RATE	COUNTY AND STATE		PERCENT CHANGE 60S/50S	PERCENT CHANGE 70S/60S	PERCENT CHANGE 70S/50S	PERCENTILE %CHANGE 70S/50S	PERCENTILE RATES 1970S
27	.3	36	.3	24	.2	LOUISIANA	LA	0	-50	-51		
1	.7					ACADIA	LA					
		1*	1.7			ALLEN	LA					
		1*	1.8			ASSUMPTION	LA					
		1	.8			BOSSIER	LA					
2	.3	2	.2	1	.1	CADDO	LA	-18	-58	-66	14	70
1	.6	2	.5	2	.4	CALCASIEU	LA	-15	-34	-44	35	83
				1*	2.5	CATAHOULA	LA					98
1*	1.4					DE SOTO	LA					
				1	.1	EAST BATON ROUGE	LA					71
		1*	4.5			EAST CARROLL	LA					
		1	.6			EVANGELINE	LA					
		1*	1.0			FRANKLIN	LA					
		1*	1.6			GRANT	LA					
2*	1.6	2*	1.1	1	.5	IBERIA	LA	-30	-58	-70	11	86
		1*	1.7			JACKSON	LA					
1	.3	1	.1	2	.1	JEFFERSON	LA	-76	87	-56	23	71
1*	1.1					JEFFERSON DAVIS	LA					
1	.6	1	.4			LAFAYETTE	LA	-34				
1*	1.7	1*	1.9	1*	.8	LA SALLE	LA	12	-56	-51	28	93
2*	2.3					LINCOLN	LA					
		1*	.7			LIVINGSTON	LA					
		1*	1.4			MOREHOUSE	LA					
1*	1.0					NATCHITOCHES	LA					
9	.4	5	.2	1	.0	ORLEANS	LA	-50	-83	-91	00	69
		1	.3			OUACHITA	LA					
		1*	1.6			POINTE COUPEE	LA					
		1	.3	2	.4	RAPIDES	LA		18			83
		1*	1.6			RICHLAND	LA					
1*	1.1			1*	.7	SABINE	LA			-36	45	91
		1	.5			ST BERNARD	LA					
				1*	1.9	ST JAMES	LA					98
1*	3.7					ST JOHN THE BAPT	LA					
		1	.6	1	.4	ST LANDRY	LA		-38			83
1*	.9	1	.9			ST MARY	LA	-6				
		2*	1.2	1	.4	ST TAMMANY	LA		-68			84
		1	.5	1	.4	TANGIPAHOA	LA		-20			83
				2*	.9	TERREBONNE	LA					94
				2*	.7	VERMILION	LA					92
		1*	1.2			VERNON	LA					
				2*	1.0	WASHINGTON	LA					95
1	1.3			1	.5	WEBSTER	LA			-62	19	86
		1*	1.3			WINN	LA					
8	.2	12	.2	12	.1	MAINE	ME	25	-23	-4		
1	.3	2	.3			ANDROSCOGGIN	ME	30				
		1	.2	1	.2	AROOSTOOK	ME		-10			76
1	.1	1	.1	2	.1	CUMBERLAND	ME	-23	57	21	77	71
				1	.5	FRANKLIN	ME					86
		1	.4			HANCOCK	ME					
1	.2	3	.6	1	.1	KENNEBEC	ME	149	-82	-55	24	70
1	.5					KNOX	ME					
1	.7					LINCOLN	ME					
				1	.2	OXFORD	ME					73
1	.2	2	.3	2	.3	PENOBSCOT	ME	74	8	88	91	80
				1	.5	SAGADAHOC	ME					88
1	.5	1	.3	1	.4	SOMERSET	ME	-36	17	-25	55	83
				1	.5	WALDO	ME					87
1	.4					WASHINGTON	ME					
		1	.2	1	.1	YORK	ME		-27			71
30	.3	35	.3	36	.2	MARYLAND	MD	-11	-23	-32		
		3	.6	1	.1	ALLEGANY	MD		-76			72
3*	.8			6*	.5	ANNE ARUNDEL	MD			-32	49	87
2	.2	2	.1	4	.1	BALTIMORE	MD	-42	14	-34	47	71
		1	.3	1	.2	CARROLL	MD		-22			78
1*	2.0					CHARLES	MD					
1	.6					DORCHESTER	MD					
3*	1.0			1	.2	FREDERICK	MD			-79	05	75
1*	1.0	1	.6			HOWARD	MD	-44				
2	.2	4	.3	8	.3	MONTGOMERY	MD	49	14	70	88	81
7*	.9	4	.3	2	.1	PRINCE GEORGE'S	MD	-69	-71	-91	00	70
		1*	.8			QUEEN ANNE'S	MD					
		1*	1.1	1	.3	ST MARY'S	MD		-68			83
1*	1.1	1*	1.2			SOMERSET	MD	3				
		1	.2	2	.3	WASHINGTON	MD		35			81
1	.5	2*	.8	1	.3	WICOMICO	MD	53	-57	-35	47	82
1	1.2					WORCESTER	MD					
7	.2	14	.3	9	.3	BALTIMORE CITY	MD	88	-15	60	86	81
79	.3	69	.2	73	.2	MASSACHUSETTS	MA	-28	-8	-33		
		1	.1	2	.2	BARNSTABLE	MA		15			73
1	.1	2	.2	4	.4	BERKSHIRE	MA	79	67	198	97	83
9	.4	9	.3	5	.2	BRISTOL	MA	-17	-48	-57	23	74
7	.2	8	.2	4	.1	ESSEX	MA	-12	-49	-55	24	70
3*	.9	1	.2	2	.4	FRANKLIN	MA	-76	87	-55	24	84
6	.3	7	.2	5	.1	HAMPDEN	MA	-16	-45	-54	26	72
2	.4			2	.2	HAMPSHIRE	MA			-38	42	77
19	.3	14	.2	16	.2	MIDDLESEX	MA	-41	-1	-42	38	74
5	.2	5	.1	10	.3	NORFOLK	MA	-23	89	46	84	80
4	.3	4	.2	6	.3	PLYMOUTH	MA	-33	22	-18	59	80
12	.2	10	.2	7	.1	SUFFOLK	MA	-7	-43	-46	34	71
11	.3	8	.2	10	.2	WORCESTER	MA	-36	-2	-37	43	76
93*	.3	78	.2	69	.2	MICHIGAN	MI	-33	-27	-51		

* SIGNIFICANTLY MORE DEATHS THAN EXPECTED

WHITE FEMALE — CANCER OF THE SALIVARY GLAND — ICD 142

COUNTY AND STATE	1950-1959 DEATHS	1950-1959 RATE	1960-1969 DEATHS	1960-1969 RATE	1970-1979 DEATHS	1970-1979 RATE	PERCENT CHANGE 60S/50S	PERCENT CHANGE 70S/60S	PERCENT CHANGE 70S/50S	PERCENTILE %CHANGE 70S/50S	PERCENTILE RATES 1970S
ALLEGAN MI	1	.4	1	.3	2	.5	-24	84	41	83	87
ALPENA MI	1	.9			2*	1.0			15	76	95
BARRY MI			1	.5							
BAY MI	1	.2			2	.3			32	80	82
BERRIEN MI	3	.5	2	.3	2	.3	-43	-10	-49	32	79
BRANCH MI					1	.5					88
CALHOUN MI	1	.2									
CHEBOYGAN MI	2*	2.9									
CLARE MI			1*	1.2	1	.5		-61			86
CLINTON MI	1	.5	2*	1.1			93				
DICKINSON MI					1	.7					90
EATON MI			2	.7							
EMMET MI	1*	1.0									
GENESEE MI	4	.3	3	.2	1	.1	-33	-73	-82	04	69
GOGEBIC MI	2*	1.3	1	.6			-55				
HILLSDALE MI			1	.4							
HOUGHTON MI	1	.4			1	.3			-28	53	80
INGHAM MI	1	.1	2	.2	1	.1	56	-53	-27	53	70
IONIA MI					1	.3					80
IRON MI			1	.8							
ISABELLA MI	1	1.0	1	.7			-35				
JACKSON MI	2	.3	2	.3			-10				
KALAMAZOO MI	5*	.8	2	.2	1	.1	-71	-59	-88	01	70
KENT MI	5	.3	4	.2	4	.1	-39	-24	-54	27	72
LEELANAU MI					2*	3.2					99
LENAWEE MI	1	.3	2	.4			57				
LIVINGSTON MI					1	.4					83
LUCE MI					1*	1.9					98
MACOMB MI	3	.5	2	.1	6	.2	-77	94	-55	25	76
MANISTEE MI			1	1.0							
MARQUETTE MI	1	.4			1	.4			-8	64	83
MENOMINEE MI	1	1.0									
MIDLAND MI			1	.6							
MISSAUKEE MI					1*	1.8					98
MONROE MI	1	.3	1	.2			-20				
MONTCALM MI					1	.4					83
MUSKEGON MI	1	.2	2	.3	1	.1	49	-56	-34	47	71
NEWAYGO MI			1	.7							
OAKLAND MI	7	.4	9	.4	7	.1	2	-60	-59	21	72
ONTONAGON MI			1*	2.0							
OTTAWA MI			1	.2							
PRESQUE ISLE MI	1*	1.4									
ROSCOMMON MI					1*	.8					92
SAGINAW MI	2	.3	2	.2	1	.1	-22	-55	-65	16	70
ST CLAIR MI	1	.3									
ST JOSEPH MI			2*	.8							
SHIAWASSEE MI	1	.4	2	.7	1	.2	72	-66	-41	38	78
TUSCOLA MI	2*	1.1									
VAN BUREN MI	4*	1.5			2	.5			-66	14	87
WAYNE MI	35*	.4	24	.2	22	.2	-40	-7	-44	35	76
WEXFORD MI			1	.9	1	.5		-43			87
MINNESOTA MN	31	.2	49	.2	50	.2	26	-17	4		
AITKIN MN			1*	.8							
ANOKA MN			1	.5							
BELTRAMI MN					1	.7					91
BENTON MN			2*	2.2							
BLUE EARTH MN			1	.2	1	.4		69			84
CARLTON MN			1	.8	1	.4		-43			85
CASS MN			1	.8							
CLAY MN					1	.3					83
COTTONWOOD MN			1*	.9	1*	.6		-38			88
CROW WING MN	1	.5									
DAKOTA MN			1	.3							
DODGE MN			1*	.8							
DOUGLAS MN					1	.4					84
FARIBAULT MN			2*	1.1							
FREEBORN MN					1	.2					77
GOODHUE MN	1	.4	3*	1.5			234				
HENNEPIN MN	5	.1	10	.2	11	.2	70	-9	56	85	75
HOUSTON MN	1*	1.2									
ISANTI MN					1*	1.1					95
ITASCA MN	1	.5			2*	.8			49	84	93
JACKSON MN	1*	1.4									
LYON MN					1	.9					94
MAHNOMEN MN					1*	3.1					99
MOWER MN			1	.4	1	.4		14			84
MURRAY MN			1*	1.4	1*	1.0		-31			94
NOBLES MN					1	.9					94
OLMSTED MN			1	.2							
OTTER TAIL MN	1	.4			5*	1.1			180	96	95
PENNINGTON MN					1*	.7					91
PINE MN			1*	1.2							
PIPESTONE MN	1*	1.5									
RAMSEY MN	8	.4	11*	.5	6	.2	21	-53	-44	36	77
REDWOOD MN			1	.8							
RENVILLE MN			1	.5	1	.4		-14			85
RICE MN	2*	1.0			2*	.7			-34	48	90
ROCK MN	1*	2.0	1*	1.3			-35				
ROSEAU MN	1*	1.9									
ST LOUIS MN	1	.1	1	.1	2	.2	-37	122	39	82	73
SHERBURNE MN	1*	1.7			1*	.5			-73	09	86
SIBLEY MN			1*	.9							
STEARNS MN					1	.2					74
SWIFT MN	1*	1.3	1*	1.1			-19				
TODD MN			1	.6							
WADENA MN					1*	.9					93
WASECA MN					1*	1.2					96
WASHINGTON MN	1	.7									
WATONWAN MN			1*	1.2	1*	1.1		-6			95

* SIGNIFICANTLY MORE DEATHS THAN EXPECTED

WHITE FEMALE — CANCER OF THE SALIVARY GLAND — ICD 142

1950-1959 DEATHS	1950-1959 RATE	1960-1969 DEATHS	1960-1969 RATE	1970-1979 DEATHS	1970-1979 RATE	COUNTY AND STATE	PERCENT CHANGE 60S/50S	PERCENT CHANGE 70S/60S	PERCENT CHANGE 70S/50S	PERCENTILE %CHANGE 70S/50S	PERCENTILE RATES 1970S
1*	2.0	1*	1.6			WILKIN MN	-21				
				2*	.5	WINONA MN					87
1	.6					WPIGHT MN					
1*	1.3					YELLOW MEDICINE MN					
23*	.4	26*	.4	17	.2	MISSISSIPPI MS	-8	-58	-61		
2*	2.9			1*	.6	ADAMS MS			-78	06	90
1	.8					ALCORN MS					
		1*	1.4			CHICKASAW MS					
1*	2.1					CHOCTAW MS					
1*	1.4			1*	.8	COPIAH MS			-42	38	92
		1	.5	1	.3	FOPPEST MS		-35			82
		1*	1.8			GEORGE MS					
2	.7	2	.7	1	.2	HARRISON MS	-6	-70	-72	10	75
4*	1.1	3	.5	1	.1	HINDS MS	-54	-77	-90	00	71
		1*	1.5			HOLMES MS					
3*	2.2			2*	.5	JONES MS			-76	07	87
				1*	1.5	KEMPER MS					97
		1*	1.7			LAMAR MS					
		3*	1.0			LAUDERDALE MS					
1	.6					LEE MS					
				1	.7	LINCOLN MS					91
1*	1.0	2*	1.8	1	.4	LOWNDES MS	80	-80	-64	16	83
1*	1.8			1*	1.4	MADISON MS			-23	57	97
		1*	1.2			MARION MS					
		1*	1.7			MARSHALL MS					
1	.8					MONFOE MS					
		1*	2.0			MONTGOMERY MS					
				1*	.9	NESHOBA MS					93
1*	1.3					NEWTON MS					
		1*	3.7			PERRY MS					
		1	.6			PIKE MS					
				1	.6	PRENTISS MS					90
				1*	.7	SIMPSON MS					90
1*	2.4					SMITH MS					
1*	1.9					TALLAHATCHIE MS					
		1*	1.2	2*	1.5	TISHOMINGO MS		24			97
		1*	5.4			TUNICA MS					
				1	1.0	UNION MS					94
1	.9	2*	1.6			WAPPEN MS	69				
		1	.6			WASHINGTON MS					
		1*	1.7			WEBSTER MS					
1*	1.9					YAZOO MS					
52	.2	49	.2	50	.2	MISSOURI MO	-23	-12	-32		
				2*	1.0	ADAIR MO					94
		1*	.9			ANDREW MO					
1	.6	1	.8	1	.4	AUDRAIN MO	46	-56	-36	44	83
2*	1.5					BATES MO					
				1	.2	BOONE MO					77
2	.4	3	.4	3	.5	BUCHANAN MO	0	27	27	79	86
		1*	1.7	1*	.8	CAMDEN MO		-53			93
				2	.6	CAPE GIRAPDEAU MO					90
1	.8					CASS MO					
		1	.2	1	.2	CLAY MO		-10			74
				2*	1.1	CLINTON MO					95
1	.5					COLE MO					
1*	1.2					DAVIESS MO					
				1	.4	DUNKLIN MO					84
1*	1.0					GENTRY MO					
3	.4			1	.1	GPEENE MO			-76	07	71
		1	.5			HENRY MO					
9	.3	9	.3	5	.1	JACKSON MO	-16	-47	-55	25	72
1	.2	1	.1	3	.5	JASPER MO	-17	278	213	98	87
		2	.6			JEFFERSON MO					
				1	.3	JOHNSON MO					80
		1*	1.5	1*	1.9	KNOX MO		26			98
1	1.4	1	.3			LAFAYETTE MO	-75				
		1	.5	1	.4	LAWPENCE MO		-14			85
				1	.6	LINCOLN MO					89
				1	.3	LINN MO					82
		1	1.0			MACON MO					
1	.4			2*	.5	MARION MO			24	78	87
				1	1.2	MILLER MO					96
		1	.4			NEWTON MO					
		1*	1.4	1*	.8	PERRY MO		-42			93
				1	.3	PETTIS MO					82
				1	.5	PHELPS MO					87
		2*	1.9			PLATTE MO					
1	1.0	1	.6			POLK MO	-38				
				1*	.8	PULASKI MO					93
				1	.3	RANDOLPH MO					80
		1*	2.6			REYNOLDS MO					
		1*	1.4			RIPLEY MO					
1	.6					ST CHARLES MO					
		1*	1.0			ST CLAIR MO					
		1	.4			ST FRANCOIS MO					
6	.2	3	.1	8	.2	ST LOUIS COUNTY MO	-70	117	-35	46	73
		1	.4			SALINE MO					
1	.7					SCOTT MO					
		2*	1.7			TEXAS MO					
				1*	1.9	WARREN MO					98
		2*	2.1			WEBSTER MO					
1*	.9					WRIGHT MO					
18*	.4	8	.2	5	.1	ST LOUIS CITY MO	-51	-50	-75	07	71
7	.3	9	.3	3	.1	MONTANA MT	2	-75	-74		
		1*	2.8			BEAVERHEAD MT					

* SIGNIFICANTLY MORE DEATHS THAN EXPECTED

WHITE FEMALE — CANCER OF THE SALIVARY GLAND — ICD 142

1950-1959 DEATHS	1950-1959 RATE	1960-1969 DEATHS	1960-1969 RATE	1970-1979 DEATHS	1970-1979 RATE	COUNTY AND STATE	PERCENT CHANGE 60S/50S	PERCENT CHANGE 70S/60S	PERCENT CHANGE 70S/50S	PERCENTILE %CHANGE 70S/50S	PERCENTILE RATES 1970S
		1	.3			CASCADE MT					
1*	1.7					CUSTER MT					
		1*	.9			DEER LODGE MT					
1	.7	2*	1.0			FLATHEAD MT	49				
				1*	14.1	GARFIELD MT					99
		1*	1.2			LAKE MT					
1	.7					MISSOULA MT					
1*	1.5					PARK#MT					
1*	1.6	1*	1.6			RAVALLI MT	2				
1	.3	1	.3	2*	.6	SILVER BOW MT	-2	74	71	89	89
1*	4.3					TETON MT					
		1	.3			YELLOWSTONE MT					
22	.3	26	.3	33*	.3	NEBRASKA NB	0	0	-1		
		1	.6	1	.2	ADAMS NB		-60			78
		1*	1.3			ANTELOPE NB					
1*	2.1					BOONE NB					
				1*	2.8	BROWN NB					99
				1	.5	BUFFALO NB					87
1	1.0	2*	2.0			CASS NB	101				
		1*	2.8	1*	2.5	CHASE NB		-11			99
				1*	1.7	CHEYENNE NB					97
		1*	1.1	1*	.9	COLFAX NB		-18			94
				1	.5	CUSTER NB					88
		1*	1.5			DIXON NB					
				1	.5	DODGE NB					86
6	.4	6	.3	7	.3	DOUGLAS NB	-17	-14	-29	52	80
				1*	1.2	FILLMORE NB					96
				1*	.6	FURNAS NB					88
				1	.3	GAGE NB					81
		2*	.7	4*	1.1	HALL NB		57			95
				1*	.7	HAMILTON NB					92
1*	3.2					HARLAN NB					
		1*	3.1	2*	5.4	HITCHCOCK NB		74			99
1*	1.0					JEFFERSON NB					
		1*	12.4			KEYA PAHA NB					
1*	1.2					KNOX NB					
1	.1	1	.1	5*	.4	LANCASTER NB	-20	362	270	99	85
		1	.5			LINCOLN NB					
		1*	.7	1*	1.4	NEMAHA NB		90			97
1*	2.0			1*	1.2	NUCKOLLS NB			-39	41	96
1	.9					OTOE NB					
2*	3.5					PHELPS NB					
				1	.4	PLATTE NB					85
1*	1.3					SALINE NB					
1*	1.6					SARPY NB					
1	.8	1	.5			SCOTTS BLUFF NB	-31				
1*	1.1					SEWARD NB					
		1*	1.8			SHERIDAN NB					
		2*	6.2			STANTON NB					
1*	1.3					WASHINGTON NB					
		1*	1.8			WAYNE NB					
1*	1.0	1*	.8			YORK NB	-24				
3	.3	1	.0	5	.2	NEVADA NV	-84	316	-34		
1	.2	1	.1	3	.2	CLARK NV	-44	102	14	75	75
2	.6			2	.3	WASHOE NV			-52	28	81
6	.2	8	.2	12	.2	NEW HAMPSHIRE NH	6	27	34		
		1	.4			BELKNAP NH					
				2*	.9	CARROLL NH					94
1	.5					CHESHIRE NH					
				2	.5	GRAFTON NH					87
3	.3	3	.3	6	.4	HILLSBOROUGH NH	-16	65	39	82	84
1	.1	2	.3			MERRIMACK NH	109				
1	.2			1	.1	ROCKINGHAM NH			-53	27	71
		2	.6	1	.3	STRAFFORD NH		-54			80
69	.3	89	.3	78	.2	NEW JERSEY NJ	-5	-29	-32		
		1	.1	2	.1	ATLANTIC NJ		79			72
5	.2	13	.3	4	.1	BERGEN NJ	96	-78	-57	23	69
2	.3			3	.2	BURLINGTON NJ			-38	42	76
9*	.6	3	.2	6	.2	CAMDEN NJ	-76	59	-62	19	78
3*	1.0			1	.1	CAPE MAY NJ			-87	01	71
1	.2			1	.1	CUMBERLAND NJ			-39	41	71
14	.3	17	.4	8	.2	ESSEX NJ	14	-51	-44	35	74
2	.4	1	.1	4	.5	GLOUCESTER NJ	-68	262	15	76	87
11	.3	14*	.4	4	.1	HUDSON NJ	15	-70	-65	15	71
				1	.2	HUNTERDON NJ					77
3	.3	9*	.7	2	.1	MERCER NJ	133	-82	-59	22	71
6	.6	5	.2	6	.2	MIDDLESEX NJ	-69	16	-64	16	77
2	.1	6	.3	7	.3	MONMOUTH NJ	89	-4	82	91	79
2	.1	1	.1	3	.1	MORRIS NJ	-47	55	-18	60	71
		3	.3	6	.3	OCEAN NJ		4			81
2	.1	4	.2	7	.3	PASSAIC NJ	87	55	190	97	80
		1	.4			SALEM NJ					
1	.3	2	.3	2	.2	SOMERSET NJ	6	-23	-19	59	77
2*	1.0	3*	1.1			SUSSEX NJ	8				
4	.2	5	.2	9	.3	UNION NJ	-15	61	36	81	80
		1	.2	2	.4	WARREN NJ		95			83
5	.2	2	.1	5	.1	NEW MEXICO NM	-72	101	-43		
1	.2	1	.1	2	.1	BERNALILLO NM	-34	43	-6	65	72
1	.5					CHAVES NM					
1*	1.2			1	.4	CURRY NM			-66	15	84
1*	2.6					LINCOLN NM					

* SIGNIFICANTLY MORE DEATHS THAN EXPECTED

\# COUNTIES COMBINED; SEE TECHNICAL NOTES

WHITE FEMALE — CANCER OF THE SALIVARY GLAND — COUNTY AND STATE — ICD 142

1950-1959 DEATHS	1950-1959 RATE	1960-1969 DEATHS	1960-1969 RATE	1970-1979 DEATHS	1970-1979 RATE	COUNTY	STATE	PERCENT CHANGE 60S/50S	PERCENT CHANGE 70S/60S	PERCENT CHANGE 70S/50S	PERCENTILE %CHANGE 70S/50S	PERCENTILE RATES 1970S
1*	1.1			1*	.8	SAN MIGUEL	NM			-32	49	92
		1	.4	1	.3	SANTA FE	NM		-23			83
178	.2	224	.2	243*	.2	NEW YORK	NY	3	-3	1		
5	.4	6	.4	6	.3	ALBANY	NY	3	-8	-5	67	82
1	.3	2	.8	1	.3	ALLEGANY	NY	136	-67	-23	57	79
23	.3	12	.1	23*	.3	BRONX	NY	-53	108	-2	69	81
2	.2	1	.1	3	.2	BROOME	NY	-53	99	-6	66	74
1	.2	4*	.8			CATTARAUGUS	NY	292				
1	.2	1	.2	5*	.8	CAYUGA	NY	-8	315	284	99	93
1	.1	2	.2	5*	.4	CHAUTAUQUA	NY	81	89	243	98	85
2	.4			1	.1	CHEMUNG	NY			-72	10	71
2	.9	1	.2			CHENANGO	NY	-75				
1	.4	2	.6			CLINTON	NY	48				
				2	.5	COLUMBIA	NY					88
		3*	1.0	2*	.6	CORTLAND	NY		-39			90
		1	.3			DELAWARE	NY					
3	.3	3	.3	2	.1	DUTCHESS	NY	-22	-48	-59	21	72
13	.3	17	.3	19	.3	ERIE	NY	4	5	9	73	81
		1	.5	1	.4	ESSEX	NY		-13			84
1	.4	1	.3	3*	1.0	FRANKLIN	NY	-18	183	132	95	94
				1	.3	FULTON	NY					81
				3*	.9	GENESEE	NY					94
		3*	1.2			GREENE	NY					
				1	.2	HERKIMER	NY					78
1	.2			1	.2	JEFFERSON	NY			-10	63	75
26	.2	18	.1	22	.2	KINGS(BROOKLYN)	NY	-34	16	-24	56	73
1	.4	1	.4	1	.4	LIVINGSTON	NY	-5	-1	-6	66	83
2	.1	10	.3	10	.3	MONROE	NY	244	1	247	98	79
1	.2	1	.3	1	.2	MONTGOMERY	NY	19	-30	-17	60	76
11	.3	13	.2	18	.2	NASSAU	NY	-38	6	-34	47	77
20	.2	34*	.4	20	.2	NEW YORK(MANHATTAN)	NY	64	-36	5	71	77
1	.1	1	.1	3	.2	NIAGARA	NY	-18	169	122	94	76
3	.2	2	.1	4	.2	ONEIDA	NY	-47	58	-16	61	74
5	.2	7	.3	12*	.5	ONONDAGA	NY	25	61	101	92	86
		1	.2	1	.2	ONTARIO	NY		-4			77
1	.1	4	.4	1	.1	ORANGE	NY	278	-77	-13	62	70
1	.5					ORLEANS	NY					
1	.2	2	.4	1	.2	OSWEGO	NY	91	-58	-20	58	73
2	.7	1	.4			OTSEGO	NY	-51				
				2	.4	PUTNAM	NY					85
16	.2	26	.2	17	.1	QUEENS	NY	0	-45	-45	34	71
2	.2	7*	.8	3	.2	RENSSELAER	NY	214	-71	-8	65	77
1	.1	3	.2	5	.3	RICHMOND	NY	135	31	206	98	80
2	.4			2	.1	ROCKLAND	NY			-58	22	72
		2	.4	1	.1	ST LAWRENCE	NY		-73			70
2	.4	2	.4	1	.1	SARATOGA	NY	-16	-74	-78	06	70
4	.5	3	.3	1	.1	SCHENECTADY	NY	-44	-78	-88	01	69
1	.6	1	.5	2*	1.3	SCHOHARIE	NY	-14	143	110	93	96
		1	.9	2*	.7	SENECA	NY		-22			91
		1	.2			STEUBEN	NY					
6	.3	6	.2	14	.2	SUFFOLK	NY	-37	51	-5	67	78
		1	.3	1	.2	SULLIVAN	NY		-35			77
		1	.5	1	.4	TIOGA	NY		-12			84
2	.3					ULSTER	NY					
1	.3	1	.3	1	.2	WARREN	NY	-7	-17	-23	57	78
		1	.4	2	.7	WASHINGTON	NY		75			91
		2	.5	1	.2	WAYNE	NY		-49			78
9	.2	11	.2	14	.2	WESTCHESTER	NY	-10	10	-1	70	78
33	.3	48	.3	42	.2	NORTH CAROLINA	NC	11	-40	-33		
		1	.3	1	.2	ALAMANCE	NC		-24			76
		1*	1.7	1*	1.1	ANSON	NC		-33			96
				1*	.9	BRUNSWICK	NC					93
		8*	1.1	2	.2	BUNCOMBE	NC		-81			75
1	.3					BURKE	NC					
				2	.5	CABARRUS	NC					87
1	.5	1	.4			CALDWELL	NC	-22				
		1	.8	2*	1.1	CARTERET	NC		28			95
		1*	1.8			CASWELL	NC					
				1	.2	CATAWBA	NC					76
		1*	1.1			CHATHAM	NC					
		1*	1.1			CHEROKEE	NC					
1	.5	2	.7	1	.3	CLEVELAND	NC	51	-54	-30	51	82
1	.5	3*	2.6			COLUMBUS	NC	372				
1	1.1	1	.5	1	.6	CRAVEN	NC	-49	4	-47	33	89
				1	.2	CUMBERLAND	NC					75
		1*	3.7			DARE	NC					
1	.6	2	.6			DAVIDSON	NC	-1				
				1	.6	DUPLIN	NC					89
1	.2			1	.2	DURHAM	NC			-16	61	75
1	.7			1	.6	EDGECOMBE	NC			-16	61	89
1	.3			3	.3	FORSYTH	NC			20	76	82
1	.3					GASTON	NC					
1*	.9					GRANVILLE	NC					
				1*	2.4	GREENE	NC					98
2	.4	2	.2	1	.1	GUILFORD	NC	-48	-65	-82	03	69
				1	.4	HARNETT	NC					85
		2*	.9	3*	.6	HENDERSON	NC		-33			90
				1*	1.2	HERTFORD	NC					96
		1	.5	1	.2	IREDELL	NC		-59			76
3*	1.9			1	.4	JOHNSTON	NC			-80	05	83
1	1.2					LENOIR	NC					
				2*	1.0	LINCOLN	NC					94
5*	.7			2	.1	MECKLENBURG	NC			-81	04	72
		1	.6	1	.7	MOORE	NC		15			90
		2	.7	1	.2	NEW HANOVER	NC		-69			77
		1*	1.1			NORTHAMPTON	NC					
1*	1.6					PASQUOTANK	NC					

* SIGNIFICANTLY MORE DEATHS THAN EXPECTED

WHITE FEMALE — CANCER OF THE SALIVARY GLAND — ICD 142

1950-1959 DEATHS	1950-1959 RATE	1960-1969 DEATHS	1960-1969 RATE	1970-1979 DEATHS	1970-1979 RATE	COUNTY AND STATE	PERCENT CHANGE 60S/50S	PERCENT CHANGE 70S/60S	PERCENT CHANGE 70S/50S	PERCENTILE %CHANGE 70S/50S	PERCENTILE RATES 1970S
1*	2.9					PERSON NC					
		1	.5	1	.4	PITT NC		-22			84
1*	1.6					POLK NC					
2*	1.2					RANDOLPH NC					
		2*	1.3			RICHMOND NC					
3*	1.6					ROBESON NC					
		2	.8			ROCKINGHAM NC					
		1	.2			ROWAN NC					
1	.5	1	.4			RUTHERFORD NC	-17				
		1	.7	1	.5	SAMPSON NC		-24			87
				1	.4	STANLY NC					85
		1*	1.4	1*	.8	TRANSYLVANIA NC		-41			93
		1	.6	1	.4	UNION NC		-34			84
		3	.4	1	.1	WAKE NC		-74			71
1*	2.8					WARREN NC					
		1*	2.5			WASHINGTON NC					
				1	.8	WATAUGA NC					92
1	.6					WAYNE NC					
		1	1.2			YADKIN NC					
				1*	1.1	YANCEY NC					95
9	.4	7	.2	11	.3	NORTH DAKOTA ND	-41	29	-24		
1*	1.1					BARNES ND					
				1*	1.7	BOTTINEAU ND					98
				1*	2.5	BURKE ND					99
1	.3					CASS ND					
		1*	2.0			CAVALIER ND					
1*	3.9					EDDY ND					
1	.5	1	.5	2*	.6	GRAND FORKS ND	-2	8	5	72	88
1*	4.3					HETTINGER ND					
		2*	4.8			LA MOURE ND					
1*	10.6					LOGAN ND					
				1*	3.7	MC KENZIE ND					99
				1*	1.9	MC LEAN ND					98
				2*	3.5	MOUNTPAIL ND					99
1*	2.9					NELSON ND					
				1*	2.5	SARGENT ND					99
1*	5.1	1*	9.3			SHERIDAN ND	81				
		1*	1.1			STARK ND					
1	.8			2*	.6	WARD ND			-27	54	90
		1*	.9			WILLIAMS ND					
88	.2	118	.2	137	.2	OHIO OH	12	-1	11		
1	.2	6*	1.0			ALLEN OH	486				
1	.5			1	.4	ASHLAND OH			-20	58	84
1	.2	1	.2			ASHTABULA OH	-5				
		2	.8	1	.3	ATHENS OH		-59			82
		2*	.9	2*	.9	AUGLAIZE OH		8			94
1	.2	2	.4	2	.3	BELMONT OH	108	-27	53	84	81
				3*	1.9	BROWN OH					98
				2	.2	BUTLER OH					72
1	.9					CARROLL OH					
		2*	1.1	1	.3	CHAMPAIGN OH		-74			80
2	.4	1	.1	4	.5	CLARK OH	-67	278	25	79	86
1	.4	1	.2	2	.4	CLERMONT OH	-43	98	12	74	85
1	.6			1	.6	CLINTON OH			10	74	90
1	.2			2	.2	COLUMBIANA OH			16	76	77
1	.7	1	.5	1	.4	COSHOCTON OH	-29	-7	-34	48	85
		1	.4	1	.3	CRAWFORD OH		-17			81
14	.2	21	.3	16	.2	CUYAHOGA OH	16	-28	-16	60	74
1	.4	1	.3	1	.2	DARKE OH	-21	-41	-54	26	74
		1	.5			DEFIANCE OH					
		2*	.8			DELAWARE OH					
				1	.2	ERIE OH					77
		2	.4			FAIRFIELD OH					
6	.2	6	.2	5	.1	FRANKLIN OH	-15	-31	-41	39	72
				1	.6	FULTON OH					90
1	.8			2*	.6	GEAUGA OH			-29	52	89
		2	.7			GREENE OH					
5	.1	9	.2	13	.3	HAMILTON OH	40	39	95	92	79
1	.3					HANCOCK OH					
		1	.4			HARDIN OH					
				2*	1.5	HARRISON OH					97
2*	1.5			1	.5	HENRY OH			-64	16	88
1	.4					HIGHLAND OH					
1	1.1					HOCKING OH					
		1	1.1			HOLMES OH					
1	.2	1	.2			JEFFERSON OH	39				
		1	.3	2*	.8	KNOX OH		171			93
		2	.3	4	.5	LAKE OH		49			87
1	.2			1	.1	LICKING OH			-35	46	72
2	.3	2	.2	4	.3	LOPAIN OH	-39	63	0	71	82
6	.3	4	.2	5	.2	LUCAS OH	-44	14	-36	44	74
		1	.8			MADISON OH					
		4	.3	7*	.4	MAHONING OH		36			84
		2	.6			MARION OH					
		2	.7	1	.3	MEDINA OH		-51			82
		1	.7	1	.4	MERCER OH		-43			84
1	.3			1	.2	MIAMI OH			-30	51	75
				1*	.5	MONROE OH					88
9*	.5	6	.3	4	.1	MONTGOMERY OH	-47	-43	-70	12	72
1*	.9					MORROW OH					
1	.2	1	.2	1	.1	MUSKINGUM OH	6	-41	-38	43	71
1	.5	1	.5			OTTAWA OH	-4				
1*	1.1	1*	1.2			PAULDING OH	15				
1	.6	1	.4			PERRY OH	-21				
		2*	1.1			PICKAWAY OH					
		1*	1.0	1	.6	PIKE OH		-43			89
				2	.4	PORTAGE OH					84

* SIGNIFICANTLY MORE DEATHS THAN EXPECTED

WHITE FEMALE — CANCER OF THE SALIVARY GLAND — ICD 142

1950-1959 DEATHS	1950-1959 RATE	1960-1969 DEATHS	1960-1969 RATE	1970-1979 DEATHS	1970-1979 RATE	COUNTY AND STATE	PERCENT CHANGE 60S/50S	PERCENT CHANGE 70S/60S	PERCENT CHANGE 70S/50S	PERCENTILE %CHANGE 70S/50S	PERCENTILE RATES 1970S
				1	.7	PUTNAM OH					91
		2	.3	5*	.7	RICHLAND OH		112			92
				1	.3	ROSS OH					79
2	.7	1	.2	1	.3	SANDUSKY OH	-70	36	-59	22	81
2	.5	1	.2	3*	.6	SCIOTO OH	-53	167	26	79	89
2	.6	1	.3			SENECA OH	-53				
		1	.5	2*	.8	SHELBY OH		59			93
2	.1	2	.1	4	.2	STARK OH	-13	60	39	82	74
2	.1	5	.2	5	.2	SUMMIT OH	128	-21	79	91	73
1	.1	2	.2	2	.2	TRUMBULL OH	142	-20	94	92	74
3*	.7	2	.4	3*	.5	TUSCARAWAS OH	-47	31	-31	50	87
				2*	1.3	UNION OH					96
1	.7	1	.7			VAN WERT OH	-12				
2*	1.0			1	.3	WARREN OH			-68	13	82
		1	.2	2	.4	WASHINGTON OH		76			85
2	.6			2	.4	WAYNE OH			-24	56	85
1	.3	1	.3	2	.5	WOOD OH	8	56	69	88	86
				1	.4	WYANDOT OH					83
28	.3	39	.3	38	.2	OKLAHOMA OK	8	-20	-14		
				2*	3.7	ADAIR OK					99
				1*	1.2	ALFALFA OK					96
				2*	2.4	ATOKA OK					98
		1*	2.9			BEAVER OK					
		2*	2.1	1*	.6	BLAINE OK		-72			89
1	.6			1	.9	CADDO OK			64	86	94
		3*	1.9			CANADIAN OK					
				1	.5	CARTER OK					86
				1	.2	CLEVELAND OK					78
2*	.8	2*	.9			COMANCHE OK	7				
1*	1.8					COTTON OK					
		1	.5			CREEK OK					
				1	.9	CUSTER OK					93
				1	.8	DELAWARE OK					93
1	.4	2	.7	1	.2	GARFIELD OK	89	-63	-31	50	78
		2*	1.1			GARVIN OK					
1	.6	1	.4			GRADY OK	-30				
1*	1.8					GRANT OK					
		1*	.9			GREER OK					
		2*	3.1			HASKELL OK					
1*	.9	1	.9			HUGHES OK	6				
2*	1.9					JACKSON OK					
1*	1.7					JEFFERSON OK					
		1	.3			KAY OK					
1*	1.4					KINGFISHER OK					
				1*	1.5	LATIMER OK					97
		1	.3	2*	.9	LE FLORE OK		162			93
		2*	1.1	1	.8	LOGAN OK		-25			93
1*	3.0	1*	2.2			LOVE OK	-27				
		1*	1.0			MC CLAIN OK					
				2*	1.3	MC CURTAIN OK					96
1*	2.1					MARSHALL OK					
1*	1.1					MAYES OK					
		1*	1.0			MURRAY OK					
		1	.3	1	.2	MUSKOGEE OK		-42			73
		1*	1.5			NOWATA OK					
4	.2	2	.1	7	.2	OKLAHOMA OK	-68	192	-6	66	78
1	.5	1	.5			OKMULGEE OK	18				
				1	.5	OSAGE OK					86
1	.7	2*	1.3			OTTAWA OK	89				
1	.6			2*	.6	PAYNE OK			0	70	89
		1	.4	1	.3	PITTSBURG OK		-24			80
		1	.6			SEMINOLE OK					
				1	.5	STEPHENS OK					88
		1*	1.6			TEXAS OK					
1*	1.5					TILLMAN OK					
3	.2	3	.2	5	.2	TULSA OK	-12	14	0	70	76
1*	1.7			1	.8	WAGONER OK			-52	28	93
1*	1.4	1*	1.1			WOODS OK	-19				
				1*	.8	WOODWARD OK					92
23	.3	26	.3	21	.2	OREGON OR	-15	-37	-47		
				1*	.7	BAKER OR					92
2	.4			2	.2	CLACKAMAS OR			-48	32	77
		1	.4			CLATSOP OR					
		1	.5	1	.3	COOS OR		-38			82
		1*	2.1			CROOK OR					
				2*	.8	DESCHUTES OR					93
		3*	1.1			DOUGLAS OR					
1	.3	1	.2			JACKSON OR	-43				
1*	3.5					JEFFERSON OR					
3*	2.1					JOSEPHINE OR					
		1	.5			KLAMATH OR					
		3	.3	4	.3	LANE OR		3			83
		1*	1.0			MALHEUR OR					
1	.2	2	.3	3	.2	MARION OR	26	-23	-3	69	76
11	.4	9	.3	4	.1	MULTNOMAH OR	-27	-59	-70	12	71
1*	1.2					TILLAMOOK OR					
1	.5	2*	.9	2*	.6	UMATILLA OR	88	-30	31	80	89
1*	1.0					UNION OR					
				1	.1	WASHINGTON OR					72
1	.8	1	.5	1	.2	YAMHILL OR	-37	-50	-68	13	78
120	.2	128	.2	147	.2	PENNSYLVANIA PA	-12	-4	-16		
13	.2	17	.2	18	.2	ALLEGHENY PA	3	-7	-5	67	74
1	.1	4	.4	2	.2	BEAVER PA	193	-56	30	80	74
				3*	.9	BEDFORD PA					93
4	.3	2	.1	2	.1	BERKS PA	-58	-40	-75	08	69

* SIGNIFICANTLY MORE DEATHS THAN EXPECTED

WHITE FEMALE — CANCER OF THE SALIVARY GLAND — ICD 142

1950-1959 DEATHS	1950-1959 RATE	1960-1969 DEATHS	1960-1969 RATE	1970-1979 DEATHS	1970-1979 RATE	COUNTY AND STATE	PERCENT CHANGE 60S/50S	PERCENT CHANGE 70S/60S	PERCENT CHANGE 70S/50S	PERCENTILE %CHANGE 70S/50S	PERCENTILE RATES 1970S
4	.5	1	.1			BLAIR PA	-77				
2	.7			1	.3	BRADFORD PA			-64	17	79
2	.2	2	.1	3	.1	BUCKS PA	-46	4	-43	36	72
2	.4	2	.3	1	.2	BUTLER PA	-11	-54	-59	21	72
1	.1			2	.1	CAMBRIA PA			8	73	72
				2	.7	CARBON PA					91
1	.3			1	.2	CENTRE PA			-39	41	76
2	.2	2	.2	8*	.5	CHESTER PA	-14	178	139	95	88
2	.5					CLEARFIELD PA					
		1	.3			COLUMBIA PA					
1	.2	2	.4	2	.4	CRAWFORD PA	106	-6	93	92	84
3	.6	1	.1	3	.3	CUMBERLAND PA	-81	175	-47	33	81
2	.2	2	.2	3	.2	DAUPHIN PA	-20	31	5	72	75
3	.1	5	.2	7	.2	DELAWARE PA	16	5	22	78	73
1	.1	4	.3	3	.2	ERIE PA	184	-45	56	85	73
2	.2	2	.2			FAYETTE PA	-8				
1	.2	2	.4			FRANKLIN PA	110				
1	.5					GREENE PA					
		1	.5			HUNTINGDON PA					
2	.6					INDIANA PA					
1	.4	1	.3	1	.6	JEFFERSON PA	-31	107	43	83	89
4	.3	6	.4	2	.1	LACKAWANNA PA	30	-75	-67	13	70
2	.2	2	.1	4	.2	LANCASTER PA	-25	42	7	72	73
2	.4	1	.2	1	.1	LAWRENCE PA	-62	-37	-76	07	70
1	.2	3	.6	2	.3	LEBANON PA	144	-50	22	77	81
2	.2	6	.4	5	.3	LEHIGH PA	144	-29	73	89	80
4	.2	4	.2	9	.3	LUZERNE PA	11	52	69	88	82
3	.5	1	.1	1	.1	LYCOMING PA	-71	-4	-72	10	71
1	.3					MC KEAN PA					
		3	.4	1	.1	MERCER PA		-68			72
				2*	.8	MIFFLIN PA					92
		2*	.8	1	.2	MONROE PA		-70			78
8	.4	7	.2	3	.1	MONTGOMERY PA	-41	-66	-80	04	70
				1	.8	MONTOUR PA					92
		3	.2	5	.4	NORTHAMPTON PA		42			83
2	.3			4*	.7	NORTHUMBERLAND PA			128	95	91
		1	.7			PERRY PA					
24	.3	25	.3	20	.2	PHILADELPHIA PA	2	-23	-21	58	76
1	.1	5	.5	3	.2	SCHUYLKILL PA	486	-58	146	96	77
1	.2	1	.2	2	.3	SOMERSET PA	1	42	43	83	82
		1	.5	1	.4	SUSQUEHANNA PA		-24			84
		1	.6	1	.6	TIOGA PA		-4			89
2	.5	1	.2			VENANGO PA	-55				
1	.3			1	.4	WARREN PA			13	75	83
1	.1	1	.1	1	.1	WASHINGTON PA	-23	21	-6	65	70
1	.6			1	.3	WAYNE PA			-43	36	82
5	.4	2	.1	5	.2	WESTMORELAND PA	-75	87	-53	28	76
1*	1.1					WYOMING PA					
3	.3	1	.1	9*	.5	YORK PA	-70	486	76	90	86
7	.2	8	.2	14	.2	RHODE ISLAND RI	-1	39	37		
1	.3	2	.3	2	.2	KENT RI	4	-30	-27	54	78
1	.4			1	.2	NEWPORT RI			-41	39	77
5	.1	5	.1	10	.2	PROVIDENCE RI	-9	84	68	88	78
		1	.3	1	.1	WASHINGTON RI		-61			71
6	.1	22	.3	19	.2	SOUTH CAROLINA SC	181	-40	70		
1*	1.5			1*	1.0	ABBEVILLE SC			-35	46	94
				2	.6	AIKEN SC					90
		1	.3	1	.2	ANDERSON SC		-24			75
		1*	.6			BERKELEY SC					
		1	.1	1	.1	CHARLESTON SC		-10			71
		1	.6			CHEROKEE SC					
		1*	1.1	1	.6	CHESTERFIELD SC		-43			90
		2*	.8	1	.4	FLORENCE SC		-46			85
		2	.2	1	.1	GREENVILLE SC		-57			70
		1*	1.1	1	.7	KERSHAW SC		-40			90
		1	.5	1	.4	LAURENS SC		-23			85
		1*	2.4			LEE SC					
		1	.5	1	.2	LEXINGTON SC		-66			73
1*	5.8			1*	4.0	MC CORMICK SC			-30	51	99
				1	.6	NEWBERRY SC					90
1	.6	1	.6			OCONEE SC	-8				
		1	.6			ORANGEBURG SC					
2	.5	3	.5	2	.3	RICHLAND SC	-1	-49	-50	30	79
		2	.5	2	.2	SPARTANBURG SC		-56			75
				1	.6	SUMTER SC					89
		1	.9			UNION SC					
1	.5	1	.4	1	.2	YORK SC	-25	-46	-59	21	76
3	.1	8	.2	11	.3	SOUTH DAKOTA SD	90	32	149		
				1	.8	BROOKINGS SD					92
				2*	.9	BROWN SD					93
1*	3.5					BRULE SD					
1*	5.3					BUTTE SD					
		1	.8	1	.8	CODINGTON SD		6			93
				1	1.1	DAVISON SD					95
				1*	1.8	FALL RIVER SD					98
1*	2.5					HUGHES SD					
				1*	1.6	HUTCHINSON SD					97
		1*	5.0			JERAULD SD					
				1*	2.4	KINGSBURY SD					98
		1*	1.2			LAKE SD					
				1*	1.1	LAWRENCE SD					96
				1*	.4	LINCOLN SD					85
		2	.4			MINNEHAHA SD					
		1*	2.2			TRIPP SD					
		1*	1.8	1*	1.3	TURNER SD		-29			96

* SIGNIFICANTLY MORE DEATHS THAN EXPECTED

1950-1959 DEATHS	1950-1959 RATE	1960-1969 DEATHS	1960-1969 RATE	1970-1979 DEATHS	1970-1979 RATE	COUNTY AND STATE	PERCENT CHANGE 60S/50S	PERCENT CHANGE 70S/60S	PERCENT CHANGE 70S/50S	PERCENTILE %CHANGE 70S/50S	PERCENTILE RATES 1970S
		1*	2.1			WALWORTH SD					
31	.2	48	.3	43	.2	TENNESSEE TN	17	-35	-24		
				1	.3	ANDERSON TN					79
1	.9					BEDFORD TN					
1	.3	1	.4	2	.5	BLOUNT TN	33	33	77	90	87
				2*	1.0	CAMPBELL TN					94
		1*	1.8	1*	1.0	CHEATHAM TN		-44			94
		1*	1.7			CHESTER TN					
		2*	1.8			COCKE TN					
1*	1.7	1*	1.0			CROCKETT TN	-39				
				3*	1.6	CUMBERLAND TN					97
6	.5	4	.2	6	.2	DAVIDSON TN	-54	12	-49	31	78
		1	.8			GILES TN					
1	.4	1	.4			GREENE TN	10				
		2*	1.5			HAMBLEN TN					
		5	.5	3	.2	HAMILTON TN		-59			75
1	1.4	2*	1.5	2*	1.0	HAWKINS TN	6	-30	-26	54	95
1*	1.8					HAYWOOD TN					
1*	2.0			2*	1.8	HUMPHREYS TN			-9	64	98
		2*	1.9	1	.6	JEFFERSON TN		-67			90
5	.7	3	.2	3	.2	KNOX TN	-66	-20	-73	10	74
		1	.7	1	.5	LINCOLN TN		-23			88
				1	.7	LOUDON TN					91
		1	.5	1	.6	MC MINN TN		21			89
				1	.6	MC NAIRY TN					90
		1*	1.4			MACON TN					
		1*	1.2			MARION TN					
1	1.3			2*	.7	MONTGOMERY TN			-47	34	91
		1	.4			OBION TN					
1*	1.3					OVERTON TN					
		1*	1.8			POLK TN					
				2*	1.3	ROBERTSON TN					96
		1	.6	1	.3	RUTHERFORD TN		-57			79
4	.3	7	.3	4	.2	SHELBY TN	14	-50	-43	37	73
		1*	1.2	1*	1.1	SMITH TN		-9			95
1	.3	1	.2			SULLIVAN TN	-25				
1	.6					SUMNER TN					
		1*	1.1			TIPTON TN					
1	.9					WARREN TN					
3*	1.1	2	.6	2	.4	WASHINGTON TN	-44	-41	-67	14	84
1	.5					WEAKLEY TN					
		1	.7			WILLIAMSON TN					
		2*	.9	1	.3	WILSON TN		-69			81
68	.2	107	.2	110	.2	TEXAS TX	9	-24	-17		
1	.7			1	.3	ANDERSON TX			-60	20	80
1	.7			1	.3	ANGELINA TX			-51	29	82
		1*	1.2	1*	1.8	AUSTIN TX		52			98
				1*	1.5	BANDERA TX					97
1*	2.8					BAYLOR TX					
		2	.5			BELL TX					
2	.1	6	.2	7	.2	BEXAR TX	73	0	73	89	74
		2*	5.8			BLANCO TX					
1	.4					BOWIE TX					
1	1.0					BRAZOS TX					
		1*	.8			BURNET TX					
		1	.8	1	.7	CALDWELL TX		-9			91
1	.2	2	.4	2	.3	CAMERON TX	94	-27	42	83	82
1*	3.8					CAMP TX					
		1	.8			CASS TX					
2*	1.0	1	.5			CHEROKEE TX	-51				
				1*	.9	CLAY TX					94
				1*	6.1	COCHRAN TX					99
		2*	3.0			COLEMAN TX					
1	.4			1	.2	COLLIN TX			-49	31	76
				1*	.4	COLORADO TX					86
				2*	1.1	COMAL TX					95
1*	5.6					CONCHO TX					
1	.8	1	.7	1	.4	COOKE TX	-14	-40	-48	32	85
		1	1.0			CORYELL TX					
				1*	5.0	CROCKETT TX					99
5	.2	12	.3	10	.2	DALLAS TX	55	-38	-3	68	74
1*	2.9			1*	1.3	DEAF SMITH TX			-55	25	96
		1*	1.5			DELTA TX					
				1	.4	DE WITT TX					84
1	.6	2*	1.7	1	.4	EASTLAND TX	202	-76	-27	53	84
1	.7					ECTOR TX					
		1*	8.3			EDWARDS TX					
1	.5	1	.6			ELLIS TX	14				
2	.3	3	.3	5	.4	EL PASO TX	25	25	56	85	84
		1	.4			ERATH TX					
1	1.1			1	.5	FALLS TX			-53	27	86
				1	.3	FANNIN TX					82
		1	.6			FAYETTE TX					
1*	2.3					FISHER TX					
1*	1.8					FLOYD TX					
				2*	.8	FORT BEND TX					93
		1	.2	2	.3	GALVESTON TX		29			80
		1*	3.3			GARZA TX					
				1	1.2	GILLESPIE TX					96
1	.3	2	.5			GRAYSON TX	96				
				1*	1.0	GRIMES TX					94
1	.9					GUADALUPE TX					
				1	.5	HALE TX					88
				1*	4.2	HAMILTON TX					99
				1*	3.9	HARDEMAN TX					99
1*	1.3	1	1.0			HARDIN TX	-27				
8	.3	11	.3	9	.2	HARRIS TX	2	-41	-40	40	72

* SIGNIFICANTLY MORE DEATHS THAN EXPECTED

WHITE FEMALE — CANCER OF THE SALIVARY GLAND — ICD 142

1950-1959 DEATHS	1950-1959 RATE	1960-1969 DEATHS	1960-1969 RATE	1970-1979 DEATHS	1970-1979 RATE	COUNTY AND STATE	PERCENT CHANGE 60S/50S	PERCENT CHANGE 70S/60S	PERCENT CHANGE 70S/50S	PERCENTILE %CHANGE 70S/50S	PERCENTILE RATES 1970S
		1	.7	2*	.9	HARRISON TX		29			94
2*	2.1			1	.7	HAYS TX			-66	15	92
		1	.1	2	.3	HIDALGO TX		97			79
				1	.3	HILL TX					83
1*	3.1					HOCKLEY TX					
1*	2.5	1*	1.8			HOOD TX	-28				
		1	.6	1	.4	HOPKINS TX		-40			83
1*	1.6					HOUSTON TX					
		2*	1.3	1	.6	HOWARD TX		-56			88
		1	.3	2	.7	HUNT TX		138			91
				1	.8	JASPER TX					92
		2	.3	2	.2	JEFFERSON TX		-29			74
		1	.3	1	.3	JOHNSON TX		-10			79
2*	1.5			1	.6	KAUFMAN TX			-63	18	88
				1*	1.0	KLEBERG TX					95
2*	1.0	1	.4			LAMAR TX	-58				
1	.9			1	.4	LAVACA TX			-51	29	85
1*	1.5	1*	1.2	1*	.7	MC CULLOCH TX	-20	-38	-51	29	92
		1*	2.6			MASON TX					
1	.6	1	.2			MIDLAND TX	-59				
		1*	1.9			MILLS TX					
1	1.1					MONTAGUE TX					
				1*	1.4	MOORE TX					97
		1*	1.8	1*	1.6	MORRIS TX		-8			97
		1	.4	2*	.5	NAVARRO TX		30			88
		1	.1	2	.2	NUECES TX		95			75
1	.8					PALO PINTO TX					
1*	2.1					PANOLA TX					
1*	1.8					POLK TX					
		2	.5			POTTER TX					
		1	.5			RANDALL TX					
1*	2.3	1*	2.3			ROBERTSON TX	2				
				1*	2.0	SAN AUGUSTINE TX					98
		2*	1.3	1	.4	SAN PATRICIO TX		-70			83
				1	.5	SHELBY TX					86
		2	.5	2	.3	SMITH TX		-28			83
1*	4.2					SOMERVELL TX					
1*	1.6	1*	1.2			STEPHENS TX	-26				
4	.2	5	.2	6	.2	TARRANT TX	-16	-19	-32	49	73
1	.5	4*	.9	3	.5	TAYLOR TX	84	-39	12	75	88
		1*	1.6			TERRY TX					
1	.4	1	.3	1	.2	TOM GREEN TX	-24	-35	-50	30	75
1	.2	2	.2	1	.1	TRAVIS TX	4	-62	-60	20	69
		1	.8			UPSHUR TX					
				1*	.5	UVALDE TX					87
		2*	1.2	2*	.8	VICTORIA TX		-29			93
				1*	1.3	WALKER TX					96
		1	.9	1	.6	WASHINGTON TX		-28			90
3*	1.8			1	.3	WEBB TX			-86	02	79
		2*	1.4			WHARTON TX					
				2	.3	WICHITA TX					79
				1	1.7	WILBARGER TX					98
		1*	1.2			WILSON TX					
		1	.6			WOOD TX					
				1*	3.3	ZAPATA TX					99
10	.4	9	.3	9	.2	UTAH UT	-28	-32	-51		
		1*	2.4			CARBON UT					
1	.4			1	.3	DAVIS UT			-29	52	81
				1*	2.6	DUCHESNE UT					99
		1*	3.4			IRON UT					
6	.5	4	.2	6	.3	SALT LAKE UT	-55	11	-50	30	79
1*	1.4					SANPETE UT					
		1	.3			UTAH UT					
2	.5	2	.5	1	.2	WEBER UT	-8	-60	-63	18	75
3	.1	3	.1	6	.2	VERMONT VT	12	59	78		
1	1.0					ADDISON VT					
1	.6	1	.9			CALEDONIA VT	40				
		1	.7			FRANKLIN VT					
				1*	1.6	LAMOILLE VT					97
				1	.5	ORANGE VT					86
				1	.9	ORLEANS VT					94
1	.3			1	.2	RUTLAND VT			-28	53	78
				1	.3	WASHINGTON VT					81
				1	.7	WINDHAM VT					91
		1	.4			WINDSOR VT					
25	.2	34	.2	33	.2	VIRGINIA VA	1	-27	-26		
		1	.8			ACCOMACK VA					
				2	.5	ALBEMARLE#VA					88
				1*	1.5	APPOMATTOX VA					97
		1	.2			ARLINGTON#VA					
3*	.9					AUGUSTA#VA					
1	.2	2	.3	4*	.5	BEDFORD#VA	30	70	121	94	88
		1*	1.2			BOTETOURT VA					
		1*	1.1			BUCHANAN VA					
1*	2.9					CAROLINE VA					
1	.6					CARROLL#VA					
3	.3	3	.2	1	.1	CHESTERFIELD#VA	-16	-66	-72	11	70
1*	1.6			1*	1.2	DICKENSON VA			-28	52	96
				1	.2	DINWIDDIE#VA					76
				5	.2	FAIRFAX#VA					75
1*	1.1	1	.9	1	.7	FRANKLIN VA	-23	-18	-36	44	91
1	.6	1	.5			FREDERICK#VA	-12				
		2	.3	1	.1	HENRICO VA		-56			72
		1	.7			HENRY#VA					
		1*	2.5			ISLE OF WIGHT VA					

* SIGNIFICANTLY MORE DEATHS THAN EXPECTED

\# COUNTIES COMBINED; SEE TECHNICAL NOTES

WHITE FEMALE — CANCER OF THE SALIVARY GLAND — ICD 142

1950-1959 DEATHS	1950-1959 RATE	1960-1969 DEATHS	1960-1969 RATE	1970-1979 DEATHS	1970-1979 RATE	COUNTY AND STATE	PERCENT CHANGE 60S/50S	PERCENT CHANGE 70S/60S	PERCENT CHANGE 70S/50S	PERCENTILE %CHANGE 70S/50S	PERCENTILE RATES 1970S
		1	.5			JAMES CITY#VA					
		1*	7.4			KING WILLIAM VA					
1	1.5					LEE VA					
				1*	1.3	LOUISA VA					96
		1	.6	1	.3	MONTGOMERY#VA		-52			80
				1*	1.5	NORTHUMBERLAND VA					97
1*	1.7					ORANGE VA					
1*	1.7			1*	.8	PATRICK VA			-54	26	92
1	.3	2	.5			PITTSYLVANIA#VA	56				
				1	.6	PULASKI VA					89
				1*	3.6	RICHMOND VA					99
2	.3	2	.2	4	.4	ROANOKE#VA	-27	66	21	77	84
1	.5	2	.7			ROCKINGHAM#VA	36				
		1	.6			SHENANDOAH VA					
				2*	.9	SMYTH VA					94
1*	1.1					SPOTSYLVANIA#VA					
1	.4			1	.4	TAZEWELL VA			18	76	84
3	.2	8*	.5	2	.1	CHESAPEAKE CITY#VA	185	-84	-55	24	70
		1	.3			HAMPTON CITY VA					
1	.5					NEWPORT NEWS CITY VA					
				1	.7	SUFFOLK CITY VA					91
21	.2	35	.2	43	.2	WASHINGTON WA	31	-10	18		
				2*	1.0	CHELAN WA					94
2	.5			3	.4	CLARK WA			-22	57	84
		1	.3	1	.3	COWLITZ WA		-14			79
		1*	1.1			DOUGLAS WA					
		1*	.9	1*	.7	FRANKLIN WA		-23			92
				1	.4	GRANT WA					85
		1	.3			GRAYS HARBOR WA					
1*	2.5	1*	1.0			ISLAND WA	-58				
1*	2.1					JEFFERSON WA					
8	.2	12	.2	12	.2	KING WA	26	-25	-6	66	74
		2	.4	2	.3	KITSAP WA		-31			80
		1	.9			KITTITAS WA					
1*	1.1					MASON WA					
		1	.7	1	.5	OKANOGAN WA		-36			86
				1	.5	PACIFIC WA					88
4	.3	4	.2	6	.3	PIERCE WA	-22	22	-4	68	80
1	.4					SKAGIT WA					
		2	.2	5	.3	SNOHOMISH WA		32			81
2	.2	3	.2	5	.3	SPOKANE WA	40	19	66	87	79
				1	.2	THURSTON WA					77
				2	.5	WALLA WALLA WA					86
		3	.7			WHATCOM WA					
		1	.6			WHITMAN WA					
1	.1	1	.1			YAKIMA WA	8				
20	.3	17	.2	27	.2	WEST VIRGINIA WV	-35	29	-15		
		1	.8	1	.4	BERKELEY WV		-48			85
		1	.8			BOONE WV					
		1	.8	1	.5	BROOKE WV		-31			88
1	.2			4*	.5	CABELL WV			192	97	87
		1*	2.2			CLAY WV					
2*	.8					FAYETTE WV					
				1	.3	GREENBRIER WV					82
		1	.9			HANCOCK WV					
1	.3	2	.4	1	.2	HARRISON WV	72	-62	-35	46	73
		2	.2	4	.3	KANAWHA WV		73			81
				1	.6	LEWIS WV					90
1	1.1			1	.5	MC DOWELL WV			-56	23	87
2	.8					MARION WV					
1	1.0					MASON WV					
1	.4	1	.3	2	.6	MERCER WV	-34	120	46	84	88
				1	.8	MINERAL WV					92
1	.4					MINGO WV					
2*	.8	1	.4	2	.5	MONONGALIA WV	-53	27	-40	40	86
2	.5			1	.2	OHIO WV			-64	17	74
1	.7	1	.3			RALEIGH WV	-56				
		1	.8	2*	1.2	RANDOLPH WV		58			96
1*	2.0			1*	.9	ROANE WV			-55	25	94
1*	1.1					TAYLOR WV					
		1*	2.0			TYLER WV					
		1	1.0	1	.6	UPSHUR WV		-34			90
1*	1.7					WEBSTER WV					
		1*	5.3	1*	3.0	WIRT WV		-42			99
2	.6	1	.2	1	.2	WOOD WV	-64	-17	-70	12	74
				1	.7	WYOMING WV					91
41	.2	64	.3	52	.2	WISCONSIN WI	31	-35	-15		
1	.6					BARRON WI					
1*	1.4	1*	1.5			BAYFIELD WI	2				
		3	.4	2	.3	BROWN WI		-40			79
		1	.8			CALUMET WI					
1	.5					COLUMBIA WI					
		1*	1.3			CRAWFORD WI					
3	.4	2	.2	3	.2	DANE WI	-46	6	-43	37	76
				1	.2	DODGE WI					75
				2*	1.0	DOOR WI					95
2*	.8	1	.4	1	.3	DOUGLAS WI	-48	-25	-61	20	81
1	.6					DUNN WI					
				3*	.6	EAU CLAIRE WI					89
1	.3	5*	1.0	2	.2	FOND DU LAC WI	308	-76	-3	69	78
1	.4	3*	1.0			GRANT WI	142				
				1	.4	GREEN WI					84
		1	.3	1	.2	JEFFERSON WI		-23			78
1	1.0					JUNEAU WI					
1	.2	1	.2	1	.1	KENOSHA WI	13	-47	-40	40	71

* SIGNIFICANTLY MORE DEATHS THAN EXPECTED

COUNTIES COMBINED; SEE TECHNICAL NOTES

WHITE FEMALE — CANCER OF THE SALIVARY GLAND — ICD 142

1950-1959 DEATHS	1950-1959 RATE	1960-1969 DEATHS	1960-1969 RATE	1970-1979 DEATHS	1970-1979 RATE	COUNTY AND STATE	PERCENT CHANGE 60S/50S	PERCENT CHANGE 70S/60S	PERCENT CHANGE 70S/50S	PERCENTILE %CHANGE 70S/50S	PERCENTILE RATES 1970S
		1	.2			LA CROSSE WI					
				4*	2.2	LINCOLN WI					98
1	.2	1	.2	1	.3	MANITOWOC WI	-3	13	9	73	79
				2	.3	MARATHON WI					80
				1	.4	MARINETTE WI					85
		1*	1.4			MARQUETTE WI					
		1	.3	1	.2	MENOMINEE#WI		-25			75
13	.3	15	.3	15	.2	MILWAUKEE WI	0	-9	-9	63	78
		1	.5			MONROE WI					
		2*	1.4	1	.4	ONEIDA WI		-70			85
		1	.2			OUTAGAMIE WI					
1	.8					OZAUKEE WI					
1	1.0					PIERCE WI					
		2*	1.1			POLK WI					
		2*	.9			PORTAGE WI					
		6*	.8	2	.2	RACINE WI		-74			76
3	.5	2	.3			ROCK WI	-46				
1*	1.5					RUSK WI					
		2*	.9			SAUK WI					
2	.5	2	.4			SHEBOYGAN WI	-23				
1*	1.5					TAYLOR WI					
		1	.6			TREMPEALEAU WI					
1	.5			1	.3	VERNON WI			-32	50	82
				1*	1.5	VILAS WI					97
				1*	1.0	WASHBURN WI					95
		1	.3			WASHINGTON WI					
2	.4	1	.1	3	.2	WAUKESHA WI	-64	84	-34	49	78
1	.2	1	.1	1	.1	WINNEBAGO WI	-29	-6	-34	48	71
1	.3	2	.6	1	.2	WOOD WI	100	-67	-34	48	75
5	.5	3	.2	4	.2	WYOMING WY	-53	-4	-54		
				1*	.7	ALBANY WY					92
				1*	2.8	CONVERSE WY					99
1*	3.0	1*	.9			FREMONT WY	-71				
1*	1.4					GOSHEN WY					
2*	.9	1	.7			LARAMIE WY	-27				
		1	.5			NATRONA WY					
1	.9			1	.6	SHERIDAN WY			-35	46	89
				1*	1.0	SWEETWATER WY					95

* SIGNIFICANTLY MORE DEATHS THAN EXPECTED

COUNTIES COMBINED; SEE TECHNICAL NOTES

NON-WHITE MALE

CANCER OF THE SALIVARY GLAND — ICD 142

1950-1959 DEATHS	1950-1959 RATE	1960-1969 DEATHS	1960-1969 RATE	1970-1979 DEATHS	1970-1979 RATE	COUNTY AND STATE	PERCENT CHANGE 60S/50S	PERCENT CHANGE 70S/60S	PERCENT CHANGE 70S/50S	PERCENTILE %CHANGE 70S/50S	PERCENTILE RATES 1970S
198	.3	294	.4	309	.3	UNITED STATES	20	-17	0		
9	.3	10	.3	9	.2	ALABAMA AL	4	-11	-7		
				1*	3.0	BUTLER AL					98
		1*	1.4	1*	1.7	CALHOUN AL		21			97
1*	3.0			1*	2.5	COLBERT AL			-18	47	98
1	1.0					DALLAS AL					
4	.5	2	.2	5	.5	JEFFERSON AL	-64	179	0	58	95
1	1.1	1	1.2			MACON AL	11				
		1*	1.5			MARENGO AL					
1	.4	3	.7			MOBILE AL	80				
		1	.5	1	.4	MONTGOMERY AL		-19			95
1*	2.5					PERRY AL					
		1*	5.8			WASHINGTON AL					
		1	.3	1	.5	ALASKA AK		61			
2	.4	2	.5	3	.4	ARIZONA AZ	3	-23	-20		
1	.7	1	.7	1	.2	MARICOPA AZ	-1	-66	-67	12	94
1*	2.0					NAVAJO AZ					
				2*	1.5	PIMA AZ					97
		1*	3.1			PINAL AZ					
4	.2	11	.6	6	.3	ARKANSAS AR	202	-49	53		
		1*	3.7			ARKANSAS AR					
1*	1.5	3*	5.5			CHICOT AR	274				
		1*	18.6			CLEVELAND AR					
		1*	5.6			CONWAY AR					
				2*	1.7	CRITTENDEN AR					97
		2*	4.0			DESHA AR					
1*	31.3					GRANT AR					
1	.5	1	.7			JEFFERSON AR	30				
				1*	1.7	LEE AR					97
				1*	3.6	LONOKE AR					98
1	.9					MISSISSIPPI AR					
		2*	1.4	1	.9	PHILLIPS AR		-40			96
				1	.4	PULASKI AR					95
11	.2	26	.5	19	.2	CALIFORNIA CA	91	-63	-30		
1	.1	4	.8			ALAMEDA CA	557				
				1	.8	CONTRA COSTA CA					96
		1	.7			FRESNO CA					
		1	.9			KERN CA					
4	.1	12	.5	11	.2	LOS ANGELES CA	239	-52	63	79	94
				1*	2.4	MERCED CA					98
				1	.1	ORANGE CA					94
		2	.7			SACRAMENTO CA					
				1	.8	SAN BERNARDINO CA					96
		1	.2	1	.0	SAN DIEGO CA		-79			94
3	.5	4	.4	1	.1	SAN FRANCISCO CA	-11	-77	-79	08	94
2*	1.6			2	.7	SAN JOAQUIN CA			-55	16	96
1*	6.1					SANTA CRUZ CA					
		1*	10.3			YUBA CA					
		1	.3	2	.4	COLORADO CO		23			
		1	.4	2	.6	DENVER-ARAPAHOE#CO		47			95
3*	1.4	1	.2	3	.3	CONNECTICUT CT	-87	80	-76		
1*	2.5	1	.6	1	.2	FAIRFIELD CT	-78	-57	-91	02	94
2*	2.2			1	.4	NEW HAVEN CT			-80	06	95
				1*	2.5	NEW LONDON CT					98
				3*	1.1	DELAWARE DE					
				3*	5.0	SUSSEX DE					99
2	.2	7	.5	6	.4	DIST. OF COLUMBIA DC	214	-22	146		
2	.2	7	.5	6	.4	WASHINGTON DC	214	-22	146	93	95
11	.4	9	.2	10	.2	FLORIDA FL	-38	-12	-45		
		1	1.2	1	1.2	ALACHUA FL		6			97
1*	2.3	1*	2.1	1*	1.9	BREVARD FL	-10	-10	-18	45	97
				1	.3	BROWARD FL					95
		1*	22.3			CHARLOTTE FL					
				2	.2	DADE FL					94
		2	.4	1	.2	DUVAL FL		-56			94
		2*	1.5			ESCAMBIA FL					
1	.4	1	.4			HILLSBOROUGH FL	8				
1*	4.7					JEFFERSON FL					
2*	5.4					LAKE FL					
1*	4.1					MADISON FL					
				2*	4.5	MANATEE FL					99
1	.5	1	.7			ORANGE FL	21				
1	.4			1	.3	PALM BEACH FL			-27	35	94
				1	.6	PINELLAS FL					95
1*	3.3					PUTNAM FL					
1*	4.8					SARASOTA FL					
1*	1.3					VOLUSIA FL					

* SIGNIFICANTLY MORE DEATHS THAN EXPECTED

COUNTIES COMBINED; SEE TECHNICAL NOTES

NON-WHITE MALE — CANCER OF THE SALIVARY GLAND — ICD 142

1950-1959 DEATHS	1950-1959 RATE	1960-1969 DEATHS	1960-1969 RATE	1970-1979 DEATHS	1970-1979 RATE	COUNTY AND STATE	PERCENT CHANGE 60S/50S	PERCENT CHANGE 70S/60S	PERCENT CHANGE 70S/50S	PERCENTILE %CHANGE 70S/50S	PERCENTILE RATES 1970S
11	.3	8	.2	15	.3	GEORGIA GA	-33	75	17		
				1*	10.8	BLECKLEY GA					99
1	.4	2	.8			CHATHAM GA	79				
		1	.3			DE KALB GA					
		1*	6.0			DODGE GA					
		1*	16.7			DOUGLAS GA					
1*	2.8					FLOYD GA					
2	.3	1	.1	3	.3	FULTON GA	-60	105	-19	43	94
				1*	1.6	GLYNN GA					97
				1*	7.7	LAMAR GA					99
1*	2.5					LOWNDES GA					
1*	3.6					MC DUFFIE GA					
				1*	5.7	MC INTOSH GA					99
		1*	2.5			MERIWETHER GA					
		1*	5.7			MORGAN GA					
				3*	1.6	MUSCOGEE#GA					97
				1*	3.7	PEACH GA					98
				1*	17.1	PIERCE GA					99
1*	3.2					RANDOLPH GA					
				1	.6	RICHMOND GA					95
1*	11.0					SEMINOLE GA					
				1*	2.6	SPALDING GA					98
1*	1.7					THOMAS GA					
1*	1.7			1*	2.1	TROUP GA			21	68	98
1*	14.9					WAYNE GA					
		9	.6	8	.3	HAWAII HI		-41			
		2	.9			HAWAII HI					
		6	.6	7	.5	HONOLULU HI		-20			95
		1	.5	1	.4	MAUI#HI		-17			95
8	.3	20	.5	12	.2	ILLINOIS IL	78	-58	-25		
1*	12.8					ADAMS IL					
		1*	5.7			CHAMPAIGN IL					
6	.2	15	.4	9	.2	COOK IL	67	-57	-29	33	94
1*	5.1					JACKSON IL					
		1*	2.3			LAKE IL					
				1*	1.9	MADISON IL					97
		1*	5.3			PULASKI IL					
		1	.5	1	.3	ST CLAIR IL		-41			95
				1*	14.1	SALINE IL					99
		1*	3.9			WINNEBAGO IL					
4	.6	5	.5	8	.6	INDIANA IN	-25	30	-3		
		1*	7.6			GRANT IN					
1	.8	2	.5	2	.7	LAKE IN	-39	28	-22	39	96
2	.8	1	.3	5*	.9	MARION IN	-64	215	15	64	96
		1*	34.4			MONROE IN					
1*	70.7					SPENCER IN					
				1*	1.8	VANDERBURGH IN					97
1	.3	1	.2	3	.6	KANSAS KS	-29	156	81		
		1	1.4	2*	2.8	SEDGWICK KS		96			98
				1*	1.8	SHAWNEE KS					97
1	1.3					WYANDOTTE KS					
4	.4	4	.4	1	.1	KENTUCKY KY	-7	-71	-73		
2*	14.1					DAVIESS KY					
		1	.9			FAYETTE KY					
1*	18.6					FRANKLIN KY					
1	.3	1	.3			JEFFERSON KY	0				
		2*	12.4			KENTON KY					
				1*	4.7	MC CRACKEN KY					99
10	.3	8	.2	12	.3	LOUISIANA LA	-35	55	1		
		1*	3.0			ACADIA LA					
				1*	6.3	BEAUREGARD LA					99
1	.3			2	.7	CADDO LA			109	87	96
1	1.4					CALCASIEU LA					
				3*	1.1	EAST BATON ROUGE LA					97
1*	2.7					IBERIA LA					
		1	.3			JEFFERSON LA					
1*	1.2			2*	2.7	MOREHOUSE LA			117	89	98
1*	1.7	1*	1.4			NATCHITOCHES LA	-17				
5	.7	3	.4	1	.1	ORLEANS LA	-39	-76	-86	04	94
				1	.9	OUACHITA LA					96
				1	.7	RAPIDES LA					96
		1	.8			ST LANDRY LA					
		1*	3.9			UNION LA					
				1*	12.1	WEST CARROLL LA					99
6	.3	8	.4	11	.4	MARYLAND MD	13	8	22		
				1	.6	ANNE ARUNDEL MD					95
1*	4.4					CECIL MD					
				1*	2.2	DORCHESTER MD					98
				1*	4.0	FREDERICK MD					98
		1*	4.7			KENT MD					
1	.8	1	.7	2	1.0	PRINCE GEORGE'S MD	-10	41	27	70	96
4	.5	6	.5	6	.4	BALTIMORE CITY MD	-8	-13	-19	41	95
1	.2	3	.7	1	.2	MASSACHUSETTS MA	171	-77	-37		

* SIGNIFICANTLY MORE DEATHS THAN EXPECTED

\# COUNTIES COMBINED; SEE TECHNICAL NOTES

1950-1959 DEATHS	1950-1959 RATE	1960-1969 DEATHS	1960-1969 RATE	1970-1979 DEATHS	1970-1979 RATE	COUNTY AND STATE	PERCENT CHANGE 60S/50S	PERCENT CHANGE 70S/60S	PERCENT CHANGE 70S/50S	PERCENTILE %CHANGE 70S/50S	PERCENTILE RATES 1970S
1*	4.6					BRISTOL MA					
		3*	1.6	1	.3	SUFFOLK MA		-79			95
3	.2	11	.3	15	.4	MICHIGAN MI	54	29	99		
1*	5.2					CALHOUN MI					
		1	.6	3*	1.7	GENESEE MI		179			97
		1*	1.6			KENT MI					
				1*	4.2	LAKE MI					98
				1	.7	SAGINAW MI					96
2	.2	9	.3	10	.4	WAYNE MI	67	15	93	85	95
				1	.4	MINNESOTA MN					
				1*	1.7	RAMSEY MN					97
8	.2	15	.4	11	.3	MISSISSIPPI MS	127	-26	68		
1*	2.1					AMITE MS					
				1	.7	BOLIVAR MS					96
1*	2.4	1*	4.0			CHICKASAW MS	70				
1*	2.7					CLAIBORNE MS					
		1	.8	1	.7	COAHOMA MS		-7			96
		1*	1.7			COPIAH MS					
		1*	5.6			COVINGTON MS					
1*	1.3					FORREST MS					
		1*	3.2			GRENADA MS					
		1*	1.7			HARRISON MS					
		1	.3	2	.7	HINDS MS		94			96
				1*	1.3	HOLMES MS					97
				1*	3.5	JASPER MS					98
		1*	2.7			JEFFERSON MS					
				1*	1.5	JONES MS					97
				1*	3.0	LAFAYETTE MS					98
1	1.2					LAUDERDALE MS					
				1	1.0	LEFLORE MS					96
		1*	1.7			LOWNDES MS					
				1*	1.8	MADISON MS					97
		1*	3.2			MARION MS					
		1*	5.8			PEARL RIVER MS					
1*	2.8	1*	2.7			TATE MS	-4				
1	1.1	1	1.1			WARREN MS	-2				
1	.4	2*	1.2	1	.6	WASHINGTON MS	170	-52	31	72	95
7	.5	9	.5	6	.3	MISSOURI MO	10	-40	-34		
1*	19.2					HOWARD MO					
1	.2	1	.3	1	.2	JACKSON MO	2	-14	-13	52	94
1*	10.5	1*	13.5			LAFAYETTE MO	29				
1*	1.1			1	.5	ST LOUIS COUNTY MO			-51	20	95
3	.4	7*	.8	4	.5	ST LOUIS CITY MO	95	-40	17	66	95
		1	.7			NEBRASKA NB					
		1*	19.5			THURSTON NB					
2	.1	5	.3	11	.4	NEW JERSEY NJ	117	15	150		
				1	.6	ATLANTIC NJ					95
		1	1.3			CAMDEN NJ					
1	.1	2	.3	3	.3	ESSEX NJ	93	24	140	91	95
		1*	2.8	1*	1.2	GLOUCESTER NJ		-56			97
				1	.5	HUDSON NJ					95
		1	.6			MERCER NJ					
1	1.3					MONMOUTH NJ					
				1*	2.1	MORRIS NJ					98
				2*	1.1	PASSAIC NJ					96
				2	.7	UNION NJ					96
1	.3			1	.1	NEW MEXICO NM			-54		
				1	.8	BERNALILLO NM					96
1*	1.2					MC KINLEY NM					
16	.5	33*	.6	23	.3	NEW YORK NY	21	-50	-39		
				1*	1.6	ALBANY NY					97
2	.4	2	.3	2	.2	BRONX NY	-22	-31	-46	25	94
				1*	11.9	CATTARAUGUS NY					99
1*	14.9					CHEMUNG NY					
2*	1.4	4*	1.0	2	.9	ERIE NY	-27	-15	-38	29	96
4	.8	1	.1	3	.2	KINGS(BROOKLYN) NY	-89	106	-77	10	94
		1	2.4			NASSAU NY					
3	.3	17*	.9	9	.4	NEW YORK(MANHATTAN) NY	222	-55	46	75	95
1*	5.9					ORANGE NY					
		4	.6			QUEENS NY					
1*	7.5			1*	7.2	RENSSELAER NY			-4	56	99
				1*	1.6	RICHMOND NY					97
1	.7	1	.9			SUFFOLK NY	35				
1	.6	3*	1.0	3*	.8	WESTCHESTER NY	74	-16	46	77	96
8	.3	14	.4	12	.3	NORTH CAROLINA NC	42	-27	4		
		1*	2.2			ALAMANCE NC					
1*	2.9					BERTIE NC					
		1*	1.9	1*	1.8	BUNCOMBE NC		-8			97
				1*	7.1	CALDWELL NC					99
		1*	1.8			CRAVEN NC					
1*	5.4	1*	2.9			DAVIDSON NC	-46				
1	1.1					DURHAM NC					
1	1.0					FORSYTH NC					

* SIGNIFICANTLY MORE DEATHS THAN EXPECTED

NON-WHITE MALE — CANCER OF THE SALIVARY GLAND — ICD 142

1950-1959 DEATHS	1950-1959 RATE	1960-1969 DEATHS	1960-1969 RATE	1970-1979 DEATHS	1970-1979 RATE	COUNTY AND STATE	PERCENT CHANGE 60S/50S	PERCENT CHANGE 70S/60S	PERCENT CHANGE 70S/50S	PERCENTILE %CHANGE 70S/50S	PERCENTILE RATES 1970S
		1*	2.5			FRANKLIN NC					
				1*	5.9	GATES NC					99
				1*	4.5	GREENE NC					99
				2	.9	GUILFORD NC					96
		1*	2.0			HERTFORD NC					
1*	2.1					IREDELL NC					
1*	2.7					JOHNSTON NC					
1*	3.7					MARTIN NC					
		3*	1.3			MECKLENBURG NC					
				1*	5.3	MONTGOMERY NC					99
		1	1.1			NEW HANOVER NC					
		2*	2.3	1	1.0	PITT NC		-56			96
		1*	4.7			RANDOLPH NC					
				1*	2.3	RICHMOND NC					98
				1*	1.4	SAMPSON NC					97
		1*	18.2	1*	12.8	SWAIN NC		-29			99
				1	.5	WAKE NC					95
1*	1.4					WILSON NC					
7	.3	6	.2	14	.4	OHIO OH	-44	140	35		
1*	2.8					BUTLER OH					
3	.3	4	.3	5	.5	CUYAHOGA OH	18	41	66	83	95
1	.3			4*	1.2	FRANKLIN OH			232	97	97
2	.8	2	.3			HAMILTON OH	-57				
				1	.2	MONTGOMERY OH					94
				3*	1.9	SUMMIT OH					97
				1*	2.0	TRUMBULL OH					98
5	.6	3	.3			OKLAHOMA OK	-53				
1*	14.9					LE FLORE OK					
1	1.2					MUSKOGEE OK					
1*	24.4					NOBLE OK					
1	1.0					OKLAHOMA OK					
		1*	2.8			SEMINOLE OK					
		1*	27.3			STEPHENS OK					
1	.9	1	.6			TULSA OK	-33				
1	1.3			2	.5	OREGON OR			-58		
1*	2.0			2*	1.1	MULTNOMAH OR			-46	22	96
10	.3	19	.5	12	.3	PENNSYLVANIA PA	44	-43	-18		
3	.4	5	.8	2	.3	ALLEGHENY PA	96	-65	-31	31	95
		2*	2.3			CHESTER PA					
				1	.5	DELAWARE PA					95
		1*	2.4			FAYETTE PA					
1*	1.1	1	.7			MONTGOMERY PA	-37				
6	.4	10	.4	9	.4	PHILADELPHIA PA	3	-17	-14	50	95
				1	1.0	RHODE ISLAND RI					
				1	1.3	PROVIDENCE RI					97
7	.3	7	.3	1	.0	SOUTH CAROLINA SC	-2	-87	-87		
		1*	4.2			BAMBERG SC					
		2*	7.9			BARNWELL SC					
		1*	2.8			BEAUFORT SC					
1*	6.8					CALHOUN SC					
		1*	2.2	1*	3.0	CHESTER SC		34			98
1*	2.4					CHESTERFIELD SC					
1*	1.9					COLLETON SC					
1*	1.0					DARLINGTON SC					
1	.6					GREENVILLE SC					
1*	1.8					LAURENS SC					
		1*	2.5			LEE SC					
1*	6.1					SALUDA SC					
		1*	1.3			WILLIAMSBURG SC					
1	1.2	1	1.1			SOUTH DAKOTA SD	-6				
1*	19.1					DEWEY SD					
		1*	5.0			SHANNON SD					
11	.5	7	.3	12	.5	TENNESSEE TN	-34	40	-8		
				1*	37.8	CARTER TN					99
2	.6	1	.5			DAVIDSON TN	-24				
1*	1.9					FAYETTE TN					
1*	2.9			1*	2.2	GIBSON TN			-24	37	98
1	.5	1	.5			HAMILTON TN	11				
		1*	2.6	1*	3.2	HAYWOOD TN		24			98
				1	.8	KNOX TN					96
1*	10.7					LAKE TN					
1*	1.4					MADISON TN					
1*	2.6					MAURY TN					
				1*	2.1	MONTGOMERY TN					98
		1*	9.9			OBION TN					
1*	3.9					RUTHERFORD TN					
2	.2	3	.4	6	.6	SHELBY TN	69	59	168	95	95
				1*	4.4	WILSON TN					99
13	.3	18	.4	25	.4	TEXAS TX	9	4	14		
		3*	1.8	3	1.3	BEXAR TX		-27			97
				1*	1.9	BRAZOS TX					98
		1*	5.7			CALDWELL TX					
				2*	2.5	CAMERON TX					98

* SIGNIFICANTLY MORE DEATHS THAN EXPECTED

NON-WHITE MALE — CANCER OF THE SALIVARY GLAND — ICD 142

1950-1959 DEATHS	1950-1959 RATE	1960-1969 DEATHS	1960-1969 RATE	1970-1979 DEATHS	1970-1979 RATE	COUNTY AND STATE	PERCENT CHANGE 60S/50S	PERCENT CHANGE 70S/60S	PERCENT CHANGE 70S/50S	PERCENTILE %CHANGE 70S/50S	PERCENTILE RATES 1970S
		1*	38.3			COLLINGSWORTH TX					
3	.6	1	.2	3	.2	DALLAS TX	-71	35	-61	14	94
1*	70.1					EPATH TX					
		1*	4.5			FAYETTE TX					
3	.8	3	.3	5	.4	HAFRIS TX	-59	39	-42	27	95
1	1.1					HARRISON TX					
1*	3.5					HUNT TX					
1	.8			2	.8	JEFFERSON TX			7	62	96
		1*	3.6			LIBERTY TX					
				1	.8	MC LENNAN TX					96
		1*	3.5			MILAM TX					
		1*	7.9			NEWTON TX					
1*	4.7					POLK TX					
				1*	2.2	ROBERTSON TX					98
1	1.0					SMITH TX					
		2	.8	4*	1.3	TARRANT TX		71			97
1*	6.6					TOM GREEN TX					
		1	1.0	1	.5	TRAVIS TX		-43			95
				1*	2.5	VICTORIA TX					98
		1*	3.5			WALLER TX					
				1*	17.5	WILBARGER TX					99
		1*	11.8			WOOD TX					
				1	.5	UTAH UT					
				1*	1.1	SALT LAKE UT					96
7	.2	8	.2	20*	.5	VIRGINIA VA	-10	149	124		
				1*	1.8	ACCOMACK VA					97
1	.8			1	.8	BEDFOPD#VA			4	60	96
1*	3.0			2*	4.9	BRUNSWICK VA			63	81	99
		1*	4.4			CHARLOTTE VA					
				1	.2	CHESTERFIELD#VA					94
				1	.8	FAIPFAX#VA					96
				3*	11.8	FAUQUIER VA					99
		1*	6.6			FRANKLIN VA					
				1*	2.4	GREENSVILLE#VA					98
1*	1.4					HALIFAX#VA					
		1*	1.5	1*	1.6	JAMES CITY#VA		5			97
1*	1.9					MECKLENBURG VA					
				1*	5.9	NELSON VA					99
1*	2.5			1*	2.3	NORTHAMPTON VA			-9	54	98
		1	.8			PITTSYLVANIA#VA					
				1*	12.4	PCWHATAN VA					99
				1*	9.9	ROCKBRIDGE#VA					99
				1*	10.8	WASHINGTON#VA					99
2	.4	3	.4	1	.2	CHESAPEAKE CITY#VA	-2	-53	-54	18	94
		1	1.1	2*	1.5	HAMPTON CITY VA		44			97
				1	.9	SUFFOLK CITY VA					96
		1	.3	3	.7	WASHINGTON WA		116			
				1*	60.6	MASON WA					99
				1*	1.1	PIERCE WA					96
				1*	6.3	SPOKANE WA					99
		1*	4.8			YAKIMA WA					
2	.3	2	.5	3	.9	WEST VIRGINIA WV	51	77	167		
				1*	2.2	FAYETTE WV					98
				1*	2.7	MC DOWELL WV					98
1*	1.6	1*	3.1			RALEIGH WV	97				
				1*	81.8	RANDOLPH WV					99
1*	6.1	1*	31.5			WYOMING WV	412				
1	.4			2	.5	WISCONSIN WI			7		
1*	8.8					MENOMINEE#WI					
				1*	114.7	ONEIDA WI					99
				1*	4.0	RACINE WI					98
1*	4.2					WYOMING WY					
1*	28.3					LARAMIE WY					

* SIGNIFICANTLY MORE DEATHS THAN EXPECTED

\# COUNTIES COMBINED; SEE TECHNICAL NOTES

NON-WHITE FEMALE — CANCER OF THE SALIVARY GLAND — ICD 142

1950-1959 DEATHS	1950-1959 RATE	1960-1969 DEATHS	1960-1969 RATE	1970-1979 DEATHS	1970-1979 RATE	COUNTY AND STATE	PERCENT CHANGE 60S/50S	PERCENT CHANGE 70S/60S	PERCENT CHANGE 70S/50S	PERCENTILE %CHANGE 70S/50S	PERCENTILE RATES 1970S
135	.2	178	.2	197	.2	UNITED STATES	1	-22	-21		
13	.3	7	.2	4	.1	ALABAMA AL	-54	-51	-78		
				1*	2.0	BUTLER AL					98
1*	1.7					CALHOUN AL					
		1*	3.7			COFFEE AL					
				1	.7	DALLAS AL					97
5*	.5	2	.2	1	.1	JEFFERSON AL	-59	-60	-84	08	95
1*	1.1					MACON AL					
1*	1.2					MADISON AL					
2	.5	3*	.6			MOBILE AL	21				
2*	.9	1	.4			MONTGOMERY AL	-60				
1*	2.5					PICKENS AL					
				1*	.8	SUMTER AL					98
		2*	1.7	3*	.8	ALASKA AK		-53			
2*	.6			1	.1	ARIZONA AZ			-81		
				1*	.7	APACHE AZ					97
1*	2.7					NAVAJO AZ					
1*	1.4					PIMA AZ					
2	.1	2	.1	4	.2	ARKANSAS AR	-5	92	83		
1*	1.0					CRITTENDEN AR					
1*	2.5					CROSS AR					
				1*	2.5	HEMPSTEAD AR					99
		1*	6.7			HOT SPRING AR					
		1*	9.4			LITTLE RIVER AR					
				1*	1.6	MILLER AR					98
				2*	.7	PULASKI AR					97
8	.2	8	.2	13	.1	CALIFORNIA CA	-12	-30	-38		
2	.4	1	.2	2	.2	ALAMEDA CA	-58	-15	-64	23	96
				1	.5	CONTRA COSTA CA					97
1*	.9					KERN CA					
				1*	5.5	KINGS CA					99
3	.2	5	.2	3	.1	LOS ANGELES CA	28	-67	-58	26	95
				1*	5.9	PLACER CA					99
1*	1.4			1	.2	SACRAMENTO CA			-87	05	96
		1*	1.1			SAN BERNARDINO CA					
				1	.1	SAN DIEGO CA					96
1	.1	1	.4	3	.3	SAN FRANCISCO CA	200	-33	100	88	96
		1	.6			COLORADO CO					
		1	.9			DENVER-ARAPAHOE#CO					
		1	.2	2	.2	CONNECTICUT CT		-8			
		1	.7			HARTFORD CT					
				1	.3	NEW HAVEN CT					96
				1*	2.0	NEW LONDON CT					98
		2*	.9			DELAWARE DE					
		1	.5			NEW CASTLE DE					
		1*	2.5			SUSSEX DE					
3	.2	8*	.4	4	.2	DIST. OF COLUMBIA DC	83	-66	-39		
3	.2	8*	.4	4	.2	WASHINGTON DC	83	-66	-39	58	96
3	.1	7	.2	6	.1	FLORIDA FL	65	-44	-8		
1	.4	1	.2			DADE FL	-31				
		1	.2	2	.3	DUVAL FL		31			96
				1	.4	ESCAMBIA FL					97
1*	1.2					GADSDEN FL					
		1	.3	1	.3	HILLSBOROUGH FL		0			96
		1*	3.6			MADISON FL					
				1*	1.6	MANATEE FL					98
		1*	4.9			NASSAU FL					
1	.5	2*	1.2			PALM BEACH FL	142				
				1	.4	PINELLAS FL					97
8	.2	12	.2	14	.2	GEORGIA GA	17	-1	16		
1*	1.3					BALDWIN GA					
		1	.6	1	.3	BIBB GA		-37			97
				1*	1.8	BURKE GA					98
				1*	1.9	CARROLL GA					98
				1*	1.4	CLARKE GA					98
				1*	2.7	COBB GA					99
		1*	3.4	1*	2.4	COLQUITT GA		-29			99
1*	3.2	1*	1.9			CRISP GA	-39				
1*	4.2					DOOLY GA					
		1*	.8			DOUGHERTY GA					
		1*	3.7			ELBERT GA					
1	.3	4	.4	2	.2	FULTON GA	49	-54	-31	61	96
1*	5.6					HANCOCK GA					
1*	3.2					HENRY GA					
				1*	3.2	LIBERTY GA					99
				1*	2.1	MACON GA					99
1	.4			2*	.9	MUSCOGEE#GA			137	94	98

* SIGNIFICANTLY MORE DEATHS THAN EXPECTED

\# COUNTIES COMBINED; SEE TECHNICAL NOTES

1950-1959 DEATHS	RATE	1960-1969 DEATHS	RATE	1970-1979 DEATHS	RATE	COUNTY AND STATE	PERCENT CHANGE 60S/50S	70S/60S	70S/50S	PERCENTILE %CHANGE 70S/50S	RATES 1970S
		1*	3.2			PEACH GA					
1	.6	1	.4	1	.3	RICHMOND GA	-27	-23	-43	44	97
				1*	1.3	THOMAS GA					98
		1*	1.3			TROUP GA					
		1	.0	5	.2	HAWAII HI		407			
				1	.4	HAWAII HI					97
		1	.1	4	.2	HONOLULU HI		289			96
10	.3	9	.2	13	.2	ILLINOIS IL	-43	27	-28		
8	.3	8	.2	10	.2	COOK IL	-44	5	-41	52	96
		1*	10.7			DU PAGE IL					
1*	182.1					FRANKLIN IL					
				1*	3.2	JACKSON IL					99
1*	3.5					MACON IL					
				1	.3	ST CLAIR IL					96
				1*	14.1	WARREN IL					99
2	.3	1	.1	2	.1	INDIANA IN	-72	34	-62		
		1*	8.1			CLARK IN					
1*	34.4					JEFFERSON IN					
				1	.2	LAKE IN					96
1	.3			1	.1	MARION IN			-56	32	96
2*	1.9					IOWA IA					
2*	4.7					POLK IA					
3*	.7					KANSAS KS					
1*	6.2					MONTGOMERY KS					
2*	1.3					WYANDOTTE KS					
3	.3	4	.4	6*	.4	KENTUCKY KY	19	18	40		
1*	23.8					ADAIR KY					
				1*	6.3	BARREN KY					99
1*	74.3					CLINTON KY					
		2*	1.8	1*	.7	FAYETTE KY		-60			97
1*	12.5					GRAVES KY					
				1*	3.9	HENDERSON KY					99
		1	.2	2	.4	JEFFERSON KY		68			97
		1*	15.7			MASON KY					
				1*	10.2	MERCER KY					99
6	.1	10	.2	8	.2	LOUISIANA LA	58	-28	14		
				1*	3.5	ASSUMPTION LA					99
1*	2.3			1*	1.8	AVOYELLES LA			-22	67	98
				1	.2	CADDO LA					96
1*	1.3					CALCASIEU LA					
		1*	2.8			CLAIBORNE LA					
		1*	2.5			CONCORDIA LA					
				1*	1.4	IBERVILLE LA					98
				1*	4.2	JEFFERSON DAVIS LA					99
				1*	1.2	LAFAYETTE LA					98
		1*	2.0			MOREHOUSE LA					
3	.3	4	.3	2	.2	ORLEANS LA	30	-54	-40	55	96
		1	.6			OUACHITA LA					
1*	.9					ST LANDRY LA					
		1*	3.3			WEST BATON ROUGE LA					
		1*	3.2			WINN LA					
5	.3	4	.2	4	.1	MARYLAND MD	-45	-33	-63		
		1*	1.2			BALTIMORE MD					
1*	3.7					MONTGOMERY MD					
1*	1.6					PRINCE GEORGE'S MD					
3	.2	3	.2	4	.2	BALTIMORE CITY MD	19	-1	17	76	96
1	.2	3	.5			MASSACHUSETTS MA	113				
		1*	39.3			HAMPSHIRE MA					
1	.4	2	.7			SUFFOLK MA	74				
3	.1	3	.1	2	.1	MICHIGAN MI	-21	-42	-54		
				1*	37.2	BARAGA MI					99
		1	.8			GENESEE MI					
3	.1	2	.1	1	.0	WAYNE MI	-49	-53	-76	11	95
				2*	.7	MINNESOTA MN					
				1*	1.3	RAMSEY MN					98
				1*	8.3	ST LOUIS MN					99
5	.2	11	.3	7	.2	MISSISSIPPI MS	107	-51	2		
1	.7					BOLIVAR MS					
		1*	3.0			CLARKE MS					
		1	1.0			COAHOMA MS					
		1*	1.8			COPIAH MS					
1*	2.0					DE SOTO MS					
		2*	2.9	1*	.8	FORREST MS		-71			98
		1*	4.0			FRANKLIN MS					
				1*	.9	HARRISON MS					98
1	.4					HINDS MS					
		1*	1.3			JACKSON MS					

* SIGNIFICANTLY MORE DEATHS THAN EXPECTED

NON-WHITE FEMALE — CANCER OF THE SALIVARY GLAND — ICD 142

1950-1959 DEATHS	RATE	1960-1969 DEATHS	RATE	1970-1979 DEATHS	RATE	COUNTY AND STATE	PERCENT CHANGE 60S/50S	70S/60S	70S/50S	PERCENTILE %CHANGE 70S/50S	RATES 1970S
				1*	4.2	KEMPER MS					99
		1*	3.3			LAFAYETTE MS					
				2*	1.7	LAUDERDALE MS					98
1*	2.8					MONROE MS					
		1*	3.1			NOXUBEE MS					
		1*	2.5			OKTIBBEHA MS					
		1*	1.5			PANOLA MS					
				1*	1.5	TALLAHATCHIE MS					98
1*	2.3					TATE MS					
				1	.4	WASHINGTON MS					97
3	.2	3	.2	2	.1	MISSOURI MO	8	-61	-58		
		1*	3.7			BUCHANAN MO					
1	.3					JACKSON MO					
		1*	5.1			LAFAYETTE MO					
		1*	7.6	1*	5.9	SCOTT MO		-22			99
2	.2			1	.1	ST LOUIS CITY MO			-67	20	95
				2*	.9	NEBRASKA NB					
				2*	1.2	DOUGLAS NB					98
				1	.5	NEVADA NV					
				1*	.7	CLARK NV					97
4	.2	4	.1	12*	.3	NEW JERSEY NJ	-37	173	71		
1	.6	1	.6			ATLANTIC NJ	8				
				1*	1.1	BURLINGTON NJ					98
1*	3.9					CUMBERLAND NJ					
1	.1	2	.2	4	.4	ESSEX NJ	105	84	277	97	97
				1*	1.0	GLOUCESTER NJ					98
				1	.4	HUDSON NJ					97
1*	.5					MERCER NJ					
				1*	.7	MIDDLESEX NJ					97
		1	.4	1	.5	MONMOUTH NJ		21			97
				1	.2	PASSAIC NJ					96
				1*	2.2	SALEM NJ					99
				1	.2	UNION NJ					96
10	.3	8	.1	9	.1	NEW YORK NY	-60	-35	-74		
1	.1	1	.1	2	.1	BRONX NY	49	-16	26	79	95
		2*	1.2	1	.5	ERIE NY		-62			97
2	.4			3	.1	KINGS(BROOKLYN) NY			-70	17	96
		1*	54.6			MADISON NY					
6	.5	3	.2	1	.0	NEW YORK(MANHATTAN) NY	-67	-77	-93	02	95
				2	.1	QUEENS NY					96
		1*	5.8			ULSTER NY					
1	.8					WESTCHESTER NY					
5	.1	12	.3	11	.2	NORTH CAROLINA NC	82	-25	37		
1*	2.6					BEAUFORT NC					
		1*	1.3			BUNCOMBE NC					
		1*	1.9			CABARRUS NC					
1	.5					DURHAM NC					
		1	.4			FORSYTH NC					
				1*	1.6	GRANVILLE NC					98
		1*	6.4			GREENE NC					
				2*	.7	GUILFORD NC					97
				1*	.6	HALIFAX NC					97
1*	1.7			1*	1.7	HARNETT NC			4	73	98
1*	1.5	1*	1.5			LENOIR NC	2				
				1*	10.6	LINCOLN NC					99
				1	.3	MECKLENBURG NC					96
1*	2.3					NASH NC					
		1*	2.1			NORTHAMPTON NC					
		1*	2.8			PENDER NC					
				1*	3.8	RANDOLPH NC					99
		1*	1.8			RICHMOND NC					
				1	.5	ROBESON NC					97
				1*	1.4	ROWAN NC					98
		1*	3.0			UNION NC					
				1*	1.5	VANCE NC					98
		2*	1.1			WAKE NC					
		1*	.7			WAYNE NC					
8	.3	5	.2	9	.2	OHIO OH	-48	27	-34		
1*	2.5					BUTLER OH					
2	.3	2	.3	2	.2	CUYAHOGA OH	10	-50	-44	41	96
				1	.2	FRANKLIN OH					96
1	.2	2	.3	2	.3	HAMILTON OH	41	27	79	85	96
				1*	9.6	LICKING OH					99
1*	.7					MAHONING OH					
				1*	26.5	MEDINA OH					99
1	.6	1	.3	1	.2	MONTGOMERY OH	-52	-12	-58	29	96
1*	10.9					POSS OH					
1*	.7			1	.3	SUMMIT OH			-54	35	96
1	.1	3	.3	2	.1	OKLAHOMA OK	138	-57	2		
		1*	3.9			CARTER OK					
		1*	6.2			CHEROKEE OK					
1*	1.2					MUSKOGEE OK					
		1	.4			OKLAHOMA OK					
				1*	4.4	POTTAWATOMIE OK					99
				1	.4	TULSA OK					97

* SIGNIFICANTLY MORE DEATHS THAN EXPECTED

1950-1959 DEATHS	RATE	1960-1969 DEATHS	RATE	1970-1979 DEATHS	RATE	COUNTY AND STATE	PERCENT CHANGE 60S/50S	70S/60S	70S/50S	PERCENTILE %CHANGE 70S/50S	RATES 1970S
		1	.5			OREGON OR					
		1*	.7			MULTNOMAH OR					
4	.1	12	.3	14*	.3	PENNSYLVANIA PA	181	-10	152		
				2	.3	ALLEGHENY PA					96
1*	2.1					FAYETTE PA					
		1*	37.7			HUNTINGDON PA					
3	.1	9	.4	11*	.3	PHILADELPHIA PA	148	-5	136	91	97
				1*	2.4	WASHINGTON PA					99
		1*	2.9			WESTMORELAND PA					
		1*	4.8			YORK PA					
5	.2	8	.3	6	.1	SOUTH CAROLINA SC	29	-44	-28		
		1*	1.1	1*	.9	AIKEN SC		-18			98
		1*	1.3			BEAUFORT SC					
1*	1.8					BERKELEY SC					
				2*	.4	CHARLESTON SC					97
		1*	2.0			CHESTER SC					
				1*	1.0	DARLINGTON SC					98
1*	4.0					EDGEFIELD SC					
1*	.5					GREENVILLE SC					
		1*	1.8			GREENWOOD SC					
		1*	2.2	1*	2.0	NEWBERRY SC		-10			98
		2*	1.2			ORANGEBURG SC					
				1*	4.9	SALUDA SC					99
1*	1.6	1	.7			SPARTANBURG SC	-57				
1*	1.5					YORK SC					
1*	1.4					SOUTH DAKOTA SD					
1*	12.4					TODD SD					
2	.1	8	.3	4	.1	TENNESSEE TN	214	-62	18		
1	.3			1	.2	DAVIDSON TN			-41	50	96
		2*	1.0			HAMILTON TN					
		1*	3.0			LAUDERDALE TN					
		1*	1.1			MADISON TN					
				1*	26.8	RHEA TN					99
1	.1	3	.3	2	.1	SHELBY TN	114	-53	1	70	96
		1*	4.2			WILSON TN					
8	.2	8	.1	7	.1	TEXAS TX	-19	-48	-58		
		1	.4			BEXAR TX					
		1*	1.2			BOWIE TX					
2	.5			3	.3	DALLAS TX			-43	47	96
		2*	1.6			GALVESTON TX					
2	.2	3	.2	2	.1	HARRIS TX	-13	-45	-52	38	95
2*	1.5					JEFFERSON TX					
				1*	16.1	MC CULLOCH TX					99
				1	.6	MC LENNAN TX					97
		1*	.8			SMITH TX					
1*	4.2					WALLER TX					
1*	2.6					WICHITA TX					
3	.1	6	.2	14*	.3	VIRGINIA VA	68	103	241		
				1*	1.5	ACCOMACK VA					98
1*	1.5			1*	1.9	ARLINGTON#VA			26	82	98
		2	.4	5*	.8	CHESTERFIELD#VA		96			97
				1	.6	DINWIDDIE#VA					97
				1*	2.8	GREENSVILLE#VA					99
1*	1.8					HALIFAX#VA					
		1*	2.0			MECKLENBURG VA					
		1*	4.5			MIDDLESEX VA					
		1*	13.3			ROCKINGHAM#VA					
		1	.1	4*	.5	CHESAPEAKE CITY#VA		276			97
1*	2.5			1*	.7	HAMPTON CITY VA			-72	14	97
1	.5	1	.4			WASHINGTON WA	-20				
		1	.7			KING WA					
1*	5.8					PIERCE WA					
1	.2	2	.4	1	.2	WEST VIRGINIA WV	99	-58	-17		
		1*	12.6			HARRISON WV					
1*	1.3			1*	1.0	KANAWHA WV			-25	64	98
		1*	2.5			MC DOWELL WV					
		1	.3	3*	.6	WISCONSIN WI		126			
				1*	26.5	ASHLAND WI					99
		1*	8.8			MENOMINEE#WI					
				2*	.6	MILWAUKEE WI					97

* SIGNIFICANTLY MORE DEATHS THAN EXPECTED

COUNTIES COMBINED; SEE TECHNICAL NOTES

WHITE MALE — CANCER OF THE NASOPHARYNX — ICD 147

1950-1959 DEATHS	1950-1959 RATE	1960-1969 DEATHS	1960-1969 RATE	1970-1979 DEATHS	1970-1979 RATE	COUNTY AND STATE	PERCENT CHANGE 60S/50S	PERCENT CHANGE 70S/60S	PERCENT CHANGE 70S/50S	PERCENTILE %CHANGE 70S/50S	PERCENTILE RATES 1970S
2500	.4	3263	.4	3587	.4	UNITED STATES	15	-2	13		
38	.4	53	.5	56	.4	ALABAMA AL	27	-13	10		
				3	1.1	BALDWIN AL					90
		1	.8			BLOUNT AL					
1	1.4					BUTLER AL					
3*	1.1	3	.9	1	.3	CALHOUN AL	-15	-66	-71	03	70
		1	.7			CHAMBERS AL					
				1	1.3	CHEROKEE AL					93
1	1.0			2*	1.2	CHILTON AL			28	66	92
		1*	2.2			CLEBURNE AL					
		2*	1.9			COFFEE AL					
1	.8			1	.5	COLBERT AL			-44	16	75
1*	1.7					CONECUH AL					
1	.7	1	.7	1	.5	COVINGTON AL	-2	-20	-21	33	78
				1*	1.5	CRENSHAW AL					94
		1	.4	1	.4	CULLMAN AL		-10			73
1	1.5	2*	2.7			DALE AL	74				
				1	.5	DE KALB AL					76
		1	.8			ELMORE AL					
1	.9	1	1.0	2*	1.6	ESCAMBIA AL	8	59	71	80	94
1	.3	1	.3	1	.3	ETOWAH AL	-1	14	12	58	69
				1	1.2	FAYETTE AL					92
1	1.1	1	1.0	1	.9	FRANKLIN AL	-7	-13	-19	34	87
				2*	1.7	GENEVA AL					95
				1	.5	HOUSTON AL					77
		1	.6	1	.5	JACKSON AL		-16			77
10	.5	13	.7	8	.4	JEFFERSON AL	29	-46	-30	25	72
				1	1.6	LAMAR AL					95
				1	.5	LAUDERDALE AL					76
				1	.8	LAWRENCE AL					85
1	1.0	1	.8	2	1.1	LIMESTONE AL	-13	30	13	59	91
1	.5	2	.7			MADISON AL	50				
1*	2.2					MARENGO AL					
2	1.0			1	.4	MARSHALL AL			-63	06	71
5	.9	5	.6	9*	.8	MOBILE AL	-32	41	-4	45	86
1*	1.9			1	1.6	MONROE AL			-15	38	95
3	1.0	2	.5	5*	1.1	MONTGOMERY AL	-52	123	8	55	90
		4*	1.6	3	1.1	MORGAN AL		-32			90
		1	1.4			PICKENS AL					
1	1.2					PIKE AL					
				1	1.0	RANDOLPH AL					90
		2*	2.1	1	.8	RUSSELL AL		-63			85
		3*	2.2			SHELBY AL					
				1	.6	TALLAPOOSA AL					79
		1	.3			TUSCALOOSA AL					
1	.5	2	.9			WALKER AL	94				
		4	.9	6	1.0	ALASKA AK		8			
9	.2	25	.4	26	.3	ARIZONA AZ	94	-39	18		
1	.3	1	.5	1	.4	COCHISE AZ	45	-23	11	57	72
		2*	1.6			COCONINO AZ					
1	1.0			1	.6	GILA AZ			-42	17	80
6	.3	13	.4	18	.3	MARICOPA AZ	44	-29	3	52	70
		1	.7			MOHAVE AZ					
				1	2.0	NAVAJO AZ					96
1	.1	6	.5	3	.1	PIMA AZ	458	-73	50	76	65
		1	.5	1	.3	PINAL AZ		-37			70
				1	1.6	SANTA CRUZ AZ					95
		1	.4			YUMA AZ					
19	.3	22	.3	29	.3	ARKANSAS AR	3	12	15		
				1	.7	BAXTER AR					83
		1	.3	1	.4	BENTON AR		30			73
				1*	5.4	CALHOUN AR					99
		1	1.2	1	1.1	CLARK AR		-7			90
				1*	2.5	CLEVELAND AR					98
		1	1.3			CONWAY AR					
		1	.5			CRAIGHEAD AR					
				1	.7	CRAWFORD AR					83
1	1.0					FAULKNER AR					
2	.7	1	.2	1	.3	GARLAND AR	-71	20	-65	06	68
1	.8					GREENE AR					
				1	.7	HOT SPRING AR					84
1*	2.0					IZARD AR					
1	1.1					JACKSON AR					
				2	.7	JEFFERSON AR					84
1*	2.9					LAFAYETTE AR					
		1	.9			LONOKE AR					
				1	1.0	MARION AR					90
1	.9			1	.9	MILLER AR			1	51	87
		1	.4			MISSISSIPPI AR					
1*	2.7					MONTGOMERY AR					
		1*	2.5			PERRY AR					
				2*	2.3	PHILLIPS AR					97
2*	3.2					POLK AR					
2	.2	9*	1.0	8	.7	PULASKI AR	313	-25	212	94	83
1	1.4					RANDOLPH AR					
1*	1.4			1	1.4	ST FRANCIS AR			-5	44	93
				1*	1.6	SEARCY AR					95
		2	.6	1	.3	SEBASTIAN AR		-59			68
				1	1.4	SEVIER AR					93
1	.6					UNION AR					
3*	1.0	3	1.0	2	.5	WASHINGTON AR	1	-51	-50	13	77

* SIGNIFICANTLY MORE DEATHS THAN EXPECTED

1950-1959 DEATHS	1950-1959 RATE	1960-1969 DEATHS	1960-1969 RATE	1970-1979 DEATHS	1970-1979 RATE	COUNTY AND STATE	PERCENT CHANGE 60S/50S	PERCENT CHANGE 70S/60S	PERCENT CHANGE 70S/50S	PERCENTILE %CHANGE 70S/50S	PERCENTILE RATES 1970S
				1*	2.0	WOODRUFF AR					96
239*	.4	284	.4	384*	.5	CALIFORNIA CA	-6	15	7		
9	.3	13	.4	15	.4	ALAMEDA CA	43	10	58	78	73
		1	.2	4	.7	BUTTE CA		230			82
1*	1.4					COLUSA CA					
4	.3	13*	.7	14	.6	CONTRA COSTA CA	140	-18	98	84	79
1*	1.1					DEL NORTE CA					
		1	.6	2	.6	EL DORADO CA		12			82
5	.4	6	.4	6	.3	FRESNO CA	-6	-10	-15	38	70
1	1.1					GLENN CA					
2	.4	1	.2	1	.2	HUMBOLDT CA	-53	9	-49	14	67
				1	.3	IMPERIAL CA					71
6	.7	2	.1	8	.6	KERN CA	-80	304	-21	33	78
				1	1.0	LASSEN CA					89
106*	.5	113	.4	130	.5	LOS ANGELES CA	-11	13	1	51	76
		1	.6	1	.5	MADERA CA		-15			76
1	.2	2	.2	5	.5	MARIN CA	39	113	195	93	77
		1	.4			MENDOCINO CA					
2	.9	1	.3	3	.8	MERCED CA	-68	158	-18	35	85
4	.7	3	.4	4	.4	MONTEREY CA	-38	0	-38	18	73
2	.6					NAPA CA					
1	.8	1	.5			NEVADA CA	-35				
6	.3	19	.5	24	.4	ORANGE CA	43	-19	16	61	74
1	.3	3	.8	2	.4	PLACER CA	166	-50	32	68	74
3	.3	5	.3	8	.3	RIVERSIDE CA	5	1	6	54	69
2	.1	11	.5	19*	.7	SACRAMENTO CA	307	29	426	99	82
1	1.3	1	1.3			SAN BENITO CA	-1				
5	.3	11	.4	12	.4	SAN BERNARDINO CA	24	-12	10	56	72
12	.4	10	.2	19	.3	SAN DIEGO CA	-47	34	-30	26	69
24*	.6	19	.5	14	.5	SAN FRANCISCO CA	-6	-4	-10	41	77
4	.4	3	.2	5	.4	SAN JOAQUIN CA	-37	55	-1	49	71
3	.8	2	.4	3	.4	SAN LUIS OBISPO CA	-49	-2	-50	14	74
8	.6	9	.5	14	.6	SAN MATEO CA	-7	9	1	51	79
2	.4	3	.4	7	.6	SANTA BARBARA CA	1	69	70	80	81
8	.4	9	.3	22	.6	SANTA CLARA CA	-27	114	57	78	81
2	.4	2	.4	3	.4	SANTA CRUZ CA	-8	-7	-15	37	72
		1	.3			SHASTA CA					
1	.5			3*	1.5	SISKIYOU CA			169	91	94
1	.2	1	.2	5	.7	SOLANO CA	5	190	204	94	83
2	.3	2	.2	5	.4	SONOMA CA	-17	59	32	69	72
3	.4	1	.1	3	.3	STANISLAUS CA	-73	147	-33	22	69
2*	1.5			2	1.0	SUTTER CA			-33	22	89
		1	.7	3*	1.7	TEHAMA CA		126			95
2	.3	6	.8	5	.6	TULARE CA	192	-27	112	86	79
				2	1.2	TUOLUMNE CA					92
1	.2	2	.1	7	.4	VENTURA CA	-15	144	108	86	72
		4*	1.3	2	.5	YOLO CA		-61			77
1	.8					YUBA CA					
23	.4	27	.3	31	.3	COLORADO CO	-6	-4	-10		
		2	.7	6*	.9	ADAMS CO		32			88
				1*	2.5	ALAMOSA CO					98
		1	.3	1	.2	BOULDER CO		-19			67
				1*	1.9	CHAFFEE CO					96
1	.9					DELTA CO					
12	.6	9	.4	7	.2	DENVER-ARAPAHOE#CO	-40	-30	-58	09	67
		1*	4.8			ELBERT CO					
		3	.6	3	.2	EL PASO CO		-65			66
				1	.5	FREMONT CO					76
		2*	97.4			HINSDALE CO					
				1	.1	JEFFERSON CO					65
2*	2.8	2*	2.2			LA PLATA CO	-20				
1	.4	2	.7	1	.2	LARIMER CO	49	-63	-45	16	67
1*	2.6					MOFFAT CO					
		1	1.1			MOPGAN CO					
		1	.8			OTEPO CO					
1*	1.6					PROWERS CO					
4	.8	2	.4	4	.8	PUEBLO CO	-55	114	-4	46	84
		1*	2.1			RIO GRANDE CO					
				1*	3.3	WASHINGTON CO					99
1	.3			4	.9	WELD CO			188	92	88
46	.4	51	.4	48	.3	CONNECTICUT CT	-5	-16	-20		
13	.5	14	.4	13	.3	FAIPFIELD CT	-13	-21	-32	24	71
14	.4	10	.3	18	.5	HARTFOPD CT	-29	47	5	54	76
3	.6			3	.4	LITCHFIELD CT			-26	28	75
		1	.2	1	.2	MIDDLESEX CT		-22			65
9	.3	21*	.6	8	.2	NEW HAVEN CT	100	-65	-30	25	67
4	.5	4	.5	3	.3	NEW LONDON CT	-2	-35	-36	21	70
		1	.4	2	.6	TOLLAND CT		49			80
3	.9					WINDHAM CT					
6	.4	9	.4	8	.4	DELAWARE DE	2	-13	-12		
		1	.3			KENT DE					
5	.5	7	.5	6	.4	NEW CASTLE DE	-6	-15	-19	34	73
1	.4	1	.3	2	.5	SUSSEX DE	-16	60	34	69	77
11	.5	11	.7	6	.5	DIST. OF COLUMBIA DC	33	-32	-9		
11	.5	11	.7	6	.5	WASHINGTON DC	33	-32	-9	42	76
61	.3	144*	.5	213*	.5	FLORIDA FL	54	-9	40		
		3*	1.6	2	.7	ALACHUA FL		-54			84
2*	1.1	1	.4	7*	2.3	BAY FL	-64	483	110	86	97
		1*	2.4			BRADFORD FL					

* SIGNIFICANTLY MORE DEATHS THAN EXPECTED

\# COUNTIES COMBINED; SEE TECHNICAL NOTES

1950-1959 DEATHS	1950-1959 RATE	1960-1969 DEATHS	1960-1969 RATE	1970-1979 DEATHS	1970-1979 RATE	COUNTY AND STATE	PERCENT CHANGE 60S/50S	PERCENT CHANGE 70S/60S	PERCENT CHANGE 70S/50S	PERCENTILE %CHANGE 70S/50S	PERCENTILE RATES 1970S
		5	.8	2	.1	BREVARD FL		-81			65
3	.2	13	.5	31*	.6	BROWARD FL	110	25	163	91	82
				1	.2	CHARLOTTE FL					66
				1	.1	CITRUS FL					65
		1	.8	3*	1.6	CLAY FL		89			95
				4	1.6	COLLIER FL					95
17	.5	32*	.6	37	.5	DADE FL	21	-15	2	52	77
1*	5.2					DIXIE FL					
1	.1	20*	1.3	10	.6	DUVAL FL	1769	-55	739	99	80
1	.2	1	.2	4	.6	ESCAMBIA FL	-25	295	198	93	80
				2*	1.8	GADSDEN FL					96
				1*	6.9	GILCHRIST FL					99
		1*	5.3	1*	2.1	GULF FL		-60			97
1*	2.0					HARDEE FL					
		1	.6	1	.4	HERNANDO FL		-25			74
				1	.3	HIGHLANDS FL					68
6	.4	14*	.8	15	.6	HILLSBOROUGH FL	74	-21	37	71	81
				1	1.1	HOLMES FL					90
2*	2.7					INDIAN RIVER FL					
1	.9	1	.9			JACKSON FL	-4				
2	.9	4*	1.3	2	.3	LAKE FL	49	-79	-68	04	68
1	.4	3	.7	6	.5	LEE FL	76	-26	31	68	78
1	.6					LEON FL					
				1*	2.0	MADISON FL					96
		2	.4	2	.2	MANATEE FL		-43			68
		1	.4			MARION FL					
1	.8	1	.6	2	.6	MONROE FL	-28	6	-24	29	80
1*	2.2			2*	3.2	NASSAU FL			43	73	98
		1	.4	1	.2	OKALOOSA FL		-36			67
1	.1	8	.7	4	.2	ORANGE FL	578	-67	125	88	67
				2	.8	OSCEOLA FL					85
4	.4	3	.2	15	.6	PALM BEACH FL	-47	179	48	75	81
				4	.2	PASCO FL					67
11	.5	13	.4	27	.5	PINELLAS FL	-18	14	-7	43	77
		6	.7	4	.3	POLK FL		-48			71
				1	.6	PUTNAM FL					80
				2	.8	ST LUCIE FL					85
1	.1	1	.1	7	.4	SARASOTA FL	-7	307	278	96	75
		1	.5	5*	1.0	SEMINOLE FL		109			88
		1	1.8			SUMTER FL					
3	.5	4	.6	2	.1	VOLUSIA FL	34	-87	-82	00	64
		1	1.2			WALTON FL					
49	.5	60	.5	90*	.6	GEORGIA GA	-3	20	16		
1	.8					BALDWIN GA					
1*	1.5			2*	3.5	BARROW GA			140	89	99
1	1.4			3*	2.7	BARTOW GA			95	83	98
1*	2.6					BEN HILL GA					
2	.6	1	.3	2	.4	BIBB GA	-52	41	-33	23	74
1*	2.2	1	1.0			BULLOCH GA	-54				
				1*	2.9	BUTTS GA					98
				1	.4	CARROLL GA					75
				2*	1.6	CATOOSA GA					95
		1*	8.1			CHARLTON GA					
3	.9	3	.7	6*	1.0	CHATHAM GA	-25	45	9	56	89
				1	1.0	CHATTOOGA GA					90
1	1.0			2	1.2	CLARKE GA			22	64	91
1*	11.8					CLAY GA					
1	1.5	1	.8	1	.3	CLAYTON GA	-46	-61	-79	01	70
1	.8	4	.7	6	.8	COBB GA	-11	25	12	57	86
				2*	2.8	COFFEE GA					98
1	.8	2*	1.9			COLQUITT GA	138				
				1*	2.1	COOK GA					97
1	1.6	1	1.1	1	.7	COWETA GA	-32	-32	-54	12	84
1*	1.6					DECATUR GA					
5	.9	2	.2	11*	.8	DE KALB GA	-81	349	-16	37	84
1*	3.6					DOOLY GA					
		3*	1.5			DOUGHERTY GA					
		1	1.1			DOUGLAS GA					
		4*	6.4			ELBERT GA					
		1	.4	2	.6	FLOYD GA		59			80
11*	.8	6	.4	10	.7	FULTON GA	-54	91	-13	39	82
		1*	2.2	1*	2.2	GILMER GA		0			97
		1	.8	3*	1.8	GLYNN GA		109			96
		1	1.3	1	.9	GORDON GA		-29			87
1*	2.0	1	1.6			GRADY GA	-19				
		3*	1.8	3	.6	GWINNETT GA		-66			81
		1	1.1	1	1.2	HABERSHAM GA		6			91
1	.5			1	.3	HALL GA			-38	18	70
		1	1.7			HARALSON GA					
				1	.6	HOUSTON GA					79
		1*	9.9			LEE GA					
1	1.1			1	.6	LOWNDES GA			-46	15	80
				1*	1.7	MERIWETHER GA					95
1*	2.5					MURRAY GA					
		3	1.1	3	.7	MUSCOGEE#GA		-39			83
				1	1.1	PAULDING GA					90
		1*	2.6			PICKENS GA					
		1*	2.8			PIERCE GA					
		1*	4.2			PIKE GA					
		1*	5.2			PUTNAM GA					
1*	4.5					RANDOLPH GA					
2	.5	3	.7			RICHMOND GA	35				
				2*	3.3	ROCKDALE GA					99
				1*	2.2	SCREVEN GA					97
1	.8	1	.7			SPALDING GA	-9				
				1	1.0	STEPHENS GA					89
		1*	1.8	1	1.4	SUMTER GA		-25			93
				1*	10.8	TALIAFERRO GA					99
		1	1.6			TATTNALL GA					

* SIGNIFICANTLY MORE DEATHS THAN EXPECTED

COUNTIES COMBINED; SEE TECHNICAL NOTES

WHITE MALE — CANCER OF THE NASOPHARYNX — ICD 147

1950-1959 DEATHS	1950-1959 RATE	1960-1969 DEATHS	1960-1969 RATE	1970-1979 DEATHS	1970-1979 RATE	COUNTY AND STATE	PERCENT CHANGE 60S/50S	PERCENT CHANGE 70S/60S	PERCENT CHANGE 70S/50S	PERCENTILE %CHANGE 70S/50S	PERCENTILE RATES 1970S
				1*	5.3	TAYLOR GA					99
		1*	2.2	1*	3.1	TELFAIR GA		38			98
				1*	4.5	TERRELL GA					99
1	.8			1	.9	THOMAS GA			2	51	86
				2*	1.7	TIFT GA					95
				2	1.0	TROUP GA					90
				1	1.2	UPSON GA					92
				3*	1.3	WALKER GA					93
				1	1.1	WALTON GA					91
		2*	1.7			WARE GA					
2*	11.8					WARREN GA					
3*	2.6	1	.6			WHITFIELD GA	-75				
		1*	3.7			WILKES GA					
1*	2.2	1*	2.3			WORTH GA	3				
		15*	2.5	7	.5	HAWAII HI		-79			
		1	4.6	1	1.0	HAWAII HI		-79			88
		11*	2.1	6	.6	HONOLULU HI		-72			79
		3*	5.1			MAUI#HI					
12	.4	5	.2	18	.5	IDAHO ID	-59	204	24		
3	.8	1	.2	3	.5	ADA ID	-70	112	-36	20	77
1	.5					BANNOCK ID					
1*	1.9					BEAR LAKE ID					
2*	2.1					BINGHAM ID					
		1	1.1	1	.9	BONNER ID		-17			87
				1	.3	CANYON ID					68
				1	1.1	CASSIA ID					91
				1*	1.8	CLEARWATER ID					96
				1*	1.6	GEM ID					95
1*	1.6			1	1.4	IDAHO ID			-16	37	93
1*	2.7					JEFFERSON ID					
				3*	1.2	KOOTENAI ID					92
		1	1.0			LATAH ID					
1	.7	1	.7	3*	2.0	NEZ PERCE ID	-3	206	196	93	97
1*	2.7					OWYHEE ID					
				1	1.2	PAYETTE ID					92
1	.5	1	.4	2	.8	TWIN FALLS ID	-17	85	54	77	85
187*	.4	183	.4	196	.4	ILLINOIS IL	-6	4	-2		
3	.7	1	.4	1	.3	ADAMS IL	-48	-26	-61	07	68
				2*	2.0	BOND IL					97
1	.5	2	1.1			BUREAU IL	113				
1	.8	1	.9			CARROLL IL	2				
1	.3	2	.5	1	.2	CHAMPAIGN IL	63	-52	-22	31	67
1	.5			1	.5	CHRISTIAN IL			-5	45	76
		2*	2.0			CLARK IL					
1	.9					CLINTON IL					
				1	.5	COLES IL					75
101*	.5	104	.5	99	.5	COOK IL	3	-5	-2	48	76
		1*	1.9	1*	2.4	CUMBERLAND IL		31			98
2	.9	2	.8	1	.4	DE KALB IL	-14	-54	-61	08	72
1	.9					DE WITT IL					
6	.7	5	.3	11	.5	DU PAGE IL	-54	47	-33	24	76
		1	.7			EDGAR IL					
1	.9	1	.8			EFFINGHAM IL	-12				
1	.7					FAYETTE IL					
				1	1.1	FORD IL					91
1	.3	1	.5	2	.7	FRANKLIN IL	82	46	166	91	84
1	.4	1	.4	1	.5	FULTON IL	5	21	27	66	75
1	1.6			1	1.0	GREENE IL			-36	20	89
				1	.8	GRUNDY IL					86
		1	2.5			HAMILTON IL					
1	.8	1	.7			HANCOCK IL	-4				
				1*	2.5	HARDIN IL					98
				1	.4	HENRY IL					71
1	.5					IROQUOIS IL					
1	.5			1	.5	JACKSON IL			-2	48	78
1	.6	3*	1.9	1	.5	JEFFERSON IL	216	-71	-7	43	78
				2*	2.1	JERSEY IL					97
2	.2	3	.3	6	.6	KANE IL	33	99	165	91	79
2	.4	1	.2	2	.4	KANKAKEE IL	-52	103	-2	48	74
1	1.0					KENDALL IL					
2	.6	1	.3			KNOX IL	-51				
3	.3	5	.3	6	.4	LAKE IL	6	15	22	64	73
2	.3	1	.2	2	.3	LA SALLE IL	-47	97	5	53	70
				1	.9	LAWRENCE IL					87
		1	.5	2	.9	LIVINGSTON IL		89			88
1	.6			1	.6	MC DONOUGH IL			10	56	81
		3	.7	4	.7	MC HENRY IL		7			83
2	.5			1	.2	MC LEAN IL			-56	11	67
2	.4	2	.4	3	.5	MACON IL	-7	53	43	73	78
		2	.9			MACOUPIN IL					
2	.2	4	.4	3	.3	MADISON IL	63	-28	18	62	68
1	1.1	1	.9			MASON IL	-17				
		2*	3.4			MENARD IL					
				1	1.0	MONROE IL					89
1	.6			1	.7	MONTGOMERY IL			20	63	82
1	.5					MORGAN IL					
				1	.4	OGLE IL					74
4	.4	3	.4	5	.6	PEORIA IL	-19	58	28	67	79
1	.6	2*	1.5	1	.7	PIKE IL	129	-54	6	54	83
				1	.8	RANDOLPH IL					84
				1	.8	RICHLAND IL					86
4	.6	2	.3	2	.3	ROCK ISLAND IL	-54	1	-53	12	68
6	.6	2	.2	4	.4	ST CLAIR IL	-73	134	-37	20	73
				1	.5	SALINE IL					78
2	.5	3	.4			SANGAMON IL	-11				

* SIGNIFICANTLY MORE DEATHS THAN EXPECTED

COUNTIES COMBINED; SEE TECHNICAL NOTES

WHITE MALE — CANCER OF THE NASOPHARYNX — ICD 147

1950-1959 DEATHS	1950-1959 RATE	1960-1969 DEATHS	1960-1969 RATE	1970-1979 DEATHS	1970-1979 RATE	COUNTY AND STATE	PERCENT CHANGE 60S/50S	PERCENT CHANGE 70S/60S	PERCENT CHANGE 70S/50S	PERCENTILE %CHANGE 70S/50S	PERCENTILE RATES 1970S
2*	1.8			1	1.0	SHELBY IL			-44	16	89
		1*	2.0			STARK IL					
				1	.4	STEPHENSON IL					74
4*	.9	3	.6	1	.2	TAZEWELL IL	-41	-68	-81	00	66
3	.6	2	.4	4	.9	VERMILION IL	-31	105	41	72	87
1	1.4					WABASH IL					
1	1.0			1	.7	WASHINGTON IL			-31	25	83
				1	.8	WAYNE IL					85
		1	.3			WHITESIDE IL					
5	.7	4	.6	4	.4	WILL IL	-17	-36	-47	15	72
1	.4			1	.4	WILLIAMSON IL			-2	49	74
4	.5	5	.6	2	.2	WINNEBAGO IL	9	-67	-64	06	66
				1	.7	WOODFORD IL					83
44	.2	71	.3	89	.4	INDIANA IN	45	17	69		
1	1.1			1	.8	ADAMS IN			-22	32	86
		4	.4	4	.4	ALLEN IN		-9			71
		1	.6			BARTHOLOMEW IN					
				2*	3.6	BENTON IN					99
				2*	2.7	BLACKFORD IN					98
1	.8	1	.7			BOONE IN	-9				
		3*	1.3	3*	1.5	CASS IN		10			94
				2	.6	CLARK IN					81
		1	.6	1	.8	CLAY IN		27			85
				1	.7	CLINTON IN					82
		1*	2.3	1*	2.0	CRAWFORD IN		-13			96
1	.8	1	1.0			DECATUR IN	18				
		1	.7			DE KALB IN					
		1	.2	4	.9	DELAWARE IN		327			87
1	.8	1	.7	1	.7	DUBOIS IN	-13	0	-13	39	83
1	.2	1	.2	1	.2	ELKHART IN	3	-25	-23	31	65
				1	1.0	FAYETTE IN					89
				1	.4	FLOYD IN					72
				1	1.0	FOUNTAIN IN					89
1	.6	1	.5			GIBSON IN	-16				
1	.7					GREENE IN					
1	.6	1	.5			HAMILTON IN	-4				
		1	.9			HANCOCK IN					
		1	1.7			HARRISON IN					
				2	1.0	HENDRICKS IN					89
		1	.4	1	.4	HENRY IN		-7			73
		2	.8	1	.3	HOWARD IN		-55			71
		1	.6			HUNTINGTON IN					
1	.8	1	.6	2	.7	JOHNSON IN	-17	7	-11	40	83
1	.5	1	.4			KNOX IN	-19				
4	.2	5	.3	3	.2	LAKE IN	33	-44	-25	28	65
3	1.3			3	.6	LA PORTE IN			-55	11	80
		1	.5	2	1.0	LAWRENCE IN		93			88
1	.2	2	.3	4	.6	MADISON IN	48	87	176	92	81
8	.4	10	.4	15	.6	MARION IN	11	48	63	79	79
		1	.6	1	.6	MARSHALL IN		-6			79
		1*	2.8			MARTIN IN					
				1	.7	MIAMI IN					82
				1	.6	MONTGOMERY IN					81
				1	.6	MORGAN IN					80
				1	1.5	NEWTON IN					94
		1	.7			NOBLE IN					
				1	1.4	PARKE IN					94
2*	2.6					PERRY IN					
		1	1.9			PIKE IN					
		2	.8			PORTER IN					
		2*	1.9	1	.9	POSEY IN		-54			87
		1	1.2			PULASKI IN					
1	.8					PUTNAM IN					
				1	.8	RIPLEY IN					86
3	.3	4	.4	4	.4	ST JOSEPH IN	19	-2	17	62	72
				1	1.4	SCOTT IN					93
				1*	2.3	SWITZERLAND IN					97
2	.7			3	.8	TIPPECANOE IN			27	66	86
		1	1.3			TIPTON IN					
3	.5	7*	.9	4	.5	VANDERBURGH IN	78	-46	-4	46	77
		1	.8			VERMILLION IN					
3	.5	3	.5	4	.8	VIGO IN	3	47	51	76	85
1	.5			1	.6	WABASH IN			31	68	81
		1*	2.1			WARREN IN					
1	.2			3	.9	WAYNE IN			275	96	87
		1	1.0			WELLS IN					
				1	1.1	WHITE IN					90
2*	2.0					WHITLEY IN					
45	.3	51	.4	40	.3	IOWA IA	8	-22	-16		
				1	.8	ALLAMAKEE IA					86
2	.4	1	.2	5	1.0	BLACK HAWK IA	-40	317	152	90	88
1	.5			1	.6	BOONE IA			13	58	81
				1	.9	BUCHANAN IA					87
		1	1.0	1	.8	BUENA VISTA IA		-19			85
1	.8					CARROLL IA					
		2	.8	1	.4	CERRO GORDO IA		-52			73
1	.8	1	1.1			CHEROKEE IA	39				
1*	1.5					CLARKE IA					
		1	1.0			CLAY IA					
1	.9	1	.7	1	1.2	CLAYTON IA	-17	68	39	71	92
2	.7	2	.7			CLINTON IA	1				
1	.9	1	1.4	1	1.0	CRAWFORD IA	52	-30	7	55	89
				1*	1.5	DECATUR IA					94
1	.5					DES MOINES IA					
1	.3	3	.9	1	.3	DUBUQUE IA	205	-69	-4	45	68
1	1.5					EMMET IA					
		1	.8			FLOYD IA					

* SIGNIFICANTLY MORE DEATHS THAN EXPECTED

WHITE MALE — CANCER OF THE NASOPHARYNX — ICD 147

1950-1959 DEATHS	1950-1959 RATE	1960-1969 DEATHS	1960-1969 RATE	1970-1979 DEATHS	1970-1979 RATE	COUNTY AND STATE	PERCENT CHANGE 60S/50S	PERCENT CHANGE 70S/60S	PERCENT CHANGE 70S/50S	PERCENTILE %CHANGE 70S/50S	PERCENTILE RATES 1970S
1	1.4					GRUNDY IA					
		1	.7	2*	1.8	HARDIN IA		165			96
		2*	1.8			HARRISON IA					
1	.9	1	.8			HENRY IA	-3				
		3*	3.3	1	1.1	IOWA IA		-67			90
1	.8	1	.8			JACKSON IA	-2				
2*	1.6					JASPER IA					
2	.7					JOHNSON IA					
1	1.1					JONES IA					
		1	.9			KEOKUK IA					
		1	.9			KOSSUTH IA					
		1	.5	1	.6	LEE IA		22			79
5*	.9			4	.6	LINN IA			-31	25	80
1	1.6	1	1.6			LUCAS IA	-1				
1	.8					MAHASKA IA					
1	.8	3*	1.5			MARSHALL IA	94				
				1	1.1	MONONA IA					91
				1*	1.2	MONROE IA					92
		1	.6	1	.5	MUSCATINE IA		-7			78
1	.9			1	1.0	O'BRIEN IA			15	60	89
1	1.4					PALO ALTO IA					
		2*	1.5			PLYMOUTH IA					
3	.3	3	.3	3	.3	POLK IA	-16	5	-11	39	69
		1	.3	3	.8	POTTAWATTAMIE IA		186			85
		1	1.1			POWESHIEK IA					
		1	.9			SAC IA					
3	.6	3	.5	1	.2	SCOTT IA	-8	-65	-68	05	66
		3*	3.9	1	1.5	SHELBY IA		-62			94
1	.8					SIOUX IA					
				1	.5	STORY IA					77
1	.8					TAMA IA					
		1	1.0			UNION IA					
2	.8			1	.4	WAPELLO IA			-52	13	73
		1	.4	3*	1.3	WEBSTER IA		184			92
1	.9					WINNESHIEK IA					
3	.6	4	.8	1	.2	WOODBURY IA	38	-75	-65	06	66
		1	1.0			WRIGHT IA					
33	.3	39	.4	31	.3	KANSAS KS	13	-25	-15		
1	1.0					ALLEN KS					
2*	1.8					BOURBON KS					
		1	1.2			BROWN KS					
1	1.4	2*	2.3			CLOUD KS	67				
2	.9					COWLEY KS					
		1	.5			CRAWFORD KS					
1	.8	1	.6	1	.6	DICKINSON KS	-18	-12	-28	26	79
		2*	4.2			DONIPHAN KS					
				2	.8	DOUGLAS KS					85
				1*	2.2	ELLSWORTH KS					97
		1	1.2			FINNEY KS					
1	.8			1	.8	FRANKLIN KS			0	50	86
1*	6.2					GRAHAM KS					
		1*	1.6			HARPER KS					
1*	1.5	1	2.9			JACKSON KS	100				
2	.4	3	.4	3	.3	JOHNSON KS	7	-31	-26	27	69
		1*	7.1			KEARNY KS					
		1*	3.4			KIOWA KS					
1	.4	1	.4	1	.5	LEAVENWORTH KS	17	10	29	67	76
				1	.5	LYON KS					78
		1	.6			MC PHERSON KS					
		1	.9	1	.9	MARSHALL KS		8			88
				1	.9	MIAMI KS					87
		2	1.1			MONTGOMERY KS					
1*	1.6					MORRIS KS					
1	1.1	1	1.0			NEMAHA KS	-6				
1	.9					NEOSHO KS					
1	1.1					OSAGE KS					
1*	1.7					OTTAWA KS					
1*	1.7					PHILLIPS KS					
1*	1.5			2*	1.9	REPUBLIC KS			32	69	96
1	.9					RILEY KS					
		1*	2.0	1*	3.0	ROOKS KS		49			98
		1	1.6			RUSSELL KS					
				2	.9	SALINE KS					88
8*	.8	9	.7	3	.2	SEDGWICK KS	-12	-74	-77	02	66
1	.2	2	.3	3	.4	SHAWNEE KS	75	30	128	88	75
		1*	2.3			STAFFORD KS					
				2*	1.5	SUMNER KS					94
				1*	2.6	THOMAS KS					98
		1	1.4			WILSON KS					
		1*	3.9	3*	7.3	WOODSON KS		88			99
3	.4	2	.3	2	.3	WYANDOTTE KS	-26	-6	-30	25	69
32	.3	61	.5	68	.5	KENTUCKY KY	79	1	80		
				2*	4.1	BALLARD KY					99
				1	.6	BARREN KY					82
		1*	2.1			BATH KY					
				1	.6	BELL KY					80
				2*	1.4	BOONE KY					94
				3*	3.7	BOURBON KY					99
				1	.9	BOYLE KY					87
		1*	2.0			BRACKEN KY					
1	1.3			1	1.4	BRECKINRIDGE KY			12	58	94
				1	.6	BULLITT KY					80
				1*	1.8	BUTLER KY					96
		2*	3.4			CALDWELL KY					
1	.9	1	1.0	2*	1.2	CALLOWAY KY	5	21	28	66	92
2	.5	7*	1.7	3	.8	CAMPBELL KY	219	-55	44	74	85
				1	1.1	CASEY KY					91

* SIGNIFICANTLY MORE DEATHS THAN EXPECTED

WHITE MALE | CANCER OF THE NASOPHARYNX | ICD 147

1950-1959 DEATHS	1950-1959 RATE	1960-1969 DEATHS	1960-1969 RATE	1970-1979 DEATHS	1970-1979 RATE	COUNTY AND STATE	PERCENT CHANGE 60S/50S	PERCENT CHANGE 70S/60S	PERCENT CHANGE 70S/50S	PERCENTILE %CHANGE 70S/50S	PERCENTILE RATES 1970S
1	.6	1	.6	1	.6	CHRISTIAN KY	-5	-5	-9	42	79
1	.4	1	.4			DAVIESS KY	0				
				1	1.4	ESTILL KY					93
		3	.6	1	.1	FAYETTE KY		-77			65
				1	1.5	FLEMING KY					94
1	.6					FRANKLIN KY					
				1*	2.0	GRANT KY					96
1	.8	1	.5	3*	1.7	GRAVES KY	-37	241	114	87	96
		1	.8	1	.9	GRAYSON KY		6			86
				1	.7	GREENUP KY					83
		1*	4.6			HANCOCK KY					
				1	.6	HARDIN KY					82
1*	2.2					HART KY					
2*	3.3					HENRY KY					
1	.6	2	1.1			HOPKINS KY	67				
1*	2.1	1*	2.1			JACKSON KY	-4				
6	.3	15	.6	19*	.7	JEFFERSON KY	150	13	182	92	84
		1*	1.8	2*	2.5	JESSAMINE KY		36			98
1	.8			1	1.0	JOHNSON KY			18	63	89
3	.6	1	.2	1	.2	KENTON KY	-69	1	-69	04	66
		1	1.0			LAWRENCE KY					
		1	1.3			LINCOLN KY					
		1*	2.7			LIVINGSTON KY					
				1	1.1	LOGAN KY					91
1*	3.4					LYON KY					
		1	.4	3	1.1	MC CRACKEN KY		222			91
				1	.6	MADISON KY					81
		2*	3.1			MARION KY					
		1	1.8			MARSHALL KY					
				1*	.9	MEADE KY					87
1*	5.6					MENIFEE KY					
		1	1.3	1	1.3	MERCER KY		-4			92
		1*	1.8			MORGAN KY					
2*	2.0			3*	2.4	MUHLENBERG KY			21	64	97
		1	1.1	1	.9	NELSON KY		-14			88
		1	1.3			OLDHAM KY					
				1*	1.6	PENDLETON KY					95
		1	.7			PERRY KY					
1	.4	1	.4			PIKE KY	1				
				1	1.5	ROWAN KY					94
		1	1.4			SCOTT KY					
		1	1.2			SHELBY KY					
1*	4.3					SPENCER KY					
1*	3.7					TRIMBLE KY					
		1	1.3			UNION KY					
1	.5	1	.4			WARREN KY	-10				
		1	1.5	1	1.1	WAYNE KY		-30			90
		1	1.7			WEBSTER KY					
		1	.7	1	.7	WHITLEY KY		12			84
1*	2.1	1*	1.8			WOODFORD KY	-15				
51*	.6	54	.5	58	.5	LOUISIANA LA	-16	-8	-23		
		1	1.0	1	.4	ASCENSION LA		-62			73
1*	1.7					ASSUMPTION LA					
		1	1.3	1	1.0	BEAUREGARD LA		-26			89
1	1.7	1	.7			BOSSIER LA	-61				
2	.4	3	.5	2	.3	CADDO LA	29	-40	-23	30	69
2	.7	2	.6	3	.5	CALCASIEU LA	-13	-11	-22	31	77
		1*	1.5			CLAIBORNE LA					
		3*	5.3	1*	1.6	CONCORDIA LA		-71			95
				1	1.4	DE SOTO LA					94
1	.2			6	.7	EAST BATON ROUGE LA			207	94	82
		1*	3.9			EAST CARROLL LA					
1*	1.7	1*	1.9	2*	3.1	GRANT LA	8	67	80	82	98
1	.8	1	.6			IBERIA LA	-23				
2*	2.8	1	1.4	1	1.2	IBERVILLE LA	-51	-14	-58	10	92
1	.2	3	.3	5	.4	JEFFERSON LA	70	17	100	85	72
1	.9	2	.8			LAFAYETTE LA	-13				
				1	.3	LAFOURCHE LA					70
1*	1.6			2*	3.2	LA SALLE LA			106	85	98
2*	3.2			1	1.3	LINCOLN LA			-60	08	93
				2	1.0	LIVINGSTON LA					89
1*	1.7					MOREHOUSE LA					
2*	2.4					NATCHITOCHES LA					
25*	1.3	15*	.8	15*	.9	ORLEANS LA	-40	13	-33	24	87
1	.5	1	.4	2	.5	OUACHITA LA	-30	37	-4	45	76
1	.3					RAPIDES LA					
		1*	2.7			RED RIVER LA					
1	1.3			1	1.3	SABINE LA			-2	48	93
				1	.7	ST BERNARD LA					83
1*	4.5					ST JAMES LA					
		1*	2.5	1*	1.8	ST JOHN THE BAPT LA		-28			96
		6*	3.3	1	.4	ST LANDRY LA		-88			73
1	1.1			3*	1.9	ST MARY LA			76	81	96
1	1.0	1	.5	1	.4	ST TAMMANY LA	-54	-15	-60	08	73
1	.7	1	.6	1	.4	TANGIPAHOA LA	-11	-35	-42	17	73
		3*	1.6			TERREBONNE LA					
				1	1.0	VERNON LA					90
		3*	2.1			WASHINGTON LA					
		1	.7	1	.6	WEBSTER LA		-8			81
				1*	2.3	WEST BATON ROUGE LA					97
12	.3	12	.2	21	.4	MAINE ME	-2	64	61		
2	.5	1	.2	1	.2	ANDROSCOGGIN ME	-54	-5	-57	10	66
1	.2			1	.3	AROOSTOOK ME			14	60	69
1	.1			3	.3	CUMBERLAND ME			183	92	70
3	.6	2	.4	3	.6	KENNEBEC ME	-29	36	-3	47	80
		1	.7			KNOX ME					
				2*	1.1	LINCOLN ME					91

* SIGNIFICANTLY MORE DEATHS THAN EXPECTED

WHITE MALE | CANCER OF THE NASOPHARYNX | ICD 147

1950-1959 DEATHS	1950-1959 RATE	1960-1969 DEATHS	1960-1969 RATE	1970-1979 DEATHS	1970-1979 RATE	COUNTY AND STATE	PERCENT CHANGE 60S/50S	PERCENT CHANGE 70S/60S	PERCENT CHANGE 70S/50S	PERCENTILE %CHANGE 70S/50S	PERCENTILE RATES 1970S
1	.4	1	.4	4*	1.6	OXFORD ME	15	263	318	97	95
		1	.2	3	.5	PENOBSCOT ME		172			77
1	.9	2*	1.7			SAGADAHOC ME	96				
				2	1.0	SOMERSET ME					89
1	.9	1	.8	1	1.0	WALDO ME	-15	23	5	53	88
		1	.7			WASHINGTON ME					
2	.4	2	.4	1	.2	YORK ME	-15	-60	-66	05	65
46	.5	60	.5	80*	.6	MARYLAND MD	2	11	13		
				2	.4	ALLEGANY MD					75
		2	.3	3	.2	ANNE ARUNDEL MD		-12			67
7	.5	2	.2	18	.7	BALTIMORE MD	-66	300	36	70	82
		1*	1.6	1	1.1	CALVERT MD		-32			91
				1	.9	CAROLINE MD					88
		4*	1.5			CARROLL MD					
1	.5	1	.5	1	.6	CECIL MD	-3	14	11	57	80
1	1.0			1	1.0	CHARLES MD			-2	47	88
				1	.6	DORCHESTER MD					82
3*	1.0	3	.9			FREDERICK MD	-10				
1	.6	5*	1.9	5*	1.2	HARFORD MD	206	-35	100	85	92
1	1.0			4*	1.9	HOWARD MD			96	84	96
1*	1.7					KENT MD					
3	.4	4	.2	8	.3	MONTGOMERY MD	-42	58	-7	43	71
1	.2	4	.3	13*	.7	PRINCE GEORGE'S MD	89	135	343	98	83
				1	.7	ST MARY'S MD					82
				1	.8	TALBOT MD					86
2	.6	3	.7	1	.2	WASHINGTON MD	12	-73	-70	04	66
		1	.7	3*	1.3	WICOMICO MD		95			93
		1	1.1			WORCESTER MD					
25*	.8	29*	1.0	16	.7	BALTIMORE CITY MD	26	-33	-15	37	82
74	.3	118	.5	109	.4	MASSACHUSETTS MA	46	-11	30		
		2	.4	2	.3	BARNSTABLE MA		-39			68
		1	.1	1	.1	BERKSHIRE MA		-1			65
6	.3	13	.6	8	.4	BRISTOL MA	112	-41	25	65	71
10	.4	11	.4	17	.5	ESSEX MA	-3	47	42	72	78
1	.4			2	.6	FRANKLIN MA			60	79	79
3	.1	7	.3	15*	.7	HAMPDEN MA	121	112	369	98	83
				3	.5	HAMPSHIRE MA					78
20	.4	32	.5	22	.4	MIDDLESEX MA	40	-31	-4	46	72
6	.3	17*	.7	13	.5	NORFOLK MA	180	-35	81	82	76
2	.3	6	.4	6	.4	PLYMOUTH MA	30	-12	14	59	72
16	.4	19	.5	13	.5	SUFFOLK MA	32	-14	14	59	75
10	.4	10	.3	7	.2	WORCESTER MA	-7	-32	-37	20	67
96	.3	142	.4	131	.4	MICHIGAN MI	31	-12	16		
		2	.7	1	.3	ALLEGAN MI		-60			68
				1	.4	BARRY MI					72
1	.3	5*	1.1	3	.7	BAY MI	348	-37	183	92	83
2	.3			3	.4	BERRIEN MI			26	65	74
1	.1	3	.5	3	.5	CALHOUN MI	216	10	248	95	77
1	.7	1	.6			CASS MI	-11				
		2*	2.8			CHARLEVOIX MI					
		1	1.3			CHEBOYGAN MI					
		1	.7			CHIPPEWA MI					
				1	.5	CLINTON MI					78
1	.5					DELTA MI					
1	.6					DICKINSON MI					
				2	.6	EATON MI					80
5	.4	5	.4	4	.3	GENESEE MI	-15	-26	-38	19	68
1	.6					GRAND TRAVERSE MI					
		1	.4	1	.4	HOUGHTON MI		0			74
1	.1	9*	1.1			INGHAM MI	1159				
1	.5	1	.5	1	.5	IONIA MI	12	-2	9	56	76
		1	1.3			IOSCO MI					
		1	1.1			IRON MI					
		3	.5	3	.5	JACKSON MI		-13			76
2	.4	6	.8	3	.4	KALAMAZOO MI	113	-46	14	60	75
				1*	1.9	KALKASKA MI					96
4	.3	5	.3	4	.2	KENT MI	11	-22	-13	38	67
1	.5			1	.3	LAPEER MI			-34	22	71
		1	.3	1	.3	LENAWEE MI		-19			68
1	.7					LIVINGSTON MI					
				1*	1.6	MACKINAC MI					95
8*	1.0	3	.2	12	.4	MACOMB MI	-81	149	-53	12	75
1	.4			1	.4	MARQUETTE MI			-10	41	72
		1	.7			MASON MI					
1	.8					MENOMINEE MI					
				2	.7	MIDLAND MI					84
1	.3	2	.5	2	.4	MONROE MI	92	-25	45	74	74
				1	.6	MONTCALM MI					79
3	.6	1	.2	3	.5	MUSKEGON MI	-75	210	-23	30	76
1	1.0			2	1.2	NEWAYGO MI			16	61	91
5	.2	9	.3	15	.4	OAKLAND MI	27	36	73	81	73
1	1.2			1	.9	OCEANA MI			-21	32	87
1	.3	2	.4	1	.2	OTTAWA MI	61	-58	-32	24	66
1	.1	4	.5	2	.3	SAGINAW MI	246	-50	74	81	68
		2	.4	7*	1.2	ST CLAIR MI		225			91
		4*	1.8	2	.8	ST JOSEPH MI		-57			85
1*	2.2					SCHOOLCRAFT MI					
		1	.4			SHIAWASSEE MI					
		1	.5			TUSCOLA MI					
				1	.4	VAN BUREN MI					73
2	.4	4	.7	2	.3	WASHTENAW MI	83	-59	-25	29	69
46	.4	60*	.6	43	.5	WAYNE MI	33	-14	15	60	76
1	1.1					WEXFORD MI					

* SIGNIFICANTLY MORE DEATHS THAN EXPECTED

WHITE MALE | CANCER OF THE NASOPHARYNX | ICD 147

1950-1959 DEATHS	1950-1959 RATE	1960-1969 DEATHS	1960-1969 RATE	1970-1979 DEATHS	1970-1979 RATE	COUNTY AND STATE	PERCENT CHANGE 60S/50S	PERCENT CHANGE 70S/60S	PERCENT CHANGE 70S/50S	PERCENTILE %CHANGE 70S/50S	PERCENTILE RATES 1970S
43	.3	64	.4	48	.3	MINNESOTA MN	38	-29	-3		
1	.9					AITKIN MN					
		3	.7	1	.2	ANOKA MN		-69			66
1	.8					BECKER MN					
		1	1.0			BELTRAMI MN					
				1	.5	BLUE EARTH MN					77
		1	.7			BROWN MN					
		2*	1.6	1	.6	CASS MN		-61			81
1	.6	2	1.2	1	.5	CLAY MN	90	-56	-17	36	77
2*	1.2	1	.4	1	.4	CROW WING MN	-65	-11	-69	04	72
		1	.2	2	.3	DAKOTA MN		14			68
				2*	1.3	DOUGLAS MN					92
				2*	1.5	FARIBAULT MN					94
1	.6					FILLMORE MN					
		1	.5			FREEBORN MN					
		1	.5			GOODHUE MN					
13	.3	20	.5	13	.3	HENNEPIN MN	52	-38	-6	44	70
		1	1.0			HOUSTON MN					
		1	.5	1	.5	ITASCA MN		-1			77
1	1.2			1	1.2	JACKSON MN			0	50	92
2*	1.2					KANDIYOHI MN					
1	1.1					LAC QUI PARLE MN					
		1*	4.2			LAKE OF THE WOODS MN					
		1	.7	1	.8	LE SUEUR MN		13			86
		1	.7			LYON MN					
				1	.6	MC LEOD MN					80
		1	1.4	1	1.1	MARSHALL MN		-23			90
		1	.7			MARTIN MN					
1	.4	1	.4	1	.4	MOWER MN	10	3	12	58	74
1	.9			1	.9	NOBLES MN			-2	48	87
1	1.3					NORMAN MN					
		1	.4	1	.2	OLMSTED MN		-56			65
1	.3			2	.7	OTTER TAIL MN			104	85	83
				1	.5	POLK MN					78
7	.4	4	.2	2	.1	RAMSEY MN	-46	-53	-75	02	64
				2*	5.9	RED LAKE MN					99
				1	1.0	RENVILLE MN					88
				1	.4	RICE MN					75
4	.3	14*	1.1	2	.2	ST LOUIS MN	257	-85	-48	15	65
1*	2.0					SHERBURNE MN					
				3	.7	STEARNS MN					83
				1	.8	TODD MN					85
				1	1.5	WADENA MN					94
		1	.5			WASHINGTON MN					
1*	1.9	1*	1.9			WILKIN MN	3				
2	1.1	2	.9			WINONA MN	-17				
1	.6					WRIGHT MN					
21	.4	29	.5	45*	.6	MISSISSIPPI MS	24	31	63		
1*	1.3	1	1.1			ADAMS MS	-12				
				1	.8	ALCORN MS					85
		1*	2.9			AMITE MS					
1	1.2					BOLIVAR MS					
		2*	3.2	3*	5.8	CALHOUN MS		80			99
1*	3.3					CARROLL MS					
		1*	1.6			CHICKASAW MS					
				1*	1.9	CLAY MS					96
				1	1.5	COAHOMA MS					94
				2*	3.7	COVINGTON MS					99
				2*	2.7	DE SOTO MS					98
1	.7	2	1.1	2	1.1	FORREST MS	55	-5	47	75	90
1*	2.2					GRENADA MS					
1	.3	2	.5	5*	1.1	HARRISON MS	87	101	276	96	91
3	.7	2	.4	1	.1	HINDS MS	-51	-60	-80	01	65
1*	12.3					ISSAQUENA MS					
		5*	2.4	1	.3	JACKSON MS		-86			70
1*	4.0					JEFFERSON DAVIS MS					
1	.7			3*	1.3	JONES MS			85	82	93
				2*	2.8	LAMAR MS					98
1	.5	1	.4	6*	2.4	LAUDERDALE MS	-21	493	369	98	98
				1	2.0	LEAKE MS					97
1	.6	1	.7			LEE MS	10				
1	1.5					LEFLORE MS					
1	1.4	1	1.1			LINCOLN MS	-20				
1	1.3					LOWNDES MS					
		2*	1.8	2*	1.8	MONROE MS		3			96
				1	1.2	NEWTON MS					91
1*	6.0					NOXUBEE MS					
		1	1.1	1	.9	PEARL RIVER MS		-12			88
		1	1.1	1	.8	PIKE MS		-26			86
				1	1.4	PONTOTOC MS					93
		1	1.1			PRENTISS MS					
				2	.9	RANKIN MS					87
1*	1.4					SCOTT MS					
				1	1.2	SUNFLOWER MS					92
1*	1.5	1	1.3			TIPPAH MS	-10				
				2*	1.6	WARREN MS					95
1	1.1	1	.6			WASHINGTON MS	-43				
		1*	2.0	1*	1.7	WAYNE MS		-18			96
				1	1.2	WINSTON MS					92
		1*	3.1			YALOBUSHA MS					
		1	1.5	1	1.2	YAZOO MS		-21			92
77	.4	76	.4	83	.4	MISSOURI MO	-7	7	0		
				1	.8	AUDRAIN MO					85
		1	1.0			BARRY MO					
1	.7	1	1.0			BATES MO	42				

* SIGNIFICANTLY MORE DEATHS THAN EXPECTED

WHITE MALE — CANCER OF THE NASOPHARYNX — ICD 147

1950-1959 DEATHS	1950-1959 RATE	1960-1969 DEATHS	1960-1969 RATE	1970-1979 DEATHS	1970-1979 RATE	COUNTY AND STATE	PERCENT CHANGE 60S/50S	PERCENT CHANGE 70S/60S	PERCENT CHANGE 70S/50S	PERCENTILE %CHANGE 70S/50S	PERCENTILE RATES 1970S
				1	1.0	BENTON MO					89
		1	.4			BOONE MO					
4	.7	4	.9			BUCHANAN MO	22				
1	.4	1	.5	1	.4	BUTLER MO	20	-15	2	52	74
				1*	1.7	CALDWELL MO					96
				1	.9	CALLAWAY MO					86
				1*	5.4	CARTER MO					99
1	.7			2	1.0	CASS MO			41	72	89
1*	1.6					CEDAR MO					
				1	1.2	CHRISTIAN MO					92
1	.2	2	.5	1	.2	CLAY MO	225	-55	47	75	67
1	.7					COLE MO					
1	.9	1	1.6	1	1.4	CRAWFORD MO	70	-16	43	73	93
1*	1.5			1	2.1	DALLAS MO			44	73	97
				1	2.5	DOUGLAS MO					98
				1	.3	FRANKLIN MO					70
		1	1.0			GASCONADE MO					
6*	1.1			1	.1	GREENE MO			-89	00	65
1	1.4					HARRISON MO					
		2*	1.5			HENRY MO					
		1	1.3			HOWARD MO					
1	.8					HOWELL MO					
15*	.6	12	.5	17*	.7	JACKSON MO	-12	39	22	65	83
2	.5	5*	1.2	1	.2	JASPER MO	139	-80	-52	13	67
		1	.3	2	.4	JEFFERSON MO		12			72
1	.7					JOHNSON MO					
		1	1.0	1	.7	LACLEDE MO		-27			83
				2	1.4	LAFAYETTE MO					93
		1	1.4			LEWIS MO					
		1	.8			LINN MO					
		2	1.3			MARION MO					
		1	1.4	1	1.5	MISSISSIPPI MO		7			94
1	1.4					MONROE MO					
1	1.5					MONTGOMERY MO					
				1	.9	MORGAN MO					88
1	.9					NEW MADRID MO					
				1	.4	NEWTON MO					74
		1	1.3			PERRY MO					
				2	1.3	PETTIS MO					93
1	.9					PHELPS MO					
		1	1.1			PIKE MO					
1	.2			1	1.1	PULASKI MO			380	99	90
				1	1.1	RANDOLPH MO					91
		1	1.2			RIPLEY MO					
1	.9	3*	1.5			ST CHARLES MO	61				
				1	1.2	ST CLAIR MO					91
		1	1.5	1	2.4	STE GENEVIEVE MO		58			98
				1	.6	ST FRANCOIS MO					79
9	.3	13	.4	12	.4	ST LOUIS COUNTY MO	11	-6	5	53	72
1	.7			1	.6	SALINE MO			-23	30	79
		3*	2.1			SCOTT MO					
				1	1.2	STONE MO					92
		1	2.1			TANEY MO					
				2*	1.4	TEXAS MO					93
				1	.9	VERNON MO					88
1*	1.4					WARREN MO					
				1*	3.5	WORTH MO					99
				2*	2.2	WRIGHT MO					97
22*	.6	13	.4	15*	.9	ST LOUIS CITY MO	-27	103	48	75	87
5	.2	13	.4	7	.2	MONTANA MT	163	-50	31		
				1*	1.6	CARBON MT					94
		1*	6.7			CARTER MT					
		1	.4	2	.6	CASCADE MT		67			81
		2*	2.0			DEER LODGE MT					
		1	1.1			FERGUS MT					
1	.5					FLATHEAD MT					
		1	.3			GALLATIN MT					
		1*	18.1			GOLDEN VALLEY MT					
		1	1.8			HILL MT					
		1	1.1			LINCOLN MT					
		1	.4	1	.4	MISSOULA MT		-10			73
				1*	3.4	PHILLIPS MT					99
1*	1.8					RICHLAND MT					
		1*	4.3			ROSEBUD MT					
		1	.4			SILVER BOW MT					
3*	1.0	1	.3	2	.5	YELLOWSTONE MT	-67	41	-54	11	75
15	.2	18	.2	25	.3	NEBRASKA NB	15	40	61		
				1*	1.6	BOX BUTTE NB					95
1*	1.4					BUTLER NB					
				1	.9	CASS NB					88
1*	1.6					CHEYENNE NB					
				1*	1.8	CLAY NB					96
1	1.1					CUSTER NB					
6	.4	6	.4	11*	.7	DOUGLAS NB	1	79	81	82	84
		1*	1.5	1*	1.5	FURNAS NB		-1			94
				1	.5	HALL NB					76
				1*	3.5	HITCHCOCK NB					99
1*	3.2					KEARNEY NB					
				1*	1.9	KEITH NB					96
		3	.5	2	.3	LANCASTER NB		-37			69
1	.7					LINCOLN NB					
1	1.1					MADISON NB					
1*	2.6					MORRILL NB					
				2*	5.3	NUCKOLLS NB					99
		1*	1.9			PHELPS NB					
				1	.8	PLATTE NB					86
		1	.7			RICHARDSON NB					

* SIGNIFICANTLY MORE DEATHS THAN EXPECTED

WHITE MALE | CANCER OF THE NASOPHARYNX | ICD 147

1950-1959 DEATHS	1950-1959 RATE	1960-1969 DEATHS	1960-1969 RATE	1970-1979 DEATHS	1970-1979 RATE	COUNTY AND STATE	PERCENT CHANGE 60S/50S	PERCENT CHANGE 70S/60S	PERCENT CHANGE 70S/50S	PERCENTILE %CHANGE 70S/50S	PERCENTILE RATES 1970S
				1	1.3	SALINE NB					92
		3*	3.5			SARPY NB					
		1	1.2			SAUNDERS NB					
1	.7			1	.6	SCOTTS BLUFF NB			-18	36	79
		1*	2.5			SHERMAN NB					
		1*	2.6			VALLEY NB					
1*	1.6					WASHINGTON NB					
3	.2	5	.3	9	.4	NEVADA NV	67	5	76		
1	.1	3	.5	5	.4	CLARK NV	215	-17	161	90	73
2	.5	2	.5	2	.3	WASHOE NV	6	-41	-38	18	69
				2*	2.5	ORMSBY-CARSON CITY NV					98
10	.4	7	.2	13	.3	NEW HAMPSHIRE NH	-39	52	-7		
				1	.4	BELKNAP NH					74
1	.5					CHESHIRE NH					
				1	.5	COOS NH					75
1	.5			1	.4	GRAFTON NH			-16	37	73
5	.7	7	.8	7	.7	HILLSBOROUGH NH	15	-17	-5	44	82
1	.3					MERRIMACK NH					
2	.8			1	.3	STRAFFORD NH			-61	08	70
				2	1.2	SULLIVAN NH					91
95	.4	145*	.5	149	.5	NEW JERSEY NJ	26	-6	18		
2	.3	6	.8	4	.4	ATLANTIC NJ	148	-44	40	71	75
8	.3	23	.6	17	.4	BERGEN NJ	118	-39	33	69	72
2	.3	4	.5	5	.5	BURLINGTON NJ	77	-8	63	79	76
2	.2	8	.4	7	.4	CAMDEN NJ	186	-21	126	88	71
				3	.6	CAPE MAY NJ					82
2	.4	3	.6	1	.2	CUMBERLAND NJ	45	-69	-55	11	66
14	.4	18	.5	18	.5	ESSEX NJ	31	17	53	77	78
1	.3	3	.4	2	.2	GLOUCESTER NJ	51	-36	-4	46	68
14	.4	19	.7	14	.5	HUDSON NJ	54	-24	17	62	77
1	.4					HUNTERDON NJ					
5	.5	5	.4	11*	.8	MERCER NJ	-14	116	85	83	86
5	.3	14*	.7	18*	.7	MIDDLESEX NJ	125	2	129	89	84
8	.7	6	.3	9	.4	MONMOUTH NJ	-48	27	-34	22	75
5	.6	7	.5	11	.6	MORRIS NJ	-16	24	4	53	80
1	.2	5	.7	2	.1	OCEAN NJ	203	-89	-66	05	64
13*	.7	15*	.7	6	.3	PASSAIC NJ	4	-60	-59	09	69
2	.9	1	.4			SALEM NJ	-59				
		1	.1	6	.7	SOMERSET NJ		349			82
2	.9			2	.5	SUSSEX NJ			-45	16	76
3	.1	6	.3	12	.5	UNION NJ	91	93	268	95	77
5*	1.5	1	.3	1	.2	WARREN NJ	-81	-28	-86	00	66
10	.4	13	.4	10	.2	NEW MEXICO NM	-6	-30	-34		
5	.9	8*	.8	4	.3	BERNALILLO NM	-13	-59	-64	06	70
1*	6.9					CATRON NM					
				1	.5	CHAVES NM					76
1	1.2					COLFAX NM					
1	.9			1	.9	CURRY NM			-3	47	87
		1	.5	1	.5	EDDY NM		6			78
		1	1.0			GRANT NM					
				1	1.2	LUNA NM					92
1	1.2	1	1.3			RIO ARRIBA NM	2				
				1	1.2	SANDOVAL NM					92
				1	.5	SAN JUAN NM					77
		1	.6			SANTA FE NM					
1*	2.6					SOCORRO NM					
		1	.5			VALENCIA NM					
347*	.5	345	.4	359*	.5	NEW YORK NY	-4	6	2		
6	.4	7	.6	4	.3	ALBANY NY	40	-54	-36	21	68
35	.5	25	.4	29*	.6	BRONX NY	-24	68	28	67	81
4	.4	3	.3	6	.5	BROOME NY	-33	90	26	65	78
1	.3	2	.5	4	1.1	CATTARAUGUS NY	85	99	268	96	90
		1	.3	5*	1.3	CAYUGA NY		414			93
4	.5	4	.5	5	.7	CHAUTAUQUA NY	-2	31	28	67	83
		4	.9	2	.4	CHEMUNG NY		-53			74
1	.5	1	.5	1	.4	CHENANGO NY	2	-4	-2	49	75
1	.4					CLINTON NY					
1	.4	2	.6			COLUMBIA NY	61				
2	1.3	1	.4			CORTLAND NY	-67				
1	.4	1	.5	1	.4	DELAWARE NY	15	-28	-17	36	71
3	.4	3	.3	3	.3	DUTCHESS NY	-10	-10	-19	35	69
24*	.5	19	.4	26	.6	ERIE NY	-22	36	6	54	79
				1	.5	ESSEX NY					78
		3*	1.5	1	.5	FRANKLIN NY		-66			77
		1	.4			FULTON NY					
		2	.8			GENESEE NY					
				1	.6	GREENE NY					82
1	.3	1	.3			HERKIMER NY	-6				
1	.2			2	.5	JEFFERSON NY			114	87	75
68*	.5	43	.4	49	.6	KINGS(BROOKLYN) NY	-30	52	7	55	79
				1	.3	LIVINGSTON NY					70
				2	.7	MADISON NY					83
8	.3	10	.3	13	.4	MONROE NY	25	26	58	78	74
				1	.3	MONTGOMERY NY					70
24*	.7	27	.5	35	.5	NASSAU NY	-32	7	-27	27	76
50*	.6	81*	1.1	22	.4	NEW YORK(MANHATTAN) NY	89	-66	-35	21	73
6	.6	5	.5	4	.4	NIAGARA NY	-21	-25	-41	17	72
5	.4	11*	.9	2	.2	ONEIDA NY	130	-83	-61	08	65
10	.5	7	.3	9	.4	ONONDAGA NY	-35	26	-18	36	75
1	.2	2	.5	2	.5	ONTARIO NY	106	4	113	86	77
4	.6	5	.5	5	.5	ORANGE NY	-11	-11	-21	33	75

* SIGNIFICANTLY MORE DEATHS THAN EXPECTED

1950-1959 DEATHS	1950-1959 RATE	1960-1969 DEATHS	1960-1969 RATE	1970-1979 DEATHS	1970-1979 RATE	COUNTY AND STATE		PERCENT CHANGE 60S/50S	PERCENT CHANGE 70S/60S	PERCENT CHANGE 70S/50S	PERCENTILE %CHANGE 70S/50S	PERCENTILE RATES 1970S
				1	.6	ORLEANS	NY					82
		1	.3	3	.7	OSWEGO	NY		162			82
1	.7			3	1.0	PUTNAM	NY			43	73	89
33	.4	30	.3	41	.4	QUEENS	NY	-15	40	20	63	75
3	.4	3	.4	2	.3	RENSSELAER	NY	8	-31	-25	28	70
8*	.8	3	.4	14*	1.1	RICHMOND	NY	-56	194	31	68	90
		2	.4	4	.4	ROCKLAND	NY		-3			71
1	.2			1	.2	ST LAWRENCE	NY			7	55	66
4*	1.0	1	.3	3	.6	SARATOGA	NY	-68	82	-42	17	79
2	.2	4	.5			SCHENECTADY	NY	100				
				1	.8	SCHOHARIE	NY					85
1	.6					SENECA	NY					
2	.4	2	.4			STEUBEN	NY	0				
5	.3	7	.2	22	.4	SUFFOLK	NY	-37	129	45	74	75
1	.3					SULLIVAN	NY					
				1	.5	TIOGA	NY					76
1	.3	1	.4	1	.3	TOMPKINS	NY	39	-15	18	62	71
2	.3	4	.6	5	.7	ULSTER	NY	78	11	97	84	83
		1	.4			WARREN	NY					
2	.7					WASHINGTON	NY					
1	.3	1	.3	4*	1.1	WAYNE	NY	0	289	289	97	90
18	.6	14	.3	16	.4	WESTCHESTER	NY	-46	14	-39	18	72
1	.6					WYOMING	NY					
				1	1.1	YATES	NY					91
40	.3	69	.5	80	.4	NORTH CAROLINA	NC	53	-7	42		
		1	.2	3	.9	ALAMANCE	NC		298			88
				1	1.1	ASHE	NC					90
2*	2.3	1	.8			BEAUFORT	NC	-65				
2*	3.6					BLADEN	NC					
2	.4	4	.6	6	.8	BUNCOMBE	NC	60	33	113	87	86
		2	1.1	1	.5	BURKE	NC		-58			75
		1	.3	2	.6	CABARRUS	NC		98			80
		2*	2.7			CARTERET	NC					
1*	2.1			1*	1.9	CASWELL	NC			-9	42	96
2*	2.3					CHATHAM	NC					
		1*	2.9			CHOWAN	NC					
		1	.5			CLEVELAND	NC					
4*	2.7	1	.9			COLUMBUS	NC	-64				
2*	2.3					CRAVEN	NC					
		1	.3	2	.4	CUMBERLAND	NC		58			74
1	.3			3	.8	DAVIDSON	NC			161	90	85
				2*	2.1	DAVIE	NC					97
1	.8			1	.7	DUPLIN	NC			-19	35	82
2	.5			4*	1.0	DURHAM	NC			85	83	89
1	.2	3	.5	1	.2	FORSYTH	NC	149	-63	-8	43	66
		2*	2.3	1	1.0	FRANKLIN	NC		-56			89
				2	.4	GASTON	NC					73
		1*	4.4			GATES	NC					
		1*	3.3			GREENE	NC					
1	.2	5	.6	4	.4	GUILFORD	NC	252	-31	144	89	73
3*	2.8	1	.7			HALIFAX	NC	-76				
1	.8					HARNETT	NC					
		1	.5	1	.4	HAYWOOD	NC		-8			75
		1	.5			HENDERSON	NC					
1*	2.2			1*	1.8	HERTFORD	NC			-20	33	96
		1*	5.2			HYDE	NC					
1	.6			5*	2.1	JOHNSTON	NC			272	96	97
		1*	3.8			JONES	NC					
		1	.5	1	.8	LENOIR	NC		56			86
1	.8					MC DOWELL	NC					
				1	.8	MACON	NC					85
		1	2.2			MARTIN	NC					
5	.6	6	.8	9*	.8	MECKLENBURG	NC	32	-8	21	64	84
		1	1.4			MONTGOMERY	NC					
		1	.9	1	.7	MOORE	NC		-30			82
		1	.6			NASH	NC					
		1	.5	1	.4	NEW HANOVER	NC		-19			73
1*	2.3	1*	3.2			NORTHAMPTON	NC	44				
		1	.4			ONSLOW	NC					
		1	.8	1	.5	ORANGE	NC		-37			78
				1	1.1	PASQUOTANK	NC					90
		1*	2.0			PENDER	NC					
		1	1.1	1	1.0	PERSON	NC		-6			89
		3*	2.0			PITT	NC					
				1	1.2	POLK	NC					91
		2	.6	2	.5	RANDOLPH	NC		-19			76
1	.7	2*	1.6			RICHMOND	NC	128				
		1	.6			ROBESON	NC					
		2	.7	1	.3	ROCKINGHAM	NC		-59			69
1	.2			3	.8	ROWAN	NC			208	94	84
1	.5			1	.4	RUTHERFORD	NC			-11	40	74
		1	.8			SAMPSON	NC					
				1	1.1	SCOTLAND	NC					91
1	.7			1	.4	STANLY	NC			-40	17	74
				2	.9	SURRY	NC					87
		1*	2.5			SWAIN	NC					
		1	1.5			TRANSYLVANIA	NC					
2*	1.3					UNION	NC					
		1	1.0			VANCE	NC					
		3	.5	2	.4	WAKE	NC		-8			75
		1*	2.6			WARREN	NC					
				1*	2.1	WASHINGTON	NC					97
		1	1.1	3*	2.9	WATAUGA	NC		161			98
				1	.3	WAYNE	NC					70
		1	.4	2	.8	WILKES	NC		103			85
				2*	2.9	YANCEY	NC					98
10	.3	8	.3	11	.4	NORTH DAKOTA	ND	-22	31	2		

* SIGNIFICANTLY MORE DEATHS THAN EXPECTED

WHITE MALE

CANCER OF THE NASOPHARYNX

ICD 147

1950-1959 DEATHS	1950-1959 RATE	1960-1969 DEATHS	1960-1969 RATE	1970-1979 DEATHS	1970-1979 RATE	COUNTY AND STATE	PERCENT CHANGE 60S/50S	PERCENT CHANGE 70S/60S	PERCENT CHANGE 70S/50S	PERCENTILE %CHANGE 70S/50S	PERCENTILE RATES 1970S
				1	1.1	BARNES ND					91
1*	2.1			1*	3.0	BENSON ND			42	72	98
				1*	2.4	BUPKE ND					97
3*	2.4	1	.6	1	.5	BURLEIGH ND	-76	-14	-79	01	77
				1	.3	CASS ND					70
				1*	1.6	CAVALIER ND					95
1*	3.0					DUNN ND					
		1*	3.3			FOSTER ND					
1	.6	3*	1.4	2	.7	GRAND FORKS ND	144	-46	31	68	84
				1*	2.5	GRANT ND					98
		1*	3.3			KIDDER ND					
1*	1.5					PEMBINA ND					
3*	4.4			1	1.2	STARK ND			-73	03	91
				1	1.1	TRAILL ND					91
		1	.5			WARD ND					
		1	1.1			WILLIAMS ND					
112	.3	197	.5	173	.4	OHIO OH	61	-16	35		
1	.2	4	.9	1	.2	ALLEN OH	323	-75	4	53	67
		2	.4	1	.2	ASHTABULA OH		-48			66
		1	.5			ATHENS OH					
1	.5	2	1.2	2	1.1	AUGLAIZE OH	122	-12	95	84	90
1	.2	2	.4	1	.3	BELMONT OH	126	-36	45	74	68
		1	.6	1	.7	BROWN OH		17			84
2	.3	3	.4	4	.3	BUTLER OH	24	-6	16	61	71
		2*	1.5			CHAMPAIGN OH					
1	.2	2	.4	3	.5	CLARK OH	103	35	174	91	77
2	.9	2	.6	1	.2	CLERMONT OH	-36	-65	-77	01	66
1	.8	1	.7			CLINTON OH	-16				
1	.2	2	.4	4	.7	COLUMBIANA OH	118	81	293	97	84
				1	.4	CRAWFORD OH					73
28	.4	39	.6	39	.6	CUYAHOGA OH	41	7	51	76	80
1	.4	2	1.0	2	.7	DARKE OH	135	-27	71	80	84
3*	.9	3	.9	1	.3	ERIE OH	0	-67	-67	05	70
2	.7	2	.6			FAIRFIELD OH	-9				
2*	1.5			1	.8	FAYETTE OH			-50	14	85
11	.5	12	.5	10	.3	FRANKLIN OH	1	-27	-26	27	71
				1	.6	FULTON OH					80
		1	.8			GALLIA OH					
1	.7			3	1.1	GEAUGA OH			54	77	90
		3	1.3	2	.5	GPEENE OH		-60			78
		1	.7			GUERNSEY OH					
11	.4	19	.5	28*	.8	HAMILTON OH	42	55	122	88	86
				1	.4	HANCOCK OH					72
		2*	1.4			HARDIN OH					
		2	1.2			HIGHLAND OH					
				1	.4	HURON OH					72
		2*	1.5	2*	1.4	JACKSON OH		-7			93
		3	.6			JEFFERSON OH					
1	.4					KNOX OH					
		3	.7	1	.1	LAKE OH		-85			65
1	.4	2	.8			LAWRENCE OH	81				
		1	.2	3	.6	LICKING OH		163			79
		3*	1.7			LOGAN OH					
		5	.6			LORAIN OH					
3	.1	13	.6	13	.6	LUCAS OH	335	3	348	98	81
8	.6	9	.7	9	.7	MAHONING OH	8	2	10	56	83
		1	.9			MEIGS OH					
1	.2					MIAMI OH					
1	1.9					MONROE OH					
6	.3	5	.2	8	.3	MONTGOMERY OH	-32	52	3	52	71
		2	.5	3	.8	MUSKINGUM OH		55			86
				1*	1.8	NOBLE OH					96
				1	.5	OTTAWA OH					78
1	.7					PICKAWAY OH					
		4*	.9	2	.4	PORTAGE OH		-54			74
		1	.9			PUTNAM OH					
		2	.4	3	.5	RICHLAND OH		41			78
1	.4	1	.3	1	.3	ROSS OH	-14	-9	-22	31	70
1	.3	2	.5	1	.2	SCIOTO OH	58	-50	-21	32	67
2	.7	1	.4			SENECA OH	-48				
				1	.5	SHELBY OH					78
4	.3	8	.5	2	.1	STARK OH	51	-75	-62	07	65
4	.3	12	.5	6	.3	SUMMIT OH	75	-51	-15	38	68
4	.5	4	.4	5	.5	TRUMBULL OH	-22	13	-11	40	76
1	.3					TUSCAPAWAS OH					
				1	.8	UNION OH					85
1	.8					VAN WERT OH					
1*	2.1					VINTON OH					
		2	.9	1	.2	WARREN OH		-81			66
				1	.2	WAYNE OH					67
2*	1.4					WILLIAMS OH					
		1	1.0			WYANDOT OH					
25	.3	35	.3	34	.3	OKLAHOMA OK	22	-13	7		
		1	2.0			ADAIR OK					
				1*	2.3	BEAVER OK					97
1*	1.5					BLAINE OK					
		1	.6			BRYAN OK					
1	.6	1	1.1			CADDO OK	77				
		1	.8			CANADIAN OK					
		1	.5			CAPTER OK					
				1	1.0	CHOCTAW OK					89
				1	.3	CLEVELAND OK					71
				3	1.0	COMANCHE OK					88
		2	1.1			CREEK OK					
1	1.3					DELAWARE OK					
				1*	2.3	ELLIS OK					97
1	.5	1	.4			GARFIELD OK	-7				

* SIGNIFICANTLY MORE DEATHS THAN EXPECTED

WHITE MALE — CANCER OF THE NASOPHARYNX — ICD 147

1950-1959 DEATHS	1950-1959 RATE	1960-1969 DEATHS	1960-1969 RATE	1970-1979 DEATHS	1970-1979 RATE	COUNTY AND STATE		PERCENT CHANGE 60S/50S	PERCENT CHANGE 70S/60S	PERCENT CHANGE 70S/50S	PERCENTILE %CHANGE 70S/50S	PERCENTILE RATES 1970S
				1	.6	GARVIN	OK					81
1	.6	1	.6			GRADY	OK	-13				
				1*	1.4	GREER	OK					93
				1	.8	HUGHES	OK					86
		1	.9			JACKSON	OK					
		1*	1.9			JOHNSTON	OK					
		1	.4			KAY	OK					
				1	1.5	KINGFISHER	OK					94
		1	1.1			LE FLORE	OK					
1	1.0	1	.9	1	.9	LOGAN	OK	-15	6	-10	40	88
1*	2.8			1*	3.4	LOVE	OK			22	64	99
				1	1.2	MC CLAIN	OK					91
				2*	2.0	MC INTOSH	OK					97
2	.7					MUSKOGEE	OK					
				1*	1.7	OKFUSKEE	OK					95
8	.7	8	.4	7	.4	OKLAHOMA	OK	-41	-15	-50	13	71
		1	.5			OKMULGEE	OK					
		1	.6			OSAGE	OK					
				1	.6	OTTAWA	OK					81
1	.5					PITTSBURG	OK					
1	.8	1	.6	1	.6	PONTOTOC	OK	-14	-9	-22	32	80
1	.5	1	.4	2	.8	POTTAWATOMIE	OK	-20	98	58	78	85
				1	.5	ROGERS	OK					78
1	.6	1	1.1			SEMINOLE	OK	75				
1	.6			1	.4	STEPHENS	OK			-20	34	75
1	1.4					TILLMAN	OK					
2	.2	6	.4	4	.3	TULSA	OK	133	-35	50	76	68
		1	1.3			WAGONER	OK					
		1	.5			WASHINGTON	OK					
18	.2	23	.2	35	.3	OREGON	OR	6	31	39		
		2	1.4			BENTON	OR					
1	.2	1	.2	5	.6	CLACKAMAS	OR	-13	264	218	94	79
1	.6	1	.6			CLATSOP	OR	-10				
				1	.5	COLUMBIA	OR					78
				1	.2	DOUGLAS	OR					66
1*	1.3					HOOD RIVER	OR					
		2	.4	1	.1	JACKSON	OR		-65			65
				3	1.0	JOSEPHINE	OR					89
				1	.6	KLAMATH	OR					80
2	.4	4	.5	3	.3	LANE	OR	44	-45	-20	34	69
		1	.8			MALHEUR	OR					
1	.2	2	.3	1	.1	MARION	OR	57	-53	-25	28	65
9	.4	8	.3	17*	.6	MULTNOMAH	OR	-23	120	69	80	81
1	1.1					TILLAMOOK	OR					
1	.4	1	.5			UMATILLA	OR	13				
1	1.1					WASCO	OR					
		1	.2	1	.1	WASHINGTON	OR		-28			65
				1	.4	YAMHILL	OR					73
200	.4	204	.4	195	.3	PENNSYLVANIA	PA	-5	-9	-14		
				1	.3	ADAMS	PA					69
34	.4	26	.3	31	.4	ALLEGHENY	PA	-25	21	-9	42	74
		1	.2	1	.2	ARMSTRONG	PA		-6			67
1	.1	5	.6	5	.4	BEAVER	PA	409	-22	298	97	75
3	.2	6	.4	4	.2	BERKS	PA	85	-40	12	57	67
2	.3	3	.4	2	.3	BLAIR	PA	52	-34	0	50	68
1	.4					BRADFORD	PA					
2	.2	2	.1	3	.2	BUCKS	PA	-51	137	17	61	66
		6*	1.1	4	.7	BUTLER	PA		-43			82
3	.3	4	.4	3	.3	CAMBRIA	PA	42	-35	-7	43	68
1*	3.8					CAMERON	PA					
		6*	1.9	1	.3	CARBON	PA		-85			69
2	.8			2	.5	CENTRE	PA			-33	23	77
		3	.2	7	.6	CHESTER	PA		151			81
1	.5					CLARION	PA					
1	.2			1	.2	CLEARFIELD	PA			10	57	68
1	.6	2	1.1			CLINTON	PA	90				
1	.4	1	.4	1	.3	COLUMBIA	PA	-11	-13	-22	31	69
1	.3			2	.5	CRAWFORD	PA			77	81	76
1	.2			3	.4	CUMBERLAND	PA			86	83	72
9*	1.0	5	.5	3	.3	DAUPHIN	PA	-48	-45	-71	03	68
12	.6	9	.4	6	.2	DELAWARE	PA	-25	-47	-60	09	67
2*	1.2			2	1.0	ELK	PA			-18	35	89
8*	.8	3	.3	5	.4	ERIE	PA	-69	51	-53	12	72
1	.1	6	.7	2	.2	FAYETTE	PA	524	-65	119	87	67
		1*	3.6			FOREST	PA					
1	.3	2	.5			FRANKLIN	PA	77				
1*	1.8					FULTON	PA					
1	.4					GREENE	PA					
3	.8					INDIANA	PA					
7	.6	2	.1	6	.4	LACKAWANNA	PA	-74	213	-19	34	75
2	.2	4	.3	5	.3	LANCASTER	PA	84	12	106	85	70
		2	.4	2	.4	LAWRENCE	PA		-16			71
		1	.2	3	.6	LEBANON	PA		167			80
6	.6	4	.3	8	.6	LEHIGH	PA	-47	87	-2	49	79
4	.3	10	.5	9	.4	LUZERNE	PA	108	-20	67	79	74
2	.4	1	.2	3	.5	LYCOMING	PA	-35	114	39	71	78
		1	.5	1	.5	MC KEAN	PA		-8			75
1	.2	3	.5	2	.3	MERCER	PA	162	-30	83	82	71
		1	.5	2	.8	MIFFLIN	PA		63			86
1	.4	2	.9	3	.9	MONROE	PA	97	0	97	84	87
6	.3	12	.5	9	.3	MONTGOMERY	PA	54	-38	-4	45	69
1	1.1					MONTOUR	PA					
3	.3	5	.5	9*	.8	NORTHAMPTON	PA	63	58	157	90	84
4	.7	2	.4			NORTHUMBERLAND	PA	-43				
1	1.0	1	.7			PERRY	PA	-30				
54*	.7	42	.6	25	.4	PHILADELPHIA	PA	-21	-31	-46	15	73
				1	1.5	PIKE	PA					94

* SIGNIFICANTLY MORE DEATHS THAN EXPECTED

WHITE MALE CANCER OF THE NASOPHARYNX ICD 147

1950-1959 DEATHS	RATE	1960-1969 DEATHS	RATE	1970-1979 DEATHS	RATE	COUNTY AND STATE	PERCENT CHANGE 60S/50S	70S/60S	70S/50S	PERCENTILE %CHANGE 70S/50S	PERCENTILE RATES 1970S
2	.2	4	.4	3	.3	SCHUYLKILL PA	68	-20	35	70	70
		1	.7			SNYDER PA					
1	.3			1	.2	SOMERSET PA			-26	27	67
		1	.6	1	.6	SUSQUEHANNA PA		2			79
1	.6					TIOGA PA					
1	.3	1	.3			VENANGO PA	-1				
1	.4			1	.4	WARREN PA			-2	47	74
4	.4	1	.1	4	.3	WASHINGTON PA	-74	247	-10	41	71
4	.2	8	.4	6	.3	WESTMORELAND PA	99	-33	34	70	69
1	.1	4	.3	2	.1	YORK PA	223	-54	50	76	65
8	.2	16	.4	20	.4	RHODE ISLAND RI	86	17	119		
1	.6	1	.6	1	.5	BRISTOL RI	-11	-11	-21	33	77
		3	.5	1	.2	KENT RI		-69			65
3*	1.1	2	.6	1	.3	NEWPORT RI	-46	-52	-74	03	69
4	.2	10	.4	15	.5	PROVIDENCE RI	133	47	242	95	78
				2	.5	WASHINGTON RI					76
13	.2	30	.4	38	.5	SOUTH CAROLINA SC	74	4	81		
		1	1.6			ABBEVILLE SC					
		1	.3			AIKEN SC					
				1	.2	ANDERSON SC					66
		1*	3.0	1*	2.5	BAMBERG SC		-19			98
				1	.4	BERKELEY SC					71
2	.5	3	.9	2	.3	CHARLESTON SC	62	-67	-47	15	69
				1	1.1	CHESTER SC					91
		1*	2.0			CLARENDON SC					
1	1.0			1	1.1	DARLINGTON SC			7	55	90
				1	1.1	DILLON SC					91
		1	.9			DORCHESTER SC					
		1*	2.0	1*	2.3	FAIRFIELD SC		14			97
		3*	1.4	1	.3	FLORENCE SC		-78			70
2	.5	1	.1	8*	.8	GREENVILLE SC	-83	903	75	81	86
		1	.5	2	1.0	GREENWOOD SC		82			88
		1	.4			HORRY SC					
		1	.8	2*	1.6	KERSHAW SC		85			95
1	.6	2	.8	1	.2	LEXINGTON SC	20	-80	-76	02	65
1*	1.4					MARION SC					
1	1.0			2*	1.6	NEWBERRY SC			55	78	95
1	.6	1	.6			OCONEE SC	-7				
				1	.5	ORANGEBURG SC					76
		1	.4	1	.4	PICKENS SC		-14			72
2	.5	6*	1.4	4	.7	RICHLAND SC	207	-50	54	77	83
1*	2.4	1*	1.8			SALUDA SC	-25				
		3	.5	2	.3	SPARTANBURG SC		-36			70
				2	1.9	SUMTER SC					96
				3*	3.1	UNION SC					98
1	.7					YORK SC					
6	.2	15	.4	6	.2	SOUTH DAKOTA SD	160	-55	18		
		1	1.0			BEADLE SD					
				1	1.5	BROOKINGS SD					94
1	.5	2	1.2	1	.7	BROWN SD	134	-42	36	70	83
1*	6.1					CAMPBELL SD					
1*	2.1					CLAY SD					
		2*	1.8			CODINGTON SD					
		1*	5.0			CORSON SD					
		1*	2.8			EDMUNDS SD					
		1*	1.6			GRANT SD					
		1*	2.3			GREGORY SD					
		2*	6.7			HANSON SD					
1*	2.3					HUGHES SD					
		1	1.2			LINCOLN SD					
				1	1.3	MEADE SD					93
1	.3			2	.5	MINNEHAHA SD			86	83	77
				1	1.2	ROBERTS SD					92
		1*	4.9			SANBORN SD					
		1*	6.8			SULLY SD					
1*	2.0					TRIPP SD					
		1*	20.2			ZIEBACH SD					
43	.3	58	.4	88*	.5	TENNESSEE TN	18	28	51		
				2	.6	ANDERSON TN					81
1	.8	2*	1.8	1	1.0	BEDFORD TN	113	-44	19	63	89
1	.5	1	.4	2	.7	BLOUNT TN	-34	93	27	66	83
		2	1.0	1	.4	BRADLEY TN		-57			75
		1	.9			CARROLL TN					
				2	.9	CARTER TN					87
		1*	1.7			CHEATHAM TN					
		1	1.0	3*	2.6	COCKE TN		162			98
		1	.9	2	1.1	COFFEE TN		32			91
1	1.4			1	.7	CUMBERLAND TN			-49	14	83
6	.6	10	.7	8	.5	DAVIDSON TN	25	-28	-10	41	78
				1	.9	DICKSON TN					87
1*	2.3					FAYETTE TN					
				2*	1.5	FRANKLIN TN					94
		1	.4			GIBSON TN					
2*	1.9					GILES TN					
		1	1.6			GRAINGER TN					
		1	.6	1	.4	GREENE TN		-29			73
		1*	2.1			GRUNDY TN					
				2	.9	HAMBLEN TN					86
1	.1	4	.4	6	.6	HAMILTON TN	196	41	318	97	81
		1	2.0			HARDEMAN TN					
				1	.5	HAWKINS TN					78
		1	1.3			HENDERSON TN					
1	.9			1	.7	HENRY TN			-23	30	82

* SIGNIFICANTLY MORE DEATHS THAN EXPECTED

WHITE MALE

CANCER OF THE NASOPHARYNX — ICD 147

1950-1959 DEATHS	RATE	1960-1969 DEATHS	RATE	1970-1979 DEATHS	RATE	COUNTY AND STATE	PERCENT CHANGE 60S/50S	70S/60S	70S/50S	PERCENTILE %CHANGE 70S/50S	RATES 1970S
1*	1.7	1*	1.6	1	1.6	HICKMAN TN	-8	2	-6	44	95
		2*	7.9			HOUSTON TN					
		1*	1.8			JACKSON TN					
				2*	1.6	JEFFERSON TN					95
2	.2	3	.3	13*	1.1	KNOX TN	19	303	381	99	90
1	1.6					LAUDERDALE TN					
				1	.6	LAWRENCE TN					81
				1*	2.2	LEWIS TN					97
1	1.0			1	.7	LINCOLN TN			-24	29	84
1	.9	1	.8	1	.7	LOUDON TN	-14	-6	-19	35	84
3*	2.2					MC MINN TN					
1	1.2					MC NAIRY TN					
1	.5	2	.9	1	.4	MADISON TN	102	-55	-10	41	74
		1	1.5	1	1.0	MARSHALL TN		-30			89
		1	.6			MAURY TN					
1	1.2					MONROE TN					
				3*	1.6	MONTGOMERY TN					95
1	.8	1	.6			OBION TN	-25				
				1	1.3	OVERTON TN					93
1	.6			2	1.0	PUTNAM TN			50	75	88
		1	.6			ROANE TN					
				1	.3	RUTHERFORD TN					71
1	1.0					SEVIER TN					
11*	.7	9	.5	13	.6	SHELBY TN	-26	20	-11	40	82
				2*	3.9	STEWART TN					99
1	.3			1	.2	SULLIVAN TN			-39	18	65
		3*	1.8	1	.3	SUMNER TN		-83			69
1	1.2	1	1.2			TIPTON TN	2				
				1*	2.2	UNION TN					97
				1	.7	WARREN TN					82
				3	.7	WASHINGTON TN					84
		1	.7			WEAKLEY TN					
1	.8	1	.7	1	.6	WILSON TN	-18	-7	-24	29	82
102	.3	177	.5	196	.4	TEXAS TX	47	-7	38		
				2*	1.5	ANDERSON TX					94
				1	.5	ANGELINA TX					77
		2*	2.1			ATASCOSA TX					
1	1.4					AUSTIN TX					
		1*	2.2			BANDERA TX					
1	1.5			1	1.1	BASTROP TX			-28	27	90
2*	3.0					BEE TX					
		3	.9	1	.3	BELL TX		-69			68
6	.3	16	.7	19	.6	BEXAR TX	169	-10	143	89	81
2	1.0	1	.4	2	.6	BOWIE TX	-65	78	-38	19	81
				4	.8	BRAZORIA TX					86
		2*	1.1	1	.5	BRAZOS TX		-53			78
				1	.5	BROWN TX					76
		2*	6.1			BURLESON TX					
1*	2.0					BURNET TX					
1	1.2					CALDWELL TX					
				1	1.3	CALLAHAN TX					92
2	.4			1	.2	CAMERON TX			-56	11	66
1	.9					CASS TX					
				2	1.3	CHEROKEE TX					92
1*	3.5					COCHRAN TX					
				1	.8	COLEMAN TX					86
1	.4	2	.9			COLLIN TX	107				
				1*	4.4	COLLINGSWORTH TX					99
				1	1.0	COLORADO TX					89
		1	1.0			COMAL TX					
		1	1.2	2*	4.0	COMANCHE TX		227			99
1	1.0					COOKE TX					
1*	4.7					COTTLE TX					
1*	2.4					CROSBY TX					
15	.5	27*	.8	22	.6	DALLAS TX	39	-26	4	52	79
				1	1.3	DAWSON TX					93
1*	2.4			1	2.0	DEAF SMITH TX			-17	36	96
		1	.5	2	.6	DENTON TX		25			81
1	1.0					DE WITT TX					
1	.9					EASTLAND TX					
		2	.4			ECTOR TX					
1	.5	2	1.0			ELLIS TX	82				
2	.2	5	.5	3	.2	EL PASO TX	135	-59	-3	47	67
				1	.6	FANNIN TX					80
		1	.8			FAYETTE TX					
2*	5.7					FLOYD TX					
1*	2.1			1	1.3	FREESTONE TX			-37	20	93
1*	2.2					GAINES TX					
3	.8	3	.5	3	.6	GALVESTON TX	-36	18	-25	28	80
		1*	3.6	1*	5.0	GARZA TX		37			99
		1	1.3			GILLESPIE TX					
1	.9					GRAY TX					
1	.4	1	.3			GREGG TX	-4				
		1	.6			GUADALUPE TX					
		1	.8			HALE TX					
		1*	5.2			HALL TX					
		2*	2.7			HAMILTON TX					
		2*	1.9			HARDIN TX					
16	.4	27*	.6	30	.6	HARRIS TX	40	0	41	71	79
1	.8			1	.6	HARRISON TX			-24	30	80
		1	.2	1	.1	HIDALGO TX		-30			65
				1	.6	HILL TX					80
				1	.9	HOCKLEY TX					88
1	.9					HOPKINS TX					
		1	1.3			HOUSTON TX					
1	.6					HOWARD TX					
				1	.9	HUTCHINSON TX					88
		1*	14.3			IRION TX					
		1*	2.8			JACKSON TX					

* SIGNIFICANTLY MORE DEATHS THAN EXPECTED

1950-1959 DEATHS	1950-1959 RATE	1960-1969 DEATHS	1960-1969 RATE	1970-1979 DEATHS	1970-1979 RATE	COUNTY AND STATE	PERCENT CHANGE 60S/50S	PERCENT CHANGE 70S/60S	PERCENT CHANGE 70S/50S	PERCENTILE %CHANGE 70S/50S	PERCENTILE RATES 1970S
		1	.9			JASPER TX					
1*	10.5					JEFF DAVIS TX					
4	.5	6	.9	7	.9	JEFFERSON TX	78	-3	72	80	87
1	1.4					JIM WELLS TX					
		1	.9			JONES TX					
		1	1.6			KARNES TX					
				1	.6	KAUFMAN TX					80
				1*	1.6	KENDALL TX					95
				1	1.5	KERR TX					94
		1	1.2	1	1.1	KLEBERG TX		-8			90
		1*	2.1			KNOX TX					
				2	1.3	LAMAR TX					93
				1	2.2	LAMB TX					97
		1*	1.9			LAMPASAS TX					
				1*	3.7	LA SALLE TX					99
2*	1.8	2*	1.7			LAVACA TX	-7				
1	.8					LIBERTY TX					
1	1.1					LIMESTONE TX					
		1	.3	1	.2	LUBBOCK TX		-40			66
				1*	2.1	LYNN TX					97
2	.4	2	.3	8*	1.2	MC LENNAN TX	-11	293	249	95	92
		1*	11.8			MC MULLEN TX					
				1	.8	MATAGORDA TX					85
		2	.7	1	.4	MIDLAND TX		-41			73
		1	1.0			MILAM TX					
				1	1.2	MONTAGUE TX					92
				1	.3	MONTGOMERY TX					70
				1	1.4	MOORE TX					93
		2*	1.7	1	.9	NACOGDOCHES TX		-47			87
		2	1.0			NAVARRO TX					
		1*	2.1			NEWTON TX					
		1	1.1			NOLAN TX					
2	.6	1	.1	2	.2	NUECES TX	-81	98	-62	07	66
		1	.5			ORANGE TX					
				1	.9	PALO PINTO TX					87
		2*	1.4			PARKER TX					
				1	1.0	POLK TX					89
1	.2	1	.2	3	.7	POTTER TX	-10	226	193	93	82
		1	.3	1	.5	RANDALL TX		39			76
		1	.9			RED RIVER TX					
				1	1.6	ROBERTSON TX					95
1	.7			3*	1.8	RUSK TX			159	90	96
		1*	2.8	1*	2.5	SAN AUGUSTINE TX		-12			98
		1	.6			SAN PATRICIO TX					
		1	1.1	1	1.3	SCURRY TX		18			93
		1	.8	2*	1.3	SHELBY TX		60			93
1	.3	1	.3	3	.6	SMITH TX	-1	118	116	87	81
				1*	5.6	SUTTON TX					99
6	.3	8	.3	15	.5	TARRANT TX	-9	70	55	77	78
2	.7			1	.2	TAYLOR TX			-65	05	67
1	1.9			1	.8	TITUS TX			-58	09	85
		2	.7	1	.3	TOM GREEN TX		-49			71
1	.1	4	.6	4	.4	TRAVIS TX	635	-38	353	98	72
		2*	2.4			UVALDE TX					
				2	1.1	VAN ZANDT TX					90
4*	3.3	3*	1.3	2	.9	VICTORIA TX	-61	-27	-72	03	88
				1*	1.6	WALLER TX					95
				1	.6	WHARTON TX					82
		4	1.0	1	.2	WICHITA TX		-77			67
				1	1.9	WILLACY TX					96
		1	.5			WILLIAMSON TX					
				1	1.4	WILSON TX					93
				1	1.0	WOOD TX					89
				1*	4.6	YOAKUM TX					99
		1	1.0	1	.9	YOUNG TX		-5			88
4	.1	7	.2	7	.2	UTAH UT	79	-23	38		
		1	.4			DAVIS UT					
4	.3	2	.2	5	.3	SALT LAKE UT	-48	76	-8	42	68
				1	1.1	SANPETE UT					90
				1	.3	UTAH UT					68
		4	1.0			WEBER UT					
2	.1	8	.4	2	.1	VERMONT VT	295	-77	-11		
		1	1.0			ADDISON VT					
				1	.6	BENNINGTON VT					80
		2*	1.7			CALEDONIA VT					
		1	.4	1	.3	CHITTENDEN VT		-15			69
		1	1.0			ORLEANS VT					
1	.4					RUTLAND VT					
1	.5	1	.6			WASHINGTON VT	26				
		2	1.3			WINDHAM VT					
37	.3	54	.4	68	.4	VIRGINIA VA	13	8	22		
2*	1.5	1	.8	1	.8	ACCOMACK VA	-48	-7	-52	13	84
1	.5			2	.7	ALBEMARLE#VA			42	72	82
		1	.7			ALLEGHANY#VA					
				1	1.0	AMHERST VA					89
3	.7	4	.7	5	.9	ARLINGTON#VA	-3	30	27	65	87
		1	.3	3	.8	AUGUSTA#VA		213			83
		2*	8.9			BATH VA					
3	.8	2	.4	1	.2	BEDFORD#VA	-45	-58	-77	02	66
		1	.4			BUCHANAN VA					
				1	.3	CARROLL#VA					71
4	.4	2	.2	1	.1	CHESTERFIELD#VA	-47	-56	-77	02	64
		1	1.5			CULPEPER VA					
1	.5			1	.3	DINWIDDIE#VA			-43	16	69
1*	6.1					ESSEX VA					

* SIGNIFICANTLY MORE DEATHS THAN EXPECTED

COUNTIES COMBINED; SEE TECHNICAL NOTES

WHITE MALE — CANCER OF THE NASOPHARYNX — ICD 147

1950-1959 DEATHS	1950-1959 RATE	1960-1969 DEATHS	1960-1969 RATE	1970-1979 DEATHS	1970-1979 RATE	COUNTY AND STATE	PERCENT CHANGE 60S/50S	PERCENT CHANGE 70S/60S	PERCENT CHANGE 70S/50S	PERCENTILE %CHANGE 70S/50S	PERCENTILE RATES 1970S
2	.4	3	.2	5	.2	FAIRFAX#VA	-53	43	-33	23	68
		1	1.1			FAUQUIER VA					
		1*	1.7			FLOYD VA					
				1*	3.4	FLUVANNA VA					99
		1	.8	1	.7	FRANKLIN VA		-11			84
				2	.9	FREDERICK#VA					87
		1*	1.7	1	1.5	GLOUCESTER VA		-13			94
		1	.9	2*	1.5	HALIFAX#VA		65			94
		3	.5	4	.6	HENRICO VA		19			79
1*	2.3					ISLE OF WIGHT VA					
		2	1.1			JAMES CITY#VA					
				1*	2.8	LANCASTER VA					98
		1	.7			LOUDOUN VA					
1	1.4					MECKLENBURG VA					
		1	.6	1	.4	MONTGOMERY#VA		-31			74
				1	1.4	ORANGE VA					94
1*	1.2					PAGE VA					
1	.3	7*	2.1	1	.2	PITTSYLVANIA#VA	505	-90	-37	19	67
1	1.6			1	.3	PRINCE WILLIAM#VA			-79	01	71
				1	.8	PULASKI VA					86
2	.7	1	.1	5	.6	ROANOKE#VA	-79	308	-14	38	79
		1	.3	2	.6	ROCKINGHAM#VA		78			80
1	.8			1	.7	SCOTT VA			-4	46	84
		3*	1.7	2	1.1	TAZEWELL VA		-35			90
				1	1.8	WARREN VA					96
2	1.1	1	.5	1	.3	WISE#VA	-55	-30	-69	04	71
				1	.8	WYTHE VA					86
9*	.7	11	.7	13*	.7	CHESAPEAKE CITY#VA	-12	13	0	49	84
				3	1.2	HAMPTON CITY VA					92
1	.5			1	.2	NEWPORT NEWS CITY VA			-49	14	67
37	.3	46	.3	48	.3	WASHINGTON WA	7	-12	-6		
		1	1.7			ASOTIN WA					
1	.3	1	.3	1	.3	BENTON WA	-7	8	1	51	69
1	.4	1	.4			CHELAN WA	2				
				1	.3	CLALLAM WA					68
3	.9	2	.4	3	.4	CLARK WA	-58	12	-53	12	75
2	.8			1	.3	COWLITZ WA			-57	10	70
		1	1.5			DOUGLAS WA					
1*	4.8					FERRY WA					
		1	.8	2*	1.6	FRANKLIN WA		112			95
		1*	5.6			GARFIELD WA					
				1	.3	GRAYS HARBOR WA					68
1*	1.0			1	.4	ISLAND WA			-60	09	73
		1*	2.0			JEFFERSON WA					
16	.4	16	.4	16	.3	KING WA	-17	-13	-28	26	70
		1	.2	1	.2	KITSAP WA		-4			66
				1	.9	KITTITAS WA					87
1	.3			1	.3	LEWIS WA			-5	44	70
3	.2	3	.2	3	.2	PIERCE WA	-11	-24	-33	23	65
		1	.4	3	.9	SKAGIT WA		144			87
3	.4	8*	.9	2	.2	SNOHOMISH WA	111	-82	-63	07	65
2	.2	2	.2	5	.4	SPOKANE WA	-6	135	121	88	72
		1	.3	1	.2	THURSTON WA		-22			67
				1	.4	WALLA WALLA WA					73
2	.6			3	.7	WHATCOM WA			16	61	83
1	.8	1	.8			WHITMAN WA	-1				
		4	.6	1	.1	YAKIMA WA		-76			65
25	.3	25	.3	32	.4	WEST VIRGINIA WV	-4	22	16		
1	.7	1	.5			BERKELEY WV	-37				
1	.8			1	.8	BOONE WV			0	50	85
1	.8					BROOKE WV					
2	.4	3	.6	1	.2	CABELL WV	38	-68	-56	10	66
		1*	2.2			CALHOUN WV					
2	.7	1	.4			FAYETTE WV	-40				
				1	.7	GREENBRIER WV					83
2*	1.4	1	.5	2	.9	HANCOCK WV	-66	99	-33	24	87
2	.5			1	.2	HARRISON WV			-57	10	67
1*	1.6			2*	2.3	JEFFERSON WV			45	74	97
4	.4	2	.2	9*	.9	KANAWHA WV	-58	402	108	86	86
1	.6					LEWIS WV					
				1	.5	LOGAN WV					77
2	.8	2	1.0			MC DOWELL WV	25				
		3	1.0	2	.6	MARION WV		-43			79
		1	.5			MARSHALL WV					
1	.4	1	.4			MERCER WV	-13				
		1	.4	1	.4	MONONGALIA WV		-7			72
1	1.1			1	.9	NICHOLAS WV			-12	39	88
1	.3	3	.9	1	.3	OHIO WV	230	-61	29	67	71
		1*	2.3			PENDLETON WV					
				1	.6	PRESTON WV					81
		1	.9			PUTNAM WV					
1	.2	1	.3			RALEIGH WV	22				
				2*	1.7	RANDOLPH WV					95
		1	1.3			RITCHIE WV					
1	1.0			1	1.2	TAYLOR WV			20	63	92
				1*	1.7	TUCKER WV					95
				1	1.0	UPSHUR WV					88
		1	.6			WAYNE WV					
1	.3			3	.8	WOOD WV			151	89	84
54	.3	73	.4	55	.3	WISCONSIN WI	21	-30	-15		
1	.2	2	.4	1	.2	BROWN WI	64	-59	-33	23	65
1*	1.9	1	1.4			BURNETT WI	-26				
		1	.4			CHIPPEWA WI					
1	.5					CLARK WI					
		1	1.0			CRAWFORD WI					

* SIGNIFICANTLY MORE DEATHS THAN EXPECTED

\# COUNTIES COMBINED; SEE TECHNICAL NOTES

WHITE MALE — CANCER OF THE NASOPHARYNX — ICD 147

1950-1959 DEATHS	1950-1959 RATE	1960-1969 DEATHS	1960-1969 RATE	1970-1979 DEATHS	1970-1979 RATE	COUNTY AND STATE	PERCENT CHANGE 60S/50S	PERCENT CHANGE 70S/60S	PERCENT CHANGE 70S/50S	PERCENTILE %CHANGE 70S/50S	PERCENTILE RATES 1970S
1	.1	5	.6	4	.4	DANE WI	369	-32	219	95	72
		1	.3			DODGE WI					
				1	.9	DOOR WI					88
1	.4					DOUGLAS WI					
1	.7			1	.8	DUNN WI			13	59	85
		3	1.2	1	.3	EAU CLAIRE WI		-75			69
1*	5.4	1*	5.0			FLORENCE WI	-9				
2	.5			2	.5	FOND DU LAC WI			0	50	77
		1	.4			GRANT WI					
1	1.2			1	.7	GREEN LAKE WI			-37	19	84
		1	.8			IOWA WI					
		2*	4.1			IRON WI					
1	.4	2	.7			JEFFERSON WI	58				
				1	.6	JUNEAU WI					81
1	.2	1	.2			KENOSHA WI	-1				
1	.3	1	.3			LA CROSSE WI	-10				
1	.9	2*	2.1			LAFAYETTE WI	131				
1	1.0			1	1.4	LANGLADE WI			36	70	94
		1	.6	1	.7	LINCOLN WI		14			84
1	.3	1	.2	1	.2	MANITOWOC WI	-11	-12	-22	32	66
1	.3	1	.2	1	.2	MARATHON WI	-24	-7	-29	26	66
		1	.5	1	.4	MARINETTE WI		-13			74
		1	.4	1	.3	MENOMINEE#WI		-29			69
23	.5	18	.4	15	.3	MILWAUKEE WI	-25	-11	-33	22	71
		1	.4			MONROE WI					
1	1.1			1	.8	ONEIDA WI			-28	26	85
1	.3	3	.7	4	.8	OUTAGAMIE WI	173	11	204	93	84
		1	.6			OZAUKEE WI					
2*	1.4	1	.6			POLK WI	-58				
		1	.5			PORTAGE WI					
2	.3	2	.3	3	.4	RACINE WI	-13	53	33	69	74
		2	.4	1	.2	ROCK WI		-57			65
1	.6	2	1.2			ST CROIX WI	88				
		2	.9			SAUK WI					
		1	1.1	1	1.2	SAWYER WI		5			91
2	.6			2	.4	SHEBOYGAN WI			-36	21	74
1	.6	1	.7	1	.7	VERNON WI	14	-2	12	58	83
				1	1.2	VILAS WI					92
				1	.3	WALWORTH WI					69
2	1.0	1	.4	1	.4	WASHINGTON WI	-60	-7	-63	07	72
1	.2	1	.2	1	.1	WAUKESHA WI	-13	-26	-36	21	65
		2	.8	1	.5	WAUPACA WI		-31			78
				2*	1.3	WAUSHARA WI					93
		2	.4	1	.2	WINNEBAGO WI		-51			66
1	.4	2	.7	1	.3	WOOD WI	64	-54	-24	29	69
5	.3	2	.1	6	.4	WYOMING WY	-58	184	19		
1	1.1			1	1.3	CARBON WY			17	62	93
				1*	2.4	CONVERSE WY					98
		1	.9			FREMONT WY					
				1*	2.3	HOT SPRINGS WY					97
1	.5					LARAMIE WY					
1	.3			1	.4	NATRONA WY			15	60	73
1	.8			1	.9	PARK-TETON#WY			6	54	86
		1	.7			SHERIDAN WY					
				1*	5.1	SUBLETTE WY					99
1	.9					SWEETWATER WY					

* SIGNIFICANTLY MORE DEATHS THAN EXPECTED

COUNTIES COMBINED; SEE TECHNICAL NOTES

WHITE FEMALE — CANCER OF THE NASOPHARYNX — ICD 147

1950-1959 DEATHS	1950-1959 RATE	1960-1969 DEATHS	1960-1969 RATE	1970-1979 DEATHS	1970-1979 RATE	COUNTY AND STATE	PERCENT CHANGE 60S/50S	PERCENT CHANGE 70S/60S	PERCENT CHANGE 70S/50S	PERCENTILE %CHANGE 70S/50S	PERCENTILE RATES 1970S
827	.1	1161	.1	1641	.2	UNITED STATES	16	19	39		
13	.1	18	.1	23	.1	ALABAMA AL	21	2	23		
				1	.3	BALDWIN AL					89
				1*	1.1	BARBOUR AL					97
		2*	1.4			BLOUNT AL					
		1*	1.8			BUTLER AL					
		1	.3			CALHOUN AL					
				1	.4	CHAMBERS AL					91
1*	1.9					CLAY AL					
				1*	1.8	CLEBURNE AL					99
1*	.4					COVINGTON AL					
1*	.4					CULLMAN AL					
		1	.5			DE KALB AL					
		1*	.8			ESCAMBIA AL					
1	.3	1	.2			ETOWAH AL	-28				
		1*	1.1			FAYETTE AL					
3	.1	6	.3	8	.3	JEFFERSON AL	88	7	101	79	88
				1*	.9	LAWRENCE AL					97
				1	.5	LEE AL					92
1*	.6					LIMESTONE AL					
1	.2			1	.1	MADISON AL			-32	28	80
				1	.7	MARION AL					95
1	.2	2	.2	4	.4	MOBILE AL	3	62	67	72	90
2*	.5					MONTGOMERY AL					
				1	.3	MORGAN AL					88
				1*	.7	RUSSELL AL					94
		1*	.8			ST CLAIR AL					
		1	.2	1	.2	TUSCALOOSA AL		-25			83
1	.5					WALKER AL					
		2	.4	1	.0	ALASKA AK		-90			
4	.1	10	.2	14	.1	ARIZONA AZ	46	-21	16		
		1	.5	1	.3	COCHISE AZ		-32			89
		1*	.8			GILA AZ					
1*	2.3			1*	.9	GRAHAM AZ			-59	09	97
2	.1	5	.1	7	.1	MARICOPA AZ	54	-28	11	51	80
				1*	.8	NAVAJO AZ					96
1	.1	3	.2	2	.1	PIMA AZ	79	-67	-41	19	79
				1	.3	YAVAPAI AZ					89
				1	.4	YUMA AZ					90
7	.1	10	.1	17	.2	ARKANSAS AR	19	24	48		
1*	1.8					BRADLEY AR					
		1*	.9			CARROLL AR					
1*	.6	1	.5			CRAIGHEAD AR	-20				
				2*	1.3	CRAWFORD AR					98
		1*	2.1			DREW AR					
				1	.6	FAULKNER AR					93
		1*	.8			INDEPENDENCE AR					
		1	.4	2*	.7	JEFFERSON AR		59			95
		1*	2.0	1*	2.4	LEE AR		20			99
		1*	3.7			MARION AR					
1*	1.0					MILLER AR					
				2*	1.4	OUACHITA AR					98
		1*	1.0			PHILLIPS AR					
		1	.1	3	.2	PULASKI AR		110			85
				1*	1.6	SCOTT AR					99
2*	.6			2	.3	SEBASTIAN AR			-51	15	88
				1*	.6	SHARP AR					94
		1	.3	2	.4	WASHINGTON AR		59			91
1*	.7					WHITE AR					
1*	1.6					YELL AR					
82	.1	120	.1	171	.2	CALIFORNIA CA	5	20	26		
7	.2	7	.1	10	.2	ALAMEDA CA	-20	55	24	56	84
				2	.3	BUTTE CA					88
1	.1	2	.1	6	.2	CONTRA COSTA CA	-15	157	117	81	85
		1*	.5	2*	.4	EL DORADO CA		-18			92
2	.1	3	.2	4	.2	FRESNO CA	28	18	50	68	83
				1*	.9	GLENN CA					96
		2*	.5			HUMBOLDT CA					
				1	.3	KINGS CA					89
1*	1.0					LAKE CA					
31	.1	47	.1	57	.2	LOS ANGELES CA	21	14	38	63	82
				1	.3	MADERA CA					88
1	.1	3*	.3	3	.3	MARIN CA	167	-14	129	82	87
		1	.1			MONTEREY CA					
				1	.2	NAPA CA					84
2*	1.4			1	.6	NEVADA CA			-60	09	94
1	.1	4	.1	15	.2	ORANGE CA	-8	215	190	88	84
				1	.2	PLACER CA					86
1	.1	3	.1	6	.2	RIVERSIDE CA	76	17	106	80	82
1	.1	2	.1	1	.0	SACRAMENTO CA	40	-55	-37	24	78
7*	.4	3	.1	8	.2	SAN BERNARDINO CA	-66	66	-44	19	84
6	.2	13*	.2	9	.1	SAN DIEGO CA	31	-51	-35	25	80
13*	.3	6	.1	8	.2	SAN FRANCISCO CA	-55	73	-22	35	85
		1	.1	3	.2	SAN JOAQUIN CA		153			84
		1	.1	1	.2	SAN LUIS OBISPO CA		8			82
2	.1	5	.2	5	.2	SAN MATEO CA	58	-10	43	65	84
		4*	.3	2	.2	SANTA BARBARA CA		-55			82
2	.1	7	.2	12	.2	SANTA CLARA CA	88	7	100	78	86
		1	.3	4*	.4	SANTA CRUZ CA		62			91

* SIGNIFICANTLY MORE DEATHS THAN EXPECTED

WHITE FEMALE | CANCER OF THE NASOPHARYNX | ICD 147

1950-1959 DEATHS	1950-1959 RATE	1960-1969 DEATHS	1960-1969 RATE	1970-1979 DEATHS	1970-1979 RATE	COUNTY AND STATE	PERCENT CHANGE 60S/50S	PERCENT CHANGE 70S/60S	PERCENT CHANGE 70S/50S	PERCENTILE %CHANGE 70S/50S	PERCENTILE RATES 1970S
		1	.1	1	.1	SONOMA CA		-20			79
1	.1			1	.1	STANISLAUS CA			-41	19	79
2	.2	1	.1	2	.2	TULARE CA	-52	47	-30	30	82
		2	.2	2	.1	VENTURA CA		-34			80
1*	.7			1	.2	YOLO CA			-62	07	86
13	.2	19*	.2	12	.1	COLORADO CO	20	-50	-40		
		1	.2			ADAMS CO					
1*	1.2					DELTA CO					
7*	.2	8	.2	7	.2	DENVER-ARAPAHOE#CO	-1	-23	-23	34	84
1	.2	3*	.4	1	.1	EL PASO CO	116	-78	-53	13	79
1*	1.9					HUERFANO CO					
		2	.3			JEFFERSON CO					
		1*	3.4			LINCOLN CO					
				2*	1.9	LOGAN CO					99
				1	.3	MESA CO					87
1*	1.1	1*	1.1			MONTROSE CO	-4				
1	.2	3*	.5	1	.1	PUEBLO CO	137	-69	-27	32	82
1	.4					WELD CO					
17	.1	13	.1	28	.2	CONNECTICUT CT	-38	69	4		
5	.2	4	.1	5	.1	FAIRFIELD CT	-39	10	-34	27	80
4	.1	4	.1	5	.1	HARTFORD CT	-24	-13	-33	27	79
				4*	.4	LITCHFIELD CT					91
				2	.3	MIDDLESEX CT					87
7	.2	4	.1	10	.2	NEW HAVEN CT	-49	88	-5	46	84
		1	.1	2	.2	NEW LONDON CT		77			82
1	.3					WINDHAM CT					
3	.2	4	.2			DELAWARE DE	-5				
		1	.4			KENT DE					
3	.3	3	.2			NEW CASTLE DE	-34				
3	.1	2	.1	3	.1	DIST. OF COLUMBIA DC	1	24	26		
3	.1	2	.1	3	.1	WASHINGTON DC	1	24	26	57	81
24	.1	45	.2	98	.2	FLORIDA FL	15	34	55		
		1	.3	2*	.4	BAY FL		35			92
		2	.3	3	.2	BREVARD FL		-29			86
2	.1	5	.1	13	.2	BROWARD FL	57	42	123	81	84
		1*	.5	2	.4	COLLIER FL		-21			91
5	.1	11	.2	17	.2	DADE FL	51	4	57	70	84
2	.1	1	.0	7*	.3	DUVAL FL	-52	569	221	92	89
		2	.4	5*	.5	ESCAMBIA FL		22			93
2	.1	1	.1	11*	.4	HILLSBOROUGH FL	-40	421	214	91	90
				3*	.6	LAKE FL					94
1*	.5	1	.3			LEE FL	-39				
1*	.8					LEON FL					
		1*	2.4			LEVY FL					
1*	2.3					MADISON FL					
				1	.1	MANATEE FL					80
		1*	.8	1	.5	MARTIN FL		-36			92
		1*	.6	1	.3	MONROE FL		-57			87
1*	2.5					NASSAU FL					
				1	.5	OKALOOSA FL					92
3*	.4	4	.3	3	.2	ORANGE FL	-29	-39	-56	11	83
		4	.3	7	.3	PALM BEACH FL		0			87
				4	.3	PASCO FL					87
3	.1	5	.1	6	.1	PINELLAS FL	6	-13	-8	44	80
1	.2	1	.1	4	.3	POLK FL	-51	247	72	74	88
1	.2	1	.1	3	.3	SARASOTA FL	-45	184	57	70	89
		1*	2.3			SUMTER FL					
1	.3	1	.1	3	.2	VOLUSIA FL	-78	261	-19	37	85
				1*	.9	WALTON FL					96
15	.1	17	.1	33	.2	GEORGIA GA	-11	50	34		
		1*	.7			BARTOW GA					
		1	.2	2	.3	CHATHAM GA		61			88
				3*	.7	CLAYTON GA					95
1	.3			3	.5	COBB GA			37	62	92
1	.1	2	.2	1	.0	DE KALB GA	22	-72	-66	06	78
				1	.2	FLOYD GA					86
4	.2	6*	.3	7*	.3	FULTON GA	30	15	50	68	89
				1*	.8	HABERSHAM GA					96
1*	1.0			1	.5	HOUSTON GA			-53	13	92
				1*	.9	JACKSON GA					96
				1*	2.1	JEFF DAVIS GA					99
		1*	.9	1*	.6	LAURENS GA		-37			93
1*	6.6					LIBERTY GA					
				1*	2.0	MC DUFFIE GA					99
1*	1.4					MERIWETHER GA					
1	.2			1	.2	MUSCOGEE#GA			-8	43	82
				1*	1.0	PAULDING GA					97
				1	.8	POLK GA					96
1*	3.9					RANDOLPH GA					
		1	.2	3*	.6	RICHMOND GA		157			94
				2*	3.6	SCREVEN GA					99
1*	1.0					SPALDING GA					
		1*	.9			SUMTER GA					
				1*	1.1	TATTNALL GA					97
				1*	.7	THOMAS GA					95
1*	.6					TROUP GA					
1*	.5	1	.4			WALKER GA	-37				
		2*	1.6			WARE GA					
1*	2.4					WAYNE GA					

* SIGNIFICANTLY MORE DEATHS THAN EXPECTED

COUNTIES COMBINED; SEE TECHNICAL NOTES

1950-1959 DEATHS	1950-1959 RATE	1960-1969 DEATHS	1960-1969 RATE	1970-1979 DEATHS	1970-1979 RATE	COUNTY AND STATE	PERCENT CHANGE 60S/50S	PERCENT CHANGE 70S/60S	PERCENT CHANGE 70S/50S	PERCENTILE %CHANGE 70S/50S	PERCENTILE RATES 1970S
		1*	3.1			WILKINSON GA					
		5*	.6	4	.4	HAWAII HI		-31			
				1*	1.0	HAWAII HI					97
		5*	.7	2	.3	HONOLULU HI		-59			88
				1*	2.8	KAUAI HI					99
2	.1	3	.1	4	.1	IDAHO ID	18	28	51		
				1	.5	BANNOCK ID					92
		1*	15.7			BOISE ID					
1	.3					CANYON ID					
				1*	1.4	GEM ID					98
		1*	2.5			JEFFERSON ID					
		1*	.8	2*	1.5	NEZ PERCE ID		88			98
1*	.6					TWIN FALLS ID					
56	.1	62	.1	84	.2	ILLINOIS IL	-3	28	25		
2*	.5					ADAMS IL					
		1*	1.4			BROWN IL					
				2*	.7	BUREAU IL					95
				1	.2	CHAMPAIGN IL					83
1*	.4					CHRISTIAN IL					
1*	.8	2*	1.1			CLARK IL	29				
		1*	.7			CLAY IL					
				1	.4	CLINTON IL					90
29	.1	31	.1	40	.2	COOK IL	-1	38	37	62	83
4*	.5	2	.1	3	.1	DU PAGE IL	-69	-17	-74	02	80
1	.3					FRANKLIN IL					
1*	.6			1	.5	HANCOCK IL			-13	39	93
2*	1.0					IROQUOIS IL					
1*	.6			1	.4	JACKSON IL			-32	29	90
2	.2	2	.2			KANE IL	-15				
				2	.4	KANKAKEE IL					89
1	.3					KNOX IL					
1	.1	1	.1	3	.2	LAKE IL	25	61	101	78	83
		1	.1	1	.1	LA SALLE IL		2			81
1*	.7			1*	1.0	LAWRENCE IL			44	66	97
				1	.4	LEE IL					91
1*	.6					MC DONOUGH IL					
		1	.2	2	.3	MC HENRY IL		47			88
		2	.5	2	.4	MC LEAN IL		-15			90
		1	.1			MACON IL					
		1	.4	1	.3	MACOUPIN IL		-25			89
2	.3	2	.2	2	.1	MADISON IL	-35	-25	-51	14	81
				1	.3	MARION IL					87
				1*	1.0	MONROE IL					97
		1	.3			MONTGOMERY IL					
		2*	.7	1	.5	MORGAN IL		-23			93
1*	1.2					MOULTRIE IL					
				1	.2	OGLE IL					85
1	.2	2	.2	1	.1	PEORIA IL	21	-44	-33	28	80
		1*	.8			PERRY IL					
				1	.8	RANDOLPH IL					95
				1*	.9	RICHLAND IL					96
		1	.1			ROCK ISLAND IL					
2	.2	1	.1	1	.1	ST CLAIR IL	-64	-17	-70	05	79
		1*	.4	1	.8	SALINE IL		129			96
		1	.1	2	.2	SANGAMON IL		34			82
				1	.2	STEPHENSON IL					86
		2	.4			TAZEWELL IL					
1	.2					VERMILION IL					
		1*	.8			WASHINGTON IL					
				1*	.6	WAYNE IL					94
1	.1	1	.1	3	.3	WILL IL	-23	140	84	77	86
				3	.3	WINNEBAGO IL					87
				1	.5	WOODFORD IL					93
11	.1	27	.1	37	.1	INDIANA IN	112	23	160		
				3	.2	ALLEN IN					84
				1*	1.8	BENTON IN					99
				1*	.8	BLACKFORD IN					96
1*	.7			1	.2	BOONE IN			-72	02	83
1*	3.2					BROWN IN					
		1	.4	2*	.7	CASS IN		66			95
		1*	.4			CLINTON IN					
				1	.2	DELAWARE IN					83
1	.2	1	.2	1	.1	ELKHART IN	-14	-21	-32	29	62
				1*	.7	HARRISON IN					95
		1	.5			HENDRICKS IN					
1*	.8					JEFFERSON IN					
				2*	.8	KNOX IN					95
				1	.4	KOSCIUSKO IN					90
1	.0	6*	.4	2	.1	LAKE IN	790	-73	136	84	79
		4*	.8			LA PORTE IN					
		1	.5			LAWRENCE IN					
2	.1	3	.1	4	.1	MARION IN	38	20	67	72	80
				1	.4	MORGAN IN					91
				1	.3	NOBLE IN					88
				1*	1.3	ORANGE IN					98
				2	.4	PORTER IN					91
2	.2	2	.2	4	.3	ST JOSEPH IN	-12	79	57	71	88
		1*	.4	1	.3	SHELBY IN		-35			88
				1*	.7	STEUBEN IN					95
1*	2.3					SWITZERLAND IN					
		1	.2	1	.2	TIPPECANOE IN		1			84
1*	1.1			1*	1.2	TIPTON IN			8	50	98
		1*	3.1			UNION IN					

* SIGNIFICANTLY MORE DEATHS THAN EXPECTED

1950-1959 DEATHS	1950-1959 RATE	1960-1969 DEATHS	1960-1969 RATE	1970-1979 DEATHS	1970-1979 RATE	COUNTY AND STATE	PERCENT CHANGE 60S/50S	PERCENT CHANGE 70S/60S	PERCENT CHANGE 70S/50S	PERCENTILE %CHANGE 70S/50S	PERCENTILE RATES 1970S
		2	.2	1	.1	VANDERBURGH IN		-46			79
		2	.4	2	.3	VIGO IN		-20			88
				1	.2	WAYNE IN					86
12	.1	20	.1	25	.1	IOWA IA	27	26	60		
1*	2.3					ADAIR IA					
		1*	2.0			ADAMS IA					
				2*	.9	BOONE IA					96
		1*	1.0	2*	1.0	BUENA VISTA IA		-3			97
				2*	1.7	BUTLER IA					99
				1*	.8	CEDAR IA					96
		1	.3			CERRO GORDO IA					
1*	1.5					CLARKE IA					
				2*	2.1	CLAY IA					99
1*	1.0					CRAWFORD IA					
		2*	3.9			DAVIS IA					
1*	1.4					GRUNDY IA					
				1*	.4	GUTHRIE IA					91
				1*	.8	HENRY IA					96
		1*	1.0			HUMBOLDT IA					
				1*	1.4	IOWA IA					98
1*	.5	1	.3			JOHNSON IA	-44				
				1*	.6	JONES IA					94
		1	.3			LEE IA					
		2	.2	1	.1	LINN IA		-42			81
		1*	1.5	1*	2.0	LUCAS IA		38			99
		1*	.7	1*	1.0	MADISON IA		49			97
				1*	.4	MAHASKA IA					91
2*	2.3	1*	.5			PAGE IA	-79				
2	.2	2	.1	2	.1	POLK IA	-7	-5	-12	39	81
1	.3	1	.2	1	.2	POTTAWATTAMIE IA	-26	-13	-35	25	82
		1*	3.0			RINGGOLD IA					
		1	.2	1	.1	SCOTT IA		-37			80
				2*	.8	STORY IA					96
		1*	.5	1*	.8	TAMA IA		58			95
		1*	.6			WARREN IA					
1*	1.3					WAYNE IA					
1	.2			1	.2	WOODBURY IA			11	52	84
8	.1	18	.1	16	.1	KANSAS KS	87	-25	40		
		1*	1.2			ATCHISON KS					
		1*	.5	1*	.5	CHEROKEE KS		-9			92
1*	.9					CLOUD KS					
		1	.6			COWLEY KS					
		2*	.8	1	.4	CRAWFORD KS		-50			91
		1*	.6			GEARY KS					
		1*	6.7			HODGEMAN KS					
1*	1.8					JEWELL KS					
				2	.2	JOHNSON KS					83
		1*	3.2			KIOWA KS					
				1	.3	LABETTE KS					88
				1	.5	LYON KS					93
		1*	.7			MARSHALL KS					
		1*	1.1			MITCHELL KS					
1	.4			2*	.6	MONTGOMERY KS			53	69	93
		1*	1.0			OSAGE KS					
		1*	1.0			OTTAWA KS					
				1	.3	SALINE KS					88
3	.3			6*	.3	SEDGWICK KS			35	61	89
1	.2			1	.1	SHAWNEE KS			-23	34	80
		2*	1.2			SUMNER KS					
1*	3.7					WOODSON KS					
		4*	.5			WYANDOTTE KS					
9	.1	19	.1	30	.2	KENTUCKY KY	81	35	144		
				2*	1.2	BARREN KY					98
		1*	.6			BELL KY					
				1	.6	BOONE KY					93
				1*	1.3	BOURBON KY					98
				1*	1.1	CALDWELL KY					97
1	.2	1	.2	1	.2	CAMPBELL KY	-5	-16	-20	36	83
1*	.6			1	.4	CHRISTIAN KY			-24	33	91
		1*	1.5			CLAY KY					
		2*	4.6			CRITTENDEN KY					
		2*	.6	1	.2	DAVIESS KY		-59			86
		1*	1.2	1*	1.9	ESTILL KY		59			99
2*	.4			3	.4	FAYETTE KY			-9	42	90
		1*	.7			FRANKLIN KY					
				1	.5	HARDIN KY					93
1*	.5					HARLAN KY					
		1*	.7			HENDERSON KY					
		1	.4			HOPKINS KY					
2	.1	3	.1	7	.2	JEFFERSON KY	14	126	156	86	85
				1*	1.1	JESSAMINE KY					97
		1*	1.0			JOHNSON KY					
		1	.2	4*	.6	KENTON KY		264			93
1*	.8					LAUREL KY					
		1*	3.0			LYON KY					
		1	.5	2*	.5	MC CRACKEN KY		1			93
				1*	.5	MARSHALL KY					93
				1*	.6	MERCER KY					94
				1	.3	PIKE KY					87
		1*	1.2			SHELBY KY					
1*	.5					WARREN KY					
10	.1	17	.1	21	.1	LOUISIANA LA	28	-2	26		
				3*	1.2	ACADIA LA					98

* SIGNIFICANTLY MORE DEATHS THAN EXPECTED

WHITE FEMALE | CANCER OF THE NASOPHARYNX | ICD 147

1950-1959 DEATHS	1950-1959 RATE	1960-1969 DEATHS	1960-1969 RATE	1970-1979 DEATHS	1970-1979 RATE	COUNTY AND STATE	PERCENT CHANGE 60S/50S	PERCENT CHANGE 70S/60S	PERCENT CHANGE 70S/50S	PERCENTILE %CHANGE 70S/50S	PERCENTILE RATES 1970S
		1*	1.9			ASSUMPTION LA					
				1	.6	AVOYELLES LA					94
		1*	1.0			BOSSIER LA					
1	.2					CADDO LA					
		1	.3			CALCASIEU LA					
1*	1.5					CLAIBORNE LA					
1*	1.7					DE SOTO LA					
		2	.2	2	.2	EAST BATON ROUGE LA		6			85
1*	1.5					FRANKLIN LA					
				1*	1.0	IBERVILLE LA					97
		1*	1.4	1*	1.1	JACKSON LA		-24			97
1	.1			2	.2	JEFFERSON LA			33	58	82
				1*	.8	JEFFERSON DAVIS LA					96
				2	.4	LAFAYETTE LA					91
		1*	.7			LINCOLN LA					
3	.1	8*	.4	3	.1	ORLEANS LA	168	-63	-1	48	81
				1	.2	OUACHITA LA					83
1	.3	1	.2	1	.2	RAPIDES LA	-25	-18	-39	21	84
1*	1.3					SABINE LA					
				1	.5	ST BERNARD LA					93
				1	.7	ST MARY LA					94
				1	.4	VERMILION LA					90
		1*	.6			WEBSTER LA					
6	.1	7	.1	15	.2	MAINE ME	6	77	88		
1	.2	1	.2	1	.2	ANDROSCOGGIN ME	9	-17	-9	42	83
1	.2			1	.3	AROOSTOOK ME			39	64	87
1	.1	4*	.3	6*	.5	CUMBERLAND ME	255	46	418	96	92
				1	.5	HANCOCK ME					93
				1	.2	KNOX ME					86
1	.2	1	.2	2	.2	PENOBSCOT ME	2	40	42	65	86
1*	.9			1*	.6	SAGADAHOC ME			-34	26	94
				1	.2	SOMERSET ME					84
1*	.6			1*	.6	WALDO ME			-6	45	93
		1	.1			YORK ME					
10	.1	21	.1	37	.2	MARYLAND MD	56	40	118		
2*	.5			1	.2	ALLEGANY MD			-52	14	85
		3*	.3	2	.1	ANNE ARUNDEL MD		-55			81
		1	.0	6	.2	BALTIMORE MD		436			83
		1	.2	1	.2	CARROLL MD		-24			83
1*	.8			1	.4	CECIL MD			-49	16	91
3*	2.2			1*	.8	DORCHESTER MD			-64	06	96
		2*	.6	1	.2	FREDERICK MD		-65			84
				1	.2	HARFORD MD					85
		1*	1.2			KENT MD					
				3	.1	MONTGOMERY MD					80
1	.1	2	.1	11*	.6	PRINCE GEORGE'S MD	63	425	754	99	94
1	.2					WASHINGTON MD					
				1	.3	WICOMICO MD					88
2	.0	11*	.3	8	.3	BALTIMORE CITY MD	461	9	513	98	88
23	.1	33	.1	53	.1	MASSACHUSETTS MA	26	45	83		
				3	.3	BERKSHIRE MA					88
3	.1	4	.1	7	.2	BRISTOL MA	13	45	64	71	85
3	.1	3	.1	10	.2	ESSEX MA	18	160	208	90	86
				1	.2	FRANKLIN MA					82
1	.0			2	.1	HAMPDEN MA			102	79	79
1	.2	2	.3			HAMPSHIRE MA	53				
1	.0	11	.1	7	.1	MIDDLESEX MA	965	-47	462	96	79
4	.2	8*	.2	6	.1	NORFOLK MA	55	-41	-9	43	81
1	.1	1	.1	4	.2	PLYMOUTH MA	-4	227	215	91	86
5	.1	1	.0	10	.2	SUFFOLK MA	-82	1246	147	86	86
4	.1	3	.1	3	.1	WORCESTER MA	-41	6	-37	23	79
29	.1	50	.1	53	.1	MICHIGAN MI	40	-6	31		
				1*	1.1	ALCONA MI					97
		1	.3	2*	.5	ALLEGAN MI		70			93
				1	.5	BARRY MI					92
1	.2	3*	.6			BAY MI	132				
1	.1					BERRIEN MI					
		1*	.5			BRANCH MI					
				1	.1	CALHOUN MI					80
		1*	.5			CASS MI					
		1*	1.1			EMMET MI					
1	.1			2	.1	GENESEE MI			71	74	80
				3*	1.1	GRAND TRAVERSE MI					97
				2	.2	INGHAM MI					84
1*	1.1			2*	1.4	IRON MI			35	61	98
				3	.4	JACKSON MI					91
		2	.2			KALAMAZOO MI					
		3	.2	2	.1	KENT MI		-34			80
		1	.5			LAPEER MI					
		1*	2.1			LEELANAU MI					
2*	.6					LENAWEE MI					
				1	.3	LIVINGSTON MI					89
1	.0	3	.2	5	.2	MACOMB MI	224	28	315	94	84
1*	.9					MASON MI					
1*	.7					MENOMINEE MI					
		2*	1.1			MIDLAND MI					
1	.3	1	.2			MONROE MI	-19				
		1	.1	1	.1	MUSKEGON MI		-11			80
				1	.8	NEWAYGO MI					95
3	.2	7	.2	5	.1	OAKLAND MI	-12	-44	-51	15	80
				1*	1.5	ONTONAGON MI					98
		1	.2	3*	.5	OTTAWA MI		141			92
		1	.1			SAGINAW MI					

* SIGNIFICANTLY MORE DEATHS THAN EXPECTED

WHITE FEMALE — CANCER OF THE NASOPHARYNX — ICD 147

1950-1959 DEATHS	1950-1959 RATE	1960-1969 DEATHS	1960-1969 RATE	1970-1979 DEATHS	1970-1979 RATE	COUNTY AND STATE	PERCENT CHANGE 60S/50S	PERCENT CHANGE 70S/60S	PERCENT CHANGE 70S/50S	PERCENTILE %CHANGE 70S/50S	PERCENTILE RATES 1970S
				1	.2	ST CLAIR MI					82
				1	.2	ST JOSEPH MI					85
		1	.4			VAN BUREN MI					
1	.2			2	.2	WASHTENAW MI			33	59	85
14	.1	19	.2	13	.1	WAYNE MI	27	-23	-3	47	80
1*	1.1					WEXFORD MI					
17	.1	14	.1	22	.1	MINNESOTA MN	-31	32	-9		
				1	.1	ANOKA MN					80
				1*	.6	BENTON MN					93
		1	.4	1	.2	BLUE EARTH MN		-64			82
				1*	.7	CARVER MN					95
1*	1.2					CHISAGO MN					
				2*	.8	CLAY MN					96
				1*	1.3	COTTONWOOD MN					98
7	.2	5	.1	3	.0	HENNEPIN MN	-35	-56	-71	03	79
1*	1.5					ISANTI MN					
				1*	.7	JACKSON MN					94
				1*	.7	MARSHALL MN					95
				1	.4	MOWER MN					90
				1*	1.0	NICOLLET MN					97
		1	.3			OLMSTED MN					
		1	.4	1	.3	POLK MN		-30			88
3	.1	3	.1	1	.0	RAMSEY MN	-24	-69	-77	01	78
				1*	2.5	RED LAKE MN					99
		1*	.9			REDWOOD MN					
2	.2	2	.1			ST LOUIS MN	-19				
1*	1.3					SIBLEY MN					
				1	.2	STEARNS MN					85
				1*	.5	TODD MN					93
1*	.5			1	.3	WASHINGTON MN			-39	20	89
1*	1.4			1*	1.3	WATONWAN MN			-7	45	98
				1	.4	WRIGHT MN					90
3	.0	12	.2	7	.1	MISSISSIPPI MS	262	-52	75		
1*	.8					ADAMS MS					
1*	1.3	1*	1.1			BOLIVAR MS	-17				
1*	3.1					CARROLL MS					
				1*	1.0	DE SOTO MS					97
		1	.5			FORREST MS					
				1*	2.9	GREENE MS					99
		1*	1.3			GRENADA MS					
		1	.2	1	.1	HINDS MS		-11			81
				1	.3	JONES MS					89
		1*	.7			LOWNDES MS					
		1*	1.0			MARION MS					
		1*	.8	1*	.6	PIKE MS		-27			93
		1*	1.0			PRENTISS MS					
				1	.4	RANKIN MS					90
		1*	1.3			SCOTT MS					
		1*	1.0			SUNFLOWER MS					
		1*	1.0			TIPPAH MS					
				1*	1.4	WAYNE MS					98
		1*	2.6			WEBSTER MS					
18	.1	31	.1	45	.2	MISSOURI MO	50	35	102		
				1*	.7	BARTON MO					95
1*	.5			2*	.5	BOONE MO			4	49	92
		1	.2			BUCHANAN MO					
2*	1.2	1*	.6			BUTLER MO	-51				
		1*	1.9			CALDWELL MO					
1*	2.3			1*	1.0	CAMDEN MO			-55	12	97
		1	.3	2*	.6	CAPE GIRARDEAU MO		68			93
		1*	1.0			CEDAR MO					
1	.2					CLAY MO					
		1	.4			COLE MO					
		1*	1.2			COOPER MO					
		1	.6			DUNKLIN MO					
		1	.4			FRANKLIN MO					
		1	.1	3	.3	GREENE MO		106			87
6	.2	3	.1	5	.2	JACKSON MO	-59	116	-12	40	83
1	.2			2	.4	JASPER MO			157	87	91
				1	.2	JEFFERSON MO					85
				1*	1.6	MC DONALD MO					99
				1	.7	MARION MO					95
				2*	1.5	MILLER MO					98
				1	.3	NEWTON MO					89
				2*	1.6	PEMISCOT MO					99
				2*	1.6	PHELPS MO					99
				1*	.9	PIKE MO					96
		1*	1.4			PULASKI MO					
				1	.7	RANDOLPH MO					95
1*	.9					RAY MO					
		2*	.8			ST FRANCOIS MO					
		8	.2	12	.2	ST LOUIS COUNTY MO		17			86
				1*	.7	TANEY MO					95
1*	.5					VERNON MO					
4	.1	7	.2	4	.1	ST LOUIS CITY MO	63	-39	0	48	80
1	.0	4	.1	3	.1	MONTANA MT	325	-56	87		
		1	.3			CASCADE MT					
				1*	.5	CUSTER MT					92
1*	.5			1	.2	MISSOULA MT			-59	09	84
		1*	10.1			POWDER RIVER MT					
		1*	2.1			ROOSEVELT MT					
		1*	3.5			TETON MT					
				1	.2	YELLOWSTONE MT					85

* SIGNIFICANTLY MORE DEATHS THAN EXPECTED

WHITE FEMALE — CANCER OF THE NASOPHARYNX — ICD 147

1950-1959 DEATHS	1950-1959 RATE	1960-1969 DEATHS	1960-1969 RATE	1970-1979 DEATHS	1970-1979 RATE	COUNTY AND STATE	PERCENT CHANGE 60S/50S	PERCENT CHANGE 70S/60S	PERCENT CHANGE 70S/50S	PERCENTILE %CHANGE 70S/50S	PERCENTILE RATES 1970S
8	.1	8	.1	13	.2	NEBRASKA NB	-15	62	38		
1*	.5	1	.7			ADAMS NB	56				
1*	1.3					ANTELOPE NB					
				1	.6	BUFFALO NB					94
1*	1.4					CUMING NB					
1*	.8	1	.4			DODGE NB	-43				
1	.1	2	.1	7*	.4	DOUGLAS NB	52	285	485	97	90
		1*	.7			FURNAS NB					
1*	.5					HALL NB					
				1*	.8	HOLT NB					96
				1*	2.1	JEFFERSON NB					99
		1*	1.4			KNOX NB					
		1	.1	1	.1	LANCASTER NB		10			81
		1*	.7			LINCOLN NB					
1*	.7					MADISON NB					
1*	1.7					NEMAHA NB					
				1*	.8	POLK NB					95
				1*	1.2	WAYNE NB					98
1	.1	5*	.3	7	.3	NEVADA NV	259	-12	217		
		3*	.4	4	.3	CLARK NV		-14			88
				1*	1.5	ELKO NV					98
1*	2.8					MINERAL NV					
		2*	.4	1	.1	WASHOE NV		-64			81
				1*	.8	ORMSBY-CARSON CITY NV					96
5	.2	5	.1	5	.1	NEW HAMPSHIRE NH	-15	-19	-31		
1	.5	1	.3			CHESHIRE NH	-37				
		1	.5	2*	.9	COOS NH		71			96
		1	.4			GRAFTON NH					
1	.1	1	.1	2	.1	HILLSBOROUGH NH	-6	47	39	64	81
1	.2					MERRIMACK NH					
2*	.4					ROCKINGHAM NH					
		1	.3	1	.2	STRAFFORD NH		-21			86
38	.1	57	.2	78*	.2	NEW JERSEY NJ	18	22	44		
		1	.1	2	.2	ATLANTIC NJ		233			85
5	.2	5	.1	13	.2	BERGEN NJ	-29	118	55	69	86
1	.1	1	.1	2	.1	BURLINGTON NJ	-12	25	10	51	81
1	.1	2	.1	4	.2	CAMDEN NJ	58	85	192	89	82
		1	.2			CAPE MAY NJ					
6	.1	8	.2	9	.2	ESSEX NJ	50	15	73	74	84
2*	.5	1	.1			GLOUCESTER NJ	-69				
6	.2	7	.2	4	.1	HUDSON NJ	12	-37	-30	31	80
2*	.7			2*	.5	HUNTERDON NJ			-32	29	92
2	.2	5*	.3	1	.0	MERCER NJ	116	-86	-70	04	78
4*	.2	6*	.3	5	.2	MIDDLESEX NJ	32	-36	-15	38	83
1	.1	5	.3	6	.2	MONMOUTH NJ	323	-5	299	93	86
				2	.1	MORRIS NJ					79
		1	.1	7*	.4	OCEAN NJ		385			90
4	.2	7*	.3	7	.3	PASSAIC NJ	52	-12	35	60	87
				1	.4	SALEM NJ					90
1	.2	2	.3	3	.3	SOMERSET NJ	57	16	82	76	89
1*	.9					SUSSEX NJ					
2	.1	5	.2	10*	.3	UNION NJ	65	95	222	92	89
3	.1	3	.1	4	.1	NEW MEXICO NM	-38	26	-21		
2	.3	2	.1	2	.1	BERNALILLO NM	-60	-5	-62	06	81
				1*	1.2	COLFAX NM					98
1*	.3			1	.3	DONA ANA NM			-23	35	87
		1	.4			SANTA FE NM					
123*	.1	139	.1	160	.2	NEW YORK NY	3	9	13		
3	.2	3	.2	3	.2	ALBANY NY	17	-21	-8	44	83
		1	.3			ALLEGANY NY					
18*	.2	10	.1	13	.2	BRONX NY	-41	60	-5	46	85
1	.1					BROOME NY					
		1	.2			CATTARAUGUS NY					
1	.1	3	.3	4*	.4	CHAUTAUQUA NY	149	14	184	88	90
		2	.3	2	.3	CHEMUNG NY		6			89
				1	.4	CLINTON NY					89
1	.3			1	.3	COLUMBIA NY			-10	41	87
1	.1	1	.1	2	.1	DUTCHESS NY	-8	30	20	55	81
4	.1	10	.2	3	.0	ERIE NY	118	-73	-41	20	79
				1	.4	ESSEX NY					90
2*	.6	1	.2			FULTON NY	-64				
1	.2					HERKIMER NY					
1	.2	1	.2	1	.2	JEFFERSON NY	28	-22	0	48	83
24*	.2	23	.2	23	.2	KINGS(BROOKLYN) NY	7	16	24	57	84
		1	.3	1	.3	MADISON NY		-15			87
3	.1	3	.1	9	.2	MONROE NY	-11	161	131	83	85
		2*	.6			MONTGOMERY NY					
6	.2	10	.1	14	.2	NASSAU NY	-7	22	13	52	83
19*	.2	22*	.3	18	.3	NEW YORK(MANHATTAN) NY	38	5	45	67	87
1	.1	1	.1	3	.2	NIAGARA NY	-1	136	134	83	85
		2	.1	2	.1	ONEIDA NY		-12			80
1	.1	4	.2	3	.1	ONONDAGA NY	250	-33	136	83	80
2*	.6					ONTARIO NY					
2	.2			6*	.4	ORANGE NY			108	80	91
1*	.4					ORLEANS NY					
				2*	.7	OTSEGO NY					94
1*	.6	1	.4	1	.3	PUTNAM NY	-26	-25	-45	17	89
16	.2	11	.1	12	.1	QUEENS NY	-41	-6	-44	18	79
1	.1			1	.1	RENSSELAER NY			8	50	81

* SIGNIFICANTLY MORE DEATHS THAN EXPECTED

WHITE FEMALE — CANCER OF THE NASOPHARYNX — ICD 147

1950-1959 DEATHS	1950-1959 RATE	1960-1969 DEATHS	1960-1969 RATE	1970-1979 DEATHS	1970-1979 RATE	COUNTY AND STATE	PERCENT CHANGE 60S/50S	PERCENT CHANGE 70S/60S	PERCENT CHANGE 70S/50S	PERCENTILE %CHANGE 70S/50S	PERCENTILE RATES 1970S
		3	.2	7*	.4	RICHMOND NY		95			91
1	.2	2	.3	2	.2	ROCKLAND NY	56	-25	16	53	84
		1	.2			ST LAWRENCE NY					
1	.2			1	.2	SARATOGA NY			-34	25	82
1	.1	2	.2	2	.2	SCHENECTADY NY	64	-12	45	67	83
		1*	.7			SCHOHARIE NY					
1	.2	1	.2	1	.2	STEUBEN NY	-12	-6	-17	37	83
2	.1	7	.2	8	.1	SUFFOLK NY	137	-24	80	76	81
1	.2	1	.1	1	.1	ULSTER NY	-27	11	-19	36	81
		1	.4	1	.4	WARREN NY		1			91
				2*	.5	WASHINGTON NY					92
1	.3	1	.3			WAYNE NY	0				
5	.1	6	.1	9	.2	WESTCHESTER NY	1	18	19	54	82
23*	.2	20	.1	34	.1	NORTH CAROLINA NC	-33	33	-11		
1	.4	1	.3			ALAMANCE NC	-26				
		1*	1.6			AVERY NC					
		1	.1	2	.2	BUNCOMBE NC		71			84
				2	.4	CABARRUS NC					90
				1	.5	CARTERET NC					92
2*	.9			1	.2	CATAWBA NC			-75	01	85
		1*	.9			CHATHAM NC					
1	.4					CLEVELAND NC					
				2*	1.0	COLUMBUS NC					97
1*	1.1	1*	.9	2*	.8	CRAVEN NC	-17	-9	-24	33	96
				2	.3	CUMBERLAND NC					89
		1*	2.2			DARE NC					
		2*	.5			DAVIDSON NC					
				1*	1.3	DAVIE NC					98
1	.4	1	.2	2	.4	DURHAM NC	-38	60	-1	47	89
1	.1			4*	.4	FORSYTH NC			213	90	91
				2	.3	GASTON NC					87
1*	4.5					GATES NC					
1*	.6					HARNETT NC					
				1	.3	HAYWOOD NC					89
1*	4.1					HOKE NC					
1	.4					IREDELL NC					
		1	.4			JOHNSTON NC					
				2*	1.4	LEE NC					98
				1	.5	LENOIR NC					92
				1	.6	LINCOLN NC					94
2	.2	1	.1	1	.1	MECKLENBURG NC	-60	0	-60	08	79
				1	.6	MOORE NC					94
				1	.3	NEW HANOVER NC					87
1*	1.9					NORTHAMPTON NC					
		2*	1.0			PITT NC					
1*	.6					RANDOLPH NC					
1*	.8	1*	.7			RICHMOND NC	-11				
		1*	.5	1	.4	ROBESON NC		-33			89
				1	.2	ROCKINGHAM NC					86
2*	.9			1	.3	RUTHERFORD NC			-62	08	89
1*	.6					SAMPSON NC					
1*	1.1					SCOTLAND NC					
2*	1.0					SURRY NC					
				1	.1	WAKE NC					80
		2*	.9	1	.4	WAYNE NC		-61			90
		1*	.6			WILSON NC					
1*	1.1	1*	1.0			YADKIN NC	-11				
		1*	1.4			YANCEY NC					
6	.2	5	.2			NORTH DAKOTA ND	-30				
1*	1.1					BARNES ND					
1*	1.6					BURLEIGH ND					
1*	4.3					DUNN ND					
		1*	3.0			EMMONS ND					
		1	.4			GRAND FORKS ND					
		1*	3.0			GRIGGS ND					
1*	1.3	1*	.9			MORTON ND	-32				
		1*	3.4			SARGENT ND					
1*	3.2					SHERIDAN ND					
1*	.8					STUTSMAN ND					
40	.1	68	.1	68	.1	OHIO OH	49	-10	34		
1*	.5			1	.4	ASHLAND OH			-6	45	91
		1	.5	1	.3	ATHENS OH		-30			89
				1	.5	AUGLAIZE OH					93
1	.2	1	.2			BELMONT OH	-23				
2	.3	1	.1			BUTLER OH	-69				
				1*	.8	CARROLL OH					95
		1	.1	2	.3	CLARK OH		73			86
		1*	.7			CLINTON OH					
				1	.1	COLUMBIANA OH					81
9	.1	14	.2	15	.2	CUYAHOGA OH	52	11	69	73	84
		1*	.5			DELAWARE OH					
				1	.3	ERIE OH					87
				1*	.8	FAYETTE OH					96
1	.0	1	.0	5	.1	FRANKLIN OH	-12	291	243	93	81
		1	.6	1	.3	GEAUGA OH		-48			88
4	.1	6	.1	10	.2	HAMILTON OH	10	72	90	77	86
1*	.7					HARDIN OH					
1	.2	1	.3	2	.4	JEFFERSON OH	6	38	47	67	90
				2	.2	LAKE OH					84
				1	.3	LAWRENCE OH					87
		3*	.6			LICKING OH					
1	.1			2	.2	LORAIN OH			21	56	82
4	.2	6	.3	3	.1	LUCAS OH	50	-54	-31	30	80
1*	1.0					MADISON OH					
1	.1	7*	.5	3	.2	MAHONING OH	283	-63	41	64	83

* SIGNIFICANTLY MORE DEATHS THAN EXPECTED

WHITE FEMALE 1950-1959 DEATHS	RATE	1960-1969 DEATHS	RATE	1970-1979 DEATHS	RATE	CANCER OF THE NASOPHARYNX COUNTY AND STATE	PERCENT CHANGE 60S/50S	70S/60S	70S/50S	ICD 147 PERCENTILE %CHANGE 70S/50S	RATES 1970S
1	.3					MARION OH					
1*	.9					MEIGS OH					
1	.3	1	.3			MIAMI OH	-14				
2	.1	5	.2	7	.3	MONTGOMERY OH	130	35	210	90	87
1	.2	1	.3	2	.4	MUSKINGUM OH	34	33	78	75	90
		1	.3			PORTAGE OH					
		1	.2	3*	.4	RICHLAND OH		173			92
		1	.3			ROSS OH					
		1	.4			SANDUSKY OH					
1	.3	2*	.5			SCIOTO OH	91				
1	.1	2	.1	3	.2	STARK OH	67	56	159	87	83
3	.1	4	.2			SUMMIT OH	12				
2*	.5					TUSCARAWAS OH					
		1*	.4			VAN WERT OH					
		1	.3			WARREN OH					
		2*	.6			WOOD OH					
9	.1	6	.0	21	.1	OKLAHOMA OK	-47	224	71		
		1*	2.6			ALFALFA OK					
1*	.8					CANADIAN OK					
				1	.3	CLEVELAND OK					87
1*	2.7					COAL OK					
				1	.3	COMANCHE OK					88
		1*	.7			CRAIG OK					
1*	.4					CREEK OK					
		1	.3			GARFIELD OK					
				1	.7	GRADY OK					94
1*	2.1					JOHNSTON OK					
1	.4			1	.3	KAY OK			-38	22	87
				1	.2	MUSKOGEE OK					86
1	.1	1	.0	2	.1	OKLAHOMA OK	-37	67	4	49	79
				1	1.0	OKMULGEE OK					97
				1	.4	OSAGE OK					90
				1	.4	PONTOTOC OK					90
				2*	.6	POTTAWATOMIE OK					94
1*	.6	1	.5			STEPHENS OK	-15				
				1*	.9	TILLMAN OK					96
2	.1	1	.1	8*	.4	TULSA OK	-62	624	176	87	90
8	.1	10	.1	17	.1	OREGON OR	-3	41	36		
		1	.1	1	.1	CLACKAMAS OR		-21			79
1*	.7					CLATSOP OR					
1*	.7	1	.4	1	.3	COOS OR	-40	-24	-54	12	89
				2*	.7	DESCHUTES OR					95
				1	.2	DOUGLAS OR					85
1	.3	1	.3			JACKSON OR	2				
				1	.5	JOSEPHINE OR					92
				1	.3	KLAMATH OR					88
		1	.1	2	.2	LANE OR		52			82
1*	.8					LINCOLN OR					
1	.4	1	.2			LINN OR	-46				
		2	.3	2	.2	MARION OR		-19			84
1	.0	3	.1	4	.1	MULTNOMAH OR	131	63	276	93	81
1*	.5					UMATILLA OR					
1	.2			2	.2	WASHINGTON OR			35	60	86
59	.1	69	.1	101	.1	PENNSYLVANIA PA	4	29	34		
				2*	.7	ADAMS PA					95
8	.1	16	.2	16	.2	ALLEGHENY PA	90	-11	69	73	83
2*	.5	1	.2			ARMSTRONG PA	-46				
1	.1			1	.1	BEAVER PA			-38	22	79
2	.1	1	.1	1	.1	BERKS PA	-53	-8	-57	10	79
1	.3					BRADFORD PA					
1	.2			3	.1	BUCKS PA			-14	38	81
2*	.4					BUTLER PA					
2	.2	2	.2	1	.1	CAMBRIA PA	-5	-60	-62	07	79
				2	.3	CENTRE PA					88
		1	.1			CHESTER PA					
2*	.5					CRAWFORD PA					
		2	.3			CUMBERLAND PA					
3	.3	3	.2	5*	.3	DAUPHIN PA	-6	28	20	54	88
		1	.0	6	.2	DELAWARE PA		471			82
1	.1	1	.1	1	.1	ERIE PA	-31	-11	-39	21	79
2	.2	1	.1	2	.2	FAYETTE PA	-44	60	-11	41	83
				1	.4	HUNTINGDON PA					90
		1	.2			INDIANA PA					
3	.2	4	.2	1	.0	LACKAWANNA PA	12	-80	-77	00	78
		1	.0	2	.1	LANCASTER PA		98			79
		1	.2	1	.2	LAWRENCE PA		2			82
				2	.4	LEBANON PA					90
1	.1	1	.1	3	.2	LEHIGH PA	31	81	138	84	84
1	.0	3	.1	6	.2	LUZERNE PA	254	95	590	98	86
2	.3	3*	.5			LYCOMING PA	42				
				1	.2	MC KEAN PA					85
				1	.1	MERCER PA					81
1	.0	4	.1	7	.2	MONTGOMERY PA	241	49	406	96	83
2	.2	1	.1	4	.3	NORTHAMPTON PA	-57	263	56	70	88
		1	.2	1	.2	NORTHUMBERLAND PA		45			86
1*	.8					PERRY PA					
12	.1	12	.1	13	.1	PHILADELPHIA PA	-1	15	14	53	81
		1*	1.3	1*	.7	PIKE PA		-48			95
2	.2	1	.1	1	.1	SCHUYLKILL PA	-62	-22	-70	04	79
2*	.5					SOMERSET PA					
				1	.4	TIOGA PA					91
1*	1.0			1*	.5	UNION PA			-49	16	93
		1	.3	2*	.6	WARREN PA		99			94
1	.1	1	.1	5*	.4	WASHINGTON PA	17	419	505	97	91
		1*	.5			WAYNE PA					

* SIGNIFICANTLY MORE DEATHS THAN EXPECTED

1950-1959 DEATHS	1950-1959 RATE	1960-1969 DEATHS	1960-1969 RATE	1970-1979 DEATHS	1970-1979 RATE	COUNTY AND STATE	PERCENT CHANGE 60S/50S	PERCENT CHANGE 70S/60S	PERCENT CHANGE 70S/50S	PERCENTILE %CHANGE 70S/50S	PERCENTILE RATES 1970S
1	.1	2	.1	5	.2	WESTMORELAND PA	123	97	339	95	85
2	.2	1	.1	2	.1	YORK PA	-50	32	-34	26	80
3	.1	5	.1	10	.2	RHODE ISLAND RI	45	98	188		
				1	.4	BRISTOL RI					90
2*	.4			1	.1	KENT RI			-72	03	81
1	.0	5	.1	8	.2	PROVIDENCE RI	376	74	727	99	86
3	.0	5	.1	21	.2	SOUTH CAROLINA SC	24	222	299		
				1*	.9	ABBEVILLE SC					96
				2*	1.4	CHESTERFIELD SC					98
				1	.5	DARLINGTON SC					92
1	.1	2	.2	1	.1	GREENVILLE SC	45	-70	-57	10	79
				1	.5	GREENWOOD SC					92
		1	.4	2*	.7	HORRY SC		86			95
				1*	.8	KERSHAW SC					96
		1*	.6			LANCASTER SC					
1*	.6					LEXINGTON SC					
		1*	5.8			MC CORMICK SC					
				1*	.7	NEWBERRY SC					95
				1	.3	OCONEE SC					89
				1	.3	PICKENS SC					87
				4*	.6	RICHLAND SC					93
1	.2			1	.1	SPARTANBURG SC			-47	17	79
				1	.4	SUMTER SC					91
				1*	.7	UNION SC					95
				2*	.5	YORK SC					92
2	.1	1	.0	4	.1	SOUTH DAKOTA SD	-57	250	49		
1*	.6	1*	.5			BROWN SD	-16				
				1*	6.1	HANSON SD					99
1*	4.0					JERAULD SD					
				1*	1.2	LINCOLN SD					98
				1*	2.3	MC COOK SD					99
				1*	.6	YANKTON SD					94
16	.1	25	.2	35	.2	TENNESSEE TN	29	11	43		
				1*	.6	BEDFORD TN					93
1*	.5					BLOUNT TN					
1*	.6	1*	.6			BRADLEY TN	-5				
		1*	.8			CAMPBELL TN					
				1*	.7	CHEATHAM TN					95
				1*	1.2	CHESTER TN					97
		1*	1.0			COCKE TN					
		1*	.6			COFFEE TN					
4*	.3	3	.2	4	.2	DAVIDSON TN	-39	18	-28	31	84
1*	.8					DYER TN					
1*	.4					GIBSON TN					
		1*	.7	1*	.7	GILES TN		5			95
				1*	1.2	GRAINGER TN					98
		1	.4			GREENE TN					
		1*	1.7			GRUNDY TN					
2	.2	2	.2			HAMILTON TN	-22				
				1	.5	HAWKINS TN					92
				1	.9	HENRY TN					96
				1*	1.3	HICKMAN TN					98
1	.1	3	.3	2	.2	KNOX TN	103	-40	21	55	82
				1*	1.0	LAUDERDALE TN					97
				3*	1.7	LAWRENCE TN					99
		2*	.7	1	.2	MADISON TN		-66			86
1*	1.1	1*	1.1			MARSHALL TN	-1				
				1	.4	MAURY TN					91
				2*	.8	MONTGOMERY TN					96
				2*	1.1	OBION TN					97
		1*	1.2			RHEA TN					
		1*	.7			ROBERTSON TN					
1*	.6					RUTHERFORD TN					
		1*	.8			SEVIER TN					
1	.1	3	.1	6	.2	SHELBY TN	147	73	327	94	86
				2	.3	SULLIVAN TN					87
		1*	1.2			UNICOI TN					
				1*	1.6	UNION TN					99
1	.4			1	.2	WASHINGTON TN			-38	22	86
				1	.6	WEAKLEY TN					94
1*	1.0					WILSON TN					
32	.1	49	.1	94	.2	TEXAS TX	27	48	88		
		1*	.9			ANDERSON TX					
				1	.5	ANGELINA TX					92
		1*	.9			AUSTIN TX					
				1*	1.3	BASTROP TX					98
1	.2					BELL TX					
3	.1	6	.2	6	.2	BEXAR TX	27	-5	20	54	82
		1	.4			BOWIE TX					
		1	.5	1	.2	BRAZORIA TX		-55			85
				1	.3	BROWN TX					89
		1*	2.7			BURLESON TX					
1	.2			1	.1	CAMERON TX			-56	11	80
				1	.3	CHEROKEE TX					87
1*	1.6					CLAY TX					
				1	.2	COLLIN TX					83
		1*	1.3	1	.7	COMAL TX		-40			95
				1*	.7	COOKE TX					95
2	.1	6	.1	13	.2	DALLAS TX	174	66	353	95	86
		1*	.8			DAWSON TX					
				1	.2	DENTON TX					85

* SIGNIFICANTLY MORE DEATHS THAN EXPECTED

WHITE FEMALE

CANCER OF THE NASOPHARYNX

ICD 147

1950-1959 DEATHS	1950-1959 RATE	1960-1969 DEATHS	1960-1969 RATE	1970-1979 DEATHS	1970-1979 RATE	COUNTY AND STATE	PERCENT CHANGE 60S/50S	PERCENT CHANGE 70S/60S	PERCENT CHANGE 70S/50S	PERCENTILE %CHANGE 70S/50S	PERCENTILE RATES 1970S
		1*	.6			EASTLAND TX					
				1	.3	ECTOR TX					87
2*	.9					ELLIS TX					
1	.1	2	.2	4	.3	EL PASO TX	50	50	125	82	89
				1*	1.9	GAINES TX					99
2*	.5	2	.4	3	.4	GALVESTON TX	-24	-1	-24	32	90
1*	3.3					GOLIAD TX					
		1	.4			GREGG TX					
		1*	2.3			GRIMES TX					
3	.1	5	.1	13	.2	HARRIS TX	37	113	192	89	84
				1*	1.3	HOOD TX					98
1*	.8			2*	1.1	HOWARD TX			38	63	97
		1	.6			HUNT TX					
1	.1	1	.1	2	.2	JEFFERSON TX	-7	92	79	75	83
1*	1.0			2*	1.3	JONES TX			28	58	98
				1*	1.9	KARNES TX					99
4*	1.0			2	.3	LUBBOCK TX			-72	03	87
1	.2	1	.1	1	.1	MC LENNAN TX	-31	-25	-48	16	79
				1*	1.2	MARION TX					98
		1	.3			MIDLAND TX					
		2*	1.5	2*	.6	MONTGOMERY TX		-63			93
		1*	1.0			NOLAN TX					
1	.1	2	.2	1	.1	NUECES TX	102	-46	9	51	81
				1*	10.3	OLDHAM TX					99
				3*	1.1	ORANGE TX					97
				1	.4	PARKER TX					92
				1	.3	RANDALL TX					88
				1*	1.9	REEVES TX					99
				1	.4	RUSK TX					90
		1*	.5	1	.4	SAN PATRICIO TX		-9			91
				1*	1.1	SCURRY TX					97
1	.3					SMITH TX					
1*	2.3					STARR TX					
3	.2	2	.1	9	.3	TARRANT TX	-50	190	44	66	87
				3*	.6	TAYLOR TX					94
		1*	1.1			TITUS TX					
		1	.1	1	.1	TRAVIS TX		-12			79
		1*	.8			UPSHUR TX					
				1*	1.0	UVALDE TX					97
				1*	.9	VAL VERDE TX					96
1*	.7					VAN ZANDT TX					
		1*	.4			VICTORIA TX					
				1	.3	WEBB TX					88
		1	.5	1	.3	WILLIAMSON TX		-45			88
		1*	1.1			WISE TX					
				1*	.4	YOUNG TX					91
1	.0	1	.0	7	.1	UTAH UT	-34	395	229		
				1	.2	DAVIS UT					83
				1*	4.3	KANE UT					99
1	.1	1	.1	4	.2	SALT LAKE UT	-34	221	112	80	83
				1	.1	WEBER UT					81
4	.2	3	.1	4	.1	VERMONT VT	-42	26	-27		
1*	.7			1	.4	BENNINGTON VT			-36	24	91
2*	.7			1	.1	CHITTENDEN VT			-83	00	80
1*	3.6					ESSEX VT					
		1*	.5			FRANKLIN VT					
		1*	1.3			LAMOILLE VT					
				1	.4	RUTLAND VT					91
				1	.5	WINDHAM VT					93
		1	.4			WINDSOR VT					
8	.1	18	.1	26	.1	VIRGINIA VA	92	4	100		
		1*	.6	1	.6	ACCOMACK VA		2			94
				1	.3	ALBEMARLE#VA					89
		1	.1	4*	.4	ARLINGTON#VA		192			91
				1*	3.5	BATH VA					99
1	.2	3*	.6	1	.1	BEDFORD#VA	149	-75	-37	23	81
1*	2.9					BUCKINGHAM VA					
				1	.1	CHESTERFIELD#VA					79
				1*	.6	CULPEPER VA					94
1	.3	1	.3			DINWIDDIE#VA	21				
1	.1	1	.1	1	.0	FAIRFAX#VA	64	-79	-66	05	78
1*	.6					FREDERICK#VA					
				1*	1.2	GILES VA					98
		3*	.6	3	.4	HENRICO VA		-37			90
		1*	.6			JAMES CITY#VA					
		1*	1.0			PAGE VA					
1	.3	1	.3			PITTSYLVANIA#VA	-12				
1*	.7			1	.6	SMYTH VA			-13	38	94
				1*	1.2	SOUTHAMPTON#VA					98
				1	.5	SPOTSYLVANIA#VA					93
		1*	.9			WYTHE VA					
1	.1	4	.2	3	.1	CHESAPEAKE CITY#VA	224	-43	84	77	81
				2*	.6	HAMPTON CITY VA					94
				3*	.7	NEWPORT NEWS CITY VA					95
12	.1	22	.1	34	.2	WASHINGTON WA	48	21	80		
		1	.4	2*	.5	BENTON WA		24			93
		1	.4			CHELAN WA					
		1*	.4			CLALLAM WA					
		1	.2			CLARK WA					
		1	.3	2*	.5	COWLITZ WA		52			92
				1	.2	GRAYS HARBOR WA					83
				1	.5	ISLAND WA					92
5	.1	4	.1	10	.2	KING WA	-33	101	35	59	82

* SIGNIFICANTLY MORE DEATHS THAN EXPECTED

COUNTIES COMBINED; SEE TECHNICAL NOTES

WHITE FEMALE — CANCER OF THE NASOPHARYNX — ICD 147

1950-1959 DEATHS	1950-1959 RATE	1960-1969 DEATHS	1960-1969 RATE	1970-1979 DEATHS	1970-1979 RATE	COUNTY AND STATE	PERCENT CHANGE 60S/50S	PERCENT CHANGE 70S/60S	PERCENT CHANGE 70S/50S	PERCENTILE %CHANGE 70S/50S	PERCENTILE RATES 1970S
		2*	.5	1	.1	KITSAP WA		-70			81
				1*	.9	KITTITAS WA					97
				1	.3	LEWIS WA					87
2	.1	1	.1	7*	.3	PIERCE WA	-52	406	144	85	89
		3*	1.1	1	.3	SKAGIT WA		-75			87
2	.4	1	.1	2	.2	SNOHCMISH WA	-65	30	-54	12	82
2	.1	2	.1	2	.1	SPOKANE WA	-15	5	-11	41	81
				1*	.9	STEVENS WA					96
		1	.3			THURSTON WA					
1	.2	2*	.5	2	.6	WHATCOM WA	135	4	144	85	93
		1	.1			YAKIMA WA					
6	.1	10	.1	18	.2	WEST VIRGINIA WV	55	52	136		
				1	.5	BERKELEY WV					92
				1*	.6	BOONE WV					94
		1	.1	2	.2	CABELL WV		78			85
				1*	1.7	GRANT WV					99
				1	.4	GREENBRIER WV					90
1*	.5					HANCOCK WV					
		1	.2			HARRISON WV					
		3	.3	2	.2	KANAWHA WV		-42			82
				1*	.6	LEWIS WV					94
				1*	.9	LINCOLN WV					96
				1	.5	MC DOWELL WV					92
1	.3	2*	.5	1	.2	MARION WV	70	-67	-44	18	82
2*	1.2					MARSHALL WV					
		1	.3	1	.2	MERCER WV		-16			86
1*	.7					MINERAL WV					
		1	.4	1	.3	MONONGALIA WV		-22			88
				1	.5	PRESTON WV					93
				1	.2	RALEIGH WV					85
1*	.7			1*	.7	RANDOLPH WV			-12	40	94
		1*	.9			WETZEL WV					
				1	.2	WOOD WV					84
20	.1	23	.1	23	.1	WISCONSIN WI	2	-17	-15		
		1*	1.4			BAYFIELD WI					
				1	.2	BROWN WI					82
1*	1.4					BUFFALO WI					
				1*	1.1	BURNETT WI					97
		1	.3			CHIPPEWA WI					
		1*	.6			CLARK WI					
1	.1			1	.1	DANE WI			-28	32	79
		1	.4			DOUGLAS WI					
		1	.2			FOND DU LAC WI					
		1	.5			GRANT WI					
		1*	.8			GREEN WI					
1	.4					JEFFERSON WI					
		1*	1.0			JUNEAU WI					
1	.2	1	.2			KENOSHA WI	-17				
		1	.2			LA CROSSE WI					
		1	.2	1	.2	MANITOWOC WI		1			85
				1	.2	MENOMINEE#WI					85
8	.2	3	.1	12	.2	MILWAUKEE WI	-66	274	26	58	84
		1*	.6			ONEIDA WI					
2*	.5					OUTAGAMIE WI					
1	.2	2	.2	2	.2	RACINE WI	53	-11	36	61	85
1	.2	1	.2	1	.2	ROCK WI	-1	-19	-20	35	82
		1	.4	1	.5	SAUK WI		32			92
1*	.7	1*	.8			TREMPEALEAU WI	3				
2*	1.3					VERNON WI					
		1	.3			WALWORTH WI					
		1*	1.3			WASHBURN WI					
1*	.5					WASHINGTON WI					
				1	.1	WAUKESHA WI					80
		1	.2	1	.1	WINNEBAGO WI		-38			80
1	.1	1	.1	1	.1	WYOMING WY	-15	-21	-33		
				1*	.8	SHERIDAN WY					96
1*	1.5	1*	1.4			SWEETWATER WY	-7				

* SIGNIFICANTLY MORE DEATHS THAN EXPECTED

\# COUNTIES COMBINED; SEE TECHNICAL NOTES

NON-WHITE MALE — CANCER OF THE NASOPHARYNX — ICD 147

1950-1959 DEATHS	1950-1959 RATE	1960-1969 DEATHS	1960-1969 RATE	1970-1979 DEATHS	1970-1979 RATE	COUNTY AND STATE	PERCENT CHANGE 60S/50S	PERCENT CHANGE 70S/60S	PERCENT CHANGE 70S/50S	PERCENTILE %CHANGE 70S/50S	PERCENTILE RATES 1970S
358	.5	549	.6	724	.6	UNITED STATES	30	2	33		
10	.3	14	.4	14	.4	ALABAMA AL	31	1	32		
				1*	4.7	AUTAUGA AL					98
		1*	1.9	1*	5.2	BULLOCK AL		180			99
1*	3.7					BUTLER AL					
				1*	5.8	CHILTON AL					99
				1*	2.3	CHOCTAW AL					97
		1	.9			DALLAS AL					
				1	2.0	ESCAMBIA AL					96
		1	1.6			ETOWAH AL					
				1	2.8	HALE AL					97
5	.6	5	.6	1	.1	JEFFERSON AL	0	-78	-78	05	91
		1*	1.3			LOWNDES AL					
		1	1.2	2	2.5	MACON AL		107			97
1	1.8					MADISON AL					
1	1.7					MARENGO AL					
1	.4	2	.5	1	.3	MOBILE AL	29	-48	-33	34	91
		1*	2.8			MONROE AL					
1	.6			1	.3	MONTGOMERY AL			-43	30	92
		1*	4.4	2*	4.1	PICKENS AL		-7			98
				1*	3.8	PIKE AL					98
		8*	5.3	8*	3.5	ALASKA AK		-33			
3	.9	3	.6	6	.5	ARIZONA AZ	-29	-17	-41		
1	1.7			1	1.3	APACHE AZ			-26	44	95
		1*	2.9			COCONINO AZ					
				1*	2.6	GILA AZ					97
		2	1.2	1	.3	MARICOPA AZ		-78			91
1	2.0					NAVAJO AZ					
1	2.5					PIMA AZ					
				3*	3.2	PINAL AZ					98
2	.1	8	.4	8	.6	ARKANSAS AR	301	34	435		
				1*	5.1	ARKANSAS AR					99
				2*	6.5	ASHLEY AR					99
		1	1.7	1	1.5	CHICOT AR		-9			95
		1	1.6			DESHA AR					
				1*	4.8	HEMPSTEAD AR					98
		1	.5			JEFFERSON AR					
				1*	9.8	LONOKE AR					99
1*	2.1					MILLER AR					
		2	1.4			PHILLIPS AR					
1	.5			1	.5	PULASKI AR			-5	55	92
		2*	3.0			ST FRANCIS AR					
		1*	4.3			SEBASTIAN AR					
				1*	11.5	WHITE AR					99
67*	1.5	95*	1.4	137*	1.1	CALIFORNIA CA	-4	-21	-24		
8*	1.4	6	.7	14*	1.3	ALAMEDA CA	-46	69	-8	52	95
				1*	3.9	BUTTE CA					98
		1	.5	2	.5	CONTRA COSTA CA		1			92
2*	1.7	1	.7	3	1.2	FRESNO CA	-56	56	-31	43	95
		1	.9	1	.7	KERN CA		-27			93
1*	6.7			1	1.8	KINGS CA			-73	08	96
12	.6	32*	1.2	46	.9	LOS ANGELES CA	108	-24	59	73	94
				1	2.4	MARIN CA					97
1	1.4	4*	3.9	1	.4	MONTEREY CA	186	-89	-68	09	92
				1	.0	ORANGE CA					91
1	.9	5*	2.1	3	.5	SACRAMENTO CA	141	-74	-37	31	92
		1	1.0			SAN BERNARDINO CA					
1	.5	3	1.5	6	.8	SAN DIEGO CA	171	-46	46	72	94
34*	5.9	32*	4.0	41*	3.9	SAN FRANCISCO CA	-32	-3	-35	33	98
2	1.3	4*	2.2	2	.9	SAN JOAQUIN CA	67	-60	-33	36	94
1	1.0	2	1.5	4	1.6	SAN MATEO CA	59	7	70	77	96
				1	2.0	SANTA BARBARA CA					96
4*	5.4	1	.5	5	1.1	SANTA CLARA CA	-91	120	-79	04	94
		1*	4.2			SANTA CRUZ CA					
				1	.5	SOLANO CA					92
				2	1.3	TULARE CA					95
		1*	54.6			TUOLUMNE CA					
				1	.4	VENTURA CA					92
1	.4			1	.1	COLORADO CO			-81		
1	.6					DENVER-ARAPAHOE#CO					
				1	.7	EL PASO CO					93
2	.7	1	.1	5	.9	CONNECTICUT CT	-83	635	23		
1	.8			4*	1.9	FAIRFIELD CT			135	90	96
1	1.7					HARTFORD CT					
				1	.8	NEW HAVEN CT					94
		1*	1.9			NEW LONDON CT					
1	.4	1	.3	2	.5	DELAWARE DE	-10	48	33		
		1	.5	2	.8	NEW CASTLE DE		48			94
1	1.5					SUSSEX DE					
15*	1.2	7	.5	14	.7	DIST. OF COLUMBIA DC	-58	27	-47		
15*	1.2	7	.5	14	.7	WASHINGTON DC	-58	27	-47	22	93

* SIGNIFICANTLY MORE DEATHS THAN EXPECTED

\# COUNTIES COMBINED; SEE TECHNICAL NOTES

NON-WHITE MALE — CANCER OF THE NASOPHARYNX — ICD 147

1950-1959 DEATHS	1950-1959 RATE	1960-1969 DEATHS	1960-1969 RATE	1970-1979 DEATHS	1970-1979 RATE	COUNTY AND STATE	PERCENT CHANGE 60S/50S	PERCENT CHANGE 70S/60S	PERCENT CHANGE 70S/50S	PERCENTILE %CHANGE 70S/50S	PERCENTILE RATES 1970S
7	.2	19	.5	33	.7	FLORIDA FL	108	38	186		
				1	1.2	ALACHUA FL					95
1*	3.1			1	2.5	BAY FL			-20	47	97
1*	2.2			1	1.2	BREVARD FL			-43	29	95
		1	.4	4	1.1	BROWARD FL		153			94
1	.2	3	.5	11*	1.1	DADE FL	178	129	538	98	94
		1*	8.7			DE SOTO FL					
1	.2	5	1.1			DUVAL FL	561				
		1	.7	1	.6	ESCAMBIA FL		-22			93
		1	.7			GADSDEN FL					
				1*	8.6	HERNANDO FL					99
		1	.6	3	1.0	HILLSBOROUGH FL		77			94
				1*	2.7	INDIAN RIVER FL					97
1*	2.1			1	2.1	LAKE FL			1	58	96
		1	.4	1	.4	PALM BEACH FL		-16			92
		2*	11.9			PASCO FL					
				1	.6	PINELLAS FL					93
1	1.0			1	.5	POLK FL			-47	23	92
				2*	3.0	SEMINOLE FL					97
		1*	4.6			SUWANNEE FL					
		2*	2.3	3*	3.6	VOLUSIA FL		57			98
1*	11.6					WASHINGTON FL					
12	.3	9	.2	20	.4	GEORGIA GA	-32	112	44		
				1*	2.6	BUTTS GA					97
1*	8.3					CANDLER GA					
1	.4	1	.4	1	.4	CHATHAM GA	-8	15	6	59	92
		1*	3.6			COLQUITT GA					
		1*	2.7			COWETA GA					
				1*	2.7	DECATUR GA					97
				1	.9	DE KALB GA					94
				1	1.0	DOUGHERTY GA					94
1*	4.9					EMANUEL GA					
1*	2.3					FLOYD GA					
2	.4	1	.1	7	.8	FULTON GA	-76	714	97	83	94
1*	1.6			1	1.8	GLYNN GA			10	61	96
				1*	5.3	HALL GA					99
		1*	4.3			HARRIS GA					
				1	1.4	LOWNDES GA					95
				1*	4.0	MONROE GA					98
				1*	5.3	MORGAN GA					99
		2	1.1			MUSCOGEE#GA					
2	1.2			1	.6	RICHMOND GA			-52	18	93
				1	2.2	SUMTER GA					96
1	1.6					THOMAS GA					
		1*	4.4			TIFT GA					
				1	1.4	TROUP GA					95
		1*	9.1			WALKER GA					
1*	5.9					WAYNE GA					
1*	16.1					WHITFIELD GA					
		45*	2.2	45*	1.8	HAWAII HI		-18			
		6*	2.2	3	1.0	HAWAII HI		-53			94
		35*	2.4	42*	2.3	HONOLULU HI		-4			97
		3*	1.7			KAUAI HI					
		1	.4			MAUI#HI					
11	.3	21	.4	40	.6	ILLINOIS IL	29	57	102		
				1	3.0	CHAMPAIGN IL					97
11	.4	16	.4	33	.6	COOK IL	-9	78	61	75	93
		1*	3.5			KANE IL					
				1	1.5	KANKAKEE IL					95
		1*	9.4			KNOX IL					
		1	3.0			LAKE IL					
				2*	6.1	MACON IL					99
		1	.5			ST CLAIR IL					
		1	1.1	1	1.6	WILL IL		41			95
				2*	2.2	WINNEBAGO IL					97
3	.3	6	.5	7	.4	INDIANA IN	68	-11	50		
				1	1.3	ALLEN IN					95
				1*	3.4	DELAWARE IN					98
				1*	10.7	FLOYD IN					99
1	.3	2	.6	2	.5	LAKE IN	75	-18	43	69	92
1	.3	4	.8	1	.2	MARION IN	230	-79	-31	41	91
				1*	6.2	TIPPECANOE IN					99
1*	2.0					VANDERBURGH IN					
		4	1.0	5	.9	KANSAS KS		-5			
		1*	20.2	1*	5.2	GEARY KS		-74			99
		1	1.2	2	1.4	SEDGWICK KS		23			95
				2*	3.7	SHAWNEE KS					98
		2	1.5			WYANDOTTE KS					
4	.4	6	.6	4	.4	KENTUCKY KY	50	-32	2		
				1*	38.3	BATH KY					99
3	.9	4	1.0	1	.2	JEFFERSON KY	9	-76	-74	06	91
		1*	6.3			KENTON KY					
		1*	34.4			LARUE KY					
1*	3.0					MC CRACKEN KY					
				1*	17.5	MARION KY					99
				1*	10.6	TRIGG KY					99

* SIGNIFICANTLY MORE DEATHS THAN EXPECTED

COUNTIES COMBINED; SEE TECHNICAL NOTES

NON-WHITE MALE — CANCER OF THE NASOPHARYNX — ICD 147

1950-1959 DEATHS	1950-1959 RATE	1960-1969 DEATHS	1960-1969 RATE	1970-1979 DEATHS	1970-1979 RATE	COUNTY AND STATE	PERCENT CHANGE 60S/50S	PERCENT CHANGE 70S/60S	PERCENT CHANGE 70S/50S	PERCENTILE %CHANGE 70S/50S	PERCENTILE RATES 1970S
15	.4	17	.4	19	.4	LOUISIANA LA	7	5	12		
				1*	3.0	ACADIA LA					97
		1*	5.1	1*	5.0	ALLEN LA		-1			98
		1	1.3	1	1.5	BOSSIER LA		12			95
		1	.3	2	.4	CADDO LA		12			92
		1	.5			CALCASIEU LA					
				1	3.1	CLAIBORNE LA					97
				1*	3.1	CONCORDIA LA					98
1	.5	1	.3			EAST BATON ROUGE LA	-35				
				1*	2.2	FRANKLIN LA					96
1*	5.0					GRANT LA					
				2*	3.5	IBERIA LA					98
1	2.0					IBERVILLE LA					
1	1.1	1	1.4			LAFAYETTE LA	28				
				1*	1.6	LAFOURCHE LA					95
		1	1.7			MOREHOUSE LA					
		1	2.1			NATCHITOCHES LA					
7	.9	5	.6	3	.3	ORLEANS LA	-34	-46	-65	13	91
				1	.6	OUACHITA LA					93
		1	.5	1	.7	RAPIDES LA		35			93
		1*	2.9			ST MARTIN LA					
1	1.6					ST MARY LA					
				1	1.7	TANGIPAHOA LA					96
				1*	6.0	TENSAS LA					99
		2*	5.3			TERREBONNE LA					
2*	4.0					WASHINGTON LA					
1*	4.4					WEST CARROLL LA					
				1*	4.0	WEST FELICIANA LA					98
		1*	6.4			MAINE ME					
		1*	35.1			CUMBERLAND ME					
13	.7	6	.3	14	.4	MARYLAND MD	-65	70	-40		
1	.9					ANNE ARUNDEL MD					
1	1.3					BALTIMORE MD					
				2*	3.5	CHARLES MD					98
1*	2.5					DORCHESTER MD					
				1	3.1	HARFORD MD					98
1*	5.0					HOWARD MD					
1*	1.4					MONTGOMERY MD					
				1	1.8	WICOMICO MD					96
8	.8	6	.4	10	.6	BALTIMORE CITY MD	-50	37	-32	38	93
6*	1.4	10*	1.8	14*	1.9	MASSACHUSETTS MA	34	4	39		
				1	2.3	BRISTOL MA					97
		1*	4.7	2*	6.7	ESSEX MA		44			99
				1	1.1	HAMPDEN MA					94
2*	3.2			2	2.5	MIDDLESEX MA			-24	45	97
1*	13.8					NORFOLK MA					
1*	5.5					PLYMOUTH MA					
2	.8	9*	3.1	8*	2.1	SUFFOLK MA	275	-31	157	91	96
11	.5	13	.4	18	.4	MICHIGAN MI	-14	1	-13		
				1	1.4	BERRIEN MI					95
				1	.6	INGHAM MI					93
1*	3.8	1	1.4			KENT MI	-64				
1*	4.2					MUSKEGON MI					
1*	11.8					ST CLAIR MI					
8	.4	12	.5	16	.5	WAYNE MI	36	1	38	68	92
		1	.7	1	.3	MINNESOTA MN		-59			
				1*	9.2	DAKOTA MN					99
		1	1.9			HENNEPIN MN					
10	.3	12	.4	22	.7	MISSISSIPPI MS	31	98	160		
1	1.7			1	1.7	ADAMS MS			0	56	96
		1*	3.1			AMITE MS					
		1	1.0	2	1.6	BOLIVAR MS		64			95
		1*	4.4			CARROLL MS					
1*	2.6					CLAY MS					
		1	1.8			FORREST MS					
2	.6			4	1.4	HINDS MS			134	87	95
				1	2.0	JACKSON MS					96
				1*	6.2	JEFFERSON DAVIS MS					99
				1	1.4	JONES MS					95
		1	1.4	2*	2.4	LAUDERDALE MS		74			97
				1	.6	LEFLORE MS					93
2*	3.9	1	2.1	1	1.9	LOWNDES MS	-47	-9	-52	19	96
				1*	3.0	MARION MS					97
				1	2.0	MARSHALL MS					96
				1*	4.6	NESHOBA MS					98
1	2.2					PIKE MS					
		1	1.8			QUITMAN MS					
1	1.5			1	1.8	RANKIN MS			19	63	96
		1*	3.1			SCOTT MS					
1*	2.4					SHARKEY MS					
				1*	4.6	SIMPSON MS					98
		1*	5.0			SMITH MS					
		1	1.1			SUNFLOWER MS					
1	1.5	1	1.1			TALLAHATCHIE MS	-28				
		1	2.0	1	.9	WARREN MS		-52			94
				2	1.5	WASHINGTON MS					95

* SIGNIFICANTLY MORE DEATHS THAN EXPECTED

NON-WHITE MALE — CANCER OF THE NASOPHARYNX — ICD 147

1950-1959 DEATHS	1950-1959 RATE	1960-1969 DEATHS	1960-1969 RATE	1970-1979 DEATHS	1970-1979 RATE	COUNTY AND STATE	PERCENT CHANGE 60S/50S	PERCENT CHANGE 70S/60S	PERCENT CHANGE 70S/50S	PERCENTILE %CHANGE 70S/50S	PERCENTILE RATES 1970S
6	.4	7	.4	9	.4	MISSOURI MO	3	15	19		
1*	6.0					BOONE MO					
		1*	13.2			DUNKLIN MO					
1	.4			1	.1	JACKSON MO			-66	11	91
1*	149.1					LINN MO					
				1*	6.0	PEMISCOT MO					99
		1	.9	1	.4	ST LOUIS COUNTY MO		-59			92
3	.3	5	.6	6	.7	ST LOUIS CITY MO	61	18	89	81	93
				1	1.3	MONTANA MT					
				1*	44.6	LEWIS AND CLARK MT					99
1	1.4	1	.8			NEBRASKA NB	-48				
		1	1.0			DOUGLAS NB					
1*	112.4					KEITH NB					
				2	1.3	NEVADA NV					
				2	2.0	CLARK NV					96
2	.1	16	.7	19	.5	NEW JERSEY NJ	869	-22	658		
		1	.7	1	.5	ATLANTIC NJ		-19			93
				1	.7	BERGEN NJ					93
				1	.4	CAMDEN NJ					92
		7	.8	8	.7	ESSEX NJ		-20			93
		1	2.0	1	1.4	GLOUCESTER NJ		-29			95
1	.3			1	.5	HUDSON NJ			76	79	92
		3*	1.8	1	.3	MERCER NJ		-81			92
				1	.3	MIDDLESEX NJ					92
1	.8	1	.6			MONMOUTH NJ	-25				
				1	.6	MORRIS NJ					93
				2	2.3	PASSAIC NJ					97
		1	2.4			SALEM NJ					
		2	1.6	1	.3	UNION NJ		-82			91
1	.3	1	.5			NEW MEXICO NM	45				
1*	663.2					TORRANCE NM					
		1*	5.7			VALENCIA NM					
73*	1.6	90*	1.4	92*	.9	NEW YORK NY	-12	-35	-43		
		1	1.6			ALBANY NY					
12*	3.0	15*	1.6	9	.6	BRONX NY	-49	-64	-82	01	93
				1*	9.1	CHEMUNG NY					99
		4	1.3	7*	1.6	ERIE NY		22			96
14*	1.6	14	.9	19	.7	KINGS(BROOKLYN) NY	-46	-16	-55	15	93
				1	.4	MONROE NY					92
2*	1.4	4*	1.3	1	.3	NASSAU NY	-4	-79	-79	02	91
43*	2.3	39*	2.0	36*	1.5	NEW YORK(MANHATTAN) NY	-12	-22	-32	40	95
1	.2	4	.7	12	.8	QUEENS NY	290	26	392	97	94
1*	9.3	1*	4.3	1*	6.2	RENSSELAER NY	-54	45	-33	37	99
				1	.7	RICHMOND NY					93
				1	1.1	ROCKLAND NY					94
		2	.9	1	.4	SUFFOLK NY		-52			92
		1*	11.3			WAYNE NY					
		5*	1.8	2	.5	WESTCHESTER NY		-71			92
9	.2	16	.3	18	.4	NORTH CAROLINA NC	53	19	83		
				1	1.5	BEAUFORT NC					95
1*	2.5					BLADEN NC					
		1	2.3			CABARRUS NC					
				1*	4.5	CARTERET NC					98
				1*	2.9	CASWELL NC					97
1*	3.9					CATAWBA NC					
		1*	2.4			CHATHAM NC					
		3*	1.9	2	1.0	CUMBERLAND NC		-48			94
1	.6	2	1.6	2	1.4	DURHAM NC	159	-11	130	86	95
1	.9	1	1.3			EDGECOMBE NC	52				
1	.7	1	.5	1	.4	GUILFORD NC	-36	-11	-43	27	92
				1	1.0	HALIFAX NC					94
				1	2.6	HERTFORD NC					97
				1	2.1	IREDELL NC					96
1	.3	1	.3	1	.2	MECKLENBURG NC	-7	-11	-18	48	91
		1	.7			NASH NC					
1	1.0					NORTHAMPTON NC					
		1*	1.6			ORANGE NC					
				1	2.8	PASQUOTANK NC					97
		2*	10.4			PERQUIMANS NC					
1	1.1			2	2.1	PITT NC			81	80	96
1	.8			1	.7	ROBESON NC			-10	51	93
				2*	3.9	UNION NC					98
		1	.3			WAKE NC					
		1	1.2			WILSON NC					
8	.4	18	.5	20	.5	OHIO OH	49	-14	28		
				1*	2.1	ALLEN OH					96
		1	2.2			BUTLER OH					
				1	1.8	CLARK OH					96
3	.4	8	.9	9	.6	CUYAHOGA OH	109	-31	44	70	93
1	.4	1	.2	2	.6	FRANKLIN OH	-44	133	30	66	93
		1*	3.5	1*	3.1	GREENE OH		-13			97
2	.6	3	.5	2	.3	HAMILTON OH	-4	-43	-45	26	92
1*	2.9					LORAIN OH					

* SIGNIFICANTLY MORE DEATHS THAN EXPECTED

NON-WHITE MALE — CANCER OF THE NASOPHARYNX — ICD 147

1950-1959 DEATHS	1950-1959 RATE	1960-1969 DEATHS	1960-1969 RATE	1970-1979 DEATHS	1970-1979 RATE	COUNTY AND STATE	PERCENT CHANGE 60S/50S	PERCENT CHANGE 70S/60S	PERCENT CHANGE 70S/50S	PERCENTILE %CHANGE 70S/50S	PERCENTILE RATES 1970S
		2	1.0			LUCAS OH					
1	1.0			1	.6	MAHONING OH			-46	25	93
		1	.4	2	.5	MONTGOMERY OH		24			92
		1	1.2			STARK OH					
				1	.3	SUMMIT OH					92
3	.4	2	.2	1	.1	OKLAHOMA OK	-49	-60	-80		
		1*	6.6			CHEROKEE OK					
1*	3.0					MC CURTAIN OK					
		1*	7.6			OSAGE OK					
1*	4.9					PITTSBURG OK					
				1*	6.8	POTTAWATOMIE OK					99
1	1.0					TULSA OK					
		3	1.5	3	1.2	OREGON OR		-21			
		1*	54.6			CLATSOP OR					
		2	1.6	3*	1.9	MULTNOMAH OR		22			96
10	.3	15	.4	23	.5	PENNSYLVANIA PA	22	37	67		
6	.9	1	.2	3	.5	ALLEGHENY PA	-83	193	-50	20	92
				1	1.8	BEAVER PA					96
				1	3.9	BUCKS PA					98
		1	.7	2	2.0	CHESTER PA		187			96
				1	.5	DELAWARE PA					92
		2*	8.4	1*	3.3	FAYETTE PA		-60			98
		1*	5.7			LANCASTER PA					
				1*	4.5	LEHIGH PA					98
				2	1.4	MONTGOMERY PA					95
1*	10.2					NORTHAMPTON PA					
3	.2	9	.4	11	.4	PHILADELPHIA PA	112	7	127	84	92
		1*	5.5			YORK PA					
				1	.6	RHODE ISLAND RI					
				1	.7	PROVIDENCE RI					93
5	.2	11	.4	16	.6	SOUTH CAROLINA SC	125	44	224		
2*	6.0	1*	5.3			ALLENDALE SC	-12				
				1	1.4	ANDERSON SC					95
				1*	2.9	BARNWELL SC					97
		1	1.9	1	1.0	BEAUFORT SC		-47			94
1	.3			2	.7	CHARLESTON SC			174	94	93
				1*	4.1	CHEROKEE SC					98
				1	1.9	CLARENDON SC					96
				1	2.1	COLLETON SC					96
		1*	2.8			DILLON SC					
		2	1.5			FLORENCE SC					
		1	1.9			GEORGETOWN SC					
		1	.7			GREENVILLE SC					
1*	3.2	1*	2.5			LEXINGTON SC	-22				
				1	2.4	MARLBORO SC					97
		1	.6	2	1.7	ORANGEBURG SC		165			96
				1*	4.5	PICKENS SC					98
1	.6	1	.3	1	.5	RICHLAND SC	-48	67	-12	50	92
				1	.8	SPARTANBURG SC					94
		1*	3.8	1*	3.3	UNION SC		-12			98
				1	1.1	YORK SC					94
8	.3	11	.4	13	.5	TENNESSEE TN	29	21	57		
		1*	133.9			BLEDSOE TN					
		1	.3	1	.3	DAVIDSON TN		-9			91
		1*	14.5			FRANKLIN TN					
		1*	4.6			GILES TN					
				1*	9.9	HAMBLEN TN					99
1	.4					HAMILTON TN					
				1	1.0	KNOX TN					94
		1	.8			MADISON TN					
				1*	5.8	OBION TN					99
		1*	5.1			ROBERTSON TN					
		1*	3.1			RUTHERFORD TN					
6	.7	4	.4	8	.8	SHELBY TN	-47	115	13	62	94
				1*	5.5	SUMNER TN					99
1*	2.1					TIPTON TN					
22	.5	27	.5	32	.4	TEXAS TX	15	-16	-4		
1*	6.5					AUSTIN TX					
2*	4.8					BELL TX					
1	.6	3	1.6	5	1.5	BEXAR TX	176	-3	167	93	95
1	1.4					BOWIE TX					
		1*	27.1			CALHOUN TX					
1*	7.6					CAMP TX					
				1	2.5	CHEROKEE TX					97
3	.7	1	.2	5	.6	DALLAS TX	-78	314	-8	54	93
1*	5.5					EL PASO TX					
		1*	2.0			FALLS TX					
				2*	18.8	FAYETTE TX					99
				1	1.6	FORT BEND TX					96
		1	.7	1	.6	GALVESTON TX		-12			93
9*	1.0	9	1.1	7	.5	HARRIS TX	5	-57	-54	16	92
		1*	4.0			HUNT TX					
		2	.8	1	.2	JEFFERSON TX		-69			91
				1*	2.3	KAUFMAN TX					97
1*	4.2	1*	3.3			LIMESTONE TX	-22				
				1	.5	LUBBOCK TX					92
		1	1.1	1	.8	MC LENNAN TX		-21			94

* SIGNIFICANTLY MORE DEATHS THAN EXPECTED

NON-WHITE MALE — CANCER OF THE NASOPHARYNX — ICD 147

1950-1959 DEATHS	1950-1959 RATE	1960-1969 DEATHS	1960-1969 RATE	1970-1979 DEATHS	1970-1979 RATE	COUNTY AND STATE	PERCENT CHANGE 60S/50S	PERCENT CHANGE 70S/60S	PERCENT CHANGE 70S/50S	PERCENTILE %CHANGE 70S/50S	PERCENTILE RATES 1970S
		1*	4.5			MATAGORDA TX					
1*	4.4					MIDLAND TX					
1*	4.7					POLK TX					
		1*	25.8			REFUGIO TX					
				1	1.5	RUSK TX					95
		1	.5	2	.6	TARRANT TX		30			93
		2	1.3	2	1.1	TRAVIS TX		-18			94
				1*	4.0	WASHINGTON TX					98
		1*	5.1			WILLIAMSON TX					
1	2.2			1	.9	UTAH UT			-61		
1*	5.4					SALT LAKE UT					
				1*	4.5	WEBER UT					98
9	.3	16	.5	20	.5	VIRGINIA VA	78	2	82		
				1	1.6	ARLINGTON#VA					95
		1*	5.0			AUGUSTA#VA					
		1	.8	1	1.1	BEDFORD#VA		28			94
1*	7.4					CHARLES CITY VA					
				1*	5.1	CHARLOTTE VA					99
1	.2	4	1.0	4	.8	CHESTERFIELD#VA	322	-26	213	95	94
				1	.6	DINWIDDIE#VA					93
				1	.2	FAIRFAX#VA					91
1*	2.4			1*	3.1	GREENSVILLE#VA			29	65	98
				1	1.4	HALIFAX#VA					95
		1*	2.6			HENRICO VA					
1*	2.5					ISLE OF WIGHT VA					
		1*	6.0			KING WILLIAM VA					
		1*	5.0			LOUISA VA					
1*	6.9					ORANGE VA					
1*	5.6					PRINCE WILLIAM#VA					
		1*	15.1			RICHMOND VA					
3	.5	4	.6	6	.8	CHESAPEAKE CITY#VA	28	27	62	76	94
		1	.7	1	1.2	HAMPTON CITY VA		70			95
				1	.3	NEWPORT NEWS CITY VA					91
		1	1.1	1	1.0	SUFFOLK CITY VA		-14			94
3	1.0	3	.6	12*	1.6	WASHINGTON WA	-42	187	66		
		1*	68.3			CHELAN WA					
2	1.1	1	.3	10*	2.5	KING WA	-70	689	135	88	97
1*	4.8	1	1.6			PIERCE WA	-67				
				1*	2.1	SNOHOMISH WA					96
				1	2.3	YAKIMA WA					97
		1	.2	1	.3	WEST VIRGINIA WV		29			
		1	1.5			KANAWHA WV					
				1*	4.0	MERCER WV					98
4*	1.3	4	1.0	3	.5	WISCONSIN WI	-25	-49	-62		
4*	2.0	4	1.4	3	.7	MILWAUKEE WI	-31	-50	-66	12	93

* SIGNIFICANTLY MORE DEATHS THAN EXPECTED

COUNTIES COMBINED; SEE TECHNICAL NOTES

NON-WHITE FEMALE — CANCER OF THE NASOPHARYNX — ICD 147

1950-1959 DEATHS	1950-1959 RATE	1960-1969 DEATHS	1960-1969 RATE	1970-1979 DEATHS	1970-1979 RATE	COUNTY AND STATE	PERCENT CHANGE 60S/50S	PERCENT CHANGE 70S/60S	PERCENT CHANGE 70S/50S	PERCENTILE %CHANGE 70S/50S	PERCENTILE RATES 1970S
88	.1	169	.2	275	.2	UNITED STATES	56	17	83		
6	.2	6	.1	2	.0	ALABAMA AL	-10	-71	-74		
		1*	2.1			CONECUH AL					
1*	2.8					HENRY AL					
2	.3	1	.1			JEFFERSON AL	-70				
1*	1.3	1*	1.7			MARENGO AL	35				
1	.3	1	.3	2	.4	MOBILE AL	-3	24	20	60	97
1	.4					MONTGOMERY AL					
		1*	2.1			SUMTER AL					
		1*	.7			TUSCALOOSA AL					
		2*	2.3	5*	2.6	ALASKA AK		13			
2*	.4	1	.1	2	.1	ARIZONA AZ	-70	-14	-74		
2*	2.9			1*	.5	APACHE AZ			-84	03	97
		1	.4			MARICOPA AZ					
				1	.4	PIMA AZ					97
2	.1	1	.0	6	.3	ARKANSAS AR	-65	554	130		
1*	1.6					CHICOT AR					
				1*	.8	CRITTENDEN AR					98
				3*	1.2	JEFFERSON AR					98
1*	1.4					MISSISSIPPI AR					
		1	.3	1	.3	PULASKI AR		-6			96
				1*	1.6	ST FRANCIS AR					98
11*	.3	20*	.3	52*	.3	CALIFORNIA CA	2	22	25		
1	.3			7*	.5	ALAMEDA CA			108	75	97
		1	.7	3*	1.0	CONTRA COSTA CA		38			98
		1*	.8			FRESNO CA					
				1*	1.3	IMPERIAL CA					98
3	.1	5	.1	18	.2	LOS ANGELES CA	10	90	110	78	96
				1	.5	ORANGE CA					97
		1*	.8			RIVERSIDE CA					
				1	.3	SACRAMENTO CA					96
				1	.1	SAN BERNARDINO CA					95
1*	.5	2*	1.0	3	.3	SAN DIEGO CA	96	-68	-38	39	96
6*	1.8	7*	1.0	10*	.8	SAN FRANCISCO CA	-45	-21	-57	21	98
		2*	1.9	2	1.3	SAN MATEO CA		-34			98
				1*	1.7	SANTA BARBARA CA					98
		1*	1.0	2	.4	SANTA CLARA CA		-59			97
				1	.5	SOLANO CA					97
				1	.8	VENTURA CA					98
				1	.1	COLORADO CO					
				1	.2	DENVER-ARAPAHOE#CO					96
1	.2			3	.3	CONNECTICUT CT			46		
				1	.2	FAIRFIELD CT					96
				1	.4	HARTFORD CT					97
1*	.7					NEW HAVEN CT					
				1*	14.9	WINDHAM CT					99
1	.6			1	.3	DELAWARE DE			-46		
1*	3.6					KENT DE					
				1	.5	NEW CASTLE DE					97
4	.3	3	.2	3	.1	DIST. OF COLUMBIA DC	-36	-43	-63		
4	.3	3	.2	3	.1	WASHINGTON DC	-36	-43	-63	17	95
		8	.2	7	.1	FLORIDA FL		-37			
		2*	1.2			BROWARD FL					
				3	.2	DADE FL					96
		2	.3	1	.2	DUVAL FL		-48			96
				1*	3.9	HIGHLANDS FL					99
		1	.4			HILLSBOROUGH FL					
				1*	1.7	LEE FL					98
		1*	.9			LEON FL					
		1	.4			PALM BEACH FL					
				1*	1.5	ST LUCIE FL					98
		1*	.9			VOLUSIA FL					
2	.0	6	.1	13	.2	GEORGIA GA	187	99	470		
1	.3	2*	.6	1	.2	CHATHAM GA	60	-57	-31	42	96
		1*	1.6			COWETA GA					
				1	.4	DE KALB GA					97
				1	.7	DOUGHERTY GA					98
		1*	1.7			FLOYD GA					
		1	.1	4	.3	FULTON GA		214			96
				1*	4.0	HANCOCK GA					99
				1*	3.3	HENRY GA					99
				1*	1.5	HOUSTON GA					98
				1*	5.2	MC INTOSH GA					99
				1	.4	MUSCOGEE#GA					96
1*	2.9					TIFT GA					
		1*	3.2			UPSON GA					
				1*	8.6	WHITFIELD GA					99

* SIGNIFICANTLY MORE DEATHS THAN EXPECTED

COUNTIES COMBINED; SEE TECHNICAL NOTES

NON-WHITE FEMALE — CANCER OF THE NASOPHARYNX — ICD 147

1950-1959 DEATHS	1950-1959 RATE	1960-1969 DEATHS	1960-1969 RATE	1970-1979 DEATHS	1970-1979 RATE	COUNTY AND STATE	PERCENT CHANGE 60S/50S	PERCENT CHANGE 70S/60S	PERCENT CHANGE 70S/50S	PERCENTILE %CHANGE 70S/50S	PERCENTILE RATES 1970S
		15*	.9	15*	.6	HAWAII HI		-27			
		2*	1.2			HAWAII HI					
		13*	.9	14*	.7	HONOLULU HI		-20			98
				1	1.0	KAUAI HI					98
6	.1	9	.2	11	.2	ILLINOIS IL	34	-7	24		
6	.1	8	.2	10	.2	COOK IL	22	-3	19	57	96
		1*	1.8			KANKAKEE IL					
				1	.2	ST CLAIR IL					96
1	.1	3	.2	1	.1	INDIANA IN	102	-74	-47		
		1*	8.9			HOWARD IN					
		1	.2			LAKE IN					
1	.3	1	.2	1	.1	MARION IN	-28	-22	-44	35	95
1	.1	2	.2	2	.2	KENTUCKY KY	97	4	104		
				1	.9	FAYETTE KY					98
1*	3.9	1*	4.7			HARLAN KY	18				
				1	.2	JEFFERSON KY					96
		1*	4.0			KENTON KY					
1	.0	8	.2	7	.1	LOUISIANA LA	570	-29	378		
				1*	3.9	BIENVILLE LA					99
		2*	3.1	1*	.9	BOSSIER LA		-72			98
				1	.5	CALCASIEU LA					97
				1*	4.3	CATAHOULA LA					99
		1*	1.0			JEFFERSON LA					
		1*	1.3			LAFAYETTE LA					
		2	.2	3	.2	ORLEANS LA		26			96
1*	2.7	1*	1.9			TERREBONNE LA	-30				
		1*	5.6			WEST CARROLL LA					
5*	.2	6	.3	8	.2	MARYLAND MD	23	-28	-12		
				1	.4	ANNE ARUNDEL MD					97
1*	.9					BALTIMORE MD					
		1*	1.3			MONTGOMERY MD					
				2	.3	PRINCE GEORGE'S MD					96
4*	.3	5	.4	5	.2	BALTIMORE CITY MD	32	-39	-20	50	96
1	.2	1	.1	3	.3	MASSACHUSETTS MA	-10	119	96		
				1*	4.6	NORFOLK MA					99
1	.3	1	.3	2	.3	SUFFOLK MA	-4	30	25	64	96
2	.1	4	.1	10	.2	MICHIGAN MI	40	83	157		
		1*	3.2	1*	2.7	JACKSON MI		-15			99
2	.1	3	.1	9	.2	WAYNE MI	16	118	154	82	96
		1	.6			MINNESOTA MN					
		1*	2.0			RAMSEY MN					
2	.1	5	.1	7	.2	MISSISSIPPI MS	96	54	202		
1*	1.8					DE SOTO MS					
				1*	2.5	GRENADA MS					99
		1*	.8			HARRISON MS					
				1	.2	HINDS MS					96
		1*	.9			HOLMES MS					
				1*	1.3	JACKSON MS					98
		1*	.8			LEFLORE MS					
1*	3.4					LINCOLN MS					
		1*	1.4			LOWNDES MS					
				1*	4.5	NESHOBA MS					99
				1*	2.1	OKTIBBEHA MS					99
				1*	1.7	RANKIN MS					99
		1*	.7			WARREN MS					
				1	.4	WASHINGTON MS					97
4	.2	1	.1	3	.1	MISSOURI MO	-76	146	-41		
2*	.6			1	.2	JACKSON MO			-68	10	96
1*	.9					ST LOUIS COUNTY MO					
1	.1	1	.1	2	.2	ST LOUIS CITY MO	9	82	98	71	96
				1	.5	NEVADA NV					
				1*	2.8	WASHOE NV					99
3	.2	1	.0	6	.1	NEW JERSEY NJ	-81	368	-12		
1*	1.4			1	.4	BERGEN NJ			-73	07	97
2	.3			2	.2	ESSEX NJ			-50	25	95
				2*	2.4	GLOUCESTER NJ					99
		1	.5			MERCER NJ					
				1	.4	PASSAIC NJ					97
		1	.5			NEW MEXICO NM					
		1*	30.7			CURRY NM					
8	.1	14	.2	43*	.3	NEW YORK NY	26	101	153		

* SIGNIFICANTLY MORE DEATHS THAN EXPECTED

NON-WHITE FEMALE — CANCER OF THE NASOPHARYNX — ICD 147

1950-1959 DEATHS	RATE	1960-1969 DEATHS	RATE	1970-1979 DEATHS	RATE	COUNTY AND STATE	PERCENT CHANGE 60S/50S	70S/60S	70S/50S	PERCENTILE %CHANGE 70S/50S	PERCENTILE RATES 1970S
3*	.4	2	.1	3	.1	BRONX NY	-63	-5	-64	14	95
1	.0	4	.2	10	.3	KINGS(BROOKLYN) NY	319	75	634	96	96
				3*	.6	NASSAU NY					97
4	.2	8	.3	14*	.5	NEW YORK(MANHATTAN) NY	79	60	186	89	97
				1*	.9	ORANGE NY					98
				6	.2	QUEENS NY					96
				1*	.8	RICHMOND NY					98
				1	.1	SUFFOLK NY					95
				1*	9.9	WAYNE NY					99
				3	.5	WESTCHESTER NY					97
1	.0	7	.1	10	.2	NORTH CAROLINA NC	557	4	582		
				1*	1.4	ANSON NC					98
		1*	7.7			CARTERET NC					
				1*	2.7	CHATHAM NC					99
		1*	3.4			CHOWAN NC					
				1*	1.5	CLEVELAND NC					98
				1	.5	CUMBERLAND NC					97
				1	.4	DURHAM NC					97
				1	.8	EDGECOMBE NC					98
				1*	.8	GASTON NC					98
				1	.7	HALIFAX NC					97
1	.4			1	.2	MECKLENBURG NC			-48	28	96
		1*	1.7			NORTHAMPTON NC					
		1*	2.3			ORANGE NC					
		1*	.9			PITT NC					
		1	.7			ROBESON NC					
		1*	1.4			VANCE NC					
				1*	1.9	WARREN NC					99
4	.1	6	.1	7	.1	OHIO OH	-5	2	-3		
		4	.3	3	.2	CUYAHOGA OH		-32			96
1	.3			1	.2	FRANKLIN OH			-47	32	96
1	.2					HAMILTON OH					
1*	2.0	1*	1.0	1*	1.6	LORAIN OH	-48	52	-21	46	98
1*	1.2					LUCAS OH					
				1	.5	MAHONING OH					97
				1*	14.4	MIAMI OH					99
		1	.5			SUMMIT OH					
1	.1	2	.2	2	.1	OKLAHOMA OK	42	-29	1		
				1*	3.2	CADDO OK					99
				1	.3	OKLAHOMA OK					96
1*	16.1	1*	10.3			OTTAWA OK	-36				
		1*	12.5			PUSHMATAHA OK					
				1	.4	OREGON OR					
				1*	62.9	LINN OR					99
4	.1	7	.1	7	.1	PENNSYLVANIA PA	26	-8	16		
		1	.1			ALLEGHENY PA					
		1*	5.0			CAMBRIA PA					
				1*	.7	CHESTER PA					97
				1	.8	MONTGOMERY PA					98
3	.1	5	.2	5	.1	PHILADELPHIA PA	18	-11	5	53	95
1*	2.5					WASHINGTON PA					
2	.1	9	.2	10	.2	SOUTH CAROLINA SC	365	-1	361		
1*	2.6					BARNWELL SC					
		1*	.9	1*	1.3	BEAUFORT SC		40			98
1	.3			3*	.8	CHARLESTON SC			188	92	98
				1*	2.4	CHESTER SC					99
		1*	.6	1	.6	FLORENCE SC		-8			97
		2*	3.1			GEORGETOWN SC					
				1*	2.6	JASPER SC					99
		1*	2.0			KERSHAW SC					
		1	.4	1	.4	ORANGEBURG SC		-3			97
		2*	.6	1	.1	RICHLAND SC		-80			95
		1	.6	1	.5	SPARTANBURG SC		-19			97
1*	1.5					SOUTH DAKOTA SD					
1*	37.8					BENNETT SD					
2	.1	12*	.4	4	.1	TENNESSEE TN	529	-72	75		
				1	.2	DAVIDSON TN					96
		1*	.8			KNOX TN					
1*	2.6					LAUDERDALE TN					
				1*	3.1	RUTHERFORD TN					99
1	.1	10*	.9	2	.1	SHELBY TN	912	-86	45	67	95
		1*	4.5			SUMNER TN					
1	.0	4	.1	10	.1	TEXAS TX	323	29	446		
				1	.3	BEXAR TX					96
				1*	1.7	BRAZORIA TX					99
				1*	7.8	CALHOUN TX					99
				1*	3.7	DE WITT TX					99
				1*	8.7	FAYETTE TX					99
		1	.1	4	.2	HARRIS TX		46			95
1*	4.0					JASPER TX					
		1	.3			JEFFERSON TX					
				1	.5	MC LENNAN TX					97
		1	.7			TRAVIS TX					

* SIGNIFICANTLY MORE DEATHS THAN EXPECTED

NON-WHITE FEMALE — CANCER OF THE NASOPHARYNX — ICD 147

1950-1959 DEATHS	1950-1959 RATE	1960-1969 DEATHS	1960-1969 RATE	1970-1979 DEATHS	1970-1979 RATE	COUNTY AND STATE	PERCENT CHANGE 60S/50S	PERCENT CHANGE 70S/60S	PERCENT CHANGE 70S/50S	PERCENTILE %CHANGE 70S/50S	PERCENTILE RATES 1970S
		1*	2.3			WICHITA TX					
7	.2	3	.1	9	.2	VIRGINIA VA	-61	126	-11		
1*	2.1					ALBEMARLE#VA					
1*	17.0					BOTETOURT VA					
1	.2			3	.5	CHESTERFIELD#VA			178	85	97
				1	.7	DINWIDDIE#VA					97
				1	.2	FAIRFAX#VA					96
1*	4.7					FAUQUIER VA					
				1*	1.5	HENRICO VA					98
		1*	6.1			NELSON VA					
		1*	8.2			ORANGE VA					
				2*	1.2	PITTSYLVANIA#VA					98
1*	3.9					PRINCE EDWARD VA					
1*	24.6					ROCKINGHAM#VA					
		1*	2.0			SOUTHAMPTON#VA					
1*	6.4					WESTMORELAND VA					
				1*	.7	SUFFOLK CITY VA					98
		1	.7	2	.3	WASHINGTON WA		-53			
				2	.6	KING WA					97
		1*	10.2			PIERCE WA					
				1	.2	WEST VIRGINIA WV					
				1*	16.9	MONONGALIA WV					99
2*	.9					WISCONSIN WI					
2*	1.5					MILWAUKEE WI					

* SIGNIFICANTLY MORE DEATHS THAN EXPECTED

COUNTIES COMBINED; SEE TECHNICAL NOTES

WHITE MALE — CANCER OF THE ESOPHAGUS — ICD 150

1950-1959 DEATHS	1950-1959 RATE	1960-1969 DEATHS	1960-1969 RATE	1970-1979 DEATHS	1970-1979 RATE	COUNTY AND STATE	PERCENT CHANGE 60S/50S	PERCENT CHANGE 70S/60S	PERCENT CHANGE 70S/50S	PERCENTILE %CHANGE 70S/50S	PERCENTILE RATES 1970S
28381	4.4	32417	4.3	37897	4.4	UNITED STATES	-3	2	-1		
174	2.1	270	2.7	407	3.3	ALABAMA AL	26	24	57		
1	2.4	3	6.0	2	2.4	AUTAUGA AL	152	-61	-1	38	31
4	2.5	3	1.4	13	4.4	BALDWIN AL	-45	224	77	71	68
3	5.9	1	1.6	4	7.2	BARBOUR AL	-73	344	22	51	94
1	2.1	4	7.6	1	2.0	BIBB AL	257	-74	-6	35	26
		1	.7			BLOUNT AL					
1	4.8					BULLOCK AL					
1	1.4	3	3.8	2	2.5	BUTLER AL	174	-35	79	72	34
4	1.5	9	3.1	5	1.4	CALHOUN AL	109	-54	-3	36	19
1	.9	6	4.5	2	1.5	CHAMBERS AL	377	-68	54	64	20
1	1.6			1	1.3	CHEROKEE AL			-19	26	18
1	.9	2	1.7	11*	8.8	CHILTON AL	96	425	930	99	97
				1	2.0	CHOCTAW AL					26
1	1.5	3	4.3	4	5.3	CLARKE AL	181	25	250	93	81
1	2.5	1	2.0			CLEBURNE AL	-18				
2	1.8	1	.9	6	4.4	COFFEE AL	-50	377	139	83	68
3	2.3	7	4.5	9	4.1	COLBERT AL	91	-9	74	71	62
1	1.7			1	2.0	CONECUH AL			15	48	26
2	1.7	1	.6	5	2.8	COVINGTON AL	-64	367	68	69	39
1	1.8	2	3.8	1	1.2	CRENSHAW AL	106	-67	-33	16	18
4	2.3	5	2.5	9	3.2	CULLMAN AL	6	28	35	57	46
		2	1.9	3	2.1	DALE AL		11			28
		7	7.4	6	6.0	DALLAS AL		-19			87
1	.5	4	1.9	8	3.2	DE KALB AL	269	71	532	98	46
1	1.0	5	4.1	5	3.4	ELMORE AL	331	-17	256	93	50
		1	1.0	7	6.1	ESCAMBIA AL		534			88
6	2.0	9	2.8	9	2.1	ETOWAH AL	36	-23	4	41	28
2	2.7	1	1.4	4	4.2	FAYETTE AL	-51	206	51	63	64
1	1.1			6	4.5	FRANKLIN AL			312	95	69
		4	4.0	4	3.8	GENEVA AL		-4			57
		1	5.9	1	6.8	GREENE AL		15			93
1	3.1			3	8.2	HALE AL			168	86	96
1	2.8	1	2.5			HENRY AL	-12				
4	2.6	3	2.2	10	4.8	HOUSTON AL	-16	119	85	74	75
2	1.5	5	3.2	4	2.2	JACKSON AL	106	-30	43	61	29
49	3.5	70	4.1	88	4.4	JEFFERSON AL	16	8	25	53	68
		2	2.8	1	1.1	LAMAR AL		-62			16
3	1.6	6	2.6	2	.7	LAUDERDALE AL	60	-72	-56	06	15
1	2.6			5	4.9	LAWRENCE AL			89	75	75
		2	1.9	11	6.8	LEE AL		258			93
1	.8	3	2.4	1	.6	LIMESTONE AL	179	-75	-29	19	14
				1	5.7	LOWNDES AL					84
				1	2.8	MACON AL					39
2	.9	5	1.4	10	2.3	MADISON AL	61	61	160	86	30
1	2.2			4	7.6	MARENGO AL			253	93	95
		3	2.5	2	1.5	MARION AL		-38			20
1	.6	1	.4	6	2.1	MARSHALL AL	-25	380	261	93	28
30	5.3	35	4.6	33	3.3	MOBILE AL	-13	-28	-37	14	48
		1	1.8	2	3.4	MONROE AL		94			50
10	3.1	11	2.6	20	4.4	MONTGOMERY AL	-17	72	43	61	69
4	2.1	4	2.1	10	3.0	MORGAN AL	3	38	42	61	41
		1	2.7	2	5.1	PERRY AL		91			79
				5	7.2	PICKENS AL					94
2	2.8	1	1.3	2	2.4	PIKE AL	-53	89	-11	31	33
				3	3.4	RANDOLPH AL					49
		4	5.3	10*	9.6	RUSSELL AL		82			97
		1	1.0	3	2.0	ST CLAIR AL		104			26
1	.9	3	2.4	4	2.4	SHELBY AL	177	-1	173	87	31
		1	3.8	1	3.7	SUMTER AL		-2			56
4	3.3	6	3.3	3	1.2	TALLADEGA AL	0	-63	-63	04	18
				4	2.8	TALLAPOOSA AL					39
6	1.9	10	2.8	14	3.5	TUSCALOOSA AL	46	25	82	73	52
4	2.2	1	.4	5	1.8	WALKER AL	-80	314	-19	26	24
2	5.4	2	4.3			WASHINGTON AL	-20				
1	4.0	1	4.0	3	11.6	WILCOX AL	-1	193	190	88	99
		1	1.3	4	4.0	WINSTON AL		209			62
		14	2.9	23	4.5	ALASKA AK		53			
76	2.3	194	3.5	298	3.2	ARIZONA AZ	49	-8	38		
		2	6.9	3	9.9	APACHE AZ		42			98
4	2.7	11	5.0	13	4.4	COCHISE AZ	87	-12	65	68	68
2	5.3	4	3.0	5	5.4	COCONINO AZ	-42	79	3	41	82
		7	6.3	5	3.5	GILA AZ		-44			51
				1	1.2	GRAHAM AZ					17
		1	1.8	3	9.6	GREENLEE AZ		443			97
38	2.3	96	3.2	166	3.2	MARICOPA AZ	37	1	38	59	46
1	2.2	7	10.2	9	2.9	MOHAVE AZ	364	-72	31	55	40
3	5.3			2	2.7	NAVAJO AZ			-49	09	38
22	3.4	42	3.4	54	2.9	PIMA AZ	0	-16	-16	28	40
1	.8	5	2.7	9	3.0	PINAL AZ	252	11	289	94	42
1	3.1	2	5.2			SANTA CRUZ AZ	65				
2	.9	9	3.8	16	5.0	YAVAPAI AZ	305	30	424	97	76
2	1.6	8	4.5	12	3.7	YUMA AZ	176	-18	128	81	56
136	2.0	218	2.8	245	2.6	ARKANSAS AR	38	-8	27		
2	2.6	1	1.6	4	3.9	ARKANSAS AR	-39	140	46	62	58
2	2.5	1	1.3	1	.9	ASHLEY AR	-47	-36	-66	03	15
		5	4.7	4	1.9	BAXTER AR		-59			25
3	1.3	7	2.6	5	1.2	BENTON AR	94	-52	-7	35	18
				5	3.6	BOONE AR					54
2	4.0			3	3.9	BRADLEY AR			-2	37	60
1	3.5					CALHOUN AR					

* SIGNIFICANTLY MORE DEATHS THAN EXPECTED

WHITE MALE — CANCER OF THE ESOPHAGUS — ICD 150

1950-1959 DEATHS	1950-1959 RATE	1960-1969 DEATHS	1960-1969 RATE	1970-1979 DEATHS	1970-1979 RATE	COUNTY AND STATE	PERCENT CHANGE 60S/50S	PERCENT CHANGE 70S/60S	PERCENT CHANGE 70S/50S	PERCENTILE %CHANGE 70S/50S	PERCENTILE RATES 1970S
2	2.2	3	3.6	2	1.6	CARROLL AR	62	-56	-29	19	21
3	7.8					CHICOT AR					
1	1.3	1	1.3	3	3.3	CLARK AR	0	160	159	86	48
2	1.9	3	2.9	4	3.0	CLAY AR	55	2	58	66	42
		1	1.8			CLEBURNE AR					
1	2.5			1	2.5	CLEVELAND AR			-2	37	34
1	1.2	2	2.3	1	1.0	COLUMBIA AR	93	-54	-12	31	16
		3	5.4	2	2.1	CONWAY AR		-60			28
3	1.6	3	1.4	9	3.8	CRAIGHEAD AR	-12	162	130	81	56
1	.9	2	1.6	7	4.8	CRAWFORD AR	78	200	434	97	74
2	3.0	4	5.8	3	3.3	CRITTENDEN AR	94	-43	11	46	49
2	3.7	2	3.8	3	4.3	CROSS AR	4	14	18	50	67
1	2.3					DESHA AR					
		1	2.1			DREW AR					
1	.9	6	4.8	4	2.2	FAULKNER AR	432	-54	143	84	29
1	1.7	1	1.2	3	4.0	FRANKLIN AR	-29	231	133	82	60
				3	4.6	FULTON AR					71
11	3.9	14	4.0	15	3.4	GARLAND AR	5	-15	-11	32	51
1	2.2	1	1.8	3	5.1	GRANT AR	-20	190	133	82	78
3	2.3	2	1.6	4	2.4	GREENE AR	-32	54	5	42	33
3	3.5	3	3.2	1	.9	HEMPSTEAD AR	-9	-73	-75	01	15
		2	1.9	4	3.3	HOT SPRING AR		72			48
		1	1.5	5	6.5	HOWARD AR		320			91
		3	2.3	1	.7	INDEPENDENCE AR		-69			15
				1	2.8	IZARD AR					39
2	2.3	5	5.1	3	2.7	JACKSON AR	127	-48	19	50	37
5	4.0	7	3.8	5	1.9	JEFFERSON AR	-5	-51	-53	07	25
		3	3.6	3	2.8	JOHNSON AR		-24			39
1	2.7	2	6.1			LAFAYETTE AR	121				
1	1.1	1	.9	4	3.3	LAWRENCE AR	-16	253	196	89	48
		1	2.5	1	2.7	LEE AR		9			38
		1	2.9	1	2.4	LINCOLN AR		-18			33
		1	3.0	2	4.5	LITTLE RIVER AR		49			69
4	4.0	3	2.3	1	1.0	LOGAN AR	-44	-57	-76	01	16
2	2.1	5	4.6	2	1.7	LONOKE AR	118	-63	-19	25	22
2	3.0			1	1.5	MADISON AR			-51	08	20
2	4.8	1	2.3	3	3.7	MARION AR	-52	63	-22	23	56
2	2.5	7	5.7	9	6.1	MILLER AR	123	8	140	83	88
3	1.7	5	3.2	4	2.1	MISSISSIPPI AR	92	-34	27	54	28
		8*	17.5	5*	9.8	MONROE AR		-44			98
				2	4.2	MONTGOMERY AR					64
		1	2.5	1	1.5	NEVADA AR		-40			20
1	2.4					NEWTON AR					
1	1.2	3	2.9	2	1.6	OUACHITA AR	142	-44	36	58	22
		3	8.5	2	5.7	PERRY AR		-33			84
2	3.3	1	1.5	5	6.6	PHILLIPS AR	-56	353	100	77	92
1	1.8					PIKE AR					
2	1.8	3	2.6	4	3.0	POINSETT AR	49	12	67	68	41
		4	4.2	4	3.3	POLK AR		-23			47
1	1.2	2	1.5	2	1.2	POPE AR	21	-20	-3	36	17
1	1.9	1	1.7			PRAIRIE AR	-11				
28	4.5	32	3.5	33	3.3	PULASKI AR	-21	-7	-26	21	48
3	4.1	2	2.3	2	1.9	RANDOLPH AR	-44	-18	-55	07	24
1	1.4	2	3.2	1	1.3	ST FRANCIS AR	125	-58	-5	36	19
1	.8	1	.6	2	1.3	SALINE AR	-24	109	58	65	19
		2	3.7			SCOTT AR					
1	2.1	1	2.8	1	1.6	SEARCY AR	31	-41	-23	23	22
6	2.2	16	5.4	6	1.6	SEBASTIAN AR	149	-71	-28	20	21
				2	2.7	SEVIER AR					38
1	1.8	2	4.7	1	.7	SHARP AR	156	-86	-63	04	14
		1	2.3			STONE AR					
5	2.9	7	4.0	6	2.9	UNION AR	35	-27	-1	38	40
		1	1.5	2	1.7	VAN BUREN AR		16			23
2	.8	6	1.9	9	2.4	WASHINGTON AR	122	26	180	88	32
1	.6	3	1.5	5	2.1	WHITE AR	136	43	238	92	28
1	3.4			3	6.0	WOODRUFF AR			79	72	88
4	6.0	2	2.4	5	5.2	YELL AR	-61	123	-13	30	80
2156	4.2	2864	4.3	3439	4.2	CALIFORNIA CA	2	-1	1		
163	5.0	186*	5.2	161	4.4	ALAMEDA CA	5	-16	-12	31	68
7*	12.5	8*	12.7	3	3.5	AMADOR CA	1	-72	-72	02	52
8	1.7	17	3.1	19	2.8	BUTTE CA	87	-11	66	68	39
6	8.8	3	3.8	5	3.6	CALAVERAS CA	-57	-4	-59	06	54
		3	3.9	2	2.7	COLUSA CA		-30			38
50	5.0	72	5.2	88	4.3	CONTRA COSTA CA	2	-17	-15	28	66
2	4.3	4	6.3	4	4.2	DEL NORTE CA	46	-33	-1	37	65
3	2.3	9	5.0	14	4.3	EL DORADO CA	112	-12	85	74	67
36	3.1	55	3.8	68	5.1	FRESNO CA	23	35	66	68	79
2	2.2	5	5.2	3	2.6	GLENN CA	139	-50	20	51	36
22	7.0	23	5.5	24	5.8	HUMBOLDT CA	-20	5	-16	27	86
5	2.5	11	3.6	12	4.0	IMPERIAL CA	44	12	62	67	62
		2	2.6	4	4.0	INYO CA		51			61
27	3.6	31	2.9	51	3.7	KERN CA	-20	29	3	40	55
8	4.2	8	4.1	12	5.5	KINGS CA	-2	34	32	56	83
4	3.2	8	5.6	9	3.1	LAKE CA	74	-44	-3	37	44
7*	10.2	2	3.2	8*	10.8	LASSEN CA	-68	234	6	43	98
699	3.5	1002	4.2	1088	4.2	LOS ANGELES CA	18	1	19	50	65
6	5.2	6	3.3	10	4.5	MADERA CA	-37	36	-14	29	69
27*	7.1	38	6.4	41	5.1	MARIN CA	-10	-21	-29	19	77
		2	4.4	2	2.6	MARIPOSA CA		-41			35
10	4.3	10	3.9	23*	7.4	MENDOCINO CA	-9	88	71	70	95
9	3.6	10	2.9	18	4.1	MERCED CA	-19	44	16	49	63
				1	1.9	MODOC CA					24
				1	4.6	MONO CA					71
24	4.9	27	4.1	44	5.3	MONTEREY CA	-17	29	7	43	81
18	4.5	21	4.7	20	3.5	NAPA CA	5	-25	-21	24	52
6	4.7	9	5.4	11	4.5	NEVADA CA	16	-17	-3	36	70
28	1.9	106	3.4	221	4.1	ORANGE CA	80	21	118	80	64
15	5.3	16	4.8	16	3.7	PLACER CA	-11	-23	-31	17	55
1	2.5	2	2.8	5	5.8	PLUMAS CA	11	107	130	81	85

* SIGNIFICANTLY MORE DEATHS THAN EXPECTED

WHITE MALE — CANCER OF THE ESOPHAGUS — ICD 150

1950-1959 DEATHS	1950-1959 RATE	1960-1969 DEATHS	1960-1969 RATE	1970-1979 DEATHS	1970-1979 RATE	COUNTY AND STATE	PERCENT CHANGE 60S/50S	PERCENT CHANGE 70S/60S	PERCENT CHANGE 70S/50S	PERCENTILE %CHANGE 70S/50S	PERCENTILE RATES 1970S
29	2.7	66	3.6	103	3.6	RIVERSIDE CA	33	0	33	57	54
76	5.1	75	3.8	119	4.9	SACRAMENTO CA	-26	29	-5	36	76
2	2.4	3	3.7	6	6.6	SAN BENITO CA	51	79	170	86	92
47	2.7	92	3.7	108	3.4	SAN BERNARDINO CA	37	-9	24	52	50
72	2.7	149	3.5	244	4.0	SAN DIEGO CA	30	12	46	62	60
317*	8.2	286*	7.7	187*	6.1	SAN FRANCISCO CA	-6	-21	-25	21	88
50	5.1	58	4.9	45	3.4	SAN JOAQUIN CA	-4	-30	-32	16	51
23*	6.7	16	3.1	20	2.8	SAN LUIS OBISPO CA	-54	-7	-58	06	40
74*	6.0	87	4.9	111	5.1	SAN MATEO CA	-19	4	-16	28	78
28	5.0	33	4.1	50	4.5	SANTA BARBARA CA	-19	11	-10	33	71
69	4.4	96	4.0	120	3.5	SANTA CLARA CA	-9	-13	-20	25	52
18	3.9	25	3.5	37	4.5	SANTA CRUZ CA	-10	28	16	48	71
7	3.4	11	3.8	22	4.8	SHASTA CA	9	27	39	59	74
7	4.2	7	4.0	9	4.4	SISKIYOU CA	-4	11	6	43	69
21	7.0	17	3.4	28	4.7	SOLANO CA	-51	39	-32	17	74
50*	7.2	38	4.0	58	4.5	SONOMA CA	-45	13	-37	13	70
16	2.4	30	3.5	39	3.7	STANISLAUS CA	47	6	56	65	56
6	5.0	4	3.3	6	3.1	SUTTER CA	-33	-9	-39	13	43
5	4.0	4	2.8	8	4.5	TEHAMA CA	-31	62	11	46	70
1	2.6	3	6.1	2	3.8	TRINITY CA	133	-37	47	62	58
12	1.9	17	2.2	24	2.6	TULARE CA	16	18	38	59	37
4	4.7	6	4.8	5	3.2	TUOLUMNE CA	2	-34	-32	17	45
19	3.3	26	3.0	56	3.7	VENTURA CA	-9	23	12	46	56
8	3.6	12	4.3	16	4.3	YOLO CA	19	1	20	51	67
2	1.6	7	5.0	8	4.9	YUBA CA	216	-3	208	90	76
139	2.1	209	2.7	290	3.1	COLORADO CO	27	14	44		
4	1.9	12	4.0	24	4.9	ADAMS CO	117	21	162	86	75
1	2.4			1	2.5	ALAMOSA CO			2	40	34
				1	2.8	BACA CO					39
1	3.0					BENT CO					
3	1.3	11	3.3	11	2.4	BOULDER CO	148	-29	77	71	32
		1	2.3	7*	13.3	CHAFFEE CO		488			99
				1	7.2	CHEYENNE CO					94
1	5.8			1	4.0	CLEAR CREEK CO			-32	17	60
1	4.0			2	6.4	CONEJOS CO			60	66	91
1	3.6					CROWLEY CO					
		2	2.2	4	2.7	DELTA CO		26			38
75	3.4	84	3.3	102	3.7	DENVER-ARAPAHOE#CO	-2	12	9	45	56
		1	5.0			DOUGLAS CO					
				1	4.2	EAGLE CO					65
1	4.7			1	4.4	ELBERT CO			-7	35	68
4	1.0	18	3.6	30	3.9	EL PASO CO	253	8	283	94	60
1	1.1	4	3.0	4	1.9	FREMONT CO	179	-39	71	70	24
1	1.5	3	4.1			GARFIELD CO	177				
1	1.9			1	2.2	HUERFANO CO			18	50	30
		1	10.0	1	13.3	JACKSON CO		32			99
5	1.9	16	3.3	31	2.7	JEFFERSON CO	77	-20	42	61	37
				2	4.0	KIT CARSON CO					61
				1	4.3	LAKE CO					66
1	1.4			4	4.1	LA PLATA CO			187	88	64
2	.7	5	1.5	4	.9	LARIMER CO	114	-39	30	55	15
5	4.2	1	1.0	6	7.1	LAS ANIMAS CO	-76	608	68	69	94
		1	3.3			LINCOLN CO					
3	3.7	5	5.7	4	4.3	LOGAN CO	55	-25	17	49	66
2	.9	9	3.5	6	1.9	MESA CO	270	-45	103	77	25
1	3.1	1	3.6			MOFFAT CO	15				
2	4.1	1	1.8	1	1.5	MONTEZUMA CO	-55	-18	-63	04	20
		3	3.4	3	3.0	MONTROSE CO		-11			43
		1	1.2	4	3.5	MORGAN CO		199			51
2	1.6	1	.8			OTERO CO	-51				
2*	15.0					OURAY CO					
1	18.7					PARK CO					
1	3.9	1	2.9	1	3.1	PHILLIPS CO	-25	4	-22	24	44
		1	5.6			PITKIN CO					
1	2.0	1	1.7			PROWERS CO	-12				
6	1.3	9	1.8	16	3.0	PUEBLO CO	41	68	137	83	42
		2	9.5			RIO BLANCO CO					
				1	2.1	RIO GRANDE CO					27
		2	5.4	2	6.2	ROUTT CO		16			89
1	4.0			1	4.8	SAGUACHE CO			23	52	75
				1	14.1	SAN MIGUEL CO					99
		1	6.2	1	4.8	SEDGWICK CO		-23			74
1	15.1					SUMMIT CO					
1	5.2			1	3.3	TELLER CO			-37	14	47
		1	2.5	1	2.8	WASHINGTON CO		12			39
6	2.0	8	2.3	6	1.5	WELD CO	18	-36	-25	22	20
1	1.5	2	2.8	1	2.3	YUMA CO	91	-18	56	65	30
741*	7.6	773*	6.6	784*	5.9	CONNECTICUT CT	-13	-11	-23		
178*	7.2	218*	7.3	218*	6.2	FAIRFIELD CT	2	-15	-13	30	89
229*	8.8	224*	7.2	205*	5.8	HARTFORD CT	-18	-19	-33	16	86
46*	8.6	48*	7.5	43	5.8	LITCHFIELD CT	-13	-22	-32	17	86
16	4.4	27	5.7	28	4.9	MIDDLESEX CT	30	-15	11	46	76
195*	7.5	186*	6.1	169*	5.6	NEW HAVEN CT	-19	-8	-26	21	83
49*	7.0	44	5.5	51	5.5	NEW LONDON CT	-21	-1	-22	24	83
12	5.7	16	5.7	25*	7.1	TOLLAND CT	2	24	26	53	94
16	5.6	10	2.9	25	6.0	WINDHAM CT	-48	109	8	44	88
63	4.8	70	4.2	83	4.1	DELAWARE DE	-14	-2	-15		
11	6.2	6	2.8	9	3.6	KENT DE	-54	26	-42	11	53
44	4.9	61	5.3	56	4.1	NEW CASTLE DE	8	-22	-16	28	64
8	3.5	3	.9	18	4.6	SUSSEX DE	-73	388	32	56	71
163*	8.0	125*	7.6	82*	7.0	DIST. OF COLUMBIA DC	-5	-7	-12		
163*	8.0	125*	7.6	82*	7.0	WASHINGTON DC	-5	-7	-12	31	94

* SIGNIFICANTLY MORE DEATHS THAN EXPECTED

\# COUNTIES COMBINED; SEE TECHNICAL NOTES

WHITE MALE

CANCER OF THE ESOPHAGUS

ICD 150

1950-1959 DEATHS	1950-1959 RATE	1960-1969 DEATHS	1960-1969 RATE	1970-1979 DEATHS	1970-1979 RATE	COUNTY AND STATE	PERCENT CHANGE 60S/50S	PERCENT CHANGE 70S/60S	PERCENT CHANGE 70S/50S	PERCENTILE %CHANGE 70S/50S	PERCENTILE RATES 1970S
525	3.4	1067	3.7	1794	3.9	FLORIDA FL	11	3	14		
5	4.3	3	1.8	13	4.0	ALACHUA FL	-57	117	-7	34	61
2	11.1			1	2.0	BAKEP FL			-82	00	26
2	2.4	9	3.8	18	5.8	BAY FL	58	56	147	84	86
		1	1.9	1	1.2	BRADFOPD FL		-36			18
7	3.2	15	3.7	49	4.9	BREVARD FL	15	33	53	64	75
27	3.6	100	3.6	201	3.5	BROWAPD FL	-2	0	-2	37	52
		1	2.9			CALHOUN FL					
2	2.5	10	4.5	18	4.5	CHARLOTTE FL	84	0	83	73	70
1	6.2	5	3.9	19	4.1	CITRUS FL	-37	5	-34	15	62
3	5.9	2	2.5	12	8.0	CLAY FL	-58	219	35	57	96
		9	7.0	16	3.6	COLLIER FL		-48			55
1	3.2	3	4.1	4	3.8	COLUMBIA FL	28	-9	16	49	56
134	4.8	228	4.3	298	4.0	DADE FL	-9	-8	-17	27	60
4	6.5	1	1.3	4	5.5	DE SOTO FL	-80	317	-16	28	83
1	7.5	1	6.0			DIXIE FL	-21				
37	3.9	59	4.6	86	5.3	DUVAL FL	18	16	37	58	81
13	4.6	24	4.8	27	3.8	ESCAMBIA FL	4	-20	-17	27	57
1	4.8					FLAGLER FL					
				2	5.4	FRANKLIN FL					82
3	3.2	1	1.1	1	.9	GADSDEN FL	-65	-22	-72	02	15
				1	6.1	GILCHRIST FL					88
1	12.4					GULF FL					
				2	6.6	HAMILTON FL					92
1	1.8	3	4.7	4	6.2	HARDEE FL	166	31	249	93	89
1	6.0	1	4.5	2	3.7	HENDRY FL	-24	-19	-38	13	55
		3	5.0	8	2.9	HEPNANDO FL		-43			40
1	1.5	2	1.2	6	3.1	HIGHLANDS FL	-20	151	100	77	44
52	3.9	91	5.1	102	4.4	HILLSBOROUGH FL	31	-14	12	46	67
1	1.5	2	3.2	2	2.9	HOLMES FL	113	-9	94	75	41
3	4.7	3	2.0	10	3.4	INDIAN RIVER FL	-58	72	-29	19	49
1	.9	1	.8	2	1.5	JACKSON FL	-10	80	63	67	20
		1	4.6			JEFFERSON FL					
		2	11.1			LAFAYETTE FL					
3	1.1	9	2.2	20	3.0	LAKE FL	101	38	177	87	42
4	2.2	18	3.3	46	4.4	LEE FL	48	32	96	76	68
7	7.3	5	2.6	7	2.3	LEON FL	-65	-12	-69	02	31
		1	2.0	2	2.3	LEVY FL		11			30
		1	8.0	1	5.7	LIBERTY FL		-28			85
		1	2.3			MADISON FL					
9	3.3	28	4.1	42	3.9	MANATEE FL	25	-6	17	49	58
5	3.1	3	.9	22	4.3	MARION FL	-71	367	37	58	66
1	1.6	8	4.2	10	3.1	MAPTIN FL	165	-27	94	75	44
5	5.4	5	2.4	9	3.4	MONROE FL	-55	40	-37	14	50
1	2.8			5	5.2	NASSAU FL			83	73	79
2	3.2	5	3.0	10	4.2	OKALOOSA FL	-6	39	31	56	65
1	7.3	3	8.3	1	1.2	OKEECHOBEE FL	15	-86	-83	00	17
23	3.1	44	3.8	69	4.3	OPANGE FL	21	13	37	58	66
3	3.6	1	.6	6	2.4	OSCEOLA FL	-84	316	-34	15	32
26	3.4	57	3.8	125	4.0	PALM BEACH FL	9	6	15	48	60
2	1.0	14	3.2	55	3.8	PASCO FL	218	21	285	94	58
64	3.1	139	3.4	219	3.9	PINELLAS FL	8	16	25	53	60
15	2.4	34	3.7	51	3.9	POLK FL	50	5	58	66	58
2	2.1	5	3.9	5	1.8	PUTNAM FL	87	-53	-12	31	24
5	5.1	6	4.8	8	4.1	ST JOHNS FL	-5	-15	-19	25	63
5	4.2	12	4.7	18	4.8	ST LUCIE FL	14	1	15	48	74
2	3.0	3	2.4	8	4.4	SANTA ROSA FL	-22	84	44	61	68
13	3.3	35	4.3	60	3.9	SARASOTA FL	27	-9	16	49	59
6	4.4	5	2.0	11	2.1	SEMINOLE FL	-55	4	-53	07	27
1	1.8	3	4.7	4	3.1	SUMTER FL	159	-34	71	70	44
1	1.8	2	3.3	1	1.3	SUWANNEE FL	84	-59	-25	22	19
				4	6.2	TAYLOR FL					89
13	2.5	34	3.3	57	3.6	VOLUSIA FL	31	7	40	60	53
		1	4.2	3	8.4	WAKULLA FL		99			96
1	1.4	2	2.4	4	3.4	WALTON FL	70	45	147	84	51
1	2.2	2	3.8	2	2.4	WASHINGTON FL	75	-36	11	46	33
267	3.0	403	3.5	526	3.6	GEORGIA GA	18	3	22		
1	3.4	1	1.8	2	4.0	APPLING GA	-47	125	20	51	62
		1	5.2			ATKINSON GA					
2	9.2			3	10.7	BACON GA			17	49	98
2	1.7			3	2.8	BALDWIN GA			68	69	40
		1	2.8			BANKS GA					
1	1.8	2	3.8	4	5.6	BARPOW GA	106	49	207	90	84
4	5.5	7	6.3	6	3.8	BAPTOW GA	14	-39	-31	18	58
				3	6.5	BEN HILL GA					91
2	5.2	1	2.4			BERRIEN GA	-54				
15	7.3	14	4.2	21	5.2	BIBB GA	-43	24	-29	19	79
		2	7.5			BRANTLEY GA					
1	2.1	1	2.0	1	2.6	BROOKS GA	-5	27	20	51	35
3*	24.8			1	4.5	BPYAN GA			-82	00	71
2	2.8	1	1.3	4	3.7	BULLOCH GA	-55	189	30	55	55
3	8.8					BUPKE GA					
1	3.8					BUTTS GA					
1	7.0					CALHOUN GA					
		1	9.9	2	15.5	CAMDEN GA		56			99
		1	3.8	2	6.2	CANDLEP GA		63			89
1	1.0	7	5.5	10	5.1	CAPPOLL GA	465	-7	427	97	78
		3	3.3	5	5.2	CATCOSA GA		56			79
1	7.2			1	7.4	CHARLTON GA			4	41	95
13	3.6	27*	6.5	28	6.0	CHATHAM GA	81	-8	67	68	87
1	1.6	1	1.4	2	2.0	CHATTOOGA GA	-12	42	25	53	26
		2	1.9	6	3.8	CHEROKEE GA		101			56
		3	2.8	8	6.3	CLAPKE GA		125			90
				1	9.0	CLAY GA					97
2	2.8	5	3.0	7	2.1	CLAYTON GA	5	-29	-25	22	28
6	3.0	10	2.8	27	4.4	COBB GA	-5	57	49	63	69
				3	3.6	COFFEE GA					52

* SIGNIFICANTLY MORE DEATHS THAN EXPECTED

1950-1959 DEATHS	RATE	1960-1969 DEATHS	RATE	1970-1979 DEATHS	RATE	COUNTY AND STATE	PERCENT CHANGE 60S/50S	70S/60S	70S/50S	PERCENTILE %CHANGE 70S/50S	RATES 1970S
1	1.1	3	2.4	7	5.4	COLQUITT GA	115	122	377	96	82
3*	12.0			1	2.2	COLUMBIA GA			-82	00	29
				2	4.6	COOK GA					71
		3	4.0	4	3.4	COWETA GA		-16			49
				1	4.8	CRAWFOPD GA					74
1	2.6	1	2.2	3	6.7	CRISP GA	-17	206	153	85	92
		1	6.4			DAWSON GA					
1	2.0	1	1.4	3	4.5	DECATUR GA	-30	216	121	80	70
15	2.8	36	4.5	46	3.9	DE KALB GA	60	-15	36	58	59
1	3.0	2	3.7	1	1.6	DODGE GA	22	-56	-46	10	22
2	6.9			2	5.4	DOOLY GA			-22	23	82
3	3.8	6	4.1	5	2.4	DOUGHERTY GA	10	-41	-36	14	33
				4	3.3	DOUGLAS GA					49
		1	2.4	3	7.1	EARLY GA		193			94
2	7.2	4*	12.1	1	1.7	EFFINGHAM GA	67	-86	-77	01	22
1	1.6	1	1.8	1	1.3	ELBERT GA	7	-23	-18	26	19
		1	1.3	2	3.1	EMANUEL GA		147			45
		1	5.2	2	6.8	EVANS GA		32			93
		2	3.3	5	6.0	FANNIN GA		84			68
		1	3.7	2	3.5	FAYETTE GA		-5			52
8	3.9	10	4.4	10	3.5	FLOYD GA	13	-21	-11	32	51
		2	4.1	3	3.5	FORSYTH GA		-13			52
1	1.9	2	3.6	1	1.4	FRANKLIN GA	89	-61	-26	21	19
60	5.1	71	4.8	66	4.5	FULTON GA	-6	-8	-13	30	69
				3	5.7	GILMER GA					85
		2*	21.2			GLASCOCK GA					
6	8.3	3	2.5	4	3.1	GLYNN GA	-70	22	-63	04	43
1	1.6			2	2.0	GORDON GA			26	53	26
1	1.7	1	1.6	3	4.2	GPADY GA	-9	167	143	84	64
1	3.5			1	2.9	GREENE GA			-18	26	40
4	3.2	1	.8	15	4.2	GWINNETT GA	-75	429	34	57	66
1	1.8			2	1.9	HABEPSHAM GA			8	44	25
3	2.2	4	2.2	6	2.5	HALL GA	-2	18	16	48	35
				1	5.6	HANCOCK GA					83
1	2.3	1	1.4	1	1.4	HARALSON GA	-37	-2	-38	13	19
				3	6.1	HARRIS GA					89
				1	3.9	HEAPD GA					60
1	2.9	6*	8.4	2	2.3	HENRY GA	191	-72	-19	25	31
1	4.2	3	6.0	8	7.1	HOUSTON GA	42	20	70	69	94
				2	5.8	IRWIN GA					86
1	1.2			2	2.2	JACKSON GA			78	72	29
1	4.3			3	8.8	JEFF DAVIS GA			104	77	97
2	5.7	1	2.9	2	4.2	JEFFERSON GA	-48	42	-27	21	65
1	6.2	3*	12.7			JENKINS GA	104				
1	3.8	2	9.1			JOHNSON GA	138				
		1	3.9	1	2.4	JONES GA		-39			33
1	3.9					LAMAR GA					
				2	8.7	LANIER GA					97
		5	7.1	4	3.7	LAURENS GA		-48			55
1	9.7			1	5.7	LEE GA			-41	12	85
		1	3.7	2	12.0	LIBERTY GA		223			99
		1	10.6			LINCOLN GA					
1	9.2					LONG GA					
4	5.0	2	1.1	4	3.1	LOWNDES GA	-77	179	-37	14	45
1	4.9	2	7.0	2	6.3	LUMPKIN GA	44	-10	29	54	90
1	5.9	1	3.3	2	3.7	MC DUFFIE GA	-45	12	-38	13	55
1	8.8			1	4.4	MC INTOSH GA			-51	08	67
		2	6.9	1	3.0	MACON GA		-56			43
3	7.8	2	4.2	1	1.6	MADISON GA	-46	-62	-80	01	21
1	7.5					MARION GA					
		1	1.7	2	3.7	MERIWETHER GA		115			55
4	9.0	3	5.9	1	1.7	MITCHELL GA	-35	-72	-82	00	22
		1	4.0	1	2.3	MONROE GA		-42			31
		1	4.7	1	4.7	MONTGOMERY GA		1			73
				1	3.9	MORGAN GA					59
		2	6.6	1	1.7	MURRAY GA		-74			23
7	3.2	14	4.1	17	4.0	MUSCOGEE#GA	30	-3	27	54	61
1	2.2	6	8.3	3	3.3	NEWTON GA	268	-60	48	63	48
		1	3.8	2	5.5	OCONEE GA		42			83
1	3.9	2	7.3	2	6.7	OGLETHORPE GA	87	-8	72	70	92
		3	4.9	4	4.8	PAULDING GA		-2			74
				2	5.5	PEACH GA					83
		1	3.5	2	3.6	PICKENS GA		1			53
1	3.1	1	2.8			PIERCE GA	-12				
		2	8.7	1	3.2	PIKE GA		-63			47
3	3.3	3	3.4	5	3.9	POLK GA	4	12	17	49	58
2*	16.2	2	8.8			PUTNAM GA	-46				
				1	12.5	QUITMAN GA					99
		3	7.3	1	2.2	PABUN GA		-70			29
1	4.4	2	8.7	1	5.2	RANDOLPH GA	98	-40	19	50	79
10	4.1	19	5.8	11	2.8	RICHMOND GA	43	-51	-30	18	40
		1	1.8	4	5.1	ROCKDALE GA		185			78
		1	3.2	1	2.2	SCPEVEN GA		-32			29
4	6.6	1	.7	10	6.7	SPALDING GA	-90	907	2	40	92
1	2.1	1	1.4	1	1.2	STEPHENS GA	-30	-19	-44	11	17
2	3.9			2	3.1	SUMTER GA			-20	25	44
		1	9.4			TALBOT GA					
1	1.7			1	1.3	TATTNALL GA			-25	22	18
		4*	18.0	1	2.4	TELFAIR GA		-87			33
		2	8.6			TERRELL GA					
2	2.0	5	4.8	2	1.9	THOMAS GA	136	-61	-8	34	24
3	4.6	4	7.0	2	2.1	TIFT GA	54	-70	-54	07	28
1	2.3			2	3.0	TOOMBS GA			27	54	41
		1	4.6	1	5.9	TREUTLEN GA		29			86
3	2.8	2	1.6	4	2.4	TROUP GA	-43	49	-15	29	32
1	4.8			2	5.6	TURNER GA			17	49	84
		2	12.6			TWIGGS GA					
		2	3.0	2	2.0	UPSON GA		-33			26
2	1.3	2	1.2	4	1.8	WALKER GA	-7	47	37	58	23
4	6.5			2	2.5	WALTON GA			-62	05	34
1	.9	4	3.2	6	4.5	WARE GA	254	42	403	97	70

* SIGNIFICANTLY MORE DEATHS THAN EXPECTED

COUNTIES COMBINED; SEE TECHNICAL NOTES

WHITE MALE | CANCER OF THE ESOPHAGUS | ICD 150

1950-1959 DEATHS	1950-1959 RATE	1960-1969 DEATHS	1960-1969 RATE	1970-1979 DEATHS	1970-1979 RATE	COUNTY AND STATE	PERCENT CHANGE 60S/50S	PERCENT CHANGE 70S/60S	PERCENT CHANGE 70S/50S	PERCENTILE %CHANGE 70S/50S	PERCENTILE RATES 1970S
				1	7.1	WARREN GA					94
				1	2.0	WASHINGTON GA					27
		2	3.2	1	1.3	WAYNE GA		-59			18
2*	28.3					WEBSTER GA					
1	4.9	1	4.5			WHEELER GA	-9				
		3	8.9	1	2.0	WHITE GA		-77			27
5	4.5	10	6.3	5	2.2	WHITFIELD GA	39	-65	-51	08	29
		1	3.7	2	6.2	WILCOX GA		69			89
1	3.2	1	3.6	1	3.5	WILKES GA	10	-1	9	44	52
2	7.4			1	3.7	WILKINSON GA			-50	08	56
		1	2.1			WORTH GA					
		36	7.1	38	4.0	HAWAII HI		-43			
		3	7.9	6	5.7	HAWAII HI		-28			84
		29*	7.2	30	4.1	HONOLULU HI		-43			63
				1	3.3	KAUAI HI					49
		4	9.5	1	1.5	MAUI#HI		-84			20
41	1.6	59	1.9	89	2.5	IDAHO ID	22	29	58		
9	2.6	10	2.3	18	3.3	ADA ID	-13	45	26	54	47
3	1.7	3	1.6	5	2.6	BANNOCK ID	-7	63	51	63	36
				1	2.2	BEAR LAKE ID					29
1	3.0					BENEWAH ID					
				2	2.0	BINGHAM ID					26
2	1.8	1	1.1	4	3.7	BONNER ID	-38	237	110	78	56
1	.8	4	2.8	3	1.6	BONNEVILLE ID	235	-41	98	76	22
1	3.3			2	5.5	BOUNDARY ID			68	69	83
4	1.8	6	1.9	5	1.5	CANYON ID	8	-24	-18	26	20
2	10.3	1	7.4	1	5.9	CARIBOU ID	-28	-21	-43	11	86
				1	1.1	CASSIA ID					16
		1	2.1			CLEARWATER ID					
1	5.6	1	6.2			CUSTER ID	12				
1	2.5			1	1.8	ELMORE ID			-30	18	23
				1	2.6	FRANKLIN ID					36
				1	2.2	FREMONT ID					30
		1	2.2	1	1.5	GEM ID		-32			20
		1	1.9	1	1.7	GOODING ID		-11			23
1	1.7	1	1.4	1	1.2	IDAHO ID	-18	-17	-31	17	17
				1	1.5	JEFFERSON ID					21
		3	6.8	3	5.3	JEROME ID		-23			80
2	1.4	8	4.6	8	3.5	KOOTENAI ID	238	-23	159	86	52
2	2.7	1	1.1	7	7.0	LATAH ID	-59	528	158	85	93
2	5.6	3	8.9			LEMHI ID	61				
				2	9.7	LEWIS ID					98
2	11.1					LINCOLN ID					
				1	2.5	MADISON ID					33
1	1.4	1	1.3	1	1.6	MINIDOKA ID	-13	30	14	47	22
		1	.7	2	1.2	NEZ PERCE ID		73			17
2	10.4					ONEIDA ID					
				1	2.3	OWYHEE ID					30
				1	1.4	PAYETTE ID					19
				1	3.8	POWER ID					58
1	1.4	7	7.3	6	6.4	SHOSHONE ID	423	-12	360	96	91
2	1.2	3	1.5	5	2.3	TWIN FALLS ID	25	56	94	75	30
		1	4.2			VALLEY ID					
1	1.9	1	2.0	2	3.6	WASHINGTON ID	7	75	86	74	53
2482*	6.2	2230*	5.1	2189*	4.8	ILLINOIS IL	-17	-5	-22		
21	5.3	17	4.1	20	5.0	ADAMS IL	-23	23	-6	35	77
1	2.3	3	4.3	4	8.3	ALEXANDER IL	91	94	271	94	96
4	4.3	5	4.7	4	4.7	BOND IL	10	0	10	45	73
3	3.0	4	3.7	5	4.0	BOONE IL	23	8	32	56	61
1	2.8	1	2.1	1	2.0	BROWN IL	-23	-5	-27	20	27
5	2.2	11	5.1	4	2.0	BUREAU IL	129	-60	-9	33	27
		1	3.4	1	1.9	CALHOUN IL		-42			25
8	6.8	1	.9	6	5.4	CARROLL IL	-87	500	-22	24	81
1	1.1	5	5.2	2	2.4	CASS IL	387	-54	125	81	33
19	5.9	10	2.7	14	3.2	CHAMPAIGN IL	-53	18	-45	10	47
10	4.3	9	4.1	9	4.4	CHRISTIAN IL	-6	6	0	39	67
2	1.7	1	.8	2	1.8	CLARK IL	-53	134	11	46	24
		3	3.0	3	3.4	CLAY IL		13			51
7	5.6	6	4.6	7	4.7	CLINTON IL	-18	4	-15	29	74
6	2.8	7	3.4	6	2.7	COLES IL	22	-18	0	38	38
1661*	9.0	1315*	6.6	1178*	6.0	COOK IL	-27	-9	-34	16	87
1	.7	3	2.5	3	2.4	CRAWFORD IL	233	-3	222	91	32
1	1.6			2	3.0	CUMBERLAND IL			84	73	43
5	2.4	7	2.9	5	1.9	DE KALB IL	22	-35	-21	24	25
4	3.7	6	6.2	1	.9	DE WITT IL	64	-85	-76	01	15
		2	2.2	1	1.0	DOUGLAS IL		-56			16
28	4.1	49	4.2	74	4.3	DU PAGE IL	2	3	6	42	66
3	2.3	3	2.2	5	4.4	EDGAR IL	-4	99	90	75	68
7	6.7	4	3.0	7	4.7	EFFINGHAM IL	-54	56	-29	19	74
4	2.9	6	4.4	3	2.3	FAYETTE IL	54	-49	-21	24	30
3	3.3	1	1.1	2	2.1	FORD IL	-67	89	-37	13	28
13	4.5	9	3.1	9	3.0	FRANKLIN IL	-30	-3	-32	17	43
6	2.2	6	2.4	5	1.9	FULTON IL	10	-20	-12	31	26
		1	2.2	2	3.4	GALLATIN IL		52			50
		7	6.1			GREENE IL					
6	5.6	4	3.5	4	3.0	GRUNDY IL	-37	-15	-46	10	42
2	2.4	2	3.2			HAMILTON IL	37				
6	3.6	6	4.3	9	6.7	HANCOCK IL	21	53	85	74	92
4*	12.2	2	7.6	1	3.3	HARDIN IL	-37	-56	-73	02	49
1	2.4	3	6.6	1	1.9	HENDERSON IL	173	-71	-22	23	25
12	4.5	4	1.4	5	1.7	HENRY IL	-68	22	-61	05	23
4	2.2	6	3.5	6	3.2	IROQUOIS IL	57	-9	43	61	45
2	1.0	4	2.1	6	2.9	JACKSON IL	107	41	192	89	41
1	1.3	1	1.4	2	2.4	JASPER IL	6	73	84	73	32
3	1.8	7	3.7	6	3.3	JEFFERSON IL	110	-10	88	74	48

* SIGNIFICANTLY MORE DEATHS THAN EXPECTED

COUNTIES COMBINED; SEE TECHNICAL NOTES

WHITE MALE — CANCER OF THE ESOPHAGUS — ICD 150

1950-1959 DEATHS	1950-1959 RATE	1960-1969 DEATHS	1960-1969 RATE	1970-1979 DEATHS	1970-1979 RATE	COUNTY AND STATE	PERCENT CHANGE 60S/50S	PERCENT CHANGE 70S/60S	PERCENT CHANGE 70S/50S	PERCENTILE %CHANGE 70S/50S	PERCENTILE RATES 1970S
1	1.0	3	3.4	3	3.3	JERSEY IL	223	-1	221	91	49
2	1.7	10*	7.9	5	4.0	JO DAVIESS IL	370	-50	135	82	60
2	4.0	5*	12.5	1	1.5	JOHNSON IL	212	-88	-62	05	21
28	3.4	33	3.6	40	4.0	KANE IL	7	10	18	50	61
9	2.3	18	4.1	18	4.4	KANKAKEE IL	75	7	87	74	67
2	2.9	3	3.2	4	3.7	KENDALL IL	11	14	27	54	55
12	4.2	11	3.5	8	2.4	KNOX IL	-18	-31	-43	11	32
46	6.6	52	5.0	62	4.8	LAKE IL	-24	-5	-27	20	74
39*	7.2	27	4.7	35	5.8	LA SALLE IL	-35	24	-20	25	85
1	.8	4	3.8	3	2.5	LAWRENCE IL	384	-35	217	91	34
3	1.6	10	5.3	5	2.7	LEE IL	237	-49	73	70	38
4	2.0	10	4.5	5	2.4	LIVINGSTON IL	127	-46	22	52	33
2	1.2	3	1.8	11	6.5	LOGAN IL	51	263	448	97	91
4	2.2	3	1.8	3	2.1	MC DONOUGH IL	-21	20	-6	35	28
14	4.2	22	5.2	17	3.4	MC HENRY IL	25	-35	-20	25	49
7	1.8	8	2.0	13	3.1	MC LEAN IL	7	59	70	69	44
16	3.2	10	1.9	25	4.5	MACON IL	-41	135	38	59	70
12	4.1	15	4.6	8	2.7	MACOUPIN IL	13	-42	-34	15	37
45	5.9	37	3.8	53	4.9	MADISON IL	-35	28	-16	27	76
6	2.7	7	3.3	14	5.9	MARION IL	20	82	118	80	87
8*	10.5	4	5.3	4	4.6	MARSHALL IL	-49	-12	-56	06	72
3	3.3	6	6.5	4	4.1	MASON IL	94	-37	23	52	63
4	6.2	1	1.5	1	1.4	MASSAC IL	-75	-11	-78	01	19
2	3.5	2	4.3	2	3.6	MENARD IL	24	-17	3	41	54
		3	3.1	3	3.1	MERCER IL		1			45
4	4.7	3	3.2	4	4.0	MONROE IL	-31	24	-14	29	61
6	2.8	11	5.5	8	4.0	MONTGOMERY IL	96	-28	42	61	61
8	3.5	4	2.1	2	1.1	MORGAN IL	-41	-47	-69	03	17
4	5.4	2	2.5	3	4.0	MOULTRIE IL	-53	58	-26	21	61
6	3.7	3	1.6	10	4.7	OGLE IL	-56	186	27	54	72
30	3.7	31	3.6	40	4.6	PEORIA IL	-2	28	25	53	72
2	1.8	6	5.5	3	2.5	PERRY IL	208	-55	40	60	34
2	2.4	8*	9.6	4	5.1	PIATT IL	305	-47	114	79	78
5	3.5	1	.8	2	1.3	PIKE IL	-76	60	-62	05	19
				2	6.7	POPE IL					92
1	2.0	2	4.0	1	2.3	PULASKI IL	105	-44	15	48	30
4*	13.4	1	3.0	1	3.9	PUTNAM IL	-78	33	-71	02	60
7	4.0	8	4.6	9	5.0	RANDOLPH IL	15	8	24	53	76
1	1.2			5	5.5	RICHLAND IL			363	96	83
31	4.8	33	4.6	40	5.6	ROCK ISLAND IL	-3	19	16	49	83
59*	7.2	50	5.5	39	4.2	ST CLAIR IL	-24	-24	-42	11	65
6	2.9	7	3.8	7	4.1	SALINE IL	34	8	45	62	64
31	4.8	50*	7.0	40	5.3	SANGAMON IL	48	-25	11	45	80
2	3.7	1	1.8	1	1.3	SCHUYLER IL	-50	-28	-64	04	18
		1	2.8	1	2.0	SCOTT IL		-27			26
3	2.0	7	4.5	4	2.6	SHELBY IL	124	-42	30	55	36
4	8.5	2	5.0	5*	12.0	STARK IL	-41	141	41	60	99
5	2.2	9	3.7	9	3.7	STEPHENSON IL	74	-2	70	69	55
10	3.3	22	5.5	14	2.6	TAZEWELL IL	66	-52	-20	25	36
4	3.1	3	2.3	1	.5	UNION IL	-25	-79	-85	00	14
16	3.4	16	3.2	20	4.1	VERMILION IL	-6	28	20	51	64
1	1.4	4	4.9	2	2.7	WABASH IL	251	-44	96	76	38
2	1.7	1	.9	10*	8.5	WARREN IL	-46	829	400	97	96
2	2.4	4	4.0	4	3.8	WASHINGTON IL	67	-5	58	65	58
2	1.8	2	2.0	3	3.0	WAYNE IL	14	51	72	70	43
		2	1.9	5	3.9	WHITE IL		109			59
5	2.0	7	2.7	14	4.8	WHITESIDE IL	32	78	136	83	74
38	6.0	25	3.7	43	4.7	WILL IL	-38	26	-22	24	73
6	2.1	11	3.9	11	3.8	WILLIAMSON IL	84	-3	79	72	58
31	4.1	40	4.3	41	4.1	WINNEBAGO IL	4	-5	-1	38	63
7	6.2	4	3.1	9	6.2	WOODFORD IL	-49	98	1	39	89
609	3.3	704	3.5	886	4.1	INDIANA IN	6	16	24		
1	.9	2	1.7	4	3.0	ADAMS IN	83	81	230	92	42
28	3.4	46	4.7	43	4.1	ALLEN IN	38	-12	21	51	64
2	1.1	4	2.3	7	3.5	BARTHOLOMEW IN	106	52	214	91	52
		1	1.7			BENTON IN					
2	2.8	3	4.3	3	3.8	BLACKFORD IN	56	-12	38	59	58
3	2.3	3	2.1	7	4.5	BOONE IN	-9	114	95	75	69
				1	2.0	BROWN IN					26
4	4.5	2	2.3	4	4.2	CARROLL IN	-50	87	-7	34	66
7	3.1	2	.9	5	2.4	CASS IN	-70	157	-23	23	32
8	4.0	4	1.5	7	2.4	CLARK IN	-62	59	-40	12	32
2	1.4	3	2.2	5	3.8	CLAY IN	57	76	177	87	57
2	1.1	7	4.1	5	3.1	CLINTON IN	260	-23	176	87	45
1	1.8	1	2.2			CRAWFORD IN	22				
1	.6	3	2.3	3	2.0	DAVIESS IN	268	-13	220	91	27
5	3.9	12*	9.1	14*	9.5	DEARBORN IN	131	5	143	84	97
2	1.7	2	1.4	4	3.6	DECATUR IN	-17	149	106	78	53
2	1.3	4	2.8	4	2.7	DE KALB IN	110	-2	105	77	38
11	2.8	12	2.8	28	6.1	DELAWARE IN	-1	119	116	79	88
7	6.0	6	4.8	12*	8.4	DUBOIS IN	-21	76	40	60	96
14	3.4	9	2.0	19	3.7	ELKHART IN	-42	85	8	44	55
7	6.9	9*	8.2	3	2.5	FAYETTE IN	20	-69	-63	04	35
6	3.1	5	2.2	15	5.9	FLOYD IN	-30	167	86	74	86
		4	4.4	2	2.1	FOUNTAIN IN		-52			27
2	2.9	3	4.1	2	2.3	FRANKLIN IN	42	-44	-21	25	31
				1	1.0	FULTON IN					16
4	2.6	12	7.0	5	2.8	GIBSON IN	163	-60	5	42	39
8	2.6	4	1.1	10	2.8	GRANT IN	-56	144	7	43	39
2	1.2	1	.6	3	2.1	GREENE IN	-46	230	77	71	27
1	.6	3	1.7	4	1.7	HAMILTON IN	166	-2	160	86	22
3	2.7	3	2.5	5	3.1	HANCOCK IN	-7	24	16	49	45
2	2.1	1	1.1	3	2.7	HARRISON IN	-49	149	28	54	38
2	1.4	6	3.6	10	4.0	HENDRICKS IN	153	12	183	88	62
4	1.8	5	2.2	6	2.5	HENRY IN	22	13	38	59	34
5	2.2	8	2.6	14	4.5	HOWARD IN	20	73	107	78	70
7	4.2	4	2.4	7	4.2	HUNTINGTON IN	-42	74	1	39	65
4	2.9	7	4.9	4	2.4	JACKSON IN	66	-51	-19	26	32
4	5.7	5	6.3	4	4.1	JASPER IN	11	-36	-29	19	62

* SIGNIFICANTLY MORE DEATHS THAN EXPECTED

WHITE MALE

CANCER OF THE ESOPHAGUS — ICD 150

1950-1959 DEATHS	1950-1959 RATE	1960-1969 DEATHS	1960-1969 RATE	1970-1979 DEATHS	1970-1979 RATE	COUNTY AND STATE	PERCENT CHANGE 60S/50S	PERCENT CHANGE 70S/60S	PERCENT CHANGE 70S/50S	PERCENTILE %CHANGE 70S/50S	PERCENTILE RATES 1970S
1	.7			1	.8	JAY IN			13	47	15
7	5.2	5	3.9	3	2.4	JEFFERSON IN	-25	-38	-53	07	33
2	2.2	2	2.3			JENNINGS IN	3				
1	.8	6	3.8	5	2.6	JOHNSON IN	395	-32	234	92	36
6	2.9	5	2.2	12	5.2	KNOX IN	-24	133	78	72	80
3	1.5	2	.9	4	1.6	KOSCIUSKO IN	-37	64	3	41	21
3	3.5	4	4.6	3	3.3	LAGRANGE IN	31	-27	-4	36	49
92*	7.0	71	4.8	88	5.1	LAKE IN	-32	8	-27	20	79
16	4.2	20	4.7	20	4.5	LA PORTE IN	14	-6	7	43	69
3	1.7	7	3.8	5	2.4	LAWRENCE IN	130	-37	46	62	33
14	3.0	14	2.7	31	5.5	MADISON IN	-8	100	85	74	83
98	4.5	132*	5.3	147*	5.6	MARION IN	18	5	24	52	84
6	3.7	3	1.7	7	3.9	MARSHALL IN	-54	130	6	42	59
				3	5.9	MARTIN IN					87
5	3.1	5	3.2	6	3.8	MIAMI IN	6	17	24	52	57
4	2.0	9	4.1	6	2.3	MONROE IN	100	-43	14	48	31
5	3.3	5	3.0	4	2.3	MONTGOMERY IN	-8	-22	-28	20	31
3	2.5	3	1.9	4	2.0	MORGAN IN	-24	4	-21	24	26
2	3.6	2	3.2	5	7.4	NEWTON IN	-11	130	106	78	95
4	2.8	4	3.0	10	6.7	NOBLE IN	7	126	141	83	92
1	3.5	3*	12.1			OHIO IN	243				
1	.9	1	1.1	1	1.1	ORANGE IN	21	-3	17	49	17
1	1.9			4	4.8	OWEN IN			150	84	74
		2	2.1	4	4.6	PARKE IN		121			71
3	3.4	6	6.2	3	3.2	PERRY IN	86	-49	-6	35	45
9	5.2	5	1.9	13	4.1	PORTER IN	-63	119	-20	25	64
3	3.1	5	4.8	3	2.7	POSEY IN	55	-44	-14	29	37
1	2.0	3	4.3	4	5.3	PULASKI IN	120	22	169	86	80
2	1.7	2	1.7	5	3.7	PUTNAM IN	1	117	118	80	55
1	.5	5	3.4	5	3.4	RANDOLPH IN	537	0	537	98	50
2	1.9	2	1.8	6	5.4	RIPLEY IN	-1	193	191	89	82
1	1.1	3	2.6	3	3.0	RUSH IN	143	15	180	88	43
56*	6.4	41	4.1	45	4.2	ST JOSEPH IN	-35	2	-34	15	65
1	1.6			3	3.8	SCOTT IN			133	82	56
5	3.2	7	4.5	1	.7	SHELBY IN	39	-85	-79	01	14
3	3.4	4	4.1			SPENCER IN	21				
7	7.6	1	1.1	6	5.8	STARKE IN	-85	404	-24	22	85
3	3.2	3	3.4	4	4.1	STEUBEN IN	5	20	26	54	63
3	1.9	3	2.2	8	6.4	SULLIVAN IN	11	196	227	92	90
9	3.0	19	5.8	20	5.2	TIPPECANOE IN	95	-10	75	71	80
1	1.1	3	3.7			TIPTON IN	233				
1	3.3					UNION IN					
24	3.7	28	3.9	44	5.7	VANDERBURGH IN	7	44	54	64	84
4	3.1	4	3.5	4	4.2	VERMILLION IN	10	23	35	57	65
15	2.9	20	3.7	20	3.9	VIGO IN	27	5	33	57	59
2	1.4	3	2.0	3	1.8	WABASH IN	39	-7	29	55	24
1	1.9			2	4.4	WARREN IN			134	82	69
1	.9	1	.9	5	3.8	WARRICK IN	2	315	323	95	56
1	1.1					WASHINGTON IN					
6	1.8	14	4.1	16	4.5	WAYNE IN	131	10	155	85	70
1	.8	2	2.0	6	5.3	WELLS IN	139	171	550	98	81
1	.9	4	3.6	7	6.0	WHITE IN	314	65	585	98	87
1	1.1	5	4.8	4	3.4	WHITLEY IN	336	-29	208	90	49
393	2.9	511	3.6	583	4.0	IOWA IA	25	12	39		
		4	5.5	3	4.0	ADAIR IA		-28			60
		2	4.5			ADAMS IA					
1	1.1	5	5.1	5	4.4	ALLAMAKEE IA	374	-15	305	95	67
2	1.9	2	1.4	4	4.4	APPANOOSE IA	-27	209	125	81	68
		1	1.3	1	1.5	AUDUBON IA		11			20
8	6.5	2	1.5	4	3.1	BENTON IA	-77	107	-52	08	45
8	2.1	17	3.4	21	3.9	BLACK HAWK IA	64	15	89	75	60
		3	1.7	7	4.4	BOONE IA		155			68
6	5.9	1	1.0	1	.9	BREMER IA	-84	-7	-85	00	15
		2	1.8	3	2.9	BUCHANAN IA		61			40
1	.8	2	1.6	10*	8.1	BUENA VISTA IA	89	404	851	99	96
1	.9	7	7.1	6	5.3	BUTLER IA	660	-25	469	97	81
1	.9	4	4.0	4	4.4	CALHOUN IA	324	12	374	96	69
3	2.5	3	2.5	5	4.1	CARROLL IA	-2	66	62	67	64
4	3.7	5	4.5	5	4.7	CASS IA	21	4	26	53	73
2	2.5	3	3.0	5	4.5	CEDAR IA	20	50	80	72	70
9	3.9	8	3.4	15	5.7	CERRO GORDO IA	-13	68	46	62	84
4	4.0	4	3.7	5	5.2	CHEROKEE IA	-8	42	30	55	80
2	2.3			3	3.8	CHICKASAW IA			66	68	57
2	3.0	1	2.2	2	3.8	CLARKE IA	-28	73	25	53	57
1	1.3	3	3.2	8	8.1	CLAY IA	142	152	511	98	96
4	3.1	3	2.3	4	3.3	CLAYTON IA	-24	41	7	44	47
7	2.7	9	3.1	17	6.0	CLINTON IA	16	94	124	80	87
3	3.0	4	3.6	2	1.9	CRAWFORD IA	20	-46	-36	14	25
5	3.7	4	2.7	4	3.0	DALLAS IA	-27	10	-19	26	42
2	3.6					DAVIS IA					
2	2.8	3	5.0	2	3.8	DECATUR IA	79	-25	34	57	57
2	2.2	5	5.3	1	1.0	DELAWARE IA	142	-81	-53	07	16
5	2.5	8	3.5	12	5.1	DES MOINES IA	39	46	103	77	78
1	1.3	4	4.7	4	4.1	DICKINSON IA	256	-13	211	91	63
22*	6.7	21	6.1	15	4.2	DUBUQUE IA	-9	-32	-38	13	64
3	4.0	2	2.6	1	1.4	EMMET IA	-35	-48	-66	03	19
4	2.7	3	1.9	4	2.5	FAYETTE IA	-28	32	-5	36	34
				6	5.2	FLOYD IA					79
3	3.8	1	1.2	2	2.9	FRANKLIN IA	-69	145	-24	22	40
2	3.1	1	1.6			FREMONT IA	-49				
		3	3.4	2	2.8	GREENE IA		-17			39
		3	3.9	6	7.0	GRUNDY IA		78			93
2	2.3	2	2.1			GUTHRIE IA	-7				
3	3.1	2	1.8	6	5.5	HAMILTON IA	-40	200	79	72	83
1	1.3	3	3.6			HANCOCK IA	190				
5	3.6	9	6.7	4	2.9	HARDIN IA	85	-57	-20	25	40
1	.8			4	4.0	HARRISON IA			387	97	61
4	3.4	4	3.9	1	.8	HENRY IA	15	-79	-76	01	15
		3	4.1	2	2.7	HUMBOLDT IA		-34			38

* SIGNIFICANTLY MORE DEATHS THAN EXPECTED

1950-1959 DEATHS	1950-1959 RATE	1960-1969 DEATHS	1960-1969 RATE	1970-1979 DEATHS	1970-1979 RATE	COUNTY AND STATE		PERCENT CHANGE 60S/50S	PERCENT CHANGE 70S/60S	PERCENT CHANGE 70S/50S	PERCENTILE %CHANGE 70S/50S	PERCENTILE RATES 1970S
5	7.5	1	1.9	1	1.2	IDA	IA	-74	-38	-84	00	17
3	3.1	9*	10.5	5	5.5	IOWA	IA	234	-47	76	71	83
2	1.7	7	5.6	3	2.4	JACKSON	IA	236	-57	46	62	33
2	1.2	4	2.3	4	2.1	JASPER	IA	92	-7	79	72	28
3	3.4	3	3.2	2	2.0	JEFFERSON	IA	-7	-36	-40	12	27
8	4.6	4	1.9	11	5.2	JOHNSON	IA	-59	175	14	47	80
1	.9	3	2.8	4	3.7	JONES	IA	194	32	288	94	55
2	1.5	1	.8	3	3.5	KEOKUK	IA	-51	369	130	81	52
3	2.5	5	3.8	5	3.8	KOSSUTH	IA	55	0	54	65	57
6	2.8	9	3.9	12	5.6	LEE	IA	38	42	96	76	83
29	5.5	25	4.1	30	4.5	LINN	IA	-26	9	-19	26	69
1	1.6			2	3.0	LOUISA	IA			95	75	43
2	2.9	4	5.6	5	6.8	LUCAS	IA	94	21	135	82	93
3	4.3	1	1.2	4	5.3	LYON	IA	-71	324	25	53	81
		4	4.5			MADISON	IA					
5	3.5	4	2.6	5	3.6	MAHASKA	IA	-26	40	4	41	54
4	2.8	7	4.3	1	.5	MARION	IA	56	-88	-81	00	14
2	1.1	7	3.1	5	2.3	MARSHALL	IA	200	-26	122	80	31
2	2.6	1	1.4	7*	10.3	MILLS	IA	-44	615	298	94	98
2	2.4	5	6.3	4	4.7	MITCHELL	IA	168	-26	99	76	73
2	2.4	6	6.3	2	2.3	MONONA	IA	159	-64	-7	34	30
		1	1.2	6	10.7	MONPOE	IA		826			98
1	1.0	5	5.6	4	4.0	MONTGOMERY	IA	439	-29	285	94	62
11	5.9	9	4.8	5	2.6	MUSCATINE	IA	-19	-45	-55	06	37
3	2.9	3	3.0	6	5.8	O'BRIEN	IA	3	91	97	76	85
1	2.2	1	2.5	1	2.1	OSCEOLA	IA	10	-15	-6	35	28
1	.6	4	2.9	6	4.7	PAGE	IA	372	62	664	99	73
1	1.3	4	4.9	8*	10.0	PALO ALTO	IA	263	105	644	99	98
		3	2.4	5	3.7	PLYMOUTH	IA		52			55
2	3.1	4	4.6	4	5.4	POCAHONTAS	IA	50	17	76	71	82
34	3.3	64*	5.6	59	5.1	POLK	IA	70	-10	53	64	78
6	1.6	18	4.8	17	4.3	POTTAWATTAMIE	IA	196	-11	164	86	66
		1	.9	3	3.1	POWESHIEK	IA		257			44
		1	1.8	2	3.7	RINGGOLD	IA		108			56
2	1.9	4	4.4	4	4.6	SAC	IA	128	6	142	84	72
20	4.0	19	3.8	30	5.2	SCOTT	IA	-5	39	32	56	80
3	3.6	1	1.0	3	3.2	SHELBY	IA	-72	212	-12	31	46
4	3.3	7	5.4	4	3.0	SIOUX	IA	67	-45	-8	34	42
5	2.7	7	3.5	4	1.8	STORY	IA	30	-48	-33	16	24
8	6.2	5	3.5	3	2.5	TAMA	IA	-43	-29	-60	05	34
5	7.2			1	2.5	TAYLOR	IA			-66	03	33
2	2.2	2	2.5	7	8.7	UNION	IA	17	243	303	95	97
1	1.1	1	1.2			VAN BUREN	IA	10				
7	2.9	8	3.3	9	3.6	WAPELLO	IA	16	7	25	53	53
		1	.9	6	5.3	WARPEN	IA		458			81
1	.9	1	.8	1	.8	WASHINGTON	IA	-10	3	-7	34	15
1	1.4			2	2.8	WAYNE	IA			98	76	39
11	5.0	12	5.1	13	5.3	WEBSTEP	IA	2	4	6	42	80
5	6.8	5	6.9	1	1.1	WINNEBAGO	IA	2	-84	-84	00	17
2	1.7	7	6.3	7	6.3	WINNESHIEK	IA	258	1	263	93	90
28	5.6	24	4.5	25	4.9	WOODBURY	IA	-21	10	-13	30	76
1	1.6	2	3.3			WORTH	IA	103				
4	3.9	1	.9	6	5.8	WRIGHT	IA	-78	564	47	62	85
192	2.0	269	2.6	358	3.3	KANSAS	KS	29	27	64		
1	.8	2	1.8	3	3.2	ALLEN	KS	130	77	307	95	46
2	3.2			2	2.6	ANDERSON	KS			-20	25	35
4	3.5	2	2.0	2	2.2	ATCHISON	KS	-44	11	-38	13	30
		1	2.2	4	9.3	BARBER	KS		320			97
1	.8	5	4.0	5	3.3	BARTON	KS	379	-17	299	94	48
2	2.2	3	2.7	2	2.2	BOURBON	KS	22	-18	0	39	29
		2	2.0			BPOWN	KS					
6	3.5	3	1.5	5	2.5	BUTLER	KS	-56	61	-29	19	34
1	3.4	1	3.5			CHASE	KS	5				
				2	5.0	CHAUTAUQUA	KS					77
1	.7	5	3.6	3	2.2	CHEROKEE	KS	423	-40	216	91	29
1	2.5			2	7.3	CHEYENNE	KS			195	89	95
		2	9.1	2	9.6	CLARK	KS		5			97
1	1.5			1	1.2	CLAY	KS			-25	22	17
3	2.8	3	3.1	1	1.1	CLOUD	KS	13	-65	-61	05	16
		2	2.4	4	7.4	COFFEY	KS		204			95
				1	5.9	COMANCHE	KS					87
4	2.0	1	.4	7	3.5	COWLEY	KS	-79	736	74	70	51
1	.3	13	5.2	9	4.2	CRAWFORD	KS	1455	-19	1157	99	65
				1	3.0	DECATUR	KS					42
2	1.7	1	.7	5	4.0	DICKINSON	KS	-62	514	135	82	61
2	3.2	1	2.5	1	1.7	DONIPHAN	KS	-21	-30	-45	10	23
1	.7	5	2.9	5	2.6	DOUGLAS	KS	332	-12	282	94	36
2	6.0			2	7.1	EDWARDS	KS			19	50	94
2	5.4	2	4.9	1	2.2	ELK	KS	-10	-55	-60	05	29
2	3.2	1	1.1	3	3.0	ELLIS	KS	-65	175	-5	36	42
3	6.9	2	3.5	1	1.7	ELLSWORTH	KS	-48	-53	-76	01	22
1	1.7	3	4.6	3	3.8	FINNEY	KS	169	-17	124	80	58
4	4.9	3	2.9	6	5.3	FORD	KS	-41	84	9	45	81
1	.9			3	2.3	FRANKLIN	KS			171	87	31
1	2.1	3	4.2	3	3.8	GEARY	KS	98	-10	79	72	57
				1	4.3	GOVE	KS					67
		1	3.2	1	3.0	GRAHAM	KS		-5			42
				1	6.9	GRANT	KS					93
		1	3.5	1	3.6	GRAY	KS		3			54
3	4.1			6	6.3	GREENWOOD	KS			54	64	90
				2	10.7	HAMILTON	KS					98
				1	1.6	HARPER	KS					21
2	1.7	4	3.1	6	4.0	HARVEY	KS	85	30	140	83	61
				1	6.4	HASKELL	KS					91
		1	7.3			HODGEMAN	KS					
2	2.9	1	1.1			JACKSON	KS	-64				
1	1.5					JEWELL	KS					
7	2.5	13	2.4	32	4.0	JOHNSON	KS	-3	64	59	66	61
				1	6.2	KEARNY	KS					89

* SIGNIFICANTLY MORE DEATHS THAN EXPECTED

WHITE MALE | CANCER OF THE ESOPHAGUS | ICD 150

1950-1959 DEATHS	1950-1959 RATE	1960-1969 DEATHS	1960-1969 RATE	1970-1979 DEATHS	1970-1979 RATE	COUNTY AND STATE	PERCENT CHANGE 60S/50S	PERCENT CHANGE 70S/60S	PERCENT CHANGE 70S/50S	PERCENTILE %CHANGE 70S/50S	PERCENTILE RATES 1970S
1	1.8	2	2.7	2	3.0	KINGMAN KS	54	11	71	69	42
				1	3.2	KIOWA KS					47
3	2.1	2	1.2	6	3.9	LABETTE KS	-43	228	88	74	59
				2	11.3	LANE KS					99
7	3.7	12	5.2	11	4.8	LEAVENWORTH KS	41	-7	31	56	75
		1	2.4	2	7.7	LINCOLN KS		225			95
		2	3.1	3	7.0	LINN KS		127			93
		1	5.3			LOGAN KS					
1	.5	4	2.5	3	2.6	LYON KS	349	6	374	96	36
		1	.8	3	1.7	MC PHERSON KS		113			23
4	3.9	1	1.0			MARION KS	-75				
2	1.9	2	1.9	3	4.0	MARSHALL KS	-1	113	111	78	60
3	2.5	1	.7	2	1.7	MIAMI KS	-71	138	-32	17	22
2	4.0	2	3.8	5	9.7	MITCHELL KS	-6	157	141	83	98
4	1.6	7	2.9	15	6.3	MONTGOMERY KS	80	115	287	94	90
		1	2.0			MORRIS KS					
				1	6.6	MORTON KS					92
1	1.2			1	1.1	NEMAHA KS			-7	34	17
2	1.7	1	1.0			NEOSHO KS	-45				
		1	2.7			NESS KS					
		1	1.8	1	1.6	NORTON KS		-14			21
1	1.0	1	.9	4	4.5	OSAGE KS	-10	411	358	96	71
				2	4.4	OSBORNE KS					68
2	3.9			3	5.0	OTTAWA KS			29	55	77
1	1.5	1	1.6	1	1.7	PAWNEE KS	8	6	14	47	23
1	1.7	1	2.3	2	4.7	PHILLIPS KS	35	107	180	88	74
4	4.8	2	2.5	1	1.1	POTTAWATOMIE KS	-47	-58	-78	01	16
		1	1.3	1	1.4	PRATT KS		11			19
1	3.9	1	3.0	1	2.6	RAWLINS KS	-23	-12	-32	17	36
9	3.2	3	1.1	5	1.6	RENO KS	-67	55	-49	09	22
1	1.5	1	.9	3	3.7	REPUBLIC KS	-36	294	153	85	56
		1	1.1	1	1.1	RICE KS		-3			16
3	2.8	5	4.1	1	.8	RILEY KS	46	-79	-70	02	15
3	9.5	1	1.7	4	7.0	ROOKS KS	-82	310	-26	21	94
		3	6.8	1	2.1	RUSH KS		-70			27
1	2.1	4	6.1	2	3.2	RUSSELL KS	189	-47	52	63	46
6	3.7	7	3.9	5	2.4	SALINE KS	5	-39	-35	14	32
1	6.6	1	4.5	1	3.0	SCOTT KS	-32	-33	-54	07	43
18	2.0	38	3.4	47	3.5	SEDGWICK KS	68	3	72	70	52
		2	3.6	1	1.4	SEWARD KS		-62			19
9	1.8	19	3.3	17	2.7	SHAWNEE KS	88	-18	54	64	38
		1	5.2			SHERIDAN KS					
		3	8.5			SHERMAN KS					
		1	1.4	1	2.3	SMITH KS		63			30
		3	7.1			STAFFORD KS					
		2	9.5			STEVENS KS					
4	3.0	4	2.7	5	3.2	SUMNER KS	-9	21	9	45	47
1	3.2	3	9.3	1	3.0	THOMAS KS	187	-68	-7	34	42
2	7.0	1	3.5			TREGO KS	-50				
		1	2.1	3	6.0	WABAUNSEE KS		184			88
1	8.4	1	9.6	1	7.2	WALLACE KS	15	-26	-15	29	94
2	2.7	2	3.0	4	4.8	WASHINGTON KS	9	61	75	71	74
		1	7.4			WICHITA KS					
3	3.0	3	3.6	3	3.3	WILSON KS	19	-9	8	44	47
1	2.1			2	6.3	WOODSON KS			195	89	90
23	3.7	22	3.4	32	5.0	WYANDOTTE KS	-8	47	35	58	77
308	2.6	403	3.1	569	3.9	KENTUCKY KY	17	27	48		
		1	1.4	2	2.2	ADAIR KY		63			30
1	1.5	1	1.5			ALLEN KY	-2				
		1	2.0	1	1.8	ANDERSON KY		-8			24
1	1.9			1	2.5	BALLARD KY			32	56	34
2	1.5	4	2.9	6	3.6	BARREN KY	86	25	133	82	53
		4	8.2	4	7.5	BATH KY		-9			95
5	3.5	4	2.9	5	3.4	BELL KY	-18	17	-4	36	49
		5	5.3	6	4.6	BOONE KY		-13			72
2	2.8	2	2.6	1	1.3	BOURBON KY	-6	-51	-54	07	18
5	2.3	10	4.5	10	3.9	BOYD KY	95	-14	68	69	58
		5	6.1	5	5.2	BOYLE KY		-15			79
				1	2.3	BRACKEN KY					31
		1	1.4	4	6.0	BREATHITT KY		322			87
1	1.2	3	4.2	1	1.0	BRECKINRIDGE KY	251	-77	-19	26	16
3	5.7	2	2.7	4	3.5	BULLITT KY	-52	30	-38	13	52
2	3.0			2	2.8	BUTLER KY			-7	34	39
1	1.4	3	4.1	4	5.7	CALDWELL KY	189	38	298	94	84
		2	1.6	3	2.1	CALLOWAY KY		31			27
19	5.2	20	5.1	28*	7.2	CAMPBELL KY	-2	43	40	60	94
1	3.0	1	2.9	2	5.9	CARLISLE KY	-1	104	101	77	87
2	5.1	2	5.1	2	4.5	CARROLL KY	0	-12	-12	31	69
1	1.1	4	4.4	4	4.3	CARTER KY	305	-3	295	94	67
1	1.4			3	3.8	CASEY KY			179	87	57
4	2.7			4	2.4	CHRISTIAN KY			-9	33	33
1	1.4	2	2.2	3	3.0	CLARK KY	52	40	113	79	43
1	1.3	2	2.3	2	2.1	CLAY KY	70	-6	60	66	29
				1	1.7	CLINTON KY					22
				1	2.1	CRITTENDEN KY					28
1	2.3			3	7.0	CUMBERLAND KY			207	90	93
5	2.2	8	2.8	13	3.9	DAVIESS KY	26	37	73	70	59
1	2.1	1	1.9			EDMONSON KY	-9				
2	3.2	2	2.7	3	4.8	ESTILL KY	-17	79	48	63	74
13	3.7	8	1.8	19	3.3	FAYETTE KY	-51	79	-12	31	47
1	1.4	2	3.1	3	4.5	FLEMING KY	124	46	227	92	71
1	.7	2	1.1	4	2.1	FLOYD KY	59	92	206	90	28
4	3.7	3	2.3	9	6.0	FRANKLIN KY	-37	156	60	66	87
2	3.3	2	3.9	1	2.0	FULTON KY	19	-50	-40	12	26
1	2.3					GARRARD KY					
				2	3.3	GRANT KY					47
2	1.3	2	1.2	4	2.3	GRAVES KY	-8	93	79	72	30
		2	2.2	4	4.1	GRAYSON KY		89			62
1	2.0	1	1.7	1	1.5	GREEN KY	-14	-11	-24	23	20

* SIGNIFICANTLY MORE DEATHS THAN EXPECTED

1950-1959 DEATHS	1950-1959 RATE	1960-1969 DEATHS	1960-1969 RATE	1970-1979 DEATHS	1970-1979 RATE	COUNTY AND STATE		PERCENT CHANGE 60S/50S	PERCENT CHANGE 70S/60S	PERCENT CHANGE 70S/50S	PERCENTILE %CHANGE 70S/50S	PERCENTILE RATES 1970S
		2	2.0	11	7.1	GREENUP	KY		257			94
1	7.4			1	2.6	HANCOCK	KY			-64	04	37
3	1.4	3	1.8	1	.5	HARDIN	KY	28	-75	-68	03	14
4	2.1	3	1.8	6	3.5	HARLAN	KY	-12	93	69	69	51
4	5.0	3	3.4	2	2.3	HARRISON	KY	-33	-32	-54	07	31
1	1.3	2	2.6	1	1.3	HART	KY	97	-52	-6	35	18
1	.7	1	.6	7	4.3	HENDERSON	KY	-13	572	486	98	67
		2	3.1			HENRY	KY					
1	3.2					HICKMAN	KY					
1	.5	7	3.6	6	2.9	HOPKINS	KY	644	-20	495	98	40
		1	2.0	1	1.8	JACKSON	KY		-13			23
100*	5.5	123*	5.6	163*	6.6	JEFFERSON	KY	1	18	19	50	91
1	1.7			1	1.4	JESSAMINE	KY			-16	28	20
1	1.1	2	2.1	3	2.5	JOHNSON	KY	93	22	135	82	34
29*	6.5	44*	8.6	36*	6.7	KENTON	KY	33	-23	3	40	92
		1	1.4	2	3.1	KNOTT	KY		112			44
1	1.0			2	2.0	KNOX	KY			108	78	26
2	3.4			1	1.6	LARUE	KY			-55	07	21
2	2.4	1	.7	4	2.4	LAUREL	KY	-69	220	0	38	32
		1	1.6	2	3.3	LAWRENCE	KY		107			48
1	2.9	2	6.1	1	1.8	LEE	KY	109	-70	-37	14	24
1	2.9			5*	9.9	LESLIE	KY			244	92	98
		1	.9	2	1.9	LETCHER	KY		121			25
				2	2.4	LEWIS	KY					32
		1	1.0	1	1.0	LINCOLN	KY		-4			16
				1	2.4	LIVINGSTON	KY					32
4	3.6	4	3.8	3	2.7	LOGAN	KY	4	-29	-25	21	38
2	5.5	1	2.9			LYON	KY	-47				
5	2.2	11	4.3	14	4.7	MC CRACKEN	KY	96	10	115	79	73
		2	3.8	2	3.1	MC CREARY	KY		-19			44
1	1.9			1	1.8	MC LEAN	KY			-5	35	24
		5	3.7	6	3.6	MADISON	KY		-1			54
1	2.3			2	3.8	MAGOFFIN	KY			63	67	57
1	1.7	1	1.5	3	4.2	MARION	KY	-8	176	155	85	65
2	2.4	2	1.9	4	3.0	MARSHALL	KY	-19	54	24	53	42
3	3.3	4	4.5	3	3.3	MASON	KY	35	-25	1	39	49
		2	4.4	2	3.5	MEADE	KY		-20			52
		2	9.8			MENIFEE	KY					
		5	6.5	1	1.0	MERCER	KY		-85			16
				3	4.5	MONROE	KY					70
		2	3.3	2	2.6	MONTGOMERY	KY		-20			37
2	3.8					MORGAN	KY					
4	2.6	6	3.8	2	1.4	MUHLENBERG	KY	47	-64	-47	09	19
1	1.6	3	3.7	9*	9.3	NELSON	KY	128	152	476	98	97
		2	5.8			NICHOLAS	KY					
2	1.6			5	4.2	OHIO	KY			156	85	65
2	4.0	1	2.1	2	3.0	OLDHAM	KY	-49	46	-26	21	42
4	7.4			2	4.1	OWEN	KY			-44	11	64
				2	6.6	OWSLEY	KY					91
				3	5.9	PENDLETON	KY					86
3	2.5	2	1.6	2	1.5	PERRY	KY	-33	-11	-40	12	20
3	1.2	3	1.2	11	3.6	PIKE	KY	-3	212	202	90	54
1	3.0	2	5.8	3	7.3	POWELL	KY	93	26	143	84	94
		1	.5	5	2.5	PULASKI	KY		414			35
1	1.8	2	2.9	2	3.0	ROCKCASTLE	KY	59	3	65	68	43
1	1.9	1	1.8	2	3.0	ROWAN	KY	-6	65	54	65	42
2	3.9			2	2.4	RUSSELL	KY			-38	13	33
1	1.2	2	2.7	3	3.6	SCOTT	KY	121	32	192	89	53
4	4.6	1	.9	3	3.0	SHELBY	KY	-81	250	-34	15	43
1	1.7	2	3.4	1	1.4	SIMPSON	KY	103	-57	-13	30	20
1	3.7			1	4.0	SPENCER	KY			7	43	61
3	4.9	1	1.5	2	2.5	TAYLOR	KY	-69	63	-50	08	34
		2	3.5	3	5.1	TODD	KY		48			78
2	4.6	1	1.9			TRIGG	KY	-58				
1	3.7	1	3.5			TRIMBLE	KY	-6				
		1	1.2	3	4.2	UNION	KY		255			66
2	1.1	8	3.8	6	2.8	WARREN	KY	256	-27	160	86	39
		1	1.8	2	4.2	WASHINGTON	KY		124			64
1	1.6	1	1.6			WAYNE	KY	5				
1	1.2	2	2.6	2	2.4	WEBSTER	KY	125	-10	103	77	31
4	3.3	4	3.1	5	3.7	WHITLEY	KY	-6	19	11	46	55
				1	2.4	WOLFE	KY					32
1	2.3			1	1.2	WOODFORD	KY			-47	09	18
251	3.5	342	3.9	457	4.1	LOUISIANA	LA	12	5	17		
3	1.8	8	5.5	14	7.2	ACADIA	LA	199	31	292	94	94
3	5.1	1	1.4	4	4.9	ALLEN	LA	-72	238	-5	36	75
1	1.4	1	1.4	3	2.6	ASCENSION	LA	-4	92	84	73	36
1	2.9	1	3.0			ASSUMPTION	LA	2				
3	2.3	1	.7	3	1.7	AVOYELLES	LA	-70	156	-24	23	23
		2	2.5	6	5.9	BEAUREGARD	LA		139			87
		4	7.0			BIENVILLE	LA					
3	6.3	5	4.9	6	4.2	BOSSIER	LA	-22	-15	-34	15	65
17	4.1	26	4.2	29	4.2	CADDO	LA	3	-1	2	40	64
7	3.3	16	4.8	19	4.0	CALCASIEU	LA	46	-16	23	52	62
1	2.3	1	2.6	1	2.2	CALDWELL	LA	14	-17	-6	35	29
		1	4.1	1	2.7	CAMERON	LA		-35			37
		1	2.5	1	2.1	CATAHOULA	LA		-15			28
2	3.4	2	2.7			CLAIBORNE	LA	-19				
3	8.6	1	2.5	1	1.8	CONCORDIA	LA	-71	-26	-79	01	24
		1	1.9	2	2.8	DE SOTO	LA		49			39
13	3.8	32*	6.5	39	5.2	EAST BATON ROUGE	LA	72	-19	39	59	80
				4*	18.7	EAST CARROLL	LA					99
		2	2.8	1	1.6	EAST FELICIANA	LA		-42			21
1	.7	2	1.6	8	6.0	EVANGELINE	LA	120	264	703	99	87
1	1.5	4	4.9	2	2.3	FRANKLIN	LA	233	-53	55	65	30
1	1.9					GRANT	LA					
3	3.5	1	.8	5	3.4	IBERIA	LA	-78	347	-1	38	51
1	1.7	1	1.3	2	3.3	IBERVILLE	LA	-25	160	96	76	48
1	3.1			2	3.1	JACKSON	LA			0	39	44

* SIGNIFICANTLY MORE DEATHS THAN EXPECTED

WHITE MALE

CANCER OF THE ESOPHAGUS

ICD 150

1950-1959 DEATHS	1950-1959 RATE	1960-1969 DEATHS	1960-1969 RATE	1970-1979 DEATHS	1970-1979 RATE	COUNTY AND STATE		PERCENT CHANGE 60S/50S	PERCENT CHANGE 70S/60S	PERCENT CHANGE 70S/50S	PERCENTILE %CHANGE 70S/50S	PERCENTILE RATES 1970S
12	3.4	22	5.0	43	3.9	JEFFERSON	LA	48	-22	15	48	59
2	2.3	1	1.6	6	5.3	JEFFERSON DAVIS	LA	-29	232	136	83	81
6	3.4	5	1.7	11	3.1	LAFAYETTE	LA	-49	77	-9	33	44
1	.9	5	3.2	6	2.7	LAFOURCHE	LA	264	-14	212	91	38
2	3.7	5	8.4	3	3.8	LA SALLE	LA	125	-54	3	41	58
		2	2.4			LINCOLN	LA					
		4	4.8	6	4.1	LIVINGSTON	LA		-14			63
1	9.3	1	3.2	1	4.2	MADISON	LA	-66	31	-55	06	65
1	1.2	3	3.7	7	6.8	MOPEHOUSE	LA	203	86	464	97	93
		2	2.9			NATCHITOCHES	LA					
87	5.3	84	4.7	96*	6.1	OPLEANS	LA	-12	30	15	48	89
8	3.5	11	4.5	16	4.6	OUACHITA	LA	28	4	33	57	72
		1	1.5	2	3.2	PLAQUEMINES	LA		107			45
				3	3.8	POINTE COUPEE	LA					57
14	5.1	13	4.2	23	6.0	RAPIDES	LA	-18	44	18	50	87
2	6.3	2	6.2			RED RIVER	LA	-1				
2	3.8			1	1.1	RICHLAND	LA			-71	02	16
3	4.7	5	5.7	3	2.4	SABINE	LA	22	-59	-49	09	32
1	2.2			4	2.3	ST BERNARD	LA			7	43	31
				1	2.5	ST CHARLES	LA					34
				1	3.3	ST HELENA	LA					49
				2	4.7	ST JAMES	LA					73
1	4.6	3	11.3	2	3.8	ST JOHN THE BAPT	LA	145	-66	-17	27	57
7	5.1	7	4.0	7	3.4	ST LANDRY	LA	-21	-17	-34	15	49
1	1.6	3	4.5	5	5.3	ST MARTIN	LA	185	16	231	92	80
4	3.9	3	3.3	2	1.4	ST MARY	LA	-15	-57	-63	04	19
2	1.8	4	2.8	5	1.4	ST TAMMANY	LA	55	-49	-21	24	19
4	2.9	10	6.0	5	2.5	TANGIPAHOA	LA	111	-59	-14	29	33
				1	3.7	TENSAS	LA					56
4	4.1	5	3.1	8	4.0	TERREBONNE	LA	-25	29	-3	37	61
		2	2.8	4	5.0	UNION	LA		78			77
7	5.3	8	4.7	9	4.9	VERMILION	LA	-11	4	-8	33	76
1	1.2	3	3.1	5	4.2	VERNON	LA	159	35	250	93	65
7	6.0	4	3.0	5	4.0	WASHINGTON	LA	-50	34	-32	17	62
3	3.3	5	4.0	2	1.2	WEBSTER	LA	22	-71	-64	04	17
		2	7.8	3	9.6	WEST BATON ROUGE	LA		23			97
1	1.8	1	1.5	3	5.3	WEST CARROLL	LA	-17	250	192	89	81
1	5.9					WEST FELICIANA	LA					
1	1.6	2	2.9	3	4.0	WINN	LA	82	38	152	85	61
164	3.6	216	4.5	249	4.8	MAINE	ME	27	7	36		
19	5.0	31*	7.3	23	5.0	ANDROSCOGGIN	ME	45	-31	0	39	77
14	3.8	16	4.3	26*	6.7	AROOSTOOK	ME	12	58	77	71	92
33	4.0	44	5.0	48	5.0	CUMBERLAND	ME	28	-2	25	53	76
5	4.3	2	1.9	4	3.3	FRANKLIN	ME	-56	73	-25	22	47
2	.9	7	3.4	13	6.1	HANCOCK	ME	256	81	545	98	88
19	4.4	16	3.5	28	5.6	KENNEBEC	ME	-22	60	26	53	83
8	4.8	9	5.1	6	3.6	KNOX	ME	7	-30	-25	22	53
2	1.8	8	6.6	7	4.9	LINCOLN	ME	271	-25	179	87	76
7	3.0	12	5.0	15	6.0	OXFORD	ME	70	19	102	77	87
16	3.0	21	4.0	26	4.6	PENOBSCOT	ME	33	15	53	64	72
4	3.6	2	1.9	5	5.1	PISCATAQUIS	ME	-48	166	39	60	77
5	4.3	5	4.2	6	4.8	SAGADAHOC	ME	-2	15	13	47	75
7	3.1	5	2.4	5	2.3	SOMERSET	ME	-21	-6	-26	21	30
		3	2.5	4	3.1	WALDO	ME		23			43
5	2.5	4	2.1	9	5.0	WASHINGTON	ME	-14	136	103	77	77
18	4.0	31	5.9	24	3.9	YORK	ME	48	-34	-2	37	59
429*	5.3	580*	5.3	766*	5.7	MARYLAND	MD	-1	9	8		
15	4.1	22	5.4	27	6.0	ALLEGANY	MD	31	12	46	62	87
20	5.7	38	6.2	62*	6.2	ANNE ARUNDEL	MD	9	0	9	44	89
50	5.1	76	4.3	126	4.6	BALTIMORE	MD	-15	8	-9	33	72
		3	5.8	7	7.2	CALVERT	MD		24			94
2	2.1	5	5.4	4	4.1	CAROLINE	MD	158	-24	96	76	63
3	1.5	10	3.4	18	5.2	CARROLL	MD	135	50	253	93	79
8	5.0	7	4.0	9	3.8	CECIL	MD	-21	-4	-24	22	58
2	3.9	5	4.9	6	4.9	CHARLES	MD	27	-1	26	53	76
1	.9	5	4.1	7	4.9	DORCHESTER	MD	370	19	460	97	76
5	1.7	7	2.3	23	6.1	FREDERICK	MD	35	162	253	93	88
3	2.9	2	1.7	3	2.4	GARRETT	MD	-40	41	-15	28	33
6	3.5	8	2.6	14	3.8	HARFORD	MD	-26	48	10	45	58
4	5.3	3	2.6	10	4.1	HOWARD	MD	-51	58	-23	23	63
		1	1.2	8*	10.4	KENT	MD		756			98
24	3.3	55	4.8	100	5.5	MONTGOMERY	MD	44	14	64	67	83
25	4.4	66*	6.3	100*	6.8	PRINCE GEORGE'S	MD	42	9	54	65	93
2	3.0	3	4.1	6	5.9	QUEEN ANNE'S	MD	39	42	97	76	86
2	2.6	5	6.3	7	5.7	ST MARY'S	MD	139	-9	116	79	84
2	2.8	2	2.5	5	6.6	SOMERSET	MD	-8	160	140	83	92
4	5.0	6	6.1	8	6.5	TALBOT	MD	20	6	30	55	91
15	4.2	18	4.3	19	3.9	WASHINGTON	MD	3	-11	-8	33	58
4	2.3	9	4.3	9	4.0	WICOMICO	MD	89	-8	74	70	61
4	4.5	6	6.3	7	6.3	WORCESTER	MD	39	-1	38	59	90
228*	8.0	218*	7.2	181*	7.2	BALTIMORE CITY	MD	-10	0	-10	32	94
1495*	6.6	1478*	6.0	1446*	5.6	MASSACHUSETTS	MA	-9	-7	-15		
10	3.3	31*	6.5	49	5.7	BARNSTABLE	MA	99	-12	74	71	84
60*	9.1	49*	6.8	47*	6.1	BERKSHIRE	MA	-25	-11	-33	16	88
81	4.4	103	5.0	111	5.1	BRISTOL	MA	14	1	15	48	78
		2	5.4	1	1.8	DUKES	MA		-66			24
170*	6.4	167*	5.7	164*	5.3	ESSEX	MA	-10	-7	-16	28	81
5	1.9	16	5.2	17	5.2	FRANKLIN	MA	179	-1	175	87	79
126*	7.2	146*	7.3	102	4.9	HAMPDEN	MA	1	-33	-32	17	76
19	4.3	26	5.4	19	3.6	HAMPSHIRE	MA	24	-33	-17	27	53
308*	6.2	313*	5.7	326*	5.7	MIDDLESEX	MA	-9	1	-8	33	85
2	11.6	2	10.9	1	3.3	NANTUCKET	MA	-6	-69	-71	02	49
115*	5.9	103	4.5	144*	5.4	NORFOLK	MA	-25	20	-10	33	82
48	4.4	67	5.4	68	4.5	PLYMOUTH	MA	24	-17	3	41	71
411*	10.5	293*	8.2	217*	7.4	SUFFOLK	MA	-22	-9	-29	19	95

* SIGNIFICANTLY MORE DEATHS THAN EXPECTED

WHITE MALE CANCER OF THE ESOPHAGUS ICD 150

1950-1959 DEATHS	1950-1959 RATE	1960-1969 DEATHS	1960-1969 RATE	1970-1979 DEATHS	1970-1979 RATE	COUNTY AND STATE	PERCENT CHANGE 60S/50S	PERCENT CHANGE 70S/60S	PERCENT CHANGE 70S/50S	PERCENTILE %CHANGE 70S/50S	PERCENTILE RATES 1970S
140	5.1	160*	5.4	180*	5.8	WORCESTER MA	7	7	15	48	65
1289*	4.7	1507*	4.8	1672*	4.9	MICHIGAN MI	2	2	4		
2	5.1			2	3.1	ALCONA MI			-39	13	44
3	5.2	3	6.0	3	5.7	ALGER MI	15	-5	9	45	85
7	2.7	5	1.8	15	4.6	ALLEGAN MI	-33	157	73	70	72
		5	4.4	11*	7.7	ALPENA MI		76			95
1	1.1			6	6.3	ANTRIM MI			462	97	90
6*	11.1	1	1.9			ARENAC MI	-83				
1	3.3	5*	10.4	1	3.3	BARAGA MI	214	-68	0	38	48
1	.6	4	2.3	6	3.0	BARRY MI	277	28	383	97	41
26*	6.7	19	4.5	25	5.2	BAY MI	-34	15	-23	23	79
3	6.5	5*	9.2	2	3.6	BENZIE MI	42	-61	-45	10	54
25	4.3	25	4.0	27	4.0	BERRIEN MI	-7	0	-7	34	62
6	3.4	4	2.4	15*	7.9	BRANCH MI	-29	232	135	82	96
13	2.7	27	4.7	22	3.6	CALHOUN MI	73	-23	34	57	54
5	3.6	10	5.7	8	4.0	CASS MI	59	-31	10	45	61
		2	2.2	6	6.1	CHARLEVOIX MI		172			88
5	6.7	2	2.5	10*	8.6	CHEBOYGAN MI	-63	246	28	54	97
6	5.1	6	4.6	6	4.7	CHIPPEWA MI	-9	2	-6	35	74
1	1.6	1	.8	5	3.2	CLARE MI	-50	290	96	76	45
3	2.0	10	6.4	6	3.4	CLINTON MI	214	-47	68	69	50
2	7.4			2	4.7	CRAWFORD MI			-37	14	73
10	5.7	8	4.7	15*	8.0	DELTA MI	-17	70	41	60	96
11*	7.6	6	3.6	10	6.1	DICKINSON MI	-53	69	-20	25	89
6	2.9	9	3.8	11	4.0	EATON MI	34	6	41	60	62
7	8.7	2	2.2	3	3.0	EMMET MI	-75	37	-65	03	43
40	4.2	51	4.3	59	4.2	GENESEE MI	3	-3	0	39	65
2	3.5	2	2.5	3	3.1	GLADWIN MI	-27	21	-12	31	44
5	4.0	11	7.1	4	2.6	GOGEBIC MI	77	-63	-34	16	37
3	1.7	2	1.0	7	3.4	GRAND TRAVERSE MI	-39	226	97	76	50
7	4.1	5	2.9	3	1.7	GRATIOT MI	-27	-43	-58	06	23
3	1.8	4	2.1	4	1.9	HILLSDALE MI	19	-9	8	44	25
11	4.5	11	5.0	9	3.9	HOUGHTON MI	9	-22	-15	29	59
7	4.0	13	7.5	9	4.5	HURON MI	86	-40	12	46	70
29	4.5	32	4.4	33	4.1	INGHAM MI	-3	-6	-9	33	64
6	2.8	10	5.1	7	3.6	IONIA MI	81	-30	26	54	53
3	3.6	2	1.8	3	2.6	IOSCO MI	-49	43	-28	20	36
2	2.0	3	3.0	4	3.2	IRON MI	49	8	60	66	46
9	6.9	2	1.6	2	1.5	ISABELLA MI	-77	-6	-79	01	20
23	4.4	16	2.9	23	3.8	JACKSON MI	-34	32	-13	30	58
19	3.1	22	3.5	19	2.6	KALAMAZOO MI	12	-27	-18	26	35
2	6.7			1	4.4	KALKASKA MI			-35	15	67
87*	6.8	67	4.6	85	5.1	KENT MI	-33	13	-24	23	78
1	5.6	1	4.5	1	3.7	KEWEENAW MI	-19	-19	-34	15	55
1	3.0	1	2.7	3	7.6	LAKE MI	-12	185	150	84	95
3	1.9	10	5.3	6	2.6	LAPEER MI	184	-52	38	59	35
2	3.8	3	5.4	2	2.8	LEELANAU MI	41	-47	-26	21	40
6	1.9	13	4.0	13	3.6	LENAWEE MI	108	-11	84	73	53
4	2.7	6	3.1	8	2.9	LIVINGSTON MI	17	-7	8	44	40
				4	9.3	LUCE MI					97
2	4.4	1	1.6	2	4.3	MACKINAC MI	-65	178	-2	37	67
32	3.6	77*	5.3	121*	6.1	MACOMB MI	47	15	69	69	88
4	3.7	2	2.1	4	3.2	MANISTEE MI	-45	57	-13	30	47
14	6.1	12	4.6	21*	7.9	MARQUETTE MI	-25	73	30	55	96
2	1.7	5	3.9	4	2.5	MASON MI	125	-35	47	62	35
		5	4.5	3	2.5	MECOSTA MI		-44			34
5	3.5	5	3.4	10	6.4	MENOMINEE MI	-3	86	80	73	90
6	6.0	6	3.6	7	4.1	MIDLAND MI	-40	16	-31	18	64
1	2.5	2	4.5	1	1.9	MISSAUKEE MI	80	-58	-25	22	25
15	4.5	17	4.1	24	5.3	MONROE MI	-10	31	18	50	81
1	.5	5	2.7	12	5.3	MONTCALM MI	439	99	972	99	81
1	5.6	1	2.4	2	4.1	MONTMORENCY MI	-57	72	-27	20	63
16	3.3	31	5.7	35	5.7	MUSKEGON MI	75	0	75	71	85
5	4.1	2	1.6	9	5.4	NEWAYGO MI	-61	239	32	56	82
47	2.8	96	3.9	170	5.1	OAKLAND MI	40	31	85	74	79
1	1.1	2	1.9	9*	8.6	OCEANA MI	72	347	667	99	97
1	1.3	1	1.2	5	4.9	OGEMAW MI	-9	305	269	94	75
1	2.0	1	1.5			ONTONAGON MI	-25				
		1	1.1	2	2.7	OSCEOLA MI		150			37
		1	3.1	2	4.1	OSCODA MI		35			64
		1	2.5	6*	10.1	OTSEGO MI		304			98
8	2.3	17	4.2	23	4.2	OTTAWA MI	80	-1	78	72	65
1	1.7	4	5.9	3	4.0	PRESQUE ISLE MI	259	-32	143	84	62
2	5.4	3	6.0	4	3.7	ROSCOMMON MI	11	-37	-30	18	56
27	4.2	25	3.6	43	5.6	SAGINAW MI	-16	58	33	57	84
9	2.1	24	4.7	34	5.9	ST CLAIR MI	126	24	181	88	86
5	2.5	5	2.5	7	2.9	ST JOSEPH MI	-1	19	18	50	41
3	1.6	10	5.3	10	5.0	SANILAC MI	229	-6	209	90	77
3	6.1	1	1.9	5	10.0	SCHOOLCRAFT MI	-69	417	63	67	98
2	.8	8	3.4	4	1.7	SHIAWASSEE MI	340	-49	126	81	23
4	1.8	9	4.1	10	4.5	TUSCOLA MI	129	9	149	84	69
8	3.3	10	4.1	12	4.5	VAN BUREN MI	22	11	35	58	70
17	3.5	15	2.8	34	5.7	WASHTENAW MI	-21	106	62	67	85
618*	7.0	652*	6.8	503*	5.8	WAYNE MI	-2	-14	-16	28	86
3	2.9	4	4.0	2	1.7	WEXFORD MI	39	-57	-40	12	23
555	3.8	659	3.9	791	4.3	MINNESOTA MN	4	11	15		
3	3.5	5	6.0	3	2.9	AITKIN MN	68	-51	-18	26	41
4	2.9	6	2.6	13	3.7	ANOKA MN	-9	41	28	54	55
4	3.3	3	2.2	5	3.7	BECKER MN	-33	70	14	47	55
1	.7	6	4.8	5	4.0	BELTRAMI MN	540	-17	430	97	60
2	3.2	2	2.2	5	4.9	BENTON MN	-29	121	56	65	76
				2	3.0	BIG STONE MN					42
7	3.7	8	3.9	8	3.6	BLUE EARTH MN	6	-9	-4	36	53
5	3.6	12*	8.0	5	3.2	BROWN MN	122	-60	-12	31	46
6	4.3	7	5.0	6	4.0	CARLTON MN	14	-19	-7	34	62
4	4.1	2	1.7	3	2.2	CARVER MN	-58	26	-47	10	29
2	2.1	3	2.2	2	1.2	CASS MN	3	-45	-43	11	17

* SIGNIFICANTLY MORE DEATHS THAN EXPECTED

WHITE MALE

CANCER OF THE ESOPHAGUS

ICD 150

1950-1959 DEATHS	1950-1959 RATE	1960-1969 DEATHS	1960-1969 RATE	1970-1979 DEATHS	1970-1979 RATE	COUNTY AND STATE	PERCENT CHANGE 60S/50S	PERCENT CHANGE 70S/60S	PERCENT CHANGE 70S/50S	PERCENTILE %CHANGE 70S/50S	PERCENTILE RATES 1970S
2	2.5	6	6.6	3	2.7	CHIPPEWA MN	162	-59	7	43	37
		3	2.7	5	4.3	CHISAGO MN		61			67
1	1.1	3	1.9	10	5.3	CLAY MN	69	184	381	96	81
				1	1.3	CLEARWATER MN					18
		2	13.4	1	4.2	COOK MN		-68			65
3	3.6	3	3.1	4	3.6	COTTONWOOD MN	-14	16	0	38	53
6	3.7	9	4.8	17*	7.4	CROW WING MN	31	54	101	77	95
5	1.9	8	2.2	19	3.9	DAKOTA MN	15	77	103	77	60
1	1.5	3	4.0	3	4.0	DODGE MN	171	0	172	87	62
2	1.5	6	4.3	6	3.5	DOUGLAS MN	184	-19	131	82	52
6	5.3	6	4.5	4	2.7	FARIBAULT MN	-17	-39	-49	09	38
7	5.2	4	2.6	3	1.6	FILLMORE MN	-50	-36	-68	03	22
2	1.1	8	4.3	8	3.8	FREEBORN MN	289	-13	238	92	57
4	2.1	6	3.0	9	4.0	GOODHUE MN	43	33	90	75	62
		2	3.0	1	1.9	GRANT MN		-35			25
152	5.0	185*	5.1	204*	5.4	HENNEPIN MN	2	7	9	44	82
2	2.1	5	5.3	2	2.1	HOUSTON MN	158	-60	2	40	28
1	1.5	3	3.4	4	3.8	HUBBARD MN	123	14	153	85	57
1	1.6	4	4.5	5	4.9	ISANTI MN	180	7	200	90	75
8	4.3	7	3.5	7	3.2	ITASCA MN	-17	-8	-24	22	47
		3	3.5	5	5.1	JACKSON MN		46			78
1	1.6	1	1.9	3	4.9	KANABEC MN	14	161	197	89	75
3	2.2	3	1.7	10	4.7	KANDIYOHI MN	-21	175	116	79	73
		1	2.2	2	4.0	KITTSON MN		80			62
3	2.9	2	2.0	5	5.2	KOOCHICHING MN	-31	159	80	72	80
1	1.5	2	2.3	4	4.1	LAC QUI PARLE MN	53	79	174	87	63
5*	10.9	2	3.1	10*	17.0	LAKE MN	-71	446	56	65	99
				1	5.2	LAKE OF THE WOODS MN					79
8	6.6	4	3.3	4	3.2	LE SUEUR MN	-49	-5	-51	08	46
1	2.5	1	1.9	4	6.4	LINCOLN MN	-24	236	156	85	90
3	2.9	3	2.5	4	3.1	LYON MN	-13	25	9	44	45
4	3.1	7	4.8	5	3.0	MC LEOD MN	56	-37	-1	37	43
2	6.6	1	2.9			MAHNOMEN MN	-55				
1	1.0	2	2.3	3	3.3	MARSHALL MN	124	43	219	91	47
		2	1.7	2	1.2	MARTIN MN		-26			17
4	3.9	4	3.5	2	1.6	MEEKER MN	-10	-55	-60	05	21
1	1.2	1	1.1	4	3.2	MILLE LACS MN	-8	197	174	87	47
5	3.8	5	3.8	6	3.9	MORRISON MN	1	4	5	42	60
6	3.4	4	2.1	7	2.9	MOWER MN	-38	35	-17	27	40
1	1.5	3	4.0	1	1.2	MURRAY MN	170	-71	-23	23	17
2	1.6	3	2.7			NICOLLET MN	66				
5	4.5	3	2.5	9	7.2	NOBLES MN	-45	191	60	66	94
4	5.7	4	6.3	6	8.3	NORMAN MN	9	33	45	62	96
11	4.8	9	3.3	13	4.2	OLMSTED MN	-32	26	-14	29	64
5	1.6	7	2.1	11	3.0	OTTER TAIL MN	29	44	86	74	42
		1	1.5	4	4.6	PENNINGTON MN		200			72
		4	3.2	7	6.2	PINE MN		96			89
		4	4.8	6	6.6	PIPESTONE MN		37			92
2	1.2	6	3.1	9	4.3	POLK MN	167	40	275	94	67
		1	1.4	2	2.2	POPE MN		59			29
80	5.2	97*	5.4	101	5.3	RAMSEY MN	5	-2	3	41	81
2	5.6			1	2.7	RED LAKE MN			-52	08	38
2	2.0	3	2.5	6	4.5	REDWOOD MN	26	78	124	80	70
6	4.6	7	5.1	4	3.4	RENVILLE MN	12	-34	-25	21	50
2	1.1	4	2.2	6	3.0	RICE MN	87	37	157	85	41
3	5.1	3	4.6	1	1.6	ROCK MN	-9	-65	-69	03	21
5	7.4					ROSEAU MN					
69*	6.0	68*	5.6	67	5.7	ST LOUIS MN	-7	2	-4	36	84
4	4.7	5	4.6	3	2.1	SCOTT MN	-2	-53	-54	07	29
1	2.3	1	1.3	2	2.1	SHERBURNE MN	-43	64	-7	34	28
5	5.8	1	1.0	3	3.3	SIBLEY MN	-82	221	-43	11	48
15	5.2	6	1.5	24	5.6	STEARNS MN	-71	274	7	43	84
5	4.6	4	3.2	3	2.3	STEELE MN	-31	-29	-51	08	30
3	6.1	2	3.0	1	2.0	STEVENS MN	-51	-34	-68	03	26
1	1.1	3	3.2	2	2.1	SWIFT MN	182	-34	85	74	27
2	1.6			4	2.7	TODD MN			67	68	37
2	4.9	1	2.9			TRAVERSE MN	-40				
3	3.3	5	4.7	1	.9	WABASHA MN	42	-81	-73	02	15
3	4.5	1	1.2	1	1.0	WADENA MN	-74	-18	-79	01	16
1	1.2	3	3.3	3	2.5	WASECA MN	182	-25	112	79	34
4	2.0	5	1.8	6	2.1	WASHINGTON MN	-10	15	4	41	28
4	5.0	5	6.0	3	3.3	WATONWAN MN	19	-45	-35	15	48
				3	5.1	WILKIN MN					77
5	2.6	12	5.9	13	5.8	WINONA MN	126	-3	119	80	85
7	4.1	1	.5	9	4.2	WRIGHT MN	-87	676	2	40	65
2	2.1	2	2.4	2	2.1	YELLOW MEDICINE MN	11	-12	-3	37	27
134	2.7	188	3.2	243	3.4	MISSISSIPPI MS	17	9	27		
3	4.7	1	1.3	2	4.0	ADAMS MS	-73	216	-15	29	61
2	2.2	5	4.3	4	3.1	ALCORN MS	94	-27	41	60	44
1	2.9			4	9.5	AMITE MS			234	92	97
1	1.2	1	1.5	3	3.6	ATTALA MS	23	141	196	89	54
8*	10.2	4	5.4	2	2.5	BOLIVAR MS	-47	-54	-76	01	34
3	5.6	1	2.1	1	1.1	CALHOUN MS	-62	-50	-81	00	16
1	3.6	1	4.6	2	5.4	CARROLL MS	30	16	52	63	82
1	2.5			1	1.6	CHICKASAW MS			-37	13	21
1	2.1					CHOCTAW MS					
				4	6.8	CLARKE MS					93
1	2.7	1	1.9	1	1.9	CLAY MS	-29	-1	-30	18	25
1	1.6			3	4.7	COAHOMA MS			195	89	73
5	6.5	3	4.2	1	1.7	COPIAH MS	-35	-59	-74	02	23
1	2.1					COVINGTON MS					
1	2.3			6	5.1	DE SOTO MS			120	80	78
2	1.4	5	3.0	5	2.6	FORREST MS	112	-13	85	74	37
		4	10.2	4	7.2	GEORGE MS		-30			94
		2	6.1			GREENE MS					
		1	2.9	2	3.7	GRENADA MS		26			55
1	1.7	1	1.4	4	5.0	HANCOCK MS	-19	252	184	88	76
9	3.0	25*	7.2	27	5.7	HARRISON MS	142	-21	90	75	84
10	3.7	21	5.0	23	4.4	HINDS MS	38	-14	19	50	67

* SIGNIFICANTLY MORE DEATHS THAN EXPECTED

WHITE MALE | CANCER OF THE ESOPHAGUS | ICD 150

1950-1959 DEATHS	1950-1959 RATE	1960-1969 DEATHS	1960-1969 RATE	1970-1979 DEATHS	1970-1979 RATE	COUNTY AND STATE	PERCENT CHANGE 60S/50S	PERCENT CHANGE 70S/60S	PERCENT CHANGE 70S/50S	PERCENTILE %CHANGE 70S/50S	PERCENTILE RATES 1970S
1	2.3	1	2.2			HOLMES MS	-5				
1	4.1			1	3.4	HUMPHREYS MS			-17	27	50
2	1.8	5	3.1	10	2.9	JACKSON MS	71	-8	58	66	40
2	4.3					JASPER MS					
		2	11.7	1	6.4	JEFFERSON MS		-46			90
1	12.4	3	8.7	1	2.4	JEFFERSON DAVIS MS	-30	-73	-81	00	32
5	3.6	7	3.7	4	1.7	JONES MS	3	-54	-52	08	23
		1	3.3			KEMPER MS					
		3	4.9	3	4.1	LAFAYETTE MS		-16			64
2	4.4			4	5.9	LAMAR MS			33	57	86
9	4.8	8	3.7	6	2.3	LAUDERDALE MS	-24	-37	-52	08	31
		2	5.6	2	4.8	LAWRENCE MS		-14			75
1	2.6			3	4.0	LEAKE MS			57	65	62
1	.8	1	.6	6	3.1	LEE MS	-21	383	283	94	44
1	1.5	2	2.8	4	4.5	LEFLORE MS	88	60	202	90	70
		1	1.1	2	2.2	LINCOLN MS		104			29
4	5.2	1	.9	5	3.7	LOWNDES MS	-83	327	-28	20	56
1	2.1					MADISON MS					
1	1.5	7*	8.7	3	3.5	MARION MS	476	-60	132	82	52
				3	5.1	MARSHALL MS					78
5	4.8	3	2.6	5	4.2	MONROE MS	-47	66	-12	31	65
		2	4.0			MONTGOMERY MS					
		5	6.9			NESHOBA MS					
1	1.4	2	2.7	2	2.7	NEWTON MS	97	0	96	76	38
2	10.6	1	3.6			NOXUBEE MS	-66				
1	2.4			2	3.4	OKTIBBEHA MS			39	59	50
1	1.4	1	1.7	1	1.1	PANOLA MS	24	-37	-22	24	17
3	4.1	5	5.7	3	2.5	PEARL RIVER MS	38	-56	-40	12	34
		1	2.8	2	8.0	PERRY MS		186			96
3	3.8	4	4.1	4	3.6	PIKE MS	8	-13	-6	35	54
2	2.7	2	2.8	4	4.7	PONTOTOC MS	4	71	77	71	74
		4	4.8	3	2.9	PRENTISS MS		-41			40
1	3.9			3	8.3	QUITMAN MS			111	78	96
2	2.3	2	1.5	3	2.2	RANKIN MS	-35	47	-5	36	29
1	1.7	3	4.9	3	3.9	SCOTT MS	180	-20	124	80	59
1	5.3			1	6.1	SHARKEY MS			14	48	88
2	2.8			4	5.2	SIMPSON MS			87	74	80
1	2.9	2	3.6	4	6.4	SMITH MS	25	77	121	80	91
1	4.1	2	7.6	1	3.4	STONE MS	84	-55	-17	27	51
1	1.2	2	4.3	4	5.6	SUNFLOWER MS	261	30	369	96	84
2	5.0	3	7.1	2	4.7	TALLAHATCHIE MS	41	-33	-6	35	73
		1	2.6	3	6.5	TATE MS		150			91
2	3.3	1	1.3	2	2.7	TIPPAH MS	-62	118	-17	27	38
2	2.9	1	1.3	5	5.9	TISHOMINGO MS	-55	342	100	77	86
		2	2.5			UNION MS					
1	2.5	2	5.0	3	6.6	WALTHALL MS	95	32	157	85	91
2	2.4	4	5.2	1	.7	WARREN MS	118	-86	-69	02	15
6	6.2	1	.9	10	7.5	WASHINGTON MS	-85	719	20	51	95
1	2.4	3	5.6	4	6.9	WAYNE MS	131	23	185	88	93
		2	4.4	1	1.7	WEBSTER MS		-61			23
		1	4.5	1	4.7	WILKINSON MS		4			73
3	4.9	1	1.5	2	3.1	WINSTON MS	-69	100	-37	14	44
		4	8.5	2	3.4	YALOBUSHA MS		-60			51
3	5.2	1	1.5	1	1.2	YAZOO MS	-70	-19	-76	01	18
720	3.7	731	3.5	821	3.7	MISSOURI MO	-6	7	1		
4	3.3	6	5.6	4	3.6	ADAIR MO	72	-37	9	44	53
		3	3.8	2	2.5	ANDREW MO		-35			34
		1	1.8			ATCHISON MO					
3	3.2	3	2.4	4	3.2	AUDRAIN MO	-24	33	1	40	46
3	2.2			3	1.8	BARRY MO			-19	26	24
		1	.9	2	2.2	BARTON MO		141			29
1	.8	2	2.3	4	3.5	BATES MO	199	49	345	96	51
1	1.2			3	3.6	BENTON MO			210	90	54
		3	4.4	1	1.2	BOLLINGER MO		-72			18
6	3.1	7	3.3	8	3.1	BOONE MO	5	-7	-2	37	43
13	2.6	20	4.1	26	5.6	BUCHANAN MO	60	36	118	79	83
1	.6	4	2.3	13	6.4	BUTLER MO	302	183	1036	99	91
		5	6.5	1	1.4	CALDWELL MO		-78			19
1	.7	4	3.2	6	4.1	CALLAWAY MO	334	31	469	98	64
2	3.1	4	4.7	5	4.6	CAMDEN MO	49	-1	47	62	72
1	.6	7	3.3	4	1.6	CAPE GIRARDEAU MO	471	-53	171	87	21
2	2.0			2	2.9	CARROLL MO			48	63	40
2	6.9			2	6.2	CARTER MO			-11	32	89
		1	.7	3	1.5	CASS MO		128			21
1	1.1			2	1.8	CEDAR MO			60	66	24
2	1.9			2	2.2	CHARITON MO			16	49	29
1	1.4	2	2.3	4	3.8	CHRISTIAN MO	60	65	164	86	58
		1	1.8	1	1.4	CLARK MO		-25			19
6	2.5	10	3.1	20	4.1	CLAY MO	24	31	62	67	64
1	1.5	4	5.8	1	1.2	CLINTON MO	297	-80	-19	26	17
4	2.8	7	3.8	9	4.1	COLE MO	38	7	48	63	63
		4	4.4	1	1.1	COOPER MO		-74			17
1	1.3	4	3.9	4	3.4	CRAWFORD MO	213	-14	171	87	50
		1	1.3	3	3.9	DADE MO		190			58
1	1.4	1	1.1	1	1.0	DALLAS MO	-21	-10	-29	19	16
4	4.2	2	2.1	4	5.8	DAVIESS MO	-50	175	38	59	86
1	1.4	1	1.7			DE KALB MO	23				
		1	1.5	2	2.2	DENT MO		44			29
1	1.7	1	1.4	2	3.0	DOUGLAS MO	-18	116	77	71	42
6	3.7	9	4.8	5	2.6	DUNKLIN MO	30	-46	-30	18	36
11	5.0	10	3.9	18	5.7	FRANKLIN MO	-22	45	13	47	84
1	1.1	3	3.0	4	4.1	GASCONADE MO	169	38	271	94	63
3	3.5			3	5.1	GENTRY MO			45	62	77
16	3.0	18	2.8	20	2.8	GREENE MO	-6	-2	-7	34	39
1	.9	2	2.0	2	2.8	GRUNDY MO	137	36	223	91	39
2	2.1	1	1.2			HARRISON MO	-41				
3	2.0	2	1.5	4	2.5	HENRY MO	-24	62	24	52	34
		1	2.0			HICKORY MO					
				2	3.2	HOLT MO					46

* SIGNIFICANTLY MORE DEATHS THAN EXPECTED

WHITE MALE — CANCER OF THE ESOPHAGUS — ICD 150

1950-1959 DEATHS	1950-1959 RATE	1960-1969 DEATHS	1960-1969 RATE	1970-1979 DEATHS	1970-1979 RATE	COUNTY AND STATE	PERCENT CHANGE 60S/50S	PERCENT CHANGE 70S/60S	PERCENT CHANGE 70S/50S	PERCENTILE %CHANGE 70S/50S	PERCENTILE RATES 1970S
		3	4.0	2	3.2	HOWARD MO		-22			46
4	3.0	1	.7	6	3.4	HOWELL MO	-76	360	12	46	50
1	1.7			2	3.7	IRON MO			115	79	56
126*	5.4	112	4.4	92	3.8	JACKSON MO	-18	-15	-30	18	57
14	3.4	15	3.4	28*	6.7	JASPER MO	3	95	100	77	92
8	3.7	6	2.0	18	4.4	JEFFERSON MO	-45	117	19	50	67
4	2.9	7	5.7	5	3.6	JOHNSON MO	98	-36	26	53	54
1	1.0	1	.8	4	3.2	LACLEDE MO	-15	277	219	91	46
4	2.6	10	6.4	3	2.0	LAFAYETTE MO	142	-68	-23	23	27
4	2.6	5	2.8	4	2.1	LAWRENCE MO	6	-24	-20	25	28
1	1.1	3	5.6	1	1.5	LEWIS MO	410	-73	36	58	20
2	2.3	1	.8	6	5.3	LINCOLN MO	-64	533	127	81	80
2	1.8	3	2.2	2	1.8	LINN MO	27	-20	1	39	24
		4	4.3	5	5.1	LIVINGSTON MO		17			78
2	2.8	1	1.0	4	5.5	MC DONALD MO	-63	424	93	75	83
2	1.5	4	3.5	1	1.1	MACON MO	138	-69	-25	22	16
2	3.8	1	1.4	2	3.2	MADISON MO	-63	132	-14	29	47
4	2.5	6	3.7	6	3.4	MARION MO	51	-7	40	60	51
1	2.8			1	2.3	MERCER MO			-16	28	31
		6	6.2	2	1.9	MILLER MO		-70			25
2	3.0	1	1.3	2	2.3	MISSISSIPPI MO	-56	73	-25	22	31
				3	3.2	MONITEAU MO					45
1	1.4	1	1.1			MONROE MO	-22				
1	1.3	2	2.6	4	6.3	MONTGOMERY MO	93	144	372	96	90
1	1.2			1	.8	MORGAN MO			-36	14	15
6	5.4	1	.9	3	3.4	NEW MADRID MO	-83	284	-37	14	50
1	.6	3	2.0	10	5.3	NEWTON MO	239	172	822	99	81
4	3.0	2	1.4	9	6.7	NODAWAY MO	-52	367	125	81	92
2	3.8					OREGON MO					
		2	2.6	4	5.5	OSAGE MO		114			83
3	8.1	3	5.4	2	3.1	OZARK MO	-33	-43	-61	05	45
1	.9	2	1.8	3	2.7	PEMISCOT MO	107	49	209	90	37
4	4.9	5	5.6	1	.9	PERRY MO	13	-85	-83	00	15
5	2.8	2	.9	7	3.8	PETTIS MO	-69	335	36	58	58
2	2.1	3	2.6	6	3.5	PHELPS MO	23	34	66	68	51
1	1.0	5	5.3	3	3.0	PIKE MO	402	-43	185	88	42
1	1.3	3	2.9	3	2.0	PLATTE MO	119	-30	54	64	27
1	.8	2	2.0	3	2.3	POLK MO	137	14	170	87	30
		4	5.6	1	1.0	PULASKI MO		-81			16
1	2.6	1	1.7			RALLS MO	-32				
2	1.1	4	3.3	5	3.3	RANDOLPH MO	197	0	198	89	48
3	2.9			7	7.0	RAY MO			141	83	93
2	3.4	3	4.2	1	1.2	RIPLEY MO	23	-70	-64	04	18
9	5.6	9	4.6	15	5.1	ST CHARLES MO	-18	11	-9	33	78
1	1.2	2	2.3	2	3.2	ST CLAIR MO	91	40	168	86	47
1	1.7	6*	8.7	4	5.2	STE GENEVIEVE MO	403	-40	202	90	80
8	4.5	5	2.2	10	4.0	ST FRANCOIS MO	-51	78	-13	30	60
79	4.3	110	4.0	135	3.8	ST LOUIS COUNTY MO	-8	-5	-12	31	57
3	1.9	1	.6	3	1.6	SALINE MO	-67	153	-17	27	21
1	2.6	1	2.2	1	1.7	SCHUYLER MO	-17	-20	-34	16	23
1	2.4					SCOTLAND MO					
5	3.8	3	1.9	6	4.0	SCOTT MO	-49	109	7	43	61
1	2.3	1	2.1			SHANNON MO	-8				
1	1.2			2	2.9	SHELBY MO			135	83	41
4	2.7	1	.8	3	1.8	STODDARD MO	-70	127	-33	16	24
		1	1.0			STONE MO					
1	1.5					SULLIVAN MO					
		1	.8	7	4.6	TANEY MO		452			72
2	1.7	2	1.5	1	.8	TEXAS MO	-14	-49	-56	06	15
2	1.1	2	1.2	5	3.2	VERNON MO	10	159	184	88	46
1	1.8	1	1.5	4	5.5	WARREN MO	-16	254	196	89	82
2	3.0			2	2.6	WASHINGTON MO			-15	29	36
		2	2.5	3	3.7	WAYNE MO		47			56
1	1.0	1	1.2			WEBSTER MO	23				
		1	2.9	2	6.4	WORTH MO		121			91
		7	7.8	3	3.2	WRIGHT MO		-59			47
265*	8.0	180*	6.5	134*	6.8	ST LOUIS CITY MO	-19	4	-16	28	93
81	2.8	88	2.7	122	3.5	MONTANA MT	-2	30	27		
1	1.8			2	4.0	BEAVERHEAD MT			127	81	62
1	2.9	1	2.6			BIG HORN MT	-11				
1	3.0	1	3.1	2	6.5	BLAINE MT	4	106	115	79	91
		1	5.6			BROADWATER MT					
1	1.6	1	1.9	1	1.4	CARBON MT	13	-23	-13	30	19
4*	27.7			1	9.1	CARTER MT			-67	03	97
6	2.2	10	3.3	11	3.6	CASCADE MT	47	8	59	66	53
2	3.0	2	2.7	2	3.3	CUSTER MT	-10	22	10	45	49
		1	4.1			DANIELS MT					
		1	2.0	2	4.3	DAWSON MT		116			67
4	3.8	5	5.4	5	5.8	DEER LODGE MT	42	8	54	64	85
		2	11.2	1	5.0	FALLON MT		-56			76
4	4.4			4	4.4	FERGUS MT			0	39	68
5	2.8	4	2.1	7	3.0	FLATHEAD MT	-25	42	7	43	42
5	5.5	1	.8	3	2.1	GALLATIN MT	-85	153	-62	05	28
				1	3.0	GLACIER MT					42
1	8.1					GOLDEN VALLEY MT					
1	8.6					GRANITE MT					
2	3.6	4	6.0			HILL MT	68				
				2	6.6	JEFFERSON MT					92
1	4.4					JUDITH BASIN MT					
2	2.3	1	1.1	2	2.4	LAKE MT	-52	116	4	41	32
3	2.1	5	3.6	8	4.7	LEWIS AND CLARK MT	69	30	120	80	73
		1	1.5	3	5.2	LINCOLN MT		251			79
				1	6.6	MC CONE MT					91
		1	2.3			MADISON MT					
				1	7.2	MEAGHER MT					94
		2*	11.8	1	6.2	MINERAL MT		-47			89
7	4.2	6	2.8	8	3.3	MISSOULA MT	-32	15	-22	24	47
1	2.2	2	7.0	2	7.0	MUSSELSHELL MT	210	0	212	91	93
3	4.7			3	4.1	PARK#MT			-13	30	63

* SIGNIFICANTLY MORE DEATHS THAN EXPECTED

COUNTIES COMBINED; SEE TECHNICAL NOTES

WHITE MALE | CANCER OF THE ESOPHAGUS | ICD 150

1950-1959 DEATHS	1950-1959 RATE	1960-1969 DEATHS	1960-1969 RATE	1970-1979 DEATHS	1970-1979 RATE	COUNTY AND STATE	PERCENT CHANGE 60S/50S	PERCENT CHANGE 70S/60S	PERCENT CHANGE 70S/50S	PERCENTILE %CHANGE 70S/50S	PERCENTILE RATES 1970S
		1	1.8			PHILLIPS MT					
2	8.9	1	3.2	4*	10.9	PONDERA MT	-64	242	22	51	98
		1	7.9	1	7.6	POWDER RIVER MT		-3			95
1	3.8			1	2.7	POWELL MT			-28	20	38
				2*	14.6	PRAIRIE MT					99
2	2.2	1	.9	2	2.0	RAVALLI MT	-58	121	-8	33	27
		1	1.4	1	1.8	RICHLAND MT		26			24
		1	1.9			ROOSEVELT MT					
2	7.1	1	3.1	2	5.9	ROSEBUD MT	-56	90	-17	27	87
2	4.5			1	2.0	SANDERS MT			-55	06	27
1	2.0			3	9.3	SHERIDAN MT			361	96	97
10	5.8	10	4.1	11	5.2	SILVER BOW MT	-29	27	-10	32	80
1	2.5			1	4.6	STILLWATER MT			82	73	71
		1	2.6	3*	12.0	SWEET GRASS MT		372			99
		1	2.1			TETON MT					
				1	2.5	TOOLE MT					34
		1	2.0	2	4.1	VALLEY MT		103			62
		2	11.5	2	9.5	WHEATLAND MT		-17			97
5	1.8	14	4.4	12	3.2	YELLOWSTONE MT	146	-26	81	73	47
224	3.3	263	3.6	315	4.2	NEBRASKA NB	9	16	26		
5	3.1	5	3.0	5	3.1	ADAMS NB	-4	5	1	39	44
		1	1.4	3	5.6	ANTELOPE NB		304			83
				1*	28.8	ARTHUR NB					99
		1*	16.3			BANNER NB					
1	1.5	2	3.1	2	4.7	BOONE NB	103	49	203	90	73
1	1.5	3	5.1	3	5.2	BOX BUTTE NB	235	2	242	92	79
1	3.9	3	10.1	1	4.0	BOYD NB	157	-60	2	40	61
1	2.8					BROWN NB					
4	2.8	3	2.2	5	3.1	BUFFALO NB	-24	42	7	43	43
2	3.0	1	1.2	4	6.0	BURT NB	-60	399	98	76	87
6	9.6	3	3.6	2	2.8	BUTLER NB	-62	-22	-71	02	39
3	3.2					CASS NB					
2	2.8	1	1.4	3	3.2	CEDAR NB	-49	123	13	47	46
1	8.3	2	6.5			CHASE NB	-22				
		1	1.6	3	5.4	CHEYENNE NB		236			82
3	4.8	2	4.1			CLAY NB	-16				
4	5.5	2	2.9	3	4.7	COLFAX NB	-48	64	-15	28	73
3	5.4	2	2.7	2	2.6	CUMING NB	-49	-5	-51	08	36
2	1.6	4	3.4	4	3.6	CUSTER NB	104	6	117	79	53
1	2.0	3	5.4	4	6.3	DAKOTA NB	173	16	215	91	90
1	1.6	1	1.6	2	3.6	DAWES NB	-3	129	122	80	54
1	.9	3	2.5	8	7.1	DAWSON NB	173	186	680	99	94
		1	5.6	2	10.3	DEUEL NB		82			98
1	1.9	2	3.8	1	1.4	DIXON NB	98	-62	-25	22	19
6	4.9	6	3.5	4	2.1	DODGE NB	-28	-40	-57	06	28
72	5.9	93*	6.6	97*	6.6	DOUGLAS NB	13	0	13	47	92
		1	2.5	2	3.6	FILLMORE NB		43			53
3	7.1	2	4.5	2	6.5	FRANKLIN NB	-37	45	-8	33	91
2	6.9					FRONTIER NB					
1	1.7	4	7.0	2	5.1	FURNAS NB	322	-27	209	90	78
6	4.4	2	1.3	8	5.1	GAGE NB	-70	284	16	49	78
1	4.6					GARDEN NB					
2	12.9	1	4.8			GARFIELD NB	-63				
1	7.3					GOSPER NB					
1*	31.5					GRANT NB					
		2	6.1	2	6.1	GREELEY NB		0			88
8	5.0	12	5.9	7	3.3	HALL NB	19	-45	-34	15	48
1	1.6			2	3.9	HAMILTON NB			140	83	59
				2	4.8	HARLAN NB					75
		1	16.1			HAYES NB					
		1	2.6	1	4.9	HITCHCOCK NB		91			76
1	1.2	1	1.1	7	8.4	HOLT NB	-13	692	589	98	96
2	6.7	3	6.6	1	1.9	HOWARD NB	0	-72	-72	02	25
		1	1.6			JEFFERSON NB					
				3	6.4	JOHNSON NB					91
1	1.9	2	4.0	1	2.0	KEARNEY NB	113	-51	4	41	26
		1	2.6	2	4.4	KEITH NB		70			68
				1*	94.4	KEYA PAHA NB					99
		1	3.5	1	3.4	KIMBALL NB		-2			51
2	2.3	1	1.1	3	4.1	KNOX NB	-52	273	78	72	62
13	2.2	21	3.2	20	2.9	LANCASTER NB	50	-11	34	57	41
1	.6	3	2.0	5	3.0	LINCOLN NB	206	53	368	96	43
				1*	56.2	LOUP NB					99
3	2.0	3	2.2	4	2.5	MADISON NB	9	16	26	54	34
2	4.0	1	1.6	4	7.2	MERRICK NB	-59	341	81	73	94
				2	5.4	MORRILL NB					82
1	2.8			1	2.1	NANCE NB			-25	22	27
2	4.2	3	5.2	1	1.6	NEMAHA NB	23	-70	-62	05	21
4	7.3			5	10.3	NUCKOLLS NB			41	60	98
3	2.8	3	3.0	6	6.4	OTOE NB	6	112	124	80	90
1	3.0	1	2.0			PAWNEE NB	-33				
				1	3.6	PERKINS NB					53
2	3.5	1	2.3	3	5.9	PHELPS NB	-34	158	71	70	87
1	1.8			3	5.8	PIERCE NB			212	91	85
4	3.9	6	4.8	8	6.0	PLATTE NB	21	25	52	64	87
		2	4.8			POLK NB					
		3	4.1	2	3.0	RED WILLOW NB		-28			41
3	3.2	2	2.0	5	5.9	RICHARDSON NB	-37	190	83	73	86
4	4.8	3	3.2	3	3.1	SALINE NB	-34	0	-34	15	45
3	5.3	2	2.4	6	4.6	SARPY NB	-54	91	-13	30	71
3	2.6	4	3.6	3	3.1	SAUNDERS NB	39	-14	20	51	45
2	2.0	2	1.7	9	5.4	SCOTTS BLUFF NB	-18	223	165	86	81
1	1.2	3	4.0	2	3.1	SEWARD NB	236	-23	160	86	45
3	6.7			3	6.0	SHERIDAN NB			-11	32	88
				1	2.4	SHERMAN NB					32
1	8.6	1	6.9			SIOUX NB	-20				
2	6.2	1	3.0			STANTON NB	-52				
2	3.7	2	2.8	2	3.0	THAYER NB	-23	6	-18	26	42
1	3.0	5*	17.7	1	2.1	THURSTON NB	480	-88	-32	17	27

* SIGNIFICANTLY MORE DEATHS THAN EXPECTED

WHITE MALE — CANCER OF THE ESOPHAGUS — ICD 150

1950-1959 DEATHS	1950-1959 RATE	1960-1969 DEATHS	1960-1969 RATE	1970-1979 DEATHS	1970-1979 RATE	COUNTY AND STATE		PERCENT CHANGE 60S/50S	PERCENT CHANGE 70S/60S	PERCENT CHANGE 70S/50S	PERCENTILE %CHANGE 70S/50S	PERCENTILE RATES 1970S
1	1.8	3	8.3	2	3.7	VALLEY	NB	354	-55	104	77	56
2	3.0	3	4.3	1	1.5	WASHINGTON	NB	44	-64	-49	09	20
2	3.5	2	3.8	3	6.2	WAYNE	NB	8	64	76	71	89
2	5.1			1	2.0	WEBSTER	NB			-60	05	27
1	24.8					WHEELER	NB					
1	1.3	2	2.4	1	1.0	YORK	NB	89	-59	-22	24	16
20	2.0	65	4.5	120	5.1	NEVADA	NV	119	13	148		
				1	1.7	CHURCHILL	NV					22
6	1.7	31	4.8	54	4.4	CLARK	NV	189	-7	167	86	69
				5	8.9	DOUGLAS	NV					97
		5	9.2	1	2.7	ELKO	NV		-71			37
				1	12.3	EUREKA	NV					99
		2	8.8	2	5.8	HUMBOLDT	NV		-34			86
		1	8.5			LANDER	NV					
		3	9.2	2	3.3	LYON	NV		-64			48
1	3.7					MINERAL	NV					
		1	2.8			NYE	NV					
				2	10.9	PERSHING	NV					98
9	3.0	19	4.4	45*	7.2	WASHOE	NV	47	65	143	84	94
2	3.6	2	4.2	1	2.8	WHITE PINE	NV	17	-35	-24	23	39
2	9.5	1	2.3	6	4.6	ORMSBY-CARSON CITY	NV	-76	103	-51	08	72
140	5.0	167*	5.3	226*	6.2	NEW HAMPSHIRE	NH	6	15	23		
4	2.8	10	6.2	10	5.3	BELKNAP	NH	120	-15	88	75	81
4	3.8	4	3.8	8	6.3	CARPOLL	NH	-1	68	67	68	90
9	4.9	13	5.6	19*	6.9	CHESHIPE	NH	15	24	42	61	93
11	5.5	17*	8.7	12	6.5	COOS	NH	60	-25	20	51	91
12	4.7	13	5.1	18	6.4	GRAFTON	NH	7	27	36	58	91
51*	6.9	48	5.6	79*	7.7	HILLSBOROUGH	NH	-20	39	11	46	96
10	2.8	23	5.8	25	5.7	MERRIMACK	NH	110	-2	106	78	85
20	5.0	19	4.0	35	5.5	ROCKINGHAM	NH	-18	36	11	46	83
14	5.8	12	4.5	13	4.1	STRAFFORD	NH	-24	-8	-30	18	62
5	3.5	8	5.2	7	3.8	SULLIVAN	NH	47	-27	8	44	57
1482*	6.8	1511*	5.6	1597*	5.2	NEW JERSEY	NJ	-17	-8	-24		
44*	6.6	43	5.3	47	5.1	ATLANTIC	NJ	-20	-3	-22	24	78
151*	6.0	195*	5.9	216*	5.3	BERGEN	NJ	-2	-9	-11	32	81
26	4.7	33	4.8	55	5.1	BURLINGTON	NJ	2	7	9	44	78
67	5.5	66	4.0	103*	5.8	CAMDEN	NJ	-27	44	5	42	86
17	6.9	16	4.2	19	3.7	CAPE MAY	NJ	-39	-13	-47	09	55
18	4.1	21	4.5	20	3.9	CUMBERLAND	NJ	9	-14	-6	35	59
253*	7.1	210*	5.5	190*	5.7	ESSEX	NJ	-22	4	-19	26	85
21	4.8	24	4.6	27	3.9	GLOUCESTER	NJ	-3	-16	-19	26	58
273*	9.7	235*	7.9	162*	6.1	HUDSON	NJ	-18	-23	-38	13	88
6	2.2	16	5.5	15	4.3	HUNTERDON	NJ	152	-23	94	75	66
76*	7.8	76*	6.7	66	5.4	MERCER	NJ	-14	-19	-30	18	82
108*	9.1	127*	8.0	131*	6.1	MIDDLESEX	NJ	-12	-24	-34	16	88
45	4.2	72	4.6	91	4.6	MONMOUTH	NJ	10	2	12	46	72
42	4.6	48	4.2	60	4.2	MORRIS	NJ	-10	0	-10	32	65
22	4.6	43	5.1	91	4.9	OCEAN	NJ	9	-3	6	43	76
126*	7.7	114*	5.9	116*	5.7	PASSAIC	NJ	-24	-3	-26	21	84
15*	7.6	5	2.6	7	2.8	SALEM	NJ	-66	7	-63	04	39
23	4.6	38	6.1	37	5.0	SOMERSET	NJ	32	-18	8	44	76
13	6.7	13	5.3	15	4.0	SUSSEX	NJ	-21	-26	-41	12	60
117*	7.3	98	4.4	106	4.4	UNION	NJ	-39	0	-39	12	69
19	6.2	18	5.6	23	5.8	WARREN	NJ	-9	4	-5	35	86
50	2.3	78	2.7	112	2.8	NEW MEXICO	NM	18	5	24		
13	2.7	25	3.4	31	2.6	BERNALILLO	NM	28	-22	-1	38	36
1	6.3	1	8.1			CATRON	NM	28				
3	2.3			7	2.8	CHAVES	NM			23	52	40
2	2.5	6	8.6	2	3.5	COLFAX	NM	243	-59	40	60	52
4	4.4	1	1.0	2	1.5	CURRY	NM	-78	56	-65	03	21
				1	4.3	DE BACA	NM					66
5	4.9	2	1.7	7	3.2	DONA ANA	NM	-65	88	-34	16	47
4	3.7	6	4.5	8	4.2	EDDY	NM	23	-7	14	48	65
2	3.3	2	3.2	5	4.9	GRANT	NM	-5	54	46	62	76
		1	4.0	1	8.3	HIDALGO	NM		106			96
1	.5	8	5.5	7	3.8	LEA	NM	922	-31	607	98	58
1	2.2					LINCOLN	NM					
1	37.5	1	9.9	3	14.0	LOS ALAMOS	NM	-74	41	-63	04	99
1	3.3	3	5.9	6	7.0	LUNA	NM	79	18	111	78	93
1	2.2	2	2.7	2	3.5	MC KINLEY	NM	23	32	63	67	52
		2	3.3	3	2.1	OTERO	NM		-36			28
1	2.3			1	1.5	QUAY	NM			-36	14	20
		1	1.5	1	1.6	ROOSEVELT	NM		6			21
		1	1.9	4	5.1	SANDOVAL	NM		170			77
1	1.5	3	4.2	3	1.6	SAN JUAN	NM	178	-61	8	44	22
1	.9	1	1.1			SAN MIGUEL	NM	23				
1	.8	6	3.5	11	4.9	SANTA FE	NM	350	38	521	98	75
1	3.6			1	2.3	SIERRA	NM			-37	14	31
1	1.8	1	1.2			TAOS	NM	-32				
1	4.1			1	2.7	TORRANCE	NM			-35	15	37
1	2.9					UNION	NM					
3	3.8	5	5.0	5	4.3	VALENCIA	NM	31	-14	13	47	66
4271*	6.3	4169*	5.5	4028*	5.2	NEW YORK	NY	-13	-5	-17		
54	4.6	76*	5.8	80*	5.8	ALBANY	NY	28	-1	27	54	85
4	1.9	5	2.3	6	2.7	ALLEGANY	NY	19	18	41	60	38
411*	6.7	309	4.8	290*	5.9	BRONX	NY	-29	21	-13	30	86
60*	6.8	52	5.3	57	5.4	BROOME	NY	-22	3	-20	25	82
15	3.7	18	4.5	20	4.8	CATTARAUGUS	NY	22	6	30	55	75
16	4.1	17	4.4	16	4.0	CAYUGA	NY	8	-9	-2	37	62
40	5.4	35	4.5	35	4.5	CHAUTAUQUA	NY	-18	1	-17	27	70
18	4.4	32*	7.1	23	5.0	CHEMUNG	NY	64	-30	14	47	76

* SIGNIFICANTLY MORE DEATHS THAN EXPECTED

WHITE MALE — CANCER OF THE ESOPHAGUS — ICD 150

1950-1959 DEATHS	RATE	1960-1969 DEATHS	RATE	1970-1979 DEATHS	RATE	COUNTY AND STATE		PERCENT CHANGE 60S/50S	70S/60S	70S/50S	PERCENTILE %CHANGE 70S/50S	RATES 1970S
6	2.7	6	2.9	14	6.3	CHENANGO	NY	7	114	130	81	90
11	4.4	6	2.4	12	4.5	CLINTON	NY	-46	90	2	40	70
11	4.0	8	2.7	9	2.7	COLUMBIA	NY	-32	-2	-34	16	37
7	3.5	8	4.1	7	3.6	CORTLAND	NY	15	-11	2	40	53
8	3.3	9	3.6	7	2.6	DELAWARE	NY	8	-29	-23	23	35
38	5.0	42	4.7	45	4.5	DUTCHESS	NY	-6	-3	-9	33	71
310*	7.7	306*	6.7	297*	6.4	ERIE	NY	-12	-5	-16	27	90
5	2.8	8	4.7	8	4.3	ESSEX	NY	71	-8	58	66	67
2	.9	9	4.6	13	6.2	FRANKLIN	NY	416	37	605	98	89
14	5.0	9	3.1	13	4.4	FULTON	NY	-38	43	-11	32	68
9	3.6	6	2.2	10	3.6	GENESEE	NY	-38	63	1	39	54
16*	8.4	5	2.4	10	4.4	GREENE	NY	-72	87	-47	09	69
				2	5.2	HAMILTON	NY					80
18	5.0	23	6.3	17	4.8	HERKIMER	NY	26	-24	-5	36	74
12	2.7	14	3.1	26	5.9	JEFFERSON	NY	15	93	122	80	87
816*	7.8	482	4.3	509*	5.3	KINGS(BROOKLYN)	NY	-44	22	-32	17	80
5	3.8	2	1.8	4	3.4	LEWIS	NY	-53	89	-11	32	50
4	1.9	6	2.7	8	3.4	LIVINGSTON	NY	43	25	79	72	49
5	2.1	6	2.4	12	4.5	MADISON	NY	15	87	114	79	69
138*	5.3	142	5.2	155*	5.4	MONROE	NY	-1	3	2	40	82
20	6.3	14	4.1	17	4.9	MONTGOMERY	NY	-35	19	-22	23	75
173*	5.9	237	5.0	291	5.2	NASSAU	NY	-14	2	-12	31	79
697*	8.6	880*	11.7	388*	6.6	NEW YORK(MANHATTAN)	NY	36	-43	-23	23	92
49	5.6	58*	6.1	55	5.3	NIAGARA	NY	9	-13	-5	35	81
68	5.8	35	2.7	59	4.5	ONEIDA	NY	-52	64	-22	24	70
90	5.5	98*	5.6	100	5.1	ONONDAGA	NY	1	-8	-7	34	79
12	3.4	8	2.2	15	3.7	ONTARIO	NY	-34	66	10	45	56
50	6.0	38	4.2	56	5.3	ORANGE	NY	-29	25	-11	32	81
7	3.7	8	4.5	2	1.2	ORLEANS	NY	21	-73	-68	03	17
19	5.0	17	4.1	21	4.6	OSWEGO	NY	-18	14	-7	34	72
10	3.3	5	1.7	9	3.0	OTSEGO	NY	-49	77	-10	32	41
5	3.9	12	6.6	17	6.2	PUTNAM	NY	69	-6	58	66	89
435*	6.4	418	4.7	449	4.7	QUEENS	NY	-27	1	-27	21	74
33	5.0	46*	6.5	50*	7.0	RENSSELAER	NY	30	7	39	59	93
67*	7.4	57	5.3	79*	6.2	RICHMOND	NY	-28	16	-17	27	89
21	5.4	30	5.3	37	4.5	ROCKLAND	NY	-1	-15	-16	28	70
12	2.5	16	3.4	25	5.2	ST LAWRENCE	NY	34	54	105	78	79
13	3.3	12	2.8	23	4.1	SARATOGA	NY	-16	47	24	52	62
41	5.5	36	4.7	42	5.2	SCHENECTADY	NY	-15	10	-6	35	79
4	2.9	2	1.4	4	2.5	SCHOHARIE	NY	-53	86	-13	30	35
2	2.6	1	1.2	4	4.5	SCHUYLER	NY	-54	266	70	69	69
2	1.2	9	5.1	6	3.4	SENECA	NY	320	-34	176	87	49
21	4.5	26	5.0	28	5.5	STEUBEN	NY	13	9	23	52	83
84	4.6	144	4.5	214	5.1	SUFFOLK	NY	-3	12	9	45	77
11	4.2	14	4.7	21	5.9	SULLIVAN	NY	10	27	40	60	86
8	4.9	7	4.1	6	3.4	TIOGA	NY	-16	-17	-31	18	49
6	2.7	7	2.9	16	6.2	TOMPKINS	NY	8	111	129	81	89
30	5.5	25	3.9	35	4.9	ULSTER	NY	-30	27	-11	32	76
7	3.2	16*	7.6	11	4.4	WARREN	NY	136	-42	36	58	68
9	3.7	11	4.6	5	1.9	WASHINGTON	NY	27	-59	-48	09	25
7	2.1	14	4.1	16	4.4	WAYNE	NY	95	7	109	78	68
191*	6.5	216*	6.0	212*	5.2	WESTCHESTER	NY	-8	-15	-21	24	79
9	4.9	7	3.9	5	2.7	WYOMING	NY	-20	-31	-45	10	38
5	4.9	4	4.0	5	4.3	YATES	NY	-17	5	-13	30	66
279	2.6	362	2.6	637	3.6	NORTH CAROLINA	NC	1	37	38		
6	2.8	3	1.1	15	4.5	ALAMANCE	NC	-61	313	60	66	71
2	4.6	1	1.4	2	2.4	ALEXANDER	NC	-70	74	-48	09	32
1	2.8	3	7.9	1	1.8	ALLEGHANY	NC	184	-78	-37	14	23
1	1.6	3	5.1	5	6.4	ANSON	NC	224	24	301	95	90
		4	3.8	3	2.6	ASHE	NC		-33			35
2	4.9	3	5.3	2	3.7	AVERY	NC	9	-30	-24	22	55
2	1.9	3	2.6	6	4.5	BEAUFORT	NC	36	75	138	83	71
1	1.9			1	1.9	BERTIE	NC			1	39	25
1	1.4	3	4.2	2	3.1	BLADEN	NC	204	-26	126	81	44
		2	3.5	7	5.6	BRUNSWICK	NC		60			84
11	2.2	17	2.9	19	2.7	BUNCOMBE	NC	34	-8	23	52	37
2	1.2	7	3.6	3	1.0	BURKE	NC	194	-73	-20	25	16
5	2.2	7	2.5	7	2.4	CABARRUS	NC	12	-4	7	44	32
2	1.3	4	2.3	9	3.5	CALDWELL	NC	73	53	163	86	51
1	4.8			1	6.0	CAMDEN	NC			25	53	87
3	3.1	2	1.6	14*	9.4	CARTERET	NC	-47	477	207	90	97
1	3.6	1	2.0			CASWELL	NC	-45				
2	1.4	5	2.1	13	3.3	CATAWBA	NC	49	58	135	82	49
1	1.1	5	4.7	5	4.1	CHATHAM	NC	314	-13	259	93	63
3	3.9	1	1.1	4	4.1	CHEROKEE	NC	-72	268	5	42	63
2	7.1			4	9.7	CHOWAN	NC			36	58	98
1	3.5			1	1.9	CLAY	NC			-47	09	24
4	3.8	1	.4	12	4.0	CLEVELAND	NC	-88	809	5	42	60
2	1.9	3	1.9	4	2.5	COLUMBUS	NC	0	32	32	56	35
3	3.2	8	7.2	7	4.5	CRAVEN	NC	124	-37	41	60	70
6	3.2	7	4.3	18	5.3	CUMBERLAND	NC	33	23	64	68	80
1	4.5			1	2.8	CURRITUCK	NC			-38	13	39
		2	5.4	4	7.6	DARE	NC		41			95
		5	1.5	9	2.4	DAVIDSON	NC		58			33
		1	1.8			DAVIE	NC					
2	2.3	1	.7	3	2.6	DUPLIN	NC	-70	263	11	45	36
8	3.5	13	4.0	22	6.5	DURHAM	NC	14	64	88	74	91
5	4.9	5	4.8	3	2.5	EDGECOMBE	NC	-1	-48	-48	09	35
11	2.8	18	3.7	20	3.0	FORSYTH	NC	34	-21	6	43	41
		3	3.5	3	3.8	FRANKLIN	NC		7			56
5	1.7	12	2.5	19	3.5	GASTON	NC	49	41	110	78	51
1	4.9	1	4.0	2	6.8	GATES	NC	-17	70	40	60	93
		1	3.3	1	2.6	GRAHAM	NC		-21			37
1	1.1	2	2.2	6	6.9	GRANVILLE	NC	99	209	513	98	93
				2	5.1	GREENE	NC					78
15	2.9	12	1.4	30	3.3	GUILFORD	NC	-53	141	13	47	47
2	1.7	10*	9.1	4	2.7	HALIFAX	NC	441	-70	62	67	38
4	2.9	1	.5	9	5.3	HARNETT	NC	-82	880	81	73	81
4	3.0	1	.6	4	1.7	HAYWOOD	NC	-81	197	-44	10	22

* SIGNIFICANTLY MORE DEATHS THAN EXPECTED

WHITE MALE — CANCER OF THE ESOPHAGUS — ICD 150

1950-1959 DEATHS	1950-1959 RATE	1960-1969 DEATHS	1960-1969 RATE	1970-1979 DEATHS	1970-1979 RATE	COUNTY AND STATE	PERCENT CHANGE 60S/50S	PERCENT CHANGE 70S/60S	PERCENT CHANGE 70S/50S	PERCENTILE %CHANGE 70S/50S	PERCENTILE RATES 1970S
2	1.5	8	3.9	10	3.4	HENDERSON NC	166	-13	131	82	49
1	2.5	2	4.0	1	2.0	HERTFORD NC	61	-50	-19	26	26
1	3.8			5*	18.9	HOKE NC			396	97	99
1	3.9	1	4.0			HYDE NC	4				
3	1.6	2	.8	9	2.7	IREDELL NC	-52	257	73	70	39
		1	1.1			JACKSON NC					
4	3.2	8	5.0	10	4.0	JOHNSTON NC	54	-20	23	52	60
1	3.7	1	4.3	2	6.2	JONES NC	15	45	67	68	89
1	1.2	2	2.2	3	3.5	LEE NC	81	59	187	88	51
3	2.3	6	4.6	10	6.6	LENOIR NC	101	45	191	88	92
3	3.3			3	2.3	LINCOLN NC			-31	17	30
4	3.9	2	1.6	4	3.0	MC DOWELL NC	-59	93	-21	24	43
1	1.2	1	1.3	4	3.0	MACON NC	2	141	145	84	43
6*	11.3	1	1.4	2	2.7	MARTIN NC	-68	98	-76	01	37
19	4.0	26	4.0	45	4.3	MECKLENBURG NC	-1	10	9	45	67
				1	1.2	MITCHELL NC					17
1	1.7	2	3.6	2	2.3	MONTGOMERY NC	112	-35	38	59	31
1	.9			8	4.3	MOORE NC			375	96	66
3	1.9	3	1.8	11	6.3	NASH NC	-5	253	237	92	90
11	7.5	7	2.9	15	5.4	NEW HANOVER NC	-62	90	-28	20	82
1	2.2	1	2.1	2	4.3	NORTHAMPTON NC	-5	107	97	76	66
2	2.4	1	.6	9	4.9	ONSLOW NC	-75	723	105	77	76
2	3.1	4	3.3	3	1.9	ORANGE NC	7	-42	-38	13	25
1	4.1			1	2.4	PAMLICO NC			-42	11	32
1	1.6	2	2.8	4	4.6	PASQUOTANK NC	70	63	179	87	71
1	4.4	1	4.0			PERQUIMANS NC	-9				
4	5.9	2	3.6	2	1.9	PERSON NC	-40	-46	-67	03	25
8	9.1	8	5.6	12	6.3	PITT NC	-38	11	-31	18	90
2	3.9	3	5.5	1	1.1	POLK NC	42	-80	-71	02	17
4	2.8	7	2.6	10	3.0	RANDOLPH NC	-7	15	7	43	41
2	1.4	3	2.6	8	6.2	RICHMOND NC	87	136	341	95	89
2	1.4	4	2.9	5	3.3	ROBESON NC	104	13	131	82	47
5	2.4	6	2.6	11	3.9	ROCKINGHAM NC	8	50	62	67	59
5	1.9	8	2.8	11	2.7	ROWAN NC	50	-5	43	61	37
6	3.8	2	1.0	3	1.3	RUTHERFORD NC	-74	37	-65	04	19
3	2.9	4	3.2	3	1.7	SAMPSON NC	11	-46	-40	12	23
3	5.9	4	8.0			SCOTLAND NC	36				
4	3.2			8	3.8	STANLY NC			16	49	57
3	3.6			3	2.5	STOKES NC			-32	17	33
1	.6	4	2.1	7	2.6	SURRY NC	255	24	341	95	36
		3	8.0			SWAIN NC					
2	3.4	1	1.3	5	4.4	TRANSYLVANIA NC	-63	248	29	55	68
3	2.6	3	1.7	3	1.5	UNION NC	-35	-11	-42	11	20
1	2.5	5	7.1	5	6.2	VANCE NC	192	-13	154	85	89
12	3.6	15	3.0	30	4.5	WAKE NC	-17	50	25	53	69
1	3.0	1	2.8	1	2.1	WARREN NC	-7	-24	-29	19	28
		3	8.3	3	7.5	WASHINGTON NC		-9			95
1	2.0			5	4.6	WATAUGA NC			136	83	72
5	4.0	4	2.5	6	2.8	WAYNE NC	-36	12	-29	19	39
		5	2.5	8	3.7	WILKES NC		47			56
4	4.6	3	2.1	10	6.0	WILSON NC	-55	189	30	55	87
		1	.9	5	3.9	YADKIN NC		317			60
				1	1.3	YANCEY NC					18
69	2.5	92	3.1	124	3.9	NORTH DAKOTA ND	23	27	55		
1	4.6	2	9.6			ADAMS ND	109				
3	3.5	2	2.6	10*	9.8	BARNES ND	-25	273	181	88	98
1	2.1			2	4.0	BENSON ND			94	75	62
1	1.6	1	1.4	1	1.3	BOTTINEAU ND	-13	-8	-20	25	18
		1	5.1	2	8.4	BOWMAN ND		66			96
3	3.0	2	1.4	7	4.4	BURLEIGH ND	-54	218	47	62	67
12	4.3	7	2.3	11	3.4	CASS ND	-47	48	-21	24	50
		1	1.6	3	5.8	CAVALIER ND		256			85
		2	4.2	1	1.7	DICKEY ND		-61			22
				2	6.4	DIVIDE ND					91
2	9.4			1	4.5	DUNN ND			-53	08	69
		1	3.1	1	3.2	EDDY ND		1			45
				1	1.9	EMMONS ND					25
		1	4.1			FOSTER ND					
4	2.8	10	5.2	8	3.9	GRAND FORKS ND	86	-25	39	59	59
1	5.9	1	4.0			GRANT ND	-32				
1	3.3	1	3.5			GRIGGS ND	8				
				2	6.4	HETTINGER ND					91
		1	7.4			KIDDER ND					
2	4.1	2	4.1	1	2.4	LA MOURE ND	0	-42	-42	11	32
1	5.0	1	4.8	1	5.4	LOGAN ND	-4	12	8	44	82
1	1.5	1	1.7	2	3.5	MC HENRY ND	16	103	135	82	51
2	6.1	1	3.9	1	2.8	MC INTOSH ND	-36	-28	-54	07	39
1	2.5			4*	14.6	MC KENZIE ND			480	98	99
		2	2.9	5	7.6	MC LEAN ND		160			95
				2	4.6	MERCER ND					72
4	4.4	2	2.1	3	2.9	MORTON ND	-53	40	-35	15	41
2	4.0	3	6.5	3	5.9	MOUNTRAIL ND	61	-9	46	62	87
1	2.3	1	1.8	2	3.2	NELSON ND	-24	82	39	60	46
				1	6.6	OLIVER ND					92
1	2.0	2	2.4	2	2.6	PEMBINA ND	19	9	29	55	37
1	3.4			3	6.6	PIERCE ND			95	75	92
1	1.5	3	3.5	3	4.0	RAMSEY ND	141	11	168	86	60
1	2.0	1	1.3	3	5.8	RANSOM ND	-35	345	191	89	85
1	3.4	2	5.8			RENVILLE ND	73				
1	.8	2	1.6	3	2.2	RICHLAND ND	95	44	182	88	30
		1	2.8			ROLETTE ND					
		2	6.1	3	8.3	SARGENT ND		34			96
1	6.6	2	15.5			SHERIDAN ND	135				
1	1.4	3	4.0	2	2.5	STARK ND	186	-37	80	72	35
				2	8.5	STEELE ND					96
3	2.5	6	4.6	4	3.0	STUTSMAN ND	84	-34	21	51	43
		1	3.0			TOWNER ND					
3	4.5	3	4.4	2	2.7	TRAILL ND	-1	-39	-40	12	38
1	1.0	6	6.5	4	3.3	WALSH ND	548	-49	231	92	48

* SIGNIFICANTLY MORE DEATHS THAN EXPECTED

WHITE MALE CANCER OF THE ESOPHAGUS ICD 150

1950-1959 DEATHS	1950-1959 RATE	1960-1969 DEATHS	1960-1969 RATE	1970-1979 DEATHS	1970-1979 RATE	COUNTY AND STATE	PERCENT CHANGE 60S/50S	PERCENT CHANGE 70S/60S	PERCENT CHANGE 70S/50S	PERCENTILE %CHANGE 70S/50S	PERCENTILE RATES 1970S
8	5.1	5	2.8	4	2.0	WARD ND	-44	-31	-61	05	26
1	1.7	2	4.0	2	3.9	WELLS ND	144	-3	136	83	59
2	4.3	5	5.2	10*	10.3	WILLIAMS ND	21	97	139	83	98
1726*	4.7	1978*	4.9	2217*	5.1	OHIO OH	5	5	10		
1	.9	1	1.0	4	2.8	ADAMS OH	6	198	217	91	40
15	3.6	12	2.7	20	4.3	ALLEN OH	-26	60	19	50	66
3	1.6	2	1.0	11	5.3	ASHLAND OH	-37	409	223	91	81
16	3.8	16	3.6	12	2.6	ASHTABULA OH	-5	-28	-32	17	36
4	1.7	8	3.6	15*	7.2	ATHENS OH	110	98	316	95	94
7	4.6	6	3.5	8	4.1	AUGLAIZE OH	-24	17	-11	32	63
21	4.5	32*	6.9	14	3.0	BELMONT OH	53	-56	-32	16	43
5	3.9	2	1.5	7	4.7	BROWN OH	-61	208	20	51	73
24	4.0	29	4.1	59*	6.8	BUTLER OH	1	66	67	68	92
2	2.0	2	2.0	2	1.7	CARROLL OH	0	-15	-15	29	22
1	.8	5	3.7	4	2.6	CHAMPAIGN OH	361	-29	225	92	36
15	3.1	25	4.5	27	4.4	CLARK OH	47	-3	43	61	67
10	4.5	15	5.5	16	4.4	CLERMONT OH	21	-19	-2	37	69
3	2.0	5	3.5	5	3.4	CLINTON OH	76	-2	72	70	50
9	2.0	18	3.5	24	4.4	COLUMBIANA OH	78	27	126	81	68
		4	2.4	6	3.4	COSHOCTON OH		40			50
6	2.7	9	4.2	12	5.3	CRAWFORD OH	53	28	96	76	81
502*	8.4	487*	7.5	410*	6.3	CUYAHOGA OH	-11	-16	-25	21	90
8	3.5	3	1.2	10	4.0	DARKE OH	-67	246	13	47	61
4	3.2	4	3.0	7	4.7	DEFIANCE OH	-4	55	49	63	73
4	2.4	5	3.1	7	3.7	DELAWARE OH	29	21	57	65	56
16	5.1	24*	7.4	27*	8.2	ERIE OH	46	10	61	67	96
6	2.1	8	2.5	19	5.3	FAIRFIELD OH	19	113	154	85	81
4	3.2	2	1.5	4	3.2	FAYETTE OH	-53	111	-1	38	46
92	4.5	142*	6.2	135	5.3	FRANKLIN OH	36	-14	16	49	80
4	3.3			8	5.3	FULTON OH			61	66	80
1	.8	4	3.3	2	1.6	GALLIA OH	285	-51	89	75	21
2	1.5	5	2.3	7	2.5	GEAUGA OH	56	8	69	69	34
10	4.5	6	2.1	13	3.4	GREENE OH	-54	65	-24	23	50
6	2.3	3	1.2	5	2.1	GUERNSEY OH	-48	75	-9	33	27
180*	5.9	197*	6.0	186*	5.6	HAMILTON OH	1	-6	-6	35	84
5	2.3	4	1.5	14	5.1	HANCOCK OH	-34	236	121	80	78
2	1.3	4	2.9	10	6.7	HARDIN OH	115	134	402	97	92
		1	.8	4	4.1	HARRISON OH		425			63
1	.7	7	5.1	14*	10.5	HENRY OH	614	104	1355	99	98
3	2.1	2	1.2	4	2.1	HIGHLAND OH	-43	69	-3	37	27
1	1.1	3	2.8	4	3.4	HOCKING OH	159	21	215	91	50
4	4.3	3	2.7	2	1.7	HOLMES OH	-38	-34	-59	05	23
2	1.0	7	3.3	10	4.6	HURON OH	219	39	343	96	71
4	2.7	5	3.7	11	7.3	JACKSON OH	37	101	175	87	95
21	5.1	19	4.1	27	6.0	JEFFERSON OH	-19	46	18	50	87
7	3.6	4	1.9	6	3.0	KNOX OH	-46	54	-17	27	42
13	3.6	26	5.4	40	5.6	LAKE OH	48	4	54	64	84
2	.9	11	4.5	8	3.0	LAWRENCE OH	426	-33	252	93	42
14	3.7	16	3.8	20	4.1	LICKING OH	3	8	11	45	62
1	.5	7	3.8	9	5.0	LOGAN OH	650	32	890	99	76
41	6.0	46	5.9	66*	7.2	LORAIN OH	-2	22	20	51	94
115*	6.2	116*	5.9	124*	6.3	LUCAS OH	-5	7	2	40	90
1	.6	5	4.5	8	6.7	MADISON OH	614	50	969	99	92
80*	6.7	71	5.7	82*	6.1	MAHONING OH	-15	8	-8	33	88
14	5.6	10	4.2	15	5.4	MARION OH	-25	29	-4	36	82
8	3.6	8	3.2	17	5.5	MEDINA OH	-12	72	52	64	83
1	.7	4	3.5	7	6.3	MEIGS OH	409	80	818	99	90
2	1.4	4	2.6	9	5.4	MERCER OH	94	104	297	94	82
4	1.4	17	5.2	9	2.6	MIAMI OH	277	-51	84	73	35
2	1.7	1	1.3	5	5.9	MONROE OH	-21	344	249	93	86
65	3.9	93	5.0	96	4.6	MONTGOMERY OH	27	-7	19	50	72
3	4.5	1	1.5	4	5.9	MORGAN OH	-66	291	31	56	86
3	3.0					MORROW OH					
10	2.7	12	3.2	20	5.4	MUSKINGUM OH	20	66	99	76	82
1	1.2	4	7.2	3	4.4	NOBLE OH	488	-39	260	93	68
7	4.8	5	3.0	15*	7.9	OTTAWA OH	-38	163	65	68	96
1	1.2	1	1.2	3	3.6	PAULDING OH	3	197	206	90	53
1	.5	8	5.5	5	3.4	PERRY OH	933	-38	545	98	50
2	1.4	5	3.3	4	2.4	PICKAWAY OH	131	-28	67	68	32
2	2.6	1	1.2	4	3.5	PIKE OH	-53	195	38	59	52
15	5.4	15	4.7	24	5.6	PORTAGE OH	-13	19	4	41	83
3	2.0	4	2.8	3	1.8	PREBLE OH	36	-35	-11	32	24
6	4.3	4	3.0	5	3.6	PUTNAM OH	-31	20	-17	27	54
11	2.4	16	3.4	20	3.8	RICHLAND OH	42	12	59	66	57
4	1.6	9	3.1	17	5.5	ROSS OH	93	79	246	93	83
7	3.1	11	4.3	15	5.6	SANDUSKY OH	38	30	80	72	84
11	3.2	13	3.3	15	3.6	SCIOTO OH	4	8	12	46	54
9	3.4	6	2.2	10	3.7	SENECA OH	-35	71	11	45	55
3	2.0	5	3.3	7	4.3	SHELBY OH	61	31	110	78	66
61	4.7	84*	5.5	81	4.9	STARK OH	17	-12	4	41	75
91	5.0	89	4.3	124*	5.7	SUMMIT OH	-15	33	13	47	84
31	4.3	33	4.0	41	4.2	TRUMBULL OH	-7	6	-1	38	65
16	4.2	18	4.6	13	3.1	TUSCARAWAS OH	8	-33	-27	20	44
2	1.7	4	3.4	9	7.5	UNION OH	99	123	344	96	95
3	2.0	7	4.8	7	4.7	VAN WERT OH	141	-2	135	82	73
		1	1.8	3	5.9	VINTON OH		221			86
4	2.7	7	3.2	10	2.9	WARREN OH	18	-10	6	43	40
6	2.6	8	3.1	12	4.4	WASHINGTON OH	19	44	71	70	68
5	1.7	9	2.8	10	2.6	WAYNE OH	63	-6	54	64	37
2	1.4	8	5.1	5	2.9	WILLIAMS OH	264	-42	111	79	41
10	3.3	13	4.2	14	3.9	WOOD OH	26	-9	15	48	58
2	1.8	2	1.8	7	6.3	WYANDOT OH	-1	252	249	93	90
219	2.2	279	2.5	402	3.2	OKLAHOMA OK	13	28	44		
1	1.5	1	1.1	1	1.3	ADAIR OK	-28	17	-16	28	18
1	1.4	4	6.0	1	1.6	ALFALFA OK	342	-74	15	48	21
1	1.7	1	1.6			ATOKA OK	-6				
1	2.6			1	2.5	BEAVER OK			-2	37	34

* SIGNIFICANTLY MORE DEATHS THAN EXPECTED

WHITE MALE

CANCER OF THE ESOPHAGUS

ICD 150

1950-1959 DEATHS	1950-1959 RATE	1960-1969 DEATHS	1960-1969 RATE	1970-1979 DEATHS	1970-1979 RATE	COUNTY AND STATE	PERCENT CHANGE 60S/50S	PERCENT CHANGE 70S/60S	PERCENT CHANGE 70S/50S	PERCENTILE %CHANGE 70S/50S	PERCENTILE RATES 1970S
		5	4.2	2	1.5	BECKHAM OK		-64			20
3	4.3	3	4.2	2	2.2	BLAINE OK	-3	-47	-48	09	30
2	1.3	5	2.9	6	3.9	BRYAN OK	126	33	201	90	59
2	1.3	3	2.0	3	1.7	CADDO OK	57	-18	29	55	22
3	2.7	5	3.7	2	1.1	CANADIAN OK	39	-71	-59	05	17
5	2.8	3	1.7	3	1.6	CARTER OK	-41	-5	-44	10	21
1	1.6			5	4.3	CHEROKEE OK			180	87	67
3	2.8	2	2.0	3	3.4	CHOCTAW OK	-28	69	21	51	50
4	2.3	2	.9	9	3.4	CLEVELAND OK	-61	287	50	63	50
		2	5.1	1	2.2	COAL OK		-56			30
5	2.9	9	4.2	11	3.6	COMANCHE OK	46	-16	23	52	53
2	4.1	1	2.1			COTTON OK	-48				
2	1.9			1	1.0	CRAIG OK			-45	10	16
4	1.9	4	1.9	12	5.4	CREEK OK	1	181	184	88	81
1	1.0	1	1.0	3	2.5	CUSTER OK	-1	139	138	83	34
2	2.3	1	.9	6	3.7	DELAWARE OK	-63	327	58	66	55
				1	1.9	DEWEY OK					25
		3	8.5	1	2.5	ELLIS OK		-71			33
5	2.1	9	3.5	6	2.1	GARFIELD OK	63	-39	-1	38	28
2	1.5	4	2.8	3	1.7	GARVIN OK	87	-41	11	46	22
2	1.1	6	3.6	2	1.4	GRADY OK	225	-62	22	52	19
1	1.7	2	3.7			GRANT OK	123				
2	3.5			3	6.1	GREER OK			75	71	88
		1	2.5			HARMON OK					
		1	3.0			HARPER OK					
1	1.7			1	1.7	HASKELL OK			1	40	22
3	2.8			1	1.3	HUGHES OK			-53	07	19
3	3.0	4	3.5	3	2.8	JACKSON OK	17	-21	-8	34	39
3	4.9			1	1.5	JEFFERSON OK			-69	02	21
		1	2.2	3	4.4	JOHNSTON OK		98			68
		5	1.9	6	2.1	KAY OK		12			29
1	1.8	2	3.2	1	1.3	KINGFISHER OK	75	-60	-29	18	18
2	2.3			2	2.9	KIOWA OK			28	54	41
1	2.5	1	1.4	2	4.5	LATIMER OK	-44	218	80	72	69
		4	2.2	4	2.0	LE FLORE OK		-9			26
5	4.1	2	1.5	1	.6	LINCOLN OK	-64	-58	-85	00	14
3	3.1	4	3.9	2	1.8	LOGAN OK	24	-54	-43	11	24
1	2.4			1	2.9	LOVE OK			22	51	41
3	4.3	4	5.4	1	1.2	MC CLAIN OK	26	-77	-71	02	18
1	.9			5	3.6	MC CURTAIN OK			302	95	54
1	1.4	1	1.8	5	6.1	MC INTOSH OK	35	232	347	96	88
		1	2.0	2	3.4	MAJOR OK		69			51
				4	4.8	MARSHALL OK					75
5	4.8	3	2.4	7	4.5	MAYES OK	-50	90	-5	35	70
		2	3.1	2	2.5	MURRAY OK		-20			34
7	2.8	7	2.7	12	4.2	MUSKOGEE OK	-4	60	54	64	65
2	3.2	2	2.9	2	3.2	NOBLE OK	-9	10	0	39	46
1	1.8					NOWATA OK					
1	1.5	1	1.4	2	3.9	OKFUSKEE OK	-9	184	158	85	58
40	3.1	41	2.5	83	4.1	OKLAHOMA OK	-20	65	32	56	64
1	.5	7	3.8	9	4.7	OKMULGEE OK	704	24	896	99	72
2	1.5	2	1.2	6	3.5	OSAGE OK	-14	181	140	83	51
		4	2.7	5	3.1	OTTAWA OK		14			44
3	4.0			4	5.3	PAWNEE OK			33	57	80
4	2.4	8	4.5	9	4.4	PAYNE OK	84	-1	82	73	68
3	1.5	8	4.0	4	2.0	PITTSBURG OK	161	-51	28	54	26
6	4.1	4	2.4	5	2.9	PONTOTOC OK	-41	21	-29	19	41
7	3.2	10	4.4	6	2.4	POTTAWATOMIE OK	38	-46	-26	21	32
1	1.4	1	1.3	5	6.9	PUSHMATAHA OK	-10	436	382	96	93
		1	3.3	1	4.1	ROGER MILLS OK		23			63
2	2.4	3	2.4	6	4.0	ROGERS OK	1	67	69	69	61
3	2.1	2	1.4	6	4.2	SEMINOLE OK	-35	207	99	76	66
3	3.9	3	3.1	3	2.5	SEQUOYAH OK	-20	-20	-36	14	34
6	3.9	7	3.6	6	2.6	STEPHENS OK	-8	-27	-33	16	36
3	5.1			2	2.8	TEXAS OK			-46	10	39
2	2.7	4	4.8	4	5.5	TILLMAN OK	79	15	106	78	83
24	2.5	41	3.4	62	4.1	TULSA OK	34	21	63	67	63
1	1.2			2	1.8	WAGONER OK			47	62	24
4	2.9	6	3.6	9	4.5	WASHINGTON OK	23	25	54	64	70
2	2.5	1	1.2	3	3.2	WASHITA OK	-52	168	29	55	47
1	1.5	2	2.7	1	1.1	WOODS OK	74	-60	-30	18	16
2	2.6	2	2.2	3	3.4	WOODWARD OK	-12	49	31	56	49
220	2.7	318	3.4	455	4.0	OREGON OR	22	19	45		
4	3.9	1	1.0	4	3.8	BAKER OR	-74	268	-2	37	57
1	.8	1	.6	3	1.6	BENTON OR	-35	182	84	73	21
15	2.8	20	3.1	23	2.5	CLACKAMAS OR	11	-19	-10	32	35
10	7.5	10	5.8	14*	7.5	CLATSOP OR	-23	30	0	39	95
4	2.9	2	1.3	2	1.3	COLUMBIA OR	-54	-1	-54	07	19
3	1.5	11	4.1	19	6.1	COOS OR	167	49	299	94	88
		2	3.6			CROOK OR					
				4	4.2	CURRY OR					65
		3	1.9	7	3.0	DESCHUTES OR		59			43
7	2.8	15	4.8	15	3.6	DOUGLAS OR	75	-25	32	56	54
1	6.2					GILLIAM OR					
		2	5.1	1	2.0	GRANT OR		-60			26
		2	5.8	1	2.4	HARNEY OR		-58			33
1	1.4	2	3.0	2	2.4	HOOD RIVER OR	124	-21	77	71	32
12	3.7	14	3.1	27	4.4	JACKSON OR	-18	44	19	50	69
				2	3.9	JEFFERSON OR					60
4	2.6	4	1.9	9	2.7	JOSEPHINE OR	-27	41	4	41	38
5	2.8	8	3.7	13	5.0	KLAMATH OR	33	35	80	72	77
1	2.3	1	5.1	5*	12.4	LAKE OR	126	143	450	97	99
12	2.1	19	2.6	32	3.2	LANE OR	25	23	54	64	46
2	1.3	3	1.7	11	4.8	LINCOLN OR	39	173	281	94	74
5	2.1	11	3.4	14	3.6	LINN OR	61	6	70	69	53
2	1.9	3	2.6	5	4.0	MALHEUR OR	39	52	113	79	60
9	1.5	17	2.5	28	3.0	MARION OR	72	21	108	78	43
				4*	15.2	MORROW OR					99
93	3.6	118	4.3	136	5.1	MULTNOMAH OR	21	18	42	61	78

* SIGNIFICANTLY MORE DEATHS THAN EXPECTED

WHITE MALE — CANCER OF THE ESOPHAGUS — ICD 150

1950-1959 DEATHS	1950-1959 RATE	1960-1969 DEATHS	1960-1969 RATE	1970-1979 DEATHS	1970-1979 RATE	COUNTY AND STATE		PERCENT CHANGE 60S/50S	PERCENT CHANGE 70S/60S	PERCENT CHANGE 70S/50S	PERCENTILE %CHANGE 70S/50S	PERCENTILE RATES 1970S
1	1.2	4	2.7	8	3.9	POLK	OR	129	43	227	92	59
				1	11.2	SHERMAN	OR					99
1	1.2	9*	8.5	1	.6	TILLAMOOK	OR	634	-93	-46	10	14
12	5.7	8	3.2	14	5.1	UMATILLA	OR	-43	59	-9	33	79
2	2.0	5	4.5	4	3.7	UNION	OR	131	-19	87	74	55
1	2.6	1	2.6	3	7.2	WALLOWA	OR	-3	181	172	87	94
		6	6.1	7	5.8	WASCO	OR		-5			85
10	2.9	13	2.6	29	3.8	WASHINGTON	OR	-8	45	33	57	57
1	7.5					WHEELER	OR					
1	.5	3	1.3	7	2.8	YAMHILL	OR	156	115	450	97	40
2517*	5.4	2573*	5.0	2710*	4.9	PENNSYLVANIA	PA	-7	-2	-9		
		5	2.0	17	5.9	ADAMS	PA		193			87
458*	7.0	451*	6.5	449*	6.2	ALLEGHENY	PA	-8	-5	-12	31	89
12	3.0	14	3.5	15	3.5	ARMSTRONG	PA	14	2	16	49	52
58*	7.7	57*	6.8	47	4.7	BEAVER	PA	-12	-31	-39	12	73
3	1.7	3	1.4	7	2.9	BEDFORD	PA	-20	110	67	68	40
43	3.4	72	4.7	72	4.3	BERKS	PA	37	-8	27	54	67
35	5.0	26	3.6	31	4.3	BLAIR	PA	-28	19	-14	29	66
8	2.8	13	4.7	6	2.0	BRADFORD	PA	67	-57	-27	20	27
30	4.2	46	4.4	82	5.2	BUCKS	PA	5	18	24	53	79
20	4.1	20	3.7	41*	6.4	BUTLER	PA	-10	73	55	65	91
58*	6.0	55	6.0	38	3.9	CAMBRIA	PA	0	-35	-35	15	59
		2	5.3	1	3.9	CAMERON	PA		-27			59
12	4.1	12	4.0	16	5.1	CARBON	PA	-3	27	24	52	77
5	2.1	9	3.3	12	3.6	CENTRE	PA	57	9	71	70	53
23	3.1	37	4.0	59	5.4	CHESTER	PA	29	33	72	70	82
4	2.1	6	3.2	2	.9	CLARION	PA	54	-71	-55	07	15
16	4.0	18	4.6	17	4.0	CLEARFIELD	PA	13	-12	-1	38	61
6	3.7	4	2.1	10	5.1	CLINTON	PA	-45	148	37	59	78
10	4.1	9	3.3	14	4.5	COLUMBIA	PA	-20	36	9	45	70
16	4.1	11	2.8	23	5.5	CRAWFORD	PA	-32	96	33	57	82
8	1.8	21	3.6	27	3.8	CUMBERLAND	PA	100	5	111	78	57
28	3.3	36	3.6	49	4.8	DAUPHIN	PA	10	31	44	61	74
90	5.5	118*	5.6	127	5.0	DELAWARE	PA	2	-10	-8	33	77
8	4.9	14*	8.4	8	4.5	ELK	PA	70	-47	-9	33	70
63*	6.2	59	5.5	68	5.8	ERIE	PA	-12	6	-7	34	85
41	4.8	34	4.0	43	5.0	FAYETTE	PA	-17	26	5	42	77
3	10.5	1	2.6			FOREST	PA	-76				
6	1.7	13	3.1	19	3.9	FRANKLIN	PA	81	25	126	81	59
1	1.9	1	1.7	2	3.4	FULTON	PA	-13	105	79	72	51
1	.4	3	1.3	10	4.6	GREENE	PA	220	240	988	99	71
8	5.0	3	1.5	10	5.0	HUNTINGDON	PA	-69	224	1	39	77
12	3.4	8	2.1	18	4.4	INDIANA	PA	-39	115	30	55	69
6	2.5	15	5.9	14	5.5	JEFFERSON	PA	136	-7	120	80	83
		3	3.9	1	.9	JUNIATA	PA		-75			16
93*	8.0	78*	6.2	79*	5.9	LACKAWANNA	PA	-23	-4	-26	21	87
31	2.8	41	3.1	61	3.9	LANCASTER	PA	13	26	43	61	60
24	4.7	28	5.5	33	6.0	LAWRENCE	PA	16	11	28	54	88
18	4.7	17	3.8	24	4.6	LEBANON	PA	-19	22	-1	37	71
54	5.4	59	5.1	70	5.2	LEHIGH	PA	-6	3	-3	36	80
126*	7.4	93	5.3	95	5.1	LUZERNE	PA	-28	-3	-31	18	78
18	3.7	22	4.1	28	5.0	LYCOMING	PA	13	22	38	59	77
11	4.3	7	2.5	9	3.1	MC KEAN	PA	-41	24	-27	20	45
18	3.4	24	4.0	29	4.4	MERCER	PA	20	10	32	56	69
5	3.4	7	3.5	5	2.2	MIFFLIN	PA	2	-37	-36	14	29
7	3.8	11	5.4	10	3.2	MONROE	PA	41	-41	-16	27	46
55	3.1	80	3.3	117	4.0	MONTGOMERY	PA	8	19	29	55	60
		4	4.3	3	3.3	MONTOUR	PA		-23			48
51	5.6	42	4.2	54	5.1	NORTHAMPTON	PA	-26	21	-10	32	78
31	6.3	24	4.3	20	3.5	NORTHUMBERLAND	PA	-32	-18	-44	11	52
3	2.8	4	2.9	4	2.6	PERRY	PA	5	-10	-5	35	36
611*	8.1	552*	7.3	405*	6.1	PHILADELPHIA	PA	-10	-17	-25	22	88
6*	10.7	5	6.5	9*	7.6	PIKE	PA	-39	16	-29	19	95
		4	5.1	2	2.1	POTTER	PA		-59			28
54*	5.5	56*	6.2	58*	5.9	SCHUYLKILL	PA	12	-5	7	43	87
2	2.5			1	.8	SNYDER	PA			-68	03	15
14	3.5	14	3.4	16	3.7	SOMERSET	PA	-4	11	6	43	56
6*	17.5			2	5.3	SULLIVAN	PA			-69	02	81
6	3.4	5	2.8	3	1.6	SUSQUEHANNA	PA	-15	-44	-53	08	21
4	2.2	10	5.4	4	2.0	TIOGA	PA	149	-63	-7	34	26
1	.9	2	1.8	5	3.7	UNION	PA	104	103	313	95	55
9	3.2	10	3.2	18	5.6	VENANGO	PA	1	74	76	71	83
6	2.4	6	2.3	9	3.4	WARREN	PA	-2	47	45	61	51
54	5.5	61*	6.0	53	4.9	WASHINGTON	PA	9	-18	-11	32	75
2	1.1	3	1.7	11	5.4	WAYNE	PA	50	226	391	97	82
82*	5.5	78	4.6	70	3.6	WESTMORELAND	PA	-17	-21	-35	15	54
1	1.1	2	2.1	3	3.2	WYOMING	PA	94	51	194	89	45
22	2.2	25	2.1	45	3.3	YORK	PA	-6	57	47	62	48
171	4.9	220*	5.5	278*	6.2	RHODE ISLAND	RI	12	13	26		
7	5.5	7	4.1	13	5.7	BRISTOL	RI	-25	41	5	42	85
14	3.5	29	6.3	51*	7.8	KENT	RI	82	24	126	81	96
10	4.1	16	6.2	16	5.0	NEWPORT	RI	50	-19	21	51	77
132	5.2	160*	5.6	177*	6.0	PROVIDENCE	RI	8	7	15	48	87
8	4.1	8	3.0	21	6.4	WASHINGTON	RI	-28	116	56	65	91
156	3.3	214	3.7	262	3.3	SOUTH CAROLINA	SC	10	-10	0		
		3	4.8	2	2.3	ABBEVILLE	SC		-52			30
9	5.8	7	3.7	13	4.4	AIKEN	SC	-35	18	-24	23	68
1	6.7					ALLENDALE	SC					
6	2.1	8	2.3	9	2.1	ANDERSON	SC	12	-11	0	38	27
1	5.1			1	2.0	BAMBERG	SC			-62	05	26
2	6.2	2	6.7	1	2.0	BARNWELL	SC	7	-71	-69	03	26
1	3.7	4	5.9	1	.5	BEAUFORT	SC	58	-92	-87	00	14
1	1.8	1	1.3	5	3.4	BERKELEY	SC	-28	164	90	75	50
1	4.5			1	3.3	CALHOUN	SC			-25	21	49
17	4.9	28*	6.6	17	3.0	CHARLESTON	SC	35	-54	-38	13	42

* SIGNIFICANTLY MORE DEATHS THAN EXPECTED

WHITE MALE — CANCER OF THE ESOPHAGUS — ICD 150

1950-1959 DEATHS	1950-1959 RATE	1960-1969 DEATHS	1960-1969 RATE	1970-1979 DEATHS	1970-1979 RATE	COUNTY AND STATE	PERCENT CHANGE 60S/50S	PERCENT CHANGE 70S/60S	PERCENT CHANGE 70S/50S	PERCENTILE %CHANGE 70S/50S	PERCENTILE RATES 1970S
3	3.0	3	2.5	7	5.0	CHEROKEE SC	-15	100	69	69	77
4	7.6	4	4.3	1	1.2	CHESTER SC	-44	-73	-85	00	17
2	5.9	1	1.0	3	2.5	CHESTERFIELD SC	-83	152	-58	06	34
2	3.3	6*	12.0	3	3.6	COLLETON SC	260	-70	10	45	54
		4	3.4	4	2.6	DARLINGTON SC		-23			35
3	5.6			3	3.9	DILLON SC			-31	18	59
2	4.6	1	1.2	3	4.4	DORCHESTER SC	-74	265	-4	36	68
		1	2.9	2	5.6	EDGEFIELD SC		97			84
1	2.5			2	4.1	FAIRFIELD SC			66	68	63
4	2.2	8	3.9	20*	9.3	FLORENCE SC	78	139	326	95	97
1	2.4	1	1.3	5	6.1	GEORGETOWN SC	-45	368	158	85	88
16	3.5	21	3.5	31	3.5	GREENVILLE SC	2	-1	1	40	52
5	4.7	2	1.4	4	2.6	GREENWOOD SC	-71	95	-44	10	36
1	2.7	1	2.5	2	3.9	HAMPTON SC	-7	53	42	61	59
2	1.1	5	2.3	10	3.6	HORRY SC	103	58	221	91	53
		1	3.8	2	6.7	JASPER SC		77			92
2	3.7	3	4.2	4	5.1	KERSHAW SC	13	22	38	59	78
2	2.3	6	4.7	4	2.4	LANCASTER SC	106	-49	5	42	33
6	5.6	8	6.3	6	3.2	LAURENS SC	13	-49	-42	11	47
1	3.1	1	2.8			LEE SC	-10				
6	4.4	6	2.2	5	1.3	LEXINGTON SC	-49	-43	-71	02	18
3	6.7	2	3.1	6	11.0	MARION SC	-53	254	65	68	98
4	9.1	1	2.0	2	2.5	MARLBORO SC	-78	29	-72	02	35
		2	1.9	5	4.1	NEWBERRY SC		120			63
4	3.2	5	3.9	3	1.7	OCONEE SC	19	-56	-48	09	23
5	5.2	3	3.0	6	3.5	ORANGEBURG SC	-42	18	-32	17	52
6	4.2	8	5.1	5	1.9	PICKENS SC	20	-62	-55	07	25
14	3.3	21	5.7	23	4.3	RICHLAND SC	71	-23	31	56	67
				1	1.5	SALUDA SC					20
8	2.1	15	3.2	14	2.2	SPARTANBURG SC	53	-31	5	42	29
3	3.2	1	1.3	7	4.5	SUMTER SC	-60	256	42	61	70
1	1.6	9*	9.9	3	3.2	UNION SC	527	-68	101	77	45
1	1.9	3	4.9	2	3.1	WILLIAMSBURG SC	156	-37	61	67	44
5	3.1	8	3.2	14	5.1	YORK SC	4	57	63	67	77
95	3.1	93	2.7	117	3.4	SOUTH DAKOTA SD	-11	25	11		
1	4.2	1	4.0	2	7.0	AURORA SD	-4	74	68	68	93
2	1.8	1	.8	2	1.9	BEADLE SD	-54	124	3	41	25
				1	8.1	BENNETT SD					96
3	5.8	1	1.6	2	3.1	BON HOMME SD	-73	102	-46	10	45
5	6.2	3	2.9	6	6.5	BROOKINGS SD	-53	121	5	42	91
5	3.5	3	1.7	2	1.1	BROWN SD	-52	-32	-67	03	17
1	2.4	2	7.1	1	3.1	BRULE SD	189	-56	28	54	45
1	2.2			2	4.8	BUTTE SD			115	79	75
1	4.9			1	5.7	CAMPBELL SD			15	48	84
		1	1.8	2	3.5	CHARLES MIX SD		102			52
2	4.3	2	5.1			CLARK SD	20				
1	2.2	1	2.4	3	6.4	CLAY SD	7	173	193	89	91
3	3.2	1	1.0	9*	8.6	CODINGTON SD	-68	756	170	86	97
1	3.7			1	3.3	CUSTER SD			-12	31	48
1	1.2	1	1.0	2	2.5	DAVISON SD	-15	140	104	77	34
2	3.0	2	2.9	1	2.0	DAY SD	-3	-33	-35	15	26
1	2.6	1	2.6			DEUEL SD	-1				
1	7.5			1	8.2	DEWEY SD			9	45	96
1	3.9			1	2.9	DOUGLAS SD			-27	20	40
3	3.1	2	2.3	2	3.5	FALL RIVER SD	-25	51	13	47	52
1	1.6	3	4.8	1	1.6	GRANT SD	194	-66	-1	38	22
1	2.6	2	4.3			GREGORY SD	68				
				1	7.8	HAAKON SD					96
1	2.0	3	7.3	2	4.9	HAMLIN SD	257	-33	139	83	75
1	4.4	1	2.8	1	4.1	HAND SD	-37	48	-7	34	63
		1	4.1			HANSON SD					
2	5.4	3	6.7	4	7.5	HUGHES SD	23	11	37	58	95
3	5.2	2	2.7	2	2.7	HUTCHINSON SD	-49	3	-47	09	38
				1	7.8	JERAULD SD					96
1	8.2					JONES SD					
3	5.0	2	3.4	2	3.1	KINGSBURY SD	-32	-10	-39	13	44
3	4.9	4	5.4			LAKE SD	12				
4	5.5	3	3.4	3	3.4	LAWRENCE SD	-38	-1	-38	13	50
1	1.1	3	4.0	2	2.5	LINCOLN SD	257	-39	117	79	33
1	10.6	1	4.9			LYMAN SD	-54				
2	4.0			1	2.1	MC COOK SD			-46	10	28
1	3.9	1	3.2			MC PHERSON SD	-18				
1	2.1					MARSHALL SD					
		1	1.4	1	1.4	MEADE SD		0			19
		1	3.2	1	3.9	MINER SD		22			59
19	5.7	20	5.2	14	3.2	MINNEHAHA SD	-10	-37	-43	11	47
2	3.6			1	2.0	MOODY SD			-44	10	27
2	1.7	2	1.0	11	4.7	PENNINGTON SD	-37	345	180	88	73
				1	3.1	PERKINS SD					45
				1	3.8	POTTER SD					58
1	1.3	4	5.8	5	6.7	ROBERTS SD	350	16	423	97	92
		1	4.1			SANBORN SD					
1	1.7	5	7.2	4	7.1	SPINK SD	328	-1	323	95	94
				1	11.2	STANLEY SD					99
1	28.1	1	6.5			SULLY SD	-77				
				1	12.7	TODD SD					99
		1	2.3	4	8.4	TRIPP SD		265			96
2	2.9	1	1.1	4	5.7	TURNER SD	-63	438	99	76	84
				4	5.7	UNION SD					85
4*	14.7	3	6.2	3	5.7	WALWORTH SD	-58	-8	-61	05	84
2	1.7	2	2.0			YANKTON SD	15				
302	2.7	375	2.8	514	3.1	TENNESSEE TN	3	12	16		
3	1.8	8	4.1	10	4.3	ANDERSON TN	122	6	136	83	66
5	5.2	1	.9	3	2.7	BEDFORD TN	-82	195	-48	09	37
3	5.1	2	3.1	4	4.7	BENTON TN	-39	52	-8	34	73
1	2.6					BLEDSOE TN					
2	1.0	7	3.1	9	2.8	BLOUNT TN	223	-9	194	89	40

* SIGNIFICANTLY MORE DEATHS THAN EXPECTED

WHITE MALE — CANCER OF THE ESOPHAGUS — ICD 150

1950-1959 DEATHS	1950-1959 RATE	1960-1969 DEATHS	1960-1969 RATE	1970-1979 DEATHS	1970-1979 RATE	COUNTY AND STATE	PERCENT CHANGE 60S/50S	PERCENT CHANGE 70S/60S	PERCENT CHANGE 70S/50S	PERCENTILE %CHANGE 70S/50S	PERCENTILE RATES 1970S
4	3.1			8	3.7	BRADLEY TN			22	51	56
4	3.5	2	1.6	4	2.8	CAMPBELL TN	-54	76	-19	25	39
1	3.0			2	4.3	CANNON TN			41	60	66
1	.9	1	.7			CARROLL TN	-20				
		8	4.6	3	1.5	CARTER TN		-68			20
		2	3.7	1	1.5	CHEATHAM TN		-60			20
1	2.3			1	2.1	CHESTER TN			-7	34	28
4	4.3	1	1.1	8	6.9	CLAIBORNE TN	-73	503	62	67	93
		2	5.7			CLAY TN					
6	6.2	2	1.9	6	4.9	COCKE TN	-70	161	-22	24	75
1	.9	4	3.6	2	1.3	COFFEE TN	323	-64	50	63	18
1	2.2	2	2.8	3	4.1	CROCKETT TN	30	47	91	75	64
		2	2.2	1	.8	CUMBERLAND TN		-62			15
45	4.3	32	2.4	60	3.8	DAVIDSON TN	-44	55	-13	30	57
1	1.9					DECATUR TN					
1	2.5					DE KALB TN					
2	2.1	4	3.9	8	6.2	DICKSON TN	86	58	193	89	89
3	2.3	2	1.5	5	3.6	DYER TN	-37	146	56	65	54
				1	1.3	FAYETTE TN					18
3	6.1			2	3.0	FENTRESS TN			-51	08	41
		6	5.4	5	4.3	FRANKLIN TN		-21			66
5	2.7	3	1.5	6	2.8	GIBSON TN	-44	85	5	42	39
1	1.1	3	2.8			GILES TN	165				
		1	1.7	1	1.7	GRAINGER TN		-4			22
5	5.0	5	2.7	6	2.4	GREENE TN	-46	-11	-52	08	32
				3	5.1	GRUNDY TN					78
3	4.1	2	1.3	3	1.8	HAMBLEN TN	-67	36	-55	06	24
20	3.1	24	3.0	42	4.4	HAMILTON TN	-3	48	44	61	68
		1	2.7			HANCOCK TN					
				4	4.4	HARDEMAN TN					68
		2	2.3	4	4.6	HARDIN TN		101			72
		2	1.6	5	2.6	HAWKINS TN		60			36
		1	1.9	2	4.6	HAYWOOD TN		139			72
		2	2.6	3	3.3	HENDERSON TN		27			48
2	1.8	4	3.4	5	3.6	HENRY TN	91	6	102	77	54
1	1.9	3	4.8	4	6.1	HICKMAN TN	158	28	230	92	88
1	3.7	2	7.9			HOUSTON TN	111				
		2	3.3	4	5.3	HUMPHREYS TN		61			81
1	1.8	1	1.9	1	1.8	JACKSON TN	9	-6	3	40	24
		2	2.1	6	4.7	JEFFERSON TN		121			73
1	1.9	2	3.1			JOHNSON TN	62				
26	3.0	30	3.0	44	3.7	KNOX TN	0	23	24	52	56
		2	5.9	1	2.7	LAKE TN		-55			37
		1	1.3	3	3.3	LAUDERDALE TN		154			49
1	.9	3	2.1	4	2.6	LAWRENCE TN	132	24	188	88	35
2	8.1	1	3.2			LEWIS TN	-60				
1	1.0			9	7.3	LINCOLN TN			615	98	95
		1	.8	8	6.2	LOUDON TN		655			89
1	.7	2	1.4	4	2.1	MC MINN TN	90	58	200	90	29
1	1.2			1	.8	MC NAIRY TN			-29	19	15
		2	2.6	2	2.4	MACON TN		-7			33
6	3.4	9	4.1	6	2.6	MADISON TN	21	-36	-23	23	36
3	4.5	3	4.1	2	2.1	MARION TN	-8	-49	-53	08	28
3	3.8	2	2.2	4	4.4	MARSHALL TN	-41	97	16	49	68
2	1.3	3	1.9	6	3.1	MAURY TN	50	63	145	84	45
1	1.2	4	3.7	2	1.7	MONROE TN	218	-53	49	63	23
3	2.9			7	3.5	MONTGOMERY TN			21	51	51
		2	9.3	2	9.3	MOORE TN		-1			97
1	1.5	4	6.7			MORGAN TN	344				
4	2.9	9	6.0	7	4.3	OBION TN	103	-28	46	62	66
4	6.0	1	1.2	2	2.2	OVERTON TN	-80	87	-63	04	29
				1	3.2	PICKETT TN					46
2	3.5	1	1.7	1	1.5	POLK TN	-51	-10	-56	06	21
3	2.3	1	.7	5	2.6	PUTNAM TN	-68	262	15	48	36
		2	2.7	1	.9	RHEA TN		-65			16
4	3.2	3	1.7	9	4.4	ROANE TN	-47	156	36	58	68
1	1.7	1	.7	4	2.7	ROBERTSON TN	-58	267	56	65	38
2	1.2	4	2.3	7	2.9	RUTHERFORD TN	91	30	149	84	41
1	1.6	1	1.4	3	4.0	SCOTT TN	-14	191	150	84	61
		1	4.0	1	3.4	SEQUATCHIE TN		-15			49
2	2.1	1	.9	6	3.4	SEVIER TN	-56	270	65	68	50
61	5.3	83*	5.9	71	4.1	SHELBY TN	12	-31	-23	23	63
2	2.9					SMITH TN					
		1	1.9	2	4.4	STEWART TN		128			69
1	.4	9	2.2	14	2.4	SULLIVAN TN	411	6	440	97	31
3	2.0	5	2.4	7	2.6	SUMNER TN	24	6	32	56	36
1	1.4	1	1.2	2	2.3	TIPTON TN	-16	95	64	67	31
		2	8.0			TROUSDALE TN					
2	3.6	1	1.3	2	2.3	UNICOI TN	-65	79	-37	14	31
1	2.7	1	2.0	1	2.3	UNION TN	-26	14	-15	29	31
1	4.4			1	4.7	VAN BUREN TN			7	43	73
4	3.8	3	2.5	2	1.4	WARREN TN	-35	-45	-64	04	19
6	2.2	13	4.0	7	1.9	WASHINGTON TN	80	-53	-16	28	25
1	1.7	2	3.1			WAYNE TN	82				
3	1.8	2	1.1	2	1.0	WEAKLEY TN	-36	-9	-42	11	16
1	2.6					WHITE TN					
3	3.2	2	1.7	1	.5	WILLIAMSON TN	-46	-73	-86	00	14
6	4.9	6	4.5	5	2.7	WILSON TN	-7	-41	-45	10	38
823	2.9	1124	3.2	1597	3.6	TEXAS TX	8	14	24		
4	3.5	4	3.0	2	1.3	ANDERSON TX	-14	-57	-63	04	18
				1	2.1	ANDREWS TX					27
3	2.1	5	2.7	7	3.0	ANGELINA TX	29	8	40	60	41
				4	5.2	ARANSAS TX					80
1	3.1	1	3.0	1	2.3	ARCHER TX	-5	-24	-27	20	30
				1	6.4	ARMSTRONG TX					90
2	2.5	4	4.2	3	2.5	ATASCOSA TX	70	-39	4	41	35
2	2.5	2	2.3			AUSTIN TX	-10				
				1	4.6	BAILEY TX					72
1	3.8	1	2.6	3	5.0	BANDERA TX	-32	96	32	56	77

* SIGNIFICANTLY MORE DEATHS THAN EXPECTED

WHITE MALE — CANCER OF THE ESOPHAGUS — ICD 150

1950-1959 DEATHS	1950-1959 RATE	1960-1969 DEATHS	1960-1969 RATE	1970-1979 DEATHS	1970-1979 RATE	COUNTY AND STATE	PERCENT CHANGE 60S/50S	PERCENT CHANGE 70S/60S	PERCENT CHANGE 70S/50S	PERCENTILE %CHANGE 70S/50S	PERCENTILE RATES 1970S
		1	1.0	5	4.6	BASTROP TX		355			71
1	2.5	1	2.3	1	2.0	BAYLOR TX	-10	-10	-18	26	27
4	6.4	3	4.4			BEE TX	-31				
5	1.8	8	2.3	14	3.3	BELL TX	28	45	86	74	48
84	4.7	119	5.2	141	5.0	BEXAR TX	9	-3	6	42	77
1	3.6	1	5.8	3	12.2	BLANCO TX	61	108	236	92	99
2	2.3	2	1.6	1	.7	BOSQUE TX	-29	-54	-67	03	15
8	3.9	11	4.7	10	3.3	BOWIE TX	22	-30	-15	29	48
7	4.5	6	3.0	17	4.0	BRAZORIA TX	-35	35	-12	31	61
5	5.1	4	2.8	6	3.1	BRAZOS TX	-45	10	-40	12	44
		2	6.7	1	2.9	BREWSTER TX		-57			40
		1	10.2			BRISCOE TX					
8	5.4	2	1.1	5	2.9	BROWN TX	-80	166	-47	09	40
		2	3.6	4	6.3	BURLESON TX		74			90
3	5.4	2	2.4	1	1.4	BURNET TX	-55	-42	-74	01	19
		3	3.7	4	4.2	CALDWELL TX		14			65
3	12.7	1	2.7	7*	8.6	CALHOUN TX	-78	213	-33	16	97
1	1.9	2	3.6			CALLAHAN TX	88				
11	3.2	16	3.4	33	5.1	CAMERON TX	4	53	59	66	79
1	3.5	1	2.9			CAMP TX	-16				
3	10.8			1	2.7	CARSON TX			-75	01	37
4	4.2	1	.9	4	2.9	CASS TX	-79	243	-29	19	41
		1	2.6			CHAMBERS TX					
3	2.0	4	2.8	2	1.2	CHEROKEE TX	36	-55	-39	12	18
1	2.1	1	2.1			CHILDRESS TX	3				
1	2.0	1	2.1	1	1.9	CLAY TX	5	-6	-2	37	25
1	5.0					COKE TX					
1	1.1	1	.9	2	2.1	COLEMAN TX	-12	121	95	75	27
5	2.9	5	2.1	8	2.5	COLLIN TX	-27	19	-13	30	35
1	1.4	5	6.2	1	1.0	COLOPADO TX	343	-83	-26	21	16
1	1.2	2	1.7	6	3.4	COMAL TX	46	98	189	88	50
		1	1.2	1	1.1	COMANCHE TX		-9			17
		1	3.2			CONCHO TX					
2	1.9	4	3.7	1	.9	COOKE TX	100	-77	-53	07	15
3	3.6			5	4.4	CORYELL TX			24	53	69
		1	3.7			CRANE TX					
1	4.5	1	5.1	1	5.8	CROCKETT TX	13	15	31	55	86
2	5.1	3	7.2			CROSBY TX	41				
1	3.2					DALLAM TX					
68	3.1	109	3.4	177	4.7	DALLAS TX	11	36	52	63	72
1	1.7	2	2.9	3	3.7	DAWSON TX	66	28	113	79	56
				1	1.3	DEAF SMITH TX					19
1	2.1			2	4.4	DELTA TX			113	79	68
		7	3.1	8	2.6	DENTON TX		-16			36
3	3.2	1	.9	3	2.8	DE WITT TX	-74	232	-12	31	40
		1	2.8			DICKENS TX					
2	5.9			1	2.3	DIMMIT TX			-60	05	31
		1	2.8	1	5.2	DONLEY TX		85			79
3	7.1	2	3.8	3	4.1	DUVAL TX	-46	8	-42	11	64
3	2.1	1	1.1	4	2.3	EASTLAND TX	-47	115	13	47	31
4	2.3	4	2.3	15	4.2	ECTOR TX	2	80	85	73	65
				2*	20.5	EDWARDS TX					99
2	1.2	11	5.4	8	3.5	ELLIS TX	367	-36	199	89	51
18	2.8	30	4.0	46	4.7	EL PASO TX	44	16	67	68	73
2	1.7	3	2.4	2	1.4	ERATH TX	40	-40	-16	28	20
4	4.0	4	4.2	4	3.2	FALLS TX	5	-24	-20	25	46
4	2.5	2	1.0	2	1.1	FANNIN TX	-61	11	-57	06	16
2	1.3	5	3.8	9	5.7	FAYETTE TX	186	52	335	95	85
1	2.5	1	2.2	1	2.1	FISHER TX	-13	-3	-16	28	28
2	3.8	3	5.2			FLOYD TX	37				
1	5.4					FOARD TX					
3	2.8	8	5.0	7	3.3	FORT BEND TX	77	-33	18	50	49
		1	2.7	1	2.6	FPANKLIN TX		-4			36
1	1.9	2	3.1	1	1.8	FREESTONE TX	67	-43	-5	36	23
2	6.7	2	4.3	2	4.1	FPIO TX	-35	-6	-39	12	62
16	4.6	17	3.9	30	5.0	GALVESTON TX	-16	29	9	45	77
1	3.4			2	7.6	GARZA TX			123	80	95
2	4.6	1	1.6	4	5.2	GILLESPIE TX	-66	234	14	47	80
		1	3.9	2	10.0	GOLIAD TX		159			98
1	1.3	2	2.2	1	.9	GONZALES TX	69	-61	-35	15	15
2	2.3	2	1.7	1	.6	GRAY TX	-28	-64	-74	02	14
7	2.1	12	3.2	14	3.1	GRAYSON TX	52	-1	50	63	45
5	3.4	8	3.1	12	3.7	GREGG TX	-7	18	9	45	55
		2	4.0			GRIMES TX					
2	1.9	5	3.6	6	3.6	GUADALUPE TX	88	-1	86	74	53
4	3.2	4	3.0	2	1.2	HALE TX	-8	-59	-62	05	18
2	4.6			1	2.0	HALL TX			-55	07	27
		1	1.3	1	1.2	HAMILTON TX		-7			18
1	4.4			1	3.2	HANSFORD TX			-27	20	46
1	2.0			1	2.5	HARDEMAN TX			29	55	35
2	3.4	4	4.7	3	2.4	HARDIN TX	37	-48	-28	19	33
96	3.8	155	4.6	199	4.1	HARRIS TX	23	-11	9	44	63
		1	.7	6	3.8	HAPPISCN TX		460			58
2	3.4	1	1.7	1	1.5	HASKELL TX	-51	-13	-57	06	20
		1	1.1	3	3.1	HAYS TX		185			44
		1	5.5	1	5.4	HEMPHILL TX		-2			81
3	3.1	6	4.6	8	4.2	HENDERSON TX	49	-8	37	58	65
13	2.8	20	3.5	35	4.3	HIDALGO TX	21	26	52	64	67
3	2.1	7	5.1	4	2.5	HILL TX	149	-50	23	52	35
1	1.0	3	5.2	3	3.2	HOCKLEY TX	393	-38	205	90	46
		1	3.1	1	1.8	HOOD TX		-40			24
3	2.6	2	1.5	2	1.2	HOPKINS TX	-42	-23	-55	06	17
1	1.6	1	1.4	4	4.3	HOUSTON TX	-15	216	170	86	66
2	1.8	2	1.7	4	3.1	HCWAPD TX	-9	87	71	69	44
4	1.9	3	1.3	6	2.6	HUNT TX	-30	94	35	58	35
1	1.2	5	6.4	4	3.0	HUTCHINSON TX	443	-54	152	85	42
3	6.2	1	1.8			JACK TX	-71				
		2	5.7	3	5.0	JACKSON TX		-12			76
				2	1.8	JASPER TX					23
1	12.2					JEFF DAVIS TX					
26	5.8	20	2.8	29	3.3	JEFFERSON TX	-52	18	-43	11	48

* SIGNIFICANTLY MORE DEATHS THAN EXPECTED

WHITE MALE — CANCER OF THE ESOPHAGUS — ICD 150

1950-1959 DEATHS	1950-1959 RATE	1960-1969 DEATHS	1960-1969 RATE	1970-1979 DEATHS	1970-1979 RATE	COUNTY AND STATE	PERCENT CHANGE 60S/50S	PERCENT CHANGE 70S/60S	PERCENT CHANGE 70S/50S	PERCENTILE %CHANGE 70S/50S	PERCENTILE RATES 1970S
2	14.4					JIM HOGG TX					
5	5.9	6	6.0	6	4.8	JIM WELLS TX	2	-21	-20	25	74
3	1.7	4	1.9	7	2.8	JOHNSON TX	8	50	61	67	39
3	3.0	3	2.8	2	1.8	JONES TX	-4	-36	-39	12	24
4	6.3	3	4.3	2	3.8	KARNES TX	-32	-12	-40	12	57
3	2.2	2	1.4	5	3.2	KAUFMAN TX	-37	128	43	61	46
		2	4.5	3	6.6	KENDALL TX		46			92
		1	14.9			KENT TX					
4	3.9	1	.6	12	5.9	KERR TX	-84	839	50	63	86
2	9.9			1	3.5	KIMBLE TX			-64	04	52
		1	7.0			KINNEY TX					
2	2.4	6	7.4	4	4.1	KLEBERG TX	212	-44	74	70	64
2	5.2	1	3.0	1	2.2	KNOX TX	-43	-25	-58	06	29
1	.5	3	1.9	7	3.5	LAMAR TX	260	86	568	98	52
1	1.7	1	1.1	3	2.9	LAMB TX	-38	173	68	69	41
1	1.9	4	6.6	3	4.3	LAMPASAS TX	248	-35	126	81	66
		1	3.9	2	10.4	LA SALLE TX		167			98
1	.9	7	6.0	4	2.7	LAVACA TX	593	-55	214	91	38
				4	7.6	LEE TX					95
		1	1.8	3	5.6	LEON TX		205			83
2	1.6	6	6.0	11	6.9	LIBERTY TX	277	16	335	95	93
4	3.9	2	2.7	3	3.1	LIMESTONE TX	-30	15	-20	25	44
		1	5.6	1	4.6	LIPSCOMB TX		-17			72
1	7.4	1	3.9	2	3.9	LIVE OAK TX	-48	0	-48	09	59
1	2.9	2	4.1	3	2.2	LLANO TX	43	-48	-25	22	29
6	1.4	13	3.0	19	3.3	LUBBOCK TX	123	9	144	84	48
		1	2.3	2	4.5	LYNN TX		95			70
1	1.5					MC CULLOCH TX					
21	4.5	23	3.8	23	3.4	MC LENNAN TX	-17	-10	-25	22	50
		1	15.0			MC MULLEN TX					
		1	3.2	1	2.3	MARION TX		-30			30
1	5.2					MARTIN TX					
1	3.8			1	4.6	MASON TX			21	51	71
1	1.3	3	3.8	3	2.3	MATAGORDA TX	201	-40	82	73	31
2	4.4			1	1.3	MAVERICK TX			-70	02	19
1	1.8	2	2.0	4	3.7	MEDINA TX	11	86	107	78	55
		1	4.6			MENARD TX					
7	8.3			8	2.9	MIDLAND TX			-65	04	41
2	1.6	3	3.0	6	5.4	MILAM TX	79	84	230	92	82
		2	4.0	1	1.7	MILLS TX		-56			23
		3	6.2	1	1.8	MITCHELL TX		-71			24
2	2.4	5	5.2	10*	8.6	MONTAGUE TX	119	67	266	93	97
3	3.0	1	.6	6	2.5	MONTGOMERY TX	-80	324	-16	28	34
1	2.6	2	4.4	1	1.8	MORRIS TX	71	-59	-30	18	24
		2	1.6	2	1.3	NACOGDOCHES TX		-17			19
2	1.3	9	5.4	3	1.6	NAVARRO TX	323	-70	27	54	22
		3	6.9	2	4.0	NEWTON TX		-42			61
1	1.3	1	1.0	1	.9	NOLAN TX	-22	-9	-29	19	15
20	4.4	17	3.1	43	5.2	NUECES TX	-30	69	18	50	80
		1	2.8			OCHILTREE TX					
				1	9.8	OLDHAM TX					98
5	3.0	2	.8	7	3.2	ORANGE TX	-74	308	5	42	46
				3	2.3	PALO PINTO TX					31
1	1.6	3	4.7			PANOLA TX	186				
1	.9	3	2.1	7	3.6	PARKER TX	149	67	316	95	53
1	1.4	1	2.1	1	1.7	PECOS TX	45	-21	15	48	22
2	3.4	3	4.5	4	3.4	POLK TX	32	-24	1	39	51
10	3.9	9	2.3	12	3.3	POTTER TX	-41	41	-17	27	47
				2	8.9	PRESIDIO TX					97
1	3.1	1	5.2	1	3.6	RAINS TX	67	-32	14	47	53
		2	1.5	7	4.6	RANDALL TX		217			72
				2*	22.5	REAGAN TX					99
				1	5.8	REAL TX					85
2	2.5	1	1.4	2	1.8	RED RIVER TX	-43	28	-27	20	24
1	2.8	2	5.7	4	7.0	REEVES TX	105	23	153	85	93
1	3.1			1	2.2	REFUGIO TX			-28	19	29
				1	9.4	ROBERTS TX					97
1	1.4	1	1.3	3	4.8	ROBERTSON TX	-9	279	244	92	75
		1	3.0			ROCKWALL TX					
2	2.6	2	2.3	2	2.3	RUNNELS TX	-12	1	-12	31	31
2	1.3	6	3.8	5	2.6	RUSK TX	203	-33	104	77	35
				1	2.4	SAN AUGUSTINE TX					31
8	6.5	4	2.8	11	5.7	SAN PATRICIO TX	-56	99	-12	30	84
				1	1.7	SAN SABA TX					23
2*	15.2					SCHLEICHER TX					
1	2.1			1	1.2	SCURRY TX			-42	12	17
				1	3.0	SHACKELFORD TX					43
2	1.8			5	4.1	SHELBY TX			123	80	63
		1	5.2			SHERMAN TX					
6	2.3	10	3.3	13	3.0	SMITH TX	42	-8	31	56	43
		2	10.1	1	3.9	SOMERVELL TX		-61			60
1	2.1	2	3.3	3	3.8	STARR TX	52	16	77	71	57
		1	1.3			STEPHENS TX					
		1	5.6			STONEWALL TX					
1	12.5	1	8.0	1	7.5	SUTTON TX	-36	-7	-40	12	95
		1	2.2	1	2.4	SWISHER TX		7			32
46	3.5	65	3.5	76	3.1	TARRANT TX	-2	-10	-11	31	45
9	3.9	8	2.3	18	4.5	TAYLOR TX	-41	91	13	47	69
		1	1.6	3	5.7	TERRY TX		251			84
				1	4.3	THROCKMORTON TX					67
2	2.5	4	4.7	3	3.6	TITUS TX	91	-24	45	62	54
5	2.0	12	4.4	7	2.0	TOM GREEN TX	116	-54	0	39	26
20	3.5	21	3.0	33	3.7	TRAVIS TX	-16	24	4	41	55
				1	1.9	TRINITY TX					25
1	1.7			5	4.6	TYLER TX			172	87	71
4	4.5	1	1.0	4	3.1	UPSHUR TX	-78	220	-31	18	45
2	3.0	7*	8.7	4	4.2	UVALDE TX	187	-52	37	58	65
1	1.4			4	4.1	VAL VERDE TX			197	89	64
1	.9			3	1.3	VAN ZANDT TX			52	64	18
8	6.1	4	3.8	7	3.6	VICTORIA TX	-39	-5	-42	11	53
				3	3.8	WALKER TX					57

* SIGNIFICANTLY MORE DEATHS THAN EXPECTED

WHITE MALE — CANCER OF THE ESOPHAGUS — ICD 150

1950-1959 DEATHS	1950-1959 RATE	1960-1969 DEATHS	1960-1969 RATE	1970-1979 DEATHS	1970-1979 RATE	COUNTY AND STATE	PERCENT CHANGE 60S/50S	PERCENT CHANGE 70S/60S	PERCENT CHANGE 70S/50S	PERCENTILE %CHANGE 70S/50S	PERCENTILE RATES 1970S
		3	8.4	1	1.6	WALLER TX		-81			22
1	2.5	1	1.6	1	1.7	WARD TX	-37	9	-31	17	23
1	1.2	4	3.6	5	5.1	WASHINGTON TX	193	40	311	95	78
8	4.5	10	4.4	16	6.0	WEBB TX	-2	37	34	57	86
5	5.5	8	6.1	4	2.3	WHARTON TX	10	-63	-59	06	30
1	2.5			1	1.6	WHEELER TX			-37	14	21
7	1.8	10	2.2	22	4.9	WICHITA TX	19	122	164	86	75
3	4.4	2	2.2	2	1.9	WILBARGER TX	-50	-15	-57	06	25
2	4.5	1	1.3	1	1.9	WILLACY TX	-71	44	-59	06	25
5	2.9	1	.6	9	3.8	WILLIAMSON TX	-80	567	32	56	57
1	1.9	4	5.3	5	6.5	WILSON TX	185	22	246	93	91
				4	11.1	WINKLER TX					98
2	2.1	3	2.7	1	.7	WISE TX	28	-76	-69	03	14
2	2.0	4	3.0	4	2.4	WOOD TX	49	-20	18	50	33
				1	3.0	YOAKUM TX					43
		4	4.0	5	4.7	YOUNG TX		17			73
1	4.1	3*	11.5			ZAPATA TX	184				
1	3.0	2	4.1	3	7.4	ZAVALA TX	34	81	142	84	95
35	1.5	63	1.9	120	2.8	UTAH UT	28	52	94		
1	5.5	1	4.3			BEAVER UT	-22				
				5	2.9	CACHE UT					40
3	5.4	3	3.2	1	.9	CARBON UT	-41	-71	-83	00	15
		6	4.3	7	2.6	DAVIS UT		-40			35
				1	3.0	DUCHESNE UT					41
				2	6.2	EMERY UT					89
1	5.3			1	4.3	GARFIELD UT			-18	26	66
		1	6.7	1	3.6	GRAND UT		-47			53
		2	4.6	3	4.9	IRON UT		7			75
		2	8.3			JUAB UT					
				3	7.5	MILLARD UT					95
20	2.0	34	2.3	57	3.1	SALT LAKE UT	14	35	55	65	45
1	2.1			2	3.0	SEVIER UT			43	61	42
				1	3.5	SUMMIT UT					51
1	2.2	1	1.0	4	5.2	TOOELE UT	-56	451	144	84	80
				2	3.6	UINTAH UT					54
1	.4	4	1.0	11	2.3	UTAH UT	164	120	481	98	30
1	8.3			1	2.9	WASATCH UT			-65	04	41
				2	1.5	WASHINGTON UT					20
6	1.8	9	2.2	16	3.2	WEBER UT	21	47	77	71	45
69	3.7	65	3.4	93	4.4	VERMONT VT	-7	27	19		
1	1.1	2	2.2	6	5.2	ADDISON VT	98	139	374	96	79
7	5.6	6	3.9	10	6.3	BENNINGTON VT	-29	61	14	47	90
7	5.8	8	6.7	4	3.0	CALEDONIA VT	16	-54	-47	09	43
10	4.0	12	4.9	19	5.4	CHITTENDEN VT	24	11	37	58	82
2	5.7	2	7.0			ESSEX VT	22				
4	2.7	4	2.8	3	2.0	FRANKLIN VT	1	-29	-28	19	26
1	5.8					GRAND ISLE VT					
2	3.4	1	1.8	2	3.2	LAMOILLE VT	-48	84	-5	36	47
1	1.0	4	4.5	2	2.1	ORANGE VT	335	-54	99	77	27
5	4.6	2	1.9	4	3.6	ORLEANS VT	-58	89	-21	24	54
7	2.9	8	3.1	14	5.2	RUTLAND VT	6	70	80	72	80
9	4.6	5	2.5	9	3.9	WASHINGTON VT	-47	59	-15	29	59
10	6.3	6	3.7	8	4.6	WINDHAM VT	-41	24	-27	20	71
3	1.4	5	2.3	12	4.8	WINDSOR VT	65	109	246	93	74
330	3.2	500	3.8	725	4.5	VIRGINIA VA	20	17	40		
3	2.6	4	3.0	7	5.1	ACCOMACK VA	12	71	93	75	78
9	5.1	9	4.2	9	3.3	ALBEMARLE#VA	-17	-21	-35	15	49
4	4.4	3	2.9	6	4.4	ALLEGHANY#VA	-33	50	0	39	67
		1	5.3	1	4.1	AMELIA VA		-23			63
1	1.5	3	3.6	1	1.3	AMHERST VA	145	-64	-11	32	18
1	2.8	5*	12.4	1	2.4	APPOMATTOX VA	337	-81	-15	29	32
15	3.6	32	5.4	39	6.1	ARLINGTON#VA	52	13	72	70	88
9	3.6	8	2.3	16	4.0	AUGUSTA#VA	-36	74	11	46	62
1	4.4	1	3.3	2	7.4	BATH VA	-25	123	68	69	95
10	2.9	21	4.8	15	2.8	BEDFORD#VA	66	-41	-3	37	39
4	5.7	4	5.0	3	3.1	BOTETOURT VA	-12	-39	-46	10	44
				1	2.5	BRUNSWICK VA					33
3	4.2	5	5.5	3	2.4	BUCHANAN VA	29	-57	-45	10	31
		1	2.6	1	2.6	BUCKINGHAM VA		0			36
				4	11.3	CAROLINE VA					99
2	1.0	6	2.8	6	2.2	CARROLL#VA	194	-22	130	81	29
				1	21.3	CHARLES CITY VA					99
3	10.0					CHARLOTTE VA					
55*	6.7	54*	5.9	68*	6.8	CHESTERFIELD#VA	-12	16	2	40	93
2	6.6	1	2.6	4	10.1	CLARKE VA	-61	289	52	63	98
				1	4.0	CRAIG VA					61
		1	1.3	4	4.9	CULPEPER VA		268			75
1	7.0			1	4.1	CUMBERLAND VA			-42	11	63
1	1.5	2	2.8	1	1.5	DICKENSON VA	87	-47	-1	38	20
9	4.8	8	3.9	9	3.2	DINWIDDIE#VA	-18	-18	-33	16	47
		1	4.8			ESSEX VA					
20	4.0	43	3.8	88	5.6	FAIRFAX#VA	-6	48	39	60	84
4	5.6	4	4.8	6	6.0	FAUQUIER VA	-14	26	8	44	87
2	4.9	4	7.6	4	6.4	FLOYD VA	56	-16	31	56	90
1	1.0	4	3.9	2	1.5	FRANKLIN VA	275	-61	45	62	20
8	5.4	9	5.1	4	1.7	FREDERICK#VA	-6	-66	-68	03	23
		1	1.3			GILES VA					
1	2.2	3	6.5			GLOUCESTER VA	190				
		2	8.4	2	5.8	GOOCHLAND VA		-31			86
				1	3.9	GREENE VA					60
		1	4.0	1	2.7	GREENSVILLE#VA		-34			37
4	4.1	10*	9.2	7	5.4	HALIFAX#VA	123	-42	30	55	82
4	5.8	3	2.8	14*	9.1	HANOVER VA	-51	220	58	66	97
5	1.7	11	2.7	27	4.8	HENRICO VA	60	78	184	88	74
1	.6	6	2.8	12	5.2	HENRY#VA	361	88	767	99	79

* SIGNIFICANTLY MORE DEATHS THAN EXPECTED

COUNTIES COMBINED; SEE TECHNICAL NOTES

WHITE MALE — CANCER OF THE ESOPHAGUS — ICD 150

1950-1959 DEATHS	1950-1959 RATE	1960-1969 DEATHS	1960-1969 RATE	1970-1979 DEATHS	1970-1979 RATE	COUNTY AND STATE	PERCENT CHANGE 60S/50S	PERCENT CHANGE 70S/60S	PERCENT CHANGE 70S/50S	PERCENTILE %CHANGE 70S/50S	PERCENTILE RATES 1970S
2	5.0			3	5.8	ISLE OF WIGHT VA			16	49	86
8	7.4	2	1.2	7	3.3	JAMES CITY#VA	-84	185	-55	07	48
		1	6.1	3*	11.7	KING AND QUEEN VA		93			99
1	4.3	1	3.3			KING GEORGE VA	-24				
2	11.5	2	14.6	1	4.0	KING WILLIAM VA	27	-72	-65	04	61
1	2.8			4	7.1	LANCASTER VA			152	85	94
3	2.4	2	1.5	2	1.6	LEE VA	-38	8	-33	16	21
4	4.7	10*	9.5	7	5.2	LOUDOUN VA	103	-45	11	46	79
1	2.3	1	1.9	3	4.8	LOUISA VA	-17	151	108	78	74
1	3.7	1	2.2	1	2.7	LUNENBURG VA	-40	20	-29	19	37
1	2.8	1	2.7	1	1.9	MADISON VA	-6	-27	-31	17	25
1	2.6	1	2.4	2	4.8	MATHEWS VA	-8	99	83	73	75
1	1.3	5	5.5	4	3.6	MECKLENBURG VA	317	-35	172	87	54
		1	3.9	1	2.4	MIDDLESEX VA		-40			32
1	.8	3	1.7	4	2.3	MONTGOMERY#VA	109	36	185	88	31
1	1.9			1	1.6	NELSON VA			-13	30	22
		2	4.0	2	5.0	NORTHAMPTON VA		25			76
1	2.3	2	5.0	5	8.3	NORTHUMBERLAND VA	123	65	268	93	96
2	4.3	4	8.5	2	3.8	NOTTOWAY VA	100	-55	-10	32	58
2	4.2	1	2.0	3	4.4	ORANGE VA	-54	125	4	41	68
1	1.6	2	2.6	2	2.2	PAGE VA	64	-12	44	61	30
1	1.9			5	5.8	PATRICK VA			202	90	86
10	3.5	10	3.1	17	4.5	PITTSYLVANIA#VA	-10	44	29	55	70
				1	6.2	POWHATAN VA					89
		2	4.6	2	4.5	PRINCE EDWARD VA		-2			71
1	1.4	3	3.0	11	5.5	PRINCE WILLIAM#VA	118	86	307	95	83
2	2.1	3	2.2	8	5.5	PULASKI VA	5	143	155	85	83
1	4.5					RAPPAHANNOCK VA					
1	5.1			2	7.0	RICHMOND VA			38	59	93
12	2.3	25	3.6	36	4.5	ROANOKE#VA	52	27	93	75	71
1	1.1	1	.8	4	3.2	ROCKBRIDGE#VA	-26	302	196	89	46
2	1.0	8	3.5	13	4.6	ROCKINGHAM#VA	236	32	345	96	72
1	1.1	4	3.9	1	.8	RUSSELL VA	267	-80	-27	20	15
1	.9			1	1.0	SCOTT VA			20	50	16
2	1.7	2	2.1	4	2.6	SHENANDOAH VA	28	23	57	65	36
		2	1.2	3	1.7	SMYTH VA		36			22
1	1.9					SOUTHAMPTON#VA					
5	6.7	3	3.1	6	4.2	SPOTSYLVANIA#VA	-54	35	-38	13	65
2	4.4			3	2.9	STAFFORD VA			-33	16	41
1	5.8			2	10.5	SURRY VA			80	72	98
		1	4.4	1	3.8	SUSSEX VA		-12			58
2	1.4	6	3.7	5	2.3	TAZEWELL VA	165	-39	62	67	30
		1	1.7	5	4.8	WARREN VA		186			75
6	3.1	7	3.1	13	4.7	WASHINGTON#VA	1	49	50	63	73
				2	2.9	WESTMORELAND VA					40
7	4.0	8	4.3	8	4.1	WISE#VA	6	-3	3	40	64
3	3.5	2	2.0	4	3.7	WYTHE VA	-41	82	7	43	56
36	3.4	77*	6.2	101*	6.3	CHESAPEAKE CITY#VA	85	0	86	74	90
5	3.0	11	4.1	19	6.8	HAMPTON CITY VA	36	67	127	81	93
6	2.6	10	4.6	14	4.7	NEWPORT NEWS CITY VA	81	0	81	73	72
4	5.0	1	1.0	4	3.5	SUFFOLK CITY VA	-81	265	-30	18	52
388	3.3	519	3.8	647	4.0	WASHINGTON WA	14	7	22		
		3	8.5	5	10.5	ADAMS WA		24			98
		2	2.3	4	4.9	ASOTIN WA		111			75
3	2.4	8	3.8	10	3.6	BENTON WA	58	-5	50	63	54
3	1.5	11	5.0	12	4.7	CHELAN WA	243	-7	219	91	73
3	2.0	9	5.0	7	2.7	CLALLAM WA	149	-46	33	57	37
6	1.3	10	2.0	19	2.9	CLARK WA	52	40	113	79	40
		1	2.5	2	6.7	COLUMBIA WA		172			92
7	2.9	9	3.2	15	4.4	COWLITZ WA	9	37	48	63	67
1	2.0	4	5.8	1	1.0	DOUGLAS WA	194	-83	-50	08	16
				1	6.6	FERRY WA					92
2	5.4	5	7.3	4	4.0	FRANKLIN WA	34	-45	-27	20	60
1	4.6	2	11.4			GARFIELD WA	149				
2	3.8	5	3.1	7	3.9	GRANT WA	-21	29	2	40	60
17	6.1	16	5.0	22	6.5	GRAYS HARBOR WA	-19	29	5	42	91
1	1.5	1	1.0	4	2.5	ISLAND WA	-31	147	71	70	34
5	9.8	5	7.7	2	2.2	JEFFERSON WA	-22	-71	-77	01	30
154	4.3	182	4.4	217	4.6	KING WA	3	5	8	44	72
2	.5	17	4.1	25	4.5	KITSAP WA	736	10	822	99	70
6	5.0	4	3.5	5	4.4	KITTITAS WA	-31	26	-13	30	67
		1	1.5	5	7.5	KLICKITAT WA		387			95
6	2.4	3	1.1	9	3.2	LEWIS WA	-55	198	33	57	46
1	1.8	1	1.5	1	1.3	LINCOLN WA	-20	-13	-30	18	18
1	1.0	3	3.4	5	3.4	MASON WA	221	1	222	91	50
4	3.5	1	.9	7	4.8	OKANOGAN WA	-75	448	36	58	74
1	1.2	2	1.6	6	5.3	PACIFIC WA	39	225	352	96	81
				3	5.8	PEND OREILLE WA					86
47	3.7	52	3.5	85	5.0	PIERCE WA	-6	43	34	57	77
		1	2.6	2	5.0	SAN JUAN WA		94			77
5	1.9	8	2.8	13	4.3	SKAGIT WA	45	55	124	80	66
		2	5.5	2	5.9	SKAMANIA WA		7			86
24	3.6	33	3.9	35	3.1	SNOHOMISH WA	8	-20	-14	29	44
42	3.4	50	3.7	48	3.3	SPOKANE WA	7	-10	-4	36	48
1	1.1	3	3.0	1	.8	STEVENS WA	169	-74	-30	18	15
7	2.9	14	4.6	16	3.9	THURSTON WA	57	-16	32	56	59
9	4.5	9	3.9	4	1.7	WALLA WALLA WA	-14	-57	-63	05	23
11	2.9	8	2.0	10	2.2	WHATCOM WA	-31	10	-24	23	30
2	1.4	4	3.0	8	5.7	WHITMAN WA	113	93	311	95	85
14	2.3	30	4.2	25	3.2	YAKIMA WA	82	-25	37	58	46
184	2.4	246	3.0	364	4.1	WEST VIRGINIA WV	25	37	71		
3	3.7	3	2.9	5	6.5	BARBOUR WV	-21	125	77	71	91
3	1.9	7	3.7	9	4.5	BERKELEY WV	92	22	133	82	69
1	.9	2	1.7	6	4.6	BOONE WV	86	167	396	97	72
2	2.5			1	1.2	BRAXTON WV			-54	07	17
4	3.9	4	3.1	5	3.4	BROOKE WV	-22	10	-14	29	50
16	3.5	17	3.3	25	4.9	CABELL WV	-7	48	37	59	75

* SIGNIFICANTLY MORE DEATHS THAN EXPECTED

COUNTIES COMBINED; SEE TECHNICAL NOTES

WHITE MALE — CANCER OF THE ESOPHAGUS — ICD 150

1950-1959 DEATHS	1950-1959 RATE	1960-1969 DEATHS	1960-1969 RATE	1970-1979 DEATHS	1970-1979 RATE	COUNTY AND STATE	PERCENT CHANGE 60S/50S	PERCENT CHANGE 70S/60S	PERCENT CHANGE 70S/50S	PERCENTILE %CHANGE 70S/50S	PERCENTILE RATES 1970S
		3	5.7	2	4.8	CALHOUN WV		-16			74
				1	2.0	CLAY WV					26
		2	4.2	2	5.2	DODDRIDGE WV		23			79
6	2.4	5	1.9	13	5.0	FAYETTE WV	-22	163	106	78	77
		2	5.4			GILMER WV					
				2	3.8	GRANT WV					57
4	2.7	8	4.6	5	2.7	GREENBRIER WV	70	-41	0	38	38
		1	1.5	2	2.5	HAMPSHIRE WV		69			35
2	1.4	5	3.4	7	3.7	HANCOCK WV	140	8	160	86	55
1	2.4	3	6.4	1	1.6	HARDY WV	171	-75	-33	16	21
14	3.7	21	5.3	13	3.1	HARRISON WV	43	-41	-15	29	44
1	1.4			3	3.0	JACKSON WV			117	79	42
		1	1.2	8	7.8	JEFFERSON WV		528			96
27	3.2	38	4.2	64*	6.2	KANAWHA WV	33	48	96	76	89
1	.7	2	1.8	3	2.5	LEWIS WV	163	40	268	93	35
		2	2.1	2	1.9	LINCOLN WV		-9			25
4	2.3	6	3.1	6	3.1	LOGAN WV	30	2	34	57	45
1	.7	8	4.3	12	6.2	MC DOWELL WV	498	44	764	99	89
14	4.7	8	2.6	14	4.1	MARION WV	-46	60	-13	30	63
3	1.7	6	3.5	8	4.0	MARSHALL WV	103	16	135	82	62
2	2.0	2	1.8	1	.8	MASON WV	-9	-55	-59	05	15
8	3.2	6	2.1	9	2.8	MERCER WV	-34	34	-12	31	39
3	3.3	4	3.6	6	4.9	MINERAL WV	9	37	48	63	76
6	4.1	2	1.4	4	2.7	MINGO WV	-67	95	-35	15	37
4	1.6	9	3.7	8	3.0	MONONGALIA WV	124	-17	85	74	43
1	1.1	1	1.8	2	3.5	MONROE WV	56	94	203	90	51
1	2.4			2	4.0	MORGAN WV			71	70	62
1	1.4	2	2.0	6	4.9	NICHOLAS WV	40	146	244	92	76
21	6.1	23*	6.6	22	6.8	OHIO WV	9	2	11	46	93
1	2.4	1	3.1	1	2.6	PLEASANTS WV	28	-16	7	43	36
1	1.9			2	3.0	POCAHONTAS WV			60	66	42
2	1.4	6	4.1	5	3.4	PRESTON WV	197	-17	148	84	51
1	1.2	2	1.7	4	3.0	PUTNAM WV	36	77	140	83	42
2	.6	9	3.2	9	2.4	RALEIGH WV	407	-25	283	94	33
2	1.6	2	1.4	2	1.3	RANDOLPH WV	-11	-7	-18	26	18
1	1.5			4	6.3	RITCHIE WV			317	95	90
		3	3.6			ROANE WV					
		2	2.3	5	6.6	SUMMERS WV		180			92
3	3.6	1	1.4	7	8.5	TAYLOR WV	-61	493	134	82	96
2	4.3			2	3.6	TUCKER WV			-16	28	54
1	2.0			3	5.1	TYLER WV			151	84	77
		2	2.0	5	5.2	UPSHUR WV		165			79
2	1.3	3	1.7	6	3.0	WAYNE WV	29	79	130	81	43
2	2.9			3	5.6	WEBSTER WV			97	76	84
3	3.2	1	1.0	4	3.9	WETZEL WV	-70	309	23	52	59
		1	4.1	2	6.1	WIRT WV		49			88
7	2.2	9	2.5	19	4.5	WOOD WV	18	77	108	78	70
		1	.6	2	1.4	WYOMING WV		122			19
995*	5.8	911*	4.7	939	4.4	WISCONSIN WI	-19	-6	-24		
2	4.8	1	1.3	4	5.4	ADAMS WI	-73	317	14	47	82
8	7.4	2	1.6	5	4.8	ASHLAND WI	-78	194	-35	14	74
4	2.2	7	3.1	8	3.4	BARRON WI	43	11	59	66	50
6	6.7	4	4.4	3	4.1	BAYFIELD WI	-34	-7	-39	13	63
27	6.8	29	5.5	22	3.4	BROWN WI	-20	-37	-50	09	51
1	1.5	1	1.3	1	1.6	BUFFALO WI	-16	24	5	42	22
1	1.4	1	1.5	4	4.8	BURNETT WI	9	221	250	93	74
7	7.8	1	.9	6	4.7	CALUMET WI	-89	441	-39	12	74
8	3.7	9	3.8	9	3.6	CHIPPEWA WI	4	-7	-4	36	53
7	3.8	4	2.2	2	.9	CLARK WI	-42	-60	-77	01	15
7	3.2	10	4.4	11	5.0	COLUMBIA WI	40	13	57	65	76
2	2.1	1	1.5	1	1.0	CRAWFORD WI	-27	-34	-52	08	16
28	3.6	35	4.0	35	3.5	DANE WI	10	-12	-3	37	52
15	4.6	15	4.3	14	3.8	DODGE WI	-5	-13	-18	27	57
4	3.2	4	3.2	5	4.1	DOOR WI	-1	29	28	54	64
18	6.6	10	4.0	16	6.8	DOUGLAS WI	-39	69	3	40	93
3	2.3	1	.7	3	1.9	DUNN WI	-70	177	-16	28	25
8	3.5	5	1.8	13	4.0	EAU CLAIRE WI	-48	123	16	49	61
1	3.4	3*	13.3			FLORENCE WI	291				
17	4.8	14	3.6	12	2.9	FOND DU LAC WI	-25	-18	-39	13	41
				3	5.0	FOREST WI					77
2	.9	11	5.0	5	2.0	GRANT WI	440	-61	112	79	26
3	2.3	6	3.9	4	2.5	GREEN WI	71	-36	9	45	33
3	3.4	1	1.2	5	4.2	GREEN LAKE WI	-64	237	22	51	65
3	2.6			2	2.1	IOWA WI			-17	27	28
2	3.5	4	7.3	2	5.4	IRON WI	109	-27	53	64	82
6	6.5	2	1.9	5	4.7	JACKSON WI	-70	145	-28	20	73
12	4.6	9	3.2	15	4.6	JEFFERSON WI	-30	42	0	38	72
1	.7	2	1.7	7	5.4	JUNEAU WI	129	217	625	98	82
24	6.9	22	4.6	26	5.1	KENOSHA WI	-34	12	-26	21	78
7	7.8	5	5.1	3	2.7	KEWAUNEE WI	-35	-46	-65	04	38
10	2.9	21	5.9	18	4.5	LA CROSSE WI	104	-24	55	65	70
3	3.2	2	2.0	5	5.1	LAFAYETTE WI	-37	154	61	67	79
9	7.6	3	2.5	5	3.7	LANGLADE WI	-67	46	-52	08	55
6	4.8	3	2.3	5	3.7	LINCOLN WI	-52	62	-22	24	56
20	6.2	24	6.2	27	6.2	MANITOWOC WI	1	0	1	39	89
23	5.9	14	3.1	18	3.9	MARATHON WI	-47	23	-35	15	58
11	5.7	10	4.3	10	4.6	MARINETTE WI	-26	8	-19	26	72
1	1.6	4	7.0	4	4.1	MARQUETTE WI	331	-41	154	85	64
13	3.9	10	2.8	14	3.6	MENOMINEE#WI	-29	29	-8	33	54
419*	10.7	350*	7.9	265*	6.0	MILWAUKEE WI	-26	-24	-44	10	87
5	2.7	6	3.0	4	1.6	MONROE WI	11	-48	-42	11	21
10*	8.9	6	5.0	15*	8.8	ONEIDA WI	-44	77	-1	38	97
17	5.3	11	2.6	22	4.6	OUTAGAMIE WI	-51	78	-14	30	71
6	5.2	12	7.7	13	5.8	OZAUKEE WI	48	-25	11	46	85
1	2.3					PEPIN WI					
2	1.4	3	2.1	4	3.2	PIERCE WI	54	50	130	81	46
5	3.4	2	1.2	9	4.8	POLK WI	-64	290	42	61	74
8	4.8	6	3.1	7	3.3	PORTAGE WI	-36	8	-31	17	48
5	5.7	2	2.2	3	2.4	PRICE WI	-61	9	-57	06	33

* SIGNIFICANTLY MORE DEATHS THAN EXPECTED

COUNTIES COMBINED; SEE TECHNICAL NOTES

1950-1959 DEATHS	1950-1959 RATE	1960-1969 DEATHS	1960-1969 RATE	1970-1979 DEATHS	1970-1979 RATE	COUNTY AND STATE	PERCENT CHANGE 60S/50S	PERCENT CHANGE 70S/60S	PERCENT CHANGE 70S/50S	PERCENTILE %CHANGE 70S/50S	PERCENTILE RATES 1970S
39*	7.7	27	4.5	36	5.2	RACINE WI	-42	17	-33	16	80
		2	2.0	4	3.9	RICHLAND WI		98			59
23	5.2	28	5.5	22	3.9	ROCK WI	6	-29	-25	22	59
4	4.2	3	3.5	4	4.3	RUSK WI	-15	22	3	41	66
7	4.8	3	1.8	3	1.6	ST CROIX WI	-62	-12	-67	03	21
6	2.9	8	3.7	8	3.3	SAUK WI	28	-9	16	49	48
4	5.6	2	4.2	1	1.1	SAWYER WI	-25	-73	-80	01	17
20	5.0	26	5.6	24	4.6	SHEBOYGAN WI	11	-17	-8	33	71
2	1.8	2	2.0	2	1.8	TAYLOR WI	12	-10	0	38	24
5	3.4	6	3.9	5	3.3	TREMPEALEAU WI	13	-14	-3	37	48
4	2.9	6	3.6	6	3.3	VERNON WI	25	-8	15	48	49
1	1.4	2	2.0	2	1.7	VILAS WI	40	-18	14	48	22
8	3.0	8	2.7	15	4.5	WALWORTH WI	-10	71	53	64	71
3	4.3	4	5.5	3	4.3	WASHBURN WI	28	-22	-1	38	66
6	3.6	10	4.3	13	4.5	WASHINGTON WI	19	4	23	52	69
17	3.9	16	2.8	38	4.3	WAUKESHA WI	-28	56	12	46	67
5	2.4	15	5.8	11	3.8	WAUPACA WI	140	-35	57	65	57
2	1.8	4	4.3	4	3.4	WAUSHAPA WI	135	-22	83	73	49
21	4.7	24	4.5	29	4.9	WINNEBAGO WI	-3	8	5	42	76
7	3.1	7	2.5	10	3.2	WOOD WI	-20	28	2	40	46
33	2.9	24	1.7	48	3.1	WYOMING WY	-42	85	7		
2	3.2	2	2.9	3	3.3	ALBANY WY	-8	12	3	41	48
1	2.3	2	3.1	1	1.5	BIG HORN WY	36	-51	-34	16	20
3	5.8	1	1.7	5	7.0	CARBON WY	-71	326	22	52	94
		2	6.2			CONVERSE WY					
				1	4.5	CROOK WY					71
1	1.2			5	4.8	FREMONT WY			311	95	75
1	2.4	1	1.6			GOSHEN WY	-34				
				2	5.1	JOHNSON WY					78
2	2.8	4	2.1	11	4.7	LARAMIE WY	-22	120	70	69	73
1	2.2					LINCOLN WY					
1	1.4	3	1.7	9	4.6	NATRONA WY	26	169	238	92	71
1	4.2	1	4.8	1	5.8	NIOBRARA WY	14	22	38	59	85
3	3.8	2	2.1	1	.7	PARK-TETON#WY	-44	-66	-81	00	15
		1	2.4			PLATTE WY					
6	4.3	2	1.3	3	2.6	SHERIDAN WY	-69	90	-41	12	35
1	10.2					SUBLETTE WY					
9*	10.5	1	1.1	2	2.1	SWEETWATER WY	-90	97	-80	00	28
		1	2.3	3	7.3	UINTA WY		214			95
1	3.4	1	3.3			WASHAKIE WY	-1				

* SIGNIFICANTLY MORE DEATHS THAN EXPECTED

\# COUNTIES COMBINED; SEE TECHNICAL NOTES

WHITE FEMALE — CANCER OF THE ESOPHAGUS — ICD 150

1950-1959 DEATHS	1950-1959 RATE	1960-1969 DEATHS	1960-1969 RATE	1970-1979 DEATHS	1970-1979 RATE	COUNTY AND STATE	PERCENT CHANGE 60S/50S	PERCENT CHANGE 70S/60S	PERCENT CHANGE 70S/50S	PERCENTILE %CHANGE 70S/50S	PERCENTILE RATES 1970S
8004	1.1	10223	1.1	14031	1.2	UNITED STATES	-2	10	7		
108	1.2	120	1.0	171	1.0	ALABAMA AL	-18	6	-12		
1	2.6					AUTAUGA AL					
2	1.3	1	.4	3	1.0	BALDWIN AL	-66	120	-26	40	59
1	1.6	1	1.4			BARBOUR AL	-10				
		2	3.2	3*	4.3	BIBB AL		35			98
1	1.0					BLOUNT AL					
1	4.4					BULLOCK AL					
1	1.2	1	1.2	1	.8	BUTLER AL	6	-34	-31	36	53
4	1.6	3	.8	6	1.2	CALHOUN AL	-48	47	-24	41	68
2	1.6	2	1.3	1	.5	CHAMBERS AL	-16	-61	-67	08	42
2	3.7	1	1.6	1	1.1	CHEROKEE AL	-57	-29	-69	07	66
1	.8	3	2.4	4	2.4	CHILTON AL	184	1	186	96	92
1	2.7					CHOCTAW AL					
				1	.9	CLARKE AL					55
				1	1.0	CLAY AL					61
1	2.1					CLEBURNE AL					
		1	.8			COFFEE AL					
2	1.5	1	.4	3	1.3	COLBERT AL	-70	197	-11	50	73
1	1.5			1	1.2	CONECUH AL			-20	44	67
		1	2.3	1	2.1	COOSA AL		-8			90
		2	1.0	4	1.4	COVINGTON AL		43			76
				2	2.2	CRENSHAW AL					90
2	1.1			1	.3	CULLMAN AL			-73	05	37
1	1.2	2	2.0	1	.8	DALE AL	71	-60	-31	36	53
		3	2.4	3	2.0	DALLAS AL		-18			89
5*	2.8	2	.9	3	1.0	DE KALB AL	-70	14	-65	10	60
3*	2.8	3	2.1			ELMORE AL	-26				
1	1.1	2	1.7	2	1.2	ESCAMBIA AL	63	-28	16	64	69
2	.6	2	.5	6	1.1	ETOWAH AL	-15	131	96	88	64
3*	4.4					FAYETTE AL					
1	1.1	1	.7			FRANKLIN AL	-40				
2	2.3	3	2.6	1	.5	GENEVA AL	10	-80	-78	03	42
				1	3.9	GREENE AL					97
1	2.1			1	1.6	HENRY AL			-21	43	82
2	1.2	2	1.1	2	.7	HOUSTON AL	-6	-38	-42	27	48
1	.8	2	1.1	3	1.4	JACKSON AL	42	21	72	83	74
20	1.2	26	1.1	32	1.1	JEFFERSON AL	-6	-4	-10	51	64
				2	1.8	LAMAR AL					86
3	1.5	2	.8	1	.2	LAUDERDALE AL	-44	-70	-83	01	36
1	1.4	1	.9	1	.7	LAWRENCE AL	-35	-30	-54	18	47
		5*	4.3	5	2.8	LEE AL		-36			94
1	1.2			2	.8	LIMESTONE AL			-35	33	53
4	1.8	3	.9	7	1.1	MADISON AL	-53	25	-41	28	64
1	2.2	1	2.0	2	2.7	MARENGO AL	-8	37	26	68	94
2	2.3	3	2.2			MARION AL	-8				
2	1.3			8	2.3	MARSHALL AL			83	85	91
10	1.5	10	1.1	15	1.2	MOBILE AL	-30	11	-22	42	68
		3*	4.3	1	1.2	MONROE AL		-73			67
4	.9	7	1.4	14	1.9	MONTGOMERY AL	47	43	111	90	88
1	.4			1	.3	MORGAN AL			-29	38	37
				1	2.1	PERRY AL					90
				1	.9	PICKENS AL					55
1	1.2	3*	3.2	2	1.8	PIKE AL	171	-42	57	80	86
1	1.1	1	.8	2	1.3	RUSSELL AL	-29	63	17	64	73
		1	.9	1	.6	ST CLAIR AL		-25			46
				5	2.4	SHELBY AL					92
2	1.0	4	1.7	4	1.3	TALLADEGA AL	69	-23	30	70	73
2	2.1	2	1.4	1	.4	TALLAPOOSA AL	-33	-70	-80	02	39
3	1.0	3	.7	3	.6	TUSCALOOSA AL	-32	-5	-36	33	45
2	.8	1	.3	2	.5	WALKER AL	-61	48	-43	26	40
2*	6.7	2*	4.8	2	3.4	WASHINGTON AL	-29	-28	-49	21	96
1	1.4	1	1.5			WINSTON AL	8				
		7*	2.7	11*	2.8	ALASKA AK		2			
29	.9	52	.9	126	1.1	ARIZONA AZ	-2	30	28		
3	1.8			6	1.6	COCHISE AZ			-13	48	81
				1	.7	COCONINO AZ					47
4*	5.0	2	1.6	2	1.5	GILA AZ	-68	-9	-71	07	78
1	2.2					GRAHAM AZ					
				1	3.3	GREENLEE AZ					96
11	.6	33	.9	69	1.0	MARICOPA AZ	54	12	73	84	62
		2	4.3	1	.3	MOHAVE AZ		-92			38
2*	3.7	1	1.7	1	.8	NAVAJO AZ	-53	-55	-79	02	52
4	.6	8	.5	30	1.3	PIMA AZ	-16	139	100	89	71
1	1.0	2	1.0	4	1.3	PINAL AZ	2	35	38	74	74
				1	1.2	SANTA CRUZ AZ					67
1	.9	2	1.0	3	.6	YAVAPAI AZ	10	-35	-28	38	46
2	1.6	2	1.2	7	2.3	YUMA AZ	-24	90	44	76	91
53	.8	75	.8	96	.8	ARKANSAS AR	5	-3	2		
1	1.7					ARKANSAS AR					
		1	1.3	2	.6	BAXTER AR		-53			45
1	.5	2	.6	2	.4	BENTON AR	30	-39	-21	44	38
		3	1.9			BOONE AR					
1	2.4			2	2.1	BRADLEY AR			-13	48	89
1	4.5					CALHOUN AR					
1	1.2					CARROLL AR					
		2*	4.2	3*	5.6	CHICOT AR		32			99
1	1.7	1	.8	1	.9	CLARK AR	-50	7	-46	23	56
		1	.7	2	1.2	CLAY AR		63			69
1	.5			1	.4	CRAIGHEAD AR			-13	48	39
		2	1.6			CRAWFORD AR					

* SIGNIFICANTLY MORE DEATHS THAN EXPECTED

WHITE FEMALE — CANCER OF THE ESOPHAGUS — ICD 150

1950-1959 DEATHS	1950-1959 RATE	1960-1969 DEATHS	1960-1969 RATE	1970-1979 DEATHS	1970-1979 RATE	COUNTY AND STATE	PERCENT CHANGE 60S/50S	PERCENT CHANGE 70S/60S	PERCENT CHANGE 70S/50S	PERCENTILE %CHANGE 70S/50S	PERCENTILE RATES 1970S
1	1.4	3*	3.7	1	.8	CRITTENDEN AR	169	-77	-38	30	54
				2	2.1	CROSS AR					90
1	2.4					DALLAS AR					
1	2.2	1	1.6			DESHA AR	-26				
		1	1.5			DREW AR					
				2	1.0	FAULKNER AR					61
2	.8	7	1.9	9	1.6	GARLAND AR	156	-16	114	90	81
1	2.7	1	1.8	1	1.0	GRANT AR	-33	-45	-63	12	60
1	.8	1	.6	1	.6	GREENE AR	-34	7	-29	38	44
1	1.0			1	.7	HEMPSTEAD AR			-29	37	48
1	1.7			1	.9	HOWARD AR			-47	22	56
				1	.3	INDEPENDENCE AR					38
				1	1.0	IZARD AR					61
3*	3.6	3*	3.1	5*	4.0	JACKSON AR	-15	28	9	60	98
4	3.0	2	.8	1	.4	JEFFERSON AR	-71	-58	-88	00	38
		2	1.8	2	1.6	JOHNSON AR		-9			81
1	1.0	1	.8	2	1.5	LAWRENCE AR	-18	73	42	76	77
1	2.7					LEE AR					
		1	2.8			LINCOLN AR					
1	2.8					LITTLE RIVER AR					
1	.9	1	.7	1	.6	LOGAN AR	-21	-15	-33	35	44
		1	.8	1	.7	LONOKE AR		-15			47
		1	1.2			MADISON AR					
1	.9			3	1.5	MILLER AR			60	81	78
2	1.2	1	.4	3	1.2	MISSISSIPPI AR	-68	218	2	57	69
		1	3.4			NEWTON AR					
		2	1.8	1	.7	OUACHITA AR		-62			48
				1	4.2	PERRY AR					98
1	1.4	1	1.1	1	.8	PHILLIPS AR	-22	-31	-46	23	52
1	.9	2	1.5	2	1.2	POINSETT AR	58	-19	27	69	68
1	1.2					POLK AR					
1	.9	1	.7	1	.5	POPE AR	-29	-18	-42	26	42
1	2.1	1	2.7			PRAIRIE AR	28				
7	.9	7	.6	20	1.5	PULASKI AR	-31	133	60	81	78
		1	1.1			RANDOLPH AR					
		1	1.2			ST FRANCIS AR					
2	1.3			5	2.5	SALINE AR			88	86	93
		1	2.1			SCOTT AR					
		1	2.2	1	1.6	SEARCY AR		-28			81
8*	2.6	4	1.0	5	.9	SEBASTIAN AR	-61	-15	-67	09	55
		1	2.2			STONE AR					
		3	1.5	2	.7	UNION AR		-51			49
		2	3.9			VAN BUREN AR					
		5	1.5	2	.3	WASHINGTON AR		-81			37
1	.7	1	.5	2	.7	WHITE AR	-31	51	4	57	48
				1	1.5	WOODRUFF AR					79
		1	1.1	1	.7	YELL AR		-31			50
724*	1.2	1043*	1.2	1653*	1.5	CALIFORNIA CA	3	25	28		
37	.9	81*	1.7	80	1.6	ALAMEDA CA	82	-6	70	83	81
1	2.1			4*	2.8	AMADOR CA			33	71	94
4	.9	4	.7	10	1.1	BUTTE CA	-31	66	15	63	64
1	1.7			3	1.8	CALAVERAS CA			4	57	85
				1	.7	COLUSA CA					50
14	1.3	24	1.4	47*	1.7	CONTRA COSTA CA	7	23	31	70	83
		1	1.6	1	.9	DEL NORTE CA		-45			56
1	.6	1	.5	2	.6	EL DORADO CA	-26	39	3	57	45
9	.8	18	1.0	25	1.1	FRESNO CA	35	9	47	77	66
		1	1.1	1	.7	GLENN CA		-33			50
4	1.2	6	1.4	12*	2.3	HUMBOLDT CA	19	62	93	87	91
2	.9	2	.7	7*	2.4	IMPERIAL CA	-21	245	172	95	92
2*	4.3	1	1.8	2	1.8	INYO CA	-57	-3	-58	14	85
8	1.1	10	.8	23	1.4	KERN CA	-24	78	35	72	77
2	1.3	3	1.4	2	.7	KINGS CA	6	-52	-49	20	47
				4	1.8	LAKE CA					85
1	3.0					LASSEN CA					
310	1.2	417*	1.3	543*	1.5	LOS ANGELES CA	4	20	25	67	79
		3	1.6	2	.8	MADERA CA		-52			52
5	1.0	8	1.0	24*	2.2	MARIN CA	3	121	128	92	90
2	1.0	5	1.8	2	.6	MENDOCINO CA	88	-66	-37	31	45
		3	.9	8	1.7	MERCED CA		86			83
		1	3.4	1	1.8	MODOC CA		-48			85
1*	23.6					MONO CA					
5	.9	11	1.3	25*	2.1	MONTEREY CA	49	57	134	92	89
3	.9	4	1.0	11	1.7	NAPA CA	7	71	83	85	83
2	1.5			7	2.2	NEVADA CA			41	75	90
16	.8	37	.8	117*	1.6	ORANGE CA	3	91	97	88	81
3	1.1	8*	2.2	12*	2.4	PLACER CA	91	9	109	89	92
		2*	3.8	1	.9	PLUMAS CA		-75			58
12	1.1	16	.7	50	1.3	RIVERSIDE CA	-35	82	19	65	72
18	1.2	19	.9	31	1.0	SACRAMENTO CA	-30	9	-23	42	59
		1	.8	3	2.9	SAN BENITO CA		249			95
16	.8	34	1.2	55	1.4	SAN BERNARDINO CA	40	23	72	84	77
43	1.3	59	1.1	119	1.5	SAN DIEGO CA	-14	34	15	63	79
80*	1.8	96*	2.1	116*	2.8	SAN FRANCISCO CA	16	35	58	80	95
7	.8	11	1.0	29*	1.8	SAN JOAQUIN CA	12	90	113	90	86
2	.5	6	1.4	7	.8	SAN LUIS OBISPO CA	153	-41	50	78	53
19	1.5	27	1.2	55*	1.9	SAN MATEO CA	-16	53	29	70	87
9	1.4	11	1.1	23	1.6	SANTA BARBARA CA	-23	45	12	61	81
33*	1.7	35	1.1	70	1.5	SANTA CLARA CA	-35	36	-12	49	79
9	1.4	10	1.0	9	.8	SANTA CRUZ CA	-26	-19	-40	28	55
3	1.4	4	1.4	3	.6	SHASTA CA	-4	-54	-56	16	45
1	.8	1	.7	3	1.3	SISKIYOU CA	-17	85	54	79	71
6	1.6	7	1.3	13	1.7	SOLANO CA	-18	32	9	59	84
7	1.0	15	1.5	22	1.3	SONOMA CA	51	-9	37	73	74
9	1.3	10	1.0	14	1.1	STANISLAUS CA	-17	7	-12	49	65
		5*	3.2	4	1.9	SUTTER CA		-40			87
		3	2.0	4	2.1	TEHAMA CA		5			89
7	1.2	5	.6	7	.6	TULARE CA	-47	7	-44	25	46
2	2.5	3	2.8	5*	2.5	TUOLUMNE CA	13	-12	0	55	93

* SIGNIFICANTLY MORE DEATHS THAN EXPECTED

WHITE FEMALE — CANCER OF THE ESOPHAGUS — ICD 150

1950-1959 DEATHS	1950-1959 RATE	1960-1969 DEATHS	1960-1969 RATE	1970-1979 DEATHS	1970-1979 RATE	COUNTY AND STATE	PERCENT CHANGE 60S/50S	PERCENT CHANGE 70S/60S	PERCENT CHANGE 70S/50S	PERCENTILE %CHANGE 70S/50S	PERCENTILE RATES 1970S
7	1.2	10	.9	25	1.3	VENTURA CA	-20	41	14	62	73
1	.7	1	.3	7	1.5	YOLO CA	-57	435	129	92	79
		3	2.3	2	1.1	YUBA CA		-50			66
48	.7	70	.8	88	.7	COLORADO CO	13	-4	8		
		2	.6	5	.9	ADAMS CO		47			55
				1	2.3	ALAMOSA CO					91
2	.6	3	.6	3	.5	BOULDER CO	-6	-19	-24	42	41
				1	1.2	CHAFFEE CO					68
		1	3.7			CONEJOS CO					
		1	3.4			CROWLEY CO					
24	.9	41	1.3	38	1.0	DENVER-ARAPAHOE#CO	42	-22	11	61	60
1*	7.4					DOUGLAS CO					
1*	8.0					EAGLE CO					
4	.7	5	.7	9	.8	EL PASO CO	-2	12	9	60	53
				2	.7	FREMONT CO					48
				1	1.1	GARFIELD CO					65
				1	4.2	GUNNISON CO					98
1	2.0					HUERFANO CO					
2	.6	4	.6	13	1.2	JEFFERSON CO	15	87	115	90	68
				1	2.2	KIT CARSON CO					91
2	2.6					LA PLATA CO					
1	.3	1	.3	2	.3	LARIMER CO	-11	23	10	60	38
1	.9			1	.9	LAS ANIMAS CO			-9	51	55
				1	.8	LOGAN CO					53
1	.4	3	1.0	1	.2	MESA CO	170	-75	-33	35	36
1	2.1					MONTEZUMA CO					
2	3.1					MONTROSE CO					
		1	.9			MORGAN CO					
1	.9	1	.6	2	1.1	OTERO CO	-25	77	32	71	66
		1*	4.1			PITKIN CO					
				1	1.6	PROWERS CO					81
3	.6	4	.6	2	.3	PUEBLO CO	10	-52	-47	22	37
		1	1.9			RIO GRANDE CO					
				1*	47.2	SAN JUAN CO					99
				1	3.8	WASHINGTON CO					97
1	.4	1	.3	1	.2	WELD CO	-9	-43	-48	22	36
126	1.1	194*	1.3	275*	1.5	CONNECTICUT CT	17	15	34		
31	1.1	52	1.3	73	1.6	FAIRFIELD CT	19	17	40	75	80
35	1.2	47	1.2	63	1.3	HARTFORD CT	4	5	9	60	71
9	1.4	11	1.3	17	1.6	LITCHFIELD CT	-4	22	18	64	82
4	.9	6	1.0	16*	2.2	MIDDLESEX CT	11	112	136	92	90
33	1.1	57	1.5	81*	1.7	NEW HAVEN CT	35	13	53	79	83
9	1.1	14	1.3	15	1.2	NEW LONDON CT	25	-8	16	63	69
1	.5	1	.3	2	.4	TOLLAND CT	-44	40	-21	43	39
4	1.1	6	1.3	8	1.7	WINDHAM CT	18	32	55	79	82
10	.6	27	1.3	41	1.5	DELAWARE DE	111	13	137		
1	.5	4	1.7	4	1.0	KENT DE	258	-40	114	90	61
5	.4	18	1.3	32*	1.8	NEW CASTLE DE	210	38	327	98	85
4	1.5	5	1.2	5	1.0	SUSSEX DE	-16	-17	-30	37	61
34	1.2	37	1.5	36	1.9	DIST. OF COLUMBIA DC	29	23	59		
34	1.2	37	1.5	36	1.9	WASHINGTON DC	29	23	59	81	87
148	.8	364	1.1	746	1.3	FLORIDA FL	36	13	53		
		3	1.3	7	1.8	ALACHUA FL		45			86
		1	3.3			BAKER FL					
2	1.8	2	.7	4	1.1	BAY FL	-59	44	-41	28	64
1	2.4			1	1.3	BRADFORD FL			-45	24	72
		8	1.4	19	1.5	BREVARD FL		12			80
10	1.1	26	.9	80	1.1	BROWARD FL	-20	26	1	56	65
		1	.2	2	.7	CHARLOTTE FL		223			49
1	5.3			5	1.2	CITRUS FL			-77	03	70
		1	1.1	1	.5	CLAY FL		-50			42
				10	2.1	COLLIER FL					90
		1	1.1	4*	3.1	COLUMBIA FL		184			95
40	1.2	73	1.2	141	1.5	DADE FL	1	24	25	68	77
		1	2.0	1	1.0	DE SOTO FL		-47			62
12	1.0	25	1.5	28	1.3	DUVAL FL	51	-18	24	67	71
6	1.5	11*	2.0	9	1.0	ESCAMBIA FL	31	-51	-35	33	60
				1	1.7	FLAGLER FL					82
		1	3.1	1	1.6	FRANKLIN FL		-48			82
		3	2.4	1	.8	GADSDEN FL		-68			52
				2	2.1	HARDEE FL					90
1	3.5			2	1.3	HERNANDO FL			-65	11	70
1	1.1	1	.6	5	.9	HIGHLANDS FL	-43	49	-15	46	59
10	.7	26	1.3	42	1.4	HILLSBOROUGH FL	68	13	90	87	76
1	1.7	1	1.9	1	.9	HOLMES FL	9	-51	-47	23	57
1	.7	3	1.3	8	2.2	INDIAN RIVER FL	82	71	211	96	90
		1	.8	2	1.0	JACKSON FL		17			60
		1	4.5	1	5.0	JEFFERSON FL		13			99
1	.4			11	1.1	LAKE FL			146	93	63
1	.4	1	.1	16	1.4	LEE FL	-75	1204	230	97	75
		2	.7	5	1.4	LEON FL		89			76
		1	2.3			LEVY FL					
1	2.3	1	1.7			MADISON FL	-24				
1	.3	4	.4	9	.5	MANATEE FL	40	11	55	79	40
2	1.1	5	2.0	6	1.0	MARION FL	81	-50	-9	51	60
1	1.2	2	1.2	3	.4	MARTIN FL	4	-69	-67	08	39
1	1.0	4	2.6	9*	2.5	MONROE FL	161	-5	149	93	93
		1	1.8	2	1.9	NASSAU FL		5			87
		2	2.2	2	.7	OKALOOSA FL		-69			48
				2	2.2	OKEECHOBEE FL					90
6	.7	18	1.3	36	1.7	ORANGE FL	79	31	135	92	84

* SIGNIFICANTLY MORE DEATHS THAN EXPECTED

\# COUNTIES COMBINED; SEE TECHNICAL NOTES

WHITE FEMALE — CANCER OF THE ESOPHAGUS — ICD 150

1950-1959 DEATHS	1950-1959 RATE	1960-1969 DEATHS	1960-1969 RATE	1970-1979 DEATHS	1970-1979 RATE	COUNTY AND STATE	PERCENT CHANGE 60S/50S	PERCENT CHANGE 70S/60S	PERCENT CHANGE 70S/50S	PERCENTILE %CHANGE 70S/50S	PERCENTILE RATES 1970S
				5	1.1	OSCEOLA FL					65
5	.6	21	1.1	48	1.3	PALM BEACH FL	79	20	115	90	72
		2	.4	14	.8	PASCO FL		110			55
10	.3	55	1.2	91	1.2	PINELLAS FL	295	3	306	98	68
6	.8	12	1.3	12	.7	POLK FL	61	-45	-11	50	48
3*	3.1	4*	2.8	4	2.0	PUTNAM FL	-10	-26	-34	34	89
4*	3.8	3	1.9	6*	3.5	ST JOHNS FL	-50	85	-8	52	97
2	2.0	3	1.3	9	1.7	ST LUCIE FL	-37	33	-16	46	84
1	1.2	1	1.1			SANTA ROSA FL	-3				
		14	1.6	35*	1.8	SARASOTA FL		12			85
1	.7	4	1.5	8	1.3	SEMINOLE FL	102	-11	79	85	73
3*	8.3	1	2.7	2	1.9	SUMTER FL	-68	-29	-77	03	87
				1	.9	SUWANNEE FL					57
2*	5.3			1	1.4	TAYLOR FL			-73	05	77
9	1.4	11	.9	24	1.2	VOLUSIA FL	-39	36	-17	46	67
2	2.7	1	1.0	2	1.5	WALTON FL	-62	49	-43	26	79
		1	1.6	3*	3.1	WASHINGTON FL		95			95
180*	1.7	184	1.3	216	1.1	GEORGIA GA	-26	-15	-37		
		1	2.8			BACON GA					
		1	.6	1	.7	BALDWIN GA		24			49
1	1.5	1	1.3	1	.8	BARROW GA	-14	-39	-48	22	51
2	2.0	3	2.6	2	1.4	BARTOW GA	33	-47	-30	37	75
1	2.2			1	1.6	BEN HILL GA			-25	41	82
1	9.3	1	1.9			BERRIEN GA	-79				
8*	2.2	6	1.3	8	1.3	BIBB GA	-41	1	-40	29	73
				1	1.7	BLECKLEY GA					84
1	1.3	1	1.2	1	.7	BULLOCH GA	-2	-42	-44	25	49
		1	1.9	1	2.5	BURKE GA		33			93
				1	2.0	BUTTS GA					88
				1	2.7	CAMDEN GA					94
1	.7	1	.6			CARROLL GA	-12				
		1	1.2	1	.6	CATOOSA GA		-50			44
6	1.5	9	1.6	5	.7	CHATHAM GA	6	-54	-51	19	49
3*	3.8					CHATTOOGA GA					
1	1.0			2	1.0	CHEROKEE GA			1	56	60
3	2.1	1	.5	3	1.4	CLARKE GA	-75	161	-34	33	74
		1*	8.1			CLAY GA					
1	1.1	6*	3.9	1	.2	CLAYTON GA	265	-94	-78	02	36
		1*	5.3			CLINCH GA					
5	2.0	4	1.1	2	.3	COBB GA	-45	-72	-85	00	37
				1	1.1	COFFEE GA					65
4*	3.5	2	1.9	2	1.1	COLQUITT GA	-44	-42	-68	08	65
				1	1.5	COLUMBIA GA					79
1	1.1	4*	3.6	2	1.2	COWETA GA	241	-68	9	59	67
		1	1.6			CRISP GA					
1	1.8			1	1.3	DECATUR GA			-29	38	71
11	1.8	9	.8	23	1.5	DE KALB GA	-52	76	-16	46	78
1	3.1	1	1.2			DODGE GA	-61				
		3*	8.9	1	2.1	DOOLY GA		-77			90
5*	5.1	2	1.8	5	2.2	DOUGHERTY GA	-65	23	-57	15	90
				2	1.4	DOUGLAS GA					74
		3*	7.2			EARLY GA					
				2	3.1	EFFINGHAM GA					96
		1	1.2	2	1.8	ELBERT GA		51			85
1	1.5	1	1.3			EMANUEL GA	-15				
1*	5.5			1	2.3	EVANS GA			-58	14	91
				1	.9	FANNIN GA					59
		1	2.8			FAYETTE GA					
7*	3.3	4	1.1	10*	2.2	FLOYD GA	-65	93	-33	35	90
1	2.3	1	1.3	1	1.4	FORSYTH GA	-45	9	-40	28	76
1	1.5			1	.9	FRANKLIN GA			-41	27	57
35*	2.1	37*	1.7	33	1.4	FULTON GA	-17	-19	-33	35	75
		2*	4.4			GILMER GA					
				1*	5.9	GLASCOCK GA					99
2	1.9	1	.7	1	.5	GLYNN GA	-60	-39	-76	04	40
		1	1.1			GORDON GA					
2*	4.4	2	3.6	1	1.8	GRADY GA	-18	-50	-59	14	86
1	3.2			1	2.2	GREENE GA			-30	37	90
1	.7	2	1.2	3	.8	GWINNETT GA	76	-32	20	65	53
2	3.7					HABERSHAM GA					
2	1.3	1	.5	3	.9	HALL GA	-63	91	-29	37	56
				1	3.8	HANCOCK GA					97
1	1.4	1	1.2			HARALSON GA	-14				
2*	4.7					HART GA					
		1	4.6			HEARD GA					
		1	1.8	4*	3.7	HENRY GA		107			97
2*	6.4			1	.5	HOUSTON GA			-92	00	41
		1	3.2			IRWIN GA					
1	1.9					JACKSON GA					
3*	7.6	2*	3.3			JEFFERSON GA	-56				
1	2.9	1	2.7			JOHNSON GA	-7				
1	2.9					LAMAR GA					
		3	2.6			LAURENS GA					
				1	3.5	LEE GA					97
2*	15.8					LIBERTY GA					
3*	3.3	1	.7	3	1.3	LOWNDES GA	-80	102	-59	14	73
1	3.0	1	2.8			LUMPKIN GA	-6				
		2*	6.1			MC DUFFIE GA					
		1*	6.6			MARION GA					
1	1.5	2	3.0	2	1.7	MERIWETHER GA	91	-44	8	59	83
				1	3.1	MILLER GA					96
		1	1.9	1	1.1	MITCHELL GA		-43			65
		1	4.1			MONTGOMERY GA					
2*	6.8			2	3.0	MURRAY GA			-56	16	95
7*	2.2	3	.6	9	1.5	MUSCOGEE#GA	-74	163	-31	36	79
		1	1.3	2	1.5	NEWTON GA		22			80
				1	.9	PAULDING GA					56
				2*	4.3	PEACH GA					98
				1	2.0	PICKENS GA					89

* SIGNIFICANTLY MORE DEATHS THAN EXPECTED

\# COUNTIES COMBINED; SEE TECHNICAL NOTES

WHITE FEMALE

CANCER OF THE ESOPHAGUS

ICD 150

1950-1959 DEATHS	1950-1959 RATE	1960-1969 DEATHS	1960-1969 RATE	1970-1979 DEATHS	1970-1979 RATE	COUNTY AND STATE	PERCENT CHANGE 60S/50S	PERCENT CHANGE 70S/60S	PERCENT CHANGE 70S/50S	PERCENTILE %CHANGE 70S/50S	PERCENTILE RATES 1970S
		1	2.4			PIERCE GA					
				1	2.7	PIKE GA					94
1	.9	3	2.1	1	.5	POLK GA	144	-77	-45	24	41
				1	2.9	PUTNAM GA					95
1	3.7	1	2.5			RANDOLPH GA	-32				
10*	3.2	6	1.5	10	1.7	RICHMOND GA	-52	14	-45	24	85
1	2.8	1	1.9	4*	3.8	ROCKDALE GA	-33	101	36	72	97
2*	5.6			1	2.3	SCREVEN GA			-59	14	91
				2*	4.5	SEMINOLE GA					98
1	1.2			3	1.3	SPALDING GA			13	61	72
2	2.4	1	1.2			STEPHENS GA	-51				
		2*	13.5			STEWART GA					
		2	2.6	1	.9	SUMTER GA		-65			58
1	1.8	1	2.3			TATTNALL GA	27				
1	3.3					TAYLOR GA					
				3*	5.5	TELFAIR GA					99
		1	2.4	1	1.7	TERRELL GA		-31			83
2	2.4	2	1.3	1	.6	THOMAS GA	-43	-57	-75	04	44
		1	1.0	2	1.4	TIFT GA		43			76
1	1.8					TOOMBS GA					
1*	5.2			1	2.5	TOWNS GA			-52	19	93
1*	7.0	1*	4.0			TREUTLEN GA	-43				
		3	1.6	2	.9	TROUP GA		-43			56
				1	1.8	TURNER GA					86
2	5.8	2	2.4	3	2.0	UPSON GA	-58	-16	-65	11	89
1	.6	3	1.6	3	1.0	WALKER GA	148	-36	58	80	60
2	2.7	2	2.3	2	1.3	WALTON GA	-15	-42	-51	19	72
2	2.3	3	2.3	4	2.3	WAPE GA	-2	-1	-3	54	91
				2*	8.8	WARREN GA					99
1	3.5					WAYNE GA					
		1*	8.3			WHEELER GA					
3	2.1	4	2.1	2	.9	WHITFIELD GA	1	-58	-58	15	56
2*	6.6	1	2.8			WILCOX GA	-58				
1	3.2			1	1.9	WILKES GA			-39	29	88
		1	3.3			WILKINSON GA					
1	2.3			2	3.0	WORTH GA			28	70	95
		9	1.4	18	1.9	HAWAII HI		35			
		1	1.3	3	2.7	HAWAII HI		117			94
		8	1.6	13	1.7	HONOLULU HI		6			85
				2*	8.3	KAUAI HI					99
16	.7	18	.6	30	.7	IDAHO ID	-25	25	-6		
4	1.2	3	.5	8	1.1	ADA ID	-57	105	-12	49	63
1	.5	2	1.0			BANNOCK ID	96				
				1	3.3	BEAR LAKE ID					96
		2*	9.6			BLAINE ID					
				1	.8	BONNER ID					50
1	2.4	1	.8	2	.8	BONNEVILLE ID	-69	13	-65	11	54
2*	10.1			1	3.1	BOUNDARY ID			-69	07	96
		2	.6	3	.6	CANYON ID		16			46
		1	1.4			CASSIA ID					
1*	10.5					CUSTER ID					
2*	6.8	1	3.0			ELMORE ID	-55				
				1	2.3	FREMONT ID					91
1	3.0			2	2.8	GEM ID			-9	51	94
		2	3.2			IDAHO ID					
				1	.2	KOOTENAI ID					36
		1	1.0	1	.8	LATAH ID		-21			53
1	4.6			1	2.9	LEMHI ID			-38	30	95
				1	4.5	LEWIS ID					98
1	2.2					MINIDOKA ID					
				2	1.1	NEZ PERCE ID					66
				1	2.6	ONEIDA ID					93
1	1.7	1	1.1			PAYETTE ID	-32				
				2	1.9	SHOSHONE ID					87
		1	.5	2	.6	TWIN FALLS ID		30			45
1*	8.2	1*	5.4			VALLEY ID	-34				
475	1.1	558	1.0	738	1.2	ILLINOIS IL	-4	15	10		
3	.7	6	1.2	6	1.0	ADAMS IL	62	-19	31	70	59
3*	4.1	2	2.2	1	1.0	ALEXANDER IL	-46	-57	-77	04	59
2	2.1	2	2.2			BOND IL	6				
1	1.1			2	1.4	BOONE IL			27	69	75
1	2.6					BROWN IL					
		2	.8	2	.6	BUREAU IL		-24			44
1	1.1			1	1.0	CARROLL IL			-4	53	62
1	.9	1	1.2	1	.7	CASS IL	30	-44	-27	39	47
4	.9	3	.6	5	.7	CHAMPAIGN IL	-36	20	-23	42	49
3	1.4	3	1.3	2	.6	CHRISTIAN IL	-5	-54	-56	15	44
		1	.4	1	.8	CLARK IL		104			53
1	.9	5*	3.6	1	.3	CLAY IL	312	-91	-61	13	38
		1	.6	1	.4	CLINTON IL		-42			38
2	.9	1	.3	5	1.3	COLES IL	-60	260	43	76	70
303*	1.5	323*	1.3	403*	1.5	COOK IL	-12	15	1	55	79
				1	.5	CRAWFORD IL					42
				1	1.0	CUMBERLAND IL					59
3	1.3	3	1.2	3	.7	DE KALB IL	-8	-44	-49	21	47
2	1.8	2	1.2			DE WITT IL	-30				
				2	1.3	DOUGLAS IL					73
7	.8	13	.9	29	1.3	DU PAGE IL	11	42	58	80	71
4	2.5	1	.5	1	.6	EDGAR IL	-81	30	-76	04	45
1	2.7					EDWARDS IL					
3	2.4	3	2.2	4	2.0	EFFINGHAM IL	-7	-9	-15	47	89
		1	.7	2	.9	FAYETTE IL		36			57
1	1.1			1	.9	FORD IL			-12	49	58
7*	2.7	1	.3	2	.4	FRANKLIN IL	-90	48	-86	00	39
3	1.1	4	1.1	1	.4	FULTON IL	0	-66	-66	09	38

* SIGNIFICANTLY MORE DEATHS THAN EXPECTED

WHITE FEMALE — CANCER OF THE ESOPHAGUS — ICD 150

1950-1959 DEATHS	1950-1959 RATE	1960-1969 DEATHS	1960-1969 RATE	1970-1979 DEATHS	1970-1979 RATE	COUNTY AND STATE	PERCENT CHANGE 60S/50S	PERCENT CHANGE 70S/60S	PERCENT CHANGE 70S/50S	PERCENTILE %CHANGE 70S/50S	PERCENTILE RATES 1970S
1	2.4					GALLATIN IL					
1	1.0			1	.6	GRUNDY IL			-34	34	46
1	1.2			1	.8	HAMILTON IL			-34	34	52
				3	1.1	HANCOCK IL					65
		1	2.3			HENDEPSON IL					
1	.4	2	.6	4	1.0	HENRY IL	65	65	173	95	59
2	1.0			2	.6	IROQUOIS IL			-40	29	45
3	1.6	1	.5	1	.3	JACKSON IL	-67	-39	-80	02	37
1	.5	2	1.0	5	1.7	JEFFERSON IL	112	74	271	98	84
1	1.6					JERSEY IL					
				2	1.6	JO DAVIESS IL					82
1	2.1	1	2.0			JOHNSON IL	-6				
3	.4	5	.4	12	.8	KANE IL	23	94	139	92	55
5	1.1	4	.8	3	.5	KANKAKEE IL	-32	-38	-58	15	40
		2	2.1	2	1.5	KENDALL IL		-29			78
2	.5	3	.6	5	.9	KNOX IL	31	33	73	84	55
8	1.1	12	1.0	33*	2.0	LAKE IL	-11	108	85	86	88
6	1.0	4	.6	5	.6	LA SALLE IL	-42	6	-39	30	45
3	2.3	2	1.2	1	.8	LAWRENCE IL	-48	-33	-65	10	52
1	.6	1	.5	2	.7	LEE IL	-7	37	28	70	49
		3	1.0	1	.3	LIVINGSTON IL		-72			37
1	.5	2	.8			LOGAN IL	61				
		1	.4	1	.7	MC DONOUGH IL		56			46
2	.6	5	.9	6	.8	MC HENRY IL	55	-11	39	74	54
2	.4	2	.4	7	1.0	MC LEAN IL	-4	160	150	93	60
8	1.5	6	.8	11	1.5	MACON IL	-42	73	1	56	78
1	.3			3	.7	MACOUPIN IL			122	91	50
7	.8	14	1.2	15	1.0	MADISON IL	45	-13	27	69	63
2	1.0	4	1.3	1	.3	MAPION IL	27	-77	-71	06	37
		1	1.1			MAPSHALL IL					
1	1.1	2	1.8	4*	3.8	MASON IL	66	114	256	97	97
		1	.9	1	1.0	MASSAC IL		10			62
1	2.1					MENARD IL					
		2	1.8	1	.7	MERCER IL		-61			48
		1	1.0	1	.9	MONROE IL		-13			55
1	.4	4	2.0	4	1.2	MONTGOMERY IL	347	-37	180	95	70
1	.4	2	.6	1	.4	MORGAN IL	77	-31	22	66	40
3	1.6	3	1.4	2	.7	OGLE IL	-13	-47	-54	18	50
2	.2	14	1.2	10	.8	PEORIA IL	465	-31	292	98	54
2	1.5	2	1.4			PERRY IL	-7				
		2	1.9			PIATT IL					
1	.6			1	.9	PIKE IL			45	76	55
1	2.7			1	3.8	POPE IL			44	76	97
1	2.2	2*	4.7			PULASKI IL	113				
1	.5			3	1.4	RANDOLPH IL			169	94	77
		1	.7			RICHLAND IL					
1	.1	8	.9	10	1.0	ROCK ISLAND IL	744	5	790	99	59
7	.9	7	.6	16	1.2	ST CLAIR IL	-28	89	36	72	68
2	1.0	2	.8	2	.6	SALINE IL	-27	-19	-40	28	45
2	.3	9	.9	14	1.2	SANGAMON IL	248	37	376	99	69
1	2.5					SCOTT IL					
3	1.7			1	.5	SHELBY IL			-72	06	40
1	1.7	2*	3.2			STAPK IL	89				
4	1.4	2	.6	5	1.2	STEPHENSON IL	-57	98	-15	47	67
3	.8	4	.8	7	1.1	TAZEWELL IL	3	36	39	75	66
1	.7			2	1.3	UNION IL			79	85	70
2	.4	4	.7	9	1.3	VEPMILION IL	56	102	215	97	73
2	2.5			1	1.4	WABASH IL			-44	25	76
		2	1.3	1	.5	WARREN IL		-61			42
1	1.0			1	.6	WASHINGTON IL			-41	28	43
		2	1.5	2	.9	WHITE IL		-42			56
		2	.5	2	.7	WHITESIDE IL		30			47
5	.7	6	.6	11	.9	WILL IL	-9	46	32	71	58
		2	.5	3	.8	WILLIAMSON IL		70			54
5	.8	11	1.0	16	1.2	WINNEBAGO IL	31	15	52	78	68
1	.7	3	1.9	1	.3	WOODFORD IL	189	-86	-58	14	37
178	.9	213	.9	313	1.1	INDIANA IN	-1	25	23		
		1	.7			ADAMS IN					
9	.9	11	.9	15	1.0	ALLEN IN	-3	17	14	62	63
		2	.9	6	2.0	BARTHOLOMEW IN		125			89
		1	.9			BENTON IN					
1	1.1			1	.8	BLACKFORD IN			-29	38	53
3	1.8					BOONE IN					
2	2.0			1	.6	CARROLL IN			-72	06	43
				2	.5	CASS IN					42
5	2.4	1	.3	3	.7	CLARK IN	-86	119	-69	07	50
1	.6	1	.5	2	1.0	CLAY IN	0	89	89	87	62
2	1.0	1	.6			CLINTON IN	-37				
1	2.0					CRAWFORD IN					
		2	.9			DAVIESS IN					
1	.7					DEARBORN IN					
				1	.3	DECATUR IN					38
2	1.2	3	2.1	2	.8	DE KALB IN	66	-64	-39	29	51
3	.7	5	.9	14*	2.1	DELAWARE IN	30	137	208	96	90
4*	3.5	2	1.4	2	1.2	DUBOIS IN	-61	-14	-66	09	67
2	.4	5	.8	11	1.3	ELKHART IN	81	66	199	96	73
		3	2.0	3	2.0	FAYETTE IN		3			89
2	.8	3	1.1	5	1.4	FLOYD IN	28	29	64	82	74
1	.9	2	1.9	2	1.5	FOUNTAIN IN	102	-22	56	79	77
		1	1.2			FPANKLIN IN					
		1	.7	3	2.0	FULTON IN		172			88
				1	.7	GIBSON IN					48
3	.9	5	1.3	3	.7	GPANT IN	44	-50	-28	38	46
1	.5	2	.9	2	1.0	GPEENE IN	79	9	96	88	59
2	1.1			1	.4	HAMILTON IN			-62	13	39
		1	.6	2	1.0	HANCOCK IN		63			60
1	1.0	2	1.8	2	1.5	HARPISON IN	84	-12	61	81	80
1	.8	1	.6	2	.7	HENDRICKS IN	-31	31	-10	51	50
6*	2.7	1	.3	5	1.2	HENRY IN	-87	261	-54	18	70

* SIGNIFICANTLY MORE DEATHS THAN EXPECTED

WHITE FEMALE — CANCER OF THE ESOPHAGUS — ICD 150

1950-1959 DEATHS	1950-1959 RATE	1960-1969 DEATHS	1960-1969 RATE	1970-1979 DEATHS	1970-1979 RATE	COUNTY AND STATE	PERCENT CHANGE 60S/50S	PERCENT CHANGE 70S/60S	PERCENT CHANGE 70S/50S	PERCENTILE %CHANGE 70S/50S	PERCENTILE RATES 1970S
1	.4	1	.3	6	1.5	HOWARD IN	-19	376	286	98	78
1	.4	1	.3	2	.9	HUNTINGTON IN	-23	163	102	89	55
2	1.3	1	.7	1	.3	JACKSON IN	-48	-52	-75	04	38
		2	1.9			JASPER IN					
3	2.6	1	.7			JAY IN	-73				
				4	2.5	JEFFERSON IN					92
		1	1.2			JENNINGS IN					
4	1.9	1	.3	3	1.2	JOHNSON IN	-86	317	-40	28	67
		1	.3	2	.5	KNOX IN		53			40
1	.6			2	.7	KOSCIUSKO IN			27	69	50
2	2.5					LAGRANGE IN					
6	.4	27	1.6	33	1.6	LAKE IN	336	5	355	99	82
4	1.1	1	.2	8	1.2	LA PORTE IN	-82	547	14	62	70
3	1.5	2	.9	1	.3	LAWRENCE IN	-39	-67	-80	02	37
3	.5	9	1.4	8	1.0	MADISON IN	172	-27	98	88	62
32	1.2	40	1.2	59	1.6	MARION IN	3	31	36	72	81
1	.6	1	.4			MARSHALL IN	-23				
2*	4.5	1	1.6			MARTIN IN	-65				
1	.6	1	.7	2	.8	MIAMI IN	15	28	47	77	54
2	1.1	1	.4	3	.9	MONROE IN	-65	148	-12	49	58
1	.4			2	.9	MONTGOMERY IN			96	88	56
1	.7					MORGAN IN					
1	1.8			1	1.4	NEWTON IN			-22	42	75
2	1.3	3	1.3			NOBLE IN	4				
				1	2.4	OHIO IN					92
4*	3.9	1	.9			ORANGE IN	-77				
		1	1.3			PARKE IN					
				1	.9	PERRY IN					56
2	2.3	1	1.3			PIKE IN	-43				
1	.6	1	.4	3	.8	PORTER IN	-37	130	45	76	54
4*	3.2					POSEY IN					
1	1.9			1	1.1	PULASKI IN			-44	24	63
1	.9	1	.6	1	.3	PUTNAM IN	-34	-42	-62	13	38
				1	.4	RANDOLPH IN					39
1	.9			1	.8	RIPLEY IN			-4	53	54
		1	.8			RUSH IN					
9	1.0	11	.9	12	.8	ST JOSEPH IN	-7	-12	-18	45	53
		1	2.3	4*	4.5	SCOTT IN		99			98
2	1.0	2	.9	1	.4	SHELBY IN	-13	-56	-62	12	39
				1	1.1	SPENCER IN					63
1	1.0	2	2.0	1	.7	STARKE IN	102	-66	-32	35	47
				3	2.5	STEUBEN IN					93
		2	1.1	2	1.2	SULLIVAN IN		10			70
1	1.7					SWITZERLAND IN					
3	.8	5	1.1	9	1.7	TIPPECANOE IN	29	53	98	88	83
1	1.3			1	1.2	TIPTON IN			-14	47	67
				2*	4.5	UNION IN					98
5	.7	11	1.2	15	1.4	VANDERBURGH IN	70	18	100	89	75
1	.9	4*	2.7	1	.6	VERMILLION IN	219	-80	-35	33	43
4	.6	8	1.2	13	1.7	VIGO IN	89	41	166	94	83
1	.5					WABASH IN					
1	2.0			2	2.5	WARREN IN			30	70	93
2	1.7	4*	2.9	1	.6	WARRICK IN	64	-80	-67	09	44
2	.5	4	.9	1	.2	WAYNE IN	73	-80	-66	09	36
1	.9			2	1.1	WELLS IN			26	68	64
2	1.9	1	.7			WHITE IN	-63				
2	2.0			5*	3.3	WHITLEY IN			65	82	96
110	.7	144	.8	180	.9	IOWA IA	12	13	26		
		1	.9	1	1.3	ADAIR IA		39			71
		1	.8	1	.7	ALLAMAKEE IA		-20			48
		2	1.2			APPANOOSE IA					
1	1.5	2	3.6			AUDUBON IA	136				
1	.7			3	1.3	BENTON IA			78	85	73
5	1.0	2	.3	8	1.0	BLACK HAWK IA	-69	223	0	55	61
1	.5	1	.4	2	.8	BOONE IA	-7	99	85	86	54
2	1.6	3	2.5	2	.9	BREMER IA	55	-62	-41	27	58
3	2.6	1	.5	1	.8	BUENA VISTA IA	-79	56	-67	09	54
2	2.2	1	.7	2	1.6	BUTLER IA	-67	130	-25	41	82
		2	1.6			CALHOUN IA					
2	1.6					CARROLL IA					
2	1.7					CASS IA					
4	1.5	2	.6	3	.8	CERRO GORDO IA	-59	23	-50	20	51
		4*	3.3	1	1.1	CHEROKEE IA		-68			63
				1	.5	CHICKASAW IA					40
1	1.4					CLARKE IA					
1	1.1	1	1.0	2	1.1	CLAY IA	-9	2	-7	52	63
				1	.4	CLAYTON IA					38
4	1.4	5	1.5	6	1.4	CLINTON IA	11	-9	1	56	75
3	3.5	1	.8			CRAWFORD IA	-77				
1	.5	3	1.4	2	.8	DALLAS IA	158	-44	43	76	51
				1	1.2	DAVIS IA					67
		1	.6			DELAWARE IA					
2	.7	4	1.3	5	1.6	DES MOINES IA	69	28	116	91	82
				3	2.4	DICKINSON IA					92
2	.4	4	.9	4	.8	DUBUQUE IA	105	-3	100	88	54
1	1.6			1	.7	EMMET IA			-56	16	48
		1	.4	1	.2	FAYETTE IA		-36			36
2	1.6	3	1.9	3	1.7	FLOYD IA	22	-10	10	60	84
1	1.3					FREMONT IA					
		1	.6	2	1.5	GREENE IA		163			78
		2	2.0			GRUNDY IA					
		1	.7	1	.5	GUTHRIE IA		-33			41
		1	.8			HAMILTON IA					
2	2.4					HANCOCK IA					
1	.7	2	.9			HARDIN IA	22				
		1	.8			HARRISON IA					
3	2.4	1	.6	4	2.9	HENRY IA	-75	395	22	67	95
		1	1.2	1	.9	HUMBOLDT IA		-27			55
2	3.5	1	1.5	1	.8	IDA IA	-58	-45	-77	04	53

* SIGNIFICANTLY MORE DEATHS THAN EXPECTED

WHITE FEMALE — CANCER OF THE ESOPHAGUS — ICD 150

1950-1959 DEATHS	1950-1959 RATE	1960-1969 DEATHS	1960-1969 RATE	1970-1979 DEATHS	1970-1979 RATE	COUNTY AND STATE		PERCENT CHANGE 60S/50S	PERCENT CHANGE 70S/60S	PERCENT CHANGE 70S/50S	PERCENTILE %CHANGE 70S/50S	PERCENTILE RATES 1970S
1	.9	1	1.1	1	.7	IOWA	IA	26	-38	-22	43	49
1	.8	1	.8	3	2.0	JACKSON	IA	-3	171	161	94	89
				3	1.1	JASPER	IA					64
				1	1.5	JEFFERSON	IA					78
4	1.7	2	.7	4	1.1	JOHNSON	IA	-56	53	-32	35	66
				1	.5	JONES	IA					42
1	.9			1	1.0	KEOKUK	IA			2	56	59
1	.8			2	1.7	KOSSUTH	IA			100	88	83
		3	1.2	2	.9	LEE	IA		-27			55
6	.9	5	.7	18*	2.0	LINN	IA	-28	208	120	91	89
				1	1.2	LOUISA	IA					68
1	.9	1	.8	2	2.0	LUCAS	IA	-8	144	126	91	89
1	1.5	1	1.2	2	2.0	LYON	IA	-16	58	33	71	88
1	1.0					MADISON	IA					
2	.9	1	.4	3	1.8	MAHASKA	IA	-53	298	89	86	85
1	.6	1	.5	1	.6	MARION	IA	-8	24	13	62	46
		1	.5	5	2.0	MARSHALL	IA		314			89
1	1.1	1	1.2			MILLS	IA	6				
				2	1.2	MITCHELL	IA					70
				1	1.1	MONONA	IA					65
				1	.6	MONROE	IA					43
1	.9	1	.8	1	1.0	MONTGOMERY	IA	-11	29	15	63	61
3	1.2	3	1.1	4	1.4	MUSCATINE	IA	-7	25	16	63	76
2	1.8			2	1.7	O'BRIEN	IA			-6	53	83
1	2.3					OSCEOLA	IA					
		2	1.2	4	3.0	PAGE	IA		161			95
				1	1.0	PALO ALTO	IA					60
1	.7			1	.7	PLYMOUTH	IA			1	56	48
		2	1.5			POCAHONTAS	IA					
8	.6	22	1.5	11	.7	POLK	IA	133	-56	1	56	47
2	.5	5	1.1	1	.1	POTTAWATTAMIE	IA	124	-90	-77	04	36
3	2.3			2	.9	POWESHIEK	IA			-60	14	58
		1	2.1			RINGGOLD	IA					
				1	.4	SAC	IA					39
2	.4	9	1.4	7	.9	SCOTT	IA	215	-36	102	89	56
1	1.1			3	2.4	SHELBY	IA			120	91	92
1	1.0	3	1.9	1	.8	SIOUX	IA	92	-61	-25	41	50
		2	.8	2	.4	STORY	IA		-53			38
3	2.2					TAMA	IA					
				3	4.1	TAYLOR	IA					98
		1	.8	1	.5	UNION	IA		-37			42
		1	.9	2	1.5	VAN BUREN	IA		57			78
5	1.9			2	.7	WAPELLO	IA			-66	10	47
		1	.5	2	1.1	WARREN	IA		109			64
1	.7					WASHINGTON	IA					
1	.8	1	1.2	2	2.5	WAYNE	IA	37	117	197	96	93
2	.8	6	1.7	3	.9	WEBSTER	IA	108	-45	13	62	58
		1	1.4			WINNEBAGO	IA					
		1	.6	1	.8	WINNESHIEK	IA		51			55
2	.4	7	1.1	4	.7	WOODBURY	IA	175	-40	66	82	46
1	.9	1	.6	2	1.5	WRIGHT	IA	-31	128	59	80	77
85	.8	86	.7	118	.8	KANSAS	KS	-14	18	1		
2	1.5			1	1.0	ALLEN	KS			-35	33	60
		1	1.0	1	.7	ATCHISON	KS		-35			47
2	2.0			1	.4	BARTON	KS			-80	02	39
2	1.3	1	1.4	2	1.3	BOURBON	KS	6	-8	-3	54	70
		1	.4	1	.4	BUTLER	KS		-2			39
1	.6	1	.7	1	.3	CHEROKEE	KS	10	-53	-48	21	38
2	2.0	1	.8	1	.5	CLAY	KS	-57	-36	-73	06	42
2	1.3	1	.6	1	.8	CLOUD	KS	-58	45	-39	29	54
				1	1.3	COFFEY	KS					71
1	.4	2	.6	3	.9	COWLEY	KS	41	46	106	89	56
1	.3	1	.3	7	1.7	CRAWFORD	KS	3	461	480	99	84
2	1.3	2	1.3	2	.8	DICKINSON	KS	1	-39	-39	30	51
1	1.4					DONIPHAN	KS					
1	.6	1	.6	3	1.4	DOUGLAS	KS	-12	151	122	91	76
2*	6.2	1	3.6			EDWARDS	KS	-43				
		2*	4.4			ELK	KS					
1	2.6	2	1.9			ELLIS	KS	-25				
				1	2.8	ELLSWORTH	KS					95
1	1.6					FINNEY	KS					
1	1.4					FORD	KS					
1	.5			2	.8	FRANKLIN	KS			48	77	51
1	1.4	1	1.1			GEARY	KS	-20				
		1	4.6			GRAHAM	KS					
1	1.1	1	1.3			HARPER	KS	18				
1	.6					HARVEY	KS					
1*	8.0					HODGEMAN	KS					
		1	1.5			JACKSON	KS					
3*	3.2	1	1.2	1	.9	JEFFERSON	KS	-61	-25	-71	07	57
2	3.4					JEWELL	KS					
2	.6	2	.3	10	.9	JOHNSON	KS	-44	183	59	80	57
		2	2.4	1	1.7	KINGMAN	KS		-29			83
3	1.6	2	.7	3	1.7	LABETTE	KS	-56	137	4	57	83
3	1.6	1	.3	2	.7	LEAVENWORTH	KS	-79	90	-59	14	47
		1	1.3	1	1.5	LINCOLN	KS		8			77
				1	1.2	LINN	KS					69
1	.6			1	.3	LYON	KS			-50	20	37
2	1.3			1	.5	MC PHERSON	KS			-66	10	40
2	1.6	1	.7	1	.7	MARSHALL	KS	-59	4	-57	15	48
1	3.3					MEADE	KS					
1	.8			2	1.1	MIAMI	KS			35	72	65
1	1.1					MITCHELL	KS					
2	.6	3	.9	4	1.2	MONTGOMERY	KS	43	40	100	89	69
		1*	6.7			MORTON	KS					
				2	1.1	NEMAHA	KS					64
2	1.4	2	1.1	1	.5	NEOSHO	KS	-16	-59	-66	10	40
		1	1.1	1	.8	NORTON	KS		-32			51
		1	.9			OSAGE	KS					

* SIGNIFICANTLY MORE DEATHS THAN EXPECTED

WHITE FEMALE — CANCER OF THE ESOPHAGUS — ICD 150

1950-1959 DEATHS	1950-1959 RATE	1960-1969 DEATHS	1960-1969 RATE	1970-1979 DEATHS	1970-1979 RATE	COUNTY AND STATE		PERCENT CHANGE 60S/50S	PERCENT CHANGE 70S/60S	PERCENT CHANGE 70S/50S	PERCENTILE %CHANGE 70S/50S	PERCENTILE RATES 1970S
				1	1.6	PAWNEE	KS					81
1	1.1	2	2.4	2	1.9	POTTAWATOMIE	KS	113	-22	67	83	87
2	3.3	1	1.3			PRATT	KS	-62				
2	.7	5	1.4	7	1.7	RENO	KS	94	21	135	92	84
				1	1.0	RICE	KS					62
2	1.1	2	.8	3	.8	SALINE	KS	-24	-6	-28	39	53
14	1.2	18	1.2	17	.9	SEDGWICK	KS	-4	-22	-25	41	58
		1	1.8			SEWARD	KS					
4	.6	6	.8	8	1.0	SHAWNEE	KS	42	18	68	83	60
		1	2.0	1	1.8	SHERMAN	KS		-9			85
		2	3.3			SMITH	KS					
1	1.9					STAFFORD	KS					
				1	3.6	STEVENS	KS					97
		1	.5	3	1.1	SUMNER	KS		127			64
1	1.0	1	.9	1	1.4	WASHINGTON	KS	-3	47	42	75	75
		1	2.9			WOODSON	KS					
6	.9	6	.7	13	1.6	WYANDOTTE	KS	-16	113	78	85	80
111	.9	161	1.0	174	.9	KENTUCKY	KY	15	-11	3		
1	1.2	1	1.1			ADAIR	KY	-13				
				1	1.3	BALLAPD	KY					72
2	1.8	1	.6	1	.4	BARREN	KY	-67	-31	-77	03	39
				1	1.5	BATH	KY					80
4*	2.6	3	2.1	2	.9	BELL	KY	-20	-57	-65	10	57
1	1.4	1	1.1			BOONE	KY	-24				
				1	1.1	BOURBON	KY					65
1	.5	1	.4	5	1.3	BOYD	KY	-30	254	149	93	72
1	1.1	2	1.7	3	2.4	BOYLE	KY	58	38	118	91	92
		1	2.8	1	1.8	BRACKEN	KY		-35			86
				1	1.4	BREATHITT	KY					75
				3	3.0	BRECKINRIDGE	KY					95
2*	3.4	2	3.2			BUTLER	KY	-6				
		1	1.4	1	.8	CALDWELL	KY		-46			51
		2	1.3	2	1.0	CALLOWAY	KY		-21			61
5	1.1	8	1.6	7	1.3	CAMPBELL	KY	44	-21	13	62	71
1	2.4	1	2.2	1	1.7	CAPROLL	KY	-8	-22	-28	38	83
		1	.8	2	1.7	CARTER	KY		118			84
2	3.6					CASEY	KY					
2	1.2	1	.5	2	.8	CHRISTIAN	KY	-63	76	-34	34	53
1	1.1	3	2.6	1	.5	CLARK	KY	146	-80	-50	20	42
1	3.0	1	2.7			CUMBERLAND	KY	-7				
2	.7	7	1.9	4	.9	DAVIESS	KY	168	-51	30	70	57
1	3.7			1	2.4	EDMONSON	KY			-35	33	92
				1*	3.4	ELLIOTT	KY					97
5	1.1	8	1.4	12	1.6	FAYETTE	KY	20	15	39	74	80
1	1.4			1	1.3	FLEMING	KY			-9	51	71
2	1.6					FLOYD	KY					
		3	1.8	3	1.5	FRANKLIN	KY		-16			80
1	1.6			1	.8	FULTON	KY			-50	20	52
1	2.0	1	1.8			GARRARD	KY	-9				
1	1.9			1	1.8	GRANT	KY			-5	53	86
		1	.4			GRAVES	KY					
1	1.0					GRAYSON	KY					
		1	1.6			GREEN	KY					
		1	.7	1	.5	GREENUP	KY		-24			42
1	2.7					HANCOCK	KY					
		3	1.7			HAPDIN	KY					
1	.6	1	.5	1	.5	HARLAN	KY	-22	3	-19	44	42
				2	2.0	HARRISON	KY					88
				1	.9	HART	KY					56
1	.6	2	1.1	1	.5	HENDERSON	KY	83	-59	-25	40	40
		1	1.1	2	2.7	HENRY	KY		146			94
				1	2.7	HICKMAN	KY					94
4	2.3	3	1.3			HOPKINS	KY	-41				
24	1.1	43	1.5	52	1.4	JEFFERSON	KY	38	-2	35	72	77
		1	1.7			JESSAMINE	KY					
1	.9					JOHNSON	KY					
5	.8	13*	1.9	12	1.6	KENTON	KY	137	-19	91	87	80
		1	1.8	1	1.2	KNOTT	KY		-31			69
1	.8	1	.7	4	2.7	KNOX	KY	-16	288	225	97	94
1	2.0			1	1.3	LARUE	KY			-37	32	72
				1	.5	LAUREL	KY					41
				1	1.4	LAWRENCE	KY					74
1	2.5	2*	4.4	1	3.4	LEE	KY	73	-23	33	71	96
		1	1.1	2	1.3	LETCHER	KY		22			71
		2	3.4	1	1.5	LEWIS	KY		-57			77
1	1.5	2	2.1	1	.9	LINCOLN	KY	39	-57	-40	28	58
				1	1.6	LIVINGSTON	KY					82
1	.8	1	.7	1	.6	LOGAN	KY	-13	-17	-27	39	44
1	3.5			1	3.8	LYON	KY			10	60	97
4	1.6	5	1.5	3	.8	MC CRACKEN	KY	-7	-50	-53	18	50
		1	1.8			MC CREARY	KY					
1	3.0					MC LEAN	KY					
		4	2.2	3	1.5	MADISON	KY		-31			79
1	1.6					MARION	KY					
1	1.7	1	.9	1	.6	MARSHALL	KY	-51	-26	-63	12	46
				1	2.4	MARTIN	KY					92
2	2.4	1	.9			MASON	KY	-64				
1	1.3	1	1.0			MERCER	KY	-25				
		1	1.5			MONTGOMERY	KY					
				1	1.5	MORGAN	KY					80
2	1.4	3	1.5	1	.3	MUHLENBERG	KY	6	-79	-78	03	37
				1	.8	NELSON	KY					51
1	.7	2	1.5	1	.5	OHIO	KY	123	-67	-26	40	41
1	2.2					OLDHAM	KY					
2	2.1			2	1.2	PERRY	KY			-42	27	70
2	1.4	1	.5	2	.6	PIKE	KY	-66	28	-57	15	44
		2*	6.0			POWELL	KY					
1	.7					PULASKI	KY					
1*	4.9					ROBERTSON	KY					

* SIGNIFICANTLY MORE DEATHS THAN EXPECTED

WHITE FEMALE | CANCER OF THE ESOPHAGUS | ICD 150

1950-1959 DEATHS	1950-1959 RATE	1960-1969 DEATHS	1960-1969 RATE	1970-1979 DEATHS	1970-1979 RATE	COUNTY AND STATE		PERCENT CHANGE 60S/50S	PERCENT CHANGE 70S/60S	PERCENT CHANGE 70S/50S	PERCENTILE %CHANGE 70S/50S	PERCENTILE RATES 1970S
		1	1.5	2	2.6	ROCKCASTLE	KY		71			93
		1	1.5			RUSSELL	KY					
1	1.2	2	2.1			SCOTT	KY	79				
1	1.2			1	.8	SHELBY	KY			-32	36	52
1	1.6					TAYLOR	KY					
				1	1.1	TODD	KY					64
		1	3.3			TRIMBLE	KY					
		2	1.9	1	1.4	UNION	KY		-29			74
1	.5			1	.3	WARREN	KY			-33	34	38
				1	1.0	WASHINGTON	KY					60
2	2.6	1	.9			WEBSTER	KY	-66				
1	.8	2	1.1	3	1.4	WHITLEY	KY	46	18	73	84	74
1	5.0					WOLFE	KY					
1	2.2					WOODFORD	KY					
68	.8	113	1.0	148	1.0	LOUISIANA	LA	18	2	20		
		1	.4	2	.9	ACADIA	LA		101			56
				1	.9	ALLEN	LA					57
		1	.9	1	.9	ASCENSION	LA		-5			56
2	1.9	1	.6	2	.9	AVOYELLES	LA	-67	43	-53	19	57
1	1.2	3*	3.6	1	1.0	BEAUREGARD	LA	192	-71	-16	46	63
1	1.9			4*	4.6	BIENVILLE	LA			146	93	98
		2	2.1	1	.5	BOSSIER	LA		-76			41
2	.5	10	1.3	13	1.3	CADDO	LA	177	2	182	95	71
3	.9	9*	2.1	6	1.0	CALCASIEU	LA	139	-52	15	62	60
				1	1.7	CALDWELL	LA					82
		1	3.5	1	2.8	CAMERON	LA		-21			95
1	2.9					CATAHOULA	LA					
				2	1.9	CLAIBORNE	LA					87
		2*	3.9	1	1.6	CONCORDIA	LA		-59			82
1	1.9			1	.9	DE SOTO	LA			-55	17	56
4	1.1	13*	2.0	12	1.3	EAST BATON ROUGE	LA	85	-35	20	66	72
1*	6.4					EAST CARROLL	LA					
1	.8			1	.6	EVANGELINE	LA			-24	41	45
				3*	3.6	GRANT	LA					97
		1	.6	3	1.4	IBERIA	LA		131			74
1	1.5			1	1.3	IBERVILLE	LA			-12	49	72
1	1.6	2	2.9	1	1.0	JACKSON	LA	81	-67	-40	29	59
3	.8	5	.6	10	.7	JEFFERSON	LA	-26	25	-8	52	50
1	1.1	2	1.8	2	1.7	JEFFERSON DAVIS	LA	57	-8	45	77	82
		2	.6	5	1.2	LAFAYETTE	LA		84			67
1	.7	1	.5	1	.3	LAFOURCHE	LA	-20	-42	-54	18	37
		1	1.4			LA SALLE	LA					
		3	2.7	4	2.3	LIVINGSTON	LA		-15			91
1	1.1	2	1.8			NATCHITOCHES	LA	67				
23	1.1	21	.9	21	1.0	ORLEANS	LA	-19	19	-4	54	62
1	.4	4	1.2	7	1.5	OUACHITA	LA	198	23	267	98	78
				2	2.8	PLAQUEMINES	LA					95
		1	1.5			POINTE COUPEE	LA					
5	1.7	4	1.0	7	1.4	RAPIDES	LA	-40	37	-18	45	74
		1	3.5			RED RIVER	LA					
		1	1.3			RICHLAND	LA					
1	2.1			1	.7	SABINE	LA			-67	08	48
1	1.7			2	.7	ST BERNARD	LA			-57	15	50
1	2.5	1	1.6			ST CHARLES	LA	-38				
				1	3.1	ST HELENA	LA					96
2	1.5	3	1.4	5	1.8	ST LANDRY	LA	-9	26	15	63	85
		2	2.5	1	1.0	ST MARTIN	LA		-61			59
		1	.6	1	.7	ST MARY	LA		8			46
				1	.3	ST TAMMANY	LA					37
3	2.0	1	.5	2	.8	TANGIPAHOA	LA	-77	71	-60	13	52
		3	1.9	2	.9	TERREBONNE	LA		-56			55
3	2.4	2	1.1	1	.4	VERMILION	LA	-56	-59	-82	01	40
				3	1.9	VERNON	LA					87
				4	2.3	WASHINGTON	LA					91
		3	2.0	4	2.0	WEBSTER	LA		4			89
1*	4.4	2*	5.9			WEST BATON ROUGE	LA	33				
				3*	4.0	WEST CARROLL	LA					97
		1*	11.2			WEST FELICIANA	LA					
2	3.3					WINN	LA					
72	1.4	73	1.2	103	1.4	MAINE	ME	-14	18	2		
8	1.8	6	1.1	8	1.2	ANDROSCOGGIN	ME	-38	9	-32	35	68
9*	3.1	3	.8	8	1.5	AROOSTOOK	ME	-73	78	-52	19	78
10	1.0	21*	1.7	25	1.9	CUMBERLAND	ME	73	10	90	87	87
				3	1.8	FRANKLIN	ME					86
3	1.3	2	.8	8*	3.0	HANCOCK	ME	-37	265	129	92	95
8	1.6	5	.9	8	1.2	KENNEBEC	ME	-47	41	-25	40	68
1	.4	2	.9	8*	2.8	KNOX	ME	136	205	620	99	94
4*	3.1			1	.5	LINCOLN	ME			-83	01	42
4	1.8	3	1.1	2	.5	OXFORD	ME	-40	-55	-73	05	40
9	1.6	8	1.1	3	.4	PENOBSCOT	ME	-29	-66	-76	04	39
1	.9	2	1.9	1	.7	PISCATAQUIS	ME	119	-61	-15	47	50
3	2.3	1	.8	2	1.3	SAGADAHOC	ME	-63	55	-42	26	72
1	.5	7*	3.0	3	1.2	SOMERSET	ME	569	-61	158	94	67
1	.8	1	.6	2	1.2	WALDO	ME	-23	87	45	76	67
5	2.3	2	.8	7*	2.8	WASHINGTON	ME	-67	261	19	65	94
5	.9	10	1.6	14	1.7	YORK	ME	80	4	87	86	83
101	1.0	159	1.1	247	1.4	MARYLAND	MD	15	21	40		
4	.9	9	1.7	5	.8	ALLEGANY	MD	84	-51	-10	50	53
2	.3	9	1.1	28*	2.5	ANNE ARUNDEL	MD	275	123	736	99	92
14	1.1	23	1.0	45	1.3	BALTIMORE	MD	-10	35	21	66	73
		1	1.5	2	2.1	CALVERT	MD		36			90
1	1.2			2	1.5	CAROLINE	MD			21	66	78
1	.3	2	.6	2	.5	CARROLL	MD	92	-15	63	82	41
1	.6	4	2.1	1	.3	CECIL	MD	238	-84	-45	24	38
		2	1.9	2	1.2	CHARLES	MD		-37			69

* SIGNIFICANTLY MORE DEATHS THAN EXPECTED

WHITE FEMALE — CANCER OF THE ESOPHAGUS — ICD 150

1950-1959 DEATHS	1950-1959 RATE	1960-1969 DEATHS	1960-1969 RATE	1970-1979 DEATHS	1970-1979 RATE	COUNTY AND STATE	PERCENT CHANGE 60S/50S	PERCENT CHANGE 70S/60S	PERCENT CHANGE 70S/50S	PERCENTILE %CHANGE 70S/50S	PERCENTILE RATES 1970S
		2	1.0	1	.5	DORCHESTER MD		-52			41
		2	.5	5	1.0	FREDERICK MD		104			62
		1	1.0	1	.5	GARRETT MD		-49			42
2	1.0	2	.3	5	1.1	HARFORD MD	-67	219	7	58	65
2	1.7	1	.9	4	1.4	HOWARD MD	-50	62	-18	45	75
1	1.9	2	2.9			KENT MD	58				
9	.9	27*	1.7	34	1.3	MONTGOMERY MD	81	-24	38	74	72
9	1.1	21	1.4	35*	1.9	PRINCE GEORGE'S MD	26	30	63	82	87
				2	1.7	QUEEN ANNE'S MD					85
1	1.9	3*	3.5			ST MARY'S MD	83				
2	2.8	1	1.0	1	.9	SOMERSET MD	-65	-5	-67	08	58
1	1.3	1	.8	1	.6	TALBOT MD	-39	-28	-56	16	43
3	.7	4	.8	9	1.4	WASHINGTON MD	19	76	110	90	76
1	.5	2	.8	2	.7	WICOMICO MD	49	-7	38	74	48
1	.8	2	1.7	1	.6	WORCESTER MD	102	-63	-26	40	45
46	1.2	38	.9	59	1.5	BALTIMORE CITY MD	-22	63	27	69	80
336	1.2	479*	1.4	554*	1.4	MASSACHUSETTS MA	20	1	21		
5	1.8	8	1.1	20	1.8	BARNSTABLE MA	-37	63	3	57	86
12	1.5	12	1.2	27*	2.3	BERKSHIRE MA	-20	92	54	79	91
23	1.0	38	1.3	32	1.0	BRISTOL MA	34	-27	-3	54	59
				2	4.0	DUKES MA					98
37	1.1	52	1.3	79*	1.7	ESSEX MA	17	36	59	80	84
3	.7	5	1.3	7	1.6	FRANKLIN MA	78	23	118	91	81
25	1.1	36	1.4	37	1.2	HAMPDEN MA	19	-14	2	56	67
1	.2	3	.4	9	1.2	HAMPSHIRE MA	107	185	490	99	68
76	1.2	108*	1.4	120	1.3	MIDDLESEX MA	19	-2	16	63	74
		1	3.2	1	2.7	NANTUCKET MA		-18			94
21	.8	42	1.2	61	1.5	NORFOLK MA	56	21	88	86	79
12	.9	30*	1.9	30	1.5	PLYMOUTH MA	102	-21	59	80	78
92*	1.9	106*	2.2	77*	1.7	SUFFOLK MA	15	-23	-11	50	83
29	.8	38	1.0	52	1.1	WORCESTER MA	14	12	27	69	64
322	1.2	401	1.1	565	1.3	MICHIGAN MI	-5	16	10		
1	3.3			1	1.1	ALCONA MI			-66	10	65
2	.8	4	1.3	7	1.9	ALLEGAN MI	65	49	146	93	87
1	.9	2	1.4	1	.6	ALPENA MI	54	-59	-37	31	43
1	1.4					ANTRIM MI					
1	1.8			1	1.0	ARENAC MI			-41	28	62
1	.6	3	1.7	4	1.6	BARRY MI	163	-1	159	94	82
4	1.0	8	1.5	3	.4	BAY MI	44	-71	-59	14	39
1	2.0	1	1.7			BENZIE MI	-17				
6	1.0	3	.5	14	1.5	BERRIEN MI	-57	229	42	75	78
1	.5	2	.8	3	1.5	BRANCH MI	62	78	189	96	77
4	.7	3	.4	6	.8	CALHOUN MI	-35	75	14	62	51
3	1.8	1	.6	2	.8	CASS MI	-69	48	-54	18	54
2	2.3			1	.8	CHARLEVOIX MI			-64	11	53
1	1.4					CHEBOYGAN MI					
5*	4.3	5*	3.9	4	2.5	CHIPPEWA MI	-10	-35	-42	27	93
2*	3.6			1	.8	CLARE MI			-78	03	52
1	.7			2	1.0	CLINTON MI			35	72	59
1*	4.2			1	2.1	CRAWFORD MI			-49	21	90
3	1.9	2	1.0	5	2.0	DELTA MI	-50	105	2	57	88
2	1.9	1	.8	2	.9	DICKINSON MI	-59	13	-54	18	55
		1	.4	4	1.2	EATON MI		212			70
		1	.8	4*	3.6	EMMET MI		333			97
9	1.0	12	.8	21	1.2	GENESEE MI	-21	51	19	65	68
2	1.7	1	.6	1	.5	GOGEBIC MI	-67	-13	-71	06	41
		1	.4	3	1.1	GRAND TRAVERSE MI		181			65
1	.5	4	1.6	2	.5	GRATIOT MI	236	-70	2	56	41
		1	.4			HILLSDALE MI					
5	2.2	5	2.1	2	.7	HOUGHTON MI	-6	-64	-66	09	50
2	1.0	1	.5	2	.7	HURON MI	-50	42	-28	38	50
9	1.2	15	1.6	19	1.6	INGHAM MI	37	0	37	73	81
2	1.0	1	.4	2	.8	IONIA MI	-67	118	-27	39	51
				1	.6	IOSCO MI					43
1	.9	2	1.6	3	2.5	IRON MI	81	55	180	95	93
		3	2.0	1	.4	ISABELLA MI		-80			39
7	1.3	11	1.7	6	.8	JACKSON MI	30	-50	-35	33	54
9	1.3	7	.8	8	.8	KALAMAZOO MI	-37	-3	-38	30	53
28*	1.8	18	.9	19	.8	KENT MI	-47	-17	-57	15	52
1*	6.4					KEWEENAW MI					
1	5.3			2*	4.8	LAKE MI			-11	50	99
		2	1.1	7*	2.6	LAPEER MI		142			93
		1	1.6	3*	3.4	LEELANAU MI		119			96
3	.8	5	1.2	3	.7	LENAWEE MI	54	-44	-13	48	48
1	.8	1	.4	3	.9	LIVINGSTON MI	-45	114	18	65	58
				1	1.3	MACKINAC MI					72
6	.8	13	.9	43	1.7	MACOMB MI	6	89	100	89	63
2	2.0	3	2.1	2	1.2	MANISTEE MI	2	-41	-40	28	69
3	1.4	6	2.2	6	2.0	MARQUETTE MI	54	-11	37	73	88
1	.8			3	1.5	MASON MI			86	86	78
		3	2.5	5*	3.3	MECOSTA MI		31			96
2	1.4	1	.6	1	.7	MENOMINEE MI	-55	13	-49	20	49
2	1.5	1	.6	6	2.3	MIDLAND MI	-63	324	56	79	91
1	3.2	1	3.0	1	1.7	MISSAUKEE MI	-8	-41	-46	23	84
3	.9	5	1.2	7	1.2	MONROE MI	27	6	34	71	69
3	1.4					MONTCALM MI					
		1	2.7	2*	5.4	MONTMORENCY MI		102			99
3	.5	9	1.4	15	1.8	MUSKEGON MI	162	30	240	97	86
1	.8	2	1.3	2	1.0	NEWAYGO MI	73	-23	34	71	61
12	.8	39	1.3	62	1.4	OAKLAND MI	64	7	75	84	75
3*	3.2	1	.7	3	2.7	OCEANA MI	-77	262	-17	46	94
				1	1.0	OGEMAW MI					62
		1	1.7	1	1.4	ONTONAGON MI		-17			77
2	2.8	1	1.1	2	1.8	OSCEOLA MI	-61	61	-38	30	85
				1	2.2	OSCODA MI					90
				2	2.4	OTSEGO MI					92
4	1.1	3	.7	8	1.0	OTTAWA MI	-41	61	-6	53	63

* SIGNIFICANTLY MORE DEATHS THAN EXPECTED

WHITE FEMALE — CANCER OF THE ESOPHAGUS — ICD 150

1950-1959 DEATHS	1950-1959 RATE	1960-1969 DEATHS	1960-1969 RATE	1970-1979 DEATHS	1970-1979 RATE	COUNTY AND STATE		PERCENT CHANGE 60S/50S	PERCENT CHANGE 70S/60S	PERCENT CHANGE 70S/50S	PERCENTILE %CHANGE 70S/50S	PERCENTILE RATES 1970S
				1	1.5	PRESQUE ISLE	MI					79
				1	1.4	ROSCOMMON	MI					75
8	1.3	11	1.3	11	1.0	SAGINAW	MI	1	-23	-22	43	62
5	1.0	12*	2.0	9	1.2	ST CLAIR	MI	108	-42	20	66	68
3	1.3	2	.7	2	.6	ST JOSEPH	MI	-47	-12	-54	18	44
4	2.3	4	1.9	3	1.0	SANILAC	MI	-16	-46	-55	17	62
				3*	4.7	SCHOOLCRAFT	MI					98
1	.3	3	1.1	4	1.2	SHIAWASSEE	MI	220	10	251	97	68
1	.6	3	1.5			TUSCOLA	MI	159				
2	.9	2	.6	5	1.6	VAN BUREN	MI	-27	146	80	85	81
		3	.4	7	.8	WASHTENAW	MI		74			51
125*	1.4	141	1.3	170*	1.5	WAYNE	MI	-8	19	9	60	80
1	1.1	2	1.9	1	.7	WEXFORD	MI	75	-63	-35	33	49
159	1.0	171	.9	274	1.2	MINNESOTA	MN	-16	34	12		
		1	.4	4	.9	ANOKA	MN		118			56
2	1.8	3	2.3	2	1.1	BECKER	MN	30	-52	-37	31	66
1	1.1			3	2.3	BELTRAMI	MN			111	90	91
		1	1.8	1	1.3	BIG STONE	MN		-26			74
2	1.0	2	.7	4	.8	BLUE EARTH	MN	-29	21	-14	47	54
2	1.4			2	1.0	BROWN	MN			-24	42	62
3	2.7	2	1.4	3	1.7	CARLTON	MN	-47	22	-36	32	84
				2	.9	CARVER	MN					57
1	1.2			2	1.3	CASS	MN			8	59	70
1	1.2			2	1.6	CHIPPEWA	MN			39	74	81
		2	1.4	2	.8	CLAY	MN		-44			51
		1	2.2			CLEARWATER	MN					
1	1.1	3	2.5	1	.6	COTTONWOOD	MN	119	-78	-51	19	43
		2	1.1	6	2.7	CROW WING	MN		157			94
3	1.2	2	.6	11	2.0	DAKOTA	MN	-51	247	71	83	88
1	1.4	1	.9	1	1.4	DODGE	MN	-33	52	2	57	76
1	.8	1	.7	1	.5	DOUGLAS	MN	-8	-31	-36	32	42
1	.9	2	1.1	1	.5	FARIBAULT	MN	29	-52	-38	30	42
				1	.6	FILLMORE	MN					46
				2	.8	FREEBORN	MN					51
4	2.0			1	.4	GOODHUE	MN			-83	01	38
		1	1.7			GRANT	MN					
44	1.2	54	1.1	80	1.4	HENNEPIN	MN	-2	24	22	66	76
		1	1.3			HUBBARD	MN					
				2	1.7	ISANTI	MN					83
1	.7	3	1.6	3	1.3	ITASCA	MN	131	-19	87	86	72
				1	.6	JACKSON	MN					45
1	1.9			1	1.7	KANABEC	MN			-14	47	83
4	2.5	2	1.1	2	.8	KANDIYOHI	MN	-55	-32	-69	07	51
				1	.7	KITTSON	MN					47
1	1.8	1	1.3	1	1.0	KOOCHICHING	MN	-26	-22	-42	27	63
		2	1.9	1	.5	LAC QUI PARLE	MN		-72			42
		1	1.9			LAKE	MN					
2	1.7			1	.8	LE SUEUR	MN			-55	17	51
1	2.8	1	1.5			LINCOLN	MN	-47				
3	2.9			1	.6	LYON	MN			-78	03	46
4*	3.2			3	1.7	MC LEOD	MN			-46	23	84
		1	4.2			MAHNOMEN	MN					
2	3.3	1	1.2			MARSHALL	MN	-65				
		1	.7	1	.3	MARTIN	MN		-64			37
1	.9	2	1.7	1	.5	MEEKER	MN	80	-69	-43	26	42
1	1.1	1	.8	1	.6	MILLE LACS	MN	-27	-30	-49	21	43
				1	.5	MORRISON	MN					41
		3	1.2	6	1.9	MOWER	MN		62			88
				2	1.8	MURRAY	MN					85
		1	.6			NICOLLET	MN					
2	2.0	1	.7			NOBLES	MN	-66				
2	3.1					NORMAN	MN					
		3	.9	1	.2	OLMSTED	MN		-82			36
2	.8	3	.7	4	1.3	OTTER TAIL	MN	-7	87	73	84	74
				4*	4.4	PENNINGTON	MN					98
				1	.6	PINE	MN					43
1	1.5	1	1.2	2	1.4	PIPESTONE	MN	-18	16	-5	53	76
1	.6			2	.6	POLK	MN			9	60	44
29	1.5	30	1.2	44	1.5	RAMSEY	MN	-24	31	0	55	80
1	1.1			1	.3	REDWOOD	MN			-68	08	38
2	1.7	2	1.2	1	.6	RENVILLE	MN	-33	-47	-64	11	45
2	1.0			4	1.3	RICE	MN			26	69	72
1	2.1					ROCK	MN					
		1	1.5	4*	4.6	ROSEAU	MN		216			98
13	1.3	20	1.5	20	1.3	ST LOUIS	MN	15	-8	6	58	74
		1	.9	3	2.1	SCOTT	MN		149			90
		1	.8	1	1.0	SHERBURNE	MN		16			59
				3	2.0	SIBLEY	MN					89
2	.7	1	.3	2	.5	STEARNS	MN	-60	73	-31	36	40
1	.9			4	2.8	STEELE	MN			214	96	94
1	1.9			1	1.6	STEVENS	MN			-16	46	81
1	1.1					SWIFT	MN					
1	.8	1	.5	3	1.9	TODD	MN	-39	271	127	91	87
1	2.7					TRAVERSE	MN					
4*	4.0	1	.9			WABASHA	MN	-78				
		3*	4.0			WADENA	MN					
1	.6			5	1.5	WASHINGTON	MN			163	94	77
				1	1.2	WATONWAN	MN					67
				1	1.1	WILKIN	MN					66
3	1.4	1	.4			WINONA	MN	-70				
		1	.6	1	.6	WRIGHT	MN		1			43
		1	.9			YELLOW MEDICINE	MN					
69	1.3	97	1.4	92	1.0	MISSISSIPPI	MS	6	-28	-24		
				2	1.5	ADAMS	MS					78
4*	3.4			2	1.3	ALCORN	MS			-63	12	71
1	1.1	2	2.2	1	.9	ATTALA	MS	105	-60	-17	45	56
3*	5.9	2	2.2	1	1.0	BOLIVAR	MS	-63	-55	-84	01	60

* SIGNIFICANTLY MORE DEATHS THAN EXPECTED

WHITE FEMALE — CANCER OF THE ESOPHAGUS — ICD 150

1950-1959 DEATHS	1950-1959 RATE	1960-1969 DEATHS	1960-1969 RATE	1970-1979 DEATHS	1970-1979 RATE	COUNTY AND STATE	PERCENT CHANGE 60S/50S	PERCENT CHANGE 70S/60S	PERCENT CHANGE 70S/50S	PERCENTILE %CHANGE 70S/50S	PERCENTILE RATES 1970S
1	1.5			2	1.9	CALHOUN MS			23	67	87
1	2.6					CHOCTAW MS					
1*	6.1					CLAIBORNE MS					
		3*	4.8	1	1.0	CLARKE MS		-79			61
		2	3.2	1	1.3	CLAY MS		-59			73
1	1.9					COAHOMA MS					
		3*	3.4			COPIAH MS					
		1	1.7	1	1.6	COVINGTON MS		-3			82
		1	2.0			DE SOTO MS					
2	1.2	9*	4.4	5	1.7	FORREST MS	263	-61	43	76	85
		3*	6.7	2	3.2	GEORGE MS		-53			96
1	2.6	1	1.5			GRENADA MS	-42				
		3*	4.5	1	.7	HANCOCK MS		-85			47
3	1.2	8*	2.0	11	2.2	HARRISON MS	71	11	89	87	90
3	.7	7	1.3	7	.9	HINDS MS	99	-31	38	74	57
				1	1.3	HOLMES MS					71
1	1.5	1	1.3	2	1.6	ITAWAMBA MS	-9	21	10	60	81
2	2.2	4	2.1	6	2.3	JACKSON MS	-1	10	9	59	92
1	2.0			2	3.1	JASPER MS			51	78	95
2	1.7			3	1.0	JONES MS			-39	29	60
		1	3.2	2*	5.2	KEMPER MS		65			99
1	2.0					LAFAYETTE MS					
2*	4.1	1	3.5	1	1.1	LAMAR MS	-13	-68	-73	05	65
5	2.0	2	.7	1	.2	LAUDERDALE MS	-67	-65	-88	00	36
		1	2.0			LAWRENCE MS					
		1	1.3			LEAKE MS					
1	.7	2	1.0			LEE MS	41				
				1	.9	LEFLORE MS					55
2	2.3	1	.9	1	.8	LINCOLN MS	-59	-18	-66	10	52
1	1.0	2	1.5	1	.5	LOWNDES MS	48	-69	-55	17	40
		1	1.3			MADISON MS					
				2	2.3	MARION MS					91
1	2.0	2*	3.5	1	1.9	MARSHALL MS	80	-45	-2	55	87
				1	.5	MONROE MS					42
1	2.1	1	2.4			MONTGOMERY MS	15				
1	1.1	3*	3.3	1	.7	NESHOBA MS	200	-79	-37	31	48
1	2.1			2	1.9	NEWTON MS			-13	48	86
				3*	9.8	NOXUBEE MS					99
1	1.9					OKTIBBEHA MS					
1	1.6					PANOLA MS					
1	1.3	1	1.2	2	1.5	PEARL RIVER MS	-12	27	11	61	78
1	4.2			1	2.3	PERRY MS			-44	25	91
		2	1.3	1	.5	PIKE MS		-57			42
2	2.4			1	.6	PONTOTOC MS			-75	04	44
1	1.2	2	1.8	1	1.1	PRENTISS MS	60	-39	-2	54	66
1	3.0	2*	4.8	1	2.4	QUITMAN MS	61	-50	-20	44	92
2	2.2	2	1.7	2	.9	RANKIN MS	-25	-47	-60	14	56
2	3.8	2	2.4			SCOTT MS	-36				
2	3.1	2	2.4	2	1.9	SIMPSON MS	-21	-20	-37	31	88
1	2.2					SMITH MS					
1*	4.3					STONE MS					
		1	1.1	2	2.2	SUNFLOWER MS		94			90
2*	4.9	2*	3.7	1	1.4	TALLAHATCHIE MS	-24	-63	-71	06	75
		1	2.2	2	2.6	TATE MS		19			93
		2	2.4	1	.8	TISHOMINGO MS		-66			53
1	1.7					UNION MS					
2*	5.7	1	2.8			WALTHALL MS	-50				
1	.8	1	.8	1	.6	WARREN MS	5	-30	-27	39	44
2	2.1	2	1.1	1	.6	WASHINGTON MS	-49	-46	-72	06	44
		1	1.8			WAYNE MS					
				2	1.9	WEBSTER MS					87
				2	2.4	WINSTON MS					92
		4*	6.8			YALOBUSHA MS					
2	2.8	1	1.2	1	.9	YAZOO MS	-56	-24	-66	10	58
188	.9	227	.8	260	.8	MISSOURI MO	-1	0	-1		
1	.9	1	.6			ADAIR MO	-35				
1	1.4					ANDREW MO					
2	3.3	1	1.4	1	1.7	ATCHISON MO	-56	21	-47	22	84
1	.8	2	1.2			AUDRAIN MO	54				
3	2.3					BARRY MO					
		1	.5	1	.4	BATES MO		-9			40
		2	2.8	1	1.2	BOLLINGER MO		-58			67
		2	.7	3	.8	BOONE MO		18			54
5	.8	10	1.6	8	1.1	BUCHANAN MO	95	-32	33	71	63
1	.9	3	1.4	1	.3	BUTLER MO	63	-78	-64	11	37
		1	.8			CALDWELL MO					
		2	1.3	3	1.5	CALLAWAY MO		10			78
1	2.2			4	4.0	CAMDEN MO			81	85	98
4	2.2	2	.8	2	.7	CAPE GIRARDEAU MO	-64	-8	-66	09	50
1	1.1			1	.6	CARROLL MO			-49	21	43
1	3.8					CARTER MO					
1	.7	1	.5	2	.9	CASS MO	-20	69	36	72	57
		1	1.3	1	.8	CEDAR MO		-40			53
				1	.6	CHARITON MO					45
		1	1.1	1	1.1	CHRISTIAN MO		-2			63
		1	1.3			CLARK MO					
5	2.0	2	.6	2	.3	CLAY MO	-71	-39	-82	01	38
2	1.0	1	.4	3	1.1	COLE MO	-61	167	5	58	64
1	.9			2	1.3	COOPER MO			48	77	73
2	2.5	1	1.1			CRAWFORD MO	-55				
1	1.8	3*	3.8	2	1.2	DADE MO	117	-67	-29	38	70
1	2.1	1	1.3	3	3.1	DALLAS MO	-37	137	50	78	96
				1	.7	DAVIESS MO					49
				1	.9	DE KALB MO					55
1	1.2			1	1.6	DENT MO			35	72	82
4	2.2	2	.8	1	.6	DUNKLIN MO	-62	-31	-74	05	44
		3	1.0	4	.9	FRANKLIN MO		-7			58
				1	.7	GASCONADE MO					49
				2	2.0	GENTRY MO					89

* SIGNIFICANTLY MORE DEATHS THAN EXPECTED

WHITE FEMALE — CANCER OF THE ESOPHAGUS — ICD 150

1950-1959 DEATHS	1950-1959 RATE	1960-1969 DEATHS	1960-1969 RATE	1970-1979 DEATHS	1970-1979 RATE	COUNTY AND STATE	PERCENT CHANGE 60S/50S	PERCENT CHANGE 70S/60S	PERCENT CHANGE 70S/50S	PERCENTILE %CHANGE 70S/50S	PERCENTILE RATES 1970S
5	.8	6	.7	5	.5	GREENE MO	-16	-24	-36	32	41
1	1.2	3	2.4			GRUNDY MO	101				
		1	.6	1	.7	HARRISON MO		13			49
1	.7					HENRY MO					
2	2.4					HOWARD MO					
1	.7	2	1.4	1	.4	HOWELL MO	91	-69	-42	27	40
		2	3.8			IRON MO					
20	.7	38	1.1	51	1.4	JACKSON MO	64	30	112	90	76
8	1.5	3	.4	10	1.3	JASPER MO	-69	199	-8	51	73
2	1.0	1	.3	6	1.2	JEFFERSON MO	-72	307	16	63	68
1	.9			1	.6	JOHNSON MO			-34	34	43
1	1.6					KNOX MO					
1	.7	3	1.8	2	.6	LAFAYETTE MO	154	-68	-19	44	44
1	.6	2	.9	2	.9	LAWRENCE MO	44	8	55	79	59
				1	1.8	LEWIS MO					86
1	.5	2	1.4	1	.5	LINN MO	172	-66	-7	52	40
		2	1.7	1	1.1	LIVINGSTON MO		-34			66
1	1.1					MC DONALD MO					
2	1.2	1	.5			MACON MO	-59				
				1	.8	MADISON MO					53
1	.5	3	1.2	2	1.4	MARION MO	164	14	201	96	74
1	1.2					MILLER MO					
		1	1.0			MISSISSIPPI MO					
		1	.6	2	2.0	MONTGOMERY MO		221			88
1	1.6	3*	4.3	2	1.4	MORGAN MO	168	-67	-12	49	76
5*	5.3	1	1.4	1	.7	NEW MADRID MO	-74	-51	-87	00	48
2	1.3	1	.6	1	.4	NEWTON MO	-56	-34	-71	07	38
				2	.7	NODAWAY MO					48
1	.9			1	.8	PEMISCOT MO			-19	44	51
		1	.7	3	2.7	PERRY MO		268			94
1	.4			1	.5	PETTIS MO			24	67	41
		1	.7	3	1.8	PHELPS MO		168			86
		1	.6	2	1.1	PIKE MO		88			65
		1	.9	1	.6	PLATTE MO		-36			43
2	1.7	1	.8			POLK MO	-57				
		1	1.3	2	1.8	PULASKI MO		36			85
1	1.5					PUTNAM MO					
1	2.8					RALLS MO					
1	.5	3	1.3	2	1.1	RANDOLPH MO	162	-16	120	91	65
1	1.0	2	1.4	1	.7	RAY MO	41	-53	-34	33	47
3	1.9			6	1.3	ST CHARLES MO			-30	36	72
1	1.8	1	1.2	1	.9	ST CLAIR MO	-31	-30	-52	19	55
5	2.3	3	1.0	4	1.1	ST FRANCOIS MO	-57	6	-55	17	63
15	.7	33	.9	42	.8	ST LOUIS COUNTY MO	28	-6	20	66	54
		1	.4			SALINE MO					
		1	1.5	1	1.2	SCOTLAND MO		-21			68
		1	.5			SCOTT MO					
1	1.0	1	1.1			SHELBY MO	7				
				1	.6	STODDARD MO					45
		1	1.7	1	.8	STONE MO		-54			52
1	1.1	1	1.0	2	1.8	SULLIVAN MO	-13	78	54	79	85
		2	2.2			TANEY MO					
2	1.9	2	1.5	1	.6	TEXAS MO	-25	-56	-67	08	46
1	.6			1	.4	VERNON MO			-37	32	38
1	.9			1	1.2	WRIGHT MO			27	69	67
53	1.3	49	1.2	37	1.1	ST LOUIS CITY MO	-8	-7	-14	47	65
29	1.1	20	.6	45	1.1	MONTANA MT	-44	80	0		
				1	2.6	BIG HORN MT					93
		1	3.7			BLAINE MT					
1	1.6			1	1.4	CARBON MT			-13	48	76
		1*	6.8			CARTER MT					
4	1.8	2	.7	6	1.5	CASCADE MT	-64	125	-18	45	78
1	1.2					CUSTER MT					
				1	1.5	DAWSON MT					78
1	1.1					DEER LODGE MT					
1*	5.8	1*	4.7			FALLON MT	-19				
				1	1.0	FERGUS MT					60
1	.7	1	.5	2	.8	FLATHEAD MT	-19	45	18	64	52
1	.9			1	.7	GALLATIN MT			-28	39	48
1*	4.3					GLACIER MT					
				1	.8	HILL MT					54
				1	3.5	JEFFERSON MT					97
2	3.1	1	1.2	2	1.7	LAKE MT	-61	46	-44	25	84
1	.8	2	1.2	1	.6	LEWIS AND CLARK MT	54	-52	-27	39	43
				1*	7.8	LIBERTY MT					99
				1	1.4	LINCOLN MT					74
				1	1.6	MADISON MT					81
				1*	6.8	MINERAL MT					99
2	1.1	1	.4	2	.7	MISSOULA MT	-60	64	-34	34	49
1	3.4			2*	5.2	MUSSELSHELL MT			52	78	99
1	1.3	1	1.2	1	1.4	PARK#MT	-9	19	8	59	75
1	3.5					PHILLIPS MT					
				1	2.3	PONDERA MT					91
				1	.8	RAVALLI MT					51
		2*	3.4			RICHLAND MT					
2*	6.8	1	3.2	1	2.7	ROOSEVELT MT	-53	-18	-61	13	94
		1	3.7			ROSEBUD MT					
				1	2.3	SANDERS MT					92
3	1.2	2	.8	5	1.6	SILVER BOW MT	-36	114	37	73	82
				2*	4.3	TETON MT					98
				2*	6.4	TOOLE MT					99
1	3.3			2	2.8	VALLEY MT			-15	47	95
4	1.8	3	.8	3	.7	YELLOWSTONE MT	-54	-20	-63	12	46
66	.9	71	.8	75	.7	NEBRASKA NB	-14	-4	-17		
				1	.6	ADAMS NB					45
				1	.9	ANTELOPE NB					55
1	1.8					BOX BUTTE NB					

* SIGNIFICANTLY MORE DEATHS THAN EXPECTED

\# COUNTIES COMBINED; SEE TECHNICAL NOTES

WHITE FEMALE — CANCER OF THE ESOPHAGUS — ICD 150

1950-1959 DEATHS	1950-1959 RATE	1960-1969 DEATHS	1960-1969 RATE	1970-1979 DEATHS	1970-1979 RATE	COUNTY AND STATE	PERCENT CHANGE 60S/50S	PERCENT CHANGE 70S/60S	PERCENT CHANGE 70S/50S	PERCENTILE %CHANGE 70S/50S	PERCENTILE RATES 1970S
		1	2.9			BOYD NB					
		1	4.0			BROWN NB					
3	1.8	2	1.2	1	.4	BUFFALO NB	-32	-66	-77	03	39
		4*	5.0			BURT NB					
3*	4.2	1	.9			BUTLER NB	-77				
				1	1.0	CASS NB					62
		1	.9			CEDAR NB					
		1	2.1			CHERRY NB					
1	2.3	1	1.4			CHEYENNE NB	-38				
1	1.6					COLFAX NB					
		1	1.2			CUMING NB					
1	2.6			2	4.0	DAWES NB			58	80	98
1	.9					DAWSON NB					
1*	5.9					DEUEL NB					
1	2.0					DIXON NB					
		2	1.0	3	.7	DODGE NB		-24			50
18	1.3	21	1.1	26	1.3	DOUGLAS NB	-13	12	-3	54	71
		1	1.0			FILLMORE NB					
2*	4.8					FRANKLIN NB					
				1	1.0	FURNAS NB					59
2	1.0	2	1.0	2	.9	GAGE NB	-4	-9	-12	48	57
1*	4.0					GARDEN NB					
1	4.3					GREELEY NB					
2	.9	4	1.5	2	.6	HALL NB	65	-37	-29	38	46
		1	2.0			HAMILTON NB					
1	1.3			1	1.5	HOLT NB			10	60	78
1*	18.6					HOOKER NB					
1	2.5					HOWARD NB					
				1	.7	JEFFERSON NB					50
1	1.8					JOHNSON NB					
				2	2.5	KEITH NB					93
				1	1.0	KNOX NB					60
7	.9	5	.5	6	.6	LANCASTER NB	-48	13	-41	28	43
		3	1.9			LINCOLN NB					
1*	16.7					LOUP NB					
		1	.6			MADISON NB					
1	2.3			1	.9	MERRICK NB			-59	14	59
1	2.7					NANCE NB					
1	1.3	2	3.1	1	.6	NEMAHA NB	143	-82	-56	16	43
		1	1.3	1	2.3	NUCKOLLS NB		67			91
				2	1.0	OTOE NB					60
				1	1.2	PAWNEE NB					69
1	2.1	1	1.7			PHELPS NB	-21				
		1	1.5			PIERCE NB					
1	1.0	4*	2.8	2	1.2	PLATTE NB	189	-57	25	68	68
1	2.0			1	1.3	RED WILLOW NB			-37	32	71
1	1.0			2	2.0	RICHARDSON NB			96	88	88
		1*	5.0			ROCK NB					
1	1.0	1	.7	1	.5	SALINE NB	-24	-35	-50	20	41
1	1.2	1	.7	4	3.5	SAUNDERS NB	-38	377	194	96	97
1	.7			1	.4	SCOTTS BLUFF NB			-45	24	39
		1	1.2			SEWARD NB					
1	3.1					STANTON NB					
		1	1.3			THAYER NB					
1	4.7					THURSTON NB					
		1	.7	2	2.7	WASHINGTON NB		290			94
		1	1.5	4*	5.9	WAYNE NB		289			99
3*	5.3	1	1.1	1	1.6	WEBSTER NB	-79	40	-70	07	81
		1	.8			YORK NB					
7	.9	15	1.0	44	1.7	NEVADA NV	21	61	95		
3	1.6	8	1.3	16	1.2	CLARK NV	-15	-12	-25	40	67
		1*	2.6	3*	7.5	DOUGLAS NV		184			99
		1	1.7			ELKO NV					
				1*	8.1	LANDER NV					99
				1	1.4	LYON NV					76
				1	4.4	MINERAL NV					98
4	1.3	4	.9	20*	2.7	WASHOE NV	-30	194	106	89	94
		1	2.4	2	1.3	ORMSBY-CARSON CITY NV		-45			72
38	1.1	45	1.1	75	1.5	NEW HAMPSHIRE NH	-4	37	32		
2	1.2	4	1.8	2	.8	BELKNAP NH	48	-55	-33	34	54
2	1.5	1	.6	4	2.3	CARROLL NH	-64	305	48	77	91
5	2.3	3	1.1	5	1.1	CHESHIRE NH	-54	7	-51	20	66
		5	2.4	4	1.5	COOS NH		-35			80
5	1.7	4	1.2	11*	2.8	GRAFTON NH	-31	131	61	81	94
10	1.0	17	1.4	23	1.6	HILLSBOROUGH NH	44	10	59	81	80
5	.9	2	.4	6	1.0	MERRIMACK NH	-59	152	3	57	59
7	1.5	4	.6	12	1.5	ROCKINGHAM NH	-61	155	-1	55	79
2	.7	4	1.1	7	1.6	STRAFFORD NH	70	43	143	93	82
		1	.5	1	.4	SULLIVAN NH		-13			39
357*	1.4	451*	1.3	595*	1.4	NEW JERSEY NJ	-6	5	-1		
11	1.4	17	1.7	16	1.1	ATLANTIC NJ	29	-34	-16	46	66
36	1.2	56	1.2	77	1.4	BERGEN NJ	4	11	16	63	75
5	.8	8	.9	19	1.4	BURLINGTON NJ	8	60	73	84	75
24	1.6	22	1.1	41	1.6	CAMDEN NJ	-35	54	0	55	82
4	1.6	4	.9	6	.9	CAPE MAY NJ	-44	4	-42	27	59
4	.9	8	1.4	9	1.2	CUMBERLAND NJ	61	-13	41	75	69
75*	1.7	68	1.4	71	1.5	ESSEX NJ	-18	8	-12	49	79
9	2.0	3	.4	12	1.5	GLOUCESTER NJ	-78	247	-25	41	78
55*	2.0	67*	1.9	67*	1.8	HUDSON NJ	-5	-5	-10	50	85
		4	1.2	5	1.1	HUNTERDON NJ		-9			65
12	1.0	20	1.4	24	1.4	MERCER NJ	38	0	38	74	76
17	1.3	31*	1.7	30	1.1	MIDDLESEX NJ	27	-36	-19	44	64
17	1.3	32	1.6	53*	1.9	MONMOUTH NJ	28	15	46	77	87
13	1.2	12	.7	29	1.5	MORRIS NJ	-39	102	22	67	79
12*	2.8	8	.9	31	1.5	OCEAN NJ	-69	76	-46	23	80

* SIGNIFICANTLY MORE DEATHS THAN EXPECTED

WHITE FEMALE | CANCER OF THE ESOPHAGUS | ICD 150

1950-1959 DEATHS	1950-1959 RATE	1960-1969 DEATHS	1960-1969 RATE	1970-1979 DEATHS	1970-1979 RATE	COUNTY AND STATE	PERCENT CHANGE 60S/50S	PERCENT CHANGE 70S/60S	PERCENT CHANGE 70S/50S	PERCENTILE %CHANGE 70S/50S	PERCENTILE RATES 1970S
33*	1.9	34	1.4	38	1.3	PASSAIC NJ	-25	-6	-29	37	73
3	1.4	1	.4	1	.3	SALEM NJ	-71	-20	-77	03	38
6	1.3	9	1.3	13	1.4	SOMERSET NJ	0	4	4	58	75
3	1.4	4	1.3	5	1.1	SUSSEX NJ	-6	-17	-22	43	63
15	.7	41	1.5	42	1.3	UNION NJ	98	-15	68	83	70
3	.9	2	.5	6	1.0	WAPREN NJ	-44	101	12	61	61
25	1.2	24	.7	49	1.1	NEW MEXICO NM	-40	48	-11		
7	1.4	7	.7	13	.9	BERNALILLO NM	-50	22	-39	30	55
4*	3.5	2	1.1	2	.7	CHAVES NM	-68	-35	-79	02	49
		1	1.5			COLFAX NM					
1	1.0	1	1.0	3	1.4	CURRY NM	2	40	44	76	76
1	.7	3	1.6	2	.6	DONA ANA NM	125	-58	-6	53	46
		3	1.5	3	1.4	EDDY NM		-7			75
1	1.6	1	1.2	2	1.9	GPANT NM	-23	59	23	67	88
1*	5.0					HIDALGO NM					
				2	.9	LEA NM					56
1	2.9					LINCOLN NM					
				2	1.7	LUNA NM					84
				1	1.4	MC KINLEY NM					77
		1	4.5	1*	3.3	MORA NM		-27			96
				1	.7	OTERO NM					50
				1	1.4	QUAY NM					74
		1	1.2	3*	4.3	RIO ARRIBA NM		248			98
2	3.6			2	2.1	ROOSEVELT NM			-43	26	89
				2	3.6	SANDOVAL NM					97
1	1.4			2	1.7	SAN MIGUEL NM			22	67	85
1	.8			2	.7	SANTA FE NM			-9	51	50
		1	1.1	1	1.0	SIERRA NM		-7			61
				1	2.3	SOCORPO NM					92
		2	3.3	1	1.5	TAOS NM		-54			79
5*	8.0	1	1.5	2	1.5	VALENCIA NM	-81	1	-81	01	79
1234*	1.6	1388*	1.4	1703*	1.6	NEW YORK NY	-9	10	-1		
24	1.7	20	1.1	28	1.4	ALBANY NY	-32	25	-15	47	76
2	.7	1	.3	3	.8	ALLEGANY NY	-63	191	8	59	53
166*	2.4	125*	1.6	186*	2.5	BRONX NY	-37	60	1	56	93
8	.8	21*	1.7	17	1.0	BROOME NY	119	-39	34	71	63
8	2.0	4	.8	6	1.1	CATTARAUGUS NY	-60	30	-48	22	64
7	1.5	6	1.3	8	1.8	CAYUGA NY	-15	41	19	65	86
8	.9	7	.7	14	1.3	CHAUTAUQUA NY	-23	91	46	77	73
8	1.5	5	.9	16*	2.4	CHEMUNG NY	-41	181	65	82	92
2	.7	2	.6	1	.4	CHENANGO NY	-16	-33	-43	26	39
4	1.5	2	.7	2	.7	CLINTON NY	-57	-2	-58	15	47
6	1.9	6	1.7	5	1.2	COLUMBIA NY	-10	-31	-38	30	67
2	1.0	5	2.0	4	1.2	CORTLAND NY	102	-39	23	67	69
5	1.9	1	.3	5	1.4	DELAWARE NY	-82	319	-24	42	76
13	1.4	12	1.1	18	1.5	DUTCHESS NY	-21	33	6	58	78
53	1.2	60	1.0	105*	1.6	ERIE NY	-11	54	38	74	82
2	.9	3	1.4	3	1.4	ESSEX NY	64	-1	63	82	76
6*	2.5	3	1.3	3	1.0	FRANKLIN NY	-49	-22	-60	13	60
6	1.6	5	1.4	3	.6	FULTON NY	-13	-56	-62	12	45
1	.4	2	.6	2	.6	GENESEE NY	43	-1	42	75	44
2	1.0	1	.4	6	1.9	GREENE NY	-62	401	91	87	87
				1	2.3	HAMILTON NY					91
3	.8	7	1.6	2	.4	HERKIMER NY	93	-72	-46	23	40
7	1.2	8	1.1	11	1.7	JEFFERSON NY	-9	52	39	74	84
234*	2.1	201*	1.5	232*	1.7	KINGS(BROOKLYN) NY	-29	16	-17	45	84
2	1.5	2	1.2	2	1.0	LEWIS NY	-18	-18	-33	34	61
4	1.7	3	1.2	2	.7	LIVINGSTON NY	-31	-44	-61	13	48
4	1.4	1	.3	1	.2	MADISON NY	-78	-36	-86	00	36
40	1.4	37	1.0	55	1.3	MONROE NY	-26	25	-7	52	71
7	1.8	3	.6	5	.9	MONTGOMERY NY	-65	41	-51	19	57
47	1.4	85*	1.4	128*	1.6	NASSAU NY	2	15	17	64	82
214*	2.3	321*	3.4	196*	2.3	NEW YORK(MANHATTAN) NY	47	-32	0	55	91
11	1.3	16	1.5	15	1.0	NIAGAPA NY	14	-29	-19	44	62
11	.8	17	1.1	20	1.1	ONEIDA NY	40	1	42	75	66
27	1.4	29	1.2	37	1.4	ONONDAGA NY	-15	12	-5	53	74
2	.5	4	.9			ONTARIO NY	70				
11	1.1	14	1.2	18	1.3	OPANGE NY	12	8	21	66	72
				4	1.5	ORLEANS NY					79
5	1.2	5	.9	12	2.0	OSWEGO NY	-22	118	71	83	88
1	.3	5	1.4	4	1.0	OTSEGO NY	397	-31	244	97	59
1	.8	10*	4.6	2	.6	PUTNAM NY	506	-86	-18	45	45
111	1.4	126	1.1	200*	1.4	QUEENS NY	-23	36	5	58	77
14	1.6	9	1.0	19	1.8	RENSSELAER NY	-39	81	10	60	86
13	1.3	14	1.0	30*	1.7	RICHMOND NY	-18	67	37	73	84
4	.7	9	1.1	13	1.1	ROCKLAND NY	50	2	53	79	66
5	1.0	4	.7	8	1.1	ST LAWRENCE NY	-31	73	18	65	66
8	1.8	6	1.1	8	1.1	SAPATOGA NY	-38	2	-37	31	66
9	1.1	12	1.2	14	1.2	SCHENECTADY NY	9	7	17	64	70
1	.7	3	2.2	1	.5	SCHOHARIE NY	224	-78	-28	39	41
		1	.9			SCHUYLER NY					
		3	1.0	3	1.2	SENECA NY		17			69
7	1.4	6	1.0	8	1.3	STEUBEN NY	-23	19	-8	52	70
22	1.0	49	1.3	91*	1.6	SUFFOLK NY	30	26	65	82	82
2	.8	1	.3	6	1.3	SULLIVAN NY	-62	326	61	81	74
7*	3.7	2	1.0	4	1.7	TIOGA NY	-72	61	-55	17	83
4	1.4	2	.6	2	.6	TOMPKINS NY	-56	1	-55	17	45
7	1.1	8	1.1	14	1.3	ULSTER NY	-6	17	11	61	71
3	1.1	10*	3.4	2	.8	WARREN NY	204	-78	-32	36	51
3	1.0	3	.9	5	1.3	WASHINGTON NY	-4	39	34	71	73
5	1.3	2	.4	3	.6	WAYNE NY	-66	44	-52	19	46
41	1.2	56	1.2	85	1.5	WESTCHESTER NY	-2	33	31	70	80
1	.6	2	.8	1	.4	WYOMING NY	47	-54	-33	35	39
3	1.9	1	.8	4	3.6	YATES NY	-61	371	85	86	97
152	1.3	203	1.2	261	1.1	NORTH CAROLINA NC	-5	-8	-13		

* SIGNIFICANTLY MORE DEATHS THAN EXPECTED

WHITE FEMALE — CANCER OF THE ESOPHAGUS — ICD 150

1950-1959 DEATHS	1950-1959 RATE	1960-1969 DEATHS	1960-1969 RATE	1970-1979 DEATHS	1970-1979 RATE	COUNTY AND STATE	PERCENT CHANGE 60S/50S	PERCENT CHANGE 70S/60S	PERCENT CHANGE 70S/50S	PERCENTILE %CHANGE 70S/50S	PERCENTILE RATES 1970S
4	1.6	1	.3	2	.4	ALAMANCE NC	-78	17	-75	04	39
1	1.4			2	1.8	ALEXANDER NC			29	70	86
		1	1.7			ALLEGHANY NC					
2	2.9	1	1.4	1	.8	ANSON NC	-52	-45	-74	05	51
		1	1.0	2	1.9	ASHE NC		101			87
1	1.7			1	1.5	AVERY NC			-14	47	77
		3	2.2	6*	3.5	BEAUFORT NC		60			97
2*	4.8			1	1.3	BERTIE NC			-74	05	71
		1	1.7	1	.9	BLADEN NC		-48			55
2*	4.1			1	.8	BRUNSWICK NC			-79	02	55
2	.4	13	1.8	17	1.6	BUNCOMBE NC	300	-10	260	97	81
		2	.9			BURKE NC					
3	1.3	3	1.0	2	.5	CABARRUS NC	-24	-48	-60	13	42
1	.8	1	.5	4	1.4	CALDWELL NC	-39	178	69	83	75
1	1.0	1	.8	1	.5	CARTERET NC	-24	-41	-55	17	40
2	1.2	6	1.8	6	1.3	CATAWBA NC	52	-30	6	58	71
2	2.1	3	2.8	2	1.6	CHATHAM NC	37	-42	-21	43	82
		1	1.0	1	.8	CHEPOKEE NC		-17			52
1	3.2	1	2.8			CHOWAN NC	-11				
1	4.1	1	2.8			CLAY NC	-32				
3	1.2	1	.3	2	.5	CLEVELAND NC	-74	68	-56	16	42
		2	1.4	3	1.4	COLUMBUS NC		0			76
1	.9			1	.5	CRAVEN NC			-46	23	41
4	2.3	2	.7	6	1.3	CUMBERLAND NC	-70	90	-44	25	72
				1	2.3	CURRITUCK NC					91
		1	3.0			DAPE NC					
2	.9	5	1.4	3	.6	DAVIDSON NC	60	-55	-27	39	46
2	2.9	1	1.1	1	1.0	DAVIE NC	-63	-6	-65	10	62
1	.9			2	1.3	DUPLIN NC			38	74	70
2	.7	15*	3.5	3	.6	DURHAM NC	431	-84	-16	46	43
5*	5.3	2	1.3	1	.5	EDGECOMBE NC	-75	-63	-91	00	41
7	1.4	11	1.8	11	1.1	FOPSYTH NC	22	-37	-24	42	65
2	3.9	1	1.1	3	2.2	FRANKLIN NC	-72	98	-44	25	90
3	1.0	7	1.2	7	.9	GASTON NC	18	-21	-7	52	58
1	4.0			1	2.5	GATES NC			-39	30	93
				1	.8	GRANVILLE NC					53
10	1.6	13	1.3	16	1.2	GUILFORD NC	-21	-7	-26	40	68
2	1.5	2	1.3	3	1.6	HALIFAX NC	-14	23	5	58	80
3	2.8	3	1.9	5	1.9	HARNETT NC	-31	1	-30	36	88
1	.6	1	.5	2	.7	HAYWOOD NC	-22	43	12	61	48
1	.7	3	1.4	2	.5	HENDEPSON NC	96	-61	-23	42	42
				1	1.3	HERTFORD NC					72
				1	2.2	HYDE NC					90
1	.6	1	.4	4	.9	IREDELL NC	-32	145	67	83	58
				2	1.7	JACKSON NC					84
3	1.7	2	.9	4	1.2	JOHNSTON NC	-48	44	-25	40	70
1*	4.3			1	2.9	JONES NC			-33	34	95
1	1.4	1	.8	2	1.3	LEE NC	-42	70	-2	54	74
2	1.7	3	2.2	4	2.0	LENOIR NC	30	-12	15	63	88
1	1.0	2	1.8			LINCOLN NC	86				
1	.8	1	.7	3	1.7	MC DOWELL NC	-15	151	113	90	83
		2	2.4	1	.7	MACON NC		-70			49
1	1.5	1	1.0	1	.9	MADISON NC	-35	-3	-37	31	58
				1	1.0	MARTIN NC					61
7	1.0	7	.6	18	1.2	MECKLENBURG NC	-38	94	21	66	70
2	3.4	1	1.2	1	.9	MITCHELL NC	-65	-27	-74	05	56
1	1.8			1	.9	MONTGOMERY NC			-51	19	56
1	.8	2	1.2	6	2.4	MOORE NC	55	99	208	96	92
3	2.1	4	2.0	3	1.1	NASH NC	-4	-46	-49	21	63
6*	3.4	2	.7	7	1.6	NEW HANOVER NC	-79	122	-53	18	81
		1	1.7			NORTHAMPTON NC					
1	.3	1	1.9	1	.8	ONSLOW NC	525	-59	155	94	52
		1	.6	1	.5	ORANGE NC		0			43
1	2.9					PAMLICO NC					
		1	1.2	3	2.6	PASQUOTANK NC		124			93
1	2.5					PENDER NC					
		1	3.5			PERQUIMANS NC					
1	1.5					PERSON NC					
4*	3.7	4	2.6	3	1.1	PITT NC	-29	-57	-69	07	66
				1	.9	POLK NC					58
4	2.2	4	1.4	2	.5	RANDOLPH NC	-35	-61	-75	04	43
1	.8	1	1.2	3	1.5	RICHMOND NC	52	29	97	88	79
1	1.6	3	1.6	3	1.4	ROBESON NC	-2	-14	-16	46	74
2	1.1	2	.6	3	.7	ROCKINGHAM NC	-40	11	-33	34	50
4	1.4	4	1.1	1	.2	ROWAN NC	-27	-82	-87	00	36
2	1.2	2	.9	3	1.0	RUTHERFORD NC	-20	12	-11	50	63
2	1.5	4	2.5	5	2.5	SAMPSON NC	74	1	75	84	93
3*	7.0	2	2.9	4*	4.5	SCOTLAND NC	-59	54	-36	32	98
		1	.6	4	1.6	STANLY NC		177			81
1	1.2	1	.9	2	1.4	STOKES NC	-21	46	15	63	74
1	.6	5	2.3	3	1.0	SURRY NC	271	-55	66	82	61
		1	2.6			SWAIN NC					
				1	.8	TRANSYLVANIA NC					53
1	.7	1	.6	2	.8	UNION NC	-16	33	11	61	52
3*	3.2	2	2.0	3	2.1	VANCE NC	-37	6	-33	35	90
9*	2.1	8	1.2	11	1.2	WAKE NC	-42	-2	-43	26	69
				1	2.1	WARREN NC					89
2	1.3	4	2.0	4	1.4	WAYNE NC	55	-31	7	58	75
1	.5	2	.9	1	.3	WILKES NC	65	-63	-39	29	38
1	1.3	5*	2.9	6*	2.5	WILSON NC	130	-14	98	88	93
		3	2.8	3	1.8	YADKIN NC		-35			85
				2	2.1	YANCEY NC					90
13	.6	19	.6	43	1.2	NORTH DAKOTA ND	8	91	106		
1*	14.9			1	4.4	ADAMS ND			-70	07	98
1	1.2					BAPNES ND					
				1	1.8	BENSON ND					86
1	1.9			1	1.2	BOTTINEAU ND			-39	29	67
		2*	8.3	1	3.8	BOWMAN ND		-55			97
				1	2.5	BURKE ND					93

* SIGNIFICANTLY MORE DEATHS THAN EXPECTED

WHITE FEMALE — CANCER OF THE ESOPHAGUS — ICD 150

1950-1959 DEATHS	1950-1959 RATE	1960-1969 DEATHS	1960-1969 RATE	1970-1979 DEATHS	1970-1979 RATE	COUNTY AND STATE		PERCENT CHANGE 60S/50S	PERCENT CHANGE 70S/60S	PERCENT CHANGE 70S/50S	PERCENTILE %CHANGE 70S/50S	PERCENTILE RATES 1970S
2	1.8	1	.6	3	1.5	BURLEIGH	ND	-66	135	-20	44	77
1	.6			7	1.6	CASS	ND			179	95	80
				1	1.7	CAVALIER	ND					83
				1	1.1	DICKEY	ND					64
		1*	5.0			DUNN	ND					
				1	3.6	FOSTER	ND					97
1	.5	1	.5	2	.9	GRAND FORKS	ND	4	71	78	85	55
				2*	6.9	GRIGGS	ND					99
				1	2.0	LA MOURE	ND					88
1	3.6					MC INTOSH	ND					
				2	2.7	MC LEAN	ND					94
		3*	2.9			MORTON	ND					
1	2.7	1	2.2			MOUNTRAIL	ND	-17				
		1	1.7	3*	3.9	NELSON	ND		126			97
1	1.7	2	2.5			PEMBINA	ND	49				
		2	2.5	1	1.1	RAMSEY	ND		-58			63
		1	1.5			RANSOM	ND					
1*	3.9			1	3.3	RENVILLE	ND			-13	48	96
				1	2.4	ROLETTE	ND					92
				1	2.5	SARGENT	ND					93
				2*	8.0	STEELE	ND					99
				3	2.1	STUTSMAN	ND					89
1	3.9			1	4.3	TOWNER	ND			11	61	98
				1	.5	TRAILL	ND					41
		2	1.8	2	1.8	WALSH	ND		2			86
1	.7	1	.4	2	.8	WARD	ND	-35	75	14	62	51
		1	.8			WILLIAMS	ND					
375	.9	474	.9	672	1.1	OHIO	OH	3	20	23		
		1	.7			ADAMS	OH					
4	.8	6	1.1	7	1.2	ALLEN	OH	27	15	45	77	69
				2	.8	ASHLAND	OH					51
3	.7	3	.5	6	1.1	ASHTABULA	OH	-24	99	51	78	63
3	1.3	1	.4	3	1.1	ATHENS	OH	-67	169	-11	50	66
2	.9	2	1.1	1	.3	AUGLAIZE	OH	20	-77	-72	06	37
4	.8	5	.9	9	1.5	BELMONT	OH	12	64	85	86	79
1	.8			2	.9	BROWN	OH			18	64	58
4	.6	7	.8	16	1.4	BUTLER	OH	40	77	148	93	75
2	1.9	1	.7			CARROLL	OH	-62				
1	.7	2	1.2	4	2.0	CHAMPAIGN	OH	59	73	174	95	89
7	1.2	9	1.3	7	.7	CLARK	OH	2	-45	-44	25	48
1	.2	5	1.6	3	.7	CLERMONT	OH	632	-58	207	96	47
3	2.0			3	1.4	CLINTON	OH			-26	40	77
5	1.0	4	.7	9	1.3	COLUMBIANA	OH	-28	88	36	72	72
2	1.2	2	1.0	3	1.4	COSHOCTON	OH	-12	35	19	65	75
3	1.2	3	1.2	3	.9	CRAWFORD	OH	2	-27	-26	40	56
72	1.1	103	1.3	139*	1.5	CUYAHOGA	OH	16	19	38	73	79
2	.7			3	1.0	DARKE	OH			52	78	62
1	.6			2	1.1	DEFIANCE	OH			83	85	66
2	1.1			3	1.2	DELAWARE	OH			9	60	68
		2	.6	1	.2	ERIE	OH		-66			36
2	.6	2	.5	6	1.3	FAIRFIELD	OH	-25	168	102	89	72
1	.6	1	.5	1	.7	FAYETTE	OH	-18	47	21	66	48
23	.9	31	1.0	48	1.2	FRANKLIN	OH	8	27	37	73	70
		2	1.2	4	1.7	FULTON	OH		42			83
1	.7	1	.6			GALLIA	OH	-12				
		1	.5	3	.9	GEAUGA	OH		66			56
2	.9	5	1.3	2	.5	GREENE	OH	55	-64	-44	25	41
2	.7	2	.6	2	.8	GUERNSEY	OH	-8	31	21	66	54
40	1.0	63	1.4	61	1.2	HAMILTON	OH	36	-13	19	65	69
3	1.0			6	1.6	HANCOCK	OH			61	81	82
		2	1.3	3	1.2	HARDIN	OH		-11			67
		1	.8	1	.7	HARRISON	OH		-10			49
4	2.2	1	.5	1	.3	HIGHLAND	OH	-77	-31	-84	00	38
1	.9	1	.7	5*	3.4	HOCKING	OH	-13	362	302	98	96
				1	.9	HOLMES	OH					57
2	.8					HURON	OH					
2	1.2	1	.7	1	.3	JACKSON	OH	-41	-56	-74	05	37
10*	2.6	4	.8	11	1.8	JEFFERSON	OH	-68	121	-30	37	86
1	.6	3	1.2	4	1.4	KNOX	OH	109	14	138	92	74
4	1.0	5	.8	8	.9	LAKE	OH	-26	21	-11	50	57
1	.5			5	1.4	LAWRENCE	OH			172	95	76
3	.7	5	.9	7	1.1	LICKING	OH	29	23	58	80	65
		1	.4	3	1.1	LOGAN	OH		195			63
4	.7	6	.6	18	1.5	LORAIN	OH	-11	146	118	91	79
18	.9	26	1.0	42	1.4	LUCAS	OH	22	38	68	83	77
2	2.1	1	.9			MADISON	OH	-58				
14	1.3	16	1.0	17	1.0	MAHONING	OH	-19	-4	-22	43	60
4	1.3	3	.8	1	.3	MARION	OH	-37	-62	-76	04	38
		2	.5	7	1.7	MEDINA	OH		239			85
		2	1.8	1	.6	MEIGS	OH		-65			46
		2	1.0	1	.3	MERCER	OH		-67			38
9*	2.7	3	.7	5	1.0	MIAMI	OH	-73	42	-62	13	61
1	.8					MONROE	OH					
21	1.2	19	.8	20	.7	MONTGOMERY	OH	-33	-13	-42	27	49
3	3.0					MORROW	OH					
2	.5	4	1.0	6	1.0	MUSKINGUM	OH	104	4	112	90	63
1	1.2	1	1.2			NOBLE	OH	5				
1	.6	1	.4	2	.8	OTTAWA	OH	-22	80	40	75	53
		1	.9			PAULDING	OH					
1	.5					PERRY	OH					
1	.7	1	.6	2	1.0	PICKAWAY	OH	-18	78	46	77	63
		2	2.2	2	1.5	PIKE	OH		-33			77
2	.7	3	.7	7	1.3	PORTAGE	OH	5	85	95	87	71
1	.7			5	2.3	PREBLE	OH			221	97	91
4	.9	4	.7	8	1.3	RICHLAND	OH	-19	67	36	73	70
4	1.4	3	1.0			ROSS	OH	-29				
1	.3	5	1.6	3	1.0	SANDUSKY	OH	404	-36	222	97	62
3	.8	2	.4	4	.6	SCIOTO	OH	-49	56	-19	44	45
		1	.4	5	1.3	SENECA	OH		260			71

* SIGNIFICANTLY MORE DEATHS THAN EXPECTED

WHITE FEMALE — CANCER OF THE ESOPHAGUS — ICD 150

1950-1959 DEATHS	1950-1959 RATE	1960-1969 DEATHS	1960-1969 RATE	1970-1979 DEATHS	1970-1979 RATE	COUNTY AND STATE	PERCENT CHANGE 60S/50S	PERCENT CHANGE 70S/60S	PERCENT CHANGE 70S/50S	PERCENTILE %CHANGE 70S/50S	PERCENTILE RATES 1970S
				1	.5	SHELBY OH					40
16	1.0	24	1.3	29	1.2	STARK OH	27	-8	18	64	69
17	1.0	24	1.0	40	1.3	SUMMIT OH	2	36	39	74	74
7	1.3	9	.9	10	.8	TRUMBULL OH	-28	-15	-39	29	51
2	.5	3	.7	6	1.2	TUSCARAWAS OH	61	63	163	94	68
3	1.9			4	2.4	UNION OH			28	69	92
2	1.0	4	2.0	3	1.3	VAN WERT OH	92	-35	25	68	72
				1	1.7	VINTON OH					84
2	.9	5	1.9	1	.3	WARREN OH	115	-82	-62	12	38
1	.3	5	1.4	1	.3	WASHINGTON OH	345	-82	-20	44	37
2	.6			5	1.0	WAYNE OH			59	80	61
		3	1.5			WILLIAMS OH					
1	.3			6	1.3	WOOD OH			289	98	72
1	.6	1	.6			WYANDOT OH	-4				
97	.9	106	.8	142	.9	OKLAHOMA OK	-17	11	-7		
		1	1.9			ADAIR OK					
		1	1.2			ALFALFA OK					
				1	1.5	ATOKA OK					79
3	2.5	4	2.9			BECKHAM OK	16				
		2	3.0			BLAINE OK					
1	.5	1	.5	2	.8	BRYAN OK	-7	57	46	77	52
		1	.4			CADDO OK					
		2	1.3	1	.3	CANADIAN OK		-79			37
3	1.5	1	.4	1	.3	CARTER OK	-75	-16	-79	02	37
1	1.5	2	2.1			CHEROKEE OK	41				
2	2.1					CHOCTAW OK					
3	1.8	1	.4	1	.3	CLEVELAND OK	-79	-12	-81	01	38
1	2.6	3*	6.0	1	1.2	COAL OK	128	-80	-53	18	69
1	.7	2	.8	3	.9	COMANCHE OK	11	19	31	70	57
1	1.8					COTTON OK					
1	1.0	4*	3.4	1	1.3	CRAIG OK	253	-62	33	71	71
2	.9	2	.7	1	.2	CREEK OK	-23	-63	-72	06	36
1	1.1	1	.9	1	.6	CUSTER OK	-22	-29	-45	24	44
				1	.6	DELAWARE OK					45
		2*	2.8	1	1.5	DEWEY OK		-46			80
5	1.7	2	.5	3	1.0	GARFIELD OK	-70	96	-41	28	60
1	.7			1	.5	GARVIN OK			-22	43	42
2	1.1	3	1.2	1	.4	GRADY OK	9	-65	-62	12	39
				1	.9	GRANT OK					57
				2	2.6	GREER OK					94
		1	1.2			HASKELL OK					
		1	.7			HUGHES OK					
		2	1.5			JACKSON OK					
1	2.0	1	1.2	1	.7	JEFFERSON OK	-41	-44	-67	09	46
1	2.2			1	1.2	JOHNSTON OK			-43	26	69
2	.8	2	.7	8	2.1	KAY OK	-16	199	152	93	90
				1	.8	KINGFISHER OK					51
1	.9			1	.8	KIOWA OK			-11	50	52
1	2.4					LATIMER OK					
1	.6	2	1.0	2	.6	LE FLORE OK	73	-34	14	62	46
1	.9					LINCOLN OK					
				1	2.0	LOVE OK					89
2	3.8	1	1.3			MC CLAIN OK	-66				
1	.9	1	.6	2	1.1	MC CURTAIN OK	-42	97	15	63	64
				2	3.2	MAJOR OK					96
				2	1.2	MAYES OK					69
1	1.3			1	.7	MURRAY OK			-43	26	50
2	.7	2	.5	7	1.9	MUSKOGEE OK	-24	265	178	95	87
				1	.8	NOWATA OK					54
1	1.4			2	3.1	OKFUSKEE OK			112	90	95
16	1.1	24	1.1	28	1.0	OKLAHOMA OK	5	-9	-4	54	61
1	.5	2	.7	5	1.6	OKMULGEE OK	42	128	224	97	82
				4	1.7	OSAGE OK					83
2	1.5	2	.9			OTTAWA OK	-36				
1	1.2			1	.6	PAWNEE OK			-51	20	44
4	2.1			1	.4	PAYNE OK			-81	01	39
3	1.5	4	1.6	4	1.6	PITTSBURG OK	5	3	8	59	82
3	1.9	1	.5	2	.6	PONTOTOC OK	-76	37	-67	09	45
		2	1.1	5	1.4	POTTAWATOMIE OK		31			75
		1	1.4			PUSHMATAHA OK					
1	.9			2	1.2	ROGERS OK			35	72	68
1	.8	1	.6	1	.4	SEMINOLE OK	-18	-36	-47	22	39
1	1.1	3	2.8			SEQUOYAH OK	155				
2	1.2	2	.9	2	.8	STEPHENS OK	-29	-12	-38	30	52
1	1.6			1	1.3	TEXAS OK			-21	43	71
1	1.1	1	.7	1	1.1	TILLMAN OK	-32	51	3	57	64
14	1.3	8	.5	25	1.2	TULSA OK	-63	142	-10	51	67
				1	.7	WAGONER OK					47
		5	2.2	2	.8	WASHINGTON OK		-64			52
1	1.2	1	.8	1	.7	WASHITA OK	-31	-18	-44	25	46
2	2.0					WOODS OK					
		1	1.0	1	.7	WOODWARD OK		-29			49
82	1.1	93	.9	155	1.1	OREGON OR	-18	28	6		
				1	1.1	BAKER OR					66
				4	1.7	BENTON OR					84
3	.7	6	.8	9	.8	CLACKAMAS OR	20	-4	15	62	52
1	.6	1	.6	4	1.9	CLATSOP OR	2	203	209	96	87
				2	1.0	COLUMBIA OR					62
2	1.4	1	.4	6	1.7	COOS OR	-74	361	18	64	83
				1	1.5	CROOK OR					77
				1	.7	CURRY OR					48
				1	.4	DESCHUTES OR					39
		5	1.7	3	.7	DOUGLAS OR		-60			47
		2*	5.0			GRANT OR					
1	1.7			2	2.6	HOOD RIVER OR			50	78	93
1	.2	5	1.1	7	1.0	JACKSON OR	382	-4	361	99	61
1*	4.0			1	3.6	JEFFERSON OR			-11	50	97

* SIGNIFICANTLY MORE DEATHS THAN EXPECTED

WHITE FEMALE | CANCER OF THE ESOPHAGUS | ICD 150

1950-1959 DEATHS	1950-1959 RATE	1960-1969 DEATHS	1960-1969 RATE	1970-1979 DEATHS	1970-1979 RATE	COUNTY AND STATE	PERCENT CHANGE 60S/50S	PERCENT CHANGE 70S/60S	PERCENT CHANGE 70S/50S	PERCENTILE %CHANGE 70S/50S	PERCENTILE RATES 1970S
		1	.5	5	1.2	JOSEPHINE OR		147			69
2	1.5	2	1.1	2	.6	KLAMATH OR	-28	-40	-56	16	46
		1	3.4			LAKE OR					
6	1.1	8	.9	11	.9	LANE OR	-11	-5	-16	46	57
1	.9	1	.6	5	2.4	LINCOLN OR	-34	307	167	94	92
3	1.3			8	1.7	LINN OR			32	71	65
10	1.6	3	.4	10	.9	MARION OR	-78	147	-45	24	56
		2*	7.0			MORROW OR					
39	1.4	39	1.1	48	1.3	MULTNOMAH OR	-19	19	-3	54	72
				1	.3	POLK OR					37
		1	.8	5*	4.1	TILLAMOOK OR		390			98
4	2.0	2	.7	4	1.4	UMATILLA OR	-65	90	-33	35	74
		1	2.0			WALLOWA OR					
		3	2.7	2	1.5	WASCO OR		-46			78
7	2.3	6	1.0	10	1.1	WASHINGTON OR	-56	7	-53	19	63
1	.4	3	1.1	2	.7	YAMHILL OR	161	-43	49	78	47
535	1.0	719	1.1	817	1.1	PENNSYLVANIA PA	11	-5	6		
				2	.4	ADAMS PA					40
81	1.2	106	1.2	125	1.2	ALLEGHENY PA	3	3	6	58	70
		4	1.0	5	.8	ARMSTRONG PA		-14			54
7	1.2	8	.8	12	.9	BEAVER PA	-33	18	-21	43	59
1	.5	2	.8	1	.4	BEDFORD PA	58	-54	-27	39	38
8	.5	12	.7	18	.8	BERKS PA	34	24	65	82	54
9	1.1	5	.6	6	.5	BLAIR PA	-51	-4	-53	18	42
5	1.8	3	1.0	5	1.1	BRADFORD PA	-46	18	-36	32	66
7	.7	17	1.3	25	1.3	BUCKS PA	97	-3	90	87	72
2	.5	5	.8	10	1.3	BUTLER PA	78	55	175	95	71
4	.4	5	.5	9	.7	CAMBRIA PA	6	51	60	81	48
1	.7	2	.6	4	.9	CARBON PA	-7	51	41	75	58
2	.7	1	.3	2	.5	CENTRE PA	-59	58	-35	33	40
8	1.0	11	1.0	13	.9	CHESTER PA	7	-10	-4	54	58
1	.8	3	1.5	2	.9	CLARION PA	93	-39	18	65	57
2	.5	7	1.5	5	.9	CLEARFIELD PA	209	-42	79	85	55
		3	1.4			CLINTON PA					
1	.3	1	.3	7	1.8	COLUMBIA PA	-21	601	452	99	85
5	1.1	5	.9	9	1.5	CRAWFORD PA	-14	60	37	73	79
1	.2	6	.8	13	1.3	CUMBERLAND PA	360	71	687	99	74
2	.2	14	1.0	15	1.0	DAUPHIN PA	507	0	507	99	62
18	.9	35	1.2	54	1.5	DELAWARE PA	41	24	74	84	79
		2	1.0	4	1.7	ELK PA		75			85
14	1.2	14	1.0	20	1.2	ERIE PA	-17	14	-6	53	67
4	.5	12	1.3	7	.7	FAYETTE PA	150	-50	25	68	47
		4	.7	6	.9	FRANKLIN PA		37			58
2	.8	4	1.5	4	1.5	GREENE PA	82	-2	77	85	78
2	1.0	1	.4	2	.8	HUNTINGDON PA	-59	87	-24	41	52
		4	1.0	5	1.0	INDIANA PA		1			60
3	1.2	3	1.0	1	.2	JEFFERSON PA	-16	-77	-81	02	36
		2	2.2			JUNIATA PA					
18	1.4	20	1.2	21	1.1	LACKAWANNA PA	-10	-8	-18	45	65
12	.9	18	1.0	19	.9	LANCASTER PA	19	-11	6	58	57
11*	2.1	4	.7	10	1.3	LAWRENCE PA	-67	93	-37	31	73
5	1.2	3	.5	3	.3	LEBANON PA	-55	-38	-72	06	38
7	.6	13	.9	17	.9	LEHIGH PA	57	2	60	81	57
18	.9	40*	1.7	25	.9	LUZERNE PA	85	-46	0	55	57
5	.8	6	.9	6	.8	LYCOMING PA	8	-10	-2	54	52
7*	2.2	2	.6	3	.8	MC KEAN PA	-75	36	-66	10	51
8	1.3	10	1.4	6	.7	MERCER PA	10	-50	-45	24	49
1	.5	1	.4	1	.3	MIFFLIN PA	-11	-31	-39	29	37
		3	1.1	3	.9	MONROE PA		-16			58
23	1.1	29	.9	50	1.2	MONTGOMERY PA	-16	28	8	59	67
1	.8	1	.8	1	.8	MONTOUR PA	9	-9	-1	55	51
9	.9	14	1.2	21	1.5	NORTHAMPTON PA	39	17	63	82	77
3	.5	4	.6	6	.7	NORTHUMBERLAND PA	24	32	64	82	50
				1	.6	PERRY PA					43
153*	1.6	177*	1.8	141	1.4	PHILADELPHIA PA	10	-20	-13	48	76
1	1.3	1	1.0			PIKE PA	-28				
		1	.9	1	1.0	POTTER PA		18			61
8	.8	9	.8	15	1.1	SCHUYLKILL PA	4	26	31	70	63
1	.8	2	1.3	1	.7	SNYDER PA	61	-44	-10	51	50
4	1.1	3	.7	4	.6	SOMERSET PA	-42	-10	-48	22	44
1	2.4					SULLIVAN PA					
		2	.9	4	1.5	SUSQUEHANNA PA		70			78
1	.5	2	.9			TIOGA PA	81				
				1	.6	UNION PA					46
2	.6	3	.8	6	1.4	VENANGO PA	47	69	148	93	75
5	1.8			3	.9	WARREN PA			-51	20	56
10	1.0	13	1.1	13	.9	WASHINGTON PA	11	-17	-8	52	58
3	1.7			4	1.6	WAYNE PA			-9	51	80
19	1.4	23	1.2	21	.9	WESTMORELAND PA	-11	-29	-37	31	55
				1	.7	WYOMING PA					47
9	.8	14	1.0	18	1.0	YORK PA	17	7	25	68	62
51	1.1	65	1.2	96	1.5	RHODE ISLAND RI	6	27	34		
		2	.9	3	1.1	BRISTOL RI		26			64
4	1.1	7	1.1	11	1.2	KENT RI	-2	14	11	61	69
9*	3.0	4	1.2	9	2.0	NEWPORT RI	-61	73	-33	34	88
36	1.1	48	1.2	64	1.5	PROVIDENCE RI	15	20	38	74	77
2	.8	4	1.2	9	2.0	WASHINGTON RI	48	64	144	93	89
73	1.4	102*	1.4	114	1.1	SOUTH CAROLINA SC	-2	-23	-24		
1	1.4	3*	3.7			ABBEVILLE SC	159				
3	2.2	4	1.6	2	.5	AIKEN SC	-28	-70	-78	03	41
		1	4.6			ALLENDALE SC					
9*	3.1	4	.9	4	.7	ANDERSON SC	-69	-25	-77	03	49
		1	2.7	1	2.1	BAMBERG SC		-21			90
1	3.4	1	2.0			BARNWELL SC	-42				
1	2.3	1	1.4	3	2.6	BEAUFORT SC	-38	83	13	61	93

* SIGNIFICANTLY MORE DEATHS THAN EXPECTED

WHITE FEMALE — CANCER OF THE ESOPHAGUS — ICD 150

1950-1959 DEATHS	1950-1959 RATE	1960-1969 DEATHS	1960-1969 RATE	1970-1979 DEATHS	1970-1979 RATE	COUNTY AND STATE		PERCENT CHANGE 60S/50S	PERCENT CHANGE 70S/60S	PERCENT CHANGE 70S/50S	PERCENTILE %CHANGE 70S/50S	PERCENTILE RATES 1970S
1	1.6					BERKELEY	SC					
6	1.4	9	2.0	9	1.2	CHARLESTON	SC	39	-42	-19	45	67
1	.9	1	.7	2	1.1	CHEROKEE	SC	-22	51	17	64	64
2	2.6	4*	3.8			CHESTER	SC	44				
2	1.9	1	1.0	1	.7	CHESTERFIELD	SC	-50	-28	-64	11	48
		1	1.2			COLLETON	SC					
2	2.5	1	.8	1	.6	DARLINGTON	SC	-70	-24	-77	03	44
3*	4.9			3	2.7	DILLON	SC			-45	24	94
1	2.2					DORCHESTER	SC					
1	.6	7*	2.9	2	.7	FLORENCE	SC	396	-75	25	67	50
		2	2.4			GEORGETOWN	SC					
9	1.9	13	1.5	14	1.1	GREENVILLE	SC	-21	-27	-42	27	66
1	.8	3	1.6	1	.4	GREENWOOD	SC	105	-73	-45	24	40
		2	1.0	5	1.3	HORRY	SC		28			72
				1	2.6	JASPER	SC					93
1	1.6	1	1.0	6*	4.8	KERSHAW	SC	-35	368	204	96	99
		2	1.7	4	2.3	LANCASTER	SC		34			91
1	.7			3	1.0	LAURENS	SC			43	76	62
1	2.4	1	2.3	1	2.0	LEE	SC	-1	-16	-17	45	88
1	.6	2	.9	3	.7	LEXINGTON	SC	51	-20	20	66	49
		1*	5.7	1	3.2	MC CORMICK	SC		-44			96
1	1.6			2	1.7	MARION	SC			7	58	85
2	4.1	1	1.4			MARLBORO	SC	-67				
		2	1.4			NEWBERRY	SC					
		2	1.3	3	1.4	OCONEE	SC		11			77
1	1.0	4	2.3	2	.9	ORANGEBURG	SC	127	-63	-17	45	55
		3	1.4	5	1.5	PICKENS	SC		12			80
9*	1.8	8	1.3	14	1.8	RICHLAND	SC	-28	37	-2	55	86
		1	1.7			SALUDA	SC					
7	1.5	7	1.1	10	1.1	SPARTANBURG	SC	-24	-2	-26	40	66
1	1.7	2	1.4	2	1.2	SUMTER	SC	-17	-15	-29	38	68
1	.8	2	1.9	3	1.9	UNION	SC	131	3	137	92	87
		1	1.4	1	1.2	WILLIAMSBURG	SC		-13			67
3	1.7	3	1.1	5	1.3	YORK	SC	-34	18	-22	43	73
27	1.0	20	.6	29	.6	SOUTH DAKOTA	SD	-42	13	-35		
		2	1.7	1	.7	BEADLE	SD		-61			47
		1*	9.4			BENNETT	SD					
1	2.1	1	1.6	1	.9	BON HOMME	SD	-23	-43	-56	16	57
4*	5.5	1	.7			BROOKINGS	SD	-87				
2	1.6			1	.6	BROWN	SD			-64	12	43
				1	1.6	BRULE	SD					81
				1	1.8	BUTTE	SD					86
1	3.7			2*	3.4	CLARK	SD			-8	52	96
		1	1.8			CLAY	SD					
		1	3.4			CUSTER	SD					
1	2.2	1	1.4	1	1.0	DAY	SD	-38	-25	-54	18	61
				1	1.6	DOUGLAS	SD					81
1	4.9					EDMUNDS	SD					
		1	2.2			FALL RIVER	SD					
				1*	5.7	HAAKON	SD					99
1	4.7			1	1.9	HAND	SD			-60	13	87
		1	1.2			HUTCHINSON	SD					
		1*	6.3			HYDE	SD					
1	1.7					LAKE	SD					
		1	1.2			LAWRENCE	SD					
				1	.8	LINCOLN	SD					53
1	2.0					MEADE	SD					
3*	12.1			1*	2.6	MINER	SD			-79	02	93
6	1.7	2	.4	5	.9	MINNEHAHA	SD	-74	104	-48	22	56
1	2.3					MOODY	SD					
		3	1.7	3	1.0	PENNINGTON	SD		-37			62
				1	3.4	PERKINS	SD					96
1	1.2					SPINK	SD					
1	2.1			1	2.0	TRIPP	SD			-5	53	89
		2	2.0	3*	3.3	TURNER	SD		62			96
				1	1.6	UNION	SD					81
2*	6.4					WALWORTH	SD					
		1	.7	2	1.4	YANKTON	SD		109			77
131	1.0	192	1.2	230	1.0	TENNESSEE	TN	13	-11	0		
2	1.0	1	.5	3	1.0	ANDERSON	TN	-50	85	-8	52	59
		1	.7	1	.6	BEDFORD	TN		-18			44
		2	2.3	2	1.8	BENTON	TN		-19			86
		1	3.1			BLEDSOE	TN					
1	.5	6	2.5	6	1.5	BLOUNT	TN	373	-41	180	95	78
2	1.7			3	1.1	BRADLEY	TN			-36	32	64
3	2.8	1	.7	4	1.9	CAMPBELL	TN	-76	189	-30	37	88
2	1.6	3	1.8	2	.8	CARROLL	TN	14	-54	-47	22	54
1	.7	1	.5	2	.7	CARTER	TN	-39	47	-11	50	47
2*	5.0					CHEATHAM	TN					
				2	2.7	CHESTER	TN					94
				1	.8	CLAIBORNE	TN					52
				1	1.9	CLAY	TN					87
2	2.2	2	1.8	4	2.8	COCKE	TN	-20	58	27	69	95
2	1.9			1	.6	COFFEE	TN			-67	09	46
1	1.5			1	.7	CROCKETT	TN			-51	19	50
				1	.7	CUMBERLAND	TN					49
14	1.0	27	1.5	20	.9	DAVIDSON	TN	48	-40	-11	49	57
		1	2.0	2	4.1	DECATUR	TN		108			98
		2	2.5			DE KALB	TN					
		2	1.9	2	1.1	DICKSON	TN		-42			64
2	1.3	2	1.2	2	1.1	DYER	TN	-6	-8	-13	48	66
				2	3.6	FAYETTE	TN					97
				1	1.4	FENTRESS	TN					77
2	2.0	3	2.5	3	2.0	FRANKLIN	TN	24	-20	-1	55	88
2	.9	1	.3	3	.8	GIBSON	TN	-64	164	-6	53	54
		1	.7	2	1.1	GILES	TN		62			65
				1	1.1	GRAINGER	TN					63

* SIGNIFICANTLY MORE DEATHS THAN EXPECTED

1950-1959 DEATHS	1950-1959 RATE	1960-1969 DEATHS	1960-1969 RATE	1970-1979 DEATHS	1970-1979 RATE	COUNTY AND STATE	PERCENT CHANGE 60S/50S	PERCENT CHANGE 70S/60S	PERCENT CHANGE 70S/50S	PERCENTILE %CHANGE 70S/50S	PERCENTILE RATES 1970S
		3	1.5	5	1.7	GREENE TN		14			64
1	2.0	1	1.7	1	1.5	GRUNDY TN	-13	-9	-21	44	80
		2	1.5	1	.5	HAMBLEN TN		-68			41
13	1.6	16	1.5	19	1.4	HAMILTON TN	-9	-10	-17	45	74
				1	2.8	HANCOCK TN					94
1	1.4	1	1.1			HARDEMAN TN	-18				
		1	1.0			HARDIN TN					
3	2.5	4	2.7	2	.8	HAWKINS TN	10	-70	-67	08	53
2*	4.2			1	1.5	HAYWOOD TN			-64	11	79
2	2.6	1	1.0			HENDERSON TN	-62				
		1	2.1			HICKMAN TN					
1	2.6					HUMPHREYS TN					
2	2.1	3	2.6	2	1.0	JEFFERSON TN	25	-59	-49	21	63
				2	2.3	JOHNSON TN					91
6	.5	18	1.3	22	1.3	KNOX TN	148	-4	137	92	71
		1	2.2	1	2.5	LAKE TN		15			93
1	1.2			1	1.2	LAUDERDALE TN			2	57	69
		2	1.3	1	.7	LAWRENCE TN		-47			47
		1	3.3	1	1.8	LEWIS TN		-45			86
2	1.8			1	.5	LINCOLN TN			-74	05	41
1	1.0	1	.7	2	1.3	LOUDON TN	-35	92	25	68	71
4*	3.0	3	1.9	3	1.3	MC MINN TN	-35	-35	-58	15	71
		1	.8	2	1.1	MC NAIRY TN		33			63
		2	.7	5	1.2	MADISON TN		83			69
1	1.3			3	2.4	MARION TN			86	86	92
2	2.2	1	1.2			MARSHALL TN	-46				
2	1.4	3	1.7	2	.7	MAURY TN	24	-59	-49	21	49
2	2.1	1	.8	2	1.1	MONROE TN	-61	39	-46	23	65
3	2.4	2	1.2	3	1.0	MONTGOMERY TN	-48	-16	-56	16	62
1	1.8					MORGAN TN					
1	.7			3	1.4	OBION TN			97	88	75
				1	1.5	POLK TN					77
2	1.5	1	.6	1	.3	PUTNAM TN	-60	-47	-79	02	37
		1	1.2	1	.7	RHEA TN		-37			50
		1	.5	3	1.1	ROANE TN		115			66
		2	1.3	2	1.1	ROBERTSON TN		-12			65
2	1.4	4	1.9			RUTHERFORD TN	38				
				2	1.9	SCOTT TN					87
		3	2.4	1	.9	SEVIER TN		-64			55
18	1.3	28	1.4	41	1.6	SHELBY TN	9	15	25	68	81
4*	6.1	1	1.7	2	1.9	SMITH TN	-72	8	-69	07	87
		1	2.5			STEWART TN					
7	2.2	5	1.1	6	.8	SULLIVAN TN	-50	-24	-62	12	54
		4	1.9	1	.3	SUMNER TN		-84			37
1	1.2	1	1.1			TIPTON TN	-8				
		1	2.9	2*	5.6	TROUSDALE TN		93			99
		1	1.4	1	.7	UNICOI TN		-51			47
				1	1.6	UNION TN					82
1	1.1	3	2.2			WARREN TN	110				
5	2.3	3	.9	3	.7	WASHINGTON TN	-60	-21	-68	08	49
		1	.8			WEAKLEY TN					
		1	1.2			WHITE TN					
		1	.7	1	.5	WILLIAMSON TN		-34			40
2	1.5	3	1.8	3	1.2	WILSON TN	22	-35	-21	43	67
409*	1.4	436	1.0	622	1.1	TEXAS TX	-26	8	-20		
3	2.1	1	.6	2	.9	ANDERSON TX	-70	52	-54	17	58
				1	2.0	ANDREWS TX					88
		1	.5	7	2.6	ANGELINA TX		397			93
				1	1.4	ARANSAS TX					75
1	1.8	1	1.1	1	.6	ATASCOSA TX	-38	-50	-69	08	43
				1	2.4	BAILEY TX					92
1	1.1	1	1.1	1	.8	BASTROP TX	2	-31	-29	37	51
2	3.2					BEE TX					
2	.7	3	.8	3	.6	BELL TX	16	-24	-12	49	44
31	1.4	45	1.5	55	1.4	BEXAR TX	4	-8	-4	54	75
		1	2.7			BLANCO TX					
				4	2.1	BOSQUE TX					89
6*	2.9	1	.3	1	.3	BOWIE TX	-88	-24	-91	00	37
1	.4	5	1.5	6	1.5	BRAZORIA TX	257	-3	247	97	78
2	2.4			3	1.2	BRAZOS TX			-51	19	68
				1	2.5	BROOKS TX					92
3	1.8	2	.9			BROWN TX	-52				
1	2.2					BURLESON TX					
1	1.8	3*	4.0			BURNET TX	122				
				1	.7	CALDWELL TX					48
1	3.2					CALHOUN TX					
		1	1.2	2	2.3	CALLAHAN TX		97			91
1	.3	8	1.5	13	1.7	CAMERON TX	445	17	538	99	84
2	3.3	1	.9	1	.6	CASS TX	-72	-36	-82	01	43
		1	7.4			CASTRO TX					
1	.6			1	.3	CHEROKEE TX			-49	21	37
		2	3.3			CHILDRESS TX					
		1	1.2			CLAY TX					
				2*	7.9	COCHRAN TX					99
1	1.0	1	.8			COLEMAN TX	-17				
3	1.2	5	1.6	4	1.0	COLLIN TX	32	-41	-22	43	59
				2	1.6	COLORADO TX					81
1	1.0			2	1.0	COMAL TX			2	56	62
		2	1.4	2	1.0	COMANCHE TX		-24			62
1	.9	4*	2.5	3	1.5	COOKE TX	187	-41	68	83	78
1	.9	1	.8	1	.7	CORYELL TX	-12	-17	-28	39	47
		1	2.5	1	1.7	CROSBY TX		-31			84
		1	3.0			DALLAM TX					
44*	1.7	38	.9	64	1.2	DALLAS TX	-48	30	-32	35	67
1	3.4					DEAF SMITH TX					
		1	1.7			DELTA TX					
1	.4	4	1.5	4	.9	DENTON TX	236	-36	115	90	58
		3	2.4	2	1.4	DE WITT TX		-40			76
2*	5.2	1	1.7	1	1.7	DUVAL TX	-67	-2	-68	08	83

* SIGNIFICANTLY MORE DEATHS THAN EXPECTED

WHITE FEMALE — CANCER OF THE ESOPHAGUS — ICD 150

1950-1959 DEATHS	1950-1959 RATE	1960-1969 DEATHS	1960-1969 RATE	1970-1979 DEATHS	1970-1979 RATE	COUNTY AND STATE	PERCENT CHANGE 60S/50S	PERCENT CHANGE 70S/60S	PERCENT CHANGE 70S/50S	PERCENTILE %CHANGE 70S/50S	PERCENTILE RATES 1970S
1	.6	1	1.1	2	1.0	EASTLAND TX	83	-14	57	79	60
1	1.7	2	1.1	3	.7	ECTOR TX	-34	-35	-57	15	49
				1*	8.2	EDWARDS TX					99
				5	1.3	ELLIS TX					70
9	1.2	8	.8	14	1.1	EL PASO TX	-33	38	-7	52	65
4*	3.4	3	1.7			ERATH TX	-49				
				2	1.4	FALLS TX					76
4	2.1	1	.5	4	1.5	FANNIN TX	-78	207	-31	36	77
1	.8	1	.5	1	.4	FAYETTE TX	-34	-19	-47	23	40
2*	4.7					FISHER TX					
1	2.5			2	3.4	FLOYD TX			36	73	96
		1	.7	1	.5	FORT BEND TX		-28			42
		1	2.0			FRANKLIN TX					
1	1.2			1	.7	FREESTONE TX			-46	23	48
1	2.3					FRIO TX					
		1	3.9			GAINES TX					
5	1.5	7	1.5	10	1.3	GALVESTON TX	-5	-8	-13	48	73
				1	3.0	GARZA TX					95
		1	.9	2	1.2	GILLESPIE TX		34			68
		2*	6.3	1	3.2	GOLIAD TX		-49			96
1	1.0					GONZALES TX					
1	1.5	1	.7	2	1.0	GRAY TX	-55	53	-31	36	61
4	1.0	9	1.6	2	.3	GRAYSON TX	57	-83	-73	05	37
3	2.1	4	1.2	8	1.9	GREGG TX	-40	51	-9	51	87
1	2.4			1	1.3	GRIMES TX			-48	22	70
		3	1.9	4	1.8	GUADALUPE TX		-3			86
1	.7			2	1.0	HALE TX			40	75	61
				2	4.2	HALL TX					98
		2	3.8			HAMILTON TX					
				2	2.3	HARDEMAN TX					91
		1	.8	3	1.7	HARDIN TX		109			84
46*	1.7	56	1.2	83	1.3	HARRIS TX	-29	8	-24	42	73
5*	4.1	1	.6	2	1.0	HARRISON TX	-85	59	-75	04	61
		1	1.2	3*	4.4	HASKELL TX		264			98
1	1.1			3	2.1	HAYS TX			84	86	90
				1	4.6	HEMPHILL TX					98
2	1.6	2	1.6	3	1.0	HENDERSON TX	-3	-34	-35	33	63
5	1.0	15*	2.4	11	1.2	HIDALGO TX	136	-50	18	64	68
3	1.8	2	.7			HILL TX	-62				
1	1.0	1	1.0	1	1.1	HOCKLEY TX	10	8	18	65	66
				2	2.3	HOOD TX					91
1	.7	1	.7	1	.5	HOPKINS TX	5	-28	-24	41	42
1	1.1	1	1.0	4*	3.0	HOUSTON TX	-8	199	175	95	95
3	2.6	2	1.4			HOWARD TX	-47				
1*	16.7					HUDSPETH TX					
3	1.1	1	.3	2	.5	HUNT TX	-77	88	-56	16	41
2	3.3	1	.6			HUTCHINSON TX	-80				
1	2.2			2	2.9	JACK TX			36	72	95
1	2.2	2	3.6			JACKSON TX	62				
1	1.5	1	1.0	3	1.7	JASPER TX	-31	70	18	64	85
10	1.7	5	.6	11	1.0	JEFFERSON TX	-66	80	-39	29	62
1*	11.2					JIM HOGG TX					
1	1.5	5*	4.8			JIM WELLS TX	207				
3	1.7	1	.5	2	.5	JOHNSON TX	-73	15	-68	08	42
4*	4.0					JONES TX					
2	1.3					KAUFMAN TX					
				3*	4.4	KENDALL TX					98
1	.8	2	1.3	2	.6	KERR TX	63	-53	-23	42	45
1	1.3	3*	3.8	2	1.7	KLEBERG TX	185	-56	25	68	83
1	2.5	2*	4.2			KNOX TX	66				
8*	3.9	2	.7	5	1.8	LAMAR TX	-83	163	-54	18	86
1	1.5	1	1.6			LAMPASAS TX	9				
				1	3.5	LA SALLE TX					97
1	.8	1	.6			LAVACA TX	-22				
1	2.4					LEE TX					
		1	1.5	1	1.1	LEON TX		-26			66
1	1.0	2	1.5	1	.5	LIBERTY TX	51	-66	-48	21	41
		1	.7	2	1.1	LIMESTONE TX		62			66
2*	5.9			1	1.9	LIVE OAK TX			-68	08	87
1	2.8			1	1.0	LLANO TX			-65	11	60
4	1.5	6	1.2	5	.6	LUBBOCK TX	-22	-46	-58	14	46
		1	2.3	1	1.7	LYNN TX		-25			83
				1	.6	MC CULLOCH TX					44
8	1.5	10	1.3	3	.3	MC LENNAN TX	-18	-77	-81	01	37
				1*	9.0	MC MULLEN TX					99
				1	1.7	MARION TX					84
1*	6.4					MARTIN TX					
1	3.1					MASON TX					
1	1.5	1	.9			MATAGORDA TX	-40				
1	3.0			1	1.0	MAVERICK TX			-67	09	60
1	1.1	2	2.0	1	.8	MEDINA TX	83	-59	-26	40	53
		1	.7	3	1.1	MIDLAND TX		45			63
1	.9	1	.6			MILAM TX	-29				
1	2.5					MILLS TX					
1	2.3					MITCHELL TX					
1	1.1	2	1.4	1	.4	MONTAGUE TX	27	-72	-65	11	39
				2	.6	MONTGOMERY TX					46
		1	3.0			MOORE TX					
				1	1.8	MORRIS TX					85
1*	5.3					MOTLEY TX					
		1	.7	1	.6	NACOGDOCHES TX		-9			45
5*	2.3	2	.8	4	1.3	NAVARRO TX	-65	60	-44	25	72
1	1.0	1	1.0	1	.6	NOLAN TX	6	-41	-37	31	45
10*	2.0	9	1.0	13	1.3	NUECES TX	-48	23	-36	32	71
1	.8	1	.3	3	1.1	ORANGE TX	-57	217	35	72	64
3	2.7	1	.9	3	1.7	PALO PINTO TX	-68	92	-38	30	83
1	1.9	3*	3.4			PANOLA TX	82				
2	1.4	1	.6	1	.5	PARKER TX	-55	-21	-65	10	41
				1	2.3	PARMER TX					91
				1	1.9	PECOS TX					87
1	1.7	1	2.0	4*	2.5	POLK TX	17	28	50	78	93

* SIGNIFICANTLY MORE DEATHS THAN EXPECTED

WHITE FEMALE — CANCER OF THE ESOPHAGUS — ICD 150

1950-1959 DEATHS	1950-1959 RATE	1960-1969 DEATHS	1960-1969 RATE	1970-1979 DEATHS	1970-1979 RATE	COUNTY AND STATE	PERCENT CHANGE 60S/50S	PERCENT CHANGE 70S/60S	PERCENT CHANGE 70S/50S	PERCENTILE %CHANGE 70S/50S	PERCENTILE RATES 1970S
3	1.0	5	1.1	3	.5	POTTER TX	11	-53	-48	21	41
				1	3.2	PRESIDIO TX					96
1	.9			2	1.0	RANDALL TX			8	59	59
		1*	6.7			REAL TX					
3*	3.4			2	2.1	RED RIVER TX			-37	31	90
1	2.1			2	3.8	REEVES TX			81	85	97
		1	2.8			REFUGIO TX					
1	1.6			1	1.1	ROBERTSON TX			-30	37	65
				3	3.7	RUNNELS TX					97
		4	2.0			RUSK TX					
				1	3.3	SABINE TX					96
				1	1.7	SAN AUGUSTINE TX					84
1	1.0			5	2.4	SAN PATRICIO TX			134	92	92
				1	1.1	SAN SABA TX					66
1	1.9	1	1.1	1	1.3	SCURRY TX	-42	24	-28	39	73
2	1.7	2	1.2	1	.7	SHELBY TX	-30	-45	-61	13	47
3	1.2	4	1.1	8	1.3	SMITH TX	-13	25	9	59	73
		1	2.2			SOMERVELL TX					
1	2.0	1	1.6	2	2.8	STARR TX	-20	75	40	75	95
1	1.7	1	1.5	1	1.0	STEPHENS TX	-10	-35	-42	27	61
		1*	15.8			STERLING TX					
22	1.5	21	.9	37	1.1	TARRANT TX	-43	25	-29	38	64
3	.9	3	.6	4	.7	TAYLOR TX	-30	5	-27	39	47
				1	3.1	THROCKMORTON TX					96
		1	.9	4	2.8	TITUS TX		192			94
2	.8	2	.6	5	1.2	TOM GREEN TX	-24	106	57	80	68
7	1.2	9	.9	18	1.2	TRAVIS TX	-23	32	2	56	70
1	2.0					TRINITY TX					
				2	1.8	TYLER TX					86
1	1.2					UPSHUR TX					
				4*	3.7	UVALDE TX					97
3*	4.9	3*	3.6	2	1.8	VAL VERDE TX	-27	-48	-62	12	86
2	1.8	2	1.1	4	1.6	VAN ZANDT TX	-39	44	-12	49	80
6*	5.2	2	1.1	1	.4	VICTORIA TX	-79	-64	-92	00	39
1	3.3	1	1.8	2	3.3	WALLER TX	-45	83	1	56	96
1	1.9	1	2.3	2	3.2	WARD TX	19	41	68	83	96
				1	1.0	WASHINGTON TX					62
6*	2.8	6	2.2	5	1.2	WEBB TX	-22	-44	-56	16	69
		1	.6	1	.4	WHARTON TX		-19			40
		1	1.4	1	2.0	WHEELER TX		45			89
4	1.1	5	.9	11	1.6	WICHITA TX	-12	74	53	79	82
1	1.0	2	2.1	1	.6	WILBARGER TX	111	-72	-40	28	44
1	2.3	1	1.5			WILLACY TX	-35				
		2	.8	2	.7	WILLIAMSON TX		-16			48
				1	2.8	WINKLER TX					95
3	3.0	1	.8	2	1.7	WISE TX	-75	131	-42	26	85
2	2.1					WOOD TX					
2*	17.9					YOAKUM TX					
3*	3.9	1	1.1	1	.9	YOUNG TX	-71	-19	-77	03	57
1*	6.6					ZAPATA TX					
2*	6.7	2*	6.2			ZAVALA TX	-8				
7	.2	11	.3	24	.5	UTAH UT	20	59	91		
1	1.3					BOX ELDER UT					
				3	.8	DAVIS UT					54
				1	2.7	EMERY UT					94
3	.2	8	.5	13	.6	SALT LAKE UT	101	23	147	93	43
1	3.9					SUMMIT UT					
		1	2.7			UINTAH UT					
1	.4	1	.2	3	.6	UTAH UT	-42	140	40	75	43
				1	.8	WASHINGTON UT					54
1	.2	1	.2	3	.5	WEBER UT	-15	130	97	88	41
33	1.5	30	1.1	40	1.3	VERMONT VT	-23	18	-9		
2	2.1	1	.8	2	1.8	ADDISON VT	-60	124	-11	49	86
3	2.0	2	.9	3	1.4	BENNINGTON VT	-53	47	-31	36	75
1	.6	1	.6	3	1.6	CALEDONIA VT	-7	177	159	94	82
3	.9	9*	2.3	2	.4	CHITTENDEN VT	155	-82	-55	17	39
1	3.5	1	2.8	1	3.4	ESSEX VT	-20	21	-3	54	96
4	2.3	1	.5	2	1.0	FRANKLIN VT	-78	99	-57	15	60
		1*	5.1			GRAND ISLE VT					
1	1.6					LAMOILLE VT					
		1	1.1	2	1.8	ORANGE VT		65			86
2	1.7			3	2.0	ORLEANS VT			19	65	89
9*	2.9	6	1.4	5	1.3	RUTLAND VT	-52	-9	-56	16	72
3	1.2	2	.6	6	1.7	WASHINGTON VT	-52	211	49	78	84
3	1.6	2	.8	7*	2.1	WINDHAM VT	-48	147	27	69	89
1	.4	3	1.0	4	1.5	WINDSOR VT	168	46	291	98	78
125	1.1	160	1.0	263	1.2	VIRGINIA VA	-7	20	11		
2	1.2	3	1.8	2	1.1	ACCOMACK VA	43	-37	-11	50	65
4	1.7	3	1.1	4	1.1	ALBEMARLE#VA	-35	-3	-37	31	64
2	1.8	2	1.7	1	.6	ALLEGHANY#VA	-4	-63	-65	11	45
		1	5.4			AMELIA VA					
				2	1.6	AMHERST VA					81
8	1.5	9	1.1	21*	2.0	ARLINGTON#VA	-26	83	36	73	89
2	.6			3	.5	AUGUSTA#VA			-13	48	41
				2*	5.3	BATH VA					99
5	1.1	8	1.4	7	1.0	BEDFORD#VA	22	-31	-16	46	59
		1	3.0	1	2.6	BLAND VA		-12			94
		2	2.3			BOTETOURT VA					
				1	1.6	BRUNSWICK VA					81
		1	2.6	1	2.4	BUCKINGHAM VA		-8			92
				2*	4.0	CAROLINE VA					98
1	.4	2	.9	3	1.0	CARROLL#VA	92	18	127	91	61
15	1.3	18	1.3	24	1.6	CHESTERFIELD#VA	-2	19	17	64	80
1	1.3			2	1.1	CULPEPER VA			-15	47	65
		1	1.7	1	1.3	DICKENSON VA		-25			70

* SIGNIFICANTLY MORE DEATHS THAN EXPECTED

\# COUNTIES COMBINED; SEE TECHNICAL NOTES

WHITE FEMALE — CANCER OF THE ESOPHAGUS — ICD 150

1950-1959 DEATHS	1950-1959 RATE	1960-1969 DEATHS	1960-1969 RATE	1970-1979 DEATHS	1970-1979 RATE	COUNTY AND STATE	PERCENT CHANGE 60S/50S	PERCENT CHANGE 70S/60S	PERCENT CHANGE 70S/50S	PERCENTILE %CHANGE 70S/50S	PERCENTILE RATES 1970S
5	1.9	1	.4	5	1.4	DINWIDDIE#VA	-80	289	-24	41	76
				1	2.3	ESSEX VA					91
5	.9	11	.9	20	1.0	FAIRFAX#VA	1	13	15	62	62
1	1.3	1	.9	1	.4	FAUQUIER VA	-27	-51	-64	11	40
1	1.9	2	2.9			FLOYD VA	53				
1	3.1					FLUVANNA VA					
1	.8	3	1.5	6	2.3	FREDERICK#VA	79	55	177	95	92
		2	2.3	1	.9	GILES VA		-61			57
1	3.0	1	1.9			GLOUCESTER VA	-37				
1*	5.0					GREENE VA					
3*	10.8					GREENSVILLE#VA					
4*	4.0			2	1.2	HALIFAX#VA			-70	07	68
		3	2.9			HANOVER VA					
4	1.5	7	1.3	17*	2.0	HENRICO VA	-9	47	35	71	88
1	.6	1	.7	3	1.1	HENRY#VA	30	46	91	87	64
				1	5.5	HIGHLAND VA					99
1	.7	4	2.3	4	1.4	JAMES CITY#VA	238	-40	104	89	75
				1	2.8	KING GEORGE VA					95
				1	.7	LANCASTER VA					49
1	.9					LEE VA					
		2	1.7	2	1.2	LOUDOUN VA		-27			69
		2*	3.3			LOUISA VA					
1	2.5	1	2.5			LUNENBURG VA	-2				
				2	3.2	MADISON VA					96
		1	2.6	2	3.2	MATHEWS VA		23			96
				2	2.0	MECKLENBURG VA					89
1	.7			2	.5	MONTGOMERY#VA			-26	40	42
1	1.8			1	1.6	NELSON VA			-14	47	81
1	1.9	1	1.9	2	2.6	NORTHAMPTON VA	-1	39	38	73	93
		1	2.1	1	1.4	NORTHUMBERLAND VA		-36			75
1	1.6					NOTTOWAY VA					
1	1.6			1	1.1	ORANGE VA			-32	36	63
		1	1.2			PAGE VA					
		1	1.3			PATRICK VA					
7*	2.2	4	1.1	7	1.4	PITTSYLVANIA#VA	-51	25	-39	30	74
		2*	4.0	1	2.1	PRINCE EDWARD VA		-47			90
1	1.5	1	.8	2	.4	PRINCE WILLIAM#VA	-46	-47	-71	06	39
2	2.3					PULASKI VA					
		2*	7.7			RICHMOND VA					
4	.6	4	.5	15	1.3	ROANOKE#VA	-15	183	140	92	73
		1	.7	2	1.3	ROCKBRIDGE#VA		93			74
1	.5	1	.3	2	.6	ROCKINGHAM#VA	-37	94	23	67	44
1	1.0			1	.8	RUSSELL VA			-20	44	51
1	.9	2	1.5	2	1.2	SCOTT VA	57	-19	27	69	68
4*	3.3			2	1.2	SHENANDOAH VA			-64	12	69
2	2.5	4	2.6			SMYTH VA	2				
				1	1.4	SOUTHAMPTON#VA					76
1	1.1	3	2.3	5	2.7	SPOTSYLVANIA#VA	112	18	151	93	94
1	2.0					STAFFORD VA					
				1*	4.5	SURRY VA					98
2	2.0			3	1.2	TAZEWELL VA			-37	32	70
4*	7.6	1	1.6	3	2.9	WARREN VA	-79	79	-62	12	95
1	.5	1	.3	1	.2	WASHINGTON#VA	-32	-26	-50	20	36
				2	3.1	WESTMORELAND VA					96
1	.5			2	.8	WISE#VA			48	77	52
		1	.8	2	1.3	WYTHE VA		58			73
12	1.2	26	1.7	42*	1.9	CHESAPEAKE CITY#VA	39	14	58	80	87
		5	1.3	4	1.1	HAMPTON CITY VA		-18			64
4	2.1	3	1.0	6	1.5	NEWPORT NEWS CITY VA	-55	54	-30	37	79
		2	1.8	2	1.2	SUFFOLK CITY VA		-31			69
132	1.1	188	1.2	236	1.2	WASHINGTON WA	7	-3	3		
				1	.5	ASOTIN WA					40
1	1.3	2	.9	1	.2	BENTON WA	-32	-76	-83	01	36
		2	.8	3	1.1	CHELAN WA		30			64
2	2.2	3	1.8	2	.8	CLALLAM WA	-20	-55	-64	11	53
7	1.9	6	1.0	6	.8	CLARK WA	-47	-24	-60	14	5[illegible]
2	.9	2	.7	7	1.7	COWLITZ WA	-18	133	90	87	85
		1	1.4	1	1.3	DOUGLAS WA		-7			71
		1*	7.1			FERRY WA					
1	1.6					FRANKLIN WA					
				2	1.0	GRANT WA					59
2	.7	3	1.1	8	2.0	GRAYS HARBOR WA	46	86	171	94	88
1	2.7	1	1.2	2	1.3	ISLAND WA	-56	12	-51	20	73
2*	4.0					JEFFERSON WA					
53	1.4	95*	1.8	80	1.2	KING WA	34	-34	-12	49	68
3	.8	2	.4	7	1.1	KITSAP WA	-48	176	42	76	64
		1	1.0			KITTITAS WA					
1	2.0	2	2.8	1	1.0	KLICKITAT WA	38	-64	-50	20	61
1	.4	2	.7	4	1.3	LEWIS WA	94	86	260	97	72
				2	2.8	LINCOLN WA					95
		1	1.0	1	.6	MASON WA		-44			44
1	1.2			1	.6	OKANOGAN WA			-47	22	45
2	2.5	1	1.2	2	1.4	PACIFIC WA	-53	18	-44	25	75
15	1.2	23	1.4	37	1.7	PIERCE WA	16	26	46	77	84
		1	3.2	1	1.4	SAN JUAN WA		-56			76
4	1.5	2	.6	3	.7	SKAGIT WA	-61	17	-54	17	48
9	1.3	10	1.1	17	1.3	SNOHOMISH WA	-20	18	-6	53	71
9	.7	8	.5	16	.9	SPOKANE WA	-35	90	23	67	56
2	2.4			1	.8	STEVENS WA			-67	09	52
3	1.3	4	1.2	4	.7	THURSTON WA	-8	-42	-47	22	47
				3*	14.1	WAHKIAKUM WA					99
2	1.1	2	.7	1	.3	WALLA WALLA WA	-40	-56	-74	05	37
5	1.3	4	.9	9	1.6	WHATCOM WA	-28	65	20	65	80
		1	.7	1	.6	WHITMAN WA		-8			46
4	.7	8	1.1	12	1.3	YAKIMA WA	59	22	94	87	73
67	.9	93	1.0	121	1.0	WEST VIRGINIA WV	15	2	18		
2	2.6			1	1.0	BARBOUR WV			-60	13	62

* SIGNIFICANTLY MORE DEATHS THAN EXPECTED

COUNTIES COMBINED; SEE TECHNICAL NOTES

WHITE FEMALE — CANCER OF THE ESOPHAGUS — ICD 150

1950-1959 DEATHS	1950-1959 RATE	1960-1969 DEATHS	1960-1969 RATE	1970-1979 DEATHS	1970-1979 RATE	COUNTY AND STATE	PERCENT CHANGE 60S/50S	PERCENT CHANGE 70S/60S	PERCENT CHANGE 70S/50S	PERCENTILE %CHANGE 70S/50S	PERCENTILE RATES 1970S
1	.6	1	.5	2	.8	BERKELEY WV	-16	47	23	67	52
1	1.1	2	1.9	1	.6	BOONE WV	76	-67	-42	27	46
1	1.3			1	.9	BRAXTON WV			-29	38	57
2	2.1	1	.8	1	.6	BROOKE WV	-63	-23	-72	06	44
7	1.3	8	1.3	5	.7	CABELL WV	-3	-43	-45	24	49
		1	2.1			CALHOUN WV					
2*	3.8					DODDRIDGE WV					
		3	1.1	7	2.0	FAYETTE WV		87			89
1	2.6	1	3.0			GILMER WV	17				
1	2.5					GRANT WV					
1	.7			2	.7	GREENBRIER WV			10	60	49
1	1.8					HAMPSHIRE WV					
				2	.8	HANCOCK WV					54
2	.5	7	1.4	5	.9	HARRISON WV	185	-34	87	86	58
1	1.3			2	1.2	JACKSON WV			-8	52	67
				1	.8	JEFFERSON WV					51
7	.9	14	1.4	22	1.6	KANAWHA WV	49	17	75	84	81
2	1.4	1	.5			LEWIS WV	-64				
		1	1.1			LINCOLN WV					
1	.4	1	.5	1	.4	LOGAN WV	14	-17	-5	53	39
2	1.6			2	1.0	MC DOWELL WV			-39	29	60
3	1.0	4	1.0	7	1.5	MARION WV	-3	55	50	78	79
1	.5	3	1.4	2	.8	MARSHALL WV	161	-45	44	76	53
1	1.0	1	.9	1	.6	MASON WV	-17	-33	-44	24	43
4	1.4	3	.9	7	1.5	MERCER WV	-34	63	8	59	79
1	.9	1	.8	2	1.4	MINERAL WV	-18	89	55	79	76
2	2.1	2	1.6	4	2.5	MINGO WV	-27	55	14	62	92
2	1.0	4	1.4	6	1.7	MONONGALIA WV	47	17	73	84	84
		2	2.1	1	.9	MONROE WV		-59			56
		1	2.5			MORGAN WV					
1	1.2	2	1.8			NICHOLAS WV	52				
3	.7	11*	2.5	10	2.0	OHIO WV	229	-19	167	94	88
				1	1.9	PLEASANTS WV					87
1	.7	1	.7	4	1.7	PRESTON WV	1	123	126	91	83
1	1.2	1	1.1	1	.6	PUTNAM WV	-9	-50	-55	17	43
2	.9	2	.7	3	.6	RALEIGH WV	-22	-14	-33	34	44
1	.7	1	.8	1	.5	RANDOLPH WV	5	-39	-36	32	40
				1	1.1	RITCHIE WV					66
		1	1.0	1	.8	ROANE WV		-21			53
				2	1.7	SUMMERS WV					83
2	1.9	1	1.1	1	.8	TAYLOR WV	-42	-25	-56	16	53
				1	1.0	TYLER WV					59
2	1.8	1	.8	1	1.0	UPSHUR WV	-54	18	-46	23	60
1	.7	1	.6	2	.9	WAYNE WV	-23	64	26	68	58
		1	2.2			WEBSTER WV					
1	.9	1	.8	1	.6	WETZEL WV	-14	-25	-35	33	45
3	.8	7	1.6	5	.8	WOOD WV	101	-47	6	58	54
				1	.8	WYOMING WV					52
180	1.0	208	.9	294	1.1	WISCONSIN WI	-11	18	5		
		1	1.2	1	1.0	ADAMS WI		-19			59
2	2.6	5*	4.9	4*	2.9	ASHLAND WI	89	-42	11	61	95
3	1.7	1	.5	4	1.2	BARRON WI	-73	150	-32	35	68
1	1.4	3*	3.7	1	1.0	BAYFIELD WI	160	-74	-32	35	60
8	1.7	8	1.2	4	.5	BROWN WI	-27	-59	-70	07	41
				1	.7	BUFFALO WI					50
1	1.2					CALUMET WI					
1	.5	1	.5			CLARK WI	-8				
		1	.4	3	.9	COLUMBIA WI		119			55
1	1.3			2	1.6	CRAWFORD WI			26	68	81
6	.6	7	.6	14	1.0	DANE WI	-7	71	59	81	61
1	.3	2	.5	5	1.2	DODGE WI	74	126	294	98	69
				3	1.4	DOOR WI					77
1	.4	6	2.1	4	1.5	DOUGLAS WI	425	-26	291	98	80
2	1.5	2	1.3	2	1.2	DUNN WI	-18	-8	-25	41	67
5	1.8	2	.5	2	.5	EAU CLAIRE WI	-75	14	-71	06	42
7	1.5	6	1.2	4	.9	FOND DU LAC WI	-25	-24	-43	26	56
2	.8	1	.3	3	.8	GRANT WI	-66	193	0	55	51
3	1.8	1	.7	2	1.0	GREEN WI	-63	50	-44	25	61
1	1.1			1	.8	GREEN LAKE WI			-29	37	52
				2	1.5	IOWA WI					80
2*	6.4	2*	4.8			IRON WI	-25				
1	1.2	2	2.2	1	.9	JACKSON WI	75	-57	-25	41	58
2	.7			5	1.1	JEFFERSON WI			66	82	65
				2	1.4	JUNEAU WI					74
4	1.0	4	.7	4	.6	KENOSHA WI	-28	-19	-42	27	44
		2	1.8	1	.6	KEWAUNEE WI		-66			45
3	.8	10*	1.9	8	1.3	LA CROSSE WI	133	-31	61	81	73
1	1.0			1	.6	LAFAYETTE WI			-37	32	45
2	2.1	1	.6			LANGLADE WI	-71				
2	2.0	1	.8	2	1.5	LINCOLN WI	-59	86	-23	42	80
1	.4	5	1.1	4	.7	MANITOWOC WI	215	-36	102	89	49
2	.5	3	.6	6	1.0	MARATHON WI	22	57	92	87	61
1	.5	4	1.7	2	.6	MARINETTE WI	226	-63	21	66	46
		1	1.2	1	.9	MARQUETTE WI		-23			56
2	.7			2	.6	MENOMINEE#WI			-17	45	43
54	1.3	56	1.0	85	1.3	MILWAUKEE WI	-22	33	4	57	73
		3	1.4	3	1.5	MONROE WI		14			80
		2	1.4	3	1.9	ONEIDA WI		38			88
7	1.9	1	.2	8	1.3	OUTAGAMIE WI	-89	551	-30	37	72
1	.7	3	1.6	3	1.0	OZAUKEE WI	121	-41	29	70	59
		1	.7	5*	2.5	PIERCE WI		250			92
2	1.8			1	.3	POLK WI			-83	01	37
2	1.2	2	.8			PORTAGE WI	-31				
1	1.2	1	1.0			PRICE WI	-13				
5	1.0	8	1.0	14	1.5	RACINE WI	7	49	59	81	80
		2	1.6	1	.4	RICHLAND WI		-74			39
9	1.7	10	1.6	12	1.5	ROCK WI	-10	-3	-13	48	79
		1	1.5			RUSK WI					
3	2.0	2	1.0	2	.9	ST CROIX WI	-49	-12	-55	17	57

* SIGNIFICANTLY MORE DEATHS THAN EXPECTED

\# COUNTIES COMBINED; SEE TECHNICAL NOTES

WHITE FEMALE

CANCER OF THE ESOPHAGUS

ICD 150

1950-1959 DEATHS	1950-1959 RATE	1960-1969 DEATHS	1960-1969 RATE	1970-1979 DEATHS	1970-1979 RATE	COUNTY AND STATE	PERCENT CHANGE 60S/50S	PERCENT CHANGE 70S/60S	PERCENT CHANGE 70S/50S	PERCENTILE %CHANGE 70S/50S	PERCENTILE RATES 1970S
1	.5	3	.9	6	1.9	SAUK WI	94	107	302	98	88
		1	1.4			SAWYER WI					
5	1.2	2	.4	7	1.1	SHEBOYGAN WI	-66	168	-10	51	65
1	1.1					TAYLOR WI					
2	1.5			2	1.3	TREMPEALEAU WI			-17	46	70
3	2.2	6*	3.5	1	.5	VERNON WI	60	-87	-79	02	40
		3*	3.9	1	.6	VILAS WI		-86			43
2	.6	1	.3	5	1.1	WALWORTH WI	-58	308	72	84	64
		1	1.7			WASHBURN WI					
1	.5	2	.9	3	.9	WASHINGTON WI	68	1	69	83	58
3	.7	5	.6	8	.7	WAUKESHA WI	-5	17	11	61	50
1	.4	2	.7	6	1.4	WAUPACA WI	81	120	299	98	77
		2	1.7	3	2.8	WAUSHARA WI		65			95
6	1.1	3	.4	6	.8	WINNEBAGO WI	-62	98	-24	41	54
3	1.6	3	1.0	8	2.0	WOOD WI	-36	97	27	69	88
9	.8	8	.5	13	.7	WYOMING WY	-29	25	-11		
		1	1.1			ALBANY WY					
1	1.7	2	3.0			CARBON WY	77				
		2*	5.1	1	1.7	CONVERSE WY		-67			83
				1	.9	FREMONT WY					57
				1	3.5	HOT SPRINGS WY					97
1	3.9					JOHNSON WY					
2	1.2			4	1.3	LARAMIE WY			8	59	70
		1	3.0			LINCOLN WY					
2	1.4			2	.7	NATRONA WY			-45	23	50
1	1.4			1	.6	PARK-TETON#WY			-61	13	43
1	.9					SHERIDAN WY					
				1*	6.0	SUBLETTE WY					99
		2	2.4	2	1.8	SWEETWATER WY		-25			86
1*	2.4					WASHAKIE WY					

* SIGNIFICANTLY MORE DEATHS THAN EXPECTED

COUNTIES COMBINED; SEE TECHNICAL NOTES

NON-WHITE MALE — CANCER OF THE ESOPHAGUS — ICD 150

1950-1959 DEATHS	1950-1959 RATE	1960-1969 DEATHS	1960-1969 RATE	1970-1979 DEATHS	1970-1979 RATE	COUNTY AND STATE	PERCENT CHANGE 60S/50S	PERCENT CHANGE 70S/60S	PERCENT CHANGE 70S/50S	PERCENTILE %CHANGE RATES 70S/50S	PERCENTILE 1970S
5142	8.0	8716	10.9	12642	12.3	UNITED STATES	36	12	53		
112	3.3	176	5.0	348	9.7	ALABAMA AL	51	94	192		
				2	7.1	AUTAUGA AL					72
2	6.7	2	6.6	3	8.0	BALDWIN AL	-1	20	19	38	73
		1	2.3	3	8.4	BARBOUR AL		262			74
2	5.0	2	6.8	2	10.1	BULLOCK AL	35	48	99	66	78
		1	2.4			BUTLER AL					
2	4.0			5	7.1	CALHOUN AL			77	59	72
		2	4.6	4	8.4	CHAMBERS AL		84			74
				1	6.8	CHILTON AL					71
		1	3.1	3	10.8	CHOCTAW AL		250			79
				7	15.6	CLARKE AL					87
				3*	51.6	CLAY AL					98
1	4.1			2	9.8	COFFEE AL			136	72	77
		1	3.1	2	5.6	COLBERT AL		84			69
2	6.0			2	7.0	CONECUH AL			18	38	71
				3	20.1	COOSA AL					92
				1	5.4	COVINGTON AL					68
		2	10.5			CRENSHAW AL					
		1	5.0	2	9.7	DALE AL		96			77
2	1.5	3	2.9	1	1.3	DALLAS AL	98	-56	-13	22	63
				1	30.7	DE KALB AL					96
				2	5.9	ELMORE AL					69
1	3.1	1	2.0	5	10.3	ESCAMBIA AL	-35	410	229	85	78
5	9.7	7	11.5	11	20.2	ETOWAH AL	19	75	107	66	92
		1	11.5	2	21.6	GENEVA AL		88			93
4	8.3			1	4.0	GREENE AL			-52	09	66
2	3.3	2	3.6	2	4.7	HALE AL	10	31	43	49	67
1	3.9			2	10.0	HENRY AL			155	74	77
		2	4.1	6	12.1	HOUSTON AL		194			81
		1	9.8	3	35.2	JACKSON AL		258			97
34	4.3	55	6.5	103	11.4	JEFFERSON AL	52	75	166	76	80
				2	24.3	LAMAR AL					94
		4	12.9	1	3.0	LAUDERDALE AL		-76			65
		1	4.4	2	13.3	LAWRENCE AL		206			84
4	7.9	2	3.0	6	10.0	LEE AL	-62	228	25	41	77
2	6.7	1	3.9	2	8.1	LIMESTONE AL	-42	108	21	40	73
		2	5.2	1	3.6	LOWNDES AL		-31			66
3	3.8	3	2.9	4	4.3	MACON AL	-24	47	12	34	66
		2	2.8	11	11.7	MADISON AL		317			81
3	3.9	5	7.5	6	10.4	MARENGO AL	92	40	168	77	79
20	7.4	26	7.4	41	10.5	MOBILE AL	0	43	43	49	79
				3	8.8	MONROE AL					75
13	6.6	23	10.8	32	14.0	MONTGOMERY AL	62	30	110	67	85
		2	6.1	2	6.8	MORGAN AL		11			71
1	1.8			1	1.8	PERRY AL			2	30	63
		1	2.9	4	13.6	PICKENS AL		372			84
		1	2.8	3	8.1	PIKE AL		186			73
1	1.3	5	9.2	8	11.3	RUSSELL AL	602	24	768	99	80
				1	6.4	ST CLAIR AL					70
		1	3.9	2	8.3	SHELBY AL		112			74
2	2.9	2	3.8	4	10.9	SUMTER AL	33	187	281	90	80
1	2.2			6	9.0	TALLADEGA AL			313	92	75
				3	8.9	TALLAPOOSA AL					75
2	2.3	5	4.9	9	7.7	TUSCALOOSA AL	113	58	237	86	73
1	3.2	2	5.5	2	8.7	WALKER AL	69	59	169	77	75
1	6.8	1	4.5	3	15.2	WASHINGTON AL	-35	241	123	70	87
		2	2.9	5	10.6	WILCOX AL		261			79
		11	9.2	6	3.8	ALASKA AK		-59			
10	3.3	26	6.0	22	3.1	ARIZONA AZ	82	-48	-6		
		2*	43.7	1	4.9	COCHISE AZ		-89			68
				1	5.6	GILA AZ					69
				1	12.8	GRAHAM AZ					83
5	6.3	18	12.5	15	5.4	MARICOPA AZ	99	-57	-14	22	68
4	8.0	6	10.0	2	2.4	PIMA AZ	25	-76	-70	05	64
1	4.7			2	4.4	PINAL AZ			-7	25	67
70	3.9	74	4.1	110	7.1	ARKANSAS AR	5	72	80		
		2	7.1	3	14.5	ARKANSAS AR		104			86
				2	5.5	ASHLEY AR					68
		1	3.2	1	6.3	BRADLEY AR		100			70
2	3.2	1	1.6	2	3.1	CHICOT AR	-50	96	-2	28	65
		1	4.4	1	3.1	CLARK AR		-29			65
		2	4.1	2	6.0	COLUMBIA AR		45			69
				2	18.7	CONWAY AR					91
		1	17.7	1	15.8	CRAIGHEAD AR		-10			87
				2*	54.9	CRAWFORD AR					98
5	3.5	1	.7	7	8.7	CRITTENDEN AR	-81	1168	146	73	75
5	15.5	1	4.3	2	9.0	CROSS AR	-72	111	-42	12	76
1	5.4			2	8.5	DALLAS AR			58	54	74
3	6.6	5	8.9	1	1.5	DESHA AR	36	-83	-77	03	63
		1	5.8	1	3.0	DREW AR		-48			65
1	3.7	4	13.4			GARLAND AR	266				
1*	34.4					GRANT AR					
1	2.6	2	4.9	1	3.8	HEMPSTEAD AR	92	-23	49	51	66
		1	6.4	1	6.5	HOT SPRING AR		2			70
1	17.2					INDEPENDENCE AR					
2	12.0			1	5.6	JACKSON AR			-54	09	69
7	4.6	12	7.6	9	5.6	JEFFERSON AR	66	-27	21	39	69
				1*	56.2	JOHNSON AR					98
				1	9.1	LAFAYETTE AR					76
1	1.7	2	3.3	3	4.3	LEE AR	97	28	151	74	66

* SIGNIFICANTLY MORE DEATHS THAN EXPECTED

NON-WHITE MALE — CANCER OF THE ESOPHAGUS — ICD 150

1950-1959 DEATHS	1950-1959 RATE	1960-1969 DEATHS	1960-1969 RATE	1970-1979 DEATHS	1970-1979 RATE	COUNTY AND STATE	PERCENT CHANGE 60S/50S	PERCENT CHANGE 70S/60S	PERCENT CHANGE 70S/50S	PERCENTILE %CHANGE 70S/50S	PERCENTILE RATES 1970S
3	8.2	1	3.6	1	2.5	LINCOLN AR	-57	-30	-69	05	64
1	4.2			2	10.6	LONOKE AR			151	74	79
4	9.4	1	3.4	5	14.2	MILLER AR	-64	319	51	52	85
5	5.5	3	3.5	3	4.0	MISSISSIPPI AR	-37	15	-28	17	66
3	7.7	2	4.3	2	4.9	MONROE AR	-45	15	-36	14	68
3	5.7	1	1.5	1	2.3	OUACHITA AR	-74	60	-58	07	64
1*	70.1					PERRY AR					
4	3.8	3	2.2	2	1.5	PHILLIPS AR	-44	-31	-61	07	63
		1	5.6			POINSETT AR					
				1	14.4	PRAIRIE AR					86
11	5.4	17	7.5	33	14.1	PULASKI AR	38	89	160	75	85
1	1.0	2	2.2	2	2.5	ST FRANCIS AR	130	13	161	75	64
2	9.0	2	8.6	7*	34.1	SEBASTIAN AR	-5	299	278	89	96
1	1.5	2	3.1	4	5.5	UNION AR	107	76	263	88	69
		1	22.3	1	16.1	WHITE AR		-28			88
1	3.2	1	2.9			WOODRUFF AR	-11				
222	7.6	441	8.2	801	7.8	CALIFORNIA CA	8	-4	3		
23	7.5	59	10.1	108	11.4	ALAMEDA CA	35	13	53	52	80
		2	21.2			BUTTE CA					
		1*	39.0			COLUSA CA					
		6	4.2	19	7.9	CONTRA COSTA CA		89			73
12	10.8	12	9.0	10	3.6	FRESNO CA	-17	-60	-66	06	66
1	3.9	1	3.9	2	2.8	IMPERIAL CA	-1	-28	-29	16	64
5	6.2	8	7.8	18	9.7	KERN CA	26	24	56	53	77
				1	2.7	KINGS CA					64
93	7.6	220	9.6	390	9.0	LOS ANGELES CA	26	-7	17	37	75
1	5.7			4	12.2	MADERA CA			114	67	82
				3	16.2	MARIN CA					88
		6	21.4	3	6.2	MERCED CA		-71			70
3	9.0	5	8.1	3	1.8	MONTEREY CA	-10	-78	-80	02	63
1	11.2					NAPA CA					
1	5.5	3	5.2	9	3.6	ORANGE CA	-6	-31	-35	14	66
1	7.3			2	10.4	PLACER CA			43	49	79
3	6.0	4	4.2	11	6.1	RIVERSIDE CA	-29	44	3	30	70
17*	21.1	14	8.1	13	4.6	SACRAMENTO CA	-62	-43	-78	02	67
1	24.9	1	15.7			SAN BENITO CA	-37				
4	8.4	3	3.4	14	6.3	SAN BERNARDINO CA	-60	85	-25	18	70
3	2.8	16	9.9	26	6.8	SAN DIEGO CA	256	-31	144	73	71
29	9.4	53	7.5	101	9.9	SAN FRANCISCO CA	-21	32	5	31	77
6	6.9	6	3.0	21	8.2	SAN JOAQUIN CA	-56	175	20	39	74
1	14.3	1	3.6	2	6.4	SAN LUIS OBISPO CA	-75	80	-55	08	70
2	5.2	4	6.5	13	7.3	SAN MATEO CA	27	11	41	48	72
		2	5.9			SANTA BARBARA CA					
5	10.5	6	7.0	6	1.3	SANTA CLARA CA	-33	-82	-88	00	63
		1	4.7	2	3.7	SANTA CRUZ CA		-21			66
				1	9.6	SHASTA CA					77
				1	9.6	SISKIYOU CA					76
3	8.6	3	4.4	7	6.0	SOLANO CA	-49	36	-31	16	69
1	10.3			1	3.1	SONOMA CA			-70	04	65
2	19.0	1	6.9	1	2.9	STANISLAUS CA	-64	-58	-85	01	65
				1	12.5	SUTTER CA					82
2	6.6	3	5.7	3	3.0	TULARE CA	-14	-47	-55	08	65
2	14.4			3	2.8	VENTURA CA			-80	02	64
				2	6.8	YOLO CA					71
4	2.8	14	6.6	22	5.1	COLORADO CO	141	-23	84		
				1	3.2	ADAMS CO					65
3	2.5	10	6.8	18	7.6	DENVER-ARAPAHOE#CO	170	11	199	81	73
		3	19.1	2	6.2	EL PASO CO		-68			70
				1	7.4	JEFFERSON CO					72
1	7.9	1	7.3			PUEBLO CO	-8				
50*	19.4	79*	19.5	136*	21.4	CONNECTICUT CT	0	10	10		
14*	16.4	31*	22.5	38*	19.8	FAIRFIELD CT	37	-12	21	39	92
16*	24.1	21*	18.6	46*	22.8	HARTFORD CT	-23	23	-5	27	94
1	22.9	1	10.8	4*	34.5	MIDDLESEX CT	-53	221	50	52	97
15*	17.7	24*	21.4	41*	20.9	NEW HAVEN CT	20	-2	18	37	92
4*	30.5	2	7.5	7	29.6	NEW LONDON CT	-75	294	-3	27	96
19	9.0	48*	19.9	61*	20.0	DELAWARE DE	120	0	121		
5	13.3	2	5.0	8	17.0	KENT DE	-63	241	28	43	89
11	8.9	38*	27.1	44*	23.6	NEW CASTLE DE	205	-13	167	76	94
3	7.5	8	13.4	9	12.4	SUSSEX DE	78	-7	64	56	82
240*	21.8	421*	26.3	540*	28.8	DIST. OF COLUMBIA DC	21	9	32		
240*	21.8	421*	26.3	540*	28.8	WASHINGTON DC	21	9	32	45	96
205	7.8	412*	12.3	605*	13.7	FLORIDA FL	59	11	77		
6	9.0	10	13.2	14	14.7	ALACHUA FL	47	12	64	55	86
		2	28.1			BAKER FL					
2	6.6	5	16.9	6	15.1	BAY FL	155	-11	127	70	86
				4	27.9	BRADFORD FL					95
1	3.1	3	6.7	13	19.4	BREVARD FL	119	190	536	97	91
3	2.4	20	11.9	35	11.9	BROWARD FL	387	0	388	94	81
		2*	77.4	1	21.0	CHARLOTTE FL		-73			92
				1	6.6	CITRUS FL					71
		4*	39.2	1	9.4	CLAY FL		-76			76
2	15.3	1	6.4	2	7.3	COLLIER FL	-58	14	-53	09	72
1	4.1	1	3.7	3	10.1	COLUMBIA FL	-12	177	144	73	78
37*	13.4	95*	18.9	108	12.8	DADE FL	41	-32	-4	27	83
2	21.3			5*	39.4	DE SOTO FL			85	61	97
47*	12.9	73*	17.4	97*	20.1	DUVAL FL	35	16	56	53	92
9	11.7	12	10.6	25	16.4	ESCAMBIA FL	-10	55	40	47	88
1	13.3			2	34.7	FRANKLIN FL			161	75	97

* SIGNIFICANTLY MORE DEATHS THAN EXPECTED

\# COUNTIES COMBINED; SEE TECHNICAL NOTES

NON-WHITE MALE — CANCER OF THE ESOPHAGUS — ICD 150

1950-1959 DEATHS	1950-1959 RATE	1960-1969 DEATHS	1960-1969 RATE	1970-1979 DEATHS	1970-1979 RATE	COUNTY AND STATE	PERCENT CHANGE 60S/50S	PERCENT CHANGE 70S/60S	PERCENT CHANGE 70S/50S	PERCENTILE %CHANGE 70S/50S	PERCENTILE RATES 1970S
1	1.1	6	6.1	6	7.5	GADSDEN FL	446	24	579	98	72
1	53.0					GLADES FL					
				3	26.7	GULF FL					95
		1	5.9	2	16.9	HAMILTON FL		185			89
		1	37.2			HARDEE FL					
		2	18.4	1	5.4	HENDRY FL		-71			68
1	4.2	1	5.7	1	3.2	HIGHLANDS FL	33	-43	-24	18	65
20	10.9	38*	16.7	44	14.7	HILLSBOROUGH FL	53	-12	35	46	86
				1*	60.6	HOLMES FL					98
1	5.1	3	12.4	4	13.9	INDIAN RIVER FL	143	12	173	78	85
2	4.2	2	5.3	5	13.1	JACKSON FL	27	146	211	82	83
1	5.9					JEFFERSON FL					
4	11.2	1	1.6	3	5.3	LAKE FL	-85	226	-53	09	68
1	3.7	1	2.4	6	11.9	LEE FL	-36	397	217	84	81
		10	13.6	11	11.0	LEON FL		-19			80
				1	6.3	LEVY FL					70
		1	4.1	4	15.8	MADISON FL		289			87
4	12.4	6	14.2	6	13.0	MANATEE FL	14	-8	5	31	83
3	5.3	5	6.6	4	4.9	MARION FL	24	-25	-7	25	68
3*	24.7	7*	50.0	2	9.2	MARTIN FL	102	-82	-63	06	76
3	14.4	5*	28.5	2	11.5	MONROE FL	98	-60	-20	20	80
1	9.9	9*	55.2	6*	31.6	NASSAU FL	456	-43	218	84	96
		2	13.5	1	4.2	OKALOOSA FL		-69			66
		1	24.6	1	14.0	OKEECHOBEE FL		-43			85
4	2.8	10	7.5	19	9.1	ORANGE FL	163	22	222	85	76
1	10.5			2	17.3	OSCEOLA FL			65	56	89
5	3.7	16	7.7	40	16.3	PALM BEACH FL	108	112	342	92	88
3*	20.1					PASCO FL					
6	6.2	16	11.1	32*	17.7	PINELLAS FL	78	60	184	80	90
6	5.6	9	6.6	20	11.1	POLK FL	16	69	97	65	80
6	14.8	2	5.8	2	4.8	PUTNAM FL	-61	-18	-68	05	67
2	7.1	6	20.1	7	21.1	ST JOHNS FL	181	5	196	81	92
1	2.7	4	7.0	9	14.1	ST LUCIE FL	164	101	432	95	85
				2	37.0	SANTA ROSA FL					97
		3	11.5	4	11.6	SARASOTA FL		0			81
4	9.2	4	8.3	6	8.7	SEMINOLE FL	-10	5	-5	26	75
1	6.1	1	6.3	2	10.4	SUWANNEE FL	3	65	70	58	78
2	14.0	1	5.6	1	6.5	TAYLOR FL	-60	17	-54	09	70
				4*	63.6	UNION FL					98
7	8.7	9	10.9	17	17.3	VOLUSIA FL	26	58	99	65	89
				2	27.5	WAKULLA FL					95
		1	17.2	3	39.8	WALTON FL		132			97
				2	20.9	WASHINGTON FL					92
163	4.7	342	8.8	547	12.8	GEORGIA GA	86	45	169		
		2	24.3			APPLING GA					
				1	22.9	ATKINSON GA					94
		3*	97.0			BACON GA					
		1	1.9	6	12.2	BALDWIN GA		547			82
				1	9.2	BARROW GA					76
1	7.2			2	12.6	BARTOW GA			76	59	82
				2	9.4	BEN HILL GA					76
		1	56.2			BERRIEN GA					
8	6.6	15	10.1	19	10.3	BIBB GA	53	2	55	53	78
				1	12.6	BLECKLEY GA					82
		1	44.6			BRANTLEY GA					
		1	3.6	4	17.2	BROOKS GA		380			89
				1	13.3	BRYAN GA					84
1	4.1	1	2.8	1	2.7	BULLOCH GA	-32	-1	-33	15	64
		1	2.4	7	18.5	BURKE GA		658			91
				1	8.2	BUTTS GA					74
		3	16.6	2	8.9	CALHOUN GA		-46			75
2	11.5			4	22.2	CAMDEN GA			93	64	93
				2	19.0	CANDLER GA					91
		3	16.0	3	13.2	CARROLL GA		-17			84
				1	16.4	CHARLTON GA					88
31*	13.3	52*	20.7	56*	21.8	CHATHAM GA	55	6	64	55	93
		1	18.9	1	16.6	CHATTOOGA GA		-12			88
2	6.6	3	7.2	5	11.9	CLARKE GA	9	65	79	60	81
		1	5.8	2	10.0	CLAYTON GA		72			78
1	12.8					CLINCH GA					
1	4.7	2	8.1	8*	21.2	COBB GA	72	162	350	93	93
2	10.8	1	5.2	1	4.6	COFFEE GA	-52	-11	-57	07	67
				2	8.2	COLQUITT GA					74
		1	6.2	1	7.5	COLUMBIA GA		21			73
		1	7.9			COOK GA					
1	2.9	1	2.6	5	13.1	COWETA GA	-10	400	352	93	84
1	10.9			1	11.2	CRAWFORD GA			3	31	80
2	7.4	1	4.7	2	8.5	CRISP GA	-37	82	14	36	74
3	9.7	1	2.9	4	12.1	DECATUR GA	-70	317	25	41	82
1	2.2	11	10.4	26	11.6	DE KALB GA	381	12	438	96	81
				1	9.9	DODGE GA					77
		1	5.6	2	13.5	DOOLY GA		141			84
3	3.9	4	5.4	14	15.8	DOUGHERTY GA	39	192	305	91	87
				2	12.7	DOUGLAS GA					83
1	4.9	2	8.7			EARLY GA	78				
		2	21.2	1	7.2	EFFINGHAM GA		-66			72
1	5.8	3	13.5			ELBERT GA	130				
		1	4.0	1	4.5	EMANUEL GA		12			67
				1	10.2	EVANS GA					78
				1*	267.9	FANNIN GA					99
1	20.2					FAYETTE GA					
1	2.3	3	7.6	5	13.9	FLOYD GA	232	82	504	97	85
		1	17.5			FRANKLIN GA					
41	7.9	94*	13.4	147*	18.4	FULTON GA	69	38	133	72	90
3	8.4	8	17.0	9	16.7	GLYNN GA	102	-2	97	65	88
1	4.5	1	5.2	4	18.1	GREENE GA	15	251	305	91	90
		2	18.1	1	10.1	GWINNETT GA		-44			78
		3	17.8	4	22.0	HALL GA		24			93
		2	8.0	1	4.5	HANCOCK GA		-44			67

* SIGNIFICANTLY MORE DEATHS THAN EXPECTED

NON-WHITE MALE — CANCER OF THE ESOPHAGUS — ICD 150

1950-1959 DEATHS	1950-1959 RATE	1960-1969 DEATHS	1960-1969 RATE	1970-1979 DEATHS	1970-1979 RATE	COUNTY AND STATE	PERCENT CHANGE 60S/50S	PERCENT CHANGE 70S/60S	PERCENT CHANGE 70S/50S	PERCENTILE %CHANGE 70S/50S	PERCENTILE RATES 1970S
		1	4.9	2	8.7	HARRIS GA		76			75
				1	8.8	HART GA					75
1	5.3	1	4.0	4	17.0	HENRY GA	-24	322	222	85	89
1	3.5	1	3.3	4	13.1	HOUSTON GA	-6	298	273	69	83
		1	8.8			IRWIN GA					
1	10.3			3	31.2	JACKSON GA			203	82	96
1	2.7			2	5.9	JEFFERSON GA			120	69	69
1	6.9					JENKINS GA					
		1	14.9			JOHNSON GA					
				2	11.5	JONES GA					81
		1	2.2	5	11.0	LAURENS GA		389			80
				1	9.9	LEE GA					77
		4	22.0	2	8.5	LIBERTY GA		-62			74
				1	12.8	LINCOLN GA					83
		1	16.4			LONG GA					
2	3.4	6	10.6	6	9.3	LOWNDES GA	212	-12	174	78	76
				2	10.8	MC DUFFIE GA					79
1	8.0	1	7.6			MC INTOSH GA	-6				
1	3.7	1	5.6	6	23.8	MACON GA	48	329	536	97	94
		1	9.6	1	10.9	MARION GA		14			79
1	2.4			4	12.9	MERIWETHER GA			434	96	83
		3	10.8	3	9.6	MITCHELL GA		-11			77
		1	4.9	1	4.6	MONROE GA		-6			67
1	10.4			2	36.3	MONTGOMERY GA			248	87	97
		1	5.9	1	5.8	MORGAN GA		-2			69
11	11.3	20	13.4	30*	17.6	MUSCOGEE#GA	18	32	56	53	90
1	4.0			2	7.3	NEWTON GA			81	60	72
				2	7.1	PEACH GA					71
1	11.2					PIERCE GA					
				1	10.9	PULASKI GA					79
				2	14.6	PUTNAM GA					86
1	28.3			1	24.8	QUITMAN GA			-12	22	95
		1	4.2			RANDOLPH GA					
6	4.9	21	13.9	20	10.7	RICHMOND GA	185	-23	119	68	79
1	21.0	2	22.0	3	27.2	ROCKDALE GA	5	24	30	43	95
				3	14.8	SCREVEN GA					86
1	11.5					SEMINOLE GA					
2	8.5	3	9.6	8	22.9	SPALDING GA	14	137	170	77	94
				2	23.4	STEPHENS GA					94
		1	6.4	3	19.1	STEWART GA		196			91
5	11.5	3	7.6	3	7.5	SUMTER GA	-34	0	-35	14	72
1	4.8			2	10.8	TALBOT GA			126	70	79
		1	4.0	2	9.9	TATTNALL GA		147			77
				2	16.1	TAYLOR GA					88
				2	15.1	TELFAIR GA					86
		2	9.0			TERRELL GA					
3	7.1	6	11.2	8	13.5	THOMAS GA	58	21	91	64	84
		1	4.9	4	16.3	TIFT GA		235			88
		1	5.0	3	17.7	TOOMBS GA		254			90
2	3.0	4	6.9	6	11.1	TROUP GA	130	61	271	89	80
		1	8.9	2	23.9	TURNER GA		168			94
		1	6.8			TWIGGS GA					
				5	19.0	UPSON GA					91
				2	22.5	WALKER GA					93
				1	5.3	WALTON GA					68
4	12.6	6	18.7	1	3.0	WARE GA	49	-84	-77	03	65
1	3.4	3	9.3			WASHINGTON GA	174				
1	9.6	3	22.0	1	6.2	WAYNE GA	130	-72	-35	14	70
				1	10.4	WHITFIELD GA					79
		1	13.5	1	17.5	WILCOX GA		30			90
		1	3.8	1	4.5	WILKES GA		18			67
				3	19.1	WILKINSON GA					91
2	7.9			1	5.5	WORTH GA			-31	16	68
		120	7.5	136	5.7	HAWAII HI		-24			
		14	5.8	22	6.9	HAWAII HI		19			71
		92	8.7	89	5.2	HONOLULU HI		-40			68
		7	5.6	8	5.0	KAUAI HI		-11			68
		7	3.7	17	7.1	MAUI#HI		92			72
1	2.8	5	12.3	1	.9	IDAHO ID	336	-93	-67		
		1	67.0			ADA ID					
		2	42.9			BANNOCK ID					
		1*	94.4			MADISON ID					
1	22.9					NEZ PERCE ID					
				1*	36.4	PAYETTE ID					97
		1*	21.0			WASHINGTON ID					
382*	13.4	569*	13.9	864*	15.8	ILLINOIS IL	4	13	18		
1	9.8	2	20.3	1	15.6	ADAMS IL	107	-23	59	55	87
1	3.2	3	8.9	2	6.6	ALEXANDER IL	175	-25	106	66	71
1	7.5	3	17.0	3	7.9	CHAMPAIGN IL	127	-54	5	31	73
1*	106.0			1*	109.3	CHRISTIAN IL			3	31	99
332*	14.8	514*	15.5	756*	16.8	COOK IL	5	8	14	35	89
1*	62.9					DE KALB IL					
		1	35.1			HENRY IL					
3	23.3	2	14.0	3	19.8	JACKSON IL	-40	42	-15	21	92
				2	36.1	JEFFERSON IL					97
6*	45.9	1	3.9	3	9.2	KANE IL	-92	135	-80	02	76
1	2.2	5	10.3	4	7.0	KANKAKEE IL	363	-33	212	83	71
2	23.2	1	10.4	1	9.2	KNOX IL	-55	-12	-60	07	76
		1	1.2	8	12.6	LAKE IL		926			82
				1	60.7	LEE IL					98
3	15.6	4	15.2	3	8.6	MACON IL	-3	-43	-45	11	75
3	7.6	2	4.5	7	12.7	MADISON IL	-41	183	67	57	83
		2	26.4	3*	51.1	MARION IL		93			98
				2	39.1	MASSAC IL					97
				1*	141.3	MONTGOMERY IL					99

* SIGNIFICANTLY MORE DEATHS THAN EXPECTED

\# COUNTIES COMBINED; SEE TECHNICAL NOTES

NON-WHITE MALE — CANCER OF THE ESOPHAGUS — ICD 150

1950-1959 DEATHS	1950-1959 RATE	1960-1969 DEATHS	1960-1969 RATE	1970-1979 DEATHS	1970-1979 RATE	COUNTY AND STATE		PERCENT CHANGE 60S/50S	PERCENT CHANGE 70S/60S	PERCENT CHANGE 70S/50S	PERCENTILE %CHANGE 70S/50S	PERCENTILE RATES 1970S
1	13.5	1	13.4	1	14.3	MORGAN	IL	-1	7	6	32	85
2	8.0	3	8.3	6	11.0	PEORIA	IL	5	32	39	47	80
1	14.1	1	4.8	2	8.2	ROCK ISLAND	IL	-66	69	-42	12	74
11	8.7	16	8.0	29	12.9	ST CLAIR	IL	-8	61	48	50	83
		1	16.3	1	28.9	SALINE	IL		77			96
7*	26.2	2	6.5	7	21.1	SANGAMON	IL	-75	226	-19	20	92
		1	12.3	2	23.9	STEPHENSON	IL		95			94
3	13.6	1	4.5	3	12.7	VERMILION	IL	-67	183	-6	26	83
1*	45.9					WARREN	IL					
				8	10.9	WILL	IL					79
1	5.3	2	6.7	4	7.4	WINNEBAGO	IL	27	11	41	48	72
77	8.8	139	12.5	171	12.3	INDIANA	IN	42	-2	39		
4	12.4	3	8.5	14*	24.0	ALLEN	IN	-32	183	93	64	94
1*	78.1					BARTHOLOMEW	IN					
				1*	81.8	CASS	IN					99
1	8.7	3	20.6			CLARK	IN	137				
1*	141.3					DAVIESS	IN					
				1*	84.8	DEARBORN	IN					99
2	7.7	2	7.0	7	24.2	DELAWARE	IN	-8	244	216	83	94
				2	16.2	ELKHART	IN					88
				1	45.9	FAYETTE	IN					98
1	18.2	1	10.5			FLOYD	IN	-42				
		1	26.5			GIBSON	IN					
		5*	32.8	1	5.1	GRANT	IN		-85			68
		1	34.2			HENRY	IN					
				1	7.5	HOWARD	IN					72
				2*	126.2	JEFFERSON	IN					99
18	9.1	27	8.6	39	9.0	LAKE	IN	-6	5	-1	28	75
		3	15.8	1	4.4	LA PORTE	IN		-72			67
		3	19.6	4	19.3	MADISON	IN		-2			91
37	11.0	66*	14.9	69	13.0	MARION	IN	35	-13	18	37	83
		1	44.6			MONROE	IN					
		1*	54.5			POSEY	IN					
3	7.4	9	17.1	6	8.2	ST JOSEPH	IN	132	-52	11	34	74
				1	49.7	SHELBY	IN					98
3	6.8	8	16.4	12*	25.0	VANDERBURGH	IN	141	53	268	88	95
4	16.0	4	14.3	5	20.1	VIGO	IN	-11	41	26	42	92
				1*	68.8	WARRICK	IN					99
2	13.0	1	5.9	3	16.0	WAYNE	IN	-54	170	23	40	88
15*	13.3	17	14.1	17	11.3	IOWA	IA	6	-20	-15		
1	6.5	3	16.4	3	14.4	BLACK HAWK	IA	154	-12	123	70	86
1*	344.1					CASS	IA					
		1*	91.1			CERRO GORDO	IA					
1*	172.0			1	40.9	CLINTON	IA			-76	03	97
		1*	112.4			DAVIS	IA					
1*	57.3					DES MOINES	IA					
1	14.0	2	34.9	2	33.8	LEE	IA	149	-3	141	72	96
				1	20.6	LINN	IA					92
1*	490.8					LUCAS	IA					
9*	21.1	8	16.9	5	10.6	POLK	IA	-20	-37	-50	10	79
				1	21.9	POTTAWATTAMIE	IA					93
				2	12.6	SCOTT	IA					82
				1	54.2	WEBSTER	IA					98
		2	27.8	1	15.2	WOODBURY	IA		-45			87
26	6.6	46	10.7	60	12.1	KANSAS	KS	64	13	85		
		1	9.3			ATCHISON	KS					
		2*	76.8			BARTON	KS					
1	30.7					BOURBON	KS					
				2*	147.6	DONIPHAN	KS					99
1	10.6	1	7.2			DOUGLAS	KS	-32				
1*	188.7					EDWARDS	KS					
1*	56.2					FINNEY	KS					
		1	8.8	1	7.2	GEARY	KS		-18			72
		1	49.1			HARVEY	KS					
		1*	163.6			JEFFERSON	KS					
		2	27.6	2	50.0	LABETTE	KS		81			98
3	10.6	3	9.6	3	12.9	LEAVENWORTH	KS	-9	34	22	40	83
		1	18.7			LYON	KS					
1	18.2	2*	50.0			MIAMI	KS	175				
2	10.2	2	12.6			MONTGOMERY	KS	23				
				1*	47.2	PRATT	KS					98
		1	15.6	1	15.3	RENO	KS		-2			87
				1	18.9	RILEY	KS					91
		2*	48.8			SALINE	KS					
3	5.7	8	9.9	7	7.1	SEDGWICK	KS	73	-28	25	41	72
3	7.4	4	8.3	9	17.2	SHAWNEE	KS	12	107	132	71	89
10	7.0	14	9.3	33*	21.3	WYANDOTTE	KS	33	128	203	82	93
90	8.9	150*	14.7	177*	17.7	KENTUCKY	KY	66	20	99		
		1	33.5			ADAIR	KY					
		1*	73.7	1*	68.3	ANDERSON	KY		-7			98
1	9.3			1	10.4	BARREN	KY			13	35	78
1	14.0	1	22.8	1	22.3	BELL	KY	63	-2	59	54	93
1	6.5	1	5.6	1	9.1	BOURBON	KY	-14	64	41	48	76
		1	22.8	1	19.5	BOYD	KY		-14			92
1	7.8	2	11.3	1	7.0	BOYLE	KY	44	-38	-10	23	71
1*	245.4					BRACKEN	KY					
1	14.6					CALDWELL	KY					
		1	22.3			CALLOWAY	KY					
1	14.0					CAMPBELL	KY					
		2*	162.9	1*	68.3	CARROLL	KY		-58			98
3	4.9	4	7.7	4	8.3	CHRISTIAN	KY	56	8	69	57	74
1	11.7	2	26.6	1	8.6	CLARK	KY	127	-68	-26	17	75
				1	53.6	CUMBERLAND	KY					98

* SIGNIFICANTLY MORE DEATHS THAN EXPECTED

NON-WHITE MALE — CANCER OF THE ESOPHAGUS — ICD 150

1950-1959 DEATHS	1950-1959 RATE	1960-1969 DEATHS	1960-1969 RATE	1970-1979 DEATHS	1970-1979 RATE	COUNTY AND STATE		PERCENT CHANGE 60S/50S	PERCENT CHANGE 70S/60S	PERCENT CHANGE 70S/50S	PERCENTILE %CHANGE 70S/50S	PERCENTILE RATES 1970S
		2	14.2	3	20.4	DAVIESS	KY		44			92
9	9.6	16	16.7	19	19.4	FAYETTE	KY	74	17	103	66	92
		1*	73.7			FLEMING	KY					
		4*	43.7	1	8.0	FRANKLIN	KY		-82			73
		1	8.1	1	11.5	FULTON	KY		41			80
		1	8.9			HARLAN	KY					
1	22.3	2*	71.8			HARRISON	KY	221				
				2	24.5	HART	KY					95
		2	10.6	1	5.5	HENDERSON	KY		-49			68
1	19.1					HENRY	KY					
1	4.8	1	6.1	3	16.2	HOPKINS	KY	29	164	241	87	88
49*	15.3	80*	21.5	96*	24.0	JEFFERSON	KY	41	11	57	54	94
5*	24.0	6*	31.6	5*	34.4	KENTON	KY	32	9	44	50	97
				1*	172.0	LEWIS	KY					99
1	11.2	2	13.3	1	8.6	LOGAN	KY	18	-35	-23	19	75
4	11.9	2	6.8	6	22.6	MC CRACKEN	KY	-43	230	90	63	94
1	5.6	1	7.1	1	7.6	MADISON	KY	26	6	34	45	73
		1*	74.3			MEADE	KY					
				1	29.8	MERCER	KY					96
				1	18.4	MONTGOMERY	KY					90
		1	10.7	1	15.7	MUHLENBERG	KY		47			87
1	18.8	3*	58.3	3*	48.8	NELSON	KY	210	-16	159	75	98
		1*	182.1			PENDLETON	KY					
				1	32.7	PERRY	KY					96
2*	45.6					PIKE	KY					
				1	27.0	PULASKI	KY					95
				2	31.7	SCOTT	KY					96
		1	9.2			SHELBY	KY					
				1	19.2	SIMPSON	KY					91
				2	40.8	TAYLOR	KY					97
1	10.3					TODD	KY					
				1	13.2	UNION	KY					84
		4	18.2	3	15.6	WARREN	KY		-15			87
		1	21.5	2*	40.5	WASHINGTON	KY		88			97
		1	8.6	3*	106.0	WEBSTER	KY		1132			99
3*	30.3			2	22.7	WOODFORD	KY			-25	18	94
198	5.8	303	7.9	441	10.8	LOUISIANA	LA	36	37	86		
1	3.3	1	2.6	5	14.8	ACADIA	LA	-23	472	343	92	86
1	4.7	3	15.0	2	9.0	ALLEN	LA	218	-40	90	63	75
3	9.4	1	3.7	1	2.9	ASCENSION	LA	-60	-24	-70	05	64
2	5.7	1	3.3	3	12.5	ASSUMPTION	LA	-42	281	120	69	82
1	3.1	1	2.7	8	21.2	AVOYELLES	LA	-13	680	579	98	93
				1	3.6	BIENVILLE	LA					66
1	2.1	2	4.5	6	10.3	BOSSIER	LA	113	126	382	94	78
14	5.2	27	8.9	38	11.5	CADDO	LA	71	29	121	69	81
2	2.5	7	6.8	10	8.2	CALCASIEU	LA	166	21	222	85	74
1	6.4			1	12.6	CALDWELL	LA			98	65	82
1	5.8	2	10.5	1	7.5	CATAHOULA	LA	82	-28	31	44	73
		1	2.7			CLAIBORNE	LA					
6	16.1	1	3.0	6	15.4	CONCORDIA	LA	-82	422	-4	27	87
1	2.0	3	5.2	2	3.0	DE SOTO	LA	163	-43	49	51	65
6	3.0	21	8.2	39	12.8	EAST BATON ROUGE	LA	173	56	326	92	83
1	2.2	1	2.1	6	17.2	EAST CARROLL	LA	-2	701	682	99	89
3	5.7	1	2.3	1	2.4	EAST FELICIANA	LA	-59	7	-57	08	64
		1	6.7	3	11.1	EVANGELINE	LA		64			80
		1	2.6			FRANKLIN	LA					
2	3.8			10	17.4	IBERIA	LA			358	93	89
2	3.8	4	7.2	9	17.0	IBERVILLE	LA	88	136	344	93	89
		1	5.4	1	5.4	JACKSON	LA		1			68
6	12.7	11	10.6	20	12.6	JEFFERSON	LA	-16	19	0	29	82
2	10.1	2	10.7	9*	40.5	JEFFERSON DAVIS	LA	6	279	302	91	97
3	6.1	5	6.8	9	11.8	LAFAYETTE	LA	11	74	93	64	81
1	4.8	3	13.8	3	12.7	LAFOURCHE	LA	186	-8	164	76	83
				1	16.4	LA SALLE	LA					88
3	9.3	2	4.5	3	6.4	LINCOLN	LA	-52	41	-32	16	70
		2	15.8	1	6.9	LIVINGSTON	LA		-56			71
2	4.2	2	3.7	3	6.5	MADISON	LA	-13	77	55	53	70
2	3.5	2	3.2	1	1.4	MOREHOUSE	LA	-7	-57	-60	07	63
2	2.9	4	6.2	7	11.2	NATCHITOCHES	LA	110	81	281	90	80
89*	13.0	121*	13.9	121	12.4	ORLEANS	LA	7	-11	-5	27	82
5	4.1	7	5.4	12	9.9	OUACHITA	LA	31	82	139	72	77
				3	16.6	PLAQUEMINES	LA					88
2	3.1	3	7.0	6	14.7	POINTE COUPEE	LA	126	112	379	94	86
6	4.8	14	10.1	25*	18.3	RAPIDES	LA	109	82	280	90	90
		1	4.3	1	4.1	RED RIVER	LA		-4			66
1	2.8			1	2.2	RICHLAND	LA			-22	19	64
		1	4.7	2	11.7	SABINE	LA		149			81
		1	4.2	2	6.5	ST CHARLES	LA		54			70
		1	5.6	3	13.4	ST HELENA	LA		140			84
3	9.6	1	3.5	2	7.0	ST JAMES	LA	-64	100	-27	17	71
		3	8.8	3	7.4	ST JOHN THE BAPT	LA		-16			72
2	1.9	6	5.8	11	9.8	ST LANDRY	LA	203	69	412	95	77
4	13.6	4	12.3	1	2.6	ST MARTIN	LA	-9	-79	-81	01	64
2	3.9	3	4.7	6	10.3	ST MARY	LA	18	120	161	75	78
2	5.6	2	5.0	5	10.4	ST TAMMANY	LA	-11	107	85	62	78
3	4.0	3	4.0	7	10.7	TANGIPAHOA	LA	-1	170	167	76	79
1	2.0	2	6.7			TENSAS	LA	244				
5	13.2	4	11.7	8	17.6	TERREBONNE	LA	-12	51	33	45	90
1	4.7	1	3.6			UNION	LA	-23				
				2	9.8	VERMILION	LA					77
		1	10.8	1	9.4	VERNON	LA		-13			76
1	2.0	2	3.6	3	5.4	WASHINGTON	LA	82	48	169	77	68
		3	4.9	3	5.7	WEBSTER	LA		16			69
1	4.1	4	14.9	1	4.3	WEST BATON ROUGE	LA	263	-71	4	31	66
		1	7.5			WEST CARROLL	LA					
1	4.2	1	4.6	1	3.8	WEST FELICIANA	LA	10	-17	-9	24	66
				1	5.1	WINN	LA					68

* SIGNIFICANTLY MORE DEATHS THAN EXPECTED

NON-WHITE MALE — CANCER OF THE ESOPHAGUS — ICD 150

1950-1959 DEATHS	1950-1959 RATE	1960-1969 DEATHS	1960-1969 RATE	1970-1979 DEATHS	1970-1979 RATE	COUNTY AND STATE	PERCENT CHANGE 60S/50S	PERCENT CHANGE 70S/60S	PERCENT CHANGE 70S/50S	PERCENTILE %CHANGE 70S/50S	PERCENTILE RATES 1970S
1	9.0			2	9.9	MAINE ME			10		
				1	27.3	AROOSTOOK ME					95
				1	23.6	CUMBERLAND ME					94
1	31.5					PENOBSCOT ME					
195*	12.2	372*	17.6	544*	20.0	MARYLAND MD	45	14	64		
1	20.2	2*	37.6			ALLEGANY MD	86				
3	2.4	12	10.7	23	17.5	ANNE ARUNDEL MD	350	64	636	98	89
5	9.1	7	8.7	18	15.2	BALTIMORE MD	-4	74	67	57	87
		2	10.4	4	15.4	CALVERT MD		48			87
1	4.7	1	5.4	3	15.9	CAROLINE MD	14	196	238	86	87
		1	8.8	1	7.0	CARROLL MD		-20			71
1	7.2	2	11.9	2	13.1	CECIL MD	66	10	83	60	84
4	14.0	8*	22.7	4	8.4	CHARLES MD	61	-63	-40	12	74
6	17.1	5	11.0	5	10.0	DORCHESTER MD	-35	-9	-41	12	77
1	7.2	4	18.4	6	26.0	FREDERICK MD	154	41	259	68	95
4	17.8	3	10.1	5	14.4	HARFORD MD	-43	42	-19	20	86
1	4.8	1	5.6	4	14.8	HOWARD MD	16	166	210	82	86
		2	9.4	4	19.1	KENT MD		103			91
4	9.0	15*	24.6	25	20.4	MONTGOMERY MD	172	-17	126	70	92
6	7.1	24	18.3	55	18.5	PRINCE GEORGE'S MD	158	1	160	75	91
		3	13.7	3	13.4	QUEEN ANNE'S MD		-2			84
1	5.1	6*	23.0	4	11.2	ST MARY'S MD	349	-52	118	68	80
4	12.7	5	15.7	5	16.7	SOMERSET MD	23	6	31	44	89
2	6.7	2	6.3	7	23.6	TALBOT MD	-7	277	252	87	94
1	10.5	7*	76.1	1	12.0	WASHINGTON MD	624	-84	14	35	81
3	7.5	6	13.3	11	20.5	WICOMICO MD	78	54	174	78	92
1	3.0	6	20.2	4	12.9	WORCESTER MD	581	-36	335	92	83
146*	16.0	248*	19.4	350*	23.5	BALTIMORE CITY MD	21	21	47	50	94
50*	13.3	64	12.7	103	13.8	MASSACHUSETTS MA	-4	8	4		
		3*	32.6	4	26.6	BARNSTABLE MA		-19			95
1	20.2	2	26.2	2	26.6	BERKSHIRE MA	30	1	31	44	95
5*	21.5	3	15.0	3	6.2	BRISTOL MA	-30	-59	-71	04	70
2	17.4			2	7.6	ESSEX MA			-57	08	73
1*	172.0					FRANKLIN MA					
3	10.2	3	5.8	8	10.8	HAMPDEN MA	-43	85	6	32	79
1*	37.5					HAMPSHIRE MA					
2	4.1	8	12.4	6	5.5	MIDDLESEX MA	205	-56	35	46	69
				1	3.5	NORFOLK MA					65
2	10.8	3	10.1	3	7.3	PLYMOUTH MA	-7	-28	-33	15	72
32*	16.2	41	14.8	71*	18.9	SUFFOLK MA	-8	27	17	37	91
1	8.2	1	6.5	3	10.0	WORCESTER MA	-20	52	22	40	77
169	8.6	354	12.2	522	12.9	MICHIGAN MI	41	6	50		
		1*	44.6	1*	34.2	BARAGA MI		-24			96
6	15.9	11*	21.5	11	15.8	BERRIEN MI	36	-26	0	28	87
4	13.2	3	9.4	8	18.2	CALHOUN MI	-29	94	38	46	90
				1	5.6	CASS MI					69
		1	67.0			EMMET MI					
6	7.7	12	10.7	10	7.9	GENESEE MI	39	-26	3	30	73
3	19.8	1	3.2	8	17.5	INGHAM MI	-84	450	-11	23	90
				1	30.7	ISABELLA MI					96
3	21.4	5	24.2	7	22.0	JACKSON MI	13	-9	3	30	93
		2	6.4	1	3.1	KALAMAZOO MI		-52			65
5	13.9	10	20.6	7	8.4	KENT MI	48	-59	-39	13	74
1	14.0			1	26.0	LAKE MI			86	62	95
		3	14.8	3	7.5	MACOMB MI		-49			72
1*	81.8					MASON MI					
1	16.7	1	9.4	2	17.0	MONROE MI	-44	80	1	29	89
		6	18.3	6	8.4	MUSKEGON MI		-54			74
2	3.8	13	14.8	11	7.3	OAKLAND MI	290	-50	93	64	72
4	20.7	3	4.6	9	8.1	SAGINAW MI	-78	77	-61	07	73
1	8.2	1	12.6	4	30.9	ST CLAIR MI	54	146	278	89	96
				2	54.7	ST JOSEPH MI					98
		1*	424.0			SANILAC MI					
				2	6.8	VAN BUREN MI					71
3	10.7	8	19.6	6	10.0	WASHTENAW MI	83	-49	-7	26	77
129	8.8	272	12.3	421*	14.2	WAYNE MI	40	15	62	55	85
18*	14.9	17	10.9	15	6.0	MINNESOTA MN	-26	-45	-60		
7	16.2	12	20.1	9	9.8	HENNEPIN MN	24	-51	-40	13	77
9*	28.7	4	9.7	6	8.4	RAMSEY MN	-66	-13	-71	04	74
2*	48.4	1	10.6			ST LOUIS MN	-78				
100	2.9	138	4.2	251	8.1	MISSISSIPPI MS	43	92	176		
2	3.5	5	6.8	16*	24.3	ADAMS MS	96	256	597	98	94
1	7.0			2	13.2	ALCORN MS			88	62	84
		1	2.8			ATTALA MS					
10	6.6	9	6.4	6	6.2	BOLIVAR MS	-4	-2	-6	26	70
				1	6.9	CALHOUN MS					71
				1	4.0	CARROLL MS					66
		1	5.1			CHICKASAW MS					
				1	9.9	CHOCTAW MS					77
		1	3.5			CLAIBORNE MS					
				4	17.6	CLARKE MS					90
1	3.2	1	3.1	1	3.0	CLAY MS	-5	-1	-6	26	65
9	6.9	8	5.9	12	12.9	COAHOMA MS	-14	117	87	62	83
		1	1.7	2	5.6	COPIAH MS		232			69
				1	6.9	COVINGTON MS					71
2	3.3	1	2.3	2	4.9	DE SOTO MS	-31	116	48	51	68
		2	3.0	3	5.1	FORREST MS		71			68
1	5.6					FRANKLIN MS					
		1	18.6			GEORGE MS					

* SIGNIFICANTLY MORE DEATHS THAN EXPECTED

NON-WHITE MALE — CANCER OF THE ESOPHAGUS — ICD 150

1950-1959 DEATHS	1950-1959 RATE	1960-1969 DEATHS	1960-1969 RATE	1970-1979 DEATHS	1970-1979 RATE	COUNTY AND STATE	PERCENT CHANGE 60S/50S	PERCENT CHANGE 70S/60S	PERCENT CHANGE 70S/50S	PERCENTILE %CHANGE 70S/50S	PERCENTILE RATES 1970S
1	2.5	6	17.4	2	5.6	GRENADA MS	586	-68	121	69	69
3*	29.9	2	18.9	3	33.1	HANCOCK MS	-37	75	11	34	96
5	10.5	10	14.5	18*	20.0	HARRISON MS	38	38	91	63	92
5	2.1	18	7.2	26	8.7	HINDS MS	241	21	313	91	75
2	2.5	2	2.4	4	7.8	HOLMES MS	-4	225	213	83	73
3	6.6	1	2.4	3	6.8	HUMPHREYS MS	-64	187	2	30	71
2	11.1					ISSAQUENA MS					
2	8.3	5	16.1	9	18.3	JACKSON MS	94	13	120	69	90
1	3.6			2	6.8	JASPER MS			87	62	71
		1	3.2	3	13.9	JEFFERSON MS		340			85
				2	4.9	JEFFERSON DAVIS MS					68
1	1.6	2	4.0	2	3.0	JONES MS	145	-25	83	61	65
1	4.0					KEMPER MS					
				2	8.9	LAFAYETTE MS					75
3	4.1	2	2.3	13	16.3	LAUDERDALE MS	-43	595	296	91	88
		1	6.1	1	7.7	LAWRENCE MS		26			73
				1	3.5	LEAKE MS					65
1	2.9	1	3.2	4	11.0	LEE MS	10	249	283	90	80
4	3.7	7	7.0	3	3.4	LEFLORE MS	87	-52	-9	24	65
1	3.2	1	2.5			LINCOLN MS	-23				
				7	12.8	LOWNDES MS					83
		2	2.6	4	5.3	MADISON MS		108			68
1	3.8	2	6.7	3	11.0	MARION MS	78	63	190	81	80
2	3.7	3	4.2	2	4.2	MARSHALL MS	15	-1	14	36	66
				3	8.7	MONROE MS					75
1	5.2					MONTGOMERY MS					
		1	12.5	1	4.4	NESHOBA MS		-65			67
1	2.4			3	9.2	NOXUBEE MS			289	90	76
1	2.2	2	5.4	1	2.8	OKTIBBEHA MS	140	-48	26	42	64
2	3.2	1	1.6	1	1.3	PANOLA MS	-50	-22	-61	07	63
		1	3.9	2	10.7	PEARL RIVER MS		174			79
2	19.4					PERRY MS					
2	4.0	2	4.0	4	7.5	PIKE MS	1	87	89	63	72
				2	17.9	PONTOTOC MS					90
				1	10.7	PRENTISS MS					79
				2	4.1	QUITMAN MS					66
1	1.7	1	2.2	5	9.8	RANKIN MS	31	337	473	96	77
		1	3.7	5	15.3	SCOTT MS		316			87
1	3.2			1	5.8	SHARKEY MS			80	60	69
1	4.3			3	11.6	SIMPSON MS			169	77	81
				3*	51.3	STONE MS					98
2	1.7	4	4.3	4	5.6	SUNFLOWER MS	159	29	235	86	69
		1	2.1	1	1.6	TALLAHATCHIE MS		-21			63
		1	4.0	1	5.5	TATE MS		38			69
		2*	33.3			TISHOMINGO MS					
2	4.0	4	7.1	3	7.2	TUNICA MS	75	2	79	59	72
				1	6.3	UNION MS					70
5	5.2	7	7.4	12	13.5	WARREN MS	43	82	160	75	84
12	6.3	6	3.5	18	10.5	WASHINGTON MS	-43	196	68	57	79
2	9.8	2	8.6			WAYNE MS	-13				
				1	2.7	WILKINSON MS					64
		3	11.0	2	5.3	WINSTON MS		-52			68
1	1.3	2	2.9	5	7.6	YAZOO MS	127	161	492	96	73
196*	12.8	251*	14.4	305*	16.0	MISSOURI MO	12	11	24		
1*	112.4					ANDREW MO					
1	18.7	1	12.3	2	31.2	AUDRAIN MO	-34	154	66	56	96
3	18.8			1	6.3	BOONE MO			-66	06	70
		2	13.1	1	8.0	BUCHANAN MO		-39			73
1	8.8			1	6.9	BUTLER MO			-22	19	71
2	11.3	1	10.9	2	18.4	CALLAWAY MO	-3	69	63	55	90
1	10.5	1	16.7	3*	65.4	CAPE GIRARDEAU MO	60	290	524	97	98
3*	69.6	1	21.2			CLAY MO	-70				
		1*	47.2			CLINTON MO					
1	11.1	1	6.6			COLE MO	-40				
1	10.3	1	12.8	2	30.4	COOPER MO	25	137	195	81	96
1*	490.8					CRAWFORD MO					
				1	23.4	FRANKLIN MO					94
		1	8.4	2	15.1	GREENE MO		81			87
1*	28.7	1*	56.2			HENRY MO	96				
2*	34.0	1	36.4	1	28.8	HOWARD MO	7	-21	-15	21	96
43*	12.8	54	13.9	66	14.5	JACKSON MO	9	4	14	35	86
1	19.5					JASPER MO					
1	28.3	2*	49.7			JEFFERSON MO	76				
1	22.3					JOHNSON MO					
		2	32.2	1	15.8	LAFAYETTE MO		-51			87
		1*	98.2			LAWRENCE MO					
				2*	129.5	LINN MO					99
				1*	89.3	MACON MO					99
2	20.0	2	25.1	1	13.7	MARION MO	25	-45	-32	16	65
1	3.6			3	23.0	MISSISSIPPI MO			535	97	94
		1*	86.0			MONITEAU MO					
		2	4.5	1	2.7	PEMISCOT MO		-40			64
1	8.4	3*	30.3			PETTIS MO	261				
		2	16.7	1	12.5	PIKE MO		-25			82
1	10.7	1	16.1			RANDOLPH MO	50				
				1	37.7	RAY MO					97
1	23.6					ST CHARLES MO					
				1*	546.4	ST CLAIR MO					99
1*	172.0					ST FRANCOIS MO					
11	13.3	14	11.8	31	14.1	ST LOUIS COUNTY MO	-12	20	6	32	85
		2	19.9	1	11.6	SALINE MO		-42			81
				1	9.4	SCOTT MO					76
		1*	37.7			SHELBY MO					
1*	344.1					STONE MO					
1*	114.7			2*	169.8	WASHINGTON MO			48	51	99
112*	15.1	152*	16.9	176*	19.4	ST LOUIS CITY MO	12	15	29	43	92
6	9.4	3	4.1	1	.7	MONTANA MT	-56	-84	-93		

* SIGNIFICANTLY MORE DEATHS THAN EXPECTED

NON-WHITE MALE — CANCER OF THE ESOPHAGUS — ICD 150

1950-1959 DEATHS	1950-1959 RATE	1960-1969 DEATHS	1960-1969 RATE	1970-1979 DEATHS	1970-1979 RATE	COUNTY AND STATE	PERCENT CHANGE 60S/50S	PERCENT CHANGE 70S/60S	PERCENT CHANGE 70S/50S	PERCENTILE %CHANGE 70S/50S	PERCENTILE RATES 1970S
				1	7.2	CASCADE MT					72
1	27.0					HILL MT					
1*	70.1	1*	54.2			LEWIS AND CLARK MT	-23				
		1	13.5			ROOSEVELT MT					
		1*	94.4			SANDERS MT					
4*	193.7					YELLOWSTONE MT					
15	14.1	20	14.8	28	17.0	NEBRASKA NB	5	15	21		
				1*	62.9	ADAMS NB					98
15*	19.3	17	17.1	26*	22.1	DOUGLAS NB	-11	29	14	36	93
		1	6.9	1	5.5	LANCASTER NB		-20			68
		1	22.3			THURSTON NB					
		1*	344.1			YORK NB					
2	4.5	3	2.7	12	6.5	NEVADA NV	-39	136	45		
1	4.6	3	4.7	9	7.3	CLARK NV	2	57	59	55	72
				1	23.6	MINERAL NV					94
1	17.0			2	5.6	WASHOE NV			-67	05	69
		2	54.7			NEW HAMPSHIRE NH					
		1*	212.0			CHESHIRE NH					
		1*	188.7			HILLSBOROUGH NH					
210*	15.1	352*	17.3	503*	17.2	NEW JERSEY NJ	15	0	14		
18*	14.5	18	12.0	21	15.3	ATLANTIC NJ	-17	27	6	32	87
9*	22.6	23*	26.0	19	11.3	BERGEN NJ	15	-57	-50	10	80
5	16.9	9	17.8	9	14.2	BURLINGTON NJ	5	-20	-16	21	85
11	10.1	18	12.0	31	15.1	CAMDEN NJ	19	26	49	51	86
4*	22.7	3	12.8	4	17.7	CAPE MAY NJ	-44	38	-22	19	90
4	10.3	4	7.2	1	1.1	CUMBERLAND NJ	-30	-85	-90	00	63
82*	17.7	140*	20.8	177*	19.6	ESSEX NJ	18	-6	11	34	92
2	3.8	8	12.7	7	9.0	GLOUCESTER NJ	233	-30	135	72	75
16*	16.0	23	19.3	39	17.1	HUDSON NJ	21	-11	7	32	89
		1	21.2			HUNTERDON NJ					
10	15.1	14	10.7	43*	23.5	MERCER NJ	-29	120	56	53	94
5	17.5	14*	21.4	16	17.7	MIDDLESEX NJ	22	-17	1	29	90
14*	14.6	17	13.7	37*	22.0	MONMOUTH NJ	-7	61	50	52	93
4*	18.1	6	21.6	3	8.8	MORRIS NJ	19	-59	-51	10	75
				7*	26.6	OCEAN NJ					95
6	11.4	18*	18.8	29*	16.8	PASSAIC NJ	65	-10	47	50	89
3	9.5	3	7.8	7	17.5	SALEM NJ	-18	124	83	61	89
2	27.1	3	16.9	2	5.1	SOMERSET NJ	-38	-70	-81	01	68
				2*	212.0	SUSSEX NJ					99
15*	11.4	30*	18.7	49*	20.2	UNION NJ	64	8	77	59	92
3	1.7	4	2.0	7	1.4	NEW MEXICO NM	15	-28	-18		
2	15.6	1	4.0	1	2.0	BERNALILLO NM	-74	-51	-87	00	63
				1	6.4	CHAVES NM					70
		1	21.2			EDDY NM					
		1	9.4	2	18.9	LEA NM		100			91
				2	2.4	MC KINLEY NM					64
				1	2.3	RIO ARRIBA NM					64
		1	4.9			SANDOVAL NM					
1	2.5					SAN JUAN NM					
637*	15.8	1098*	18.8	1374*	16.0	NEW YORK NY	19	-15	1		
2	9.9	5	13.7	16*	27.4	ALBANY NY	38	101	176	79	95
58*	14.6	114*	15.1	163	13.8	BRONX NY	3	-9	-5	26	85
3*	42.0	1	13.2	1	11.5	BROOME NY	-68	-13	-73	03	80
1	17.0					CATTARAUGUS NY					
				1	8.4	CHAUTAUQUA NY					74
1	31.5	3*	37.3	2	15.3	CHEMUNG NY	19	-59	-51	10	87
				1	13.3	CLINTON NY					84
2*	27.1					COLUMBIA NY					
2	4.4	4	7.2	7	12.1	DUTCHESS NY	63	68	175	79	82
27*	12.3	42	16.0	63	17.0	ERIE NY	30	6	38	46	89
1*	85.2	1*	65.5			ESSEX NY	-23				
		1	37.2			FRANKLIN NY					
		2*	34.7	1	11.4	GENESEE NY		-67			80
				1	25.8	GREENE NY					95
103*	12.0	202*	15.0	337*	16.2	KINGS(BROOKLYN) NY	25	8	35	45	88
3	5.7	9	11.3	22	14.1	MONROE NY	99	25	149	74	85
13*	13.7	32*	25.7	42	17.0	NASSAU NY	88	-34	24	41	89
318*	19.8	521*	27.5	378*	18.0	NEW YORK(MANHATTAN) NY	39	-34	-9	24	90
1	3.4	2	7.1	2	3.7	NIAGARA NY	111	-48	10	34	66
1	14.9	3	11.2	1	3.7	ONEIDA NY	-25	-67	-75	03	66
4	17.1	7	24.3	16*	21.8	ONONDAGA NY	42	-10	27	42	93
		1	62.9	1	15.8	ONTARIO NY		-75			87
3	16.8	1	2.2	8	13.3	ORANGE NY	-87	494	-21	19	84
		1	34.4	1	13.8	ORLEANS NY		-60			85
		1*	86.0			OTSEGO NY					
39*	11.7	66	10.7	158	12.9	QUEENS NY	-8	20	10	33	83
2*	62.9	1	16.7	1	4.4	RENSSELAER NY	-73	-74	-93	00	67
5*	20.5	3	9.4	9	14.8	RICHMOND NY	-54	57	-28	17	86
2	13.1	3	8.3	5	10.3	ROCKLAND NY	-36	24	-21	19	78
2*	41.5	1	13.8			SARATOGA NY	-67				
1	7.6	1	8.4	3	17.3	SCHENECTADY NY	11	107	130	71	89
1*	133.9					SCHOHARIE NY					
		3*	54.6	1	13.7	STEUBEN NY		-75			85
5	5.9	11	6.3	43*	18.6	SUFFOLK NY	8	193	217	84	91
		2	41.4	4	34.2	SULLIVAN NY		-17			97
		1*	172.0			TIOGA NY					
		3*	53.1	3*	49.6	TOMPKINS NY		-7			98
1	5.0	2	11.8	4	16.5	ULSTER NY	138	40	232	85	88
		1	7.7	2	13.2	WAYNE NY		73			84

* SIGNIFICANTLY MORE DEATHS THAN EXPECTED

NON-WHITE MALE | CANCER OF THE ESOPHAGUS | ICD 150

1950-1959 DEATHS	1950-1959 RATE	1960-1969 DEATHS	1960-1969 RATE	1970-1979 DEATHS	1970-1979 RATE	COUNTY AND STATE	PERCENT CHANGE 60S/50S	PERCENT CHANGE 70S/60S	PERCENT CHANGE 70S/50S	PERCENTILE %CHANGE 70S/50S	PERCENTILE RATES 1970S
36*	23.4	47*	17.4	76*	20.5	WESTCHESTER NY	-26	18	-13	22	92
				1	70.7	WYOMING NY					99
154	4.7	277	7.2	519	11.4	NORTH CAROLINA NC	52	60	142		
		2	3.0	5	6.6	ALAMANCE NC		122			71
		1	17.5			ALEXANDER NC					
1	2.3			1	2.5	ANSON NC			9	33	64
3	6.1	3	6.6	7	13.8	BEAUFORT NC	8	111	127	70	85
1	1.8	5	10.3	6	12.2	BERTIE NC	467	18	570	98	82
		2	4.5	5	11.8	BLADEN NC		164			81
1	4.8	1	4.1	1	3.6	BRUNSWICK NC	-15	-13	-26	17	65
7	10.1	10	15.0	14*	21.1	BUNCOMBE NC	48	41	109	67	93
				4	18.7	BURKE NC					91
		4	11.3	6	13.6	CABARRUS NC		21			84
1	11.8	1	3.7	2	16.3	CALDWELL NC	-69	346	38	46	88
		1	12.4			CAMDEN NC					
2	18.6	2	17.5			CARTERET NC	-6				
2	8.5	1	3.1	1	2.4	CASWELL NC	-64	-24	-72	04	64
				2	6.7	CATAWBA NC					71
1	2.6	4	15.8	2	5.6	CHATHAM NC	510	-65	115	68	69
		2*	135.7			CHEROKEE NC					
2	9.2	2	12.3	1	6.1	CHOWAN NC	33	-50	-33	15	70
		1	2.0	4	8.8	CLEVELAND NC		345			75
2	3.4	2	4.1	2	3.2	COLUMBUS NC	19	-22	-8	25	65
4	7.6	3	5.0	7	12.1	CRAVEN NC	-35	144	59	54	81
1	1.1	7	7.4	16	10.3	CUMBERLAND NC	606	38	874	99	78
				1	9.6	CURRITUCK NC					76
		1*	60.6			DARE NC					
		1	2.9	2	4.9	DAVIDSON NC		68			68
				2	21.6	DAVIE NC					93
		3	6.6	6	11.3	DUPLIN NC		70			80
8	10.7	13	8.3	26	14.7	DURHAM NC	-23	78	37	46	86
		4	4.6	11	12.4	EDGECOMBE NC		171			82
8	5.8	12	7.1	37*	18.9	FORSYTH NC	21	167	223	85	91
		3	7.6	5	12.3	FRANKLIN NC		63			82
2	4.3	1	1.5	7	9.9	GASTON NC	-66	582	129	71	77
				1	4.6	GATES NC					67
2	4.2	2	3.8	2	3.5	GRANVILLE NC	-9	-8	-16	21	65
1	2.9			1	3.9	GREENE NC			34	45	66
21*	17.1	25	12.7	27	11.3	GUILFORD NC	-26	-11	-34	15	80
1	1.1	5	5.3	10	10.4	HALIFAX NC	384	98	859	99	78
		2	4.6	8	17.1	HARNETT NC		269			89
		1	22.8			HAYWOOD NC					
		3*	34.5	3	28.0	HENDERSON NC		-19			96
1	2.8	1	1.7	4	7.7	HERTFORD NC	-37	345	178	79	73
				5	18.5	HOKE NC					91
3	13.8	4	9.1	7	16.0	IREDELL NC	-34	77	16	37	88
				1	11.0	JACKSON NC					80
2	4.6	2	4.2	9	17.7	JOHNSTON NC	-9	323	286	90	90
				1	7.2	JONES NC					72
		4	19.1	3	10.0	LEE NC		-48			77
7	12.3	3	5.4	8	10.6	LENOIR NC	-56	97	-14	22	79
1	13.3	1	8.8	2	16.3	LINCOLN NC	-34	86	23	40	88
		1	17.7	1	12.6	MC DOWELL NC		-29			82
2	5.6	3	7.0	4	9.0	MARTIN NC	24	29	60	55	76
12	6.1	15	6.9	35	12.3	MECKLENBURG NC	12	78	100	66	82
1*	281.2					MITCHELL NC					
1	7.3			5	26.4	MONTGOMERY NC			260	88	95
2	12.2	3	8.7	1	2.3	MOORE NC	-29	-73	-81	01	64
		9	12.6	12	14.5	NASH NC		15			86
11*	16.4	10	11.9	15	18.4	NEW HANOVER NC	-27	55	13	35	91
2	4.1	2	3.8	9	15.0	NORTHAMPTON NC	-6	292	269	89	86
		2	8.7	7	20.6	ONSLOW NC		138			92
1	4.5	3	8.2	9	19.7	ORANGE NC	84	139	341	92	92
				3	22.1	PAMLICO NC					93
2	6.9	3	8.0	8	19.2	PASQUOTANK NC	17	139	179	79	91
2	6.1	6	20.2	2	5.7	PENDER NC	231	-72	-7	25	69
1	12.4	1	6.0	1	4.0	PERQUIMANS NC	-52	-33	-68	05	66
		2	7.8	2	5.5	PERSON NC		-30			69
5	6.3	10	11.9	9	8.5	PITT NC	89	-29	34	45	74
		1	15.8			POLK NC					
		1	5.3	1	4.7	RANDOLPH NC		-11			67
		1	2.3	4	8.2	RICHMOND NC		249			74
4	4.8	8	5.9	8	4.6	ROBESON NC	23	-22	-4	27	67
3	6.2	3	5.9	7	11.8	ROCKINGHAM NC	-6	101	89	63	81
3	5.9	2	4.3	6	10.1	ROWAN NC	-28	135	70	58	78
		1	5.2	2	8.1	RUTHERFORD NC		55			73
		1	2.1	3	3.8	SAMPSON NC		81			66
1	2.9			6	14.0	SCOTLAND NC			383	94	65
				1	4.4	STANLY NC					67
				2	24.8	STOKES NC					95
1	9.4	3*	29.0	1	8.6	SURRY NC	208	-70	-8	25	75
				1	31.3	TRANSYLVANIA NC					96
				1	13.3	TYRRELL NC					84
1	3.8	1	5.9	2	5.5	UNION NC	54	-7	42	49	68
1	1.9			5	8.5	VANCE NC			354	93	74
4	3.6	21	13.7	36*	17.4	WAKE NC	281	27	384	94	89
4	9.2	2	4.5	6	15.1	WARREN NC	-51	235	65	56	87
1	4.1	4	20.8	1	4.7	WASHINGTON NC	405	-78	13	35	67
		1*	273.2			WATAUGA NC					
3	2.7	5	4.7	10	10.0	WAYNE NC	72	113	266	88	77
		1	10.4	3	23.5	WILKES NC		125			94
1	1.2	5	6.3	14	17.8	WILSON NC	410	184	1350	99	90
		1	18.4			YADKIN NC					
				1*	78.1	YANCEY NC					99
1	3.4			2	4.5	NORTH DAKOTA ND			32		
				1*	53.0	MC LEAN ND					98
1*	112.4					WARD ND					

* SIGNIFICANTLY MORE DEATHS THAN EXPECTED

NON-WHITE MALE — CANCER OF THE ESOPHAGUS — ICD 150

1950-1959 DEATHS	1950-1959 RATE	1960-1969 DEATHS	1960-1969 RATE	1970-1979 DEATHS	1970-1979 RATE	COUNTY AND STATE	PERCENT CHANGE 60S/50S	PERCENT CHANGE 70S/60S	PERCENT CHANGE 70S/50S	PERCENTILE %CHANGE 70S/50S	PERCENTILE RATES 1970S
				1*	98.2	WILLIAMS ND					99
254*	10.4	406*	12.7	572*	14.6	OHIO OH	22	15	40		
2	9.9	3	13.8	6	17.3	ALLEN OH	40	25	75	58	89
				1*	68.3	ASHLAND OH					98
1	9.7	1	9.9	1	6.0	ASHTABULA OH	2	-40	-38	13	69
		1	28.9			ATHENS OH					
				1	10.7	BELMONT OH					79
4	12.3	7	18.8	5	10.4	BUTLER OH	53	-45	-15	22	79
		1*	172.0			CARPOLL OH					
				1	16.7	CHAMPAIGN OH					88
6	11.9	8	15.4	8	13.6	CLARK OH	30	-12	15	36	84
				1	21.3	CLERMONT OH					93
		1	70.1	1	31.3	CLINTON OH		-55			96
1	14.9			1	10.1	COLUMBIANA OH			-32	15	78
65	9.4	124	12.6	203*	16.0	CUYAHOGA OH	34	27	70	58	88
		1	24.9			DELAWARE OH					
2	13.7	3	17.4	4	16.5	ERIE OH	27	-5	20	39	88
		1	18.6			FAYETTE OH					
18	6.5	30	9.1	55	14.0	FRANKLIN OH	39	54	114	67	85
		1	12.6			GALLIA OH					
				3*	56.8	GEAUGA OH					98
3	14.6	4	16.2	1	2.9	GREENE OH	11	-82	-80	02	65
		1	20.2			GUERNSEY OH					
64*	15.1	100*	18.8	109*	17.9	HAMILTON OH	24	-5	18	38	90
				1	44.6	HANCOCK OH					97
1	18.6	1	24.4			HIGHLAND OH	31				
				1*	57.3	JACKSON OH					98
2	11.1	1	2.8	3	13.0	JEFFERSON OH	-75	371	17	37	83
1	11.2	1	11.5	2	18.3	LAKE OH	3	60	65	56	90
1	9.9					LAWRENCE OH					
		1	17.0	1	10.5	LICKING OH		-38			79
5	17.2	4	9.0	4	6.6	LORAIN OH	-48	-26	-62	06	71
16	13.8	18	10.7	35	16.7	LUCAS OH	-23	57	21	39	88
13	9.6	25*	16.5	18	10.2	MAHONING OH	73	-38	7	32	78
		1	13.6	1	16.3	MARION OH		20			88
		1	62.9			MEDINA OH					
17	9.0	26	10.0	42	12.8	MONTGOMERY OH	12	28	43	49	83
3*	22.8			4	28.9	MUSKINGUM OH			27	42	96
				1	9.0	PORTAGE OH					76
2	37.4	1	4.5	5	17.7	RICHLAND OH	-88	293	-53	09	90
				1	3.5	ROSS OH					65
		1	9.3			SCIOTO OH					
1*	67.0					SENECA OH					
		1*	70.1			SHELBY OH					
7	13.6	8	11.9	12	13.5	STARK OH	-13	14	-1	28	84
15	11.2	22	13.8	31	15.6	SUMMIT OH	24	13	39	47	87
2	4.2	5	9.8	5	9.4	TRUMBULL OH	132	-4	122	69	76
		1	19.3			TUSCARAWAS OH					
1*	62.9					VAN WEPT OH					
1	21.3					WAPREN OH					
		1	19.7	1	19.1	WASHINGTON OH		-3			91
				3*	77.8	WAYNE OH					99
41	4.7	63	6.1	88	6.9	OKLAHOMA OK	31	14	49		
		1	5.4			ADAIR OK					
		1	11.2	1	12.4	BRYAN OK		11			82
				2	14.6	CADDO OK					86
		4	20.4	1	6.6	CARTER OK		-67			71
				1	4.0	CHEROKEE OK					66
1	4.5			2	8.0	CHOCTAW OK			78	59	73
1	5.6	4	15.3			COMANCHE OK	174				
		2	23.2	2	11.4	CPAIG OK		-51			80
1	3.2	2	8.7	1	3.3	CREEK OK	169	-62	2	29	65
2*	33.9			1	9.9	GARFIELD OK			-71	04	77
				1	9.0	GARVIN OK					75
1	14.9					GRADY OK					
				1	19.1	GREER OK					91
				1	10.1	HUGHES OK					78
				1*	298.1	JEFFERSON OK					99
		1	8.4	1	10.3	KAY OK		23			78
		1	16.3			KIOWA OK					
2*	30.1					LINCOLN OK					
		1	6.7	2	11.5	LOGAN OK		70			81
3	8.9	2	4.7	2	5.9	MC CUPTAIN OK	-47	25	-34	14	69
				3	17.9	MC INTOSH OK					90
		1	12.2			MAYES OK					
		1	24.6			MUPRAY OK					
3	4.2	1	1.0	7	9.7	MUSKOGEE OK	-76	886	132	72	77
		1	21.5			NOBLE OK					
		1	11.5			NOWATA OK					
		1	4.3	1	4.6	OKFUSKEE OK		9			67
12	9.1	15	8.9	22	8.9	OKLAHOMA OK	-2	0	-2	28	75
1	2.1	2	6.9	2	6.5	OKMULGEE OK	229	-5	212	83	70
1	9.1	1	8.3			OSAGE OK	-9				
		1	15.7	1	11.7	PAYNE OK		-26			81
1	4.5					PITTSBURG OK					
1	8.6	1	8.1			POTTAWATOMIE OK	-6				
				2	36.3	PUSHMATAHA OK					97
1	18.9					ROGEPS OK					
				2	8.2	SEMINOLE OK					74
		1	37.5	1	12.6	STEPHENS OK		-66			82
3*	34.6			2	21.6	TILLMAN OK			-38	13	93
7	6.9	14	8.8	24	12.1	TULSA OK	28	38	76	59	82
		1	6.5			WAGONER OK					
		1	6.4	1	7.9	WASHINGTON OK		23			73
		1*	424.0			WASHITA OK					

* SIGNIFICANTLY MORE DEATHS THAN EXPECTED

NON-WHITE MALE — CANCER OF THE ESOPHAGUS — ICD 150

1950-1959 DEATHS	1950-1959 RATE	1960-1969 DEATHS	1960-1969 RATE	1970-1979 DEATHS	1970-1979 RATE	COUNTY AND STATE	PERCENT CHANGE 60S/50S	PERCENT CHANGE 70S/60S	PERCENT CHANGE 70S/50S	PERCENTILE %CHANGE 70S/50S	PERCENTILE RATES 1970S
9	7.3	12	7.0	17	6.5	OREGON OR	-4	-8	-12		
1*	114.7					JACKSON OR					
1	32.1			1	18.4	MALHEUR OR			-43	12	90
7	9.1	11	10.0	15	9.7	MULTNOMAH OR	10	-3	7	32	77
		1	17.0			UMATILLA OR					
				1	19.1	WASHINGTON OR					91
354*	13.0	600*	16.4	674*	16.0	PENNSYLVANIA PA	26	-2	23		
				1	40.9	ADAMS PA					97
56	10.2	84	12.6	96	14.3	ALLEGHENY PA	23	13	39	47	85
1	74.3			1	15.7	ARMSTRONG PA			-79	02	87
3	8.5	6	12.1	13*	26.8	BEAVER PA	42	122	215	83	95
		1*	47.2			BEDFORD PA					
3	30.3	1	5.0	1	3.4	BERKS PA	-84	-31	-89	00	65
2*	31.2	1	15.6	1	28.1	BLAIR PA	-50	80	-10	23	96
3*	27.5	3	11.3	4	13.6	BUCKS PA	-59	20	-51	10	84
1	5.9	2	18.1	3	19.0	CAMBRIA PA	206	5	222	85	91
1*	546.4					CARBON PA					
4	6.0	10	14.2	16	18.6	CHESTER PA	138	31	212	82	91
1	21.9					CUMBERLAND PA					
12*	17.3	16*	19.3	10	9.5	DAUPHIN PA	11	-50	-45	11	76
14	13.8	22	13.6	36*	19.6	DELAWARE PA	-1	44	42	48	92
		2	7.2	2	5.5	ERIE PA		-23			69
2	4.5	1	2.9	1	3.1	FAYETTE PA	-37	7	-32	15	65
		1	13.3	1	13.7	FRANKLIN PA		3			85
				1*	212.0	FULTON PA					99
1*	42.4	1	37.7			GREENE PA	-11				
1	20.6	1	18.1			LACKAWANNA PA	-12				
1	7.3	1	6.7	5	23.1	LANCASTER PA	-8	245	216	84	94
		3	26.3	2	16.6	LAWRENCE PA		-37			88
1*	54.6			1	33.5	LEBANON PA			-39	13	96
		4*	55.4			LEHIGH PA					
2*	40.1	3*	41.0	1	11.1	LUZERNE PA	2	-73	-72	04	80
1	15.6	2	48.4			LYCOMING PA	210				
		3	20.0	3	12.5	MERCER PA		-37			82
				1*	89.3	MIFFLIN PA					99
		1	47.2			MONROE PA					
13*	20.2	13	15.4	16	12.8	MONTGOMERY PA	-24	-17	-37	13	83
1	8.9	3	29.3	4	22.4	NORTHAMPTON PA	229	-23	152	74	93
				1*	49.2	NORTHUMBERLAND PA					98
222*	14.6	403*	18.6	436*	17.2	PHILADELPHIA PA	27	-8	17	37	89
		3*	145.0	1	84.8	SCHUYLKILL PA		-42			99
		1*	344.1			SUSQUEHANNA PA					
				1*	18.7	VENANGO PA					91
2	3.7	3	5.8	2	4.6	WASHINGTON PA	55	-20	24	41	67
4	13.4	4	11.0	7	22.1	WESTMORELAND PA	-18	101	65	56	93
2	9.5	1	4.0	6	22.0	YORK PA	-58	452	130	71	93
8	13.0	14*	20.1	22*	19.5	RHODE ISLAND RI	55	-3	50		
				1	22.3	KENT RI					93
1	9.9			2	16.8	NEWPORT RI			70	58	89
7	14.9	14*	26.3	19*	21.5	PROVIDENCE RI	76	-18	44	49	93
193	8.2	215	8.5	395*	13.7	SOUTH CAROLINA SC	4	62	68		
				2	7.9	ABBEVILLE SC					73
4	5.5	7	8.9	17*	19.6	AIKEN SC	64	119	259	87	92
3	10.8	3	12.7	3	14.2	ALLENDALE SC	17	12	31	44	85
4	6.5	2	3.2	5	6.6	ANDERSON SC	-52	108	1	29	70
1	3.6	1	4.0	2	7.0	BAMBERG SC	11	76	96	65	71
1	3.3					BARNWELL SC					
8*	17.6	10*	19.1	19*	32.5	BEAUFORT SC	8	70	85	61	96
10*	20.2	12*	24.2	10	14.6	BERKELEY SC	20	-40	-28	17	86
2	8.6			2	7.8	CALHOUN SC			-9	24	73
67*	34.4	52*	21.5	88*	31.5	CHARLESTON SC	-37	46	-8	25	96
1	4.7	1	4.5	2	8.6	CHEROKEE SC	-5	93	83	61	75
		3	8.2	2	5.8	CHESTER SC		-30			69
2	5.7	2	6.5	2	4.6	CHESTERFIELD SC	14	-29	-20	20	67
6	10.6	2	3.8	5	11.3	CLARENDON SC	-64	201	7	33	80
5	11.2	6	13.0	4	8.3	COLLETON SC	16	-36	-26	17	74
		8	12.3	8	11.5	DARLINGTON SC		-7			81
1	2.7			1	2.7	DILLON SC			1	29	64
1	2.5	2	5.2	11*	22.4	DORCHESTER SC	113	330	815	99	93
1	3.9	2	9.0	3	11.9	EDGEFIELD SC	130	32	203	82	81
1	3.3	1	2.3	5	12.3	FAIRFIELD SC	-29	424	270	89	82
6	6.0	6	5.9	15	12.8	FLORENCE SC	-1	116	113	67	83
8*	18.5	6	12.7	12	20.1	GEORGETOWN SC	-31	58	9	33	92
3	2.3	8	6.8	21	14.2	GREENVILLE SC	190	109	505	97	85
1	2.5	3	6.2	1	2.0	GREENWOOD SC	146	-68	-20	20	64
6*	25.7	3	12.2	4	13.2	HAMPTON SC	-52	8	-49	11	84
2	4.5	5	7.8	7	12.8	HORRY SC	73	63	183	80	83
1	4.0			6	20.0	JASPER SC			405	95	92
		1	2.7	7	15.4	KERSHAW SC		481			87
2	5.6			3	7.3	LANCASTER SC			31	44	72
2	3.8	4	9.6	4	9.4	LAURENS SC	152	-3	145	73	76
1	1.7			1	3.2	LEE SC			84	61	65
		2	5.3	7	17.2	LEXINGTON SC		225			89
		1	5.4	2	14.9	MC CORMICK SC		179			86
1	2.2	3	4.7	7	11.6	MARION SC	118	147	438	96	81
2	4.1	1	2.9	5	12.4	MARLBORO SC	-31	335	202	81	82
2	5.3	1	3.1	3	9.1	NEWBERRY SC	-41	196	73	58	76
		1	5.9	3	17.1	OCONEE SC		189			89
12	12.2	14	11.9	18	13.2	ORANGEBURG SC	-3	11	8	33	84
				3	14.0	PICKENS SC					85
10	6.0	15	7.0	31	13.1	RICHLAND SC	17	88	119	68	83
		1	5.0	1	4.9	SALUDA SC		-3			68
4	4.1	7	5.4	9	6.5	SPARTANBURG SC	32	19	58	54	70

* SIGNIFICANTLY MORE DEATHS THAN EXPECTED

NON-WHITE MALE — CANCER OF THE ESOPHAGUS — ICD 150

1950-1959 DEATHS	1950-1959 RATE	1960-1969 DEATHS	1960-1969 RATE	1970-1979 DEATHS	1970-1979 RATE	COUNTY AND STATE	PERCENT CHANGE 60S/50S	PERCENT CHANGE 70S/60S	PERCENT CHANGE 70S/50S	PERCENTILE %CHANGE 70S/50S	PERCENTILE RATES 1970S
4	4.9	9	8.7	11	9.0	SUMTER SC	78	3	83	60	76
2	6.7	1	3.6	5	18.0	UNION SC	-46	397	167	77	90
5	8.5	4	8.0	9	11.8	WILLIAMSBURG SC	-6	47	38	47	81
1	2.0	5	7.2	9	12.0	YORK SC	268	68	517	97	81
2	2.5	2	1.9	4	3.7	SOUTH DAKOTA SD	-25	97	48		
				1	37.5	BUFFALO SD					97
				1*	66.2	HUGHES SD					98
1*	94.4					JACKSON SD					
1	30.3					MELLETTE SD					
		1*	62.9			MINNEHAHA SD					
		1	4.5	1	3.7	SHANNON SD		-17			66
				1	4.8	TODD SD					67
126	5.6	216	8.9	312	12.2	TENNESSEE TN	58	37	117		
		1	28.7	2	21.4	ANDERSON TN		-25			93
1	7.9	2	14.3	1	10.1	BEDFORD TN	81	-29	28	42	78
				3	25.8	BLOUNT TN					95
		3*	51.5	1	12.8	BRADLEY TN		-75			83
		1*	98.2			CAMPBELL TN					
1	7.2	1	5.7	1	8.6	CARROLL TN	-20	51	20	39	75
				1	44.6	CARTER TN					97
				1	32.6	CHEATHAM TN					96
		1	16.1	1	11.8	CHESTER TN		-27			81
1	6.9			2	15.6	CROCKETT TN			127	70	87
22	7.0	54*	16.4	70*	19.3	DAVIDSON TN	134	18	176	79	91
		1	15.0	1	12.7	DICKSON TN		-15			83
				1	4.3	DYER TN					66
1	2.0	1	1.6	4	6.0	FAYETTE TN	-17	274	209	82	69
1	10.7	1	10.4	1	13.9	FRANKLIN TN	-2	33	30	43	85
		1	2.3	3	6.7	GIBSON TN		196			71
		1	22.5	1	19.3	GREENE TN		-14			91
				3*	36.4	HAMBLEN TN					97
12	6.5	28	14.2	38*	18.7	HAMILTON TN	118	31	187	80	91
2	6.6			1	2.4	HARDEMAN TN			-63	06	64
1	14.9					HAWKINS TN					
				2	4.4	HAYWOOD TN					67
				2	22.3	HENDERSON TN					93
				2	10.8	HENRY TN					79
		1*	53.0	1*	49.7	HOUSTON TN		-6			98
				1*	94.4	JOHNSON TN					99
5	4.7	5	5.2	9	8.9	KNOX TN	12	70	90	63	75
				1	21.3	LAKE TN					93
1	2.6	4	11.8	2	4.7	LAUDERDALE TN	353	-60	80	60	67
		1	7.2			LINCOLN TN					
				1	10.1	MC MINN TN					78
1	16.7			1	21.3	MC NAIRY TN			27	42	93
3	4.8	8	10.0	7	7.1	MADISON TN	109	-29	48	51	72
				2	34.6	MARION TN					97
1	11.9			1	13.3	MARSHALL TN			12	34	84
3	7.8	6	15.4	6	16.9	MAURY TN	97	10	117	68	89
		1	23.4			MONROE TN					
1	2.8	3	8.2	6	16.4	MONTGOMERY TN	188	100	476	96	88
1	7.7					OBION TN					
				1	38.3	RHEA TN					97
1	14.9	1	11.9	3*	39.5	ROANE TN	-20	232	165	76	97
		1	4.8			ROBERTSON TN					
1	3.8	1	2.9	8*	24.5	RUTHERFORD TN	-24	748	540	98	95
56	7.2	80	8.6	108	10.7	SHELBY TN	19	25	48	50	79
		1	24.8	1	12.8	SULLIVAN TN		-49			83
1	5.3			3	13.1	SUMNER TN			148	73	83
1	1.8	2	4.9	1	2.7	TIPTON TN	174	-44	54	52	64
2	7.1	2	5.6	2	16.1	WASHINGTON TN	-21	189	128	71	88
		2	21.4	1	10.6	WEAKLEY TN		-50			79
				1	21.5	WHITE TN					93
				1	5.8	WILLIAMSON TN					69
6*	44.6	1	6.1	2	9.7	WILSON TN	-86	60	-78	02	77
248	5.8	337	6.7	611	8.9	TEXAS TX	16	32	53		
2	4.5	2	4.7	3	7.6	ANDERSON TX	4	62	69	57	73
		1	2.9	1	2.3	ANGELINA TX		-22			64
2	12.9			2	13.2	AUSTIN TX			2	30	84
		1*	60.7			BAILEY TX					
1	3.7	1	5.8	5	18.4	BASTROP TX	57	217	398	95	91
				1*	112.4	BAYLOR TX					99
1*	57.3					BEE TX					
4	14.1	2	6.8	5	9.4	BELL TX	-51	38	-33	15	76
10	7.0	21	10.5	26	6.9	BEXAR TX	52	-35	-1	28	71
		1*	53.0			BOSQUE TX					
3	4.9	4	6.2	6	8.8	BOWIE TX	28	42	81	60	75
5	15.1	3	7.7	2	3.9	BRAZORIA TX	-49	-49	-74	03	66
2	4.7	2	4.6	2	4.3	BRAZOS TX	-1	-8	-9	24	66
1	5.0			2	17.5	BURLESON TX			253	87	90
1	16.6	1	22.3			CAMERON TX	35				
				3	18.0	CAMP TX					90
1	3.2			1	2.4	CASS TX			-26	18	64
1	2.3			2	7.9	CHEROKEE TX			240	86	73
1*	212.0					COCHRAN TX					
1*	49.7					COLEMAN TX					
1	5.3	1	6.0	2	10.2	COLLIN TX	12	72	92	64	78
		1	4.4	3	14.5	COLORADO TX		232			86
		1*	39.0			COMAL TX					
1	28.1					COOKE TX					
				1	2.7	CORYELL TX					64
		1	32.1			CROSBY TX					
30	6.8	40	7.2	115	14.8	DALLAS TX	5	106	116	68	86
		1	8.4			DENTON TX					
2	10.0	2	13.2	2	10.3	DE WITT TX	33	-23	3	30	78

* SIGNIFICANTLY MORE DEATHS THAN EXPECTED

NON-WHITE MALE

CANCER OF THE ESOPHAGUS

ICD 150

COUNTY AND STATE	1950-1959 DEATHS	1950-1959 RATE	1960-1969 DEATHS	1960-1969 RATE	1970-1979 DEATHS	1970-1979 RATE	PERCENT CHANGE 60S/50S	PERCENT CHANGE 70S/60S	PERCENT CHANGE 70S/50S	PERCENTILE %CHANGE 70S/50S	PERCENTILE RATES 1970S
ECTOR TX			2	9.3	2	7.3		-22			72
ELLIS TX	2	4.5	2	4.6	7	16.4	1	260	264	88	68
EL PASO TX	2	13.5	1	5.1	3	1.2	-62	-77	-91	00	63
FALLS TX	1	2.6	1	2.4	2	7.1	-6	192	174	78	72
FANNIN TX	2	13.3			1	7.9			-41	12	73
FISHER TX			1*	29.8							
FORT BEND TX	2	5.3	2	4.6	7	14.3	-12	209	170	78	85
FREESTONE TX	1	3.0	1	3.2	1	2.7	7	-18	-13	22	64
GALVESTON TX	7	7.7	9	8.5	22	14.5	11	70	88	62	86
GOLIAD TX			1	32.7							
GONZALES TX					1	4.5					67
GRAYSON TX	1	3.4	3	10.4	2	5.4	205	-48	57	54	68
GREGG TX			6	8.6	4	6.1		-29			70
GRIMES TX			1	6.9	1	4.0		-41			66
GUADALUPE TX					1	3.5					65
HARDEMAN TX	1	21.5									
HARDIN TX					2	12.0					81
HARRIS TX	65	9.6	98	10.3	162	11.5	7	12	19	38	80
HARRISON TX	2	2.1	3	3.2	2	2.6	51	-18	24	40	64
HASKELL TX	1*	21.1									
HAYS TX	1	16.4									
HENDERSON TX					1	8.2					74
HILL TX	1	5.6	1	7.0	2	10.3	27	46	85	62	78
HOPKINS TX			1	9.6							
HOUSTON TX	3	8.9			2	4.5			-49	11	67
HOWARD TX					1	5.6					69
HUNT TX	1	4.3			3	10.8			148	73	79
HUTCHINSON TX			1	29.8							
JACK TX	1*	182.1									
JACKSON TX	1	10.5	1	11.2			6				
JASPER TX					2	6.8					71
JEFFERSON TX	3	1.8	24	10.5	21	8.0	470	-24	335	92	73
JOHNSON TX			1	14.1							
JONES TX	1	15.8									
KAUFMAN TX					1	3.4					65
KENDALL TX			1*	589.6							
KERR TX	2*	92.5									
KLEBERG TX					2	50.2					98
LAMAR TX					3	9.8					77
LAMB TX	1	26.5									
LAVACA TX			1	8.5	1	12.0		40			81
LEE TX	1	37.2									
LEON TX					1	3.5					65
LIBERTY TX			1	2.7	3	9.4		242			76
LIMESTONE TX	1	3.3	1	3.0	3	10.4	-6	242	221	84	78
LLANO TX					1*	74.3					99
LUBBOCK TX	5*	15.3	1	6.2	3	5.2	-59	-16	-66	06	68
MC LENNAN TX	8	8.4	11	10.2	12	10.0	22	-2	19	38	78
MADISON TX			1	9.6	1	11.2		16			80
MARION TX			2	9.1	1	3.3		-64			65
MATAGORDA TX	2	9.1	5	22.3	4	13.4	145	-40	47	50	84
MIDLAND TX	1	9.8			2	8.3			-16	21	74
MILAM TX	1	5.9	1	6.5	1	7.8	10	20	31	44	73
MONTGOMERY TX	1	3.0			1	3.7			22	40	66
MORRIS TX					2	13.5					84
NACOGDOCHES TX	1	3.6			5	14.0			293	90	85
NAVARRO TX			1	1.7	1	4.7		173			67
NEWTON TX	1	7.0			1	6.3			-11	23	70
NOLAN TX			1	35.3	1	28.7		-19			96
NUECES TX	1	2.0	1	2.7	9	10.3	32	290	414	95	78
ORANGE TX	3	15.7	1	3.5			-78				
POLK TX	2	9.6									
POTTER TX	1	4.3	2	9.1	7	20.0	110	119	359	94	92
RED RIVER TX	2	10.6									
REEVES TX					1	8.8					75
REFUGIO TX					1	18.2					90
ROBERTSON TX	4	11.9	3	8.3	1	6.0	-31	-27	-50	11	69
ROCKWALL TX					1	54.2					98
RUNNELS TX	1*	37.5									
RUSK TX					4	8.9					75
SABINE TX	1	15.8									
SAN AUGUSTINE TX	1	7.9									
SAN JACINTO TX					1	14.4					86
SHELBY TX	1	3.4	1	4.8	2	9.6	44	98	185	80	76
SMITH TX	5	6.8	2	2.2	7	6.5	-68	194	-4	27	70
TARRANT TX	15	8.1	27	11.0	46	13.7	35	25	68	57	85
TAYLOR TX	3*	30.1	2	18.3	1	4.8	-39	-74	-84	01	67
TITUS TX			1	7.3	1	9.4		29			76
TOM GREEN TX			2	14.7	2	6.8		-53			71
TRAVIS TX	5	5.2	7	6.5	12	7.2	26	11	40	47	72
TRINITY TX					4*	28.1					96
TYLER TX			1	9.4							
UPSHUR TX	1	5.7									
VAN ZANDT TX					1	10.5					79
VICTORIA TX			4	20.6	2	8.1		-61			73
WALKER TX					6	12.5					82
WALLER TX			2	9.1	4	14.4		58			86
WASHINGTON TX	3	7.8			2	12.0			54	52	81
WHARTON TX	3	8.7	5	12.0	1	3.8	38	-68	-56	08	66
WICHITA TX	1	3.0	1	3.0	5	13.3	-1	347	344	93	84
WILLIAMSON TX	1	5.0	1	4.0	1	3.6	-20	-10	-28	16	66
WILSON TX			1*	86.0							
WINKLER TX					1	39.0					97
WOOD TX					2	15.8					87
UTAH UT	5	10.1	3	4.9			-52				
DAVIS UT	1	15.7									
SALT LAKE UT	2	10.1	1	2.1			-79				
WEBER UT	2	23.9	2	22.0			-8				

* SIGNIFICANTLY MORE DEATHS THAN EXPECTED

NON-WHITE MALE — CANCER OF THE ESOPHAGUS — ICD 150

1950-1959 DEATHS	1950-1959 RATE	1960-1969 DEATHS	1960-1969 RATE	1970-1979 DEATHS	1970-1979 RATE	COUNTY AND STATE	PERCENT CHANGE 60S/50S	PERCENT CHANGE 70S/60S	PERCENT CHANGE 70S/50S	PERCENTILE %CHANGE 70S/50S	PERCENTILE RATES 1970S
1	26.5			1	18.1	VERMONT VT			-32		
1*	424.0			1*	344.1	BENNINGTON VT			-19	20	99
209	7.2	400*	12.4	560*	14.9	VIRGINIA VA	73	20	107		
2	4.4	4	8.5	1	1.9	ACCOMACK VA	94	-78	-57	08	63
6	14.3	8	16.2	10	18.4	ALBEMARLE#VA	13	14	29	43	90
1	9.4	1	5.9	4	29.9	ALLEGHANY#VA	-37	404	217	84	96
1	6.5	1	5.7	2	13.9	AMELIA VA	-12	143	113	67	85
				1	4.3	AMHERST VA					67
1	10.5					APPOMATTOX VA					
5*	17.4	7	25.3	15*	26.6	ARLINGTON#VA	45	5	53	52	95
1	5.4			3	10.5	AUGUSTA#VA			96	65	79
10	12.5	14	13.8	19	17.3	BEDFORD#VA	10	26	39	47	89
		1*	267.9			BLAND VA					
4	10.8	2	5.3	6	14.5	BRUNSWICK VA	-51	175	35	45	86
1	5.0	1	5.0			BUCKINGHAM VA	-2				
2	8.8			4	12.4	CAROLINE VA			42	48	82
				2*	39.5	CARROLL#VA					97
		1	7.7			CHARLES CITY VA					
2	11.7	1	4.9			CHARLOTTE VA	-58				
31	8.4	92*	22.2	108*	23.2	CHESTERFIELD#VA	164	5	176	79	94
		2	34.4	2	31.4	CLARKE VA		-9			96
1	6.0			1	5.4	CULPEPER VA			-11	23	68
1	5.5					CUMBERLAND VA					
6	4.1	11	6.2	21	13.0	DINWIDDIE#VA	50	110	215	83	83
		1	5.7	3	18.1	ESSEX VA		218			90
9	11.3	20*	19.7	30*	24.3	FAIRFAX#VA	74	24	115	67	95
2	7.8	1	3.8	7*	25.9	FAUQUIER VA	-51	573	233	86	95
		1	7.7	2	14.6	FLUVANNA VA		89			86
1	8.2	2	21.5	1	6.1	FRANKLIN VA	163	-71	-25	18	70
1	15.1	1	11.9	1	12.0	FREDERICK#VA	-22	1	-21	20	81
		2	15.0	3	14.5	GLOUCESTER VA		-4			86
		3	17.6	1	4.8	GOOCHLAND VA		-73			67
2	7.3	2	8.0	2	6.1	GREENSVILLE#VA	9	-23	-16	21	70
		2	3.3	4	6.3	HALIFAX#VA		89			70
1	2.5	4	15.0	4	12.9	HANOVER VA	507	-14	423	95	83
		4	12.1	7	10.2	HENRICO VA		-16			78
2	4.4	4	6.1	8	11.8	HENRY#VA	40	93	170	78	81
1	4.3	4	14.6	3	11.1	ISLE OF WIGHT VA	239	-24	158	74	80
5	13.0	10*	18.7	10	16.7	JAMES CITY#VA	43	-10	29	43	89
				1	8.4	KING AND QUEEN VA					74
		1	16.4			KING GEORGE VA					
				2	13.2	KING WILLIAM VA					84
		1	5.1	5*	25.8	LANCASTER VA		410			95
2	11.3	3	15.2	5	26.8	LOUDOUN VA	35	76	136	72	95
				2	7.1	LOUISA VA					72
		3	16.2	2	8.4	LUNENBURG VA		-48			74
		1	9.3			MADISON VA					
1	9.3			1	15.7	MATHEWS VA			69	57	87
2	4.0	2	3.7	5	9.2	MECKLENBURG VA	-7	147	130	71	76
1	7.8	2	10.2	3	22.8	MIDDLESEX VA	30	124	191	81	94
				1	9.1	MONTGOMERY#VA					76
1	5.7	1	6.9	1	6.6	NELSON VA	21	-4	16	36	71
2	4.7	2	5.0	1	2.4	NORTHAMPTON VA	7	-52	-49	11	64
1	5.0	2	11.6	1	4.6	NORTHUMBERLAND VA	134	-60	-7	25	67
1	3.6			3	10.0	NOTTOWAY VA			180	80	77
1	6.8	1	6.1			ORANGE VA	-10				
1*	47.2			1	23.6	PAGE VA			-50	10	94
		1	21.9			PATRICK VA					
5	5.8	7	6.8	18	15.5	PITTSYLVANIA#VA	19	127	170	77	87
1	9.9			1	6.3	POWHATAN VA			-37	14	70
2	7.2	2	7.7	1	3.5	PRINCE EDWARD VA	7	-54	-51	10	65
2	16.0	1	9.0	4	22.6	PRINCE WILLIAM#VA	-44	151	41	48	94
1	17.1			2	20.3	PULASKI VA			19	38	92
		1	8.2	2	21.1	RICHMOND VA		159			93
9	9.8	14	15.1	12	12.1	ROANOKE#VA	55	-20	24	41	82
		2	21.7	1	11.5	ROCKBRIDGE#VA		-47			80
1	19.6	1	12.3			ROCKINGHAM#VA	-37				
1*	109.3					SCOTT VA					
				1	35.3	SMYTH VA					97
1	1.7	5	11.6	4	6.6	SOUTHAMPTON#VA	592	-43	295	91	71
		2	8.4	5	15.8	SPOTSYLVANIA#VA		88			87
		4*	57.1	3	27.7	STAFFORD VA		-52			95
		2	13.7	2	13.6	SURRY VA		-1			84
1	3.8	1	3.4			SUSSEX VA	-9				
1	8.8	1	9.1	1	11.2	TAZEWELL VA	3	23	26	42	80
		1	18.9	4*	81.6	WARREN VA		332			99
1	7.7			2	25.6	WASHINGTON#VA			233	86	95
1	4.9	1	5.0	5	25.7	WESTMORELAND VA	4	408	427	95	95
		2	33.1	1	15.8	WISE#VA		-52			87
44	8.6	82*	14.8	112*	16.6	CHESAPEAKE CITY#VA	74	12	94	65	88
11*	14.0	20*	24.1	12	10.4	HAMPTON CITY VA	72	-57	-26	18	79
9	7.3	17	14.9	33*	20.5	NEWPORT NEWS CITY VA	105	38	183	80	92
6	7.4	7	7.2	15	14.9	SUFFOLK CITY VA	-3	106	100	66	86
17	6.0	33	8.5	48	7.3	WASHINGTON WA	41	-14	21		
				1	22.9	BENTON WA					94
				1	8.2	CLARK WA					74
		1	13.9			FRANKLIN WA					
1	44.6	2*	35.6			GRANT WA	-20				
10	7.3	19	8.1	31	8.3	KING WA	11	2	13	35	74
		2	21.9	2	10.4	KITSAP WA		-53			78
3	11.0	4	12.8	7	9.7	PIERCE WA	17	-25	-12	23	77
2	9.4	2	9.6	2	8.5	SPOKANE WA	2	-12	-10	24	74
		1	74.3			WALLA WALLA WA					
		1*	188.7			WHITMAN WA					
1	5.6	1	3.2	4	8.3	YAKIMA WA	-43	158	48	50	74

* SIGNIFICANTLY MORE DEATHS THAN EXPECTED

\# COUNTIES COMBINED; SEE TECHNICAL NOTES

NON-WHITE MALE

CANCER OF THE ESOPHAGUS — ICD 150

1950-1959 DEATHS	1950-1959 RATE	1960-1969 DEATHS	1960-1969 RATE	1970-1979 DEATHS	1970-1979 RATE	COUNTY AND STATE	PERCENT CHANGE 60S/50S	PERCENT CHANGE 70S/60S	PERCENT CHANGE 70S/50S	PERCENTILE %CHANGE 70S/50S	PERCENTILE RATES 1970S
19	3.9	35	7.8	45	12.8	WEST VIRGINIA WV	98	65	227		
		1	6.4			BERKELEY WV					
				1	85.2	BOONE WV					99
6*	28.1	2	6.3	2	8.9	CABELL WV	-77	40	-68	05	75
		3	7.4	1	5.7	FAYETTE WV		-23			69
				1	12.1	GREENBRIER WV					81
		1*	172.0			HAMPSHIRE WV					
		2*	162.9			HARDY WV					
		1	13.4			HARRISON WV					
1	6.7					JEFFERSON WV					
2	2.7	5	7.6	5	7.7	KANAWHA WV	177	1	181	80	73
		2	14.1	1	5.0	LOGAN WV		-65			68
5	7.1	5	8.5	12*	24.3	MC DOWELL WV	19	187	243	87	94
1	5.2	1	3.8	2	13.6	MARION WV	-27	260	162	76	84
		2	5.5	4	12.3	MERCER WV		122			82
				1	45.9	MINERAL WV					98
1	9.0	1	9.4	1	10.3	MINGO WV	4	9	14	36	78
		1	8.4			MONONGALIA WV					
		2	18.1	7*	78.5	OHIO WV		333			99
3	5.0	4	8.7	6	16.9	RALEIGH WV	75	95	240	87	89
		1	18.7			SUMMERS WV					
		1	38.3	1	27.0	WYOMING WV		-30			95
14	6.6	30	9.6	40	7.6	WISCONSIN WI	46	-21	15		
				1	38.3	ASHLAND WI					97
1	54.5			1	9.2	DANE WI			-83	01	76
		1*	106.0			DOUGLAS WI					
12	9.0	24	10.5	36	9.7	MILWAUKEE WI	17	-8	8	33	77
		2*	39.3			OUTAGAMIE WI					
		1	3.9			RACINE WI					
1	12.0	1	9.6	1	7.2	ROCK WI	-20	-25	-40	12	72
				1	25.9	WALWORTH WI					95
		1*	94.4			WAUKESHA WI					
1	4.7	2	7.2			WYOMING WY	53				
		1*	99.4			CARBON WY					
		1	30.4			NATRONA WY					
1*	133.9					WASHAKIE WY					

* SIGNIFICANTLY MORE DEATHS THAN EXPECTED

NON-WHITE FEMALE — CANCER OF THE ESOPHAGUS — ICD 150

1950-1959 DEATHS	1950-1959 RATE	1960-1969 DEATHS	1960-1969 RATE	1970-1979 DEATHS	1970-1979 RATE	COUNTY AND STATE	PERCENT CHANGE 60S/50S	PERCENT CHANGE 70S/60S	PERCENT CHANGE 70S/50S	PERCENTILE %CHANGE 70S/50S	PERCENTILE RATES 1970S
1262	1.8	2284	2.5	3869	3.0	UNITED STATES	35	23	65		
45	1.2	60	1.4	108	2.2	ALABAMA AL	19	58	89		
		1	2.4	2	4.1	BALDWIN AL		72			90
1	1.7					BARBOUR AL					
1	1.8					BULLOCK AL					
				1	1.1	BUTLER AL					79
		1	1.5	3	3.3	CALHOUN AL		119			87
		1	1.3			CHAMBERS AL					
				1	7.0	CHILTON AL					95
				1	2.3	CLARKE AL					83
				1	3.0	COFFEE AL					86
				1	1.8	COLBERT AL					81
1	2.3			1	1.9	CONECUH AL			-16	30	82
		1	7.6			COOSA AL					
				1	2.6	COVINGTON AL					84
		2*	8.9	1	1.8	CRENSHAW AL		-80			81
		4	2.5	2	1.4	DALLAS AL		-44			80
1	2.0	1	2.6			ELMORE AL	31				
				2	4.6	ESCAMBIA AL					91
		1	1.4			ETOWAH AL					
1*	21.0					FRANKLIN AL					
				1	2.6	GREENE AL					85
1	3.0					HENRY AL					
1	2.4	1	1.7	1	1.3	HOUSTON AL	-28	-27	-47	13	80
				1	8.1	JACKSON AL					96
15	1.7	15	1.3	42	3.5	JEFFERSON AL	-22	159	102	75	88
				1	10.9	LAMAR AL					97
				2	2.5	LEE AL					84
				1	4.1	LIMESTONE AL					90
		1	.9	1	1.2	MACON AL		30			79
2	2.4	3	3.6	2	1.8	MADISON AL	47	-50	-27	24	81
1	1.3	1	1.5	1	.8	MARENGO AL	13	-47	-40	17	79
				1*	17.5	MARSHALL AL					98
9	2.9	14	3.4	21	4.0	MOBILE AL	18	15	36	56	89
4	1.7	5	1.7	3	.9	MONTGOMERY AL	2	-49	-48	13	79
1	1.3			1	1.4	PERRY AL			12	45	80
1	1.9					PICKENS AL					
2	4.2	1	1.8			PIKE AL	-57				
		2	1.9	1	1.0	RUSSELL AL		-44			79
				1	2.7	SHELBY AL					85
1	1.5	2	3.0	2	2.0	TALLADEGA AL	100	-33	33	55	82
				2	4.0	TALLAPOOSA AL					89
2	1.9	3	2.1	6	4.0	TUSCALOOSA AL	9	88	106	75	89
		4	4.6	2	1.2	ALASKA AK		-74			
		4	.9	7	.9	ARIZONA AZ		-4			
		1*	20.2			COCHISE AZ					
		2	1.3	6	2.0	MARICOPA AZ		51			82
		1	1.5	1	.7	PIMA AZ		-56			78
29	1.6	21	1.1	41	1.9	ARKANSAS AR	-33	81	21		
				1	5.5	ARKANSAS AR					93
1	3.1	1	3.0	1	2.4	ASHLEY AR	-2	-20	-22	27	84
1	1.5	1	1.5			CHICOT AR	-3				
				1	2.8	CLARK AR					85
				1	1.4	COLUMBIA AR					80
3	2.4	1	.8	2	1.6	CRITTENDEN AR	-67	103	-33	20	81
				2	11.9	DALLAS AR					97
3*	8.7					GARLAND AR					
		2	4.4	1	3.5	HEMPSTEAD AR		-21			87
				2*	23.1	HOWARD AR					99
4	2.4	2	.9	6	2.9	JEFFERSON AR	-62	211	20	49	86
1	1.7	2	4.0			LEE AR	132				
				1	4.4	LINCOLN AR					90
				1	5.3	LITTLE RIVER AR					92
				1	3.0	LONOKE AR					86
1	2.6					MILLER AR					
1	1.0			2	2.0	MISSISSIPPI AR			101	74	82
3*	7.4			1	1.2	MONROE AR			-84	00	79
				1	3.5	NEVADA AR					87
		1	2.7	1	1.0	OUACHITA AR		-62			79
3	2.6	3	2.2	5	3.6	PHILLIPS AR	-18	67	37	57	88
				2*	15.4	POINSETT AR					98
6	2.5	6	2.2	8	2.5	PULASKI AR	-12	14	0	40	84
1	1.2			1	1.0	ST FRANCIS AR			-14	31	79
1*	22.5					SEVIER AR					
		1	1.2			UNION AR					
		1	3.2			WOODRUFF AR					
44	1.6	116	2.1	257	2.1	CALIFORNIA CA	32	-2	30		
6	1.8	17	2.8	27	2.4	ALAMEDA CA	56	-14	35	56	84
		1*	18.6	1	5.9	BUTTE CA		-68			94
1	1.0	2	1.1	4	1.6	CONTRA COSTA CA	5	45	52	63	81
1	2.2	1	.6	3	.8	FRESNO CA	-73	27	-66	07	79
1*	182.1			1*	15.7	GLENN CA			-91	00	98
				3	5.0	IMPERIAL CA					92
1	1.7	1	1.1	2	1.2	KERN CA	-36	6	-33	21	79
		1	11.2			KINGS CA					
		1*	37.5			LAKE CA					
23	1.8	59	2.1	144	2.5	LOS ANGELES CA	18	19	41	60	84
1*	8.4					MADERA CA					
				1	9.1	MENDOCINO CA					96

* SIGNIFICANTLY MORE DEATHS THAN EXPECTED

NON-WHITE FEMALE

CANCER OF THE ESOPHAGUS

ICD 150

1950-1959 DEATHS	1950-1959 RATE	1960-1969 DEATHS	1960-1969 RATE	1970-1979 DEATHS	1970-1979 RATE	COUNTY AND STATE	PERCENT CHANGE 60S/50S	PERCENT CHANGE 70S/60S	PERCENT CHANGE 70S/50S	PERCENTILE %CHANGE 70S/50S	PERCENTILE RATES 1970S
				1	2.7	MERCED CA					65
		1	1.5			MONTEREY CA					
1	1.1	2	2.3	3	1.8	RIVERSIDE CA	97	-21	56	64	81
1	.8	3	1.7	5	1.5	SACRAMENTO CA	110	-11	87	71	80
				5	2.1	SAN BERNARDINO CA					83
1	.2	8	4.2	11	1.9	SAN DIEGO CA	1585	-55	650	97	82
7	2.3	14	2.4	31	2.6	SAN FRANCISCO CA	6	8	14	46	85
		3	3.5	3	1.4	SAN JOAQUIN CA		-62			80
				2	.7	SAN MATEO CA					78
		2	1.6	1	.2	SANTA CLARA CA		-86			78
				1	3.1	SANTA CRUZ CA					86
				3	3.0	SOLANO CA					86
				1	1.9	SONOMA CA					82
				1	1.9	STANISLAUS CA					82
				1*	84.8	TRINITY CA					99
				1	.6	VENTURA CA					78
				1	3.4	YUBA CA					87
3	2.4	3	1.4	5	.8	COLORADO CO	-41	-46	-68		
2	2.3	3	2.0	4	1.1	DENVER-ARAPAHOE#CO	-13	-43	-51	12	79
1*	8.3					EL PASO CO					
				1	2.2	PUEBLO CO					83
6	2.1	20*	4.5	34	3.9	CONNECTICUT CT	110	-13	83		
1	.8	9*	5.9	17*	6.1	FAIRFIELD CT	639	3	660	97	94
		3	2.4	4	1.5	HARTFORD CT		-37			80
				1*	12.0	LITCHFIELD CT					97
4*	4.3	7	5.2	11	4.3	NEW HAVEN CT	21	-17	0	40	90
		1	4.7	1	1.9	NEW LONDON CT		-60			82
1*	62.9					WINDHAM CT					
5	2.5	11	3.9	15	4.0	DELAWARE DE	57	4	63		
				2	2.9	KENT DE					86
3	2.7	9*	5.0	9	3.8	NEW CASTLE DE	83	-23	41	60	89
2	3.8	2	3.4	4	5.6	SUSSEX DE	-11	64	45	62	93
36	2.7	91*	4.3	142*	5.6	DIST. OF COLUMBIA DC	62	30	110		
36	2.7	91*	4.3	142*	5.6	WASHINGTON DC	62	30	110	77	93
50	1.7	115*	3.0	163	3.1	FLORIDA FL	73	3	78		
		1	1.1	6	5.5	ALACHUA FL		382			93
		1*	14.0			BAKER FL					
				2	3.3	BAY FL					87
1	4.9	2	2.8	2	3.1	BREVARD FL	-42	11	-36	18	87
		6	3.1	8	2.2	BROWARD FL		-28			83
		1*	35.3			CHARLOTTE FL					
				1	10.5	CITRUS FL					97
1*	11.2	1	7.3			CLAY FL	-35				
		1*	13.8			COLLIER FL					
4	1.0	23*	3.8	28	2.5	DADE FL	274	-34	146	83	84
13*	2.8	26*	4.8	30*	4.8	DUVAL FL	70	0	69	68	92
3	2.6	1	.6	7	3.3	ESCAMBIA FL	-76	438	27	52	87
				1*	15.8	FLAGLER FL					98
				1*	12.3	FRANKLIN FL					97
1	.9					GADSDEN FL					
				1	2.5	HIGHLANDS FL					84
5	2.2	13*	4.8	12	3.5	HILLSBOROUGH FL	114	-27	56	63	88
		1	4.2	1	3.0	INDIAN RIVER FL		-30			86
		1	2.0	2	4.2	JACKSON FL		111			90
2*	5.2	1	3.9			LAKE FL	-25				
1	5.5	1	2.9			LEE FL	-46				
				2	1.7	LEON FL					81
		1	2.8			MADISON FL					
2*	5.3	3*	7.5			MANATEE FL	40				
3	5.3	2	2.4	2	2.3	MARION FL	-55	-5	-58	10	83
		1	5.8	2	8.6	MONROE FL		48			96
		1	14.1	1	6.4	OKALOOSA FL		-55			94
		1*	23.4			OKEECHOBEE FL					
1	.7	3	1.8	6	2.8	ORANGE FL	157	61	315	94	85
				1	8.5	OSCEOLA FL					96
1	.3	8	4.0	8	2.6	PALM BEACH FL	1035	-34	646	97	85
3	3.1	6	4.0	11	4.7	PINELLAS FL	29	18	52	63	91
3	2.4	4	2.8	8	4.6	POLK FL	18	62	90	72	91
				2	3.9	PUTNAM FL					89
		1	2.7			ST JOHNS FL					
				5*	7.4	ST LUCIE FL					95
2*	7.5	1	4.2	5*	10.4	SARASOTA FL	-44	148	38	58	97
				4	4.7	SEMINOLE FL					91
				1	6.8	SUMTER FL					95
2*	10.2	1	3.8			SUWANNEE FL	-63				
		1	8.0	1	4.1	TAYLOR FL		-48			90
2	2.2	1	.8	2	1.4	VOLUSIA FL	-65	79	-36	18	80
75	1.7	123	2.4	148	2.6	GEORGIA GA	42	5	50		
		1	8.9			APPLING GA					
				1*	14.9	ATKINSON GA					98
		1*	23.8			BACON GA					
		1	1.2	5*	7.6	BALDWIN GA		523			95
		1	5.9			BARTOW GA					
				1	5.4	BEN HILL GA					93
		1*	11.7			BERRIEN GA					
6	3.5	6	2.7	7	2.6	BIBB GA	-23	-5	-26	26	85
2	3.4			1	1.9	BURKE GA			-45	15	82
		1	4.7			BUTTS GA					
		1	3.5			CALHOUN GA					

* SIGNIFICANTLY MORE DEATHS THAN EXPECTED

COUNTIES COMBINED; SEE TECHNICAL NOTES

NON-WHITE FEMALE — CANCER OF THE ESOPHAGUS — ICD 150

COUNTY AND STATE	1950-1959 DEATHS	1950-1959 RATE	1960-1969 DEATHS	1960-1969 RATE	1970-1979 DEATHS	1970-1979 RATE	PERCENT CHANGE 60S/50S	PERCENT CHANGE 70S/60S	PERCENT CHANGE 70S/50S	PERCENTILE %CHANGE 70S/50S	PERCENTILE RATES 1970S
CAMDEN GA			1	5.8							
CANDLER GA					1	12.2					97
CARROLL GA					2	5.7					94
CHATHAM GA	16*	5.5	21*	6.6	21*	5.8	20	-12	5	43	94
CHATTOOGA GA			1*	9.6							
CLARKE GA			3	5.3	1	1.5		-72			80
COBB GA					1	2.0					82
COFFEE GA	1	3.5			1	3.9			12	44	89
COOK GA	1*	9.9									
COWETA GA			2	4.3	1	2.3		-46			84
CRISP GA			1	2.3							
DECATUR GA	1	2.2									
DE KALB GA	1	.8	3	2.2	7	3.3	174	49	309	93	87
DOUGHERTY GA	2	1.7	3	2.5	2	1.3	50	-48	-22	28	80
EFFINGHAM GA	1*	9.8									
ELBERT GA					1	4.4					90
FLOYD GA	1	2.3			1	1.8			-21	28	81
FRANKLIN GA			1*	16.6							
FULTON GA	17	2.6	26	2.6	35	3.0	1	15	16	48	86
GLYNN GA	1	2.2	2	3.1	6*	9.0	40	190	306	93	96
GREENE GA	2*	5.7									
HARRIS GA	2*	6.4									
HART GA					2*	15.6					98
HOUSTON GA	2*	8.0	1	3.1			-61				
JASPER GA					1	7.8					96
JEFF DAVIS GA					1	12.6					97
JOHNSON GA			2*	20.1							
LAMAR GA			1	6.1							
LANIER GA					1*	14.3					98
LAURENS GA					1	2.1					83
LIBERTY GA	2*	11.0			3*	11.2			2	41	97
LINCOLN GA			1*	9.6							
LOWNDES GA	1	1.3	3	4.2	2	1.9	215	-56	38	57	82
MC DUFFIE GA					1	3.0					86
MC INTOSH GA	1	6.7			1	4.9			-26	26	92
MITCHELL GA					2	5.4					93
MONROE GA			1	4.4	1	3.9		-11			89
MONTGOMERY GA	1*	11.1									
MUSCOGEE#GA	3	2.5	3	1.6	9	3.6	-34	120	45	61	88
NEWTON GA			1	4.1							
PEACH GA			1	3.5							
PIERCE GA	1*	13.3									
PUTNAM GA					1	6.6					95
RANDOLPH GA	1	3.2			1	2.3			-27	24	84
RICHMOND GA	2	1.2	9	4.5	10	4.0	279	-13	231	90	89
SCREVEN GA			1	2.2							
SPALDING GA			2	4.7	2	4.1		-12			90
STEWART GA			2*	9.1							
SUMTER GA			1	1.9	1	1.3		-28			80
TALBOT GA					1	4.9					92
TALIAFERRO GA			1*	9.4							
TATTNALL GA	1*	6.4	1	7.2	1	6.1	13	-15	-4	37	94
TELFAIR GA			1	5.6	2*	9.9		78			96
TERRELL GA			1	3.0	1	3.4		12			87
THOMAS GA			1	1.5	1	1.7		10			81
TIFT GA			1	3.7							
TOOMBS GA	1	4.4	1	4.3	1	3.2	-4	-25	-27	23	87
TREUTLEN GA					1	9.9					96
TROUP GA			3	4.0	2	2.6		-36			85
TWIGGS GA	1	6.4	1	6.5			3				
UPSON GA			1	3.2							
WALTON GA			1	4.4							
WARREN GA	1	10.2									
WASHINGTON GA	1	1.8	1	2.5	2	3.9	37	55	112	78	89
WAYNE GA	2*	12.8									
WHEELER GA			1*	16.1	1*	17.3		8			98
WILKES GA			1	4.5							
WILKINSON GA			1	5.8							
HAWAII HI			11	.8	22	.9		16			
HAWAII HI			2	1.3							
HONOLULU HI			7	.7	15	.8		16			79
KAUAI HI			1	.8	3	2.6		212			85
MAUI#HI			1	.6	4	1.9		229			82
ILLINOIS IL	87*	2.6	177*	3.5	277*	4.1	32	17	55		
ADAMS IL	1*	13.0									
ALEXANDER IL	1	3.1			1	7.3			132	81	95
CHAMPAIGN IL			1	2.5	1	2.1		-16			83
COLES IL					1*	44.6					99
COOK IL	76*	2.9	154*	3.7	255*	4.5	27	24	57	65	91
JACKSON IL	1	5.7			1	4.0			-30	22	90
KANE IL			1	3.6							
KANKAKEE IL			2	3.5	3	5.2		49			92
LAKE IL	1	2.9	3	7.0	1	.8	139	-88	-71	04	79
MACON IL	1	7.3	2	8.2			12				
MADISON IL	1	1.9									
MARION IL			1*	9.8							
MORGAN IL			1*	22.1							
PEORIA IL			2	4.2	3	5.5		29			93
PULASKI IL					1	3.8					89
RANDOLPH IL					1*	27.1					99
ROCK ISLAND IL	1*	9.6			1	2.3			-76	02	83
ST CLAIR IL	4	2.9	4	1.6	2	.8	-44	-54	-74	02	79
SALINE IL			1*	15.1							
SANGAMON IL			2	5.1	3	7.2		41			95
STEPHENSON IL			1*	15.6							
WILL IL			1	3.1	1	1.1		-65			79
WINNEBAGO IL			1	2.4	2	4.3		77			90

* SIGNIFICANTLY MORE DEATHS THAN EXPECTED

COUNTIES COMBINED; SEE TECHNICAL NOTES

NON-WHITE FEMALE

CANCER OF THE ESOPHAGUS

ICD 150

1950-1959 DEATHS	1950-1959 RATE	1960-1969 DEATHS	1960-1969 RATE	1970-1979 DEATHS	1970-1979 RATE	COUNTY AND STATE		PERCENT CHANGE 60S/50S	PERCENT CHANGE 70S/60S	PERCENT CHANGE 70S/50S	PERCENTILE %CHANGE 70S/50S	PERCENTILE RATES 1970S
16	1.8	40	3.1	59	3.6	INDIANA	IN	74	16	102		
2*	7.5	2	3.2	4	7.2	ALLEN	IN	-57	123	-4	36	95
		1	3.1	1	3.5	DELAWARE	IN		13			88
1*	45.5					FAYETTE	IN					
		1*	35.1	1*	16.1	GIBSON	IN		-54			98
		1	7.1			HOWARD	IN					
5	2.0	8	2.4	19	3.8	LAKE	IN	15	59	84	70	89
				1	4.4	LA PORTE	IN					91
				1	3.6	MADISON	IN					88
8	2.2	20*	3.6	29*	4.4	MARION	IN	65	23	103	75	91
				1*	21.3	MONROE	IN					99
		2	4.5	1	1.0	ST JOSEPH	IN		-78			79
		3	5.3			VANDERBURGH	IN					
		1	3.8	1	2.5	VIGO	IN		-35			84
		1	6.7			WAYNE	IN					
4	3.7	5	3.7	3	1.4	IOWA	IA	1	-61	-61		
		1	5.2			BLACK HAWK	IA					
2	4.4	4*	7.8	3	4.0	POLK	IA	75	-48	-9	34	90
1*	15.3					SCOTT	IA					
1*	14.2					WOODBURY	IA					
7	1.7	10	2.2	18	3.0	KANSAS	KS	30	36	77		
1*	39.3					BARTON	KS					
				1*	8.6	COWLEY	KS					96
				1*	20.2	CRAWFORD	KS					98
		1	5.9			DOUGLAS	KS					
1*	172.0					ELLSWORTH	KS					
1*	10.5			1	13.0	LABETTE	KS			24	50	97
				1	6.6	LEAVENWORTH	KS					95
				1*	22.3	MIAMI	KS					99
		1	6.1			MONTGOMERY	KS					
1*	61.3					NEOSHO	KS					
1*	26.5					SALINE	KS					
1	2.7	5*	6.9	1	.8	SEDGWICK	KS	161	-89	-70	05	79
				2	3.3	SHAWNEE	KS					87
1	.5	3	1.8	10	5.0	WYANDOTTE	KS	233	177	820	99	92
29*	2.7	40	3.5	61*	4.9	KENTUCKY	KY	28	40	80		
		1*	10.3			BELL	KY					
				3*	13.4	BOURBON	KY					98
		2*	13.2	1	5.4	BOYLE	KY		-59			93
		3	5.6	2	3.7	CHRISTIAN	KY		-33			88
				1	10.9	CLARK	KY					97
2*	11.0	1	5.6	1	6.5	DAVIESS	KY	-49	15	-41	16	94
7*	7.4	4	3.4	6	4.5	FAYETTE	KY	-53	31	-39	17	91
		1*	13.3			FULTON	KY					
		1	4.5	1	16.4	HARLAN	KY		263			98
				1	6.4	HENDERSON	KY					94
				1	3.0	HOPKINS	KY					86
14*	4.0	23*	5.0	34*	6.5	JEFFERSON	KY	27	30	65	66	94
4*	18.6			1	5.6	KENTON	KY			-70	05	93
1*	21.8					LINCOLN	KY					
				1	3.4	MC CRACKEN	KY					87
				1*	14.5	MARION	KY					98
		1*	9.4			MASON	KY					
		1	8.8	2*	19.5	SHELBY	KY		121			98
				1	6.2	TODD	KY					94
				2*	21.2	UNION	KY					99
				1	5.1	WARREN	KY					92
1*	14.4	1*	20.2	1*	12.8	WEBSTER	KY	41	-37	-11	33	97
		1*	11.4			WOODFORD	KY					
67	1.8	103	2.2	129	2.5	LOUISIANA	LA	24	11	37		
				3	6.3	ACADIA	LA					94
1	4.7			2	7.8	ALLEN	LA			64	65	95
1	1.7	3*	8.1	2	4.2	ASCENSION	LA	384	-48	154	84	90
		1	3.7			ASSUMPTION	LA					
				1	1.8	AVOYELLES	LA					82
1	5.4	1	6.0	1	4.6	BEAUREGARD	LA	11	-23	-14	30	91
		1	2.6			BIENVILLE	LA					
		1	1.7			BOSSIER	LA					
4	1.3	6	1.6	13	2.8	CADDO	LA	20	79	115	78	85
		6*	5.6	6	4.1	CALCASIEU	LA		-27			90
		1	8.0			CALDWELL	LA					
1	2.1	2	4.3			CLAIBORNE	LA	104				
1	2.3	3	6.4	2	3.7	CONCORDIA	LA	183	-42	64	66	88
3	1.0	8	2.7	8	2.1	EAST BATON ROUGE	LA	183	-23	118	79	82
3*	6.8	1	1.4			EAST CARROLL	LA	-80				
		1	2.0	2	3.9	EAST FELICIANA	LA		95			89
				1	4.1	EVANGELINE	LA					90
				1	1.8	FRANKLIN	LA					81
3*	5.8			4	5.6	IBERIA	LA			-3	37	93
				1	1.5	IBERVILLE	LA					80
				1	3.7	JACKSON	LA					88
		4	3.0	1	.6	JEFFERSON	LA		-81			78
				1	3.9	JEFFERSON DAVIS	LA					89
1	2.2	1	1.2	2	2.2	LAFAYETTE	LA	-45	86	1	41	83
				1	3.4	LAFOURCHE	LA					87
				2*	22.7	LA SALLE	LA					99
		1	1.6	1	2.1	LINCOLN	LA		35			83
		1	2.0	1	1.3	MADISON	LA		-36			80
		2	2.0	1	1.1	MOREHOUSE	LA		-43			79
		1	1.0			NATCHITOCHES	LA					
32*	3.8	31	2.7	34	2.7	ORLEANS	LA	-28	-2	-30	22	85

* SIGNIFICANTLY MORE DEATHS THAN EXPECTED

NON-WHITE FEMALE — CANCER OF THE ESOPHAGUS — ICD 150

1950-1959 DEATHS	1950-1959 RATE	1960-1969 DEATHS	1960-1969 RATE	1970-1979 DEATHS	1970-1979 RATE	COUNTY AND STATE	PERCENT CHANGE 60S/50S	PERCENT CHANGE 70S/60S	PERCENT CHANGE 70S/50S	PERCENTILE %CHANGE 70S/50S	PERCENTILE RATES 1970S
2	2.1	3	2.0	2	1.3	OUACHITA LA	-4	-33	-36	19	80
1	7.3					PLAQUEMINES LA					
2	5.0	1	2.3	2	4.5	POINTE COUPEE LA	-54	97	-10	33	91
1	.6	3	1.8	6	3.1	RAPIDES LA	196	70	402	95	86
		1	4.6	1	4.7	RED RIVER LA		1			91
1	3.5			1	2.3	RICHLAND LA			-35	19	83
		1*	15.7			ST BERNARD LA					
				2	8.3	ST HELENA LA					96
1	3.4	1	2.6	1	2.7	ST JAMES LA	-24	7	-19	29	85
		1	2.8	4*	9.9	ST JOHN THE BAPT LA		248			96
3	2.5	3	2.5	3	2.4	ST LANDRY LA	1	-6	-6	36	84
				1	2.5	ST MARTIN LA					84
		4*	6.2	2	3.3	ST MARY LA		-46			87
				2	3.7	ST TAMMANY LA					88
		3	3.6	3	3.1	TANGIPAHOA LA		-13			86
2	4.5	1	3.3			TENSAS LA	-26				
2	6.1	2	3.9			TERREBONNE LA	-37				
1	4.9	2*	9.8			VERMILION LA	99				
				1	6.6	VERNON LA					95
		1	1.3	4	6.5	WASHINGTON LA		404			94
				2	6.7	WEST BATON ROUGE LA					95
39	2.4	72	2.9	141*	4.3	MARYLAND MD	22	47	80		
3	2.7	4	3.5	4	2.5	ANNE ARUNDEL MD	28	-28	-8	35	84
1	2.3	3	3.1	5	3.9	BALTIMORE MD	37	27	73	68	89
				2	9.7	CALVERT MD					96
		1	8.0			CECIL MD					
				3	6.2	CHARLES MD					94
				2	4.5	DORCHESTER MD					91
				1	3.8	FREDERICK MD					89
2*	11.6			2	4.4	HARFORD MD			-62	09	91
		2*	11.4	1	5.9	KENT MD		-48			94
				3	1.4	MONTGOMERY MD					80
3	3.9	3	1.9	13	4.5	PRINCE GEORGE'S MD	-51	131	13	45	91
		1	3.5	1	3.8	QUEEN ANNE'S MD		10			89
				1	4.7	ST MARY'S MD					91
1	3.1	2	6.1	2	4.3	SOMERSET MD	95	-29	38	58	90
1	2.9			4	7.1	WICOMICO MD			146	83	95
				3	8.1	WORCESTER MD					96
28*	2.7	56*	3.6	94*	4.9	BALTIMORE CITY MD	35	34	82	69	92
8	2.0	20	3.4	21	2.2	MASSACHUSETTS MA	68	-35	8		
		1*	11.7	1	9.0	BERKSHIRE MA		-23			96
		1	3.4			BRISTOL MA					
				1	1.4	ESSEX MA					80
1	3.3	1	1.4	3	2.5	HAMPDEN MA	-56	72	-25	27	84
1*	74.3					HAMPSHIRE MA					
1	1.5	2	2.7	4	2.9	MIDDLESEX MA	84	5	94	72	86
1	8.6			1	2.9	PLYMOUTH MA			-67	06	86
3	1.3	12	3.6	11	2.4	SUFFOLK MA	171	-33	83	70	84
1*	6.7	3*	17.7			WORCESTER MA	164				
29	1.4	84	2.6	164	3.5	MICHIGAN MI	86	32	146		
				1	13.0	ALLEGAN MI					97
1	2.4	2	3.9	2	2.1	BERRIEN MI	66	-46	-11	33	83
				1	1.7	CALHOUN MI					81
		1	4.2	1	2.2	CASS MI		-48			83
1	1.0	5	2.9	5	2.4	GENESEE MI	188	-20	131	81	84
				3	4.8	INGHAM MI					92
		2*	7.9	2	4.5	KALAMAZOO MI		-43			91
		1	1.2	5	5.3	KENT MI		328			93
				1	5.8	LAKE MI					94
				1	1.8	MACOMB MI					81
		1	1.4	1	2.6	MUSKEGON MI		80			84
		2	1.5	5	3.4	OAKLAND MI		120			87
		3	3.6	6	4.8	SAGINAW MI		34			92
		2*	16.6			ST CLAIR MI					
1	4.8	1	3.3	4	5.2	WASHTENAW MI	-32	57	7	44	92
26	1.6	64	2.7	126	3.6	WAYNE MI	72	34	130	81	88
		4	2.6	11	4.0	MINNESOTA MN		55			
				1*	106.0	CROW WING MN					99
		1	2.0	5	4.1	HENNEPIN MN		107			90
				1*	546.4	PIPESTONE MN					99
		2	4.2	3	4.2	RAMSEY MN		1			90
		1*	11.6	1	8.3	ST LOUIS MN		-29			96
39	1.1	45	1.2	65	1.6	MISSISSIPPI MS	10	33	47		
1	.9	3	3.4	5	4.5	ADAMS MS	266	31	381	94	91
3	1.5	1	.6	3	2.9	BOLIVAR MS	-63	414	93	72	85
				1	2.4	CLAIBORNE MS					84
		1	2.2	1	1.7	CLAY MS		-23			81
1	.7	3	1.7	4	2.4	COAHOMA MS	136	38	227	90	84
1	2.0					COPIAH MS					
1	1.6	2	3.8	1	1.9	DE SOTO MS	134	-51	14	46	82
				1	1.2	FORREST MS					79
		2*	9.9			FRANKLIN MS					
		1	1.3	2	1.8	HARRISON MS		41			81
6	2.6	2	.6	3	.8	HINDS MS	-75	22	-69	06	79
1	1.2	2	2.1	2	2.7	HOLMES MS	76	27	124	80	65
1	3.7			1	1.8	HUMPHREYS MS			-51	12	81
				1*	14.0	ISSAQUENA MS					98
		2	6.0	3	5.1	JACKSON MS		-14			92
		2	4.2	2	5.3	JEFFERSON MS		28			93
1	3.2					JEFFERSON DAVIS MS					
1	1.5	1	1.3	2	2.2	JONES MS	-10	65	48	62	83

* SIGNIFICANTLY MORE DEATHS THAN EXPECTED

NON-WHITE FEMALE

CANCER OF THE ESOPHAGUS

ICD 150

1950-1959 DEATHS	1950-1959 RATE	1960-1969 DEATHS	1960-1969 RATE	1970-1979 DEATHS	1970-1979 RATE	COUNTY AND STATE	PERCENT CHANGE 60S/50S	PERCENT CHANGE 70S/60S	PERCENT CHANGE 70S/50S	PERCENTILE %CHANGE 70S/50S	PERCENTILE RATES 1970S
1	.7	1	.7	1	.9	LAUDERDALE MS	-2	26	23	50	79
1	6.5					LAWRENCE MS					
				1	3.4	LEAKE MS					87
		1	1.9			LEE MS					
		1	.8	4	3.0	LEFLORE MS		271			86
1	3.1			1	2.2	LINCOLN MS			-26	25	83
		2	2.8	1	1.4	LOWNDES MS		-50			80
		1	1.1	3	3.0	MADISON MS		167			86
2*	5.0	1	3.7			MARION MS	-25				
		1	1.3			MONROE MS					
2*	7.1					NEWTON MS					
1	1.9	1	2.7			NOXUBEE MS	41				
				1	2.2	OKTIBBEHA MS					83
2	2.7			1	1.5	PANOLA MS			-46	14	80
1	6.0					PEARL RIVER MS					
2	3.8			3	4.1	PIKE MS			7	43	90
				1	5.7	PONTOTOC MS					94
1	2.8					QUITMAN MS					
		1	1.9			RANKIN MS					
				2	4.6	SCOTT MS					91
1	2.7					SHARKEY MS					
		1	1.0	2	2.2	SUNFLOWER MS		130			83
		1	1.8			TALLAHATCHIE MS					
				2	5.1	TATE MS					92
		1	1.8	2	5.1	TUNICA MS		184			92
		1	5.9			UNION MS					
		1	4.7			WALTHALL MS					
2	2.0	4	3.3	4	3.7	WARREN MS	62	11	79	69	88
3	1.4			3	1.5	WASHINGTON MS			1	40	80
1	3.9	1	4.8	1	5.3	WAYNE MS	23	11	36	56	93
		1	11.2			WEBSTER MS					
1	1.2	2	2.2			YAZOO MS	82				
35	2.3	58	2.8	101*	4.2	MISSOURI MO	22	48	79		
				1	7.3	BUCHANAN MO					95
				1	4.4	BUTLER MO					90
				1	6.7	CALLAWAY MO					95
1*	13.7					CAPE GIRARDEAU MO					
				1*	32.8	CHARITON MO					99
		1*	13.7			COLE MO					
1*	16.1					COOPER MO					
				1	5.4	GREENE MO					93
				1*	85.2	HOWELL MO					99
4	1.3	15	3.3	26*	4.6	JACKSON MO	159	37	256	91	91
		1*	12.0			JASPER MO					
				1*	12.8	JEFFERSON MO					97
1*	24.6					JOHNSON MO					
				1*	29.8	LIVINGSTON MO					99
1*	9.0	1*	7.3	1	14.0	MARION MO	-19	93	56	64	98
		1	5.7			MISSISSIPPI MO					
				2*	10.7	NEW MADRID MO					97
		1	2.2	1	2.1	PEMISCOT MO		-5			83
				1*	18.2	RANDOLPH MO					98
2	2.2	3	1.8	10	4.0	ST LOUIS COUNTY MO	-18	127	85	70	90
		2*	20.3	1	15.6	SALINE MO		-23			98
		1*	106.0			WRIGHT MO					
25*	3.2	32	2.9	51*	4.2	ST LOUIS CITY MO	-8	44	32	55	90
2	1.7	8*	4.8	11	5.1	NEBRASKA NB	179	6	195		
2	2.2	8*	6.2	11*	7.0	DOUGLAS NB	179	13	215	88	95
1	1.5	2	3.3	3	1.8	NEVADA NV	124	-46	21		
1	3.0	2	5.4	3	2.7	CLARK NV	79	-50	-11	32	85
33	2.1	54	2.1	142*	3.6	NEW JERSEY NJ	-1	69	68		
3	2.6	3	1.7	10	5.3	ATLANTIC NJ	-33	209	106	76	93
		2	1.6	9	4.3	BERGEN NJ		165			90
1	3.8			1	1.3	BURLINGTON NJ			-65	08	80
3	2.3	3	1.3	13	5.0	CAMDEN NJ	-46	303	119	80	92
1	4.0	3	3.9	2	2.7	CUMBERLAND NJ	-2	-32	-33	20	85
13	2.4	16	1.7	52	3.9	ESSEX NJ	-30	136	65	66	89
1	2.2	1	1.5	1	1.3	GLOUCESTER NJ	-34	-13	-42	16	80
3	4.5	9*	6.3	13	3.7	HUDSON NJ	38	-40	-17	30	89
		4	2.4	8	3.4	MERCER NJ		45			87
1	1.9	2	4.0	6	4.8	MIDDLESEX NJ	111	21	155	85	92
2	1.5	8*	5.6	4	1.9	MONMOUTH NJ	268	-66	25	51	82
		1	2.5			MORRIS NJ					
				1	2.3	OCEAN NJ					84
		1	.6	11	4.8	PASSAIC NJ		673			92
3*	8.8			1	2.1	SALEM NJ			-76	01	82
				3	5.4	SOMERSET NJ					93
2	1.3	1	.7	7	2.3	UNION NJ	-51	253	74	69	84
1	.9	1	.5	2	.5	NEW MEXICO NM	-52	10	-47		
				1	7.6	LEA NM					95
				1*	26.8	LUNA NM					99
		1	1.5			MC KINLEY NM					
1	7.6					SAN JUAN NM					
125*	2.6	283*	3.7	473*	3.8	NEW YORK NY	42	3	46		
		3	5.5	5*	6.6	ALBANY NY		20			94
11	2.9	20	1.8	75	4.0	BRONX NY	-36	117	39	59	90
1*	14.1					CATTARAUGUS NY					
		1*	35.1			CAYUGA NY					
				2*	16.3	CHEMUNG NY					98

* SIGNIFICANTLY MORE DEATHS THAN EXPECTED

NON-WHITE FEMALE — CANCER OF THE ESOPHAGUS — ICD 150

1950-1959 DEATHS	1950-1959 RATE	1960-1969 DEATHS	1960-1969 RATE	1970-1979 DEATHS	1970-1979 RATE	COUNTY AND STATE		PERCENT CHANGE 60S/50S	PERCENT CHANGE 70S/60S	PERCENT CHANGE 70S/50S	PERCENTILE %CHANGE 70S/50S	PERCENTILE RATES 1970S
1	2.3	2	5.5	1	1.7	DUTCHESS	NY	140	-69	-26	26	81
8*	3.6	10	2.9	13	3.0	ERIE	NY	-20	2	-19	29	86
				1	13.6	FRANKLIN	NY					98
23	1.7	59	3.1	111	3.5	KINGS(BROOKLYN)	NY	88	12	111	77	88
1*	844.2					LEWIS	NY					
1	2.2	2	2.0	6	2.5	MONROE	NY	-8	22	13	45	84
2	3.6	4	1.6	12	3.3	NASSAU	NY	-54	100	-9	34	87
62*	3.3	138*	5.9	140*	4.7	NEW YORK(MANHATTAN)	NY	81	-20	45	61	91
		1	2.0			NIAGARA	NY					
		1	5.5			ONEIDA	NY					
1	2.4	3	4.2	3	3.2	ONONDAGA	NY	75	-25	32	54	87
		1	4.8	2	3.7	ORANGE	NY		-23			88
		1*	10.7	1*	26.5	ORLEANS	NY		147			99
7	1.8	17	2.1	56	3.0	QUEENS	NY	15	44	65	67	86
		2*	24.2	1	5.7	RENSSELAER	NY		-76			94
1	3.1			2	2.9	RICHMOND	NY			-6	35	86
		1	4.2	2	2.5	ROCKLAND	NY		-41			84
				1	7.0	SCHENECTADY	NY					95
2	2.6	5	2.5	9	3.6	SUFFOLK	NY	-1	40	38	58	88
				1	10.9	TOMPKINS	NY					97
1*	11.9			1	3.5	ULSTER	NY			-71	04	87
3	1.5	12	2.9	28*	5.0	WESTCHESTER	NY	86	75	225	89	92
60	1.7	70	1.6	117	2.0	NORTH CAROLINA	NC	-7	30	21		
2	5.7	1	2.2			ALAMANCE	NC	-61				
1*	34.4					ALEXANDER	NC					
1	2.7					ANSON	NC					
1	2.8	1	1.9			BEAUFORT	NC	-31				
1	3.1					BERTIE	NC					
		1	2.8	1	2.0	BLADEN	NC		-28			82
				2	5.0	BRUNSWICK	NC					92
2	2.2	3	3.6	2	2.2	BUNCOMBE	NC	65	-40	-2	38	83
1*	5.7					BURKE	NC					
2	4.7			3	5.4	CABARRUS	NC			15	47	93
		1	8.6	2*	11.0	CALDWELL	NC		28			97
				1	3.0	CATAWBA	NC					86
3*	12.1					CHATHAM	NC					
1	1.8	2	3.6	4	7.4	CLEVELAND	NC	103	105	315	93	95
				1	1.8	COLUMBUS	NC					81
2	3.0	1	1.1	1	1.4	CRAVEN	NC	-63	24	-55	10	80
		2	1.7	2	1.0	CUMBERLAND	NC		-43			79
		2	7.2			DAVIDSON	NC					
		1	1.7			DUPLIN	NC					
		4	2.5	3	1.3	DURHAM	NC		-49			80
		1	.9	1	.9	EDGECOMBE	NC		0			79
1	.4	3	1.2	10	3.8	FORSYTH	NC	233	216	953	99	89
2	4.8	2	4.0			FRANKLIN	NC	-17				
1	1.2	1	1.7	3	2.9	GASTON	NC	33	73	130	80	85
				1	1.0	GRANVILLE	NC					79
1	6.3					GREENE	NC					
3	1.7	5	2.3	7	2.3	GUILFORD	NC	31	-1	30	53	83
2	2.3					HALIFAX	NC					
		1	2.1	2	3.4	HARNETT	NC		67			87
		2*	16.8			HENDERSON	NC					
				2	2.6	HERTFORD	NC					84
		1	4.4	1	3.4	HOKE	NC		-23			87
				1	1.8	IREDELL	NC					81
		2	6.6	1	2.4	LEE	NC		-63			84
1	1.6			3	2.6	LENOIR	NC			61	65	84
1	2.8	1	2.3	1	1.9	MARTIN	NC	-18	-14	-29	23	82
2	1.4	1	1.0	12	3.0	MECKLENBURG	NC	-33	213	111	77	86
				1	4.1	MONTGOMERY	NC					90
1	2.1	1	2.0	3	5.5	MOORE	NC	-1	169	167	86	93
		1	1.1	4	4.3	NASH	NC		286			90
1	1.6	3	2.9	5	3.9	NEW HANOVER	NC	89	33	150	83	89
4*	6.8	1	1.6	1	1.5	NORTHAMPTON	NC	-76	-8	-78	01	80
				1	1.7	ONSLOW	NC					81
1	4.9			1	1.6	ORANGE	NC			-66	07	81
1	4.4					PASQUOTANK	NC					
				1	2.8	PENDER	NC					85
				1	3.3	PERQUIMANS	NC					87
1	.8	5	4.5			PITT	NC	465				
1	9.4	3*	14.0	2	6.9	RANDOLPH	NC	48	-51	-27	25	95
2	3.8	2	3.7			RICHMOND	NC	-2				
3	1.7	2	1.3	4	1.7	ROBESON	NC	-24	37	4	42	81
		1	1.6	3	4.3	ROCKINGHAM	NC		172			90
1	3.4	2	3.1	3	4.0	ROWAN	NC	-7	26	17	48	89
		1	1.9	1	.9	SAMPSON	NC		-50			79
		1	2.4	2	3.5	SCOTLAND	NC		51			88
2*	14.0					STANLY	NC					
1*	10.5					SURRY	NC					
				1*	21.5	TRANSYLVANIA	NC					99
1	2.2	1	2.8			UNION	NC	24				
		1	1.5	1	1.5	VANCE	NC		-3			80
3	2.1	3	1.6	7	2.6	WAKE	NC	-25	65	23	50	85
2	5.2			1	1.9	WARREN	NC			-63	08	82
4	3.7	2	1.7	3	2.1	WAYNE	NC	-54	22	-44	15	82
		1	1.0	4	3.8	WILSON	NC		281			89
57	2.1	99	2.8	198*	4.1	OHIO	OH	31	48	93		
1*	8.9					BELMONT	OH					
2*	5.4			5*	9.1	BUTLER	OH			68	68	96
				2	2.7	CLARK	OH					85
				1*	13.8	CLERMONT	OH					98
22*	2.5	27	2.3	58	3.7	CUYAHOGA	OH	-7	57	46	62	88
		1*	21.5			FAYETTE	OH					
7	2.5	15	4.2	18	3.5	FRANKLIN	OH	66	-16	39	59	88
				2*	30.8	GEAUGA	OH					99
		2	6.9	3*	9.1	GREENE	OH		31			96

* SIGNIFICANTLY MORE DEATHS THAN EXPECTED

NON-WHITE FEMALE — CANCER OF THE ESOPHAGUS — ICD 150

1950-1959 DEATHS	1950-1959 RATE	1960-1969 DEATHS	1960-1969 RATE	1970-1979 DEATHS	1970-1979 RATE	COUNTY AND STATE	PERCENT CHANGE 60S/50S	PERCENT CHANGE 70S/60S	PERCENT CHANGE 70S/50S	PERCENTILE %CHANGE 70S/50S	PERCENTILE RATES 1970S
13	2.8	26*	3.9	53*	6.7	HAMILTON OH	39	72	139	82	95
		1	3.6	1	3.2	JEFFERSON OH		-11			87
				1	7.5	LAKE OH					95
		1*	15.1			LICKING OH					
1*	33.1					LOGAN OH					
		1	2.2	5	7.8	LORAIN OH		253			96
3	2.6	9	5.7	12	4.4	LUCAS OH	116	-23	67	67	91
1	1.0			8	3.8	MAHONING OH			271	92	89
				1	7.9	MIAMI OH					96
3	1.0	5	1.8	11	2.9	MONTGOMERY OH	78	57	180	87	85
1*	6.2	1	4.6			MUSKINGUM OH	-25				
1*	12.9					PORTAGE OH					
				1	3.4	RICHLAND OH					87
				1*	17.2	SANDUSKY OH					98
				1	13.6	SCIOTO OH					98
		1	1.1	2	1.8	STARK OH		63			81
1	.5	6	3.6	11	4.5	SUMMIT OH	570	24	733	98	91
1	5.2	1	1.1	1	1.4	TRUMBULL OH	-78	24	-73	03	80
		1*	42.0			TUSCARAWAS OH					
		1*	22.3			WARREN OH					
8	.9	11	.9	32	2.0	OKLAHOMA OK	3	128	135		
				1	4.6	ADAIR OK					91
				1*	11.7	ATOKA OK					97
				1	13.7	CANADIAN OK					98
		1	5.7			CARTER OK					
1	8.3					CHEROKEE OK					
				1	1.6	COMANCHE OK					81
				1	2.9	CREEK OK					86
1*	10.1					DELAWARE OK					
		2*	16.4	1	9.8	GARFIELD OK		-40			96
1*	10.7			1	10.2	GRADY OK			-5	36	96
				1	7.7	HUGHES OK					95
				1	11.6	JACKSON OK					97
				2*	15.7	KAY OK					98
				1*	28.9	MURRAY OK					99
3	3.3			5	4.7	MUSKOGEE OK			42	61	91
		2	.9	6	2.0	OKLAHOMA OK		116			82
				2	3.9	OKMULGEE OK					89
				1	3.7	OTTAWA OK					88
		1	3.1			PITTSBURG OK					
1*	8.4					ROGERS OK					
		1	6.8	1	6.3	SEQUOYAH OK		-8			94
1	.9	4	2.0	5	1.8	TULSA OK	117	-10	96	73	81
		2	1.5	1	.4	OREGON OR		-77			
		2	2.4	1	.6	MULTNOMAH OR		-75			78
59	1.8	128	2.9	269*	4.9	PENNSYLVANIA PA	59	70	170		
9	1.7	14	2.0	34	4.0	ALLEGHENY PA	19	101	139	82	90
1	1.9	1	3.3	3	4.7	BEAVER PA	76	43	152	84	91
		1	3.4	5*	13.2	BERKS PA		284			97
1*	14.1					BUCKS PA					
1*	7.6					CAMBRIA PA					
				1*	163.6	CARBON PA					99
		1	1.3	1	1.4	CHESTER PA		9			80
3	4.3	6*	6.0	7	5.2	DAUPHIN PA	40	-14	20	49	92
3	2.1	1	.4	7	2.7	DELAWARE PA	-81	566	27	52	85
		1	3.3	2	4.0	ERIE PA		19			89
1	2.2					FAYETTE PA					
				1	5.6	FRANKLIN PA					93
				1*	13.8	LACKAWANNA PA					98
		1	6.3	4*	14.7	LANCASTER PA		133			98
		1	8.4			LAWRENCE PA					
1	1.3	2	1.9	5	3.6	MONTGOMERY PA	41	94	173	86	88
2*	35.6			3*	20.4	NORTHAMPTON PA			-43	16	98
37	1.7	99*	3.6	190*	5.5	PHILADELPHIA PA	105	55	217	89	93
				1	2.1	WASHINGTON PA					83
				3	8.4	WESTMORELAND PA					96
				1	2.7	YORK PA					85
2	3.3			3	2.3	RHODE ISLAND RI			-30		
1	2.2			3	2.9	PROVIDENCE RI			30	54	86
1*	24.6					WASHINGTON RI					
107*	3.8	90	2.9	143*	3.8	SOUTH CAROLINA SC	-23	30	1		
		1	3.3			ABBEVILLE SC					
2	2.0	3	3.4	4	3.9	AIKEN SC	69	15	94	73	89
2*	6.9	1	3.3	1	4.5	ALLENDALE SC	-52	35	-35	19	91
		1	1.2	1	1.1	ANDERSON SC		-1			79
				1	2.5	BAMBERG SC					84
1	2.2	1	2.4			BARNWELL SC	6				
9*	14.2	8*	12.7	11*	15.0	BEAUFORT SC	-11	18	5	43	98
6*	9.8	2	3.3	4	5.4	BERKELEY SC	-66	61	-45	14	93
1	3.8	2	6.5	3*	8.3	CALHOUN SC	73	26	119	79	96
28*	10.3	19*	6.6	29*	7.8	CHARLESTON SC	-36	18	-24	27	96
1	3.2			1	2.8	CHEROKEE SC			-12	32	85
1	2.0	1	2.1	1	1.9	CHESTER SC	6	-8	-3	37	82
3*	8.9	1	2.3			CHESTERFIELD SC	-75				
4*	6.8	1	1.6	4	4.9	CLARENDON SC	-77	214	-27	23	92
3*	6.1	4*	7.5	2	3.1	COLLETON SC	23	-59	-50	13	86
1	1.4	2	2.8	2	1.9	DARLINGTON SC	106	-32	39	59	82
1	2.5	1	3.0	1	2.2	DILLON SC	20	-28	-14	31	83
1	2.7	4*	10.5	2	3.1	DORCHESTER SC	291	-70	15	47	86
				1	3.2	EDGEFIELD SC					87
7*	7.3	2	1.5	3	1.7	FLORENCE SC	-79	12	-77	01	81

* SIGNIFICANTLY MORE DEATHS THAN EXPECTED

NON-WHITE FEMALE — CANCER OF THE ESOPHAGUS — ICD 150

1950-1959 DEATHS	1950-1959 RATE	1960-1969 DEATHS	1960-1969 RATE	1970-1979 DEATHS	1970-1979 RATE	COUNTY AND STATE		PERCENT CHANGE 60S/50S	PERCENT CHANGE 70S/60S	PERCENT CHANGE 70S/50S	PERCENTILE %CHANGE 70S/50S	PERCENTILE RATES 1970S
7*	14.1	6*	9.8	4	5.7	GEORGETOWN	SC	-31	-42	-60	09	93
		4	2.4	4	2.2	GREENVILLE	SC		-6			83
		1	1.7	3	4.3	GREENWOOD	SC		149			90
3*	7.7	1	2.8	2	3.7	HAMPTON	SC	-64	33	-53	11	88
1	2.2			1	1.3	HORRY	SC			-39	18	80
2*	8.2			4*	10.6	JASPER	SC			30	54	97
		1	2.5			LANCASTER	SC					
1	1.6	1	1.8	4	5.9	LAURENS	SC	8	233	258	91	94
		2	5.1	1	1.8	LEXINGTON	SC		-65			81
		1	5.2	1	5.1	MC CORMICK	SC		-2			92
		1	1.4	3	4.1	MARION	SC		196			90
3*	5.4	1	3.0	1	1.6	MARLBORO	SC	-44	-47	-70	05	81
		2	3.8	1	2.4	NEWBERRY	SC		-37			84
				3*	12.4	OCONEE	SC					97
10*	6.6	3	1.9	12*	6.6	ORANGEBURG	SC	-72	255	0	39	95
2	.9	6	2.0	10	3.2	RICHLAND	SC	124	63	266	92	87
				1	4.0	SALUDA	SC					89
		1	.6	9	5.1	SPARTANBURG	SC		778			92
3	2.4	3	2.1	4	2.8	SUMTER	SC	-14	35	16	48	85
2	2.6	1	1.2	3	2.9	WILLIAMSBURG	SC	-53	145	15	47	86
2	3.2	1	1.2	1	.8	YORK	SC	-62	-32	-74	02	79
				1	1.0	SOUTH DAKOTA	SD					
				1	4.6	SHANNON	SD					91
36	1.4	83	2.8	97	3.0	TENNESSEE	TN	96	5	105		
				1	9.0	ANDERSON	TN					96
				2*	119.2	BENTON	TN					99
		1	6.9			BLOUNT	TN					
1*	32.6					COCKE	TN					
9	2.5	20*	4.8	23*	5.0	DAVIDSON	TN	95	3	102	74	92
		1	4.3	1	2.3	DYER	TN		-46			83
1	1.9					FAYETTE	TN					
1	2.5			2	5.1	GIBSON	TN			108	76	92
				1	6.2	GILES	TN					94
2	1.1	7	2.7	10	3.6	HAMILTON	TN	148	33	230	90	88
1	2.8					HARDEMAN	TN					
1	1.6					HAYWOOD	TN					
				1	3.8	HENRY	TN					89
				1*	38.5	HICKMAN	TN					99
		1*	37.8			HUMPHREYS	TN					
		1*	94.4			JOHNSON	TN					
1	.9	8*	6.1	4	2.8	KNOX	TN	559	-54	204	88	85
				1*	54.6	LOUDON	TN					99
2	2.1	1	.8	3	2.6	MADISON	TN	-63	239	26	52	85
1*	14.4	1*	14.5	1*	14.8	MARION	TN	1	2	2	41	98
		1*	21.3			MONROE	TN					
		1	2.2	2	3.7	MONTGOMERY	TN		65			88
				1	3.9	OBION	TN					89
		1*	24.8			RHEA	TN					
1*	20.2	1*	11.5			ROANE	TN	-43				
12	1.4	32	2.9	40	3.0	SHELBY	TN	103	6	115	79	86
1	5.7					SUMNER	TN					
1	2.3					TIPTON	TN					
		1*	21.1			TROUSDALE	TN					
1*	7.1			1	6.9	WASHINGTON	TN			-3	38	95
				1	8.0	WEAKLEY	TN					96
		1	4.8	1	4.8	WILLIAMSON	TN		0			91
		4*	20.5			WILSON	TN					
65	1.5	101	1.8	172	2.1	TEXAS	TX	20	19	43		
		2	3.7			ANDERSON	TX					
1	2.8	1	2.5	1	2.2	ANGELINA	TX	-10	-10	-19	29	83
2*	7.1					BASTROP	TX					
		1*	27.0	1	10.3	BEE	TX		-62			97
				1	2.1	BELL	TX					82
1	1.1	6	2.5	12	3.0	BEXAR	TX	137	20	185	87	86
2	3.1			2	2.1	BOWIE	TX			-31	22	83
		1	3.1	2	3.5	BRAZORIA	TX		13			88
2*	13.6			1	5.5	CALDWELL	TX			-60	09	93
1	7.5					CAMP	TX					
				1	1.9	CASS	TX					82
1*	23.4	1*	12.5			COOKE	TX	-47				
1*	80.4					CROCKETT	TX					
6	1.5	9	1.3	29	3.2	DALLAS	TX	-16	146	108	76	87
1	4.9	1	6.0			DE WITT	TX	21				
				1*	24.6	EASTLAND	TX					99
				2	7.6	ECTOR	TX					95
4*	9.1			2	4.4	ELLIS	TX			-52	11	90
				1	.3	EL PASO	TX					78
				2	5.8	FALLS	TX					94
1*	344.1					FOARD	TX					
		1	3.2			FREESTONE	TX					
2	2.1	2	1.7	13*	7.2	GALVESTON	TX	-17	316	247	91	95
		1*	19.9			GRAY	TX					
1	2.2	3*	8.5			GRIMES	TX	279				
1	5.4					GUADALUPE	TX					
1	6.0	1	5.6	1	5.5	HARDIN	TX	-6	-2	-8	34	93
11	1.7	31	2.7	35	2.1	HARRIS	TX	61	-23	25	51	82
1	1.0	1	1.1	3	2.8	HARRISON	TX	6	152	167	86	85
		1	5.6			HENDERSON	TX					
				1	5.5	HILL	TX					93
				1	3.5	HOUSTON	TX					88
				1	5.2	HOWARD	TX					92
		2	8.9	2	7.4	HUNT	TX		-17			95
4	1.6	11*	5.0	12	4.0	JEFFERSON	TX	221	-21	154	84	89
				2*	23.0	JOHNSON	TX					99
				1	1.8	KAUFMAN	TX					81

* SIGNIFICANTLY MORE DEATHS THAN EXPECTED

NON-WHITE FEMALE — CANCER OF THE ESOPHAGUS — ICD 150

1950-1959 DEATHS	1950-1959 RATE	1960-1969 DEATHS	1960-1969 RATE	1970-1979 DEATHS	1970-1979 RATE	COUNTY AND STATE	PERCENT CHANGE 60S/50S	PERCENT CHANGE 70S/60S	PERCENT CHANGE 70S/50S	PERCENTILE %CHANGE 70S/50S	PERCENTILE RATES 1970S
				1*	21.8	KERR TX					99
				1	10.8	LAMB TX					97
		1	3.5	1	3.2	LIBERTY TX		-10			87
		2	4.9	2	5.6	LIMESTONE TX		14			93
1	3.2			1	1.0	LUBBOCK TX			-69	06	79
5*	5.4	2	1.5	3	1.6	MC LENNAN TX	-73	8	-71	04	81
		1	3.9	1	2.7	MATAGORDA TX		-32			85
1	3.0	2*	12.4			MILAM TX	307				
		1	2.4	2	6.0	MONTGOMERY TX		150			94
				1	2.2	NACOGDOCHES TX					83
1	1.6	1	2.6	4	3.2	NUECES TX	60	22	95	73	87
				1	1.9	POTTER TX					82
		1	3.4			ROBERTSON TX					
1	1.6	1	1.9	2	4.4	RUSK TX	24	127	181	87	91
				1	17.0	SAN PATRICIO TX					98
		1	3.7			SHELBY TX					
1	1.3	2	1.7	1	.9	SMITH TX	31	-47	-31	21	79
2	1.0	3	1.0	12	2.9	TARRANT TX	-2	197	191	88	86
				1	6.1	TITUS TX					94
		2	1.3	3	1.3	TRAVIS TX		-1			80
2*	9.7					UPSHUR TX					
		3*	13.5	1	2.1	VICTORIA TX		-84			83
1	3.0	1	2.7	1	2.0	WALKER TX	-11	-25	-34	20	82
1	4.8			1	5.3	WALLER TX			11	44	93
1	1.6					WASHINGTON TX					
		1*	73.7			WHEELER TX					
2*	5.1					WICHITA TX					
2*	7.7					WILLIAMSON TX					
				1*	37.8	WINKLER TX					99
				1	.6	UTAH UT					
				1	1.3	SALT LAKE UT					80
				1*	9.0	VERMONT VT					
				1*	546.4	ADDISON VT					99
41	1.4	95	2.7	176*	4.0	VIRGINIA VA	94	46	185		
1	1.9	2	5.8	1	1.9	ACCOMACK VA	202	-68	-2	38	82
2	4.4	1	1.9	2	2.5	ALBEMARLE#VA	-57	32	-44	15	84
				1	3.9	ALLEGHANY#VA					89
1	5.9	1	5.0			AMHERST VA	-16				
				1	9.6	APPOMATTOX VA					96
1	1.0	1	.8	8*	6.2	BEDFORD#VA	-25	724	518	95	94
		1	2.4			BRUNSWICK VA					
		1	4.4			BUCKINGHAM VA					
		2*	9.2	1	3.5	CAROLINE VA		-62			88
				1	3.7	CHARLES CITY VA					88
5	.9	16	3.2	32*	5.3	CHESTERFIELD#VA	254	64	479	95	93
				1	5.3	CULPEPER VA					93
1	6.6	1	8.3			CUMBERLAND VA	25				
1	.7	8	4.4	12*	6.1	DINWIDDIE#VA	545	39	797	98	94
		1	6.2			ESSEX VA					
1	.7	5	5.5	8	4.4	FAIRFAX#VA	708	-20	546	96	91
		2*	8.5	3*	10.3	FAUQUIER VA		21			97
				1	5.7	FLUVANNA VA					94
				2*	10.5	GOOCHLAND VA					97
1	2.6					GREENSVILLE#VA					
3	5.0	2	3.0	2	2.4	HALIFAX#VA	-40	-19	-51	12	84
				2	6.1	HANOVER VA					94
1	4.5	1	3.1	1	2.1	HENRICO VA	-31	-31	-52	11	83
2	4.4	1	1.5			HENRY#VA	-66				
		1	3.7	2	5.3	ISLE OF WIGHT VA		43			92
1	4.2			4	5.5	JAMES CITY#VA			33	55	93
1*	6.9					KING AND QUEEN VA					
				1	10.5	KING GEORGE VA					97
				1	6.3	KING WILLIAM VA					94
				2*	10.2	LANCASTER VA					97
		1	4.9			LUNENBURG VA					
1*	12.8					MADISON VA					
		3*	32.7			MATHEWS VA					
				1	1.6	MECKLENBURG VA					81
				2*	8.3	MIDDLESEX VA					96
				1	8.0	MONTGOMERY#VA					96
1	7.3					NELSON VA					
1	2.5	1	3.7	1	2.2	NORTHAMPTON VA	46	-41	-14	31	83
				1	2.3	NOTTOWAY VA					83
1*	5.7	2*	14.3			ORANGE VA	150				
2	2.2	2	1.7	1	.6	PITTSYLVANIA#VA	-23	-66	-73	03	78
1	3.3			1	3.5	PRINCE EDWARD VA			5	42	88
		1*	12.4			PULASKI VA					
		1*	10.9			RICHMOND VA					
1	.9	3	2.7	9*	7.6	ROANOKE#VA	197	186	748	98	95
				1	8.7	ROCKBRIDGE#VA					96
		1*	14.6	1	12.5	ROCKINGHAM#VA		-15			97
1	1.8			3	4.8	SOUTHAMPTON#VA			163	85	92
				1	3.0	SPOTSYLVANIA#VA					86
				2	7.5	SUSSEX VA					95
		1*	7.9			TAZEWELL VA					
1	5.6					WESTMORELAND VA					
4	.8	18	2.6	36	4.1	CHESAPEAKE CITY#VA	209	60	394	94	90
		6*	7.2	9*	6.8	HAMPTON CITY VA		-6			95
4	4.4	6	3.8	11*	5.6	NEWPORT NEWS CITY VA	-13	47	28	53	93
1	.8	2	1.7	6	4.8	SUFFOLK CITY VA	124	180	529	96	92
2	1.2	4	1.2	8	1.2	WASHINGTON WA	1	1	2		
				2*	69.4	CHELAN WA					99
		1*	32.6			CLALLAM WA					

* SIGNIFICANTLY MORE DEATHS THAN EXPECTED

\# COUNTIES COMBINED; SEE TECHNICAL NOTES

NON-WHITE FEMALE — CANCER OF THE ESOPHAGUS — ICD 150

1950-1959 DEATHS	1950-1959 RATE	1960-1969 DEATHS	1960-1969 RATE	1970-1979 DEATHS	1970-1979 RATE	COUNTY AND STATE	PERCENT CHANGE 60S/50S	PERCENT CHANGE 70S/60S	PERCENT CHANGE 70S/50S	PERCENTILE %CHANGE 70S/50S	PERCENTILE RATES 1970S
1	1.2	1	.5	4	1.2	KING WA	-53	112	-1	39	79
		1*	12.4			OKANOGAN WA					
				1	1.8	PIERCE WA					81
1	8.9			1	3.3	SPOKANE WA			-63	08	87
		1	3.5			YAKIMA WA					
4	.8	7	1.6	14	3.1	WEST VIRGINIA WV	101	96	293		
2*	7.0			3*	8.8	CABELL WV			26	51	96
				2	5.0	FAYETTE WV					92
1*	7.1					GREENBRIER WV					
		4	5.1	3	3.5	KANAWHA WV		-31			88
		1	1.4	1	1.3	MC DOWELL WV		-2			80
				1*	117.9	MARSHALL WV					99
		2	4.8			MERCER WV					
				1	12.9	MINGO WV					97
				1*	56.2	MONROE WV					99
				1	6.5	OHIO WV					94
1	1.9			1	3.5	RALEIGH WV			88	71	88
5	2.6	9	3.5	11	2.1	WISCONSIN WI	37	-41	-19		
5*	4.1	9	5.3	11	3.0	MILWAUKEE WI	31	-44	-27	25	86
1	5.3					WYOMING WY					
1*	40.9					NATRONA WY					

* SIGNIFICANTLY MORE DEATHS THAN EXPECTED

WHITE MALE

CANCER OF THE STOMACH

ICD 151

COUNTY AND STATE	1950-1959 DEATHS	1950-1959 RATE	1960-1969 DEATHS	1960-1969 RATE	1970-1979 DEATHS	1970-1979 RATE	PERCENT CHANGE 60S/50S	PERCENT CHANGE 70S/60S	PERCENT CHANGE 70S/50S	PERCENTILE %CHANGE 70S/50S	PERCENTILE RATES 1970S
UNITED STATES	125640	20.2	96672	13.2	75604	9.0	-35	-32	-56		
ALABAMA AL	1138	15.1	949	10.0	750	6.5	-33	-35	-57		
AUTAUGA AL	7	15.1	6	11.4	5	7.4	-25	-35	-51	60	49
BALDWIN AL	33	21.9	17	8.1	16	5.7	-63	-30	-74	16	29
BARBOUR AL	5	10.2	1	1.4	1	1.3	-86	-7	-87	03	05
BIBB AL	7	15.2	7	14.4	7	12.5	-6	-13	-18	89	88
BLOUNT AL	13	12.3	9	8.1	4	2.4	-34	-70	-80	08	07
BULLOCK AL	2	8.7	3	17.0	1	4.5	95	-74	-49	65	19
BUTLER AL	7	9.6	7	10.6	6	7.1	10	-33	-26	86	46
CALHOUN AL	29	13.7	28	10.3	22	6.4	-25	-38	-53	56	38
CHAMBERS AL	11	11.8	10	9.1	7	5.6	-22	-39	-52	57	28
CHEROKEE AL	8	13.5	6	8.3	6	7.3	-39	-11	-46	69	48
CHILTON AL	23	22.2	12	13.4	9	7.8	-40	-41	-65	32	54
CHOCTAW AL	6	22.2	4	9.4	5	11.7	-58	25	-47	67	85
CLARKE AL	9	15.3	6	8.5	1	1.4	-45	-84	-91	01	05
CLAY AL	4	6.7	14	21.8	8	11.0	224	-50	63	98	81
CLEBURNE AL	6	13.5	7	15.8	4	8.6	17	-46	-36	80	63
COFFEE AL	11	12.9	8	8.3	6	4.4	-36	-47	-66	30	18
COLBERT AL	9	9.3	16	10.2	12	5.8	9	-43	-38	79	30
CONECUH AL	11	22.8	4	7.0			-69				
COOSA AL	7	19.1	1	2.5	4	8.5	-87	243	-55	52	62
COVINGTON AL	14	11.2	12	7.6	9	5.3	-32	-30	-52	57	26
CRENSHAW AL	8	14.6	6	10.7	1	1.7	-27	-84	-88	02	05
CULLMAN AL	30	16.8	22	9.9	16	6.1	-41	-38	-63	35	34
DALE AL	7	10.4	1	1.4	5	5.9	-87	337	-43	73	32
DALLAS AL	7	8.6	8	9.1	8	7.2	6	-21	-16	90	47
DE KALB AL	26	16.3	19	9.9	26	11.2	-39	13	-32	83	82
ELMORE AL	12	12.0	14	13.1	11	7.5	9	-43	-38	79	50
ESCAMBIA AL	9	11.6	9	9.2	9	8.4	-21	-9	-27	85	61
ETOWAH AL	42	15.8	40	11.3	28	6.8	-28	-40	-57	49	42
FAYETTE AL	7	11.9	5	6.7	5	6.6	-44	-2	-45	71	40
FRANKLIN AL	17	18.9	22*	21.2	4	3.5	13	-83	-81	07	12
GENEVA AL	14	16.1	8	9.0	5	4.6	-44	-49	-71	20	20
GREENE AL	1	7.9	1	10.6	1	8.9	35	-16	13	95	66
HALE AL	2	6.6	5	16.5	5	12.2	148	-26	83	98	87
HENRY AL	5	13.3	1	2.7	4	7.1	-80	163	-47	68	46
HOUSTON AL	14	13.8	13	8.7	10	5.4	-37	-37	-61	41	27
JACKSON AL	24	17.5	17	11.4	11	6.5	-35	-43	-63	36	38
JEFFERSON AL	178	14.2	174	10.5	131	6.9	-26	-34	-51	60	43
LAMAR AL	8	14.1	5	7.6	8	10.3	-46	36	-27	85	77
LAUDERDALE AL	15	9.0	35	17.0	14	4.5	89	-74	-50	62	19
LAWRENCE AL	14	21.8	9	10.8	1	.9	-50	-92	-96	00	04
LEE AL	7	15.2	10	9.5	4	2.1	-38	-77	-86	03	06
LIMESTONE AL	23	24.8	18	15.0	5	3.1	-39	-79	-88	02	09
LOWNDES AL	4	22.5	2	13.3			-41				
MACON AL	6	24.0	2	7.3	3	8.8	-69	20	-63	35	65
MADISON AL	22	11.9	21	7.6	18	4.8	-36	-37	-59	44	22
MARENGO AL	5	12.0	3	6.6	3	5.9	-45	-10	-51	61	32
MARION AL	16	17.6	8	7.2	4	3.4	-59	-52	-81	08	11
MARSHALL AL	28	17.1	33	16.5	21	7.4	-3	-55	-57	49	49
MOBILE AL	89	17.7	51	7.5	76	9.2	-57	23	-48	66	69
MONROE AL	5	9.7	4	7.6	5	7.9	-22	4	-19	89	55
MONTGOMERY AL	32	12.8	24	7.3	24	5.7	-43	-22	-56	52	29
MORGAN AL	24	12.9	30	14.0	13	4.3	9	-69	-67	28	17
PERRY AL	1	4.1	1	3.0	3	7.8	-25	157	92	98	54
PICKENS AL	11	19.1	12	18.4	3	3.9	-4	-79	-80	09	14
PIKE AL	15	24.8	8	11.2	5	6.4	-55	-43	-74	16	38
RANDOLPH AL	11	14.7	7	8.7	6	6.4	-41	-26	-56	50	38
RUSSELL AL	9	13.2	11	12.8	15	13.3	-3	4	1	94	91
ST CLAIR AL	11	12.5	13	12.6	15	11.4	1	-10	-9	92	83
SHELBY AL	21	22.3	7	7.3	5	3.9	-67	-46	-82	06	15
SUMTER AL	5	24.4	2	9.9	3	10.0	-60	1	-59	44	75
TALLADEGA AL	29	22.0	19	11.2	18	8.3	-49	-26	-62	37	59
TALLAPOOSA AL	15	15.0	7	5.7	4	2.6	-62	-54	-82	06	08
TUSCALOOSA AL	20	8.3	20	6.7	26	6.6	-19	0	-20	89	40
WALKER AL	39	20.9	20	8.3	16	5.6	-60	-32	-73	18	29
WASHINGTON AL	8	20.3	7	15.2	5	9.8	-25	-35	-52	59	74
WILCOX AL	4	15.2	2	7.7	2	7.0	-49	-9	-54	55	44
WINSTON AL	16	23.5	15	20.6	12	13.4	-12	-35	-43	73	91
ALASKA AK			41	13.8	37	9.8		-29			
ARIZONA AZ	527	17.1	623	11.7	690	7.6	-31	-35	-56		
APACHE AZ	5	22.3	6	27.9			25				
COCHISE AZ	26	16.7	23	11.7	22	7.6	-30	-35	-55	54	51
COCONINO AZ	15	27.4	10	10.1	10	10.4	-63	3	-62	38	78
GILA AZ	14	17.8	10	10.5	14	10.1	-41	-4	-43	73	76
GRAHAM AZ	10	22.3	8	13.5	6	7.9	-39	-41	-64	33	55
GREENLEE AZ	5	18.2	3	10.4	5	18.2	-43	74	0	94	98
MARICOPA AZ	241	15.0	327	11.0	391	7.5	-26	-32	-50	62	50
MOHAVE AZ	6	13.7	5	6.6	12	4.6	-52	-31	-67	29	19
NAVAJO AZ	11	24.6	10	17.5	7	6.0	-29	-66	-75	15	33
PIMA AZ	118	20.8	136	12.6	135	7.4	-40	-41	-65	32	49
PINAL AZ	15	11.0	34	16.9	23	8.2	53	-51	-26	86	58
SANTA CRUZ AZ	6	19.2	9	20.0	6	9.7	4	-52	-50	63	73
YAVAPAI AZ	29	18.2	23	10.0	30	7.6	-45	-24	-58	47	52
YUMA AZ	26	24.9	19	11.6	29	9.8	-53	-16	-61	41	74
ARKANSAS AR	954	14.5	810	10.4	678	7.3	-28	-31	-50		
ARKANSAS AR	14	17.1	9	9.9	4	3.6	-42	-64	-79	09	12
ASHLEY AR	7	11.4	5	8.1	4	5.1	-29	-37	-55	53	24
BAXTER AR	11	16.7	5	4.2	13	6.8	-75	63	-59	45	43
BENTON AR	29	12.9	22	8.1	27	7.0	-37	-14	-46	69	45
BOONE AR	10	12.0	9	7.8	13	9.0	-35	15	-25	87	67

* SIGNIFICANTLY MORE DEATHS THAN EXPECTED

WHITE MALE

CANCER OF THE STOMACH — COUNTY AND STATE

ICD 151

COUNTY AND STATE	1950-1959 DEATHS	1950-1959 RATE	1960-1969 DEATHS	1960-1969 RATE	1970-1979 DEATHS	1970-1979 RATE	PERCENT CHANGE 60S/50S	PERCENT CHANGE 70S/60S	PERCENT CHANGE 70S/50S	PERCENTILE %CHANGE 70S/50S	PERCENTILE RATES 1970S
BRADLEY AR	8	16.5	9	15.1	5	7.3	-8	-52	-56	51	48
CALHOUN AR	5	20.4			1	3.0			-85	04	09
CARROLL AR	8	9.4	9	9.3	11	10.0	-1	7	6	95	76
CHICOT AR	5	13.4	5	11.4	3	6.6	-15	-42	-50	62	40
CLARK AR	7	8.8	7	8.5	14*	15.7	-3	85	79	98	95
CLAY AR	11	9.6	19	15.9	13	9.2	66	-43	-4	93	69
CLEBURNE AR	10	18.1	9	13.5	2	2.2	-26	-84	-88	02	06
CLEVELAND AR	4	11.3	5	15.2	1	2.5	35	-84	-78	11	07
COLUMBIA AR	9	11.8	9	9.3	10	9.4	-21	0	-21	88	70
CONWAY AR	12	19.5	6	8.1	5	5.9	-59	-27	-70	23	32
CRAIGHEAD AR	32	17.2	23	11.4	22	8.9	-34	-22	-48	66	66
CRAWFORD AR	8	7.4	12	9.2	10	6.4	24	-30	-13	91	38
CRITTENDEN AR	8	19.5	6	10.1	10	13.3	-48	31	-32	83	91
CROSS AR	8	15.1	9	13.3	6	8.2	-12	-38	-45	70	59
DALLAS AR	5	13.3	4	8.5	2	4.3	-36	-49	-67	27	18
DESHA AR	5	9.2	10	20.2	2	3.7	119	-82	-60	43	13
DREW AR	10	19.2	5	8.8	4	7.0	-54	-21	-64	34	44
FAULKNER AR	24	26.0	17	14.5	8	5.0	-44	-66	-81	08	22
FRANKLIN AR	6	10.3	12	16.6	7	8.8	61	-47	-15	90	65
FULTON AR	5	10.5	4	7.4	4	5.0	-29	-33	-52	57	23
GARLAND AR	47	18.5	40	10.6	38	8.0	-43	-25	-57	49	56
GRANT AR	3	6.1	2	3.6	3	4.9	-40	35	-19	89	22
GREENE AR	11	9.3	9	6.9	8	5.1	-26	-26	-45	70	24
HEMPSTEAD AR	12	14.3	6	7.6	6	5.6	-47	-26	-61	41	29
HOT SPRING AR	11	11.3	13	14.1	5	4.2	25	-70	-63	36	17
HOWARD AR	7	13.2	7	11.2	3	4.4	-15	-60	-66	29	18
INDEPENDENCE AR	13	12.2	8	6.8	11	7.1	-44	5	-41	75	46
IZARD AR	6	13.1	4	7.8	7	8.6	-41	11	-34	82	63
JACKSON AR	17	19.2	13	14.3	14	13.4	-25	-7	-30	84	91
JEFFERSON AR	30	20.0	25	12.5	12	5.2	-38	-58	-74	17	25
JOHNSON AR	7	9.1	20*	20.2	7	6.1	121	-70	-33	82	34
LAFAYETTE AR	4	10.8			2	5.6			-48	66	29
LAWRENCE AR	23	27.9	4	3.8	12	11.4	-87	204	-59	45	84
LEE AR	5	14.8	3	8.2	4	10.6	-44	28	-29	85	79
LINCOLN AR	3	9.7	4	11.8	5	13.6	22	15	40	97	92
LITTLE RIVER AR	8	21.0	6	15.2	5	11.1	-27	-27	-47	67	82
LOGAN AR	13	11.5	13	10.8	7	5.8	-6	-46	-49	64	30
LONOKE AR	11	11.8	10	9.4	4	3.4	-20	-64	-71	20	11
MADISON AR	7	10.8	6	8.4	6	7.9	-22	-6	-26	86	55
MARION AR	7	22.1	5	10.4	4	5.8	-53	-44	-74	17	31
MILLER AR	16	15.9	12	9.7	17	12.0	-39	25	-24	87	87
MISSISSIPPI AR	28	16.6	19	11.2	15	8.5	-33	-24	-49	65	62
MONROE AR	7	15.4	2	4.1	6	13.0	-73	216	-16	90	90
MONTGOMERY AR	8	21.9	3	6.3	2	3.6	-71	-43	-84	05	12
NEVADA AR	9	20.6	3	7.8	3	4.6	-62	-41	-78	11	20
NEWTON AR	8	20.4			3	8.8			-57	49	65
OUACHITA AR	12	15.8	6	6.4	9	8.8	-59	38	-44	71	65
PERRY AR	3	9.5	6	18.1	2	5.1	90	-72	-46	69	24
PHILLIPS AR	8	12.1	11	15.6	5	6.5	29	-58	-46	69	39
PIKE AR	6	13.8	7	12.3	10	17.8	-11	44	29	96	97
POINSETT AR	17	16.0	15	13.5	10	7.7	-16	-43	-52	58	52
POLK AR	9	11.9	7	8.8	12	11.0	-26	26	-7	92	81
POPE AR	18	19.2	26*	21.2	13	8.2	10	-61	-57	48	58
PRAIRIE AR	10	21.2	4	7.6	4	8.2	-64	9	-61	39	58
PULASKI AR	87	14.0	65	8.7	59	6.3	-38	-28	-55	53	36
RANDOLPH AR	11	14.9	9	11.8	7	7.1	-21	-40	-53	57	45
ST FRANCIS AR	17*	29.7	6	9.6	8	12.1	-68	26	-59	44	87
SALINE AR	16	12.3	11	7.9	10	5.3	-35	-33	-57	49	26
SCOTT AR	8	16.6	5	9.4	1	2.7	-43	-72	-84	05	08
SEARCY AR	2	3.7	5	9.8	7	13.0	165	32	250	99	90
SEBASTIAN AR	35	13.5	29	9.2	27	7.4	-32	-20	-45	70	49
SEVIER AR	9	17.8	9	13.3	2	2.7	-25	-80	-85	04	08
SHARP AR	3	9.7	11	19.3	8	9.5	100	-51	-2	93	72
STONE AR	5	13.7	3	7.2	7	12.7	-47	76	-7	92	89
UNION AR	17	11.5	16	8.7	15	7.3	-24	-17	-37	79	48
VAN BUREN AR	6	12.6	7	13.4	9	9.8	6	-27	-22	88	74
WASHINGTON AR	39	16.0	39	13.2	13	3.5	-18	-74	-78	10	11
WHITE AR	14	8.9	25	12.6	22	9.3	42	-26	4	94	70
WOODRUFF AR	7	15.7	7	17.0	2	3.8	8	-78	-76	14	13
YELL AR	13	17.3	14	15.7	6	6.8	-9	-57	-61	40	42
CALIFORNIA CA	9392	19.0	8499	13.1	7606*	9.6	-31	-26	-49		
ALAMEDA CA	581	18.9	464	13.5	366*	10.3	-29	-23	-45	70	77
ALPINE CA	2*	110.1									
AMADOR CA	3	5.0	9	13.1	5	4.5	162	-65	-10	92	19
BUTTE CA	81	18.7	61	11.2	42	5.8	-40	-48	-69	25	31
CALAVERAS CA	12	17.5	15	17.6	11	9.5	1	-46	-46	70	72
COLUSA CA	19	28.3	13	18.2	6	9.1	-36	-50	-68	26	68
CONTRA COSTA CA	174	19.1	179	12.6	180	8.6	-34	-32	-55	52	63
DEL NORTE CA	9	18.1	10	13.8	10	13.4	4	-29	-26	86	91
EL DORADO CA	22	20.4	10	5.4	23	9.9	-73	81	-52	59	74
FRESNO CA	239	21.6	168	13.6	165	9.8	-37	-28	-55	53	74
GLENN CA	18	20.9	10	10.7	8	7.3	-49	-32	-65	31	48
HUMBOLDT CA	104*	32.4	64	16.4	31	7.0	-49	-57	-78	11	45
IMPERIAL CA	25	11.0	26	9.2	37*	13.1	-17	43	19	96	90
INYO CA	10	16.6	14	17.9	4	3.5	8	-81	-79	10	11
KERN CA	109	13.4	114	10.6	114	9.1	-21	-15	-32	83	68
KINGS CA	35	20.7	22	11.0	21	9.4	-47	-15	-55	53	71
LAKE CA	13	12.2	18	11.2	22	9.1	-8	-19	-26	87	68
LASSEN CA	14	26.8	11	14.4	6	8.0	-46	-45	-70	22	56
LOS ANGELES CA	3689	19.6	3108	13.4	2437*	9.8	-32	-26	-50	63	74
MADERA CA	24	18.2	25	13.4	24	10.6	-26	-21	-42	74	79
MARIN CA	72	18.7	62	11.7	70	9.0	-37	-24	-52	59	66
MARIPOSA CA	5	13.0	9	19.6	2	3.4	51	-83	-74	16	11
MENDOCINO CA	37	15.6	35	13.9	27	9.7	-11	-30	-38	79	73
MERCED CA	44	16.2	33	10.6	39	9.4	-34	-12	-42	74	70
MODOC CA	10	28.9	6	14.4	1	1.9	-50	-87	-93	00	06
MONO CA	4	38.0	2	27.8	2	10.9	-27	-61	-71	20	81
MONTEREY CA	89	19.7	97	15.0	67	7.9	-24	-48	-60	42	55
NAPA CA	57	15.3	40	8.4	46	8.4	-45	0	-45	70	60

* SIGNIFICANTLY MORE DEATHS THAN EXPECTED

WHITE MALE — CANCER OF THE STOMACH — ICD 151

1950-1959 DEATHS	1950-1959 RATE	1960-1969 DEATHS	1960-1969 RATE	1970-1979 DEATHS	1970-1979 RATE	COUNTY AND STATE	PERCENT CHANGE 60S/50S	PERCENT CHANGE 70S/60S	PERCENT CHANGE 70S/50S	PERCENTILE %CHANGE 70S/50S	PERCENTILE RATES 1970S
26	20.2	21	12.7	24	9.4	NEVADA CA	-37	-26	-53	56	71
210	15.0	288	9.5	478	9.0	ORANGE CA	-37	-5	-40	76	67
44	17.0	37	11.1	40	9.2	PLACER CA	-35	-17	-46	69	69
10	21.3	7	15.2	7	10.2	PLUMAS CA	-28	-33	-52	58	77
151	14.0	195	10.7	262	9.1	RIVERSIDE CA	-24	-15	-35	81	68
274	21.2	283*	16.0	246*	10.9	SACRAMENTO CA	-25	-31	-48	66	81
15	21.0	12	16.0	8	9.1	SAN BENITO CA	-24	-43	-57	50	68
242	15.7	253	10.8	256	8.5	SAN BERNARDINO CA	-31	-22	-46	69	62
400	15.4	474	11.8	483	8.1	SAN DIEGO CA	-23	-32	-48	67	57
893*	24.6	696*	19.2	430*	13.9	SAN FRANCISCO CA	-22	-28	-43	72	93
211	21.2	193*	16.3	146*	11.1	SAN JOAQUIN CA	-23	-32	-48	66	82
63	18.2	47	9.5	36	5.5	SAN LUIS OBISPO CA	-48	-42	-70	22	27
213	19.2	229	14.4	231*	11.3	SAN MATEO CA	-25	-22	-41	75	83
114	21.4	95	11.7	118	10.8	SANTA BARBARA CA	-45	-8	-49	64	80
250	16.4	290	12.1	349*	10.6	SANTA CLARA CA	-26	-12	-35	81	79
96	19.3	83	12.2	75	8.9	SANTA CRUZ CA	-37	-27	-54	55	66
27	14.8	40	14.8	41	9.7	SHASTA CA	0	-35	-35	81	73
1	5.0	1	5.1	1	9.4	SIERRA CA	2	83	87	98	71
28	18.7	22	13.0	15	7.3	SISKIYOU CA	-30	-44	-61	40	48
58	17.2	50	11.8	58	10.4	SOLANO CA	-32	-11	-39	77	78
151	22.1	125	13.3	125	9.3	SONOMA CA	-40	-30	-58	48	70
114	17.7	107	13.1	90	9.1	STANISLAUS CA	-26	-31	-49	65	68
16	12.3	14	9.0	12	6.5	SUTTER CA	-26	-29	-47	67	38
25	22.0	19	13.5	11	5.9	TEHAMA CA	-39	-56	-73	18	32
6	17.7	6	19.5	5	10.9	TRINITY CA	11	-44	-38	78	81
105	16.8	83	10.7	89	10.2	TULARE CA	-36	-4	-39	77	77
10	11.4	17	16.9	18	11.7	TUOLUMNE CA	48	-31	3	94	85
70	13.8	104	12.1	134	9.8	VENTURA CA	-12	-19	-29	84	74
44	20.2	49	18.3	34	9.7	YOLO CA	-9	-47	-52	58	73
24	20.2	14	11.2	18	11.5	YUBA CA	-45	3	-43	73	84
1232	19.9	853	11.3	718	7.9	COLORADO CO	-43	-30	-60		
27	15.9	29	10.8	35	7.0	ADAMS CO	-32	-35	-56	50	44
10	30.1	9	20.6	7	16.8	ALAMOSA CO	-32	-18	-44	71	96
2	17.3	3	26.2	3	24.3	ARCHULETA CO	52	-7	41	97	99
6	18.3	3	10.9	3	10.4	BACA CO	-40	-5	-43	72	78
4	8.6	2	3.8	2	4.7	BENT CO	-55	23	-45	70	21
41	16.5	27	8.5	36	7.9	BOULDER CO	-48	-7	-52	59	55
15*	37.6	7	13.8	3	5.3	CHAFFEE CO	-63	-61	-86	03	26
3	23.5			1	7.3	CHEYENNE CO			-69	24	48
2	12.2	2	10.9			CLEAR CREEK CO	-11				
11	31.0	14*	39.2	4	14.5	CONEJOS CO	27	-63	-53	56	94
8*	52.4	3	16.4	3	17.6	COSTILLA CO	-69	8	-66	29	97
4	15.8	7*	28.9	3	14.5	CROWLEY CO	83	-50	-8	92	94
6*	60.7	1	10.1	1	15.2	CUSTER CO	-83	50	-75	15	95
16	15.9	15	13.9	9	6.6	DELTA CO	-13	-52	-58	47	40
409	19.5	287	11.5	217	8.1	DENVER-ARAPAHOE#CO	-41	-29	-59	46	57
				2	3.3	DOUGLAS CO					11
7	42.7	4	15.3	4	15.0	EAGLE CO	-64	-2	-65	31	95
2	10.6			1	4.6	ELBERT CO			-56	51	20
68	17.5	35	6.7	47	6.4	EL PASO CO	-62	-4	-63	34	37
29	22.4	15	10.4	14	8.3	FREMONT CO	-54	-20	-63	36	60
5	8.2	7	9.8	8	9.6	GARFIELD CO	19	-2	16	95	72
1	14.3	1	14.1	1	12.6	GILPIN CO	-2	-10	-12	91	89
3	19.9	3	27.8	3	14.0	GRAND CO	40	-50	-29	84	93
5	24.5	2	10.9	2	16.5	GUNNISON CO	-56	52	-33	83	96
16	27.6	9	15.6	6	14.1	HUERFANO CO	-44	-9	-49	65	93
2	32.1	1	12.3			JACKSON CO	-62				
55	21.4	53	10.5	64	7.9	JEFFERSON CO	-51	-25	-63	35	54
		2	12.4	3	27.7	KIOWA CO		123			99
11	29.0	2	5.1	3	6.8	KIT CARSON CO	-82	32	-77	13	42
6	33.3	1	6.4	1	5.1	LAKE CO	-81	-21	-85	04	23
14	19.5	7	8.9	14*	16.1	LA PLATA CO	-55	81	-18	90	96
45	18.4	33	10.4	30	6.9	LARIMER CO	-43	-34	-62	37	43
41*	32.4	28*	24.1	11	10.9	LAS ANIMAS CO	-26	-55	-66	29	81
1	3.2	3	10.6	4	13.4	LINCOLN CO	230	27	318	99	92
33*	40.4	13	15.3	6	6.6	LOGAN CO	-62	-57	-84	05	40
38	19.1	23	8.7	14	4.4	MESA CO	-54	-49	-77	12	18
9	33.7	1	2.8	1	3.5	MOFFAT CO	-92	27	-90	01	12
4	9.9	5	10.0	6	10.4	MONTEZUMA CO	1	4	6	95	78
22	29.0	11	13.0	11	10.6	MONTROSE CO	-55	-19	-64	34	79
13	14.2	8	7.8	9	8.4	MORGAN CO	-45	9	-41	76	61
28	23.7	13	10.8	9	7.5	OTERO CO	-55	-30	-68	25	51
3	28.0	2	35.9			OURAY CO	29				
1	8.0	1	10.5			PARK CO	31				
4	13.3	1	2.9	2	5.5	PHILLIPS CO	-78	90	-59	45	27
2	35.5					PITKIN CO					
18	31.2	6	9.7	2	3.7	PROWERS CO	-69	-62	-88	02	13
87	19.1	75	14.5	56	10.6	PUEBLO CO	-24	-27	-45	71	79
1	4.5	4	29.1			RIO BLANCO CO	551				
6	12.3	7	12.8	8	13.7	RIO GRANDE CO	4	8	12	95	92
10	29.6	2	5.8	4	12.0	ROUTT CO	-80	106	-59	43	86
5	26.2	6*	30.3	4	20.7	SAGUACHE CO	16	-32	-21	88	98
4	47.2	1	15.7	2	18.5	SAN MIGUEL CO	-67	18	-61	41	98
4	17.6	2	8.8	1	5.5	SEDGWICK CO	-50	-38	-69	24	27
2	90.7	3*	135.4			SUMMIT CO	49				
5	32.3	1	6.3	3	12.5	TELLER CO	-81	100	-61	39	88
2	5.1	3	7.9	4	11.7	WASHINGTON CO	56	48	131	99	85
48	16.0	46	13.5	31	7.8	WELD CO	-16	-42	-51	60	54
8	14.4	4	6.0			YUMA CO	-58				
2126*	22.3	1922*	17.1	1501*	11.5	CONNECTICUT CT	-24	-32	-48		
552*	22.8	492*	17.2	383*	11.3	FAIRFIELD CT	-24	-34	-50	62	83
541*	21.8	559*	18.9	434*	12.7	HARTFORD CT	-13	-33	-42	74	89
114	21.5	94	15.4	93*	12.8	LITCHFIELD CT	-28	-16	-40	76	90
70	18.7	64	14.2	51	9.5	MIDDLESEX CT	-24	-33	-49	64	72
606*	23.8	485*	16.3	378*	11.3	NEW HAVEN CT	-32	-31	-52	57	83
114	16.7	122	15.1	97	10.8	NEW LONDON CT	-10	-28	-35	80	80
47	22.8	35	13.9	22	6.6	TOLLAND CT	-39	-52	-71	21	40

* SIGNIFICANTLY MORE DEATHS THAN EXPECTED

COUNTIES COMBINED; SEE TECHNICAL NOTES

WHITE MALE

CANCER OF THE STOMACH — ICD 151

COUNTY AND STATE	1950-1959 DEATHS	1950-1959 RATE	1960-1969 DEATHS	1960-1969 RATE	1970-1979 DEATHS	1970-1979 RATE	PERCENT CHANGE 60S/50S	PERCENT CHANGE 70S/60S	PERCENT CHANGE 70S/50S	PERCENTILE %CHANGE 70S/50S	PERCENTILE RATES 1970S
WINDHAM CT	82*	27.2	71*	20.9	43	10.8	-23	-48	-60	42	80
DELAWARE DE	203	15.9	192	12.2	162	8.4	-23	-31	-47		
KENT DE	27	14.3	14	6.6	14	5.4	-54	-18	-62	38	27
NEW CASTLE DE	141	16.7	152	14.2	128	10.1	-15	-29	-39	77	76
SUSSEX DE	35	13.6	26	8.8	20	5.1	-35	-42	-62	37	24
DIST. OF COLUMBIA DC	311	16.2	179	10.9	99	8.6	-32	-22	-47		
WASHINGTON DC	311	16.2	179	10.9	99	8.6	-32	-22	-47	68	63
FLORIDA FL	2049	13.4	2854	9.9	3626	7.6	-26	-23	-43		
ALACHUA FL	13	9.9	18	10.3	17	5.8	5	-44	-41	75	31
BAKER FL	2	24.8	3	13.0	2	4.0	-48	-69	-84	05	15
BAY FL	15	11.5	13	6.6	26	8.8	-43	34	-24	87	65
BRADFORD FL	6	15.8	6	12.5	7	11.7	-21	-6	-26	86	85
BREVARD FL	20	9.1	55	13.6	53	5.6	49	-59	-39	77	28
BROWARD FL	99	10.2	263	9.6	460	8.0	-5	-17	-21	88	56
CALHOUN FL	4	14.4	2	6.6	1	3.0	-54	-55	-79	09	09
CHARLOTTE FL	8	18.2	22	9.7	41	8.6	-47	-12	-53	56	62
CITRUS FL	8	13.5	12	11.2	24	7.5	-17	-33	-44	71	50
CLAY FL	4	7.6	4	4.6	16	9.0	-40	95	18	96	66
COLLIER FL	4	7.5	10	6.6	26	6.5	-13	-2	-14	91	38
COLUMBIA FL	14	28.5	3	4.7	8	7.4	-83	58	-74	17	50
DADE FL	485	16.9	621	11.8	756*	9.7	-30	-17	-42	74	74
DE SOTO FL	4	8.1	3	4.4	5	5.4	-45	22	-33	82	27
DIXIE FL	1	7.3	1	7.0			-4				
DUVAL FL	115	14.2	109	9.6	104	6.8	-32	-29	-52	58	42
ESCAMBIA FL	29	9.9	34	7.3	42	7.1	-27	-2	-28	85	46
FLAGLER FL	3	21.8	3	11.8	2	6.2	-46	-48	-72	20	35
FRANKLIN FL	1	4.3	4	10.2	6	17.9	138	77	321	99	97
GADSDEN FL	7	9.0	7	6.8	4	4.6	-25	-32	-49	64	20
GILCHRIST FL	2	81.3	1	7.3	1	7.4	-91	2	-91	01	50
GLADES FL			1	8.9	3	13.1		47			90
GULF FL	5	49.9	6	24.5	4	10.2	-51	-58	-80	09	77
HAMILTON FL	3	13.6	3	13.2	3	11.7	-3	-11	-14	91	85
HARDEE FL	6	11.2	5	9.1	4	5.0	-19	-45	-56	52	23
HENDRY FL	2	12.5	1	2.9	4	7.4	-77	158	-41	75	49
HERNANDO FL	4	8.6	9	8.7	31	12.2	1	40	41	97	87
HIGHLANDS FL	9	13.9	10	6.1	21	6.9	-56	13	-50	62	43
HILLSBOROUGH FL	177	14.1	189	10.6	165	7.2	-25	-33	-49	64	46
HOLMES FL	6	10.7	1	1.4	1	1.8	-87	23	-83	06	06
INDIAN RIVER FL	10	10.5	17	9.4	22	6.2	-10	-34	-41	76	35
JACKSON FL	14	15.7	8	7.2	4	3.0	-54	-58	-81	08	09
JEFFERSON FL	3	14.4	1	5.2	3	12.7	-63	143	-11	91	89
LAFAYETTE FL			1	6.4							
LAKE FL	37	15.7	53	11.2	46	7.0	-29	-38	-56	51	44
LEE FL	27	15.9	48	9.1	79	6.6	-43	-27	-58	46	40
LEON FL	11	14.6	11	8.0	13	5.9	-45	-26	-59	43	32
LEVY FL	7	18.8	1	1.8	8	7.7	-90	322	-59	44	52
LIBERTY FL			1	5.3	1	7.5		42			50
MADISON FL	5	13.6	3	7.0	2	4.1	-49	-41	-70	23	16
MANATEE FL	30	7.4	54	7.0	74	6.8	-5	-3	-8	92	42
MARION FL	20	14.2	24	10.1	30	6.1	-29	-39	-57	48	34
MARTIN FL	5	6.8	5	3.7	32	9.0	-46	147	32	97	67
MONROE FL	18	18.0	17	10.4	17	6.2	-42	-41	-66	30	35
NASSAU FL	4	12.0	5	11.7	3	4.3	-2	-63	-64	33	17
OKALOOSA FL	10	13.6	5	3.5	14	6.0	-74	71	-56	50	33
OKEECHOBEE FL	1	6.2	2	13.5	6	7.3	117	-46	17	95	48
ORANGE FL	97	14.9	86	8.0	104	6.7	-46	-16	-55	52	41
OSCEOLA FL	21	17.4	17	9.2	25	9.8	-47	7	-44	72	74
PALM BEACH FL	87	11.5	162	9.6	235	6.9	-17	-27	-40	77	44
PASCO FL	25	13.4	51	10.4	116	6.7	-23	-35	-50	63	41
PINELLAS FL	283	12.0	459	10.3	496	8.0	-14	-22	-33	82	56
POLK FL	77	13.4	77	8.3	81	6.2	-38	-26	-54	55	34
PUTNAM FL	8	9.0	9	7.5	16	7.9	-17	5	-13	91	54
ST JOHNS FL	14	13.9	10	7.8	16	8.2	-44	5	-41	75	58
ST LUCIE FL	18	18.6	35	16.5	24	5.3	-11	-68	-71	20	26
SANTA ROSA FL	11	15.8	10	11.4	4	3.3	-28	-71	-79	10	10
SARASOTA FL	35	10.3	107	12.2	119	8.0	18	-35	-23	88	56
SEMINOLE FL	8	6.8	26	11.6	32	7.3	70	-37	8	95	48
SUMTER FL	10	23.7	8	12.0	9	9.0	-50	-24	-62	38	67
SUWANNEE FL	8	16.6	3	5.1			-69				
TAYLOR FL	5	14.4	4	9.4	6	11.4	-35	22	-21	89	83
UNION FL	1	5.5	1	5.7	2	9.4	4	66	73	98	71
VOLUSIA FL	73	11.2	103	9.1	122	7.2	-19	-21	-36	80	46
WAKULLA FL	1	6.2	1	4.5	1	2.3	-28	-49	-64	34	06
WALTON FL	6	8.3	4	5.0	4	3.6	-40	-27	-57	49	12
WASHINGTON FL	3	6.3	6	11.5	3	4.4	82	-62	-30	84	18
GEORGIA GA	1203	14.3	918	8.4	829	6.0	-41	-29	-58		
APPLING GA	3	9.4	2	4.8	4	7.7	-49	60	-18	90	53
ATKINSON GA	6	34.2	2	17.4	1	4.2	-49	-76	-88	02	16
BACON GA	4	19.1	1	2.6	1	2.5	-86	-4	-87	03	07
BAKER GA	2	23.9	2	31.6	1	7.6	32	-76	-68	26	52
BALDWIN GA	2	1.7	4	3.8	6	5.6	117	49	223	99	28
BANKS GA	4	14.4	1	5.1	2	5.1	-65	-1	-65	32	24
BARROW GA	9	21.1	2	3.5	3	4.2	-84	21	-80	08	16
BARTOW GA	8	10.1	8	7.0	11	9.2	-31	32	-9	92	69
BEN HILL GA	5	12.8	9	20.4	2	3.7	59	-82	-71	21	13
BERRIEN GA	6	14.3	8	18.8	3	7.2	31	-62	-50	63	47
BIBB GA	38	16.2	30	9.5	28	9.0	-41	-5	-44	71	67
BLECKLEY GA	1	3.6	2	6.9	1	2.3	92	-66	-35	81	07
BRANTLEY GA	2	10.4	2	12.5	2	5.7	20	-55	-46	70	29
BROOKS GA	4	9.5	3	6.5	2	4.0	-31	-39	-58	47	15
BRYAN GA	1	10.2			3	10.3			1	94	77
BULLOCH GA	7	12.4	5	6.2	4	3.8	-50	-38	-69	24	14
BURKE GA	4	11.8	4	11.6	3	6.6	-2	-42	-44	72	40

* SIGNIFICANTLY MORE DEATHS THAN EXPECTED

WHITE MALE

CANCER OF THE STOMACH

ICD 151

1950-1959 DEATHS	1950-1959 RATE	1960-1969 DEATHS	1960-1969 RATE	1970-1979 DEATHS	1970-1979 RATE	COUNTY AND STATE	PERCENT CHANGE 60S/50S	PERCENT CHANGE 70S/60S	PERCENT CHANGE 70S/50S	PERCENTILE %CHANGE 70S/50S	PERCENTILE RATES 1970S
2	8.4	2	7.1			BUTTS GA	-16				
3	19.4	1	10.6	3	20.6	CALHOUN GA	-45	94	7	95	98
3	21.3	1	7.0	1	8.6	CAMDEN GA	-67	23	-59	44	63
7	44.9	1	3.6	2	6.2	CANDLER GA	-92	70	-86	03	35
24	20.5	16	11.8	13	7.3	CARROLL GA	-42	-38	-64	33	48
2	3.6	5	7.0	6	4.6	CATOOSA GA	92	-34	26	96	20
1	7.5	3	24.9			CHARLTON GA	233				
48	17.1	35	8.9	35	7.5	CHATHAM GA	-48	-16	-56	50	50
13	21.8	7	8.1	4	4.6	CHATTOOGA GA	-63	-44	-79	10	19
18	23.9	9	8.6	10	7.3	CHEROKEE GA	-64	-15	-69	23	48
10	12.6	7	6.6	9	6.4	CLARKE GA	-47	-4	-49	64	37
				1	18.6	CLAY GA					98
5	6.6	9	6.9	11	4.2	CLAYTON GA	5	-39	-36	80	16
1	16.1					CLINCH GA					
38	20.1	33	9.9	29	5.3	COBB GA	-51	-46	-74	17	26
6	10.8	9	15.3	6	8.0	COFFEE GA	41	-47	-26	86	57
10	14.2	7	6.3	12	9.4	COLQUITT GA	-55	49	-33	82	71
3	17.0	4	16.3	2	5.4	COLUMBIA GA	-4	-67	-68	25	27
6	18.6	6	16.5	5	11.0	COOK GA	-12	-33	-41	76	81
23*	32.9	6	7.4	2	1.6	COWETA GA	-77	-79	-95	00	05
3	20.7	1	5.8	1	12.4	CRAWFORD GA	-72	112	-40	76	88
6	14.3	2	6.3	5	9.7	CRISP GA	-56	53	-33	83	73
3	9.7	5	14.8	4	10.3	DADE GA	52	-30	6	95	78
4	28.9			1	4.3	DAWSON GA			-85	04	17
11	18.9	3	4.8	6	9.0	DECATUR GA	-75	88	-53	57	67
52	13.7	66	8.8	55	4.9	DE KALB GA	-36	-44	-64	34	22
6	11.3	4	6.6	5	7.4	DODGE GA	-42	13	-34	81	50
6	17.6	2	7.1	1	3.4	DOOLY GA	-60	-53	-81	08	11
14	19.3	17	16.5	13	7.2	DOUGHERTY GA	-14	-56	-63	36	47
6	17.3	4	5.2	3	2.8	DOUGLAS GA	-70	-46	-84	05	09
6	19.2			3	9.4	EARLY GA			-51	61	71
		1	14.5			ECHOLS GA					
7	28.9	2	10.3	3	12.2	EFFINGHAM GA	-64	18	-58	47	87
7	17.0	6	9.3	4	6.1	ELBERT GA	-45	-34	-64	34	34
5	10.1	5	8.0	1	1.3	EMANUEL GA	-21	-84	-87	03	05
2	13.9			1	3.2	EVANS GA			-77	12	10
6	12.2	5	7.6	8	10.3	FANNIN GA	-38	36	-15	90	77
		3	9.0	2	3.4	FAYETTE GA		-62			11
19	10.5	13	5.3	16	5.6	FLOYD GA	-49	4	-47	67	28
9	24.9	7	13.3	12	14.2	FORSYTH GA	-47	6	-43	73	93
6	10.9	7	12.0	4	7.1	FRANKLIN GA	10	-41	-35	81	46
156	13.0	119	8.7	89	6.3	FULTON GA	-33	-28	-51	60	36
3	8.9	1	2.4	8	15.1	GILMER GA	-73	519	70	98	95
		1	7.5	1	8.1	GLASCOCK GA		9			57
9	15.5	11	12.0	7	4.2	GLYNN GA	-23	-64	-73	19	17
11	15.1	11	14.7	4	3.5	GORDON GA	-2	-76	-77	13	12
10	24.8	3	4.9	1	1.2	GRADY GA	-80	-76	-95	00	04
7	24.1	7	23.9	1	2.7	GREENE GA	-1	-89	-89	02	08
14	11.6	15	9.9	15	6.2	GWINNETT GA	-15	-37	-46	68	35
9	14.6	5	6.8	3	3.1	HABERSHAM GA	-53	-54	-79	10	10
19	12.9	12	6.8	11	5.2	HALL GA	-47	-24	-60	42	24
3	21.6	3	25.8			HANCOCK GA	20				
9	16.1	4	6.3	2	3.0	HARALSON GA	-61	-52	-82	07	09
		5	16.2	1	1.7	HARRIS GA		-90			05
4	10.9	3	5.5	1	1.4	HART GA	-50	-74	-87	03	05
2	7.5	1	4.1	3	11.6	HEARD GA	-46	186	55	98	85
7	21.1	4	7.3	3	3.5	HENRY GA	-66	-51	-83	06	12
5	7.5	6	9.5	5	4.5	HOUSTON GA	27	-53	-40	76	19
8	31.5	3	10.0	2	7.5	IRWIN GA	-68	-25	-76	14	51
9	13.8	7	10.4	4	4.3	JACKSON GA	-25	-59	-69	24	17
5	32.2			2	10.0	JASPER GA			-69	24	75
2	11.6	2	12.0			JEFF DAVIS GA	3				
5	15.0	5	15.1	2	3.9	JEFFERSON GA	0	-74	-74	16	14
1	9.0	2	7.7			JENKINS GA	-14				
3	9.7	4	14.2	1	3.2	JOHNSON GA	46	-77	-67	28	10
4	31.8	1	6.2	2	6.9	JONES GA	-81	12	-78	11	43
2	9.5	3	11.6	2	5.9	LAMAR GA	22	-49	-38	78	31
		1	6.7			LANIER GA					
5	5.9	3	2.7	5	3.8	LAURENS GA	-54	43	-35	81	14
1	10.4					LEE GA					
1	9.8	2	9.6	2	8.0	LIBERTY GA	-3	-16	-19	89	56
		1	6.6	2	7.4	LINCOLN GA		12			49
3	43.7	2	17.6	2	14.2	LONG GA	-60	-19	-68	27	93
12	14.7	7	6.8	12	8.6	LOWNDES GA	-54	27	-42	75	63
2	9.1	1	3.8			LUMPKIN GA	-59				
5	18.4	1	2.9	4	8.8	MC DUFFIE GA	-84	203	-52	58	65
1	10.5	1	7.0	1	5.8	MC INTOSH GA	-33	-17	-44	71	31
2	7.5	2	7.8	2	5.9	MACON GA	4	-25	-21	88	32
6	13.3	1	2.1	5	8.7	MADISON GA	-84	309	-35	81	64
3	22.6			1	11.1	MARION GA			-51	61	82
8	20.7	7	13.0	2	3.3	MERIWETHER GA	-37	-75	-84	05	10
1	4.0	1	5.0	3	9.7	MILLER GA	24	95	142	99	73
3	6.5	5	11.1	5	10.2	MITCHELL GA	71	-8	58	98	77
2	7.9	2	9.0	4	10.0	MONROE GA	14	11	27	96	75
3	16.0	1	4.7	1	4.1	MONTGOMERY GA	-71	-12	-75	16	15
4	15.8	1	3.3	1	2.6	MORGAN GA	-79	-19	-83	06	08
6	16.7	3	8.5	4	6.2	MURRAY GA	-49	-26	-63	36	35
28	12.8	22	7.6	30	9.2	MUSCOGEE#GA	-40	21	-28	85	69
6	10.3	6	9.0	7	7.8	NEWTON GA	-13	-13	-24	87	54
3	12.1	1	3.4	1	3.0	OCONEE GA	-72	-13	-76	14	09
4	15.6	1	3.7	3	9.0	OGLETHORPE GA	-76	144	-42	74	67
7	17.2	6	10.4	5	6.9	PAULDING GA	-40	-34	-60	42	43
		1	3.8	2	5.5	PEACH GA		47			27
9	27.7	1	3.7	1	1.7	PICKENS GA	-87	-54	-94	00	05
4	13.2	4	11.4	2	4.9	PIERCE GA	-14	-57	-63	35	22
3	13.4	3	12.7	1	3.7	PIKE GA	-5	-71	-72	19	13
8	11.0	7	6.3	3	2.3	POLK GA	-43	-64	-79	09	07
		1	3.7	1	3.3	PULASKI GA		-13			10
5	35.2	1	4.2	2	8.6	PUTNAM GA	-88	103	-76	14	63
3*	62.0	1	13.8	1	12.5	QUITMAN GA	-78	-9	-80	09	88
7	23.7	5	12.8	5	9.7	RABUN GA	-46	-24	-59	45	74

* SIGNIFICANTLY MORE DEATHS THAN EXPECTED

COUNTIES COMBINED; SEE TECHNICAL NOTES

1950-1959 DEATHS	1950-1959 RATE	1960-1969 DEATHS	1960-1969 RATE	1970-1979 DEATHS	1970-1979 RATE	COUNTY AND STATE	PERCENT CHANGE 60S/50S	PERCENT CHANGE 70S/60S	PERCENT CHANGE 70S/50S	PERCENTILE %CHANGE 70S/50S	PERCENTILE RATES 1970S
4	20.5	1	4.4	1	3.0	RANDOLPH GA	-79	-32	-85	04	09
28	15.7	27	10.1	18	5.4	RICHMOND GA	-36	-46	-65	30	27
5	16.8	6	15.2	4	3.7	ROCKDALE GA	-10	-76	-78	11	13
1	12.8			1	10.2	SCHLEY GA			-20	89	77
3	14.8	2	5.3	2	3.9	SCREVEN GA	-64	-26	-74	17	14
		1	4.2	2	7.4	SEMINOLE GA		75			49
15	20.3	10	8.5	10	7.7	SPALDING GA	-58	-10	-62	38	53
9	19.8	7	10.2	8	9.0	STEPHENS GA	-49	-11	-55	53	67
2	13.3	2	14.0	1	8.0	STEWART GA	5	-43	-40	77	56
4	9.8			3	4.5	SUMTER GA			-54	54	19
2	14.8	1	6.7			TALBOT GA	-55				
1	13.2	1	8.6			TALIAFERRO GA	-35				
4	7.9	1	1.3	8	13.4	TATTNALL GA	-84	969	70	98	92
5	20.8	3	13.8	2	7.3	TAYLOR GA	-34	-47	-65	32	48
4	11.4	3	13.2	1	3.1	TELFAIR GA	16	-76	-73	18	10
3	12.6	3	13.4	5	17.8	TERRELL GA	7	33	42	97	97
9	8.7	8	9.1	6	5.4	THOMAS GA	5	-41	-38	78	26
5	8.7	5	6.7	4	5.1	TIFT GA	-23	-23	-41	75	24
3	7.0	2	4.4	3	4.0	TOOMBS GA	-37	-8	-42	74	15
3	17.7					TOWNS GA					
2	12.6	2	12.2	4	17.3	TREUTLEN GA	-3	42	38	97	97
14	12.2	15	11.9	7	5.7	TROUP GA	-2	-52	-53	56	29
4	12.8	3	11.4	1	2.9	TURNER GA	-11	-75	-78	11	09
		2	15.8	2	12.3	TWIGGS GA		-22			88
5	16.8	2	4.9	6	11.8	UNION GA	-71	140	-30	84	85
12	23.3	4	4.9	9	9.4	UPSON GA	-79	91	-60	43	71
16	13.8	17	10.6	17	8.2	WALKER GA	-24	-23	-41	75	58
9	15.0	7	9.7	5	5.2	WALTON GA	-36	-46	-65	30	24
19	26.7	9	8.0	5	4.0	WARE GA	-70	-50	-85	04	15
4	25.7	1	5.6			WARREN GA	-78				
1	2.7			2	4.9	WASHINGTON GA			86	98	22
4	13.6	6	11.7	1	2.4	WAYNE GA	-14	-80	-82	06	07
1	17.5					WEBSTER GA					
1	7.0	1	5.0	1	4.4	WHEELER GA	-29	-11	-37	79	18
4	15.6	2	6.2	2	4.1	WHITE GA	-60	-34	-74	17	16
17	14.7	14	9.9	11	5.2	WHITFIELD GA	-33	-47	-64	32	25
1	3.9	1	3.4	2	6.0	WILCOX GA	-12	79	57	98	33
3	11.2	4	13.4	4	12.0	WILKES GA	20	-10	8	95	86
2	11.6	2	10.9	3	10.9	WILKINSON GA	-6	0	-6	92	80
5	13.4	4	8.8	2	5.2	WORTH GA	-34	-41	-61	39	24
		83	15.7	105*	12.1	HAWAII HI		-23			
		13*	24.7	11	10.8	HAWAII HI		-56			80
		59	14.1	74*	11.5	HONOLULU HI		-19			84
		5	22.2	7*	17.7	KAUAI HI		-20			97
		6	14.8	13*	17.4	MAUI#HI		17			97
515	20.2	326	10.7	314	8.8	IDAHO ID	-47	-17	-56		
47	14.8	45	10.7	37	7.0	ADA ID	-28	-34	-53	57	44
24	19.0	23	13.1	17	8.5	BANNOCK ID	-31	-36	-55	52	61
6	22.4	1	3.0	4	10.3	BEAR LAKE ID	-87	249	-54	55	78
9	20.8	3	7.0	3	7.9	BENEWAH ID	-66	13	-62	38	55
24	29.3	13	12.9	13	11.9	BINGHAM ID	-56	-8	-60	43	86
6	33.0	3	11.4	4	18.9	BLAINE ID	-66	67	-43	73	98
15	17.0	8	8.4	7	6.1	BONNER ID	-51	-27	-64	34	34
21	18.5	19	13.6	21	11.3	BONNEVILLE ID	-26	-17	-39	77	83
5	16.8	4	13.2	4	13.0	BOUNDARY ID	-22	-1	-23	88	90
3	33.3			1	11.8	BUTTE ID			-65	32	85
2	36.7					CAMAS ID					
31	12.4	29	9.6	36	10.1	CANYON ID	-22	5	-19	89	76
4	24.8	1	3.4	2	7.9	CARIBOU ID	-86	130	-68	25	54
14	26.4	6	9.0	10	14.1	CASSIA ID	-66	56	-47	68	93
		1	18.9			CLARK ID					
9	18.6	6	14.5	3	6.3	CLEARWATER ID	-22	-56	-66	29	36
5	33.0			2	11.8	CUSTER ID			-64	33	85
16*	53.5	6	14.3	1	3.5	ELMORE ID	-73	-75	-93	00	12
7	21.3	3	7.6	2	4.8	FRANKLIN ID	-64	-37	-78	11	21
9	34.3	6	18.1			FREMONT ID	-47				
7	14.4	6	12.4	7	12.0	GEM ID	-14	-3	-17	90	86
11	23.7	6	9.5	5	8.3	GOODING ID	-60	-13	-65	31	59
8	13.3	4	5.9	3	4.6	IDAHO ID	-56	-23	-66	30	19
7	23.3	6	13.5	1	1.5	JEFFERSON ID	-42	-89	-94	00	05
8	24.9	5	8.6	8	12.2	JEROME ID	-66	42	-51	61	87
42	27.1	18	9.9	18	7.6	KOOTENAI ID	-63	-23	-72	20	52
15	17.8	10	10.2	11	11.0	LATAH ID	-43	7	-38	78	81
7	20.4	3	9.2	5	15.0	LEMHI ID	-55	63	-27	86	95
5	22.9	1	4.6	1	4.1	LEWIS ID	-80	-12	-82	07	15
2	13.0	3	18.5	1	4.2	LINCOLN ID	43	-77	-68	26	16
9	37.6	3	8.5	9*	23.5	MADISON ID	-78	178	-38	79	99
12	30.1	7	11.4	7	11.6	MINIDOKA ID	-62	2	-62	38	84
19	16.7	17	12.8	6	3.9	NEZ PERCE ID	-23	-69	-76	13	15
5	29.1	2	9.3	3	12.7	ONEIDA ID	-68	36	-56	50	89
4	14.1			4	9.9	OWYHEE ID			-29	84	75
11	16.9	12	16.3	5	5.2	PAYETTE ID	-4	-68	-69	24	25
2	11.3	2	10.2	4	18.8	POWER ID	-10	85	66	98	98
34*	40.6	7	8.6	15*	17.5	SHOSHONE ID	-79	103	-57	49	97
		4*	37.4	4*	31.7	TETON ID		-15			99
36	19.6	24	11.3	25	10.4	TWIN FALLS ID	-42	-8	-47	68	78
2	12.7	2	10.5	1	4.1	VALLEY ID	-17	-61	-68	26	15
12	25.6	7	13.7	4	6.7	WASHINGTON ID	-47	-51	-74	17	42
8678*	22.4	6032*	14.3	4241*	9.6	ILLINOIS IL	-36	-33	-57		
91	23.7	42	10.9	26	6.4	ADAMS IL	-54	-41	-73	18	37
14	20.1	9	15.4	5	9.4	ALEXANDER IL	-23	-39	-53	56	71
20	21.9	10	12.2	6	7.2	BOND IL	-44	-41	-67	27	46
16	16.9	14	14.1	7	6.5	BOONE IL	-17	-54	-61	39	39
10	19.7	6	11.7	3	7.4	BROWN IL	-41	-37	-62	37	49
43	20.1	24	11.1	20	8.9	BUREAU IL	-45	-20	-56	51	65

* SIGNIFICANTLY MORE DEATHS THAN EXPECTED

COUNTIES COMBINED; SEE TECHNICAL NOTES

1950-1959 DEATHS	RATE	1960-1969 DEATHS	RATE	1970-1979 DEATHS	RATE	COUNTY AND STATE		PERCENT CHANGE 60S/50S	70S/60S	70S/50S	PERCENTILE %CHANGE 70S/50S	RATES 1970S
9	20.9	3	6.4	1	4.1	CALHOUN	IL	-69	-37	-80	08	16
24	22.3	19	16.3	15	13.1	CARROLL	IL	-27	-20	-42	75	90
15	18.9	6	6.2	2	2.3	CASS	IL	-67	-63	-88	02	07
40	12.2	26	7.1	20	5.0	CHAMPAIGN	IL	-42	-30	-59	44	23
29	12.9	35	15.9	18	8.6	CHRISTIAN	IL	23	-46	-33	82	63
12	10.2	14	11.0	8	7.5	CLARK	IL	8	-32	-26	86	50
15	16.0	6	6.0	10	9.8	CLAY	IL	-63	64	-39	77	74
24	20.7	18	14.2	8	5.6	CLINTON	IL	-31	-61	-73	18	28
35	16.3	10	4.6	19	8.6	COLES	IL	-72	88	-47	67	63
4761*	26.4	3369*	17.7	2270*	12.1	COOK	IL	-33	-32	-54	54	87
19	15.2	10	7.8	6	5.0	CRAWFORD	IL	-49	-36	-67	28	23
10	17.4	8	14.4	6	8.9	CUMBERLAND	IL	-17	-38	-49	65	66
39	18.7	27	12.1	26	10.4	DE KALB	IL	-35	-14	-44	71	78
15	14.5	11	11.5	5	5.0	DE WITT	IL	-21	-57	-66	30	23
11	12.1	7	7.6	5	5.3	DOUGLAS	IL	-37	-30	-56	51	26
160	24.2	134	12.6	162	10.2	DU PAGE	IL	-48	-20	-58	47	76
22	16.2	11	7.9	4	2.9	EDGAR	IL	-51	-64	-82	06	09
11	22.4	4	10.0	4	7.7	EDWARDS	IL	-55	-23	-66	30	52
17	15.3	5	4.0	7	5.4	EFFINGHAM	IL	-74	37	-65	32	26
21	14.7	10	7.4	8	5.9	FAYETTE	IL	-50	-20	-60	42	31
18	19.7	17	17.7	7	7.4	FORD	IL	-10	-58	-62	37	49
61	21.5	31	9.8	29	10.1	FRANKLIN	IL	-54	3	-53	57	76
38	15.6	22	8.8	10	3.7	FULTON	IL	-44	-57	-76	14	13
10	20.0	10	21.6	8	17.8	GALLATIN	IL	8	-18	-11	91	97
22	19.1	12	9.1	6	5.1	GREENE	IL	-53	-44	-74	17	23
24	24.8	16	14.7	12	9.5	GRUNDY	IL	-41	-35	-62	38	72
11	14.5	9	12.6	4	5.8	HAMILTON	IL	-13	-54	-60	42	31
13	8.1	10	6.5	12	8.1	HANCOCK	IL	-20	26	0	94	57
10	30.5	4	11.9	2	5.7	HARDIN	IL	-61	-53	-81	07	29
2	4.2	3	5.9	7	14.2	HENDERSON	IL	41	141	239	99	93
61	23.1	46	16.6	19	6.5	HENRY	IL	-28	-61	-72	20	39
37	21.6	18	9.9	13	7.4	IROQUOIS	IL	-54	-26	-66	30	49
29	16.8	16	8.3	12	6.0	JACKSON	IL	-50	-28	-64	34	33
15	19.6	7	10.7	7	9.5	JASPER	IL	-46	-11	-52	59	72
29	16.4	26	14.2	13	6.8	JEFFERSON	IL	-13	-52	-58	46	42
18	21.3	9	10.4	3	3.4	JERSEY	IL	-51	-67	-84	05	11
24	19.9	23	18.2	15	12.5	JO DAVIESS	IL	-9	-31	-37	79	88
3	5.9	1	1.9	7	13.0	JOHNSON	IL	-68	593	120	98	90
163	22.0	116	13.4	76	8.0	KANE	IL	-39	-40	-63	35	57
65	15.4	48	11.4	30	7.6	KANKAKEE	IL	-26	-33	-51	61	52
16	25.7	11	15.0	5	4.7	KENDALL	IL	-42	-69	-82	07	21
56	19.3	34	10.6	23	7.0	KNOX	IL	-45	-35	-64	34	44
185*	24.4	155	15.6	105	8.7	LAKE	IL	-36	-44	-64	32	64
124	24.8	69	12.8	63	10.9	LA SALLE	IL	-48	-15	-56	51	81
27	25.7	8	6.9	1	.8	LAWRENCE	IL	-73	-88	-97	00	04
33	19.5	23	13.1	14	7.9	LEE	IL	-33	-40	-60	43	55
30	14.2	20	9.3	14	6.3	LIVINGSTON	IL	-35	-33	-56	51	36
25	15.8	13	7.8	11	6.4	LOGAN	IL	-51	-18	-59	44	38
25	16.0	13	8.0	9	5.7	MC DONOUGH	IL	-50	-28	-64	33	30
71	24.4	57	13.7	51	10.1	MC HENRY	IL	-44	-26	-59	46	76
43	11.2	31	7.6	37	9.0	MC LEAN	IL	-32	19	-19	89	67
65	14.2	49	9.7	33	6.1	MACON	IL	-32	-36	-57	49	34
48	15.9	40	13.1	24	7.8	MACOUPIN	IL	-18	-40	-51	61	54
125	16.7	100	10.9	69	7.0	MADISON	IL	-35	-36	-58	47	45
42	19.9	27	11.7	22	9.0	MARION	IL	-41	-23	-55	53	67
14	18.5	7	9.4	5	5.9	MARSHALL	IL	-49	-38	-68	25	31
15	18.2	7	7.8	8	7.7	MASON	IL	-57	-2	-58	47	52
20	27.3	8	10.5	4	4.9	MASSAC	IL	-62	-54	-82	07	22
5	9.3	6	10.2	5	7.2	MENARD	IL	9	-29	-23	88	47
16	16.2	10	10.0	6	5.7	MERCER	IL	-38	-43	-65	32	30
22	29.5	16	17.7	10	9.7	MONROE	IL	-40	-45	-67	28	73
50	23.6	32	16.0	11	5.7	MONTGOMERY	IL	-32	-64	-76	14	30
22	9.3	15	7.0	9	4.8	MORGAN	IL	-25	-31	-48	65	21
10	12.3	11	14.3	8	10.1	MOULTRIE	IL	16	-29	-18	90	76
24	13.3	22	11.5	15	7.1	OGLE	IL	-13	-38	-46	69	46
122	16.0	87	10.5	76	9.0	PEORIA	IL	-34	-14	-44	72	67
25	20.5	12	10.1	10	8.1	PERRY	IL	-51	-20	-61	41	57
14	18.8	4	4.7	5	6.1	PIATT	IL	-75	30	-67	27	34
25	18.5	13	9.8	12	8.5	PIKE	IL	-47	-13	-54	55	62
5	12.4	3	9.1			POPE	IL	-27				
10	22.7	4	8.5	1	2.5	PULASKI	IL	-63	-70	-89	02	07
6	18.4	11*	36.5	2	6.5	PUTNAM	IL	98	-82	-65	31	38
25	15.7	13	7.4	11	6.0	RANDOLPH	IL	-53	-18	-62	39	33
19	20.9	10	10.1	8	8.7	RICHLAND	IL	-52	-14	-58	46	64
150	24.0	86	12.3	63	8.7	ROCK ISLAND	IL	-49	-29	-64	34	64
159	20.2	90	10.2	76	8.2	ST CLAIR	IL	-50	-19	-59	44	59
39	21.8	33	18.8	8	4.4	SALINE	IL	-14	-76	-80	09	18
105	16.5	69	10.0	58	7.9	SANGAMON	IL	-39	-21	-52	58	55
17	28.2	5	8.9	4	6.1	SCHUYLER	IL	-69	-31	-78	11	34
6	15.1	4	9.3			SCOTT	IL	-38				
21	13.7	24	15.2	10	7.1	SHELBY	IL	11	-54	-49	65	45
6	12.0	6	12.8	3	6.6	STARK	IL	7	-49	-45	70	39
37	16.8	29	13.0	14	5.7	STEPHENSON	IL	-23	-56	-66	30	30
56	21.9	33	9.7	39	8.7	TAZEWELL	IL	-56	-10	-60	42	64
15	11.5	4	3.1	8	6.5	UNION	IL	-73	110	-44	72	38
94	20.7	48	9.8	41	8.6	VERMILION	IL	-53	-12	-58	46	63
8	9.9	9	11.4	6	7.8	WABASH	IL	15	-32	-22	88	53
15	12.9	14	11.8	8	7.4	WARREN	IL	-8	-37	-43	73	49
20	19.7	9	7.9	5	4.5	WASHINGTON	IL	-60	-42	-77	12	19
16	13.8	10	8.7	8	6.5	WAYNE	IL	-37	-25	-53	56	39
21	16.8	10	8.2	8	6.6	WHITE	IL	-51	-19	-61	40	40
32	13.3	27	10.1	23	7.9	WHITESIDE	IL	-24	-22	-41	76	55
151*	24.6	113	15.4	88	10.4	WILL	IL	-37	-33	-58	47	78
55	20.5	38	13.4	17	5.8	WILLIAMSON	IL	-35	-57	-72	20	30
132	18.3	131	16.4	95	9.9	WINNEBAGO	IL	-11	-39	-46	69	75
23	20.9	17	13.2	9	6.1	WOODFORD	IL	-37	-54	-71	21	34
3111	17.4	2118	10.8	1505	7.1	INDIANA	IN	-38	-34	-59		
11	10.7	10	8.6	9	7.1	ADAMS	IN	-19	-18	-34	82	46
120	16.2	95	11.4	68	6.6	ALLEN	IN	-30	-42	-59	44	40

* SIGNIFICANTLY MORE DEATHS THAN EXPECTED

WHITE MALE

CANCER OF THE STOMACH

ICD 151

1950-1959 DEATHS	1950-1959 RATE	1960-1969 DEATHS	1960-1969 RATE	1970-1979 DEATHS	1970-1979 RATE	COUNTY AND STATE	PERCENT CHANGE 60S/50S	PERCENT CHANGE 70S/60S	PERCENT CHANGE 70S/50S	PERCENTILE %CHANGE 70S/50S	PERCENTILE RATES 1970S
26	16.1	21	11.1	12	5.9	BARTHOLOMEW IN	-31	-47	-63	36	32
6	11.5	2	3.1	2	3.3	BENTON IN	-73	6	-71	21	11
11	16.3	4	5.6	4	6.0	BLACKFORD IN	-66	9	-63	36	33
23	16.3	7	5.0	13	8.7	BOONE IN	-69	75	-46	68	64
6	17.1	2	5.2	2	4.0	BPOWN IN	-70	-23	-77	12	15
12	12.5	9	9.6	8	8.1	CARROLL IN	-23	-15	-35	81	57
31	14.5	12	5.6	10	5.0	CASS IN	-61	-10	-65	30	23
23	13.7	20	8.8	19	6.6	CLARK IN	-35	-25	-52	59	40
24	15.9	10	6.4	8	5.4	CLAY IN	-59	-16	-66	29	26
24	15.7	16	9.8	14	8.9	CLINTON IN	-37	-9	-43	73	66
8	12.9	2	3.7	2	3.5	CRAWFORD IN	-71	-6	-73	18	12
24	17.3	18	12.5	6	4.1	DAVIESS IN	-27	-67	-76	13	16
22	18.3	8	5.7	10	6.6	DEARBOPN IN	-69	15	-64	33	40
18	15.8	10	9.0	6	5.3	DECATUR IN	-43	-41	-66	29	25
27	19.2	8	5.6	6	4.2	DE KALB IN	-71	-25	-78	11	16
38	9.8	34	8.1	18	3.9	DELAWARE IN	-18	-52	-61	41	14
12	10.9	22	16.8	10	7.1	DUBOIS IN	55	-58	-35	81	46
63	15.8	49	10.9	53	10.8	ELKHART IN	-31	-1	-31	83	80
19	18.6	7	6.5	11	8.7	FAYETTE IN	-65	35	-53	56	64
32	16.7	23	10.8	22	9.0	FLOYD IN	-35	-17	-46	69	67
15	14.2	9	8.9	2	1.9	FOUNTAIN IN	-37	-79	-87	03	06
8	9.9	4	5.1	4	4.6	FRANKLIN IN	-49	-10	-54	55	19
14	14.8	11	10.1	6	6.0	FULTON IN	-32	-40	-59	44	33
32	19.8	19	10.7	20	11.5	GIBSON IN	-46	8	-42	74	84
37	12.1	36	11.1	15	4.3	GPANT IN	-8	-61	-64	32	17
29	16.0	34*	19.6	12	6.8	GREENE IN	22	-65	-58	48	42
24	14.9	22	12.4	19	7.9	HAMILTON IN	-16	-37	-47	67	55
14	12.6	6	4.8	6	3.9	HANCOCK IN	-62	-18	-69	24	14
17	16.5	9	8.2	10	9.0	HARRISON IN	-50	9	-46	70	67
19	14.8	17	9.8	14	6.7	HENDRICKS IN	-34	-32	-55	53	41
26	12.2	16	7.0	12	5.1	HENRY IN	-43	-28	-58	46	23
39	17.0	26	10.1	20	6.7	HOWARD IN	-40	-34	-60	41	41
28	16.7	20	11.8	7	4.0	HUNTINGTON IN	-29	-66	-76	14	15
17	11.8	19	12.5	12	7.6	JACKSON IN	5	-39	-36	80	51
15	18.2	7	8.0	7	6.9	JASPER IN	-56	-14	-62	38	43
16	13.1	10	8.1	6	5.2	JAY IN	-38	-36	-60	42	25
19	15.0	10	7.6	4	3.1	JEFFERSON IN	-49	-59	-79	10	10
12	14.0	4	4.5	6	6.8	JENNINGS IN	-68	50	-52	59	42
27	18.7	12	6.9	9	4.8	JOHNSON IN	-63	-31	-74	16	21
57	24.5	25	10.8	11	4.8	KNOX IN	-56	-55	-80	08	22
34	19.2	18	9.1	22	9.4	KOSCIUSKO IN	-53	3	-51	60	71
24	29.0	6	7.3	10	11.0	LAGRANGE IN	-75	51	-62	38	81
398*	31.7	258*	17.4	177*	10.9	LAKE IN	-45	-37	-66	30	80
87	24.5	56	13.3	34	7.6	LA PORTE IN	-46	-42	-69	25	52
20	11.8	28	15.9	6	3.0	LAWRENCE IN	35	-81	-75	15	09
65	14.2	40	8.5	36	6.5	MADISON IN	-40	-23	-54	54	39
321	15.5	232	10.1	160	6.4	MARION IN	-35	-37	-59	45	37
27	17.8	15	9.2	10	5.6	MARSHALL IN	-45	-39	-69	25	28
4	7.4	4	8.3	5	9.3	MARTIN IN	12	12	25	96	69
35	21.7	15	9.1	5	3.5	MIAMI IN	-58	-62	-84	05	12
31	16.5	25	12.7	15	5.9	MONROE IN	-23	-53	-64	34	32
25	16.1	14	8.6	8	4.7	MONTGOMERY IN	-47	-45	-71	22	21
16	12.4	17	11.2	14	7.7	MORGAN IN	-9	-31	-38	79	53
11	20.0	5	8.8	4	5.7	NEWTON IN	-56	-35	-72	20	29
30	19.9	10	7.4	10	6.8	NOBLE IN	-63	-7	-66	30	42
4	15.3	2	7.7	2	7.0	OHIO IN	-50	-9	-54	55	45
13	13.6	11	12.0	4	4.0	OPANGE IN	-12	-66	-70	22	15
17	22.6	12	15.3	9	11.0	OWEN IN	-32	-28	-51	60	81
16	15.6	9	8.6	3	3.0	PAPKE IN	-45	-66	-81	08	09
16	18.3	10	11.0	3	3.1	PERRY IN	-40	-72	-83	06	10
16	19.4	8	10.6	2	2.3	PIKE IN	-45	-78	-88	02	07
33	17.0	36	15.4	36	12.1	PORTER IN	-10	-21	-29	85	87
13	13.4	14	14.0	8	7.2	POSEY IN	4	-48	-46	69	47
18	27.3	4	5.4	5	6.6	PULASKI IN	-80	22	-76	14	40
16	12.8	15	11.5	7	4.8	PUTNAM IN	-10	-58	-62	37	22
19	12.0	13	8.4	5	3.3	RANDOLPH IN	-30	-61	-73	18	10
17	14.5	11	9.5	7	5.4	RIPLEY IN	-35	-43	-63	35	26
12	10.4	14	12.3	9	8.4	RUSH IN	19	-31	-19	89	61
196*	24.6	121	13.0	115*	11.1	ST JOSEPH IN	-47	-15	-55	53	82
1	1.9	12	17.7	6	6.9	SCOTT IN	841	-61	267	99	43
23	14.8	13	8.3	11	6.4	SHELBY IN	-44	-23	-57	49	37
17	17.8	10	9.4	7	7.1	SPENCER IN	-47	-25	-60	41	45
27*	28.8	34*	32.9	12	11.4	STARKE IN	14	-65	-60	41	83
15	15.8	14	15.0	5	4.8	STEUBEN IN	-5	-68	-70	23	21
33	22.4	23	16.1	15	11.9	SULLIVAN IN	-28	-26	-47	68	86
7	15.5	7	15.4	4	8.5	SWITZERLAND IN	0	-45	-45	70	61
43	14.9	34	10.6	20	5.4	TIPPECANOE IN	-29	-49	-64	34	26
7	8.6	3	3.8	5	6.2	TIPTON IN	-56	63	-28	85	35
4	10.7	4	11.4			UNION IN	6				
106	16.9	81	11.2	53	6.8	VANDERBURGH IN	-33	-40	-60	43	42
24	19.5	5	4.5	8	7.6	VERMILLION IN	-77	70	-61	40	52
85	15.7	61	11.3	45	9.0	VIGO IN	-28	-20	-42	73	67
23	16.1	15	10.1	9	5.7	WABASH IN	-37	-43	-64	33	30
5	10.5	5	10.7	3	6.0	WARREN IN	2	-44	-44	72	32
21	19.7	15	12.4	14	10.6	WARRICK IN	-37	-14	-46	69	79
6	6.6	7	7.5	9	8.2	WASHINGTON IN	13	10	24	96	59
39	12.6	19	5.7	19	5.6	WAYNE IN	-55	-2	-55	52	28
13	11.9	10	9.5	4	3.5	WELLS IN	-20	-63	-71	21	12
22	22.1	9	8.3	6	4.9	WHITE IN	-62	-41	-78	11	22
11	11.4	9	8.6	4	3.5	WHITLEY IN	-25	-59	-69	23	12
2507	18.7	1608	11.2	1172	7.9	IOWA IA	-40	-29	-58		
11	16.2	12	17.0	6	9.0	ADAIR IA	5	-47	-45	71	66
9	17.8			7	14.8	ADAMS IA			-17	90	95
13	13.6	13	13.9	9	8.3	ALLAMAKEE IA	2	-41	-39	77	59
28	20.6	8	7.3	7	5.3	APPANOOSE IA	-65	-27	-74	16	25
14	21.7	15*	23.0	10	15.7	AUDUBON IA	6	-32	-28	85	95
26	21.3	14	10.4	8	5.8	BENTON IA	-51	-44	-73	18	31
73	17.4	48	9.9	46	8.9	BLACK HAWK IA	-43	-10	-49	64	66
23	14.4	23	13.0	4	2.4	BOONE IA	-10	-82	-83	06	07

* SIGNIFICANTLY MORE DEATHS THAN EXPECTED

WHITE MALE — CANCER OF THE STOMACH — ICD 151

1950-1959 DEATHS	1950-1959 RATE	1960-1969 DEATHS	1960-1969 RATE	1970-1979 DEATHS	1970-1979 RATE	COUNTY AND STATE		PERCENT CHANGE 60S/50S	PERCENT CHANGE 70S/60S	PERCENT CHANGE 70S/50S	PERCENTILE %CHANGE 70S/50S	PERCENTILE RATES 1970S
16	15.5	11	9.9	16	13.3	BREMER	IA	-36	35	-14	91	91
17	13.6	12	9.8	5	4.5	BUCHANAN	IA	-28	-54	-67	28	19
29	24.8	19	15.2	12	9.0	BUENA VISTA	IA	-39	-41	-64	34	67
13	13.5	15	15.2	6	6.1	BUTLER	IA	13	-60	-55	53	34
17	19.3	6	6.3	10	8.6	CALHOUN	IA	-67	36	-55	52	63
22	19.7	14	11.0	13	10.0	CARROLL	IA	-44	-10	-49	64	75
17	15.4	16	14.0	6	5.5	CASS	IA	-9	-61	-64	33	27
13	13.1	11	10.7	5	4.5	CEDAR	IA	-19	-58	-66	29	19
43	19.0	38	14.7	33*	12.4	CERRO GORDO	IA	-23	-15	-35	81	88
13	13.3	16	15.2	5	5.4	CHEROKEE	IA	14	-64	-59	44	27
15	17.7	14	16.1	10	10.8	CHICKASAW	IA	-9	-33	-39	77	80
8	13.2	6	10.5	1	2.0	CLARKE	IA	-21	-81	-85	04	06
17	19.7	11	10.7	7	6.2	CLAY	IA	-46	-42	-68	25	35
25	18.9	17	12.4	14	9.7	CLAYTON	IA	-34	-22	-49	65	73
61	24.3	36	13.3	27	9.7	CLINTON	IA	-45	-27	-60	42	73
23	25.1	22*	22.1	5	4.6	CRAWFORD	IA	-12	-79	-81	07	20
25	18.3	14	9.1	10	6.7	DALLAS	IA	-50	-26	-63	35	41
8	13.9	3	4.4	3	5.5	DAVIS	IA	-69	27	-60	42	28
16	19.3	5	5.8	3	4.7	DECATUR	IA	-70	-20	-76	14	20
19	19.9	11	11.2	8	8.2	DELAWARE	IA	-44	-26	-58	46	59
43	20.2	25	10.6	14	5.5	DES MOINES	IA	-47	-48	-73	19	27
15	21.1	8	10.0	10	11.0	DICKINSON	IA	-53	10	-48	66	81
77	24.0	48	14.3	25	7.1	DUBUQUE	IA	-40	-50	-70	22	46
18	27.3	13	18.0	9	10.8	EMMET	IA	-34	-40	-60	41	80
44	27.9	11	6.8	14	8.9	FAYETTE	IA	-76	31	-68	26	66
18	15.8	14	12.3	10	9.0	FLOYD	IA	-22	-27	-43	73	67
17	21.6	10	11.2	2	2.3	FRANKLIN	IA	-48	-80	-90	02	06
9	12.5	11	15.4	4	5.6	FREMONT	IA	23	-63	-55	53	29
19	23.8	12	14.5	3	3.4	GREENE	IA	-39	-77	-86	03	11
12	19.7	3	3.8	10	11.9	GRUNDY	IA	-81	211	-40	77	86
12	14.3	10	11.8	6	7.1	GUTHRIE	IA	-18	-40	-50	62	46
21	21.8	7	6.4	9	8.4	HAMILTON	IA	-70	31	-61	39	61
18	23.3	14	18.1	12	14.2	HANCOCK	IA	-22	-21	-39	78	93
35	26.8	20	14.3	8	5.7	HARDIN	IA	-47	-60	-79	10	30
13	12.6	11	10.3	4	3.8	HARRISON	IA	-18	-63	-70	22	13
18	14.6	10	8.4	11	10.7	HENRY	IA	-42	27	-26	86	80
15	19.7	17*	20.5	5	6.0	HOWARD	IA	4	-71	-69	23	33
18	26.0	9	12.0	5	6.0	HUMBOLDT	IA	-54	-50	-77	12	33
12	21.7	11	18.0	9	14.6	IDA	IA	-17	-19	-33	82	94
18	20.8	11	12.5	11	11.6	IOWA	IA	-40	-8	-44	71	84
20	17.0	12	10.4	11	8.8	JACKSON	IA	-39	-16	-48	66	65
33	20.4	24	13.3	7	3.7	JASPER	IA	-35	-72	-82	07	13
14	15.6	4	4.2	4	4.3	JEFFERSON	IA	-73	2	-73	19	17
20	11.3	13	6.4	12	5.5	JOHNSON	IA	-44	-13	-51	60	27
22	21.0	11	10.3	3	2.7	JONES	IA	-51	-73	-87	03	08
18	17.2	10	9.0	14	13.4	KEOKUK	IA	-48	48	-23	88	91
27	22.7	19	14.9	9	6.8	KOSSUTH	IA	-34	-54	-70	23	42
30	14.0	18	8.1	13	6.0	LEE	IA	-42	-25	-57	49	33
77	14.6	51	8.7	51	8.1	LINN	IA	-40	-8	-45	71	57
9	14.4	5	8.1	2	2.9	LOUISA	IA	-44	-65	-80	08	09
16	20.0	8	10.0	4	5.9	LUCAS	IA	-50	-42	-71	21	31
20	31.2	12	16.1	7	9.2	LYON	IA	-48	-43	-70	22	69
9	11.4	7	8.3	6	7.6	MADISON	IA	-27	-8	-33	82	52
23	16.2	7	4.6	12	8.2	MAHASKA	IA	-72	80	-49	64	58
24	15.3	14	8.0	14	7.8	MARION	IA	-48	-2	-49	65	54
29	15.4	11	5.1	11	5.0	MARSHALL	IA	-67	-3	-68	26	23
7	10.3	7	9.7	9	13.3	MILLS	IA	-6	37	29	96	91
16	19.3	13	14.8	5	5.7	MITCHELL	IA	-23	-61	-70	22	29
9	10.3	15	16.2	7	7.5	MONONA	IA	58	-54	-27	86	51
20	25.4	12	14.6	3	3.8	MONROE	IA	-43	-74	-85	04	13
16	16.9	8	7.4	8	8.0	MONTGOMERY	IA	-56	8	-53	57	56
31	16.6	18	9.7	16	8.5	MUSCATINE	IA	-41	-13	-49	65	62
23	24.2	18	16.6	9	8.1	O'BRIEN	IA	-32	-51	-66	29	57
9	17.5	8	16.2	7	12.6	OSCEOLA	IA	-7	-22	-28	85	89
19	13.0	18	12.2	10	7.2	PAGE	IA	-6	-41	-45	71	47
11	14.6	15	18.1	4	5.2	PALO ALTO	IA	24	-71	-64	32	24
30	25.8	33*	26.1	18	13.2	PLYMOUTH	IA	1	-49	-49	65	91
13	19.0	17	20.7	6	7.6	POCAHONTAS	IA	9	-63	-60	42	51
163	17.1	110	10.6	104	9.3	POLK	IA	-38	-12	-46	69	70
57	16.8	36	10.0	25	6.4	POTTAWATTAMIE	IA	-40	-36	-62	37	37
15	13.7	10	9.8	4	3.9	POWESHIEK	IA	-29	-60	-72	20	14
9	14.3	3	6.0	4	10.1	RINGGOLD	IA	-58	68	-29	84	76
14	13.9	8	7.3	9	8.6	SAC	IA	-47	18	-38	78	63
104	23.4	60	11.6	42	7.6	SCOTT	IA	-51	-35	-68	26	51
12	14.7	12	12.2	5	4.6	SHELBY	IA	-17	-62	-69	25	20
39*	32.9	23	17.6	18	12.1	SIOUX	IA	-46	-31	-63	35	87
27	14.7	22	10.6	16	7.2	STORY	IA	-28	-32	-51	61	47
22	17.0	17	12.8	10	7.7	TAMA	IA	-25	-40	-55	53	53
10	11.7	6	6.9	3	3.8	TAYLOR	IA	-41	-44	-67	27	14
16	15.3	5	5.2	9	9.1	UNION	IA	-66	73	-41	75	68
18	22.5	8	10.9	6	8.4	VAN BUREN	IA	-51	-23	-62	37	61
24	10.4	18	7.4	11	4.4	WAPELLO	IA	-28	-42	-58	47	18
18	18.5	10	9.5	8	6.4	WARREN	IA	-49	-32	-65	30	38
16	13.3	9	7.5	5	4.3	WASHINGTON	IA	-44	-43	-68	26	17
13	16.2	6	7.4	7	8.8	WAYNE	IA	-54	19	-45	70	65
53	23.9	32	13.5	24	9.8	WEBSTER	IA	-43	-27	-59	45	74
15	21.5	12	15.5	13	15.5	WINNEBAGO	IA	-28	0	-28	85	95
28	24.3	15	12.7	16	12.1	WINNESHIEK	IA	-48	-5	-50	62	87
129*	25.8	50	9.5	52	10.0	WOODBURY	IA	-63	4	-61	39	75
18	30.3	7	11.5	8	12.2	WORTH	IA	-62	6	-60	43	87
26	25.8	16	14.5	14	12.7	WRIGHT	IA	-44	-12	-51	61	89
1488	15.6	1036	9.9	776	7.0	KANSAS	KS	-37	-29	-55		
21	17.1	5	4.8	8	7.8	ALLEN	KS	-72	62	-54	54	53
10	14.4	7	9.6	4	6.2	ANDERSON	KS	-33	-36	-57	49	35
13	12.0	5	5.3	11	11.5	ATCHISON	KS	-56	116	-4	93	84
5	10.6	7	12.4	2	4.0	BARBER	KS	17	-68	-63	36	15
26	26.1	18	13.5	13	8.3	BARTON	KS	-48	-39	-68	25	59
16	14.5	16	13.0	8	6.2	BOURBON	KS	-11	-52	-57	48	35
14	13.4	8	8.5	6	5.3	BROWN	KS	-37	-37	-60	41	26

* SIGNIFICANTLY MORE DEATHS THAN EXPECTED

WHITE MALE

CANCER OF THE STOMACH — COUNTY AND STATE — ICD 151

1950-1959 DEATHS	1950-1959 RATE	1960-1969 DEATHS	1960-1969 RATE	1970-1979 DEATHS	1970-1979 RATE	COUNTY AND STATE	PERCENT CHANGE 60S/50S	PERCENT CHANGE 70S/60S	PERCENT CHANGE 70S/50S	PERCENTILE %CHANGE 70S/50S	PERCENTILE RATES 1970S
32	19.0	21	11.7	17	8.3	BUTLER KS	-39	-29	-56	50	60
6	20.4	2	6.7	2	6.9	CHASE KS	-67	3	-66	30	44
6	12.6	4	7.5	4	9.0	CHAUTAUQUA KS	-40	20	-28	85	67
22	16.4	12	9.1	13	9.8	CHEROKEE KS	-45	8	-40	76	74
6	22.7	1	3.5			CHEYENNE KS	-85				
5	25.5	3	14.1	2	9.3	CLARK KS	-45	-34	-64	34	70
11	13.9	4	4.6	7	9.3	CLAY KS	-67	102	-33	82	69
17	16.7	10	10.0	11	11.3	CLOUD KS	-40	13	-33	83	83
9	11.6	10	13.3	3	6.2	COFFEY KS	14	-54	-47	68	34
3	14.2	1	4.8	1	5.6	COMANCHE KS	-67	18	-61	41	29
26	12.1	18	8.3	12	6.0	COWLEY KS	-31	-28	-50	62	33
49	18.3	26	10.4	24	9.2	CRAWFORD KS	-43	-11	-50	63	69
7	17.3	5	13.2	3	8.1	DECATUR KS	-24	-38	-53	56	58
25	19.4	16	11.0	9	6.2	DICKINSON KS	-44	-43	-68	26	35
14	22.3	7	12.2	3	5.0	DONIPHAN KS	-45	-59	-78	11	23
19	12.7	9	5.7	10	5.7	DOUGLAS KS	-55	1	-55	53	30
11	30.6	2	6.8	3	12.7	EDWARDS KS	-78	85	-58	46	89
9	18.9	4	7.1	4	10.2	ELK KS	-62	44	-46	69	77
16	29.1	14	20.7	12	12.1	ELLIS KS	-29	-41	-58	46	87
10	19.1	5	8.9	4	8.9	ELLSWORTH KS	-53	0	-53	56	66
8	14.3	9	13.3	5	6.7	FINNEY KS	-7	-50	-53	56	41
17	19.1	15	14.8	5	4.5	FORD KS	-22	-69	-76	13	19
20	16.0	14	10.2	4	2.7	FRANKLIN KS	-36	-73	-83	06	08
12	19.4	9	12.5	6	8.3	GEARY KS	-36	-34	-57	48	59
2	10.4	1	4.7	2	8.5	GOVE KS	-54	80	-18	89	62
		4	13.3	3	10.7	GRAHAM KS		-19			80
4	34.2			1	3.5	GRANT KS			-90	01	12
				1	3.6	GRAY KS					12
2	58.1	2	19.9			GREELEY KS	-66				
16	18.5	11	13.6	5	7.2	GREENWOOD KS	-27	-47	-61	40	47
1	7.2	1	6.6	1	5.8	HAMILTON KS	-8	-12	-19	89	31
14	22.2	7	11.1	4	7.1	HARPER KS	-50	-36	-68	26	46
15	13.4	13	9.8	15	10.0	HARVEY KS	-27	2	-25	87	76
		3	30.6	2	12.8	HASKELL KS		-58			89
		2	11.7			HODGEMAN KS					
9	11.6	7	9.5	13*	17.3	JACKSON KS	-18	83	49	97	97
14	17.3	9	10.7	6	7.0	JEFFERSON KS	-38	-34	-60	43	45
13	18.9	3	4.8			JEWELL KS	-75				
38	14.2	32	6.9	53	6.8	JOHNSON KS	-51	-2	-52	58	42
2	17.9			1	8.6	KEARNY KS			-52	59	63
5	10.2	7	11.9	2	3.3	KINGMAN KS	18	-72	-67	27	11
2	9.0	3	10.1	2	8.8	KIOWA KS	12	-13	-2	93	65
25	15.5	22	13.6	12	7.7	LABETTE KS	-13	-43	-50	62	53
3	22.8			1	4.3	LANE KS			-81	08	17
33	16.1	21	9.4	17	8.0	LEAVENWORTH KS	-41	-16	-50	62	56
11	24.5	4	9.3	5	12.8	LINCOLN KS	-62	38	-48	67	90
11	13.3	8	13.3	5	7.0	LINN KS	0	-47	-47	67	45
5	25.5	1	5.9			LOGAN KS	-77				
20	12.8	10	6.9	16	10.6	LYON KS	-46	54	-17	90	79
20	15.3	24	16.3	14	8.9	MC PHERSON KS	6	-45	-42	74	66
17	17.6	10	9.5	8	6.3	MARION KS	-46	-34	-64	33	36
21	18.8	16	14.7	7	6.0	MARSHALL KS	-22	-59	-68	25	33
		6	21.6	4	12.2	MEADE KS		-43			87
14	11.7	13	10.3	8	6.2	MIAMI KS	-12	-40	-47	68	35
9	15.0	5	6.9	2	3.3	MITCHELL KS	-54	-52	-78	11	10
36	14.1	29	10.9	19	8.4	MONTGOMERY KS	-23	-23	-41	75	60
2	4.1	6	10.0	2	3.2	MORRIS KS	143	-68	-23	88	10
1	14.9	1	12.4			MORTON KS	-17				
7	8.4	7	9.1	6	6.9	NEMAHA KS	8	-24	-18	89	43
17	14.1	19	14.8	7	5.8	NEOSHO KS	5	-61	-59	44	30
8	25.4	5	13.9	2	5.2	NESS KS	-45	-63	-80	09	25
9	16.3	2	3.3	3	5.6	NORTON KS	-79	67	-66	30	28
19	20.7	9	8.6	7	6.6	OSAGE KS	-59	-24	-68	25	39
10	17.2	2	4.2	6	10.3	OSBORNE KS	-75	143	-40	76	77
6	11.4	8	13.4	5	9.7	OTTAWA KS	17	-27	-14	91	74
5	7.6	7	11.3	3	6.5	PAWNEE KS	49	-43	-15	90	38
6	10.5	4	6.0	4	7.7	PHILLIPS KS	-43	29	-27	86	53
20	22.4	6	6.8	6	7.4	POTTAWATOMIE KS	-70	9	-67	28	49
8	11.9	4	6.0	3	4.2	PRATT KS	-49	-31	-65	32	16
8	29.0	3	9.3	1	4.4	RAWLINS KS	-68	-53	-85	04	18
32	12.1	35	12.0	18	5.6	RENO KS	-1	-54	-54	55	28
19	22.5	10	14.2	3	5.3	REPUBLIC KS	-37	-63	-77	13	25
16	22.1	7	8.0	10	13.3	RICE KS	-64	66	-40	77	91
17	14.8	14	11.3	8	6.2	RILEY KS	-24	-45	-58	47	35
6	12.6	1	1.9	4	8.0	ROOKS KS	-85	326	-36	80	56
11	28.9	5	12.4			RUSH KS	-57				
17	28.2	10	15.8	6	8.8	RUSSELL KS	-44	-44	-69	25	65
21	13.0	18	9.8	17	8.5	SALINE KS	-24	-13	-35	81	62
2	13.1	1	4.0	2	8.2	SCOTT KS	-69	105	-37	79	59
120	14.4	100	9.9	76	6.2	SEDGWICK KS	-31	-38	-57	48	35
8	21.4	5	11.8	3	5.7	SEWARD KS	-45	-52	-73	17	29
81	16.4	57	10.0	31	4.9	SHAWNEE KS	-39	-51	-70	22	22
3	18.6	3	14.7	1	5.0	SHERIDAN KS	-21	-66	-73	18	23
6	19.3	2	5.0	2	5.3	SHERMAN KS	-74	6	-72	19	26
10	16.7	3	4.6	3	3.3	SMITH KS	-72	-29	-80	08	10
6	12.3	2	4.3	1	1.7	STAFFORD KS	-65	-61	-86	03	05
1	11.2	1	9.9			STANTON KS	-11				
				1	4.5	STEVENS KS					19
21	14.2	12	7.4	9	5.6	SUMNER KS	-48	-25	-61	40	28
11	33.5	3	10.5	4	9.2	THOMAS KS	-69	-12	-72	19	69
3	11.1	1	3.4	5	16.4	TREGO KS	-69	384	48	97	96
4	8.0	5	9.8	7	11.4	WABAUNSEE KS	22	16	42	97	83
1	8.9			2	13.0	WALLACE KS			45	97	90
11	13.9	11	15.1	3	3.8	WASHINGTON KS	9	-75	-73	18	13
1	8.0			1	5.7	WICHITA KS			-29	84	29
10	11.3	4	4.2	6	6.9	WILSON KS	-63	66	-39	78	44
9	18.3	5	9.4	3	8.3	WOODSON KS	-49	-12	-55	54	60
88	15.6	77	12.5	50	8.1	WYANDOTTE KS	-20	-35	-48	66	57
1779	15.5	1361	10.5	1074	7.4	KENTUCKY KY	-32	-29	-52		

* SIGNIFICANTLY MORE DEATHS THAN EXPECTED

WHITE MALE | CANCER OF THE STOMACH | ICD 151

1950-1959 DEATHS	1950-1959 RATE	1960-1969 DEATHS	1960-1969 RATE	1970-1979 DEATHS	1970-1979 RATE	COUNTY AND STATE	PERCENT CHANGE 60S/50S	PERCENT CHANGE 70S/60S	PERCENT CHANGE 70S/50S	PERCENTILE %CHANGE 70S/50S	PERCENTILE RATES 1970S
10	13.4	2	2.5	8	10.0	ADAIR KY	-81	301	-25	87	76
7	9.1	6	7.1	6	8.3	ALLEN KY	-22	17	-9	92	59
11	24.3	3	6.7	4	7.9	ANDERSON KY	-73	19	-67	27	55
4	8.3	4	7.0	4	7.3	BALLARD KY	-15	3	-12	91	48
15	11.5	13	9.5	7	4.3	BARREN KY	-18	-54	-62	37	18
11	25.6	4	9.0	5	9.4	BATH KY	-65	4	-63	35	71
22	16.7	12	8.0	14	9.4	BELL KY	-52	18	-43	72	71
9	11.8	4	4.9	7	5.8	BOONE KY	-59	20	-51	61	31
11	17.8	7	9.2	4	4.8	BOURBON KY	-48	-48	-73	18	21
28	16.1	31	14.2	19	7.7	BOYD KY	-12	-46	-52	58	53
8	9.3	8	8.5	12	12.2	BOYLE KY	-8	43	32	96	87
6	13.0	2	4.7	9*	18.4	BRACKEN KY	-64	292	42	97	98
12	19.0	6	9.4	6	8.4	BREATHITT KY	-50	-11	-56	51	61
11	14.9	4	4.6	8	9.1	BRECKINRIDGE KY	-69	101	-39	78	68
9	19.3	11	15.9	5	4.7	BULLITT KY	-18	-70	-76	14	21
8	14.4	5	8.2	6	9.0	BUTLER KY	-43	10	-37	79	67
9	13.0	8	10.2	5	7.5	CALDWELL KY	-22	-26	-43	73	50
17	16.2	16	13.0	11	7.2	CALLOWAY KY	-20	-44	-56	52	47
66	19.0	35	9.1	28	7.2	CAMPBELL KY	-52	-21	-62	38	47
1	2.6	7	18.1	1	2.3	CARLISLE KY	602	-87	-11	92	07
3	5.7	2	4.3			CARROLL KY	-25				
24	26.6	17	18.3	23*	22.0	CARTER KY	-31	20	-17	90	98
8	10.6	5	6.7	6	8.0	CASEY KY	-37	19	-25	87	56
11	7.4	20	12.0	6	3.6	CHRISTIAN KY	62	-70	-52	59	12
11	14.2	7	7.7	8	7.8	CLARK KY	-46	0	-45	70	54
10	14.6	13	17.3	12	13.8	CLAY KY	19	-21	-6	93	92
11	28.6	5	12.9	4	8.0	CLINTON KY	-55	-38	-72	19	56
9	18.5	6	11.2	5	9.4	CRITTENDEN KY	-39	-16	-49	64	71
5	12.3	6	14.8	2	3.8	CUMBERLAND KY	20	-74	-69	24	14
31	13.7	30	10.4	18	5.4	DAVIESS KY	-24	-48	-61	40	26
7	18.2	5	10.9	2	3.5	EDMONSON KY	-40	-68	-81	08	11
7	26.5	9*	28.2	2	14.9	ELLIOTT KY	6	-47	-44	72	95
14	24.7	7	10.2	7	9.0	ESTILL KY	-59	-11	-63	34	67
52	14.6	51	11.8	45	8.5	FAYETTE KY	-19	-28	-42	74	61
12	17.5	10	16.0	3	4.4	FLEMING KY	-9	-72	-75	15	18
25	16.9	17	11.3	15	8.2	FLOYD KY	-33	-27	-52	59	58
9	8.2	11	8.5	11	7.6	FRANKLIN KY	4	-11	-8	92	51
14	23.9	8	15.1	3	5.6	FULTON KY	-37	-63	-77	13	28
6	28.9	1	4.7	2	8.5	GALLATIN KY	-84	81	-71	21	62
6	13.2	2	4.1	4	7.0	GARRARD KY	-69	70	-47	67	44
12	20.6	7	11.0	3	4.2	GRANT KY	-47	-62	-80	09	16
18	11.1	16	8.9	12	6.5	GRAVES KY	-20	-27	-41	75	39
14	16.9	4	3.5	5	4.2	GRAYSON KY	-79	20	-75	15	16
10	20.2	4	7.2	1	1.6	GREEN KY	-64	-78	-92	01	05
25	27.1	18	15.4	11	7.6	GREENUP KY	-43	-51	-72	20	52
8	22.9	3	7.6	3	9.4	HANCOCK KY	-67	24	-59	46	71
16	12.1	21	13.6	16	8.0	HARDIN KY	13	-41	-34	82	56
30	17.9	34*	20.9	19	11.2	HARLAN KY	17	-47	-37	79	82
10	11.5	9	10.8	6	6.8	HARRISON KY	-6	-37	-40	76	42
9	11.8	9	12.4	6	7.6	HART KY	5	-39	-36	80	51
26	20.7	17	12.4	7	4.4	HENDERSON KY	-40	-65	-79	10	18
10	15.3	3	4.0	4	5.7	HENRY KY	-74	42	-63	35	29
5	11.9	5	11.9	3	7.6	HICKMAN KY	0	-36	-36	80	51
20	11.2	20	10.4	9	4.4	HOPKINS KY	-7	-58	-61	39	18
5	9.2	3	5.9	7	11.4	JACKSON KY	-36	94	23	96	83
248	14.3	198	9.8	167	7.1	JEFFERSON KY	-32	-27	-50	62	46
6	11.2	6	10.0	8	11.9	JESSAMINE KY	-10	18	6	95	86
13	13.9	15	16.5	7	7.0	JOHNSON KY	18	-57	-49	64	45
56	13.2	54	10.9	35	6.5	KENTON KY	-17	-41	-51	61	38
14	23.2	4	5.6	6	9.0	KNOTT KY	-76	60	-61	39	66
18	19.5	18	15.1	11	8.4	KNOX KY	-23	-44	-57	49	60
4	9.0	6	10.7	2	3.0	LARUE KY	20	-72	-67	28	09
17	15.4	13	9.8	12	7.4	LAUREL KY	-37	-24	-52	59	50
24*	34.9	10	15.2	11	14.0	LAWRENCE KY	-57	-8	-60	43	93
5	12.5	6	16.1	4	10.2	LEE KY	28	-37	-19	89	77
14*	36.8	4	10.9	1	1.9	LESLIE KY	-70	-83	-95	00	06
21	21.2	10	9.5	11	8.9	LETCHER KY	-55	-7	-58	47	65
12	19.8	11	17.4	4	6.1	LEWIS KY	-12	-65	-69	24	34
15	18.9	8	8.8	9	9.5	LINCOLN KY	-53	7	-50	63	72
7	17.8	4	9.3	1	2.0	LIVINGSTON KY	-48	-79	-89	02	06
16	15.1	10	9.0	6	5.3	LOGAN KY	-41	-41	-65	31	25
10	34.1	2	6.2	2	6.2	LYON KY	-82	1	-82	07	36
22	10.0	15	5.9	23	8.3	MC CRACKEN KY	-41	40	-17	90	60
9	18.7	11	19.7	8	11.8	MC CREARY KY	5	-40	-37	79	86
7	12.2	5	9.0	1	1.5	MC LEAN KY	-26	-83	-88	02	05
18	15.2	13	9.3	18	11.1	MADISON KY	-39	19	-27	85	82
5	10.6	7	14.2			MAGOFFIN KY	34				
14	25.3	5	8.2	6	9.0	MARION KY	-68	10	-64	33	67
22	29.7	10	10.8	1	.7	MARSHALL KY	-64	-93	-98	00	04
8	25.8	6	16.6	6	12.9	MARTIN KY	-36	-22	-50	63	90
17	23.6	13	15.8	6	7.0	MASON KY	-33	-56	-70	22	44
3	8.1	5	11.9	1	1.7	MEADE KY	47	-85	-78	10	05
6	25.0	3	11.1	6*	27.3	MENIFEE KY	-55	146	9	95	99
12	16.8	8	10.9	7	8.0	MERCER KY	-35	-26	-52	58	56
13	29.4	1	1.8	8	14.7	METCALFE KY	-94	718	-50	62	94
5	9.3	9	15.1	1	1.3	MONROE KY	63	-92	-86	03	05
5	9.5	9	14.2	6	8.3	MONTGOMERY KY	49	-42	-13	91	59
7	12.5	9	15.7	5	8.6	MORGAN KY	26	-45	-31	83	64
26	17.5	6	3.7	12	8.1	MUHLENBERG KY	-79	119	-54	55	58
11	14.3	6	7.0	3	3.1	NELSON KY	-51	-56	-78	11	10
6	13.9	6	15.0	6	15.9	NICHOLAS KY	8	6	15	95	96
18	15.6	10	8.1	7	5.6	OHIO KY	-48	-31	-64	33	28
4	8.0	1	1.7	5	7.8	OLDHAM KY	-79	361	-2	93	54
2	3.9	7	14.4	2	3.9	OWEN KY	267	-73	0	93	14
3	10.2	2	4.9	1	2.7	OWSLEY KY	-51	-45	-73	18	08
12	21.7	7	12.0	5	8.4	PENDLETON KY	-45	-30	-61	40	61
22	19.0	21	17.5	14	11.2	PERRY KY	-8	-36	-41	75	82
27	12.6	33	13.4	27	9.4	PIKE KY	6	-30	-25	87	71
3	11.4	5	17.0	3	7.0	POWELL KY	49	-58	-38	78	45
27	16.6	26	14.1	14	6.4	PULASKI KY	-15	-55	-61	39	38
1	4.8	2	12.6			ROBERTSON KY	163				

* SIGNIFICANTLY MORE DEATHS THAN EXPECTED

1950-1959 DEATHS	1950-1959 RATE	1960-1969 DEATHS	1960-1969 RATE	1970-1979 DEATHS	1970-1979 RATE	COUNTY AND STATE	PERCENT CHANGE 60S/50S	PERCENT CHANGE 70S/60S	PERCENT CHANGE 70S/50S	PERCENTILE %CHANGE 70S/50S	PERCENTILE RATES 1970S
8	13.5	4	6.7	6	8.3	ROCKCASTLE KY	-50	23	-38	78	60
4	8.3	6	10.5	8	12.2	ROWAN KY	26	17	46	97	87
6	11.0	5	10.3	10	13.3	RUSSELL KY	-6	29	21	96	91
7	9.9	6	8.4	3	3.5	SCOTT KY	-15	-58	-64	33	12
12	14.1	8	8.7	6	5.7	SHELBY KY	-38	-34	-60	43	29
9	17.3	7	11.3	4	5.7	SIMPSON KY	-35	-50	-67	27	29
5	19.4	2	7.4	3	11.5	SPENCER KY	-62	56	-41	76	84
3	5.4	10	14.4	6	7.0	TAYLOR KY	168	-51	31	96	45
13	25.3	3	5.5	1	1.8	TODD KY	-78	-67	-93	00	06
13	31.5	3	6.1	10*	20.2	TRIGG KY	-81	232	-36	80	98
3	10.8	1	3.4	2	7.3	TRIMBLE KY	-69	117	-32	83	48
11	16.7	3	4.2	3	4.3	UNION KY	-75	3	-74	16	18
17	9.2	20	9.9	14	6.0	WARREN KY	7	-39	-35	81	33
3	5.8	13*	24.7	3	5.8	WASHINGTON KY	325	-77	0	93	31
6	8.6	5	6.7	5	6.4	WAYNE KY	-23	-4	-26	86	38
17	21.8	10	11.0	5	5.3	WEBSTER KY	-50	-52	-76	14	25
24	18.6	15	11.1	13	9.3	WHITLEY KY	-41	-16	-50	63	70
6	19.0	2	6.4	7*	19.3	WOLFE KY	-66	203	1	94	98
9	21.8	5	10.4	7	11.5	WOODFORD KY	-52	11	-47	67	84
1256	19.0	950	11.3	828	8.0	LOUISIANA LA	-41	-29	-58		
42*	32.1	21	13.0	20	11.3	ACADIA LA	-59	-14	-65	31	83
7	12.5	7	11.4	4	5.3	ALLEN LA	-8	-53	-57	48	26
12	19.4	14	17.4	9	8.6	ASCENSION LA	-11	-50	-56	51	63
3	6.6	7	16.1	5	7.7	ASSUMPTION LA	145	-53	16	95	52
28	25.9	17	12.8	9	5.7	AVOYELLES LA	-51	-55	-78	11	30
17	24.5	10	13.6	5	5.0	BEAUREGARD LA	-45	-63	-80	09	23
8	16.5	8	13.8	11*	18.1	BIENVILLE LA	-17	32	10	95	97
13	20.1	8	12.4	15	10.0	BOSSIER LA	-38	-20	-50	62	75
64	14.5	58	9.6	49	7.4	CADDO LA	-34	-22	-49	65	49
50	27.3	27	8.9	33	7.9	CALCASIEU LA	-67	-11	-71	21	55
7	16.9	3	6.8	2	4.3	CALDWELL LA	-60	-37	-75	16	17
3	14.3	2	6.0	2	7.7	CAMERON LA	-58	28	-46	69	53
2	6.2	4	11.5	3	6.3	CATAHOULA LA	84	-45	1	94	36
7	13.0	9	14.5	6	8.9	CLAIBORNE LA	12	-39	-31	83	66
6	22.6	3	6.8	4	8.3	CONCORDIA LA	-70	22	-63	35	60
12	21.0	10	14.7	5	6.3	DE SOTO LA	-30	-57	-70	22	36
37	13.8	36	8.4	41	6.3	EAST BATON ROUGE LA	-39	-25	-54	54	36
3	18.3	4	17.9	2	15.2	EAST CARROLL LA	-3	-15	-17	90	95
2	4.2	5	9.5	4	8.2	EAST FELICIANA LA	126	-13	96	98	59
19	22.3	18	18.3	7	5.9	EVANGELINE LA	-18	-68	-73	17	32
16	25.1	10	12.5	8	8.5	FRANKLIN LA	-50	-32	-66	29	62
7	11.8	8	14.5	4	5.3	GRANT LA	23	-64	-55	52	25
17	19.3	13	11.3	13	8.8	IBERIA LA	-41	-22	-54	54	65
17	31.5	8	10.6	8	11.5	IBERVILLE LA	-66	8	-64	34	84
6	12.6	11	18.9	4	5.5	JACKSON LA	50	-71	-56	50	27
57	22.6	64	13.2	73	7.9	JEFFERSON LA	-42	-40	-65	31	55
20	26.3	17	21.0	8	7.4	JEFFERSON DAVIS LA	-20	-65	-72	19	48
28	21.1	20	11.2	33	11.9	LAFAYETTE LA	-47	6	-44	72	86
23	19.8	23	17.4	14	7.3	LAFOURCHE LA	-12	-58	-63	36	48
9	18.0	1	1.5	5	7.4	LA SALLE LA	-91	378	-59	45	49
13	23.5	9	12.0	8	9.3	LINCOLN LA	-49	-22	-60	42	70
10	16.6	8	8.3	6	4.2	LIVINGSTON LA	-50	-50	-75	15	16
5	29.1	2	8.4	1	2.8	MADISON LA	-71	-66	-90	01	09
13	23.3	3	4.0	3	3.4	MOREHOUSE LA	-83	-16	-86	04	11
11	12.9	10	9.3	7	5.9	NATCHITOCHES LA	-28	-37	-55	53	31
268	17.9	168	10.2	117	7.8	ORLEANS LA	-43	-23	-56	50	54
35	18.0	34	12.7	22	6.4	OUACHITA LA	-29	-50	-65	32	37
5	21.9	4	6.4	6	10.0	PLAQUEMINES LA	-71	55	-55	53	75
6	15.6	4	9.6	8	13.4	POINTE COUPEE LA	-39	40	-14	91	92
45	19.4	25	8.1	24	6.4	RAPIDES LA	-58	-21	-67	27	37
3	10.4	4	14.3	1	2.3	RED RIVER LA	37	-84	-77	12	07
10	17.6	5	8.9	8	10.2	RICHLAND LA	-50	15	-42	74	77
9	12.1	6	7.1	6	5.8	SABINE LA	-42	-19	-53	57	30
7	19.3	9	10.3	22*	14.4	ST BERNARD LA	-47	39	-26	87	94
4	8.8	5	12.4	6	8.1	ST CHARLES LA	41	-35	-8	92	57
1	4.6	1	4.5	4	15.9	ST HELENA LA	-3	255	245	99	96
9	32.6	1	2.5	6	14.1	ST JAMES LA	-92	457	-57	49	93
7	30.7	6	19.1	5	11.6	ST JOHN THE BAPT LA	-38	-39	-62	37	85
36	23.9	24	15.2	26	12.5	ST LANDRY LA	-36	-18	-47	67	88
14	27.4	17*	22.7	14*	14.8	ST MARTIN LA	-17	-35	-46	69	95
15	17.1	16	16.0	7	5.2	ST MARY LA	-6	-67	-69	24	25
14	13.4	12	7.5	11	4.4	ST TAMMANY LA	-44	-42	-67	27	18
28	19.1	18	11.4	18	10.0	TANGIPAHOA LA	-41	-12	-48	66	75
3	17.4			1	5.6	TENSAS LA			-68	25	28
32*	32.6	19	14.8	26*	14.1	TERREBONNE LA	-54	-5	-57	49	93
9	13.9	8	11.4	3	4.1	UNION LA	-18	-64	-71	21	16
29	23.2	16	10.3	24	13.8	VERMILION LA	-56	35	-41	76	92
20	25.4	17	18.6	7	6.0	VERNON LA	-27	-68	-76	13	33
17	17.8	12	10.0	5	3.8	WASHINGTON LA	-44	-62	-79	10	13
10	10.1	17	14.4	8	5.3	WEBSTER LA	42	-63	-48	66	25
6	23.3	6	25.9	5	16.0	WEST BATON ROUGE LA	11	-38	-31	83	96
12	21.5	8	14.0	3	5.1	WEST CARROLL LA	-35	-64	-76	13	23
		2	11.8	3	23.0	WEST FELICIANA LA		95			99
8	13.0	8	12.2	1	1.2	WINN LA	-6	-90	-91	01	04
1008*	22.2	693*	14.6	496	9.8	MAINE ME	-34	-33	-56		
114*	29.5	66	15.5	53	11.7	ANDROSCOGGIN ME	-47	-25	-60	41	85
114*	32.6	63	17.3	56*	14.8	AROOSTOOK ME	-47	-14	-54	54	95
158	19.4	97	11.1	82	8.7	CUMBERLAND ME	-42	-22	-55	52	64
21	20.0	15	13.8	12	10.1	FRANKLIN ME	-31	-27	-50	63	76
37	20.1	35	17.3	14	6.4	HANCOCK ME	-14	-63	-68	26	38
68	20.4	68	15.1	42	8.4	KENNEBEC ME	-26	-44	-59	45	61
41	22.6	29	16.0	16	8.7	KNOX ME	-29	-45	-61	39	64
23	20.1	24	18.5	18	12.0	LINCOLN ME	-8	-36	-41	76	86
34	14.1	37	15.8	15	5.9	OXFORD ME	12	-63	-58	47	32
104	20.9	61	11.6	50	9.2	PENOBSCOT ME	-44	-20	-56	51	69
29	26.0	18	17.0	8	8.6	PISCATAQUIS ME	-35	-50	-67	27	62
21	19.1	18	15.1	16	13.3	SAGADAHOC ME	-21	-12	-30	84	91

* SIGNIFICANTLY MORE DEATHS THAN EXPECTED

WHITE MALE — CANCER OF THE STOMACH — ICD 151

1950-1959 DEATHS	1950-1959 RATE	1960-1969 DEATHS	1960-1969 RATE	1970-1979 DEATHS	1970-1979 RATE	COUNTY AND STATE	PERCENT CHANGE 60S/50S	PERCENT CHANGE 70S/60S	PERCENT CHANGE 70S/50S	PERCENTILE %CHANGE 70S/50S	PERCENTILE RATES 1970S
51	24.4	27	12.6	16	7.4	SOMERSET ME	-48	-42	-70	23	49
16	11.9	31*	25.0	13	10.4	WALDO ME	110	-59	-13	91	78
54	27.0	24	12.3	19	9.6	WASHINGTON ME	-54	-22	-64	33	72
103	22.5	80	15.4	66	10.9	YORK ME	-32	-29	-51	60	81
1444	18.4	1115	11.2	1051	8.4	MARYLAND MD	-39	-25	-54		
74	20.9	47	12.0	36	8.4	ALLEGANY MD	-42	-30	-60	43	61
66	20.0	70	12.6	65	8.3	ANNE ARUNDEL MD	-37	-34	-59	46	59
167	16.8	165	10.5	225	9.5	BALTIMORE MD	-37	-10	-44	72	72
8	19.9	9	17.7	8	9.5	CALVERT MD	-11	-47	-53	57	72
14	17.5	6	6.1	2	1.9	CAROLINE MD	-65	-69	-89	02	06
27	11.8	23	8.4	14	4.5	CARROLL MD	-29	-46	-61	39	19
28	19.6	24	13.7	12	6.1	CECIL MD	-30	-56	-69	24	33
11	16.8	10	15.7	11	7.8	CHARLES MD	-7	-50	-53	56	54
23	19.5	17	13.1	13	10.3	DORCHESTER MD	-33	-22	-47	67	77
36	14.4	29	10.1	32	9.0	FREDERICK MD	-29	-11	-37	79	67
24	25.5	15	14.5	11	8.9	GARRETT MD	-43	-38	-65	31	66
31	18.5	35	13.9	22	5.8	HARFORD MD	-25	-59	-69	24	30
9	11.7	5	3.9	12	5.6	HOWARD MD	-67	44	-52	58	29
15	25.7	5	8.5	3	3.5	KENT MD	-67	-59	-86	03	12
85	13.4	99	9.3	130	7.7	MONTGOMERY MD	-31	-18	-43	73	52
70	13.9	92	10.3	119	9.8	PRINCE GEORGE'S MD	-26	-5	-29	84	74
11	17.0	6	8.7	5	6.6	QUEEN ANNE'S MD	-49	-25	-62	39	39
13	20.6	3	3.6	8	7.1	ST MARY'S MD	-83	100	-65	31	46
12	15.0	10	12.7	8	9.9	SOMERSET MD	-16	-22	-34	82	75
19	23.8	12	12.6	12	9.4	TALBOT MD	-47	-25	-60	41	71
46	13.4	28	7.2	36	7.4	WASHINGTON MD	-46	4	-44	71	49
22	13.3	15	8.4	11	5.1	WICOMICO MD	-37	-39	-62	38	24
17	18.3	9	9.4	5	4.1	WORCESTER MD	-48	-56	-77	12	16
616	21.4	381	13.5	251	10.1	BALTIMORE CITY MD	-37	-25	-53	57	76
4996*	22.6	3786*	15.8	2958*	11.6	MASSACHUSETTS MA	-30	-27	-49		
44	15.1	55	11.7	59	7.0	BARNSTABLE MA	-22	-40	-53	56	45
174*	27.3	108	15.5	63	8.3	BERKSHIRE MA	-43	-46	-69	23	60
502*	27.5	379*	19.4	341*	15.9	BRISTOL MA	-30	-18	-42	74	96
8	25.3	6	16.7	3	7.1	DUKES MA	-34	-57	-72	20	46
589*	22.2	502*	17.5	350*	11.5	ESSEX MA	-21	-34	-48	66	84
45	15.8	38	12.5	36	11.1	FRANKLIN MA	-21	-11	-30	84	82
418*	24.5	316*	16.5	225*	11.1	HAMPDEN MA	-33	-33	-55	53	82
77	18.3	71	15.7	58	11.5	HAMPSHIRE MA	-14	-26	-37	79	84
1073*	22.2	762*	14.3	634*	11.2	MIDDLESEX MA	-36	-22	-50	63	82
3	17.7	2	12.0	1	4.1	NANTUCKET MA	-32	-66	-77	12	16
343	18.3	357*	15.7	281*	10.6	NORFOLK MA	-14	-32	-42	74	79
213	20.1	168	13.9	158*	11.1	PLYMOUTH MA	-31	-20	-45	71	82
919*	24.6	593*	16.7	406*	13.8	SUFFOLK MA	-32	-18	-44	72	92
588*	21.8	429*	14.8	343*	11.2	WORCESTER MA	-32	-24	-49	65	82
6288*	24.2	4558*	15.3	3176*	9.7	MICHIGAN MI	-37	-37	-60		
14*	37.2	11	22.8	5	6.6	ALCONA MI	-39	-71	-82	07	40
25*	49.9	17*	31.8	5	10.8	ALGER MI	-36	-66	-78	10	80
67	26.2	42	15.2	35	11.0	ALLEGAN MI	-42	-27	-58	48	82
46*	41.8	23	19.2	14	10.4	ALPENA MI	-54	-46	-75	15	78
21	28.4	12	14.9	7	7.2	ANTRIM MI	-47	-52	-75	16	47
17	29.2	11	17.6	7	7.9	ARENAC MI	-40	-55	-73	18	55
13	29.0	14*	29.3	4	7.3	BARAGA MI	1	-75	-75	15	48
23	16.2	16	9.8	16	8.0	BARRY MI	-39	-18	-50	62	56
98*	26.5	79*	18.9	53	11.4	BAY MI	-29	-40	-57	49	83
5	10.1	9	18.2	4	6.7	BENZIE MI	80	-63	-34	82	41
136*	26.0	105*	17.2	52	8.0	BERRIEN MI	-34	-53	-69	24	56
32	20.0	16	9.5	10	5.8	BRANCH MI	-52	-39	-71	21	30
86	17.3	62	11.4	36	6.2	CALHOUN MI	-34	-46	-64	33	35
45*	31.0	18	10.7	12	6.4	CASS MI	-66	-40	-79	09	37
15	19.1	12	14.1	10	10.4	CHARLEVOIX MI	-26	-26	-45	70	78
21	27.6	14	17.7	10	10.3	CHEBOYGAN MI	-36	-42	-63	37	77
29	22.3	21	16.1	13	9.1	CHIPPEWA MI	-28	-43	-59	45	68
6	10.3	11	15.2	6	5.0	CLARE MI	47	-67	-51	60	23
22	15.4	14	9.2	15	8.3	CLINTON MI	-41	-9	-46	69	59
6	31.2	3	10.0	6	13.6	CRAWFORD MI	-68	36	-56	50	92
76*	43.1	46*	24.8	18	9.1	DELTA MI	-43	-63	-79	10	68
45*	34.3	30*	18.6	22	13.1	DICKINSON MI	-46	-30	-62	38	90
37	18.1	19	8.8	18	6.9	EATON MI	-51	-23	-62	37	43
16	18.7	12	13.6	8	8.4	EMMET MI	-27	-38	-55	52	61
204	20.5	160	14.2	99	7.6	GENESEE MI	-31	-46	-63	36	52
11	20.0	15	21.8	10	11.5	GLADWIN MI	9	-47	-43	73	84
65*	42.0	38*	23.2	19	12.9	GOGEBIC MI	-45	-45	-69	23	90
20	11.6	21	11.4	14	6.9	GRAND TRAVERSE MI	-1	-40	-41	76	43
33	18.6	16	8.9	8	4.4	GRATIOT MI	-52	-50	-76	13	18
25	13.8	21	11.8	18	8.8	HILLSDALE MI	-14	-25	-36	80	65
101*	41.2	43*	19.3	37*	17.6	HOUGHTON MI	-53	-9	-57	48	97
32	17.7	29	15.2	29*	13.3	HURON MI	-14	-12	-25	87	91
136	22.0	98	14.2	49	6.7	INGHAM MI	-35	-53	-70	23	41
38	18.6	20	10.0	24	12.5	IONIA MI	-46	25	-33	83	88
14	19.1	13	13.8	16	11.9	IOSCO MI	-28	-14	-38	79	86
61*	58.0	23*	20.4	29*	26.3	IRON MI	-65	29	-55	53	99
22	18.0	17	13.4	7	5.1	ISABELLA MI	-25	-62	-72	20	24
79	15.7	54	10.2	33	5.9	JACKSON MI	-35	-43	-63	36	31
136	24.0	72	11.4	67	9.4	KALAMAZOO MI	-52	-17	-61	41	71
5	16.8	8*	24.0	5	13.3	KALKASKA MI	43	-45	-21	89	91
276	22.1	188	12.7	139	8.6	KENT MI	-42	-32	-61	40	63
10*	48.1	3	17.7	1	3.7	KEWEENAW MI	-63	-79	-92	00	13
8	27.4	1	2.4	5	11.6	LAKE MI	-91	375	-58	48	84
33	18.6	22	11.8	18	8.6	LAPEER MI	-36	-27	-53	56	63
5	10.4	6	11.2	4	6.5	LEELANAU MI	8	-42	-37	79	39
56	17.9	32	9.7	30	8.7	LENAWEE MI	-46	-10	-51	60	64
29	18.8	24	12.8	16	5.9	LIVINGSTON MI	-32	-54	-69	24	31
5	9.7	4	7.6	5	14.0	LUCE MI	-22	85	45	97	93
11	24.5	5	10.8	7	11.2	MACKINAC MI	-56	4	-54	54	82
191*	26.1	204*	17.0	207*	10.9	MACOMB MI	-35	-36	-58	47	81

* SIGNIFICANTLY MORE DEATHS THAN EXPECTED

WHITE MALE — CANCER OF THE STOMACH — ICD 151

1950-1959 DEATHS	1950-1959 RATE	1960-1969 DEATHS	1960-1969 RATE	1970-1979 DEATHS	1970-1979 RATE	COUNTY AND STATE	PERCENT CHANGE 60S/50S	PERCENT CHANGE 70S/60S	PERCENT CHANGE 70S/50S	PERCENTILE %CHANGE 70S/50S	PERCENTILE RATES 1970S
33*	32.4	13	11.0	14	11.4	MANISTEE MI	-66	4	-65	32	84
74*	31.4	58*	24.0	32	12.7	MARQUETTE MI	-24	-47	-59	44	89
29	24.3	18	14.4	9	6.1	MASON MI	-41	-58	-75	15	34
19	17.9	19	17.6	6	4.8	MECOSTA MI	-1	-73	-73	18	21
50*	38.0	30*	21.6	13	8.7	MENOMINEE MI	-43	-60	-77	12	64
36	29.1	21	13.3	19	10.0	MIDLAND MI	-54	-25	-66	30	75
13*	35.5	4	10.4	6	12.6	MISSAUKEE MI	-71	21	-65	32	89
83	26.4	49	12.6	37	8.6	MONROE MI	-52	-32	-68	26	62
32	17.4	24	12.4	16	6.9	MONTCALM MI	-29	-44	-60	42	43
6	27.9	5	17.5	3	6.7	MONTMORENCY MI	-37	-62	-76	13	41
118*	26.6	69	13.2	48	8.4	MUSKEGON MI	-50	-37	-68	25	60
26	20.7	23	16.4	8	4.9	NEWAYGO MI	-20	-70	-76	13	22
310	21.7	321	14.2	292	9.3	OAKLAND MI	-35	-34	-57	48	70
26	27.5	12	12.6	5	4.3	OCEANA MI	-54	-66	-84	05	17
12	22.3	12	16.7	14	14.3	OGEMAW MI	-25	-15	-36	80	94
30*	46.4	12	19.7	5	6.5	ONTONAGON MI	-58	-67	-86	03	39
20	24.0	9	10.9	6	6.7	OSCEOLA MI	-55	-38	-72	19	41
4	21.6	5	15.7	2	4.4	OSCODA MI	-27	-72	-80	09	18
8	24.6	5	10.9	7	11.8	OTSEGO MI	-56	9	-52	58	85
93*	27.6	67	16.8	49	9.9	OTTAWA MI	-39	-41	-64	33	74
17	30.1	10	15.1	18*	23.7	PRESQUE ISLE MI	-50	57	-21	88	99
8	18.4	8	11.3	5	8.0	ROSCOMMON MI	-38	-30	-57	49	55
144	24.2	87	13.4	63	8.6	SAGINAW MI	-45	-36	-64	33	63
84	19.1	77	15.6	42	7.5	ST CLAIR MI	-18	-52	-61	40	50
44	22.1	24	11.2	24	10.4	ST JOSEPH MI	-49	-8	-53	56	78
33	17.0	26	14.1	18	8.5	SANILAC MI	-17	-40	-50	63	62
14	29.9	3	6.2	7	14.5	SCHOOLCRAFT MI	-79	134	-52	60	94
58	25.0	28	12.0	16	6.5	SHIAWASSEE MI	-52	-46	-74	16	39
46	22.6	39	18.1	14	6.7	TUSCOLA MI	-20	-63	-70	22	41
49	20.2	28	10.5	24	8.7	VAN BUREN MI	-48	-18	-57	48	64
78	17.5	62	12.2	52	8.7	WASHTENAW MI	-30	-29	-51	61	64
2168*	25.5	1618*	17.6	1014*	12.2	WAYNE MI	-31	-31	-52	57	87
16	15.0	18	17.5	6	5.1	WEXFORD MI	17	-71	-66	29	24
3672*	25.1	2718*	16.2	1882*	10.2	MINNESOTA MN	-35	-37	-59		
19	20.8	18	16.7	16*	14.6	AITKIN MN	-20	-13	-30	84	94
25	14.7	33	15.0	23	6.9	ANOKA MN	2	-54	-53	56	43
49*	37.1	28*	18.9	10	6.6	BECKER MN	-49	-65	-82	06	39
34	27.9	25*	20.1	10	7.4	BELTRAMI MN	-28	-63	-74	17	49
22	32.4	17	19.2	10	8.9	BENTON MN	-41	-54	-73	18	66
14	26.9	10	18.2	9	14.6	BIG STONE MN	-33	-20	-46	69	94
40	21.1	30	13.9	20	9.1	BLUE EARTH MN	-34	-35	-57	48	68
31	24.7	22	14.3	18	11.0	BROWN MN	-42	-23	-56	52	81
39	28.4	21	14.6	18	12.4	CARLTON MN	-48	-15	-56	50	88
16	16.1	16	14.5	14	9.8	CARVER MN	-10	-32	-39	77	74
20	20.1	16	13.2	8	6.9	CASS MN	-34	-48	-66	30	43
23	25.9	18	18.6	12	11.5	CHIPPEWA MN	-28	-38	-55	52	84
21	21.0	17	14.9	17	13.6	CHISAGO MN	-29	-9	-35	81	92
30	24.0	29	17.5	28*	13.8	CLAY MN	-27	-21	-42	74	93
21*	34.9	10	16.1	10	13.5	CLEARWATER MN	-54	-16	-61	39	92
4	21.3	3	15.9	3	11.3	COOK MN	-25	-29	-47	68	83
21	27.3	12	13.8	10	8.6	COTTONWOOD MN	-49	-37	-68	25	64
40	22.7	33	16.8	20	8.7	CROW WING MN	-26	-48	-61	39	64
51	23.9	45	14.9	31	7.1	DAKOTA MN	-38	-52	-70	22	46
11	15.9	9	12.3	8	9.7	DODGE MN	-23	-21	-39	77	73
35	27.9	23	15.4	13	7.6	DOUGLAS MN	-45	-51	-73	18	52
30	25.5	17	12.6	12	8.3	FARIBAULT MN	-51	-35	-68	26	59
24	16.5	19	12.2	19	11.6	FILLMORE MN	-26	-5	-29	84	85
35	22.0	39*	20.7	14	6.6	FREEBORN MN	-6	-68	-70	23	40
43	22.5	31	14.7	20	8.3	GOODHUE MN	-35	-43	-63	36	59
10	19.0	6	9.7	5	6.8	GRANT MN	-49	-29	-64	34	42
758*	25.1	535*	15.4	366	9.9	HENNEPIN MN	-39	-36	-61	41	75
16	18.6	15	14.4	10	9.2	HOUSTON MN	-22	-36	-51	61	69
17	25.1	15	22.1	11	11.7	HUBBARD MN	-12	-47	-53	56	85
15	19.5	13	13.3	13	12.9	ISANTI MN	-32	-3	-34	82	90
43	24.5	37	20.0	26	12.2	ITASCA MN	-18	-39	-50	63	87
18	24.6	16	18.5	8	8.4	JACKSON MN	-25	-54	-66	30	61
16	28.7	8	13.2	5	6.3	KANABEC MN	-54	-52	-78	11	37
45	26.6	29	15.8	26	12.9	KANDIYOHI MN	-41	-18	-51	60	90
14	27.0	12	22.1	2	3.5	KITTSON MN	-18	-84	-87	03	12
34*	34.0	23*	25.2	7	7.7	KOOCHICHING MN	-26	-69	-77	12	53
14	16.9	12	14.7	14*	15.1	LAC QUI PARLE MN	-13	2	-11	92	95
20*	42.9	13*	25.0	7	10.6	LAKE MN	-42	-57	-75	15	79
5	22.2	3	12.6	4	12.6	LAKE OF THE WOODS MN	-43	0	-43	73	89
38*	33.8	16	13.1	20*	14.5	LE SUEUR MN	-61	11	-57	48	94
18*	34.2	20*	35.2	9	13.3	LINCOLN MN	3	-62	-61	40	91
30	27.6	20	15.9	22*	16.3	LYON MN	-42	3	-41	75	96
29	22.0	20	13.8	12	7.0	MC LEOD MN	-37	-49	-68	26	45
5	16.3	6	16.4	4	10.5	MAHNOMEN MN	1	-36	-35	80	78
24	29.5	13	14.1	7	7.7	MARSHALL MN	-52	-45	-74	17	53
30	25.6	17	12.6	6	4.0	MARTIN MN	-51	-69	-85	05	15
31*	30.2	13	11.0	10	6.9	MEEKER MN	-64	-37	-77	12	43
23	24.4	14	13.0	12	9.9	MILLE LACS MN	-47	-24	-59	44	75
24	18.5	19	12.6	12	7.2	MORRISON MN	-32	-43	-61	40	47
40	23.2	41*	18.9	30	12.4	MOWER MN	-18	-34	-46	68	88
14	23.0	12	16.8	7	8.9	MURRAY MN	-27	-47	-61	39	66
21	18.2	16	14.0	5	4.4	NICOLLET MN	-23	-69	-76	14	18
26	24.1	25*	20.7	15	12.4	NOBLES MN	-14	-40	-48	65	88
27*	40.9	18*	23.7	7	9.8	NORMAN MN	-42	-59	-76	13	74
46	21.0	44	16.5	27	8.7	OLMSTED MN	-22	-47	-58	46	64
74	24.4	53	15.2	36	9.5	OTTER TAIL MN	-38	-37	-61	40	72
20	27.5	18*	21.7	15*	16.1	PENNINGTON MN	-21	-26	-41	75	96
29	27.9	26*	21.9	17	13.2	PINE MN	-22	-39	-52	57	91
16	23.5	16	20.6	7	8.0	PIPESTONE MN	-12	-61	-66	30	56
38	19.9	32	14.4	17	7.5	POLK MN	-27	-48	-62	37	50
23*	32.8	13	16.5	9	10.8	POPE MN	-50	-35	-67	27	80
335	22.4	255	14.6	180	9.7	RAMSEY MN	-35	-34	-57	49	73
8	27.0	6	16.6	2	5.6	RED LAKE MN	-39	-66	-79	09	28
21	20.0	15	12.7	10	7.4	REDWOOD MN	-37	-42	-63	35	49
28	22.2	22	15.3	7	4.4	RENVILLE MN	-31	-71	-80	09	19

* SIGNIFICANTLY MORE DEATHS THAN EXPECTED

WHITE MALE — CANCER OF THE STOMACH — ICD 151

1950-1959 DEATHS	1950-1959 RATE	1960-1969 DEATHS	1960-1969 RATE	1970-1979 DEATHS	1970-1979 RATE	COUNTY AND STATE	PERCENT CHANGE 60S/50S	PERCENT CHANGE 70S/60S	PERCENT CHANGE 70S/50S	PERCENTILE %CHANGE 70S/50S	PERCENTILE RATES 1970S
28	16.8	32	17.3	18	8.8	RICE MN	2	-49	-48	66	65
17	30.4	8	12.0	9	13.1	ROCK MN	-61	10	-57	49	91
28*	41.0	18*	23.1	10	12.4	ROSEAU MN	-44	-46	-70	23	88
388*	33.5	279*	22.9	193*	16.2	ST LOUIS MN	-32	-29	-52	59	96
23	28.9	15	14.5	17	12.8	SCOTT MN	-50	-12	-56	51	89
16	26.2	9	11.5	9	9.4	SHERBURNE MN	-56	-18	-64	34	71
20	23.0	11	11.1	18*	16.8	SIBLEY MN	-52	52	-27	85	96
76	25.0	63	17.0	47	11.0	STEARNS MN	-32	-35	-56	51	81
31	29.2	14	11.2	12	8.3	STEELE MN	-62	-26	-72	20	59
17	31.6	5	8.1	9	13.8	STEVENS MN	-74	70	-56	50	92
22	29.6	21*	23.7	12	12.2	SWIFT MN	-20	-49	-59	45	87
36	27.4	32*	21.8	19	11.2	TODD MN	-20	-49	-59	45	82
11	30.2	7	17.5	4	7.8	TRAVERSE MN	-42	-55	-74	16	54
23	21.9	30*	26.9	9	7.7	WABASHA MN	22	-71	-65	31	53
11	17.4	9	13.4	9	10.3	WADENA MN	-23	-23	-41	75	77
25	30.4	6	7.0	9	9.2	WASECA MN	-77	30	-70	23	69
32	19.1	39	19.6	26	8.8	WASHINGTON MN	2	-55	-54	55	65
12	15.9	13	15.8	7	7.1	WATONWAN MN	0	-55	-55	52	45
14	30.1	4	8.1	5	9.1	WILKIN MN	-73	12	-70	22	68
43	21.8	20	9.4	21	9.1	WINONA MN	-57	-3	-58	46	68
33	19.7	24	12.4	24	10.7	WRIGHT MN	-37	-14	-46	69	79
25	28.2	11	11.6	13	12.0	YELLOW MEDICINE MN	-59	3	-58	48	86
676	14.2	591	10.4	495	7.3	MISSISSIPPI MS	-27	-30	-49		
8	18.9	10	19.2	9	10.9	ADAMS MS	1	-43	-42	74	81
13	13.2	17	15.3	8	6.2	ALCORN MS	17	-59	-53	57	35
5	17.4	5	14.0	4	8.0	AMITE MS	-20	-43	-54	55	56
10	15.1	7	9.0	7	7.5	ATTALA MS	-40	-17	-50	62	50
5	21.1	4	18.9	3	9.0	BENTON MS	-10	-53	-57	48	66
11	18.0	13	16.4	8	9.6	BOLIVAR MS	-9	-41	-47	68	72
10	19.0	11	16.3	4	5.8	CALHOUN MS	-14	-64	-69	24	31
5	17.4	3	9.3	1	4.3	CARROLL MS	-47	-54	-75	15	17
3	6.4	10	19.2	3	5.5	CHICKASAW MS	200	-71	-14	91	28
3	7.5	4	10.7	5	14.3	CHOCTAW MS	43	34	91	98	94
4	25.0	1	8.3	2	11.4	CLAIBORNE MS	-67	38	-54	54	83
12	22.2	8	14.0	8	13.2	CLARKE MS	-37	-6	-41	76	91
3	8.2	2	4.6	3	6.2	CLAY MS	-44	34	-24	87	35
11	22.2	7	11.8	3	4.7	COAHOMA MS	-47	-60	-79	10	20
7	11.5	7	9.9	5	7.5	COPIAH MS	-14	-24	-35	81	50
9	20.8	12*	30.3			COVINGTON MS	46				
5	12.7	2	3.7	5	4.5	DE SOTO MS	-71	24	-64	33	19
12	10.0	19	13.6	14	7.8	FORREST MS	36	-43	-22	88	54
7	23.3	6	18.0	3	8.8	FRANKLIN MS	-23	-51	-62	37	65
4	11.3	5	13.0	5	10.3	GEORGE MS	15	-21	-9	92	77
4	13.0	7	21.8	6	17.2	GREENE MS	68	-21	33	97	97
5	16.0	5	9.7	4	6.6	GRENADA MS	-39	-32	-59	46	40
9	18.4	6	8.6	2	2.2	HANCOCK MS	-53	-75	-88	02	06
33	12.1	38	11.9	40	9.4	HARRISON MS	-2	-21	-23	88	70
26	9.1	32	8.5	34	6.4	HINDS MS	-7	-25	-30	84	37
7	15.6	2	5.2	4	8.7	HOLMES MS	-67	68	-45	71	64
5	19.7	10*	40.7	5	20.7	HUMPHREYS MS	106	-49	5	94	98
1	22.3					ISSAQUENA MS					
15	22.1	4	5.5	7	7.3	ITAWAMBA MS	-75	33	-67	28	48
12	12.6	11	6.5	22	9.5	JACKSON MS	-48	46	-24	87	72
3	6.6	6	12.8	2	3.6	JASPER MS	95	-72	-45	70	12
2	9.7	1	7.8	1	6.2	JEFFERSON MS	-19	-21	-36	80	35
5	16.9	1	4.6	6	17.0	JEFFERSON DAVIS MS	-73	265	0	94	97
21	17.7	16	8.0	14	6.3	JONES MS	-55	-21	-64	33	36
5	18.6	3	9.7	3	8.3	KEMPER MS	-48	-14	-55	52	60
10	18.2	2	3.5			LAFAYETTE MS	-81				
9	23.6	6	11.7	1	1.9	LAMAR MS	-51	-84	-92	01	06
24	13.4	18	8.7	15	7.0	LAUDERDALE MS	-35	-20	-48	66	44
5	15.9	3	9.8	3	6.9	LAWRENCE MS	-38	-29	-56	50	44
8	12.7	11	16.4	6	8.8	LEAKE MS	30	-46	-30	84	65
20	15.8	12	8.5	15	8.8	LEE MS	-46	4	-44	71	65
9	13.9	14	22.3	4	4.9	LEFLORE MS	61	-78	-65	31	22
13	16.6	9	10.4	1	1.0	LINCOLN MS	-38	-90	-94	00	04
8	9.8	10	10.4	8	6.6	LOWNDES MS	7	-37	-33	83	39
10	19.8	3	6.2	3	5.0	MADISON MS	-69	-19	-75	16	23
9	16.1	9	13.5	13*	16.8	MARION MS	-16	24	4	94	96
5	15.2	3	7.2	5	9.2	MARSHALL MS	-53	27	-40	77	69
18	19.7	17	15.9	7	5.7	MONROE MS	-19	-64	-71	21	30
1	3.0	4	8.0	6	11.9	MONTGOMERY MS	171	49	303	99	86
12	15.6	4	5.4	5	5.5	NESHOBA MS	-66	3	-65	32	27
9	13.2	6	8.0	3	3.7	NEWTON MS	-40	-54	-72	19	13
2	11.2	3	15.4			NOXUBEE MS	37				
5	11.2	3	7.0	4	6.5	OKTIBBEHA MS	-38	-7	-42	74	39
11	17.3	6	8.9	5	6.6	PANOLA MS	-49	-26	-62	38	40
3	4.4	12	13.8	4	4.2	PEARL RIVER MS	211	-70	-6	93	16
2	7.8	2	4.9	1	2.6	PERRY MS	-37	-47	-67	28	07
10	11.6	7	7.5	11	9.6	PIKE MS	-36	29	-17	90	72
14	18.6	8	9.6	8	8.3	PONTOTOC MS	-48	-14	-55	52	60
8	11.7	5	5.8	5	5.0	PRENTISS MS	-50	-15	-58	48	22
6	18.5	3	9.8	1	2.7	QUITMAN MS	-47	-73	-86	04	08
9	13.7	4	4.1	12	7.7	RANKIN MS	-70	87	-43	72	53
7	13.5	2	3.4	2	2.7	SCOTT MS	-75	-22	-80	08	08
1	8.6			3	19.0	SHARKEY MS			120	98	98
10	16.1	3	4.1	6	7.0	SIMPSON MS	-75	71	-56	50	45
10	16.3	1	1.6	6	9.9	SMITH MS	-90	534	-40	77	74
1	5.2	2	5.5	1	3.2	STONE MS	6	-41	-38	79	10
7	12.1	7	9.7	3	4.0	SUNFLOWER MS	-20	-58	-67	28	15
10	23.7	8	16.6	6	13.3	TALLAHATCHIE MS	-30	-20	-44	72	91
4	12.1	3	7.1	5	9.3	TATE MS	-41	31	-23	87	70
9	14.4	8	11.1	6	6.8	TIPPAH MS	-23	-39	-53	57	42
7	12.1	7	8.7	9	10.0	TISHOMINGO MS	-28	15	-18	90	75
1	5.6			2	14.9	TUNICA MS			167	99	95
5	6.6	8	9.1	6	5.9	UNION MS	36	-34	-10	92	32
5	17.1	2	4.6	2	4.1	WALTHALL MS	-73	-11	-76	13	15
11	14.2	6	7.4	6	5.5	WARREN MS	-48	-25	-61	40	27
8	8.5	16	18.7	11	8.6	WASHINGTON MS	119	-54	1	94	63

* SIGNIFICANTLY MORE DEATHS THAN EXPECTED

WHITE MALE CANCER OF THE STOMACH ICD 151

1950-1959 DEATHS	1950-1959 RATE	1960-1969 DEATHS	1960-1969 RATE	1970-1979 DEATHS	1970-1979 RATE	COUNTY AND STATE	PERCENT CHANGE 60S/50S	PERCENT CHANGE 70S/60S	PERCENT CHANGE 70S/50S	PERCENTILE %CHANGE 70S/50S	PERCENTILE RATES 1970S
8	19.9	1	1.7	6	10.0	WAYNE MS	-91	478	-50	63	75
4	8.7	5	10.4	5	8.3	WEBSTER MS	19	-20	-5	93	60
5	22.5	3	17.1	4	18.4	WILKINSON MS	-24	8	-18	89	98
8	14.0	10	16.6	2	2.7	WINSTON MS	18	-84	-81	08	08
5	10.8	14*	27.3	2	3.4	YALOBUSHA MS	152	-88	-69	25	11
10	16.9	6	10.7	6	9.3	YAZOO MS	-37	-13	-45	70	70
3292	17.2	2150	10.4	1472	6.7	MISSOURI MO	-40	-36	-61		
15	13.3	14	12.3	3	2.7	ADAIR MO	-7	-78	-80	08	08
16	20.0	10	12.8	4	4.8	ANDREW MO	-36	-62	-76	14	22
6	10.1	7	11.0	5	8.3	ATCHISON MO	9	-25	-18	90	59
20	16.6	8	6.4	8	5.9	AUDRAIN MO	-61	-8	-65	32	32
26	21.6	22	16.2	6	3.8	BARRY MO	-25	-77	-82	06	14
16	17.0	10	11.4	6	5.9	BARTON MO	-33	-49	-65	30	31
21	17.1	10	7.6	7	5.4	BATES MO	-56	-29	-68	25	27
16	26.7	2	2.8	10	10.2	BENTON MO	-90	269	-62	38	77
6	9.7	6	8.1	1	1.4	BOLLINGER MO	-17	-83	-86	04	05
27	14.3	23	10.5	19	7.5	BOONE MO	-27	-28	-47	67	51
70	14.3	55	11.2	33	7.0	BUCHANAN MO	-22	-38	-51	60	44
27	15.7	20	10.8	13	6.5	BUTLER MO	-31	-40	-59	46	39
12	15.1	4	4.5	5	5.6	CALDWELL MO	-70	24	-63	36	28
14	9.2	17	12.3	9	6.2	CALLAWAY MO	34	-49	-32	83	36
8	14.6	11	14.6	9	6.6	CAMDEN MO	0	-55	-55	53	40
32	19.4	15	6.8	15	5.9	CAPE GIRARDEAU MO	-65	-14	-70	23	31
15	15.0	11	11.1	9	9.5	CARROLL MO	-26	-14	-36	80	72
9	37.2	4	13.2	1	2.4	CARTER MO	-65	-82	-94	00	07
21	15.8	25	16.8	7	3.9	CASS MO	7	-76	-75	15	15
13	17.8	4	4.9	16*	18.0	CEDAR MO	-73	270	1	94	97
16	15.4	9	11.1	5	8.3	CHARITON MO	-28	-25	-46	69	59
18	24.6	10	12.4	6	6.7	CHRISTIAN MO	-50	-46	-73	18	41
12	17.2	4	6.2	3	5.6	CLARK MO	-64	-10	-68	26	28
42	18.3	28	9.4	21	4.7	CLAY MO	-48	-50	-74	16	20
8	8.5	1	.9	2	2.3	CLINTON MO	-89	145	-73	18	07
21	13.7	14	7.9	10	4.7	COLE MO	-42	-40	-65	30	21
12	12.9	11	12.1	4	4.3	COOPER MO	-7	-65	-67	28	17
15	18.2	12	11.0	9	8.1	CRAWFORD MO	-40	-26	-56	51	57
9	12.8	6	8.3	6	7.6	DADE MO	-35	-8	-40	76	52
10	15.2	8	9.8	7	7.7	DALLAS MO	-36	-21	-50	63	53
17	18.6	9	11.1	10	14.0	DAVIESS MO	-40	26	-25	87	93
10	15.3	4	7.1	3	7.6	DE KALB MO	-53	6	-50	62	51
9	12.8	10	13.3	4	5.2	DENT MO	4	-61	-60	43	24
14	21.7	6	8.9	7	9.3	DOUGLAS MO	-59	5	-57	49	70
30	18.6	16	8.6	15	7.4	DUNKLIN MO	-54	-14	-60	42	49
52	23.7	36	14.9	25	8.4	FRANKLIN MO	-37	-44	-65	32	61
18	22.1	13	15.5	7	7.4	GASCONADE MO	-30	-52	-66	29	49
14	18.3	6	8.4	7	8.8	GENTRY MO	-54	5	-52	58	64
89	16.7	48	7.5	56	7.7	GREENE MO	-55	2	-54	55	52
18	17.7	17	16.9	10	10.7	GRUNDY MO	-5	-36	-39	77	79
23	22.7	11	10.5	4	6.2	HARRISON MO	-54	-41	-73	18	35
27	18.5	11	7.0	7	4.1	HENRY MO	-62	-41	-78	11	16
6	16.1	2	4.1	1	2.2	HICKORY MO	-74	-46	-86	03	06
16	22.6	8	10.9	2	3.8	HOLT MO	-52	-65	-83	06	14
9	12.5	3	3.8			HOWARD MO	-70				
20	15.3	15	9.9	12	8.0	HOWELL MO	-35	-20	-48	66	56
12	25.7	10	17.3	2	2.8	IRON MO	-33	-84	-89	02	09
356	15.9	225	9.3	166	6.9	JACKSON MO	-42	-26	-57	50	43
66	16.3	43	10.2	19	4.3	JASPER MO	-37	-57	-73	17	18
32	15.8	26	9.4	33	7.6	JEFFERSON MO	-41	-19	-52	58	51
16	11.5	13	9.3	7	4.8	JOHNSON MO	-19	-48	-58	47	21
9	17.7	3	6.0	3	5.8	KNOX MO	-66	-4	-67	27	30
23	21.2	15	12.6	8	5.8	LACLEDE MO	-41	-53	-72	19	31
23	15.5	20	12.4	12	7.0	LAFAYETTE MO	-20	-43	-55	53	45
32	20.8	24	14.7	12	6.2	LAWRENCE MO	-29	-58	-70	22	35
7	9.6	7	9.0	3	5.6	LEWIS MO	-6	-38	-42	75	28
12	12.7	5	5.0	4	3.4	LINCOLN MO	-61	-31	-73	18	11
14	10.4	22	16.9	8	6.1	LINN MO	62	-64	-42	74	34
13	12.5	12	11.4	4	3.8	LIVINGSTON MO	-9	-67	-70	23	13
12	15.6	4	4.8	2	2.2	MC DONALD MO	-69	-54	-86	04	06
25	18.6	12	9.3	6	4.9	MACON MO	-50	-47	-74	17	22
6	11.1	8	13.0	1	1.2	MADISON MO	17	-91	-89	02	04
6	15.3	4	8.7	1	1.7	MARIES MO	-43	-81	-89	02	05
20	12.1	11	6.6	10	5.8	MARION MO	-46	-12	-52	58	30
7	12.8	6	12.7	2	5.4	MERCER MO	-1	-58	-58	47	26
14	16.5	16	16.5	8	7.2	MILLER MO	0	-56	-57	50	47
12	17.5	9	12.1	5	6.4	MISSISSIPPI MO	-31	-47	-63	34	37
13	17.8	4	5.5	10	12.4	MONITEAU MO	-69	124	-30	84	88
9	11.3	2	2.7	4	5.6	MONROE MO	-76	103	-51	61	28
18	21.1	10	10.7	6	7.2	MONTGOMERY MO	-50	-32	-66	30	47
9	11.5	7	9.3	7	6.4	MORGAN MO	-19	-32	-44	71	37
22	22.0	10	9.7	7	6.8	NEW MADRID MO	-56	-30	-69	24	42
19	12.1	15	9.8	13	6.6	NEWTON MO	-20	-32	-45	70	40
28	19.2	17	12.2	11	9.5	NODAWAY MO	-36	-23	-51	61	72
20	29.7	5	6.5	2	2.4	OREGON MO	-78	-64	-92	01	07
14	21.6	4	5.9	1	1.2	OSAGE MO	-73	-80	-94	00	04
3	6.4	5	9.0	1	1.4	OZARK MO	40	-85	-79	10	05
23	21.7	12	9.6	6	5.4	PEMISCOT MO	-56	-44	-75	15	26
12	15.5	12	13.0	5	4.4	PERRY MO	-16	-66	-72	20	18
37	19.7	20	10.0	11	5.9	PETTIS MO	-49	-41	-70	22	32
25	21.6	13	9.7	10	6.6	PHELPS MO	-55	-32	-70	23	40
18	17.2	5	5.6	7	6.9	PIKE MO	-68	23	-60	42	43
15	17.8	8	8.0	10	7.6	PLATTE MO	-55	-4	-57	48	52
17	16.2	6	5.3	8	6.6	POLK MO	-67	24	-59	45	40
12	18.6	7	10.4	6	6.4	PULASKI MO	-44	-39	-66	30	37
7	10.7	11	17.4	5	11.2	PUTNAM MO	63	-36	5	94	82
6	10.5	6	9.4	5	10.1	RALLS MO	-10	7	-4	93	76
25	16.0	21	13.7	8	6.0	RANDOLPH MO	-15	-56	-63	36	32
22	19.4	10	9.3	2	2.0	RAY MO	-52	-79	-90	01	06
6	17.2	2	6.2	3	6.7	REYNOLDS MO	-64	8	-61	40	41
14	21.1	3	3.4	8	9.6	RIPLEY MO	-84	182	-54	54	73
22	14.0	28	14.5	21	7.2	ST CHARLES MO	4	-50	-49	65	47

* SIGNIFICANTLY MORE DEATHS THAN EXPECTED

WHITE MALE — CANCER OF THE STOMACH — ICD 151

1950-1959 DEATHS	RATE	1960-1969 DEATHS	RATE	1970-1979 DEATHS	RATE	COUNTY AND STATE	PERCENT CHANGE 60S/50S	70S/60S	70S/50S	PERCENTILE %CHANGE 70S/50S	RATES 1970S
7	8.4	6	9.9	4	5.0	ST CLAIR MO	18	-49	-40	76	23
17	33.2	7	11.4	6	7.8	STE GENEVIEVE MO	-66	-31	-76	13	54
38	20.1	31	14.0	15	6.5	ST FRANCOIS MO	-30	-53	-67	27	39
281	17.6	259	10.3	239	6.8	ST LOUIS COUNTY MO	-42	-34	-62	39	42
31	20.3	15	9.5	11	7.0	SALINE MO	-53	-26	-65	31	45
10	22.9	5	10.9	2	6.8	SCHUYLER MO	-53	-37	-70	22	42
4	7.4	6	12.5	2	4.7	SCOTLAND MO	69	-62	-36	80	21
17	14.6	14	9.8	13	8.4	SCOTT MO	-33	-13	-42	74	61
8	16.9	3	7.0	2	4.0	SHANNON MO	-59	-43	-77	13	15
12	15.7	5	6.3	1	.9	SHELBY MO	-60	-85	-94	00	04
24	17.7	18	12.0	11	6.3	STODDARD MO	-32	-48	-65	32	36
13	27.2	4	7.2	7	6.3	STONE MO	-74	-12	-77	13	37
9	12.0	11	14.0	10	14.4	SULLIVAN MO	16	3	20	96	94
7	12.1	9	9.2	8	5.9	TANEY MO	-24	-36	-51	60	32
14	12.1	16	13.4	11	8.8	TEXAS MO	11	-34	-27	86	65
23	14.4	18	11.7	10	6.5	VERNON MO	-19	-44	-55	53	39
11	18.3	4	7.8	2	2.7	WARREN MO	-58	-65	-85	04	08
11	14.2	12	16.4	8	9.2	WASHINGTON MO	16	-44	-35	81	69
13	21.9	9	11.9	7	8.5	WAYNE MO	-45	-29	-61	39	62
15	17.1	8	8.3	8	6.9	WEBSTER MO	-51	-17	-60	43	43
4	12.7	3	7.8			WORTH MO	-38				
19	20.0	13	12.9	8	7.6	WRIGHT MO	-36	-41	-62	37	51
646	20.6	340	12.4	165	7.9	ST LOUIS CITY MO	-40	-36	-62	38	54
610	20.4	419	13.1	283	8.3	MONTANA MT	-36	-36	-59		
12	28.5	11*	23.5	8*	17.5	BEAVERHEAD MT	-17	-26	-39	78	97
4	12.2	7	21.6	4	11.6	BIG HORN MT	77	-46	-5	93	85
5	12.4	1	2.9	3	11.3	BLAINE MT	-77	294	-9	92	83
2	8.8	2	11.9	4*	22.1	BROADWATER MT	35	86	150	99	98
9	15.4	8	13.4	5	8.1	CARBON MT	-13	-39	-47	67	57
		5*	37.3	3	22.8	CARTER MT		-39			99
52	20.5	43	14.9	30	9.4	CASCADE MT	-27	-37	-54	54	71
10	25.5	6	16.6	2	5.7	CHOUTEAU MT	-35	-66	-78	12	30
19	26.3	10	13.8	4	5.8	CUSTER MT	-48	-58	-78	11	30
5	58.5	2	7.3	2	11.8	DANIELS MT	-87	60	-80	09	85
11	25.1	4	8.3	9*	18.0	DAWSON MT	-67	117	-28	85	97
16	16.4	9	11.2	7	9.7	DEER LODGE MT	-32	-13	-41	76	73
4	23.3	2	11.5	1	4.8	FALLON MT	-51	-58	-79	09	22
15	16.5	11	12.7	7	7.9	FERGUS MT	-23	-38	-52	57	55
36	20.0	20	10.1	28	12.8	FLATHEAD MT	-50	27	-36	80	89
24	24.5	8	7.1	6	4.3	GALLATIN MT	-71	-38	-82	07	18
1	11.2	3	27.1	1	14.9	GARFIELD MT	141	-45	32	97	95
4	16.7	5	15.7	3	11.0	GLACIER MT	-6	-30	-34	82	81
2	16.7	2	35.6	1	20.6	GOLDEN VALLEY MT	114	-42	24	96	98
1	4.5	4	27.9	1	5.6	GRANITE MT	520	-80	24	96	28
15	22.5	14	18.7	3	3.7	HILL MT	-17	-80	-84	05	13
4	20.5			2	6.3	JEFFERSON MT			-69	24	36
1	6.2	5	30.2			JUDITH BASIN MT	384				
14	19.1	8	9.5	6	6.2	LAKE MT	-50	-35	-68	27	35
22	18.2	11	8.3	8	5.0	LEWIS AND CLARK MT	-55	-39	-72	19	23
		2	14.8			LIBERTY MT					
4	8.0	6	12.0	7	11.3	LINCOLN MT	50	-6	41	97	83
4	19.3	5	29.9			MC CONE MT	55				
7	18.2	7	19.1	3	8.4	MADISON MT	5	-56	-54	55	60
2	10.0	1	5.3			MEAGHER MT	-47				
3	22.4	1	6.9			MINERAL MT	-69				
32	19.0	27	13.5	17	7.0	MISSOULA MT	-29	-48	-63	35	44
16*	38.3	3	8.9	3	9.7	MUSSELSHELL MT	-77	8	-75	15	73
12	16.9	10	12.7	5	7.0	PARK#MT	-25	-45	-59	46	44
2	30.1	1	74.3	1	14.6	PETROLEUM MT	147	-80	-51	60	94
12	34.6	5	13.6	2	4.1	PHILLIPS MT	-61	-70	-88	02	16
3	9.1	4	13.0	3	9.0	PONDERA MT	44	-31	0	93	67
3	21.0	2	15.0			POWDER RIVER MT	-29				
12	36.3	6	19.6	2	6.3	POWELL MT	-46	-68	-83	06	36
5	39.0	1	8.6	1	11.2	PRAIRIE MT	-78	31	-71	20	83
13	17.4	8	8.7	7	6.4	RAVALLI MT	-50	-26	-63	35	38
17*	34.8	8	14.4	4	8.5	RICHLAND MT	-59	-41	-76	14	62
10	26.0	7	16.9	2	4.8	ROOSEVELT MT	-35	-72	-82	07	21
3	9.2	4	15.8	1	3.1	ROSEBUD MT	72	-80	-66	29	10
13	26.8	8	18.5	5	11.0	SANDERS MT	-31	-40	-59	45	82
5	15.1	4	8.5	3	6.7	SHERIDAN MT	-44	-22	-56	51	41
57	24.7	38	16.4	17	8.3	SILVER BOW MT	-34	-49	-66	29	60
4	13.0	1	2.6	4	10.7	STILLWATER MT	-80	307	-17	90	79
5	21.4	5	24.4	4	15.9	SWEET GRASS MT	14	-35	-26	86	96
8	20.8	4	10.0	2	3.8	TETON MT	-52	-62	-82	07	13
10	39.2	5	17.8	2	7.7	TOOLE MT	-55	-57	-80	08	52
1	11.8	1	15.1			TREASURE MT	28				
16	28.6	7	12.0	4	7.1	VALLEY MT	-58	-41	-75	15	45
6	29.7	3	16.7	3	17.6	WHEATLAND MT	-44	5	-41	75	97
1	11.1	1	8.6	1	7.5	WIBAUX MT	-22	-13	-32	83	50
36	14.8	33	10.6	32	8.7	YELLOWSTONE MT	-29	-18	-41	75	64
1496*	22.2	982	13.3	619	7.9	NEBRASKA NB	-40	-41	-64		
29	17.6	18	9.9	17	9.7	ADAMS NB	-44	-2	-45	70	73
18	28.3	10	14.7	3	5.5	ANTELOPE NB	-48	-63	-81	08	27
		1	47.2			ARTHUR NB					
1	18.1			1	12.2	BLAINE NB			-33	82	87
14	24.2	4	6.7	5	9.3	BOONE NB	-72	38	-62	39	70
13	26.5	7	11.7	3	4.9	BOX BUTTE NB	-56	-58	-81	07	22
7	26.4	7	24.5	4	11.9	BOYD NB	-7	-51	-55	53	86
12*	37.5	8*	26.1			BROWN NB	-30				
29	20.1	20	12.3	11	6.7	BUFFALO NB	-39	-45	-67	28	41
15	20.3	9	11.7	7	8.5	BURT NB	-42	-28	-58	46	61
20	30.0	11	15.1	10	12.1	BUTLER NB	-50	-20	-60	43	87
18	18.0	15	14.4	14	13.2	CASS NB	-20	-8	-26	86	91
13	18.9	12	16.0	8	10.7	CEDAR NB	-15	-33	-43	72	80
1	3.5	3	9.4			CHASE NB	169				
10	27.1	5	11.1	2	5.5	CHERRY NB	-59	-50	-80	09	27
14	26.5	9	16.5	9	14.6	CHEYENNE NB	-38	-11	-45	71	94

* SIGNIFICANTLY MORE DEATHS THAN EXPECTED

\# COUNTIES COMBINED; SEE TECHNICAL NOTES

WHITE MALE | CANCER OF THE STOMACH | ICD 151

1950-1959 DEATHS	1950-1959 RATE	1960-1969 DEATHS	1960-1969 RATE	1970-1979 DEATHS	1970-1979 RATE	COUNTY AND STATE		PERCENT CHANGE 60S/50S	PERCENT CHANGE 70S/60S	PERCENT CHANGE 70S/50S	PERCENTILE %CHANGE 70S/50S	PERCENTILE RATES 1970S
14	23.2	7	12.6	4	6.7	CLAY	NB	-46	-47	-71	21	41
22*	32.4	7	9.7	2	2.1	COLFAX	NB	-70	-78	-94	00	06
19	32.4	16*	21.8	7	8.5	CUMING	NB	-33	-61	-74	17	62
24	21.8	23*	21.1	10	9.6	CUSTER	NB	-3	-54	-56	51	73
16*	32.8	15*	26.5	4	6.3	DAKOTA	NB	-19	-76	-81	08	36
18*	37.5	12	21.2	3	5.2	DAWES	NB	-43	-75	-86	03	25
26	29.0	9	7.8	2	1.6	DAWSON	NB	-73	-80	-95	00	05
4	26.7	1	7.1	2	7.7	DEUEL	NB	-73	8	-71	20	53
17*	31.8	19*	32.4	11*	17.9	DIXON	NB	2	-45	-44	72	97
22	14.6	19	10.8	15	7.5	DODGE	NB	-26	-31	-49	65	50
289*	24.1	180	13.3	116	8.0	DOUGLAS	NB	-45	-40	-67	28	57
1	5.1	1	3.9	1	4.9	DUNDY	NB	-23	24	-5	93	22
10	13.9	5	8.0	10	15.4	FILLMORE	NB	-43	92	10	95	95
7	16.3	2	5.4	2	3.7	FRANKLIN	NB	-67	-31	-77	12	13
5	19.4	2	6.5			FRONTIER	NB	-66				
5	8.2	10	13.1	8	9.3	FURNAS	NB	61	-29	15	95	70
23	14.5	26	15.7	9	5.7	GAGE	NB	9	-64	-60	42	30
7	30.9	1	6.8	4*	34.0	GARDEN	NB	-78	403	10	95	99
5	28.5	2	10.9	1	4.4	GARFIELD	NB	-62	-60	-85	05	18
1	7.5	3	25.9			GOSPER	NB	245				
		3*	50.5			GRANT	NB					
4	12.9	6	19.3	2	7.2	GREELEY	NB	49	-63	-44	72	47
33	21.3	22	11.1	15	6.5	HALL	NB	-48	-41	-70	23	39
16	28.8	2	3.2	3	6.4	HAMILTON	NB	-89	101	-78	11	37
6	16.2	7	18.5	2	4.8	HARLAN	NB	14	-74	-70	22	21
1	14.9			2*	138.1	HAYES	NB			828	99	99
8	27.7	2	5.9	2	6.1	HITCHCOCK	NB	-79	5	-78	11	34
19	24.9	15	18.1	5	5.3	HOLT	NB	-27	-71	-79	10	25
1	15.0			1	13.5	HOOKER	NB			-10	92	92
12	26.6	7	15.5	4	8.4	HOWARD	NB	-42	-46	-68	25	61
10	13.6	10	13.1	2	3.6	JEFFERSON	NB	-4	-73	-74	17	12
9	20.8	6	12.3	4	9.2	JOHNSON	NB	-41	-25	-56	52	69
14	39.5	4	8.5			KEARNEY	NB	-79				
10	27.8	7	15.6	5	10.0	KEITH	NB	-44	-36	-64	34	76
3	31.7			1	18.9	KEYA PAHA	NB			-40	76	98
10*	46.9	1	4.3	2	7.8	KIMBALL	NB	-91	82	-83	06	54
14	15.8	8	8.8	2	2.0	KNOX	NB	-44	-78	-88	02	06
111	19.2	86	13.5	61	8.9	LANCASTER	NB	-30	-34	-54	55	66
20	16.4	15	10.5	15	9.1	LINCOLN	NB	-36	-13	-45	71	68
2	24.3	1	17.7	2*	96.7	LOGAN	NB	-27	447	299	99	99
3	41.3	2	42.3			LOUP	NB	3				
1	37.5					MC PHERSON	NB					
43	29.3	21	12.4	13	6.9	MADISON	NB	-58	-45	-76	13	43
5	9.6	10	18.0	2	3.7	MERRICK	NB	88	-80	-61	39	13
2	4.3	6	14.1	3	6.4	MORRILL	NB	232	-55	49	97	37
4	10.0	9	21.3	4	12.7	NANCE	NB	112	-40	27	96	89
10	16.9	9	15.3	5	9.6	NEMAHA	NB	-9	-37	-43	73	73
8	15.7	4	7.3	1	1.6	NUCKOLLS	NB	-54	-78	-90	01	05
22	21.8	13	12.1	10	8.6	OTOE	NB	-45	-29	-61	41	63
7	14.5	4	9.7	2	3.8	PAWNEE	NB	-34	-61	-74	17	14
3	14.8	2	9.1	3	10.7	PERKINS	NB	-38	17	-28	85	79
13	22.5	10	14.9	6	9.7	PHELPS	NB	-34	-35	-57	48	73
16*	33.3	7	14.5	3	5.6	PIERCE	NB	-56	-62	-83	06	28
27	26.1	13	11.3	10	7.5	PLATTE	NB	-57	-33	-71	21	51
15	28.8	4	7.7	6	13.3	POLK	NB	-73	72	-54	55	91
15	25.5	8	11.7	3	3.7	RED WILLOW	NB	-54	-68	-85	04	13
15	15.5	6	5.7	8	8.2	RICHARDSON	NB	-63	43	-47	68	58
5	31.1	5*	33.3	2	14.4	ROCK	NB	7	-57	-54	55	94
21	21.6	17	16.6	6	5.1	SALINE	NB	-23	-69	-76	13	24
11	17.7	9	11.2	12	8.7	SARPY	NB	-37	-22	-51	61	64
26	24.2	18	15.1	11	9.3	SAUNDERS	NB	-38	-39	-62	38	70
34	27.8	16	10.9	19	10.5	SCOTTS BLUFF	NB	-61	-3	-62	38	78
15	19.5	15	17.8	7	7.8	SEWARD	NB	-8	-56	-60	43	54
15	32.4	3	6.6	6	11.6	SHERIDAN	NB	-80	75	-64	33	85
8	24.3	5	15.6	6	16.1	SHERMAN	NB	-36	3	-34	82	96
1	9.9			1	9.2	SIOUX	NB			-7	92	69
7	22.1	9*	25.1	3	8.5	STANTON	NB	13	-66	-61	39	62
11	16.0	8	11.4	6	9.6	THAYER	NB	-29	-16	-40	76	72
2	30.9					THOMAS	NB					
14*	42.0	2	5.3	4	10.8	THURSTON	NB	-87	103	-74	16	80
9	20.6	6	11.1	8	18.7	VALLEY	NB	-46	69	-9	92	98
13	20.9	14	19.2	2	2.8	WASHINGTON	NB	-8	-85	-87	03	08
15	30.6	8	15.4			WAYNE	NB	-50				
13	26.6	5	8.7	4	7.3	WEBSTER	NB	-67	-16	-72	19	48
2	30.5	1	16.1	1	16.1	WHEELER	NB	-47	0	-47	67	96
13	15.2	12	14.0	7	7.4	YORK	NB	-7	-47	-51	61	50
179	21.3	153	12.5	181	8.1	NEVADA	NV	-41	-35	-62		
10	26.8	4	9.0	3	4.3	CHURCHILL	NV	-66	-52	-84	05	17
50	21.8	62	12.9	101	8.9	CLARK	NV	-41	-31	-59	44	65
4	26.4	1	6.4	1	3.0	DOUGLAS	NV	-76	-53	-89	02	09
9	17.5	4	6.4	8	16.4	ELKO	NV	-63	155	-6	93	96
1	18.6	1	16.4	2*	44.1	ESMERALDA	NV	-12	170	137	99	99
		2	40.4			EUREKA	NV					
8	27.5	6	22.3	1	4.2	HUMBOLDT	NV	-19	-81	-85	05	16
3	27.6					LANDER	NV					
6*	43.0			1	4.6	LINCOLN	NV			-89	02	20
5	19.4	4	13.5	5	9.3	LYON	NV	-30	-31	-52	58	70
4	15.6			2	9.8	MINERAL	NV			-37	79	74
5	20.3	5	22.7	4	10.9	NYE	NV	11	-52	-46	68	81
11*	72.3	7*	43.8	1	4.5	PERSHING	NV	-39	-90	-94	00	19
2	40.4					STOREY	NV					
43	15.4	45	11.4	49	8.0	WASHOE	NV	-26	-30	-48	66	56
14	31.8	11*	24.1	1	2.7	WHITE PINE	NV	-24	-89	-91	01	08
4	18.4	1	3.7	2	1.7	ORMSBY-CARSON CITY	NV	-80	-53	-91	01	05
604	21.8	448*	14.7	351	9.8	NEW HAMPSHIRE	NH	-33	-33	-55		
24	16.8	20	13.0	16	8.6	BELKNAP	NH	-23	-34	-49	65	63
15	14.6	11	10.5	12	8.6	CARROLL	NH	-28	-18	-41	75	63

* SIGNIFICANTLY MORE DEATHS THAN EXPECTED

WHITE MALE — CANCER OF THE STOMACH — ICD 151

1950-1959 DEATHS	1950-1959 RATE	1960-1969 DEATHS	1960-1969 RATE	1970-1979 DEATHS	1970-1979 RATE	COUNTY AND STATE	PERCENT CHANGE 60S/50S	PERCENT CHANGE 70S/60S	PERCENT CHANGE 70S/50S	PERCENTILE %CHANGE 70S/50S	PERCENTILE RATES 1970S
37	18.6	34	16.0	26	9.9	CHESHIPE NH	-14	-38	-47	68	75
44	23.8	32	16.8	21	11.0	COOS NH	-30	-35	-54	55	81
48	19.5	40	15.7	29	10.1	GRAFTON NH	-19	-36	-48	66	76
199*	25.5	132	15.5	121*	12.4	HILLSBOPOUGH NH	-39	-20	-51	60	88
64	17.7	49	12.8	32	7.3	MERRIMACK NH	-28	-43	-59	45	48
79	20.6	72	15.3	55	8.9	ROCKINGHAM NH	-26	-42	-57	49	66
68*	28.5	42	16.4	32	10.4	STPAFFORD NH	-42	-36	-63	35	78
26	20.0	16	10.2	7	4.0	SULLIVAN NH	-49	-61	-80	08	15
5360*	**25.0**	**4225***	**16.3**	**3510***	**11.8**	**NEW JERSEY NJ**	**-35**	**-27**	**-53**		
170*	25.2	125	14.0	117*	12.7	ATLANTIC NJ	-44	-9	-49	64	89
618*	25.1	535*	16.1	463*	11.9	BERGEN NJ	-36	-26	-53	57	86
120	22.0	100	14.1	109	10.8	BURLINGTON NJ	-36	-23	-51	61	80
287*	22.7	225	14.7	173	10.4	CAMDEN NJ	-35	-29	-54	54	78
67	25.3	52	14.8	41	7.5	CAPE MAY NJ	-42	-49	-70	22	50
81	20.4	78*	17.1	60	12.3	CUMBERLAND NJ	-16	-28	-40	77	88
880*	24.2	531*	14.5	369*	11.4	ESSEX NJ	-40	-22	-53	56	83
84	21.3	72	14.3	68	10.8	GLOUCESTER NJ	-33	-24	-49	64	80
818*	30.4	546*	19.5	353*	13.6	HUDSON NJ	-36	-30	-55	52	92
56	22.4	46	16.5	27	8.5	HUNTERDON NJ	-26	-48	-62	38	62
230*	24.2	205*	19.1	153*	12.7	MERCER NJ	-21	-34	-47	67	89
347*	30.3	287*	19.6	288*	14.6	MIDDLESEX NJ	-35	-25	-52	59	94
234	20.5	246*	16.0	224*	11.6	MONMOUTH NJ	-22	-28	-43	72	85
177	21.6	150	13.4	143	10.6	MORRIS NJ	-38	-21	-51	61	79
88	21.5	119	14.2	192*	9.7	OCEAN NJ	-34	-31	-55	53	74
465*	29.0	358*	18.9	285*	14.2	PASSAIC NJ	-35	-25	-51	61	93
35	17.9	29	13.4	18	7.0	SALEM NJ	-25	-48	-61	41	45
116*	25.5	88	15.2	70	10.0	SOMERSET NJ	-40	-34	-61	41	76
37	18.6	40	14.7	43	12.2	SUSSEX NJ	-21	-17	-35	81	87
402*	24.8	343*	17.0	289*	12.5	UNION NJ	-32	-26	-49	64	88
48	15.8	50	14.8	25	6.9	WARPEN NJ	-6	-53	-56	51	44
494*	**22.8**	**475***	**16.7**	**371***	**10.2**	**NEW MEXICO NM**	**-27**	**-39**	**-55**		
128*	26.6	96	12.3	108	10.1	BERNALILLO NM	-54	-18	-62	37	76
3	18.9	3	21.9	1	8.5	CATPON NM	16	-61	-55	53	62
13	11.0	18	11.5	18	8.7	CHAVES NM	4	-24	-21	89	64
18	31.7	12	15.7	8	12.1	COLFAX NM	-51	-23	-62	38	87
10	10.4	13	11.9	6	4.7	CURPY NM	14	-61	-55	52	20
2	10.5	1	5.9			DE BACA NM	-44				
30	24.8	25	16.3	27	11.9	DONA ANA NM	-34	-27	-52	58	86
15	14.0	27	18.6	16	8.5	EDDY NM	34	-54	-39	78	62
13	20.2	11	15.1	11	12.2	GPANT NM	-25	-19	-40	77	87
8	37.8	5	20.4	5*	19.7	GUADALUPE NM	-46	-3	-48	66	98
5	32.9	2	31.8			HAPDING NM	-3				
2	9.6	9*	43.2	1	4.3	HIDALGO NM	349	-90	-55	52	17
9	11.4	15	10.4	10	5.9	LEA NM	-8	-44	-48	66	32
11	40.4	7	16.4	6	12.2	LINCOLN NM	-59	-26	-70	23	87
1	3.1	2	3.8	1	1.3	LOS ALAMOS NM	23	-66	-58	47	05
7	21.4	8	18.2	1	3.2	LUNA NM	-15	-82	-85	04	10
8	21.1	9	20.3	7	15.5	MC KINLEY NM	-4	-23	-27	86	95
8	22.8	6	20.3	2	6.1	MORA NM	-11	-70	-73	18	34
11	26.3	12	17.1	7	4.7	OTERO NM	-35	-73	-82	07	20
12	22.6	16*	26.7	5	8.2	QUAY NM	18	-69	-64	34	58
17	23.1	26*	33.1	14*	23.6	RIO ARPIBA NM	43	-29	2	94	99
18	28.0	9	11.8	4	5.4	ROOSEVELT NM	-58	-55	-81	08	26
10	33.4	2	5.2	20*	29.7	SANDOVAL NM	-84	465	-11	91	99
10	20.7	7	7.9	5	4.4	SAN JUAN NM	-62	-44	-79	10	19
28	27.8	25*	23.9	15*	15.1	SAN MIGUEL NM	-14	-37	-46	70	95
24	19.7	41*	24.7	30*	14.7	SANTA FE NM	26	-41	-26	87	94
16	30.7	11	15.3	4	4.2	SIERRA NM	-50	-73	-86	03	16
4	10.4	7	19.1	3	6.3	SOCOPRO NM	83	-67	-39	77	36
10	18.8	15*	22.3	9	17.0	TAOS NM	19	-24	-9	92	97
6	24.5	9*	31.8	5	14.0	TORPANCE NM	30	-56	-43	73	93
8	23.0	4	10.6	3	8.5	UNION NM	-54	-19	-63	36	62
29*	46.1	22*	24.5	19*	13.5	VALENCIA NM	-47	-45	-71	21	92
16278*	**24.2**	**12159***	**16.3**	**8670***	**11.4**	**NEW YORK NY**	**-33**	**-30**	**-53**		
236	19.5	178	13.3	122	9.1	ALBANY NY	-31	-32	-53	56	68
40	18.2	26	11.6	12	5.3	ALLEGANY NY	-37	-54	-71	21	26
1857*	30.7	944*	14.9	697*	13.5	BRONX NY	-52	-9	-56	51	92
183	21.4	130	13.7	123*	12.0	BROOME NY	-36	-13	-44	72	86
87	22.3	57	14.3	33	8.1	CATTARAUGUS NY	-36	-43	-63	35	58
66	17.1	48	12.6	23	6.0	CAYUGA NY	-26	-53	-65	31	33
151	20.4	98	12.8	76	9.7	CHAUTAUQUA NY	-37	-24	-53	57	73
67	17.4	58	13.4	40	9.3	CHEMUNG NY	-23	-30	-46	69	70
41	19.6	20	9.1	19	8.2	CHENANGO NY	-53	-10	-58	47	58
60	25.3	31	13.2	35*	14.3	CLINTON NY	-48	8	-44	72	94
49	18.0	46	15.9	28	8.6	COLUMBIA NY	-12	-46	-52	57	62
35	19.1	18	9.1	15	7.4	CORTLAND NY	-52	-18	-61	40	49
37	15.7	16	6.7	13	5.0	DELAWARE NY	-57	-25	-68	25	23
124	16.5	107	12.5	93	9.7	DUTCHESS NY	-24	-22	-41	75	73
949*	23.8	749*	17.1	532*	11.6	ERIE NY	-28	-32	-51	60	85
45	26.0	22	12.3	12	6.3	ESSEX NY	-53	-49	-76	14	36
37	17.3	29	13.7	17	8.1	FRANKLIN NY	-21	-41	-53	56	57
53	18.9	31	10.6	32	10.8	FULTON NY	-44	2	-43	73	80
41	16.5	27	10.5	18	6.7	GENESEE NY	-37	-36	-59	44	41
44	25.0	27	13.2	27	12.8	GREENE NY	-47	-3	-49	65	89
7	28.3	6	19.5	1	2.8	HAMILTON NY	-31	-86	-90	01	08
88	24.6	46	12.6	31	8.2	HEPKIMER NY	-49	-35	-67	29	58
80	17.5	50	10.9	32	7.3	JEFFERSON NY	-38	-33	-58	47	48
3064*	28.3	1584*	14.5	1281*	13.2	KINGS(BROOKLYN) NY	-49	-9	-53	56	91
23	18.2	17	14.5	10	8.6	LEWIS NY	-21	-41	-53	57	63
26	12.6	25	11.7	19	8.1	LIVINGSTON NY	-7	-31	-36	80	57
51	20.8	31	12.5	22	8.4	MADISON NY	-40	-33	-59	44	61
469	19.0	380	14.0	267	9.4	MONROE NY	-26	-33	-51	61	71
91*	28.2	53	15.5	29	8.5	MONTGOMERY NY	-45	-45	-70	23	62
620	21.6	719*	16.1	607*	11.3	NASSAU NY	-26	-30	-48	66	83
2400*	30.0	2603*	34.7	762*	12.7	NEW YORK(MANHATTAN) NY	15	-63	-58	47	89

* SIGNIFICANTLY MORE DEATHS THAN EXPECTED

WHITE MALE — CANCER OF THE STOMACH — ICD 151

1950-1959 DEATHS	1950-1959 RATE	1960-1969 DEATHS	1960-1969 RATE	1970-1979 DEATHS	1970-1979 RATE	COUNTY AND STATE	PERCENT CHANGE 60S/50S	PERCENT CHANGE 70S/60S	PERCENT CHANGE 70S/50S	PERCENTILE %CHANGE 70S/50S	PERCENTILE RATES 1970S
181	21.4	143	14.8	102	9.9	NIAGARA NY	-31	-33	-54	55	74
227	19.1	157	12.3	132	10.3	ONEIDA NY	-35	-17	-46	69	77
321	19.4	235	13.5	204*	10.7	ONONDAGA NY	-31	-21	-45	70	79
63	18.7	35	9.8	31	8.0	ONTARIO NY	-48	-18	-57	49	57
147	17.8	158*	17.2	113*	11.0	ORANGE NY	-3	-36	-38	78	81
28	16.8	29	16.3	16	9.6	ORLEANS NY	-3	-41	-43	73	73
67	16.6	32	7.4	40	9.1	OSWEGO NY	-55	22	-45	70	68
64	21.5	32	10.6	24	7.5	OTSEGO NY	-51	-29	-65	31	51
36	28.7	19	10.9	29	11.5	PUTNAM NY	-62	5	-60	43	84
1738*	26.2	1247*	14.2	1179*	12.6	QUEENS NY	-46	-12	-52	58	88
139	21.7	87	12.4	75	10.6	RENSSELAER NY	-43	-14	-51	61	79
255*	28.6	138	13.1	176*	14.7	RICHMOND NY	-54	13	-48	65	94
103*	24.5	85	15.1	86	11.4	ROCKLAND NY	-38	-25	-54	56	83
74	15.5	56	11.7	50	10.4	ST LAWRENCE NY	-25	-11	-33	82	78
78	19.9	53	12.3	43	8.3	SARATOGA NY	-38	-32	-58	47	60
130	18.0	115	15.3	94*	11.9	SCHENECTADY NY	-15	-23	-34	81	86
28	20.9	15	11.1	12	8.0	SCHOHARIE NY	-47	-28	-62	38	56
9	11.2	8	10.1	4	4.5	SCHUYLER NY	-10	-55	-60	43	19
31	16.5	14	7.7	11	5.9	SENECA NY	-54	-23	-64	33	32
84	16.7	57	11.0	34	6.7	STEUBEN NY	-34	-39	-60	42	41
352	18.6	404	13.0	478*	11.7	SUFFOLK NY	-30	-11	-37	79	85
67	26.1	48	16.5	38	10.9	SULLIVAN NY	-37	-34	-58	47	81
16	9.6	30	18.4	20	10.9	TIOGA NY	92	-41	14	95	81
46	20.4	20	8.4	22	8.8	TOMPKINS NY	-59	4	-57	48	65
122	22.7	102	16.2	77	10.5	ULSTER NY	-28	-35	-54	55	78
40	18.6	24	10.8	23	9.4	WARREN NY	-42	-13	-50	63	71
49	19.8	25	10.0	25	10.0	WASHINGTON NY	-49	0	-49	64	75
73	22.8	43	12.7	34	9.5	WAYNE NY	-44	-25	-58	46	72
550	18.8	546*	15.7	451*	11.4	WESTCHESTER NY	-16	-27	-39	77	84
22	12.3	18	9.9	9	4.9	WYOMING NY	-19	-50	-60	43	22
17	15.4	8	7.3	10	9.1	YATES NY	-52	24	-41	75	68
1367	13.4	1120	8.7	1084	6.5	NORTH CAROLINA NC	-35	-26	-52		
24	13.4	21	9.0	15	4.6	ALAMANCE NC	-33	-48	-65	31	20
2	3.8	5	9.4	2	2.1	ALEXANDER NC	144	-78	-46	68	06
9	23.7	4	9.6	1	1.9	ALLEGHANY NC	-59	-80	-92	01	06
7	16.1	7	10.7	3	5.0	ANSON NC	-34	-53	-69	25	23
15	17.2	14	14.8	17*	15.0	ASHE NC	-14	1	-13	91	95
3	6.5	7	12.8	5	7.5	AVERY NC	98	-41	16	95	51
15	16.3	10	9.4	9	7.6	BEAUFORT NC	-42	-20	-54	56	51
9	20.5	11	23.4	8	17.3	BERTIE NC	14	-26	-15	90	97
6	10.1	2	2.9	2	2.7	BLADEN NC	-72	-5	-73	18	08
4	7.8	5	8.1	3	2.8	BRUNSWICK NC	4	-65	-64	34	08
60	12.8	61	10.5	50	7.1	BUNCOMBE NC	-18	-32	-44	72	46
17	11.6	12	5.7	15	6.4	BURKE NC	-51	12	-45	71	38
15	9.1	15	6.8	21	7.3	CABARRUS NC	-25	7	-19	89	48
14	11.4	13	7.6	16	7.4	CALDWELL NC	-34	-2	-35	80	49
1	6.0	1	6.2	3	17.3	CAMDEN NC	4	178	190	99	97
12	14.4	15	13.9	10	6.7	CARTERET NC	-4	-52	-54	55	41
2	5.9	4	10.2	4	8.1	CASWELL NC	72	-20	37	97	57
20	12.0	17	7.4	22	6.7	CATAWBA NC	-38	-10	-44	72	41
8	10.4	9	10.8	1	.9	CHATHAM NC	4	-92	-92	01	04
21	28.7	3	3.4	8	7.4	CHEROKEE NC	-88	119	-74	16	49
2	8.3	3	8.8	4	14.4	CHOWAN NC	6	64	74	98	94
2	7.2	3	9.4	3	5.8	CLAY NC	31	-38	-19	89	31
25	14.3	20	9.3	18	6.7	CLEVELAND NC	-35	-28	-53	56	41
9	9.6	6	4.9	17	11.8	COLUMBUS NC	-49	141	23	96	86
16	17.5	13	12.1	11	8.9	CRAVEN NC	-31	-27	-50	64	65
16	9.6	21	7.4	23	6.3	CUMBERLAND NC	-23	-15	-34	82	37
4	17.1	1	3.7	3	11.2	CURRITUCK NC	-78	201	-34	81	82
3	10.4	5	13.4	3	6.4	DARE NC	29	-52	-39	78	37
26	14.6	19	8.5	14	3.4	DAVIDSON NC	-42	-59	-76	13	11
6	10.8	5	7.4	5	6.3	DAVIE NC	-32	-15	-42	74	36
14	17.3	7	8.9	3	2.7	DUPLIN NC	-49	-70	-84	05	08
28	12.4	24	8.5	26	7.8	DURHAM NC	-32	-8	-37	79	54
8	10.1	10	9.8	14	11.4	EDGECOMBE NC	-3	17	13	95	84
47	13.1	46	10.1	33	4.8	FORSYTH NC	-23	-52	-63	35	21
14	19.8	7	10.5	3	3.5	FRANKLIN NC	-47	-67	-83	06	11
39	13.8	32	8.0	40	7.2	GASTON NC	-42	-10	-48	67	47
2	8.4			2	8.7	GATES NC			4	94	64
8	33.3	2	7.2	1	3.3	GRAHAM NC	-78	-54	-90	01	11
3	3.6	8	9.6	4	4.6	GRANVILLE NC	165	-52	27	96	19
5	17.0	4	15.2	3	6.9	GREENE NC	-11	-55	-60	43	43
53	10.7	48	7.1	54	6.2	GUILFORD NC	-34	-13	-42	73	35
9	9.0	14	12.7	8	6.0	HALIFAX NC	40	-53	-34	82	33
11	8.4	17	15.4	13	8.9	HARNETT NC	84	-42	6	95	66
24	17.8	18	9.7	21	9.4	HAYWOOD NC	-46	-2	-47	67	71
15	10.4	17	8.2	26	8.6	HENDERSON NC	-21	5	-17	90	63
4	11.5	6	16.3	2	3.4	HERTFORD NC	42	-79	-71	21	11
1	3.7	1	3.4	1	2.2	HOKE NC	-9	-34	-40	76	06
8	33.7	1	5.1			HYDE NC	-85				
21	12.8	17	8.1	22	8.2	IREDELL NC	-36	1	-36	80	58
16	22.0	10	12.3	7	7.2	JACKSON NC	-44	-41	-67	27	47
31	21.0	26	14.9	19	9.2	JOHNSTON NC	-29	-38	-56	50	69
7	32.8			1	2.9	JONES NC			-91	01	09
10	17.6	5	6.5	9	8.0	LEE NC	-63	25	-54	54	56
14	20.8	4	4.2	13	10.5	LENOIR NC	-80	148	-49	64	78
7	8.1	8	7.6	8	5.7	LINCOLN NC	-6	-25	-30	84	29
9	10.4	15	14.0	9	6.7	MC DOWELL NC	35	-52	-35	80	41
10	15.9	6	6.9	9	7.6	MACON NC	-57	10	-52	58	51
9	12.1	11	12.7	7	7.4	MADISON NC	5	-42	-39	78	49
4	8.2	5	7.8	5	7.5	MARTIN NC	-4	-5	-9	92	50
55	12.7	53	8.5	58	6.3	MECKLENBURG NC	-33	-27	-51	62	36
10	17.4	7	9.7	10	12.8	MITCHELL NC	-44	32	-26	86	90
10	19.9	8	11.8	2	2.6	MONTGOMERY NC	-41	-78	-87	03	07
7	7.0	7	5.4	5	2.3	MOORE NC	-24	-57	-67	28	07
19	16.1	16	11.8	12	7.0	NASH NC	-26	-41	-56	50	45
22	13.0	16	7.9	21	8.6	NEW HANOVER NC	-40	9	-34	81	62
5	9.9	9	19.7	4	7.1	NORTHAMPTON NC	100	-64	-28	85	46
9	12.0	10	8.2	8	4.6	ONSLOW NC	-32	-43	-62	38	20

* SIGNIFICANTLY MORE DEATHS THAN EXPECTED

WHITE MALE

CANCER OF THE STOMACH

ICD 151

1950-1959 DEATHS	1950-1959 RATE	1960-1969 DEATHS	1960-1969 RATE	1970-1979 DEATHS	1970-1979 RATE	COUNTY AND STATE	PERCENT CHANGE 60S/50S	PERCENT CHANGE 70S/60S	PERCENT CHANGE 70S/50S	PERCENTILE %CHANGE 70S/50S	PERCENTILE RATES 1970S
6	8.9	6	6.0	2	1.5	ORANGE NC	-33	-75	-84	05	05
3	9.2	7	22.9	5	12.1	PAMLICO NC	149	-47	32	97	87
4	8.4	6	9.1	6	7.4	PASQUOTANK NC	8	-19	-12	91	50
7	23.9	5	12.1	4	6.0	PENDER NC	-49	-51	-75	15	33
2	7.8	2	7.4	3	10.0	PERQUIMANS NC	-5	35	29	96	75
10	18.2	7	10.3	4	4.9	PERSON NC	-44	-52	-73	18	22
10	10.7	14	10.9	10	5.5	PITT NC	3	-49	-48	66	28
5	10.6	1	1.4	7	8.1	POLK NC	-87	500	-23	87	58
23	14.3	18	8.1	25	7.7	RANDOLPH NC	-44	-5	-46	69	53
16	16.4	6	6.3	8	7.4	RICHMOND NC	-62	19	-55	53	50
17	15.5	12	9.3	12	7.3	ROBESON NC	-40	-21	-53	57	48
17	10.4	18	8.8	24	8.1	ROCKINGHAM NC	-15	-8	-22	88	57
23	11.4	18	6.7	16	3.8	ROWAN NC	-41	-43	-67	29	13
24	17.3	19	9.8	9	4.2	RUTHERFORD NC	-43	-57	-76	14	17
18	18.8	7	5.4	12	8.1	SAMPSON NC	-71	50	-57	49	58
4	9.3	6	13.4	5	9.4	SCOTLAND NC	45	-30	2	94	71
12	10.4	5	3.3	8	4.4	STANLY NC	-69	34	-58	47	18
14	17.4	3	3.4	4	4.5	STOKES NC	-81	35	-74	16	19
22	16.8	17	8.7	10	4.3	SURRY NC	-48	-50	-74	16	17
6	18.0	2	5.3	2	4.7	SWAIN NC	-70	-12	-74	16	20
9	17.2	5	8.8	2	2.1	TRANSYLVANIA NC	-49	-77	-88	02	06
1	7.1	2	12.1	2	10.5	TYRRELL NC	71	-13	49	97	78
22	19.2	4	3.3	10	5.1	UNION NC	-83	54	-73	17	24
11	15.3	2	2.6	5	4.7	VANCE NC	-83	82	-70	23	20
41	12.4	25	5.9	29	5.3	WAKE NC	-52	-10	-57	48	25
12*	42.1	5	12.6	3	7.5	WARREN NC	-70	-40	-82	07	50
6	18.5	5	17.8	2	4.5	WASHINGTON NC	-3	-75	-76	14	19
8	13.1	6	7.3	8	7.9	WATAUGA NC	-45	9	-40	76	55
16	14.8	10	7.4	9	4.6	WAYNE NC	-50	-37	-69	24	20
23	15.2	17	9.7	10	5.3	WILKES NC	-36	-46	-65	31	25
12	11.4	15	11.3	11	6.6	WILSON NC	-1	-42	-42	74	40
12	14.6	7	7.6	6	4.7	YADKIN NC	-48	-38	-68	26	21
10	16.7	11	16.6	12*	15.5	YANCEY NC	-1	-7	-7	92	95
748*	26.9	556*	18.4	363*	11.2	NORTH DAKOTA ND	-31	-39	-58		
8	33.7	8*	34.9	3	14.1	ADAMS ND	4	-60	-58	47	93
30*	34.3	18	19.3	15*	15.6	BARNES ND	-44	-19	-55	53	95
7	14.4	7	14.7	7	14.2	BENSON ND	2	-3	-1	93	93
1	15.8			1	49.7	BILLINGS ND			215	99	99
12	18.6	11	16.1	5	7.0	BOTTINEAU ND	-13	-57	-62	37	44
4	18.8	4	22.1			BOWMAN ND	18				
14*	32.8	9*	26.9	7*	23.2	BURKE ND	-18	-14	-29	84	99
24	23.9	22	16.3	17	10.3	BURLEIGH ND	-32	-37	-57	49	77
63	24.3	39	13.1	32	9.9	CASS ND	-46	-24	-59	44	75
9	16.9	12	25.5	3	6.2	CAVALIER ND	51	-76	-63	35	35
15*	36.5	7	15.4			DICKEY ND	-58				
15*	44.0	6	15.7	5	15.3	DIVIDE ND	-64	-3	-65	31	95
3	11.5	5	20.5	5*	20.0	DUNN ND	79	-3	74	98	98
7	26.6	5	16.0	7*	21.7	EDDY ND	-40	35	-18	89	98
7	20.3	7	20.6	5	11.8	EMMONS ND	2	-43	-42	74	86
10*	40.6	4	13.4	1	3.0	FOSTER ND	-67	-78	-93	00	09
5	29.4	6*	39.9	1	5.7	GOLDEN VALLEY ND	36	-86	-81	08	30
35	20.3	36	18.7	24	11.5	GRAND FORKS ND	-8	-38	-43	73	84
4	17.0	5	17.6	5	22.6	GRANT ND	4	29	33	97	99
9	28.0	6	17.8	7*	19.5	GRIGGS ND	-36	9	-30	84	98
5	19.6	6	23.4	2	6.8	HETTINGER ND	19	-71	-65	31	42
7	26.5	4	15.9	2	8.8	KIDDER ND	-40	-45	-67	28	65
9	19.4	11	22.7	7	13.4	LA MOURE ND	17	-41	-31	83	91
9*	43.2	7*	30.2	1	6.3	LOGAN ND	-30	-79	-86	04	36
15	25.1	12	21.4	4	6.3	MC HENRY ND	-15	-70	-75	15	37
8	28.0	4	14.7	8*	20.9	MC INTOSH ND	-47	42	-25	87	98
7	19.2	13*	33.3	2	6.8	MC KENZIE ND	73	-79	-64	33	42
14	22.5	13	17.9	8	11.3	MC LEAN ND	-20	-37	-50	63	83
7	25.1	12*	40.5	6	16.7	MERCER ND	61	-59	-33	82	96
26*	31.8	15	16.5	13	13.1	MORTON ND	-48	-21	-59	45	90
18*	31.2	14*	24.0	5	10.2	MOUNTRAIL ND	-23	-57	-67	27	77
15	30.1	8	16.2	1	2.4	NELSON ND	-46	-85	-92	01	07
4	36.1	2	16.0	1	7.6	OLIVER ND	-56	-52	-79	10	52
21*	32.4	15	20.1	7	8.4	PEMBINA ND	-38	-58	-74	16	61
15*	48.6	4	10.4	7	21.5	PIERCE ND	-79	107	-56	51	98
19	26.6	13	16.9	12	14.0	RAMSEY ND	-36	-18	-47	67	93
14	30.3	8	13.4	7	11.1	RANSOM ND	-56	-17	-63	35	82
4	16.7	5	20.3			RENVILLE ND	22				
33*	33.4	15	13.0	10	9.5	RICHLAND ND	-61	-27	-72	20	72
10	36.0	8*	24.8	8*	28.8	ROLETTE ND	-31	16	-20	89	99
9	24.7	2	5.3	5	13.8	SARGENT ND	-79	160	-44	71	92
2	10.8	2	10.9			SHERIDAN ND	1				
2	35.7	4*	52.0			SIOUX ND	46				
3	36.3	1	17.9	1	141.3	SLOPE ND	-51	691	290	99	99
21*	36.6	19*	26.1	8	9.2	STARK ND	-29	-65	-75	15	69
9	34.2	9*	33.7	1	4.4	STEELE ND	-1	-87	-87	03	18
20	15.9	21	15.9	17	12.8	STUTSMAN ND	0	-20	-19	89	89
8	28.1	1	2.9	3	10.7	TOWNER ND	-90	273	-62	38	79
10	18.1	10	13.1	11	15.7	TRAILL ND	-28	20	-13	91	96
32*	36.6	15	13.9	6	5.3	WALSH ND	-62	-62	-85	04	26
47*	28.5	35*	19.6	31*	14.8	WARD ND	-31	-24	-48	66	94
17*	35.6	8	16.5	6	11.0	WELLS ND	-54	-34	-69	24	81
26	27.0	23*	22.0	13	13.1	WILLIAMS ND	-19	-40	-52	60	90
7061	19.9	5195	13.3	3636	8.7	OHIO OH	-33	-35	-56		
17	13.8	11	9.4	5	4.1	ADAMS OH	-32	-57	-71	21	15
69	17.2	48	10.9	38	8.3	ALLEN OH	-37	-24	-52	59	59
24	13.5	25	13.1	13	6.4	ASHLAND OH	-3	-51	-52	58	38
132*	32.9	67	15.5	43	9.3	ASHTABULA OH	-53	-40	-72	20	70
36	15.2	32	14.2	12	5.8	ATHENS OH	-7	-59	-62	37	30
27	15.9	18	10.2	15	8.0	AUGLAIZE OH	-36	-22	-50	63	56
97	21.0	63	13.6	37	8.3	BELMONT OH	-35	-39	-60	41	60
27	19.9	12	8.1	18	12.0	BROWN OH	-59	48	-40	77	86
70	12.0	62	9.2	43	4.9	BUTLER OH	-23	-47	-59	44	22

* SIGNIFICANTLY MORE DEATHS THAN EXPECTED

WHITE MALE

CANCER OF THE STOMACH

ICD 151

1950-1959 DEATHS	RATE	1960-1969 DEATHS	RATE	1970-1979 DEATHS	RATE	COUNTY AND STATE	PERCENT CHANGE 60S/50S	70S/60S	70S/50S	PERCENTILE %CHANGE 70S/50S	RATES 1970S
22	21.9	16	15.0	13	10.9	CARROLL OH	-32	-27	-50	62	81
16	12.4	14	9.8	9	6.0	CHAMPAIGN OH	-20	-39	-52	59	32
89	18.7	41	7.5	45	7.3	CLARK OH	-60	-3	-61	40	48
42	18.2	18	6.3	36	9.7	CLERMONT OH	-65	54	-47	68	73
19	13.4	15	10.3	15	10.4	CLINTON OH	-23	0	-23	88	78
107	21.7	66	13.6	43	8.2	COLUMBIANA OH	-37	-40	-62	37	58
25	15.2	15	8.8	10	5.7	COSHOCTON OH	-42	-35	-63	36	29
37	17.2	17	7.7	10	4.5	CRAWFORD OH	-55	-42	-74	17	19
1519*	26.7	1155*	18.5	808*	12.9	CUYAHOGA OH	-31	-30	-52	59	90
35	15.0	21	8.7	18	7.0	DARKE OH	-42	-20	-54	55	44
31	23.0	15	10.4	12	8.0	DEFIANCE OH	-55	-23	-65	31	57
14	8.8	15	9.5	15	8.1	DELAWARE OH	8	-15	-8	92	57
46	16.9	39	12.7	32	9.5	ERIE OH	-25	-25	-44	72	72
36	12.9	30	9.7	17	4.8	FAIRFIELD OH	-25	-51	-63	35	21
20	16.0	10	7.9	3	2.4	FAYETTE OH	-51	-70	-85	04	07
299	15.7	251	11.6	195	7.8	FRANKLIN OH	-26	-32	-50	62	54
20	14.3	18	12.6	9	5.9	FULTON OH	-12	-53	-59	45	31
17	12.9	9	7.1	17	13.3	GALLIA OH	-45	86	3	94	91
28	21.5	27	15.8	23	10.4	GEAUGA OH	-27	-34	-52	59	78
31	15.6	21	7.6	27	8.2	GREENE OH	-51	8	-47	67	59
44	17.7	26	9.2	14	6.3	GUERNSEY OH	-48	-31	-64	33	37
452	15.6	376	12.2	253	7.8	HAMILTON OH	-22	-36	-50	62	53
39	16.9	19	7.4	16	5.9	HANCOCK OH	-56	-20	-65	32	32
25	15.8	12	7.9	6	4.2	HARDIN OH	-50	-47	-73	17	16
17	17.5	11	11.2	8	8.3	HARRISON OH	-36	-26	-52	57	59
35	27.4	16	12.0	11	8.1	HENRY OH	-56	-32	-70	22	57
20	11.4	14	8.0	16	9.0	HIGHLAND OH	-30	12	-21	88	67
22	20.9	12	11.1	7	6.1	HOCKING OH	-47	-45	-71	21	34
13	14.1	16	15.2	3	2.5	HOLMES OH	8	-84	-83	06	07
33	15.4	28	13.3	16	7.5	HURON OH	-14	-43	-51	60	51
21	14.4	22	14.3	14	9.4	JACKSON OH	-1	-34	-34	81	71
102	23.2	81*	18.2	44	9.5	JEFFERSON OH	-22	-48	-59	44	72
23	11.6	25	12.6	8	3.8	KNOX OH	8	-70	-67	27	14
97*	29.3	64	13.6	67	10.7	LAKE OH	-54	-21	-63	35	79
36	16.6	37	15.8	19	7.2	LAWRENCE OH	-5	-55	-57	49	47
57	14.3	37	8.6	37	7.8	LICKING OH	-40	-9	-46	69	54
26	15.6	9	4.9	17	8.7	LOGAN OH	-68	77	-44	72	64
176*	26.9	132*	17.2	74	8.6	LORAIN OH	-36	-50	-68	26	63
367	21.2	256	13.7	160	8.3	LUCAS OH	-35	-39	-61	41	60
24	23.6	7	6.4	7	6.3	MADISON OH	-73	-1	-73	18	37
341*	29.8	251*	20.0	161*	12.3	MAHONING OH	-33	-38	-59	46	88
42	17.0	26	10.0	18	6.6	MARION OH	-41	-34	-61	39	39
41	19.1	23	8.4	26	7.9	MEDINA OH	-56	-6	-58	46	55
19	14.1	14	10.9	7	6.3	MEIGS OH	-23	-43	-56	51	36
35	23.3	12	8.2	12	7.5	MERCER OH	-65	-8	-68	27	51
47	16.8	28	9.5	14	3.8	MIAMI OH	-43	-60	-77	12	14
23	24.2	12	13.1	3	3.1	MONROE OH	-46	-77	-87	02	09
245	16.2	167	9.6	121	6.2	MONTGOMERY OH	-40	-36	-62	38	35
6	7.4	5	6.3	3	3.9	MORGAN OH	-14	-39	-47	67	14
12	12.2	6	6.2	2	1.8	MORROW OH	-49	-71	-85	04	06
57	15.2	36	9.9	22	6.0	MUSKINGUM OH	-35	-39	-61	40	32
16	20.8	6	8.8	6	9.1	NOBLE OH	-58	4	-56	50	68
37	24.3	24	14.4	11	6.0	OTTAWA OH	-41	-58	-75	15	33
13	16.3	5	6.3	6	7.0	PAULDING OH	-61	11	-57	49	45
29	17.7	22	14.3	11	7.5	PERRY OH	-19	-48	-58	48	50
24	17.0	11	7.3	6	3.5	PICKAWAY OH	-57	-53	-79	09	12
16	20.9	10	11.1	13	12.0	PIKE OH	-47	8	-43	73	86
67	23.0	34	10.5	29	6.9	PORTAGE OH	-54	-34	-70	23	44
16	11.0	11	7.2	12	7.2	PREBLE OH	-35	1	-35	81	47
12	8.3	12	9.1	6	4.3	PUTNAM OH	9	-52	-48	66	17
73	18.6	45	10.1	28	5.6	RICHLAND OH	-46	-44	-70	23	29
43	16.3	23	7.9	28	9.0	ROSS OH	-52	14	-45	70	67
31	13.9	27	10.7	27	10.2	SANDUSKY OH	-23	-5	-27	86	77
78	21.6	52	13.0	29	7.2	SCIOTO OH	-40	-45	-67	28	47
48	18.1	34	12.6	21	7.8	SENECA OH	-30	-38	-57	49	54
28	20.0	16	10.6	7	4.1	SHELBY OH	-47	-61	-79	09	16
261	20.1	214	14.4	159	10.0	STARK OH	-28	-31	-50	62	75
317	19.3	317*	16.2	202	9.2	SUMMIT OH	-16	-43	-52	58	69
158	22.5	155*	19.0	103	10.8	TRUMBULL OH	-15	-43	-52	59	80
61	16.1	51	12.9	40	9.8	TUSCARAWAS OH	-20	-24	-39	78	74
16	14.9	7	6.1	6	5.3	UNION OH	-59	-13	-65	32	25
20	12.9	20	13.4	13	8.4	VAN WERT OH	4	-37	-35	81	61
8	15.6	10	16.5	9	15.4	VINTON OH	6	-7	-1	93	95
33	18.7	23	10.6	21	8.3	WARREN OH	-43	-22	-56	52	60
38	15.1	37	14.1	20	7.2	WASHINGTON OH	-7	-49	-52	58	47
54	18.5	33	10.2	23	6.4	WAYNE OH	-45	-37	-65	31	38
30	20.8	22	14.0	13	7.7	WILLIAMS OH	-33	-45	-63	36	52
65	22.8	37	12.2	18	5.0	WOOD OH	-46	-59	-78	11	23
20	17.7	6	5.4	2	1.8	WYANDOT OH	-70	-67	-90	01	05
1666	17.5	1161	10.7	788	6.4	OKLAHOMA OK	-39	-40	-63		
8	11.9	8	11.5	7	10.1	ADAIR OK	-3	-12	-15	90	76
11	16.1	7	9.6	5	9.3	ALFALFA OK	-40	-4	-42	74	70
11	18.0	8	12.6	3	5.1	ATOKA OK	-30	-60	-72	20	24
4	13.0	6	16.8	2	6.6	BEAVER OK	29	-61	-49	64	40
28	26.2	16	13.8	10	7.5	BECKHAM OK	-47	-45	-71	21	51
15	21.4	15	19.4	6	8.0	BLAINE OK	-9	-59	-63	36	55
19	13.2	18	11.6	9	5.6	BRYAN OK	-12	-51	-57	48	29
27	17.3	20	12.8	6	3.7	CADDO OK	-26	-71	-78	10	13
22	18.9	11	8.0	8	4.6	CANADIAN OK	-57	-43	-76	14	20
25	15.3	25	12.7	10	4.6	CARTER OK	-17	-64	-70	22	20
13	18.5	13	15.4	8	7.5	CHEROKEE OK	-16	-51	-59	43	50
19	20.7	10	10.1	6	7.3	CHOCTAW OK	-51	-28	-65	31	48
3	14.6			1	4.8	CIMARRON OK			-67	28	21
17	9.7	10	5.0	13	4.3	CLEVELAND OK	-49	-14	-56	51	17
12	32.0	4	10.4	6	16.1	COAL OK	-68	55	-50	63	96
29	17.4	21	10.6	9	3.6	COMANCHE OK	-39	-66	-79	09	12
8	15.0	8	14.8	4	8.7	COTTON OK	-1	-41	-42	74	64
23	19.9	8	8.1	3	3.1	CRAIG OK	-59	-61	-84	05	10
36	18.3	19	9.0	21	9.0	CREEK OK	-51	-1	-51	60	66

* SIGNIFICANTLY MORE DEATHS THAN EXPECTED

WHITE MALE — CANCER OF THE STOMACH — ICD 151

1950-1959 DEATHS	RATE	1960-1969 DEATHS	RATE	1970-1979 DEATHS	RATE	COUNTY AND STATE	PERCENT CHANGE 60S/50S	70S/60S	70S/50S	PERCENTILE %CHANGE 70S/50S	RATES 1970S
19	19.4	13	12.5	3	2.8	CUSTER OK	-36	-78	-86	04	08
13	15.2	7	7.9	4	2.7	DELAWARE OK	-48	-65	-82	07	08
7	16.2	5	12.6	3	7.5	DEWEY OK	-22	-40	-54	56	51
4	10.5	4	10.6	1	2.4	ELLIS OK	2	-77	-77	12	07
27	12.0	27	10.3	14	4.9	GARFIELD OK	-14	-52	-59	45	22
23	18.1	16	10.6	17	9.4	GARVIN OK	-41	-12	-48	66	71
33	19.0	18	10.1	14	7.7	GRADY OK	-47	-24	-60	43	52
12	20.4	6	10.7	3	5.2	GRANT OK	-48	-52	-75	16	24
14	22.3	13	19.1	6	10.0	GREER OK	-14	-48	-55	52	75
5	15.4	5	12.9	1	2.4	HARMON OK	-16	-81	-84	05	07
2	5.5	6	18.8	3	8.6	HARPER OK	240	-54	56	98	63
16	28.2	8	12.7	3	4.8	HASKELL OK	-55	-62	-83	06	21
11	10.6	14	14.3	6	6.1	HUGHES OK	36	-57	-42	74	34
18	19.0	10	9.3	7	5.8	JACKSON OK	-51	-38	-69	23	31
10	16.5	3	5.1	3	5.0	JEFFERSON OK	-69	-2	-70	23	23
6	13.0	9	13.5	3	4.3	JOHNSTON OK	4	-68	-67	27	17
51	22.3	30	11.7	21	7.3	KAY OK	-48	-37	-67	27	48
8	12.7	9	13.0	4	5.5	KINGFISHER OK	3	-58	-56	50	28
21	22.7	13	13.4	10	11.8	KIOWA OK	-41	-12	-48	66	85
4	8.3	2	6.0	4	7.5	LATIMER OK	-28	24	-10	92	50
26	15.5	25	13.3	19	9.3	LE FLORE OK	-15	-30	-40	76	70
31	27.2	22	17.5	11	9.1	LINCOLN OK	-36	-48	-67	28	68
20	19.2	9	8.5	5	4.1	LOGAN OK	-56	-52	-79	10	15
8	20.3	6	15.3	3	6.4	LOVE OK	-25	-58	-69	25	37
13	19.0	5	7.7	12	13.4	MC CLAIN OK	-59	74	-29	84	92
20	16.4	12	9.2	7	4.6	MC CURTAIN OK	-44	-50	-72	19	20
10	15.3	8	13.6	7	7.7	MC INTOSH OK	-11	-44	-50	63	52
8	15.2	4	7.9	3	5.1	MAJOR OK	-48	-36	-67	28	23
3	6.2	5	9.8	7	9.9	MARSHALL OK	59	0	59	98	74
25	24.2	20	16.7	8	5.3	MAYES OK	-31	-68	-78	11	26
16	23.0	6	8.1	5	5.2	MURRAY OK	-65	-36	-77	12	25
43	17.4	32	11.9	20	7.2	MUSKOGEE OK	-32	-39	-58	46	47
13	21.4	4	5.5	5	7.6	NOBLE OK	-74	37	-64	32	51
11	16.7	4	6.0	4	6.5	NOWATA OK	-64	9	-61	41	39
7	11.6	3	5.0	3	4.8	OKFUSKEE OK	-57	-5	-59	45	21
177	15.2	152	9.8	136	7.0	OKLAHOMA OK	-35	-29	-54	55	44
39	21.4	17	8.8	8	4.3	OKMULGEE OK	-59	-51	-80	09	17
33	22.8	14	8.9	11	6.7	OSAGE OK	-61	-24	-70	22	42
28	20.0	21	14.4	16	10.1	OTTAWA OK	-28	-30	-49	64	76
20	23.9	9	10.1	1	1.6	PAWNEE OK	-58	-84	-93	00	05
32	18.9	16	8.6	13	6.0	PAYNE OK	-55	-30	-68	26	33
27	14.1	15	7.3	21	10.1	PITTSBURG OK	-48	38	-29	85	76
23	15.9	8	5.6	6	3.3	PONTOTOC OK	-65	-42	-79	09	10
45	20.9	29	12.2	16	6.1	POTTAWATOMIE OK	-42	-50	-71	21	34
7	11.5	8	13.4	2	3.2	PUSHMATAHA OK	16	-76	-72	19	10
4	12.4	5	16.2			ROGER MILLS OK	30				
23	21.9	14	12.1	6	4.1	ROGERS OK	-45	-66	-81	07	15
35	27.4	21	15.4	7	4.7	SEMINOLE OK	-44	-69	-83	06	21
17	20.6	10	9.9	8	6.8	SEQUOYAH OK	-52	-32	-67	27	42
23	14.7	24	13.1	17	7.8	STEPHENS OK	-10	-41	-47	68	54
6	9.3	5	7.8	8	11.0	TEXAS OK	-17	42	18	96	82
17	22.8	6	8.2	4	5.2	TILLMAN OK	-64	-37	-77	12	24
156	16.9	127	10.8	91	6.4	TULSA OK	-36	-41	-62	37	37
11	16.8	9	10.6	8	7.4	WAGONER OK	-37	-30	-56	51	49
21	15.6	17	10.7	7	3.8	WASHINGTON OK	-31	-65	-76	14	13
9	10.9	8	9.6	7	8.3	WASHITA OK	-12	-14	-24	87	59
6	8.2	9	12.1	5	6.1	WOODS OK	47	-50	-26	86	34
19	21.8	7	7.7	4	4.6	WOODWARD OK	-65	-41	-79	10	19
1441	19.1	1141	12.5	880	8.0	OREGON OR	-34	-36	-58		
17	17.7	8	8.1	4	3.8	BAKER OR	-54	-54	-79	10	13
13	11.9	17	11.7	10	4.9	BENTON OR	-2	-58	-59	46	22
78	17.6	83	14.0	77	9.7	CLACKAMAS OR	-21	-31	-45	70	73
37	25.6	30	16.4	15	8.5	CLATSOP OR	-36	-48	-67	28	62
31	26.3	15	10.6	7	4.4	COLUMBIA OR	-60	-59	-83	06	18
35	19.4	26	11.3	28	9.5	COOS OR	-42	-16	-51	60	72
9	29.8	5	13.3	5	7.9	CROOK OR	-55	-40	-73	17	55
4	10.0	6	9.0	11	10.8	CURRY OR	-11	20	8	95	80
20	18.6	11	7.7	18	8.4	DESCHUTES OR	-59	9	-55	53	60
39	17.1	34	11.2	34	8.8	DOUGLAS OR	-35	-21	-49	65	65
4	25.3	2	16.8	2	14.4	GILLIAM OR	-34	-14	-43	73	94
1	3.0	6	14.9	6	14.1	GRANT OR	396	-5	372	99	93
5	16.2	7	23.8	3	8.1	HARNEY OR	47	-66	-50	63	58
8	12.9	8	11.9	3	3.9	HOOD RIVER OR	-8	-67	-70	23	15
51	15.4	37	8.1	53	8.9	JACKSON OR	-47	10	-42	74	66
4	24.8	3	12.0	1	4.7	JEFFERSON OR	-52	-61	-81	08	20
43	27.2	19	9.1	27	9.0	JOSEPHINE OR	-66	-1	-67	28	67
40	24.1	31	17.8	17	8.0	KLAMATH OR	-26	-55	-67	28	56
5	19.9	5	16.2			LAKE OR	-18				
85	15.9	87	12.4	69	7.3	LANE OR	-22	-41	-54	54	47
18	12.5	17	11.1	18	9.2	LINCOLN OR	-11	-18	-26	86	69
43	19.6	36	12.0	24	6.4	LINN OR	-38	-47	-67	27	38
18	20.5	16	14.9	11	8.4	MALHEUR OR	-28	-43	-59	45	61
104	18.8	72	10.2	54	6.1	MARION OR	-46	-41	-68	26	34
4	16.1	5	21.7	3	10.7	MORROW OR	35	-51	-34	82	79
487	19.8	375	13.9	247	9.1	MULTNOMAH OR	-30	-35	-54	55	68
22	18.4	27	17.8	10	5.2	POLK OR	-3	-71	-72	20	25
2	13.5	1	8.6			SHERMAN OR	-36				
23	28.7	11	10.4	7	6.2	TILLAMOOK OR	-64	-41	-78	10	35
21	10.7	27	11.3	22	8.4	UMATILLA OR	6	-26	-22	88	60
22	22.6	16	14.9	2	1.7	UNION OR	-34	-88	-92	01	05
10	29.5	5	13.6	2	4.8	WALLOWA OR	-54	-65	-84	05	21
20	23.1	12	12.6	6	5.5	WASCO OR	-46	-56	-76	14	28
71	20.5	53	11.7	56	7.9	WASHINGTON OR	-43	-33	-62	39	55
3	34.8	2	31.3	1	12.2	WHEELER OR	-10	-61	-65	31	87
44	23.7	26	12.7	27	10.8	YAMHILL OR	-47	-14	-54	54	80
10298*	22.7	7302*	14.7	5191*	9.7	PENNSYLVANIA PA	-35	-34	-57		
25	11.9	17	7.3	20	6.5	ADAMS PA	-39	-11	-45	70	39

* SIGNIFICANTLY MORE DEATHS THAN EXPECTED

WHITE MALE — CANCER OF THE STOMACH — ICD 151

1950-1959 DEATHS	1950-1959 RATE	1960-1969 DEATHS	1960-1969 RATE	1970-1979 DEATHS	1970-1979 RATE	COUNTY AND STATE	PERCENT CHANGE 60S/50S	PERCENT CHANGE 70S/60S	PERCENT CHANGE 70S/50S	PERCENTILE %CHANGE 70S/50S	PERCENTILE RATES 1970S
1609*	25.4	1161*	17.1	796*	11.3	ALLEGHENY PA	-33	-34	-56	51	83
66	17.6	64	16.0	33	7.9	ARMSTRONG PA	-9	-50	-55	52	55
219*	28.4	141*	16.8	117*	12.4	BEAVER PA	-41	-26	-56	50	88
47	24.5	19	8.8	20	9.1	BEDFORD PA	-64	3	-63	36	68
205	16.1	156	10.9	148	9.6	BERKS PA	-33	-12	-40	76	72
132	19.8	100	14.2	66	9.1	BLAIR PA	-28	-36	-54	55	68
52	19.2	31	11.1	27	9.2	BRADFORD PA	-42	-18	-52	58	69
151	19.9	138	13.2	132	9.4	BUCKS PA	-33	-29	-53	57	71
69	18.1	64	12.1	38	6.1	BUTLER PA	-33	-49	-66	29	34
220*	24.4	180*	19.1	129*	13.5	CAMBRIA PA	-22	-29	-44	71	92
5	15.9	6*	22.7			CAMERON PA	43				
74*	27.2	40	14.6	29	9.4	CARBON PA	-46	-36	-65	30	71
52	22.2	37	13.8	23	7.4	CENTRE PA	-38	-46	-66	29	50
109	15.8	92	11.0	84	7.9	CHESTER PA	-31	-28	-50	63	55
27	14.3	26	13.7	19	9.4	CLARION PA	-4	-31	-34	82	71
80	19.7	60	14.8	39	9.5	CLEARFIELD PA	-25	-36	-52	58	72
30	20.1	19	10.9	7	3.7	CLINTON PA	-46	-66	-82	07	13
45	17.7	30	11.0	22	7.2	COLUMBIA PA	-38	-35	-59	43	46
76	20.9	56	14.0	43	10.2	CRAWFORD PA	-33	-27	-51	60	77
62	15.5	43	9.6	43	6.7	CUMBERLAND PA	-38	-30	-57	50	41
135	15.5	103	10.6	78	7.9	DAUPHIN PA	-32	-26	-49	64	55
332	21.0	254	12.1	246	10.1	DELAWARE PA	-42	-16	-52	59	76
43	26.7	34*	20.6	26*	14.2	ELK PA	-23	-31	-47	68	93
252*	26.3	166*	15.2	118	10.3	ERIE PA	-42	-32	-61	41	77
236*	28.6	146*	17.4	78	9.1	FAYETTE PA	-39	-48	-68	25	68
6	22.8	12*	38.0	2	6.8	FOREST PA	67	-82	-70	22	43
32	9.7	41	10.5	19	3.9	FRANKLIN PA	8	-63	-60	42	14
12	27.6	3	6.1	6	10.8	FULTON PA	-78	77	-61	40	80
49	23.7	34	14.6	19	8.8	GREENE PA	-38	-40	-63	36	65
27	15.9	28	15.5	13	7.1	HUNTINGDON PA	-2	-54	-55	52	46
98*	27.2	62	16.6	46	11.4	INDIANA PA	-39	-31	-58	47	84
53	21.0	42	15.8	31	11.3	JEFFERSON PA	-25	-29	-46	68	83
4	6.3	8	10.9	5	5.5	JUNIATA PA	74	-49	-12	91	27
361*	30.8	218*	17.9	163*	12.6	LACKAWANNA PA	-42	-29	-59	45	89
163	14.8	103	8.0	98	6.6	LANCASTER PA	-46	-18	-56	51	39
103	20.6	76	14.1	55	10.1	LAWRENCE PA	-31	-29	-51	60	76
60	1ø.0	51	11.8	29	5.8	LEBANON PA	-26	-51	-64	34	31
201	21.1	154	13.5	118	9.3	LEHIGH PA	-36	-31	-56	51	70
474*	29.1	318*	17.8	210*	11.1	LUZERNE PA	-39	-37	-62	38	82
76	15.2	44	8.5	45	8.0	LYCOMING PA	-44	-6	-47	67	56
50	19.2	47	17.3	30	10.6	MC KEAN PA	-10	-39	-45	71	79
129	23.7	85	14.7	53	8.5	MERCER PA	-38	-42	-64	33	62
23	12.6	20	10.6	7	3.3	MIFFLIN PA	-16	-69	-74	17	10
42	23.1	30	14.0	29	10.1	MONROE PA	-39	-28	-56	50	76
257	15.0	270	12.5	219	8.1	MONTGOMERY PA	-17	-35	-46	69	57
8	9.1	6	6.5	13	14.9	MONTOUR PA	-28	129	64	98	95
207*	23.9	157*	16.4	108	10.1	NORTHAMPTON PA	-32	-38	-58	48	76
103	20.7	54	10.2	45	8.0	NORTHUMBERLAND PA	-51	-22	-62	39	56
16	14.7	16	12.7	10	7.1	PERRY PA	-14	-44	-52	59	46
1930*	25.4	1225*	16.6	770*	11.6	PHILADELPHIA PA	-35	-30	-54	54	85
11	22.1	9	12.1	12	11.7	PIKE PA	-45	-3	-47	67	85
17	17.1	8	9.0	4	4.3	POTTER PA	-47	-53	-75	15	17
257*	29.3	146*	16.0	105*	11.6	SCHUYLKILL PA	-45	-28	-61	41	84
11	9.9	8	6.8	7	4.9	SNYDER PA	-32	-28	-51	61	22
122*	31.2	79*	18.8	37	8.8	SOMERSET PA	-40	-53	-72	20	65
5	13.9	7	20.1	2	5.1	SULLIVAN PA	45	-75	-63	35	24
49*	28.9	21	12.2	22	11.8	SUSQUEHANNA PA	-58	-3	-59	44	85
45	25.3	20	10.9	15	7.8	TIOGA PA	-57	-29	-69	24	53
14	13.3	7	6.6	7	5.5	UNION PA	-51	-16	-59	46	27
55	18.4	33	11.3	18	5.8	VENANGO PA	-39	-48	-68	25	31
35	14.6	24	9.7	20	8.1	WARREN PA	-33	-16	-44	71	58
253*	26.4	218*	21.0	115	10.8	WASHINGTON PA	-21	-48	-59	44	80
35	20.2	22	12.4	16	8.3	WAYNE PA	-38	-33	-59	45	59
367*	25.3	283*	17.1	171	9.3	WESTMORELAND PA	-32	-45	-63	36	70
18	19.6	15	16.8	16*	15.2	WYOMING PA	-15	-9	-23	88	95
147	16.6	108	9.7	100	7.4	YORK PA	-42	-23	-55	52	49
881*	24.9	673*	17.1	559*	12.7	RHODE ISLAND RI	-31	-26	-49		
38	27.4	34*	19.6	30*	13.6	BRISTOL RI	-28	-31	-50	62	92
93*	25.1	78	16.2	76*	12.5	KENT RI	-36	-22	-50	63	88
45	20.1	24	9.1	38	12.4	NEWPORT RI	-55	37	-38	78	88
664*	25.6	506*	18.2	379*	13.0	PROVIDENCE RI	-29	-29	-49	64	90
41	19.9	31	12.9	36	10.7	WASHINGTON RI	-35	-17	-47	68	79
542	13.0	464	8.5	464	6.4	SOUTH CAROLINA SC	-34	-25	-51		
9	15.3	6	9.2	2	2.9	ABBEVILLE SC	-40	-69	-81	07	09
13	9.9	14	7.4	27	10.4	AIKEN SC	-25	40	5	94	78
5	33.2	1	5.9	1	5.1	ALLENDALE SC	-82	-13	-85	05	24
39	17.5	31	9.8	27	6.4	ANDERSON SC	-44	-34	-63	35	38
6	17.0	4	10.7	1	2.6	BAMBERG SC	-37	-76	-85	04	07
4	13.0	4	10.4	3	6.3	BARNWELL SC	-20	-40	-52	59	36
3	7.1	4	6.1	7	7.0	BEAUFORT SC	-14	14	-1	93	45
7	15.8	5	12.0	7	5.8	BERKELEY SC	-24	-52	-64	34	30
3	15.2	1	3.9	3	11.1	CALHOUN SC	-75	187	-27	85	82
43	16.4	42	12.2	29	6.0	CHARLESTON SC	-25	-51	-63	35	33
15	17.3	12	11.0	8	5.8	CHEROKEE SC	-36	-47	-66	29	31
10	15.0	4	7.2	8	8.7	CHESTER SC	-52	21	-42	75	64
7	10.4	7	9.7	3	2.6	CHESTERFIELD SC	-7	-73	-75	16	08
1	3.2	3	6.7	3	7.0	CLARENDON SC	114	4	123	99	45
7	13.6	6	11.3	2	3.1	COLLETON SC	-17	-73	-77	12	09
10	12.5	13	12.5	7	5.1	DARLINGTON SC	0	-59	-60	43	23
7	14.4	5	8.1	6	10.2	DILLON SC	-43	26	-29	84	77
3	9.6	6	11.0	2	2.1	DORCHESTER SC	15	-81	-78	11	06
1	3.1	2	6.6	3	7.0	EDGEFIELD SC	115	6	129	99	45
10	32.9	1	2.9	2	4.6	FAIRFIELD SC	-91	61	-86	03	20
20	16.0	19	11.5	13	6.4	FLORENCE SC	-28	-44	-60	42	38
5	16.1	8	14.1	3	4.2	GEORGETOWN SC	-12	-71	-74	16	16
48	10.6	48	7.4	55	7.0	GREENVILLE SC	-30	-6	-34	81	44
11	10.4	19	15.0	7	4.0	GREENWOOD SC	44	-73	-62	38	15

* SIGNIFICANTLY MORE DEATHS THAN EXPECTED

WHITE MALE

CANCER OF THE STOMACH — COUNTY AND STATE — ICD 151

1950-1959 DEATHS	1950-1959 RATE	1960-1969 DEATHS	1960-1969 RATE	1970-1979 DEATHS	1970-1979 RATE	COUNTY AND STATE		PERCENT CHANGE 60S/50S	PERCENT CHANGE 70S/60S	PERCENT CHANGE 70S/50S	PERCENTILE %CHANGE 70S/50S	PERCENTILE RATES 1970S
4	13.1	3	15.5	4	9.8	HAMPTON	SC	19	-37	-25	87	74
13	17.5	12	6.5	18	6.3	HORRY	SC	-63	-4	-64	33	36
		2	9.4	2	8.5	JASPER	SC		-10			62
5	11.9			5	4.6	KERSHAW	SC			-61	40	20
6	7.1	5	6.1	6	4.4	LANCASTER	SC	-14	-28	-38	79	18
11	10.0	14	10.2	16	10.4	LAURENS	SC	2	2	4	94	78
4	12.2	4	12.5			LEE	SC	2				
13	8.7	13	6.8	17	5.2	LEXINGTON	SC	-22	-23	-40	77	25
		4	23.2	2	11.7	MC CORMICK	SC		-50			85
7	16.8	5	7.9	1	1.1	MARION	SC	-53	-86	-93	00	04
7	14.4	3	4.9	2	3.0	MARLBORO	SC	-66	-38	-79	10	09
12	14.9	6	6.2	7	6.6	NEWBERRY	SC	-58	6	-56	51	40
15	12.7	12	9.2	7	3.8	OCONEE	SC	-28	-59	-70	22	13
11	12.2	15	13.1	12	8.4	ORANGEBURG	SC	8	-36	-31	83	61
14	11.0	9	5.2	18	9.7	PICKENS	SC	-53	85	-12	91	73
36	11.1	25	6.9	42	7.9	RICHLAND	SC	-38	15	-29	84	55
1	2.7	4	8.9	3	4.7	SALUDA	SC	228	-47	73	98	20
55	15.5	26	6.4	38	6.4	SPARTANBURG	SC	-58	-1	-59	45	37
9	12.0	8	8.8	5	2.9	SUMTER	SC	-27	-67	-76	14	09
5	6.7	4	4.2	5	5.7	UNION	SC	-38	37	-15	90	30
8	24.4	8	16.4	10*	16.7	WILLIAMSBURG	SC	-33	2	-31	83	96
19	13.5	17	8.7	15	6.6	YORK	SC	-35	-25	-51	60	39
677	22.0	492	14.4	307	8.6	SOUTH DAKOTA	SD	-35	-41	-61		
1	3.7	6	22.6	1	4.0	AURORA	SD	511	-82	7	95	15
22	21.2	12	10.7	10	8.6	BEADLE	SD	-49	-20	-59	44	63
6*	62.8					BENNETT	SD					
10	18.8	7	11.5	3	4.7	BON HOMME	SD	-39	-60	-75	15	20
26*	30.1	19	20.0	7	7.0	BROOKINGS	SD	-34	-65	-77	12	44
24	18.1	17	10.9	18	10.7	BROWN	SD	-40	-2	-41	75	79
4	13.0	4	13.7	1	3.4	BRULE	SD	5	-75	-74	17	11
5	11.5	10	23.2	3	6.2	BUTTE	SD	101	-73	-47	68	34
5	39.1	4	23.9	2	11.5	CAMPBELL	SD	-39	-52	-70	22	84
8	14.6	7	12.5	6	10.0	CHARLES MIX	SD	-14	-20	-32	83	75
6	14.5	6	12.9	4	9.9	CLARK	SD	-11	-23	-32	83	75
15	27.7	11	21.0	2	4.4	CLAY	SD	-24	-79	-84	05	18
18	19.3	13	12.5	11	9.9	CODINGTON	SD	-36	-20	-49	65	75
3	36.3	7*	37.3	3	20.9	CORSON	SD	3	-44	-43	73	98
8	27.3	2	7.4	3	10.1	CUSTER	SD	-73	37	-63	36	76
13	16.2	14	15.4	8	7.8	DAVISON	SD	-5	-49	-52	59	54
17	27.5	13	20.9	5	6.5	DAY	SD	-24	-69	-76	13	38
9	22.3	15*	34.7	2	5.2	DEUEL	SD	55	-85	-77	13	25
3	16.8	2	13.2	5*	35.6	DEWEY	SD	-21	169	113	98	99
7	34.0	2	6.7	6*	21.3	DOUGLAS	SD	-80	217	-37	79	98
7	22.1	1	4.6	2	5.3	EDMUNDS	SD	-79	15	-76	14	26
12	13.8	4	3.8	6	8.6	FALL RIVER	SD	-72	125	-38	78	62
6	25.4	4	15.8	1	4.1	FAULK	SD	-38	-74	-84	05	16
10	16.2	11	18.8	9	14.3	GRANT	SD	3	-24	-22	88	94
18*	41.7	7	14.7	3	5.8	GREGORY	SD	-65	-60	-86	03	31
4	23.4	2	10.9	1	7.3	HAAKON	SD	-53	-34	-69	24	48
8	19.2	6	12.7	4	12.6	HAMLIN	SD	-34	-1	-34	82	89
6	21.5	4	11.9	1	2.9	HAND	SD	-44	-75	-86	03	09
4	22.8	2	10.7	1	3.6	HANSON	SD	-53	-67	-84	05	12
3	24.4	2	18.1			HARDING	SD	-26				
3	8.2	9	19.8	1	2.1	HUGHES	SD	140	-89	-75	15	06
8	13.0	5	7.1	1	.9	HUTCHINSON	SD	-46	-87	-93	00	04
2	13.3	1	4.8	1	5.8	HYDE	SD	-64	20	-56	50	31
		3	29.1	2	17.4	JACKSON	SD		-40			97
6	23.0	7	23.7	1	5.0	JERAULD	SD	3	-79	-78	11	23
3	27.9	2	24.4	2	17.3	JONES	SD	-13	-29	-38	78	97
15	26.6	7	11.3	6	8.7	KINGSBURY	SD	-58	-23	-67	27	64
15	25.0	7	10.9	11	16.3	LAKE	SD	-56	50	-35	81	96
18	23.4	11	13.7	6	6.6	LAWRENCE	SD	-42	-51	-72	20	40
23	30.7	16	19.0	6	6.7	LINCOLN	SD	-38	-65	-78	11	41
7	33.7	1	3.3	2	11.4	LYMAN	SD	-90	251	-66	29	84
8	18.0	2	3.4	8	16.6	MC COOK	SD	-81	383	-8	92	96
9	35.1	6	21.2	2	5.6	MC PHERSON	SD	-40	-74	-84	05	28
13	31.9	10	20.9	6	13.8	MARSHALL	SD	-34	-34	-57	49	93
10	14.6	3	4.1	5	6.7	MEADE	SD	-72	64	-54	55	41
5*	48.4	2	18.4			MELLETTE	SD	-62				
7	25.5	5	13.2	2	14.4	MINER	SD	-48	9	-44	72	94
74	23.5	60	16.0	49	11.4	MINNEHAHA	SD	-32	-29	-52	59	83
11	23.8	8	15.7	4	8.3	MOODY	SD	-34	-47	-65	31	60
25	19.4	20	10.9	26	11.7	PENNINGTON	SD	-44	7	-40	77	85
8	29.2	6	19.7	1	2.8	PERKINS	SD	-33	-86	-90	01	09
10*	46.9	2	7.1			POTTER	SD	-85				
21	30.1	17*	23.3	3	3.9	ROBERTS	SD	-23	-83	-87	03	14
3	10.6	1	3.2	3	11.8	SANBORN	SD	-70	273	11	95	85
2	98.2			1	22.3	SHANNON	SD			-77	12	99
14	23.8	10	14.9	5	7.6	SPINK	SD	-37	-49	-68	26	52
2	16.4	3	25.3			STANLEY	SD	54				
2	30.8	1	7.9	1	6.3	SULLY	SD	-74	-20	-80	09	36
		2	45.2	1	8.2	TODD	SD		-82			58
10	24.8	13*	26.3	1	1.5	TRIPP	SD	6	-94	-94	00	05
14	20.2	12	16.7	4	4.1	TURNER	SD	-17	-75	-79	09	16
12	19.3	15*	24.5	3	5.3	UNION	SD	27	-78	-72	19	26
14*	41.3	2	5.5	6	13.0	WALWORTH	SD	-87	135	-68	25	90
13	12.8	7	6.7	8	8.4	YANKTON	SD	-48	26	-34	81	61
2	30.4	2	31.9	1	22.3	ZIEBACH	SD	5	-30	-27	86	99
1594	15.1	1213	9.4	1068	6.8	TENNESSEE	TN	-38	-27	-55		
16	12.8	8	4.7	10	3.4	ANDERSON	TN	-63	-29	-74	17	11
15	15.4	8	7.9	8	6.1	BEDFORD	TN	-49	-23	-61	41	34
15	28.1	7	10.5	3	3.4	BENTON	TN	-63	-68	-88	02	11
4	11.3	5	14.0			BLEDSOE	TN	23				
22	14.1	13	5.4	15	5.4	BLOUNT	TN	-62	0	-62	38	26
18	16.9	12	8.0	17	8.4	BRADLEY	TN	-52	4	-50	62	60
19	17.0	14	10.8	11	6.9	CAMPBELL	TN	-36	-36	-59	44	43
14*	37.8	3	6.6	2	4.3	CANNON	TN	-82	-35	-89	02	18

* SIGNIFICANTLY MORE DEATHS THAN EXPECTED

WHITE MALE — CANCER OF THE STOMACH — ICD 151

1950-1959 DEATHS	1950-1959 RATE	1960-1969 DEATHS	1960-1969 RATE	1970-1979 DEATHS	1970-1979 RATE	COUNTY AND STATE		PERCENT CHANGE 60S/50S	PERCENT CHANGE 70S/60S	PERCENT CHANGE 70S/50S	PERCENTILE %CHANGE 70S/50S	PERCENTILE RATES 1970S
16	14.1	11	9.1	8	5.3	CARROLL	TN	-36	-41	-62	37	26
13	10.8	13	7.8	20	8.7	CARTER	TN	-28	10	-20	89	64
10	25.5	6	13.0	6	8.2	CHEATHAM	TN	-49	-37	-68	26	59
6	16.3	2	4.7	3	6.5	CHESTER	TN	-71	37	-60	42	38
19	22.0	11	12.2	10	9.6	CLAIBORNE	TN	-44	-21	-56	50	73
8	28.0	5	14.2	3	7.2	CLAY	TN	-49	-50	-74	16	47
15	17.0	13	14.3	12	9.6	COCKE	TN	-16	-33	-44	72	72
15	15.3	11	10.7	11	7.3	COFFEE	TN	-30	-32	-52	58	48
10	16.4	11	18.4	6	7.9	CROCKETT	TN	12	-57	-52	59	55
12	15.3	12	13.6	8	6.2	CUMBERLAND	TN	-11	-55	-60	42	34
121	12.4	114	9.3	85	5.7	DAVIDSON	TN	-25	-39	-55	54	29
8	19.7	9	16.9	4	6.3	DECATUR	TN	-14	-62	-68	26	37
8	15.9	6	9.8	4	5.5	DE KALB	TN	-38	-44	-65	31	27
15	16.9	9	8.8	8	6.4	DICKSON	TN	-45	-28	-62	37	37
28	22.8	14	10.5	12	7.7	DYER	TN	-54	-26	-66	29	53
6	16.7	5	11.4	3	5.2	FAYETTE	TN	-32	-55	-69	24	24
11	24.4	7	13.1	3	4.6	FENTRESS	TN	-46	-65	-81	08	20
14	16.5	10	9.6	4	2.7	FRANKLIN	TN	-42	-72	-83	06	08
20	10.9	13	6.1	13	5.5	GIBSON	TN	-44	-10	-50	63	27
16	17.1	16	15.3	4	3.5	GILES	TN	-11	-77	-80	09	11
3	6.1	2	3.0	4	5.4	GRAINGER	TN	-50	77	-12	91	26
23	16.0	19	11.2	15	8.0	GREENE	TN	-30	-28	-50	63	57
4	8.5	6	12.6	7	11.8	GRUNDY	TN	49	-7	39	97	85
7	9.1	9	8.0	11	7.4	HAMBLEN	TN	-12	-7	-18	89	49
76	12.6	52	6.6	67	7.3	HAMILTON	TN	-47	9	-42	74	48
3	9.0	5	14.5	1	2.8	HANCOCK	TN	62	-80	-68	25	09
10	13.3	12	14.7	7	7.6	HARDEMAN	TN	11	-49	-43	73	51
2	2.6	5	6.0	4	4.1	HARDIN	TN	127	-31	56	98	16
16	14.6	12	9.4	9	6.0	HAWKINS	TN	-36	-36	-59	46	33
8	20.8	8	17.3	3	5.4	HAYWOOD	TN	-17	-69	-74	17	27
12	16.5	4	5.0	5	5.3	HENDERSON	TN	-69	5	-68	26	25
19	16.5	11	9.6	6	3.9	HENRY	TN	-42	-60	-77	13	14
9	17.3	6	10.5	5	6.5	HICKMAN	TN	-40	-38	-62	37	39
7	24.7	2	6.7	1	2.5	HOUSTON	TN	-73	-63	-90	01	07
10	19.4	3	5.3	7	9.7	HUMPHREYS	TN	-73	82	-50	62	73
6	12.6	7	13.7	2	3.3	JACKSON	TN	8	-76	-74	16	10
14	18.6	10	10.7	12	10.0	JEFFERSON	TN	-42	-7	-46	69	75
7	12.9	4	6.5	4	5.9	JOHNSON	TN	-50	-9	-54	54	32
84	11.4	96	10.3	75	6.8	KNOX	TN	-10	-35	-41	75	42
3	11.2	3	11.0	4	14.4	LAKE	TN	-2	31	28	96	94
14	20.5	11	15.9	6	8.4	LAUDERDALE	TN	-22	-47	-59	45	61
13	10.9	15	11.5	13	8.4	LAWRENCE	TN	5	-27	-23	88	60
10*	37.7	2	6.2	5	13.3	LEWIS	TN	-84	116	-65	32	91
10	10.6	17	17.5	8	7.6	LINCOLN	TN	65	-56	-28	85	52
5	5.9	10	10.8	10	8.4	LOUDON	TN	81	-22	42	97	61
22	19.2	13	9.1	14	8.1	MC MINN	TN	-53	-11	-55	47	57
20	25.8	9	9.1	10	9.5	MC NAIRY	TN	-65	5	-63	36	72
14	22.9	10	14.9	7	8.9	MACON	TN	-35	-40	-61	40	66
15	8.3	21	9.8	16	6.6	MADISON	TN	17	-33	-21	88	39
11	24.3	5	6.4	5	6.3	MARION	TN	-74	-1	-74	16	36
14	21.9	9	12.4	7	8.1	MARSHALL	TN	-43	-34	-63	36	58
12	8.7	14	9.5	12	6.4	MAURY	TN	9	-33	-27	86	37
4	17.9	1	5.0	1	3.2	MEIGS	TN	-72	-35	-82	07	10
12	13.1	9	8.6	6	5.1	MONROE	TN	-34	-41	-61	40	24
14	12.6	13	9.7	15	8.1	MONTGOMERY	TN	-23	-17	-36	80	57
2	11.6					MOORE	TN					
9	15.8	5	8.7	1	1.4	MORGAN	TN	-45	-84	-91	01	05
21	15.7	17	11.2	18	11.2	OBION	TN	-29	0	-29	65	82
16	22.9	7	8.9	3	3.6	OVERTON	TN	-61	-59	-84	05	12
4	12.4	2	8.4	1	2.2	PERRY	TN	-32	-74	-82	07	06
3	14.7	1	9.4	1	4.2	PICKETT	TN	-36	-55	-71	21	17
11	23.8	8	15.7	2	3.9	POLK	TN	-34	-75	-84	05	14
11	8.3	11	7.6	13	7.1	PUTNAM	TN	-8	-7	-15	90	46
9	15.4	7	9.6	6	7.1	RHEA	TN	-38	-26	-54	54	45
21	19.2	15	11.6	11	5.8	ROANE	TN	-40	-50	-70	23	30
20	18.9	6	4.7	5	3.5	ROBERTSON	TN	-75	-26	-82	07	12
11	8.9	10	5.9	17	6.8	RUTHERFORD	TN	-33	15	-23	88	43
11	18.6	3	4.2	2	2.7	SCOTT	TN	-77	-35	-85	04	08
13*	61.6	1	3.7	2	5.9	SEQUATCHIE	TN	-94	61	-90	01	32
12	13.2	15	13.4	15	9.3	SEVIER	TN	2	-31	-30	84	69
149	14.0	137	9.9	118	7.5	SHELBY	TN	-30	-24	-46	68	51
18	25.8	4	5.5	1	1.5	SMITH	TN	-79	-72	-94	00	05
3	6.1	6	12.3	7	13.5	STEWART	TN	103	10	123	99	92
46	16.8	34	8.7	45	9.3	SULLIVAN	TN	-48	6	-45	70	69
20	14.6	14	7.3	20	7.8	SUMNER	TN	-50	6	-47	68	53
12	18.1	10	13.2	12	13.1	TIPTON	TN	-27	-1	-28	85	90
5	23.1			1	3.8	TROUSDALE	TN			-83	05	14
11	18.1	4	5.5	8	9.7	UNICOI	TN	-69	76	-46	69	74
5	12.1	2	5.1	4	8.0	UNION	TN	-58	56	-34	82	56
4	26.7	1	5.3			VAN BUREN	TN	-80				
15	15.6	14	12.9	5	3.4	WARREN	TN	-17	-73	-78	11	11
34	13.3	22	6.5	30	8.3	WASHINGTON	TN	-51	27	-38	79	59
2	3.8	5	8.4	7	10.6	WAYNE	TN	118	27	177	99	79
31	21.0	20	12.7	10	6.1	WEAKLEY	TN	-39	-52	-71	21	34
10	14.2	6	7.8	6	7.2	WHITE	TN	-45	-8	-50	63	47
32*	39.7	15	14.3	11	6.9	WILLIAMSON	TN	-64	-52	-83	06	43
16	13.7	8	5.7	9	5.1	WILSON	TN	-58	-10	-62	37	24
4195	15.7	3759	11.0	3271	7.7	TEXAS	TX	-30	-30	-51		
18	17.2	11	8.4	4	3.3	ANDERSON	TX	-51	-61	-81	08	10
		2	6.1	4	11.4	ANDREWS	TX		86			83
15	12.3	23	13.3	16	7.1	ANGELINA	TX	7	-46	-42	74	46
3	10.2	3	6.9	4	6.6	ARANSAS	TX	-32	-4	-35	81	40
4	13.0	3	8.1	1	2.5	ARCHER	TX	-38	-69	-81	08	07
3	25.1					ARMSTRONG	TX					
16	23.3	13	14.3	12	11.2	ATASCOSA	TX	-39	-22	-52	59	82
16	26.0	12	13.4	9	8.6	AUSTIN	TX	-49	-36	-67	28	63
5	16.4	7	22.3	2	5.3	BAILEY	TX	36	-76	-68	26	26
2	7.6	1	2.6	1	2.2	BANDERA	TX	-66	-14	-71	20	06
8	10.5	9	10.7	5	4.3	BASTROP	TX	2	-60	-59	45	17

* SIGNIFICANTLY MORE DEATHS THAN EXPECTED

WHITE MALE

CANCER OF THE STOMACH

ICD 151

1950-1959 DEATHS	1950-1959 RATE	1960-1969 DEATHS	1960-1969 RATE	1970-1979 DEATHS	1970-1979 RATE	COUNTY AND STATE	PERCENT CHANGE 60S/50S	PERCENT CHANGE 70S/60S	PERCENT CHANGE 70S/50S	PERCENTILE %CHANGE 70S/50S	PERCENTILE RATES 1970S
5	14.6	1	2.2	3	8.2	BAYLOR TX	-85	266	-44	71	58
12	22.4	8	10.0	7	8.3	BEE TX	-55	-17	-63	36	60
38	13.7	27	8.0	22	5.3	BELL TX	-41	-34	-61	40	26
289	16.6	307	13.8	337*	12.2	BEXAR TX	-17	-12	-27	86	87
7	30.0	3	12.1	4	12.2	BLANCO TX	-59	0	-59	44	87
13	15.6	6	5.5	6	5.3	BOSQUE TX	-65	-4	-66	29	25
34	18.8	22	9.1	23	8.2	BOWIE TX	-51	-10	-56	50	58
29	22.3	29	12.3	26	8.1	BRAZORIA TX	-45	-34	-64	34	57
15	15.6	11	9.4	12	7.4	BRAZOS TX	-40	-20	-52	58	50
8	31.1	5	19.5	3	9.7	BREWSTER TX	-38	-50	-69	24	73
		3	27.9			BRISCOE TX					
6	23.1	2	9.6	7*	22.0	BROOKS TX	-58	129	-4	93	98
24	15.4	13	7.7	9	5.1	BROWN TX	-50	-33	-67	29	24
6	11.3	5	10.0	4	6.7	BURLESON TX	-12	-33	-41	75	41
9	18.0	8	11.3	7	5.3	BURNET TX	-37	-53	-71	21	25
12	15.1	8	9.2	10	9.2	CALDWELL TX	-39	0	-39	77	69
7	26.3	3	6.0	5	7.7	CALHOUN TX	-77	28	-71	21	53
11	18.5	8	12.3	4	5.1	CALLAHAN TX	-33	-58	-72	19	24
54	15.1	52	11.1	69	10.7	CAMERON TX	-26	-4	-29	84	80
9	30.8	3	8.7	5	14.2	CAMP TX	-72	63	-54	55	93
3	11.7	4	12.4			CARSON TX	6				
10	12.2	9	9.0	6	5.0	CASS TX	-27	-44	-59	45	23
1	7.0	3	21.6	2	11.3	CASTRO TX	207	-48	60	98	83
8	35.2	4	10.0	1	4.1	CHAMBERS TX	-72	-59	-88	02	16
13	8.2	14	7.9	10	5.4	CHEROKEE TX	-3	-31	-34	82	27
11	22.3	7	13.0	3	5.3	CHILDRESS TX	-41	-60	-76	13	25
5	9.6	3	5.4	6	9.9	CLAY TX	-43	83	4	94	75
4	26.8	3	12.7			COCHRAN TX	-53				
1	6.8	3	16.8	1	7.1	COKE TX	148	-58	5	94	46
7	8.0	11	12.2	4	4.1	COLEMAN TX	52	-66	-49	64	16
23	11.4	25	10.9	15	5.3	COLLIN TX	-5	-51	-54	55	25
14*	42.0	6	17.7	1	2.5	COLLINGSWORTH TX	-58	-86	-94	00	07
15	21.5	10	12.9	4	4.4	COLORADO TX	-40	-66	-80	09	18
13	16.4	18	17.6	14	9.0	COMAL TX	8	-49	-45	70	67
22	22.5	8	8.6	5	4.5	COMANCHE TX	-62	-48	-80	09	19
4	21.6	4	16.0	3	15.5	CONCHO TX	-26	-3	-28	85	95
19	18.2	16	14.1	10	7.8	COOKE TX	-22	-45	-57	49	54
9	10.9	6	6.5	3	3.0	CORYELL TX	-40	-55	-73	18	09
3	14.3	4	20.8	1	3.3	COTTLE TX	46	-84	-77	12	11
4*	52.3			2	14.6	CRANE TX			-72	19	94
1	21.0	3	20.2	1	8.2	CROCKETT TX	-3	-59	-61	40	58
4	11.0	1	2.1	4	10.8	CROSBY TX	-81	409	-2	93	80
2	43.6	1	6.6	1	8.6	CULBERSON TX	-85	30	-80	08	63
6	18.3	6	17.4	2	6.6	DALLAM TX	-5	-62	-64	34	40
262	13.8	253	8.5	247	6.5	DALLAS TX	-38	-24	-53	57	39
7	14.6	12	16.8	11	14.5	DAWSON TX	15	-14	-1	93	94
8	23.6	5	10.2	7	10.7	DEAF SMITH TX	-57	5	-55	53	80
8	18.4	5	12.2	2	4.2	DELTA TX	-34	-65	-77	12	17
24	13.5	15	6.9	22	7.3	DENTON TX	-49	6	-46	69	48
8	9.0	5	4.4	11	9.0	DE WITT TX	-51	106	0	94	67
6	20.6	3	9.4	1	3.1	DICKENS TX	-55	-67	-85	04	10
11	33.1	2	5.8	6	14.0	DIMMIT TX	-83	143	-58	48	93
5	15.3	5	15.4	3	13.8	DONLEY TX	0	-10	-10	92	92
10	23.1	10	20.5	7	13.6	DUVAL TX	-11	-34	-41	75	92
21	15.0	16	10.3	10	6.0	EASTLAND TX	-31	-42	-60	42	33
16	16.0	29	15.4	15	6.4	ECTOR TX	-3	-58	-60	43	38
		2	19.9	1	10.5	EDWARDS TX		-47			78
37	20.4	12	6.0	11	5.0	ELLIS TX	-71	-16	-75	15	23
112	19.3	122	15.3	114*	11.9	EL PASO TX	-21	-22	-38	78	86
11	9.5	19	13.7	7	4.9	ERATH TX	44	-64	-48	66	22
13	14.2	14	12.8	9	9.8	FALLS TX	-10	-23	-31	83	74
15	8.9	16	9.4	11	6.0	FANNIN TX	6	-36	-32	83	33
21	16.7	13	9.7	14	10.6	FAYETTE TX	-42	9	-37	79	79
3	6.7	4	9.6	3	6.4	FISHER TX	43	-34	-5	93	37
8	16.8	9	20.0	5	11.3	FLOYD TX	19	-43	-33	83	83
5	26.5					FOARD TX					
29*	32.0	20	16.7	20	10.4	FORT BEND TX	-48	-38	-68	27	78
1	2.6	7	17.1	1	2.6	FRANKLIN TX	563	-85	0	94	08
11	17.2	4	5.9	3	3.7	FREESTONE TX	-65	-38	-78	10	13
10	26.6	7	16.6	2	3.8	FRIO TX	-38	-77	-86	04	14
2	8.6	4	11.5	5	9.0	GAINES TX	34	-22	4	94	67
43	14.0	56	13.9	52	9.4	GALVESTON TX	-1	-32	-33	83	71
8	33.4	3	10.9			GARZA TX	-67				
17	28.0	6	7.0	5	4.5	GILLESPIE TX	-75	-36	-84	05	19
				1	74.3	GLASSCOCK TX					99
2	8.9	3	13.8	4	14.2	GOLIAD TX	54	3	59	98	93
20	25.4	11	12.3	6	6.2	GONZALES TX	-52	-50	-76	14	35
7	7.1	9	9.1	13	10.2	GRAY TX	29	11	43	97	77
34	11.2	29	7.9	23	5.4	GRAYSON TX	-29	-31	-51	60	27
31	21.8	19	9.7	23	7.5	GREGG TX	-56	-22	-66	30	50
7	13.9	7	12.0	7	12.0	GRIMES TX	-14	0	-14	91	86
16	16.1	23	17.1	15	8.2	GUADALUPE TX	6	-52	-49	64	59
12	11.4	11	8.0	10	6.9	HALE TX	-30	-14	-40	77	43
5	11.4	4	12.0	1	3.1	HALL TX	6	-74	-73	18	10
12	16.8	9	11.6	4	5.1	HAMILTON TX	-31	-56	-69	23	24
5	53.9	2	14.6	1	3.8	HANSFORD TX	-73	-74	-93	00	13
7	14.8	5	10.2	4	7.5	HARDEMAN TX	-31	-26	-49	64	51
7	9.2	9	8.2	13	9.3	HARDIN TX	-11	14	1	94	69
351	15.8	388	11.5	334	7.5	HARRIS TX	-27	-35	-52	57	51
14	13.7	7	5.6	10	6.4	HARRISON TX	-60	16	-53	56	38
1	7.6	1	9.2			HARTLEY TX	21				
8	13.4	7	11.2	5	7.6	HASKELL TX	-16	-33	-43	73	51
10	14.2	9	12.2	14	13.9	HAYS TX	-14	14	-2	93	93
2	11.8	2	12.4			HEMPHILL TX	6				
11	10.7	13	9.2	8	4.3	HENDERSON TX	-14	-54	-60	42	17
93	20.7	84	15.0	78	9.6	HIDALGO TX	-27	-36	-53	56	73
16	10.4	9	5.7	19	13.1	HILL TX	-45	131	26	96	90
9	15.3	8	10.8	3	3.8	HOCKLEY TX	-29	-65	-75	15	13
3	7.7	11*	26.5	3	4.8	HOOD TX	245	-82	-38	79	21
22	18.0	10	7.9	10	8.2	HOPKINS TX	-56	3	-54	54	58
11	15.2	3	3.7	2	2.7	HOUSTON TX	-75	-29	-82	06	08

* SIGNIFICANTLY MORE DEATHS THAN EXPECTED

WHITE MALE

CANCER OF THE STOMACH — ICD 151

1950-1959 DEATHS	1950-1959 RATE	1960-1969 DEATHS	1960-1969 RATE	1970-1979 DEATHS	1970-1979 RATE	COUNTY AND STATE	PERCENT CHANGE 60S/50S	PERCENT CHANGE 70S/60S	PERCENT CHANGE 70S/50S	PERCENTILE %CHANGE 70S/50S	PERCENTILE RATES 1970S
13	13.4	9	6.4	7	4.6	HOWARD TX	-52	-28	-66	30	20
		3	37.1	1	9.3	HUDSPETH TX		-75			70
23	11.9	13	6.4	12	4.9	HUNT TX	-46	-23	-59	45	22
18	26.7	6	7.1	6	7.4	HUTCHINSON TX	-73	4	-72	19	49
1	14.3	2	23.3			IRION TX	62				
4	8.7	8	17.4	5	8.9	JACK TX	100	-49	2	94	66
3	7.2	10	20.9	4	9.0	JACKSON TX	191	-57	26	96	67
10	13.5	4	4.0	7	5.2	JASPER TX	-70	30	-61	39	25
1	12.8	1	14.1			JEFF DAVIS TX	10				
81	15.2	72	11.1	74	9.3	JEFFERSON TX	-27	-17	-39	77	69
3	22.0	3	17.9	3	10.9	JIM HOGG TX	-19	-39	-51	61	80
21	27.9	21*	20.0	12	9.9	JIM WELLS TX	-28	-50	-64	32	75
13	7.8	15	7.7	12	4.7	JOHNSON TX	-2	-39	-40	76	21
17	18.3	16	14.8	7	6.5	JONES TX	-19	-56	-64	33	39
7	10.7	11	17.1	8	13.3	KARNES TX	59	-22	25	96	91
17	12.1	13	9.0	11	6.7	KAUFMAN TX	-26	-25	-44	71	41
7	18.9	4	9.9	8	14.1	KENDALL TX	-48	43	-25	87	93
1	38.5					KENEDY TX					
		2	18.9			KENT TX					
10	10.2	14	8.9	6	2.9	KERR TX	-12	-68	-72	20	09
9*	42.1	8*	31.3	4	12.1	KIMBLE TX	-26	-61	-71	20	87
1	9.9	3	24.8	1	8.3	KINNEY TX	150	-66	-16	90	60
18*	33.2	3	3.2	5	5.9	KLEBERG TX	-90	86	-82	07	32
10	24.9	6	13.2			KNOX TX	-47				
21	11.9	23	12.4	17	8.8	LAMAR TX	4	-29	-26	86	65
21*	30.9	13	14.9	4	4.5	LAMB TX	-52	-70	-85	04	19
7	13.9	5	9.3	4	6.9	LAMPASAS TX	-33	-25	-50	62	44
10*	39.4	5	17.6	3	12.7	LA SALLE TX	-55	-28	-68	26	89
20	18.4	17	14.0	6	5.1	LAVACA TX	-24	-63	-72	19	24
7	18.1	4	8.5	3	5.7	LEE TX	-53	-32	-68	25	30
7	14.7	3	6.5	2	4.1	LEON TX	-56	-38	-72	19	15
15	16.6	11	8.9	10	6.0	LIBERTY TX	-46	-33	-64	33	32
7	7.0	8	7.6	8	6.9	LIMESTONE TX	8	-10	-3	93	43
3	19.9	1	5.2	1	4.9	LIPSCOMB TX	-74	-7	-76	14	22
7	21.5	6	17.2	2	6.3	LIVE OAK TX	-20	-63	-71	21	36
5	12.8	8	13.8	6	7.6	LLANO TX	8	-45	-40	76	51
48	18.2	41	9.3	32	6.0	LUBBOCK TX	-49	-36	-67	27	32
4	9.1	4	10.5	5	11.6	LYNN TX	15	10	26	96	84
10	16.1	7	11.6	8	13.1	MC CULLOCH TX	-28	13	-19	89	90
57	11.6	64	10.9	40	5.9	MC LENNAN TX	-6	-46	-49	64	32
1	15.5	1	14.1			MC MULLEN TX	-9				
6	15.6	4	10.7	4	7.7	MADISON TX	-31	-28	-50	62	53
4	17.4	4	14.0	5	12.0	MARION TX	-20	-14	-31	83	87
3	19.3	3	17.4			MARTIN TX	-10				
1	3.9	4	17.4	1	3.9	MASON TX	348	-77	1	94	15
15	21.5	11	12.0	8	7.1	MATAGORDA TX	-44	-41	-67	28	46
10	24.2	10	24.2	4	6.0	MAVERICK TX	0	-75	-75	15	33
11	14.4	7	7.5	12	10.9	MEDINA TX	-48	45	-25	87	80
8	41.8	4	17.9	3	13.0	MENARD TX	-57	-28	-69	24	90
17	29.9	15	13.1	10	5.0	MIDLAND TX	-56	-62	-83	06	23
22	19.3	15	13.5	11	8.5	MILAM TX	-30	-37	-56	51	61
2	5.8	4	7.7	4	8.4	MILLS TX	33	9	45	97	60
5	10.2	3	6.0	4	6.3	MITCHELL TX	-41	4	-38	78	36
7	7.5	9	8.2	4	3.3	MONTAGUE TX	8	-59	-56	51	11
12	12.4	9	6.2	23	8.2	MONTGOMERY TX	-50	33	-33	82	58
1	1.9	5	14.0	3	5.8	MOORE TX	635	-59	203	99	30
3	9.1	4	8.6	3	5.4	MORRIS TX	-6	-37	-41	75	26
4	27.0	2	17.2	1	3.9	MOTLEY TX	-36	-78	-86	04	14
20	18.1	8	6.3	9	5.6	NACOGDOCHES TX	-65	-13	-69	24	28
23	13.7	12	6.7	7	3.6	NAVARRO TX	-51	-47	-74	16	12
3	8.1	9	21.2	3	5.8	NEWTON TX	162	-72	-28	85	31
16	22.0	6	7.4	7	7.4	NOLAN TX	-66	0	-66	29	49
75	14.8	89	15.3	77	10.8	NUECES TX	3	-30	-27	85	80
2	7.5	2	7.7	5	15.3	OCHILTREE TX	3	98	104	98	95
1	12.8			1	10.3	OLDHAM TX			-19	89	77
18	21.8	20	10.8	16	6.8	ORANGE TX	-51	-37	-69	24	42
23	22.4	11	9.4	9	7.1	PALO PINTO TX	-58	-25	-68	25	45
11	17.8	4	5.4	12	13.1	PANOLA TX	-69	140	-27	86	90
17	13.4	13	9.3	6	3.2	PARKER TX	-31	-65	-76	14	10
2	10.5	4	11.3	6	15.8	PARMER TX	8	39	50	97	96
8	22.9	3	8.4	4	10.6	PECOS TX	-63	26	-54	55	79
7	14.5	4	6.2	5	5.1	POLK TX	-57	-17	-65	32	24
43	17.5	39	11.8	24	6.8	POTTER TX	-33	-42	-61	40	43
4	19.4	5	19.8	1	4.7	PRESIDIO TX	2	-76	-76	14	21
3	19.7	2	7.1			RAINS TX	-64				
5	9.8	9	10.3	15	9.2	RANDALL TX	5	-10	-6	93	69
4	55.3	2	11.7	1	18.9	REAGAN TX	-79	62	-66	30	98
		2	15.1	2	13.5	REAL TX		-11			92
10	11.2	7	7.3	7	7.8	RED RIVER TX	-34	7	-30	84	54
9	29.6	5	10.4	2	5.1	REEVES TX	-65	-51	-83	06	24
1	3.2	3	7.6	5	10.3	REFUGIO TX	139	35	223	99	77
				1	14.6	ROBERTS TX					94
6	9.4	6	8.8	5	6.9	ROBERTSON TX	-6	-21	-26	86	43
5	19.8	6	19.6	1	2.4	ROCKWALL TX	-1	-88	-88	02	07
11	15.6	12	13.5	12	14.2	RUNNELS TX	-13	5	-9	92	93
20	16.9	15	10.1	12	5.6	RUSK TX	-40	-45	-67	27	28
8	21.1	4	10.6	6	14.1	SABINE TX	-50	33	-33	82	93
5	16.2	2	6.4	5	11.9	SAN AUGUSTINE TX	-61	87	-27	86	86
4	17.5	2	6.9	3	7.0	SAN JACINTO TX	-60	1	-60	42	44
22	17.4	19	12.0	23	12.6	SAN PATRICIO TX	-31	5	-28	85	89
5	9.4	5	9.7	2	3.1	SAN SABA TX	3	-68	-67	28	10
2	19.3	2	17.5	1	7.1	SCHLEICHER TX	-9	-59	-63	35	46
10	17.8	10	14.0	6	8.3	SCURRY TX	-21	-41	-53	56	59
10*	42.4	4	17.4	2	6.3	SHACKELFORD TX	-59	-64	-85	04	36
8	7.9	11	9.6	6	5.5	SHELBY TX	22	-42	-30	84	28
1	14.9	1	6.9	1	4.9	SHERMAN TX	-54	-28	-67	28	22
37	17.7	27	8.7	31	7.1	SMITH TX	-50	-19	-60	43	46
		3	15.5	1	3.5	SOMERVELL TX		-78			12
15*	34.5	5	7.7	13*	17.6	STARR TX	-78	130	-49	64	97
3	5.3	5	8.7	4	6.1	STEPHENS TX	63	-30	14	95	34
1	22.9			3*	50.9	STERLING TX			122	99	99

* SIGNIFICANTLY MORE DEATHS THAN EXPECTED

WHITE MALE | CANCER OF THE STOMACH | ICD 151

1950-1959 DEATHS	1950-1959 RATE	1960-1969 DEATHS	1960-1969 RATE	1970-1979 DEATHS	1970-1979 RATE	COUNTY AND STATE	PERCENT CHANGE 60S/50S	PERCENT CHANGE 70S/60S	PERCENT CHANGE 70S/50S	PERCENTILE %CHANGE 70S/50S	PERCENTILE RATES 1970S
2	12.3	1	5.4	2	8.0	STONEWALL TX	-56	48	-36	80	56
2	18.6			4*	23.4	SUTTON TX			26	96	99
3	8.8	2	4.5	2	4.5	SWISHER TX	-49	1	-49	65	19
201	16.5	161	8.9	150	6.6	TARRANT TX	-46	-26	-60	41	39
30	12.7	30	9.6	27	6.7	TAYLOR TX	-25	-30	-47	67	41
1	9.8	2	19.8			TERRELL TX	101				
9	22.0	9	18.5	3	6.0	TERRY TX	-16	-67	-73	19	33
2	8.8			2	7.5	THROCKMORTON TX			-14	91	51
11	18.3	9	10.6	4	3.9	TITUS TX	-42	-64	-79	10	14
39	18.8	25	9.4	30	9.5	TOM GREEN TX	-50	2	-49	64	72
66	12.3	92	12.9	81	9.3	TRAVIS TX	5	-28	-24	87	70
4	10.6	1	2.2			TRINITY TX	-79				
11	24.7	5	7.9	5	6.2	TYLER TX	-68	-22	-75	15	35
11	12.6	12	12.2	8	6.8	UPSHUR TX	-3	-44	-46	69	42
2	29.8	5*	43.4	3	16.7	UPTON TX	46	-61	-44	72	96
9	13.7	10	12.4	10	10.7	UVALDE TX	-10	-14	-22	88	79
20*	34.9	11	15.4	7	7.9	VAL VERDE TX	-56	-49	-77	12	54
23	19.1	13	9.2	19	10.6	VAN ZANDT TX	-52	16	-44	71	79
21	20.9	21	15.3	15	8.4	VICTORIA TX	-27	-45	-60	42	60
3	5.3	8	13.6	4	6.1	WALKER TX	157	-55	15	95	34
8	26.2	3	13.4	8	16.2	WALLER TX	-49	21	-38	78	96
9	21.8	4	12.1	5	12.6	WARD TX	-44	4	-42	74	89
12	14.3	20*	20.9	8	6.9	WASHINGTON TX	47	-67	-51	60	44
36	19.4	29	12.8	30	11.0	WEBB TX	-34	-14	-43	73	82
21	19.8	17	13.1	9	6.0	WHARTON TX	-33	-54	-70	23	33
6	12.9	7	14.3	2	3.6	WHEELER TX	11	-75	-72	19	12
41	12.1	30	7.2	25	5.5	WICHITA TX	-41	-23	-54	54	28
15	17.9	12	12.5	8	7.1	WILBARGER TX	-30	-44	-60	41	45
12	25.8	12	19.7	4	7.0	WILLACY TX	-24	-64	-73	18	44
33	22.8	23	11.1	10	4.0	WILLIAMSON TX	-51	-64	-82	06	15
13	18.7	11	15.2	7	9.7	WILSON TX	-19	-37	-48	66	73
4	27.1	6	18.9	2	5.1	WINKLER TX	-30	-73	-81	07	23
14	13.6	7	6.3	11	8.4	WISE TX	-54	34	-38	78	61
9	9.8	8	7.0	9	6.9	WOOD TX	-29	-1	-30	84	43
		4	19.2			YOAKUM TX					
6	6.5	16	15.3	7	5.6	YOUNG TX	135	-63	-13	91	29
1	4.7	2	10.4	2	7.2	ZAPATA TX	121	-31	52	98	47
10	26.0	4	9.6	6	16.0	ZAVALA TX	-63	66	-39	78	96
454	18.4	363	11.5	315	8.0	UTAH UT	-37	-30	-56		
6	43.8	3	15.3	1	3.7	BEAVER UT	-65	-76	-92	01	13
15	20.8	13	15.1	8	7.9	BOX ELDER UT	-27	-47	-62	38	55
26	21.7	15	10.2	15	8.6	CACHE UT	-53	-15	-60	42	63
21	33.1	16	20.2	16*	18.1	CARBON UT	-39	-10	-45	70	97
14	14.6	18	12.8	20	8.2	DAVIS UT	-12	-36	-44	71	58
7	34.1	4	14.8	6	27.4	DUCHESNE UT	-57	85	-20	89	99
5	20.4	3	11.9	3	7.7	EMERY UT	-42	-35	-62	37	53
2	15.8	3	18.5	1	9.4	GARFIELD UT	18	-49	-41	76	71
		2	17.8	2	9.3	GRAND UT		-48			70
4	13.1	5	12.3	5	10.4	IRON UT	-6	-16	-21	88	78
4	18.0	2	10.1	1	3.3	JUAB UT	-44	-67	-82	07	11
1	10.9	1	10.0			KANE UT	-8				
11	34.6	3	9.2			MILLARD UT	-74				
1	9.6	1	7.4			MORGAN UT	-22				
		1	19.1			PIUTE UT					
2	21.5	2	21.9			RICH UT	2				
187	17.3	150	10.9	139	8.2	SALT LAKE UT	-37	-24	-53	57	58
2	13.5	1	7.4			SAN JUAN UT	-45				
15	24.0	15	22.3	8	12.0	SANPETE UT	-7	-46	-50	63	86
17*	40.0	5	9.7	2	2.9	SEVIER UT	-76	-71	-93	00	09
4	15.9	3	11.4	7*	23.7	SUMMIT UT	-29	109	49	97	99
6	10.5	9	17.0	7	13.3	TOOELE UT	62	-22	27	96	91
9	32.8	4	12.0	3	6.1	UINTAH UT	-63	-49	-81	07	34
42	16.1	23	6.7	34	7.8	UTAH UT	-59	17	-52	59	54
1	5.0	4	14.5	3	11.7	WASATCH UT	192	-19	135	99	85
3	7.7	4	7.6	5	4.8	WASHINGTON UT	-2	-37	-38	78	21
				3*	22.6	WAYNE UT					99
49	15.7	53	13.4	26	5.6	WEBER UT	-14	-58	-64	33	28
346	18.8	210	11.0	155	7.4	VERMONT VT	-42	-33	-61		
19	20.3	6	6.1	7	6.0	ADDISON VT	-70	-1	-70	22	33
24	19.3	19	13.0	7	4.3	BENNINGTON VT	-32	-67	-78	11	17
31	23.7	15	12.0	8	6.4	CALEDONIA VT	-49	-47	-73	17	37
33	13.2	28	10.4	21	6.8	CHITTENDEN VT	-22	-34	-48	65	42
9	26.8	6	20.0	7*	21.2	ESSEX VT	-25	6	-21	88	98
31	21.2	16	11.1	8	5.2	FRANKLIN VT	-47	-53	-75	15	25
7	39.2	1	5.6	2	11.0	GRAND ISLE VT	-86	99	-72	19	81
10	17.1	6	10.2	3	4.4	LAMOILLE VT	-41	-57	-74	16	18
17	18.4	6	6.4	4	4.4	ORANGE VT	-65	-31	-76	14	18
22	21.2	8	7.8	10	9.1	ORLEANS VT	-63	17	-57	49	68
49	20.1	32	12.8	36*	13.7	RUTLAND VT	-36	7	-32	83	92
33	17.7	22	10.7	13	5.7	WASHINGTON VT	-39	-47	-68	26	29
29	18.2	13	7.6	11	6.3	WINDHAM VT	-58	-17	-66	30	36
32	15.6	32	14.3	18	7.2	WINDSOR VT	-8	-50	-54	55	47
1402	15.0	1166	9.8	1014	6.7	VIRGINIA VA	-35	-31	-55		
16	12.8	17	13.2	5	4.0	ACCOMACK VA	3	-70	-69	24	15
20	12.4	14	6.7	15	5.7	ALBEMARLE#VA	-45	-16	-54	54	29
15	18.1	12	11.3	12	9.1	ALLEGHANY#VA	-38	-19	-50	63	68
5	23.4	4	16.9			AMELIA VA	-28				
9	13.6	4	4.9	10	10.3	AMHERST VA	-64	111	-24	87	77
4	10.7	4	11.2	1	2.0	APPOMATTOX VA	5	-82	-81	08	06
56	15.7	52	10.6	52	9.0	ARLINGTON#VA	-33	-15	-43	73	66
27	11.3	27	8.6	19	5.1	AUGUSTA#VA	-24	-41	-55	53	24
7	38.2	3	13.7	2	7.2	BATH VA	-64	-47	-81	08	47
37	10.3	42	9.5	34	6.4	BEDFORD#VA	-7	-33	-38	78	37
5	41.1	1	3.4	2	10.1	BLAND VA	-92	198	-75	15	76
15	22.3	7	9.0	10	11.4	BOTETOURT VA	-60	28	-49	65	84

* SIGNIFICANTLY MORE DEATHS THAN EXPECTED

\# COUNTIES COMBINED; SEE TECHNICAL NOTES

WHITE MALE — CANCER OF THE STOMACH — ICD 151

1950-1959 DEATHS	1950-1959 RATE	1960-1969 DEATHS	1960-1969 RATE	1970-1979 DEATHS	1970-1979 RATE	COUNTY AND STATE	PERCENT CHANGE 60S/50S	PERCENT CHANGE 70S/60S	PERCENT CHANGE 70S/50S	PERCENTILE %CHANGE 70S/50S	PERCENTILE RATES 1970S
5	18.4	6	19.4	1	2.4	BRUNSWICK VA	5	-88	-87	03	07
15	18.1	20*	19.8	12	11.5	BUCHANAN VA	9	-42	-36	80	84
2	5.8	6	15.8	3	8.0	BUCKINGHAM VA	173	-49	38	97	56
5	15.5	4	12.9	3	7.6	CAROLINE VA	-17	-41	-51	61	52
36	18.3	22	10.3	4	1.5	CARROLL#VA	-44	-85	-92	01	05
1	16.4					CHARLES CITY VA					
8	18.8	5	11.1	6	13.6	CHARLOTTE VA	-41	22	-28	85	92
106	13.6	88	10.5	67	7.2	CHESTERFIELD#VA	-23	-32	-47	67	46
3	10.6	5	15.3	5	13.4	CLARKE VA	45	-12	27	96	92
1	10.2	2	13.8	2	9.6	CRAIG VA	35	-31	-6	93	72
4	7.8	6	9.7	7	8.8	CULPEPER VA	24	-9	13	95	65
2	12.8	1	12.5			CUMBERLAND VA	-2				
10	16.5	14	22.1	4	4.8	DICKENSON VA	34	-78	-71	21	21
18	10.7	16	7.6	9	3.4	DINWIDDIE#VA	-29	-56	-69	25	11
2	9.4	3	14.3	2	6.8	ESSEX VA	53	-52	-27	85	42
62	14.9	87	10.0	83	6.2	FAIRFAX#VA	-33	-38	-58	46	35
12	18.4	11	12.8	9	9.1	FAUQUIER VA	-31	-29	-51	61	68
13	25.2	7	13.3	2	3.0	FLOYD VA	-47	-77	-88	02	09
3	11.4	2	7.4	2	5.4	FLUVANNA VA	-35	-27	-52	58	27
8	8.6	6	5.6	9	7.0	FRANKLIN VA	-35	24	-19	89	44
11	8.6	11	6.9	5	2.4	FREDERICK#VA	-20	-65	-72	19	07
9	19.0	7	9.4	9	10.0	GILES VA	-51	7	-47	67	76
6	18.4	4	8.7	3	4.6	GLOUCESTER VA	-53	-47	-75	15	20
		2	7.9	4	13.4	GOOCHLAND VA		68			91
4	22.2	1	7.0	3	12.7	GREENE VA	-69	81	-43	73	89
4	15.2	2	6.6	1	3.1	GREENSVILLE#VA	-56	-53	-80	09	10
24	28.9	8	7.9	8	6.9	HALIFAX#VA	-73	-12	-76	14	43
9	13.8	5	5.3	11	8.8	HANOVER VA	-62	66	-36	80	65
21	9.0	43	11.6	41	7.9	HENRICO VA	28	-31	-12	91	55
14	14.7	10	5.5	5	2.2	HENRY#VA	-63	-59	-85	04	06
8	38.5	2	12.0			HIGHLAND VA	-69				
2	7.5	3	8.9	2	3.2	ISLE OF WIGHT VA	18	-64	-58	47	10
13	13.7	8	6.7	10	5.7	JAMES CITY#VA	-51	-15	-58	46	29
3	14.1			1	5.5	KING AND QUEEN VA			-61	40	27
6	31.7	4	18.0			KING GEORGE VA	-43				
6	32.0	2	8.9	1	4.2	KING WILLIAM VA	-72	-53	-87	03	16
4	13.7	4	10.8	1	1.4	LANCASTER VA	-21	-87	-90	01	05
23	20.0	17	13.3	12	9.1	LEE VA	-33	-32	-55	54	68
12	13.5	6	5.7	8	6.0	LOUDOUN VA	-58	6	-55	52	33
7	20.5	1	1.9	8	13.8	LOUISA VA	-91	615	-32	83	93
6	15.5	4	11.0	2	4.7	LUNENBURG VA	-29	-58	-70	22	20
8	23.6	4	11.9	7	17.9	MADISON VA	-49	50	-24	87	97
5	12.1	3	6.4	2	3.5	MATHEWS VA	-47	-46	-71	20	11
10	16.9	8	11.2	11	10.7	MECKLENBURG VA	-34	-4	-37	80	80
1	4.8	1	3.6			MIDDLESEX VA	-24				
23	19.2	8	5.8	14	7.3	MONTGOMERY#VA	-70	25	-62	37	48
4	7.2	9	17.3	4	6.8	NELSON VA	141	-61	-6	93	42
7	15.4	3	6.0	4	7.7	NORTHAMPTON VA	-61	29	-50	62	52
5	16.4	3	7.9	2	3.9	NORTHUMBERLAND VA	-52	-50	-76	14	15
6	12.6	5	11.2	6	10.9	NOTTOWAY VA	-11	-3	-14	91	81
1	1.6	3	5.0	7	9.1	ORANGE VA	215	81	472	99	68
12	18.4	3	4.3	6	6.8	PAGE VA	-77	59	-63	36	42
9	14.0	3	4.2	9	11.5	PATRICK VA	-70	171	-18	89	84
35	15.8	34	11.0	21	5.7	PITTSYLVANIA#VA	-30	-48	-64	34	29
2	13.1	2	9.9	4	16.0	POWHATAN VA	-24	61	22	96	96
8	21.3	6	14.3	5	12.6	PRINCE EDWARD VA	-33	-12	-41	76	89
10	14.4	15	17.3	17	9.5	PRINCE WILLIAM#VA	20	-45	-34	82	72
15	21.5	20*	18.1	6	4.8	PULASKI VA	-16	-74	-78	11	21
6	25.1	1	3.2	1	3.9	RAPPAHANNOCK VA	-87	21	-85	05	14
5	24.7			1	5.7	RICHMOND VA			-77	12	30
66	13.4	65	10.5	52	6.6	ROANOKE#VA	-21	-37	-51	61	39
25	27.6	10	9.2	8	6.2	ROCKBRIDGE#VA	-67	-32	-78	12	35
29	16.2	18	8.5	20	8.0	ROCKINGHAM#VA	-47	-7	-51	61	56
16	17.9	8	7.5	7	6.7	RUSSELL VA	-58	-11	-63	36	41
9	8.6	15	12.7	15	12.0	SCOTT VA	47	-6	39	97	86
12	11.2	12	10.2	5	3.7	SHENANDOAH VA	-9	-63	-67	29	13
20	17.7	12	8.9	15	9.0	SMYTH VA	-49	1	-49	64	66
5	10.3	3	5.8	4	5.8	SOUTHAMPTON#VA	-43	0	-43	72	31
9	13.4	7	7.4	11	8.1	SPOTSYLVANIA#VA	-45	9	-39	77	57
6	14.5	1	1.0	6	5.6	STAFFORD VA	-93	478	-61	40	29
		1	7.2	1	5.0	SURRY VA		-31			23
6	30.0					SUSSEX VA					
21	15.3	22	14.5	22	11.9	TAZEWELL VA	-5	-18	-22	88	86
14	23.3	2	3.1	11	12.0	WARREN VA	-87	283	-49	65	86
36	18.9	19	8.4	18	6.3	WASHINGTON#VA	-55	-26	-67	28	36
4	12.4	2	7.3	7	10.6	WESTMORELAND VA	-41	46	-14	91	79
33	21.2	29	16.8	13	7.0	WISE#VA	-21	-58	-67	28	44
14	17.0	8	9.0	5	4.9	WYTHE VA	-47	-46	-71	20	22
127	14.8	119	10.3	108	7.0	CHESAPEAKE CITY#VA	-31	-31	-53	57	45
20	12.2	20	9.9	12	4.2	HAMPTON CITY VA	-19	-57	-65	31	17
23	15.8	14	7.2	14	4.2	NEWPORT NEWS CITY VA	-54	-41	-73	18	17
11	13.5	6	6.2	6	6.8	SUFFOLK CITY VA	-54	11	-49	64	42
2438	20.9	1760	13.1	1338	8.6	WASHINGTON WA	-37	-35	-59		
6	19.5	9	27.3	3	6.2	ADAMS WA	39	-77	-68	26	35
15	21.8	11	14.3	12	13.5	ASOTIN WA	-34	-5	-38	79	92
28	20.2	22	11.8	19	6.0	BENTON WA	-42	-49	-70	22	33
26	13.6	22	10.0	15	5.9	CHELAN WA	-26	-41	-57	50	31
25	19.6	24	14.3	25	10.1	CLALLAM WA	-27	-29	-48	66	76
70	17.6	70	14.7	59	9.5	CLARK WA	-16	-35	-46	69	72
7	22.2	3	10.1	3	9.9	COLUMBIA WA	-55	-2	-55	52	75
38	18.3	31	12.6	23	7.5	COWLITZ WA	-31	-41	-59	44	50
9	17.3	8	12.0	8	8.6	DOUGLAS WA	-30	-29	-50	62	63
3	15.2	1	5.3	3	15.7	FERRY WA	-65	196	3	94	95
7	12.7	9	11.8	5	4.7	FRANKLIN WA	-7	-60	-63	36	21
5	29.1	2	11.8			GARFIELD WA	-60				
12	11.8	13	11.2	21	13.0	GRANT WA	-6	16	10	95	90
83*	26.7	61*	19.2	51*	15.4	GRAYS HARBOR WA	-28	-20	-42	74	95
13	22.2	9	10.2	9	6.1	ISLAND WA	-54	-40	-73	19	34
10	21.0	5	8.8	7	7.7	JEFFERSON WA	-58	-12	-63	35	53

* SIGNIFICANTLY MORE DEATHS THAN EXPECTED

\# COUNTIES COMBINED; SEE TECHNICAL NOTES

1950-1959 DEATHS	1950-1959 RATE	1960-1969 DEATHS	1960-1969 RATE	1970-1979 DEATHS	1970-1979 RATE	COUNTY AND STATE	PERCENT CHANGE 60S/50S	PERCENT CHANGE 70S/60S	PERCENT CHANGE 70S/50S	PERCENTILE %CHANGE 70S/50S	PERCENTILE RATES 1970S
845*	24.2	547	13.6	401	8.9	KING WA	-44	-35	-63	35	66
45	13.1	57	14.0	43	8.4	KITSAP WA	7	-40	-35	80	61
19	17.8	13	11.4	8	7.4	KITTITAS WA	-36	-35	-59	46	49
12	18.6	6	9.3	3	4.0	KLICKITAT WA	-50	-57	-79	10	15
57	22.8	39	14.4	30	10.1	LEWIS WA	-37	-30	-56	52	76
9	15.7	7	9.7	5	6.9	LINCOLN WA	-38	-29	-56	51	43
17	23.3	9	9.7	8	5.3	MASON WA	-58	-45	-77	12	26
16	13.2	14	10.3	6	3.8	OKANOGAN WA	-22	-63	-71	21	14
18	22.2	4	4.0	11	9.3	PACIFIC WA	-82	134	-58	47	70
14	32.2	4	8.6	6	12.3	PEND OREILLE WA	-73	43	-62	39	88
256	20.1	189	13.3	141	8.5	PIERCE WA	-34	-36	-58	48	62
6	25.0	1	2.6	6	15.4	SAN JUAN WA	-90	492	-38	78	95
34	12.1	34	11.5	18	5.5	SKAGIT WA	-5	-52	-55	54	27
5	21.5	1	3.4	2	6.6	SKAMANIA WA	-84	93	-69	24	40
191*	29.3	123	14.9	103	9.7	SNOHOMISH WA	-49	-34	-67	28	74
228	18.6	164	12.0	133	9.1	SPOKANE WA	-35	-24	-51	61	68
19	17.8	15	14.2	7	6.0	STEVENS WA	-20	-58	-67	29	32
49	20.7	41	13.9	29	7.0	THURSTON WA	-33	-50	-66	29	45
5	23.1	3	14.6	1	4.1	WAHKIAKUM WA	-37	-72	-82	07	16
26	13.0	31	13.7	9	3.8	WALLA WALLA WA	5	-72	-70	22	14
77	20.3	51	12.6	37	8.3	WHATCOM WA	-38	-34	-59	45	59
23	17.0	16	12.4	13	9.0	WHITMAN WA	-27	-27	-47	68	67
110	18.3	91	13.1	55	7.2	YAKIMA WA	-28	-45	-61	41	47
1317	17.8	1081	13.4	701	8.1	WEST VIRGINIA WV	-25	-39	-55		
10	11.0	10	10.6	7	8.7	BARBOUR WV	-4	-18	-21	88	64
16	10.5	10	6.1	10	4.9	BERKELEY WV	-41	-19	-53	57	22
25	28.1	19	16.5	12	10.0	BOONE WV	-41	-40	-65	32	75
17	18.6	11	12.7	12	14.4	BRAXTON WV	-32	14	-23	88	94
16	13.6	15	11.8	16	11.2	BROOKE WV	-13	-5	-17	90	83
54	12.5	45	9.5	33	6.7	CABELL WV	-24	-29	-46	68	41
6	12.7	4	9.1	5	9.8	CALHOUN WV	-28	7	-23	87	74
8	15.9	9	17.8	5	9.4	CLAY WV	12	-47	-41	75	71
8	15.6	2	4.2	3	6.3	DODDRIDGE WV	-73	49	-59	44	36
29	12.0	38	16.0	22	8.7	FAYETTE WV	34	-46	-27	85	64
12	26.2	8	17.4	7	15.8	GILMER WV	-34	-9	-40	77	96
5	13.1	5	13.1	6	12.9	GRANT WV	0	-1	-1	93	90
26	18.9	32*	19.3	10	5.8	GREENBRIER WV	2	-70	-69	24	30
11	18.4	6	9.5	3	3.9	HAMPSHIRE WV	-48	-59	-79	10	14
36	28.8	27	18.5	20	11.7	HANCOCK WV	-36	-37	-59	44	85
8	18.0	5	9.9	6	11.2	HARDY WV	-45	14	-38	79	82
72	18.5	57	14.2	43	10.2	HARRISON WV	-23	-28	-45	70	77
18	23.4	12	14.6	8	7.7	JACKSON WV	-38	-47	-67	27	53
10	17.3	6	9.0	11	11.9	JEFFERSON WV	-48	32	-31	83	86
123	17.5	115	13.2	65	6.6	KANAWHA WV	-25	-50	-63	36	39
24	19.0	19	13.7	16	13.3	LEWIS WV	-28	-3	-30	84	91
21	23.9	16	17.4	6	5.5	LINCOLN WV	-27	-69	-77	12	27
33	21.6	28	15.0	14	7.8	LOGAN WV	-31	-48	-64	34	54
38	19.2	29	17.4	15	9.3	MC DOWELL WV	-9	-47	-52	59	70
75	25.6	59*	18.5	41	12.3	MARION WV	-28	-34	-52	58	88
39	22.9	24	12.6	23	12.0	MARSHALL WV	-45	-5	-48	67	86
10	9.9	11	10.2	3	2.6	MASON WV	3	-75	-74	17	08
37	16.3	34	12.5	24	8.5	MERCER WV	-23	-32	-48	66	62
16	17.7	11	11.2	4	3.9	MINERAL WV	-37	-65	-78	11	14
25	18.9	18	12.1	13	9.2	MINGO WV	-36	-24	-51	60	69
50	22.7	36	14.6	24	9.1	MONONGALIA WV	-36	-37	-60	43	68
10	14.7	13	19.0	8	11.2	MONROE WV	29	-41	-24	87	82
8	20.1	1	2.3	1	1.9	MORGAN WV	-89	-17	-91	01	06
15	15.3	23*	21.1	16	13.0	NICHOLAS WV	38	-38	-15	90	90
62	18.7	52	15.1	29	9.2	OHIO WV	-20	-39	-51	60	69
4	9.3	4	8.2	1	2.4	PENDLETON WV	-12	-71	-74	16	07
5	13.3	7	20.6	1	2.8	PLEASANTS WV	55	-86	-79	10	08
11	18.5	5	8.0	7	10.4	POCAHONTAS WV	-57	30	-44	72	78
26	18.3	21	15.8	11	7.3	PRESTON WV	-14	-54	-60	42	48
9	10.3	14	12.7	12	8.5	PUTNAM WV	23	-33	-17	90	62
42	16.5	34	11.6	22	6.5	RALEIGH WV	-30	-44	-61	40	38
17	13.0	22	14.8	8	5.4	RANDOLPH WV	14	-63	-58	47	27
9	11.3	9	10.9	6	8.2	RITCHIE WV	-4	-24	-27	86	58
13	13.1	17	17.1	12	12.2	ROANE WV	30	-29	-7	92	87
12	15.1	12	15.0	3	3.5	SUMMERS WV	-1	-76	-77	13	12
14	16.3	9	9.6	8	9.0	TAYLOR WV	-41	-6	-45	70	66
12	23.7	7	15.4	6	11.4	TUCKER WV	-35	-26	-52	58	83
13	18.3	5	8.8	4	6.5	TYLER WV	-52	-27	-65	32	38
17	17.3	11	11.1	12	11.2	UPSHUR WV	-36	0	-36	80	82
39	25.4	17	9.7	10	5.3	WAYNE WV	-62	-45	-79	09	25
13	21.4	8	12.8	2	2.8	WEBSTER WV	-40	-78	-87	03	08
18	17.6	12	12.0	8	7.5	WETZEL WV	-32	-37	-57	48	50
4	14.7	1	3.1	4	13.0	WIRT WV	-79	313	-11	91	90
52	16.9	51	14.6	18	4.5	WOOD WV	-14	-69	-73	17	19
14	16.2	5	4.2	5	4.5	WYOMING WV	-74	7	-72	19	19
4118*	24.6	3033*	15.8	2125*	10.0	WISCONSIN WI	-36	-37	-60		
12	21.8	6	10.0	7	7.7	ADAMS WI	-54	-23	-65	32	53
36*	34.9	17	15.1	20*	17.9	ASHLAND WI	-57	19	-49	65	97
48	24.6	42*	19.4	33*	12.7	BARRON WI	-21	-35	-48	65	89
28*	31.2	9	10.0	11	12.2	BAYFIELD WI	-68	23	-61	41	88
80	19.5	85*	17.3	53	8.9	BROWN WI	-11	-48	-54	54	66
15	18.8	15	18.6	6	6.6	BUFFALO WI	-1	-64	-65	32	40
16	25.2	14	18.4	12	13.9	BURNETT WI	-27	-25	-45	71	93
18	19.8	15	14.9	10	7.9	CALUMET WI	-25	-47	-60	42	55
44	21.8	25	11.0	20	7.7	CHIPPEWA WI	-50	-30	-65	31	52
36	19.5	39*	19.6	27	12.8	CLARK WI	0	-35	-34	81	89
41	19.6	27	11.7	17	7.0	COLUMBIA WI	-40	-40	-64	33	44
18	20.1	11	11.5	14	13.5	CRAWFORD WI	-43	17	-33	83	92
159	22.4	101	11.6	76	7.8	DANE WI	-48	-33	-65	31	53
64	20.4	50	14.8	31	8.3	DODGE WI	-28	-44	-60	43	59
32	27.8	12	9.5	19	13.8	DOOR WI	-66	44	-50	62	92
78*	30.9	43	16.8	41*	16.5	DOUGLAS WI	-46	-2	-47	68	96
30	22.6	33*	20.8	16	9.5	DUNN WI	-8	-54	-58	47	72

* SIGNIFICANTLY MORE DEATHS THAN EXPECTED

WHITE MALE — CANCER OF THE STOMACH — ICD 151

1950-1959 DEATHS	1950-1959 RATE	1960-1969 DEATHS	1960-1969 RATE	1970-1979 DEATHS	1970-1979 RATE	COUNTY AND STATE	PERCENT CHANGE 60S/50S	PERCENT CHANGE 70S/60S	PERCENT CHANGE 70S/50S	PERCENTILE %CHANGE 70S/50S	PERCENTILE RATES 1970S
50	20.6	42	14.6	41*	13.0	EAU CLAIRE WI	-29	-10	-37	80	90
11*	54.7	4	14.9	3	15.0	FLORENCE WI	-73	1	-73	19	95
74	22.0	51	13.5	33	8.1	FOND DU LAC WI	-39	-40	-63	35	57
19*	40.1	13*	23.9	6	11.0	FOREST WI	-40	-54	-73	18	81
45	19.6	23	9.8	14	5.7	GRANT WI	-50	-41	-71	21	30
30	23.0	19	12.8	20	12.3	GREEN WI	-45	-4	-47	68	88
20	21.2	17	15.7	12	9.4	GREEN LAKE WI	-26	-40	-56	51	71
22	20.0	14	12.0	5	4.7	IOWA WI	-40	-61	-77	13	21
21*	41.6	14*	31.4	12*	22.1	IRON WI	-25	-30	-47	68	98
22	22.4	16	16.0	11	10.9	JACKSON WI	-29	-32	-51	60	81
52	20.6	26	8.9	24	7.7	JEFFERSON WI	-57	-13	-63	37	53
28	25.1	13	11.1	14	11.5	JUNEAU WI	-56	4	-54	54	84
81	22.6	95*	21.8	58	11.3	KENOSHA WI	-4	-48	-50	63	83
30*	32.1	16	15.8	13	12.3	KEWAUNEE WI	-51	-22	-62	38	88
73	23.8	46	12.9	30	7.7	LA CROSSE WI	-46	-41	-68	26	53
22	21.9	16	15.9	10	10.6	LAFAYETTE WI	-28	-33	-52	59	79
23	20.2	20	15.5	24*	18.7	LANGLADE WI	-23	21	-8	92	98
35*	28.7	33*	24.2	23*	14.8	LINCOLN WI	-16	-39	-49	65	94
86*	28.0	58	15.9	28	6.5	MANITOWOC WI	-43	-59	-77	12	38
88	23.5	87*	20.8	59*	12.4	MARATHON WI	-11	-40	-47	67	88
61*	32.7	41*	19.2	27	11.4	MARINETTE WI	-41	-41	-65	31	83
16	24.0	10	16.3	9	9.8	MARQUETTE WI	-32	-40	-59	45	74
97*	30.5	60	16.1	40	9.9	MENOMINEE#WI	-47	-38	-67	27	75
1022*	27.0	765*	18.0	449*	10.2	MILWAUKEE WI	-33	-43	-62	38	77
35	19.5	23	11.8	17	8.4	MONROE WI	-39	-29	-57	48	60
25	24.2	31*	24.5	12	7.0	ONEIDA WI	2	-71	-71	21	45
80	24.5	60	14.9	42	8.9	OUTAGAMIE WI	-39	-41	-64	34	66
30	27.6	31*	18.9	13	6.6	OZAUKEE WI	-32	-65	-76	13	40
13	28.4	2	4.0	2	3.4	PEPIN WI	-86	-15	-88	02	11
32	25.6	18	13.7	9	6.0	PIERCE WI	-47	-56	-77	13	33
42	28.4	26	15.4	25	12.4	POLK WI	-46	-19	-56	50	88
45	26.9	34	18.4	23	10.4	PORTAGE WI	-32	-43	-61	39	78
28	28.3	14	12.9	10	8.1	PRICE WI	-55	-37	-71	20	57
125	24.4	104*	16.8	66	9.7	RACINE WI	-31	-42	-60	42	73
21	19.9	11	10.7	10	9.3	RICHLAND WI	-47	-13	-54	56	69
99	22.4	72	14.6	46	8.4	ROCK WI	-35	-42	-62	37	61
22	25.9	21*	22.7	12	13.3	RUSK WI	-12	-41	-49	65	91
28	19.1	23	14.7	15	7.8	ST CROIX WI	-23	-47	-59	44	53
38	19.1	28	12.6	18	7.2	SAUK WI	-34	-43	-63	37	46
15	21.2	13	17.9	10	13.4	SAWYER WI	-15	-25	-37	80	92
110*	27.0	52	10.9	54	10.4	SHEBOYGAN WI	-60	-5	-61	39	78
27	30.3	24*	22.1	14	12.5	TAYLOR WI	-27	-44	-59	45	88
33	21.9	26	16.0	14	8.8	TREMPEALEAU WI	-27	-45	-60	43	65
45	27.7	19	12.0	17	9.4	VERNON WI	-57	-22	-66	29	70
10	17.6	5	7.5	14	12.8	VILAS WI	-58	71	-27	85	89
58	23.9	30	9.8	38	11.4	WALWORTH WI	-59	17	-52	58	84
20	27.0	7	8.6	15*	16.9	WASHBURN WI	-68	97	-37	79	97
31	19.7	29	13.8	36	12.8	WASHINGTON WI	-30	-8	-35	81	89
106	24.2	92	15.2	92	10.9	WAUKESHA WI	-37	-28	-55	53	81
54	23.8	43	14.9	24	8.0	WAUPACA WI	-37	-46	-66	29	56
24	22.6	21	20.2	10	8.0	WAUSHARA WI	-11	-61	-65	31	55
84	18.7	84	16.3	53	9.3	WINNEBAGO WI	-13	-43	-50	63	70
55	24.2	47	17.1	36	11.0	WOOD WI	-29	-36	-54	54	81
229	19.7	157	11.4	119	7.7	WYOMING WY	-42	-32	-61		
13	19.3	6	7.1	7	8.4	ALBANY WY	-63	17	-57	50	60
10	21.8	7	11.9	4	5.7	BIG HORN WY	-45	-52	-74	17	29
4	18.6	3	7.8	2	3.5	CAMPBELL WY	-58	-55	-81	07	12
10	15.4	8	11.7	5	7.6	CARBON WY	-24	-35	-51	61	52
4	12.7	1	2.7	4	10.8	CONVERSE WY	-79	305	-15	90	80
5	23.1			4	17.3	CROOK WY			-25	87	97
10	12.6	6	6.7	7	7.8	FREMONT WY	-47	17	-38	78	54
11	19.8	7	11.8	2	2.6	GOSHEN WY	-40	-78	-87	03	08
6	20.7	3	7.5	2	3.7	HOT SPRINGS WY	-64	-51	-82	06	12
4	13.0	5	14.1	4	11.5	JOHNSON WY	8	-19	-12	91	84
22	12.3	28	15.2	20	8.6	LARAMIE WY	24	-44	-30	84	63
17*	49.4	10*	25.1	4	10.7	LINCOLN WY	-49	-57	-78	11	79
19	14.3	21	12.6	11	5.2	NATRONA WY	-12	-58	-63	35	25
6	27.4	4	19.1	1	5.9	NIOBRARA WY	-30	-69	-79	10	32
9	15.2	8	9.6	6	5.9	PARK-TETON#WY	-37	-38	-61	40	32
8	19.3	4	9.3	3	6.6	PLATTE WY	-51	-29	-66	30	40
16	14.8	11	8.8	11	9.1	SHERIDAN WY	-40	3	-38	78	68
5	40.0	3	23.8	2	29.8	SUBLETTE WY	-40	25	-25	87	99
32*	42.7	15	19.2	9	9.7	SWEETWATER WY	-55	-49	-77	12	74
7	19.1	2	4.8	4	11.1	UINTA WY	-75	129	-42	74	82
4	19.3	2	8.9	6	16.3	WASHAKIE WY	-54	84	-15	90	96

* SIGNIFICANTLY MORE DEATHS THAN EXPECTED

\# COUNTIES COMBINED; SEE TECHNICAL NOTES

WHITE FEMALE | CANCER OF THE STOMACH | ICD 151

1950-1959 DEATHS	1950-1959 RATE	1960-1969 DEATHS	1960-1969 RATE	1970-1979 DEATHS	1970-1979 RATE	COUNTY AND STATE	PERCENT CHANGE 60S/50S	PERCENT CHANGE 70S/60S	PERCENT CHANGE 70S/50S	PERCENTILE %CHANGE 70S/50S	PERCENTILE RATES 1970S
75224	10.7	61985	6.7	51456	4.3	UNITED STATES	-38	-35	-60		
745	8.7	577	4.8	553	3.4	ALABAMA AL	-45	-29	-61		
2	3.9	3	4.7	6	6.3	AUTAUGA AL	20	34	61	97	88
9	6.9	10	4.2	17	4.4	BALDWIN AL	-38	4	-36	81	66
6	10.8	3	3.7	8	7.6	BARBOUR AL	-66	107	-30	85	93
5	9.3	1	1.6	1	1.3	BIBB AL	-83	-21	-86	05	13
8	7.0	6	4.8	4	1.9	BLOUNT AL	-32	-59	-72	26	20
5	20.3	1	5.2			BULLOCK AL	-74				
4	8.0	2	2.2	2	2.3	BUTLER AL	-73	4	-72	27	24
26	12.5	12	3.5	23	4.6	CALHOUN AL	-72	34	-63	43	70
9	7.9	8	5.0	8	3.9	CHAMBERS AL	-36	-22	-50	67	58
3	6.5	5	6.8	1	1.1	CHEROKEE AL	4	-83	-83	10	12
13	12.5	5	4.0	12	6.8	CHILTON AL	-68	68	-46	73	90
3	7.6	2	4.7	2	3.9	CHOCTAW AL	-39	-15	-48	69	58
5	7.4	4	4.9			CLARKE AL	-34				
10	16.2	4	5.4			CLAY AL	-67				
7	15.1			2	2.7	CLEBURNE AL			-82	10	31
4	4.8	3	2.4	2	1.2	COFFEE AL	-51	-50	-75	21	12
10	8.2	11	5.8	6	2.5	COLBERT AL	-28	-58	-70	30	28
8	16.2	3	4.1			CONECUH AL	-74				
3	7.2	2	4.1	2	4.1	COOSA AL	-43	2	-42	76	62
7	5.6	9	4.7	14	6.1	COVINGTON AL	-15	28	8	93	85
2	3.1	3	4.3			CRENSHAW AL	38				
21	13.0	9	3.5	12	3.4	CULLMAN AL	-73	-4	-74	23	46
6	8.0	5	4.2	1	.5	DALE AL	-47	-89	-94	00	09
3	2.9	9	6.9	8	5.0	DALLAS AL	138	-28	71	97	75
13	7.4	15	6.3	8	2.5	DE KALB AL	-14	-61	-66	37	28
9	10.0	7	5.9	2	1.3	ELMORE AL	-41	-78	-87	04	13
9	10.0	5	4.3	3	2.0	ESCAMBIA AL	-57	-52	-80	14	21
20	6.7	18	4.6	24	4.3	ETOWAH AL	-30	-6	-35	82	65
7	10.4	6	6.9	4	3.7	FAYETTE AL	-34	-46	-64	41	53
11	11.9	18*	15.3	9	6.1	FRANKLIN AL	28	-60	-49	68	86
7	8.1	5	4.0	1	.9	GENEVA AL	-51	-78	-89	03	10
1	7.3			2	4.5	GREENE AL			-39	79	67
2	6.1	2	3.7	3	5.7	HALE AL	-39	54	-6	91	83
2	4.2			2	3.2	HENRY AL			-24	87	42
18	11.8	3	1.4	3	1.1	HOUSTON AL	-88	-24	-91	02	11
14	10.4	12	7.1	7	2.9	JACKSON AL	-32	-58	-72	27	36
124	7.7	81	3.6	100	3.4	JEFFERSON AL	-54	-4	-56	58	46
1	1.7	3	3.9	10*	10.0	LAMAR AL	124	160	482	99	97
13	7.0	11	4.3	9	2.2	LAUDERDALE AL	-39	-49	-69	32	23
3	4.4	5	4.9	6	4.7	LAWRENCE AL	11	-4	7	93	71
7	7.7	4	2.7	7	3.7	LEE AL	-65	36	-52	64	53
13	12.8	7	5.0	4	2.1	LIMESTONE AL	-61	-58	-84	08	21
2	9.8	1	5.6			LOWNDES AL	-43				
1	4.3	6*	18.8	3	9.3	MACON AL	336	-51	114	98	96
15	5.6	17	4.4	21	3.5	MADISON AL	-21	-20	-37	81	49
1	2.3	3	5.4	1	1.2	MARENGO AL	134	-77	-47	72	12
8	8.6	12	9.4	5	2.6	MARION AL	9	-73	-70	29	29
17	9.6	11	4.3	9	2.8	MARSHALL AL	-55	-35	-71	28	33
60	9.8	49	5.4	51	4.0	MOBILE AL	-45	-26	-59	52	60
3	6.3	6	9.2	2	2.4	MONROE AL	46	-74	-62	45	26
25	7.1	20	3.8	26	3.5	MONTGOMERY AL	-47	-8	-51	66	48
15	7.8	13	4.7	13	3.1	MORGAN AL	-39	-34	-60	50	41
		2	4.2	3	6.5	PERRY AL		53			89
4	6.6	4	5.2	1	.6	PICKENS AL	-22	-87	-90	02	10
9	11.0	5	5.0	3	2.4	PIKE AL	-54	-52	-78	16	27
5	6.6	6	6.4	5	3.8	RANDOLPH AL	-3	-41	-43	76	55
7	8.1	6	4.7	2	1.4	RUSSELL AL	-42	-70	-83	09	13
9	10.5	6	5.4	5	3.0	ST CLAIR AL	-49	-45	-72	27	37
11	12.9	4	3.0	4	1.9	SHELBY AL	-77	-36	-85	06	19
5	18.8	2	4.1	1	2.5	SUMTER AL	-78	-40	-87	04	28
12	8.5	18	8.3	9	3.0	TALLADEGA AL	-2	-63	-64	40	38
12	11.9	10	6.5	10	5.3	TALLAPOOSA AL	-45	-19	-55	58	79
14	5.1	17	4.2	18	3.5	TUSCALOOSA AL	-18	-16	-31	84	48
37*	21.2	15	5.9	19	5.1	WALKER AL	-72	-14	-76	19	77
5	15.3	2	4.0			WASHINGTON AL	-74				
3	9.7			1	1.8	WILCOX AL			-82	11	17
12*	17.7	20*	26.7	6	5.5	WINSTON AL	51	-79	-69	32	81
		12	7.5	16	4.7	ALASKA AK		-38			
254	8.3	371	6.3	394	3.6	ARIZONA AZ	-24	-43	-57		
1	4.8	1	5.6			APACHE AZ	16				
6	4.4	22	10.4	27*	8.1	COCHISE AZ	136	-22	84	98	94
5	13.9	5	5.5	4	2.9	COCONINO AZ	-60	-48	-79	14	34
6	7.4	10	9.7	5	2.8	GILA AZ	32	-71	-61	46	34
8	17.8	6	10.9	4	4.8	GRAHAM AZ	-39	-56	-73	25	73
5	16.3	2	6.3			GREENLEE AZ	-61				
133	8.2	189	5.7	221	3.4	MARICOPA AZ	-31	-39	-58	53	47
		7	11.0	5	1.9	MOHAVE AZ		-82			19
4	7.1	6	7.7	4	3.6	NAVAJO AZ	10	-54	-50	68	50
50	7.8	89	6.7	75	3.2	PIMA AZ	-15	-52	-59	51	42
10	7.2	8	4.0	15	5.6	PINAL AZ	-44	39	-22	87	82
6	19.0	4	7.8	5	7.1	SANTA CRUZ AZ	-59	-9	-62	44	92
14	10.6	12	5.3	12	2.7	YAVAPAI AZ	-50	-49	-75	22	31
6	5.3	10	5.1	17	6.1	YUMA AZ	-3	19	15	94	86
605	9.2	439	4.9	421	3.5	ARKANSAS AR	-46	-29	-62		
11	14.7	8	7.8	4	3.0	ARKANSAS AR	-47	-62	-80	14	37
1	1.5	6	7.2	6	5.7	ASHLEY AR	392	-20	292	99	83
8	15.6	1	.6	7	2.8	BAXTER AR	-96	346	-82	10	33
12	4.9	13	3.9	14	3.0	BENTON AR	-21	-23	-39	79	38
5	6.2	2	1.3	8	4.3	BOONE AR	-79	218	-32	83	64

* SIGNIFICANTLY MORE DEATHS THAN EXPECTED

WHITE FEMALE — CANCER OF THE STOMACH — ICD 151

1950-1959 DEATHS	1950-1959 RATE	1960-1969 DEATHS	1960-1969 RATE	1970-1979 DEATHS	1970-1979 RATE	COUNTY AND STATE	PERCENT CHANGE 60S/50S	PERCENT CHANGE 70S/60S	PERCENT CHANGE 70S/50S	PERCENTILE %CHANGE 70S/50S	PERCENTILE RATES 1970S
3	5.9	4	6.3	2	1.9	BRADLEY AR	7	-70	-67	35	19
		2	7.7	2	5.2	CALHOUN AR		-33			78
12	14.3	6	5.6	5	3.4	CARROLL AR	-61	-39	-76	19	47
2	4.9	2	4.5	4	7.6	CHICOT AR	-9	69	54	97	93
8	9.9	5	4.6	5	3.4	CLARK AR	-53	-27	-66	37	46
16	14.4	9	6.6	6	3.6	CLAY AR	-54	-45	-75	22	51
3	6.0	2	3.0	4	3.8	CLEBURNE AR	-50	28	-36	81	55
4	13.5			1	2.1	CLEVELAND AR			-84	07	22
8	9.5	5	5.0	3	1.8	COLUMBIA AR	-48	-63	-81	12	18
10	14.6	6	7.0	1	.8	CONWAY AR	-52	-88	-94	00	10
23	13.2	8	3.3	18	5.8	CRAIGHEAD AR	-75	73	-56	57	83
6	5.9	7	4.7	5	2.4	CRAWFORD AR	-21	-49	-60	50	26
6	10.2	3	4.4	4	3.8	CRITTENDEN AR	-56	-14	-62	45	56
4	6.6	2	2.2	7	7.1	CROSS AR	-66	214	7	93	92
3	8.8	6	12.1	1	1.6	DALLAS AR	38	-87	-82	10	15
4	9.7	3	5.7	2	2.3	DESHA AR	-41	-60	-77	18	24
4	9.1	1	1.4	3	3.8	DREW AR	-85	178	-59	52	54
12	11.5	9	6.8	5	2.1	FAULKNER AR	-41	-69	-82	11	22
6	8.9	3	3.7	3	3.7	FRANKLIN AR	-58	0	-58	53	54
2	4.6	3	5.7	2	3.3	FULTON AR	24	-42	-28	85	44
24	11.2	12	3.6	19	3.8	GARLAND AR	-68	5	-66	36	54
3	15.1	1	1.8	5	6.9	GRANT AR	-88	271	-54	60	91
10	9.0	10	6.6	8	4.4	GREENE AR	-27	-34	-51	65	66
10	12.2	3	2.7	3	2.2	HEMPSTEAD AR	-77	-18	-82	11	24
6	6.0	6	5.2	8	4.4	HOT SPRING AR	-14	-15	-27	85	66
4	7.2	1	1.5	4	4.7	HOWARD AR	-79	216	-35	82	71
8	8.1	10	7.2	6	3.1	INDEPENDENCE AR	-11	-57	-62	45	39
7	15.6	3	5.5	4	3.9	IZARD AR	-65	-28	-75	22	58
11	15.4	7	7.6	5	3.9	JACKSON AR	-51	-48	-74	22	58
19	11.1	10	4.5	9	3.0	JEFFERSON AR	-59	-34	-73	25	37
12	15.2	13*	12.6	5	4.2	JOHNSON AR	-17	-66	-72	26	63
3	8.6					LAFAYETTE AR					
11	12.4	4	3.9	5	3.2	LAWRENCE AR	-68	-19	-74	23	41
1	2.6	2	5.0	4	8.4	LEE AR	97	68	230	99	95
2	6.1	2	5.5	2	4.2	LINCOLN AR	-10	-24	-31	84	63
1	2.2	1	2.2			LITTLE RIVER AR	-3				
10	9.9	3	2.5	7	4.1	LOGAN AR	-75	63	-59	52	61
15	16.3	3	2.4	8	4.6	LONOKE AR	-85	90	-71	27	70
9	15.8	5	7.0	4	5.8	MADISON AR	-56	-18	-63	42	84
1	3.3			1	1.1	MARION AR			-68	34	11
6	5.0	7	4.8	2	1.0	MILLER AR	-4	-79	-80	14	11
18	11.0	10	4.8	7	2.8	MISSISSIPPI AR	-57	-42	-75	21	32
3	7.7	2	4.0	1	1.6	MONROE AR	-48	-61	-80	14	15
1	3.5	1	2.3	2	4.0	MONTGOMERY AR	-36	78	14	94	59
7	12.9	2	3.5	1	1.4	NEVADA AR	-73	-61	-89	02	13
1	3.0	2	5.8	1	2.3	NEWTON AR	89	-60	-25	87	25
7	11.0	4	3.6	6	4.0	OUACHITA AR	-67	9	-64	41	59
2	7.5	2	6.1	1	2.2	PERRY AR	-19	-64	-71	28	23
5	7.9	4	4.6	5	5.1	PHILLIPS AR	-41	10	-36	81	77
4	10.9	5	8.7	8*	9.4	PIKE AR	-20	7	-14	90	96
11	13.8	7	6.2	8	5.6	POINSETT AR	-55	-9	-59	51	82
7	9.0	2	1.9	7	4.8	POLK AR	-79	156	-47	71	72
13	11.5	16*	11.6	12	6.2	POPE AR	1	-46	-46	72	87
		11*	24.0	4	6.5	PRAIRIE AR		-73			88
61	8.2	53	4.9	51	3.7	PULASKI AR	-40	-26	-55	59	51
6	9.9	5	5.7	1	1.4	RANDOLPH AR	-43	-74	-85	06	14
4	7.7	4	6.0	6	6.5	ST FRANCIS AR	-22	9	-15	89	89
7	5.7	4	2.5	3	1.4	SALINE AR	-55	-45	-75	21	13
4	9.0	4	9.7	2	3.3	SCOTT AR	8	-65	-63	43	45
2	4.2	1	2.2	1	2.2	SEARCY AR	-49	3	-47	71	24
21	6.8	34	8.0	14	2.4	SEBASTIAN AR	17	-70	-64	40	27
4	6.6	5	6.7	2	2.6	SEVIER AR	2	-61	-60	48	30
5	13.3	2	3.7	9*	8.7	SHARP AR	-72	137	-35	82	96
4	13.8	7*	20.0	5	8.5	STONE AR	45	-58	-38	79	95
9	5.1	6	3.1	7	2.3	UNION AR	-40	-24	-54	61	26
6	14.6	1	2.8	2	2.0	VAN BUREN AR	-81	-29	-87	05	20
24	9.8	9	2.5	9	1.9	WASHINGTON AR	-75	-22	-80	13	19
14	8.4	10	4.8	8	2.8	WHITE AR	-43	-43	-67	35	32
4	9.0	3	6.1	3	4.9	WOODRUFF AR	-33	-19	-45	73	74
5	7.4	4	4.3	4	3.9	YELL AR	-42	-9	-47	71	57
5634	9.6	5611	6.6	5097*	4.5	CALIFORNIA CA	-31	-32	-53		
385	9.6	345	7.0	274*	5.0	ALAMEDA CA	-28	-29	-48	69	75
1*	136.6					ALPINE CA					
8	17.7	2	3.3	8	7.0	AMADOR CA	-81	111	-60	48	92
20	4.9	27	4.7	42	4.3	BUTTE CA	-5	-9	-14	90	64
2	4.1	7	8.7	4	3.5	CALAVERAS CA	110	-59	-15	89	49
7	13.3	4	6.8			COLUSA CA	-49				
80	9.1	106	5.9	131	4.8	CONTRA COSTA CA	-34	-20	-47	71	72
5	15.4	4	6.9			DEL NORTE CA	-55				
3	3.0	4	2.5	11	3.5	EL DORADO CA	-17	42	17	94	48
110	9.5	115	6.8	135*	6.1	FRESNO CA	-28	-11	-36	81	86
5	6.4	4	4.1	4	3.5	GLENN CA	-36	-15	-45	73	48
39	13.8	31	7.7	24	4.8	HUMBOLDT CA	-44	-38	-65	38	72
14	9.9	13	6.0	13	4.7	IMPERIAL CA	-40	-22	-53	63	70
2	5.2	6	11.1	6	5.2	INYO CA	114	-53	0	92	78
64	8.6	75	6.5	62	4.0	KERN CA	-25	-38	-54	61	59
17	11.6	14	6.7	13	4.8	KINGS CA	-42	-29	-59	51	72
2	2.3	8	5.3	9	2.9	LAKE CA	134	-45	29	95	36
8	16.2	2	2.9	2	1.8	LASSEN CA	-82	-38	-89	03	17
2482	9.8	2335	6.9	1789*	4.7	LOS ANGELES CA	-29	-32	-52	65	71
7	5.6	13	7.3	13	5.1	MADERA CA	29	-30	-10	90	76
47	10.5	47	6.3	44	3.5	MARIN CA	-40	-44	-66	37	49
2	9.2			1	2.3	MARIPOSA CA			-75	21	24
22	11.3	14	5.3	20	5.7	MENDOCINO CA	-53	8	-49	68	83
25	11.7	18	5.5	32*	6.8	MERCED CA	-53	25	-42	76	91
2	8.1	2	4.6			MODOC CA	-43				
43	7.9	58	7.0	64	5.2	MONTEREY CA	-12	-25	-34	82	78
21	6.0	34	6.8	33	5.0	NAPA CA	12	-26	-16	89	76
12	10.3	8	4.3	10	3.1	NEVADA CA	-58	-29	-70	29	39

* SIGNIFICANTLY MORE DEATHS THAN EXPECTED

WHITE FEMALE

CANCER OF THE STOMACH

ICD 151

1950-1959 DEATHS	1950-1959 RATE	1960-1969 DEATHS	1960-1969 RATE	1970-1979 DEATHS	1970-1979 RATE	COUNTY AND STATE	PERCENT CHANGE 60S/50S	PERCENT CHANGE 70S/60S	PERCENT CHANGE 70S/50S	PERCENTILE %CHANGE 70S/50S	PERCENTILE RATES 1970S
138	8.1	210	5.1	305	3.9	ORANGE CA	-37	-24	-52	64	57
18	7.6	17	4.7	20	3.4	PLACER CA	-38	-27	-55	59	47
3	8.6	3	6.6	1	1.6	PLUMAS CA	-23	-76	-81	12	15
85	7.5	112	5.1	128	3.4	RIVERSIDE CA	-32	-34	-55	60	46
120	8.6	129	6.1	152	4.8	SACRAMENTO CA	-29	-21	-44	74	72
7	10.6	5	5.7	7	6.0	SAN BENITO CA	-47	6	-43	75	85
154	9.4	152	5.3	158	3.8	SAN BERNARDINO CA	-43	-29	-60	49	54
250	7.9	296	5.6	305	3.6	SAN DIEGO CA	-29	-36	-54	60	50
547*	12.8	408*	8.7	300*	6.3	SAN FRANCISCO CA	-32	-27	-51	66	87
81	9.0	83	6.6	68	4.1	SAN JOAQUIN CA	-27	-39	-55	59	60
35	10.0	24	4.7	32	4.0	SAN LUIS OBISPO CA	-53	-15	-60	49	59
149	11.7	152	7.2	164*	5.3	SAN MATEO CA	-38	-26	-54	60	79
73	11.1	58	5.3	78	4.8	SANTA BARBARA CA	-53	-8	-57	56	73
147	7.8	218	6.9	215	4.6	SANTA CLARA CA	-11	-34	-42	77	69
54	9.4	59	6.7	62	4.9	SANTA CRUZ CA	-29	-27	-48	69	73
18	10.5	21	7.5	22	4.2	SHASTA CA	-29	-45	-60	48	63
		1	7.6			SIERRA CA					
18	13.6	9	5.9	7	3.4	SISKIYOU CA	-57	-41	-75	22	47
38	11.7	43	8.6	27	3.6	SOLANO CA	-26	-58	-69	32	51
63	8.9	75	6.6	92	5.3	SONOMA CA	-25	-20	-40	78	79
46	6.7	68	7.1	55	4.2	STANISLAUS CA	6	-40	-37	81	64
9	8.7	8	5.1	8	3.7	SUTTER CA	-41	-28	-58	55	52
13	12.4	15	10.4	7	2.8	TEHAMA CA	-16	-73	-78	17	33
		2	6.9	4	9.0	TRINITY CA		30			96
53	8.9	46	5.6	50	4.5	TULARE CA	-37	-20	-49	68	68
6	8.7	6	5.5	6	3.4	TUOLUMNE CA	-37	-38	-61	46	46
41	6.9	67	6.2	61	3.2	VENTURA CA	-10	-48	-53	62	43
16	8.4	19	6.5	15	3.3	YOLO CA	-22	-50	-61	46	43
17*	17.4	9	6.4	4	2.2	YUBA CA	-63	-66	-87	04	23
727	10.5	502	5.3	479	3.8	COLORADO CO	-49	-29	-64		
12	10.9	27	8.1	18	3.3	ADAMS CO	-25	-60	-70	30	43
5	18.1	3	7.2	4	7.5	ALAMOSA CO	-60	3	-59	52	93
1	8.5	1	7.2			ARCHULETA CO	-16				
				3	8.4	BACA CO					95
2	5.7	1	3.0	2	5.6	BENT CO	-47	85	-2	92	82
21	7.0	19	4.2	25	3.3	BOULDER CO	-40	-20	-52	64	45
4	9.6	6	10.3	3	4.7	CHAFFEE CO	7	-54	-51	66	71
		2	10.0			CHEYENNE CO					
3	18.3	2	12.3			CLEAR CREEK CO	-33				
10*	30.6	2	4.4			CONEJOS CO	-86				
1	4.2	5*	31.9	4*	23.9	COSTILLA CO	660	-25	469	99	99
2	8.6	1	3.2	1	3.6	CROWLEY CO	-63	14	-58	54	51
1	14.1					CUSTER CO					
13	14.0	9	7.6	8	4.9	DELTA CO	-46	-36	-65	38	73
269	9.9	204	5.8	172	4.0	DENVER-ARAPAHOE# CO	-41	-31	-59	50	60
1	23.6	1	14.1			DOLORES CO	-40				
1	5.2	2	6.7	2	2.5	DOUGLAS CO	30	-62	-51	66	29
3	23.0	1	3.6	1	5.2	EAGLE CO	-84	46	-77	18	78
2	9.7	2	9.9	1	3.5	ELBERT CO	2	-64	-64	41	49
58	11.3	25	3.5	32	2.9	EL PASO CO	-69	-17	-75	22	34
14	9.4	8	4.0	9	4.3	FREMONT CO	-57	6	-54	61	65
4	6.5	4	5.6	1	1.1	GARFIELD CO	-14	-80	-82	10	12
1	6.1	1	9.3			GRAND CO	51				
1	5.5	1	6.6	4*	15.6	GUNNISON CO	21	137	187	99	99
11*	24.0	9*	17.9	2	3.3	HUERFANO CO	-25	-81	-86	05	44
1	16.4					JACKSON CO					
24	8.8	25	4.3	40	3.3	JEFFERSON CO	-51	-24	-63	44	44
4*	37.2			1	6.1	KIOWA CO			-84	08	86
7	25.5	3	7.8	1	1.2	KIT CARSON CO	-70	-85	-95	00	12
3	20.9	3	20.8			LAKE CO	0				
5	7.8	2	2.3	6	5.4	LA PLATA CO	-71	137	-32	84	79
25	9.2	16	4.3	20	3.4	LARIMER CO	-54	-19	-63	44	47
18*	17.1	8	6.7	5	3.2	LAS ANIMAS CO	-61	-53	-81	11	41
4	14.8	1	3.4	2	5.0	LINCOLN CO	-77	46	-66	36	75
9	11.8	2	1.6	3	2.5	LOGAN CO	-87	59	-79	16	29
26	12.8	13	4.3	13	3.3	MESA CO	-67	-22	-74	23	44
1	3.7	1	3.3			MOFFAT CO	-11				
3	6.5			3	4.2	MONTEZUMA CO			-36	81	63
12	19.3			2	1.9	MONTROSE CO			-90	02	19
13	15.2	2	2.1	4	3.1	MORGAN CO	-87	50	-80	14	40
10	8.2	10	6.8	6	3.3	OTERO CO	-18	-51	-60	49	43
3*	26.4					OURAY CO					
1	37.2	1	15.8	1	4.5	PARK CO	-58	-71	-88	04	68
3	11.1	1	2.4	2	4.6	PHILLIPS CO	-79	95	-59	52	69
1	13.5			1	14.9	PITKIN CO			10	94	99
9	14.4	6	6.8	3	2.1	PROWERS CO	-53	-68	-85	07	22
46	9.2	33	5.5	43*	6.4	PUEBLO CO	-40	16	-30	84	88
2	12.6			1	4.4	RIO BLANCO CO			-65	39	67
9	19.3	7	13.8	6*	10.7	RIO GRANDE CO	-28	-23	-45	74	97
3	10.4	2	7.0			ROUTT CO	-32				
		1	8.0			SAGUACHE CO					
1	19.2					SAN JUAN CO					
2	34.3					SAN MIGUEL CO					
		2	7.8	1	4.9	SEDGWICK CO		-37			74
				1	37.2	SUMMIT CO					99
2	12.2	1	6.2	1	3.5	TELLER CO	-49	-43	-71	28	49
6	22.0	1	2.8	1	1.8	WASHINGTON CO	-87	-36	-92	01	18
26	8.5	23	5.7	19	3.5	WELD CO	-33	-39	-59	51	48
8	14.9	2	3.3	1	2.2	YUMA CO	-78	-34	-85	06	23
1329*	11.9	1142*	7.8	1000*	5.2	CONNECTICUT CT	-34	-33	-56		
347*	12.2	285	7.5	289*	5.8	FAIRFIELD CT	-39	-22	-52	64	84
343*	12.0	284	7.4	248	4.9	HARTFORD CT	-38	-33	-59	52	74
70	11.4	61	7.4	50	4.8	LITCHFIELD CT	-35	-35	-58	54	72
43	9.7	49	8.3	44	5.9	MIDDLESEX CT	-15	-29	-39	79	85
348	11.6	330*	8.6	258*	5.1	NEW HAVEN CT	-26	-40	-56	58	77
97	11.7	70	6.8	59	4.6	NEW LONDON CT	-42	-33	-61	47	69
28	12.1	20	6.3	20	4.5	TOLLAND CT	-48	-29	-63	43	67

* SIGNIFICANTLY MORE DEATHS THAN EXPECTED

COUNTIES COMBINED; SEE TECHNICAL NOTES

WHITE FEMALE — CANCER OF THE STOMACH — ICD 151

1950-1959 DEATHS	1950-1959 RATE	1960-1969 DEATHS	1960-1969 RATE	1970-1979 DEATHS	1970-1979 RATE	COUNTY AND STATE	PERCENT CHANGE 60S/50S	PERCENT CHANGE 70S/60S	PERCENT CHANGE 70S/50S	PERCENTILE %CHANGE 70S/50S	PERCENTILE RATES 1970S
53	13.9	43	9.1	32	5.0	WINDHAM CT	-35	-45	-64	41	76
147	9.8	131	6.4	101	3.6	DELAWARE DE	-35	-43	-63		
18	8.6	9	3.5	17	4.9	KENT DE	-59	38	-44	75	73
100	9.9	103	7.4	65	3.6	NEW CASTLE DE	-26	-52	-64	40	50
29	9.9	19	4.5	19	3.1	SUSSEX DE	-55	-31	-69	33	40
232	8.2	145	5.2	83	3.8	DIST. OF COLUMBIA DC	-36	-28	-54		
232	8.2	145	5.2	83	3.8	WASHINGTON DC	-36	-28	-54	61	54
1089	6.8	1596	4.9	2092	3.5	FLORIDA FL	-29	-29	-49		
9	6.7	8	3.4	8	1.9	ALACHUA FL	-50	-44	-72	27	19
1	6.0	2	11.5			BAKER FL	93				
4	2.5	10	4.6	10	2.7	BAY FL	81	-41	6	93	31
2	6.4	3	6.1	2	2.9	BRADFORD FL	-5	-53	-55	59	35
18	9.7	24	5.1	32	3.0	BREVARD FL	-47	-42	-69	31	37
55	7.7	129	4.4	270	3.8	BROWARD FL	-42	-14	-50	66	55
		2	7.1	2	3.9	CALHOUN FL		-44			58
1	1.8	10	4.2	15	2.3	CHARLOTTE FL	128	-45	26	95	26
2	3.8	7	7.2	10	2.3	CITRUS FL	88	-68	-40	77	25
1	2.3	3	2.7	3	1.6	CLAY FL	16	-39	-30	85	16
		3	2.2	12	3.0	COLLIER FL		41			39
1	1.6			2	1.4	COLUMBIA FL			-8	91	14
282	9.4	382	6.3	475	4.5	DADE FL	-32	-29	-52	64	68
4	10.4	3	4.0	4	3.1	DE SOTO FL	-62	-21	-70	30	40
2	18.1					DIXIE FL					
65	6.5	75	4.7	83	3.8	DUVAL FL	-28	-20	-42	76	54
18	5.2	23	3.9	22	2.6	ESCAMBIA FL	-24	-34	-50	67	30
1	8.5	2	9.0	2	4.0	FLAGLER FL	7	-55	-52	63	60
1	3.5	2	7.7	2	5.5	FRANKLIN FL	120	-29	56	97	81
4	3.5	5	3.7	2	1.6	GADSDEN FL	4	-57	-55	59	15
1	9.2			1	3.8	GILCHRIST FL			-59	51	54
1	28.1					GLADES FL					
1	5.1	4	14.6	1	1.9	GULF FL	189	-87	-62	45	19
3	14.2	1	3.9	1	2.4	HAMILTON FL	-73	-37	-83	09	27
2	5.4	2	3.0	1	1.1	HARDEE FL	-44	-64	-80	13	11
2	7.7	3	11.4	1	1.7	HENDRY FL	48	-85	-78	16	16
1	2.8	6	5.5	3	.9	HERNANDO FL	94	-84	-69	32	11
3	2.8	11	5.6	9	2.1	HIGHLANDS FL	101	-63	-26	86	21
68	5.3	122	5.8	114	3.8	HILLSBOROUGH FL	10	-36	-29	85	54
2	3.9	3	4.8	1	.9	HOLMES FL	23	-81	-76	19	11
3	3.0	13	6.6	10	2.6	INDIAN RIVER FL	117	-61	-14	90	30
2	2.1	6	4.2	6	3.5	JACKSON FL	105	-16	71	97	49
2	8.6	1	5.3	2	5.1	JEFFERSON FL	-39	-3	-41	77	77
19	7.0	16	3.2	27	3.2	LAKE FL	-54	1	-54	61	42
9	5.0	22	4.8	38	3.0	LEE FL	-4	-36	-39	79	39
5	3.3	8	3.9	5	1.4	LEON FL	18	-64	-58	54	13
6	17.5	1	1.6	2	2.3	LEVY FL	-91	47	-87	05	25
				2	12.1	LIBERTY FL					98
		3	7.0	2	2.8	MADISON FL		-59			34
22	6.5	37	4.9	53	3.7	MANATEE FL	-24	-26	-44	75	52
6	3.5	8	3.1	15	2.9	MARION FL	-12	-7	-18	89	35
2	3.7	2	1.0	15	4.0	MARTIN FL	-73	284	5	93	58
6	6.8	7	3.7	6	2.2	MONROE FL	-46	-39	-67	35	24
				7	7.5	NASSAU FL					93
9	8.5	3	2.0	9	3.5	OKALOOSA FL	-76	76	-58	53	49
1	8.2	1	5.3	4	6.1	OKEECHOBEE FL	-35	16	-25	86	87
55	7.0	79	5.5	80	3.7	ORANGE FL	-21	-33	-47	70	51
8	7.6	9	5.0	12	2.7	OSCEOLA FL	-34	-46	-65	39	31
42	5.8	83	4.9	154	3.7	PALM BEACH FL	-17	-24	-37	80	52
11	6.5	29	6.7	62	4.0	PASCO FL	3	-40	-38	80	60
149	6.0	224	4.2	246	3.0	PINELLAS FL	-30	-27	-49	68	39
35	5.4	38	3.5	57	3.2	POLK FL	-35	-9	-41	77	41
9	9.4	7	5.0	5	2.1	PUTNAM FL	-46	-59	-78	17	21
14	11.9	13	7.7	8	2.7	ST JOHNS FL	-35	-66	-77	17	31
7	8.3	15	7.1	7	1.5	ST LUCIE FL	-14	-79	-82	11	15
6	10.5	5	4.5	1	.5	SANTA ROSA FL	-57	-89	-95	00	09
18	4.8	33	3.7	61	3.3	SARASOTA FL	-23	-10	-31	84	44
7	5.5	8	2.7	18	3.3	SEMINOLE FL	-51	23	-40	78	44
2	8.1	3	5.0	3	2.4	SUMTER FL	-38	-52	-71	29	26
6	9.7	3	4.3			SUWANNEE FL	-55				
5	17.0			4	5.3	TAYLOR FL			-69	32	79
1	8.6	2	10.9			UNION FL	26				
60	8.8	69	5.4	83	3.8	VOLUSIA FL	-39	-28	-56	57	56
2	12.9					WAKULLA FL					
4	6.0	2	2.6			WALTON FL	-57				
1	2.5	1	1.6			WASHINGTON FL	-36				
722	7.1	656	4.5	643	3.2	GEORGIA GA	-37	-29	-55		
1	3.0	1	2.1			APPLING GA	-32				
2	10.3					ATKINSON GA					
2	6.9					BACON GA					
		1	8.6			BAKER GA					
3	1.9	4	2.2	5	2.8	BALDWIN GA	12	32	48	96	34
1	3.0	5	14.5	1	2.4	BANKS GA	375	-83	-20	88	27
2	3.2	5	6.1	4	3.8	BARROW GA	90	-37	19	95	56
6	6.3	3	2.4	9	5.1	BARTOW GA	-62	114	-19	88	77
5	10.6	4	5.8	5	6.4	BEN HILL GA	-46	11	-40	78	88
2	4.6			3	4.3	BERRIEN GA			-8	91	64
20	5.5	18	3.7	19	3.0	BIBB GA	-34	-19	-46	72	37
2	10.6	4	10.7			BLECKLEY GA	2				
1	5.9	1	5.3	2	5.5	BRANTLEY GA	-10	3	-7	91	81
				7*	8.7	BROOKS GA					96
2	16.0			2	6.6	BRYAN GA			-59	51	89
5	8.8	3	3.5	1	.8	BULLOCH GA	-61	-77	-91	01	10
3	9.6			5	7.0	BURKE GA			-27	85	91
6	19.6	1	2.0	1	1.8	BUTTS GA	-90	-10	-91	01	17

* SIGNIFICANTLY MORE DEATHS THAN EXPECTED

WHITE FEMALE

CANCER OF THE STOMACH

ICD 151

1950-1959 DEATHS	1950-1959 RATE	1960-1969 DEATHS	1960-1969 RATE	1970-1979 DEATHS	1970-1979 RATE	COUNTY AND STATE	PERCENT CHANGE 60S/50S	PERCENT CHANGE 70S/60S	PERCENT CHANGE 70S/50S	PERCENTILE %CHANGE 70S/50S	PERCENTILE RATES 1970S
1	6.2	1	6.7	1	3.8	CALHOUN GA	8	-44	-39	78	54
1	7.5	3	14.3	1	3.5	CAMDEN GA	92	-75	-53	63	49
		1	3.8	1	2.9	CANDLER GA		-22			36
18	13.9	9	5.1	4	1.6	CARROLL GA	-63	-68	-88	03	15
2	2.5	3	3.4	7	4.6	CATOOSA GA	35	34	80	98	69
31	8.1	27	4.6	16	2.2	CHATHAM GA	-43	-53	-73	25	23
9	13.6	12*	13.9	3	2.6	CHATTOOGA GA	2	-82	-81	12	29
15	17.8	5	3.8	5	2.7	CHEROKEE GA	-79	-29	-85	06	31
4	3.2	3	1.8	3	1.5	CLARKE GA	-44	-19	-55	60	14
1	8.4	2	15.8	1	6.5	CLAY GA	89	-59	-22	87	89
4	5.6	7	4.0	12	4.1	CLAYTON GA	-30	4	-27	85	61
				1	3.9	CLINCH GA					58
11	5.5	8	1.8	26	3.3	COBB GA	-67	85	-39	79	45
4	9.9	2	2.4	5	5.4	COFFEE GA	-75	121	-46	73	80
4	4.0	5	3.8	3	1.9	COLQUITT GA	-5	-51	-53	62	18
1	4.5	1	3.3	3	3.3	COLUMBIA GA	-28	2	-27	86	45
5	14.6	2	4.5	2	3.5	COOK GA	-69	-22	-76	19	48
9	10.8	8	6.8	1	.6	COWETA GA	-37	-91	-94	00	10
2	16.6	1	7.3	1	4.1	CRAWFORD GA	-56	-43	-75	21	62
3	8.2	4	7.0			CRISP GA	-14				
2	8.3			1	2.1	DADE GA			-74	23	22
				2	8.8	DAWSON GA					96
2	4.1	4	5.1	7	8.2	DECATUR GA	26	61	103	98	95
21	3.3	62	5.9	65	4.0	DE KALB GA	80	-32	22	95	59
3	6.1	1	1.8	4	4.9	DODGE GA	-70	171	-19	88	74
3	9.1	2	5.9	1	1.7	DOOLY GA	-35	-71	-81	12	16
5	5.4	7	4.2	7	2.6	DOUGHERTY GA	-23	-37	-52	64	30
4	9.5	6	6.2	4	2.4	DOUGLAS GA	-35	-61	-74	23	27
1	2.8	3	7.4	1	1.8	EARLY GA	169	-76	-35	82	18
1	15.1	1	11.6	1	10.0	ECHOLS GA	-23	-14	-34	82	97
4	14.5	1	2.7	3	5.6	EFFINGHAM GA	-81	106	-62	46	82
5	8.4	4	4.9	4	5.7	ELBERT GA	-42	18	-32	83	83
1	1.9	2	2.7	1	1.5	EMANUEL GA	40	-43	-20	88	15
2	10.0			2	5.5	EVANS GA			-45	73	81
6	10.0	4	5.4	2	1.5	FANNIN GA	-46	-72	-85	06	14
1	4.9			3	4.5	FAYETTE GA			-8	91	68
16	7.7	9	2.9	8	2.1	FLOYD GA	-62	-30	-73	24	21
10*	19.1	2	2.9	4	3.8	FORSYTH GA	-85	31	-80	13	55
2	3.3	3	4.3	7	8.3	FRANKLIN GA	30	94	152	99	95
102	6.4	95	4.5	76	3.2	FULTON GA	-30	-28	-50	67	42
5	12.6	5	10.4	4	6.2	GILMER GA	-17	-41	-51	66	87
1	13.5			1	5.3	GLASCOCK GA			-61	47	78
6	8.6	6	4.5	7	3.8	GLYNN GA	-47	-16	-56	58	54
4	7.1	1	.9	3	2.1	GORDON GA	-87	123	-70	29	22
6	11.7	2	3.2	2	2.4	GRADY GA	-73	-25	-80	14	26
6	16.2	1	2.2	3	6.3	GREENE GA	-86	186	-61	47	87
8	6.9	7	3.9	14	3.7	GWINNETT GA	-43	-5	-46	73	53
4	5.7	8	8.9	2	1.6	HABERSHAM GA	54	-82	-72	27	15
9	5.4	9	4.0	10	2.9	HALL GA	-26	-28	-47	72	35
1	3.7	1	5.5			HANCOCK GA	48				
5	10.1	2	2.5	2	2.1	HARALSON GA	-75	-16	-79	15	21
4	12.3	3	7.5	2	3.7	HARRIS GA	-39	-50	-70	30	53
1	2.3	3	4.7	3	3.3	HART GA	102	-30	42	96	44
		3	10.3			HEARD GA					
4	8.1	7	10.7	2	1.5	HENRY GA	32	-86	-81	12	15
1	3.1	2	.9	6	2.7	HOUSTON GA	-72	205	-13	90	32
6	19.3	4	11.7	2	4.5	IRWIN GA	-39	-61	-77	18	68
6	7.5	6	6.2	7	5.7	JACKSON GA	-18	-7	-24	87	83
				2	6.7	JASPER GA					90
4	16.0	1	3.2	1	1.8	JEFF DAVIS GA	-80	-44	-89	03	18
2	4.5	7*	14.5	1	1.7	JEFFERSON GA	223	-88	-62	45	16
		1	4.5			JENKINS GA					
2	7.7	2	7.2			JOHNSON GA	-7				
2	10.7	2	7.5	3	8.3	JONES GA	-30	11	-23	87	95
2	5.7	2	4.9	2	3.6	LAMAR GA	-14	-26	-37	80	50
2	13.8					LANIER GA					
4	4.6	2	1.6	5	2.8	LAURENS GA	-65	78	-38	80	34
				1	4.7	LEE GA					70
				1	3.3	LIBERTY GA					43
2	18.6	1	3.7	1	4.2	LINCOLN GA	-80	13	-77	17	63
5	5.1	7	5.6	5	2.5	LOWNDES GA	10	-56	-52	64	28
2	6.2	2	4.8	1	2.5	LUMPKIN GA	-23	-47	-59	50	28
5	18.9	2	5.4	1	1.9	MC DUFFIE GA	-71	-65	-90	02	19
1	7.6	1	7.9			MC INTOSH GA	4				
6	22.0					MACON GA					
2	4.3	5	9.2	2	2.7	MADISON GA	112	-71	-38	80	31
3	5.9	2	3.0	4	4.7	MERIWETHER GA	-49	56	-20	88	71
2	10.7	3	10.7	1	3.2	MILLER GA	0	-70	-70	30	42
2	4.1	2	3.5	2	2.2	MITCHELL GA	-15	-37	-47	72	23
4	13.5	1	1.5	5	10.5	MONROE GA	-89	594	-22	87	97
2	9.0	1	4.0	1	3.1	MONTGOMERY GA	-55	-22	-65	38	40
4	13.9	1	2.9	3	6.9	MORGAN GA	-79	140	-51	66	91
1	2.6	5	10.1	3	5.7	MURRAY GA	297	-44	124	98	83
10	3.3	30	7.0	22	3.9	MUSCOGEE#GA	111	-45	17	94	56
2	2.9	7	8.8	5	3.7	NEWTON GA	202	-58	28	95	53
3	13.7	1	2.9	1	1.9	OCONEE GA	-79	-33	-86	05	19
1	3.7	4	11.9	3	6.8	OGLETHORPE GA	225	-43	87	98	91
3	6.6	2	3.1	4	3.8	PAULDING GA	-53	22	-43	75	54
2	7.5	1	3.4	3	6.4	PEACH GA	-54	88	-14	90	88
5	13.3	4	9.1	2	3.6	PICKENS GA	-32	-61	-73	25	50
4	16.3	4	10.9			PIEPCE GA	-33				
2	7.5	2	6.2			PIKE GA	-18				
4	4.2	8	6.2	5	3.2	POLK GA	45	-48	-24	87	42
		1	3.7	1	2.7	PULASKI GA		-26			32
				1	2.8	PUTNAM GA					33
1	28.1					QUITMAN GA					
2	5.9	3	7.3	2	2.1	RABUN GA	23	-71	-65	39	21
4	13.8	1	4.8	1	1.7	RANDOLPH GA	-65	-64	-87	04	17
18	6.9	11	2.5	17	3.0	RICHMOND GA	-63	18	-57	56	37
4	21.0			3	3.9	ROCKDALE GA			-81	12	58
1	10.6			1	7.3	SCHLEY GA			-31	84	92

* SIGNIFICANTLY MORE DEATHS THAN EXPECTED

COUNTIES COMBINED; SEE TECHNICAL NOTES

WHITE FEMALE — CANCER OF THE STOMACH — ICD 151

1950-1959 DEATHS	1950-1959 RATE	1960-1969 DEATHS	1960-1969 RATE	1970-1979 DEATHS	1970-1979 RATE	COUNTY AND STATE	PERCENT CHANGE 60S/50S	PERCENT CHANGE 70S/60S	PERCENT CHANGE 70S/50S	PERCENTILE %CHANGE 70S/50S	PERCENTILE RATES 1970S
				1	1.6	SCREVEN GA					15
1	5.2			2	5.1	SEMINOLE GA			-1	92	77
5	5.1	5	3.4	5	2.4	SPALDING GA	-33	-31	-54	61	26
3	4.9	4	4.7	6	5.0	STEPHENS GA	-4	5	1	92	75
		2	10.5	1	4.4	STEWART GA		-58			66
3	4.9	2	2.4	3	3.9	SUMTER GA	-50	63	-20	88	58
1	7.7	1	8.5	1	4.7	TALBOT GA	11	-45	-39	79	71
1	5.8					TALIAFERRO GA					
6	12.7	1	1.7	4	5.1	TATTNALL GA	-87	210	-59	50	77
4	15.1	3	9.9	1	2.4	TAYLOR GA	-34	-76	-84	07	26
7	25.0	3	5.8	2	3.5	TELFAIR GA	-77	-39	-86	05	48
		3	6.1	2	4.8	TERRELL GA		-21			73
4	4.2	3	2.6	2	1.1	THOMAS GA	-39	-57	-74	24	12
3	6.3	1	1.2	1	.8	TIFT GA	-81	-30	-87	04	10
2	3.4	1	1.3	2	1.8	TOOMBS GA	-63	41	-47	71	17
1	5.2	1	5.6			TOWNS GA	9				
4	33.2					TREUTLEN GA					
8	6.6	6	2.9	8	3.7	TROUP GA	-55	26	-43	75	53
2	6.5	2	6.3			TURNER GA	-3				
		1	5.2	1	4.0	TWIGGS GA		-23			59
6	20.9	5	14.5			UNION GA	-31				
11	14.1	5	5.4	6	4.9	UPSON GA	-62	-9	-65	38	73
12	11.5	12	6.2	13	4.4	WALKER GA	-46	-30	-62	45	65
4	6.0	2	2.1	5	3.8	WALTON GA	-65	81	-37	81	55
2	3.0	6	4.5	3	1.7	WARE GA	54	-64	-44	74	16
4	21.4			1	4.8	WARREN GA			-78	17	72
4	11.6	2	2.7	1	1.3	WASHINGTON GA	-77	-54	-89	02	12
3	6.4	2	3.7	4	4.5	WAYNE GA	-42	20	-30	84	67
2	32.6					WEBSTER GA					
3	31.0			1	5.3	WHEELER GA			-83	09	79
2	7.0	1	2.6	1	2.2	WHITE GA	-63	-15	-69	33	23
18	14.7	13	6.5	10	3.6	WHITFIELD GA	-56	-44	-75	20	50
1	3.5	1	2.7	3	8.1	WILCOX GA	-22	196	131	99	94
5	14.6	3	7.3	1	2.3	WILKES GA	-50	-68	-84	08	25
1	3.6	2	7.8	3	8.5	WILKINSON GA	114	9	134	99	95
8	19.7	4	8.5			WORTH GA	-57				
		47	7.9	62*	6.4	HAWAII HI		-19			
		6	11.3	9*	10.5	HAWAII HI		-7			97
		38	7.9	49*	6.4	HONOLULU HI		-18			88
				1	3.1	KAUAI HI					41
		3	7.4	3	4.2	MAUI#HI		-44			62
232	10.0	190	5.9	152	3.5	IDAHO ID	-41	-40	-65		
27	7.7	26	5.0	26	3.6	ADA ID	-35	-28	-53	62	50
1	11.2			1	6.7	ADAMS ID			-40	78	90
15	13.9	15	8.0	9	3.5	BANNOCK ID	-43	-56	-75	21	48
4	17.7	3	11.7	3	7.7	BEAR LAKE ID	-34	-34	-57	56	93
5	21.4	1	3.7			BENEWAH ID	-83				
6	6.9	7	7.0	9	6.7	BINGHAM ID	1	-4	-2	92	90
3	16.9	1	4.0			BLAINE ID	-76				
4	5.3	5	5.8	3	2.5	BONNER ID	9	-56	-52	63	29
15	13.7	11	6.6	8	3.9	BONNEVILLE ID	-52	-40	-71	27	58
1	3.2	2	6.9	1	1.8	BOUNDARY ID	113	-74	-44	75	18
1	22.5					BUTTE ID					
19	7.5	11	3.4	16	4.0	CANYON ID	-54	15	-47	70	59
2	11.8	1	3.0			CARIBOU ID	-75				
7	15.6	10*	13.9	1	.8	CASSIA ID	-11	-94	-95	00	10
1	26.8					CLARK ID					
2	4.1	1	2.8	1	1.6	CLEARWATER ID	-31	-44	-62	46	15
		2	16.2			CUSTER ID					
6*	25.9	5	14.4			ELMORE ID	-44				
3	8.6	6*	14.8	2	4.1	FRANKLIN ID	72	-72	-53	63	61
1	3.4	3	9.0	3	6.5	FREMONT ID	167	-28	93	98	89
3	7.1	2	3.4	1	1.4	GEM ID	-52	-59	-80	13	14
5	10.9	3	5.1	1	1.2	GOODING ID	-53	-76	-89	03	12
4	9.0	4	8.5	1	1.5	IDAHO ID	-5	-82	-83	08	14
5	21.3	3	7.0	1	2.0	JEFFERSON ID	-67	-71	-91	02	20
4	8.0	6	11.1	1	2.1	JEROME ID	39	-81	-74	23	21
10	7.1	7	3.4	12	3.9	KOOTENAI ID	-52	13	-46	73	57
17*	22.1	11	10.0	4	2.5	LATAH ID	-55	-74	-88	03	29
5	25.0	1	3.9	3	8.1	LEMHI ID	-84	107	-67	35	94
2	10.2	1	5.0	3*	11.3	LEWIS ID	-51	127	10	94	98
1	9.9	1	8.6			LINCOLN ID	-13				
2	7.6			4	7.9	MADISON ID			4	93	94
5	11.5			5	7.0	MINIDOKA ID			-40	78	91
7	6.2	16*	10.8	11	5.8	NEZ PERCE ID	75	-47	-6	91	83
2	9.2			2	7.8	ONEIDA ID			-15	89	94
3	16.6	3	14.8			OWYHEE ID	-11				
6	10.5	1	.8	2	1.8	PAYETTE ID	-92	114	-83	09	17
1	6.4	1	6.5			POWER ID	3				
9	22.4	5	6.5	7	8.3	SHOSHONE ID	-71	26	-63	43	95
13	8.1	14	5.2	9	2.4	TWIN FALLS ID	-35	-54	-70	29	27
1	4.3					VALLEY ID					
4	8.5	1	1.9	2	4.0	WASHINGTON ID	-78	112	-52	63	60
5091*	11.7	3954*	7.4	2876*	4.5	ILLINOIS IL	-37	-39	-62		
33	7.4	18	3.4	18	3.4	ADAMS IL	-55	2	-54	61	47
6	7.7	4	4.7	1	2.0	ALEXANDER IL	-39	-58	-74	23	20
10	10.0	6	5.0	7	4.9	BOND IL	-50	-2	-51	65	73
12	10.7	13	10.4	6	3.6	BOONE IL	-2	-65	-66	37	51
4	6.8	4	6.6	2	2.2	BROWN IL	-4	-67	-68	33	23
21	8.8	13	4.9	11	3.4	BUREAU IL	-44	-32	-62	45	45
1	3.2			2	6.1	CALHOUN IL			90	98	86
7	5.6	10	7.0	3	1.7	CARPOLL IL	25	-75	-69	32	17
11	11.2	7	6.3	6	3.5	CASS IL	-44	-44	-69	33	48
39	9.3	14	2.5	18	2.7	CHAMPAIGN IL	-73	9	-71	28	31
16	6.6	17	6.1	10	3.2	CHRISTIAN IL	-9	-46	-51	65	43

* SIGNIFICANTLY MORE DEATHS THAN EXPECTED

COUNTIES COMBINED; SEE TECHNICAL NOTES

WHITE FEMALE

CANCER OF THE STOMACH

ICD 151

COUNTY AND STATE	1950-1959 DEATHS	1950-1959 RATE	1960-1969 DEATHS	1960-1969 RATE	1970-1979 DEATHS	1970-1979 RATE	PERCENT CHANGE 60S/50S	PERCENT CHANGE 70S/60S	PERCENT CHANGE 70S/50S	PERCENTILE %CHANGE 70S/50S	PERCENTILE RATES 1970S
CLARK IL	7	5.3	13	9.2	3	1.7	75	-81	-68	35	16
CLAY IL	11	11.0	7	5.6	2	1.3	-49	-78	-89	03	12
CLINTON IL	13	12.4	7	4.3	5	2.3	-65	-46	-81	11	25
COLES IL	16	6.2	19	6.6	7	2.0	6	-70	-68	33	20
COOK IL	2897*	14.5	2256*	9.3	1606*	5.8	-36	-38	-60	49	84
CRAWFORD IL	20	14.6	9	5.6	4	2.2	-61	-62	-85	06	22
CUMBERLAND IL	6	9.9	4	5.6	3	4.2	-44	-24	-57	55	63
DE KALB IL	22	9.4	18	6.5	11	2.9	-31	-56	-70	30	34
DE WITT IL	4	3.6	6	4.6	3	2.1	27	-55	-43	76	21
DOUGLAS IL	12	12.4	5	4.3	4	2.3	-65	-47	-82	11	25
DU PAGE IL	84	10.1	98	6.8	99	4.2	-33	-39	-59	52	63
EDGAR IL	16	10.0	9	4.7	5	2.5	-53	-46	-74	22	29
EDWARDS IL	4	7.1			2	3.8			-46	72	55
EFFINGHAM IL	4	4.1	6	3.9	11	5.4	-5	37	30	95	79
FAYETTE IL	26*	20.0	6	3.7	5	2.5	-82	-31	-87	04	28
FORD IL	4	4.2	11	9.5	3	2.4	127	-74	-42	77	27
FRANKLIN IL	25	9.9	23	7.2	16	4.2	-28	-42	-58	53	62
FULTON IL	22	8.1	10	3.3	9	2.6	-59	-22	-68	34	30
GALLATIN IL	6	12.0	7*	14.8	2	3.2	23	-79	-74	24	41
GREENE IL	13	8.9	12	8.3	3	2.1	-6	-75	-77	18	21
GRUNDY IL	21*	20.0	11	8.6	6	3.7	-57	-57	-81	11	53
HAMILTON IL	7	8.4	6	7.6	5	4.9	-10	-35	-41	77	74
HANCOCK IL	13	6.8	5	2.5	2	1.0	-63	-62	-86	05	11
HARDIN IL	4	11.6	3	7.2	2	5.1	-38	-30	-57	56	76
HENDERSON IL	5	11.1	3	6.6			-40				
HENRY IL	39	13.6	22	6.1	13	3.5	-55	-42	-74	24	49
IROQUOIS IL	16	9.5	11	4.8	10	3.8	-50	-20	-60	50	56
JACKSON IL	21	10.7	14	5.9	7	2.6	-45	-56	-76	20	30
JASPER IL	12	15.2	7	7.9	4	3.2	-48	-60	-79	15	42
JEFFERSON IL	22	12.3	15	7.0	22*	8.3	-43	18	-33	83	95
JERSEY IL	6	7.7	6	6.7	7	6.3	-13	-6	-19	89	87
JO DAVIESS IL	14	10.1	12	7.5	2	1.1	-25	-86	-89	02	11
JOHNSON IL	7	14.3	3	5.3	2	2.1	-63	-60	-85	06	22
KANE IL	105	10.9	50	4.3	65	4.5	-61	5	-59	52	68
KANKAKEE IL	34	7.7	21	3.9	16	2.8	-49	-28	-63	43	34
KENDALL IL	16*	22.3	13*	13.5	7	5.1	-40	-62	-77	18	77
KNOX IL	37	10.4	38	8.5	12	2.1	-19	-75	-80	14	21
LAKE IL	78	10.2	74	6.3	66	4.1	-38	-35	-60	49	60
LA SALLE IL	54	9.3	64*	9.1	29	3.2	-2	-65	-65	38	42
LAWRENCE IL	9	7.3	8	5.0	4	2.0	-32	-60	-73	25	20
LEE IL	22	11.1	10	4.4	9	2.8	-60	-37	-75	22	33
LIVINGSTON IL	13	5.9	13	4.7	10	3.2	-19	-32	-45	73	43
LOGAN IL	12	6.5	3	1.4	9	3.4	-78	137	-48	69	45
MC DONOUGH IL	18	9.5	7	2.7	8	2.9	-71	7	-69	31	36
MC HENRY IL	40	12.6	37	7.7	32	4.9	-39	-36	-61	47	74
MC LEAN IL	37	7.8	26	4.5	25	3.8	-42	-15	-51	66	55
MACON IL	51	9.1	30	4.5	22	2.7	-51	-40	-71	28	31
MACOUPIN IL	25	8.0	17	4.4	19	4.3	-45	-4	-47	71	64
MADISON IL	60	7.0	52	4.6	48	3.2	-35	-30	-54	61	42
MARION IL	35*	14.9	24	7.8	17	4.8	-48	-38	-68	34	72
MARSHALL IL	11	14.5	4	3.8	4	3.7	-74	-2	-74	23	53
MASON IL	5	5.3	9	8.0	3	1.8	51	-78	-67	36	17
MASSAC IL	11	15.7	6	6.2	2	1.8	-61	-71	-89	03	17
MENARD IL	6	9.9	1	.9	1	.6	-91	-36	-94	00	10
MERCER IL	11	10.0	7	5.6	8	5.4	-44	-4	-46	72	80
MONROE IL	12	15.1	6	5.7	3	2.7	-62	-52	-82	11	32
MONTGOMERY IL	23	9.7	14	5.0	8	2.2	-49	-57	-78	17	22
MORGAN IL	19	6.8	13	4.2	13	4.8	-39	16	-29	85	72
MOULTRIE IL	9	8.9	6	4.3	7	5.1	-51	19	-42	76	77
OGLE IL	14	7.9	14	6.2	8	2.9	-21	-53	-63	44	36
PEORIA IL	69	7.4	50	4.4	39	3.0	-40	-33	-60	49	37
PERRY IL	13	10.7	7	4.8	3	1.9	-55	-61	-82	10	19
PIATT IL	5	4.9	7	6.6	2	2.1	36	-69	-58	55	21
PIKE IL	7	5.4	8	4.8	3	2.4	-12	-50	-56	57	26
POPE IL	6	18.7									
PULASKI IL	4	8.7	4	8.2			-6				
PUTNAM IL	2	7.8	4	14.2	1	3.0	81	-79	-61	46	38
RANDOLPH IL	15	9.2	11	5.4	10	3.4	-41	-38	-63	43	46
RICHLAND IL	7	7.1	7	5.4	4	3.6	-24	-33	-49	68	50
ROCK ISLAND IL	73	10.6	54	6.2	54	5.2	-41	-17	-51	65	78
ST CLAIR IL	85	9.4	72	6.4	45	3.4	-33	-46	-64	42	47
SALINE IL	15	9.0	19	8.1	4	1.5	-10	-82	-84	08	14
SANGAMON IL	49	6.6	55	5.5	41	3.7	-17	-32	-44	75	53
SCHUYLER IL	6	9.5	2	2.6	2	2.3	-73	-10	-76	20	25
SCOTT IL	5	10.5	3	5.8	1	1.7	-45	-70	-83	08	17
SHELBY IL	15	8.5	8	4.6	9	4.4	-46	-5	-49	68	65
STARK IL	8	13.6	1	1.3	4	5.5	-90	312	-60	49	81
STEPHENSON IL	32	11.4	21	6.6	20	5.5	-42	-17	-52	64	81
TAZEWELL IL	20	6.2	22	4.9	19	3.0	-21	-39	-52	64	36
UNION IL	5	3.4	5	3.4	3	1.5	0	-55	-55	59	15
VERMILION IL	46	9.3	36	6.0	25	3.7	-36	-39	-61	47	51
WABASH IL	12	13.7	7	7.0	5	5.3	-49	-23	-61	47	79
WARREN IL	3	1.8	6	3.9	3	1.7	113	-55	-5	92	17
WASHINGTON IL	10	9.5	7	5.6	8	4.9	-42	-12	-49	69	74
WAYNE IL	9	8.4	9	7.2	5	3.8	-14	-47	-55	60	55
WHITE IL	8	6.2	8	5.7	7	3.8	-8	-34	-39	79	54
WHITESIDE IL	18	6.7	15	4.6	4	1.0	-32	-77	-84	07	11
WILL IL	84	12.6	97*	10.9	58	4.7	-14	-57	-63	44	71
WILLIAMSON IL	27	9.2	20	6.0	17	3.5	-35	-41	-62	46	49
WINNEBAGO IL	96	12.0	64	6.0	49	3.6	-50	-40	-70	29	49
WOODFORD IL	11	8.2	8	4.0	7	3.3	-51	-17	-59	51	45
INDIANA IN	1830	9.1	1316	5.2	1038	3.4	-43	-34	-62		
ADAMS IN	7	5.8	6	4.2	11	6.0	-29	44	3	93	85
ALLEN IN	85	9.6	60	4.9	47	3.1	-49	-37	-68	34	40
BARTHOLOMEW IN	14	7.4	12	4.9	10	3.4	-34	-31	-54	61	46
BENTON IN	5	7.6	4	5.3	5	6.3	-30	20	-16	89	68
BLACKFORD IN	6	7.2	7	7.4	1	.8	3	-90	-90	02	10
BOONE IN	15	9.0	7	3.3	9	3.4	-63	2	-63	44	45
BROWN IN	3	8.7	1	2.1	1	2.5	-76	21	-71	27	28

* SIGNIFICANTLY MORE DEATHS THAN EXPECTED

WHITE FEMALE — CANCER OF THE STOMACH — ICD 151

1950-1959 DEATHS	1950-1959 RATE	1960-1969 DEATHS	1960-1969 RATE	1970-1979 DEATHS	1970-1979 RATE	COUNTY AND STATE	PERCENT CHANGE 60S/50S	PERCENT CHANGE 70S/60S	PERCENT CHANGE 70S/50S	PERCENTILE %CHANGE 70S/50S	PERCENTILE RATES 1970S
3	2.7	2	1.5	1	.9	CARROLL IN	-43	-44	-68	33	10
22	8.0	12	3.8	13	4.6	CASS IN	-53	23	-42	76	70
12	5.9	15	5.3	22	5.6	CLARK IN	-10	5	-6	92	82
13	7.5	15	7.3	13	6.8	CLAY IN	-3	-7	-10	91	91
14	7.5	15	6.9	11	3.5	CLINTON IN	-8	-49	-53	62	48
6	10.7	2	4.0	2	2.8	CRAWFORD IN	-63	-31	-74	23	33
14	9.3	9	5.0	6	3.1	DAVIESS IN	-47	-37	-66	36	40
13	9.8	11	6.6	12	6.5	DEARBORN IN	-33	-2	-34	82	88
5	3.8	10	6.9	5	2.8	DECATUR IN	83	-60	-27	86	33
16	9.9	8	4.6	6	2.9	DE KALB IN	-54	-37	-71	28	35
31	7.6	14	2.5	20	3.1	DELAWARE IN	-67	24	-59	51	40
9	7.6	5	3.4	6	3.8	DUBOIS IN	-56	14	-49	68	55
39	8.4	21	3.4	15	2.0	ELKHART IN	-60	-40	-76	20	20
16	17.2	11	8.0	6	4.2	FAYETTE IN	-54	-47	-75	20	63
19	7.6	9	3.0	15	4.1	FLOYD IN	-61	37	-46	72	62
11	9.3	9	6.6	6	5.6	FOUNTAIN IN	-29	-15	-40	78	82
7	8.0	3	3.0	3	2.4	FRANKLIN IN	-62	-19	-70	31	27
7	6.3	6	5.3	5	3.3	FULTON IN	-15	-38	-47	70	44
23	12.8	10	4.6	5	2.1	GIBSON IN	-64	-55	-84	08	21
33	10.4	17	4.4	13	2.6	GRANT IN	-58	-40	-75	22	30
20	11.1	20	8.9	5	2.2	GREENE IN	-20	-75	-80	13	23
14	8.2	14	6.1	10	3.2	HAMILTON IN	-25	-48	-61	47	42
5	3.6	8	5.0	4	2.0	HANCOCK IN	41	-61	-45	74	20
8	8.9	3	2.6	2	1.5	HARRISON IN	-71	-44	-83	08	14
9	5.9	5	2.4	8	2.7	HENDRICKS IN	-59	11	-55	59	31
18	7.8	12	4.3	4	1.0	HENRY IN	-46	-76	-87	04	11
21	7.6	15	4.3	11	2.4	HOWARD IN	-43	-43	-68	34	27
22	9.8	9	3.5	9	2.3	HUNTINGTON IN	-64	-34	-76	19	26
9	6.0	12	6.8	10	4.2	JACKSON IN	12	-38	-30	84	63
2	2.2	8	8.0	2	1.8	JASPER IN	266	-77	-17	89	18
7	5.9	5	3.4	7	4.5	JAY IN	-42	31	-24	87	67
14	9.8	5	3.1	3	1.5	JEFFERSON IN	-68	-53	-85	07	14
5	6.9	2	2.1	6	5.8	JENNINGS IN	-70	184	-15	89	84
16	8.8	9	4.0	5	1.3	JOHNSON IN	-55	-68	-86	06	13
29	11.3	18	6.1	11	3.1	KNOX IN	-46	-49	-73	25	40
25	12.0	11	4.6	6	2.1	KOSCIUSKO IN	-62	-55	-83	09	21
6	6.8	3	3.2	4	3.1	LAGRANGE IN	-54	-3	-55	59	39
173*	15.1	137*	8.5	120*	6.0	LAKE IN	-44	-30	-60	48	85
43	11.6	42	8.4	27	4.4	LA PORTE IN	-27	-47	-62	46	66
17	9.5	9	4.1	5	1.9	LAWRENCE IN	-57	-54	-80	13	19
42	8.2	37	5.7	25	3.0	MADISON IN	-30	-47	-63	42	38
207	7.9	161	4.9	121	3.1	MARION IN	-38	-37	-61	47	39
15	8.8	5	2.3	5	1.8	MARSHALL IN	-74	-23	-80	14	17
3	6.3	3	5.3	2	3.3	MARTIN IN	-17	-37	-48	70	44
11	6.0	12	6.0	6	2.4	MIAMI IN	-1	-60	-60	50	27
14	7.7	16	6.2	11	3.0	MONROE IN	-20	-52	-61	46	37
19	9.8	12	5.4	9	3.9	MONTGOMERY IN	-46	-26	-60	49	58
13	9.5	8	4.2	5	2.3	MORGAN IN	-55	-46	-76	19	24
5	8.3			2	2.3	NEWTON IN			-72	27	25
18	10.1	9	4.9	6	2.6	NOBLE IN	-51	-46	-74	24	30
4	12.9	4	11.9	1	2.6	OHIO IN	-8	-79	-80	13	29
9	9.3	3	3.3	6	4.8	ORANGE IN	-64	43	-48	69	72
11	13.5	4	4.0	5	5.0	OWEN IN	-70	25	-63	43	76
9	8.2	6	4.8	6	4.9	PARKE IN	-42	2	-41	77	74
10	10.4	6	6.2	5	3.2	PERRY IN	-40	-49	-69	31	42
11	12.4	8	8.7	4	3.8	PIKE IN	-29	-56	-69	32	55
10	5.2	17	6.1	21	5.0	PORTER IN	18	-18	-3	92	76
17	14.4	6	4.6	11	7.9	POSEY IN	-68	73	-45	73	94
7	10.3	3	3.6	5	4.3	PULASKI IN	-65	18	-58	53	64
9	6.8	9	5.9	6	2.7	PUTNAM IN	-13	-54	-60	49	32
10	5.8	6	3.1	5	2.4	RANDOLPH IN	-47	-22	-59	52	26
15	12.4	10	6.6	4	2.6	RIPLEY IN	-46	-61	-79	15	29
5	4.3	10	6.7	10	6.8	RUSH IN	56	2	59	97	90
113	12.6	78	6.4	66	4.5	ST JOSEPH IN	-49	-30	-64	40	67
1	1.9	3	4.2	2	2.2	SCOTT IN	118	-46	16	94	24
14	8.0	7	3.0	5	2.1	SHELBY IN	-62	-31	-74	23	21
14	14.9	8	7.7	5	4.7	SPENCER IN	-48	-39	-68	33	71
7	8.4	8	6.1	6	4.6	STARKE IN	-27	-24	-45	73	69
8	8.9	2	1.5	6	4.1	STEUBEN IN	-83	176	-54	61	62
20	12.4	10	5.5	4	3.6	SULLIVAN IN	-55	-36	-71	27	49
8	15.0	4	7.1			SWITZERLAND IN	-52				
21	5.7	19	4.3	10	2.0	TIPPECANOE IN	-24	-54	-65	39	20
6	6.9	8	8.2	6	5.1	TIPTON IN	18	-38	-26	86	77
6	14.9	2	3.8	2	3.7	UNION IN	-74	-4	-75	20	52
49	6.3	47	4.9	35	2.9	VANDERBURGH IN	-23	-40	-53	62	36
11	10.0	11	8.0	7	4.0	VERMILLION IN	-20	-50	-60	50	60
61	9.8	39	5.4	24	3.1	VIGO IN	-45	-42	-68	33	40
24	13.5	14	5.3	6	2.3	WABASH IN	-61	-57	-83	09	24
5	11.0	4	9.8	2	3.8	WARREN IN	-11	-61	-65	39	56
11	9.0	12	8.8	4	2.3	WARRICK IN	-3	-74	-74	23	25
4	4.0	3	2.4	4	3.1	WASHINGTON IN	-41	30	-23	87	39
29	7.4	21	4.6	20	3.8	WAYNE IN	-38	-18	-49	68	54
4	3.0	5	3.8			WELLS IN	28				
12	11.7	2	1.4	7	5.0	WHITE IN	-88	248	-58	54	74
17	16.3	6	4.1	8	5.3	WHITLEY IN	-74	29	-67	35	79
1459	9.5	1050	5.6	767	3.5	IOWA IA	-41	-38	-63		
4	5.1	9	9.9	3	2.1	ADAIR IA	92	-78	-58	53	22
2	3.8	4	6.7	6*	11.5	ADAMS IA	78	71	204	99	98
13	13.2	14*	11.3	6	3.5	ALLAMAKEE IA	-14	-69	-73	25	49
33*	23.2	8	4.6	11	6.1	APPANOOSE IA	-80	32	-74	24	86
3	4.9	5	7.1	5	6.5	AUDUBON IA	43	-8	31	96	89
11	8.0	10	6.3	7	3.3	BENTON IA	-21	-48	-59	51	44
44	8.9	38	5.9	24	3.1	BLACK HAWK IA	-34	-48	-66	37	39
12	6.3	10	4.9	8	3.3	BOONE IA	-22	-33	-48	69	43
10	8.7	4	2.4	14*	8.1	BREMER IA	-72	231	-7	91	94
6	5.0	8	6.0	2	1.3	BUCHANAN IA	21	-79	-74	23	13
13	10.9	4	3.0	8	4.8	BUENA VISTA IA	-72	60	-56	58	73
7	6.4	5	4.3	3	1.8	BUTLER IA	-32	-59	-72	27	17
12	11.8	4	3.2	4	1.9	CALHOUN IA	-73	-42	-84	07	18

* SIGNIFICANTLY MORE DEATHS THAN EXPECTED

WHITE FEMALE | CANCER OF THE STOMACH | ICD 151

1950-1959 DEATHS	1950-1959 RATE	1960-1969 DEATHS	1960-1969 RATE	1970-1979 DEATHS	1970-1979 RATE	COUNTY AND STATE	PERCENT CHANGE 60S/50S	PERCENT CHANGE 70S/60S	PERCENT CHANGE 70S/50S	PERCENTILE %CHANGE 70S/50S	PERCENTILE RATES 1970S
12	10.1	14	9.6	9	5.0	CARROLL IA	-5	-48	-50	67	76
10	8.7	12	8.0	6	3.6	CASS IA	-8	-55	-59	52	50
12	11.1	6	4.2	7	5.7	CEDAR IA	-62	36	-49	69	83
31	12.2	29	8.2	14	3.1	CERRO GORDO IA	-32	-62	-74	23	41
11	9.0	11	8.3	2	1.8	CHEROKEE IA	-8	-78	-80	13	18
10	10.8	4	3.5	7	6.3	CHICKASAW IA	-68	79	-42	76	87
10	15.1	2	2.2			CLARKE IA	-85				
9	10.0	5	4.6	6	3.9	CLAY IA	-54	-14	-61	47	57
17	12.4	8	5.2	1	.5	CLAYTON IA	-58	-91	-96	00	09
26	9.1	27	8.1	20	4.8	CLINTON IA	-11	-41	-48	70	72
15	15.9	9	7.9	5	3.2	CRAWFORD IA	-50	-59	-80	14	42
7	4.3	19	8.9	5	2.7	DALLAS IA	105	-70	-38	80	31
5	7.4	4	5.6	3	3.2	DAVIS IA	-24	-43	-57	55	41
4	5.2	8	9.5	2	2.1	DECATUR IA	81	-78	-61	48	21
7	6.9	6	4.4	3	2.4	DELAWARE IA	-36	-46	-65	38	26
21	7.9	19	5.5	16	4.6	DES MOINES IA	-31	-16	-42	76	69
9	13.3	4	4.2	6	4.4	DICKINSON IA	-69	7	-67	36	67
43	10.1	28	5.3	22	3.9	DUBUQUE IA	-47	-28	-62	46	56
12	16.5	5	5.3	9	7.3	EMMET IA	-68	39	-56	58	92
12	7.2	4	2.2	8	3.1	FAYETTE IA	-70	42	-57	56	40
7	5.7	3	2.8	2	.8	FLOYD IA	-51	-70	-85	06	10
8	8.7	5	5.1	4	3.5	FRANKLIN IA	-41	-31	-59	51	49
8	10.6	3	4.3	3	5.1	FREMONT IA	-59	18	-52	64	76
13	13.2	6	5.6	3	2.5	GREENE IA	-58	-55	-81	12	28
13	17.0	7	6.8	3	1.9	GRUNDY IA	-60	-72	-89	03	19
10	10.5	6	5.2	1	.8	GUTHRIE IA	-51	-85	-93	01	10
14	12.1	5	4.0	5	3.4	HAMILTON IA	-67	-15	-72	26	45
8	11.5	3	3.4	5	4.3	HANCOCK IA	-70	25	-63	43	64
11	7.8	6	3.2	7	3.3	HARDIN IA	-59	2	-58	53	43
11	9.1	6	3.7	7	4.9	HARRISON IA	-59	33	-46	72	74
13	8.0	8	4.3	4	2.6	HENRY IA	-46	-39	-67	35	30
7	8.1	7	7.9	6	4.9	HOWARD IA	-3	-39	-40	78	74
12	17.1	10	11.1	5	4.5	HUMBOLDT IA	-35	-59	-74	24	68
7	11.4	2	2.8	3	2.7	IDA IA	-75	-5	-77	18	31
8	7.5	9	7.4	7	4.5	IOWA IA	-2	-40	-41	77	67
16	14.0	11	7.7	9	5.1	JACKSON IA	-45	-35	-64	41	76
14	8.3	9	3.9	8	2.7	JASPER IA	-52	-32	-68	35	31
4	4.3	3	2.3	3	2.3	JEFFERSON IA	-47	0	-48	70	24
20	9.3	15	5.7	22*	6.3	JOHNSON IA	-39	11	-33	83	87
10	9.1	11	8.3	2	1.2	JONES IA	-9	-86	-87	05	12
7	5.6	8	5.1	6	5.1	KEOKUK IA	-9	0	-10	90	77
16	13.5	7	4.7	6	3.2	KOSSUTH IA	-65	-32	-76	19	42
24	9.5	17	5.7	9	2.8	LEE IA	-40	-51	-71	29	33
50	7.8	46	5.4	38	3.8	LINN IA	-31	-29	-51	66	56
2	3.4	2	2.5	4	5.6	LOUISA IA	-29	129	64	97	82
14	14.6	13*	12.5	3	2.7	LUCAS IA	-14	-78	-81	11	32
8	12.1	8	9.8	2	2.4	LYON IA	-19	-76	-80	13	26
12	12.4	3	2.5	3	2.5	MADISON IA	-80	0	-80	14	28
15	8.1	7	4.1	3	1.7	MAHASKA IA	-49	-60	-79	14	16
19	12.3	8	4.6	5	2.5	MARION IA	-63	-45	-80	14	29
17	7.5	10	3.8	6	1.8	MARSHALL IA	-49	-53	-76	19	17
6	7.6	1	1.1	1	.8	MILLS IA	-86	-26	-90	02	10
14	15.9	13*	14.3	6	3.9	MITCHELL IA	-10	-73	-76	20	56
7	9.0	5	4.9	2	2.8	MONONA IA	-45	-43	-69	32	34
6	7.8	7	6.9	2	3.8	MONROE IA	-10	-45	-51	66	55
4	3.3	4	3.2	8	5.4	MONTGOMERY IA	-2	69	65	97	80
21	9.2	24	9.5	10	3.1	MUSCATINE IA	3	-67	-66	37	40
16	15.0	7	6.2	7	3.5	O'BRIEN IA	-59	-44	-77	18	47
8	17.9	7	11.7	2	3.4	OSCEOLA IA	-35	-71	-81	12	45
5	2.9	6	2.4	2	1.1	PAGE IA	-17	-55	-63	43	11
6	7.7	7	7.6	7	5.0	PALO ALTO IA	-1	-34	-34	82	76
12	9.9	10	6.4	11	4.8	PLYMOUTH IA	-36	-24	-51	65	73
9	12.1	7	6.8	6	4.1	POCAHONTAS IA	-44	-39	-66	37	62
103	8.4	65	4.3	61	3.4	POLK IA	-49	-19	-59	52	47
20	5.2	23	4.8	13	2.3	POTTAWATTAMIE IA	-7	-52	-56	58	25
2	1.4	7	5.4	4	2.7	POWESHIEK IA	286	-49	97	98	32
9	14.4	2	2.6	4	6.8	RINGGOLD IA	-82	157	-53	62	90
12	11.4	4	3.5	3	1.9	SAC IA	-69	-45	-83	09	19
47	8.7	29	4.1	18	2.1	SCOTT IA	-53	-48	-76	20	22
11	12.9	15*	14.7	13*	8.0	SHELBY IA	14	-45	-38	80	94
15	12.3	13	7.9	11	5.3	SIOUX IA	-36	-33	-57	55	79
28	12.4	11	4.1	11	3.3	STORY IA	-67	-19	-73	24	45
15	11.7	17*	10.4	7	3.8	TAMA IA	-11	-63	-67	35	55
12	13.7	4	3.1	3	2.9	TAYLOR IA	-77	-8	-79	15	35
5	4.7	7	4.8	6	4.0	UNION IA	2	-17	-16	89	59
12	13.9	5	5.4			VAN BUREN IA	-61				
19	7.2	16	5.0	19	4.9	WAPELLO IA	-31	-1	-31	84	74
11	10.4	6	4.9	8	5.4	WARREN IA	-53	12	-48	70	80
14	9.7	13	8.0	6	3.6	WASHINGTON IA	-18	-56	-63	42	49
7	6.8	9	8.0	4	4.4	WAYNE IA	18	-46	-36	81	66
32	12.7	20	5.7	13	3.6	WEBSTER IA	-55	-37	-72	27	49
14*	18.1	8	7.3	5	3.4	WINNEBAGO IA	-60	-53	-81	12	46
24*	18.9	13	9.7	9	4.9	WINNESHIEK IA	-49	-50	-74	23	74
62	11.3	36	5.5	27	3.4	WOODBURY IA	-51	-38	-70	30	46
5	8.4	9*	13.9	5	5.3	WORTH IA	65	-62	-38	80	78
15	13.7	9	6.6	7	3.7	WRIGHT IA	-51	-44	-73	25	53
849	7.9	631	4.7	489	3.1	KANSAS KS	-41	-34	-61		
9	6.5	4	2.3	6	3.6	ALLEN KS	-65	60	-45	74	51
2	2.5	3	2.7			ANDERSON KS	10				
14	11.3	7	4.5	3	2.8	ATCHISON KS	-60	-38	-75	21	33
4	8.6	1	1.4	3	4.2	BARBER KS	-84	203	-51	65	63
14	11.1	9	5.7	6	2.7	BARTON KS	-49	-52	-75	21	32
7	5.2	13	9.6	9	7.0	BOURBON KS	85	-27	35	96	91
10	7.8	3	2.9	3	2.0	BROWN KS	-62	-33	-75	22	20
9	5.6	12	5.2	7	2.5	BUTLER KS	-7	-53	-56	57	27
3	9.5	1	2.5	1	1.6	CHASE KS	-74	-34	-83	09	16
3	5.3	2	3.7	2	3.7	CHAUTAUQUA KS	-31	-1	-31	84	52
14	9.0	13	6.9	9	4.3	CHEROKEE KS	-23	-37	-51	65	65
3	13.6	2	7.3			CHEYENNE KS	-46				

* SIGNIFICANTLY MORE DEATHS THAN EXPECTED

WHITE FEMALE — CANCER OF THE STOMACH — ICD 151

1950-1959 DEATHS	1950-1959 RATE	1960-1969 DEATHS	1960-1969 RATE	1970-1979 DEATHS	1970-1979 RATE	COUNTY AND STATE	PERCENT CHANGE 60S/50S	PERCENT CHANGE 70S/60S	PERCENT CHANGE 70S/50S	PERCENTILE %CHANGE 70S/50S	PERCENTILE RATES 1970S
3	14.8	2	9.8			CLARK KS	-33				
6	7.3	5	5.3	3	2.9	CLAY KS	-27	-45	-60	50	36
10	7.5	7	4.6	5	1.9	CLOUD KS	-39	-59	-75	21	19
6	7.4	7	9.7	5	5.6	COFFEY KS	30	-42	-24	87	82
4	24.7					COMANCHE KS					
14	5.6	10	3.4	6	1.7	COWLEY KS	-40	-49	-69	32	16
19	6.8	12	3.4	11	2.9	CRAWFORD KS	-50	-14	-57	55	35
4	10.4	2	4.3	3	5.6	DECATUR KS	-58	29	-46	72	82
17	10.7	12	6.7	7	4.0	DICKINSON KS	-37	-41	-63	44	59
7	10.4	6	9.9			DONIPHAN KS	-5				
20	10.7	13	5.7	6	2.2	DOUGLAS KS	-47	-61	-79	15	24
2	5.0	2	5.4			EDWARDS KS	7				
3	5.9	5	9.8	2	3.7	ELK KS	66	-62	-37	81	53
10	14.9	9	10.6	6	5.3	ELLIS KS	-29	-49	-64	40	79
3	4.7	3	4.2	3	4.0	ELLSWORTH KS	-10	-4	-14	90	60
2	2.8	4	5.2	2	1.3	FINNEY KS	84	-75	-54	61	13
8	8.2	4	3.5	5	3.1	FORD KS	-57	-13	-63	43	39
10	6.7	7	4.3	7	4.3	FRANKLIN KS	-35	0	-36	81	65
6	8.1	5	5.1	3	2.9	GEARY KS	-37	-44	-64	40	35
				1	5.6	GOVE KS					82
4	17.0			2	5.4	GRAHAM KS			-68	34	80
		1	3.7			GRANT KS					
1	4.5	1	3.6	1	2.4	GRAY KS	-20	-33	-46	72	26
11	12.5	6	7.0	3	3.1	GREENWOOD KS	-44	-55	-75	21	40
2	15.5			1	4.2	HAMILTON KS			-73	24	62
3	4.7	5	6.4	3	4.1	HARPER KS	36	-36	-12	90	62
10	7.1	7	4.0	5	2.0	HARVEY KS	-44	-49	-71	28	21
		1	8.6			HASKELL KS					
		3	23.0	1	3.4	HODGEMAN KS		-85			46
4	4.8	4	4.0	4	4.2	JACKSON KS	-16	6	-11	90	64
12	13.8	5	5.5	3	3.0	JEFFERSON KS	-60	-45	-78	16	38
9	13.3	2	2.8	3	4.7	JEWELL KS	-79	71	-64	40	71
23	6.7	20	3.2	29	2.7	JOHNSON KS	-51	-17	-60	50	31
		1	5.3			KEARNY KS					
5	7.5	2	2.4	3	3.2	KINGMAN KS	-68	34	-57	56	43
		2	5.8			KIOWA KS					
14	7.0	15	6.3	7	2.9	LABETTE KS	-10	-53	-58	54	36
2	13.8					LANE KS					
12	6.3	9	4.3	6	2.6	LEAVENWORTH KS	-32	-40	-59	52	30
4	9.3	3	5.5	3	5.0	LINCOLN KS	-42	-9	-47	71	75
5	9.0	5	7.2	2	2.1	LINN KS	-19	-71	-76	19	22
1	5.4					LOGAN KS					
12	6.8	12	5.4	9	4.5	LYON KS	-20	-17	-34	82	68
17	11.3	10	5.1	8	2.8	MC PHERSON KS	-55	-45	-75	20	33
14	13.9	3	2.3	9	5.9	MARION KS	-84	158	-58	54	84
6	4.9	6	4.4	6	6.0	MARSHALL KS	-11	36	21	95	85
1	5.9	1	2.6	3	9.4	MEADE KS	-56	261	59	97	96
6	4.8	8	5.0	3	1.7	MIAMI KS	6	-66	-64	40	16
6	9.0	4	4.3			MITCHELL KS	-52				
21	7.3	17	4.7	9	2.7	MONTGOMERY KS	-35	-44	-64	41	31
3	5.2	3	4.0	2	2.1	MORRIS KS	-23	-47	-59	51	22
1	12.5					MORTON KS					
6	5.9	3	2.5	6	4.0	NEMAHA KS	-58	62	-32	83	59
13	9.2	7	4.0	2	1.4	NEOSHO KS	-57	-66	-85	06	13
4	11.6	5	11.0			NESS KS	-5				
6	10.2	3	4.2	2	2.1	NORTON KS	-58	-50	-79	15	22
11	11.9	8	7.0	1	1.3	OSAGE KS	-42	-81	-89	03	13
3	5.3	3	2.9	1	.9	OSBORNE KS	-46	-69	-83	09	10
16*	27.9	2	2.7	3	4.4	OTTAWA KS	-90	66	-84	07	66
4	5.4	2	2.8	3	4.2	PAWNEE KS	-49	51	-24	87	62
5	8.6	4	7.1	2	3.7	PHILLIPS KS	-17	-48	-57	55	52
8	8.8	6	5.9	4	3.1	POTTAWATOMIE KS	-34	-47	-65	39	40
6	10.8	6	6.8	3	3.0	PRATT KS	-37	-56	-72	26	38
4	16.2	2	6.5	3	8.7	RAWLINS KS	-60	34	-46	72	96
26	9.1	9	2.2	15	3.1	RENO KS	-76	45	-66	37	40
9	11.1	4	4.2	3	3.6	REPUBLIC KS	-62	-15	-68	34	50
4	4.8	6	5.9	5	3.8	RICE KS	22	-35	-21	88	56
7	4.9	10	6.1	5	2.3	RILEY KS	24	-62	-54	61	25
2	4.3			4	7.4	ROOKS KS			73	97	93
4	11.4	2	3.5	1	2.4	RUSH KS	-69	-32	-79	15	26
5	9.9	7	10.1	5	5.5	RUSSELL KS	2	-45	-44	74	81
13	6.9	18	7.0	11	3.8	SALINE KS	2	-47	-46	73	54
1	5.1					SCOTT KS					
75	7.2	70	4.7	57	3.0	SEDGWICK KS	-34	-36	-58	53	37
1	2.0	3	4.9	2	2.3	SEWARD KS	149	-53	16	94	25
48	7.5	30	3.6	31	3.3	SHAWNEE KS	-52	-9	-56	57	43
3	21.2			1	4.0	SHERIDAN KS			-81	11	58
2	5.9	1	2.2	2	3.9	SHERMAN KS	-63	79	-34	82	57
9	16.1	5	6.6	5	5.9	SMITH KS	-59	-11	-63	42	85
2	4.0	4	6.7	2	2.3	STAFFORD KS	67	-66	-44	75	24
1	74.3	2*	22.0			STANTON KS	-70				
1	37.2			2	7.4	STEVENS KS			-80	13	93
11	6.5	12	5.5	12	5.6	SUMNER KS	-15	2	-13	90	82
1	2.9	1	2.8	1	2.1	THOMAS KS	-2	-25	-27	86	21
2	8.2	3	8.4	2	3.6	TREGO KS	2	-57	-56	57	50
5	9.3	1	1.5	3	3.6	WABAUNSEE KS	-84	148	-61	47	51
				1	7.4	WALLACE KS					93
9	10.4	6	6.0	6	5.4	WASHINGTON KS	-42	-11	-48	69	80
1	10.3	1	8.8	1	5.5	WICHITA KS	-15	-38	-47	71	81
6	5.9	3	2.9	1	.5	WILSON KS	-50	-82	-91	01	09
4	8.2	6	11.9	4	6.8	WOODSON KS	45	-43	-18	89	90
56	8.3	35	4.3	28	3.1	WYANDOTTE KS	-48	-27	-62	44	40
1166	9.4	964	6.2	805	4.1	KENTUCKY KY	-34	-33	-56		
4	6.1	3	3.8			ADAIR KY	-37				
12	15.2	7	7.6	3	3.3	ALLEN KY	-50	-57	-78	16	43
5	11.4	6	12.0	5	7.9	ANDERSON KY	5	-34	-31	84	94
2	4.0	4	6.0	2	2.8	BALLARD KY	50	-53	-29	85	34
13	10.0	8	5.2	7	3.2	BARREN KY	-48	-39	-68	33	41
5	11.2	6	11.1	2	3.5	BATH KY	-1	-68	-69	32	49

* SIGNIFICANTLY MORE DEATHS THAN EXPECTED

WHITE FEMALE — CANCER OF THE STOMACH — ICD 151

1950-1959		1960-1969		1970-1979		COUNTY AND STATE		PERCENT CHANGE			PERCENTILE	
											%CHANGE	RATES
DEATHS	RATE	DEATHS	RATE	DEATHS	RATE			60S/50S	70S/60S	70S/50S	70S/50S	1970S
9	6.1	7	4.8	9	4.3	BELL	KY	-21	-11	-29	85	65
4	5.1	10	10.2	7	4.8	BOONE	KY	99	-53	-6	91	73
5	6.3	3	3.4	6	5.0	BOURBON	KY	-46	47	-21	88	75
20	10.0	23	8.7	11	3.0	BOYD	KY	-13	-66	-70	29	37
8	8.5	2	1.6	6	4.6	BOYLE	KY	-81	182	-45	73	70
5	9.9	6	10.0	1	2.2	BRACKEN	KY	1	-78	-78	17	23
4	6.0	4	6.4	8*	10.5	BREATHITT	KY	7	65	76	98	97
14	17.9	8	9.3	12*	11.6	BRECKINRIDGE	KY	-48	25	-35	82	98
4	8.0	10*	14.8	4	3.7	BULLITT	KY	85	-75	-53	62	53
6	11.6	3	5.5	3	4.6	BUTLER	KY	-52	-17	-60	49	69
10	12.0	7	6.9	5	4.0	CALDWELL	KY	-43	-41	-66	36	60
8	6.9	11	7.2	8	4.6	CALLOWAY	KY	4	-36	-33	83	69
43	10.1	34	6.3	12	2.1	CAMPBELL	KY	-37	-67	-79	15	21
1	3.2	1	2.5	1	2.5	CARLISLE	KY	-23	2	-21	88	28
4	8.4	8*	14.1	2	3.7	CARROLL	KY	67	-74	-56	57	52
7	8.3	9	9.1	14*	10.8	CARTER	KY	10	19	31	96	98
5	7.6	6	8.0	4	4.2	CASEY	KY	5	-47	-44	74	64
15	9.0	15	7.2	13	5.0	CHRISTIAN	KY	-20	-31	-45	74	75
14	16.0	7	6.0	5	3.4	CLARK	KY	-62	-43	-79	16	47
3	4.8	7	8.9	2	2.6	CLAY	KY	86	-71	-46	73	30
8*	27.0	3	5.9	7*	13.7	CLINTON	KY	-78	132	-49	68	99
7	10.2	6	7.6	1	1.1	CRITTENDEN	KY	-26	-85	-89	02	12
1	2.7	7*	15.2	8*	15.1	CUMBERLAND	KY	468	-1	464	99	99
24	9.2	22	5.9	10	2.0	DAVIESS	KY	-35	-66	-78	16	20
4	13.2	2	4.2	2	4.2	EDMONSON	KY	-68	2	-68	34	64
4	18.3	3	10.2	3*	70.2	ELLIOTT	KY	-44	586	283	99	99
8	15.6	3	3.8	3	3.9	ESTILL	KY	-76	3	-75	21	57
38	8.0	30	4.9	34	4.2	FAYETTE	KY	-39	-14	-48	70	63
6	9.2	4	5.6	1	1.3	FLEMING	KY	-40	-76	-85	06	13
19	17.4	9	5.2	9	4.1	FLOYD	KY	-70	-21	-76	19	61
8	6.6	12	7.3	8	4.0	FRANKLIN	KY	11	-45	-38	79	60
6	9.5	6	7.6	3	2.9	FULTON	KY	-20	-62	-69	31	36
1	5.2	2	7.1			GALLATIN	KY	38				
2	4.2	3	5.5	1	1.4	GARRARD	KY	32	-74	-66	37	13
8	14.6	2	3.0	3	4.8	GRANT	KY	-80	62	-67	35	72
20	12.1	12	5.6	11	5.0	GRAVES	KY	-54	-11	-59	52	75
9	10.3	8	7.3	9	6.4	GRAYSON	KY	-29	-12	-38	80	88
8	15.6	5	8.9			GREEN	KY	-43				
10	10.4	15*	11.7	11	6.5	GREENUP	KY	12	-44	-38	80	89
2	6.1	2	7.8	2	6.1	HANCOCK	KY	28	-22	0	92	86
12	9.3	19*	11.5	6	2.2	HARDIN	KY	23	-81	-76	19	24
8	4.3	14	7.5	10	4.5	HARLAN	KY	76	-41	4	93	67
11	12.2	7	8.4	2	1.9	HARRISON	KY	-31	-77	-84	07	19
8	13.3	4	5.5	1	.9	HART	KY	-59	-84	-93	00	11
6	4.3	11	5.7	8	3.4	HENDERSON	KY	32	-40	-22	88	46
4	6.0	2	3.3	2	3.0	HENRY	KY	-44	-11	-50	67	37
3	7.8	5	11.0	5	9.1	HICKMAN	KY	41	-17	17	94	96
22	12.2	12	4.9	14	4.6	HOPKINS	KY	-59	-8	-63	44	68
3	6.2	2	3.9	4	5.6	JACKSON	KY	-38	46	-9	91	82
169	7.6	131	4.4	144	3.9	JEFFERSON	KY	-41	-12	-49	69	57
5	9.5	2	2.9	3	3.5	JESSAMINE	KY	-69	20	-63	42	48
8	9.5	7	7.4	2	1.4	JOHNSON	KY	-22	-81	-85	07	14
41	7.6	41	5.8	29	3.4	KENTON	KY	-24	-41	-55	59	47
7	13.6			3	3.9	KNOTT	KY			-71	27	57
14	13.8	8	6.4	13*	8.6	KNOX	KY	-53	34	-38	80	96
1	2.0	3	5.3	1	1.8	LARUE	KY	158	-66	-14	90	17
13	12.9	6	4.2	6	3.0	LAUREL	KY	-67	-29	-77	18	37
8	11.7	4	6.1	6	7.0	LAWRENCE	KY	-48	14	-40	78	91
1	2.5	4	9.7	6*	12.3	LEE	KY	293	27	398	99	98
3	9.4	2	4.0	2	3.8	LESLIE	KY	-57	-7	-60	50	54
12	11.6	9	8.3	12*	9.3	LETCHER	KY	-29	12	-20	88	96
3	7.0	4	7.2	1	1.1	LEWIS	KY	3	-84	-84	08	12
9	12.0	7	8.4	7	6.1	LINCOLN	KY	-30	-27	-49	69	86
8	20.0	2	4.0	3	4.0	LIVINGSTON	KY	-80	-1	-80	14	60
15	13.9	11	8.9	4	2.9	LOGAN	KY	-36	-67	-79	15	36
2	6.8	3	10.6			LYON	KY	56				
26	11.1	19	6.0	13	3.0	MC CRACKEN	KY	-46	-49	-73	26	39
6	14.1	5	8.2	5	7.0	MC CREARY	KY	-42	-15	-50	66	91
6	11.8	1	1.5	3	3.8	MC LEAN	KY	-88	159	-68	33	54
6	4.8	11	6.5	14	6.1	MADISON	KY	37	-6	28	95	86
5	11.8	1	2.3	1	1.6	MAGOFFIN	KY	-81	-28	-86	05	16
8	11.0	7	7.6	5	5.2	MARION	KY	-30	-32	-53	63	78
11	14.8	7	6.2	2	1.3	MARSHALL	KY	-58	-79	-91	01	13
8*	30.8	5	13.4	3	6.5	MARTIN	KY	-56	-51	-79	16	89
3	3.5	7	7.1	8	6.1	MASON	KY	103	-14	76	97	86
1	2.7	2	4.6	2	3.1	MEADE	KY	73	-33	15	94	39
1	3.6	2	10.7	2	8.1	MENIFEE	KY	194	-24	124	98	94
12	15.2	7	8.3	2	1.8	MERCER	KY	-45	-78	-88	03	18
7	17.2	3	6.4	5	7.1	METCALFE	KY	-63	10	-59	51	92
7	13.5	5	8.6	2	2.8	MONROE	KY	-36	-68	-79	15	33
3	5.7	4	4.8	6	6.6	MONTGOMERY	KY	-15	37	17	94	89
2	3.4			3	4.4	MORGAN	KY			30	96	67
13	9.8	12	7.3	9	4.4	MUHLENBERG	KY	-26	-40	-56	58	65
2	2.4	9	8.9	2	1.4	NELSON	KY	269	-84	-41	77	13
3	6.7					NICHOLAS	KY					
16	13.3	16*	13.2	3	2.2	OHIO	KY	-1	-83	-83	09	24
5	9.6	7	12.1	7*	8.7	OLDHAM	KY	27	-28	-9	91	96
2	4.3	3	6.6	2	3.0	OWEN	KY	54	-54	-30	85	38
4	16.2	1	3.2	2	6.7	OWSLEY	KY	-80	108	-59	52	90
11	19.4	3	5.7	2	2.9	PENDLETON	KY	-71	-49	-85	07	35
20*	19.2	15*	12.5	7	4.5	PERRY	KY	-35	-64	-77	19	68
19	10.2	21	9.0	25*	8.0	PIKE	KY	-11	-11	-22	88	94
7*	27.1	2	5.9	5*	11.3	POWELL	KY	-78	92	-58	53	98
13	8.8	9	4.8	7	2.5	PULASKI	KY	-46	-46	-71	28	29
1	4.9			1	4.8	ROBERTSON	KY			-1	92	73
3	5.6	7	11.1	2	2.3	ROCKCASTLE	KY	99	-79	-58	54	26
4	9.5	3	4.8	8*	10.9	ROWAN	KY	-49	126	15	94	98
5	9.8	3	4.4	5	5.9	RUSSELL	KY	-55	34	-40	78	84
6	9.4	5	6.5	5	4.8	SCOTT	KY	-31	-27	-49	68	72
6	7.1	4	3.8	1	.5	SHELBY	KY	-46	-88	-94	00	09
3	5.4	3	3.6	1	1.3	SIMPSON	KY	-33	-63	-75	21	13

* SIGNIFICANTLY MORE DEATHS THAN EXPECTED

WHITE FEMALE 1950-1959 DEATHS	RATE	1960-1969 DEATHS	RATE	1970-1979 DEATHS	RATE	CANCER OF THE STOMACH COUNTY AND STATE	PERCENT CHANGE 60S/50S	70S/60S	70S/50S	ICD 151 PERCENTILE %CHANGE 70S/50S	RATES 1970S
7*	27.5			2	6.1	SPENCER KY			-78	17	86
6	8.4	4	4.9	4	3.5	TAYLOR KY	-42	-28	-58	53	48
5	9.1	4	5.6	4	5.0	TODD KY	-38	-11	-45	73	75
5	12.8	5	9.6	4	6.8	TRIGG KY	-25	-29	-47	72	91
2	8.4	2	6.5			TRIMBLE KY	-23				
9	12.5	4	4.5	6	6.9	UNION KY	-64	51	-45	73	91
18	8.1	20	7.6	13	4.1	WARREN KY	-5	-47	-50	67	60
4	7.3	4	6.4	4	5.0	WASHINGTON KY	-12	-22	-31	84	76
8	13.0	2	2.7	6	5.8	WAYNE KY	-79	113	-55	59	84
10	11.4	4	3.3	6	5.3	WEBSTER KY	-71	59	-53	62	79
9	7.6	10	6.6	9	4.8	WHITLEY KY	-13	-28	-37	80	72
3	9.9	2	5.5	2	6.1	WOLFE KY	-45	11	-39	79	86
6	12.4	3	4.8	3	3.7	WOODFORD KY	-61	-24	-70	29	52
717	9.3	617	5.7	571	3.9	LOUISIANA LA	-39	-32	-59		
19	15.0	22*	11.9	11	4.2	ACADIA LA	-21	-65	-72	27	63
8	13.4	5	6.6	3	3.2	ALLEN LA	-51	-51	-76	19	42
5	8.0	11*	12.4	6	4.6	ASCENSION LA	56	-63	-42	76	70
3	6.2	1	1.5	1	1.2	ASSUMPTION LA	-76	-19	-81	12	12
18	15.9	8	5.2	6	2.7	AVOYELLES LA	-67	-48	-83	09	32
8	12.1	3	3.7	4	3.1	BEAUREGARD LA	-69	-16	-74	23	40
6	9.8	5	7.1	1	1.4	BIENVILLE LA	-28	-81	-86	05	13
4	5.4	7	5.9	5	2.4	BOSSIER LA	11	-60	-55	58	26
40	8.2	38	4.8	43	4.2	CADDO LA	-42	-13	-49	68	63
13	5.0	17	4.9	16	2.9	CALCASIEU LA	-1	-40	-41	77	36
5	13.7			2	3.0	CALDWELL LA			-78	17	38
2	10.7	1	2.6	1	2.0	CAMERON LA	-76	-25	-82	11	20
2	10.8	2	7.2	5*	9.9	CATAHOULA LA	-33	36	-9	91	97
2	2.6	7	9.1	3	2.5	CLAIBORNE LA	247	-73	-5	92	28
1	5.5	1	2.8	1	1.5	CONCORDIA LA	-49	-48	-73	24	14
8	13.0	5	5.7	2	1.9	DE SOTO LA	-57	-67	-86	06	18
25	7.2	26	4.4	34	3.9	EAST BATON ROUGE LA	-39	-13	-46	72	56
2	10.5	1	5.1	2	7.1	EAST CARROLL LA	-51	38	-33	83	92
2	3.2	3	4.9			EAST FELICIANA LA	50				
7	8.3	3	2.3	7	4.2	EVANGELINE LA	-72	80	-49	68	63
6	9.0	4	4.2	8	7.2	FRANKLIN LA	-53	70	-20	88	92
6	12.9	3	4.4	2	2.1	GRANT LA	-66	-52	-84	08	22
9	8.6	18*	11.0	8	3.8	IBERIA LA	28	-65	-56	58	55
8	14.7	1	1.3	1	1.0	IBERVILLE LA	-91	-20	-93	01	11
2	4.4	5	7.0	1	1.0	JACKSON LA	57	-85	-77	18	11
29	8.2	32	4.7	55	4.0	JEFFERSON LA	-43	-15	-51	65	60
10	12.7	8	7.2	4	2.8	JEFFERSON DAVIS LA	-43	-62	-78	16	33
17	9.6	20	7.6	16	3.8	LAFAYETTE LA	-21	-51	-61	47	54
15	11.6	15	7.7	18	6.8	LAFOURCHE LA	-33	-12	-41	77	91
5	9.5	6	9.1	2	1.8	LA SALLE LA	-4	-80	-81	12	17
9	11.7	4	3.4	1	.7	LINCOLN LA	-71	-81	-94	00	10
3	3.7	5	5.7			LIVINGSTON LA	54				
2	14.2	2	7.3			MADISON LA	-48				
2	3.8	2	2.0	6	4.6	MOREHOUSE LA	-46	125	22	95	69
9	10.7	2	1.8	3	2.5	NATCHITOCHES LA	-83	35	-77	18	28
188	9.0	129	5.2	123	4.9	ORLEANS LA	-42	-6	-45	73	74
16	6.9	15	4.6	12	2.4	OUACHITA LA	-34	-46	-65	40	27
3	5.4	3	4.4	2	4.3	PLAQUEMINES LA	-17	-3	-19	88	65
8	20.0	5	9.5	2	2.7	POINTE COUPEE LA	-53	-71	-86	05	31
28	11.0	9	2.3	14	2.7	RAPIDES LA	-79	15	-76	20	31
7*	30.9	4	10.3	2	4.9	RED PIVER LA	-67	-52	-84	07	74
12*	22.7	4	5.6	6	6.2	RICHLAND LA	-75	9	-73	25	87
5	7.2	5	4.8	3	2.3	SABINE LA	-33	-52	-68	34	25
3	10.7	5	4.6	10	4.8	ST BERNARD LA	-57	6	-55	59	73
4	17.8	10*	18.0	5	6.0	ST CHARLES LA	1	-67	-66	37	85
2	10.7					ST HELENA LA					
4	11.0	1	2.1	4	7.1	ST JAMES LA	-81	242	-35	81	92
4	12.2	3	7.6	5	8.5	ST JOHN THE BAPT LA	-38	12	-31	84	95
19	12.4	21	9.9	14	4.7	ST LANDRY LA	-20	-53	-62	45	71
3	5.0	5	5.7	10*	8.4	ST MARTIN LA	15	47	70	97	95
7	8.4	8	7.0	5	2.9	ST MARY LA	-17	-58	-65	38	36
7	6.6	11	6.6	14	4.3	ST TAMMANY LA	0	-35	-35	82	65
17	11.3	11	5.4	8	2.9	TANGIPAHOA LA	-52	-46	-74	23	36
1	6.3			1	5.1	TENSAS LA			-19	88	77
9	7.3	8	5.1	13	5.8	TERREBONNE LA	-30	13	-21	88	83
7	10.6	4	4.9	5	4.5	UNION LA	-54	-9	-58	53	67
15	12.3	19*	10.5	13	5.3	VERMILION LA	-15	-50	-57	55	79
7	10.1	9	10.0	6	4.3	VERNON LA	0	-57	-57	56	65
9	7.3	12	7.8	4	2.2	WASHINGTON LA	6	-72	-70	30	23
10	10.9	10	7.0	8	4.0	WEBSTER LA	-36	-43	-64	42	59
2	8.2	2	6.5	2	5.7	WEST BATON ROUGE LA	-20	-13	-30	84	83
5	9.0	3	4.9	2	3.0	WEST CARPOLL LA	-45	-39	-67	36	37
2	18.9	2	17.0			WEST FELICIANA LA	-10				
1	2.3	6	7.7			WINN LA	231				
576	11.1	439	7.3	388*	5.3	MAINE ME	-34	-27	-52		
52	11.0	53*	9.7	42	6.1	ANDROSCOGGIN ME	-11	-37	-44	74	86
50*	15.5	41*	11.0	29	6.3	AROOSTOOK ME	-29	-43	-60	50	87
92	8.6	95	7.8	65	4.2	CUMBERLAND ME	-10	-45	-51	66	64
13	11.3	15*	12.3	8	5.1	FRANKLIN ME	10	-59	-55	60	77
26	12.7	21	9.0	11	3.8	HANCOCK ME	-29	-57	-70	30	56
55	11.0	44	7.8	42	5.8	KENNEBEC ME	-29	-25	-47	71	84
29	12.9	22	8.4	16	5.8	KNOX ME	-34	-31	-55	59	84
17	13.0	12	7.2	7	3.0	LINCOLN ME	-44	-58	-77	19	38
26	11.0	14	5.0	19	6.0	OXFORD ME	-55	21	-45	73	85
67	11.6	39	5.9	42	5.4	PENOBSCOT ME	-49	-9	-54	61	79
17	16.0	7	6.0	9	7.3	PISCATAQUIS ME	-63	22	-54	60	92
8	6.2	7	5.3	9	5.5	SAGADAHOC ME	-14	5	-10	90	81
26	12.0	14	5.6	14	5.0	SOMERSET ME	-53	-11	-58	53	75
14	9.9	1	.9	7	4.4	WALDO ME	-91	403	-56	58	66
24	11.9	15	6.7	13	5.3	WASHINGTON ME	-43	-21	-55	59	79
60	11.0	39	5.8	55*	6.4	YORK ME	-47	10	-42	76	88

* SIGNIFICANTLY MORE DEATHS THAN EXPECTED

1950-1959 DEATHS	1950-1959 RATE	1960-1969 DEATHS	1960-1969 RATE	1970-1979 DEATHS	1970-1979 RATE	COUNTY AND STATE	PERCENT CHANGE 60S/50S	PERCENT CHANGE 70S/60S	PERCENT CHANGE 70S/50S	PERCENTILE %CHANGE 70S/50S	PERCENTILE RATES 1970S
964	9.9	862	6.4	712	3.9	MARYLAND MD	-36	-39	-61		
64*	16.1	45	8.7	29	4.5	ALLEGANY MD	-46	-48	-72	26	68
38	11.5	55	8.3	46	4.2	ANNE ARUNDEL MD	-28	-49	-64	42	63
117	9.7	149	6.8	136	3.9	BALTIMORE MD	-30	-42	-60	50	57
8*	22.5	5	8.6	4	4.2	CALVERT MD	-62	-51	-81	11	63
6	5.8	10	9.0	2	1.5	CAROLINE MD	55	-84	-74	22	14
16	5.4	9	2.7	8	1.8	CARROLL MD	-50	-33	-66	36	18
21	14.6	10	5.4	17	7.2	CECIL MD	-63	35	-51	66	92
2	6.3	3	3.1	2	1.2	CHARLES MD	-51	-61	-81	12	12
15	11.6	10	5.7	6	3.0	DORCHESTER MD	-50	-47	-74	24	39
34	11.0	17	4.5	19	3.7	FREDERICK MD	-59	-19	-66	36	52
5	5.7	6	5.1	11	6.6	GARRETT MD	-11	31	16	94	89
24	12.7	17	6.0	11	2.2	HARFORD MD	-52	-63	-82	10	24
10	10.6	6	4.2	11	4.2	HOWARD MD	-61	1	-60	48	63
8	12.2	6	7.5	1	.8	KENT MD	-38	-89	-93	01	10
68	8.3	87	5.5	104	3.8	MONTGOMERY MD	-34	-30	-54	60	55
54	8.4	84	6.7	74	4.0	PRINCE GEORGE'S MD	-20	-40	-52	64	60
7	11.1	3	3.5	1	.6	QUEEN ANNE'S MD	-69	-84	-95	00	10
4	7.4	3	2.6	4	2.9	ST MARY'S MD	-65	10	-61	47	35
6	7.0	6	6.1	5	4.2	SOMERSET MD	-13	-31	-40	78	63
9	9.7	5	3.5	5	2.9	TALBOT MD	-64	-16	-70	30	35
35	8.7	32	6.1	15	2.2	WASHINGTON MD	-30	-63	-74	23	24
15	7.9	13	5.0	14	3.9	WICOMICO MD	-37	-22	-51	65	57
8	8.0	4	3.1	6	3.8	WORCESTER MD	-61	22	-53	63	55
390	10.3	277	6.9	181	4.6	BALTIMORE CITY MD	-33	-34	-56	58	68
3596*	12.5	2703*	7.8	2236*	5.5	MASSACHUSETTS MA	-37	-30	-56		
27	7.3	42	6.5	53	4.6	BARNSTABLE MA	-10	-29	-36	81	70
122*	15.2	76	7.7	62	5.3	BERKSHIRE MA	-49	-32	-65	38	79
333*	14.5	259*	9.2	240*	7.1	BRISTOL MA	-36	-23	-51	66	92
4	9.4	4	6.2	3	5.1	DUKES MA	-34	-17	-45	73	77
415*	11.9	345*	8.3	260*	5.2	ESSEX MA	-31	-37	-56	57	78
39	10.9	26	6.3	28	5.9	FRANKLIN MA	-42	-6	-46	72	85
278*	12.9	214*	7.8	155	4.8	HAMPDEN MA	-39	-38	-63	44	72
69	12.7	57	8.6	42	5.4	HAMPSHIRE MA	-32	-37	-58	55	80
760*	11.7	620*	7.8	502*	5.4	MIDDLESEX MA	-33	-31	-53	62	80
1	3.2	2	5.1			NANTUCKET MA	58				
292	11.3	224	6.6	218*	5.0	NORFOLK MA	-42	-24	-56	58	76
172*	12.8	116	6.8	123*	5.4	PLYMOUTH MA	-47	-20	-58	54	80
674*	13.9	416*	8.2	285*	5.5	SUFFOLK MA	-41	-33	-61	48	81
410*	12.3	302*	7.6	265*	5.8	WORCESTER MA	-39	-24	-53	62	84
3180*	11.7	2563*	7.2	1998	4.5	MICHIGAN MI	-38	-38	-61		
4	12.8			3	4.7	ALCONA MI			-63	43	71
4	10.5	4	9.6	3	5.8	ALGER MI	-8	-40	-45	74	84
23	9.2	26	8.4	16	3.9	ALLEGAN MI	-8	-54	-58	54	57
13	12.6	7	5.4	11	6.6	ALPENA MI	-57	21	-48	70	89
9	14.3	5	6.3	2	1.5	ANTRIM MI	-56	-77	-90	02	14
6	12.3	4	7.3	3	3.9	ARENAC MI	-41	-47	-68	33	57
4	11.1	4	9.8	4	7.3	BARAGA MI	-11	-26	-34	82	92
6	3.7	8	4.2	7	3.3	BARRY MI	15	-21	-9	91	44
53	13.1	47	9.1	28	4.3	BAY MI	-31	-53	-67	35	65
4	8.4	5	9.5	3	4.4	BENZIE MI	13	-53	-47	71	66
61	10.6	46	6.1	37	3.9	BERRIEN MI	-43	-37	-64	41	56
12	6.4	9	4.1	9	3.9	BRANCH MI	-36	-4	-39	79	58
51	8.8	41	5.7	27	3.3	CALHOUN MI	-35	-41	-62	45	45
25*	15.9	8	4.4	12	4.8	CASS MI	-72	8	-70	30	72
6	8.7	5	5.2	8	6.8	CHARLEVOIX MI	-40	29	-22	87	90
10	13.8	6	7.3	9	7.8	CHEBOYGAN MI	-47	7	-44	75	94
10	7.4	10	7.9	8	5.0	CHIPPEWA MI	7	-37	-33	83	75
5	9.5	5	6.6	5	3.6	CLARE MI	-31	-45	-62	46	51
13	8.7	5	2.8	9	4.1	CLINTON MI	-67	42	-53	62	60
6*	31.0	2	6.5	1	2.9	CRAWFORD MI	-79	-55	-91	02	35
36*	24.1	23*	11.4	15	6.0	DELTA MI	-53	-47	-75	21	85
22*	17.2	18*	10.8	11	5.4	DICKINSON MI	-37	-50	-69	32	80
14	6.2	15	5.6	12	3.7	EATON MI	-11	-34	-41	77	52
7	7.5	4	3.4	5	3.3	EMMET MI	-54	-5	-56	57	43
107	10.1	94	6.7	76	4.3	GENESEE MI	-33	-36	-58	54	64
2	4.1	4	6.6	5	4.9	GLADWIN MI	60	-26	19	94	74
24*	17.8	33*	20.9	17*	8.4	GOGEBIC MI	17	-60	-53	62	95
14	7.4	9	4.3	10	3.7	GRAND TRAVERSE MI	-42	-13	-49	68	54
22	11.0	9	3.8	4	1.9	GRATIOT MI	-66	-48	-82	10	19
11	5.9	12	5.7	8	3.1	HILLSDALE MI	-3	-46	-47	71	40
65*	27.2	18	7.1	14	4.9	HOUGHTON MI	-74	-31	-82	11	74
24	14.1	17	8.4	11	3.9	HURON MI	-40	-54	-72	26	57
66	8.8	64	6.8	46	3.9	INGHAM MI	-22	-43	-56	58	57
31	14.2	23	8.9	13	4.4	IONIA MI	-37	-50	-69	32	66
5	7.6	6	6.2	4	3.0	IOSCO MI	-18	-52	-61	48	37
13	13.7	4	3.5	11*	8.1	IRON MI	-75	133	-41	77	94
14	11.4	10	6.8	4	2.0	ISABELLA MI	-40	-70	-82	10	20
36	6.4	47	6.9	31	3.8	JACKSON MI	8	-45	-40	78	55
72	10.8	27	3.2	43	4.2	KALAMAZOO MI	-70	30	-61	46	63
3	11.8	3	14.3	1	1.9	KALKASKA MI	22	-87	-84	08	19
176	11.4	126	6.3	81	3.3	KENT MI	-44	-47	-71	29	45
3	23.6	1	5.7			KEWEENAW MI	-76				
2	10.6	1	4.1	2	5.7	LAKE MI	-62	39	-46	72	83
18	10.5	18	8.5	12	4.7	LAPEER MI	-19	-45	-56	58	70
1	2.8	2	3.3	2	2.4	LEELANAU MI	19	-25	-11	90	27
29	8.4	16	3.5	16	3.3	LENAWEE MI	-58	-7	-61	48	43
16	10.2	17	7.8	10	2.8	LIVINGSTON MI	-24	-64	-72	26	34
2	4.6	3	6.5	2	4.9	LUCE MI	43	-25	8	93	74
4	9.8	4	7.4	6*	11.2	MACKINAC MI	-25	51	14	94	98
93*	13.9	102	8.0	135*	5.3	MACOMB MI	-43	-33	-62	46	79
16	13.2	5	3.9	7	4.2	MANISTEE MI	-71	8	-68	33	62
27	12.7	29*	10.6	21	6.6	MARQUETTE MI	-16	-38	-48	69	89
25*	23.0	16*	10.3	7	3.8	MASON MI	-55	-63	-83	09	56
12	11.7	7	5.7	3	1.8	MECOSTA MI	-51	-69	-85	07	18

* SIGNIFICANTLY MORE DEATHS THAN EXPECTED

WHITE FEMALE — CANCER OF THE STOMACH

1950-1959 DEATHS	1950-1959 RATE	1960-1969 DEATHS	1960-1969 RATE	1970-1979 DEATHS	1970-1979 RATE	COUNTY AND STATE	PERCENT CHANGE 60S/50S	PERCENT CHANGE 70S/60S	PERCENT CHANGE 70S/50S	PERCENTILE %CHANGE 70S/50S	PERCENTILE RATES 1970S
26*	21.8	20*	13.3	11	5.5	MENOMINEE MI	-39	-59	-75	22	81
10	7.9	15	8.5	9	3.4	MIDLAND MI	7	-60	-57	56	46
5	16.5	3	9.7	4	7.1	MISSAUKEE MI	-42	-26	-57	56	92
30	9.3	27	6.4	23	4.1	MONROE MI	-32	-36	-56	57	61
23	11.7	8	3.4	8	2.6	MONTCALM MI	-71	-21	-77	17	30
2	16.1	2	6.9	2	2.9	MONTMORENCY MI	-57	-58	-82	11	35
48	9.9	49	7.6	37	4.5	MUSKEGON MI	-24	-41	-55	59	67
12	10.9	5	3.7	12	5.7	NEWAYGO MI	-67	56	-48	70	83
160	10.1	201	7.3	188	4.2	OAKLAND MI	-28	-43	-59	52	62
12	12.9	3	2.8	5	3.9	OCEANA MI	-78	41	-70	31	58
5	11.3	3	3.9	6	6.1	OGEMAW MI	-65	55	-46	72	86
11*	22.0	7*	13.5	8*	13.0	ONTONAGON MI	-39	-3	-41	77	98
5	6.9	5	5.3	3	2.4	OSCEOLA MI	-23	-55	-65	38	27
		1	5.4	1	2.7	OSCODA MI		-49			32
5	16.6	2	5.3	4	4.6	OTSEGO MI	-68	-13	-72	26	70
41	11.9	46*	9.3	30	4.4	OTTAWA MI	-22	-53	-63	42	66
3	7.0	9*	13.1	5	5.3	PRESQUE ISLE MI	86	-59	-24	87	79
2	6.1			2	1.5	ROSCOMMON MI			-75	22	15
88	12.9	54	6.3	30	2.7	SAGINAW MI	-51	-57	-79	15	32
48	10.6	43	7.2	27	3.8	ST CLAIR MI	-32	-48	-64	40	54
25	11.5	11	4.2	13	3.9	ST JOSEPH MI	-64	-7	-66	37	57
22	13.2	14	7.0	12	4.6	SANILAC MI	-47	-34	-65	39	70
10*	24.7	8*	16.0	2	3.2	SCHOOLCRAFT MI	-35	-80	-87	04	42
25	10.0	23	7.9	11	3.2	SHIAWASSEE MI	-21	-59	-68	34	42
30	14.8	15	6.0	16	5.6	TUSCOLA MI	-59	-7	-62	44	82
24	9.2	15	4.9	17	4.6	VAN BUREN MI	-46	-6	-50	68	70
49	8.3	37	5.2	35	3.9	WASHTENAW MI	-37	-26	-53	62	56
1118*	13.2	898*	8.4	617*	5.4	WAYNE MI	-36	-36	-59	51	80
4	4.4	5	4.1	10	6.7	WEXFORD MI	-9	65	51	97	90
2000*	13.1	1616*	8.1	1171*	4.6	MINNESOTA MN	-38	-43	-65		
5	6.7	7	8.0	5	4.6	AITKIN MN	19	-42	-31	84	70
13	7.7	18	8.1	18	4.1	ANOKA MN	4	-49	-47	71	61
12	11.6	11	8.1	9	5.0	BECKER MN	-30	-39	-57	56	75
17*	18.1	16*	13.2	5	3.6	BELTRAMI MN	-27	-73	-80	13	50
9	13.6	8	9.3	9	7.1	BENTON MN	-32	-24	-48	69	92
7	15.1	9*	15.2	5	5.5	BIG STONE MN	1	-64	-64	41	81
27	11.7	14	4.8	14	4.4	BLUE EARTH MN	-59	-8	-62	45	67
17	11.7	13	7.0	9	4.0	BROWN MN	-40	-43	-66	37	59
14	11.9	13	9.2	16*	8.4	CARLTON MN	-23	-9	-30	84	95
16	16.8	13	10.5	7	3.9	CARVER MN	-38	-63	-77	18	57
10	11.9	13	11.8	6	4.4	CASS MN	-1	-63	-63	42	66
15*	18.6	6	5.4	5	2.9	CHIPPEWA MN	-71	-46	-85	07	35
9	9.9	15*	13.9	8	5.0	CHISAGO MN	41	-64	-50	67	75
19	15.6	16	8.7	12	4.4	CLAY MN	-44	-50	-72	27	65
7	17.1	6	12.5	2	3.4	CLEARWATER MN	-27	-73	-80	13	46
1	12.4			1	5.9	COOK MN			-52	64	85
8	11.1	7	6.8	3	2.2	COTTONWOOD MN	-39	-68	-81	12	22
19	12.9	22*	11.0	13	4.8	CROW WING MN	-15	-56	-63	44	73
20	9.6	32	9.5	20	3.6	DAKOTA MN	-1	-62	-62	45	51
6	8.1	5	5.7	1	.5	DODGE MN	-30	-91	-94	00	09
17	14.9	12	7.7	11	5.5	DOUGLAS MN	-48	-28	-63	43	81
14	11.2	17*	11.1	3	1.3	FARIBAULT MN	-1	-88	-88	03	13
23	14.9	15	8.5	7	2.7	FILLMORE MN	-43	-68	-82	11	32
14	8.3	14	6.6	10	4.1	FREEBOPN MN	-20	-38	-51	65	61
23	11.1	16	6.3	3	.9	GOODHUE MN	-43	-85	-92	01	11
5	11.5	5	7.4	2	2.7	GRANT MN	-35	-64	-77	18	31
456*	12.1	329	6.6	276	4.4	HENNEPIN MN	-45	-33	-63	42	66
7	8.9	14*	12.5	7	5.7	HOUSTON MN	40	-54	-36	81	83
9	17.3	8	12.5	2	2.0	HUBBAPD MN	-28	-84	-88	03	20
16*	23.9	9	10.9	4	3.7	ISANTI MN	-54	-66	-84	07	53
20	16.2	18*	11.0	17*	6.8	ITASCA MN	-32	-38	-58	53	90
13	17.4	9	9.7	4	2.7	JACKSON MN	-44	-72	-84	07	32
9*	23.2	3	5.2	1	1.6	KANABEC MN	-78	-70	-93	01	15
23	15.3	11	5.7	13	5.5	KANDIYOHI MN	-63	-2	-64	41	81
6	14.4	5	10.5	1	2.2	KITTSON MN	-27	-79	-85	07	23
5	12.1	4	5.1	4	4.1	KOOCHICHING MN	-58	-20	-66	36	61
15*	21.6	8	9.0	6	5.3	LAC QUI PARLE MN	-58	-41	-76	20	78
9*	23.7	5	9.6	2	3.1	LAKE MN	-60	-68	-87	04	40
2	8.7			1	3.9	LAKE OF THE WOODS MN			-55	60	58
11	9.6	11	7.8	5	3.0	LE SUEUR MN	-19	-62	-69	31	37
5	10.6	3	5.7	4	5.1	LINCOLN MN	-47	-11	-52	64	76
18*	16.8	14	10.1	15*	8.4	LYON MN	-40	-17	-50	67	95
18	13.3	18*	10.9	6	3.3	MC LEOD MN	-19	-70	-75	20	43
5	21.6	3	13.1	2	5.8	MAHNOMEN MN	-39	-56	-73	24	83
12*	21.4	7	9.6	6	6.6	MARSHALL MN	-55	-31	-69	31	89
19	15.5	7	4.6	10	4.6	MARTIN MN	-70	0	-70	30	70
11	11.9	11	7.9	8	5.0	MEEKER MN	-34	-37	-58	53	75
10	12.9	9	8.9	6	4.2	MILLE LACS MN	-31	-53	-68	34	63
15	12.3	19*	13.2	5	2.1	MORRISON MN	7	-84	-83	09	22
24	11.9	24	9.3	8	2.2	MOWER MN	-22	-76	-81	11	23
7	11.9	3	3.7	7	7.4	MURRAY MN	-69	102	-38	80	93
14	11.1	10	7.1	4	3.3	NICOLLET MN	-36	-54	-71	29	43
11	10.0	13	9.6	5	3.0	NOBLES MN	-5	-68	-70	30	38
20*	30.2	8	10.0	10*	9.7	NORMAN MN	-67	-3	-68	34	97
29	10.5	26	6.6	21	4.0	OLMSTED MN	-37	-39	-61	46	60
46*	16.4	22	6.5	22	5.1	OTTER TAIL MN	-61	-21	-69	32	77
13*	20.7	14*	15.3	3	3.7	PENNINGTON MN	-26	-76	-82	10	53
10	11.1	12	11.4	5	4.4	PINE MN	3	-62	-61	48	65
8	11.6	13*	15.3	3	2.6	PIPESTONE MN	32	-83	-78	17	29
19	10.7	15	6.5	14	5.4	POLK MN	-39	-18	-50	67	79
9	14.9	9	9.7	9*	7.1	POPE MN	-35	-27	-53	63	92
255*	13.4	188	7.5	137	4.4	RAMSEY MN	-44	-41	-67	35	66
8*	30.0	4	11.9	3	7.1	RED LAKE MN	-60	-40	-76	19	92
8	7.2	12	8.1	11	7.3	REDWOOD MN	12	-10	0	92	92
15	13.1	14	9.2	8	3.9	RENVILLE MN	-29	-58	-70	29	56
16	8.4	16	7.0	10	3.6	RICE MN	-17	-48	-57	56	51
9	16.5	2	2.3	5	6.4	ROCK MN	-86	181	-61	46	88
12*	23.0	11*	16.8	3	4.6	ROSEAU MN	-27	-73	-80	13	69
157*	15.0	146*	11.0	133*	8.3	ST LOUIS MN	-27	-24	-44	74	95

* SIGNIFICANTLY MORE DEATHS THAN EXPECTED

WHITE FEMALE — CANCER OF THE STOMACH — ICD 151

1950-1959 DEATHS	1950-1959 RATE	1960-1969 DEATHS	1960-1969 RATE	1970-1979 DEATHS	1970-1979 RATE	COUNTY AND STATE	PERCENT CHANGE 60S/50S	PERCENT CHANGE 70S/60S	PERCENT CHANGE 70S/50S	PERCENTILE %CHANGE 70S/50S	PERCENTILE RATES 1970S
12	13.3	12	10.2	6	3.7	SCOTT MN	-23	-64	-72	26	53
4	6.6	2	2.5	7	6.0	SHERBURNE MN	-62	143	-8	91	85
7	8.5	8	7.8	4	3.1	SIBLEY MN	-9	-60	-63	42	40
44	14.9	43*	11.1	21	4.6	STEARNS MN	-26	-59	-69	31	69
15	13.6	19*	12.6	6	3.5	STEELE MN	-7	-72	-74	23	49
2	4.4	6	10.2	2	2.0	STEVENS MN	131	-80	-54	60	20
16*	22.2	11*	11.3	5	3.7	SWIFT MN	-49	-67	-83	09	53
24*	20.6	12	8.4	11	6.0	TODD MN	-59	-28	-71	28	85
3	8.9	1	2.1	4	6.8	TRAVERSE MN	-77	226	-24	87	90
13	12.4	16*	11.8	6	3.5	WABASHA MN	-5	-70	-72	27	48
3	4.8	5	6.5	7	7.0	WADENA MN	36	7	45	96	91
10	12.1	4	3.8	7	4.4	WASECA MN	-69	17	-63	42	66
16	9.8	18	7.9	18	5.2	WASHINGTON MN	-19	-35	-47	71	78
11	14.3	8	8.0	5	4.5	WATONWAN MN	-44	-44	-69	33	68
10*	22.9	7	11.7	3	3.0	WILKIN MN	-49	-74	-87	05	38
33	14.7	8	2.8	14	4.1	WINONA MN	-81	48	-72	26	61
16	10.5	15	8.1	10	3.9	WRIGHT MN	-23	-52	-63	43	57
10	13.1	11	11.0	5	3.8	YELLOW MEDICINE MN	-16	-66	-71	27	54
484	9.3	391	5.5	327	3.5	MISSISSIPPI MS	-41	-37	-63		
3	5.2	7	8.7	8	6.8	ADAMS MS	68	-22	31	96	90
11	9.8	13	10.4	8	4.1	ALCORN MS	6	-60	-58	54	62
5	12.6	3	6.1	1	1.7	AMITE MS	-52	-73	-87	05	16
10	14.3	8	8.2	4	3.1	ATTALA MS	-42	-62	-78	17	41
2	11.1					BENTON MS					
12	18.2	6	6.5	2	2.0	BOLIVAR MS	-65	-69	-89	03	20
12*	19.4	5	6.6	3	3.4	CALHOUN MS	-66	-49	-83	09	46
4	12.5	2	5.8	3	7.3	CARROLL MS	-54	25	-42	76	92
2	3.6	2	2.5	5	5.2	CHICKASAW MS	-30	106	45	96	78
4	10.2	4	10.0	2	3.2	CHOCTAW MS	-2	-68	-69	32	42
11*	21.3	6	9.5	3	3.1	CLARKE MS	-55	-67	-85	06	40
3	6.9	3	5.1	3	3.8	CLAY MS	-26	-25	-45	74	55
4	5.9	6	7.3	3	3.2	COAHOMA MS	24	-57	-47	71	41
9	13.4	4	4.2	2	1.7	COPIAH MS	-69	-59	-87	04	16
6	14.4	1	1.4	2	3.2	COVINGTON MS	-90	123	-78	17	42
3	7.0	3	4.2	4	2.7	DE SOTO MS	-39	-36	-61	47	31
10	6.9	6	2.8	12	4.6	FORREST MS	-60	66	-33	83	69
6	21.0	3	10.9	1	1.7	FRANKLIN MS	-48	-85	-92	01	16
4	15.5	3	6.7	3	5.0	GEORGE MS	-57	-25	-68	34	75
3	12.5	2	6.8	4	9.6	GREENE MS	-45	41	-23	87	97
4	11.4	5	8.3	1	1.1	GRENADA MS	-27	-87	-90	02	11
4	8.2	4	6.2	5	4.6	HANCOCK MS	-25	-25	-44	74	69
17	7.1	13	3.7	17	3.2	HARRISON MS	-47	-14	-54	60	42
26	6.4	30	4.9	32	3.9	HINDS MS	-23	-22	-40	78	56
5	9.2	2	4.1	1	1.5	HOLMES MS	-56	-64	-84	07	14
2	6.4			1	2.6	HUMPHREYS MS			-60	49	29
6	8.9	6	7.2	3	2.9	ITAWAMBA MS	-19	-60	-68	35	35
7	6.6	6	3.7	5	1.6	JACKSON MS	-44	-56	-75	20	15
5	11.3	8*	15.1	4	5.5	JASPER MS	34	-63	-51	65	81
2	10.0	1	5.1			JEFFERSON MS	-49				
6	21.9	4	11.5	5*	10.5	JEFFERSON DAVIS MS	-47	-9	-52	64	97
15	8.8	10	4.3	7	2.4	JONES MS	-51	-44	-73	26	27
1	3.1	1	1.9	2	6.2	KEMPER MS	-38	224	100	98	87
4	6.7	3	3.6	7	7.7	LAFAYETTE MS	-46	112	15	94	93
2	4.9	5	8.3	1	1.4	LAMAR MS	69	-83	-72	27	13
19	8.5	14	4.6	13	3.7	LAUDERDALE MS	-46	-20	-57	56	52
7*	20.8	6*	16.8	2	4.3	LAWRENCE MS	-19	-74	-79	15	65
13*	23.6	4	5.1	3	3.4	LEAKE MS	-78	-33	-86	06	46
13	9.5	4	2.1	9	3.7	LEE MS	-77	74	-61	48	53
8	11.0	5	4.8	5	4.0	LEFLORE MS	-57	-16	-64	42	59
2	2.2	3	2.7	7	4.5	LINCOLN MS	19	69	101	98	68
6	6.2	12	9.0	7	4.0	LOWNDES MS	45	-55	-35	81	59
8	17.1	1	1.5	2	2.8	MADISON MS	-91	92	-83	08	34
3	4.3	8	9.3	2	1.7	MARION MS	115	-82	-62	46	16
4	9.5	2	3.3	4	6.1	MAPSHALL MS	-65	84	-35	81	86
7	6.8	6	4.8	4	2.0	MONROE MS	-29	-58	-70	29	20
5	11.6	2	3.8			MONTGOMERY MS	-67				
2	2.0	11	11.7	3	2.7	NESHOBA MS	469	-77	33	96	32
8	11.9	5	5.3	5	4.8	NEWTON MS	-55	-10	-59	50	72
5	19.2	1	3.1	2	4.2	NOXUBEE MS	-84	36	-78	16	63
3	6.4	1	2.0	3	3.4	OKTIBBEHA MS	-69	72	-47	71	46
11	17.2	10	13.2			PANOLA MS	-23				
2	2.5	3	3.8	3	2.6	PEARL RIVER MS	52	-32	3	93	30
4	13.2	3	9.2	1	2.7	PEPRY MS	-30	-71	-79	14	31
9	8.7	6	4.2	6	3.2	PIKE MS	-52	-23	-63	43	42
5	6.2	4	4.4	11*	8.5	PONTOTOC MS	-28	93	39	96	96
4	5.0	6	5.8	4	2.8	PRENTISS MS	16	-51	-43	75	33
2	7.7					QUITMAN MS					
6	8.3	6	4.7	8	3.7	RANKIN MS	-43	-22	-56	58	52
8	15.7	2	2.4	3	3.0	SCOTT MS	-85	24	-81	12	38
2	16.7	4*	21.8	1	5.3	SHARKEY MS	31	-76	-68	33	78
3	6.3	6	7.4	3	2.8	SIMPSON MS	18	-62	-55	59	34
5	8.6	4	6.0	3	3.5	SMITH MS	-30	-42	-60	50	47
2	7.7	1	2.7	1	2.4	STONE MS	-65	-11	-68	33	27
9	14.2	4	4.9	3	2.9	SUNFLOWER MS	-66	-40	-79	14	36
5	11.4	1	1.5			TALLAHATCHIE MS	-86				
4	9.8	9*	17.1	2	3.1	TATE MS	74	-82	-68	33	40
8	11.6	5	5.9	6	6.1	TIPPAH MS	-49	2	-47	70	86
6	9.6	4	4.4	3	2.5	TISHOMINGO MS	-54	-43	-74	23	28
7	8.6	8	7.9	4	2.6	UNION MS	-8	-67	-69	31	30
2	5.5			3	5.1	WALTHALL MS			-7	91	77
5	5.3	9	6.9	6	3.6	WARREN MS	30	-48	-32	83	50
3	3.0	7	4.9	4	2.2	WASHINGTON MS	66	-55	-25	86	23
1	3.1	3	5.2			WAYNE MS	67				
8	19.2	5	9.4	2	3.9	WEBSTER MS	-51	-58	-79	14	58
4	17.5	1	4.0	4*	11.4	WILKINSON MS	-77	187	-35	82	98
7	14.1	1	1.5	6	5.9	WINSTON MS	-89	291	-58	53	84
5	10.0	7	11.0			YALOBUSHA MS	9				
4	7.2	2	2.8	5	5.5	YAZOO MS	-62	100	-23	87	81

* SIGNIFICANTLY MORE DEATHS THAN EXPECTED

WHITE FEMALE

CANCER OF THE STOMACH

ICD 151

COUNTY AND STATE	1950-1959 DEATHS	1950-1959 RATE	1960-1969 DEATHS	1960-1969 RATE	1970-1979 DEATHS	1970-1979 RATE	PERCENT CHANGE 60S/50S	PERCENT CHANGE 70S/60S	PERCENT CHANGE 70S/50S	PERCENTILE %CHANGE 70S/50S	PERCENTILE RATES 1970S
MISSOURI MO	2095	9.5	1486	5.4	1034	3.2	-43	-41	-66		
ADAIR MO	9	6.4	11	7.2	4	2.3	12	-69	-65	39	24
ANDREW MO	10	11.9	5	4.5	7	5.2	-62	14	-57	56	77
ATCHISON MO	8	11.8	7	9.7			-19				
AUDRAIN MO	14	10.2	9	5.2	2	.7	-49	-86	-93	01	10
BARRY MO	11	9.1	7	4.2	10	5.2	-55	24	-44	75	77
BARTON MO	10	9.6	6	5.4	11*	10.7	-44	97	11	94	97
BATES MO	10	7.2	6	3.8	2	.9	-47	-77	-88	03	10
BENTON MO	2	3.1	2	2.4	2	1.7	-21	-29	-44	74	16
BOLLINGER MO	8	15.0	8	12.5	4	7.4	-16	-41	-51	66	93
BOONE MO	20	8.6	12	3.8	10	3.0	-56	-23	-66	37	36
BUCHANAN MO	59	9.1	35	5.1	19	2.4	-44	-52	-73	24	27
BUTLER MO	20	12.8	14	7.1	11	4.8	-44	-33	-63	44	72
CALDWELL MO	7	8.2	4	5.2	3	2.8	-36	-46	-65	38	34
CALLAWAY MO	10	6.7	9	5.5	6	3.2	-17	-43	-53	63	41
CAMDEN MO	2	4.6	2	2.5	2	1.5	-45	-42	-68	33	14
CAPE GIRARDEAU MO	15	7.3	18	6.6	11	3.2	-10	-51	-55	59	43
CARROLL MO	16	15.9	8	6.8	3	3.5	-57	-49	-78	16	48
CARTER MO	1	5.4	2	6.5	2	4.7	22	-29	-13	90	70
CASS MO	14	9.7	16	8.9	8	3.5	-8	-61	-64	41	48
CEDAR MO	8	9.8	5	5.1	6	4.8	-48	-5	-51	65	72
CHARITON MO	8	7.8	11	11.2	5	4.0	42	-64	-49	69	60
CHRISTIAN MO	12	16.7	11	11.8	3	2.4	-30	-80	-86	06	26
CLARK MO	2	3.4	5	5.6	4	5.8	65	4	72	97	84
CLAY MO	22	8.8	21	5.3	17	3.1	-39	-42	-65	38	39
CLINTON MO	10	9.6	3	3.0	2	1.9	-69	-37	-80	13	19
COLE MO	11	7.0	11	5.0	5	1.6	-28	-68	-77	18	15
COOPER MO	8	7.2	7	6.5	3	1.8	-10	-72	-74	22	18
CRAWFORD MO	6	9.1	4	3.5	4	2.8	-62	-21	-70	30	32
DADE MO	7	9.8	6	8.8	1	.6	-10	-93	-94	00	10
DALLAS MO	9	15.6	3	4.5	5	4.6	-71	2	-71	29	69
DAVIESS MO	13	13.7	9	8.8	4	4.1	-35	-53	-70	30	62
DE KALB MO	6	9.0	3	4.5	2	3.1	-49	-32	-66	37	39
DENT MO	11	15.5	2	2.1	1	.8	-87	-60	-95	00	10
DOUGLAS MO	4	7.0	7	11.9	5	5.7	70	-52	-18	89	83
DUNKLIN MO	20	13.1	14	6.5	9	3.6	-51	-44	-73	26	50
FRANKLIN MO	23	10.8	19	6.9	16	4.4	-36	-36	-59	51	66
GASCONADE MO	20*	24.7	6	6.4	1	.7	-74	-88	-97	00	10
GENTRY MO	8	9.1	3	2.3	3	2.0	-74	-13	-78	17	20
GREENE MO	65	10.2	44	5.0	29	2.6	-51	-49	-75	21	29
GRUNDY MO	11	8.8	9	6.3	8	6.4	-29	3	-27	85	88
HARRISON MO	7	7.0	9	6.7	4	3.0	-5	-55	-57	56	38
HENRY MO	10	5.9	13	6.9	7	3.6	17	-48	-39	78	49
HICKORY MO	2	4.9	2	5.2	1	1.5	5	-70	-69	33	15
HOLT MO	9	11.7	8	10.4	3	3.2	-11	-69	-73	25	41
HOWARD MO	3	3.7	1	.8	3	3.8	-80	399	2	93	54
HOWELL MO	15	11.4	6	3.4	8	4.2	-70	25	-63	44	64
IRON MO	7	12.7	7	11.7	1	.5	-8	-96	-96	00	09
JACKSON MO	209	7.1	155	4.3	117	2.9	-40	-33	-60	50	35
JASPER MO	54	9.9	40	6.3	27	4.0	-37	-36	-60	49	59
JEFFERSON MO	19	9.3	19	6.1	24	4.9	-35	-19	-47	71	74
JOHNSON MO	14	8.9	9	5.1	5	2.4	-42	-53	-73	25	27
KNOX MO	3	5.7	4	7.7	1	1.5	37	-80	-73	25	15
LACLEDE MO	10	10.1	7	5.5	6	3.5	-45	-37	-65	38	48
LAFAYETTE MO	13	7.9	10	4.7	9	3.7	-40	-21	-53	63	53
LAWRENCE MO	16	9.5	13	6.0	14	5.9	-37	-1	-38	80	84
LEWIS MO	4	5.9	4	3.7			-37				
LINCOLN MO	10	10.0	7	5.8	2	1.5	-42	-74	-85	06	14
LINN MO	19	12.3	15	9.2	3	1.1	-26	-88	-91	02	12
LIVINGSTON MO	17	13.8	12	9.3	4	2.4	-32	-74	-83	09	26
MC DONALD MO	10	13.6	9	9.6	6	6.2	-30	-36	-55	60	87
MACON MO	19	11.9	12	7.2	6	2.7	-39	-62	-77	18	32
MADISON MO	5	9.8	5	8.9	1	1.0	-9	-89	-90	02	11
MARIES MO	8*	22.6	4	8.9			-61				
MARION MO	12	5.6	7	2.5	9	2.9	-55	14	-48	69	35
MERCER MO	10*	21.1	1	1.8	6*	10.5	-92	495	-50	67	97
MILLER MO	13	16.0	8	8.0	2	1.4	-50	-82	-91	01	13
MISSISSIPPI MO	12	20.2	2	2.1	2	1.9	-89	-9	-90	02	19
MONITEAU MO	16*	18.6	8	8.0	3	2.1	-57	-74	-89	03	21
MONROE MO	5	6.1	2	3.5	3	2.9	-43	-17	-52	63	35
MONTGOMERY MO	4	5.1	8	6.8	4	2.9	35	-58	-43	76	35
MORGAN MO	8	12.5	6	7.4	3	2.2	-41	-70	-82	10	23
NEW MADRID MO	14	14.8	7	5.6	3	2.6	-62	-54	-83	10	29
NEWTON MO	11	7.3	8	3.8	9	3.3	-48	-14	-55	59	43
NODAWAY MO	13	7.0	7	3.7	6	2.6	-48	-29	-63	43	30
OREGON MO	10	17.2	2	3.2	3	2.4	-81	-24	-86	05	27
OSAGE MO	8	12.7	1	1.2	6	8.5	-90	581	-33	83	95
OZARK MO	2	5.2	2	4.1	5	8.5	-21	106	64	97	95
PEMISCOT MO	10	9.1	7	5.9	4	2.5	-35	-58	-73	25	28
PERRY MO	7	9.1	4	3.8	5	3.6	-58	-6	-61	48	50
PETTIS MO	18	8.3	12	4.1	13	4.0	-51	-1	-51	65	60
PHELPS MO	9	7.5	10	6.8	8	3.6	-9	-47	-52	64	51
PIKE MO	9	7.8	7	5.7	2	1.0	-27	-82	-87	05	11
PLATTE MO	8	8.5	7	5.8	5	3.2	-32	-45	-62	45	42
POLK MO	11	9.1	9	6.4	2	1.2	-29	-82	-87	04	12
PULASKI MO	8	11.8	8	9.4	9	7.9	-20	-15	-33	83	94
PUTNAM MO	9	14.0	17*	22.1	8*	9.7	58	-56	-30	84	97
RALLS MO	7	12.5	1	2.8	3	3.5	-77	23	-72	26	48
RANDOLPH MO	18	9.6	9	4.0	5	2.1	-58	-48	-78	16	22
RAY MO	9	8.8	12	8.4	6	3.4	-5	-59	-61	47	47
REYNOLDS MO	3	8.8			1	2.7			-70	30	31
RIPLEY MO	8	14.8	4	4.9	5	5.3	-67	7	-64	40	78
ST CHARLES MO	6	3.6	10	3.7	12	2.9	2	-19	-18	89	36
ST CLAIR MO	5	6.6	1	1.5	4	4.0	-78	173	-39	78	59
STE GENEVIEVE MO	8	15.4	3	4.7	3	3.6	-70	-22	-76	19	51
ST FRANCOIS MO	18	8.9	20	7.5	13	4.0	-16	-46	-55	60	59
ST LOUIS COUNTY MO	189	8.7	205	5.8	188	3.6	-33	-37	-58	53	51
SALINE MO	17	9.4	12	6.2	3	1.5	-34	-76	-84	07	14
SCHUYLER MO	5	9.2	3	4.6	1	2.5	-50	-44	-72	26	29

* SIGNIFICANTLY MORE DEATHS THAN EXPECTED

WHITE FEMALE

CANCER OF THE STOMACH

ICD 151

1950-1959 DEATHS	1950-1959 RATE	1960-1969 DEATHS	1960-1969 RATE	1970-1979 DEATHS	1970-1979 RATE	COUNTY AND STATE	PERCENT CHANGE 60S/50S	PERCENT CHANGE 70S/60S	PERCENT CHANGE 70S/50S	PERCENTILE %CHANGE 70S/50S	PERCENTILE RATES 1970S
5	8.1	4	5.1	4	5.6	SCOTLAND MO	-37	10	-31	84	82
12	10.5	10	6.3	6	2.4	SCOTT MO	-39	-62	-77	18	27
5	12.5	5	12.3	2	4.2	SHANNON MO	-2	-65	-66	37	64
6	7.1	3	3.7	2	1.8	SHELBY MO	-48	-51	-74	23	18
10	8.1	8	4.8	5	2.2	STODDARD MO	-41	-54	-73	25	23
4	8.0	1	1.7	2	2.2	STONE MO	-78	27	-73	26	23
9	10.3	7	7.3	8*	9.3	SULLIVAN MO	-29	27	-9	91	96
9	19.2	6	7.9	2	1.4	TANEY MO	-59	-82	-93	01	14
11	11.7	9	7.5	3	1.7	TEXAS MO	-36	-78	-86	06	16
12	6.2	14	6.2	3	1.9	VERNON MO	0	-69	-70	31	19
6	10.7	2	2.8	5	6.6	WARREN MO	-73	134	-38	80	89
9	14.3	5	6.7	5	5.6	WASHINGTON MO	-53	-17	-61	48	82
2	3.2	4	6.5	3	3.7	WAYNE MO	103	-44	13	94	51
8	9.6	6	6.3	1	.6	WEBSTER MO	-34	-91	-94	00	10
4	10.7	1	2.3			WORTH MO	-79				
10	10.3	7	5.5	2	1.8	WRIGHT MO	-46	-68	-83	09	17
438	10.6	209	4.9	115	3.0	ST LOUIS CITY MO	-54	-39	-72	27	38
288	11.2	226	6.8	162	3.9	MONTANA MT	-39	-43	-66		
5	16.6	4	9.6			BEAVERHEAD MT	-42				
3	11.3	1	4.2	1	2.5	BIG HORN MT	-63	-41	-78	16	28
3	11.1	2	5.3			BLAINE MT	-52				
		2	17.1	2	10.7	BROADWATER MT		-37			97
5	9.6	3	6.2	1	1.1	CARBON MT	-36	-82	-89	03	11
1	8.6	2	14.5			CARTER MT	67				
28	12.0	29	8.7	18	4.2	CASCADE MT	-27	-53	-65	38	62
6	19.1	3	8.7			CHOUTEAU MT	-55				
6	11.2	5	6.5	4	4.8	CUSTER MT	-42	-26	-57	55	72
3	18.8	1	4.1	1	5.0	DANIELS MT	-78	21	-73	24	75
8*	19.7	6	13.8	1	1.4	DAWSON MT	-30	-90	-93	01	14
3	3.4	5	5.8	1	.9	DEER LODGE MT	70	-84	-74	24	11
1	5.8			1	5.1	FALLON MT			-13	90	77
9	15.1	2	2.3	5	3.0	FERGUS MT	-85	33	-80	13	38
11	8.3	13	7.0	14.	5.7	FLATHEAD MT	-16	-19	-32	83	83
11	10.2	7	5.1	7	4.7	GALLATIN MT	-50	-7	-54	61	71
1	2.1	2	7.1	3	10.8	GLACIER MT	230	53	404	99	98
		1	13.5			GOLDEN VALLEY MT					
2	16.2					GRANITE MT					
4	9.2	6	8.0	1	1.5	HILL MT	-13	-81	-84	08	14
4	29.4	1	4.8			JEFFERSON MT	-84				
3	19.2	1	7.2	1	5.9	JUDITH BASIN MT	-63	-18	-69	31	85
4	5.8	4	4.9	2	1.7	LAKE MT	-17	-64	-70	30	17
12	9.5	10	6.2	9	4.3	LEWIS AND CLARK MT	-34	-31	-54	60	65
6	17.1	4	10.2	1	1.5	LINCOLN MT	-40	-85	-91	01	14
		2	14.2	1	7.9	MC CONE MT		-44			94
4	13.8			5*	8.5	MADISON MT			-38	79	95
2	28.3					MEAGHER MT					
1	12.6					MINERAL MT					
23	13.0	18	7.6	16	5.6	MISSOULA MT	-42	-26	-57	56	82
6*	22.0	4	12.5	4*	12.3	MUSSELSHELL MT	-43	-2	-44	74	98
8	14.1	8	10.5	1	1.2	PARK#MT	-25	-89	-92	01	12
7*	25.6	3	8.9	4*	9.5	PHILLIPS MT	-65	6	-63	43	97
4	18.8	3	10.0	1	1.7	PONDERA MT	-47	-83	-91	02	17
2	24.3	1	9.1			POWDER RIVER MT	-63				
2	7.0	3	8.3	4	10.2	POWELL MT	18	22	45	96	97
		1	8.7			PRAIRIE MT					
6	9.1	6	7.2	7	5.4	RAVALLI MT	-21	-26	-41	77	79
7	20.3	3	5.6			RICHLAND MT	-73				
3	7.9	2	5.2			ROOSEVELT MT	-35				
2	10.4	4	14.9	2	8.5	ROSEBUD MT	43	-43	-19	89	95
5	15.5					SANDERS MT					
2	10.0	4	10.3	4	8.5	SHERIDAN MT	3	-17	-14	90	96
30	12.2	23	8.5	12	4.0	SILVER BOW MT	-30	-53	-67	35	60
				2	4.9	STILLWATER MT					73
		3	13.1			SWEET GRASS MT					
4	12.6	3	9.4			TETON MT	-26				
		4	11.5	2	6.5	TOOLE MT		-44			88
3	6.4	3	5.7	1	2.3	VALLEY MT	-11	-60	-64	40	24
		1	11.8	1	17.1	WIBAUX MT		45			99
28	11.6	13	3.7	22	4.5	YELLOWSTONE MT	-68	23	-61	48	68
786	10.8	617	6.7	452	4.1	NEBRASKA NB	-38	-39	-62		
16	8.9	7	3.0	9	2.2	ADAMS NB	-67	-26	-75	20	23
8	11.1	8	9.2	5	6.4	ANTELOPE NB	-18	-30	-42	76	88
				1*	74.3	BANNER NB					99
1	28.3					BLAINE NB					
7	12.1	6	9.1	5	7.7	BOONE NB	-25	-16	-36	81	93
7	13.6	5	8.1	5	7.4	BOX BUTTE NB	-40	-9	-45	73	93
10*	42.3	4	13.5	2	5.0	BOYD NB	-68	-63	-88	03	76
4	15.2	3	9.4	1	2.0	BROWN NB	-38	-79	-87	05	20
16	10.7	9	5.3	11	5.5	BUFFALO NB	-51	3	-49	68	81
9	14.1	10*	11.6	2	1.4	BURT NB	-18	-88	-90	02	14
12	17.4	7	7.8	5	7.2	BUTLER NB	-55	-8	-59	52	92
14	13.9	12	9.1	5	3.6	CASS NB	-34	-60	-74	24	51
4	5.6	9	10.6	5	4.2	CEDAR NB	90	-61	-25	86	63
5	20.3	2	5.5	1	2.0	CHASE NB	-73	-63	-90	02	20
4	11.8	4	8.8	1	1.1	CHERRY NB	-25	-88	-91	02	11
5	12.5	2	2.7	2	2.3	CHEYENNE NB	-78	-14	-81	11	25
3	4.1	5	7.2	5	6.3	CLAY NB	76	-12	54	97	87
9	12.9	6	6.9	5	5.5	COLFAX NB	-47	-20	-57	55	81
6	8.7	4	4.4	5	4.6	CUMING NB	-50	6	-47	71	70
15	13.8	10	7.5	6	3.5	CUSTER NB	-46	-54	-75	21	47
4	10.3	5	8.7	8*	9.7	DAKOTA NB	-16	12	-6	91	97
8	16.9	3	4.8	3	4.0	DAWES NB	-72	-15	-76	19	59
11	11.8	7	5.3	10	5.6	DAWSON NB	-55	5	-53	62	82
1	5.3	1	3.9	2	12.0	DEUEL NB	-26	209	128	98	98
10*	19.6	14*	22.2	8*	11.9	DIXON NB	13	-46	-39	79	98
20	11.9	20	8.7	7	2.4	DODGE NB	-27	-73	-80	13	26
169	12.3	107	5.8	93	4.1	DOUGLAS NB	-53	-30	-67	35	61

* SIGNIFICANTLY MORE DEATHS THAN EXPECTED

\# COUNTIES COMBINED; SEE TECHNICAL NOTES

WHITE FEMALE

CANCER OF THE STOMACH — ICD 151

1950-1959 DEATHS	1950-1959 RATE	1960-1969 DEATHS	1960-1969 RATE	1970-1979 DEATHS	1970-1979 RATE	COUNTY AND STATE		PERCENT CHANGE 60S/50S	PERCENT CHANGE 70S/60S	PERCENT CHANGE 70S/50S	PERCENTILE %CHANGE 70S/50S	PERCENTILE RATES 1970S
1	4.5					DUNDY	NB					
6	8.2	4	5.7	3	3.3	FILLMORE	NB	-31	-41	-59	50	45
5	12.6	5	12.3	2	3.4	FRANKLIN	NB	-2	-72	-73	25	47
1	3.8	1	3.5	1	2.1	FRONTIER	NB	-6	-40	-44	75	22
5	7.6	2	3.6	2	4.4	FURNAS	NB	-53	25	-41	77	67
13	8.1	14	6.5	9	4.0	GAGE	NB	-19	-38	-50	67	60
		1	5.2	1	11.2	GARDEN	NB		118			98
1	4.3	3	12.3	1	4.1	GARFIELD	NB	185	-67	-5	92	61
		1	7.3			GOSPER	NB					
				1	56.2	GRANT	NB					99
6*	28.7	4	13.2	1	2.2	GREELEY	NB	-54	-83	-92	01	23
12	7.0	13	5.3	12	3.4	HALL	NB	-24	-36	-52	65	46
6	9.7	6	8.7	3	3.5	HAMILTON	NB	-10	-60	-64	41	49
5	13.0	2	4.0	1	1.9	HARLAN	NB	-69	-53	-86	06	18
		2*	19.4			HAYES	NB					
2	7.6	3	10.2	1	3.5	HITCHCOCK	NB	35	-66	-54	61	48
4	5.3	12*	12.7	8	6.8	HOLT	NB	142	-47	28	95	90
1	12.5			1	11.2	HOOKER	NB			-11	90	98
9*	24.9	5	10.1	1	1.7	HOWARD	NB	-60	-83	-93	01	17
10	11.9	5	5.2	1	1.2	JEFFERSON	NB	-56	-76	-90	02	12
5	10.9	5	8.7	4	7.1	JOHNSON	NB	-20	-18	-35	82	92
3	7.0	3	4.9	5	8.4	KEARNEY	NB	-31	73	20	95	95
5	18.4	4	9.9	2	3.4	KEITH	NB	-46	-66	-82	11	46
		1	13.4			KEYA PAHA	NB					
		3	11.1	2	2.7	KIMBALL	NB		-75			32
9	12.8	5	6.1	5	5.0	KNOX	NB	-52	-19	-61	47	75
57	7.4	67	7.2	33	3.0	LANCASTER	NB	-2	-58	-59	51	38
14	11.2	7	4.0	9	4.6	LINCOLN	NB	-64	13	-59	51	69
11	7.0	15	7.9	9	4.4	MADISON	NB	13	-44	-37	81	66
4	7.2	3	5.2	2	3.4	MERRICK	NB	-28	-35	-53	63	46
4	12.3	4	9.3	2	2.8	MORRILL	NB	-24	-70	-77	17	33
8*	21.4	3	6.5	2	3.6	NANCE	NB	-70	-45	-83	08	49
5	6.9	3	4.3	4	4.3	NEMAHA	NB	-37	-1	-37	80	65
4	6.6			2	3.2	NUCKOLLS	NB			-51	65	42
20*	18.3	10	7.5	5	2.3	OTOE	NB	-59	-70	-88	04	24
9	17.0	2	3.7	3	4.3	PAWNEE	NB	-78	16	-75	22	65
1	5.7	1	4.5	1	3.1	PERKINS	NB	-22	-31	-46	72	40
14*	24.2	5	5.0	4	2.9	PHELPS	NB	-79	-42	-88	03	36
3	6.6	3	4.9	1	.8	PIERCE	NB	-26	-84	-88	03	10
17	17.3	6	4.4	6	3.1	PLATTE	NB	-75	-30	-82	10	39
6	11.4	2	3.0	6	7.6	POLK	NB	-74	157	-33	83	93
8	12.1	6	7.2	6	5.0	RED WILLOW	NB	-41	-30	-59	52	76
9	8.6	6	4.1	5	2.9	RICHARDSON	NB	-52	-30	-66	37	35
4*	26.9	3	16.5			ROCK	NB	-39				
8	7.7	8	7.0	7	4.3	SALINE	NB	-9	-39	-44	74	64
6	10.4			4	2.6	SARPY	NB			-75	22	30
17*	16.9	6	4.4	7	5.4	SAUNDERS	NB	-74	24	-68	34	80
16	12.6	10	5.8	10	4.5	SCOTTS BLUFF	NB	-54	-23	-64	40	67
7	8.3	9	8.1	2	2.8	SEWARD	NB	-3	-66	-67	36	33
4	8.7	2	3.9	3	4.6	SHERIDAN	NB	-55	18	-47	71	69
3	9.0	3	9.6	4	10.9	SHERMAN	NB	7	13	21	95	98
3	10.6	1	4.6	4	9.2	STANTON	NB	-56	98	-13	90	96
7	9.5	11*	14.0	5	4.7	THAYER	NB	48	-67	-51	66	70
		2*	29.9	1	16.4	THOMAS	NB		-45			99
		7*	20.9	3	5.8	THURSTON	NB		-72			84
2	4.3	7	12.5	1	1.4	VALLEY	NB	190	-89	-68	34	13
12	16.7	12*	12.5	5	6.0	WASHINGTON	NB	-25	-51	-64	41	85
3	6.1	5	7.8	10*	12.1	WAYNE	NB	29	55	100	98	98
7	14.1	7	10.1	5	4.5	WEBSTER	NB	-28	-55	-68	34	68
6	7.0	7	6.1	2	1.7	YORK	NB	-13	-71	-75	21	17
63	9.5	69	5.7	82	3.4	NEVADA	NV	-40	-41	-64		
3	11.2			2	3.4	CHURCHILL	NV			-70	30	46
18	8.8	34	6.8	46	3.7	CLARK	NV	-23	-45	-58	54	53
		1	4.1	1	1.2	DOUGLAS	NV		-71			12
1	2.6	1	2.1			ELKO	NV	-20				
1	56.2					ESMERALDA	NV					
1	26.5					EUREKA	NV					
1	5.7	1	4.3	3	10.3	HUMBOLDT	NV	-24	138	81	98	97
2	28.6					LANDER	NV					
3*	27.8	1	8.4			LINCOLN	NV	-70				
		2	6.8	2	5.4	LYON	NV		-21			80
1	5.1			3	11.7	MINERAL	NV			131	98	98
4*	29.8	2	12.9			NYE	NV	-57				
				1	6.8	PERSHING	NV					91
1	31.3					STOREY	NV					
24	9.5	24	5.6	16	2.3	WASHOE	NV	-41	-58	-75	21	25
		2	7.2	4	9.3	WHITE PINE	NV		29			96
3	16.0	1	4.9	4	3.3	ORMSBY-CARSON CITY	NV	-70	-32	-79	15	44
389	11.6	297	7.2	250	4.7	NEW HAMPSHIRE	NH	-38	-35	-60		
14	8.1	9	4.1	8	2.9	BELKNAP	NH	-49	-30	-64	41	36
8	7.3	11	8.2	4	2.1	CARROLL	NH	12	-74	-71	28	22
23	9.6	18	6.4	20	5.4	CHESHIRE	NH	-34	-15	-44	75	80
27	14.7	22*	9.8	18*	6.6	COOS	NH	-33	-33	-55	59	89
31	10.7	17	5.0	24	5.8	GRAFTON	NH	-54	17	-46	73	84
154*	16.0	103*	8.3	92*	5.8	HILLSBOROUGH	NH	-48	-30	-63	42	84
44	8.7	33	5.8	21	3.0	MERRIMACK	NH	-33	-48	-65	39	39
45	9.6	36	5.9	37	4.2	ROCKINGHAM	NH	-39	-28	-56	57	64
34	11.7	36*	10.4	19	4.2	STRAFFORD	NH	-11	-60	-64	40	62
9	5.8	12	6.3	7	2.4	SULLIVAN	NH	9	-62	-59	52	27
3385*	13.5	2896*	8.6	2514*	5.9	NEW JERSEY	NJ	-36	-32	-56		
94	11.3	81	7.4	67	4.7	ATLANTIC	NJ	-35	-37	-59	52	71
436*	14.7	387*	9.0	378*	6.7	BERGEN	NJ	-39	-25	-54	60	90
71	11.4	65	6.8	69	5.0	BURLINGTON	NJ	-40	-27	-56	57	75
187*	12.7	126	6.5	126	5.1	CAMDEN	NJ	-49	-21	-60	49	77
28	10.1	36	8.1	27	3.7	CAPE MAY	NJ	-20	-55	-64	41	52

* SIGNIFICANTLY MORE DEATHS THAN EXPECTED

WHITE FEMALE

CANCER OF THE STOMACH — ICD 151

1950-1959 DEATHS	1950-1959 RATE	1960-1969 DEATHS	1960-1969 RATE	1970-1979 DEATHS	1970-1979 RATE	COUNTY AND STATE		PERCENT CHANGE 60S/50S	PERCENT CHANGE 70S/60S	PERCENT CHANGE 70S/50S	PERCENTILE %CHANGE 70S/50S	PERCENTILE RATES 1970S
52	11.2	40	7.0	41	5.7	CUMBERLAND	NJ	-37	-19	-49	68	83
586*	13.1	440*	8.9	283*	5.9	ESSEX	NJ	-32	-35	-55	59	84
47	10.5	34	5.5	38	4.5	GLOUCESTER	NJ	-47	-19	-57	55	68
477*	15.9	390*	11.1	284*	7.4	HUDSON	NJ	-30	-33	-53	62	93
31	11.5	25	7.0	24	5.6	HUNTERDON	NJ	-39	-20	-52	64	82
152*	13.2	101	7.1	116*	6.6	MERCER	NJ	-46	-8	-50	67	89
207*	17.2	201*	10.6	181*	6.7	MIDDLESEX	NJ	-38	-37	-61	47	90
149	10.6	184*	9.0	145	4.8	MONMOUTH	NJ	-15	-46	-54	61	73
112	11.2	87	5.9	98	5.0	MORRIS	NJ	-47	-16	-55	58	75
51	11.3	72	7.0	103	4.2	OCEAN	NJ	-38	-40	-63	42	62
265*	14.9	263*	11.0	181*	6.4	PASSAIC	NJ	-26	-42	-57	55	88
19	9.0	14	5.6	11	3.2	SALEM	NJ	-38	-43	-65	39	41
60	12.3	55	7.4	51	5.0	SOMERSET	NJ	-40	-32	-59	52	76
23	10.4	32*	9.8	30	6.5	SUSSEX	NJ	-6	-33	-37	80	89
301*	15.4	235*	8.5	230*	6.8	UNION	NJ	-45	-20	-56	58	91
37	10.9	28	6.1	31	5.6	WARREN	NJ	-44	-9	-49	68	82
291*	14.1	268*	8.6	285*	6.3	NEW MEXICO	NM	-39	-27	-56		
69	13.7	64	7.2	90*	6.1	BERNALILLO	NM	-48	-15	-56	58	86
1	11.1	1	8.6			CATRON	NM	-22				
10	7.8	12	6.4	15	5.5	CHAVES	NM	-18	-15	-30	85	81
7	11.4	9*	12.4	8*	11.2	COLFAX	NM	8	-10	-2	92	98
11	12.3	8	5.9	7	3.6	CURRY	NM	-52	-38	-70	29	51
2	12.2	2	11.1			DE BACA	NM	-9				
19	14.3	9	5.2	12	4.2	DONA ANA	NM	-64	-20	-71	28	62
8	6.0	6	3.5	10	3.9	EDDY	NM	-41	10	-35	82	57
10	16.3	7	9.8	6	5.4	GRANT	NM	-40	-45	-67	36	80
5*	30.6	5*	24.5	1	4.0	GUADALUPE	NM	-20	-84	-87	05	59
2	24.7	2*	43.5			HARDING	NM	76				
		4*	21.0	1	3.7	HIDALGO	NM		-83			52
7	6.6	10	6.7	5	2.2	LEA	NM	1	-68	-67	35	23
4	11.7	2	5.3	2	4.4	LINCOLN	NM	-54	-17	-62	45	66
2	24.4	1	6.3			LOS ALAMOS	NM	-74				
3	10.1	2	3.8	3	3.8	LUNA	NM	-62	0	-62	45	55
5	17.9	2	2.4	5	8.6	MC KINLEY	NM	-86	255	-52	64	96
4	19.3	3	11.5	2	7.7	MORA	NM	-40	-33	-60	49	94
4	12.3	14*	18.5	7	5.9	OTERO	NM	50	-68	-52	63	84
7	13.6	9*	13.0	9*	12.9	QUAY	NM	-5	-1	-6	92	98
7	10.8	18*	25.0	10*	13.9	RIO ARRIBA	NM	132	-44	29	95	99
8	13.5	2	2.5	5	6.6	ROOSEVELT	NM	-81	161	-51	65	89
8*	39.5	4	12.1	4	7.6	SANDOVAL	NM	-70	-37	-81	12	93
7	13.0	3	2.2	9	5.6	SAN JUAN	NM	-83	156	-57	56	82
15	17.7	13*	12.6	14*	13.1	SAN MIGUEL	NM	-29	4	-26	86	99
20	16.5	22*	11.0	22*	8.0	SANTA FE	NM	-33	-27	-51	65	94
3	7.5	7	12.7	1	1.1	SIERRA	NM	69	-91	-85	06	12
6	20.4	3	9.4	7*	15.7	SOCORRO	NM	-54	67	-23	87	99
12*	24.7	13*	19.3	12*	18.5	TAOS	NM	-22	-4	-25	86	99
5	23.7	2	6.0	2	5.8	TORRANCE	NM	-75	-3	-75	20	84
11*	36.7	2	5.6	3	8.4	UNION	NM	-85	51	-77	18	95
9	21.5	7	8.6	13*	8.8	VALENCIA	NM	-60	3	-59	51	96
10353*	13.3	8092*	8.5	6498*	5.9	NEW YORK	NY	-37	-31	-56		
177	12.3	114	6.6	98	4.7	ALBANY	NY	-47	-28	-62	45	71
16	6.1	17	6.5	10	3.2	ALLEGANY	NY	5	-50	-47	70	42
1260*	18.8	660*	8.2	622*	8.1	BRONX	NY	-56	-1	-57	56	94
92	9.0	80	6.4	78	5.1	BROOME	NY	-29	-20	-43	75	77
42	10.0	31	6.2	39*	6.9	CATTARAUGUS	NY	-38	11	-31	84	91
51	11.7	37	7.6	14	2.6	CAYUGA	NY	-35	-66	-78	17	30
103	11.9	60	6.0	43	3.6	CHAUTAUQUA	NY	-50	-40	-70	31	51
54	10.2	39	6.3	35	5.2	CHEMUNG	NY	-38	-18	-49	68	78
19	7.5	16	5.3	13	4.3	CHENANGO	NY	-29	-18	-42	76	65
41*	16.4	25	8.6	19	5.4	CLINTON	NY	-47	-37	-67	36	80
33	10.7	35*	9.8	21	5.1	COLUMBIA	NY	-8	-48	-52	64	77
14	6.2	11	4.2	10	3.7	CORTLAND	NY	-33	-12	-41	77	51
26	10.0	11	3.9	11	3.2	DELAWARE	NY	-61	-19	-68	33	41
79	8.3	88	7.7	67	5.0	DUTCHESS	NY	-8	-35	-40	78	75
564*	12.3	490*	8.6	382*	5.7	ERIE	NY	-30	-34	-53	62	83
14	7.0	16	6.9	8	2.7	ESSEX	NY	-1	-60	-61	48	32
28	12.1	28*	10.5	18	6.4	FRANKLIN	NY	-14	-39	-47	71	88
24	6.6	12	2.9	16	3.8	FULTON	NY	-56	34	-42	77	56
30	10.8	20	6.4	18	4.8	GENESEE	NY	-41	-25	-56	58	72
17	8.2	27*	10.6	14	4.4	GREENE	NY	30	-58	-46	72	66
3	14.2	3	10.0	1	2.5	HAMILTON	NY	-30	-75	-82	10	28
40	10.4	25	5.6	25	4.6	HERKIMER	NY	-46	-19	-56	57	69
36	6.7	28	4.5	22	3.5	JEFFERSON	NY	-33	-23	-48	70	48
2118*	18.0	1068*	7.9	947*	6.9	KINGS(BROOKLYN)	NY	-56	-12	-61	46	91
11	8.6	7	4.7	6	3.5	LEWIS	NY	-45	-26	-59	51	48
23	10.0	4	1.4	7	2.2	LIVINGSTON	NY	-86	55	-78	16	23
19	7.1	15	4.7	17	4.8	MADISON	NY	-34	2	-33	83	72
327	10.9	251	6.8	198	4.3	MONROE	NY	-38	-36	-60	48	65
51	13.1	37	7.9	22	4.5	MONTGOMERY	NY	-39	-44	-66	37	67
437*	12.5	512*	8.9	503*	6.5	NASSAU	NY	-29	-27	-48	70	89
1427*	15.4	1708*	17.7	596*	6.4	NEW YORK(MANHATTAN)	NY	15	-64	-58	53	88
87	9.7	76	6.6	86*	6.1	NIAGARA	NY	-32	-7	-37	81	86
120	8.8	72	4.4	96	5.3	ONEIDA	NY	-50	20	-40	78	78
159	8.4	143	6.0	123	4.3	ONONDAGA	NY	-28	-29	-50	68	64
27	7.2	22	4.9	22	3.8	ONTARIO	NY	-31	-23	-47	71	55
114	11.5	93	7.8	82*	5.7	ORANGE	NY	-32	-27	-50	67	83
14	7.5	13	6.3	19*	7.4	ORLEANS	NY	-16	17	-2	92	93
26	5.6	33	6.5	24	4.0	OSWEGO	NY	14	-39	-30	85	59
28	8.2	15	3.8	8	1.8	OTSEGO	NY	-54	-51	-78	17	18
21	14.3	18	7.8	15	4.4	PUTNAM	NY	-45	-44	-69	32	66
1163*	15.0	825*	7.3	902*	6.6	QUEENS	NY	-51	-10	-56	57	89
91	10.8	71	7.5	44	3.9	RENSSELAER	NY	-31	-48	-64	40	56
124*	13.4	95	7.4	116*	6.8	RICHMOND	NY	-45	-8	-49	68	91
53	9.7	49	6.2	67*	5.5	ROCKLAND	NY	-37	-11	-44	75	81
42	7.8	37	6.1	36	5.2	ST LAWRENCE	NY	-21	-15	-33	83	78
34	7.9	40	7.3	35	4.9	SARATOGA	NY	-7	-33	-38	80	73
100	12.2	68	6.6	60	4.8	SCHENECTADY	NY	-46	-27	-61	48	72

* SIGNIFICANTLY MORE DEATHS THAN EXPECTED

WHITE FEMALE | CANCER OF THE STOMACH | ICD 151

1950-1959 DEATHS	1950-1959 RATE	1960-1969 DEATHS	1960-1969 RATE	1970-1979 DEATHS	1970-1979 RATE	COUNTY AND STATE		PERCENT CHANGE 60S/50S	PERCENT CHANGE 70S/60S	PERCENT CHANGE 70S/50S	PERCENTILE %CHANGE 70S/50S	PERCENTILE RATES 1970S
10	6.7	10	6.2	12	7.1	SCHOHARIE	NY	-8	15	6	93	92
4	4.4	1	1.1	3	2.6	SCHUYLER	NY	-74	127	-41	77	30
17	7.6	14	6.0	8	2.8	SENECA	NY	-21	-54	-64	41	33
53	10.1	32	5.3	26	3.9	STEUBEN	NY	-48	-26	-62	46	57
231	10.7	280	7.3	312*	5.4	SUFFOLK	NY	-32	-26	-49	68	80
33	13.3	28	9.5	25	6.6	SULLIVAN	NY	-28	-31	-51	66	89
13	7.0	9	4.0	8	3.4	TIOGA	NY	-43	-16	-52	64	46
32	10.8	26	7.4	11	2.9	TOMPKINS	NY	-32	-60	-73	25	36
70	10.9	52	6.6	47	4.6	ULSTER	NY	-40	-29	-57	55	70
19	7.3	20	6.4	11	2.7	WARREN	NY	-12	-58	-63	43	31
34	12.0	28	8.3	12	3.0	WASHINGTON	NY	-31	-64	-75	21	37
37	9.7	29	6.4	25	4.8	WAYNE	NY	-34	-25	-50	67	73
398	11.1	398*	8.4	367*	6.3	WESTCHESTER	NY	-24	-25	-43	75	87
15	8.2	8	3.7	9	3.7	WYOMING	NY	-56	1	-55	59	53
8	6.2	12	7.9	5	3.3	YATES	NY	29	-58	-46	72	45
870	7.4	796	4.7	727	3.0	NORTH CAROLINA	NC	-36	-36	-59		
11	4.5	15	4.3	17	3.4	ALAMANCE	NC	-5	-21	-24	87	47
5	9.5			4	3.0	ALEXANDER	NC			-69	33	37
12*	29.5	4	8.3	4	5.1	ALLEGHANY	NC	-72	-38	-83	09	77
7	11.3	3	2.9	2	2.0	ANSON	NC	-74	-33	-83	09	20
3	3.4	3	2.5	6	4.5	ASHE	NC	-26	77	32	96	68
		4	7.2	2	3.0	AVERY	NC		-58			38
5	5.7	8	5.9	6	3.3	BEAUFORT	NC	2	-44	-43	76	44
6	10.9	4	6.1	3	4.6	BERTIE	NC	-44	-25	-58	54	69
2	3.7	2	3.1	4	4.2	BLADEN	NC	-16	34	13	94	63
1	4.6	5	7.5	5	3.4	BRUNSWICK	NC	62	-54	-26	86	47
39	7.1	34	4.7	33	3.4	BUNCOMBE	NC	-34	-28	-53	63	45
11	7.1	7	3.0	8	2.5	BURKE	NC	-58	-17	-65	39	28
13	6.1	12	3.9	10	2.3	CABARRUS	NC	-35	-41	-62	45	25
12	9.3	10	5.0	11	3.7	CALDWELL	NC	-46	-27	-61	47	51
1	5.0			1	3.9	CAMDEN	NC			-21	88	58
5	6.2	11	8.7	8	4.5	CARTERET	NC	41	-49	-28	85	67
2	4.2	3	5.8	2	2.8	CASWELL	NC	38	-52	-33	83	33
9	4.0	10	3.1	14	3.0	CATAWBA	NC	-21	-4	-24	87	38
6	7.9	4	3.9	5	3.5	CHATHAM	NC	-51	-10	-56	58	49
7	10.1	8	9.0	5	4.2	CHEROKEE	NC	-10	-53	-58	54	64
3	10.7	4	12.5			CHOWAN	NC	16				
4	13.7	1	2.7	1	3.1	CLAY	NC	-80	13	-77	17	40
18	8.9	12	4.7	14	3.6	CLEVELAND	NC	-47	-23	-59	51	51
6	5.3	5	3.2	4	2.0	COLUMBUS	NC	-40	-37	-62	45	20
13	14.4	5	3.6	4	2.2	CRAVEN	NC	-75	-37	-84	07	24
15	8.0	12	4.5	20	4.7	CUMBERLAND	NC	-45	5	-42	76	70
2	7.6	2	6.6	4	8.4	CURRITUCK	NC	-13	27	10	94	95
5	17.8	1	2.5	4	6.4	DARE	NC	-86	151	-64	41	88
20	9.2	11	3.3	7	1.4	DAVIDSON	NC	-64	-58	-85	07	13
3	4.6	2	2.3	5	4.1	DAVIE	NC	-50	80	-11	90	61
5	5.6	6	5.0	5	3.0	DUPLIN	NC	-11	-41	-47	70	37
17	5.7	20	5.1	17	3.0	DURHAM	NC	-10	-41	-47	71	39
9	8.6	6	4.7	7	3.7	EDGECOMBE	NC	-45	-22	-57	56	52
29	7.2	30	4.6	34	3.3	FORSYTH	NC	-36	-28	-54	61	44
3	4.2	6	6.4	3	2.6	FRANKLIN	NC	51	-60	-40	78	29
22	6.1	24	4.5	18	2.4	GASTON	NC	-27	-46	-61	48	27
1	3.3					GATES	NC					
1	7.4	3	11.3			GRAHAM	NC	52				
4	4.5			1	.9	GRANVILLE	NC			-80	14	11
1	5.1	3	6.7	3	6.7	GREENE	NC	30	1	31	96	90
41	6.8	35	3.6	37	2.7	GUILFORD	NC	-46	-27	-61	48	31
4	3.7	6	4.3	9	4.7	HALIFAX	NC	18	10	29	95	71
15	12.6	13	8.0	11	4.5	HARNETT	NC	-36	-44	-65	40	67
8	5.2	11	5.1	9	3.1	HAYWOOD	NC	0	-40	-41	77	39
13	8.0	10	4.2	16	4.3	HENDERSON	NC	-47	1	-47	71	64
4	13.7	4	7.3	4	4.7	HERTFORD	NC	-47	-36	-66	37	70
		3	9.1	1	1.7	HOKE	NC		-81			16
3	13.6	2	6.4	1	2.3	HYDE	NC	-53	-64	-83	09	25
16	8.2	15	5.2	6	1.5	IREDELL	NC	-37	-71	-81	11	15
10	13.4	4	4.6	4	3.7	JACKSON	NC	-66	-20	-73	25	51
20	11.3	20	8.7	16	5.0	JOHNSTON	NC	-23	-43	-56	57	75
3	34.2			2	5.2	JONES	NC			-85	07	78
6	11.2	2	1.9	4	2.6	LEE	NC	-83	37	-77	18	29
12	11.4	8	6.3	3	1.5	LENOIR	NC	-45	-76	-87	04	14
4	4.3	3	2.3	5	2.9	LINCOLN	NC	-47	27	-32	83	36
12	15.3	5	3.8	9	4.7	MC DOWELL	NC	-75	24	-69	31	71
3	4.0	5	5.5	5	4.0	MACON	NC	38	-26	1	92	60
10	12.4	8	8.8	7	5.9	MADISON	NC	-29	-33	-52	63	84
3	5.7	4	6.6	3	3.1	MARTIN	NC	16	-53	-45	73	40
43	6.0	44	4.3	43	2.9	MECKLENBURG	NC	-28	-34	-53	63	34
4	6.8	7	9.2	8	8.5	MITCHELL	NC	35	-7	26	95	96
6	9.2	5	6.8	4	4.0	MONTGOMERY	NC	-26	-42	-57	57	59
8	7.3	5	3.3	8	3.0	MOORE	NC	-55	-10	-59	50	37
15	12.9	10	6.2	6	2.5	NASH	NC	-52	-60	-81	12	28
17	8.0	15	5.5	7	1.7	NEW HANOVER	NC	-31	-68	-78	16	17
3	5.1	5	7.2	3	4.1	NORTHAMPTON	NC	41	-44	-20	88	60
4	3.8	5	5.3	2	1.1	ONSLOW	NC	37	-80	-72	27	11
4	5.0	5	3.2	6	2.7	ORANGE	NC	-36	-17	-47	71	31
				3	5.8	PAMLICO	NC					84
2	3.2	5	5.2	6	4.6	PASQUOTANK	NC	64	-13	43	96	68
1	2.6	1	2.3	6	7.9	PENDER	NC	-10	243	208	99	94
		1	3.2	2	4.5	PERQUIMANS	NC		39			67
8	13.5	5	6.1	2	2.3	PERSON	NC	-55	-61	-83	10	26
9	8.4	13	7.6	8	3.0	PITT	NC	-9	-60	-64	40	38
2	3.9	6	8.0			POLK	NC	106				
13	8.1	14	4.7	12	3.0	RANDOLPH	NC	-42	-38	-64	42	36
6	7.0	6	4.3	4	1.9	RICHMOND	NC	-39	-55	-72	26	19
8	5.7	3	1.5	5	2.1	ROBESON	NC	-73	38	-63	42	21
13	6.5	19	6.7	10	2.4	ROCKINGHAM	NC	2	-63	-63	44	27
25	9.5	22	6.1	9	1.7	ROWAN	NC	-36	-72	-82	11	16
16	10.2	8	3.6	12	3.8	RUTHERFORD	NC	-65	5	-63	43	54
14	11.7	12	8.1	4	1.9	SAMPSON	NC	-31	-76	-84	08	19
5	9.9	2	2.8	3	3.6	SCOTLAND	NC	-72	27	-64	40	50

* SIGNIFICANTLY MORE DEATHS THAN EXPECTED

1950-1959 DEATHS	1950-1959 RATE	1960-1969 DEATHS	1960-1969 RATE	1970-1979 DEATHS	1970-1979 RATE	COUNTY AND STATE	PERCENT CHANGE 60S/50S	PERCENT CHANGE 70S/60S	PERCENT CHANGE 70S/50S	PERCENTILE %CHANGE 70S/50S	PERCENTILE RATES 1970S
9	6.5	10	5.8	11	4.2	STANLY NC	-11	-27	-35	82	64
4	5.9	3	3.7	5	3.6	STOKES NC	-37	-3	-39	79	50
16	10.5	8	3.7	7	2.2	SURRY NC	-64	-41	-79	15	23
1	3.3	2	4.9			SWAIN NC	49				
9	17.6	2	2.7	2	2.0	TRANSYLVANIA NC	-85	-26	-88	03	20
		1	5.6			TYRRELL NC					
8	6.3	12	6.8	4	1.5	UNION NC	7	-78	-76	19	15
7	9.1	9	8.3	2	1.7	VANCE NC	-9	-79	-81	12	16
17	4.2	21	3.2	23	2.6	WAKE NC	-24	-19	-38	80	29
2	4.9	3	5.9	1	2.1	WARREN NC	21	-65	-58	54	21
1	4.0	3	8.5	3	6.3	WASHINGTON NC	115	-26	60	97	88
12	17.2	6	6.5	4	2.7	WATAUGA NC	-62	-58	-84	07	32
9	7.6	13	6.2	10	3.6	WAYNE NC	-19	-42	-53	63	50
8	5.4	9	4.2	9	3.3	WILKES NC	-22	-23	-40	78	43
6	4.0	8	5.0	7	3.0	WILSON NC	24	-40	-26	86	37
3	3.5	4	3.3	9	5.8	YADKIN NC	-6	75	64	97	84
7	11.7	6	8.5	4	4.6	YANCEY NC	-27	-46	-61	47	69
417*	17.0	297*	9.6	231*	6.0	NORTH DAKOTA ND	-44	-38	-65		
6*	33.9	1	2.6	3	7.0	ADAMS ND	-92	174	-79	15	91
13	15.6	12*	11.9	8	7.7	BARNES ND	-24	-35	-51	66	93
6	16.6	3	6.8	4	8.1	BENSON ND	-59	19	-51	65	94
8	16.5	10*	17.3	1	.9	BOTTINEAU ND	4	-95	-95	00	10
6*	34.3	3	13.8	1	2.1	BOWMAN ND	-60	-85	-94	00	22
8*	27.4	5*	15.3	1	2.8	BURKE ND	-44	-82	-90	02	33
13	11.7	12	7.0	17*	7.6	BURLEIGH ND	-40	8	-35	81	93
37	14.4	21	6.0	23	4.9	CASS ND	-59	-18	-66	36	73
11*	24.5	5	11.1	3	4.5	CAVALIER ND	-55	-59	-82	11	68
5	15.6	1	1.5	4	5.6	DICKEY ND	-91	281	-64	41	82
2	16.4	2	6.5	3	10.0	DIVIDE ND	-60	54	-39	79	97
2	11.8	1	5.0	1	3.9	DUNN ND	-57	-22	-67	36	57
3	16.6	1	5.1	2	5.1	EDDY ND	-69	-1	-70	31	76
3	9.3	5	13.6	6*	14.4	EMMONS ND	46	6	55	97	99
3	15.7	4	13.2	1	2.8	FOSTER ND	-16	-79	-82	10	33
4*	26.3	3	16.3	1	3.3	GOLDEN VALLEY ND	-38	-80	-87	04	44
32*	17.6	17	7.3	14	5.3	GRAND FORKS ND	-59	-27	-70	30	78
1	6.6	1	3.6	1	4.6	GRANT ND	-45	28	-30	85	70
6*	25.0	2	7.2			GRIGGS ND	-71				
5	22.1	3	12.8	3	10.0	HETTINGER ND	-42	-22	-55	59	97
6*	28.1	4*	20.5	2	8.0	KIDDER ND	-27	-61	-72	27	94
4	10.5	9*	19.7	1	2.1	LA MOURE ND	87	-89	-80	14	22
4	19.6	4*	18.2	2*	79.0	LOGAN ND	-7	334	304	99	99
9	17.5	6	11.3	7*	11.0	MC HENRY ND	-35	-3	-37	80	98
3	12.2	1	2.8	4	10.2	MC INTOSH ND	-77	264	-16	89	97
8*	40.4	2	8.1	3	8.2	MC KENZIE ND	-80	1	-80	14	95
10*	19.5	9*	17.4	12*	14.9	MC LEAN ND	-10	-15	-24	87	99
3	11.5	4	12.6	4	8.2	MERCER ND	9	-35	-29	85	95
8	9.6	14*	14.8	7	5.5	MORTON ND	54	-63	-43	76	81
9*	22.1	3	6.4	4	5.8	MOUNTRAIL ND	-71	-9	-74	24	84
9*	23.0	2	4.0	4	8.4	NELSON ND	-82	107	-64	42	95
3*	41.7					OLIVER ND					
10	16.8	3	4.5	3	4.1	PEMBINA ND	-73	-9	-75	20	62
6	23.1	3	7.7	4	11.2	PIERCE ND	-67	45	-52	64	98
11	19.9	7	8.8	8	8.2	RAMSEY ND	-56	-7	-59	52	95
5	10.2	6	10.5	3	5.7	RANSOM ND	2	-46	-45	74	83
4	20.6	2	13.3	3*	12.9	RENVILLE ND	-36	-3	-38	80	98
25*	28.3	13*	11.3	3	1.5	RICHLAND ND	-60	-86	-95	00	15
2	7.8			1	3.2	ROLETTE ND			-59	52	42
5	15.5	4	11.6	2	6.6	SARGENT ND	-25	-43	-58	55	89
1	7.0			2	10.5	SHERIDAN ND			49	97	97
1	13.8	1	28.1			SLOPE ND	104				
14*	25.4	15*	18.3	3	2.8	STARK ND	-28	-85	-89	03	33
6*	32.6	2	9.5	3*	11.7	STEELE ND	-71	23	-64	40	98
11	9.4	18*	13.5	8	5.0	STUTSMAN ND	44	-63	-46	72	76
10*	40.4	3	12.0	4*	11.7	TOWNER ND	-70	-3	-71	28	98
10	17.6	5	6.1	6	5.1	TRAILL ND	-66	-15	-71	28	77
15*	20.1	10	9.6	5	4.1	WALSH ND	-52	-57	-79	14	62
22	14.6	16	7.6	15	5.7	WARD ND	-48	-25	-61	47	83
2	4.3	6	13.0	3	4.9	WELLS ND	204	-63	13	94	73
7	9.0	13*	12.5	8	6.7	WILLIAMS ND	39	-47	-26	86	90
4197	10.4	3189	6.4	2334	3.9	OHIO OH	-39	-39	-63		
7	5.6	6	4.8	7	5.2	ADAMS OH	-15	8	-8	91	78
30	6.6	30	5.3	13	2.1	ALLEN OH	-20	-61	-69	32	21
29	13.4	9	3.7	15	5.1	ASHLAND OH	-72	38	-62	45	77
70*	15.5	42	7.9	23	3.5	ASHTABULA OH	-49	-56	-77	17	48
18	6.9	16	5.6	10	3.6	ATHENS OH	-19	-36	-48	70	51
22	11.8	11	4.8	11	4.2	AUGLAIZE OH	-59	-14	-65	39	62
44	9.0	26	4.8	22	3.8	BELMONT OH	-47	-20	-58	54	55
8	5.4	7	4.5	4	2.3	BROWN OH	-17	-49	-58	54	25
53	7.8	45	5.1	44	3.7	BUTLER OH	-35	-27	-53	63	53
11	10.4	6	5.0	8	6.1	CARROLL OH	-52	22	-42	76	86
12	7.8	13	7.0	4	2.5	CHAMPAIGN OH	-9	-64	-67	35	29
43	7.5	37	4.9	29	3.2	CLARK OH	-34	-36	-58	54	41
20	8.6	14	4.4	15	3.5	CLERMONT OH	-49	-20	-60	50	48
13	7.7	8	4.3	4	1.9	CLINTON OH	-44	-56	-75	21	19
61	11.6	50	7.9	21	2.9	COLUMBIANA OH	-32	-63	-75	22	35
28*	15.4	13	6.5	5	1.9	COSHOCTON OH	-58	-70	-87	04	19
40*	16.4	21	7.4	10	2.8	CRAWFORD OH	-55	-61	-83	09	34
902*	13.9	741*	9.1	529*	5.6	CUYAHOGA OH	-34	-39	-60	49	82
15	5.6	15	4.8	10	2.3	DARKE OH	-13	-53	-59	50	24
9	5.6	5	2.8	13	5.7	DEFIANCE OH	-50	101	1	92	83
12	6.8	9	4.5	6	2.3	DELAWARE OH	-35	-48	-66	37	25
26	9.2	17	5.0	15	3.4	ERIE OH	-45	-32	-63	43	47
27	9.2	19	4.8	17	3.2	FAIRFIELD OH	-47	-33	-65	39	43
10	7.0	4	2.6	4	2.1	FAYETTE OH	-63	-18	-70	30	21
216	8.6	151	4.8	130	3.2	FRANKLIN OH	-45	-32	-62	45	43
15	10.0	5	2.9	3	1.2	FULTON OH	-71	-59	-88	04	12
8	6.2	5	3.6	4	2.5	GALLIA OH	-42	-30	-60	50	28

* SIGNIFICANTLY MORE DEATHS THAN EXPECTED

WHITE FEMALE

CANCER OF THE STOMACH

ICD 151

1950-1959 DEATHS	1950-1959 RATE	1960-1969 DEATHS	1960-1969 RATE	1970-1979 DEATHS	1970-1979 RATE	COUNTY AND STATE	PERCENT CHANGE 60S/50S	PERCENT CHANGE 70S/60S	PERCENT CHANGE 70S/50S	PERCENTILE %CHANGE 70S/50S	PERCENTILE RATES 1970S
16	12.1	10	5.5	8	2.9	GEAUGA OH	-55	-47	-76	20	36
15	6.2	12	4.0	5	1.2	GREENE OH	-35	-70	-81	12	12
34	12.4	10	3.8	13	3.9	GUERNSEY OH	-69	4	-68	34	58
328	8.5	257	5.6	197	3.7	HAMILTON OH	-35	-34	-57	56	52
33	11.7	15	4.4	8	1.9	HANCOCK OH	-63	-56	-83	08	19
18	10.7	12	5.5	6	2.8	HARDIN OH	-48	-49	-74	24	34
7	6.3	9	7.9	1	.6	HARRISON OH	25	-93	-91	01	10
19	14.1	16	10.1	6	3.3	HENRY OH	-29	-68	-77	18	43
11	5.4	12	5.1	7	2.9	HIGHLAND OH	-7	-42	-46	72	36
12	10.8	11	9.0	5	3.8	HOCKING OH	-17	-58	-65	39	54
15	16.1	7	5.0	4	3.0	HOLMES OH	-69	-40	-81	12	39
21	8.7	16	5.5	9	2.8	HURON OH	-37	-49	-68	34	34
23	15.0	19*	10.8	9	3.4	JACKSON OH	-28	-68	-77	18	47
50	12.7	32	6.8	20	3.4	JEFFERSON OH	-46	-49	-73	25	47
14	5.9	13	4.8	6	2.0	KNOX OH	-18	-58	-66	37	20
45	12.5	40	6.8	46	5.2	LAKE OH	-46	-23	-58	53	78
21	9.4	19	6.6	6	1.8	LAWRENCE OH	-30	-72	-81	12	18
29	6.7	26	4.7	18	2.8	LICKING OH	-29	-41	-59	53	33
18	8.8	13	5.2	11	4.2	LOGAN OH	-41	-20	-53	62	62
102*	15.0	72	7.8	53	4.4	LORAIN OH	-48	-43	-70	29	67
198	9.9	138	5.5	110	3.8	LUCAS OH	-45	-31	-62	45	55
6	5.3	8	6.3	12*	7.6	MADISON OH	20	20	44	96	93
167*	14.3	147*	10.0	95	5.4	MAHONING OH	-31	-46	-63	44	80
28	9.5	20	5.7	12	3.1	MARION OH	-40	-47	-68	34	39
21	8.8	15	4.6	20	4.8	MEDINA OH	-48	4	-46	72	72
12	9.1	11	7.3	6	4.2	MEIGS OH	-20	-43	-54	61	63
15	10.1	12	6.6	8	3.1	MERCER OH	-35	-52	-69	32	41
34	9.8	13	2.9	18	3.4	MIAMI OH	-70	18	-65	39	47
15	14.2	4	4.2	6	5.3	MONROE OH	-71	26	-63	43	78
168	9.6	108	4.7	74	2.6	MONTGOMERY OH	-51	-45	-73	25	30
6	6.6	5	5.1	4	5.1	MORGAN OH	-23	-1	-24	87	76
3	2.9	4	3.6	2	1.5	MORROW OH	24	-58	-48	70	14
47	10.9	20	4.2	22	3.8	MUSKINGUM OH	-61	-10	-65	39	55
8	10.3	6	6.6	3	3.2	NOBLE OH	-36	-52	-69	31	41
16	10.0	10	5.1	18*	7.5	OTTAWA OH	-49	48	-25	86	93
7	8.1	4	4.4	1	1.0	PAULDING OH	-46	-77	-87	04	11
7	4.0	12	6.3	10	4.7	PERRY OH	58	-25	18	94	71
11	7.3	7	4.2	8	4.1	PICKAWAY OH	-43	-1	-43	75	62
10	15.9	5	6.0	3	2.8	PIKE OH	-62	-54	-83	09	33
29	9.6	31	8.2	14	2.7	PORTAGE OH	-15	-67	-72	27	32
23	14.5	8	4.0	7	3.6	PPEBLE OH	-72	-11	-75	21	50
15	10.4	8	4.8	6	3.0	PUTNAM OH	-54	-38	-71	27	37
34	7.5	26	4.6	28	4.1	RICHLAND OH	-39	-10	-46	73	61
34	12.9	17	5.8	10	3.0	ROSS OH	-55	-48	-77	19	38
30	10.7	13	3.8	18	4.5	SANDUSKY OH	-65	19	-58	54	67
49	12.1	26	5.3	17	2.9	SCIOTO OH	-56	-45	-76	20	36
23	7.4	29	8.4	13	3.3	SENECA OH	13	-60	-55	59	45
11	6.7	10	5.4	8	3.8	SHELBY OH	-20	-29	-43	75	55
134	9.5	113	6.1	76	3.5	STARK OH	-35	-43	-63	42	48
195	11.0	185	7.7	118	3.9	SUMMIT OH	-30	-50	-65	40	57
80	11.2	80	8.5	72*	5.9	TRUMBULL OH	-24	-31	-48	70	84
32	8.2	29	5.9	19	3.3	TUSCARAWAS OH	-28	-45	-60	49	43
8	6.2	12	7.0	5	2.7	UNION OH	13	-62	-57	55	31
17	9.2	16	7.1	11	4.8	VAN WERT OH	-22	-33	-48	69	72
5	9.1	5	8.6	3	4.3	VINTON OH	-5	-50	-53	63	65
16	7.9	9	3.5	10	2.5	WARREN OH	-56	-27	-68	33	28
22	7.1	11	3.7	11	3.0	WASHINGTON OH	-49	-18	-58	54	38
21	6.6	22	5.3	11	2.2	WAYNE OH	-19	-58	-66	37	24
15	8.2	12	5.8	11	4.7	WILLIAMS OH	-29	-19	-43	76	71
34	10.9	24	6.2	14	2.6	WOOD OH	-44	-57	-76	20	30
11	8.0	8	5.4	6	3.4	WYANDOT OH	-32	-37	-57	56	47
901	8.8	763	5.6	613	3.6	OKLAHOMA OK	-36	-37	-59		
2	3.4	5	6.2	3	3.7	ADAIR OK	81	-40	9	93	53
7	9.3	5	7.7	4	3.8	ALFALFA OK	-18	-50	-59	51	55
3	5.4	4	6.6	2	2.3	ATOKA OK	22	-65	-58	54	24
3	8.6	4	9.5	1	1.8	BEAVER OK	10	-81	-79	16	18
12	10.0	14	8.4	3	1.8	BECKHAM OK	-16	-79	-82	10	17
12	17.2	9	9.3	6	5.3	BLAINE OK	-46	-43	-69	31	79
13	8.6	9	4.4	7	3.3	BRYAN OK	-48	-25	-61	46	45
14	8.4	10	5.9	9	4.6	CADDO OK	-29	-22	-45	73	69
6	4.2	8	5.1	6	3.2	CANADIAN OK	21	-38	-25	86	41
11	5.7	14	5.5	12	4.6	CARTER OK	-5	-16	-20	88	69
10	16.6	8	8.4	4	3.3	CHEROKEE OK	-49	-61	-80	13	44
11	12.2	9	8.7	7	6.3	CHOCTAW OK	-29	-27	-48	69	87
1	5.4	2	8.7	2	7.4	CIMARRON OK	61	-15	37	96	93
9	4.8	9	3.3	16	3.9	CLEVELAND OK	-30	16	-19	88	57
10*	36.7	4	9.6			COAL OK	-74				
13	7.6	12	5.1	11	3.5	COMANCHE OK	-34	-30	-54	61	49
9	17.5	3	4.6	3	4.0	COTTON OK	-74	-14	-77	17	59
11	9.8	2	1.9	3	2.2	CRAIG OK	-81	15	-78	16	22
22	11.4	21	7.9	7	2.6	CREEK OK	-31	-67	-77	17	29
9	8.7	3	2.2	9	5.1	CUSTER OK	-75	131	-41	77	77
5	8.7	2	1.9	8	4.5	DELAWARE OK	-78	136	-48	70	68
3	6.3	2	3.8	5	8.3	DEWEY OK	-39	116	31	96	95
2	4.4	3	5.2			ELLIS OK	19				
14	4.9	21	5.9	17	3.5	GARFIELD OK	20	-40	-28	85	49
11	7.9	9	5.0	5	2.2	GARVIN OK	-37	-57	-73	26	22
15	8.2	7	3.5	3	1.1	GRADY OK	-57	-70	-87	04	11
4	6.0	6	8.1	6	5.6	GRANT OK	36	-31	-7	91	82
4	6.5	7	7.0	3	5.2	GREER OK	8	-25	-19	88	78
2	4.8	4	10.2	3	4.4	HARMON OK	110	-57	-9	91	66
		2	5.3	2	4.6	HARPER OK		-14			69
11*	19.9	7	11.2	4	5.0	HASKELL OK	-44	-55	-75	21	75
7	7.9	9	7.0	8	4.9	HUGHES OK	-12	-30	-38	80	74
9	9.2	7	5.4	7	4.7	JACKSON OK	-41	-13	-49	68	71
6	10.4	7	8.5	3	3.1	JEFFERSON OK	-19	-63	-70	29	40
8	14.5	4	7.4	3	3.6	JOHNSTON OK	-49	-51	-75	21	51
31	12.0	22	6.7	21	4.8	KAY OK	-44	-28	-60	49	73
6	9.5	5	5.9	5	3.8	KINGFISHER OK	-37	-36	-60	50	55

* SIGNIFICANTLY MORE DEATHS THAN EXPECTED

WHITE FEMALE — CANCER OF THE STOMACH — ICD 151

1950-1959 DEATHS	1950-1959 RATE	1960-1969 DEATHS	1960-1969 RATE	1970-1979 DEATHS	1970-1979 RATE	COUNTY AND STATE	PERCENT CHANGE 60S/50S	PERCENT CHANGE 70S/60S	PERCENT CHANGE 70S/50S	PERCENTILE %CHANGE 70S/50S	PERCENTILE RATES 1970S
5	4.9	7	6.6	4	2.4	KIOWA OK	34	-64	-51	65	26
3	7.9	1	2.6	5	7.7	LATIMER OK	-67	198	-2	92	94
17	10.4	11	5.2	12	4.1	LE FLORE OK	-49	-22	-60	49	61
16	13.9	17*	11.1	10	5.3	LINCOLN OK	-20	-52	-62	45	79
19	15.3	8	5.8	7	3.8	LOGAN OK	-62	-34	-75	21	55
7	22.2	2	6.4			LOVE OK	-71				
6	8.7	7	7.6	3	3.0	MC CLAIN OK	-13	-60	-65	38	38
10	9.3	13	9.1	6	3.9	MC CURTAIN OK	-3	-57	-58	53	56
2	3.5	5	6.4	9*	8.7	MC INTOSH OK	85	35	150	99	96
2	3.9	6	10.1	1	1.1	MAJOR OK	162	-90	-73	26	11
4	10.2	2	3.6	5	5.9	MARSHALL OK	-65	63	-42	76	85
10	10.3	6	4.7	4	2.0	MAYES OK	-55	-57	-80	13	21
2	2.8			3	3.0	MURRAY OK			8	93	38
30	11.2	24	7.2	19	4.9	MUSKOGEE OK	-36	-32	-56	57	74
8	12.4	2	2.6	2	2.2	NOBLE OK	-79	-14	-82	10	23
7	11.3	1	1.3	1	1.3	NOWATA OK	-88	-3	-89	03	13
4	8.6	3	3.7	3	3.7	OKFUSKEE OK	-57	-1	-58	54	52
99	6.8	107	5.0	94	3.3	OKLAHOMA OK	-26	-34	-51	65	44
23	12.4	20	7.9	11	4.5	OKMULGEE OK	-36	-43	-63	42	68
13	9.1	14	8.0	8	3.1	OSAGE OK	-11	-61	-65	38	41
7	4.3	17	8.9	9	3.9	OTTAWA OK	106	-56	-10	90	56
14*	18.6	8	8.6	6	5.7	PAWNEE OK	-54	-34	-69	31	83
27	14.2	14	5.6	14	4.8	PAYNE OK	-60	-14	-66	37	73
11	5.8	14	5.6	8	2.9	PITTSBURG OK	-4	-49	-51	66	34
8	5.3	6	3.2	7	2.9	PONTOTOC OK	-40	-9	-46	72	35
15	6.3	11	3.9	9	2.4	POTTAWATOMIE OK	-38	-40	-62	44	26
5	9.6	5	7.6	3	4.1	PUSHMATAHA OK	-21	-46	-58	54	61
2	6.9	3	9.0	2	5.0	ROGER MILLS OK	29	-45	-29	85	75
10	9.8	6	5.2	4	2.4	ROGERS OK	-47	-54	-76	20	26
23*	16.8	9	5.3	5	2.6	SEMINOLE OK	-69	-51	-85	07	29
14	18.2	6	6.4	8	5.4	SEQUOYAH OK	-65	-17	-71	29	79
13	8.3	15	6.6	10	3.3	STEPHENS OK	-20	-50	-60	49	44
		5	6.9	2	1.9	TEXAS OK		-72			19
9	10.5	9	9.0	5	4.4	TILLMAN OK	-14	-51	-58	53	66
91	8.2	64	3.9	73	3.4	TULSA OK	-52	-14	-59	51	45
7	11.4	4	4.6	5	4.0	WAGONER OK	-59	-14	-65	38	58
13	8.3	9	4.4	9	3.0	WASHINGTON OK	-47	-31	-63	42	38
2	2.4	9	8.4	3	2.4	WASHITA OK	254	-71	1	92	26
9	11.2	10	10.1	2	1.6	WOODS OK	-10	-84	-85	06	16
8	9.1	4	3.9	4	3.3	WOODWARD OK	-57	-15	-64	41	44
745	9.7	597	5.6	537	3.6	OREGON OR	-42	-35	-62		
9	11.1	3	3.0	5	3.9	BAKER OR	-73	31	-65	38	56
8	6.0	7	3.9	9	3.9	BENTON OR	-35	0	-35	82	57
43	9.2	29	4.0	38	3.4	CLACKAMAS OR	-56	-15	-63	44	47
19	13.5	20*	10.2	8	4.1	CLATSOP OR	-25	-60	-70	30	61
17*	19.2	11	8.6	10	5.6	COLUMBIA OR	-55	-35	-71	28	82
17	10.6	13	5.1	17	4.9	COOS OR	-51	-4	-53	62	74
		2	4.1	1	1.9	CROOK OR		-53			19
4	11.6			1	1.0	CURRY OR			-91	01	11
8	10.3	5	3.7	12	5.1	DESCHUTES OR	-64	37	-51	66	76
20	10.2	13	4.5	12	2.6	DOUGLAS OR	-56	-42	-75	22	30
2	15.2	2	16.3			GILLIAM OR	7				
2	10.1	3	8.8			GRANT OR	-13				
1	5.9	2	7.1			HARNEY OR	20				
4	5.8	4	5.5	3	3.7	HOOD RIVER OR	-5	-34	-37	80	52
24	7.6	29	6.2	25	3.2	JACKSON OR	-19	-48	-57	55	43
		2	9.9			JEFFERSON OR					
10	6.8	16	8.0	16	4.0	JOSEPHINE OR	17	-50	-42	76	59
9	8.0	13	7.1	9	3.2	KLAMATH OR	-12	-55	-61	48	41
		3	11.1			LAKE OR					
34	6.5	42	5.0	46	3.9	LANE OR	-22	-23	-40	78	56
9	7.8	5	3.1	16	6.9	LINCOLN OR	-61	124	-12	90	91
22	9.9	16	5.0	11	2.3	LINN OR	-49	-54	-77	18	25
8	11.9	4	3.6	8	5.6	MALHEUR OR	-70	54	-53	62	82
47	7.7	44	5.0	46	3.7	MARION OR	-35	-26	-52	63	52
2	12.4	2	6.6	1	4.1	MORROW OR	-47	-38	-67	35	61
298	10.3	213	5.8	167	3.8	MULTNOMAH OR	-43	-34	-63	44	56
17	14.6	13	7.1	8	3.3	POLK OR	-52	-53	-77	17	44
1	8.4					SHERMAN OR					
8	10.7	3	2.7	7	4.7	TILLAMOOK OR	-74	73	-56	58	71
20	11.0	9	3.7	8	2.5	UMATILLA OR	-66	-32	-77	18	28
13	14.5	6	5.8	7	5.3	UNION OR	-60	-9	-64	41	78
7*	29.8	2	5.4	2	3.9	WALLOWA OR	-82	-27	-87	05	58
10	12.0	9	8.0	7	4.8	WASCO OR	-33	-40	-60	49	72
30	10.1	26	4.8	28	2.7	WASHINGTON OR	-52	-43	-73	25	32
1	9.6			2*	18.7	WHEELER OR			95	98	99
21	11.9	26*	10.6	7	2.0	YAMHILL OR	-11	-81	-83	08	20
6073*	11.8	4722*	7.5	3548*	4.6	PENNSYLVANIA PA	-36	-38	-61		
19	8.2	16	5.2	6	1.6	ADAMS PA	-37	-69	-80	13	15
950*	13.7	713*	8.3	541*	5.3	ALLEGHENY PA	-40	-36	-61	47	79
43	11.7	28	6.4	25	4.5	ARMSTRONG PA	-46	-29	-62	46	68
99*	13.7	86*	9.3	56	4.7	BEAVER PA	-33	-49	-66	37	71
24	13.1	13	5.8	13	5.0	BEDFORD PA	-56	-15	-62	45	75
166	11.7	129	7.2	93	4.2	BERKS PA	-38	-42	-64	40	63
65	8.2	50	5.5	43	4.0	BLAIR PA	-33	-26	-51	66	60
38	12.0	34*	9.7	13	3.5	BRADFORD PA	-19	-64	-71	28	47
68	8.1	99	7.7	83	4.2	BUCKS PA	-5	-46	-48	69	63
53	10.6	40	6.3	23	2.7	BUTLER PA	-40	-57	-74	23	32
136*	15.3	100*	9.5	60	4.7	CAMBRIA PA	-38	-51	-70	31	70
3	9.4	2	4.8	2	3.7	CAMERON PA	-49	-21	-60	49	53
46*	16.4	25	7.2	23	5.6	CARBON PA	-56	-23	-66	37	82
24	9.0	24	7.5	12	2.9	CENTRE PA	-16	-62	-68	33	34
67	8.3	51	4.7	59	4.3	CHESTER PA	-44	-8	-48	69	65
16	8.6	10	4.5	9	3.4	CLARION PA	-47	-26	-61	47	45
41	10.5	37	8.0	12	2.2	CLEARFIELD PA	-24	-72	-79	15	24
8	4.7	12	5.9	10	4.1	CLINTON PA	27	-30	-11	90	62
28	9.4	18	5.1	10	2.4	COLUMBIA PA	-46	-52	-74	23	27

* SIGNIFICANTLY MORE DEATHS THAN EXPECTED

WHITE FEMALE — CANCER OF THE STOMACH — ICD 151

1950-1959 DEATHS	1950-1959 RATE	1960-1969 DEATHS	1960-1969 RATE	1970-1979 DEATHS	1970-1979 RATE	COUNTY AND STATE		PERCENT CHANGE 60S/50S	PERCENT CHANGE 70S/60S	PERCENT CHANGE 70S/50S	PERCENTILE %CHANGE 70S/50S	PERCENTILE RATES 1970S
54	12.3	26	5.1	26	4.2	CRAWFORD	PA	-58	-17	-65	38	64
41	8.3	38	5.2	25	2.5	CUMBERLAND	PA	-37	-53	-70	29	27
91	8.6	69	5.3	47	3.0	DAUPHIN	PA	-39	-43	-65	39	38
207	9.9	203	7.0	185	4.9	DELAWARE	PA	-30	-29	-50	67	74
30*	19.8	23*	12.0	14	6.2	ELK	PA	-39	-49	-69	32	87
124	11.1	99	7.3	78	4.9	ERIE	PA	-34	-34	-56	57	73
113*	14.5	78	8.5	66*	5.9	FAYETTE	PA	-42	-30	-59	51	85
2	9.0	3	10.9			FOREST	PA	21				
25	6.1	29	5.4	23	3.3	FRANKLIN	PA	-11	-40	-47	71	43
7	13.8	3	5.9	2	2.6	FULTON	PA	-57	-56	-81	11	29
24	11.6	23	10.0	8	2.6	GREENE	PA	-13	-74	-78	17	30
17	9.3	15	6.9	9	3.5	HUNTINGDON	PA	-26	-50	-63	44	48
48	13.4	39	9.2	21	4.1	INDIANA	PA	-31	-56	-70	31	61
22	8.3	22	7.2	11	3.7	JEFFERSON	PA	-14	-49	-56	57	51
2	2.8	8	8.3	1	.6	JUNIATA	PA	195	-93	-79	16	10
257*	18.2	181*	10.9	125*	6.1	LACKAWANNA	PA	-40	-44	-66	36	86
107	7.9	77	4.3	77	3.4	LANCASTER	PA	-46	-21	-57	55	46
62	12.4	43	6.9	43	5.9	LAWRENCE	PA	-45	-14	-52	63	85
36	8.3	41	7.7	31	4.5	LEBANON	PA	-7	-41	-45	73	68
116	10.4	104	7.3	98	5.4	LEHIGH	PA	-29	-27	-48	69	80
296*	15.3	223*	9.6	180*	6.2	LUZERNE	PA	-37	-35	-59	51	87
48	7.9	39	5.5	27	3.4	LYCOMING	PA	-31	-39	-58	54	45
25	8.5	20	5.8	14	3.6	MC KEAN	PA	-31	-38	-57	55	51
55	9.4	34	5.0	38	4.4	MERCER	PA	-47	-11	-53	63	67
13	6.4	9	3.7	8	2.5	MIFFLIN	PA	-42	-34	-62	46	28
21	10.3	30*	11.7	17	4.7	MONROE	PA	14	-60	-54	60	71
193	9.1	212	6.8	164	3.8	MONTGOMERY	PA	-25	-44	-58	54	55
16	13.7	4	2.7	5	3.7	MONTOUR	PA	-80	35	-73	24	52
109	11.5	93	7.7	88*	5.9	NORTHAMPTON	PA	-32	-24	-48	69	85
65	10.6	41	5.9	41	4.8	NORTHUMBERLAND	PA	-44	-20	-55	59	72
18	15.3	10	6.6	3	1.7	PERRY	PA	-57	-75	-89	03	16
1158*	12.4	799*	8.0	515*	5.0	PHILADELPHIA	PA	-36	-37	-60	50	76
7	11.2	4	5.7	6	5.3	PIKE	PA	-50	-7	-53	62	78
9	8.5	7	6.6	3	3.0	POTTER	PA	-22	-54	-65	39	38
136*	14.0	94*	8.7	89*	6.7	SCHUYLKILL	PA	-38	-23	-52	63	90
10	9.4	5	3.4	3	1.7	SNYDER	PA	-64	-50	-82	10	16
61*	15.7	33	7.0	26	4.6	SOMERSET	PA	-55	-34	-71	28	69
8*	21.0	4	9.8	3	6.3	SULLIVAN	PA	-53	-36	-70	30	87
25	13.1	25*	11.6	8	3.1	SUSQUEHANNA	PA	-11	-74	-77	18	39
26	13.1	14	6.3	8	2.9	TIOGA	PA	-52	-54	-78	16	35
7	6.1	4	2.7	3	1.2	UNION	PA	-56	-53	-80	14	12
27	7.8	26	6.7	17	4.1	VENANGO	PA	-14	-38	-47	71	62
28	9.8	20	6.3	15	4.4	WARREN	PA	-35	-30	-55	60	67
124*	13.6	100*	8.7	76	5.4	WASHINGTON	PA	-36	-38	-60	49	80
16	8.3	10	4.6	14	5.5	WAYNE	PA	-45	21	-33	83	81
204*	14.5	166*	9.0	129*	5.4	WESTMORELAND	PA	-38	-41	-63	43	79
14	14.3	9	9.1	5	3.4	WYOMING	PA	-37	-62	-76	19	47
75	7.1	76	5.3	60	3.3	YORK	PA	-24	-39	-54	61	43
562*	12.7	491*	9.0	380*	5.6	RHODE ISLAND	RI	-29	-38	-56		
21	13.8	18	8.2	19	6.8	BRISTOL	RI	-40	-17	-50	66	91
51	11.7	62*	9.8	47	5.0	KENT	RI	-16	-49	-57	55	75
22	7.5	30	8.3	21	4.6	NEWPORT	RI	10	-45	-39	79	69
446*	13.4	362*	9.3	277*	6.0	PROVIDENCE	RI	-31	-35	-56	58	85
22	9.1	19	5.9	16	3.4	WASHINGTON	RI	-36	-42	-63	44	46
368	7.4	313	4.4	312	3.0	SOUTH CAROLINA	SC	-41	-31	-59		
4	8.7	3	3.4	6	5.9	ABBEVILLE	SC	-61	74	-33	83	84
7	4.1	8	3.5	6	1.5	AIKEN	SC	-15	-57	-63	43	15
1	6.6	1	4.0			ALLENDALE	SC	-40				
22	7.8	13	3.3	16	2.6	ANDERSON	SC	-57	-21	-66	36	30
2	5.2	1	2.3	1	1.6	BAMBERG	SC	-55	-31	-69	32	15
2	5.4	2	4.5	4	6.4	BARNWELL	SC	-15	41	20	95	88
3	9.0	1	2.2	2	1.8	BEAUFORT	SC	-76	-18	-80	13	17
2	4.0			3	2.9	BERKELEY	SC			-27	85	35
1	3.9	2	7.8			CALHOUN	SC	101				
23	6.0	25	5.0	21	2.8	CHARLESTON	SC	-17	-43	-53	62	34
10	10.0	6	4.8	5	2.5	CHEROKEE	SC	-52	-48	-75	21	28
3	3.8	7	6.8	4	2.9	CHESTER	SC	78	-58	-25	86	34
5	5.7	5	6.1	8	5.8	CHESTERFIELD	SC	7	-4	3	93	84
1	2.3	5	9.8	1	1.2	CLARENDON	SC	327	-88	-48	70	12
8	16.3	3	3.6	3	2.7	COLLETON	SC	-78	-25	-83	09	32
10	9.9	3	2.4	10	4.5	DARLINGTON	SC	-76	89	-54	61	68
4	8.0	1	1.4	1	1.1	DILLON	SC	-82	-22	-86	05	12
4	8.5	4	5.8	3	2.9	DORCHESTER	SC	-33	-50	-66	36	35
1	3.1					EDGEFIELD	SC					
1	2.6	4	7.6	2	2.7	FAIRFIELD	SC	191	-65	3	93	31
12	7.2	11	5.3	10	3.1	FLORENCE	SC	-27	-41	-57	55	40
1	2.6	2	2.6	2	2.1	GEORGETOWN	SC	-2	-19	-21	88	21
46	8.6	46	5.6	34	2.8	GREENVILLE	SC	-35	-49	-67	35	34
7	6.0	7	3.6	6	2.6	GREENWOOD	SC	-40	-28	-57	56	30
3	8.7	4	10.0	2	3.7	HAMPTON	SC	15	-63	-57	55	53
5	4.1	5	2.7	12	4.1	HORRY	SC	-35	54	0	92	62
3	17.7			1	2.7	JASPER	SC			-85	07	32
7	9.8	8	10.2	1	.7	KERSHAW	SC	4	-93	-93	01	10
5	4.8	4	2.8	5	2.7	LANCASTER	SC	-42	-3	-43	75	32
9	6.6	10	5.7	11	5.0	LAURENS	SC	-14	-12	-25	86	75
5	21.5	1	2.4	1	1.8	LEE	SC	-89	-26	-92	01	18
9	5.9	5	2.4	12	2.7	LEXINGTON	SC	-59	13	-54	61	31
2	14.2	1	5.2			MC CORMICK	SC	-63				
4	7.9	3	4.2			MARION	SC	-48				
6	9.3	8	10.1	4	4.3	MARLBORO	SC	9	-58	-54	61	64
6	6.0	5	4.1	4	2.7	NEWBERRY	SC	-31	-34	-55	59	31
7	5.8	5	3.1	3	1.3	OCONEE	SC	-46	-60	-78	16	12
7	5.7	3	2.0	5	2.2	ORANGEBURG	SC	-65	9	-62	44	22
8	5.9	7	3.1	15	5.1	PICKENS	SC	-48	66	-15	90	76
25	6.4	25	4.3	25	3.3	RICHLAND	SC	-33	-24	-49	68	43
2	4.8	1	1.7	1	1.5	SALUDA	SC	-64	-16	-70	31	14
45	10.4	34	5.6	35	4.1	SPARTANBURG	SC	-46	-28	-61	47	60

* SIGNIFICANTLY MORE DEATHS THAN EXPECTED

WHITE FEMALE

CANCER OF THE STOMACH

ICD 151

1950-1959 DEATHS	1950-1959 RATE	1960-1969 DEATHS	1960-1969 RATE	1970-1979 DEATHS	1970-1979 RATE	COUNTY AND STATE	PERCENT CHANGE 60S/50S	PERCENT CHANGE 70S/60S	PERCENT CHANGE 70S/50S	PERCENTILE %CHANGE 70S/50S	PERCENTILE RATES 1970S
10	12.1	8	5.6	4	1.7	SUMTER SC	-53	-69	-86	06	16
6	7.5	4	3.9	7	4.9	UNION SC	-48	26	-35	82	73
1	2.1	6	10.5	4	4.4	WILLIAMSBURG SC	406	-59	110	98	65
13	8.3	6	2.5	12	3.3	YORK SC	-70	32	-60	49	44
362*	12.6	255	6.8	187	3.9	SOUTH DAKOTA SD	-46	-42	-69		
1	3.9	2	6.4			AURORA SD	61				
14	14.8	7	5.4	8	4.6	BEADLE SD	-64	-14	-69	32	70
1	7.7	1	9.2			BENNETT SD	21				
8	14.6	3	5.4	5	6.6	BON HOMME SD	-63	23	-54	60	90
8	10.5	10	9.3	6	3.6	BROOKINGS SD	-12	-61	-65	38	51
9	6.2	12	6.4	6	2.5	BROWN SD	2	-60	-59	50	29
3	10.3	1	3.3	1	1.7	BRULE SD	-68	-49	-84	08	16
6	19.4	3	5.8	2	4.1	BUTTE SD	-70	-29	-79	15	61
		2	16.4			CAMPBELL SD					
6	13.5	4	7.0	5	6.5	CHARLES MIX SD	-48	-7	-52	64	89
1	2.7	4	9.4	2	3.3	CLARK SD	249	-65	22	95	44
7	12.9	6	9.3	4	4.4	CLAY SD	-28	-53	-66	37	66
11	12.7	6	4.3	5	3.1	CODINGTON SD	-66	-28	-76	20	39
1	11.1	2	12.2			CORSON SD	10				
2	9.1	3	14.0			CUSTER SD	54				
11	11.7	2	1.6	8	6.4	DAVISON SD	-86	291	-46	73	88
11*	20.2	3	3.7	5	4.0	DAY SD	-81	6	-80	13	59
9*	27.6	1	2.4	4	6.8	DEUEL SD	-91	177	-76	20	90
3	13.1	2	5.8	1	2.8	DOUGLAS SD	-56	-52	-79	16	33
5	21.8	4	13.0	3	6.4	EDMUNDS SD	-40	-51	-71	29	88
6	12.2	8*	13.1	1	1.5	FALL RIVER SD	7	-88	-88	04	15
2	10.3	2	6.6			FAULK SD	-36				
8	16.8	2	2.7	4	6.4	GRANT SD	-84	139	-62	45	88
4	9.4	3	6.5	4	6.1	GREGORY SD	-32	-6	-35	81	86
2	13.3	4*	22.7			HAAKON SD	71				
5	14.2	3	7.5	3	5.0	HAMLIN SD	-47	-34	-65	39	76
9*	35.2	1	4.4	3	4.9	HAND SD	-88	12	-86	05	74
1	4.7	3	18.3			HANSON SD	291				
2	19.0	1	14.9			HARDING SD	-22				
3	8.2	4	7.3	4	5.5	HUGHES SD	-12	-24	-33	83	81
5	8.7	4	5.8	6	6.2	HUTCHINSON SD	-34	8	-29	85	87
2	13.4					HYDE SD					
2	17.6	3*	23.7			JACKSON SD	35				
4	17.9	2	5.4	5*	16.9	JERAULD SD	-70	215	-5	92	99
		1	8.5			JONES SD					
8	17.0	3	5.0	2	2.3	KINGSBURY SD	-70	-55	-87	05	24
9	15.7	7	8.9	3	4.0	LAKE SD	-43	-55	-74	23	60
7	8.9	5	4.7	5	4.1	LAWRENCE SD	-47	-15	-55	60	60
9	12.6	7	7.8	2	1.2	LINCOLN SD	-38	-85	-91	02	12
2	12.6	2	10.7			LYMAN SD	-15				
5	12.3	2	3.1	3	7.3	MC COOK SD	-75	133	-41	77	92
4	15.6	4	12.0	1	2.3	MC PHERSON SD	-23	-81	-85	06	25
2	6.1	2	5.5	3	7.4	MARSHALL SD	-10	35	22	95	93
5	10.4	7	11.7	1	1.4	MEADE SD	13	-88	-87	05	13
1	13.2	2*	27.2			MELLETTE SD	105				
3	10.9	3	9.3	2	7.1	MINER SD	-15	-24	-35	81	92
40	11.6	42	8.8	26	4.1	MINNEHAHA SD	-24	-53	-64	40	62
4	9.8	4	7.6	3	4.4	MOODY SD	-22	-43	-56	58	65
24*	16.8	8	4.2	10	3.4	PENNINGTON SD	-75	-20	-80	13	45
4	11.1	4	12.4	1	2.4	PERKINS SD	12	-81	-78	16	26
3	16.7	1	3.1	1	2.1	POTTER SD	-81	-33	-88	04	21
6	9.6	6	9.6	4	3.7	ROBERTS SD	0	-62	-62	46	52
2	7.6					SANBORN SD					
9	15.1	2	2.4	1	1.8	SPINK SD	-84	-23	-88	03	18
2	21.0					STANLEY SD					
2	40.5					SULLY SD					
1	13.3	1	14.9	1	12.2	TODD SD	12	-18	-8	91	98
3	8.1	2	3.8			TRIPP SD	-53				
11	16.8	7	9.4	3	1.9	TURNER SD	-44	-80	-89	03	18
11*	19.5	4	6.1	7	9.2	UNION SD	-69	51	-52	63	96
1	2.9	5	11.4	3	8.2	WALWORTH SD	291	-28	182	99	95
12	11.1	6	4.6	10	6.9	YANKTON SD	-58	48	-38	80	91
1013	8.5	833	5.1	735	3.3	TENNESSEE TN	-40	-34	-61		
16	10.3	11	4.7	15	4.6	ANDERSON TN	-55	-3	-56	58	69
4	3.8	6	4.3	7	4.7	BEDFORD TN	14	10	25	95	71
7	12.1	5	6.4	1	1.0	BENTON TN	-48	-84	-92	01	11
4	12.5	2	5.5	1	2.3	BLEDSOE TN	-56	-58	-81	11	25
15	8.5	10	4.0	13	3.0	BLOUNT TN	-53	-25	-65	39	37
13	10.0	18	10.4	8	3.2	BRADLEY TN	5	-69	-68	34	42
13	12.0	11	7.5	5	2.4	CAMPBELL TN	-38	-68	-80	14	27
1	2.2	3	5.3			CANNON TN	140				
12	10.0	8	4.8	10	5.1	CARROLL TN	-52	7	-49	68	77
5	3.7	6	3.3	13	4.5	CARTER TN	-11	35	20	95	67
7	18.5	5	9.9	2	2.4	CHEATHAM TN	-46	-75	-87	05	27
3	7.4	1	1.7	1	1.8	CHESTER TN	-78	6	-76	19	17
6	8.1	8	8.0	7	5.1	CLAIBORNE TN	-1	-36	-37	81	77
1	2.7	3	7.7	3	6.5	CLAY TN	191	-16	144	99	89
20*	24.6	6	5.6	9	6.2	COCKE TN	-77	11	-75	22	87
9	9.6	5	3.6	2	1.2	COFFEE TN	-62	-67	-87	04	12
3	4.8	4	5.3	2	2.0	CROCKETT TN	12	-63	-59	52	20
9	13.5	4	4.0	3	2.2	CUMBERLAND TN	-71	-44	-84	08	24
84	6.5	65	3.6	71	3.1	DAVIDSON TN	-44	-16	-53	63	39
10*	24.9	4	8.0	6	7.8	DECATUR TN	-68	-3	-69	32	94
10	17.9	2	3.0	3	3.0	DE KALB TN	-83	1	-83	09	38
5	6.0	6	4.9	3	1.9	DICKSON TN	-19	-61	-69	32	19
12	9.8	7	4.2	7	3.3	DYER TN	-57	-21	-66	36	44
2	4.9	5	11.2	1	1.3	FAYETTE TN	127	-88	-73	26	13
5	11.0	5	9.5	2	3.0	FENTRESS TN	-14	-69	-73	25	37
9	8.6	5	3.6	4	2.4	FRANKLIN TN	-58	-34	-73	26	26
25	12.4	16	6.3	12	3.2	GIBSON TN	-49	-50	-75	22	41
5	4.8	9	6.9	5	3.7	GILES TN	44	-46	-22	87	53
5	9.5	2	3.5			GRAINGER TN	-63				

* SIGNIFICANTLY MORE DEATHS THAN EXPECTED

WHITE FEMALE — CANCER OF THE STOMACH — ICD 151

1950-1959 DEATHS	1950-1959 RATE	1960-1969 DEATHS	1960-1969 RATE	1970-1979 DEATHS	1970-1979 RATE	COUNTY AND STATE	PERCENT CHANGE 60S/50S	PERCENT CHANGE 70S/60S	PERCENT CHANGE 70S/50S	PERCENTILE %CHANGE 70S/50S	PERCENTILE RATES 1970S
12	9.0	11	5.2	10	3.4	GREENE TN	-42	-35	-62	45	47
4	8.1	4	6.4	5	5.9	GRUNDY TN	-21	-8	-27	85	85
9	9.2	11	7.4	5	2.2	HAMBLEN TN	-19	-70	-76	20	23
60	7.7	40	3.9	44	3.2	HAMILTON TN	-49	-18	-58	53	43
4	16.1	2	5.1	2	4.6	HANCOCK TN	-68	-10	-71	27	70
8	8.7	6	5.8	4	3.6	HARDEMAN TN	-34	-39	-59	51	50
2	2.4	5	5.6	4	3.1	HARDIN TN	134	-44	30	95	40
5	5.7	11	7.2	4	2.0	HAWKINS TN	26	-73	-66	37	20
5	10.5	5	8.8	4	4.0	HAYWOOD TN	-16	-54	-62	46	59
3	5.2	4	4.5	5	3.6	HENDERSON TN	-14	-19	-31	84	51
17	15.1	12	7.7	9	5.4	HENRY TN	-49	-31	-64	40	79
7	18.1	1	1.2	3	3.2	HICKMAN TN	-93	159	-83	10	41
3	11.6	2	5.4			HOUSTON TN	-53				
5	10.4	3	4.7	3	3.9	HUMPHREYS TN	-54	-18	-63	44	57
3	6.8	5	8.5	4	6.2	JACKSON TN	25	-28	-10	91	87
6	7.0	7	6.6	8	4.6	JEFFERSON TN	-5	-31	-34	82	69
5	10.5	7	11.2	2	2.3	JOHNSON TN	7	-79	-78	16	25
50	5.6	55	4.4	52	3.1	KNOX TN	-22	-28	-44	74	41
		1	3.3	3	7.3	LAKE TN		121			92
5	7.0	7	7.0	4	4.3	LAUDERDALE TN	0	-38	-38	79	65
14	14.0	13	8.1	7	3.3	LAWRENCE TN	-42	-59	-77	18	43
1	3.6	2	5.5	3	6.7	LEWIS TN	55	21	87	98	90
11	10.8	8	6.2	9	5.4	LINCOLN TN	-43	-12	-50	67	80
5	5.3	5	4.1	6	3.4	LOUDON TN	-21	-19	-36	81	45
12	8.7	8	4.4	8	3.5	MC MINN TN	-49	-20	-59	51	49
11	12.9	7	6.2	1	.7	MC NAIRY TN	-52	-88	-94	00	10
10	17.7	4	5.4	3	2.6	MACON TN	-70	-52	-85	06	30
21	9.2	11	3.7	6	1.5	MADISON TN	-60	-59	-84	08	15
3	3.7	3	3.2	5	4.3	MARION TN	-13	32	14	94	64
8	8.9	2	1.5	9	6.1	MARSHALL TN	-84	321	-31	84	86
12	8.9	5	2.5	6	2.3	MAURY TN	-72	-8	-74	23	25
2	11.5	1	6.8			MEIGS TN	-41				
7	9.7	6	5.1	3	2.3	MONROE TN	-48	-55	-76	19	25
7	5.6	14	8.5	7	2.8	MONTGOMERY TN	51	-67	-50	67	33
		3	15.2	1	5.1	MOORE TN		-66			77
3	5.2	4	6.4	5	6.4	MORGAN TN	24	-1	23	95	88
14	9.8	7	3.9	11	4.7	OBION TN	-61	23	-52	65	71
3	4.5	6	6.9	6	5.0	OVERTON TN	53	-27	12	94	76
3	10.7	1	2.5	3	6.7	PERRY TN	-77	170	-38	80	90
6*	31.0	3	12.5	2	8.7	PICKETT TN	-60	-30	-72	27	96
6	13.4	5	8.7	1	1.4	POLK TN	-35	-84	-90	02	13
8	7.8	11	6.4	11	4.4	PUTNAM TN	-18	-31	-44	75	66
4	5.9	7	7.7	3	3.2	RHEA TN	31	-59	-47	71	41
5	4.9	8	4.5	10	4.1	ROANE TN	-9	-9	-17	89	60
9	7.3	7	4.8	4	2.1	ROBERTSON TN	-34	-55	-71	28	22
22	13.8	11	5.4	12	3.7	RUTHERFORD TN	-61	-32	-73	24	53
7	15.3	3	4.2	4	4.1	SCOTT TN	-73	-2	-73	25	62
4	21.4	1	3.1	1	2.7	SEQUATCHIE TN	-85	-15	-87	04	31
8	9.2	13	10.7	4	2.2	SEVIER TN	16	-80	-76	19	23
103	7.1	100	4.9	93	3.6	SHELBY TN	-31	-27	-50	67	50
6	8.3	3	3.7			SMITH TN	-56				
5	12.4	3	6.5			STEWART TN	-47				
29	8.9	24	4.7	36	5.0	SULLIVAN TN	-47	8	-43	75	76
17	11.8	11	5.6	11	3.8	SUMNER TN	-52	-33	-68	34	54
7	8.7	7	7.8	3	2.4	TIPTON TN	-10	-69	-72	26	27
2	7.9	1	3.0	3	8.0	TROUSDALE TN	-62	171	2	92	94
2	4.4	5	6.1	2	2.1	UNICOI TN	37	-65	-52	64	22
		1	2.2	3	5.4	UNION TN		148			80
1	9.4					VAN BUREN TN					
16	16.1	4	3.2	5	2.5	WARREN TN	-80	-23	-85	07	27
16	7.3	21	6.4	17	3.8	WASHINGTON TN	-12	-40	-48	70	56
7	15.1	1	2.1	2	3.0	WAYNE TN	-86	46	-80	14	39
21	13.0	11	5.9	5	2.3	WEAKLEY TN	-55	-61	-82	10	24
3	3.6	8	8.2	4	3.8	WHITE TN	131	-54	5	93	54
12	12.7	10	7.7	5	2.7	WILLI MSON TN	-40	-64	-79	16	32
7	6.1	3	1.8	4	1.5	WILSON TN	-71	-18	-76	19	14
2552	8.4	2568	6.0	2396	4.1	TEXAS TX	-29	-31	-51		
5	3.6	6	3.4	12	6.3	ANDERSON TX	-7	85	72	97	87
		1	1.1			ANDREWS TX					
8	5.9	4	2.0	10	3.5	ANGELINA TX	-65	70	-41	77	48
1	3.1	5	11.8	7*	10.2	ARANSAS TX	280	-13	229	99	97
1	6.8	1	2.7	1	2.2	ARCHER TX	-60	-17	-67	35	23
1	6.5					ARMSTRONG TX					
4	5.3	16*	17.2	9	6.7	ATASCOSA TX	228	-61	28	95	90
9	12.2	9	8.0	4	3.7	AUSTIN TX	-34	-54	-70	30	53
5	18.4			2	4.4	BAILEY TX			-76	19	66
		1	3.8			BANDERA TX					
6	7.2	7	6.9	7	5.4	BASTROP TX	-4	-21	-24	87	80
3	7.9	2	4.1	4	5.3	BAYLOR TX	-49	30	-33	83	79
2	3.0	7	7.3	7	6.5	BEE TX	141	-11	114	98	89
23	8.6	21	5.7	20	3.9	BELL TX	-34	-31	-54	60	58
246	11.5	250*	8.3	235*	5.9	BEXAR TX	-28	-28	-48	69	85
2	7.2	3	11.4			BLANCO TX	58				
1	33.5					BORDEN TX					
7	8.3	6	5.2	10	5.8	BOSQUE TX	-37	11	-31	84	83
23	10.5	20	6.7	19	4.8	BOWIE TX	-36	-28	-54	61	73
13	11.2	21	8.0	18	3.9	BRAZORIA TX	-29	-52	-65	38	56
5	5.6	10	6.6	9	3.3	BRAZOS TX	19	-50	-40	77	44
3	11.5	3	8.2	3	7.5	BREWSTER TX	-29	-9	-35	81	93
1	6.0					BRISCOE TX					
2	10.5	2	5.9	2	5.1	BROOKS TX	-44	-14	-52	64	77
11	6.5	15	6.3	14	5.4	BROWN TX	-3	-15	-17	89	80
4	9.5	5	8.5	2	1.8	BURLESON TX	-10	-79	-81	11	17
9	15.8	4	4.7	7	5.1	BURNET TX	-70	8	-67	35	77
10	11.1	7	8.1	3	1.8	CALDWELL TX	-27	-78	-84	08	17
3	8.4	2	4.6	3	4.6	CALHOUN TX	-45	0	-45	74	69
3	5.1	3	4.2	5	5.4	CALLAHAN TX	-18	30	6	93	80
34	8.1	40	7.5	44	6.0	CAMERON TX	-8	-20	-26	86	85
4	13.2	4	8.7	2	3.6	CAMP TX	-34	-59	-73	25	50

* SIGNIFICANTLY MORE DEATHS THAN EXPECTED

WHITE FEMALE

CANCER OF THE STOMACH — ICD 151

COUNTY AND STATE	1950-1959 DEATHS	1950-1959 RATE	1960-1969 DEATHS	1960-1969 RATE	1970-1979 DEATHS	1970-1979 RATE	PERCENT CHANGE 60S/50S	PERCENT CHANGE 70S/60S	PERCENT CHANGE 70S/50S	PERCENTILE %CHANGE 70S/50S	PERCENTILE RATES 1970S
CARSON TX	3	12.7	2	5.9	3	6.4	-54	8	-50	67	88
CASS TX	14	15.3	6	4.9	10	6.7	-68	36	-56	57	90
CASTRO TX	1	3.4	1	2.6	1	2.6	-23	-2	-25	86	29
CHAMBERS TX	2	8.3	1	3.1			-63				
CHEROKEE TX	11	6.7	13	6.5	12	5.0	-4	-22	-25	86	76
CHILDRESS TX	6	11.7	2	2.6	2	2.4	-78	-7	-79	15	27
CLAY TX	2	3.8	4	6.2	3	3.1	61	-50	-20	88	39
COCHRAN TX	1	4.7			4*	17.7			280	99	99
COKE TX	1	4.9	2	9.8	2	7.1	101	-27	46	96	92
COLEMAN TX	6	6.5	11	9.0	4	2.4	38	-74	-64	41	26
COLLIN TX	14	6.1	20	7.1	17	4.0	17	-44	-35	82	59
COLLINGSWORTH TX	6	15.2	1	2.0			-87				
COLORADO TX	7	9.1	2	1.9	2	2.1	-79	9	-77	18	21
COMAL TX	8	9.7	6	4.6	10	4.8	-52	3	-51	66	72
COMANCHE TX	9	8.7	9	6.9	3	3.4	-20	-51	-61	47	46
CONCHO TX	5	24.6			1	3.4			-86	05	46
COOKE TX	6	5.3	11	6.9	12	7.0	29	1	30	96	91
CORYELL TX	4	3.8	4	3.3	3	1.6	-12	-53	-59	51	15
COTTLE TX	2	9.9									
CRANE TX					1	4.5					67
CROCKETT TX	2	16.7									
CROSBY TX	5	14.2	1	1.8	2	3.5	-87	92	-76	20	47
DALLAM TX	1	3.1	1	2.8	2	5.5	-8	95	79	98	81
DALLAS TX	178	6.8	206	5.1	202	3.6	-26	-30	-48	70	50
DAWSON TX	10	15.7	5	7.1	1	1.0	-55	-86	-94	00	11
DEAF SMITH TX	2	5.1	2	4.0	3	3.4	-22	-15	-34	82	46
DELTA TX	8	19.7	2	3.5	3	5.0	-82	43	-75	22	75
DENTON TX	17	9.2	18	6.8	18	4.1	-26	-40	-56	58	61
DE WITT TX	12	11.1	8	5.8	16*	8.9	-48	54	-20	88	96
DICKENS TX	3	10.2	2	5.7	3	7.7	-44	33	-25	86	93
DIMMIT TX	9*	26.8	1	2.6	2	4.5	-90	72	-83	09	68
DONLEY TX	1	3.0			4	9.9			229	99	97
DUVAL TX	4	7.4	3	4.7	1	1.5	-36	-69	-80	13	14
EASTLAND TX	21	14.3	14	6.2	9	4.7	-56	-25	-67	35	70
ECTOR TX	10	13.0	13	5.5	13	3.7	-58	-31	-71	28	54
EDWARDS TX					1	5.6					81
ELLIS TX	14	6.6	16	5.8	9	3.3	-12	-43	-50	67	44
EL PASO TX	70	10.6	87*	8.5	83*	6.6	-20	-23	-38	79	89
ERATH TX	5	3.6	6	3.0	4	1.7	-16	-44	-53	62	16
FALLS TX	11	10.4	4	2.7	7	4.3	-74	58	-59	52	64
FANNIN TX	11	5.8	15	6.6	10	3.9	14	-40	-32	83	58
FAYETTE TX	14	10.3	13	7.6	12	7.9	-26	4	-23	87	94
FISHER TX	2	5.4	3	6.1	2	3.2	14	-49	-41	77	41
FLOYD TX	2	4.0	5	7.9	3	5.9	97	-26	47	96	84
FOARD TX			1	4.2	4*	18.7		347			99
FORT BEND TX	10	11.9	13	9.0	9	3.7	-25	-58	-69	32	53
FRANKLIN TX	2	6.3	2	3.1	3	4.5	-51	46	-29	85	68
FREESTONE TX	3	5.8	4	3.8	3	3.2	-34	-16	-45	74	41
FRIO TX	6	14.8	2	3.5	5	7.7	-76	119	-48	69	93
GAINES TX			5	13.2	2	3.9		-71			56
GALVESTON TX	30	9.0	28	5.2	31	4.4	-42	-15	-51	65	66
GARZA TX	1	5.1	1	3.4	1	2.9	-33	-16	-44	74	34
GILLESPIE TX	6	7.1	7	8.6	3	2.6	20	-69	-63	42	30
GLASSCOCK TX					1*	18.9					99
GOLIAD TX	2	8.7	4	10.5	2	6.8	21	-35	-22	88	90
GONZALES TX	14	14.9	4	3.8	5	5.0	-74	30	-66	36	75
GRAY TX	8	8.9	1	1.5	8	4.2	-83	175	-53	62	63
GRAYSON TX	32	8.3	29	5.7	26	3.8	-31	-33	-54	61	55
GREGG TX	17	8.6	8	3.0	15	3.8	-65	26	-56	57	54
GRIMES TX	3	5.1	5	7.0	7	9.5	38	35	86	98	97
GUADALUPE TX	16	15.6	14	9.2	10	4.7	-41	-49	-70	30	70
HALE TX	5	3.9	8	5.0	9	4.9	30	-2	26	95	74
HALL TX	2	4.4	1	2.6			-41				
HAMILTON TX	1	1.4	9	10.1	3	2.7	635	-74	94	98	31
HANSFORD TX			1	5.3	1	3.9		-26			58
HARDEMAN TX	3	6.4	2	3.2	3	3.7	-50	14	-43	75	52
HARDIN TX	2	4.3	3	2.6	4	2.2	-39	-15	-49	69	23
HARRIS TX	229	8.8	264	6.1	243	3.8	-30	-38	-56	57	55
HARRISON TX	12	10.3	3	1.6	9	4.4	-84	174	-57	55	66
HARTLEY TX	1	5.9									
HASKELL TX	5	7.9	7	9.5	3	4.8	21	-50	-39	79	73
HAYS TX	5	6.7	1	1.1	6	4.4	-83	298	-34	82	66
HEMPHILL TX	2	15.2	1	5.3	2	7.7	-65	45	-50	67	93
HENDERSON TX	14	13.5	6	4.2	13	5.4	-69	30	-60	50	80
HIDALGO TX	55	12.2	52	8.1	48	5.0	-34	-39	-59	50	75
HILL TX	13	7.4	12	5.3	4	1.4	-28	-73	-81	12	14
HOCKLEY TX	4	7.0	4	5.2	4	4.1	-25	-22	-41	77	61
HOOD TX	2	5.5	10*	21.7	5	7.7	290	-64	39	96	94
HOPKINS TX	5	3.9	7	4.5	6	2.8	16	-38	-28	85	33
HOUSTON TX	8	11.6	4	4.9	2	2.1	-58	-57	-82	10	21
HOWARD TX	3	3.1	7	4.5	7	3.9	45	-13	26	95	57
HUDSPETH TX	1	7.7			3*	27.1			249	99	99
HUNT TX	16	7.0	15	5.6	12	3.8	-20	-32	-46	72	54
HUTCHINSON TX	6	8.5	8	8.2	4	2.5	-3	-70	-71	28	28
JACK TX	3	7.2	2	4.4	1	1.6	-40	-62	-77	17	16
JACKSON TX	4	10.0	2	3.5	5	6.1	-65	73	-39	79	86
JASPER TX	4	4.9	4	4.1	4	2.9	-15	-29	-40	78	36
JEFFERSON TX	59	10.0	47	5.5	49	4.4	-45	-20	-56	57	66
JIM HOGG TX	1	4.6	2	9.0	1	3.9	98	-57	-15	89	57
JIM WELLS TX	8	9.6	11	9.8	8	5.0	1	-48	-48	70	76
JOHNSON TX	8	4.1	10	3.8	13	3.3	-7	-13	-19	88	44
JONES TX	6	5.6	14	10.5	8	6.8	88	-35	22	95	90
KARNES TX	13*	20.0	7	9.2	2	1.8	-54	-80	-91	01	18
KAUFMAN TX	16	10.0	10	5.4	10	4.1	-46	-23	-59	52	62
KENDALL TX	4	10.4	3	5.7	3	3.9	-45	-32	-63	43	56
KENEDY TX			1*	28.1							
KENT TX			1	7.3	2*	12.9		77			98
KERR TX	6	6.2	10	5.3	6	2.2	-14	-59	-65	39	23
KIMBLE TX	3	11.3	3	9.9	5*	21.7	-13	119	92	98	99
KINNEY TX	1	10.9	1	14.1	2*	19.1	29	36	76	98	99
KLEBERG TX	5	7.1	8	7.8	2	1.9	10	-76	-73	24	19

* SIGNIFICANTLY MORE DEATHS THAN EXPECTED

WHITE FEMALE — CANCER OF THE STOMACH — ICD 151

1950-1959 DEATHS	1950-1959 RATE	1960-1969 DEATHS	1960-1969 RATE	1970-1979 DEATHS	1970-1979 RATE	COUNTY AND STATE	PERCENT CHANGE 60S/50S	PERCENT CHANGE 70S/60S	PERCENT CHANGE 70S/50S	PERCENTILE %CHANGE 70S/50S	PERCENTILE RATES 1970S
5	13.3	5	10.3	1	1.8	KNOX TX	-23	-83	-86	05	18
15	7.8	15	5.7	14	4.7	LAMAR TX	-27	-17	-39	78	71
5	7.4	6	6.1	4	3.3	LAMB TX	-18	-46	-55	58	44
10	18.1	5	6.7	5	5.5	LAMPASAS TX	-63	-17	-69	31	81
4	14.8	1	3.4	1	3.4	LA SALLE TX	-77	-1	-77	18	45
12	10.0	11	6.6	4	2.0	LAVACA TX	-34	-70	-80	14	20
2	4.8	3	5.7	2	2.3	LEE TX	18	-59	-52	64	25
4	8.3	4	7.7	1	1.1	LEON TX	-8	-86	-87	04	11
10	13.3	15*	11.3	10	5.3	LIBERTY TX	-15	-53	-60	48	78
11	8.6	7	5.3	3	2.0	LIMESTONE TX	-39	-62	-77	18	20
1	4.9	2	9.2			LIPSCOMB TX	88				
3	8.9	2	4.8	1	2.1	LIVE OAK TX	-46	-55	-76	20	22
3	7.6	2	3.2	2	2.3	LLANO TX	-58	-29	-70	30	25
1*	119.2	1*	98.2			LOVING TX	-18				
25	6.9	30	5.5	20	2.6	LUBBOCK TX	-22	-52	-62	45	30
2	4.6	1	2.1	4	7.6	LYNN TX	-54	258	65	97	93
9	15.2	3	3.6	6	5.5	MC CULLOCH TX	-76	53	-64	41	81
37	6.5	29	3.8	32	3.2	MC LENNAN TX	-41	-18	-52	64	41
				1	15.1	MC MULLEN TX					99
5	14.4	5	9.2	3	4.0	MADISON TX	-36	-57	-72	26	59
		2	7.2	4	7.2	MARION TX		-1			92
6*	36.0	1	4.4	2	10.2	MAPTIN TX	-88	131	-72	27	97
3	9.6	4	14.3	1	2.1	MASON TX	49	-86	-79	16	21
12	15.9	4	3.6	8	6.0	MATAGORDA TX	-78	67	-62	44	85
3	8.9	8*	15.2	4	5.2	MAVERICK TX	70	-66	-42	76	78
6	6.7	5	5.5	8	6.1	MEDINA TX	-17	10	-9	91	86
		4	15.0	1	3.4	MENARD TX		-77			46
8	11.3	8	4.4	8	2.8	MIDLAND TX	-61	-37	-75	21	33
9	7.2	10	7.8	5	2.6	MILAM TX	9	-66	-63	42	30
5	12.9	4	6.8	2	6.8	MILLS TX	-47	0	-47	70	90
2	3.3	5	8.2	2	2.3	MITCHELL TX	150	-72	-30	85	25
5	4.9	7	4.5	4	2.4	MONTAGUE TX	-8	-46	-51	66	26
4	5.6	9	6.8	6	1.6	MONTGOMERY TX	22	-76	-71	29	16
1	1.4	4	9.6	4	7.4	MOORE TX	569	-24	411	99	93
2	4.5	4	6.7	2	2.5	MORRIS TX	50	-63	-45	74	28
4	23.2	3	22.2			MOTLEY TX	-4				
10	8.5	10	6.5	3	1.2	NACOGDOCHES TX	-23	-82	-86	05	12
19	9.8	10	4.0	9	2.9	NAVARRO TX	-59	-29	-71	28	34
1	3.3	5	11.5	1	2.5	NEWTON TX	253	-79	-25	87	27
14	16.5	1	1.1	5	3.5	NOLAN TX	-93	206	-79	15	48
49	9.9	31	4.4	57*	5.6	NUECES TX	-55	25	-44	75	81
1	4.9	1	4.1	1	2.2	OCHILTREE TX	-16	-46	-55	59	23
2*	34.0			2*	23.4	OLDHAM TX			-31	84	99
11	9.9	10	6.0	10	3.5	ORANGE TX	-39	-42	-65	39	48
7	6.8	9	5.9	7	4.1	PALO PINTO TX	-13	-30	-40	78	61
4	5.4	5	5.9	5	4.9	PANOLA TX	10	-18	-10	90	74
4	2.9	4	2.4	8	3.3	PARKER TX	-19	41	14	94	45
1	4.8	1	3.2	3	6.4	PARMER TX	-34	103	34	96	88
		4	7.6	1	2.6	PECOS TX		-66			30
3	5.5	8	11.8	4	3.4	POLK TX	115	-72	-39	79	45
23	7.6	18	4.1	26	4.6	POTTER TX	-46	12	-39	78	69
2	6.6	1	3.1	2	7.0	PRESIDIO TX	-52	122	7	93	91
2	10.1			1	2.4	RAINS TX			-76	19	26
3	4.6	7	6.5	4	1.7	RANDALL TX	41	-75	-64	41	16
		1	18.6	1	8.3	REAGAN TX		-56			95
		1	6.5			REAL TX					
11	11.7	5	4.6	4	2.6	RED RIVER TX	-60	-45	-78	16	29
10*	28.5	6	12.2	3	5.7	REEVES TX	-57	-54	-80	13	83
4	10.4					REFUGIO TX					
6	10.4	4	4.9	2	3.0	ROBERTSON TX	-53	-38	-71	28	38
3	9.1	2	3.9			ROCKWALL TX	-57				
5	5.7	3	2.5	7	6.2	RUNNELS TX	-55	143	8	93	87
12	8.5	8	3.8	9	3.9	RUSK TX	-55	3	-53	62	58
2	6.3	6*	16.7	4	6.1	SABINE TX	165	-64	-3	92	86
1	3.0	2	4.5	1	1.8	SAN AUGUSTINE TX	48	-60	-41	77	18
2	10.7			2	4.2	SAN JACINTO TX			-60	48	64
12	10.9	10	6.6	10	4.6	SAN PATRICIO TX	-40	-30	-57	55	70
5	9.3	2	3.7	4	5.2	SAN SABA TX	-61	41	-45	74	78
				1	3.5	SCHLEICHER TX					49
3	3.7	7	7.6	4	4.0	SCURRY TX	106	-48	7	93	58
4	14.5	2	8.3	2	4.0	SHACKELFORD TX	-43	-52	-73	25	58
6	6.2	8	5.7	8	4.3	SHELBY TX	-7	-25	-30	84	64
2	111.5	2	15.2	1	6.5	SHERMAN TX	-86	-57	-94	00	88
14	5.3	18	4.6	21	3.6	SMITH TX	-14	-22	-33	83	50
1	3.9	1	5.2			SOMERVELL TX	36				
4	6.4	3	4.2	9*	10.0	STARR TX	-35	138	56	97	97
2	3.7	3	3.7	6	7.4	STEPHENS TX	-1	101	99	98	93
1	13.1	1	15.7			STERLING TX	20				
3	20.1	2	10.3			STONEWALL TX	-49				
		1	6.4			SUTTON TX					
2	5.7	4	8.7	4	5.4	SWISHER TX	53	-38	-5	92	80
132	8.2	116	4.8	108	3.1	TARRANT TX	-42	-36	-63	44	39
11	4.2	20	4.6	11	1.9	TAYLOR TX	11	-59	-54	60	19
2	22.2					TERRELL TX					
3	8.5	2	3.8	3	4.3	TERRY TX	-55	14	-49	68	65
		2	7.7	1	2.6	THROCKMORTON TX		-67			29
3	4.3	7	7.4	5	4.0	TITUS TX	70	-45	-7	91	60
10	4.1	16	4.8	18	3.8	TOM GREEN TX	17	-21	-8	91	54
41	6.1	72	7.6	56	4.0	TRAVIS TX	24	-47	-34	82	59
3	7.0	2	4.3	1	1.8	TRINITY TX	-38	-57	-74	24	18
4	8.2	2	2.7	3	2.5	TYLER TX	-67	-7	-70	31	28
4	4.3	9	8.1	5	3.9	UPSHUR TX	90	-52	-8	91	58
		1	5.0			UPTON TX					
4	5.7	2	2.2	6	4.7	UVALDE TX	-61	111	-17	89	71
6	9.9	5	5.8	9*	8.2	VAL VERDE TX	-41	40	-18	89	94
13	10.6	10	6.7	9	3.9	VAN ZANDT TX	-37	-42	-64	41	56
15	13.3	17	9.6	9	3.5	VICTORIA TX	-28	-63	-73	24	49
3	6.3	1	1.8	2	2.5	WALKER TX	-72	40	-61	48	28
3	9.1	4	8.9	1	1.7	WALLER TX	-2	-81	-81	12	16
		3	8.1	3	5.1	WARD TX		-38			76
11	11.7	10	7.7	9	6.6	WASHINGTON TX	-35	-14	-43	75	89

* SIGNIFICANTLY MORE DEATHS THAN EXPECTED

WHITE FEMALE

CANCER OF THE STOMACH

ICD 151

COUNTY AND STATE	1950-1959 DEATHS	1950-1959 RATE	1960-1969 DEATHS	1960-1969 RATE	1970-1979 DEATHS	1970-1979 RATE	PERCENT CHANGE 60S/50S	PERCENT CHANGE 70S/60S	PERCENT CHANGE 70S/50S	PERCENTILE %CHANGE 70S/50S	PERCENTILE RATES 1970S
WEBB TX	31*	14.7	18	6.5	18	4.6	-56	-30	-69	32	68
WHARTON TX	11	9.0	14	9.2	6	3.5	2	-62	-61	47	48
WHEELER TX	2	4.5	5	7.7	3	6.6	71	-14	47	96	89
WICHITA TX	21	5.8	39	7.1	19	3.0	23	-58	-49	69	37
WILBARGER TX	6	6.5	1	1.2	10	6.9	-82	497	7	93	91
WILLACY TX	9	20.0	3	4.6	7*	10.1	-77	118	-50	67	97
WILLIAMSON TX	13	7.3	12	4.9	10	3.0	-33	-38	-59	52	38
WILSON TX	8	12.8	10*	12.9	7	8.7	1	-33	-32	83	96
WINKLER TX			1	1.4	2	6.3		341			87
WISE TX	5	4.6	5	3.9	3	1.9	-15	-51	-59	52	18
WOOD TX	6	5.5	4	3.0	12	6.0	-45	99	9	93	85
YOAKUM TX	1	18.7	2	9.2			-51				
YOUNG TX	8	9.0	5	4.2	4	2.7	-53	-36	-70	29	31
ZAPATA TX	1	6.4	3	13.4			109				
ZAVALA TX	3	7.7	5	11.6	3	7.1	50	-39	-9	91	92
UTAH UT	267	10.0	232	6.2	184	3.5	-38	-43	-65		
BEAVER UT			2	8.8	2	8.6		-2			96
BOX ELDER UT	7	9.5	8	8.0	4	2.8	-15	-66	-71	28	33
CACHE UT	17	11.8	14	7.4	7	3.0	-37	-60	-75	22	37
CARBON UT	9	18.7	5	7.5	3	3.2	-60	-58	-83	09	41
DAVIS UT	18*	16.7	11	6.4	16	5.4	-62	-15	-67	35	80
DUCHESNE UT	4	18.3			1	2.8			-84	07	34
EMERY UT			3	12.4	2	6.0		-51			85
GARFIELD UT			1	7.5	1	6.2		-17			87
GRAND UT	1	24.8	1	2.5			-90				
IRON UT	3	9.7	1	2.3	2	4.1	-77	81	-58	54	62
JUAB UT	1	2.8			1	3.0			4	93	37
KANE UT			1	10.6							
MILLARD UT	4	12.0	7*	18.7	2	3.6	55	-81	-70	29	50
MORGAN UT	3	27.5			1	6.1			-78	17	86
PIUTE UT	1	19.1	1	12.4	1	12.7	-35	3	-34	82	98
RICH UT	1	15.3			1	10.6			-31	84	97
SALT LAKE UT	109	9.1	113	6.5	74	3.2	-29	-51	-65	39	42
SAN JUAN UT			1	7.4							
SANPETE UT	5	8.0	3	3.7	3	2.9	-54	-23	-64	40	35
SEVIER UT	11*	25.5	2	3.7	4	4.5	-86	23	-82	10	68
SUMMIT UT	4	15.7	2	7.3	2	9.4	-53	29	-40	78	96
TOOELE UT	5	10.5	3	4.6	4	4.9	-56	7	-53	62	74
UINTAH UT	6*	22.5	2	4.8	4	7.0	-79	44	-69	32	91
UTAH UT	19	6.7	21	5.1	20	3.3	-24	-36	-51	66	43
WASATCH UT	3	16.8	2	8.4			-50				
WASHINGTON UT	2	4.3	1	1.8	6	5.6	-58	215	31	96	82
WAYNE UT	1	16.3									
WEBER UT	33	10.4	27	5.5	23	3.6	-47	-35	-65	38	50
VERMONT VT	190	8.5	155	6.1	110	3.6	-28	-41	-57		
ADDISON VT	16	15.5	4	3.1	5	3.2	-80	1	-80	14	42
BENNINGTON VT	9	5.8	12	6.6	8	3.2	15	-52	-45	74	42
CALEDONIA VT	12	8.4	10	6.3	4	3.1	-24	-52	-64	42	39
CHITTENDEN VT	21	6.5	19	4.9	21	4.5	-24	-9	-31	84	67
ESSEX VT	1	3.3	2	5.9			77				
FRANKLIN VT	17	11.0	12	6.9	5	2.3	-37	-67	-79	15	24
GRAND ISLE VT			3	19.4							
LAMOILLE VT	8	13.1	9*	14.1	3	3.3	7	-76	-75	22	45
ORANGE VT	3	2.8	3	2.8	5	3.6	2	26	28	95	50
ORLEANS VT	10	8.2	12	9.5	7	4.7	16	-51	-43	76	70
RUTLAND VT	29	9.0	20	5.8	15	3.4	-35	-42	-62	45	46
WASHINGTON VT	33	12.9	22	7.0	20	5.5	-45	-22	-57	55	81
WINDHAM VT	13	5.8	18	6.9	7	2.9	17	-57	-50	67	36
WINDSOR VT	18	7.5	9	3.2	10	3.0	-57	-7	-61	48	37
VIRGINIA VA	905	8.1	793	5.0	719	3.3	-38	-34	-59		
ACCOMACK VA	7	4.6	8	4.1	8	3.5	-11	-15	-24	87	48
ALBEMARLE#VA	15	7.1	14	4.8	14	3.5	-32	-26	-50	67	49
ALLEGHANY#VA	17	16.9	4	2.9	5	2.7	-83	-6	-84	08	32
AMELIA VA	3	14.6	2	12.5			-14				
AMHERST VA	9	14.2	5	5.6	1	.9	-60	-85	-94	00	10
APPOMATTOX VA	2	5.8	1	2.1	1	1.7	-63	-18	-70	29	17
ARLINGTON#VA	38	7.3	37	4.8	28	2.8	-34	-42	-62	46	33
AUGUSTA#VA	24	7.8	18	4.3	17	3.4	-44	-22	-57	56	46
BATH VA	2	7.7									
BEDFORD#VA	34	7.7	21	3.6	20	2.8	-54	-21	-63	42	34
BLAND VA	1	3.8									
BOTETOURT VA	11	16.6	8	9.4	5	4.3	-44	-54	-74	23	64
BRUNSWICK VA	2	7.3	2	4.1	1	1.5	-44	-63	-79	15	15
BUCHANAN VA	12*	22.8	11*	13.7	6	4.7	-40	-65	-79	15	71
BUCKINGHAM VA	2	7.3	1	2.5			-65				
CAROLINE VA	4	11.0	2	5.2	4	9.2	-53	77	-16	89	96
CARROLL#VA	20	10.3	9	3.7	11	3.3	-64	-9	-68	35	44
CHARLOTTE VA	5	14.5	3	7.4	1	1.8	-49	-76	-88	04	17
CHESTERFIELD#VA	89	7.8	75	5.2	58	3.7	-33	-30	-53	62	52
CLARKE VA	3	9.9			1	1.8			-82	11	18
CRAIG VA	1	5.9	1	4.1			-30				
CULPEPER VA	4	5.4	5	5.2	6	4.4	-4	-15	-18	89	66
CUMBERLAND VA	1	6.2	1	5.7	2	8.4	-9	49	36	96	95
DICKENSON VA	5	11.9	3	3.8	4	4.7	-68	23	-60	48	71
DINWIDDIE#VA	13	7.4	9	3.2	13	3.4	-57	8	-54	61	47
ESSEX VA	2	9.3	2	6.6			-29				
FAIRFAX#VA	35	6.5	54	4.6	75	3.5	-29	-24	-46	72	48
FAUQUIER VA	4	4.8	6	5.4	9	7.2	11	35	50	97	92
FLOYD VA	12*	21.4	11*	16.4	1	1.2	-23	-93	-94	00	12
FLUVANNA VA	2	7.1			1	3.1			-56	57	40
FRANKLIN VA	9	10.1	8	7.1	5	3.1	-30	-56	-69	32	41
FREDERICK#VA	9	6.1	10	4.9	7	2.6	-20	-48	-58	53	29
GILES VA	6	10.9	2	3.0	4	4.1	-72	34	-63	44	60
GLOUCESTER VA	3	7.2	3	5.1	2	2.1	-29	-60	-71	28	21
GOOCHLAND VA	2	8.2	1	3.0			-63				

* SIGNIFICANTLY MORE DEATHS THAN EXPECTED

COUNTIES COMBINED; SEE TECHNICAL NOTES

WHITE FEMALE — CANCER OF THE STOMACH — ICD 151

1950-1959 DEATHS	1950-1959 RATE	1960-1969 DEATHS	1960-1969 RATE	1970-1979 DEATHS	1970-1979 RATE	COUNTY AND STATE	PERCENT CHANGE 60S/50S	PERCENT CHANGE 70S/60S	PERCENT CHANGE 70S/50S	PERCENTILE %CHANGE 70S/50S	PERCENTILE RATES 1970S
5*	26.1	2	7.4	1	3.7	GREENE VA	-72	-50	-86	06	53
2	5.9	1	2.4	1	2.1	GREENSVILLE#VA	-60	-11	-64	40	22
9	8.5	9	6.7	5	2.9	HALIFAX#VA	-21	-57	-66	36	34
2	2.4	1	1.0	10	5.9	HANOVER VA	-59	482	140	99	84
16	6.6	33	7.0	21	2.5	HENRICO VA	6	-64	-62	44	28
8	6.4	10	6.2	12	4.2	HENRY#VA	-2	-32	-34	83	63
1	4.3	2	9.0	1	3.7	HIGHLAND VA	109	-59	-15	89	52
5	15.1	4	7.8	2	2.9	ISLE OF WIGHT VA	-48	-63	-81	12	35
5	4.5	4	2.2	11	4.5	JAMES CITY#VA	-51	104	0	92	67
1	6.2					KING AND QUEEN VA					
1	5.6	1	3.7	1	3.4	KING GEORGE VA	-33	-8	-39	79	47
		1	4.2			KING WILLIAM VA					
4	13.5			2	2.7	LANCASTER VA			-80	13	32
11	9.3	17*	12.6	10	6.1	LEE VA	35	-51	-34	82	86
3	2.9	5	3.8	8	4.8	LOUDOUN VA	31	28	67	97	73
3	6.4			3	3.4	LOUISA VA			-48	70	46
5	14.6	4	7.8	3	4.1	LUNENBURG VA	-46	-48	-72	27	61
3	8.3			3	7.0	MADISON VA			-16	89	91
6	11.9	3	4.6	5	5.0	MATHEWS VA	-61	7	-58	53	74
5	7.7	2	2.4	6	4.4	MECKLENBURG VA	-69	84	-44	75	65
5	19.0	3	7.5	1	1.5	MIDDLESEX VA	-61	-81	-92	01	14
12	8.1	7	3.8	7	2.6	MONTGOMERY#VA	-53	-31	-68	34	30
6	12.9	2	3.1	4	5.7	NELSON VA	-76	85	-56	57	83
8	15.5	6	9.5	5	6.2	NORTHAMPTON VA	-38	-35	-60	49	87
2	4.2	3	5.5	2	2.5	NORTHUMBERLAND VA	31	-54	-40	78	29
3	6.2	3	5.2			NOTTOWAY VA	-16				
4	7.4	3	3.9	3	3.2	ORANGE VA	-47	-18	-57	56	42
4	5.4	4	4.4	2	1.9	PAGE VA	-19	-57	-65	38	19
3	4.9	2	2.6	1	.9	PATRICK VA	-47	-67	-83	10	10
21	6.9	20	4.8	16	3.0	PITTSYLVANIA#VA	-31	-38	-57	56	37
2	10.8	1	5.0	1	2.8	POWHATAN VA	-54	-44	-74	22	32
4	9.0	2	3.4	3	4.3	PRINCE EDWARD VA	-62	27	-52	64	65
7	10.3	9	6.4	9	3.7	PRINCE WILLIAM#VA	-38	-42	-64	41	54
8	9.0	9	6.8	4	2.5	PULASKI VA	-24	-63	-72	26	28
3	13.6	2	6.9			RAPPAHANNOCK VA	-50				
1	5.2			4	15.7	RICHMOND VA			205	99	99
38	6.5	40	4.6	24	2.0	ROANOKE#VA	-28	-57	-69	32	20
13	11.4	4	2.8	12	6.9	ROCKBRIDGE#VA	-75	147	-39	78	91
16	7.1	18	6.3	14	3.5	ROCKINGHAM#VA	-12	-44	-50	67	49
5	5.9	8	7.8	7	5.0	RUSSELL VA	33	-36	-15	89	76
6	5.4	8	6.0	6	3.3	SCOTT VA	12	-44	-38	80	45
14	11.1	9	6.5	5	2.8	SHENANDOAH VA	-41	-57	-75	21	33
12	10.3	8	5.1	6	3.1	SMYTH VA	-50	-39	-70	30	40
5	11.0			3	3.7	SOUTHAMPTON#VA			-67	36	52
8	8.7	4	3.3	4	2.3	SPOTSYLVANIA#VA	-61	-32	-74	24	24
3	8.5	5	8.6	3	2.9	STAFFORD VA	1	-66	-66	37	36
2	15.0	1	5.9	2	8.7	SURRY VA	-61	47	-42	76	96
		2	7.5	1	3.0	SUSSEX VA		-61			36
16	10.9	13	6.7	9	3.8	TAZEWELL VA	-38	-43	-65	39	55
5	8.8	5	6.8	2	1.6	WARREN VA	-23	-77	-82	10	15
27	13.2	22	7.9	20	5.6	WASHINGTON#VA	-40	-30	-58	54	82
2	8.3	1	1.7	2	2.3	WESTMORELAND VA	-80	37	-72	26	25
14	8.7	16	7.5	16	6.0	WISE#VA	-13	-20	-31	84	85
10	9.8	12	10.2	7	3.6	WYTHE VA	4	-64	-63	43	51
79	7.1	74	4.4	64	2.9	CHESAPEAKE CITY#VA	-37	-36	-60	49	34
13	8.9	15	6.5	14	4.0	HAMPTON CITY VA	-27	-38	-55	59	60
16	9.4	10	3.6	19	4.6	NEWPORT NEWS CITY VA	-62	27	-52	64	68
5	6.4	6	5.0	2	1.3	SUFFOLK CITY VA	-23	-73	-79	15	13
1208	10.3	1001	6.2	807	3.8	WASHINGTON WA	-39	-39	-63		
3	12.1	4	9.5	6*	9.7	ADAMS WA	-22	2	-20	88	97
4	3.9	4	4.0	4	3.4	ASOTIN WA	4	-16	-13	90	45
10	7.7	3	1.9	15	4.8	BENTON WA	-76	155	-38	80	72
23	12.2	10	4.0	12	3.5	CHELAN WA	-68	-10	-71	28	49
13	11.4	7	3.6	12	4.1	CLALLAM WA	-68	13	-64	41	61
45	11.2	29	5.0	29	3.9	CLARK WA	-55	-23	-65	38	57
22	10.7	17	6.4	18	4.6	COWLITZ WA	-40	-28	-57	55	69
4	9.5	11*	15.5			DOUGLAS WA	62				
2	19.2			1	5.1	FERRY WA			-74	24	76
4	7.6	2	2.5	5	3.9	FRANKLIN WA	-67	57	-48	70	58
				1	5.7	GARFIELD WA					83
6	11.1	5	3.9	6	3.0	GRANT WA	-65	-23	-73	25	38
24	10.0	29	9.3	20	4.7	GRAYS HARBOR WA	-7	-49	-53	63	71
4	7.8	6	6.7	4	2.5	ISLAND WA	-15	-63	-69	33	28
4	9.9			1	1.2	JEFFERSON WA			-88	04	12
465*	12.0	347	6.5	268	4.0	KING WA	-46	-39	-67	36	58
42	13.7	23	4.8	31	4.7	KITSAP WA	-65	-2	-65	38	71
10	10.4	3	2.6	5	3.1	KITTITAS WA	-75	19	-71	28	39
3	5.7	12*	18.6	2	2.5	KLICKITAT WA	227	-87	-56	57	28
23	9.0	24	7.8	13	3.2	LEWIS WA	-14	-59	-65	39	42
6	11.2	4	6.8	1	1.3	LINCOLN WA	-39	-81	-89	03	13
8	13.7	12*	12.2	5	3.6	MASON WA	-11	-71	-74	23	50
8	7.9	4	3.0	2	1.3	OKANOGAN WA	-62	-58	-84	08	13
10	12.7	5	5.1	8	5.9	PACIFIC WA	-60	16	-53	62	85
3	9.0	4	11.8	3	5.8	PEND OREILLE WA	31	-51	-36	81	83
126	9.6	129	7.5	98	4.4	PIERCE WA	-22	-42	-55	60	66
3	14.4	4	12.9	2	4.8	SAN JUAN WA	-10	-63	-67	36	72
17	6.9	13	4.2	17	4.0	SKAGIT WA	-40	-4	-42	76	59
3	12.5			2	6.2	SKAMANIA WA			-50	67	87
75	11.5	64	6.7	45	3.1	SNOHOMISH WA	-42	-53	-73	25	40
110	8.7	91	5.5	84	4.0	SPOKANE WA	-37	-27	-54	60	59
4	5.0	10	10.5	6	4.5	STEVENS WA	109	-57	-10	90	68
19	8.4	26	7.6	14	2.3	THURSTON WA	-9	-70	-72	26	25
1	5.7	2	8.6			WAHKIAKUM WA	51				
17	8.3	15	5.3	12	4.1	WALLA WALLA WA	-36	-23	-51	65	61
26	6.5	27	5.7	23	3.7	WHATCOM WA	-12	-36	-44	75	52
16	12.4	12	7.5	3	2.2	WHITMAN WA	-40	-71	-82	10	23
45	7.9	43	5.6	29	3.1	YAKIMA WA	-30	-45	-61	47	39

* SIGNIFICANTLY MORE DEATHS THAN EXPECTED

COUNTIES COMBINED; SEE TECHNICAL NOTES

WHITE FEMALE — CANCER OF THE STOMACH — ICD 151

1950-1959 DEATHS	1950-1959 RATE	1960-1969 DEATHS	1960-1969 RATE	1970-1979 DEATHS	1970-1979 RATE	COUNTY AND STATE	PERCENT CHANGE 60S/50S	PERCENT CHANGE 70S/60S	PERCENT CHANGE 70S/50S	PERCENTILE %CHANGE 70S/50S	PERCENTILE RATES 1970S
733	9.9	610	6.6	470	4.2	WEST VIRGINIA WV	-34	-37	-58		
11	13.4	4	4.0	3	2.9	BARBOUR WV	-70	-27	-78	16	35
16	10.6	7	4.4	5	2.2	BERKELEY WV	-58	-49	-79	15	24
7	9.6	6	5.2	5	3.7	BOONE WV	-46	-29	-61	46	53
8	9.7	3	4.0	6	5.9	BRAXTON WV	-59	50	-39	79	85
11	10.1	13	9.9	10	6.1	BROOKE WV	-1	-38	-39	79	86
38	7.4	41	6.3	21	2.9	CABELL WV	-14	-55	-61	47	35
6	15.7	3	6.6	3	5.6	CALHOUN WV	-58	-15	-65	40	82
3	7.9	4	7.8	2	3.3	CLAY WV	0	-58	-58	53	44
4	8.6	4	6.5	4	6.4	DODDRIDGE WV	-24	-2	-26	86	88
29	14.4	16	5.9	25*	8.2	FAYETTE WV	-59	39	-43	75	95
6	13.9	5	10.8	3	5.8	GILMER WV	-22	-46	-58	53	84
7	17.7	11*	24.4	1	1.9	GRANT WV	38	-92	-89	02	18
13	9.3	10	5.7	9	4.2	GREENBRIER WV	-39	-25	-55	60	63
4	8.0	5	7.4	1	1.4	HAMPSHIRE WV	-8	-81	-82	10	14
14	11.5	15	9.4	13	6.2	HANCOCK WV	-18	-34	-46	73	87
4	9.1	4	7.8	3	4.6	HARDY WV	-14	-40	-49	68	70
47	11.9	29	5.9	22	3.8	HARRISON WV	-50	-35	-68	34	56
11	13.2	10	10.9	4	3.0	JACKSON WV	-18	-73	-77	17	37
9	14.5	11*	11.6	4	3.6	JEFFERSON WV	-20	-69	-75	21	50
76	10.2	63	6.0	57	4.2	KANAWHA WV	-41	-31	-59	51	63
21	15.3	14	8.8	11	6.1	LEWIS WV	-43	-31	-60	49	86
6	7.8	6	6.3	7	6.4	LINCOLN WV	-19	1	-19	89	88
11	7.5	13	6.9	11	5.3	LOGAN WV	-8	-22	-29	85	79
14	10.0	16	8.7	13	6.7	MC DOWELL WV	-13	-24	-33	83	90
32	10.5	25	6.4	25	5.2	MARION WV	-39	-19	-51	66	78
20	12.0	12	5.9	10	3.9	MARSHALL WV	-51	-33	-67	35	58
7	6.7	11	9.2	5	3.6	MASON WV	37	-61	-47	71	50
15	6.7	22	6.8	7	1.8	MERCER WV	2	-74	-73	24	18
12	12.7	10	9.1	7	5.0	MINERAL WV	-28	-45	-61	48	75
12	9.2	11	7.7	6	3.7	MINGO WV	-16	-52	-59	50	53
25	11.1	19	6.8	14	4.1	MONONGALIA WV	-39	-40	-63	43	61
6	9.3	3	3.9	3	3.6	MONROE WV	-58	-8	-62	46	50
3	6.5	5	10.5			MORGAN WV	62				
9	9.2	8	7.2	11*	7.8	NICHOLAS WV	-22	9	-15	89	94
31	7.5	21	4.5	18	3.7	OHIO WV	-40	-17	-50	67	54
1	2.2	1	2.5	4	7.1	PENDLETON WV	12	186	220	99	92
1	3.0	3	7.5	2	4.0	PLEASANTS WV	155	-47	35	96	59
5	9.6	4	7.3	2	3.6	POCAHONTAS WV	-23	-51	-62	45	51
18	14.2	12	7.8	9	4.6	PRESTON WV	-45	-41	-68	34	69
4	4.4	9	8.3	7	4.9	PUTNAM WV	86	-40	11	94	74
20	9.9	23	7.3	24	5.5	RALEIGH WV	-27	-24	-44	74	81
8	6.6	7	4.6	5	2.9	RANDOLPH WV	-30	-38	-56	57	34
9	10.1	2	2.2	4	4.9	RITCHIE WV	-78	125	-51	65	74
9	11.2	6	6.4	5	4.1	ROANE WV	-43	-36	-64	42	61
12	15.7	8	8.8	3	3.0	SUMMERS WV	-44	-66	-81	12	37
12	14.0	4	3.3	4	2.8	TAYLOR WV	-77	-14	-80	14	33
6	12.1	5	8.5	4	5.3	TUCKER WV	-30	-38	-57	56	78
7	9.7	5	5.4	4	5.4	TYLER WV	-44	0	-44	74	80
7	6.8	7	5.9	6	3.3	UPSHUR WV	-13	-44	-52	65	44
8	5.9	8	4.2	5	2.1	WAYNE WV	-28	-50	-64	41	22
4	7.8	5	8.4	4	5.9	WEBSTER WV	7	-30	-25	86	84
11	11.6	8	6.9	4	3.0	WETZEL WV	-40	-57	-75	22	36
		4	12.8	2	5.7	WIRT WV		-55			83
37	10.5	22	4.9	15	2.6	WOOD WV	-54	-48	-76	20	29
6	8.6	7	7.6	8	6.5	WYOMING WV	-12	-14	-25	86	88
2472*	13.9	1828*	7.9	1353*	4.6	WISCONSIN WI	-43	-42	-67		
6	13.1	5	8.1	4	3.9	ADAMS WI	-38	-52	-70	29	57
19*	19.6	15*	12.7	7	6.5	ASHLAND WI	-35	-49	-67	36	89
23	14.1	16	7.0	9	3.0	BARRON WI	-51	-57	-79	15	37
20*	28.6	7	9.3	3	2.9	BAYFIELD WI	-67	-69	-90	02	35
61	13.9	49	7.7	38	4.5	BROWN WI	-44	-42	-68	34	67
8	10.2	7	7.6	7	6.6	BUFFALO WI	-25	-13	-35	81	89
3	5.1	5	6.7	5	5.6	BURNETT WI	30	-17	8	93	81
13	15.0	6	5.5	4	2.6	CALUMET WI	-63	-53	-83	09	30
33*	16.6	13	5.0	20	6.1	CHIPPEWA WI	-70	21	-63	43	86
26*	15.5	17	8.4	11	4.5	CLARK WI	-46	-47	-71	28	67
23	10.3	10	3.9	10	3.1	COLUMBIA WI	-62	-21	-70	30	40
9	11.0	5	5.3	2	1.5	CRAWFORD WI	-51	-72	-86	05	15
100	11.8	69	5.9	70	4.6	DANE WI	-50	-22	-61	47	69
39	12.1	27	6.6	21	3.7	DODGE WI	-45	-44	-69	31	52
23*	21.7	14	9.9	7	4.2	DOOR WI	-54	-57	-81	12	64
43*	18.1	34*	12.7	15	4.7	DOUGLAS WI	-30	-63	-74	23	71
13	9.5	18*	10.7	12	6.2	DUNN WI	12	-42	-35	82	87
32	12.1	33	9.0	19	4.1	EAU CLAIRE WI	-25	-55	-66	36	61
5*	28.1	5*	31.2	4*	17.3	FLORENCE WI	11	-45	-38	79	99
53	12.5	30	5.5	18	3.0	FOND DU LAC WI	-56	-45	-76	20	39
2	5.3	4	9.0	1	1.8	FOREST WI	69	-80	-67	36	17
27	10.1	23	7.6	15	4.3	GRANT WI	-25	-43	-57	56	65
19	12.1	14	7.4	10	3.9	GREEN WI	-39	-48	-68	34	57
10	10.6	15*	12.5	5	3.4	GREEN LAKE WI	18	-73	-68	33	45
10	8.4	16*	10.9	6	4.8	IOWA WI	29	-56	-43	76	73
7	16.8	3	5.5	6*	9.0	IRON WI	-67	63	-47	72	96
15*	18.3	10	10.5	3	2.3	JACKSON WI	-43	-78	-87	04	25
34	12.0	26	7.1	18	3.7	JEFFERSON WI	-41	-48	-69	31	52
12	12.0	7	5.8	6	3.7	JUNEAU WI	-51	-37	-69	31	52
51	14.8	40	7.7	33	4.8	KENOSHA WI	-48	-38	-68	34	72
13	15.3	7	7.0	8	6.1	KEWAUNEE WI	-55	-12	-60	49	86
40	10.4	34	7.1	26	4.1	LA CROSSE WI	-32	-42	-61	48	61
9	9.2	7	5.8	9	6.8	LAFAYETTE WI	-36	16	-26	86	90
17	16.5	7	6.1	5	3.2	LANGLADE WI	-63	-47	-81	12	42
16	14.8	13	8.9	14*	7.8	LINCOLN WI	-40	-13	-47	70	94
51	14.7	33	7.2	20	3.3	MANITOWOC WI	-51	-54	-77	18	45
53*	15.7	36	7.5	34	5.4	MARATHON WI	-52	-28	-66	38	80
40*	21.5	18	7.6	11	3.5	MARINETTE WI	-65	-54	-84	08	48
7	12.4	5	6.6	2	1.9	MARQUETTE WI	-47	-72	-85	06	18
47*	16.7	32	8.7	27	5.8	MENOMINEE#WI	-48	-34	-66	38	83

* SIGNIFICANTLY MORE DEATHS THAN EXPECTED

\# COUNTIES COMBINED; SEE TECHNICAL NOTES

WHITE FEMALE — CANCER OF THE STOMACH — ICD 151

1950-1959 DEATHS	1950-1959 RATE	1960-1969 DEATHS	1960-1969 RATE	1970-1979 DEATHS	1970-1979 RATE	COUNTY AND STATE	PERCENT CHANGE 60S/50S	PERCENT CHANGE 70S/60S	PERCENT CHANGE 70S/50S	PERCENTILE %CHANGE 70S/50S	PERCENTILE RATES 1970S
638*	14.7	481*	8.7	321*	4.8	MILWAUKEE WI	-41	-44	-67	35	73
24	14.7	15	7.5	10	3.9	MONPOE WI	-49	-48	-74	24	57
13	13.5	14	10.1	5	2.8	ONEIDA WI	-25	-72	-79	15	33
50	13.6	34	6.6	23	3.7	OUTAGAMIE WI	-52	-44	-73	25	52
13	11.3	19	10.0	17	6.3	OZAUKEE WI	-11	-37	-45	74	87
2	4.4	3	5.9	1	1.4	PEPIN WI	34	-76	-67	35	14
10	8.2	11	7.4	4	2.1	PIERCE WI	-10	-72	-74	22	22
29*	24.1	18*	10.7	12	5.0	POLK WI	-56	-54	-79	15	75
25	14.5	20	9.8	20*	7.1	PORTAGE WI	-32	-28	-51	65	92
7	8.9	12*	13.2	12*	9.3	PRICE WI	48	-30	4	93	96
84*	14.9	50	6.7	50	5.3	RACINE WI	-55	-21	-64	40	79
12	10.9	9	7.9	8	6.7	RICHLAND WI	-27	-16	-38	79	90
68	13.1	45	7.1	40	4.9	ROCK WI	-46	-31	-63	43	74
6	10.1	7	7.9	9	7.6	RUSK WI	-22	-4	-25	86	93
20	14.2	12	6.9	12	4.3	ST CROIX WI	-51	-37	-69	31	65
24	11.1	14	4.7	6	1.6	SAUK WI	-58	-65	-85	06	16
4	8.4	5	8.0	6	8.0	SAWYER WI	-4	-1	-4	92	94
63	14.1	66*	11.5	32	4.5	SHEBOYGAN WI	-19	-61	-68	33	67
10	12.5	8	8.3	6	4.3	TAYLOR WI	-34	-47	-65	38	65
26*	18.7	17	10.4	8	3.5	TREMPEALEAU WI	-45	-67	-82	11	47
17	11.4	16	8.5	9	4.1	VERNON WI	-25	-52	-64	41	61
8	22.4	2	5.5	5	4.2	VILAS WI	-76	-22	-81	12	64
39	14.4	21	6.3	20	4.2	WALWORTH WI	-56	-34	-71	28	62
4	6.3	5	6.5	2	2.9	WASHBURN WI	2	-55	-54	61	36
22	11.7	21	8.2	19	5.3	WASHINGTON WI	-30	-35	-55	60	79
71*	14.6	61	8.4	68*	6.2	WAUKESHA WI	-42	-26	-57	55	87
39*	17.2	16	5.2	12	2.8	WAUPACA WI	-70	-46	-84	08	34
12	13.4	8	7.1	5	4.2	WAUSHAPA WI	-47	-41	-68	33	63
68	13.2	46	7.1	39	4.6	WINNEBAGO WI	-46	-35	-65	39	70
32	14.5	37*	11.3	27*	6.6	WOOD WI	-22	-42	-54	60	89
111	10.9	66	4.6	65	3.5	WYOMING WY	-58	-24	-68		
6	9.3	4	4.5	2	2.1	ALBANY WY	-52	-53	-77	17	22
4	8.0	3	5.0	1	1.3	BIG HORN WY	-38	-74	-84	08	13
3	14.1	1	2.7			CAMPBELL WY	-81				
1	2.5	2	4.1	3	4.3	CARBON WY	69	4	75	97	64
3	10.4	1	2.5			CONVERSE WY	-76				
1	4.6	1	5.0	1	5.2	CROOK WY	8	6	15	94	78
10	22.7	8	8.8	2	1.6	FREMONT WY	-61	-82	-93	01	15
5	9.6	1	1.5	5	6.9	GOSHEN WY	-85	366	-28	85	91
2	11.2	3	7.2	1	2.3	HOT SPRINGS WY	-36	-69	-80	13	24
1	4.1	3	9.1			JOHNSON WY	123				
16	10.6	9	4.0	16	5.4	LARAMIE WY	-62	34	-49	69	80
2	7.4	3	8.6	5*	11.4	LINCOLN WY	16	33	53	97	98
13	11.0	9	4.9	8	3.3	NATRONA WY	-55	-33	-70	29	43
4	7.5	3	3.8	5	3.9	PARK-TETON#WY	-50	2	-49	69	56
5	15.4	2	5.5	2	3.9	PLATTE WY	-65	-29	-75	22	57
13	13.0	6	4.7	8	5.6	SHERIDAN WY	-64	19	-57	56	82
2	31.3	1	6.4			SUBLETTE WY	-79				
10	18.8	2	2.2	3	2.9	SWEETWATER WY	-89	35	-85	07	35
5	16.2	2	5.3	2	4.8	UINTA WY	-67	-9	-70	29	73
4	20.8	1	3.5	1	3.4	WASHAKIE WY	-83	-3	-84	08	46

* SIGNIFICANTLY MORE DEATHS THAN EXPECTED

COUNTIES COMBINED; SEE TECHNICAL NOTES

NON-WHITE MALE | CANCER OF THE STOMACH | ICD 151

1950-1959 DEATHS	RATE	1960-1969 DEATHS	RATE	1970-1979 DEATHS	RATE	COUNTY AND STATE	PERCENT CHANGE 60S/50S	70S/60S	70S/50S	PERCENTILE %CHANGE 70S/50S	RATES 1970S
16598	28.0	17045	23.1	15112	15.8	UNITED STATES	-17	-32	-44		
725	21.8	692	19.9	597	16.0	ALABAMA AL	-9	-20	-27		
3	8.8	5	17.3	2	7.4	AUTAUGA AL	97	-58	-16	71	62
3	9.8	5	24.2	10	24.3	BALDWIN AL	146	0	147	97	90
3	7.4	14	31.3	11	27.3	BARBOUR AL	325	-13	272	99	92
5	27.3	5	29.6	2	16.5	BIBB AL	9	-44	-39	51	79
		2	62.9			BLOUNT AL					
11	25.0	6	17.2	3	7.4	BULLOCK AL	-31	-57	-70	14	62
6	16.6	7	16.5	3	8.3	BUTLER AL	-1	-50	-50	39	63
12	23.5	10	20.1	6	9.2	CALHOUN AL	-14	-54	-61	25	64
9	18.7	4	8.8	10	20.6	CHAMBERS AL	-53	135	10	82	86
1	17.2	2	45.4	2	35.6	CHEROKEE AL	164	-21	107	95	95
3	23.2	2	12.8	1	6.6	CHILTON AL	-45	-48	-71	13	61
9	24.7	5	14.8	5	15.0	CHOCTAW AL	-40	1	-39	51	76
9	17.9	10	20.9	5	9.3	CLARKE AL	17	-56	-48	41	64
		2	26.9	2	26.8	CLAY AL		0			92
				1	24.8	CLEBURNE AL					90
		4	23.0	6	24.6	COFFEE AL		7			90
4	14.8	7	23.6	10	26.0	COLBERT AL	60	10	76	92	91
10	32.1	9	28.1	4	10.9	CONECUH AL	-13	-61	-66	19	67
5	34.6	2	13.9	1	5.7	COOSA AL	-60	-59	-84	04	60
4	20.9	7	31.3	5	28.3	COVINGTON AL	49	-9	35	87	93
2	9.7	4	21.6	2	11.3	CRENSHAW AL	124	-48	17	84	68
1	44.6					CULLMAN AL					
5	33.8	2	11.0	3	13.8	DALE AL	-67	25	-59	28	73
24	18.4	18	14.6	18	15.1	DALLAS AL	-21	3	-18	70	76
		1	33.5			DE KALB AL					
4	10.9	12	32.7	2	4.4	ELMORE AL	199	-86	-59	27	58
13	34.4	10	23.1	9	19.7	ESCAMBIA AL	-33	-15	-43	47	85
7	17.2	7	19.4	8	15.4	ETOWAH AL	12	-20	-11	74	76
2	20.6	2	24.7	1	7.2	FAYETTE AL	20	-71	-65	20	61
1	13.4	1	15.1	1	18.7	FRANKLIN AL	13	24	40	88	83
1	10.6	2	17.1	1	5.4	GENEVA AL	61	-69	-50	40	59
11	23.1	4	8.2	6	18.2	GREENE AL	-64	121	-21	67	83
10	20.6	12	23.8	6	9.2	HALE AL	16	-61	-55	33	64
7	32.3	8	37.5	1	3.5	HENRY AL	16	-91	-89	02	58
12	25.6	13	25.7	6	11.1	HOUSTON AL	0	-57	-57	30	67
		1	10.8	3	35.3	JACKSON AL		226			95
198	25.9	167	19.6	158	17.2	JEFFERSON AL	-24	-13	-34	56	80
2	22.4	1	10.8	1	10.6	LAMAR AL	-52	-1	-53	36	67
7	23.6	6	22.0	4	11.2	LAUDERDALE AL	-7	-49	-52	36	68
3	15.1	4	19.4	3	17.4	LAWRENCE AL	29	-11	15	83	81
12	20.4	15	25.4	8	12.9	LEE AL	25	-49	-37	53	71
10	37.4	6	20.0	2	6.9	LIMESTONE AL	-46	-66	-82	06	61
9	19.3	5	11.3	4	10.3	LOWNDES AL	-41	-9	-47	43	66
16	14.0	11	10.6	27*	25.4	MACON AL	-25	140	81	93	91
6	8.5	12	15.7	7	8.1	MADISON AL	85	-48	-5	77	63
16	21.0	15	22.2	11	16.2	MARENGO AL	6	-27	-23	66	78
1	37.3	2	149.6			MARION AL	301				
2	50.9	1	21.0			MARSHALL AL	-59				
60	25.2	76	22.9	65	18.1	MOBILE AL	-9	-21	-28	62	82
10	24.2	7	16.8	7	16.0	MONROE AL	-31	-5	-34	56	78
45	22.0	41	20.1	41	17.7	MONTGOMERY AL	-9	-12	-20	68	81
5	16.1	6	16.8	4	11.3	MORGAN AL	4	-33	-30	60	68
11	19.9	8	15.9	4	9.4	PERRY AL	-20	-41	-53	35	65
8	23.3	4	12.4	7	18.7	PICKENS AL	-47	51	-20	68	83
8	21.6	13	33.6	12*	34.1	PIKE AL	55	2	58	90	95
7	35.7	9*	52.6	3	16.7	RANDOLPH AL	47	-68	-53	34	80
7	12.9	15	24.8	10	13.8	RUSSELL AL	92	-44	7	81	73
1	11.2	4	24.9	1	6.9	ST CLAIR AL	122	-72	-38	51	61
7	28.2	2	8.1	4	16.0	SHELBY AL	-71	98	-43	47	78
14	23.2	10	15.7	10	16.5	SUMTER AL	-32	5	-29	61	79
16	28.6	10	17.5	13	20.3	TALLADEGA AL	-39	16	-29	61	86
4	12.3	2	6.0	4	13.8	TALLAPOOSA AL	-52	131	12	82	73
21	22.3	17	17.8	18	16.8	TUSCALOOSA AL	-20	-5	-25	65	80
3	9.7	10	39.9	3	11.1	WALKER AL	311	-72	14	83	67
3	17.3			3	18.9	WASHINGTON AL			9	82	84
16	28.4	7	12.7	6	12.2	WILCOX AL	-55	-4	-57	30	70
		1*	344.1	1*	344.1	WINSTON AL		0			99
		17	14.1	24	13.9	ALASKA AK		-1			
63	18.1	92	22.1	81	10.3	ARIZONA AZ	22	-53	-43		
7	15.1	19	35.3	15	16.3	APACHE AZ	134	-54	8	81	79
		1	17.0	1	4.7	COCHISE AZ		-72			59
5	18.6	5	19.1	5	9.8	COCONINO AZ	3	-48	-47	42	65
1	10.3	5	43.8	3	20.3	GILA AZ	325	-54	97	94	86
		1	26.8			GRAHAM AZ					
24	23.3	27	19.7	23	7.8	MARICOPA AZ	-15	-60	-66	18	62
6	13.1	9	16.4	14	18.0	NAVAJO AZ	25	10	38	87	82
9	14.8	12	18.5	12	10.0	PIMA AZ	25	-46	-32	58	66
7	24.4	9	27.3	5	11.5	PINAL AZ	12	-58	-53	35	68
2	50.1					YAVAPAI AZ					
2	24.1	4	25.3	3	9.5	YUMA AZ	5	-62	-61	26	65
369	20.7	358	20.2	269	15.3	ARKANSAS AR	-3	-24	-26		
9	36.5	12*	44.1	7	23.9	ARKANSAS AR	21	-46	-35	55	89
8	23.2	5	14.4	5	16.0	ASHLEY AR	-38	11	-31	59	78
1*	188.7					BOONE AR					
5	23.0	6	23.3	6	24.4	BRADLEY AR	2	5	6	80	90
1	10.5	4	39.6	2	14.6	CALHOUN AR	277	-63	39	88	74
22	33.3	19	29.5	12	19.9	CHICOT AR	-11	-33	-40	49	85
2	7.5	5	20.8	2	7.4	CLARK AR	177	-64	-1	78	62
2	24.1	2	28.4			CLEVELAND AR	18				

* SIGNIFICANTLY MORE DEATHS THAN EXPECTED

COUNTY AND STATE	1950-1959 DEATHS	1950-1959 RATE	1960-1969 DEATHS	1960-1969 RATE	1970-1979 DEATHS	1970-1979 RATE	PERCENT CHANGE 60S/50S	PERCENT CHANGE 70S/60S	PERCENT CHANGE 70S/50S	PERCENTILE %CHANGE 70S/50S	PERCENTILE RATES 1970S
COLUMBIA AR	14	33.1	11	23.9	6	14.0	-28	-41	-58	29	73
CONWAY AR	6	34.7	3	30.3	1	4.2	-13	-86	-88	02	58
CRAIGHEAD AR	3	31.8	1	10.4	1	20.6	-67	97	-35	55	86
CRAWFORD AR	2	57.8	1	19.1	1	16.1	-67	-16	-72	13	78
CRITTENDEN AR	15	11.9	24	20.4	16	14.8	72	-27	25	85	75
CROSS AR	4	13.1	7	30.3	4	15.9	131	-48	21	84	77
DALLAS AR	7	40.3	5	27.3	4	22.9	-32	-16	-43	47	89
DESHA AR	7	15.6	8	19.1	9	22.7	22	19	45	89	88
DREW AR	8	30.0	1	4.7	3	14.5	-84	212	-51	37	74
FAULKNER AR			1	8.5	1	8.0		-5			62
GARLAND AR	1	3.4	8	30.9	4	15.2	817	-51	351	99	76
HEMPSTEAD AR	7	16.8	8	19.9	4	10.8	19	-46	-36	54	67
HOT SPRING AR	6	58.9	2	14.7	2	15.5	-75	5	-74	11	76
HOWARD AR	1	8.5	6	51.3			505				
INDEPENDENCE AR	1	74.3			1	54.5			-27	63	97
JACKSON AR	4	24.7	3	19.7	6*	31.7	-20	61	28	86	94
JEFFERSON AR	44	26.7	22	13.5	28	18.1	-49	34	-32	57	82
JOHNSON AR			1	67.0	1*	60.7		-9			98
LAFAYETTE AR	7	32.2	1	6.1	4	15.6	-81	157	-51	37	77
LEE AR	6	10.0	18	29.2	7	12.3	190	-58	22	85	70
LINCOLN AR	6	15.5	9	29.4	4	18.2	89	-38	17	84	82
LITTLE RIVER AR			2	11.9	2	14.8		25			75
LONOKE AR	4	15.6	5	18.3	5	18.8	18	3	21	84	84
MARION AR					1*	490.8					99
MILLER AR	8	23.0	7	21.7	8	20.3	-6	-6	-12	73	86
MISSISSIPPI AR	9	9.8	16	20.2	16	21.4	107	6	119	96	87
MONROE AR	10	24.3	8	20.6	4	10.8	-15	-47	-56	32	67
NEVADA AR	5	22.7	2	8.7	4	22.4	-62	157	-1	78	88
OUACHITA AR	9	17.1	5	8.8	6	9.9	-49	13	-42	48	66
PHILLIPS AR	23	20.0	17	14.5	9	7.8	-28	-46	-61	25	62
PIKE AR			1	28.1							
POINSETT AR	4	28.9	2	15.1	2	25.0	-48	66	-13	73	91
POPE AR	1	39.3	1	36.4			-7				
PRAIRIE AR			4	46.6							
PULASKI AR	57	26.4	42	18.3	36	14.4	-31	-22	-46	44	74
ST FRANCIS AR	15	18.3	21	27.6	13	19.8	50	-28	8	81	85
SALINE AR	1	11.2	1	10.8	2	23.7	-4	120	111	95	89
SEBASTIAN AR	1	5.6	5	22.1	1	5.0	298	-78	-10	74	59
SEVIER AR			4*	104.3							
UNION AR	11	19.5	13	22.0	10	15.2	13	-31	-22	67	76
WASHINGTON AR	1	54.2	1	53.0	1	18.1	-2	-66	-67	18	82
WHITE AR	1	16.1	1	17.2	2	21.2	6	24	32	87	87
WOODRUFF AR	9	32.1	7	24.7	6	24.1	-23	-3	-25	64	90
YELL AR	1	74.3									
CALIFORNIA CA	935*	34.0	1038	22.0	1178	12.4	-35	-44	-63		
ALAMEDA CA	80	29.8	98	19.1	117	12.9	-36	-32	-57	30	71
BUTTE CA	3	38.7	2	14.2	3	16.5	-63	16	-57	29	79
COLUSA CA	3*	78.0	2*	166.0	1	21.3	113	-87	-73	12	87
CONTRA COSTA CA	14	20.8	23	27.0	23	9.1	30	-66	-56	31	64
DEL NORTE CA			1	34.9							
FRESNO CA	45*	41.7	48*	35.4	29	12.3	-15	-65	-70	14	70
GLENN CA	1	188.7			1	23.6			-88	02	89
HUMBOLDT CA	2	23.3	1	6.8	3	18.7	-71	174	-20	68	83
IMPERIAL CA	4	14.4	6	21.7	2	2.7	50	-88	-81	06	57
KERN CA	21	36.8	16	19.3	13	7.9	-48	-59	-79	08	62
KINGS CA	5	37.3	3	19.8	2	6.5	-47	-67	-83	05	60
LOS ANGELES CA	414*	36.5	492	23.8	562	14.4	-35	-40	-61	26	74
MADERA CA	4	26.4	2	11.3	4	12.7	-57	12	-52	37	71
MARIN CA	1	15.1									
MARIPOSA CA			1*	68.8							
MENDOCINO CA	1	8.9									
MERCED CA	6	29.1	6	21.9	7	13.0	-25	-40	-55	33	71
MODOC CA	1	68.3									
MONO CA	1*	188.7									
MONTEREY CA	12	36.3	13	26.5	8	7.1	-27	-73	-80	07	61
ORANGE CA	7	47.7	13	22.7	17	7.8	-52	-66	-84	04	62
PLACER CA	6	46.9			1	5.3			-89	02	59
RIVERSIDE CA	14	37.0	14	19.7	22	15.0	-47	-24	-59	28	76
SACRAMENTO CA	35	37.9	35	26.6	43	14.8	-30	-45	-61	25	75
SAN BENITO CA	1	12.7									
SAN BERNARDINO CA	11	24.8	7	7.6	12	6.7	-69	-12	-73	12	61
SAN DIEGO CA	25	29.8	29	18.4	35	9.9	-38	-46	-67	18	66
SAN FRANCISCO CA	90	28.0	113	21.8	134	13.4	-22	-39	-52	36	72
SAN JOAQUIN CA	36	34.1	31	20.5	34	13.3	-40	-35	-61	25	72
SAN LUIS OBISPO CA	5	62.6	4	25.4	5	15.6	-59	-38	-75	11	77
SAN MATEO CA	9	24.4	15	28.4	22	9.6	17	-66	-60	26	65
SANTA BARBARA CA	9	43.5	1	3.0	10	15.0	-93	402	-65	20	76
SANTA CLARA CA	30*	56.1	32*	36.5	28	12.8	-35	-65	-77	09	71
SANTA CRUZ CA	7	46.8	7	36.7	4	8.9	-21	-76	-81	06	64
SHASTA CA			3	73.3	2	20.3		-72			86
SISKIYOU CA					2	25.8					91
SOLANO CA	2	2.8	4	4.9	7	6.2	72	27	118	96	60
SONOMA CA	3	36.0	4	28.6	8	34.4	-21	20	-4	77	95
STANISLAUS CA	1	8.7	3	19.1	5	15.7	120	-18	82	93	77
SUTTER CA	3	37.8	1	12.5			-67				
TULARE CA	7	15.7	3	7.8	4	6.1	-50	-22	-61	25	60
TUOLUMNE CA	1	74.3									
VENTURA CA	3	30.8	3	16.4	5	6.3	-47	-62	-80	07	60
YOLO CA	4	41.8	1	11.1	2	3.9	-73	-65	-91	01	58
YUBA CA	8*	127.8	1	11.5	1	8.8	-91	-23	-93	00	64
COLORADO CO	51	38.3	29	15.1	29	6.7	-61	-55	-82		
ADAMS CO	2	46.9	1	21.2	3	11.5	-55	-46	-75	10	68
BOULDER CO	2*	150.7									
CROWLEY CO	2*	1080.3									
DENVER-APAPAHOE#CO	29	34.9	21	15.7	22	9.3	-55	-41	-73	12	65
EAGLE CO	1*	344.1									
EL PASO CO	2	24.2	2	11.4	2	4.5	-53	-61	-81	06	59

* SIGNIFICANTLY MORE DEATHS THAN EXPECTED

COUNTIES COMBINED; SEE TECHNICAL NOTES

NON-WHITE MALE — CANCER OF THE STOMACH — ICD 151

1950-1959 DEATHS	1950-1959 RATE	1960-1969 DEATHS	1960-1969 RATE	1970-1979 DEATHS	1970-1979 RATE	COUNTY AND STATE	PERCENT CHANGE 60S/50S	PERCENT CHANGE 70S/60S	PERCENT CHANGE 70S/50S	PERCENTILE %CHANGE 70S/50S	PERCENTILE RATES 1970S
2*	256.5					HUERFANO CO					
		1	74.3	1	9.0	JEFFERSON CO		-88			64
		1*	56.2			LAS ANIMAS CO					
1*	86.0	1*	112.4			LOGAN CO	31				
1*	344.1					MORGAN CO					
2	101.7	1	33.5			OTERO CO	-67				
2*	735.1					PROWERS CO					
3	25.3	1	6.2	1	2.4	PUEBLO CO	-76	-61	-90	01	57
1*	133.9					SEDGWICK CO					
1	22.5					WELD CO					
78	37.6	81	23.7	104	18.8	CONNECTICUT CT	-37	-21	-50		
32*	52.4	24	23.1	35	21.3	FAIRFIELD CT	-56	-8	-59	27	87
20	32.0	16	13.8	35	18.8	HARTFORD CT	-57	36	-41	48	84
1	31.5	1	18.7	1	12.7	LITCHFIELD CT	-40	-32	-59	27	71
1	23.4	1	22.5	2	16.2	MIDDLESEX CT	-4	-28	-31	59	78
20	30.3	32	30.0	29	19.8	NEW HAVEN CT	-1	-34	-35	55	85
2	17.5	7	44.4	2	4.9	NEW LONDON CT	154	-89	-72	13	59
2*	157.1					WINDHAM CT					
83*	43.3	58	24.8	49	17.3	DELAWARE DE	-43	-30	-60		
10	27.6	10	24.9	8	15.9	KENT DE	-10	-36	-42	47	77
61*	57.3	39	27.5	34	20.0	NEW CASTLE DE	-52	-27	-65	20	85
12	26.0	9	17.6	7	13.3	SUSSEX DE	-33	-24	-49	40	72
323*	34.5	362*	25.9	360*	22.6	DIST. OF COLUMBIA DC	-25	-13	-34		
323*	34.5	362*	25.9	360*	22.6	WASHINGTON DC	-25	-13	-34	56	88
595	23.8	708	22.3	662	16.1	FLORIDA FL	-6	-28	-32		
12	18.9	19	25.2	13	14.3	ALACHUA FL	33	-43	-24	65	74
2	33.9	3	47.3			BAKER FL	40				
6	26.0	5	18.3	7	17.8	BAY FL	-30	-2	-31	59	82
		2	18.0	1	6.8	BRADFORD FL		-62			61
12	41.4	9	23.3	8	12.4	BREVARD FL	-44	-47	-70	15	70
23	23.6	38	24.3	44	17.1	BROWARD FL	3	-30	-28	62	80
		2	58.7			CALHOUN FL					
1	26.8			1	26.8	CHARLOTTE FL			0	78	92
3	41.9	2	27.2			CITRUS FL	-35				
1	9.1	4	41.0			CLAY FL	352				
1	6.1	3	22.8	3	13.0	COLLIER FL	272	-43	112	95	71
4	15.2	4	17.1	4	16.3	COLUMBIA FL	12	-5	7	81	79
58	23.5	100	24.2	107	14.7	DADE FL	3	-39	-37	52	75
		1	8.7	1	7.5	DE SOTO FL		-14			62
				1	24.5	DIXIE FL					90
77	21.9	105	26.2	90*	19.4	DUVAL FL	20	-26	-11	74	85
14	15.6	17	15.2	20	13.3	ESCAMBIA FL	-2	-12	-14	72	72
		3	47.9			FLAGLER FL					
3	44.1	1	17.6	1	18.9	FRANKLIN FL	-60	7	-57	29	84
14	19.6	15	16.8	10	11.6	GADSDEN FL	-14	-31	-41	49	69
1	47.2					GILCHRIST FL					
1	44.6	3*	84.5			GLADES FL	89				
3	30.4	2	17.2			GULF FL	-44				
5	29.5	2	13.3	3	17.6	HAMILTON FL	-55	33	-40	50	81
		2	37.1	3	34.5	HARDEE FL		-7			95
4	58.9	2	16.8	5	29.9	HENDRY FL	-71	78	-49	40	94
4	55.7			3	28.5	HERNANDO FL			-49	41	93
3	21.2	2	7.8	7	27.8	HIGHLANDS FL	-63	257	31	86	92
46	23.2	48	21.1	45	16.2	HILLSBOROUGH FL	-9	-23	-30	60	78
1	42.1					HOLMES FL					
2	10.2	7	43.1	6	21.6	INDIAN RIVER FL	323	-50	113	96	87
4	10.1	11	27.6	7	18.0	JACKSON FL	173	-35	78	92	82
6	27.3	5	24.2	2	10.7	JEFFERSON FL	-11	-56	-61	26	67
				2*	142.2	LAFAYETTE FL					99
11	28.7	7	15.0	5	9.7	LAKE FL	-48	-35	-66	19	65
7	26.4	12	32.2	4	10.5	LEE FL	22	-67	-60	27	67
14	20.1	16	21.8	12	14.9	LEON FL	8	-32	-26	63	75
4	26.3	1	10.8	2	11.4	LEVY FL	-59	6	-56	31	68
		1	40.9			LIBERTY FL					
7	29.6	7	27.7	8*	30.6	MADISON FL	-6	10	3	80	94
7	21.0	4	10.0	2	4.2	MANATEE FL	-52	-58	-80	07	58
25	38.9	21	29.3	15	18.1	MARION FL	-25	-38	-54	34	82
1	8.4	2	14.0	2	14.3	MARTIN FL	67	2	70	92	74
4	27.0	4	26.5	3	15.8	MONROE FL	-2	-40	-42	48	77
4	28.6	9*	56.9	1	8.6	NASSAU FL	99	-85	-70	15	63
4	65.0	1	7.6	1	10.2	OKALOOSA FL	-88	35	-84	04	66
2	54.9			1	11.9	OKEECHOBEE FL			-78	08	69
22	23.8	25	18.6	24	13.4	ORANGE FL	-22	-28	-44	46	72
		2	18.6	3	28.9	OSCEOLA FL		55			93
29	20.5	42	22.0	37	15.5	PALM BEACH FL	7	-30	-25	65	76
2	16.3	2	16.5	1	5.2	PASCO FL	1	-68	-68	16	59
18	23.5	30	23.3	32	19.6	PINELLAS FL	-1	-16	-17	71	85
22	19.6	16	12.0	33	19.3	POLK FL	-39	60	-2	78	84
8	21.0	12	31.0	7	14.6	PUTNAM FL	48	-53	-30	59	75
8	31.4	6	19.8	7	20.7	ST JOHNS FL	-37	5	-34	56	86
11	37.3	10	24.9	13	22.4	ST LUCIE FL	-33	-10	-40	50	88
2	24.2	1	20.2	2	47.2	SANTA ROSA FL	-16	134	96	94	96
6	26.8	6	25.9	6	18.8	SARASOTA FL	-4	-27	-30	60	84
19	38.8	20*	36.2	17	24.9	SEMINOLE FL	-7	-31	-36	54	91
		6	43.3	4	24.1	SUMTER FL		-44			90
5	26.9	4	24.7	2	13.1	SUWANNEE FL	-8	-47	-51	38	71
7	41.5	3	23.8	5	29.0	TAYLOR FL	-43	22	-30	60	93
2	27.6	1	13.3	1	7.4	UNION FL	-52	-44	-73	12	62
29*	39.0	17	19.8	17	16.6	VOLUSIA FL	-49	-16	-57	29	80
1	16.9	2	32.4			WAKULLA FL	92				
2	20.6	1	11.5			WALTON FL	-44				
1	7.8			1	10.2	WASHINGTON FL			31	86	66

* SIGNIFICANTLY MORE DEATHS THAN EXPECTED

NON-WHITE MALE — CANCER OF THE STOMACH — ICD 151

1950-1959 DEATHS	1950-1959 RATE	1960-1969 DEATHS	1960-1969 RATE	1970-1979 DEATHS	1970-1979 RATE	COUNTY AND STATE	PERCENT CHANGE 60S/50S	PERCENT CHANGE 70S/60S	PERCENT CHANGE 70S/50S	PERCENTILE %CHANGE 70S/50S	PERCENTILE RATES 1970S
807	23.7	762	20.6	610	14.7	GEORGIA GA	-13	-29	-38		
1	9.6	1	11.8	1	12.8	APPLING GA	23	8	33	87	71
3	51.6	1	21.9			ATKINSON GA	-58				
2	61.0	1	33.5			BACON GA	-45				
3	32.4	3	36.8			BAKER GA	14				
12	25.2	4	6.8	6	12.9	BALDWIN GA	-73	90	-49	41	71
1	14.1	2	21.0			BARROW GA	49				
6	47.8	3	20.3			BARTOW GA	-57				
2	11.6	3	13.4	2	8.4	BEN HILL GA	16	-37	-28	62	63
3	51.3	2	31.3			BERRIEN GA	-39				
32	23.9	25	15.6	35	19.8	BIBB GA	-35	27	-17	70	85
1	8.2					BLECKLEY GA					
2	50.2					BRANTLEY GA					
1	3.7	5	20.3	5	20.5	BROOKS GA	445	1	453	99	86
2	24.1	6*	81.4	1	14.4	BRYAN GA	238	-82	-40	50	74
2	7.1	7	23.8	3	10.2	BULLOCH GA	236	-57	44	88	66
11	20.0	7	16.0	4	10.3	BURKE GA	-20	-35	-48	41	66
3	19.9			3	23.1	BUTTS GA			16	83	89
3	13.2	4	21.7	4	22.1	CALHOUN GA	64	2	67	91	88
2	23.0	3	18.1	2	11.3	CAMDEN GA	-21	-38	-51	38	68
1	9.7					CANDLER GA					
5	24.0	1	6.1	4	16.1	CARROLL GA	-75	166	-33	57	78
1	19.1	2	42.6	1	15.1	CHARLTON GA	123	-64	-21	68	76
79*	37.8	79*	35.8	58*	22.8	CHATHAM GA	-5	-36	-40	50	88
4*	87.4	1	15.0			CHATTOOGA GA	-83				
		2	47.7	2	61.7	CHEROKEE GA		30			98
11	29.6	8	22.7	12*	28.5	CLARKE GA	-24	26	-4	77	93
1	18.7	1	10.6	1	9.2	CLAY GA	-43	-13	-51	38	64
1	9.9	5	47.0	1	6.7	CLAYTON GA	374	-86	-33	57	61
1	13.7					CLINCH GA					
9	63.0	8	29.4	1	4.5	COBB GA	-53	-85	-93	00	59
4	24.3	3	15.1	5	24.5	COFFEE GA	-38	62	1	79	90
4	18.0	2	8.6	4	14.7	COLQUITT GA	-53	72	-18	70	75
2	10.0			1	6.9	COLUMBIA GA			-32	59	61
		2	19.6	1	8.0	COOK GA		-59			62
4	9.1	6	16.6	5	14.1	COWETA GA	83	-15	56	90	73
4	39.8	3	30.1			CRAWFORD GA	-25				
3	11.1	4	16.5	2	9.7	CRISP GA	48	-41	-13	73	65
				1*	424.0	DADE GA					99
10	30.3	13	36.1	4	11.9	DECATUR GA	19	-67	-61	26	69
14	30.9	19	21.5	16	8.1	DE KALB GA	-30	-62	-74	12	63
3	20.4	1	6.2	5	31.2	DODGE GA	-69	403	53	90	94
4	21.0	1	5.5	1	5.9	DOOLY GA	-74	8	-72	13	60
16	24.8	15	19.7	11	11.4	DOUGHERTY GA	-21	-42	-54	34	68
		2	21.7			DOUGLAS GA					
6	20.8	4	15.5	3	12.6	EARLY GA	-25	-19	-39	51	70
3	29.5	2	17.1			EFFINGHAM GA	-42				
1	4.8	3	16.7	1	4.9	ELBERT GA	248	-71	2	79	59
5	23.7	4	17.7	3	12.5	EMANUEL GA	-25	-29	-47	42	70
5	60.0	3	31.3	2	16.4	EVANS GA	-48	-47	-73	13	79
1	14.0	1	12.9	1	19.9	FAYETTE GA	-8	54	42	88	85
9	27.3	5	15.0	8	21.1	FLOYD GA	-45	41	-23	66	87
				4*	96.4	FRANKLIN GA					98
132	25.5	143	21.7	103	13.5	FULTON GA	-15	-38	-47	42	72
1	49.1					GLASCOCK GA					
11	31.6	6	12.5	11	20.9	GLYNN GA	-60	68	-34	56	87
2	52.7	1	31.5	1	21.0	GORDON GA	-40	-33	-60	27	87
3	13.8	4	17.3	1	4.7	GRADY GA	25	-73	-66	19	59
3	15.6	2	9.1	2	8.7	GREENE GA	-42	-4	-44	46	63
3	41.9	2	19.6	2	14.7	GWINNETT GA	-53	-25	-65	21	75
1	28.1					HABERSHAM GA					
5	52.1	2	14.2	1	3.3	HALL GA	-73	-76	-94	00	58
4	19.4	3	15.6	2	8.7	HANCOCK GA	-20	-44	-55	32	63
				1	24.8	HARALSON GA					90
5	24.4	4	19.1	5	25.0	HARRIS GA	-22	31	2	79	91
3	20.2			3	22.4	HART GA			11	82	88
		2	63.2			HEARD GA					
5	24.7	1	4.4	2	8.1	HENRY GA	-82	83	-67	18	63
7	27.1	4	15.4	3	10.5	HOUSTON GA	-43	-32	-61	25	67
2	14.0	1	12.8	1	8.8	IRWIN GA	-8	-31	-37	53	64
3	37.7	3	34.1	2	29.6	JACKSON GA	-10	-13	-22	67	93
2	16.4	4	45.8	2	22.1	JASPER GA	180	-52	35	87	88
1	19.3	1	13.3			JEFF DAVIS GA	-31				
4	11.6	4	11.8	3	8.8	JEFFERSON GA	2	-26	-24	65	64
1	6.7	2	19.4	1	9.3	JENKINS GA	192	-52	39	88	64
3	34.5	1	11.9	1	12.3	JOHNSON GA	-66	4	-64	21	70
10*	87.9	4	34.1	1	6.2	JONES GA	-61	-82	-93	00	60
3	23.3	1	9.4	2	14.2	LAMAR GA	-60	52	-39	51	74
		2	35.8			LANIER GA					
5	12.1	13	30.5	7	14.7	LAURENS GA	153	-52	22	84	75
3	17.8	2	17.6	1	11.1	LEE GA	-1	-37	-38	52	67
3	16.2	5	25.6	6	27.4	LIBERTY GA	58	7	69	91	92
		1	14.1	2	18.1	LINCOLN GA		28			82
4*	65.8					LONG GA					
14	26.7	9	16.4	8	12.3	LOWNDES GA	-38	-25	-54	34	70
3	22.1	2	14.0	3	14.9	MC DUFFIE GA	-37	6	-33	57	75
4	31.0	6	41.5	3	19.8	MC INTOSH GA	34	-52	-36	54	85
8	27.7	4	17.5	2	8.7	MACON GA	-37	-50	-68	16	64
		1	13.3			MADISON GA					
4	37.7			3	37.9	MARION GA			0	79	95
11	35.0	8	24.5	7	19.4	MERIWETHER GA	-30	-21	-45	45	85
3	38.9	2	57.8	2	26.6	MILLER GA	48	-54	-32	58	92
7	22.6	7	23.0	6	18.8	MITCHELL GA	1	-18	-17	71	84
2	12.7	2	14.7	1	5.9	MONROE GA	16	-60	-54	34	60
1	11.5	1	18.7			MONTGOMERY GA	63				
		2	12.2	4	19.0	MORGAN GA		56			84
19	17.8	25	25.7	25	17.5	MUSCOGEE#GA	44	-32	-2	77	81
3	15.9	1	4.4	4	13.5	NEWTON GA	-73	211	-15	72	72
4	39.3	2	16.8			OGLETHORPE GA	-57				

* SIGNIFICANTLY MORE DEATHS THAN EXPECTED

\# COUNTIES COMBINED; SEE TECHNICAL NOTES

NON-WHITE MALE — CANCER OF THE STOMACH — ICD 151

1950-1959 DEATHS	1950-1959 RATE	1960-1969 DEATHS	1960-1969 RATE	1970-1979 DEATHS	1970-1979 RATE	COUNTY AND STATE	PERCENT CHANGE 60S/50S	PERCENT CHANGE 70S/60S	PERCENT CHANGE 70S/50S	PERCENTILE %CHANGE 70S/50S	PERCENTILE RATES 1970S
1	31.3					PAULDING GA					
5	22.2	5	19.9	5	17.7	PEACH GA	-11	-11	-20	68	82
		1	119.2			PICKENS GA					
1	12.2	1	28.1	1	20.6	PIERCE GA	131	-27	69	91	86
4	37.4	2	18.5			PIKE GA	-51				
1	7.4	5	30.8	5	34.5	POLK GA	314	12	363	99	95
3	26.9	1	6.0	3	26.9	PULASKI GA	-78	352	0	79	92
4	34.8			1	6.5	PUTNAM GA			-81	06	61
1	21.9					QUITMAN GA					
4	15.1	5	20.7	4	15.5	RANDOLPH GA	38	-25	3	79	77
29	25.3	29	21.5	34	20.1	RICHMOND GA	-15	-7	-21	68	85
3	41.2	1	14.5			ROCKDALE GA	-65				
1	14.4	1	37.5	1	18.9	SCHLEY GA	160	-50	31	86	84
4	15.7	7	27.6	5	19.5	SCREVEN GA	76	-29	24	85	85
2	22.7	2	20.9	1	11.2	SEMINOLE GA	-8	-47	-51	38	67
5	20.0	5	17.5	4	13.6	SPALDING GA	-12	-22	-32	58	72
		4	47.4	2	18.0	STEPHENS GA		-62			82
9	47.3	1	5.8	6*	41.9	STEWART GA	-88	618	-11	74	96
5	11.8	4	10.1	13*	28.5	SUMTER GA	-15	182	141	97	93
5	36.5	3	23.7	1	5.3	TALBOT GA	-35	-78	-86	03	59
2	16.8	3	36.7	2	18.6	TALIAFERRO GA	119	-49	11	82	83
2	11.4	5	38.1	2	9.7	TATTNALL GA	235	-74	-14	72	65
2	13.7	2	14.5	1	8.6	TAYLOR GA	6	-41	-37	52	63
1	6.6	3	20.5	2	13.3	TELFAIR GA	212	-35	101	95	72
4	14.7	5	19.8	4	15.2	TERRELL GA	35	-23	4	80	76
4	8.1	10	18.5	7	14.5	THOMAS GA	127	-22	78	92	74
4	18.9	3	15.1	5	17.8	TIFT GA	-20	18	-6	76	82
5	35.9	7	40.0	3	14.8	TOOMBS GA	12	-63	-59	28	75
				1*	546.6	TOWNS GA					99
4	62.3	2	32.4			TREUTLEN GA	-48				
8	16.5	18	33.3	7	13.6	TROUP GA	102	-59	-17	70	72
3	32.7	1	10.3			TURNER GA	-69				
2	14.0	2	14.5	4	24.9	TWIGGS GA	4	72	79	93	91
3	13.3	7	32.9	1	3.4	UPSON GA	147	-90	-75	11	58
3	18.1	5	31.7	2	9.6	WALTON GA	76	-70	-47	42	65
13	42.2	14*	40.8	7	22.9	WARE GA	-3	-44	-46	44	89
3	17.6	3	25.5	2	12.0	WARREN GA	45	-53	-32	58	69
12	35.4	5	14.7	8	23.7	WASHINGTON GA	-58	61	-33	57	89
2	16.7	6*	60.7	3	22.6	WAYNE GA	264	-63	35	87	88
3	48.9					WEBSTER GA					
2	36.8	1	6.2			WHEELER GA	-83				
1	81.8					WHITE GA					
2	38.1	1	13.7			WHITFIELD GA	-64				
1	8.6	2	17.1			WILCOX GA	97				
10	43.0	6	27.9	1	5.1	WILKES GA	-35	-82	-88	02	59
5	36.3	3	25.9	2	10.4	WILKINSON GA	-29	-60	-71	14	66
4	18.6	4	18.8	2	9.6	WORTH GA	1	-49	-48	41	65
		532*	33.1	539*	23.5	HAWAII HI		-29			
		88*	33.9	77*	25.0	HAWAII HI		-26			91
		337*	31.6	363*	22.6	HONOLULU HI		-28			88
		40	32.5	36	23.3	KAUAI HI		-28			89
		67*	40.3	63*	28.2	MAUI#HI		-30			93
11	31.1	8	21.3	6	11.7	IDAHO ID	-31	-45	-63		
1	53.6	1	74.3	1	16.1	ADA ID	39	-78	-70	15	78
2	34.9	3	50.1	1	16.4	BANNOCK ID	44	-67	-53	35	79
1	10.6					BINGHAM ID					
1	94.4			1	47.2	BONNEVILLE ID			-50	39	96
2	125.8	2	97.8	1	8.6	CANYON ID	-22	-91	-93	00	63
				1*	112.4	JEROME ID					98
1*	188.7					MADISON ID					
1*	188.7	1*	112.4			MINIDOKA ID	-40				
1*	67.0					PAYETTE ID					
1	27.0	1	172.0	1*	53.6	WASHINGTON ID	538	-69	99	94	97
846*	32.6	911*	25.2	840*	17.2	ILLINOIS IL	-23	-32	-47		
1	13.3			1	11.1	ADAMS IL			-16	71	67
8	22.9	8	22.7	2	14.1	ALEXANDER IL	-1	-38	-38	51	73
		1	18.9			BOND IL					
6	38.5	5	28.7	4	16.0	CHAMPAIGN IL	-25	-44	-58	28	78
		1*	53.6			CLINTON IL					
688*	34.7	767*	26.6	719*	18.2	COOK IL	-23	-32	-48	42	83
		1	47.2			DE KALB IL					
1	25.8	1	12.5	3	12.3	DU PAGE IL	-52	-1	-52	36	70
1	188.7					EDGAR IL					
1*	344.1					EDWARDS IL					
1*	98.2					FAYETTE IL					
		1*	74.3			GALLATIN IL					
1	57.3	1	34.4			HENRY IL	-40				
4	24.3	3	18.3	7*	41.6	JACKSON IL	-25	127	71	92	96
1	17.2					JEFFERSON IL					
2	21.6	4	17.9	5	14.8	KANE IL	-17	-17	-32	58	75
7	20.5	4	8.1	12	21.0	KANKAKEE IL	-60	159	2	79	87
2	19.7	1	15.8	1	6.7	KNOX IL	-20	-57	-66	19	61
9	32.8	7	20.8	6	6.0	LAKE IL	-36	-71	-82	05	60
				1	21.3	LEE IL					87
		1*	98.2			LIVINGSTON IL					
		1	37.7			LOGAN IL					
				1	112.4	MC HENRY IL					98
1	27.3	1	17.1			MC LEAN IL	-37				
1	5.9	2	11.1	4	12.7	MACON IL	88	15	115	96	71
				1	62.9	MACOUPIN IL					98
9	21.6	8	16.5	7	11.9	MADISON IL	-23	-28	-45	45	69
3	55.5	4*	64.5	1	14.9	MARION IL	16	-77	-73	12	75
		1	10.5	2	27.4	MASSAC IL		161			92
3	41.4	1	11.2			MORGAN IL	-73				
5	22.3	5	15.8	8	16.3	PEORIA IL	-29	3	-27	63	79

* SIGNIFICANTLY MORE DEATHS THAN EXPECTED

COUNTIES COMBINED; SEE TECHNICAL NOTES

1950-1959 DEATHS	1950-1959 RATE	1960-1969 DEATHS	1960-1969 RATE	1970-1979 DEATHS	1970-1979 RATE	COUNTY AND STATE	PERCENT CHANGE 60S/50S	PERCENT CHANGE 70S/60S	PERCENT CHANGE 70S/50S	PERCENTILE %CHANGE 70S/50S	PERCENTILE RATES 1970S
		2	107.6	1	38.2	PERRY IL		-64			95
4	18.8	7	30.1	2	8.2	PULASKI IL	60	-73	-57	30	63
3	38.1	1	18.9	1	16.4	RANDOLPH IL	-51	-13	-57	30	79
1	9.4	3	17.1	5	17.5	ROCK ISLAND IL	81	2	86	93	81
51	37.3	46	24.3	25	11.4	ST CLAIR IL	-35	-53	-69	15	68
1	14.1	2	25.9	2	50.0	SALINE IL	84	93	255	98	97
7	28.5	4	13.1	3	9.0	SANGAMON IL	-54	-32	-68	16	64
		1*	267.9			SCHUYLER IL					
		1	29.8			STEPHENSON IL					
1	32.1					UNION IL					
9	44.7	2	7.9	2	8.8	VERMILION IL	-82	12	-80	07	64
1	53.0	2*	85.4			WARREN IL	61				
5	28.7	6	22.3	5	14.6	WILL IL	-23	-35	-49	40	75
3	45.7	1	12.6			WILLIAMSON IL	-72				
5	45.0	4	15.1	9	21.5	WINNEBAGO IL	-66	42	-52	36	87
241	31.8	222	21.6	173	13.3	INDIANA IN	-32	-39	-58		
5	28.2	1	3.3	6	9.0	ALLEN IN	-88	178	-68	17	64
1	344.1					BARTHOLOMEW IN					
1	73.2	1	70.7	1	47.2	CASS IN	-4	-33	-36	55	96
2	21.5	2	14.9	2	11.9	CLARK IN	-31	-20	-45	45	69
3	12.0	5	17.6	7	23.5	DELAWARE IN	47	33	95	94	89
		1	8.5	2	14.0	ELKHART IN		66			73
1	11.5	2	27.7	2	22.7	FLOYD IN	142	-18	98	94	88
2	40.7					GIBSON IN					
1	8.4	3	24.7	2	11.6	GRANT IN	194	-53	39	88	69
1	33.5	1	31.3			HAMILTON IN	-7				
				1*	74.3	HARRISON IN					98
1	34.4					HENRY IN					
5*	77.1	3	40.8	1	7.1	HOWARD IN	-47	-83	-91	01	61
		1*	112.4			HUNTINGTON IN					
1*	163.6					JACKSON IN					
1*	424.0					JASPER IN					
		1*	294.8			JAY IN					
1	212.0	1	56.2			JOHNSON IN	-73				
1	70.1			1	24.4	KNOX IN			-65	20	90
62	34.4	72	25.7	57	14.6	LAKE IN	-25	-43	-58	29	74
4	31.9	3	14.4	3	14.1	LA PORTE IN	-55	-2	-56	32	73
1*	84.8					LAWRENCE IN					
4	45.1	4	31.6	3	15.8	MADISON IN	-30	-50	-65	21	77
105	35.5	90	22.0	64	12.9	MARION IN	-38	-42	-64	22	71
2	67.0					MONROE IN					
1	53.6					ORANGE IN					
		1	74.5			POSEY IN					
				1*	133.9	PULASKI IN					99
1	70.7			1*	42.0	PUTNAM IN			-41	49	96
1*	188.7	1*	112.4			RANDOLPH IN	-40				
11	26.6	9	17.6	8	12.7	ST JOSEPH IN	-34	-28	-52	37	71
1	38.3					SPENCER IN					
1	44.6					TIPPECANOE IN					
13	32.4	11	24.5	4	8.1	VANDERBURGH IN	-24	-67	-75	11	63
5	19.1	3	11.8	5	18.0	VIGO IN	-38	53	-6	76	82
2	13.8	6	35.0	2	12.2	WAYNE IN	153	-65	-12	73	70
30	28.2	26	21.8	12	8.1	IOWA IA	-23	-63	-71		
4	29.3	4	41.3	4	19.7	BLACK HAWK IA	41	-52	-33	57	85
1*	344.1					CEDAR IA					
		1	38.3			CERRO GORDO IA					
1	38.3					CLINTON IA					
1	30.3					DES MOINES IA					
2	29.0	2	29.1	1	15.8	LEE IA	0	-46	-46	44	77
1	18.1	1	27.0			LINN IA	49				
				1	29.8	MARSHALL IA					94
				1*	424.0	MONTGOMERY IA					99
14	34.4	10	23.2	3	6.4	POLK IA	-33	-72	-81	06	60
1	49.1	1	22.3			POTTAWATTAMIE IA	-55				
		1	12.5	1	3.5	SCOTT IA		-72			58
		1	54.5			TAMA IA					
1	27.0	1	37.5			WAPELLO IA	39				
1	70.1					WEBSTER IA					
3	49.6	4	56.6	1	11.8	WOODBURY IA	14	-79	-76	10	69
97	25.5	83	19.6	46	9.4	KANSAS KS	-23	-52	-63		
4	39.8	1	18.6	2	28.8	ATCHISON KS	-53	54	-28	62	93
		1	32.6			BARTON KS					
		1	19.1			BOURBON KS					
1	38.2	1	18.6			BROWN KS	-51				
		1*	74.3			CHAUTAUQUA KS					
1	18.9			1	16.4	COWLEY KS			-13	73	79
3*	83.9					CRAWFORD KS					
				1	28.1	DONIPHAN KS					93
3	26.1	2	19.3	2	18.7	DOUGLAS KS	-26	-3	-28	62	83
1	89.3	1	136.6			FINNEY KS	53				
		1	47.1			FRANKLIN KS					
2	34.9					GEARY KS					
				1*	344.1	GREENWOOD KS					99
		1	24.4			HARVEY KS					
3*	122.9	2	59.5			JACKSON KS	-52				
1	24.6	1	23.4			JOHNSON KS	-5				
1	10.5	1	14.1	2	33.6	LABETTE KS	35	138	220	98	94
4	15.3	4	19.0	1	4.6	LEAVENWORTH KS	24	-76	-70	15	59
1	29.8					LYON KS					
		1*	112.4			MARION KS					
1	20.6					MIAMI KS					
10*	56.9	4	19.5	1	5.7	MONTGOMERY KS	-66	-71	-90	01	60
1*	596.2					NORTON KS					
1	34.4					PRATT KS					
2	56.3					RENO KS					

* SIGNIFICANTLY MORE DEATHS THAN EXPECTED

NON-WHITE MALE — CANCER OF THE STOMACH — ICD 151

1950-1959 DEATHS	1950-1959 RATE	1960-1969 DEATHS	1960-1969 RATE	1970-1979 DEATHS	1970-1979 RATE	COUNTY AND STATE	PERCENT CHANGE 60S/50S	PERCENT CHANGE 70S/60S	PERCENT CHANGE 70S/50S	PERCENTILE %CHANGE 70S/50S	PERCENTILE RATES 1970S
1	34.2					RILEY KS					
1	25.8			2	33.0	SALINE KS			28	85	94
9	22.0	7	10.3	8	9.8	SEDGWICK KS	-53	-4	-55	32	65
				1	54.5	SEWARD KS					97
11	26.8	6	14.1	7	12.4	SHAWNEE KS	-47	-12	-54	34	70
		1	70.7			SUMNER KS					
1	89.3					WABAUNSEE KS					
1*	149.1					WILSON KS					
33	23.1	46	30.6	17	10.5	WYANDOTTE KS	33	-66	-55	33	67
251	25.1	229	22.2	164	15.7	KENTUCKY KY	-12	-29	-38		
1	15.6	1	38.3			ADAIR KY	145				
3*	56.3					BALLARD KY					
1	8.6	1	10.5			BARREN KY	21				
1	24.8	2*	76.5	1	47.2	BATH KY	209	-38	90	93	96
3	38.2	2	23.7			BELL KY	-38				
1	74.3					BOONE KY					
4	20.5	4	26.0	6*	50.4	BOURBON KY	26	94	146	97	97
		1	18.1	1	20.6	BOYD KY		14			86
3	22.0	5	27.8	2	11.7	BOYLE KY	26	-58	-47	43	69
1*	114.7	1*	62.9			BREATHITT KY	-45				
1	49.2					BULLITT KY					
2	23.4	1	13.4			CALDWELL KY	-43				
1	37.5			2*	47.7	CALLOWAY KY			27	85	96
1	22.9	1	31.3			CAMPBELL KY	36				
6	10.9	10	20.1	6	12.0	CHRISTIAN KY	84	-40	10	82	69
2	15.3	6*	45.5			CLARK KY	198				
				1*	49.2	CLAY KY					97
3*	122.9					CUMBERLAND KY					
7	46.6	5	27.6	1	7.2	DAVIESS KY	-41	-74	-85	03	61
1*	56.2					EDMONSON KY					
1*	141.3					ESTILL KY					
25	26.6	19	19.4	24*	23.5	FAYETTE KY	-27	21	-12	73	89
		1	49.2	1	49.2	FLEMING KY		0			97
1	31.3			1	86.0	FLOYD KY			175	97	98
3	45.9	1	11.2			FRANKLIN KY	-76				
2	21.7	2	20.6	2	31.2	FULTON KY	-5	52	44	88	94
1	18.6					GARRARD KY					
1	11.6	1	6.2			GRAVES KY	-47				
				1*	188.7	GRAYSON KY					99
1	26.8	3*	100.5	1	56.2	GREEN KY	275	-44	110	95	97
1	4.4	2	37.3	1	2.2	HARDIN KY	742	-94	-50	39	57
3	21.1	5	20.8	2	10.0	HARLAN KY	-2	-52	-53	36	66
		2	64.8			HARRISON KY					
		1	15.6			HART KY					
8	45.8	4	26.0	2	16.3	HENDERSON KY	-43	-37	-64	21	78
1	37.7	3*	76.3			HENRY KY	102				
1	37.5	1	17.5	1	29.8	HICKMAN KY	-53	70	-21	68	94
6	24.3	3	12.6	4	19.1	HOPKINS KY	-48	52	-21	67	84
89	30.3	72	21.5	62	15.8	JEFFERSON KY	-29	-27	-48	41	77
1	16.6			1	11.2	JESSAMINE KY			-33	57	68
3	14.1	9*	49.0	4	24.4	KENTON KY	248	-50	73	92	90
		1*	344.1	1*	245.4	KNOTT KY		-29			99
1	31.5					LARUE KY					
1*	112.4					LESLIE KY					
1	14.4	2	30.3	3*	76.7	LETCHER KY	110	153	431	99	98
1	15.7	4*	80.8			LINCOLN KY	414				
1	9.4	2	14.9	4	31.9	LOGAN KY	58	114	239	98	94
2	40.8					LYON KY					
4	13.7	9	31.1	4	12.3	MC CRACKEN KY	127	-60	-10	74	70
2	11.5	6	36.7	2	10.2	MADISON KY	219	-72	-11	74	66
1	20.2			1	21.0	MARION KY			4	80	87
		2	22.7	1	10.5	MASON KY		-54			66
		1	89.3			MEADE KY					
3	54.0	3	51.4			MERCER KY	-5				
		1	67.0			METCALFE KY					
1	28.1	1	22.3	1	57.3	MONROE KY	-21	157	104	95	97
3	32.8	2	36.8	2	48.2	MONTGOMERY KY	12	31	47	89	97
2	23.6	1	14.1			MUHLENBERG KY	-40				
4	55.8	2	28.2	2	25.1	NELSON KY	-49	-11	-55	32	91
1	29.8					OHIO KY					
1	23.6					PERRY KY					
				2*	61.2	PIKE KY					98
1*	136.6			1*	212.0	POWELL KY			55	90	99
				1	23.6	PULASKI KY					89
3	26.2	1	8.4	2	23.6	SCOTT KY	-68	181	-10	75	89
3	27.9	4	37.6	1	9.3	SHELBY KY	35	-75	-67	18	65
1	16.4	1	13.5			SIMPSON KY	-18				
1	22.3	4*	98.2			TAYLOR KY	340				
3	29.0	1	12.1	2	25.4	TODD KY	-58	109	-12	73	91
		1	18.4			TRIGG KY					
6*	75.5	2	30.8	1	24.8	UNION KY	-59	-19	-67	18	90
3	18.9	4	19.1	1	5.8	WARREN KY	1	-70	-70	15	60
1	21.5	1	37.5	1	20.2	WASHINGTON KY	74	-46	-6	76	86
				1	28.1	WAYNE KY					93
5	55.1	2	27.5	1	26.5	WEBSTER KY	-50	-4	-52	37	92
5	57.0	2	23.2	2	24.8	WOODFORD KY	-59	7	-56	31	90
1235*	36.9	1145*	30.5	800*	19.4	LOUISIANA LA	-17	-36	-47		
16*	52.4	5	14.1	5	12.9	ACADIA LA	-73	-8	-75	11	71
6	34.7	9*	45.3	3	15.1	ALLEN LA	31	-67	-57	31	76
8	23.0	4	12.1	14*	41.2	ASCENSION LA	-47	241	79	93	96
13*	51.0	14*	54.7	4	17.5	ASSUMPTION LA	7	-68	-66	19	81
16	44.8	10	27.9	7	17.4	AVOYELLES LA	-38	-38	-61	25	81
7	41.0	5	30.5	2	11.3	BEAUREGARD LA	-26	-63	-73	13	68
8	23.3	8	21.7	3	6.0	BIENVILLE LA	-7	-73	-74	11	60
14	27.5	18	29.5	9	17.2	BOSSIER LA	7	-42	-37	52	81
80	30.3	84	27.1	68	19.6	CADDO LA	-11	-28	-35	55	85
41*	53.3	31*	34.0	22	20.6	CALCASIEU LA	-36	-39	-61	24	86

* SIGNIFICANTLY MORE DEATHS THAN EXPECTED

NON-WHITE MALE | CANCER OF THE STOMACH | ICD 151

1950-1959 DEATHS	1950-1959 RATE	1960-1969 DEATHS	1960-1969 RATE	1970-1979 DEATHS	1970-1979 RATE	COUNTY AND STATE	PERCENT CHANGE 60S/50S	PERCENT CHANGE 70S/60S	PERCENT CHANGE 70S/50S	PERCENTILE %CHANGE 70S/50S	PERCENTILE RATES 1970S
6	46.6	2	14.2	1	6.6	CALDWELL LA	-69	-54	-86	03	61
1	3.9	7	37.2	2	12.6	CATAHOULA LA	861	-66	227	98	71
4	8.8	8	17.2	9	17.6	CLAIBORNE LA	95	2	100	94	81
5	14.8	8	20.3	4	9.4	CONCORDIA LA	37	-54	-36	54	65
22	39.2	18	26.8	10	16.4	DE SOTO LA	-32	-39	-58	29	79
55	31.0	60	25.7	51	18.7	EAST BATON ROUGE LA	-17	-27	-40	50	83
9	21.7	8	21.7	8	22.5	EAST CARROLL LA	0	4	4	80	88
10	20.5	8	16.8	6	13.3	EAST FELICIANA LA	-18	-21	-35	55	72
8	36.9	9	44.5	6	25.8	EVANGELINE LA	21	-42	-30	60	91
16	46.1	7	18.1	6	14.9	FRANKLIN LA	-61	-18	-68	17	75
3	19.7	7*	46.8	4	25.5	GRANT LA	138	-46	30	86	91
22*	45.0	27*	54.1	22*	40.4	IBERIA LA	20	-25	-10	74	96
21	36.1	28*	48.7	9	15.5	IBERVILLE LA	35	-68	-57	30	77
4	19.9	5	22.0	2	10.5	JACKSON LA	10	-52	-47	42	67
35*	46.2	32*	29.4	24	17.4	JEFFERSON LA	-36	-41	-62	24	81
7	34.7	7	35.8	8*	37.6	JEFFERSON DAVIS LA	3	5	8	81	95
24*	53.1	22*	38.6	17	26.8	LAFAYETTE LA	-27	-30	-49	40	92
16*	73.5	7	32.5	5	23.2	LAFOURCHE LA	-56	-29	-68	16	89
3	34.8	2	21.7	2	18.9	LA SALLE LA	-38	-13	-46	44	84
8	26.4	10	24.3	4	7.8	LINCOLN LA	-8	-68	-70	14	62
4	33.5	2	20.7	4	28.4	LIVINGSTON LA	-38	38	-15	72	93
14	28.4	16	34.9	6	14.9	MADISON LA	23	-57	-48	42	75
15	26.3	14	23.5	12	20.3	MOREHOUSE LA	-11	-14	-23	66	86
16	23.5	9	14.3	7	10.6	NATCHITOCHES LA	-39	-26	-55	33	67
316*	45.8	273*	33.0	205*	22.2	ORLEANS LA	-28	-33	-52	37	88
30	29.1	36	28.6	19	14.8	OUACHITA LA	-2	-48	-49	40	75
7	38.2	9*	45.7	3	14.2	PLAQUEMINES LA	20	-69	-63	23	74
24*	50.0	12	27.1	9	21.6	POINTE COUPEE LA	-46	-20	-57	30	87
43	33.0	44	32.5	18	13.0	RAPIDES LA	-1	-60	-61	26	71
6	26.2	6	25.1	1	4.3	RED RIVER LA	-4	-83	-84	04	58
13	34.2	7	15.4	2	6.2	RICHLAND LA	-55	-60	-82	05	60
4	21.9	7	33.5	1	4.6	SABINE LA	53	-86	-79	08	59
6*	94.9	4*	66.9	5*	61.2	ST BERNARD LA	-30	-8	-35	55	98
12*	67.1	3	15.9	3	12.8	ST CHARLES LA	-76	-19	-81	06	71
3	22.7	5	25.8	1	4.2	ST HELENA LA	13	-84	-82	06	58
19*	64.6	9	30.9	14*	48.7	ST JAMES LA	-52	57	-25	64	97
13	40.8	17*	58.3	9	24.9	ST JOHN THE BAPT LA	43	-57	-39	51	91
35	37.7	46*	43.2	36*	29.7	ST LANDRY LA	15	-31	-21	67	93
11	37.2	13*	43.2	8	22.0	ST MARTIN LA	16	-49	-41	49	88
18	30.8	23*	40.4	21*	35.4	ST MARY LA	31	-12	15	83	95
11	33.7	5	13.0	7	16.0	ST TAMMANY LA	-61	23	-52	36	78
29*	45.2	28*	39.7	15	18.4	TANGIPAHOA LA	-12	-54	-59	28	83
11	31.2	7	18.6	4	11.6	TENSAS LA	-40	-38	-63	23	68
13	35.2	18*	51.2	14*	34.5	TERREBONNE LA	46	-33	-2	78	95
4	17.9	5	18.5	1	3.1	UNION LA	4	-83	-83	05	58
9*	53.4	5	25.8	4	19.8	VERMILION LA	-52	-23	-63	23	85
2	17.4	2	16.3	1	8.2	VERNON LA	-6	-50	-53	35	63
19	38.2	6	11.9	5	8.9	WASHINGTON LA	-69	-25	-77	09	64
17	37.4	18	31.9	11	16.0	WEBSTER LA	-15	-50	-57	30	78
4	12.7	8	29.1	11*	42.0	WEST BATON ROUGE LA	128	44	230	98	96
5	38.9	5	35.7	1	7.9	WEST CARROLL LA	-8	-78	-80	07	62
5	19.0	6	27.4	2	9.0	WEST FELICIANA LA	44	-67	-53	36	64
9	34.2	7	28.4	3	11.6	WINN LA	-17	-59	-66	18	68
2	15.1	1	5.8	1	5.2	MAINE ME	-61	-10	-65		
1	133.9					AROOSTOOK ME					
1	35.1					CUMBERLAND ME					
				1*	114.7	HANCOCK ME					99
		1	133.9			WASHINGTON ME					
480*	33.3	465	25.6	440*	18.3	MARYLAND MD	-23	-29	-45		
1	17.5	2	43.0	1	17.2	ALLEGANY MD	145	-60	-2	78	80
25	29.4	27	23.6	20	16.4	ANNE ARUNDEL MD	-20	-31	-44	46	79
16	29.8	12	20.9	10	9.3	BALTIMORE MD	-30	-55	-69	16	65
9*	61.7	6	33.2	4	16.7	CALVERT MD	-46	-50	-73	12	60
11*	61.3	2	11.1			CAROLINE MD	-82				
3	37.7					CARROLL MD					
4	38.7	9*	62.6	1	5.8	CECIL MD	62	-91	-85	03	60
8	32.5	5	14.0	7	19.9	CHARLES MD	-57	42	-39	51	85
2	5.5	8	17.8	8	17.4	DORCHESTER MD	222	-2	214	98	81
8	46.3	8	37.2	5	21.9	FREDERICK MD	-20	-41	-53	36	88
3	11.4	7	25.4	2	5.9	HARFORD MD	123	-77	-48	41	60
4	24.2	5	33.0	2	9.3	HOWARD MD	36	-72	-62	24	65
5	27.2	6	29.5	1	4.5	KENT MD	9	-85	-84	04	59
7	16.2	16	33.9	15	13.4	MONTGOMERY MD	109	-61	-17	70	72
18	25.5	27	27.4	42	18.5	PRINCE GEORGE'S MD	7	-32	-27	62	83
8	43.1	5	21.9	1	5.1	QUEEN ANNE'S MD	-49	-77	-88	02	59
2	12.7	8	37.3	5	17.3	ST MARY'S MD	193	-54	36	87	81
11	32.9	7	20.9	8	24.5	SOMERSET MD	-37	17	-26	64	90
6	21.1	6	22.0	5	15.6	TALBOT MD	4	-29	-26	63	77
5	60.2	4	43.0			WASHINGTON MD	-29				
10	29.3	11	23.5	8	16.3	WICOMICO MD	-20	-31	-44	45	78
8	28.8	6	18.7	3	9.1	WORCESTER MD	-35	-51	-68	16	64
306*	36.4	278	26.7	292*	21.9	BALTIMORE CITY MD	-27	-18	-40	50	88
113	32.7	102	22.3	101	14.5	MASSACHUSETTS MA	-32	-35	-56		
5	49.6	1	22.3	2	10.7	BARNSTABLE MA	-55	-52	-78	08	67
1	22.5	2	31.6	1	10.9	BERKSHIRE MA	40	-66	-52	37	67
11	47.5	7	28.3	5	12.0	BRISTOL MA	-40	-58	-75	11	69
5	47.6	2	16.1	1	3.1	ESSEX MA	-66	-81	-93	00	58
		2*	215.7			FRANKLIN MA					
7	28.7	10	28.3	11	16.0	HAMPDEN MA	-1	-44	-44	46	78
14	31.3	11	17.5	10	9.8	MIDDLESEX MA	-44	-44	-69	16	65
1*	67.0					NANTUCKET MA					
1	9.0	1	4.6	2	10.9	NORFOLK MA	-48	135	21	84	67
6	25.6	5	18.4	3	6.9	PLYMOUTH MA	-28	-63	-73	12	61
59	32.6	61	25.4	60	17.7	SUFFOLK MA	-22	-30	-46	44	82
3	31.5			6	25.4	WORCESTER MA			-20	69	91

* SIGNIFICANTLY MORE DEATHS THAN EXPECTED

NON-WHITE MALE CANCER OF THE STOMACH ICD 151

1950-1959 DEATHS	1950-1959 RATE	1960-1969 DEATHS	1960-1969 RATE	1970-1979 DEATHS	1970-1979 RATE	COUNTY AND STATE	PERCENT CHANGE 60S/50S	PERCENT CHANGE 70S/60S	PERCENT CHANGE 70S/50S	PERCENTILE %CHANGE 70S/50S	PERCENTILE RATES 1970S
492	31.1	607	23.5	550	15.2	MICHIGAN MI	-25	-35	-51		
		1*	490.8			ALGER MI					
1	22.5	2	25.9	2	32.4	ALLEGAN MI	15	25	44	89	94
1	57.3					BARAGA MI					
		1	23.6			BAY MI					
12	39.3	12	28.2	15	22.7	BERRIEN MI	-28	-19	-42	47	88
4	15.9	3	14.1	2	3.2	CALHOUN MI	-11	-77	-80	07	58
8	44.1	4	15.2	3	11.9	CASS MI	-66	-22	-73	12	69
2*	360.7					DELTA MI					
				1	44.6	EMMET MI					96
20	37.7	18	24.1	18	13.3	GENESEE MI	-36	-45	-65	21	72
1*	344.1					GOGEBIC MI					
		1	38.2			GRAND TRAVERSE MI					
5	45.5	6	23.2	6	15.7	INGHAM MI	-49	-32	-66	20	77
		1	18.2			IONIA MI					
1*	344.1					IRON MI					
3	14.6	3	8.9	1	3.0	JACKSON MI	-39	-66	-79	07	58
6	45.0	3	18.4	6	21.2	KALAMAZOO MI	-59	16	-53	35	87
10	40.5	11	24.6	15	20.8	KENT MI	-39	-15	-49	41	86
		4	50.7	4	21.2	LAKE MI		-58			87
1	56.2			1	112.4	LAPEER MI			100	94	98
		1	54.5			LEELANAU MI					
		2*	205.2	2	56.3	LENAWEE MI		-73			97
5	38.3	2	10.4	1	2.8	MACOMB MI	-73	-73	-93	00	58
1	37.5					MECOSTA MI					
1*	344.1			2*	124.3	MIDLAND MI			-64	22	99
1	17.2	2	19.2	2	16.3	MONROE MI	11	-15	-5	77	79
1*	74.3					MONTCALM MI					
10	43.9	11	22.5	4	7.1	MUSKEGON MI	-49	-69	-84	04	61
1	20.4	1	13.8	2	28.3	NEWAYGO MI	-33	106	39	88	93
10	20.6	19	36.6	14	11.3	OAKLAND MI	78	-69	-45	44	68
1*	133.9					OTTAWA MI					
4	8.0	14	27.1	16	17.4	SAGINAW MI	237	-36	117	96	81
		1	8.5			ST CLAIR MI					
		3	37.6	2	16.3	ST JOSEPH MI		-57			79
		1*	267.9			TUSCOLA MI					
5	43.1	5	20.9	9	27.2	VAN BUREN MI	-51	30	-37	53	92
6	25.4	8	19.5	12	23.3	WASHTENAW MI	-23	20	-8	75	89
371	31.5	467	24.1	410	15.5	WAYNE MI	-23	-36	-51	39	77
31	27.2	22	14.4	19	7.7	MINNESOTA MN	-47	-46	-72		
3	75.8	3*	71.3			BECKER MN	-6				
1	12.3			2	23.7	BELTRAMI MN			93	94	89
		1	37.2			CARLTON MN					
1	13.0	1	19.1			CASS MN	47				
1	114.7					CLEARWATER MN					
9	21.3	11	20.0	10	8.3	HENNEPIN MN	-6	-59	-61	26	63
1	24.8					ITASCA MN					
1	28.1	1	21.2			MAHNOMEN MN	-25				
				1*	67.0	OLMSTED MN					98
1*	188.7					POLK MN					
13	42.7	5	11.9	5	9.3	RAMSEY MN	-72	-21	-78	08	65
				1	11.5	ST LOUIS MN					68
846	25.2	747	22.1	508	14.8	MISSISSIPPI MS	-12	-33	-41		
15	23.8	19	28.6	15	19.5	ADAMS MS	20	-32	-18	70	85
3	17.0	3	18.2			ALCORN MS	7				
6	19.7	6	16.5	2	7.4	AMITE MS	-17	-55	-63	23	62
8	23.8	11	33.4	5	11.9	ATTALA MS	41	-64	-50	39	69
4	31.2	1	9.3	1	6.9	BENTON MS	-70	-26	-78	09	61
45	29.2	27	19.2	25	20.7	BOLIVAR MS	-34	8	-29	61	86
1	9.9	2	12.1	1	6.3	CALHOUN MS	22	-48	-36	54	60
6	19.1	4	20.1	3	14.2	CARROLL MS	5	-29	-26	64	74
7	23.5	5	18.4	4	16.6	CHICKASAW MS	-21	-10	-29	61	79
3	35.3			3	23.6	CHOCTAW MS			-33	57	89
3	9.8	15*	48.9	3	7.9	CLAIBORNE MS	399	-84	-19	69	62
8	30.2	4	14.2	7	28.4	CLARKE MS	-53	99	-6	76	93
7	19.5	9	24.8	5	14.2	CLAY MS	27	-43	-27	62	73
33	24.6	30	24.2	14	13.8	COAHOMA MS	-1	-43	-44	46	73
12	23.3	11	21.7	9	17.5	COPIAH MS	-7	-19	-25	64	81
5	24.3	4	24.6	3	15.4	COVINGTON MS	1	-37	-37	53	76
10	17.3	7	13.2	2	4.0	DE SOTO MS	-24	-70	-77	09	58
18	34.9	11	21.8	8	13.8	FORREST MS	-38	-37	-61	26	73
7	47.9	4	26.2	4	30.8	FRANKLIN MS	-45	18	-36	54	94
2	52.4	2	53.8	3*	42.0	GEORGE MS	3	-22	-20	68	96
2	24.5	1	11.2	1	10.5	GREENE MS	-54	-6	-57	29	66
12	32.2	13	32.3	5	14.6	GRENADA MS	0	-55	-55	33	75
3	30.6	2	17.8	2	18.1	HANCOCK MS	-42	2	-41	49	82
12	23.2	11	19.3	15	18.0	HARRISON MS	-17	-7	-22	66	82
42	19.8	50	20.4	55	18.1	HINDS MS	3	-11	-9	75	82
26	31.1	17	19.7	5	6.8	HOLMES MS	-37	-65	-78	09	61
13	27.7	14	30.5	6	14.2	HUMPHREYS MS	10	-54	-49	40	73
2	11.2	1	6.5	1	14.3	ISSAQUENA MS	-42	120	28	86	74
		2	44.6			ITAWAMBA MS					
10	43.1	8	25.8	4	9.4	JACKSON MS	-40	-64	-78	08	65
7	23.1	5	17.6	2	4.6	JASPER MS	-24	-74	-80	07	59
10	31.5	10	32.6	8	25.5	JEFFERSON MS	3	-22	-19	69	91
10	38.7	6	24.1	6	25.4	JEFFERSON DAVIS MS	-38	5	-34	56	91
13	26.8	12	22.0	11	17.3	JONES MS	-18	-21	-35	55	81
5	17.5	2	8.0	1	3.2	KEMPER MS	-54	-60	-82	05	58
5	18.9	3	12.0	3	9.1	LAFAYETTE MS	-36	-24	-52	37	64
1	11.6	5*	60.3	1	14.4	LAMAR MS	422	-76	24	85	74
31	39.4	21	25.1	16	18.8	LAUDERDALE MS	-36	-25	-52	36	84
8	52.8	4	26.9	2	13.0	LAWRENCE MS	-49	-52	-75	10	71
10	38.8	9	30.6	4	14.1	LEAKE MS	-21	-54	-64	22	73
7	17.5	15*	42.0	6	13.6	LEE MS	141	-68	-22	67	72
24	20.3	34	32.7	11	11.2	LEFLORE MS	61	-66	-45	45	68

* SIGNIFICANTLY MORE DEATHS THAN EXPECTED

NON-WHITE MALE — CANCER OF THE STOMACH — ICD 151

1950-1959 DEATHS	1950-1959 RATE	1960-1969 DEATHS	1960-1969 RATE	1970-1979 DEATHS	1970-1979 RATE	COUNTY AND STATE	PERCENT CHANGE 60S/50S	PERCENT CHANGE 70S/60S	PERCENT CHANGE 70S/50S	PERCENTILE %CHANGE 70S/50S	PERCENTILE RATES 1970S
7	20.9	5	15.6	3	7.8	LINCOLN MS	-25	-50	-63	23	62
6	9.4	9	14.4	6	8.9	LOWNDES MS	54	-38	-5	76	64
26	31.3	14	17.4	14	18.9	MADISON MS	-45	9	-40	50	84
5	17.2	8	27.8	10*	33.6	MARION MS	62	21	95	94	94
13	24.1	4	8.0	4	8.4	MARSHALL MS	-67	5	-65	20	63
10	21.2	9	20.8	4	9.4	MONROE MS	-2	-55	-56	32	65
6	27.0	2	8.4	5	22.4	MONTGOMERY MS	-69	168	-17	70	88
5	32.9	2	7.4	4	19.2	NESHOBA MS	-78	161	-42	48	84
8	31.7	3	11.2	4	13.3	NEWTON MS	-65	19	-58	29	72
9	20.2	7	14.8	7	15.5	NOXUBEE MS	-27	5	-23	65	77
9	24.9	13	38.1	7	18.8	OKTIBBEHA MS	53	-51	-24	65	84
18	28.0	18	28.0	17*	26.0	PANOLA MS	0	-7	-7	76	91
2	8.9	1	5.7	5	27.9	PEARL RIVER MS	-36	388	214	98	92
1	10.5			3	23.3	PERRY MS			123	96	89
14	26.7	8	14.1	10	16.2	PIKE MS	-47	15	-39	51	78
2	18.0	2	17.5	1	7.0	PONTOTOC MS	-3	-60	-61	25	61
1	12.3					PRENTISS MS					
17	30.4	11	22.7	5	10.3	QUITMAN MS	-26	-54	-66	19	66
7	15.7	9	16.8	4	6.5	RANKIN MS	7	-61	-59	28	60
5	17.6	3	9.0	3	7.7	SCOTT MS	-49	-15	-56	31	62
12	38.9	11*	40.1	2	7.9	SHARKEY MS	3	-80	-80	07	62
8	38.7	2	7.2	3	10.2	SIMPSON MS	-82	42	-74	12	66
4	44.6	3	34.6			SMITH MS	-23				
2	27.5	1	14.9	2	27.1	STONE MS	-46	82	-1	78	92
26	22.6	25	23.3	11	12.6	SUNFLOWER MS	3	-46	-44	46	71
13	22.1	13	24.6	5	9.2	TALLAHATCHIE MS	11	-62	-58	29	64
6	20.4	3	8.3	5	14.3	TATE MS	-59	72	-30	60	74
1	11.2	4	31.1	3	17.9	TIPPAH MS	177	-42	60	90	82
2	51.3	1	15.7			TISHOMINGO MS	-69				
23	36.7	14	26.0	4	9.2	TUNICA MS	-29	-65	-75	11	64
2	15.6	1	8.8			UNION MS	-43				
5	25.4	5	23.6	8*	41.4	WALTHALL MS	-7	75	63	91	96
20	23.2	23	26.1	7	8.7	WARREN MS	13	-67	-62	24	64
63	37.1	45	26.9	28	17.7	WASHINGTON MS	-27	-34	-52	36	82
3	14.2	7	36.9	7	31.1	WAYNE MS	159	-16	118	96	94
1	14.9	1	8.0	2	19.2	WEBSTER MS	-46	140	29	86	84
4	11.0	7	19.1	7	17.5	WILKINSON MS	73	-8	59	90	81
4	16.1	10	31.1	4	10.0	WINSTON MS	94	-68	-38	52	66
5	19.1	3	14.7	3	16.7	YALOBUSHA MS	-23	14	-13	73	80
15	18.7	20	28.5	9	13.0	YAZOO MS	52	-55	-31	59	71
417	29.5	399	23.7	252	13.3	MISSOURI MO	-20	-44	-55		
1	18.7	2	21.6			AUDRAIN MO	15				
1*	188.7					BARRY MO					
1*	114.7					BATES MO					
3	18.7	5	29.7	4	21.2	BOONE MO	59	-28	13	83	87
2	10.2	2	16.0	3	17.6	BUCHANAN MO	57	10	72	92	81
3	15.4	2	12.9	1	19.5	BUTLER MO	-16	51	27	85	85
		2	11.1	2	15.4	CALLAWAY MO		39			76
1	17.9	1	22.3	1	17.2	CAPE GIRARDEAU MO	25	-23	-4	77	80
1	68.8					CARROLL MO					
6*	81.9	1	10.7			CHARITON MO	-87				
4*	75.7	1	14.9			CLAY MO	-80				
		1	24.8			CLINTON MO					
1	10.3	3	25.3	3	35.0	COLE MO	146	38	240	98	95
3	32.3					COOPER MO					
2	28.4	3	41.4	1	14.9	DUNKLIN MO	46	-64	-48	42	75
1	53.0			1	19.1	FRANKLIN MO			-64	22	84
2	14.8	2	15.4			GREENE MO	5				
				1*	68.8	HENRY MO					98
7*	76.5	1	10.8	1	14.3	HOWARD MO	-86	33	-81	06	74
83	29.2	84	23.4	55	12.8	JACKSON MO	-20	-45	-56	31	71
4	74.0	1	22.5			JASPER MO	-70				
		3*	75.2	1	18.4	JEFFERSON MO		-75			83
3	56.0			1	39.0	JOHNSON MO			-30	59	96
3	41.5	1	15.7	1	14.1	LAFAYETTE MO	-62	-10	-66	19	73
1	26.5	2	44.2			LINCOLN MO	67				
2*	135.7					LINN MO					
				1*	60.6	LIVINGSTON MO					97
1	11.4	3	29.0	1	14.1	MARION MO	154	-51	24	85	73
		1*	112.4			MILLER MO					
5	23.0	4	21.3	2	8.1	MISSISSIPPI MO	-8	-62	-65	21	63
1	26.8					MONROE MO					
3	71.1	2	36.7			MONTGOMERY MO	-48				
		2*	172.5			MORGAN MO					
8	36.5	2	12.9	1	4.9	NEW MADRID MO	-65	-62	-87	03	59
		1	62.9			NEWTON MO					
11	27.8	15	34.1	4	8.9	PEMISCOT MO	23	-74	-68	17	64
3	29.6	1	13.3	1	11.1	PETTIS MO	-55	-17	-62	24	67
4	40.5	2	15.4	1	14.1	PIKE MO	-62	-8	-65	21	73
				1*	74.3	PLATTE MO					98
1	112.4					RALLS MO					
4	45.4	1	14.0			RANDOLPH MO	-69				
1*	589.6					RIPLEY MO					
1	14.3			1	16.1	ST CHARLES MO			12	82	78
1*	245.4					STE GENEVIEVE MO					
15	18.3	25	25.3	27	14.7	ST LOUIS COUNTY MO	38	-42	-20	69	75
5	42.5	1	7.5	1	16.6	SALINE MO	-82	122	-61	26	79
2	14.5	4	33.7	4	32.3	SCOTT MO	132	-4	122	96	94
2*	278.3					SHELBY MO					
1	17.6	1	28.1	1	29.8	STODDARD MO	59	6	69	91	94
1*	424.0					TANEY MO					
1	74.3					WARREN MO					
1	212.0					WRIGHT MO					
210	31.0	217	25.9	130	14.4	ST LOUIS CITY MO	-17	-44	-54	34	74
25	36.7	8	11.1	8	8.5	MONTANA MT	-70	-23	-77		
5*	47.3			1	11.2	BIG HORN MT			-76	10	68
3	56.8					BLAINE MT					

* SIGNIFICANTLY MORE DEATHS THAN EXPECTED

NON-WHITE MALE 1950-1959		1960-1969		1970-1979		CANCER OF THE STOMACH COUNTY AND STATE		PERCENT CHANGE			ICD 151 PERCENTILE	
											%CHANGE	RATES
DEATHS	RATE	DEATHS	RATE	DEATHS	RATE			60S/50S	70S/60S	70S/50S	70S/50S	1970S
2	57.9			1	7.7	CASCADE	MT			-87	03	62
1*	344.1					CHOUTEAU	MT					
1*	344.1	1*	112.4			CUSTER	MT	-67				
		1	60.6			DEER LODGE	MT					
1	8.4			1	8.3	GLACIER	MT			-1	78	63
1	28.1			1	21.5	HILL	MT			-23	65	87
4*	90.3					LAKE	MT					
1	62.9	1	44.6			LEWIS AND CLARK	MT	-29				
		1	57.3			MISSOULA	MT					
1	56.2					PHILLIPS	MT					
				1*	163.6	POWELL	MT					99
1	20.2	2	23.4	3	41.1	ROOSEVELT	MT	16	76	103	95	96
		1	67.0			SANDERS	MT					
1*	74.3					TETON	MT					
3*	183.0					VALLEY	MT					
		1	28.4			YELLOWSTONE	MT					
37	36.9	29	22.6	22	13.9	NEBRASKA	NB	-39	-39	-62		
1	172.0			1*	89.3	BOX BUTTE	NB			-48	41	98
		1*	212.0			CHERRY	NB					
29	40.0	25	25.9	17	15.0	DOUGLAS	NB	-35	-42	-62	24	76
3	44.2			3	43.3	LANCASTER	NB			-2	78	96
1	67.0					LINCOLN	NB					
		1*	112.4			MADISON	NB					
		1*	112.4			RICHARDSON	NB					
		1	42.4	1	17.2	SCOTTS BLUFF	NB		-60			80
1*	344.1					SIOUX	NB					
2	29.5					THURSTON	NB					
14	28.7	12	12.8	28	17.9	NEVADA	NV	-55	40	-38		
1	60.7					CHURCHILL	NV					
7	28.6	7	12.5	19	21.3	CLARK	NV	-56	71	-25	64	87
		1	26.5			ELKO	NV					
				1	24.6	HUMBOLDT	NV					90
		1*	114.7			LANDER	NV					
1	56.2	1	24.6	1	23.4	MINERAL	NV	-56	-5	-58	28	89
1	62.9					NYE	NV					
1*	98.3					PERSHING	NV					
2	19.0	2	11.2	4	13.5	WASHOE	NV	-41	20	-29	61	72
1	61.3					WHITE PINE	NV					
				3*	115.2	ORMSBY-CARSON CITY	NV					99
3	75.3	2	49.4	1	7.8	NEW HAMPSHIRE	NH	-34	-84	-90		
		1*	188.7			GRAFTON	NH					
1*	267.9					MERRIMACK	NH					
2*	225.6	1	38.3	1	24.6	ROCKINGHAM	NH	-83	-36	-89	01	90
434*	34.6	520*	29.7	483*	19.3	NEW JERSEY	NJ	-14	-35	-44		
41	34.7	29	20.3	27	18.3	ATLANTIC	NJ	-42	-10	-47	42	83
17	42.0	9	12.2	15	15.0	BERGEN	NJ	-71	23	-64	22	75
17*	49.6	8	15.1	18	23.4	BURLINGTON	NJ	-70	55	-53	35	89
31	29.9	52*	41.3	29	16.1	CAMDEN	NJ	38	-61	-46	43	78
4	25.3	2	10.3	4	16.7	CAPE MAY	NJ	-59	62	-34	56	80
10	26.5	13	27.8	13	19.9	CUMBERLAND	NJ	5	-28	-25	64	85
149*	38.7	165*	28.6	144	17.1	ESSEX	NJ	-26	-40	-56	31	80
12	28.2	12	26.8	11	17.9	GLOUCESTER	NJ	-5	-33	-36	53	82
23	24.3	32	27.5	38	20.4	HUDSON	NJ	13	-26	-16	71	86
1	57.3					HUNTERDON	NJ					
17	27.7	36	41.0	26	16.3	MERCER	NJ	48	-60	-41	48	79
6	21.5	31*	59.9	16	16.2	MIDDLESEX	NJ	179	-73	-25	64	78
27	28.4	39	32.3	45*	31.9	MONMOUTH	NJ	14	-1	12	83	94
6	37.0	1	6.2	6	20.8	MORRIS	NJ	-83	232	-44	46	86
4	42.4	6	43.3	6	26.6	OCEAN	NJ	2	-38	-37	52	92
10	22.2	26*	37.1	24	18.2	PASSAIC	NJ	67	-51	-18	70	83
16*	66.5	14*	47.0	9	21.9	SALEM	NJ	-29	-53	-67	18	87
3	15.3	2	11.5	4	15.8	SOMERSET	NJ	-24	37	3	80	77
				2*	75.7	SUSSEX	NJ					98
40*	45.8	43	31.9	46	25.8	UNION	NJ	-30	-19	-44	46	91
24	14.3	31	14.8	53	12.2	NEW MEXICO	NM	4	-18	-14		
		1	8.0	5	5.6	BERNALILLO	NM		-31			59
2	54.3	4*	138.9	1	9.4	CURRY	NM	156	-93	-83	05	65
1	33.5	1	33.5	1	2.8	DONA ANA	NM	0	-91	-91	01	58
1	38.2			1	8.2	EDDY	NM			-79	08	63
1	9.4	2	18.9	3	22.3	LEA	NM	100	18	136	97	88
				1	33.5	LUNA	NM					94
7	16.1	3	4.8	20*	28.2	MC KINLEY	NM	-70	485	75	92	93
3	68.0	1	13.2	1	11.7	OTERO	NM	-81	-12	-83	05	69
		3	39.1	1	1.6	RIO ARRIBA	NM		-96			57
3	15.5	2	10.6	5	16.6	SANDOVAL	NM	-32	57	7	81	79
3	11.0	7	19.6	8	15.3	SAN JUAN	NM	78	-22	39	88	76
		1	20.4	1	18.7	SANTA FE	NM		-8			83
1*	490.8					SIERRA	NM					
				1	16.6	SOCORRO	NM					79
		1	13.8	1	3.7	TAOS	NM		-73			58
2	12.9	5	24.9	3	7.6	VALENCIA	NM	94	-69	-41	49	62
1192*	35.9	1283*	25.7	1336*	17.8	NEW YORK	NY	-28	-31	-51		
6	18.7	5	20.5	10	19.9	ALBANY	NY	10	-3	6	80	85
133*	48.4	119	21.1	161	15.5	BRONX	NY	-56	-27	-68	17	77
3	40.9	2	25.5			BROOME	NY	-37				
1	17.9			1	14.1	CATTARAUGUS	NY			-21	67	73
2	106.5					CAYUGA	NY					
		2	28.2	1	9.2	CHAUTAUQUA	NY		-67			64
1	13.8	2	25.3	2	28.9	CHEMUNG	NY	84	14	110	95	93

* SIGNIFICANTLY MORE DEATHS THAN EXPECTED

NON-WHITE MALE — CANCER OF THE STOMACH — ICD 151

1950-1959 DEATHS	1950-1959 RATE	1960-1969 DEATHS	1960-1969 RATE	1970-1979 DEATHS	1970-1979 RATE	COUNTY AND STATE		PERCENT CHANGE 60S/50S	PERCENT CHANGE 70S/60S	PERCENT CHANGE 70S/50S	PERCENTILE %CHANGE 70S/50S	PERCENTILE RATES 1970S
				2*	175.3	CHENANGO	NY					99
				1	13.0	CLINTON	NY					71
3	57.9	2	24.8	2	41.1	COLUMBIA	NY	-57	66	-29	61	96
5	24.7	8	19.1	6	13.3	DUTCHESS	NY	-23	-31	-46	43	72
49	32.0	60	24.3	68	20.3	ERIE	NY	-24	-17	-37	53	86
		1	13.4			FRANKLIN	NY					
1	21.2	2	37.3	2	36.2	GENESEE	NY	76	-3	71	92	95
1	94.4	1	23.6	2	55.7	GREENE	NY	-75	136	-41	49	97
		1	84.8			JEFFERSON	NY					
246*	37.0	225	20.2	312*	18.4	KINGS(BROOKLYN)	NY	-45	-9	-50	39	83
1	133.9					LIVINGSTON	NY					
2*	224.8					MADISON	NY					
9	32.6	12	23.6	27	24.2	MONROE	NY	-28	3	-26	63	90
18	32.8	28	35.0	33	15.9	NASSAU	NY	6	-54	-52	37	77
527*	38.5	611*	36.2	420*	21.0	NEW YORK(MANHATTAN)	NY	-6	-42	-45	44	87
1	9.3	3	11.4	4	8.5	NIAGARA	NY	23	-26	-8	75	63
1	4.5	2	9.2	4	22.8	ONEIDA	NY	104	148	408	99	88
5	19.1	10	32.0	5	11.3	ONONDAGA	NY	67	-65	-41	48	68
2	139.0	1	32.1			ONTARIO	NY	-77				
10	52.3	12	32.0	5	10.3	ORANGE	NY	-39	-68	-80	07	66
2*	73.0	1	19.6			ORLEANS	NY	-73				
1*	198.7					OTSEGO	NY					
67	26.6	77	14.0	149	14.7	QUEENS	NY	-47	5	-45	45	75
		1	8.8	1	5.3	RENSSELAER	NY		-40			59
6	25.0	6	24.6	8	13.4	RICHMOND	NY	-2	-46	-47	43	72
5	34.9	5	17.1	10	20.6	ROCKLAND	NY	-51	20	-41	48	86
1	37.5					SARATOGA	NY					
1	31.5			2	14.3	SCHENECTADY	NY			-55	33	74
1*	424.0					SCHOHARIE	NY					
		1	22.8			STEUBEN	NY					
13	13.0	26	15.4	37	18.2	SUFFOLK	NY	18	18	39	88	82
2	59.6	2	33.5	2	11.9	SULLIVAN	NY	-44	-65	-80	07	69
1	74.3					TIOGA	NY					
2	36.3	1	12.8			TOMPKINS	NY	-65				
5	35.4	2	16.4	5	24.1	ULSTER	NY	-54	47	-32	58	90
		1*	122.7			WASHINGTON	NY					
		2	32.6	1	18.1	WAYNE	NY		-44			82
58*	41.9	49	23.8	52	16.4	WESTCHESTER	NY	-43	-31	-61	26	79
				1*	56.2	YATES	NY					97
715	23.2	690	19.4	681	16.0	NORTH CAROLINA	NC	-17	-18	-31		
10	28.5	11	22.3	11	15.5	ALAMANCE	NC	-22	-30	-45	44	77
		1	29.8			ALEXANDER	NC					
1	84.8					ALLEGHANY	NC					
14	36.6	4	12.1	3	6.7	ANSON	NC	-67	-45	-82	06	61
1	89.3			1*	273.2	ASHE	NC			206	98	99
18	38.5	13	28.2	9	16.9	BEAUFORT	NC	-27	-40	-56	31	80
7	16.4	15	33.4	7	14.4	BERTIE	NC	104	-57	-12	73	74
8	26.6	3	8.6	4	9.0	BLADEN	NC	-68	5	-66	18	64
3	13.3	2	7.8	7	24.0	BRUNSWICK	NC	-42	209	80	93	89
11	18.6	21	33.8	19*	30.2	BUNCOMBE	NC	81	-11	62	90	94
2	15.6	4	31.6	2	12.1	BURKE	NC	103	-62	-22	66	69
5	15.7	3	14.0	9	20.3	CABARRUS	NC	-11	45	29	86	86
7*	78.0	2	19.4	3	22.7	CALDWELL	NC	-75	17	-71	14	88
1	14.9	1	10.3	1	12.6	CAMDEN	NC	-31	22	-15	71	70
4	35.1	3	31.2			CARTERET	NC	-11				
3	13.7	3	8.0	2	5.7	CASWELL	NC	-42	-29	-58	28	60
5	24.7	9*	57.8	4	16.9	CATAWBA	NC	134	-71	-31	59	80
2	10.4	4	17.4	6	16.7	CHATHAM	NC	67	-4	60	90	80
1	60.7	1	44.6			CHEROKEE	NC	-26				
8	50.6	3	18.3	5	28.1	CHOWAN	NC	-64	53	-44	45	92
9	23.4	7	16.8	7	14.2	CLEVELAND	NC	-28	-16	-39	51	74
7	16.3	6	11.6	4	6.5	COLUMBUS	NC	-29	-44	-60	27	60
10	19.7	11	20.0	12	19.1	CRAVEN	NC	1	-4	-3	77	84
20	35.5	20	19.6	26	18.8	CUMBERLAND	NC	-45	-4	-47	42	83
		2	20.9	1	11.1	CURRITUCK	NC		-47			67
		1	133.9			DARE	NC					
4	17.7	5	19.6	4	11.6	DAVIDSON	NC	11	-41	-34	56	68
2	26.3					DAVIE	NC					
13	29.2	7	13.7	11	20.5	DUPLIN	NC	-53	50	-30	60	86
23	21.8	19	14.7	32	18.5	DURHAM	NC	-33	26	-15	72	83
18	25.5	24	33.0	24*	27.7	EDGECOMBE	NC	29	-16	9	81	92
43	31.7	35	22.5	36	19.9	FORSYTH	NC	-29	-12	-37	52	85
7	18.8	9	23.0	8	18.3	FRANKLIN	NC	22	-20	-2	77	83
14	33.2	13	29.6	9	13.9	GASTON	NC	-11	-53	-58	29	73
3	17.2	2	11.8	3	18.7	GATES	NC	-31	58	9	82	83
		1*	98.2			GRAHAM	NC					
4	12.9	7	12.8	3	5.7	GRANVILLE	NC	-1	-56	-56	31	60
3	23.9	4	24.3	4	15.1	GREENE	NC	2	-38	-37	53	76
29	26.5	37	21.9	33	14.7	GUILFORD	NC	-17	-33	-45	45	75
19	23.7	12	14.8	12	12.1	HALIFAX	NC	-37	-18	-49	40	69
10	34.6	6	17.1	11	22.2	HARNETT	NC	-51	30	-36	54	88
1	38.3	1	26.8	2	31.5	HAYWOOD	NC	-30	17	-18	70	94
2	28.6	1	12.6	2	21.6	HENDERSON	NC	-56	72	-25	64	87
11	28.1	3	6.3	5	10.0	HERTFORD	NC	-77	57	-65	21	66
5	19.2	2	8.7	2	8.6	HOKE	NC	-55	-1	-55	32	63
1	9.6	5*	58.3			HYDE	NC	506				
4	16.5	10	25.7	4	7.8	IREDELL	NC	56	-70	-53	35	62
1	22.8			2	23.8	JACKSON	NC			4	80	89
7	21.4	6	12.1	9	19.0	JOHNSTON	NC	-43	57	-11	74	84
1	8.4	1	6.0	1	5.9	JONES	NC	-28	-2	-30	61	60
8	48.8	4	26.8	5	16.6	LEE	NC	-45	-38	-66	19	79
10	15.3	9	12.6	10	12.0	LENOIR	NC	-18	-5	-22	67	69
3	29.1	3	30.5	1	9.8	LINCOLN	NC	5	-68	-66	19	65
1	18.4	1	14.9	1	13.7	MC DOWELL	NC	-19	-8	-26	63	72
4	12.7	13	44.3	8	18.8	MARTIN	NC	249	-58	48	89	84
36	23.5	44	20.0	39	15.1	MECKLENBURG	NC	-15	-25	-36	54	76
4	32.0	3	21.5	2	11.6	MONTGOMERY	NC	-33	-46	-64	22	68
9	33.2	5	13.3	12*	26.9	MOORE	NC	-60	102	-19	69	92
13	20.3	18	27.5	12	16.0	NASH	NC	36	-42	-21	67	78

* SIGNIFICANTLY MORE DEATHS THAN EXPECTED

NON-WHITE MALE — CANCER OF THE STOMACH — ICD 151

COUNTY AND STATE	1950-1959 DEATHS	1950-1959 RATE	1960-1969 DEATHS	1960-1969 RATE	1970-1979 DEATHS	1970-1979 RATE	PERCENT CHANGE 60S/50S	PERCENT CHANGE 70S/60S	PERCENT CHANGE 70S/50S	PERCENTILE %CHANGE 70S/50S	PERCENTILE RATES 1970S
NEW HANOVER NC	16	26.1	24*	35.3	10	12.0	35	-66	-54	34	69
NORTHAMPTON NC	6	11.8	8	14.3	6	12.0	22	-16	2	79	69
ONSLOW NC	3	13.9	3	18.2	8	26.8	31	47	93	93	92
ORANGE NC	3	11.4	3	10.0	8	18.0	-13	81	57	90	82
PAMLICO NC	2	18.2			3	21.3			17	84	87
PASQUOTANK NC	12	39.3	6	16.4	7	19.1	-58	16	-51	38	84
PENDER NC	2	6.1	6	20.1	4	13.9	228	-31	128	96	73
PERQUIMANS NC	3	18.2			1	3.9			-79	08	58
PERSON NC	3	10.6	6	19.0	5	16.1	79	-15	52	89	78
PITT NC	21	27.9	22	25.6	24*	26.3	-8	3	-6	76	92
POLK NC	1	15.6			2	25.5			63	91	91
RANDOLPH NC	3	29.5	4	23.7	3	12.6	-20	-47	-57	29	71
RICHMOND NC	8	27.1	8	21.2	7	16.7	-21	-22	-38	51	80
ROBESON NC	17	14.7	20	14.1	25	15.4	-4	9	5	80	76
ROCKINGHAM NC	12	28.2	9	20.3	6	13.6	-28	-33	-52	37	72
ROWAN NC	9	18.9	14	27.0	9	14.6	43	-46	-23	66	74
RUTHERFORD NC	2	11.7	3	15.9	2	9.7	36	-39	-17	70	65
SAMPSON NC	8	15.3	10	19.6	7	11.7	28	-40	-23	66	69
SCOTLAND NC	7	23.4	4	13.5	8	22.7	-42	69	-3	77	88
STANLY NC	7	54.8	3	15.9	2	14.0	-71	-11	-74	11	73
STOKES NC	1	20.2									
SURRY NC	2	22.0	2	21.6	3	25.3	-2	17	15	83	91
SWAIN NC			1	11.5	2	24.5		114			90
TRANSYLVANIA NC	2	69.1	1	24.6	1	32.1	-64	31	-54	34	94
TYRRELL NC	1	16.1									
UNION NC	3	10.7	3	11.2	2	5.5	5	-50	-48	41	59
VANCE NC	12	28.8	12	26.1	9	16.9	-9	-35	-41	48	80
WAKE NC	28	23.1	16	11.0	31	16.6	-52	51	-28	62	80
WARREN NC	11	22.2	10	21.7	7	15.5	-2	-29	-30	60	77
WASHINGTON NC	4	33.4	3	17.4	3	14.4	-48	-17	-57	30	74
WAYNE NC	21	25.1	17	16.6	18	18.5	-34	11	-26	63	83
WILKES NC					4	46.5					96
WILSON NC	15	21.8	12	16.8	4	5.1	-23	-69	-76	10	59
NORTH DAKOTA ND	7	26.0	5	15.2	2	3.0	-41	-80	-88		
CASS ND					1	15.6					77
DUNN ND	2*	456.6									
GRAND FORKS ND	1*	267.9									
MOUNTRAIL ND			1	44.6							
ROLETTE ND	4	33.8	3	19.8			-41				
SIOUX ND					1	20.4					86
WILLIAMS ND			1	112.4							
OHIO OH	682*	32.5	672	23.3	560	15.5	-28	-34	-52		
ALLEN OH	1	3.2	1	3.6	4	19.1	14	426	498	99	84
ASHTABULA OH			2	18.3	1	16.1		-12			78
ATHENS OH			1	16.3	2	48.0		194			96
BELMONT OH	4	36.2	1	8.6			-76				
BROWN OH	1	22.9									
BUTLER OH	8	25.6	6	18.9	7	16.6	-26	-12	-35	55	79
CARROLL OH					1*	78.1					98
CHAMPAIGN OH			1	15.7	2	34.9		122			95
CLARK OH	15	35.0	11	19.5	9	15.1	-44	-23	-57	30	76
CLERMONT OH			2	83.0	1	14.1		-83			73
COLUMBIANA OH	3	40.8	1	28.1	3	61.4	-31	118	50	89	98
COSHOCTON OH					1*	57.3					97
CRAWFORD OH	1	45.9	3*	167.0			264				
CUYAHOGA OH	200	34.6	222	25.2	190	16.3	-27	-35	-53	35	79
DELAWARE OH	3*	73.6	3*	77.9			6				
ERIE OH	1	24.8	1	6.9			-72				
FAYETTE OH	1	28.7	2	52.7			84				
FRANKLIN OH	80	35.0	68	22.1	49	12.8	-37	-42	-63	22	71
GALLIA OH	1	11.8	3	45.1	1	18.8	283	-58	60	90	84
GEAUGA OH	1	22.7									
GREENE OH	8	41.5	3	13.1	3	10.3	-68	-21	-75	11	66
GUERNSEY OH	1	20.6			2*	49.3			139	97	97
HAMILTON OH	120	32.0	119	24.4	98	16.7	-24	-32	-48	41	80
HIGHLAND OH	2	45.7			1	38.3			-16	71	95
HURON OH	2*	110.0									
JEFFERSON OH	12*	74.5	7	32.8	2	7.0	-56	-79	-91	01	61
KNOX OH	2	62.1									
LAKE OH	3	70.7	2	35.8	2	11.3	-49	-68	-84	04	68
LAWRENCE OH	5	58.9	3	29.5			-50				
LICKING OH	3	77.8									
LORAIN OH	8	26.5	8	21.3	8	17.0	-20	-20	-36	54	80
LUCAS OH	39	37.6	41	28.0	29	14.6	-26	-48	-61	25	75
MADISON OH	1	16.7			1	12.4			-26	64	70
MAHONING OH	35	33.2	34	23.6	30	18.0	-29	-24	-46	44	82
MARION OH	1	26.5									
MEDINA OH	1	68.8	1	27.3			-60				
MEIGS OH					1	53.6					97
MIAMI OH			3	49.3	1	12.1		-75			69
MONTGOMERY OH	50	32.5	50	22.9	44	15.9	-30	-30	-51	38	77
MORGAN OH	1	67.0	1	37.5			-44				
MORROW OH			1*	74.3							
MUSKINGUM OH	4	29.4	2	14.2	5	34.4	-52	141	17	84	95
OTTAWA OH	1	61.3			1	56.2			-8	75	97
PERRY OH			1	62.9							
PICKAWAY OH	2*	73.3	1	53.6			-27				
PIKE OH	1	68.8									
PORTAGE OH	1	12.8	2	19.9	1	9.3	55	-53	-28	62	65
RICHLAND OH	5	45.1	6	29.7	5	19.7	-34	-34	-56	31	85
ROSS OH	2	11.8	1	11.2			-4				
SANDUSKY OH	1	28.3	1	36.4			29				
SCIOTO OH	4	40.9			3	34.1			-16	71	95
STARK OH	6	11.7	17	26.5	17	19.9	127	-25	70	92	85
SUMMIT OH	28	31.1	30	22.8	23	12.5	-27	-45	-60	27	70
TRUMBULL OH	8	26.7	7	15.9	10	20.5	-40	29	-23	66	86
TUSCARAWAS OH	1	26.8	3*	70.1	1	24.6	162	-65	-8	75	90

* SIGNIFICANTLY MORE DEATHS THAN EXPECTED

NON-WHITE MALE — CANCER OF THE STOMACH — ICD 151

1950-1959 DEATHS	1950-1959 RATE	1960-1969 DEATHS	1960-1969 RATE	1970-1979 DEATHS	1970-1979 RATE	COUNTY AND STATE	PERCENT CHANGE 60S/50S	PERCENT CHANGE 70S/60S	PERCENT CHANGE 70S/50S	PERCENTILE %CHANGE 70S/50S	PERCENTILE RATES 1970S
1*	112.4					VAN WERT OH					
				1	24.8	WARREN OH					90
3	64.9					WASHINGTON OH					
196	22.7	175	17.0	155	12.3	OKLAHOMA OK	-25	-28	-46		
3	25.3	2	11.9	2	9.8	ADAIR OK	-53	-18	-61	25	65
1	21.3	1	13.4	1	12.5	ATOKA OK	-37	-7	-42	48	70
3	54.6	1	10.8			BLAINE OK	-80				
1	15.7	1	10.1	2	15.6	BRYAN OK	-36	54	-1	78	77
3	22.3	7*	47.4			CADDO OK	112				
		3	53.9	2	28.6	CANADIAN OK		-47			93
5	27.2	4	19.2	3	12.3	CARTER OK	-29	-36	-55	33	70
8*	58.5	1	8.6	2	7.7	CHEROKEE OK	-85	-11	-87	02	62
2	8.3	2	8.1	3	18.4	CHOCTAW OK	-2	127	122	96	83
		1*	424.0			CIMARRON OK					
2	97.8	1	21.5			COAL OK	-78				
5	26.5	7	30.2	4	10.5	COMANCHE OK	14	-65	-60	27	67
1	89.3					COTTON OK					
1	16.6			2	12.3	CRAIG OK			-26	63	70
6	28.7	10	40.4	2	9.4	CREEK OK	41	-77	-67	18	65
1	14.9					CUSTER OK					
1	11.2			1	5.1	DELAWARE OK			-54	33	59
		1*	267.9	1*	106.0	DEWEY OK		-60			98
3	59.3	2	30.9			GARFIELD OK	-48				
2	38.0	3	39.7	2	18.9	GARVIN OK	5	-53	-50	39	84
		2	22.7			GRADY OK					
1	73.7					GREER OK					
		1	89.3	1	54.5	HARMON OK		-39			97
		1	18.7			HASKELL OK					
5	58.7					HUGHES OK					
2	84.4	1	14.9	2	25.1	JACKSON OK	-82	69	-70	14	91
1	98.2					JEFFERSON OK					
		1	14.9			JOHNSTON OK					
1	12.7	1	11.2	4	29.9	KAY OK	-12	168	135	97	94
2	37.0	1	16.1			KINGFISHER OK	-57				
1	15.7					KIOWA OK					
1	19.1					LATIMER OK					
		2	17.0			LE FLORE OK					
4	50.5					LINCOLN OK					
5	24.1	5	21.7	1	3.2	LOGAN OK	-10	-85	-87	03	58
2	103.4					LOVE OK					
4	12.4	3	9.1	3	7.2	MC CURTAIN OK	-27	-21	-42	48	61
3	19.1	1	5.4	1	11.4	MC INTOSH OK	-72	112	-40	50	68
1	49.2					MARSHALL OK					
3	39.4	2	20.6	4	31.7	MAYES OK	-48	54	-19	69	94
17	21.9	18	21.0	9	12.0	MUSKOGEE OK	-4	-43	-45	45	69
		1	21.5			NOBLE OK					
1	9.9	2	19.6	1	11.5	NOWATA OK	97	-41	15	83	68
4	15.2	3	17.3	2	11.1	OKFUSKEE OK	14	-36	-27	62	67
32	26.4	31	18.2	38	16.9	OKLAHOMA OK	-31	-7	-36	54	80
10	25.2	8	19.8	9	22.1	OKMULGEE OK	-22	12	-13	73	88
1	10.7	2	14.0	1	3.9	OSAGE OK	31	-72	-64	22	58
1	31.5	2	39.9	1	11.8	PAWNEE OK	27	-70	-63	23	69
2	34.9	1	18.2			PAYNE OK	-48				
8	41.7	3	16.1	6	22.9	PITTSBURG OK	-62	43	-45	45	89
6*	89.8			1	6.9	PONTOTOC OK			-92	00	61
1	8.9					POTTAWATOMIE OK					
3	79.9	4*	97.1	1	12.4	PUSHMATAHA OK	22	-87	-84	04	70
2	31.7	1	18.7	1	5.7	ROGERS OK	-41	-69	-82	05	60
6	28.9	9	36.9	8*	28.4	SEMINOLE OK	28	-23	-2	78	93
2	23.9			1	6.7	SEQUOYAH OK			-72	13	61
				2	20.9	STEPHENS OK					87
3	35.3	2	25.2	1	8.3	TILLMAN OK	-29	-67	-77	09	63
8	7.5	14	9.4	30	15.7	TULSA OK	25	66	108	95	77
4	19.7	3	12.3			WAGONER OK	-38				
1	12.8	3	21.4			WASHINGTON OK	67				
35	30.4	23	15.1	21	9.2	OREGON OR	-50	-39	-70		
1*	245.4					BENTON OR					
		1	17.9	2	20.1	CLACKAMAS OR		12			85
				2*	63.2	HOOD RIVER OR					98
1	53.0					JEFFERSON OR					
				1	16.3	KLAMATH OR					79
1	31.4					LANE OR					
2	44.5	2	16.4			MALHEUR OR	-63				
				1*	163.6	MORROW OR					99
29*	41.4	17	17.7	14	10.7	MULTNOMAH OR	-57	-40	-74	11	67
		1	18.6	1	26.8	UMATILLA OR		44			92
1	67.0					WASHINGTON OR					
		2*	164.1			YAMHILL OR					
893*	36.5	911*	27.5	691*	17.6	PENNSYLVANIA PA	-25	-36	-52		
173*	38.0	172*	27.6	118	17.3	ALLEGHENY PA	-27	-37	-54	33	81
		1	18.1			ARMSTRONG PA					
5	13.8	9	26.3	8	15.4	BEAVER PA	90	-41	11	82	76
2	17.7	3	13.5	3	9.7	BERKS PA	-24	-28	-45	45	65
1	18.7	4*	135.8	1	15.7	BLAIR PA	625	-88	-16	71	77
1	8.2	5	25.2	3	13.7	BUCKS PA	209	-46	68	91	73
4	62.5	4	31.8	4	25.3	CAMBRIA PA	-49	-20	-60	27	91
				1	37.2	CENTRE PA					95
22	36.9	20	31.5	14	17.4	CHESTER PA	-15	-45	-53	35	81
1	19.1	1	26.8			CRAWFORD PA	40				
3	60.8	1	26.8	1	21.5	CUMBERLAND PA	-56	-20	-65	21	87
24	39.7	14	16.0	14	14.0	DAUPHIN PA	-60	-12	-65	21	73
35	33.3	30	17.2	29	16.1	DELAWARE PA	-48	-7	-52	37	78
6	53.7	7	27.1	4	15.7	ERIE PA	-50	-42	-71	14	77
16	40.7	16*	40.8	8	19.0	FAYETTE PA	0	-53	-53	34	84
2	44.0	3	37.4	1	13.5	FRANKLIN PA	-15	-64	-69	16	72

* SIGNIFICANTLY MORE DEATHS THAN EXPECTED

NON-WHITE MALE — CANCER OF THE STOMACH — ICD 151

1950-1959 DEATHS	1950-1959 RATE	1960-1969 DEATHS	1960-1969 RATE	1970-1979 DEATHS	1970-1979 RATE	COUNTY AND STATE	PERCENT CHANGE 60S/50S	PERCENT CHANGE 70S/60S	PERCENT CHANGE 70S/50S	PERCENTILE %CHANGE 70S/50S	PERCENTILE RATES 1970S
1	37.2	1	37.7			GREENE PA	2				
		1	37.8			HUNTINGDON PA					
1	35.1					INDIANA PA					
2	53.0	3	145.6	1	15.8	LACKAWANNA PA	175	-89	-70	15	77
4	30.9	3	17.8	5	26.1	LANCASTER PA	-42	47	-15	72	91
6	59.5	2	15.5	1	9.2	LAWRENCE PA	-74	-41	-85	04	64
		2*	70.9	2*	47.4	LEBANON PA		-33			96
3	75.4	1	17.2	2	22.2	LUZERNE PA	-77	30	-71	14	88
2	48.4	1	37.5			LYCOMING PA	-23				
1*	91.1					MC KEAN PA					
4	27.3	5	29.7	5	24.4	MERCER PA	9	-18	-11	74	90
		1*	114.7			MIFFLIN PA					
1	49.7			1	17.9	MONROE PA			-64	22	82
18	31.9	12	17.6	17	17.3	MONTGOMERY PA	-45	-2	-46	44	81
				2*	384.0	MONTOUR PA					99
3	58.3	2	23.4	1	5.5	NORTHAMPTON PA	-60	-76	-91	01	59
1	114.7			1*	188.7	NORTHUMBERLAND PA			65	91	99
514*	36.4	543*	28.1	427*	18.3	PHILADELPHIA PA	-23	-35	-50	39	83
1	106.0	4*	128.2	1	188.7	SCHUYLKILL PA	21	47	78	93	99
				1*	24.8	SOMERSET PA					90
		1*	267.9			SUSQUEHANNA PA					
1	47.2	2*	81.7			VENANGO PA	73				
15	33.7	13	28.0	5	11.0	WASHINGTON PA	-17	-61	-67	17	67
		1	28.3			WAYNE PA					
12	49.3	12	36.0	4	10.8	WESTMORELAND PA	-27	-70	-78	08	67
8	62.7	11*	49.7	6	24.0	YORK PA	-21	-52	-62	24	90
19	33.0	14	19.4	13	12.1	RHODE ISLAND RI	-41	-38	-63		
1	38.2	1	44.6			KENT RI	17				
4	42.2	2	24.6	1	6.0	NEWPORT RI	-42	-76	-86	03	60
14	34.2	10	17.7	10	12.2	PROVIDENCE RI	-48	-31	-64	22	70
		1	15.1	2	30.8	WASHINGTON RI		103			94
607	27.0	527	21.9	509*	18.4	SOUTH CAROLINA SC	-19	-16	-32		
7	26.9	3	13.4	4	16.1	ABBEVILLE SC	-50	21	-40	50	78
23	33.6	11	16.7	8	10.2	AIKEN SC	-50	-39	-70	15	66
9	43.9	9	40.2	6	29.7	ALLENDALE SC	-8	-26	-32	58	93
8	12.6	7	11.1	12	18.6	ANDERSON SC	-12	67	47	89	83
13*	51.4	2	6.6	13*	45.7	BAMBERG SC	-87	591	-11	74	96
3	10.7	9	35.1			BARNWELL SC	230				
15	30.0	8	15.9	11	19.5	BEAUFORT SC	-47	23	-35	55	85
21*	39.4	14	28.7	16*	23.4	BERKELEY SC	-27	-18	-41	49	89
7	26.3	3	14.0	8*	33.8	CALHOUN SC	-47	142	29	86	95
66*	36.4	58	27.1	87*	32.7	CHARLESTON SC	-26	21	-10	74	94
4	19.0	6	26.5	3	12.1	CHEROKEE SC	39	-54	-36	54	70
9	20.7	10	24.9	5	13.6	CHESTER SC	20	-45	-34	56	72
8	23.2	8	23.2	2	5.0	CHESTERFIELD SC	0	-79	-79	08	59
12	23.1	9	17.8	9	16.1	CLARENDON SC	-23	-9	-30	59	78
11	30.2	10	21.4	10	18.8	COLLETON SC	-29	-12	-38	52	84
9	16.7	10	16.4	14	20.5	DARLINGTON SC	-1	24	23	85	86
10	28.8	4	10.8	4	11.7	DILLON SC	-63	9	-59	28	69
15*	43.5	4	9.5	8	18.7	DORCHESTER SC	-78	97	-57	30	83
8	26.6	3	12.1	2	8.5	EDGEFIELD SC	-54	-30	-68	17	63
13	34.6	5	12.9	6	13.5	FAIRFIELD SC	-63	5	-61	26	72
33*	40.0	30*	31.3	22	17.9	FLORENCE SC	-22	-43	-55	32	82
11	26.7	17*	36.9	13	21.6	GEORGETOWN SC	38	-41	-19	69	87
21	23.3	25	24.1	17	13.1	GREENVILLE SC	4	-46	-44	46	72
7	17.0	13	28.0	10	19.3	GREENWOOD SC	65	-31	13	83	84
6	23.6	6	21.9	8	25.8	HAMPTON SC	-7	18	9	82	91
3	8.7	10	25.6	4	5.9	HORRY SC	194	-77	-32	58	60
6	25.6	10	35.8	4	14.5	JASPER SC	40	-60	-43	47	74
10	21.7	8	18.8	9	19.0	KERSHAW SC	-13	1	-12	73	84
14*	45.7	5	17.2	4	10.7	LANCASTER SC	-62	-38	-77	10	67
11	24.4	6	13.6	8	17.1	LAURENS SC	-44	26	-30	60	80
5	29.8	6	20.9	9	26.9	LEE SC	-30	29	-10	75	92
4	12.5	9	25.7	9	22.9	LEXINGTON SC	105	-11	83	93	89
8	48.2	4	26.4	3	18.6	MC CORMICK SC	-45	-29	-61	25	83
11	21.9	18*	40.1	6	11.0	MARION SC	83	-73	-50	39	67
13	36.7	4	11.5	7	17.2	MARLBORO SC	-69	50	-53	35	81
6	16.5	4	10.9	5	13.3	NEWBERRY SC	-34	22	-20	68	72
1	5.6	3	18.3	1	5.2	OCONEE SC	224	-72	-9	75	59
36	33.6	26	24.3	26	19.0	ORANGEBURG SC	-28	-22	-44	46	84
5	32.8	3	19.0	2	7.3	PICKENS SC	-42	-61	-78	09	62
32	20.1	43	20.4	47	19.1	RICHLAND SC	1	-6	-5	76	84
6	24.2	4	23.0	3	17.3	SALUDA SC	-5	-25	-29	61	81
26	26.5	20	18.8	13	10.3	SPARTANBURG SC	-29	-45	-61	25	66
17	19.0	28	28.2	18	15.3	SUMTER SC	48	-46	-19	69	76
10	36.1	8	27.5	5	17.0	UNION SC	-24	-38	-53	35	80
17	26.1	12	16.9	15	20.6	WILLIAMSBURG SC	-35	22	-21	67	86
17	27.2	12	19.1	13	20.7	YORK SC	-30	8	-24	65	86
16	20.5	19	22.0	14	13.1	SOUTH DAKOTA SD	8	-41	-36		
		1*	490.8			BEADLE SD					
2	61.5	1	42.4			BENNETT SD	-31				
1	62.9					CHARLES MIX SD					
		1	37.5			CORSON SD					
1	86.0					DAY SD					
2	37.4	3	67.4	2	28.7	DEWEY SD	80	-57	-23	66	93
				1*	74.3	GREGORY SD					98
1	54.5			2*	79.6	JACKSON SD			46	89	98
				1*	596.2	JONES SD					99
		1	44.6			LYMAN SD					
		1*	62.9			MARSHALL SD					
1	21.5	2	65.9			MELLETTE SD	206				
				1*	267.9	POTTER SD					99
4*	94.3	1	21.2	2	29.5	ROBERTS SD	-78	39	-69	16	93
3	15.3	6	30.9	4	16.3	SHANNON SD	102	-47	6	81	78
1	21.0	1	7.4			TODD SD	-65				

* SIGNIFICANTLY MORE DEATHS THAN EXPECTED

NON-WHITE MALE — CANCER OF THE STOMACH — ICD 151

1950-1959 DEATHS	1950-1959 RATE	1960-1969 DEATHS	1960-1969 RATE	1970-1979 DEATHS	1970-1979 RATE	COUNTY AND STATE	PERCENT CHANGE 60S/50S	PERCENT CHANGE 70S/60S	PERCENT CHANGE 70S/50S	PERCENTILE %CHANGE 70S/50S	PERCENTILE RATES 1970S
				1*	89.3	WALWORTH SD					98
		1	31.3			ZIEBACH SD					
533	24.9	446	18.7	407	15.6	TENNESSEE TN	-25	-17	-38		
1	21.5	2	37.3	1	11.5	ANDERSON TN	73	-69	-47	43	68
4	30.8	1	6.5	3	26.0	BEDFORD TN	-79	298	-16	71	91
				2*	104.7	BENTON TN					98
4	30.7			1	9.3	BLOUNT TN			-70	15	64
				1	18.7	BRADLEY TN					83
2*	137.3					CAMPBELL TN					
		2	11.4	4	26.6	CARROLL TN		133			92
				1	23.6	CHEATHAM TN					89
1	14.9	2	29.2			CHESTER TN	97				
1	40.9					COCKE TN					
2	82.9	1	38.5			COFFEE TN	-54				
1	6.4	2	13.5			CROCKETT TN	111				
76	27.6	44	14.4	49	13.5	DAVIDSON TN	-48	-6	-51	38	72
4*	207.0					DECATUR TN					
2*	85.4			1	57.3	DE KALB TN			-33	57	97
3	49.0	1	14.1	1	13.5	DICKSON TN	-71	-4	-72	13	72
10	44.9	5	21.5	2	16.3	DYER TN	-52	-25	-64	22	78
7	11.1	8	13.5	9	16.9	FAYETTE TN	22	25	52	89	80
3	32.1	1	11.6	2	20.2	FRANKLIN TN	-64	74	-37	52	86
6	15.4	8	18.6	7	14.6	GIBSON TN	21	-22	-5	76	74
3	13.8	6	28.8	5	27.2	GILES TN	109	-6	97	94	92
1	34.4	1	25.6			GREENE TN	-25				
1	16.4	1	14.9			HAMBLEN TN	-9				
38	23.8	33	18.1	30	14.7	HAMILTON TN	-24	-18	-38	52	75
5	16.5	3	8.8	10	24.8	HARDEMAN TN	-47	180	50	89	90
		1	15.8	1	17.7	HARDIN TN		12			82
2	48.3	2	44.0	2	38.9	HAWKINS TN	-9	-12	-20	69	95
8	15.6	3	6.3	4	7.2	HAYWOOD TN	-60	14	-54	34	61
1	11.8	2	29.1	1	14.8	HENDERSON TN	147	-49	25	85	75
4	21.0	2	11.6	1	7.8	HENRY TN	-45	-32	-63	23	62
1	45.5			1	62.9	HICKMAN TN			38	88	98
1	38.3	1	89.3	1*	49.2	HOUSTON TN	133	-45	28	86	97
				1	112.4	HUMPHREYS TN					98
		2	45.9			JEFFERSON TN					
1	67.0					JOHNSON TN					
24	25.3	16	18.1	19	17.3	KNOX TN	-28	-4	-32	58	81
4	49.7	4	46.8	3	39.8	LAKE TN	-6	-15	-20	68	96
8	22.0	8	20.3	8	23.3	LAUDERDALE TN	-7	14	6	80	89
		1	44.6	1	39.0	LAWRENCE TN		-13			96
		1*	267.9			LEWIS TN					
2	15.2			2	15.6	LINCOLN TN			3	79	77
				2*	116.8	LOUDON TN					99
4	56.8	3	40.0			MC MINN TN	-30				
3	58.0	1	16.7			MC NAIRY TN	-71				
17	22.9	18	20.5	13	15.3	MADISON TN	-11	-25	-33	57	76
2	27.3			1	13.4	MARION TN			-51	38	72
		2	24.7	1	11.5	MARSHALL TN		-54			68
5	13.5	10	25.7	3	8.1	MAURY TN	91	-68	-40	50	63
				1*	114.7	MEIGS TN					99
1	37.5					MONROE TN					
9	21.4	7	19.0	6	14.9	MONTGOMERY TN	-11	-22	-31	59	75
		2	11.5	5	38.4	OBION TN		233			95
2	112.1					PUTNAM TN					
		2*	69.3			RHEA TN					
		1	15.8	1	10.1	ROANE TN		-36			66
6	26.8	1	5.1	2	9.9	ROBERTSON TN	-81	94	-63	23	66
9	29.1	7	23.0	4	12.8	RUTHERFORD TN	-21	-44	-56	31	71
1*	84.8					SEVIER TN					
193	25.7	198	22.0	168	16.3	SHELBY TN	-14	-26	-36	53	79
1	37.5			2*	61.8	SMITH TN			65	91	98
7*	79.5	1	9.9	4*	35.2	SULLIVAN TN	-88	255	-56	32	95
12*	76.6	9*	42.7	5	23.1	SUMNER TN	-44	-46	-70	15	89
11	27.7	5	11.6	8	18.5	TIPTON TN	-58	59	-33	56	83
1	22.5	1	15.6			TROUSDALE TN	-30				
1	14.1			1	28.4	WARREN TN			102	95	93
6	33.5	3	17.4	1	7.7	WASHINGTON TN	-48	-56	-77	09	62
1	67.0	1	86.0			WAYNE TN	28				
2	20.6	2	21.4			WEAKLEY TN	4				
		1	53.0			WHITE TN					
5	24.3	4	20.1	4	19.5	WILLIAMSON TN	-17	-3	-20	69	85
3	20.3	3	16.2	1	4.4	WILSON TN	-20	-73	-79	08	58
964	23.8	1037	21.4	869	13.0	TEXAS TX	-10	-39	-45		
13	31.9	9	20.1	14*	27.4	ANDERSON TX	-37	37	-14	72	92
6	28.6	7	19.8	6	15.2	ANGELINA TX	-31	-23	-47	43	76
1	7.4	6	43.3	7*	47.3	AUSTIN TX	483	9	536	99	96
7	21.2	7	23.4	5	14.5	BASTROP TX	11	-38	-31	59	74
4	11.5	5	12.5	2	4.0	BELL TX	8	-68	-66	20	58
27	24.5	40	23.2	38	12.0	BEXAR TX	-5	-48	-51	38	69
				1*	133.9	BLANCO TX					99
13	21.4	15	21.2	5	5.9	BOWIE TX	-1	-72	-72	13	60
7	21.4	11	27.5	6	10.5	BRAZORIA TX	28	-62	-51	38	66
12	30.0	7	17.1	5	8.9	BRAZOS TX	-43	-48	-70	14	64
				1	42.4	BREWSTER TX					96
				1	14.5	BROWN TX					74
6	34.1	5	23.7	1	4.5	BURLESON TX	-30	-81	-87	02	59
6	41.1	4	23.3	4	20.4	CALDWELL TX	-43	-12	-50	39	86
		2*	89.9			CALHOUN TX					
2	37.3	2	35.4	1	1.8	CAMERON TX	-5	-95	-95	00	57
4	24.8	2	10.7	2	10.2	CAMP TX	-57	-4	-59	28	66
6	19.7	3	10.3	8	22.7	CASS TX	-48	120	15	83	88
1	15.8	2	22.0	1	11.2	CHAMBERS TX	40	-49	-29	61	68
3	7.7	7	14.8	7	14.1	CHEROKEE TX	92	-5	84	93	73
		1	32.7			CHILDRESS TX					
		1*	424.0			CLAY TX					

* SIGNIFICANTLY MORE DEATHS THAN EXPECTED

NON-WHITE MALE — CANCER OF THE STOMACH — ICD 151

COUNTY AND STATE	1950-1959 DEATHS	1950-1959 RATE	1960-1969 DEATHS	1960-1969 RATE	1970-1979 DEATHS	1970-1979 RATE	PERCENT CHANGE 60S/50S	PERCENT CHANGE 70S/60S	PERCENT CHANGE 70S/50S	PERCENTILE %CHANGE 70S/50S	PERCENTILE RATES 1970S
COLLIN TX	5	25.1	2	9.5	2	10.1	-62	6	-60	27	66
COLLINGSWORTH TX	1	31.3	1	31.3			0				
COLORADO TX	8	38.5	4	20.5	3	11.3	-47	-45	-71	14	68
COMAL TX			1	172.0							
COOKE TX			1	23.8							
COTTLE TX					1	98.2					98
CROCKETT TX			1*	61.3							
CROSBY TX			2*	70.7							
DALLAS TX	89	24.3	116	23.2	116	16.0	-5	-31	-34	56	78
DAWSON TX			2	38.6	1	15.3		-60			76
DELTA TX			1	26.5	1	59.6		125			97
DENTON TX	6	57.1	1	8.3	2	14.0	-86	69	-76	10	73
DE WITT TX	5	27.2	1	5.6	3	12.3	-80	121	-55	33	70
DICKENS TX			1	56.2							
DONLEY TX	1	98.2	1	37.2			-62				
EASTLAND TX			1	60.6							
ECTOR TX	1	14.6	5	32.7	4	30.4	124	-7	108	95	94
ELLIS TX	11	24.4	7	14.6	3	7.3	-40	-50	-70	15	62
EL PASO TX	5	25.1	3	12.6	3	3.7	-50	-70	-85	03	58
FALLS TX	8	20.8	5	14.8	1	2.1	-29	-86	-90	01	57
FANNIN TX	2	9.7	2	11.7	3	22.3	21	90	130	96	88
FAYETTE TX	7	34.3	3	25.5	2	22.2	-26	-13	-35	55	88
FISHER TX			1	56.2							
FORT BEND TX	24*	69.3	10	27.0	4	6.4	-61	-76	-91	01	60
FRANKLIN TX	1	37.5			1	37.8			1	79	95
FREESTONE TX	2	7.2	6	20.1	4	11.0	179	-45	52	89	67
GALVESTON TX	24	30.5	27	22.1	25	16.5	-28	-26	-46	44	79
GARZA TX	1	84.8									
GOLIAD TX	1	18.7									
GONZALES TX	3	15.6	1	6.5	3	24.9	-58	282	59	90	91
GRAY TX			1	32.6							
GRAYSON TX	6	19.2	5	14.2	1	3.5	-26	-75	-82	06	58
GREGG TX	15	25.6	17	26.0	8	11.3	1	-56	-56	32	68
GRIMES TX	3	10.0	7	21.4	2	7.4	114	-65	-26	63	62
GUADALUPE TX	3	17.6	3	17.4	4	20.1	-1	15	14	83	85
HALE TX			2	42.0	1	6.2		-85			60
HARDEMAN TX			1	24.9							
HARDIN TX	2	13.6	3	19.4	3	14.7	43	-24	9	81	75
HARRIS TX	162	26.5	206	24.6	212	16.8	-7	-32	-37	53	80
HARRISON TX	10	10.5	18	19.7	12	14.0	88	-29	34	87	73
HASKELL TX			1	53.6	1	18.1		-66			82
HAYS TX			2	40.5	1	16.1		-60			78
HENDERSON TX	4	22.3	7	31.7	4	16.8	42	-47	-24	65	80
HIDALGO TX	2	58.0	1	29.8			-49				
HILL TX	1	5.7	2	12.0	4	23.0	109	92	301	99	89
HOCKLEY TX					1	12.4					70
HOOD TX			1*	188.7							
HOPKINS TX	2	21.5	1	11.2	3	28.4	-48	155	32	87	93
HOUSTON TX	6	15.3	3	8.6	2	5.8	-44	-33	-62	24	60
HOWARD TX					1	8.4					63
HUNT TX	2	9.2	9	37.2	6	24.1	306	-35	163	97	90
IRION TX			1*	188.7							
JACKSON TX	2	31.4	3	36.2	3	30.4	15	-16	-3	77	94
JASPER TX	6	26.2	8	29.6			13				
JEFFERSON TX	80*	44.7	67*	32.4	45	17.2	-27	-47	-62	24	81
JIM WELLS TX			1	53.0							
JOHNSON TX	1	12.9	1	12.0	3	34.3	-7	187	166	97	95
JONES TX	2	34.7	2	31.8	1	11.8	-8	-63	-66	19	69
KARNES TX	2	96.0	1	49.2			-49				
KAUFMAN TX	7	15.8	8	19.4	3	9.8	23	-50	-38	52	65
KERR TX			1	74.3	1	17.9		-76			82
KINNEY TX			1*	188.7	1*	54.6		-71			97
KLEBERG TX			1	23.6							
KNOX TX			1	31.3							
LAMAR TX	6	19.6	3	8.7	4	9.8	-55	13	-50	39	65
LAMB TX			4*	101.9							
LAVACA TX	2	13.7	3	25.1			83				
LEE TX	5	36.0	1	6.9			-81				
LEON TX	3	12.8	5	26.4	2	14.9	107	-44	17	84	75
LIBERTY TX	8	27.7	1	3.6	8	25.2	-87	609	-9	75	91
LIMESTONE TX	5	14.4	2	5.8	3	13.1	-59	125	-9	75	72
LUBBOCK TX	7	28.2	7	20.9	10	15.2	-26	-27	-46	43	76
MC LENNAN TX	17	19.0	20	19.3	16	13.4	2	-31	-29	61	72
MADISON TX	2	17.0	3	30.2			78				
MARION TX	7	30.2	4	16.8	1	5.2	-44	-69	-83	05	59
MARTIN TX					1	86.0					98
MATAGORDA TX	5	21.6	6	24.0	6	16.5	11	-31	-24	65	79
MEDINA TX			1	38.2							
MIDLAND TX	1	37.2	1	4.9	1	3.7	-87	-24	-90	01	58
MILAM TX	5	21.2	4	18.1	1	3.6	-15	-80	-83	04	58
MITCHELL TX	1	28.1	1	62.9	1	15.7	124	-75	-44	46	77
MONTGOMERY TX	11	40.4	5	20.4	4	12.7	-50	-38	-69	16	71
MORRIS TX					2	12.7					71
NACOGDOCHES TX	8	26.7	7	21.0	6	15.3	-21	-27	-43	47	76
NAVARRO TX	10	24.5	11	23.5	7	18.0	-4	-23	-26	63	82
NEWTON TX	4	22.1	5	30.8	2	14.0	40	-55	-37	53	73
NOLAN TX	1	74.3	1	24.5	1	20.2	-67	-18	-73	12	86
NUECES TX	5	25.4	5	12.6	7	8.3	-50	-34	-67	17	63
ORANGE TX	6	35.6	4	22.5	2	8.7	-37	-62	-76	10	63
PALO PINTO TX	1	37.7			3*	113.9			202	98	99
PANOLA TX	1	4.6	3	13.1	5	19.7	186	50	329	99	85
POLK TX	10	47.0	4	18.7	2	7.0	-60	-63	-85	03	61
POTTER TX	1	4.3	4	15.8	3	9.6	265	-40	120	96	65
REAGAN TX			1*	81.8							
RED RIVER TX			2	11.7	3	17.2		46			80
REEVES TX					1	38.3					95
REFUGIO TX	2	49.5									
ROBERTSON TX	5	13.4	5	15.6	6	14.6	16	-6	9	82	75
ROCKWALL TX			2	52.1							
RUNNELS TX	1	74.3			1	17.2			-77	09	80
RUSK TX	14	29.1	9	16.2	7	15.3	-44	-6	-47	42	76

* SIGNIFICANTLY MORE DEATHS THAN EXPECTED

NON-WHITE MALE — CANCER OF THE STOMACH — ICD 151

1950-1959 DEATHS	1950-1959 RATE	1960-1969 DEATHS	1960-1969 RATE	1970-1979 DEATHS	1970-1979 RATE	COUNTY AND STATE	PERCENT CHANGE 60S/50S	PERCENT CHANGE 70S/60S	PERCENT CHANGE 70S/50S	PERCENTILE %CHANGE 70S/50S	PERCENTILE RATES 1970S
				1	9.9	SABINE TX					65
3	30.7	2	16.2			SAN AUGUSTINE TX	-47				
5	28.9	3	21.9	2	10.6	SAN JACINTO TX	-24	-51	-63	23	67
		1	37.7			SAN PATRICIO TX					
				1	53.0	SCURRY TX					97
2	9.6	4	15.4	1	3.6	SHELBY TX	60	-76	-62	24	58
18	21.4	25	24.6	24	20.9	SMITH TX	15	-15	-2	77	87
		1*	344.1			SUTTON TX					
47	28.2	59	25.7	42	14.3	TARRANT TX	-9	-44	-49	40	74
3	39.5	1	7.6			TAYLOR TX	-81				
		1	37.2			TERRY TX					
3	27.8	3	17.5	1	5.0	TITUS TX	-37	-72	-82	05	59
3	24.6			2	8.1	TOM GREEN TX			-67	18	63
18	20.5	21	19.8	11	6.7	TRAVIS TX	-4	-66	-67	17	61
2	18.4	6	47.8	5	30.1	TRINITY TX	160	-37	64	91	94
3	31.6	1	8.2	2	21.0	TYLER TX	-74	156	-33	57	87
5	22.4	5	19.8	7	23.9	UPSHUR TX	-12	21	7	81	89
		1*	133.9			UPTON TX					
		1*	114.7			UVALDE TX					
1	11.8	2	29.3			VAN ZANDT TX	148				
5	25.7	3	14.9	1	8.3	VICTORIA TX	-42	-44	-68	17	63
7	19.5	5	12.2	2	4.4	WALKER TX	-37	-64	-77	09	58
4	18.5	3	14.3	4	14.3	WALLER TX	-23	0	-23	66	74
				1	28.7	WARD TX					93
5	13.7	3	7.7	6	16.9	WASHINGTON TX	-43	119	24	85	80
8	23.6	14	39.4	7	16.0	WHARTON TX	67	-60	-32	58	78
1	112.4					WHEELER TX					
8	35.0	6	20.2	5	12.8	WICHITA TX	-42	-37	-63	22	71
1	22.9	1	10.8	1	17.6	WILBARGER TX	-53	64	-23	66	81
1*	112.4					WILLACY TX					
12*	52.3	3	12.5	3	9.3	WILLIAMSON TX	-76	-25	-82	05	65
				1	21.0	WILSON TX					87
1*	344.1					WINKLER TX					
3	20.3			3	16.4	WOOD TX			-20	69	79
30*	56.9	16	26.8	9	8.1	UTAH UT	-53	-70	-86		
4*	114.7	1	28.1	1	12.5	BOX ELDER UT	-76	-56	-89	02	70
1	27.0	1	112.4			CARBON UT	317				
2	46.3			1	27.0	DAVIS UT			-42	48	92
16*	67.2	4	17.2	3	4.5	SALT LAKE UT	-74	-74	-93	00	59
1	23.8	2	21.3	1	12.2	SAN JUAN UT	-10	-43	-49	40	70
1*	74.3					SEVIER UT					
1	94.4	1	94.4			UTAH UT	0				
4	44.2	7*	67.2	3	15.1	WEBER UT	52	-77	-66	20	76
1	17.9					VERMONT VT					
1*	89.3					WINDSOR VT					
771	29.2	690	22.9	626*	18.1	VIRGINIA VA	-21	-21	-38		
13	26.9	7	14.1	9	18.2	ACCOMACK VA	-48	29	-32	58	83
14	36.8	13	28.6	4	7.4	ALBEMARLE#VA	-22	-74	-80	07	62
5	36.2	6	54.1	2	12.2	ALLEGHANY#VA	49	-78	-66	18	70
3	17.3	1	5.7	2	10.2	AMELIA VA	-67	78	-41	49	66
6	30.0	4	18.2	2	8.4	AMHERST VA	-39	-54	-72	13	63
2	22.1	1	8.6			APPOMATTOX VA	-61				
8	38.4	3	12.0	6	15.2	ARLINGTON#VA	-69	27	-60	27	76
6	29.9	3	13.4	2	6.7	AUGUSTA#VA	-55	-50	-78	09	61
3*	118.2	3*	121.7			BATH VA	3				
11	12.3	23	24.7	19	17.2	BEDFORD#VA	100	-30	39	88	80
1*	188.7					BLAND VA					
3	33.5					BOTETOURT VA					
6	17.0	6	16.4	6	14.6	BRUNSWICK VA	-4	-11	-14	72	75
				1*	596.2	BUCHANAN VA					99
12*	58.6	4	20.3	2	8.1	BUCKINGHAM VA	-65	-60	-86	03	62
9	36.9	4	15.9	4	13.9	CAROLINE VA	-57	-12	-62	24	73
				1	23.6	CARROLL#VA					89
4	33.7	3	22.3	5	24.3	CHARLES CITY VA	-34	9	-28	62	90
4	20.0	5	22.5	3	12.3	CHARLOTTE VA	12	-45	-39	51	70
89	28.0	97	26.0	71	16.2	CHESTERFIELD#VA	-7	-38	-42	48	78
2	35.1	1	21.0			CLARKE VA	-40				
4	26.4	3	19.1	3	17.3	CULPEPER VA	-28	-10	-35	55	81
4	26.7	4	20.3	2	12.1	CUMBERLAND VA	-24	-40	-55	33	69
28	19.3	34	23.1	41*	26.3	DINWIDDIE#VA	20	14	36	87	92
3	37.4	1	9.3	4	26.2	ESSEX VA	-75	181	-30	60	91
16	32.9	22	31.0	24	22.9	FAIRFAX#VA	-6	-26	-30	59	89
5	19.8	10	49.9	7	24.4	FAUQUIER VA	152	-51	23	85	90
1	56.2					FLOYD VA					
3	24.6	1	10.2	1	8.6	FLUVANNA VA	-58	-16	-65	20	63
5	47.7	5	45.9	1	6.6	FRANKLIN VA	-4	-86	-86	03	61
5*	83.2	2	25.0	4*	44.8	FREDERICK#VA	-70	79	-46	43	96
5	31.2	3	17.4	6	29.1	GLOUCESTER VA	-44	67	-7	76	93
2	11.3	4	26.5	2	13.4	GOOCHLAND VA	134	-49	18	84	72
		1	31.3			GREENE VA					
7	27.2	11*	36.8	11*	35.4	GREENSVILLE#VA	35	-4	30	86	95
21	37.4	9	16.4	8	13.7	HALIFAX#VA	-56	-16	-63	23	73
11	50.9	6	25.4	7	28.9	HANOVER VA	-50	13	-43	47	93
8	43.0	5	19.4	4	13.8	HENRICO VA	-55	-29	-68	17	73
9	33.0	9	20.4	15*	26.3	HENRY#VA	-38	29	-20	68	92
10	40.6	5	16.6	4	11.5	ISLE OF WIGHT VA	-59	-30	-72	13	68
8	22.9	10	21.1	18*	39.3	JAMES CITY#VA	-8	87	72	92	96
9*	59.5	3	20.5	1	6.9	KING AND QUEEN VA	-66	-66	-88	02	61
2	26.4	2	23.3	2	25.8	KING GEORGE VA	-12	11	-2	77	91
6	39.4	4	30.8	6*	42.5	KING WILLIAM VA	-22	38	8	81	96
7	44.0	2	11.0	3	19.5	LANCASTER VA	-75	77	-56	32	85
5	28.8	6	32.8	2	10.2	LOUDOUN VA	14	-69	-65	21	66
8	37.8	2	8.6	5	18.3	LOUISA VA	-77	113	-51	37	83
2	9.4	2	10.6	5	33.8	LUNENBURG VA	14	218	261	99	94
1	14.1	2	27.9	1	11.6	MADISON VA	98	-58	-18	70	69

* SIGNIFICANTLY MORE DEATHS THAN EXPECTED

\# COUNTIES COMBINED; SEE TECHNICAL NOTES

NON-WHITE MALE — CANCER OF THE STOMACH — ICD 151

1950-1959 DEATHS	1950-1959 RATE	1960-1969 DEATHS	1960-1969 RATE	1970-1979 DEATHS	1970-1979 RATE	COUNTY AND STATE	PERCENT CHANGE 60S/50S	PERCENT CHANGE 70S/60S	PERCENT CHANGE 70S/50S	PERCENTILE %CHANGE 70S/50S	PERCENTILE RATES 1970S
3	36.6	2	26.2	2	27.1	MATHEWS VA	-28	3	-26	63	92
18	35.6	9	16.9	7	12.2	MECKLENBURG VA	-53	-28	-66	20	70
5	35.0	6	46.3	3	22.2	MIDDLESEX VA	32	-52	-37	53	88
1	8.4	2	26.1			MONTGOMERY#VA	211				
6	39.4	2	11.6	5	29.5	NELSON VA	-71	154	-25	64	93
11	27.7	9	23.2	6	16.5	NORTHAMPTON VA	-16	-29	-41	49	79
5	30.8	7	46.0	2	12.8	NORTHUMBERLAND VA	49	-72	-58	28	71
7	28.3	5	18.0	3	10.5	NOTTOWAY VA	-36	-42	-63	23	66
		3	23.3	2	11.5	ORANGE VA		-51			68
1	37.5					PAGE VA					
1	22.5					PATRICK VA					
22	28.4	15	16.4	11	9.2	PITTSYLVANIA#VA	-42	-44	-68	17	64
1	6.1	1	6.8	1	5.7	POWHATAN VA	10	-16	-8	76	59
6	21.8	5	20.0	6	22.3	PRINCE EDWARD VA	-8	11	2	79	88
4	34.5	3	20.0	5	25.2	PRINCE WILLIAM#VA	-42	26	-27	62	91
1	12.5			2	42.5	PULASKI VA			240	98	96
		1	22.5	1	24.8	RAPPAHANNOCK VA		10			90
1	17.2	1	9.9	1	14.5	RICHMOND VA	-42	46	-15	72	74
23	29.3	18	22.2	17	16.7	ROANOKE#VA	-24	-25	-43	47	80
5	52.8	1	8.2	3	28.3	ROCKBRIDGE#VA	-84	244	-46	43	93
3	60.0	1	15.8			ROCKINGHAM#VA	-74				
		1	47.2			RUSSELL VA					
				1*	196.5	SCOTT VA					99
1	44.6	1	31.3			SHENANDOAH VA	-30				
2	86.1					SMYTH VA					
20*	44.6	13	24.9	11	18.6	SOUTHAMPTON#VA	-44	-25	-58	29	83
9	50.8	9	42.5	4	17.6	SPOTSYLVANIA#VA	-16	-59	-65	20	81
4*	57.3					STAFFORD VA					
3	20.1	1	10.2	2	11.9	SURRY VA	-49	17	-41	49	69
4	15.6	3	10.4	9*	37.2	SUSSEX VA	-33	256	138	97	95
2	18.0	2	19.2			TAZEWELL VA	6				
1	22.9					WARREN VA					
1	12.5	3	31.1	1	11.2	WASHINGTON#VA	149	-64	-10	74	68
12*	70.7	1	5.0	5	20.8	WESTMORELAND VA	-93	318	-71	14	87
3	26.4	3	44.6	2	44.8	WISE#VA	69	0	70	91	96
1	22.5					WYTHE VA					
115	27.5	147*	27.9	131*	21.3	CHESAPEAKE CITY#VA	1	-24	-23	66	87
19	28.7	16	18.9	17	17.4	HAMPTON CITY VA	-34	-8	-40	50	81
33	35.0	23	17.5	24	19.0	NEWPORT NEWS CITY VA	-50	8	-46	44	84
21	32.6	24	28.2	16	17.4	SUFFOLK CITY VA	-13	-38	-47	43	81
99	39.3	92	26.1	79	12.9	WASHINGTON WA	-34	-51	-67		
1*	267.9					ASOTIN WA					
		1	56.2			BENTON WA					
		1	94.4			CHELAN WA					
				1	10.3	CLARK WA					66
1	28.3	3	55.0	1	12.8	FRANKLIN WA	94	-77	-55	33	71
2	137.2			1	10.4	GRANT WA			-92	00	66
2	115.9					GRAYS HARBOR WA					
		1*	89.3			ISLAND WA					
1	112.4					JEFFERSON WA					
62*	46.2	63	29.8	55	15.8	KING WA	-36	-47	-66	19	77
2	22.3	2	14.3			KITSAP WA	-36				
1*	67.0					LEWIS WA					
1	10.1					OKANOGAN WA					
5	26.9	5	14.8	11	20.5	PIERCE WA	-45	38	-24	65	86
				1	27.0	SKAGIT WA					92
1	21.3	1	23.6			SNOHOMISH WA	11				
11*	61.4	10*	49.1	3	14.4	SPOKANE WA	-20	-71	-77	10	74
1	44.6			1	38.5	STEVENS WA			-14	72	95
				1	7.4	THURSTON WA					62
		1	18.2			WALLA WALLA WA					
1	20.6					WHATCOM WA					
7	35.8	4	28.8	4	8.7	YAKIMA WA	-20	-70	-76	10	63
131	29.2	106	24.1	84*	20.1	WEST VIRGINIA WV	-17	-17	-31		
1	67.0					BARBOUR WV					
1	10.6					BERKELEY WV					
6	22.9	9	34.4	6	24.0	CABELL WV	50	-30	5	80	90
5	9.6	10	25.8	10*	27.0	FAYETTE WV	169	5	182	97	92
4	44.7	1	9.4	1	10.7	GREENBRIER WV	-79	14	-76	10	67
1	15.7			1	14.3	HANCOCK WV			-9	75	74
				1	81.8	HARDY WV					98
		1	23.8	1	12.2	HARRISON WV		-49			70
3	25.8	3	23.6	6*	50.0	JEFFERSON WV	-9	112	94	94	97
22	37.6	17	24.0	14	21.4	KANAWHA WV	-36	-11	-43	47	87
6	23.5	6	30.5	1	5.9	LOGAN WV	30	-81	-75	11	60
16	25.0	13	21.3	16*	29.2	MC DOWELL WV	-15	37	17	84	93
3	17.6	7	32.3	2	12.3	MARION WV	84	-62	-30	60	70
1	49.2	1	62.9			MARSHALL WV	28				
1	24.6					MASON WV					
15	49.2	11	28.1	2	6.0	MERCER WV	-43	-79	-88	02	60
1	28.9			1	20.6	MINERAL WV			-29	61	86
4	32.3	1	5.9	3	34.5	MINGO WV	-82	483	7	81	95
2	21.8	6*	89.9	1	18.6	MONONGALIA WV	312	-79	-15	72	83
2	86.3					MONROE WV					
3	20.1	1	8.6	2	16.5	OHIO WV	-57	91	-18	70	79
3*	99.4	1	54.6	1	56.2	POCAHONTAS WV	-45	3	-43	47	97
25*	50.2	14	29.8	12	24.7	RALEIGH WV	-41	-17	-51	38	90
1	68.3					RANDOLPH WV					
1	28.1	1	24.8	2*	49.1	SUMMERS WV	-12	98	75	92	97
1*	112.4					TUCKER WV					
2	56.1	1	22.3	1	28.8	WOOD WV	-60	29	-49	40	93
1	10.3	2	50.7			WYOMING WV	390				
42	28.6	40	15.8	46	11.5	WISCONSIN WI	-45	-27	-60		
2	41.2	1	49.7	1	56.2	ASHLAND WI	21	13	36	87	97
1	57.3					BAYFIELD WI					

* SIGNIFICANTLY MORE DEATHS THAN EXPECTED

COUNTIES COMBINED; SEE TECHNICAL NOTES

NON-WHITE MALE — CANCER OF THE STOMACH — ICD 151

1950-1959 DEATHS	1950-1959 RATE	1960-1969 DEATHS	1960-1969 RATE	1970-1979 DEATHS	1970-1979 RATE	COUNTY AND STATE	PERCENT CHANGE 60S/50S	PERCENT CHANGE 70S/60S	PERCENT CHANGE 70S/50S	PERCENTILE %CHANGE 70S/50S	PERCENTILE RATES 1970S
2	64.3					BROWN WI					
				1*	60.3	BURNETT WI					97
		3	44.5	3	15.8	DANE WI		-65			77
1	245.4			1	38.3	DOUGLAS WI			-84	04	95
		1	24.5	1	12.6	KENOSHA WI		-49			70
1*	163.6					MANITOWOC WI					
2	22.3			2	17.5	MENOMINEE#WI			-22	67	81
29	37.6	31	19.2	31	11.8	MILWAUKEE WI	-49	-39	-69	16	69
				1*	273.2	ONEIDA WI					99
2	25.2	2	17.1	1	7.4	RACINE WI	-32	-57	-70	14	62
1	20.6	1	12.8	2	17.3	ROCK WI	-38	34	-16	71	81
		1	26.5	1	28.1	SAWYER WI		6			93
1	57.3					VILAS WI					
				1	35.1	WAUKESHA WI					95
7	28.1	1	6.6	1	2.7	WYOMING WY	-76	-60	-91		
1	133.9					CARBON WY					
3	30.1					FREMONT WY					
1*	589.6					HOT SPRINGS WY					
1*	267.9					JOHNSON WY					
1	59.6			1	30.3	NATRONA WY			-49	40	94
		1	56.2			SWEETWATER WY					

* SIGNIFICANTLY MORE DEATHS THAN EXPECTED

\# COUNTIES COMBINED; SEE TECHNICAL NOTES

NON-WHITE FEMALE

CANCER OF THE STOMACH

ICD 151

1950-1959 DEATHS	1950-1959 RATE	1960-1969 DEATHS	1960-1969 RATE	1970-1979 DEATHS	1970-1979 RATE	COUNTY AND STATE	PERCENT CHANGE 60S/50S	PERCENT CHANGE 70S/60S	PERCENT CHANGE 70S/50S	PERCENTILE %CHANGE 70S/50S	PERCENTILE RATES 1970S
8176	13.2	8503	10.0	8577	6.9	UNITED STATES	-24	-31	-48		
475	12.4	370	8.6	344	6.7	ALABAMA AL	-30	-23	-46		
3	8.9	5	13.9	1	1.9	AUTAUGA AL	57	-86	-78	12	65
3	8.8	3	7.6	2	3.6	BALDWIN AL	-14	-53	-60	35	69
6	9.5	7	11.7	8	12.3	BARBOUR AL	23	5	30	89	91
5	22.6	1	4.7			BIBB AL	-79				
13*	26.5	5	8.6	4	7.7	BULLOCK AL	-68	-11	-71	22	82
1	1.9	2	4.3	4	7.9	BUTLER AL	124	85	315	99	83
10	17.3	5	7.3	3	3.6	CALHOUN AL	-58	-50	-79	11	69
7	13.8	4	6.3	5	6.8	CHAMBERS AL	-54	8	-51	45	79
2	49.5			1	11.8	CHEROKEE AL			-76	15	91
6*	41.6	1	6.5	1	7.2	CHILTON AL	-84	10	-83	07	80
4	10.8	3	8.7	3	6.6	CHOCTAW AL	-20	-24	-39	56	78
6	12.5	3	6.1	4	5.5	CLARKE AL	-51	-10	-56	40	74
2	23.1	1	20.2			CLAY AL	-13				
2	10.6	2	8.0	2	6.1	COFFEE AL	-24	-24	-42	54	76
2	6.0	1	2.2	4	7.6	COLBERT AL	-63	242	27	89	82
6	16.2	4	8.3	4	8.7	CONECUH AL	-49	4	-46	49	85
		2	18.0	1	7.5	COOSA AL		-58			81
1	3.7	2	6.5	1	4.4	COVINGTON AL	75	-32	19	88	71
3	15.5	2	8.3			CRENSHAW AL	-47				
3	15.0	1	5.5			DALE AL	-63				
19	12.0	13	7.9	7	3.2	DALLAS AL	-34	-59	-73	19	68
				1	17.2	DE KALB AL					95
9	21.0	3	7.8	4	8.6	ELMORE AL	-63	10	-59	36	85
3	7.8	9	16.7	3	5.5	ESCAMBIA AL	115	-67	-29	66	75
4	8.7	9	16.4	6	8.7	ETOWAH AL	89	-47	0	81	85
				1	6.7	FAYETTE AL					79
1	17.1					FRANKLIN AL					
1	8.5			1	5.2	GENEVA AL			-38	57	74
8	15.2	4	7.9	2	4.7	GREENE AL	-48	-40	-69	25	72
6	8.5	5	9.9	2	2.5	HALE AL	16	-75	-71	22	66
4	17.9	3	9.7			HENRY AL	-46				
5	7.5	7	10.4	3	4.1	HOUSTON AL	38	-61	-46	50	70
		1	8.7			JACKSON AL					
113	12.4	95	9.4	94	7.4	JEFFERSON AL	-24	-21	-40	55	81
		1	8.2			LAMAR AL					
1	3.5	2	5.3	1	1.9	LAUDERDALE AL	51	-64	-45	51	65
2	7.8	5	20.6	1	3.3	LAWRENCE AL	165	-84	-58	37	68
9	13.3	13*	17.8	7	8.6	LEE AL	34	-52	-35	59	85
2	6.2					LIMESTONE AL					
3	5.9	2	3.8	2	4.4	LOWNDES AL	-36	16	-25	68	71
9	10.1	6	6.4	17*	14.7	MACON AL	-37	129	45	91	93
7	8.5	2	2.1	5	4.0	MADISON AL	-76	96	-52	44	70
7	8.2	4	4.7	7	7.9	MARENGO AL	-42	68	-3	80	83
		1	33.5	2*	62.3	MARION AL		86			98
38	12.9	37	9.6	40	7.8	MOBILE AL	-25	-20	-40	55	82
3	6.6	2	5.0	5	9.7	MONROE AL	-24	95	48	92	88
34	13.3	26	8.5	25	7.0	MONTGOMERY AL	-36	-18	-47	48	80
3	8.5	2	5.4	2	3.8	MORGAN AL	-37	-28	-55	41	69
4	6.1	3	4.6	2	3.9	PERRY AL	-24	-16	-36	59	69
3	6.0	2	3.9	2	3.2	PICKENS AL	-35	-18	-47	49	68
6	13.9	3	5.8	6	8.6	PIKE AL	-59	49	-39	56	85
		2	12.5	3	16.2	RANDOLPH AL		29			94
9	10.4	12	14.0	3	2.8	RUSSELL AL	35	-80	-73	19	66
3	17.2	2	11.0	1	5.2	ST CLAIR AL	-36	-53	-70	24	73
3	13.8	2	7.1	2	5.8	SHELBY AL	-49	-18	-58	37	75
15	22.5	8	10.8	10*	11.6	SUMTER AL	-52	8	-48	47	90
12	19.1	9	11.7	5	5.6	TALLADEGA AL	-39	-52	-71	23	75
3	7.0	4	8.9	2	4.0	TALLAPOOSA AL	27	-56	-43	52	70
28*	26.4	10	8.2	16	9.8	TUSCALOOSA AL	-69	20	-63	31	88
6	20.7	2	5.5	1	2.0	WALKER AL	-74	-64	-90	01	65
				1	3.7	WASHINGTON AL					69
7	10.3	5	7.8	4	5.5	WILCOX AL	-24	-30	-47	48	74
		6	6.7	9	5.2	ALASKA AK		-22			
27	10.2	37	10.5	38	4.4	ARIZONA AZ	3	-58	-57		
3	7.1	7	15.5	6	5.3	APACHE AZ	120	-66	-25	68	74
2*	85.4			1	4.1	COCHISE AZ			-95	00	70
		5	23.9	4	8.0	COCONINO AZ		-67			83
		2	21.4	1	7.0	GILA AZ		-67			80
10	15.4	11	9.0	11	3.6	MARICOPA AZ	-42	-60	-77	14	69
5	10.3	2	5.0	7	7.8	NAVAJO AZ	-51	56	-24	69	82
5	9.9	5	8.0	5	4.3	PIMA AZ	-19	-47	-57	38	71
1	7.2	4	28.7	3	6.9	PINAL AZ	296	-76	-5	78	79
1	29.8					YAVAPAI AZ					
		1	10.4			YUMA AZ					
202	11.6	165	8.4	142	6.4	ARKANSAS AR	-28	-24	-45		
2	10.6	2	8.5	2	6.9	ARKANSAS AR	-20	-18	-35	60	79
2	5.8	1	2.4	2	3.2	ASHLEY AR	-59	35	-44	51	68
1	4.5	7*	32.3	2	8.2	BRADLEY AR	611	-75	81	95	84
				1	5.6	CALHOUN AR					75
9	14.1	10	14.1	5	5.4	CHICOT AR	0	-62	-62	31	74
3	11.9	2	6.6	4	12.8	CLARK AR	-45	94	8	85	92
2	16.5	1	13.3			CLEVELAND AR	-19				
3	6.9	3	6.0	3	4.5	COLUMBIA AR	-13	-26	-35	59	71
4	21.6	2	10.5	2	8.1	CONWAY AR	-51	-23	-63	31	83
		2	18.9			CRAIGHEAD AR					
18	17.2	8	7.1	16*	14.2	CRITTENDEN AR	-59	100	-18	72	93
1	3.5	2	9.4	5*	15.8	CROSS AR	169	69	355	99	94
1	3.4	4	15.6	1	3.1	DALLAS AR	358	-80	-8	78	68
4	8.6	2	4.1	8*	15.2	DESHA AR	-53	272	76	95	94

* SIGNIFICANTLY MORE DEATHS THAN EXPECTED

NON-WHITE FEMALE — CANCER OF THE STOMACH — ICD 151

1950-1959 DEATHS	1950-1959 RATE	1960-1969 DEATHS	1960-1969 RATE	1970-1979 DEATHS	1970-1979 RATE	COUNTY AND STATE	PERCENT CHANGE 60S/50S	PERCENT CHANGE 70S/60S	PERCENT CHANGE 70S/50S	PERCENTILE %CHANGE 70S/50S	PERCENTILE RATES 1970S
4	14.5	1	3.1	1	4.3	DREW AR	-79	40	-70	23	71
1	10.6	2	15.5			FAULKNER AR	46				
4	14.6			2	6.9	GARLAND AR			-53	43	79
1	133.9					GRANT AR					
4	9.8	4	7.8	3	7.5	HEMPSTEAD AR	-20	-4	-24	69	81
3	28.7	3	24.1	1	3.7	HOT SPRING AR	-16	-85	-87	04	69
2	20.7			1	2.2	HOWARD AR			-89	02	65
				1	9.9	INDEPENDENCE AR					88
2	13.7	2	11.2	3	15.4	JACKSON AR	-18	38	12	86	94
15	9.3	9	4.6	7	3.3	JEFFERSON AR	-51	-28	-64	29	68
9*	40.0	2	8.2	3	9.5	LAFAYETTE AR	-79	15	-76	15	87
4	6.7	3	4.5	1	.7	LEE AR	-32	-84	-89	02	64
5	16.4	3	11.5	4	12.9	LINCOLN AR	-30	12	-22	71	92
1	6.4	2	8.8	2	10.2	LITTLE RIVER AR	36	16	59	93	89
3	13.1	3	8.7	3	9.2	LONOKE AR	-33	5	-30	65	86
10*	25.8	4	8.4	5	8.6	MILLER AR	-68	3	-67	27	85
10	13.0	14	16.6	6	6.7	MISSISSIPPI AR	28	-60	-48	47	79
3	7.8	3	7.4	2	4.2	MONROE AR	-4	-43	-46	50	71
		2	10.5			NEVADA AR					
7	12.6	3	4.6	2	2.7	OUACHITA AR	-64	-41	-78	12	66
14	11.5	11	9.1	6	4.2	PHILLIPS AR	-21	-54	-64	30	70
1	7.2	2	15.3			POINSETT AR	113				
1	28.3			1	19.1	POPE AR			-32	62	95
		1	6.2			PRAIRIE AR					
26	11.4	22	7.9	16	4.8	PULASKI AR	-31	-39	-58	37	73
9	10.9	6	7.6	8	10.2	ST FRANCIS AR	-30	34	-7	78	89
3	12.4	3	10.7	4	14.5	SEBASTIAN AR	-14	35	17	87	93
				1	22.3	SEVIER AR					96
5	11.8	8	11.8	6	7.8	UNION AR	0	-33	-33	61	82
1	74.3					WASHINGTON AR					
		2	33.3	1	11.2	WHITE AR		-66			90
4	13.1	3	11.0	1	3.3	WOODRUFF AR	-16	-70	-75	17	68
		1	49.1			YELL AR					
325	13.3	462	9.5	711	6.4	CALIFORNIA CA	-28	-33	-52		
30	10.8	44	7.5	74	7.2	ALAMEDA CA	-31	-3	-33	62	81
1*	112.4			1*	49.1	AMADOR CA			-56	39	98
		1*	94.4			COLUSA CA					
3	13.0	10	15.6	15	5.6	CONTRA COSTA CA	21	-64	-57	39	75
				1	25.9	DEL NORTE CA					96
3	3.6	15	14.4	12	4.5	FRESNO CA	297	-69	25	88	72
		1*	78.1			GLENN CA					
2	12.6			4	6.0	IMPERIAL CA			-53	44	76
		1	14.3			INYO CA					
7	19.7	7	11.1	5	2.8	KERN CA	-44	-75	-86	04	67
		5*	44.0	2	5.9	KINGS CA		-87			75
178	14.0	228	9.2	361	6.8	LOS ANGELES CA	-34	-26	-51	45	79
		2	11.7	1	2.4	MADERA CA		-79			66
1	34.4	1	11.8			MARIN CA	-66				
		2	37.6			MENDOCINO CA					
3	17.7	2	9.1	2	3.3	MERCED CA	-49	-63	-81	09	68
		1*	89.3			MODOC CA					
1	9.6	4	15.1	2	1.5	MONTEREY CA	58	-90	-84	05	64
1	15.7	5	11.0	14	4.5	ORANGE CA	-30	-59	-71	21	71
		3*	82.3	3	15.9	PLACER CA		-81			94
5	18.3	3	4.2	6	2.9	RIVERSIDE CA	-77	-31	-84	06	67
12	19.9	23*	22.1	26	9.9	SACRAMENTO CA	11	-55	-50	46	88
1*	81.8					SAN BENITO CA					
5	11.8	8	7.5	9	3.4	SAN BERNARDINO CA	-36	-55	-71	21	68
6	5.5	11	5.9	28	5.5	SAN DIEGO CA	6	-6	0	81	74
34	17.1	38	8.2	71	7.1	SAN FRANCISCO CA	-52	-14	-59	37	80
5	10.5	7	6.8	11	6.2	SAN JOAQUIN CA	-35	-8	-41	55	76
1	25.8	1	9.4			SAN LUIS OBISPO CA	-63				
2	5.4	7	8.4	13	6.6	SAN MATEO CA	57	-22	23	88	78
		1	4.1	7	10.6	SANTA BARBARA CA		156			89
13*	32.3	14	12.8	21	6.3	SANTA CLARA CA	-60	-51	-81	09	77
2	41.4	3	29.7	1	2.1	SANTA CRUZ CA	-28	-93	-95	00	65
1	20.6	1	13.3	1	11.1	SHASTA CA	-36	-16	-46	50	90
2	6.9	5	18.3	8	7.0	SOLANO CA	165	-62	1	82	79
2	48.5	3	35.1	2	8.2	SONOMA CA	-28	-77	-83	06	83
1	23.6			2	2.9	STANISLAUS CA			-88	03	67
1	12.4	2	9.9	2	2.7	TULARE CA	-20	-73	-78	12	66
1	3.1	1	10.7	2	1.7	VENTURA CA	249	-84	-45	51	64
1	18.2			3	17.4	YOLO CA			-4	79	95
		2*	42.5	1	3.3	YUBA CA		-92			68
16	12.3	27	12.5	23	4.6	COLORADO CO	2	-63	-62		
1	13.7	2	38.2			ADAMS CO	180				
				1*	74.3	ARCHULETA CO					99
		1*	21.8			BOULDER CO					
8	9.8	22	14.7	18	6.3	DENVER-ARAPAHOE#CO	50	-57	-36	59	77
3	23.4			2	1.6	EL PASO CO			-93	00	64
1*	62.9					OTERO CO					
2	22.5			1	4.4	PUEBLO CO			-80	09	71
1	26.5	2*	45.9	1	11.2	WELD CO	73	-76	-58	37	90
19	8.5	31	7.3	45	6.1	CONNECTICUT CT	-14	-17	-29		
7	11.3	13	10.3	15	6.5	FAIRFIELD CT	-9	-37	-43	53	78
5	7.8	8	5.7	10	4.6	HARTFORD CT	-26	-21	-42	54	72
		1	24.4			LITCHFIELD CT					
		1	20.2			MIDDLESEX CT					
6	6.8	6	4.2	19	8.4	NEW HAVEN CT	-38	99	23	88	84
1	12.4	2	11.6	1	3.7	NEW LONDON CT	-6	-68	-70	23	69
21	12.6	29	12.0	37*	11.2	DELAWARE DE	-4	-7	-11		
4	13.4	2	4.9	8	13.5	KENT DE	-63	175	1	81	92
13	12.8	19	13.1	24*	12.2	NEW CASTLE DE	2	-7	-5	79	91

* SIGNIFICANTLY MORE DEATHS THAN EXPECTED

COUNTIES COMBINED; SEE TECHNICAL NOTES

NON-WHITE FEMALE — CANCER OF THE STOMACH — ICD 151

1950-1959 DEATHS	1950-1959 RATE	1960-1969 DEATHS	1960-1969 RATE	1970-1979 DEATHS	1970-1979 RATE	COUNTY AND STATE	PERCENT CHANGE 60S/50S	PERCENT CHANGE 70S/60S	PERCENT CHANGE 70S/50S	PERCENTILE %CHANGE 70S/50S	PERCENTILE RATES 1970S
4	10.9	8	15.7	5	7.1	SUSSEX DE	44	-55	-35	60	80
159	14.2	166	9.8	164	7.0	DIST. OF COLUMBIA DC	-31	-29	-51		
159	14.2	166	9.8	164	7.0	WASHINGTON DC	-31	-29	-51	45	80
254	10.2	285	8.1	320	6.1	FLORIDA FL	-20	-25	-40		
6	7.9	5	5.7	3	2.1	ALACHUA FL	-28	-64	-74	18	65
3	22.8	2	5.5	2	3.4	BAY FL	-76	-37	-85	05	68
		2	18.9	1	24.8	BRADFORD FL		31			96
2	10.9	3	6.6	4	5.6	BREVARD FL	-40	-15	-48	47	75
11	16.8	13	7.2	17	5.9	BROWARD FL	-57	-19	-65	28	75
1	62.9					CHARLOTTE FL					
1	23.6	3*	47.5	1	14.9	CLAY FL	102	-69	-37	58	93
		2	17.2	1	7.6	COLLIER FL		-55			82
		4	13.0	2	4.4	COLUMBIA FL		-66			71
24	8.2	52	11.8	48	4.5	DADE FL	43	-62	-46	50	71
		1	56.2	1	8.5	DE SOTO FL		-85			85
1	56.2					DIXIE FL					
39	10.3	42	8.4	45	7.0	DUVAL FL	-19	-16	-32	63	80
12	12.5	6	3.5	16	8.2	ESCAMBIA FL	-72	138	-34	61	84
				2	15.3	FRANKLIN FL					94
8	9.4	13	11.8	7	5.5	GADSDEN FL	25	-54	-42	54	74
1	27.1					GLADES FL					
1	14.3			1	9.4	GULF FL			-34	61	87
2	13.9	1	7.7	2	8.2	HAMILTON FL	-44	6	-41	55	84
		2	20.9			HERNANDO FL					
1	7.0	3	20.9	2	6.8	HIGHLANDS FL	199	-68	-3	80	79
16	7.4	15	6.0	22	6.3	HILLSBOROUGH FL	-19	4	-16	73	77
14*	26.5	2	3.8	3	4.6	JACKSON FL	-86	22	-83	07	72
2	7.3	2	8.1	2	6.5	JEFFERSON FL	11	-19	-11	75	78
4	12.2	2	4.0	1	1.3	LAKE FL	-67	-67	-89	02	64
2	8.4	5	13.3	4	8.1	LEE FL	58	-40	-5	79	83
7	9.1	8	8.1	8	6.0	LEON FL	-11	-25	-34	61	76
1	6.0	1	9.4	1	5.9	LEVY FL	57	-37	-1	81	76
1	133.9			1*	24.6	LIBERTY FL			-82	08	96
1	3.5	4	13.3	2	5.2	MADISON FL	275	-61	46	91	73
1	2.9	2	5.4	4	7.3	MANATEE FL	85	36	152	98	81
10	13.6	7	9.3	4	3.9	MARION FL	-32	-58	-71	21	70
				2	9.5	MARTIN FL					87
3	21.0	2	11.9	1	3.7	MONROE FL	-44	-69	-82	07	69
3	18.7	2	9.9	4*	16.4	NASSAU FL	-47	65	-12	75	94
1	11.9					OKALOOSA FL					
		1	56.2			OKEECHOBEE FL					
8	8.2	9	6.4	14	5.3	ORANGE FL	-22	-17	-35	60	74
2	35.1			1	10.7	OSCEOLA FL			-69	24	89
10	6.0	12	6.2	19	7.1	PALM BEACH FL	3	15	19	87	80
1	8.4	1	4.7	2	8.9	PASCO FL	-44	89	6	84	86
5	6.9	11	7.9	16	7.3	PINELLAS FL	14	-7	7	85	81
8	9.4	8	5.4	15	8.5	POLK FL	-43	59	-10	76	85
2	4.2	7	13.9	3	6.3	PUTNAM FL	228	-55	48	92	77
9*	30.1	6	15.9	8*	18.6	ST JOHNS FL	-47	17	-38	57	95
3	10.1	2	4.9	5	8.4	ST LUCIE FL	-51	70	-17	73	84
1	13.6					SANTA ROSA FL					
3	14.9	4	11.6	2	4.9	SARASOTA FL	-22	-57	-67	27	73
11*	21.6	6	9.4	7	7.7	SEMINOLE FL	-56	-19	-64	29	82
				1	7.0	SUMTER FL					80
3	12.3	2	8.4	1	4.9	SUWANNEE FL	-32	-42	-61	34	73
2	16.5			2	11.6	TAYLOR FL			-29	65	90
4	4.1	9	9.4	8	5.7	VOLUSIA FL	128	-39	40	90	75
				2	19.7	WAKULLA FL					95
2	18.9					WALTON FL					
1	10.0	1	10.9			WASHINGTON FL	9				
455	11.1	419	8.5	384	6.3	GEORGIA GA	-23	-26	-43		
				2	20.4	APPLING GA					95
		1	22.5			ATKINSON GA					
1	28.7			1	25.8	BACON GA			-10	76	96
2	19.7			1	12.7	BAKER GA			-35	59	92
9	12.4	4	4.8	6	9.4	BALDWIN GA	-62	97	-24	68	87
3*	35.4	1	8.4	1	4.4	BARROW GA	-76	-48	-88	03	71
1	5.1			3	9.9	BARTOW GA			94	96	88
2	9.9	4	17.3			BEN HILL GA	75				
3*	53.2	1	15.7	3*	35.0	BERRIEN GA	-70	123	-34	61	98
19	11.1	18	7.8	24	8.6	BIBB GA	-30	11	-22	70	85
2	14.7	2	20.5	1	8.6	BLECKLEY GA	39	-58	-41	54	85
		1*	28.9			BRANTLEY GA					
3	10.1	4	11.9	2	6.0	BROOKS GA	18	-49	-41	55	76
		2	30.6	2*	26.5	BRYAN GA		-14			97
1	3.5	3	8.0	2	4.8	BULLOCH GA	131	-40	38	90	72
6	8.9	3	5.2	2	2.7	BURKE GA	-42	-48	-70	24	66
		3	21.3	2	7.7	BUTTS GA		-64			82
5	21.8			1	4.0	CALHOUN GA			-82	08	70
		2	12.4	1	4.2	CAMDEN GA		-66			70
		1	9.6			CANDLER GA					
4	15.5	1	3.9	2	6.0	CARROLL GA	-75	54	-61	32	76
1*	106.0					CATOOSA GA					
1	13.6					CHARLTON GA					
41	17.2	28	9.2	28	7.4	CHATHAM GA	-47	-19	-57	39	81
				1	8.2	CHATTOOGA GA					83
5	10.4	2	3.5	6	8.4	CLARKE GA	-66	136	-19	72	84
		1	8.8	1	6.6	CLAY GA		-25			78
		2	11.1	3	12.4	CLAYTON GA		11			91
5	19.3	3	8.1	5	11.0	COBB GA	-58	35	-43	53	90
		1	3.8	2	9.2	COFFEE GA		143			86
4	11.9	1	3.2	1	2.9	COLQUITT GA	-73	-9	-76	16	67
1	5.9	1	5.1	2	13.5	COLUMBIA GA	-15	167	128	97	92
1	8.3	1	7.1			COOK GA	-15				
2	5.2	1	2.4	4	8.2	COWETA GA	-55	248	58	93	84

* SIGNIFICANTLY MORE DEATHS THAN EXPECTED

NON-WHITE FEMALE — CANCER OF THE STOMACH — ICD 151

1950-1959 DEATHS	1950-1959 RATE	1960-1969 DEATHS	1960-1969 RATE	1970-1979 DEATHS	1970-1979 RATE	COUNTY AND STATE	PERCENT CHANGE 60S/50S	PERCENT CHANGE 70S/60S	PERCENT CHANGE 70S/50S	PERCENTILE %CHANGE 70S/50S	PERCENTILE RATES 1970S
		2	16.5	1	9.6	CRAWFORD GA		-42			87
5	15.7			2	4.8	CRISP GA			-69	25	73
4	8.1	6	13.1	5	10.2	DECATUR GA	60	-22	25	89	89
4	7.5	9	6.3	14	4.8	DE KALB GA	-16	-24	-36	58	72
2	9.7	3	21.2	1	3.8	DODGE GA	119	-82	-60	34	69
4	16.3	4	14.5			DOOLY GA	-11				
6	7.7	9	7.7	7	4.5	DOUGHERTY GA	-1	-41	-41	54	72
1	15.8	1	10.6			DOUGLAS GA	-33				
1	3.4	4	13.2	1	3.6	EARLY GA	283	-73	4	83	69
2	18.2					EFFINGHAM GA					
		2	8.1	4	17.2	ELBERT GA		112			95
		2	7.1			EMANUEL GA					
2	22.1			1	5.3	EVANS GA			-76	15	74
3*	44.4	2	27.9			FAYETTE GA	-37				
1	2.9	3	6.4	2	3.0	FLOYD GA	119	-53	2	82	67
67	10.3	87	9.8	76	6.7	FULTON GA	-5	-32	-35	60	78
4	9.4	7	12.2	3	4.2	GLYNN GA	29	-66	-55	40	70
		1	16.1	1	21.9	GORDON GA		36			96
3	12.2	4	14.9	2	7.0	GRADY GA	23	-53	-43	53	79
3	11.7	4	13.6	2	5.7	GREENE GA	17	-58	-51	45	75
1	9.2	2	18.5			GWINNETT GA	100				
				1	26.5	HABERSHAM GA					97
1	6.6	2	9.4	1	3.0	HALL GA	42	-68	-55	41	67
1	3.9	4	14.3	3	8.2	HANCOCK GA	263	-43	108	97	84
2	30.6	1	14.3			HARALSON GA	-53				
1	5.1	3	14.9	1	3.8	HARRIS GA	194	-74	-24	69	69
1	8.1	2	18.9	1	7.7	HART GA	132	-59	-6	78	82
5	24.5			2	7.6	HENRY GA			-69	25	82
1	7.3	1	3.3	1	3.2	HOUSTON GA	-55	-4	-56	39	68
1	8.2			1	9.4	IRWIN GA			15	87	87
2	22.5			1	8.5	JACKSON GA			-62	31	84
4	26.1			1	6.7	JASPER GA			-74	17	79
2	28.2			2	18.2	JEFF DAVIS GA			-35	59	95
3	9.8			5	9.4	JEFFERSON GA			-5	79	87
		2	10.1	2	9.6	JENKINS GA		-5			87
1	10.5	2	18.5	2	19.0	JOHNSON GA	76	3	81	95	95
		2	9.7			JONES GA					
		1	4.7	1	3.4	LAMAR GA		-28			68
1	14.9	1	19.6	1	22.3	LANIER GA	32	14	50	92	96
3	5.3	4	7.5	6	7.6	LAURENS GA	42	2	45	91	82
1	4.1	1	7.6			LEE GA	83				
5	23.7	2	8.1	3	9.9	LIBERTY GA	-66	22	-58	37	88
1	7.8	3*	36.9			LINCOLN GA	372				
				2*	34.2	LONG GA					97
3	6.1	4	5.5	4	4.2	LOWNDES GA	-10	-25	-32	63	70
1	6.2	3	11.4	1	2.9	MC DUFFIE GA	83	-74	-53	43	67
2	11.0	3	17.7	2	10.0	MC INTOSH GA	61	-43	-9	76	88
3	9.1	2	4.4	4	9.3	MACON GA	-51	112	3	83	86
2	28.6	1	21.0	1	11.8	MADISON GA	-27	-44	-59	36	91
1	7.8					MARION GA					
5	12.0	5	12.1	6	13.6	MERIWETHER GA	1	12	13	86	92
2	24.9					MILLER GA					
2	5.1	5	12.8	5	10.4	MITCHELL GA	150	-19	102	97	89
3	18.0	3	16.8	2	7.7	MONROE GA	-6	-54	-57	38	82
2	19.2	1	12.1			MONTGOMERY GA	-37				
1	4.4	3	14.3	1	3.9	MORGAN GA	225	-73	-11	75	70
17	14.2	16	8.7	13	5.6	MUSCOGEE#GA	-39	-35	-61	34	75
2	11.6	3	11.4	5	15.3	NEWTON GA	-2	34	32	90	94
		1	7.6			OGLETHORPE GA					
		1	23.8			PAULDING GA					
1	3.0			1	2.3	PEACH GA			-24	69	65
		1*	38.3	1*	27.0	PICKENS GA		-30			97
2	21.3					PIERCE GA					
1	10.7					PIKE GA					
3	18.4			1	6.1	POLK GA			-67	26	76
6*	39.1	1	6.1	1	4.2	PULASKI GA	-85	-30	-89	02	71
4	28.9	2	10.5	1	4.8	PUTNAM GA	-64	-55	-83	06	72
2	30.3					QUITMAN GA					
1*	147.4					RABUN GA					
3	9.0			2	4.3	RANDOLPH GA			-53	43	71
21	14.6	22	11.3	13	4.8	RICHMOND GA	-23	-57	-67	26	72
1	8.8					ROCKDALE GA					
				1	17.9	SCHLEY GA					95
4	11.1	1	4.5	3	8.2	SCREVEN GA	-59	83	-26	67	84
2	19.1					SEMINOLE GA					
3	8.4	5	12.1	4	7.1	SPALDING GA	45	-42	-16	74	80
		1	8.6			STEPHENS GA					
1	5.1			2	7.9	STEWART GA			55	93	83
6	11.1	3	4.7	6	8.4	SUMTER GA	-58	78	-25	68	84
4	23.6	1	5.2			TALBOT GA	-78				
		3	25.2	2	10.2	TATTNALL GA		-59			89
3	22.6	2	10.7			TAYLOR GA	-52				
2	10.7	2	10.2	2	12.9	TELFAIR GA	-4	26	21	88	92
5	15.9	3	8.8	3	8.5	TERRELL GA	-45	-4	-47	49	85
14*	25.0	7	10.4	7	9.2	THOMAS GA	-59	-11	-63	30	86
4	14.0	1	3.5	1	3.2	TIFT GA	-75	-8	-77	14	68
1	4.4	3	12.4	1	4.6	TOOMBS GA	181	-63	4	83	72
8	12.7	4	5.4	1	.9	TROUP GA	-57	-84	-93	00	64
1	10.6					TURNER GA					
1	7.7			1	6.5	TWIGGS GA			-15	74	78
1	3.1	4	14.0	3	8.9	UPSON GA	349	-37	183	98	85
1	9.1					WALKER GA					
4	21.2	3	14.9	1	4.9	WALTON GA	-30	-67	-77	14	73
7	18.3	5	10.9	1	2.0	WARE GA	-40	-82	-89	02	65
5	30.9					WARREN GA					
6	14.4	5	12.2	6	12.9	WASHINGTON GA	-15	5	-10	76	92
		2	12.2	2	13.0	WAYNE GA		6			92
2	27.3			1	15.0	WEBSTER GA			-45	51	93
2	29.4	1	12.5			WHITFIELD GA	-58				
4*	34.1			1	5.9	WILCOX GA			-83	07	76
		4	14.4	1	2.9	WILKES GA		-80			67

* SIGNIFICANTLY MORE DEATHS THAN EXPECTED

COUNTIES COMBINED; SEE TECHNICAL NOTES

NON-WHITE FEMALE — CANCER OF THE STOMACH — ICD 151

1950-1959 DEATHS	1950-1959 RATE	1960-1969 DEATHS	1960-1969 RATE	1970-1979 DEATHS	1970-1979 RATE	COUNTY AND STATE	PERCENT CHANGE 60S/50S	PERCENT CHANGE 70S/60S	PERCENT CHANGE 70S/50S	PERCENTILE %CHANGE 70S/50S	PERCENTILE RATES 1970S
2	16.4	2	11.4	2	9.9	WILKINSON GA	-30	-13	-40	56	88
		5	20.6			WORTH GA					
		324*	22.3	299*	13.5	HAWAII HI		-39			
		39*	21.3	33*	13.3	HAWAII HI		-38			92
		238*	22.7	234*	14.2	HONOLULU HI		-38			93
		22*	26.1	12	9.9	KAUAI HI		-62			88
		25*	18.1	20	10.1	MAUI#HI		-44			88
11*	40.8	5	15.6	2	2.7	IDAHO ID	-62	-83	-93		
3*	59.5	1	37.5			BANNOCK ID	-37				
1*	74.3					BENEWAH ID					
2*	143.6	1*	29.5			BONNEVILLE ID	-79				
2*	106.1	1*	28.7	1	9.8	CANYON ID	-73	-66	-91	01	88
				1*	74.3	GEM ID					99
		1*	94.4			KOOTENAI ID					
3*	98.3					NEZ PERCE ID					
		1*	114.7			WASHINGTON ID					
410	14.4	475*	11.5	489*	7.8	ILLINOIS IL	-20	-32	-46		
				3*	37.4	ADAMS IL					98
8	24.9	3	8.9	4	11.3	ALEXANDER IL	-64	27	-55	41	90
		1*	267.9			CASS IL					
		2	8.9	1	2.8	CHAMPAIGN IL		-69			66
1*	188.7					CHRISTIAN IL					
				2*	237.3	CLINTON IL					99
345*	15.0	409*	12.1	425*	8.3	COOK IL	-20	-31	-45	51	84
		1*	112.4			DE KALB IL					
				1	1.8	DU PAGE IL					64
2	14.2			2	8.2	JACKSON IL			-42	54	84
		1	19.3			JEFFERSON IL					
1	8.8			1	3.0	KANE IL			-66	27	67
1	3.4	2	3.9	1	1.8	KANKAKEE IL	13	-53	-47	48	65
1	11.1					KNOX IL					
2	6.4	4	10.2	3	5.9	LAKE IL	60	-42	-7	78	76
				1	29.8	LA SALLE IL					97
1*	56.2	1*	47.1			LOGAN IL	-16				
1*	133.9					MC DONOUGH IL					
1	6.5			3	7.5	MACON IL			15	87	81
5	13.2	6	13.0	2	3.0	MADISON IL	-1	-77	-78	13	67
		1	15.2			MARION IL					
		2	28.5			MASSAC IL					
				1*	44.6	MONTGOMERY IL					98
		1	9.0			MORGAN IL					
2	9.1	5	13.2	1	1.9	PEORIA IL	46	-86	-79	11	65
5	19.1	5	18.9	1	3.2	PULASKI IL	-1	-83	-83	06	68
2*	61.0	1	18.6			RANDOLPH IL	-70				
1	14.5	2	11.0			ROCK ISLAND IL	-24				
20	13.7	17	8.9	26	9.1	ST CLAIR IL	-35	3	-33	61	86
3	11.1	3	9.0	1	2.5	SANGAMON IL	-19	-72	-77	13	66
1	47.2					STEPHENSON IL					
1*	196.5					TAZEWELL IL					
2	8.4	3	14.9	2	7.6	VERMILION IL	77	-49	-10	76	82
2	9.3	4	19.8	5	10.2	WILL IL	113	-49	10	85	89
		1	11.6			WILLIAMSON IL					
2	10.7			3	9.1	WINNEBAGO IL			-15	74	86
98	12.9	85	7.4	113	7.0	INDIANA IN	-43	-5	-46		
		1	2.2	4	6.6	ALLEN IN		204			78
1*	94.4					BOONE IN					
1	8.6			2	9.1	CLARK IN			5	84	86
6*	33.4	2	7.4	1	2.8	DELAWARE IN	-78	-62	-91	01	67
				1	4.8	ELKHART IN					72
		1	22.3			FAYETTE IN					
2	26.2					FLOYD IN					
1	10.6	2	16.5			GRANT IN	55				
1	16.1	1	10.6	2	16.2	HOWARD IN	-34	53	1	82	94
				1*	37.2	JAY IN					98
1*	67.0					JENNINGS IN					
				1*	112.4	KNOX IN					99
24	12.2	23	7.3	41	9.1	LAKE IN	-41	26	-25	68	86
1	6.7	1	3.7	2	9.7	LA PORTE IN	-44	160	45	91	88
42	12.4	42	8.8	42	6.4	MARION IN	-29	-27	-48	47	78
		1*	43.0	1*	141.3	MONTGOMERY IN		229			99
1*	182.1					OHIO IN					
		1*	86.0			PUTNAM IN					
8	32.7	1	1.4	7	8.5	ST JOSEPH IN	-96	492	-74	18	85
1*	70.7					SPENCER IN					
1	38.2					TIPPECANOE IN					
2	5.4	4	8.5	3	4.6	VANDERBURGH IN	58	-46	-15	74	72
3	12.3	3	11.6	4	9.3	VIGO IN	-5	-20	-25	68	86
		2*	137.6			WARRICK IN					
2	19.2			1	5.6	WAYNE IN			-71	21	75
14	13.8	11	9.0	11	5.7	IOWA IA	-35	-37	-59		
		2	18.5	2	6.1	BLACK HAWK IA		-67			76
1*	344.1					BUCHANAN IA					
1	24.8	1	33.5			DES MOINES IA	35				
2*	186.8	2*	380.3			DUBUQUE IA	104				
1*	424.0					HUMBOLDT IA					
				2*	200.4	JOHNSON IA					99
2	36.2			2*	25.4	LEE IA			-30	65	96
4	10.3	5	11.4	2	2.9	POLK IA	11	-75	-72	20	67
1	33.1					POTTAWATTAMIE IA					
		1	14.3	1	6.4	SCOTT IA		-56			77
				2*	121.5	TAMA IA					99

* SIGNIFICANTLY MORE DEATHS THAN EXPECTED

\# COUNTIES COMBINED; SEE TECHNICAL NOTES

NON-WHITE FEMALE | CANCER OF THE STOMACH | ICD 151

1950-1959 DEATHS	1950-1959 RATE	1960-1969 DEATHS	1960-1969 RATE	1970-1979 DEATHS	1970-1979 RATE	COUNTY AND STATE	PERCENT CHANGE 60S/50S	PERCENT CHANGE 70S/60S	PERCENT CHANGE 70S/50S	PERCENTILE %CHANGE 70S/50S	PERCENTILE RATES 1970S
2	37.3					WOODBURY IA					
41	11.5	31	6.6	33	5.0	KANSAS KS	-43	-24	-57		
2*	77.3					BARTON KS					
		1	18.7			CHEROKEE KS					
2	36.4					COWLEY KS					
				1	11.2	CRAWFORD KS					90
1	37.5					DONIPHAN KS					
1	8.6					DOUGLAS KS					
2*	353.9					FORD KS					
				1*	38.3	FRANKLIN KS					98
1	31.5	1	11.2			GEARY KS	-65				
		1*	344.1			KIOWA KS					
		1	7.0			LABETTE KS					
2	12.7	1	5.9			LEAVENWORTH KS	-53				
1	28.1					MIAMI KS					
5	31.2	2	8.3			MONTGOMERY KS	-73				
1*	94.4					MORRIS KS					
				1*	14.9	PRATT KS					93
1	12.4	1	18.9	4*	44.5	RENO KS	52	136	259	98	98
1	23.6	1	24.5			RILEY KS	4				
		2*	30.5	1	5.9	SALINE KS		-81			76
5	11.8	3	4.5	6	5.3	SEDGWICK KS	-62	17	-55	40	74
3	6.3	5	9.1	5	6.6	SHAWNEE KS	44	-28	5	84	78
				1*	74.3	SHERMAN KS					99
13	9.9	12	7.4	13	6.0	WYANDOTTE KS	-25	-19	-39	56	76
124	12.0	106	9.0	93	6.8	KENTUCKY KY	-25	-24	-43		
1	19.1					ADAIR KY					
1	9.1					BARREN KY					
				1	10.3	BELL KY					89
		6*	34.7	2	6.9	BOURBON KY		-80			79
				1	22.8	BOYD KY					96
1	5.8	1	5.2	1	5.7	BOYLE KY	-10	9	-2	80	75
1*	56.2					BRACKEN KY					
1	37.7	1	26.5			BRECKINRIDGE KY	-30				
1*	94.4					BULLITT KY					
1	20.4					CALDWELL KY					
1	45.5	1	14.1			CALLOWAY KY	-69				
1	21.2			1	21.9	CAMPBELL KY			3	83	96
4	6.1	1	1.5	5	6.3	CHRISTIAN KY	-76	330	3	83	77
4*	38.8	1	6.2	1	3.9	CLARK KY	-84	-37	-90	01	70
1*	188.7					CLAY KY					
				2*	70.3	CUMBERLAND KY					99
4	23.9	5*	24.4	2	9.5	DAVIESS KY	2	-61	-60	34	87
15	16.0	9	8.4	9	6.5	FAYETTE KY	-48	-23	-60	35	78
		1	29.8			FLEMING KY					
				1*	34.4	FLOYD KY					97
				3*	20.2	FRANKLIN KY					95
2	21.0			1	15.2	FULTON KY			-28	66	94
1	18.2			1	30.4	GARRARD KY			67	94	97
2	35.0					GRAVES KY					
1	47.2					GREEN KY					
				1*	74.3	GREENUP KY					99
		1	2.8	3*	32.7	HARDIN KY		1049			97
2	15.4	4*	28.7	1	5.3	HARLAN KY	86	-82	-66	27	74
				1	22.5	HARRISON KY					96
1	25.8					HART KY					
4	18.6	3	14.6			HENDERSON KY	-22				
				1	20.6	HENRY KY					96
				1	18.9	HICKMAN KY					95
3	12.5			2	5.1	HOPKINS KY			-59	35	73
40	11.9	44	10.3	34	6.5	JEFFERSON KY	-14	-37	-46	50	78
1	11.5	3*	44.9			JESSAMINE KY	292				
				2*	33.0	LETCHER KY					97
1	14.3					LINCOLN KY					
1	8.2	2	13.4	1	8.5	LOGAN KY	63	-37	3	83	85
6	15.3	4	9.1	2	4.3	MC CRACKEN KY	-41	-53	-72	20	71
2	15.7	2	15.7	2	13.6	MADISON KY	0	-13	-13	75	92
		1	13.5			MARION KY					
1	14.5					MASON KY					
1*	53.6	1*	24.8	1*	49.2	MONROE KY	-54	98	-8	77	98
3	30.7	1	14.1	2*	20.8	MONTGOMERY KY	-54	47	-32	63	96
1	7.9					MUHLENBERG KY					
		1	8.6	1	14.9	NELSON KY		73			93
		1	18.6			OHIO KY					
2*	42.1					OLDHAM KY					
1	21.3					PERRY KY					
1	112.4					PIKE KY					
		1	14.9	2*	23.6	PULASKI KY		59			96
2	19.3	2	11.9	1	11.5	SCOTT KY	-38	-4	-41	55	90
1	10.5	2	15.5			SHELBY KY	48				
				1	7.9	SIMPSON KY					83
		2*	55.6			TAYLOR KY					
1	19.3	1	17.3			TRIGG KY	-10				
				1	15.8	UNION KY					94
2	8.9	1	3.6			WARREN KY	-59				
1	10.9					WEBSTER KY					
		1*	99.4			WHITLEY KY					
3	25.2	2	22.8	2	26.3	WOODFORD KY	-9	15	4	84	96
574*	15.5	586*	13.1	491*	9.1	LOUISIANA LA	-15	-31	-42		
3	6.8	6	15.6	2	3.5	ACADIA LA	131	-78	-48	47	69
1	5.5	6*	29.6	7*	26.3	ALLEN LA	442	-11	381	99	96
8	21.4	5	12.0	8*	19.4	ASCENSION LA	-44	62	-9	76	95
3	11.4	2	6.6	5*	15.0	ASSUMPTION LA	-43	128	31	90	93
6	17.1	4	9.7	6	11.5	AVOYELLES LA	-43	18	-33	62	90
1	6.5	1	4.7			BEAUREGARD LA	-28				

* SIGNIFICANTLY MORE DEATHS THAN EXPECTED

NON-WHITE FEMALE | CANCER OF THE STOMACH | ICD 151

1950-1959 DEATHS	1950-1959 RATE	1960-1969 DEATHS	1960-1969 RATE	1970-1979 DEATHS	1970-1979 RATE	COUNTY AND STATE		PERCENT CHANGE 60S/50S	PERCENT CHANGE 70S/60S	PERCENT CHANGE 70S/50S	PERCENTILE %CHANGE 70S/50S	PERCENTILE RATES 1970S
3	8.3	5	14.1	1	3.0	BIENVILLE	LA	69	-79	-64	29	67
4	6.0	7	11.0	3	5.2	BOSSIER	LA	85	-53	-13	74	73
43	13.8	58*	14.6	40	7.9	CADDO	LA	6	-46	-43	53	83
18*	25.0	8	8.5	10	7.9	CALCASIEU	LA	-66	-7	-68	25	83
		1	8.2	1	8.3	CALDWELL	LA		1			84
3	16.3	5*	21.9	3	10.6	CATAHOULA	LA	34	-52	-35	60	89
7	12.4	3	5.5	5	8.4	CLAIBORNE	LA	-56	53	-32	62	84
4	6.8	8	18.0	3	5.7	CONCORDIA	LA	163	-68	-17	73	75
8	11.4	15*	18.8	12*	15.7	DE SOTO	LA	64	-17	37	90	94
26	12.6	28	9.7	25	6.7	EAST BATON ROUGE	LA	-23	-31	-47	49	79
5	12.3	7	12.7	5	9.2	EAST CARROLL	LA	3	-28	-26	67	86
2	3.4	5	9.1	5	9.5	EAST FELICIANA	LA	168	4	180	98	87
2	13.8	3	14.5	3	8.6	EVANGELINE	LA	5	-40	-37	57	85
5	13.7			3	7.3	FRANKLIN	LA			-46	49	81
8	17.2	13*	20.7	20*	28.7	IBERIA	LA	21	38	67	94	97
12	19.2	6	8.1	9	12.5	IBERVILLE	LA	-58	54	-35	60	91
3	20.0	1	4.1	1	2.2	JACKSON	LA	-80	-45	-89	02	65
6	7.8	26*	23.5	10	6.0	JEFFERSON	LA	201	-74	-23	70	76
6*	36.4	2	9.7	1	4.0	JEFFERSON DAVIS	LA	-73	-58	-89	02	70
10	19.5	9	13.9	7	7.5	LAFAYETTE	LA	-28	-46	-61	32	81
6*	32.7	7*	28.4	3	10.4	LAFOURCHE	LA	-13	-63	-68	26	89
8	23.7	3	7.5	2	3.2	LINCOLN	LA	-69	-58	-87	04	68
				2	13.3	LIVINGSTON	LA					92
7	14.3	4	7.9	3	5.4	MADISON	LA	-45	-31	-62	31	74
13	21.5	5	7.3	8	8.8	MOREHOUSE	LA	-66	21	-59	36	85
9	13.2	3	3.6	6	7.1	NATCHITOCHES	LA	-72	95	-46	50	80
133*	15.8	142*	13.7	154*	11.9	ORLEANS	LA	-13	-13	-24	68	91
25*	20.3	18	11.8	12	6.9	OUACHITA	LA	-42	-42	-66	27	79
4	28.6	1	4.9	1	5.2	PLAQUEMINES	LA	-83	7	-82	08	74
16*	33.0	8	14.9	4	7.2	POINTE COUPEE	LA	-55	-52	-78	12	80
26*	19.0	27*	17.9	9	5.1	RAPIDES	LA	-6	-71	-73	19	73
4	17.1	2	8.5	1	2.9	RED RIVER	LA	-50	-66	-83	06	67
3	8.1	5	9.3	3	4.7	RICHLAND	LA	14	-49	-42	54	72
5	25.1	2	15.0			SABINE	LA	-40				
3*	44.4			1	7.8	ST BERNARD	LA			-82	07	82
8*	38.2	4	13.7	5*	18.1	ST CHARLES	LA	-64	33	-53	43	95
1	8.0	2	10.6			ST HELENA	LA	32				
7	18.5	11*	29.2	2	5.0	ST JAMES	LA	58	-83	-73	19	73
8*	26.3	9*	23.1	1	2.1	ST JOHN THE BAPT	LA	-12	-91	-92	01	65
18	16.8	21*	17.8	18*	13.0	ST LANDRY	LA	6	-27	-23	70	92
4	14.1	5	13.9	3	8.2	ST MARTIN	LA	-2	-41	-42	53	84
6	9.7	11	16.9	11*	15.0	ST MARY	LA	74	-11	55	93	93
4	11.8	4	10.0	7	13.6	ST TAMMANY	LA	-15	35	15	87	92
12	21.2	14*	17.7	4	4.3	TANGIPAHOA	LA	-16	-76	-80	10	71
7	13.9	1	2.9	4	7.6	TENSAS	LA	-79	164	-45	51	82
7	19.1	5	11.2	7	13.6	TERREBONNE	LA	-41	22	-28	66	92
5	22.9	2	7.3	1	3.3	UNION	LA	-68	-55	-86	04	68
2	9.2	1	5.0	1	4.0	VERMILION	LA	-46	-19	-56	39	70
		3	22.8			VERNON	LA					
3	5.7	9	14.5	3	3.7	WASHINGTON	LA	154	-74	-35	60	69
11	24.9	14*	20.3	8	10.0	WEBSTER	LA	-18	-51	-60	35	88
2	7.5	5	17.2	7*	20.2	WEST BATON ROUGE	LA	130	17	170	98	95
2	16.9	2	11.6	1	8.6	WEST CARROLL	LA	-32	-25	-49	46	85
4	17.1			3	16.0	WEST FELICIANA	LA			-6	78	94
4	19.6	2	8.2	1	2.4	WINN	LA	-58	-71	-88	03	65
2	13.4	1	8.8	1	2.7	MAINE	ME	-34	-69	-80		
				1*	49.7	ANDROSCOGGIN	ME					98
2*	89.6					CUMBERLAND	ME					
		1*	109.3			YORK	ME					
236*	16.8	196	9.6	245*	7.9	MARYLAND	MD	-43	-18	-53		
17	22.0	10	8.6	12	8.2	ANNE ARUNDEL	MD	-61	-5	-63	30	83
10	18.8	12	17.9	12	9.7	BALTIMORE	MD	-5	-46	-49	47	88
5*	31.8	4	20.0	5*	20.1	CALVERT	MD	-37	0	-37	58	95
4	26.5	2	10.6	1	4.0	CAROLINE	MD	-60	-62	-85	05	70
		1	12.8	1	6.1	CARROLL	MD		-52			76
1	9.1	2	17.9			CECIL	MD	97				
2	10.4	2	6.1	1	2.2	CHARLES	MD	-41	-64	-79	11	65
1	2.5	3	7.2	7*	13.8	DORCHESTER	MD	192	92	460	99	93
1	5.5	2	10.8	5*	19.1	FREDERICK	MD	95	76	244	98	95
1	6.2	2	8.0	1	2.5	HARFORD	MD	27	-69	-61	34	66
1	9.1	3	28.0	4	11.8	HOWARD	MD	210	-58	31	89	91
3	20.2	1	6.7	3	14.5	KENT	MD	-67	115	-28	66	93
3	8.8	2	2.3	13	8.0	MONTGOMERY	MD	-73	241	-9	76	83
9	13.8	8	8.0	19	4.9	PRINCE GEORGE'S	MD	-42	-39	-65	28	73
3	18.0	1	4.8			QUEEN ANNE'S	MD	-73				
1	4.8	2	10.9			ST MARY'S	MD	128				
2	8.5	2	5.5	4	10.1	SOMERSET	MD	-35	82	18	87	88
3	10.7	3	8.9	3	7.2	TALBOT	MD	-17	-19	-33	62	80
1	10.6			2	14.7	WASHINGTON	MD			38	90	93
5	17.1	6	12.8	2	3.3	WICOMICO	MD	-25	-74	-81	09	68
5	21.1	2	7.1	2	5.1	WORCESTER	MD	-66	-29	-76	15	73
158*	18.5	126	9.9	148*	8.3	BALTIMORE CITY	MD	-46	-16	-55	40	84
48	13.4	58	10.3	66	7.0	MASSACHUSETTS	MA	-23	-33	-48		
		2	16.7	1	3.3	BARNSTABLE	MA		-80			68
				1	8.0	BERKSHIRE	MA					83
5	29.6	2	8.3	1	3.1	BRISTOL	MA	-72	-63	-90	02	67
				1	12.8	DUKES	MA					92
1	7.8	2	12.3	2	6.3	ESSEX	MA	58	-49	-20	71	77
		1*	188.7			FRANKLIN	MA					
3	9.8	4	8.8	7	7.2	HAMPDEN	MA	-10	-18	-27	66	81
9	16.0	6	6.8	14	10.4	MIDDLESEX	MA	-58	54	-35	59	89
1	28.1	1	6.1	2	3.3	NORFOLK	MA	-78	-46	-88	03	68
1	4.8	1	4.5			PLYMOUTH	MA	-6				
28	14.8	38	12.8	33	6.8	SUFFOLK	MA	-13	-47	-54	42	79
		1	5.8	4	14.2	WORCESTER	MA		146			93

* SIGNIFICANTLY MORE DEATHS THAN EXPECTED

NON-WHITE FEMALE — CANCER OF THE STOMACH — ICD 151

1950-1959 DEATHS	1950-1959 RATE	1960-1969 DEATHS	1960-1969 RATE	1970-1979 DEATHS	1970-1979 RATE	COUNTY AND STATE	PERCENT CHANGE 60S/50S	PERCENT CHANGE 70S/60S	PERCENT CHANGE 70S/50S	PERCENTILE %CHANGE 70S/50S	PERCENTILE RATES 1970S
237	15.8	258	9.1	290	6.7	MICHIGAN MI	-42	-27	-58		
1	37.7	1	17.9			ALLEGAN MI	-53				
1*	172.0					ANTRIM MI					
				1*	35.1	BARAGA MI					98
4	13.3	5	9.8	6	7.5	BERRIEN MI	-26	-23	-44	52	82
3	11.6	4	14.2	3	5.7	CALHOUN MI	22	-60	-51	45	75
4	28.7	1	5.4			CASS MI	-81				
1*	112.4					CHIPPEWA MI					
5	5.3	10	8.6	10	4.1	GENESEE MI	63	-52	-21	71	70
1	8.6	3	14.1	3	4.6	INGHAM MI	63	-67	-46	50	72
3	21.2			3	9.4	JACKSON MI			-56	40	87
3	24.5	2	6.9	1	3.6	KALAMAZOO MI	-72	-48	-85	05	69
3	9.9	4	7.6	3	2.7	KENT MI	-23	-65	-73	19	66
1*	98.2					LAPEER MI					
1*	114.7			1*	188.7	LEELANAU MI			65	94	99
1*	86.0					LENAWEE MI					
				3	10.9	MACOMB MI					90
		1*	74.3			MECOSTA MI					
2*	28.3			1	5.7	MONROE MI			-80	10	75
3	18.8	4	10.0	4	6.3	MUSKEGON MI	-47	-37	-67	26	77
3	4.1	8	10.6	6	4.1	OAKLAND MI	160	-61	0	81	70
1*	188.7					OTTAWA MI					
10*	44.6	6	10.8	8	8.7	SAGINAW MI	-76	-19	-80	10	85
2	29.6	1	7.3	1	6.4	ST CLAIR MI	-75	-13	-78	12	77
		4	17.0	3	9.0	VAN BUREN MI		-47			86
3	12.0	2	3.3	12*	17.6	WASHTENAW MI	-72	430	47	92	95
180	16.2	202	9.5	221	7.0	WAYNE MI	-41	-27	-57	38	79
1*	490.8					WEXFORD MI					
10	8.3	14	8.7	16	5.9	MINNESOTA MN	5	-32	-29		
1*	196.5					ANOKA MN					
		2	21.4	2*	29.6	BELTRAMI MN		38			97
1*	98.2					CAPLTON MN					
1	14.3	1	20.2	1	13.2	CASS MN	41	-34	-8	77	92
1*	114.7					CLEARWATER MN					
1	5.0	5	7.0	5	3.9	HENNEPIN MN	41	-45	-22	70	69
		1*	112.4			MILLE LACS MN					
1*	94.7					POLK MN					
3	8.4	3	7.9	7	10.3	RAMSEY MN	-6	31	23	88	89
1	20.2	2*	28.8	1	7.9	ST LOUIS MN	42	-72	-61	33	83
503	14.3	464*	12.0	319	6.9	MISSISSIPPI MS	-16	-42	-51		
16	19.7	17*	18.7	10	8.5	ADAMS MS	-5	-55	-57	38	85
2	12.3			2	9.2	ALCORN MS			-25	67	86
2	8.5	6	19.7	5	16.0	AMITE MS	132	-19	89	96	94
4	12.3	5	12.6	4	8.0	ATTALA MS	3	-37	-35	60	83
3	32.2					BENTON MS					
31*	19.4	19	11.2	12	7.2	BOLIVAR MS	-42	-36	-63	30	80
2	25.2	1	7.0			CALHOUN MS	-72				
		5	20.8	3	14.9	CARROLL MS		-28			93
3	10.2	3	10.5	1	3.0	CHICKASAW MS	3	-71	-71	22	67
1	9.9			1	7.4	CHOCTAW MS			-25	68	81
5	15.0	6	16.0	3	7.0	CLAIBORNE MS	7	-56	-53	43	80
7	25.3	2	5.5	2	7.1	CLARKE MS	-78	30	-72	21	80
7	20.1	2	5.7	5	9.7	CLAY MS	-72	70	-52	44	87
21	15.9	14	10.4	8	6.1	COAHOMA MS	-35	-41	-61	32	76
8	13.9	8	14.0	4	5.3	COPIAH MS	1	-63	-62	31	74
3	18.1			2	7.5	COVINGTON MS			-59	36	81
5	9.0	10*	18.8	4	7.0	DE SOTO MS	108	-63	-22	70	80
5	8.9	5	7.9	7	9.1	FORREST MS	-11	15	3	82	86
3	12.1	2	10.3			FRANKLIN MS	-15				
2*	33.1	2*	40.9			GEORGE MS	23				
		2	27.4			GREENE MS					
14*	35.7	12*	28.7	6	11.0	GRENADA MS	-20	-62	-69	25	90
2	21.4	1	10.2	1	6.2	HANCOCK MS	-52	-39	-71	22	77
4	7.2	10	13.8	1	1.2	HARRISON MS	91	-91	-84	06	64
17	7.0	22	7.6	31	7.2	HINDS MS	9	-5	4	83	81
18	19.6	12	14.1	15*	14.4	HOLMES MS	-28	2	-26	66	93
4	9.9	6	12.8	4	6.7	HUMPHREYS MS	30	-48	-32	63	78
2	17.6	1	7.5	1	5.3	ISSAQUENA MS	-57	-30	-70	24	74
9*	35.9	6	16.8	4	9.7	JACKSON MS	-53	-42	-73	19	88
4	11.6	8*	22.9	2	5.4	JASPER MS	98	-77	-54	42	74
6	14.8	4	11.4	7*	16.5	JEFFERSON MS	-23	45	11	86	94
2	6.8	6	18.4	6*	15.3	JEFFERSON DAVIS MS	169	-17	124	97	94
7	13.4	7	13.1	4	4.2	JONES MS	-2	-68	-69	25	71
8*	23.8	3	10.5	2	6.7	KEMPER MS	-56	-36	-72	21	79
3	12.3	1	3.3			LAFAYETTE MS	-73				
2	18.4	1	10.9	1	7.9	LAMAR MS	-41	-28	-57	38	83
12	12.6	18*	15.8	9	6.1	LAUDERDALE MS	25	-61	-52	44	76
7*	43.4	5*	27.0	2	12.3	LAWRENCE MS	-38	-54	-72	21	91
9*	39.2	5	16.4	7*	16.1	LEAKE MS	-58	-2	-59	36	94
6	15.8	3	6.4	2	3.8	LEE MS	-59	-40	-76	16	69
14	12.5	19*	16.3	7	5.5	LEFLOPE MS	30	-66	-56	40	74
7	20.3	6	14.8	3	7.9	LINCOLN MS	-27	-47	-61	33	83
9	14.2	5	6.9	4	3.7	LOWNDES MS	-51	-47	-74	18	69
18	20.7	18*	19.4	5	4.4	MADISON MS	-6	-77	-79	12	71
2	8.3	6	19.6	3	6.4	MAPION MS	136	-67	-23	70	77
4	6.9	3	4.3	1	1.2	MARSHALL MS	-38	-71	-82	08	64
4	7.4	4	7.7	4	6.9	MONROE MS	4	-11	-8	77	79
1	6.7	3	13.2	2	7.9	MONTGOMERY MS	96	-40	17	87	83
3	15.8	2	10.1			NESHOBA MS	-36				
7*	27.5	4	13.5	1	3.7	NEWTON MS	-51	-73	-87	04	69
5	10.1	3	5.4	3	6.4	NOXUBEE MS	-46	18	-36	59	77
3	6.7	8	19.6	4	6.2	OKTIBBEHA MS	192	-68	-8	78	76
15*	24.3	13*	18.6	2	2.0	PANOLA MS	-23	-89	-92	01	65
1	5.4	1	5.4			PEARL RIVER MS	-1				
2	19.9	1	8.1			PERRY MS	-59				

* SIGNIFICANTLY MORE DEATHS THAN EXPECTED

NON-WHITE FEMALE — CANCER OF THE STOMACH — ICD 151

1950-1959 DEATHS	1950-1959 RATE	1960-1969 DEATHS	1960-1969 RATE	1970-1979 DEATHS	1970-1979 RATE	COUNTY AND STATE	PERCENT CHANGE 60S/50S	PERCENT CHANGE 70S/60S	PERCENT CHANGE 70S/50S	PERCENTILE %CHANGE 70S/50S	PERCENTILE RATES 1970S
8	13.7	12*	18.8	2	2.3	PIKE MS	37	-88	-83	06	65
1	7.5			2	14.2	PONTOTOC MS			88	96	93
		1	9.6			PRENTISS MS					
7	15.4	4	7.6	6	11.2	QUITMAN MS	-51	48	-27	66	90
2	5.2	3	4.5	2	3.0	RANKIN MS	-13	-34	-43	53	67
5	18.8	8*	24.1	4	9.4	SCOTT MS	28	-61	-50	46	87
1	3.6			4	15.1	SHARKEY MS			323	99	94
4	17.6	6	22.1	2	5.2	SIMPSON MS	25	-76	-70	23	74
2	26.2					SMITH MS					
1	18.9	2	21.9			STONE MS	16				
21	19.4	5	4.4	9	7.4	SUNFLOWER MS	-77	69	-62	32	81
5	8.8	3	4.8	4	7.8	TALLAHATCHIE MS	-46	62	-12	75	82
4	13.1	9*	25.4	2	5.8	TATE MS	93	-77	-55	40	75
1	13.7	1	8.6			TIPPAH MS	-37				
11	19.7	8	16.9	9*	18.0	TUNICA MS	-14	6	-8	77	95
1	5.9	2	11.8	2	9.5	UNION MS	100	-20	60	94	87
2	8.7	1	4.9	1	5.9	WALTHALL MS	-44	21	-33	62	76
10	8.5	10	7.9	14	10.5	WARREN MS	-6	33	24	88	89
29	15.3	43*	20.9	20	8.9	WASHINGTON MS	37	-57	-42	54	86
1	6.0			1	2.8	WAYNE MS			-53	42	66
1	12.6	1	14.9			WEBSTER MS	18				
6	18.7	2	5.4	2	4.4	WILKINSON MS	-71	-18	-76	14	71
4	18.2	3	9.5	6*	14.8	WINSTON MS	-48	55	-19	72	93
1	3.9	3	13.9	5*	16.9	YALOBUSHA MS	252	22	328	99	94
14	16.0	1	1.3	5	4.8	YAZOO MS	-92	285	-70	24	73
215	14.8	187	9.5	153	6.2	MISSOURI MO	-36	-35	-58		
5	29.2			5*	21.7	BOONE MO			-25	67	96
4	21.8			2	5.8	BUCHANAN MO			-73	19	75
3	24.6	2	13.8			BUTLER MO	-44				
1	6.8	2	15.8			CALLAWAY MO	134				
3*	38.5					CAPE GIRARDEAU MO					
2*	98.1					CARROLL MO					
1	29.8			2*	26.6	CLAY MO			-11	75	97
1	78.1					CLINTON MO					
1	22.5					COLE MO					
1	10.2					COOPER MO					
				1	8.6	DUNKLIN MO					85
1	21.0					FRANKLIN MO					
		1	7.5	1	4.6	GREENE MO		-38			72
4*	46.4					HOWARD MO					
1*	212.0					HOWELL MO					
1*	114.7					IRON MO					
49	18.2	42	10.1	33	5.8	JACKSON MO	-45	-43	-68	25	75
				1	21.0	JEFFERSON MO					96
				1	12.6	LAFAYETTE MO					91
1	23.6	1	21.0			LINCOLN MO	-11				
1	9.0					MARION MO					
3	17.8	3	14.9	1	9.8	MISSISSIPPI MO	-16	-34	-45	51	88
		1*	98.2			MONITEAU MO					
1	27.0	2*	51.3			MONTGOMERY MO	90				
		1*	74.3			MORGAN MO					
1	5.3			2	10.4	NEW MADRID MO			95	96	89
1*	106.0	1*	38.2			NEWTON MO	-64				
9*	27.3	8*	23.1	5	10.5	PEMISCOT MO	-15	-55	-61	32	89
2	13.9					PETTIS MO					
3	34.8	1	8.6	1	7.5	PIKE MO	-75	-13	-78	12	82
				1*	56.2	PLATTE MO					98
2	26.3					RANDOLPH MO					
3*	90.1					RAY MO					
1	20.2	1	19.3			ST CHARLES MO	-5				
7	8.0	4	2.8	11	3.9	ST LOUIS COUNTY MO	-65	41	-51	44	70
1	6.6	3	30.8			SALINE MO	366				
2	21.0	1	7.2	1	8.5	SCOTT MO	-66	18	-60	35	84
2*	37.8					STODDARD MO					
1*	344.1					TEXAS MO					
				1*	24.8	WARREN MO					96
96	13.0	113	10.9	84	6.8	ST LOUIS CITY MO	-16	-38	-48	47	79
6	10.9	7	11.3	11	10.5	MONTANA MT	4	-7	-3		
1	14.1	1	6.7	1	4.1	BIG HORN MT	-53	-39	-71	22	70
		1	43.0	2*	28.9	CASCADE MT		-33			97
2*	685.2			1*	163.6	GALLATIN MT			-76	15	99
1	9.1					GLACIER MT					
				1	27.0	HILL MT					97
				1	9.3	LAKE MT					86
				1*	188.7	LINCOLN MT					99
		1*	56.2			PHILLIPS MT					
2	37.1	2	29.4	1	9.0	ROOSEVELT MT	-21	-69	-76	16	86
				1	13.7	ROSEBUD MT					92
		1*	62.9	1*	114.7	SANDERS MT		82			99
				1*	182.1	TETON MT					99
		1*	133.9			VALLEY MT					
14	14.1	12	7.9	22*	10.1	NEBRASKA NB	-44	27	-28		
				1*	182.1	CHERRY NB					99
11	14.1	9	7.7	17*	10.9	DOUGLAS NB	-46	43	-23	70	90
				1	6.4	LANCASTER NB					78
1*	344.1					POLK NB					
				2*	9.3	SARPY NB					86
		1	24.8	1	8.3	SCOTTS BLUFF NB		-67			84
2	49.1	2*	40.6			THURSTON NB	-17				
6	15.9	7	11.1	11	7.0	NEVADA NV	-30	-36	-56		
2	18.6	6	13.2	9	9.6	CLARK NV	-29	-28	-49	47	87
1*	62.9					DOUGLAS NV					
		1*	188.7			HUMBOLDT NV					

* SIGNIFICANTLY MORE DEATHS THAN EXPECTED

NON-WHITE FEMALE — CANCER OF THE STOMACH — ICD 151

1950-1959 DEATHS	1950-1959 RATE	1960-1969 DEATHS	1960-1969 RATE	1970-1979 DEATHS	1970-1979 RATE	COUNTY AND STATE	PERCENT CHANGE 60S/50S	PERCENT CHANGE 70S/60S	PERCENT CHANGE 70S/50S	PERCENTILE %CHANGE 70S/50S	PERCENTILE RATES 1970S
1	19.7					MINERAL NV					
2	28.6			2	6.7	WASHOE NV			-77	14	78
1	27.3	1	15.8	2	6.8	NEW HAMPSHIRE NH	-42	-57	-75		
				1*	37.2	CARROLL NH					98
1*	109.3	1*	67.0			HILLSBOROUGH NH	-39				
				1	10.5	ROCKINGHAM NH					89
236*	17.2	201	9.1	265	7.9	NEW JERSEY NJ	-47	-14	-54		
21	14.9	15	8.8	21	9.1	ATLANTIC NJ	-41	4	-39	56	86
11	17.0	7	5.8	20*	11.7	BERGEN NJ	-66	102	-31	64	91
6	15.3	2	4.1	3	1.9	BURLINGTON NJ	-73	-54	-88	03	65
21*	20.8	19	11.9	25	10.6	CAMDEN NJ	-43	-11	-49	46	89
6*	36.7	3	12.3	1	3.2	CAPE MAY NJ	-67	-74	-91	01	68
8	27.8	4	6.9	5	6.9	CUMBERLAND NJ	-75	-1	-75	17	79
79*	17.5	73	9.6	72	6.9	ESSEX NJ	-45	-28	-61	34	79
4	11.9	6	12.0	5	6.3	GLOUCESTER NJ	0	-48	-48	48	77
17	18.6	12	8.0	21	7.7	HUDSON NJ	-57	-4	-59	36	82
9	11.4	11	8.6	13	6.8	MERCER NJ	-24	-21	-40	55	79
3	6.3	5	9.3	11	10.5	MIDDLESEX NJ	49	13	68	94	89
13	15.6	10	8.2	19	9.7	MONMOUTH NJ	-48	19	-38	57	87
3	11.9	2	7.8	2	2.7	MORRIS NJ	-34	-66	-78	13	66
1	11.7	1	6.9	3	11.5	OCEAN NJ	-41	66	-2	80	90
5	9.8	7	5.7	17	9.5	PASSAIC NJ	-42	66	-3	79	87
6	21.7	6	20.8	4	8.9	SALEM NJ	-4	-57	-59	36	86
3	50.7			1	3.5	SOMERSET NJ			-93	00	69
20	17.8	18	10.8	22	7.8	UNION NJ	-39	-28	-56	40	82
12	9.5	15	8.6	14	2.8	NEW MEXICO NM	-10	-67	-71		
1	11.6			2	2.2	BERNALILLO NM			-81	09	65
		1	19.6			CURRY NM					
1	38.3					DONA ANA NM					
		3*	30.7	2	13.2	LEA NM		-57			92
				1*	42.6	LINCOLN NM					98
5	15.5	5	10.3	2	2.9	MC KINLEY NM	-34	-72	-81	08	67
1	34.4	1	12.6			OTERO NM	-63				
				1	6.4	RIO ARRIBA NM					78
1	12.6					SANDOVAL NM					
3	11.0	4	15.0	4	5.2	SAN JUAN NM	37	-65	-53	43	73
		1	7.5	2	6.7	VALENCIA NM		-11			79
641*	17.2	759*	12.0	860*	7.9	NEW YORK NY	-30	-34	-54		
4	17.5	2	3.0	3	4.4	ALBANY NY	-83	47	-75	17	71
72*	21.8	80	9.8	138	8.5	BRONX NY	-55	-14	-61	33	84
1	17.9					BROOME NY					
		1	14.9	1	11.0	CATTARAUGUS NY		-26			90
		1*	56.2			CAYUGA NY					
1	20.6					CHEMUNG NY					
1	31.5					COLUMBIA NY					
1*	196.5					CORTLAND NY					
1	4.0	3	7.6	4	7.1	DUTCHESS NY	92	-6	81	95	80
19	13.3	20	8.0	44*	10.5	ERIE NY	-40	32	-21	71	89
4*	57.1			1	13.5	FRANKLIN NY			-76	14	92
				1	12.6	GENESEE NY					91
				1	26.5	GREENE NY					97
137*	17.9	145	9.4	204	8.2	KINGS(BROOKLYN) NY	-47	-13	-54	41	84
1	112.4			1	22.1	LIVINGSTON NY			-80	10	96
7	26.9	8	12.5	15	10.5	MONROE NY	-53	-16	-61	33	89
		1*	106.0			MONTGOMERY NY					
6	5.9	13	5.7	31	9.4	NASSAU NY	-4	66	60	93	87
304*	19.9	356*	17.8	235*	8.4	NEW YORK(MANHATTAN) NY	-11	-53	-58	37	84
2	32.4					NIAGARA NY					
1	7.5	3	18.9	1	4.5	ONEIDA NY	152	-76	-40	55	71
2	8.8	1	3.2	11*	14.8	ONONDAGA NY	-63	357	69	94	93
2	7.3			1	1.8	ORANGE NY			-75	17	65
33	10.4	67	8.9	103	6.6	QUEENS NY	-15	-26	-37	57	78
1	17.5	1	8.8	2	12.3	RENSSELAER NY	-50	41	-30	65	91
1	4.7	2	5.3	4	6.5	RICHMOND NY	14	23	39	90	78
2	10.8	3	6.9	4	8.5	ROCKLAND NY	-36	24	-21	71	85
4*	81.5					SARATOGA NY					
1	13.3	1	6.0	2	10.7	SCHENECTADY NY	-55	80	-20	72	90
11	9.8	14	8.1	14	5.5	SUFFOLK NY	-17	-32	-44	52	74
		1	10.9			SULLIVAN NY					
				1	7.4	TOMPKINS NY					81
		1	16.1			ULSTER NY					
22	12.3	35	11.5	38	7.5	WESTCHESTER NY	-6	-35	-39	56	81
379	11.2	349	8.1	382	6.6	NORTH CAROLINA NC	-28	-18	-41		
7	16.7	6	11.8	6	7.1	ALAMANCE NC	-29	-40	-58	37	80
2*	165.0					ALLEGHANY NC					
3	6.6	2	4.4	4	6.5	ANSON NC	-33	46	-2	80	78
10	20.6	5	8.9	6	8.9	BEAUFORT NC	-57	0	-57	38	86
6	12.1	3	5.5	3	5.5	BERTIE NC	-55	0	-55	41	74
2	8.1	7	16.5	1	2.1	BLADEN NC	105	-87	-74	18	65
2	10.6	2	6.8	1	3.1	BRUNSWICK NC	-36	-54	-70	23	68
5	7.1	5	6.8	8	8.0	BUNCOMBE NC	-4	19	14	86	83
3	23.8	3	15.1			BURKE NC	-37				
5	14.0	8	16.3	7	12.1	CABARRUS NC	17	-26	-13	75	91
2	15.9	1	12.5			CALDWELL NC	-21				
		1	10.7	1	12.5	CAMDEN NC		17			91
1	8.2	3	23.9			CARTERET NC	193				
2	10.5			1	2.4	CASWELL NC			-77	13	65
2	9.6	1	4.8	5	12.2	CATAWBA NC	-50	155	27	89	91
3	15.9	1	4.9	3	7.4	CHATHAM NC	-69	51	-54	42	81
2	9.5			1	2.9	CHOWAN NC			-70	23	67
6	15.1	6	10.5	6	9.5	CLEVELAND NC	-31	-9	-37	58	87

* SIGNIFICANTLY MORE DEATHS THAN EXPECTED

NON-WHITE FEMALE — CANCER OF THE STOMACH — ICD 151

1950-1959 DEATHS	1950-1959 RATE	1960-1969 DEATHS	1960-1969 RATE	1970-1979 DEATHS	1970-1979 RATE	COUNTY AND STATE	PERCENT CHANGE 60S/50S	PERCENT CHANGE 70S/60S	PERCENT CHANGE 70S/50S	PERCENTILE %CHANGE 70S/50S	PERCENTILE RATES 1970S
4	8.8	3	6.6	5	6.7	COLUMBUS NC	-25	2	-24	69	79
10	18.2	5	7.7	6	7.9	CRAVEN NC	-58	3	-57	39	83
7	10.0	5	5.0	11	5.6	CUMBERLAND NC	-51	14	-44	52	75
		3*	34.3	1	12.8	CURRITUCK NC		-63			92
1*	188.7					DARE NC					
2	7.7	5	15.3	2	4.4	DAVIDSON NC	99	-72	-43	52	71
3*	35.7	1	8.4	1	9.9	DAVIE NC	-76	18	-72	20	88
5	12.3	5	9.7	2	3.2	DUPLIN NC	-22	-67	-74	18	68
20	15.9	17	9.4	16	6.8	DURHAM NC	-41	-27	-57	38	79
11	14.1	9	9.8	10	7.9	EDGECOMBE NC	-30	-20	-44	52	83
16	10.6	19	8.3	22	7.5	FORSYTH NC	-22	-9	-29	65	81
4	10.7	6	13.6	6	9.9	FRANKLIN NC	27	-27	-8	78	88
2	3.0	4	5.1	5	5.1	GASTON NC	74	-1	72	95	73
1	7.3	2	11.4	2	6.0	GATES NC	58	-48	-18	73	76
8	18.1	1	2.1	6	8.4	GRANVILLE NC	-88	295	-54	42	84
1	8.6	2	13.9	3	15.2	GREENE NC	62	9	77	95	94
13	7.8	14	6.9	17	6.2	GUILFORD NC	-11	-11	-21	71	76
7	7.6	7	6.8	9	6.6	HALIFAX NC	-11	-3	-14	74	78
1	2.7	3	5.9	4	6.2	HARNETT NC	124	5	135	97	77
		1	18.1			HAYWOOD NC					
5*	66.9			1	5.4	HENDERSON NC			-92	01	74
4	15.1	2	3.7	7	11.5	HERTFORD NC	-76	213	-24	69	90
3	13.0			1	2.7	HOKE NC			-80	11	66
4	9.7	5	12.1	6	10.5	IREDELL NC	24	-13	8	85	89
		4*	57.7			JACKSON NC					
2	7.1	9	16.5	8	12.1	JOHNSTON NC	132	-26	71	94	91
		1	7.7			JONES NC					
4	20.6			1	2.7	LEE NC			-87	04	66
7	13.8	3	5.2	3	2.7	LENOIR NC	-62	-49	-80	09	66
2	37.5	2	37.7			MC DOWELL NC	0				
1*	37.5			1*	37.2	MADISON NC			-1	81	98
3	7.0	5	13.7	7	13.8	MARTIN NC	96	1	99	96	93
28	14.5	19	7.3	35	10.0	MECKLENBURG NC	-49	37	-31	64	88
1	12.6	1	6.4	1	4.6	MONTGOMERY NC	-49	-28	-63	30	72
4	12.1	2	3.7	4	6.3	MOORE NC	-70	72	-48	48	77
15*	21.9	6	7.1	11	11.6	NASH NC	-67	63	-47	49	90
7	7.0	5	5.0	9	7.2	NEW HANOVER NC	-29	45	3	82	81
4	9.4	7	12.9	4	6.4	NORTHAMPTON NC	36	-50	-32	63	77
2	5.7			1	3.4	ONSLOW NC			-40	55	68
3	13.4	2	6.0	4	8.6	ORANGE NC	-56	44	-36	58	85
1	8.4	2	18.7	1	4.4	PAMLICO NC	123	-77	-48	48	71
1	4.1	4	10.5	3	5.8	PASQUOTANK NC	154	-45	40	91	75
4	12.5	7	19.0	2	3.7	PENDER NC	52	-81	-71	23	69
2	14.9	1	4.2	2	12.1	PERQUIMANS NC	-72	186	-19	72	91
3	13.3	4	15.8	2	4.8	PERSON NC	19	-70	-64	29	72
7	6.8	12	9.9	11	7.3	PITT NC	46	-26	8	85	81
2	42.4					POLK NC					
4	35.0	2	8.1	2	7.1	RANDOLPH NC	-77	-13	-80	10	80
9	21.9	3	6.3	6	8.9	RICHMOND NC	-71	40	-59	35	86
8	6.8	13	8.3	14	7.2	ROBESON NC	22	-13	6	84	81
2	5.1	4	6.1	4	5.0	ROCKINGHAM NC	20	-18	-1	80	73
2	4.4	3	5.1	5	6.4	ROWAN NC	16	27	47	91	77
1	4.8	3	13.3	1	2.3	RUTHERFORD NC	176	-83	-53	42	65
4	7.1	6	9.8	1	1.1	SAMPSON NC	37	-89	-85	05	64
6	18.4	2	4.2	4	7.8	SCOTLAND NC	-77	87	-57	38	82
2	13.1	2	7.4	1	3.0	STANLY NC	-44	-59	-77	13	67
				1	12.2	STOKES NC					91
2	23.0	1	8.2	1	8.0	SURRY NC	-64	-2	-65	28	83
				2*	64.9	TRANSYLVANIA NC					99
		2	33.9	1	10.3	TYRRELL NC		-69			89
2	7.6	2	5.3	3	6.2	UNION NC	-30	18	-18	72	77
1	2.6	4	7.3	3	3.9	VANCE NC	181	-46	51	92	70
15	11.3	15	8.7	8	2.9	WAKE NC	-23	-66	-74	18	67
3	6.6	1	2.1	2	4.0	WARREN NC	-68	93	-39	57	70
1	4.8	4	22.5	2	7.1	WASHINGTON NC	374	-69	49	92	80
13	12.8	11	8.9	9	6.1	WAYNE NC	-30	-32	-53	43	76
				1	6.1	WILKES NC					76
4	7.0	2	2.0	5	4.8	WILSON NC	-71	136	-32	64	72
2*	35.4					YADKIN NC					
5	16.5	2	6.3	3	5.0	NORTH DAKOTA ND	-62	-21	-70		
1*	81.8			1*	23.8	GRAND FORKS ND			-71	22	96
3	28.3			1	5.7	ROLETTE ND			-80	10	75
		1	22.3	1	24.8	SIOUX ND		11			96
1*	74.3					STARK ND					
		1*	114.7			WILLIAMS ND					
300	14.2	306	9.9	314	6.9	OHIO OH	-30	-30	-52		
1	4.8	1	6.3	1	2.7	ALLEN OH	33	-58	-44	51	66
				2	17.7	ASHTABULA OH					95
1	8.2			1	5.6	BELMONT OH			-32	63	75
2	8.6			4	7.3	BUTLER OH			-15	74	81
1	17.9	1	14.3			CHAMPAIGN OH	-20				
9	20.4	7	11.2	3	4.0	CLARK OH	-45	-65	-81	09	70
1	27.3					CLERMONT OH					
				1	31.3	CLINTON OH					97
3*	48.2	1	9.2			COLUMBIANA OH	-81				
1*	74.3			1*	39.0	CRAWFORD OH			-48	48	98
97	14.4	97	9.8	102	6.9	CUYAHOGA OH	-32	-29	-52	44	79
1*	56.2					DARKE OH					
2*	41.7					DELAWARE OH					
				2	12.5	ERIE OH					91
1	18.7					FAYETTE OH					
32	14.1	37	11.2	30	6.3	FRANKLIN OH	-20	-44	-55	40	77
				1	28.1	GALLIA OH					97
3	13.7	3	9.3	1	3.4	GREENE OH	-32	-64	-75	16	68
		1	14.1			GUERNSEY OH					
49	12.6	59	10.8	61	7.8	HAMILTON OH	-15	-27	-38	57	82
		1*	49.2			HANCOCK OH					

* SIGNIFICANTLY MORE DEATHS THAN EXPECTED

NON-WHITE FEMALE — CANCER OF THE STOMACH — ICD 151

1950-1959 DEATHS	1950-1959 RATE	1960-1969 DEATHS	1960-1969 RATE	1970-1979 DEATHS	1970-1979 RATE	COUNTY AND STATE	PERCENT CHANGE 60S/50S	PERCENT CHANGE 70S/60S	PERCENT CHANGE 70S/50S	PERCENTILE %CHANGE 70S/50S	PERCENTILE RATES 1970S
		1*	62.9			HARDIN OH					
1	60.6					HARRISON OH					
				2*	43.1	HURON OH					98
3	16.8			5*	16.7	JEFFERSON OH			-1	81	94
1	8.7	2	21.7	1	3.4	LAKE OH	150	-84	-61	33	68
2	46.7	2	21.4			LICKING OH	-54				
		1	28.1			LOGAN OH					
5	19.1	4	8.8	8	12.2	LORAIN OH	-54	38	-36	58	91
8	8.1	16	9.6	23	9.1	LUCAS OH	19	-6	13	86	86
1	24.6					MADISON OH					
21*	19.5	18	14.2	12	6.5	MAHONING OH	-27	-55	-67	26	78
		1	24.6			MARION OH					
				1	28.9	MEDINA OH					97
		1	28.1			MEIGS OH					
1	15.1	2	26.6			MIAMI OH	75				
21	17.2	15	5.9	24	6.7	MONTGOMERY OH	-66	14	-61	33	78
2	20.4	3	18.4			MUSKINGUM OH	-10				
				1*	25.8	OTTAWA OH					96
				1*	86.0	PAULDING OH					99
				1*	20.2	PICKAWAY OH					95
		1*	98.2			PIKE OH					
2	44.4			1	10.7	PORTAGE OH			-76	16	90
2	15.8	2	14.2	1	3.2	RICHLAND OH	-10	-77	-80	11	68
1	9.8	1	7.5	2	18.4	ROSS OH	-23	144	87	96	95
		1	37.2			SENECA OH					
3	5.6	4	6.0	4	4.0	STARK OH	7	-34	-29	65	70
18	21.9	14	10.3	11	4.8	SUMMIT OH	-53	-54	-78	12	72
3	11.6	6	17.0	6	12.5	TRUMBULL OH	46	-26	8	85	91
		1	37.2			WARREN OH					
		1	11.2			WASHINGTON OH					
1*	37.2	1*	94.4			WOOD OH	154				
127	14.5	120	10.3	100	5.9	OKLAHOMA OK	-29	-42	-59		
2	23.3	6*	44.9	2	8.0	ADAIR OK	93	-82	-66	27	83
1	15.0			1	12.6	ATOKA OK			-16	73	91
1*	44.6			1*	37.2	BECKHAM OK			-17	73	98
1	14.0	4*	48.0	2*	14.9	BLAINE OK	242	-69	6	84	93
1	9.8	2	18.6			BRYAN OK	89				
1	6.0			5*	21.9	CADDO OK			263	99	96
2	40.3			1	11.1	CANADIAN OK			-72	20	90
2	9.4	1	6.0	2	5.1	CARTER OK	-36	-16	-46	50	73
4	29.4	2	10.4	1	3.1	CHEROKEE OK	-65	-70	-89	02	68
4	19.5	3	12.2	2	6.2	CHOCTAW OK	-37	-49	-68	26	76
1*	49.7					CLEVELAND OK					
2*	51.0					COAL OK					
5	25.1	6*	26.5	1	2.9	COMANCHE OK	5	-89	-88	03	67
		1*	49.2			COTTON OK					
		3	24.7	1	3.1	CRAIG OK		-87			68
3	12.3	5	18.0	1	3.0	CREEK OK	47	-83	-76	16	67
				2*	27.9	CUSTER OK					97
2	27.4	1	9.4	2	10.2	DELAWARE OK	-66	9	-63	31	89
1	10.5	2	15.3	2	16.2	GARFIELD OK	46	6	55	93	94
1	22.3	2	30.1	1	7.9	GARVIN OK	35	-74	-65	28	83
				1	11.2	GRADY OK					90
		1	37.5	1	36.4	HASKELL OK		-3			98
3	38.8	3	22.8			HUGHES OK	-41				
		1	12.4	1	11.7	JACKSON OK		-6			91
				1*	28.1	JEFFERSON OK					97
		1	31.4			JOHNSTON OK					
3*	53.1	2	16.6			KAY OK	-69				
2	37.0			1	34.2	KINGFISHER OK			-8	77	97
				2*	35.7	KIOWA OK					98
2	18.5	1	8.6	1	4.1	LE FLORE OK	-53	-52	-78	13	70
1	16.4	1	12.4	2	18.9	LINCOLN OK	-24	52	15	87	95
4	16.3	3	13.5	4	18.5	LOGAN OK	-17	36	13	86	95
1	44.6					LOVE OK					
5	17.9	2	5.3	4	8.1	MC CURTAIN OK	-71	53	-55	41	83
1	5.6					MC INTOSH OK					
1	13.5	3*	22.8			MAYES OK	69				
				1	16.7	MURRAY OK					94
14	15.3	7	6.6	6	4.6	MUSKOGEE OK	-57	-30	-70	24	72
		1	28.1			NOBLE OK					
1	9.1			1	12.9	NOWATA OK			43	91	92
1	5.1	4	18.8	2	7.0	OKFUSKEE OK	268	-63	38	90	80
17	18.0	19	9.4	17	5.6	OKLAHOMA OK	-48	-41	-69	25	75
6	14.3	7	14.4	1	1.8	OKMULGEE OK	1	-87	-87	03	65
2	32.8	1	4.5	3	12.6	OSAGE OK	-86	177	-62	32	91
1	21.3	1	18.9	1	15.8	PAWNEE OK	-11	-16	-26	67	94
1	37.2			2	28.3	PAYNE OK			-24	69	97
1	5.7	3	14.4	1	4.3	PITTSBURG OK	151	-70	-24	68	71
2	26.6	2	21.1			PONTOTOC OK	-21				
2	18.6	2	15.9			POTTAWATOMIE OK	-15				
1	23.6	1	21.5			PUSHMATAHA OK	-9				
1	18.6					ROGERS OK					
3	12.4	4	18.1	2	6.6	SEMINOLE OK	45	-63	-47	49	78
2	20.2	1	7.0	1	4.9	SEQUOYAH OK	-65	-30	-76	16	73
				1	8.2	TILLMAN OK					84
9	8.7	10	6.0	14	5.5	TULSA OK	-31	-8	-37	58	74
6*	27.9	1	4.2			WAGONER OK	-85				
				2	8.3	WASHINGTON OK					84
12	14.9	12	8.1	21	8.2	OREGON OR	-46	2	-45		
1*	70.7					COOS OR					
		1*	112.4			HOOD RIVER OR					
				2*	35.0	JEFFERSON OR					98
1	28.7					KLAMATH OR					
		1	27.1			LANE OR					
1	22.9			4*	36.6	MALHEUR OR			59	93	98
				1	4.6	MARION OR					72

* SIGNIFICANTLY MORE DEATHS THAN EXPECTED

NON-WHITE FEMALE

CANCER OF THE STOMACH

ICD 151

1950-1959 DEATHS	1950-1959 RATE	1960-1969 DEATHS	1960-1969 RATE	1970-1979 DEATHS	1970-1979 RATE	COUNTY AND STATE		PERCENT CHANGE 60S/50S	PERCENT CHANGE 70S/60S	PERCENT CHANGE 70S/50S	PERCENTILE %CHANGE 70S/50S	PERCENTILE RATES 1970S
7	14.0	10	10.6	14	9.7	MULTNOMAH	OR	-24	-8	-30	64	88
1	25.8					UMATILLA	OR					
1*	37.2					YAMHILL	OR					
374	14.3	359	9.4	349	6.6	PENNSYLVANIA	PA	-34	-30	-54		
71	15.5	51	8.1	52	6.3	ALLEGHENY	PA	-48	-22	-59	35	77
1	19.1	2*	39.9	1	20.2	ARMSTRONG	PA	109	-49	6	84	95
4	9.7	2	4.7	1	2.0	BEAVER	PA	-51	-57	-79	11	65
1	12.4	2	11.8	2	5.8	BERKS	PA	-5	-51	-53	43	75
1	15.8					BLAIR	PA					
1	14.9	1	5.9	3	11.3	BUCKS	PA	-60	91	-24	69	90
				1	27.0	BUTLER	PA					97
3	24.1	2	15.8	2	12.2	CAMBRIA	PA	-34	-23	-49	46	91
12	27.2	6	8.8	5	5.3	CHESTER	PA	-68	-40	-80	10	74
				2*	63.2	CLEARFIELD	PA					98
				1	5.7	CRAWFORD	PA					75
2	23.0					CUMBERLAND	PA					
8	13.8	8	10.5	11	8.0	DAUPHIN	PA	-24	-24	-42	54	83
16	13.7	14	9.3	22	9.7	DELAWARE	PA	-32	5	-29	65	88
				1*	89.3	ELK	PA					99
2	13.6	2	9.3	1	3.5	ERIE	PA	-32	-62	-74	18	69
2	5.7	2	4.9	5	11.1	FAYETTE	PA	-14	125	94	96	90
2	31.4			2	14.5	FRANKLIN	PA			-54	42	93
1*	267.9					FULTON	PA					
1	57.3	1*	34.4			GREENE	PA	-40				
				1*	33.5	HUNTINGDON	PA					97
2*	40.0			1	7.5	LACKAWANNA	PA			-81	09	81
3	26.4	1	6.1	2	8.5	LANCASTER	PA	-77	39	-68	26	85
1	9.6					LAWRENCE	PA					
2*	146.6					LEBANON	PA					
		2*	28.5			LEHIGH	PA					
1	24.8			1	9.6	LUZERNE	PA			-61	33	87
2	31.8					LYCOMING	PA					
2	16.5					MERCER	PA					
1*	49.2	1	44.6			MONROE	PA	-9				
11	16.9	12	12.9	7	5.2	MONTGOMERY	PA	-23	-60	-69	25	73
		1	9.4	2	16.5	NORTHAMPTON	PA		75			94
				1*	68.8	NORTHUMBERLAND	PA					99
210	13.3	240	10.0	215	6.6	PHILADELPHIA	PA	-24	-35	-50	45	78
2*	814.3					PIKE	PA					
				1*	163.6	SUSQUEHANNA	PA					99
1	32.7					VENANGO	PA					
1*	141.3					WARREN	PA					
3	8.8	3	7.2	3	6.3	WASHINGTON	PA	-18	-13	-29	65	77
				1*	112.4	WAYNE	PA					99
3	8.9	5	14.1	1	2.7	WESTMORELAND	PA	59	-81	-70	23	66
1	6.9	1	4.1	1	4.0	YORK	PA	-40	-3	-42	53	70
8	12.8	7	8.7	15	10.6	RHODE ISLAND	RI	-32	22	-17		
				1*	54.2	BRISTOL	RI					98
		1*	61.3			KENT	RI					
2	16.7	2	19.1			NEWPORT	RI	15				
4	8.4	4	6.7	14*	12.9	PROVIDENCE	RI	-20	92	53	92	92
2*	51.5					WASHINGTON	RI					
358	13.5	325	10.6	275	7.1	SOUTH CAROLINA	SC	-21	-33	-47		
4	15.0	4	13.8	3	10.1	ABBEVILLE	SC	-8	-27	-32	62	89
13	16.9	5	6.0	3	2.8	AIKEN	SC	-65	-53	-83	06	66
1	3.4	5	17.7	4	13.0	ALLENDALE	SC	422	-27	282	99	92
7	9.4	6	7.6	3	3.1	ANDERSON	SC	-19	-59	-67	26	68
3	10.4	1	3.2	3	7.1	BAMBERG	SC	-69	119	-32	63	80
3	8.4	4	12.6			BARNWELL	SC	51				
6	8.7	9	14.4	5	6.5	BEAUFORT	SC	65	-55	-26	67	78
17*	31.2	3	4.2	6	7.3	BERKELEY	SC	-87	76	-77	14	81
7	23.2	5	15.7	3	9.5	CALHOUN	SC	-33	-39	-59	36	87
50*	20.0	45*	16.1	36	9.4	CHARLESTON	SC	-19	-42	-53	43	86
4	20.2	4	13.1	1	2.6	CHEROKEE	SC	-35	-80	-87	03	66
14*	27.3	8	15.0	5	10.2	CHESTER	SC	-45	-32	-63	31	89
6	14.9	2	4.5	2	4.1	CHESTERFIELD	SC	-70	-9	-73	20	70
9	13.8	4	6.0	4	6.2	CLARENDON	SC	-57	4	-55	41	76
5	10.8	4	7.3	10*	15.4	COLLETON	SC	-33	112	43	91	94
9	13.3	5	6.7	5	4.9	DARLINGTON	SC	-50	-26	-63	30	73
		2	5.1	6	10.3	DILLON	SC		103			89
8	19.7	4	9.6	4	6.0	DORCHESTER	SC	-52	-38	-70	24	76
4	16.5			1	3.0	EDGEFIELD	SC			-82	08	67
6	13.5	7	15.3	3	5.8	FAIRFIELD	SC	13	-62	-57	39	75
15	14.0	15	12.8	12	7.7	FLORENCE	SC	-9	-40	-45	51	82
8	14.7	11*	17.0	8	10.2	GEORGETOWN	SC	16	-40	-31	64	89
9	9.2	20	14.4	7	3.5	GREENVILLE	SC	57	-76	-62	32	69
13*	26.8	10	17.0	5	6.6	GREENWOOD	SC	-36	-61	-75	16	78
4	12.6	4	10.6	4	9.9	HAMPTON	SC	-15	-7	-21	71	88
1	2.5	8	16.6	4	5.4	HORRY	SC	560	-68	115	97	74
5	19.3	8*	24.9	1	2.8	JASPER	SC	29	-89	-85	05	66
2	4.7	6	11.0	3	5.0	KERSHAW	SC	133	-54	7	85	73
2	3.7	2	4.1	4	7.3	LANCASTER	SC	11	77	97	96	81
4	6.1	4	7.2	7	9.8	LAURENS	SC	18	37	61	94	88
5	13.5	2	4.7	2	4.7	LEE	SC	-65	1	-65	28	72
5	14.9	2	4.9	7	14.7	LEXINGTON	SC	-67	203	-1	80	93
4	24.5	1	4.5			MC CORMICK	SC	-82				
5	8.9	4	6.8	5	6.2	MARION	SC	-23	-10	-30	64	76
6	13.7	3	5.4	7	10.1	MARLBORO	SC	-60	86	-26	67	88
8	19.9	8	17.4	2	3.7	NEWBERRY	SC	-12	-79	-82	08	69
4	23.3	1	5.2	2	8.3	OCONEE	SC	-78	61	-65	29	84
17	12.1	18	12.6	8	4.4	ORANGEBURG	SC	4	-65	-64	29	71
		2	7.9	1	3.1	PICKENS	SC		-61			67
20	10.3	28	11.6	36*	10.8	RICHLAND	SC	12	-7	5	84	90
1	6.8	1	5.3	2	9.9	SALUDA	SC	-22	88	47	92	88
8	7.3	7	4.7	13	7.1	SPARTANBURG	SC	-35	51	-3	80	80

* SIGNIFICANTLY MORE DEATHS THAN EXPECTED

NON-WHITE FEMALE — CANCER OF THE STOMACH — ICD 151

1950-1959 DEATHS	1950-1959 RATE	1960-1969 DEATHS	1960-1969 RATE	1970-1979 DEATHS	1970-1979 RATE	COUNTY AND STATE	PERCENT CHANGE 60S/50S	PERCENT CHANGE 70S/60S	PERCENT CHANGE 70S/50S	PERCENTILE %CHANGE 70S/50S	PERCENTILE RATES 1970S
19	17.0	7	5.2	10	6.3	SUMTER SC	-69	21	-63	30	77
3	8.3	3	8.8	4	9.7	UNION SC	6	10	16	87	87
4	6.0	16*	19.3	6	5.7	WILLIAMSBURG SC	220	-70	-5	79	75
10	14.0	7	9.5	8	7.9	YORK SC	-32	-17	-44	52	83
13	20.2	12	13.8	10	8.7	SOUTH DAKOTA SD	-32	-37	-57		
2*	79.3	1	38.5	3*	108.1	BENNETT SD	-51	180	36	90	99
		1	18.1			CHARLES MIX SD					
		1*	546.4			CODINGTON SD					
1	112.4	2*	137.2			CORSON SD	22				
		2*	34.2	1	11.9	DEWEY SD		-65			91
		1*	109.3			GREGORY SD					
				1*	84.8	HUGHES SD					99
1*	490.8					MARSHALL SD					
				1*	29.5	MELLETTE SD					97
1*	84.8					MOODY SD					
3*	111.2			1	7.4	PENNINGTON SD			-93	00	81
2*	64.0	2*	38.8			ROBERTS SD	-39				
1	8.1	1	5.5			SHANNON SD	-32				
1	13.0			2	13.0	TODD SD			0	81	92
		1*	49.7			TRIPP SD					
1	37.2			1*	37.7	ZIEBACH SD			2	82	98
260	11.2	254	9.0	215	6.3	TENNESSEE TN	-19	-31	-44		
1	30.7					ANDERSON TN					
		2	13.2			BEDFORD TN					
				3*	157.1	BLEDSOE TN					99
1	10.1			2	16.3	BRADLEY TN			61	94	94
2	15.0			1	7.2	CARROLL TN			-52	44	80
1	20.2					CHESTER TN					
				1	26.5	COCKE TN					97
2*	47.3					COFFEE TN					
4*	32.9	1	7.9			CROCKETT TN	-76				
32	10.4	25	6.2	38	7.9	DAVIDSON TN	-40	27	-24	69	83
		1	18.7			DECATUR TN					
1	43.0	1*	43.0			DE KALB TN	0				
1	37.2	1	14.5	1	14.9	DICKSON TN	-61	2	-60	34	93
2	11.1	2	9.0	1	4.4	DYER TN	-19	-51	-61	34	71
6	11.5	2	3.3	3	4.4	FAYETTE TN	-71	34	-61	33	71
6*	67.7			1	11.8	FRANKLIN TN			-83	07	91
5	10.3	5	10.3	4	7.0	GIBSON TN	0	-32	-32	63	80
		3	12.9			GILES TN					
		1*	74.3			GRAINGER TN					
1	24.8			2*	22.8	GREENE TN			-8	77	96
15	7.6	21	8.7	27	9.5	HAMILTON TN	14	9	25	88	87
6	14.7	4	9.5			HARDEMAN TN	-35				
1	1.8	7	13.2	2	3.7	HAYWOOD TN	636	-72	108	97	69
		1	13.6	1	8.4	HENDERSON TN		-39			84
3	16.1	1	4.4			HENRY TN	-73				
1	24.4	1	16.7			HICKMAN TN	-31				
		1*	74.3	2*	63.2	HOUSTON TN		-15			98
1	30.3					HUMPHREYS TN					
		1	23.6	1	20.2	JEFFERSON TN		-14			95
16	17.1	14	11.4	7	4.6	KNOX TN	-33	-60	-73	19	72
2	23.4	2	23.8	2*	19.0	LAKE TN	2	-20	-19	72	95
2	6.1	8*	20.8			LAUDERDALE TN	240				
		2	20.0			LINCOLN TN					
1	18.9	2	23.4			MC MINN TN	24				
1	17.0	1	14.9			MC NAIRY TN	-12				
7	8.6	7	6.6	10	8.2	MADISON TN	-24	25	-5	79	84
1	14.1	1	9.6	1	9.0	MARSHALL TN	-32	-6	-36	58	86
6	13.4	7	15.4	5	9.7	MAURY TN	14	-37	-28	66	87
5	12.2			3	6.2	MONTGOMERY TN			-49	46	76
2	10.5	9*	51.9	2	12.6	OBION TN	396	-76	20	88	91
		2*	51.7	1	24.6	RHEA TN		-53			96
1	19.3					ROANE TN					
1	4.1	1	3.2	2	6.3	ROBERTSON TN	-21	95	55	93	77
1	3.1	1	2.7	1	2.7	RUTHERFORD TN	-13	0	-14	74	66
98	11.6	105	9.9	84	6.1	SHELBY TN	-15	-38	-47	48	76
1	24.6					SMITH TN					
2	21.4	3	24.5	1	6.3	SULLIVAN TN	15	-74	-71	22	77
3	15.0			1	4.2	SUMNER TN			-72	20	71
10*	26.5			3	6.8	TIPTON TN			-74	17	79
1	5.9	1	6.7	1	5.3	WASHINGTON TN	14	-22	-11	75	74
2	17.8	2	23.3			WEAKLEY TN	31				
2	10.9	3	12.4			WILLIAMSON TN	14				
3	18.3	2	10.1	1	3.2	WILSON TN	-45	-68	-82	07	68
471	11.2	552	10.1	463	5.5	TEXAS TX	-10	-45	-51		
4	9.7	5	9.1	3	4.8	ANDERSON TX	-6	-48	-51	45	72
2	9.7	1	2.0	1	3.0	ANGELINA TX	-79	46	-70	24	67
4	23.7	1	7.7	1	4.2	AUSTIN TX	-68	-46	-82	07	70
3	10.2	1	6.2	2	4.4	BASTROP TX	-39	-29	-57	38	71
3	10.6	1	3.3	2	2.2	BELL TX	-69	-32	-79	11	65
16	10.6	25	10.2	38	7.1	BEXAR TX	-3	-30	-33	62	80
1*	162.1					BLANCO TX					
9	11.8	7	8.8	4	4.3	BOWIE TX	-25	-51	-64	29	71
3	12.5	8*	23.5	5	9.7	BRAZORIA TX	89	-59	-22	70	88
5	12.6	4	7.4	5	6.4	BRAZOS TX	-41	-14	-49	46	77
		1	17.2			BROWN TX					
2	8.8	1	3.1	4	26.1	BURLESON TX	-65	743	195	98	96
2	13.0	1	5.4			CALDWELL TX	-59				
1	4.0	1	6.2			CAMP TX	57				
1	2.9	4	10.7	3	10.1	CASS TX	263	-6	242	98	88
		1*	344.1			CASTRO TX					
				2	19.4	CHAMBERS TX					95
6	13.0	1	1.8	3	5.1	CHEROKEE TX	-86	190	-61	33	73
1	7.3	1	5.7			COLLIN TX	-22				

* SIGNIFICANTLY MORE DEATHS THAN EXPECTED

NON-WHITE FEMALE — CANCER OF THE STOMACH — ICD 151

1950-1959 DEATHS	1950-1959 RATE	1960-1969 DEATHS	1960-1969 RATE	1970-1979 DEATHS	1970-1979 RATE	COUNTY AND STATE	PERCENT CHANGE 60S/50S	PERCENT CHANGE 70S/60S	PERCENT CHANGE 70S/50S	PERCENTILE %CHANGE 70S/50S	PERCENTILE RATES 1970S
		1*	54.5			COLLINGSWORTH TX					
1	2.8					COLORADO TX					
		2*	54.3	1	9.3	COOKE TX		-83			86
		1*	16.1			CORYELL TX					
58	15.9	66	10.9	51	5.6	DALLAS TX	-31	-49	-65	28	75
		2	16.4			DENTON TX					
3	12.1	3	12.6			DE WITT TX	5				
2*	87.4					EASTLAND TX					
		2	12.8	1	8.4	ECTOR TX		-35			84
6	13.4	2	3.6	5	9.0	ELLIS TX	-73	152	-33	62	86
2	14.2	3	22.3	3	2.2	EL PASO TX	57	-90	-85	05	65
4	10.0	5	12.5	1	2.0	FALLS TX	24	-84	-80	10	65
2	14.6	2	13.3			FANNIN TX	-8				
3	12.8	5	19.1	3	13.2	FAYETTE TX	49	-31	3	83	92
9*	25.2	4	9.1	4	5.3	FORT BEND TX	-64	-42	-79	11	74
3	10.7	3	9.2	1	2.6	FREESTONE TX	-14	-72	-76	15	66
8	8.0	11	10.3	15	8.7	GALVESTON TX	30	-15	10	86	85
1	22.5					GOLIAD TX					
1	5.8					GONZALES TX					
4	11.0	3	5.8	1	2.0	GRAYSON TX	-47	-66	-82	08	65
2	2.4	11	14.3	4	4.5	GREGG TX	489	-68	87	95	72
4	11.7	6	16.2	1	2.6	GRIMES TX	39	-84	-78	13	66
2	10.3	3	13.2	1	3.6	GUADALUPE TX	29	-73	-65	28	69
1	11.8	2*	69.5			HALE TX	489				
		1*	44.6	2*	87.9	HALL TX		97			99
		1	22.7			HARDEMAN TX					
2	15.5	1	4.8			HARDIN TX	-69				
93	15.0	116*	12.2	110	6.8	HARRIS TX	-18	-44	-54	41	79
14	15.1	4	4.1	10	8.4	HARRISON TX	-73	105	-44	52	84
1	9.1	1	11.8			HAYS TX	30				
3	13.4	1	3.6	5	14.1	HENDERSON TX	-73	289	6	84	93
		1	38.3			HIDALGO TX					
2	12.8	1	6.2	2	11.2	HILL TX	-52	80	-13	75	90
1	12.1	2	18.0			HOPKINS TX	48				
		3	7.4	1	1.9	HOUSTON TX		-74			65
1	4.0	5	17.6	3	9.4	HUNT TX	345	-46	138	97	87
2	22.9	1	8.6			JACKSON TX	-62				
3	17.6	2	7.3	5	14.6	JASPER TX	-59	101	-17	73	93
13	6.8	30	14.6	26	8.5	JEFFERSON TX	114	-42	25	89	85
2	20.8			1	7.2	JOHNSON TX			-65	28	80
		1	24.4			JONES TX					
4	10.2	1	2.1	1	2.6	KAUFMAN TX	-79	22	-75	17	66
1	49.2			1	14.1	KERR TX			-71	21	93
2*	483.5					KING TX					
				1	6.3	KLEBERG TX					77
		1*	70.7			KNOX TX					
2	5.4	1	2.5	3	5.0	LAMAR TX	-55	102	-8	77	73
1	15.3	1	23.8			LAMB TX	55				
				1*	37.5	LAMPASAS TX					98
3	25.9	4*	25.3	1	7.6	LAVACA TX	-2	-70	-71	22	82
2	14.1	1	5.6	3	13.9	LEE TX	-60	148	-1	80	93
3	11.2	1	7.4	1	3.1	LEON TX	-34	-58	-72	20	67
2	11.0	4	12.3	2	5.5	LIBERTY TX	12	-55	-50	46	74
5	14.7	5	13.2	3	9.0	LIMESTONE TX	-10	-32	-39	56	86
3	22.3	2	5.9	1	.5	LUBBOCK TX	-74	-92	-98	00	64
				1	22.3	LYNN TX					96
9	9.2	11	8.5	7	4.9	MC LENNAN TX	-8	-42	-46	50	73
				1	7.2	MADISON TX					80
1	3.7	3	12.7	3	9.5	MARION TX	240	-25	155	98	87
4	16.3	3	10.4	2	5.4	MATAGORDA TX	-36	-48	-67	27	74
				1	4.5	MIDLAND TX					72
1	4.2	2	8.3	2	6.5	MILAM TX	99	-22	56	93	78
7	22.2	2	5.6	1	1.4	MONTGOMERY TX	-75	-74	-94	00	64
		1	6.3	2	12.7	MORRIS TX		103			92
		1*	133.9			MOTLEY TX					
4	13.7	1	2.4	1	1.9	NACOGDOCHES TX	-82	-23	-86	04	65
7	12.1	6	11.5	2	2.8	NAVARRO TX	-5	-76	-77	14	66
3	25.3					NEWTON TX					
		1	23.8			NOLAN TX					
		5	10.2	6	5.3	NUECES TX		-48			74
		5*	21.7	1	3.1	ORANGE TX		-86			67
2*	44.9					PALO PINTO TX					
1	4.9	2	7.1	1	3.6	PANOLA TX	47	-49	-25	67	69
1	38.2					PARKER TX					
2	8.7	7*	34.0			POLK TX	293				
1	11.8	1	6.1	1	2.9	POTTER TX	-48	-52	-75	16	67
1*	74.3					PRESIDIO TX					
		1*	37.2			RAINS TX					
		1	5.3	2	7.5	RED RIVER TX		41			82
5	10.7	1	1.9	2	3.9	ROBERTSON TX	-82	108	-63	30	70
2	3.7	6	10.4	3	3.7	RUSK TX	183	-64	1	82	69
		2	23.8	2	34.2	SABINE TX		44			97
4*	33.0			1	8.7	SAN AUGUSTINE TX			-74	18	85
1	6.5	4	19.2			SAN JACINTO TX	197				
		1	188.7			SAN PATRICIO TX					
		1*	49.1			SCURRY TX					
3	13.1	2	7.4	1	3.7	SHELBY TX	-43	-50	-72	20	69
17	20.9	8	7.2	6	4.6	SMITH TX	-65	-37	-78	12	72
				1*	29.8	STEPHENS TX					97
12	6.2	29	10.6	33	7.8	TARRANT TX	72	-26	27	89	82
		1	9.3	4	16.9	TAYLOR TX		82			94
1	5.5	3	17.3			TITUS TX	217				
1	3.7	1	8.3	1	3.9	TOM GREEN TX	121	-53	4	83	69
8	6.9	11	7.5	11	4.7	TRAVIS TX	9	-37	-31	64	72
2	14.4			2	14.8	TRINITY TX			3	82	93
		1	3.9	1	2.6	UPSHUR TX		-33			66
1*	68.3					UVALDE TX					
1	11.1	1	12.8			VAN ZANDT TX	16				
4	22.0					VICTORIA TX					
1	3.8	3	8.6	1	1.5	WALKER TX	127	-82	-60	35	64
2	6.6	4	16.7			WALLER TX	153				

* SIGNIFICANTLY MORE DEATHS THAN EXPECTED

NON-WHITE FEMALE — CANCER OF THE STOMACH — ICD 151

1950-1959 DEATHS	1950-1959 RATE	1960-1969 DEATHS	1960-1969 RATE	1970-1979 DEATHS	1970-1979 RATE	COUNTY AND STATE	PERCENT CHANGE 60S/50S	PERCENT CHANGE 70S/60S	PERCENT CHANGE 70S/50S	PERCENTILE %CHANGE 70S/50S	PERCENTILE RATES 1970S
1*	65.5					WARD TX					
6	15.1	8	22.0	1	2.2	WASHINGTON TX	46	-90	-85	05	65
1*	490.8					WEBB TX					
6	16.7	5	11.4	2	4.0	WHARTON TX	-32	-65	-76	15	70
		3	9.7	8*	14.1	WICHITA TX		45			93
2	7.0	4	12.3	1	2.5	WILLIAMSON TX	75	-79	-64	29	66
		3*	323.4			WINKLER TX					
2	16.5	3	17.6			WOOD TX	7				
		1*	267.9			YOAKUM TX					
10*	29.9	5	10.5	11	9.0	UTAH UT	-65	-14	-70		
2*	58.1			2*	40.4	BOX ELDER UT			-30	64	98
				1	14.1	DAVIS UT					93
2*	360.7					DUCHESNE UT					
2	11.8	3	15.0	5	10.9	SALT LAKE UT	27	-28	-8	77	90
		1	14.9			SAN JUAN UT					
4*	139.9					UINTAH UT					
		1	10.5	3	11.5	WEBER UT		10			90
1	62.9			1	9.4	VERMONT VT			-85		
1*	188.7					ADDISON VT					
				1*	94.4	WINDSOR VT					99
346	12.8	290	8.7	273	6.2	VIRGINIA VA	-32	-29	-52		
3	6.6	4	7.7			ACCOMACK VA	16				
7	14.1	6	10.8	3	3.9	ALBEMARLE#VA	-24	-64	-72	20	70
7*	55.9	1	7.5			ALLEGHANY#VA	-87				
		2	12.3			AMELIA VA					
6*	31.7			3	13.0	AMHERST VA			-59	36	92
1	12.2			1	9.4	APPOMATTOX VA			-23	70	87
5	28.3	3	9.8	2	2.5	ARLINGTON#VA	-66	-75	-91	01	66
4	16.7	2	6.3	3	7.0	AUGUSTA#VA	-63	12	-58	37	80
15	14.8	10	9.0	9	6.5	BEDFORD#VA	-39	-29	-56	39	78
		1	10.8			BOTETOURT VA					
10*	31.5	1	3.2	2	4.9	BRUNSWICK VA	-90	54	-85	05	73
1	5.2	1	5.3			BUCKINGHAM VA	4				
7*	33.4			2	6.3	CAROLINE VA			-81	09	77
				1	20.6	CAPROLL#VA					96
1	6.9	2	12.1			CHAPLES CITY VA	75				
		2	8.3			CHARLOTTE VA					
40	10.4	40	8.8	36	5.9	CHESTERFIELD#VA	-16	-32	-43	53	76
1	16.6	2	36.0			CLARKE VA	117				
1	6.4	2	10.3	1	6.1	CULPEPER VA	62	-41	-4	79	76
		3	20.9	1	4.5	CUMBERLAND VA		-79			71
1*	112.4					DICKENSON VA					
12	8.0	5	2.9	13	6.5	DINWIDDIE#VA	-64	122	-19	72	78
3	19.4			1	9.4	ESSEX VA			-51	45	87
10	21.7	6	8.2	7	2.9	FAIPFAX#VA	-62	-65	-87	04	67
2	9.4	4	16.6	1	2.7	FAUQUIER VA	76	-84	-71	22	66
		1	8.3	2	13.5	FLUVANNA VA		63			92
2	15.2	2	19.2	1	5.1	FRANKLIN VA	26	-73	-66	27	73
1	14.0			1	8.9	FREDERICK#VA			-36	58	86
		1*	37.8			GILES VA					
1	7.4	3	16.5			GLOUCESTER VA	122				
1	10.6	1	6.4	2	7.2	GOOCHLAND VA	-39	12	-32	63	80
				1*	18.6	GREENE VA					95
3	9.8	1	2.9	4	9.7	GREENSVILLE#VA	-70	233	-1	81	88
7	11.5	7	11.2	6	7.5	HALIFAX#VA	-3	-33	-34	61	82
6	28.1	1	4.4	2	6.4	HANOVER VA	-84	46	-77	13	77
1	4.8	1	6.2	1	1.0	HENPICO VA	30	-84	-80	11	64
6	15.9	5	12.5	9	13.0	HENRY#VA	-21	3	-18	72	92
4	21.6	4	13.2	3	8.6	ISLE OF WIGHT VA	-39	-35	-60	34	85
2	9.5	5	13.6	5	7.5	JAMES CITY#VA	43	-45	-21	71	81
1	5.8	2	12.0	1	8.2	KING AND QUEEN VA	106	-32	41	91	84
1	13.5					KING GEORGE VA					
2	13.9	1	6.9	2	10.2	KING WILLIAM VA	-51	49	-27	66	89
2	14.4	2	12.7	2	9.5	LANCASTER VA	-12	-25	-34	61	87
1	7.7	2	12.4	1	4.9	LOUDOUN VA	62	-60	-36	59	73
2	7.8	3	12.2	1	4.8	LOUISA VA	55	-60	-39	57	72
8*	50.7			2	7.4	LUNENBURG VA			-85	04	81
1	10.4	1	16.1			MADISON VA	55				
		1	11.1	2	15.1	MATHEWS VA		36			94
8	18.8	1	3.1	6	9.3	MECKLENBURG VA	-83	199	-50	45	86
2	15.9					MIDDLESEX VA					
2	19.0	3	24.8	1	8.8	MONTGOMERY#VA	30	-64	-54	42	85
2	14.0	1	6.6	1	7.5	NELSON VA	-53	13	-46	49	81
6	15.4	4	11.8	8*	17.1	NORTHAMPTON VA	-24	45	11	86	95
1	6.3			2	11.5	NORTHUMBEPLAND VA			82	95	90
3	13.9	2	6.8	1	2.5	NOTTOWAY VA	-51	-64	-82	08	66
1	9.4	1	5.8			ORANGE VA	-38				
1	31.3					PAGE VA					
14	13.8	7	6.1	11	7.7	PITTSYLVANIA#VA	-56	27	-44	52	82
				3*	26.6	POWHATAN VA					97
3	11.4	4	15.3	3	10.4	PRINCE EDWARD VA	34	-32	-9	77	89
1	9.8	2	14.6			PRINCE WILLIAM#VA	49				
1	8.6					PULASKI VA					
		1	22.9			RAPPAHANNOCK VA					
		2	18.3	1	7.9	RICHMOND VA		-57			83
4	6.0	14	13.2	11	7.7	ROANOKE#VA	120	-42	28	89	82
1	9.3					ROCKBRIDGE#VA					
2	46.2					ROCKINGHAM#VA					
1*	44.6	1*	24.8			RUSSELL VA	-44				
1	31.5					SMYTH VA					
6	16.6	5	9.7	4	6.3	SOUTHAMPTON#VA	-42	-35	-62	32	77
1	5.6	1	4.8	3	9.5	SPOTSYLVANIA#VA	-13	97	72	95	87
		1	7.3			SUPRY VA					
3	15.8			3	9.6	SUSSEX VA			-40	56	87
3	39.6			3*	26.3	TAZEWELL VA			-34	61	96

* SIGNIFICANTLY MORE DEATHS THAN EXPECTED

\# COUNTIES COMBINED; SEE TECHNICAL NOTES

NON-WHITE FEMALE — CANCER OF THE STOMACH — ICD 151

1950-1959 DEATHS	1950-1959 RATE	1960-1969 DEATHS	1960-1969 RATE	1970-1979 DEATHS	1970-1979 RATE	COUNTY AND STATE	PERCENT CHANGE 60S/50S	PERCENT CHANGE 70S/60S	PERCENT CHANGE 70S/50S	PERCENTILE %CHANGE 70S/50S	PERCENTILE RATES 1970S
2	21.5	1	9.3	1	5.6	WASHINGTON#VA	-57	-40	-74	18	75
1	6.1	1	5.1	2	7.0	WESTMORELAND VA	-18	38	13	86	79
1	11.7	2	21.1	1	10.6	WISE#VA	80	-50	-9	76	89
1	18.9	2*	22.9			WYTHE VA	21				
59	13.2	65	11.0	51	6.4	CHESAPEAKE CITY#VA	-17	-42	-52	44	77
4	6.8	6	8.1	4	2.5	HAMPTON CITY VA	18	-69	-63	30	66
10	10.4	10	7.7	13	7.4	NEWPORT NEWS CITY VA	-26	-4	-29	65	81
12	14.7	12	12.3	6	4.8	SUFFOLK CITY VA	-16	-61	-68	26	72
41*	19.6	44	14.6	49	7.6	WASHINGTON WA	-25	-48	-61		
		1*	28.1			COWLITZ WA					
		1	34.4			FERRY WA					
1	9.9					FRANKLIN WA					
		1	19.9			GRAYS HARBOR WA					
				1*	38.5	ISLAND WA					98
26*	25.4	35*	20.3	30	8.5	KING WA	-20	-58	-66	27	85
1	16.1	1	15.1	1	3.9	KITSAP WA	-6	-74	-76	15	69
1*	201.4					LEWIS WA					
				1	12.5	OKANOGAN WA					91
				1*	133.9	PEND OREILLE WA					99
7*	32.9	1	3.1	5	5.4	PIERCE WA	-90	72	-84	06	74
		1	53.6	1	11.1	SKAGIT WA		-79			90
1	31.4			2	6.9	SNOHOMISH WA			-78	13	79
		1	6.5	5*	16.6	SPOKANE WA		153			94
		1	18.6			STEVENS WA					
2*	45.1					WHATCOM WA					
2	11.4	1	4.8	2	5.4	YAKIMA WA	-58	13	-52	44	74
55	14.8	40	8.9	33	6.2	WEST VIRGINIA WV	-40	-30	-58		
1	15.1					BERKELEY WV					
				1*	114.7	BRAXTON WV					99
2*	162.1			1*	38.5	BROOKE WV			-76	15	98
1	9.4	2	5.6	3	7.9	CABELL WV	-40	40	-16	73	83
2	5.1	3	9.2	2	4.7	FAYETTE WV	79	-49	-9	76	72
1*	84.8					GRANT WV					
		1	5.7			GREENBRIER WV					
1	7.0	3	25.0			JEFFERSON WV	255				
12	17.7	8	9.8	6	6.3	KANAWHA WV	-45	-35	-64	29	77
1*	74.3					LEWIS WV					
1	4.6	1	4.0	1	4.4	LOGAN WV	-14	12	-4	79	71
7	11.8	7	12.9	6	9.7	MC DOWELL WV	9	-25	-17	73	88
4	24.2	2	10.2	5*	26.5	MARION WV	-58	160	10	85	97
		1	26.5			MASON WV					
6	20.4	3	7.0	3	5.1	MERCER WV	-66	-27	-75	17	73
1	37.7	1	30.7			MINERAL WV	-19				
2	24.6	2	12.5	1	12.9	MINGO WV	-49	3	-47	48	92
		1	18.7			MONONGALIA WV					
4*	30.0			1	8.0	OHIO WV			-73	19	83
5	17.2	5	12.9	3	5.2	RALEIGH WV	-25	-60	-70	23	73
1	21.0					SUMMERS WV					
2	38.8					WOOD WV					
1	14.4					WYOMING WV					
19	15.1	26	10.4	38	7.7	WISCONSIN WI	-31	-26	-49		
1	37.2			2*	76.0	ASHLAND WI			104	97	99
1*	57.3					BAYFIELD WI					
1*	112.4					CALUMET WI					
		1	14.3	2	17.7	DANE WI		23			95
		2*	188.7			FOREST WI					
		3*	35.4			MENOMINEE#WI					
12	16.9	15	7.5	29	8.6	MILWAUKEE WI	-56	16	-49	47	85
1	20.2			1	9.3	OUTAGAMIE WI			-54	41	86
1*	74.3					PIERCE WI					
		2	18.1	2	6.6	RACINE WI		-64			78
1	10.6	1	9.6	1	6.5	ROCK WI	-10	-32	-39	56	78
1	47.2					VILAS WI					
		1*	37.2			WALWORTH WI					
		1*	94.4			WAUKESHA WI					
				1*	29.8	WINNEBAGO WI					97
5*	30.8			2	7.0	WYOMING WY			-77		
2	31.2			1	11.6	FREMONT WY			-63	31	90
1	38.3					LARAMIE WY					
1*	546.4					PARK-TETON#WY					
1*	60.7					SWEETWATER WY					
				1*	56.2	WASHAKIE WY					98

* SIGNIFICANTLY MORE DEATHS THAN EXPECTED

\# COUNTIES COMBINED; SEE TECHNICAL NOTES

WHITE MALE

CANCER OF THE LARGE INTESTINE

ICD 153,159.0

1950-1959 DEATHS	1950-1959 RATE	1960-1969 DEATHS	1960-1969 RATE	1970-1979 DEATHS	1970-1979 RATE	COUNTY AND STATE	PERCENT CHANGE 60S/50S	PERCENT CHANGE 70S/60S	PERCENT CHANGE 70S/50S	PERCENTILE %CHANGE 70S/50S	PERCENTILE RATES 1970S
107014	17.1	134848	18.3	169905	20.0	UNITED STATES	7	10	17		
839	11.0	1250	13.3	1726	14.8	ALABAMA AL	21	11	34		
3	6.6	8	16.5	8	11.6	AUTAUGA AL	151	-30	77	75	18
19	12.8	26	12.6	48	16.3	BALDWIN AL	-1	29	28	48	47
5	12.0	13	25.5	10	15.3	BARBOUR AL	113	-40	28	49	40
2	4.3	5	12.0	10	17.7	BIBB AL	177	48	310	96	56
11	11.1	22	18.8	20	13.3	BLOUNT AL	70	-29	20	42	27
		2	8.6	2	10.2	BULLOCK AL		19			13
4	6.2	4	5.7	11	13.6	BUTLER AL	-8	140	121	85	29
21	9.9	34	13.5	46	13.6	CALHOUN AL	36	0	36	54	29
7	6.8	18	16.4	12	9.9	CHAMBERS AL	141	-40	45	61	12
5	9.5	5	7.1	11	12.8	CHEROKEE AL	-25	80	35	53	24
6	6.2	17	15.5	14	11.3	CHILTON AL	152	-27	83	77	17
5	14.8	2	4.9	9	21.7	CHOCTAW AL	-67	339	47	61	80
7	11.1	8	11.2	16	19.9	CLARKE AL	1	78	80	76	71
4	6.5	5	7.7	5	6.9	CLAY AL	18	-9	6	30	06
4	8.7	4	8.4	3	5.2	CLEBURNE AL	-3	-38	-40	06	04
7	7.3	8	9.0	14	11.1	COFFEE AL	23	23	52	65	16
8	8.9	19	12.5	23	11.9	COLBERT AL	40	-4	33	52	19
6	15.5	4	6.9	13	23.7	CONECUH AL	-56	244	52	65	88
4	8.7	4	14.3	8	19.4	COOSA AL	64	35	122	85	67
9	7.6	20	13.2	21	12.5	COVINGTON AL	75	-5	66	71	23
5	10.7	5	9.0	8	12.0	CRENSHAW AL	-16	33	12	35	19
15	9.3	16	7.8	40	14.5	CULLMAN AL	-17	87	56	66	34
6	8.6	11	12.1	14	12.4	DALE AL	41	3	45	60	22
9	11.3	11	13.4	18	16.0	DALLAS AL	18	19	41	58	45
12	7.5	29	14.6	29	12.0	DE KALB AL	93	-17	60	68	20
10	10.7	12	10.7	16	10.8	ELMORE AL	0	2	1	25	15
10	15.6	12	12.2	10	9.0	ESCAMBIA AL	-22	-26	-42	05	10
20	7.7	35	10.2	52	12.4	ETOWAH AL	33	22	62	69	22
8	12.0	8	10.9	13	16.1	FAYETTE AL	-8	47	35	53	45
16	16.3	12	11.6	14	10.9	FRANKLIN AL	-29	-6	-33	08	15
7	8.2	10	11.0	14	13.2	GENEVA AL	35	19	61	69	26
1	9.4	3	21.0	1	8.6	GREENE AL	125	-59	-8	19	09
3	10.1	7	21.4	5	14.5	HALE AL	112	-32	44	59	35
3	7.2	3	9.5	2	4.3	HENRY AL	32	-54	-39	06	03
11	9.2	26	16.8	27	12.9	HOUSTON AL	83	-23	40	57	25
13	9.3	13	8.3	28	15.8	JACKSON AL	-10	91	71	73	44
192	14.7	273	16.5	339	18.1	JEFFERSON AL	12	9	23	44	59
6	9.7	12	18.0	14	18.0	LAMAR AL	86	0	86	78	58
18	11.5	21	10.5	34	12.3	LAUDERDALE AL	-8	17	7	30	21
9	13.5	8	10.6	14	14.6	LAWRENCE AL	-21	38	9	32	35
8	12.0	23	24.4	18	13.3	LEE AL	104	-46	11	33	27
7	8.1	15	12.5	25	16.6	LIMESTONE AL	54	32	105	82	49
1	5.0	1	6.0	2	11.9	LOWNDES AL	20	98	138	88	19
4	16.5	3	9.9	5	14.6	MACON AL	-40	48	-11	17	35
20	8.9	42	14.7	84	18.2	MADISON AL	65	24	104	82	60
3	6.8	5	10.4	3	5.9	MARENGO AL	53	-43	-12	16	04
5	5.0	10	9.7	16	11.7	MARION AL	93	20	133	87	19
8	4.7	35	16.3	34	12.6	MARSHALL AL	245	-23	166	91	23
68	13.3	104	15.0	176	20.0	MOBILE AL	12	34	50	64	71
1	2.2	11	19.9	10	14.5	MONROE AL	816	-27	570	98	35
36	14.4	52	16.2	73	18.5	MONTGOMERY AL	12	15	29	49	62
23	12.2	29	13.6	36	12.7	MORGAN AL	12	-7	4	28	23
10*	37.3			9	25.7	PERRY AL			-31	09	93
3	4.6	7	10.3	10	13.6	PICKENS AL	125	32	196	93	29
3	4.8	6	9.8	10	11.2	PIKE AL	102	15	132	87	17
6	7.4	10	11.9	7	7.4	RANDOLPH AL	61	-38	0	24	06
10	14.9	9	10.7	8	6.1	RUSSELL AL	-28	-43	-59	02	04
7	7.4	10	10.2	12	9.3	ST CLAIR AL	38	-9	25	46	11
12	12.9	10	9.0	23	14.3	SHELBY AL	-31	59	11	33	33
11*	51.1	3	14.5	4	13.7	SUMTER AL	-72	-5	-73	00	30
13	11.2	21	11.9	29	13.0	TALLADEGA AL	7	9	16	39	25
14	13.2	13	11.5	36	24.8	TALLAPOOSA AL	-13	115	87	78	91
23	8.2	38	12.0	50	13.3	TUSCALOOSA AL	47	11	63	70	27
20	10.0	23	9.8	28	9.5	WALKER AL	-2	-3	-5	21	11
4	10.4	5	10.7	6	11.2	WASHINGTON AL	3	5	8	32	17
4	23.9	2	8.3	8	36.5	WILCOX AL	-65	341	53	65	99
4	6.5	8	10.8	8	8.5	WINSTON AL	65	-21	30	50	09
		64	19.1	74	13.5	ALASKA AK		-29			
364	11.3	708	12.9	1445	15.7	ARIZONA AZ	14	22	39		
1	5.8	1	2.7	5	14.0	APACHE AZ	-54	426	141	88	31
25	18.2	14	7.1	45	15.3	COCHISE AZ	-61	117	-16	15	40
6	10.0	6	6.6	19	16.0	COCONINO AZ	-34	142	60	69	45
14	16.6	5	4.9	22	14.8	GILA AZ	-70	203	-11	17	36
3	5.9	8	14.2	13	17.3	GRAHAM AZ	139	22	191	92	54
4	12.2	5	16.6	2	5.5	GREENLEE AZ	36	-67	-55	02	04
188	11.2	426	14.0	864	16.6	MARICOPA AZ	25	19	48	62	49
1	2.1	5	6.5	37	16.7	MOHAVE AZ	204	157	681	99	50
2	6.1	5	8.0	15	17.5	NAVAJO AZ	32	118	188	92	55
74	10.8	166	15.0	273	14.5	PIMA AZ	39	-3	35	53	35
10	9.6	18	9.1	45	15.7	PINAL AZ	-5	72	63	70	43
6	17.6	4	10.0	8	15.1	SANTA CPUZ AZ	-43	50	-15	15	38
14	9.4	26	11.6	65	16.2	YAVAPAI AZ	23	40	73	74	46
16	15.7	19	9.4	32	9.8	YUMA AZ	-40	4	-38	06	12
678	10.3	947	12.1	1428	15.2	ARKANSAS AR	17	26	47		
7	9.7	5	5.1	16	15.6	ARKANSAS AR	-47	203	61	69	42
5	12.0	6	8.0	11	13.3	ASHLEY AR	-34	68	11	34	27
8	12.7	13	9.6	34	15.0	BAXTER AR	-24	56	19	41	38
23	9.9	44	16.1	52	13.1	BENTON AR	62	-18	32	52	26
5	4.8	9	7.7	25	17.4	BOONE AR	59	126	260	95	54

* SIGNIFICANTLY MORE DEATHS THAN EXPECTED

WHITE MALE | CANCER OF THE LARGE INTESTINE | ICD 153,159.0

1950-1959 DEATHS	1950-1959 RATE	1960-1969 DEATHS	1960-1969 RATE	1970-1979 DEATHS	1970-1979 RATE	COUNTY AND STATE		PERCENT CHANGE 60S/50S	PERCENT CHANGE 70S/60S	PERCENT CHANGE 70S/50S	PERCENTILE %CHANGE 70S/50S	PERCENTILE RATES 1970S
8	16.2	10	16.9	13	21.3	BRADLEY	AR	4	26	31	51	78
2	8.1	2	8.6	3	10.7	CALHOUN	AR	6	24	32	52	14
5	5.6	14	16.1	15	14.0	CARROLL	AR	189	-13	152	89	31
11	36.0	2	4.5	7	16.3	CHICOT	AR	-87	260	-55	03	47
9	11.7	8	9.7	12	12.5	CLARK	AR	-17	29	6	30	22
9	7.6	10	7.9	20	15.1	CLAY	AR	3	91	98	81	38
8	15.5	3	4.0	16	16.3	CLEBURNE	AR	-74	310	5	28	47
1	3.2	5	15.3	4	10.3	CLEVELAND	AR	376	-33	221	94	13
6	7.4	12	12.6	14	12.7	COLUMBIA	AR	71	1	73	74	24
9	14.3	12	16.7	11	13.4	CONWAY	AR	17	-20	-7	20	28
17	9.2	22	10.6	35	15.4	CRAIGHEAD	AR	16	44	68	72	40
7	5.7	15	11.2	17	10.2	CRAWFORD	AR	96	-8	79	76	13
5	9.6	8	9.9	17	18.8	CRITTENDEN	AR	3	90	97	81	64
8	13.0	3	4.6	8	10.5	CROSS	AR	-65	131	-19	13	14
2	4.5	2	3.9	10	23.6	DALLAS	AR	-13	503	422	98	88
4	9.1	3	5.3	9	15.4	DESHA	AR	-41	189	69	72	40
4	10.2	4	7.1	8	11.5	DREW	AR	-30	61	12	35	18
16	17.1	16	13.6	26	16.6	FAULKNER	AR	-20	22	-3	23	49
5	8.0	10	12.9	11	12.5	FRANKLIN	AR	61	-3	56	66	22
1	1.7	6	10.7	5	7.7	FULTON	AR	546	-28	362	97	07
37	14.6	52	14.3	91	20.3	GARLAND	AR	-2	42	40	57	73
1	2.4	8	15.8	13	22.0	GRANT	AR	547	39	802	99	82
14	11.6	14	10.7	19	11.8	GREENE	AR	-7	10	2	25	19
9	11.7	15	16.5	22	22.5	HEMPSTEAD	AR	41	36	92	80	83
5	6.5	9	9.2	22	18.8	HOT SPRING	AR	43	103	191	92	64
4	6.6	7	12.4	9	13.0	HOWARD	AR	87	4	96	80	25
7	5.8	20	15.8	19	13.2	INDEPENDENCE	AR	171	-17	126	86	27
3	6.5	7	12.0	15	17.5	IZARD	AR	86	45	170	91	55
16	23.4	4	4.5	22	21.5	JACKSON	AR	-81	382	-8	19	79
20	12.7	31	17.3	38	17.4	JEFFERSON	AR	37	0	37	55	54
5	6.2	12	14.6	19	18.0	JOHNSON	AR	135	23	190	92	59
3	7.7	2	5.0	6	12.8	LAFAYETTE	AR	-36	158	66	71	24
6	6.4	8	8.5	15	12.9	LAWRENCE	AR	34	52	103	82	25
3	7.6	7	18.5	7	17.7	LEE	AR	145	-4	134	87	56
8	25.6	6	18.2	7	17.2	LINCOLN	AR	-29	-5	-33	08	53
4	18.1	5	12.4	7	14.2	LITTLE RIVER	AR	-32	15	-21	12	33
11	9.7	17	14.6	23	20.1	LOGAN	AR	50	38	106	83	71
7	7.6	14	13.4	23	18.6	LONOKE	AR	76	39	145	89	62
3	4.6	3	4.8	8	10.3	MADISON	AR	4	115	125	86	13
3	6.3	10	19.6	11	14.5	MARION	AR	210	-26	129	86	35
9	8.0	6	4.6	21	14.6	MILLER	AR	-43	220	83	77	35
16	9.3	27	15.3	30	16.3	MISSISSIPPI	AR	65	7	76	75	47
3	5.7	3	6.4	8	15.7	MONROE	AR	12	145	176	91	43
3	8.9	6	17.5	4	7.8	MONTGOMERY	AR	96	-55	-12	16	07
9	21.4	13	23.0	3	5.0	NEVADA	AR	8	-78	-77	00	03
1	2.1	5	13.6	9	21.1	NEWTON	AR	554	55	914	99	77
11	15.4	13	13.2	16	14.4	OUACHITA	AR	-14	9	-6	20	34
		3	11.3	6	14.8	PERRY	AR		32			36
10	17.1	7	10.7	13	15.6	PHILLIPS	AR	-37	45	-9	18	42
5	10.2	9	15.8	6	9.8	PIKE	AR	56	-38	-3	22	12
8	7.5	11	9.7	22	17.6	POINSETT	AR	29	82	135	87	56
9	11.3	16	15.2	21	20.8	POLK	AR	35	37	85	78	75
12	11.5	11	8.7	14	9.3	POPE	AR	-24	7	-19	13	10
6	12.2	4	10.4	7	13.3	PRAIRIE	AR	-14	28	9	33	27
73	11.9	130	16.3	167	17.9	PULASKI	AR	37	10	51	64	58
8	10.7	7	8.6	12	12.0	RANDOLPH	AR	-20	40	12	35	20
9	18.0	7	11.2	6	9.6	ST FRANCIS	AR	-38	-14	-47	04	11
6	5.1	15	10.3	18	9.7	SALINE	AR	102	-6	89	79	12
3	7.1	3	5.6	10	15.5	SCOTT	AR	-22	176	116	85	41
6	12.3	7	14.4	5	8.0	SEARCY	AR	17	-44	-35	07	08
28	10.7	43	13.7	54	14.1	SEBASTIAN	AR	28	3	32	51	32
8	13.4	8	12.4	8	10.3	SEVIER	AR	-7	-17	-23	11	14
3	6.5	6	10.2	19	18.4	SHARP	AR	57	81	185	92	61
4	13.4	1	3.1	3	5.4	STONE	AR	-77	71	-60	02	04
19	14.0	18	10.5	38	18.9	UNION	AR	-25	80	35	53	64
4	8.0	8	14.5	11	16.1	VAN BUREN	AR	82	11	102	82	46
28	11.1	30	9.7	51	13.5	WASHINGTON	AR	-13	39	21	42	28
12	7.1	18	8.8	30	13.0	WHITE	AR	24	48	83	77	25
6	12.4	6	13.2	6	11.8	WOODRUFF	AR	7	-11	-4	21	19
5	6.7	11	13.1	15	15.1	YELL	AR	94	16	125	86	38
7470	15.0	10676	16.3	14891	18.8	CALIFORNIA	CA	9	16	26		
517	17.0	655	19.0	750	21.2	ALAMEDA	CA	12	11	25	46	77
		1	112.4			ALPINE	CA					
13	23.3	16	23.5	12	12.3	AMADOR	CA	1	-48	-47	04	21
59	14.5	77	13.9	124	16.4	BUTTE	CA	-4	18	13	36	48
4	5.1	17	20.1	19	16.7	CALAVERAS	CA	291	-17	225	94	50
13	22.6	13	18.6	12	18.2	COLUSA	CA	-18	-2	-20	12	60
143	14.8	244	17.0	407	20.2	CONTRA COSTA	CA	15	19	37	54	72
5	9.9	12	19.4	16	20.2	DEL NORTE	CA	96	4	104	82	72
25	24.1	26	17.6	59	20.7	EL DORADO	CA	-27	18	-14	15	75
153	13.8	203	14.4	289	17.1	FRESNO	CA	4	18	23	44	52
13	16.1	14	14.6	21	20.0	GLENN	CA	-9	37	24	45	71
44	12.7	73	19.3	75	17.1	HUMBOLDT	CA	52	-11	35	53	53
12	5.7	38	13.1	37	13.7	IMPERIAL	CA	128	5	139	88	30
5	9.6	9	11.8	15	13.8	INYO	CA	22	17	43	59	30
80	10.3	136	13.0	213	16.3	KERN	CA	26	25	57	67	47
22	13.3	30	14.6	29	13.9	KINGS	CA	10	-5	5	29	31
9	7.4	26	16.3	46	18.4	LAKE	CA	119	13	147	89	61
7	9.3	22*	33.2	10	13.1	LASSEN	CA	257	-60	41	58	26
3058	15.9	3906	16.6	4766	19.4	LOS ANGELES	CA	4	17	22	43	67
18	13.2	15	8.3	43	19.3	MADERA	CA	-37	132	46	61	67
61	15.6	97	16.5	162	20.2	MARIN	CA	6	22	29	50	72
7	16.2	5	12.1	9	16.8	MARIPOSA	CA	-25	38	3	27	50
26	11.3	30	11.3	46	15.2	MENDOCINO	CA	0	34	34	53	39
27	10.7	56	17.3	80	20.2	MERCED	CA	62	16	88	79	72
6	14.8	8	17.4	7	15.3	MODOC	CA	18	-12	3	27	39
		2	16.9	3	13.0	MONO	CA		-23			25
51	10.2	98	14.4	169	20.2	MONTEREY	CA	42	40	98	81	72
44	12.3	60	12.8	91	16.9	NAPA	CA	5	31	37	55	51

* SIGNIFICANTLY MORE DEATHS THAN EXPECTED

WHITE MALE — CANCER OF THE LARGE INTESTINE — ICD 153,159.0

1950-1959 DEATHS	1950-1959 RATE	1960-1969 DEATHS	1960-1969 RATE	1970-1979 DEATHS	1970-1979 RATE	COUNTY AND STATE	PERCENT CHANGE 60S/50S	PERCENT CHANGE 70S/60S	PERCENT CHANGE 70S/50S	PERCENTILE %CHANGE 70S/50S	PERCENTILE RATES 1970S
13	10.6	29	17.9	38	15.3	NEVADA CA	69	-14	45	60	40
191	13.0	461	15.2	1018	19.4	ORANGE CA	16	28	49	63	67
34	13.1	44	13.2	72	15.5	PLACER CA	1	17	18	40	41
2	3.7	12	17.6	14	20.8	PLUMAS CA	379	18	465	98	75
113	10.7	274	14.3	463	15.4	RIVERSIDE CA	34	7	43	59	40
182	13.4	299	16.8	470	20.1	SACRAMENTO CA	25	20	50	64	72
13	17.9	19	24.1	12	14.2	SAN BENITO CA	34	-41	-21	12	33
189	12.1	361	15.4	498	16.3	SAN BERNARDINO CA	28	5	35	53	47
370	13.9	624	15.3	1083	17.9	SAN DIEGO CA	10	17	29	49	58
715*	19.9	767*	21.3	758*	24.3	SAN FRANCISCO CA	7	14	22	43	90
132	13.8	155	13.3	237	18.0	SAN JOAQUIN CA	-3	35	31	51	58
40	12.1	72	14.6	101	14.6	SAN LUIS OBISPO CA	20	0	20	42	35
170	15.6	315	19.0	433	20.9	SAN MATEO CA	22	10	34	53	76
86	17.7	128	15.7	220	20.2	SANTA BARBARA CA	-11	29	15	37	72
258	16.3	396	16.5	652	20.0	SANTA CLARA CA	2	21	23	44	71
65	12.4	114	15.8	148	17.5	SANTA CRUZ CA	27	11	41	57	55
29	17.8	28	9.8	68	16.0	SHASTA CA	-45	63	-10	18	45
		3	20.2	7	38.3	SIERRA CA		90			99
33	20.5	21	12.4	29	14.6	SISKIYOU CA	-40	18	-29	09	35
48	12.9	70	16.1	105	18.1	SOLANO CA	24	12	40	57	59
107	15.1	162	17.2	241	17.9	SONOMA CA	14	4	19	41	58
75	11.0	93	11.5	164	16.4	STANISLAUS CA	4	42	48	62	47
17	13.6	20	12.9	33	17.5	SUTTER CA	-5	35	29	49	55
16	13.1	35	24.9	37	19.2	TEHAMA CA	91	-23	47	62	66
1	3.0			4	7.8	TRINITY CA			162	90	07
67	10.7	87	11.3	114	13.2	TULARE CA	5	17	23	44	26
2	2.6	22	19.3	26	16.9	TUOLUMNE CA	640	-13	547	98	51
41	8.6	126	14.0	240	17.2	VENTURA CA	63	23	100	81	53
24	10.9	34	12.2	64	18.4	YOLO CA	12	50	68	72	61
15	12.5	16	12.5	32	19.9	YUBA CA	-1	59	58	68	70
812	12.9	1017	13.3	1451	16.0	COLORADO CO	4	20	25		
20	12.6	34	10.9	79	16.7	ADAMS CO	-14	54	33	52	50
2	4.7	8	17.9	6	13.7	ALAMOSA CO	281	-23	192	93	30
				1	12.4	ARCHULETA CO					21
7	26.8	4	20.6	4	11.3	BACA CO	-23	-45	-58	02	17
1	1.8	1	4.1	3	7.1	BENT CO	132	73	301	96	06
22	8.8	42	12.8	75	16.2	BOULDER CO	45	26	83	77	46
4	8.8	4	7.6	12	20.8	CHAFFEE CO	-14	174	135	87	75
1	6.8					CHEYENNE CO					
3	20.8	4	19.1	1	4.4	CLEAR CREEK CO	-8	-77	-79	00	03
1	2.2	3	8.2	1	2.5	CONEJOS CO	279	-69	17	40	02
1	5.1	4	24.5	1	4.9	COSTILLA CO	380	-80	-4	21	03
4	14.1	4	15.8	4	16.0	CROWLEY CO	12	2	14	36	45
1	11.6	1	12.6	1	14.5	CUSTER CO	8	15	25	45	35
9	9.6	14	12.3	17	13.0	DELTA CO	29	6	36	54	26
348	16.5	393	15.8	498	18.6	DENVER-ARAPAHOE# CO	-4	17	13	36	62
		1	13.8			DOLORES CO					
2	10.7	4	14.2	8	15.3	DOUGLAS CO	32	8	43	59	40
2	8.1			3	11.8	EAGLE CO			46	61	19
1	4.5	3	13.0	1	4.2	ELBERT CO	188	-68	-7	19	03
52	13.1	65	13.2	123	17.0	EL PASO CO	1	29	30	50	52
16	12.7	17	10.2	17	10.1	FREMONT CO	-20	-1	-20	12	13
12	19.2	6	8.6	9	10.4	GARFIELD CO	-55	21	-46	04	14
		1	27.0	1	22.5	GILPIN CO		-17			84
4	19.1			2	25.5	GRAND CO			33	52	92
		5	25.1	1	4.1	GUNNISON CO		-83			02
				1*	98.2	HINSDALE CO					99
11	18.9	7	11.9	9	19.8	HUERFANO CO	-37	67	5	28	70
		3	43.8			JACKSON CO					
40	15.1	78	14.6	159	19.1	JEFFERSON CO	-4	31	26	47	65
		1	10.4	2	13.3	KIOWA CO		28			27
4	10.5	11	26.8	7	15.0	KIT CARSON CO	156	-44	43	59	38
1	4.3	2	17.4	2	27.3	LAKE CO	307	57	540	98	95
7	9.9	4	4.9	13	14.7	LA PLATA CO	-50	199	49	63	36
24	9.6	45	13.6	65	15.2	LARIMER CO	43	11	59	68	39
10	8.1	12	10.7	20	18.5	LAS ANIMAS CO	31	74	128	86	62
4	12.2	5	16.1	4	12.7	LINCOLN CO	32	-21	4	28	24
9	11.6	7	8.2	15	15.8	LOGAN CO	-29	93	37	54	43
15	7.2	29	10.6	53	16.1	MESA CO	48	51	125	86	45
4	17.3	1	2.8	4	12.2	MOFFAT CO	-84	337	-29	09	20
4	9.2	9	17.5	9	15.4	MONTEZUMA CO	91	-12	68	72	40
11	14.0	9	10.3	10	9.6	MONTROSE CO	-26	-7	-32	08	11
13	14.3	11	10.9	12	10.7	MORGAN CO	-24	-2	-25	10	15
7	6.0	14	11.5	12	9.9	OTERO CO	92	-14	65	70	12
4	42.0	1	10.6			OURAY CO	-75				
		2	22.7	4	50.5	PARK CO		122			99
4	15.6	3	11.5	2	4.6	PHILLIPS CO	-26	-60	-70	00	03
2	19.8	2	37.0	1	3.0	PITKIN CO	87	-92	-85	00	02
5	7.9	7	11.2	8	10.9	PROWERS CO	42	-3	38	56	15
46	9.7	52	10.1	84	15.7	PUEBLO CO	4	55	62	69	43
3	14.0	2	14.8	6	35.6	RIO BLANCO CO	6	141	154	69	98
7	14.8	9	17.6	8	13.3	RIO GRANDE CO	19	-25	-10	17	27
6	14.8	5	14.8	3	7.1	ROUTT CO	0	-52	-52	03	06
		1	4.4			SAGUACHE CO					
1	16.3	1	22.8			SAN JUAN CO	40				
3	38.2			4*	60.6	SAN MIGUEL CO			59	68	99
7	31.3	2	9.3	1	4.8	SEDGWICK CO	-70	-49	-85	00	03
1	22.5					SUMMIT CO					
2	14.8	4	31.1	5	22.2	TELLER CO	110	-29	50	64	83
6	17.1	2	4.8	9	25.2	WASHINGTON CO	-72	421	47	62	92
32	10.6	47	14.1	42	10.2	WELD CO	33	-27	-3	23	13
6	10.4	11	17.1	9	14.4	YUMA CO	64	-16	38	56	34
2027*	21.4	2518*	22.4	3086*	23.7	CONNECTICUT CT	5	6	11		
544*	23.2	657*	22.8	806*	24.3	FAIRFIELD CT	-2	7	5	28	90
504*	21.0	646*	22.3	780*	22.6	HARTFORD CT	6	2	8	31	84
89	17.2	132	21.0	164	22.4	LITCHFIELD CT	22	7	30	50	83
62	16.4	86	18.9	113	20.7	MIDDLESEX CT	15	10	26	47	75

* SIGNIFICANTLY MORE DEATHS THAN EXPECTED

COUNTIES COMBINED; SEE TECHNICAL NOTES

WHITE MALE — CANCER OF THE LARGE INTESTINE — ICD 153,159.0

1950-1959 DEATHS	1950-1959 RATE	1960-1969 DEATHS	1960-1969 RATE	1970-1979 DEATHS	1970-1979 RATE	COUNTY AND STATE	PERCENT CHANGE 60S/50S	PERCENT CHANGE 70S/60S	PERCENT CHANGE 70S/50S	PERCENTILE %CHANGE 70S/50S	PERCENTILE RATES 1970S
625*	24.3	701*	23.7	853*	25.3	NEW HAVEN CT	-2	7	4	28	92
116	16.4	175*	22.1	212*	23.4	NEW LONDON CT	35	6	43	59	87
32	16.0	45	16.5	69	21.9	TOLLAND CT	3	33	37	55	81
55	18.0	76	22.6	89	22.7	WINDHAM CT	25	1	26	47	85
239	18.6	319	20.2	433*	22.6	DELAWARE DE	9	12	22		
27	14.4	30	14.1	36	13.6	KENT DE	-2	-3	-6	21	29
186*	21.9	244*	23.0	336*	26.5	NEW CASTLE DE	5	15	21	42	94
26	10.2	45	14.9	61	15.9	SUSSEX DE	46	7	56	67	44
463*	23.8	330	20.0	274*	23.4	DIST. OF COLUMBIA DC	-16	17	-2		
463*	23.8	330	20.0	274*	23.4	WASHINGTON DC	-16	17	-2	23	87
2171	13.6	4730	16.0	9272	19.0	FLORIDA FL	17	19	39		
13	9.3	24	12.1	52	18.1	ALACHUA FL	30	50	95	80	59
		5	20.4	9	21.5	BAKER FL		6			79
12	10.7	19	10.0	58	19.9	BAY FL	-6	99	87	78	70
3	10.7	2	4.7	11	16.0	BRADFORD FL	-56	240	49	63	45
20	11.0	60	12.5	203	19.5	BREVARD FL	14	56	78	75	68
128	13.5	511	16.9	1292	20.6	BROWARD FL	25	22	53	65	74
3	12.5	4	11.0	5	13.1	CALHOUN FL	-12	20	5	29	26
5	7.8	46	16.5	92	15.8	CHARLOTTE FL	110	-4	101	81	43
3	6.5	10	7.0	52	15.3	CITRUS FL	8	119	136	87	40
2	2.1	6	7.4	36	22.3	CLAY FL	246	201	942	99	83
2	6.2	15	11.0	79	17.8	COLLIER FL	77	62	188	92	57
5	9.5	10	14.9	15	14.4	COLUMBIA FL	57	-4	52	65	34
526	18.1	1004	18.5	1698*	21.7	DADE FL	2	17	20	42	80
2	4.6	6	9.1	14	15.6	DE SOTO FL	98	72	240	95	42
3	19.6	1	4.6	4	14.2	DIXIE FL	-76	206	-27	10	33
110	12.4	217	18.0	263	17.2	DUVAL FL	45	-4	38	56	53
23	7.3	48	11.3	80	13.3	ESCAMBIA FL	55	18	82	77	27
3	35.3	5	21.3	9	18.8	FLAGLER FL	-40	-12	-47	04	63
3	11.3	7	24.5	8	23.1	FRANKLIN FL	117	-6	105	82	86
8	9.0	8	8.4	12	12.5	GADSDEN FL	-6	48	39	56	22
1	11.8	1	5.9	4	15.4	GILCHRIST FL	-50	161	30	51	40
		1	7.1	3	16.0	GLADES FL		126			45
3	14.1	1	3.7	7	21.6	GULF FL	-74	489	53	65	79
2	17.0	1	3.5	6	23.4	HAMILTON FL	-79	567	37	55	87
4	7.2	10	18.1	8	11.4	HARDEE FL	149	-37	58	67	18
1	4.9	4	15.0	6	15.2	HENDRY FL	205	1	210	93	39
2	4.1	13	16.3	47	18.8	HERNANDO FL	298	15	358	97	64
13	14.2	23	13.2	62	17.4	HIGHLANDS FL	-7	32	23	44	55
146	11.5	272	15.2	387	16.4	HILLSBOROUGH FL	33	8	43	59	48
3	5.3	5	9.2	13	17.6	HOLMES FL	73	90	229	94	56
16	22.7	37	19.2	64	19.4	INDIAN RIVER FL	-15	1	-15	15	67
11	13.0	9	7.6	17	12.8	JACKSON FL	-41	67	-2	23	24
4	19.5	1	4.3	5	19.4	JEFFERSON FL	-78	349	0	24	67
1	6.4	1	7.9	1	3.8	LAFAYETTE FL	24	-52	-40	06	02
32	11.1	64	14.4	123	16.7	LAKE FL	30	16	50	64	50
21	12.1	71	13.0	215	17.3	LEE FL	8	32	43	59	54
8	9.9	14	8.6	43	18.7	LEON FL	-13	118	89	79	63
2	6.1	6	11.7	12	12.9	LEVY FL	93	10	112	84	24
2	18.9	3	19.2	1	6.1	LIBERTY FL	1	-68	-68	01	04
2	5.6	7	17.4	9	18.1	MADISON FL	210	4	221	94	59
46	12.6	126	15.0	233	19.2	MANATEE FL	19	27	52	65	66
18	11.5	27	10.4	86	15.9	MARION FL	-9	54	39	56	45
5	6.8	32	17.7	75	19.5	MARTIN FL	159	10	185	92	68
26*	26.7	30	18.7	48	15.1	MONROE FL	-30	-19	-43	05	38
3	8.9	10	25.4	12	13.7	NASSAU FL	184	-46	53	65	30
2	3.3	16	12.0	33	13.1	OKALOOSA FL	263	9	297	96	26
2	14.5	1	2.2	13	16.1	OKEECHOBEE FL	-85	643	11	34	46
120	16.5	177	15.6	302	19.3	ORANGE FL	-5	23	16	39	67
9	6.1	26	13.6	51	18.9	OSCEOLA FL	123	39	211	93	64
117	14.2	257	15.4	675	18.8	PALM BEACH FL	9	22	33	52	64
31	14.9	83	13.5	302	20.3	PASCO FL	-9	50	36	54	73
362	16.0	771	16.9	1262	19.7	PINELLAS FL	6	17	23	44	69
62	13.6	123	13.1	217	16.0	POLK FL	-4	23	18	40	45
8	9.2	13	10.6	33	16.5	PUTNAM FL	15	55	78	76	48
20	20.1	27	20.9	34	17.9	ST JOHNS FL	4	-14	-11	17	58
19	15.6	40	16.9	68	17.3	ST LUCIE FL	8	2	11	34	54
4	6.8	10	11.4	16	11.4	SANTA ROSA FL	67	0	68	72	17
41	11.6	136	13.9	290	15.9	SARASOTA FL	20	15	38	55	44
9	5.8	34	16.2	66	13.1	SEMINOLE FL	177	-19	124	86	26
5	11.6	5	7.5	14	12.9	SUMTER FL	-35	71	11	34	25
6	13.7	9	15.2	19	22.7	SUWANNEE FL	11	49	65	71	84
2	5.7	7	16.2	17	28.5	TAYLOR FL	185	77	403	98	97
2	10.1	2	12.5	5	21.8	UNION FL	23	75	116	84	80
72	10.9	201	18.2	345	20.7	VOLUSIA FL	67	14	90	79	75
3	17.5	2	9.3	5	14.2	WAKULLA FL	-47	52	-19	13	32
5	7.4	6	8.1	12	11.1	WALTON FL	10	37	50	64	16
4	8.2	7	13.6	14	20.9	WASHINGTON FL	66	54	154	89	76
890	10.5	1396	13.0	1938	14.1	GEORGIA GA	24	9	35		
2	4.8	3	7.4	2	5.2	APPLING GA	53	-29	9	32	04
		5	32.7	1	4.8	ATKINSON GA		-85			03
		4	12.0	4	14.1	BACON GA		18			32
				2	18.9	BAKER GA					65
3	3.0	5	5.1	4	3.9	BALDWIN GA	70	-24	30	50	02
2	6.0	3	9.5	4	10.9	BANKS GA	59	15	82	77	15
1	2.0	6	11.6	13	17.7	BARROW GA	485	53	795	99	56
7	8.1	7	6.9	21	17.9	BARTOW GA	-15	159	119	85	58
9	22.9	7	16.3	7	15.0	BEN HILL GA	-29	-8	-35	07	37
5	15.4	6	14.3	9	16.4	BERRIEN GA	-7	15	7	30	48
29	10.8	44	14.2	51	14.7	BIBB GA	31	4	36	54	36
2	10.7	2	5.5	4	13.7	BLECKLEY GA	-49	149	27	48	29
1	5.5	3	14.9	2	8.1	BRANTLEY GA	173	-46	48	62	08
2	4.4	1	2.2	1	2.3	BROOKS GA	-49	4	-46	04	02

* SIGNIFICANTLY MORE DEATHS THAN EXPECTED

WHITE MALE — CANCER OF THE LARGE INTESTINE — ICD 153,159.0

1950-1959 DEATHS	1950-1959 RATE	1960-1969 DEATHS	1960-1969 RATE	1970-1979 DEATHS	1970-1979 RATE	COUNTY AND STATE	PERCENT CHANGE 60S/50S	PERCENT CHANGE 70S/60S	PERCENT CHANGE 70S/50S	PERCENTILE %CHANGE 70S/50S	PERCENTILE RATES 1970S
		5	31.1	3	16.8	BRYAN GA		-46			51
1	1.6	8	11.4	5	5.7	BULLOCH GA	613	-50	259	95	04
6	19.8	3	8.9	5	13.9	BURKE GA	-55	56	-30	09	31
3	14.2	6	20.2	8	22.3	BUTTS GA	42	10	56	67	83
				3	20.2	CALHOUN GA					73
1	14.9			4	13.8	CAMDEN GA			-7	19	30
2	13.2	1	3.9	5	16.1	CANDLER GA	-71	316	22	43	46
15	12.2	17	12.2	18	10.7	CARROLL GA	0	-12	-12	16	15
6	9.7	12	17.0	18	18.2	CATOOSA GA	76	7	88	79	59
1	8.6	1	7.5	5	24.7	CHARLTON GA	-13	227	186	92	91
46	15.2	64	16.3	78	17.5	CHATHAM GA	7	7	15	38	55
2	3.5	8	10.8	12	13.3	CHATTOOGA GA	212	23	285	96	27
11	13.1	10	9.9	22	15.6	CHEROKEE GA	-25	57	19	41	42
8	9.0	19	19.7	18	13.5	CLARKE GA	119	-31	50	63	28
		1	14.1			CLAY GA					
9	19.4	10	6.8	31	11.7	CLAYTON GA	-65	72	-40	06	18
1	8.9	2	12.4			CLINCH GA	39				
20	11.5	42	12.8	66	11.7	COBB GA	12	-9	2	25	19
6	9.9	5	8.4	10	12.0	COFFEE GA	-16	44	21	42	20
6	6.0	7	6.5	14	11.0	COLQUITT GA	8	70	83	77	16
2	8.1	4	16.8	8	14.1	COLUMBIA GA	107	-16	74	74	32
4	18.3	8	19.0	9	20.8	COOK GA	4	10	14	37	75
4	6.0	12	13.8	17	15.4	COWETA GA	130	12	157	90	41
2	13.9			1	4.5	CRAWFORD GA			-68	01	03
7	18.1	12	28.5	8	15.0	CRISP GA	58	-47	-17	14	38
2	10.8	5	18.5	12	29.8	DADE GA	71	61	175	91	97
2	12.7	2	12.0	3	16.5	DAWSON GA	-6	38	30	50	48
7	12.1	4	7.4	12	17.9	DECATUR GA	-39	143	48	62	58
64	12.4	110	15.0	174	16.7	DE KALB GA	21	12	35	53	50
3	6.3	3	9.2	9	16.6	DODGE GA	48	80	165	91	49
1	3.0	3	9.4	8	26.2	DOOLY GA	209	179	761	99	94
7	7.4	15	14.5	22	15.7	DOUGHERTY GA	96	8	112	84	43
1	1.7	7	10.3	14	13.8	DOUGLAS GA	491	34	694	99	30
2	6.9			2	4.6	EARLY GA			-33	08	03
		1	37.2			ECHOLS GA					
4	17.0	3	13.6	6	10.6	EFFINGHAM GA	-20	-22	-38	06	14
5	9.5	4	6.1	9	13.8	ELBERT GA	-36	126	45	61	30
2	2.9	3	6.8	7	8.9	EMANUEL GA	133	32	206	93	10
		3	32.9	1	4.5	EVANS GA		-86			03
3	5.4	4	6.1	10	13.2	FANNIN GA	14	115	145	88	27
3	12.0	3	9.7	8	13.7	FAYETTE GA	-19	41	14	37	29
21	10.5	32	13.6	44	16.1	FLOYD GA	30	18	54	66	46
4	12.5	6	13.8	17	19.8	FORSYTH GA	11	43	59	68	70
6	10.5	10	18.7	5	9.1	FRANKLIN GA	79	-52	-13	16	10
160	14.2	225	16.9	250	17.8	FULTON GA	19	5	25	46	57
6	20.5	5	11.6	7	13.6	GILMER GA	-43	17	-34	08	29
		1	22.5			GLASCOCK GA					
6	8.3	9	9.9	25	16.7	GLYNN GA	20	68	102	81	50
4	4.8	10	12.1	13	14.4	GORDON GA	154	19	203	93	34
6	13.8	4	7.0	7	10.5	GRADY GA	-49	50	-24	11	14
2	7.0	3	8.8	4	11.5	GREENE GA	25	31	63	70	18
15	13.0	26	15.7	33	11.4	GWINNETT GA	21	-27	-12	16	17
6	10.5	7	9.2	14	15.3	HABERSHAM GA	-12	66	46	61	39
7	5.1	23	13.1	26	11.3	HALL GA	158	-13	123	86	17
		3	18.1	4	29.3	HANCOCK GA		62			97
5	9.1	4	6.7	6	8.1	HARALSON GA	-26	21	-11	17	08
6	24.6	5	15.1	5	9.7	HARRIS GA	-39	-36	-61	01	12
2	5.0	5	9.4	11	14.0	HART GA	87	49	179	91	32
4	18.3	1	4.2	5	18.9	HEARD GA	-77	354	3	27	64
2	4.6	3	5.2	9	9.9	HENRY GA	14	89	115	84	12
3	9.5	8	12.6	23	17.9	HOUSTON GA	32	42	88	78	58
2	6.5	5	18.9	3	8.8	IRWIN GA	189	-53	35	54	09
5	7.4	5	7.3	9	10.7	JACKSON GA	-1	46	45	60	14
2	10.6	1	5.1	3	19.1	JASPER GA	-52	272	80	76	65
		3	11.1	1	3.9	JEFF DAVIS GA		-65			02
5	22.1	4	11.1	3	8.3	JEFFERSON GA	-50	-25	-63	01	08
1	6.7			2	7.2	JENKINS GA			7	31	06
2	7.8	3	11.8	3	9.9	JOHNSON GA	51	-16	26	47	12
1	5.5	3	16.3	5	12.0	JONES GA	197	-27	117	85	19
6	21.7	3	11.3	8	23.2	LAMAR GA	-48	106	7	31	86
1	18.6			2	13.1	LANIER GA			-29	09	26
5	6.7	13	15.6	15	12.3	LAURENS GA	134	-21	85	78	21
1	22.5	2	42.2	1	37.2	LEE GA	88	-12	65	71	99
1	7.4	3	24.6	5	21.3	LIBERTY GA	231	-13	186	92	78
5	37.3	1	6.6	2	9.9	LINCOLN GA	-82	49	-74	00	12
		1	8.5	4	34.5	LONG GA		307			98
6	6.6	17	14.5	19	13.7	LOWNDES GA	120	-5	108	83	30
2	11.7	4	15.8	11	26.7	LUMPKIN GA	35	69	129	86	95
4	15.3			4	10.9	MC DUFFIE GA			-28	09	16
		4	37.2	2	9.0	MC INTOSH GA		-76			10
2	7.5	4	15.7	4	12.3	MACON GA	108	-22	63	70	21
4	9.7	6	11.6	7	13.1	MADISON GA	19	14	35	54	26
1	8.9	2	15.1	2	16.8	MARION GA	70	11	88	79	50
5	16.0	6	12.0	6	10.6	MERIWETHER GA	-25	-11	-34	08	14
2	10.1	1	3.6	2	7.7	MILLER GA	-64	113	-24	11	07
5	12.4	6	14.2	9	18.2	MITCHELL GA	14	29	47	61	60
1	3.5	11*	41.2	7	18.7	MONPOE GA	1061	-55	428	98	63
1	5.9	4	20.2	3	13.7	MONTGOMERY GA	242	-32	133	87	30
6	23.4	5	16.3	7	21.5	MORGAN GA	-31	32	-8	19	79
7	19.7	7	19.4	5	8.6	MUPRAY GA	-1	-56	-57	02	09
24	12.1	55	18.1	51	13.8	MUSCOGEE#GA	50	-24	14	37	30
2	2.6	11	16.2	13	14.1	NEWTON GA	514	-13	434	98	32
1	5.6	2	8.2	2	6.4	OCONEE GA	46	-22	14	37	05
2	6.9			4	13.3	OGLETHORPE GA			94	80	27
5	12.2	9	14.5	9	13.0	PAULDING GA	19	-10	7	31	25
2	9.7	3	11.6	6	18.1	PEACH GA	20	56	87	78	59
6	18.7	2	4.5	3	6.2	PICKENS GA	-76	36	-67	01	04
1	3.1	3	9.7	5	12.6	PIERCE GA	216	30	310	96	23
4	18.0	3	11.9	8	25.9	PIKE GA	-34	117	43	59	93
9	10.1	19	18.9	15	12.0	POLK GA	87	-37	18	41	19
4	29.4	3	20.7	6	22.5	PULASKI GA	-29	9	-23	11	84

* SIGNIFICANTLY MORE DEATHS THAN EXPECTED

COUNTIES COMBINED; SEE TECHNICAL NOTES

WHITE MALE — CANCER OF THE LARGE INTESTINE — ICD 153,159.0

1950-1959 DEATHS	1950-1959 RATE	1960-1969 DEATHS	1960-1969 RATE	1970-1979 DEATHS	1970-1979 RATE	COUNTY AND STATE	PERCENT CHANGE 60S/50S	PERCENT CHANGE 70S/60S	PERCENT CHANGE 70S/50S	PERCENTILE %CHANGE 70S/50S	PERCENTILE RATES 1970S
2	15.6	2	16.5	6	26.3	PUTNAM GA	5	60	68	72	94
1	26.8			1	15.8	QUITMAN GA			-41	05	44
2	6.8	3	7.2	2	3.4	RABUN GA	6	-53	-50	03	02
		4	20.0			RANDOLPH GA					
25	11.6	40	16.0	44	12.3	RICHMOND GA	38	-23	6	30	21
4	17.6	2	5.3	7	7.7	ROCKDALE GA	-70	45	-56	02	07
3	40.2					SCHLEY GA					
		3	9.4	3	9.1	SCREVEN GA		-3			10
		2	9.8	3	10.3	SEMINOLE GA		6			14
3	4.8	14	14.9	15	10.9	SPALDING GA	212	-27	128	86	15
2	3.3	9	12.3	16	18.7	STEPHENS GA	270	52	462	98	63
4	34.4			1	9.4	STEWART GA			-73	00	11
7	13.2	9	15.4	13	19.7	SUMTER GA	16	29	49	63	69
2	19.4	2	12.5	2	14.9	TALBOT GA	-35	19	-23	11	37
		1	26.5			TALIAFERRO GA					
2	3.4	3	5.0	16	26.1	TATTNALL GA	44	425	658	99	94
1	3.8	2	8.0	1	3.7	TAYLOR GA	109	-53	-3	23	02
2	5.7	5	12.6	7	17.0	TELFAIR GA	121	35	197	93	52
2	8.1	3	15.6	4	17.0	TERRELL GA	92	9	109	83	52
10	10.1	10	9.9	14	13.0	THOMAS GA	-2	31	29	49	25
4	7.6	8	13.5	16	18.9	TIFT GA	77	39	147	89	64
5	10.8	3	6.3	5	7.9	TOOMBS GA	-42	26	-27	10	07
2	8.8	3	12.1	4	12.6	TOWNS GA	38	4	43	59	23
		1	4.9	2	10.1	TREUTLEN GA		107			13
11	9.6	13	10.5	34	21.6	TROUP GA	9	107	125	86	79
4	14.1	4	13.8	8	24.3	TURNER GA	-2	76	72	74	90
1	14.1			3	18.3	TWIGGS GA			31	51	61
1	3.7	1	3.1	6	12.4	UNION GA	-17	299	230	94	21
3	5.0	11	14.5	21	24.1	UPSON GA	190	67	383	97	89
8	7.3	23	13.7	30	14.1	WALKER GA	87	3	92	80	32
3	4.3	6	8.7	8	8.5	WALTON GA	106	-3	100	81	09
10	14.0	17	17.3	18	16.1	WARE GA	23	-7	15	38	46
2	11.6	2	14.4	1	4.6	WARREN GA	24	-68	-60	02	03
4	13.9	2	6.7	5	9.6	WASHINGTON GA	-52	43	-31	09	11
3	5.8	6	10.5	4	6.5	WAYNE GA	82	-38	13	35	05
		2	29.6			WEBSTER GA					
		2	9.8	1	4.4	WHEELER GA		-55			03
1	4.8	2	6.6	5	10.9	WHITE GA	39	65	130	86	16
4	3.6	13	8.8	24	11.2	WHITFIELD GA	143	28	211	93	17
1	4.1	2	6.5	3	10.4	WILCOX GA	58	59	152	89	14
6	27.0	7	22.2	5	17.8	WILKES GA	-18	-20	-34	07	57
1	4.3	1	3.9	2	5.8	WILKINSON GA	-9	51	37	55	04
4	12.3	6	16.5	4	8.3	WORTH GA	34	-49	-32	08	08
		66	12.2	131	15.1	HAWAII HI		24			
		6	10.7	16	14.4	HAWAII HI		34			34
		55	13.4	103	15.8	HONOLULU HI		17			43
		2	5.1	4	11.4	KAUAI HI		124			17
		3	8.4	8	11.7	MAUI#HI		39			19
317	12.3	390	12.9	501	14.0	IDAHO ID	5	9	14		
37	11.6	68	16.1	50	9.5	ADA ID	38	-41	-18	13	11
		1	5.3	2	13.1	ADAMS ID		149			26
18	15.1	19	11.1	32	16.2	BANNOCK ID	-27	46	7	31	46
3	9.4	1	2.7	3	8.6	BEAR LAKE ID	-71	218	-9	18	09
6	16.4	6	14.9	5	12.5	BENEWAH ID	-9	-16	-24	11	22
6	8.9	12	12.6	16	13.9	BINGHAM ID	41	11	56	66	31
1	2.9	3	13.1	6	27.5	BLAINE ID	352	109	847	99	96
1	12.4	1	12.2			BOISE ID	-2				
16	17.4	11	11.7	18	17.8	BONNER ID	-33	52	2	26	57
11	10.5	18	12.9	28	15.8	BONNEVILLE ID	23	22	50	64	43
4	12.9	4	14.5	2	5.3	BOUNDARY ID	13	-64	-59	02	04
		2	23.5	3	17.9	BUTTE ID		-24			58
1	37.2			1	17.7	CAMAS ID			-52	03	56
23	8.6	47	15.2	46	12.4	CANYON ID	77	-19	44	60	21
4	18.6	1	4.6	5	23.9	CARIBOU ID	-75	417	28	49	89
4	8.2	7	10.6	11	15.2	CASSIA ID	29	44	86	78	39
1	37.7	1	18.2	1	17.0	CLARK ID	-52	-7	-55	02	52
6	11.4	8	17.0	9	19.4	CLEARWATER ID	49	14	69	73	67
4	27.8			2	12.0	CUSTER ID			-57	02	20
3	11.5	8	22.4	12	26.9	ELMORE ID	95	20	134	87	95
4	11.2	4	11.4	10	21.9	FRANKLIN ID	2	93	96	80	81
3	9.1	8	20.9	6	16.3	FREMONT ID	128	-22	78	76	47
7	16.2	7	13.6	6	10.4	GEM ID	-16	-24	-36	07	14
5	11.8	6	9.8	5	7.9	GOODING ID	-18	-19	-33	08	08
16	26.4	9	15.0	13	18.7	IDAHO ID	-43	25	-29	09	63
1	3.1	2	4.6	4	6.6	JEFFERSON ID	48	45	114	84	05
8	19.1	4	11.0	8	13.3	JEROME ID	-42	20	-30	09	27
19	12.3	21	11.2	51	21.9	KOOTENAI ID	-9	95	77	75	81
13	14.9	13	14.2	14	13.2	LATAH ID	-5	-7	-11	17	26
8	25.8	5	15.4	6	15.4	LEMHI ID	-40	0	-40	06	40
2	8.3	2	9.1	3	13.0	LEWIS ID	10	44	57	67	26
3	16.3	2	11.3	5	24.9	LINCOLN ID	-30	119	53	65	91
3	10.8	3	12.4	5	11.5	MADISON ID	15	-8	6	30	18
7	15.5	8	14.7	8	12.8	MINIDOKA ID	-5	-13	-18	14	24
18	15.3	22	15.9	35	22.9	NEZ PERCE ID	4	44	50	63	85
2	10.6	2	9.4	3	13.2	ONEIDA ID	-11	41	25	46	27
1	3.4	4	12.0	2	5.3	OWYHEE ID	248	-56	55	66	04
9	13.2	8	10.6	6	7.6	PAYETTE ID	-19	-28	-42	05	07
3	17.7	2	9.1	3	15.5	POWER ID	-48	70	-12	16	41
12	19.9	9	11.1	13	13.9	SHOSHONE ID	-44	25	-30	09	31
		4	61.2	2	14.6	TETON ID		-76			35
19	10.1	24	11.3	32	12.7	TWIN FALLS ID	12	12	25	47	23
		1	5.4	3	11.1	VALLEY ID		106			16
5	10.0	2	4.2	6	11.6	WASHINGTON ID	-58	176	16	39	18
7816*	20.0	8771*	20.7	9768*	22.1	ILLINOIS IL	4	7	11		
69	17.9	84	21.2	100	24.7	ADAMS IL	18	16	37	55	91

* SIGNIFICANTLY MORE DEATHS THAN EXPECTED

COUNTIES COMBINED; SEE TECHNICAL NOTES

WHITE MALE — CANCER OF THE LARGE INTESTINE — ICD 153,159.0

1950-1959 DEATHS	1950-1959 RATE	1960-1969 DEATHS	1960-1969 RATE	1970-1979 DEATHS	1970-1979 RATE	COUNTY AND STATE		PERCENT CHANGE 60S/50S	PERCENT CHANGE 70S/60S	PERCENT CHANGE 70S/50S	PERCENTILE %CHANGE 70S/50S	PERCENTILE RATES 1970S
12	17.5	11	17.3	8	13.6	ALEXANDER	IL	-1	-22	-22	11	29
12	12.2	14	15.5	18	19.0	BOND	IL	27	23	56	67	65
23	24.7	17	17.2	25	22.8	BOONE	IL	-30	32	-8	19	85
6	12.8	3	5.9	13	30.2	BROWN	IL	-54	414	136	87	97
37	17.9	52	23.9	43	18.7	BUREAU	IL	33	-22	4	28	63
3	6.4	7	14.7	10	23.6	CALHOUN	IL	130	60	268	95	88
18	16.3	13	10.8	18	15.8	CARROLL	IL	-34	47	-3	22	43
21	24.0	21	23.6	19	21.5	CASS	IL	-2	-9	-10	18	79
48	14.4	74	19.5	103	23.9	CHAMPAIGN	IL	35	22	66	71	89
36	16.7	31	14.2	47	21.5	CHRISTIAN	IL	-15	51	29	49	79
17	14.4	19	16.7	28	25.2	CLARK	IL	15	52	75	75	92
13	13.1	7	6.9	21	20.0	CLAY	IL	-47	189	52	65	71
18	15.6	24	18.2	31	21.0	CLINTON	IL	17	15	35	53	76
30	14.2	37	17.2	50	22.9	COLES	IL	21	34	61	69	85
4184*	23.3	4405*	23.2	4542*	24.2	COOK	IL	0	4	4	27	90
27	21.7	30	23.5	28	21.6	CRAWFORD	IL	8	-8	-1	24	79
10	17.3	7	10.6	9	14.3	CUMBERLAND	IL	-39	35	-17	14	33
33	15.3	31	13.7	45	18.0	DE KALB	IL	-10	32	18	40	59
12	13.0	15	15.5	16	17.0	DE WITT	IL	19	9	30	51	52
13	13.8	18	19.8	13	13.4	DOUGLAS	IL	44	-32	-2	23	28
140	20.0	242*	22.1	372*	23.2	DU PAGE	IL	10	5	16	38	86
15	11.1	30	22.2	28	20.5	EDGAR	IL	100	-8	85	78	74
6	10.8	6	11.5	15	27.9	EDWARDS	IL	7	143	159	90	96
19	16.9	20	15.7	29	19.8	EFFINGHAM	IL	-7	27	18	40	70
14	9.6	24	18.0	20	14.1	FAYETTE	IL	87	-22	46	61	32
9	10.4	11	12.0	20	21.5	FORD	IL	16	79	108	83	79
46	16.8	60	21.1	61	21.8	FRANKLIN	IL	26	3	30	50	80
31	12.6	39	15.9	60	23.7	FULTON	IL	26	49	88	79	88
13	28.0	11	24.9	7	16.9	GALLATIN	IL	-11	-32	-40	06	51
15	13.3	17	13.9	30	25.7	GREENE	IL	5	84	93	80	93
16	16.4	26	23.0	24	18.3	GRUNDY	IL	41	-20	12	35	61
4	5.3	16	22.8	11	15.3	HAMILTON	IL	327	-33	188	92	40
33	20.3	25	17.2	29	19.7	HANCOCK	IL	-15	14	-3	22	69
5	18.4	2	6.4	10	30.5	HARDIN	IL	-65	374	66	71	98
3	6.3	3	6.4	10	19.6	HENDERSON	IL	2	207	212	94	69
54	20.1	50	18.1	54	18.6	HENRY	IL	-10	3	-7	19	62
29	16.3	36	19.9	38	19.8	IROQUOIS	IL	22	0	22	43	70
24	13.5	28	14.1	30	14.6	JACKSON	IL	5	3	8	32	35
9	12.5	7	12.0	10	14.0	JASPER	IL	-4	17	13	35	32
29	16.4	31	16.6	40	20.0	JEFFERSON	IL	1	20	22	43	71
16	20.5	15	17.0	15	15.7	JERSEY	IL	-17	-8	-23	11	43
28	22.8	20	16.5	25	20.4	JO DAVIESS	IL	-28	23	-11	17	73
2	3.4	9	18.0	11	19.3	JOHNSON	IL	437	7	475	98	67
157*	21.1	185	21.4	180	18.9	KANE	IL	1	-12	-10	17	64
63	15.2	84	20.1	86	21.5	KANKAKEE	IL	33	7	42	58	79
12	17.8	17	21.6	19	18.3	KENDALL	IL	21	-15	3	26	60
44	14.4	63	19.3	54	16.6	KNOX	IL	34	-14	16	38	49
158*	20.1	195	19.3	218	18.0	LAKE	IL	-4	-7	-11	17	59
101	20.2	124*	22.5	139	23.4	LA SALLE	IL	11	4	16	38	87
15	14.7	24	21.7	25	21.9	LAWRENCE	IL	47	1	49	62	81
29	17.0	45	24.6	37	20.8	LEE	IL	45	-15	23	44	75
32	15.5	56*	25.5	45	20.9	LIVINGSTON	IL	64	-18	35	53	76
31	20.4	37	22.4	26	15.4	LOGAN	IL	10	-31	-25	11	40
28	17.1	30	18.8	26	16.5	MC DONOUGH	IL	10	-12	-3	22	48
55	18.1	88	22.2	110	21.3	MC HENRY	IL	23	-4	18	40	78
66	17.4	61	15.1	80	19.1	MC LEAN	IL	-13	26	10	33	66
76	16.8	100	19.5	115	21.1	MACON	IL	16	8	26	47	77
54	17.6	56	18.2	61	21.1	MACOUPIN	IL	4	16	20	42	77
119	15.7	186	20.3	234	22.1	MADISON	IL	29	9	41	57	82
29	13.8	50	22.3	54	22.3	MARION	IL	62	0	62	69	83
17	22.6	15	19.3	8	9.4	MARSHALL	IL	-14	-51	-58	02	11
15	16.9	22	23.2	28	25.2	MASON	IL	38	9	49	63	92
10	14.5	12	15.7	17	20.3	MASSAC	IL	8	29	40	57	73
9	15.3	9	14.1	12	19.8	MENARD	IL	-8	41	30	50	70
20	19.9	19	18.8	18	17.9	MERCER	IL	-5	-5	-10	18	58
24*	31.0	18	20.1	19	19.2	MONROE	IL	-35	-4	-38	06	66
27	12.9	42	20.7	45	22.9	MONTGOMERY	IL	61	11	78	76	85
33	14.9	34	16.1	45	23.6	MORGAN	IL	8	47	58	67	88
4	5.3	21	26.1	14	16.4	MOULTRIE	IL	397	-37	211	93	47
41	23.3	34	18.0	46	22.1	OGLE	IL	-23	23	-5	21	82
137	17.4	167	20.1	197*	23.2	PEORIA	IL	15	15	33	52	86
18	16.0	19	15.2	33	27.1	PERRY	IL	-5	79	70	73	95
20	26.0	14	16.6	18	22.6	PIATT	IL	-36	36	-13	16	84
26	19.1	21	15.7	19	13.9	PIKE	IL	-18	-12	-27	10	31
6	20.2	2	6.5	5	16.5	POPE	IL	-68	154	-19	13	48
4	8.3	5	11.7	13	31.1	PULASKI	IL	40	166	273	95	98
2	7.7	8	25.1	5	17.2	PUTNAM	IL	225	-32	122	85	53
22	13.3	22	12.6	32	17.2	RANDOLPH	IL	-5	36	29	50	53
15	17.2	19	21.1	15	16.6	RICHLAND	IL	23	-21	-3	22	49
152*	24.1	142	20.0	131	18.0	ROCK ISLAND	IL	-17	-10	-25	10	59
161	19.8	194*	21.4	230*	24.8	ST CLAIR	IL	8	16	25	46	91
14	7.1	29	15.8	43	23.8	SALINE	IL	124	50	236	95	88
112	17.4	134	19.2	185*	25.0	SANGAMON	IL	10	30	44	60	92
10	17.5	9	14.7	14	25.9	SCHUYLER	IL	-16	76	49	62	93
7	17.1	4	10.0	5	13.5	SCOTT	IL	-42	35	-21	12	28
21	12.8	23	15.4	21	14.3	SHELBY	IL	20	-7	12	35	33
10	20.0	8	16.5	12	27.1	STARK	IL	-18	64	35	54	95
40	18.7	46	19.8	56	23.3	STEPHENSON	IL	6	18	25	45	87
74*	24.4	61	15.9	90	19.5	TAZEWELL	IL	-35	23	-20	12	68
12	9.1	9	6.8	18	14.1	UNION	IL	-25	106	56	66	32
67	14.2	83	17.3	106	22.3	VERMILION	IL	22	29	57	67	83
20	25.3	14	17.9	6	8.2	WABASH	IL	-29	-54	-68	01	08
16	14.0	19	16.3	24	19.9	WARREN	IL	17	22	42	58	70
13	13.3	18	18.4	25	22.7	WASHINGTON	IL	38	23	70	73	84
16	14.5	25	21.5	14	11.8	WAYNE	IL	48	-45	-19	13	19
17	14.9	15	12.7	22	18.1	WHITE	IL	-15	43	22	43	59
44	18.6	63	23.2	82*	27.8	WHITESIDE	IL	24	20	49	63	96
101	16.2	116	15.6	177	20.0	WILL	IL	-4	28	23	44	71
37	13.1	51	18.7	63	21.7	WILLIAMSON	IL	43	16	66	71	80
129	18.8	136	16.1	192	20.0	WINNEBAGO	IL	-14	24	7	30	71
22	19.0	20	15.7	18	12.8	WOODFORD	IL	-18	-19	-33	08	24

* SIGNIFICANTLY MORE DEATHS THAN EXPECTED

WHITE MALE

CANCER OF THE LARGE INTESTINE

ICD 153,159.0

1950-1959 DEATHS	1950-1959 RATE	1960-1969 DEATHS	1960-1969 RATE	1970-1979 DEATHS	1970-1979 RATE	COUNTY AND STATE	PERCENT CHANGE 60S/50S	PERCENT CHANGE 70S/60S	PERCENT CHANGE 70S/50S	PERCENTILE %CHANGE 70S/50S	PERCENTILE RATES 1970S
3013	16.7	3714	18.8	4205	19.8	INDIANA IN	13	6	19		
17	16.0	16	13.8	24	18.9	ADAMS IN	-14	38	18	40	64
135	17.5	161	18.3	206	20.2	ALLEN IN	4	11	15	38	72
22	12.2	30	15.1	34	15.0	BARTHOLOMEW IN	24	-1	22	43	37
5	9.7	14	23.4	13	22.7	BENTON IN	140	-3	133	87	84
4	5.6	13	18.0	12	16.0	BLACKFORD IN	223	-11	188	92	45
23	17.0	31	21.7	27	17.3	BOONE IN	27	-20	2	25	54
2	5.4	7	17.4	12	23.6	BROWN IN	221	36	335	97	88
17	18.0	24	26.0	19	19.1	CARROLL IN	44	-26	6	30	65
32	14.2	43	19.9	33	16.7	CASS IN	40	-16	17	40	50
27	15.0	43	19.7	45	15.7	CLARK IN	31	-20	5	29	43
28	20.7	26	17.0	33	21.8	CLAY IN	-18	29	6	29	81
20	12.8	37	22.4	41	25.2	CLINTON IN	75	13	97	80	92
4	7.0	6	10.3	10	18.4	CRAWFORD IN	46	79	162	90	61
19	14.1	21	15.0	33	22.8	DAVIESS IN	6	52	61	69	85
32*	27.2	23	17.6	26	17.8	DEARBORN IN	-35	1	-35	07	57
13	11.7	16	13.3	18	16.1	DECATUR IN	14	21	38	56	46
21	15.0	35	24.6	26	18.3	DE KALB IN	64	-26	22	43	61
53	14.2	61	14.3	93	21.1	DELAWARE IN	1	47	49	63	77
19	16.5	24	18.0	39	26.8	DUBOIS IN	9	49	62	70	95
61	15.3	75	16.5	99	20.1	ELKHART IN	8	22	31	51	71
20	20.0	22	20.5	31	26.9	FAYETTE IN	2	31	34	53	95
39	21.3	55	24.3	55	22.0	FLOYD IN	14	-9	3	27	82
18	17.1	19	19.2	14	14.4	FOUNTAIN IN	12	-25	-16	14	33
15	17.7	13	15.8	12	13.7	FRANKLIN IN	-11	-13	-23	11	29
22	22.5	19	18.2	23	23.4	FULTON IN	-19	29	4	28	87
31	19.1	37	21.8	39	21.7	GIBSON IN	14	0	13	36	80
31	9.9	54	16.5	56	16.4	GRANT IN	67	0	66	71	48
33	20.0	23	13.5	36	20.6	GREENE IN	-33	53	3	26	74
15	9.9	30	16.1	40	17.1	HAMILTON IN	64	6	74	74	53
13	12.4	20	15.9	21	14.3	HANCOCK IN	28	-10	15	38	33
7	6.8	13	12.8	25	22.2	HARRISON IN	87	74	225	94	82
21	16.2	30	17.7	49	24.1	HENDRICKS IN	9	36	49	63	89
31	14.5	41	17.8	45	18.2	HENRY IN	23	2	26	47	60
34	14.6	45	16.9	47	16.3	HOWARD IN	16	-4	11	34	47
33	19.5	39	22.4	38	21.2	HUNTINGTON IN	14	-5	8	32	78
16	11.1	29	20.0	39	23.9	JACKSON IN	79	20	115	84	89
14	15.5	12	13.2	17	17.8	JASPER IN	-14	34	15	38	57
18	15.0	24	19.9	23	18.7	JAY IN	33	-6	25	46	63
20	15.4	15	12.0	13	10.3	JEFFERSON IN	-22	-14	-33	08	13
16	20.0	18	21.7	13	14.3	JENNINGS IN	9	-34	-29	09	33
24	16.2	36	21.3	52	23.9	JOHNSON IN	31	12	47	62	89
30	12.7	49	21.1	36	15.6	KNOX IN	66	-26	23	44	42
30	16.0	30	14.7	46	19.5	KOSCIUSKO IN	-8	33	22	43	68
5	5.9	18	21.9	7	7.4	LAGRANGE IN	271	-66	26	47	06
277*	21.2	324*	21.7	366*	23.1	LAKE IN	2	7	9	32	86
78	21.5	77	18.7	100	22.5	LA PORTE IN	-13	21	4	28	84
20	11.7	37	20.5	30	14.8	LAWRENCE IN	75	-27	27	48	37
60	14.3	103	21.1	137*	25.8	MADISON IN	48	22	81	76	93
386	18.7	511*	21.7	503	20.6	MARION IN	16	-5	10	33	74
24	15.6	23	13.5	37	20.3	MARSHALL IN	-14	50	30	50	73
9	18.6	9	16.7	15	28.8	MARTIN IN	-10	73	55	66	97
24	15.4	24	14.7	23	14.6	MIAMI IN	-5	-1	-5	21	35
32	17.6	38	18.6	44	16.6	MONROE IN	6	-11	-6	21	49
28	17.2	40	23.8	39	22.8	MONTGOMERY IN	38	-4	33	52	85
22	17.8	21	13.9	18	10.1	MORGAN IN	-22	-28	-43	05	13
10	18.4	15	23.7	20	29.2	NEWTON IN	29	23	59	68	97
24	16.8	29	19.8	24	16.3	NOBLE IN	18	-18	-3	22	47
4	17.4	5	23.1	6	26.8	OHIO IN	33	16	54	66	95
24	26.4	21	20.6	15	14.7	ORANGE IN	-22	-28	-44	05	36
14	19.0	23*	31.3	18	21.8	OWEN IN	64	-30	15	38	80
14	14.0	24	24.1	17	18.3	PARKE IN	72	-24	31	51	61
15	18.8	18	20.0	21	21.8	PERRY IN	7	9	16	39	81
14	16.1	18	22.3	23	28.5	PIKE IN	39	28	77	75	96
32	16.6	47	20.0	57	18.3	PORTER IN	21	-8	11	33	61
16	15.4	22	21.4	29	26.4	POSEY IN	39	23	71	73	94
7	10.9	11	16.4	20	28.0	PULASKI IN	50	71	156	90	96
17	13.3	21	16.6	18	13.4	PUTNAM IN	25	-19	1	25	28
28	18.4	20	13.5	35	24.1	RANDOLPH IN	-26	78	31	51	89
20	18.0	22	19.1	24	19.0	RIPLEY IN	6	-1	5	29	65
19	17.3	16	15.3	18	17.4	RUSH IN	-12	14	0	25	54
169*	20.2	195	20.9	221	21.3	ST JOSEPH IN	4	2	6	30	78
5	8.8	11	16.3	15	19.2	SCOTT IN	84	18	118	85	66
23	15.0	20	12.3	28	17.3	SHELBY IN	-18	41	15	38	54
20	21.4	17	19.1	18	17.1	SPENCER IN	-11	-11	-20	12	52
17	18.6	18	18.4	22	20.7	STARKE IN	-1	12	11	34	74
17	18.0	15	15.8	18	16.5	STEUBEN IN	-13	4	-9	18	48
18	13.0	20	14.7	20	14.7	SULLIVAN IN	14	0	13	36	35
5	9.9	7	15.9	7	16.4	SWITZERLAND IN	61	4	67	71	48
40	13.9	58	17.3	70	19.3	TIPPECANOE IN	24	12	38	56	67
10	11.5	14	16.5	16	20.0	TIPTON IN	43	21	73	74	71
8	22.1	4	12.3			UNION IN	-44				
115	18.5	135	19.0	148	19.3	VANDERBURGH IN	3	2	4	28	66
11	9.3	25	23.0	20	18.7	VERMILLION IN	147	-19	100	81	63
97	18.3	84	16.2	113	22.1	VIGO IN	-12	37	21	42	82
30	20.6	18	12.0	31	19.2	WABASH IN	-42	60	-7	20	66
6	12.7	1	2.2	6	12.0	WARREN IN	-83	454	-6	21	20
15	13.5	25	20.7	19	14.1	WARRICK IN	54	-32	4	28	32
11	12.4	15	15.6	22	19.8	WASHINGTON IN	26	27	59	68	70
32	9.9	67	20.3	68	20.1	WAYNE IN	104	-1	102	82	72
14	13.0	13	11.2	16	14.0	WELLS IN	-14	26	8	31	31
22	22.0	17	15.5	26	21.0	WHITE IN	-30	36	-4	21	77
20	20.2	19	18.4	24	21.0	WHITLEY IN	-9	14	4	27	76
2261	16.7	2553	17.8	3134*	21.2	IOWA IA	7	19	27		
11	15.4	12	17.7	13	18.7	ADAIR IA	15	6	21	42	63
5	10.2	10	21.0	7	15.1	ADAMS IA	107	-28	48	62	38
15	15.7	17	16.6	19	17.9	ALLAMAKEE IA	6	8	14	37	58

* SIGNIFICANTLY MORE DEATHS THAN EXPECTED

WHITE MALE | CANCER OF THE LARGE INTESTINE | ICD 153,159.0

1950-1959 DEATHS	1950-1959 RATE	1960-1969 DEATHS	1960-1969 RATE	1970-1979 DEATHS	1970-1979 RATE	COUNTY AND STATE		PERCENT CHANGE 60S/50S	PERCENT CHANGE 70S/60S	PERCENT CHANGE 70S/50S	PERCENTILE %CHANGE 70S/50S	PERCENTILE RATES 1970S
16	13.2	24	19.6	14	12.4	APPANOOSE	IA	49	-37	-6	20	21
10	16.6	8	11.9	11	16.8	AUDUBON	IA	-28	41	2	25	51
26	21.5	28	21.8	25	18.4	BENTON	IA	1	-16	-14	15	61
73	16.7	68	13.9	123	23.8	BLACK HAWK	IA	-17	71	42	59	88
19	11.5	35	20.3	26	15.9	BOONE	IA	77	-21	39	56	44
13	12.6	13	11.8	22	18.0	BREMER	IA	-7	53	42	58	58
18	14.9	19	17.1	31	28.2	BUCHANAN	IA	15	65	89	79	96
20	16.9	25	20.3	26	19.1	BUENA VISTA	IA	21	-6	14	36	66
11	11.8	13	12.3	30	25.9	BUTLER	IA	4	111	119	85	93
16	17.1	17	16.3	20	20.7	CALHOUN	IA	-5	27	21	42	74
20	17.4	27	22.3	29	22.6	CARROLL	IA	28	1	30	50	84
22	20.1	26	22.8	24	19.3	CASS	IA	13	-15	-4	21	67
15	15.2	23	21.3	24	21.8	CEDAR	IA	40	3	44	60	81
44	19.8	37	14.5	63	24.3	CERRO GORDO	IA	-27	68	23	44	90
18	16.4	20	20.0	24	23.8	CHEROKEE	IA	22	19	45	61	88
14	17.8	11	12.6	14	16.1	CHICKASAW	IA	-29	28	-10	18	45
11	18.6	5	9.7	14	26.0	CLARKE	IA	-48	167	40	57	93
16	18.1	17	17.3	28	26.3	CLAY	IA	-4	52	45	61	94
27	20.2	24	18.5	29	21.5	CLAYTON	IA	-8	16	6	30	79
61*	22.5	48	17.2	69	24.3	CLINTON	IA	-23	41	8	32	90
22	21.4	20	18.6	28	25.0	CRAWFORD	IA	-13	34	17	39	92
22	15.9	31	21.7	39	26.2	DALLAS	IA	36	21	64	70	94
5	7.4	11	18.5	13	21.0	DAVIS	IA	149	14	184	92	77
15	17.4	6	7.9	18	28.3	DECATUR	IA	-55	260	63	70	96
11	11.6	19	19.4	16	16.9	DELAWARE	IA	67	-13	45	61	51
45	21.0	46	19.4	54	22.1	DES MOINES	IA	-7	14	5	29	82
12	16.2	9	11.1	19	19.9	DICKINSON	IA	-31	79	23	44	71
58	17.9	58	17.3	78	21.9	DUBUQUE	IA	-4	27	22	43	81
11	16.0	17	22.9	14	17.5	EMMET	IA	43	-23	10	33	55
31	19.7	28	17.7	29	17.6	FAYETTE	IA	-10	-1	-11	17	56
20	17.3	22	18.4	25	21.8	FLOYD	IA	6	19	26	47	80
14	17.1	11	13.0	31*	33.9	FRANKLIN	IA	-24	160	99	81	98
14	20.0	11	17.5	17	25.2	FREMONT	IA	-12	44	26	47	92
19	23.9	15	17.9	20	23.8	GREENE	IA	-25	33	0	24	88
9	12.5	11	15.1	21	25.7	GRUNDY	IA	22	69	106	83	93
17	18.7	18	19.7	16	17.4	GUTHRIE	IA	5	-12	-7	19	54
19	18.6	22	21.0	22	20.5	HAMILTON	IA	12	-2	10	33	74
12	16.9	26*	32.0	14	17.2	HANCOCK	IA	89	-46	2	25	53
20	15.2	24	17.3	34	25.0	HARDIN	IA	14	44	64	70	92
15	14.4	18	16.7	18	17.1	HARRISON	IA	16	2	19	41	53
20	16.7	19	15.5	16	14.3	HENRY	IA	-7	-8	-14	15	33
12	15.7	17	21.5	14	16.5	HOWARD	IA	37	-23	5	29	49
5	7.6	13	18.4	15	18.6	HUMBOLDT	IA	143	1	144	88	62
12	20.9	12	18.5	15	21.7	IDA	IA	-11	17	4	28	80
9	10.3	20	21.5	20	20.9	IOWA	IA	108	-3	102	82	76
21	18.6	19	16.2	27	21.0	JACKSON	IA	-13	29	13	35	76
28	17.2	33	18.4	45	23.8	JASPER	IA	7	30	39	56	88
14	15.4	22	23.0	18	20.4	JEFFERSON	IA	49	-11	32	52	73
30	16.8	39	19.3	46	21.4	JOHNSON	IA	15	11	28	49	79
26	25.2	19	18.1	25	23.0	JONES	IA	-28	27	-9	18	86
21	18.9	16	14.6	10	8.7	KEOKUK	IA	-23	-40	-54	03	09
12	10.5	24	18.8	25	19.4	KOSSUTH	IA	78	3	84	77	67
54*	25.2	47	21.4	56	26.2	LEE	IA	-15	22	4	28	94
89	17.6	113	19.3	136	21.1	LINN	IA	10	9	20	41	77
10	15.6	22*	34.5	17	26.1	LOUISA	IA	122	-24	68	72	93
7	8.6	14	18.5	8	10.8	LUCAS	IA	114	-42	25	46	15
11	15.3	27*	36.4	25*	31.6	LYON	IA	138	-13	107	83	98
16	19.9	18	22.2	18	20.5	MADISON	IA	12	-8	3	27	74
24	17.1	36	24.7	44*	29.5	MAHASKA	IA	44	19	72	74	97
27	17.1	23	12.5	27	14.9	MARION	IA	-27	19	-12	16	37
25	12.5	34	15.7	34	14.5	MARSHALL	IA	26	-8	16	39	34
10	14.0	10	14.1	12	18.4	MILLS	IA	1	31	32	51	61
7	9.0	14	16.9	15	19.3	MITCHELL	IA	87	14	114	84	67
5	5.8	9	10.6	20	22.3	MONONA	IA	85	110	288	96	83
8	10.6	8	9.8	10	15.4	MONROE	IA	-8	57	45	61	40
14	14.0	17	16.7	17	19.2	MONTGOMERY	IA	19	15	37	55	66
34	18.6	43	23.1	40	21.1	MUSCATINE	IA	24	-9	14	36	77
19	18.9	13	11.5	19	16.9	O'BRIEN	IA	-39	47	-10	18	51
3	6.8	7	12.9	11	19.6	OSCEOLA	IA	88	52	186	92	68
22	15.1	24	18.0	28	20.2	PAGE	IA	19	12	34	52	72
14	18.8	7	9.0	12	13.8	PALO ALTO	IA	-52	54	-26	10	30
24	21.9	20	15.6	23	17.0	PLYMOUTH	IA	-29	9	-22	11	52
17	20.9	16	19.4	19	22.3	POCAHONTAS	IA	-7	15	7	30	83
164	16.6	221	20.6	262*	23.5	POLK	IA	24	14	41	58	87
51	15.0	58	15.8	73	18.7	POTTAWATTAMIE	IA	5	19	25	46	63
19	18.5	13	13.1	34*	33.8	POWESHIEK	IA	-29	158	83	77	98
8	14.0	6	11.3	10	17.7	RINGGOLD	IA	-19	57	26	47	57
16	17.5	15	14.1	29	27.8	SAC	IA	-20	97	59	68	96
83	16.9	99	18.7	138*	24.2	SCOTT	IA	11	29	43	59	90
16	21.9	8	8.5	15	14.8	SHELBY	IA	-61	75	-32	08	36
28	23.4	28	21.3	40	26.8	SIOUX	IA	-9	26	15	37	95
29	15.1	38	18.3	47	21.3	STORY	IA	21	16	41	58	78
20	16.0	18	13.5	30	21.6	TAMA	IA	-16	60	35	53	79
9	11.5	21	26.0	11	15.6	TAYLOR	IA	126	-40	35	54	42
15	15.1	15	17.1	16	16.1	UNION	IA	14	-6	7	30	45
11	16.1	9	13.1	20	31.2	VAN BUREN	IA	-18	138	94	80	98
29	12.8	42	17.5	49	19.7	WAPELLO	IA	37	12	54	65	69
12	12.2	14	13.2	23	19.2	WARREN	IA	9	45	58	67	66
19	17.4	21	17.5	24	19.6	WASHINGTON	IA	1	12	13	35	69
11	13.4	11	14.5	6	6.9	WAYNE	IA	8	-53	-49	03	06
33	14.6	39	16.4	51	20.4	WEBSTER	IA	12	25	40	57	73
14	19.9	12	14.7	17	22.0	WINNEBAGO	IA	-26	49	10	33	81
19	16.1	15	13.0	24	19.8	WINNESHIEK	IA	-19	52	24	44	70
95	18.8	90	16.7	111	21.6	WOODBURY	IA	-12	30	15	37	79
4	6.3	18	28.9	11	17.6	WORTH	IA	355	-39	178	91	56
14	13.2	17	15.3	15	13.9	WRIGHT	IA	16	-9	6	29	31
1257	13.0	1560	14.9	1972	17.8	KANSAS	KS	14	20	36		
20	18.3	18	19.4	12	10.1	ALLEN	KS	6	-48	-45	05	13
6	8.8	11	15.5	15	22.0	ANDERSON	KS	77	42	151	89	82

* SIGNIFICANTLY MORE DEATHS THAN EXPECTED

WHITE MALE

CANCER OF THE LARGE INTESTINE

ICD 153,159.0

COUNTY AND STATE	1950-1959 DEATHS	1950-1959 RATE	1960-1969 DEATHS	1960-1969 RATE	1970-1979 DEATHS	1970-1979 RATE	PERCENT CHANGE 60S/50S	PERCENT CHANGE 70S/60S	PERCENT CHANGE 70S/50S	PERCENTILE %CHANGE 70S/50S	PERCENTILE RATES 1970S
ATCHISON KS	16	14.3	20	19.8	21	21.3	39	7	49	63	78
BARBER KS	4	7.8	9	19.2	6	11.8	147	-38	52	65	19
BARTON KS	21	19.7	25	17.6	49*	31.5	-11	79	60	68	98
BOURBON KS	9	8.4	15	12.4	18	16.5	47	33	96	80	48
BROWN KS	11	11.6	11	13.0	13	12.9	12	-1	11	34	24
BUTLER KS	19	11.0	30	16.1	32	15.2	46	-6	38	55	39
CHASE KS	3	9.4	1	3.7	4	14.5	-60	291	55	66	34
CHAUTAUQUA KS	8	17.5	8	17.8	10	21.4	2	20	22	43	78
CHEROKEE KS	22	17.2	16	12.5	27	21.1	-27	69	22	43	77
CHEYENNE KS	6	26.4	2	6.9	3	10.5	-74	53	-60	02	14
CLARK KS	2	10.9	1	4.8	5	22.5	-56	370	107	83	83
CLAY KS	6	7.8	12	16.2	14	15.9	108	-2	104	82	44
CLOUD KS	10	10.3	14	13.3	12	12.3	29	-7	20	41	21
COFFEY KS	11	15.1	13	18.2	8	13.3	21	-27	-12	16	27
COMANCHE KS	5	22.2	4	26.6	3	14.3	20	-46	-36	07	33
COWLEY KS	27	12.8	26	12.4	45	21.3	-4	72	66	71	78
CRAWFORD KS	31	12.3	40	15.5	53	20.8	27	34	70	73	75
DECATUR KS	9	22.2	9	21.5	7	17.9	-3	-17	-19	13	58
DICKINSON KS	26	18.9	15	11.0	26	17.9	-42	63	-5	21	58
DONIPHAN KS	13	19.7	6	9.9	12	18.8	-50	90	-5	21	64
DOUGLAS KS	15	9.8	24	14.0	24	13.1	42	-6	33	52	26
EDWARDS KS	3	11.2	3	9.2	8	26.7	-18	192	139	88	95
ELK KS	6	12.1	4	8.0	7	17.5	-34	119	45	60	55
ELLIS KS	6	8.7	11	14.6	15	16.1	68	10	86	78	45
ELLSWORTH KS	7	13.5	10	20.6	6	12.1	52	-41	-10	17	20
FINNEY KS	4	7.1	7	11.5	10	13.2	62	15	86	78	27
FORD KS	14	15.9	12	11.1	17	14.9	-30	34	-6	20	37
FRANKLIN KS	26	20.2	16	12.0	27	19.6	-41	63	-3	22	69
GEARY KS	7	11.9	9	12.7	15	19.4	7	53	64	70	68
GOVE KS	2	9.9	4	16.6	1	3.9	69	-76	-60	01	02
GRAHAM KS	1	4.2	4	14.2	5	18.7	242	32	350	97	63
GRANT KS	1	8.4	2	19.2	3	9.6	129	-50	15	37	11
GRAY KS	1	4.7	4	15.4	11*	47.3	226	207	903	99	99
GREELEY KS	2	34.9	2	40.3	5*	58.7	16	46	68	72	99
GREENWOOD KS	16	18.8	12	14.6	17	21.8	-22	49	16	38	80
HAMILTON KS	1	7.5			3	16.3			117	85	47
HARPER KS	5	8.6	6	9.7	7	10.5	13	8	22	43	14
HARVEY KS	13	11.0	12	8.5	29	18.6	-23	120	69	72	62
HASKELL KS	1	9.8	2	15.2	2	16.9	54	11	72	74	51
JACKSON KS	6	7.8	12	14.9	19	26.4	91	77	238	95	94
JEFFERSON KS	9	12.8	12	15.3	13	15.3	19	0	19	41	39
JEWELL KS	8	13.4	6	10.4	7	12.6	-22	21	-6	20	23
JOHNSON KS	38	11.3	82	17.1	120	15.9	51	-7	41	57	44
KEARNY KS	1	11.8	3	21.7	1	6.9	84	-68	-42	05	05
KINGMAN KS	8	13.7	11	18.6	10	16.6	35	-11	21	42	49
KIOWA KS	5	19.3	3	11.8	6	20.9	-39	77	8	31	76
LABETTE KS	15	8.6	27	16.9	30	19.6	96	16	127	86	68
LANE KS			2	12.2	4	25.6		110			93
LEAVENWORTH KS	31	13.6	34	15.2	40	18.9	12	25	40	57	65
LINCOLN KS	6	14.4	5	12.5	6	15.5	-13	24	8	32	42
LINN KS	7	8.2	5	7.4	12	17.5	-11	138	112	84	55
LOGAN KS	1	5.4	4	18.0	5	21.4	231	19	293	96	78
LYON KS	12	7.7	22	13.6	18	10.7	76	-21	39	56	15
MC PHERSON KS	14	10.6	20	13.1	28	16.6	24	26	56	67	49
MARION KS	16	17.3	17	16.1	19	16.4	-7	2	-5	21	48
MARSHALL KS	15	13.2	14	12.4	20	19.1	-6	54	44	60	65
MEADE KS	4	17.4	2	7.1	3	9.0	-59	28	-48	04	10
MIAMI KS	11	9.3	15	11.7	19	14.8	26	26	59	68	36
MITCHELL KS	6	9.9	15	22.8	12	22.7	131	-1	130	86	84
MONTGOMERY KS	40	16.2	39	15.7	41	16.7	-3	6	3	26	50
MORRIS KS	4	7.8	6	10.3	7	15.7	32	53	102	82	42
MORTON KS	1	8.8	3	21.4	6	38.7	143	81	339	97	99
NEMAHA KS	8	10.3	10	11.6	11	12.4	13	7	20	42	22
NEOSHO KS	24	19.9	18	13.9	14	10.5	-30	-24	-47	04	14
NESS KS	5	17.3	7	19.9	4	12.6	15	-36	-27	10	23
NORTON KS	8	13.6	7	13.3	5	8.2	-3	-38	-40	06	08
OSAGE KS	12	12.8	13	13.2	15	14.3	3	9	12	34	33
OSBORNE KS	8	15.1	4	7.6	8	12.0	-50	59	-20	12	20
OTTAWA KS	19*	34.6	7	12.8	9	18.3	-63	42	-47	04	60
PAWNEE KS	2	4.3	10	17.0	15	27.5	298	62	545	98	96
PHILLIPS KS	4	6.7	7	12.7	9	14.7	90	15	120	85	36
POTTAWATOMIE KS	15	17.7	16	17.8	16	18.4	1	3	4	28	61
PRATT KS	4	8.0	14	22.8	10	14.4	183	-37	80	76	34
RAWLINS KS	2	8.5	5	15.2	9	26.0	79	71	206	93	93
RENO KS	24	8.9	38	12.7	61	19.2	42	51	114	84	66
REPUBLIC KS	10	14.6	15	18.6	15	18.5	27	-1	26	47	62
RICE KS	8	11.8	15	19.3	10	12.4	63	-36	5	29	22
RILEY KS	12	10.5	22	17.4	21	16.6	66	-4	59	68	49
ROOKS KS	5	11.7	10	21.3	8	15.2	82	-29	30	50	39
RUSH KS	12*	34.8	5	13.3	12	28.1	-62	112	-19	13	96
RUSSELL KS	3	5.7	12	18.5	13	20.1	227	8	254	95	71
SALINE KS	23	13.8	40	21.2	48	23.1	53	9	68	72	86
SCOTT KS	2	10.8	4	17.8	1	3.8	65	-79	-65	01	02
SEDGWICK KS	112	12.2	182	16.9	211	17.2	38	2	41	58	53
SEWARD KS	7	22.8	8	15.7	4	7.0	-31	-55	-69	01	06
SHAWNEE KS	62	12.2	72	12.3	125	20.1	1	63	64	70	71
SHERIDAN KS	3	18.6	4	17.6	2	10.0	-5	-44	-46	04	12
SHERMAN KS	4	12.6	7	22.0	7	17.2	75	-22	37	55	53
SMITH KS	6	8.8	7	11.8	8	14.7	34	25	68	72	36
STAFFORD KS	7	14.4	7	13.3	7	12.4	-7	-6	-13	15	22
STANTON KS	1	13.2	1	37.7			185				
STEVENS KS			2	10.0	1	3.9		-61			02
SUMNER KS	28	20.1	30	19.5	34	22.3	-3	14	11	34	83
THOMAS KS	4	18.1	5	13.4	14	32.3	-26	141	79	76	98
TREGO KS			10*	35.1	8	24.1		-32			89
WABAUNSEE KS	10	19.7	12	21.5	4	9.3	9	-57	-53	03	10
WALLACE KS			1	8.6	2	17.5		103			55
WASHINGTON KS	7	8.6	9	11.6	10	13.2	35	14	53	65	27
WICHITA KS			2	14.7							
WILSON KS	17	18.0	11	11.7	13	15.9	-35	36	-12	16	44
WOODSON KS	4	7.3	2	3.7	9	18.6	-50	404	153	89	62

* SIGNIFICANTLY MORE DEATHS THAN EXPECTED

WHITE MALE | CANCER OF THE LARGE INTESTINE | ICD 153,159.0

1950-1959 DEATHS	1950-1959 RATE	1960-1969 DEATHS	1960-1969 RATE	1970-1979 DEATHS	1970-1979 RATE	COUNTY AND STATE	PERCENT CHANGE 60S/50S	PERCENT CHANGE 70S/60S	PERCENT CHANGE 70S/50S	PERCENTILE %CHANGE 70S/50S	PERCENTILE RATES 1970S
101	16.7	100	16.5	146	23.9	WYANDOTTE KS	-1	45	43	59	89
1512	13.2	1958	15.1	2609	18.0	KENTUCKY KY	14	19	36		
5	6.7	9	11.9	11	13.6	ADAIR KY	77	14	102	82	29
13	17.9	7	9.1	10	12.2	ALLEN KY	-49	34	-32	08	20
1	2.0	8	19.9	9	17.9	ANDERSON KY	904	-10	799	99	57
10	21.5	6	11.6	10	17.1	BALLARD KY	-46	48	-21	12	53
16	12.9	19	14.2	26	15.9	BARREN KY	10	13	24	45	45
7	15.3	5	9.8	7	12.7	BATH KY	-36	30	-17	14	24
4	2.9	13	9.3	20	14.3	BELL KY	224	55	400	98	33
12	16.6	20	20.8	30	24.0	BOONE KY	25	15	44	60	89
15	23.1	10	13.1	24	29.0	BOURBON KY	-43	122	26	47	97
29	16.4	28	12.9	51	20.6	BOYD KY	-21	59	25	46	74
8	9.3	8	8.9	17	17.8	BOYLE KY	-4	99	91	80	57
4	8.9	5	10.5	11	24.2	BRACKEN KY	19	130	174	91	90
5	8.3	9	14.4	13	18.0	BREATHITT KY	72	26	116	85	59
7	9.0	9	10.8	12	13.0	BRECKINRIDGE KY	19	20	44	60	25
2	3.2	6	9.0	19	17.8	BULLITT KY	181	97	454	98	57
4	6.2	11	19.2	12	17.9	BUTLER KY	209	-7	188	92	58
14	20.9	18	23.2	12	13.5	CALDWELL KY	11	-42	-35	07	28
7	7.0	13	11.3	15	9.6	CALLOWAY KY	60	-15	37	54	11
92*	26.7	108*	28.4	127*	33.1	CAMPBELL KY	7	16	24	45	98
3	8.7	4	10.0	9	22.3	CARLISLE KY	15	123	157	90	83
8	18.9	5	10.4	6	13.0	CARROLL KY	-45	25	-31	08	25
9	9.8	10	10.9	19	17.3	CARTER KY	11	59	77	75	54
9	16.1	6	8.3	11	13.7	CASEY KY	-49	65	-15	15	29
14	9.2	23	14.0	23	13.7	CHRISTIAN KY	52	-3	48	62	29
10	13.0	15	16.3	15	14.4	CLARK KY	25	-11	11	34	34
3	4.4	4	5.2	11	11.8	CLAY KY	17	130	170	91	19
4	11.8	3	7.3	6	11.1	CLINTON KY	-38	52	-6	20	16
8	15.8	9	15.7	10	17.5	CRITTENDEN KY	-1	12	11	34	55
3	8.7	1	2.0	3	6.4	CUMBERLAND KY	-77	222	-26	10	05
34	14.6	43	14.9	69	21.9	DAVIESS KY	2	47	50	64	81
4	10.4	2	4.3	10	21.9	EDMONSON KY	-59	409	110	83	81
1	4.0	3	9.7	4	38.2	ELLIOTT KY	144	295	863	99	99
4	7.2	7	10.3	8	10.8	ESTILL KY	43	4	49	63	15
40	11.1	68	15.6	102	18.7	FAYETTE KY	41	19	68	72	63
3	4.2	8	12.2	12	17.8	FLEMING KY	186	46	319	96	57
13	9.7	12	7.3	22	11.5	FLOYD KY	-25	59	19	41	18
8	6.7	14	10.6	31	21.6	FRANKLIN KY	58	103	221	94	79
8	14.4	6	10.8	10	20.0	FULTON KY	-25	85	39	56	71
5	24.6	2	8.2	3	13.3	GALLATIN KY	-66	61	-46	04	27
8	20.5	5	10.4	10	18.2	GARRARD KY	-49	75	-11	17	60
11	19.2	9	14.4	14	21.0	GRANT KY	-25	46	10	33	76
15	9.4	28	15.7	29	14.6	GRAVES KY	66	-7	55	66	35
11	12.0	12	12.6	14	12.5	GRAYSON KY	5	-1	4	28	22
5	9.8	7	12.3	11	16.5	GREEN KY	25	34	68	72	48
17	16.0	12	10.2	22	15.5	GREENUP KY	-37	53	-3	22	41
4	12.7	2	5.9	7	19.5	HANCOCK KY	-54	232	54	65	68
12	8.7	19	12.1	40	20.3	HARDIN KY	39	68	133	87	73
11	6.4	20	13.0	28	16.1	HARLAN KY	102	24	151	89	46
14	18.3	13	16.1	17	19.4	HARRISON KY	-12	20	5	29	67
5	6.7	10	13.3	12	15.1	HART KY	99	14	127	86	38
23	18.0	27	19.6	29	18.7	HENDERSON KY	9	-5	4	27	62
7	11.3	4	6.0	8	12.0	HENRY KY	-47	100	6	30	20
3	7.9	10	25.8	13	31.4	HICKMAN KY	225	21	295	96	98
17	10.2	29	15.0	31	15.3	HOPKINS KY	47	2	50	63	39
4	8.6	8	15.4	10	15.2	JACKSON KY	79	-1	77	75	39
347*	20.4	440*	21.6	538*	23.1	JEFFERSON KY	6	7	13	36	86
6	12.7	6	10.1	13	16.9	JESSAMINE KY	-20	67	33	52	51
7	9.2	6	7.1	16	14.8	JOHNSON KY	-23	110	61	69	36
91	20.4	115*	23.2	146*	27.6	KENTON KY	14	19	35	54	96
2	4.1	5	7.3	6	8.6	KNOTT KY	76	18	107	83	09
9	8.2	16	13.5	18	14.7	KNOX KY	64	9	79	76	36
5	10.5	7	12.9	21*	33.0	LARUE KY	22	157	215	94	98
6	6.0	13	10.5	15	9.4	LAUREL KY	74	-11	55	66	11
4	5.2	10	15.3	11	14.0	LAWRENCE KY	195	-8	170	91	31
4	10.6	4	8.6	2	5.5	LEE KY	-19	-37	-49	03	04
		3	7.4	5	10.0	LESLIE KY		34			12
9	7.9	18	15.6	19	15.5	LETCHER KY	99	-1	97	81	41
2	3.1	8	12.8	6	9.1	LEWIS KY	314	-29	195	93	10
6	7.5	12	14.2	9	8.9	LINCOLN KY	88	-37	18	40	10
2	5.1	10	21.8	9	18.4	LIVINGSTON KY	327	-15	262	95	61
15	15.5	12	10.5	23	18.2	LOGAN KY	-32	73	18	40	60
		3	9.5	10	25.3	LYON KY		167			92
31	14.2	52	21.5	60	21.2	MC CRACKEN KY	52	-1	50	63	78
5	8.9	4	7.2	8	12.9	MC CREARY KY	-19	80	46	61	25
2	3.5	7	13.0	10	18.3	MC LEAN KY	276	41	431	98	61
15	13.2	22	16.0	23	14.0	MADISON KY	21	-12	6	30	31
4	8.1	5	9.6	2	3.7	MAGOFFIN KY	18	-62	-55	03	02
		5	7.9	8	11.5	MARION KY		47			18
13	19.8	9	8.7	20	14.7	MARSHALL KY	-56	69	-26	10	36
3	8.5	2	5.6	4	8.6	MARTIN KY	-34	54	1	25	09
13	15.6	19	22.7	16	19.0	MASON KY	45	-16	22	43	65
4	9.7	5	12.9	8	15.2	MEADE KY	33	18	57	67	39
3	11.4	1	5.1	4	16.0	MENIFEE KY	-56	217	40	57	45
18	25.8	10	13.2	13	15.6	MERCER KY	-49	18	-40	06	42
3	6.2	3	6.2	3	5.2	METCALFE KY	0	-16	-16	14	04
7	11.4	7	11.9	11	17.0	MONROE KY	5	43	50	64	52
8	15.4	6	11.1	13	17.9	MONTGOMERY KY	-28	62	16	39	58
6	10.2	5	9.0	12	19.0	MORGAN KY	-11	111	87	78	65
17	12.8	18	11.7	22	13.8	MUHLENBERG KY	-8	17	7	31	30
9	12.2	8	9.3	20	20.1	NELSON KY	-24	117	64	70	72
6	13.8	8	18.8	14*	31.9	NICHOLAS KY	36	70	131	86	98
9	7.4	22	17.2	19	14.8	OHIO KY	132	-14	100	81	36
9	18.0	7	13.3	15	21.2	OLDHAM KY	-26	60	18	40	77
8	15.1	6	11.6	11	20.3	OWEN KY	-23	75	35	53	73
5	16.4	5	15.3	3	9.7	OWSLEY KY	-7	-37	-41	05	12
9	16.3	13	23.7	11	19.0	PENDLETON KY	46	-20	17	39	65
4	3.8	7	5.7	11	8.3	PERRY KY	50	45	117	85	08

* SIGNIFICANTLY MORE DEATHS THAN EXPECTED

WHITE MALE

CANCER OF THE LARGE INTESTINE

ICD 153,159.0

1950-1959 DEATHS	1950-1959 RATE	1960-1969 DEATHS	1960-1969 RATE	1970-1979 DEATHS	1970-1979 RATE	COUNTY AND STATE		PERCENT CHANGE 60S/50S	PERCENT CHANGE 70S/60S	PERCENT CHANGE 70S/50S	PERCENTILE %CHANGE 70S/50S	PERCENTILE RATES 1970S
15	7.3	22	8.6	35	12.3	PIKE	KY	18	43	69	72	21
2	7.1	8	23.7	5	12.5	POWELL	KY	233	-47	76	75	23
21	13.0	20	10.5	28	13.1	PULASKI	KY	-19	24	0	25	26
		3	15.6	1	5.7	ROBERTSON	KY		-63			04
5	8.6	5	8.6	13	19.0	ROCKCASTLE	KY	0	120	121	85	65
8	16.1	6	10.2	9	13.2	ROWAN	KY	-37	30	-18	13	26
2	7.6	2	5.4	6	7.8	RUSSELL	KY	-29	43	2	26	07
5	7.4	11	14.8	11	13.5	SCOTT	KY	100	-8	84	77	29
10	11.9	7	7.7	17	16.3	SHELBY	KY	-35	112	37	55	47
12	22.9	8	13.8	13	19.9	SIMPSON	KY	-40	45	-13	16	71
3	10.5	4	15.1	1	3.9	SPENCER	KY	44	-75	-63	01	02
9	15.3	15	22.0	12	14.4	TAYLOR	KY	44	-35	-6	20	34
5	9.8	17*	30.1	9	16.3	TODD	KY	208	-46	67	71	47
2	4.4	9	23.1	10	19.1	TRIGG	KY	426	-17	335	97	66
7	24.1	6	20.8	3	10.0	TRIMBLE	KY	-14	-52	-58	02	13
7	10.7	12	16.7	11	15.8	UNION	KY	56	-5	48	62	44
26	14.4	34	17.0	44	19.2	WARREN	KY	18	13	34	53	66
7	14.7	4	7.3	8	15.2	WASHINGTON	KY	-50	109	4	27	39
4	7.5	7	10.3	8	9.4	WAYNE	KY	37	-9	26	47	11
8	9.9	13	15.0	13	14.8	WEBSTER	KY	51	-1	49	63	36
11	8.7	18	14.0	21	15.2	WHITLEY	KY	61	9	75	75	39
1	2.8	2	4.8	1	3.1	WOLFE	KY	72	-36	9	33	02
3	7.1	9	17.8	7	11.4	WOODFORD	KY	151	-36	61	69	17
812	12.0	1246	14.5	1824	17.4	LOUISIANA	LA	21	20	45		
19	13.9	16	9.5	27	15.1	ACADIA	LA	-32	59	9	32	38
4	7.2	10	14.7	6	7.5	ALLEN	LA	104	-49	5	29	06
10	16.8	12	14.5	15	14.1	ASCENSION	LA	-14	-3	-16	14	32
3	10.0	6	13.4	10	20.6	ASSUMPTION	LA	34	54	105	83	74
21	20.4	19	13.7	25	16.9	AVOYELLES	LA	-33	23	-17	14	51
7	9.7	12	15.1	16	16.3	BEAUREGARD	LA	55	8	67	71	47
6	9.9	8	13.2	9	15.7	BIENVILLE	LA	34	19	59	68	43
7	8.9	14	13.4	32	22.3	BOSSIER	LA	51	67	151	89	83
58	12.2	103	18.7	139	20.8	CADDO	LA	54	11	71	73	75
19	7.4	35	9.3	59	14.0	CALCASIEU	LA	25	51	89	79	31
2	4.5	8	19.2	9	17.3	CALDWELL	LA	327	-10	286	96	54
		2	12.5	2	5.2	CAMERON	LA		-58			04
5	19.1	4	11.1	8	17.8	CATAHOULA	LA	-42	60	-7	19	57
4	6.5	9	13.4	11	15.8	CLAIBORNE	LA	105	18	142	88	43
3	7.5	9	25.6	7	13.5	CONCORDIA	LA	242	-47	80	76	28
9	15.7	8	12.5	11	14.7	DE SOTO	LA	-21	18	-6	20	36
38	13.6	71	16.6	99	15.2	EAST BATON ROUGE	LA	22	-8	12	35	39
4	16.1	2	11.9	1	4.8	EAST CARROLL	LA	-26	-60	-70	01	03
1	2.5	3	5.1	10	19.7	EAST FELICIANA	LA	104	286	689	99	69
11	12.7	12	10.7	20	16.2	EVANGELINE	LA	-15	51	28	48	46
6	10.2	13	16.6	19	20.6	FRANKLIN	LA	63	24	101	81	74
3	7.0	8	12.1	10	13.0	GRANT	LA	75	7	88	78	26
11	14.0	21	18.1	23	14.6	IBERIA	LA	29	-19	4	28	35
7	12.5	6	10.3	11	15.7	IBERVILLE	LA	-18	53	26	47	43
4	8.9	7	11.9	11	17.5	JACKSON	LA	33	47	96	80	55
35	12.7	93	13.6	211	22.8	JEFFERSON	LA	7	67	79	76	65
11	14.5	14	16.0	28	27.5	JEFFERSON DAVIS	LA	10	72	90	79	96
12	8.0	20	10.8	35	12.4	LAFAYETTE	LA	34	15	54	65	21
12	11.7	20	13.0	16	7.0	LAFOURCHE	LA	11	-46	-41	06	06
3	5.2	4	7.0	11	14.1	LA SALLE	LA	35	102	173	91	32
3	4.8	6	7.8	15	16.4	LINCOLN	LA	62	110	241	95	48
4	6.2	7	7.5	19	13.5	LIVINGSTON	LA	21	79	118	85	28
		4	17.6	3	10.7	MADISON	LA		-39			15
1	2.3	11	14.7	15	17.1	MOREHOUSE	LA	551	16	658	99	52
5	5.2	16	15.8	13	11.1	NATCHITOCHES	LA	202	-30	111	84	16
247	16.1	312	19.0	353*	23.8	ORLEANS	LA	18	25	48	62	88
18	9.7	30	11.4	50	15.1	OUACHITA	LA	17	33	55	66	38
1	2.6	5	17.4	9	19.8	PLAQUEMINES	LA	576	14	671	99	70
4	11.4	9	19.3	12	19.1	POINTE COUPEE	LA	68	-1	67	71	66
32	12.5	40	13.1	59	15.7	RAPIDES	LA	4	21	26	47	43
1	3.3	1	3.6	3	7.9	RED RIVER	LA	9	122	142	88	08
9	17.0	1	1.5	10	12.4	RICHLAND	LA	-91	703	-27	10	21
10	13.0	19	22.1	15	13.4	SABINE	LA	70	-39	3	27	28
6	15.1	14	19.9	45*	29.9	ST BERNARD	LA	32	50	98	81	97
4	14.6	7	11.1	15	23.3	ST CHARLES	LA	-24	110	60	69	87
		1	5.6	2	16.0	ST HELENA	LA		187			45
3	10.4	5	14.5	12	30.1	ST JAMES	LA	40	108	191	92	97
1	3.1	6	16.0	9	18.3	ST JOHN THE BAPT	LA	412	15	487	98	60
12	9.2	19	12.3	30	15.8	ST LANDRY	LA	33	29	72	74	44
7	19.5	4	4.9	8	8.3	ST MARTIN	LA	-75	71	-58	02	08
6	7.2	16	14.7	18	13.9	ST MARY	LA	105	-6	93	80	31
13	12.1	21	15.1	35	12.4	ST TAMMANY	LA	24	-18	2	26	22
22	14.2	32	19.1	27	12.9	TANGIPAHOA	LA	35	-32	-9	18	25
3	16.4	3	15.8	3	17.1	TENSAS	LA	-4	8	4	28	53
9	8.0	14	9.8	31	16.9	TERREBONNE	LA	23	72	111	84	51
5	8.6	7	10.0	11	12.5	UNION	LA	16	24	44	60	22
8	6.7	13	8.2	29	15.4	VERMILION	LA	23	87	130	86	41
14	17.0	3	3.3	12	11.1	VERNON	LA	-81	239	-35	07	16
7	6.5	15	11.8	10	7.3	WASHINGTON	LA	81	-38	12	34	06
8	9.3	17	12.8	24	14.9	WEBSTER	LA	38	17	61	69	37
3	21.3	4	16.4	9	25.6	WEST BATON ROUGE	LA	-23	57	20	42	93
5	10.7	8	15.4	10	15.4	WEST CARROLL	LA	44	0	44	60	40
2	18.3	1	5.6	5	42.9	WEST FELICIANA	LA	-69	661	134	87	99
4	5.8	6	9.3	12	16.5	WINN	LA	60	77	183	91	48
876*	19.5	938*	19.8	1051	20.5	MAINE	ME	1	4	5		
99*	25.8	103*	25.1	87	19.5	ANDROSCOGGIN	ME	-3	-22	-24	11	68
60	17.1	55	15.6	73	19.2	AROOSTOOK	ME	-9	23	13	35	66
143	17.7	198*	22.8	210	22.1	CUMBERLAND	ME	29	-3	25	45	82
26	23.5	25	23.5	21	17.7	FRANKLIN	ME	0	-25	-25	11	56
46	24.5	42	21.8	37	16.9	HANCOCK	ME	-11	-23	-31	09	51
89	21.5	83	18.5	71	14.3	KENNEBEC	ME	-14	-23	-34	08	33
33	19.0	28	15.3	36	19.3	KNOX	ME	-20	26	1	25	67
24	19.4	26	20.7	28	19.4	LINCOLN	ME	7	-6	0	25	68

* SIGNIFICANTLY MORE DEATHS THAN EXPECTED

WHITE MALE

CANCER OF THE LARGE INTESTINE

ICD 153,159.0

1950-1959 DEATHS	1950-1959 RATE	1960-1969 DEATHS	1960-1969 RATE	1970-1979 DEATHS	1970-1979 RATE	COUNTY AND STATE	PERCENT CHANGE 60S/50S	PERCENT CHANGE 70S/60S	PERCENT CHANGE 70S/50S	PERCENTILE %CHANGE 70S/50S	PERCENTILE RATES 1970S
39	16.7	44	18.5	55	21.9	OXFORD ME	11	18	31	51	81
91	18.2	92	17.4	119	21.8	PENOBSCOT ME	-4	26	20	42	81
19	16.9	19	17.6	17	18.1	PISCATAQUIS ME	5	3	7	31	59
17	15.8	23	19.7	34	27.5	SAGADAHOC ME	25	40	74	74	95
35	17.5	43	20.5	45	20.4	SOMERSET ME	17	0	16	39	73
22	17.1	28	21.9	24	18.2	WALDO ME	28	-17	7	31	60
44	22.1	34	17.8	44	22.7	WASHINGTON ME	-19	28	3	26	84
89	19.7	95	18.6	150*	25.0	YORK ME	-6	35	27	48	92
1439	18.0	2096*	20.6	2938*	23.8	MARYLAND MD	14	15	32		
58	14.8	76	18.4	105	24.2	ALLEGANY MD	24	32	64	70	90
55	15.7	112	20.5	226*	26.6	ANNE ARUNDEL MD	31	30	69	73	94
198	19.5	313	19.0	603*	25.3	BALTIMORE MD	-3	33	30	50	92
10	27.1	9	15.9	17	19.6	CALVERT MD	-41	23	-28	10	68
13	15.0	11	11.2	17	16.7	CAROLINE MD	-25	48	11	34	50
32	13.6	47	16.9	70	21.1	CARROLL MD	24	25	55	66	77
20	14.3	26	15.8	44	22.7	CECIL MD	10	44	59	68	84
6	10.4	9	10.7	27	24.5	CHARLES MD	3	129	136	87	90
19	16.0	22	16.7	36	26.7	DORCHESTER MD	4	61	68	72	95
38	14.4	72*	24.5	95*	26.6	FREDERICK MD	70	9	85	78	94
10	9.9	8	7.4	20	16.4	GARRETT MD	-26	122	65	70	48
26	15.5	45	18.5	85	23.7	HARFORD MD	19	28	53	65	88
15	20.8	24	18.4	50	25.7	HOWARD MD	-11	40	24	45	93
9	14.2	12	16.9	12	15.4	KENT MD	19	-9	8	32	40
132*	21.8	230	20.6	363	21.0	MONTGOMERY MD	-5	2	-3	22	77
90	18.1	191	20.7	251	21.1	PRINCE GEORGE'S MD	14	2	16	39	77
6	9.8	6	8.0	13	15.5	QUEEN ANNE'S MD	-18	93	59	68	42
3	5.5	19	24.4	27	23.0	ST MARY'S MD	341	-6	315	96	86
12	14.6	9	11.8	12	13.9	SOMERSET MD	-19	18	-5	21	31
10	12.1	15	14.7	22	17.3	TALBOT MD	21	18	43	59	54
38	10.4	64	15.9	117*	24.1	WASHINGTON MD	53	51	131	86	89
20	12.3	27	14.0	45	21.3	WICOMICO MD	14	52	73	74	78
16	18.1	13	13.0	23	19.5	WORCESTER MD	-28	50	8	31	68
603*	21.4	736*	26.7	658*	26.6	BALTIMORE CITY MD	25	0	24	45	94
5063*	23.2	5265*	22.2	6138*	24.1	MASSACHUSETTS MA	-4	9	4		
53	18.2	92	20.6	189	23.3	BARNSTABLE MA	13	13	28	48	87
121	18.7	137	20.1	165	22.0	BERKSHIRE MA	8	9	18	40	81
494*	27.5	475*	24.5	541*	25.3	BRISTOL MA	-11	4	-8	19	92
6	18.7	12*	35.1	11	25.6	DUKES MA	88	-27	37	55	93
587*	22.3	614*	21.5	723*	23.8	ESSEX MA	-4	11	7	30	88
68*	24.4	72	23.3	63	19.0	FRANKLIN MA	-4	-18	-22	12	65
368*	21.8	379	20.1	473*	23.2	HAMPDEN MA	-8	15	6	30	86
81	19.6	89	19.3	131*	26.0	HAMPSHIRE MA	-1	35	33	52	93
1149*	24.1	1208*	23.1	1409*	25.0	MIDDLESEX MA	-4	8	4	27	92
3	22.2	3	15.7	6	31.9	NANTUCKET MA	-29	103	44	60	98
404*	21.7	525*	23.2	629*	23.8	NORFOLK MA	7	3	10	33	88
225*	21.3	254	20.9	361*	25.0	PLYMOUTH MA	-2	20	18	40	92
935*	25.4	853*	24.5	808*	27.4	SUFFOLK MA	-4	12	8	31	95
569*	21.4	552	19.1	629	20.6	WORCESTER MA	-11	8	-4	22	74
4843*	18.4	5782*	19.2	6843*	20.8	MICHIGAN MI	4	8	13		
3	9.0	9	18.7	16	23.8	ALCONA MI	107	27	163	90	88
9	18.2	4	7.1	14	25.2	ALGER MI	-61	253	38	56	92
40	16.2	45	16.3	51	16.0	ALLEGAN MI	1	-1	-1	24	45
16	14.5	22	17.7	34	24.8	ALPENA MI	22	40	71	73	91
16	26.5	9	12.1	26	30.5	ANTRIM MI	-54	153	15	38	98
6	10.6	10	16.6	17	22.3	ARENAC MI	56	34	110	83	83
8	20.3	7	15.3	8	15.5	BARAGA MI	-25	1	-24	11	41
21	14.0	36	22.0	35	18.1	BARRY MI	56	-17	29	50	59
71	18.5	75	18.1	99	22.0	BAY MI	-2	21	19	41	82
6	13.5	6	11.4	11	17.8	BENZIE MI	-16	56	32	51	57
87	16.7	126	20.9	124	18.8	BERRIEN MI	25	-10	13	36	64
12	7.3	30	17.4	42	22.7	BRANCH MI	139	31	213	94	85
88	17.3	97	17.6	117	19.9	CALHOUN MI	1	13	15	37	70
33	21.9	31	18.2	42	22.3	CASS MI	-17	22	2	25	83
9	10.6	5	6.4	19	19.7	CHARLEVOIX MI	-40	210	85	78	69
7	8.9	9	10.9	18	15.9	CHEBOYGAN MI	23	46	80	76	45
22	16.9	19	14.4	31	22.1	CHIPPEWA MI	-14	53	31	51	82
4	6.7	18	23.5	26	18.3	CLARE MI	248	-22	171	91	60
20	13.3	23	13.4	34	19.7	CLINTON MI	1	47	48	62	69
1	3.5	5	17.8	13	28.8	CRAWFORD MI	406	62	720	99	97
23	13.8	31	17.0	38	18.9	DELTA MI	23	11	37	55	64
26	19.6	23	15.8	36	20.2	DICKINSON MI	-19	27	3	26	72
33	15.7	50	22.7	53	19.5	EATON MI	44	-14	24	45	68
16	18.7	16	17.9	19	19.6	EMMET MI	-5	10	4	28	69
169	16.6	237	20.4	300*	23.1	GENESEE MI	23	13	39	56	86
10	19.3	13	17.6	20	18.9	GLADWIN MI	-9	8	-2	23	65
23	16.0	26	15.7	20	13.0	GOGEBIC MI	-2	-17	-18	13	26
18	10.9	27	14.3	47	23.2	GRAND TRAVERSE MI	31	62	113	84	86
31	17.3	20	12.0	36	20.3	GRATIOT MI	-31	70	18	40	73
31	16.9	33	18.1	29	14.1	HILLSDALE MI	7	-22	-17	14	32
49	20.3	47	21.0	46	19.5	HOUGHTON MI	4	-7	-4	22	68
36	20.7	31	16.1	38	18.0	HURON MI	-22	12	-13	16	59
105	16.6	130	18.5	170	22.8	INGHAM MI	12	23	38	55	85
36	17.8	36	18.3	31	15.8	IONIA MI	3	-14	-11	16	43
10	15.5	10	10.7	29	20.0	IOSCO MI	-31	88	29	49	71
21	20.6	23	21.5	15	13.2	IRON MI	4	-38	-36	07	27
16	13.0	19	14.8	25	16.9	ISABELLA MI	14	14	30	51	51
94	19.6	83	15.8	124	21.8	JACKSON MI	-19	37	11	34	80
75	13.1	116	18.2	134	19.6	KALAMAZOO MI	39	8	50	63	69
2	6.7	8	27.2	8	17.3	KALKASKA MI	304	-36	157	90	54
207	15.9	296	20.0	268	16.4	KENT MI	26	-18	3	26	48
4	25.1	1	5.0	1	3.4	KEWEENAW MI	-80	-32	-87	00	02
5	20.9	7	22.0	8	17.1	LAKE MI	5	-22	-18	13	53
29	16.4	26	13.5	39	18.5	LAPEER MI	-18	37	13	35	62
11	21.5	10	17.0	13	18.8	LEELANAU MI	-21	10	-13	16	63
54	17.5	52	15.8	56	16.2	LENAWEE MI	-10	2	-8	19	46

* SIGNIFICANTLY MORE DEATHS THAN EXPECTED

WHITE MALE

CANCER OF THE LARGE INTESTINE

ICD 153,159.0

1950-1959 DEATHS	1950-1959 RATE	1960-1969 DEATHS	1960-1969 RATE	1970-1979 DEATHS	1970-1979 RATE	COUNTY AND STATE	PERCENT CHANGE 60S/50S	PERCENT CHANGE 70S/60S	PERCENT CHANGE 70S/50S	PERCENTILE %CHANGE 70S/50S	PERCENTILE RATES 1970S
14	9.2	25	13.6	61	22.7	LIVINGSTON MI	48	67	147	89	84
6	12.5	4	8.3	7	18.3	LUCE MI	-33	120	47	62	60
8	16.5	4	8.0	9	15.5	MACKINAC MI	-51	93	-6	20	41
130	18.2	242	18.9	442*	22.2	MACOMB MI	4	17	22	43	82
13	11.8	20	18.0	22	17.0	MANISTEE MI	53	-6	44	60	52
33	14.7	52	22.2	47	18.4	MARQUETTE MI	51	-17	25	46	61
19	16.0	31	24.5	25	17.7	MASON MI	53	-28	10	33	56
21	20.1	13	12.1	23	18.6	MECOSTA MI	-40	53	-7	19	62
22	16.8	26	18.4	34	23.9	MENOMINEE MI	10	30	43	59	89
22	16.9	22	11.3	42	20.0	MIDLAND MI	-33	77	18	40	71
5	13.4	11	28.5	6	11.4	MISSAUKEE MI	114	-60	-15	15	17
47	13.8	69	18.4	85	19.4	MONROE MI	33	5	40	57	67
37	21.2	34	17.6	38	17.2	MONTCALM MI	-17	-2	-19	13	53
5	21.2	5	14.7	12	23.2	MONTMORENCY MI	-30	57	9	33	86
86	18.4	86	16.4	115	19.4	MUSKEGON MI	-11	18	5	29	67
23	18.5	17	12.3	26	15.3	NEWAYGO MI	-34	25	-17	14	40
273	18.5	445	19.1	683	21.2	OAKLAND MI	4	11	15	38	78
19	20.0	26*	26.5	19	18.4	OCEANA MI	32	-31	-8	19	61
9	18.7	15	20.8	24	24.4	OGEMAW MI	11	17	30	51	90
12	19.0	4	5.7	14	21.3	ONTONAGON MI	-70	272	12	35	78
12	15.3	17	21.8	16	19.2	OSCEOLA MI	42	-12	25	46	66
3	14.7	4	15.3	8	15.7	OSCODA MI	4	3	7	31	43
4	10.1	6	11.4	13	20.8	OTSEGO MI	13	82	106	83	75
58	17.2	61	14.9	119	23.5	OTTAWA MI	-13	57	36	54	87
9	14.4	14	21.0	13	15.5	PRESQUE ISLE MI	46	-26	8	32	41
11	27.5	7	11.6	31*	27.7	ROSCOMMON MI	-58	138	1	25	96
125	20.5	133	18.9	162	21.6	SAGINAW MI	-8	14	5	29	79
77	16.5	94	18.6	146*	26.0	ST CLAIR MI	12	40	58	67	93
33	16.5	38	17.8	45	18.7	ST JOSEPH MI	8	5	14	36	63
37	20.5	28	15.1	46	20.8	SANILAC MI	-27	38	1	25	75
8	17.5	16*	34.5	14	25.9	SCHOOLCRAFT MI	97	-25	48	62	93
37	16.1	54	22.9	49	20.1	SHIAWASSEE MI	42	-12	25	45	72
33	15.5	43	20.8	39	17.8	TUSCOLA MI	34	-14	15	37	57
43	17.4	43	16.8	49	17.8	VAN BUREN MI	-4	6	2	26	57
98*	22.0	95	18.3	111	18.1	WASHTENAW MI	-17	-1	-18	13	59
1820*	22.0	1997*	21.8	1844*	22.2	WAYNE MI	-1	2	1	25	82
18	17.8	24	23.9	24	20.2	WEXFORD MI	35	-16	14	36	72
2369	15.9	2917	17.3	3531	19.1	MINNESOTA MN	9	10	20		
14	16.4	14	15.0	22	20.4	AITKIN MN	-9	36	24	45	73
20	13.6	28	12.3	62	16.8	ANOKA MN	-9	36	24	45	50
15	11.5	23	16.3	21	12.9	BECKER MN	41	-20	12	35	25
12	10.6	19	15.5	29	22.8	BELTRAMI MN	46	47	115	84	85
12	16.0	8	9.2	19	17.2	BENTON MN	-42	87	8	31	53
8	15.1	4	7.7	13	20.8	BIG STONE MN	-49	171	38	56	75
30	15.1	42	19.6	49	21.2	BLUE EARTH MN	30	8	40	57	78
27	19.9	31	20.5	41	24.4	BROWN MN	3	19	22	43	90
15	11.0	18	12.6	22	15.1	CARLTON MN	14	20	37	55	38
13	13.5	28	23.8	26	18.9	CARVER MN	76	-21	40	57	64
12	11.3	23	17.9	24	14.4	CASS MN	59	-19	28	49	34
10	11.1	16	16.9	21	20.9	CHIPPEWA MN	52	23	87	78	76
17	18.0	19	17.6	27	22.5	CHISAGO MN	-2	28	25	46	83
20	15.6	22	13.1	30	15.4	CLAY MN	-16	18	-1	24	40
4	6.7	12	19.6	11	16.6	CLEARWATER MN	192	-16	146	89	49
5	25.5	2	9.2	1	5.1	COOK MN	-64	-44	-80	00	03
16	19.5	17	18.4	17	17.1	COTTONWOOD MN	-6	-7	-12	16	53
26	15.1	24	12.2	49	20.8	CROW WING MN	-19	70	38	56	75
37	17.2	59	20.8	94	21.6	DAKOTA MN	21	4	25	46	79
10	13.8	12	16.3	16	19.9	DODGE MN	18	22	44	60	71
19	15.5	26	17.2	26	15.0	DOUGLAS MN	11	-13	-3	22	37
15	11.7	24	19.6	30	20.7	FARIBAULT MN	67	6	76	75	75
26	18.4	22	14.7	25	14.4	FILLMORE MN	-20	-2	-22	12	34
23	15.4	27	14.8	33	15.3	FREEBORN MN	-4	3	-1	24	39
25	12.8	33	16.2	47	20.7	GOODHUE MN	27	28	62	70	75
11	19.6	13	21.3	12	17.8	GRANT MN	8	-16	-9	18	57
505	16.3	682	19.4	783	21.0	HENNEPIN MN	19	8	29	49	76
12	14.5	11	11.5	22	20.4	HOUSTON MN	-21	78	40	57	73
12	16.9	12	15.8	15	17.6	HUBBARD MN	-6	12	4	28	56
14	17.9	12	13.8	15	15.0	ISANTI MN	-23	8	-16	14	37
17	10.1	27	13.5	36	16.5	ITASCA MN	34	22	64	70	49
13	16.3	9	10.5	18	19.6	JACKSON MN	-35	86	21	42	69
8	15.3	10	14.9	22*	30.7	KANABEC MN	-2	105	100	81	98
11	6.7	30	16.1	31	14.9	KANDIYOHI MN	138	-8	120	85	37
8	15.5	8	15.2	12	21.8	KITTSON MN	-2	43	40	57	80
11	13.1	17	17.3	19	20.7	KOOCHICHING MN	32	20	59	68	75
9	11.4	11	13.1	19	22.8	LAC QUI PARLE MN	15	74	100	81	85
7	17.4	8	13.2	6	9.3	LAKE MN	-24	-30	-47	04	10
2	7.0	4	13.7	5	20.6	LAKE OF THE WOODS MN	97	50	195	93	74
24	21.2	27	21.7	28	20.2	LE SUEUR MN	3	-7	-4	21	72
10	20.8	7	12.3	8	12.6	LINCOLN MN	-41	3	-39	06	23
13	11.6	23	18.7	21	15.4	LYON MN	62	-18	33	52	41
19	14.3	29	19.5	35	21.3	MC LEOD MN	36	10	49	63	78
4	15.1	2	5.7	8	22.0	MAHNOMEN MN	-62	283	45	61	82
9	11.7	18	20.4	8	8.7	MARSHALL MN	75	-57	-25	10	09
16	12.6	19	13.3	25	16.1	MARTIN MN	6	21	27	48	45
15	15.3	24	19.4	23	17.5	MEEKER MN	27	-10	15	37	55
15	16.6	14	14.2	20	17.7	MILLE LACS MN	-14	24	7	30	56
23	17.9	29	18.7	25	15.9	MORRISON MN	5	-15	-11	17	44
26	14.0	39	18.8	36	15.0	MOWER MN	35	-20	8	31	38
12	18.0	17	23.5	15	18.7	MURRAY MN	30	-20	4	27	63
18	15.5	19	16.5	20	18.0	NICOLLET MN	6	9	16	39	59
16	15.6	18	15.1	26	19.1	NOBLES MN	-4	27	22	43	66
7	10.3	18	22.0	20	24.8	NORMAN MN	113	13	141	88	91
40	17.5	36	13.6	51	16.4	OLMSTED MN	-22	21	-6	20	48
36	12.0	48	14.5	74	20.1	OTTER TAIL MN	22	38	68	72	72
11	14.5	12	15.2	13	16.1	PENNINGTON MN	5	7	11	34	46
13	11.0	16	14.5	39*	32.5	PINE MN	32	124	195	93	98
5	7.3	17	22.2	20	23.9	PIPESTONE MN	203	7	225	94	89
19	9.5	28	13.0	40	17.8	POLK MN	37	37	87	78	57
11	14.4	13	16.2	15	16.2	POPE MN	12	0	12	35	46

* SIGNIFICANTLY MORE DEATHS THAN EXPECTED

WHITE MALE — CANCER OF THE LARGE INTESTINE — ICD 153,159.0

1950-1959 DEATHS	1950-1959 RATE	1960-1969 DEATHS	1960-1969 RATE	1970-1979 DEATHS	1970-1979 RATE	COUNTY AND STATE	PERCENT CHANGE 60S/50S	PERCENT CHANGE 70S/60S	PERCENT CHANGE 70S/50S	PERCENTILE %CHANGE 70S/50S	PERCENTILE RATES 1970S
317*	20.4	363*	21.0	396	21.0	RAMSEY MN	3	0	3	26	76
1	2.5	9	27.2	4	10.0	RED LAKE MN	969	-63	292	96	12
20	17.9	17	14.1	23	16.9	REDWOOD MN	-21	20	-5	21	51
21	16.7	22	15.9	27	18.3	RENVILLE MN	-4	15	10	33	60
25	15.3	23	12.4	36	17.3	RICE MN	-19	40	13	36	54
8	13.8	10	16.4	14	19.6	ROCK MN	18	20	42	58	69
11	15.2	14	18.4	15	17.3	ROSEAU MN	21	-6	14	37	54
204	17.5	185	15.3	200	17.0	ST LOUIS MN	-13	11	-3	22	52
16	18.0	26	24.5	31	23.6	SCOTT MN	37	-4	31	51	88
10	17.6	14	17.8	24	25.1	SHERBURNE MN	1	41	43	59	92
9	10.1	12	12.3	24	22.8	SIBLEY MN	22	86	127	86	85
52	17.0	58	16.0	98	22.9	STEARNS MN	-6	43	35	53	85
24	22.4	17	13.7	30	20.4	STEELE MN	-39	49	-9	18	73
11	20.5	13	22.8	5	7.8	STEVENS MN	11	-66	-62	01	07
9	11.4	28*	32.0	18	19.2	SWIFT MN	179	-40	68	72	66
16	12.3	30	20.1	26	17.6	TODD MN	64	-13	43	59	56
7	18.6	8	17.5	4	8.4	TRAVERSE MN	-6	-52	-55	02	08
16	16.3	24	23.5	17	15.0	WABASHA MN	44	-36	-8	18	37
7	11.8	11	14.8	15	18.3	WADENA MN	26	24	56	66	61
15	17.6	12	13.2	17	17.2	WASECA MN	-25	31	-2	23	54
33	18.4	39	18.6	37	13.4	WASHINGTON MN	1	-28	-27	10	28
15	20.1	10	11.7	15	18.0	WATONWAN MN	-42	53	-11	17	59
7	15.6	10	19.7	7	12.6	WILKIN MN	26	-36	-19	13	23
40	20.1	36	16.6	54	23.5	WINONA MN	-18	41	16	39	87
28	17.0	34	17.8	47	20.3	WRIGHT MN	5	14	19	41	73
14	15.2	12	11.8	7	8.6	YELLOW MEDICINE MN	-22	-28	-43	05	09
505	10.7	773	13.5	1050	15.4	MISSISSIPPI MS	27	14	44		
6	23.2	12	20.6	10	13.6	ADAMS MS	-11	-34	-42	05	29
3	3.1	20	16.7	26	19.4	ALCORN MS	446	17	536	98	68
7	19.8	4	8.3	6	13.9	AMITE MS	-58	67	-30	09	31
9	13.6	5	7.9	16	19.0	ATTALA MS	-42	141	39	57	65
1	4.7	1	3.6	6	28.9	BENTON MS	-22	697	520	98	97
9	14.7	7	9.5	12	13.7	BOLIVAR MS	-35	44	-6	20	30
3	5.1	11	17.2	7	10.0	CALHOUN MS	239	-42	96	80	12
3	9.4	2	9.0	4	13.4	CARROLL MS	-3	48	43	59	28
7	14.3	1	1.9	10	16.8	CHICKASAW MS	-87	799	18	40	50
5	11.6	2	4.9	5	12.5	CHOCTAW MS	-58	156	8	32	22
1	6.0	1	7.3	1	7.5	CLAIBORNE MS	22	3	26	47	06
4	8.9	3	5.4	6	10.4	CLARKE MS	-39	93	17	40	14
2	4.5	6	15.7	5	9.6	CLAY MS	246	-39	112	84	11
10	18.9	14	23.1	5	7.7	COAHOMA MS	22	-67	-59	02	07
10	14.3	15	19.6	13	17.0	COPIAH MS	37	-13	19	41	52
5	10.7	5	13.7	7	12.3	COVINGTON MS	29	-11	15	38	20
4	10.9	8	11.6	25	23.9	DE SOTO MS	7	106	120	85	89
14	11.1	28	18.4	29	15.2	FORREST MS	67	-18	37	55	39
3	10.0	5	15.6	4	13.9	FRANKLIN MS	56	-11	38	56	31
4	13.5	1	1.9	4	6.8	GEORGE MS	-86	260	-50	03	05
4	13.1	4	14.7	5	14.2	GREENE MS	12	-3	8	32	33
4	11.7	12	28.6	12	20.5	GRENADA MS	146	-28	76	75	74
4	7.1	7	10.2	11	12.9	HANCOCK MS	43	27	81	77	25
18	6.2	48	14.7	64	14.7	HARRISON MS	137	0	137	88	36
28	9.5	67	18.0	106	21.3	HINDS MS	89	18	123	85	78
5	11.7	7	15.8	4	8.8	HOLMES MS	35	-44	-25	11	09
1	6.8	5	19.3	5	20.0	HUMPHREYS MS	186	4	196	93	71
1	21.5	1	17.5	2	36.7	ISSAQUENA MS	-18	109	70	73	99
5	7.8	6	7.9	12	12.5	ITAWAMBA MS	1	59	61	69	23
13	11.8	18	12.1	30	11.5	JACKSON MS	2	-5	-3	23	18
4	9.2	5	10.9	9	17.5	JASPER MS	19	61	91	79	55
3	15.7	2	12.4	1	6.5	JEFFERSON MS	-21	-47	-59	02	05
5	22.3	2	6.0	7	19.9	JEFFERSON DAVIS MS	-73	232	-11	17	70
17	10.7	22	11.5	27	13.1	JONES MS	8	13	22	43	26
2	7.5	4	12.5	5	13.9	KEMPER MS	66	11	85	78	31
7	14.0	10	17.7	10	13.2	LAFAYETTE MS	27	-26	-6	20	27
4	9.8	4	7.2	8	12.0	LAMAR MS	-26	67	23	44	20
13	7.6	20	9.8	40	17.6	LAUDERDALE MS	29	81	132	87	56
4	11.8	6	15.7	6	15.3	LAWRENCE MS	33	-2	29	50	39
4	6.1	12	19.3	15	19.8	LEAKE MS	214	3	222	94	70
11	11.8	23	15.5	32	17.8	LEE MS	31	15	51	64	57
10	18.2	16	24.0	16	18.4	LEFLORE MS	32	-23	1	25	61
6	9.3	9	10.3	13	12.4	LINCOLN MS	12	20	34	53	21
12	15.2	9	8.8	18	16.2	LOWNDES MS	-42	84	6	30	46
2	4.6	5	11.0	12	20.4	MADISON MS	139	86	344	97	73
4	7.1	8	10.8	10	12.1	MARION MS	53	12	72	73	20
6	18.1	6	15.2	6	10.6	MARSHALL MS	-16	-30	-42	05	14
5	5.4	5	4.4	17	13.7	MONROE MS	-18	210	155	89	30
7	15.7	7	18.1	8	18.5	MONTGOMERY MS	16	2	18	40	62
12	17.2	9	9.6	9	9.2	NESHOBA MS	-44	-4	-47	04	10
5	7.2	10	13.9	9	10.8	NEWTON MS	94	-23	50	64	15
3	16.2	1	3.6	4	17.4	NOXUBEE MS	-78	388	7	31	55
4	9.4	8	16.0	11	16.5	OKTIBBEHA MS	69	3	75	75	48
4	6.5	12	16.7	14	17.1	PANOLA MS	158	2	164	90	52
11	17.0	11	13.4	9	7.4	PEARL RIVER MS	-21	-45	-56	02	06
2	6.2	4	15.0	3	7.8	PERRY MS	142	-48	25	46	07
14	16.1	18	19.1	12	11.7	PIKE MS	18	-39	-27	10	19
4	6.4	8	10.4	14	14.3	PONTOTOC MS	62	37	123	85	33
4	10.1	6	6.9	11	10.8	PRENTISS MS	-32	58	7	31	15
5	18.2	6	19.3	5	15.6	QUITMAN MS	6	-19	-14	15	42
6	10.2	12	12.4	16	10.4	RANKIN MS	21	-16	2	25	14
5	7.8	7	10.0	17	22.6	SCOTT MS	29	126	191	92	84
		2	14.9	2	10.0	SHARKEY MS		-33			12
6	10.0	11	15.4	14	19.0	SIMPSON MS	55	23	90	79	65
5	8.9	7	11.2	5	7.8	SMITH MS	26	-30	-12	16	07
3	15.8	3	11.1	3	8.9	STONE MS	-30	-19	-44	05	10
10	17.8	11	17.2	14	22.1	SUNFLOWER MS	-3	28	24	45	82
2	4.7	5	10.5	9	19.5	TALLAHATCHIE MS	124	85	314	96	68
4	13.1	10	24.1	9	16.7	TATE MS	84	-30	28	49	50
6	9.0	10	13.8	15	18.7	TIPPAH MS	52	36	107	83	63
4	6.2	10	14.3	12	13.4	TISHOMINGO MS	132	-6	117	85	28
6*	49.1	4	28.0	2	16.1	TUNICA MS	-43	-43	-67	01	45

* SIGNIFICANTLY MORE DEATHS THAN EXPECTED

WHITE MALE | CANCER OF THE LARGE INTESTINE | ICD 153,159.0

1950-1959 DEATHS	1950-1959 RATE	1960-1969 DEATHS	1960-1969 RATE	1970-1979 DEATHS	1970-1979 RATE	COUNTY AND STATE	PERCENT CHANGE 60S/50S	PERCENT CHANGE 70S/60S	PERCENT CHANGE 70S/50S	PERCENTILE %CHANGE 70S/50S	PERCENTILE RATES 1970S
2	2.3	15	18.1	11	11.5	UNION MS	700	-37	407	98	18
5	14.0	2	4.6	11	27.0	WALTHALL MS	-67	487	93	80	95
16	23.0	13	14.1	23	23.6	WARREN MS	-39	68	3	26	88
8	11.1	21	20.8	20	18.6	WASHINGTON MS	86	-10	68	71	62
2	6.0	3	6.3	11	20.5	WAYNE MS	6	225	243	95	74
6	14.7	5	9.2	5	7.9	WEBSTER MS	-38	-14	-46	04	08
5	28.9	7	32.1	2	10.5	WILKINSON MS	11	-67	-64	01	14
7	12.0	5	8.1	13	19.2	WINSTON MS	-33	138	60	69	66
5	10.3	5	10.4	10	18.7	YALOBUSHA MS	1	80	81	77	63
5	10.4	11	18.8	13	20.9	YAZOO MS	81	11	101	81	76
3112	16.1	3603	17.3	4365	19.7	MISSOURI MO	7	14	22		
15	12.4	12	9.8	24	21.1	ADAIR MO	-21	114	70	73	77
11	14.4	8	10.6	13	15.4	ANDREW MO	-26	45	7	30	41
13	22.4	10	18.0	7	12.9	ATCHISON MO	-20	-28	-43	05	24
14	11.4	27	20.3	29	20.9	AUDRAIN MO	78	3	83	77	76
11	9.0	23	16.7	25	16.4	BARRY MO	86	-2	82	77	47
7	7.9	7	7.5	11	11.9	BARTON MO	-4	58	51	64	19
14	10.7	17	13.1	23	19.8	BATES MO	22	51	84	78	70
11	14.0	8	8.6	15	13.5	BENTON MO	-38	56	-4	22	28
7	10.8	8	12.8	9	13.4	BOLLINGER MO	19	4	24	45	28
21	10.7	23	10.9	50	19.1	BOONE MO	2	75	78	76	65
66	12.8	97	20.5	99	21.7	BUCHANAN MO	60	6	69	73	80
23	12.9	21	11.0	19	9.6	BUTLER MO	-14	-13	-25	11	11
13	16.9	8	11.3	21	28.3	CALDWELL MO	-33	150	68	72	96
10	7.2	22	15.7	20	13.8	CALLAWAY MO	118	-12	92	80	30
5	9.4	8	11.7	19	14.0	CAMDEN MO	25	20	50	63	31
19	11.3	26	12.2	36	14.3	CAPE GIRARDEAU MO	8	17	27	48	33
9	9.4	8	7.3	13	14.4	CARROLL MO	-22	96	54	66	34
6	19.4	4	11.5	6	19.9	CARTER MO	-41	72	3	26	71
16	12.4	20	13.4	22	12.8	CASS MO	8	-5	3	26	24
13	18.7	15	17.0	15	17.4	CEDAR MO	-9	2	-7	20	55
14	14.5	17	18.6	23	26.6	CHARITON MO	28	44	84	77	94
8	10.5	11	13.7	14	14.9	CHRISTIAN MO	30	8	41	58	37
13	19.6	13	19.2	10	16.6	CLARK MO	-2	-14	-15	15	49
35	15.1	61	20.7	81	19.3	CLAY MO	37	-7	28	49	67
7	9.1	9	11.2	13	16.9	CLINTON MO	24	51	87	78	51
21	13.3	20	10.8	43	20.2	COLE MO	-19	88	53	65	72
12	13.0	11	12.5	16	18.2	COOPER MO	-4	45	40	57	60
14	18.3	13	14.8	17	14.4	CRAWFORD MO	-19	-3	-21	12	34
9	13.8	6	8.4	9	15.5	DADE MO	-39	85	12	35	41
8	10.7	8	8.5	7	8.9	DALLAS MO	-21	5	-17	14	10
9	10.1	15	18.4	10	12.4	DAVIESS MO	83	-33	22	43	21
5	8.3	9	16.4	13	19.3	DE KALB MO	98	18	134	87	67
16	22.7	6	8.1	14	14.8	DENT MO	-64	83	-35	07	36
3	4.8	6	8.7	12	16.9	DOUGLAS MO	82	93	252	95	51
20	11.7	22	11.6	29	15.3	DUNKLIN MO	0	32	32	51	40
35	16.5	42	16.4	52	16.9	FRANKLIN MO	0	3	3	26	51
14	17.6	20	22.3	30*	30.7	GASCONADE MO	27	37	75	74	98
12	15.1	10	12.3	10	13.5	GENTRY MO	-19	10	-11	17	28
57	10.2	80	12.5	129	17.2	GREENE MO	23	37	69	72	53
14	14.3	19	19.8	22	25.2	GRUNDY MO	38	27	76	75	92
14	13.8	16	18.8	14	16.6	HARRISON MO	36	-12	20	42	49
24	16.7	23	15.6	23	15.9	HENRY MO	-6	2	-5	21	44
4	12.0	6	13.6	5	7.9	HICKORY MO	13	-42	-34	07	07
15	20.3	7	11.4	6	9.8	HOLT MO	-44	-14	-52	03	12
10	13.7	7	10.0	15	22.5	HOWARD MO	-28	126	64	70	84
18	14.7	17	12.5	33	19.1	HOWELL MO	-15	53	30	50	65
8	16.8	5	9.1	15	22.7	IRON MO	-46	150	35	54	84
467*	20.3	466	19.1	487	20.3	JACKSON MO	-6	6	0	24	73
59	14.4	65	15.3	84	19.2	JASPER MO	6	26	33	52	66
28	13.9	39	13.1	79	19.6	JEFFERSON MO	-6	49	41	58	69
24	18.2	25	17.7	19	13.5	JOHNSON MO	-3	-24	-26	10	28
9	17.6	4	7.5	11	25.0	KNOX MO	-58	234	42	58	91
10	8.9	17	14.6	26	19.0	LACLEDE MO	63	30	113	84	65
21	14.2	25	15.7	29	16.2	LAFAYETTE MO	11	3	14	37	46
21	13.0	21	12.5	23	12.8	LAWRENCE MO	-4	3	-2	23	24
14	18.8	11	15.9	11	18.2	LEWIS MO	-15	14	-3	22	60
15	15.4	10	9.0	26	21.9	LINCOLN MO	-42	144	42	58	81
12	8.5	31	23.2	27	21.8	LINN MO	174	-6	157	90	80
19	18.2	10	8.5	22	21.2	LIVINGSTON MO	-53	149	17	39	78
12	13.8	13	15.1	10	12.6	MC DONALD MO	9	-16	-8	19	23
22	15.7	11	8.4	14	10.7	MACON MO	-47	27	-32	08	15
4	8.0	6	10.1	10	12.0	MADISON MO	26	19	49	63	20
4	9.3	5	12.1	5	9.0	MARIES MO	30	-26	-4	22	10
25	15.5	33	19.8	27	16.2	MARION MO	28	-19	4	28	46
4	7.8	7	15.6	3	5.9	MERCER MO	100	-62	-24	11	04
18	22.1	11	11.1	11	10.0	MILLER MO	-50	-10	-55	03	13
7	10.1	10	13.7	15	20.0	MISSISSIPPI MO	36	46	99	81	71
10	13.6	20	26.0	8	11.2	MONITEAU MO	91	-57	-18	13	17
7	10.1	13	16.7	21	29.0	MONROE MO	65	74	188	92	97
11	13.0	8	10.4	14	16.6	MONTGOMERY MO	-20	60	28	48	49
9	11.6	10	11.7	22	22.2	MORGAN MO	0	90	91	79	82
9	10.7	13	13.0	17	17.9	NEW MADRID MO	22	38	68	72	58
14	9.0	20	12.2	25	12.5	NEWTON MO	35	2	38	56	22
28	19.0	30	21.8	22	17.9	NODAWAY MO	15	-18	-6	21	58
8	11.9	9	12.3	8	11.4	OREGON MO	4	-7	-4	22	18
8	13.7	12	17.6	15	20.5	OSAGE MO	29	16	50	64	74
7	18.9	7	15.7	7	11.8	OZARK MO	-17	-25	-38	06	19
10	8.6	13	11.7	12	10.8	PEMISCOT MO	36	-8	26	47	15
12	14.4	19	20.8	16	14.9	PERRY MO	44	-28	3	27	37
28	15.0	35	16.9	48	22.9	PETTIS MO	13	35	52	65	85
19	16.1	22	17.0	29	18.9	PHELPS MO	6	11	17	39	64
15	15.1	17	16.5	15	14.8	PIKE MO	9	-11	-3	23	36
15	16.8	15	13.2	20	15.1	PLATTE MO	-21	14	-10	18	38
15	14.0	10	9.6	14	11.3	POLK MO	-32	17	-20	12	17
7	10.7	10	12.3	15	16.8	PULASKI MO	15	36	57	67	50
9	16.0	10	17.4	3	4.7	PUTNAM MO	9	-73	-71	00	03
5	8.5	4	6.5	13	24.0	RALLS MO	-24	270	182	91	89
14	8.8	28	16.8	34	23.9	RANDOLPH MO	91	42	172	91	89

* SIGNIFICANTLY MORE DEATHS THAN EXPECTED

WHITE MALE CANCER OF THE LARGE INTESTINE ICD 153,159.0

1950-1959 DEATHS	1950-1959 RATE	1960-1969 DEATHS	1960-1969 RATE	1970-1979 DEATHS	1970-1979 RATE	COUNTY AND STATE	PERCENT CHANGE 60S/50S	PERCENT CHANGE 70S/60S	PERCENT CHANGE 70S/50S	PERCENTILE %CHANGE 70S/50S	PERCENTILE RATES 1970S
14	13.1	6	5.3	20	19.5	RAY MO	-60	270	49	62	68
6	15.1	4	11.1	6	15.4	REYNOLDS MO	-26	39	2	26	41
6	10.1	5	5.5	12	13.1	RIPLEY MO	-45	137	30	50	26
24	15.4	39	19.5	64	20.6	ST CHARLES MO	27	6	34	53	74
18	25.5	10	13.0	13	18.9	ST CLAIR MO	-49	46	-26	10	64
9	14.7	9	13.5	17	22.2	STE GENEVIEVE MO	-8	64	51	64	82
19	9.9	38	18.4	33	14.7	ST FRANCOIS MO	85	-20	48	62	36
328	18.4	610*	23.1	918*	26.6	ST LOUIS COUNTY MO	25	15	44	60	94
33	21.7	37	23.6	48*	30.2	SALINE MO	9	28	39	57	97
4	13.4	3	7.2	6	13.5	SCHUYLER MO	-46	88	1	25	29
8	14.6	8	17.4	8	15.9	SCOTLAND MO	19	-9	8	32	44
15	12.1	21	14.5	31	19.1	SCOTT MO	19	32	57	67	66
3	6.3	8	19.0	2	4.8	SHANNON MO	202	-75	-23	11	03
14	16.6	6	8.6	7	11.0	SHELBY MO	-48	28	-34	08	16
18	13.5	22	14.3	27	15.5	STODDARD MO	5	9	15	37	41
6	12.9	6	10.4	10	12.7	STONE MO	-19	22	-1	24	23
15	21.0	10	14.6	9	12.2	SULLIVAN MO	-30	-16	-42	05	20
13	24.9	15	15.5	22	14.8	TANEY MO	-38	-4	-40	06	37
13	12.3	20	17.2	29	21.9	TEXAS MO	40	27	78	76	81
10	6.1	24	15.4	19	13.4	VERNON MO	151	-13	119	85	28
8	14.0	11	16.9	11	14.5	WARREN MO	20	-14	3	27	34
9	13.6	7	9.9	10	13.5	WASHINGTON MO	-27	37	-1	24	29
10	15.5	10	15.7	11	13.2	WAYNE MO	2	-16	-14	15	27
12	13.8	18	17.2	16	15.4	WEBSTER MO	24	-10	12	34	41
6	19.9	3	8.6	5	21.2	WORTH MO	-57	146	6	30	77
8	8.8	11	10.9	13	12.8	WRIGHT MO	25	17	46	61	24
705*	22.3	646*	24.0	540*	26.5	ST LOUIS CITY MO	7	10	19	41	94
380	12.7	513	15.9	582	17.0	MONTANA MT	25	7	34		
6	13.9	7	15.4	6	13.5	BEAVERHEAD MT	11	-12	-3	23	29
3	10.3	2	6.9	4	11.8	BIG HORN MT	-33	72	15	37	19
2	5.1	8	21.6	2	7.0	BLAINE MT	322	-68	36	54	06
2	9.6	3	18.3	7	37.1	BROADWATER MT	90	103	285	96	99
4	7.2	7	11.6	9	15.1	CARBON MT	60	30	108	83	38
1	8.3	1	7.7	2	8.3	CARTER MT	-8	8	-1	24	08
39	17.0	54	18.4	57	17.6	CASCADE MT	8	-4	4	27	56
7	19.0	10	25.8	6	15.9	CHOUTEAU MT	36	-38	-17	14	44
4	4.7	2	2.6	13	15.1	CUSTER MT	-46	484	217	94	38
2	6.6	6	43.0	2	6.4	DANIELS MT	547	-85	-4	22	05
7	14.4	3	6.0	11	23.3	DAWSON MT	-59	291	61	69	87
11	11.2	13	13.3	17	22.9	DEER LODGE MT	19	72	105	82	85
4	31.9	6	31.6	1	6.0	FALLON MT	-1	-81	-81	00	04
10	10.8	15	17.9	15	18.3	FERGUS MT	66	2	70	73	60
23	13.2	24	12.3	44	20.1	FLATHEAD MT	-7	64	53	65	72
11	10.1	25	22.5	23	17.1	GALLATIN MT	122	-24	68	72	53
		1	14.4	2	58.0	GARFIELD MT		302			99
4	11.4	1	3.4	3	10.0	GLACIER MT	-70	194	-12	16	13
		3	45.0			GOLDEN VALLEY MT					
2	10.0	4	20.8	3	16.9	GRANITE MT	107	-19	68	72	51
13	22.8	12	16.6	14	20.4	HILL MT	-27	23	-11	17	73
2	9.1			7	28.5	JEFFERSON MT			212	94	96
				1	5.2	JUDITH BASIN MT					04
8	10.2	13	15.4	12	12.4	LAKE MT	50	-19	22	43	22
20	16.8	29	21.1	36	22.9	LEWIS AND CLARK MT	26	9	37	55	85
		2	37.5			LIBERTY MT					
1	2.4	7	12.4	9	10.9	LINCOLN MT	418	-12	354	97	15
4	20.8	3	16.1	1	8.4	MC CONE MT	-23	-48	-60	02	08
5	12.7	4	11.7	8	22.3	MADISON MT	-8	90	75	75	83
3	16.5			2	17.0	MEAGHER MT			3	27	52
4	29.2	2	14.2	2	15.4	MINERAL MT	-51	8	-47	04	40
20	11.5	33	16.4	34	14.4	MISSOULA MT	42	-12	25	46	34
2	4.0	8	27.3	3	10.2	MUSSELSHELL MT	586	-63	157	90	13
6	9.8	9	12.1	10	13.0	PARK#MT	23	7	32	51	25
3	7.8	9	24.7	6	15.8	PHILLIPS MT	215	-36	102	82	43
3	10.9	4	12.4	9	27.3	PONDERA MT	14	120	151	89	95
1	6.9	4	40.9			POWDER RIVER MT	495				
7	22.6	8	24.1	9	23.9	POWELL MT	6	0	6	30	89
1	10.7	1	7.5	4	24.8	PRAIRIE MT	-30	230	132	87	91
9	12.7	13	14.1	24	21.9	RAVALLI MT	11	55	72	74	81
5	8.9	7	13.5	12	23.1	RICHLAND MT	52	71	160	90	86
7	18.3	3	6.5	6	14.9	ROOSEVELT MT	-65	131	-19	13	37
6	21.6	3	9.0	4	17.1	ROSEBUD MT	-59	90	-21	12	53
3	5.9	8	15.1	12	24.8	SANDERS MT	154	64	317	96	91
6	17.5	6	18.2	4	11.0	SHERIDAN MT	4	-39	-37	07	16
32	13.3	45	20.8	43	20.0	SILVER BOW MT	56	-3	51	64	71
5	21.6	5	16.0	3	8.5	STILLWATER MT	-26	-47	-61	01	09
1	7.0	7	24.4	3	9.9	SWEET GRASS MT	247	-60	40	57	12
5	14.0	6	15.9	1	1.8	TETON MT	14	-89	-87	00	02
5	13.6	4	12.1	4	14.8	TOOLE MT	-11	22	9	33	36
7	12.1	13	22.4	7	13.2	VALLEY MT	84	-41	9	32	27
4	18.7	3	14.9	5	29.0	WHEATLAND MT	-20	94	55	66	97
				1	11.6	WIBAUX MT					18
40	17.0	47	15.9	59	15.9	YELLOWSTONE MT	-6	0	-6	20	44
1232	18.2	1336	18.1	1528	19.5	NEBRASKA NB	-1	8	7		
33	20.7	27	15.1	40	22.4	ADAMS NB	-27	49	8	32	83
11	18.2	10	14.3	9	12.9	ANTELOPE NB	-22	-9	-29	09	25
		1	22.5			ARTHUR NB					
		1	23.6			BANNER NB					
1	15.6			3	44.4	BLAINE NB			184	91	99
12	20.1	9	15.7	10	18.5	BOONE NB	-22	18	-8	19	62
2	3.3	8	13.2	9	15.1	BOX BUTTE NB	304	14	360	97	38
4	14.2	7	24.1	4	11.4	BOYD NB	70	-53	-19	13	18
4	14.1	3	11.7	5	17.7	BROWN NB	-17	51	26	47	57
18	12.8	29	18.9	25	14.7	BUFFALO NB	47	-22	14	37	36
6	8.2	7	9.1	10	11.8	BURT NB	10	30	44	60	19
24*	35.9	18	23.8	23*	29.7	BUTLER NB	-34	25	-17	14	97
13	13.2	24	23.4	28	27.0	CASS NB	77	15	104	82	95
20*	28.4	6	8.0	7	8.7	CEDAR NB	-72	8	-69	01	09

* SIGNIFICANTLY MORE DEATHS THAN EXPECTED

COUNTIES COMBINED; SEE TECHNICAL NOTES

WHITE MALE

CANCER OF THE LARGE INTESTINE

ICD 153,159.0

1950-1959 DEATHS	1950-1959 RATE	1960-1969 DEATHS	1960-1969 RATE	1970-1979 DEATHS	1970-1979 RATE	COUNTY AND STATE		PERCENT CHANGE 60S/50S	PERCENT CHANGE 70S/60S	PERCENT CHANGE 70S/50S	PERCENTILE %CHANGE 70S/50S	PERCENTILE RATES 1970S
2	7.4	4	19.4	5	17.7	CHASE	NB	161	-8	139	88	57
5	12.5	6	15.1	8	20.2	CHERRY	NB	21	34	62	70	72
6	12.6	9	16.2	5	8.5	CHEYENNE	NB	29	-47	-32	08	09
10	15.8	11	18.8	11	20.9	CLAY	NB	19	11	32	51	75
18*	28.6	18	24.3	16	19.1	COLFAX	NB	-15	-21	-33	08	66
14	21.2	12	16.9	9	12.5	CUMING	NB	-20	-26	-41	05	23
10	9.1	16	15.0	15	14.4	CUSTER	NB	65	-4	58	67	33
9	17.2	10	16.6	15	23.1	DAKOTA	NB	-3	39	34	53	86
9	18.8	8	14.4	13	25.0	DAWES	NB	-23	74	33	52	91
10	10.8	27	24.7	17	13.7	DAWSON	NB	130	-44	28	48	30
3	19.9	2	8.0	6	23.6	DEUEL	NB	-60	196	19	41	88
5	10.4	8	15.5	5	7.9	DIXON	NB	49	-49	-24	11	07
26	17.4	21	13.0	39	19.5	DODGE	NB	-25	50	12	35	68
266*	21.9	268	19.9	309	21.4	DOUGLAS	NB	-9	7	-2	23	78
1	5.1	1	3.5	2	8.5	DUNDY	NB	-31	143	67	71	09
10	17.0	16	25.3	15	23.0	FILLMORE	NB	49	-9	35	54	85
5	13.4	8	17.6	3	6.9	FRANKLIN	NB	31	-61	-49	03	05
4	24.0	6	20.6	8	31.3	FRONTIER	NB	-14	52	30	51	98
11	17.7	14	23.4	6	10.2	FURNAS	NB	32	-56	-42	05	13
34	21.9	29	18.3	29	16.9	GAGE	NB	-16	-7	-23	11	51
1	5.3	1	3.4	9*	147.7	GARDEN	NB	-35	4200	2712	99	99
1	5.9	3	14.6	1	9.6	GARFIELD	NB	147	-34	63	70	11
1	6.7			1	6.3	GOSPER	NB			-7	20	05
1	15.1	2	32.9			GRANT	NB	117				
7	24.9	10*	35.1	7	23.0	GREELEY	NB	41	-34	-8	19	85
31	17.2	44	21.3	46	20.4	HALL	NB	23	-4	19	41	73
11	20.5	6	10.8	13	24.2	HAMILTON	NB	-47	123	18	40	89
3	7.7	9	23.6	4	17.1	HARLAN	NB	205	-28	121	85	52
3	23.9	1	8.9	2	86.1	HAYES	NB	-63	865	261	95	99
5	16.9	5	18.1	3	8.4	HITCHCOCK	NB	7	-53	-50	03	09
7	9.4	12	13.9	14	17.1	HOLT	NB	48	23	81	77	52
2	32.3	1	12.6	2	26.6	HOOKER	NB	-61	111	-18	13	94
6	12.8	5	10.4	11	21.9	HOWARD	NB	-19	111	71	73	81
8	10.6	10	12.1	14	18.8	JEFFERSON	NB	15	55	78	75	64
8	17.9	14*	30.4	11	22.3	JOHNSON	NB	70	-27	25	45	83
6	14.3	3	5.5	6	14.2	KEARNEY	NB	-61	157	0	24	33
8	21.8	10	24.3	5	11.3	KEITH	NB	12	-54	-48	04	17
2	18.8			2	27.2	KEYA PAHA	NB			44	60	95
4	17.9	5	22.4	3	11.5	KIMBALL	NB	25	-49	-36	07	18
14	17.1	19	21.8	18	18.4	KNOX	NB	28	-15	8	31	61
122*	20.7	148*	23.1	149	21.6	LANCASTER	NB	12	-6	5	28	80
25	19.2	18	12.3	29	17.3	LINCOLN	NB	-36	41	-10	18	54
1	24.8					LOGAN	NB					
1	17.2			1	18.6	LOUP	NB			8	32	62
24	15.7	21	12.7	37	21.1	MADISON	NB	-19	66	34	53	77
13	25.8	9	16.7	9	16.2	MERRICK	NB	-35	-3	-37	06	46
2	5.1	8	19.9	5	14.7	MORRILL	NB	291	-26	188	92	35
5	13.3	12	31.2	6	14.4	NANCE	NB	135	-54	9	32	34
10	15.7	14	22.8	10	17.8	NEMAHA	NB	45	-22	13	36	57
12	22.4	13	23.8	11	18.5	NUCKOLLS	NB	6	-22	-18	14	62
20	19.8	16	15.5	30	27.8	OTOE	NB	-21	79	41	58	96
10	21.4	6	12.2	8	22.3	PAWNEE	NB	-43	82	4	28	83
2	13.2	7	34.9	4	16.1	PERKINS	NB	164	-54	22	43	45
13	23.6	8	11.2	11	16.2	PHELPS	NB	-53	45	-32	08	46
5	11.6	6	11.7	16	28.9	PIERCE	NB	0	148	149	89	97
18	17.5	19	16.5	30	22.5	PLATTE	NB	-6	36	28	49	83
14	27.5	9	16.8	13	24.4	POLK	NB	-39	46	-11	16	90
9	15.2	9	13.3	14	17.8	RED WILLOW	NB	-12	34	17	39	57
19	20.2	19	20.5	26	25.3	RICHARDSON	NB	2	23	25	46	92
3	18.2	2	15.9	2	15.1	ROCK	NB	-13	-5	-17	14	38
27*	27.7	23	21.9	24	22.7	SALINE	NB	-21	4	-18	13	84
10	17.6	13	16.1	18	12.4	SARPY	NB	-9	-23	-30	09	22
16	15.2	21	18.6	29	24.3	SAUNDERS	NB	22	31	60	68	90
20	15.4	30	19.7	28	15.7	SCOTTS BLUFF	NB	28	-21	2	25	42
9	11.2	16	19.6	20	23.0	SEWARD	NB	76	17	105	83	85
11	23.0	7	14.5	11	22.4	SHERIDAN	NB	-37	54	-3	23	83
6	18.4	2	5.7	5	13.5	SHERMAN	NB	-69	138	-27	10	28
		1	8.6	1	18.6	SIOUX	NB		115			62
3	10.5	3	10.4	7	19.2	STANTON	NB	-1	86	84	77	66
10	14.0	18	27.4	17	28.4	THAYER	NB	96	4	103	82	96
1	15.1	1	12.6			THOMAS	NB	-17				
9	29.2	6	17.7	10	28.7	THURSTON	NB	-39	62	-2	23	97
10	22.9	6	15.0	9	20.8	VALLEY	NB	-35	39	-9	18	75
11	17.1	16	21.5	15	19.4	WASHINGTON	NB	26	-10	14	36	68
4	9.0	8	15.2	16	28.6	WAYNE	NB	70	89	220	94	97
10	21.5	4	6.9	11	24.8	WEBSTER	NB	-68	256	15	38	91
1	12.8	2	31.7	2	42.4	WHEELER	NB	147	34	230	94	99
11	13.2	11	13.2	13	15.4	YORK	NB	0	17	17	39	40
83	10.3	212	16.0	362	15.7	NEVADA	NV	54	-2	52		
2	5.4	7	17.8	16	24.2	CHURCHILL	NV	233	36	352	97	90
19	8.3	99	17.7	175	15.1	CLARK	NV	114	-14	83	77	39
1	6.4	1	4.6	12	23.8	DOUGLAS	NV	-28	416	273	95	88
2	4.2	2	3.7	9	15.9	ELKO	NV	-13	333	276	95	44
1	19.1	1	14.1			ESMERALDA	NV	-27				
1	18.7					EUREKA	NV					
		1	2.8	5	14.6	HUMBOLDT	NV		418			35
1	24.8	3	26.4	2	26.0	LANDER	NV	7	-2	5	29	93
1	6.7					LINCOLN	NV					
1	4.1	3	9.2	7	11.8	LYON	NV	125	28	187	92	19
1	2.8	3	8.5	5	21.6	MINERAL	NV	206	153	676	99	79
6	24.9	2	6.9			NYE	NV	-72				
4	18.3	7*	33.2	2	11.1	PERSHING	NV	81	-67	-40	06	16
		1	26.8	1	29.8	STOREY	NV		11			97
33	12.4	71	17.1	111	18.0	WASHOE	NV	38	5	45	60	58
6	16.0	7	16.9	6	12.9	WHITE PINE	NV	6	-24	-19	13	25
4	17.2	4	8.9	11	13.7	ORMSBY-CARSON CITY	NV	-48	53	-21	12	29
568*	20.6	635*	20.6	801*	22.2	NEW HAMPSHIRE	NH	0	8	7		

* SIGNIFICANTLY MORE DEATHS THAN EXPECTED

WHITE MALE — CANCER OF THE LARGE INTESTINE — ICD 153,159.0

1950-1959 DEATHS	1950-1959 RATE	1960-1969 DEATHS	1960-1969 RATE	1970-1979 DEATHS	1970-1979 RATE	COUNTY AND STATE	PERCENT CHANGE 60S/50S	PERCENT CHANGE 70S/60S	PERCENT CHANGE 70S/50S	PERCENTILE %CHANGE 70S/50S	PERCENTILE RATES 1970S
27	18.4	19	11.7	32	15.8	BELKNAP NH	-37	35	-14	15	44
16	15.7	21	19.8	27	19.4	CARROLL NH	26	-2	23	44	67
33	16.6	48	22.3	70*	26.9	CHESHIRE NH	34	21	62	70	95
38	20.8	43	23.1	37	19.7	COOS NH	11	-15	-5	21	69
46	18.2	48	18.2	42	14.8	GRAFTON NH	0	-19	-19	13	36
165*	21.8	189*	21.8	247*	24.8	HILLSBOROUGH NH	0	14	14	36	91
65	18.4	60	15.8	97	22.7	MERRIMACK NH	-14	44	24	45	85
92*	23.7	118*	24.7	142	23.1	ROCKINGHAM NH	4	-7	-3	23	86
61*	25.7	61	23.8	77	25.0	STRAFFORD NH	-8	5	-3	23	92
25	18.3	28	17.7	30	17.0	SULLIVAN NH	-4	-4	-7	19	52
4990*	23.5	6329*	24.4	7844*	26.3	NEW JERSEY NJ	4	8	12		
163*	23.2	200*	22.7	224*	23.2	ATLANTIC NJ	-2	2	0	25	87
553*	23.2	786*	23.8	1027*	25.7	BERGEN NJ	3	8	11	34	93
133*	24.3	179*	24.0	224	21.7	BURLINGTON NJ	-1	-9	-11	17	80
297*	24.1	430*	27.5	512*	29.9	CAMDEN NJ	14	9	24	45	97
64*	25.1	85	22.5	124	24.3	CAPE MAY NJ	-10	8	-3	22	90
70	17.8	98	21.5	148*	29.4	CUMBERLAND NJ	21	37	66	71	97
921*	25.9	980*	27.4	889*	27.3	ESSEX NJ	6	0	5	29	95
80	20.1	111	21.3	151*	24.7	GLOUCESTER NJ	6	16	23	44	91
713*	26.7	700*	25.2	703*	27.3	HUDSON NJ	-6	8	2	26	95
37	15.9	49	16.6	75	23.0	HUNTERDON NJ	5	38	44	60	85
223*	23.2	290*	25.9	305*	25.8	MERCER NJ	12	-1	11	34	93
266*	23.6	398*	27.1	538*	27.6	MIDDLESEX NJ	15	2	17	39	96
242*	20.9	410*	27.1	526*	27.2	MONMOUTH NJ	30	0	30	50	95
159	19.3	226	20.2	341*	24.3	MORRIS NJ	5	20	26	47	90
79	19.4	186	21.1	500*	26.2	OCEAN NJ	9	24	35	54	94
387*	24.4	391*	21.5	538*	27.5	PASSAIC NJ	-12	28	13	35	96
32	16.9	43	19.4	65	26.5	SALEM NJ	14	37	57	67	94
76	17.3	128	21.2	161	22.4	SOMERSET NJ	23	5	30	50	83
42	20.3	42	15.0	72	19.1	SUSSEX NJ	-26	27	-6	20	65
384*	23.9	504*	24.6	643*	27.5	UNION NJ	3	12	15	38	96
69*	23.0	93*	27.7	78	20.9	WARREN NJ	20	-24	-9	18	76
182	8.2	316	11.0	578	15.4	NEW MEXICO NM	35	40	88		
50	9.1	98	13.2	193	17.3	BERNALILLO NM	44	31	90	79	54
		1	10.6	2	10.7	CATRON NM		1			15
10	7.3	21	13.4	56	25.0	CHAVES NM	84	87	244	95	92
10	14.3	7	11.7	16	21.8	COLFAX NM	-19	87	52	65	80
11	12.9	12	10.2	9	6.6	CURRY NM	-21	-35	-49	03	05
2	24.8	4	22.1	4	33.1	DE BACA NM	-11	50	33	52	98
13	12.4	17	10.9	31	14.1	DONA ANA NM	-12	30	14	37	32
8	7.1	20	13.1	23	12.8	EDDY NM	85	-3	80	76	24
11	16.5	6	7.3	10	11.0	GRANT NM	-55	50	-33	08	16
		1	4.2	5	25.3	GUADALUPE NM		509			92
1	9.0			1	9.6	HARDING NM			6	30	11
1	5.0	6	36.9	3	18.4	HIDALGO NM	639	-50	270	95	61
6	8.6	10	8.4	21	12.5	LEA NM	-2	49	45	61	23
2	6.0	3	7.8	7	16.9	LINCOLN NM	30	116	179	91	51
2	19.7	2	2.5	5	4.7	LOS ALAMOS NM	-87	89	-76	00	03
		5	10.8	20	22.7	LUNA NM		110			85
8	29.6	8	24.3	4	8.8	MC KINLEY NM	-18	-64	-70	01	09
1	4.0	4	12.9	2	7.8	MORA NM	222	-39	96	80	07
2	4.5	7	8.8	20	20.8	OTERO NM	97	137	367	97	75
8	15.5	4	7.5	8	13.0	QUAY NM	-52	75	-16	14	26
3	3.8	4	5.3	11	15.8	RIO ARRIBA NM	39	196	311	96	43
2	3.0	10	13.2	10	13.5	ROOSEVELT NM	336	2	346	97	29
1	2.7	6	13.6	8	9.1	SANDOVAL NM	396	-33	233	94	10
3	4.4	5	5.5	18	14.4	SAN JUAN NM	25	162	228	94	34
5	5.0	7	7.4	8	8.3	SAN MIGUEL NM	46	13	66	71	08
5	3.4	18	10.8	32	14.6	SANTA FE NM	217	35	328	97	35
6	10.0	9	11.0	16	18.9	SIERRA NM	10	72	89	79	64
1	2.6	2	5.6	7	17.1	SOCORRO NM	118	206	568	98	52
		8	11.4	5	8.6	TAOS NM		-24			09
2	7.2			5	17.0	TORRANCE NM			136	87	52
3	8.7	4	10.8	8	24.5	UNION NM	24	127	181	91	90
5	6.7	7	7.3	10	6.3	VALENCIA NM	8	-13	-6	20	05
15083*	22.4	18000*	24.2	19140*	25.2	NEW YORK NY	8	4	12		
239*	20.7	303*	23.6	350*	26.1	ALBANY NY	14	10	26	47	93
33	15.2	30	13.7	59	26.1	ALLEGANY NY	-10	90	72	73	93
1671*	28.0	1394*	22.2	1432*	27.6	BRONX NY	-21	24	-2	23	96
182*	20.8	227*	23.8	247*	24.6	BROOME NY	14	3	18	40	91
75	18.8	82	20.4	93	22.6	CATTARAUGUS NY	8	11	20	42	84
64	17.2	67	17.6	79	20.0	CAYUGA NY	2	14	16	39	71
132	17.6	157	20.5	166	21.2	CHAUTAUQUA NY	17	3	20	42	77
70	17.7	86	20.0	107	23.4	CHEMUNG NY	13	17	32	51	87
33	15.4	37	17.5	57	24.4	CHENANGO NY	14	39	59	68	90
24	10.3	38	16.2	47	18.8	CLINTON NY	57	16	82	77	64
52	19.1	48	16.4	80	24.2	COLUMBIA NY	-14	47	27	48	90
31	16.2	34	17.7	59*	29.4	CORTLAND NY	10	66	82	77	97
33	13.9	33	14.0	35	13.0	DELAWARE NY	1	-7	-6	20	26
134	17.7	173	20.0	212	21.7	DUTCHESS NY	13	9	23	44	80
842*	21.1	1004*	22.6	1174*	25.9	ERIE NY	7	15	23	44	93
26	15.2	26	14.8	44	23.0	ESSEX NY	-3	56	51	64	86
29	14.0	39	18.6	46	22.6	FRANKLIN NY	33	21	61	69	84
54	18.8	77*	26.2	72	23.7	FULTON NY	39	-10	26	47	88
48	19.6	40	15.3	59	21.9	GENESEE NY	-22	44	12	35	81
33	18.9	44	21.9	67*	28.7	GREENE NY	15	31	51	64	97
1	3.3	7	24.0	8	23.8	HAMILTON NY	637	-1	632	99	88
75	21.4	73	20.0	93*	25.6	HERKIMER NY	-6	27	19	41	92
88	19.3	82	18.3	110*	25.2	JEFFERSON NY	-5	38	31	51	92
2769*	25.9	2466*	22.4	2503*	25.8	KINGS(BROOKLYN) NY	-13	15	0	24	93
22	17.9	23	19.8	28	24.0	LEWIS NY	10	21	34	52	89
33	15.7	49	23.2	39	16.5	LIVINGSTON NY	48	-29	6	29	49
34	13.9	44	17.7	56	21.8	MADISON NY	27	23	57	67	80
465*	18.8	599*	22.1	682*	23.9	MONROE NY	17	8	27	48	89
83*	26.8	76	22.4	71	20.7	MONTGOMERY NY	-16	-7	-23	11	75

* SIGNIFICANTLY MORE DEATHS THAN EXPECTED

WHITE MALE

CANCER OF THE LARGE INTESTINE — ICD 153,159.0

1950-1959 DEATHS	1950-1959 RATE	1960-1969 DEATHS	1960-1969 RATE	1970-1979 DEATHS	1970-1979 RATE	COUNTY AND STATE		PERCENT CHANGE 60S/50S	70S/60S	70S/50S	PERCENTILE %CHANGE 70S/50S	PERCENTILE RATES 1970S
642*	21.6	1112*	25.8	1542*	28.3	NASSAU	NY	20	10	31	51	96
1960*	25.2	3323*	45.1	1531*	25.7	NEW YORK(MANHATTAN)	NY	79	-43	2	25	93
146	16.9	194	20.1	242*	23.2	NIAGARA	NY	19	16	37	55	87
205	17.3	261*	20.6	279	21.8	ONEIDA	NY	19	6	26	47	80
336*	20.5	375*	21.0	458*	24.3	ONONDAGA	NY	3	16	19	41	90
53	15.7	83*	23.4	84	21.4	ONTARIO	NY	48	-8	36	54	78
174*	21.3	212*	22.6	249*	24.3	ORANGE	NY	6	7	14	37	90
49*	28.8	41	22.7	26	14.7	ORLEANS	NY	-21	-35	-49	03	36
72	17.7	73	17.2	70	15.6	OSWEGO	NY	-3	-10	-12	16	42
56	19.1	45	15.5	61	19.0	OTSEGO	NY	-19	23	0	24	65
32*	24.2	47*	24.9	65	25.3	PUTNAM	NY	3	2	5	28	92
1626*	24.1	1860*	20.9	2501*	26.6	QUEENS	NY	-13	27	10	33	94
133*	20.4	177*	25.8	183*	25.9	RENSSELAER	NY	27	0	27	48	93
207*	23.0	188	17.9	343*	27.9	RICHMOND	NY	-22	56	21	42	96
85	20.3	111	19.7	198*	25.8	ROCKLAND	NY	-3	31	27	48	93
63	14.3	84	17.7	95	19.8	ST LAWRENCE	NY	23	12	38	56	70
74	19.7	89	20.9	116	22.1	SARATOGA	NY	6	6	12	35	82
183*	25.5	186*	24.2	182	22.7	SCHENECTADY	NY	-5	-6	-11	17	84
26	19.3	14	10.2	31	20.8	SCHOHARIE	NY	-47	103	8	31	75
18	22.4	16	20.9	9	10.9	SCHUYLER	NY	-7	-48	-51	03	16
20	11.0	34	18.0	29	16.7	SENECA	NY	63	-7	51	64	50
84	16.5	108	21.6	96	18.5	STEUBEN	NY	31	-15	12	35	62
297	15.2	603*	19.7	1065*	26.2	SUFFOLK	NY	30	33	72	74	94
54	21.7	71*	23.9	78	22.5	SULLIVAN	NY	10	-6	4	27	84
33	19.8	33	20.3	30	16.7	TIOGA	NY	3	-18	-16	14	50
37	16.5	57	24.2	53	20.6	TOMPKINS	NY	47	-15	25	46	74
124*	22.1	145*	22.8	181*	24.8	ULSTER	NY	3	8	12	35	91
49	22.7	46	20.6	60	23.2	WARREN	NY	-9	13	2	26	86
49	19.6	54	22.1	45	17.5	WASHINGTON	NY	13	-21	-11	17	55
57	17.2	68	20.1	76	21.5	WAYNE	NY	17	7	25	46	79
715*	25.1	889*	24.7	1025*	26.1	WESTCHESTER	NY	-1	5	4	28	93
31	17.8	24	13.2	41	22.1	WYOMING	NY	-25	67	25	45	82
13	12.3	19	18.2	21	19.1	YATES	NY	48	5	55	66	66
969	9.3	1471	11.2	2460	14.6	NORTH CAROLINA	NC	20	30	57		
31	17.1	24	9.2	54	16.0	ALAMANCE	NC	-46	74	-6	20	45
3	6.4	5	7.4	10	12.4	ALEXANDER	NC	15	68	94	80	22
		5	11.1	7	13.9	ALLEGHANY	NC		24			31
11	21.1	3	5.1	6	8.3	ANSON	NC	-76	64	-61	01	08
9	10.1	10	10.3	9	7.9	ASHE	NC	2	-23	-22	12	07
1	2.0	6	10.4	4	6.2	AVERY	NC	407	-40	203	93	04
6	7.3	17	15.9	18	13.8	BEAUFORT	NC	119	-13	90	79	30
5	10.8	9	18.4	10	20.7	BERTIE	NC	70	12	91	79	75
1	1.8	3	5.2	8	9.7	BLADEN	NC	184	87	433	98	12
1	1.6	2	2.8	5	4.3	BRUNSWICK	NC	77	55	174	91	03
41	9.6	69	12.1	96	13.9	BUNCOMBE	NC	26	15	44	60	31
9	5.9	13	7.1	25	10.3	BURKE	NC	21	44	73	74	13
15	8.1	32	14.3	46	14.7	CABARRUS	NC	77	3	82	77	36
6	5.4	16	9.3	28	12.5	CALDWELL	NC	72	34	130	86	22
1	4.9	1	6.2	5	22.0	CAMDEN	NC	27	253	350	97	81
7	7.4	12	12.6	21	12.9	CARTERET	NC	71	2	75	75	25
3	6.0	8	17.0	9	18.4	CASWELL	NC	183	8	207	93	61
17	7.5	20	7.3	53	16.9	CATAWBA	NC	-3	133	124	86	51
6	10.4	14	16.5	12	11.4	CHATHAM	NC	59	-31	9	33	17
5	7.9	6	7.0	9	8.6	CHEROKEE	NC	-12	24	9	33	09
5	16.0	2	6.4	11	30.2	CHOWAN	NC	-60	375	89	79	97
2	6.9	5	14.9	3	8.4	CLAY	NC	115	-44	21	42	08
17	10.3	26	12.9	54	20.9	CLEVELAND	NC	24	63	102	82	76
8	7.6	10	8.7	10	8.4	COLUMBUS	NC	16	-4	11	33	08
7	8.6	9	7.8	28	21.0	CRAVEN	NC	-10	170	144	88	76
20	11.8	13	4.6	55	17.4	CUMBERLAND	NC	-61	280	48	62	55
1	3.8	3	10.7	2	4.5	CURRITUCK	NC	179	-57	19	41	03
		4	12.8	9	15.9	DARE	NC		24			44
21	11.4	32	13.0	46	12.4	DAVIDSON	NC	14	-5	9	32	22
6	12.0	3	4.2	11	12.9	DAVIE	NC	-65	205	8	31	25
5	6.0	3	3.6	11	8.7	DUPLIN	NC	-39	140	46	61	09
27	13.3	55	20.3	62	17.7	DURHAM	NC	53	-13	33	52	56
9	10.3	13	11.6	20	16.2	EDGECOMBE	NC	13	40	58	67	46
37	10.0	70	14.7	126	19.8	FORSYTH	NC	47	35	99	81	70
7	10.8	7	8.4	10	12.5	FRANKLIN	NC	-22	49	16	39	22
23	8.4	46	11.9	78	15.3	GASTON	NC	41	29	82	77	39
		1	3.7	9	34.7	GATES	NC		846			98
4	15.0	2	6.2	2	5.1	GRAHAM	NC	-58	-17	-66	01	04
6	7.0	5	5.9	9	12.2	GRANVILLE	NC	-16	109	76	75	20
1	3.4	3	11.4	3	12.8	GREENE	NC	238	13	281	96	24
53	9.4	104	14.8	132	14.7	GUILFORD	NC	57	-1	56	66	36
3	4.8	12	10.3	24	18.1	HALIFAX	NC	112	76	273	95	59
10	7.8	18	17.0	23	15.3	HARNETT	NC	118	-10	96	80	39
8	5.6	12	6.6	26	11.5	HAYWOOD	NC	18	74	106	83	18
20	13.9	26	13.0	48	16.2	HENDERSON	NC	-7	24	16	39	46
2	7.6	9	21.8	10	19.6	HERTFORD	NC	187	-10	158	90	69
2	8.8	3	12.0	7	23.3	HOKE	NC	37	93	166	91	87
5	22.6	4	15.9	1	5.9	HYDE	NC	-30	-63	-74	00	04
16	8.9	30	12.4	48	16.7	IREDELL	NC	39	35	88	79	50
6	9.2	8	9.5	13	12.7	JACKSON	NC	3	33	38	55	24
15	10.2	17	8.6	32	14.9	JOHNSTON	NC	-16	74	46	61	37
1	4.2	2	9.4	1	7.0	JONES	NC	124	-25	67	71	06
5	8.4	6	8.7	23	21.4	LEE	NC	4	146	156	90	78
5	6.6	7	6.5	24	19.4	LENOIR	NC	-2	199	194	93	67
7	10.4	17	17.3	20	15.5	LINCOLN	NC	67	-10	49	63	41
7	9.5	12	10.9	20	15.4	MC DOWELL	NC	14	42	61	69	41
2	3.5	8	9.6	5	4.0	MACON	NC	173	-59	13	36	02
3	4.2	7	8.8	7	7.1	MADISON	NC	108	-19	68	72	06
2	4.2	7	13.0	10	15.8	MARTIN	NC	206	22	273	95	43
69	13.6	105	15.4	170	17.7	MECKLENBURG	NC	13	15	30	50	57
5	8.6	3	4.7	7	9.0	MITCHELL	NC	-45	92	5	28	10
2	4.0	9	13.6	5	6.3	MONTGOMERY	NC	238	-53	57	67	05
9	8.7	16	12.8	24	12.5	MOORE	NC	47	-2	43	59	22
10	10.0	15	11.3	23	12.8	NASH	NC	13	14	29	49	24
10	5.9	18	8.3	42	15.7	NEW HANOVER	NC	42	88	168	91	43

* SIGNIFICANTLY MORE DEATHS THAN EXPECTED

WHITE MALE — CANCER OF THE LARGE INTESTINE — ICD 153,159.0

1950-1959 DEATHS	1950-1959 RATE	1960-1969 DEATHS	1960-1969 RATE	1970-1979 DEATHS	1970-1979 RATE	COUNTY AND STATE		PERCENT CHANGE 60S/50S	PERCENT CHANGE 70S/60S	PERCENT CHANGE 70S/50S	PERCENTILE %CHANGE 70S/50S	PERCENTILE RATES 1970S
3	6.9	10	20.1	9	17.3	NORTHAMPTON	NC	194	-14	152	89	54
3	2.6	7	5.7	17	12.7	ONSLOW	NC	122	124	398	97	24
10	11.0	10	9.1	22	14.1	ORANGE	NC	-17	55	28	49	32
2	7.2	2	6.0	3	7.0	PAMLICO	NC	-17	18	-2	23	06
9	19.4	5	8.1	14	18.1	PASQUOTANK	NC	-58	123	-7	19	59
3	7.4	6	12.7	9	15.0	PENDER	NC	72	18	102	82	37
1	4.3	2	6.8	4	11.5	PERQUIMANS	NC	59	69	169	91	18
3	4.6	6	7.3	8	10.9	PERSON	NC	58	49	136	87	15
8	7.0	19	14.9	21	12.2	PITT	NC	112	-18	73	74	20
3	8.8	3	4.4	17	20.8	POLK	NC	-50	376	136	87	75
7	3.4	29	12.7	47	14.7	RANDOLPH	NC	268	16	326	96	35
11	17.9	15	13.6	21	16.1	RICHMOND	NC	-24	18	-10	17	45
7	6.4	12	8.3	20	12.6	ROBESON	NC	29	53	97	81	23
16	8.9	30	14.0	43	15.9	ROCKINGHAM	NC	56	14	78	76	44
25	10.4	30	10.3	53	13.0	ROWAN	NC	-1	26	25	45	26
22	15.4	19	10.2	45	19.8	RUTHERFORD	NC	-34	94	28	49	70
14	12.7	12	8.9	14	9.7	SAMPSON	NC	-29	8	-24	11	12
5	10.3	3	6.9	8	11.1	SCOTLAND	NC	-33	60	8	31	16
12	12.3	16	9.7	24	13.4	STANLY	NC	-21	38	9	33	28
8	10.4	6	6.9	14	12.6	STOKES	NC	-34	84	22	43	23
13	8.1	12	6.5	27	11.3	SURRY	NC	-19	74	40	57	17
4	12.9	6	15.1	4	8.0	SWAIN	NC	17	-47	-38	06	08
7	14.7	5	8.1	7	7.8	TRANSYLVANIA	NC	-45	-4	-47	04	07
		1	14.9	2	15.3	TYRRELL	NC		3			39
20	16.5	10	8.2	25	12.2	UNION	NC	-50	48	-26	10	20
5	7.7	10	11.9	19	23.1	VANCE	NC	53	95	199	93	86
25	8.2	54	11.5	96	15.9	WAKE	NC	40	38	94	80	44
1	2.4	3	7.8	7	16.7	WARREN	NC	228	113	598	98	50
		3	8.3	8	20.1	WASHINGTON	NC		142			72
11	16.0	4	5.4	11	10.0	WATAUGA	NC	-66	86	-37	07	13
13	9.4	13	8.9	34	19.5	WAYNE	NC	-6	120	107	83	68
10	6.4	22	12.5	16	7.6	WILKES	NC	95	-39	18	41	07
13	11.8	14	12.0	21	14.9	WILSON	NC	2	24	26	47	37
8	9.1	6	5.8	20	15.4	YADKIN	NC	-36	163	70	73	40
6	9.8	4	6.0	11	13.8	YANCEY	NC	-39	129	40	57	30
319	11.4	361	12.0	525	16.2	NORTH DAKOTA	ND	5	35	42		
2	9.5	3	13.4	2	9.0	ADAMS	ND	40	-33	-6	20	10
10	11.5	6	5.9	19	19.5	BARNES	ND	-49	229	69	73	68
8	16.0	2	4.0	10	20.7	BENSON	ND	-75	414	29	50	75
		1	21.5	2*	91.3	BILLINGS	ND		325			99
8	13.1	9	14.2	17	24.2	BOTTINEAU	ND	8	70	84	78	90
2	13.9	3	12.2	3	13.7	BOWMAN	ND	-12	12	-1	24	29
3	7.5	3	11.4	3	10.4	BURKE	ND	52	-9	39	56	14
18	17.0	14	10.5	32	18.9	BURLEIGH	ND	-38	80	11	34	64
23	8.7	47	15.7	70	21.5	CASS	ND	80	37	147	89	79
5	8.9	6	12.4	8	15.7	CAVALIER	ND	39	26	75	75	42
10	21.0	7	13.2	7	13.0	DICKEY	ND	-37	-1	-38	06	25
6	15.9	4	12.9	5	17.7	DIVIDE	ND	-19	38	11	34	57
2	14.3	3	13.0			DUNN	ND	-9				
1	3.7	3	11.2	3	9.0	EDDY	ND	207	-20	146	89	10
3	10.1	3	9.5	5	13.9	EMMONS	ND	-6	46	37	55	31
1	3.9	6	21.7	8	24.7	FOSTER	ND	460	14	536	98	91
1	5.6	1	3.9	6	45.1	GOLDEN VALLEY	ND	-29	1053	713	99	99
16	9.2	19	9.6	40	19.1	GRAND FORKS	ND	5	98	108	83	65
11*	50.2	3	10.9	3	9.5	GRANT	ND	-78	-13	-81	00	11
6	21.7	6	17.1	5	13.3	GRIGGS	ND	-21	-22	-39	06	27
3	12.1	2	7.2	1	5.1	HETTINGER	ND	-41	-29	-58	02	03
3	13.6	1	4.5	3	12.4	KIDDER	ND	-67	175	-9	18	21
4	8.9	5	10.7	8	15.6	LA MOURE	ND	21	46	76	75	42
3	21.9	1	7.4	2	36.2	LOGAN	ND	-66	387	65	70	99
6	12.6	8	14.0	8	15.0	MC HENRY	ND	11	7	19	41	38
4	15.2	8	27.7	7	16.8	MC INTOSH	ND	82	-39	10	33	50
5	11.9	2	6.2	2	5.0	MC KENZIE	ND	-47	-21	-58	02	03
9	15.5	7	9.3	12	15.3	MC LEAN	ND	-40	65	-1	24	39
1	2.6	1	3.0	4	12.0	MERCER	ND	14	303	361	97	20
8	11.0	13	14.4	18	17.1	MORTON	ND	31	19	55	66	52
5	11.6	5	9.5	6	11.1	MOUNTRAIL	ND	-18	17	-4	21	16
2	4.6	8	16.5	6	11.2	NELSON	ND	256	-32	143	88	17
1	11.8	2	46.1	4	40.2	OLIVER	ND	291	-13	241	95	99
5	7.2	7	9.3	15	20.5	PEMBINA	ND	29	121	185	92	74
6	18.7	2	6.6	7	21.1	PIERCE	ND	-65	219	13	35	77
6	8.7	7	9.3	8	9.4	RAMSEY	ND	7	1	8	32	11
5	9.2	7	13.3	8	13.0	RANSOM	ND	44	-2	41	58	25
3	9.4	3	10.4	4	13.7	RENVILLE	ND	11	32	46	61	30
12	12.9	15	13.5	18	14.5	RICHLAND	ND	5	7	12	35	34
1	3.2	2	6.0	3	8.3	ROLETTE	ND	87	40	162	90	08
6	14.9	2	4.7	4	10.7	SARGENT	ND	-69	129	-28	09	14
		3	15.5	4	21.7	SHERIDAN	ND		40			80
1	12.0	1	14.5	1	56.2	SIOUX	ND	21	287	370	97	99
1	8.9	1	24.8	1	14.9	SLOPE	ND	178	-40	67	71	37
9	15.2	10	14.1	13	15.5	STARK	ND	-8	10	1	25	41
3	10.9	1	7.4			STEELE	ND	-32				
11	8.5	15	11.5	18	14.0	STUTSMAN	ND	35	22	65	70	31
6	21.2	2	6.6	4	11.4	TOWNER	ND	-69	73	-46	04	17
5	8.6	10	14.1	14	18.1	TRAILL	ND	65	29	112	84	59
14	15.6	20	20.3	22	19.6	WALSH	ND	30	-3	25	46	69
20	11.5	22	11.6	32	15.6	WARD	ND	1	34	35	54	42
8	16.8	8	16.9	8	14.9	WELLS	ND	1	-12	-11	17	37
7	7.4	11	9.9	12	12.6	WILLIAMS	ND	34	27	69	73	23
6810*	19.0	8339*	21.2	9281*	22.1	OHIO	OH	12	5	17		
21	15.7	22	15.8	29	21.7	ADAMS	OH	1	37	38	56	80
79	19.3	89	20.6	86	18.9	ALLEN	OH	7	-8	-2	23	64
22	12.4	31	16.0	53	25.9	ASHLAND	OH	29	62	109	83	93
80	19.6	94	22.0	80	17.1	ASHTABULA	OH	12	-22	-13	16	53
44	18.7	46	20.3	40	19.2	ATHENS	OH	8	-6	2	26	66
47*	27.4	47*	26.8	47	24.6	AUGLAIZE	OH	-2	-8	-10	18	91
83	18.2	90	19.5	86	18.8	BELMONT	OH	7	-3	3	27	64

* SIGNIFICANTLY MORE DEATHS THAN EXPECTED

WHITE MALE

CANCER OF THE LARGE INTESTINE

ICD 153,159.0

1950-1959 DEATHS	1950-1959 RATE	1960-1969 DEATHS	1960-1969 RATE	1970-1979 DEATHS	1970-1979 RATE	COUNTY AND STATE	PERCENT CHANGE 60S/50S	PERCENT CHANGE 70S/60S	PERCENT CHANGE 70S/50S	PERCENTILE %CHANGE 70S/50S	PERCENTILE RATES 1970S
19	13.3	25	17.0	32	21.6	BROWN OH	28	27	62	70	79
90	15.2	140	19.5	183	22.0	BUTLER OH	29	13	45	61	81
22	21.0	17	15.7	21	17.9	CARROLL OH	-25	14	-15	15	58
18	12.7	22	15.7	18	12.3	CHAMPAIGN OH	24	-22	-3	22	21
81	16.7	119	21.3	116	19.1	CLARK OH	27	-11	14	36	65
42	18.1	53	18.7	72	20.4	CLERMONT OH	3	9	13	36	74
15	10.3	25	17.2	27	18.3	CLINTON OH	67	7	78	75	61
85	18.0	98	19.4	112	21.4	COLUMBIANA OH	8	10	19	41	78
27	16.0	42	24.8	48	27.0	COSHOCTON OH	55	9	68	72	95
35	16.9	51	23.4	55	24.8	CRAWFORD OH	39	6	47	62	91
1321*	23.0	1542*	24.8	1513*	24.2	CUYAHOGA OH	8	-2	5	29	90
24	10.2	33	14.0	62	24.1	DARKE OH	38	72	137	87	89
28	21.9	29	20.5	33	21.7	DEFIANCE OH	-6	6	-1	24	80
25	15.8	26	15.7	45	25.3	DELAWARE OH	-1	61	59	68	92
53	19.2	50	16.5	74	23.1	ERIE OH	-14	40	20	42	86
42	15.0	53	17.0	85	24.1	FAIRFIELD OH	13	42	60	69	89
24	19.8	19	14.9	28	20.8	FAYETTE OH	-25	40	5	29	75
385*	20.0	560*	25.2	595*	24.5	FRANKLIN OH	26	-3	23	44	90
11	7.7	17	11.9	29	19.3	FULTON OH	56	62	152	89	67
19	14.9	25	19.9	14	10.6	GALLIA OH	34	-47	-29	09	14
21	13.5	36	19.2	50	21.7	GEAUGA OH	43	13	61	69	80
25	12.4	48	17.9	76	21.7	GREENE OH	45	21	76	75	80
39	15.8	37	14.9	45	19.7	GUERNSEY OH	-6	33	25	46	69
708*	23.8	802*	25.5	893*	27.3	HAMILTON OH	7	7	15	38	95
35	14.6	44	17.1	49	17.9	HANCOCK OH	17	4	22	43	58
25	15.7	23	15.4	36	24.6	HARDIN OH	-2	59	57	67	91
14	13.9	19	18.3	16	16.7	HARRISON OH	32	-9	20	42	50
25	19.0	15	11.6	29	21.4	HENRY OH	-39	85	13	36	79
17	9.6	29	16.3	46	25.5	HIGHLAND OH	69	56	165	91	92
20	18.6	26	23.3	23	20.1	HOCKING OH	25	-14	8	31	71
14	14.8	19	19.5	22	19.5	HOLMES OH	31	0	31	51	68
40	18.8	49	22.9	36	16.7	HURON OH	22	-27	-11	17	50
16	10.8	11	7.2	25	16.2	JACKSON OH	-34	126	51	64	46
62	14.1	99	22.4	78	17.1	JEFFERSON OH	59	-23	21	42	53
37	19.6	36	18.4	39	18.0	KNOX OH	-6	-2	-8	19	59
76	22.2	106	23.3	151	23.2	LAKE OH	5	0	5	28	87
34	15.6	45	19.0	64	24.6	LAWRENCE OH	21	30	57	67	91
66	16.6	82	19.4	91	19.2	LICKING OH	16	-1	15	38	66
32	17.5	38	20.6	39	20.1	LOGAN OH	17	-2	14	37	72
121	18.6	136	17.5	201	23.4	LORAIN OH	-6	33	26	47	87
348*	20.1	430*	22.4	424*	22.2	LUCAS OH	12	-1	11	33	82
12	11.7	21	19.6	22	18.5	MADISON OH	68	-6	59	68	62
269*	23.7	330*	26.2	345*	26.5	MAHONING OH	10	1	12	34	94
27	11.1	38	14.8	53	20.1	MARION OH	33	36	81	77	72
34	15.5	47	18.3	53	17.2	MEDINA OH	18	-6	11	34	53
19	14.2	20	17.0	14	12.0	MEIGS OH	20	-29	-15	15	20
23	15.7	27	18.2	35	21.5	MERCER OH	16	18	37	55	79
43	14.7	48	15.4	75	21.3	MIAMI OH	5	38	45	60	78
20	18.4	26*	26.5	22	23.3	MONROE OH	44	-12	27	48	87
294*	20.2	360*	20.5	429	22.0	MONTGOMERY OH	2	7	9	32	81
16	19.1	14	17.7	11	12.6	MORGAN OH	-8	-29	-34	07	23
11	11.6	13	13.0	14	12.6	MORROW OH	12	-3	8	32	23
57	15.6	72	19.5	81	21.9	MUSKINGUM OH	25	12	41	57	81
11	14.2	11	15.6	12	17.5	NOBLE OH	10	12	24	45	56
22	14.0	32	19.7	37	20.3	OTTAWA OH	41	3	45	61	73
12	14.1	8	10.4	20	24.5	PAULDING OH	-27	136	73	74	91
33	19.8	39	24.2	31	21.0	PERRY OH	22	-13	6	30	76
22	15.4	27	18.0	35	21.0	PICKAWAY OH	17	16	36	54	76
8	9.9	9	8.8	15	12.8	PIKE OH	-11	45	30	50	24
59	20.9	72	21.9	71	18.0	PORTAGE OH	5	-18	-14	15	59
19	13.3	26	17.1	30	18.7	PREBLE OH	28	9	41	57	63
20	15.8	34*	25.7	29	21.2	PUTNAM OH	63	-18	34	53	77
82	20.4	99	21.7	108	21.6	RICHLAND OH	6	0	6	29	79
33	13.2	42	14.2	58	18.9	ROSS OH	7	33	43	59	64
47	19.4	35	13.7	52	19.9	SANDUSKY OH	-29	45	2	26	70
57	15.9	85	21.8	82	20.0	SCIOTO OH	37	-8	25	46	71
49	18.0	47	17.3	56	20.6	SENECA OH	-4	19	14	37	74
20	13.9	17	11.5	36	21.6	SHELBY OH	-17	87	56	66	79
230	17.9	283	19.2	350	21.9	STARK OH	7	14	22	43	81
346*	20.6	430*	21.5	456	21.3	SUMMIT OH	5	-1	3	27	78
130	18.2	183*	22.1	215	22.7	TRUMBULL OH	22	3	25	46	84
48	12.6	75	19.3	74	18.0	TUSCARAWAS OH	53	-7	43	59	59
19	15.9	14	12.3	33*	27.7	UNION OH	-23	125	74	74	96
24	16.6	30	19.7	34	21.6	VAN WERT OH	19	9	30	50	79
8	12.5	4	6.7	11	19.3	VINTON OH	-46	186	54	66	67
24	13.1	35	15.9	44	16.1	WARREN OH	22	1	23	44	46
41	16.5	52	19.7	52	18.6	WASHINGTON OH	19	-6	12	35	62
26	9.0	56	17.1	62	16.7	WAYNE OH	89	-2	85	78	50
21	14.1	19	12.3	31	18.5	WILLIAMS OH	-13	51	32	51	62
42	14.6	48	15.6	58	16.1	WOOD OH	7	3	11	34	46
15	14.3	16	14.2	19	16.4	WYANDOT OH	-1	16	15	37	48
1162	12.1	1483	13.6	2075	16.8	OKLAHOMA OK	12	23	38		
2	2.7	9	12.0	10	12.8	ADAIR OK	350	7	381	97	24
6	9.4	9	12.6	13	19.9	ALFALFA OK	33	58	110	83	70
3	4.5	8	12.3	2	2.3	ATOKA OK	172	-81	-49	03	02
5	12.5	9	26.3	3	7.4	BEAVER OK	110	-72	-41	06	06
12	10.8	15	13.8	17	14.3	BECKHAM OK	27	4	32	52	33
15	23.5	14	18.5	10	13.9	BLAINE OK	-21	-25	-41	05	31
18	12.2	20	12.9	26	14.8	BRYAN OK	6	15	21	42	37
21	14.2	19	12.7	26	15.9	CADDO OK	-10	26	12	35	44
22	18.2	19	13.2	26	15.5	CANADIAN OK	-27	17	-15	15	41
15	8.8	25	13.0	36	16.2	CARTER OK	48	24	83	77	46
7	9.5	12	13.0	20	18.0	CHEROKEE OK	36	39	89	79	59
9	9.4	10	10.9	12	13.3	CHOCTAW OK	16	21	41	58	27
1	4.8	1	4.6	4	16.0	CIMARRON OK	-5	246	230	94	45
13	7.3	22	9.7	31	10.8	CLEVELAND OK	33	12	48	62	15
2	4.8	7	18.8	9	21.1	COAL OK	292	12	341	97	77
18	12.2	31	13.3	38	14.9	COMANCHE OK	9	12	22	43	37
10	20.2	17*	33.8	7	13.7	COTTON OK	67	-59	-32	08	29

* SIGNIFICANTLY MORE DEATHS THAN EXPECTED

WHITE MALE — CANCER OF THE LARGE INTESTINE — ICD 153,159.0

1950-1959 DEATHS	1950-1959 RATE	1960-1969 DEATHS	1960-1969 RATE	1970-1979 DEATHS	1970-1979 RATE	COUNTY AND STATE	PERCENT CHANGE 60S/50S	PERCENT CHANGE 70S/60S	PERCENT CHANGE 70S/50S	PERCENTILE %CHANGE 70S/50S	PERCENTILE RATES 1970S
8	8.1	18	18.3	16	17.7	CRAIG OK	125	-3	117	85	56
22	11.0	35	17.4	44	18.8	CREEK OK	58	8	71	73	64
14	14.6	13	12.3	22	17.6	CUSTER OK	-15	43	21	42	56
11	11.6	7	6.3	16	11.5	DELAWARE OK	-46	83	-1	24	18
3	7.0	4	13.3	2	4.2	DEWEY OK	90	-68	-39	06	03
2	5.2	5	11.6	2	5.0	ELLIS OK	123	-56	-3	23	03
26	11.1	30	11.4	46	15.9	GARFIELD OK	3	40	44	60	44
8	5.8	18	12.1	22	12.8	GARVIN OK	107	6	119	85	24
23	13.4	26	14.0	33	18.3	GRADY OK	4	31	37	55	60
4	6.5	10	18.2	11	19.3	GRANT OK	180	6	197	93	67
6	10.1	13	19.4	15	21.3	GREER OK	92	10	111	84	78
5	13.0	4	9.6	6	14.8	HARMON OK	-26	54	13	36	36
4	14.2	6	18.2	8	23.6	HARPER OK	28	29	66	71	88
6	10.9	6	11.0	11	15.9	HASKELL OK	1	44	45	61	44
8	8.4	13	15.4	8	7.8	HUGHES OK	84	-49	-7	19	07
10	11.6	9	8.2	16	13.7	JACKSON OK	-29	66	18	40	29
8	13.0	5	8.3	11	20.0	JEFFERSON OK	-36	140	54	66	71
6	10.7	6	10.6	6	9.9	JOHNSTON OK	-1	-7	-8	19	12
33	14.4	21	8.2	44	15.7	KAY OK	-43	92	9	32	43
5	8.5	5	8.9	15	19.6	KINGFISHER OK	5	120	132	87	69
12	13.3	12	14.7	21	22.3	KIOWA OK	11	51	67	71	83
2	4.3	4	8.8	10	17.9	LATIMER OK	103	104	312	96	58
12	7.4	26	14.0	36	17.2	LE FLORE OK	89	23	133	87	54
17	15.1	21	16.0	24	16.9	LINCOLN OK	6	6	12	35	52
13	13.3	17	15.5	16	12.9	LOGAN OK	16	-17	-3	23	25
4	11.9	4	9.9	4	8.7	LOVE OK	-17	-12	-27	10	09
9	13.8	9	12.4	18	20.6	MC CLAIN OK	-10	66	50	63	74
15	13.2	15	12.2	17	11.3	MC CURTAIN OK	-8	-7	-15	15	17
6	9.0	2	2.7	14	14.4	MC INTOSH OK	-71	441	59	68	34
8	14.5	6	10.7	6	10.3	MAJOR OK	-26	-4	-29	09	13
6	14.0	5	9.9	7	9.3	MARSHALL OK	-30	-6	-34	07	10
9	9.4	15	12.0	24	15.3	MAYES OK	27	28	62	70	39
8	11.1	7	8.0	8	9.0	MURRAY OK	-28	12	-19	13	10
28	11.9	36	13.5	45	16.3	MUSKOGEE OK	13	21	37	55	47
11	17.5	10	15.9	8	12.0	NOBLE OK	-9	-25	-32	08	19
10	14.4	8	12.7	9	14.7	NOWATA OK	-12	15	2	26	35
11	20.3	10	15.4	4	7.5	OKFUSKEE OK	-24	-51	-63	01	06
179	15.6	200	12.8	386	20.4	OKLAHOMA OK	-18	59	31	51	73
21	11.7	33	17.0	30	15.0	OKMULGEE OK	45	-12	28	49	37
12	9.0	22	14.0	22	13.0	OSAGE OK	55	-7	43	59	25
19	13.3	26	17.3	27	17.7	OTTAWA OK	30	2	33	52	56
15	18.8	14	16.5	9	10.2	PAWNEE OK	-12	-38	-46	04	13
21	12.6	30	15.9	40	18.5	PAYNE OK	26	17	47	62	62
25	13.1	24	12.1	41	18.6	PITTSBURG OK	-8	54	42	58	62
10	7.0	18	11.4	24	14.6	PONTOTOC OK	63	28	109	83	35
28	13.3	27	11.6	42	16.3	POTTAWATOMIE OK	-13	41	23	44	47
5	9.5	11	19.0	13	19.1	PUSHMATAHA OK	100	1	101	81	65
7	22.0	3	9.8	8	29.9	ROGER MILLS OK	-56	206	36	54	97
7	7.1	15	12.6	25	18.2	ROGERS OK	78	44	157	90	59
17	12.1	19	12.3	34	23.4	SEMINOLE OK	2	90	94	80	87
6	7.2	9	10.3	19	15.4	SEQUOYAH OK	44	50	116	84	41
20	12.6	31	16.6	44	20.1	STEPHENS OK	32	21	60	68	72
7	11.6	9	14.6	8	10.7	TEXAS OK	26	-26	-8	19	15
10	12.5	8	10.6	10	12.4	TILLMAN OK	-16	18	-1	24	22
128	13.7	182	15.3	273	18.8	TULSA OK	12	23	37	55	63
6	8.4	11	13.1	23	21.2	WAGONER OK	55	62	152	89	77
18	13.1	37	21.8	40	19.5	WASHINGTON OK	67	-10	50	63	68
7	8.4	20	24.1	17	20.2	WASHITA OK	189	-16	142	88	73
15	21.1	13	17.4	11	14.1	WOODS OK	-17	-19	-33	08	32
7	7.8	13	14.5	14	15.9	WOODWARD OK	86	10	104	82	44
1060	13.8	1386	15.0	1746	15.7	OREGON OR	9	5	14		
8	8.3	16	16.6	13	12.8	BAKER OR	99	-23	54	65	24
13	11.1	19	12.8	26	13.8	BENTON OR	15	8	25	46	31
69	14.7	92	15.1	135	16.2	CLACKAMAS OR	3	7	10	33	46
28	17.4	17	10.6	27	14.7	CLATSOP OR	-39	38	-16	14	35
10	8.4	11	8.4	22	13.6	COLUMBIA OR	0	62	62	69	29
34	19.8	29	11.2	44	15.2	COOS OR	-44	36	-23	11	39
3	10.8	9	19.4	7	10.4	CROOK OR	79	-46	-4	22	14
6	14.2	6	9.8	9	9.0	CURRY OR	-31	-8	-37	07	10
10	9.1	22	16.5	27	11.6	DESCHUTES OR	81	-29	28	48	18
29	12.0	48	15.6	48	12.4	DOUGLAS OR	30	-20	4	27	22
1	8.6	2	16.1			GILLIAM OR	88				
8	19.8	8	19.9	4	10.1	GRANT OR	1	-50	-49	03	13
2	7.7	4	11.5	8	22.2	HARNEY OR	50	93	190	92	82
13	23.2	12	17.0	19	24.4	HOOD RIVER OR	-27	43	5	29	90
37	10.5	76	17.2	99	16.6	JACKSON OR	65	-4	58	68	49
3	22.1	2	8.4	8	21.1	JEFFERSON OR	-62	151	-5	21	77
24	14.7	38	17.7	52	16.9	JOSEPHINE OR	21	-5	15	37	51
17	10.0	25	12.8	41	17.6	KLAMATH OR	27	38	75	75	56
3	8.7	2	7.0	7	18.3	LAKE OR	-20	162	111	84	60
71	12.4	131	17.7	148	15.3	LANE OR	42	-13	23	44	40
11	8.6	9	5.5	23	11.0	LINCOLN OR	-36	100	28	48	16
35	15.4	40	13.6	49	13.2	LINN OR	-12	-2	-14	15	27
10	11.7	12	11.0	21	16.3	MALHEUR OR	-6	49	40	57	47
83	15.1	82	11.6	139	15.6	MARION OR	-23	34	4	27	42
1	3.9	5	17.7	2	6.6	MORROW OR	354	-63	70	73	05
359	14.3	471	17.2	463	17.1	MULTNOMAH OR	21	-1	20	41	52
8	6.5	28	17.5	29	14.1	POLK OR	168	-19	116	84	32
		2	18.7	2	15.6	SHERMAN OR		-16			42
6	7.5	10	10.3	21	15.4	TILLAMOOK OR	36	50	105	82	41
34	16.0	39	16.2	52	20.4	UMATILLA OR	1	26	28	48	73
19	18.9	10	9.7	19	16.6	UNION OR	-48	70	-12	16	49
1	2.7	8	20.3	3	6.6	WALLOWA OR	652	-67	145	88	05
13	15.4	13	12.9	29	24.8	WASCO OR	-17	93	61	69	91
47	14.2	66	13.5	112	15.7	WASHINGTON OR	-5	16	11	34	43
3	17.7	3	24.4	1	8.6	WHEELER OR	38	-65	-51	03	09
41	23.1	19	9.4	37	14.8	YAMHILL OR	-59	57	-36	07	36

WHITE MALE — CANCER OF THE LARGE INTESTINE — ICD 153,159.0

1950-1959 DEATHS	1950-1959 RATE	1960-1969 DEATHS	1960-1969 RATE	1970-1979 DEATHS	1970-1979 RATE	COUNTY AND STATE		PERCENT CHANGE 60S/50S	PERCENT CHANGE 70S/60S	PERCENT CHANGE 70S/50S	PERCENTILE %CHANGE 70S/50S	PERCENTILE RATES 1970S
9347*	20.7	10848*	21.8	12343*	23.0	PENNSYLVANIA	PA	5	6	11		
35	16.8	33	13.7	51	18.2	ADAMS	PA	-18	33	8	32	60
1399*	22.6	1602*	23.4	1641*	23.3	ALLEGHENY	PA	3	0	3	26	87
54	14.6	57	14.5	65	15.4	ARMSTRONG	PA	-1	6	6	29	41
156*	21.5	189*	21.9	219*	22.7	BEAVER	PA	2	4	5	29	84
30	15.6	26	12.3	30	13.1	BEDFORD	PA	-21	6	-16	14	26
243	18.7	275	19.2	367*	23.0	BERKS	PA	2	20	23	44	86
106	15.8	137	19.7	140	19.5	BLAIR	PA	25	-1	24	45	68
47	17.4	38	13.9	63	21.9	BRADFORD	PA	-20	58	26	47	81
137	18.4	206	18.6	309	20.8	BUCKS	PA	1	12	13	36	75
76	16.0	92	17.1	129	21.2	BUTLER	PA	7	23	32	52	77
163	18.9	177	19.1	207	21.4	CAMBRIA	PA	1	12	13	36	78
10*	32.9	9	33.5	9	24.0	CAMERON	PA	2	-28	-27	10	89
47	17.9	62	22.7	63	21.2	CARBON	PA	27	-7	18	40	77
45	18.4	38	14.3	64	20.1	CENTRE	PA	-22	41	9	33	72
122	18.2	160	18.6	230	22.2	CHESTER	PA	2	19	22	43	82
34	18.2	28	14.4	38	18.8	CLARION	PA	-21	30	3	26	63
68	17.0	61	14.8	89	21.0	CLEARFIELD	PA	-13	42	23	44	76
34	21.1	26	15.3	43	22.7	CLINTON	PA	-27	49	8	31	84
46	17.5	55	20.0	61	20.4	COLUMBIA	PA	14	2	16	39	73
84*	22.0	75	19.2	81	18.8	CRAWFORD	PA	-13	-2	-14	15	64
76	18.6	99	18.5	154	23.2	CUMBERLAND	PA	-1	25	25	46	87
147	16.7	232*	24.7	230*	23.5	DAUPHIN	PA	48	-5	41	58	88
345*	20.1	518*	24.5	669*	26.9	DELAWARE	PA	22	10	34	52	95
21	13.8	41	24.4	39	20.9	ELK	PA	77	-14	52	65	76
205*	21.1	208	19.7	259	22.6	ERIE	PA	-7	15	7	31	84
145	17.6	178	21.1	157	18.5	FAYETTE	PA	20	-12	5	29	62
		6	19.1	10	24.4	FOREST	PA		28			90
56	16.1	70	17.9	101	20.6	FRANKLIN	PA	11	15	28	48	74
7	18.7	12	23.3	14	23.8	FULTON	PA	25	2	27	48	88
25	11.8	32	14.3	29	13.4	GREENE	PA	22	-6	14	37	28
25	14.3	35	19.6	27	13.8	HUNTINGDON	PA	38	-30	-4	22	30
43	12.3	70	18.4	79	19.9	INDIANA	PA	50	8	62	69	70
33	13.1	44	16.3	48	17.4	JEFFERSON	PA	25	7	33	52	55
12	17.5	16	22.0	17	19.1	JUNIATA	PA	26	-13	9	32	65
335*	28.5	304*	24.8	351*	27.6	LACKAWANNA	PA	-13	11	-3	22	96
189	16.9	238	18.5	356*	23.4	LANCASTER	PA	10	26	38	56	87
70	14.2	107	20.4	117	21.8	LAWRENCE	PA	43	7	53	65	80
77	21.2	71	16.7	97	19.7	LEBANON	PA	-21	18	-7	20	70
183	20.3	227	20.0	304*	23.5	LEHIGH	PA	-1	17	16	39	88
345*	21.2	390*	22.2	520*	27.5	LUZERNE	PA	5	24	30	50	95
103	20.8	110	21.1	140*	24.9	LYCOMING	PA	1	18	19	41	91
46	16.5	60	22.2	52	18.0	MC KEAN	PA	35	-19	9	32	58
97	18.4	112	19.5	112	17.9	MERCER	PA	6	-8	-2	23	58
25	13.1	35	17.5	40	18.3	MIFFLIN	PA	34	4	39	57	60
33	18.8	43	20.1	50	17.8	MONROE	PA	7	-11	-5	21	57
393*	23.5	484*	21.6	648*	23.4	MONTGOMERY	PA	-8	8	-1	24	87
15	17.9	8	9.3	21	23.0	MONTOUR	PA	-48	148	28	49	86
181*	21.2	205*	21.4	243*	23.2	NORTHAMPTON	PA	1	8	9	33	86
71	13.7	135*	25.2	134*	23.9	NORTHUMBERLAND	PA	83	-5	74	74	89
31*	28.3	19	14.7	35	23.6	PERRY	PA	-48	60	-17	14	88
1919*	25.8	2119*	28.6	1871*	28.1	PHILADELPHIA	PA	11	-2	9	32	96
7	11.7	11	14.6	13	10.0	PIKE	PA	25	-32	-14	15	13
16	16.0	20	21.0	20	21.6	POTTER	PA	31	3	35	53	79
194*	22.5	184	19.8	191	19.8	SCHUYLKILL	PA	-12	0	-12	16	70
9	7.9	21	18.3	26	19.1	SNYDER	PA	132	4	142	88	65
64	15.9	63	15.1	62	14.4	SOMERSET	PA	-5	-5	-10	18	33
7	19.0	7	19.9	7	17.2	SULLIVAN	PA	5	-14	-10	18	53
38	22.8	40	23.1	34	18.3	SUSQUEHANNA	PA	1	-21	-20	12	61
22	12.2	26	14.3	28	14.3	TIOGA	PA	17	0	17	39	33
21	19.4	17	15.0	16	12.6	UNION	PA	-23	-16	-35	07	23
65	21.7	64	21.0	60	19.4	VENANGO	PA	-3	-8	-11	17	67
34	13.8	44	17.3	46	18.0	WARREN	PA	25	4	30	50	58
173	17.7	201	19.9	224	20.8	WASHINGTON	PA	12	5	17	40	75
39	22.2	31	18.0	46	23.4	WAYNE	PA	-19	30	5	29	87
263	18.4	303	18.2	402	21.4	WESTMORELAND	PA	-1	18	16	38	78
13	14.5	15	16.3	23	21.8	WYOMING	PA	12	34	50	64	80
193*	20.4	227	19.8	322*	24.0	YORK	PA	-3	21	18	40	89
809*	23.6	1007*	25.5	1171*	26.5	RHODE ISLAND	RI	8	4	12		
27	21.1	36	21.3	59*	26.1	BRISTOL	RI	1	23	24	45	93
80*	23.4	127*	26.0	169*	27.3	KENT	RI	11	5	16	39	95
47	21.1	54	20.5	65	21.1	NEWPORT	RI	-3	3	0	24	77
609*	24.0	732*	26.4	812*	27.6	PROVIDENCE	RI	10	5	15	38	96
46	23.7	58	22.8	66	20.0	WASHINGTON	RI	-4	-12	-16	14	71
410	9.8	677	12.8	1155	16.0	SOUTH CAROLINA	SC	32	25	64		
4	6.5	9	17.0	7	9.9	ABBEVILLE	SC	163	-42	54	65	12
12	7.1	24	13.5	37	12.9	AIKEN	SC	91	-4	83	77	25
2	11.8	1	5.9	4	21.0	ALLENDALE	SC	-50	255	78	75	77
20	8.8	48	17.2	69	17.3	ANDERSON	SC	96	1	97	81	54
		3	9.5			BAMBERG	SC					
4	14.7	5	12.1	5	12.0	BARNWELL	SC	-18	0	-18	13	20
3	7.1	7	14.5	16	10.5	BEAUFORT	SC	104	-28	47	62	14
4	11.8			11	8.7	BERKELEY	SC			-26	10	09
		3	22.8	4	15.5	CALHOUN	SC		-32			42
32	11.6	76	20.2	91	18.0	CHARLESTON	SC	75	-11	56	66	59
15	15.6	13	12.3	28	22.7	CHEROKEE	SC	-21	84	45	61	84
7	11.8	11	16.3	21	25.2	CHESTER	SC	38	54	114	84	92
8	9.9	10	12.5	16	16.0	CHESTERFIELD	SC	26	28	61	69	45
1	3.2	6	15.8	7	13.1	CLARENDON	SC	400	-17	315	96	26
3	7.2	8	13.8	11	14.6	COLLETON	SC	91	6	102	82	35
8	9.1	12	13.1	26	18.7	DARLINGTON	SC	44	43	105	83	62
1	2.5	7	12.2	12	17.8	DILLON	SC	391	46	617	99	57
4	11.3	3	8.2	12	15.4	DORCHESTER	SC	-28	88	37	54	40
1	4.5	4	12.9	4	11.1	EDGEFIELD	SC	187	-14	148	89	17
5	24.4	3	7.3	8	16.4	FAIRFIELD	SC	-70	124	-33	08	48

* SIGNIFICANTLY MORE DEATHS THAN EXPECTED

1950-1959 DEATHS	1950-1959 RATE	1960-1969 DEATHS	1960-1969 RATE	1970-1979 DEATHS	1970-1979 RATE	COUNTY AND STATE		PERCENT CHANGE 60S/50S	PERCENT CHANGE 70S/60S	PERCENT CHANGE 70S/50S	PERCENTILE %CHANGE 70S/50S	PERCENTILE RATES 1970S
10	8.6	19	11.3	45	21.2	FLORENCE	SC	32	87	148	89	78
1	4.9	2	2.9	9	11.5	GEORGETOWN	SC	-42	301	135	87	18
52	11.5	84	14.4	131	16.3	GREENVILLE	SC	25	13	41	58	47
3	2.7	15	13.1	38	24.1	GREENWOOD	SC	390	83	799	99	89
4	12.6	3	7.4	7	16.6	HAMPTON	SC	-41	124	32	51	49
10	7.8	15	9.3	28	10.7	HORRY	SC	20	15	37	55	15
1	4.7	4	18.0	2	7.6	JASPER	SC	283	-58	62	70	07
5	8.7	15	23.0	21	23.6	KERSHAW	SC	165	3	173	91	88
4	6.0	8	7.5	17	11.8	LANCASTER	SC	26	57	98	81	19
14	13.8	17	13.4	20	11.6	LAURENS	SC	-2	-13	-15	15	18
2	6.9	4	12.0	3	9.7	LEE	SC	75	-19	41	58	12
16	12.0	27	13.6	42	14.0	LEXINGTON	SC	13	3	17	39	31
1	6.7	2	11.9	1	5.7	MC CORMICK	SC	78	-52	-15	15	04
		3	6.0	12	21.8	MARION	SC		266			80
5	14.1	3	6.6	9	12.5	MARLBORO	SC	-53	89	-11	16	22
9	12.7	9	9.2	25	20.9	NEWBERRY	SC	-27	126	64	70	76
15	16.5	14	12.3	16	9.0	OCONEE	SC	-25	-27	-46	04	10
3	2.6	16	14.2	31	20.5	ORANGEBURG	SC	444	44	686	99	74
11	8.7	15	9.9	23	11.1	PICKENS	SC	13	12	27	48	16
27	8.4	46	11.8	86	17.7	RICHLAND	SC	40	50	110	83	56
6	14.0	5	11.3	11	19.6	SALUDA	SC	-19	73	40	57	68
42	10.6	54	11.4	93	16.5	SPARTANBURG	SC	8	45	56	67	48
7	11.2	11	11.3	13	9.3	SUMTER	SC	1	-18	-17	14	11
7	9.7	6	8.7	12	13.6	UNION	SC	-10	57	41	58	29
1	1.9	6	11.3	13	22.3	WILLIAMSBURG	SC	490	96	1057	99	83
20	15.2	21	10.9	58	23.0	YORK	SC	-28	111	52	65	85
449	14.9	507	14.9	651	18.2	SOUTH DAKOTA	SD	0	23	23		
2	9.4	9	32.1	1	4.0	AURORA	SD	241	-88	-57	02	02
8	7.9	18	15.9	29	26.3	BEADLE	SD	102	65	234	94	94
1	16.1	3	24.4	1	6.6	BENNETT	SD	52	-73	-59	02	05
5	11.6	5	9.6	18	30.1	BON HOMME	SD	-17	213	160	90	97
20	23.1	20	20.3	28	27.2	BROOKINGS	SD	-12	34	18	40	95
29	20.2	19	11.8	34	19.4	BROWN	SD	-42	65	-4	22	67
4	14.4	10	31.4	5	15.9	BRULE	SD	118	-49	10	33	44
2*	100.1			2*	79.9	BUFFALO	SD			-20	12	99
4	10.2	7	15.5	6	10.2	BUTTE	SD	52	-34	0	24	13
1	7.4	1	7.7	1	7.0	CAMPBELL	SD	3	-9	-6	20	06
6	10.9	6	10.4	8	14.7	CHARLES MIX	SD	-5	41	35	54	36
8	20.9	12	29.7	6	17.0	CLARK	SD	42	-43	-19	13	52
11	20.9	12	24.4	16	32.5	CLAY	SD	17	33	56	66	98
12	12.5	12	11.4	18	16.2	CODINGTON	SD	-9	42	30	50	46
4	21.4	2	11.7	2	11.1	CORSON	SD	-45	-5	-48	04	16
1	3.7	2	7.9	3	11.2	CUSTER	SD	116	41	205	93	17
9	10.9	9	9.2	20	22.0	DAVISON	SD	-15	138	102	82	81
7	9.8	7	10.9	13	19.9	DAY	SD	11	84	103	82	71
5	13.1	7	17.6	11	27.8	DEUEL	SD	34	59	113	84	96
2	13.2	1	14.9	4	31.3	DEWEY	SD	13	111	137	88	98
		2	8.0	4	13.2	DOUGLAS	SD		66			27
2	8.1	3	15.6	5	12.5	EDMUNDS	SD	93	-20	55	66	22
12	17.5	7	8.1	10	12.2	FALL RIVER	SD	-54	52	-30	09	20
2	11.9	2	8.6	2	6.2	FAULK	SD	-28	-28	-48	04	05
9	17.0	11	19.6	2	3.2	GRANT	SD	16	-83	-81	00	02
9	19.3	4	9.9	5	11.0	GREGORY	SD	-48	10	-43	05	16
1	5.6	1	5.3	2	12.6	HAAKON	SD	-6	137	124	86	23
3	6.1	6	12.9	10	23.1	HAMLIN	SD	112	78	277	96	86
6	21.3	7	20.4	11	31.7	HAND	SD	-4	55	49	63	98
		5	26.7	5	18.9	HANSON	SD		-29			64
3	22.3	3	29.6	2	15.7	HARDING	SD	33	-47	-29	09	42
3	7.3	3	6.5	9	18.4	HUGHES	SD	-10	181	153	89	61
13	24.1	11	16.5	18	26.4	HUTCHINSON	SD	-31	59	9	33	94
2	17.0	2	13.5	5	30.8	HYDE	SD	-21	128	81	77	98
3	26.9	1	11.2	3	26.8	JACKSON	SD	-58	138	-1	24	95
2	11.9	5	20.7	3	12.3	JERAULD	SD	74	-40	4	27	21
1	8.6	6*	58.1	1	8.4	JONES	SD	575	-86	-3	23	08
7	13.1	10	15.8	18	34.4	KINGSBURY	SD	21	117	163	90	98
5	7.8	12	17.1	5	7.6	LAKE	SD	119	-55	-2	23	07
11	14.2	13	16.1	14	17.4	LAWRENCE	SD	13	9	22	43	55
21*	28.3	16	20.1	22	26.6	LINCOLN	SD	-29	32	-6	20	94
1	10.6	2	9.5	2	10.2	LYMAN	SD	-10	8	-4	22	13
5	10.2	6	12.1	14	30.2	MC COOK	SD	19	150	198	93	97
2	7.4	3	9.3	4	12.3	MC PHERSON	SD	26	32	66	71	21
2	5.3	6	18.6	11	24.2	MARSHALL	SD	251	31	359	97	90
7	12.0	5	7.1	9	12.0	MEADE	SD	-41	68	0	24	19
3	30.1	1	12.1	1	8.2	MELLETTE	SD	-60	-32	-73	00	08
7	22.5	8	22.8	7	93.7	MINER	SD	1	311	316	96	99
56	18.0	52	13.5	96	22.2	MINNEHAHA	SD	-25	64	24	45	82
9	18.1	8	16.4	3	7.0	MOODY	SD	-9	-57	-62	01	06
20	14.7	27	14.6	36	16.3	PENNINGTON	SD	0	12	11	34	47
6	23.5	3	10.4	4	12.5	PERKINS	SD	-56	20	-47	04	23
4	21.3	2	7.8	5	18.5	POTTER	SD	-63	137	-13	16	62
5	7.4	12	16.0	11	15.7	ROBERTS	SD	116	-2	111	84	42
6	19.8	5	19.7	2	7.4	SANBORN	SD	0	-63	-63	01	06
10	15.3	4	6.5	14	23.0	SPINK	SD	-57	253	50	64	85
2	19.8	1	9.6	3	30.4	STANLEY	SD	-52	218	54	65	97
		2	13.6			SULLY	SD					
2	20.9			1	9.9	TODD	SD			-52	03	12
6	16.3	8	17.9	4	8.9	TRIPP	SD	10	-51	-46	04	10
14	21.9	20	25.3	11	13.7	TURNER	SD	15	-46	-37	06	30
9	15.6	12	19.3	10	16.6	UNION	SD	24	-14	7	30	49
5	17.1	4	9.7	6	12.2	WALWORTH	SD	-43	26	-29	09	20
12	11.7	23	23.1	15	15.0	YANKTON	SD	97	-35	28	49	38
		1	18.7			ZIEBACH	SD					
1228	11.3	1628	12.4	2393	15.0	TENNESSEE	TN	10	21	33		
17	14.0	32	15.9	36	13.4	ANDERSON	TN	14	-16	-4	21	28
9	9.3	15	14.1	15	12.6	BEDFORD	TN	51	-11	35	53	23
5	8.6	8	12.5	21	25.5	BENTON	TN	45	104	197	93	92
4	11.7	4	11.1	3	7.6	BLEDSOE	TN	-5	-31	-35	07	07

* SIGNIFICANTLY MORE DEATHS THAN EXPECTED

WHITE MALE

CANCER OF THE LARGE INTESTINE

ICD 153,159.0

1950-1959 DEATHS	1950-1959 RATE	1960-1969 DEATHS	1960-1969 RATE	1970-1979 DEATHS	1970-1979 RATE	COUNTY AND STATE	PERCENT CHANGE 60S/50S	PERCENT CHANGE 70S/60S	PERCENT CHANGE 70S/50S	PERCENTILE %CHANGE 70S/50S	PERCENTILE RATES 1970S
19	11.9	27	11.1	50	15.7	BLOUNT TN	-6	41	33	52	43
7	8.0	13	8.2	26	13.0	BRADLEY TN	2	59	62	70	25
4	3.2	5	4.0	15	9.4	CAMPBELL TN	25	134	191	92	11
1	3.3	5	10.9	6	11.1	CANNON TN	229	2	237	95	16
14	12.3	24	19.4	21	14.1	CARROLL TN	57	-27	15	37	32
23	16.2	15	8.5	32	14.3	CAPTER TN	-48	68	-12	16	33
4	10.8	6	11.2	6	9.8	CHEATHAM TN	4	-13	-9	18	12
8	18.9	4	8.8	12	22.8	CHESTER TN	-53	159	21	42	85
9	11.5	5	4.9	13	12.3	CLAIBORNE TN	-58	152	7	30	20
1	3.1	8	24.4	3	9.3	CLAY TN	692	-62	202	93	11
10	10.6	15	14.8	21	18.9	COCKE TN	39	28	78	75	64
7	7.6	13	11.7	28	17.2	COFFEE TN	54	47	126	86	53
6	9.3	6	9.7	16	23.5	CROCKETT TN	5	141	154	89	87
9	12.2	3	3.3	17	13.1	CUMBEPLAND TN	-73	293	8	31	26
136	13.4	197	16.3	235	16.4	DAVIDSON TN	22	1	23	44	48
8	19.3	4	8.0	7	11.4	DECATUR TN	-59	43	-41	05	17
4	8.2	7	12.7	6	8.9	DE KALB TN	55	-29	9	33	10
7	8.3	26	26.0	17	13.8	DICKSON TN	213	-47	66	71	30
18	14.2	18	14.2	19	12.9	DYER TN	0	-9	-9	18	25
4	10.2	3	7.5	11	19.3	FAYETTE TN	-26	157	90	79	67
8	16.9	3	5.0	8	12.5	FENTRESS TN	-70	150	-26	10	22
6	5.9	7	6.6	20	15.1	FRANKLIN TN	11	131	155	90	38
25	14.1	28	13.4	39	16.5	GIBSON TN	-5	24	17	39	49
6	6.0	17	16.2	12	10.7	GILES TN	170	-34	79	76	15
4	7.3	6	9.9	8	10.8	GRAINGER TN	36	9	49	63	15
13	8.9	16	9.1	24	11.5	GREENE TN	2	27	30	50	18
2	4.2	9	18.5	5	9.3	GRUNDY TN	344	-50	123	85	10
7	7.1	12	9.6	15	10.2	HAMBLEN TN	35	6	43	59	13
89	13.7	117	15.6	158	17.4	HAMILTON TN	13	12	27	48	55
2	5.7	4	11.6	6	17.6	HANCOCK TN	104	51	209	93	56
5	6.5	11	13.5	18	21.0	HARDEMAN TN	108	55	224	94	76
5	7.0	4	4.3	12	10.8	HARDIN TN	-38	150	54	65	15
13	12.2	15	11.9	18	10.9	HAWKINS TN	-3	-8	-11	17	16
5	12.9	5	11.3	10	18.9	HAYWOOD TN	-12	67	47	62	64
6	9.8	9	10.5	13	14.0	HENDERSON TN	8	33	43	59	31
7	6.7	24	19.3	32	21.6	HENRY TN	188	12	222	94	79
6	11.1	7	13.4	7	11.2	HICKMAN TN	20	-16	1	25	17
4	17.8	6	18.1	2	6.2	HOUSTON TN	2	-66	-65	01	04
4	7.3	6	10.5	10	15.3	HUMPHREYS TN	45	46	111	84	40
2	4.4	1	2.2	1	2.4	JACKSON TN	-50	10	-45	04	02
8	11.0	15	17.0	18	14.1	JEFFERSON TN	54	-17	28	49	32
7	13.4	9	15.4	7	10.6	JOHNSON TN	15	-31	-21	12	14
74	9.9	110	11.5	171	15.0	KNOX TN	15	31	51	64	38
3	10.0	2	6.2	1	2.8	LAKE TN	-38	-55	-72	00	02
4	6.3	8	10.5	12	17.2	LAUDERDALE TN	67	63	173	91	53
7	6.4	11	8.3	19	12.0	LAWRENCE TN	29	44	86	78	19
7	24.9	2	6.1	6	16.2	LEWIS TN	-76	167	-35	07	46
6	5.7	13	13.6	17	13.6	LINCOLN TN	136	1	138	88	29
7	8.9	8	7.0	19	15.8	LOUDON TN	-21	124	78	76	43
9	7.0	11	7.5	22	13.0	MC MINN TN	6	75	85	78	26
10	11.3	15	15.6	21	19.7	MC NAIRY TN	39	26	75	74	69
3	5.1	8	10.7	10	12.3	MACON TN	110	15	141	88	21
24	12.8	28	13.1	40	16.3	MADISON TN	2	24	27	48	47
3	4.1	13	16.7	14	14.9	MARION TN	310	-11	265	95	37
20*	29.6	12	15.0	18	18.8	MARSHALL TN	-49	25	-37	07	63
13	9.4	12	7.8	29	16.0	MAURY TN	-17	104	70	73	45
		1	4.1	5	20.0	MEIGS TN		385			71
5	5.6	13	12.7	13	10.7	MONROE TN	126	-16	90	79	15
13	11.8	18	11.5	28	15.0	MONTGOMERY TN	-3	30	27	48	38
6	48.1	1	4.6	2	9.6	MOORE TN	-90	108	-80	00	11
7	13.4	6	10.4	4	5.4	MORGAN TN	-22	-48	-60	02	04
10	6.9	16	10.6	29	17.6	OBION TN	54	66	155	89	56
7	9.2	4	4.8	16	18.2	OVERTON TN	-48	278	98	81	60
3	10.6	2	4.7	3	12.3	PERRY TN	-55	160	16	38	21
4	24.5	1	5.0	4	17.3	PICKETT TN	-80	249	-29	09	54
3	6.1	5	9.8	11	21.7	POLK TN	60	121	255	95	80
10	7.7	16	11.8	20	11.1	PUTNAM TN	55	-7	44	60	16
12	22.1	8	12.1	16	17.5	RHEA TN	-45	44	-21	12	55
23	18.8	16	11.5	19	9.6	ROANE TN	-39	-17	-49	03	11
11	10.9	14	11.2	21	14.0	ROBERTSON TN	3	26	29	49	32
17	12.0	18	10.4	25	9.8	RUTHERFORD TN	-13	-6	-18	13	12
4	6.6	10	15.6	15	18.3	SCOTT TN	135	17	176	91	60
2	12.1	2	9.1	3	9.7	SEQUATCHIE TN	-24	6	-20	12	12
7	7.0	6	5.6	22	13.5	SEVIER TN	-20	142	94	80	28
166	16.0	229	16.6	315	19.8	SHELBY TN	4	19	24	45	70
6	8.9	5	6.6	10	12.9	SMITH TN	-26	96	45	60	25
5	12.5	3	6.8	8	14.9	STEWART TN	-45	117	19	41	37
26	9.0	42	10.3	79	14.8	SULLIVAN TN	14	44	64	70	37
15	10.4	25	14.0	35	13.6	SUMNER TN	34	-3	30	50	29
8	11.6	10	11.9	19	19.8	TIPTON TN	2	67	70	73	70
				3	11.7	TROUSDALE TN					18
4	6.3	8	10.8	9	10.9	UNICOI TN	70	1	71	73	15
4	15.9			3	6.3	UNION TN			-61	01	05
3	18.4	2	12.3			VAN BUREN TN	-33				
14	13.6	12	10.1	23	16.0	WARREN TN	-26	58	18	40	45
29	11.8	33	10.1	53	14.6	WASHINGTON TN	-14	44	24	45	35
6	14.7	4	8.0	6	9.5	WAYNE TN	-45	18	-35	07	11
14	9.3	23	14.2	30	16.6	WEAKLEY TN	52	17	78	76	49
11	16.0	4	4.9	12	12.2	WHITE TN	-69	147	-24	11	20
10	13.0	8	6.8	25	14.8	WILLIAMSON TN	-48	118	14	37	36
5	4.3	16	11.3	29	15.4	WILSON TN	163	36	256	95	40
3106	11.5	4686	13.6	6906	16.2	TEXAS TX	18	19	40		
20	17.5	17	14.2	31	19.9	ANDERSON TX	-19	40	14	36	70
2	51.7	2	4.4	6	18.4	ANDREWS TX	-92	320	-64	01	61
18	13.7	15	8.5	32	13.9	ANGELINA TX	-38	63	1	25	31
1	2.6	3	4.5	9	12.8	ARANSAS TX	75	185	400	98	24
2	5.5	7	24.1	7	17.4	ARCHER TX	340	-28	217	94	54
1	8.5	3	29.2	1	5.9	ARMSTRONG TX	243	-80	-31	09	04
6	8.8	7	7.5	18	16.8	ATASCOSA TX	-15	125	91	79	50

* SIGNIFICANTLY MORE DEATHS THAN EXPECTED

WHITE MALE — CANCER OF THE LARGE INTESTINE — ICD 153,159.0

1950-1959 DEATHS	1950-1959 RATE	1960-1969 DEATHS	1960-1969 RATE	1970-1979 DEATHS	1970-1979 RATE	COUNTY AND STATE	PERCENT CHANGE 60S/50S	PERCENT CHANGE 70S/60S	PERCENT CHANGE 70S/50S	PERCENTILE %CHANGE 70S/50S	PERCENTILE RATES 1970S
9	11.9	18	20.2	27	27.7	AUSTIN TX	69	37	132	87	96
3	12.9	2	9.7	4	10.8	BAILEY TX	-25	11	-16	14	15
2	7.2	3	9.1	7	16.3	BANDERA TX	27	79	127	86	47
12	16.8	8	9.8	12	11.1	BASTROP TX	-42	14	-34	08	16
4	12.1	5	14.6	10	23.2	BAYLOR TX	20	59	91	79	86
8	13.8	7	9.1	8	9.4	BEE TX	-35	4	-32	08	11
18	6.7	38	11.1	57	14.1	BELL TX	65	27	109	83	32
214	12.7	352	15.8	512	18.6	BEXAR TX	24	18	46	61	62
4	18.8	4	16.0	13*	36.4	BLANCO TX	-15	128	94	80	99
1	37.2					BORDEN TX					
8	10.1	13	12.2	20	17.9	BOSQUE TX	22	46	77	75	58
25	14.8	37	15.1	40	13.7	BOWIE TX	2	-9	-8	19	30
17	11.4	28	12.2	58	16.8	BRAZORIA TX	7	38	48	62	51
7	7.4	10	7.7	28	17.0	BRAZOS TX	5	120	130	86	52
1	5.1	1	3.2	4	12.7	BREWSTER TX	-38	300	149	89	24
3	19.4	1	6.5	1	5.9	BRISCOE TX	-67	-9	-70	01	04
3	10.1	2	9.8	8	22.4	BROOKS TX	-3	129	122	85	83
17	11.1	20	11.1	35	17.9	BROWN TX	0	60	61	69	58
5	9.8	6	13.3	8	12.1	BURLESON TX	36	-9	24	45	20
9	17.0	10	11.8	15	14.4	BURNET TX	-31	23	-15	15	34
7	9.0	10	11.1	15	15.7	CALDWELL TX	24	42	75	75	43
4	17.8	7	18.7	14	26.4	CALHOUN TX	5	41	48	62	94
4	8.0	8	11.5	11	16.2	CALLAHAN TX	44	41	103	82	46
40	11.0	48	10.4	67	11.0	CAMERON TX	-5	6	0	25	16
2	5.9	3	8.9	5	14.4	CAMP TX	51	62	144	88	34
1	3.6	4	12.6	6	18.1	CARSON TX	255	43	407	98	59
4	4.2	13	12.2	27	21.6	CASS TX	188	78	412	98	79
2	11.4	2	10.6	4	13.5	CASTRO TX	-7	27	19	41	28
3	12.5	2	5.5	7	12.6	CHAMBERS TX	-56	128	1	25	23
14	10.0	18	11.6	33	17.4	CHEROKEE TX	16	49	74	74	54
10	20.5	2	3.7	9	18.8	CHILDRESS TX	-82	408	-8	18	63
9	17.0	10	21.0	12	19.6	CLAY TX	24	-7	15	38	69
				1	3.7	COCHRAN TX					02
1	7.0	5	24.5	4	15.9	COKE TX	249	-35	126	86	44
3	3.4	11	13.7	15	15.9	COLEMAN TX	299	16	363	97	44
24	12.1	18	7.8	41	12.4	COLLIN TX	-36	60	3	26	22
6	18.4	7	16.7	6	17.5	COLLINGSWORTH TX	-10	5	-5	21	56
11	17.3	6	8.3	21	22.1	COLORADO TX	-52	167	27	48	82
6	7.0	9	9.5	18	11.0	COMAL TX	34	16	56	67	16
9	8.6	11	12.8	13	12.5	COMANCHE TX	49	-3	45	60	22
1	3.9	5	24.9	1	4.2	CONCHO TX	540	-83	8	31	03
10	9.3	20	17.4	26	19.3	COOKE TX	87	11	107	83	66
8	9.4	17	17.7	12	10.9	CORYELL TX	88	-39	15	38	15
1	6.6	2	9.2	2	6.7	COTTLE TX	40	-28	1	25	05
1	9.8			3	22.3	CRANE TX			127	86	83
		2	15.7	1	6.6	CROCKETT TX		-58			05
9	23.7	7	17.9	6	14.4	CROSBY TX	-25	-20	-39	06	34
1	17.9	1	7.2	2	16.9	CULBERSON TX	-60	135	-6	21	51
3	8.8			3	9.5	DALLAM TX			8	31	11
296	15.1	488	17.1	655	17.8	DALLAS TX	13	4	18	40	57
7	11.3	3	4.5	17	22.7	DAWSON TX	-60	409	101	81	84
4	17.2	4	8.9	9	14.4	DEAF SMITH TX	-48	61	-16	14	34
6	15.9	1	1.9	4	7.4	DELTA TX	-88	297	-53	03	06
22	12.3	31	14.1	37	12.6	DENTON TX	14	-11	2	26	23
9	9.2	16	15.4	26	21.4	DE WITT TX	68	39	132	87	78
3	11.3	1	3.9	6	20.4	DICKENS TX	-65	419	81	76	73
4	13.0	1	2.8	3	6.8	DIMMIT TX	-78	142	-48	04	05
4	12.1	2	6.2	5	16.8	DONLEY TX	-48	170	39	56	50
4	7.8	2	4.9	9	15.9	DUVAL TX	-37	223	105	82	44
20	14.9	25	16.8	34	22.2	EASTLAND TX	12	32	49	63	82
18	19.2	26	12.9	42	13.8	ECTOR TX	-33	7	-28	10	30
2	18.9			1	10.2	EDWARDS TX			-46	04	13
19	10.1	28	14.1	59	26.3	ELLIS TX	39	87	160	90	94
72	13.0	108	13.1	153	16.6	EL PASO TX	1	26	28	48	49
12	10.4	20	14.6	15	9.1	ERATH TX	40	-38	-13	16	10
13	12.8	16	15.7	17	15.6	FALLS TX	23	-1	22	43	42
14	9.6	24	14.0	27	14.2	FANNIN TX	46	2	48	62	33
11	9.9	18	12.8	25	17.8	FAYETTE TX	29	39	80	76	57
6	13.5	2	5.0	2	6.7	FISHER TX	-63	35	-50	03	05
8	16.3	5	9.3	7	13.3	FLOYD TX	-43	44	-18	13	28
1	5.2	2	9.6	4	24.6	FOARD TX	86	157	378	97	91
12	13.3	24	19.3	30	15.9	FORT BEND TX	45	-18	19	41	44
4	12.9	5	14.6	10	23.1	FRANKLIN TX	13	58	79	76	86
9	14.7	11	17.6	13	17.1	FREESTONE TX	20	-3	16	38	52
6	15.8	3	6.0	8	14.6	FRIO TX	-62	144	-8	19	35
2	9.5	5	16.9	4	7.6	GAINES TX	77	-55	-20	12	07
51	13.6	67	14.9	125	23.0	GALVESTON TX	10	54	69	72	86
6	24.9	3	12.8	2	6.3	GARZA TX	-49	-51	-75	00	05
6	9.4	12	16.5	15	14.7	GILLESPIE TX	75	-11	56	67	36
1	74.3			1	112.4	GLASSCOCK TX			51	64	99
2	8.9	2	10.4	3	11.4	GOLIAD TX	17	9	28	49	17
7	8.8	6	7.0	16	16.4	GONZALES TX	-20	134	87	78	48
9	12.3	14	13.6	21	15.4	GRAY TX	11	13	25	46	40
41	13.4	54	14.8	61	13.7	GRAYSON TX	11	-7	3	26	30
29	19.2	35	13.8	48	15.6	GREGG TX	-28	13	-18	13	42
4	8.1	4	6.8	13	23.0	GRIMES TX	-16	239	185	92	85
11	11.0	20	15.4	19	11.5	GUADALUPE TX	40	-25	5	29	18
13	11.9	12	8.7	24	16.4	HALE TX	-27	89	39	56	48
5	12.5	3	7.3	5	11.6	HALL TX	-42	60	-7	20	18
9	13.3	12	18.0	11	13.7	HAMILTON TX	36	-23	4	27	30
1	8.2	2	15.0	2	7.2	HANSFORD TX	83	-52	-12	16	06
7	15.7	10	19.7	8	19.7	HARDEMAN TX	25	0	25	46	69
10	12.0	7	6.8	8	6.6	HARDIN TX	-44	-2	-45	05	05
271	11.3	582	17.9	792	18.3	HARRIS TX	58	3	62	69	60
12	12.9	18	13.9	21	14.1	HARRISON TX	7	2	9	32	32
				2	11.5	HARTLEY TX					18
6	10.5	11	17.6	12	22.0	HASKELL TX	68	25	110	84	82
4	6.2	7	8.0	12	12.3	HAYS TX	29	53	97	81	21
2	11.6	1	6.2			HEMPHILL TX	-46				
13	14.2	16	12.7	23	12.5	HENDERSON TX	-11	-1	-12	16	23
44	10.6	49	8.5	67	8.1	HIDALGO TX	-20	-5	-24	11	08

* SIGNIFICANTLY MORE DEATHS THAN EXPECTED

WHITE MALE

CANCER OF THE LARGE INTESTINE

ICD 153,159.0

1950-1959 DEATHS	1950-1959 RATE	1960-1969 DEATHS	1960-1969 RATE	1970-1979 DEATHS	1970-1979 RATE	COUNTY AND STATE	PERCENT CHANGE 60S/50S	PERCENT CHANGE 70S/60S	PERCENT CHANGE 70S/50S	PERCENTILE %CHANGE 70S/50S	PERCENTILE RATES 1970S
18	13.1	13	8.6	17	8.7	HILL TX	-34	1	-34	07	09
4	9.1	6	8.7	12	16.8	HOCKLEY TX	-4	93	86	78	51
7	19.1	4	9.9	20	26.0	HOOD TX	-48	163	36	54	93
9	8.1	20	15.8	23	16.9	HOPKINS TX	96	7	110	83	51
4	5.2	1	1.2	13	12.3	HOUSTON TX	-76	889	136	87	20
12	10.5	14	10.2	24	17.4	HOWARD TX	-3	70	66	71	54
1	11.6	1	13.5			HUDSPETH TX	16				
26	13.7	34	15.7	45	18.4	HUNT TX	15	17	34	53	61
6	11.5	17	28.3	18	14.1	HUTCHINSON TX	147	-50	23	44	32
6	14.6	8	17.1	8	14.5	JACK TX	17	-15	-1	24	35
5	10.6	9	19.0	6	10.4	JACKSON TX	80	-45	-2	23	14
1	1.0	12	12.6	17	13.7	JASPER TX	1164	9	1273	99	30
				3	35.4	JEFF DAVIS TX					98
81	16.8	114	16.9	143	18.2	JEFFERSON TX	1	8	8	32	60
2	11.2	2	9.4	2	7.2	JIM HOGG TX	-16	-24	-36	07	06
7	7.4	10	9.2	17	12.8	JIM WELLS TX	25	39	74	74	24
19	10.3	23	11.7	44	16.6	JOHNSON TX	14	41	61	69	49
13	12.9	8	7.2	15	13.2	JONES TX	-44	85	3	26	27
13	22.5	10	14.3	13	17.7	KARNES TX	-36	23	-22	12	56
10	7.4	26	17.1	31	18.9	KAUFMAN TX	131	10	155	90	64
6	20.2	9	21.5	10	17.9	KENDALL TX	7	-17	-11	17	58
1	10.3					KENT TX					
10	10.7	17	11.0	36	18.6	KERR TX	2	70	74	74	62
1	3.5	2	6.5	3	10.0	KIMBLE TX	87	55	189	92	13
		1	12.5	1	8.2	KINNEY TX		-34			08
4	8.3	4	7.2	14	19.0	KLEBERG TX	-14	165	128	86	65
7	18.5	9	18.8	9	20.8	KNOX TX	2	10	12	35	75
31	18.2	33	18.6	26	12.4	LAMAR TX	2	-33	-32	08	21
3	3.4	11	12.5	15	15.7	LAMB TX	272	25	365	97	42
6	10.3	8	14.0	9	14.1	LAMPASAS TX	37	1	37	55	32
5	19.5	5	19.6	3	11.5	LA SALLE TX	1	-41	-41	05	18
9	8.1	25	20.0	25	17.7	LAVACA TX	146	-12	118	85	56
2	4.8	7	16.1	8	15.4	LEE TX	236	-4	222	94	40
5	10.0	4	9.1	4	8.4	LEON TX	-9	-7	-15	15	09
10	9.9	7	5.6	28	17.7	LIBERTY TX	-44	218	79	76	56
9	9.1	18	17.8	18	16.6	LIMESTONE TX	96	-6	83	77	49
2	12.6	3	16.8	6	30.1	LIPSCOMB TX	34	79	140	88	97
4	11.6	2	5.0	7	14.9	LIVE OAK TX	-57	197	29	49	37
		6	10.0	10	9.1	LLANO TX		-9			10
1*	84.4	1*	112.4			LOVING TX	33				
33	10.7	53	11.5	86	15.7	LUBBOCK TX	7	36	47	61	42
3	7.6	4	10.7	2	4.5	LYNN TX	41	-58	-40	06	03
6	9.7	9	13.4	11	17.5	MC CULLOCH TX	38	30	79	76	55
54	10.8	80	14.2	136	20.6	MC LENNAN TX	32	45	91	79	74
1	3.1	3	7.8	13	31.0	MADISON TX	153	297	904	99	98
2	8.3	2	6.3	5	14.4	MARION TX	-25	130	73	74	34
1	8.0	4	23.5	2	8.1	MARTIN TX	193	-65	1	25	08
9*	34.8	1	3.2	4	14.1	MASON TX	-91	341	-60	02	32
4	5.0	10	11.4	13	11.5	MATAGORDA TX	130	1	131	87	18
3	8.6	6	12.5	5	7.6	MAVERICK TX	46	-39	-11	17	07
13	18.6	8	8.8	15	13.4	MEDINA TX	-53	52	-28	10	28
2	8.1	2	8.8	1	4.6	MENARD TX	9	-47	-43	05	03
9	13.4	18	12.8	31	15.7	MIDLAND TX	-4	22	17	39	42
11	10.9	14	11.3	23	17.3	MILAM TX	3	54	58	68	54
5	12.8	1	1.9	3	6.5	MILLS TX	-85	239	-49	03	05
8	16.9	5	9.1	3	4.6	MITCHELL TX	-46	-49	-73	00	03
12	11.8	14	13.3	27	21.9	MONTAGUE TX	13	65	86	78	81
6	6.3	10	7.9	41	13.9	MONTGOMERY TX	26	77	122	85	31
2	3.9	5	11.5	7	14.8	MOORE TX	196	29	283	96	37
5	12.7	4	8.5	15	26.9	MORRIS TX	-33	217	112	84	95
5	33.2	1	8.3	3	22.8	MOTLEY TX	-75	176	-31	08	85
15	14.5	15	11.0	23	15.0	NACOGDOCHES TX	-24	36	3	27	37
15	8.9	22	12.3	28	14.1	NAVARRO TX	39	14	58	67	32
2	5.9			1	1.4	NEWTON TX			-76	00	01
10	13.8	10	11.9	14	15.7	NOLAN TX	-14	32	13	36	42
37	8.3	76	12.2	115	15.8	NUECES TX	48	29	91	79	43
4	21.5	6	18.3	3	9.5	OCHILTREE TX	-15	-48	-56	02	11
		1	9.3			OLDHAM TX					
12	10.0	25	15.5	46	19.6	ORANGE TX	55	26	95	80	69
22	20.5	19	16.5	14	10.9	PALO PINTO TX	-19	-34	-47	04	15
7	13.5	3	3.8	12	14.0	PANOLA TX	-72	267	3	27	31
8	6.4	20	13.7	21	11.2	PARKER TX	113	-18	74	74	17
3	9.1	7	25.9	10	27.0	PARMER TX	184	4	196	93	95
1	3.6	5	14.2	5	10.4	PECOS TX	292	-27	188	92	14
6	10.1	4	6.2	11	9.7	POLK TX	-38	56	-4	22	12
35	14.2	38	10.8	63	17.1	POTTER TX	-24	58	20	42	52
3	11.6	2	7.8	1	3.3	PRESIDIO TX	-33	-57	-71	00	02
1	3.2	2	6.7	1	2.3	RAINS TX	110	-65	-27	10	02
3	3.9	9	10.1	32	17.4	RANDALL TX	161	73	352	97	55
1	11.5			2	27.5	REAGAN TX			139	88	95
		1	6.5	1	7.8	REAL TX		20			07
8	11.1	16	17.6	18	18.2	RED RIVER TX	59	3	65	70	60
2	6.1	1	2.2	4	7.6	REEVES TX	-64	246	23	44	07
1	2.0	3	9.1	5	11.4	REFUGIO TX	357	25	469	98	17
				1	17.0	ROBERTS TX					52
10	17.6	9	15.1	8	10.1	ROBERTSON TX	-14	-33	-43	05	13
2	9.4	3	8.9	1	2.3	ROCKWALL TX	-5	-74	-75	00	02
7	10.2	21	26.2	12	14.0	RUNNELS TX	156	-47	37	55	31
14	9.7	20	11.8	23	11.4	RUSK TX	21	-4	17	39	17
2	5.5	4	9.4	6	14.4	SABINE TX	70	53	160	90	34
3	12.5	5	18.3	6	15.4	SAN AUGUSTINE TX	47	-16	23	44	40
2	10.5	2	7.9	3	7.1	SAN JACINTO TX	-26	-10	-33	08	06
5	5.4	19	12.2	19	10.9	SAN PATRICIO TX	123	-10	100	81	16
7	15.5	9	17.2	6	11.3	SAN SABA TX	11	-34	-27	10	17
1	6.3	5	50.2	6	49.9	SCHLEICHER TX	694	-1	690	99	99
6	9.9	6	7.1	10	12.9	SCURRY TX	-29	82	30	50	25
1	4.6	4	18.6	5	17.2	SHACKELFORD TX	303	-8	273	95	53
7	6.3	13	12.0	22	18.7	SHELBY TX	91	55	196	93	63
1	11.8	1	11.1	2	12.3	SHERMAN TX	-6	11	5	28	21
20	9.0	38	12.4	56	13.4	SMITH TX	39	7	49	63	28
2	11.7	3	13.8	5	21.9	SOMERVELL TX	17	59	87	78	81

* SIGNIFICANTLY MORE DEATHS THAN EXPECTED

WHITE MALE — CANCER OF THE LARGE INTESTINE — ICD 153,159.0

1950-1959 DEATHS	1950-1959 RATE	1960-1969 DEATHS	1960-1969 RATE	1970-1979 DEATHS	1970-1979 RATE	COUNTY AND STATE	PERCENT CHANGE 60S/50S	PERCENT CHANGE 70S/60S	PERCENT CHANGE 70S/50S	PERCENTILE %CHANGE 70S/50S	PERCENTILE RATES 1970S
2	5.6	3	4.8	6	7.9	STARR TX	-13	63	42	58	07
5	9.1	6	9.6	9	14.4	STEPHENS TX	6	50	58	67	34
1	15.5			1	37.5	STERLING TX			142	88	99
2	11.6	3	17.8	6	34.4	STONEWALL TX	54	93	197	93	98
		1	4.8	2	12.3	SUTTON TX		154			21
3	9.5	5	11.1	11	23.0	SWISHER TX	17	107	141	88	85
166	12.4	280	15.6	445	19.5	TARRANT TX	26	25	58	67	68
28	11.3	50	15.3	64	16.0	TAYLOR TX	36	5	42	58	45
		2	19.0	1	11.1	TERRELL TX		-42			16
3	6.0	11	19.4	4	6.7	TERRY TX	221	-65	11	34	05
1	5.2	2	9.2	2	7.9	THROCKMORTON TX	78	-15	52	65	07
7	8.8	14	17.2	17	17.8	TITUS TX	96	4	103	82	57
18	8.6	40	14.6	41	12.3	TOM GREEN TX	69	-16	42	58	21
59	11.6	82	11.3	165	18.4	TRAVIS TX	-3	63	59	68	61
		6	14.2	8	15.4	TRINITY TX		8			41
8	15.8	4	6.9	12	11.8	TYLER TX	-56	72	-25	11	19
5	5.7	14	14.5	19	15.1	UPSHUR TX	154	4	164	90	38
3	21.4			1	4.8	UPTON TX			-78	00	03
7	10.4	8	10.8	11	12.0	UVALDE TX	4	11	15	38	19
2	3.4	7	11.7	13	13.6	VAL VERDE TX	246	17	304	96	29
16	14.4	12	8.4	23	12.4	VAN ZANDT TX	-41	47	-13	15	22
11	11.2	19	12.6	30	16.0	VICTORIA TX	12	27	43	59	45
3	4.2	6	9.2	6	7.6	WALKER TX	119	-18	79	76	07
5	18.8	8	24.2	14	28.1	WALLER TX	28	16	49	63	96
2	4.1	4	11.3	7	17.5	WARD TX	179	54	330	97	55
10	13.1	11	11.6	25	23.3	WASHINGTON TX	-11	102	79	76	87
9	5.6	13	5.7	30	10.9	WEBB TX	1	93	95	80	16
9	7.3	18	13.5	28	19.7	WHARTON TX	85	46	169	91	69
3	6.2	6	12.5	5	8.4	WHEELER TX	102	-33	36	54	09
40	10.9	63	15.6	77	16.9	WICHITA TX	44	8	56	66	51
9	9.8	6	5.9	18	16.4	WILBARGER TX	-40	180	68	72	48
2	4.7	2	3.0	9	15.3	WILLACY TX	-36	408	225	94	39
16	9.5	18	8.8	33	13.2	WILLIAMSON TX	-7	50	39	56	27
5	8.0	10	15.2	9	11.4	WILSON TX	90	-25	42	58	17
		3	11.7	9	36.2	WINKLER TX		209			99
17	17.4	21	18.6	17	12.5	WISE TX	7	-33	-28	09	22
13	13.5	17	14.5	32	23.3	WOOD TX	8	61	73	74	87
2	10.3	2	13.8	5	17.8	YOAKUM TX	34	29	72	74	57
6	7.0	17	16.2	27	22.6	YOUNG TX	131	39	222	94	84
1	5.4	2	11.9	2	6.9	ZAPATA TX	122	-42	28	49	06
1	2.0			3	7.4	ZAVALA TX			276	96	06
261	10.6	401	12.7	544	13.3	UTAH UT	20	5	26		
2	25.0	3	13.5	2	7.4	BEAVER UT	-46	-45	-70	00	06
7	11.5	10	12.0	15	12.9	BOX ELDER UT	4	8	13	35	25
9	6.7	18	12.2	24	13.4	CACHE UT	81	10	99	81	28
10	17.3	15	17.9	10	10.2	CARBON UT	4	-43	-41	06	13
6	5.5	12	7.6	30	12.7	DAVIS UT	39	67	133	87	24
1	4.6	5	17.1	4	12.3	DUCHESNE UT	271	-28	166	91	21
2	10.1	3	11.7	7	24.0	EMERY UT	16	105	138	88	89
		2	16.8	1	10.2	GARFIELD UT		-39			13
		2	22.6	2	10.1	GRAND UT		-55			13
3	8.8	5	13.0	9	18.1	IRON UT	48	39	106	83	59
				5	22.3	JUAB UT					83
3	35.4	1	9.3	3	18.6	KANE UT	-74	101	-48	04	62
2	5.7	5	12.5	5	10.5	MILLARD UT	121	-16	86	78	14
		1	12.2	3	16.5	MORGAN UT		35			48
1	14.0					PIUTE UT					
		2	22.4	1	18.7	RICH UT		-16			63
135	12.7	200	14.6	244	14.1	SALT LAKE UT	15	-3	11	34	32
1	2.6					SAN JUAN UT					
8	14.2	8	11.9	6	9.2	SANPETE UT	-16	-22	-35	07	10
5	12.2	5	10.3	3	4.2	SEVIER UT	-15	-60	-66	01	02
2	7.6			2	6.5	SUMMIT UT			-15	15	05
1	2.1	11	19.7	12	19.9	TOOELE UT	852	1	860	99	70
5	18.1	3	9.5	4	7.1	UINTAH UT	-47	-26	-61	01	06
22	8.4	41	11.7	58	11.8	UTAH UT	38	1	39	57	19
1	6.6	1	4.0	7	24.2	WASATCH UT	-39	498	266	95	90
7	16.6	2	2.4	7	6.3	WASHINGTON UT	-86	166	-62	01	05
1	13.4			1	7.9	WAYNE UT			-41	05	07
27	8.5	46	11.8	79	16.6	WEBER UT	39	40	95	80	49
310	16.9	332	17.4	400	19.0	VERMONT VT	3	9	12		
12	13.1	23	24.6	16	15.0	ADDISON VT	87	-39	15	37	38
24	19.3	15	11.0	31	19.2	BENNINGTON VT	-43	74	0	24	66
20	15.2	16	13.7	21	16.7	CALEDONIA VT	-10	22	10	33	50
28	11.2	49	18.1	49	15.6	CHITTENDEN VT	61	-14	39	56	42
7	19.8	3	9.9	7	19.0	ESSEX VT	-50	92	-4	21	65
27	19.1	24	16.8	39	26.4	FRANKLIN VT	-12	57	38	56	94
		4	21.3	4	21.6	GRAND ISLE VT		1			79
9	16.4	12	20.4	14	21.7	LAMOILLE VT	24	7	32	52	80
18	19.2	14	15.8	16	15.4	ORANGE VT	-18	-2	-20	12	41
23	23.0	13	12.7	26	23.6	ORLEANS VT	-45	86	3	26	88
49	21.7	67*	27.2	43	15.8	RUTLAND VT	25	-42	-27	10	43
24	12.9	42	20.3	50	22.1	WASHINGTON VT	57	9	71	73	82
33	20.3	26	15.4	33	18.7	WINDHAM VT	-24	22	-8	19	63
36	17.5	24	10.6	51	20.4	WINDSOR VT	-39	92	17	39	73
1156	12.0	1718	14.3	2692	17.9	VIRGINIA VA	19	26	49		
15	11.4	24	18.3	30	23.2	ACCOMACK VA	61	27	104	82	86
22	13.5	24	11.8	36	14.2	ALBEMARLE#VA	-12	21	6	29	33
12	12.5	25	21.3	16	13.1	ALLEGHANY#VA	71	-38	5	29	26
2	10.6	4	16.9	4	16.9	AMELIA VA	60	0	59	68	51
4	5.9	8	9.6	11	12.6	AMHERST VA	64	31	114	84	23
3	9.0	4	11.0	5	11.6	APPOMATTOX VA	22	6	29	49	18
63	16.4	97	18.0	143*	24.9	ARLINGTON#VA	10	39	52	65	91
24	9.1	39	12.5	56	15.4	AUGUSTA#VA	37	24	70	73	41
9*	41.4	5	19.0	6	23.9	BATH VA	-54	26	-42	05	89

* SIGNIFICANTLY MORE DEATHS THAN EXPECTED

\# COUNTIES COMBINED; SEE TECHNICAL NOTES

WHITE MALE

CANCER OF THE LARGE INTESTINE

ICD 153,159.0

1950-1959 DEATHS	1950-1959 RATE	1960-1969 DEATHS	1960-1969 RATE	1970-1979 DEATHS	1970-1979 RATE	COUNTY AND STATE	PERCENT CHANGE 60S/50S	PERCENT CHANGE 70S/60S	PERCENT CHANGE 70S/50S	PERCENTILE %CHANGE 70S/50S	PERCENTILE RATES 1970S
33	8.7	52	11.8	94	17.5	BEDFORD#VA	35	48	100	81	55
		3	19.3	2	6.3	BLAND VA		-67			05
3	4.6	10	12.5	10	11.8	BOTETOURT VA	172	-5	159	90	19
6	18.7	4	10.7	5	16.5	BRUNSWICK VA	-43	54	-12	16	48
7	9.7	6	7.0	12	10.1	BUCHANAN VA	-28	45	4	27	13
1	2.5	6	16.1	4	10.6	BUCKINGHAM VA	548	-34	327	96	14
1	2.7	4	12.4	7	19.5	CAROLINE VA	353	58	615	98	68
29	15.6	29	13.6	36	14.4	CARROLL#VA	-13	6	-8	19	34
				1	17.1	CHARLES CITY VA					52
8	23.8	5	10.4	7	15.2	CHARLOTTE VA	-56	45	-36	07	39
147	19.7	165	18.9	185	20.4	CHESTERFIELD#VA	-4	8	4	27	73
4	14.8	6	18.2	8	21.2	CLARKE VA	23	16	43	59	77
		4	25.7	4	20.7	CRAIG VA		-20			74
11	21.5	11	16.8	19	24.3	CULPEPER VA	-22	44	13	36	90
4	22.4	1	5.4	2	9.3	CUMBERLAND VA	-76	73	-59	02	10
4	7.7	8	13.0	10	13.2	DICKENSON VA	69	1	71	73	27
16	10.6	32	15.6	49	20.2	DINWIDDIE#VA	47	30	91	79	72
3	17.2	5	22.4	7	24.1	ESSEX VA	30	8	40	57	89
78	15.5	127	15.9	285	21.0	FAIRFAX#VA	2	33	35	54	76
13	16.1	12	14.5	20	20.7	FAUQUIER VA	-10	42	28	49	74
2	3.8	6	10.8	7	9.9	FLOYD VA	182	-8	159	90	12
1	4.0	9	31.0	5	14.5	FLUVANNA VA	675	-53	262	95	35
6	6.4	6	5.7	14	11.0	FRANKLIN VA	-12	93	70	73	16
12	8.7	37	22.3	48	23.0	FREDERICK#VA	156	3	164	90	86
7	12.1	4	6.2	14	17.1	GILES VA	-49	176	42	58	53
7	16.4	4	8.4	12	19.3	GLOUCESTER VA	-49	128	18	40	67
2	8.3			5	15.9	GOOCHLAND VA			91	79	44
2	10.7	1	4.3			GREENE VA	-60				
2	7.5	5	14.7	7	16.8	GREENSVILLE#VA	95	14	123	86	50
14	15.1	6	5.8	12	9.7	HALIFAX#VA	-62	68	-36	07	12
3	4.9	16	16.4	31	21.9	HANOVER VA	234	34	347	97	81
18	8.1	84	24.9	124	23.1	HENRICO VA	208	-7	186	92	86
5	4.6	15	8.5	31	13.1	HENRY#VA	85	55	187	92	26
5	25.1	2	9.2	3	17.4	HIGHLAND VA	-63	88	-31	09	54
1	2.2	7	18.8	7	14.2	ISLE OF WIGHT VA	738	-25	533	98	33
8	8.3	14	9.3	25	14.0	JAMES CITY#VA	11	51	68	72	31
1	4.5	3	24.5	3	13.6	KING AND QUEEN VA	445	-45	202	93	29
1	6.2	4	23.9	5	23.9	KING GEORGE VA	282	0	282	96	89
2	13.4	7	40.1	6	32.2	KING WILLIAM VA	200	-20	141	88	98
4	10.7	6	13.2	12	20.9	LANCASTER VA	23	58	95	80	76
13	11.9	9	7.2	23	18.4	LEE VA	-39	154	54	66	61
16	19.0	11	11.2	26	19.7	LOUDOUN VA	-41	77	4	27	69
2	4.4	5	10.5	10	16.4	LOUISA VA	140	57	277	96	48
1	2.2	6	16.9	4	9.9	LUNENBURG VA	683	-42	358	97	12
5	15.5	4	11.8	6	12.8	MADISON VA	-24	9	-17	14	24
3	7.3	7	14.6	11	20.5	MATHEWS VA	99	40	180	91	74
2	5.1	9	11.4	15	16.2	MECKLENBURG VA	121	42	215	94	46
4	15.4	2	7.1	7	19.5	MIDDLESEX VA	-54	174	27	47	68
6	5.6	11	7.9	23	13.1	MONTGOMERY#VA	42	66	136	87	26
4	9.8	5	10.0	8	15.1	NELSON VA	2	51	54	66	38
10	21.2	4	8.8	14	28.5	NORTHAMPTON VA	-59	225	34	53	96
7	18.2	6	12.5	6	10.9	NORTHUMBERLAND VA	-31	-13	-40	06	15
5	11.2	9	18.2	9	16.5	NOTTOWAY VA	63	-9	47	62	48
9	18.3	8	15.3	16	22.9	ORANGE VA	-16	49	25	46	85
5	7.7	8	11.2	16	19.6	PAGE VA	46	76	157	90	69
8	12.2	7	9.7	7	8.3	PATRICK VA	-21	-15	-32	08	08
21	8.4	49	16.5	69	17.7	PITTSYLVANIA#VA	95	7	110	83	56
2	14.2	5	25.4	2	7.3	POWHATAN VA	78	-71	-49	03	06
4	11.5	5	12.9	8	16.9	PRINCE EDWARD VA	12	31	47	61	51
5	11.3	19	18.3	34	23.0	PRINCE WILLIAM#VA	62	26	104	82	86
10	11.5	9	9.2	15	11.7	PULASKI VA	-20	27	2	25	19
1	5.1	3	13.7	4	18.7	RAPPAHANNOCK VA	167	37	265	95	63
5	27.2	5	23.5	2	6.1	RICHMOND VA	-14	-74	-78	00	04
67	14.1	81	13.0	155	19.8	ROANOKE#VA	-7	52	41	58	70
15	16.3	12	11.4	19	14.9	ROCKBRIDGE#VA	-30	31	-9	18	37
22	12.1	36	16.6	42	15.1	ROCKINGHAM#VA	37	-9	25	46	38
8	9.1	9	8.5	17	14.1	RUSSELL VA	-6	66	56	66	32
6	6.3	14	12.2	19	15.8	SCOTT VA	94	30	151	89	43
16	15.9	18	15.3	14	9.5	SHENANDOAH VA	-4	-38	-40	06	11
10	8.9	10	6.9	14	8.8	SMYTH VA	-22	28	-1	24	09
5	10.0	6	11.1	7	13.3	SOUTHAMPTON#VA	12	19	33	52	27
9	11.6	15	16.3	24	19.4	SPOTSYLVANIA#VA	41	19	67	71	68
6	12.6	11	18.8	14	18.2	STAFFORD VA	49	-3	44	60	60
2	11.2	2	12.9	3	19.7	SURRY VA	14	53	75	75	69
2	7.4	2	8.1	5	21.2	SUSSEX VA	9	161	184	92	77
14	9.3	11	6.5	25	13.0	TAZEWELL VA	-30	99	39	56	25
4	10.2	11	17.4	18	22.4	WARREN VA	71	28	120	85	83
11	5.9	19	8.4	42	15.2	WASHINGTON#VA	44	80	159	90	39
5	14.1	8	18.3	13	21.1	WESTMORELAND VA	30	15	50	64	77
10	7.5	22	13.3	24	12.6	WISE#VA	78	-5	69	73	23
7	8.0	13	14.3	10	10.3	WYTHE VA	80	-28	29	50	14
116	13.4	200	16.7	299	19.5	CHESAPEAKE CITY#VA	25	17	46	61	68
23	15.7	21	9.8	52	20.7	HAMPTON CITY VA	-38	112	32	51	75
22	12.6	28	13.2	70	24.8	NEWPORT NEWS CITY VA	4	88	96	80	91
3	3.2	7	9.9	26	23.4	SUFFOLK CITY VA	204	138	622	99	87
1742	14.9	2088	15.5	2832	18.0	WASHINGTON WA	4	16	21		
4	14.4	4	8.6	7	16.3	ADAMS WA	-40	89	13	36	47
11	14.4	11	14.0	15	16.6	ASOTIN WA	-2	18	16	38	49
16	11.4	34	18.7	57	21.9	BENTON WA	63	17	92	80	81
31	16.4	26	12.0	39	14.8	CHELAN WA	-27	24	-10	18	36
15	10.6	23	13.8	48	18.6	CLALLAM WA	29	35	75	75	62
60	15.3	59	12.1	80	12.5	CLARK WA	-21	3	-18	13	22
2	6.4	5	16.9	2	6.2	COLUMBIA WA	162	-63	-4	22	04
34	16.9	37	14.1	62	19.5	COWLITZ WA	-16	38	15	38	68
5	8.5	9	14.2	14	14.3	DOUGLAS WA	67	0	68	72	33
1	5.3	2	11.3	3	20.7	FERRY WA	113	84	291	96	75
5	9.4	10	14.1	20	19.6	FRANKLIN WA	49	39	108	83	69
2	14.1	5	30.2	1	6.5	GARFIELD WA	113	-78	-54	03	05
13	15.6	24	20.4	25	12.8	GRANT WA	31	-37	-18	13	24

COUNTIES COMBINED; SEE TECHNICAL NOTES

WHITE MALE

CANCER OF THE LARGE INTESTINE

ICD 153,159.0

1950-1959 DEATHS	1950-1959 RATE	1960-1969 DEATHS	1960-1969 RATE	1970-1979 DEATHS	1970-1979 RATE	COUNTY AND STATE	PERCENT CHANGE 60S/50S	PERCENT CHANGE 70S/60S	PERCENT CHANGE 70S/50S	PERCENTILE %CHANGE 70S/50S	PERCENTILE RATES 1970S
50	16.7	50	16.1	52	15.4	GRAYS HARBOR WA	-3	-5	-8	19	41
8	12.3	11	14.1	24	15.6	ISLAND WA	15	10	27	48	42
9	18.2	6	10.0	15	17.8	JEFFERSON WA	-45	78	-2	23	57
583	16.7	695	17.2	879	19.3	KING WA	3	12	16	38	66
48	14.8	63	15.1	104	19.8	KITSAP WA	2	31	34	53	70
16	14.8	9	8.1	24	22.5	KITTITAS WA	-45	177	52	64	84
5	9.5	5	7.4	13	16.6	KLICKITAT WA	-22	124	76	75	49
36	14.0	56	20.6	44	15.1	LEWIS WA	47	-27	7	31	38
5	8.4	11	16.9	15	21.4	LINCOLN WA	101	27	156	90	78
10	12.1	11	10.7	29	20.2	MASON WA	-12	89	66	71	72
18	13.9	20	15.0	28	18.7	OKANOGAN WA	8	24	34	53	63
10	10.3	9	9.1	18	15.6	PACIFIC WA	-12	73	51	64	42
5	10.5	7	16.6	7	16.7	PEND OREILLE WA	58	1	59	68	50
190	14.6	252	17.7	292	17.6	PIERCE WA	21	-1	20	42	56
1	3.7	9	30.9	6	12.3	SAN JUAN WA	735	-60	233	94	21
38	14.6	30	10.6	44	13.4	SKAGIT WA	-27	26	-8	18	28
2	8.6	1	6.8	5	15.7	SKAMANIA WA	-21	133	84	77	43
109	16.6	146	17.0	227	21.0	SNOHOMISH WA	2	24	27	47	76
192	15.3	190	14.0	266	18.3	SPOKANE WA	-8	30	20	41	60
10	9.5	12	11.1	17	15.3	STEVENS WA	17	38	62	69	39
30	13.1	47	16.2	73	17.3	THURSTON WA	24	7	32	52	54
2	11.5	4	17.8	5	20.6	WAHKIAKUM WA	54	16	79	76	74
16	7.6	34	15.3	38	15.4	WALLA WALLA WA	100	1	101	81	40
51	13.5	58	14.2	77	17.4	WHATCOM WA	5	23	29	49	55
22	15.8	26	19.1	26	18.7	WHITMAN WA	21	-2	18	41	63
77	12.8	77	11.3	131	17.1	YAKIMA WA	-12	52	34	53	53
870	11.6	1095	13.6	1438	16.5	WEST VIRGINIA WV	17	22	42		
9	9.7	9	9.5	13	13.7	BARBOUR WV	-2	45	42	58	30
15	10.7	32	18.3	35	17.9	BERKELEY WV	71	-2	67	71	58
7	7.8	11	9.5	15	11.8	BOONE WV	21	24	51	64	19
13	15.3	5	5.4	17	18.9	BRAXTON WV	-65	249	23	44	64
13	13.2	19	15.8	23	16.3	BROOKE WV	20	3	23	44	47
69	16.5	81	17.1	113	22.4	CABELL WV	3	32	36	54	83
5	10.5	3	6.0	5	10.0	CALHOUN WV	-43	67	-5	21	12
4	7.7	2	3.3	10	18.4	CLAY WV	-58	461	137	88	61
7	14.3	3	5.8	10	23.9	DODDRIDGE WV	-59	309	68	71	89
20	8.3	21	8.9	24	9.2	FAYETTE WV	6	4	11	34	10
6	13.9	5	11.1	7	14.7	GILMER WV	-21	32	5	29	35
6	19.6	12	27.1	4	8.3	GRANT WV	38	-69	-57	02	08
11	6.8	15	9.3	15	8.3	GREENBRIER WV	37	-11	22	43	08
8	13.4	8	12.5	12	17.1	HAMPSHIRE WV	-7	37	28	49	53
19	14.2	28	19.6	36	20.6	HANCOCK WV	38	5	44	60	74
7	15.2	8	16.6	5	8.7	HARDY WV	9	-48	-43	05	09
52	13.3	58	14.5	79	19.1	HARRISON WV	9	32	44	60	66
11	14.4	9	10.8	16	15.4	JACKSON WV	-25	42	7	31	40
7	11.4	8	10.6	19	19.6	JEFFERSON WV	-7	85	72	74	69
85	11.5	128	15.1	184	19.0	KANAWHA WV	32	25	65	71	65
13	9.8	12	9.4	16	13.8	LEWIS WV	-4	47	41	58	30
7	8.6	14	15.6	15	14.5	LINCOLN WV	82	-8	68	72	34
17	8.4	20	10.7	22	11.6	LOGAN WV	28	9	39	56	18
12	6.1	24	12.9	33	19.7	MC DOWELL WV	111	52	220	94	69
33	11.2	42	13.7	68	20.2	MARION WV	22	48	81	76	72
33	18.9	41	21.8	46	23.8	MARSHALL WV	15	9	26	47	88
12	11.4	15	13.8	22	18.2	MASON WV	22	32	60	69	60
27	10.3	38	14.1	35	11.5	MERCER WV	36	-18	11	34	18
9	12.1	18	18.1	21	19.3	MINERAL WV	50	6	60	68	66
9	6.1	8	5.6	12	8.8	MINGO WV	-9	56	43	59	09
30	13.4	30	12.1	41	15.3	MONONGALIA WV	-10	27	14	37	39
4	5.7	11	14.9	3	4.2	MONROE WV	163	-72	-27	10	03
4	10.4	4	10.7	13	26.5	MORGAN WV	2	148	154	89	94
13	12.1	6	5.5	17	14.3	NICHOLAS WV	-54	159	19	41	33
68	19.9	61	17.8	88*	26.6	OHIO WV	-11	50	34	53	94
		5	11.5	2	4.4	PENDLETON WV		-62			03
4	12.8	4	12.4	9	24.3	PLEASANTS WV	-4	96	89	79	90
2	3.3	7	11.2	12	19.0	POCAHONTAS WV	239	70	476	98	65
16	11.4	14	9.9	24	15.5	PRESTON WV	-13	56	36	54	41
4	4.9	16	15.8	22	16.6	PUTNAM WV	221	5	237	95	49
30	11.5	26	8.8	41	12.0	RALEIGH WV	-23	36	4	28	20
5	4.1	13	9.5	26	17.5	RANDOLPH WV	133	85	331	97	55
16	20.5	7	10.2	10	13.5	RITCHIE WV	-50	32	-34	07	29
6	6.7	11	10.9	12	13.5	ROANE WV	63	24	103	82	28
7	8.7	11	14.6	8	10.4	SUMMERS WV	66	-29	19	41	14
12	13.8	14	16.7	15	17.4	TAYLOR WV	21	4	27	48	55
6	12.6	7	14.4	9	18.9	TUCKER WV	14	32	51	64	65
5	7.6	13	21.7	10	16.4	TYLER WV	186	-24	116	85	48
5	4.7	14	14.0	15	13.4	UPSHUR WV	195	-4	183	91	28
9	5.8	18	10.6	24	12.3	WAYNE WV	83	17	114	84	21
3	5.3	3	5.1	3	4.9	WEBSTER WV	-2	-4	-6	20	03
11	10.8	23	22.4	16	14.9	WETZEL WV	108	-34	38	56	37
3	9.6	2	6.7	4	15.0	WIRT WV	-30	123	56	67	38
57	18.3	73	20.7	73	18.3	WOOD WV	13	-12	0	24	60
4	3.8	5	5.5	9	8.0	WYOMING WV	43	46	109	83	08
3210*	18.9	3720*	19.2	4164	19.4	WISCONSIN WI	1	1	3		
6	10.0	8	11.5	20	24.6	ADAMS WI	15	113	145	88	91
15	14.0	21	20.3	24	22.2	ASHLAND WI	45	9	58	67	82
23	11.7	39	18.3	42	17.4	BARRON WI	57	-5	49	63	54
14	17.0	14	14.7	23	24.8	BAYFIELD WI	-14	68	45	61	91
82	20.3	88	16.7	118	19.3	BROWN WI	-18	15	-5	21	67
16	20.9	20	23.9	17	20.3	BUFFALO WI	15	-15	-3	22	73
15	24.2	9	11.9	12	12.7	BURNETT WI	-51	7	-47	04	24
15	16.9	22	20.5	18	14.1	CALUMET WI	21	-31	-17	14	32
39	18.0	46	20.7	44	17.0	CHIPPEWA WI	15	-18	-6	20	52
36	19.5	31	15.9	36	17.3	CLARK WI	-18	9	-11	17	54
40	19.1	33	13.9	49	20.1	COLUMBIA WI	-27	44	5	29	71
22	24.4	7	7.5	18	17.1	CRAWFORD WI	-69	129	-30	09	53
103	14.3	160	19.0	163	16.3	DANE WI	32	-14	13	36	47
47	15.3	80*	23.2	84	21.8	DODGE WI	52	-6	42	58	80

* SIGNIFICANTLY MORE DEATHS THAN EXPECTED

WHITE MALE

CANCER OF THE LARGE INTESTINE

ICD 153,159.0

1950-1959 DEATHS	1950-1959 RATE	1960-1969 DEATHS	1960-1969 RATE	1970-1979 DEATHS	1970-1979 RATE	COUNTY AND STATE	PERCENT CHANGE 60S/50S	PERCENT CHANGE 70S/60S	PERCENT CHANGE 70S/50S	PERCENTILE %CHANGE 70S/50S	PERCENTILE RATES 1970S
19	16.4	19	14.5	24	15.5	DOOR WI	-12	7	-6	21	41
39	15.7	45	18.0	41	16.8	DOUGLAS WI	15	-7	7	30	50
27	20.0	23	14.7	27	15.5	DUNN WI	-27	5	-23	11	41
37	14.7	38	13.0	67	21.0	EAU CLAIRE WI	-11	61	43	59	77
3	21.0	3	12.2	5	18.7	FLORENCE WI	-42	53	-11	17	63
67	18.9	74	19.5	83	20.2	FOND DU LAC WI	3	3	7	30	72
9	20.0	2	3.5	7	12.4	FOREST WI	-82	250	-38	06	21
35	15.1	47	20.1	39	15.8	GRANT WI	33	-21	5	28	43
31	21.5	24	16.3	25	15.0	GREEN WI	-24	-8	-30	09	38
15	16.5	20	19.3	30	25.5	GREEN LAKE WI	17	32	54	66	92
21	18.6	14	13.8	18	17.3	IOWA WI	-26	26	-7	19	54
7	15.2	7	12.7	8	16.2	IRON WI	-16	27	7	30	46
10	10.5	13	12.8	16	14.3	JACKSON WI	21	12	35	54	33
55	21.0	69	23.8	58	18.1	JEFFERSON WI	13	-24	-14	15	59
13	12.4	20	18.0	24	19.8	JUNEAU WI	45	10	60	68	70
70	17.6	82	17.4	113	21.9	KENOSHA WI	-1	26	24	45	81
14	17.0	18	18.6	24	23.2	KEWAUNEE WI	9	25	37	55	87
45	14.4	81	22.4	64	16.3	LA CROSSE WI	56	-27	14	36	47
19	19.5	16	16.3	13	13.8	LAFAYETTE WI	-17	-15	-29	09	30
16	13.9	16	12.8	16	12.2	LANGLADE WI	-8	-4	-12	16	20
21	17.6	38*	28.2	34	22.1	LINCOLN WI	60	-22	25	46	82
88*	27.1	87	23.0	106	24.7	MANITOWOC WI	-15	7	-9	18	91
57	14.6	89	20.7	85	18.1	MARATHON WI	42	-13	24	45	59
31	17.5	43	20.4	41	16.9	MARINETTE WI	17	-17	-3	22	51
11	20.0	17	25.4	9	11.3	MARQUETTE WI	27	-56	-44	05	17
59	18.2	80	22.3	75	18.8	MENOMINEE#WI	23	-16	3	27	63
846*	22.9	974*	22.7	1018*	23.3	MILWAUKEE WI	-1	3	2	26	87
28	15.5	35	17.8	30	14.2	MONROE WI	15	-21	-9	18	33
12	10.9	26	20.0	37	22.1	ONEIDA WI	83	11	103	82	82
79*	23.6	63	15.7	79	16.7	OUTAGAMIE WI	-33	7	-29	09	50
26	21.7	30	18.2	29	12.3	OZAUKEE WI	-16	-32	-43	05	21
12	27.1	4	8.4	6	12.8	PEPIN WI	-69	52	-53	03	24
18	13.8	20	15.8	22	16.3	PIERCE WI	14	3	18	40	47
26	17.0	22	12.1	38	19.3	POLK WI	-29	60	14	36	67
26	15.7	42	22.2	49	22.4	PORTAGE WI	42	1	43	59	83
20	20.0	14	14.3	18	15.5	PRICE WI	-28	8	-22	12	42
115*	22.4	107	17.1	120	17.6	RACINE WI	-24	2	-22	12	56
25	24.0	9	8.9	19	18.3	RICHLAND WI	-63	107	-23	11	61
80	17.4	114*	22.9	112	20.2	ROCK WI	31	-12	16	38	72
11	12.4	18	20.2	18	18.7	RUSK WI	63	-8	51	64	63
28	19.4	25	15.6	29	15.2	ST CROIX WI	-20	-3	-22	12	39
34	16.9	51	23.5	35	14.6	SAUK WI	39	-38	-14	15	35
9	13.3	6	7.3	16	17.5	SAWYER WI	-45	138	32	51	55
111*	26.9	108*	22.3	104	20.1	SHEBOYGAN WI	-17	-10	-25	10	72
11	10.7	15	14.4	18	16.1	TAYLOR WI	34	11	50	63	45
20	14.3	35	21.7	39	22.6	TREMPEALEAU WI	52	4	59	68	84
20	13.2	20	12.2	30	16.8	VERNON WI	-7	37	27	48	50
5	11.5	15	19.9	23	21.0	VILAS WI	73	5	82	77	76
38	15.2	49	16.6	71	20.9	WALWORTH WI	9	26	38	55	76
10	14.0	14	20.8	17	20.6	WASHBURN WI	48	-1	47	62	74
28	15.4	46	20.3	58	20.4	WASHINGTON WI	32	1	33	52	73
83	18.4	107	17.3	171	21.0	WAUKESHA WI	-6	21	14	37	76
46	20.5	45	16.5	46	16.2	WAUPACA WI	-20	-2	-21	12	47
13	13.3	19	19.7	17	14.2	WAUSHARA WI	49	-28	7	30	33
96	21.2	79	15.3	120	20.9	WINNEBAGO WI	-28	37	-2	23	76
42	17.9	45	16.2	55	16.9	WOOD WI	-10	4	-6	20	51
130	11.3	173	12.4	216	14.2	WYOMING WY	10	15	26		
7	12.5	17	21.3	15	20.1	ALBANY WY	71	-6	61	69	72
6	12.5	4	7.0	11	18.2	BIG HORN WY	-44	159	45	61	60
1	5.6	5	15.3	4	13.3	CAMPBELL WY	175	-13	139	88	27
8	12.5	4	5.8	7	9.5	CARBON WY	-54	65	-23	11	11
3	9.7	4	11.4	1	2.8	CONVERSE WY	17	-75	-71	00	02
				4	16.4	CROOK WY					47
9	12.7	12	13.5	12	11.9	FREMONT WY	6	-12	-6	20	19
4	6.7	7	11.0	10	14.4	GOSHEN WY	62	32	114	84	34
3	9.2	4	16.3	10	26.3	HOT SPRINGS WY	78	62	187	92	94
6	21.3	4	10.7	8	21.3	JOHNSON WY	-50	99	0	24	78
21	13.1	28	14.5	51	22.7	LARAMIE WY	11	57	74	74	84
4	13.5	4	9.6	3	5.6	LINCOLN WY	-29	-41	-58	02	04
13	11.8	28	15.9	23	10.9	NATRONA WY	35	-31	-7	19	16
1	8.0	1	7.5	1	3.7	NIOBRARA WY	-7	-50	-54	03	02
8	12.8	11	13.7	10	10.7	PARK-TETON#WY	7	-22	-17	14	15
7	18.5	8	19.5	8	18.2	PLATTE WY	6	-7	-2	23	60
13	9.9	13	10.5	15	12.0	SHERIDAN WY	6	14	21	42	20
1	10.2	2	10.0	2	8.6	SUBLETTE WY	-3	-14	-16	14	09
8	11.0	11	13.8	13	13.7	SWEETWATER WY	25	-1	24	45	29
4	9.7	4	10.1	3	7.7	UINTA WY	5	-24	-21	12	07
1	3.0	2	6.3	3	9.3	WASHAKIE WY	110	47	208	93	11

* SIGNIFICANTLY MORE DEATHS THAN EXPECTED

COUNTIES COMBINED; SEE TECHNICAL NOTES

WHITE FEMALE — CANCER OF THE LARGE INTESTINE — ICD 153,159.0

1950-1959 DEATHS	1950-1959 RATE	1960-1969 DEATHS	1960-1969 RATE	1970-1979 DEATHS	1970-1979 RATE	COUNTY AND STATE	PERCENT CHANGE 60S/50S	PERCENT CHANGE 70S/60S	PERCENT CHANGE 70S/50S	PERCENTILE %CHANGE RATES 70S/50S	PERCENTILE 1970S
128567	18.0	159080	17.1	194585	16.4	UNITED STATES	-5	-4	-9		
1132	12.9	1544	12.7	2139	12.9	ALABAMA AL	-2	2	0		
5	11.0	7	11.6	10	11.2	AUTAUGA AL	5	-3	2	52	22
15	9.5	31	13.1	48	12.8	BALDWIN AL	38	-3	35	76	32
7	11.9	11	13.9	11	12.8	BARBOUR AL	17	-8	7	58	32
8	17.2	4	6.8	5	6.0	BIBB AL	-60	-12	-65	02	05
13	12.3	11	9.0	25	12.9	BLOUNT AL	-27	44	5	56	34
5	19.0	4	11.2	3	11.6	BULLOCK AL	-41	3	-39	11	24
12	15.8	8	8.4	10	7.1	BUTLER AL	-47	-15	-55	05	07
28	10.5	44	12.9	71	14.5	CALHOUN AL	23	12	38	78	47
15	12.6	21	13.2	23	11.9	CHAMBERS AL	5	-10	-6	44	26
4	6.7	9	11.3	13	13.8	CHEROKEE AL	70	22	108	91	42
12	10.8	15	11.7	18	10.2	CHILTON AL	8	-13	-6	44	16
6	16.5	4	9.8	4	7.5	CHOCTAW AL	-41	-23	-55	05	07
3	4.6	8	10.3	14	13.3	CLARKE AL	125	30	191	96	37
6	10.5	12	14.7	11	10.5	CLAY AL	39	-28	0	51	18
5	11.1	5	9.5	7	10.1	CLEBURNE AL	-14	6	-8	41	16
8	9.2	9	6.5	18	9.8	COFFEE AL	-30	51	6	57	15
18	13.4	18	10.2	25	10.4	COLBERT AL	-24	1	-23	26	17
6	11.8	8	13.0	8	10.9	CONECUH AL	9	-16	-8	42	20
14*	37.6	5	11.6	2	3.3	COOSA AL	-69	-72	-91	00	02
19	12.3	27	14.0	33	14.1	COVINGTON AL	14	0	14	64	44
8	12.9	7	10.1	8	9.3	CRENSHAW AL	-21	-9	-28	20	12
33	18.3	26	10.4	39	10.9	CULLMAN AL	-43	4	-41	11	19
7	9.0	15	14.4	13	7.8	DALE AL	61	-46	-13	36	08
14	12.7	13	9.5	14	9.3	DALLAS AL	-25	-3	-27	21	12
24	14.0	28	12.2	35	11.7	DE KALB AL	-13	-5	-17	32	24
11	12.6	18	14.1	18	9.5	ELMORE AL	12	-32	-25	23	13
16	19.3	19	16.6	13	8.2	ESCAMBIA AL	-14	-51	-57	04	09
43	14.7	59	14.2	94	16.4	ETOWAH AL	-4	16	11	62	62
12	18.9	8	9.5	13	12.4	FAYETTE AL	-49	30	-34	15	30
19	23.3	21	17.9	18	10.3	FRANKLIN AL	-23	-42	-56	05	17
7	8.1	9	7.6	14	9.6	GENEVA AL	-6	27	19	68	14
3	19.4	3	14.9	5	17.9	GREENE AL	-23	20	-8	42	73
2	6.5	2	4.6	12	20.9	HALE AL	-29	354	222	96	89
3	6.5	6	10.4	11	16.2	HENRY AL	60	56	150	94	61
23	15.5	25	12.9	29	10.0	HOUSTON AL	-17	-22	-35	14	16
19	13.5	20	11.3	36	15.2	JACKSON AL	-17	35	12	62	53
241	15.3	333	14.5	442	14.7	JEFFERSON AL	-5	1	-4	47	49
7	10.5	8	10.5	13	12.9	LAMAR AL	0	23	23	70	34
13	7.3	22	8.1	52	13.5	LAUDERDALE AL	10	67	84	88	38
8	11.2	8	8.4	15	11.6	LAWRENCE AL	-25	38	4	54	24
11	11.9	18	13.6	21	11.2	LEE AL	15	-18	-6	44	21
20	18.6	15	10.2	24	11.0	LIMESTONE AL	-45	8	-41	11	20
2	11.6	2	10.3	5	17.4	LOWNDES AL	-12	70	50	82	69
2	6.7	7	24.7	11*	32.6	MACON AL	271	32	390	98	98
32	14.3	61	17.0	73	11.8	MADISON AL	19	-31	-17	31	25
5	10.7	7	12.5	7	10.2	MARENGO AL	16	-18	-5	45	16
12	11.3	17	12.9	18	10.5	MARION AL	14	-18	-7	43	18
23	14.1	26	10.2	54	15.7	MARSHALL AL	-27	54	12	62	57
83	13.4	151	16.6	172	13.7	MOBILE AL	24	-18	2	52	40
6	12.3	7	10.0	9	9.9	MONROE AL	-19	-1	-20	29	15
39	10.1	83	15.7	116	16.1	MONTGOMERY AL	55	3	59	84	59
20	10.2	39	14.3	46	11.8	MORGAN AL	40	-17	16	65	25
1	2.9	1	2.9	2	5.2	PERRY AL	0	79	79	87	04
7	11.1	8	10.9	14	16.7	PICKENS AL	-1	53	51	82	64
8	10.0	13	13.3	9	7.7	PIKE AL	34	-42	-23	25	08
4	4.8	6	5.8	14	11.6	RANDOLPH AL	22	99	143	93	24
16	20.0	13	11.2	23	14.4	RUSSELL AL	-44	29	-28	20	47
6	6.5	5	4.4	18	10.7	ST CLAIR AL	-32	141	65	85	19
6	6.6	14	10.8	26	13.3	SHELBY AL	63	23	101	90	37
2	7.5	7	20.3	5	10.2	SUMTER AL	171	-50	36	77	16
19	12.3	22	10.1	39	13.0	TALLADEGA AL	-18	28	5	56	34
17	16.3	29	18.5	35	16.8	TALLAPOOSA AL	14	-10	3	53	65
30	9.9	34	8.1	75	14.5	TUSCALOOSA AL	-18	79	46	81	48
26	12.9	27	10.1	44	11.9	WALKER AL	-22	18	-7	42	26
4	11.0	6	12.7	6	10.4	WASHINGTON AL	16	-18	-5	45	18
6	23.2	5	15.4	7	18.5	WILCOX AL	-33	20	-20	28	77
3	5.2	10	12.8	13	11.5	WINSTON AL	146	-11	120	92	23
		43	18.3	54	11.3	ALASKA AK		-39			
426	13.1	826	13.4	1491	13.1	ARIZONA AZ	2	-3	0		
4	22.5	3	13.7	2	7.8	APACHE AZ	-39	-43	-65	02	08
33	20.6	27	12.8	45	13.2	COCHISE AZ	-38	3	-36	14	36
7	17.7	8	10.6	15	12.1	COCONINO AZ	-41	14	-32	17	27
9	11.4	11	10.5	25	14.9	GILA AZ	-8	42	30	74	50
1	2.0	7	11.2	11	13.0	GRAHAM AZ	452	16	539	99	35
		1	2.9	3	10.2	GREENLEE AZ		253			17
237	13.6	489	14.0	889	13.4	MARICOPA AZ	3	-4	-2	49	37
1	3.9	4	4.2	30	12.5	MOHAVE AZ	7	200	221	96	31
7	16.9	8	10.0	13	12.3	NAVAJO AZ	-41	23	-27	22	29
92	13.2	186	13.6	327	13.4	PIMA AZ	3	-1	1	52	38
3	2.1	27	14.6	46	14.5	PINAL AZ	587	-1	581	99	47
		1	1.6	12	14.8	SANTA CRUZ AZ		830			50
22	17.0	37	17.3	49	11.9	YAVAPAI AZ	2	-31	-30	19	26
10	11.8	17	10.7	24	6.7	YUMA AZ	-9	-38	-43	09	06
817	12.1	1156	12.8	1737	14.1	ARKANSAS AR	5	11	17		
6	7.9	19	18.7	22	15.3	ARKANSAS AR	137	-18	94	89	54
8	12.3	7	8.9	16	13.5	ASHLEY AR	-28	52	10	60	38
7	13.1	13	9.8	32	14.1	BAXTER AR	-25	43	7	58	44
31	12.7	45	13.5	45	8.7	BENTON AR	6	-36	-32	17	10
8	10.1	11	8.6	28	13.5	BOONE AR	-15	56	33	75	38

* SIGNIFICANTLY MORE DEATHS THAN EXPECTED

WHITE FEMALE — CANCER OF THE LARGE INTESTINE — ICD 153,159.0

1950-1959 DEATHS	1950-1959 RATE	1960-1969 DEATHS	1960-1969 RATE	1970-1979 DEATHS	1970-1979 RATE	COUNTY AND STATE	PERCENT CHANGE 60S/50S	PERCENT CHANGE 70S/60S	PERCENT CHANGE 70S/50S	PERCENTILE %CHANGE 70S/50S	PERCENTILE RATES 1970S
9	17.3	19*	28.0	23*	28.0	BRADLEY AR	62	0	62	85	98
		6	21.1	7	21.8	CALHOUN AR		3			91
6	8.0	13	12.5	14	10.6	CARROLL AR	56	-16	32	75	18
3	9.4	4	8.7	5	9.6	CHICOT AR	-7	10	2	53	14
15	17.7	16	16.8	15	12.6	CLARK AR	-5	-25	-29	20	32
9	7.6	20	14.6	28	15.5	CLAY AR	92	7	105	90	56
8	14.6	8	12.0	15	15.5	CLEBURNE AR	-17	29	6	57	56
4	13.9	3	8.1	8	18.3	CLEVELAND AR	-42	126	32	75	75
10	12.3	20	17.8	17	11.9	COLUMBIA AR	45	-33	-3	48	26
7	11.0	13	15.9	18	15.9	CONWAY AR	45	0	44	80	58
22	11.9	35	14.5	50	14.4	CRAIGHEAD AR	21	0	21	69	47
13	12.2	12	8.6	22	11.9	CRAWFORD AR	-30	39	-3	48	26
4	9.8	9	11.5	12	10.7	CRITTENDEN AR	18	-7	10	60	19
10	16.9	11	13.7	7	7.1	CROSS AR	-19	-48	-58	04	07
6	14.0	7	13.3	12	18.8	DALLAS AR	-5	41	34	76	78
6	15.4	10	19.4	5	7.6	DESHA AR	26	-61	-51	06	07
7	11.2	7	10.3	11	12.8	DREW AR	-8	25	15	64	33
15	14.1	21	14.4	30	14.6	FAULKNER AR	2	2	4	54	48
10	15.3	9	10.7	9	9.0	FRANKLIN AR	-30	-16	-41	10	11
5	11.9	12	17.5	8	9.3	FULTON AR	47	-47	-22	27	13
39	17.0	39	10.8	82	14.3	GARLAND AR	-36	33	-16	33	46
5	13.2	6	11.7	7	8.6	GRANT AR	-12	-26	-35	15	10
16	12.8	16	9.5	32	15.2	GREENE AR	-26	61	19	67	53
16	18.4	23	21.4	26	21.0	HEMPSTEAD AR	16	-2	14	64	89
6	7.7	12	10.6	22	12.7	HOT SPRING AR	37	20	66	86	32
5	9.0	12	16.0	18	19.3	HOWARD AR	78	20	113	91	81
4	3.7	19	13.7	29	14.9	INDEPENDENCE AR	268	8	300	98	50
8	17.4	5	9.8	14	19.5	IZARD AR	-44	100	13	62	83
17	20.3	17	17.6	19	14.7	JACKSON AR	-13	-16	-27	21	49
21	12.1	26	11.3	45	14.2	JEFFERSON AR	-6	25	18	67	45
8	10.4	14	15.1	20	14.2	JOHNSON AR	45	-6	36	77	45
6	15.8	4	9.6	4	7.2	LAFAYETTE AR	-39	-25	-54	05	07
7	10.0	17	15.4	28	21.0	LAWRENCE AR	55	36	111	91	89
2	6.3	5	11.7	7	13.7	LEE AR	87	17	118	92	40
3	9.7	4	11.4	9	20.0	LINCOLN AR	18	75	107	91	85
2	8.8	7	16.0	6	9.9	LITTLE RIVER AR	82	-38	13	63	15
16	15.0	18	13.4	12	8.8	LOGAN AR	-10	-35	-41	10	11
8	8.3	14	11.0	19	11.0	LONOKE AR	31	1	32	75	21
7	11.6	7	10.0	9	11.4	MADISON AR	-14	13	-2	49	23
5	19.3	7	15.7	13	14.5	MARION AR	-18	-8	-25	24	47
11	9.4	20	13.1	23	12.2	MILLER AR	39	-7	29	74	28
16	10.7	27	12.7	34	13.8	MISSISSIPPI AR	19	8	29	74	41
4	11.7	7	14.0	10	17.4	MONROE AR	19	24	48	81	69
2	5.8	3	8.4	9	15.1	MONTGOMERY AR	45	79	160	94	52
12	23.1	6	9.3	10	12.8	NEVADA AR	-60	38	-45	09	33
2	6.7	2	7.1			NEWTON AR	6				
14	16.0	15	12.3	27	17.8	OUACHITA AR	-23	44	11	61	72
3	11.3	3	8.2	6	13.2	PERRY AR	-28	61	17	66	36
2	3.2	16	17.3	28*	25.3	PHILLIPS AR	445	46	696	99	96
9	19.7	11	18.2	23*	27.1	PIKE AR	-7	48	37	77	97
6	5.9	8	6.7	23	15.2	POINSETT AR	14	127	159	94	53
14	18.0	18	18.2	18	13.5	POLK AR	1	-26	-25	23	39
18	16.8	15	9.1	33	16.1	POPE AR	-46	76	-4	46	60
5	11.3	5	9.9	7	11.4	PRAIRIE AR	-13	15	0	51	23
106	13.7	147	13.7	196	13.7	PULASKI AR	0	0	0	51	40
4	5.6	13	18.0	23	21.1	RANDOLPH AR	220	17	275	97	90
11	22.7	7	10.2	11	12.7	ST FRANCIS AR	-55	24	-44	09	32
10	7.8	10	6.2	40	18.1	SALINE AR	-21	194	133	93	74
3	6.7	6	9.9	11	16.6	SCOTT AR	47	69	148	94	63
3	7.0	3	6.6	5	6.7	SEARCY AR	-6	1	-5	46	06
36	11.2	50	12.0	80	14.1	SEBASTIAN AR	7	17	26	72	44
10	17.5	12	16.6	14	15.0	SEVIER AR	-5	-10	-14	35	51
1	2.0	8	11.4	8	9.2	SHARP AR	476	-20	363	98	12
3	9.1			8	14.9	STONE AR			64	85	51
21	13.8	28	12.8	45	15.0	UNION AR	-7	17	9	59	51
3	6.3	5	8.1	7	7.1	VAN BUREN AR	29	-13	13	63	07
26	10.0	44	12.5	69	14.0	WASHINGTON AR	25	13	41	79	43
14	8.2	24	11.0	48	15.6	WHITE AR	34	42	90	89	56
5	13.4	3	6.8	14	23.4	WOODRUFF AR	-50	247	74	87	94
13	18.8	18	19.0	27	20.1	YELL AR	1	6	7	57	85
9255	15.4	12923	15.1	16862	15.0	CALIFORNIA CA	-2	-1	-3		
688	17.1	822	16.8	896	16.4	ALAMEDA CA	-2	-2	-4	47	62
9	17.7	11	15.4	11	9.3	AMADOR CA	-13	-40	-48	07	12
51	13.0	79	12.7	132	14.4	BUTTE CA	-2	13	11	61	47
6	10.8	11	12.5	18	14.7	CALAVERAS CA	17	17	37	77	49
8	15.3	10	15.8	14	17.7	COLUSA CA	3	12	16	65	72
168	16.7	255	14.4	429	15.4	CONTRA COSTA CA	-14	7	-8	42	55
7	17.3	8	13.1	9	10.7	DEL NORTE CA	-24	-18	-38	12	19
15	16.8	26	16.3	34	9.7	EL DORADO CA	-3	-41	-43	10	14
169	14.5	250	14.9	278	12.0	FRESNO CA	3	-19	-17	32	27
7	10.0	12	12.1	20	14.4	GLENN CA	21	20	45	80	47
46	14.6	55	13.2	82	15.7	HUMBOLDT CA	-10	19	8	58	57
18	12.3	29	14.2	28	10.3	IMPERIAL CA	16	-27	-16	33	17
5	14.4	9	12.9	16	14.7	INYO CA	-11	14	2	53	49
97	12.4	131	10.9	235	14.6	KERN CA	-12	34	18	67	48
17	11.0	24	11.3	38	14.0	KINGS CA	2	24	27	72	43
5	4.3	17	10.3	37	12.6	LAKE CA	140	22	192	96	31
4	8.1	8	11.9	18	20.9	LASSEN CA	48	76	160	94	89
4155	16.2	5238	15.5	5783	15.3	LOS ANGELES CA	-4	-1	-5	45	54
13	9.3	25	13.4	33	12.5	MADERA CA	43	-6	34	76	31
83	17.9	122	16.1	210	18.5	MARIN CA	-10	15	3	53	76
1	3.3	3	7.9	11	18.2	MARIPOSA CA	139	130	449	99	75
33	16.3	37	14.0	39	11.5	MENDOCINO CA	-14	-18	-29	20	23
29	12.3	56	16.2	79	16.8	MERCED CA	32	3	36	77	65
2	6.8	3	6.8	7	12.3	MODOC CA	0	82	81	88	29
1	15.6	1	37.2	2	12.6	MONO CA	138	-66	-19	29	31
68	12.5	104	12.6	180	15.1	MONTEREY CA	1	19	21	69	52
54	15.6	61	12.5	88	13.1	NAPA CA	-20	5	-16	33	36
9	6.7	32	17.6	30	10.0	NEVADA CA	164	-43	50	82	16

* SIGNIFICANTLY MORE DEATHS THAN EXPECTED

WHITE FEMALE — CANCER OF THE LARGE INTESTINE — ICD 153,159.0

1950-1959 DEATHS	1950-1959 RATE	1960-1969 DEATHS	1960-1969 RATE	1970-1979 DEATHS	1970-1979 RATE	COUNTY AND STATE	PERCENT CHANGE 60S/50S	PERCENT CHANGE 70S/60S	PERCENT CHANGE 70S/50S	PERCENTILE %CHANGE 70S/50S	PERCENTILE RATES 1970S
235	13.1	570	13.5	1192	15.1	ORANGE CA	3	12	16	65	52
28	11.5	48	12.9	76	13.6	PLACER CA	12	5	18	67	39
3	7.7	7	12.8	20*	24.0	PLUMAS CA	65	88	210	96	95
139	12.1	281	13.1	452	11.7	RIVERSIDE CA	8	-10	-3	48	24
232	16.4	350	16.4	509	15.7	SACRAMENTO CA	0	-4	-4	47	57
5	7.6	16	19.6	16	14.9	SAN BENITO CA	156	-24	96	89	51
236	13.7	395	13.8	564	13.8	SAN BERNARDINO CA	1	0	1	52	41
508	15.6	759	14.3	1167	13.9	SAN DIEGO CA	-8	-3	-10	39	42
765	17.8	880*	18.4	845*	17.8	SAN FRANCISCO CA	3	-3	0	51	73
116	13.0	150	12.0	231	13.9	SAN JOAQUIN CA	-8	17	8	58	42
39	11.6	64	12.3	111	13.6	SAN LUIS OBISPO CA	6	11	17	66	40
222	16.8	357	16.4	494	16.5	SAN MATEO CA	-3	1	-2	49	63
85	12.9	173	15.4	235	14.2	SANTA BARBARA CA	19	-8	10	61	45
280	14.5	473	14.6	731	15.1	SANTA CLAPA CA	0	4	4	55	53
75	12.8	146	16.5	172	14.5	SANTA CRUZ CA	28	-12	13	63	47
29	16.2	47	16.2	65	12.6	SHASTA CA	0	-22	-22	26	32
		1	5.8	1	5.5	SIERPA CA		-5			04
20	16.3	23	14.5	35	16.2	SISKIYOU CA	-11	11	-1	50	60
59	16.3	63	12.6	135	18.0	SOLANO CA	-23	43	11	61	74
97	13.7	154	13.5	273	15.6	SONOMA CA	-2	16	13	63	56
100	14.8	121	12.3	198	14.6	STANISLAUS CA	-17	19	-1	50	48
18	19.3	36	22.3	30	12.6	SUTTER CA	16	-44	-35	15	31
14	13.4	30	19.7	30	13.3	TEHAMA CA	47	-33	0	50	37
2	8.5	6	22.0	6	12.3	TRINITY CA	158	-44	44	80	29
68	11.3	117	14.1	146	13.0	TULARE CA	25	-8	15	65	35
8	11.1	16	13.5	26	13.7	TUOLUMNE CA	21	2	23	70	40
61	10.5	142	13.0	257	13.4	VENTURA CA	24	3	28	73	38
31	16.9	37	12.4	61	14.1	YOLO CA	-26	14	-16	32	45
12	13.3	22	16.9	27	14.4	YUBA CA	28	-15	8	59	47
1039	14.7	1367	14.5	1765	14.2	COLORADO CO	-2	-2	-4		
21	11.0	40	10.4	85	13.7	ADAMS CO	-5	31	25	71	40
6	15.2	5	10.9	3	5.4	ALAMOSA CO	-29	-50	-65	02	04
1	9.0					ARCHULETA CO					
3	11.3	2	7.1	8	20.4	BACA CO	-37	189	81	88	87
4	11.8	7	16.2	8	20.4	BENT CO	38	26	73	87	87
55	17.5	85	18.0	96	13.2	BOULDER CO	3	-27	-25	23	36
4	9.4	3	5.8	6	7.8	CHAFFEE CO	-38	33	-17	31	08
1	7.1	2	14.4	3	19.9	CHEYENNE CO	103	38	180	95	84
2	11.9	4	21.0	4	19.5	CLEAR CREEK CO	76	-7	63	85	82
3	9.1	7	19.5	7	22.4	CONEJOS CO	116	15	147	94	92
		3	20.1	2	12.5	COSTILLA CO		-38			31
1	3.9	5	19.9	7	23.9	CROWLEY CO	408	20	511	99	95
1	12.2	1	10.5	2	24.0	CUSTER CO	-14	129	97	90	95
13	14.2	14	12.5	24	14.4	DELTA CO	-12	15	1	52	46
481	17.5	532	15.2	574	13.8	DENVER-ARAPAHOE#CO	-13	-9	-21	27	41
2	55.3					DOLORES CO					
3	17.2	8	27.0	3	6.1	DOUGLAS CO	57	-78	-65	02	05
1	5.8	5	31.7	4	22.0	EAGLE CO	443	-30	278	98	92
2	10.2	1	4.8	4	14.7	ELBERT CO	-53	208	44	80	49
66	12.6	88	12.0	171	16.3	EL PASO CO	-4	35	29	74	61
16	11.3	16	8.7	38	16.2	FREMONT CO	-23	87	44	79	60
5	9.1	19	24.0	13	13.2	GARFIELD CO	165	-45	45	80	36
2	49.1	1	12.8	1	12.4	GILPIN CO	-74	-3	-75	01	30
3	32.3			1	11.2	GRAND CO			-65	02	22
		2	8.2	3	12.5	GUNNISON CO		53			31
9	19.2	7	13.3	5	9.2	HUERFANO CO	-31	-31	-52	06	12
		1	11.2	6*	85.8	JACKSON CO		669			99
46	14.7	105	16.6	175	15.6	JEFFERSON CO	13	-6	6	57	56
2	17.0	3	20.9	2	16.0	KIOWA CO	23	-23	-6	44	59
2	6.3	7	17.5	9	19.4	KIT CARSON CO	177	11	208	96	82
6	32.4	6	37.3	1	6.2	LAKE CO	15	-83	-81	00	05
7	11.0	10	10.9	13	10.7	LA PLATA CO	-1	-2	-3	48	19
37	13.9	52	13.4	95	16.1	LARIMER CO	-4	20	16	65	60
12	11.0	17	14.9	12	9.4	LAS ANIMAS CO	35	-37	-14	35	13
2	9.9	2	5.3	5	10.6	LINCOLN CO	-46	100	8	58	18
9	10.9	20	19.6	10	8.0	LOGAN CO	79	-59	-27	21	08
29	13.3	40	13.5	57	13.1	MESA CO	1	-4	-2	49	35
		1	35.3			MINERAL CO					
3	13.0	5	15.9	8	20.2	MOFFAT CO	23	26	55	83	86
3	6.6	4	7.5	12	15.9	MONTEZUMA CO	13	114	141	93	59
10	14.5	14	15.1	15	11.7	MONTROSE CO	4	-22	-19	30	25
5	5.5	19	16.7	20	13.5	MORGAN CO	201	-19	144	93	39
21	16.4	23	13.8	22	13.3	OTERO CO	-16	-4	-19	30	37
1	10.2	3	34.4	4	43.7	OURAY CO	236	27	328	98	99
		1	18.9	1	10.6	PARK CO		-44			18
5	18.7	5	13.8	5	11.4	PHILLIPS CO	-26	-17	-39	12	23
1	18.6			2	17.7	PITKIN CO			-5	45	71
8	12.5	9	12.7	15	19.6	PROWERS CO	1	55	57	83	83
55	11.0	79	13.2	90	13.3	PUEBLO CO	20	1	21	69	37
2	14.5	2	10.6	1	3.1	RIO BLANCO CO	-27	-71	-78	00	02
5	10.5	7	13.5	13	21.1	RIO GRANDE CO	29	56	100	90	89
7	22.7	8	28.4	5	15.3	ROUTT CO	25	-46	-33	16	53
2	11.0	1	5.2	4	18.2	SAGUACHE CO	-53	254	66	86	75
2	58.5	2*	45.9	1	37.8	SAN JUAN CO	-22	-18	-35	14	99
2	21.5	1	14.1	2	26.0	SAN MIGUEL CO	-34	85	21	69	97
2	9.7	4	17.2	7	28.6	SEDGWICK CO	77	66	194	96	98
1	18.8	1	29.8	1	12.2	SUMMIT CO	58	-59	-35	14	28
4	27.1	2	10.1	5	19.8	TELLER CO	-63	96	-27	21	84
4	14.0	1	3.2	5	12.2	WASHINGTON CO	-77	280	-13	37	28
31	10.0	50	13.4	55	10.4	WELD CO	34	-22	4	55	17
8	15.4	5	6.9	15	20.7	YUMA CO	-55	201	34	76	88
2492*	22.0	2917*	19.7	3488*	18.4	CONNECTICUT CT	-11	-6	-16		
618*	21.7	804*	21.0	940*	19.3	FAIRFIELD CT	-3	-8	-11	38	81
633*	21.8	752*	19.3	889*	17.6	HARTFORD CT	-11	-9	-19	29	71
141*	23.2	147	17.8	157	15.0	LITCHFIELD CT	-24	-16	-35	14	51
73	16.0	110	18.6	116	15.0	MIDDLESEX CT	16	-20	-7	43	51
745*	24.4	752*	19.5	925*	18.8	NEW HAVEN CT	-20	-3	-23	26	79

* SIGNIFICANTLY MORE DEATHS THAN EXPECTED

COUNTIES COMBINED; SEE TECHNICAL NOTES

WHITE FEMALE — CANCER OF THE LARGE INTESTINE — ICD 153,159.0

1950-1959 DEATHS	1950-1959 RATE	1960-1969 DEATHS	1960-1969 RATE	1970-1979 DEATHS	1970-1979 RATE	COUNTY AND STATE	PERCENT CHANGE 60S/50S	PERCENT CHANGE 70S/60S	PERCENT CHANGE 70S/50S	PERCENTILE %CHANGE 70S/50S	PERCENTILE RATES 1970S
152	17.9	210*	20.2	272*	21.0	NEW LONDON CT	13	4	17	66	89
50	21.0	64	20.3	79	18.0	TOLLAND CT	-4	-11	-15	35	73
80	20.9	78	16.3	110	19.3	WINDHAM CT	-22	18	-8	42	81
289	18.5	421*	20.3	505*	18.2	DELAWARE DE	9	-10	-2		
25	12.0	47	18.0	63	17.9	KENT DE	50	0	50	81	73
226*	21.3	312*	22.1	360*	19.4	NEW CASTLE DE	4	-12	-9	41	82
38	13.9	62	15.5	82	14.6	SUSSEX DE	11	-6	5	55	48
595*	20.6	491	18.3	359	16.5	DIST. OF COLUMBIA DC	-11	-9	-20		
595*	20.6	491	18.3	359	16.5	WASHINGTON DC	-11	-9	-20	29	63
2213	13.4	4635	13.8	9010	14.7	FLORIDA FL	3	7	10		
21	14.5	34	14.2	62	15.8	ALACHUA FL	-2	11	9	59	57
1	4.4	1	3.5	6	14.2	BAKER FL	-21	307	223	96	45
16	14.2	27	11.8	58	15.7	BAY FL	-17	33	10	61	57
1	2.3	6	11.2	7	9.8	BRADFORD FL	382	-13	320	98	15
25	12.1	76	15.1	184	15.0	BREVARD FL	24	0	24	71	52
123	12.8	434	13.9	1098	14.8	BROWARD FL	9	6	16	65	49
		3	8.7	1	1.5	CALHOUN FL		-82			01
10	17.8	26	11.0	88	13.3	CHARLOTTE FL	-38	21	-25	23	37
4	6.8	21	18.6	46	10.3	CITRUS FL	175	-44	53	82	17
8	13.5	10	9.4	30	14.6	CLAY FL	-31	56	8	59	48
2	4.7	10	9.7	61	12.9	COLLIER FL	106	33	173	95	34
12	24.6	7	8.5	13	9.9	COLUMBIA FL	-65	17	-60	03	15
492	15.5	937	14.9	1666	16.2	DADE FL	-3	9	5	55	60
10	16.5	11	15.5	21	18.7	DE SOTO FL	-6	21	13	63	78
2	21.2	1	5.0	5	14.8	DIXIE FL	-77	198	-30	18	50
140	13.1	249	15.5	331	14.5	DUVAL FL	18	-6	11	61	48
37	11.0	78	14.4	86	10.3	ESCAMBIA FL	31	-28	-6	44	17
2	22.3	2	8.2	7	12.9	FLAGLER FL	-63	58	-42	10	34
4	22.7	2	7.9	5	11.7	FRANKLIN FL	-65	48	-49	07	24
14	13.2	12	8.4	11	6.7	GADSDEN FL	-37	-20	-49	06	06
		1	6.1	5	23.3	GILCHRIST FL		279			94
1	15.1			5	21.9	GLADES FL			45	80	91
2	10.8	3	9.7	8	17.7	GULF FL	-11	83	64	85	72
4	20.4	1	3.8	7	17.0	HAMILTON FL	-81	344	-17	32	66
4	10.2	8	10.7	15	15.8	HARDEE FL	5	47	54	83	57
3	20.4	4	16.6	6	12.9	HENDRY FL	-18	-22	-37	13	34
7	17.8	17	17.0	37	13.3	HERNANDO FL	-5	-21	-25	23	37
10	13.0	30	18.0	43	11.7	HIGHLANDS FL	38	-35	-10	39	25
173	13.3	281	13.4	431	13.9	HILLSBOROUGH FL	0	3	4	54	42
6	10.9	11	16.6	6	6.4	HOLMES FL	52	-61	-41	11	05
8	7.0	22	12.0	73	18.3	INDIAN RIVER FL	71	52	160	94	76
16	16.5	12	9.3	31	17.3	JACKSON FL	-44	86	5	55	68
1	4.3	2	7.9	5	15.2	JEFFERSON FL	83	92	252	97	53
2	22.5	1	7.3	3	16.5	LAFAYETTE FL	-68	128	-26	22	63
26	10.6	82	16.6	115	13.5	LAKE FL	57	-19	27	72	38
18	11.6	59	11.5	187	13.6	LEE FL	-1	19	17	66	40
21	15.0	32	15.4	49	13.7	LEON FL	3	-12	-9	41	40
8	25.2	7	12.2	18	17.2	LEVY FL	-52	41	-32	17	68
		1	6.1	1	5.7	LIBERTY FL		-7			04
7	18.5	4	7.6	3	4.4	MADISON FL	-59	-41	-76	01	03
46	13.4	90	10.9	224	14.9	MANATEE FL	-19	37	12	62	51
21	13.3	29	11.5	69	12.2	MARION FL	-14	6	-8	41	28
7	13.7	17	10.3	56	12.4	MARTIN FL	-25	20	-10	40	30
17	17.7	27	14.6	46	16.1	MONROE FL	-17	10	-9	40	60
3	9.0	3	5.9	17	19.0	NASSAU FL	-34	220	111	91	80
7	8.1	22	15.5	40	13.9	OKALOOSA FL	90	-10	70	86	42
1	5.3	4	14.2	16	20.9	OKEECHOBEE FL	170	47	298	98	89
111	13.8	211	14.7	341	15.4	ORANGE FL	7	4	12	62	55
8	5.7	32	14.9	49	13.7	OSCEOLA FL	162	-8	140	93	40
101	12.0	225	11.6	656	16.0	PALM BEACH FL	-3	38	33	75	59
21	9.5	68	11.7	235	14.2	PASCO FL	23	21	49	81	45
351	14.2	770	14.7	1272	15.2	PINELLAS FL	4	3	7	57	53
60	9.6	122	11.3	224	12.6	POLK FL	17	12	31	74	31
7	7.1	21	14.7	39	16.9	PUTNAM FL	108	15	139	93	66
14	11.2	34	19.8	46	16.9	ST JOHNS FL	77	-15	50	82	65
12	12.0	29	13.2	58	12.9	ST LUCIE FL	10	-2	7	58	34
2	2.7	11	10.3	17	9.2	SANTA ROSA FL	286	-11	245	97	12
47	12.4	132	13.6	293	13.9	SARASOTA FL	10	2	13	62	42
11	7.6	37	13.8	89	15.1	SEMINOLE FL	82	9	99	90	52
6	17.3	10	17.0	15	13.0	SUMTER FL	-2	-23	-25	24	35
7	15.3	10	14.1	13	11.6	SUWANNEE FL	-7	-18	-24	24	24
1	5.7	6	14.1	6	8.4	TAYLOR FL	146	-40	47	81	09
2	12.0	2	10.8	4	14.3	UNION FL	-10	32	19	67	46
96	14.6	187	14.4	330	15.3	VOLUSIA FL	-1	6	5	56	54
1	6.1	1	5.4	9	22.6	WAKULLA FL	-12	321	273	97	93
7	10.7	6	6.5	6	5.6	WALTON FL	-39	-14	-48	07	04
5	11.1	4	5.9	6	7.7	WASHINGTON FL	-47	31	-31	18	08
1272	12.2	1814	12.4	2535	12.4	GEORGIA GA	1	0	1		
1	2.1	3	5.8	3	4.6	APPLING GA	176	-21	118	92	03
6	31.5	1	4.8	1	4.4	ATKINSON GA	-85	-9	-86	00	03
4	16.7	1	3.1	5	11.1	BACON GA	-81	259	-33	16	21
3	35.6	2	19.3			BAKER GA	-46				
4	2.9	8	4.8	18	11.3	BALDWIN GA	67	136	294	98	22
4	13.3	3	8.4	3	7.1	BANKS GA	-37	-15	-47	08	07
4	6.2	8	9.4	13	11.3	BARROW GA	52	20	82	88	22
18	18.6	14	12.0	19	11.6	BARTOW GA	-35	-3	-37	13	24
6	11.4	7	10.1	4	4.5	BEN HILL GA	-11	-55	-60	03	03
2	4.6	3	5.7	4	6.1	BERRIEN GA	26	7	34	76	05
48	14.7	59	12.4	84	13.6	BIBB GA	-15	9	-8	42	39
6	22.1	4	10.0	4	6.7	BLECKLEY GA	-55	-33	-70	02	06
1	9.4	3	14.1	7	20.7	BRANTLEY GA	51	47	121	92	88
5	10.1	5	8.1	10	12.6	BROOKS GA	-20	56	25	71	31
2	20.6	1	4.1	8	32.9	BRYAN GA	-80	694	60	84	98

* SIGNIFICANTLY MORE DEATHS THAN EXPECTED

WHITE FEMALE

CANCER OF THE LARGE INTESTINE

ICD 153,159.0

1950-1959 DEATHS	1950-1959 RATE	1960-1969 DEATHS	1960-1969 RATE	1970-1979 DEATHS	1970-1979 RATE	COUNTY AND STATE	PERCENT CHANGE 60S/50S	PERCENT CHANGE 70S/60S	PERCENT CHANGE 70S/50S	PERCENTILE %CHANGE 70S/50S	PERCENTILE RATES 1970S
3	4.5	8	8.0	13	9.5	BULLOCH GA	77	18	109	91	13
2	6.2	3	7.7	8	12.8	BURKE GA	25	67	107	91	33
2	6.8			7	9.7	BUTTS GA			44	79	14
2	14.5	1	5.6	2	10.7	CALHOUN GA	-61	91	-26	22	19
1	12.5	2	8.0	1	2.7	CAMDEN GA	-36	-66	-78	01	02
3	18.9	5	20.1	1	3.1	CANDLER GA	7	-85	-84	00	02
22	17.1	24	12.9	25	9.4	CARROLL GA	-24	-27	-45	09	13
5	7.6	11	13.6	28	20.7	CATOOSA GA	78	53	172	95	88
2	13.8	1	8.6	2	8.7	CHARLTON GA	-37	1	-37	13	10
61	15.0	95	16.5	116	15.5	CHATHAM GA	10	-6	4	54	56
8	12.3	19	22.6	17	14.1	CHATTOOGA GA	84	-38	15	64	44
11	12.1	14	11.5	19	10.1	CHEROKEE GA	-5	-12	-17	32	16
18	14.6	22	13.3	34	13.4	CLARKE GA	-9	1	-8	42	38
				1	6.7	CLAY GA					06
7	8.4	19	11.5	50	14.8	CLAYTON GA	37	28	75	87	49
1	11.2	3	19.3			CLINCH GA	72				
27	12.3	47	10.8	119	14.4	COBB GA	-12	33	17	66	47
5	8.3	8	10.6	11	10.9	COFFEE GA	29	3	32	75	20
12	11.5	20	16.2	24	13.8	COLQUITT GA	42	-15	21	69	42
2	15.2	5	13.4	6	6.9	COLUMBIA GA	-12	-48	-55	05	06
8	26.4	2	4.6	6	10.2	COOK GA	-83	123	-61	03	16
9	10.3	18	15.8	20	11.6	COWETA GA	54	-27	13	63	24
2	15.2			3	13.3	CRAWFORD GA			-12	37	37
7	15.4	3	5.0	13	16.0	CRISP GA	-67	221	4	55	59
5	23.3	6	17.4	6	11.4	DADE GA	-25	-34	-51	06	23
		2	10.8	3	17.7	DAWSON GA		64			71
6	11.6	15	20.3	15	16.4	DECATUR GA	75	-19	41	79	62
76	11.6	145	13.4	208	12.5	DE KALB GA	15	-7	7	58	31
2	4.8	5	6.9	10	12.3	DODGE GA	45	78	158	94	29
6	19.9	2	4.7	3	7.6	DOOLY GA	-77	63	-62	03	07
6	6.4	18	10.9	29	12.1	DOUGHERTY GA	70	11	88	89	27
3	5.2	8	9.3	15	9.5	DOUGLAS GA	78	2	82	88	13
2	5.8	6	13.0	10	18.4	EARLY GA	125	41	217	96	76
2	42.7	2	24.1			ECHOLS GA	-43				
5	21.0	5	14.6	5	8.7	EFFINGHAM GA	-31	-40	-58	04	10
10	18.0	15	19.5	10	9.9	ELBERT GA	8	-49	-45	09	15
3	6.6	8	12.0	6	5.9	EMANUEL GA	83	-51	-11	39	04
1	5.2	1	4.3	2	6.0	EVANS GA	-17	41	17	66	05
7	12.5	8	9.9	11	9.6	FANNIN GA	-21	-3	-23	25	14
5	20.1	5	13.4	10	14.4	FAYETTE GA	-34	8	-28	20	47
27	12.3	42	13.4	52	12.3	FLOYD GA	9	-8	0	51	29
6	12.2	7	12.2	9	8.7	FORSYTH GA	0	-29	-29	20	10
12	20.9	6	9.0	12	12.0	FRANKLIN GA	-57	34	-42	10	27
235	14.5	331	15.7	339	14.0	FULTON GA	9	-11	-3	48	43
4	10.7	3	6.4	9	12.9	GILMER GA	-40	101	21	69	34
1	9.1	1	5.5	3	21.7	GLASCOCK GA	-40	296	138	93	91
7	7.3	22	16.0	19	9.1	GLYNN GA	119	-43	25	71	12
13	16.5	16	15.8	18	12.2	GORDON GA	-4	-23	-26	22	28
3	4.7	8	11.9	11	12.4	GRADY GA	152	4	163	94	30
4	12.9	5	12.6	7	13.4	GREENE GA	-3	7	4	55	38
12	8.5	18	9.0	52	13.3	GWINNETT GA	5	48	55	83	37
6	9.5	8	9.2	9	7.4	HABERSHAM GA	-3	-19	-22	27	07
10	6.7	21	9.5	41	12.3	HALL GA	42	30	84	88	29
3	12.4	1	4.2	3	10.3	HANCOCK GA	-66	145	-17	32	17
5	9.2	18	24.0	7	6.4	HARALSON GA	162	-74	-31	18	05
6	20.4	5	12.1	5	10.3	HARRIS GA	-41	-14	-49	06	17
6	12.6	9	13.4	15	16.3	HART GA	6	21	29	74	61
5	22.7	3	9.8	5	14.0	HEARD GA	-57	43	-38	12	43
6	14.6	7	9.5	13	10.0	HENRY GA	-35	6	-31	17	16
9	20.8	4	3.6	24	14.0	HOUSTON GA	-83	288	-33	16	43
1	3.9	5	13.7	10	19.5	IRWIN GA	256	42	406	98	82
10	13.6	12	11.6	13	9.9	JACKSON GA	-15	-15	-27	21	15
3	18.3	2	7.3	3	6.6	JASPER GA	-60	-9	-64	03	06
1	3.5	3	10.0	3	7.2	JEFF DAVIS GA	187	-28	106	90	07
7	17.0	5	10.2	10	15.4	JEFFERSON GA	-40	50	-10	40	55
		2	7.6	4	14.4	JENKINS GA		90			47
2	6.0	2	4.9	3	8.0	JOHNSON GA	-17	62	34	76	08
2	11.2	3	9.7	2	3.9	JONES GA	-13	-60	-65	02	02
5	14.7	1	2.9	9	17.0	LAMAR GA	-81	494	15	65	66
2	15.8	1	5.0	1	5.4	LANIER GA	-68	8	-66	02	04
13	17.2	15	13.6	21	13.9	LAURENS GA	-21	2	-20	29	42
2	26.1	1	8.6	2	8.3	LEE GA	-67	-4	-68	02	09
				8	23.2	LIBERTY GA					94
6	37.0	2	8.0	1	4.7	LINCOLN GA	-78	-41	-87	00	03
1	11.2					LONG GA					
14	16.9	21	14.2	21	10.2	LOWNDES GA	-16	-28	-40	11	16
3	11.7	5	13.2	3	4.7	LUMPKIN GA	13	-64	-60	03	03
1	3.0	4	9.4	9	14.1	MC DUFFIE GA	214	50	372	98	44
2	22.7	2	11.8	5	27.8	MC INTOSH GA	-48	136	23	70	98
2	9.5	9	23.0	3	8.1	MACON GA	143	-65	-14	35	09
4	8.3	9	16.4	4	5.3	MADISON GA	99	-67	-35	14	04
1	7.5	2	16.9	5	28.8	MARION GA	123	71	282	98	98
8	15.4	4	6.0	14	15.0	MERIWETHER GA	-61	148	-3	48	51
1	3.9	4	13.8	7	19.1	MILLER GA	258	39	397	98	80
3	5.2	3	4.4	4	5.2	MITCHELL GA	-16	18	-1	50	03
4	13.8	13*	34.9	6	8.1	MONROE GA	152	-77	-41	10	09
3	13.1	4	17.7	2	7.9	MONTGOMERY GA	35	-55	-39	11	08
5	17.0	4	11.3	6	11.7	MORGAN GA	-34	4	-31	17	24
1	2.3	7	14.0	14	20.9	MURRAY GA	508	49	804	99	89
25	8.5	51	11.7	81	14.1	MUSCOGEE# GA	38	20	67	86	44
8	12.7	19	22.1	16	11.9	NEWTON GA	74	-46	-6	44	26
3	13.1			4	9.7	OCONEE GA			-25	23	14
3	9.8	4	12.3	9	21.5	OGLETHORPE GA	26	74	120	92	91
8	19.9	7	10.0	10	10.1	PAULDING GA	-50	2	-49	07	16
3	11.2	2	5.9	8	17.2	PEACH GA	-48	193	53	83	67
5	13.9	10	21.1	9	13.1	PICKENS GA	51	-38	-6	44	36
2	7.9	3	8.0	4	9.4	PIERCE GA	1	16	18	67	13
2	9.1	4	13.1	5	12.0	PIKE GA	44	-9	31	74	26
5	4.1	14	10.8	26	13.5	POLK GA	161	25	226	97	39
10*	63.5	5	18.8	10	25.5	PULASKI GA	-70	36	-60	03	97
2	10.5	1	4.8	3	9.2	PUTNAM GA	-55	93	-13	37	12

* SIGNIFICANTLY MORE DEATHS THAN EXPECTED

COUNTIES COMBINED; SEE TECHNICAL NOTES

WHITE FEMALE — CANCER OF THE LARGE INTESTINE — ICD 153,159.0

1950-1959 DEATHS	1950-1959 RATE	1960-1969 DEATHS	1960-1969 RATE	1970-1979 DEATHS	1970-1979 RATE	COUNTY AND STATE	PERCENT CHANGE 60S/50S	PERCENT CHANGE 70S/60S	PERCENT CHANGE 70S/50S	PERCENTILE %CHANGE 70S/50S	PERCENTILE RATES 1970S
				3	42.1	QUITMAN GA					99
		3	6.4	4	7.1	RABUN GA		12			07
6	24.2	3	8.9	5	12.3	RANDOLPH GA	-63	38	-49	07	29
33	11.2	53	12.9	65	11.7	RICHMOND GA	16	-9	5	55	24
3	9.8	4	8.3	12	12.1	ROCKDALE GA	-15	45	24	71	27
4	50.0			2	13.8	SCHLEY GA			-72	01	41
1	2.6	4	9.2	5	9.0	SCREVEN GA	259	-3	249	97	11
		1	5.7	3	7.4	SEMINOLE GA		30			07
14	12.4	21	13.1	24	11.7	SPALDING GA	6	-11	-6	45	24
8	12.2	10	11.7	12	10.1	STEPHENS GA	-5	-13	-17	31	16
2	12.4	4	24.8	2	8.9	STEWART GA	100	-64	-28	20	11
11	19.1	12	12.0	14	12.1	SUMTER GA	-37	1	-37	13	28
2	11.8	3	14.0	2	14.1	TALBOT GA	19	1	20	68	44
1	9.6	1	5.6	1	12.1	TALIAFERRO GA	-42	116	26	72	27
1	2.6	4	6.6	8	9.7	TATTNALL GA	152	46	270	97	14
4	16.0	5	21.0	6	15.5	TAYLOR GA	32	-26	-3	48	55
7	17.1	8	15.5	6	8.5	TELFAIR GA	-9	-45	-50	06	10
4	12.8	4	13.6	4	9.0	TERRELL GA	6	-34	-30	19	11
12	11.8	16	11.8	11	7.2	THOMAS GA	0	-39	-39	11	07
8	14.5	6	6.9	15	12.1	TIFT GA	-52	74	-16	33	28
5	9.3	5	6.5	8	6.6	TOOMBS GA	-30	2	-28	20	06
4	21.1			2	5.0	TOWNS GA			-76	01	03
1	4.4	2	10.3	4	15.3	TREUTLEN GA	137	49	252	97	54
20	14.2	23	12.8	43	17.5	TROUP GA	-9	36	24	71	70
4	15.5	8	24.1	2	5.5	TURNER GA	56	-77	-65	03	04
1	7.3			3	13.8	TWIGGS GA			90	89	41
4	14.0	3	7.7	4	7.5	UNION GA	-45	-2	-46	08	07
4	4.9	10	12.0	17	13.5	UPSON GA	146	13	177	95	39
13	9.0	23	11.3	38	13.6	WALKER GA	26	20	50	82	39
7	10.7	11	11.2	16	11.0	WALTON GA	4	-1	3	53	20
8	8.1	21	15.8	14	8.4	WARE GA	95	-47	3	54	09
1	5.3	3	13.7	4	17.7	WARREN GA	159	29	235	97	72
4	8.0	3	4.5	14	19.0	WASHINGTON GA	-44	326	139	93	80
2	3.7	7	12.0	8	8.9	WAYNE GA	224	-26	139	93	11
2	39.7	3	41.0	1	11.6	WEBSTER GA	3	-72	-71	01	24
2	13.4	3	15.6	4	15.0	WHEELER GA	16	-4	12	62	52
3	12.1	1	2.5	5	8.9	WHITE GA	-80	263	-27	22	11
12	10.3	14	7.3	34	12.1	WHITFIELD GA	-29	66	17	66	27
6	26.0	2	6.0	5	13.2	WILCOX GA	-77	121	-49	06	36
2	5.4	4	10.1	5	8.9	WILKES GA	87	-12	64	85	11
1	4.3			10	28.4	WILKINSON GA			556	99	98
9	22.2	3	5.8	7	10.1	WORTH GA	-74	72	-55	05	16
		80	13.3	139	14.3	HAWAII HI		8			
		5	9.4	13	13.2	HAWAII HI		40			36
		71	14.6	110	14.5	HONOLULU HI		-1			47
		3	11.1	7	19.0	KAUAI HI		71			80
		1	2.7	9	10.9	MAUI#HI		308			20
318	13.1	416	12.9	528	12.2	IDAHO ID	-1	-5	-7		
42	11.3	73	13.9	88	12.1	ADA ID	23	-13	8	58	28
1	7.5	3	27.5	4	22.8	ADAMS ID	268	-17	205	96	94
9	6.2	35	18.7	47	18.5	BANNOCK ID	200	-1	197	96	77
3	16.2	3	7.9	4	9.1	BEAR LAKE ID	-51	14	-44	09	12
11*	52.6	2	6.8	2	5.6	BENEWAH ID	-87	-18	-89	00	04
9	11.2	10	9.2	8	6.0	BINGHAM ID	-18	-35	-47	08	05
2	11.5	1	6.2	4	13.5	BLAINE ID	-46	118	17	67	39
				1	9.4	BOISE ID					13
10	13.8	6	6.9	14	12.0	BONNER ID	-50	72	-13	36	26
10	8.2	12	7.3	21	9.3	BONNEVILLE ID	-12	29	14	63	13
6	23.2	2	7.2	3	7.0	BOUNDARY ID	-69	-3	-70	01	06
		1	7.6	1	5.4	BUTTE ID		-29			04
34	12.9	48	14.2	62	13.0	CANYON ID	10	-9	1	52	35
3	13.0	5	20.8	1	3.3	CARIBOU ID	60	-84	-75	01	02
11	20.8	4	6.0	15	16.7	CASSIA ID	-71	178	-20	29	64
1	26.8			1	37.2	CLARK ID			39	78	99
4	13.1	3	8.8	3	6.2	CLEARWATER ID	-33	-30	-53	05	05
1	6.8	3	25.5			CUSTER ID	273				
5	21.2	4	9.4	7	12.0	ELMORE ID	-56	28	-43	09	27
2	6.6	7	17.3	7	16.2	FRANKLIN ID	162	-7	144	93	60
3	10.3	5	13.5	3	6.5	FREMONT ID	31	-52	-37	13	06
5	12.3	7	13.3	6	9.3	GEM ID	8	-30	-24	24	13
6	15.8	7	12.7	7	10.9	GOODING ID	-19	-15	-31	18	20
10	21.5	6	10.8	9	11.2	IDAHO ID	-50	3	-48	07	21
6	21.2	7	16.9	2	3.3	JEFFERSON ID	-20	-81	-85	00	02
3	7.6	6	10.5	13	17.8	JEROME ID	38	69	133	93	72
17	12.2	28	14.8	37	13.2	KOOTENAI ID	22	-11	8	59	36
13	15.8	15	14.7	26	18.5	LATAH ID	-7	26	17	66	76
4	18.6	5	17.7	3	8.4	LEMHI ID	-4	-53	-55	05	09
3	15.0	1	4.7	3	14.5	LEWIS ID	-69	207	-4	47	47
1	7.0	2	10.2			LINCOLN ID	46				
5	16.9	6	14.5	6	11.7	MADISON ID	-14	-19	-31	18	24
8	17.1	4	8.3	13	15.9	MINIDOKA ID	-51	91	-7	43	58
18	15.6	27	18.4	27	14.5	NEZ PERCE ID	18	-21	-7	43	48
1	5.1	4	14.2	5	25.4	ONEIDA ID	182	79	403	98	96
2	8.7	5	17.5	3	8.7	OWYHEE ID	101	-50	0	51	10
11	17.7	9	10.7	9	7.3	PAYETTE ID	-40	-31	-59	04	07
2	16.1	1	4.8	1	4.3	POWER ID	-70	-11	-73	01	03
10	11.9	12	15.5	19	20.4	SHOSHONE ID	31	31	72	87	87
1	14.5	1	9.4	1	8.6	TETON ID	-35	-9	-41	11	10
17	9.5	25	10.0	28	9.9	TWIN FALLS ID	5	0	4	55	15
2	18.5			2	9.6	VALLEY ID			-48	07	14
6	13.8	11	19.2	12	17.7	WASHINGTON ID	39	-8	28	73	71
8770*	19.8	9983*	18.6	11167*	17.7	ILLINOIS IL	-6	-5	-11		
75	16.6	98	20.4	86	14.7	ADAMS IL	23	-28	-11	38	49
19	25.1	14	19.1	24*	26.3	ALEXANDER IL	-24	38	5	56	97
25	26.2	19	16.7	24	18.8	BOND IL	-36	12	-28	20	78

* SIGNIFICANTLY MORE DEATHS THAN EXPECTED

COUNTIES COMBINED; SEE TECHNICAL NOTES

WHITE FEMALE — CANCER OF THE LARGE INTESTINE — ICD 153,159.0

COUNTY AND STATE

COUNTY AND STATE	1950-1959 DEATHS	1950-1959 RATE	1960-1969 DEATHS	1960-1969 RATE	1970-1979 DEATHS	1970-1979 RATE	PERCENT CHANGE 60S/50S	PERCENT CHANGE 70S/60S	PERCENT CHANGE 70S/50S	PERCENTILE %CHANGE 70S/50S	PERCENTILE RATES 1970S
BOONE IL	22	21.1	31*	26.1	38*	24.9	24	-5	18	67	96
BROWN IL	7	13.9	15	28.4	5	8.6	105	-70	-38	12	10
BUREAU IL	57*	24.4	42	16.1	54	18.2	-34	13	-25	23	75
CALHOUN IL	5	16.4	4	8.6	10	21.9	-48	156	34	76	91
CARROLL IL	20	17.0	25	17.3	28	17.7	2	3	5	55	72
CASS IL	22	22.0	18	15.8	22	18.5	-28	18	-16	33	77
CHAMPAIGN IL	89	20.4	83	15.7	101	15.7	-23	0	-23	25	57
CHRISTIAN IL	34	14.8	40	15.1	48	14.2	2	-6	-4	47	45
CLARK IL	32*	25.3	22	15.3	27	17.1	-39	12	-32	17	67
CLAY IL	19	19.8	17	14.8	15	9.3	-25	-37	-53	05	13
CLINTON IL	29	24.7	23	15.6	34	16.4	-37	5	-33	16	62
COLES IL	43	17.5	42	14.5	57	17.1	-17	18	-2	49	67
COOK IL	4364*	21.1	4818*	19.7	4912*	18.0	-7	-9	-15	34	73
CRAWFORD IL	29	21.3	27	18.0	28	15.0	-16	-17	-30	19	51
CUMBERLAND IL	11	16.3	23*	33.1	14	16.9	103	-49	4	54	66
DE KALB IL	42	17.4	57	20.0	59	17.0	15	-15	-2	49	66
DE WITT IL	27	24.9	16	12.1	24	15.6	-51	29	-37	13	56
DOUGLAS IL	19	17.3	22	17.5	23	15.8	1	-10	-9	40	58
DU PAGE IL	170	20.4	241	16.3	404	17.4	-20	7	-15	34	69
EDGAR IL	24	14.9	36	21.3	35	18.3	43	-14	23	70	76
EDWARDS IL	15	26.9	15	20.4	13	19.3	-24	-5	-28	20	81
EFFINGHAM IL	21	17.7	36	23.4	30	15.2	32	-35	-14	35	53
FAYETTE IL	17	12.5	9	5.8	28	15.6	-53	169	25	72	56
FORD IL	15	14.5	20	17.2	19	15.3	18	-11	5	56	54
FRANKLIN IL	46	17.1	60	17.5	74	18.5	2	6	8	59	76
FULTON IL	54	20.1	47	15.8	54	16.0	-22	2	-20	28	59
GALLATIN IL	14*	34.7	10	20.1	8	12.6	-42	-37	-64	03	31
GREENE IL	25	19.4	32	25.0	35	22.8	29	-9	18	67	93
GRUNDY IL	25	23.0	23	17.7	27	16.6	-23	-6	-28	21	64
HAMILTON IL	11	14.1	8	9.7	16	15.3	-32	58	8	58	53
HANCOCK IL	42	22.3	31	15.9	36	18.9	-29	19	-15	34	79
HARDIN IL	2	7.9	7	20.1	9	20.2	154	1	156	94	86
HENDERSON IL	5	13.0	11	21.0	8	13.1	62	-38	1	52	35
HENRY IL	52	18.2	70	20.8	68	15.1	14	-28	-17	31	52
IROQUOIS IL	42	22.9	48	21.5	43	16.4	-6	-24	-28	20	62
JACKSON IL	28	14.6	38	16.3	39	13.6	12	-17	-7	43	40
JASPER IL	20	26.6	12	17.8	20	23.7	-33	33	-11	38	95
JEFFERSON IL	38	20.9	30	13.6	58	20.6	-35	51	-2	49	88
JERSEY IL	22	28.4	14	14.0	19	14.5	-51	4	-49	07	48
JO DAVIESS IL	29	23.1	22	16.3	35	23.0	-29	40	0	50	94
JOHNSON IL	7	14.4	9	15.0	10	15.7	4	5	9	59	57
KANE IL	190	19.6	217	18.3	226	15.8	-7	-13	-19	29	58
KANKAKEE IL	77	16.6	79	15.0	101	18.3	-10	22	10	60	75
KENDALL IL	15	20.8	19	19.9	27	19.4	-4	-2	-7	43	82
KNOX IL	67	18.5	69	16.8	67	14.1	-9	-16	-24	25	44
LAKE IL	139	17.8	211	17.7	259	15.9	-1	-10	-11	39	58
LA SALLE IL	121	20.4	144	20.1	176*	20.9	-2	4	2	53	89
LAWRENCE IL	15	12.5	18	11.9	24	14.5	-5	22	16	66	48
LEE IL	33	16.4	32	14.4	48	18.5	-13	29	12	62	76
LIVINGSTON IL	55	23.7	52	19.3	74*	22.6	-18	17	-5	45	93
LOGAN IL	46	24.6	34	15.1	50	19.0	-39	26	-23	25	80
MC DONOUGH IL	38	20.8	42	19.6	45	19.3	-6	-1	-7	43	82
MC HENRY IL	64	20.1	79	17.5	102	15.2	-13	-13	-25	24	53
MC LEAN IL	78	16.6	76	13.4	99	15.0	-19	12	-9	40	52
MACON IL	97	17.2	130	18.6	151	18.0	8	-4	4	55	73
MACOUPIN IL	51	16.3	81*	23.3	72	16.6	43	-29	1	52	63
MADISON IL	119	14.0	213	18.3	282*	19.3	30	6	38	77	82
MARION IL	45	19.6	49	18.0	65	19.9	-8	11	2	53	85
MARSHALL IL	9	11.6	11	11.5	16	16.2	-1	40	40	78	60
MASON IL	16	16.4	23	20.5	19	13.8	25	-33	-16	33	41
MASSAC IL	23*	32.1	21	23.1	15	13.0	-28	-44	-59	03	35
MENARD IL	18	28.3	18	27.9	10	9.9	-2	-65	-65	02	15
MERCER IL	20	20.0	17	14.8	20	15.3	-26	3	-23	25	54
MONROE IL	16	19.8	7	7.1	24	17.7	-64	150	-11	39	71
MONTGOMERY IL	33	13.6	52	21.2	63	23.0	56	8	69	86	94
MORGAN IL	36	12.9	47	16.8	39	12.9	31	-24	0	51	33
MOULTRIE IL	16	16.1	18	17.3	22	16.7	7	-4	3	54	64
OGLE IL	42	22.4	60*	26.0	54	19.3	16	-26	-14	36	82
PEORIA IL	190	20.2	219	19.6	222	17.3	-3	-12	-14	35	68
PERRY IL	13	10.9	23	17.3	32	19.1	58	11	75	87	80
PIATT IL	15	17.5	18	18.3	21	19.8	5	8	13	63	84
PIKE IL	18	10.9	28	18.0	32	17.3	66	-4	59	84	68
POPE IL	6	17.5	7	23.2	3	8.8	32	-62	-50	06	11
PULASKI IL	8	16.6	11	22.5	8	12.9	36	-43	-23	26	33
PUTNAM IL	5	21.3	6	20.0	8	23.1	-6	15	8	59	94
RANDOLPH IL	45*	26.9	34	16.8	32	14.7	-37	-13	-45	08	49
RICHLAND IL	22	21.7	19	15.1	24	18.3	-30	21	-16	33	75
ROCK ISLAND IL	140	19.6	152	17.5	171	16.7	-11	-5	-15	34	64
ST CLAIR IL	189	20.6	223	19.4	270*	20.3	-6	4	-1	50	86
SALINE IL	27	14.4	41	17.6	55	19.0	22	8	32	75	80
SANGAMON IL	166*	21.1	177	18.0	235*	20.0	-15	11	-5	45	85
SCHUYLER IL	13	19.7	16	22.7	14	15.0	15	-34	-24	24	51
SCOTT IL	10	20.2	6	12.7	12	25.1	-37	98	24	71	96
SHELBY IL	39	23.0	28	14.3	34	18.4	-38	29	-20	28	76
STARK IL	3	5.0	8	14.6	11	18.5	189	27	267	97	76
STEPHENSON IL	54	19.8	65	21.0	71	19.9	6	-5	0	51	84
TAZEWELL IL	59	18.5	90	19.4	122	19.3	5	-1	4	54	81
UNION IL	19	13.9	12	8.3	27	19.9	-40	141	44	79	84
VERMILION IL	103	21.0	113	19.6	136*	20.1	-7	3	-4	46	86
WABASH IL	29*	35.7	18	18.6	21	22.3	-48	20	-37	13	92
WARREN IL	23	16.4	21	13.3	31	17.5	-19	32	7	57	70
WASHINGTON IL	9	8.9	21	19.9	35*	27.3	125	37	208	96	97
WAYNE IL	23	20.8	30	22.7	27	19.2	9	-15	-8	42	81
WHITE IL	20	16.6	21	14.5	43*	28.8	-13	99	73	87	98
WHITESIDE IL	41	15.3	69	21.2	78	20.2	38	-5	31	75	86
WILL IL	95	13.9	146	16.4	196	16.3	17	-1	17	66	61
WILLIAMSON IL	39	14.1	62	18.4	70	17.3	30	-6	23	70	68
WINNEBAGO IL	142	17.4	170	15.9	207	15.5	-8	-3	-11	39	56
WOODFORD IL	12	9.2	25	15.2	48	23.7	66	56	159	94	95

* SIGNIFICANTLY MORE DEATHS THAN EXPECTED

WHITE FEMALE — CANCER OF THE LARGE INTESTINE — ICD 153,159.0

1950-1959 DEATHS	1950-1959 RATE	1960-1969 DEATHS	1960-1969 RATE	1970-1979 DEATHS	1970-1979 RATE	COUNTY AND STATE	PERCENT CHANGE 60S/50S	PERCENT CHANGE 70S/60S	PERCENT CHANGE 70S/50S	PERCENTILE %CHANGE 70S/50S	PERCENTILE RATES 1970S
3828*	18.7	4687*	18.7	5359*	17.8	INDIANA IN	0	-5	-5		
19	16.0	17	10.8	30	17.5	ADAMS IN	-32	61	9	60	70
174	18.1	245*	19.7	264	17.3	ALLEN IN	9	-12	-4	46	69
32	16.8	45	19.7	49	16.4	BARTHOLOMEW IN	18	-17	-2	49	62
12	18.2	11	14.1	11	14.1	BENTON IN	-23	0	-22	26	44
12	15.8	23	25.4	19	18.6	BLACKFORD IN	61	-27	18	67	77
44*	27.2	43	21.6	42	18.8	BOONE IN	-21	-13	-31	18	78
5	15.0	7	18.5	9	14.9	BROWN IN	23	-20	-1	50	50
22	21.7	25	21.8	31*	24.2	CAPPOLL IN	0	11	11	62	95
44	16.6	51	17.4	54	18.4	CASS IN	4	6	11	61	76
41	19.6	43	15.0	61	15.6	CLARK IN	-23	4	-20	28	56
27	16.4	47*	23.3	46	21.4	CLAY IN	42	-8	31	74	90
41	21.4	39	19.0	59*	22.0	CLINTON IN	-11	16	3	53	92
6	10.8	13	23.3	13	20.3	CRAWFORD IN	116	-13	88	89	87
29	20.0	29	16.6	39	19.3	DAVIESS IN	-17	16	-4	47	81
42*	31.1	28	17.1	44	24.2	DEARBORN IN	-45	42	-22	26	95
16	12.9	32	21.9	33	20.6	DECATUR IN	70	-6	60	84	88
18	11.4	31	17.9	38	19.8	DE KALB IN	58	10	74	87	84
81	19.0	91	16.5	106	15.9	DELAWARE IN	-13	-4	-16	33	59
24	19.9	31	20.6	32	16.9	DUBOIS IN	3	-18	-15	34	66
91	18.9	101	17.1	112	15.7	ELKHART IN	-9	-9	-17	31	57
22	20.4	21	14.8	28	17.1	FAYETTE IN	-27	16	-16	33	67
38	16.5	57	19.7	72	20.1	FLOYD IN	20	2	22	70	86
28	25.8	28	22.4	31	21.5	FOUNTAIN IN	-13	-4	-17	32	90
20	22.6	17	19.1	14	13.3	FRANKLIN IN	-16	-30	-41	10	37
28	25.3	26	23.0	18	13.5	FULTON IN	-9	-41	-46	08	39
36	20.2	45	20.8	52	19.9	GIBSON IN	3	-4	-1	50	84
50	15.2	77	19.5	75	15.7	GRANT IN	28	-20	3	53	57
38	21.5	34	17.6	44	17.6	GREENE IN	-18	0	-18	30	71
37	21.1	51	21.7	70*	20.5	HAMILTON IN	3	-5	-2	49	87
24	17.9	29	18.2	29	13.6	HANCOCK IN	1	-25	-24	24	39
17	17.5	24	21.9	27	20.8	HARRISON IN	25	-5	19	67	88
36	24.7	36	17.1	49	17.3	HENDRICKS IN	-31	1	-30	18	68
42	18.1	55	19.5	53	14.8	HENRY IN	8	-24	-18	30	50
48	17.4	51	14.3	74	17.4	HOWARD IN	-18	22	0	50	69
38	17.5	51	18.3	63*	21.3	HUNTINGTON IN	5	16	22	70	90
32	21.1	33	18.4	46	20.3	JACKSON IN	-13	10	-4	47	86
17	19.9	25*	25.1	19	15.4	JASPER IN	26	-39	-23	26	55
25	18.0	32	20.8	30	18.1	JAY IN	16	-13	1	52	74
12	8.8	24	14.7	37	21.2	JEFFERSON IN	68	44	142	93	90
19	25.2	21	24.2	17	16.2	JENNINGS IN	-4	-33	-35	14	61
31	17.3	40	15.5	60	16.8	JOHNSON IN	-11	8	-3	48	65
33	12.8	49	16.5	71	21.0	KNOX IN	28	27	63	85	89
40	19.0	55	22.0	54	17.3	KOSCIUSKO IN	16	-21	-9	40	68
6	7.7	22	23.5	23	20.0	LAGRANGE IN	207	-15	161	94	85
252*	20.5	333*	19.9	381*	18.9	LAKE IN	-3	-5	-8	42	79
80	20.3	98	20.0	116	19.3	LA PORTE IN	-1	-4	-5	45	81
36	19.4	50	22.3	49	18.5	LAWRENCE IN	15	-17	-4	46	77
85	16.8	108	16.7	151	19.5	MADISON IN	-1	17	16	65	82
552*	20.5	683*	20.5	662	17.2	MARION IN	0	-16	-16	32	67
46*	27.4	26	13.3	50	18.1	MARSHALL IN	-51	36	-34	15	74
10	23.6	7	13.8	20*	34.2	MARTIN IN	-41	147	45	80	99
42	22.8	44	21.8	42	19.6	MIAMI IN	-5	-10	-14	35	83
34	17.1	34	13.5	60	16.6	MONROE IN	-21	23	-3	48	63
37	19.1	43	19.8	43	17.4	MONTGOMERY IN	4	-12	-9	41	69
24	17.8	28	16.1	42	17.7	MORGAN IN	-10	10	0	51	72
7	10.9	18	23.9	13	16.2	NEWTON IN	120	-32	49	81	60
30	18.4	24	13.3	37	16.1	NOBLE IN	-28	21	-13	37	59
4	16.5	5	18.4	7	19.7	OHIO IN	11	7	19	68	83
20	22.1	20	18.8	32*	25.3	ORANGE IN	-15	35	14	64	96
20	26.5	25*	28.5	18	19.3	OWEN IN	7	-32	-27	21	81
31*	30.5	23	19.5	26	20.5	PARKE IN	-36	5	-33	16	87
20	22.9	26	23.6	23	17.2	PERRY IN	3	-27	-25	24	68
9	10.9	24*	26.1	21	19.1	PIKE IN	140	-27	75	87	80
46	23.2	50	18.0	66	15.9	PORTER IN	-22	-12	-31	17	59
22	18.9	22	17.8	31	20.6	POSEY IN	-6	16	9	59	88
23*	32.3	13	17.0	17	17.7	PULASKI IN	-47	4	-45	08	72
17	12.4	23	14.2	29	15.7	PUTNAM IN	14	10	26	72	57
32	18.8	30	16.8	44	22.5	RANDOLPH IN	-11	34	19	68	93
28	23.7	20	15.0	28	17.3	RIPLEY IN	-36	15	-27	21	68
28	23.9	18	14.4	32	21.3	RUSH IN	-40	47	-11	38	90
156	17.0	204	17.0	213	14.9	ST JOSEPH IN	0	-13	-13	37	50
7	12.4	17	23.7	15	14.7	SCOTT IN	91	-38	18	67	49
19	11.1	37	18.4	41	18.4	SHELBY IN	65	0	65	85	76
20	22.0	15	14.9	18	13.5	SPENCER IN	-32	-9	-39	12	39
14	17.1	21	17.7	25	18.4	STARKE IN	3	4	7	58	76
19	20.5	17	14.6	24	17.0	STEUBEN IN	-29	16	-17	31	66
20	12.6	34	20.2	35	17.4	SULLIVAN IN	61	-14	38	78	69
12	22.1	11	23.4	5	9.6	SWITZERLAND IN	6	-59	-57	04	14
54	15.2	83	18.9	95	17.2	TIPPECANOE IN	24	-9	13	63	67
10	10.7	17	16.7	28*	23.8	TIPTON IN	57	42	124	92	95
3	9.3	9	22.1	9	20.1	UNION IN	138	-9	116	92	85
143	18.0	183	18.6	219	18.4	VANDERBURGH IN	4	-1	3	53	76
21	19.3	26	20.5	45*	30.0	VERMILLION IN	6	46	56	83	98
103	16.5	116	16.4	133	17.1	VIGO IN	-1	4	3	54	67
39	20.7	53*	24.1	51	20.8	WABASH IN	16	-14	0	51	88
7	15.9	11	22.8	15	27.0	WARREN IN	44	18	70	86	97
15	12.2	29	20.8	24	13.3	WARRICK IN	70	-36	8	59	37
21	20.9	15	12.9	22	16.2	WASHINGTON IN	-38	25	-23	26	60
46	11.5	69	14.7	76	15.1	WAYNE IN	28	3	31	75	52
26	21.3	18	13.3	20	12.8	WELLS IN	-38	-4	-40	11	33
24	21.6	31	25.1	25	14.8	WHITE IN	16	-41	-31	18	50
17	15.8	21	17.1	33	20.2	WHITLEY IN	9	18	28	73	86
2723	18.0	3184	17.6	3797*	18.2	IOWA IA	-2	3	1		
8	12.0	13	15.5	19	20.5	ADAIR IA	29	33	71	86	87
9	17.1	8	14.1	9	15.0	ADAMS IA	-18	6	-13	37	51
16	16.4	28	23.2	17	11.7	ALLAMAKEE IA	41	-49	-29	20	25

* SIGNIFICANTLY MORE DEATHS THAN EXPECTED

WHITE FEMALE

CANCER OF THE LARGE INTESTINE

ICD 153,159.0

1950-1959 DEATHS	1950-1959 RATE	1960-1969 DEATHS	1960-1969 RATE	1970-1979 DEATHS	1970-1979 RATE	COUNTY AND STATE	PERCENT CHANGE 60S/50S	PERCENT CHANGE 70S/60S	PERCENT CHANGE 70S/50S	PERCENTILE %CHANGE 70S/50S	PERCENTILE RATES 1970S
24	18.0	32	18.4	19	10.3	APPANOOSE IA	2	-44	-43	10	17
11	18.8	9	12.7	16	19.7	AUDUBON IA	-32	55	5	55	83
19	14.8	31	21.3	34	18.2	BENTON IA	44	-14	23	71	75
74	15.0	109	16.8	125	16.5	BLACK HAWK IA	12	-2	10	60	62
26	13.9	36	16.6	47	18.9	BOONE IA	19	14	35	77	79
24	22.3	23	16.8	28	16.9	BREMER IA	-25	0	-24	24	65
20	16.1	25	19.0	27	17.8	BUCHANAN IA	18	-6	10	61	72
20	17.1	30	20.2	18	8.9	BUENA VISTA IA	18	-56	-48	07	11
21	19.6	18	15.2	21	14.8	BUTLER IA	-23	-3	-25	24	49
9	9.1	27	23.0	16	11.3	CALHOUN IA	153	-51	24	71	22
22	17.8	15	9.9	40	21.2	CARROLL IA	-45	115	19	68	90
22	17.9	34	25.5	20	13.7	CASS IA	42	-46	-24	25	41
28*	27.6	21	17.0	26	18.6	CEDAR IA	-38	9	-33	16	77
49	18.5	42	12.7	76	19.7	CERRO GORDO IA	-32	56	6	57	84
21	18.2	24	19.5	21	15.1	CHEROKEE IA	7	-23	-17	31	52
10	11.7	14	14.2	11	10.9	CHICKASAW IA	21	-23	-7	43	20
19*	29.2	17	26.2	18	21.4	CLARKE IA	-10	-18	-27	21	90
19	19.3	22	18.2	35*	25.3	CLAY IA	-6	39	31	74	96
29	22.0	28	17.8	25	16.5	CLAYTON IA	-19	-7	-25	23	63
73*	25.3	74	22.1	77	19.1	CLINTON IA	-13	-13	-24	24	81
24	23.3	26	21.5	22	16.2	CRAWFORD IA	-8	-25	-31	18	60
20	12.3	27	14.1	44	22.2	DALLAS IA	14	58	80	87	92
14	21.8	13	17.0	17	26.7	DAVIS IA	-22	57	22	70	97
17	17.9	15	19.1	19	18.9	DECATUR IA	7	-1	5	56	79
18	19.2	14	12.6	22	18.5	DELAWARE IA	-35	47	-4	47	77
49	18.3	53	17.3	60	16.5	DES MOINES IA	-6	-4	-10	40	63
14	20.1	19	21.4	24	21.7	DICKINSON IA	6	1	8	58	91
72	17.3	107*	21.2	135*	22.7	DUBUQUE IA	23	7	32	75	93
15	20.0	25*	25.5	19	16.9	EMMET IA	28	-34	-15	34	65
31	18.7	34	18.0	41	18.9	FAYETTE IA	-4	5	1	52	79
21	16.9	29	17.9	33	21.0	FLOYD IA	6	17	24	71	89
11	13.3	17	17.2	15	12.5	FRANKLIN IA	30	-27	-6	44	31
11	14.3	17	21.2	21	26.2	FREMONT IA	48	24	83	88	97
21	23.0	19	17.0	26	20.4	GREENE IA	-26	20	-11	38	87
13	17.0	18	17.2	21	16.1	GRUNDY IA	1	-6	-5	45	60
22	24.6	15	13.2	21	14.0	GUTHRIE IA	-46	6	-43	09	43
14	12.6	22	18.0	33	22.8	HAMILTON IA	43	27	82	88	94
15	19.4	15	16.1	23	17.6	HANCOCK IA	-17	9	-9	40	70
21	14.3	31	17.6	36	18.9	HARDIN IA	24	7	32	75	79
22	19.0	37*	27.3	21	13.5	HARRISON IA	44	-51	-29	20	38
22	15.2	24	15.6	26	16.0	HENRY IA	3	3	5	56	59
19	24.4	11	11.3	19	16.4	HOWARD IA	-54	45	-33	16	62
21*	29.8	12	13.9	26*	26.4	HUMBOLDT IA	-53	90	-11	38	97
6	10.4	11	16.5	16	19.1	IDA IA	58	16	83	88	81
25	25.5	18	16.6	28	21.6	IOWA IA	-35	30	-16	34	91
33*	29.2	15	10.9	29	20.0	JACKSON IA	-63	85	-31	18	85
32	17.9	41	18.2	39	14.7	JASPER IA	2	-19	-18	30	48
12	11.1	20	16.3	19	13.9	JEFFERSON IA	46	-15	24	71	42
45	20.7	44	16.7	84*	26.2	JOHNSON IA	-20	57	26	72	97
26	24.7	27	20.5	29	21.0	JONES IA	-17	2	-15	34	89
27	21.6	24	17.6	26	19.6	KEOKUK IA	-18	11	-9	40	83
22	17.9	19	13.0	26	15.0	KOSSUTH IA	-27	15	-16	32	51
58	22.9	63	21.8	78*	24.9	LEE IA	-5	14	9	59	96
110	17.5	160	19.8	189*	19.5	LINN IA	13	-2	11	61	82
11	18.0	12	14.7	19	23.9	LOUISA IA	-19	63	33	75	95
10	11.1	24	23.9	11	13.5	LUCAS IA	115	-44	22	69	39
13	19.6	13	16.2	12	11.4	LYON IA	-17	-30	-42	10	23
25*	29.0	20	21.3	21	18.7	MADISON IA	-26	-12	-35	14	78
42	24.3	28	15.2	36	17.7	MAHASKA IA	-38	16	-27	21	71
23	14.7	37	20.3	36	17.2	MARION IA	38	-15	17	66	68
28	13.0	47	17.7	48	16.3	MARSHALL IA	37	-8	25	72	61
8	11.4	13	16.8	21	23.7	MILLS IA	47	41	108	91	95
9	9.9	21	19.9	19	14.6	MITCHELL IA	102	-27	48	81	48
17	19.9	19	18.7	14	11.3	MONONA IA	-6	-39	-43	09	22
15	19.3	11	12.8	20	23.3	MONROE IA	-33	81	21	69	94
19	18.4	17	14.7	20	15.3	MONTGOMERY IA	-20	4	-17	32	54
29	14.3	51	21.8	60*	20.6	MUSCATINE IA	52	-5	44	79	88
28*	25.9	31	22.8	18	10.3	O'BRIEN IA	-12	-55	-60	03	17
8	16.7	8	13.3	25*	35.3	OSCEOLA IA	-20	166	112	91	99
18	11.5	29	16.4	34	16.3	PAGE IA	42	0	42	79	61
15	19.8	15	15.6	19	17.4	PALO ALTO IA	-21	12	-12	37	69
24	21.8	22	14.7	29	16.5	PLYMOUTH IA	-32	12	-24	24	62
15	19.1	27*	29.2	23	18.8	POCAHONTAS IA	53	-36	-1	50	79
217	17.5	259	17.1	298	16.7	POLK IA	-2	-3	-5	45	64
61	15.5	80	17.4	97	17.5	POTTAWATTAMIE IA	12	0	13	62	70
23	20.1	22	16.4	28	20.9	POWESHIEK IA	-19	27	4	54	89
10	16.1	6	8.9	11	15.9	RINGGOLD IA	-45	79	-1	50	58
15	14.3	16	12.4	32	24.3	SAC IA	-13	97	70	86	95
96	17.2	120	17.5	145	18.9	SCOTT IA	2	8	10	60	79
20	24.3	13	10.1	23	19.3	SHELBY IA	-58	90	-21	28	81
30	24.7	28	17.7	36	18.7	SIOUX IA	-29	6	-24	24	78
30	13.9	39	14.5	57	18.0	STORY IA	5	24	30	74	74
21	16.3	27	18.1	32	17.9	TAMA IA	11	-1	10	60	73
15	16.2	12	10.7	19	18.7	TAYLOR IA	-34	75	15	65	78
19	15.8	25	20.8	27	19.0	UNION IA	32	-9	20	68	80
15	16.9	16	19.6	20	22.7	VAN BUREN IA	16	16	34	76	93
43	15.7	53	16.5	72	19.6	WAPELLO IA	5	19	25	72	83
17	15.6	27	21.7	29	17.1	WARREN IA	40	-22	10	60	66
30	23.0	26	15.4	32	20.0	WASHINGTON IA	-33	30	-13	36	85
17	20.3	18	19.2	19	17.1	WAYNE IA	-5	-11	-16	33	67
41	16.7	59	18.5	67	16.5	WEBSTER IA	11	-11	-1	50	62
9	11.9	17	17.7	21	16.6	WINNEBAGO IA	49	-6	40	78	64
23	19.2	19	13.8	29	17.5	WINNESHIEK IA	-28	27	-9	41	70
113	20.2	104	15.7	132	17.4	WOODBURY IA	-22	11	-14	35	69
4	7.0	10	13.7	19	20.2	WORTH IA	95	47	187	95	86
21	19.8	21	14.9	35	22.1	WRIGHT IA	-25	49	12	62	92
1616	14.9	2076	15.7	2416	15.5	KANSAS KS	5	-2	4		
19	15.1	33	21.9	19	11.2	ALLEN KS	44	-49	-26	22	21
11	16.5	11	13.8	13	11.0	ANDERSON KS	-16	-21	-33	16	20

* SIGNIFICANTLY MORE DEATHS THAN EXPECTED

WHITE FEMALE — CANCER OF THE LARGE INTESTINE — ICD 153,159.0

1950-1959 DEATHS	1950-1959 RATE	1960-1969 DEATHS	1960-1969 RATE	1970-1979 DEATHS	1970-1979 RATE	COUNTY AND STATE	PERCENT CHANGE 60S/50S	PERCENT CHANGE 70S/60S	PERCENT CHANGE 70S/50S	PERCENTILE %CHANGE 70S/50S	PERCENTILE RATES 1970S
25	19.4	17	11.3	27	21.3	ATCHISON KS	-41	87	10	60	90
9	19.5	9	15.7	10	13.9	BARBER KS	-20	-11	-29	20	42
11	9.2	27	16.5	32	15.7	BARTON KS	78	-5	70	86	57
11	8.5	21	14.4	25	14.4	BOURBON KS	68	0	69	86	47
28	24.9	17	15.5	20	12.4	BROWN KS	-38	-20	-50	06	30
19	11.5	38	17.5	50	17.6	BUTLER KS	52	0	53	82	70
3	9.8	7	19.9	6	14.3	CHASE KS	103	-28	46	80	46
10	19.3	9	15.1	11	16.7	CHAUTAUQUA KS	-22	11	-14	36	64
24	15.6	31	17.8	25	11.5	CHEROKEE KS	14	-35	-26	22	23
3	10.8	7	18.9	3	9.9	CHEYENNE KS	75	-48	-9	41	15
5	21.4	5	19.8	1	2.7	CLARK KS	-8	-86	-87	00	02
13	13.8	17	17.0	20	17.9	CLAY KS	23	5	30	74	73
20	17.0	21	16.4	24	14.1	CLOUD KS	-4	-14	-17	31	44
12	16.0	13	16.8	20	22.1	COFFEY KS	5	31	38	78	92
3	15.2	4	11.6	3	7.7	COMANCHE KS	-24	-34	-50	06	08
31	12.3	41	14.7	48	15.1	COWLEY KS	19	3	22	70	52
38	12.7	41	12.3	58	17.4	CRAWFORD KS	-4	42	37	77	69
4	10.2	7	16.5	10	20.2	DECATUR KS	62	22	98	90	86
23	15.8	28	14.3	30	13.6	DICKINSON KS	-10	-5	-14	35	39
8	12.6	15	24.1	6	8.6	DONIPHAN KS	91	-64	-32	17	10
29	15.8	28	12.6	35	14.0	DOUGLAS KS	-20	11	-11	38	43
5	13.6	3	6.7	9	21.1	EDWARDS KS	-51	214	55	83	89
8	16.1	7	11.0	7	11.8	ELK KS	-32	7	-27	21	25
9	15.4	18	20.1	26	20.0	ELLIS KS	31	-1	30	74	85
8	15.3	9	14.5	8	10.6	ELLSWORTH KS	-5	-27	-31	18	18
12	21.8	14	18.8	13	13.3	FINNEY KS	-14	-29	-39	12	37
24	25.0	15	12.3	25	16.3	FORD KS	-51	33	-35	15	61
25	17.6	23	14.2	21	11.2	FRANKLIN KS	-20	-21	-37	13	21
11	13.7	11	11.8	24	23.3	GEARY KS	-14	97	70	86	94
2	10.1	2	8.2	1	3.6	GOVE KS	-19	-57	-65	02	02
1	4.1	4	15.2	4	13.0	GRAHAM KS	268	-14	216	96	35
1	14.1	4	18.2	5	23.1	GRANT KS	29	27	64	85	94
		1	3.4	3	7.2	GRAY KS		112			07
		1	10.3	4	37.9	GREELEY KS		266			99
15	16.2	18	16.0	16	15.1	GREENWOOD KS	-1	-6	-7	43	53
1	7.0			5	24.8	HAMILTON KS			254	97	96
12	17.0	12	15.0	9	8.4	HARPER KS	-12	-44	-50	06	09
17	12.0	25	12.1	38	16.1	HARVEY KS	1	33	35	76	60
1	18.6	1	7.0	3	17.7	HASKELL KS	-62	151	-5	45	71
				5	21.6	HODGEMAN KS					91
15	18.8	13	14.9	21	24.8	JACKSON KS	-20	66	32	75	96
11	12.9	15	18.8	10	9.0	JEFFERSON KS	46	-52	-30	19	11
15	22.8	2	2.4	10	15.5	JEWELL KS	-90	559	-32	17	56
50	14.0	126	19.1	145	13.5	JOHNSON KS	37	-29	-3	47	39
1	7.5	4	30.0			KEARNY KS	301				
11	18.9	13	17.0	9	12.1	KINGMAN KS	-10	-29	-36	14	28
2	7.7	10	29.9	10	25.6	KIOWA KS	290	-14	234	97	97
25	13.3	36	15.9	43	19.2	LABETTE KS	19	21	44	80	81
2	16.8	1	7.3	2	8.7	LANE KS	-57	19	-49	07	10
36	18.5	43	20.0	48	18.7	LEAVENWORTH KS	8	-6	1	52	78
7	17.3	10	22.8	12	26.3	LINCOLN KS	32	15	52	82	97
9	12.3	12	19.2	9	9.0	LINN KS	57	-53	-27	21	11
2	11.7	2	9.5	7	27.7	LOGAN KS	-18	191	137	93	98
34	20.0	28	13.8	41	16.8	LYON KS	-31	22	-16	33	65
16	10.6	30	17.2	33	13.9	MC PHERSON KS	62	-19	31	74	42
16	16.5	24	18.8	18	11.1	MARION KS	14	-41	-33	16	21
25	21.5	30	23.0	22	15.7	MARSHALL KS	7	-31	-27	22	57
2	8.2	6	16.6	7	13.5	MEADE KS	103	-19	64	85	38
18	12.6	21	13.5	39	22.9	MIAMI KS	7	70	83	88	94
11	15.1	9	10.3	15	17.4	MITCHELL KS	-32	70	16	65	69
33	11.6	54	15.5	65	18.0	MONTGOMERY KS	34	16	56	83	74
13	21.9	11	19.3	12	16.2	MORRIS KS	-12	-16	-26	22	60
2	17.5			2	14.3	MORTON KS			-18	30	46
20	22.2	18	16.4	22	20.5	NEMAHA KS	-26	25	-8	42	87
17	12.1	36	21.2	25	14.1	NEOSHO KS	75	-33	17	66	44
6	16.5	6	12.9	4	7.7	NESS KS	-22	-40	-53	05	08
11	18.5	9	13.2	18	22.0	NORTON KS	-29	67	19	67	92
17	16.4	20	17.3	21	16.7	OSAGE KS	5	-3	2	53	64
10	16.7	9	14.1	7	9.6	OSBORNE KS	-16	-32	-43	10	14
5	7.8	11	18.5	12	16.3	OTTAWA KS	137	-12	108	91	61
5	6.7	11	15.2	9	15.7	PAWNEE KS	128	3	135	93	57
9	16.7	9	15.4	9	12.4	PHILLIPS KS	-8	-20	-26	22	29
14	17.6	7	7.2	24	23.2	POTTAWATOMIE KS	-59	221	32	75	94
3	4.2	9	11.5	8	6.9	PRATT KS	176	-40	64	85	06
7	25.5	2	7.1	2	3.9	RAWLINS KS	-72	-45	-85	00	02
34	11.2	44	11.8	76	16.4	RENO KS	5	40	46	81	62
11	15.0	25*	26.8	17	18.3	REPUBLIC KS	78	-32	22	70	76
16	19.1	18	19.1	27	23.0	RICE KS	0	20	20	68	94
19	13.9	31	19.5	15	8.5	RILEY KS	41	-56	-38	12	10
3	6.4	8	15.2	12	17.9	ROOKS KS	138	18	181	95	73
15*	42.6	2	3.9	7	12.4	RUSH KS	-91	215	-71	01	29
13	23.3	10	14.0	15	18.9	RUSSELL KS	-40	35	-19	30	79
37	19.4	36	14.1	26	8.6	SALINE KS	-27	-39	-55	05	10
3	17.7	3	12.3	5	14.1	SCOTT KS	-31	15	-20	28	44
151	13.2	239	16.1	279	15.1	SEDGWICK KS	22	-6	14	64	52
5	11.0	9	14.8	7	7.7	SEWARD KS	35	-48	-30	19	08
101	15.8	124	15.6	155	16.7	SHAWNEE KS	-1	7	6	56	64
1	4.2	2	8.6	12*	49.8	SHERIDAN KS	107	480	1099	99	99
2	5.1	9	20.0	6	13.1	SHERMAN KS	293	-35	157	94	36
12	19.4	17	22.7	8	9.1	SMITH KS	17	-60	-53	05	12
7	13.7	4	8.4	11	18.8	STAFFORD KS	-39	124	37	77	78
		1	14.1			STANTON KS					
				4	15.1	STEVENS KS					53
29	17.0	33	16.0	38	16.8	SUMNER KS	-6	5	-1	50	65
2	6.9	12	26.7	8	14.3	THOMAS KS	285	-46	106	90	46
2	8.1	3	9.7	4	13.1	TREGO KS	20	34	62	84	35
4	8.5	8	12.9	4	5.0	WABAUNSEE KS	52	-62	-42	10	03
3	37.1	2	16.4	6*	36.0	WALLACE KS	-56	120	-3	48	99
16	18.4	9	9.8	18	18.1	WASHINGTON KS	-47	85	-2	49	74
3	33.2	4	30.9	1	5.0	WICHITA KS	-7	-84	-85	00	03
23	21.8	21	18.0	25	23.5	WILSON KS	-18	31	8	58	94

* SIGNIFICANTLY MORE DEATHS THAN EXPECTED

WHITE FEMALE | CANCER OF THE LARGE INTESTINE | ICD 153,159.0

1950-1959 DEATHS	1950-1959 RATE	1960-1969 DEATHS	1960-1969 RATE	1970-1979 DEATHS	1970-1979 RATE	COUNTY AND STATE	PERCENT CHANGE 60S/50S	PERCENT CHANGE 70S/60S	PERCENT CHANGE 70S/50S	PERCENTILE %CHANGE 70S/50S	PERCENTILE RATES 1970S
9	19.9	5	9.5	9	15.0	WOODSON KS	-52	58	-25	24	51
104	15.1	135	16.6	158	17.3	WYANDOTTE KS	10	4	15	64	68
2058	16.5	2468	15.8	3340*	17.1	KENTUCKY KY	-4	8	4		
6	8.6	19	22.5	16	17.0	ADAIR KY	161	-25	97	90	66
5	6.7	6	6.9	17	15.3	ALLEN KY	2	122	126	92	54
8	18.7	8	15.0	16	22.7	ANDERSON KY	-20	52	21	69	93
7	13.8	10	16.3	11	14.0	BALLARD KY	18	-14	1	52	43
28	21.8	19	12.0	41	19.7	BARREN KY	-45	64	-10	39	83
5	11.4	9	16.8	9	14.9	BATH KY	47	-11	31	74	51
23	17.8	26	15.5	33	16.6	BELL KY	-13	7	-6	43	63
10	15.1	20	19.2	33	21.7	BOONE KY	28	13	44	80	91
11	13.6	16	16.6	16	12.3	BOURBON KY	22	-26	-10	39	29
31	14.6	49	18.0	70	20.1	BOYD KY	24	11	38	77	85
17	16.8	18	16.1	30	21.6	BOYLE KY	-4	34	29	73	91
8	18.7	5	10.8	17*	31.1	BRACKEN KY	-42	188	67	86	98
5	8.3	5	7.7	7	9.1	BREATHITT KY	-7	19	10	60	12
13	16.0	12	13.3	15	13.7	BRECKINRIDGE KY	-17	3	-14	35	40
9	18.7	9	12.2	23	22.4	BULLITT KY	-35	84	20	68	92
7	16.7	14	22.8	21*	27.3	BUTLER KY	36	20	63	85	97
10	13.0	24	24.4	22	19.0	CALDWELL KY	87	-22	46	80	80
23	20.6	25	16.9	30	14.7	CALLOWAY KY	-18	-13	-29	20	49
98	22.6	131*	24.8	133*	22.5	CAMPBELL KY	10	-9	0	51	93
9	27.1	6	14.4	8	15.7	CARLISLE KY	-47	9	-42	10	57
10	20.7	13	25.1	10	17.3	CARROLL KY	21	-31	-17	32	68
12	14.6	16	16.3	19	15.5	CARTER KY	11	-5	6	56	55
8	14.7	10	13.0	11	12.8	CASEY KY	-12	-1	-13	36	33
26	15.2	28	13.2	50	19.7	CHRISTIAN KY	-13	49	29	74	83
16	18.0	15	12.6	26	18.7	CLARK KY	-30	49	4	54	78
3	5.8	9	12.3	14	15.3	CLAY KY	113	24	165	94	54
3	7.1	6	14.4	8	14.3	CLINTON KY	102	-1	100	90	45
9	13.7	12	19.9	8	11.8	CRITTENDEN KY	46	-41	-14	35	25
1	2.5	2	4.8	7	14.0	CUMBERLAND KY	93	192	465	99	43
45	16.8	49	13.6	82	17.7	DAVIESS KY	-19	30	5	56	71
		5	10.1	8	14.5	EDMONSON KY		43			48
3	12.2	4	13.3	3	19.2	ELLIOTT KY	9	44	57	83	81
4	6.3	8	11.1	14	14.4	ESTILL KY	75	30	127	92	46
62	13.1	86	13.9	159*	19.7	FAYETTE KY	6	42	51	82	84
12	19.1	5	7.7	10	12.3	FLEMING KY	-60	61	-36	14	29
25	21.1	17	10.3	23	10.4	FLOYD KY	-51	1	-51	06	17
20	15.8	32	19.3	33	15.8	FRANKLIN KY	22	-18	0	51	58
10	15.0	14	18.3	19	24.7	FULTON KY	22	35	65	85	96
3	16.7	2	8.1	6	23.6	GALLATIN KY	-51	190	41	79	95
4	8.4	5	9.9	6	10.0	GARRARD KY	17	1	19	67	16
13	25.6	15	23.9	10	13.3	GRANT KY	-7	-44	-48	07	37
18	10.3	27	11.6	34	12.4	GRAVES KY	13	7	21	69	30
18	20.6	12	11.8	16	11.8	GRAYSON KY	-43	0	-43	10	25
6	11.7	8	14.1	15	21.6	GREEN KY	21	53	85	88	91
17	18.9	19	14.5	18	9.8	GREENUP KY	-23	-33	-48	07	14
8	21.4	4	10.6	8	19.6	HANCOCK KY	-51	86	-8	41	83
30	23.4	23	12.7	40	16.7	HARDIN KY	-46	32	-29	20	64
12	7.5	13	6.9	24	11.2	HARLAN KY	-8	64	50	82	22
16	18.1	23	22.7	26	21.6	HARRISON KY	26	-5	19	68	91
7	10.5	10	13.4	13	14.2	HART KY	27	6	35	76	45
11	6.9	30	16.3	43	18.9	HENDERSON KY	136	16	175	95	79
8	11.4	6	8.3	12	14.2	HENRY KY	-28	71	24	71	45
12	29.5	11	23.5	15*	30.7	HICKMAN KY	-20	31	4	54	98
27	14.6	20	7.8	46	16.0	HOPKINS KY	-47	107	10	60	59
5	13.7	4	8.3	13	18.7	JACKSON KY	-40	127	36	77	78
480*	21.0	521	17.6	704*	19.3	JEFFERSON KY	-16	9	-8	41	81
9	16.6	12	18.4	14	16.0	JESSAMINE KY	11	-13	-4	47	59
8	9.8	12	12.9	16	12.7	JOHNSON KY	31	-1	29	74	32
109	18.9	161*	23.5	217*	26.2	KENTON KY	24	12	38	78	97
4	8.2	5	7.8	12	14.7	KNOTT KY	-5	88	79	87	49
12	11.9	20	15.9	15	9.5	KNOX KY	34	-40	-20	28	13
11	22.6	17*	29.1	4	4.9	LARUE KY	29	-83	-78	01	03
11	10.6	22	15.5	19	9.8	LAUREL KY	46	-37	-7	42	15
13	18.6	6	7.7	13	14.1	LAWRENCE KY	-58	83	-24	24	45
4	11.2	5	11.3	10	17.7	LEE KY	1	57	58	84	72
1	3.6	2	5.0	4	7.5	LESLIE KY	39	48	107	91	07
7	8.5	16	16.0	9	7.0	LETCHER KY	89	-56	-17	31	06
9	15.8	14	23.5	9	11.8	LEWIS KY	49	-50	-25	23	25
10	12.9	14	16.0	22	20.1	LINCOLN KY	25	26	56	83	86
6	15.0	9	17.7	13	20.7	LIVINGSTON KY	18	17	38	78	88
29*	26.9	15	11.9	35*	22.6	LOGAN KY	-56	89	-16	33	93
5	19.4	6	20.2	9	21.7	LYON KY	4	8	12	62	91
30	12.0	61	19.1	66	15.8	MC CRACKEN KY	59	-17	32	75	58
1	2.2	9	15.9	14	17.6	MC CREARY KY	610	11	688	99	70
4	7.9	6	10.0	16	22.2	MC LEAN KY	26	122	179	95	92
13	10.1	21	12.5	25	11.2	MADISON KY	24	-10	12	62	22
3	7.8	3	6.5			MAGOFFIN KY	-17				
22*	30.2	10	12.4	12	11.6	MARION KY	-59	-7	-62	03	24
16	22.5	19	20.3	24	15.3	MARSHALL KY	-10	-25	-32	17	54
7	22.7	5	13.8	8	15.9	MARTIN KY	-39	15	-30	18	58
16	18.3	18	18.0	16	13.5	MASON KY	-2	-25	-26	22	39
2	4.4	7	14.0	5	8.7	MEADE KY	214	-38	96	89	10
1	5.3	2	8.1	4	17.7	MENIFEE KY	53	119	236	97	72
21	26.0	16	17.1	17	15.3	MERCER KY	-34	-10	-41	10	54
5	15.9	11	25.8	6	9.3	METCALFE KY	63	-64	-41	11	13
9	17.1	12	18.7	11	13.8	MONROE KY	9	-26	-20	29	41
17	26.4	10	13.0	26*	25.3	MONTGOMERY KY	-51	94	-4	47	96
3	5.4	9	15.5	12	16.8	MORGAN KY	186	8	209	96	65
22	16.4	24	13.8	37	17.6	MUHLENBERG KY	-16	27	7	58	70
20	24.9	16	15.3	18	13.3	NELSON KY	-39	-13	-47	08	37
4	8.9	6	12.0	9	19.4	NICHOLAS KY	36	61	118	92	82
16	14.4	32*	24.8	35*	22.6	OHIO KY	72	-9	57	83	93
3	6.0	10	16.7	11	12.6	OLDHAM KY	177	-24	109	91	32
7	13.9	6	10.8	14	22.3	OWEN KY	-22	107	60	84	92
		3	7.8	2	6.5	OWSLEY KY		-17			06
16	28.5	13	21.1	13	20.0	PENDLETON KY	-26	-5	-30	19	85

* SIGNIFICANTLY MORE DEATHS THAN EXPECTED

WHITE FEMALE

CANCER OF THE LARGE INTESTINE

ICD 153,159.0

1950-1959 DEATHS	1950-1959 RATE	1960-1969 DEATHS	1960-1969 RATE	1970-1979 DEATHS	1970-1979 RATE	COUNTY AND STATE	PERCENT CHANGE 60S/50S	PERCENT CHANGE 70S/60S	PERCENT CHANGE 70S/50S	PERCENTILE %CHANGE 70S/50S	PERCENTILE RATES 1970S
6	6.6	10	8.2	12	7.8	PERRY KY	24	-5	18	67	08
15	8.5	33	12.4	37	11.2	PIKE KY	46	-10	31	75	22
6	22.0	8	24.1	8	17.5	POWELL KY	10	-27	-20	28	70
23	14.9	16	7.4	34	12.0	PULASKI KY	-50	62	-20	29	27
5	27.6	3	15.3	6	31.6	ROBERTSON KY	-44	106	14	64	98
6	10.6	7	11.4	14	17.4	ROCKCASTLE KY	8	52	64	85	69
8	17.6	5	8.0	12	15.0	ROWAN KY	-54	86	-15	34	51
7	16.2	6	10.1	9	9.4	RUSSELL KY	-38	-7	-42	10	13
12	16.9	15	18.5	19	18.3	SCOTT KY	9	-1	8	58	75
14	16.0	13	13.1	16	12.3	SHELBY KY	-18	-6	-23	25	29
12	18.8	12	16.7	19	19.0	SIMPSON KY	-11	13	1	52	80
7	29.8	3	11.2	5	17.7	SPENCER KY	-62	57	-41	11	71
10	14.8	12	13.4	21	17.2	TAYLOR KY	-9	28	16	65	68
8	13.7	11	18.2	13	18.0	TODD KY	33	-1	32	75	74
5	12.4	13	26.5	11	18.2	TRIGG KY	113	-31	46	81	75
2	8.2	4	13.5	5	15.4	TRIMBLE KY	64	14	87	89	55
12	16.9	9	10.6	24*	24.3	UNION KY	-37	130	44	79	95
35	17.0	45	18.0	51	15.7	WARREN KY	6	-13	-7	42	57
4	8.0	9	16.4	6	8.9	WASHINGTON KY	106	-46	12	62	11
4	6.6	7	8.9	13	14.3	WAYNE KY	34	61	115	92	46
20	21.9	15	13.1	22	17.5	WEBSTER KY	-40	34	-20	29	70
16	12.5	19	12.3	20	10.4	WHITLEY KY	-2	-16	-17	32	17
6	21.2	3	10.3	5	14.0	WOLFE KY	-51	36	-34	15	43
12	23.8	10	16.3	15	17.0	WOODFORD KY	-32	4	-29	20	66
1112	14.1	1588	14.4	2085	13.9	LOUISIANA LA	2	-3	-1		
20	14.2	24	12.4	33	13.0	ACADIA LA	-12	4	-9	41	34
5	9.4	12	18.7	14	13.1	ALLEN LA	98	-30	39	78	36
9	14.4	16	16.7	24	17.6	ASCENSION LA	16	5	22	70	70
3	6.1	8	13.9	10	14.8	ASSUMPTION LA	129	6	143	93	49
20	17.5	22	13.8	42	18.3	AVOYELLES LA	-21	32	5	55	75
11	17.8	15	16.7	11	9.3	BEAUREGARD LA	-6	-45	-48	07	12
3	6.0	8	11.2	16	18.1	BIENVILLE LA	86	61	200	96	74
4	7.1	15	12.4	30	15.2	BOSSIER LA	76	22	114	91	53
84	15.4	119	14.9	153	14.3	CADDO LA	-3	-4	-7	42	46
40	17.3	48	12.0	80	13.9	CALCASIEU LA	-31	16	-20	29	42
8	23.8	15*	30.5	4	6.1	CALDWELL LA	28	-80	-74	01	05
2	8.2	2	6.4			CAMERON LA	-22				
4	18.7	5	12.8	8	16.5	CATAHOULA LA	-32	29	-12	38	63
10	16.0	10	12.4	18	19.1	CLAIBORNE LA	-23	54	19	68	80
1	5.1	6	18.1	14	23.6	CONCORDIA LA	255	30	362	98	94
9	14.1	12	15.0	14	13.4	DE SOTO LA	7	-11	-5	45	38
45	12.4	96	15.4	140	15.0	EAST BATON ROUGE LA	25	-2	22	69	52
2	11.1	3	12.6	6	22.4	EAST CARROLL LA	13	78	102	90	92
10	14.4	5	6.4	5	7.9	EAST FELICIANA LA	-56	24	-45	09	08
15	15.8	23	19.4	19	11.3	EVANGELINE LA	23	-42	-28	20	22
8	12.1	12	12.9	16	14.1	FRANKLIN LA	7	9	16	65	44
5	9.1	9	13.0	17	20.2	GRANT LA	44	55	123	92	86
19	18.2	15	8.6	17	7.9	IBERIA LA	-53	-9	-57	04	08
9	16.4	8	10.5	12	12.8	IBERVILLE LA	-36	22	-22	27	33
6	10.9	8	11.4	6	6.5	JACKSON LA	5	-43	-40	11	06
57	16.2	99	14.0	185	13.3	JEFFERSON LA	-14	-5	-18	31	37
17	20.5	17	15.8	23	16.1	JEFFERSON DAVIS LA	-23	2	-22	27	59
25	13.6	38	14.7	35	8.4	LAFAYETTE LA	8	-43	-38	12	09
21	15.7	8	4.5	29	11.0	LAFOURCHE LA	-71	143	-30	19	20
8	16.0	4	5.8	9	8.2	LA SALLE LA	-64	41	-49	07	09
11	13.6	13	11.9	18	12.8	LINCOLN LA	-13	8	-6	43	33
7	10.7	10	11.0	19	11.4	LIVINGSTON LA	3	3	6	57	22
2	9.3	3	11.6	4	9.8	MADISON LA	25	-16	6	56	15
6	11.0	15	17.2	15	10.5	MOREHOUSE LA	56	-39	-4	46	18
9	9.0	16	13.4	22	14.9	NATCHITOCHES LA	49	12	66	86	51
345	16.3	426	17.2	423	16.7	ORLEANS LA	6	-3	3	53	64
23	9.7	64	18.5	71	14.4	OUACHITA LA	91	-22	49	81	47
8	25.8	5	14.8	7	14.3	PLAQUEMINES LA	-43	-3	-45	09	46
5	11.7	9	15.7	9	12.3	POINTE COUPEE LA	35	-22	5	56	29
25	8.7	53	13.3	58	11.1	RAPIDES LA	52	-16	27	73	21
3	9.9	7	20.2	7	13.2	RED RIVER LA	104	-35	33	75	36
7	12.2	11	14.4	13	12.6	RICHLAND LA	18	-13	4	54	32
8	10.6	6	6.4	19	13.5	SABINE LA	-40	113	28	73	39
1	2.2	21	21.8	40	20.1	ST BERNARD LA	897	-8	821	99	86
1	2.1	6	12.9	7	7.7	ST CHARLES LA	516	-41	266	97	08
1	7.4	4	16.1	1	4.1	ST HELENA LA	116	-74	-44	09	03
6	15.9	10	22.1	9	16.5	ST JAMES LA	39	-25	4	55	63
7	20.3	8	20.4	15	24.6	ST JOHN THE BAPT LA	1	20	21	69	96
11	7.4	24	11.2	35	12.1	ST LANDRY LA	52	7	63	85	27
7	11.2	11	12.8	11	9.9	ST MARTIN LA	14	-23	-12	38	15
12	12.7	21	16.9	23	13.2	ST MARY LA	33	-22	4	55	36
12	11.3	23	13.7	57	18.9	ST TAMMANY LA	22	37	68	86	79
29	19.5	34	17.2	40	13.7	TANGIPAHOA LA	-12	-20	-29	19	41
3	19.2	2	9.5	1	3.9	TENSAS LA	-50	-59	-80	00	03
8	7.4	13	8.6	30	12.6	TERREBONNE LA	17	46	72	87	32
7	11.9	11	13.5	11	9.5	UNION LA	14	-30	-20	29	13
14	11.0	14	8.1	27	11.0	VERMILION LA	-26	37	1	52	21
13	17.3	12	11.9	13	8.9	VERNON LA	-31	-26	-49	07	11
14	17.0	18	12.0	24	12.1	WASHINGTON LA	-29	0	-29	20	27
6	6.5	13	8.8	35	16.2	WEBSTER LA	37	84	151	94	60
2	9.6	4	13.3	3	8.5	WEST BATON ROUGE LA	38	-36	-12	38	09
8	16.6	11	18.3	10	13.2	WEST CARROLL LA	10	-28	-21	28	36
1	9.9					WEST FELICIANA LA					
10	17.6	8	11.5	18	19.9	WINN LA	-35	73	13	63	84
1234*	23.7	1202*	20.0	1354*	18.7	MAINE ME	-16	-6	-21		
145*	30.8	101	18.2	113	16.6	ANDROSCOGGIN ME	-41	-9	-46	08	63
64	21.5	82	21.3	66	13.9	AROOSTOOK ME	-1	-35	-35	14	42
226*	21.2	239	19.4	291*	20.2	CUMBERLAND ME	-8	4	-5	45	86
33*	29.5	28	21.4	25	15.4	FRANKLIN ME	-28	-28	-48	07	55
46	21.7	53	21.4	66	19.9	HANCOCK ME	-1	-7	-8	41	84
127*	25.5	110	18.9	109	15.5	KENNEBEC ME	-26	-18	-39	12	56
46	19.0	55	22.6	46	15.8	KNOX ME	19	-30	-17	32	58

* SIGNIFICANTLY MORE DEATHS THAN EXPECTED

WHITE FEMALE

CANCER OF THE LARGE INTESTINE

ICD 153,159.0

1950-1959 DEATHS	1950-1959 RATE	1960-1969 DEATHS	1960-1969 RATE	1970-1979 DEATHS	1970-1979 RATE	COUNTY AND STATE	PERCENT CHANGE 60S/50S	PERCENT CHANGE 70S/60S	PERCENT CHANGE 70S/50S	PERCENTILE %CHANGE 70S/50S	PERCENTILE RATES 1970S
45*	34.6	41*	27.7	27	14.3	LINCOLN ME	-20	-48	-59	04	46
62*	25.8	48	17.5	73*	21.6	OXFORD ME	-32	23	-16	32	91
113	19.6	125	19.0	159*	21.0	PENOBSCOT ME	-3	10	7	58	89
20	18.2	28	24.8	27	21.2	PISCATAQUIS ME	36	-15	17	66	90
27	19.7	20	14.0	35	22.6	SAGADAHOC ME	-29	61	15	64	93
45	21.0	27	11.4	50	17.7	SOMERSET ME	-46	55	-15	34	72
36	24.4	33	22.3	34	19.2	WALDO ME	-8	-14	-21	27	81
53*	26.3	53*	23.5	61*	23.6	WASHINGTON ME	-11	1	-10	39	95
146*	26.5	159*	23.5	172*	19.7	YORK ME	-11	-16	-26	23	83
1928*	19.4	2582*	18.8	3436*	18.8	MARYLAND MD	-3	0	-3		
71	16.7	94	18.2	115	17.9	ALLEGANY MD	9	-1	7	58	73
80	20.3	106	15.4	201	17.6	ANNE ARUNDEL MD	-24	14	-13	36	71
263	20.9	405	18.2	704*	20.4	BALTIMORE MD	-13	12	-2	49	87
8	20.5	10	15.0	16	16.0	CALVERT MD	-27	6	-22	26	59
13	13.2	17	15.5	28	22.7	CAROLINE MD	18	47	72	87	93
42	15.3	63	18.2	91	19.9	CARROLL MD	19	10	30	74	85
33	22.7	34	18.4	43	17.3	CECIL MD	-19	-6	-24	25	69
6	10.5	24*	29.7	16	10.4	CHARLES MD	184	-65	0	51	18
25	20.0	21	12.6	37	20.5	DORCHESTER MD	-37	63	2	53	87
57	18.2	85*	22.9	95	19.1	FREDERICK MD	26	-17	5	56	80
9	10.1	14	12.2	25	15.5	GARRETT MD	21	27	54	83	56
31	17.3	67*	24.1	78	17.4	HAPFORD MD	39	-28	0	51	69
12	12.3	35	22.5	48	18.0	HOWARD MD	83	-20	46	81	73
11	17.1	11	13.8	14	14.2	KENT MD	-19	3	-17	32	45
150	18.0	314	18.8	491*	18.7	MONTGOMERY MD	5	-1	4	54	78
98	14.4	222	17.4	333	18.1	PRINCE GEORGE'S MD	21	4	26	72	74
2	2.4	8	9.1	21	20.3	QUEEN ANNE'S MD	277	122	737	99	86
11	21.1	19	19.7	28	17.7	ST MARY'S MD	-6	-10	-16	33	71
14	16.4	17	17.7	26	22.9	SOMERSET MD	8	29	39	78	94
26	25.5	33*	26.8	34	18.2	TALBOT MD	5	-32	-28	20	75
75	18.0	104	19.6	147*	20.9	WASHINGTON MD	9	6	16	65	89
29	14.4	26	10.3	60	17.4	WICOMICO MD	-28	69	21	69	69
10	9.7	18	13.9	24	15.3	WORCESTER MD	43	10	58	84	54
852*	22.4	835*	20.6	761*	18.8	BALTIMORE CITY MD	-8	-9	-16	33	79
6292*	21.9	6853*	20.1	7498*	18.6	MASSACHUSETTS MA	-9	-7	-15		
107*	28.9	112	19.2	222	18.9	BARNSTABLE MA	-33	-2	-34	15	79
190*	23.4	210*	22.1	215	18.7	BERKSHIRE MA	-6	-15	-20	28	78
543*	23.4	632*	22.7	656*	19.2	BRISTOL MA	-3	-15	-18	31	81
9	21.8	5	9.7	14	19.0	DUKES MA	-56	96	-13	37	80
772*	22.3	865*	21.0	908*	18.6	ESSEX MA	-6	-11	-16	32	77
72	20.3	82	19.8	96	20.7	FRANKLIN MA	-2	5	2	53	88
454*	20.9	518*	19.5	632*	19.8	HAMPDEN MA	-7	2	-5	45	84
113	21.1	126	19.4	144	19.6	HAMPSHIRE MA	-8	1	-7	43	83
1582*	24.2	1560*	19.8	1719*	18.7	MIDDLESEX MA	-18	-6	-23	26	77
5	18.2	4	12.0	10	22.9	NANTUCKET MA	-34	92	26	72	94
525*	20.1	662*	19.6	767*	18.0	NORFOLK MA	-2	-8	-10	39	74
284*	21.7	315	19.1	428*	19.7	PLYMOUTH MA	-12	3	-10	40	83
1037*	21.4	1069*	21.4	922*	18.7	SUFFOLK MA	0	-13	-13	37	77
599	18.1	693	17.4	765	16.6	WORCESTER MA	-4	-5	-8	41	63
5406*	19.4	6329*	17.7	7085	16.0	MICHIGAN MI	-9	-9	-17		
5	18.1	5	10.5	12	16.9	ALCONA MI	-42	61	-7	43	65
4	10.0	3	6.7	11	20.5	ALGER MI	-34	208	105	90	87
56	21.2	56	18.0	56	14.3	ALLEGAN MI	-15	-21	-33	16	46
22	21.7	26	19.6	30	17.4	ALPENA MI	-10	-11	-20	29	69
18*	31.6	8	11.1	18	16.6	ANTRIM MI	-65	50	-48	07	63
8	15.5	8	14.0	8	9.6	ARENAC MI	-10	-31	-38	12	14
4	14.8	6	17.0	7	13.4	BARAGA MI	14	-21	-9	40	38
34	21.6	23	11.8	27	10.8	BARRY MI	-46	-8	-50	06	19
76	18.2	103	19.8	76	12.1	BAY MI	8	-39	-34	16	27
7	17.3	6	10.6	16	21.1	BENZIE MI	-39	99	22	70	90
115	20.3	131	17.8	157	17.2	BERRIEN MI	-12	-4	-15	34	67
40	21.5	36	17.6	36	14.7	BRANCH MI	-18	-17	-32	17	48
103	17.8	108	15.6	133	16.4	CALHOUN MI	-13	5	-8	42	62
37	23.6	30	15.8	29	12.4	CASS MI	-33	-22	-47	08	30
20	26.4	12	12.7	17	16.3	CHARLEVOIX MI	-52	29	-38	12	61
15	21.2	17	18.5	12	9.1	CHEBOYGAN MI	-13	-51	-57	04	12
18	15.6	19	14.6	31	19.6	CHIPPEWA MI	-6	34	26	72	83
14	26.4	10	13.1	19	12.7	CLARE MI	-51	-3	-52	06	32
26	17.0	31	17.3	36	16.4	CLINTON MI	2	-5	-3	48	62
3	12.8	7	22.2	14*	31.9	CRAWFORD MI	73	44	148	94	98
23	14.7	39	19.6	39	15.8	DELTA MI	33	-19	8	59	58
29	24.5	28	17.3	44	21.5	DICKINSON MI	-30	24	-12	37	90
36	15.4	54	20.4	65	18.8	EATON MI	33	-8	22	70	79
14	16.4	20	19.1	20	15.1	EMMET MI	16	-21	-8	41	52
191	18.2	234	15.9	312	17.2	GENESEE MI	-13	8	-5	45	67
7	13.6	7	9.6	13	12.9	GLADWIN MI	-30	34	-5	45	34
26	21.7	21	13.8	34	18.7	GOGEBIC MI	-37	36	-14	36	78
17	9.0	35	15.6	38	13.3	GRAND TRAVERSE MI	73	-15	47	81	37
41	21.6	37	17.8	40	16.9	GRATIOT MI	-18	-5	-22	27	65
30	15.9	36	17.5	40	15.5	HILLSDALE MI	10	-11	-3	49	56
50	22.9	40	15.9	44	17.6	HOUGHTON MI	-31	11	-23	25	71
38	23.2	34	16.7	34	12.2	HURON MI	-28	-27	-48	07	28
126	16.0	180	18.8	196	17.2	INGHAM MI	17	-9	7	58	67
47	22.0	43	18.7	52	18.2	IONIA MI	-15	-3	-17	31	74
12	20.4	20	21.1	27	18.7	IOSCO MI	3	-11	-9	41	78
15	17.3	15	14.5	18	13.0	IRON MI	-16	-10	-25	24	35
26	20.1	27	18.2	21	11.2	ISABELLA MI	-9	-38	-44	09	21
92	16.4	139	20.4	143	18.2	JACKSON MI	24	-11	11	61	75
114	16.6	165	19.6	156	15.2	KALAMAZOO MI	18	-23	-9	41	53
5	22.0	3	10.5	6	12.4	KALKASKA MI	-52	18	-44	09	30
314	20.1	346	17.5	337	14.1	KENT MI	-13	-19	-30	19	45
1	4.8			1	5.2	KEWEENAW MI			8	58	03
8	37.0	3	9.5	5	12.9	LAKE MI	-74	36	-65	02	33
39	22.5	37	17.7	49	18.7	LAPEER MI	-21	5	-17	32	78
7	16.7	11	17.5	14	17.9	LEELANAU MI	5	2	8	58	73

* SIGNIFICANTLY MORE DEATHS THAN EXPECTED

WHITE FEMALE

CANCER OF THE LARGE INTESTINE

ICD 153,159.0

1950-1959 DEATHS	1950-1959 RATE	1960-1969 DEATHS	1960-1969 RATE	1970-1979 DEATHS	1970-1979 RATE	COUNTY AND STATE	PERCENT CHANGE 60S/50S	PERCENT CHANGE 70S/60S	PERCENT CHANGE 70S/50S	PERCENTILE %CHANGE 70S/50S	PERCENTILE RATES 1970S
68	19.0	76	18.2	87	17.1	LENAWEE MI	-4	-6	-10	39	67
43*	28.2	43	20.3	57	16.7	LIVINGSTON MI	-28	-18	-41	11	64
6	14.3	2	4.3	7	14.8	LUCE MI	-70	241	3	54	49
12	28.2	11	20.7	8	11.3	MACKINAC MI	-26	-46	-60	03	22
145	19.3	248	18.0	441	17.3	MACOMB MI	-7	-4	-11	39	68
17	15.5	17	13.6	21	12.0	MANISTEE MI	-12	-12	-23	26	27
39	16.7	45	16.5	40	12.2	MARQUETTE MI	-2	-26	-27	21	28
23	18.1	20	13.4	27	14.4	MASON MI	-26	7	-21	28	46
23	24.0	24	20.1	20	14.0	MECOSTA MI	-16	-30	-42	10	43
18	14.5	29	19.6	34	18.3	MENOMINEE MI	35	-7	26	72	75
25	19.3	40	21.6	47	17.4	MIDLAND MI	12	-20	-10	39	69
6	19.7	7	20.2	4	9.0	MISSAUKEE MI	3	-55	-54	05	11
57	17.8	72	16.5	78	13.3	MONROE MI	-7	-20	-25	23	37
37	19.8	33	14.7	44	15.7	MONTCALM MI	-26	7	-21	28	57
4	19.6	9	34.9	7	12.5	MONTMORENCY MI	78	-64	-36	14	31
102	20.6	105	16.2	155	18.7	MUSKEGON MI	-21	15	-9	40	78
33*	29.9	22	15.9	41	21.4	NEWAYGO MI	-47	34	-29	20	90
326	19.0	476	16.5	686	15.3	OAKLAND MI	-13	-7	-19	29	54
13	13.8	11	11.0	11	9.2	OCEANA MI	-20	-17	-34	15	12
3	6.6	8	11.8	22	20.2	OGEMAW MI	80	71	208	96	86
15*	35.8	12	23.5	8	13.1	ONTONAGON MI	-34	-44	-63	03	35
17	22.6	23*	26.6	20	17.9	OSCEOLA MI	18	-33	-21	27	73
3	19.9			7	15.1	OSCODA MI			-24	24	52
2	6.5	5	11.1	2	3.0	OTSEGO MI	70	-73	-53	05	02
80	22.1	97	19.9	95	13.7	OTTAWA MI	-10	-31	-38	12	41
11	20.5	9	13.5	23*	24.4	PRESQUE ISLE MI	-34	81	19	67	96
6	18.2	10	20.7	21	19.5	ROSCOMMON MI	14	-6	7	58	82
105	15.1	146	16.9	146	14.2	SAGINAW MI	12	-16	-6	44	45
106*	22.7	111	18.9	121	16.7	ST CLAIR MI	-17	-11	-26	22	64
29	13.5	42	16.2	55	17.5	ST JOSEPH MI	21	8	30	74	70
26	14.5	46	22.2	45	17.7	SANILAC MI	53	-20	22	70	71
11	26.0	5	9.7	10	14.9	SCHOOLCRAFT MI	-63	54	-43	10	51
52	20.5	50	17.7	54	16.2	SHIAWASSEE MI	-14	-9	-21	27	60
42	19.7	51	21.7	51	19.0	TUSCOLA MI	10	-12	-3	48	80
49	19.8	52	17.3	53	14.8	VAN BUREN MI	-13	-15	-25	23	49
111	18.8	124	17.2	138	15.4	WASHTENAW MI	-9	-10	-18	30	54
1787*	20.3	2005*	18.5	1879	16.7	WAYNE MI	-9	-10	-18	31	64
21	21.7	24	20.3	27	17.7	WEXFORD MI	-7	-13	-18	30	72
2578	16.5	3154	15.8	3686	14.9	MINNESOTA MN	-4	-6	-9		
15	22.4	12	15.9	11	10.7	AITKIN MN	-29	-33	-52	05	19
25	14.4	40	15.9	67	14.4	ANOKA MN	10	-9	0	51	47
10	9.8	23	16.7	21	11.4	BECKER MN	70	-31	17	66	23
11	11.9	21	17.0	21	13.4	BELTRAMI MN	43	-21	13	62	38
9	14.6	13	15.7	19	16.6	BENTON MN	7	6	13	63	63
7	14.5	6	9.9	9	12.2	BIG STONE MN	-32	23	-16	33	28
38	16.1	41	15.2	43	13.7	BLUE EARTH MN	-6	-10	-15	34	40
20	13.5	27	14.1	44	19.7	BROWN MN	5	40	47	81	83
23	18.9	22	14.9	22	11.7	CARLTON MN	-21	-22	-38	12	24
16	17.0	14	11.4	19	12.0	CARVER MN	-33	6	-29	19	27
8	9.1	23	19.9	24	15.7	CASS MN	119	-21	73	87	57
6	6.8	14	13.9	16	11.4	CHIPPEWA MN	106	-19	68	86	22
11	12.9	19	19.6	34*	25.6	CHISAGO MN	52	31	99	90	97
12	8.9	27	14.7	40	15.9	CLAY MN	65	8	79	87	59
4	10.2	8	16.0	6	10.4	CLEARWATER MN	56	-35	2	52	18
2	15.6	2	15.4	3	13.0	COOK MN	-1	-16	-17	32	34
17	22.0	20	18.6	18	14.8	COTTONWOOD MN	-16	-21	-33	16	49
26	16.1	41	20.5	45	15.8	CROW WING MN	27	-23	-2	49	58
35	14.9	44	12.6	88	15.3	DAKOTA MN	-16	21	2	53	53
14	19.3	21	27.0	14	14.9	DODGE MN	40	-45	-23	26	50
15	12.9	20	12.9	24	11.0	DOUGLAS MN	0	-15	-15	34	20
25	19.1	35	22.9	31	17.6	FARIBAULT MN	20	-23	-8	42	71
26	17.6	19	11.8	35	19.6	FILLMORE MN	-33	66	12	62	83
23	13.6	25	11.3	24	8.8	FREEBORN MN	-17	-22	-35	14	11
38	19.1	35	14.3	44	14.6	GOODHUE MN	-25	2	-24	25	48
3	6.6	10	18.0	11	16.0	GRANT MN	174	-11	143	93	59
701	18.1	826	16.5	920	15.5	HENNEPIN MN	-9	-6	-14	35	56
13	16.4	17	18.7	18	14.0	HOUSTON MN	14	-25	-14	35	43
8	16.8	11	16.6	11	11.6	HUBBARD MN	-1	-30	-31	18	24
15	22.2	17	20.5	17	14.5	ISANTI MN	-8	-29	-35	15	47
23	19.0	24	14.0	32	13.9	ITASCA MN	-26	-1	-27	21	42
11	14.9	14	13.9	26	22.9	JACKSON MN	-7	65	53	82	94
10	24.4	6	9.7	11	16.4	KANABEC MN	-60	68	-33	16	62
26	17.0	31	15.3	29	12.6	KANDIYOHI MN	-10	-18	-26	22	31
9	20.9	8	14.6	6	12.6	KITTSON MN	-30	-13	-39	11	32
3	4.1	10	13.0	22	24.3	KOOCHICHING MN	218	87	493	99	96
8	11.1	15	16.9	18	17.2	LAC QUI PARLE MN	53	1	55	83	68
12	30.3	8	15.1	13	19.7	LAKE MN	-50	30	-35	14	83
1	4.5	2	10.4	3	12.9	LAKE OF THE WOODS MN	130	24	184	95	34
21	19.0	25	19.2	27	17.0	LE SUEUR MN	1	-11	-11	39	66
6	12.4	8	12.7	6	4.8	LINCOLN MN	2	-62	-61	03	03
11	11.1	24	18.6	23	14.2	LYON MN	68	-24	28	73	45
22	16.6	33	20.0	32	17.1	MC LEOD MN	20	-14	3	54	67
2	7.1	3	11.9	9	28.1	MAHNOMEN MN	67	135	293	98	98
8	15.4	11	14.7	9	9.8	MARSHALL MN	-5	-33	-36	13	15
19	16.7	22	13.5	40	19.2	MARTIN MN	-19	42	15	65	81
12	12.7	12	10.0	27	17.3	MEEKER MN	-21	73	36	77	68
16	19.5	12	13.3	28	24.0	MILLE LACS MN	-32	80	23	70	95
21	18.3	27	16.8	24	12.3	MORRISON MN	-8	-27	-33	16	29
33	16.3	38	14.9	42	13.5	MOWER MN	-9	-9	-17	31	39
8	14.0	13	17.2	19	20.2	MURRAY MN	23	17	44	80	86
17	14.1	21	15.6	17	12.7	NICOLLET MN	11	-18	-10	40	32
17	16.5	22	16.0	22	13.2	NOBLES MN	-3	-18	-20	29	36
12	17.9	17	19.4	15	16.1	NORMAN MN	8	-17	-10	40	60
44	15.2	41	10.5	63	12.8	OLMSTED MN	-31	22	-16	33	33
46	16.3	42	12.7	67	15.2	OTTER TAIL MN	-23	20	-7	43	53
7	11.5	11	13.4	13	14.6	PENNINGTON MN	16	9	27	72	48
15	17.1	22	20.2	14	12.0	PINE MN	18	-41	-30	19	27
7	10.1	11	12.5	15	14.5	PIPESTONE MN	24	16	43	79	47
31	17.4	26	12.2	32	12.1	POLK MN	-30	0	-30	18	27

* SIGNIFICANTLY MORE DEATHS THAN EXPECTED

WHITE FEMALE

CANCER OF THE LARGE INTESTINE

ICD 153,159.0

COUNTY AND STATE	1950-1959 DEATHS	1950-1959 RATE	1960-1969 DEATHS	1960-1969 RATE	1970-1979 DEATHS	1970-1979 RATE	PERCENT CHANGE 60S/50S	PERCENT CHANGE 70S/60S	PERCENT CHANGE 70S/50S	PERCENTILE %CHANGE 70S/50S	PERCENTILE RATES 1970S
POPE MN	5	7.9	11	12.5	9	5.9	58	-53	-26	22	04
RAMSEY MN	343	17.2	411	16.2	453	14.9	-6	-8	-13	36	50
RED LAKE MN	6	21.2	4	12.5	8	25.4	-41	104	20	68	96
REDWOOD MN	19	17.9	27	20.3	17	9.1	13	-55	-49	06	12
RENVILLE MN	16	13.7	28	18.1	28	15.4	32	-15	12	62	55
RICE MN	24	12.6	34	14.7	52	19.1	16	30	51	82	80
ROCK MN	13	27.1	9	14.2	14	14.9	-47	4	-45	09	50
ROSEAU MN	9	16.8	9	13.9	10	10.5	-17	-25	-38	12	18
ST LOUIS MN	143	13.2	210	15.9	211	13.7	20	-14	3	54	40
SCOTT MN	18	20.7	26	23.6	25	15.7	14	-33	-24	24	57
SHERBURNE MN	4	7.8	13	16.6	18	15.0	113	-10	92	89	52
SIBLEY MN	11	13.0	15	14.9	28*	22.7	15	53	75	87	93
STEARNS MN	65	20.8	65	16.7	74	14.8	-19	-12	-29	20	50
STEELE MN	17	13.9	19	13.3	21	10.2	-4	-23	-26	22	17
STEVENS MN	3	6.3	12	21.8	11	12.3	247	-44	95	89	29
SWIFT MN	13	17.3	11	11.6	23	21.1	-33	82	22	70	89
TODD MN	27	22.1	28	18.7	25	14.2	-15	-24	-35	14	45
TRAVERSE MN	9	23.5	2	5.7	11	19.7	-76	247	-16	33	83
WABASHA MN	16	15.8	24	19.2	25	17.1	22	-11	8	59	66
WADENA MN	13	23.3	11	15.1	18	17.4	-35	15	-26	23	69
WASECA MN	16	17.1	22	19.5	21	15.1	14	-22	-12	38	52
WASHINGTON MN	38	21.3	39	16.8	54	15.7	-21	-7	-26	22	57
WATONWAN MN	9	11.2	18	18.9	17	15.0	68	-21	34	76	51
WILKIN MN	10	23.1	8	14.9	14	18.0	-36	21	-22	26	74
WINONA MN	43	19.5	40	15.5	48	14.4	-21	-7	-26	22	46
WRIGHT MN	27	17.1	33	17.2	34	12.0	0	-30	-30	19	27
YELLOW MEDICINE MN	7	8.9	13	13.1	14	12.2	47	-6	38	77	28

COUNTY AND STATE	1950-1959 DEATHS	1950-1959 RATE	1960-1969 DEATHS	1960-1969 RATE	1970-1979 DEATHS	1970-1979 RATE	PERCENT CHANGE 60S/50S	PERCENT CHANGE 70S/60S	PERCENT CHANGE 70S/50S	PERCENTILE %CHANGE 70S/50S	PERCENTILE RATES 1970S
MISSISSIPPI MS	701	13.2	962	13.5	1325	13.9	2	3	5		
ADAMS MS	9	15.6	14	18.5	19	16.4	19	-12	5	56	61
ALCORN MS	16	15.1	22	15.9	30	16.6	5	4	10	60	63
AMITE MS	8	19.0	5	11.2	11	22.6	-41	101	19	68	93
ATTALA MS	10	14.5	12	13.5	15	13.5	-7	-1	-7	42	38
BENTON MS	5	22.6	8	27.5	5	12.5	22	-54	-44	09	31
BOLIVAR MS	9	12.4	9	9.8	17	14.5	-21	49	17	66	47
CALHOUN MS	7	9.5	4	5.3	10	10.0	-44	89	6	56	16
CARROLL MS	4	12.4	6	17.2	6	16.8	39	-2	35	77	65
CHICKASAW MS	10	18.0	11	15.6	12	14.1	-14	-10	-22	27	44
CHOCTAW MS	4	10.4	8	18.1	6	13.7	74	-24	31	74	40
CLAIBORNE MS	2	12.5	2	11.2	1	4.8	-11	-57	-61	03	03
CLARKE MS	4	7.3	6	9.1	7	9.8	25	8	35	76	14
CLAY MS	8	19.7	5	8.7	18	20.7	-56	138	5	56	88
COAHOMA MS	12	18.8	8	9.2	13	13.3	-51	44	-29	19	37
COPIAH MS	10	12.6	18	19.4	14	14.1	54	-27	12	62	44
COVINGTON MS	5	11.9	7	12.4	8	10.8	4	-13	-9	40	19
DE SOTO MS	5	11.6	7	12.6	14	13.5	8	7	16	66	38
FORREST MS	15	10.4	29	13.8	36	12.3	32	-11	18	67	29
FRANKLIN MS	4	14.4	5	13.9	3	8.6	-4	-38	-40	11	10
GEORGE MS	6	19.1	7	18.3	9	14.6	-4	-20	-23	25	48
GREENE MS	3	11.4	5	22.7	11	28.4	99	25	149	94	98
GRENADA MS	6	16.4	5	6.5	10	12.2	-60	87	-26	22	28
HANCOCK MS	2	4.5	14	20.8	13	12.7	362	-39	183	95	32
HARRISON MS	37	15.5	55	15.0	82	15.0	-3	0	-3	47	51
HINDS MS	54	14.4	78	13.4	114	14.1	-7	5	-2	49	44
HOLMES MS	14	29.9	10	16.3	12	20.7	-46	28	-31	18	88
HUMPHREYS MS	3	12.0	7	22.6	3	8.8	88	-61	-27	22	11
ISSAQUENA MS					2	41.8					99
ITAWAMBA MS	9	14.7	4	4.6	15	12.7	-69	176	-14	35	32
JACKSON MS	18	16.1	20	11.4	39	11.8	-29	4	-27	22	25
JASPER MS	2	4.1	8	15.3	8	12.4	274	-19	203	96	29
JEFFERSON MS	1	4.6	5	23.9	3	14.9	418	-38	223	96	50
JEFFERSON DAVIS MS	6	21.9	3	8.0	6	12.8	-63	60	-42	10	33
JONES MS	24	15.9	26	11.0	49	16.1	-31	46	1	52	60
KEMPER MS	5	13.6	4	9.2	4	8.5	-33	-7	-37	13	10
LAFAYETTE MS	7	12.2	12	16.1	10	8.6	32	-47	-30	19	10
LAMAR MS	2	4.0	14	25.7	3	3.1	544	-88	-23	26	02
LAUDERDALE MS	24	10.2	44	14.7	36	9.9	44	-33	-3	48	15
LAWRENCE MS	3	10.2	12*	29.7	12	24.6	191	-17	141	93	96
LEAKE MS	5	10.2	14	19.4	13	13.0	91	-33	28	73	35
LEE MS	22	16.3	24	12.7	36	14.0	-22	11	-14	36	43
LEFLORE MS	7	9.0	12	11.0	32*	24.8	22	125	174	95	96
LINCOLN MS	6	7.6	14	12.7	21	13.7	67	9	81	88	41
LOWNDES MS	7	6.6	24	18.2	17	9.2	175	-50	39	78	12
MADISON MS	8	17.8	5	8.2	11	12.4	-54	51	-30	18	30
MARION MS	10	18.8	11	13.6	17	14.4	-28	6	-23	25	47
MARSHALL MS	5	11.9	8	15.4	6	9.2	30	-40	-22	26	12
MONROE MS	15	14.2	17	12.8	29	16.3	-10	27	14	64	61
MONTGOMERY MS	6	12.2	8	15.2	8	12.2	25	-20	0	51	28
NESHOBA MS	6	7.1	13	13.8	11	8.1	95	-41	15	64	09
NEWTON MS	10	14.1	17	18.3	10	8.2	30	-55	-42	10	09
NOXUBEE MS	2	6.9	5	14.7	4	10.6	115	-28	54	83	18
OKTIBBEHA MS	7	13.5	14	22.0	13	15.5	62	-30	14	64	55
PANOLA MS	12	17.6	9	10.9	10	8.8	-38	-19	-50	06	11
PEARL RIVER MS	6	9.3	12	12.9	20	15.0	39	16	62	85	52
PERRY MS	1	3.5	2	6.1	1	2.0	75	-67	-42	10	02
PIKE MS	18	16.5	8	5.6	26	16.8	-66	199	2	53	65
PONTOTOC MS	6	7.3	9	9.0	19	15.8	24	75	118	92	58
PRENTISS MS	8	11.9	19	17.9	20	13.5	51	-25	13	63	38
QUITMAN MS	9	28.1	4	11.2	5	11.4	-60	2	-59	04	23
RANKIN MS	12	15.1	7	6.2	26	14.0	-59	127	-8	42	43
SCOTT MS	3	4.9	9	10.9	17	16.5	122	52	238	97	63
SHARKEY MS	1	7.5	4	25.3	6	33.6	238	33	350	98	99
SIMPSON MS	6	9.3	13	17.2	21	20.2	85	17	117	92	86
SMITH MS	2	4.7	9	15.9	11	13.5	236	-15	186	95	38
STONE MS	2	7.4	6	20.6	6	15.4	176	-25	106	91	55
SUNFLOWER MS	6	9.6	11	13.1	14	14.5	36	11	51	82	47
TALLAHATCHIE MS	2	4.5	8	15.6	10	15.5	249	0	248	97	56
TATE MS	12	31.3	6	13.0	7	9.8	-59	-24	-69	02	15
TIPPAH MS	4	5.6	8	10.2	22	20.6	84	101	270	97	88
TISHOMINGO MS	9	14.3	14	16.0	21	18.4	12	15	29	73	76

* SIGNIFICANTLY MORE DEATHS THAN EXPECTED

WHITE FEMALE

CANCER OF THE LARGE INTESTINE

ICD 153,159.0

1950-1959 DEATHS	1950-1959 RATE	1960-1969 DEATHS	1960-1969 RATE	1970-1979 DEATHS	1970-1979 RATE	COUNTY AND STATE	PERCENT CHANGE 60S/50S	PERCENT CHANGE 70S/60S	PERCENT CHANGE 70S/50S	PERCENTILE %CHANGE 70S/50S	PERCENTILE RATES 1970S
2	18.7	3	16.4	1	3.7	TUNICA MS	-12	-77	-80	00	02
10	13.5	11	10.8	20	14.3	UNION MS	-20	32	6	56	46
4	12.8	5	12.1	4	6.4	WALTHALL MS	-5	-47	-50	06	05
20	19.8	14	10.6	30	18.3	WARREN MS	-47	73	-7	42	75
12	11.4	22	16.7	30	17.1	WASHINGTON MS	46	3	50	82	67
4	10.9	9	16.2	3	4.4	WAYNE MS	49	-73	-59	04	03
6	13.1	12	19.2	9	13.8	WEBSTER MS	46	-28	5	56	42
5	21.9	2	8.6	7	22.6	WILKINSON MS	-61	163	4	54	93
11	20.5	9	11.8	14	15.9	WINSTON MS	-42	35	-22	26	58
9	16.3	5	9.2	11	14.3	YALOBUSHA MS	-43	55	-12	38	46
10	15.4	7	8.4	20	18.2	YAZOO MS	-46	117	18	67	75
3939	17.7	4620	17.2	5009	15.9	MISSOURI MO	-3	-8	-10		
22	16.8	24	17.0	19	11.3	ADAIR MO	1	-34	-33	16	22
9	11.2	11	10.4	26*	23.6	ANDREW MO	-7	127	111	91	05
9	14.4	12	17.7	24*	33.1	ATCHISON MO	23	87	129	93	99
23	16.0	34	19.0	36	18.4	AUDRAIN MO	19	-3	15	65	76
23	17.6	21	12.4	26	12.1	BARRY MO	-30	-2	-31	18	28
18	19.7	18	17.9	16	12.8	BARTON MO	-9	-28	-35	15	33
35*	29.4	25	16.9	19	12.2	BATES MO	-43	-28	-59	04	28
11	16.0	16	18.8	20	19.5	BENTON MO	18	4	22	70	82
13	22.6	12	19.2	10	14.3	BOLLINGER MO	-15	-25	-36	13	46
41	17.5	40	13.5	65	16.4	BOONE MO	-23	22	-6	44	62
82	12.9	124	17.8	108	14.9	BUCHANAN MO	38	-16	16	65	50
17	11.2	25	12.4	27	10.8	BUTLER MO	11	-13	-4	47	19
21	26.2	17	17.2	13	16.8	CALDWELL MO	-34	-2	-36	14	65
14	9.4	18	10.3	26	15.6	CALLAWAY MO	10	51	66	86	56
7	16.2	17	23.7	15	11.5	CAMDEN MO	47	-52	-29	20	23
26	12.3	38	13.6	55	15.4	CAPE GIRARDEAU MO	11	13	25	72	55
24	20.6	21	16.6	17	13.0	CARROLL MO	-20	-21	-37	13	35
9*	43.7	4	16.3	2	5.6	CARTER MO	-63	-66	-87	00	04
24	16.7	30	16.5	40	17.4	CASS MO	-1	5	4	54	69
11	13.9	10	13.8	16	13.0	CEDAR MO	-1	-6	-7	43	34
14	14.4	16	16.0	15	13.9	CHARITON MO	11	-13	-4	47	42
9	12.0	7	8.1	19	16.7	CHRISTIAN MO	-33	107	39	78	64
19*	29.3	20*	29.9	16	19.4	CLARK MO	2	-35	-34	15	82
52	20.2	68	16.6	100	17.6	CLAY MO	-18	6	-13	37	70
21	22.7	33*	30.7	20	15.0	CLINTON MO	35	-51	-34	15	52
25	14.5	41	17.3	33	11.0	COLE MO	19	-36	-24	24	20
6	5.9	18	16.5	20	16.4	COOPER MO	178	-1	177	95	61
13	17.6	16	15.7	20	15.2	CRAWFORD MO	-11	-3	-14	36	53
12	17.4	8	14.8	7	7.1	DADE MO	-15	-52	-59	04	07
10	16.8	5	6.3	11	10.5	DALLAS MO	-63	67	-38	12	18
18	24.3	12	14.3	17	18.9	DAVIESS MO	-41	33	-22	26	79
7	10.8	11	22.2	11	12.9	DE KALB MO	105	-42	19	67	34
12	18.5	15	17.7	18	17.3	DENT MO	-4	-3	-6	44	68
8	14.5	9	14.0	7	9.2	DOUGLAS MO	-4	-34	-37	13	12
18	10.5	15	6.7	31	11.6	DUNKLIN MO	-36	72	10	61	23
41	19.3	57	20.3	70	19.0	FRANKLIN MO	5	-6	-2	49	79
8	9.8	12	11.8	12	8.5	GASCONADE MO	21	-28	-13	36	10
24*	26.3	19	19.5	19	22.0	GENTRY MO	-26	13	-17	32	91
114	17.4	127	14.6	163	14.2	GREENE MO	-16	-3	-19	30	45
20	17.3	19	17.7	29	22.0	GRUNDY MO	2	24	27	72	92
20	19.5	23	19.6	12	11.4	HARRISON MO	1	-42	-42	10	23
27	16.4	35	18.2	26	14.7	HENRY MO	11	-19	-10	39	48
3	7.5	5	11.7	5	11.0	HICKORY MO	56	-5	48	81	21
16	24.5	10	12.9	17	26.4	HOLT MO	-47	105	8	58	97
13	15.2	18	19.6	19	19.9	HOWARD MO	30	1	31	74	84
21	18.7	21	13.0	31	14.8	HOWELL MO	-31	14	-21	28	50
17*	28.2	7	10.7	5	5.5	IRON MO	-62	-49	-81	00	04
545	18.5	590	16.7	609	15.8	JACKSON MO	-9	-6	-15	33	57
85	16.0	103	16.5	114	16.4	JASPER MO	3	0	3	53	62
42	18.1	54	16.1	77	14.9	JEFFERSON MO	-11	-8	-18	31	50
32	23.5	28	16.4	37	21.2	JOHNSON MO	-30	29	-10	39	90
11	23.5	10	16.3	14	23.3	KNOX MO	-31	43	-1	50	94
17	16.3	12	8.2	19	11.2	LACLEDE MO	-50	36	-31	17	21
26	16.4	30	14.3	35	15.2	LAFAYETTE MO	-13	7	-7	42	53
22	13.1	37	17.4	39	17.6	LAWRENCE MO	33	1	34	76	71
17	22.0	13	17.3	11	12.5	LEWIS MO	-21	-28	-43	09	30
8	8.5	18	13.9	19	13.4	LINCOLN MO	63	-4	57	83	37
28	19.0	46*	27.6	24	13.4	LINN MO	46	-52	-30	19	37
29	24.0	25	19.1	31	23.3	LIVINGSTON MO	-20	22	-3	48	94
10	12.9	17	18.8	15	13.7	MC DONALD MO	46	-27	6	56	40
20	12.9	25	14.4	20	13.0	MACON MO	12	-10	0	51	34
9	18.7	16	22.8	17	19.9	MADISON MO	22	-13	7	57	84
3	8.9	2	4.3	5	9.3	MARIES MO	-52	117	5	55	13
36	18.7	34	15.0	42	16.6	MARION MO	-20	11	-11	38	63
6	15.3	10	18.0	12	22.8	MERCER MO	17	27	49	81	93
14	16.6	17	15.7	16	11.3	MILLER MO	-5	-28	-32	17	22
9	14.2	10	11.0	9	9.3	MISSISSIPPI MO	-23	-16	-35	15	12
14	16.7	13	14.5	14	12.8	MONITEAU MO	-13	-12	-24	23	33
7	7.1	17	15.8	19	19.6	MONROE MO	122	24	175	95	83
26*	27.3	19	19.0	15	13.8	MONTGOMERY MO	-30	-27	-49	06	41
9	13.9	14	18.0	12	9.8	MORGAN MO	29	-46	-30	19	15
14	15.4	10	8.9	14	10.4	NEW MADRID MO	-42	17	-32	17	18
17	10.7	27	13.3	35	12.9	NEWTON MO	24	-3	20	68	34
32	20.3	30	15.6	29	14.2	NODAWAY MO	-23	-9	-30	19	45
11	18.1	11	14.2	11	12.1	OREGON MO	-22	-14	-33	16	28
8	13.0	12	17.8	14	17.4	OSAGE MO	37	-2	34	76	69
7	18.6	6	12.6	6	11.4	OZARK MO	-32	-10	-39	12	23
9	8.9	14	10.4	20	12.9	PEMISCOT MO	17	24	45	80	34
7	9.6	24	25.4	21	18.6	PERRY MO	165	-27	94	89	77
31	14.2	59	23.1	50	17.1	PETTIS MO	63	-26	21	69	67
23	20.4	22	14.8	26	14.3	PHELPS MO	-28	-3	-30	19	46
22	20.4	25	19.7	22	16.6	PIKE MO	-3	-16	-18	30	63
19	22.1	23	21.3	18	11.9	PLATTE MO	-4	-44	-46	08	26
13	11.2	15	11.9	16	9.7	POLK MO	6	-18	-13	36	14
16	25.3	9	9.6	13	10.4	PULASKI MO	-62	8	-59	04	17
12	20.4	9	13.4	19*	22.6	PUTNAM MO	-34	68	11	61	93
5	8.4	15	24.5	6	8.3	RALLS MO	193	-66	-1	50	09

* SIGNIFICANTLY MORE DEATHS THAN EXPECTED

WHITE FEMALE | CANCER OF THE LARGE INTESTINE | ICD 153,159.0

1950-1959 DEATHS	1950-1959 RATE	1960-1969 DEATHS	1960-1969 RATE	1970-1979 DEATHS	1970-1979 RATE	COUNTY AND STATE	PERCENT CHANGE 60S/50S	PERCENT CHANGE 70S/60S	PERCENT CHANGE 70S/50S	PERCENTILE %CHANGE 70S/50S	PERCENTILE RATES 1970S
33	18.9	43	22.9	30	13.4	RANDOLPH MO	21	-41	-29	20	38
18	17.3	20	16.9	23	17.4	RAY MO	-2	3	1	52	70
9	32.5	6	17.2	2	3.4	REYNOLDS MO	-47	-80	-90	00	02
7	15.2	5	8.3	7	6.6	RIPLEY MO	-45	-20	-56	04	06
27	15.3	39	14.4	53	12.2	ST CHARLES MO	-6	-15	-20	28	28
14	19.0	11	15.5	15	16.4	ST CLAIR MO	-18	6	-13	36	62
7	13.0	9	13.4	19	24.1	STE GENEVIEVE MO	3	80	85	88	95
36	16.6	33	12.4	59	18.3	ST FRANCOIS MO	-25	47	10	60	75
436	19.0	699*	19.2	940*	18.1	ST LOUIS COUNTY MO	1	-5	-5	46	74
35	19.8	39	21.0	53*	25.4	SALINE MO	6	21	29	73	96
10	19.8	8	13.3	3	6.9	SCHUYLER MO	-33	-48	-65	02	06
8	14.2	10	17.2	10	19.8	SCOTLAND MO	21	15	40	78	84
13	11.5	28	18.2	37	16.8	SCOTT MO	58	-8	46	80	65
3	7.6	11	23.6	6	11.4	SHANNON MO	211	-52	50	82	23
13	16.1	23	26.5	16	18.7	SHELBY MO	65	-29	16	66	78
14	11.4	17	10.2	29	13.8	STODDARD MO	-10	36	22	69	41
7	15.2	6	8.6	14	15.2	STONE MO	-43	76	0	51	53
11	12.0	16	20.1	18	21.9	SULLIVAN MO	67	9	82	88	91
11	19.3	16	18.6	25	18.0	TANEY MO	-4	-3	-7	43	74
13	12.9	16	12.9	20	12.6	TEXAS MO	0	-3	-3	48	31
27	14.8	32	16.7	25	14.1	VERNON MO	13	-16	-4	46	45
13	23.9	11	16.9	10	12.7	WARREN MO	-29	-25	-47	08	32
9	13.7	19	27.0	19	22.0	WASHINGTON MO	97	-19	60	84	92
4	6.8	11	17.7	17	19.1	WAYNE MO	159	8	179	95	80
8	8.8	16	16.7	13	10.0	WEBSTER MO	88	-40	13	62	15
10	26.4	7	23.8	5	10.9	WORTH MO	-10	-54	-59	04	19
11	11.8	15	14.6	19	15.0	WRIGHT MO	24	2	27	72	51
854*	20.1	819*	19.5	644	17.3	ST LOUIS CITY MO	-3	-11	-14	35	68
407	15.6	467	14.0	584	14.1	MONTANA MT	-11	1	-10		
5	18.3	2	4.9	9	19.8	BEAVERHEAD MT	-73	301	8	59	84
1	3.9	7	24.5	4	13.7	BIG HORN MT	531	-44	253	97	41
4	15.8	3	9.8	2	5.2	BLAINE MT	-38	-47	-67	02	04
1	9.9	1	7.4	1	5.3	BROADWATER MT	-25	-29	-47	08	04
5	9.8	6	10.3	12	19.6	CARBON MT	5	90	99	90	83
		5*	39.3	1	10.9	CARTER MT		-72			20
32	13.6	40	12.4	55	13.1	CASCADE MT	-9	6	-4	47	35
3	9.6	4	12.7	4	12.6	CHOUTEAU MT	33	-1	32	75	31
7	11.0	8	11.8	8	9.2	CUSTER MT	7	-22	-16	32	12
2	16.7	4	18.3	5	18.4	DANIELS MT	9	1	10	61	76
5	12.5	5	10.1	9	15.0	DAWSON MT	-19	49	20	68	51
14	16.4	9	9.3	22	22.5	DEER LODGE MT	-43	142	37	77	93
2	12.4	1	5.4	5	27.5	FALLON MT	-57	414	122	92	97
12	15.7	11	14.6	9	6.7	FERGUS MT	-7	-54	-57	04	06
29	21.3	26	13.2	30	12.0	FLATHEAD MT	-38	-9	-44	09	27
15	15.6	21	16.5	33	20.1	GALLATIN MT	6	22	29	74	85
		2	21.0	2	26.8	GARFIELD MT		28			97
2	12.1	2	6.4	2	6.2	GLACIER MT	-47	-2	-48	07	05
1	22.5	1	14.9	1	17.9	GOLDEN VALLEY MT	-34	20	-21	28	73
2	13.7	3	19.9	6*	39.5	GRANITE MT	46	98	189	95	99
4	7.9	10	12.6	6	7.1	HILL MT	60	-44	-10	39	07
4	22.8	4	15.3	4	14.3	JEFFERSON MT	-33	-7	-37	13	46
1	8.3			1	7.0	JUDITH BASIN MT			-16	33	06
9	13.9	9	11.4	14	12.9	LAKE MT	-18	14	-7	43	34
27	21.0	27	16.3	37	17.1	LEWIS AND CLARK MT	-23	5	-19	30	67
4*	85.7	2	19.3	1	6.5	LIBERTY MT	-78	-66	-92	00	06
5	12.7	5	11.1	12	19.2	LINCOLN MT	-13	73	51	82	81
1	5.6	1	8.7	1	6.9	MC CONE MT	55	-20	24	71	06
1	3.7	5	16.9	6	14.1	MADISON MT	356	-17	280	98	44
1	13.4	1	7.3	2	18.6	MEAGHER MT	-46	157	39	78	77
1	8.3	3	28.5			MINERAL MT	243				
28	16.8	30	12.7	42	14.9	MISSOULA MT	-24	17	-12	38	50
1	3.2	4	15.7	2	6.8	MUSSELSHELL MT	384	-56	111	91	06
10	15.8	15	18.9	15	16.0	PARK#MT	20	-16	1	52	59
		1	30.7	1	62.9	PETROLEUM MT		105			99
5	20.1	4	10.7	7	18.8	PHILLIPS MT	-47	75	-6	44	78
2	7.7	5	15.0	5	12.3	PONDERA MT	94	-18	59	84	29
1	11.6	1	8.3			POWDER RIVER MT	-29				
8	26.4	8	24.9	8	20.2	POWELL MT	-5	-19	-23	25	86
2	19.5	2	20.4	2	18.3	PRAIRIE MT	5	-10	-6	44	75
8	10.8	13	13.9	12	9.3	RAVALLI MT	28	-33	-14	36	13
7	16.7	8	15.6	15	22.2	RICHLAND MT	-6	42	33	76	92
4	10.9	3	6.4	5	12.0	ROOSEVELT MT	-41	87	10	60	26
2	10.7	1	4.6	10*	35.1	ROSEBUD MT	-57	664	229	97	99
5	15.8	2	5.2	5	9.7	SANDERS MT	-67	86	-39	12	14
4	18.9	8	20.8	2	2.0	SHERIDAN MT	10	-90	-89	00	02
62*	24.3	40	14.6	55	17.8	SILVER BOW MT	-40	22	-27	22	72
4	15.3	4	12.9	9	21.3	STILLWATER MT	-16	66	39	78	90
4	22.2	5	26.3	3	11.9	SWEET GRASS MT	19	-55	-46	08	26
		5	14.0	5	10.7	TETON MT		-24			19
4	14.8	5	14.7	7	21.6	TOOLE MT	-1	48	46	80	91
		1	24.8	1	19.1	TREASURE MT		-23			80
5	9.4	5	10.3	5	9.0	VALLEY MT	10	-13	-4	46	11
1	6.0	6	36.5	2	9.6	WHEATLAND MT	511	-74	61	84	14
		1	11.8	2	18.8	WIBAUX MT		59			79
40	17.4	62	17.3	60	12.1	YELLOWSTONE MT	0	-30	-30	18	28
1397*	19.2	1603	17.9	1828	17.0	NEBRASKA NB	-6	-5	-11		
28	15.5	38	16.5	41	15.4	ADAMS NB	6	-7	-1	50	55
12	18.9	12	18.1	12	16.3	ANTELOPE NB	-4	-10	-14	35	61
3*	59.1					BANNER NB					
1	28.1			2	28.3	BLAINE NB			1	52	98
12	21.4	14	22.7	11	14.3	BOONE NB	6	-37	-33	16	46
4	6.8	7	9.9	17	15.9	BOX BUTTE NB	46	61	135	93	58
6	24.7	5	15.5	4	14.7	BOYD NB	-37	-5	-40	11	49
3	11.2	12*	31.9	9	22.4	BROWN NB	185	-30	100	90	92
19	12.7	30	16.9	40	19.1	BUFFALO NB	34	13	51	82	80
7	11.3	10	13.5	16	21.3	BURT NB	20	58	89	89	90
25*	35.5	19	24.8	13	12.3	BUTLER NB	-30	-50	-65	02	29

* SIGNIFICANTLY MORE DEATHS THAN EXPECTED

COUNTIES COMBINED; SEE TECHNICAL NOTES

WHITE FEMALE

CANCER OF THE LARGE INTESTINE — ICD 153,159.0

1950-1959 DEATHS	1950-1959 RATE	1960-1969 DEATHS	1960-1969 RATE	1970-1979 DEATHS	1970-1979 RATE	COUNTY AND STATE		PERCENT CHANGE 60S/50S	PERCENT CHANGE 70S/60S	PERCENT CHANGE 70S/50S	PERCENTILE %CHANGE 70S/50S	PERCENTILE RATES 1970S
27	26.7	33*	26.9	21	15.4	CASS	NB	1	-43	-42	10	55
6	8.8	13	16.6	16	17.4	CEDAR	NB	88	5	97	90	69
2	7.9	1	3.7	4	8.8	CHASE	NB	-53	136	11	61	10
10	29.1	10	24.2	8	17.6	CHERRY	NB	-17	-27	-39	11	71
12	28.2	10	15.6	15	19.8	CHEYENNE	NB	-45	27	-30	19	84
11	16.2	17	20.7	9	11.7	CLAY	NB	28	-44	-28	21	24
19*	29.9	21	25.7	23	20.3	COLFAX	NB	-14	-21	-32	17	86
12	18.1	14	16.0	26*	25.8	CUMING	NB	-12	61	42	79	97
15	14.1	18	15.4	17	11.1	CUSTER	NB	9	-28	-21	27	21
9	19.0	13	20.9	10	11.7	DAKOTA	NB	10	-44	-39	12	24
12	23.2	7	13.0	8	13.0	DAWES	NB	-44	0	-44	09	34
14	14.0	24	17.7	28	15.4	DAWSON	NB	27	-13	10	61	55
3	14.4	5	24.3	7	24.1	DEUEL	NB	69	-1	67	86	95
11	20.9	12	20.7	10	14.6	DIXON	NB	-1	-29	-30	18	48
23	13.7	39	19.2	42	14.1	DODGE	NB	40	-27	2	53	44
295*	20.6	341	18.7	388	17.7	DOUGLAS	NB	-9	-6	-14	35	72
6	27.5	6	26.2	3	9.7	DUNDY	NB	-5	-63	-65	02	14
4	6.0	16	19.1	22*	24.4	FILLMORE	NB	217	28	304	98	96
11	25.2	13	28.1	7	10.5	FRANKLIN	NB	12	-63	-58	04	18
6	24.5	7	23.3	6	19.0	FRONTIER	NB	-5	-19	-22	26	80
9	12.5	13	18.1	18	23.6	FURNAS	NB	44	31	88	89	95
33	18.9	30	14.6	48	19.7	GAGE	NB	-23	35	4	54	83
4	21.4	2	6.6	3	17.0	GARDEN	NB	-69	157	-20	28	66
1	6.6	2	8.0	10*	36.4	GARFIELD	NB	21	355	451	99	99
3	30.9	1	6.9	2	11.0	GOSPER	NB	-78	59	-64	03	20
2	59.7					GRANT	NB					
3	11.5	4	16.3	4	12.2	GREELEY	NB	42	-25	6	57	29
36	19.5	39	16.1	62	19.6	HALL	NB	-18	22	1	51	83
13	23.5	9	14.4	13	19.1	HAMILTON	NB	-39	33	-19	30	80
8	20.1	6	13.1	7	17.2	HARLAN	NB	-35	31	-14	35	67
3	44.8			2	53.5	HAYES	NB			19	68	99
5	18.5	7	20.9	3	9.7	HITCHCOCK	NB	13	-53	-47	07	14
12	16.6	12	12.6	9	7.7	HOLT	NB	-24	-39	-54	05	08
				1	11.8	HOOKER	NB					25
4	10.1	6	13.6	5	9.1	HOWARD	NB	35	-33	-9	40	12
13	15.7	10	9.9	23	20.7	JEFFERSON	NB	-37	110	32	75	88
13	27.3	15*	31.6	19*	32.8	JOHNSON	NB	16	4	20	68	98
4	9.5	6	10.5	10	15.5	KEARNEY	NB	11	48	64	85	56
5	14.7	15*	33.0	16	25.2	KEITH	NB	125	-24	72	87	96
1	12.3	2	22.2	2	37.3	KEYA PAHA	NB	81	68	204	96	99
7*	40.6	3	12.7	2	7.3	KIMBALL	NB	-69	-43	-82	00	07
14	19.0	14	16.5	6	6.2	KNOX	NB	-14	-62	-67	02	05
135	18.3	162	18.0	192	18.1	LANCASTER	NB	-1	0	-1	50	74
15	11.7	27	16.6	27	12.7	LINCOLN	NB	42	-24	8	59	32
1	14.6			1	39.0	LOGAN	NB			167	94	99
		2	39.6	1	18.9	LOUP	NB		-52			79
1	30.4					MC PHERSON	NB					
36	22.3	27	13.3	45	21.0	MADISON	NB	-41	58	-6	44	89
15	28.2	9	15.1	22*	34.3	MERRICK	NB	-46	127	22	69	99
5	15.6	7	15.8	11	22.5	MORRILL	NB	1	42	44	80	93
9	25.2	6	13.6	2	3.5	NANCE	NB	-46	-74	-86	00	02
14	21.0	17	22.3	18	18.7	NEMAHA	NB	6	-16	-11	38	77
13	24.0	8	11.4	13	20.6	NUCKOLLS	NB	-52	80	-14	36	88
29*	27.9	32*	25.1	32	20.3	OTOE	NB	-10	-19	-27	21	86
12	24.6	9	22.5	8	11.9	PAWNEE	NB	-9	-47	-52	06	26
4	22.6	2	8.0	3	7.9	PERKINS	NB	-64	-1	-65	02	08
15	24.2	12	17.1	10	10.7	PHELPS	NB	-30	-37	-56	05	19
13	27.0	6	10.3	16	22.2	PIERCE	NB	-62	116	-18	31	92
23	22.8	21	15.6	26	14.3	PLATTE	NB	-32	-8	-37	13	46
4	8.1	13	22.5	13	22.0	POLK	NB	177	-3	170	95	92
10	14.2	20	26.0	22	22.2	RED WILLOW	NB	84	-15	57	83	92
33*	31.4	20	17.0	24	17.2	RICHARDSON	NB	-46	1	-45	08	67
6*	41.1	1	8.2	2	6.6	ROCK	NB	-80	-19	-84	00	06
17	17.1	23	19.1	25	18.4	SALINE	NB	12	-4	8	58	76
8	14.2	18	17.1	22	11.5	SARPY	NB	20	-33	-19	29	23
19	19.5	25	21.4	33	21.8	SAUNDERS	NB	10	2	12	62	91
28	20.7	34	18.6	36	14.5	SCOTTS BLUFF	NB	-10	-22	-30	19	48
16	19.4	19	21.2	14	10.3	SEWARD	NB	10	-51	-47	08	17
9	20.6	10	16.8	4	5.2	SHERIDAN	NB	-18	-69	-75	01	04
8	25.9	8	26.1	7	19.4	SHERMAN	NB	1	-26	-25	23	82
2	15.8	2	20.0	1	12.4	SIOUX	NB	27	-38	-21	27	30
3	9.9	4	10.0	8	20.4	STANTON	NB	1	104	107	91	87
12	17.7	11	16.3	16	14.0	THAYER	NB	-8	-14	-21	27	43
		2	34.4	1	17.2	THOMAS	NB		-50			68
7	27.4	3	8.8	8	20.3	THURSTON	NB	-68	132	-26	22	87
6	12.6	14	29.8	6	11.5	VALLEY	NB	136	-61	-9	40	23
15	19.1	14	14.1	21	20.2	WASHINGTON	NB	-26	44	6	57	86
2	3.9	6	9.2	9	12.2	WAYNE	NB	136	32	213	96	28
9	16.2	14	19.8	8	18.2	WEBSTER	NB	22	-8	12	62	75
		1	20.2	2	29.0	WHEELER	NB		43			98
22	22.0	31*	28.9	19	14.9	YORK	NB	32	-49	-32	17	50
96	13.9	182	14.5	358	14.2	NEVADA	NV	4	-2	2		
8	29.7	3	7.5	7	10.9	CHURCHILL	NV	-75	46	-63	03	20
32	16.9	68	12.4	170	13.5	CLARK	NV	-27	9	-20	29	39
4	36.2	7*	37.5	4	12.0	DOUGLAS	NV	3	-68	-67	02	27
1	2.2	7	15.0	6	9.3	ELKO	NV	585	-38	324	98	13
1	38.3	1	24.5			ESMERALDA	NV	-36				
2	12.0	4	18.1	2	6.4	HUMBOLDT	NV	51	-64	-46	08	05
		1	11.5	2	24.8	LANDER	NV		116			96
		1	7.0	1	4.9	LINCOLN	NV		-30			03
		3	12.3	11	20.4	LYON	NV		67			87
1	3.7	3	11.1	3	12.7	MINERAL	NV	199	15	243	97	32
2	14.1	10*	66.5	1	3.4	NYE	NV	371	-95	-76	01	02
4	30.6	1	8.0	1	6.2	PERSHING	NV	-74	-22	-80	00	05
		2	48.5	1	13.7	STOREY	NV		-72			40
37	14.7	60	14.6	118	16.2	WASHOE	NV	-1	11	11	61	60
2	6.4	8	24.0	8	17.9	WHITE PINE	NV	274	-26	178	95	73
2	7.5	3	7.6	23	23.8	ORMSBY-CARSON CITY	NV	1	214	218	96	95

* SIGNIFICANTLY MORE DEATHS THAN EXPECTED

WHITE FEMALE — CANCER OF THE LARGE INTESTINE — ICD 153,159.0

1950-1959 DEATHS	1950-1959 RATE	1960-1969 DEATHS	1960-1969 RATE	1970-1979 DEATHS	1970-1979 RATE	COUNTY AND STATE	PERCENT CHANGE 60S/50S	PERCENT CHANGE 70S/60S	PERCENT CHANGE 70S/50S	PERCENTILE %CHANGE 70S/50S	PERCENTILE RATES 1970S
749*	22.5	813*	19.9	1036*	19.6	NEW HAMPSHIRE NH	-11	-1	-13		
33	18.6	37	17.3	58	22.2	BELKNAP NH	-7	28	19	68	92
18	15.5	16	11.5	37	18.1	CARPOLL NH	-26	57	17	66	74
54	23.4	61	21.5	65	17.7	CHESHIPE NH	-8	-18	-24	24	72
30	16.1	47	22.0	46	17.7	COOS NH	36	-19	10	60	72
57	20.6	60	18.3	74	18.3	GRAFTON NH	-11	0	-11	38	75
237*	24.3	260*	21.7	345*	22.5	HILLSBOPOUGH NH	-11	4	-7	42	93
79	16.7	95	16.9	125	18.9	MERRIMACK NH	1	12	13	63	79
120*	26.4	115	19.0	157	17.8	ROCKINGHAM NH	-28	-7	-33	16	72
80*	27.5	81*	23.1	89	19.0	STRAFFORD NH	-16	-18	-31	18	80
41*	26.7	41	22.1	40	17.3	SULLIVAN NH	-17	-22	-35	14	68
6127*	24.1	7455*	22.0	8585*	20.1	NEW JERSEY NJ	-9	-9	-16		
213*	26.2	252*	22.3	321*	22.8	ATLANTIC NJ	-15	2	-13	37	93
723*	23.4	935*	21.3	1135*	20.0	BERGEN NJ	-9	-6	-15	35	85
151*	23.3	235*	24.9	274*	19.6	BURLINGTON NJ	7	-21	-16	33	83
371*	24.6	519*	25.8	564*	22.7	CAMDEN NJ	5	-12	-8	42	93
71*	26.0	78	17.4	137*	19.3	CAPE MAY NJ	-33	11	-26	22	81
106*	23.0	135*	23.5	144*	20.0	CUMBERLAND NJ	2	-15	-13	36	85
1093*	24.2	1123*	22.8	916*	18.9	ESSEX NJ	-6	-17	-22	26	79
96	22.2	152*	24.3	175*	20.5	GLOUCESTER NJ	9	-15	-8	42	87
794*	25.7	753*	20.8	798*	20.9	HUDSON NJ	-19	0	-19	30	89
79*	28.8	73	21.0	82	18.8	HUNTERDON NJ	-27	-11	-35	15	78
277*	24.1	339*	23.5	377*	21.3	MERCER NJ	-2	-9	-11	38	90
291*	23.5	403*	21.0	544*	20.1	MIDDLESEX NJ	-10	-4	-14	35	86
356*	24.9	487*	23.3	622*	21.4	MONMOUTH NJ	-6	-8	-14	35	90
228*	22.0	307*	20.8	395*	20.1	MOPRIS NJ	-5	-4	-9	41	85
106*	24.4	215*	21.7	469*	19.4	OCEAN NJ	-11	-11	-21	28	82
423*	23.1	513*	21.5	549*	19.4	PASSAIC NJ	-7	-10	-16	33	82
48	23.0	55	21.4	63	18.7	SALEM NJ	-7	-13	-19	30	77
111	21.9	140	18.8	156	15.7	SOMERSET NJ	-14	-16	-28	21	57
42	19.4	62	18.7	102*	20.8	SUSSEX NJ	-4	11	7	58	88
475*	24.1	602*	22.2	652*	19.6	UNION NJ	-8	-12	-19	30	83
73	21.6	77	17.6	110*	20.5	WARREN NJ	-19	16	-5	45	87
231	10.9	388	12.1	644	13.9	NEW MEXICO NM	11	15	27		
72	13.9	126	13.4	234	15.9	BERNALILLO NM	-4	18	14	64	58
10	8.1	14	7.9	37	14.4	CHAVES NM	-3	84	78	87	47
8	14.6	7	10.8	11	13.9	COLFAX NM	-26	28	-5	46	42
13	15.9	13	9.7	21	11.1	CURRY NM	-39	15	-30	18	21
1	7.0	1	5.5			DE BACA NM	-21				
11	9.4	18	10.0	39	14.0	DONA ANA NM	6	40	49	81	43
14	12.0	22	13.0	35	14.3	EDDY NM	9	10	20	68	46
16	26.5	13	16.6	14	11.2	GRANT NM	-37	-33	-58	04	21
3	16.3	1	5.1	4	19.7	GUADALUPE NM	-69	287	21	69	84
2	20.9	1	12.7	1	11.2	HARDING NM	-39	-12	-46	08	22
2	11.0	4	20.4	4	16.7	HIDALGO NM	85	-19	51	82	64
9	12.6	14	9.2	25	12.5	LEA NM	-27	35	-1	50	31
8	28.8	8	19.7	2	3.0	LINCOLN NM	-32	-85	-90	00	02
2	13.2	3	17.5	6	16.3	LOS ALAMOS NM	33	-7	23	71	61
2	5.4	4	8.0	8	8.7	LUNA NM	48	9	61	84	10
1	5.3	5	8.6	7	12.0	MC KINLEY NM	61	40	125	92	26
1	4.4	7	29.0			MORA NM	560				
7	13.9	12	12.9	16	12.8	OTERO NM	-7	-1	-8	41	32
3	6.1	5	7.5	13	17.3	QUAY NM	22	132	184	95	69
2	3.2	12	16.0	10	12.4	RIO ARRIBA NM	399	-22	288	98	30
4	6.0	14	17.5	13	13.7	ROOSEVELT NM	190	-21	127	92	41
4	13.7	3	8.9	10	10.7	SANDOVAL NM	-35	20	-22	27	19
1	1.8	9	10.1	20	12.8	SAN JUAN NM	469	27	621	99	33
4	4.5	13	13.2	17	15.4	SAN MIGUEL NM	195	17	245	97	55
16	13.6	28	13.7	41	14.3	SANTA FE NM	1	4	5	56	46
4	9.5	6	11.2	9	11.8	SIERRA NM	19	5	24	71	25
1	2.4	4	10.9	11	23.8	SOCORRO NM	361	118	905	99	95
1	2.5	8	12.0	6	9.6	TAOS NM	380	-20	285	98	14
2	7.8	1	3.7	2	5.3	TORRANCE NM	-53	42	-33	16	04
3	11.0	1	2.9	4	8.8	UNION NM	-74	202	-20	28	11
4	7.6	11	15.1	24	17.2	VALENCIA NM	98	14	127	92	68
17264*	21.8	19528*	20.3	21369*	19.5	NEW YORK NY	-7	-4	-11		
312*	21.6	381*	21.8	429*	20.9	ALBANY NY	1	-4	-3	48	89
61*	24.3	51	17.7	62	19.3	ALLEGANY NY	-27	9	-20	28	82
1779*	25.3	1418	17.5	1479*	19.3	BPONX NY	-31	10	-24	25	81
208	19.9	240	19.0	291*	19.3	BROOME NY	-4	2	-3	48	82
76	17.7	92	19.0	110	19.4	CATTARAUGUS NY	7	2	10	60	82
79	18.0	99	20.6	92	16.4	CAYUGA NY	14	-20	-9	41	62
159	18.3	172	17.9	218*	19.5	CHAUTAUQUA NY	-3	9	6	57	82
91	17.9	111	18.8	133	19.8	CHEMUNG NY	5	5	11	61	84
63*	24.7	46	15.5	53	16.3	CHENANGO NY	-37	6	-34	15	61
49	19.2	45	15.0	73	20.5	CLINTON NY	-22	37	7	57	87
54	17.5	73	20.9	88	20.1	COLUMBIA NY	19	-4	15	64	86
39	17.6	48	19.1	52	18.2	CORTLAND NY	8	-5	3	54	75
56	21.5	56	19.5	83*	24.4	DELAWARE NY	-9	25	13	63	96
166	17.7	208	18.3	231	16.9	DUTCHESS NY	4	-8	-5	46	65
881	18.8	983	17.0	1250*	18.8	ERIE NY	-10	10	0	51	78
36	17.9	41	18.6	51	19.2	ESSEX NY	4	3	7	58	81
36	15.2	50	19.3	61	21.5	FRANKLIN NY	27	11	42	79	90
96*	26.6	81	20.9	83	19.8	FULTON NY	-22	-5	-26	22	84
62	22.4	50	15.7	68	18.5	GENESEE NY	-30	18	-18	31	76
43	20.7	58*	22.4	45	14.6	GREENE NY	8	-35	-29	19	48
10*	39.2	1	3.1	4	10.2	HAMILTON NY	-92	232	-74	01	16
94*	24.6	89	19.5	103	19.9	HERKIMER NY	-21	2	-19	30	85
102	18.3	110	17.6	134*	20.8	JEFFERSON NY	-4	18	14	64	88
2980*	24.3	2417	17.5	2693*	20.0	KINGS(BROOKLYN) NY	-28	14	-18	31	85
25	19.1	24	17.2	32	21.7	LEWIS NY	-10	26	14	64	91
41	17.3	55	20.5	59	17.8	LIVINGSTON NY	19	-13	3	53	72
51	18.9	64	20.9	80*	22.6	MADISON NY	11	8	20	68	93

* SIGNIFICANTLY MORE DEATHS THAN EXPECTED

WHITE FEMALE

CANCER OF THE LARGE INTESTINE

ICD 153,159.0

1950-1959 DEATHS	1950-1959 RATE	1960-1969 DEATHS	1960-1969 RATE	1970-1979 DEATHS	1970-1979 RATE	COUNTY AND STATE		PERCENT CHANGE 60S/50S	PERCENT CHANGE 70S/60S	PERCENT CHANGE 70S/50S	PERCENTILE %CHANGE 70S/50S	PERCENTILE RATES 1970S
567	18.8	736*	19.9	801*	18.2	MONROE	NY	6	-9	-3	47	74
80	20.9	80	18.3	89	18.3	MONTGOMERY	NY	-12	0	-13	37	75
807*	21.9	1232*	21.1	1635*	20.9	NASSAU	NY	-4	-1	-4	46	89
2164*	23.4	3563*	36.9	1733*	18.8	NEW YORK(MANHATTAN)	NY	57	-49	-20	29	78
180	19.1	199	17.2	252	18.0	NIAGARA	NY	-10	4	-6	44	74
238	17.6	281	17.5	360*	19.5	ONEIDA	NY	-1	11	11	61	82
388*	20.3	420	17.8	532*	18.9	ONONDAGA	NY	-12	6	-7	43	79
76	21.0	82	19.4	105	19.8	ONTARIO	NY	-8	2	-6	44	84
214*	21.5	257*	21.7	288*	20.6	ORANGE	NY	1	-5	-4	47	88
45	22.4	50	23.4	44	19.1	ORLEANS	NY	5	-18	-14	35	81
93	20.9	98	19.5	118	20.0	OSWEGO	NY	-7	2	-4	46	85
82*	23.4	62	16.0	81	17.4	OTSEGO	NY	-32	9	-26	22	69
30	21.9	42	20.2	67	20.6	PUTNAM	NY	-8	2	-6	44	88
1845*	22.4	1951	16.7	2715*	19.7	QUEENS	NY	-25	18	-12	38	83
221*	26.6	230*	24.2	245*	23.1	RENSSELAER	NY	-9	-4	-13	36	94
245*	25.5	195	14.8	364*	20.9	RICHMOND	NY	-42	41	-18	30	89
121*	22.1	145	17.8	207	18.0	ROCKLAND	NY	-19	1	-19	30	74
90	16.5	107	18.4	124	18.3	ST LAWRENCE	NY	11	0	11	61	75
108*	24.4	103	18.9	148*	20.5	SARATOGA	NY	-23	8	-16	33	87
158	18.7	190	18.5	205	17.4	SCHENECTADY	NY	-1	-6	-7	43	69
38*	27.2	28	17.9	35	17.9	SCHOHARIE	NY	-34	0	-34	15	73
18	22.5	12	13.4	22	18.7	SCHUYLER	NY	-41	40	-17	31	77
35	15.3	42	18.5	37	14.9	SENECA	NY	21	-20	-3	48	50
108	20.7	119	20.1	122	18.4	STEUBEN	NY	-2	-9	-11	39	76
355	16.1	696	18.5	1172*	20.6	SUFFOLK	NY	15	11	28	73	88
46	18.3	79*	26.2	75	17.9	SULLIVAN	NY	43	-32	-2	49	73
39	21.5	41	19.4	59*	23.5	TIOGA	NY	-10	21	10	60	94
58	19.7	48	14.5	85*	21.6	TOMPKINS	NY	-26	48	9	60	91
139	21.4	165*	20.8	202*	20.0	ULSTER	NY	-3	-4	-7	43	85
52	19.6	56	18.6	81*	21.4	WARREN	NY	-5	15	9	59	90
61	21.4	52	16.3	67	18.7	WASHINGTON	NY	-24	15	-13	37	78
63	17.0	79	18.8	82	16.6	WAYNE	NY	10	-12	-3	49	63
779*	21.2	962*	20.2	1110*	19.1	WESTCHESTER	NY	-5	-5	-10	39	81
47	23.5	36	16.1	35	14.8	WYOMING	NY	-32	-8	-37	13	50
25	19.6	27	19.9	15	8.4	YATES	NY	2	-58	-57	04	09
1393	11.7	2014	12.0	3054	12.6	NORTH CAROLINA	NC	3	5	8		
25	9.7	40	12.2	68	13.5	ALAMANCE	NC	26	11	40	78	39
8	14.1	8	10.3	12	10.9	ALEXANDER	NC	-27	5	-23	26	20
1	2.6	15*	29.0	13	18.6	ALLEGHANY	NC	1009	-36	610	99	77
12	19.5	8	9.6	10	8.7	ANSON	NC	-51	-9	-55	05	10
11	11.9	18	15.8	14	10.0	ASHE	NC	33	-37	-16	33	15
4	7.1	2	3.5	8	10.7	AVERY	NC	-50	207	52	82	19
9	7.8	13	9.4	20	11.0	BEAUFORT	NC	20	17	40	78	20
8	14.9	11	17.1	12	17.1	BERTIE	NC	15	0	15	65	67
6	10.7	8	11.0	15	14.8	BLADEN	NC	3	35	39	78	50
2	6.2	6	10.7	6	5.2	BRUNSWICK	NC	73	-51	-15	34	04
85	15.8	73	9.7	124	11.9	BUNCOMBE	NC	-39	23	-25	24	26
17	9.8	12	4.9	28	8.2	BURKE	NC	-49	65	-16	32	09
27	12.3	36	12.3	62	14.4	CABARRUS	NC	1	16	17	66	46
14	11.0	20	10.5	27	9.6	CALDWELL	NC	-5	-8	-12	37	14
		2	10.6	4	16.9	CAMDEN	NC		59			66
10	10.5	23	17.6	17	8.6	CARTERET	NC	66	-51	-19	30	10
9	23.4	5	8.7	8	10.9	CASWELL	NC	-63	26	-53	05	20
21	10.8	33	10.3	68	14.5	CATAWBA	NC	-5	40	34	76	47
10	12.0	12	11.2	24	15.9	CHATHAM	NC	-6	42	33	75	58
12	17.0	5	5.5	19	15.4	CHEROKEE	NC	-68	181	-10	40	55
6	18.1	4	10.6	10	20.3	CHOWAN	NC	-42	92	12	62	86
1	4.1	4	11.9	4	8.4	CLAY	NC	190	-29	105	90	09
25	12.2	41	15.1	66	17.6	CLEVELAND	NC	24	17	45	80	71
6	5.8	15	10.0	18	9.8	COLUMBUS	NC	71	-2	68	86	14
9	7.8	17	12.6	34	17.0	CRAVEN	NC	63	34	119	92	66
9	5.6	31	11.2	67	15.8	CUMBERLAND	NC	98	41	180	95	58
2	9.8	1	3.2	9	18.6	CURRITUCK	NC	-67	482	91	89	77
1	3.6	4	9.1	7	11.9	DARE	NC	152	30	228	97	26
28	12.8	39	11.7	68	13.7	DAVIDSON	NC	-8	17	7	58	41
11	17.5	9	10.5	9	8.1	DAVIE	NC	-40	-23	-54	05	08
13	12.5	18	14.6	22	13.7	DUPLIN	NC	17	-6	9	60	41
40	13.4	51	12.5	74	13.1	DURHAM	NC	-7	5	-3	49	35
11	9.2	21	16.2	24	14.0	EDGECOMBE	NC	76	-13	53	82	43
63	14.4	102	15.5	130	12.8	FORSYTH	NC	8	-18	-12	38	33
10	14.5	12	13.4	14	10.8	FRANKLIN	NC	-8	-20	-26	23	19
42	12.5	59	12.1	80	10.8	GASTON	NC	-4	-10	-14	36	19
4	17.2	2	6.4	10	26.5	GATES	NC	-63	312	54	83	97
1	8.3	4	13.1	1	2.6	GRAHAM	NC	58	-80	-69	02	02
8	8.3	10	9.0	13	10.8	GRANVILLE	NC	9	20	30	74	19
3	9.2	3	8.7	6	13.5	GREENE	NC	-4	55	48	81	39
85	13.0	131	13.8	191	14.1	GUILFORD	NC	6	2	8	59	44
7	5.2	21	14.5	32	15.1	HALIFAX	NC	182	4	193	96	53
19	15.0	11	7.1	42	18.4	HARNETT	NC	-52	158	23	70	76
13	9.7	17	8.6	30	10.3	HAYWOOD	NC	-12	21	7	57	17
23	13.3	31	13.1	39	9.5	HENDERSON	NC	-2	-27	-28	21	13
11	27.7	2	3.1	8	9.1	HERTFORD	NC	-89	194	-67	02	12
1	2.2	3	11.7	5	12.1	HOKE	NC	436	3	453	99	27
4	15.1	1	2.5	5	14.3	HYDE	NC	-83	467	-5	45	46
23	10.7	38	12.7	53	13.3	IREDELL	NC	19	5	25	71	37
12	17.6	5	5.9	12	9.3	JACKSON	NC	-66	58	-47	08	13
17	11.9	20	8.8	37	11.4	JOHNSTON	NC	-26	29	-4	46	23
1	4.0	3	21.6	6	17.4	JONES	NC	436	-19	332	98	69
6	10.1	11	10.7	24	15.9	LEE	NC	6	49	59	84	59
13	13.7	26	17.4	25	11.9	LENOIR	NC	27	-32	-13	36	26
10	10.5	16	13.1	29	15.6	LINCOLN	NC	25	19	49	81	56
11	12.4	13	9.9	24	13.5	MC DOWELL	NC	-20	37	9	59	38
6	9.1	9	8.9	9	5.9	MACON	NC	-2	-34	-35	14	04
6	8.7	9	9.5	9	7.2	MADISON	NC	9	-25	-17	31	07
3	5.8	10	14.4	7	8.6	MARTIN	NC	148	-40	48	81	10
75	11.2	151	14.9	222	15.1	MECKLENBURG	NC	33	1	35	76	52
4	6.4	5	6.9	11	10.4	MITCHELL	NC	9	50	64	85	18
8	12.9	10	12.6	15	13.8	MONTGOMERY	NC	-3	10	7	58	42
13	11.9	29	18.4	31	12.3	MOORE	NC	54	-33	3	53	29

* SIGNIFICANTLY MORE DEATHS THAN EXPECTED

WHITE FEMALE — CANCER OF THE LARGE INTESTINE — ICD 153,159.0

1950-1959 DEATHS	1950-1959 RATE	1960-1969 DEATHS	1960-1969 RATE	1970-1979 DEATHS	1970-1979 RATE	COUNTY AND STATE	PERCENT CHANGE 60S/50S	PERCENT CHANGE 70S/60S	PERCENT CHANGE 70S/50S	PERCENTILE %CHANGE 70S/50S	PERCENTILE RATES 1970S
14	11.4	26	13.6	36	14.5	NASH NC	20	7	28	73	48
37	17.5	36	12.8	50	12.2	NEW HANOVER NC	-27	-5	-31	18	28
1	1.7	10	14.4	11	14.1	NORTHAMPTON NC	741	-2	724	99	44
11	13.6	11	11.3	23	13.9	ONSLOW NC	-17	23	2	52	42
6	8.0	10	6.8	29	12.8	ORANGE NC	-15	88	60	84	33
2	7.7	1	2.6	6	11.7	PAMLICO NC	-67	354	51	82	25
7	9.3	10	10.9	17	13.7	PASQUOTANK NC	17	25	47	81	41
4	10.4	6	11.3	12	16.8	PENDER NC	9	49	62	85	65
		2	5.1	6	13.8	PERQUIMANS NC		172			41
7	10.3	7	9.8	13	10.8	PERSON NC	-5	10	4	55	19
14	10.2	23	12.8	32	12.8	PITT NC	26	0	26	72	33
8	13.6	10	11.7	18	15.0	POLK NC	-14	28	10	61	51
23	12.4	26	9.8	48	11.8	RANDOLPH NC	-21	21	-4	47	25
13	11.2	19	14.9	22	11.2	RICHMOND NC	33	-25	0	51	21
17	13.2	17	9.5	30	12.4	ROBESON NC	-28	31	-6	44	30
27	13.3	39	14.2	44	10.8	ROCKINGHAM NC	6	-24	-19	29	19
33	13.7	47	12.2	75	13.5	ROWAN NC	-11	11	-1	50	39
13	7.8	23	9.4	53	16.2	RUTHERFORD NC	20	71	106	90	60
14	12.6	9	6.4	14	6.9	SAMPSON NC	-49	8	-45	09	06
5	9.3	6	10.8	14	14.8	SCOTLAND NC	17	36	59	84	49
10	7.7	22	11.8	35	13.2	STANLY NC	53	12	72	87	36
9	11.9	16	15.7	11	7.8	STOKES NC	32	-50	-34	15	08
12	7.3	28	12.1	47	14.1	SURRY NC	65	16	92	89	44
3	10.8	3	7.5	4	8.1	SWAIN NC	-30	8	-24	24	09
7	15.3	13	18.6	11	8.9	TRANSYLVANIA NC	21	-52	-42	10	11
2	12.9			5	22.8	TYRRELL NC			77	87	93
15	11.3	24	13.6	26	9.8	UNION NC	20	-28	-14	36	14
13	18.3	6	5.6	16	11.0	VANCE NC	-69	95	-40	11	20
32	8.1	75	12.1	115	12.0	WAKE NC	49	-1	48	81	27
6	14.9	9	17.6	7	10.4	WARREN NC	18	-41	-30	18	17
2	7.4	2	5.0	7	14.6	WASHINGTON NC	-33	195	99	90	48
5	7.2	18	19.8	12	8.1	WATAUGA NC	175	-59	13	63	09
16	12.7	26	12.7	31	10.1	WAYNE NC	0	-20	-20	28	16
23	14.3	20	9.3	25	8.4	WILKES NC	-35	-9	-41	11	09
12	9.9	19	10.9	28	12.3	WILSON NC	10	13	24	71	29
10	11.3	14	12.1	20	13.6	YADKIN NC	7	12	20	68	39
5	8.0	12	16.1	10	9.5	YANCEY NC	103	-41	19	67	13
345	13.8	422	13.6	505	12.9	NORTH DAKOTA ND	-2	-5	-7		
1	7.0	2	8.8	6	18.0	ADAMS ND	26	104	156	94	73
7	8.8	13	13.3	13	10.1	BARNES ND	51	-24	15	65	16
6	14.4	11	25.7	7	14.9	BENSON ND	79	-42	4	54	51
2	52.9					BILLINGS ND					
9	19.0	12	19.9	5	8.3	BOTTINEAU ND	5	-58	-56	04	09
4	21.5	2	8.0	3	10.2	BOWMAN ND	-63	27	-52	05	16
1	3.6	7	21.4	6	16.4	BURKE ND	500	-23	360	98	62
17	14.4	17	10.3	36	15.9	BURLEIGH ND	-28	54	11	61	58
49	17.8	43	12.6	64	13.6	CASS ND	-29	8	-24	25	39
9	19.9	8	16.1	12	21.2	CAVALIER ND	-19	32	6	57	90
3	7.4	7	15.7	12	18.9	DICKEY ND	111	21	155	94	79
4	12.6	1	2.7	5	12.2	DIVIDE ND	-79	361	-3	48	29
3	15.3	3	16.9	2	9.5	DUNN ND	10	-44	-38	12	13
2	8.3	7	27.2	7	18.6	EDDY ND	226	-32	123	92	77
6	20.6	5	13.6	5	12.2	EMMONS ND	-34	-10	-41	11	28
1	5.3	8	28.0	4	15.7	FOSTER ND	427	-44	195	96	57
1	6.0	4	30.6	3	9.1	GOLDEN VALLEY ND	406	-70	50	82	12
23	13.4	25	10.7	22	8.2	GRAND FORKS ND	-20	-23	-39	12	09
1	5.0	2	8.5	4	13.2	GRANT ND	72	55	166	94	36
4	16.8	2	6.3	3	7.2	GRIGGS ND	-62	14	-57	04	07
3	16.7	3	11.4	4	13.7	HETTINGER ND	-32	21	-18	31	41
1	4.9	6	31.3			KIDDER ND	536				
4	10.6	10	23.4	7	11.1	LA MOURE ND	120	-53	4	55	21
1	4.8	6	26.4	4	112.6	LOGAN ND	445	326	2223	99	99
8	15.7	11	19.5	13	23.7	MC HENRY ND	24	22	51	82	95
6	19.8	7	21.9	7	13.5	MC INTOSH ND	10	-38	-32	17	39
3	9.4	3	9.6	7	17.3	MC KENZIE ND	3	80	85	88	69
8	14.3	6	9.4	9	12.4	MC LEAN ND	-34	32	-13	36	30
3	15.6	3	10.7			MERCER ND	-31				
7	9.8	10	11.1	15	10.7	MORTON ND	13	-3	9	60	19
4	10.1	5	11.6	12	19.9	MOUNTRAIL ND	15	71	96	89	84
6	15.3	5	12.9	6	15.3	NELSON ND	-15	18	0	51	54
2	21.6			1	13.4	OLIVER ND			-38	12	38
10	16.7	11	15.7	12	12.5	PEMBINA ND	-6	-20	-25	23	30
5	14.2	4	9.6	7	13.2	PIERCE ND	-32	37	-7	43	36
8	13.0	8	11.2	15	17.9	RAMSEY ND	-14	60	38	77	73
3	5.9	8	13.6	8	14.7	RANSOM ND	129	8	148	94	49
4	15.4	6	26.6	3	10.1	RENVILLE ND	73	-62	-35	15	16
15	16.9	20	18.1	21	15.1	RICHLAND ND	7	-16	-11	39	53
5	22.6	1	3.4	6	16.3	ROLETTE ND	-85	375	-28	21	61
3	9.2	4	11.9	3	9.6	SARGENT ND	29	-19	5	55	14
1	7.3	4	22.9	1	7.0	SHERIDAN ND	216	-69	-3	48	07
		1	28.7			SIOUX ND					
2	40.9					SLOPE ND					
8	14.3	9	11.3	17	15.3	STARK ND	-21	35	7	57	54
2	14.2	2	9.9	3	12.4	STEELE ND	-30	25	-13	37	30
16	13.5	16	12.0	10	6.3	STUTSMAN ND	-11	-48	-53	05	05
3	11.0	6	21.4	9	24.3	TOWNER ND	95	14	122	92	95
4	7.7	10	13.2	17	18.8	TRAILL ND	70	43	143	93	79
15	16.3	14	13.9	12	10.8	WALSH ND	-15	-22	-34	15	19
23	15.1	21	10.0	34	12.7	WARD ND	-33	27	-15	34	32
5	12.0	11	23.8	8	13.4	WELLS ND	98	-43	12	62	38
4	5.3	12	11.5	15	12.1	WILLIAMS ND	116	6	128	93	27
7888*	19.3	9527*	19.0	10959*	18.3	OHIO OH	-1	-4	-5		
28	22.9	28	22.7	31	21.9	ADAMS OH	-1	-4	-5	46	91
76	16.3	98	17.7	125	18.5	ALLEN OH	9	4	13	63	76
25	12.1	54	22.5	59	20.1	ASHLAND OH	86	-11	66	86	85
99	22.2	94	17.5	101	16.4	ASHTABULA OH	-21	-6	-26	22	62
44	16.6	60	21.9	53	18.7	ATHENS OH	32	-15	13	63	78

* SIGNIFICANTLY MORE DEATHS THAN EXPECTED

WHITE FEMALE

CANCER OF THE LARGE INTESTINE

ICD 153,159.0

1950-1959 DEATHS	1950-1959 RATE	1960-1969 DEATHS	1960-1969 RATE	1970-1979 DEATHS	1970-1979 RATE	COUNTY AND STATE		PERCENT CHANGE 60S/50S	PERCENT CHANGE 70S/60S	PERCENT CHANGE 70S/50S	PERCENTILE %CHANGE 70S/50S	PERCENTILE RATES 1970S
28	14.7	39	17.7	34	12.2	AUGLAIZE	OH	21	-31	-17	32	28
81	16.7	115	20.9	108	17.6	BELMONT	OH	25	-16	5	56	70
26	18.0	23	15.3	35	18.8	BROWN	OH	-15	23	5	55	79
121	17.6	149	16.5	218	18.5	BUTLER	OH	-6	12	5	56	76
32*	31.2	29*	24.8	25	17.8	CARROLL	OH	-21	-28	-43	09	73
31	20.6	36	20.7	38	18.0	CHAMPAIGN	OH	1	-13	-13	37	74
120	20.3	135	18.1	160	18.5	CLARK	OH	-11	3	-9	41	77
46	19.0	60	18.4	103*	22.2	CLERMONT	OH	-3	21	17	67	92
21	12.6	27	14.6	39	18.7	CLINTON	OH	16	27	48	81	77
127*	23.9	139*	22.1	148*	20.5	COLUMBIANA	OH	-7	-8	-14	35	87
37	20.2	36	17.5	38	16.3	COSHOCTON	OH	-13	-7	-20	29	61
50	21.8	46	16.0	56	16.7	CRAWFORD	OH	-26	4	-23	25	64
1308*	19.8	1614*	19.8	1577	16.8	CUYAHOGA	OH	0	-15	-15	34	65
40	15.6	39	13.0	75	22.6	DARKE	OH	-16	74	45	80	93
37	24.3	26	15.3	37	17.7	DEFIANCE	OH	-37	16	-27	21	71
38	21.6	30	15.4	42	16.2	DELAWARE	OH	-29	6	-25	23	61
64	21.3	67	18.8	60	14.2	ERIE	OH	-12	-24	-33	16	45
62	20.1	83*	21.7	107*	21.4	FAIRFIELD	OH	8	-2	6	57	90
22	15.8	26	15.5	36	19.6	FAYETTE	OH	-2	26	24	71	83
541*	21.5	617*	19.4	735*	18.8	FRANKLIN	OH	-10	-3	-13	37	79
20	13.6	18	10.3	38	16.9	FULTON	OH	-24	64	25	71	65
18	13.5	24	16.5	30	19.0	GALLIA	OH	22	15	41	79	80
34	24.1	42	23.1	37	13.1	GEAUGA	OH	-4	-43	-46	08	36
43	17.4	47	14.5	83	17.9	GREENE	OH	-17	23	3	53	73
61	23.0	50	18.8	44	15.0	GUERNSEY	OH	-18	-20	-35	15	52
836*	21.5	1022*	22.1	1142*	21.7	HAMILTON	OH	3	-2	1	52	91
53	18.1	64	19.2	65	16.0	HANCOCK	OH	6	-17	-11	38	59
26	14.9	40	20.8	43	20.5	HARDIN	OH	39	-2	37	77	87
18	16.9	13	11.7	17	12.9	HARRISON	OH	-31	10	-24	25	33
23	18.0	23	15.6	25	14.1	HENRY	OH	-14	-9	-22	27	44
33	18.5	31	15.3	48	21.0	HIGHLAND	OH	-17	38	14	63	89
26	24.1	31*	25.7	28	20.0	HOCKING	OH	7	-22	-17	32	85
19	18.7	12	10.9	26	20.2	HOLMES	OH	-42	85	8	59	86
52	21.4	56	21.2	58	19.2	HURON	OH	-1	-9	-10	39	81
18	11.8	36	20.0	30	13.9	JACKSON	OH	69	-31	17	67	42
86	22.2	106*	21.7	125*	21.7	JEFFERSON	OH	-2	0	-2	49	91
34	15.3	38	14.4	51	17.7	KNOX	OH	-6	23	15	65	71
85	21.8	115	19.7	164	18.8	LAKE	OH	-9	-4	-14	36	79
36	16.3	66*	24.5	62	18.0	LAWRENCE	OH	50	-26	11	61	74
77	17.3	113*	21.4	115	18.0	LICKING	OH	24	-16	4	55	74
31	15.5	48	20.2	64*	24.8	LOGAN	OH	30	23	60	84	96
132	19.3	156	16.5	210	17.6	LORAIN	OH	-15	6	-9	40	70
328	15.8	424	16.9	578*	20.3	LUCAS	OH	6	20	28	73	86
20	18.1	19	14.9	42*	27.3	MADISON	OH	-17	84	51	82	97
243*	20.8	288*	19.4	330*	18.6	MAHONING	OH	-7	-4	-10	39	77
45	14.8	63	17.9	78	20.0	MARION	OH	21	12	35	77	85
36	15.6	69	22.1	68	16.4	MEDINA	OH	41	-26	5	55	62
26	20.7	22	16.0	32	20.5	MEIGS	OH	-23	28	-1	50	87
28	19.1	28	15.2	51*	22.2	MERCER	OH	-20	46	16	66	92
63	18.0	84	20.5	114*	22.7	MIAMI	OH	14	11	27	72	93
21	20.1	24	23.6	21	18.2	MONROE	OH	17	-23	-10	40	75
388*	21.1	453*	19.4	527*	18.5	MONTGOMERY	OH	-8	-5	-12	37	77
13	15.1	20	21.8	11	8.4	MORGAN	OH	44	-62	-45	09	09
15	14.7	28*	25.3	25	19.4	MORROW	OH	72	-23	32	75	82
72	16.7	83	17.7	107	20.9	MUSKINGUM	OH	6	18	25	71	89
14	19.5	11	16.0	17	22.2	NOBLE	OH	-18	39	14	64	92
32	20.5	37	18.9	40	16.4	OTTAWA	OH	-8	-13	-20	29	62
14	18.4	12	12.7	19	17.7	PAULDING	OH	-31	40	-4	47	72
40	24.5	40	23.7	44	21.4	PERRY	OH	-3	-10	-13	37	90
26	16.7	32	18.7	44	22.9	PICKAWAY	OH	12	22	37	77	94
16	22.6	21	22.6	28	22.2	PIKE	OH	0	-2	-2	49	92
65	21.2	63	16.0	84	16.4	PORTAGE	OH	-24	2	-23	26	62
27	17.8	22	12.1	38	17.0	PREBLE	OH	-32	41	-4	47	66
33	24.4	29	19.8	36	20.6	PUTNAM	OH	-19	4	-15	34	88
83	18.3	100	17.9	116	16.7	RICHLAND	OH	-2	-6	-9	41	64
45	17.0	59	19.7	71	20.5	ROSS	OH	16	4	21	69	88
60	21.0	58	17.7	58	16.1	SANDUSKY	OH	-16	-9	-24	25	59
77	20.2	103*	21.6	98	17.0	SCIOTO	OH	7	-21	-16	33	66
56	18.7	48	13.9	62	16.6	SENECA	OH	-26	19	-11	38	63
35	21.7	30	16.4	50	21.7	SHELBY	OH	-24	32	0	51	91
279	19.3	347	18.8	383	17.0	STARK	OH	-3	-9	-12	38	66
368	19.4	458	18.5	488	15.9	SUMMIT	OH	-4	-14	-18	31	59
94	13.2	155	16.3	218	17.8	TRUMBULL	OH	23	9	35	76	72
68	16.9	103*	22.2	86	15.5	TUSCARAWAS	OH	31	-30	-9	41	55
25	18.5	29	20.1	22	14.0	UNION	OH	9	-31	-25	23	42
24	13.6	41	19.3	45	20.8	VAN WERT	OH	42	8	53	82	88
19*	34.2	5	9.0	6	9.8	VINTON	OH	-74	9	-71	01	15
33	16.3	48	19.0	70	19.0	WARREN	OH	16	0	16	66	80
47	15.7	74*	22.5	76	19.9	WASHINGTON	OH	44	-12	27	72	84
57	17.3	62	15.3	82	16.6	WAYNE	OH	-12	8	-4	46	63
28	17.6	33	16.9	45	19.0	WILLIAMS	OH	-4	12	8	59	80
40	12.7	62	16.3	74	15.6	WOOD	OH	28	-4	23	70	56
20	15.5	19	13.1	30	17.8	WYANDOT	OH	-16	36	14	64	72
1430	13.8	1891	13.9	2595	14.9	OKLAHOMA	OK	1	7	8		
9	16.0	3	3.5	13	14.0	ADAIR	OK	-78	296	-13	37	43
14	20.9	16	19.3	16	20.5	ALFALFA	OK	-8	7	-2	49	87
6	11.5	10	15.5	5	6.5	ATOKA	OK	35	-58	-43	09	06
6	15.9	5	11.9	6	12.3	BEAVER	OK	-25	4	-22	26	29
19	16.6	18	11.7	23	10.5	BECKHAM	OK	-30	-10	-37	13	18
14	20.1	14	15.0	15	10.8	BLAINE	OK	-25	-28	-46	08	19
18	12.1	25	14.1	45	20.0	BRYAN	OK	17	42	66	85	85
27	16.7	33	17.6	39	16.7	CADDO	OK	6	-5	0	51	64
15	12.3	26	16.3	30	14.0	CANADIAN	OK	32	-14	14	64	43
27	13.9	37	15.3	27	9.1	CARTER	OK	10	-41	-35	14	12
7	10.0	8	7.3	20	14.0	CHEROKEE	OK	-27	91	40	78	43
4	4.5	16	16.1	16	15.3	CHOCTAW	OK	255	-5	238	97	54
1	6.2	5	24.8	5	17.0	CIMARRON	OK	297	-31	173	95	66
27	14.5	22	8.7	67	17.1	CLEVELAND	OK	-40	96	18	67	67
1	2.6	9	20.2	3	5.4	COAL	OK	666	-73	105	90	04

* SIGNIFICANTLY MORE DEATHS THAN EXPECTED

WHITE FEMALE — CANCER OF THE LARGE INTESTINE — COUNTY AND STATE — ICD 153,159.0

1950-1959 DEATHS	1950-1959 RATE	1960-1969 DEATHS	1960-1969 RATE	1970-1979 DEATHS	1970-1979 RATE	COUNTY AND STATE	PERCENT CHANGE 60S/50S	PERCENT CHANGE 70S/60S	PERCENT CHANGE 70S/50S	PERCENTILE %CHANGE 70S/50S	PERCENTILE RATES 1970S
12	6.8	41	16.5	59	16.8	COMANCHE OK	142	2	147	94	65
6	10.8	7	12.1	6	9.8	COTTON OK	12	-19	-9	40	15
8	7.3	16	13.9	20	17.7	CRAIG OK	90	27	142	93	72
40	18.8	38	14.6	59	18.0	CREEK OK	-23	24	-4	46	74
21	20.6	16	11.3	17	9.5	CUSTER OK	-45	-16	-54	05	13
14	22.4	15	15.2	19	12.8	DELAWARE OK	-32	-15	-43	10	33
7	16.4	4	8.8	10	15.8	DEWEY OK	-46	79	-3	47	58
5	12.7	5	12.4	5	8.2	ELLIS OK	-2	-33	-35	14	09
35	12.7	51	13.6	63	13.5	GARFIELD OK	7	-1	6	57	39
14	10.2	24	13.7	35	14.3	GARVIN OK	34	4	39	78	45
18	9.7	28	13.6	22	9.2	GRADY OK	41	-32	-5	46	12
9	14.8	11	15.2	10	9.9	GRANT OK	3	-35	-33	16	15
10	15.1	20	22.4	17	17.3	GREER OK	49	-23	15	65	69
4	11.8	7	14.5	7	18.4	HARMON OK	23	27	56	83	76
5	18.2	6	16.0	2	4.5	HARPER OK	-12	-72	-75	01	03
4	6.8	4	6.1	13	14.2	HASKELL OK	-10	131	108	91	45
13	13.6	19	15.1	29	21.9	HUGHES OK	11	45	60	84	91
15	14.5	17	12.1	27	15.6	JACKSON OK	-17	29	7	58	56
11	16.9	7	8.9	10	13.3	JEFFERSON OK	-47	49	-22	27	37
1	2.0	12	20.9	12	16.4	JOHNSTON OK	939	-21	715	99	62
52	20.1	46	13.8	56	13.7	KAY OK	-31	-1	-32	17	40
10	16.2	7	8.8	14	16.8	KINGFISHER OK	-46	91	4	54	65
13	13.0	10	9.2	21	16.5	KIOWA OK	-29	78	27	72	62
2	4.3	4	8.1	9	14.1	LATIMER OK	86	75	225	97	44
17	10.1	21	10.0	44	17.1	LE FLORE OK	-1	71	69	86	67
15	12.1	17	11.3	22	14.0	LINCOLN OK	-7	25	16	65	43
17	15.0	24	18.0	32	16.0	LOGAN OK	21	-11	7	58	59
2	5.0	4	10.3			LOVE OK	106				
6	8.0	12	14.2	19	17.1	MC CLAIN OK	76	20	113	91	67
17	16.1	12	8.6	21	12.1	MC CURTAIN OK	-47	41	-25	23	27
5	7.9	13	18.4	10	10.3	MC INTOSH OK	132	-44	30	74	17
8	15.8	12	19.9	13	17.6	MAJOR OK	26	-11	12	62	71
12	26.1	9	14.4	8	9.9	MARSHALL OK	-45	-31	-62	03	15
6	6.1	18	13.9	30	17.6	MAYES OK	128	27	190	95	71
8	11.9	4	4.9	24	23.9	MURRAY OK	-59	385	101	90	95
36	13.4	46	13.8	65	16.9	MUSKOGEE OK	3	23	27	72	66
13	20.5	15	19.4	12	11.8	NOBLE OK	-5	-39	-42	10	25
14	23.4	14	21.1	8	9.5	NOWATA OK	-10	-55	-59	04	13
4	7.1	13	18.2	10	13.3	OKFUSKEE OK	157	-27	87	89	37
174	11.5	285	13.3	466	16.1	OKLAHOMA OK	15	22	41	78	60
25	12.9	34	15.0	40	14.6	OKMULGEE OK	16	-3	13	62	48
16	10.8	19	10.4	19	8.8	OSAGE OK	-4	-15	-19	30	11
24	15.2	37	19.4	30	13.5	OTTAWA OK	28	-31	-12	38	38
17	23.4	8	8.1	19	18.6	PAWNEE OK	-65	130	-20	28	77
28	14.2	40	16.4	39	13.4	PAYNE OK	16	-18	-5	45	38
29	15.5	32	13.6	45	15.0	PITTSBURG OK	-13	11	-3	47	51
19	11.4	22	11.3	31	11.7	PONTOTOC OK	-1	3	3	53	24
44	18.8	49	16.2	65	17.3	POTTAWATOMIE OK	-14	7	-8	42	68
2	4.0	7	10.6	17	19.7	PUSHMATAHA OK	167	87	399	98	83
7	23.8	4	10.6	5	12.7	ROGER MILLS OK	-56	21	-46	08	32
13	14.5	17	13.0	33	18.5	ROGERS OK	-10	42	28	73	77
19	12.9	31	18.3	37	18.7	SEMINOLE OK	42	2	45	80	77
9	11.1	15	14.3	20	12.4	SEQUOYAH OK	28	-13	11	61	30
24	15.3	30	12.1	39	12.0	STEPHENS OK	-21	-1	-22	27	27
9	16.1	17	21.8	11	11.4	TEXAS OK	35	-48	-29	20	23
14	16.7	13	13.4	18	14.3	TILLMAN OK	-20	7	-14	35	46
188	16.2	237	14.3	348	15.5	TULSA OK	-11	8	-4	47	56
6	9.7	8	9.6	18	14.0	WAGONER OK	-1	46	44	79	43
20	12.5	45	20.1	52	17.2	WASHINGTON OK	61	-15	38	77	67
11	12.6	18	19.7	12	9.5	WASHITA OK	56	-52	-25	23	13
16	19.4	13	12.2	20	15.7	WOODS OK	-37	28	-19	29	57
12	13.7	14	13.5	21	16.4	WOODWARD OK	-1	21	19	68	62
1205	15.3	1562	14.5	1942	13.4	OREGON OR	-5	-8	-12		
10	12.3	16	14.7	7	5.7	BAKER OR	20	-61	-53	05	04
13	9.6	32	17.4	31	11.6	BENTON OR	83	-33	22	70	24
64	13.6	117	16.9	160	14.8	CLACKAMAS OR	24	-12	9	59	50
12	10.4	30	16.2	31	13.3	CLATSOP OR	56	-18	28	73	37
14	13.5	15	11.7	27	15.4	COLUMBIA OR	-13	31	14	64	55
27	15.8	39	16.6	44	13.3	COOS OR	5	-20	-15	34	37
1	2.4	4	9.1	13	18.2	CROOK OR	285	99	668	99	75
4	12.1	9	13.6	15	15.9	CURRY OR	13	17	32	75	59
14	15.2	14	9.6	19	7.8	DESCHUTES OR	-37	-18	-49	07	08
32	15.5	42	14.5	46	10.9	DOUGLAS OR	-6	-25	-30	19	20
		3	21.4	2	11.6	GILLIAM OR		-46			24
7	28.0	3	7.9	5	11.6	GRANT OR	-72	48	-58	04	24
1	4.7	2	6.7	4	12.2	HARNEY OR	41	83	158	94	28
9	15.9	9	12.2	15	16.9	HOOD RIVER OR	-23	38	6	57	65
45	13.8	73	15.1	107	14.5	JACKSON OR	10	-4	6	56	48
3	19.8	2	9.5	2	5.0	JEFFERSON OR	-52	-47	-75	01	03
20	14.5	33	15.9	45	12.8	JOSEPHINE OR	10	-19	-11	38	33
24	20.2	31	16.0	36	13.7	KLAMATH OR	-21	-15	-32	16	40
9*	43.5	2	6.3	4	11.0	LAKE OR	-86	75	-75	01	20
89	15.7	106	12.7	171	14.2	LANE OR	-19	12	-10	39	45
18	15.8	25	15.3	36	16.5	LINCOLN OR	-3	8	4	55	62
42	17.5	43	13.2	61	13.1	LINN OR	-25	0	-25	23	36
10	14.6	19	19.5	32	22.9	MALHEUR OR	33	18	57	83	94
106	17.1	119	13.8	166	13.7	MARION OR	-19	-1	-20	29	40
3	11.4	3	10.5	4	12.2	MORROW OR	-8	16	7	57	28
452	15.5	542	14.9	554	13.4	MULTNOMAH OR	-4	-10	-14	35	37
21	17.2	27	15.1	31	11.9	POLK OR	-12	-21	-31	18	26
3	28.4			3	22.1	SHERMAN OR			-22	26	92
8	9.2	10	9.7	13	8.8	TILLAMOOK OR	5	-9	-5	45	10
21	11.3	30	12.1	42	13.7	UMATILLA OR	7	13	21	69	40
14	14.5	15	12.9	22	16.7	UNION OR	-11	29	15	65	64
3	9.7	7	17.9	4	6.8	WALLOWA OR	84	-62	-30	19	06
11	13.1	16	12.6	24	16.1	WASCO OR	-3	28	24	71	60
51	15.4	89	16.2	122	12.7	WASHINGTON OR	5	-21	-17	31	32
44	24.7	35	14.6	44	12.6	YAMHILL OR	-41	-14	-49	07	31

* SIGNIFICANTLY MORE DEATHS THAN EXPECTED

WHITE FEMALE

CANCER OF THE LARGE INTESTINE

ICD 153,159.0

1950-1959 DEATHS	1950-1959 RATE	1960-1969 DEATHS	1960-1969 RATE	1970-1979 DEATHS	1970-1979 RATE	COUNTY AND STATE		PERCENT CHANGE 60S/50S	PERCENT CHANGE 70S/60S	PERCENT CHANGE 70S/50S	PERCENTILE %CHANGE 70S/50S	PERCENTILE RATES 1970S
11374*	21.8	12733*	20.1	14343*	18.8	PENNSYLVANIA	PA	-8	-6	-14		
48	19.8	62	20.4	58	15.2	ADAMS	PA	3	-26	-24	25	53
1538*	21.8	1714*	19.6	1805*	17.7	ALLEGHENY	PA	-10	-10	-19	30	71
68	18.8	77	17.5	86	16.1	ARMSTRONG	PA	-7	-8	-15	35	59
158*	22.2	166	17.5	201	16.7	BEAVER	PA	-21	-5	-25	23	64
45	24.5	27	12.4	43	14.8	BEDFORD	PA	-49	20	-39	11	50
288	19.7	377*	20.9	419*	19.1	BERKS	PA	6	-9	-3	48	80
172*	21.5	172	18.5	179	17.0	BLAIR	PA	-14	-9	-21	27	66
58	19.1	53	15.8	60	15.2	BRADFORD	PA	-17	-4	-20	28	53
192*	22.6	261*	19.7	391*	19.5	BUCKS	PA	-13	-1	-14	36	83
95	18.4	111	17.3	142	17.6	BUTLER	PA	-6	2	-4	47	71
145	16.7	199	18.8	228	17.8	CAMBRIA	PA	13	-6	7	57	72
8	22.7	5	12.6	6	14.1	CAMERON	PA	-44	12	-38	12	44
62	21.3	59	17.2	76	18.2	CARBON	PA	-19	6	-15	35	75
47	17.7	57	17.0	73	17.0	CENTRE	PA	-4	0	-4	46	66
189*	23.3	226*	21.2	287*	20.4	CHESTER	PA	-9	-4	-12	37	87
43	22.6	44	20.6	45	17.5	CLARION	PA	-9	-15	-23	26	70
82	20.4	92	19.8	82	15.8	CLEARFIELD	PA	-3	-21	-23	25	57
40	23.1	43	21.5	38	15.8	CLINTON	PA	-7	-26	-31	17	58
65	22.2	69	19.7	72	17.7	COLUMBIA	PA	-11	-10	-20	28	71
86	19.8	87	17.5	102	17.5	CRAWFORD	PA	-11	0	-11	38	70
94	17.9	144	20.2	215*	22.1	CUMBERLAND	PA	13	9	23	71	92
194	17.9	267*	20.4	257	17.0	DAUPHIN	PA	14	-17	-5	45	66
535*	25.6	635*	21.5	806*	21.5	DELAWARE	PA	-16	0	-16	33	90
33	20.6	47*	25.2	50	21.5	ELK	PA	22	-15	4	55	90
223	20.0	225	16.6	295	18.2	ERIE	PA	-17	9	-9	40	75
123	16.0	155	16.8	172	15.7	FAYETTE	PA	5	-7	-2	49	57
4	16.1	7	23.9	2	6.1	FOPEST	PA	48	-74	-62	03	05
79	19.8	104	20.3	138*	20.5	FRANKLIN	PA	3	1	3	54	87
7	15.4	8	15.2	19*	27.8	FULTON	PA	-1	84	81	88	98
39	18.5	34	13.9	41	15.4	GREENE	PA	-25	11	-17	32	54
36	19.5	25	11.8	42	16.1	HUNTINGDON	PA	-40	36	-18	31	59
64	18.0	57	13.3	90	18.3	INDIANA	PA	-26	37	1	52	75
52	19.1	49	15.9	57	16.6	JEFFERSON	PA	-17	4	-13	36	63
14	19.4	14	15.4	21	17.7	JUNIATA	PA	-20	15	-9	41	72
400*	28.4	400*	24.3	481*	24.1	LACKAWANNA	PA	-14	-1	-15	34	95
243	17.8	339*	19.6	459*	20.8	LANCASTER	PA	10	6	17	66	88
77	15.3	116	18.8	142	19.9	LAWRENCE	PA	22	6	30	74	84
107*	24.7	116*	21.4	114	16.7	LEBANON	PA	-14	-22	-32	17	64
216	19.3	235	16.3	308	16.8	LEHIGH	PA	-16	3	-13	37	65
464*	23.8	517*	21.8	556*	19.4	LUZERNE	PA	-8	-11	-19	30	82
117	19.3	172*	24.6	172*	21.4	LYCOMING	PA	28	-13	11	61	90
33	11.3	67	19.6	70	17.9	MC KEAN	PA	73	-9	58	83	73
103	17.7	129	18.6	138	16.6	MERCER	PA	5	-11	-6	44	63
36	18.3	52	21.7	42	13.0	MIFFLIN	PA	19	-40	-29	20	35
37	18.1	61*	22.9	72	18.9	MONROE	PA	27	-17	4	55	79
457*	20.6	607*	19.6	875*	20.9	MONTGOMERY	PA	-5	7	1	52	89
19	17.9	12	8.7	19	12.2	MONTOUR	PA	-51	40	-32	17	28
193	19.5	201	16.7	263	17.9	NORTHAMPTON	PA	-15	7	-9	41	73
125	20.9	135	19.5	146	18.2	NORTHUMBERLAND	PA	-7	-7	-13	36	75
36*	31.1	26	17.9	37	19.9	PERRY	PA	-42	12	-36	14	85
2405*	25.5	2411*	23.9	2084*	20.4	PHILADELPHIA	PA	-6	-14	-20	29	87
24*	39.1	16	21.3	28	21.0	PIKE	PA	-46	-1	-46	08	89
19	18.7	20	18.5	25	19.9	POTTER	PA	-1	7	6	57	84
214*	22.0	228*	20.5	274*	20.6	SCHUYLKILL	PA	-7	0	-7	43	88
24	20.8	29	20.7	29	16.3	SNYDER	PA	-1	-21	-22	27	61
79	20.6	74	15.6	75	13.0	SOMEPSET	PA	-24	-17	-37	13	34
3	7.4	7	19.3	7	14.9	SULLIVAN	PA	159	-23	99	90	50
40	22.2	46	21.9	40	16.8	SUSQUEHANNA	PA	-1	-23	-24	24	65
46	23.4	42	20.4	55	22.5	TIOGA	PA	-13	10	-4	47	93
28	23.7	29	19.1	14	8.4	UNION	PA	-19	-56	-65	02	09
76	21.5	84*	22.0	76	17.9	VENANGO	PA	2	-19	-17	32	73
41	14.0	47	15.3	61	17.2	WARREN	PA	9	12	23	70	68
188	20.8	181	16.1	236	16.6	WASHINGTON	PA	-23	3	-20	28	63
59*	32.1	44	21.5	66*	24.9	WAYNE	PA	-33	16	-22	26	96
280	19.9	286	15.3	372	15.3	WESTMORELAND	PA	-23	0	-23	25	54
18	18.5	28*	23.3	29	21.0	WYOMING	PA	26	-10	14	63	89
241*	22.1	304*	21.3	380*	20.9	YORK	PA	-4	-2	-6	44	89
1030*	23.2	1220*	22.5	1458*	21.7	RHODE ISLAND	RI	-3	-3	-6		
36	23.0	45	20.4	67*	22.8	BRISTOL	RI	-11	12	-1	50	93
106*	23.9	134*	21.1	191*	21.0	KENT	RI	-11	-1	-12	37	89
86*	29.7	81*	22.6	120*	25.3	NEWPORT	RI	-24	12	-15	34	96
753*	22.7	891*	22.9	974*	21.3	PROVIDENCE	RI	1	-7	-6	44	90
49	21.4	69	20.9	106*	23.2	WASHINGTON	RI	-3	11	8	59	94
582	11.4	926	12.7	1302	12.4	SOUTH CAROLINA	SC	12	-3	9		
8	13.4	12	15.1	18	16.1	ABBEVILLE	SC	13	7	21	69	60
21	15.2	30	12.6	44	11.9	AIKEN	SC	-17	-5	-21	27	26
5	26.0	4	14.4	3	8.7	ALLENDALE	SC	-45	-39	-67	02	10
31	9.7	44	10.9	70	11.9	ANDERSON	SC	13	9	23	70	26
2	7.1	5	11.7	6	12.9	BAMBERG	SC	66	10	82	88	34
3	8.1	2	4.0	15	24.3	BARNWELL	SC	-51	506	199	96	95
5	10.8	8	17.2	12	10.3	BEAUFORT	SC	60	-40	-4	46	17
2	3.3	5	6.1	11	8.3	BERKELEY	SC	85	37	154	94	09
2	8.2	3	11.2	2	6.0	CALHOUN	SC	36	-47	-27	21	05
46	12.3	69	13.1	97	13.1	CHARLESTON	SC	7	0	7	57	35
8	8.6	19	12.7	22	12.5	CHEROKEE	SC	48	-2	45	80	31
15	19.3	16	15.3	16	12.1	CHESTER	SC	-21	-21	-37	13	27
6	7.2	12	11.8	13	8.9	CHESTERFIELD	SC	65	-25	24	71	11
2	5.5	10	18.7	5	7.4	CLARENDON	SC	241	-60	35	76	07
12	21.8	10	14.6	9	9.0	COLLETON	SC	-33	-38	-59	04	11
9	9.4	22	14.9	28	13.8	DARLINGTON	SC	59	-8	46	81	41
2	2.9	4	5.3	12	11.6	DILLON	SC	82	121	302	98	24
5	9.7	5	7.5	15	12.3	DORCHESTER	SC	-22	64	27	73	29
3	9.2	3	7.5	8	13.1	EDGEFIELD	SC	-19	76	42	79	36
7	19.1	7	14.1	13	21.6	FAIRFIELD	SC	-26	53	13	63	91

* SIGNIFICANTLY MORE DEATHS THAN EXPECTED

WHITE FEMALE — CANCER OF THE LARGE INTESTINE — ICD 153,159.0

1950-1959 DEATHS	1950-1959 RATE	1960-1969 DEATHS	1960-1969 RATE	1970-1979 DEATHS	1970-1979 RATE	COUNTY AND STATE		PERCENT CHANGE 60S/50S	PERCENT CHANGE 70S/60S	PERCENT CHANGE 70S/50S	PERCENTILE %CHANGE 70S/50S	PERCENTILE RATES 1970S
16	10.3	38	16.7	41	12.9	FLORENCE	SC	63	-23	25	72	34
4	6.7	9	14.4	8	8.1	GEORGETOWN	SC	114	-43	21	69	09
56	10.1	113	13.6	161	13.1	GREENVILLE	SC	35	-3	30	74	36
15	11.3	23	13.1	29	11.8	GREENWOOD	SC	16	-10	4	55	25
4	10.5	3	7.5	6	11.4	HAMPTON	SC	-29	51	8	59	22
8	5.7	17	8.3	38	11.7	HORRY	SC	46	41	106	90	25
2	9.2	4	16.4	3	7.9	JASPER	SC	77	-52	-15	34	08
9	13.2	17	19.4	17	12.8	KERSHAW	SC	47	-34	-3	48	33
16	18.7	14	11.3	22	11.2	LANCASTER	SC	-40	-1	-40	11	22
15	10.3	19	10.5	26	11.2	LAURENS	SC	3	6	9	60	21
5	14.8	4	9.5	1	2.0	LEE	SC	-36	-79	-87	00	01
15	10.0	21	8.8	54	12.7	LEXINGTON	SC	-12	45	28	73	32
				4	16.9	MC CORMICK	SC					66
8	14.9	10	13.7	12	12.6	MARION	SC	-8	-8	-15	34	32
8	14.4	3	3.7	14	14.7	MARLBORO	SC	-74	295	2	52	48
12	12.7	24	18.3	19	10.9	NEWBERRY	SC	45	-40	-14	36	20
23	19.9	17	10.5	24	10.6	OCONEE	SC	-47	0	-47	08	18
11	10.2	22	13.2	31	12.3	ORANGEBURG	SC	30	-7	21	69	29
13	8.6	23	11.4	38	12.6	PICKENS	SC	32	10	45	80	31
58	13.8	90	15.2	116	14.4	RICHLAND	SC	9	-5	4	55	47
3	6.6	8	14.7	7	9.8	SALUDA	SC	122	-33	49	81	15
36	8.5	82	13.0	111	12.4	SPARTANBURG	SC	52	-4	46	80	30
12	15.0	15	11.4	29	15.1	SUMTER	SC	-24	33	1	51	53
12	14.5	14	13.6	9	6.4	UNION	SC	-7	-53	-56	04	05
2	3.8	9	14.5	7	8.2	WILLIAMSBURG	SC	284	-44	116	92	09
25	14.0	37	13.9	56	15.4	YORK	SC	-1	11	10	61	55
447	15.1	559	15.1	759	16.8	SOUTH DAKOTA	SD	1	11	12		
3	14.2	4	11.2	4	9.3	AURORA	SD	-21	-17	-35	15	12
13	13.5	21	15.7	30	17.2	BEADLE	SD	16	10	27	73	68
1	9.6					BENNETT	SD					
7	13.6	10	17.1	12	17.5	BON HOMME	SD	25	2	28	73	70
10	11.3	12	11.6	13	11.6	BROOKINGS	SD	3	0	3	53	24
16	10.3	20	10.2	38	15.1	BROWN	SD	-1	48	47	81	52
5	17.9	7	19.5	2	4.3	BRULE	SD	9	-78	-76	01	03
		1	37.7			BUFFALO	SD					
4	12.4	6	11.4	9	15.8	BUTTE	SD	-8	39	27	72	58
		4	29.5	1	6.0	CAMPBELL	SD		-80			05
6	12.3	6	9.9	11	19.7	CHARLES MIX	SD	-19	99	60	84	84
9	24.5	10	27.2	8	16.1	CLARK	SD	11	-41	-34	15	60
7	13.2	5	8.4	12	18.9	CLAY	SD	-37	125	42	79	79
12	12.2	16	11.8	26	16.8	CODINGTON	SD	-3	42	38	77	65
4	41.8	2	13.8	1	7.1	CORSON	SD	-67	-49	-83	00	07
3	12.7	7	23.8	2	6.6	CUSTER	SD	87	-72	-48	07	06
20	20.9	24	21.1	16	11.2	DAVISON	SD	1	-47	-47	08	21
6	10.8	14	24.1	14	16.8	DAY	SD	124	-31	55	83	64
2	4.9	2	3.7	5	10.1	DEUEL	SD	-25	174	106	90	16
2	18.0	1	6.5	6*	46.9	DEWEY	SD	-64	623	160	94	99
9*	37.4	2	7.6	7	23.9	DOUGLAS	SD	-80	216	-36	13	95
7	26.8	3	8.3	5	13.0	EDMUNDS	SD	-69	57	-52	06	34
4	7.5	10	16.8	4	5.9	FALL RIVER	SD	123	-65	-22	26	04
2	11.1	1	2.7	5	15.0	FAULK	SD	-75	447	35	76	51
7	14.8	9	14.1	19*	26.7	GRANT	SD	-4	89	81	88	97
13*	33.0	9	19.9	3	4.2	GREGORY	SD	-40	-79	-87	00	03
3	23.1	2	11.3	2	11.6	HAAKON	SD	-51	2	-50	06	24
6	18.2	10	21.1	15*	29.2	HAMLIN	SD	16	38	60	84	98
3	15.0	8	27.1	10	24.6	HAND	SD	80	-9	64	85	96
2	8.8	1	4.1	2	11.2	HANSON	SD	-53	173	27	73	22
				2	15.4	HARDING	SD					54
10	29.1	10	18.8	14	21.8	HUGHES	SD	-36	16	-25	23	91
6	11.4	8	9.9	27*	31.6	HUTCHINSON	SD	-13	218	176	95	98
1	6.9	4	21.6	1	2.6	HYDE	SD	214	-88	-63	03	02
1	8.6	3	23.3			JACKSON	SD	169				
3	11.2	5	17.2	7	22.7	JERAULD	SD	53	32	102	90	93
		2	21.8	2	17.9	JONES	SD		-18			73
12	25.9	9	19.1	15	23.8	KINGSBURY	SD	-26	24	-8	41	95
8	13.0	17	23.3	16	16.0	LAKE	SD	79	-31	23	71	59
21	25.2	15	16.6	26	23.2	LAWRENCE	SD	-34	40	-8	42	94
12	16.6	11	12.3	16	13.0	LINCOLN	SD	-26	5	-22	26	34
3	19.8	3	15.4	8*	46.4	LYMAN	SD	-22	201	135	93	99
10	23.8	7	13.1	10	19.4	MC COOK	SD	-45	48	-18	30	82
		4	12.9	6	13.7	MC PHERSON	SD		6			40
6	19.7	6	17.8	7	16.7	MARSHALL	SD	-10	-6	-15	34	64
2	4.0	5	8.4	10	11.9	MEADE	SD	113	41	199	96	25
3	9.3	3	8.3	9*	60.7	MINER	SD	-11	629	552	99	99
55	15.7	82	17.6	104	16.8	MINNEHAHA	SD	12	-5	7	57	64
5	12.1	11	23.2	10	16.4	MOODY	SD	92	-29	36	77	61
28	18.1	31	14.7	45	15.2	PENNINGTON	SD	-19	3	-16	33	53
4	16.3	4	14.1	9	27.0	PERKINS	SD	-14	92	66	85	97
5	24.5	3	13.4	5	13.1	POTTER	SD	-45	-2	-47	08	35
5	8.1	11	16.2	22*	25.9	ROBERTS	SD	101	60	221	96	97
7	27.9	5	18.4	6	18.8	SANBORN	SD	-34	2	-33	16	78
4	5.4	12	16.8	12	12.9	SPINK	SD	210	-23	137	93	34
2	25.7	4*	41.8	4	46.4	STANLEY	SD	63	11	81	88	99
1	9.4			2	19.9	SULLY	SD			112	91	85
		2	24.5	1	11.2	TODD	SD		-54			22
7	18.0	6	11.5	17*	36.4	TRIPP	SD	-36	217	102	90	99
11	16.6	18	21.5	21	23.8	TURNER	SD	30	10	43	79	95
7	12.8	5	7.6	17	19.7	UNION	SD	-41	160	54	83	84
6	18.8	8	19.5	8	12.4	WALWORTH	SD	4	-37	-34	15	30
16	15.0	18	14.2	17	12.8	YANKTON	SD	-5	-10	-15	34	33
				1	22.1	ZIEBACH	SD					92
1666	13.5	2397	14.4	3136	14.1	TENNESSEE	TN	7	-2	4		
16	11.1	30	13.3	45	13.0	ANDERSON	TN	20	-3	17	66	35
15	13.7	26	19.1	25	14.4	BEDFORD	TN	40	-25	6	56	47
10	15.5	7	8.9	16	14.0	BENTON	TN	-43	58	-9	40	43
2	6.6	4	12.7	6	15.3	BLEDSOE	TN	92	20	132	93	54
16	8.7	42	15.0	25	6.1	BLOUNT	TN	71	-59	-30	18	05

* SIGNIFICANTLY MORE DEATHS THAN EXPECTED

WHITE FEMALE

CANCER OF THE LARGE INTESTINE

ICD 153,159.0

1950-1959 DEATHS	1950-1959 RATE	1960-1969 DEATHS	1960-1969 RATE	1970-1979 DEATHS	1970-1979 RATE	COUNTY AND STATE		PERCENT CHANGE 60S/50S	PERCENT CHANGE 70S/60S	PERCENT CHANGE 70S/50S	PERCENTILE %CHANGE 70S/50S	PERCENTILE RATES 1970S
19	15.9	21	13.1	41	15.0	BRADLEY	TN	-18	14	-6	44	51
15	12.9	14	9.2	22	10.6	CAMPBELL	TN	-28	15	-18	31	18
3	7.6	3	5.8	8	12.7	CANNON	TN	-23	117	66	86	32
14	11.7	25	15.7	24	11.4	CARROLL	TN	34	-27	-3	48	23
24	18.5	28	15.2	30	10.5	CARTER	TN	-18	-31	-43	09	18
5	10.9	4	7.5	16	20.4	CHEATHAM	TN	-31	171	87	89	87
3	6.8	11	20.2	12	17.9	CHESTER	TN	197	-11	163	94	73
16	18.7	13	13.1	12	9.5	CLAIBORNE	TN	-30	-27	-49	07	13
2	5.5	6	17.9	2	3.6	CLAY	TN	223	-80	-36	14	02
13	17.3	14	14.0	18	13.0	COCKE	TN	-19	-7	-25	23	34
15	16.5	21	15.9	28	13.3	COFFEE	TN	-4	-16	-19	29	37
7	11.0	11	13.1	19	20.5	CROCKETT	TN	19	57	86	89	87
5	7.4	14	13.8	17	10.6	CUMBERLAND	TN	86	-23	43	79	18
180	13.7	279	15.6	338	14.4	DAVIDSON	TN	13	-7	5	56	47
5	12.2	7	11.9	16	19.9	DECATUR	TN	-2	67	63	85	84
4	8.7	12	17.9	8	8.5	DE KALB	TN	106	-53	-3	48	10
14	15.6	14	12.2	29	17.8	DICKSON	TN	-22	47	14	64	72
10	7.4	19	12.0	34	16.3	DYER	TN	62	37	121	92	61
5	12.7	3	6.3	11	17.1	FAYETTE	TN	-50	172	35	76	67
3	5.5	2	3.2	12	17.0	FENTRESS	TN	-42	427	207	96	66
17	16.9	11	8.3	27	16.2	FRANKLIN	TN	-51	96	-4	46	60
31	15.3	45	17.1	42	12.4	GIBSON	TN	12	-27	-19	30	30
12	10.6	22	16.5	23	13.9	GILES	TN	56	-15	32	75	42
4	6.6	15	22.4	10	11.9	GRAINGER	TN	239	-47	80	87	26
28	18.0	22	9.9	34	11.9	GREENE	TN	-45	20	-34	15	26
5	10.9	5	8.5	11	14.3	GRUNDY	TN	-21	67	31	74	45
11	10.5	13	8.8	31	15.2	HAMBLEN	TN	-16	73	46	80	53
118	15.4	167	16.0	209	14.7	HAMILTON	TN	4	-8	-4	46	49
5	19.0	5	13.1	9	21.2	HANCOCK	TN	-31	61	11	61	90
8	8.3	8	8.7	19	14.9	HARDEMAN	TN	5	72	80	88	50
9	13.3	10	10.3	19	14.4	HARDIN	TN	-22	39	8	59	46
12	10.7	17	10.8	26	12.4	HAWKINS	TN	0	15	16	65	30
6	11.3	10	16.2	16	20.8	HAYWOOD	TN	43	28	84	88	88
9	11.2	15	15.4	17	13.1	HENDERSON	TN	38	-15	17	66	35
22	18.4	30	19.5	28	15.8	HENRY	TN	6	-19	-14	35	58
8	13.8	7	10.6	10	11.1	HICKMAN	TN	-23	5	-19	29	21
4	19.0	4	12.2	5	12.4	HOUSTON	TN	-36	2	-35	15	30
9	18.7	5	7.4	17	19.1	HUMPHREYS	TN	-61	158	2	53	80
1	1.3	2	3.4	7	10.6	JACKSON	TN	163	210	715	99	18
11	12.1	18	15.9	19	11.3	JEFFERSON	TN	32	-29	-7	43	22
7	14.4	1	1.7	7	8.1	JOHNSON	TN	-88	376	-44	09	08
139	15.4	186	14.5	251	14.7	KNOX	TN	-6	1	-5	46	48
3	11.1	4	11.2	11	23.4	LAKE	TN	1	108	111	91	94
7	9.6	11	12.4	7	6.4	LAUDERDALE	TN	30	-49	-33	16	05
14	12.0	21	14.0	24	11.7	LAWRENCE	TN	16	-16	-2	49	24
8	40.7	3	8.8	6	11.3	LEWIS	TN	-78	28	-72	01	22
16	15.3	22	17.4	33	20.5	LINCOLN	TN	14	18	34	76	87
14	14.4	17	13.4	28	15.8	LOUDON	TN	-7	17	9	60	57
18	13.4	27	14.7	41	16.9	MC MINN	TN	10	15	26	72	65
12	14.7	18	16.1	25	17.6	MC NAIRY	TN	10	10	20	68	71
4	7.0	9	11.4	13	13.8	MACON	TN	63	22	98	90	41
24	10.8	39	12.3	66	17.6	MADISON	TN	13	43	62	85	70
15	21.1	17	19.9	13	10.6	MARION	TN	-5	-47	-50	06	18
15	16.7	7	6.1	15	10.2	MARSHALL	TN	-64	68	-39	12	17
25	16.3	21	10.6	53	21.3	MAURY	TN	-35	101	31	74	90
1	4.4	2	9.0	2	7.7	MEIGS	TN	104	-15	74	87	08
9	10.4	19	17.3	23	15.1	MONROE	TN	66	-12	46	80	53
16	11.9	23	13.2	39	14.8	MONTGOMERY	TN	12	11	24	71	49
2	11.0	2	9.6	4	17.8	MOORE	TN	-12	85	62	84	72
4	7.3	7	10.8	10	12.7	MORGAN	TN	47	18	73	87	32
15	10.1	37	19.0	28	11.8	OBION	TN	87	-38	16	66	25
12	17.9	13	14.7	15	13.6	OVERTON	TN	-18	-8	-24	24	40
2	8.0	4	13.1	8	17.0	PERRY	TN	64	29	113	91	66
1	5.5	1	5.9			PICKETT	TN	7				
7	13.8	10	16.1	10,	13.6	POLK	TN	17	-16	-1	50	40
20	14.7	26	14.3	32	13.1	PUTNAM	TN	-3	-9	-11	38	35
12	17.6	16	18.0	10	8.8	RHEA	TN	2	-51	-50	06	11
13	10.1	23	13.0	31	12.4	ROANE	TN	29	-4	23	71	30
14	12.0	19	13.4	19	9.7	ROBERTSON	TN	12	-27	-18	30	14
15	10.0	23	10.9	45	14.4	RUTHERFORD	TN	9	33	45	80	47
3	6.2	13	18.4	9	10.9	SCOTT	TN	198	-41	76	87	20
3	12.7	1	3.8	6	15.4	SEQUATCHIE	TN	-70	308	22	69	55
9	10.8	18	14.9	16	8.7	SEVIER	TN	38	-41	-19	29	10
240	16.3	367	17.5	429	16.4	SHELBY	TN	7	-7	0	51	62
7	10.7	11	14.4	12	11.7	SMITH	TN	34	-18	9	60	25
3	7.3	2	4.6	7	12.5	STEWART	TN	-37	173	73	87	31
48	14.1	57	10.9	99	13.5	SULLIVAN	TN	-23	24	-5	46	38
29	20.5	32	16.4	34	10.8	SUMNER	TN	-20	-34	-47	08	19
7	9.8	19	20.5	20	15.9	TIPTON	TN	110	-22	63	85	58
3	11.6	5	17.5			TROUSDALE	TN	52				
6	10.6	10	12.2	13	12.1	UNICOI	TN	15	-1	14	64	27
		7	16.3	7	11.6	UNION	TN		-29			24
		4	20.8	4	19.5	VAN BUREN	TN		-6			82
13	12.6	20	14.6	32	17.3	WARREN	TN	16	19	37	77	68
27	11.8	49	14.8	62	13.7	WASHINGTON	TN	25	-8	15	65	40
7	13.7	7	12.1	10	13.0	WAYNE	TN	-12	8	-5	45	35
19	11.8	37	19.5	36	14.3	WEAKLEY	TN	66	-27	22	70	46
7	9.2	11	11.3	16	12.7	WHITE	TN	22	13	38	77	32
12	12.5	12	9.2	28	14.2	WILLIAMSON	TN	-26	54	14	64	45
8	7.1	26	15.7	34	13.9	WILSON	TN	122	-11	98	90	42
3899	12.7	5826	13.4	8003	13.5	TEXAS	TX	5	1	6		
20	14.8	20	10.8	29	13.5	ANDERSON	TX	-27	25	-9	40	38
2	19.8	7	26.6	6	13.2	ANDREWS	TX	35	-51	-34	16	36
21	15.6	17	8.1	30	10.0	ANGELINA	TX	-48	23	-36	13	16
4	22.2	3	7.9	6	8.0	ARANSAS	TX	-65	2	-64	03	08
4	13.5	6	15.2	7	14.0	ARCHER	TX	12	-8	4	54	43
		2	11.8	1	6.2	ARMSTRONG	TX		-47			05
7	10.1	7	7.5	20	16.8	ATASCOSA	TX	-26	123	66	86	65
11	12.8	17	18.4	29	18.8	AUSTIN	TX	43	2	46	81	78

WHITE FEMALE — CANCER OF THE LARGE INTESTINE — ICD 153,159.0

1950-1959 DEATHS	1950-1959 RATE	1960-1969 DEATHS	1960-1969 RATE	1970-1979 DEATHS	1970-1979 RATE	COUNTY AND STATE	PERCENT CHANGE 60S/50S	PERCENT CHANGE 70S/60S	PERCENT CHANGE 70S/50S	PERCENTILE %CHANGE 70S/50S	PERCENTILE RATES 1970S
6	27.2	4	13.6	4	9.9	BAILEY TX	-50	-27	-63	03	15
3	11.6	4	10.4	10	19.4	BANDERA TX	-10	87	68	86	82
3	4.1	13	13.9	23	16.7	BASTROP TX	241	20	310	98	64
3	7.4	5	9.8	10	14.6	BAYLOR TX	32	49	96	89	48
16	22.8	11	12.2	10	8.9	BEE TX	-47	-27	-61	03	11
35	12.9	46	11.6	51	9.4	BELL TX	-11	-19	-28	21	13
256	12.2	389	12.9	567	14.3	BEXAR TX	6	11	17	66	46
3	14.3	5	14.8	3	5.9	BLANCO TX	3	-60	-59	04	05
		1	21.3			BORDEN TX					
10	10.2	13	11.1	21	13.5	BOSQUE TX	9	22	32	75	39
23	10.3	45	14.8	56	13.9	BOWIE TX	43	-6	34	76	42
15	10.5	38	15.5	54	12.8	BRAZORIA TX	48	-17	22	70	33
19	18.6	26	17.0	33	14.0	BRAZOS TX	-8	-18	-25	24	43
5	16.7	6	18.9	7	19.3	BREWSTER TX	13	2	16	65	81
1	6.0	4	21.5	5	30.5	BRISCOE TX	258	42	407	98	98
2	7.5	3	9.1	2	5.2	BROOKS TX	22	-43	-30	18	04
20	10.7	30	12.6	44	15.4	BROWN TX	18	22	44	79	55
6	12.3	10	15.7	20*	27.2	BURLESON TX	28	73	122	92	97
6	10.8	10	10.9	20	12.4	BURNET TX	1	14	15	65	30
9	9.7	9	7.3	17	11.2	CALDWELL TX	-25	53	15	64	21
3	9.3	8	19.9	13	19.7	CALHOUN TX	114	-1	112	91	84
7	12.2	13	16.0	17	19.6	CALLAHAN TX	31	23	60	84	83
34	8.3	38	7.4	57	7.5	CAMERON TX	-12	2	-10	39	07
6	18.2	2	4.1	5	10.4	CAMP TX	-77	152	-43	10	18
2	8.0	5	15.2	10	22.4	CARSON TX	90	47	179	95	92
7	9.1	18	13.3	24	13.7	CASS TX	47	3	51	82	41
1	8.6	1	5.2	3	11.5	CASTRO TX	-39	120	34	76	23
4	30.2	3	7.8	10	18.1	CHAMBERS TX	-74	131	-40	11	74
23	14.9	26	12.6	16	5.5	CHEROKEE TX	-15	-56	-63	03	04
5	9.4	7	10.5	8	11.0	CHILDRESS TX	12	5	17	66	21
5	8.8	13	20.7	14	18.3	CLAY TX	135	-12	108	91	75
1	7.0	3	12.5	3	11.9	COCHRAN TX	77	-4	69	86	26
4	19.4	5	18.9	2	5.8	COKE TX	-2	-69	-70	01	04
6	6.8	16	14.4	17	11.1	COLEMAN TX	110	-23	61	84	21
29	12.7	36	11.9	48	11.1	COLLIN TX	-6	-7	-13	37	21
1	2.4	6	13.7	4	14.9	COLLINGSWORTH TX	460	9	508	99	50
13	17.8	9	8.9	9	7.2	COLORADO TX	-50	-19	-59	03	07
9	10.2	17	13.7	32	14.6	COMAL TX	34	6	43	79	48
13	12.4	20	14.7	10	7.2	COMANCHE TX	19	-51	-42	10	07
		7	21.6	2	5.2	CONCHO TX		-76			03
14	12.3	16	10.7	26	14.8	COOKE TX	-13	38	20	68	50
12	13.0	14	11.1	38*	23.7	CORYELL TX	-15	114	83	88	95
3	16.3	3	13.8	3	7.9	COTTLE TX	-15	-43	-52	06	08
1	11.1	2	11.9	5	29.0	CRANE TX	7	143	161	94	98
		2	15.1	2	12.3	CROCKETT TX		-18			29
5	13.4	4	8.8	7	12.8	CROSBY TX	-34	45	-4	46	33
		1	8.3	1	10.1	CULBERSON TX		22			16
3	12.7	1	3.2	1	2.6	DALLAM TX	-75	-19	-80	00	02
399	15.0	617	15.1	834	14.5	DALLAS TX	1	-4	-3	48	48
8	14.0	10	12.2	14	14.1	DAWSON TX	-13	15	1	52	44
2	6.2	4	7.0	10	11.7	DEAF SMITH TX	12	67	87	89	25
8	14.0	8	13.0	3	4.5	DELTA TX	-7	-66	-68	02	03
28	13.2	45	16.1	43	10.4	DENTON TX	22	-35	-21	27	17
7	5.9	17	11.7	26	12.4	DE WITT TX	98	6	111	91	30
2	6.8	2	6.8	2	5.3	DICKENS TX	0	-22	-22	27	04
3	11.9	6	18.6	2	4.1	DIMMIT TX	56	-78	-66	02	03
5	15.8	4	10.7	4	8.1	DONLEY TX	-32	-25	-49	07	08
1	2.5	6	10.1	7	10.0	DUVAL TX	303	-2	297	98	15
24	15.4	22	11.5	25	13.9	EASTLAND TX	-25	20	-10	39	42
12	10.4	33	17.4	38	11.0	ECTOR TX	68	-37	6	57	20
2	17.2	2	17.1	3	24.6	EDWARDS TX	0	44	43	79	96
22	10.9	36	13.4	37	10.2	ELLIS TX	23	-24	-6	44	16
79	10.9	124	12.0	196	15.6	EL PASO TX	10	30	42	79	56
15	12.2	19	11.7	18	8.6	ERATH TX	-4	-27	-30	19	10
12	11.0	16	11.6	19	11.4	FALLS TX	6	-2	3	54	22
26	14.6	26	13.6	28	11.0	FANNIN TX	-7	-19	-24	24	21
16	11.7	26	17.0	26	13.9	FAYETTE TX	45	-18	19	67	42
3	7.9	3	6.0	9	17.8	FISHER TX	-24	196	125	92	72
11	21.9	5	7.7	9	15.0	FLOYD TX	-65	94	-31	18	52
2	7.9	3	11.8	4	11.9	FOARD TX	50	1	51	82	26
12	14.0	19	13.8	25	10.4	FORT BEND TX	-2	-25	-26	22	17
3	10.6	3	7.7	15	24.7	FRANKLIN TX	-27	220	133	93	96
4	5.3	10	10.7	8	9.9	FREESTONE TX	102	-7	88	89	15
1	2.6	9	18.4	7	12.4	FRIO TX	606	-33	376	98	30
5	14.5	7	18.9	7	13.2	GAINES TX	31	-30	-9	41	36
49	13.6	76	15.0	119	16.5	GALVESTON TX	11	10	22	70	63
2	8.4	2	6.5	4	10.2	GARZA TX	-23	57	21	69	16
3	3.8	15	14.4	17	11.1	GILLESPIE TX	275	-23	190	95	21
				1	29.8	GLASSCOCK TX					98
7	30.9	7	23.6	6	17.5	GOLIAD TX	-24	-26	-43	09	70
12	12.4	18	17.3	10	9.3	GONZALES TX	40	-46	-25	23	13
16	16.7	26	18.7	28	15.9	GRAY TX	12	-15	-5	46	58
53	13.7	64	12.8	76	12.0	GRAYSON TX	-7	-6	-12	37	26
27	14.1	55	18.3	52	11.9	GREGG TX	30	-35	-15	34	26
5	7.6	11	16.0	10	12.1	GRIMES TX	110	-24	59	84	28
10	9.5	21	13.0	42	18.1	GUADALUPE TX	37	40	91	89	74
15	12.6	22	13.6	19	9.4	HALE TX	8	-31	-25	23	13
4	8.5	6	15.6	8	12.4	HALL TX	83	-20	46	80	30
9	11.6	15	16.2	16	13.8	HAMILTON TX	40	-14	19	68	41
3	20.6	1	4.0	4	13.2	HANSFORD TX	-81	230	-36	13	36
6	11.8	5	6.3	8	10.7	HARDEMAN TX	-47	71	-9	40	19
9	11.8	12	11.1	19	11.6	HARDIN TX	-7	5	-2	49	24
417	15.0	628	14.4	932	14.8	HARRIS TX	-4	2	-1	50	49
16	13.6	28	16.8	30	13.0	HARRISON TX	24	-23	-4	46	34
1	18.9	3	28.2	1	5.5	HARTLEY TX	50	-81	-71	01	04
9	14.5	11	16.9	12	16.4	HASKELL TX	17	-3	13	63	61
14	18.0	15	14.2	16	12.0	HAYS TX	-21	-15	-33	16	27
3	17.7	1	4.8	3	12.9	HEMPHILL TX	-73	167	-27	21	34
22	21.5	20	13.4	34	13.6	HENDERSON TX	-38	2	-37	13	40
37	8.3	52	8.1	75	7.9	HIDALGO TX	-3	-2	-4	46	08
19	10.9	36	17.0	35	14.7	HILL TX	56	-14	35	76	49

* SIGNIFICANTLY MORE DEATHS THAN EXPECTED

WHITE FEMALE

CANCER OF THE LARGE INTESTINE

ICD 153,159.0

1950-1959 DEATHS	1950-1959 RATE	1960-1969 DEATHS	1960-1969 RATE	1970-1979 DEATHS	1970-1979 RATE	COUNTY AND STATE	PERCENT CHANGE 60S/50S	PERCENT CHANGE 70S/60S	PERCENT CHANGE 70S/50S	PERCENTILE %CHANGE 70S/50S	PERCENTILE RATES 1970S
5	9.6	14	17.3	11	10.7	HOCKLEY TX	80	-38	11	61	18
5	16.5	12	27.6	16	18.3	HOOD TX	68	-34	11	61	76
14	11.3	22	13.7	35	17.4	HOPKINS TX	21	27	54	83	69
13	17.3	10	10.1	15	12.5	HOUSTON TX	-42	24	-28	21	31
10	8.6	14	9.6	24	12.5	HOWARD TX	11	31	45	80	31
		2	18.2	1	14.9	HUDSPETH TX		-18			50
38	16.3	35	11.8	66	17.0	HUNT TX	-28	44	4	54	66
5	6.2	18	16.2	25	17.1	HUTCHINSON TX	160	6	174	95	66
12	26.3	8	13.7	12	16.2	JACK TX	-48	18	-38	12	60
7	17.9	10	19.2	9	12.0	JACKSON TX	7	-38	-33	16	26
14	18.7	13	11.9	26	16.4	JASPER TX	-36	37	-12	37	62
91	15.5	165	19.4	178	15.6	JEFFERSON TX	25	-20	1	52	56
3	16.4	4	17.4	1	2.9	JIM HOGG TX	6	-83	-82	00	02
5	6.5	21	16.5	16	10.3	JIM WELLS TX	153	-38	58	83	17
24	12.7	35	13.7	63	16.3	JOHNSON TX	7	19	28	73	61
22	21.5	19	13.9	27	17.0	JONES TX	-35	22	-21	27	66
11	17.5	13	15.9	11	11.4	KARNES TX	-9	-29	-35	14	23
12	7.9	24	11.6	30	12.1	KAUFMAN TX	47	5	54	83	28
4	11.0	8	16.5	11	17.1	KENDALL TX	50	4	56	83	67
		1	8.2	2	12.9	KENT TX		58			34
5	6.1	15	9.0	23	8.0	KERR TX	49	-11	32	75	08
3	12.4	4	13.5	3	9.5	KIMBLE TX	9	-30	-24	25	13
		3	24.2			KINNEY TX					
4	5.0	11	12.7	10	8.9	KLEBERG TX	153	-30	77	87	11
3	6.7	7	13.2	6	10.0	KNOX TX	96	-24	49	81	16
23	11.0	43	17.7	32	11.3	LAMAR TX	60	-36	2	53	22
7	10.9	15	16.2	15	12.3	LAMB TX	49	-24	13	63	29
2	3.5	10	15.1	10	13.1	LAMPASAS TX	328	-14	269	97	35
6	24.3	1	4.0	8	31.6	LA SALLE TX	-84	688	30	74	98
21	17.1	20	12.6	20	12.5	LAVACA TX	-27	-1	-27	21	31
6	12.9	11	20.5	9	11.9	LEE TX	59	-42	-8	42	26
3	7.1	4	10.7	7	7.8	LEON TX	51	-27	10	61	08
14	16.8	12	9.0	24	11.4	LIBERTY TX	-47	27	-32	17	23
15	12.4	20	15.2	22	13.5	LIMESTONE TX	22	-11	9	59	39
1	8.3	2	11.5	8*	31.2	LIPSCOMB TX	39	172	278	97	98
4	12.3	3	7.3	12	22.2	LIVE OAK TX	-41	205	81	88	92
9	25.1	2	4.2	15	15.3	LLANO TX	-83	265	-39	12	54
47	13.8	58	9.8	107	13.7	LUBBOCK TX	-29	39	-1	50	40
6	17.2	4	8.0	3	5.0	LYNN TX	-53	-38	-71	01	03
11	17.7	12	16.1	17	18.8	MC CULLOCH TX	-9	17	6	57	78
59	10.7	93	11.9	148	15.0	MC LENNAN TX	11	26	40	78	51
6	18.7	11	28.7	7	10.2	MADISON TX	54	-64	-45	08	17
1	3.5	6	19.7	10	20.5	MARION TX	467	4	490	99	87
2	17.2	2	8.5	5	20.7	MARTIN TX	-50	144	21	69	88
3	10.5	4	9.2	7	14.1	MASON TX	-13	53	34	76	44
9	12.3	19	18.5	16	10.9	MATAGORDA TX	51	-41	-11	38	20
6	15.8	1	1.9	8	9.8	MAVERICK TX	-88	424	-38	12	15
11	14.5	12	12.4	10	6.4	MEDINA TX	-14	-49	-56	05	05
1	4.7	2	7.2	6	26.7	MENARD TX	52	272	466	99	97
8	8.6	22	14.2	31	11.0	MIDLAND TX	66	-22	29	73	21
11	9.2	18	11.5	21	14.2	MILAM TX	25	24	55	83	45
5	13.8	5	10.3	9	12.4	MILLS TX	-25	20	-10	39	30
6	11.2	5	7.4	10	15.3	MITCHELL TX	-34	106	36	77	54
14	12.4	22	15.0	23	13.9	MONTAGUE TX	21	-7	12	62	42
7	8.7	12	8.0	36	10.8	MONTGOMERY TX	-8	35	24	71	19
1	1.4	5	12.6	10	16.8	MOORE TX	776	33	1067	99	65
3	7.9	9	16.6	10	11.0	MORRIS TX	110	-33	40	78	21
2	11.1	6	30.0	7	32.2	MOTLEY TX	169	7	189	95	98
18	16.4	29	17.5	33	14.4	NACOGDOCHES TX	6	-18	-12	37	47
26	13.0	32	13.9	52	16.7	NAVARRO TX	7	20	29	73	64
3	9.4	4	8.2	5	8.5	NEWTON TX	-13	4	-10	39	10
15	18.6	22	21.0	22	15.6	NOLAN TX	13	-26	-16	33	56
53	11.0	94	12.7	121	12.4	NUECES TX	15	-2	13	63	29
4	12.0	8	24.4	5	11.4	OCHILTREE TX	103	-53	-5	45	23
				2	30.2	OLDHAM TX					98
12	9.8	18	9.0	37	13.1	ORANGE TX	-8	46	33	76	35
23	19.6	23	16.9	28	15.9	PALO PINTO TX	-13	-6	-19	30	59
7	11.1	10	10.8	19	17.2	PANOLA TX	-2	59	55	83	67
22	16.5	25	15.7	39	16.1	PARKER TX	-4	2	-2	49	59
2	6.9	6	19.4	5	13.4	PARMER TX	181	-31	94	89	38
1	2.5	5	11.9	5	9.6	PECOS TX	367	-19	277	97	14
9	16.5	7	8.9	16	13.2	POLK TX	-46	48	-20	29	36
40	12.6	53	12.3	86	15.4	POTTER TX	-2	24	22	70	54
2	7.9	4	14.9	3	9.6	PRESIDIO TX	89	-36	22	70	14
3	15.2	2	9.0	5	12.8	RAINS TX	-41	42	-16	33	33
12	17.5	15	11.8	29	12.0	RANDALL TX	-32	1	-32	17	26
2	32.9	3	29.2	2	15.3	REAGAN TX	-11	-48	-54	05	54
2	21.8	1	8.3	3	15.3	REAL TX	-62	85	-30	19	54
10	10.9	17	15.2	24	15.2	RED RIVER TX	39	0	39	78	53
7	17.2	6	10.6	3	5.5	REEVES TX	-38	-48	-68	02	04
1	4.2	5	12.3	7	13.0	REFUGIO TX	192	6	210	96	35
		1	15.6	3	42.1	ROBERTS TX		169			99
8	12.3	12	16.5	13	16.5	ROBERTSON TX	34	0	34	76	62
2	7.8	4	8.2	7	12.6	ROCKWALL TX	5	53	61	84	31
11	12.6	12	10.5	10	7.0	RUNNELS TX	-16	-33	-44	09	06
24	16.6	20	9.8	37	14.0	RUSK TX	-41	43	-15	34	43
3	9.2	6	13.8	9	19.4	SABINE TX	50	41	110	91	82
9	32.4	5	11.0	7	19.5	SAN AUGUSTINE TX	-66	78	-40	11	82
3	19.1	1	7.4	6	15.8	SAN JACINTO TX	-61	114	-17	31	57
12	11.7	22	14.8	22	10.7	SAN PATRICIO TX	26	-28	-9	41	19
13	25.8	10	15.7	13	17.5	SAN SABA TX	-39	12	-32	17	70
2	16.6	4	22.2	3	17.4	SCHLEICHER TX	34	-21	5	56	69
7	10.9	13	15.9	18	17.5	SCURRY TX	46	10	61	84	70
3	11.5	2	5.1	9	27.4	SHACKELFORD TX	-56	441	138	93	97
13	12.1	21	14.6	20	12.8	SHELBY TX	21	-12	6	56	33
2	16.0	1	9.0	1	5.4	SHERMAN TX	-44	-40	-67	02	04
45	15.8	73	18.4	82	14.0	SMITH TX	16	-24	-12	38	43
3	13.7	2	10.5			SOMERVELL TX	-23				
		4	7.4	2	2.1	STARR TX		-71			02
10	19.0	6	7.9	12	13.4	STEPHENS TX	-59	70	-29	19	38
		2	31.3	2	25.9	STERLING TX		-17			97

* SIGNIFICANTLY MORE DEATHS THAN EXPECTED

WHITE FEMALE — CANCER OF THE LARGE INTESTINE — ICD 153,159.0

1950-1959 DEATHS	1950-1959 RATE	1960-1969 DEATHS	1960-1969 RATE	1970-1979 DEATHS	1970-1979 RATE	COUNTY AND STATE	PERCENT CHANGE 60S/50S	PERCENT CHANGE 70S/60S	PERCENT CHANGE 70S/50S	PERCENTILE %CHANGE 70S/50S	PERCENTILE RATES 1970S
1	7.4	2	12.1	4	17.1	STONEWALL TX	62	42	130	93	67
		3	18.7	2	9.7	SUTTON TX		-48			14
3	10.9	9	18.9	9	15.7	SWISHER TX	73	-17	44	79	57
207	13.1	339	14.1	538	15.5	TARRANT TX	8	10	19	68	56
47	15.8	51	11.5	66	11.3	TAYLOR TX	-27	-2	-29	20	22
		2	16.8			TERRELL TX					
5	11.1	5	9.3	13	18.7	TERRY TX	-16	102	69	86	78
1	5.3	5	21.7	1	2.7	THROCKMORTON TX	308	-88	-50	06	02
9	12.4	18	17.5	17	12.8	TITUS TX	42	-27	4	54	33
24	10.5	41	11.6	48	10.1	TOM GREEN TX	10	-12	-3	48	16
77	11.7	130	13.7	172	12.4	TRAVIS TX	17	-9	6	57	30
4	9.1	3	6.4	7	9.1	TRINITY TX	-30	43	-1	50	12
3	8.3	5	6.5	12	10.9	TYLER TX	-22	69	32	75	20
15	18.4	24	19.8	19	12.5	UPSHUR TX	7	-37	-32	17	31
1	12.5	2	9.9	3	15.3	UPTON TX	-21	54	22	70	54
10	13.9	5	5.2	13	11.5	UVALDE TX	-63	120	-18	31	23
7	11.3	11	12.7	16	13.6	VAL VERDE TX	13	7	20	68	39
15	12.8	16	10.2	24	9.9	VAN ZANDT TX	-20	-3	-23	26	15
14	13.0	25	14.7	37	15.5	VICTORIA TX	13	5	19	68	55
2	4.8	8	11.5	15	12.8	WALKER TX	141	11	169	95	33
8	26.1	5	12.1	8	12.4	WALLER TX	-54	2	-53	05	30
1	3.8	4	8.9	3	4.9	WARD TX	134	-45	28	73	03
9	10.6	19	15.4	29	17.8	WASHINGTON TX	46	15	68	86	73
17	8.2	20	7.1	27	6.9	WEBB TX	-13	-3	-15	34	06
8	6.5	19	12.6	30	15.4	WHARTON TX	93	22	136	93	55
5	11.8	11	19.8	8	11.4	WHEELER TX	68	-42	-3	48	23
42	10.6	77	14.0	102	14.9	WICHITA TX	31	6	40	78	50
10	10.8	15	11.6	27	19.7	WILBARGER TX	8	70	83	88	83
2	4.8	1	1.6	7	10.2	WILLACY TX	-66	522	110	91	16
20	10.6	30	11.6	44	13.5	WILLIAMSON TX	9	17	27	72	39
3	5.1	7	9.0	12	11.0	WILSON TX	77	22	117	92	20
2	11.2	4	13.6	6	14.2	WINKLER TX	21	5	27	72	45
12	11.6	14	11.9	22	13.1	WISE TX	2	10	13	62	35
10	11.1	24	18.0	27	13.1	WOOD TX	63	-27	19	67	36
3	13.5	1	2.5	5	17.4	YOAKUM TX	-82	603	29	74	69
13	13.4	15	11.5	26	13.9	YOUNG TX	-14	21	3	54	42
1	7.0	3	16.4			ZAPATA TX	135				
3	10.5	5	12.1	5	11.3	ZAVALA TX	15	-7	8	58	22
317	11.5	438	11.6	609	11.7	UTAH UT	1	1	2		
4	22.8	2	9.6	4	14.4	BEAVER UT	-58	49	-37	13	46
8	10.4	15	14.9	16	11.4	BOX ELDER UT	43	-23	10	60	23
20	14.6	16	9.2	25	10.6	CACHE UT	-37	15	-28	21	18
6	6.7	8	12.2	15	15.1	CARBON UT	84	23	127	92	53
15	14.4	18	8.8	27	9.1	DAVIS UT	-39	4	-37	13	12
3	12.5	3	10.9	7	20.1	DUCHESNE UT	-13	84	61	84	85
1	4.1	2	6.9	1	2.8	EMERY UT	69	-59	-31	18	02
1	6.9	2	19.5	3	17.3	GARFIELD UT	181	-11	149	94	68
		3	21.6	4	15.8	GRAND UT		-27			58
5	17.8	2	4.1	7	12.5	IRON UT	-77	207	-30	19	31
		3	10.7	1	3.1	JUAB UT		-71			02
		2	16.0	2	9.2	KANE UT		-43			12
6	17.7	10	26.8	5	11.3	MILLARD UT	51	-58	-36	13	22
1	9.0	2	14.3	2	15.0	MORGAN UT	58	5	66	86	52
		1	14.9			PIUTE UT					
1	13.6					RICH UT					
158	12.6	212	12.1	276	11.8	SALT LAKE UT	-4	-2	-6	44	25
2	20.2	1	10.2	1	4.5	SAN JUAN UT	-49	-56	-78	01	03
4	6.3	5	6.7	10	11.5	SANPETE UT	6	72	82	88	23
9	22.2	11	18.2	9	11.4	SEVIER UT	-18	-37	-49	07	22
3	11.9	4	14.2	3	8.1	SUMMIT UT	19	-43	-32	17	08
5	12.0	6	9.5	17	22.3	TOOELE UT	-21	134	85	88	92
		3	8.6	10	19.1	UINTAH UT		122			80
27	8.6	49	12.3	82	13.8	UTAH UT	43	12	60	84	41
1	4.4	1	3.9	1	3.0	WASATCH UT	-12	-23	-32	17	02
4	10.4	8	11.6	4	2.4	WASHINGTON UT	11	-80	-77	01	02
				1	11.4	WAYNE UT					23
33	10.0	49	10.2	76	11.8	WEBER UT	2	15	18	67	25
444*	20.4	477*	19.0	502	16.6	VERMONT VT	-7	-12	-18		
13	13.3	26	23.4	22	16.1	ADDISON VT	76	-31	21	69	60
37	23.7	35	18.3	35	16.1	BENNINGTON VT	-23	-12	-32	17	59
30	21.1	24	15.6	35	20.0	CALEDONIA VT	-26	29	-5	45	85
54	16.9	71	18.3	86	18.6	CHITTENDEN VT	9	1	10	60	77
6	19.5	6	20.5	8	23.2	ESSEX VT	5	13	19	68	94
34	21.9	40	22.7	31	14.2	FRANKLIN VT	4	-37	-35	14	45
4	23.8	2	17.5	3	12.1	GRAND ISLE VT	-26	-31	-49	06	27
13	20.1	10	14.6	15	14.7	LAMOILLE VT	-28	1	-27	21	49
15	15.5	18	17.1	32*	25.3	ORANGE VT	10	48	64	85	96
25	22.3	28	23.3	28	18.1	ORLEANS VT	4	-22	-19	30	74
75*	25.2	62	17.2	67	15.8	RUTLAND VT	-32	-8	-37	13	58
58	23.5	56	19.5	49	14.0	WASHINGTON VT	-17	-28	-40	11	43
45	22.1	42	17.0	42	16.7	WINDHAM VT	-23	-2	-24	24	64
35	14.7	57	20.7	49	14.2	WINDSOR VT	40	-31	-4	47	45
1745	15.2	2315	14.5	3274	15.0	VIRGINIA VA	-4	3	-1		
16	10.9	26	14.9	40	19.0	ACCOMACK VA	37	27	75	87	80
37	17.4	44	14.6	52	13.8	ALBEMARLE#VA	-16	-6	-21	28	41
19	16.5	27	18.7	31	16.5	ALLEGHANY#VA	13	-12	0	51	63
6	30.1	3	11.7	7	24.7	AMELIA VA	-61	112	-18	31	96
8	12.1	9	10.3	14	11.7	AMHERST VA	-15	14	-3	48	25
3	7.7	5	12.5	6	9.9	APPOMATTOX VA	63	-21	29	74	15
131*	23.8	140	17.4	181	17.8	ARLINGTON#VA	-27	3	-25	23	73
42	12.5	40	9.3	66	11.8	AUGUSTA#VA	-26	27	-6	44	25
11*	41.2	4	13.5	11*	30.5	BATH VA	-67	125	-26	22	98
72	16.1	91	15.3	83	10.7	BEDFORD#VA	-5	-30	-34	16	19
4	16.0	4	14.5	5	15.5	BLAND VA	-9	7	-3	48	56
12	18.0	8	9.2	11	10.2	BOTETOURT VA	-49	11	-43	09	16

* SIGNIFICANTLY MORE DEATHS THAN EXPECTED

COUNTIES COMBINED; SEE TECHNICAL NOTES

WHITE FEMALE — CANCER OF THE LARGE INTESTINE — ICD 153,159.0

1950-1959 DEATHS	1950-1959 RATE	1960-1969 DEATHS	1960-1969 RATE	1970-1979 DEATHS	1970-1979 RATE	COUNTY AND STATE	PERCENT CHANGE 60S/50S	PERCENT CHANGE 70S/60S	PERCENT CHANGE 70S/50S	PERCENTILE %CHANGE 70S/50S	PERCENTILE RATES 1970S
2	4.8	10	22.2	8	15.8	BRUNSWICK VA	359	-29	227	97	58
6	8.4	4	4.1	20	17.7	BUCHANAN VA	-50	329	112	91	72
4	10.9	6	16.1	9	17.7	BUCKINGHAM VA	48	10	63	85	71
8	22.4	5	14.7	9	17.3	CAROLINE VA	-35	18	-23	25	68
35	17.7	38	15.4	39	11.3	CARROLL#VA	-13	-27	-36	13	22
1	15.7	1	14.9			CHARLES CITY VA	-5				
8	20.0	6	13.1	13	23.1	CHARLOTTE VA	-34	76	15	65	94
202	17.5	220	15.6	268	16.3	CHESTERFIELD#VA	-11	5	-7	43	61
5	16.8	7	20.9	13	24.4	CLARKE VA	25	17	45	80	96
5	26.2	3	14.0	3	10.3	CRAIG VA	-47	-27	-61	03	17
9	15.3	15	16.5	17	14.8	CULPEPER VA	8	-10	-3	47	49
2	18.1			9*	33.5	CUMBERLAND VA			85	88	99
4	6.9	7	11.1	9	10.6	DICKENSON VA	61	-4	54	83	18
30	14.1	57	20.2	53	14.1	DINWIDDIE#VA	44	-30	0	51	44
4	20.5	5	17.9	7	21.8	ESSEX VA	-12	22	7	57	91
110	17.8	197	16.1	338	16.1	FAIRFAX#VA	-9	0	-9	40	59
12	14.6	28*	27.6	22	16.2	FAUQUIER VA	88	-41	11	61	60
7	11.7	3	4.1	7	9.4	FLOYD VA	-65	128	-20	29	13
2	7.7	3	11.0	7	17.7	FLUVANNA VA	44	61	131	93	71
6	6.0	14	12.3	10	6.8	FRANKLIN VA	105	-45	14	64	06
23	14.2	37	17.8	56	19.9	FREDERICK#VA	26	12	41	79	84
7	12.1	5	6.3	6	6.3	GILES VA	-48	0	-48	07	05
6	14.1	12	20.9	21	24.2	GLOUCESTER VA	48	16	72	87	95
4	16.9	6	21.0	7	19.1	GOOCHLAND VA	24	-9	13	63	81
5	25.2	2	10.1	2	6.7	GREENE VA	-60	-34	-74	01	06
2	5.3	3	7.3	6	10.3	GREENSVILLE#VA	38	42	97	90	17
13	14.2	18	13.8	18	11.1	HALIFAX#VA	-3	-20	-22	27	21
9	11.0	18	15.6	35	19.2	HANOVER VA	43	23	75	87	81
39	16.3	84	16.5	131	15.3	HENRICO VA	2	-8	-6	44	54
24	22.5	21	10.5	38	12.8	HENRY#VA	-53	23	-43	10	33
4	23.3	5	23.1	4	18.5	HIGHLAND VA	-1	-20	-21	28	76
4	11.0	7	15.1	13	20.2	ISLE OF WIGHT VA	37	34	84	88	86
6	4.9	13	7.8	22	8.6	JAMES CITY#VA	58	11	76	87	10
4	24.4	4	23.6	3	11.7	KING AND QUEEN VA	-4	-50	-52	06	25
4	18.8	2	7.3	6	17.8	KING GEORGE VA	-61	145	-5	45	72
4	24.9	6	24.3	3	10.9	KING WILLIAM VA	-2	-55	-56	04	20
7	20.1	7	16.8	12	15.7	LANCASTER VA	-17	-6	-22	27	57
11	9.0	17	12.1	21	12.8	LEE VA	34	6	42	79	32
20	20.5	19	15.0	35	20.5	LOUDOUN VA	-27	37	0	51	87
4	9.2	10	17.6	9	14.0	LOUISA VA	91	-20	52	82	43
6	17.1	8	16.8	7	12.1	LUNENBURG VA	-1	-28	-29	19	27
6	17.0	7	15.2	5	10.8	MADISON VA	-10	-29	-37	13	19
12	23.9	7	10.9	11	13.9	MATHEWS VA	-54	27	-42	10	42
11	15.7	10	11.3	13	10.1	MECKLENBURG VA	-28	-11	-36	14	16
5	18.1	9	28.8	8	14.2	MIDDLESEX VA	59	-51	-22	27	45
17	12.8	25	13.1	34	12.6	MONTGOMERY#VA	2	-4	-2	49	31
6	11.7	3	5.4	7	10.0	NELSON VA	-54	87	-14	35	16
8	15.5	7	10.0	14	17.5	NORTHAMPTON VA	-35	75	13	63	70
1	2.4	5	10.0	16	22.8	NORTHUMBERLAND VA	314	128	846	99	93
11	21.1	11	15.6	7	9.2	NOTTOWAY VA	-26	-41	-56	04	12
9	17.2	14	20.9	17	19.2	ORANGE VA	21	-8	11	62	81
6	9.4	19	21.5	22	18.7	PAGE VA	128	-13	99	90	78
4	7.2	8	11.2	12	11.7	PATRICK VA	55	5	62	85	25
32	10.5	42	10.0	69	12.2	PITTSYLVANIA#VA	-5	22	16	65	28
2	11.8	3	13.0	2	6.7	POWHATAN VA	10	-48	-43	09	06
6	14.1	5	8.4	10	16.2	PRINCE EDWARD VA	-40	92	15	64	60
6	8.0	23	17.6	40	14.7	PRINCE WILLIAM#VA	121	-17	85	88	49
19	20.0	21	16.6	18	9.7	PULASKI VA	-17	-41	-51	06	14
6	26.4	1	4.0	3	10.3	RAPPAHANNOCK VA	-85	162	-61	03	17
5	23.9	6	18.8	11	29.5	RICHMOND VA	-21	57	24	71	98
87	14.3	117	13.8	182	15.4	ROANOKE#VA	-4	12	8	58	55
15	13.9	20	14.7	34	20.1	ROCKBRIDGE#VA	6	37	45	80	85
28	12.3	45	15.0	62	15.1	ROCKINGHAM#VA	22	1	23	70	52
6	6.4	11	9.5	19	13.5	RUSSELL VA	49	42	111	91	38
11	10.6	13	10.1	20	12.0	SCOTT VA	-4	18	13	63	27
24	19.4	21	14.1	17	8.8	SHENANDOAH VA	-27	-37	-55	05	11
16	13.8	15	9.7	22	10.6	SMYTH VA	-30	10	-23	26	18
5	8.9	5	8.0	14	16.6	SOUTHAMPTON#VA	-10	109	87	89	63
17	18.0	15	11.2	30	15.8	SPOTSYLVANIA#VA	-38	41	-13	37	58
2	4.9	5	7.4	9	9.6	STAFFORD VA	49	31	95	89	14
2	12.9	1	5.1	4	17.6	SURRY VA	-60	244	36	77	70
3	10.9	3	11.9	4	13.6	SUSSEX VA	9	14	25	71	40
15	11.3	27	15.0	38	15.5	TAZEWELL VA	32	3	37	77	56
13	21.3	6	8.7	22	21.1	WARREN VA	-59	143	-1	50	89
20	9.3	29	10.3	44	11.3	WASHINGTON#VA	11	9	21	69	22
4	12.0	8	16.6	12	16.7	WESTMORELAND VA	38	1	39	78	64
25	15.4	24	11.1	42	16.2	WISE#VA	-27	46	6	56	61
17	16.5	20	16.0	28	18.6	WYTHE VA	-3	16	13	63	77
163	14.4	282	17.3	390	17.3	CHESAPEAKE CITY#VA	21	0	20	68	68
30	18.6	36	15.4	56	17.7	HAMPTON CITY VA	-17	15	-5	46	72
30	16.8	42	14.3	66	15.4	NEWPORT NEWS CITY VA	-15	8	-8	41	55
11	12.6	10	8.1	17	10.3	SUFFOLK CITY VA	-36	28	-18	30	17
1805	15.0	2370	14.8	3062	14.5	WASHINGTON WA	-1	-2	-3		
2	8.5	2	4.8	5	8.3	ADAMS WA	-44	74	-2	49	09
11	14.5	13	13.7	10	7.2	ASOTIN WA	-5	-48	-50	06	07
15	10.3	33	14.1	46	13.2	BENTON WA	38	-7	28	73	36
19	10.0	42	16.5	37	11.8	CHELAN WA	64	-29	17	66	25
10	9.1	21	11.4	34	12.5	CLALLAM WA	26	9	37	77	31
58	14.1	83	14.7	130	15.5	CLARK WA	4	5	10	60	55
2	6.9	5	14.9	8	19.0	COLUMBIA WA	115	27	174	95	80
26	12.7	31	10.2	66	16.3	COWLITZ WA	-20	60	29	73	61
3	6.6	15	20.2	8	8.8	DOUGLAS WA	206	-56	33	75	11
		3	21.9	1	7.4	FERRY WA		-66			07
9	20.5	10	10.7	13	10.8	FRANKLIN WA	-48	0	-47	07	19
3	16.6	1	3.2	1	6.0	GARFIELD WA	-81	87	-64	03	05
5	5.1	20	15.1	30	15.5	GRANT WA	194	3	203	96	56
56	22.3	56	17.2	61	15.0	GRAYS HARBOR WA	-23	-13	-33	16	51
8	16.4	7	7.4	20	10.9	ISLAND WA	-55	47	-34	15	20
8	18.4	7	12.4	9	10.9	JEFFERSON WA	-33	-12	-41	11	20

* SIGNIFICANTLY MORE DEATHS THAN EXPECTED

\# COUNTIES COMBINED; SEE TECHNICAL NOTES

WHITE FEMALE

CANCER OF THE LARGE INTESTINE

ICD 153,159.0

1950-1959 DEATHS	1950-1959 RATE	1960-1969 DEATHS	1960-1969 RATE	1970-1979 DEATHS	1970-1979 RATE	COUNTY AND STATE	PERCENT CHANGE 60S/50S	PERCENT CHANGE 70S/60S	PERCENT CHANGE 70S/50S	PERCENTILE %CHANGE 70S/50S	PERCENTILE RATES 1970S
651	16.2	790	14.9	983	14.7	KING WA	-8	-1	-9	40	49
45	12.6	71	14.8	89	13.2	KITSAP WA	17	-11	5	55	36
14	14.1	16	14.7	18	14.2	KITTITAS WA	4	-3	0	51	45
11	19.7	5	7.8	12	15.0	KLICKITAT WA	-60	92	-24	25	52
51	20.5	51	17.0	51	13.5	LEWIS WA	-17	-21	-34	15	38
10	19.3	7	11.5	11	13.2	LINCOLN WA	-41	15	-32	17	36
11	14.7	16	16.6	28	19.0	MASON WA	12	15	29	73	80
15	14.5	17	13.3	24	15.0	OKANOGAN WA	-9	13	3	53	51
11	12.7	18	16.7	27	20.3	PACIFIC WA	32	22	60	84	86
7	24.5	4	11.5	4	8.2	PEND OREILLE WA	-53	-29	-67	02	09
186	14.1	266	15.6	351	15.5	PIERCE WA	11	-1	10	60	56
4	19.1	1	4.3	12	21.8	SAN JUAN WA	-77	405	14	64	91
32	12.3	47	15.2	57	13.9	SKAGIT WA	24	-8	14	63	42
1	4.8	2	7.8	2	5.2	SKAMANIA WA	63	-34	8	59	03
109	17.4	162	17.1	233	16.5	SNOHOMISH WA	-2	-4	-5	45	62
171	13.3	238	14.4	292	14.4	SPOKANE WA	8	0	8	59	47
8	9.7	14	14.4	20	16.1	STEVENS WA	48	12	66	86	59
30	12.0	56	16.0	97	17.4	THURSTON WA	33	8	44	80	69
3	13.6	2	9.6	5	18.4	WAHKIAKUM WA	-30	92	35	76	76
42	20.3	48	18.1	56	16.8	WALLA WALLA WA	-11	-7	-17	32	65
51	12.7	65	13.8	70	11.6	WHATCOM WA	8	-15	-9	41	24
21	15.6	20	12.5	36	21.3	WHITMAN WA	-20	71	37	77	90
86	14.7	105	13.8	105	10.9	YAKIMA WA	-6	-21	-26	23	20
1175	15.6	1366	14.6	1663	14.5	WEST VIRGINIA WV	-6	-1	-7		
11	13.5	16	14.9	17	14.6	BARBOUR WV	11	-2	8	59	48
25	16.4	39	21.5	44	18.6	BERKELEY WV	31	-13	14	64	77
15	16.7	10	8.6	17	11.6	BOONE WV	-49	35	-30	18	24
7	9.2	14	14.9	13	12.5	BRAXTON WV	62	-16	36	77	31
19	17.0	14	10.1	38	22.6	BROOKE WV	-41	124	33	75	93
81	16.1	90	13.6	112	14.7	CABELL WV	-15	8	-9	41	49
11	25.6	3	6.5	14	26.6	CALHOUN WV	-75	310	4	54	97
1	2.9	3	5.7	3	5.4	CLAY WV	99	-6	88	89	04
11	22.9	5	11.3	17*	33.7	DODDRIDGE WV	-51	199	47	81	99
27	14.1	24	9.0	34	10.0	FAYETTE WV	-36	10	-29	19	15
12	28.7	8	17.6	4	6.5	GILMER WV	-39	-63	-78	01	05
6	15.9	4	9.4	9	15.3	GRANT WV	-41	63	-4	47	54
27	18.3	17	8.7	21	9.0	GREENBRIER WV	-52	3	-51	06	11
16	29.4	16	25.0	21*	25.8	HAMPSHIRE WV	-15	3	-12	38	97
21	16.3	36	21.2	38	17.7	HANCOCK WV	30	-17	8	59	71
6	14.1	10	20.6	9	13.8	HARDY WV	46	-33	-2	49	41
52	12.9	85	17.4	86	15.2	HARRISON WV	35	-13	18	67	53
7	9.0	16	17.0	18	14.9	JACKSON WV	88	-12	65	85	51
12	16.0	19	20.9	17	13.8	JEFFERSON WV	31	-34	-13	36	41
126	16.1	144	13.4	214	15.8	KANAWHA WV	-17	18	-2	49	58
33	25.0	24	15.4	21	11.9	LEWIS WV	-38	-23	-53	05	26
9	11.8	16	17.5	20	16.4	LINCOLN WV	49	-7	39	78	61
15	12.0	28	16.0	26	11.1	LOGAN WV	33	-30	-7	42	21
15	9.9	26	15.1	23	11.8	MC DOWELL WV	52	-22	19	67	25
43	14.2	51	13.5	64	13.2	MARION WV	-5	-2	-7	43	37
40	23.3	44	21.3	45	17.7	MARSHALL WV	-9	-17	-24	25	71
10	10.4	16	13.2	24	16.2	MASON WV	28	23	57	83	60
36	14.3	54	16.1	56	12.4	MERCER WV	12	-23	-13	36	30
21	22.0	17	14.3	27	17.5	MINERAL WV	-35	22	-21	28	70
13	12.6	15	10.8	24	14.8	MINGO WV	-14	36	17	66	49
31	13.1	34	11.7	49	13.8	MONONGALIA WV	-11	18	5	55	41
16	25.4	4	5.1	9	9.3	MONROE WV	-80	82	-64	03	12
11	28.5	8	16.4	12	18.3	MORGAN WV	-43	12	-36	14	76
17	19.9	12	10.8	27	18.9	NICHOLAS WV	-46	76	-5	45	79
69	16.7	85	18.7	99	19.8	OHIO WV	12	6	18	67	84
8	21.3	7	14.4	6	10.8	PENDLETON WV	-33	-24	-49	07	19
7	22.4	7	18.1	3	6.2	PLEASANTS WV	-19	-66	-72	01	05
6	12.3	11	19.1	12	17.5	POCAHONTAS WV	54	-8	42	79	70
13	10.2	22	14.9	19	10.4	PRESTON WV	46	-30	2	53	18
5	5.7	9	8.3	27	17.1	PUTNAM WV	46	106	200	96	67
45	19.5	30	9.2	61	13.9	RALEIGH WV	-53	51	-29	20	42
10	7.6	26	17.4	17	9.7	RANDOLPH WV	128	-44	27	72	14
11	12.8	19	22.3	11	13.5	RITCHIE WV	74	-39	5	56	39
21	24.5	15	14.7	11	10.3	ROANE WV	-40	-30	-58	04	17
15	19.2	15	16.8	13	11.8	SUMMERS WV	-13	-30	-38	12	25
14	15.0	14	12.8	19	15.4	TAYLOR WV	-15	20	3	53	55
3	5.4	9	15.9	6	8.6	TUCKER WV	196	-46	59	84	10
14	21.4	10	12.2	10	13.7	TYLER WV	-43	12	-36	14	40
11	10.7	13	10.9	23	15.1	UPSHUR WV	2	38	41	79	52
13	9.3	11	5.9	26	11.3	WAYNE WV	-36	90	21	69	22
4	7.5	11	18.7	7	10.6	WEBSTER WV	150	-43	42	79	18
20	20.9	19	16.8	16	11.4	WETZEL WV	-19	-32	-46	08	23
		8	27.6	5	15.3	WIRT WV		-44			54
75	20.2	92	20.2	89	15.4	WOOD WV	0	-24	-24	25	55
8	13.7	11	9.9	10	8.3	WYOMING WV	-28	-16	-39	11	09
3352	18.4	4111*	17.9	4586	16.2	WISCONSIN WI	-3	-10	-12		
8	18.7	21*	35.9	16	16.9	ADAMS WI	92	-53	-10	40	65
22	22.1	20	17.3	26	20.6	ASHLAND WI	-22	19	-7	43	88
24	14.5	39	18.2	42	15.4	BARRON WI	25	-15	6	57	55
8	10.7	13	17.2	12	12.6	BAYFIELD WI	61	-27	18	67	32
68	14.2	103	15.7	138	16.2	BROWN WI	10	3	14	63	60
10	13.7	11	12.8	13	12.1	BUFFALO WI	-6	-5	-11	38	28
6	12.3	21*	32.9	16	18.6	BURNETT WI	167	-43	51	82	77
23	25.3	15	13.4	21	14.0	CALUMET WI	-47	4	-45	09	43
28	13.6	42	15.7	39	12.8	CHIPPEWA WI	16	-18	-5	45	33
28	17.4	48*	24.4	40	16.3	CLARK WI	40	-33	-7	43	61
48	21.0	54	19.9	57	17.5	COLUMBIA WI	-6	-12	-17	32	70
13	16.4	17	18.0	22	20.6	CRAWFORD WI	10	14	26	72	88
167	19.1	192	16.8	209	14.1	DANE WI	-12	-16	-26	22	44
50	15.8	82	21.1	88	18.1	DODGE WI	34	-14	15	64	74
15	13.9	32	23.4	21	13.0	DOOR WI	68	-45	-7	43	34
41	17.4	48	17.2	56	17.3	DOUGLAS WI	-1	1	-1	50	68
19	13.7	33	20.4	24	13.2	DUNN WI	49	-35	-4	47	36

* SIGNIFICANTLY MORE DEATHS THAN EXPECTED

WHITE FEMALE

CANCER OF THE LARGE INTESTINE

ICD 153,159.0

1950-1959 DEATHS	1950-1959 RATE	1960-1969 DEATHS	1960-1969 RATE	1970-1979 DEATHS	1970-1979 RATE	COUNTY AND STATE	PERCENT CHANGE 60S/50S	PERCENT CHANGE 70S/60S	PERCENT CHANGE 70S/50S	PERCENTILE %CHANGE 70S/50S	PERCENTILE RATES 1970S
50	17.4	61	17.0	92	20.1	EAU CLAIRE WI	-3	18	15	65	85
3	16.9	3	16.3	3	13.4	FLORENCE WI	-4	-18	-21	27	38
74	17.6	87	17.2	102	17.1	FOND DU LAC WI	-3	-1	-3	48	66
5	15.0	10	24.7	9	17.1	FOREST WI	65	-31	14	64	67
33	13.1	66*	23.2	61	18.4	GRANT WI	77	-21	40	78	76
41*	27.2	31	18.0	36	18.0	GREEN WI	-34	0	-34	15	74
18	19.6	13	10.8	29	19.5	GREEN LAKE WI	-45	81	0	50	83
22	18.9	21	18.8	20	13.7	IOWA WI	-1	-27	-28	21	40
4	10.1	8	15.0	4	6.2	IRON WI	49	-58	-38	12	05
20	24.4	14	15.1	22	20.3	JACKSON WI	-38	34	-17	32	86
70*	23.9	60	17.2	82	18.7	JEFFERSON WI	-28	9	-22	27	78
23	24.0	26	23.3	25	15.7	JUNEAU WI	-3	-33	-35	15	57
62	16.7	76	14.0	115	16.9	KENOSHA WI	-16	21	1	52	66
13	14.9	13	12.3	23	17.9	KEWAUNEE WI	-18	46	21	69	73
74	19.5	82	17.6	106	18.3	LA CROSSE WI	-10	4	-6	44	75
14	14.9	13	11.7	23	19.2	LAFAYETTE WI	-21	63	28	73	81
20	19.1	18	14.3	22	15.1	LANGLADE WI	-25	6	-21	28	53
17	15.3	27	18.6	41	21.5	LINCOLN WI	21	15	40	78	90
80	22.2	97	21.9	80	13.5	MANITOWOC WI	-1	-38	-39	12	39
58	16.1	76	16.1	91	15.0	MARATHON WI	0	-7	-7	43	51
38	20.4	44	18.5	55	18.6	MARINETTE WI	-9	0	-9	40	77
12	19.9	14	19.5	15	18.0	MARQUETTE WI	-2	-8	-10	39	74
58	20.3	84*	23.8	67	15.0	MENOMINEE#WI	17	-37	-26	22	52
836	18.6	1060*	19.0	1070	16.6	MILWAUKEE WI	2	-12	-11	39	63
27	15.9	26	13.5	38	16.2	MONROE WI	-15	20	2	53	61
24	24.3	27	19.7	34	18.2	ONEIDA WI	-19	-8	-25	23	75
65	17.0	91	18.1	102	15.5	OUTAGAMIE WI	6	-14	-9	41	56
22	17.5	32	17.1	54	19.2	OZAUKEE WI	-2	12	10	60	81
11	26.6	7	13.2	13	22.2	PEPIN WI	-50	67	-17	32	92
16	13.3	20	12.7	21	13.4	PIERCE WI	-4	5	1	52	37
14	10.1	30	18.2	31	13.0	POLK WI	80	-29	28	73	35
30	17.4	33	15.7	44	16.5	PORTAGE WI	-10	6	-5	45	63
19	22.9	17	17.0	25	21.6	PRICE WI	-25	27	-6	44	91
117	19.7	125	16.2	141	15.0	RACINE WI	-18	-7	-24	25	52
22	21.1	24	20.7	22	14.1	RICHLAND WI	-2	-32	-33	16	44
95	18.1	121	18.5	123	15.1	ROCK WI	3	-18	-16	32	52
9	11.6	15	16.4	12	11.7	RUSK WI	41	-29	1	52	25
21	14.5	26	15.4	33	13.6	ST CROIX WI	6	-12	-6	44	40
53*	24.5	42	16.7	40	13.1	SAUK WI	-32	-22	-47	08	35
9	19.1	9	14.6	18	21.0	SAWYER WI	-23	43	10	60	89
98	21.2	108	18.7	123	17.9	SHEBOYGAN WI	-12	-4	-16	33	73
10	12.5	11	12.1	14	12.4	TAYLOR WI	-3	3	-1	50	29
20	14.1	17	9.9	41	17.3	TREMPEALEAU WI	-30	74	23	70	68
34	23.1	30	16.7	32	14.1	VERNON WI	-28	-16	-39	12	44
11	24.8	5	8.7	21	17.5	VILAS WI	-65	100	-30	19	70
64	23.2	67	18.5	84	18.3	WALWORTH WI	-20	-1	-21	27	76
12	20.3	14	18.9	12	13.0	WASHBURN WI	-7	-31	-36	14	35
32	17.6	50	20.2	47	13.3	WASHINGTON WI	14	-34	-24	24	37
90	18.1	121	16.4	165	15.3	WAUKESHA WI	-9	-7	-15	34	54
51	23.2	52	18.0	63	17.6	WAUPACA WI	-22	-2	-24	24	71
20	22.9	12	12.1	16	13.8	WAUSHARA WI	-47	14	-40	11	41
79	15.2	118	17.9	111	14.0	WINNEBAGO WI	18	-22	-8	42	43
54	23.1	66	21.3	73	16.8	WOOD WI	-8	-21	-27	21	65
166	15.4	169	11.9	260	14.1	WYOMING WY	-23	19	-8		
13	19.4	9	9.6	16	14.8	ALBANY WY	-50	54	-24	25	50
7	12.8	4	7.3	8	10.6	BIG HORN WY	-43	44	-17	31	18
5	21.9	3	10.3	8	18.7	CAMPBELL WY	-53	81	-15	34	77
10	23.1	8	12.4	9	13.7	CARBON WY	-46	11	-41	11	40
2	8.3	4	10.8	9	18.7	CONVERSE WY	30	73	126	92	78
1	5.4	2	12.5	4	20.0	CROOK WY	133	60	273	97	85
6	7.6	7	9.0	20	16.6	FREMONT WY	18	84	118	92	63
4	7.1	7	10.7	10	11.9	GOSHEN WY	51	11	68	86	26
4	13.1	3	7.9	6	12.1	HOT SPRINGS WY	-40	54	-7	42	27
2	8.9	8	27.0	5	12.9	JOHNSON WY	203	-52	44	79	34
26	17.1	32	14.1	43	14.4	LARAMIE WY	-18	2	-16	33	46
2	5.9	3	7.2	2	4.7	LINCOLN WY	23	-35	-20	29	03
26	20.8	24	12.6	37	14.8	NATRONA WY	-39	18	-29	20	50
2	9.4	2	9.8	4	18.0	NIOBRARA WY	4	83	91	89	74
10	16.5	14	16.5	17	13.1	PARK-TETON#WY	0	-21	-20	28	35
4	11.5	5	12.9	10	17.1	PLATTE WY	12	33	48	81	67
14	13.7	16	12.5	19	12.6	SHERIDAN WY	-9	1	-8	42	32
1	11.6	1	8.0	2	11.1	SUBLETTE WY	-31	38	-5	45	21
15	26.8	11	14.5	18	17.3	SWEETWATER WY	-46	19	-36	14	68
3	10.1			8	19.4	UINTA WY			92	89	82
9*	47.2	3	10.1	3	7.6	WASHAKIE WY	-79	-25	-84	00	07

* SIGNIFICANTLY MORE DEATHS THAN EXPECTED

COUNTIES COMBINED; SEE TECHNICAL NOTES

NON-WHITE MALE — CANCER OF THE LARGE INTESTINE — ICD 153,159.0

1950-1959 DEATHS	1950-1959 RATE	1960-1969 DEATHS	1960-1969 RATE	1970-1979 DEATHS	1970-1979 RATE	COUNTY AND STATE	PERCENT CHANGE 60S/50S	PERCENT CHANGE 70S/60S	PERCENT CHANGE 70S/50S	PERCENTILE %CHANGE 70S/50S	PERCENTILE RATES 1970S
6332	10.8	10393	14.0	15673	16.2	UNITED STATES	30	16	51		
259	7.7	353	10.1	482	12.7	ALABAMA AL	30	26	64		
3	9.2	4	12.3	6	26.7	AUTAUGA AL	33	118	190	84	89
1	2.4	4	10.9	5	9.8	BALDWIN AL	353	-10	309	91	66
2	4.3	5	12.0	2	5.6	BARBOUR AL	183	-53	32	46	60
1	5.0	2	9.7	2	16.6	BIBB AL	95	71	233	88	79
				2*	71.6	BLOUNT AL					97
5	12.1	3	7.6	3	7.2	BULLOCK AL	-37	-5	-41	16	62
5	13.1	4	9.8	3	7.2	BUTLER AL	-25	-26	-45	13	62
3	5.8	2	3.1	8	12.7	CALHOUN AL	-46	304	119	74	71
3	5.5	7	14.5	2	4.3	CHAMBERS AL	162	-70	-22	23	58
				1	6.2	CHILTON AL					60
3	9.4	3	8.9			CHOCTAW AL	-6				
4	7.3	5	11.6	5	11.3	CLARKE AL	59	-3	55	57	68
		1	13.8	1	9.4	CLAY AL		-32			65
2	9.1	2	11.0	2	7.6	COFFEE AL	21	-31	-16	26	62
3	8.4	6	17.6	4	12.1	COLBERT AL	109	-32	43	52	70
6	19.4	1	2.7	3	9.0	CONECUH AL	-86	237	-54	09	65
5*	44.0	2	14.1	1	5.2	COOSA AL	-68	-63	-88	00	59
1	5.4	1	4.4			COVINGTON AL	-19				
1	6.1	3	14.0	2	12.3	CRENSHAW AL	129	-13	100	69	70
				1	24.6	CULLMAN AL					88
		1	7.5	3	14.0	DALE AL		87			74
16	12.6	14	11.4	10	8.3	DALLAS AL	-10	-27	-35	18	63
2*	53.0			1	21.0	DE KALB AL			-60	07	85
3	8.5	3	8.2	9	21.0	ELMORE AL	-3	155	147	79	85
5	11.0	1	2.3	1	2.4	ESCAMBIA AL	-79	7	-78	01	56
9*	18.4	7	12.0	5	8.6	ETOWAH AL	-35	-28	-53	09	64
2	16.8	1	10.9			FAYETTE AL	-35				
2	16.7			3	25.6	GENEVA AL			54	56	89
5	10.4	4	10.1	9	23.7	GREENE AL	-3	134	127	76	88
5	10.1	6	10.4	7	16.3	HALE AL	3	57	63	60	79
3	13.7	3	10.2	5	25.6	HENRY AL	-25	150	86	65	89
		3	6.4	4	6.6	HOUSTON AL		4			61
		2	22.8			JACKSON AL					
56	7.9	111	13.5	161	17.6	JEFFERSON AL	71	30	122	75	81
				2	21.9	LAMAR AL					86
2	6.8	2	6.2	6	19.8	LAUDERDALE AL	-8	218	191	85	84
5*	24.7			2	8.3	LAWRENCE AL			-67	05	63
1	2.1	6	9.8	2	3.3	LEE AL	355	-66	53	56	57
1	3.9	2	6.7	1	3.3	LIMESTONE AL	70	-51	-17	26	57
1	1.6	3	6.1	1	4.2	LOWNDES AL	273	-32	154	80	58
4	4.4	6	6.3	9	9.0	MACON AL	41	44	104	70	65
8	11.6	5	7.0	14	16.1	MADISON AL	-40	131	39	50	78
5	7.0	6	9.3	12	15.9	MARENGO AL	33	71	128	76	78
1	20.2			1	26.8	MARSHALL AL			32	47	90
19	7.1	24	7.2	51	13.8	MOBILE AL	1	92	94	68	73
		3	6.2	2	4.0	MONROE AL		-34			58
18	8.7	25	12.0	29	12.0	MONTGOMERY AL	38	-1	37	49	70
3	10.9	7	20.4	4	10.8	MORGAN AL	87	-47	-1	33	68
3	6.0	2	3.9	5	11.8	PERRY AL	-34	200	97	68	70
1	3.0	1	3.2	1	2.7	PICKENS AL	7	-15	-8	31	57
1	2.7	6	14.4	2	4.2	PIKE AL	437	-71	56	58	58
1	6.7	1	5.3	4	19.9	RANDOLPH AL	-21	274	196	85	84
		4	7.9	12	16.4	RUSSELL AL		108			79
2	10.7	4	26.3			ST CLAIR AL	146				
		1	3.9	6	22.3	SHELBY AL		468			87
3	5.0	5	6.9	5	8.3	SUMTER AL	39	19	66	61	64
7	12.2	11	17.8	8	11.4	TALLADEGA AL	46	-36	-6	32	69
4	10.8	1	3.3	7	17.1	TALLAPOOSA AL	-69	416	58	58	80
6	5.3	7	7.1	18	14.6	TUSCALOOSA AL	34	105	174	83	75
2	7.6	5	16.4	4	14.2	WALKER AL	115	-13	87	66	74
2	11.5	3	18.8	1	4.5	WASHINGTON AL	64	-76	-60	07	58
3	4.7	2	3.3	2	4.9	WILCOX AL	-30	48	4	35	59
		13	10.5	29	16.9	ALASKA AK		61			
20	5.8	36	8.6	48	6.5	ARIZONA AZ	47	-25	11		
2	3.5	3	4.4	6	7.6	APACHE AZ	25	72	115	73	62
1	56.2	1	19.1	1	14.1	COCHISE AZ	-66	-27	-75	03	74
				1	1.5	COCONINO AZ					56
		1	9.2	1	6.4	GILA AZ		-31			61
		1*	94.4			GREENLEE AZ					
7	6.6	16	11.8	23	8.4	MARICOPA AZ	77	-29	26	43	64
		1	2.5			NAVAJO AZ					
8	15.0	8	12.7	10	7.4	PIMA AZ	-16	-42	-51	10	62
1	6.2	3	8.6	2	4.6	PINAL AZ	37	-46	-26	22	58
		1*	53.6			SANTA CRUZ AZ					
1	5.3	1	4.0	4	10.3	YUMA AZ	-23	153	94	67	67
141	8.3	206	11.9	235	12.9	ARKANSAS AR	44	8	55		
1	3.4	2	11.0	4	13.3	ARKANSAS AR	227	20	293	91	72
3	11.2	1	2.9	6	14.8	ASHLEY AR	-74	415	32	47	75
		4	18.5	4	15.1	BRADLEY AR		-18			76
1	12.1	1	8.4	1	6.7	CALHOUN AR	-31	-19	-44	14	61
5	7.8	6	13.9	6	7.2	CHICOT AR	78	-49	-9	31	61
2	8.4	3	11.6	4	13.0	CLARK AR	38	13	55	57	72
		1	8.6	1	10.3	CLEVELAND AR		19			67
2	5.0	6	14.2	4	8.0	COLUMBIA AR	184	-44	59	59	63
1	6.4	2	10.6	1	3.9	CONWAY AR	66	-63	-39	16	58
1	13.4	1	22.5			CRAIGHEAD AR	68				
6	5.3	8	7.2	11	10.6	CRITTENDEN AR	37	47	101	70	67
2	6.8	1	3.3	2	7.3	CROSS AR	-51	117	7	36	62
5*	33.2	3	13.9	2	11.8	DALLAS AR	-58	-15	-64	06	70

* SIGNIFICANTLY MORE DEATHS THAN EXPECTED

NON-WHITE MALE — CANCER OF THE LARGE INTESTINE — ICD 153,159.0

1950-1959 DEATHS	1950-1959 RATE	1960-1969 DEATHS	1960-1969 RATE	1970-1979 DEATHS	1970-1979 RATE	COUNTY AND STATE	PERCENT CHANGE 60S/50S	PERCENT CHANGE 70S/60S	PERCENT CHANGE 70S/50S	PERCENTILE %CHANGE 70S/50S	PERCENTILE RATES 1970S
1	2.5	4	8.3	4	9.3	DESHA AR	234	12	274	90	65
4	15.2	1	3.9	4	13.8	DREW AR	-74	251	-10	30	73
2	18.2					FAULKNER AR					
3	12.3	7	26.9	5	15.7	GARLAND AR	119	-41	28	45	77
2	5.5	4	10.4	10	25.5	HEMPSTEAD AR	88	145	361	93	89
1	6.3	2	15.7			HOT SPRING AR	147				
		2	15.4	3	19.7	HOWARD AR		28			84
2	12.3	4	25.0	3	16.2	JACKSON AR	103	-35	31	46	78
17	10.7	32	21.5	20	11.4	JEFFERSON AR	101	-47	7	36	69
1	4.6	2	7.0	1	4.8	LAFAYETTE AR	52	-32	4	35	59
3	5.3	2	2.7	3	4.6	LEE AR	-50	71	-14	27	58
2	6.3	1	2.8	2	9.5	LINCOLN AR	-55	239	53	56	65
1	7.4	1	6.6	2	14.4	LITTLE RIVER AR	-11	118	94	67	75
2	9.3	5	21.1	5	18.0	LONOKE AR	128	-15	95	68	81
6	17.4	4	13.0	6	15.5	MILLER AR	-25	19	-11	29	77
5	6.2	5	5.6	12	18.1	MISSISSIPPI AR	-9	224	193	85	81
		5	12.8	5	18.8	MONROE AR		47			82
4	22.9	1	4.3			NEVADA AR	-81				
6	13.2	8	15.1	2	3.0	OUACHITA AR	14	-80	-77	02	57
7	5.7	13	10.9	17	15.0	PHILLIPS AR	92	37	163	81	76
		1	60.6	2*	131.6	PIKE AR		117			99
		2	16.0	1	5.8	POINSETT AR		-64			60
		1	18.9	1	29.8	POPE AR		58			91
2	21.7	2	26.2	1	10.1	PRAIRIE AR	21	-61	-53	09	66
23	11.1	32	14.0	42	16.8	PULASKI AR	26	20	51	55	80
8	10.5	12	15.3	8	9.7	ST FRANCIS AR	46	-37	-8	31	66
1	8.6					SALINE AR					
3	14.0	4	22.1	8*	32.2	SEBASTIAN AR	58	46	131	76	92
				2	26.3	SEVIER AR					89
4	7.1	5	7.9	13	17.9	UNION AR	11	126	152	80	81
				2	51.3	WASHINGTON AR					96
		1	17.2	3	37.7	WHITE AR		119			94
2	7.5	3	15.8	2	6.7	WOODRUFF AR	111	-57	-10	29	61
		1	49.2			YELL AR					
309	10.7	614	12.4	1226	12.4	CALIFORNIA CA	16	0	16		
34	10.9	67	13.2	153	16.6	ALAMEDA CA	21	26	52	56	79
1	18.7			1	5.6	BUTTE CA			-70	04	60
1	40.9	1	112.4			COLUSA CA	175				
6	9.0	9	6.7	28	12.6	CONTRA COSTA CA	-25	88	40	50	71
1*	70.7			1	44.6	DEL NORTE CA			-37	17	95
13	12.4	10	7.5	21	8.4	FRESNO CA	-40	13	-32	19	64
1	6.8	1	8.6	3	17.3	HUMBOLDT CA	27	100	153	80	80
		2	6.6	3	4.2	IMPERIAL CA		-37			58
1	39.0	1	18.9	1	10.0	INYO CA	-52	-47	-74	03	66
7	10.5	14	15.4	22	12.1	KERN CA	46	-21	15	39	70
2	17.9	1	6.1	2	5.0	KINGS CA	-66	-19	-72	04	59
				1	10.2	LAKE CA					66
134	10.9	276	13.0	571	14.1	LOS ANGELES CA	19	9	29	45	74
3	25.7	2	10.6	2	5.0	MADERA CA	-59	-52	-80	01	59
				4	18.0	MARIN CA					81
		1*	212.0			MARIPOSA CA					
1	18.7	1	11.8	1	6.5	MENDOCINO CA	-37	-45	-65	06	61
1	5.9	3	9.9	7	14.7	MERCED CA	68	48	149	79	75
1	2.7	4	9.2	15	9.1	MONTEREY CA	246	-2	240	89	65
		2	2.7	13	3.8	ORANGE CA		42			57
		3	28.0	1	7.0	PLACER CA		-75			61
4	8.7	12	16.4	18	10.5	RIVERSIDE CA	88	-36	21	41	67
22*	25.3	20	11.4	46	16.0	SACRAMENTO CA	-55	41	-37	17	78
4	9.8	12	14.7	15	6.0	SAN BERNARDINO CA	50	-59	-38	16	60
9	12.1	14	7.7	42	10.0	SAN DIEGO CA	-36	29	-17	25	66
36	12.0	86	15.5	147	15.1	SAN FRANCISCO CA	29	-2	26	43	76
6	5.2	21	13.7	19	7.7	SAN JOAQUIN CA	164	-44	48	54	63
2	23.0	2	12.5	1	2.6	SAN LUIS OBISPO CA	-46	-79	-89	00	57
2	2.5	6	6.4	15	6.4	SAN MATEO CA	157	0	157	81	61
		2	5.3	9	10.5	SANTA BARBARA CA		97			67
6	11.0	16	14.4	22	8.4	SANTA CLARA CA	31	-42	-24	23	64
1	6.0	2	7.5	4	10.4	SANTA CRUZ CA	25	39	74	63	67
1	18.6			1	7.9	SHASTA CA			-58	07	63
		1	11.5	2	34.6	SISKIYOU CA		202			93
3	21.4	4	34.0	11	11.8	SOLANO CA	59	-65	-45	13	69
1	12.6	2	23.0	6	15.8	SONOMA CA	83	-31	26	43	78
				2	8.8	STANISLAUS CA					64
1	9.4			1	9.3	SUTTER CA			-2	33	65
1*	212.0					TEHAMA CA					
2	4.0	10	17.5	10	10.3	TULARE CA	343	-41	160	81	67
		2	6.4	4	2.8	VENTURA CA		-56			57
1	12.8	1	10.6	1	2.9	YOLO CA	-17	-73	-77	02	57
		3	32.4			YUBA CA					
20	14.3	30	15.1	36	8.8	COLORADO CO	6	-42	-39		
1	24.6	1	6.9	1	3.0	ADAMS CO	-72	-57	-88	00	57
		1*	267.9			ALAMOSA CO					
				1	15.7	BOULDER CO					77
16	18.6	25	18.5	29	13.0	DENVER-ARAPAHOE#CO	-1	-30	-30	20	72
2	28.9	1	5.5	2	6.5	EL PASO CO	-81	18	-77	01	61
		2*	112.9			OTERO CO					
				2	7.2	PUEBLO CO					62
				1*	85.2	SEDGWICK CO					98
1	16.7					WELD CO					
42*	22.6	42	12.1	100	19.0	CONNECTICUT CT	-46	57	-16		
7	9.4	13	11.2	31	20.2	FAIRFIELD CT	19	80	115	73	84
19*	40.3	9	12.0	29	19.1	HARTFORD CT	-70	59	-53	09	83
3*	88.0	1	12.8			LITCHFIELD CT	-85				
1	18.8	1	11.9			MIDDLESEX CT	-37				
10	16.2	17	14.6	35	22.2	NEW HAVEN CT	-10	52	37	48	87
2	17.5	1	6.1	5	14.1	NEW LONDON CT	-65	131	-20	24	74

* SIGNIFICANTLY MORE DEATHS THAN EXPECTED

COUNTIES COMBINED; SEE TECHNICAL NOTES

NON-WHITE MALE — CANCER OF THE LARGE INTESTINE — ICD 153,159.0

1950-1959 DEATHS	1950-1959 RATE	1960-1969 DEATHS	1960-1969 RATE	1970-1979 DEATHS	1970-1979 RATE	COUNTY AND STATE	PERCENT CHANGE 60S/50S	PERCENT CHANGE 70S/60S	PERCENT CHANGE 70S/50S	PERCENTILE %CHANGE 70S/50S	PERCENTILE RATES 1970S
21	10.9	42	18.0	54	19.5	DELAWARE DE	66	8	80		
2	5.2	4	9.5	12	25.5	KENT DE	82	170	392	94	89
16	15.9	32*	23.4	32	20.2	NEW CASTLE DE	48	-14	28	44	84
3	5.2	6	10.9	10	13.7	SUSSEX DE	110	25	163	81	73
165*	17.9	306*	23.7	415*	26.3	DIST. OF COLUMBIA DC	32	11	47		
165*	17.9	306*	23.7	415*	26.3	WASHINGTON DC	32	11	47	53	89
194	8.0	313	9.9	523	12.4	FLORIDA FL	24	25	55		
10	15.5	9	12.5	12	14.0	ALACHUA FL	-20	12	-10	30	74
1	2.5	4	10.7	5	14.9	BAY FL	337	39	506	96	76
		1	7.9	1	15.7	BRADFORD FL		99			77
2	9.3	3	5.8	6	8.1	BREVARD FL	-38	40	-13	28	63
11	13.0	16	10.1	34	13.8	BROWARD FL	-22	36	7	36	73
		1	27.3	2	39.9	CHARLOTTE FL		46			94
1	9.6	1	12.6	1	10.2	CITRUS FL	31	-19	7	36	66
1	15.3	1	8.9	1	6.9	CLAY FL	-42	-23	-55	09	61
1	7.7	1	7.0	5	20.9	COLLIER FL	-9	198	171	82	85
1	4.1	4	18.2	5	17.3	COLUMBIA FL	348	-5	327	92	80
28	11.1	49	11.0	84	10.5	DADE FL	-1	-4	-5	32	67
1	10.1	1	8.2	2	15.9	DE SOTO FL	-19	94	57	58	78
23	7.1	51	12.4	66	14.2	DUVAL FL	74	15	100	69	74
2	2.7	13	12.1	17	12.4	ESCAMBIA FL	340	3	353	93	71
2	31.1					FLAGLER FL					
				1	15.1	FRANKLIN FL					76
6	6.7	4	4.2	11	11.9	GADSDEN FL	-37	183	77	64	70
				1*	45.4	GILCHRIST FL					95
1	23.6	1	16.1	1	8.6	GULF FL	-32	-47	-63	06	64
				3	21.5	HAMILTON FL					86
		2	66.9	1	10.8	HARDEE FL		-84			68
1	11.4	1	10.9			HENDRY FL	-4				
				1	14.1	HERNANDO FL					74
		2	13.0	5	25.1	HIGHLANDS FL		94			89
12	7.8	26	11.9	42	15.0	HILLSBOROUGH FL	51	27	92	66	76
		1	44.6			HOLMES FL					
1	4.4	1	7.3	3	16.7	INDIAN RIVER FL	68	128	283	90	79
4	11.6	4	10.1	6	14.2	JACKSON FL	-13	41	22	42	74
1	5.1	2	10.3	1	5.2	JEFFERSON FL	102	-50	1	35	59
2	4.5	6	14.0	9	16.9	LAKE FL	212	21	277	90	80
3	9.2	4	11.2	7	13.0	LEE FL	23	16	42	51	72
6	7.9	7	9.8	10	11.9	LEON FL	25	21	51	55	70
1	5.0	2	13.1	2	12.9	LEVY FL	165	-1	161	81	72
4	19.6	2	9.6	1	3.9	MADISON FL	-51	-59	-80	01	57
4	12.9	5	12.6	7	15.0	MANATEE FL	-2	18	16	39	76
2	2.7	5	7.0	14	17.0	MARION FL	156	143	521	96	80
2	12.3			3	18.6	MARTIN FL			50	55	82
		5*	24.4			MONROE FL					
		1	7.9	3	23.3	NASSAU FL		196			87
				1	5.3	OKALOOSA FL					59
		1	14.4	2	26.8	OKEECHOBEE FL		87			90
5	4.6	12	8.8	20	10.2	ORANGE FL	92	17	124	75	66
		1	12.2			OSCEOLA FL					
13	8.4	9	5.2	27	11.3	PALM BEACH FL	-38	118	34	47	68
3	25.1	2	15.6			PASCO FL	-38				
9	11.2	10	8.0	28	15.4	PINELLAS FL	-28	93	38	49	77
7	8.8	12	9.6	22	13.0	POLK FL	10	35	48	54	72
5	12.4	6	14.4	7	15.5	PUTNAM FL	16	8	25	43	77
3	13.8	3	13.1	2	7.0	ST JOHNS FL	-5	-46	-49	11	61
2	4.2	1	1.7	6	9.2	ST LUCIE FL	-60	448	120	74	65
1	9.8	1	14.0	1	11.2	SANTA ROSA FL	43	-20	13	38	68
		1	4.5	3	8.1	SARASOTA FL		82			63
1	1.7	3	5.9	5	6.9	SEMINOLE FL	243	17	300	91	61
2	14.1	2	14.1	1	4.1	SUMTER FL	0	-71	-71	04	58
1	5.8	3	14.8	2	15.9	SUWANNEE FL	153	7	171	82	78
				2	11.1	TAYLOR FL					68
7	9.8	8	9.2	18	17.1	VOLUSIA FL	-5	85	75	64	80
				1	18.2	WAKULLA FL					81
		2	27.0	1	10.6	WALTON FL		-61			67
1	16.1			1	10.8	WASHINGTON FL			-33	18	68
283	8.5	373	10.0	575	13.6	GEORGIA GA	19	36	61		
		1	26.0			BACON GA					
1	8.6			3	44.3	BAKER GA			416	95	95
1	1.5	1	1.9	6	11.8	BALDWIN GA	24	527	679	99	69
				2	18.7	BARROW GA					82
2	15.1	2	12.3	4	24.8	BARTOW GA	-18	102	65	61	89
		2	12.3	3	14.4	BEN HILL GA		17			75
		1	22.3	1	16.7	BERRIEN GA		-25			79
12	10.3	20	12.2	21	12.4	BIBB GA	18	2	21	41	71
2	28.5	2	26.3			BLECKLEY GA	-8				
4	13.6	3	10.9			BROOKS GA	-20				
		2	25.6	1	16.3	BRYAN GA		-36			79
		3	7.3	4	11.2	BULLOCH GA		53			68
		4	7.9	4	9.4	BURKE GA		19			65
1	9.1			2	13.9	BUTTS GA			53	56	73
		1	4.9	1	5.6	CALHOUN GA		14			60
		1	6.8	3	17.2	CAMDEN GA		152			80
		1	11.5			CANDLER GA					
		2	9.5	2	11.8	CARROLL GA		23			69
1*	60.7	1*	53.6	2*	267.9	CATOOSA GA	-12	400	341	93	99
1	15.7	1	16.4	2	27.5	CHARLTON GA	4	68	75	63	90
27	12.5	26	10.4	53	20.6	CHATHAM GA	-17	99	65	61	85
1	17.5	2	27.7	1	11.1	CHATTOOGA GA	58	-60	-37	17	68
1	2.5	4	9.6	5	10.8	CLARKE GA	283	12	328	92	68
1	6.6					CLAY GA					
1	6.7	1	14.1	1	5.2	CLAYTON GA	108	-63	-23	23	59

* SIGNIFICANTLY MORE DEATHS THAN EXPECTED

NON-WHITE MALE — CANCER OF THE LARGE INTESTINE — ICD 153,159.0

1950-1959 DEATHS	1950-1959 RATE	1960-1969 DEATHS	1960-1969 RATE	1970-1979 DEATHS	1970-1979 RATE	COUNTY AND STATE	PERCENT CHANGE 60S/50S	PERCENT CHANGE 70S/60S	PERCENT CHANGE 70S/50S	PERCENTILE %CHANGE 70S/50S	PERCENTILE RATES 1970S
				3	36.7	CLINCH GA					93
2	9.6	2	8.9	13*	54.2	COBB GA	-7	510	467	96	96
1	6.0	2	9.3	3	17.8	COFFEE GA	56	91	199	86	81
2	9.3	1	4.1	2	8.1	COLQUITT GA	-55	96	-12	29	63
		2	14.7	2	14.5	COLUMBIA GA		-1			75
				3	35.6	COOK GA					93
4	13.3	1	2.9	2	5.4	COWETA GA	-78	89	-59	07	60
3*	28.6					CRAWFORD GA					
1	3.9	5	19.5	3	12.7	CRISP GA	398	-35	224	87	71
				1*	163.6	DADE GA					99
3	8.3	2	6.8	7	18.7	DECATUR GA	-17	174	126	76	82
4	8.0	10	10.0	23	12.4	DE KALB GA	24	25	55	57	71
1	6.5	2	16.4			DODGE GA	150				
1	4.7	1	5.6	2	10.2	DOOLY GA	19	82	117	74	66
1	1.5	7	8.5	11	11.0	DOUGHERTY GA	449	30	613	98	68
1	16.9			2	24.3	DOUGLAS GA			43	52	88
2	8.4			1	3.2	EARLY GA			-62	07	57
1	8.9	3	25.8	1	7.0	EFFINGHAM GA	189	-73	-21	24	61
2	12.3	1	4.9	1	5.6	ELBERT GA	-60	13	-55	09	60
1	3.8	5	21.2	4	20.2	EMANUEL GA	462	-5	434	95	84
		4*	46.3	3	29.1	EVANS GA		-37			91
				3*	45.8	FAYETTE GA					95
6	21.0	3	8.5	5	14.0	FLOYD GA	-60	65	-33	18	74
1	19.3	2	40.6	1	17.5	FRANKLIN GA	110	-57	-9	30	80
68*	12.9	90	14.7	135	18.0	FULTON GA	14	22	40	50	81
		1	44.6			GLASCOCK GA					
2	4.7			19*	35.1	GLYNN GA			650	98	93
				1	15.0	GORDON GA					76
		3	13.8	4	15.6	GRADY GA		13			77
2	8.9	1	5.7	2	8.1	GREENE GA	-36	41	-9	30	63
2	19.2	2	18.5			GWINNETT GA	-4				
1	8.8	2	13.8	2	12.2	HALL GA	56	-11	39	50	70
3	12.9	1	3.5	4	16.7	HANCOCK GA	-72	371	30	45	79
		1	16.1			HARALSON GA					
1	6.2			1	4.2	HARRIS GA			-33	18	58
		1	7.1	2	14.1	HART GA		97			74
				1	32.1	HEARD GA					92
2	8.6	1	4.2	4	20.2	HENRY GA	-51	380	133	77	84
3	14.5	4	15.5	6	19.6	HOUSTON GA	7	26	35	48	83
				1	10.7	JACKSON GA					67
3	24.7	1	8.2	2	14.5	JASPER GA	-67	76	-41	15	75
				1	14.8	JEFF DAVIS GA					75
2	7.4			6	16.0	JEFFERSON GA			114	73	78
				3	15.1	JENKINS GA					76
1	11.2	1	14.9			JOHNSON GA	33				
2	21.7	2	14.7	1	4.8	JONES GA	-32	-67	-78	01	59
3	21.0	3	25.2	5	30.7	LAMAR GA	20	22	46	53	92
8	18.8	4	9.9	3	7.2	LAURENS GA	-47	-27	-62	07	62
				1	7.2	LEE GA					62
				1	4.0	LIBERTY GA					58
1	9.8			1	7.5	LINCOLN GA			-23	23	62
		2	4.0	8	14.1	LOWNDES GA		250			74
				3	20.5	MC DUFFIE GA					85
3	16.8	2	15.8	2	15.1	MC INTOSH GA	-6	-4	-10	30	76
		4	14.1	3	14.3	MACON GA		1			74
1	14.9			2	47.9	MADISON GA			222	87	96
		2	19.3	2	21.9	MARION GA		13			86
2	5.7	3	9.4	2	5.2	MERIWETHER GA	67	-45	-8	31	59
				1	22.8	MILLER GA					87
1	4.1			7	22.5	MITCHELL GA			444	95	87
1	6.5	3	19.1			MONROE GA	194				
1	10.4			1	18.7	MONTGOMERY GA			80	64	82
1	7.5	4	27.1	2	10.1	MORGAN GA	262	-63	35	48	66
10	12.9	15	12.9	13	7.2	MUSCOGEE#GA	0	-44	-44	14	62
1	3.4	4	20.7	3	15.6	NEWTON GA	499	-24	353	93	77
		3	29.3			OGLETHORPE GA					
				1	12.5	PAULDING GA					71
1	3.2	3	10.6	2	5.9	PEACH GA	229	-44	84	65	60
2*	201.7					PICKENS GA					
				2	29.3	PIERCE GA					91
		2	21.2	1	10.1	PIKE GA		-52			66
		3	16.0	1	4.6	POLK GA		-71			58
		1	10.6	4	28.4	PULASKI GA		167			90
		1*	74.3			RABUN GA					
4	14.3	1	4.7	4	15.3	RANDOLPH GA	-67	224	7	37	77
8	6.3	15	10.5	23	13.5	RICHMOND GA	65	28	112	73	73
2	35.7			1	5.6	ROCKDALE GA			-84	00	60
		1	13.8			SCHLEY GA					
		1	3.0	1	4.7	SCREVEN GA		53			58
				4	11.6	SPALDING GA					69
1	26.5	2	31.3	2	18.3	STEPHENS GA	18	-42	-31	19	82
2	11.4					STEWART GA					
3	8.8	3	6.9	8	19.0	SUMTER GA	-22	176	116	74	82
		2	11.7	1	4.5	TALBOT GA		-61			58
2	20.4					TALIAFERRO GA					
				2	10.7	TATTNALL GA					67
1	9.4	4	34.3			TAYLOR GA	266				
		1	8.3	2	14.5	TELFAIR GA		75			75
6	18.8	5	16.5	4	15.0	TERRELL GA	-12	-9	-20	24	76
4	8.3	5	10.5	2	3.5	THOMAS GA	26	-66	-57	08	57
3	15.3	2	7.9	3	11.1	TIFT GA	-49	42	-27	21	68
1	6.3					TOOMBS GA					
1	15.7					TREUTLEN GA					
		5	8.7	10	18.7	TROUP GA		115			82
		1	9.5			TURNER GA					
1	7.2	3	19.7			TWIGGS GA	173				
4	22.1			1	4.0	UPSON GA			-82	01	58
1	14.9					WALKER GA					
1	4.0	2	10.3	1	3.5	WALTON GA	154	-65	-12	28	57
3	8.7	4	11.0	13*	41.0	WARE GA	26	274	370	94	95
7	22.1	2	5.7	4	11.8	WASHINGTON GA	-74	106	-47	12	69

* SIGNIFICANTLY MORE DEATHS THAN EXPECTED

COUNTIES COMBINED; SEE TECHNICAL NOTES

NON-WHITE MALE

CANCER OF THE LARGE INTESTINE

ICD 153,159.0

1950-1959 DEATHS	1950-1959 RATE	1960-1969 DEATHS	1960-1969 RATE	1970-1979 DEATHS	1970-1979 RATE	COUNTY AND STATE	PERCENT CHANGE 60S/50S	PERCENT CHANGE 70S/60S	PERCENT CHANGE 70S/50S	PERCENTILE %CHANGE 70S/50S	PERCENTILE RATES 1970S
		3	22.6			WAYNE GA					
1	18.9					WEBSTER GA					
1	19.6			1	16.3	WHEELER GA			-17	26	79
				1	9.3	WHITFIELD GA					65
1	4.7	5	26.0	2	11.8	WILKES GA	453	-54	152	80	70
		2	15.8	1	6.4	WILKINSON GA		-59			61
4	18.5			1	4.7	WORTH GA			-75	03	58
		206	12.5	348	14.5	HAWAII HI		16			
		34	13.0	41	12.9	HAWAII HI		0			72
		137	12.9	264	15.7	HONOLULU HI		22			77
		11	8.8	19	12.6	KAUAI HI		43			71
		24	13.0	24	9.8	MAUI#HI		-25			66
		8	20.4	3	3.8	IDAHO ID		-82			
		2*	66.0	2	34.4	ADA ID		-48			93
		2	35.5			BANNOCK ID					
				1	18.1	BINGHAM ID					81
		1	37.2			BONNEVILLE ID					
		1	22.5			CANYON ID					
		1	47.2			NEZ PERCE ID					
		1*	54.6			PAYETTE ID					
408*	16.3	769*	22.0	992*	20.5	ILLINOIS IL	35	-7	26		
1	15.7			4*	48.5	ADAMS IL			208	86	96
4	12.5	5	17.4	6	28.1	ALEXANDER IL	39	61	125	75	90
				2*	130.9	BOND IL					99
2	11.7	2	8.9	6	20.8	CHAMPAIGN IL	-24	135	78	64	85
339*	18.0	648*	23.5	849*	21.6	COOK IL	30	-8	20	41	86
		1*	62.9			DE KALB IL					
2*	67.8	1	24.4	2	7.0	DU PAGE IL	-64	-71	-90	00	61
2*	230.6					EDGAR IL					
		1*	74.3			FRANKLIN IL					
		1*	188.7			GALLATIN IL					
				1	62.9	HENRY IL					97
1*	44.6	1*	56.2			IROQUOIS IL	26				
1	6.4	3	19.3	4	25.2	JACKSON IL	203	31	296	91	89
		2	47.1	2	33.5	JEFFERSON IL		-29			93
		1	7.5	9*	44.2	KANE IL		486			95
2	4.9	4	7.8	4	7.2	KANKAKEE IL	58	-8	46	53	62
1	11.2	2	18.3	1	8.0	KNOX IL	64	-56	-28	21	63
1	1.3	5	15.2	8	15.0	LAKE IL	1060	-2	1040	99	76
1*	60.6			3*	158.3	LA SALLE IL			161	81	99
		1	17.6	3*	146.3	MC LEAN IL		730			99
4*	26.0	5	22.9	6	19.5	MACON IL	-12	-15	-25	22	83
3	6.9	3	5.8	5	8.6	MADISON IL	-16	49	26	43	64
		2	29.5	1	14.9	MARION IL		-50			75
1	14.0	2	26.2	1	20.2	MASSAC IL	87	-23	44	52	84
1	7.3	2	26.9	2	31.3	MORGAN IL	267	16	327	92	92
3	9.7	5	14.3	4	8.3	PEORIA IL	47	-42	-14	27	64
1*	141.3					POPE IL					
1	4.3	4	16.8	4	28.8	PULASKI IL	286	72	564	97	91
1	11.1					RANDOLPH IL					
2	19.9	4	23.6	6	20.7	ROCK ISLAND IL	19	-12	4	35	85
15	11.3	32	16.9	34	15.2	ST CLAIR IL	50	-10	35	48	76
3*	58.8	4*	56.6			SALINE IL	-4				
5	21.6	12*	44.6	7	21.3	SANGAMON IL	106	-52	-1	34	86
				1	13.8	STEPHENSON IL					73
				1*	74.3	UNION IL					97
1	4.6	4	17.6	2	6.9	VERMILION IL	281	-60	51	55	61
1*	49.7	1	61.3	2*	131.5	WARREN IL	23	114	165	82	99
		1*	74.3			WAYNE IL					
4	15.6	6	20.5	6	9.6	WILL IL	31	-53	-38	16	66
		2	29.1			WILLIAMSON IL					
5*	37.4	2	13.0	6	20.3	WINNEBAGO IL	-65	56	-46	13	84
111*	14.9	188*	18.5	277*	21.2	INDIANA IN	24	15	43		
		3	11.2	6	12.1	ALLEN IN		8			70
		1*	589.6			BLACKFORD IN					
2	19.9	1	8.0	1	14.1	CLARK IN	-60	75	-29	20	74
				1*	57.3	DEARBORN IN					97
3	13.7	7	26.0	5	16.4	DELAWARE IN	90	-37	20	41	79
		3	35.7	1	12.6	ELKHART IN		-65			71
2*	76.6	3*	107.9	1	31.3	FAYETTE IN	41	-71	-59	07	92
3*	41.9	2	20.7	3	29.4	FLOYD IN	-51	42	-30	20	91
1	18.6					GIBSON IN					
3	26.8	2	18.2	2	11.8	GRANT IN	-32	-35	-56	08	69
1	29.8	1	27.0			HAMILTON IN	-9				
2	26.5	1	8.1			HOWARD IN	-69				
1*	114.7					JACKSON IN					
		1	31.5	1	47.1	JEFFERSON IN		50			95
1*	133.9					JENNINGS IN					
				1	33.1	JOHNSON IN					93
24	15.2	48	17.5	93*	24.0	LAKE IN	15	37	58	58	88
2	10.9	2	8.9	5	27.6	LA PORTE IN	-18	209	154	80	90
3	24.6	2	12.6	2	14.0	MADISON IN	-49	11	-43	14	74
49*	15.1	86*	21.0	110*	21.8	MARION IN	39	4	44	52	86
		1	31.3			MONROE IN					
				1	89.3	PORTER IN					98
		1	24.8	2*	115.9	POSEY IN		367			99
4	20.9	8	16.7	8	11.7	ST JOSEPH IN	-20	-30	-44	14	69
				2*	103.4	SHELBY IN					98
		1	38.3			TIPPECANOE IN					
7	16.2	8	20.6	15*	30.6	VANDERBURGH IN	27	49	89	66	92
2	9.6	2	10.4	10*	41.8	VIGO IN	9	301	337	93	95
		1*	56.2			WABASH IN					
		1*	62.9	1*	70.1	WARRICK IN		11			97

* SIGNIFICANTLY MORE DEATHS THAN EXPECTED

COUNTIES COMBINED; SEE TECHNICAL NOTES

NON-WHITE MALE

CANCER OF THE LARGE INTESTINE

ICD 153,159.0

1950-1959 DEATHS	1950-1959 RATE	1960-1969 DEATHS	1960-1969 RATE	1970-1979 DEATHS	1970-1979 RATE	COUNTY AND STATE	PERCENT CHANGE 60S/50S	PERCENT CHANGE 70S/60S	PERCENT CHANGE 70S/50S	PERCENTILE %CHANGE 70S/50S	PERCENTILE RATES 1970S
1	6.2	2	11.1	6	34.2	WAYNE IN	78	209	450	95	93
15	14.3	23	18.6	25	17.2	IOWA IA	30	-8	20		
1	6.9	4	18.0	2	8.0	BLACK HAWK IA	160	-56	16	39	63
		1*	424.0			BUCHANAN IA					
		1*	188.7			CASS IA					
				1	37.2	CERRO GORDO IA					94
1*	86.0					DALLAS IA					
2*	121.5	1	28.3			DES MOINES IA	-77				
				1*	267.9	HENRY IA					99
				1	28.8	JOHNSON IA					91
1	13.3	3*	49.5	2	26.9	LEE IA	271	-46	102	70	90
				2	20.6	LINN IA					85
9*	22.3	13*	29.8	13	27.1	POLK IA	34	-9	22	42	90
				1	31.5	POTTAWATTAMIE IA					92
				1	9.4	SCOTT IA					65
1	23.6			1	34.4	WAPELLO IA			46	53	93
62*	16.5	77	18.0	88	17.5	KANSAS KS	9	-2	6		
		1	27.0	1	37.2	ALLEN KS		38			94
3*	29.2	2	28.3			ATCHISON KS	-3				
1	26.8	1	67.0	1	67.0	BARTON KS	150	0	150	79	97
		3*	95.9	2*	40.5	BOURBON KS		-58			94
		1*	112.4			BUTLER KS					
		2*	116.8	1	37.5	CHEROKEE KS		-68			94
		2	29.7	1	14.6	COWLEY KS		-51			75
1	28.7	1	37.8	1	12.4	CRAWFORD KS	32	-67	-57	08	71
				1*	94.4	DICKINSON KS					98
2*	55.6			3*	103.0	DONIPHAN KS			85	65	98
2	13.1	3	19.7	2	12.4	DOUGLAS KS	50	-37	-6	32	71
		1*	62.9			FORD KS					
		1	12.0	4*	43.8	GEARY KS		266			95
		3*	90.1			GRAHAM KS					
1	33.5					JOHNSON KS					
		1	17.5	2	28.2	LABETTE KS		61			90
4	16.3	4	12.5	3	14.0	LEAVENWORTH KS	-23	12	-14	27	74
				1*	56.2	LINN KS					96
1	26.8	1	18.7	1	49.2	LYON KS	-30	162	83	65	96
1	22.5					MIAMI KS					
3	15.5	1	4.0	5	26.1	MONTGOMERY KS	-74	551	69	62	89
		1*	112.4	1*	37.2	MORRIS KS		-67			94
				1*	74.3	NEMAHA KS					97
1*	74.3					NEOSHO KS					
		1*	27.0			PRATT KS					
		1	19.3	1	13.4	RENO KS		-31			72
				1*	188.7	RICE KS					99
1	7.1					RILEY KS					
2*	112.4	1	21.0	3*	59.6	SALINE KS	-81	184	-47	12	97
5	11.2	6	8.4	15	15.8	SEDGWICK KS	-25	88	41	50	78
		1*	70.7			SEWARD KS					
5	12.1	10	21.3	9	16.2	SHAWNEE KS	76	-24	33	47	78
				1*	74.3	STEVENS KS					97
				1*	37.2	WABAUNSEE KS					94
29*	22.4	28	18.3	26	16.1	WYANDOTTE KS	-18	-12	-28	20	78
141*	14.4	189*	18.5	233*	22.5	KENTUCKY KY	29	22	57		
2*	50.7	1	27.3			ADAIR KY	-46				
1	24.6	1	35.3			BALLARD KY	44				
		3	35.8	5*	51.9	BARREN KY		45			96
				4*	80.9	BELL KY					98
5*	38.1	3	18.3	5*	36.2	BOURBON KY	-52	98	-5	33	93
2*	44.4	1	31.5			BOYD KY	-29				
1	8.1	4	21.4	3	19.2	BOYLE KY	166	-11	138	77	83
				1	29.8	BRECKINRIDGE KY					91
2	28.4	1	16.3	1	13.5	CALDWELL KY	-43	-17	-53	10	73
		1	39.7			CALLOWAY KY					
				3*	46.9	CAMPBELL KY					95
				1*	94.4	CARLISLE KY					98
1*	94.4					CARROLL KY					
4	7.4	7	13.0	9	18.2	CHRISTIAN KY	76	40	148	79	81
2	22.5	2	22.0	2	23.4	CLARK KY	-3	6	4	35	87
1*	70.3					CLAY KY					
3	20.6	3	18.4	2	13.4	DAVIESS KY	-10	-27	-35	18	73
13	13.1	15	15.6	26*	25.3	FAYETTE KY	19	62	93	67	89
		1	37.2			FLEMING KY					
		1	44.6			FLOYD KY					
1	22.5	4*	55.9			FRANKLIN KY	148				
2	22.9	1	13.7	1	10.2	FULTON KY	-40	-25	-55	09	66
		1	24.8	2	52.8	GARRARD KY		113			96
1	12.3	1	9.3	1	10.6	GRAVES KY	-24	14	-14	28	67
		1	42.6			GREEN KY					
1*	267.9					GREENUP KY					
		1*	136.6			HANCOCK KY					
1	24.6	1	19.1	1	13.2	HARDIN KY	-22	-31	-46	13	72
		2	9.9	1	5.0	HARLAN KY		-50			59
		2*	70.9	1	112.4	HARRISON KY		59			98
		1	16.4			HART KY					
1	5.6	6*	34.5	3	20.4	HENDERSON KY	511	-41	261	89	84
		1	22.5			HICKMAN KY					
3	12.7	5	22.9	4	19.2	HOPKINS KY	79	-16	50	55	83
43	15.0	71*	21.5	103*	26.9	JEFFERSON KY	44	25	80	65	90
4	17.2	5	24.8	7*	42.2	KENTON KY	44	70	145	79	95
		2*	69.2	1	74.3	KNOX KY		7			97
		1	35.3			LARUE KY					
1	22.3	2	38.1			LETCHER KY	71				
2	26.0	1	14.0	1	23.6	LINCOLN KY	-46	68	-9	30	87
1*	188.7					LIVINGSTON KY					
2	16.4			3	24.2	LOGAN KY			48	53	88

* SIGNIFICANTLY MORE DEATHS THAN EXPECTED

NON-WHITE MALE — CANCER OF THE LARGE INTESTINE — ICD 153,159.0

1950-1959 DEATHS	1950-1959 RATE	1960-1969 DEATHS	1960-1969 RATE	1970-1979 DEATHS	1970-1979 RATE	COUNTY AND STATE	PERCENT CHANGE 60S/50S	PERCENT CHANGE 70S/60S	PERCENT CHANGE 70S/50S	PERCENTILE %CHANGE 70S/50S	PERCENTILE RATES 1970S
2	6.6	6	19.3	9	29.1	MC CRACKEN KY	194	51	344	93	91
4	26.0	1	7.0	3	17.4	MADISON KY	-73	148	-33	18	80
1	14.1			1	32.8	MARION KY			133	77	93
2	16.9			2	29.4	MASON KY			74	63	91
		2	43.8			MERCER KY					
		1	30.3			METCALFE KY					
3*	37.1	1	12.7			MONTGOMERY KY	-66				
4*	47.6			2	24.6	MUHLENBERG KY			-48	12	88
		1	14.1	2	29.5	NELSON KY		110			91
2*	65.6					NICHOLAS KY					
1	31.5					OHIO KY					
		1	11.2			OLDHAM KY					
1	27.0					PERRY KY					
		2*	64.3			PIKE KY					
		1*	188.7			POWELL KY					
1	31.5	3*	177.0			PULASKI KY	463				
1	10.4	3	25.1	3	35.5	SCOTT KY	141	41	240	88	93
4*	39.4	2	19.1	1	8.9	SHELBY KY	-52	-53	-77	02	64
1	11.5	1	11.9	2	22.9	SIMPSON KY	3	93	100	69	87
1	22.3	1	17.2			TAYLOR KY	-23				
2	17.3			1	5.7	TODD KY			-67	05	60
1	10.8	2	26.0	2	26.1	TRIGG KY	142	0	142	78	89
1	10.1	4*	69.3	2	36.1	UNION KY	585	-48	257	89	93
3	13.3	2	8.7	9*	41.2	WARREN KY	-35	376	209	86	95
1*	74.3	1	57.3			WAYNE KY	-23				
				2	22.5	WEBSTER KY					87
				1*	74.3	WHITLEY KY					97
5*	56.5	1	19.6			WOODFORD KY	-65				
298	8.9	467	12.5	592	14.2	LOUISIANA LA	41	13	60		
4	12.2	3	9.5	1	3.0	ACADIA LA	-22	-68	-75	02	57
2	12.9	1	6.2	4	21.1	ALLEN LA	-52	241	63	61	86
1	2.2	8	23.3	4	11.6	ASCENSION LA	968	-50	431	95	69
		3	12.1			ASSUMPTION LA					
3	8.7	5	12.1	3	7.1	AVOYELLES LA	40	-41	-18	25	61
2	13.3	4	21.0	5	32.3	BEAUREGARD LA	58	54	143	78	92
5	14.8	4	10.0	4	10.9	BIENVILLE LA	-32	10	-26	22	68
1	2.1	7	12.4	7	12.0	BOSSIER LA	499	-3	482	96	70
9	3.3	42	13.8	55	15.8	CADDO LA	315	14	374	94	78
6	8.0	11	11.0	15	12.5	CALCASIEU LA	37	14	55	58	71
		2	15.2	1	7.2	CALDWELL LA		-53			62
				1	112.4	CAMERON LA					98
1	5.9	5	30.6	1	5.3	CATAHOULA LA	419	-83	-10	30	59
		1	1.7	4	6.6	CLAIBORNE LA		287			61
6	17.6	5	10.8	4	10.2	CONCORDIA LA	-38	-6	-42	15	66
8	13.0	11	16.2	10	14.6	DE SOTO LA	25	-10	12	38	75
22	11.4	27	11.9	42	14.5	EAST BATON ROUGE LA	5	22	28	44	75
5	11.9	4	11.5	7	18.4	EAST CARROLL LA	-4	61	55	57	82
		3	6.9	3	7.3	EAST FELICIANA LA		7			62
2	7.5	2	11.0	2	8.2	EVANGELINE LA	47	-25	10	37	63
3	8.3	2	5.2	6	17.1	FRANKLIN LA	-37	230	107	71	80
3	18.4	5*	36.9	3	15.8	GRANT LA	101	-57	-14	27	78
5	11.6	7	14.1	2	2.9	IBERIA LA	21	-79	-75	02	57
3	4.7	6	10.7	8	15.1	IBERVILLE LA	128	42	222	87	76
		3	13.1	2	11.5	JACKSON LA		-12			69
8	10.8	8	10.2	21	14.7	JEFFERSON LA	-6	45	36	48	75
2	12.6	2	8.8	3	14.9	JEFFERSON DAVIS LA	-30	69	19	40	76
7	15.0	10	19.9	9	12.7	LAFAYETTE LA	32	-36	-15	26	71
3	15.1	3	13.9	5	20.9	LAFOURCHE LA	-7	50	38	50	85
		1	18.7			LA SALLE LA					
1	3.1	4	8.6	6	12.0	LINCOLN LA	179	41	292	91	70
1	9.9	1	9.3	1	6.2	LIVINGSTON LA	-6	-33	-37	17	60
3	6.5	10	19.8	8	18.8	MADISON LA	205	-5	191	85	82
4	6.8	8	14.4	8	13.5	MOREHOUSE LA	111	-6	97	68	73
4	6.8	2	2.8	5	8.4	NATCHITOCHES LA	-59	203	24	43	64
86	12.6	148*	18.1	196*	20.9	ORLEANS LA	44	15	66	61	85
11	9.6	9	7.6	20	15.6	OUACHITA LA	-21	104	62	60	77
1	3.6	2	11.7	2	10.9	PLAQUEMINES LA	227	-7	204	86	68
2	5.2	3	7.3	8	19.2	POINTE COUPEE LA	40	163	269	90	83
6	4.8	12	9.0	20	13.1	RAPIDES LA	87	47	175	83	72
1	3.5			1	4.7	RED RIVER LA			32	47	59
4	9.0	6	13.8	6	19.0	RICHLAND LA	53	38	111	72	83
2	14.0	4	21.8	1	6.2	SABINE LA	56	-71	-55	09	60
				1	16.4	ST BERNARD LA					79
1	4.9	1	4.0	5	18.6	ST CHARLES LA	-18	360	278	90	82
2	5.4	2	6.5	5	15.5	ST JAMES LA	20	139	187	84	77
3	14.0	5	16.6	10	27.1	ST JOHN THE BAPT LA	19	63	94	67	90
10	9.9	3	2.6	11	9.1	ST LANDRY LA	-74	253	-8	31	65
2	7.8	4	14.3	2	5.1	ST MARTIN LA	84	-64	-34	18	59
7	13.2	5	8.6	6	10.5	ST MARY LA	-35	22	-21	24	67
		6	15.5	4	7.4	ST TAMMANY LA		-52			62
6	8.5	11	14.1	2	2.5	TANGIPAHOA LA	67	-82	-71	04	56
3	6.9	5	12.6	3	8.1	TENSAS LA	83	-35	18	40	63
1	2.7	5	12.3	5	13.7	TERREBONNE LA	358	11	408	95	73
1	4.1	2	7.8	2	6.0	UNION LA	88	-23	45	52	60
2	12.9			2	8.0	VERMILION LA			-38	16	63
2	13.5	2	18.5			VERNON LA	37				
7	15.2	2	2.9	4	6.7	WASHINGTON LA	-81	132	-56	08	61
5	10.2	6	9.7	6	8.4	WEBSTER LA	-5	-14	-17	25	64
3	9.6	1	3.3	4	15.5	WEST BATON ROUGE LA	-65	363	61	60	77
				2	24.3	WEST CARROLL LA					88
5	19.7	2	10.1	1	5.1	WEST FELICIANA LA	-49	-50	-74	03	59
1	5.6	1	5.4	3	11.3	WINN LA	-4	109	101	70	69
1	7.7	2	13.2	1	2.7	MAINE ME	72	-79	-64		
		1*	67.0			AROOSTOOK ME					
				1	15.2	CUMBERLAND ME					76
		1	37.5			PENOBSCOT ME					
1*	53.6					WASHINGTON ME					

* SIGNIFICANTLY MORE DEATHS THAN EXPECTED

NON-WHITE MALE — CANCER OF THE LARGE INTESTINE — ICD 153,159.0

1950-1959 DEATHS	1950-1959 RATE	1960-1969 DEATHS	1960-1969 RATE	1970-1979 DEATHS	1970-1979 RATE	COUNTY AND STATE	PERCENT CHANGE 60S/50S	PERCENT CHANGE 70S/60S	PERCENT CHANGE 70S/50S	PERCENTILE %CHANGE 70S/50S	PERCENTILE RATES 1970S
175	12.6	327*	17.8	499*	20.6	MARYLAND MD	42	16	64		
2*	33.0	3*	55.5	4*	75.3	ALLEGANY MD	68	36	129	76	98
11	10.4	15	14.1	25	19.9	ANNE ARUNDEL MD	37	41	93	67	84
6.	9.6	12	17.2	19	20.2	BALTIMORE MD	80	17	111	72	84
		6*	36.2	4	16.9	CALVERT MD		-53			80
		3	21.1	8*	46.1	CAROLINE MD		118			95
1	10.1	3	21.9	1	7.2	CARROLL MD	116	-67	-29	20	62
1	18.6	1	9.0	2	11.8	CECIL MD	-52	31	-36	17	69
2	10.6	6	17.6	5	13.8	CHARLES MD	67	-22	31	46	73
7	19.4	4	9.4	9	20.2	DORCHESTER MD	-52	116	4	36	84
2	11.3	3	14.0	4	18.3	FREDERICK MD	24	30	62	60	82
3	16.2	7	27.0	8	27.3	HARFORD MD	66	1	68	62	90
2	17.9	5*	30.0	2	10.6	HOWARD MD	67	-65	-41	15	67
		2	10.0	2	9.8	KENT MD		-3			66
5	11.7	7	12.7	15	15.3	MONTGOMERY MD	8	21	31	46	77
8	10.2	15	13.1	45	14.1	PRINCE GEORGE'S MD	28	7	37	49	74
		2	8.2	4	15.6	QUEEN ANNE'S MD		90			77
4	21.7	4	17.3	4	12.3	ST MARY'S MD	-20	-29	-43	15	70
1	2.8	3	8.8	5	16.7	SOMERSET MD	210	91	492	96	79
2	8.1	7	22.1	5	16.6	TALBOT MD	174	-25	106	71	79
2	20.2	1	14.5	1	10.5	WASHINGTON MD	-28	-28	-48	12	67
5	15.8	5	11.4	9	17.0	WICOMICO MD	-28	49	7	37	80
2	7.6	3	10.0	2	5.9	WORCESTER MD	32	-41	-22	23	60
109*	14.2	210*	19.7	316*	23.9	BALTIMORE CITY MD	39	21	68	62	88
77*	22.4	76	15.6	120	17.4	MASSACHUSETTS MA	-30	11	-22		
1	8.0	1	9.6	3	20.6	BARNSTABLE MA	20	115	157	81	85
1	17.1	3*	57.1	2	19.5	BERKSHIRE MA	233	-66	14	39	83
3	10.8	5	21.5	7	20.5	BRISTOL MA	99	-5	90	66	85
2*	84.9					DUKES MA					
3	27.7	1	7.8	2	7.6	ESSEX MA	-72	-2	-73	03	62
7*	24.8	1	2.3	15	20.9	HAMPDEN MA	-91	802	-16	26	85
12*	23.7	8	12.3	15	16.3	MIDDLESEX MA	-48	32	-31	19	79
1*	74.3					NANTUCKET MA					
2	24.2	5*	35.4	4	17.4	NORFOLK MA	46	-51	-28	21	80
2	10.6	5	22.5	3	8.3	PLYMOUTH MA	112	-63	-22	23	64
42*	24.4	45	17.7	65	20.3	SUFFOLK MA	-28	15	-17	26	84
1	10.6	2	10.9	4	16.3	WORCESTER MA	3	49	54	57	79
218*	13.2	431*	16.7	702*	20.1	MICHIGAN MI	27	21	53		
1	22.8	3*	48.1			ALLEGAN MI	111				
				1	38.2	BAY MI					94
4	11.8	8	16.1	13	20.2	BERRIEN MI	37	26	72	63	84
4	15.3	5	14.6	9	20.4	CALHOUN MI	-4	40	34	47	85
3	14.3	5	23.4	3	12.9	CASS MI	63	-45	-10	29	72
1*	67.0					CHARLEVOIX MI					
1*	344.1					CRAWFORD MI					
		1*	212.0			DELTA MI					
8	14.5	12	14.8	43*	29.0	GENESEE MI	2	95	100	69	91
				1*	212.0	HILLSDALE MI					99
1	10.3	2	7.6	5	13.6	INGHAM MI	-26	79	32	47	73
		1	18.2			IONIA MI					
1	39.3			1	47.1	ISABELLA MI			20	41	95
1	2.0	3	12.5	4	14.7	JACKSON MI	537	17	648	98	75
3	18.2	1	7.4	3	10.2	KALAMAZOO MI	-59	38	-44	14	67
1	3.3	4	8.4	17	23.8	KENT MI	153	184	618	98	88
		3	25.7	3	15.8	LAKE MI		-38			78
1*	62.9					LEELANAU MI					
3	10.5	1	3.7	5	15.4	MACOMB MI	-64	311	46	53	77
1*	22.8	1	86.0			MARQUETTE MI	278				
1*	188.7					MISSAUKEE MI					
		1	13.4	2	17.1	MONROE MI		28			80
2	8.9	5	11.4	11	17.1	MUSKEGON MI	27	51	92	66	80
		1	13.8	2	64.6	NEWAYGO MI		369			97
2	3.1	14	17.1	13	11.1	OAKLAND MI	458	-35	262	90	68
1*	74.3					OGEMAW MI					
				1	9.2	OTTAWA MI					65
2	5.0	8	23.2	7	10.6	SAGINAW MI	367	-54	114	73	67
1	18.7	3	29.4	6*	49.4	ST CLAIR MI	57	68	163	81	96
		1	17.7	1	10.4	ST JOSEPH MI		-41			67
				1	26.5	TUSCOLA MI					89
3	16.1	5	18.3	4	11.7	VAN BUREN MI	13	-36	-28	21	69
7*	22.1	4	10.0	9	19.7	WASHTENAW MI	-55	98	-11	29	84
165*	13.9	339*	17.7	537*	21.5	WAYNE MI	27	22	55	57	86
16	12.9	26	17.8	51	22.8	MINNESOTA MN	37	29	77		
1*	89.3					ANOKA MN					
				2	23.2	BELTRAMI MN					87
				1*	267.9	BLUE EARTH MN					99
7	17.5	15*	30.1	23	24.8	HENNEPIN MN	72	-18	42	51	88
				1	27.0	MAHNOMEN MN					90
1*	112.4					PINE MN					
6	18.1	10	24.5	22*	38.3	RAMSEY MN	35	56	111	72	94
1	5.3	1	14.1	2	25.5	ST LOUIS MN	167	81	384	94	89
271	8.2	354	10.5	412	11.8	MISSISSIPPI MS	28	13	45		
5	7.9	5	6.7	19	23.9	ADAMS MS	-15	256	201	86	88
		2	14.5	2	11.4	ALCORN MS		-21			69
		4	13.0	4	11.6	AMITE MS		-11			69
6	17.8	2	4.6	3	7.6	ATTALA MS	-74	66	-57	08	62
		1	7.2	1	8.6	BENTON MS		21			64
14	8.9	15	10.9	14	10.6	BOLIVAR MS	22	-3	18	40	67
4	13.9	3	10.9			CARROLL MS	-21				
		3	8.9			CHICKASAW MS					
1	8.4	1	9.6	4	32.2	CHOCTAW MS	15	235	285	91	92

* SIGNIFICANTLY MORE DEATHS THAN EXPECTED

NON-WHITE MALE — CANCER OF THE LARGE INTESTINE — ICD 153,159.0

1950-1959 DEATHS	1950-1959 RATE	1960-1969 DEATHS	1960-1969 RATE	1970-1979 DEATHS	1970-1979 RATE	COUNTY AND STATE		PERCENT CHANGE 60S/50S	PERCENT CHANGE 70S/60S	PERCENT CHANGE 70S/50S	PERCENTILE %CHANGE 70S/50S	PERCENTILE RATES 1970S
3	9.1	1	3.3	3	7.6	CLAIBORNE	MS	-64	133	-16	26	62
1	3.1	2	8.2	3	8.8	CLARKE	MS	165	8	185	84	64
4	11.9	5	13.0	2	5.7	CLAY	MS	9	-56	-52	10	60
15	11.8	10	8.6	10	8.6	COAHOMA	MS	-27	-1	-27	21	64
7	13.5	11	23.4	9	16.8	COPIAH	MS	74	-28	25	43	80
2	10.7	2	12.6			COVINGTON	MS	17				
8	14.6	3	5.5	2	4.1	DE SOTO	MS	-62	-26	-72	04	58
3	5.4	6	9.2	3	4.7	FORREST	MS	70	-48	-12	28	59
2	11.2	3	20.0	3	21.5	FRANKLIN	MS	79	8	93	67	86
1	14.1	1	15.8			GEORGE	MS	12				
2	27.7	1	10.3	1	11.9	GREENE	MS	-63	15	-57	08	70
3	8.7	7	17.7	2	6.8	GRENADA	MS	105	-62	-21	24	61
		1	7.9	3	25.6	HANCOCK	MS		225			89
1	2.0	12	20.2	11	13.4	HARRISON	MS	910	-34	571	97	72
15	7.0	30	11.8	33	11.2	HINDS	MS	67	-5	59	59	68
6	7.3	10	13.1	9	11.2	HOLMES	MS	80	-15	54	57	68
2	4.0	5	11.1	5	12.4	HUMPHREYS	MS	180	12	213	87	71
				1	12.6	ISSAQUENA	MS					71
				1	10.7	ITAWAMBA	MS					67
6*	30.1	2	5.5	8	15.7	JACKSON	MS	-82	185	-48	12	77
4	14.3	4	13.2	3	9.2	JASPER	MS	-8	-30	-35	17	65
1	4.4	1	2.7	2	6.0	JEFFERSON	MS	-38	123	38	49	60
5	17.4	2	8.1	3	10.5	JEFFERSON DAVIS	MS	-54	30	-40	16	67
3	5.1	4	6.2	8	12.5	JONES	MS	22	102	146	79	71
4	16.1			4	13.4	KEMPER	MS			-17	26	72
5	20.5	6	23.2	2	5.2	LAFAYETTE	MS	13	-78	-75	03	59
2	21.4	1	12.4			LAMAR	MS	-42				
5	6.9	9	11.4	11	12.8	LAUDERDALE	MS	65	12	85	65	72
		5	27.2	3	17.1	LAWRENCE	MS		-37			80
		1	2.5	5	13.4	LEAKE	MS		431			73
6	12.9	3	7.8	5	13.9	LEE	MS	-39	77	8	37	73
8	7.0	13	12.0	18	20.0	LEFLORE	MS	73	66	187	84	84
5	15.1	5	15.2	3	7.9	LINCOLN	MS	0	-48	-48	12	63
3	5.8	8	14.0	2	3.2	LOWNDES	MS	141	-77	-44	13	57
4	5.3	5	6.0	12	12.6	MADISON	MS	13	111	138	77	71
1	3.3	2	6.5	3	11.3	MARION	MS	99	74	246	89	69
5	9.9	5	9.9	4	6.9	MARSHALL	MS	0	-30	-30	20	61
2	3.7	1	1.6	11	25.3	MONROE	MS	-58	1503	576	97	89
		3	10.6	3	12.9	MONTGOMERY	MS		23			72
		1	5.6	3	13.7	NESHOBA	MS		145			73
4	16.0	1	4.4			NEWTON	MS	-72				
5	11.5	5	11.7	7	19.7	NOXUBEE	MS	2	69	72	63	84
		1	2.8	5	14.1	OKTIBBEHA	MS		407			74
7	11.6	15*	23.3	11	17.3	PANOLA	MS	102	-26	50	54	80
				3	15.0	PEARL RIVER	MS					76
				1	9.1	PERRY	MS					65
7	13.6	9	16.5	9	14.5	PIKE	MS	21	-12	7	36	75
				3	21.7	PONTOTOC	MS					86
				2	18.9	PRENTISS	MS					82
2	4.7	9	17.4	2	3.8	QUITMAN	MS	273	-78	-18	25	57
2	4.9	1	2.4	2	4.4	RANKIN	MS	-51	83	-11	29	58
2	6.6	1	2.3	5	12.8	SCOTT	MS	-66	466	95	68	72
3	9.6	3	12.9	3	12.4	SHARKEY	MS	35	-4	30	45	71
2	10.6	6	22.7	9*	30.3	SIMPSON	MS	114	34	186	84	92
		1	10.2	1	11.4	SMITH	MS		12			69
		1	16.4	2	24.2	STONE	MS		48			88
12	10.4	8	7.3	9	11.9	SUNFLOWER	MS	-29	63	15	39	70
3	5.5	4	6.8	3	10.9	TALLAHATCHIE	MS	23	60	98	69	68
		2	5.9	7	20.1	TATE	MS		241			84
		3	22.4	3	21.3	TIPPAH	MS		-5			86
6	10.4	6	12.2	6	12.7	TUNICA	MS	17	4	22	42	71
		2	14.7	5	39.3	UNION	MS		168			94
1	3.8	3	14.2	1	5.9	WALTHALL	MS	272	-59	54	57	60
13	13.5	15	16.5	14	13.7	WARREN	MS	22	-17	1	35	73
14	8.5	15	9.8	23	14.5	WASHINGTON	MS	16	48	71	62	75
2	10.1	2	9.1	2	8.9	WAYNE	MS	-10	-2	-12	29	64
1	10.4			1	14.9	WEBSTER	MS			43	51	75
2	5.8	3	8.3	2	7.7	WILKINSON	MS	44	-7	33	47	63
1	4.9			3	10.3	WINSTON	MS			110	72	67
1	4.9			2	10.4	YALOBUSHA	MS			113	73	67
7	9.5	11	14.6	5	8.1	YAZOO	MS	53	-45	-15	27	63
194*	14.0	284*	16.6	380*	20.1	MISSOURI	MO	19	21	44		
2	27.8	3	29.3	3	29.6	AUDRAIN	MO	5	1	6	36	91
1*	112.4					BATES	MO					
		3	18.5	5	29.6	BOONE	MO		60			91
3	13.3	3	25.5	3	18.2	BUCHANAN	MO	92	-29	37	49	81
		3	17.8	1	12.1	BUTLER	MO		-32			70
2	16.1			3	24.3	CALLAWAY	MO			51	55	88
1	16.3	2	30.7	3*	52.1	CAPE GIRARDEAU	MO	88	70	219	87	96
				1	37.5	CARROLL	MO					94
1*	188.7					CASS	MO					
				2	26.1	CHARITON	MO					89
		1	6.6			CLAY	MO					
		1*	267.9			CLINTON	MO					
1	8.8	3	33.8	1	9.3	COLE	MO	282	-72	5	36	65
		1	14.5			COOPER	MO					
		3*	40.8			DUNKLIN	MO					
1	31.3					FRANKLIN	MO					
		4	31.4	1	8.1	GREENE	MO		-74			63
2	27.4	3*	33.2	1	31.0	HOWARD	MO	21	-7	13	38	92
		1*	344.1			HOWELL	MO					
39	14.8	67	17.7	91*	21.1	JACKSON	MO	19	19	42	51	86
		2	24.8			JASPER	MO					
3*	76.4	1	26.8	1	22.5	JEFFERSON	MO	-65	-16	-71	04	87
		1	38.5			JOHNSON	MO					
1*	74.3					KNOX	MO					
1	11.8					LAFAYETTE	MO					
		1	37.5			LEWIS	MO					
2*	51.2			2	35.0	LINCOLN	MO			-32	19	93

* SIGNIFICANTLY MORE DEATHS THAN EXPECTED

NON-WHITE MALE — CANCER OF THE LARGE INTESTINE — ICD 153,159.0

1950-1959 DEATHS	1950-1959 RATE	1960-1969 DEATHS	1960-1969 RATE	1970-1979 DEATHS	1970-1979 RATE	COUNTY AND STATE	PERCENT CHANGE 60S/50S	PERCENT CHANGE 70S/60S	PERCENT CHANGE 70S/50S	PERCENTILE %CHANGE 70S/50S	PERCENTILE RATES 1970S
		1	28.1			LINN MO					
		2*	85.4			LIVINGSTON MO					
		1	28.1			MACON MO					
5*	51.0	2	19.3			MARION MO	-62				
3	11.1	2	7.2			MISSISSIPPI MO	-35				
		1*	188.7	1*	91.1	MONITEAU MO		-52			98
		1	74.3			MONROE MO					
1	22.9					MONTGOMERY MO					
5	20.4			3	12.1	NEW MADRID MO			-41	15	70
3	7.2	1	1.8	1	3.9	PEMISCOT MO	-75	111	-46	13	57
		1	10.1	3	28.1	PETTIS MO		178			90
1	9.4	2	18.8	1	8.0	PIKE MO	101	-57	-14	27	63
2	21.0	1	11.9	3	43.2	RANDOLPH MO	-44	264	105	70	95
2*	44.0					RAY MO					
				2	25.9	ST CHARLES MO					89
1*	98.2			1*	70.7	ST FRANCOIS MO			-28	21	97
16*	22.4	10	8.9	30	16.0	ST LOUIS COUNTY MO	-60	79	-29	20	78
1	8.1	1	8.9	3	32.0	SALINE MO	10	258	294	91	92
		2	12.0	1	8.3	SCOTT MO		-31			64
				1*	18.6	SHELBY MO					82
		1	17.6			STODDARD MO					
				1*	60.7	VERNON MO					97
1*	49.2			1	54.6	WARREN MO			11	38	96
93*	14.2	152*	17.8	210*	23.0	ST LOUIS CITY MO	25	30	62	60	87
7	11.0	11	15.0	7	7.2	MONTANA MT	37	-52	-34		
		1	10.2	1	11.8	BIG HORN MT		15			69
1	18.9	1	19.1			BLAINE MT	1				
				1	12.7	CASCADE MT					71
1*	490.8	1*	56.2			GALLATIN MT	-89				
		3	25.4	1	11.1	GLACIER MT		-56			68
		2*	49.3			HILL MT					
		1	33.5			LAKE MT					
		1	94.4			LEWIS AND CLARK MT					
1*	37.8					PONDERA MT					
1*	74.3					POWELL MT					
1	14.9			1	7.3	ROOSEVELT MT			-51	10	62
				1	67.0	ROSEBUD MT					97
				1*	172.0	TETON MT					99
1*	267.9					WHEATLAND MT					
1	33.1	1	26.5	1	17.9	YELLOWSTONE MT	-20	-33	-46	13	81
18	16.8	28*	20.5	35	21.5	NEBRASKA NB	22	5	28		
		1*	43.0			ADAMS NB					
		1*	141.3			CHEYENNE NB					
1*	112.4					DAKOTA NB					
14	17.5	21	21.7	28*	23.8	DOUGLAS NB	24	9	36	48	88
		1*	78.1	1*	49.2	KNOX NB		-37			96
3*	49.6	1	4.4	1	21.0	LANCASTER NB	-91	380	-58	08	85
				1*	344.1	OTOE NB					99
				1	67.0	SARPY NB					97
				2*	206.8	SHERIDAN NB					99
		3*	45.2	1	15.0	THURSTON NB		-67			76
5	12.4	5	4.7	18	11.8	NEVADA NV	-62	151	-5		
1*	31.3					CHURCHILL NV					
2	7.8	2	3.7	14	10.3	CLARK NV	-53	181	32	46	67
		1*	424.0			ESMERALDA NV					
				1*	112.4	EUREKA NV					98
		1	26.0			MINERAL NV					
2	37.8	1	5.3	3	14.2	WASHOE NV	-86	168	-62	06	74
		1	3.8			NEW HAMPSHIRE NH					
		1	7.0			ROCKINGHAM NH					
219*	18.1	308*	17.4	536*	21.1	NEW JERSEY NJ	-4	21	16		
28*	24.3	34*	24.8	43*	29.4	ATLANTIC NJ	2	19	21	41	91
6	14.2	23*	38.3	25	21.2	BERGEN NJ	169	-45	49	54	86
9*	31.6	8	20.7	21*	26.6	BURLINGTON NJ	-35	29	-16	26	89
23*	20.9	19	14.6	42	24.2	CAMDEN NJ	-30	66	16	39	88
		3	14.8	4	16.9	CAPE MAY NJ		14			80
3	6.8	5	10.7	13	20.9	CUMBERLAND NJ	56	96	206	86	85
67*	17.8	107*	17.3	181*	23.2	ESSEX NJ	-3	34	30	46	87
2	6.1	5	8.3	9	15.1	GLOUCESTER NJ	36	82	147	79	76
18*	25.6	21	22.6	25	11.4	HUDSON NJ	-12	-50	-55	08	69
17*	29.8	18	14.3	28	16.7	MERCER NJ	-52	17	-44	14	79
4	7.7	13	26.6	16	15.1	MIDDLESEX NJ	247	-43	98	68	76
16	18.2	15	13.4	33	22.3	MONMOUTH NJ	-26	66	23	42	87
2	22.3	2	10.0	6	16.4	MORRIS NJ	-55	64	-26	22	79
1	9.8	3	16.9	7	29.3	OCEAN NJ	72	73	198	85	91
2	3.4	12	13.7	17	11.1	PASSAIC NJ	310	-19	231	88	68
1	5.1	2	5.5	13*	33.8	SALEM NJ	7	518	561	97	93
1	6.2			2	8.1	SOMERSET NJ			30	45	63
				3*	215.8	SUSSEX NJ					99
19*	18.6	18	13.6	47	22.0	UNION NJ	-27	62	18	40	87
				1	24.4	WARREN NJ					88
6	4.3	6	2.9	18	3.3	NEW MEXICO NM	-32	14	-22		
1	7.2	1	3.4	2	1.5	BERNALILLO NM	-53	-56	-79	01	56
				1	19.1	CHAVES NM					83
1*	74.3					COLFAX NM					
1	70.7			1	9.1	CURRY NM			-87	00	65
				1	3.2	DONA ANA NM					57
				2	16.2	LEA NM					78
1	2.4	2	4.5	4	4.2	MC KINLEY NM	87	-7	73	63	58

* SIGNIFICANTLY MORE DEATHS THAN EXPECTED

NON-WHITE MALE — CANCER OF THE LARGE INTESTINE — ICD 153,159.0

1950-1959 DEATHS	1950-1959 RATE	1960-1969 DEATHS	1960-1969 RATE	1970-1979 DEATHS	1970-1979 RATE	COUNTY AND STATE		PERCENT CHANGE 60S/50S	PERCENT CHANGE 70S/60S	PERCENT CHANGE 70S/50S	PERCENTILE %CHANGE 70S/50S	PERCENTILE RATES 1970S
		1*	114.7			QUAY	NM					
				2	5.3	RIO ARRIBA	NM					59
		1	5.7	1	3.7	SANDOVAL	NM		-34			57
1	4.5	1	2.0	1	1.6	SAN JUAN	NM	-56	-17	-64	06	56
1	28.1			1	3.3	SANTA FE	NM			-88	00	57
				1	1.8	TAOS	NM					56
				1	1.4	VALENCIA	NM					56
502*	15.3	911*	18.0	1385*	18.7	NEW YORK	NY	18	4	22		
1	5.6	3	9.1	11	23.1	ALBANY	NY	64	153	314	92	87
46	16.8	113*	18.1	160	15.7	BRONX	NY	8	-14	-7	31	77
1	23.6	1	13.2	1	17.2	BROOME	NY	-44	30	-27	21	80
1	24.8	3*	44.3	3	43.0	CATTARAUGUS	NY	79	-3	74	63	95
1	112.4			1	37.2	CAYUGA	NY			-67	05	94
		1	16.4	1	9.9	CHAUTAUQUA	NY		-40			66
1	14.9	4*	40.9	3	42.7	CHEMUNG	NY	175	4	187	84	95
		1	26.5			CLINTON	NY					
1	12.6	1	13.2	1	11.8	COLUMBIA	NY	5	-11	-6	32	69
		1*	188.7			CORTLAND	NY					
5	18.1	8	17.3	4	8.8	DUTCHESS	NY	-4	-49	-51	10	64
17	10.2	45	18.3	81*	24.4	ERIE	NY	81	33	140	78	88
1	12.8	1	10.5	2	26.4	FRANKLIN	NY	-18	152	105	71	89
				1	21.0	GENESEE	NY					85
		1	31.4			GREENE	NY					
97*	14.5	154	15.2	305	18.6	KINGS(BROOKLYN)	NY	5	23	29	45	82
4	6.1	4	10.0	15	11.5	MONROE	NY	63	15	88	66	69
1*	188.7					MONTGOMERY	NY					
11	16.0	13	13.1	44	21.1	NASSAU	NY	-18	61	32	47	86
222*	16.6	381*	22.7	425*	21.7	NEW YORK(MANHATTAN)	NY	36	-5	30	45	86
		2	14.5	10	26.1	NIAGARA	NY		80			89
		2	8.8	1	3.3	ONEIDA	NY		-62			57
2	10.8	8	22.2	6	11.3	ONONDAGA	NY	106	-49	5	36	68
4	21.8	6	14.2	11	20.3	ORANGE	NY	-35	43	-7	31	84
				1	114.7	OSWEGO	NY					98
1*	172.0					OTSEGO	NY					
42*	20.0	86	14.3	174	16.1	QUEENS	NY	-29	13	-19	24	78
1	16.9			3	28.3	RENSSELAER	NY			67	62	90
5	24.0	4	11.3	13	23.7	RICHMOND	NY	-53	109	-1	34	88
1	3.4	3	9.7	8	20.7	ROCKLAND	NY	184	113	505	96	85
				2	21.7	SARATOGA	NY					86
1	15.0	2	24.1	6*	41.2	SCHENECTADY	NY	61	71	175	83	95
		1	34.4			SENECA	NY					
1	28.4					STEUBEN	NY					
8	8.4	15	10.3	31	13.4	SUFFOLK	NY	23	31	60	60	73
				1	60.6	TIOGA	NY					97
		2	48.4	1	11.2	TOMPKINS	NY		-77			68
		3	12.1	3	13.6	ULSTER	NY		12			73
				1	8.0	WAYNE	NY					63
26*	16.3	42	18.9	55	18.2	WESTCHESTER	NY	16	-3	12	38	81
247	8.3	335	9.5	628	14.9	NORTH CAROLINA	NC	15	56	79		
3	6.3	5	13.8	11	17.3	ALAMANCE	NC	118	25	173	83	80
				2	48.2	ALEXANDER	NC					96
1*	70.7					ALLEGHANY	NC					
2	5.6	1	1.4	5	12.1	ANSON	NC	-76	789	115	73	70
2	4.5	2	4.5	6	11.3	BEAUFORT	NC	1	150	153	80	69
5	12.5	2	4.5	9	19.1	BERTIE	NC	-64	327	53	56	83
7*	22.6	4	8.7	2	4.3	BLADEN	NC	-62	-51	-81	01	58
				7	23.4	BRUNSWICK	NC					87
4	5.0	15*	22.3	10	16.5	BUNCOMBE	NC	350	-26	233	88	79
1	4.8	2	13.5	4	20.6	BURKE	NC	178	53	324	92	85
2	9.4	2	6.0	3	6.4	CABARRUS	NC	-36	7	-31	19	61
1	8.3	3	31.3	1	8.3	CALDWELL	NC	278	-74	0	34	64
2	30.6	1	18.6	2	14.3	CARTERET	NC	-39	-23	-53	09	74
5	21.0	2	8.2	6	19.2	CASWELL	NC	-61	135	-9	30	83
		2	7.7	4	14.2	CATAWBA	NC		84			74
2	12.2	2	8.8	2	6.1	CHATHAM	NC	-28	-30	-49	11	60
		1*	84.8	1	33.5	CHEROKEE	NC		-61			93
				1	4.2	CHOWAN	NC					58
3	8.5	3	7.5	7	18.2	CLEVELAND	NC	-13	144	113	73	81
3	6.8	5	10.1	4	5.9	COLUMBUS	NC	48	-41	-13	28	60
1	1.4	8	13.3	6	9.0	CRAVEN	NC	876	-32	563	97	65
5	6.9	11	10.2	15	13.7	CUMBERLAND	NC	48	35	99	69	73
				1	60.6	DARE	NC					97
1	2.7	3	7.5	5	17.2	DAVIDSON	NC	182	130	549	97	80
				2	20.1	DAVIE	NC					84
2	4.5	4	8.3	10	21.9	DUPLIN	NC	86	165	392	94	87
6	7.6	19	15.0	27	15.9	DURHAM	NC	97	6	108	71	78
9	17.4	3	4.9	18	20.5	EDGECOMBE	NC	-72	319	18	40	85
16	12.4	23	15.0	27	15.1	FORSYTH	NC	21	0	21	42	76
3	7.3	7	17.8	6	12.1	FRANKLIN	NC	143	-32	65	61	70
5	12.1	4	7.2	13	23.3	GASTON	NC	-40	223	93	67	87
1	10.6			3	13.8	GATES	NC			30	45	73
2	5.4	2	5.0	12	21.4	GRANVILLE	NC	-9	332	294	91	86
2	11.0					GREENE	NC					
15	15.3	25	15.0	43	20.9	GUILFORD	NC	-2	39	37	49	85
7	7.8	6	8.0	10	10.0	HALIFAX	NC	3	25	28	44	66
1	1.7	1	3.2	9	21.5	HARNETT	NC	85	582	1164	99	86
		1	28.8			HAYWOOD	NC					
4*	53.5			2	19.6	HENDERSON	NC			-63	06	83
1	2.9	10	29.9	6	12.6	HERTFORD	NC	937	-58	336	92	71
		1	4.3	3	13.2	HOKE	NC		206			72
				2	17.8	HYDE	NC					81
5	14.6	3	8.9	6	13.1	IREDELL	NC	-39	48	-10	29	72
1	26.8			2	19.4	JACKSON	NC			-28	21	83
2	5.1	2	5.8	8	17.7	JOHNSTON	NC	14	204	248	89	81
		2	13.3	1	5.6	JONES	NC		-58			60
1	5.8	2	8.7	4	15.5	LEE	NC	50	77	165	82	77
1	3.3	5	7.1	17	21.8	LENOIR	NC	120	205	570	97	86

* SIGNIFICANTLY MORE DEATHS THAN EXPECTED

NON-WHITE MALE — CANCER OF THE LARGE INTESTINE — ICD 153,159.0

1950-1959 DEATHS	1950-1959 RATE	1960-1969 DEATHS	1960-1969 RATE	1970-1979 DEATHS	1970-1979 RATE	COUNTY AND STATE	PERCENT CHANGE 60S/50S	PERCENT CHANGE 70S/60S	PERCENT CHANGE 70S/50S	PERCENTILE %CHANGE 70S/50S	PERCENTILE RATES 1970S
				2	30.5	MC DOWELL NC					92
1	3.6	2	9.1	4	8.6	MARTIN NC	150	-5	138	77	64
14	9.5	15	7.9	44	16.6	MECKLENBURG NC	-16	109	75	63	79
2	20.2			2	11.5	MONTGOMERY NC			-43	14	69
1	4.2	7	20.9	6	13.9	MOORE NC	397	-34	230	87	73
2	2.1	5	5.5	14	19.5	NASH NC	156	257	814	99	83
7	9.6	6	8.5	9	12.4	NEW HANOVER NC	-11	46	30	45	71
5	10.5	3	5.4	3	6.2	NORTHAMPTON NC	-48	14	-41	15	60
1	2.0			4	9.7	ONSLOW NC			378	94	66
6*	25.4	1	3.3	3	7.9	ORANGE NC	-87	137	-69	05	63
1	12.1					PAMLICO NC					
2	5.3	5	13.7	3	7.2	PASQUOTANK NC	158	-48	35	48	62
		3	9.7	3	9.8	PENDER NC		0			66
1	6.1	1	8.6	2	9.6	PERQUIMANS NC	42	11	57	58	65
		4	14.6	4	11.2	PERSON NC		-23			68
5	6.1	5	6.0	17	18.2	PITT NC	-2	205	200	86	81
		1	16.7			POLK NC					
		3	16.4	3	14.1	RANDOLPH NC		-14			74
1	2.5	4	10.4	4	8.4	RICHMOND NC	316	-19	238	88	64
9	6.0	6	3.7	22	14.4	RCBESON NC	-38	287	141	78	75
8	20.8	8	15.0	11	19.5	ROCKINGHAM NC	-28	30	-6	32	83
1	2.1	6	12.0	13	20.7	ROWAN NC	470	72	880	99	85
4	25.2	2	13.3	3	13.7	RUTHERFORD NC	-47	3	-46	13	73
3	5.5	6	10.6	5	7.6	SAMPSON NC	91	-28	37	49	62
2	5.9	2	5.3	4	9.6	SCOTLAND NC	-10	82	63	61	65
1	7.5	2	25.0	3	14.7	STANLY NC	234	-41	97	68	75
1	13.3					STOKES NC					
1	12.2			4	36.6	SURRY NC			201	86	93
		1	11.5			SWAIN NC					
1*	84.8					TRANSYLVANIA NC					
1	21.0			1	13.3	TYRRELL NC			-36	17	72
3	15.2	1	4.5	5	15.6	UNION NC	-71	249	2	35	77
5	11.7	4	10.4	9	16.9	VANCE NC	-11	63	44	52	80
12	10.0	22	14.9	34	18.5	WAKE NC	48	24	84	65	82
3	6.2	4	11.9	8	18.5	WARREN NC	91	56	197	85	82
1	10.2	1	5.7	3	13.4	WASHINGTON NC	-44	134	31	46	73
6	8.9	3	2.6	19	19.7	WAYNE NC	-71	652	120	74	84
				3	23.0	WILKES NC					87
2	2.9	6	9.2	8	9.8	WILSON NC	214	6	232	88	66
		1	16.1			YADKIN NC					
		1*	141.3	1*	68.3	YANCEY NC		-52			97
5	18.3	2	6.1	2	5.5	NORTH DAKOTA ND	-67	-9	-70		
1*	267.9					CASS ND					
		1*	112.4			MC LEAN ND					
1*	245.4					MOUNTRAIL ND					
1	12.8	1	5.9	2	12.8	ROLETTE ND	-54	118	1	34	72
1	24.4					SIOUX ND					
1*	68.8					WARD ND					
336*	16.3	572*	19.3	846*	23.8	OHIO OH	19	23	46		
1	9.4	5	17.1	8	29.5	ALLEN OH	82	72	213	87	91
				1*	133.9	ASHLAND OH					99
		1	15.7	1	8.0	ASHTABULA OH		-49			63
1	15.6	1	22.3	1	24.4	ATHENS OH	43	9	56	58	88
4*	36.7	2	18.6	1	10.1	BELMONT OH	-49	-46	-73	03	66
		1	33.5			BROWN OH					
4	14.1	6	15.5	6	16.4	BUTLER OH	10	5	16	40	79
				1	15.8	CHAMPAIGN OH					77
8	15.5	11	20.1	8	14.5	CLARK OH	29	-28	-7	32	75
				1	18.2	CLERMONT OH					81
1	17.7	5*	51.6	3	40.6	COLUMBIANA OH	192	-21	130	76	94
1*	68.3			1	47.1	COSHOCTON OH			-31	19	95
1*	42.4					CRAWFORD OH					
85*	13.8	167*	19.3	254*	22.5	CUYAHOGA OH	40	16	63	61	87
1*	106.0					DARKE OH					
				1*	188.7	DEFIANCE OH					99
		1	24.5			DELAWARE OH					
1	14.5	1	9.4	6	32.0	ERIE OH	-35	242	121	75	92
1*	117.9			1*	56.2	FAIRFIELD OH			-52	10	96
1	23.8	2*	49.0	1	27.0	FAYETTE OH	106	-45	13	38	90
47*	21.8	71*	23.5	94*	25.5	FRANKLIN OH	8	8	17	40	89
				1	14.9	GEAUGA OH					76
3	16.6	10*	40.0	9*	36.1	GREENE OH	141	-10	118	74	93
56*	15.6	116*	22.8	153*	26.9	HAMILTON OH	47	18	73	63	90
				1	32.7	HANCOCK OH					93
		1*	122.7	1*	133.9	HAPDIN OH		9			99
				1	24.8	HARRISON OH					88
1	16.1	1	22.3	1	42.4	HIGHLAND OH	39	90	164	82	95
				2*	41.9	HUPON OH					95
				1*	89.3	JACKSON OH					98
6*	31.7	3	12.5	10*	39.9	JEFFERSON OH	-61	219	26	44	94
2*	73.1	1	53.6	1	37.7	KNOX OH	-27	-30	-48	11	94
		2	26.1	1	10.6	LAKE OH		-59			67
2	24.8	4*	45.1	2	18.9	LAWRENCE OH	82	-58	-24	22	82
1	28.1	1	17.9	4*	44.7	LICKING OH	-36	150	59	59	95
1	27.3			2*	65.7	LOGAN OH			141	78	97
4	11.7	5	14.1	12	18.9	LOPAIN OH	20	34	61	60	82
18	15.5	27	15.0	58*	29.4	LUCAS OH	-3	95	89	66	91
2	45.0					MADISON OH					
12	16.0	24	16.9	37	22.6	MAHONING OH	5	34	41	50	87
2*	28.9	1	11.9	2	27.4	MIAMI OH	-59	130	-5	33	90
27*	18.3	41	17.7	76*	26.4	MONTGOMERY OH	-3	49	44	52	89
				1	33.6	MORGAN OH					93
2	15.1	2	14.6	1	4.9	MUSKINGUM OH	-3	-67	-68	05	59
				1	57.3	OTTAWA OH					97
				1	53.6	PICKAWAY OH					96
2	29.3	3	33.2	1	14.5	PORTAGE OH	13	-56	-50	11	75
3	14.2	2	9.0	3	9.7	RICHLAND OH	-37	8	-32	19	66

* SIGNIFICANTLY MORE DEATHS THAN EXPECTED

NON-WHITE MALE — CANCER OF THE LARGE INTESTINE — ICD 153,159.0

1950-1959 DEATHS	1950-1959 RATE	1960-1969 DEATHS	1960-1969 RATE	1970-1979 DEATHS	1970-1979 RATE	COUNTY AND STATE		PERCENT CHANGE 60S/50S	PERCENT CHANGE 70S/60S	PERCENT CHANGE 70S/50S	PERCENTILE %CHANGE 70S/50S	PERCENTILE RATES 1970S
1	14.9	1	3.8	1	5.0	ROSS	OH	-75	32	-66	05	59
				2	53.9	SANDUSKY	OH					96
2	31.6	1	9.4	5*	62.5	SCIOTO	OH	-70	563	98	69	97
		2*	71.9	1	16.4	SENECA	OH		-77			79
				1	31.6	SHELBY	OH					92
6	12.7	10	14.8	13	18.5	STARK	OH	16	25	45	52	82
17*	22.4	30*	22.4	35	19.3	SUMMIT	OH	0	-14	-14	28	83
6	22.7	5	12.4	11	21.5	TRUMBULL	OH	-45	73	-6	32	86
1	22.9	1	26.5	1	21.0	TUSCARAWAS	OH	15	-21	-9	31	85
		1*	37.2			UNION	OH					
1	20.2	2*	45.9	1	19.1	WARREN	OH	127	-58	-5	32	83
1	21.0	1	38.4	2	56.6	WASHINGTON	OH	83	48	169	82	96
				1	23.6	WAYNE	OH					87
78	9.0	114	11.1	180	14.0	OKLAHOMA	OK	23	27	56		
5*	48.4	2	15.5	1	5.2	ADAIR	OK	-68	-67	-89	00	59
		1	47.2			BECKHAM	OK					
		1	9.9			BLAINE	OK					
1	18.6	1	23.6	1	9.6	BRYAN	OK	27	-59	-49	11	65
		4	28.4	2	10.3	CADDO	OK		-64			67
1	15.6					CANADIAN	OK					
3	16.9	2	10.0	4	18.4	CARTER	OK	-41	83	9	37	82
		1	3.7	1	5.9	CHEROKEE	OK		58			60
1	4.7					CHOCTAW	OK					
				1	18.6	CLEVELAND	OK					82
2	11.5	2	9.2	5	15.8	COMANCHE	OK	-20	72	37	49	78
				2	19.0	CRAIG	OK					82
1	3.6	4	15.1	2	7.5	CREEK	OK	316	-50	106	71	62
		1	13.7			CUSTER	OK					
		1	6.5	1	7.3	DELAWARE	OK		11			62
1	15.7			2	25.1	GARFIELD	OK			60	59	89
2	35.9					GARVIN	OK					
1	10.3	2	26.0	1	12.7	GRADY	OK	152	-51	23	42	71
1*	74.3			1	37.2	GREER	OK			-50	11	94
				2*	82.6	HARMON	OK					98
		1	44.6	3*	63.6	HASKELL	OK		43			97
		1	8.4	3	27.5	HUGHES	OK		229			90
1	10.8			2	27.3	JACKSON	OK			154	80	90
				1*	53.6	JEFFERSON	OK					96
		1	43.3	2	32.3	JOHNSTON	OK		-25			92
5*	66.8	2	16.7	2	15.5	KAY	OK	-75	-8	-77	02	77
				1	16.1	KINGFISHER	OK					78
		2	28.5			KIOWA	OK					
				2	40.9	LATIMER	OK					95
1	10.9	2	13.0			LE FLORE	OK	19				
2	22.3					LINCOLN	OK					
1	9.7	3	15.1	7*	39.5	LOGAN	OK	57	161	309	91	94
1*	47.2					LOVE	OK					
				3*	115.2	MC CLAIN	OK					99
5	16.3	4	14.6	9	24.4	MC CURTAIN	OK	-10	67	50	54	88
1	6.9			1	5.2	MC INTOSH	OK			-24	22	59
		1	8.6	1	6.3	MAYES	OK		-27			60
9	10.0	13	17.3	11	14.3	MUSKOGEE	OK	73	-17	44	52	74
				2	45.2	NOBLE	OK					95
		2	43.4			NOWATA	OK					
		2	8.3	4	19.8	OKFUSKEE	OK		140			84
12	12.1	18	11.1	35	15.2	OKLAHOMA	OK	-9	37	25	43	76
3	6.7	4	9.3	5	10.7	OKMULGEE	OK	38	16	60	59	67
		1	7.1	3	15.0	OSAGE	OK		111			76
				2	19.2	OTTAWA	OK					83
1	31.5					PAWNEE	OK					
1	14.1	2	31.5			PAYNE	OK	123				
				1	3.8	PITTSBURG	OK					57
				1	12.4	PONTOTOC	OK					71
1	6.9	4	25.5	2	15.3	POTTAWATOMIE	OK	272	-40	122	75	76
1	28.3					PUSHMATAHA	OK					
		5*	64.9			ROGERS	OK					
1	4.6	4	15.7	8	32.0	SEMINOLE	OK	239	104	592	98	92
				1	4.1	SEQUOYAH	OK					58
2*	55.3			1	7.8	STEPHENS	OK			-86	00	63
		1	10.7			TILLMAN	OK					
8	8.7	16	10.9	36	19.1	TULSA	OK	25	75	119	74	83
2	10.5	2	9.5	4	21.5	WAGONER	OK	-9	127	106	71	86
1	12.6	1	12.2	1	5.6	WASHINGTON	OK	-3	-54	-56	08	60
9	8.3	14	9.1	37	15.6	OREGON	OR	9	72	88		
				1*	163.6	BAKER	OR					99
				2	11.8	CLACKAMAS	OR					69
				1	38.2	CLATSOP	OR					94
		1*	212.0			COLUMBIA	OR					
				3*	136.0	DOUGLAS	OR					99
		1	84.8			LANE	OR					
				1	67.0	LINCOLN	OR					97
2	26.5			2	12.8	MALHEUR	OR			-52	10	72
6	8.8	11	11.1	24	18.2	MULTNOMAH	OR	26	64	107	71	81
		1	23.6			UMATILLA	OR					
				1*	112.4	UNION	OR					98
1	37.2			1	54.7	WASCO	OR			47	53	96
				1	7.3	WASHINGTON	OR					62
388*	16.1	636*	19.4	981*	25.3	PENNSYLVANIA	PA	20	31	57		
1*	112.4					ADAMS	PA					
65	13.8	115*	18.8	167*	25.2	ALLEGHENY	PA	36	34	83	65	89
		2	50.5	2	38.2	ARMSTRONG	PA		-24			94
9*	33.4	5	12.6	9	18.3	BEAVER	PA	-62	46	-45	13	82
1	6.7	3	16.7	3	12.3	BERKS	PA	147	-26	82	65	70
2*	95.9	1	14.9	2	32.5	BLAIR	PA	-84	118	-66	06	92
3	23.2	2	10.9	7	17.7	BUCKS	PA	-53	63	-24	23	81

* SIGNIFICANTLY MORE DEATHS THAN EXPECTED

1950-1959 DEATHS	1950-1959 RATE	1960-1969 DEATHS	1960-1969 RATE	1970-1979 DEATHS	1970-1979 RATE	COUNTY AND STATE		PERCENT CHANGE 60S/50S	PERCENT CHANGE 70S/60S	PERCENT CHANGE 70S/50S	PERCENTILE %CHANGE 70S/50S	PERCENTILE RATES 1970S
1	36.6			4*	84.7	BUTLER	PA			131	77	98
2	12.5	1	5.9	6*	41.7	CAMBRIA	PA	-53	602	232	88	95
11	17.5	13	19.0	24*	28.1	CHESTER	PA	8	48	61	60	90
		1*	267.9			CLINTON	PA					
				1	21.5	CRAWFORD	PA					86
1	19.1	1	17.2			CUMBERLAND	PA	-10				
11	19.2	15	19.4	25*	26.0	DAUPHIN	PA	1	34	35	48	89
17	15.7	22	15.5	53*	30.2	DELAWARE	PA	-2	95	92	67	91
5*	33.3	5	19.4	10*	39.8	ERIE	PA	-42	105	19	41	94
5	13.7	8	17.5	7	19.4	FAYETTE	PA	28	11	42	51	83
2	36.6	2	20.8	5*	55.6	FRANKLIN	PA	-43	167	52	56	96
		1	26.8			GREENE	PA					
1*	28.7	1	40.9			INDIANA	PA	43				
				2	37.8	LACKAWANNA	PA					94
3	17.2	2	12.6	5	24.6	LANCASTER	PA	-27	96	43	51	88
2	16.2	2	13.6	2	18.7	LAWRENCE	PA	-16	37	16	39	82
		1	56.2	1	37.7	LEBANON	PA		-33			94
4*	66.5	3*	34.8	2	19.4	LUZERNE	PA	-48	-44	-71	04	83
		2	30.6	5*	72.1	LYCOMING	PA		136			97
				1*	94.4	MC KEAN	PA					98
1	4.7	3	14.5	3	15.1	MERCER	PA	208	4	221	87	76
1*	56.2			3*	82.1	MONROE	PA			46	53	98
9	14.3	18	25.0	31*	27.5	MONTGOMERY	PA	75	10	93	67	90
		1	12.5	4	32.5	NORTHAMPTON	PA		160			92
				1*	89.3	NORTHUMBERLAND	PA					98
1*	490.8			1*	267.9	PERRY	PA			-45	13	99
216*	16.4	382*	20.0	562*	24.6	PHILADELPHIA	PA	22	23	49	54	88
1*	112.4			1	37.2	SCHUYLKILL	PA			-67	05	94
2*	197.0	1*	106.0			SOMERSET	PA	-46				
				1*	188.7	TIOGA	PA					99
1	22.3					VENANGO	PA					
5	11.9	13*	28.0	13*	28.8	WASHINGTON	PA	134	3	141	78	91
4	11.9	6	19.3	8	24.0	WESTMORELAND	PA	63	24	103	70	88
1	5.5	4	22.8	10*	41.7	YORK	PA	314	83	657	99	95
10	16.9	17*	23.6	20	16.6	RHODE ISLAND	RI	40	-30	-2		
		1	62.9			KENT	RI					
1	10.1	2	27.5	1	7.2	NEWPORT	RI	171	-74	-29	20	62
9	20.6	13	24.0	15	15.5	PROVIDENCE	RI	17	-35	-25	22	77
		1	15.1	4*	58.6	WASHINGTON	RI		287			97
163	7.2	213	8.7	350	12.5	SOUTH CAROLINA	SC	21	44	75		
4	16.9	4	17.6	1	4.0	ABBEVILLE	SC	4	-78	-77	02	58
10	16.9	3	4.6	4	5.4	AIKEN	SC	-73	19	-68	05	59
2	9.7	3	12.2	4	17.2	ALLENDALE	SC	26	41	78	64	80
6	9.9	5	7.5	10	13.6	ANDERSON	SC	-25	83	38	49	73
1	3.3			4	12.4	BAMBEPG	SC			273	90	70
1	5.6					BARNWELL	SC					
6	11.5	4	9.1	16*	26.9	BEAUFORT	SC	-21	197	134	77	90
3	7.4	3	5.0	3	4.5	BERKELEY	SC	-32	-12	-40	16	58
3	10.8			1	3.1	CALHOUN	SC			-71	04	57
16	8.7	26	12.0	37	13.9	CHARLESTON	SC	38	16	60	59	74
		2	9.5	2	7.9	CHEROKEE	SC		-17			63
6	11.3	4	11.4	2	3.8	CHESTER	SC	1	-66	-66	06	57
4	14.9	2	5.4	2	5.2	CHESTERFIELD	SC	-64	-5	-65	06	59
		4	7.6	5	8.1	CLARENDON	SC		6			63
1	2.5	3	7.0	6	12.5	COLLETON	SC	181	78	400	94	71
2	3.3	10	14.0	17*	23.6	DARLINGTON	SC	324	69	616	98	88
		4	9.0	2	5.8	DILLON	SC		-35			60
2	6.6	5	15.7	5	9.8	DORCHESTER	SC	137	-38	48	54	66
3	8.7	5	17.4	3	10.7	EDGEFIELD	SC	101	-39	24	42	67
3	6.0	2	5.7	7	16.8	FAIRFIELD	SC	-5	194	179	84	80
5	6.3	12	11.3	16	13.5	FLOPENCE	SC	80	20	116	74	73
2	4.1	2	3.6	9	15.8	GEORGETOWN	SC	-13	343	283	90	78
8	7.4	14	13.3	27	19.0	GREENVILLE	SC	79	43	157	80	83
4	10.1	3	7.6	7	14.0	GREENWOOD	SC	-24	83	39	50	74
3	10.1	2	6.8	5	17.8	HAMPTON	SC	-33	163	76	64	81
1	4.7	3	4.4	2	5.2	HORRY	SC	-7	19	10	37	59
		3	9.5	3	10.0	JASPER	SC		5			66
7	15.0	6	14.0	6	11.0	KERSHAW	SC	-7	-21	-26	22	68
2	5.4	3	10.5	4	11.4	LANCASTER	SC	94	8	110	72	69
5	11.8	7	13.9	4	7.8	LAURENS	SC	18	-44	-34	18	63
1	2.5	1	2.5	4	11.3	LEE	SC	0	361	361	93	69
1	2.6	1	2.4	8	19.6	LEXINGTON	SC	-6	702	651	99	83
1	6.0			3	18.0	MC CORMICK	SC			201	86	81
		5	10.1	6	11.5	MARION	SC		13			69
2	3.9	3	7.5	7	14.6	MARLBORO	SC	90	96	271	90	75
2	5.4	4	11.0	4	11.2	NEWBERRY	SC	106	1	108	72	68
3	16.5	4	28.3	2	9.3	OCONEE	SC	72	-67	-43	14	65
10	9.3	3	2.5	12	9.7	ORANGEBURG	SC	-73	287	5	36	66
1	4.6	3	19.2	4	20.9	PICKENS	SC	318	8	353	93	85
9	6.6	12	6.4	31	13.4	RICHLAND	SC	-3	110	104	70	72
1	5.8			2	12.8	SALUDA	SC			121	75	72
7	7.6	10	7.8	23	17.7	SPARTANBURG	SC	2	127	132	77	81
5	5.8	11	10.8	9	8.1	SUMTER	SC	86	-25	39	50	63
4	13.7	2	6.6	4	12.0	UNION	SC	-52	81	-13	28	70
2	3.2	4	6.6	3	3.7	WILLIAMSBURG	SC	104	-44	15	39	57
4	6.8	6	9.9	14	20.2	YORK	SC	46	105	199	85	84
7	8.3	5	5.7	4	3.9	SOUTH DAKOTA	SD	-31	-31	-53		
1	47.1					BUFFALO	SD					
2*	33.8					DEWEY	SD					
				1*	44.6	LYMAN	SD					95
				1	94.4	MINNEHAHA	SD					98
1*	67.0					MOODY	SD					
1	38.2					ROBERTS	SD					
1	6.4	3	16.3	2	8.4	SHANNON	SD	156	-49	31	46	64
1*	424.0					STANLEY	SD					

* SIGNIFICANTLY MORE DEATHS THAN EXPECTED

NON-WHITE MALE — CANCER OF THE LARGE INTESTINE — ICD 153,159.0

1950-1959 DEATHS	1950-1959 RATE	1960-1969 DEATHS	1960-1969 RATE	1970-1979 DEATHS	1970-1979 RATE	COUNTY AND STATE	PERCENT CHANGE 60S/50S	PERCENT CHANGE 70S/60S	PERCENT CHANGE 70S/50S	PERCENTILE %CHANGE 70S/50S	PERCENTILE RATES 1970S
		1	7.9			TODD SD					
		1*	43.0			TRIPP SD					
214	10.0	312	12.9	456	17.3	TENNESSEE TN	29	34	73		
		1	11.6	3	30.4	ANDERSON TN		161			92
2	11.7	1	6.7	2	17.8	BEDFORD TN	-42	164	53	56	81
1*	188.7					BENTON TN					
		8*	61.2	4	31.7	BLOUNT TN		-48			92
				3	30.8	BRADLEY TN					92
2	10.9			1	5.8	CARROLL TN			-47	12	60
		1	49.7			CARTER TN					
		1	18.7			CHEATHAM TN					
				1*	74.3	CLAIBORNE TN					97
		1	44.6			COCKE TN					
				2	54.6	COFFEE TN					96
1	7.0	1	6.3	4	28.8	CROCKETT TN	-11	359	309	92	91
26	9.8	51	16.6	70	19.6	DAVIDSON TN	69	18	100	69	83
				1	28.7	DECATUR TN					91
2	32.9			1	22.5	DICKSON TN			-32	19	87
4	20.9	2	8.3	4	16.6	DYER TN	-60	101	-20	24	79
3	4.4	2	4.4	6	10.5	FAYETTE TN	0	139	138	78	67
		4*	45.2	1	9.1	FRANKLIN TN		-80			65
9*	23.6	6	14.9	9	20.0	GIBSON TN	-37	34	-15	27	84
3	15.9	5	27.3	1	4.0	GILES TN	71	-85	-75	03	58
1	34.4					GREENE TN					
1	14.2			1	13.4	HAMBLEN TN			-6	32	72
23	13.8	32	17.6	42	21.9	HAMILTON TN	27	24	58	59	86
2	6.6	1	3.0	6	16.3	HARDEMAN TN	-54	435	147	79	79
1	17.9			2	25.5	HARDIN TN			43	51	89
1	20.2					HAWKINS TN					
2	4.6	3	5.7	4	11.2	HAYWOOD TN	24	96	142	78	68
		1	10.6	4*	49.7	HENDERSON TN		368			96
3	15.9	6*	33.4	4	22.5	HENRY TN	109	-33	41	51	87
		1	14.5	1	36.4	HICKMAN TN		151			93
1	29.8			1	32.6	HUMPHREYS TN			10	37	92
2*	39.4			2	57.6	JEFFERSON TN			46	53	97
8	11.5	14	14.9	26*	25.6	KNOX TN	30	72	123	75	89
1	14.9			2	23.5	LAKE TN			58	59	87
4	10.9	2	5.2	6	16.1	LAUDERDALE TN	-52	208	49	54	78
2	16.4	1	7.4	1	6.8	LINCOLN TN	-55	-9	-59	07	61
				1	27.0	LOUDON TN					90
		3*	39.3	2	27.5	MC MINN TN		-30			90
		1	17.2	1	22.1	MC NAIRY TN		29			87
10	12.4	14	16.1	17	16.8	MADISON TN	30	4	35	48	80
2*	36.6	1	12.8	1	14.1	MARION TN	-65	11	-61	07	74
1	2.2	2	5.2	8	19.4	MAURY TN	134	275	778	99	83
		1*	84.8			MEIGS TN					
		1	16.4	2*	58.4	MONROE TN		257			97
3	6.5	3	8.0	6	13.5	MONTGOMERY TN	23	68	108	72	73
1	6.2	2	14.4	5	30.3	OBION TN	131	111	386	94	91
		1*	93.8			PERRY TN					
				1	40.9	PUTNAM TN					95
2*	55.6					RHEA TN					
2	31.3	2	27.2	2	27.2	ROANE TN	-13	0	-13	28	90
3	12.2	4	17.2	4	14.4	ROBERTSON TN	41	-16	18	40	75
1	4.1	3	9.6	8	24.7	RUTHERFORD TN	134	158	505	96	88
70	9.1	114	12.5	159	15.4	SHELBY TN	38	23	70	62	77
				1	56.2	SMITH TN					96
1	14.1	3	27.8	2	17.8	SULLIVAN TN	97	-36	27	44	81
2	11.0			2	9.0	SUMNER TN			-18	25	65
3	7.4	3	7.6	7	15.8	TIPTON TN	3	109	114	73	78
		1	20.6			TROUSDALE TN					
4	29.2	2	11.7	5	37.2	WASHINGTON TN	-60	217	27	44	94
1*	94.4			1*	68.8	WAYNE TN			-27	22	97
2	18.2	1	15.0	2	34.3	WEAKLEY TN	-18	130	89	66	93
1	5.3	2	8.8	2	8.0	WILLIAMSON TN	65	-8	51	55	63
		3	19.1	2	10.7	WILSON TN		-44			67
327	8.0	566	11.7	895	13.2	TEXAS TX	47	13	66		
5	11.1	5	12.5	6	14.5	ANDERSON TX	13	16	30	46	75
				1	67.0	ANDREWS TX					97
2	8.5	2	5.7	4	10.7	ANGELINA TX	-33	88	27	44	68
1	6.3			1	5.2	AUSTIN TX			-18	25	59
3	11.3	1	2.8	6	20.2	BASTROP TX	-76	627	78	64	84
2	6.3			5	10.0	BELL TX			58	58	66
9	5.7	31	16.1	45	13.0	BEXAR TX	180	-19	126	75	72
				1	43.0	BOSQUE TX					95
7	10.0	11	16.8	11	15.6	BOWIE TX	68	-7	56	58	77
1	2.0	10	28.6	8	14.5	BRAZORIA TX	1347	-49	632	98	75
5	12.5	4	9.0	5	9.6	BRAZOS TX	-28	6	-24	22	65
		1	17.9	1	13.6	BROWN TX		-24			73
		2	9.8	4	19.1	BURLESON TX		94			83
1	7.5	3	18.3	1	4.7	CALDWELL TX	143	-74	-38	17	59
2*	35.4					CAMERON TX					
2	11.3	4	24.9	3	14.3	CAMP TX	120	-42	27	44	74
3	9.1	3	8.0	3	10.1	CASS TX	-12	26	11	38	66
		1	12.7	1	8.6	CHAMBERS TX		-32			64
4	9.4	2	4.3	7	13.2	CHEROKEE TX	-54	206	41	51	72
				1*	182.1	CLAY TX					99
		1	6.5	3	15.6	COLLIN TX		139			77
		1	18.6			COLLINGSWORTH TX					
3	14.8	3	11.5	4	16.9	COLORADO TX	-22	47	14	39	80
1	32.1					COOKE TX					
		1	43.0			COTTLE TX					
22	6.4	68	13.1	108	14.5	DALLAS TX	105	11	127	76	75
		1	30.4	1	15.8	DAWSON TX		-48			77
1	10.3	1	16.1	4	35.3	DENTON TX	56	120	243	89	93
1	4.2	3	16.4	4	14.2	DE WITT TX	291	-13	240	88	74
				1	27.0	DICKENS TX					90

* SIGNIFICANTLY MORE DEATHS THAN EXPECTED

NON-WHITE MALE — CANCER OF THE LARGE INTESTINE — ICD 153,159.0

1950-1959 DEATHS	1950-1959 RATE	1960-1969 DEATHS	1960-1969 RATE	1970-1979 DEATHS	1970-1979 RATE	COUNTY AND STATE	PERCENT CHANGE 60S/50S	PERCENT CHANGE 70S/60S	PERCENT CHANGE 70S/50S	PERCENTILE %CHANGE 70S/50S	PERCENTILE RATES 1970S
1*	84.8					DONLEY TX					
				2*	81.1	EASTLAND TX					98
1	4.8	1	6.5	5	24.8	ECTOR TX	35	284	420	95	89
2	4.2	7	14.6	7	14.6	ELLIS TX	245	0	244	89	75
2	10.6	2	9.4	6	5.0	EL PASO TX	-11	-47	-53	09	59
3	6.4	1	2.4	5	15.4	FALLS TX	-62	538	143	79	77
3	22.4	4	27.9	1	6.6	FANNIN TX	25	-76	-70	04	61
1	5.7	3	17.1	5	24.3	FAYETTE TX	200	42	326	92	88
2	3.8	2	6.7	5	8.7	FORT BEND TX	78	30	131	76	64
4	13.1	2	31.2	1	3.1	FREESTONE TX	138	-90	-76	02	57
8	9.7	17	14.5	22	14.5	GALVESTON TX	50	0	50	54	75
		1*	57.3			GARZA TX					
				1*	267.9	GILLESPIE TX					99
		2*	61.8			GOLIAD TX					
		1	4.8	2	7.3	GONZALES TX		52			62
1	26.9	1	47.2	1	18.8	GRAY TX	76	-60	-30	20	82
1	3.1	4	11.6	7	20.9	GRAYSON TX	278	80	580	97	85
5	9.2	8	12.5	8	11.0	GREGG TX	35	-12	19	41	68
3	10.5	5	15.8	1	2.6	GRIMES TX	51	-83	-75	03	57
1	5.9	2	11.0	6	30.1	GUADALUPE TX	86	173	407	95	91
1	17.5			1	9.0	HALE TX			-49	11	64
2	12.6			2	9.9	HARDIN TX			-21	24	66
43	6.1	111	12.5	207	16.3	HARRIS TX	107	30	168	82	79
14	16.1	10	10.4	14	16.2	HARRISON TX	-35	56	1	34	79
		1	18.9	3	21.3	HAYS TX		13			86
		4	21.4	3	14.7	HENDERSON TX		-32			75
2	9.0	1	8.3	4	23.9	HILL TX	-8	189	165	82	88
		2*	45.0			HOCKLEY TX					
		1	8.9	2	19.1	HOPKINS TX		114			83
1	2.4	1	2.6	4	11.0	HOUSTON TX	8	331	364	93	68
				1	7.4	HOWARD TX					62
4	18.2	1	3.7	4	15.6	HUNT TX	-80	321	-14	27	77
		2*	54.6	1	20.2	HUTCHINSON TX		-63			84
		3	40.5	1	9.4	JACKSON TX		-77			65
1	3.4			2	7.0	JASPER TX			107	71	61
19	12.8	32	16.6	52	21.0	JEFFERSON TX	29	26	63	61	86
				2	30.9	JIM WELLS TX					92
1	10.3					JOHNSON TX					
		2	37.2			JONES TX					
		1	49.2			KARNES TX					
2	5.8	5	11.5	5	13.8	KAUFMAN TX	99	20	139	78	73
				1*	112.4	KENDALL TX					98
2*	130.6	1	24.4			KERR TX	-81				
3*	90.3					KLEBERG TX					
3	10.3	5	14.6	3	8.4	LAMAR TX	42	-42	-18	25	64
				2	20.5	LAMB TX					85
1	8.8	3	21.0	3	30.5	LAVACA TX	138	45	247	89	92
1	7.9	1	5.6	2	12.7	LEE TX	-28	126	62	60	71
2	6.0	1	9.9	1	2.9	LEON TX	66	-71	-52	10	57
3	15.4	6	20.5	3	9.7	LIBERTY TX	33	-53	-37	17	66
2	5.9	2	7.2	5	16.0	LIMESTONE TX	22	121	170	82	78
2	7.4	4	13.6	5	9.1	LUBBOCK TX	84	-33	24	43	65
1*	39.7					LYNN TX					
5	5.3	11	10.5	18	14.5	MC LENNAN TX	100	38	176	83	75
		2	21.7	1	9.2	MADISON TX		-58			65
		1	3.3	1	2.7	MARION TX		-20			57
2	8.1	2	7.6	6	17.8	MATAGORDA TX	-7	135	119	74	81
1	7.2			6	19.7	MIDLAND TX			175	83	84
2	7.9			3	15.4	MILAM TX			95	68	77
		1	40.9			MITCHELL TX					
4	13.7	1	3.3	3	7.2	MONTGOMERY TX	-76	119	-47	12	62
2	19.1			3	17.6	MORRIS TX			-8	31	81
				1*	106.0	MOTLEY TX					98
3	10.8	3	9.3	4	8.9	NACOGDOCHES TX	-14	-4	-17	25	64
4	8.6	4	9.7	5	13.2	NAVARRO TX	13	36	54	57	72
		1	5.0	1	4.2	NEWTON TX		-16			58
3	21.0	1	1.7	7	6.8	NUECES TX	-92	299	-68	05	61
		1	4.3	1	5.7	ORANGE TX		31			60
		3*	88.9			PALO PINTO TX					
4	17.1			3	13.3	PANOLA TX			-22	23	72
1*	57.3	1*	74.3			PARKER TX	30				
2	31.5	4	24.9	5	16.3	POTTER TX	-21	-35	-48	12	79
				1	53.6	REAGAN TX					96
		3	21.8	1	4.6	RED RIVER TX		-79			58
		1	18.6			REFUGIO TX					
2	5.4	2	4.8	1	2.6	ROBERTSON TX	-10	-46	-51	10	57
1	22.5	1	29.8	1	24.4	ROCKWALL TX	32	-18	8	37	88
		2*	61.8	1	74.3	RUNNELS TX		20			97
5	9.6	4	7.5	4	9.6	RUSK TX	-22	27	0	34	65
		1	9.2	1	12.6	SABINE TX		36			71
				3	21.9	SAN AUGUSTINE TX					87
1	5.9	3	14.3	3	16.3	SAN JACINTO TX	143	13	176	83	79
				1	27.0	SCURRY TX					90
		2	8.3	5	21.2	SHELBY TX		154			86
9	9.7	5	4.6	14	11.3	SMITH TX	-53	149	17	40	69
		1*	60.6			SWISHER TX					
18	10.0	33	14.7	46	15.9	TARRANT TX	47	8	59	59	78
3	23.2	4	27.5	1	6.8	TAYLOR TX	19	-75	-71	04	61
1	8.8	2	15.9	1	5.1	TITUS TX	80	-68	-43	15	59
1	6.1	2	11.7	1	5.3	TOM GREEN TX	91	-55	-15	27	59
8	9.1	18	17.6	33	20.8	TRAVIS TX	93	18	129	76	85
				2	21.3	TRINITY TX					86
1	7.0	2	19.1	2	19.5	TYLER TX	172	2	177	84	83
4	17.8	2	8.1	5	18.4	UPSHUR TX	-55	128	3	35	82
3*	49.2					VAN ZANDT TX					
		3	16.7	3	13.3	VICTORIA TX		-20			72
2	5.5	1	2.1	4	9.5	WALKER TX	-63	358	72	62	65
2	8.2	3	12.3	5	15.8	WALLER TX	51	29	94	68	78
1*	70.3					WARD TX					
1	2.8	3	8.5	5	17.0	WASHINGTON TX	201	100	501	96	80
1	2.2	2	4.3	6	12.3	WHARTON TX	93	187	454	96	70

* SIGNIFICANTLY MORE DEATHS THAN EXPECTED

NON-WHITE MALE — CANCER OF THE LARGE INTESTINE — ICD 153,159.0

1950-1959 DEATHS	1950-1959 RATE	1960-1969 DEATHS	1960-1969 RATE	1970-1979 DEATHS	1970-1979 RATE	COUNTY AND STATE	PERCENT CHANGE 60S/50S	PERCENT CHANGE 70S/60S	PERCENT CHANGE 70S/50S	PERCENTILE %CHANGE 70S/50S	PERCENTILE RATES 1970S
1*	49.2					WHEELER TX					
3	13.5	5	14.6	9	23.4	WICHITA TX	8	60	74	63	87
				3	35.2	WILBARGER TX					93
1	3.6	1	3.7	7	30.4	WILLIAMSON TX	3	722	748	99	92
1*	33.5					WILSON TX					
		1*	133.9	1*	54.5	WISE TX		-59			96
1	8.0	1	6.2			WOOD TX	-22				
8	14.7	10	13.6	9	10.4	UTAH UT	-7	-24	-30		
		1	24.8			BOX ELDER UT					
1	30.4	1	27.0			DAVIS UT	-11				
				1*	188.7	IRON UT					99
7*	28.0	7	21.8	5	14.0	SALT LAKE UT	-22	-36	-50	11	74
		1	8.2	3	14.0	WEBER UT		71			74
		2*	46.1	1	13.5	VERMONT VT		-71			
		1*	188.7			FRANKLIN VT					
				1*	188.7	WASHINGTON VT					99
		1*	56.2			WINDHAM VT					
245	9.0	411	13.7	611*	17.5	VIRGINIA VA	52	28	95		
5	10.9	5	9.8	4	7.2	ACCOMACK VA	-10	-27	-34	18	62
2	6.2	4	10.0	6	12.1	ALBEMARLE#VA	61	22	96	68	70
2	14.3	7*	53.4	4	29.3	ALLEGHANY#VA	273	-45	105	70	91
3	17.6	2	13.4	3	16.2	AMELIA VA	-24	21	-8	31	78
2	7.5	1	4.6	2	6.8	AMHERST VA	-38	47	-9	30	61
		1	9.1	3	30.2	APPOMATTOX VA		232			91
4	19.5	7	25.9	9	19.2	ARLINGTON#VA	33	-26	-1	33	83
2	9.6	3	12.1	4	13.4	AUGUSTA#VA	26	11	40	50	72
6	6.5	12	12.1	23	20.5	BEDFORD#VA	88	69	218	87	85
		2	24.3	2	30.1	BOTETOURT VA		24			91
2	5.4	1	2.2	2	6.0	BRUNSWICK VA	-58	165	11	38	60
1	4.3	2	11.3	2	8.5	BUCKINGHAM VA	162	-24	99	69	64
2	10.4	1	5.0	4	12.1	CAROLINE VA	-52	145	17	40	70
1	31.5	1	28.1			CARROLL#VA	-11				
1	7.4	2	14.5	1	4.2	CHARLES CITY VA	94	-71	-43	14	58
4	19.6	3	12.9	4	19.8	CHARLOTTE VA	-34	53	1	35	84
37	12.1	65	17.7	85	20.4	CHESTERFIELD#VA	46	16	68	62	84
3*	67.5	1	13.2	1	16.7	CLARKE VA	-80	27	-75	02	79
3	21.3	2	13.5	2	10.9	CULPEPER VA	-36	-19	-49	11	68
1	5.0	1	8.0	4	23.3	CUMBERLAND VA	61	192	369	94	87
9	6.0	16	10.1	31	20.4	DINWIDDIE#VA	67	103	239	88	85
4*	25.6			3	15.1	ESSEX VA			-41	15	76
5	9.1	8	12.2	17	15.4	FAIRFAX#VA	34	26	68	62	77
2	7.1	3	12.0	5	19.3	FAUQUIER VA	70	61	173	83	83
1*	89.3					FLOYD VA					
2	19.3	2	16.4	2	18.9	FLUVANNA VA	-15	15	-2	33	82
2	15.2			2	15.1	FRANKLIN VA			0	34	76
1	14.0					FREDERICK#VA					
1	4.8	3	16.8	6	33.8	GLOUCESTER VA	252	101	607	98	93
		2	11.6	3	18.2	GOOCHLAND VA		57			81
				1	37.5	GREENE VA					94
		2	10.2	1	4.0	GREENSVILLE#VA		-61			58
6	8.4	8	15.1	14	21.7	HALIFAX#VA	81	43	159	81	86
3	12.9	2	8.5	3	11.6	HANOVER VA	-34	36	-11	29	69
3	10.3	2	8.5	13*	31.5	HENRICO VA	-18	271	206	86	92
2	6.1	7	15.5	8	12.7	HENRY#VA	152	-18	108	71	71
2	7.4	3	11.5	7	27.7	ISLE OF WIGHT VA	56	141	275	90	90
5	14.3	6	12.2	8	12.3	JAMES CITY#VA	-14	1	-14	27	70
2	12.1	2	15.8	4	28.6	KING AND QUEEN VA	30	82	136	77	91
		2	23.7	3	37.6	KING GEORGE VA		59			94
		2	11.7	1	6.7	KING WILLIAM VA		-43			61
1	6.0	6*	34.3	3	15.9	LANCASTER VA	467	-54	164	81	78
4	26.8	5	29.3	1	5.4	LOUDOUN VA	9	-82	-80	01	59
2	7.8	10*	47.0	3	11.8	LOUISA VA	504	-75	52	55	69
1	7.4	3	14.0	3	15.0	LUNENBURG VA	88	7	101	70	76
		1	10.0	4*	54.9	MADISON VA		448			96
1	12.6	2	29.8	3	35.1	MATHEWS VA	137	18	179	84	93
5	9.9	7	13.1	7	12.2	MECKLENBURG VA	33	-7	23	42	70
		2	19.4	2	13.6	MIDDLESEX VA		-30			73
2	19.5	3	30.1	4	34.4	MONTGOMERY#VA	54	14	76	64	93
1	6.3			1	8.0	NELSON VA			28	44	63
2	6.1	4	15.1	6	17.7	NORTHAMPTON VA	149	17	191	85	81
5*	28.9	2	11.5			NORTHUMBERLAND VA	-60				
2	7.8	1	3.4	2	6.1	NOTTOWAY VA	-57	81	-22	23	60
3	19.5			4	29.6	ORANGE VA			52	56	91
		2*	67.6	1	38.3	PAGE VA		-43			94
1	21.0			1	12.3	PATRICK VA			-41	15	70
3	3.7	12	13.3	26	23.8	PITTSYLVANIA#VA	258	78	538	97	88
		1	6.5	3	27.4	POWHATAN VA		319			90
3	12.5	2	7.2	7	29.2	PRINCE EDWARD VA	-42	305	134	77	91
3	25.9	1	9.1	5	26.1	PRINCE WILLIAM#VA	-65	188	1	34	89
2	21.5	1	13.0	1	21.0	PULASKI VA	-39	61	-2	33	85
1	8.4	1	9.8	2	17.8	RICHMOND VA	16	82	112	72	81
11	13.8	13	15.8	22	20.8	ROANOKE#VA	14	32	51	55	85
1	11.2	2	22.6	2	28.2	ROCKBRIDGE#VA	103	24	152	80	90
				2	38.5	ROCKINGHAM#VA					94
1	89.3					RUSSELL VA					
5	10.7	2	3.7	6	12.0	SOUTHAMPTON#VA	-65	221	12	38	70
1	6.1	4	18.4	8	30.7	SPOTSYLVANIA#VA	202	67	405	95	92
1	14.3			1	9.4	STAFFORD VA			-34	18	65
		4	35.7			SURRY VA					
2	7.7	4	13.6	5	16.1	SUSSEX VA	77	18	110	72	78
1	9.3			3	32.6	TAZEWELL VA			252	89	92
1	14.0	1	33.1			WARREN VA	136				
1	14.3	2	34.3	3	30.4	WASHINGTON#VA	139	-11	112	72	92
		4	19.3	3	13.3	WESTMORELAND VA		-31			72
		1	12.3			WISE#VA					

* SIGNIFICANTLY MORE DEATHS THAN EXPECTED

\# COUNTIES COMBINED; SEE TECHNICAL NOTES

NON-WHITE MALE — CANCER OF THE LARGE INTESTINE — ICD 153,159.0

1950-1959 DEATHS	1950-1959 RATE	1960-1969 DEATHS	1960-1969 RATE	1970-1979 DEATHS	1970-1979 RATE	COUNTY AND STATE	PERCENT CHANGE 60S/50S	PERCENT CHANGE 70S/60S	PERCENT CHANGE 70S/50S	PERCENTILE %CHANGE 70S/50S	PERCENTILE RATES 1970S
		1	18.9			WYTHE VA					
30	6.6	77	15.0	114	19.2	CHESAPEAKE CITY#VA	128	28	192	85	83
3	3.4	9	10.3	19	17.7	HAMPTON CITY VA	205	72	424	95	81
5	6.3	15	12.1	21	13.9	NEWPORT NEWS CITY VA	92	15	121	75	74
9	13.7	13	17.3	16	16.4	SUFFOLK CITY VA	26	-5	19	41	79
33	12.0	51	13.6	81	13.1	WASHINGTON WA	14	-4	9		
				1	33.5	CLALLAM WA					93
				1	5.9	CLARK WA					60
				1	56.2	COWLITZ WA					96
		1	31.3			FERRY WA					
				1	7.1	FRANKLIN WA					61
		1	22.8			GRANT WA					
		1	22.3			GRAYS HARBOR WA					
21	13.5	28	12.1	57	16.4	KING WA	-11	36	22	42	79
				2	21.2	KITSAP WA					86
		1*	74.3			KLICKITAT WA					
1*	37.5					MASON WA					
		3*	66.6			OKANOGAN WA					
4	20.8	6	19.1	11	12.7	PIERCE WA	-8	-34	-39	16	71
		3*	87.0			SKAGIT WA					
				1*	273.3	SKAMANIA WA					99
		1	23.6			SNOHOMISH WA					
4	24.2	3	14.2			SPOKANE WA	-41				
				1	42.4	STEVENS WA					95
1*	34.2	2*	159.5	1	29.8	THURSTON WA	367	-81	-13	28	91
				1	22.5	WALLA WALLA WA					87
1	37.7			2	36.1	WHATCOM WA			-4	33	93
		1*	344.1			WHITMAN WA					
1	2.8			1	1.7	YAKIMA WA			-40	16	56
46	9.0	72	15.9	55	13.0	WEST VIRGINIA WV	77	-18	44		
2	13.9	1	6.5	2	20.5	BERKELEY WV	-53	214	48	54	85
		1	26.5			BOONE WV					
1	44.6	1	30.3			BROOKE WV	-32				
1	4.4	7	27.1	8*	31.1	CABELL WV	513	15	604	98	92
6	12.9	4	10.9	3	8.9	FAYETTE WV	-15	-18	-31	19	64
		2*	196.8			GRANT WV					
		2	28.3	1	12.6	GREENBRIER WV		-55			71
				2	90.7	HANCOCK WV					98
		1	12.2	1	18.9	HARRISON WV		55			82
1	11.2	3	22.2			JEFFERSON WV	98				
9	13.7	15	20.3	7	10.0	KANAWHA WV	48	-51	-27	21	66
3	9.0	2	10.5	2	7.4	LOGAN WV	16	-30	-19	25	62
8	8.3	14	23.9	5	9.0	MC DOWELL WV	188	-62	8	37	64
2	9.8	3	16.5	4	17.3	MARION WV	69	5	78	64	80
		1	42.4			MARSHALL WV					
3	7.4			6	20.4	MERCER WV			176	83	84
2	10.8					MINGO WV					
1	7.6	3*	33.9	3	32.7	MONONGALIA WV	344	-3	328	92	93
1	11.5					OHIO WV					
5	8.5	11	25.4	9	16.1	RALEIGH WV	200	-37	90	66	78
		1	22.5			SUMMERS WV					
1*	344.1					WETZEL WV					
				1*	546.4	WIRT WV					99
				1	18.7	WOOD WV					82
17	11.5	35	14.6	70	16.9	WISCONSIN WI	26	16	47		
		1*	344.1			ADAMS WI					
				2*	55.6	ASHLAND WI					96
		1	94.4	1	57.3	BAYFIELD WI		-39			97
		1	16.4	3	67.9	BROWN WI		315			97
		1*	53.6	1*	57.3	BURNETT WI		7			97
		1	14.5	1	11.2	DANE WI		-23			68
				1	33.5	DOUGLAS WI					93
		1*	62.0			FOREST WI					
1*	67.0			1	86.0	JACKSON WI			28	45	98
		1	30.3	1	14.9	KENOSHA WI		-51			75
		1*	112.4			MANITOWOC WI					
5*	56.1	2	20.0	7*	45.4	MENOMINEE#WI	-64	127	-19	24	95
9	10.1	20	13.0	46	16.1	MILWAUKEE WI	29	24	59	59	78
		1*	47.2			MONROE WI					
				1*	94.4	POLK WI					98
		1	12.6	3	17.3	RACINE WI		37			80
1	12.6	1	17.9	1	12.4	ROCK WI	42	-31	-2	33	71
		1	28.1			SAWYER WI					
1	37.7			1	37.8	VILAS WI			0	34	94
		1*	298.1			WASHINGTON WI					
1	4.5	1	3.4	2	6.6	WYOMING WY	-26	97	46		
				1*	56.2	BIG HORN WY					96
1	29.8					LARAMIE WY					
		1*	74.3			PARK-TETON#WY					
				1*	74.3	WASHAKIE WY					97

* SIGNIFICANTLY MORE DEATHS THAN EXPECTED

\# COUNTIES COMBINED; SEE TECHNICAL NOTES

NON-WHITE FEMALE — CANCER OF THE LARGE INTESTINE — ICD 153,159.0

COUNTY AND STATE	1950-1959 DEATHS	1950-1959 RATE	1960-1969 DEATHS	1960-1969 RATE	1970-1979 DEATHS	1970-1979 RATE	PERCENT CHANGE 60S/50S	PERCENT CHANGE 70S/60S	PERCENT CHANGE 70S/50S	PERCENTILE %CHANGE 70S/50S	PERCENTILE RATES 1970S
UNITED STATES	7622	12.0	12138	14.1	18620	14.9	18	6	24		
ALABAMA AL	301	7.9	454	10.4	693	13.1	31	27	66		
AUTAUGA AL	2	5.8	3	9.1	7	17.9	56	97	207	90	81
BALDWIN AL	1	3.2	2	4.7	7	12.2	48	160	285	93	69
BARBOUR AL	2	4.2	2	3.3	10	15.6	-22	380	274	93	77
BIBB AL	5	26.8	3	15.0	6	29.3	-44	95	9	46	92
BLOUNT AL			1	40.9							
BULLOCK AL	1	1.8	7	13.4	5	9.1	657	-33	411	97	62
BUTLER AL	3	6.3	5	11.7	3	7.8	86	-33	24	53	60
CALHOUN AL	7	11.2	12	15.7	9	9.9	40	-37	-12	34	64
CHAMBERS AL	5	11.3	6	9.6	11	15.2	-15	58	35	57	76
CHEROKEE AL			1	11.0							
CHILTON AL	2	14.8	2	9.4	1	5.6	-36	-40	-62	08	57
CHOCTAW AL	3	8.5	6	18.9	3	8.5	121	-55	0	41	61
CLARKE AL	3	5.5	8	14.9	10	16.4	171	10	198	89	79
CLAY AL	1	11.5	1	18.7			63				
COFFEE AL	1	3.9			4	12.1			213	90	69
COLBERT AL	4	10.7	5	12.7	9	16.1	19	27	51	62	78
CONECUH AL	3	8.1	3	7.8	3	5.8	-5	-26	-29	24	57
COOSA AL	2	14.7	1	10.6	2	10.8	-28	2	-26	26	66
COVINGTON AL			5	16.9	6	20.5		22			85
CRENSHAW AL	3	15.3	1	3.8	2	9.7	-75	154	-36	21	64
CULLMAN AL			1	24.8							
DALE AL					2	5.4					57
DALLAS AL	13	7.9	15	8.9	19	10.0	12	12	26	54	64
DE KALB AL	1	28.7									
ELMORE AL	3	7.4	2	4.6	6	14.2	-38	208	92	74	74
ESCAMBIA AL	3	9.4	1	1.7	10	14.5	-82	746	55	64	75
ETOWAH AL	4	7.3	9	15.2	17	21.4	109	40	194	89	87
FAYETTE AL	2	21.7	2	19.0	2	9.3	-12	-51	-57	10	63
FRANKLIN AL			1	13.4							
GENEVA AL			1	5.5	2	12.9		137			72
GREENE AL	3	6.3	6	12.0	7	11.2	90	-7	77	71	67
HALE AL	1	1.4	6	9.4	8	10.7	563	14	653	99	66
HENRY AL	2	6.5	1	4.6	2	6.9	-28	49	7	45	59
HOUSTON AL	3	7.3	7	10.2	5	7.0	39	-31	-4	39	59
JACKSON AL			1	6.8	3	20.7		206			86
JEFFERSON AL	77	8.8	119	11.7	206	16.0	32	37	81	72	78
LAMAR AL	1	9.6	1	11.1			16				
LAUDERDALE AL	1	3.0	3	7.1	5	10.7	140	51	262	93	66
LAWRENCE AL	5	23.6	4	14.2	6	23.7	-40	66	0	42	89
LEE AL	5	8.4	6	7.8	11	12.1	-8	56	44	60	69
LIMESTONE AL	3	12.7	3	8.0	2	5.3	-37	-34	-58	09	57
LOWNDES AL	3	6.1			1	1.7			-73	04	54
MACON AL	9	10.0	12	11.7	12	9.9	17	-15	-1	40	64
MADISON AL	9	12.6	13	14.9	25	20.4	18	37	62	67	85
MARENGO AL	7	9.5	11	13.3	6	5.6	40	-58	-41	18	57
MARION AL					2*	45.5					95
MARSHALL AL	1	67.0									
MOBILE AL	23	7.6	54	12.8	85	16.8	69	31	122	80	80
MONROE AL	2	5.0	12*	23.2	7	11.4	368	-51	130	81	67
MONTGOMERY AL	19	6.8	21	6.9	35	9.4	1	36	37	58	63
MORGAN AL	4	11.7	2	5.4	9	20.1	-54	275	73	70	85
PERRY AL	3	5.7	4	6.5	7	10.2	14	56	78	72	65
PICKENS AL	3	6.7	5	11.4	1	1.8	71	-84	-73	03	54
PIKE AL	1	1.5	5	9.7	6	9.1	533	-6	496	98	62
RANDOLPH AL	3	13.6	3	13.2	4	16.2	-3	23	19	51	78
RUSSELL AL	3	3.0	6	6.6	14	13.3	118	101	338	96	72
ST CLAIR AL	2	18.3	2	11.6	3	16.5	-36	42	-10	34	79
SHELBY AL	3	11.9	4	14.5	6	19.3	22	32	61	66	84
SUMTER AL	6	7.8	8	10.8	9	9.8	39	-10	25	53	64
TALLADEGA AL	6	10.1	4	5.2	10	11.2	-48	113	10	46	67
TALLAPOOSA AL	2	4.4	5	11.1	4	6.1	152	-45	39	58	58
TUSCALOOSA AL	8	7.4	10	6.9	22	13.1	-6	90	78	72	72
WALKER AL	6	20.6	4	13.9	6	16.7	-33	20	-19	30	79
WASHINGTON AL					4	16.0					78
WILCOX AL	3	4.9	6	10.0	4	5.5	101	-44	12	47	57
ALASKA AK			12	12.9	34*	23.4		82			
ARIZONA AZ	10	3.9	29	7.7	46	5.7	95	-25	46		
APACHE AZ	1	2.5	1	1.7	4	4.1	-34	148	64	67	55
COCHISE AZ	1	25.8	1	38.3	2	13.1	48	-66	-49	14	72
COCONINO AZ			1	6.2							
GRAHAM AZ			1	14.9							
MARICOPA AZ	6	7.8	15	11.5	26	8.0	48	-30	3	43	61
NAVAJO AZ	1	3.5	2	3.2	5	7.8	-8	142	122	81	60
PIMA AZ	1	2.0	6	9.3	9	6.4	353	-31	213	90	58
PINAL AZ			1	4.6							
YAVAPAI AZ			1	30.4							
ARKANSAS AR	130	7.3	224	11.5	295	12.9	57	13	78		
ARKANSAS AR			6	23.0	2	5.5		-76			57
ASHLEY AR			3	7.3	3	5.0		-32			56
BRADLEY AR	1	5.2	4	18.3	5	18.2	249	-1	247	92	82
CALHOUN AR	2	17.3	1	10.5			-40				
CHICOT AR	4	6.2	15	22.5	14	16.9	264	-25	173	86	80
CLARK AR	2	9.3	3	10.5	2	5.5	14	-48	-41	18	57
CLEVELAND AR	1	6.2	2	21.7	1	7.9	251	-64	27	54	60
COLUMBIA AR	3	8.1	2	3.4	7	12.3	-58	264	53	64	70
CONWAY AR	3	16.7	2	10.3	5	21.9	-39	113	31	56	87
CRAIGHEAD AR					4	34.7					94
CRAWFORD AR			1	28.3							
CRITTENDEN AR	7	5.9	6	5.1	11	10.0	-13	94	69	69	64

* SIGNIFICANTLY MORE DEATHS THAN EXPECTED

NON-WHITE FEMALE — CANCER OF THE LARGE INTESTINE — ICD 153,159.0

1950-1959 DEATHS	1950-1959 RATE	1960-1969 DEATHS	1960-1969 RATE	1970-1979 DEATHS	1970-1979 RATE	COUNTY AND STATE	PERCENT CHANGE 60S/50S	PERCENT CHANGE 70S/60S	PERCENT CHANGE 70S/50S	PERCENTILE %CHANGE 70S/50S	PERCENTILE RATES 1970S
1	2.3	3	10.7	2	6.6	CROSS AR	372	-38	192	88	59
3	14.5	4	15.8	4	14.9	DALLAS AR	9	-5	3	43	76
2	4.5	9	18.1	7	12.5	DESHA AR	301	-31	178	87	71
4	17.2	1	6.2			DREW AR	-64				
2	20.2	1	8.9	1	4.3	FAULKNER AR	-56	-52	-79	02	56
2	5.9	8	21.0	7	17.2	GARLAND AR	253	-18	189	88	80
2	5.1	9	22.4	5	12.0	HEMPSTEAD AR	335	-46	134	82	69
1	8.6	2	18.0	3	17.4	HOT SPRING AR	109	-4	102	77	81
1	10.0					HOWARD AR					
		3	18.9	6	27.8	JACKSON AR		47			91
6	3.5	18	9.9	20	9.7	JEFFERSON AR	182	-3	174	86	64
				1	37.7	JOHNSON AR					94
5	18.8	4	15.3	4	14.3	LAFAYETTE AR	-19	-6	-24	27	74
1*	68.8					LAWRENCE AR					
1	1.9	2	2.6	10	14.6	LEE AR	34	465	660	99	75
		2	6.2	2	6.3	LINCOLN AR		1			58
1	6.5	3	16.9			LITTLE RIVER AR	160				
1*	49.2			1	24.5	LOGAN AR			-50	13	89
2	7.6	2	7.7	4	11.5	LONOKE AR	1	50	51	62	68
7	18.4	4	8.8	7	11.5	MILLER AR	-52	30	-37	20	68
3	4.3	6	6.6	17	18.8	MISSISSIPPI AR	54	185	339	96	83
7	17.3	4	9.6	12	29.1	MONROE AR	-44	202	68	68	92
				1*	74.3	MONTGOMERY AR					97
		6	30.0	4	16.5	NEVADA AR		-45			79
4	7.3	11	16.3	11	14.9	OUACHITA AR	123	-9	104	77	76
9	8.2	8	6.2	13	8.9	PHILLIPS AR	-25	43	8	45	62
1*	38.2	1	49.7	1	112.4	PIKE AR	30	126	194	89	98
		1	5.8	1	5.4	POINSETT AR		-8			57
				2	28.2	POPE AR					92
1	10.9	1	12.5	2	22.7	PRAIRIE AR	14	82	108	78	88
14	6.1	40	14.6	53	15.7	PULASKI AR	138	8	156	84	77
14	18.9	9	10.5	12	17.0	ST FRANCIS AR	-44	62	-10	35	80
				3*	53.9	SALINE AR					96
		1*	112.4			SEARCY AR					
2	8.6	2	7.3	3	8.0	SEBASTIAN AR	-16	10	-8	36	60
				1	12.4	SEVIER AR					70
7	11.0	9	13.6	15	16.3	UNION AR	24	20	49	61	78
				1	17.2	WASHINGTON AR					80
		1	19.6	1	10.3	WHITE AR		-48			65
3	11.0	3	10.2	4	14.4	WOODRUFF AR	-7	41	31	56	75
		1	30.3			YELL AR					
285	10.8	615	12.5	1236	11.0	CALIFORNIA CA	15	-12	2		
36	12.1	68	11.3	134	12.7	ALAMEDA CA	-7	13	5	44	71
		1	16.1	2	9.3	BUTTE CA		-42			63
		1*	188.7			CALAVERAS CA					
2	3.0	8	5.4	37	14.4	CONTRA COSTA CA	79	166	376	96	75
8	10.2	8	7.1	18	8.2	FRESNO CA	-30	16	-19	30	61
1	11.5	4	37.5	1	5.5	HUMBOLDT CA	227	-85	-52	12	57
1	6.6	2	12.0			IMPERIAL CA	83				
2*	110.8					INYO CA					
4	7.9	11	18.1	15	9.0	KERN CA	128	-50	13	48	62
		1	6.9	2	5.9	KINGS CA		-15			58
		1*	39.0			LASSEN CA					
157	12.1	363	14.3	684	12.9	LOS ANGELES CA	18	-10	6	44	71
		4	23.6	3	6.5	MADERA CA		-73			58
1	23.6	1	15.7	1	5.6	MARIN CA	-33	-64	-76	03	57
				1	10.3	MENDOCINO CA					65
1	6.5	2	7.8	6	15.1	MERCED CA	20	93	131	82	76
		2	5.9	9	6.8	MONTEREY CA		15			59
1	12.0					NAPA CA					
		1*	85.2			NEVADA CA					
				14	5.9	ORANGE CA					58
		2	14.5	3	20.8	PLACER CA		43			86
4	12.4	4	4.4	13	5.9	RIVERSIDE CA	-64	32	-53	12	57
9	10.1	15	10.3	33	10.1	SACRAMENTO CA	2	-2	0	41	64
				1	26.5	SAN BENITO CA					91
5	12.6	7	10.3	16	6.2	SAN BERNARDINO CA	-18	-40	-50	13	58
10	11.6	19	11.0	44	10.6	SAN DIEGO CA	-5	-4	-8	35	66
26	11.1	61	13.3	134	13.0	SAN FRANCISCO CA	20	-2	17	50	72
3	5.6	6	7.6	7	3.5	SAN JOAQUIN CA	37	-54	-38	20	55
				1	3.6	SAN LUIS OBISPO CA					55
2	3.7	7	13.1	18	8.5	SAN MATEO CA	252	-35	130	81	61
1	5.7	1	4.6	4	7.4	SANTA BARBARA CA	-18	59	31	56	60
3	4.8	6	9.4	16	3.9	SANTA CLARA CA	96	-58	-18	31	55
		1	10.5			SANTA CRUZ CA					
				1	11.1	SHASTA CA					67
				1	18.2	SISKIYOU CA					82
5	17.1	3	10.9	5	5.4	SOLANO CA	-36	-50	-68	04	57
		1	7.5	1	5.1	SONOMA CA		-32			56
				1	1.5	STANISLAUS CA					54
				3	25.8	SUTTER CA					90
1	6.5	1	4.0	2	1.3	TULARE CA	-37	-67	-79	02	54
1	8.8	2	5.8	4	4.3	VENTURA CA	-34	-27	-52	13	55
1	17.2					YOLO CA					
		1	31.5	1	8.2	YUBA CA		-74			61
11	8.4	32	14.4	53	10.3	COLORADO CO	72	-29	23		
				1	4.8	ADAMS CO					56
9	10.2	28	17.6	35	11.7	DENVER-ARAPAHOE#CO	73	-34	14	48	68
		1	1.8	7	14.9	EL PASO CO		722			76
				1	18.9	JEFFERSON CO					83
				2	19.3	OTEPO CO					84
2	15.9	3	21.3	5	12.7	PUEBLO CO	34	-40	-20	30	71
				1*	198.7	SEDGWICK CO					99
				1	4.5	WELD CO					56
46*	19.4	64	15.5	112	16.2	CONNECTICUT CT	-20	4	-17		

* SIGNIFICANTLY MORE DEATHS THAN EXPECTED

COUNTIES COMBINED; SEE TECHNICAL NOTES

NON-WHITE FEMALE — CANCER OF THE LARGE INTESTINE — ICD 153,159.0

1950-1959 DEATHS	1950-1959 RATE	1960-1969 DEATHS	1960-1969 RATE	1970-1979 DEATHS	1970-1979 RATE	COUNTY AND STATE	PERCENT CHANGE 60S/50S	PERCENT CHANGE 70S/60S	PERCENT CHANGE 70S/50S	PERCENTILE %CHANGE 70S/50S	PERCENTILE RATES 1970S
15	17.5	22	17.6	33	14.4	FAIRFIELD CT	0	-18	-18	31	74
17*	26.1	9	8.7	27	12.4	HARTFORD CT	-66	42	-52	12	70
1	35.3	2	39.4	1	8.1	LITCHFIELD CT	12	-79	-77	02	61
		1	12.3	4	26.4	MIDDLESEX CT		115			90
10	13.4	23	16.7	40	20.7	NEW HAVEN CT	25	24	55	65	86
3	30.8	5	27.7	6	24.2	NEW LONDON CT	-10	-13	-21	29	89
				1	9.4	TOLLAND CT					63
		2*	99.6			WINDHAM CT					
24	11.9	43	17.4	72*	20.7	DELAWARE DE	46	19	74		
2	7.0	5	9.7	9	14.7	KENT DE	37	52	109	78	75
18	14.3	34*	23.1	51*	23.9	NEW CASTLE DE	61	4	67	68	89
4	8.9	4	6.7	12	18.5	SUSSEX DE	-24	176	109	78	82
211*	18.4	368*	21.6	514*	22.0	DIST. OF COLUMBIA DC	17	2	19		
211*	18.4	368*	21.6	514*	22.0	WASHINGTON DC	17	2	19	51	88
222	8.7	382	10.7	700	13.1	FLORIDA FL	23	23	51		
17*	23.7	22*	24.0	17	14.4	ALACHUA FL	2	-40	-39	19	75
				1	12.0	BAKER FL					69
		2	6.6	8	15.4	BAY FL		132			77
1	12.6	2	16.1	1	5.9	BRADFORD FL	28	-63	-53	12	58
		8	16.9	9	12.3	BREVARD FL		-27			70
12	15.3	23	13.0	37	11.9	BROWARD FL	-15	-9	-22	28	69
		1	17.9			CALHOUN FL					
1	26.5	1	26.5			CHARLOTTE FL	0				
1	14.9			1	7.0	CITRUS FL			-53	12	59
3*	28.0					CLAY FL					
				3	16.4	COLLIER FL					79
4	18.1	3	9.8	4	13.3	COLUMBIA FL	-46	36	-27	25	72
14	4.1	47	9.8	103	10.6	DADE FL	136	8	155	84	66
1	8.8	1	9.4	1	7.0	DE SOTO FL	7	-25	-20	30	59
		1	53.6			DIXIE FL					
29	7.6	58	11.8	93	14.0	DUVAL FL	54	19	84	73	74
11	9.6	17	12.6	40	18.8	ESCAMBIA FL	31	49	96	75	83
6	7.7	3	2.7	13	10.3	GADSDEN FL	-65	277	33	57	65
		1	13.3	2	17.3	GULF FL		30			80
1	5.5	2	14.9	1	3.7	HAMILTON FL	172	-75	-33	22	55
1	21.5	1	35.1			HARDEE FL	63				
1	47.2	2	31.9	5*	39.0	HENDRY FL	-32	22	-17	31	95
				5*	38.9	HERNANDO FL					95
1	5.8	2	8.8	1	4.6	HIGHLANDS FL	53	-48	-20	29	56
11	5.4	41	14.9	56	15.9	HILLSBOROUGH FL	173	7	191	88	77
				4*	116.8	HOLMES FL					98
				5	19.4	INDIAN RIVER FL					84
5	8.1	5	8.9	3	6.3	JACKSON FL	9	-29	-22	28	58
3	10.5	1	4.6	2	7.6	JEFFERSON FL	-57	67	-27	25	60
5	18.7	5	10.3	11	17.7	LAKE FL	-45	72	-5	38	81
1	24.8	3	8.4	10	23.2	LEE FL	-66	177	-7	37	88
3	4.2	12	11.3	17	12.7	LEON FL	168	12	200	89	71
1	5.7	2	13.1	2	11.5	LEVY FL	131	-12	103	77	68
1	4.3	2	5.5	2	5.6	MADISON FL	28	1	29	55	57
4	10.1	4	9.2	5	8.8	MANATEE FL	-9	-4	-13	33	62
5	7.3	5	6.4	19	17.7	MARION FL	-12	175	142	83	81
4*	35.3	1	6.5	3	22.2	MARTIN FL	-82	241	-37	20	88
1	7.0	1	3.9	1	5.3	MONROE FL	-45	35	-25	26	56
2	16.5	2	11.4	2	8.4	NASSAU FL	-31	-27	-49	14	61
1	17.2			2	16.0	OKALOOSA FL			-7	37	78
11	10.3	20	14.4	27	12.4	ORANGE FL	40	-14	20	52	70
				5*	44.5	OSCEOLA FL					95
14	16.8	15	7.7	35	12.0	PALM BEACH FL	-54	57	-28	24	69
3	26.1			1	4.5	PASCO FL			-83	01	56
9	7.6	15	11.3	37	16.5	PINELLAS FL	49	47	119	80	79
5	4.0	13	11.4	36	18.9	POLK FL	186	66	375	96	83
5	12.6	6	11.4	8	15.7	PUTNAM FL	-10	38	24	53	77
3	8.3	4	10.4	5	11.9	ST JOHNS FL	26	14	44	60	69
5	19.7	3	7.8	9	17.0	ST LUCIE FL	-60	118	-14	33	80
				1	7.7	SANTA ROSA FL					60
		1	6.2	5	11.9	SARASOTA FL		91			69
4	7.1	5	8.0	8	8.8	SEMINOLE FL	13	10	25	53	62
3	25.6	1	7.2			SUMTER FL	-72				
				4	17.3	SUWANNEE FL					80
		4	20.1	2	9.4	TAYLOR FL		-53			63
7	7.7	10	9.1	24	16.9	VOLUSIA FL	18	86	119	80	80
1	10.9	1	7.2	1	9.5	WALTON FL	-34	32	-13	33	63
1	8.8	3	34.0	3	21.4	WASHINGTON FL	286	-37	143	83	87
390	9.3	560	11.2	813	13.3	GEORGIA GA	20	19	43		
1	8.5	1	9.9			APPLING GA	17				
		1	44.6			BACON GA					
1	1.8	4	5.1	7	10.0	BALDWIN GA	186	95	457	98	64
		1	9.8	3	21.5	BARROW GA		119			87
		2	10.5	3	14.6	BARTOW GA		39			75
1	4.6	5	21.3	5	16.0	BEN HILL GA	359	-25	245	92	78
1	14.1			1	14.1	BERRIEN GA			0	41	74
15	7.8	28	12.7	42	14.9	BIBB GA	63	18	91	74	76
		1	8.6	2	12.7	BLECKLEY GA		47			71
		1	47.1			BRANTLEY GA					
1	2.3	1	3.5	3	8.6	BROOKS GA	51	149	274	93	61
2	24.1	3*	85.7	2	27.4	BRYAN GA	255	-68	14	48	91
1	3.0	4	10.3	1	2.5	BULLOCH GA	244	-76	-18	31	54
5	8.2	2	3.1	7	12.5	BURKE GA	-62	307	54	64	71
		1	6.4	6	30.5	BUTTS GA		375			93
2	12.5	4	18.7	5	20.9	CALHOUN GA	50	12	67	68	86
3	21.2	2	13.3	2	9.7	CAMDEN GA	-37	-27	-54	11	64
				1	7.4	CANDLER GA					60
		2	10.5	3	9.9	CARROLL GA		-6			64

* SIGNIFICANTLY MORE DEATHS THAN EXPECTED

NON-WHITE FEMALE

CANCER OF THE LARGE INTESTINE

ICD 153,159.0

1950-1959 DEATHS	1950-1959 RATE	1960-1969 DEATHS	1960-1969 RATE	1970-1979 DEATHS	1970-1979 RATE	COUNTY AND STATE	PERCENT CHANGE 60S/50S	PERCENT CHANGE 70S/60S	PERCENT CHANGE 70S/50S	PERCENTILE %CHANGE 70S/50S	PERCENTILE RATES 1970S
25	10.4	48	15.3	64	17.3	CHATHAM GA	47	13	66	67	80
1	14.9			3	32.9	CHATTOOGA GA			121	80	93
7	16.4	4	6.3	7	9.6	CLARKE GA	-62	53	-41	18	63
2	19.4			2	14.3	CLAY GA			-26	25	74
		1	4.8	7*	27.6	CLAYTON GA		470			91
1	24.8	1	11.1			CLINCH GA	-55				
2	6.1	5	14.1	4	8.9	COBB GA	130	-37	44	60	62
		2	26.4	4	12.4	COFFEE GA		-53			70
		5	15.9	6	18.2	COLQUITT GA		14			82
2	11.1	2	10.7	1	4.7	COLUMBIA GA	-4	-56	-58	09	56
1	6.7					COOK GA					
6	15.3	4	9.0	5	8.5	COWETA GA	-41	-6	-45	16	61
1	7.2					CRAWFORD GA					
2	5.7	7	19.9	8	19.7	CRISP GA	247	-1	243	92	84
3	5.6	7	15.2	1	1.5	DECATUR GA	170	-90	-74	03	54
5	7.1	25	20.5	45	17.1	DE KALB GA	187	-16	140	83	80
				1	5.2	DODGE GA					56
2	9.0	7	25.4	4	13.7	DOOLY GA	181	-46	51	62	73
6	6.2	10	9.5	22	14.3	DOUGHERTY GA	52	51	129	81	74
		1	8.2			DOUGLAS GA					
5	18.1	1	4.5	4	11.2	EARLY GA	-75	148	-38	19	67
		1*	67.0			ECHOLS GA					
1	10.1	1	7.2	2	16.1	EFFINGHAM GA	-29	124	59	66	78
3	12.5	2	7.1	6	19.7	EMANUEL GA	-44	179	58	65	84
2	30.2	3	27.1	1	7.2	EVANS GA	-10	-73	-76	03	59
				1	12.8	FAYETTE GA					71
3	9.7	3	6.7	10	18.9	FLOYD GA	-31	182	95	75	83
		1	17.9	1	16.1	FRANKLIN GA		-10			78
105*	14.8	126	13.8	189	16.4	FULTON GA	-7	19	11	46	79
1	34.2	1	67.0			GLASCOCK GA	96				
7	13.9	7	12.7	7	9.9	GLYNN GA	-8	-22	-29	24	64
		3*	57.2			GORDON GA					
1	4.2	3	10.9	3	8.8	GRADY GA	158	-19	109	78	62
3	10.9	3	10.6	2	8.2	GREENE GA	-3	-23	-25	27	61
1	10.1	1	7.0	3	21.1	GWINNETT GA	-31	201	109	78	86
		1	30.3	1	12.4	HABERSHAM GA		-59			70
1	6.0	4	24.1	2	6.2	HALL GA	303	-74	4	43	58
3	12.3			7	24.2	HANCOCK GA			97	76	89
1	24.4			1	16.1	HARALSON GA			-34	22	78
3	15.4	3	11.9	4	10.5	HARRIS GA	-23	-12	-32	22	65
		2	22.7	2	10.5	HART GA		-54			65
1	15.2			1	17.5	HEARD GA			15	49	81
3	15.8	3	10.3	1	3.0	HENRY GA	-35	-70	-81	01	55
3	8.8	3	9.4	4	8.5	HOUSTON GA	7	-10	-4	39	61
		2	11.6	1	8.8	IRWIN GA		-24			62
1	11.6	1	8.9	2	16.7	JACKSON GA	-23	87	43	60	79
2	13.9	1	7.4	2	14.6	JASPER GA	-47	96	5	44	75
1	15.8	1	15.3	2	18.5	JEFF DAVIS GA	-3	21	17	50	82
3	7.4	3	7.5	3	6.9	JEFFERSON GA	0	-8	-8	36	59
1	4.3	5	21.5	2	11.0	JENKINS GA	398	-49	154	84	66
		1	9.6	1	10.5	JOHNSON GA		9			65
		1	5.2	3	12.8	JONES GA		145			71
		2	11.4	3	17.5	LAMAR GA		54			81
3	5.9	6	11.5	7	11.7	LAURENS GA	93	2	96	75	68
1	7.0	3	21.2	3	19.7	LEE GA	202	-7	181	87	84
1	4.8	1	3.8	1	3.9	LIBERTY GA	-20	3	-18	30	55
2	37.2	1	12.3	1	6.6	LINCOLN GA	-67	-46	-82	01	58
2	2.7	6	8.1	10	11.7	LOWNDES GA	200	46	337	95	68
		5	20.6			MC DUFFIE GA					
1	5.8	1	5.9	6	31.0	MC INTOSH GA	2	422	432	97	93
1	3.5	3	7.9	2	4.0	MACON GA	124	-49	14	48	55
1	9.9					MADISON GA					
1	7.8			1	5.7	MARION GA			-27	25	57
5	13.5	3	7.1	4	7.3	MERIWETHER GA	-48	3	-46	16	59
1	18.9			3	45.2	MILLER GA			140	83	95
5	13.8	1	2.1	5	11.4	MITCHELL GA	-85	448	-17	31	67
		5	26.5	3	12.1	MONROE GA		-54			69
1	13.3	1	10.3	2	23.9	MONTGOMERY GA	-22	132	81	72	89
5	23.9	2	8.0	4	20.2	MORGAN GA	-66	151	-15	32	85
13	11.2	14	8.0	29	11.8	MUSCOGEE#GA	-28	48	6	44	68
		2	6.7	5	14.0	NEWTON GA		108			74
3*	54.1	2	29.1	1	25.8	OCONEE GA	-46	-11	-52	12	90
2	12.6			1	9.3	OGLETHORPE GA			-26	26	63
		1	23.8			PAULDING GA					
		3	8.5	5	11.9	PEACH GA		39			68
1*	62.9			1	38.2	PICKENS GA			-39	19	95
		2	15.1	1	6.2	PIERCE GA		-59			58
1	11.2	2	18.0	1	6.9	PIKE GA	61	-62	-38	19	59
2	13.6	2	9.4	4	12.4	POLK GA	-31	32	-9	35	70
4	22.6	2	10.8	6	30.0	PULASKI GA	-52	177	33	57	92
5*	36.8	1	4.5	1	5.8	PUTNAM GA	-88	29	-84	01	57
1	20.2					QUITMAN GA					
10	6.7	20	10.1	34	12.8	RICHMOND GA	51	27	91	74	71
2	27.6	2	19.0	2	12.3	ROCKDALE GA	-31	-35	-55	11	70
				1	9.4	SCHLEY GA					63
3	9.8	2	6.3	8	26.8	SCREVEN GA	-36	327	174	86	91
		1	7.3	1	4.8	SEMINOLE GA		-34			56
2	6.1	5	12.6	5	10.3	SPALDING GA	108	-19	69	69	65
2	20.0	2	15.6	1	7.2	STEPHENS GA	-22	-54	-64	07	59
1	3.1	1	4.4	1	3.0	STEWART GA	42	-32	-3	40	55
4	7.1	5	8.1	12	16.9	SUMTER GA	13	110	138	83	80
3	17.2	2	10.5	2	10.1	TALBOT GA	-39	-3	-41	18	65
2	16.8	2	19.4			TALIAFERRO GA	15				
		2	15.5	1	5.4	TATTNALL GA		-65			57
2	14.4					TAYLOR GA					
1	6.4	4	20.8	3	12.4	TELFAIR GA	226	-40	94	75	70
4	11.1	4	9.3	5	14.0	TERRELL GA	-17	51	26	54	74
8	15.2	6	9.1	10	14.2	THOMAS GA	-40	57	-6	37	74
5	23.0	3	10.2	4	11.2	TIFT GA	-56	9	-51	13	67
4	20.8	5	21.4	2	5.4	TOOMBS GA	3	-75	-74	03	57
		2	23.2	1	9.0	TREUTLEN GA		-61			62

* SIGNIFICANTLY MORE DEATHS THAN EXPECTED

COUNTIES COMBINED; SEE TECHNICAL NOTES

NON-WHITE FEMALE — CANCER OF THE LARGE INTESTINE — ICD 153,159.0

COUNTY AND STATE	1950-1959 DEATHS	1950-1959 RATE	1960-1969 DEATHS	1960-1969 RATE	1970-1979 DEATHS	1970-1979 RATE	PERCENT CHANGE 60S/50S	PERCENT CHANGE 70S/60S	PERCENT CHANGE 70S/50S	PERCENTILE %CHANGE 70S/50S	PERCENTILE RATES 1970S
TROUP GA	3	5.5	5	7.7	11	12.3	39	61	124	81	70
TURNER GA	5*	48.0	1	9.0	2	15.2	-81	68	-68	04	76
TWIGGS GA	2	11.1	1	5.2	1	6.4	-54	24	-43	16	58
UPSON GA	1	3.2	5	15.9	7	20.9	403	32	562	98	86
WALKER GA			2	23.0	5*	40.0		74			95
WALTON GA	1	4.4			5	19.4			340	96	84
WARE GA	1	2.3	6	13.1	5	9.3	471	-29	304	94	62
WARREN GA			2	10.8	1	5.4		-50			57
WASHINGTON GA	3	5.8	4	8.5	8	15.9	48	86	176	87	78
WAYNE GA			1	6.4	4	21.9		243			87
WEBSTER GA			1	16.4	1	15.0		-9			76
WHEELER GA	1	16.1	1	15.7			-2				
WHITE GA					1	78.1					97
WHITFIELD GA					2	21.1					86
WILCOX GA					1	9.9					64
WILKES GA			3	8.9	6	19.8		124			84
WILKINSON GA	1	4.6	3	16.4	3	16.2	254	-2	247	92	78
WORTH GA	3	11.0	7	24.2	2	9.0	121	-63	-18	30	62
HAWAII HI			130	9.3	236	10.3		11			
HAWAII HI			16	8.4	23	9.3		11			63
HONOLULU HI			98	9.8	183	10.6		8			66
KAUAI HI			9	10.7	7	6.1		-43			58
MAUI#HI			7	5.8	23	11.8		102			68
IDAHO ID			4	13.1	5	7.7		-41			
BANNOCK ID					1	10.6					66
BINGHAM ID			1	21.0	1	16.1		-24			78
BONNEVILLE ID					1	24.6					89
BOUNDARY ID			1*	112.4							
CANYON ID					1	10.3					65
KOOTENAI ID			1*	74.3							
LATAH ID			1*	490.8							
LINCOLN ID					1*	106.0					98
ILLINOIS IL	512*	18.1	817*	19.5	1107*	17.4	8	-11	-4		
ADAMS IL	2	31.2	2	22.9	3	37.2	-27	63	19	51	94
ALEXANDER IL	8	21.4	6	19.6	4	16.6	-8	-15	-22	28	79
BOONE IL					1*	49.7					96
CHAMPAIGN IL	7*	38.6	5	14.8	5	13.9	-62	-6	-64	07	73
CHRISTIAN IL			1*	196.5							
COLES IL	1*	112.4									
COOK IL	433*	19.0	703*	20.8	910*	17.5	10	-16	-8	36	81
DE KALB IL					1	44.6					95
DU PAGE IL			1	24.4	3	2.6		-89			54
EDGAR IL					1*	136.6					99
HENRY IL			1	74.3							
IROQUOIS IL					1	68.8					97
JACKSON IL	3	27.1	4	21.7	1	2.4	-20	-89	-91	00	54
JEFFERSON IL			4*	82.9	4*	45.9		-45			96
KANE IL	2	24.1			6	17.1			-29	24	80
KANKAKEE IL			3	5.5	9	14.3		158			74
KNOX IL			3	28.4	4	30.6		8			93
LAKE IL	3	16.2	7	21.1	6	8.0	30	-62	-51	13	61
LAWPENCE IL			1*	62.9	2*	112.9		79			98
MC HENRY IL					1	35.1					94
MC LEAN IL	1	16.7			2	18.7			12	47	83
MACON IL	5*	23.3	1	5.7	11*	27.3	-75	378	18	50	91
MADISON IL	4	9.8	8	15.8	18*	25.1	62	59	157	85	90
MARION IL	1	22.3			2	21.6			-3	39	87
MASSAC IL	1	8.6	1	9.4	3	42.3	9	351	393	97	95
MORGAN IL	3*	50.0	1	12.2	3*	44.7	-76	267	-11	34	95
PEORIA IL	2	8.4	6	14.3	11	21.5	71	51	157	85	87
PERRY IL	1	20.2			1	35.3			75	71	94
PULASKI IL	1	3.7	3	13.9	6	26.5	280	91	625	99	91
RANDOLPH IL			2*	50.9	3*	78.5		54			97
ROCK ISLAND IL	1	14.1	1	6.0	5	17.3	-57	186	22	52	80
ST CLAIR IL	16	11.7	23	11.0	47	16.4	-6	49	41	59	79
SALINE IL			1	14.1	1	28.8		103			92
SANGAMON IL	6	22.7	12*	36.8	13*	27.2	62	-26	20	52	91
SHELBY IL			1*	188.7							
STEPHENSON IL	2*	56.9	2	35.1			-38				
VERMILION IL	4	20.9	4	17.9	3	11.6	-14	-35	-44	16	68
WARREN IL	2*	95.5			1	47.2			-51	13	96
WILL IL	2	14.6	6	22.9	8	12.1	57	-47	-17	31	69
WILLIAMSON IL			3*	44.1	1	18.7		-58			83
WINNEBAGO IL	1	6.6	1	4.0	6	10.9	-40	172	64	67	66
INDIANA IN	136*	17.7	234*	20.2	329*	20.2	14	0	14		
ALLEN IN	4	17.8	3	10.3	11	14.4	-42	40	-19	30	75
BARTHOLOMEW IN					1	35.1					94
CLARK IN	4*	43.1	5*	39.8	5	30.0	-8	-25	-30	23	92
CLAY IN			1*	56.2	1	47.2		-16			96
DAVIESS IN			1*	94.4							
DEARBORN IN			1*	172.0	1*	133.9		-22			99
DE KALB IN					1*	74.3					97
DELAWARE IN			5	18.0	6	15.7		-13			77
ELKHART IN	2*	40.3	1	9.4	1	5.9	-77	-37	-85	01	58
FAYETTE IN	1	18.6	1	38.5	3*	76.3	107	98	310	95	97
FLOYD IN	3*	34.4	1	11.8	2	14.6	-66	24	-58	09	75
GRANT IN	2	22.7	1	6.7	3	14.9	-70	121	-34	21	76
HAMILTON IN					1	12.4					70
HENDRICKS IN			1*	344.1							
HENRY IN					1	14.9					76
HOWARD IN			1	10.6	3	22.3		110			88
JEFFERSON IN					1	20.6					86
JENNINGS IN	1*	188.7									

* SIGNIFICANTLY MORE DEATHS THAN EXPECTED

COUNTIES COMBINED; SEE TECHNICAL NOTES

NON-WHITE FEMALE

CANCER OF THE LARGE INTESTINE

ICD 153,159.0

COUNTY AND STATE	1950-1959 DEATHS	1950-1959 RATE	1960-1969 DEATHS	1960-1969 RATE	1970-1979 DEATHS	1970-1979 RATE	PERCENT CHANGE 60S/50S	PERCENT CHANGE 70S/60S	PERCENT CHANGE 70S/50S	PERCENTILE %CHANGE 70S/50S	PERCENTILE RATES 1970S
JOHNSON IN			1	44.6	1	32.7		-27			93
KOSCIUSKO IN	1*	112.4									
LAGRANGE IN					1*	112.4					98
LAKE IN	24	12.3	57*	19.3	93*	20.5	57	6	67	68	85
LA PORTE IN			2	11.3	2	8.7		-23			62
LAWRENCE IN					2*	558.0					99
MADISON IN	1	10.7	1	5.6	6	23.5	-47	316	119	80	89
MARION IN	65*	19.7	109*	22.3	135*	20.6	13	-7	5	44	86
MARSHALL IN			1*	133.9							
MIAMI IN	1*	47.2			1	74.3			58	65	97
MONROE IN	1	34.2									
MONTGOMERY IN			1*	67.0	1*	133.9		100			99
ORANGE IN					1*	62.9					97
POSEY IN					2*	115.9					98
PUTNAM IN					1	60.6					97
RANDOLPH IN			1*	94.4							
ST JOSEPH IN	4	14.6	11	17.8	10	11.3	22	-36	-23	28	67
SHELBY IN	2*	87.6	2*	83.9	1	38.3	-4	-54	-56	10	95
SULLIVAN IN			1*	112.4							
TIPPECANOE IN			2*	53.4	1	12.4		-77			70
VANDERBURGH IN	13*	31.7	9	15.4	17*	25.4	-52	65	-20	30	90
VIGO IN	3	10.8	9*	31.3	6	18.2	189	-42	68	68	82
WABASH IN	1*	112.4									
WARRICK IN			1*	56.2	1*	81.8		46			98
WAYNE IN	3	18.6	4	19.9	6	28.4	7	43	53	63	92
IOWA IA	21*	20.8	24	18.4	26	14.7	-11	-20	-29		
ADAIR IA	1*	267.9									
ALLAMAKEE IA			1*	112.4							
BLACK HAWK IA			5	34.9	2	9.5		-73			63
BUCHANAN IA			1*	267.9							
CERRO GORDO IA	2*	172.5									
CLINTON IA			2*	131.7							
DES MOINES IA	1	40.9			1	22.5			-45	16	88
DUBUQUE IA	1*	74.3									
HENRY IA	1*	47.2									
LEE IA	1	12.3	1	23.8			93				
MAHASKA IA					1*	344.1					99
POLK IA	12*	29.2	10	19.0	17*	25.9	-35	36	-12	34	90
POTTAWATTAMIE IA			1	20.2	1	14.1		-30			74
SCOTT IA	1	26.8	1	9.9	3	31.5	-63	217	17	50	93
WAPELLO IA			1	33.5							
WOODBURY IA	1	20.2	1	12.8	1	9.3	-37	-27	-54	11	63
KANSAS KS	50	13.2	80	16.5	113	17.5	25	6	33		
ALLEN KS	1	49.2									
ANDERSON KS					1*	112.4					98
ATCHISON KS			2	28.0							
BARTON KS	1	38.2	2*	63.5	1	38.2	66	-40	0	41	95
BROWN KS	1	43.0									
BUTLER KS			1*	53.6	1	27.0		-50			91
CHAUTAUQUA KS					1*	56.2					96
CHEROKEE KS			1	22.5							
COWLEY KS			1	13.7							
CRAWFORD KS			2*	58.1	1	49.7		-14			96
DONIPHAN KS					1	22.3					88
DOUGLAS KS			6*	35.8							
FINNEY KS	1*	37.2									
GEARY KS	1	19.6	1	14.9	1	11.8	-24	-21	-40	18	68
GRAHAM KS					1*	43.0					95
HARVEY KS	1	26.5			2	37.4			41	59	94
JOHNSON KS	1	31.5	2	27.4	2	16.0	-13	-41	-49	14	78
LABETTE KS			3	42.5	1	7.2		-83			59
LEAVENWORTH KS	3	17.7	2	11.5	5	27.9	-35	143	57	65	91
LYON KS	1	31.3	3*	81.5			161				
MIAMI KS	1	18.2	1	37.5	1	16.4	106	-56	-10	34	79
MONTGOMERY KS	4	21.8	3	13.3	7	28.7	-39	117	32	56	92
NEOSHO KS			1	74.3							
PAWNEE KS			1*	74.3							
PRATT KS	2*	186.8	1	38.3	1	49.1	-80	28	-74	03	96
RENO KS					2	35.6					94
RILEY KS					2	45.7					95
SALINE KS					2	29.3					92
SEDGWICK KS	5	15.7	7	9.9	25	20.6	-37	109	31	56	85
SHAWNEE KS	4	8.9	12	21.9	6	8.3	146	-62	-7	36	61
SUMNER KS	1*	109.3									
WABAUNSEE KS	1*	81.8									
WYANDOTTE KS	21	14.3	28	15.6	49*	21.6	9	38	51	62	87
KENTUCKY KY	177*	16.6	210*	17.5	288*	20.8	6	19	25		
ADAIR KY			1	27.0							
ALLEN KY	1*	44.6			2*	159.4			257	92	99
ANDERSON KY			1	37.5							
BALLARD KY					1	18.7					83
BARREN KY	1	10.1	3	26.9			166				
BATH KY	2*	101.7									
BELL KY	1	13.4	3	43.7	2	30.9	226	-29	130	82	93
BOURBON KY	4	23.2	2	9.6	5	21.6	-59	126	-7	37	87
BOYD KY			1	22.8	2	19.0		-17			83
BOYLE KY	4	24.2	3	18.9	4	17.6	-22	-7	-27	25	81
BRECKINRIDGE KY	1	60.6									
BULLITT KY	1*	188.7									
CALDWELL KY			3*	35.5	3	31.2		-12			93
CALLOWAY KY	3*	89.9									
CAMPBELL KY					2	43.2					95
CARLISLE KY	1*	805.6									
CARROLL KY			1*	74.3	1*	54.5		-27			96
CHRISTIAN KY	6	9.8	5	8.7	13	21.1	-11	142	115	79	86

* SIGNIFICANTLY MORE DEATHS THAN EXPECTED

NON-WHITE FEMALE

CANCER OF THE LARGE INTESTINE

ICD 153,159.0

1950-1959 DEATHS	RATE	1960-1969 DEATHS	RATE	1970-1979 DEATHS	RATE	COUNTY AND STATE	PERCENT CHANGE 60S/50S	70S/60S	70S/50S	PERCENTILE %CHANGE 70S/50S	RATES 1970S
2	17.4	3	24.6	5*	35.4	CLARK KY	42	44	104	77	94
1*	70.1			1	44.6	CRITTENDEN KY			-36	21	95
		1	29.8			CUMBERLAND KY					
3	15.9	1	5.7	2	11.9	DAVIESS KY	-64	110	-25	27	69
1*	212.0					EDMONSON KY					
1*	198.7					ESTILL KY					
16	15.8	28*	24.1	23	16.7	FAYETTE KY	52	-31	6	44	79
		1	112.4			FLOYD KY					
1	8.6	2	14.7	4	26.3	FRANKLIN KY	70	79	204	90	90
1	7.8	1	9.6	2	20.9	FULTON KY	22	119	168	85	86
		6*	88.8			GARRARD KY					
4*	37.8	2	19.9	1	6.3	GRAVES KY	-47	-69	-83	01	58
1	35.3	1	49.2			GREEN KY	39				
2*	19.7	4*	29.6	1	11.6	HARDIN KY	50	-61	-41	18	68
1	4.2	2	11.4	3	15.8	HARLAN KY	172	38	276	93	77
1	18.1			1	27.0	HARRISON KY			49	61	91
1	14.4	1	16.4	2	26.9	HART KY	13	64	86	73	91
1	4.9	2	9.2	6	27.3	HENDERSON KY	88	195	454	97	91
				1	18.6	HENRY KY					82
3	14.2	1	3.4	1	3.1	HOPKINS KY	-76	-8	-78	02	55
61*	17.1	81*	19.1	132*	24.3	JEFFERSON KY	11	27	42	60	89
1	15.6	2	25.6			JESSAMINE KY	64				
4	18.6	5	22.0	9*	36.9	KENTON KY	19	68	99	76	94
		1	19.1			KNOX KY					
				1	89.3	LAUREL KY					98
				1*	133.9	LAWRENCE KY					99
2	39.2			5*	81.7	LINCOLN KY			108	78	98
1*	267.9			1*	188.7	LIVINGSTON KY			-30	23	99
3	21.5	2	17.4	5	37.4	LOGAN KY	-19	115	74	70	94
1*	64.9					LYON KY					
4	11.1	5	13.3	6	15.7	MC CRACKEN KY	20	18	41	59	77
4	29.2	3	20.1	4	21.7	MADISON KY	-31	8	-26	26	87
1	18.9			1	13.2	MARION KY			-30	23	72
2	17.7	1	10.2			MASON KY	-42				
1*	141.3	1	37.7			MEADE KY	-73				
		1	24.6	2	28.4	MERCER KY		16			92
		1*	68.3			METCALFE KY					
1	12.5	1	13.4	1	15.3	MONTGOMERY KY	7	14	23	52	77
6*	59.3	1	9.9	1	21.3	MUHLENBERG KY	-83	114	-64	06	87
3*	36.3	1	11.6	1	12.6	NELSON KY	-68	8	-65	06	71
1*	68.8	2*	172.8			NICHOLAS KY	151				
1	31.5			2*	50.0	OHIO KY			59	66	96
				1	18.7	OWEN KY					83
1	22.3					PIKE KY					
		2	39.2			PULASKI KY					
2	17.3	3	31.3	2	11.9	SCOTT KY	81	-62	-31	22	69
1	9.8	2	18.9	4	30.8	SHELBY KY	93	63	215	90	93
		2	24.9	1	9.9	SIMPSON KY		-60			64
				1*	122.7	SPENCER KY					98
				1	14.5	TAYLOR KY					75
		2	17.2	2	13.5	TODD KY		-21			73
		2	29.7			TRIGG KY					
1	11.6			3	32.0	UNION KY			174	86	93
6*	24.8	8*	31.4	9*	31.8	WARREN KY	27	1	28	55	93
1	28.3			1	24.8	WASHINGTON KY			-12	34	90
1	10.0			2	23.9	WEBSTER KY			139	83	89
2	23.1	2	20.1	1	15.7	WOODFORD KY	-13	-22	-32	22	77
413	11.3	579	12.8	752	13.8	LOUISIANA LA	14	8	22		
		8	19.3	9	18.9	ACADIA LA		-2			83
3	18.6	3	14.1	4	14.6	ALLEN LA	-24	3	-22	29	75
2	5.5	6	14.1	5	10.5	ASCENSION LA	155	-25	90	74	65
2	7.6	4	15.5	2	5.6	ASSUMPTION LA	104	-64	-26	26	57
2	5.0	9	20.2	6	12.9	AVOYELLES LA	304	-36	160	85	72
1	7.4	7*	40.9	5	21.6	BEAUREGARD LA	449	-47	190	88	87
2	5.7	5	10.8	9	18.3	BIENVILLE LA	89	70	220	91	82
3	5.0	6	8.4	11	14.6	BOSSIER LA	67	74	189	88	75
35	11.0	35	8.6	62	12.7	CADDO LA	-22	48	16	49	71
8	10.1	16	15.3	18	13.0	CALCASIEU LA	52	-15	29	55	72
		3	23.2			CALDWELL LA					
5	22.1	1	3.7	4	15.9	CATAHOULA LA	-83	327	-28	24	77
6	12.4	7	13.5	13	20.5	CLAIBORNE LA	9	52	65	67	85
4	7.9	10	20.7	7	13.5	CONCORDIA LA	161	-35	70	69	73
5	7.9	10	11.6	10	10.3	DE SOTO LA	48	-11	31	56	65
26	11.1	32	10.2	53	13.3	EAST BATON ROUGE LA	-8	30	19	51	72
		6	10.8	7	17.1	EAST CARROLL LA		58			80
3	5.9	4	7.4	2	4.6	EAST FELICIANA LA	25	-38	-22	28	56
1	5.0	1	7.5	3	12.5	EVANGELINE LA	50	66	150	84	71
1	4.0	8	19.3	7	16.5	FRANKLIN LA	381	-15	311	95	79
4	23.8	3	18.0	2	8.1	GRANT LA	-24	-55	-66	06	61
8	17.6	6	9.8	20*	27.0	IBERIA LA	-45	176	53	63	91
6	8.7	3	4.2	10	13.2	IBERVILLE LA	-52	218	52	63	72
2	11.2	3	14.3	1	2.6	JACKSON LA	28	-82	-77	02	54
10	14.4	19	17.7	17	9.6	JEFFERSON LA	23	-46	-33	22	63
5	28.6	5	25.8	5	21.5	JEFFERSON DAVIS LA	-10	-17	-25	27	87
4	7.3	7	9.9	13	13.9	LAFAYETTE LA	37	40	92	74	73
6*	30.4	1	4.6	7	22.5	LAFOURCHE LA	-85	389	-26	26	88
2	32.0	1	10.9	2	31.3	LA SALLE LA	-66	187	-2	40	93
1	3.2	3	6.3	6	8.9	LINCOLN LA	99	42	183	87	62
1	14.4	2	13.3	2	14.5	LIVINGSTON LA	-8	9	1	42	75
4	6.5	7	11.4	9	11.9	MADISON LA	74	4	82	72	68
9	16.2	10	14.1	11	11.9	MOREHOUSE LA	-13	-15	-27	25	69
4	4.8	8	10.1	7	6.8	NATCHITOCHES LA	110	-33	41	59	59
129*	15.4	178*	16.6	226*	17.4	ORLEANS LA	8	5	13	48	81
9	8.2	23	15.6	16	8.9	OUACHITA LA	90	-43	8	45	62
4	25.1	1	4.7	2	9.2	PLAQUEMINES LA	-81	95	-63	07	62
2	3.7	4	7.5	6	9.9	POINTE COUPEE LA	105	32	171	86	64
12	8.8	12	8.3	22	11.8	RAPIDES LA	-5	42	34	57	68
1	4.3	3	14.5	3	11.5	RED RIVER LA	237	-20	169	86	67
5	13.4	2	3.8	11	20.6	RICHLAND LA	-72	445	55	64	86

* SIGNIFICANTLY MORE DEATHS THAN EXPECTED

NON-WHITE FEMALE

CANCER OF THE LARGE INTESTINE

ICD 153,159.0

1950-1959 DEATHS	1950-1959 RATE	1960-1969 DEATHS	1960-1969 RATE	1970-1979 DEATHS	1970-1979 RATE	COUNTY AND STATE	PERCENT CHANGE 60S/50S	PERCENT CHANGE 70S/60S	PERCENT CHANGE 70S/50S	PERCENTILE %CHANGE 70S/50S	PERCENTILE RATES 1970S
1	5.5	2	10.0	2	6.2	SABINE LA	82	-38	13	47	58
1	14.9	1	15.7	1	9.3	ST BERNARD LA	6	-41	-37	20	63
2	7.8	4	14.9	3	9.9	ST CHARLES LA	91	-34	27	54	64
				2	7.3	ST HELENA LA					59
5	15.2	8	21.3	9	23.8	ST JAMES LA	39	12	56	65	89
2	5.6	4	9.4	10	23.2	ST JOHN THE BAPT LA	67	148	313	95	88
9	9.8	13	11.0	13	8.6	ST LANDRY LA	13	-23	-12	33	61
2	7.0	2	5.2	5	11.7	ST MARTIN LA	-26	125	67	68	68
7	10.9	10	15.0	11	14.6	ST MARY LA	37	-3	34	57	75
6	20.2	6	14.6	4	7.3	ST TAMMANY LA	-28	-50	-64	07	59
10	16.7	8	10.8	6	6.3	TANGIPAHOA LA	-36	-42	-62	07	58
8	21.2	3	8.4	9	21.1	TENSAS LA	-60	151	-1	40	86
4	10.1	3	7.9	7	13.3	TERREBONNE LA	-22	68	31	56	72
2	8.1	4	14.6	6	14.0	UNION LA	80	-4	72	70	73
2	13.8	2	8.8	3	12.1	VERMILION LA	-37	38	-12	34	69
1	9.8	2	18.4	3	21.4	VERNON LA	89	16	120	80	87
6	15.1	2	2.7	8	11.1	WASHINGTON LA	-82	305	-26	25	67
4	9.4	8	12.4	4	4.4	WEBSTER LA	32	-65	-54	11	56
1	3.8	4	13.2	4	11.3	WEST BATON ROUGE LA	245	-14	195	89	67
		6*	38.6	4	27.5	WEST CARROLL LA		-29			91
1	3.1	3	14.9	3	12.6	WEST FELICIANA LA	381	-15	307	94	71
2	14.2	2	8.1			WINN LA	-43				
4*	35.5	4	26.4	5	18.8	MAINE ME	-25	-29	-47		
1	45.5	1	39.0	1	12.4	CUMBERLAND ME	-14	-68	-73	03	70
1*	109.3					KENNEBEC ME					
1*	112.4					LINCOLN ME					
1	20.2	1	40.9	2	48.4	PENOBSCOT ME	102	18	139	83	96
		1*	112.4			PISCATAQUIS ME					
				1*	273.2	SOMERSET ME					99
				1	112.4	WASHINGTON ME					98
		1*	98.2			YORK ME					
197	13.7	338*	16.7	570*	18.7	MARYLAND MD	22	12	36		
		1	17.7	6*	83.6	ALLEGANY MD		373			98
9	11.7	13	11.4	31*	21.5	ANNE ARUNDEL MD	-3	89	83	72	87
4	7.8	16	24.4	35*	29.5	BALTIMORE MD	210	21	275	93	92
2	13.7	4	20.9	6	22.0	CALVERT MD	53	5	61	66	88
2	12.9	3	16.3	5	24.8	CAROLINE MD	27	52	93	74	90
3*	46.6	2	18.0	2	10.9	CARROLL MD	-61	-39	-77	02	66
		1	9.4	3	22.0	CECIL MD		133			88
		1	2.8	7	18.1	CHARLES MD		552			81
1	3.0	4	10.1	7	15.7	DORCHESTER MD	240	56	429	97	77
6*	31.1	3	14.6	5	18.8	FREDERICK MD	-53	28	-40	19	83
6*	28.6	4	15.0	8	21.3	HARFORD MD	-48	42	-25	26	87
3	22.6	1	6.6	4	15.6	HOWARD MD	-71	136	-31	23	77
1	5.9	1	5.2			KENT MD	-11				
9	22.3	6	11.0	27	19.3	MONTGOMERY MD	-51	76	-14	32	84
6	8.6	19	16.4	57	18.8	PRINCE GEORGE'S MD	90	15	118	80	83
		3	13.6	2	7.5	QUEEN ANNE'S MD		-45			60
3	19.5	2	10.5	4	12.2	ST MARY'S MD	-46	16	-37	20	70
3	11.3	6	18.3	3	6.9	SOMERSET MD	62	-62	-39	19	59
5	19.5	5	14.3	7	19.8	TALBOT MD	-27	39	1	42	84
4*	35.9	2	18.1			WASHINGTON MD	-50				
3	10.1	3	6.7	9	12.5	WICOMICO MD	-33	85	24	53	71
2	10.5	6	23.4	6	15.2	WORCESTER MD	123	-35	46	61	76
125	13.9	232*	18.5	336*	18.7	BALTIMORE CITY MD	34	1	35	57	82
73*	19.6	78	13.5	136	14.1	MASSACHUSETTS MA	-31	4	-28		
2	25.3	2	22.2	1	3.5	BARNSTABLE MA	-12	-84	-86	00	55
3*	42.9			2	18.2	BERKSHIRE MA			-58	09	82
6*	31.9	3	10.2	7	14.0	BRISTOL MA	-68	38	-56	10	74
1*	122.7	3*	99.3			DUKES MA	-19				
5*	44.7	2	10.7	2	5.2	ESSEX MA	-76	-51	-88	00	56
1	3.0	10	17.2	17	16.2	HAMPDEN MA	474	-6	441	97	78
12	21.0	9	10.0	16	10.7	MIDDLESEX MA	-52	7	-49	14	66
		1	16.1	6	17.9	NORFOLK MA		11			81
2	13.1	6*	30.6	5	10.2	PLYMOUTH MA	133	-67	-23	28	65
38*	19.0	38	12.4	77	16.6	SUFFOLK MA	-35	34	-13	33	79
3	25.7	4	18.9	3	9.6	WORCESTER MA	-26	-49	-63	07	63
296*	17.2	458*	16.0	723*	16.4	MICHIGAN MI	-7	2	-4		
2*	48.3			3	26.4	ALLEGAN MI			-45	16	90
		1*	59.6			BARAGA MI					
		1	42.4	1	28.7	BAY MI		-32			92
		5	9.5	15	20.9	BERRIEN MI		121			86
6*	19.9	7	18.9	6	10.4	CALHOUN MI	-5	-45	-48	15	65
3	15.4	2	9.8	5	16.7	CASS MI	-36	71	9	45	79
1*	114.7	2*	227.1	1*	60.6	CHARLEVOIX MI	98	-73	-47	15	97
		1	53.0	1	17.2	CHIPPEWA MI		-68			80
				1*	245.4	DICKINSON MI					99
				1	62.9	EATON MI					97
2*	66.4	1	24.6			EMMET MI	-63				
12*	19.2	14	13.6	23	14.4	GENESEE MI	-29	5	-25	26	75
		2	6.0	10	20.4	INGHAM MI		243			85
		1*	62.9			IONIA MI					
1	6.3	4	17.5	3	7.9	JACKSON MI	177	-55	26	54	60
2	17.5	4	16.8	4	11.4	KALAMAZOO MI	-4	-32	-35	21	67
7	21.9	8	12.0	11	12.3	KENT MI	-45	2	-44	16	70
2	12.4	4	22.9	5	20.1	LAKE MI	84	-12	62	66	85
		5	20.7	6	18.7	MACOMB MI		-10			83
				1	86.0	MECOSTA MI					98
		4	7.0	11	16.8	MUSKEGON MI		139			80
		2	73.9	3	60.3	NEWAYGO MI		-18			97
7	10.7	8	12.0	26	18.2	OAKLAND MI	12	52	69	68	82
1*	136.6	1*	114.7			OCEANA MI	-16				
4	13.7	7	10.7	11	11.8	SAGINAW MI	-21	10	-14	32	68

* SIGNIFICANTLY MORE DEATHS THAN EXPECTED

NON-WHITE FEMALE — CANCER OF THE LARGE INTESTINE — ICD 153,159.0

1950-1959 DEATHS	1950-1959 RATE	1960-1969 DEATHS	1960-1969 RATE	1970-1979 DEATHS	1970-1979 RATE	COUNTY AND STATE	PERCENT CHANGE 60S/50S	PERCENT CHANGE 70S/60S	PERCENT CHANGE 70S/50S	PERCENTILE %CHANGE 70S/50S	PERCENTILE RATES 1970S
1	8.2	4	31.8	3	19.9	ST CLAIR MI	288	-37	143	84	84
1	56.2	2	25.8			ST JOSEPH MI	-54				
6*	45.7	2	6.4	6	14.9	VAN BUREN MI	-86	135	-67	05	76
4	13.3	9	15.8	12	17.6	WASHTENAW MI	18	12	32	56	81
234*	18.3	357*	16.9	554*	17.0	WAYNE MI	-8	1	-7	37	80
17	16.2	20	12.8	33	12.0	MINNESOTA MN	-21	-6	-26		
				1	24.9	ANOKA MN					90
1	14.9			1	23.6	BECKER MN			59	66	89
2	45.3	1	18.9	2	22.4	BELTRAMI MN	-58	19	-51	13	88
1	27.0	2	82.8	2	31.6	CASS MN	207	-62	17	50	93
6	18.2	7	12.1	14	11.3	HENNEPIN MN	-34	-6	-38	20	67
		1*	424.0			LAC QUI PARLE MN					
1	44.6	1	45.5			MAHNOMEN MN	2				
				1	37.2	OLMSTED MN					94
5	16.0	7	14.6	11	15.0	RAMSEY MN	-9	3	-6	37	76
		1*	74.3			RICE MN					
1	27.3					ST LOUIS MN					
				1	7.9	WASHINGTON MN					60
348	9.9	474	12.2	554	12.5	MISSISSIPPI MS	24	2	27		
12	15.0	14	15.6	15	14.1	ADAMS MS	4	-9	-6	38	74
2	13.7	1	7.0	1	5.1	ALCORN MS	-49	-27	-63	07	56
2	5.5	12*	40.8	6	18.2	AMITE MS	638	-55	229	91	82
3	8.5	4	9.6	4	9.5	ATTALA MS	12	0	12	47	63
1	10.8	1	10.3	1	5.7	BENTON MS	-4	-44	-47	15	57
12	7.2	24	14.8	25	15.1	BOLIVAR MS	105	3	110	78	76
1	5.5	1	7.5	2	16.7	CALHOUN MS	37	124	206	90	79
2	7.8	1	4.0	1	4.0	CARROLL MS	-48	0	-48	14	55
1	3.3	2	6.0	2	4.7	CHICKASAW MS	81	-21	43	60	56
		1	7.1			CHOCTAW MS					
2	6.5	1	2.6	4	9.8	CLAIBORNE MS	-59	274	52	63	64
2	7.4	5	16.4	5	14.9	CLARKE MS	122	-9	102	77	76
7	18.9	4	8.2	7	14.2	CLAY MS	-57	74	-25	27	74
16	10.8	10	7.0	21	13.6	COAHOMA MS	-35	94	26	54	73
7	11.8	8	12.5	8	11.3	COPIAH MS	6	-10	-5	39	67
3	18.1	4	20.9	2	7.8	COVINGTON MS	15	-63	-57	10	60
5	10.6	7	13.1	4	6.7	DE SOTO MS	23	-49	-37	20	59
2	3.9	13	19.6	12	13.9	FORREST MS	404	-29	257	92	73
2	12.5	2	7.8	2	8.1	FRANKLIN MS	-37	4	-35	21	61
		1	24.8			GEORGE MS					
1	18.7	1	8.6	1	7.5	GREENE MS	-54	-13	-60	09	60
3	9.3	1	2.0	3	6.7	GRENADA MS	-79	241	-28	24	59
1	12.5	1	7.3	1	8.8	HANCOCK MS	-42	21	-30	23	62
4	7.5	9	13.2	14	13.0	HARRISON MS	75	-1	73	70	72
29	11.8	29	10.1	46	11.0	HINDS MS	-15	9	-7	37	66
28*	32.5	10	10.0	21	19.8	HOLMES MS	-69	97	-39	19	84
1	1.8	10	20.1	9	16.8	HUMPHREYS MS	1005	-17	820	99	80
				2	26.2	ISSAQUENA MS					90
				1	9.9	ITAWAMBA MS					64
7*	24.3	10*	28.6	8	15.9	JACKSON MS	18	-44	-34	21	78
1	3.6	3	7.9	3	6.6	JASPER MS	122	-16	86	73	59
		5	13.5	5	10.5	JEFFERSON MS		-23			65
5	18.9	3	10.9	5	14.8	JEFFERSON DAVIS MS	-43	36	-22	29	75
3	8.3	10	15.1	11	13.7	JONES MS	81	-9	65	67	73
2	7.2	4	13.4	3	7.8	KEMPER MS	87	-42	8	45	60
2	5.9	7	23.6	6	16.3	LAFAYETTE MS	297	-31	175	86	78
1	8.6	3	30.8	1	10.2	LAMAR MS	259	-67	19	51	65
9	8.1	9	8.4	9	7.6	LAUDERDALE MS	4	-10	-6	37	60
1	7.4	2	11.7	2	9.6	LAWRENCE MS	57	-18	29	55	63
1	4.5	5	16.5	2	6.9	LEAKE MS	268	-58	54	64	59
4	11.9	7	14.1	11	20.0	LEE MS	19	42	69	69	85
9	8.7	15	12.2	15	11.8	LEFLORE MS	40	-4	35	57	68
3	7.3	3	8.0	9	18.5	LINCOLN MS	9	132	154	84	82
4	6.0	6	8.3	10	10.6	LOWNDES MS	39	28	77	71	66
10	10.8	9	10.1	12	11.0	MADISON MS	-6	9	2	42	66
2	7.0	6	16.3	3	6.9	MARION MS	133	-58	-2	40	59
5	8.8	4	5.9	4	5.5	MARSHALL MS	-33	-7	-38	19	57
2	3.4	2	3.8	3	4.5	MONROE MS	13	19	35	58	56
3	11.6	1	3.6	8	22.6	MONTGOMERY MS	-69	522	95	75	88
1	3.7	4	19.1	2	6.9	NESHOBA MS	420	-64	88	74	59
4	19.1	3	10.2	3	8.9	NEWTON MS	-46	-13	-53	12	62
3	4.8	4	7.6	4	8.3	NOXUBEE MS	58	10	74	70	61
9	21.3	10	21.6	4	7.5	OKTIBBEHA MS	2	-65	-65	06	60
11	17.5	13	20.1	6	9.4	PANOLA MS	15	-53	-46	15	63
1	5.6	3	16.3	2	9.8	PEARL RIVER MS	192	-40	75	71	64
2	15.2	2	17.5			PERRY MS	15				
4	6.4	12	17.9	11	14.2	PIKE MS	178	-20	121	80	74
1	9.3			4	24.9	PONTOTOC MS			167	85	90
		1	12.6	3	28.3	PRENTISS MS		125			92
6	11.4	6	13.3	7	13.5	QUITMAN MS	16	1	18	50	73
2	3.9	7	11.8	5	7.5	RANKIN MS	205	-36	94	75	60
2	5.3	7	21.4	4	9.8	SCOTT MS	304	-54	86	73	64
2	6.9	2	7.6	2	9.4	SHARKEY MS	11	22	36	58	63
3	10.8	4	13.8	2	5.3	SIMPSON MS	28	-62	-51	13	57
		1	13.5	2	26.5	STONE MS		97			91
10	10.3	5	4.6	13	11.9	SUNFLOWER MS	-56	162	15	49	69
5	8.7	9	15.8	10	15.6	TALLAHATCHIE MS	81	-1	79	72	77
3	8.2	6	15.6	8	18.3	TATE MS	90	17	123	81	82
		1	7.2	3	23.1	TIPPAH MS		219			88
3	4.8	6	13.0	9	20.6	TUNICA MS	169	58	326	95	85
1	6.9	5	33.6	1	6.4	UNION MS	386	-81	-7	37	58
2	12.9	3	16.2	2	8.1	WALTHALL MS	25	-50	-37	20	61
16	14.1	21	18.5	23	18.4	WARREN MS	31	0	30	55	82
20	10.8	21	10.1	45	20.1	WASHINGTON MS	-7	100	86	73	85
1	5.3	3	12.9	1	4.9	WAYNE MS	146	-62	-7	36	56
1	8.4	2	17.3	3	26.6	WEBSTER MS	106	54	217	91	91
2	6.5	3	8.2	5	11.7	WILKINSON MS	27	43	81	72	68
2	8.2	3	10.7	7	19.8	WINSTON MS	31	85	142	83	84

* SIGNIFICANTLY MORE DEATHS THAN EXPECTED

NON-WHITE FEMALE — CANCER OF THE LARGE INTESTINE — ICD 153,159.0

1950-1959 DEATHS	1950-1959 RATE	1960-1969 DEATHS	1960-1969 RATE	1970-1979 DEATHS	1970-1979 RATE	COUNTY AND STATE	PERCENT CHANGE 60S/50S	PERCENT CHANGE 70S/60S	PERCENT CHANGE 70S/50S	PERCENTILE %CHANGE 70S/50S	PERCENTILE RATES 1970S
1	3.9	2	7.7	4	16.0	YALOBUSHA MS	94	109	307	94	78
7	9.3	14	16.0	9	10.4	YAZOO MS	72	-35	12	47	65
204	13.6	332*	16.7	451*	18.0	MISSOURI MO	23	8	32		
		3	32.6	2	21.4	AUDRAIN MO		-34			87
1*	133.9					BATES MO					
3	19.4	5	27.1	4	16.4	BOONE MO	40	-40	-15	32	79
2	9.5	2	13.8	2	8.7	BUCHANAN MO	45	-37	-9	35	61
		4	24.7	1	5.5	BUTLER MO		-78			57
3	23.0	1	5.7	1	7.4	CALLAWAY MO	-75	30	-68	05	60
		2	25.3			CAPE GIRARDEAU MO					
				1	23.6	CHARITON MO					89
1	28.3			2	22.9	CLAY MO			-19	30	88
1*	38.2	1	70.7	1	38.5	CLINTON MO	85	-45	1	42	95
2	32.5	3*	45.2	2	19.0	COLE MO	39	-58	-42	17	83
				1	10.5	COOPER MO					65
		1*	112.4			DADE MO					
		1	11.8			DUNKLIN MO					
				1	24.5	FRANKLIN MO					89
		5*	32.0	4	20.7	GREENE MO		-35			86
1*	298.1					GRUNDY MO					
1*	106.0					HENRY MO					
1	7.8	4*	45.4	2	26.7	HOWARD MO	480	-41	241	92	91
39	13.4	74	17.5	115*	20.3	JACKSON MO	31	16	51	63	85
3*	41.0			3	29.0	JASPER MO			-29	23	92
1	14.9	1	31.5	1	14.1	JEFFERSON MO	112	-55	-5	38	74
		1*	70.3	1*	81.8	LACLEDE MO		16			98
5*	50.9	2	19.4	1	10.5	LAFAYETTE MO	-62	-46	-79	02	65
1*	67.0					LINN MO					
		2*	48.5			LIVINGSTON MO					
1	31.5			1	54.5	MACON MO			73	70	96
1	8.8	3	36.7	2	20.8	MARION MO	316	-43	135	82	86
4	20.7	2	9.4			MISSISSIPPI MO	-55				
1*	62.9			1	37.7	MONITEAU MO			-40	18	94
		2*	82.6			MONROE MO					
				4*	118.6	MONTGOMERY MO					98
				1*	28.1	MORGAN MO					92
1	4.7					NEW MADRID MO					
2	5.9	5	14.4	1	2.1	PEMISCOT MO	146	-85	-64	06	54
2	17.7	1	6.1	3	21.1	PETTIS MO	-66	247	20	52	86
		1*	114.7			PHELPS MO					
		2	33.5	4*	36.0	PIKE MO		7			94
1*	344.1					POLK MO					
1*	57.3	1*	117.9			RALLS MO	106				
1	9.3	3	39.9	3	43.5	RANDOLPH MO	330	9	368	96	95
1	37.5			2*	73.6	RAY MO			96	75	97
2*	43.1			1	8.2	ST CHARLES MO			-81	01	61
		1*	344.1			ST CLAIR MO					
		1*	267.9			STE GENEVIEVE MO					
		1	81.8			ST FRANCOIS MO					
9	10.2	10	9.2	49	20.1	ST LOUIS COUNTY MO	-10	119	97	76	85
1	10.9	1	10.1	6*	46.7	SALINE MO	-7	363	330	95	96
2	24.4	2	19.3	3	19.0	SCOTT MO	-21	-1	-22	29	83
		1	37.7			STODDARD MO					
1*	112.4					SULLIVAN MO					
1*	267.9					TANEY MO					
107	13.3	183*	17.5	225*	17.9	ST LOUIS CITY MO	31	2	34	57	81
6	11.1	5	7.5	10	10.5	MONTANA MT	-33	41	-6		
1	22.3	1	21.5			BLAINE MT	-4				
1*	424.0					DEER LODGE MT					
				1	67.0	GARFIELD MT					97
1	9.6	2	13.8	4	28.7	GLACIER MT	43	108	197	89	92
2	34.7			2	20.1	LAKE MT			-42	17	85
		1*	112.4			LEWIS AND CLARK MT					
				1*	85.2	RAVALLI MT					98
1	14.3	1	11.9	1	9.4	ROOSEVELT MT	-17	-21	-35	21	63
				1	13.4	YELLOWSTONE MT					72
15	14.5	27	18.3	27	13.7	NEBRASKA NB	26	-25	-5		
1*	122.7			1*	133.9	ADAMS NB			9	45	99
11	14.0	24*	20.8	22	15.3	DOUGLAS NB	49	-27	9	46	76
				1*	57.3	KNOX NB					96
1	18.6	2	26.3	1	10.9	LANCASTER NB	42	-59	-42	17	66
		1*	98.2			LINCOLN NB					
				1	12.3	SCOTTS BLUFF NB					70
1*	188.7					SHERIDAN NB					
1	20.2			1	12.5	THURSTON NB			-38	19	71
6	13.1	9	13.8	14	7.4	NEVADA NV	5	-46	-43		
				1	24.9	CHURCHILL NV					90
4	28.4	3	7.2	11	9.2	CLARK NV	-75	27	-68	05	62
		1	26.0			ELKO NV					
		1*	188.7	1	49.1	HUMBOLDT NV		-74			96
		1*	74.3			LINCOLN NV					
1*	74.5	1*	109.3			LYON NV	47				
1	6.1	1	5.2	1	2.9	WASHOE NV	-14	-45	-53	12	55
		1*	141.3			ORMSBY-CARSON CITY NV					
1	62.9	1	16.4	2	9.5	NEW HAMPSHIRE NH	-74	-42	-85		
				1*	188.7	COOS NH					99
				1	16.9	HILLSBOROUGH NH					80
1*	188.7	1	40.9			ROCKINGHAM NH	-78				
234*	16.0	345	15.4	670*	19.4	NEW JERSEY NJ	-4	26	21		

* SIGNIFICANTLY MORE DEATHS THAN EXPECTED

NON-WHITE FEMALE — CANCER OF THE LARGE INTESTINE — ICD 153,159.0

1950-1959 DEATHS	1950-1959 RATE	1960-1969 DEATHS	1960-1969 RATE	1970-1979 DEATHS	1970-1979 RATE	COUNTY AND STATE	PERCENT CHANGE 60S/50S	PERCENT CHANGE 70S/60S	PERCENT CHANGE 70S/50S	PERCENTILE %CHANGE 70S/50S	PERCENTILE RATES 1970S
22	15.5	42*	23.5	54*	24.1	ATLANTIC NJ	51	3	56	65	89
8	12.6	20	20.5	37	21.6	BERGEN NJ	63	5	72	70	87
6	18.0	10	21.8	26*	27.0	BURLINGTON NJ	21	24	50	62	91
17	16.2	32	18.5	50	20.1	CAMDEN NJ	15	9	24	53	85
4	20.9	3	12.4	4	12.1	CAPE MAY NJ	-41	-2	-42	17	69
4	12.8	7	12.9	10	15.6	CUMBERLAND NJ	1	21	22	52	77
83*	17.0	102	13.7	212*	18.9	ESSEX NJ	-19	38	11	46	83
7	15.0	5	9.6	17	20.4	GLOUCESTER NJ	-36	113	36	58	85
22*	23.4	14	8.0	47	15.5	HUDSON NJ	-66	94	-34	22	77
1	53.0	1	23.6			HUNTERDON NJ	-55				
13	19.2	33*	23.5	41	22.1	MERCER NJ	23	-6	15	48	88
5	21.4	6	7.8	15	13.8	MIDDLESEX NJ	-64	77	-36	21	73
10	8.7	22	15.8	38	18.3	MONMOUTH NJ	81	16	111	79	82
1	5.5	4	13.8	19*	42.1	MORRIS NJ	150	205	663	99	95
				10*	30.8	OCEAN NJ					93
7	12.1	11	8.1	21	11.9	PASSAIC NJ	-33	48	-1	40	69
2	9.0	8	28.0	9	21.2	SALEM NJ	210	-24	135	82	86
3	25.6	2	21.6	5	18.1	SOMERSET NJ	-16	-16	-29	24	82
				1	49.2	SUSSEX NJ					96
19	15.7	23	12.7	52	18.2	UNION NJ	-19	43	16	49	82
				2	41.8	WARREN NJ					95
6	4.6	7	4.3	15	2.4	NEW MEXICO NM	-6	-45	-49		
2	13.0	1	4.6	4	2.8	BERNALILLO NM	-64	-40	-79	02	55
1	25.9					CHAVES NM					
		1	18.8			EDDY NM					
1*	112.4					HIDALGO NM					
		2	37.7	4	26.8	LEA NM		-29			91
2	5.8	1	3.0	2	1.9	MC KINLEY NM	-49	-36	-67	05	54
		1	17.0			OTERO NM					
				2	3.9	RIO ARRIBA NM					55
		1	2.8	1	1.6	SAN JUAN NM		-44			54
				1	23.8	SOCORRO NM					89
				1	1.7	VALENCIA NM					54
680*	16.8	1103*	16.9	1751*	15.9	NEW YORK NY	1	-6	-5		
6	20.5	4	9.9	10	15.9	ALBANY NY	-52	60	-23	28	77
77*	18.5	108	12.5	237	13.8	BRONX NY	-33	11	-25	26	73
		2	24.0	2	12.5	BROOME NY		-48			71
		2	26.0			CATTARAUGUS NY					
				1	19.1	CAYUGA NY					84
		1	19.1			CHAUTAUQUA NY					
				2	12.3	CHEMUNG NY					70
2*	333.2					CHENANGO NY					
		1*	267.9			CLINTON NY					
2*	31.2			3	35.3	COLUMBIA NY			13	48	94
		1*	94.4			CORTLAND NY					
		1*	98.3			DELAWARE NY					
1	2.8	3	7.5	7	11.9	DUTCHESS NY	165	59	321	95	68
22	13.3	54*	22.5	76	18.8	ERIE NY	69	-16	41	59	83
1*	74.3					ESSEX NY					
2	32.1					FRANKLIN NY					
1*	81.8					FULTON NY					
1	33.5					GENESEE NY					
1	30.4			3*	89.1	GREENE NY			193	88	98
		1*	106.0			JEFFERSON NY					
130*	15.7	210	13.7	380	14.5	KINGS(BROOKLYN) NY	-13	6	-8	36	75
		3*	79.9			LIVINGSTON NY					
4	12.7	10	12.9	28	16.3	MONROE NY	1	27	28	55	79
11	12.4	21	11.6	59	19.1	NASSAU NY	-7	65	54	64	84
311*	19.3	460*	22.3	496*	17.6	NEW YORK(MANHATTAN) NY	16	-21	-9	35	81
3	39.9	2	7.9	6	16.8	NIAGARA NY	-80	112	-58	09	80
		3	23.0	1	4.0	ONEIDA NY		-83			55
1	5.1	6	13.2	12	15.1	ONONDAGA NY	159	14	195	89	76
				3*	47.8	ONTARIO NY					96
4	15.8	2	5.3	10	16.8	ORANGE NY	-67	219	6	45	80
		1	94.4			ORLEANS NY					
		1*	66.2			OSWEGO NY					
47	13.8	98	11.8	230	14.6	QUEENS NY	-14	24	6	44	75
		1	18.4	3	21.3	RENSSELAER NY		15			87
5	34.8	7	17.3	12	14.8	RICHMOND NY	-50	-14	-57	10	76
2	9.5	4	10.5	10	12.5	ROCKLAND NY	11	20	33	56	71
		1	38.5			ST LAWRENCE NY					
		1	16.3	2	19.0	SARATOGA NY		17			83
		2	14.6	1	3.9	SCHENECTADY NY		-74			55
8	7.3	20	12.2	42	15.3	SUFFOLK NY	68	25	110	78	76
		1	33.5	3	23.8	SULLIVAN NY		-29			89
1	27.0	2	24.0	5*	55.6	TOMPKINS NY	-11	132	106	77	96
1	10.5	2	13.7	3	10.1	ULSTER NY	31	-26	-4	39	64
1*	424.0					WASHINGTON NY					
1*	188.7			2	23.1	WAYNE NY			-88	00	88
34	17.1	67*	20.5	102*	20.6	WESTCHESTER NY	20	1	21	52	86
292	8.5	516	11.8	818	14.1	NORTH CAROLINA NC	39	19	66		
1	2.5	3	6.5	14	16.4	ALAMANCE NC	161	154	563	99	79
2*	77.5			2	28.6	ALEXANDER NC			-63	07	92
1*	57.3	1*	47.2	1	80.4	ALLEGHANY NC	-18	70	40	59	98
4	9.9	6	12.5	7	12.7	ANSON NC	26	1	27	54	71
		1*	112.4			ASHE NC					
		1*	56.2	1*	89.3	AVERY NC		59			98
4	6.9	5	9.5	9	12.9	BEAUFORT NC	37	36	86	73	71
4	9.8	9	18.4	6	9.7	BERTIE NC	88	-47	-1	40	63
3	8.7	1	2.1	5	8.8	BLADEN NC	-76	313	1	42	62
3	15.7	4	14.0	3	8.3	BRUNSWICK NC	-11	-41	-47	15	61
8	13.1	9	11.3	12	12.1	BUNCOMBE NC	-14	8	-8	36	69
		4	20.5	2	8.0	BURKE NC		-61			61
8	26.8	9	18.9	9	13.7	CABARRUS NC	-29	-27	-49	14	73
2	22.0	2	15.9	1	5.6	CALDWELL NC	-27	-65	-75	03	57

* SIGNIFICANTLY MORE DEATHS THAN EXPECTED

NON-WHITE FEMALE — CANCER OF THE LARGE INTESTINE — ICD 153,159.0

1950-1959 DEATHS	1950-1959 RATE	1960-1969 DEATHS	1960-1969 RATE	1970-1979 DEATHS	1970-1979 RATE	COUNTY AND STATE		PERCENT CHANGE 60S/50S	PERCENT CHANGE 70S/60S	PERCENT CHANGE 70S/50S	PERCENTILE %CHANGE 70S/50S	PERCENTILE RATES 1970S
1	10.4	1	13.0	1	9.8	CAMDEN	NC	25	-24	-6	38	64
1	7.5	3	20.9	4	24.7	CARTERET	NC	180	18	231	91	90
4	17.5	1	2.3	6	15.5	CASWELL	NC	-87	567	-11	34	77
1	6.5	4	13.7	5	12.8	CATAWBA	NC	109	-6	96	75	71
2	13.7	8	29.3	6	15.3	CHATHAM	NC	113	-48	12	47	77
		2	8.5	6	23.3	CHOWAN	NC		175			89
4	9.0	5	10.4	7	9.8	CLEVELAND	NC	15	-6	9	45	64
1	2.1	3	4.6	6	7.8	COLUMBUS	NC	122	69	276	93	60
3	4.5	4	6.1	13	15.9	CRAVEN	NC	36	160	255	92	78
2	2.2	7	4.9	18	8.4	CUMBERLAND	NC	124	74	290	94	61
		1	14.5			CURRITUCK	NC					
1	2.7	4	11.1	8	15.9	DAVIDSON	NC	316	44	499	98	78
		1	9.3			DAVIE	NC					
5	14.5	8	13.8	12	16.0	DUPLIN	NC	-5	16	10	46	78
13	10.5	27	14.6	36	15.8	DURHAM	NC	38	9	50	62	77
4	4.4	8	8.4	14	10.4	EDGECOMBE	NC	89	25	137	83	65
13	8.2	35	16.4	54	19.3	FORSYTH	NC	99	18	135	82	84
7	19.8	4	8.5	12	19.9	FRANKLIN	NC	-57	133	0	42	84
5	9.4	9	14.3	19	20.6	GASTON	NC	52	44	119	80	85
1	6.5			4	19.3	GATES	NC			198	89	84
4	8.0	5	9.1	10	14.8	GRANVILLE	NC	14	62	85	73	75
1	4.8	4	20.8	5	20.5	GREENE	NC	332	-1	327	95	85
19	13.8	33	15.6	50	17.0	GUILFORD	NC	13	9	24	52	80
7	8.3	8	7.2	16	12.4	HALIFAX	NC	-13	72	50	62	70
3	11.5	5	11.7	11	19.8	HARNETT	NC	2	70	72	70	84
1	12.5			3	21.2	HENDERSON	NC			70	69	87
3	7.2	10	19.9	10	14.1	HERTFORD	NC	175	-29	94	74	74
2	7.5			4	11.5	HOKE	NC			53	63	67
3	9.5	4	9.0	15*	25.8	IREDELL	NC	-5	187	173	86	90
		1	11.1	1	22.5	JACKSON	NC		103			88
		5	11.7	7	10.8	JOHNSTON	NC		-7			66
				3	14.2	JONES	NC					74
		2	8.9	4	11.8	LEE	NC		33			68
2	3.2	9	13.9	12	11.2	LENOIR	NC	342	-20	255	92	67
		2	15.0			LINCOLN	NC					
				2	19.2	MC DOWELL	NC					84
5	14.0	5	10.3	7	12.9	MARTIN	NC	-26	25	-8	35	71
13	7.0	38	13.5	69	18.6	MECKLENBURG	NC	94	38	167	85	82
4*	27.1	2	10.6	3	12.2	MONTGOMERY	NC	-61	15	-55	11	70
4	10.5	6	14.5	7	11.8	MOORE	NC	38	-18	13	47	68
		8	10.4	11	10.9	NASH	NC		5			66
16*	17.7	12	13.7	16	13.6	NEW HANOVER	NC	-23	-1	-23	27	73
1	2.0	4	6.6	15	22.1	NORTHAMPTON	NC	234	234	1018	99	88
3	20.6	4	17.8	9	24.0	ONSLOW	NC	-14	35	16	49	89
		4	10.2	9	15.6	ORANGE	NC		53			77
1	7.3	2	15.5	2	10.4	PAMLICO	NC	112	-33	43	60	65
3	8.3	7	18.7	3	6.3	PASQUOTANK	NC	126	-67	-24	27	58
2	7.2	1	4.5	6	13.3	PENDER	NC	-37	197	86	73	72
4	22.9	3	16.4	3	12.2	PERQUIMANS	NC	-28	-26	-47	15	69
6	23.7	4	13.2	2	5.5	PERSON	NC	-45	-58	-77	02	57
7	6.8	11	10.9	10	6.9	PITT	NC	61	-37	1	42	59
1	14.9			1	13.0	POLK	NC			-13	33	72
		3	11.5	2	7.8	RANDOLPH	NC		-33			60
6	14.0	6	11.9	7	9.4	RICHMOND	NC	-15	-21	-33	22	63
5	3.8	16	10.5	24	11.0	ROBESON	NC	175	4	187	87	66
2	4.6	15*	24.1	23*	30.4	ROCKINGHAM	NC	428	26	567	99	93
3	6.8	9	15.3	15	19.8	ROWAN	NC	126	29	192	88	84
1	4.6	2	9.7	3	9.5	RUTHERFORD	NC	112	-2	107	78	63
2	4.4	8	11.9	8	9.5	SAMPSON	NC	172	-20	118	80	63
3	8.4	1	2.8	7	13.1	SCOTLAND	NC	-66	363	55	65	72
1	5.5	3	18.9	2	6.5	STANLY	NC	242	-66	18	50	58
		1	17.0			STOKES	NC					
		3	23.2	4	24.8	SURRY	NC		7			90
				1	9.3	SWAIN	NC					63
2*	64.6			1	24.8	TRANSYLVANIA	NC			-62	08	90
		2	26.8			TYRRELL	NC					
5	15.1	4	9.8	7	15.6	UNION	NC	-35	60	3	43	77
4	11.1	13	23.7	9	12.8	VANCE	NC	113	-46	15	48	71
10	6.9	13	6.6	26	9.8	WAKE	NC	-5	49	41	59	64
5	11.8	5	10.3	10	16.5	WARREN	NC	-13	60	40	59	79
		1	4.4	2	8.9	WASHINGTON	NC		103			62
				1	49.7	WATAUGA	NC					96
14	15.4	12	10.3	17	12.2	WAYNE	NC	-33	19	-21	29	70
1	8.4	1	7.4	3	24.7	WILKES	NC	-11	232	194	89	89
5	6.5	9	11.3	17	15.9	WILSON	NC	73	41	143	84	78
2	9.5	4	12.0	7	13.2	NORTH DAKOTA	ND	26	10	39		
		1	24.8			BENSON	ND					
				1	84.8	MC KENZIE	ND					98
				2*	188.7	MC LEAN	ND					99
		1*	60.6			MOUNTRAIL	ND					
1	11.2	2	16.8	2	10.5	ROLETTE	ND	50	-37	-6	37	66
1*	188.7					STARK	ND					
				1	5.4	WARD	ND					57
				1*	43.0	WILLIAMS	ND					95
366*	17.1	596*	18.0	956*	20.8	OHIO	OH	5	16	22		
1*	133.9					ADAMS	OH					
1	4.7	5	21.9	6	20.9	ALLEN	OH	364	-5	343	96	86
1	26.5	1	12.6			ASHTABULA	OH	-52				
2*	96.8			2	46.1	ATHENS	OH			-52	12	96
1	8.2	1	6.4	3	24.3	BELMONT	OH	-22	282	197	89	89
1	27.0	1	37.2	1	34.2	BROWN	OH	38	-8	27	54	94
9*	29.9	3	6.6	12	19.8	BUTLER	OH	-78	201	-34	22	84
1	18.9			1	14.5	CHAMPAIGN	OH			-23	27	75
7	14.4	15	23.1	26*	31.1	CLARK	OH	61	34	116	79	93
1	37.2					CLERMONT	OH					
		1	28.3	2*	69.5	CLINTON	OH		146			97
		1	8.6	3	23.3	COLUMBIANA	OH		171			89

* SIGNIFICANTLY MORE DEATHS THAN EXPECTED

NON-WHITE FEMALE — CANCER OF THE LARGE INTESTINE — ICD 153,159.0

COUNTY AND STATE	1950-1959 DEATHS	1950-1959 RATE	1960-1969 DEATHS	1960-1969 RATE	1970-1979 DEATHS	1970-1979 RATE	PERCENT CHANGE 60S/50S	PERCENT CHANGE 70S/60S	PERCENT CHANGE 70S/50S	PERCENTILE %CHANGE 70S/50S	PERCENTILE RATES 1970S
COSHOCTON OH					4*	109.9					98
CRAWFORD OH			1	49.2							
CUYAHOGA OH	91	13.9	188*	18.1	277*	18.6	30	3	34	57	82
DARKE OH	1*	60.7	1*	188.7			211				
DELAWARE OH	1	24.4			1	17.7			-27	25	81
ERIE OH			2	20.7	5	25.9		25			90
FAIRFIELD OH					1	133.9					99
FAYETTE OH	3*	59.4	1	38.3	1	22.8	-36	-41	-62	08	88
FRANKLIN OH	66*	26.8	73*	20.8	123*	25.7	-22	23	-4	39	90
GALLIA OH	1	14.1			3*	67.9			383	97	97
GEAUGA OH					1	15.3					77
GREENE OH	7*	30.5	7	26.1	5	13.4	-14	-49	-56	10	72
GUERNSEY OH			2	35.1	1	21.0		-40			86
HAMILTON OH	67*	15.6	129*	22.2	186*	24.3	43	10	56	65	89
HANCOCK OH	1*	56.2	2*	114.1			103				
HARRISON OH			1	53.6							
HIGHLAND OH			1	21.0	1	12.4		-41			70
HOCKING OH			1*	114.7							
JEFFERSON OH	2	8.6	7*	27.1	8	28.7	216	6	234	91	92
KNOX OH	1	24.8	1	38.5			56				
LAKE OH					1	11.1					67
LAWRENCE OH	2	27.8	6*	70.0	3	26.8	152	-62	-3	39	91
LICKING OH	1	21.0	1	8.2	1	12.2	-61	49	-42	17	69
LOGAN OH					3*	74.7					97
LORAIN OH	3	11.3	4	10.3	12	17.3	-8	68	54	64	81
LUCAS OH	15	15.8	28	15.2	37	14.5	-4	-5	-8	35	75
MADISON OH	1	33.5	1	24.4	3*	64.6	-27	165	93	74	97
MAHONING OH	20*	23.8	22	15.2	46*	23.7	-36	56	0	41	89
MARION OH	1	28.7	2*	46.9			64				
MEDINA OH			1	53.6	1	22.9		-57			88
MEIGS OH			1	68.3							
MIAMI OH					2	21.6					87
MONROE OH					1*	546.6					99
MONTGOMERY OH	26	18.3	36	14.7	63	17.3	-20	18	-5	38	80
MORGAN OH					2*	80.6					98
MORROW OH			1*	424.0							
MUSKINGUM OH	1	7.2	3	23.1	4	20.1	222	-13	180	87	85
PAULDING OH					1	68.8					97
PERRY OH	1*	94.4									
PICKAWAY OH					2*	90.8					98
PIKE OH					1	38.3					95
PORTAGE OH	1	18.6	3	50.8	1	8.2	173	-84	-56	10	61
RICHLAND OH	2	7.3	1	3.3	6	19.0	-54	468	160	85	83
ROSS OH	3	32.9	4*	38.6	8*	57.9	18	50	76	71	96
SANDUSKY OH			1	28.1							
SCIOTO OH	2	32.8	3	31.5	1	10.7	-4	-66	-67	05	66
SENECA OH					1	22.3					88
SHELBY OH					1	57.3					96
STARK OH	6	13.8	10	14.5	24*	22.7	5	56	64	67	88
SUMMIT OH	12	13.3	17	10.5	44	19.3	-21	83	46	61	84
TRUMBULL OH	2	5.1	3	6.1	8	13.3	19	118	158	85	72
TUSCARAWAS OH					1	18.1					81
UNION OH	1*	112.4	1*	133.9			19				
WARREN OH			1	34.4	1	18.9		-45			83
WASHINGTON OH					2	32.9					93
WAYNE OH					2	41.7					95
WOOD OH			1	54.5							
OKLAHOMA OK	97	10.5	176	14.5	227	13.3	39	-8	28		
ADAIR OK	3	26.3	1	8.0			-69				
BLAINE OK	4*	49.2	1	12.6			-74				
BRYAN OK			2	17.6							
CADDO OK	1	13.5	1	7.4	3	12.3	-45	66	-9	35	70
CANADIAN OK	1	12.7			1	13.3			5	44	72
CARTER OK	4	18.9	6	17.1	1	2.5	-10	-85	-87	00	54
CHEROKEE OK	1	5.3	1	7.3	1	4.8	37	-34	-10	34	56
CHOCTAW OK	1	5.6	4	16.7	7	26.0	199	56	366	96	90
COAL OK	1	35.3	2*	60.7	1	13.5	72	-78	-62	08	73
COMANCHE OK	5*	29.4	7	22.2	8	17.2	-25	-23	-42	17	80
CRAIG OK			1	7.3	3	16.0		118			78
CREEK OK	2	7.2	3	9.2	4	11.4	27	24	58	66	67
CUSTER OK					2	56.1					96
DELAWARE OK			1	12.5	1	5.7		-54			57
GARFIELD OK	2	26.7	2	21.2	2	19.7	-21	-7	-26	25	84
GARVIN OK					1	14.2					74
GRADY OK	1	8.4	1	6.2	4	28.6	-25	358	242	92	92
GREER OK					2*	71.1					97
HARMON OK			1*	89.3							
HASKELL OK	1	29.8									
HUGHES OK	3	29.6	2	18.2	4	29.5	-39	62	0	41	92
JACKSON OK			1	17.2	1	7.4		-57			60
JEFFERSON OK			1*	62.9							
JOHNSTON OK					1	11.8					68
KAY OK	2	16.1	1	9.4			-42				
KINGFISHER OK	1	19.6	2	40.2			105				
KIOWA OK			1	14.5	1	28.1		94			92
LATIMER OK			1	24.4							
LE FLORE OK			2	12.4	2	9.0		-27			62
LINCOLN OK			7*	83.1	3	32.5		-61			93
LOGAN OK	5	24.4	8*	30.8	6	21.0	26	-32	-14	32	86
LOVE OK	1	112.4									
MC CLAIN OK					1	14.1					74
MC CURTAIN OK	2	7.2	5	14.4	6	11.2	101	-22	57	65	67
MC INTOSH OK			2	7.1	4	13.1		84			72
MARSHALL OK	1*	62.9	1	42.0			-33				
MAYES OK	1	24.8			2	10.2			-59	09	65
MURRAY OK	1	22.9	1	74.3			224				
MUSKOGEE OK	5	5.6	17	15.7	22	22.1	181	41	297	94	88
NOWATA OK	2	21.6			2	13.9			-36	21	73
OKFUSKEE OK			4	18.1	4	13.8		-24			73

* SIGNIFICANTLY MORE DEATHS THAN EXPECTED

NON-WHITE FEMALE — CANCER OF THE LARGE INTESTINE — ICD 153,159.0

1950-1959 DEATHS	1950-1959 RATE	1960-1969 DEATHS	1960-1969 RATE	1970-1979 DEATHS	1970-1979 RATE	COUNTY AND STATE		PERCENT CHANGE 60S/50S	PERCENT CHANGE 70S/60S	PERCENT CHANGE 70S/50S	PERCENTILE %CHANGE 70S/50S	PERCENTILE RATES 1970S
9	7.9	34	16.7	50	16.7	OKLAHOMA	OK	113	0	112	79	79
5	12.4	7	13.7	11	20.2	OKMULGEE	OK	11	48	64	67	85
		4	40.6	4	10.4	OSAGE	OK		-74			65
		1	9.9			OTTAWA	OK					
1	23.6					PAWNEE	OK					
1	22.5	2	25.7	1	9.0	PAYNE	OK	14	-65	-60	09	62
3	15.9	5	27.2	8	26.5	PITTSBURG	OK	71	-2	67	68	91
		1	9.6	2	12.1	PONTOTOC	OK		27			69
2	11.2	3	19.4	1	4.7	POTTAWATOMIE	OK	73	-76	-58	09	56
				1	2.4	ROGERS	OK					54
2	7.5	2	8.5	6	19.5	SEMINOLE	OK	14	129	161	85	84
2	18.9			2	11.0	SEQUOYAH	OK			-41	17	66
1	11.1	1	9.6	2	19.5	TILLMAN	OK	-14	104	76	71	84
17	14.4	20	11.1	33	12.1	TULSA	OK	-23	9	-16	32	69
3	18.8	1	6.1	2	6.0	WAGONER	OK	-68	-1	-68	04	58
		5*	70.5	4	17.6	WASHINGTON	OK		-75			81
10	12.7	16	10.0	31	11.5	OREGON	OR	-21	15	-9		
		1	28.7			CLACKAMAS	OR					
				1	44.6	COOS	OR					95
				1*	94.4	HARNEY	OR					98
1*	89.3	1	29.5			JEFFERSON	OR	-67				
		1*	66.2			JOSEPHINE	OR					
		1	22.9	2	35.3	KLAMATH	OR		54			94
				1	8.8	LANE	OR					62
		1*	133.9			LINCOLN	OR					
9	21.2	11	10.4	26	16.3	MULTNOMAH	OR	-51	57	-23	28	78
538*	19.7	723*	18.7	1056*	20.0	PENNSYLVANIA	PA	-5	7	2		
		1*	54.6			ADAMS	PA					
85*	17.9	132*	20.5	177*	21.4	ALLEGHENY	PA	14	4	19	51	87
1	21.3					ARMSTRONG	PA					
8*	31.9	4	11.1	8	13.6	BEAVER	PA	-65	22	-57	10	73
6*	38.5	7*	31.2	6	20.4	BERKS	PA	-19	-35	-47	15	85
1	21.0			2	27.1	BLAIR	PA			29	55	91
		4	22.1	8	18.5	BUCKS	PA		-16			82
		1	49.7	1	27.0	BUTLER	PA		-46			91
4*	70.9	2	15.1	4	24.4	CAMBRIA	PA	-79	62	-66	06	89
				1	67.0	CENTRE	PA					97
18*	35.8	13	21.9	17	17.7	CHESTER	PA	-39	-19	-51	13	81
1*	34.4	1	74.3	1	31.5	CLEARFIELD	PA	116	-58	-9	35	93
2*	44.6	1	20.2	1	21.2	CRAWFORD	PA	-55	5	-53	12	86
2	24.7	1	11.9	3	23.3	CUMBERLAND	PA	-52	96	-6	38	88
15*	24.6	13	12.8	31*	24.1	DAUPHIN	PA	-48	88	-2	40	89
19	17.7	32	17.3	49	20.0	DELAWARE	PA	-2	16	13	47	85
				2*	226.2	ELK	PA					99
3	23.0	2	12.6	1	3.0	ERIE	PA	-45	-76	-87	00	55
10*	30.2	7	19.2	14*	33.8	FAYETTE	PA	-37	76	12	47	94
2	39.5			5*	46.5	FRANKLIN	PA			18	50	96
		1	54.2			GREENE	PA					
				2*	58.3	HUNTINGDON	PA					97
		1	56.2	1	11.1	LACKAWANNA	PA		-80			67
1	7.4	3	22.2	7	29.1	LANCASTER	PA	201	31	295	94	92
1	10.9	5*	39.4	5*	33.0	LAWRENCE	PA	261	-16	202	90	93
				1	31.3	LEBANON	PA					93
		1	37.2	1	4.6	LEHIGH	PA		-88			56
				1	9.6	LUZERNE	PA					63
1	21.0			2	17.4	LYCOMING	PA			-17	31	81
				1*	424.0	MC KEAN	PA					99
3	18.5	4	23.9	6	24.9	MERCER	PA	29	4	35	57	90
7	10.0	13	12.6	39*	27.8	MONTGOMERY	PA	25	121	176	87	91
1*	163.6	2*	360.7	2*	556.1	MONTOUR	PA	121	54	240	91	99
		2	16.4	3	13.5	NORTHAMPTON	PA		-18			73
1*	134.3	1*	98.2	1	47.2	NORTHUMBERLAND	PA	-27	-52	-65	06	96
322*	19.3	449*	18.8	628*	19.0	PHILADELPHIA	PA	-2	1	-1	40	83
		1	26.5			SCHUYLKILL	PA					
1*	37.5			1	86.0	SOMERSET	PA			130	81	98
				1*	70.7	SUSQUEHANNA	PA					97
		1	44.6			VENANGO	PA					
				1*	245.4	WARREN	PA					99
10*	27.7	11*	25.0	11	21.5	WASHINGTON	PA	-10	-14	-23	28	87
		1*	490.8			WAYNE	PA					
9*	33.1	3	11.3	7	17.7	WESTMORELAND	PA	-66	56	-47	15	81
4	37.2	3	14.8	4	12.6	YORK	PA	-60	-15	-66	05	71
8	14.2	12	13.6	19	13.4	RHODE ISLAND	RI	-4	-2	-6		
2	24.1	2	14.8	1	8.5	NEWPORT	RI	-39	-42	-65	06	61
5	12.9	10	15.3	17	14.9	PROVIDENCE	RI	18	-3	15	49	76
1	19.1			1	10.2	WASHINGTON	RI			-47	15	65
198	7.4	283	9.3	440	11.4	SOUTH CAROLINA	SC	25	23	54		
3	9.5	4	13.8	1	2.4	ABBEVILLE	SC	46	-83	-75	03	54
5	6.2	11	12.6	14	12.2	AIKEN	SC	104	-3	97	76	69
2	6.0	2	9.0	1	3.8	ALLENDALE	SC	50	-58	-37	20	55
6	8.0	3	3.9	12	11.7	ANDERSON	SC	-51	199	46	61	68
2	6.3	2	7.4	2	4.4	BAMBERG	SC	17	-40	-30	23	56
2	5.2	1	2.7	4	9.3	BARNWELL	SC	-47	240	80	72	63
6	9.9	9	14.1	18	22.1	BEAUFORT	SC	43	57	124	81	88
4	6.4	1	1.5	6	7.0	BERKELEY	SC	-77	372	10	46	59
2	7.7	3	10.4	1	3.5	CALHOUN	SC	36	-66	-54	11	55
23	9.8	32	11.0	46	12.7	CHARLESTON	SC	12	15	29	55	71
		3	11.4	6	18.2	CHEROKEE	SC		60			82
7	13.0	7	12.9	10	19.1	CHESTER	SC	-1	48	47	61	83
2	6.4	4	8.1	4	7.3	CHESTERFIELD	SC	27	-11	13	48	59
6	9.5	4	7.0	6	8.7	CLARENDON	SC	-26	24	-8	36	62
7	14.8	4	7.0	8	12.3	COLLETON	SC	-53	76	-17	31	70
7	11.0	11	13.2	15	14.4	DARLINGTON	SC	20	9	31	56	75

* SIGNIFICANTLY MORE DEATHS THAN EXPECTED

NON-WHITE FEMALE — CANCER OF THE LARGE INTESTINE — ICD 153,159.0

1950-1959 DEATHS	1950-1959 RATE	1960-1969 DEATHS	1960-1969 RATE	1970-1979 DEATHS	1970-1979 RATE	COUNTY AND STATE	PERCENT CHANGE 60S/50S	PERCENT CHANGE 70S/60S	PERCENT CHANGE 70S/50S	PERCENTILE %CHANGE 70S/50S	PERCENTILE RATES 1970S
1	1.8	2	4.8	6	11.2	DILLON SC	158	135	506	98	67
1	1.8	2	5.9	7	11.0	DORCHESTER SC	227	88	515	98	66
2	6.0	3	11.3	1	3.6	EDGEFIELD SC	89	-68	-40	18	55
4	7.8	6	11.5	6	10.9	FAIRFIELD SC	48	-5	40	59	66
7	7.1	9	6.6	13	7.9	FLORENCE SC	-7	19	11	46	60
		3	4.8	7	8.7	GEORGETOWN SC		80			62
10	9.6	12	8.0	25	12.3	GREENVILLE SC	-17	53	28	55	70
2	3.2	6	10.4	7	9.9	GREENWOOD SC	223	-5	208	90	64
		3	7.8	4	10.0	HAMPTON SC		29			64
1	5.0	2	2.7	3	4.4	HORRY SC	-45	63	-11	34	56
2	9.2	6	20.3	5	14.1	JASPER SC	121	-31	53	63	74
4	7.5	5	9.4	8	12.5	KERSHAW SC	24	34	66	68	71
4	10.9	3	7.6	5	10.5	LANCASTER SC	-30	38	-3	40	66
3	7.1	4	7.2	9	12.2	LAURENS SC	2	70	73	70	70
4	12.8	2	3.9	9	19.9	LEE SC	-69	408	55	64	84
3	5.8	4	10.1	7	12.4	LEXINGTON SC	73	23	113	79	70
2	11.9	2	11.7	2	10.4	MC CORMICK SC	-2	-11	-13	33	65
4	6.4	2	3.4	6	6.8	MARION SC	-47	98	6	44	59
4	9.1	7	13.3	8	13.2	MARLBORO SC	46	-1	45	61	72
2	4.4	5	10.4	12	25.3	NEWBERRY SC	138	143	478	98	90
2	10.3	4	21.4	5	20.9	OCONEE SC	108	-2	103	77	86
9	8.5	13	9.1	22	11.9	ORANGEBURG SC	6	31	40	59	69
2	12.6	8*	32.9	7	24.6	PICKENS SC	162	-25	96	75	89
9	5.0	21	8.2	44	13.7	RICHLAND SC	65	67	177	87	73
		3	16.7	3	13.3	SALUDA SC		-21			72
8	6.9	20	13.4	22	12.0	SPARTANBURG SC	95	-10	75	71	69
9	6.9	4	2.9	11	6.5	SUMTER SC	-58	125	-6	38	58
5	12.7	3	8.2	7	13.5	UNION SC	-35	65	6	45	73
6	8.8	8	11.7	3	3.3	WILLIAMSBURG SC	33	-72	-62	07	55
4	6.1	10	12.1	12	11.9	YORK SC	99	-2	95	75	68
5	6.4	10	13.0	12	10.4	SOUTH DAKOTA SD	103	-20	63		
1	34.2	1	24.8	2*	59.5	BENNETT SD	-27	140	74	71	97
1*	344.1					BRULE SD					
				1	13.0	BUFFALO SD					72
				2	41.2	CHARLES MIX SD					95
1	13.0					CORSON SD					
		2	34.8			DEWEY SD					
				1*	172.0	HYDE SD					99
		1	62.9	1	24.8	JACKSON SD		-61			90
				1*	188.7	MOODY SD					99
1	28.7	1	19.9	1	17.2	ROBERTS SD	-31	-13	-40	18	80
1	7.7			2	8.9	SHANNON SD			17	49	62
				1*	212.0	STANLEY SD					99
		1	9.4			TODD SD					
		1*	37.7			TRIPP SD					
		1*	424.0			WALWORTH SD					
		2*	68.5			ZIEBACH SD					
238	10.2	386	13.5	545	15.5	TENNESSEE TN	33	15	52		
3*	64.1	1	29.8	1	7.1	ANDERSON TN	-54	-76	-89	00	59
3	20.3	3	22.1	3	14.8	BEDFORD TN	9	-33	-27	25	75
		4	26.8	1	4.6	BLOUNT TN		-83			56
				1	7.6	BRADLEY TN					60
		1*	37.5			CAMPBELL TN					
1	5.1	1	4.9	3	13.5	CARROLL TN	-5	177	163	85	73
		1	18.7			CARTER TN					
		1	20.6	1	31.4	CHEATHAM TN		52			93
		1	18.6	2	32.2	CHESTER TN		73			93
				1	26.8	COCKE TN					91
1	14.9	3	16.6	2	17.4	CROCKETT TN	11	5	17	50	81
38	11.4	72	18.3	80	15.8	DAVIDSON TN	61	-14	39	58	77
				1	34.4	DECATUR TN					94
1	14.5	2	24.6	1	14.9	DICKSON TN	70	-40	2	43	76
2	6.7	3	9.5	7	21.4	DYER TN	40	125	217	90	87
5	8.2	3	5.0	2	3.0	FAYETTE TN	-40	-39	-63	07	55
1	10.4	2	30.6	1	10.1	FRANKLIN TN	193	-67	-3	39	64
2	5.3	7	13.8	18*	31.9	GIBSON TN	160	130	500	98	93
2	8.6	2	8.9	3	14.6	GILES TN	4	63	70	69	75
1	10.8	1	9.8	2	12.8	HAMBLEN TN	-9	31	19	51	71
32	16.3	28	11.3	60*	20.3	HAMILTON TN	-31	80	24	53	85
1	3.2	5	12.0	11	23.9	HARDEMAN TN	273	99	640	99	89
		1	14.3	1	11.6	HARDIN TN		-19			68
		1	18.4	1	13.8	HAWKINS TN		-25			73
		8	15.4	6	10.7	HAYWOOD TN		-31			66
1	15.3	1	11.2	2	17.6	HENDERSON TN	-27	57	15	48	81
2	13.8	2	9.5	2	8.0	HENRY TN	-31	-16	-42	17	61
		1	22.5	3*	68.3	HICKMAN TN		204			97
		1	18.6			HUMPHREYS TN					
				1	20.2	JEFFERSON TN					85
7	8.3	18	13.9	28	17.9	KNOX TN	68	29	117	79	81
		1	8.9	3	37.5	LAKE TN		321			94
3	9.9	7	19.5	5	12.3	LAUDERDALE TN	98	-37	25	54	70
				1	28.1	LAWRENCE TN					92
1	8.0	2	13.6	5	29.9	LINCOLN TN	69	121	274	93	92
		1	12.1	3	28.2	MC MINN TN		133			92
		2	43.3	1	13.8	MC NAIRY TN		-68			73
13	15.2	8	7.0	26	20.7	MADISON TN	-54	197	36	58	86
				2	29.7	MARION TN					92
1	6.6			2	17.7	MARSHALL TN			168	86	81
2	6.3	2	4.1	11	20.4	MAURY TN	-35	399	227	91	85
1	53.6			1	16.7	MONROE TN			-69	04	79
2	5.0	7	14.8	8	14.3	MONTGOMERY TN	198	-4	187	88	74
		1*	67.0			MOORE TN					
2	10.1	2	14.7	7*	42.0	OBION TN	46	185	315	95	95
1	26.8					PUTNAM TN					
				1	26.5	RHEA TN					90
				1	9.3	ROANE TN					63
1	4.7			6	19.1	ROBERTSON TN			307	94	83

* SIGNIFICANTLY MORE DEATHS THAN EXPECTED

NON-WHITE FEMALE

CANCER OF THE LARGE INTESTINE

ICD 153,159.0

1950-1959 DEATHS	1950-1959 RATE	1960-1969 DEATHS	1960-1969 RATE	1970-1979 DEATHS	1970-1979 RATE	COUNTY AND STATE	PERCENT CHANGE 60S/50S	PERCENT CHANGE 70S/60S	PERCENT CHANGE 70S/50S	PERCENTILE %CHANGE 70S/50S	PERCENTILE RATES 1970S
1	3.7	3	8.7	6	12.0	RUTHERFORD TN	136	37	224	91	69
93	11.1	154	14.3	189	13.9	SHELBY TN	29	-3	25	53	73
2	18.3	2	15.5	1	5.8	SULLIVAN TN	-15	-62	-68	05	57
3	13.2	3	13.0	4	13.1	SUMNER TN	-1	1	-1	41	72
1	3.1	5	10.9	3	5.3	TIPTON TN	249	-52	69	69	57
3*	125.0	1	26.8	1	74.3	TROUSDALE TN	-79	178	-41	18	97
1	14.1	1	14.3	1	9.9	WARREN TN	2	-31	-29	23	64
4*	34.5	2	16.2	2	10.7	WASHINGTON TN	-53	-34	-69	04	66
		2	20.7	4	32.4	WEAKLEY TN		57			93
		1	31.3	1	18.6	WHITE TN		-41			82
		1	3.7	3	10.5	WILLIAMSON TN		183			65
		4	18.6	2	6.0	WILSON TN		-68			58
390	8.9	695	12.5	1120	13.1	TEXAS TX	41	5	48		
1	2.5	13*	26.2	5	10.2	ANDERSON TX	941	-61	307	94	65
2	8.3	7	19.1	7	14.1	ANGELINA TX	129	-26	69	68	74
		1*	114.7	1	32.7	ARANSAS TX		-71			93
		1	5.0	3	15.0	AUSTIN TX		201			76
2	6.9	7	16.8	5	14.7	BASTROP TX	143	-12	113	79	75
3	9.5	4	12.3	8	16.0	BELL TX	30	30	69	69	78
25	16.7	34	14.6	67	13.9	BEXAR TX	-12	-5	-16	32	73
1*	86.0					BLANCO TX					
6	9.1	8	10.2	14	15.5	BOWIE TX	11	53	70	69	77
2	6.3	2	5.6	8	15.5	BRAZORIA TX	-12	179	145	84	77
4	9.4	6	11.4	8	11.7	BRAZOS TX	21	3	24	53	68
1	19.1	1	17.0			BROWN TX	-11				
		1	5.4	2	6.9	BURLESON TX		27			59
2	11.9	3	14.8	2	9.3	CALDWELL TX	25	-37	-22	29	63
				2	4.7	CAMERON TX					56
1	7.1	3	17.0	8*	35.7	CAMP TX	140	110	403	97	94
4	13.3	3	7.0	2	6.1	CASS TX	-47	-12	-54	11	58
				1	12.2	CHAMBERS TX					69
4	7.9	8	19.3	9	18.3	CHEROKEE TX	144	-5	131	82	82
				1*	114.7	CLAY TX					98
		1*	59.6			COCHPAN TX					
				1	14.9	COLEMAN TX					76
1	7.4	2	10.6	2	6.9	COLLIN TX	42	-35	-7	36	59
6	22.4	3	9.2	4	12.5	COLORADO TX	-59	37	-44	16	71
		1	24.4			COMAL TX					
4*	82.2	1	17.2	1	14.3	COOKE TX	-79	-16	-83	01	74
		1	22.5			CORYELL TX					
		1	70.1			CROSBY TX					
33	8.7	89	15.4	144	15.5	DALLAS TX	77	1	79	72	77
1*	19.9					DAWSON TX					
				1	34.4	DEAF SMITH TX					94
1	23.6			1	54.5	DELTA TX			131	82	96
1	7.7	2	17.4	3	14.9	DENTON TX	127	-15	94	74	76
2	8.3	1	3.0	2	4.4	DE WITT TX	-64	47	-48	14	56
				1*	149.1	DONLEY TX					99
		2	19.8	3	21.9	ECTOR TX		11			87
4	10.3	5	9.3	4	7.4	ELLIS TX	-10	-21	-28	24	60
2	9.7	3	5.9	9	4.5	EL PASO TX	-39	-24	-54	11	56
2	4.7	8	16.1	7	13.0	FALLS TX	244	-19	177	87	72
2	12.8	1	4.9	4	25.5	FANNIN TX	-62	423	99	76	90
1	5.3	2	9.7	6	32.5	FAYETTE TX	83	236	514	98	93
3	8.7	3	7.0	5	8.4	FORT BEND TX	-20	20	-4	39	61
1*	43.0			1	37.5	FRANKLIN TX			-13	33	94
4	15.4	7	21.3	5	12.7	FREESTONE TX	38	-41	-18	31	71
4	4.6	22	16.0	25	14.0	GALVESTON TX	244	-13	201	90	74
				2*	56.8	GARZA TX					96
		1	19.1	1	11.2	GOLIAD TX		-41			67
		2	10.6	5	22.4	GONZALES TX		110			88
				1	16.4	GRAY TX					79
4	9.3	6	14.8	5	8.9	GRAYSON TX	59	-40	-5	38	62
8	11.6	11	12.9	11	11.9	GREGG TX	11	-8	2	43	68
3	11.8	6	22.8	5	12.9	GRIMES TX	93	-43	10	46	72
2	9.9	6	31.8	1	2.4	GUADALUPE TX	220	-93	-76	03	54
		1	40.9	1	47.1	HALL TX		15			96
				1*	149.1	HAMILTON TX					99
		2	11.9	3	14.6	HARDIN TX		23			75
61	9.2	115	11.4	227	14.0	HARRIS TX	24	23	53	63	73
14	14.2	9	8.3	27*	22.0	HARRISON TX	-42	166	55	64	88
				1	35.3	HASKELL TX					94
1	14.1	1	12.4	1	4.2	HAYS TX	-12	-66	-70	04	55
3	11.8	4	16.3	2	6.3	HENDERSON TX	39	-62	-47	15	58
1	26.5	1	30.3			HIDALGO TX	14				
2	15.3	3	15.6	5	23.3	HILL TX	3	49	53	63	89
1	12.9			1	7.9	HOPKINS TX			-39	19	60
4	9.9	1	2.1	8	14.4	HOUSTON TX	-79	589	45	61	75
1	19.9	1	31.5	2	14.1	HOWAPD TX	58	-55	-29	23	74
2	7.7	3	10.9	6	18.1	HUNT TX	42	67	136	83	81
				1*	294.8	JACK TX					99
1	11.1			3	18.9	JACKSON TX			70	69	83
3	15.7	1	4.6	3	8.8	JASPER TX	-71	91	-44	16	62
23	11.6	27	13.2	53	17.6	JEFFERSON TX	14	34	52	63	81
1*	89.3					JIM WELLS TX					
1	9.4	2	19.2	1	8.0	JOHNSON TX	103	-58	-15	32	61
				1	20.2	KAPNES TX					85
1	2.4	4	8.7	1	1.8	KAUFMAN TX	258	-79	-25	26	54
1*	344.1					KENEDY TX					
		1	18.7			KERR TX					
		1	7.7	2	11.5	KLEBERG TX		49			67
1	2.5	5	13.4	5	10.7	LAMAR TX	432	-20	326	95	66
				1	44.6	LAMPASAS TX					95
				1	23.6	LA SALLE TX					89
		1	9.1			LAVACA TX					
1	5.3	2	13.8	5	19.4	LEE TX	163	41	269	93	84
2	8.0	5	22.9	4	15.9	LEON TX	187	-30	100	76	78
2	8.3	5	16.3	4	12.0	LIBERTY TX	97	-26	45	60	69
2	8.1	3	6.6	8	16.7	LIMESTONE TX	-19	155	107	77	79

* SIGNIFICANTLY MORE DEATHS THAN EXPECTED

NON-WHITE FEMALE — CANCER OF THE LARGE INTESTINE — ICD 153,159.0

1950-1959 DEATHS	1950-1959 RATE	1960-1969 DEATHS	1960-1969 RATE	1970-1979 DEATHS	1970-1979 RATE	COUNTY AND STATE	PERCENT CHANGE 60S/50S	PERCENT CHANGE 70S/60S	PERCENT CHANGE 70S/50S	PERCENTILE %CHANGE 70S/50S	PERCENTILE RATES 1970S
1	2.2	1	2.6	5	7.5	LUBBOCK TX	16	187	233	91	60
		2*	228.7			LYNN TX					
				1	20.6	MC CULLOCH TX					86
5	5.2	18	14.4	24	14.0	MC LENNAN TX	179	-3	170	86	74
2	16.5	1	8.3	4	26.6	MADISON TX	-50	220	62	67	91
		2	6.9			MARION TX					
2	8.2	4	14.0	3	7.1	MATAGORDA TX	71	-49	-13	33	59
1	22.3	2	8.7	3	7.7	MIDLAND TX	-61	-11	-65	06	60
		4	16.6	4	11.4	MILAM TX		-31			67
2	5.9	3	9.3	7	11.9	MONTGOMERY TX	56	28	100	76	68
3	36.2	1	8.2	1	4.2	MORRIS TX	-77	-49	-88	00	55
		1*	94.4			MOTLEY TX					
3	10.8	10*	27.2	10	19.0	NACOGDOCHES TX	152	-30	75	71	83
2	3.9	8	15.7	10	20.9	NAVARRO TX	300	34	433	97	86
1	7.2	3	18.1	2	10.8	NEWTON TX	150	-40	49	62	66
2	5.3	3	5.4	6	5.1	NUECES TX	1	-5	-4	39	56
		1	4.6	1	4.0	ORANGE TX		-12			55
		1	16.4			PALO PINTO TX					
4	16.6	2	7.8	5	16.5	PANOLA TX	-53	113	0	41	79
				2*	58.4	PARKER TX					97
1	5.0			3	12.6	POLK TX			153	84	71
1	3.7	1	2.9	6	16.7	POTTER TX	-22	479	350	96	79
		1	49.2			RAINS TX					
		2	8.0	3	12.5	RED RIVER TX		56			71
1	30.7	1	27.0	1	21.0	REFUGIO TX	-12	-22	-32	22	86
4	10.0	5	9.0	4	11.1	ROBERTSON TX	-9	23	11	46	67
				2	51.5	RUNNELS TX					96
5	9.9	8	13.3	10	15.7	RUSK TX	34	18	59	66	77
1	16.1	1	8.8			SABINE TX	-45				
4*	36.4	2	18.4	3	15.9	SAN AUGUSTINE TX	-49	-13	-56	10	78
2	11.4	3	15.5	4	16.8	SAN JACINTO TX	35	8	47	61	80
		1	94.4			SAN PATRICIO TX					
		2*	107.0	1	19.1	SCURRY TX		-82			84
		1*	74.3			SHACKELFORD TX					
		5	20.3	5	16.2	SHELBY TX		-20			78
13	13.6	9	8.0	27	18.8	SMITH TX	-41	133	38	58	83
				2*	44.8	STEPHENS TX					95
		1*	212.0			STONEWALL TX					
21	9.4	41	15.3	80*	19.0	TARRANT TX	62	24	101	77	83
		1	4.4	3	11.5	TAYLOR TX		161			67
		2	13.7	2	10.7	TITUS TX		-22			66
4*	28.7			4	11.3	TOM GREEN TX			-60	08	67
7	5.4	15	10.6	28	12.4	TRAVIS TX	97	18	131	82	70
2	14.3	1	5.5			TRINITY TX	-61				
5	20.5	4	17.7	1	3.1	UPSHUR TX	-13	-83	-85	01	55
				1	9.3	VAL VERDE TX					63
		2	28.9			VAN ZANDT TX					
		4	18.2	3	9.4	VICTORIA TX		-48			63
		3	7.9	5	11.1	WALKER TX		41			67
1	5.6	3	11.5	5	10.4	WALLER TX	105	-10	85	73	65
4	10.0	5	10.0	9	22.4	WASHINGTON TX	-1	125	124	81	88
3	7.5	4	8.9	10	17.2	WHARTON TX	19	92	128	81	80
		1*	86.0			WHEELER TX					
2	17.1	5	12.4	10	18.2	WICHITA TX	-28	47	6	45	82
		2	25.1			WILBARGER TX					
2	9.1	5	16.9	3	7.5	WILLIAMSON TX	87	-56	-17	31	60
				1*	141.3	WISE TX					99
3	19.2			5	24.7	WOOD TX			29	55	90
3	8.9	2	3.3	14	14.6	UTAH UT	-63	345	64		
				1	29.6	BOX ELDER UT					92
				1	14.5	DAVIS UT					75
3	25.1			9	21.1	SALT LAKE UT			-16	32	86
		1	10.2			SAN JUAN UT					
		1	8.1	3	16.3	WEBER UT		101			79
		2	33.1			VERMONT VT					
		2*	109.8			WINDHAM VT					
357	12.5	525*	15.5	741*	16.7	VIRGINIA VA	24	8	33		
2	3.8	5	12.2	3	4.8	ACCOMACK VA	220	-61	25	54	56
9	20.5	7	12.3	16	21.6	ALBEMARLE#VA	-40	76	5	44	87
4*	28.3	6*	41.8	5	33.2	ALLEGHANY#VA	48	-21	17	50	93
3	19.8	1	6.2	2	11.4	AMELIA VA	-69	83	-43	17	67
2	10.2	3	14.4	2	7.7	AMHERST VA	42	-47	-25	27	60
1	12.8	1	10.9	2	15.3	APPOMATTOX VA	-15	40	19	51	76
7*	30.0	2	5.7	17*	30.6	ARLINGTON#VA	-81	440	2	42	93
4	17.3	4	12.3	5	13.3	AUGUSTA#VA	-29	8	-23	28	72
		1	24.6	2*	86.9	BATH VA		253			98
13	13.4	25*	21.5	22	16.0	BEDFORD#VA	60	-25	19	51	78
1	17.2	3	41.6	3	34.4	BOTETOURT VA	142	-17	101	76	94
6	17.2	6	16.5	6	13.9	BRUNSWICK VA	-4	-16	-19	30	73
				8*	33.4	BUCKINGHAM VA					94
5	25.4	5	21.9	3	8.0	CAROLINE VA	-14	-64	-69	04	60
2*	41.9	1	54.6	1	18.6	CARROLL#VA	30	-66	-56	11	82
4*	39.3	2	20.2	2	14.9	CHARLES CITY VA	-49	-26	-62	08	76
3	15.2	4	17.9	1	6.0	CHARLOTTE VA	18	-67	-61	08	58
44	11.2	78	16.8	98	16.3	CHESTERFIELD#VA	49	-3	45	60	78
		1	17.2	1	16.4	CLARKE VA		-5			79
3	20.1	4	26.2	2	10.2	CULPEPER VA	30	-61	-50	14	65
1	6.4	3	17.3	3	18.4	CUMBERLAND VA	172	6	189	88	82
15	8.7	22	12.2	28	13.9	DINWIDDIE#VA	40	14	59	66	73
2	18.3	4	28.9	7*	39.7	ESSEX VA	58	37	117	79	95
9	17.5	18*	25.3	30	18.2	FAIRFAX#VA	44	-28	4	43	82
9*	42.0	9*	38.7	3	12.2	FAUQUIER VA	-8	-68	-71	04	70
		4	31.5	3	29.0	FLUVANNA VA		-8			92
4*	29.8	2	15.9	2	15.4	FRANKLIN VA	-47	-3	-48	14	77
4*	55.4	2	30.2	3	30.9	FREDERICK#VA	-46	2	-44	16	93

* SIGNIFICANTLY MORE DEATHS THAN EXPECTED

COUNTIES COMBINED; SEE TECHNICAL NOTES

NON-WHITE FEMALE — CANCER OF THE LARGE INTESTINE — ICD 153,159.0

1950-1959 DEATHS	1950-1959 RATE	1960-1969 DEATHS	1960-1969 RATE	1970-1979 DEATHS	1970-1979 RATE	COUNTY AND STATE	PERCENT CHANGE 60S/50S	PERCENT CHANGE 70S/60S	PERCENT CHANGE 70S/50S	PERCENTILE %CHANGE 70S/50S	PERCENTILE RATES 1970S
		1*	45.5			GILES VA					
2	10.2	1	4.9	1	6.5	GLOUCESTER VA	-52	33	-37	21	58
1	8.5	2	12.3	2	10.1	GOOCHLAND VA	45	-17	20	51	65
				3*	116.2	GREENE VA					98
4	10.8	2	7.0	5	11.3	GREENSVILLE#VA	-36	63	4	43	67
8	12.1	8	13.2	13	18.8	HALIFAX#VA	10	42	55	65	83
5	20.8	5	20.1	2	5.8	HANOVER VA	-4	-71	-72	04	57
2	8.7	5	15.8	9	13.1	HENRICO VA	81	-17	51	62	72
3	6.7	6	10.3	10	14.5	HENRY#VA	54	40	116	79	75
1	6.1	8*	28.4	1	2.3	ISLE OF WIGHT VA	367	-92	-62	08	54
4	11.1	8	21.7	10	15.2	JAMES CITY#VA	95	-30	36	58	76
				5*	37.1	KING AND QUEEN VA					94
2	25.5	1	11.1			KING GEORGE VA	-56				
1	6.5	3	19.8	2	12.0	KING WILLIAM VA	206	-39	85	73	69
1	6.9	2	10.4	7*	35.0	LANCASTER VA	52	236	409	97	94
5*	34.7	3	14.7	3	12.1	LOUDOUN VA	-58	-18	-65	06	69
3	12.4	3	10.7	5	20.3	LOUISA VA	-14	89	63	67	85
		1	5.2	4	14.9	LUNENBURG VA		189			76
1	11.2	2	33.8	1	10.3	MADISON VA	203	-69	-7	36	65
1	11.1	4*	37.8	2	17.7	MATHEWS VA	241	-53	59	66	81
5	9.7	4	7.1	4	6.1	MECKLENBURG VA	-27	-14	-37	20	58
		8*	63.8	5	30.1	MIDDLESEX VA		-53			93
1	14.1	2	16.2	5	34.9	MONTGOMERY#VA	15	116	148	84	94
		1	5.4	2	9.7	NELSON VA		81			64
5	14.7	5	12.0	2	4.9	NORTHAMPTON VA	-18	-59	-66	05	56
1	7.0	1	5.0	6	27.5	NORTHUMBERLAND VA	-29	455	294	94	91
		4	14.6	9	20.8	NOTTOWAY VA		42			86
1	5.3	2	11.5	2	11.0	ORANGE VA	118	-5	107	78	66
				1	27.0	PAGE VA					91
		1	32.6	3*	50.1	PATRICK VA		53			96
14	14.3	17	14.1	32*	21.5	PITTSYLVANIA#VA	-1	52	50	62	87
3	25.0	2	16.9	3	26.2	POWHATAN VA	-33	55	5	43	90
1	3.7	4	16.0	7	21.5	PRINCE EDWARD VA	334	35	484	98	87
3	27.5	3	16.2	2	3.9	PRINCE WILLIAM#VA	-41	-76	-86	00	55
1	10.5	4*	39.8	1	9.9	PULASKI VA	279	-75	-5	38	64
2*	55.1	2	39.8			RAPPAHANNOCK VA	-28				
				3	27.0	RICHMOND VA					91
12	13.1	23*	21.8	20	15.3	ROANOKE#VA	67	-30	17	49	76
		2	17.3	2	14.8	ROCKBRIDGE#VA		-14			76
		1	14.8			ROCKINGHAM#VA					
		1*	294.8	1	74.3	SCOTT VA		-75			97
		1*	54.6			SHENANDOAH VA					
				1	17.2	SMYTH VA					80
2	3.8	7	13.2	12	18.1	SOUTHAMPTON#VA	245	37	373	96	81
3	14.3	5	22.1	9*	29.1	SPOTSYLVANIA#VA	55	32	104	77	92
		3*	45.3	3	28.8	STAFFORD VA		-36			92
4*	29.9	4	25.4	4	22.4	SURRY VA	-15	-12	-25	27	88
6	20.7	3	11.8	3	9.2	SUSSEX VA	-43	-23	-56	10	62
1	16.1	1	8.9			TAZEWELL VA	-44				
		1	19.1	3*	41.5	WARREN VA		117			95
3	27.6	2	14.3	2	10.9	WASHINGTON#VA	-48	-24	-60	08	66
3	17.8	4	18.4	5	21.5	WESTMORELAND VA	3	17	21	52	87
1	10.1			2	20.0	WISE#VA			98	76	85
3*	50.0	1	17.0	1	11.6	WYTHE VA	-66	-31	-77	02	68
58	11.9	76	12.8	144*	17.5	CHESAPEAKE CITY#VA	8	37	48	61	81
6	10.0	11	11.7	17	14.4	HAMPTON CITY VA	18	22	44	60	74
11	11.6	14	12.1	37	20.3	NEWPORT NEWS CITY VA	4	68	75	71	85
6	7.1	17	15.6	23	19.9	SUFFOLK CITY VA	121	27	181	87	84
25	13.3	39	11.3	84	13.0	WASHINGTON WA	-15	15	-2		
				2	18.3	CLARK WA					82
				1	21.3	COWLITZ WA					87
1	37.5	1	34.4	1	18.6	FERRY WA	-8	-46	-50	13	82
1	53.0	1	74.3	1	8.9	FRANKLIN WA	40	-88	-83	01	62
		1	25.8			GRANT WA					
1	47.2					GRAYS HARBOR WA					
				1	62.9	JEFFERSON WA					97
14	12.0	21	10.3	51	14.0	KING WA	-14	36	17	49	74
		3*	69.6	1	8.6	KITSAP WA		-88			61
2*	105.4					KLICKITAT WA					
1*	182.2					LEWIS WA					
		1*	39.7	1	37.2	MASON WA		-6			94
3	20.3	4	16.8	13	16.2	PIERCE WA	-17	-4	-20	29	78
		2	33.2	2	16.8	SNOHOMISH WA		-49			80
		1	3.6	1	3.3	SPOKANE WA		-10			55
		1	26.0			STEVENS WA					
				2	71.8	THURSTON WA					97
				1	25.8	WALLA WALLA WA					90
				1	16.7	WHATCOM WA					79
2	13.1	3	17.2	5	13.0	YAKIMA WA	32	-24	0	41	72
43	10.2	57	12.1	84	16.4	WEST VIRGINIA WV	18	35	60		
		1*	86.0			BARBOUR WV					
1	15.1	1	13.6			BERKELEY WV	-10				
				1	53.0	BOONE WV					96
1	49.1					BROOKE WV					
		4	12.1	2	5.1	CABELL WV		-58			56
2	6.4	5	13.3	3	7.4	FAYETTE WV	109	-44	16	49	60
1*	62.9	3*	323.4			GRANT WV	414				
2	19.0			2	13.8	GREENBRIER WV			-28	24	73
1	11.9			2	38.9	HANCOCK WV			228	91	95
				1*	112.4	HARDY WV					98
2	18.6	1	13.3	2	14.8	HARRISON WV	-28	11	-21	29	75
		3	20.1	2	13.2	JEFFERSON WV		-34			72
6	7.7	10	12.1	28*	29.9	KANAWHA WV	56	147	286	93	92
2	12.5	3	13.1	2	11.8	LOGAN WV	5	-10	-6	38	68
6	9.8	6	11.5	12	21.1	MC DOWELL WV	17	84	116	79	86
2	21.4	2	10.5	5	21.9	MARION WV	-51	109	2	43	87
3	9.1	5	10.5	3	6.5	MERCER WV	15	-38	-28	24	58

* SIGNIFICANTLY MORE DEATHS THAN EXPECTED

COUNTIES COMBINED; SEE TECHNICAL NOTES

NON-WHITE FEMALE — CANCER OF THE LARGE INTESTINE — ICD 153,159.0

1950-1959 DEATHS	1950-1959 RATE	1960-1969 DEATHS	1960-1969 RATE	1970-1979 DEATHS	1970-1979 RATE	COUNTY AND STATE	PERCENT CHANGE 60S/50S	PERCENT CHANGE 70S/60S	PERCENT CHANGE 70S/50S	PERCENTILE %CHANGE 70S/50S	PERCENTILE RATES 1970S
		2	12.1			MINGO WV					
		2	25.4	2	24.7	MONONGALIA WV		-3			90
1	38.2			1	15.8	MONROE WV			-59	09	77
2	13.4	3	22.6	2	16.6	OHIO WV	69	-26	24	53	79
9	19.6	4	7.4	12	17.0	RALEIGH WV	-62	129	-13	33	80
1*	98.2					RANDOLPH WV					
		1	24.8			SUMMERS WV					
1*	267.9					TYLER WV					
				1*	267.9	WIRT WV					99
		1	37.8			WOOD WV					
				1	14.1	WYOMING WV					74
23	16.2	33	11.8	77	14.4	WISCONSIN WI	-27	22	-11		
		2*	606.2	1*	344.1	ADAMS WI		-43			99
		1	38.3	1	23.8	ASHLAND WI		-38			89
		1*	67.0			BAYFIELD WI					
1	21.9	1	22.5	1	8.5	BROWN WI	3	-62	-61	08	61
1*	344.1					BUFFALO WI					
				3	15.8	DANE WI					77
1*	212.0	1*	344.1			DOUGLAS WI	62				
				1*	141.3	JUNEAU WI					99
		1	22.9			KENOSHA WI					
		1*	112.4			LA CROSSE WI					
1	18.6	2	18.1			MENOMINEE#WI	-3				
14	20.2	19	8.6	64	17.4	MILWAUKEE WI	-57	102	-14	32	81
1*	298.1					ONEIDA WI					
		1	22.5			OUTAGAMIE WI					
2	35.6			3	11.8	RACINE WI			-67	05	68
1	14.4	1	10.9	1	7.3	ROCK WI	-25	-33	-50	14	59
		1	47.1	1	19.6	SAWYER WI		-58			84
1	43.0					VILAS WI					
		1	47.2	1	26.5	WALWORTH WI		-44			90
1	6.6	1	4.6	4	11.2	WYOMING WY	-31	145	70		
				1	94.4	CARBON WY					98
1	18.4	1	10.2	1	18.7	FREMONT WY	-45	83	2	42	83
				2	18.8	LARAMIE WY					83

* SIGNIFICANTLY MORE DEATHS THAN EXPECTED

\# COUNTIES COMBINED; SEE TECHNICAL NOTES

WHITE MALE CANCER OF THE RECTUM ICD 154 EXCEPT 154.3

1950-1959 DEATHS	1950-1959 RATE	1960-1969 DEATHS	1960-1969 RATE	1970-1979 DEATHS	1970-1979 RATE	COUNTY AND STATE	PERCENT CHANGE 60S/50S	PERCENT CHANGE 70S/60S	PERCENT CHANGE 70S/50S	PERCENTILE %CHANGE 70S/50S	PERCENTILE RATES 1970S
56746	9.0	55437	7.5	48913	5.8	UNITED STATES	-17	-23	-36		
304	3.8	318	3.3	340	2.9	ALABAMA AL	-14	-10	-23		
1	1.9	2	4.0	2	2.3	AUTAUGA AL	108	-41	24	76	24
8	5.0	5	2.7	7	2.1	BALDWIN AL	-46	-22	-58	13	22
1	2.0	1	1.8	1	1.7	BARBOUR AL	-9	-8	-16	55	18
2	4.3	1	1.9	3	6.1	BIBB AL	-56	223	43	82	72
2	2.9	4	3.2	2	1.4	BLOUNT AL	13	-56	-50	20	17
		1	4.8			BULLOCK AL					
2	2.8	2	2.6	1	1.4	BUTLER AL	-4	-47	-50	20	16
9	4.3	4	1.7	6	1.9	CALHOUN AL	-61	13	-56	14	20
4	3.6	8	6.8	7	4.9	CHAMBERS AL	87	-27	36	80	59
2	3.6			2	2.6	CHEROKEE AL			-29	43	27
4	4.0	4	3.5	3	2.5	CHILTON AL	-12	-27	-36	35	27
1	2.2	1	2.1	1	1.9	CHOCTAW AL	-5	-10	-14	57	21
		3	4.5	2	2.5	CLARKE AL		-45			26
1	1.4	2	3.0	5	7.5	CLAY AL	112	150	430	99	85
2	4.3	1	2.0	1	1.7	CLEBURNE AL	-53	-15	-60	11	19
2	2.3	2	1.7	3	2.4	COFFEE AL	-26	42	5	69	25
4	3.2	3	1.8			COLBERT AL	-43				
2	3.7			1	1.6	CONECUH AL			-56	14	18
1	2.7					COOSA AL					
3	2.0			5	2.7	COVINGTON AL			38	81	29
1	1.9	2	3.3	1	1.7	CRENSHAW AL	72	-48	-11	60	19
6	3.5	5	2.4	9	3.3	CULLMAN AL	-31	38	-5	64	37
1	1.3	5	5.8	3	2.5	DALE AL	336	-57	87	90	26
3	4.0	2	2.4	4	3.8	DALLAS AL	-40	62	-4	65	44
3	1.6	8	3.8	4	1.6	DE KALB AL	132	-58	-2	66	18
3	3.1	4	3.6	2	1.4	ELMORE AL	15	-60	-54	16	17
2	2.0	2	2.0	9	8.0	ESCAMBIA AL	-2	311	304	98	89
12	4.4	7	2.1	12	2.8	ETOWAH AL	-51	31	-36	35	30
6	10.6	4	5.0	5	6.5	FAYETTE AL	-53	28	-39	32	76
3	3.5	2	1.9	7	5.7	FRANKLIN AL	-45	195	62	86	68
2	2.1	2	2.0	4	3.6	GENEVA AL	-2	75	72	88	40
		1	11.2			GREENE AL					
1	3.7	1	3.4	2	4.9	HALE AL	-8	44	33	79	59
1	2.5	1	2.2	3	5.0	HENRY AL	-15	135	100	91	60
4	2.8	3	2.0	6	3.4	HOUSTON AL	-28	69	21	75	37
4	3.4	7	4.6	4	2.2	JACKSON AL	33	-51	-35	36	23
76	5.5	85	5.2	71	3.8	JEFFERSON AL	-5	-27	-30	42	43
3	4.4			1	1.2	LAMAR AL			-73	06	15
6	3.7	4	2.0	8	2.8	LAUDERDALE AL	-45	38	-24	48	30
4	6.9					LAWRENCE AL					
2	3.0	2	3.2	1	.6	LEE AL	7	-81	-80	02	13
1	.9	1	.8	4	2.3	LIMESTONE AL	-10	187	159	95	24
				2	10.0	LOWNDES AL					96
1	4.3			1	2.9	MACON AL			-33	39	31
9	4.3	6	1.7	17	4.1	MADISON AL	-61	144	-4	64	48
3	7.9	1	1.9	1	1.8	MARENGO AL	-76	-7	-77	03	20
1	1.0	1	.7	1	.8	MARION AL	-28	11	-20	52	14
6	4.9	7	3.6	6	2.3	MARSHALL AL	-27	-35	-52	17	24
30	5.7	24	3.4	33	3.8	MOBILE AL	-39	10	-33	39	43
1	1.6	1	1.5			MONROE AL	-2				
10	3.4	22	5.8	11	2.8	MONTGOMERY AL	70	-52	-19	53	30
10	5.4	11	4.9	9	3.7	MORGAN AL	-9	-24	-31	42	42
2	7.8	2	6.7	2	7.4	PERRY AL	-15	11	-6	63	85
2	3.4					PICKENS AL					
2	2.9	1	1.3	2	2.4	PIKE AL	-55	88	-16	55	25
4	5.4	1	1.3	1	1.3	RANDOLPH AL	-76	0	-76	04	16
2	2.9	5	4.2	1	.7	RUSSELL AL	45	-82	-74	05	13
		4	3.7	4	2.9	ST CLAIR AL		-23			31
4	4.8	4	3.1	5	2.8	SHELBY AL	-36	-11	-43	27	30
		3	14.4			SUMTER AL					
6	3.7	7	4.2	6	3.0	TALLADEGA AL	15	-30	-20	52	32
3	2.7	1	.9	1	.5	TALLAPOOSA AL	-67	-40	-80	02	13
6	2.3	7	2.0	12	3.1	TUSCALOOSA AL	-13	53	33	79	34
2	.8	13	5.4	10	3.4	WALKER AL	540	-37	305	98	38
1	3.2	1	3.0			WASHINGTON AL	-5				
1	4.0	2	7.7	1	4.7	WILCOX AL	92	-38	18	75	56
3	4.5	2	2.7	2	2.4	WINSTON AL	-40	-11	-47	24	25
		19	5.1	18	4.2	ALASKA AK		-17			
186	6.2	298	5.5	383	4.2	ARIZONA AZ	-11	-23	-32		
2	6.3					APACHE AZ					
10	7.6	4	2.6	16	5.3	COCHISE AZ	-65	103	-30	43	64
4	5.9	4	3.6			COCONINO AZ	-39				
3	2.9	7	6.7	9	5.9	GILA AZ	136	-12	107	92	70
4	10.1	2	3.4	2	3.1	GRAHAM AZ	-66	-8	-69	07	34
1	3.0			2	5.7	GREENLEE AZ			90	90	67
95	6.2	184	6.1	212	4.1	MARICOPA AZ	-1	-33	-33	38	48
1	1.9	2	2.2	4	1.4	MOHAVE AZ	15	-35	-25	47	17
3	8.9	1	1.1	5	4.3	NAVAJO AZ	-87	284	-52	18	50
49	8.4	64	5.4	87	4.6	PIMA AZ	-36	-15	-45	25	55
6	3.5	5	2.1	9	3.0	PINAL AZ	-39	39	-15	57	32
1	3.2	3	6.5	5	9.0	SANTA CRUZ AZ	104	38	182	96	93
5	3.4	11	4.8	23	6.1	YAVAPAI AZ	42	29	82	89	73
2	1.0	11	6.4	9	2.6	YUMA AZ	535	-59	161	95	28
227	3.4	255	3.3	350	3.8	ARKANSAS AR	-4	15	10		
1	1.7			5	4.7	ARKANSAS AR			173	96	57
				2	2.5	ASHLEY AR					26
3	5.2	1	.6	6	3.5	BAXTER AR	-88	460	-33	38	39
5	2.1	15	5.1	17	4.2	BENTON AR	141	-17	99	91	49
3	3.2	5	4.2	6	4.5	BOONE AR	30	7	39	81	53

1950-1959 DEATHS	1950-1959 RATE	1960-1969 DEATHS	1960-1969 RATE	1970-1979 DEATHS	1970-1979 RATE	COUNTY AND STATE	PERCENT CHANGE 60S/50S	PERCENT CHANGE 70S/60S	PERCENT CHANGE 70S/50S	PERCENTILE %CHANGE 70S/50S	PERCENTILE RATES 1970S
3	5.4	1	1.7	1	1.5	BRADLEY AR	-70	-12	-73	05	17
1	3.7			1	3.6	CALHOUN AR			-3	65	41
2	2.2	1	1.2	6	6.0	CARROLL AR	-48	413	166	95	71
1	3.4	1	2.6	5	12.6	CHICOT AR	-25	393	269	98	98
3	3.8	4	4.8	6	6.8	CLARK AR	28	40	79	89	79
3	2.7	5	3.7	4	3.0	CLAY AR	35	-18	10	71	33
2	3.9	2	3.2	2	2.0	CLEBURNE AR	-19	-38	-49	21	21
				1	2.5	CLEVELAND AR					26
2	2.5	6	7.1	3	3.0	COLUMBIA AR	180	-57	21	75	33
		2	3.0			CONWAY AR					
7	3.9	2	.9	9	3.8	CRAIGHEAD AR	-77	321	-2	66	43
5	4.4	4	3.1	3	2.0	CRAWFORD AR	-29	-35	-53	16	22
1	1.5	6	7.1	5	5.3	CRITTENDEN AR	386	-24	268	98	64
1	5.3	1	2.2	2	4.4	DALLAS AR	-58	102	-16	55	52
1	3.1			2	3.3	DESHA AR			5	69	36
1	2.1	2	3.7	3	5.0	DREW AR	76	34	136	93	60
2	1.8	2	1.5	6	4.2	FAULKNER AR	-14	168	131	93	48
3	4.2	3	4.0	2	2.4	FRANKLIN AR	-5	-40	-43	27	25
1	2.3	1	1.8	4	6.0	FULTON AR	-23	232	157	95	71
22	8.9	14	3.9	21	4.4	GARLAND AR	-56	14	-50	19	52
1	1.8	2	3.5	1	1.7	GRANT AR	93	-50	-4	64	19
4	3.0	8	5.3	7	4.5	GREENE AR	79	-16	51	84	53
4	5.0	7	8.6	7	7.6	HEMPSTEAD AR	73	-11	53	84	86
2	1.9	4	4.3	3	2.6	HOT SPRING AR	120	-38	35	80	28
2	3.0	1	1.4	4	5.2	HOWARD AR	-54	276	72	88	62
4	3.2	4	3.1	4	2.6	INDEPENDENCE AR	-4	-15	-18	53	28
		1	2.3	5	6.1	IZARD AR		162			73
4	4.5	2	2.0	3	3.3	JACKSON AR	-56	64	-28	44	36
6	4.1	11	6.2	15	6.2	JEFFERSON AR	52	-2	50	84	73
1	1.4	4	3.9	3	2.6	JOHNSON AR	171	-34	78	88	27
1	2.8	1	2.7	1	3.0	LAFAYETTE AR	-1	9	8	70	32
4	6.0	1	.8	2	2.3	LAWRENCE AR	-86	181	-61	10	24
1	2.6	1	2.4	1	3.1	LEE AR	-5	30	23	76	34
		1	2.6			LINCOLN AR					
1	2.5	2	5.6			LITTLE RIVER AR	129				
4	3.6	6	4.5	3	3.1	LOGAN AR	23	-31	-15	57	34
		3	2.8	3	2.5	LONOKE AR		-8			27
3	4.9	1	1.3			MADISON AR	-74				
3	7.1	1	1.4	3	3.0	MARION AR	-80	110	-58	13	33
2	1.5	2	1.6	6	4.4	MILLER AR	6	171	187	96	51
6	3.1	3	1.9	9	4.9	MISSISSIPPI AR	-39	157	57	85	59
		1	2.1	3	6.2	MONROE AR		198			74
				1	2.6	MONTGOMERY AR					27
		1	3.4	2	4.3	NEWTON AR		29			51
5	6.1	3	3.1	3	2.6	OUACHITA AR	-49	-16	-57	13	28
1	2.8					PERRY AR					
3	6.0	5	6.7	5	6.5	PHILLIPS AR	11	-3	8	70	77
1	4.1	3	5.5			PIKE AR	33				
3	2.8	7	6.6	8	6.7	POINSETT AR	133	2	139	94	79
2	2.4	5	5.0	2	1.5	POLK AR	108	-70	-37	34	17
3	2.8	2	1.5	7	4.4	POPE AR	-45	186	56	85	52
2	3.9	2	4.5	4	7.9	PRAIRIE AR	13	77	100	91	88
35	5.8	33	4.1	44	4.6	PULASKI AR	-29	13	-20	51	54
6	7.8	1	1.1	3	4.3	RANDOLPH AR	-86	291	-45	26	50
2	3.0	1	2.9	2	2.8	ST FRANCIS AR	-4	-1	-4	64	30
4	3.5	4	2.7	6	3.3	SALINE AR	-25	26	-6	63	37
1	2.4	1	1.7	3	5.1	SCOTT AR	-29	206	117	92	61
		2	4.1			SEARCY AR					
16	5.8	14	4.4	9	2.3	SEBASTIAN AR	-24	-49	-61	11	23
2	3.7	3	5.7	1	1.8	SEVIER AR	55	-68	-50	20	20
				2	1.2	SHARP AR					15
		1	1.8	2	3.2	STONE AR		74			35
2	1.1	5	3.2	6	2.9	UNION AR	199	-10	170	95	31
		1	1.5	2	3.0	VAN BUREN AR		106			33
9	3.6	13	4.3	19	4.9	WASHINGTON AR	19	13	34	80	58
3	1.7	2	1.0	12	5.4	WHITE AR	-44	442	206	96	64
		1	2.1	1	2.3	WOODRUFF AR		6			23
1	1.2			1	.8	YELL AR			-31	41	14
4229	8.5	4723	7.2	4356	5.5	CALIFORNIA CA	-15	-24	-35		
281	9.1	255	7.4	222	6.3	ALAMEDA CA	-19	-15	-31	42	75
3	4.9	11*	17.1	3	2.9	AMADOR CA	253	-83	-40	31	31
32	7.5	43	7.6	28	3.8	BUTTE CA	2	-50	-49	21	43
8	12.2	8	9.5	8	6.7	CALAVERAS CA	-22	-30	-45	25	79
4	5.7	6	9.3	3	4.4	COLUSA CA	61	-52	-23	48	52
69	7.6	100	6.9	90	4.5	CONTRA COSTA CA	-9	-35	-41	30	53
2	4.6	3	5.9	3	3.8	DEL NORTE CA	28	-35	-17	55	43
2	1.8	7	5.4	12	3.9	EL DORADO CA	195	-28	113	92	45
66	5.8	91	6.3	102	6.1	FRESNO CA	8	-3	4	68	72
3	3.6	3	3.9	8	7.2	GLENN CA	8	86	102	91	83
38	10.9	30	7.9	27	6.1	HUMBOLDT CA	-28	-23	-44	27	72
9	3.9	19	6.6	14	5.4	IMPERIAL CA	71	-19	39	81	64
2	2.9	2	3.3	7	7.0	INYO CA	15	109	141	94	81
52	6.9	74	6.9	62	4.7	KERN CA	0	-32	-32	40	56
13	8.6	6	2.8	13	6.4	KINGS CA	-68	129	-26	46	75
1	.7	9	6.4	11	5.2	LAKE CA	791	-19	621	99	62
3	4.6	1	1.3	4	5.6	LASSEN CA	-71	317	22	76	67
1650	8.5	1848	7.8	1475	6.0	LOS ANGELES CA	-8	-24	-30	42	71
10	6.9	8	4.4	10	4.4	MADERA CA	-36	-1	-36	35	52
35	8.7	54	9.2	42	5.7	MARIN CA	6	-38	-35	37	67
2	4.9	3	6.5	2	3.8	MARIPOSA CA	32	-40	-21	50	44
19	7.8	17	6.7	21	7.0	MENDOCINO CA	-15	5	-11	60	81
17	7.4	25	7.7	25	6.2	MERCED CA	4	-20	-16	55	73
1	3.3	2	4.9	1	1.8	MODOC CA	49	-63	-45	25	20
1	6.9			1	6.5	MONO CA			-5	63	77
36	7.6	49	7.7	48	5.8	MONTEREY CA	1	-24	-24	48	69
24	7.0	36	7.7	24	4.4	NAPA CA	11	-43	-37	34	51
21*	15.0	5	3.6	14	6.2	NEVADA CA	-76	71	-59	12	74
94	7.0	166	5.6	247	4.8	ORANGE CA	-20	-15	-32	40	57
21	8.1	15	4.8	18	4.2	PLACER CA	-41	-12	-48	22	49

* SIGNIFICANTLY MORE DEATHS THAN EXPECTED

WHITE MALE

CANCER OF THE RECTUM

ICD 154 EXCEPT 154.3

1950-1959 DEATHS	1950-1959 RATE	1960-1969 DEATHS	1960-1969 RATE	1970-1979 DEATHS	1970-1979 RATE	COUNTY AND STATE	PERCENT CHANGE 60S/50S	PERCENT CHANGE 70S/60S	PERCENT CHANGE 70S/50S	PERCENTILE %CHANGE 70S/50S	PERCENTILE RATES 1970S
2	3.7	2	3.1	1	1.0	PLUMAS CA	-17	-68	-73	05	14
73	7.0	111	6.3	121	4.1	RIVERSIDE CA	-11	-35	-42	28	47
140	10.2	139	7.9	141	5.9	SACRAMENTO CA	-23	-25	-42	29	70
9	13.2	9	11.0	8	9.5	SAN BENITO CA	-16	-14	-28	44	95
111	7.0	131	5.5	177	5.9	SAN BERNARDINO CA	-21	7	-16	55	70
188	7.2	254	6.5	275	4.6	SAN DIEGO CA	-9	-30	-36	35	54
471*	13.2	362*	9.8	212*	6.8	SAN FRANCISCO CA	-25	-31	-49	22	79
82	8.6	89	7.6	74	5.7	SAN JOAQUIN CA	-11	-25	-34	38	68
26	7.9	33	6.8	30	4.4	SAN LUIS OBISPO CA	-13	-36	-45	26	51
118	12.0	124	7.8	137	6.6	SAN MATEO CA	-36	-15	-45	25	78
47	8.9	50	6.5	48	4.5	SANTA BARBARA CA	-27	-31	-50	20	53
138	8.9	172	7.0	167	5.3	SANTA CLARA CA	-21	-25	-41	30	63
54	10.4	42	5.8	47	5.5	SANTA CRUZ CA	-45	-4	-47	23	66
5	2.5	12	4.1	26	6.3	SHASTA CA	63	52	147	94	74
3	18.3	1	8.6	1	4.6	SIERRA CA	-53	-47	-75	04	54
11	6.3	7	4.2	8	4.3	SISKIYOU CA	-33	3	-31	41	50
17	5.1	33	7.5	41	7.0	SOLANO CA	47	-6	38	81	81
66	9.8	60	6.3	70	5.3	SONOMA CA	-35	-16	-45	25	64
51	7.5	44	5.3	43	4.2	STANISLAUS CA	-30	-20	-44	27	49
5	4.4	7	5.0	11	6.2	SUTTER CA	14	24	41	82	74
5	4.2	7	4.8	11	5.9	TEHAMA CA	14	23	41	82	70
4	11.4					TRINITY CA					
33	5.3	48	6.2	58	6.7	TULARE CA	17	6	25	77	78
4	4.3	9	9.0	8	5.3	TUOLUMNE CA	107	-41	22	76	64
23	4.6	60	7.4	64	4.5	VENTURA CA	59	-38	-2	66	54
14	7.1	16	5.8	20	5.6	YOLO CA	-18	-3	-21	50	67
10	8.4	5	3.7	12	7.8	YUBA CA	-56	114	-6	63	87
403	6.4	412	5.5	346	3.9	COLORADO CO	-14	-29	-39		
7	4.9	13	5.4	11	2.4	ADAMS CO	11	-56	-51	18	25
1	2.1	2	5.1			ALAMOSA CO	139				
		2	16.8	1	6.5	ARCHULETA CO		-61			77
1	2.6			3	8.1	BACA CO			208	96	89
		2	3.8	1	2.4	BENT CO		-37			25
14	5.7	16	4.9	14	3.1	BOULDER CO	-15	-37	-46	24	34
2	3.7	6	11.8	3	6.1	CHAFFEE CO	216	-48	63	86	73
1	5.5					CHEYENNE CO					
		2	9.6	2	8.5	CLEAR CREEK CO		-11			91
1	2.5					CONEJOS CO					
		3	17.8			COSTILLA CO					
				2	8.2	CROWLEY CO					90
2	17.3			1	11.8	CUSTER CO			-32	40	98
6	5.9	10	7.6	4	3.2	DELTA CO	30	-58	-46	24	35
175	8.2	150	6.1	113	4.3	DENVER-ARAPAHOE#CO	-26	-29	-48	23	50
2	10.4	2	8.3	1	3.9	DOUGLAS CO	-20	-53	-63	10	45
1	5.2					EAGLE CO					
1	5.7	1	3.5	2	6.2	ELBERT CO	-38	75	8	70	73
40	10.5	29	5.6	30	4.1	EL PASO CO	-47	-27	-61	11	48
4	3.1	11	7.9	7	3.6	FREMONT CO	155	-54	16	74	41
1	1.6	5	6.9			GARFIELD CO	338				
1	14.9	1	31.5			GILPIN CO	111				
		1	7.7			GRAND CO					
1	6.1	2	8.0			GUNNISON CO	32				
4	7.3	3	6.0	2	4.2	HUERFANO CO	-18	-30	-43	28	49
		1	13.2			JACKSON CO					
23	8.4	36	7.3	36	4.6	JEFFERSON CO	-14	-37	-45	25	55
1	7.1					KIOWA CO					
3	7.7			2	4.3	KIT CARSON CO			-45	26	50
2	2.7	4	5.4	6	6.4	LA PLATA CO	104	18	141	94	76
11	4.4	20	6.0	10	2.4	LARIMER CO	37	-61	-46	24	24
6	5.2	11	8.4	9	7.9	LAS ANIMAS CO	61	-6	51	84	88
1	3.4			1	2.9	LINCOLN CO			-15	56	31
1	1.1	6	6.6	6	5.7	LOGAN CO	476	-15	392	99	67
15	6.9	13	4.9	12	3.8	MESA CO	-29	-22	-45	25	43
		2	7.1			MOFFAT CO					
1	2.1	2	4.1	3	5.8	MONTEZUMA CO	90	42	170	95	69
2	2.9	3	3.3	4	3.7	MONTROSE CO	13	15	30	79	42
6	6.0	3	3.0	5	4.8	MORGAN CO	-50	60	-20	51	57
8	7.8	4	3.4	7	6.7	OTERO CO	-56	95	-14	58	78
1	10.7			1	16.7	OURAY CO			56	85	99
		1	6.5			PARK CO					
2	6.2					PHILLIPS CO					
3	4.5	3	4.1			PROWERS CO	-9				
25	5.4	20	4.0	16	3.0	PUEBLO CO	-27	-24	-45	26	33
				1	6.7	RIO BLANCO CO					79
1	2.1	1	2.1	1	1.8	RIO GRANDE CO	1	-13	-12	59	20
1	2.0	2	5.9	1	2.7	ROUTT CO	199	-54	36	80	29
1	4.9	1	4.3			SAGUACHE CO	-12				
		1	4.7			SEDGWICK CO					
		1	6.2	1	4.4	TELLER CO		-29			52
2	6.3	1	2.7			WASHINGTON CO	-57				
18	5.9	12	3.4	26	6.8	WELD CO	-42	100	16	74	80
4	7.1	3	4.8	1	1.6	YUMA CO	-32	-67	-78	03	17
1177*	12.4	1063*	9.4	905*	7.0	CONNECTICUT CT	-24	-26	-44		
308*	13.2	271*	9.5	213	6.5	FAIRFIELD CT	-28	-32	-51	19	77
278*	11.4	262*	8.8	245*	7.0	HARTFORD CT	-23	-20	-38	32	82
76*	14.7	76*	12.2	49	6.6	LITCHFIELD CT	-17	-46	-55	15	78
31	8.0	40	8.8	38	7.1	MIDDLESEX CT	9	-19	-12	60	82
352*	13.5	292*	9.9	226*	6.8	NEW HAVEN CT	-27	-31	-50	20	79
80*	11.8	67	8.1	84*	9.3	NEW LONDON CT	-31	14	-22	50	94
21	9.9	18	6.7	23	7.2	TOLLAND CT	-32	7	-27	45	83
31	10.5	37	10.5	27	6.7	WINDHAM CT	0	-36	-36	35	79
122	9.1	118	7.5	126	6.8	DELAWARE DE	-18	-9	-25		
14	7.2	18	8.5	15	5.7	KENT DE	18	-33	-21	50	68
92	10.5	87	8.1	87	7.2	NEW CASTLE DE	-22	-11	-31	42	83
16	5.9	13	4.3	24	6.2	SUSSEX DE	-28	45	4	68	74

* SIGNIFICANTLY MORE DEATHS THAN EXPECTED

COUNTIES COMBINED; SEE TECHNICAL NOTES

WHITE MALE CANCER OF THE RECTUM ICD 154 EXCEPT 154.3

1950-1959 DEATHS	1950-1959 RATE	1960-1969 DEATHS	1960-1969 RATE	1970-1979 DEATHS	1970-1979 RATE	COUNTY AND STATE	PERCENT CHANGE 60S/50S	PERCENT CHANGE 70S/60S	PERCENT CHANGE 70S/50S	PERCENTILE %CHANGE RATES 70S/50S	PERCENTILE 1970S
234*	11.7	124	7.6	71	5.9	DIST. OF COLUMBIA DC	-35	-23	-50		
234*	11.7	124	7.6	71	5.9	WASHINGTON DC	-35	-23	-50	20	70
977	6.3	1598	5.5	2125	4.6	FLORIDA FL	-12	-17	-27		
6	5.1	13	6.4	9	3.3	ALACHUA FL	25	-49	-36	35	36
		2	8.0			BAKER FL					
3	2.1	9	5.3	6	2.3	BAY FL	152	-56	11	72	24
2	6.2			3	5.0	BRADFORD FL			-18	53	60
5	3.1	20	4.5	51	5.3	BREVARD FL	47	16	70	87	63
42	4.6	171	5.9	271	4.7	BROWARD FL	30	-21	3	68	56
1	3.5					CALHOUN FL					
2	4.7	13	5.2	32	6.1	CHARLOTTE FL	11	16	28	78	72
1	2.6	4	3.2	18	4.6	CITRUS FL	26	42	79	89	54
2	2.3	6	6.4	6	3.8	CLAY FL	177	-40	65	86	43
4	11.5	6	4.1	13	2.8	COLLIER FL	-64	-33	-76	04	30
6	11.2	3	4.2	5	5.5	COLUMBIA FL	-62	32	-50	19	66
245	8.2	337	6.4	382	5.1	DADE FL	-22	-21	-38	32	61
3	7.5	4	5.0	4	4.9	DE SOTO FL	-34	-2	-35	36	59
		1	6.7			DIXIE FL					
61	7.4	57	4.8	66	4.5	DUVAL FL	-36	-6	-39	31	53
13	4.7	21	5.0	24	3.9	ESCAMBIA FL	4	-22	-19	53	44
1	10.2	1	4.0	1	3.3	FLAGLER FL	-61	-16	-68	07	37
		1	2.1	1	2.3	FRANKLIN FL		9			24
4	4.1	3	2.9	1	1.0	GADSDEN FL	-30	-64	-75	05	14
		2	12.9			GILCHRIST FL					
1	9.8	1	14.9	1	5.7	GLADES FL	51	-62	-42	28	67
1	4.1	1	6.2	1	1.8	GULF FL	51	-70	-56	14	20
2	3.3	1	1.7			HARDEE FL	-50				
1	7.5			3	6.6	HENDRY FL			-12	59	78
6	13.4	5	5.2	16	6.8	HERNANDO FL	-61	33	-49	21	80
5	5.6	10	5.1	12	3.4	HIGHLANDS FL	-10	-33	-39	31	38
87	7.1	116	6.5	111	4.9	HILLSBOROUGH FL	-9	-25	-32	40	58
2	5.8	1	1.4	2	2.9	HOLMES FL	-76	108	-50	19	31
6	7.2	7	3.4	17	5.9	INDIAN RIVER FL	-52	71	-18	54	70
1	1.0	5	4.3	4	3.2	JACKSON FL	335	-27	219	97	34
1	5.4	2	8.9	1	4.6	JEFFERSON FL	66	-48	-13	58	55
		1	12.4			LAFAYETTE FL					
14	5.3	16	3.2	23	3.1	LAKE FL	-39	-5	-42	29	33
5	2.5	29	6.0	44	3.5	LEE FL	140	-41	42	82	39
4	4.1	5	3.2	10	4.4	LEON FL	-22	39	9	71	52
2	5.1	1	3.4	2	1.9	LEVY FL	-33	-43	-62	10	21
				1	5.2	LIBERTY FL					62
1	2.6			2	3.8	MADISON FL			42	82	42
25	7.6	32	4.9	52	4.8	MANATEE FL	-35	-3	-37	34	57
9	6.6	13	5.8	22	4.5	MARION FL	-11	-23	-32	40	53
2	2.1	4	2.4	16	5.4	MARTIN FL	12	129	156	95	65
4	3.6	12	6.0	18	7.8	MONROE FL	66	31	116	92	88
				1	1.9	NASSAU FL					21
1	1.3	3	2.5	5	1.9	OKALOOSA FL	87	-24	43	82	21
				3	6.9	OKEECHOBEE FL					81
42	5.9	60	5.2	66	4.2	ORANGE FL	-13	-18	-28	44	49
7	3.9	17	9.6	13	5.2	OSCEOLA FL	143	-46	31	79	62
50	6.4	85	5.1	135	4.3	PALM BEACH FL	-19	-17	-33	39	50
13	6.1	25	4.1	58	3.9	PASCO FL	-32	-5	-36	36	45
179	7.7	246	5.8	313	5.1	PINELLAS FL	-25	-12	-34	38	60
29	4.4	53	5.6	67	5.1	POLK FL	29	-9	17	74	61
2	2.3	6	4.9	9	4.6	PUTNAM FL	113	-5	101	91	55
3	4.9	4	3.1	5	2.1	ST JOHNS FL	-38	-30	-57	14	22
4	3.4	10	3.9	27	6.7	ST LUCIE FL	12	73	94	90	78
		6	6.8	3	1.8	SANTA ROSA FL		-74			19
17	4.8	55	6.5	67	4.5	SARASOTA FL	36	-31	-6	63	53
2	1.5	14	6.1	14	3.1	SEMINOLE FL	295	-50	99	91	33
		3	5.0	6	5.1	SUMTER FL		1			61
2	3.8	1	1.7	6	7.7	SUWANNEE FL	-56	358	101	91	87
1	2.4	1	2.2			TAYLOR FL	-8				
		3	14.4			UNION FL					
38	6.3	65	5.6	72	4.3	VOLUSIA FL	-11	-22	-31	41	51
2	12.0	1	5.0	2	4.1	WAKULLA FL	-59	-18	-66	08	47
2	3.0	2	2.6	2	1.9	WALTON FL	-14	-26	-36	35	21
3	7.2	2	4.6			WASHINGTON FL	-35				
335	3.8	415	3.8	463	3.4	GEORGIA GA	-1	-11	-11		
1	2.5			2	3.9	APPLING GA			58	85	45
1	5.3	1	7.0	1	3.8	ATKINSON GA	32	-45	-28	44	43
		1	12.8			BAKER GA					
3	2.8	1	.8	3	3.0	BALDWIN GA	-71	285	10	71	33
1	3.2	3	9.5	2	5.0	BANKS GA	194	-47	55	85	60
1	1.9			3	3.8	BARROW GA			95	90	43
2	2.3	7	7.6	5	3.9	BARTOW GA	229	-49	69	87	44
2	5.1					BEN HILL GA					
		1	2.2	2	3.2	BERRIEN GA		43			35
6	2.0	11	3.6	12	3.2	BIBB GA	81	-11	61	86	36
				1	2.9	BLECKLEY GA					32
1	4.6	1	3.7			BRANTLEY GA	-21				
2	5.1	4	9.2	6	11.7	BROOKS GA	79	28	128	93	98
				1	8.6	BRYAN GA					92
2	3.5	2	2.6			BULLOCH GA	-27				
		3	8.6	2	4.6	BURKE GA		-46			55
1	7.2	1	7.2			CALHOUN GA	0				
2	12.0	1	8.6			CAMDEN GA	-28				
1	3.1					CANDLER GA					
2	1.6	4	2.7	9	5.1	CARROLL GA	66	91	217	97	61
2	3.5			2	2.1	CATOOSA GA			-38	33	22
				1	4.3	CHARLTON GA					50
22	7.5	22	5.9	20	4.1	CHATHAM GA	-21	-31	-45	25	48
1	1.4	2	2.5	1	1.1	CHATTOOGA GA	78	-55	-21	51	15
		3	2.5	2	1.2	CHEROKEE GA		-51			15

* SIGNIFICANTLY MORE DEATHS THAN EXPECTED

WHITE MALE — CANCER OF THE RECTUM — ICD 154 EXCEPT 154.3

1950-1959 DEATHS	1950-1959 RATE	1960-1969 DEATHS	1960-1969 RATE	1970-1979 DEATHS	1970-1979 RATE	COUNTY AND STATE	PERCENT CHANGE 60S/50S	PERCENT CHANGE 70S/60S	PERCENT CHANGE 70S/50S	PERCENTILE %CHANGE 70S/50S	PERCENTILE RATES 1970S
4	5.0	2	1.9	6	4.4	CLARKE GA	-63	137	-12	59	52
		3	3.0	10	5.7	CLAYTON GA		86			67
1	8.9					CLINCH GA					
10	3.7	16	4.4	15	2.4	COBB GA	18	-46	-36	35	25
		1	1.7	3	3.4	COFFEE GA		101			38
		4	4.3	2	1.6	COLQUITT GA		-62			18
		2	7.2	2	2.9	COLUMBIA GA		-59			32
		1	2.4	2	6.1	COOK GA		159			73
3	4.3	1	1.3	3	2.9	COWETA GA	-69	117	-32	40	31
1	8.0					CRAWFORD GA					
3	8.2					CRISP GA					
				3	7.4	DADE GA					84
1	7.5					DAWSON GA					
4	6.4	2	3.1	1	1.8	DECATUR GA	-52	-41	-72	06	20
24	5.1	37	4.5	41	4.4	DE KALB GA	-11	-1	-12	59	52
1	2.0			1	1.5	DODGE GA			-26	46	17
		1	2.9			DOOLY GA					
2	2.6	2	1.8	6	4.4	DOUGHERTY GA	-30	148	73	88	52
3	6.0	2	2.4	3	3.0	DOUGLAS GA	-60	24	-50	19	32
				1	2.7	EARLY GA					29
				1	11.8	ECHOLS GA					98
1	3.6			1	1.7	EFFINGHAM GA			-51	18	19
		1	2.6	1	1.6	ELBERT GA		-36			18
2	3.9	2	3.0	2	2.8	EMANUEL GA	-22	-9	-29	43	29
				3	14.1	EVANS GA					99
4	7.2	1	1.6	5	6.1	FANNIN GA	-78	292	-15	56	73
1	3.8	1	2.3	1	2.2	FAYETTE GA	-41	-1	-41	29	23
16	7.6	15	7.1	11	3.9	FLOYD GA	-7	-46	-50	20	44
2	4.6	2	3.3	1	1.6	FORSYTH GA	-29	-51	-65	08	18
1	1.7	3	5.0	2	3.5	FRANKLIN GA	195	-31	104	91	39
63	5.3	70	5.0	64	4.5	FULTON GA	-6	-9	-15	56	53
1	2.1			3	6.6	GILMER GA			208	97	77
5	6.6	6	6.2	5	3.3	GLYNN GA	-6	-47	-50	19	36
1	1.3	3	4.1	1	.9	GORDON GA	225	-79	-31	42	14
2	4.3	4	7.3	1	1.3	GRADY GA	71	-82	-69	07	16
				1	2.8	GREENE GA					30
3	2.3	6	3.5	9	4.0	GWINNETT GA	53	13	73	88	46
1	1.7	1	1.2	2	2.4	HABERSHAM GA	-28	106	47	83	26
4	3.1	4	2.5	8	3.5	HALL GA	-19	41	14	73	39
1	1.8	2	3.5	2	2.9	HARALSON GA	95	-19	59	85	31
1	3.1	1	3.9	6*	11.9	HARRIS GA	26	206	287	98	98
2	4.2					HART GA					
1	5.1	1	3.9			HEARD GA	-24				
1	1.8	1	1.9	2	2.4	HENRY GA	2	26	28	78	25
		4	8.4	4	2.1	HOUSTON GA		-75			22
1	3.4	1	3.0			IRWIN GA	-10				
1	1.6	1	1.2	3	3.4	JACKSON GA	-21	173	116	92	38
1	5.3					JASPER GA					
				3	9.1	JEFF DAVIS GA					93
1	2.3	4	9.9			JEFFERSON GA	324				
		1	4.4			JENKINS GA					
		1	4.1			JONES GA					
2	7.1					LAMAR GA					
		1	7.3			LANIER GA					
3	3.6	2	1.7	4	3.9	LAURENS GA	-54	139	10	71	46
		1	4.4			LIBERTY GA					
		2	9.9			LINCOLN GA					
				1	9.6	LONG GA					95
2	2.1	4	3.7			LOWNDES GA	76				
3	14.0	2	6.8	4	10.0	LUMPKIN GA	-52	48	-29	43	96
1	4.5					MC DUFFIE GA					
		1	7.0	2	9.7	MC INTOSH GA		37			95
1	4.4	1	4.1			MACON GA	-6				
1	2.3	1	1.8			MADISON GA	-23				
1	7.7	1	7.3			MARION GA	-5				
3	6.4	4	7.4	3	5.3	MERIWETHER GA	16	-28	-16	55	64
1	5.2			2	9.9	MILLER GA			89	90	96
2	7.8					MITCHELL GA					
		1	3.3			MONROE GA					
1	5.7			1	4.0	MONTGOMERY GA			-28	44	47
		3	10.6			MORGAN GA					
		1	1.8	1	1.7	MURRAY GA		-8			19
7	2.5	15	5.3	14	3.6	MUSCOGEE#GA	111	-31	45	83	41
2	4.3	3	3.9	1	.7	NEWTON GA	-8	-83	-84	01	13
3	11.5			1	5.3	OGLETHORPE GA			-54	16	63
				3	4.1	PAULDING GA					47
		1	3.0			PEACH GA					
2	5.4			1	1.5	PICKENS GA			-72	06	17
3	10.1			1	2.3	PIERCE GA			-77	04	24
1	3.7					PIKE GA					
5	6.6	2	1.7	1	1.0	POLK GA	-74	-42	-85	01	14
		4*	21.0	1	3.2	PULASKI GA		-85			35
2	19.6			2	8.9	PUTNAM GA			-55	15	92
				3	5.4	RABUN GA					65
1	6.6					RANDOLPH GA					
10	3.8	19	5.9	23	6.8	RICHMOND GA	55	15	78	88	80
1	3.7	1	2.8	2	2.1	ROCKDALE GA	-23	-25	-43	28	22
1	10.5	2*	20.4			SCHLEY GA	95				
		2	8.1	1	2.3	SCREVEN GA		-72			24
10	14.2	6	5.4	4	2.7	SPALDING GA	-62	-49	-81	02	29
6	10.9	2	3.2	7	7.8	STEPHENS GA	-71	146	-29	43	87
				1	7.6	STEWART GA					86
1	1.8			3	4.3	SUMTER GA			133	93	50
				2	15.7	TALBOT GA					99
		1	9.6	1	14.1	TALIAFERRO GA		48			99
		1	3.2	3	4.9	TATTNALL GA		52			59
2	9.0	2	7.5			TAYLOR GA	-16				
1	5.7	2	4.9			TELFAIR GA	-14				
		1	3.8	1	3.4	TERRELL GA		-10			37
1	.8	6	5.9	3	2.5	THOMAS GA	605	-58	194	96	26
		3	4.6	1	1.2	TIFT GA		-73			15

* SIGNIFICANTLY MORE DEATHS THAN EXPECTED

COUNTIES COMBINED; SEE TECHNICAL NOTES

WHITE MALE — CANCER OF THE RECTUM — ICD 154 EXCEPT 154.3

1950-1959 DEATHS	1950-1959 RATE	1960-1969 DEATHS	1960-1969 RATE	1970-1979 DEATHS	1970-1979 RATE	COUNTY AND STATE	PERCENT CHANGE 60S/50S	PERCENT CHANGE 70S/60S	PERCENT CHANGE 70S/50S	PERCENTILE %CHANGE 70S/50S	PERCENTILE RATES 1970S
1	1.9	1	1.8			TOOMBS GA	-4				
		1	4.2	2	5.4	TOWNS GA		30			65
		1	6.5			TREUTLEN GA					
4	3.9	3	2.8	5	2.8	TROUP GA	-29	1	-28	44	31
		1	3.5	1	3.2	TURNER GA		-9			35
2	12.3					TWIGGS GA					
2	4.3	1	1.0	2	2.2	UPSON GA	-77	122	-50	20	23
2	1.2	8	4.8	8	3.4	WALKER GA	296	-30	179	96	38
2	3.0	3	4.5	5	4.7	WALTON GA	49	4	54	85	55
3	3.1	3	2.6	7	5.7	WARE GA	-15	118	85	89	68
		1	6.1			WARREN GA					
1	2.6	3	6.8			WASHINGTON GA	162				
2	5.3	1	2.6	4	6.1	WAYNE GA	-52	137	13	73	72
				1	12.8	WEBSTER GA					99
				1	4.8	WHEELER GA					57
1	4.6	1	3.1	2	5.6	WHITE GA	-33	82	21	76	67
3	2.8	7	4.1	9	4.5	WHITFIELD GA	48	8	61	86	53
2	8.2					WILCOX GA					
1	3.2	2	5.9			WILKES GA	86				
2	8.6			1	2.9	WILKINSON GA			-66	08	31
		2	6.1	1	1.8	WORTH GA		-71			19
		32	5.3	29	3.6	HAWAII HI		-33			
		5	8.6	4	3.2	HAWAII HI		-63			35
		27	5.6	22	3.8	HONOLULU HI		-33			42
				3	3.8	MAUI#HI					43
131	5.1	147	4.9	138	3.8	IDAHO ID	-4	-22	-26		
15	4.3	15	3.7	19	3.5	ADA ID	-14	-5	-18	54	39
4	2.8	10	6.1	4	1.9	BANNOCK ID	121	-69	-31	41	20
1	2.9			1	2.7	BEAR LAKE ID			-9	61	28
4	9.8	1	2.7	1	2.3	BENEWAH ID	-73	-13	-76	04	24
4	6.2	2	2.8	7	5.9	BINGHAM ID	-54	106	-5	63	69
1	5.9			1	2.6	BLAINE ID			-55	14	28
4	4.2	6	6.3	5	4.6	BONNER ID	49	-26	10	71	55
7	7.7	5	3.4	7	3.5	BONNEVILLE ID	-55	3	-54	15	40
		2	7.0			BOUNDARY ID					
1	15.7					BUTTE ID					
1	17.0	1	24.6	1	21.5	CAMAS ID	45	-13	27	78	99
12	4.9	18	5.8	20	5.7	CANYON ID	18	-2	16	74	68
1	5.3	1	3.5			CARIBOU ID	-35				
3	4.8	5	7.7	4	4.8	CASSIA ID	61	-37	1	67	58
2	3.9	2	6.0	1	2.8	CLEARWATER ID	56	-54	-28	44	30
		1	6.4	1	4.2	CUSTER ID		-35			49
		1	2.1	3	6.0	ELMORE ID		186			71
				1	2.1	FRANKLIN ID					22
3	10.5			1	2.3	FREMONT ID			-78	03	24
3	6.5	1	1.8	6	8.7	GEM ID	-72	381	35	80	92
2	4.0	2	4.0	3	4.4	GOODING ID	0	9	9	71	51
7	12.0	4	6.7	5	7.3	IDAHO ID	-44	9	-39	32	84
2	6.5	3	6.3	3	5.2	JEFFERSON ID	-4	-17	-20	51	62
3	7.4	5	12.3	4	6.1	JEROME ID	66	-51	-18	54	72
9	6.2	7	4.0	6	2.9	KOOTENAI ID	-35	-28	-53	16	31
6	6.6	8	8.2	3	2.9	LATAH ID	25	-65	-56	14	31
1	2.7	1	3.0	1	3.7	LEMHI ID	9	26	37	81	42
		2	8.2	1	7.4	LEWIS ID		-9			85
2	12.7					LINCOLN ID					
		1	2.6	2	4.4	MADISON ID		70			52
		3	5.6	3	3.9	MINIDOKA ID		-30			44
14	11.9	15	10.6	5	3.2	NEZ PERCE ID	-11	-70	-73	05	35
1	6.3	1	4.5	2	10.5	ONEIDA ID	-29	136	67	87	97
2	5.9	1	3.1	1	2.8	OWYHEE ID	-47	-9	-52	18	31
2	3.9	6	7.4	2	2.0	PAYETTE ID	90	-73	-48	22	21
2	14.5			1	4.2	POWER ID			-71	06	49
8	10.1	4	4.7	4	5.1	SHOSHONE ID	-54	10	-49	21	61
4	2.1	7	3.2	8	3.1	TWIN FALLS ID	53	-2	50	84	34
				1	3.5	VALLEY ID					39
		6	13.9			WASHINGTON ID					
4196*	10.7	3986*	9.3	3207*	7.3	ILLINOIS IL	-13	-22	-32		
38	9.9	44*	10.9	25	6.3	ADAMS IL	10	-42	-36	35	75
4	5.4	3	4.6	4	7.1	ALEXANDER IL	-14	52	32	79	82
6	6.4	8	9.2	9	9.7	BOND IL	44	5	51	84	95
7	7.2	11	11.4	6	5.4	BOONE IL	59	-53	-25	47	64
4	8.3	3	7.0	2	5.0	BROWN IL	-16	-28	-39	31	60
19	8.5	15	6.5	12	5.1	BUREAU IL	-24	-22	-40	30	61
2	5.0	3	6.6	6*	12.4	CALHOUN IL	32	89	149	94	98
6	5.4	5	4.3	15*	13.2	CARROLL IL	-21	210	145	94	99
		7	7.5	4	3.8	CASS IL		-48			44
29	9.3	27	7.1	26	6.4	CHAMPAIGN IL	-24	-10	-31	41	75
22	9.8	26*	12.0	15	7.2	CHRISTIAN IL	22	-40	-27	45	83
8	7.5	7	6.3	4	3.3	CLARK IL	-16	-47	-56	14	37
3	3.6	13*	13.8	5	4.8	CLAY IL	286	-65	35	80	58
7	5.9	9	6.6	18*	12.7	CLINTON IL	12	92	115	92	98
11	5.1	12	5.7	17	7.5	COLES IL	12	31	47	83	85
2345*	13.1	2064*	10.7	1509*	8.1	COOK IL	-18	-25	-38	33	89
7	5.8	6	4.6	4	2.9	CRAWFCRD IL	-20	-36	-49	21	32
5	8.1	3	5.2	2	3.6	CUMBERLAND IL	-36	-30	-55	14	40
16	7.5	23	10.3	25*	10.1	DE KALB IL	37	-2	35	80	96
9	9.7	6	5.8	5	5.1	DE WITT IL	-40	-12	-47	23	61
9	10.0	6	6.7	5	5.0	DOUGLAS IL	-33	-26	-50	19	59
67	8.7	110*	9.4	115	7.2	DU PAGE IL	8	-23	-17	55	83
14	10.1	6	4.7	9	7.0	EDGAR IL	-54	50	-31	41	81
1	1.7	4	7.5	3	4.9	EDWARDS IL	349	-35	190	96	58
11	8.9	12	9.8	9	6.1	EFFINGHAM IL	10	-38	-32	40	72
8	5.5	12	8.5	5	3.2	FAYETTE IL	55	-62	-41	29	36
4	5.2	10	11.3	8	8.5	FORD IL	119	-24	66	87	91
16	6.0	17	6.2	20	7.7	FRANKLIN IL	4	23	28	78	86

* SIGNIFICANTLY MORE DEATHS THAN EXPECTED

COUNTIES COMBINED; SEE TECHNICAL NOTES

WHITE MALE

CANCER OF THE RECTUM

ICD 154 EXCEPT 154.3

1950-1959 DEATHS	1950-1959 RATE	1960-1969 DEATHS	1960-1969 RATE	1970-1979 DEATHS	1970-1979 RATE	COUNTY AND STATE	PERCENT CHANGE 60S/50S	PERCENT CHANGE 70S/60S	PERCENT CHANGE 70S/50S	PERCENTILE %CHANGE 70S/50S	PERCENTILE RATES 1970S
23	8.8	18	7.2	9	3.6	FULTON IL	-18	-51	-59	11	40
1	1.9	3	7.3	4	8.8	GALLATIN IL	284	20	361	99	92
10	8.2	7	5.9	8	7.2	GREENE IL	-29	23	-13	58	83
5	5.2	10	9.0	11	9.5	GRUNDY IL	75	5	84	89	95
4	5.2	3	4.6	2	2.8	HAMILTON IL	-11	-39	-46	24	30
11	6.9	15	10.6	4	2.7	HANCOCK IL	53	-75	-61	10	29
		2	5.3	4	11.1	HARDIN IL		109			97
3	6.7	2	3.9	2	3.8	HENDERSON IL	-42	-2	-43	28	44
22	8.6	30*	10.7	17	5.9	HENRY IL	24	-45	-31	41	70
16	8.9	16	9.1	20*	10.5	IROQUOIS IL	3	15	18	75	97
18	9.5	14	7.2	7	3.5	JACKSON IL	-24	-52	-63	09	39
3	4.0	3	3.8	4	6.4	JASPER IL	-5	67	60	85	76
13	7.3	4	2.1	10	5.0	JEFFERSON IL	-70	132	-31	41	60
7	8.1	6	6.9	6	6.8	JERSEY IL	-16	0	-16	56	80
9	7.3	7	5.7	5	4.0	JO DAVIESS IL	-22	-30	-46	25	46
2	4.1	4	8.8	5	8.6	JOHNSON IL	116	-1	112	92	92
72	9.5	62	6.8	70	7.4	KANE IL	-29	10	-22	50	85
20	4.8	26	6.1	25	6.3	KANKAKEE IL	27	3	31	79	74
4	6.1	8	9.5	6	4.6	KENDALL IL	56	-52	-25	47	54
32	10.4	21	6.3	16	4.9	KNOX IL	-39	-23	-53	16	58
71	9.5	95	9.1	92*	7.5	LAKE IL	-5	-17	-21	51	86
59	11.7	48	8.6	39	6.6	LA SALLE IL	-27	-23	-43	27	78
8	7.1	5	4.6	5	4.4	LAWRENCE IL	-36	-4	-38	33	52
22	12.1	15	8.3	8	4.4	LEE IL	-31	-47	-63	09	52
16	7.7	14	6.2	11	5.1	LIVINGSTON IL	-19	-17	-33	38	61
8	5.1	8	4.7	9	5.4	LOGAN IL	-7	13	5	69	64
7	4.4	8	5.0	9	5.6	MC DONOUGH IL	13	12	26	77	66
37	12.4	35	8.5	44*	8.9	MC HENRY IL	-31	4	-28	44	93
44	11.0	42	10.4	31	7.4	MC LEAN IL	-5	-29	-33	39	84
36	8.2	45	8.9	42	7.8	MACON IL	9	-13	-5	64	87
23	8.0	21	6.2	19	6.6	MACOUPIN IL	-22	6	-17	54	78
85	10.3	88*	9.9	78*	7.4	MADISON IL	-4	-25	-28	45	85
16	7.4	15	6.5	14	5.7	MARION IL	-12	-13	-23	49	68
2	2.6	5	6.5	4	4.9	MARSHALL IL	156	-25	91	90	58
13	16.3	1	1.0	6	5.8	MASON IL	-94	498	-65	09	69
3	4.3	5	6.7	6	7.0	MASSAC IL	55	3	60	86	81
9	15.4	7	11.7	1	1.7	MENARD IL	-24	-86	-89	00	18
4	3.7	8	7.5	6	6.1	MERCER IL	104	-19	65	86	72
13*	17.7	17*	19.5	9	9.1	MONROE IL	10	-53	-49	22	93
16	7.9	15	7.2	16	7.8	MONTGOMERY IL	-9	9	-1	66	87
19	8.8	17	8.4	13	6.9	MORGAN IL	-5	-18	-22	50	80
8	8.1	9	9.2	6	7.6	MOULTRIE IL	13	-17	-7	62	86
16	9.1	13	6.6	9	4.4	OGLE IL	-28	-34	-52	18	51
64	8.5	75	9.3	63	7.5	PEORIA IL	10	-19	-12	59	85
9	7.8	8	6.5	9	7.3	PERRY IL	-17	13	-6	63	84
8	10.8	6	7.4	4	4.7	PIATT IL	-31	-36	-56	14	57
9	7.5	16	11.9	6	4.2	PIKE IL	58	-65	-45	26	48
2	7.7	2	6.7	1	4.0	POPE IL	-13	-41	-49	22	46
2	3.9	2	4.1	2	4.1	PULASKI IL	4	0	4	68	47
2	7.0	6*	22.0	3	10.0	PUTNAM IL	216	-54	44	83	96
11	6.4	11	6.2	10	5.6	RANDOLPH IL	-3	-10	-13	58	67
5	6.0	11	11.2	8	7.6	RICHLAND IL	89	-32	27	78	86
68	10.6	64	9.2	54	7.7	ROCK ISLAND IL	-13	-17	-28	44	86
91	11.3	94*	10.3	67	7.3	ST CLAIR IL	-9	-29	-36	36	84
12	6.4	15	7.8	9	5.3	SALINE IL	22	-32	-17	55	63
70	11.2	77*	11.0	55	7.6	SANGAMON IL	-1	-31	-32	39	86
6	9.9	1	1.5	4	6.5	SCHUYLER IL	-85	332	-35	37	76
8*	19.6	1	2.9	2	5.0	SCOTT IL	-85	74	-75	05	59
10	6.3	7	4.5	4	2.7	SHELBY IL	-30	-40	-58	13	29
4	8.5	1	2.2	2	4.5	STARK IL	-74	107	-47	23	53
21	9.2	14	6.0	24*	10.3	STEPHENSON IL	-35	73	12	72	96
31	9.7	30	7.5	23	4.7	TAZEWELL IL	-23	-37	-51	18	56
6	4.8	6	5.2	2	2.0	UNION IL	9	-61	-58	13	22
37	8.2	37	7.8	46*	9.7	VERMILION IL	-5	25	19	75	95
6	8.3	4	5.2	4	5.0	WABASH IL	-37	-4	-40	31	60
10	8.6	8	7.1	5	4.1	WARREN IL	-17	-42	-52	18	48
8	7.9	6	5.4	6	4.9	WASHINGTON IL	-31	-11	-38	33	58
4	3.8	8	6.6	6	4.7	WAYNE IL	72	-29	22	76	56
6	4.9	8	6.7	8	6.9	WHITE IL	36	2	39	81	80
25	10.5	14	5.2	17	5.8	WHITESIDE IL	-50	11	-45	25	68
72	11.8	69	8.8	67*	8.0	WILL IL	-25	-10	-32	39	88
12	4.5	26	9.3	25*	8.7	WILLIAMSON IL	108	-7	94	90	92
76	10.4	84*	9.9	65	6.8	WINNEBAGO IL	-6	-31	-35	37	80
13	11.2	6	4.7	7	4.7	WOODFORD IL	-58	0	-58	13	57
1450	8.0	1435	7.3	1226	5.8	INDIANA IN	-9	-20	-27		
10	10.4	10	8.4	12	9.4	ADAMS IN	-19	11	-10	61	94
64	8.0	72	8.1	74	7.4	ALLEN IN	1	-9	-8	62	84
14	8.3	11	5.8	5	2.4	BARTHOLOMEW IN	-30	-59	-72	06	25
5	9.0	6	10.7	3	4.8	BENTON IN	19	-55	-47	24	57
2	2.9	6	9.1	4	5.3	BLACKFORD IN	209	-41	82	89	64
14	10.6	12	8.3	7	4.7	BOONE IN	-22	-43	-56	14	56
		1	1.8	3	5.3	BROWN IN		187			63
9	10.2	4	4.0	8	8.4	CARROLL IN	-61	112	-17	54	91
12	5.0	15	7.0	18*	9.1	CASS IN	40	30	82	89	93
8	4.1	16	6.7	9	3.4	CLARK IN	61	-49	-17	54	38
13	8.8	16	10.4	13	9.4	CLAY IN	18	-9	7	70	94
6	4.0	11	6.9	5	2.9	CLINTON IN	73	-57	-26	46	32
4	6.1	4	7.5	4	7.1	CRAWFORD IN	23	-5	16	74	83
7	5.8	8	5.8	4	2.6	DAVIESS IN	0	-55	-55	15	27
3	2.8	12	8.9	11	7.8	DEARBORN IN	217	-13	177	96	87
14	12.6	8	7.7	9	7.7	DECATUR IN	-39	1	-39	32	87
9	6.9	7	5.1	13	9.3	DE KALB IN	-26	82	35	80	94
28	7.1	22	5.1	25	5.6	DELAWARE IN	-27	10	-20	51	67
5	4.7	7	5.7	13	9.0	DUBOIS IN	24	57	94	90	93
32	7.9	32	7.1	25	4.9	ELKHART IN	-10	-31	-37	34	59
14	13.5	11	10.2	5	5.0	FAYETTE IN	-25	-51	-63	09	60
19	9.9	15	6.7	15	6.1	FLOYD IN	-32	-9	-38	33	73
12	10.9	7	6.9	8	8.1	FOUNTAIN IN	-37	18	-25	47	89
9	11.3	4	5.0	5	5.9	FRANKLIN IN	-56	19	-48	23	70

* SIGNIFICANTLY MORE DEATHS THAN EXPECTED

WHITE MALE

CANCER OF THE RECTUM — ICD 154 EXCEPT 154.3

1950-1959 DEATHS	1950-1959 RATE	1960-1969 DEATHS	1960-1969 RATE	1970-1979 DEATHS	1970-1979 RATE	COUNTY AND STATE	PERCENT CHANGE 60S/50S	PERCENT CHANGE 70S/60S	PERCENT CHANGE 70S/50S	PERCENTILE %CHANGE 70S/50S	PERCENTILE RATES 1970S
3	3.2	5	5.3	7	7.7	FULTON IN	68	44	143	94	87
14	8.2	8	4.8	5	3.0	GIBSON IN	-42	-38	-64	09	32
14	4.3	21	6.3	21	6.3	GRANT IN	45	1	46	83	75
11	6.3	16	9.1	8	4.4	GREENE IN	45	-52	-30	43	52
8	5.3	15	8.2	11	4.6	HAMILTON IN	55	-44	-14	58	54
9	7.8	8	6.5	5	3.5	HANCOCK IN	-16	-46	-55	15	39
5	4.6	3	3.1	3	2.8	HARRISON IN	-32	-9	-38	33	30
10	7.0	7	4.5	5	2.6	HENDRICKS IN	-36	-41	-62	10	28
15	7.4	8	3.6	16	6.8	HENRY IN	-52	91	-8	62	80
19	8.1	20	7.2	17	6.1	HOWARD IN	-10	-15	-24	48	73
14	7.8	12	6.8	12	6.9	HUNTINGTON IN	-13	2	-11	60	81
14	9.5	8	5.7	14	8.6	JACKSON IN	-40	52	-9	61	91
5	5.8	5	5.3	3	3.3	JASPER IN	-8	-38	-42	28	37
6	4.7	7	5.7	8	6.5	JAY IN	22	14	38	81	76
11	8.7	9	7.2	9	7.0	JEFFERSON IN	-18	-3	-20	52	81
4	4.8	6	7.6	5	6.0	JENNINGS IN	58	-21	26	77	71
20*	14.8	8	4.4	6	3.0	JOHNSON IN	-70	-32	-80	02	32
29	12.4	18	7.9	11	5.1	KNOX IN	-36	-36	-59	12	60
11	5.9	6	2.8	9	3.9	KOSCIUSKO IN	-53	37	-35	36	44
7	8.3	4	4.7	4	4.2	LAGRANGE IN	-44	-10	-49	21	49
145*	10.8	151*	9.9	113	6.9	LAKE IN	-9	-30	-36	35	81
29	8.0	30	7.0	30	6.7	LA PORTE IN	-13	-5	-17	55	78
9	5.1	9	5.0	11	5.6	LAWRENCE IN	-2	10	8	70	66
23	5.3	30	6.1	32	5.6	MADISON IN	15	-8	6	69	67
201	9.3	209*	8.9	154	6.2	MARION IN	-4	-31	-34	37	73
10	6.3	3	1.8	5	2.7	MARSHALL IN	-72	50	-58	13	29
5	10.6	3	5.7	1	1.8	MARTIN IN	-46	-69	-83	01	19
10	5.8	5	3.1	8	5.0	MIAMI IN	-47	60	-15	57	59
14	7.4	16	8.0	12	4.9	MONROE IN	9	-39	-33	38	59
7	4.4	12	7.2	7	4.2	MONTGOMERY IN	64	-41	-3	65	49
12	9.6	9	6.1	6	3.5	MORGAN IN	-36	-43	-64	09	39
6	10.8	3	4.7	2	2.7	NEWTON IN	-57	-41	-75	05	29
17	11.6	8	5.5	10	6.5	NOBLE IN	-53	19	-44	26	76
		1	4.5			OHIO IN					
4	3.9	5	5.1	7	6.8	ORANGE IN	33	33	76	88	80
4	5.1	4	5.4	1	1.1	OWEN IN	6	-80	-79	03	15
6	5.7	7	7.6	6	6.1	PARKE IN	34	-20	7	70	72
3	3.6	8	9.0	4	4.1	PERRY IN	147	-55	12	72	47
4	5.5	5	6.7	5	6.1	PIKE IN	21	-8	11	72	73
13	6.1	18	7.4	18	5.9	PORTER IN	22	-20	-3	65	70
9	8.7	6	6.3	5	4.5	POSEY IN	-28	-29	-49	21	53
6	8.6	7	10.5	5	7.4	PULASKI IN	22	-29	-14	57	85
6	5.0	9	7.3	6	4.5	PUTNAM IN	45	-38	-10	61	53
8	4.9	14	9.1	8	5.4	RANDOLPH IN	84	-41	9	71	64
9	8.2	3	2.4	9	7.6	RIPLEY IN	-70	211	-8	62	86
6	5.8	10	8.9	6	5.6	RUSH IN	54	-37	-3	65	67
72	9.4	83	8.4	59	5.8	ST JOSEPH IN	-11	-31	-38	33	69
2	3.5	4	5.8	3	3.6	SCOTT IN	65	-37	3	68	41
7	4.5	16	10.1	12	7.2	SHELBY IN	126	-29	60	86	83
4	4.2	5	4.8	6	6.1	SPENCER IN	14	27	45	83	72
6	6.2	9	8.0	3	2.4	STARKE IN	29	-70	-62	10	25
10	10.9	3	3.0	5	4.5	STEUBEN IN	-73	49	-59	12	53
11	6.8	10	7.4	8	6.0	SULLIVAN IN	9	-19	-12	59	71
1	2.8	4	9.3	1	3.0	SWITZERLAND IN	232	-68	6	70	32
27	9.1	24	7.3	22	6.1	TIPPECANOE IN	-19	-16	-32	40	73
6	6.8	4	5.0	4	5.1	TIPTON IN	-26	3	-24	48	62
2	5.5	1	3.0	1	3.5	UNION IN	-45	15	-37	34	39
54	8.5	59	8.5	46	6.1	VANDERBURGH IN	0	-28	-28	44	72
9	8.0	14	12.2	9	8.0	VERMILLION IN	53	-35	0	66	88
56	10.2	38	7.1	37	7.1	VIGO IN	-30	-1	-31	42	82
7	4.6	6	4.0	5	3.0	WABASH IN	-14	-25	-36	35	33
3	6.3	5	10.4	3	6.5	WARREN IN	66	-37	4	68	77
6	5.4	7	5.9	7	5.2	WARRICK IN	10	-13	-4	64	62
4	4.3	6	5.9	3	2.6	WASHINGTON IN	37	-56	-40	30	27
33	10.7	22	6.4	21	6.0	WAYNE IN	-40	-7	-44	27	71
6	5.4	7	6.1	5	4.4	WELLS IN	14	-28	-18	54	52
7	7.0	5	4.7	9	7.4	WHITE IN	-33	58	5	69	84
9	9.0	7	6.6	9	8.5	WHITLEY IN	-26	28	-6	63	91
1044	7.7	944	6.6	829	5.6	IOWA IA	-14	-15	-27		
3	4.1	6	9.7	5	8.9	ADAIR IA	135	-9	114	92	92
5	9.7	4	7.5	4	8.4	ADAMS IA	-22	12	-13	58	90
11	11.0	7	6.9	1	1.0	ALLAMAKEE IA	-38	-85	-91	00	14
8	6.7	3	2.6	6	5.5	APPANOOSE IA	-61	107	-19	53	65
		6	9.9	3	5.4	AUDUBON IA		-46			64
7	5.5	8	6.1	13	9.7	BENTON IA	11	59	77	88	95
33	7.8	29	5.9	21	4.1	BLACK HAWK IA	-24	-31	-47	23	48
13	8.1	15	9.7	10	5.9	BOONE IA	20	-40	-27	45	70
11	10.8	7	6.4	9	7.3	BREMER IA	-40	14	-32	39	84
4	4.1	6	5.5	5	4.5	BUCHANAN IA	35	-19	9	71	53
4	3.2	8	6.6	6	4.5	BUENA VISTA IA	107	-32	41	82	54
6	6.4	4	3.5	2	1.6	BUTLER IA	-45	-56	-76	04	17
8	8.1	9	8.8	6	6.1	CALHOUN IA	8	-31	-25	47	72
8	7.2	5	4.2	8	6.7	CARROLL IA	-42	60	-7	62	78
10	8.9	9	7.6	10	8.8	CASS IA	-15	16	-2	66	92
7	7.4	6	5.5	4	3.9	CEDAR IA	-26	-29	-48	22	44
14	6.1	21	8.3	13	5.1	CERRO GORDO IA	36	-39	-17	54	60
11	9.9	4	3.9	5	5.7	CHEROKEE IA	-61	47	-42	28	68
6	7.3	4	4.6	3	3.4	CHICKASAW IA	-37	-24	-53	17	38
6	10.8	5	8.4	4	7.0	CLARKE IA	-22	-16	-35	36	81
9	9.5	6	6.0	2	1.8	CLAY IA	-37	-70	-81	02	20
17	12.9	9	6.8	8	5.3	CLAYTON IA	-48	-21	-59	12	63
25	9.4	18	6.5	17	6.1	CLINTON IA	-30	-7	-35	36	72
10	9.7	5	5.0	6	5.3	CRAWFORD IA	-48	7	-45	26	64
9	6.2	5	3.3	7	4.5	DALLAS IA	-46	34	-27	45	53
4	5.9	3	5.4	6	12.9	DAVIS IA	-8	137	119	92	99
8	9.6	1	1.0	5	7.5	DECATUR IA	-89	616	-22	49	85
8	9.2	15*	15.7	9	9.7	DELAWARE IA	70	-38	6	69	95
22	9.9	23	9.7	19	7.5	DES MOINES IA	-2	-22	-24	48	86
2	3.0	6	7.4	8	9.5	DICKINSON IA	149	30	224	97	95

* SIGNIFICANTLY MORE DEATHS THAN EXPECTED

WHITE MALE — CANCER OF THE RECTUM — ICD 154 EXCEPT 154.3

1950-1959 DEATHS	1950-1959 RATE	1960-1969 DEATHS	1960-1969 RATE	1970-1979 DEATHS	1970-1979 RATE	COUNTY AND STATE		PERCENT CHANGE 60S/50S	PERCENT CHANGE 70S/60S	PERCENT CHANGE 70S/50S	PERCENTILE %CHANGE 70S/50S	PERCENTILE RATES 1970S
38	12.0	21	6.3	25	7.1	DUBUQUE	IA	-48	13	-41	29	82
4	5.7	1	1.3	4	5.1	EMMET	IA	-77	282	-11	60	61
17	11.4	19*	12.1	14	9.2	FAYETTE	IA	6	-23	-19	52	94
9	7.6	9	7.5	10	8.0	FLOYD	IA	-1	7	6	69	89
12	15.1	6	7.0	7	7.6	FRANKLIN	IA	-54	9	-50	20	86
6	8.8	3	4.7	4	5.5	FREMONT	IA	-46	17	-37	34	66
7	8.6	5	6.0	3	3.5	GREENE	IA	-30	-40	-59	12	40
		2	2.6	5	5.9	GRUNDY	IA		131			70
4	4.2	5	5.5	4	5.5	GUTHRIE	IA	32	0	32	79	66
3	2.7	5	5.4	3	3.1	HAMILTON	IA	99	-43	14	73	34
4	6.1	4	5.6	8	9.2	HANCOCK	IA	-9	65	51	84	94
11	8.0	7	5.3	7	4.7	HARDIN	IA	-34	-11	-41	30	56
9	8.3	12	10.6	12*	11.3	HARRISON	IA	29	7	37	81	98
7	5.5	7	6.1	6	5.2	HENRY	IA	11	-15	-6	63	62
7	9.5	6	7.4	7	9.0	HOWARD	IA	-22	21	-5	63	93
4	6.0	7	9.9	1	1.2	HUMBOLDT	IA	64	-88	-80	02	15
2	4.2	3	4.4	2	3.9	IDA	IA	5	-12	-7	62	44
8	8.8	6	7.1	3	3.1	IOWA	IA	-20	-56	-64	09	34
9	7.9	6	4.8	6	5.0	JACKSON	IA	-39	4	-36	35	60
8	4.8	15	8.3	11	5.9	JASPER	IA	74	-29	24	76	71
8	8.7	4	4.6	4	3.6	JEFFERSON	IA	-48	-21	-59	12	41
21	11.9	11	5.4	8	3.8	JOHNSON	IA	-54	-30	-68	07	43
2	1.9	5	5.0	10	9.3	JONES	IA	164	87	393	99	94
7	6.4	5	4.7	5	4.8	KEOKUK	IA	-27	2	-25	47	57
7	5.7	9	6.8	6	4.7	KOSSUTH	IA	19	-32	-19	53	56
16	7.4	24	10.6	13	6.1	LEE	IA	42	-43	-19	53	72
48	9.2	47	8.0	41	6.4	LINN	IA	-13	-20	-31	42	76
7	10.9	7	10.6	2	3.0	LOUISA	IA	-3	-71	-72	06	33
8	10.3	5	5.8	3	3.6	LUCAS	IA	-44	-39	-65	08	40
3	5.5	3	4.1	8	9.1	LYON	IA	-26	124	65	87	93
4	5.0	5	5.6	4	4.2	MADISON	IA	13	-25	-15	57	49
13	9.2	14	9.6	4	3.0	MAHASKA	IA	5	-69	-67	08	33
6	4.0	7	4.2	7	3.9	MARION	IA	5	-8	-3	65	44
18	8.7	21	10.0	19	8.3	MARSHALL	IA	14	-17	-5	64	90
5	6.9	4	5.6	4	5.9	MILLS	IA	-19	6	-14	57	70
8	9.9	4	5.1	3	4.2	MITCHELL	IA	-48	-18	-57	13	49
7	7.6	4	4.4	4	4.7	MONONA	IA	-42	7	-38	33	56
3	4.5	6	9.9	6	8.7	MONROE	IA	121	-12	94	90	92
2	2.2	1	1.3	3	2.7	MONTGOMERY	IA	-42	106	20	75	29
11	6.1	12	6.4	13	6.7	MUSCATINE	IA	5	4	9	71	78
9	8.9	5	4.6	4	3.3	O'BRIEN	IA	-48	-29	-63	09	36
4	8.9			4	7.1	OSCEOLA	IA			-21	51	82
8	5.2	6	4.4	8	6.2	PAGE	IA	-16	42	20	75	74
7	10.2	4	4.6	5	6.9	PALO ALTO	IA	-55	49	-33	39	80
5	4.3	5	4.0	8	5.8	PLYMOUTH	IA	-8	46	35	80	69
4	6.6	7	10.1	3	3.3	POCAHONTAS	IA	53	-67	-49	20	37
89	9.0	66	6.1	65	5.7	POLK	IA	-32	-6	-37	34	68
17	4.7	23	6.0	16	4.1	POTTAWATTAMIE	IA	28	-32	-13	58	48
5	5.0	8	8.0	4	3.6	POWESHIEK	IA	60	-55	-27	45	41
4	7.3	1	1.9	4	7.3	RINGGOLD	IA	-75	294	0	67	84
9	8.9	6	6.5	2	1.7	SAC	IA	-27	-73	-81	02	19
47	10.2	47	9.5	49*	8.7	SCOTT	IA	-7	-8	-14	57	92
13*	15.5	6	6.4	1	.7	SHELBY	IA	-58	-89	-96	00	13
5	4.0	8	6.2	5	3.3	SIOUX	IA	55	-46	-17	55	37
13	6.8	8	4.1	8	3.6	STORY	IA	-40	-11	-46	24	41
15	12.2	7	5.5	8	5.6	TAMA	IA	-55	1	-55	15	66
6	7.3	1	1.1	7	10.8	TAYLOR	IA	-85	920	48	84	97
6	6.1	8	8.2	9	8.7	UNION	IA	33	7	42	82	92
8	13.0	5	6.8	4	6.8	VAN BUREN	IA	-48	0	-48	22	79
19	8.3	18	7.8	22*	9.0	WAPELLO	IA	-6	15	9	71	93
5	4.8	4	3.8	7	5.3	WARREN	IA	-22	41	10	71	64
8	6.9	14	11.9	2	1.7	WASHINGTON	IA	72	-86	-76	04	18
3	3.5	6	7.6	3	4.4	WAYNE	IA	117	-43	24	77	51
11	5.4	12	5.0	7	2.9	WEBSTER	IA	-7	-41	-45	25	32
6	8.5	7	9.6			WINNEBAGO	IA	13				
12	10.6	16*	14.2	3	2.4	WINNESHIEK	IA	33	-83	-77	03	26
34	6.5	32	6.1	29	5.7	WOODBURY	IA	-7	-6	-13	59	68
3	4.5	5	8.2	4	7.1	WORTH	IA	81	-13	58	85	82
7	7.2	5	4.4	4	3.4	WRIGHT	IA	-39	-23	-53	17	37
563	5.8	561	5.4	483	4.4	KANSAS	KS	-8	-18	-25		
5	4.1	3	2.6	4	3.9	ALLEN	KS	-36	51	-3	65	45
3	4.3	6	8.6	4	6.0	ANDERSON	KS	98	-30	38	81	71
7	6.3	4	3.7	4	4.1	ATCHISON	KS	-42	12	-35	37	48
2	4.1	2	3.4	1	1.9	BARBER	KS	-17	-44	-53	17	21
5	5.5	6	4.5	4	2.6	BARTON	KS	-18	-43	-53	17	27
6	5.3	5	4.4	6	4.8	BOURBON	KS	-18	10	-9	61	58
4	3.8	5	5.2	9	8.6	BROWN	KS	39	65	128	93	91
11	6.5	7	3.8	5	2.5	BUTLER	KS	-42	-35	-62	10	26
4	13.2			2	8.6	CHASE	KS			-34	37	91
4	7.6	1	1.8	1	2.9	CHAUTAUQUA	KS	-76	61	-61	10	32
11	8.6	10	7.3	9	6.8	CHEROKEE	KS	-16	-6	-21	50	80
1	2.6	1	4.0	1	3.1	CHEYENNE	KS	56	-23	20	75	34
2	11.2			2	9.0	CLARK	KS			-20	52	93
7	9.3	1	1.0	4	5.5	CLAY	KS	-89	446	-41	29	65
6	5.8	4	3.8	3	3.8	CLOUD	KS	-34	1	-34	38	44
6	7.2	1	2.0	2	2.4	COFFEY	KS	-72	19	-67	08	24
		1	5.3			COMANCHE	KS					
11	5.2	9	4.2	1	.4	COWLEY	KS	-19	-90	-92	00	13
26	10.0	19	7.5	14	5.7	CRAWFORD	KS	-25	-24	-43	27	68
2	5.1	2	6.0	1	1.8	DECATUR	KS	17	-69	-64	09	20
9	6.6	8	5.6	9	7.1	DICKINSON	KS	-15	26	8	70	82
4	6.5	3	6.0	4	5.7	DONIPHAN	KS	-7	-4	-11	60	68
10	6.7	6	3.5	5	2.6	DOUGLAS	KS	-48	-24	-61	11	28
1	3.3	4	13.2	4	10.8	EDWARDS	KS	305	-19	229	97	97
		2	5.1			ELK	KS					
5	7.8	3	3.3	5	5.8	ELLIS	KS	-58	76	-26	46	68
1	1.6	4	7.0	2	3.1	ELLSWORTH	KS	344	-56	96	90	34
6	10.3	4	5.8	6	7.8	FINNEY	KS	-44	36	-24	48	87
9	10.8	8	7.6	6	5.9	FORD	KS	-29	-23	-45	25	69

* SIGNIFICANTLY MORE DEATHS THAN EXPECTED

WHITE MALE

CANCER OF THE RECTUM — ICD 154 EXCEPT 154.3

COUNTY AND STATE	1950-1959 DEATHS	1950-1959 RATE	1960-1969 DEATHS	1960-1969 RATE	1970-1979 DEATHS	1970-1979 RATE	PERCENT CHANGE 60S/50S	PERCENT CHANGE 70S/60S	PERCENT CHANGE 70S/50S	PERCENTILE %CHANGE 70S/50S	PERCENTILE RATES 1970S
FRANKLIN KS	4	3.0	5	3.6	6	5.1	19	43	69	87	61
GEARY KS	8	11.0	6	8.7	4	5.5	-21	-37	-50	20	66
GOVE KS			1	4.6	2	7.5		61			85
GRAHAM KS	3	14.0	4	14.7			5				
GRANT KS					1	5.9					70
GRAY KS			1	4.6	1	3.6		-22			40
GREELEY KS			1	11.2	1	16.1		44			99
GREENWOOD KS	6	7.8	7	8.2	5	7.1	4	-12	-9	61	83
HAMILTON KS	1	7.5									
HARPER KS	2	3.1	4	6.2	1	2.1	97	-66	-33	39	22
HARVEY KS	8	6.9	7	5.3	7	4.4	-24	-16	-37	34	52
HASKELL KS			1	6.2	1	9.6		17			95
HODGEMAN KS	2	12.6									
JACKSON KS	2	2.3	4	6.8	1	1.3	194	-81	-44	27	16
JEFFERSON KS	5	6.5	4	4.4	1	1.2	-33	-72	-82	01	15
JEWELL KS	2	2.6	4	6.7	2	3.3	152	-50	25	77	37
JOHNSON KS	13	4.3	24	4.8	33	4.3	10	-9	0	67	51
KEARNY KS	2	14.9									
KINGMAN KS	3	5.8	3	4.9	2	3.1	-16	-36	-47	23	34
KIOWA KS	1	3.6	1	3.0			-18				
LABETTE KS	9	5.2	9	5.4	12	7.4	3	37	41	82	84
LANE KS	1	11.2	2	11.8			5				
LEAVENWORTH KS	16	7.7	10	4.2	10	4.5	-45	5	-42	29	52
LINCOLN KS	1	2.0	2	4.4	1	2.6	125	-40	34	80	28
LINN KS	1	1.3	2	2.8	2	2.4	118	-16	83	89	25
LOGAN KS	1	4.0			2	11.5			190	96	98
LYON KS	11	7.2	11	7.1	4	2.8	-1	-60	-61	11	30
MC PHERSON KS	13	10.4	10	6.7	7	4.3	-35	-36	-58	12	50
MARION KS	8	8.7	5	4.5	12*	11.4	-48	151	31	79	98
MARSHALL KS	3	2.5	7	5.7	5	4.2	129	-26	69	87	49
MEADE KS	4	17.2	5*	17.1			0				
MIAMI KS	7	5.8	6	5.1	3	2.6	-12	-49	-55	15	28
MITCHELL KS	1	2.4	4	8.3	3	4.0	244	-52	65	87	46
MONTGOMERY KS	11	4.4	17	6.7	10	4.0	51	-40	-9	61	46
MORRIS KS					2	2.5					27
NEMAHA KS	3	3.7	3	3.0	2	2.6	-20	-14	-31	41	27
NEOSHO KS	10	8.1	7	5.5	5	3.9	-32	-29	-52	18	45
NESS KS	2	5.6	2	6.0	1	3.0	8	-51	-47	24	32
NORTON KS	1	1.9	1	1.9	1	1.4	-2	-25	-27	46	16
OSAGE KS	3	3.3	6	5.9	7	6.7	79	14	105	91	79
OSBORNE KS	3	5.8	1	2.4	2	4.7	-59	96	-19	53	56
OTTAWA KS	5	9.3	3	5.2	2	3.5	-44	-34	-63	10	39
PAWNEE KS			5	8.1	2	3.6		-56			40
PHILLIPS KS	1	1.8	2	2.6	2	3.0	43	17	67	87	33
POTTAWATOMIE KS	4	4.9	2	2.2	6	6.4	-55	189	30	79	75
PRATT KS	5	8.1	4	5.7	5	7.5	-29	32	-7	62	86
RENO KS	9	3.2	17	5.8	17	5.4	80	-6	68	87	65
REPUBLIC KS	4	5.3	2	2.3	5	6.4	-56	176	20	75	75
RICE KS	3	3.5	5	6.6			87				
RILEY KS	4	3.7	13	10.6	8	6.6	186	-38	78	88	78
ROOKS KS	3	6.3	2	3.9			-38				
RUSH KS	1	2.6	1	2.1	3	7.3	-20	250	179	96	84
RUSSELL KS	5	11.2	5	9.1	2	3.0	-19	-67	-73	05	32
SALINE KS	9	5.1	14	8.3	10	5.1	62	-39	-1	66	61
SCOTT KS	1	5.5	2	7.3	2	6.3	32	-14	14	73	74
SEDGWICK KS	58	6.5	41	4.0	58	4.8	-39	21	-26	46	57
SEWARD KS	2	9.0	3	5.9	3	4.6	-34	-22	-48	22	55
SHAWNEE KS	22	4.4	39	6.9	24	3.8	58	-45	-13	58	43
SHERIDAN KS	1	5.4	2	8.6	3	12.9	58	51	138	93	99
SHERMAN KS			3	7.6	2	5.1		-34			61
SMITH KS			7	8.7	1	1.2		-86			15
STAFFORD KS	4	8.7	2	4.4	2	3.5	-49	-21	-60	11	39
STANTON KS	1	11.8									
STEVENS KS					1	5.5					65
SUMNER KS	8	5.6	4	2.7	4	2.7	-53	3	-52	18	29
THOMAS KS	2	5.4	1	3.0	3	7.3	-45	144	34	80	83
TREGO KS	1	4.7	2	6.7			44				
WABAUNSEE KS	2	4.0	3	5.9	3	4.8	49	-19	21	75	57
WALLACE KS	1	10.6	1	8.0			-25				
WASHINGTON KS	10	13.2	4	5.2	3	3.1	-60	-40	-76	04	34
WICHITA KS					2	12.7					98
WILSON KS	7	7.9	1	.9	2	3.0	-89	236	-62	10	32
WOODSON KS	1	1.8	1	1.8	1	2.7	-2	54	51	84	29
WYANDOTTE KS	50	8.7	54	8.6	30	4.7	-1	-45	-46	25	57
KENTUCKY KY	577	4.9	664	5.1	634	4.4	4	-15	-11		
ADAIR KY	3	4.0	6	7.7	2	2.1	91	-73	-48	22	22
ALLEN KY	1	1.4	1	1.3	5	5.9	-5	355	333	98	69
ANDERSON KY	3	7.8	1	2.5	1	1.6	-68	-37	-80	02	18
BALLARD KY	5	12.9	1	1.8	4	7.0	-86	284	-46	25	81
BARREN KY	2	1.4	3	2.1	6	3.6	49	66	147	94	40
BATH KY	1	2.2	3	5.7	1	1.8	154	-69	-21	50	19
BELL KY	7	5.0	4	3.1	1	.6	-37	-80	-87	00	13
BOONE KY	3	3.6	5	5.1	3	3.5	39	-30	-3	65	39
BOURBON KY	6	7.8	5	6.8	6	7.1	-13	4	-10	61	82
BOYD KY	18	9.8	16	7.5	19	7.6	-23	1	-22	49	86
BOYLE KY	5	5.6	3	3.7	4	3.9	-35	8	-29	43	46
BRACKEN KY	2	4.6	4	9.4	2	4.1	107	-56	-9	61	48
BREATHITT KY			2	2.7	1	1.3		-54			16
BRECKINRIDGE KY	2	3.2	1	1.0	5	5.3	-69	443	68	87	63
BULLITT KY	2	3.6	7	9.8	2	1.5	171	-84	-58	13	17
BUTLER KY	3	5.1	3	4.7	1	1.3	-7	-73	-74	05	16
CALDWELL KY	4	6.0	2	2.4	5	5.6	-61	137	-7	62	66
CALLOWAY KY	8	7.7	6	4.8	5	2.9	-37	-41	-63	10	31
CAMPBELL KY	40	11.1	32	8.1	32*	8.3	-27	2	-25	47	90
CARLISLE KY			2	4.8							
CARROLL KY	1	3.5	2	4.8	4	8.7	37	79	145	94	92
CARTER KY	3	3.1	5	5.5	1	1.4	80	-75	-55	15	16
CASEY KY	1	1.5	1	1.3	1	1.2	-12	-11	-22	49	15
CHRISTIAN KY	5	3.1	8	4.8	9	5.5	55	16	80	89	66

* SIGNIFICANTLY MORE DEATHS THAN EXPECTED

WHITE MALE

CANCER OF THE RECTUM

ICD 154 EXCEPT 154.3

1950-1959 DEATHS	1950-1959 RATE	1960-1969 DEATHS	1960-1969 RATE	1970-1979 DEATHS	1970-1979 RATE	COUNTY AND STATE	PERCENT CHANGE 60S/50S	PERCENT CHANGE 70S/60S	PERCENT CHANGE 70S/50S	PERCENTILE %CHANGE 70S/50S	PERCENTILE RATES 1970S
8	9.7	4	4.3	7	7.0	CLARK KY	-56	62	-28	44	81
		2	2.7	1	1.4	CLAY KY		-47			17
3	7.2	1	2.0	1	1.8	CLINTON KY	-72	-13	-76	04	19
		1	2.1	1	1.6	CRITTENDEN KY		-21			18
				2	4.5	CUMBERLAND KY					53
7	3.2	7	2.4	8	2.5	DAVIESS KY	-24	4	-21	51	26
				1	2.2	EDMONSON KY					23
		2	6.3			ELLIOTT KY					
2	3.1	2	2.9	2	2.4	ESTILL KY	-7	-17	-23	49	25
24	6.1	14	3.1	28	4.9	FAYETTE KY	-49	59	-19	53	59
2	3.2	2	3.0	1	1.6	FLEMING KY	-6	-45	-48	22	18
5	3.2	2	1.4	7	3.9	FLOYD KY	-58	188	22	76	45
4	4.3	6	4.6	7	5.0	FRANKLIN KY	7	10	18	75	60
2	3.5			2	3.8	FULTON KY			8	70	44
1	3.5	2	10.1			GALLATIN KY	188				
1	2.4	3	6.2	4	7.9	GARRARD KY	157	28	230	97	88
5	8.6	5	8.3	4	5.7	GRANT KY	-3	-31	-34	38	68
11	6.9	7	3.9	4	2.0	GRAVES KY	-43	-49	-71	06	21
2	2.2	6	5.9	3	2.8	GRAYSON KY	168	-53	26	77	30
2	4.1	4	7.7	2	3.2	GREEN KY	89	-59	-22	49	35
4	4.7	2	1.6	11	7.5	GREENUP KY	-65	361	60	85	85
3	9.3			1	3.5	HANCOCK KY			-62	10	39
6	3.9	9	6.1	6	2.5	HARDIN KY	54	-58	-35	36	27
7	5.2	6	3.5	8	4.7	HARLAN KY	-33	35	-9	61	56
4	4.8	6	7.4			HARRISON KY	53				
		3	3.8	3	3.7	HART KY		-2			42
6	4.7	15	10.5	8	5.1	HENDERSON KY	123	-52	7	70	61
4	6.8	4	6.3			HENRY KY	-8				
		2	4.7	2	5.5	HICKMAN KY		17			65
8	4.7	7	4.0	13	6.5	HOPKINS KY	-15	64	39	81	77
		1	2.1	2	3.1	JACKSON KY		48			33
137	7.6	150	7.3	125	5.5	JEFFERSON KY	-4	-25	-28	44	65
3	5.9	1	1.8	2	2.9	JESSAMINE KY	-70	62	-51	19	31
5	5.3	2	1.8	1	.9	JOHNSON KY	-65	-54	-84	01	14
49	10.6	69*	14.3	44*	8.0	KENTON KY	35	-44	-25	47	88
2	3.1	1	1.5	1	1.5	KNOTT KY	-53	3	-51	19	17
1	1.1	4	3.6	4	2.9	KNOX KY	237	-19	173	96	31
2	4.2	3	5.7	3	5.7	LARUE KY	35	0	35	80	68
4	3.8	1	.7	4	2.4	LAUREL KY	-80	227	-36	35	25
1	1.6	2	3.0	5	6.9	LAWRENCE KY	82	131	320	98	81
1	3.2	2	5.8	2	4.1	LEE KY	79	-30	26	77	47
1	2.2					LESLIE KY					
1	.7	2	1.6	6	5.3	LETCHER KY	120	225	616	99	63
5	8.4	3	4.7	4	6.1	LEWIS KY	-45	30	-28	44	72
3	3.6	3	3.5	7	7.6	LINCOLN KY	-1	117	114	92	86
		4	10.1	1	2.0	LIVINGSTON KY		-80			21
5	5.5	2	1.7	6	4.6	LOGAN KY	-69	168	-18	54	54
1	3.1	1	3.2	1	3.2	LYON KY	3	1	4	68	36
4	1.7	11	4.7	14	5.2	MC CRACKEN KY	185	10	214	97	62
2	3.7	1	1.6	1	1.5	MC CREARY KY	-57	-6	-60	11	17
		4	7.0	2	3.1	MC LEAN KY		-55			34
5	4.0	3	2.3	9	5.5	MADISON KY	-44	143	36	80	66
				1	1.6	MAGOFFIN KY					18
4	8.4	1	1.7	5	7.4	MARION KY	-79	330	-11	60	85
1	1.5	10	11.0	4	3.0	MARSHALL KY	617	-73	96	90	33
5	6.1	4	4.7	3	3.3	MASON KY	-23	-30	-46	25	37
4	7.9	3	7.2	2	4.6	MEADE KY	-9	-36	-42	28	55
1	4.6	1	4.8	1	3.5	MENIFEE KY	4	-26	-23	48	39
3	4.2	4	5.3	2	2.4	MERCER KY	26	-54	-42	28	25
1	1.9	5	11.0	4	7.9	METCALFE KY	489	-28	327	98	88
2	3.0					MONROE KY					
1	1.4	1	1.7	4	5.7	MONTGOMERY KY	16	242	296	98	68
2	3.7					MORGAN KY					
7	4.7	9	6.1	9	5.5	MUHLENBERG KY	30	-10	17	74	66
2	2.9	2	2.3	1	1.2	NELSON KY	-20	-48	-59	12	15
3	7.5	2	4.1	5	12.3	NICHOLAS KY	-45	197	64	86	98
4	3.3	7	5.9	8	6.0	OHIO KY	76	2	81	89	71
		2	3.9			OLDHAM KY					
3	7.2	2	3.9	4	8.0	OWEN KY	-45	104	12	72	89
		1	2.6			OWSLEY KY					
5	8.3	5	9.3	1	1.9	PENDLETON KY	12	-80	-78	03	20
1	.7	5	4.1	3	2.3	PERRY KY	498	-45	226	97	23
3	1.2	12	4.8	6	1.8	PIKE KY	300	-63	49	84	20
1	3.4	2	6.4	1	2.3	POWELL KY	89	-65	-34	38	23
6	3.7	14	7.6	6	2.8	PULASKI KY	108	-64	-24	48	30
		1	5.6			ROBERTSON KY					
3	5.3	4	7.0	1	1.2	ROCKCASTLE KY	32	-84	-78	03	15
3	6.1	1	1.8	4	5.8	ROWAN KY	-71	228	-5	64	69
2	3.6	2	4.3	2	2.8	RUSSELL KY	18	-36	-24	47	29
3	5.1	4	5.5	2	2.4	SCOTT KY	8	-57	-54	16	25
8	9.2	6	6.0	1	1.0	SHELBY KY	-35	-84	-90	00	14
1	2.4			3	4.4	SIMPSON KY			85	89	52
1	3.0	1	3.7			SPENCER KY	25				
				4	4.6	TAYLOR KY					54
2	3.7	3	5.4	3	5.3	TODD KY	46	-2	43	83	63
1	2.0	3	7.3	3	5.3	TRIGG KY	263	-27	166	95	64
1	4.5	2	7.9	1	4.3	TRIMBLE KY	74	-46	-6	63	50
3	4.5	10*	14.4	4	5.8	UNION KY	216	-60	27	78	69
3	1.8	8	3.9	9	3.9	WARREN KY	116	-2	112	92	44
2	3.7	1	1.8	3	6.1	WASHINGTON KY	-50	230	65	86	73
		2	2.4	4	4.8	WAYNE KY		98			57
		5	6.4	4	4.7	WEBSTER KY		-27			56
4	3.9	1	.7	7	5.2	WHITLEY KY	-82	660	33	79	62
				2	4.9	WOLFE KY					58
3	8.3	1	1.7	6	10.9	WOODFORD KY	-79	532	31	79	97
424	6.2	479	5.5	501	4.8	LOUISIANA LA	-12	-13	-23		
7	5.8	12	7.9	7	3.8	ACADIA LA	36	-52	-34	37	44
		1	1.3	1	1.4	ALLEN LA		7			16
2	2.4	4	5.2	5	4.2	ASCENSION LA	114	-18	76	88	50

* SIGNIFICANTLY MORE DEATHS THAN EXPECTED

WHITE MALE

CANCER OF THE RECTUM

ICD 154 EXCEPT 154.3

1950-1959 DEATHS	1950-1959 RATE	1960-1969 DEATHS	1960-1969 RATE	1970-1979 DEATHS	1970-1979 RATE	COUNTY AND STATE	PERCENT CHANGE 60S/50S	PERCENT CHANGE 70S/60S	PERCENT CHANGE 70S/50S	PERCENTILE %CHANGE 70S/50S	PERCENTILE RATES 1970S
2	4.5	3	7.8	4	8.4	ASSUMPTION LA	71	8	85	89	90
4	3.4	7	5.3	8	5.0	AVOYELLES LA	57	-6	47	83	60
1	1.6	1	1.2	6	6.3	BEAUREGARD LA	-23	408	293	98	75
3	5.7			4	6.8	BIENVILLE LA			20	75	80
2	3.3	5	4.6	7	4.8	BOSSIER LA	40	4	45	83	58
25	5.8	22	3.9	21	3.0	CADDO LA	-33	-22	-48	23	33
13	5.7	22	7.1	15	3.4	CALCASIEU LA	25	-52	-40	30	37
1	2.3	2	4.6	4	8.6	CALDWELL LA	97	90	274	98	92
2	11.3					CAMERON LA					
				3	7.1	CATAHOULA LA					82
1	1.8	2	3.0	1	1.4	CLAIBORNE LA	69	-54	-23	49	16
		1	1.5	1	1.7	CONCORDIA LA		10			19
				2	2.9	DE SOTO LA					32
16	5.7	21	3.9	28	4.7	EAST BATON ROUGE LA	-33	21	-19	53	55
				2	3.2	EAST FELICIANA LA					35
5	6.7	2	2.2	2	1.6	EVANGELINE LA	-68	-24	-76	04	18
2	3.8	2	2.5	2	2.3	FRANKLIN LA	-36	-7	-40	30	24
2	2.8	3	4.2	1	1.0	GRANT LA	51	-76	-63	09	14
3	3.3	6	4.5	7	5.4	IBERIA LA	36	22	65	87	65
1	1.5	2	2.7			IBERVILLE LA	83				
		3	4.9	3	4.0	JACKSON LA		-18			46
23	8.4	33	5.5	64	7.4	JEFFERSON LA	-35	35	-12	59	85
3	3.8	4	4.5	7	6.4	JEFFERSON DAVIS LA	17	43	66	87	75
6	4.8	9	4.7	9	2.9	LAFAYETTE LA	-1	-39	-39	31	31
10	9.2	3	1.9	9	4.3	LAFOURCHE LA	-79	126	-53	17	51
1	1.8	2	2.7	1	1.7	LA SALLE LA	50	-37	-5	63	19
1	1.5	4	5.3	5	5.7	LINCOLN LA	246	9	277	98	68
1	2.6	4	5.0	1	.9	LIVINGSTON LA	94	-82	-65	08	14
1	3.4					MADISON LA					
1	1.5			3	3.2	MOREHOUSE LA			110	92	36
4	4.7			5	4.1	NATCHITOCHES LA			-13	58	48
185*	12.0	153*	9.2	121*	8.0	ORLEANS LA	-23	-13	-34	38	88
5	2.0	18	6.4	13	3.7	OUACHITA LA	220	-42	85	89	42
5	13.4	4	5.8	4	7.9	PLAQUEMINES LA	-57	36	-41	29	88
3	8.7			2	3.5	POINTE COUPEE LA			-59	11	39
11	4.4	12	4.0	18	4.9	RAPIDES LA	-9	22	11	72	59
		1	4.4	1	2.5	RED RIVER LA		-43			26
1	2.2	2	2.9			RICHLAND LA	29				
		2	2.1	4	3.5	SABINE LA		63			38
		9	12.7	13	9.0	ST BERNARD LA		-29			93
2	6.2	4	7.3	3	4.0	ST CHARLES LA	17	-45	-35	36	47
		3	14.9			ST HELENA LA					
1	2.2	2	5.3	2	6.2	ST JAMES LA	144	17	184	96	73
2	9.6	2	5.0	1	1.4	ST JOHN THE BAPT LA	-47	-72	-85	00	16
5	3.9	11	6.4	10	4.3	ST LANDRY LA	62	-33	9	71	50
5	9.6	5	5.8	6	7.0	ST MARTIN LA	-40	20	-27	45	81
7	8.6	9	9.0	6	5.2	ST MARY LA	5	-42	-39	31	62
6	5.6	10	6.3	9	3.7	ST TAMMANY LA	12	-41	-34	38	42
4	2.7	7	4.4	5	2.2	TANGIPAHOA LA	67	-51	-18	54	23
2	10.7	2	10.1			TENSAS LA	-6				
7	6.5	11	7.0	12	6.9	TERREBONNE LA	7	-1	6	69	81
7	10.6	6	8.3	2	2.3	UNION LA	-22	-72	-78	03	24
6	4.4	8	4.9	12	6.7	VERMILION LA	12	37	53	84	79
1	1.1	4	4.3	4	3.4	VERNON LA	304	-20	221	97	38
3	3.0	2	1.4	2	1.2	WASHINGTON LA	-54	-11	-59	12	16
7	7.5	7	5.5	11	7.2	WEBSTER LA	-27	31	-4	64	83
1	4.2					WEST BATON ROUGE LA					
3	6.8	2	3.9	1	1.3	WEST CARROLL LA	-43	-67	-81	02	16
		1	8.2			WEST FELICIANA LA					
2	4.1	2	3.3	1	1.2	WINN LA	-21	-62	-70	07	15
500*	10.9	432*	9.1	413*	8.1	MAINE ME	-16	-11	-26		
44	11.6	50*	12.0	40*	8.8	ANDROSCOGGIN ME	4	-27	-24	48	92
38	11.3	34	9.2	36*	9.6	AROOSTOOK ME	-18	4	-15	56	95
76	8.9	76	8.7	81*	8.5	CUMBERLAND ME	-3	-2	-5	64	91
21*	19.2	10	9.4	13*	11.3	FRANKLIN ME	-51	20	-41	29	98
16	8.2	25*	12.2	16	7.1	HANCOCK ME	49	-41	-12	59	83
60*	14.4	35	7.7	37	7.5	KENNEBEC ME	-47	-2	-48	23	86
18	10.6	12	7.1	11	5.9	KNOX ME	-33	-17	-44	26	70
16	13.2	7	5.6	10	6.9	LINCOLN ME	-58	24	-47	23	81
22	9.2	27*	11.3	18	7.2	OXFORD ME	24	-36	-21	50	83
37	7.3	42	8.0	42	7.8	PENOBSCOT ME	10	-3	7	70	87
12	10.5	3	2.6	10	10.8	PISCATAQUIS ME	-75	310	2	67	97
11	10.1	11	9.2	11	9.0	SAGADAHOC ME	-8	-3	-11	60	93
30*	13.7	26*	12.4	12	5.8	SOMERSET ME	-10	-53	-58	13	69
12	9.2	14	11.2	10	8.3	WALDO ME	22	-26	-10	60	90
23	12.0	14	7.4	12	5.9	WASHINGTON ME	-39	-21	-51	18	69
64*	14.0	46	9.1	54*	9.2	YORK ME	-35	2	-34	38	94
753	9.3	814	7.9	793*	6.3	MARYLAND MD	-15	-20	-32		
26	7.0	34	8.4	30	6.9	ALLEGANY MD	20	-19	-2	66	80
30	8.9	54	8.5	53	6.3	ANNE ARUNDEL MD	-5	-25	-29	43	75
106	9.8	132	8.1	159	6.6	BALTIMORE MD	-17	-18	-32	39	78
2	5.7	3	5.6	2	2.6	CALVERT MD	-3	-54	-55	15	28
4	4.5	5	6.3	2	1.8	CAROLINE MD	41	-71	-59	12	20
23	9.7	22	7.9	24	7.1	CARROLL MD	-19	-10	-27	45	82
17	11.3	14	8.4	11	5.3	CECIL MD	-26	-37	-54	16	63
2	4.1	3	3.1	2	1.2	CHARLES MD	-24	-60	-70	07	15
8	6.4	8	6.4	13	9.4	DORCHESTER MD	2	45	48	83	94
22	8.3	31	10.5	24	7.2	FREDERICK MD	28	-32	-13	58	83
2	2.2	6	5.4	7	5.9	GARRETT MD	146	9	168	95	69
17	8.7	13	5.1	23	7.1	HARFORD MD	-41	39	-18	53	82
2	1.9	10	8.8	13	5.3	HOWARD MD	369	-39	185	96	64
5	8.1	3	4.2	2	2.4	KENT MD	-48	-43	-71	06	25
48	7.6	71	6.9	73	4.4	MONTGOMERY MD	-10	-36	-43	27	51
49	9.4	73	7.9	75	6.6	PRINCE GEORGE'S MD	-16	-17	-30	42	77
2	3.0	4	5.6	5	5.6	QUEEN ANNE'S MD	89	0	88	90	66
3	5.9	5	7.1	7	6.0	ST MARY'S MD	19	-16	0	67	71
4	5.1	3	4.3	3	3.9	SOMERSET MD	-16	-10	-24	48	44

* SIGNIFICANTLY MORE DEATHS THAN EXPECTED

WHITE MALE — CANCER OF THE RECTUM — ICD 154 EXCEPT 154.3

1950-1959 DEATHS	1950-1959 RATE	1960-1969 DEATHS	1960-1969 RATE	1970-1979 DEATHS	1970-1979 RATE	COUNTY AND STATE	PERCENT CHANGE 60S/50S	PERCENT CHANGE 70S/60S	PERCENT CHANGE 70S/50S	PERCENTILE %CHANGE 70S/50S	PERCENTILE RATES 1970S
5	5.7	3	3.1	6	4.9	TALBOT MD	-45	56	-14	57	58
29	8.8	40	9.4	35	7.2	WASHINGTON MD	6	-23	-18	54	83
7	4.6	6	3.8	17	7.6	WICOMICO MD	-19	102	63	86	86
10	12.3	8	8.8	6	5.0	WORCESTER MD	-28	-43	-59	11	60
330*	11.7	263*	9.2	201*	7.9	BALTIMORE CITY MD	-21	-14	-32	40	88
2641*	12.0	2214*	9.2	2006*	7.9	MASSACHUSETTS MA	-23	-15	-35		
36	12.8	33	7.0	47	5.8	BARNSTABLE MA	-45	-17	-55	15	69
70	11.2	83*	11.8	67*	8.9	BERKSHIRE MA	5	-25	-21	51	92
233*	13.2	188*	9.7	189*	8.8	BRISTOL MA	-26	-9	-33	39	92
3	8.6	2	5.3	4	9.3	DUKES MA	-38	75	8	70	94
330*	12.4	267*	9.2	253*	8.4	ESSEX MA	-26	-9	-33	39	90
27	9.6	26	8.4	35*	10.6	FRANKLIN MA	-12	26	11	72	97
192*	10.9	192*	10.1	174*	8.5	HAMPDEN MA	-7	-16	-22	50	91
37	9.1	37	8.0	32	6.1	HAMPSHIRE MA	-12	-24	-33	39	72
600*	12.6	484*	9.1	410*	7.3	MIDDLESEX MA	-28	-20	-43	28	83
		4*	21.6	1	4.7	NANTUCKET MA		-78			56
211*	11.5	194	8.6	196*	7.5	NORFOLK MA	-25	-13	-35	37	85
113	10.6	121*	9.9	117*	8.0	PLYMOUTH MA	-7	-19	-24	48	89
501*	13.2	351*	10.0	244*	8.2	SUFFOLK MA	-24	-18	-38	33	89
288*	10.8	232	7.9	237*	7.7	WORCESTER MA	-27	-3	-29	43	87
2662*	9.9	2498*	8.3	1879	5.7	MICHIGAN MI	-17	-31	-43		
2	5.5	4	7.5	4	5.1	ALCONA MI	36	-32	-7	62	61
3	5.3	5	9.6	2	3.9	ALGER MI	81	-60	-27	45	44
18	6.9	23	8.3	12	3.8	ALLEGAN MI	22	-55	-45	25	42
8	6.7	9	7.4	9	6.4	ALPENA MI	10	-13	-5	64	76
4	5.6	1	1.2	6	6.4	ANTRIM MI	-78	413	15	73	76
6	12.2	7	12.1	3	3.6	ARENAC MI	-1	-71	-71	06	40
4	9.5			2	4.0	BARAGA MI			-57	13	46
7	4.9	13	8.0	15	7.7	BARRY MI	63	-4	56	85	87
37	9.7	38	9.4	32	6.9	BAY MI	-4	-27	-29	43	80
4	9.7	4	9.0	1	1.6	BENZIE MI	-7	-82	-84	01	18
50	9.9	38	6.3	33	5.1	BERRIEN MI	-36	-20	-49	21	60
15	8.5	19	11.3	15	8.1	BRANCH MI	32	-28	-5	64	89
44	8.7	36	6.6	31	5.3	CALHOUN MI	-23	-21	-39	32	63
15	10.0	14	8.4	6	3.0	CASS MI	-16	-65	-70	07	32
6	7.0	6	7.3	5	5.6	CHARLEVOIX MI	4	-23	-20	52	67
3	4.0	7	9.6	7	7.2	CHEBOYGAN MI	138	-25	80	89	83
9	6.7	13	9.8	7	5.3	CHIPPEWA MI	45	-46	-21	50	63
2	3.0	4	4.1	4	4.2	CLARE MI	34	3	38	81	49
19	12.3	15	9.5	6	3.5	CLINTON MI	-23	-63	-71	06	40
2	9.4	1	3.5	3	5.8	CRAWFORD MI	-63	65	-39	32	69
13	7.7	13	7.1	11	5.3	DELTA MI	-8	-25	-30	42	64
12	8.6	14	8.1	14	8.3	DICKINSON MI	-5	2	-3	65	90
12	5.9	13	5.6	15	5.6	EATON MI	-6	0	-6	63	66
8	9.1	3	3.6	9	8.5	EMMET MI	-60	133	-7	62	91
93	9.1	93	8.0	47	3.6	GENESEE MI	-12	-55	-60	11	41
		4	6.3	4	4.3	GLADWIN MI		-31			51
7	4.4	19*	12.9	16*	9.9	GOGEBIC MI	195	-23	126	93	96
12	7.0	10	5.4	7	3.7	GRAND TRAVERSE MI	-22	-32	-47	23	42
10	5.7	10	5.8	7	3.7	GRATIOT MI	2	-36	-34	37	42
9	5.1	18	9.8	14	6.9	HILLSDALE MI	90	-29	35	80	81
27	11.6	17	7.7	18	8.6	HOUGHTON MI	-34	11	-27	46	91
17	9.2	16	8.6	22*	10.3	HURON MI	-6	20	13	72	96
79*	11.7	54	7.6	32	4.0	INGHAM MI	-35	-47	-65	08	47
8	4.2	11	5.5	7	3.4	IONIA MI	32	-38	-18	54	38
10	15.3	3	3.1	4	3.9	IOSCO MI	-80	24	-75	05	44
10	8.0	10	9.4	7	5.5	IRON MI	17	-42	-32	40	65
7	5.8	7	5.7	6	3.8	ISABELLA MI	-3	-33	-34	37	43
39	8.0	40	7.7	18	3.3	JACKSON MI	-4	-57	-59	12	36
55	10.0	36	5.5	33	4.9	KALAMAZOO MI	-45	-10	-51	19	59
3	11.0	6*	18.1	4	9.4	KALKASKA MI	65	-48	-14	57	94
133	10.0	118	7.8	98	5.9	KENT MI	-22	-24	-41	30	70
2	9.7	2	12.2	2	9.2	KEWEENAW MI	26	-25	-5	63	94
4	12.3	1	2.7	3	6.1	LAKE MI	-78	124	-51	19	72
18	10.2	9	4.9	12	5.6	LAPEER MI	-51	14	-45	26	67
		4	7.1	3	3.9	LEELANAU MI		-45			45
28	9.0	31	9.5	21	6.1	LENAWEE MI	5	-36	-32	39	72
10	6.0	14	7.1	13	4.9	LIVINGSTON MI	17	-31	-19	52	58
1	2.5	2	4.3	1	2.4	LUCE MI	72	-45	-6	63	25
1	2.5	7	13.5	2	3.5	MACKINAC MI	442	-74	41	82	39
89*	11.6	120*	9.9	107	5.4	MACOMB MI	-15	-46	-54	16	64
18*	18.1	6	5.6	8	6.3	MANISTEE MI	-69	14	-65	08	75
24	10.7	16	6.1	13	4.7	MARQUETTE MI	-43	-23	-56	14	56
9	7.3	12	9.9	7	4.5	MASON MI	36	-54	-38	33	54
9	9.3	9	7.9	7	5.6	MECOSTA MI	-15	-30	-40	31	66
6	4.3	14	10.1	10	6.4	MENOMINEE MI	135	-36	49	84	76
8	6.2	17	11.4	6	2.8	MIDLAND MI	83	-76	-55	15	30
1	2.5			5	11.1	MISSAUKEE MI			341	98	97
30	9.1	29	7.7	38*	8.4	MONROE MI	-15	10	-7	62	90
15	8.1	14	7.2	13	5.8	MONTCALM MI	-11	-20	-29	43	68
		5	15.4	2	3.7	MONTMORENCY MI		-76			42
36	7.4	32	5.7	33	5.5	MUSKEGON MI	-24	-3	-26	46	66
8	7.1	7	5.3	5	3.5	NEWAYGO MI	-25	-35	-51	19	39
146	9.4	179	7.5	205	6.1	OAKLAND MI	-20	-19	-35	36	72
5	5.4	5	5.1	7	7.0	OCEANA MI	-4	37	31	79	81
5	7.7	6	8.7	6	6.5	OGEMAW MI	13	-25	-15	56	77
3	5.3			2	3.6	ONTONAGON MI			-33	39	40
4	4.7	3	3.4	3	2.8	OSCEOLA MI	-28	-16	-40	31	30
3	12.9	1	3.1	3	5.7	OSCODA MI	-76	83	-55	14	68
3	7.9	5	10.5	3	5.2	OTSEGO MI	34	-50	-34	38	62
32	9.1	27	6.7	27	5.3	OTTAWA MI	-26	-22	-42	28	63
5	8.0	4	7.3	3	3.2	PRESQUE ISLE MI	-9	-56	-60	11	35
6	13.3	1	1.3	8	7.3	ROSCOMMON MI	-91	483	-45	26	84
72	11.6	55	8.1	45	6.2	SAGINAW MI	-31	-23	-47	23	73
40	9.0	43	8.6	30	5.4	ST CLAIR MI	-3	-38	-40	31	64
17	8.1	27*	12.7	17	7.3	ST JOSEPH MI	57	-42	-10	61	84
11	6.4	7	3.9	17	7.9	SANILAC MI	-39	104	24	77	88

* SIGNIFICANTLY MORE DEATHS THAN EXPECTED

WHITE MALE

CANCER OF THE RECTUM — ICD 154 EXCEPT 154.3

1950-1959 DEATHS	1950-1959 RATE	1960-1969 DEATHS	1960-1969 RATE	1970-1979 DEATHS	1970-1979 RATE	COUNTY AND STATE	PERCENT CHANGE 60S/50S	PERCENT CHANGE 70S/60S	PERCENT CHANGE 70S/50S	PERCENTILE %CHANGE 70S/50S	PERCENTILE RATES 1970S
4	8.7	6	11.5	2	3.6	SCHOOLCRAFT MI	31	-68	-59	12	41
16	6.9	16	6.8	12	4.5	SHIAWASSEE MI	-2	-33	-35	37	53
13	6.4	12	5.6	9	4.1	TUSCOLA MI	-12	-28	-36	35	47
17	6.7	12	4.7	12	4.3	VAN BUREN MI	-30	-7	-35	36	51
33	7.1	31	5.6	29	4.5	WASHTENAW MI	-21	-19	-36	35	54
1078*	12.3	930*	10.1	530*	6.3	WAYNE MI	-18	-38	-49	21	75
10	10.1	5	4.7	2	1.7	WEXFOPD MI	-53	-64	-83	01	18
1291	8.6	1347*	8.0	1029	5.6	MINNESOTA MN	-7	-30	-36		
3	4.1	10	9.5	9	9.9	AITKIN MN	132	4	142	94	96
11	7.3	14	6.8	10	2.7	ANOKA MN	-7	-61	-63	09	29
12	8.7	13	9.3	7	4.4	BECKER MN	7	-53	-50	20	52
8	6.8	6	5.1	12	8.9	BELTRAMI MN	-26	76	31	79	93
8	10.7	9	10.6	6	6.2	BENTON MN	-1	-42	-42	28	73
12*	22.2	6	10.8	5	8.0	BIG STONE MN	-52	-26	-64	09	88
18	8.8	23	10.7	23*	10.2	BLUE EARTH MN	22	-5	15	73	96
21*	15.3	17	11.1	19*	11.5	BPOWN MN	-28	4	-25	47	98
8	6.4	10	7.1	5	3.2	CARLTON MN	11	-54	-49	21	36
9	9.4	11	9.6	10	7.4	CARVER MN	2	-23	-21	50	85
11	11.0	11	8.7	8	5.1	CASS MN	-21	-42	-54	15	61
10	12.8	5	5.4	3	3.0	CHIPPEWA MN	-58	-44	-77	04	32
3	3.1	11	11.3	3	2.1	CHISAGO MN	261	-81	-33	38	22
13	9.5	8	5.2	8	4.2	CLAY MN	-46	-20	-56	14	48
3	5.3	1	1.8	1	2.1	CLEARWATER MN	-67	18	-61	11	22
		2	8.8	3	12.2	COOK MN		39			98
1	1.2	1	1.1	3	3.4	COTTONWOOD MN	-6	201	184	96	37
5	3.0	7	3.6	9	3.9	CROW WING MN	20	8	30	78	45
20	7.9	29	10.4	19	5.1	DAKOTA MN	30	-51	-36	35	61
7	9.9	4	5.8	6	7.2	DODGE MN	-42	25	-28	45	83
10	8.1	10	7.5	9	5.1	DOUGLAS MN	-7	-33	-37	34	60
15	11.7	16	12.5	4	3.1	FAPIBAULT MN	7	-75	-74	05	34
10	6.6	10	6.3	6	3.9	FILLMORE MN	-6	-37	-41	30	45
18	10.5	17	9.0	10	4.8	FREEBOPN MN	-14	-46	-54	16	58
18	9.4	12	5.9	6	2.7	GOODHUE MN	-37	-55	-72	06	29
2	4.5	1	1.6	2	3.3	GRANT MN	-64	100	-27	45	36
279	8.9	277	7.9	201	5.4	HENNEPIN MN	-11	-31	-39	32	64
3	3.6	5	5.6	2	1.8	HOUSTON MN	54	-68	-50	20	20
4	5.7	7	9.6	4	4.2	HUBBARD MN	67	-56	-26	46	49
5	6.3	1	1.0	4	3.9	ISANTI MN	-84	300	-38	34	46
20	11.7	14	7.4	18	8.4	ITASCA MN	-37	14	-28	44	90
4	5.2	6	6.7	5	4.9	JACKSON MN	30	-28	-6	63	58
5	8.9	6	9.1	4	6.1	KANABEC MN	2	-32	-31	41	73
7	3.9	9	4.8	11	5.2	KANDIYOHI MN	23	7	32	79	62
5	9.1	5	9.1	2	4.0	KITTSON MN	0	-56	-56	14	46
5	6.2	8	8.8	5	5.4	KOOCHICHING MN	42	-38	-12	59	65
9	11.6	6	6.5	3	3.9	LAC QUI PARLE MN	-44	-40	-66	08	45
3	5.5	4	7.3	3	5.5	LAKE MN	32	-24	0	67	66
3	10.1	2	6.5			LAKE OF THE WOODS MN	-35				
8	7.2	5	4.3	11	8.0	LE SUEUR MN	-40	83	10	72	88
5	9.0	3	5.2	7	12.4	LINCOLN MN	-42	137	37	81	98
11	10.3	9	7.4	15*	11.1	LYON MN	-28	51	8	70	97
10	8.4	15	10.0	13	7.6	MC LEOD MN	20	-24	-9	61	86
5	14.5	1	3.3	1	4.1	MAHNOMEN MN	-77	25	-72	06	48
3	3.7	5	5.5	3	3.0	MARSHALL MN	48	-45	-18	53	33
11	8.2	8	5.6	8	4.8	MARTIN MN	-31	-14	-41	29	58
6	6.0	10	8.8	11	8.6	MEEKER MN	48	-3	44	83	91
7	7.5	6	6.3	5	3.6	MILLE LACS MN	-15	-44	-52	17	40
3	2.5	23*	15.2	7	4.0	MORRISON MN	518	-74	62	86	46
9	5.2	10	4.4	16	6.4	MOWER MN	-16	45	22	76	76
1	3.1	4	6.0	3	3.9	MUPPAY MN	95	-35	26	78	45
9	7.7	7	6.2	10	8.9	NICOLLET MN	-20	43	15	73	92
8	8.1	6	4.7	13*	9.5	NOBLES MN	-42	100	17	74	95
3	4.4	9	11.0	7	6.6	NORMAN MN	151	-40	49	84	77
14	6.2	20	7.4	10	3.3	OLMSTED MN	18	-55	-47	24	37
20	6.7	21	6.2	19	4.6	OTTER TAIL MN	-7	-26	-31	41	54
5	6.8	6	8.0	3	3.8	PENNINGTON MN	17	-52	-44	27	43
8	7.8	8	6.3	8	6.6	PINE MN	-19	5	-15	56	78
8	11.0	3	3.5	7	8.1	PIPESTONE MN	-68	130	-26	46	89
12	6.2	14	6.4	10	4.7	POLK MN	3	-27	-25	47	56
6	9.7	10	12.8	3	2.9	POPE MN	33	-78	-70	06	31
185*	12.4	164*	9.4	100	5.3	RAMSEY MN	-24	-44	-57	13	63
2	6.5	2	7.2	1	2.3	RED LAKE MN	12	-68	-64	09	24
9	8.5	15*	12.6	8	5.8	REDWOOD MN	48	-54	-32	40	69
10	7.8	9	6.3	8	5.2	RENVILLE MN	-20	-18	-34	38	62
12	6.9	15	8.0	13	6.7	RICE MN	16	-17	-4	64	78
2	3.2	3	4.5	1	1.5	ROCK MN	38	-67	-55	15	17
4	5.7	5	6.4	7	7.7	ROSEAU MN	14	19	36	80	86
112	9.9	124*	10.2	65	7.2	ST LOUIS MN	3	-30	-28	44	83
11	13.2	10	10.5	10	7.6	SCOTT MN	-21	-28	-43	28	86
5	7.6	8	10.5	4	4.9	SHERBUPNE MN	37	-53	-36	35	59
7	7.6	6	6.1	7	6.9	SIBLEY MN	-20	13	-9	61	81
37	12.0	32	8.8	30	7.0	STEAPNS MN	-27	-20	-42	29	81
9	8.9	11	8.7	8	5.6	STEELE MN	-3	-35	-37	34	67
4	7.3	7	11.2	4	6.2	STEVENS MN	53	-45	-16	56	73
5	7.7	6	7.3	6	5.9	SWIFT MN	-6	-18	-23	48	70
6	4.2	9	6.3	4	2.5	TODD MN	51	-61	-41	30	26
5	12.1	2	4.9	3	5.5	TRAVERSE MN	-59	12	-55	15	66
8	7.7	9	8.3	7	5.6	WABASHA MN	8	-33	-27	45	67
7	11.2	6	7.8	5	5.2	WADENA MN	-30	-34	-54	16	62
8	8.8	7	7.6	6	6.7	WASECA MN	-14	-11	-23	48	79
14	8.2	19	8.5	11	4.2	WASHINGTON MN	3	-50	-49	21	49
9	11.2	7	7.8	10*	12.6	WATONWAN MN	-31	62	12	72	98
7	15.5	4	7.5	2	3.8	WILKIN MN	-51	-49	-75	04	43
19	9.6	27*	12.4	15	6.3	WINONA MN	30	-49	-34	38	75
9	5.3	12	6.2	10	4.2	WRIGHT MN	16	-32	-21	51	49
5	5.3	11	10.6	6	4.7	YELLOW MEDICINE MN	101	-56	-12	59	56
182	3.8	206	3.5	209	3.1	MISSISSIPPI MS	-7	-13	-19		
5	10.0	8	13.6	4	4.1	ADAMS MS	35	-70	-59	11	47

* SIGNIFICANTLY MORE DEATHS THAN EXPECTED

WHITE MALE — CANCER OF THE RECTUM — ICD 154 EXCEPT 154.3

1950-1959 DEATHS	1950-1959 RATE	1960-1969 DEATHS	1960-1969 RATE	1970-1979 DEATHS	1970-1979 RATE	COUNTY AND STATE		PERCENT CHANGE 60S/50S	PERCENT CHANGE 70S/60S	PERCENT CHANGE 70S/50S	PERCENTILE %CHANGE 70S/50S	PERCENTILE RATES 1970S
1	1.0	1	.8	6	4.8	ALCORN	MS	-20	512	386	99	57
2	4.9					AMITE	MS					
1	2.9	2	2.7	3	3.5	ATTALA	MS	-5	27	21	75	38
2	9.5	1	3.2			BENTON	MS	-67				
3	4.5	5	6.1	3	3.6	BOLIVAR	MS	36	-41	-20	52	40
1	1.8	5	7.6			CALHOUN	MS	324				
1	3.6	1	3.4			CARROLL	MS	-4				
5	10.2			2	3.6	CHICKASAW	MS			-65	08	40
		2	5.4	1	2.8	CHOCTAW	MS		-49			30
1	6.0					CLAIBORNE	MS					
		1	1.7	1	2.1	CLARKE	MS		26			22
		4	11.0	2	3.8	CLAY	MS		-65			44
2	3.8	2	3.3	3	4.5	COAHOMA	MS	-15	38	17	74	53
2	5.2	5	6.3	2	2.6	COPIAH	MS	23	-58	-49	21	28
2	6.6	2	3.6	3	5.6	COVINGTON	MS	-45	57	-14	57	67
3	7.8	2	4.9	1	.9	DE SOTO	MS	-37	-81	-88	00	14
1	.8	6	3.9	4	2.2	FORREST	MS	364	-43	164	95	23
1	2.4	6	14.5	1	1.7	GEORGE	MS	499	-88	-28	44	19
2	5.9	1	3.3	1	2.5	GREENE	MS	-45	-24	-58	12	26
		1	2.4	1	1.5	GRENADA	MS		-40			17
3	5.3	2	3.7			HANCOCK	MS	-29				
18	7.3	11	3.4	22	5.4	HARRISON	MS	-54	59	-27	45	64
14	5.5	14	3.3	12	2.5	HINDS	MS	-39	-24	-54	15	26
3	7.6	1	2.2	2	4.5	HOLMES	MS	-71	106	-41	30	53
		1	3.4	2	6.6	HUMPHREYS	MS		92			78
3	4.4	2	2.6	1	1.0	ITAWAMBA	MS	-41	-62	-78	03	14
2	1.9	3	2.2	12	4.6	JACKSON	MS	12	112	138	94	55
2	4.0			3	6.9	JASPER	MS			75	88	81
1	18.6	2	11.7			JEFFERSON	MS	-37				
2	11.5	2	6.5	2	6.1	JEFFERSON DAVIS	MS	-43	-6	-47	23	72
3	1.9	5	2.5	6	3.1	JONES	MS	30	24	62	86	34
1	3.2			2	6.3	KEMPER	MS			96	90	74
1	1.7	2	3.3	2	2.8	LAFAYETTE	MS	88	-15	60	86	30
2	3.7	1	2.7	3	4.2	LAMAR	MS	-27	59	16	73	49
4	2.8	9	4.3	9	4.2	LAUDERDALE	MS	57	-3	53	84	49
		2	6.2			LAWRENCE	MS					
2	3.6	3	5.0	2	2.4	LEAKE	MS	38	-52	-34	38	25
1	1.0	8	5.3	3	1.9	LEE	MS	460	-64	104	91	21
3	4.0	5	6.7	4	4.6	LEFLORE	MS	69	-31	16	74	55
4	5.1	4	4.1	1	.9	LINCOLN	MS	-19	-78	-82	01	14
2	2.0	4	4.4	9	7.7	LOWNDES	MS	116	77	282	98	87
1	2.4	2	4.2	2	2.9	MADISON	MS	78	-30	24	77	32
1	2.4	2	3.1			MARION	MS	29				
1	2.9	2	5.0	1	1.9	MARSHALL	MS	75	-62	-34	38	20
4	4.5	5	4.6	4	3.4	MONROE	MS	3	-27	-25	47	37
2	4.8	1	3.1	2	4.1	MONTGOMERY	MS	-36	34	-14	58	48
1	1.6	2	2.1	2	2.2	NESHOBA	MS	37	5	43	82	23
				4	4.5	NEWTON	MS					54
1	3.7	1	4.2			NOXUBEE	MS	15				
2	3.3	2	6.0	3	3.9	OKTIBBEHA	MS	81	-35	17	74	45
4	6.3	2	2.9	2	2.4	PANOLA	MS	-54	-16	-61	10	26
3	5.0	3	3.5	4	3.4	PEARL RIVER	MS	-30	-2	-31	42	38
2	2.2	4	3.8	4	3.8	PIKE	MS	69	0	70	87	43
2	3.3	3	3.5	1	.9	PONTOTOC	MS	6	-76	-74	05	14
1	1.4			6	6.1	PRENTISS	MS			354	99	73
4	15.9			1	3.5	QUITMAN	MS			-78	03	39
		6	5.3	6	4.5	RANKIN	MS		-15			53
3	5.1			1	1.2	SCOTT	MS			-76	04	15
3	28.0					SHARKEY	MS					
1	1.4	2	3.0	1	1.1	SIMPSON	MS	108	-62	-21	50	15
3	5.5					SMITH	MS					
2	10.0	1	3.3			STONE	MS	-67				
2	3.7	2	2.8	2	2.5	SUNFLOWER	MS	-24	-10	-32	41	27
5	11.2	3	7.1	1	2.2	TALLAHATCHIE	MS	-37	-70	-81	02	22
1	2.7			5	9.4	TATE	MS			252	97	94
1	2.0	3	4.1			TIPPAH	MS	106				
		3	3.7	1	1.0	TISHOMINGO	MS		-73			14
1	4.8			1	7.0	TUNICA	MS			46	83	82
5	6.3	2	2.3	4	3.8	UNION	MS	-64	67	-39	31	43
1	3.4	1	2.3	1	2.6	WALTHALL	MS	-33	14	-23	48	28
6	7.3	4	4.4	3	2.6	WARREN	MS	-39	-41	-64	09	27
8	10.4	3	3.5	3	2.0	WASHINGTON	MS	-66	-45	-81	02	21
		2	3.9	1	2.6	WAYNE	MS		-35			27
1	2.0	3	5.0	2	3.2	WEBSTER	MS	143	-35	58	85	35
2	9.1					WILKINSON	MS					
		3	5.0	3	4.6	WINSTON	MS		-8			55
1	2.0	2	4.1	2	3.4	YALOBUSHA	MS	107	-18	70	87	37
1	1.9	1	1.4	1	1.8	YAZOO	MS	-28	25	-10	61	19
1488	7.7	1258	6.0	1061	4.8	MISSOURI	MO	-22	-20	-37		
8	6.6	5	4.4	4	3.2	ADAIR	MO	-34	-27	-52	18	35
5	6.3	10	12.9	2	2.3	ANDREW	MO	104	-82	-64	09	24
3	5.5	4	7.0	3	6.5	ATCHISON	MO	28	-7	19	75	77
10	8.0	12	9.3	6	4.5	AUDRAIN	MO	17	-52	-44	27	52
2	1.5	4	3.3	7	3.8	BARRY	MO	126	14	157	95	43
5	6.6	6	6.3	4	3.7	BARTON	MO	-4	-41	-43	27	42
6	4.4	8	7.5	5	3.5	BATES	MO	68	-53	-21	51	39
2	2.9	1	1.1	3	4.1	BENTON	MO	-63	275	40	81	48
1	1.3	6	8.9	7	8.8	BOLLINGER	MO	578	-1	573	99	92
15	7.6	11	5.0	12	4.7	BOONE	MO	-34	-7	-38	32	55
43	8.4	29	5.8	22	4.7	BUCHANAN	MO	-31	-20	-45	26	55
3	1.7	5	2.6	6	3.1	BUTLER	MO	53	17	78	88	33
3	3.5	3	3.6			CALDWELL	MO	3				
3	2.1	2	1.5	5	3.7	CALLAWAY	MO	-28	148	78	88	41
1	1.7	5	5.0	3	3.0	CAMDEN	MO	197	-42	74	88	32
15	8.0	19	8.7	16	6.1	CAPE GIRARDEAU	MO	9	-30	-23	48	73
7	6.6	4	4.4	6	7.1	CARROLL	MO	-33	62	9	70	83
1	3.6	3	9.3	1	4.2	CARTER	MO	157	-55	17	74	49
6	4.6	8	5.4	7	3.9	CASS	MO	16	-27	-15	57	45
2	2.2	4	4.1	3	2.8	CEDAR	MO	90	-31	31	79	30

WHITE MALE — CANCER OF THE RECTUM — ICD 154 EXCEPT 154.3

COUNTY AND STATE	1950-1959 DEATHS	1950-1959 RATE	1960-1969 DEATHS	1960-1969 RATE	1970-1979 DEATHS	1970-1979 RATE	PERCENT CHANGE 60S/50S	PERCENT CHANGE 70S/60S	PERCENT CHANGE 70S/50S	PERCENTILE %CHANGE 70S/50S	PERCENTILE RATES 1970S
CHARITON MO	6	6.2	5	4.8	3	3.7	-22	-22	-40	31	42
CHRISTIAN MO	6	8.1	4	4.5	6	5.7	-45	27	-30	43	68
CLARK MO	5	6.8	3	5.7	1	1.4	-16	-76	-80	02	16
CLAY MO	14	5.8	10	2.8	12	2.6	-52	-5	-54	15	28
CLINTON MO	6	9.9	3	4.2	3	3.2	-58	-22	-67	08	36
COLE MO	12	7.1	10	5.4	11	5.2	-25	-3	-27	45	62
COOPER MO	5	5.4	2	2.4	7	7.5	-56	219	39	81	85
CRAWFORD MO	7	9.1	5	6.8	7	6.3	-26	-6	-31	42	75
DADE MO	4	6.7	1	2.9	1	1.2	-57	-59	-82	01	15
DALLAS MO	2	3.4	4	5.2	2	2.2	50	-58	-37	34	23
DAVIESS MO	5	6.1	2	2.1	4	5.5	-66	165	-9	61	66
DE KALB MO	3	4.1	3	4.3	4	5.9	5	36	44	83	70
DENT MO			2	2.3	3	2.9		25			31
DOUGLAS MO	1	1.7	1	1.4			-19				
DUNKLIN MO	2	1.2	6	3.2	13	6.2	173	96	435	99	74
FRANKLIN MO	8	4.2	17	6.7	18	6.0	60	-11	43	82	71
GASCONADE MO	4	4.8	7	8.0	2	2.1	67	-73	-55	14	22
GENTRY MO	3	3.6	9	12.3			240				
GREENE MO	28	5.3	19	3.0	37	5.1	-44	72	-5	64	61
GRUNDY MO	7	7.0	6	7.2	4	5.5	3	-23	-21	50	66
HARRISON MO	5	5.1	4	3.7	3	3.6	-27	-5	-31	42	40
HENRY MO	7	4.3	5	3.3	2	1.1	-24	-67	-75	05	15
HICKORY MO			1	2.1	4	5.5		161			66
HOLT MO	6	9.2	2	2.6	6	9.0	-71	243	-2	66	93
HOWARD MO	1	1.4	4	5.7	5	7.8	309	36	457	99	87
HOWELL MO	6	4.9	1	.6	6	3.4	-87	444	-30	42	38
IRON MO	1	2.2	2	3.3	2	3.1	51	-8	38	81	33
JACKSON MO	199	8.6	157	6.5	101	4.2	-25	-35	-51	18	49
JASPER MO	24	5.9	26	6.2	35	7.8	5	27	34	80	87
JEFFERSON MO	18	8.5	12	3.9	24	5.8	-54	48	-32	40	68
JOHNSON MO	4	2.9	4	3.2	3	2.6	12	-19	-9	61	28
KNOX MO	1	2.4	1	3.4	2	3.0	41	-12	24	77	33
LACLEDE MO	7	6.5	7	5.8	5	3.6	-11	-38	-45	26	41
LAFAYETTE MO	9	6.3	8	5.1	9	5.2	-19	2	-17	55	62
LAWRENCE MO	6	3.7	11	6.3	13	7.1	71	12	92	90	83
LEWIS MO	1	1.4	5	6.9	4	5.7	381	-18	296	98	67
LINCOLN MO	7	7.4	3	2.9	2	1.5	-61	-47	-79	02	17
LINN MO	9	6.4	13	10.7	10	8.7	67	-19	36	80	92
LIVINGSTON MO	4	3.7	2	2.1	1	1.0	-43	-51	-73	06	14
MC DONALD MO	4	4.5	6	7.1			58				
MACON MO	11	8.4	11	8.9	8	6.1	5	-31	-27	45	73
MADISON MO	2	3.8	2	3.0	7	9.5	-21	220	152	94	95
MARIES MO	3	6.2	1	2.1			-67				
MARION MO	16	9.7	15	8.9	7	4.0	-9	-54	-58	12	47
MERCER MO	3	5.9			3	6.7			14	73	79
MILLER MO	3	3.3	3	3.2	3	2.5	-3	-21	-23	48	27
MISSISSIPPI MO	3	4.1	1	1.6	1	1.4	-61	-10	-65	09	17
MONITEAU MO	1	1.6	5	7.2	4	4.3	349	-40	170	95	51
MONROE MO	3	3.4	3	4.3	1	1.9	26	-55	-43	27	21
MONTGOMERY MO	8	9.3	4	5.7	2	2.3	-39	-59	-75	04	24
MORGAN MO	1	1.1	1	1.0	3	3.2	-3	209	201	96	35
NEW MADRID MO	4	3.7	1	1.3	1	.9	-66	-28	-75	04	14
NEWTON MO	7	4.6	5	2.8	8	3.9	-38	37	-15	57	44
NODAWAY MO	7	4.9	6	4.1	6	3.9	-16	-5	-20	52	45
OREGON MO	1	1.8	2	2.5	5	6.7	39	164	266	97	79
OSAGE MO	3	4.4	6	9.1	2	2.5	105	-73	-44	26	26
OZARK MO	2	5.0	3	7.0	2	3.4	40	-51	-31	41	38
PEMISCOT MO	6	5.9	5	4.2	2	1.6	-30	-61	-72	06	18
PERRY MO	7	8.7	8	8.8	6	6.8	2	-23	-21	50	80
PETTIS MO	15	8.0	13	6.5	9	4.5	-18	-31	-44	27	53
PHELPS MO	10	8.5	10	7.4	4	2.5	-12	-66	-70	06	26
PIKE MO	8	7.5	4	3.9	1	.9	-49	-76	-87	00	14
PLATTE MO	5	5.4	7	7.0	5	4.8	30	-32	-12	59	57
POLK MO	3	2.8	4	3.1	6	5.2	11	65	83	89	62
PULASKI MO	4	7.0	2	2.7	4	4.3	-61	59	-38	32	50
PUTNAM MO	3	4.9			2	3.1			-36	35	34
RALLS MO	1	1.5	1	1.7	2	3.5	16	105	137	93	39
RANDOLPH MO	12	8.0	7	4.9	6	4.3	-39	-13	-47	24	50
RAY MO	6	6.4	2	2.0	2	1.8	-69	-10	-72	06	19
REYNOLDS MO	1	3.0	1	2.3	2	6.3	-25	178	109	92	75
RIPLEY MO	3	3.7	2	2.8	1	1.3	-25	-54	-65	08	16
ST CHARLES MO	10	7.2	14	6.1	15	4.7	-15	-24	-35	36	56
ST CLAIR MO			3	3.7	2	3.8		3			43
STE GENEVIEVE MO	6	11.8	5	7.6	6	7.6	-35	-1	-36	35	86
ST FRANCOIS MO	7	3.8	8	3.6	11	4.9	-4	34	29	78	58
ST LOUIS COUNTY MO	184	10.8	213	8.2	180	5.1	-24	-38	-53	17	60
SALINE MO	7	4.5	5	3.0	1	.6	-33	-80	-87	00	13
SCHUYLER MO			2	4.2	3	9.0		114			93
SCOTLAND MO	5	9.1	3	4.8	1	1.6	-47	-67	-83	01	18
SCOTT MO	10	8.0	8	5.6	9	5.5	-30	-1	-31	42	66
SHANNON MO			2	4.2							
SHELBY MO	4	4.5	2	2.9	3	3.9	-37	36	-14	57	45
STODDARD MO	7	5.0	5	3.2	5	3.2	-36	-2	-37	34	35
STONE MO	2	5.3	3	5.0	1	.6	-6	-87	-88	00	13
SULLIVAN MO	5	6.6			1	1.4			-78	03	17
TANEY MO	1	1.4	5	7.3	3	2.9	437	-60	116	92	32
TEXAS MO	5	4.4	1	.8	3	2.2	-82	186	-50	20	23
VERNON MO	6	3.5	2	1.4	4	2.6	-60	89	-25	47	28
WARREN MO	6	9.8	6	10.1	4	5.2	3	-48	-47	23	62
WASHINGTON MO	3	4.7	3	3.7	4	5.7	-22	56	21	76	68
WAYNE MO	2	3.2	3	3.8	8	12.0	19	219	279	98	98
WEBSTER MO	4	4.4	4	3.9	2	1.7	-12	-55	-60	11	19
WORTH MO			2	9.0							
WRIGHT MO	3	3.6	5	5.2	3	2.6	46	-50	-26	46	28
ST LOUIS CITY MO	461*	14.4	277*	10.1	180*	8.5	-30	-16	-41	30	91
MONTANA MT	200	6.6	187	5.8	198	5.8	-11	-1	-12		
BEAVERHEAD MT	2	4.5	4	10.0	3	6.8	123	-32	51	84	80
BIG HORN MT	3	8.8	1	2.9	1	3.7	-66	25	-58	12	41
BLAINE MT	1	4.4	3	7.5	2	5.3	72	-29	22	76	63

* SIGNIFICANTLY MORE DEATHS THAN EXPECTED

WHITE MALE — CANCER OF THE RECTUM — ICD 154 EXCEPT 154.3

1950-1959 DEATHS	1950-1959 RATE	1960-1969 DEATHS	1960-1969 RATE	1970-1979 DEATHS	1970-1979 RATE	COUNTY AND STATE	PERCENT CHANGE 60S/50S	PERCENT CHANGE 70S/60S	PERCENT CHANGE 70S/50S	PERCENTILE %CHANGE 70S/50S	PERCENTILE RATES 1970S
		1	8.0	1	4.4	BROADWATER MT		-45			52
3	5.5					CARBON MT					
16	6.5	23	8.0	22	7.0	CASCADE MT	23	-13	8	70	81
2	6.7	4	9.3	4	9.9	CHOUTEAU MT	38	7	48	84	96
6	10.5	5	7.7	5	6.8	CUSTER MT	-27	-12	-36	36	79
		1	3.5	2	8.9	DANIELS MT		152			93
1	2.3	1	1.8	3	6.3	DAWSON MT	-22	245	171	95	75
3	3.7	4	3.8	5	5.4	DEER LODGE MT	2	43	45	83	65
1	6.2					FALLON MT					
11	13.8	4	4.4	2	2.4	FERGUS MT	-68	-46	-83	01	25
16	9.0	8	4.4	16	7.1	FLATHEAD MT	-51	63	-20	51	83
4	3.6	9	7.9	5	3.3	GALLATIN MT	122	-59	-8	62	36
1	6.5					GARFIELD MT					
4	20.8	1	4.3	2	5.8	GLACIER MT	-79	35	-72	06	69
1	8.0					GOLDEN VALLEY MT					
		2	10.8			GRANITE MT					
1	1.4	6	8.2	9*	13.5	HILL MT	501	65	890	99	99
2	10.5	2	8.3	1	3.7	JEFFERSON MT	-22	-55	-65	08	42
1	4.8	1	5.1			JUDITH BASIN MT	4				
7	9.1	5	6.1	5	5.6	LAKE MT	-34	-7	-39	32	67
8	6.4	10	7.8	11	7.1	LEWIS AND CLARK MT	21	-9	10	71	82
1	7.1					LIBERTY MT					
3	6.8	6	11.7	1	1.3	LINCOLN MT	71	-89	-81	02	16
2	4.6	1	2.6			MADISON MT	-42				
		1	6.0	2	13.0	MINERAL MT		117			99
11	6.0	8	4.0	18	7.5	MISSOULA MT	-33	86	26	77	85
1	2.6			2	6.8	MUSSELSHELL MT			164	95	79
4	6.0	6	8.4			PARK#MT	41				
3	7.7	2	5.1			PHILLIPS MT	-34				
1	3.7	1	2.9	2	5.1	PONDERA MT	-22	79	39	81	61
		2	15.5	1	6.9	POWDER RIVER MT		-55			81
		2	6.4			POWELL MT					
		1	7.0	2	13.6	PRAIRIE MT		92			99
8	9.5	3	3.2	6	5.1	RAVALLI MT	-66	57	-47	24	61
5	9.6	3	5.8	2	4.6	RICHLAND MT	-39	-20	-52	18	55
1	2.8	6	13.8	3	7.3	ROOSEVELT MT	388	-47	157	95	84
		2	7.7	1	3.9	ROSEBUD MT		-49			45
4	8.3	2	4.8	1	1.7	SANDERS MT	-43	-64	-80	02	19
3	8.7	3	7.1	4	12.0	SHERIDAN MT	-18	68	37	81	98
23	9.1	8	3.5	21*	10.2	SILVER BOW MT	-61	191	12	72	96
3	9.1	1	3.5	3	10.1	STILLWATER MT	-62	189	11	72	96
				1	4.5	SWEET GRASS MT					54
4	13.0	3	7.3			TETON MT	-43				
3	8.9			2	5.8	TOOLE MT			-35	36	68
1	9.4	1	16.4			TREASURE MT	75				
3	5.2	5	8.8			VALLEY MT	70				
1	3.8	1	6.1	2	10.7	WHEATLAND MT	59	76	181	96	97
1	10.3	1	11.2			WIBAUX MT	9				
20	7.6	23	7.9	25	6.8	YELLOWSTONE MT	4	-14	-11	60	79
507	7.5	480	6.5	469	6.0	NEBRASKA NB	-13	-7	-20		
10	5.7	8	4.7	16	8.8	ADAMS NB	-18	88	54	85	92
7	11.5	1	2.0	2	3.3	ANTELOPE NB	-82	64	-71	06	36
		1	11.6			BLAINE NB					
3	5.6	3	5.2	3	4.8	BOONE NB	-7	-9	-15	56	57
6	10.4	4	7.0			BOX BUTTE NB	-33				
1	3.6	2	6.4	2	6.3	BOYD NB	76	-2	73	88	74
1	4.1	4	13.9			BROWN NB	239				
5	3.4	8	4.9	7	3.9	BUFFALO NB	46	-21	16	74	45
3	3.8	7	9.4	4	5.4	BURT NB	144	-42	41	82	65
2	3.2	9	12.7	5	7.6	BUTLER NB	300	-41	138	93	86
14	14.0	9	8.7	10	9.3	CASS NB	-38	8	-33	38	94
5	7.2	4	5.5	5	5.9	CEDAR NB	-24	7	-18	53	69
2	9.7	3	11.1	3	9.5	CHASE NB	15	-15	-2	66	95
2	5.2	2	5.0	2	4.7	CHERRY NB	-4	-5	-8	62	57
2	3.2	2	3.4	5	8.3	CHEYENNE NB	7	143	161	95	90
8	11.9	2	3.0	3	5.2	CLAY NB	-74	72	-56	14	63
3	5.1	3	4.7	7	10.3	COLFAX NB	-10	121	99	91	96
6	8.3	9	11.7	6	7.0	CUMING NB	41	-40	-16	55	81
7	6.7	10	9.0	6	6.0	CUSTER NB	33	-33	-11	60	71
1	1.7	5	9.9	6	9.3	DAKOTA NB	489	-5	457	99	94
2	4.3	3	5.6	3	6.1	DAWES NB	30	10	43	82	73
5	4.8	8	7.1	5	4.0	DAWSON NB	48	-43	-15	57	47
		2	11.5			DEUEL NB					
2	3.7	3	5.9	2	3.7	DIXON NB	58	-38	-1	66	41
8	5.5	11	6.2	17	9.4	DODGE NB	13	51	71	88	94
103	8.4	113	8.3	111*	7.8	DOUGLAS NB	-1	-7	-8	62	87
1	3.1	4	18.7			DUNDY NB	505				
6	9.6			3	4.4	FILLMORE NB			-54	15	51
4	9.7	3	6.8	3	6.5	FRANKLIN NB	-29	-5	-33	39	76
				1	3.5	FRONTIER NB					39
3	5.5	3	3.8	2	2.7	FURNAS NB	-31	-30	-52	18	29
15	10.4	6	3.3	11	6.6	GAGE NB	-69	103	-36	35	78
		2	9.2			GARDEN NB					
4*	22.4	2	11.6	1	3.7	GARFIELD NB	-48	-68	-83	01	42
				1	10.6	GOSPER NB					97
4	13.6	1	3.0			GREELEY NB	-78				
18	11.7	8	3.7	19	8.3	HALL NB	-68	123	-29	43	90
		2	3.8	2	3.4	HAMILTON NB		-10			38
2	6.0	3	7.6	1	2.8	HARLAN NB	27	-64	-54	16	29
1	4.0	1	3.3	2	5.8	HITCHCOCK NB	-18	76	45	83	69
8	9.3	5	5.9	3	3.8	HOLT NB	-37	-35	-59	12	43
4	8.7	3	5.6	5	11.0	HOWARD NB	-35	95	27	78	97
5	6.5	4	5.0	2	2.2	JEFFERSON NB	-22	-56	-66	08	23
4	9.2	1	2.0	4	8.9	JOHNSON NB	-78	341	-4	64	92
1	2.0	1	2.2	2	4.3	KEARNEY NB	11	94	115	92	51
		2	4.1	2	4.6	KEITH NB		14			55
2	8.7	2	8.9			KIMBALL NB	2				
7	8.5	7	7.8	6	6.6	KNOX NB	-9	-15	-23	49	77
44	7.7	46	7.2	43	6.3	LANCASTER NB	-6	-13	-18	54	74

* SIGNIFICANTLY MORE DEATHS THAN EXPECTED

COUNTIES COMBINED; SEE TECHNICAL NOTES

WHITE MALE

CANCER OF THE RECTUM

ICD 154 EXCEPT 154.3

1950-1959 DEATHS	1950-1959 RATE	1960-1969 DEATHS	1960-1969 RATE	1970-1979 DEATHS	1970-1979 RATE	COUNTY AND STATE	PERCENT CHANGE 60S/50S	PERCENT CHANGE 70S/60S	PERCENT CHANGE 70S/50S	PERCENTILE %CHANGE 70S/50S	PERCENTILE RATES 1970S
13	11.9	11	7.6	8	5.0	LINCOLN NB	-36	-34	-58	12	59
13	8.6	10	6.0	12	6.8	MADISON NB	-31	14	-21	50	80
4	7.4	4	7.0	6	11.0	MERRICK NB	-6	56	47	83	97
1	3.9	1	1.9	1	2.0	MORRILL NB	-51	3	-50	20	21
1	2.4	3	6.4	1	2.0	NANCE NB	173	-68	-13	58	22
4	6.6	6	10.4	2	3.4	NEMAHA NB	57	-67	-48	22	38
		4	7.5	2	2.9	NUCKOLLS NB		-62			31
8	7.9	12	11.7	7	5.9	OTOE NB	47	-49	-26	46	70
2	4.4	7*	16.8	5	12.8	PAWNEE NB	279	-24	189	96	99
1	5.0			1	3.5	PERKINS NB			-30	43	39
4	6.8	1	1.4	5	7.3	PHELPS NB	-80	433	8	70	84
2	4.1	5	9.1	3	4.6	PIERCE NB	123	-49	14	73	55
9	9.2	8	6.7	6	4.7	PLATTE NB	-27	-31	-49	21	56
8	16.6	1	1.8	3	5.5	POLK NB	-89	206	-67	07	65
7	10.5			5	8.0	RED WILLOW NB			-24	48	88
11	12.0	5	5.6	6	6.8	RICHAPDSON NB	-53	21	-43	27	80
		1	7.5			POCK NB					
14	14.1	8	7.6	8	8.8	SALINE NB	-46	15	-38	33	92
3	5.2	6	7.3	5	4.3	SARPY NB	39	-40	-17	55	51
12	11.7	7	6.0	7	6.0	SAUNDERS NB	-49	1	-49	22	71
6	4.2	3	1.9	6	3.2	SCOTTS BLUFF NB	-55	72	-22	49	36
10	13.0	9	11.1	4	4.8	SEWARD NB	-15	-57	-63	10	58
1	2.2	1	2.0	2	3.6	SHERIDAN NB	-11	86	65	87	41
4	11.9	3	8.8	1	2.4	SHERMAN NB	-26	-72	-79	02	26
1	9.4					SIOUX NB					
		2	5.4			STANTON NB					
8	11.8	5	7.7	4	6.2	THAYER NB	-34	-20	-47	23	74
				1	22.5	THOMAS NB					99
5	14.4	1	3.0	3	9.8	THUPSTON NB	-79	222	-32	40	96
1	2.9	3	6.3	5	9.4	VALLEY NB	116	50	224	97	94
3	4.8	1	1.3	3	3.7	WASHINGTON NB	-72	177	-22	49	42
		1	1.9	1	1.4	WAYNE NB		-28			16
3	6.9	3	6.2	2	3.9	WEBSTER NB	-9	-37	-43	28	46
1	11.4					WHEELER NB					
6	6.9	6	6.5	1	1.2	YORK NB	-7	-81	-82	01	15
53	6.5	76	5.4	119	5.5	NEVADA NV	-18	3	-15		
2	5.0			3	4.8	CHURCHILL NV			-4	64	58
11	4.4	22	3.2	54	4.6	CLARK NV	-27	43	5	69	55
2	12.4			1	1.0	DOUGLAS NV			-92	00	15
1	1.4	5	10.2	2	4.5	ELKO NV	603	-56	212	97	53
1	37.2					EUREKA NV					
5	16.3	3	10.1	2	7.6	HUMBOLDT NV	-38	-25	-54	16	86
		1	18.7	1	7.4	LANDER NV		-60			85
1	6.4	2	13.4	1	5.2	LINCOLN NV	110	-61	-18	54	63
1	3.9	2	6.7	2	3.5	LYON NV	70	-47	-10	60	40
1	5.4	1	8.0	1	2.8	MINERAL NV	50	-65	-47	23	30
				1	2.4	NYE NV					25
1	4.5	1	5.4			PERSHING NV	19				
23	8.5	35	8.1	38	6.9	WASHOE NV	-5	-15	-19	52	80
3	7.3	3	6.6	5	12.8	WHITE PINE NV	-9	93	75	88	99
1	5.6	1	2.4	8	9.7	ORMSBY-CARSON CITY NV	-57	296	72	88	95
350*	12.6	318*	10.3	283*	7.8	NEW HAMPSHIRE NH	-18	-25	-38		
19	13.4	19*	12.3	9	4.7	BELKNAP NH	-8	-62	-65	08	56
8	7.8	10	8.3	8	5.8	CARROLL NH	6	-30	-26	46	69
23	11.4	27*	12.2	24*	9.2	CHESHIPE NH	7	-25	-20	52	94
20	10.1	19	10.0	15	7.8	COOS NH	-1	-22	-22	49	87
33*	13.6	22	8.7	29*	10.4	GRAFTON NH	-36	20	-23	49	97
111*	14.5	114*	13.1	99*	10.1	HILLSBOROUGH NH	-10	-23	-31	42	96
46*	13.0	28	7.3	28	6.6	MERRIMACK NH	-44	-10	-49	21	77
37	9.4	38	8.1	33	5.1	ROCKINGHAM NH	-14	-37	-46	25	61
28	12.3	25	10.0	31*	10.2	STRAFFORD NH	-19	3	-17	55	96
25*	17.7	16	10.5	7	4.0	SULLIVAN NH	-40	-62	-77	03	46
3033*	13.9	2974*	11.4	2379*	8.0	NEW JERSEY NJ	-18	-29	-42		
89*	13.2	90*	10.9	79*	8.2	ATLANTIC NJ	-18	-24	-38	33	90
355*	14.2	370*	11.3	310*	7.9	BERGEN NJ	-20	-30	-44	27	88
67*	12.6	66	8.8	72	7.4	BURLINGTON NJ	-30	-16	-41	29	85
156*	12.0	170*	11.3	109	6.5	CAMDEN NJ	-6	-43	-46	24	76
31	12.7	35	8.9	29	6.0	CAPE MAY NJ	-30	-33	-53	17	71
29	7.2	38	8.5	35	7.0	CUMBERLAND NJ	17	-18	-4	64	81
587*	15.7	449*	12.2	237*	7.3	ESSEX NJ	-22	-40	-54	16	84
48	11.9	46	8.4	38	6.1	GLOUCESTER NJ	-30	-27	-49	21	72
448*	16.2	396*	14.0	288*	11.2	HUDSON NJ	-13	-21	-31	41	97
25	9.9	43*	14.3	27	8.5	HUNTERDON NJ	44	-40	-14	58	91
145*	15.1	134*	12.1	117*	9.9	MERCER NJ	-20	-18	-34	37	96
158*	14.4	162*	10.7	158*	7.8	MIDDLESEX NJ	-26	-27	-46	25	87
122	10.7	159*	10.4	123	6.4	MONMOUTH NJ	-3	-39	-40	30	75
89	10.1	116*	10.1	101*	7.5	MORRIS NJ	0	-26	-26	46	85
46	11.2	81	8.9	124	6.7	OCEAN NJ	-20	-24	-40	31	79
268*	16.4	224*	11.9	188*	9.5	PASSAIC NJ	-27	-21	-42	28	95
13	6.3	24	10.8	27*	10.7	SALEM NJ	71	-1	69	87	97
49	9.8	55	9.5	60*	8.0	SOMERSET NJ	-3	-16	-18	53	88
36*	19.0	24	7.9	33*	9.1	SUSSEX NJ	-58	15	-52	17	93
232*	13.5	244*	12.1	178*	7.8	UNION NJ	-10	-35	-42	29	87
40*	12.8	48*	14.0	46*	12.5	WARREN NJ	9	-11	-3	65	98
82	3.8	103	3.6	144	3.9	NEW MEXICO NM	-6	8	2		
21	4.8	31	4.4	42	3.9	BERNALILLO NM	-9	-12	-20	52	44
1	6.1			1	8.6	CATRON NM			42	82	91
4	4.6	4	2.6	5	2.1	CHAVES NM	-44	-21	-55	14	22
2	3.2	3	4.4	3	4.2	COLFAX NM	39	-6	31	79	48
3	3.1	6	5.5	8	6.1	CURRY NM	78	11	98	91	72
2	10.9	1	6.2	1	5.1	DE BACA NM	-43	-18	-54	16	60
4	3.2	9	5.4	4	1.5	DONA ANA NM	70	-72	-53	17	17
7	6.0	4	2.6	10	5.4	EDDY NM	-56	106	-10	61	65

* SIGNIFICANTLY MORE DEATHS THAN EXPECTED

WHITE MALE

CANCER OF THE RECTUM — ICD 154 EXCEPT 154.3

1950-1959 DEATHS	1950-1959 RATE	1960-1969 DEATHS	1960-1969 RATE	1970-1979 DEATHS	1970-1979 RATE	COUNTY AND STATE	PERCENT CHANGE 60S/50S	PERCENT CHANGE 70S/60S	PERCENT CHANGE 70S/50S	PERCENTILE %CHANGE 70S/50S	PERCENTILE RATES 1970S
4	7.3	4	5.1	4	5.0	GRANT NM	-30	-3	-32	40	59
		1	5.3			HIDALGO NM					
3	1.5	7	5.2	7	3.5	LEA NM	252	-32	139	94	39
		2	4.4			LINCOLN NM					
1	2.1	1	3.1	2	33.4	LOS ALAMOS NM	46	987	1490	99	99
5	15.7	1	1.8	2	2.8	LUNA NM	-89	62	-82	01	31
		2	3.8	3	8.6	MC KINLEY NM		128			91
		2	5.3	1	3.3	MORA NM		-38			37
2	5.7	2	4.0	4	4.0	OTERO NM	-31	1	-30	42	46
6	10.4	3	4.8	3	4.9	QUAY NM	-53	0	-53	16	58
1	1.6			8	11.7	RIO ARRIBA NM			626	99	98
		1	1.5	1	1.6	ROOSEVELT NM		6			18
		3	10.1	2	3.3	SANDOVAL NM		-67			36
3	3.1	4	4.4	8	6.3	SAN JUAN NM	41	42	100	91	74
1	1.3			2	2.6	SAN MIGUEL NM			108	92	28
3	2.7	4	2.5	13	6.5	SANTA FE NM	-8	166	144	94	77
		1	1.1	2	1.8	SIERRA NM		75			20
1	2.0	1	3.0	1	2.1	SOCORRO NM	50	-31	3	68	22
2	3.0	1	1.4	3	5.2	TAOS NM	-54	277	75	88	62
		2	7.9	1	3.3	TORRANCE NM		-58			37
5	14.7	1	2.8	1	2.9	UNION NM	-81	5	-80	02	31
1	1.9	2	1.6	2	1.4	VALENCIA NM	-16	-13	-27	45	16
9550*	14.1	8105*	10.8	6150*	8.1	NEW YORK NY	-23	-25	-43		
176*	15.2	125*	9.6	96	7.2	ALBANY NY	-37	-25	-52	17	83
16	7.3	23	10.7	17	7.5	ALLEGANY NY	47	-30	3	68	85
1006*	16.9	611*	9.8	451*	8.7	BRONX NY	-42	-11	-49	21	92
89	10.2	111*	12.1	79*	7.7	BROOME NY	19	-37	-25	47	86
42	10.4	34	8.5	43*	10.4	CATTARAUGUS NY	-18	22	0	67	97
38	10.2	29	7.7	32*	8.1	CAYUGA NY	-25	6	-20	51	89
73	9.7	80*	10.3	55	7.0	CHAUTAUQUA NY	6	-32	-28	45	82
49	12.4	44	10.1	37	8.2	CHEMUNG NY	-19	-19	-34	37	89
19	8.5	15	6.7	16	7.0	CHENANGO NY	-21	4	-18	54	82
35*	14.0	17	7.3	24*	9.6	CLINTON NY	-48	31	-32	40	95
34*	12.3	20	7.2	25	7.6	COLUMBIA NY	-42	6	-38	32	86
29*	15.6	28*	14.9	15	7.7	CORTLAND NY	-5	-48	-51	19	87
24	10.4	19	7.9	24*	9.2	DELAWARE NY	-23	15	-12	60	94
90*	11.8	87*	10.0	78*	8.1	DUTCHESS NY	-15	-19	-31	41	89
579*	14.8	499*	11.3	417*	9.2	ERIE NY	-24	-19	-38	32	93
18	10.4	9	4.8	11	5.6	ESSEX NY	-54	16	-46	24	67
21	10.2	13	6.0	19*	9.0	FRANKLIN NY	-41	50	-11	60	93
42*	14.6	39*	13.1	22	7.3	FULTON NY	-10	-44	-50	20	84
17	6.9	21	8.6	23	8.6	GENESEE NY	25	0	25	77	91
20	12.2	23*	11.8	14	6.0	GREENE NY	-4	-49	-51	19	71
1	3.1	4	16.6	2	5.7	HAMILTON NY	432	-66	81	89	67
42	12.1	38	10.6	29	8.1	HERKIMER NY	-13	-23	-33	39	89
50	10.7	44	9.5	38*	8.8	JEFFERSON NY	-11	-8	-18	54	92
1759*	16.3	990*	9.1	758*	7.8	KINGS(BROOKLYN) NY	-45	-14	-52	18	87
13	10.1	12	10.4	7	6.1	LEWIS NY	3	-41	-40	31	73
21	10.0	20	9.1	24*	10.2	LIVINGSTON NY	-10	13	2	67	96
32*	13.5	22	8.9	14	5.4	MADISON NY	-34	-39	-60	11	65
302*	12.4	264*	9.6	213*	7.4	MONROE NY	-22	-23	-40	31	85
47*	14.3	24	6.9	49*	14.7	MONTGOMERY NY	-52	112	3	68	99
449*	15.2	483*	11.0	404*	7.5	NASSAU NY	-28	-32	-51	19	85
1309*	16.4	1609*	21.5	423*	7.1	NEW YORK(MANHATTAN) NY	31	-67	-57	14	82
99*	11.5	98*	10.1	100*	9.7	NIAGARA NY	-12	-4	-16	55	95
124	10.4	109	8.5	104*	8.2	ONEIDA NY	-18	-4	-21	50	90
195*	11.8	179*	10.0	147*	7.7	ONONDAGA NY	-15	-23	-34	37	87
26	7.6	25	7.0	27	7.0	ONTARIO NY	-8	0	-8	62	81
79	9.4	99*	10.4	94*	9.2	ORANGE NY	11	-12	-3	65	93
18	10.6	19	11.1	12	6.7	ORLEANS NY	4	-40	-37	34	78
60*	14.7	31	7.5	36	8.2	OSWEGO NY	-49	8	-44	26	89
32	10.6	25	8.5	33*	10.5	OTSEGO NY	-19	24	0	66	97
18	13.7	15	8.4	21	8.4	PUTNAM NY	-39	0	-39	32	90
1096*	16.1	830*	9.3	810*	8.7	QUEENS NY	-42	-7	-46	24	92
104*	16.1	90*	12.7	83*	11.7	RENSSELAER NY	-21	-8	-27	45	98
144*	16.0	92	8.7	115*	9.6	RICHMOND NY	-46	11	-40	31	95
35	8.8	40	6.8	45	6.2	ROCKLAND NY	-23	-9	-30	43	74
36	7.5	43	8.9	38	7.9	ST LAWRENCE NY	20	-11	6	69	88
58*	15.3	49*	11.5	43*	8.4	SARATOGA NY	-25	-27	-45	25	90
77	11.4	72	9.5	53	6.6	SCHENECTADY NY	-17	-31	-43	28	77
18	12.9	11	7.7	9	5.8	SCHOHARIE NY	-41	-24	-55	15	69
8	10.1	6	7.6			SCHUYLER NY	-25				
10	5.2	13	6.9	7	3.9	SENECA NY	32	-43	-25	47	45
59	12.0	36	7.1	31	6.1	STEUBEN NY	-41	-14	-50	20	72
175	9.4	236	7.8	327*	8.0	SUFFOLK NY	-17	3	-15	57	88
30	12.5	27	8.7	34*	10.1	SULLIVAN NY	-30	16	-20	52	96
13	7.8	13	7.9	9	5.2	TIOGA NY	1	-34	-34	38	62
30*	12.7	20	8.5	16	6.1	TOMPKINS NY	-33	-29	-52	18	72
62	11.4	66*	10.1	68*	9.5	ULSTER NY	-11	-6	-17	55	95
20	9.2	16	7.0	17	7.1	WARREN NY	-23	2	-22	49	83
26	10.9	26	10.6	27*	10.6	WASHINGTON NY	-3	0	-3	65	97
41	12.2	28	8.2	25	7.0	WAYNE NY	-33	-14	-42	28	82
384*	13.3	413*	12.0	335*	8.6	WESTCHESTER NY	-10	-28	-35	36	91
18	9.7	12	6.7	12	6.6	WYOMING NY	-31	-1	-32	40	77
13	11.6	9	8.6	13*	11.4	YATES NY	-26	33	-2	66	98
401	3.9	562	4.3	649	3.9	NORTH CAROLINA NC	10	-10	-1		
5	2.3	9	3.5	15	4.3	ALAMANCE NC	53	26	93	90	51
3	6.0	2	2.4	6	7.5	ALEXANDER NC	-60	209	24	76	85
1	2.4			1	2.1	ALLEGHANY NC			-13	58	22
2	3.5	3	4.7	2	2.6	ANSON NC	35	-45	-26	46	27
1	1.4	4	4.0	2	1.6	ASHE NC	187	-59	17	74	18
				3	5.3	AVERY NC					64
1	1.1	11	10.5	6	5.2	BEAUFORT NC	838	-51	364	99	62
1	2.1			4	6.7	BERTIE NC			219	97	79
		2	2.6	5	6.2	BLADEN NC		139			74
1	2.0	5	6.3	4	3.7	BRUNSWICK NC	217	-42	85	89	41
21	4.8	23	4.0	33	4.7	BUNCOMBE NC	-16	17	-3	65	56

* SIGNIFICANTLY MORE DEATHS THAN EXPECTED

WHITE MALE — CANCER OF THE RECTUM — ICD 154 EXCEPT 154.3

1950-1959 DEATHS	1950-1959 RATE	1960-1969 DEATHS	1960-1969 RATE	1970-1979 DEATHS	1970-1979 RATE	COUNTY AND STATE	PERCENT CHANGE 60S/50S	PERCENT CHANGE 70S/60S	PERCENT CHANGE 70S/50S	PERCENTILE %CHANGE 70S/50S	PERCENTILE RATES 1970S
5	4.1	6	3.4	11	4.7	BURKE NC	-16	37	16	73	56
4	3.8	10	4.4	15	4.9	CABARRUS NC	17	10	29	78	59
5	4.0	9	5.6	4	1.6	CALDWELL NC	40	-72	-61	10	17
		1	3.9			CAMDEN NC					
1	1.2	7	7.7	5	3.1	CARTERET NC	555	-59	166	95	34
2	4.5	1	3.1	1	1.7	CASWELL NC	-31	-45	-62	10	19
4	2.0	9	3.7	15	4.4	CATAWBA NC	89	19	124	93	51
2	2.8	2	2.5	7	6.5	CHATHAM NC	-11	163	133	93	77
		9	11.2			CHEROKEE NC					
				3	9.3	CHOWAN NC					94
		1	3.4	2	4.2	CLAY NC		24			49
7	5.3	8	3.6	14	5.3	CLEVELAND NC	-32	46	-1	66	63
1	1.1	4	3.4	2	1.2	COLUMBUS NC	199	-65	3	68	15
6	9.7	2	1.3	5	4.3	CRAVEN NC	-87	243	-56	14	50
4	2.4	12	7.0	14	4.9	CUMBERLAND NC	188	-30	102	91	59
3	11.8	3	12.6			CURRITUCK NC	6				
		1	2.8	1	1.8	DARE NC		-36			19
6	2.6	9	3.1	10	3.0	DAVIDSON NC	19	-3	15	73	33
1	1.5	3	4.2	6	6.7	DAVIE NC	186	58	351	98	78
1	1.0	2	2.2	4	3.5	DUPLIN NC	117	57	240	97	39
6	3.0	7	2.7	8	2.7	DURHAM NC	-13	1	-11	60	29
2	2.3	3	3.0	4	3.3	EDGECOMBE NC	30	9	42	82	36
13	3.1	31	6.3	30	4.8	FORSYTH NC	100	-24	52	84	57
3	4.3	2	3.0	4	4.7	FRANKLIN NC	-31	56	8	70	55
9	2.5	14	3.5	15	2.8	GASTON NC	36	-20	10	71	30
		3	11.7			GATES NC					
1	4.6	2	7.0			GRAHAM NC	53				
1	1.2	1	1.4	1	1.1	GRANVILLE NC	19	-23	-8	62	15
1	5.9	1	2.5	5*	12.8	GPEENE NC	-57	403	117	92	99
26	5.0	29	4.5	38	4.4	GUILFORD NC	-9	-3	-12	59	51
5	5.3	7	6.4	5	3.6	HALIFAX NC	20	-44	-33	39	40
4	3.3	5	4.4	11	6.4	HARNETT NC	31	46	91	90	75
3	2.2	8	4.7	7	3.2	HAYWOOD NC	108	-32	42	82	35
13	10.0	9	4.6	14	4.9	HENDERSON NC	-54	7	-51	19	58
1	2.4	2	3.6			HERTFORD NC	49				
1	4.2			1	2.1	HOKE NC			-49	21	22
1	4.0	2	8.1			HYDE NC	103				
9	4.6	17	7.6	12	4.1	IREDELL NC	65	-46	-10	60	48
3	4.2	2	2.8	3	3.1	JACKSON NC	-33	11	-25	47	34
6	4.5	6	3.3	13	6.4	JOHNSTON NC	-26	91	41	82	75
		2	8.7	3	10.2	JONES NC		17			96
2	3.8	2	1.9	4	4.0	LEE NC	-51	114	5	69	46
6	5.6	5	4.9	1	.5	LENOIR NC	-12	-90	-91	00	13
2	4.2	6	5.6	1	.8	LINCOLN NC	34	-86	-81	01	13
3	3.2	3	2.7	7	5.3	MC DOWELL NC	-15	96	67	87	63
1	1.6	9	10.6	4	3.9	MACON NC	576	-63	149	94	44
5	6.2	4	4.5	4	4.4	MADISON NC	-26	-2	-28	44	52
		1	2.5	2	5.5	MARTIN NC		118			65
36	7.1	36	5.5	44	4.4	MECKLENBURG NC	-22	-20	-37	34	52
		2	3.5	2	2.3	MITCHELL NC		-32			24
		4	6.0	4	5.2	MONTGOMERY NC		-12			63
10	10.2	8	5.8	5	3.1	MOORE NC	-44	-46	-69	07	34
6	7.7	6	3.8	2	1.0	NASH NC	-50	-75	-87	00	14
12	7.4	12	6.2	12	4.7	NEW HANOVER NC	-16	-24	-36	36	57
		5	10.5	1	1.6	NORTHAMPTON NC		-85			18
3	5.8	1	.6	1	.5	ONSLOW NC	-89	-19	-91	00	13
3	4.2	3	2.5	6	3.7	ORANGE NC	-41	50	-11	60	42
1	3.2	2	6.2	2	5.1	PAMLICO NC	93	-18	59	85	61
5	9.3	3	4.9	4	5.2	PASQUOTANK NC	-47	4	-45	26	62
2	6.1	1	1.7	2	3.0	PENDER NC	-72	75	-51	19	33
1	4.0	2	6.7			PERQUIMANS NC	68				
3	4.2	4	5.1	2	2.4	PERSON NC	22	-53	-42	28	25
4	5.8	6	5.5	5	2.7	PITT NC	-6	-50	-53	17	29
1	2.5			5	5.8	POLK NC			129	93	69
3	1.6	10	3.5	6	2.0	RANDOLPH NC	116	-43	23	76	21
4	4.4	3	2.2	9	6.3	RICHMOND NC	-50	180	41	82	74
5	4.0	2	1.6	9	5.5	ROBESON NC	-61	256	37	81	66
5	2.3	13	5.6	15	5.4	ROCKINGHAM NC	147	-2	142	94	65
9	3.6	9	3.1	17	4.4	ROWAN NC	-13	41	23	76	51
2	1.2	11	6.5	7	3.2	RUTHERFORD NC	436	-52	159	95	34
5	4.9	2	1.9	2	1.5	SAMPSON NC	-62	-19	-69	07	17
2	6.8	2	5.9			SCOTLAND NC	-14				
5	5.5	11	7.5	10	5.5	STANLY NC	37	-27	0	67	65
5	7.0	6	6.2	6	5.4	STOKES NC	-12	-13	-23	48	65
6	4.5	6	2.8	8	3.2	SURRY NC	-38	16	-29	44	35
		2	5.7	5	10.6	SWAIN NC		84			97
		3	4.8	5	6.1	TRANSYLVANIA NC		27			72
3	2.0	6	4.2	8	4.8	UNION NC	109	12	135	93	57
2	3.0	3	3.7	3	3.4	VANCE NC	24	-7	15	73	38
13	4.2	9	2.0	19	3.6	WAKE NC	-52	78	-14	58	40
1	3.0	1	2.5	1	2.3	WARREN NC	-16	-9	-23	49	24
4	12.1	3	10.4			WASHINGTON NC	-14				
4	5.9	4	4.9	4	4.1	WATAUGA NC	-18	-16	-31	41	47
10	8.2	6	4.1	5	3.3	WAYNE NC	-49	-21	-60	11	36
5	3.1	6	3.6	7	3.4	WILKES NC	16	-5	11	72	37
3	2.3	3	2.2	5	4.3	WILSON NC	-7	101	86	89	51
1	.7	2	2.0	2	1.5	YADKIN NC	173	-27	98	91	17
2	3.2	3	4.7	2	2.3	YANCEY NC	45	-50	-28	44	24
181	6.5	163	5.4	158	4.9	NORTH DAKOTA ND	-17	-10	-25		
		1	3.4	1	3.6	ADAMS ND		4			40
6	6.8	4	4.2	4	4.0	BARNES ND	-38	-4	-41	30	47
2	4.2	2	4.0	1	2.2	BENSON ND	-4	-44	-46	24	23
6	9.6			1	1.3	BOTTINEAU ND			-86	00	16
4	17.3	2	9.2	3	12.5	BOWMAN ND	-47	35	-28	44	98
1	2.6	2	5.9	1	2.9	BURKE ND	127	-52	10	71	31
8	8.1	6	4.9	3	1.9	BURLEIGH ND	-40	-62	-77	03	20
20	7.1	20	6.6	18	5.6	CASS ND	-7	-15	-21	51	67
1	1.7	2	5.1	2	3.5	CAVALIER ND	203	-32	107	92	39
3	6.6	4	7.8	1	1.7	DICKEY ND	19	-79	-75	05	18

* SIGNIFICANTLY MORE DEATHS THAN EXPECTED

WHITE MALE

CANCER OF THE RECTUM

ICD 154 EXCEPT 154.3

COUNTY AND STATE		1950-1959 DEATHS	1950-1959 RATE	1960-1969 DEATHS	1960-1969 RATE	1970-1979 DEATHS	1970-1979 RATE	PERCENT CHANGE 60S/50S	PERCENT CHANGE 70S/60S	PERCENT CHANGE 70S/50S	PERCENTILE %CHANGE 70S/50S	PERCENTILE RATES 1970S
DIVIDE	ND	5	10.7	1	3.8	2	5.0	-64	31	-53	17	60
DUNN	ND	1	3.9	1	3.9	2	6.5	0	68	69	87	77
EDDY	ND	3	12.8			2	7.1			-45	26	82
EMMONS	ND	4	13.1	1	3.4	4	9.7	-74	187	-26	46	95
FOSTER	ND			3	14.5	2	7.4		-49			85
GOLDEN VALLEY	ND			1	6.7							
GRAND FORKS	ND	15	8.1	15	7.7	12	5.6	-5	-27	-31	42	67
GRIGGS	ND	5	16.2	1	2.6	2	6.2	-84	140	-62	10	73
HETTINGER	ND	1	7.0	1	3.3	2	7.6	-53	130	8	70	86
KIDDER	ND			3	15.0	1	3.7		-76			41
LA MOURE	ND	2	4.3	3	5.8	2	3.9	37	-33	-9	61	44
MC HENRY	ND	5	8.8	2	3.0	3	5.6	-66	89	-37	34	67
MC INTOSH	ND	1	3.0	1	2.7	2	5.0	-10	89	70	87	60
MC KENZIE	ND	2	6.5									
MC LEAN	ND			3	3.9	1	1.3		-65			16
MERCER	ND	4	13.1	3	11.3			-13				
MORTON	ND	2	2.3	8	8.9	3	2.9	286	-67	25	77	31
MOUNTRAIL	ND	1	1.3	3	4.5	4	7.0	234	57	425	99	82
NELSON	ND	5	11.8	2	3.4	3	4.3	-71	27	-64	09	50
OLIVER	ND			1	8.0							
PEMBINA	ND	4	6.2	3	5.3	2	2.3	-14	-57	-63	10	24
PIERCE	ND	1	3.1	1	4.1			32				
RAMSEY	ND	3	4.3	4	5.4	8	9.6	26	79	125	93	95
RANSOM	ND	2	5.9									
RENVILLE	ND	1	5.7	3	13.3	1	3.5	133	-73	-38	33	40
RICHLAND	ND	6	5.9	6	5.3	8	6.6	-10	26	13	73	78
ROLETTE	ND	2	6.2	2	6.7	1	2.5	7	-63	-60	11	26
SARGENT	ND	1	2.6	3	7.2			181				
SHERIDAN	ND			2	12.6	1	4.1		-68			48
SLOPE	ND			1	11.5							
STARK	ND	8	12.6	4	5.6	5	6.7	-56	20	-47	23	78
STEELE	ND	4	19.0	1	3.6			-81				
STUTSMAN	ND	9	7.4	5	3.7	11	7.8	-50	109	5	69	87
TOWNER	ND	3	9.7	3	9.9			2				
TRAILL	ND	6	9.8	3	3.8	4	6.0	-61	57	-38	32	71
WALSH	ND	3	2.9	5	5.0	11	10.4	74	106	258	97	97
WARD	ND	16	9.6	13	7.3	20*	9.7	-23	32	1	67	95
WELLS	ND			4	8.4	2	4.2		-51			49
WILLIAMS	ND	5	4.8	4	3.7	2	2.1	-22	-44	-56	14	22
OHIO	OH	3416*	9.4	3358*	8.5	2896*	6.9	-10	-19	-27		
ADAMS	OH	5	4.0	6	4.6	4	3.0	14	-34	-24	48	33
ALLEN	OH	33	8.0	40	9.2	34	7.3	15	-21	-8	62	84
ASHLAND	OH	11	6.1	15	7.7	18	9.0	25	17	46	83	93
ASHTABULA	OH	51	11.9	33	7.7	28	6.0	-36	-21	-50	20	71
ATHENS	OH	16	6.5	13	5.5	23*	10.8	-15	95	66	87	97
AUGLAIZE	OH	11	6.2	10	5.7	6	3.1	-8	-46	-50	20	34
BELMONT	OH	34	7.0	39	8.3	29	6.4	18	-23	-9	61	76
BROWN	OH	6	4.5	14	9.6	3	1.9	114	-80	-57	13	21
BUTLER	OH	57	9.4	45	6.5	46	5.6	-31	-15	-41	29	66
CARROLL	OH	8	7.3	11	10.1	3	2.7	38	-74	-64	09	28
CHAMPAIGN	OH	11	7.9	3	2.1	10	6.7	-73	215	-15	56	78
CLARK	OH	38	7.9	47	8.4	42	6.7	6	-20	-15	56	78
CLERMONT	OH	28	12.8	34*	12.5	23	6.3	-3	-50	-51	19	74
CLINTON	OH	20*	14.4	13	9.2	6	4.2	-36	-54	-71	06	49
COLUMBIANA	OH	37	7.8	40	8.2	20	3.8	6	-54	-51	19	43
COSHOCTON	OH	17	10.3	12	6.9	11	6.3	-33	-8	-38	32	75
CRAWFORD	OH	9	4.1	14	6.5	18	8.1	59	25	99	91	89
CUYAHOGA	OH	728*	12.6	690*	11.0	540*	8.6	-12	-23	-32	40	91
DARKE	OH	16	7.2	20	8.2	12	4.9	15	-41	-32	40	58
DEFIANCE	OH	12	9.1	13	9.0	7	4.8	-1	-46	-47	24	58
DELAWARE	OH	10	6.6	5	3.1	8	4.4	-52	40	-33	39	52
ERIE	OH	27	9.3	26	8.3	29*	9.0	-10	8	-3	65	93
FAIRFIELD	OH	22	8.8	23	7.3	20	5.7	-17	-21	-35	37	68
FAYETTE	OH	6	5.1	7	5.6	3	2.4	9	-57	-53	17	25
FRANKLIN	OH	184	9.5	169	7.3	148	6.0	-23	-19	-37	34	71
FULTON	OH	4	3.1	8	5.8	7	4.6	86	-21	47	83	54
GALLIA	OH	6	4.5	7	5.4	6	5.3	20	-2	18	75	64
GEAUGA	OH	6	4.4	11	6.8	10	4.9	56	-28	12	72	58
GREENE	OH	8	3.3	15	5.3	23	6.3	61	18	89	90	74
GUERNSEY	OH	18	7.0	13	5.3	17	7.3	-24	38	4	68	84
HAMILTON	OH	376*	12.8	367*	11.8	311*	9.6	-8	-19	-25	47	95
HANCOCK	OH	13	5.6	19	7.3	13	4.8	31	-34	-13	58	58
HARDIN	OH	13	8.6	10	6.9	14	9.5	-20	38	10	71	95
HARRISON	OH	7	7.0	6	5.8	5	5.0	-16	-14	-28	44	60
HENRY	OH	7	5.8	7	5.5	9	6.7	-6	23	15	73	79
HIGHLAND	OH	7	3.8	10	5.4	18*	10.1	42	86	164	95	96
HOCKING	OH	14	14.1	9	8.6	3	2.6	-39	-70	-81	01	28
HOLMES	OH	8	8.9	7	6.9	8	6.9	-23	0	-23	49	80
HURON	OH	10	4.8	18	8.5	10	4.8	79	-44	0	67	57
JACKSON	OH	8	5.0	14	9.1	6	4.1	82	-55	-19	53	47
JEFFERSON	OH	47	10.5	40	9.5	31	6.9	-10	-28	-35	37	80
KNOX	OH	16	8.6	12	6.3	17	8.2	-27	29	-5	64	89
LAKE	OH	23	6.5	40	8.5	41	6.5	30	-23	0	67	77
LAWRENCE	OH	9	4.0	16	6.8	9	3.5	72	-49	-12	59	39
LICKING	OH	39	10.1	31	7.5	39	8.5	-26	14	-15	56	91
LOGAN	OH	15	8.0	14	7.8	17	8.5	-3	9	6	69	91
LORAIN	OH	68	10.4	77*	9.8	51	6.2	-6	-37	-40	30	74
LUCAS	OH	172	9.7	194*	10.0	143*	7.4	3	-26	-24	48	85
MADISON	OH	8	7.7	3	2.8	5	4.5	-63	58	-42	29	52
MAHONING	OH	138*	12.1	105	8.3	83	6.5	-31	-22	-46	24	76
MARION	OH	18	7.1	24	9.4	21	7.9	32	-16	11	72	88
MEDINA	OH	21	9.4	11	4.4	23	7.4	-54	69	-22	50	84
MEIGS	OH	9	6.7	7	5.6	5	4.3	-17	-23	-36	35	50
MERCER	OH	11	7.4	7	4.4	11	6.8	-41	54	-9	61	79
MIAMI	OH	31	10.8	22	7.0	16	4.5	-36	-35	-58	12	54
MONROE	OH	7	6.6	7	7.1	5	5.2	7	-26	-21	51	62
MONTGOMERY	OH	123	7.9	120	6.7	111	5.8	-15	-14	-27	45	69
MORGAN	OH	5	7.1	3	2.9	1	1.8	-59	-40	-75	04	19
MORROW	OH	5	5.5	7	6.8	5	4.7	22	-31	-16	56	55

* SIGNIFICANTLY MORE DEATHS THAN EXPECTED

CANCER OF THE RECTUM — ICD 154 EXCEPT 154.3

WHITE MALE

1950-1959 DEATHS	1950-1959 RATE	1960-1969 DEATHS	1960-1969 RATE	1970-1979 DEATHS	1970-1979 RATE	COUNTY AND STATE	PERCENT CHANGE 60S/50S	PERCENT CHANGE 70S/60S	PERCENT CHANGE 70S/50S	PERCENTILE %CHANGE 70S/50S	PERCENTILE RATES 1970S
26	7.2	25	6.8	26	7.0	MUSKINGUM OH	-6	3	-3	65	82
7	9.0	5	7.9	4	5.9	NOBLE OH	-13	-25	-35	37	70
12	8.0	9	5.4	11	5.5	OTTAWA OH	-33	2	-31	41	65
6	7.6	2	2.6			PAULDING OH	-65				
12	7.0	21*	13.3	16*	11.1	PERRY OH	91	-17	59	85	97
7	4.8	5	3.3	8	4.9	PICKAWAY OH	-32	52	4	68	59
3	3.9	5	5.8	2	1.9	PIKE OH	50	-67	-50	20	21
32	11.0	25	8.0	29	7.1	PORTAGE OH	-27	-12	-36	35	82
8	5.7	11	7.2	8	5.0	PREBLE OH	25	-30	-12	60	60
15	11.2	9	7.1	10	7.2	PUTNAM OH	-37	2	-36	36	83
31	7.8	40	8.9	31	6.3	RICHLAND OH	14	-29	-19	52	74
26	10.1	19	6.5	19	6.2	ROSS OH	-36	-4	-38	33	74
30	12.4	11	4.3	14	5.3	SANDUSKY OH	-66	24	-57	13	63
20	5.4	23	5.5	27	6.7	SCIOTO OH	3	22	25	77	79
20	7.6	14	5.2	22	8.2	SENECA OH	-31	57	9	71	90
12	8.3	6	4.2	9	5.9	SHELBY OH	-49	39	-29	43	70
125	9.3	127	8.4	121*	7.5	STARK OH	-9	-11	-19	52	85
144	8.8	168	8.4	149*	7.0	SUMMIT OH	-5	-17	-21	51	81
52	7.0	68	8.4	63	6.9	TRUMBULL OH	21	-18	-1	66	80
38	9.8	31	8.0	34*	8.3	TUSCARAWAS OH	-18	4	-15	56	90
5	4.3	8	6.9	4	3.2	UNION OH	59	-54	-26	46	35
7	5.0	14	9.3	12	7.7	VAN WEPT OH	87	-16	56	85	87
5	9.6	4	6.8	6	10.1	VINTON OH	-30	50	5	69	96
14	7.8	14	5.8	18	6.4	WARREN OH	-26	11	-18	54	76
22	8.9	18	6.9	18	6.5	WASHINGTON OH	-22	-7	-27	45	76
26	9.1	33	9.8	27	7.4	WAYNE OH	8	-24	-18	53	85
11	7.2	11	7.2	7	4.3	WILLIAMS OH	0	-41	-41	29	50
26	9.0	22	7.1	10	2.8	WOOD OH	-22	-61	-69	07	30
7	6.3	7	6.6	6	5.3	WYANDOT OH	4	-19	-16	56	64
456	4.7	418	3.8	432	3.5	OKLAHOMA OK	-18	-8	-25		
2	3.3	4	6.5	1	1.4	ALFALFA OK	96	-78	-57	13	16
4	6.3	2	2.9	2	2.8	ATOKA OK	-54	-6	-56	14	30
3	9.1	2	6.8			BEAVER OK	-25				
3	2.9	5	4.0	3	2.6	BECKHAM OK	37	-35	-11	60	27
4	5.4	6	7.6	5	5.6	BLAINE OK	40	-27	3	68	66
3	2.0	7	4.2	4	2.1	BRYAN OK	114	-51	5	69	22
4	2.8	8	4.8	7	3.9	CADDO OK	75	-19	42	82	45
4	4.0	8	6.2	1	.7	CANADIAN OK	54	-89	-83	01	13
5	3.4	7	3.5	6	3.0	CARTER OK	3	-14	-11	60	33
3	4.1	3	3.3	5	4.9	CHEROKEE OK	-18	48	21	75	59
2	1.9	2	2.7	2	2.2	CHOCTAW OK	42	-21	13	72	23
1	5.0	1	3.9			CIMARRON OK	-22				
5	3.0	5	2.1	4	1.6	CLEVELAND OK	-28	-26	-47	23	17
1	4.3	2	6.3			COAL OK	47				
7	4.2	9	4.9	11	4.1	COMANCHE OK	19	-17	-1	66	48
2	4.0			3	6.6	COTTON OK			65	86	78
5	4.7	2	2.2	3	3.1	CRAIG OK	-53	40	-35	37	33
7	3.4	9	4.2	6	2.7	CREEK OK	23	-37	-22	49	29
8	8.4	6	5.5	3	2.6	CUSTER OK	-34	-53	-69	07	27
2	2.2	5	4.3	3	2.1	DELAWARE OK	100	-50	0	66	22
4	9.6	1	2.2	2	4.5	DEWEY OK	-77	101	-54	16	52
1	4.5					ELLIS OK					
16	6.8	13	5.1	16	5.6	GARFIELD OK	-25	11	-17	54	67
6	4.5	6	4.0	9	5.6	GARVIN OK	-10	39	26	77	67
9	5.4	8	4.6	7	3.7	GRADY OK	-14	-19	-31	42	42
3	5.0	6	10.0			GRANT OK	101				
3	4.9	2	3.2	2	4.7	GREER OK	-35	48	-4	64	56
1	3.2	2	5.0			HARMON OK	55				
2	5.7	3	9.6			HARPER OK	70				
4	7.0			2	2.8	HASKELL OK			-59	11	30
4	4.4	5	5.0	4	4.0	HUGHES OK	14	-19	-8	62	46
9	9.3	3	2.7	2	1.8	JACKSON OK	-71	-33	-81	02	20
2	3.5					JEFFERSON OK					
		1	1.7	2	3.2	JOHNSTON OK		90			36
16	7.1	8	3.1	16	5.8	KAY OK	-56	85	-18	53	69
4	6.8			3	4.2	KINGFISHER OK			-38	33	49
3	3.3	1	1.1	5	5.6	KIOWA OK	-67	425	71	87	66
1	2.7	2	3.7	2	2.9	LATIMER OK	38	-21	9	71	32
3	2.6	6	3.3	9	4.3	LE FLORE OK	25	31	63	86	50
3	2.9			3	2.1	LINCOLN OK			-30	42	22
4	4.0	3	2.6	1	.8	LOGAN OK	-34	-69	-79	03	14
1	2.3			2	4.6	LOVE OK			94	90	54
3	4.1	4	5.0	4	4.4	MC CLAIN OK	21	-13	5	69	51
7	6.4	5	3.7	5	3.4	MC CURTAIN OK	-42	-8	-46	24	38
1	1.4	2	2.5	3	3.4	MC INTOSH OK	74	37	139	94	38
2	3.6	1	1.8			MAJOR OK	-51				
1	2.5			4	6.6	MARSHALL OK			167	95	78
2	1.9	3	2.4	5	3.0	MAYES OK	25	26	57	85	33
1	1.6	6	6.7	4	4.6	MURRAY OK	309	-31	183	96	55
16	6.6	17	6.5	18	6.5	MUSKOGEE OK	-1	0	-1	66	77
2	3.3	1	1.3	4	5.9	NOBLE OK	-59	341	81	89	71
6	12.2	1	2.2			NOWATA OK	-82				
3	4.9	1	1.9	2	2.6	OKFUSKEE OK	-62	39	-47	23	28
65	5.4	81	5.1	75	4.1	OKLAHOMA OK	-6	-20	-25	47	47
9	4.9	7	3.8	7	3.8	OKMULGEE OK	-24	3	-22	49	44
3	1.8	7	4.5	6	3.6	OSAGE OK	143	-19	97	91	41
12	8.8	9	6.3	12	8.4	OTTAWA OK	-29	34	-5	64	90
				2	2.2	PAWNEE OK					23
9	5.3	11	6.0	7	3.5	PAYNE OK	13	-41	-33	38	40
7	3.5	8	4.3	9	4.3	PITTSBURG OK	23	1	24	76	51
5	3.4	5	3.6	3	1.8	PONTOTOC OK	6	-50	-47	23	19
12	5.8	6	2.6	9	3.4	POTTAWATOMIE OK	-54	28	-41	29	37
3	4.5	1	1.4	3	4.8	PUSHMATAHA OK	-70	254	7	70	57
3	9.0	5	16.0	3	7.9	ROGER MILLS OK	77	-51	-12	59	88
3	3.1	5	4.5	1	.8	ROGERS OK	42	-63	-76	04	13
8	5.9	5	3.5	4	3.1	SEMINOLE OK	-41	-12	-48	22	34
6	7.6	2	2.2	2	1.8	SEQUOYAH OK	-71	-16	-76	04	20
4	2.4	3	1.6	4	1.9	STEPHENS OK	-31	14	-21	50	20
3	4.5					TEXAS OK					

* SIGNIFICANTLY MORE DEATHS THAN EXPECTED

WHITE MALE — CANCER OF THE RECTUM — ICD 154 EXCEPT 154.3

1950-1959 DEATHS	1950-1959 RATE	1960-1969 DEATHS	1960-1969 RATE	1970-1979 DEATHS	1970-1979 RATE	COUNTY AND STATE	PERCENT CHANGE 60S/50S	PERCENT CHANGE 70S/60S	PERCENT CHANGE 70S/50S	PERCENTILE %CHANGE 70S/50S	PERCENTILE RATES 1970S
2	2.4	3	3.6	1	1.4	TILLMAN OK	53	-62	-42	29	16
58	6.2	39	3.1	67	4.6	TULSA OK	-50	48	-26	46	54
4	6.2	2	2.3	2	1.9	WAGONER OK	-63	-19	-70	07	20
11	7.9	2	1.5	4	1.9	WASHINGTON OK	-81	29	-75	04	21
4	4.9	3	3.3	1	1.0	WASHITA OK	-32	-69	-79	03	14
2	2.8	3	4.7	1	1.1	WOODS OK	67	-76	-60	11	15
5	5.7	7	8.0	4	4.7	WOODWARD OK	40	-42	-18	53	55
523	6.8	544	5.9	552	5.0	OREGON OR	-13	-16	-27		
6	6.9	6	5.7	5	4.6	BAKER OR	-17	-19	-33	39	55
5	4.0	10	6.5	10	5.1	BENTON OR	64	-22	28	78	61
53	12.2	31	5.3	49	6.0	CLACKAMAS OR	-57	14	-50	19	71
11	6.6	13	7.6	13	7.4	CLATSOP OR	14	-2	11	72	84
6	5.4	12	9.0	1	.7	COLUMBIA OR	66	-93	-88	00	13
11	5.5	17	7.1	20	6.6	COOS OR	30	-7	21	75	78
2	8.1	1	2.3	1	1.4	CROOK OR	-72	-38	-82	01	16
		5	6.8	6	6.3	CURRY OR		-8			74
9	8.0	9	6.7	8	3.9	DESCHUTES OR	-17	-41	-51	18	45
15	5.2	21	7.8	21	5.2	DOUGLAS OR	50	-34	0	67	62
1	8.2					GILLIAM OR					
		2	5.2	1	2.2	GRANT OR		-57			23
5	16.6			2	5.9	HARNEY OR			-64	09	70
5	8.1	4	5.0	3	4.1	HOOD RIVER OR	-38	-19	-50	20	47
23	6.8	22	4.8	19	3.2	JACKSON OR	-29	-34	-53	17	35
7	4.6	14	6.6	25	8.2	JOSEPHINE OR	43	23	77	88	89
6	4.2	11	5.5	13	5.2	KLAMATH OR	32	-5	25	77	63
2	6.5	2	6.1	1	2.3	LAKE OR	-5	-62	-64	09	24
27	5.1	29	4.3	47	4.8	LANE OR	-16	12	-6	63	58
6	4.0	4	2.5	9	4.3	LINCOLN OR	-39	74	7	70	51
24	10.3	23	8.0	26	6.8	LINN OR	-22	-16	-34	37	79
3	2.7	5	4.4	6	5.0	MALHEUR OR	64	12	84	89	60
27	5.0	37	5.5	35	4.0	MARION OR	10	-27	-20	52	46
3	12.3			1	4.1	MORROW OR			-67	08	47
187	7.5	189	6.9	163	6.0	MULTNOMAH OR	-8	-14	-20	51	71
9	7.6	3	1.9	5	2.5	POLK OR	-74	29	-67	08	26
1	8.1					SHERMAN OR					
		5	4.4	3	2.2	TILLAMOOK OR		-50			23
12	5.7	15	6.6	16	6.3	UMATILLA OR	16	-4	12	72	75
5	5.2	8	7.6	2	1.7	UNION OR	48	-78	-67	07	19
4	10.9	2	4.8	1	2.7	WALLOWA OR	-56	-45	-76	04	28
3	4.0	7	6.7	3	2.6	WASCO OR	66	-61	-35	36	28
30	8.3	24	4.7	29	4.1	WASHINGTON OR	-44	-13	-51	19	47
3*	23.1					WHEELER OR					
12	6.7	13	6.3	8	3.2	YAMHILL OR	-5	-49	-52	18	35
5233*	11.4	4800*	9.6	3847*	7.2	PENNSYLVANIA PA	-16	-25	-37		
24	10.9	27*	11.2	23	8.2	ADAMS PA	3	-27	-25	47	89
757*	12.0	707*	10.5	531*	7.5	ALLEGHENY PA	-12	-28	-37	34	86
34	9.1	18	4.6	32	7.7	ARMSTRONG PA	-50	67	-16	56	86
94*	12.3	91*	10.8	72*	7.8	BEAVER PA	-13	-28	-37	34	87
15	7.8	16	7.6	14	6.1	BEDFORD PA	-2	-20	-22	50	72
141*	10.8	154*	10.8	109	6.8	BERKS PA	1	-37	-37	34	80
80*	11.8	64	9.1	46	6.6	BLAIR PA	-23	-28	-45	26	77
39*	14.3	19	7.1	18	6.3	BRADFORD PA	-50	-11	-56	14	75
71	10.2	92	9.0	75	5.3	BUCKS PA	-12	-41	-48	22	64
45	9.4	52	9.7	39	6.5	BUTLER PA	4	-34	-31	41	76
103*	11.7	81	8.9	60	6.5	CAMBRIA PA	-24	-26	-44	27	77
3	10.0	2	7.5	4	13.0	CAMERON PA	-25	75	31	79	99
28	9.9	30*	10.4	37*	12.4	CARBON PA	5	19	25	77	98
21	9.4	28	10.5	25	8.1	CENTRE PA	12	-23	-14	57	89
59	8.5	78	9.6	64	6.0	CHESTER PA	13	-37	-29	43	71
14	7.3	15	8.1	7	3.4	CLARION PA	12	-58	-53	17	38
35	8.5	39	9.4	36	8.1	CLEARFIELD PA	10	-14	-5	64	89
14	8.5	12	7.3	16	8.8	CLINTON PA	-15	21	3	68	92
31	11.3	22	8.1	13	4.3	COLUMBIA PA	-28	-47	-62	10	50
31	8.3	31	7.8	36	8.3	CRAWFORD PA	-6	6	-1	66	90
45	10.7	43	8.2	46	7.0	CUMBERLAND PA	-24	-15	-35	36	81
102*	11.4	78	8.3	63	6.3	DAUPHIN PA	-27	-23	-44	26	75
194*	12.2	222*	10.3	168*	6.9	DELAWARE PA	-16	-33	-44	27	80
15	9.3	18	11.1	16	9.1	ELK PA	19	-17	-2	66	93
116*	12.0	103*	9.5	80	6.9	ERIE PA	-20	-27	-42	28	81
71	8.7	84*	10.0	53	6.2	FAYETTE PA	14	-38	-29	43	74
3	12.6	1	4.1			FOREST PA	-67				
30	8.6	36	9.2	41*	8.4	FRANKLIN PA	6	-8	-2	65	90
2	4.2	1	1.7			FULTON PA	-60				
22	10.4	12	5.6	14	6.2	GREENE PA	-46	10	-41	30	73
20	11.5	19	10.4	17	8.5	HUNTINGDON PA	-9	-18	-26	46	91
30	8.1	24	6.3	21	5.2	INDIANA PA	-23	-17	-36	35	62
18	6.9	19	7.3	15	5.6	JEFFERSON PA	6	-23	-18	53	67
1	1.4	5	6.4	5	5.4	JUNIATA PA	368	-16	294	98	65
172*	14.6	152*	12.7	125*	9.6	LACKAWANNA PA	-13	-24	-34	37	95
101	9.1	107	8.4	98	6.5	LANCASTER PA	-8	-22	-29	43	77
46	8.8	54*	10.1	42	7.8	LAWRENCE PA	15	-23	-12	59	87
36	9.9	36	8.6	29	5.8	LEBANON PA	-13	-32	-41	29	69
116*	11.8	138*	12.5	103*	8.3	LEHIGH PA	6	-33	-30	43	90
259*	15.8	197*	10.8	166*	8.6	LUZERNE PA	-31	-21	-46	25	91
50	9.9	46	8.8	26	4.5	LYCOMING PA	-11	-48	-54	16	54
27	10.1	25	9.5	26*	9.0	MC KEAN PA	-6	-5	-10	60	93
53	9.7	49	8.4	42	6.7	MERCER PA	-14	-20	-31	41	79
29*	16.0	20	10.4	20*	9.9	MIFFLIN PA	-35	-5	-38	33	96
18	9.7	32*	15.1	25*	8.1	MONROE PA	56	-46	-17	55	89
181*	11.0	199*	8.7	182	6.7	MONTGOMERY PA	-21	-23	-39	32	79
7	8.2	9	11.2	12*	13.5	MONTOUR PA	36	21	64	86	99
104*	11.5	96*	9.9	74	6.9	NORTHAMPTON PA	-14	-31	-40	31	80
42	8.0	58*	11.2	46*	8.8	NORTHUMBERLAND PA	40	-22	9	71	92
24*	22.1	8	5.8	13	8.6	PERRY PA	-74	49	-61	10	91
1064*	14.2	805*	10.8	500*	7.4	PHILADELPHIA PA	-23	-32	-48	23	84
7	13.9	8	10.2	9	8.4	PIKE PA	-27	-18	-40	31	90
9	9.2	5	5.2	9	8.9	POTTER PA	-43	70	-3	65	93

* SIGNIFICANTLY MORE DEATHS THAN EXPECTED

WHITE MALE — CANCER OF THE RECTUM — ICD 154 EXCEPT 154.3

1950-1959 DEATHS	1950-1959 RATE	1960-1969 DEATHS	1960-1969 RATE	1970-1979 DEATHS	1970-1979 RATE	COUNTY AND STATE	PERCENT CHANGE 60S/50S	PERCENT CHANGE 70S/60S	PERCENT CHANGE 70S/50S	PERCENTILE %CHANGE 70S/50S	PERCENTILE RATES 1970S
122*	13.2	100*	10.9	76*	8.5	SCHUYLKILL PA	-17	-22	-36	36	91
9	8.6	7	5.7	10	6.9	SNYDER PA	-34	21	-20	52	80
27	6.5	38	9.3	34	7.9	SOMERSET PA	42	-15	21	76	88
3	8.1	2	5.8	2	4.8	SULLIVAN PA	-29	-17	-41	30	57
18	11.4	11	6.5	12	6.7	SUSQUEHANNA PA	-43	3	-41	29	78
13	7.1	13	6.7	8	4.2	TIOGA PA	-5	-38	-41	29	49
8	7.7	7	6.2	11	8.6	UNION PA	-20	39	12	72	91
26	8.2	18	5.8	25	8.0	VENANGO PA	-30	38	-2	65	89
23	9.3	13	5.0	11	4.7	WARREN PA	-46	-6	-50	20	56
64	6.3	92	8.8	79	7.3	WASHINGTON PA	41	-17	16	74	84
19	11.1	17	9.6	10	5.2	WAYNE PA	-13	-46	-53	17	63
136	9.0	108	6.6	125	6.9	WESTMORELAND PA	-26	4	-23	49	81
15*	17.4	9	9.9	8	7.9	WYOMING PA	-43	-21	-55	15	88
97	10.8	78	7.0	93	7.1	YORK PA	-35	1	-34	37	82
473*	13.7	470*	11.8	411*	9.3	RHODE ISLAND RI	-13	-21	-32		
19*	13.3	13	7.7	7	3.3	BRISTOL RI	-42	-57	-75	04	37
58*	17.1	69*	14.1	66*	11.0	KENT RI	-18	-22	-36	35	97
21	9.7	32*	12.3	29*	9.1	NEWPORT RI	27	-26	-6	63	93
361*	14.1	335*	12.0	277*	9.4	PROVIDENCE RI	-15	-21	-33	39	94
14	7.3	21	8.4	32*	9.9	WASHINGTON RI	14	19	35	80	96
159	3.8	247	4.5	242	3.2	SOUTH CAROLINA SC	20	-29	-15		
		4	6.1	2	2.8	ABBEVILLE SC		-54			30
5	4.0	11	5.7	7	2.2	AIKEN SC	42	-62	-46	25	23
8	3.7	13	5.2	17	3.9	ANDERSON SC	42	-24	7	70	45
				3	8.1	BAMBEPG SC					89
1	3.3	1	2.4	1	1.8	BARNWELL SC	-28	-25	-46	24	19
2	6.0			2	1.2	BEAUFORT SC			-80	02	15
2	6.2	2	6.2	1	.6	BERKELEY SC	0	-90	-90	00	13
3	18.6	1	3.6			CALHOUN SC	-81				
22	7.4	20	5.8	26	5.1	CHAPLESTON SC	-22	-12	-31	41	61
4	4.8	12	11.3	4	3.0	CHEROKEE SC	135	-73	-37	34	33
3	6.5	3	3.3	4	4.1	CHESTER SC	-49	23	-37	34	48
3	4.4	1	1.6	3	2.7	CHESTERFIELD SC	-63	66	-39	32	29
				2	3.6	CLARENDON SC					40
3	5.6	2	3.3	1	1.1	COLLETON SC	-42	-67	-81	02	15
4	3.9	1	1.1	7	4.7	DARLINGTON SC	-72	330	21	76	56
		2	2.8			DILLON SC					
		1	1.1			DORCHESTER SC					
2	5.7	2	5.7			EDGEFIELD SC	-1				
		1	3.0	1	2.0	FAIPFIELD SC		-35			21
4	3.4	9	6.0	10	4.2	FLORENCE SC	76	-30	23	76	49
1	4.3	1	1.2	1	.8	GEOPGETOWN SC	-73	-28	-81	02	14
18	3.8	37	6.1	36	4.5	GREENVILLE SC	60	-25	20	75	54
4	3.6	2	1.6	4	2.2	GREENWOOD SC	-56	39	-39	32	23
1	10.6	1	2.9			HAMPTON SC	-73				
2	1.3	5	3.1	6	1.8	HORRY SC	132	-42	34	80	20
1	5.5					JASPER SC					
2	3.3	8	11.5	5	4.2	KERSHAW SC	243	-63	26	77	49
3	3.8	2	1.7	2	2.6	LANCASTER SC	-53	50	-30	42	28
5	4.2	8	5.4	2	1.1	LAURENS SC	29	-80	-74	05	15
1	3.2	2	6.3	2	5.3	LEE SC	94	-16	63	86	63
2	1.5	7	3.0	10	3.5	LEXINGTON SC	107	15	138	93	38
				1	7.3	MC CORMICK SC					84
1	3.7			3	4.2	MARION SC			11	72	48
3	5.8			3	4.0	MARLBOPO SC			-32	40	46
3	4.9	8	8.6	3	2.5	NEWBERRY SC	74	-70	-48	22	27
5	3.8	5	3.7	5	2.5	OCONEE SC	-2	-33	-34	38	27
3	2.8	6	4.4	6	3.5	ORANGEBURG SC	57	-19	26	78	40
3	2.7	8	4.8	6	2.6	PICKENS SC	79	-46	-3	65	28
11	3.2	15	3.7	15	2.8	RICHLAND SC	16	-25	-13	58	30
2	4.9	2	3.8			SALUDA SC	-22				
9	3.0	27	5.6	24	4.1	SPARTANBURG SC	87	-26	38	81	48
5	5.3	5	4.4	5	4.3	SUMTER SC	-18	-1	-18	54	51
		4	4.1	3	2.9	UNION SC		-28			32
2	3.4	2	3.2	1	1.7	WILLIAMSBURG SC	-4	-48	-50	20	18
6	3.9	6	4.0	8	3.3	YORK SC	3	-18	-16	56	36
205	6.6	220	6.5	197	5.5	SOUTH DAKOTA SD	-2	-15	-17		
2	9.0	2	6.8	1	2.7	AURORA SD	-25	-61	-71	06	28
10	9.1	8	7.5	8	7.2	BEADLE SD	-17	-4	-20	51	83
				1	7.8	BENNETT SD					87
2	3.3	5	9.4	3	4.3	BON HOMME SD	183	-54	31	79	51
6	7.1	10	10.4	5	5.1	BROOKINGS SD	47	-51	-27	45	61
10	7.1	13	7.8	11	6.2	BROWN SD	10	-21	-13	58	73
2	5.8	2	5.2	2	6.1	BPULE SD	-10	17	6	69	72
				3	5.8	BUTTE SD					69
2	3.4	6	10.3	5	8.5	CHARLES MIX SD	202	-17	150	94	91
3	7.5	1	1.9			CLARK SD	-75				
3	6.9	4	8.1	3	5.1	CLAY SD	17	-37	-26	46	61
11	11.6	8	8.1	10	9.5	CODINGTON SD	-31	18	-18	54	95
		3	17.0			CORSON SD					
2	10.4	1	3.0	3	9.5	CUSTER SD	-71	214	-9	61	94
4	4.9	9	10.4	3	2.3	DAVISON SD	114	-78	-52	18	24
3	3.9	7	10.5	6	8.5	DAY SD	167	-20	115	92	91
1	2.4			2	5.4	DEUEL SD			129	93	65
				1	6.9	DEWEY SD					81
				3	11.1	DOUGLAS SD					97
1	3.8	3	7.7	1	1.7	EDMUNDS SD	104	-78	-54	15	19
7	11.4	3	3.2	2	2.4	FALL RIVER SD	-72	-24	-79	03	26
3	10.4	1	2.8	1	3.2	FAULK SD	-73	15	-69	07	36
4	8.1	2	2.9			GRANT SD	-64				
2	4.5	6	13.1	2	3.9	GREGORY SD	192	-70	-12	59	46
1	6.1			2	12.0	HAAKON SD			97	90	98
5	13.1	2	4.3	4	7.8	HAMLIN SD	-68	82	-41	30	87
3	9.9	1	4.3	1	2.9	HAND SD	-56	-32	-70	07	32
1	4.6			2	10.7	HANSON SD			132	93	97

* SIGNIFICANTLY MORE DEATHS THAN EXPECTED

WHITE MALE

CANCER OF THE RECTUM

ICD 154 EXCEPT 154.3

1950-1959 DEATHS	1950-1959 RATE	1960-1969 DEATHS	1960-1969 RATE	1970-1979 DEATHS	1970-1979 RATE	COUNTY AND STATE	PERCENT CHANGE 60S/50S	PERCENT CHANGE 70S/60S	PERCENT CHANGE 70S/50S	PERCENTILE %CHANGE 70S/50S	PERCENTILE RATES 1970S
		1	9.4	1	11.7	HARDING SD		24			98
3	6.9	4	8.9	8*	15.7	HUGHES SD	29	77	129	93	99
4	7.7	3	4.8	2	3.6	HUTCHINSON SD	-38	-24	-53	17	41
		1	9.3	1	6.7	HYDE SD		-27			79
1	3.1	3	11.6	2	6.0	JERAULD SD	278	-48	95	90	71
				2*	24.4	JONES SD					99
4	7.9	3	5.0	1	1.6	KINGSBURY SD	-37	-68	-80	02	17
5	8.6	4	5.8	5	8.4	LAKE SD	-32	44	-2	65	90
4	5.2	7	8.4	3	3.9	LAWRENCE SD	61	-54	-25	47	44
5	6.6	2	2.8	6	6.4	LINCOLN SD	-57	128	-3	65	76
1	8.3					LYMAN SD					
5	11.6	7	13.3	2	4.6	MC COOK SD	15	-66	-61	11	54
3	8.3			1	2.3	MC PHERSON SD			-72	06	24
6	14.2	3	6.1	1	1.9	MARSHALL SD	-57	-69	-87	00	21
2	5.2			5	6.9	MEADE SD			33	79	80
1	8.6	2	23.9			MELLETTE SD	177				
1	2.6	1	3.1	2	25.3	MINER SD	20	725	887	99	99
26	7.6	25	6.8	22	5.1	MINNEHAHA SD	-11	-25	-34	38	60
2	3.9	5	10.3	2	4.1	MOODY SD	167	-61	5	69	47
12	9.6	9	4.6	14	6.0	PENNINGTON SD	-52	31	-37	34	71
4	12.2			1	3.1	PERKINS SD			-75	05	33
1	7.4	3	13.1			POTTER SD	76				
1	1.7	5	7.2	3	3.9	ROBERTS SD	327	-46	129	93	44
1	3.0	3	11.8	1	3.9	SANBORN SD	289	-67	29	78	45
1	20.6	2*	57.4	1*	74.3	SHANNON SD	179	30	262	97	99
5	11.3	6	8.6	4	5.7	SPINK SD	-24	-33	-49	21	68
1	8.9					STANLEY SD					
		1	8.5			SULLY SD					
3	6.0	6	14.2	4	8.8	TRIPP SD	138	-38	48	83	92
3	3.9	5	6.2	5	6.1	TURNER SD	57	-2	54	85	72
2	3.0	4	6.5	4	5.9	UNION SD	115	-8	98	91	71
6	18.6	5	11.4	5	10.4	WALWORTH SD	-39	-9	-44	26	97
4	3.7	2	2.1	4	4.3	YANKTON SD	-44	110	18	75	51
		1	19.1			ZIEBACH SD					
463	4.2	512	3.9	583	3.7	TENNESSEE TN	-8	-6	-14		
4	2.0	9	5.1	8	2.9	ANDERSON TN	152	-44	41	82	31
5	4.9	5	4.3	7	5.5	BEDFORD TN	-11	27	13	73	66
5	11.8	4	6.7	1	1.2	BENTON TN	-43	-82	-90	00	15
		4	10.5	1	2.7	BLEDSOE TN		-74			29
2	1.2	7	2.9	9	3.1	BLOUNT TN	141	5	153	95	34
7	6.0	3	2.4	8	4.3	BRADLEY TN	-60	80	-29	44	50
4	3.8	5	4.3	3	1.7	CAMPBELL TN	11	-60	-55	15	19
3	6.7	1	2.4			CANNON TN	-64				
6	5.0	10	9.1	6	4.2	CARROLL TN	82	-53	-15	56	50
8	5.4	7	3.9	13	6.0	CARTER TN	-27	54	12	72	72
				2	3.8	CHEATHAM TN					42
4	9.0	3	7.0	1	2.6	CHESTER TN	-23	-63	-72	06	27
3	3.2	7	6.9	2	1.8	CLAIBORNE TN	114	-75	-46	25	19
		1	3.3	2	5.0	CLAY TN		48			59
5	5.8	6	6.4	4	3.1	COCKE TN	11	-51	-46	24	34
2	2.4	3	2.5	5	3.2	COFFEE TN	2	28	32	79	35
4	6.3			4	5.9	CROCKETT TN			-6	63	70
2	2.4	3	3.4	1	1.0	CUMBERLAND TN	39	-71	-59	12	14
53	5.3	62	4.8	66	4.4	DAVIDSON TN	-9	-8	-16	55	52
		3	5.9			DECATUR TN					
2	4.7	1	1.6	2	2.7	DE KALB TN	-65	65	-43	27	28
2	2.3	3	3.0	4	3.5	DICKSON TN	28	16	49	84	38
4	3.9	9	6.5	8	5.3	DYER TN	68	-19	37	81	63
				1	2.9	FAYETTE TN					31
1	2.0					FENTRESS TN					
3	3.2	4	3.9	3	2.8	FRANKLIN TN	22	-28	-12	59	30
9	5.0	9	4.2	6	2.5	GIBSON TN	-15	-40	-49	21	27
6	5.7	3	2.9	2	1.8	GILES TN	-48	-38	-68	07	20
1	1.8			2	2.3	GRAINGER TN			29	78	24
3	1.8	8	4.4	9	4.3	GREENE TN	151	-3	145	94	50
5	5.0	2	1.7	4	2.5	HAMBLEN TN	-67	51	-50	20	26
28	5.0	35	4.8	37	4.0	HAMILTON TN	-5	-17	-21	51	46
1	2.7			1	3.1	HANCOCK TN			16	74	34
2	2.5	1	1.3	1	1.5	HARDEMAN TN	-50	17	-41	29	17
3	4.2	4	5.1	4	3.6	HARDIN TN	19	-28	-14	57	41
2	1.5	6	4.6	7	4.5	HAWKINS TN	206	-2	200	96	53
2	4.3	3	7.0	3	5.6	HAYWOOD TN	62	-21	29	78	66
4	5.6	1	1.4	3	2.9	HENDERSON TN	-76	111	-49	22	31
4	3.3	8	6.4	7	5.2	HENRY TN	91	-18	56	85	62
2	3.7	2	4.1	4	5.9	HICKMAN TN	11	46	62	86	70
1	6.8			3	9.5	HOUSTON TN			40	81	95
2	3.7			3	4.3	HUMPHREYS TN			16	74	51
1	2.6	1	1.9	3	4.8	JACKSON TN	-28	161	89	90	58
2	2.5	3	3.0	2	1.6	JEFFERSON TN	20	-44	-33	39	18
2	3.6			2	3.0	JOHNSON TN			-17	55	32
44	5.9	36	3.9	56	5.0	KNOX TN	-34	31	-14	57	60
		1	4.0	1	2.5	LAKE TN		-39			26
3	3.9	2	2.5	1	1.3	LAUDERDALE TN	-38	-49	-68	07	16
6	5.4			6	3.8	LAWRENCE TN			-29	43	43
1	4.5	1	2.9	3	7.8	LEWIS TN	-35	168	75	88	87
7	8.3	4	4.6	3	2.5	LINCOLN TN	-45	-45	-70	07	27
5	4.9	2	2.0			LOUDON TN	-59				
6	4.6	3	2.0	5	2.8	MC MINN TN	-57	41	-39	32	30
5	6.5	4	3.8	3	2.7	MC NAIRY TN	-41	-29	-58	12	29
1	2.0	2	2.6	2	2.4	MACON TN	28	-8	18	75	25
6	3.0	11	5.0	12	4.9	MADISON TN	64	-1	61	86	59
2	6.3	1	2.2	5	5.5	MARION TN	-65	151	-12	59	65
8	10.6	3	3.8			MARSHALL TN	-64				
7	5.9	6	4.2	5	2.8	MAURY TN	-28	-35	-53	16	29
				1	3.5	MEIGS TN					39
2	2.0	5	4.7	3	2.5	MONROE TN	140	-48	24	77	26
6	4.9	9	5.7	6	2.7	MONTGOMERY TN	16	-53	-45	25	29
1	5.0			1	5.7	MOORE TN			13	73	68
		2	3.8	1	1.9	MORGAN TN		-49			21

* SIGNIFICANTLY MORE DEATHS THAN EXPECTED

WHITE MALE — CANCER OF THE RECTUM — ICD 154 EXCEPT 154.3

1950-1959 DEATHS	1950-1959 RATE	1960-1969 DEATHS	1960-1969 RATE	1970-1979 DEATHS	1970-1979 RATE	COUNTY AND STATE	PERCENT CHANGE 60S/50S	PERCENT CHANGE 70S/60S	PERCENT CHANGE 70S/50S	PERCENTILE %CHANGE 70S/50S	PERCENTILE RATES 1970S
2	1.4	4	2.7	9	5.3	OBION TN	93	97	281	98	64
3	5.3	2	2.5	3	3.3	OVERTON TN	-53	32	-38	33	36
2	7.3	1	2.6			PERRY TN	-65				
		1	4.4			PICKETT TN					
1	2.2	4	8.6	2	4.0	POLK TN	298	-53	87	90	47
3	2.5	8	5.3	5	2.8	PUTNAM TN	116	-47	14	73	30
		1	1.4	4	4.1	RHEA TN		201			47
5	4.2	4	3.0	6	3.3	ROANE TN	-29	9	-22	49	36
6	6.3	4	3.4	5	3.5	ROBERTSON TN	-46	4	-44	26	40
7	4.8	5	2.8	4	1.6	RUTHERFORD TN	-43	-41	-66	08	18
5	8.6	3	4.3	1	1.3	SCOTT TN	-50	-70	-85	01	16
1	4.9	1	3.9	2	5.9	SEQUATCHIE TN	-21	54	22	76	70
3	3.4	6	4.7	7	4.5	SEVIER TN	38	-3	34	80	54
53	4.6	65	4.7	76	4.6	SHELBY TN	1	-3	-2	66	54
4	5.9	3	4.1	1	1.3	SMITH TN	-30	-69	-78	03	16
2	5.5	2	3.7	2	4.5	STEWART TN	-33	22	-19	53	53
12	4.2	18	4.7	24	4.3	SULLIVAN TN	14	-9	4	68	50
2	1.2	9	4.5	3	1.1	SUMNER TN	290	-76	-6	63	15
4	5.5	1	1.2	6	6.1	TIPTON TN	-78	405	12	72	73
		1	4.3			TROUSDALE TN					
4	7.0	2	2.8	5	5.9	UNICOI TN	-60	108	-16	56	70
1	2.0					UNION TN					
1	.9	5	4.5	8	5.8	WARREN TN	376	30	517	99	69
11	4.3	9	2.7	14	3.8	WASHINGTON TN	-37	43	-10	60	43
1	1.8			1	1.6	WAYNE TN			-13	58	18
3	1.9	5	3.2	4	2.3	WEAKLEY TN	69	-30	18	75	23
1	1.2	2	2.5	2	2.2	WHITE TN	98	-9	81	89	23
3	3.9	3	2.7	4	2.5	WILLIAMSON TN	-29	-8	-34	37	27
4	3.1	5	3.8	7	3.9	WILSON TN	22	2	25	77	45
1174	4.4	1386	4.0	1494	3.5	TEXAS TX	-9	-12	-20		
3	2.4	4	3.2	3	1.8	ANDERSON TX	32	-43	-25	47	20
				2	5.6	ANDREWS TX					67
5	3.8	9	4.8	9	3.9	ANGELINA TX	27	-19	4	68	45
		1	1.4	1	1.0	ARANSAS TX		-30			14
1	4.0			1	4.6	ARCHER TX			16	74	55
				1	6.3	ARMSTRONG TX					74
2	3.2			1	1.0	ATASCOSA TX			-70	07	14
2	2.7	3	4.2	3	3.1	AUSTIN TX	58	-27	15	73	33
		1	3.4	5	10.2	BANDERA TX		202			96
7	8.4	3	3.2	1	.8	BASTROP TX	-62	-76	-91	00	13
		1	2.2	1	2.6	BAYLOR TX		15			27
1	1.6	4	5.1	3	3.7	BEE TX	214	-27	130	93	42
10	3.6	12	3.4	15	3.8	BELL TX	-7	13	6	69	43
105	6.6	118	5.3	98	3.7	BEXAR TX	-19	-31	-44	27	41
3	13.1					BLANCO TX					
		1	16.3			BORDEN TX					
4	4.5	1	2.4	3	2.4	BOSQUE TX	-46	0	-47	24	25
7	3.8	10	4.3	7	2.4	BOWIE TX	12	-45	-38	33	25
5	3.9	9	4.1	16	5.0	BRAZORIA TX	5	22	29	78	60
1	.9	6	4.2	1	.6	BRAZOS TX	362	-85	-32	40	13
2	7.5	3	12.6	1	3.2	BREWSTER TX	68	-75	-57	13	35
3	13.6	1	4.9	1	3.3	BROOKS TX	-64	-33	-76	04	36
4	3.0	7	4.1	3	1.6	BROWN TX	39	-62	-48	22	17
4	7.6	2	5.0	3	4.1	BURLESON TX	-34	-18	-46	25	48
2	3.5	4	5.1	6	4.9	BURNET TX	47	-4	41	82	59
4	5.3	4	5.4	1	.9	CALDWELL TX	3	-83	-83	01	14
1	4.7	1	1.0	2	4.2	CALHOUN TX	-79	313	-12	59	48
3	5.2	1	1.4	2	2.5	CALLAHAN TX	-73	77	-52	18	27
18	5.3	9	1.7	8	1.4	CAMERON TX	-68	-17	-74	05	16
		2	7.3			CAMP TX					
		1	3.6	1	3.1	CARSON TX		-13			34
3	4.2	3	3.0	1	.8	CASS TX	-28	-74	-82	01	13
		1	1.9	1	1.4	CHAMBERS TX		-27			17
5	3.6	7	4.1	7	3.7	CHEROKEE TX	16	-10	4	68	42
2	4.3	1	1.8	2	3.3	CHILDRESS TX	-59	89	-23	49	37
3	5.4	1	2.1	3	4.4	CLAY TX	-62	113	-18	53	52
1	3.6					COCHRAN TX					
1	5.2	1	4.0			COKE TX	-23				
5	7.6	4	5.3	2	2.0	COLEMAN TX	-30	-62	-74	05	21
2	.9	7	3.0	15	5.2	COLLIN TX	234	73	479	99	62
		1	4.6			COLLINGSWORTH TX					
1	1.4	5	6.3	6	6.6	COLORADO TX	341	5	365	99	78
5	8.1	2	1.7	5	2.8	COMAL TX	-80	70	-65	08	30
2	2.0	4	4.4	4	4.7	COMANCHE TX	119	7	133	93	57
2	8.8	1	3.2			CONCHO TX	-64				
1	.9	8	7.2	4	3.1	COOKE TX	728	-57	255	97	34
		4	4.4	1	1.0	CORYELL TX		-77			14
		1	37.2			CRANE TX					
1	21.0					CROCKETT TX					
1	3.4	1	2.7			CROSBY TX	-20				
2	5.5	2	6.8	1	3.1	DALLAM TX	24	-54	-43	27	34
107	5.5	115	3.9	156	4.1	DALLAS TX	-29	6	-25	47	48
3	5.2	2	3.1	2	2.5	DAWSON TX	-40	-21	-53	17	26
		1	1.9	2	4.0	DEAF SMITH TX		115			46
2	4.9			1	6.4	DELTA TX			30	79	75
11	6.4	12	6.0	12	3.9	DENTON TX	-6	-35	-39	32	45
4	3.9	6	5.2	4	3.5	DE WITT TX	34	-34	-11	60	38
1	4.4	3	11.0			DICKENS TX	151				
1	3.1	2	5.4	1	2.2	DIMMIT TX	73	-58	-27	45	23
		1	4.6	1	2.7	DONLEY TX		-42			29
1	2.6	1	1.9	2	4.3	DUVAL TX	-28	129	65	86	50
5	3.4	4	2.7	2	1.6	EASTLAND TX	-19	-42	-53	17	18
2	2.8	5	2.7	15	5.7	ECTOR TX	-2	109	104	91	68
9	4.8	4	2.1	3	1.3	ELLIS TX	-56	-38	-73	06	16
27	4.4	33	4.1	43	4.3	EL PASO TX	-6	4	-2	66	50
9	7.3	5	3.4	10	6.3	ERATH TX	-53	82	-14	57	74
3	3.2	6	5.7	4	4.0	FALLS TX	78	-29	26	77	47
5	2.9	7	4.2	6	3.6	FANNIN TX	41	-14	21	76	40
5	3.9	12	8.0	8	5.7	FAYETTE TX	102	-28	45	83	68

WHITE MALE

CANCER OF THE RECTUM

ICD 154 EXCEPT 154.3

1950-1959		1960-1969		1970-1979		COUNTY AND STATE		PERCENT CHANGE			PERCENTILE	
DEATHS	RATE	DEATHS	RATE	DEATHS	RATE			60S/50S	70S/60S	70S/50S	%CHANGE 70S/50S	RATES 1970S
1	2.1	5	11.5			FISHER	TX	454				
2	5.1	2	3.5			FLOYD	TX	-32				
3	2.6	2	1.4	6	3.5	FORT BEND	TX	-45	150	36	80	40
		1	3.0	1	2.3	FRANKLIN	TX		-22			24
3	4.8	2	2.8	3	3.9	FREESTONE	TX	-41	41	-17	54	45
1	2.5	1	2.5			FRIO	TX	0				
		2	7.6	1	1.9	GAINES	TX		-75			21
17	6.0	30	6.6	19	3.5	GALVESTON	TX	10	-47	-42	28	39
1	4.0					GARZA	TX					
4	6.5	6	8.5	2	1.6	GILLESPIE	TX	32	-81	-75	05	18
		1	24.6			GLASSCOCK	TX					
2	9.3			3	10.5	GOLIAD	TX			14	73	97
2	2.2	2	2.1	4	4.5	GONZALES	TX	-3	112	105	91	54
2	2.0	4	4.2	7	4.9	GRAY	TX	105	17	140	94	59
16	5.0	16	4.3	21	4.9	GRAYSON	TX	-15	14	-3	65	58
6	3.1	16	7.0	19	6.4	GREGG	TX	127	-8	108	92	76
5	9.2	5	10.8	3	4.8	GRIMES	TX	17	-56	-48	22	57
3	3.2	2	1.5	5	3.5	GUADALUPE	TX	-53	131	8	70	39
5	4.7	4	2.7	5	3.4	HALE	TX	-42	25	-27	45	38
1	2.4					HALL	TX					
2	3.2	3	3.5	2	2.5	HAMILTON	TX	7	-27	-22	50	27
1	4.2	2	8.5	1	3.7	HANSFORD	TX	105	-57	-12	59	41
2	4.5	3	6.6			HARDEMAN	TX	48				
3	5.0			7	5.2	HARDIN	TX			6	69	63
115	5.3	169	5.0	187	4.1	HARRIS	TX	-6	-18	-23	49	47
3	2.7	2	1.5	7	4.8	HARRISON	TX	-45	219	76	88	57
2	3.1	3	4.8			HASKELL	TX	54				
2	2.5	1	1.1	3	3.4	HAYS	TX	-57	219	36	80	38
2	2.1	3	2.3	2	1.0	HENDERSON	TX	8	-58	-54	15	14
20	4.2	10	1.7	20	2.5	HIDALGO	TX	-60	49	-41	30	26
6	4.3	7	4.5	3	2.0	HILL	TX	5	-55	-53	17	21
1	1.6	3	3.8	2	3.2	HOCKLEY	TX	142	-15	105	92	36
		1	2.7			HOOD	TX					
4	3.5	11	8.7	1	.6	HOPKINS	TX	148	-93	-83	01	13
1	1.4	2	2.6	1	1.7	HOUSTON	TX	83	-34	21	76	19
6	6.0	2	1.9	8	5.6	HOWARD	TX	-69	198	-8	62	66
3	1.5	12	5.8	10	4.3	HUNT	TX	286	-26	185	96	50
		2	2.0	3	2.5	HUTCHINSON	TX		29			27
				1	9.6	IRION	TX					95
1	2.2	2	4.1	1	2.2	JACK	TX	87	-46	2	67	23
1	4.1	4	7.6			JACKSON	TX	83				
3	3.9			2	1.6	JASPER	TX			-60	11	18
28	5.4	23	3.3	38	4.5	JEFFERSON	TX	-39	34	-18	54	52
2	1.8	2	2.2	3	2.4	JIM WELLS	TX	23	9	34	80	25
1	.5	2	.9	7	2.6	JOHNSON	TX	72	186	393	99	27
2	1.8	5	4.4	8	6.6	JONES	TX	145	49	266	98	78
4	5.9			2	2.9	KARNES	TX			-51	19	31
7	5.1	4	2.8	2	1.2	KAUFMAN	TX	-46	-57	-77	03	15
2	5.9	2	3.9	2	3.8	KENDALL	TX	-33	-4	-36	36	43
		1	9.2	1	9.1	KENT	TX		-2			93
3	3.2	2	1.8	6	2.6	KERR	TX	-45	49	-17	54	28
				1	3.0	KIMBLE	TX					32
1	18.6					KINNEY	TX					
3	4.1	2	1.8	3	2.5	KLEBERG	TX	-56	42	-38	33	26
11	6.7	10	5.4	9	4.1	LAMAR	TX	-20	-24	-39	31	47
2	2.1	4	5.5	5	5.6	LAMB	TX	166	2	171	95	67
4	8.0			2	3.8	LAMPASAS	TX			-52	18	43
7	5.9	10	8.7	7	4.8	LAVACA	TX	48	-45	-19	53	57
		1	2.1	1	1.6	LEE	TX		-21			18
4	10.2	3	5.6	2	3.4	LEON	TX	-46	-39	-67	08	37
3	3.4	4	3.0	8	5.0	LIBERTY	TX	-11	65	46	83	59
8	7.8	3	3.6	3	3.8	LIMESTONE	TX	-54	5	-52	18	43
1	3.0	3	8.1	2	4.7	LIVE OAK	TX	173	-43	56	85	56
1	5.3			1	1.1	LLANO	TX			-80	02	15
13	4.7	10	2.3	17	3.3	LUBBOCK	TX	-50	40	-30	42	36
4	6.7	2	5.1	1	1.4	MC CULLOCH	TX	-23	-73	-79	03	16
24	4.9	29	4.8	23	3.4	MC LENNAN	TX	-3	-29	-31	41	37
3	7.2	2	5.5	2	4.9	MADISON	TX	-23	-11	-32	41	59
		1	3.3	2	5.5	MARION	TX		70			66
		4	15.4			MASON	TX					
5	7.2	4	4.8	4	4.0	MATAGORDA	TX	-33	-17	-45	26	46
		1	1.7	2	2.5	MAVERICK	TX		47			26
5	6.1	4	4.4	4	3.6	MEDINA	TX	-28	-18	-41	30	41
		1	4.6			MENARD	TX					
		3	1.7	12	6.5	MIDLAND	TX		280			76
4	3.6	3	2.3	6	4.6	MILAM	TX	-34	98	30	79	55
		3	8.3	2	3.2	MILLS	TX		-62			35
2	3.8			1	2.2	MITCHELL	TX			-42	28	23
3	3.2	2	2.1	4	3.3	MONTAGUE	TX	-33	56	4	68	36
		2	2.1	9	3.9	MONTGOMERY	TX		87			45
1	3.2	2	5.7			MOORE	TX	78				
2	4.4	1	2.1			MORRIS	TX	-53				
				1	3.9	MOTLEY	TX					45
3	2.4	5	4.0	7	4.7	NACOGDOCHES	TX	70	17	98	91	56
5	3.1	6	3.2	10	5.1	NAVARRO	TX	5	56	64	86	60
1	2.5			1	2.3	NEWTON	TX			-8	62	24
2	2.9	4	4.7	3	3.4	NOLAN	TX	62	-28	16	74	38
17	5.1	23	3.7	26	3.2	NUECES	TX	-27	-15	-38	33	35
				1	2.4	OCHILTREE	TX					25
4	3.7	5	3.5	7	3.7	ORANGE	TX	-4	4	0	67	41
4	3.7	2	1.9	5	4.0	PALO PINTO	TX	-50	117	8	70	46
2	3.2	1	1.2	2	2.3	PANOLA	TX	-63	89	-30	42	23
6	4.9	5	3.5	6	3.2	PARKER	TX	-27	-11	-35	36	35
		1	4.5			PARMER	TX					
1	2.0					PECOS	TX					
2	3.8	2	2.9	2	1.9	POLK	TX	-25	-35	-51	19	20
11	4.3	20	5.8	13	3.4	POTTER	TX	35	-42	-22	50	37
				1	3.9	PRESIDIO	TX					45
1	3.3	1	3.5	1	3.3	RAINS	TX	5	-6	-1	66	37
5	7.5	7	6.9	5	2.8	RANDALL	TX	-9	-60	-63	09	29
3	6.0			4	3.7	RED RIVER	TX			-38	32	41

WHITE MALE — CANCER OF THE RECTUM — ICD 154 EXCEPT 154.3

1950-1959 DEATHS	1950-1959 RATE	1960-1969 DEATHS	1960-1969 RATE	1970-1979 DEATHS	1970-1979 RATE	COUNTY AND STATE	PERCENT CHANGE 60S/50S	PERCENT CHANGE 70S/60S	PERCENT CHANGE 70S/50S	PERCENTILE %CHANGE 70S/50S	PERCENTILE RATES 1970S
				1	1.8	REEVES TX					20
2	3.8			1	1.8	REFUGIO TX			-53	16	19
3	5.1	3	4.3	4	4.7	ROBERTSON TX	-15	9	-7	62	56
2	6.8	2	7.0			ROCKWALL TX	3				
5	6.0	4	5.8	6	7.0	RUNNELS TX	-4	21	17	74	82
8	5.9	7	4.1	5	2.6	RUSK TX	-30	-38	-57	13	27
2	5.6	1	2.6	3	6.6	SABINE TX	-53	152	18	74	77
3	10.6			1	1.5	SAN AUGUSTINE TX			-86	00	17
3	18.8	1	4.5	3	6.5	SAN JACINTO TX	-76	45	-65	08	77
5	4.2	5	3.1	4	2.2	SAN PATRICIO TX	-25	-30	-48	22	23
1	2.1	1	1.7			SAN SABA TX	-19				
		1	6.7			SCHLEICHER TX					
1	1.6	1	1.4	3	3.5	SCURRY TX	-14	154	119	92	39
		3	12.9			SHACKELFORD TX					
3	2.9			7	5.3	SHELBY TX			85	89	63
1	12.2	2	19.7	2	12.5	SHERMAN TX	62	-37	3	68	98
18	7.5	12	4.4	11	2.7	SMITH TX	-41	-39	-64	09	29
1	5.5			1	3.9	SOMERVELL TX			-29	43	44
1	2.0	3	4.3	5	6.2	STARR TX	113	45	209	97	74
3	6.3	3	4.6	2	3.3	STEPHENS TX	-28	-28	-48	23	37
1	6.3					STONEWALL TX					
				1	4.6	SUTTON TX					54
		1	2.4	3	6.4	SWISHER TX		163			75
75	5.8	87	4.7	83	3.4	TARRANT TX	-19	-27	-41	30	38
8	3.0	21	6.2	11	2.8	TAYLOR TX	108	-54	-4	64	31
1	2.3	1	2.9	2	3.9	TERRY TX	25	36	70	87	45
1	5.6	1	4.5	1	4.5	THROCKMORTON TX	-21	1	-20	52	53
3	4.0	3	3.8	4	4.7	TITUS TX	-3	23	19	75	56
11	4.6	19	6.7	12	3.7	TOM GREEN TX	45	-44	-19	52	42
26	4.9	24	3.4	27	3.1	TRAVIS TX	-30	-10	-37	34	34
		1	3.3	2	3.7	TRINITY TX		11			42
				3	3.3	TYLER TX					36
5	6.0	5	5.2	4	3.4	UPSHUR TX	-14	-34	-43	27	38
		1	3.8	1	6.2	UPTON TX		62			74
2	3.2	6	8.2	5	5.7	UVALDE TX	160	-31	81	89	68
3	4.8	4	5.3	1	1.3	VAL VERDE TX	9	-76	-74	05	16
3	2.3	7	4.9	4	2.3	VAN ZANDT TX	115	-52	2	67	24
4	2.8	6	4.8	8	4.5	VICTORIA TX	70	-6	59	85	53
2	4.6	2	2.9	2	2.0	WALKER TX	-36	-31	-56	14	21
2	6.8	3	8.8	1	1.7	WALLER TX	29	-81	-75	04	19
1	6.3					WARD TX					
4	5.3	10	11.4	5	4.0	WASHINGTON TX	115	-65	-24	48	46
5	2.9	7	3.4	17	6.0	WEBB TX	18	78	111	92	71
8	7.6	3	2.1	6	4.0	WHARTON TX	-72	89	-47	24	46
1	2.3	2	3.8	2	4.7	WHEELER TX	64	23	102	91	56
12	3.7	17	3.9	22	4.8	WICHITA TX	6	22	30	78	57
3	3.3	2	2.0	3	3.4	WILBARGER TX	-40	71	2	67	37
		3	5.6			WILLACY TX					
11	6.4	13	6.6	9	4.0	WILLIAMSON TX	2	-40	-38	33	46
3	4.8			2	2.8	WILSON TX			-42	28	29
1	1.9			1	9.3	WINKLER TX			400	99	94
3	2.8	8	7.4	2	1.4	WISE TX	163	-81	-49	21	17
5	5.0	5	4.4	5	3.0	WOOD TX	-11	-32	-40	31	32
				1	3.5	YOAKUM TX					40
		3	2.7	4	4.1	YOUNG TX		49			47
		1	5.1	1	2.6	ZAPATA TX		-49			28
122	5.0	139	4.5	127	3.2	UTAH UT	-9	-29	-36		
				2	8.3	BEAVER UT					90
4	5.8	3	3.3			BOX ELDER UT	-44				
4	3.3	7	4.9	5	3.2	CACHE UT	47	-35	-5	63	35
3	4.2	3	4.7	5	5.9	CARBON UT	11	25	39	81	70
5	6.3	8	5.6	6	2.3	DAVIS UT	-12	-58	-63	09	24
2	9.6	3	9.9	1	2.4	DUCHESNE UT	2	-75	-75	05	26
				1	3.3	EMERY UT					37
1	5.6			1	7.5	GARFIELD UT			33	79	85
1	12.6					GRAND UT					
		2	5.4	1	2.2	IRON UT		-59			23
		2	9.0	1	3.7	JUAB UT		-59			42
1	2.5			1	2.1	MILLARD UT			-16	56	22
1	37.5	1	17.9			PIUTE UT	-52				
59	5.5	64	4.8	68	4.1	SALT LAKE UT	-13	-15	-26	46	48
1	18.9					SAN JUAN UT					
3	4.6	3	4.4	3	4.7	SANPETE UT	-5	8	3	68	56
		3	5.3			SEVIER UT					
				1	3.3	SUMMIT UT					37
2	5.8	3	5.1	1	2.0	TOOELE UT	-11	-62	-66	08	21
1	2.6	1	3.5	1	1.9	UINTAH UT	35	-46	-27	45	20
15	5.9	14	4.0	11	2.4	UTAH UT	-32	-41	-60	11	25
		3	16.8			WASATCH UT					
2	4.9	4	8.2	1	1.3	WASHINGTON UT	67	-84	-74	05	16
17	5.5	15	3.9	17	3.5	WEBER UT	-29	-9	-35	36	40
180	9.8	193*	10.0	164*	7.8	VERMONT VT	2	-21	-20		
6	6.4	8	8.0	8	7.3	ADDISON VT	25	-10	13	73	84
8	6.6	4	2.9	10	6.6	BENNINGTON VT	-57	130	0	67	77
13	10.0	17*	14.3	4	3.2	CALEDONIA VT	44	-78	-68	07	35
29	12.3	39*	14.2	23	7.3	CHITTENDEN VT	15	-48	-40	30	84
7*	20.1	2	5.5	4	11.9	ESSEX VT	-72	114	-41	30	98
12	8.9	12	8.3	10	7.0	FRANKLIN VT	-7	-15	-21	50	82
2	10.7					GRAND ISLE VT					
4	6.8	4	7.1	6	9.9	LAMOILLE VT	4	40	46	83	96
10	11.7	9	10.3	8	8.0	ORANGE VT	-12	-22	-31	41	89
17*	17.0	8	7.8	14*	12.8	ORLEANS VT	-54	64	-25	47	99
21	9.1	30*	12.3	25*	9.3	RUTLAND VT	35	-25	1	67	94
21	10.8	12	6.1	22*	10.3	WASHINGTON VT	-43	68	-5	64	96
13	7.8	20*	11.5	14	8.2	WINDHAM VT	47	-29	5	69	90
17	7.8	28*	12.5	16	6.5	WINDSOR VT	59	-48	-17	55	77

* SIGNIFICANTLY MORE DEATHS THAN EXPECTED

WHITE MALE

CANCER OF THE RECTUM

ICD 154 EXCEPT 154.3

1950-1959 DEATHS	1950-1959 RATE	1960-1969 DEATHS	1960-1969 RATE	1970-1979 DEATHS	1970-1979 RATE	COUNTY AND STATE	PERCENT CHANGE 60S/50S	PERCENT CHANGE 70S/60S	PERCENT CHANGE 70S/50S	PERCENTILE %CHANGE 70S/50S	PERCENTILE RATES 1970S
515	5.3	650	5.3	669	4.5	VIRGINIA VA	-1	-15	-16		
3	2.2	7	5.3	11	7.9	ACCOMACK VA	138	48	251	97	88
11	6.7	6	2.9	15	6.1	ALBEMARLE#VA	-57	112	-10	61	72
7	7.8	3	2.8	9	6.7	ALLEGHANY#VA	-64	134	-15	57	78
		1	4.1			AMELIA VA					
3	4.7	5	7.3	1	1.0	AMHERST VA	57	-87	-79	03	14
3	7.9	1	2.4	1	2.0	APPOMATTOX VA	-70	-15	-74	05	22
25	6.9	41	8.6	23	4.3	ARLINGTON#VA	24	-50	-38	32	50
16	5.6	24	7.6	28	8.2	AUGUSTA#VA	36	8	46	83	90
2	7.5	1	3.3	2	6.4	BATH VA	-56	95	-14	58	76
15	4.1	17	3.8	18	3.2	BEDFORD#VA	-6	-16	-21	51	35
2	8.7	1	3.1			BLAND VA	-64				
4	6.7			3	3.1	BOTETOURT VA			-53	17	34
1	3.1	2	5.3	1	2.3	BRUNSWICK VA	72	-57	-26	47	24
3	3.4	7	6.5	3	2.8	BUCHANAN VA	91	-57	-17	54	30
2	5.2	3	7.7	1	2.4	BUCKINGHAM VA	50	-69	-54	16	25
2	5.8	2	6.1	1	2.2	CAROLINE VA	6	-64	-62	10	23
6	3.0	10	4.5	5	2.0	CARROLL#VA	49	-56	-35	37	21
				1	37.2	CHARLES CITY VA					99
				3	5.9	CHARLOTTE VA					70
51	6.8	48	5.6	34	3.7	CHESTERFIELD#VA	-17	-34	-45	25	42
2	6.4	4	12.0	2	5.4	CLARKE VA	88	-55	-16	56	64
		1	6.4			CRAIG VA					
3	5.6	1	1.3	2	2.4	CULPEPER VA	-77	85	-57	13	25
1	7.1	1	5.0	1	5.4	CUMBERLAND VA	-30	9	-24	48	65
2	2.9	6	9.2	2	2.3	DICKENSON VA	210	-74	-21	51	24
6	3.4	11	4.8	10	4.1	DINWIDDIE#VA	39	-15	18	75	47
1	5.1	2	9.2	3	10.1	ESSEX VA	81	10	100	91	96
26	6.2	44	5.1	65	4.6	FAIRFAX#VA	-18	-10	-26	46	54
4	5.3	5	5.9	5	4.9	FAUQUIER VA	13	-18	-7	62	59
2	4.3	2	4.0	4	6.4	FLOYD VA	-8	61	49	84	76
1	2.9	3	11.1	3	8.0	FLUVANNA VA	286	-28	177	96	88
5	5.9	7	6.7	6	4.7	FRANKLIN VA	14	-30	-20	51	56
9	6.7	18	11.2	14	6.8	FREDERICK#VA	69	-40	2	67	79
2	3.6	4	6.1	6	6.4	GILES VA	67	6	76	88	76
1	2.5	2	3.9	2	3.2	GLOUCESTER VA	57	-18	30	78	35
2	7.4			1	3.8	GOOCHLAND VA			-48	22	43
		1	4.7			GREENE VA					
1	3.7	3	12.0			GREENSVILLE#VA	224				
6	6.3	6	5.1	6	5.1	HALIFAX#VA	-18	-1	-19	52	60
5	6.3	6	6.3	6	5.2	HANOVER VA	0	-19	-19	53	62
9	3.5	15	4.1	25	4.9	HENRICO VA	16	21	40	81	59
5	4.2	6	3.6	13	6.8	HENRY#VA	-15	92	63	86	80
1	4.1	1	6.1	1	4.6	HIGHLAND VA	50	-25	13	72	55
		1	3.6	4	7.2	ISLE OF WIGHT VA		100			83
3	3.2	4	2.8	9	5.0	JAMES CITY#VA	-12	75	55	85	59
1	5.7			1	4.6	KING AND QUEEN VA			-20	52	54
		3	11.4			KING GEORGE VA					
1	4.1	1	5.0	3	15.5	KING WILLIAM VA	21	210	275	98	99
1	3.2			4	7.6	LANCASTER VA			138	94	86
2	1.6	11	9.2	1	.7	LEE VA	468	-93	-58	13	13
5	5.5	4	4.3	5	3.8	LOUDOUN VA	-22	-12	-32	40	43
2	4.3	3	7.0	3	5.2	LOUISA VA	63	-26	20	75	62
3	7.2	2	4.7			LUNENBURG VA	-35				
4	12.0	2	5.7	3	7.1	MADISON VA	-53	25	-41	30	82
4	10.8	1	2.0	3	7.3	MATHEWS VA	-81	265	-32	40	84
3	8.9	3	3.8	9	9.0	MECKLENBURG VA	-58	140	1	67	93
1	3.5	2	6.5			MIDDLESEX VA	85				
5	5.1	4	2.7	6	3.4	MONTGOMERY#VA	-47	25	-34	37	37
4	7.8	2	5.3	1	2.5	NELSON VA	-32	-54	-68	07	26
				2	3.6	NORTHAMPTON VA					41
1	2.2			1	1.9	NORTHUMBERLAND VA			-12	59	21
2	4.1			2	4.0	NOTTOWAY VA			-3	65	46
4	8.5	3	5.2	5	7.3	ORANGE VA	-39	40	-14	57	84
5	7.7	7	9.2			PAGE VA	20				
		3	4.2	2	2.4	PATRICK VA		-44			25
14	5.1	21	6.5	20	5.7	PITTSYLVANIA#VA	27	-11	13	73	68
		2	9.3	2	6.2	POWHATAN VA		-33			74
1	2.6	2	4.6	5	10.9	PRINCE EDWARD VA	80	134	321	98	97
3	2.7	7	5.6	7	3.8	PRINCE WILLIAM#VA	107	-32	41	82	43
5	6.2	4	3.7	3	2.1	PULASKI VA	-40	-43	-66	08	22
2	11.1					RAPPAHANNOCK VA					
2	9.6	2	8.9	3	12.5	RICHMOND VA	-7	40	30	78	98
28	5.9	47	7.6	23	3.0	ROANOKE#VA	29	-61	-49	21	33
7	8.4	6	5.3	5	3.6	ROCKBRIDGE#VA	-37	-31	-57	13	41
15	8.2	17	7.3	22	8.6	ROCKINGHAM#VA	-11	17	5	68	91
7	6.1	8	8.2	2	2.0	RUSSELL VA	33	-76	-68	07	21
5	5.0	1	.8	4	3.0	SCOTT VA	-85	296	-39	31	33
8	8.0	3	3.2	12	8.8	SHENANDOAH VA	-61	179	10	71	92
7	5.8	6	4.9	2	1.4	SMYTH VA	-17	-71	-76	04	17
4	8.9	2	4.3	1	1.7	SOUTHAMPTON#VA	-52	-60	-81	02	19
3	3.7	4	4.5	5	3.8	SPOTSYLVANIA#VA	21	-16	2	67	42
1	2.8	5	7.2	3	4.5	STAFFORD VA	154	-37	59	85	53
				1	5.1	SURRY VA					61
		4	16.0			SUSSEX VA					
3	2.1	8	4.4	11	6.4	TAZEWELL VA	110	44	202	96	75
2	3.7	5	9.4	4	4.6	WARREN VA	156	-51	26	77	55
10	5.3	10	4.2	16	5.8	WASHINGTON#VA	-20	37	10	71	69
		1	1.9	2	2.9	WESTMORELAND VA		53			31
5	3.4	8	4.3	5	2.7	WISE#VA	27	-38	-22	50	28
1	1.1	8	8.7	3	3.7	WYTHE VA	682	-57	233	97	42
54	6.2	61	5.0	63	4.2	CHESAPEAKE CITY#VA	-20	-16	-33	39	49
14	8.7	8	3.1	15	5.4	HAMPTON CITY VA	-64	73	-38	33	65
9	5.8	13	7.7	22	7.1	NEWPORT NEWS CITY VA	33	-8	22	76	82
4	4.4	3	3.7	7	7.0	SUFFOLK CITY VA	-15	88	59	85	82
989	8.5	920	6.8	825	5.3	WASHINGTON WA	-19	-23	-38		
3	8.1	3	8.5	1	1.9	ADAMS WA	5	-78	-77	03	20

COUNTIES COMBINED; SEE TECHNICAL NOTES

WHITE MALE

CANCER OF THE RECTUM

ICD 154 EXCEPT 154.3

1950-1959 DEATHS	1950-1959 RATE	1960-1969 DEATHS	1960-1969 RATE	1970-1979 DEATHS	1970-1979 RATE	COUNTY AND STATE	PERCENT CHANGE 60S/50S	PERCENT CHANGE 70S/60S	PERCENT CHANGE 70S/50S	PERCENTILE %CHANGE 70S/50S	PERCENTILE RATES 1970S
4	5.8	5	5.6	5	6.4	ASOTIN WA	-5	15	9	71	75
16	9.9	14	6.7	13	4.6	BENTON WA	-32	-31	-53	17	55
10	5.3	13	6.1	12	4.9	CHELAN WA	15	-20	-8	62	59
7	4.3	3	1.6	11	4.6	CLALLAM WA	-63	183	5	69	54
27	6.8	31	6.3	40	6.5	CLARK WA	-8	4	-4	64	77
2	7.2			2	6.8	COLUMBIA WA			-6	63	79
19	8.4	19	7.1	14	4.3	COWLITZ WA	-16	-39	-49	22	51
2	3.4	3	6.2	4	5.1	DOUGLAS WA	81	-18	49	84	61
				2	9.3	FERRY WA					94
4	8.8	5	5.9	4	4.5	FRANKLIN WA	-34	-23	-49	21	53
1	4.6	1	6.1			GARFIELD WA	33				
5	9.7	8	5.8	4	2.0	GRANT WA	-41	-65	-79	03	22
30	9.7	22	7.0	19	5.9	GRAYS HARBOR WA	-28	-16	-40	31	69
3	4.7	9	11.7	6	4.1	ISLAND WA	146	-65	-14	57	47
3	5.5			4	4.3	JEFFERSON WA			-21	50	50
315	9.1	315	7.8	232	5.1	KING WA	-14	-35	-44	27	61
24	7.0	19	4.4	20	3.7	KITSAP WA	-37	-16	-47	23	42
7	5.9	9	8.1	8	6.8	KITTITAS WA	37	-15	17	74	80
13*	22.3	6	8.7	2	2.7	KLICKITAT WA	-61	-69	-88	00	29
26	10.2	19	7.2	13	4.5	LEWIS WA	-30	-37	-56	14	53
3	4.5	4	6.6	4	5.8	LINCOLN WA	45	-12	28	78	69
13*	15.5	6	6.2	6	3.9	MASON WA	-60	-37	-75	05	44
7	5.8	7	5.3	1	.7	OKANOGAN WA	-8	-87	-88	00	13
8	9.7	4	3.6	3	3.0	PACIFIC WA	-62	-17	-69	07	33
1	2.2	2	5.2			PEND OREILLE WA	141				
117	9.2	101	7.1	110	6.6	PIERCE WA	-23	-8	-29	43	77
				1	2.1	SAN JUAN WA					22
21	7.5	19	6.8	11	3.4	SKAGIT WA	-9	-50	-54	15	38
2	8.5	1	3.8	4	12.2	SKAMANIA WA	-56	225	44	83	98
56	8.6	52	6.2	66	6.1	SNOHOMISH WA	-28	-1	-29	43	73
111	9.0	105	7.8	94	6.5	SPOKANE WA	-13	-17	-28	44	76
7	6.1	6	5.7	3	2.5	STEVENS WA	-6	-57	-60	11	26
25	10.2	26	8.6	17	4.3	THURSTON WA	-15	-50	-57	13	50
1	4.7					WAHKIAKUM WA					
11	5.2	9	3.8	11	4.5	WALLA WALLA WA	-26	17	-13	58	53
32	8.6	29	7.1	19	4.4	WHATCOM WA	-17	-38	-48	22	52
10	7.9	5	3.7	11	7.9	WHITMAN WA	-53	113	0	66	88
43	7.5	40	5.7	48	6.4	YAKIMA WA	-24	12	-15	56	76
407	5.4	476	5.9	492	5.6	WEST VIRGINIA WV	9	-4	4		
6	6.7	5	5.5	6	6.8	BARBOUR WV	-17	22	1	67	79
13	7.9	10	6.3	12	5.5	BERKELEY WV	-20	-13	-30	42	66
6	5.0	4	3.2	6	4.6	BOONE WV	-36	42	-9	61	54
5	5.2	4	4.8	2	2.9	BRAXTON WV	-8	-39	-44	26	31
11	11.0	8	7.6	3	2.0	BROOKE WV	-31	-74	-82	01	21
27	6.0	18	3.7	25	5.0	CABELL WV	-38	37	-15	56	60
		3	6.8	1	1.6	CALHOUN WV		-77			18
				5	8.6	CLAY WV					91
1	2.6	2	5.7			DODDRIDGE WV	122				
5	2.1	16	7.1	13	4.8	FAYETTE WV	235	-32	127	93	57
4	8.8	2	4.4	1	1.9	GILMER WV	-50	-58	-79	03	20
3	7.8	2	4.6	3	6.1	GRANT WV	-40	31	-22	49	72
7	5.3	3	2.1	15	8.1	GREENBRIER WV	-61	295	52	84	89
3	5.1	10*	15.9	3	4.5	HAMPSHIRE WV	209	-72	-13	58	53
17	12.3	20*	14.6	14	8.1	HANCOCK WV	19	-44	-34	38	89
1	1.8	5	10.0	4	7.3	HARDY WV	450	-27	303	98	84
24	6.4	34	8.4	36*	8.7	HARRISON WV	32	3	35	80	92
3	4.2	2	2.5	4	4.4	JACKSON WV	-40	73	3	68	51
2	2.9	6	8.1	3	3.0	JEFFERSON WV	176	-63	1	67	32
44	5.2	44	5.2	61	6.0	KANAWHA WV	-1	16	15	73	71
8	6.0	6	4.6	6	4.8	LEWIS WV	-23	4	-20	51	57
1	1.1	2	2.1	2	2.1	LINCOLN WV	91	1	93	90	22
2	1.3	11	5.9	10	5.3	LOGAN WV	365	-10	319	98	64
5	3.6	9	5.9	12	7.3	MC DOWELL WV	64	24	104	91	84
23	7.3	30	9.4	25	7.5	MARION WV	28	-20	2	68	86
16	8.8	18	9.9	14	7.2	MARSHALL WV	12	-27	-18	53	83
5	5.1	3	2.7	6	4.8	MASON WV	-48	79	-6	63	58
14	5.4	15	5.5	9	3.0	MERCER WV	2	-46	-45	26	32
4	3.6	10	9.5	8	6.5	MINERAL WV	161	-32	78	88	76
4	2.5	5	3.8	8	5.7	MINGO WV	55	49	131	93	68
16	7.5	15	5.8	21	7.9	MONONGALIA WV	-22	35	6	69	88
7	10.4	3	3.9	4	5.3	MONROE WV	-62	36	-49	22	63
		3	6.9	3	6.4	MORGAN WV		-7			76
3	3.2	8	7.4	3	2.5	NICHOLAS WV	132	-66	-20	52	27
24	7.2	27	8.1	21	6.3	OHIO WV	12	-22	-12	59	75
		2	4.4	2	4.0	PENDLETON WV		-8			46
2	5.7	2	6.2	3	8.4	PLEASANTS WV	10	36	49	84	90
1	1.9	2	3.4	4	5.9	POCAHONTAS WV	79	77	217	97	70
7	4.9	8	5.5	9	6.0	PRESTON WV	11	9	21	76	71
2	2.2	3	2.9	9	6.8	PUTNAM WV	34	135	214	97	80
14	6.6	6	2.0	14	3.8	RALEIGH WV	-70	93	-42	28	43
11	8.8	11	7.4	5	3.7	RANDOLPH WV	-16	-51	-58	12	41
1	1.0	6	8.6	1	1.1	RITCHIE WV	801	-87	17	74	15
3	2.8	9	9.6	4	4.1	ROANE WV	250	-57	50	84	48
5	6.2	4	5.7	6	7.8	SUMMERS WV	-7	37	27	78	88
4	5.0	9	10.8	7	8.0	TAYLOR WV	118	-26	62	86	89
2	3.3			3	6.5	TUCKER WV			96	90	77
1	1.4	3	5.3	4	7.4	TYLER WV	279	41	435	99	85
3	3.1	3	3.4	4	3.6	UPSHUR WV	9	6	16	74	41
6	3.8	4	2.3	4	2.0	WAYNE WV	-40	-11	-46	24	21
3	4.9			5	7.3	WEBSTER WV			49	84	84
5	5.0	9	9.0	2	1.8	WETZEL WV	82	-80	-63	10	20
1	3.2	2	5.8	2	7.6	WIRT WV	83	31	140	94	86
19	6.3	28	7.8	30	7.6	WOOD WV	24	-3	21	75	86
3	3.3	2	2.3	5	3.6	WYOMING WV	-32	61	9	71	41
1910*	11.1	1702*	8.8	1452*	6.8	WISCONSIN WI	-21	-22	-38		
1	2.0	4	7.7	5	5.1	ADAMS WI	284	-34	152	94	60
10	9.9	7	5.6	11	9.6	ASHLAND WI	-43	70	-4	65	95

* SIGNIFICANTLY MORE DEATHS THAN EXPECTED

WHITE MALE

CANCER OF THE RECTUM

ICD 154 EXCEPT 154.3

COUNTY AND STATE	1950-1959 DEATHS	1950-1959 RATE	1960-1969 DEATHS	1960-1969 RATE	1970-1979 DEATHS	1970-1979 RATE	PERCENT CHANGE 60S/50S	PERCENT CHANGE 70S/60S	PERCENT CHANGE 70S/50S	PERCENTILE %CHANGE 70S/50S	PERCENTILE RATES 1970S
BARRON WI	13	7.1	16	7.4	14	5.9	4	-20	-17	55	70
BAYFIELD WI	2	2.8	6	7.0	7	7.3	145	4	155	95	83
BROWN WI	44	10.5	45	8.9	49*	8.1	-15	-9	-23	49	89
BUFFALO WI	14*	18.3	7	8.1	4	4.1	-56	-49	-78	03	48
BURNETT WI	8	12.0	5	5.6	2	2.1	-53	-62	-82	01	22
CALUMET WI	9	9.6	11	10.0	8	6.5	4	-35	-32	40	77
CHIPPEWA WI	14	6.5	14	6.0	18	7.1	-8	18	9	71	83
CLARK WI	11	5.8	18	8.9	17	8.2	54	-8	42	82	89
COLUMBIA WI	21	9.8	16	6.8	21	8.7	-31	29	-11	60	92
CRAWFORD WI	6	6.4	7	8.6	4	3.5	33	-59	-45	26	40
DANE WI	62	8.8	60	7.1	45	4.5	-20	-36	-49	22	54
DODGE WI	49*	15.3	26	7.8	24	6.4	-49	-19	-59	12	75
DOOR WI	10	8.1	5	3.9	7	4.9	-52	26	-40	31	58
DOUGLAS WI	21	8.3	13	5.3	15	6.1	-36	16	-26	46	73
DUNN WI	8	5.2	7	4.7	5	3.2	-9	-32	-39	32	35
EAU CLAIRE WI	39*	15.2	34*	12.0	22	7.0	-21	-41	-54	16	82
FLORENCE WI	3	13.4	6*	25.5	2	6.2	90	-76	-54	16	74
FOND DU LAC WI	40	11.5	37	9.6	39*	9.6	-16	0	-17	55	95
FOREST WI	3	4.8	6	10.9	4	6.3	127	-42	31	79	75
GRANT WI	15	6.7	18	8.0	7	2.8	19	-65	-58	13	30
GREEN WI	12	8.5	10	6.8	6	3.6	-20	-48	-58	12	40
GREEN LAKE WI	10	11.2	11	11.6	11	8.9	4	-23	-20	51	93
IOWA WI	13	12.0	13	11.8	6	5.4	-1	-54	-55	15	65
IRON WI	1	2.0	6	11.6	3	6.3	468	-46	207	96	74
JACKSON WI	4	4.1	7	6.5	4	3.9	59	-41	-6	62	44
JEFFERSON WI	30	11.5	28	9.6	25	7.9	-16	-18	-31	41	88
JUNEAU WI	6	5.5	7	6.0	8	5.9	8	-3	6	69	69
KENOSHA WI	49*	13.6	43	9.4	27	5.2	-31	-45	-62	10	63
KEWAUNEE WI	6	6.4	4	4.2	4	4.0	-34	-5	-38	33	46
LA CROSSE WI	27	8.5	32	8.9	27	6.9	5	-23	-19	52	80
LAFAYETTE WI	11	11.1	11	11.0	9	9.4	0	-14	-15	57	94
LANGLADE WI	9	7.6	17*	13.4	7	5.1	76	-62	-33	39	61
LINCOLN WI	10	8.0	11	8.1	15*	9.3	2	15	17	74	94
MANITOWOC WI	27	9.2	38	10.3	23	5.3	12	-49	-42	28	63
MARATHON WI	34	9.0	38	8.6	33	6.8	-4	-21	-24	48	79
MARINETTE WI	17	7.9	10	4.3	24*	10.2	-46	137	28	78	96
MARQUETTE WI	3	5.1	3	4.0	8	9.4	-22	135	83	89	94
MENOMINEE#WI	25	7.7	25	6.9	25	6.4	-10	-8	-17	55	76
MILWAUKEE WI	664*	17.4	475*	10.9	350*	8.1	-37	-26	-54	16	89
MONROE WI	4	2.4	6	3.1	12	6.0	33	91	153	94	71
ONEIDA WI	14	12.2	13	10.7	13	8.3	-12	-22	-32	40	90
OUTAGAMIE WI	43	12.6	27	6.4	44*	9.3	-49	47	-26	46	94
OZAUKEE WI	17	14.5	24*	13.4	12	5.3	-8	-60	-63	09	63
PEPIN WI	2	4.6	3	5.9	3	5.9	29	0	29	78	70
PIERCE WI	10	7.8	10	7.5	6	4.2	-3	-45	-46	24	49
POLK WI	11	7.2	10	5.6	7	3.7	-22	-35	-49	21	41
PORTAGE WI	25*	14.7	19	10.2	19	8.9	-31	-13	-40	31	93
PRICE WI	7	7.3	3	2.7	11	9.1	-64	242	24	77	93
RACINE WI	73*	13.6	55	8.7	46	6.7	-36	-23	-51	19	78
RICHLAND WI	3	3.0	5	4.8	9	8.2	58	71	171	95	90
ROCK WI	46	10.3	44	9.1	35	6.4	-11	-30	-38	33	75
RUSK WI	10	11.1	9	10.3	5	5.4	-7	-48	-52	18	65
ST CROIX WI	9	6.1	8	5.2	6	3.2	-14	-38	-47	23	35
SAUK WI	18	9.2	15	7.1	13	5.4	-23	-23	-41	29	65
SAWYER WI	6	8.7	5	8.0	7	9.9	-7	23	14	73	96
SHEBOYGAN WI	50	11.9	50*	10.8	40	7.7	-10	-28	-35	36	87
TAYLOR WI	8	7.8	13	12.3	8	6.8	56	-45	-13	58	79
TREMPEALEAU WI	14	9.5	13	7.8	8	3.7	-18	-53	-61	10	42
VERNON WI	9	5.8	7	4.5	14	7.1	-22	58	23	76	83
VILAS WI	2	3.2	7	10.4	7	7.1	225	-32	122	93	82
WALWORTH WI	25	9.7	21	7.6	18	5.4	-21	-29	-44	27	65
WASHBURN WI	5	7.4	8	10.1	8	9.4	35	-7	26	77	94
WASHINGTON WI	13	7.0	24	11.0	17	6.0	57	-45	-14	58	71
WAUKESHA WI	34	7.3	36	5.8	53	6.5	-22	12	-12	59	76
WAUPACA WI	17	7.8	19	6.7	24	8.4	-14	25	8	70	90
WAUSHARA WI	8	8.3	8	7.0	10	7.8	-16	12	-6	63	87
WINNEBAGO WI	45	10.1	50	9.5	27	4.6	-5	-51	-54	16	55
WOOD WI	21	8.8	33*	12.0	25	8.0	37	-33	-9	61	88
WYOMING WY	68	6.0	74	5.2	61	4.0	-12	-23	-32		
ALBANY WY	7	11.7	5	6.4	4	4.9	-45	-23	-58	13	59
BIG HORN WY	3	6.2	1	1.5	3	4.8	-76	220	-23	49	57
CAMPBELL WY					1	1.6					18
CARBON WY	1	1.2	1	1.8	3	4.3	49	142	259	97	50
CONVERSE WY	2	7.8	1	2.6	1	3.9	-66	50	-49	21	46
CROOK WY			1	6.6	1	4.2		-37			49
FREMONT WY	4	6.7	2	2.6	5	5.4	-61	109	-19	52	64
GOSHEN WY	2	2.9	3	4.1	3	4.0	43	-2	39	81	47
HOT SPRINGS WY	2	7.1	4	9.4	5	14.7	32	57	107	92	99
JOHNSON WY	1	2.3									
LARAMIE WY	13	8.2	12	5.3	8	3.5	-35	-35	-58	12	38
LINCOLN WY	2	6.6	2	5.5	2	5.6	-17	2	-15	56	67
NATRONA WY	9	6.4	14	7.4	7	3.8	16	-49	-41	29	42
NIOBRARA WY	1	4.8	2	11.1	1	5.4	130	-51	12	72	64
PARK-TETON#WY	2	3.2	4	6.3	5	5.2	99	-17	65	87	62
PLATTE WY			2	5.2	1	2.6		-50			28
SHERIDAN WY	10	8.4	8	6.2	5	4.1	-26	-34	-51	18	48
SUBLETTE WY			1	11.8	1	16.1		36			99
SWEETWATER WY	5	5.8	6	6.4	4	4.1	11	-36	-28	44	48
UINTA WY	1	2.7									
WASHAKIE WY	1	7.5	2	5.1	1	3.2	-32	-37	-57	13	36

* SIGNIFICANTLY MORE DEATHS THAN EXPECTED

COUNTIES COMBINED; SEE TECHNICAL NOTES

WHITE FEMALE

CANCER OF THE RECTUM

ICD 154 EXCEPT 154.3

1950-1959 DEATHS	1950-1959 RATE	1960-1969 DEATHS	1960-1969 RATE	1970-1979 DEATHS	1970-1979 RATE	COUNTY AND STATE	PERCENT CHANGE 60S/50S	PERCENT CHANGE 70S/60S	PERCENT CHANGE 70S/50S	PERCENTILE %CHANGE 70S/50S	PERCENTILE RATES 1970S
42704	6.0	42476	4.6	39607	3.4	UNITED STATES	-23	-27	-44		
333	3.8	323	2.7	328	2.0	ALABAMA AL	-29	-25	-47		
2	4.7	3	4.8	2	1.9	AUTAUGA AL	2	-61	-60	22	35
5	3.3	7	3.3	9	2.6	BALDWIN AL	-2	-19	-21	66	52
2	3.2	2	2.4			BARBOUR AL	-23				
3	6.5	3	6.0	1	1.1	BIBB AL	-7	-82	-83	02	22
7	8.3	5	4.3	1	.6	BLOUNT AL	-48	-87	-93	00	17
1	5.0	2	7.3			BULLOCK AL	46				
2	2.6	1	1.1	1	.7	BUTLER AL	-58	-36	-73	09	18
11	4.6	9	2.6	10	2.0	CALHOUN AL	-44	-23	-56	26	37
4	3.1	4	2.9	6	2.7	CHAMBERS AL	-8	-6	-13	71	53
1	1.2	1	1.3	1	.9	CHEROKEE AL	2	-28	-27	62	20
2	1.8	3	2.4	3	1.8	CHILTON AL	32	-24	0	77	33
		1	2.4	1	1.8	CHOCTAW AL		-27			32
2	4.3	1	1.1	2	2.0	CLARKE AL	-73	73	-54	28	37
1	1.3	3	4.0	4	3.7	CLAY AL	214	-7	193	97	74
2	4.4			2	2.9	CLEBURNE AL			-35	54	57
3	3.6	1	.8	1	.5	COFFEE AL	-79	-37	-87	01	17
5	4.0	5	2.9	7	2.8	COLBERT AL	-27	-2	-29	60	56
1	1.9	1	1.6	1	1.1	CONECUH AL	-15	-29	-40	48	22
2	5.2	2	4.5			COOSA AL	-13				
1	.6	2	1.1	10	3.8	COVINGTON AL	84	256	557	99	76
2	3.9			1	1.0	CRENSHAW AL			-74	08	21
9	4.8	6	2.3	8	2.2	CULLMAN AL	-51	-5	-54	29	42
1	1.3	3	3.1	2	1.4	DALE AL	139	-56	5	79	25
3	2.9	6	4.2	3	1.9	DALLAS AL	44	-55	-36	53	34
3	1.5	2	.8	4	1.4	DE KALB AL	-46	74	-7	74	27
1	1.0	2	1.5			ELMORE AL	49				
4	7.1	2	1.7	5	2.9	ESCAMBIA AL	-76	68	-60	23	57
8	2.5	10	2.6	11	2.1	ETOWAH AL	3	-17	-15	70	39
2	2.7	3	3.3	4	3.5	FAYETTE AL	21	6	28	85	69
		2	1.4	4	2.2	FRANKLIN AL		57			42
4	4.5	4	3.5	1	.6	GENEVA AL	-23	-84	-88	01	17
1	1.8	1	1.7	1	1.1	HENRY AL	-4	-34	-37	51	22
8	6.1	4	2.0	3	1.1	HOUSTON AL	-67	-45	-82	03	22
8	5.7	4	2.6	2	.7	JACKSON AL	-54	-72	-87	01	18
83	5.1	71	3.1	69	2.4	JEFFERSON AL	-38	-24	-53	30	46
3	5.3			2	1.8	LAMAR AL			-66	16	34
8	4.2	11	3.9	17	4.7	LAUDERDALE AL	-7	21	13	81	86
3	4.2	5	6.4	5	4.1	LAWRENCE AL	53	-36	-2	76	80
3	4.4	4	3.6	6	2.9	LEE AL	-18	-20	-34	55	57
		2	1.4	2	.9	LIMESTONE AL		-33			20
1	3.8	1	2.4			MACON AL	-37				
5	1.5	17	4.4	12	2.0	MADISON AL	191	-55	32	86	38
				2	2.6	MARENGO AL					50
1	.8	3	2.2	2	1.2	MARION AL	164	-48	37	87	22
5	2.6	1	.4	6	1.7	MARSHALL AL	-85	326	-37	52	30
32	4.9	24	2.6	32	2.6	MOBILE AL	-47	1	-47	38	50
1	2.9	2	2.5	1	.9	MONROE AL	-11	-64	-68	14	20
27	7.2	28	5.2	16	2.3	MONTGOMERY AL	-27	-57	-68	13	44
6	3.1	6	2.1	7	1.8	MORGAN AL	-33	-12	-41	47	34
2	6.2	1	2.3			PERRY AL	-63				
1	1.8	3	4.1	2	2.1	PICKENS AL	132	-49	17	82	39
1	1.5					PIKE AL					
1	1.1	2	1.8	3	2.2	RANDOLPH AL	61	24	99	94	41
4	4.2	5	4.3	1	.6	RUSSELL AL	2	-87	-87	01	17
1	1.3	3	2.4	5	2.9	ST CLAIR AL	93	20	132	95	58
5	4.3	4	3.0	3	1.4	SHELBY AL	-29	-54	-68	14	26
1	3.5	2	5.6	1	2.3	SUMTER AL	59	-58	-34	55	45
7	4.8	6	2.8	4	1.3	TALLADEGA AL	-42	-52	-72	10	25
4	5.2	1	.6	1	.5	TALLAPOOSA AL	-88	-16	-90	00	17
6	2.2			9	1.7	TUSCALOOSA AL			-21	66	31
10	5.7	13	4.6	7	1.7	WALKER AL	-20	-62	-69	12	32
				1	1.9	WASHINGTON AL					35
1	1.0	3	3.8	1	1.0	WINSTON AL	268	-74	-5	75	20
		9	4.2	16	4.4	ALASKA AK		4			
160	5.3	225	3.7	280	2.6	ARIZONA AZ	-29	-31	-51		
		1	2.4	1	3.2	APACHE AZ		35			65
9	5.9	6	2.5	12	3.6	COCHISE AZ	-58	46	-39	50	73
		1	.7	5	4.9	COCONINO AZ		600			88
3	2.8	3	3.0	6	3.9	GILA AZ	7	31	40	87	77
1	2.2			3	3.3	GRAHAM AZ			47	68	66
1	4.0			1	2.0	GREENLEE AZ			-51	32	37
80	5.1	120	3.6	163	2.5	MARICOPA AZ	-30	-29	-51	32	49
2	8.5	3	3.1	1	.3	MOHAVE AZ	-63	-91	-97	00	16
2	3.8	2	1.9	5	5.5	NAVAJO AZ	-50	190	46	88	92
45	6.6	57	4.3	59	2.6	PIMA AZ	-35	-39	-61	21	51
4	3.0	6	3.2	4	1.4	PINAL AZ	6	-57	-54	28	25
		1	1.7	1	.8	SANTA CRUZ AZ		-52			19
8	7.0	16	8.4	13	3.4	YAVAPAI AZ	20	-60	-52	31	68
5	4.3	9	4.7	6	2.3	YUMA AZ	9	-50	-46	40	45
258	3.9	235	2.7	264	2.2	ARKANSAS AR	-31	-16	-43		
3	3.7	3	2.8	3	2.3	ARKANSAS AR	-24	-19	-38	50	44
3	4.7			5	4.6	ASHLEY AR			-2	76	86
2	3.2			3	1.0	BAXTER AR			-69	12	20
9	3.8	9	2.8	9	1.9	BENTON AR	-27	-31	-49	34	36
5	6.7			4	1.4	BOONE AR			-79	04	26
		3	4.5	1	.9	BRADLEY AR		-81			19
1	3.2					CALHOUN AR					
3	4.4	4	3.4	2	1.4	CARROLL AR	-22	-59	-68	14	26
2	4.8					CHICOT AR					

WHITE FEMALE

CANCER OF THE RECTUM

ICD 154 EXCEPT 154.3

1950-1959 DEATHS	1950-1959 RATE	1960-1969 DEATHS	1960-1969 RATE	1970-1979 DEATHS	1970-1979 RATE	COUNTY AND STATE	PERCENT CHANGE 60S/50S	PERCENT CHANGE 70S/60S	PERCENT CHANGE 70S/50S	PERCENTILE %CHANGE 70S/50S	PERCENTILE RATES 1970S
2	3.0	2	2.2			CLARK AR	-26				
6	5.3	3	2.4	3	1.9	CLAY AR	-55	-22	-65	17	35
2	4.0	1	1.9	4	3.5	CLEBURNE AR	-53	86	-13	71	70
		1	2.6	1	2.0	CLEVELAND AR		-22			37
4	4.9	3	2.6	2	1.0	COLUMBIA AR	-47	-62	-80	04	21
		1	1.1	3	3.3	CONWAY AR		211			67
3	1.9	7	2.7	6	1.6	CRAIGHEAD AR	40	-40	-16	69	30
4	3.5	6	4.2			CRAWFORD AR	19				
1	3.1			2	2.1	CRITTENDEN AR			-31	58	40
1	2.1	2	3.0	5	5.5	CROSS AR	43	87	167	97	92
3	8.8					DALLAS AR					
				1	1.4	DESHA AR					25
1	1.6	2	3.0	5	7.9	DREW AR	92	162	403	99	98
4	3.6	4	3.0	6	3.1	FAULKNER AR	-15	2	-13	71	62
5	8.2	2	2.2	1	.8	FRANKLIN AR	-74	-63	-90	00	19
		5	7.5	2	1.8	FULTON AR		-76			34
11	4.6	14	4.3	20	4.1	GARLAND AR	-5	-5	-9	73	80
		1	1.7	2	2.2	GRANT AR		33			42
1	.7	1	.7	8	4.4	GREENE AR	-3	521	501	99	84
4	4.9	7	6.2	4	3.2	HEMPSTEAD AR	26	-48	-34	55	65
3	3.7	2	2.1	3	2.0	HOT SPRING AR	-43	-5	-46	40	37
3	4.8	6	8.0	3	2.9	HOWARD AR	65	-63	-39	49	59
5	4.6	1	.6	2	.9	INDEPENDENCE AR	-87	50	-80	04	19
1	2.2	1	1.5	3	3.5	IZARD AR	-30	128	60	90	69
7	9.1	2	2.0	3	2.4	JACKSON AR	-78	21	-73	09	47
8	4.6	5	2.5	7	2.4	JEFFERSON AR	-46	-2	-47	38	47
5	5.7	1	1.5	1	.8	JOHNSON AR	-73	-45	-85	02	19
2	5.6	2	5.8	1	2.7	LAFAYETTE AR	3	-54	-52	30	52
1	1.2	2	1.7	2	1.7	LAWRENCE AR	44	0	44	88	32
		1	2.8	1	2.2	LEE AR		-20			42
				2	4.4	LINCOLN AR					83
2	8.0	2	5.6	1	2.3	LITTLE RIVER AR	-30	-58	-71	11	45
3	3.0	4	3.2	2	1.3	LOGAN AR	4	-59	-58	25	24
1	.9	1	1.0	5	3.3	LONOKE AR	10	215	247	98	66
2	4.1	1	1.5	1	1.1	MADISON AR	-63	-27	-73	09	22
		1	2.4	1	1.0	MARION AR		-58			21
3	2.5	3	2.0	8	4.1	MILLER AR	-21	104	61	91	80
5	2.9	7	3.7	3	1.3	MISSISSIPPI AR	28	-66	-57	26	23
2	4.8			1	1.9	MONROE AR			-61	21	34
1	2.9					MONTGOMERY AR					
1	1.9	5	8.7			NEVADA AR	352				
1	2.8	2	5.7	3	6.7	NEWTON AR	100	19	137	96	96
1	1.2	2	1.5	3	1.8	OUACHITA AR	25	18	48	89	33
2	6.7	1	3.4	4*	8.1	PERRY AR	-49	136	20	83	98
2	3.0	3	3.2	1	.9	PHILLIPS AR	7	-72	-70	12	19
		3	4.9	2	3.0	PIKE AR		-38			61
5	5.7	6	4.6	5	3.3	POINSETT AR	-20	-28	-42	45	66
2	2.6	4	4.4	1	.7	POLK AR	67	-83	-71	10	18
2	1.6	1	.9	6	3.1	POPE AR	-46	255	92	93	63
2	4.1	1	2.1	1	1.6	PRAIRIE AR	-49	-23	-61	21	29
42	5.4	28	2.5	35	2.5	PULASKI AR	-53	0	-53	29	49
3	4.3	2	2.1	5	4.7	RANDOLPH AR	-51	124	10	80	87
1	2.8	1	1.5	1	1.0	ST FRANCIS AR	-46	-35	-65	17	21
4	3.8	3	1.8	2	.9	SALINE AR	-53	-47	-75	07	20
1	2.1			1	1.8	SCOTT AR			-16	69	32
				1	1.4	SEARCY AR					26
11	3.8	14	3.5	15	2.6	SEBASTIAN AR	-8	-26	-32	57	50
1	1.9	8*	10.4			SEVIER AR	457				
2	5.4			2	1.8	SHARP AR			-67	15	33
1	3.2	2	5.4	2	4.5	STONE AR	69	-16	41	88	85
13	8.4	4	2.0	6	2.3	UNION AR	-77	15	-73	09	43
1	2.2			1	1.3	VAN BUREN AR			-42	45	24
17	6.5	11	3.3	8	1.8	WASHINGTON AR	-50	-47	-73	09	32
6	3.7	5	2.2	5	1.7	WHITE AR	-40	-23	-54	29	32
1	2.4	3	6.8			WOODRUFF AR	181				
2	3.4	3	3.2	2	1.4	YELL AR	-3	-55	-57	26	27
3522	5.9	3812	4.5	3611	3.2	CALIFORNIA CA	-24	-28	-45		
258	6.5	202	4.1	189	3.4	ALAMEDA CA	-36	-17	-47	39	68
2	3.7	3	3.9			AMADOR CA	4				
22	5.5	23	3.8	18	2.0	BUTTE CA	-32	-48	-65	17	36
3	7.4	1	1.5	4	2.5	CALAVERAS CA	-80	74	-66	16	49
3	5.8	4	5.5	1	1.3	COLUSA CA	-6	-77	-78	05	24
61	6.5	66	3.9	105	3.7	CONTRA COSTA CA	-40	-4	-42	45	74
5	14.3	6*	8.9	2	2.5	DEL NORTE CA	-38	-72	-82	03	49
4	5.0	7	4.7	4	1.1	EL DORADO CA	-5	-77	-78	05	22
57	4.9	63	3.7	54	2.5	FRESNO CA	-23	-34	-49	35	48
3	4.0	2	1.8	2	1.6	GLENN CA	-55	-12	-60	22	29
26	8.2	19	4.6	19	3.8	HUMBOLDT CA	-44	-18	-54	28	75
7	5.0	7	4.0	5	1.7	IMPERIAL CA	-20	-57	-66	16	31
1	2.3	6	8.7	2	2.7	INYO CA	276	-69	17	82	53
50	6.2	58	5.0	50	3.0	KERN CA	-19	-39	-51	32	61
6	3.6	9	4.3	5	1.9	KINGS CA	17	-55	-48	37	35
5	5.9	4	2.8	10	3.3	LAKE CA	-53	17	-45	42	66
2	3.1	4	6.9	1	1.1	LASSEN CA	121	-84	-65	17	22
1538	5.9	1565	4.7	1241	3.3	LOS ANGELES CA	-21	-30	-45	42	66
6	4.2	7	3.5	9	3.6	MADERA CA	-15	1	-15	70	71
36	8.0	34	4.3	40	3.6	MARIN CA	-47	-15	-55	28	72
2	7.0	1	2.5	1	1.8	MARIPOSA CA	-64	-27	-74	08	34
12	6.0	10	3.8	10	2.7	MENDOCINO CA	-36	-29	-55	28	53
13	5.8	15	4.4	11	2.3	MERCED CA	-24	-49	-61	21	43
4	11.8					MODOC CA					
				3*	15.5	MONO CA					99
39	8.1	44	5.3	34	2.9	MONTEREY CA	-34	-46	-64	18	57
18	4.8	17	3.7	25	3.7	NAPA CA	-22	-1	-23	65	73
7	6.3	12	7.2	11	3.3	NEVADA CA	15	-54	-47	39	67
94	5.0	169	4.2	230	3.0	ORANGE CA	-16	-29	-40	47	60
10	4.4	16	4.6	15	2.6	PLACER CA	4	-44	-42	46	50
1	2.2	3	4.5	1	1.2	PLUMAS CA	100	-73	-46	40	23
54	5.0	75	3.6	105	2.8	RIVERSIDE CA	-27	-22	-43	44	56

* SIGNIFICANTLY MORE DEATHS THAN EXPECTED

1950-1959 DEATHS	1950-1959 RATE	1960-1969 DEATHS	1960-1969 RATE	1970-1979 DEATHS	1970-1979 RATE	COUNTY AND STATE	PERCENT CHANGE 60S/50S	PERCENT CHANGE 70S/60S	PERCENT CHANGE 70S/50S	PERCENTILE %CHANGE 70S/50S	PERCENTILE RATES 1970S
78	5.5	114	5.3	113	3.5	SACRAMENTO CA	-4	-34	-36	53	70
4	6.5	6	7.4	3	3.0	SAN BENITO CA	14	-59	-54	29	60
73	4.2	101	3.5	103	2.6	SAN BERNARDINO CA	-16	-28	-39	49	50
160	5.0	227	4.4	261	3.1	SAN DIEGO CA	-13	-29	-38	51	62
354*	8.2	293*	6.1	207*	4.3	SAN FRANCISCO CA	-26	-29	-48	37	82
54	6.0	46	3.6	48	2.8	SAN JOAQUIN CA	-40	-23	-53	29	55
19	5.8	25	4.8	23	2.8	SAN LUIS OBISPO CA	-18	-41	-52	31	55
71	5.7	102	4.8	115	3.8	SAN MATEO CA	-15	-21	-33	56	76
43	6.5	47	4.4	54	3.5	SANTA BARBARA CA	-33	-20	-47	39	69
101	5.4	139	4.4	169	3.6	SANTA CLARA CA	-19	-18	-34	55	71
32	5.6	51	5.6	30	2.6	SANTA CRUZ CA	0	-54	-54	29	51
5	2.7	7	2.4	15	3.2	SHASTA CA	-9	32	19	83	64
1	11.2	1	7.0			SIERRA CA	-38				
7	5.7	6	3.7	3	1.2	SISKIYOU CA	-35	-67	-79	05	23
22	6.6	26	5.2	35*	4.7	SOLANO CA	-20	-9	-28	61	87
39	5.6	44	3.9	57	3.3	SONOMA CA	-30	-15	-41	46	66
22	3.3	28	2.9	36	2.7	STANISLAUS CA	-13	-7	-19	68	53
3	3.1	3	1.7	7	3.2	SUTTER CA	-45	93	6	79	65
3	3.4	4	2.6	6	2.9	TEHAMA CA	-22	11	-13	71	58
3	11.8					TRINITY CA					
31	5.3	31	3.7	41	3.6	TULARE CA	-30	-3	-32	58	72
4	5.7	3	2.6	7	3.7	TUOLUMNE CA	-54	41	-35	54	74
23	4.3	40	3.7	54	2.8	VENTURA CA	-14	-24	-34	55	56
17	9.2	11	4.0	10	2.5	YOLO CA	-57	-37	-73	09	49
4	4.2	5	3.8	7	4.3	YUBA CA	-10	15	3	78	83
362	5.1	299	3.2	308	2.5	COLORADO CO	-38	-23	-52		
9	5.8	13	4.4	14	2.2	ADAMS CO	-24	-50	-62	20	42
3	11.0					ALAMOSA CO					
1	5.1					BACA CO					
2	6.3	2	5.7			BENT CO	-10				
13	4.2	18	3.8	13	1.8	BOULDER CO	-8	-53	-57	25	33
3	6.7	3	5.2	1	1.4	CHAFFEE CO	-23	-73	-80	04	26
1	5.6					CHEYENNE CO					
		1	6.1			CLEAR CREEK CO					
		1	2.2			CONEJOS CO					
		1	6.8			COSTILLA CO					
1	5.2					CROWLEY CO					
1	11.1					CUSTER CO					
6	6.3	4	2.8	3	2.0	DELTA CO	-55	-30	-68	13	37
155	5.6	129	3.7	129	3.1	DENVER-ARAPAHOE#CO	-34	-17	-45	41	62
1	18.9			1	13.3	DOLORES CO			-30	59	99
1	5.7			3	6.0	DOUGLAS CO			4	78	94
2	11.9	3*	14.2			EAGLE CO	19				
		2	8.1			ELBERT CO					
27	5.3	23	2.9	25	2.3	EL PASO CO	-46	-19	-56	26	45
9	7.0	6	3.8	4	1.8	FREMONT CO	-46	-54	-75	07	32
1	1.7	1	1.2			GARFIELD CO	-28				
1	28.1					GILPIN CO					
1	8.6	1	4.8	1	4.4	GRAND CO	-44	-8	-49	35	84
5	11.6			2	3.2	HUERFANO CO			-72	10	65
2*	76.5					JACKSON CO					
18	6.2	19	3.0	25	2.2	JEFFERSON CO	-52	-26	-64	18	43
1	3.3	2	4.5	2	5.3	KIT CARSON CO	37	17	60	90	91
		2	8.8	2	6.3	LAKE CO		-28			95
4	5.5	3	3.7	2	2.1	LA PLATA CO	-33	-43	-61	21	39
6	2.1	13	3.5	12	2.0	LARIMER CO	64	-44	-7	74	36
3	2.8	3	2.5	2	1.4	LAS ANIMAS CO	-12	-41	-49	36	26
2	8.3	1	2.2	2	3.8	LINCOLN CO	-73	67	-55	27	75
6	7.4	1	1.1	5	3.7	LOGAN CO	-86	252	-50	34	74
13	6.1	6	2.0	10	2.2	MESA CO	-68	9	-65	17	40
1	4.4	1	3.7	1	2.8	MOFFAT CO	-16	-23	-35	54	56
3	6.9	1	1.8			MONTEZUMA CO	-74				
3	3.9	3	3.6	3	2.5	MONTROSE CO	-6	-30	-34	55	49
4	5.0			2	1.7	MORGAN CO			-66	16	31
4	3.7	6	4.5	5	3.2	OTERO CO	20	-29	-15	70	64
1	14.1	1	14.9	1	8.6	PARK CO	6	-42	-39	50	98
3	10.9			2	4.3	PHILLIPS CO			-60	22	83
1	1.8			1	1.2	PROWERS CO			-32	58	23
20	4.2	14	2.4	15	2.3	PUEBLO CO	-42	-7	-46	41	43
3*	19.7	1	4.1			RIO BLANCO CO	-79				
2	4.3			1	1.4	RIO GRANDE CO			-68	14	26
3	14.0			2	6.3	ROUTT CO			-55	27	95
		1	6.8			SAGUACHE CO					
				1*	37.2	SAN JUAN CO					99
2	8.3			1	4.1	SEDGWICK CO			-51	32	80
1	2.9			3	8.5	WASHINGTON CO			195	97	98
10	3.5	13	3.3	10	1.8	WELD CO	-7	-44	-48	37	34
3	5.2			2	1.6	YUMA CO			-69	13	29
833*	7.3	768*	5.2	743*	3.9	CONNECTICUT CT	-29	-25	-47		
218*	7.6	197	5.2	172	3.5	FAIRFIELD CT	-31	-33	-54	28	69
196	6.8	196	5.1	175	3.5	HARTFORD CT	-26	-31	-48	36	70
47	8.0	46	5.8	42	4.1	LITCHFIELD CT	-27	-30	-49	35	80
22	4.7	24	4.0	27	3.4	MIDDLESEX CT	-16	-15	-29	60	67
243*	8.0	207*	5.4	187	3.7	NEW HAVEN CT	-32	-31	-53	30	74
55	6.6	61	5.8	80*	6.1	NEW LONDON CT	-11	5	-7	74	94
20	7.9	15	4.8	26*	6.0	TOLLAND CT	-39	25	-24	64	94
32*	8.3	22	4.8	34*	5.9	WINDHAM CT	-42	22	-30	60	93
90	5.8	89	4.4	111	4.0	DELAWARE DE	-25	-9	-31		
6	3.1	12	4.9	15	4.1	KENT DE	57	-16	32	86	80
66	6.3	59	4.2	76	4.1	NEW CASTLE DE	-33	-3	-34	55	80
18	6.1	18	4.7	20	3.5	SUSSEX DE	-24	-25	-42	45	70
197	6.8	132	4.8	66	2.8	DIST. OF COLUMBIA DC	-30	-41	-59		
197	6.8	132	4.8	66	2.8	WASHINGTON DC	-30	-41	-59	24	56

* SIGNIFICANTLY MORE DEATHS THAN EXPECTED

\# COUNTIES COMBINED; SEE TECHNICAL NOTES

WHITE FEMALE — CANCER OF THE RECTUM — ICD 154 EXCEPT 154.3

1950-1959 DEATHS	1950-1959 RATE	1960-1969 DEATHS	1960-1969 RATE	1970-1979 DEATHS	1970-1979 RATE	COUNTY AND STATE	PERCENT CHANGE 60S/50S	PERCENT CHANGE 70S/60S	PERCENT CHANGE 70S/50S	PERCENTILE %CHANGE 70S/50S	PERCENTILE RATES 1970S
747	4.6	1190	3.7	1666	2.8	FLORIDA FL	-21	-23	-39		
11	7.2	8	3.5	11	3.0	ALACHUA FL	-52	-15	-59	24	60
1	6.7	1	3.7	4	7.5	BAKER FL	-44	101	12	81	97
8	5.5	8	3.5	5	1.2	BAY FL	-37	-65	-78	05	23
1	2.3	1	1.7	3	4.0	BRADFORD FL	-24	133	76	92	79
4	1.8	17	3.8	37	3.2	BREVARD FL	109	-16	75	92	64
31	3.8	113	3.9	186	2.5	BROWARD FL	1	-34	-33	56	49
		1	3.1	2	4.0	CALHOUN FL		31			79
		10	5.5	14	2.0	CHARLOTTE FL		-63			38
1	2.2	5	4.9	13	3.1	CITRUS FL	121	-36	41	88	63
1	1.1	1	1.0	4	1.7	CLAY FL	-5	62	53	89	31
1	1.6	2	1.7	15	3.4	COLLIER FL	5	104	113	94	68
3	5.1	1	1.4	2	1.6	COLUMBIA FL	-72	11	-69	13	29
180	5.9	265	4.4	310	3.1	DADE FL	-25	-29	-47	38	63
1	1.7	1	1.1	1	.7	DE SOTO FL	-35	-37	-59	23	18
				1	3.0	DIXIE FL					60
45	4.8	71	4.4	60	2.7	DUVAL FL	-8	-39	-44	43	53
8	2.6	16	3.3	28	3.2	ESCAMBIA FL	30	-3	26	84	65
		1	1.9	3	6.2	FRANKLIN FL		224			95
2	1.9	2	1.6	1	.4	GADSDEN FL	-17	-73	-77	06	16
1	7.3			1	3.7	GILCHRIST FL			-49	35	74
		2	5.8	3	6.7	GULF FL		15			96
				1	2.2	HAMILTON FL					41
2	5.6	2	3.6			HARDEE FL	-36				
				1	3.0	HENDRY FL					61
4	15.8	3	2.6	4	1.4	HERNANDO FL	-83	-48	-91	00	25
3	2.9	5	2.1	7	1.8	HIGHLANDS FL	-25	-18	-39	50	32
74	5.5	74	3.5	90	2.9	HILLSBOROUGH FL	-36	-18	-48	37	57
1	1.7	2	2.9	1	1.5	HOLMES FL	69	-48	-12	71	28
4	6.0	8	4.1	14	3.4	INDIAN RIVER FL	-33	-17	-44	43	67
2	2.1	1	.6	4	2.4	JACKSON FL	-71	285	11	81	46
		2	6.8			JEFFERSON FL					
1	7.0			2	9.2	LAFAYETTE FL			32	86	98
14	5.4	13	3.0	18	2.1	LAKE FL	-43	-31	-61	22	39
9	4.4	16	2.6	38	3.1	LEE FL	-40	17	-30	60	62
4	3.0	7	3.2	7	1.8	LEON FL	4	-44	-41	46	33
2	10.4	3	5.4	2	1.8	LEVY FL	-48	-68	-83	02	32
2	5.3	2	4.4			MADISON FL	-18				
20	5.6	19	2.6	35	2.5	MANATEE FL	-54	-3	-56	27	48
3	1.8	13	5.1	21	3.6	MARION FL	180	-29	99	94	73
5	6.2	2	2.5	12	3.5	MARTIN FL	-59	38	-43	44	70
4	4.4	6	3.1	6	2.4	MONROE FL	-29	-23	-45	41	47
				1	.9	NASSAU FL					20
2	4.2	5	2.8	8	3.0	OKALOOSA FL	-34	9	-28	61	61
1	2.9	2	4.7	1	1.2	OKEECHOBEE FL	61	-76	-61	22	22
28	3.3	53	3.8	64	2.9	ORANGE FL	13	-23	-14	70	57
7	4.6	7	4.3	8	2.9	OSCEOLA FL	-5	-34	-37	51	57
45	5.6	71	3.8	106	2.8	PALM BEACH FL	-32	-27	-50	33	55
4	2.2	21	3.9	42	2.3	PASCO FL	80	-42	5	79	44
111	4.4	193	3.8	230	3.0	PINELLAS FL	-14	-21	-33	57	59
33	5.2	25	2.3	41	2.6	POLK FL	-55	11	-51	33	50
1	1.3	7	4.7	4	1.7	PUTNAM FL	266	-65	29	85	30
5	4.7	7	4.2	7	2.4	ST JOHNS FL	-10	-42	-48	37	47
3	2.2	11	4.6	15	3.6	ST LUCIE FL	111	-22	64	91	71
1	1.3	1	1.2	7	3.8	SANTA ROSA FL	-8	217	193	97	76
15	4.5	27	2.2	74	3.8	SARASOTA FL	-51	72	-16	69	75
1	1.0	8	3.4	20	3.2	SEMINOLE FL	239	-6	220	98	65
4	8.3	1	1.8	3	2.2	SUMTER FL	-78	21	-74	09	41
2	3.7	2	2.8	2	1.4	SUWANNEE FL	-26	-48	-61	21	27
1	3.2					TAYLOR FL					
1	7.1					UNION FL					
28	4.3	42	3.1	65	2.8	VOLUSIA FL	-29	-9	-36	53	55
		1	1.3	1	.9	WALTON FL		-32			19
1	2.1	2	3.0			WASHINGTON FL	45				
354	3.4	406	2.8	451	2.2	GEORGIA GA	-18	-20	-34		
		2	3.6	2	3.0	APPLING GA		-19			59
2	8.7					ATKINSON GA					
2	6.4	1	2.9			BACON GA	-55				
1	.5					BALDWIN GA					
1	3.2	1	2.6	2	4.2	BANKS GA	-17	59	32	86	81
3	4.1	2	2.9			BARROW GA	-28				
3	4.2	5	3.8	3	1.8	BARTOW GA	-9	-53	-57	25	33
		1	1.8	1	1.1	BEN HILL GA		-36			22
4	8.2	3	6.3	1	1.4	BERRIEN GA	-23	-78	-83	02	26
11	3.3	9	2.0	16	2.7	BIBB GA	-41	37	-19	67	53
2	12.2					BLECKLEY GA					
1	9.4	1	4.1	1	3.1	BRANTLEY GA	-56	-23	-66	16	63
				4	5.7	BROOKS GA					93
1	13.5			1	3.9	BRYAN GA			-71	11	78
4	5.7	5	4.8	1	.6	BULLOCH GA	-17	-87	-89	00	17
3	10.5	1	2.1			BURKE GA	-80				
1	3.1			2	2.9	BUTTS GA			-8	73	57
1	6.7					CALHOUN GA					
		3*	14.7			CAMDEN GA					
4	2.8	3	1.5	4	1.6	CARROLL GA	-46	3	-44	43	29
1	5.0	4	4.5	2	1.3	CATOOSA GA	-10	-71	-74	08	24
		2	13.4			CHARLTON GA					
22	5.5	25	4.3	22	3.0	CHATHAM GA	-22	-31	-46	40	60
2	3.1	3	3.3	4	3.1	CHATTOOGA GA	4	-5	-1	76	62
		5	3.8	5	2.9	CHEROKEE GA		-23			58
3	2.5	3	1.6	7	3.0	CLARKE GA	-38	94	21	83	60
		1	10.0	1	11.2	CLAY GA		11			99
2	3.4	3	2.3	11	3.5	CLAYTON GA	-33	54	3	78	69
		1	9.4			CLINCH GA					
8	3.6	9	2.0	16	1.9	COBB GA	-43	-5	-46	40	36
2	2.8			1	1.0	COFFEE GA			-64	18	21
4	4.4	6	5.0	4	2.2	COLQUITT GA	13	-56	-50	34	42

* SIGNIFICANTLY MORE DEATHS THAN EXPECTED

WHITE FEMALE

CANCER OF THE RECTUM

ICD 154 EXCEPT 154.3

1950-1959 DEATHS	1950-1959 RATE	1960-1969 DEATHS	1960-1969 RATE	1970-1979 DEATHS	1970-1979 RATE	COUNTY AND STATE	PERCENT CHANGE 60S/50S	PERCENT CHANGE 70S/60S	PERCENT CHANGE 70S/50S	PERCENTILE %CHANGE 70S/50S	PERCENTILE RATES 1970S
		1	2.8	2	1.9	COLUMBIA GA		-31			36
3	7.3	1	2.3	2	3.3	COOK GA	-69	43	-55	27	65
5	6.1	3	2.8	4	2.5	COWETA GA	-54	-11	-59	24	48
2	7.3			1	1.4	DADE GA			-80	04	27
		1	6.2	1	4.6	DAWSON GA		-27			85
3	4.8	3	4.2	2	1.8	DECATUR GA	-14	-57	-63	20	33
15	2.4	34	3.2	41	2.6	DE KALB GA	35	-20	8	80	50
1	3.6	1	1.4			DODGE GA	-62				
		1	3.7	1	2.0	DOOLY GA		-45			38
2	1.8	5	2.7	4	1.6	DOUGHERTY GA	53	-40	-8	73	30
1	2.5	5	5.5	4	2.4	DOUGLAS GA	124	-56	-2	76	47
		2	4.5	1	1.6	EARLY GA		-65			29
				2	3.5	EFFINGHAM GA					70
3	4.8	2	2.4	3	3.6	ELBERT GA	-50	48	-25	63	72
2	3.6					EMANUEL GA					
1	4.4	1	3.8			EVANS GA	-12				
2	4.0	1	1.1			FANNIN GA	-72				
1	4.4	1	2.7			FAYETTE GA	-39				
14	5.6	15	5.0	7	1.7	FLOYD GA	-11	-67	-71	11	30
		3	4.5	2	2.2	FORSYTH GA		-50			43
1	1.7	1	1.3	5	5.0	FRANKLIN GA	-24	292	198	98	89
68	4.1	85	4.0	61	2.4	FULTON GA	-2	-39	-41	47	47
1	2.2	1	2.9	2	3.6	GILMER GA	30	24	61	91	71
				1	5.0	GLASCOCK GA					89
4	5.2	1	.6	4	1.8	GLYNN GA	-88	195	-65	17	33
2	3.2	4	4.0	1	.6	GORDON GA	25	-85	-81	03	17
1	1.5			1	1.5	GPADY GA			2	78	27
		2	5.0	2	5.6	GREENE GA		12			92
5	4.0	5	2.6	9	2.5	GWINNETT GA	-35	-2	-37	52	50
2	2.5	2	2.1	2	1.5	HABERSHAM GA	-15	-29	-40	49	28
7	4.7	2	.9	7	2.0	HALL GA	-81	133	-57	25	38
1	4.6			2	10.9	HANCOCK GA			136	96	99
1	1.7	2	2.9	7*	6.6	HAPALSON GA	70	131	294	99	96
2	6.7			1	1.5	HARRIS GA			-77	06	28
2	3.7			3	3.2	HART GA			-13	71	65
		1	1.4	2	2.1	HENRY GA		51			39
1	2.3	2	1.7	1	.6	HOUSTON GA	-27	-65	-74	08	17
1	3.3	2	5.5	4	9.6	IRWIN GA	67	74	191	97	98
3	3.3	1	1.1	2	1.7	JACKSON GA	-67	53	-50	34	30
2	9.7					JASPER GA					
		1	3.8	1	2.4	JEFF DAVIS GA		-37			46
2	5.4	3	6.1	2	2.7	JEFFERSON GA	13	-56	-50	33	54
1	7.3	1	3.8			JENKINS GA	-48				
		2	6.3			JOHNSON GA					
3	9.4	1	2.2	2	3.3	LAMAR GA	-76	48	-65	17	66
1	6.3					LANIER GA					
2	3.3	2	1.7	4	3.0	LAURENS GA	-49	80	-8	74	61
1	6.5	1	4.1	1	3.0	LINCOLN GA	-37	-26	-53	29	61
2	1.6	5	3.5	7	3.5	LOWNDES GA	122	0	121	95	69
1	4.0					LUMPKIN GA					
		1	2.7	1	1.3	MC DUFFIE GA		-51			25
1	3.5					MACON GA					
2	4.0					MADISON GA					
1	8.0					MARION GA					
3	5.5			2	2.2	MERIWETHER GA			-59	23	43
3	13.7	1	2.9			MILLER GA	-79				
2	4.5	1	1.8	2	2.3	MITCHELL GA	-61	31	-49	36	45
1	3.9	1	2.5			MONROE GA	-36				
		1	3.7			MONTGOMERY GA					
1	3.3			2	4.9	MORGAN GA			49	89	88
1	1.6					MURRAY GA					
13	3.7	12	2.7	18	3.0	MUSCOGEE#GA	-27	11	-19	68	61
1	1.5	1	1.2	1	.6	NEWTON GA	-19	-49	-59	24	17
		1	2.8			OCONEE GA					
		2	5.9	1	2.1	OGLETHORPE GA		-64			40
3	6.0	2	2.8	3	2.6	PAULDING GA	-53	-6	-56	26	52
1	3.9	1	2.9	1	2.5	PEACH GA	-24	-15	-35	53	49
		1	2.3			PICKENS GA					
		1	3.3			PIERCE GA					
2	6.6			1	2.2	PIKE GA			-67	15	41
1	1.0	5	3.6	4	2.4	POLK GA	252	-34	131	95	46
1	6.1			3*	9.2	PUTNAM GA			49	89	98
1	2.7	1	2.4	2	2.7	RABUN GA	-11	11	-1	76	53
		3	8.7	1	2.0	RANDOLPH GA		-77			37
7	2.6	13	3.2	22	4.0	RICHMOND GA	22	25	52	89	78
				1	1.6	ROCKDALE GA					30
1	11.2	3*	34.0	1	5.7	SCHLEY GA	203	-83	-49	35	92
1	2.6					SCREVEN GA					
1	6.1	1	3.5			SEMINOLE GA	-43				
3	3.0	6	4.2	4	2.0	SPALDING GA	41	-53	-34	56	37
1	1.4	2	2.3	3	2.5	STEPHENS GA	61	10	78	92	49
1	5.7					STEWART GA					
2	3.7	1	1.0	3	2.3	SUMTER GA	-72	122	-38	50	44
		2	13.2			TALBOT GA					
				1	5.9	TALIAFERRO GA					94
2	4.3	1	1.7	3	4.3	TATTNALL GA	-61	152	-1	77	82
1	3.2			1	3.5	TAYLOR GA			10	80	70
		1	1.7	1	.9	TELFAIR GA		-46			20
3	13.5			3	7.8	TERRELL GA			-42	45	97
3	3.3	4	3.0	3	1.7	THOMAS GA	-9	-44	-49	35	31
1	1.2	1	1.2	3	2.7	TIFT GA	-3	135	128	95	54
		1	1.3	3	2.7	TOOMBS GA		106			54
1	12.4			1	2.5	TOWNS GA			-80	04	49
2	16.3	1	4.3			TREUTLEN GA	-74				
6	4.6	6	3.2	8	3.1	TROUP GA	-30	-3	-33	57	62
		1	2.3	2	4.3	TURNER GA		84			82
2	6.3			1	1.4	UNION GA			-78	05	26
3	3.9	1	1.0	1	.5	UPSON GA	-73	-51	-87	01	17
4	2.6	5	2.6	6	2.0	WALKER GA	1	-25	-24	64	36
4	5.4	1	1.1			WALTON GA	-79				
1	1.0	3	2.2	6	2.7	WARE GA	123	28	184	97	54

* SIGNIFICANTLY MORE DEATHS THAN EXPECTED

\# COUNTIES COMBINED; SEE TECHNICAL NOTES

WHITE FEMALE — CANCER OF THE RECTUM — ICD 154 EXCEPT 154.3

1950-1959 DEATHS	1950-1959 RATE	1960-1969 DEATHS	1960-1969 RATE	1970-1979 DEATHS	1970-1979 RATE	COUNTY AND STATE	PERCENT CHANGE 60S/50S	PERCENT CHANGE 70S/60S	PERCENT CHANGE 70S/50S	PERCENTILE %CHANGE 70S/50S	PERCENTILE RATES 1970S
		2	9.6			WARREN GA					
		1	1.4			WASHINGTON GA					
1	2.1			3	3.6	WAYNE GA			71	92	72
1	15.8	1	11.2			WEBSTER GA	-29				
1	5.8			2	6.0	WHEELER GA			3	78	94
1	1.6	3	1.7	3	1.1	WHITFIELD GA	8	-38	-33	56	21
1	3.5					WILCOX GA					
1	3.6	2	3.7			WILKES GA	2				
		1	2.8	1	3.2	WILKINSON GA		14			64
		1	2.1	1	1.5	WORTH GA		-30			27
		25	3.9	27	2.8	HAWAII HI		-28			
		1	1.3	3	3.4	HAWAII HI		163			67
		21	4.2	22	2.8	HONOLULU HI		-32			56
		1	3.1			KAUAI HI					
		2	4.2	2	3.0	MAUI#HI		-28			61
102	4.2	109	3.4	94	2.2	IDAHO ID	-18	-37	-48		
17	4.5	15	2.7	13	1.9	ADA ID	-40	-30	-57	25	36
		1	6.1	1	7.8	ADAMS ID		30			97
2	1.5	6	3.6	7	2.7	BANNOCK ID	136	-23	81	92	54
				2	4.1	BEAR LAKE ID					79
3	12.9			2	5.5	BENEWAH ID			-57	25	92
3	3.5	1	.8	1	.8	BINGHAM ID	-76	-3	-77	06	19
		1	4.7			BLAINE ID					
		1	11.8			BOISE ID					
4	5.3	2	2.3	4	3.8	BONNER ID	-57	65	-29	60	75
1	1.0	9	6.5	4	1.9	BONNEVILLE ID	529	-71	82	93	35
		1	3.6			BOUNDARY ID					
		1	14.1			BUTTE ID					
16	6.4	11	3.1	9	1.7	CANYON ID	-51	-46	-74	09	31
		1	7.5			CARIBOU ID					
3	5.2	4	6.0	1	1.2	CASSIA ID	14	-79	-76	06	23
				1*	53.6	CLARK ID					99
1	3.7	3	8.9	1	2.0	CLEARWATER ID	140	-77	-46	40	37
				1	5.9	CUSTER ID					94
1	3.9	1	2.4	1	2.0	ELMORE ID	-38	-17	-49	35	37
		2	4.2			FRANKLIN ID					
1	5.3			1	2.4	FREMONT ID			-54	28	47
3	6.9	3	5.9	3	4.4	GEM ID	-15	-24	-36	53	84
1	2.2	2	4.3	3	4.6	GOODING ID	93	9	111	94	86
2	4.2	4	6.7			IDAHO ID	61				
1	4.3	1	2.6	3	5.3	JEFFERSON ID	-39	103	23	84	91
3	7.4	3	6.5	2	2.8	JEROME ID	-12	-57	-62	20	56
9	6.7	6	3.3	6	2.0	KOOTENAI ID	-50	-39	-70	12	38
3	3.6	4	4.1	3	2.4	LATAH ID	13	-42	-35	54	45
		1	3.5			LEMHI ID					
1	6.2	1	5.2	3*	13.7	LEWIS ID	-17	166	120	95	99
2	14.3			1	4.5	LINCOLN ID			-69	13	65
1	3.6			1	1.7	MADISON ID			-54	29	30
3	7.4					MINIDOKA ID					
9	7.7	10	6.7	5	2.6	NEZ PERCE ID	-12	-61	-66	16	52
1	3.9	1	5.5			ONEIDA ID	40				
1	6.3	3	13.7	4*	9.5	OWYHEE ID	117	-30	52	89	98
		1	1.0	3	2.4	PAYETTE ID		144			46
		1	6.5	1	4.1	POWER ID		-37			80
1	1.6	2	2.9	3	3.2	SHOSHONE ID	82	9	98	94	64
1	12.2					TETON ID					
7	3.9	4	1.6	4	1.2	TWIN FALLS ID	-59	-22	-68	14	23
1	7.0					VALLEY ID					
		2	4.4			WASHINGTON ID					
2895*	6.5	2865*	5.3	2428*	3.9	ILLINOIS IL	-18	-28	-40		
34	7.3	27	5.4	28	4.8	ADAMS IL	-27	-10	-34	55	88
3	4.4	2	2.2	2	2.3	ALEXANDER IL	-51	5	-49	35	43
5	6.0	5	5.9	3	2.8	BOND IL	-2	-53	-53	29	55
8	7.2	8	6.5	5	3.9	BOONE IL	-11	-40	-47	39	76
2	3.6			3	3.8	BROWN IL			5	79	75
18	7.8	12	5.0	8	3.0	BUREAU IL	-37	-39	-62	20	60
		1	2.2	3	6.4	CALHOUN IL		188			95
5	3.9	6	4.0	7	4.8	CARROLL IL	3	20	24	84	88
3	2.8	10*	8.5	3	2.9	CASS IL	206	-66	4	78	58
39*	9.1	24	4.5	14	2.1	CHAMPAIGN IL	-51	-52	-77	06	40
7	3.0	15	6.0	12	3.3	CHRISTIAN IL	97	-45	9	80	66
2	1.8	4	3.0	8	4.8	CLARK IL	67	60	168	97	88
3	3.1	10	9.3	7	4.3	CLAY IL	202	-53	41	88	83
5	3.9	7	4.8	2	.8	CLINTON IL	22	-82	-79	05	19
10	4.1	15	5.0	13	4.1	COLES IL	24	-19	0	77	79
1550*	7.4	1396*	5.7	1115*	4.1	COOK IL	-23	-28	-45	42	80
9	6.5	6	3.5	4	2.6	CRAWFORD IL	-46	-26	-60	22	50
4	5.9	2	2.7	2	2.4	CUMBERLAND IL	-55	-10	-59	23	46
11	4.8	23*	8.5	14	3.7	DE KALB IL	79	-57	-22	65	73
9	8.3	6	5.1	10*	6.0	DE WITT IL	-38	19	-27	62	94
3	2.8	5	3.8	9	5.8	DOUGLAS IL	39	52	111	94	93
49	5.7	62	4.2	76	3.3	DU PAGE IL	-27	-21	-42	45	66
9	6.2	9	5.3	7	3.8	EDGAR IL	-13	-28	-38	51	76
		4	8.4	2	2.6	EDWARDS IL		-69			51
3	2.7	9	5.7	11	5.6	EFFINGHAM IL	111	-2	106	94	92
9	7.5	8	5.3	5	2.5	FAYETTE IL	-29	-53	-67	15	48
9	9.2	7	6.4	1	1.3	FORD IL	-31	-80	-86	01	24
13	5.1	16	4.6	11	2.8	FRANKLIN IL	-10	-39	-46	40	54
20	7.3	8	2.5	17	4.4	FULTON IL	-66	79	-40	48	83
3	6.6	3	5.4			GALLATIN IL	-17				
3	2.1	10	6.6	1	.7	GREENE IL	214	-89	-64	18	18
6	5.6	8	6.0	3	1.9	GRUNDY IL	8	-69	-66	16	35
5	6.6	2	2.4	2	2.0	HAMILTON IL	-64	-15	-69	13	38
12	6.9	1	.4	9	4.7	HANCOCK IL	-94	973	-32	58	87
		2	5.8	1	1.9	HARDIN IL		-68			34

* SIGNIFICANTLY MORE DEATHS THAN EXPECTED

COUNTIES COMBINED; SEE TECHNICAL NOTES

1950-1959 DEATHS	1950-1959 RATE	1960-1969 DEATHS	1960-1969 RATE	1970-1979 DEATHS	1970-1979 RATE	COUNTY AND STATE	PERCENT CHANGE 60S/50S	PERCENT CHANGE 70S/60S	PERCENT CHANGE 70S/50S	PERCENTILE %CHANGE 70S/50S	PERCENTILE RATES 1970S
2	4.2	1	2.0	2	3.0	HENDERSON IL	-52	51	-28	61	61
11	3.9	21	6.2	11	2.7	HENRY IL	58	-57	-31	58	53
8	4.4	8	3.6	18*	7.0	IROQUOIS IL	-17	92	58	90	96
7	3.5	9	3.8	9	3.2	JACKSON IL	7	-15	-9	73	64
2	2.5	4	4.4	4	3.8	JASPER IL	75	-12	54	90	76
16	8.2	8	3.7	12	4.7	JEFFERSON IL	-55	27	-43	44	86
6	7.2	3	2.7	10*	7.8	JERSEY IL	-62	189	9	80	97
10	6.9	9	6.2	4	2.5	JO DAVIESS IL	-9	-60	-64	18	48
		1	1.7	3	4.0	JOHNSON IL		134			78
46	4.8	57	4.8	49	3.5	KANE IL	1	-28	-27	62	69
15	3.3	25	4.8	19	3.5	KANKAKEE IL	45	-27	5	79	70
5	7.3	5	5.3	7	5.9	KENDALL IL	-28	11	-20	67	93
20	5.7	19	4.5	17	3.5	KNOX IL	-21	-22	-38	50	69
53	6.3	57	4.9	57	3.6	LAKE IL	-23	-27	-44	43	71
34	5.9	38	5.2	38	4.8	LA SALLE IL	-13	-8	-20	67	87
5	4.0	6	4.6	6	2.8	LAWRENCE IL	15	-40	-31	58	55
10	5.2	11	4.7	8	2.9	LEE IL	-9	-39	-45	42	57
14	5.9	12	3.8	7	2.3	LIVINGSTON IL	-36	-39	-61	21	45
9	4.8	13	6.1	6	2.3	LOGAN IL	27	-63	-53	30	44
18*	9.2	12	6.4	5	2.2	MC DONOUGH IL	-31	-66	-76	06	41
25	7.5	33*	7.5	34*	5.2	MC HENRY IL	0	-31	-31	59	90
29	6.1	47*	8.7	22	3.5	MC LEAN IL	42	-59	-42	45	70
37	6.6	31	4.5	39	5.1	MACON IL	-31	11	-23	64	89
9	3.0	16	4.1	12	3.2	MACOUPIN IL	35	-22	5	79	64
52	5.9	59	5.2	57	3.9	MADISON IL	-11	-24	-33	57	78
11	4.9	17	5.9	13	4.0	MARION IL	22	-33	-18	68	78
3	3.8	4	4.0	6	5.3	MARSHALL IL	6	33	41	88	91
4	4.5	7	5.9	8	5.5	MASON IL	32	-7	22	83	92
6	7.8			4	3.5	MASSAC IL			-55	28	70
3	4.3	4	6.4	4	4.4	MENARD IL	49	-31	4	78	84
9	8.4	6	5.0	6	4.2	MERCER IL	-40	-17	-50	33	81
9*	11.3	4	3.8	7	5.5	MONROE IL	-67	47	-51	32	92
13	5.7	10	3.6	5	2.0	MONTGOMERY IL	-38	-44	-65	17	37
13	4.7	12	4.0	10	3.6	MORGAN IL	-15	-8	-22	65	72
1	1.0	2	1.3	4	3.6	MOULTRIE IL	26	180	253	98	71
13	6.9	10	4.3	12	4.2	OGLE IL	-38	-2	-39	49	81
44	4.7	61	5.5	45	3.4	PEORIA IL	17	-37	-26	62	69
6	5.3	8	5.4	8	5.1	PERRY IL	2	-5	-3	75	90
5	5.8	4	3.5	5	5.4	PIATT IL	-40	57	-6	74	92
11	6.9	12	7.1	7	3.8	PIKE IL	4	-47	-45	42	75
1	3.2	2	6.5	2	6.6	POPE IL	103	1	105	94	96
6*	13.5	1	2.1	2	2.9	PULASKI IL	-85	43	-78	05	59
2	10.0	3	10.0			PUTNAM IL	1				
9	5.3	5	2.4	9	3.9	RANDOLPH IL	-55	63	-26	63	76
6	6.0	10	7.4	7	5.5	RICHLAND IL	24	-25	-7	74	92
48	7.0	51	6.0	45	4.6	ROCK ISLAND IL	-13	-24	-34	55	85
50	5.3	59	5.2	43	3.2	ST CLAIR IL	-2	-38	-39	49	64
8	4.2	10	4.2	9	3.0	SALINE IL	0	-29	-28	61	60
57	7.4	60	6.2	52	4.3	SANGAMON IL	-17	-30	-42	45	82
1	2.2	1	.9			SCHUYLER IL	-60				
1	1.8	4	7.8			SCOTT IL	323				
13	8.4	7	3.5	7	4.0	SHELBY IL	-58	15	-52	31	79
5	8.7	3	5.0			STARK IL	-42				
14	5.1	13	4.3	10	2.7	STEPHENSON IL	-16	-36	-46	40	54
15	4.4	27	5.8	14	2.2	TAZEWELL IL	31	-62	-50	33	42
4	3.2	7	4.8	5	3.3	UNION IL	48	-31	2	78	66
26	5.3	32	5.8	21	3.0	VERMILION IL	9	-48	-44	43	60
4	4.8	4	4.6	3	2.4	WABASH IL	-5	-47	-50	34	47
7	4.6	4	2.5	5	3.5	WARREN IL	-46	41	-24	64	69
5	4.7	5	4.4	5	3.5	WASHINGTON IL	-6	-20	-24	64	70
4	3.1	2	1.4	8	5.1	WAYNE IL	-56	278	67	91	89
5	3.9	7	5.4	7	4.3	WHITE IL	38	-20	10	80	83
17	6.5	14	4.4	11	2.6	WHITESIDE IL	-33	-40	-60	22	51
61*	8.9	59*	6.5	73*	6.1	WILL IL	-27	-6	-32	58	94
9	3.1	13	3.8	14	3.6	WILLIAMSON IL	23	-5	17	82	72
45	5.9	59	5.6	50	3.8	WINNEBAGO IL	-6	-31	-35	54	76
7	5.9	4	2.7	8	3.8	WOODFORD IL	-55	43	-35	54	75
1148	5.6	1135	4.5	928	3.1	INDIANA IN	-19	-32	-45		
8	6.6	7	4.7	7	4.1	ADAMS IN	-29	-12	-37	51	80
52	5.5	62	5.1	55	3.6	ALLEN IN	-6	-29	-34	56	72
6	3.3	6	2.7	7	2.4	BARTHOLOMEW IN	-18	-13	-28	61	46
6	9.4	5	6.8	4	6.0	BENTON IN	-28	-12	-37	52	94
4	4.8	1	1.4	3	3.0	BLACKFORD IN	-72	123	-37	51	60
8	4.7	10	4.9	3	1.7	BOONE IN	3	-64	-63	19	32
1	3.1			1	1.5	BROWN IN			-52	30	27
		3	2.4	9*	6.2	CARROLL IN		157			95
9	3.8	13	4.2	6	2.1	CASS IN	11	-49	-43	44	40
11	4.9	6	2.0	16	3.9	CLARK IN	-60	100	-20	67	77
13	8.3	10	4.9	13*	5.3	CLAY IN	-40	6	-36	52	90
9	4.9	6	2.2	2	.6	CLINTON IN	-54	-72	-87	01	17
3	5.8	2	3.3	4	6.8	CRAWFORD IN	-43	105	16	82	96
1	.5	4	2.5	6	2.8	DAVIESS IN	403	8	445	99	55
2	1.3	10	6.4	12*	5.7	DEARBORN IN	398	-11	342	99	92
2	2.1	6	4.2	6	3.5	DECATUR IN	103	-17	68	91	69
4	2.5	7	3.6	4	1.7	DE KALB IN	47	-53	-31	59	31
22	5.0	22	3.9	20	3.0	DELAWARE IN	-21	-23	-39	49	60
4	3.3	9	6.1	8	4.2	DUBOIS IN	84	-31	27	84	81
26	5.6	27	4.5	32	4.5	ELKHART IN	-20	-1	-20	66	84
12*	10.8	11	8.2	7	4.1	FAYETTE IN	-24	-50	-62	20	79
18	7.5	15	5.0	11	2.9	FLOYD IN	-34	-42	-61	21	57
9	8.5	11*	8.5	2	1.6	FOUNTAIN IN	0	-82	-82	03	28
3	3.3	6	6.4	2	2.2	FRANKLIN IN	95	-65	-31	58	43
5	4.7	3	2.6	5	4.0	FULTON IN	-43	52	-14	71	79
11	6.2	8	4.8	7	3.3	GIBSON IN	-23	-32	-48	37	65
17	5.0	13	3.7	12	2.6	GRANT IN	-27	-29	-48	37	51
10	6.2	9	4.5	8	3.1	GREENE IN	-28	-30	-50	34	62
12	6.5	9	3.8	13	4.2	HAMILTON IN	-42	13	-35	54	81
3	2.6	8	4.4	4	2.1	HANCOCK IN	73	-53	-18	68	39
5	5.5	5	4.5	2	1.9	HARRISON IN	-18	-57	-65	17	36

* SIGNIFICANTLY MORE DEATHS THAN EXPECTED

WHITE FEMALE

CANCER OF THE RECTUM

ICD 154 EXCEPT 154.3

COUNTY AND STATE		1950-1959		1960-1969		1970-1979		PERCENT CHANGE			PERCENTILE	
		DEATHS	RATE	DEATHS	RATE	DEATHS	RATE	60S/50S	70S/60S	70S/50S	%CHANGE 70S/50S	RATES 1970S
HENDRICKS	IN	7	4.3	6	3.2	4	1.4	-26	-57	-68	14	26
HENRY	IN	13	5.8	10	3.7	13	3.8	-37	3	-35	53	74
HOWARD	IN	18	6.3	22	6.2	15	3.6	-2	-42	-43	44	72
HUNTINGTON	IN	10	4.7	10	4.0	6	2.4	-13	-40	-48	36	47
JACKSON	IN	4	2.6	6	3.1	7	3.1	20	0	20	83	62
JASPER	IN	7	8.1	6	6.0	4	3.3	-25	-45	-58	24	67
JAY	IN	3	1.9	6	4.0	4	2.3	105	-43	18	82	44
JEFFERSON	IN	6	3.5	4	2.3	11*	5.9	-34	157	69	92	94
JENNINGS	IN	3	4.4	4	4.4	2	1.7	1	-62	-61	21	31
JOHNSON	IN	13	6.5	16	5.2	8	2.3	-20	-56	-65	17	44
KNOX	IN	13	5.8	13	4.5	6	1.8	-22	-60	-69	13	33
KOSCIUSKO	IN	14	6.5	11	4.3	5	1.6	-34	-62	-75	07	30
LAGRANGE	IN	2	2.3	2	2.3	1	1.2	3	-51	-49	35	22
LAKE	IN	85	6.5	63	3.6	59	2.9	-45	-19	-55	27	58
LA PORTE	IN	29	7.6	25	5.0	17	2.8	-34	-44	-63	19	56
LAWRENCE	IN	5	2.6	2	.8	4	1.5	-68	76	-44	43	27
MADISON	IN	32	6.1	24	3.7	18	2.2	-39	-42	-65	18	41
MARION	IN	145	5.3	152	4.6	107	2.8	-14	-39	-47	37	55
MARSHALL	IN	11	6.6	7	3.5	6	2.3	-47	-35	-66	16	44
MARTIN	IN	3	6.2	1	2.0	1	1.4	-67	-32	-77	06	26
MIAMI	IN	14	7.4	15	7.0	11	4.7	-5	-33	-36	52	87
MONROE	IN	5	2.4	13	4.9	11	3.4	101	-30	41	88	68
MONTGOMERY	IN	7	3.8	15	6.2	8	3.9	62	-36	4	78	78
MORGAN	IN	6	4.0	6	3.4	8	3.4	-16	1	-15	70	68
NEWTON	IN	5	8.7	7*	9.3	2	2.1	6	-78	-76	06	39
NOBLE	IN	12	7.3	6	3.2	8	3.7	-56	15	-49	35	74
OHIO	IN	2	7.3			2	5.7			-21	66	93
ORANGE	IN	7	7.1	8	7.3	2	1.3	3	-82	-82	03	24
OWEN	IN	3	3.8	6	6.0			56				
PARKE	IN	3	2.7	1	1.4	2	1.6	-50	21	-39	49	30
PERRY	IN	3	3.4	6	5.8	2	1.7	70	-72	-52	31	30
PIKE	IN	4	4.4	2	1.9	4	3.9	-58	108	-12	72	76
PORTER	IN	13	7.1	15	5.5	22*	5.2	-22	-6	-27	62	90
POSEY	IN	2	1.8	9	6.9	6	4.5	288	-34	155	96	85
PULASKI	IN	5	7.7	3	3.7	2	1.8	-52	-51	-77	06	33
PUTNAM	IN	8	6.1	8	4.8	8	3.7	-21	-23	-40	49	73
RANDOLPH	IN	14	8.1	11	6.3	4	1.5	-23	-76	-81	04	28
RIPLEY	IN	3	2.7	7	5.4	3	1.8	101	-67	-33	56	33
RUSH	IN	10	8.7	9	7.4	11*	6.7	-15	-8	-22	65	96
ST JOSEPH	IN	66	7.0	63	5.3	49	3.3	-24	-38	-53	29	66
SCOTT	IN	2	3.2	3	4.0	2	1.7	27	-58	-46	39	31
SHELBY	IN	11	6.4	9	4.0	14*	5.6	-38	40	-13	71	92
SPENCER	IN	5	5.5	1	.9	2	1.7	-83	90	-68	13	31
STARKE	IN	3	3.6	2	1.5	3	2.6	-57	69	-28	61	50
STEUBEN	IN	4	3.7	8	6.7	2	1.4	83	-79	-61	21	26
SULLIVAN	IN	8	5.0	8	4.6	5	2.3	-9	-49	-54	28	45
SWITZERLAND	IN	1	2.4	6*	12.0	3	5.3	392	-55	119	95	91
TIPPECANOE	IN	31*	8.3	16	3.4	18	3.6	-59	6	-57	25	72
TIPTON	IN	4	4.6	4	4.2	4	3.5	-9	-16	-24	64	70
UNION	IN	2	4.9	1	2.9	2	4.4	-42	52	-12	72	83
VANDERBURGH	IN	52	6.5	51	5.3	42	3.5	-19	-33	-46	40	70
VERMILLION	IN	6	5.5	12*	9.0	7	5.9	65	-35	8	80	94
VIGO	IN	37	6.1	30	4.2	26	3.0	-31	-28	-50	33	61
WABASH	IN	10	5.1	6	2.9	7	2.2	-44	-24	-57	25	41
WARREN	IN	3	6.7	2	4.9	1	1.4	-27	-71	-79	05	26
WARRICK	IN	5	4.1	10	6.9	5	2.9	68	-58	-30	60	58
WASHINGTON	IN	4	4.6	4	3.4	3	2.3	-26	-31	-49	34	45
WAYNE	IN	21	5.6	22	4.8	18	3.1	-15	-34	-44	42	63
WELLS	IN	10	8.4	4	3.2	3	1.5	-62	-55	-83	02	27
WHITE	IN	10	9.3	11*	8.2	4	2.4	-12	-71	-74	08	46
WHITLEY	IN	8	7.9	11*	8.3	3	2.5	5	-69	-68	14	49
IOWA	IA	770	5.0	742	4.1	658	3.2	-18	-23	-37		
ADAIR	IA	1	1.4	4	4.9	4	3.7	258	-24	172	97	73
ADAMS	IA	1	2.3			2	3.0			32	86	61
ALLAMAKEE	IA	4	3.6	4	3.5			-3				
APPANOOSE	IA	9	6.3	9	6.0	10*	6.4	-4	6	1	77	95
AUDUBON	IA	5	8.3	3	4.0	3	2.9	-52	-26	-65	17	59
BENTON	IA	9	7.1	3	1.9	8	5.3	-73	183	-25	64	91
BLACK HAWK	IA	31	6.3	24	3.9	22	3.0	-38	-22	-52	31	60
BOONE	IA	7	3.7	8	3.5	4	2.3	-7	-35	-39	49	43
BREMER	IA	3	2.9	5	3.5	2	1.0	23	-73	-67	15	20
BUCHANAN	IA	1	1.0	2	1.6	2	1.5	53	-4	46	88	28
BUENA VISTA	IA	4	3.1	6	3.6	3	1.5	17	-60	-53	30	27
BUTLER	IA	5	5.4	5	3.7	3	2.3	-32	-38	-58	24	44
CALHOUN	IA	7	6.7	2	2.3	1	.6	-66	-75	-92	00	17
CARROLL	IA	8	6.8	4	2.7	12*	6.0	-60	117	-13	71	94
CASS	IA	7	5.7	11	7.4	6	3.1	29	-58	-46	40	62
CEDAR	IA	4	3.7	7	5.9	3	2.1	59	-64	-43	44	40
CERRO GORDO	IA	15	5.8	13	3.9	11	3.0	-33	-25	-49	34	59
CHEROKEE	IA	6	5.3	3	2.6	6	3.6	-50	36	-33	57	71
CHICKASAW	IA	6	6.7	8	7.8	1	1.0	16	-87	-85	02	21
CLARKE	IA	3	4.4	3	5.4	3	6.0	23	13	38	87	94
CLAY	IA	4	4.1	4	3.3	2	1.5	-21	-55	-64	18	27
CLAYTON	IA	7	5.5	7	4.7	5	2.4	-14	-49	-56	27	47
CLINTON	IA	21	7.5	12	3.4	15	4.2	-54	23	-44	43	81
CRAWFORD	IA	5	4.8	6	5.0	1	.4	4	-93	-92	00	16
DALLAS	IA	5	3.3	7	4.1	10	4.2	23	1	25	84	81
DAVIS	IA	5	7.1	3	4.2	4	4.3	-41	3	-40	48	82
DECATUR	IA	3	3.2	5	4.8	1	1.3	49	-73	-60	22	24
DELAWARE	IA	7	7.7	5	4.3	3	2.6	-43	-39	-66	16	52
DES MOINES	IA	9	3.0	19	6.0	12	3.3	98	-45	10	80	67
DICKINSON	IA	5	7.0	1	1.1	8*	5.7	-84	431	-17	68	93
DUBUQUE	IA	31	7.3	32	6.5	9	1.7	-11	-73	-76	06	32
EMMET	IA	3	3.9	1	1.2	5	4.6	-70	303	19	82	86
FAYETTE	IA	12	7.2	9	4.8	5	2.0	-34	-58	-72	10	38
FLOYD	IA	2	1.5	7	4.4	4	1.9	193	-56	28	85	35
FRANKLIN	IA	1	1.4	6	6.0	3	3.5	340	-42	155	96	69
FREMONT	IA	8	10.6	3	3.6	3	3.6	-66	-1	-66	16	71
GREENE	IA	6	6.2	1	.8	2	2.2	-87	183	-64	18	43

* SIGNIFICANTLY MORE DEATHS THAN EXPECTED

WHITE FEMALE

CANCER OF THE RECTUM

ICD 154 EXCEPT 154.3

1950-1959 DEATHS	1950-1959 RATE	1960-1969 DEATHS	1960-1969 RATE	1970-1979 DEATHS	1970-1979 RATE	COUNTY AND STATE	PERCENT CHANGE 60S/50S	PERCENT CHANGE 70S/60S	PERCENT CHANGE 70S/50S	PERCENTILE %CHANGE 70S/50S	PERCENTILE RATES 1970S
4	5.4	4	4.1	5	4.6	GRUNDY IA	-24	13	-14	70	86
5	4.5	5	4.5	7	6.6	GUTHRIE IA	1	45	47	88	96
2	1.8	3	2.1	6	4.7	HAMILTON IA	16	120	156	96	87
4	6.4			5	5.7	HANCOCK IA			-10	72	93
4	3.2	7	4.2	9	4.2	HARDIN IA	28	0	29	85	81
3	2.4	6	4.3	5	4.0	HARRISON IA	81	-7	68	92	78
5	4.1	3	2.9	4	2.8	HENRY IA	-30	0	-30	59	56
6	7.6	4	4.6	3	3.9	HOWARD IA	-40	-14	-48	36	78
3	4.3	4	4.7	3	2.2	HUMBOLDT IA	10	-53	-48	37	43
2	3.3	1	1.3	6	6.5	IDA IA	-61	410	97	93	96
7	6.7	5	4.9	2	1.6	IOWA IA	-27	-67	-76	07	30
4	3.4	6	4.0	7	5.5	JACKSON IA	18	38	63	91	92
8	4.3	12	5.3	11	3.9	JASPER IA	23	-27	-10	73	77
4	4.7	4	3.8	2	1.2	JEFFERSON IA	-20	-69	-75	07	22
4	1.8	21*	7.9	16	4.4	JOHNSON IA	334	-44	144	96	84
5	5.0	9	7.4	7	4.3	JONES IA	49	-42	-13	71	83
8	6.7	2	1.4	2	2.1	KEOKUK IA	-80	54	-68	13	39
5	4.1	5	3.0	4	2.0	KOSSUTH IA	-29	-31	-51	33	38
15	5.9	12	4.5	21*	6.9	LEE IA	-23	53	18	82	96
33	5.1	46	5.5	36	3.8	LINN IA	8	-32	-27	62	74
4	6.0	1	1.2	2	1.9	LOUISA IA	-79	54	-68	14	36
6	5.8	2	1.7			LUCAS IA	-71				
5	7.5	3	3.5			LYON IA	-53				
7	7.9	7	6.9	2	1.7	MADISON IA	-12	-76	-79	05	31
9	6.0	8	4.7	10	5.2	MAHASKA IA	-21	9	-13	71	90
6	3.8	7	3.9	7	3.7	MARION IA	2	-3	-1	76	74
11	4.8	15	5.8	16	5.4	MARSHALL IA	19	-5	13	81	92
8	10.6	2	2.6	3	3.5	MILLS IA	-76	36	-67	15	69
2	2.5	4	3.6	3	1.3	MITCHELL IA	45	-64	-47	38	24
5	6.4	2	2.1	5	4.0	MONONA IA	-67	87	-38	50	79
2	2.6	2	3.0	4	3.9	MONROE IA	15	28	48	89	77
5	5.1	2	1.3	9	5.4	MONTGOMERY IA	-74	312	5	79	91
11	4.9	8	3.5	9	3.2	MUSCATINE IA	-28	-11	-36	53	63
2	1.7	5	4.2	12*	7.1	O'BRIEN IA	141	69	307	99	97
		1	2.0	1	1.4	OSCEOLA IA		-27			26
4	2.6	6	2.9	8	4.6	PAGE IA	11	58	75	92	86
2	3.2	3	3.4	2	1.3	PALO ALTO IA	6	-61	-59	24	25
5	4.0	3	2.1	7	3.9	PLYMOUTH IA	-49	92	-1	76	78
4	4.7	4	4.0	5	4.2	POCAHONTAS IA	-14	4	-10	72	81
67	5.4	55	3.7	53	3.1	POLK IA	-32	-14	-42	45	63
21	5.3	22	4.8	11	2.2	POTTAWATTAMIE IA	-9	-55	-59	23	40
8	6.8	4	4.0	7	3.5	POWESHIEK IA	-41	-12	-48	36	71
		2	3.0	2	2.6	RINGGOLD IA		-13			50
6	5.8	2	1.3			SAC IA	-78				
45	8.1	30	4.6	30	3.7	SCOTT IA	-43	-19	-54	28	74
3	3.6	6	6.2	5	3.0	SHELBY IA	73	-52	-16	69	60
2	1.5	4	2.5	2	1.0	SIOUX IA	67	-60	-33	57	21
9	4.1	10	3.8	7	2.2	STORY IA	-7	-42	-46	41	43
7	5.2	8	5.4	5	3.0	TAMA IA	2	-43	-42	45	61
5	4.7	3	2.8	2	1.8	TAYLOR IA	-41	-35	-61	21	33
7	6.8	10	7.6	6	3.8	UNION IA	11	-50	-45	42	75
3	3.9	6	6.3	6	7.7	VAN BUREN IA	60	23	96	93	97
11	4.0	8	2.5	7	2.2	WAPELLO IA	-39	-12	-46	40	41
1	1.2	11*	8.8			WARREN IA	644				
8	6.7	6	3.7	7	3.5	WASHINGTON IA	-44	-5	-47	39	71
4	4.0	6	5.4	1	2.2	WAYNE IA	33	-59	-46	40	41
15	5.7	12	3.2	8	2.6	WEBSTER IA	-44	-20	-55	28	51
4	5.1	2	2.7	2	1.6	WINNEBAGO IA	-47	-39	-68	14	30
6	5.0			3	1.7	WINNESHIEK IA			-67	15	30
32	5.7	33	5.1	24	3.0	WOODBURY IA	-11	-41	-48	37	60
2	3.3	1	1.7	2	1.9	WORTH IA	-47	11	-41	46	35
5	4.0	5	3.8	4	2.4	WRIGHT IA	-3	-38	-40	48	46
447	4.1	466	3.6	426	2.7	KANSAS KS	-14	-25	-36		
2	1.7	5	3.2	1	.3	ALLEN KS	87	-90	-80	04	16
6	8.7			2	1.8	ANDERSON KS			-80	04	33
7	5.4	6	4.3	5	3.2	ATCHISON KS	-19	-26	-40	48	65
		1	2.4	1	1.3	BARBER KS		-45			24
2	1.5	3	1.9	5	2.4	BARTON KS	28	25	60	91	46
3	2.5	10	6.9	5	3.9	BOURBON KS	171	-44	52	89	76
2	1.6	11*	8.7	3	2.2	BROWN KS	426	-75	32	86	41
9	4.8	4	2.0	11	3.5	BUTLER KS	-59	80	-26	62	70
1	2.7	1	2.7			CHASE KS	-1				
2	2.9					CHAUTAUQUA KS					
7	4.4	9	4.9	7	3.4	CHEROKEE KS	10	-31	-23	64	67
2	7.2	1	2.7	2	4.7	CHEYENNE KS	-63	78	-34	55	87
				1	1.8	CLARK KS					34
3	3.2	2	1.9	4	2.2	CLAY KS	-41	18	-30	59	42
6	4.9	5	2.9	2	.6	CLOUD KS	-41	-81	-89	01	17
4	5.6			1	.7	COFFEY KS			-87	01	18
2	9.7			1	1.8	COMANCHE KS			-81	04	34
9	3.7	11	3.7	11	3.8	COWLEY KS	0	1	1	77	75
9	3.4	6	2.2	14	3.6	CRAWFORD KS	-35	62	5	79	72
		1	2.4	1	1.5	DECATUR KS		-37			28
5	2.9	3	1.3	9	3.6	DICKINSON KS	-55	175	25	84	72
5	8.2			2	2.6	DONIPHAN KS			-69	13	50
4	2.1	8	3.7	7	3.2	DOUGLAS KS	78	-12	56	90	64
		2	4.5	3	6.3	EDWARDS KS		41			95
2	5.0	1	3.1	1	1.6	ELK KS	-38	-47	-67	15	30
		1	1.2	5	4.1	ELLIS KS		249			80
4	7.7	4	6.7	3	6.0	ELLSWORTH KS	-13	-10	-21	65	94
3	5.1	3	4.2	2	1.8	FINNEY KS	-18	-57	-65	18	33
3	2.6	5	4.1	7	4.3	FORD KS	58	4	65	91	82
9	6.5	6	3.5	5	2.3	FRANKLIN KS	-45	-34	-64	18	45
1	1.4	8*	8.4	6	5.2	GEARY KS	515	-37	285	99	90
		1	4.2			GOVE KS					
				1	2.3	GRAHAM KS					45
1	11.2			1	4.0	GRANT KS			-64	19	79
		2	7.2	1	3.6	GRAY KS		-50			71
8	10.0	5	4.5	3	2.2	GREENWOOD KS	-55	-50	-78	06	42

* SIGNIFICANTLY MORE DEATHS THAN EXPECTED

WHITE FEMALE — CANCER OF THE RECTUM — ICD 154 EXCEPT 154.3

1950-1959 DEATHS	1950-1959 RATE	1960-1969 DEATHS	1960-1969 RATE	1970-1979 DEATHS	1970-1979 RATE	COUNTY AND STATE	PERCENT CHANGE 60S/50S	PERCENT CHANGE 70S/60S	PERCENT CHANGE 70S/50S	PERCENTILE %CHANGE 70S/50S	PERCENTILE RATES 1970S
3	4.7	2	2.1	3	4.3	HARPER KS	-54	99	-8	73	82
9	6.1	5	2.9	7	2.6	HARVEY KS	-53	-10	-58	24	50
1	1.2	3	3.3	3	3.0	JACKSON KS	182	-10	155	96	59
2	2.1	2	2.4	3	2.6	JEFFERSON KS	13	10	24	84	51
1	1.7	3	5.3			JEWELL KS	202				
14	3.8	19	2.9	20	1.9	JOHNSON KS	-24	-33	-49	35	36
		1	6.2			KEARNY KS					
		3	5.5			KINGMAN KS					
4*	14.6	2	5.3			KIOWA KS	-63				
8	4.1	11	5.0	7	2.5	LABETTE KS	24	-51	-40	49	48
1	10.5	1	6.5			LANE KS	-38				
8	4.2	9	3.7	15*	6.4	LEAVENWORTH KS	-12	73	53	89	95
2	4.4	2	3.1	2	6.1	LINCOLN KS	-30	99	39	87	94
2	3.0	2	2.9	3	3.9	LINN KS	-3	37	32	86	78
2	15.0	2	9.0			LOGAN KS	-40				
11	6.5	13	5.5	8	3.2	LYON KS	-15	-42	-50	33	65
6	4.4	6	3.8	3	1.5	MC PHERSON KS	-12	-62	-66	16	27
8	7.7	5	4.5	5	2.9	MARION KS	-42	-35	-62	20	58
6	5.2	2	1.2			MARSHALL KS	-76				
1	3.9	2	5.8	2	6.4	MEADE KS	48	11	64	91	95
4	2.9	5	2.6	2	1.6	MIAMI KS	-8	-40	-45	42	29
1	1.4	1	1.5	1	2.2	MITCHELL KS	5	47	55	90	41
10	3.7	6	1.8	12	3.5	MONTGOMERY KS	-52	99	-4	75	70
1	2.4	2	2.8	3	3.4	MORRIS KS	19	19	42	88	67
2*	21.9			1	6.0	MORTON KS			-73	09	94
4	4.2	3	3.1			NEMAHA KS	-27				
4	2.7	7	4.3	7	3.3	NEOSHO KS	59	-24	21	83	66
3	10.0	2	6.2	1	4.6	NESS KS	-38	-26	-54	28	86
4	6.1	2	2.3	4	4.1	NORTON KS	-63	83	-33	57	80
5	5.6	5	4.8	3	2.7	OSAGE KS	-15	-44	-52	30	53
3	4.8	1	1.1			OSBORNE KS	-77				
		1	3.7	1	1.0	OTTAWA KS		-74			20
7	9.5	2	2.3	3	4.1	PAWNEE KS	-76	82	-57	26	80
2	3.2	1	1.1	3	2.4	PHILLIPS KS	-66	116	-27	62	46
2	2.2	2	1.8	4	2.8	POTTAWATOMIE KS	-18	54	27	84	55
6	7.9	3	3.5	3	2.6	PRATT KS	-56	-27	-68	14	50
1	4.0					RAWLINS KS					
17	5.6	15	4.1	13	2.7	RENO KS	-28	-33	-51	32	54
1	1.3	3	3.0			REPUBLIC KS	132				
3	3.7	2	2.2	5	3.6	RICE KS	-40	62	-2	76	72
5	3.4	8	4.6	4	1.8	RILEY KS	35	-61	-47	39	34
2	4.1			1	1.3	ROOKS KS			-68	13	24
1	3.1	3	6.1	1	2.6	RUSH KS	95	-58	-17	68	50
3	5.1	3	4.4	3	3.7	RUSSELL KS	-13	-16	-27	62	73
6	3.2	10	4.0	6	2.0	SALINE KS	26	-49	-36	53	38
				1	3.9	SCOTT KS					77
39	3.5	52	3.5	41	2.1	SEDGWICK KS	-1	-40	-40	47	39
3	8.5	3	4.6	1	1.6	SEWARD KS	-46	-65	-81	04	29
33	5.1	32	4.3	29	2.8	SHAWNEE KS	-16	-35	-45	41	56
1	2.9	3	7.0	3	5.0	SHERMAN KS	139	-29	68	92	89
1	1.3	5	7.7	2	3.6	SMITH KS	491	-54	173	97	71
2	4.1	5	8.2	5	5.4	STAFFORD KS	100	-35	30	85	91
				1	6.5	STANTON KS					96
2	16.4	1	4.5			STEVENS KS	-73				
7	4.2	7	3.9	7	3.3	SUMNER KS	-7	-15	-21	66	67
3	11.4			3	6.3	THOMAS KS			-45	41	95
		2	6.0	2	5.8	TREGO KS		-4			93
2	3.4			1	1.3	WABAUNSEE KS			-63	20	24
3	3.4	7	6.8	1	1.4	WASHINGTON KS	99	-80	-60	22	25
1	9.3					WICHITA KS					
4	3.8	3	2.0	4	2.9	WILSON KS	-47	43	-25	64	57
4	10.5	1	2.9	1	5.3	WOODSON KS	-72	81	-50	34	90
35	5.1	38	4.7	29	3.2	WYANDOTTE KS	-8	-33	-38	50	64
512	4.1	550	3.5	543	2.8	KENTUCKY KY	-13	-20	-31		
4	6.9	1	1.5	2	1.8	ALLEN KY	-78	13	-75	07	32
2	4.4	1	2.1			ANDERSON KY	-53				
2	4.6			2	3.1	BALLARD KY			-33	56	62
8	6.0	4	2.5	4	2.1	BARREN KY	-58	-15	-64	18	40
		1	1.8	2	3.0	BATH KY		69			61
4	3.0	5	3.2	6	3.0	BELL KY	6	-6	0	77	61
3	4.2	1	.9	9	6.0	BOONE KY	-78	554	43	88	94
2	2.2	3	2.5	3	2.7	BOURBON KY	14	6	21	83	53
11	5.5	12	4.4	14	4.1	BOYD KY	-20	-8	-26	63	79
7	6.3	3	2.8	4	2.9	BOYLE KY	-55	5	-53	29	59
		2	3.6	2	3.2	BRACKEN KY		-10			65
		1	1.4	3	3.7	BREATHITT KY		166			73
2	2.4	7	7.7	3	3.1	BRECKINRIDGE KY	221	-60	29	85	62
2	3.9	2	3.1	6	5.4	BULLITT KY	-21	77	40	87	91
2	4.1	3	5.0	1	1.4	BUTLER KY	21	-73	-67	15	25
6	8.0			3	2.1	CALDWELL KY			-73	09	40
4	3.7	7	4.2	9	4.6	CALLOWAY KY	12	9	22	83	85
29	6.8	34*	6.4	25	4.1	CAMPBELL KY	-6	-35	-39	49	80
4	10.5	2	4.9			CARLISLE KY	-54				
3	5.6	3	6.3	1	1.4	CARROLL KY	12	-77	-75	08	26
		5	5.0	4	3.0	CARTER KY		-40			61
2	2.9	1	1.3	1	1.0	CASEY KY	-56	-18	-64	18	21
3	1.6	3	1.4	9	3.8	CHRISTIAN KY	-11	177	147	96	76
2	2.4	5	4.1	4	2.8	CLARK KY	74	-31	20	83	56
		3	4.1	1	1.1	CLAY KY		-73			22
		3	7.0	2	3.3	CLINTON KY		-53			67
4	5.5	1	1.9	1	1.8	CRITTENDEN KY	-66	-4	-67	14	33
				1	1.9	CUMBERLAND KY					35
7	2.6	4	1.1	12	2.4	DAVIESS KY	-57	116	-8	74	47
1	3.1	2	5.0	1	1.7	EDMONSON KY	60	-66	-45	42	31
		2	6.4			ELLIOTT KY					
2	3.1	4	5.0	1	.7	ESTILL KY	60	-85	-76	07	18
24	5.4	12	2.0	10	1.3	FAYETTE KY	-63	-33	-75	07	24
2	3.1	1	1.4	2	2.3	FLEMING KY	-55	65	-26	63	44
1	.8	4	2.4	6	2.6	FLOYD KY	196	8	218	98	51

* SIGNIFICANTLY MORE DEATHS THAN EXPECTED

WHITE FEMALE — CANCER OF THE RECTUM — ICD 154 EXCEPT 154.3

1950-1959 DEATHS	1950-1959 RATE	1960-1969 DEATHS	1960-1969 RATE	1970-1979 DEATHS	1970-1979 RATE	COUNTY AND STATE	PERCENT CHANGE 60S/50S	PERCENT CHANGE 70S/60S	PERCENT CHANGE 70S/50S	PERCENTILE %CHANGE 70S/50S	PERCENTILE RATES 1970S
6	4.4	6	3.8	5	2.4	FRANKLIN KY	-14	-36	-45	41	47
1	1.7	2	3.0	4	4.7	FULTON KY	72	57	170	97	86
2	12.5	2	10.9	2	8.6	GALLATIN KY	-13	-21	-31	58	98
2	4.3	3	6.3	1	1.8	GARRARD KY	48	-72	-58	24	33
4	6.5	2	3.4	5	5.9	GRANT KY	-48	75	-9	73	93
10	5.6	3	1.4	6	2.2	GRAVES KY	-75	58	-60	22	42
3	3.3	3	2.8	1	.9	GRAYSON KY	-15	-70	-74	08	19
2	3.9			1	1.4	GREEN KY			-65	16	25
3	2.9	5	4.2	10	5.9	GREENUP KY	44	42	104	94	94
1	3.6	1	2.2	1	2.6	HANCOCK KY	-38	17	-28	61	51
5	3.7	3	1.8	12	5.1	HARDIN KY	-50	178	40	87	89
5	3.5	5	3.1	5	2.3	HARLAN KY	-12	-26	-35	54	44
1	1.0	3	2.9	1	.8	HARRISON KY	195	-73	-19	67	19
2	3.0	1	1.2	4	3.9	HART KY	-60	229	30	86	76
9	7.1	6	3.2	5	2.3	HENDERSON KY	-54	-29	-67	14	45
4	6.4	2	2.3	8*	9.5	HENRY KY	-64	315	49	89	98
1	2.6	2	4.5	1	2.1	HICKMAN KY	72	-52	-18	68	40
9	5.1	11	4.8	7	2.5	HOPKINS KY	-5	-49	-52	31	48
		1	2.0			JACKSON KY					
123	5.3	117	3.9	116	3.2	JEFFERSON KY	-25	-19	-40	48	64
		1	1.4	2	2.3	JESSAMINE KY		65			43
4	3.8	3	2.7	2	1.7	JOHNSON KY	-30	-36	-55	27	31
28	4.8	45*	6.3	37	4.4	KENTON KY	33	-31	-9	73	83
4	9.0	1	1.4	2	2.2	KNOTT KY	-84	56	-75	07	43
6	6.1	2	1.6	4	3.0	KNOX KY	-73	81	-51	32	59
				4	5.2	LARUE KY					90
1	.9	1	.7	4	2.3	LAUREL KY	-30	248	143	96	44
1	1.6	4	5.3	2	2.0	LAWRENCE KY	230	-62	24	84	37
2	6.2	2	5.5			LEE KY	-11				
				2	4.3	LESLIE KY					82
3	2.8	1	1.4	4	2.9	LETCHER KY	-49	110	6	79	59
1	1.7	1	1.8	1	1.8	LEWIS KY	6	1	6	79	34
1	1.3	6	6.0	1	.9	LINCOLN KY	366	-84	-28	61	20
1	2.4	1	2.6	1	1.7	LIVINGSTON KY	10	-35	-29	61	32
4	3.7	4	3.4	2	1.8	LOGAN KY	-7	-48	-52	31	32
		3	9.1			LYON KY					
10	4.1	13	4.2	12	3.3	MC CRACKEN KY	2	-22	-20	66	66
		1	1.8	2	2.5	MC CREARY KY		40			49
1	1.9	3	5.2	2	2.6	MC LEAN KY	172	-51	33	86	50
4	2.9	2	1.2	3	1.5	MADISON KY	-57	19	-49	36	27
		2	4.2			MAGOFFIN KY					
		5	5.7	1	1.3	MARION KY		-78			24
4	5.0	5	5.3	3	1.7	MARSHALL KY	5	-68	-66	16	31
1	3.2					MARTIN KY					
5	5.7	6	5.8	7	5.8	MASON KY	2	-1	1	77	93
4	10.0	2	4.6	2	2.4	MEADE KY	-54	-48	-76	06	46
1	5.4	3*	12.6	1	3.7	MENIFEE KY	133	-71	-32	57	73
2	2.5	1	1.2	2	1.8	MERCER KY	-53	51	-28	61	33
				1	1.7	METCALFE KY					32
5	9.6	3	4.6	1	1.4	MONROE KY	-52	-69	-85	02	26
5	8.7	3	3.7	3	3.2	MONTGOMERY KY	-57	-14	-63	19	64
		6*	9.6	3	4.1	MORGAN KY		-58			79
4	3.2	6	3.4	3	1.6	MUHLENBERG KY	7	-53	-50	34	29
4	4.5	8	7.9	1	1.0	NELSON KY	75	-87	-78	06	21
3	7.0	2	3.6			NICHOLAS KY	-48				
4	3.2	6	5.1	6	3.8	OHIO KY	59	-25	20	83	76
2	4.3	3	5.3	1	1.3	OLDHAM KY	24	-75	-69	13	25
4	8.2	1	2.8	2	2.7	OWEN KY	-66	-3	-67	14	53
9*	15.2	2	3.1	1	2.1	PENDLETON KY	-80	-32	-86	01	39
1	.9			4	2.5	PERRY KY			169	97	49
4	1.8	7	2.8	7	2.2	PIKE KY	55	-21	22	83	40
		1	2.9	2	4.2	POWELL KY		43			81
3	1.9	5	2.6	6	2.2	PULASKI KY	39	-16	17	82	41
		1	5.2	1	5.9	ROBERTSON KY		13			94
5	8.6	6	9.3	1	1.3	ROCKCASTLE KY	8	-86	-85	02	24
		2	3.0			ROWAN KY					
1	2.4	2	3.3	1	.9	RUSSELL KY	34	-72	-63	19	19
4	5.4	4	4.9	3	2.5	SCOTT KY	-10	-50	-54	28	48
1	1.2	10*	8.8	2	1.4	SHELBY KY	613	-84	11	81	25
3	4.6	4	5.1	2	2.7	SIMPSON KY	12	-47	-40	47	53
		2	7.2			SPENCER KY					
3	4.8	1	1.3	9*	7.6	TAYLOR KY	-74	490	56	90	97
4	7.3	3	5.1	4	5.5	TODD KY	-30	9	-24	64	92
1	2.1			2	4.0	TRIGG KY			89	93	79
		1	3.4			TRIMBLE KY					
2	3.0	6	6.7	1	1.0	UNION KY	127	-85	-66	16	21
7	3.3	9	3.4	11	3.5	WARREN KY	3	4	8	80	70
		3	5.5	1	1.4	WASHINGTON KY		-75			25
2	3.3	1	1.4			WAYNE KY	-59				
4	4.5			2	1.9	WEBSTER KY			-58	24	35
3	2.8	3	1.9	6	3.2	WHITLEY KY	-31	65	13	81	63
				1	4.1	WOLFE KY					80
		1	1.3	3	3.8	WOODFORD KY		201			75
367	4.5	369	3.4	377	2.6	LOUISIANA LA	-26	-24	-44		
9	5.6	14	7.7	12	4.7	ACADIA LA	38	-39	-16	69	86
1	1.6	1	1.4	3	3.2	ALLEN LA	-15	126	92	93	64
3	4.0	1	1.2	1	.8	ASCENSION LA	-71	-33	-81	04	18
3	5.9	1	1.8			ASSUMPTION LA	-70				
7	6.8	7	4.1	1	.4	AVOYELLES LA	-40	-89	-94	00	16
1	1.7	1	1.3	5	4.2	BEAUREGARD LA	-21	211	146	96	81
1	1.5	4	5.2	1	1.1	BIENVILLE LA	246	-80	-29	60	21
4	3.9	4	4.3	4	2.0	BOSSIER LA	11	-53	-48	37	38
15	2.9	24	3.1	27	2.5	CADDO LA	8	-18	-12	72	49
11	3.6	15	3.9	19	3.4	CALCASIEU LA	8	-12	-5	75	68
				1	1.6	CATAHOULA LA					29
3	4.3	4	4.3	3	2.9	CLAIBORNE LA	0	-32	-32	58	58
		2	4.7	1	1.9	CONCORDIA LA		-60			35
		3	3.7	3	2.7	DE SOTO LA		-26			54
17	4.2	19	3.4	12	1.3	EAST BATON ROUGE LA	-20	-61	-68	13	25

* SIGNIFICANTLY MORE DEATHS THAN EXPECTED

WHITE FEMALE — CANCER OF THE RECTUM — ICD 154 EXCEPT 154.3

1950-1959 DEATHS	1950-1959 RATE	1960-1969 DEATHS	1960-1969 RATE	1970-1979 DEATHS	1970-1979 RATE	COUNTY AND STATE	PERCENT CHANGE 60S/50S	PERCENT CHANGE 70S/60S	PERCENT CHANGE 70S/50S	PERCENTILE %CHANGE 70S/50S	PERCENTILE RATES 1970S
1	5.0	1	3.9			EAST CARROLL LA	-21				
		2	2.7	2	2.7	EAST FELICIANA LA		1			53
4	4.2	4	3.5	2	1.3	EVANGELINE LA	-16	-62	-68	14	24
2	2.4	1	1.1	1	.6	FRANKLIN LA	-54	-43	-74	08	17
2	4.2	1	2.2	3	3.1	GRANT LA	-49	44	-26	63	63
2	1.5	4	2.4	8	3.8	IBERIA LA	53	59	144	96	75
1	1.7	3	3.9	3	3.5	IBERVILLE LA	123	-10	101	94	69
2	4.2			3	2.9	JACKSON LA			-31	59	58
19	5.2	25	3.5	35	2.7	JEFFERSON LA	-33	-22	-47	38	54
1	.9	2	1.7	3	2.0	JEFFERSON DAVIS LA	79	22	118	95	38
2	1.1	3	1.0	9	2.2	LAFAYETTE LA	-2	115	110	94	42
3	1.9	7	4.0	9	3.3	LAFOURCHE LA	113	-17	76	92	66
1	1.9	1	1.3			LA SALLE LA	-28				
6	7.6	4	3.2	2	1.5	LINCOLN LA	-58	-53	-80	04	28
1	1.9	5	4.9	3	2.0	LIVINGSTON LA	152	-60	1	77	36
		3	3.0	2	1.8	MOREHOUSE LA		-39			34
2	2.1	4	3.2	3	1.9	NATCHITOCHES LA	50	-40	-10	73	36
152*	7.2	117	4.8	90	3.6	ORLEANS LA	-34	-25	-50	33	72
10	4.6	6	1.6	9	1.8	OUACHITA LA	-65	12	-61	22	33
		2	5.5	1	1.6	PLAQUEMINES LA		-71			29
3	6.7	1	1.6			POINTE COUPEE LA	-76				
11	3.8	7	1.7	13	2.6	RAPIDES LA	-57	60	-31	58	52
				1	1.7	RED RIVER LA					30
1	2.0	4	5.5			RICHLAND LA	171				
2	2.6	1	1.0			SABINE LA	-63				
3	6.1	3	2.4	3	1.3	ST BERNARD LA	-61	-44	-78	05	25
1	1.9	1	1.3	2	1.9	ST CHARLES LA	-31	44	-1	77	35
1	2.6	3	6.0			ST JAMES LA	133				
4	10.9	2	4.8			ST JOHN THE BAPT LA	-56				
5	3.1	5	2.3	10	3.5	ST LANDRY LA	-25	53	14	81	70
5	7.5	1	1.1	3	2.6	ST MARTIN LA	-85	126	-66	16	50
5	6.2	9	7.3	10	6.1	ST MARY LA	18	-16	-1	76	94
4	4.5	4	2.3	10	3.6	ST TAMMANY LA	-48	54	-20	66	71
6	4.1	3	1.4	4	1.3	TANGIPAHOA LA	-67	-6	-69	13	24
1	7.4					TENSAS LA					
5	4.6	7	4.4	11	4.5	TERREBONNE LA	-5	2	-3	76	85
2	3.3	1	1.2	3	2.8	UNION LA	-65	134	-17	68	54
6	4.4	5	2.6	9	3.6	VERMILION LA	-40	35	-19	67	71
4	5.8	4	4.4	2	1.7	VERNON LA	-23	-62	-71	11	30
1	.9	4	2.6	4	1.9	WASHINGTON LA	193	-25	120	95	36
4	3.4	4	2.5	6	2.9	WEBSTER LA	-26	14	-15	70	57
4*	17.0	1	2.5	1	2.0	WEST BATON ROUGE LA	-85	-21	-88	01	37
2	4.2			3	3.6	WEST CARROLL LA			-14	70	72
1	1.6	4	5.3	1	2.0	WINN LA	233	-63	24	84	37
404*	7.8	385*	6.4	352*	4.8	MAINE ME	-17	-25	-38		
47*	9.8	43*	7.7	42*	6.2	ANDROSCOGGIN ME	-21	-19	-37	52	95
19	6.5	15	3.8	29*	6.2	AROOSTOOK ME	-42	64	-5	75	95
80	7.6	87*	7.1	69*	4.5	CUMBERLAND ME	-6	-37	-40	47	85
8	7.4	10	8.2	5	3.4	FRANKLIN ME	11	-59	-54	28	68
18	8.6	12	4.9	11	4.3	HANCOCK ME	-44	-11	-50	34	83
39	7.9	31	5.5	25	3.6	KENNEBEC ME	-31	-35	-55	27	71
15	7.4	18*	6.4	11	3.5	KNOX ME	-14	-46	-53	29	69
11	8.7	8	4.6	11	5.9	LINCOLN ME	-47	26	-33	57	93
12	4.9	16	5.7	15	4.5	OXFORD ME	16	-21	-9	73	85
35	6.0	32	4.8	36	4.6	PENOBSCOT ME	-19	-4	-23	65	86
9	8.3	7	5.6	5	3.4	PISCATAQUIS ME	-33	-40	-59	23	68
9	6.8	9	6.2	8	5.2	SAGADAHOC ME	-9	-16	-24	64	90
23*	10.8	24*	10.3	18*	6.5	SOMERSET ME	-4	-36	-39	49	96
17*	11.6	4	3.0	6	3.3	WALDO ME	-74	12	-71	11	67
16	8.1	18*	8.8	10	3.8	WASHINGTON ME	9	-56	-53	30	75
46	8.4	51*	7.6	51*	5.8	YORK ME	-9	-24	-31	58	93
622	6.3	664	4.8	705*	3.9	MARYLAND MD	-23	-19	-38		
19	4.5	31	5.9	28	4.6	ALLEGANY MD	31	-21	3	78	86
20	5.4	39	5.7	56*	4.9	ANNE ARUNDEL MD	5	-14	-9	73	88
70	5.5	108	4.9	127	3.7	BALTIMORE MD	-10	-24	-32	57	74
1	2.6	1	1.8	10*	10.3	CALVERT MD	-32	483	296	99	99
4	4.3	5	5.1	2	1.5	CAROLINE MD	18	-70	-65	17	27
15	5.2	16	4.7	19	4.3	CARROLL MD	-9	-9	-17	68	82
7	4.6	4	2.2	13	5.2	CECIL MD	-52	136	14	81	90
6	13.5	5	6.7	4	3.4	CHARLES MD	-50	-49	-75	08	68
10	7.4	10	6.2	8	4.9	DORCHESTER MD	-17	-20	-34	56	88
16	5.2	22	5.6	16	3.4	FREDERICK MD	7	-40	-35	54	67
3	3.4	1	1.1	4	2.7	GARRETT MD	-69	152	-21	66	53
11	5.9	11	3.6	11	2.5	HARFORD MD	-39	-30	-57	25	49
11*	11.0	6	4.2	16*	6.2	HOWARD MD	-62	48	-44	43	95
1	2.0	3	3.6	1	1.0	KENT MD	86	-73	-51	32	20
66	8.5	66	4.1	76	3.0	MONTGOMERY MD	-52	-27	-65	17	60
35	5.0	54	4.0	71	3.8	PRINCE GEORGE'S MD	-19	-6	-24	64	75
2	4.1	2	2.1	2	1.8	QUEEN ANNE'S MD	-49	-13	-55	27	34
		5	6.5	7	5.2	ST MARY'S MD		-20			90
1	1.4	3	3.1	4	3.5	SOMERSET MD	120	12	145	96	69
3	2.7	10	7.3	4	2.5	TALBOT MD	169	-66	-9	73	48
30	7.5	33	6.0	39*	5.8	WASHINGTON MD	-20	-4	-23	65	93
6	2.8	5	1.8	14	4.3	WICOMICO MD	-37	140	51	89	82
2	1.9	5	4.2	7	5.3	WORCESTER MD	121	26	178	97	91
283*	7.4	219*	5.5	166*	4.3	BALTIMORE CITY MD	-25	-23	-42	45	82
2120*	7.4	1878*	5.5	1672*	4.2	MASSACHUSETTS MA	-25	-24	-43		
21	5.6	30	4.8	54*	4.8	BARNSTABLE MA	-14	-1	-15	70	87
65*	8.1	72*	7.7	47	3.9	BERKSHIRE MA	-5	-49	-52	31	77
192*	8.3	167*	6.1	164*	4.9	BRISTOL MA	-26	-20	-41	46	88
2	4.5	1	1.8	2	2.6	DUKES MA	-59	41	-42	45	51
240*	6.8	225*	5.5	195*	4.0	ESSEX MA	-20	-27	-42	46	78
31*	8.8	27	6.3	23	5.1	FRANKLIN MA	-28	-19	-42	45	89
171*	7.9	144	5.4	153*	4.7	HAMPDEN MA	-32	-14	-41	46	86
31	5.8	32	4.7	20	2.6	HAMPSHIRE MA	-19	-44	-55	27	51

* SIGNIFICANTLY MORE DEATHS THAN EXPECTED

WHITE FEMALE CANCER OF THE RECTUM ICD 154 EXCEPT 154.3

1950-1959 DEATHS	1950-1959 RATE	1960-1969 DEATHS	1960-1969 RATE	1970-1979 DEATHS	1970-1979 RATE	COUNTY AND STATE		PERCENT CHANGE 60S/50S	PERCENT CHANGE 70S/60S	PERCENT CHANGE 70S/50S	PERCENTILE %CHANGE 70S/50S	PERCENTILE RATES 1970S
469*	7.2	418*	5.3	414*	4.5	MIDDLESEX	MA	-26	-16	-38	51	84
3	12.7	2	7.7	1	2.4	NANTUCKET	MA	-39	-69	-81	04	47
190*	7.3	177	5.2	163	3.8	NORFOLK	MA	-28	-27	-48	37	76
109*	8.1	88	5.3	79	3.8	PLYMOUTH	MA	-34	-28	-53	30	76
386*	8.0	300*	5.9	206*	4.4	SUFFOLK	MA	-26	-26	-45	41	84
210	6.4	195	5.1	151	3.3	WORCESTER	MA	-20	-34	-47	38	67
1736	6.2	1693	4.8	1386	3.2	MICHIGAN	MI	-23	-34	-49		
1	3.2			1	2.1	ALCONA	MI			-33	56	40
3	7.0	1	2.3			ALGER	MI	-67				
9	3.5	14	4.6	12	3.1	ALLEGAN	MI	33	-32	-10	72	63
2	1.8	4	2.9	9	4.9	ALPENA	MI	64	68	174	97	88
6	10.0	3	4.4	4	3.6	ANTRIM	MI	-56	-19	-64	18	71
		3	5.0	1	1.4	ARENAC	MI		-71			26
9	5.4	9	4.8	10	3.8	BARRY	MI	-13	-21	-31	58	74
33	7.5	37*	7.3	18	2.9	BAY	MI	-2	-60	-61	21	58
2	5.1	1	1.8	2	2.6	BENZIE	MI	-65	48	-48	36	52
40	6.8	31	4.3	18	2.0	BERRIEN	MI	-37	-52	-70	12	38
17	9.4	9	4.4	4	1.7	BPANCH	MI	-54	-60	-81	04	32
31	5.4	23	3.3	36	4.6	CALHOUN	MI	-38	38	-14	70	86
9	6.1	5	2.8	6	2.7	CASS	MI	-53	-6	-56	26	53
5	7.0	5	5.3	2	2.0	CHARLEVOIX	MI	-25	-63	-72	10	36
6	8.5			3	2.3	CHEBOYGAN	MI			-73	09	44
10	7.9	2	1.5	2	1.3	CHIPPEWA	MI	-81	-18	-84	02	23
2	4.7	5	6.0	7	5.0	CLARE	MI	28	-17	6	79	88
12	7.6	16*	9.1	6	2.5	CLINTON	MI	19	-72	-67	15	49
1	5.2	2	5.9	3	7.2	CRAWFORD	MI	12	23	38	87	97
12	8.3	15*	8.0	4	1.5	DELTA	MI	-4	-81	-82	03	28
4	2.8	6	3.8	4	1.6	DICKINSON	MI	38	-59	-44	43	28
14	6.2	13	5.2	12	3.3	EATON	MI	-15	-37	-46	39	67
9	9.9	11*	11.6	3	2.3	EMMET	MI	17	-80	-77	06	44
76	7.0	60	4.2	29	1.6	GENESEE	MI	-40	-61	-77	06	30
3	6.5	4	4.9	4	3.9	GLADWIN	MI	-24	-21	-40	48	77
10	9.0	8	4.8	9	5.4	GOGEBIC	MI	-46	11	-41	47	91
5	2.7	9	4.1	10	3.2	GRAND TRAVERSE	MI	55	-23	18	82	64
7	3.3	7	2.5	10	4.3	GPATIOT	MI	-24	69	29	85	82
16	8.2	14	6.9	8	3.0	HILLSDALE	MI	-16	-57	-63	19	60
17	7.7	12	4.9	12	4.6	HOUGHTON	MI	-36	-7	-40	47	86
4	2.4	12	5.8	10	3.8	HURON	MI	147	-35	60	90	74
55	7.0	46	4.7	42	3.4	INGHAM	MI	-32	-29	-52	31	67
10	4.7	12	5.3	4	1.3	IONIA	MI	13	-76	-73	09	24
4	7.2	1	1.4	3	1.9	IOSCO	MI	-81	40	-74	08	35
6	7.2	4	4.2	2	1.3	IRON	MI	-42	-70	-83	03	24
15*	12.7	5	3.2	3	1.7	ISABELLA	MI	-75	-46	-87	01	31
34	6.0	33	4.9	19	2.3	JACKSON	MI	-19	-54	-63	20	43
29	4.4	33	3.9	36	3.4	KALAMAZOO	MI	-10	-13	-21	65	68
2	11.3			3	8.7	KALKASKA	MI			-23	65	98
85	5.5	85	4.4	84	3.5	KENT	MI	-20	-20	-36	53	70
		1	9.3			KEWEENAW	MI					
		4*	13.4	1	2.7	LAKE	MI		-80			53
16	9.5	10	4.6	10	3.8	LAPEER	MI	-51	-18	-60	22	75
1	2.3	1	2.1	2	2.1	LEELANAU	MI	-8	0	-8	73	39
19	5.4	16	3.6	17	3.1	LENAWEE	MI	-33	-15	-43	44	62
10	6.5	7	3.2	8	2.4	LIVINGSTON	MI	-50	-25	-63	20	47
1	2.3	1	2.2			LUCE	MI	-5				
2	4.8	1	2.2	1	1.2	MACKINAC	MI	-53	-44	-74	08	23
53	6.8	82	6.3	78	3.2	MACOMB	MI	-6	-50	-53	29	63
6	5.3	9	7.1	6	3.4	MANISTEE	MI	34	-52	-35	53	68
20	8.9	7	2.7	14	4.4	MARQUETTE	MI	-69	59	-51	32	83
2	1.9	3	2.2	3	1.4	MASON	MI	16	-39	-29	60	25
6	7.1	6	5.0	4	2.7	MECOSTA	MI	-29	-47	-62	20	52
7	5.2	6	4.0	8	4.5	MENOMINEE	MI	-25	13	-15	70	85
4	3.3	11	6.2	10	3.7	MIDLAND	MI	89	-41	12	81	73
				1	2.5	MISSAUKEE	MI					48
20	5.8	19	4.4	23	4.0	MONROE	MI	-24	-9	-31	59	79
7	3.9	11	5.0	5	2.0	MONTCALM	MI	31	-60	-47	38	38
2	10.9	2	8.3	1	1.9	MONTMORENCY	MI	-24	-77	-82	03	36
22	4.8	35	5.5	28	3.3	MUSKEGON	MI	14	-41	-32	57	65
5	4.6	1	.6	10	5.1	NEWAYGO	MI	-86	706	12	81	89
90	5.3	145	5.1	138	3.1	OAKLAND	MI	-4	-39	-41	46	62
1	1.0					OCEANA	MI					
2	3.9	4	5.8	7*	7.3	OGEMAW	MI	48	26	86	93	97
		1	1.9	1	2.2	ONTONAGON	MI		17			43
6	7.4	1	1.6	2	2.1	OSCEOLA	MI	-78	32	-71	10	40
1	8.6	1	4.4			OSCODA	MI	-49				
2	7.1	5*	13.0	5	7.2	OTSEGO	MI	82	-45	1	77	97
18	4.8	23	4.8	16	2.4	OTTAWA	MI	-1	-49	-50	34	47
1	1.5	6	9.0	2	2.8	PRESQUE ISLE	MI	489	-69	82	93	55
2	5.9	2	3.6	6	6.2	ROSCOMMON	MI	-40	73	5	79	95
51	7.2	40	4.7	32	3.2	SAGINAW	MI	-35	-32	-56	27	64
35	7.5	30	5.3	32	4.3	ST CLAIR	MI	-30	-18	-43	44	83
14	6.4	19*	7.3	14	4.3	ST JOSEPH	MI	15	-41	-33	57	82
9	5.2	7	3.5	8	2.8	SANILAC	MI	-34	-19	-47	39	55
		5	9.7			SCHOOLCRAFT	MI					
13	5.3	10	3.6	8	2.6	SHIAWASSEE	MI	-31	-27	-50	33	52
14	7.0	7	2.9	5	1.5	TUSCOLA	MI	-59	-48	-79	05	27
15	5.9	17	6.3	9	2.3	VAN BUREN	MI	7	-62	-60	22	45
26	4.3	29	4.0	20	2.3	WASHTENAW	MI	-8	-44	-48	37	43
612*	6.6	543*	5.0	393	3.5	WAYNE	MI	-25	-30	-47	38	69
6	6.3	5	4.3	3	2.3	WEXFORD	MI	-32	-47	-64	18	44
805	5.2	777	3.9	747	3.1	MINNESOTA	MN	-24	-22	-40		
5	6.8	3	4.2	3	3.7	AITKIN	MN	-38	-12	-45	41	73
2	1.3	7	2.5	11	2.5	ANOKA	MN	90	-1	88	93	48
8	7.8	6	4.3	9	5.3	BECKER	MN	-45	24	-31	58	91
7	6.9	4	3.6	3	1.8	BELTRAMI	MN	-49	-48	-73	09	34
3	4.9	4	4.8	4	4.5	BENTON	MN	-3	-7	-10	73	84
4	8.7	4	6.6	3	3.7	BIG STONE	MN	-24	-44	-57	25	73
15	6.6	18	6.5	14	4.0	BLUE EARTH	MN	-1	-39	-39	49	79

* SIGNIFICANTLY MORE DEATHS THAN EXPECTED

WHITE FEMALE

CANCER OF THE RECTUM — ICD 154 EXCEPT 154.3

1950-1959 DEATHS	1950-1959 RATE	1960-1969 DEATHS	1960-1969 RATE	1970-1979 DEATHS	1970-1979 RATE	COUNTY AND STATE	PERCENT CHANGE 60S/50S	PERCENT CHANGE 70S/60S	PERCENT CHANGE 70S/50S	PERCENTILE %CHANGE 70S/50S	PERCENTILE RATES 1970S
9	6.0	8	4.4	10	5.4	BROWN MN	-26	22	-10	72	91
1	.8	2	1.4	7	4.0	CARLTON MN	79	198	432	99	79
5	5.3	2	1.8	5	3.2	CARVER MN	-67	80	-40	48	64
7	8.1	2	2.0	7	4.2	CASS MN	-75	108	-49	35	60
3	4.4	7	6.2	5	3.4	CHIPPEWA MN	40	-44	-22	65	69
5	6.0	8	8.1	10*	7.2	CHISAGO MN	35	-10	21	83	97
5	3.7	7	3.8	7	2.8	CLAY MN	4	-26	-23	65	56
1	2.3			1	1.1	CLEARWATER MN			-55	27	21
1	10.7					COOK MN					
2	2.5	1	.8	3	3.5	COTTONWOOD MN	-69	356	43	88	70
8	4.6	5	2.4	8	2.9	CROW WING MN	-48	19	-38	51	57
9	4.1	14	4.2	13	2.3	DAKOTA MN	2	-45	-43	44	45
2	2.9	1	1.6	5	4.8	DODGE MN	-47	206	63	91	87
4	3.6	10	6.2	4	1.7	DOUGLAS MN	73	-73	-53	30	31
5	3.8	9	6.0	4	2.2	FARIBAULT MN	60	-63	-41	46	42
6	4.3	5	2.6	4	1.8	FILLMORE MN	-39	-32	-59	24	32
7	4.3	17*	8.3	7	2.9	FREEBORN MN	94	-66	-33	56	56
7	3.5	14	6.2	5	1.5	GOODHUE MN	79	-75	-56	26	28
1	2.2	1	2.0	1	1.9	GRANT MN	-8	-6	-13	71	36
238	6.2	201	4.1	181	3.1	HENNEPIN MN	-34	-25	-50	33	62
4	5.0	4	3.5	5	3.2	HOUSTON MN	-30	-9	-36	52	63
2	4.1	2	2.6	1	2.1	HUBBARD MN	-37	-17	-48	37	40
1	1.3	2	2.5	2	2.0	ISANTI MN	99	-21	57	90	37
7	4.6	4	2.2	10	4.5	ITASCA MN	-52	105	-1	76	85
5	6.5	3	3.7	4	2.8	JACKSON MN	-44	-24	-57	25	56
3	6.2	4	6.8	1	1.8	KANABEC MN	9	-73	-71	11	33
5	3.5	7	3.5	10	4.7	KANDIYOHI MN	-1	36	35	87	87
3	7.0	1	2.7	1	1.6	KITTSON MN	-62	-41	-78	06	28
2	3.3	4	6.5			KOOCHICHING MN	98				
2	2.6	2	2.0	4	4.3	LAC QUI PARLE MN	-24	113	62	91	82
2	4.2	1	2.0	2	2.5	LAKE MN	-52	26	-40	48	49
				1	4.0	LAKE OF THE WOODS MN					78
5	4.5			6	3.7	LE SUEUR MN			-19	68	73
2	3.9	1	1.9	6*	9.6	LINCOLN MN	-52	416	150	96	98
5	5.0	9	6.8	4	2.9	LYON MN	36	-57	-42	45	57
2	1.3	8	5.1	7	3.9	MC LEOD MN	299	-22	210	98	77
3	10.6	5*	18.2	1	3.7	MAHNOMEN MN	72	-80	-65	17	74
4	7.0			2	2.1	MARSHALL MN			-70	12	39
5	4.1	2	1.3	8	3.4	MARTIN MN	-69	170	-16	69	68
6	6.7	2	1.8	5	3.6	MEEKER MN	-73	98	-46	40	72
3	3.8	1	1.0	3	2.5	MILLE LACS MN	-74	150	-34	55	49
10	8.2	4	2.8	3	1.6	MORRISON MN	-65	-44	-81	04	29
6	3.2	8	3.2	3	.9	MOWER MN	-2	-72	-72	10	19
1	1.8	2	2.6	2	2.5	MURRAY MN	40	-2	37	87	49
1	.9	3	2.2	3	2.5	NICOLLET MN	146	15	184	97	50
12*	11.7	1	.7	2	1.0	NOBLES MN	-94	47	-92	00	20
4	6.0	4	4.4	3	2.2	NORMAN MN	-28	-50	-64	18	41
14	5.0	14	3.8	13	3.0	OLMSTED MN	-25	-20	-40	48	61
12	4.2	9	2.5	10	2.2	OTTER TAIL MN	-40	-14	-49	36	41
3	5.0	3	4.4	4	3.6	PENNINGTON MN	-12	-18	-27	62	72
2	2.3	5	5.9	2	1.3	PINE MN	149	-77	-43	44	25
3	4.5	1	1.6	2	3.0	PIPESTONE MN	-65	84	-35	54	59
8	4.6	9	3.7	4	1.4	POLK MN	-20	-63	-71	11	25
3	5.2	1	1.1	4	3.8	POPE MN	-79	250	-27	62	75
121	6.2	111	4.4	83	2.7	RAMSEY MN	-28	-38	-56	27	54
1	3.1			1	3.5	RED LAKE MN			13	81	70
1	.9	6	4.0	5	2.6	REDWOOD MN	329	-35	181	97	51
9	7.7	5	3.9	7	3.9	RENVILLE MN	-50	2	-49	36	78
8	4.1	5	2.3	15	5.4	RICE MN	-44	136	32	86	91
3	5.6	2	3.2	2	2.0	ROCK MN	-43	-36	-64	18	38
1	1.9	5	7.8	2	2.4	ROSEAU MN	313	-68	31	86	48
53	4.9	77	5.8	63	4.2	ST LOUIS MN	17	-28	-16	69	81
3	3.1	4	3.5	2	1.1	SCOTT MN	15	-68	-64	19	22
2	3.4	3	2.5	1	.4	SHERBURNE MN	-28	-84	-89	01	16
4	5.2	2	2.0	4	3.0	SIBLEY MN	-61	47	-42	45	59
15	5.2	18	4.6	17	3.3	STEARNS MN	-10	-30	-37	52	66
3	2.3	3	2.2	9	4.5	STEELE MN	-3	106	99	94	85
1	1.8	3	4.7	3	4.3	STEVENS MN	168	-8	145	96	83
1	1.2	1	1.2	1	.6	SWIFT MN	3	-48	-47	39	17
3	2.3	2	1.4	4	2.2	TODD MN	-41	57	-7	74	41
1	2.5	2	4.2	1	3.4	TRAVERSE MN	66	-18	36	87	68
10	9.9	7	5.1	1	.5	WABASHA MN	-48	-90	-95	00	17
2	3.6	2	2.9	1	.9	WADENA MN	-18	-68	-74	08	20
3	3.7	2	1.6	4	3.7	WASECA MN	-56	126	0	77	73
8	4.7	5	2.3	10	3.1	WASHINGTON MN	-50	32	-34	55	62
5	6.5	6	6.9	6	5.9	WATONWAN MN	6	-15	-10	73	93
5	11.6	3	4.6	3	3.9	WILKIN MN	-61	-15	-66	16	77
15	6.9	5	1.8	19*	6.1	WINONA MN	-74	234	-12	72	94
6	4.1	8	4.1	6	2.4	WRIGHT MN	0	-41	-41	47	47
1	1.4	4	3.9	5	3.8	YELLOW MEDICINE MN	183	-1	181	97	76
174	3.2	192	2.7	208	2.2	MISSISSIPPI MS	-15	-19	-31		
1	1.3	5	5.9	9*	7.9	ADAMS MS	347	33	494	99	97
1	1.2	6	4.3	9	4.7	ALCORN MS	249	9	280	98	86
1	2.5	1	2.0	3	6.0	AMITE MS	-19	201	143	96	94
2	2.4	5	5.7	4	4.2	ATTALA MS	140	-25	80	92	82
				1	2.3	BENTON MS					45
9*	13.0	1	1.0	3	2.8	BOLIVAR MS	-92	171	-79	05	55
1	1.5	3	4.0	3	3.1	CALHOUN MS	165	-23	104	94	62
2	3.8	1	1.4	1	1.0	CHICKASAW MS	-64	-26	-73	09	21
1	2.9					CHOCTAW MS					
1	6.6			1	4.2	CLAIBORNE MS			-37	51	80
1	1.7			1	1.3	CLARKE MS			-23	65	25
		1	1.8	1	1.9	CLAY MS		1			34
5	6.5	6	7.0	2	2.3	COAHOMA MS	9	-68	-65	17	43
3	3.6			2	1.6	COPIAH MS			-56	26	28
2	3.9			3	3.7	COVINGTON MS			-3	75	74
2	4.7			3	2.2	DE SOTO MS			-52	31	43
9	6.1	3	1.5	8	3.2	FORREST MS	-76	121	-47	38	64
1	3.2	1	2.7			FRANKLIN MS	-15				

* SIGNIFICANTLY MORE DEATHS THAN EXPECTED

WHITE FEMALE

CANCER OF THE RECTUM

ICD 154 EXCEPT 154.3

1950-1959 DEATHS	1950-1959 RATE	1960-1969 DEATHS	1960-1969 RATE	1970-1979 DEATHS	1970-1979 RATE	COUNTY AND STATE	PERCENT CHANGE 60S/50S	PERCENT CHANGE 70S/60S	PERCENT CHANGE 70S/50S	PERCENTILE %CHANGE 70S/50S	PERCENTILE RATES 1970S
		1	2.0	1	1.2	GEORGE MS		-38			23
				2	4.3	GREENE MS					83
		1	1.5	1	1.4	GRENADA MS		-7			25
1	2.2	9*	12.2	1	1.1	HANCOCK MS	459	-91	-52	31	21
13	5.2	16	4.2	13	2.5	HARRISON MS	-18	-41	-52	31	48
18	5.0	16	2.8	19	2.4	HINDS MS	-43	-15	-51	32	47
				2	3.6	HOLMES MS					72
2	8.1	1	3.7	1	3.0	HUMPHREYS MS	-55	-17	-62	20	61
				5	4.6	ITAWAMBA MS					85
1	.8	4	2.3	11	3.5	JACKSON MS	183	55	338	99	69
1	1.9	1	1.5	3	3.6	JASPER MS	-19	135	89	93	72
2	10.7			1	5.0	JEFFERSON MS			-53	29	89
		1	2.8	1	1.7	JEFFERSON DAVIS MS		-39			30
6	3.3	4	1.6	7	2.4	JONES MS	-50	44	-28	61	46
		1	1.7			LAMAR MS					
7	3.0	9	3.0	11	2.8	LAUDERDALE MS	0	-5	-5	75	56
1	2.7					LAWRENCE MS					
		2	2.9	1	1.0	LEAKE MS		-67			20
4	2.8	4	2.1	4	1.5	LEE MS	-26	-31	-49	35	27
1	1.3			1	.7	LEFLORE MS			-45	41	18
4	4.8	1	.8	4	2.7	LINCOLN MS	-83	230	-43	44	54
10	9.6	6	4.7	3	1.7	LOWNDES MS	-51	-64	-82	03	31
1	2.0	1	1.9	1	1.2	MADISON MS	-5	-33	-36	52	23
1	2.2	4	4.5			MARION MS	107				
2	4.5	1	1.5	2	3.7	MARSHALL MS	-66	137	-18	68	73
3	2.4	5	3.9	4	2.7	MONROE MS	58	-31	10	80	53
2	3.9	4	7.4	2	3.7	MONTGOMERY MS	89	-51	-7	74	73
1	1.1	1	1.0	4	3.2	NESHOBA MS	-4	211	199	98	65
4	5.1	1	1.2	1	1.1	NEWTON MS	-76	-4	-77	06	22
		2	5.7	1	2.0	NOXUBEE MS		-66			36
3	6.1	1	1.6	4	4.3	OKTIBBEHA MS	-73	166	-29	60	83
2	3.4	2	2.6			PANOLA MS	-24				
1	2.0	3	4.5	2	1.5	PEARL RIVER MS	131	-67	-23	65	28
		1	2.7	2	4.9	PERRY MS		86			88
3	2.9	2	1.3	8	4.5	PIKE MS	-54	238	57	90	85
2	2.4	3	3.0			PONTOTOC MS	25				
4	4.6	3	2.6	3	1.9	PRENTISS MS	-45	-26	-60	23	35
		1	2.6	2	4.7	QUITMAN MS		83			87
1	.9	4	4.0	3	1.4	RANKIN MS	321	-64	50	89	26
2	4.4	1	1.3	1	.6	SCOTT MS	-71	-53	-86	01	17
		1	5.5	1	5.3	SHARKEY MS		-4			90
		3	3.7	2	1.9	SIMPSON MS		-48			36
1	1.9					SMITH MS					
1	1.5	2	3.0	1	.8	SUNFLOWER MS	94	-73	-47	39	19
1	2.3	2	3.7			TALLAHATCHIE MS	64				
3	4.3	2	2.3	1	.8	TIPPAH MS	-47	-63	-80	04	19
		1	1.2	3	2.6	TISHOMINGO MS		127			52
		1	5.7			TUNICA MS					
3	3.6	3	3.4	3	2.1	UNION MS	-5	-40	-43	44	38
		1	2.3			WALTHALL MS					
4	3.5	11*	8.2	3	1.5	WARREN MS	134	-81	-56	26	28
3	3.0	3	2.2	4	2.3	WASHINGTON MS	-24	2	-23	65	44
2	5.3	1	1.7			WAYNE MS	-67				
1	1.8	2	3.4	2	2.2	WEBSTER MS	87	-36	20	83	40
2	9.5	2	7.8	1	3.2	WILKINSON MS	-17	-59	-66	16	65
3	5.9	3	4.6			WINSTON MS	-22				
3	5.7	1	1.9			YALOBUSHA MS	-67				
2	3.2	3	4.0	1	.5	YAZOO MS	27	-87	-84	02	17
1136	5.1	1057	3.9	1004	3.2	MISSOURI MO	-23	-19	-37		
3	2.2	5	3.7	6	4.0	ADAIR MO	67	8	81	92	79
4	4.4	3	3.6	2	2.1	ANDREW MO	-19	-41	-52	30	39
6	9.4	1	1.9			ATCHISON MO	-79				
10	6.8	4	2.3	5	2.3	AUDRAIN MO	-66	0	-66	16	45
5	3.8	6	3.5	6	3.8	BARRY MO	-9	10	0	77	76
5	5.7	4	3.0	2	2.1	BARTON MO	-47	-32	-64	18	39
4	2.8	5	3.2	3	1.3	BATES MO	14	-58	-52	31	25
3	5.0	2	2.0	4	3.5	BENTON MO	-60	70	-31	58	69
2	3.4	2	3.2	3	5.1	BOLLINGER MO	-8	60	48	89	89
9	4.2	13	4.7	5	1.2	BOONE MO	12	-74	-72	10	23
39	6.2	36	5.5	23	2.9	BUCHANAN MO	-11	-47	-53	30	58
5	3.0	8	3.8	9	3.9	BUTLER MO	29	0	29	85	76
1	1.0	1	.8	1	1.5	CALDWELL MO	-25	93	44	88	28
3	2.1	5	3.3	5	2.7	CALLAWAY MO	58	-17	31	66	54
3	7.2			2	2.7	CAMDEN MO			-63	19	53
6	2.9	10	3.7	12	3.7	CAPE GIRARDEAU MO	25	2	28	85	74
6	5.0	7	5.3	5	3.3	CARROLL MO	6	-38	-34	55	66
2	9.0			1	2.6	CARTER MO			-71	11	50
4	2.7	9	5.2	5	2.1	CASS MO	92	-59	-21	66	40
1	1.3	4	6.5	8*	7.3	CEDAR MO	412	12	473	99	97
5	4.8	4	3.9	2	2.3	CHARITON MO	-19	-41	-52	31	44
6	7.1	3	3.5	4	3.7	CHRISTIAN MO	-52	8	-48	37	74
2	3.1	2	2.9	2	2.5	CLARK MO	-7	-14	-20	67	48
4	1.5	12	3.2	14	2.6	CLAY MO	113	-17	78	92	[illegible]
2	2.2	4	2.9	5	3.7	CLINTON MO	30	29	68	91	74
9	4.8	7	3.0	6	2.1	COLE MO	-36	-30	-55	27	40
1	1.0	7	5.6	3	1.6	COOPER MO	448	-71	61	91	30
4	5.4	4	3.6	6	3.8	CRAWFORD MO	-34	8	-29	60	76
3	5.3	5	6.1	3	3.3	DADE MO	17	-46	-37	52	67
		2	2.6	2	2.8	DALLAS MO		5			55
1	1.2	1	1.8	3	3.7	DAVIESS MO	49	104	203	98	74
3	5.3	1	1.0	2	1.7	DE KALB MO	-82	71	-69	13	30
1	1.8	2	2.1			DENT MO	13				
1	1.8	1	1.6	4	6.4	DOUGLAS MO	-12	289	244	98	95
5	2.8	8	3.9	4	1.4	DUNKLIN MO	39	-65	-51	32	26
9	4.2	7	2.5	11	2.9	FRANKLIN MO	-40	14	-31	58	57
2	2.7	2	1.9	6	4.7	GASCONADE MO	-30	149	73	92	87
4	5.2	9*	7.4	5	3.6	GENTRY MO	43	-52	-31	58	71
31	4.8	29	3.2	41	3.8	GREENE MO	-34	19	-21	65	75
5	4.4	1	.9	3	2.6	GRUNDY MO	-80	197	-41	47	51

* SIGNIFICANTLY MORE DEATHS THAN EXPECTED

WHITE FEMALE — CANCER OF THE RECTUM — ICD 154 EXCEPT 154.3

1950-1959 DEATHS	1950-1959 RATE	1960-1969 DEATHS	1960-1969 RATE	1970-1979 DEATHS	1970-1979 RATE	COUNTY AND STATE	PERCENT CHANGE 60S/50S	PERCENT CHANGE 70S/60S	PERCENT CHANGE 70S/50S	PERCENTILE %CHANGE 70S/50S	PERCENTILE RATES 1970S
5	5.3	5	4.2	7	6.9	HARRISON MO	-21	67	32	86	96
5	3.7	4	2.4	4	2.2	HENRY MO	-35	-10	-41	46	41
1	3.6	1	1.9	5*	8.9	HICKORY MO	-47	365	149	96	98
1	1.3	2	3.1	2	2.6	HOLT MO	146	-17	105	94	50
1	1.3	5	7.1	4	4.7	HOWARD MO	450	-34	264	98	87
8	6.2	2	1.4	6	2.8	HOWELL MO	-77	99	-55	28	56
3	4.4	1	1.0	2	4.3	IRON MO	-78	331	-3	75	82
153	5.2	161	4.5	112	2.9	JACKSON MO	-14	-34	-44	43	58
20	3.8	31	5.0	28	4.3	JASPER MO	33	-15	14	81	82
9	3.9	11	3.2	16	3.3	JEFFERSON MO	-17	3	-14	70	67
7	4.6	7	4.1	4	1.6	JOHNSON MO	-11	-62	-67	15	28
3	6.2	3	4.9	1	1.0	KNOX MO	-22	-80	-85	02	20
2	1.8	8	6.9	3	2.4	LACLEDE MO	277	-65	33	86	47
4	2.3	5	2.8	6	2.6	LAFAYETTE MO	23	-5	17	82	51
10	6.2	3	1.8	8	3.6	LAWRENCE MO	-72	103	-43	44	71
2	2.8	2	1.8	4	5.4	LEWIS MO	-36	200	93	93	91
2	1.9	3	2.0	5	3.8	LINCOLN MO	8	85	100	94	75
9	6.8	4	1.8	9	5.9	LINN MO	-73	226	-12	71	94
2	1.4	5	3.8	4	2.6	LIVINGSTON MO	162	-30	82	93	51
4	5.2	4	4.4	3	2.9	MC DONALD MO	-15	-34	-44	43	58
8	5.5	5	3.1	4	1.6	MACON MO	-45	-47	-71	11	30
2	3.3	6	7.5	5	5.0	MADISON MO	129	-33	53	90	89
2	5.4			3	5.2	MARIES MO			-5	75	90
5	2.6	12	5.5	10	3.8	MARION MO	114	-30	49	89	75
				1	1.2	MERCER MO					22
4	4.8			4	2.8	MILLER MO			-41	46	56
1	1.5	1	1.1	1	.9	MISSISSIPPI MO	-24	-20	-39	49	20
2	1.9	5	6.9	2	3.0	MONITEAU MO	257	-56	56	90	61
1	1.0	4	3.7	5	5.1	MONROE MO	289	38	435	99	90
3	3.8	4	4.5	4	4.1	MONTGOMERY MO	20	-8	10	80	80
2	3.9	3	3.7	1	1.0	MORGAN MO	-3	-73	-74	08	21
4	4.5	4	3.3	2	1.3	NEW MADRID MO	-27	-62	-72	10	23
8	5.0	6	3.4	10	4.1	NEWTON MO	-32	22	-17	69	80
8	4.8	5	2.3	4	2.3	NODAWAY MO	-52	1	-52	31	45
1	2.2			3	3.2	OREGON MO			42	88	64
5	9.1	2	2.7	4	4.4	OSAGE MO	-71	62	-52	31	83
				1	1.8	OZARK MO					33
3	3.8	2	1.6	8	5.2	PEMISCOT MO	-57	217	37	87	90
6	8.0	6	6.7			PERRY MO	-16				
8	3.6	9	3.6	4	1.1	PETTIS MO	1	-68	-68	14	22
9	7.5	1	.7	6	2.9	PHELPS MO	-91	307	-61	21	58
6	5.7	5	4.1	6	4.2	PIKE MO	-29	2	-27	62	80
1	1.2	1	.9	4	2.8	PLATTE MO	-22	208	140	96	56
5	4.7	4	3.2	1	.4	POLK MO	-32	-87	-91	00	16
1	1.7			3	2.8	PULASKI MO			61	91	55
3	4.5			1	1.8	PUTNAM MO			-59	23	34
4	7.5	1	1.9	4	5.2	RALLS MO	-75	173	-31	58	90
7	3.9	12	5.7	3	2.3	RANDOLPH MO	46	-59	-41	47	45
1	1.0	1	.6	3	2.2	RAY MO	-37	249	120	95	41
		1	1.2	1	1.7	RIPLEY MO		38			32
5	2.8	16	6.0	5	1.0	ST CHARLES MO	111	-83	-65	17	21
1	1.2	1	1.0	1	.8	ST CLAIR MO	-20	-20	-36	53	18
2	3.9	5	7.1	3	3.4	STE GENEVIEVE MO	80	-52	-15	70	67
7	3.5	10	3.9	14	4.5	ST FRANCOIS MO	14	15	30	86	85
127	5.7	172	4.7	205*	4.0	ST LOUIS COUNTY MO	-17	-16	-30	60	79
4	2.3	4	1.8	6	2.7	SALINE MO	-19	47	19	82	53
		1	1.8			SCHUYLER MO					
5	8.1	1	1.2			SCOTLAND MO	-85				
8	5.4	4	2.3	10	4.6	SCOTT MO	-57	99	-15	69	85
4	12.3	1	2.0			SHANNON MO	-83				
1	.9	4	4.2	3	3.5	SHELBY MO	363	-17	284	99	69
3	2.7	6	3.4	5	2.2	STODDARD MO	29	-37	-18	68	41
1	2.8	1	1.9			STONE MO	-31				
1	1.8	5	5.2			SULLIVAN MO	183				
1	2.1	4	5.4	5	3.5	TANEY MO	163	-35	71	92	70
7	6.7	4	3.1	5	2.9	TEXAS MO	-53	-7	-57	25	58
7	4.1	1	.5	2	.9	VERNON MO	-89	88	-79	05	19
3	5.2	7*	10.2	2	2.3	WARREN MO	96	-77	-55	27	45
5	7.8	1	1.4	3	3.1	WASHINGTON MO	-82	122	-59	23	63
4	7.4	5	7.6	4	4.7	WAYNE MO	2	-38	-37	52	87
2	2.3	6	5.3	3	2.1	WEBSTER MO	132	-60	-6	74	40
1	3.4					WORTH MO					
		1	.8	3	3.7	WRIGHT MO		345			73
357*	8.4	194	4.7	151	4.2	ST LOUIS CITY MO	-44	-10	-50	33	81
114	4.3	142	4.3	137	3.2	MONTANA MT	-1	-24	-25		
1	3.1			2	4.4	BEAVERHEAD MT			42	88	84
2	8.6					BIG HORN MT					
		1	2.5			BLAINE MT					
		1	6.5			BROADWATER MT					
2	4.1	1	1.6	2	3.8	CARBON MT	-61	139	-8	73	75
1	8.4	2*	17.9			CARTER MT	113				
13	5.1	18	5.8	17	4.0	CASCADE MT	13	-31	-22	65	78
2	6.3					CHOUTEAU MT					
		10*	13.6	6	5.0	CUSTER MT		-63			89
		1	6.8			DANIELS MT					
1	3.1	1	2.0			DAWSON MT	-36				
3	3.8	3	3.2	3	2.9	DEER LODGE MT	-17	-9	-24	64	58
7	10.3			3	3.7	FERGUS MT			-65	17	73
8	5.5	3	1.5	15*	5.7	FLATHEAD MT	-72	271	5	79	93
1	1.2	4	2.9	3	1.6	GALLATIN MT	142	-45	34	87	29
1	4.9					GLACIER MT					
		2*	30.3			GOLDEN VALLEY MT					
1	8.2	2	15.2			GRANITE MT	86				
4	6.6	4	4.8	6	6.9	HILL MT	-28	44	4	78	96
3*	14.4	2	8.6	3*	8.9	JEFFERSON MT	-41	3	-39	50	98
		3	3.9	1	1.1	LAKE MT		-72			22
4	3.2	8	4.7	3	1.5	LEWIS AND CLARK MT	48	-68	-53	29	27
3	12.0	3	5.4	2	3.8	LINCOLN MT	-55	-30	-68	14	75
				1	7.9	MC CONE MT					97

* SIGNIFICANTLY MORE DEATHS THAN EXPECTED

1950-1959 DEATHS	1950-1959 RATE	1960-1969 DEATHS	1960-1969 RATE	1970-1979 DEATHS	1970-1979 RATE	COUNTY AND STATE	PERCENT CHANGE 60S/50S	PERCENT CHANGE 70S/60S	PERCENT CHANGE 70S/50S	PERCENTILE %CHANGE 70S/50S	PERCENTILE RATES 1970S
1	3.5	2	6.1			MADISON MT	74				
		1	7.0			MINERAL MT					
11	6.4	8	3.7	12	3.5	MISSOULA MT	-42	-6	-46	40	69
		1	4.2	3	7.8	MUSSELSHELL MT		85			97
1	1.6	4	5.0	2	2.3	PARK#MT	218	-54	47	88	45
		3	12.4	1	1.9	PHILLIPS MT		-85			35
2	7.9	3	9.7	1	3.0	PONDERA MT	23	-70	-63	20	59
		1	9.6			POWDER RIVER MT					
		1	2.8	1	2.7	POWELL MT		-4			53
		2*	14.4	1	8.4	PRAIRIE MT		-42			98
5	6.1	4	3.9	6	4.1	RAVALLI MT	-37	5	-33	56	79
		1	1.4			RICHLAND MT					
1	5.7	2	4.3	5*	10.4	ROOSEVELT MT	-24	139	82	93	99
		1	3.2			ROSEBUD MT					
2	6.1			2	3.6	SANDERS MT			-40	48	72
1	4.3	1	1.8	1	3.6	SHERIDAN MT	-58	103	-15	69	72
19	7.4	13	4.8	12	4.4	SILVER BOW MT	-35	-10	-41	47	83
1	4.4	1	3.5			STILLWATER MT	-19				
1	5.8	1	4.8			SWEET GRASS MT	-19				
1	3.0	2	5.0			TETON MT	68				
		1	3.5	2	4.8	TOOLE MT		38			87
				1*	10.6	TREASURE MT					99
1	2.2			1	2.0	VALLEY MT			-8	74	37
		2	12.9			WHEATLAND MT					
10	4.0	18	5.2	19	4.0	YELLOWSTONE MT	30	-24	-1	76	78
349	4.8	346	3.9	323	3.0	NEBRASKA NB	-19	-23	-38		
7	3.8	10	4.5	5	1.6	ADAMS NB	19	-65	-58	24	29
3	4.4	1	1.9	1	1.2	ANTELOPE NB	-56	-38	-73	09	23
1	1.6	3	4.9	2	2.3	BOONE NB	204	-52	46	88	45
2	3.8	6	8.4	2	2.0	BOX BUTTE NB	121	-76	-47	39	38
2	7.4	1	2.8	2	5.0	BOYD NB	-62	79	-32	57	89
4	13.4	1	1.8	2	5.5	BROWN NB	-87	202	-59	23	92
6	3.9	4	2.4	4	1.7	BUFFALO NB	-38	-32	-57	25	30
1	1.4			3	3.0	BURT NB			111	94	60
4	5.7	4	5.8	5	7.2	BUTLER NB	3	24	28	85	97
6	6.3	1	.8	2	1.9	CASS NB	-88	141	-71	11	34
3	4.3	4	5.4	1	.6	CEDAR NB	26	-88	-85	02	17
1	3.8	3	11.3	1	1.1	CHASE NB	198	-90	-70	11	22
5*	14.6	2	5.1	1	1.4	CHERRY NB	-65	-73	-91	00	25
1	1.4	2	3.4	3	3.7	CHEYENNE NB	150	9	173	97	74
2	2.7	3	3.8			CLAY NB	41				
2	3.1	4	4.0	4	3.6	COLFAX NB	32	-11	16	82	71
4	6.1	4	5.3	5	5.2	CUMING NB	-13	-2	-14	70	90
3	2.5	5	3.8	4	3.3	CUSTER NB	53	-15	31	86	65
2	4.2	5	8.5	6*	7.1	DAKOTA NB	102	-17	68	91	97
2	3.5	2	2.5	4	7.7	DAWES NB	-28	206	120	95	97
2	1.9	4	2.9	8	5.7	DAWSON NB	50	97	195	98	93
		1	4.7			DEUEL NB					
3	5.9			4	7.0	DIXON NB			19	82	96
7	4.1	13	6.2	13	5.3	DODGE NB	50	-16	27	84	90
87	6.0	90	5.0	76	3.6	DOUGLAS NB	-16	-29	-41	47	71
3	16.6			1	2.4	DUNDY NB			-86	01	46
6	9.3	2	2.4	2	2.3	FILLMORE NB	-75	-4	-76	07	43
1	2.5	3	6.9	1	1.1	FRANKLIN NB	178	-84	-56	26	22
		2	7.1			FRONTIER NB					
2	3.4	3	4.3	1	.8	FURNAS NB	29	-81	-76	07	19
10	5.8	10	5.0	10	3.6	GAGE NB	-14	-28	-38	50	71
1	4.7	1	4.0	2*	19.5	GARDEN NB	-14	385	316	99	99
2	12.5	1	4.4			GARFIELD NB	-65				
1	17.7	1*	16.6			GRANT NB	-6				
1	3.6	2	8.6			GREELEY NB	136				
10	6.0	12	4.9	17*	5.3	HALL NB	-18	10	-11	72	91
2	3.3	1	1.4	2	2.9	HAMILTON NB	-59	112	-13	71	57
1	2.8	2	4.4	1	1.7	HARLAN NB	56	-61	-40	48	31
				1*	23.6	HAYES NB					99
1	4.2					HITCHCOCK NB					
6	7.8	4	4.5	1	.8	HOLT NB	-43	-83	-90	00	18
1	16.1					HOOKER NB					
1	2.1	1	2.3			HOWARD NB	10				
3	3.5	4	3.4	2	2.0	JEFFERSON NB	-4	-41	-43	43	37
3	6.2	1	1.7	2	3.3	JOHNSON NB	-73	95	-47	39	66
3	6.1	2	2.9	4	4.4	KEARNEY NB	-53	51	-28	61	84
2	7.8	1	2.8	1	1.7	KEITH NB	-64	-40	-79	05	31
1	10.3	2*	19.1			KEYA PAHA NB	85				
		1	3.9	2	5.2	KIMBALL NB		33			90
7	9.8	7	8.2	1	.7	KNOX NB	-17	-91	-93	00	18
31	4.1	34	3.7	38	3.5	LANCASTER NB	-11	-6	-16	69	69
4	3.2	5	3.2	4	2.0	LINCOLN NB	0	-38	-39	49	36
				1*	37.7	LOGAN NB					99
				1*	22.8	LOUP NB					99
7	4.0	4	2.3	7	2.6	MADISON NB	-43	16	-34	55	52
2	3.4	1	1.5	2	1.7	MERRICK NB	-55	9	-51	33	31
		1	2.5			MORRILL NB					
1	2.6	1	2.0	1	1.4	NANCE NB	-24	-28	-45	42	26
1	1.6	1	1.8	1	.6	NEMAHA NB	12	-67	-63	19	17
3	5.4	1	1.2	2	2.3	NUCKOLLS NB	-78	93	-57	25	44
11	11.1	5	3.7	5	2.7	OTOE NB	-66	-27	-75	07	54
7*	15.5	3	5.2			PAWNEE NB	-67				
2	10.7	1	3.9			PERKINS NB	-64				
2	3.3	2	2.3	2	2.2	PHELPS NB	-28	-6	-32	57	42
1	2.1	1	2.1	2	2.8	PIERCE NB	0	31	30	85	55
5	4.6	4	2.7	4	1.9	PLATTE NB	-41	-28	-58	24	36
1	2.0	2	2.2			POLK NB	11				
		1	1.0	2	1.3	RED WILLOW NB		26			24
3	2.7	5	3.7	2	1.7	RICHARDSON NB	36	-55	-38	30	31
6	6.4	5	4.2	1	.6	SALINE NB	-35	-86	-91	00	17
1	2.0	1	.4	3	2.1	SARPY NB	-81	447	6	79	40
8	8.5	5	4.7	4	2.5	SAUNDERS NB	-45	-47	-71	11	49
5	3.4	2	1.1	6	2.7	SCOTTS BLUFF NB	-66	139	-19	67	54

* SIGNIFICANTLY MORE DEATHS THAN EXPECTED

COUNTIES COMBINED; SEE TECHNICAL NOTES

WHITE FEMALE — CANCER OF THE RECTUM — ICD 154 EXCEPT 154.3

1950-1959 DEATHS	1950-1959 RATE	1960-1969 DEATHS	1960-1969 RATE	1970-1979 DEATHS	1970-1979 RATE	COUNTY AND STATE	PERCENT CHANGE 60S/50S	PERCENT CHANGE 70S/60S	PERCENT CHANGE 70S/50S	PERCENTILE %CHANGE 70S/50S	PERCENTILE RATES 1970S
2	2.5	3	2.7	1	.7	SEWARD NB	10	-76	-74	09	18
2	4.6	2	3.6	2	2.9	SHERIDAN NB	-23	-18	-37	52	58
		2	5.7			SHERMAN NB					
1	8.5					SIOUX NB					
1	4.1	1	4.0	1	2.1	STANTON NB	-2	-49	-50	34	39
4	5.9	2	3.2	2	2.7	THAYER NB	-46	-15	-54	28	53
				1*	21.0	THOMAS NB					99
2	7.2	3	9.0	3	6.4	THURSTON NB	25	-29	-12	72	95
1	2.0	3	5.3	3	5.9	VALLEY NB	171	10	198	98	93
4	5.4			4	3.9	WASHINGTON NB			-27	62	78
1	1.7	1	1.9	5	6.2	WAYNE NB	10	234	268	98	95
1	1.7	3	4.8	2	2.0	WEBSTER NB	184	-59	17	82	36
1	1.0	5	4.6	2	1.2	YORK NB	367	-75	17	82	22
43	5.5	55	4.4	76	3.2	NEVADA NV	-21	-28	-43		
		1	3.1	2	2.7	CHURCHILL NV		-14			53
15	4.6	18	3.2	35	2.8	CLARK NV	-31	-13	-40	48	54
1	16.1			1	1.6	DOUGLAS NV			-90	00	29
2	4.4			1	1.6	ELKO NV			-64	19	29
1	5.1	2	8.1			HUMBOLDT NV	58				
		1	7.2	1	4.7	LINCOLN NV		-35			87
1	6.7	2	9.3			MINERAL NV	38				
1	8.4	1	7.4			NYE NV	-11				
				1*	17.9	STOREY NV					99
19	7.1	25	6.0	30	4.5	WASHOE NV	-15	-25	-36	53	85
1	3.7			2	4.4	WHITE PINE NV			17	82	84
2	7.9	5*	13.3	3	2.7	ORMSBY-CARSON CITY NV	67	-80	-66	16	53
230*	6.9	236*	5.8	238*	4.5	NEW HAMPSHIRE NH	-16	-22	-35		
9	5.2	13	6.3	8	2.7	BELKNAP NH	21	-56	-47	38	54
1	.8	13*	9.6	7	3.4	CARROLL NH	1146	-64	345	99	68
15	6.4	13	4.5	15	3.8	CHESHIRE NH	-30	-15	-41	46	75
12	6.4	14	6.6	17*	6.9	COOS NH	2	5	8	79	96
17	6.1	22	6.4	19	4.6	GRAFTON NH	4	-28	-25	64	86
77*	7.8	79*	6.7	83*	5.5	HILLSBOROUGH NH	-14	-18	-29	60	92
30	6.8	26	4.8	23	3.6	MERRIMACK NH	-29	-24	-47	39	72
36	7.9	31	5.2	31	3.6	ROCKINGHAM NH	-35	-30	-55	28	72
18	6.2	17	5.0	22	4.7	STRAFFORD NH	-19	-6	-24	64	87
15	9.4	8	4.5	13	5.5	SULLIVAN NH	-53	23	-42	45	92
2147*	8.4	2226*	6.6	2094*	4.9	NEW JERSEY NJ	-21	-26	-41		
75*	9.2	85*	7.5	70*	5.0	ATLANTIC NJ	-19	-33	-46	40	89
254*	8.5	289*	6.6	280*	5.0	BERGEN NJ	-22	-25	-42	46	89
52	8.0	40	4.3	54	3.8	BURLINGTON NJ	-46	-11	-52	31	75
122*	8.0	142*	7.0	104*	4.2	CAMDEN NJ	-13	-39	-47	38	81
17	6.0	28	6.0	31	4.5	CAPE MAY NJ	-1	-25	-26	63	84
29	6.0	39*	6.9	30	4.2	CUMBERLAND NJ	14	-39	-31	59	81
420*	9.2	316*	6.4	244*	5.0	ESSEX NJ	-30	-22	-45	41	89
38	8.2	44*	7.2	36	4.2	GLOUCESTER NJ	-12	-42	-49	35	81
310*	9.8	268*	7.5	242*	6.4	HUDSON NJ	-24	-15	-35	54	95
24	8.8	25*	7.1	25*	5.7	HUNTERDON NJ	-20	-20	-36	53	92
96*	8.3	89*	6.3	87*	4.8	MERCER NJ	-24	-24	-42	45	87
90	7.0	128*	6.7	141*	5.2	MIDDLESEX NJ	-5	-22	-25	63	90
115*	8.0	124*	6.0	154*	5.3	MONMOUTH NJ	-25	-12	-34	56	90
68	6.5	102*	6.9	95*	4.8	MORRIS NJ	5	-30	-27	62	87
28	6.1	59	6.2	109*	4.4	OCEAN NJ	2	-29	-27	62	84
160*	8.8	169*	7.1	126*	4.4	PASSAIC NJ	-19	-38	-50	33	84
13	6.2	18	6.9	20*	6.0	SALEM NJ	12	-13	-3	75	94
35	7.0	37	5.1	44	4.3	SOMERSET NJ	-28	-14	-38	51	83
14	6.5	26*	7.9	40*	8.2	SUSSEX NJ	22	4	26	84	98
156*	7.7	174*	6.4	129	3.8	UNION NJ	-16	-40	-50	33	76
31*	9.3	24	5.5	33*	6.4	WARREN NJ	-41	15	-32	58	95
73	3.5	114	3.7	94	2.1	NEW MEXICO NM	7	-44	-41		
22	4.6	36	4.2	41	2.8	BERNALILLO NM	-9	-32	-38	50	56
				1	6.4	CATRON NM					95
3	2.4	2	1.0	4	1.5	CHAVES NM	-58	50	-37	52	28
		5	6.9			COLFAX NM					
		4	3.0	3	1.6	CURRY NM		-46			29
2	1.6	5	3.0	2	.9	DONA ANA NM	88	-68	-40	47	20
6	5.6	4	2.3	4	1.6	EDDY NM	-59	-29	-71	11	29
2	4.1			3	2.6	GRANT NM			-36	53	51
3*	17.3			1	2.3	HIDALGO NM			-87	01	43
4	4.5	7	4.4	4	1.8	LEA NM	-2	-59	-60	22	33
2	8.1	6*	14.8			LINCOLN NM	83				
1	3.1	2	4.6			LOS ALAMOS NM	51				
2	6.6	2	4.0	1	.9	LUNA NM	-39	-78	-87	01	19
				1	2.0	MC KINLEY NM					38
		1	3.7	2	9.9	MORA NM		166			99
5	10.4	3	4.0			OTERO NM	-62				
2	3.5	1	1.6	1	2.0	QUAY NM	-54	23	-43	44	37
1	1.1	2	2.2	3	3.8	RIO ARRIBA NM	108	73	261	98	76
5	7.8	4	5.0	4	4.0	ROOSEVELT NM	-36	-21	-49	34	78
1	3.4	1	2.0	2	2.0	SANDOVAL NM	-40	0	-40	47	37
		2	2.7	2	1.4	SAN JUAN NM		-48			26
2	2.5	3	2.8	1	.8	SAN MIGUEL NM	10	-70	-67	15	19
5	4.0	10	4.9	7	2.4	SANTA FE NM	23	-50	-38	50	48
1	3.5	2	3.1	2	2.8	SIERRA NM	-13	-9	-21	66	56
		1	2.7	1	1.8	SOCORRO NM		-35			32
1	1.2	1	1.4	3	4.8	TAOS NM	18	240	302	99	88
1	4.3	1	3.9	1	2.5	TORRANCE NM	-10	-35	-41	46	49
		4	11.5			UNION NM					
2	2.4	5	8.3			VALENCIA NM	245				
6741*	8.5	5951*	6.2	5028*	4.6	NEW YORK NY	-27	-26	-46		
139*	9.7	93	5.3	108*	5.3	ALBANY NY	-45	0	-45	41	91

* SIGNIFICANTLY MORE DEATHS THAN EXPECTED

WHITE FEMALE — CANCER OF THE PECTUM — ICD 154 EXCEPT 154.3

1950-1959 DEATHS	1950-1959 RATE	1960-1969 DEATHS	1960-1969 RATE	1970-1979 DEATHS	1970-1979 RATE	COUNTY AND STATE		PERCENT CHANGE 60S/50S	PERCENT CHANGE 70S/60S	PERCENT CHANGE 70S/50S	PERCENTILE %CHANGE 70S/50S	PERCENTILE RATES 1970S
9	3.9	5	1.6	21*	5.8	ALLEGANY	NY	-60	275	51	89	93
750*	10.6	497*	6.1	376*	4.9	BRONX	NY	-42	-20	-54	28	88
71	6.9	83*	6.7	52	3.5	BROOME	NY	-3	-48	-50	34	69
32	7.3	26	5.4	23	4.0	CATTARAUGUS	NY	-26	-26	-46	40	78
34	7.8	25	5.3	22	4.0	CAYUGA	NY	-31	-24	-48	37	79
48	5.6	45	4.8	42	3.9	CHAUTAUQUA	NY	-14	-19	-30	59	76
38	7.2	42*	6.9	33*	4.7	CHEMUNG	NY	-4	-32	-34	55	86
11	4.2	20	7.3	19*	6.1	CHENANGO	NY	72	-16	44	88	94
14	5.4	20	6.9	22*	5.9	CLINTON	NY	28	-15	9	80	93
23	7.3	22	6.2	20	4.6	COLUMBIA	NY	-15	-27	-38	51	85
12	5.3	29*	11.6	20*	6.7	CORTLAND	NY	118	-42	27	84	96
25*	9.4	11	3.7	18	5.3	DELAWARE	NY	-61	45	-44	43	91
58	6.2	56	4.9	69*	5.1	DUTCHESS	NY	-21	5	-17	68	90
348*	7.5	358*	6.3	302*	4.5	ERIE	NY	-16	-28	-40	49	85
20*	10.3	19*	8.2	18*	7.4	ESSEX	NY	-20	-10	-28	61	97
14	6.1	11	4.4	22*	7.8	FRANKLIN	NY	-27	76	28	85	97
19	5.4	24	5.8	20	4.5	FULTON	NY	8	-23	-17	69	85
14	5.3	17	5.5	21*	5.9	GENESEE	NY	6	7	12	81	93
17	8.6	16	6.5	10	3.3	GREENE	NY	-25	-49	-61	21	66
2	13.2			4*	11.1	HAMILTON	NY			-16	69	99
26	6.7	23	5.1	27*	5.4	HERKIMER	NY	-25	6	-20	66	91
48*	8.7	45*	7.4	39*	6.2	JEFFERSON	NY	-15	-17	-30	60	95
1204*	9.8	752*	5.5	580*	4.3	KINGS(BROOKLYN)	NY	-44	-22	-56	26	82
8	6.6	11*	8.9	9	5.9	LEWIS	NY	35	-34	-11	72	93
15	6.4	17	6.4	20*	5.7	LIVINGSTON	NY	-1	-10	-11	72	92
19	6.8	10	3.2	22*	5.9	MADISON	NY	-53	87	-12	71	94
196	6.5	193	5.3	154	3.6	MONROE	NY	-19	-32	-45	41	71
27	7.3	28	6.5	31*	6.1	MONTGOMERY	NY	-11	-6	-16	69	94
364*	10.2	319*	5.4	346*	4.5	NASSAU	NY	-46	-18	-56	26	84
900*	9.7	1078*	11.2	442*	4.7	NEW YORK(MANHATTAN)	NY	15	-58	-51	32	87
59	6.2	67	5.8	71*	5.0	NIAGARA	NY	-7	-13	-19	68	89
104*	7.7	106*	6.5	87*	4.9	ONEIDA	NY	-15	-24	-36	53	88
124	6.5	97	4.0	111	3.9	ONONDAGA	NY	-38	-3	-40	48	77
17	4.6	27	6.0	26	5.1	ONTARIO	NY	29	-14	10	80	89
74	7.5	82*	7.0	75*	5.4	ORANGE	NY	-6	-23	-28	61	91
12	6.3	15	6.9	8	3.4	ORLEANS	NY	10	-51	-46	40	68
32	7.0	23	4.7	25	4.4	OSWEGO	NY	-33	-7	-38	50	84
21	6.4	25	6.6	22*	5.1	OTSEGO	NY	3	-22	-19	67	69
11	7.7	13	5.8	6	1.9	PUTNAM	NY	-25	-68	-76	07	35
726*	8.7	600*	5.2	625*	4.5	QUEENS	NY	-40	-13	-47	38	85
76*	9.3	68*	7.1	62*	5.8	RENSSELAER	NY	-24	-18	-37	51	93
107*	10.9	59	4.5	82*	4.8	RICHMOND	NY	-58	5	-56	26	87
37	6.8	44	5.7	42	3.8	ROCKLAND	NY	-16	-33	-44	42	75
32	6.1	36	6.1	31	4.6	ST LAWRENCE	NY	0	-26	-26	63	85
36	8.1	31	5.6	32	4.6	SARATOGA	NY	-32	-17	-43	44	86
68*	8.0	52	5.2	55*	4.6	SCHENECTADY	NY	-35	-12	-43	44	86
3	2.2	6	3.9	9	5.1	SCHOHARIE	NY	76	33	134	95	89
5	5.9	4	4.3	4	3.1	SCHUYLER	NY	-27	-28	-47	38	63
8	3.3	5	2.3	7	2.8	SENECA	NY	-29	19	-15	70	56
31	5.9	32	5.6	28	4.3	STEUBEN	NY	-4	-25	-28	61	82
141	6.5	199	5.3	240*	4.4	SUFFOLK	NY	-19	-17	-33	56	83
11	4.7	12	3.9	19	4.6	SULLIVAN	NY	-18	20	-2	76	86
19*	10.5	16*	7.6	10	3.9	TIOGA	NY	-28	-49	-63	19	76
18	6.1	13	3.9	16	4.3	TOMPKINS	NY	-36	10	-30	60	82
57*	9.0	45	5.9	57*	6.0	ULSTER	NY	-35	2	-34	56	94
22	8.6	17	5.6	18	5.1	WARREN	NY	-35	-8	-40	48	89
18	6.5	17	5.2	25*	6.3	WASHINGTON	NY	-21	21	-4	75	95
25	6.9	23	5.2	22	4.5	WAYNE	NY	-25	-12	-34	55	85
317*	8.8	300*	6.4	278*	4.8	WESTCHESTER	NY	-27	-25	-45	41	88
17	8.4	10	4.8	11	4.9	WYOMING	NY	-44	4	-41	46	88
8	5.9	12*	8.6	12*	7.7	YATES	NY	46	-11	31	86	97
435	3.7	513	3.0	564	2.3	NORTH CAROLINA	NC	-18	-22	-36		
9	3.8	9	2.8	11	2.3	ALAMANCE	NC	-27	-19	-41	46	43
4	6.9	2	2.9	2	2.2	ALEXANDER	NC	-58	-25	-68	13	41
		1	1.7	3	4.4	ALLEGHANY	NC		164			84
3	4.9	1	1.3	1	1.2	ANSON	NC	-73	-6	-75	07	23
1	1.0	5	4.5	3	1.9	ASHE	NC	352	-58	88	93	35
1	2.5	2	3.5	2	2.6	AVERY	NC	39	-27	2	78	50
2	1.3	5	3.6	5	2.6	BEAUFORT	NC	185	-28	104	94	51
				2	2.9	BERTIE	NC					59
1	2.3	3	5.0	2	2.2	BLADEN	NC	117	-56	-5	75	41
2	4.7	6	7.9	3	2.6	BRUNSWICK	NC	68	-68	-45	41	50
31	5.6	25	3.3	31	2.9	BUNCOMBE	NC	-41	-13	-49	36	58
1	.7	5	2.2	5	1.4	BURKE	NC	216	-34	108	94	26
7	3.4	12	4.1	9	2.2	CABARRUS	NC	21	-46	-35	54	42
3	2.7	6	3.2	5	1.6	CALDWELL	NC	19	-50	-41	46	29
1	6.7	2	12.6	1	3.7	CAMDEN	NC	87	-70	-45	42	74
4	4.8	4	3.2	3	1.4	CARTERET	NC	-34	-57	-72	10	25
2	3.9	1	1.9	1	1.5	CASWELL	NC	-52	-23	-62	20	27
13	7.1	9	2.8	15	3.1	CATAWBA	NC	-61	13	-56	26	62
4	5.3	6	5.7	5	3.7	CHATHAM	NC	7	-34	-30	59	74
		3	3.3	2	1.6	CHEROKEE	NC		-52			29
2	6.3	2	5.7	4	7.9	CHOWAN	NC	-9	38	25	84	98
1	3.4			2	4.0	CLAY	NC			17	82	78
7	3.8	10	3.8	5	1.3	CLEVELAND	NC	0	-67	-67	15	23
2	2.0	4	2.6	5	2.5	COLUMBUS	NC	28	-4	23	83	49
2	2.0	4	2.5	4	2.1	CRAVEN	NC	27	-16	7	79	39
8	5.4	5	2.2	11	2.7	CUMBERLAND	NC	-60	27	-49	34	54
3	18.4	1	3.0	2	3.3	CURRITUCK	NC	-84	11	-82	03	67
2	6.9	2	6.0	2	2.3	DARE	NC	-12	-63	-67	15	43
7	3.2	9	2.8	10	2.0	DAVIDSON	NC	-13	-28	-37	51	38
1	1.5	4	4.7	3	2.6	DAVIE	NC	205	-43	73	92	52
3	3.4	2	1.5	3	1.9	DUPLIN	NC	-55	28	-43	44	36
8	2.4	12	2.8	19	3.5	DURHAM	NC	19	23	46	88	69
6	7.2	8	5.8	3	1.7	EDGECOMBE	NC	-19	-70	-76	07	31
17	3.6	16	2.3	30	3.2	FORSYTH	NC	-36	36	-12	71	63
3	4.0			1	.8	FRANKLIN	NC			-81	04	18
9	2.6	9	1.7	17	2.3	GASTON	NC	-34	37	-10	72	44
		3	10.2	2	4.9	GRAHAM	NC		-52			88

* SIGNIFICANTLY MORE DEATHS THAN EXPECTED

WHITE FEMALE

CANCER OF THE RECTUM — ICD 154 EXCEPT 154.3

1950-1959 DEATHS	1950-1959 RATE	1960-1969 DEATHS	1960-1969 RATE	1970-1979 DEATHS	1970-1979 RATE	COUNTY AND STATE	PERCENT CHANGE 60S/50S	PERCENT CHANGE 70S/60S	PERCENT CHANGE 70S/50S	PERCENTILE %CHANGE 70S/50S	PERCENTILE RATES 1970S
3	3.1	2	2.0	1	.7	GRANVILLE NC	-34	-68	-79	05	18
2	7.1	1	2.6	1	1.8	GREENE NC	-63	-32	-75	07	33
22	3.7	20	2.1	24	1.8	GUILFORD NC	-42	-17	-52	30	32
4	3.3	6	3.8	4	1.7	HALIFAX NC	16	-55	-48	36	31
5	3.7	8	4.7	7	2.7	HARNETT NC	25	-43	-29	60	52
4	2.6	5	2.4	4	1.4	HAYWOOD NC	-7	-44	-48	37	25
6	3.5	5	2.3	10	2.2	HENDERSON NC	-34	-5	-37	51	41
		2	3.1	5	5.6	HERTFORD NC		80			92
1	3.8					HOKE NC					
1	5.9	1	3.5	1	4.0	HYDE NC	-41	15	-32	58	79
8	3.7	8	3.0	8	2.0	IREDELL NC	-20	-33	-46	39	37
3	5.5	1	1.3			JACKSON NC	-77				
4	2.1	6	2.9	10	3.1	JOHNSTON NC	39	10	53	89	63
2	7.4	2	8.4	1	3.0	JONES NC	13	-65	-60	22	60
3	4.5	4	5.0	1	.8	LEE NC	10	-84	-82	03	19
4	3.5	5	3.5	2	.9	LENOIR NC	-1	-73	-73	09	20
4	4.0	8	6.5	6	3.2	LINCOLN NC	63	-51	-20	67	64
2	2.5	7	5.0	3	1.6	MC DOWELL NC	102	-68	-35	54	30
		5	5.5	6	4.7	MACON NC		-14			87
2	2.7	1	.9	3	2.7	MADISON NC	-68	219	1	77	54
2	5.1	1	1.4	1	1.1	MARTIN NC	-73	-21	-78	05	22
23	3.9	42	4.3	31	2.2	MECKLENBURG NC	10	-49	-44	42	41
5	9.2	1	1.1	2	2.2	MITCHELL NC	-88	90	-76	06	40
2	2.9	4	5.1	4	3.7	MONTGOMERY NC	73	-27	26	84	74
9	8.1	4	2.6	6	2.4	MOORE NC	-68	-9	-71	11	46
7	4.6	8	4.1	6	2.2	NASH NC	-10	-46	-51	32	42
7	3.2	14	4.8	19	4.5	NEW HANOVER NC	49	-6	40	87	85
4	6.6	1	1.2	1	1.0	NORTHAMPTON NC	-82	-11	-84	02	21
5	8.1	1	1.1	4	2.3	ONSLOW NC	-87	115	-72	10	44
5	6.0	5	3.0	5	2.4	ORANGE NC	-50	-22	-61	22	46
		1	2.6	3	6.7	PAMLICO NC		160			96
2	3.4	1	1.1	2	1.6	PASQUOTANK NC	-68	49	-53	30	29
3	7.2	1	2.0	2	2.4	PENDER NC	-72	17	-67	14	45
4*	17.9	2	5.7			PERQUIMANS NC	-68				
1	1.6	1	1.0	2	1.6	PERSON NC	-36	55	-1	76	29
3	2.7	6	3.2	2	.8	PITT NC	18	-75	-70	11	19
3	5.3	2	2.3	3	3.6	POLK NC	-56	57	-31	58	72
9	4.6	9	3.2	10	2.4	RANDOLPH NC	-29	-26	-47	38	47
3	2.1	5	3.2	1	.5	RICHMOND NC	57	-86	-77	06	17
7	4.7	6	3.3	6	2.5	ROBESON NC	-30	-23	-46	39	49
6	2.7	8	2.7	12	2.8	ROCKINGHAM NC	-1	4	3	78	55
13	5.6	13	3.2	19	3.4	ROWAN NC	-43	6	-39	49	68
8	4.6	4	1.8	8	2.5	RUTHERFORD NC	-60	38	-45	41	49
1	.9	4	2.7	3	1.5	SAMPSON NC	219	-46	72	92	27
		2	3.0	2	2.0	SCOTLAND NC		-33			38
2	1.5	3	1.4	2	1.0	STANLY NC	-7	-31	-37	52	20
2	2.5	1	.9	2	1.3	STOKES NC	-65	49	-48	37	24
1	.5	10	4.5	3	.9	SURRY NC	855	-79	102	94	20
1	5.0	1	1.7	1	1.6	SWAIN NC	-66	-8	-69	13	28
4	8.1	2	2.3	5	4.1	TRANSYLVANIA NC	-71	78	-49	35	80
4	2.9	6	3.3	4	1.5	UNION NC	15	-56	-49	35	27
4	4.6	7	6.4	6	4.0	VANCE NC	38	-38	-14	70	79
17	4.0	26	4.0	26	2.8	WAKE NC	0	-30	-30	60	56
		1	2.1	3	7.1	WARREN NC		238			96
1	2.9	2	5.9	1	1.9	WASHINGTON NC	100	-68	-35	54	35
2	3.0	2	2.3	5	3.8	WATAUGA NC	-25	68	27	84	76
1	.8	2	1.1	5	1.8	WAYNE NC	34	61	116	95	33
9	6.0	1	.4	4	1.4	WILKES NC	-93	241	-76	07	26
1	1.2	3	1.7	6	3.0	WILSON NC	43	79	155	96	60
2	2.1	3	3.0	4	2.3	YADKIN NC	47	-25	11	81	44
2	3.2	2	2.7	2	2.2	YANCEY NC	-14	-19	-31	59	42
102	4.2	154	5.0	106	2.7	NORTH DAKOTA ND	19	-45	-35		
		1	5.0			ADAMS ND					
6	7.4	2	2.2	2	1.3	BARNES ND	-70	-39	-82	03	25
2	5.6	3	7.1			BENSON ND	25				
		2	3.5	1	1.5	BOTTINEAU ND		-59			27
		1	3.3			BOWMAN ND					
3	11.6	1	2.7			BURKE ND	-77				
2	2.4	11	6.3	7	3.0	BURLEIGH ND	161	-53	23	83	59
9	3.2	18	5.2	12	2.4	CASS ND	65	-54	-25	63	46
1	2.1			2	2.8	CAVALIER ND			31	86	54
1	2.2			3	4.8	DICKEY ND			119	95	88
4*	13.7	2	6.1			DIVIDE ND	-55				
		2	8.5	1	4.7	DUNN ND		-45			87
1	6.2			2	3.1	EDDY ND			-50	33	62
2	5.9			1	2.1	EMMONS ND			-64	18	40
		2	5.6	1	3.8	FOSTER ND		-32			75
11	6.6	10	4.5	11	4.0	GRAND FORKS ND	-32	-11	-40	48	78
		1	3.3			GRANT ND					
		2	8.5			GRIGGS ND					
		2	7.6	1	3.0	HETTINGER ND		-61			60
		2	9.9	1	4.2	KIDDER ND		-57			81
3	9.0	5*	12.0	1	1.6	LA MOURE ND	32	-87	-83	02	28
2	9.4					LOGAN ND					
2	4.2	4	7.4	3	5.1	MC HENRY ND	74	-31	20	83	89
2	7.8	4*	11.5	2	5.7	MC INTOSH ND	47	-50	-26	62	93
1	4.2					MC KENZIE ND					
3	6.3	3	5.0	2	2.2	MC LEAN ND	-21	-55	-64	18	43
1	3.6	6*	20.8			MERCER ND	478				
		7	7.2	6	4.8	MORTON ND		-33			88
1	2.6	3	6.6	1	2.2	MOUNTRAIL ND	155	-67	-15	70	41
1	3.7	1	3.1	2	4.6	NELSON ND	-17	48	23	84	86
		1	24.8			OLIVER ND					
1	1.7	1	1.4	2	3.1	PEMBINA ND	-17	120	82	93	62
2	9.1	1	3.2	1	2.8	PIERCE ND	-64	-14	-70	12	55
3	4.7	3	4.7	6	7.4	RAMSEY ND	0	55	55	90	97
		2	3.5	2	2.6	RANSOM ND		-25			52
1	3.5			1	4.3	RENVILLE ND			24	84	83
2	2.1	6	5.6	2	2.1	RICHLAND ND	171	-61	5	79	40

* SIGNIFICANTLY MORE DEATHS THAN EXPECTED

WHITE FEMALE

CANCER OF THE RECTUM

ICD 154 EXCEPT 154.3

1950-1959 DEATHS	1950-1959 RATE	1960-1969 DEATHS	1960-1969 RATE	1970-1979 DEATHS	1970-1979 RATE	COUNTY AND STATE	PERCENT CHANGE 60S/50S	PERCENT CHANGE 70S/60S	PERCENT CHANGE 70S/50S	PERCENTILE %CHANGE 70S/50S	PERCENTILE RATES 1970S
		2	8.8			ROLETTE ND					
		2	6.6	1	2.3	SARGENT ND		-65			44
		1	10.6			SHERIDAN ND					
		1*	11.2			SIOUX ND					
4	6.8	1	1.3	4	3.3	STARK ND	-81	152	-52	30	65
3	16.3	2	8.8	3*	11.4	STEELE ND	-46	29	-30	59	99
9	7.9	2	1.6	6	3.4	STUTSMAN ND	-80	113	-57	25	68
		1	3.8			TOWNER ND					
3	5.4	6	7.9	2	1.7	TRAILL ND	47	-78	-68	14	32
3	3.8	3	2.8	3	2.0	WALSH ND	-27	-27	-47	39	38
7	5.2	17*	8.4	8	3.2	WARD ND	61	-62	-40	49	63
3	7.1	2	3.9	1	2.0	WELLS ND	-45	-48	-72	10	37
3	3.7	5	5.1	2	1.4	WILLIAMS ND	38	-72	-61	21	26
2562*	6.3	2497*	5.0	2238*	3.8	OHIO OH	-20	-24	-40		
9	7.7	8	6.5	4	2.9	ADAMS OH	-15	-56	-62	20	58
23	5.0	35	6.2	30	4.2	ALLEN OH	25	-32	-15	69	81
11	5.4	12	4.6	9	3.0	ASHLAND OH	-14	-34	-43	43	61
40*	9.0	25	4.8	18	3.0	ASHTABULA OH	-46	-37	-66	16	61
16	6.1	4	1.5	9	3.7	ATHENS OH	-76	151	-40	48	73
8	4.3	9	4.2	3	1.0	AUGLAIZE OH	-2	-77	-78	06	20
23	4.7	23	4.3	26	4.5	BELMONT OH	-9	5	-5	75	85
9	6.8	9	5.5	2	1.2	BROWN OH	-19	-79	-83	02	23
36	5.2	37	4.1	38	3.3	BUTLER OH	-21	-21	-38	50	65
7	6.7	1	.7	7	5.3	CARROLL OH	-89	614	-22	65	90
4	2.5	5	3.3	9	4.6	CHAMPAIGN OH	31	40	83	93	86
30	5.2	33	4.4	18	2.1	CLARK OH	-17	-53	-61	21	38
16	6.7	14	4.3	12	2.7	CLERMONT OH	-36	-38	-60	22	53
6	3.3	13	6.7	5	2.4	CLINTON OH	101	-65	-29	60	46
32	5.8	31	5.0	28	3.9	COLUMBIANA OH	-14	-22	-33	56	76
9	4.9	6	2.9	6	2.6	COSHOCTON OH	-40	-9	-46	40	52
16	6.3	12	4.1	10	2.9	CRAWFORD OH	-35	-30	-55	27	57
511*	7.6	447*	5.5	407*	4.3	CUYAHOGA OH	-27	-21	-43	44	83
21	8.0	11	3.8	12	3.3	DARKE OH	-53	-13	-59	23	66
4	2.6	14*	8.1	4	2.3	DEFIANCE OH	212	-72	-13	71	43
7	4.3	6	3.0	7	2.9	DELAWARE OH	-29	-3	-31	59	59
13	4.1	15	4.3	14	3.2	ERIE OH	5	-25	-21	66	65
26	8.5	17	4.6	7	1.6	FAIRFIELD OH	-46	-66	-82	03	28
4	2.7	8	4.8	4	2.4	FAYETTE OH	74	-51	-14	70	46
175	7.0	147	4.6	136	3.5	FRANKLIN OH	-34	-24	-50	34	70
7	5.1	8	4.5	4	1.7	FULTON OH	-11	-62	-67	15	31
9	7.0	5	3.3	5	3.1	GALLIA OH	-52	-5	-55	27	63
9	6.7	11	5.7	9	3.3	GEAUGA OH	-14	-43	-51	33	66
10	4.7	10	3.2	11	2.4	GREENE OH	-31	-26	-49	34	46
12	4.3	15	5.7	10	3.2	GUERNSEY OH	33	-43	-25	64	65
307*	7.9	294*	6.4	246*	4.7	HAMILTON OH	-19	-27	-41	47	86
14	4.9	12	3.6	14	3.1	HANCOCK OH	-27	-14	-37	52	63
11	7.3	9	4.9	4	1.8	HARDIN OH	-33	-63	-75	07	33
1	.9	10*	8.8	3	2.7	HARRISON OH	907	-70	205	98	52
9	6.8	7	4.6	5	2.9	HENRY OH	-32	-38	-57	25	57
10	5.1	9	5.2	14*	6.4	HIGHLAND OH	2	22	25	84	95
4	3.7	13*	11.2	5	3.6	HOCKING OH	206	-68	-2	76	71
7	7.8	5	4.2	5	3.9	HOLMES OH	-46	-6	-50	34	77
9	3.7	14	5.3	8	2.6	HURON OH	43	-51	-29	60	51
9	6.1	10	6.0	5	2.0	JACKSON OH	-2	-66	-67	15	38
30	7.1	24	4.9	25	4.2	JEFFERSON OH	-31	-14	-41	47	81
7	2.7	19*	7.0	13	4.8	KNOX OH	157	-31	77	92	88
15	4.1	48*	8.2	46*	5.3	LAKE OH	100	-36	29	85	91
12	5.4	11	4.0	9	2.6	LAWRENCE OH	-26	-34	-52	32	51
33	7.5	23	4.4	25	3.9	LICKING OH	-41	-11	-48	37	77
15	7.4	12	5.2	9	3.5	LOGAN OH	-30	-32	-53	30	69
38	5.4	46	4.8	54	4.6	LORAIN OH	-11	-3	-14	70	86
114	5.5	111	4.5	94	3.3	LUCAS OH	-19	-27	-41	47	66
4	3.6	4	3.3	5	3.6	MADISON OH	-8	7	-1	76	71
87	7.2	62	4.2	61	3.4	MAHONING OH	-42	-18	-52	30	68
16	5.6	20	5.8	15	3.9	MARION OH	3	-32	-30	59	77
22*	9.9	21	6.8	10	2.5	MEDINA OH	-31	-64	-75	07	48
10	7.9	5	3.4	5	3.5	MEIGS OH	-57	5	-55	27	71
8	5.4	5	2.6	8	4.2	MERCER OH	-52	65	-21	66	81
29	8.1	18	4.5	15	3.0	MIAMI OH	-45	-33	-63	19	60
5	4.8	5	5.2			MONROE OH	9				
98	5.6	92	4.0	94	3.4	MONTGOMERY OH	-28	-16	-40	49	67
3	3.5	8	9.0	2	1.9	MORGAN OH	155	-79	-47	38	34
7	6.9	6	5.7	3	2.6	MORROW OH	-18	-55	-63	19	50
26	5.8	19	3.8	20	3.8	MUSKINGUM OH	-34	-2	-36	53	74
1	1.2	3	4.2	6*	8.3	NOBLE OH	254	97	598	99	98
6	3.5	7	3.6	13	5.5	OTTAWA OH	3	52	56	90	92
2	2.6	3	3.1	2	1.8	PAULDING OH	16	-43	-33	56	32
12	6.9	13	7.3	14*	7.6	PERRY OH	6	3	9	80	97
7	4.8	7	4.3	9	4.7	PICKAWAY OH	-10	8	-3	75	86
2	2.8	8	9.2	3	1.8	PIKE OH	232	-80	-34	56	34
18	6.0	24	5.9	23	4.8	PORTAGE OH	-3	-18	-20	67	88
10	6.6	9	5.0	9	4.3	PREBLE OH	-25	-13	-35	54	83
11	8.3	10	6.1	8	4.5	PUTNAM OH	-26	-27	-46	39	84
26	5.9	28	5.0	38*	5.6	RICHLAND OH	-16	12	-5	75	92
20	7.9	10	3.5	11	3.2	ROSS OH	-55	-9	-59	23	64
9	3.4	9	2.7	7	1.8	SANDUSKY OH	-21	-32	-46	39	34
13	3.1	21	4.3	30*	5.0	SCIOTO OH	38	17	61	91	89
16	5.2	16	4.8	13	3.5	SENECA OH	-8	-27	-32	57	69
10	6.5	8	4.3	10	3.4	SHELBY OH	-34	-21	-48	37	68
70	4.9	89	4.9	93	4.4	STARK OH	0	-9	-9	73	84
105	5.6	128	5.2	90	3.0	SUMMIT OH	-7	-41	-46	40	61
36	5.4	35	3.7	37	3.0	TRUMBULL OH	-32	-19	-45	42	60
23	5.9	30	6.5	26	4.8	TUSCARAWAS OH	10	-26	-19	67	88
7	5.9	8	5.3	5	2.8	UNION OH	-11	-48	-53	29	54
11	6.5	9	4.5	19*	8.4	VAN WERT OH	-31	85	29	85	98
		4	7.8	4	6.1	VINTON OH		-21			95
13	6.3	14	5.2	12	3.2	WARREN OH	-17	-38	-49	35	64
27*	9.2	18	5.5	13	3.2	WASHINGTON OH	-41	-42	-65	17	64
24	7.5	23	5.7	18	3.8	WAYNE OH	-24	-33	-49	35	75

* SIGNIFICANTLY MORE DEATHS THAN EXPECTED

WHITE FEMALE

CANCER OF THE RECTUM — ICD 154 EXCEPT 154.3

1950-1959 DEATHS	1950-1959 RATE	1960-1969 DEATHS	1960-1969 RATE	1970-1979 DEATHS	1970-1979 RATE	COUNTY AND STATE	PERCENT CHANGE 60S/50S	PERCENT CHANGE 70S/60S	PERCENT CHANGE 70S/50S	PERCENTILE %CHANGE 70S/50S	PERCENTILE RATES 1970S
4	2.3	11	5.0	9	4.2	WILLIAMS OH	119	-16	83	93	81
10	3.2	13	3.4	21	4.2	WOOD OH	7	24	33	86	81
6	4.5	6	4.3	9	6.2	WYANDOT OH	-6	45	36	87	95
363	3.5	425	3.2	405	2.3	OKLAHOMA OK	-10	-26	-34		
		3	5.0	2	2.2	ADAIR OK		-56			42
3	4.3			2	2.0	ALFALFA OK			-53	29	37
2	4.5	4	7.7	1	1.4	ATOKA OK	70	-82	-70	12	25
2	6.0	1	2.6	1	2.2	BEAVER OK	-56	-16	-63	19	41
2	1.7	5	3.2	7	3.8	BECKHAM OK	89	17	122	95	75
2	2.7	3	2.6	4	3.7	BLAINE OK	-4	42	36	87	73
2	1.4	9	5.3	7	3.0	BRYAN OK	280	-44	114	95	59
3	1.8	8	4.5	4	2.2	CADDO OK	149	-51	23	83	42
4	3.2	9	6.4	6	2.8	CANADIAN OK	104	-57	-13	71	54
6	3.2	12	5.1	11	4.0	CARTER OK	60	-22	24	84	78
5	6.8	1	.9	3	2.5	CHEROKEE OK	-86	166	-63	19	48
		2	1.8			CHOCTAW OK					
4	2.3	5	1.9	6	1.7	CLEVELAND OK	-19	-9	-26	63	32
5	2.8	13	5.2	9	2.8	COMANCHE OK	83	-47	-3	76	55
		3	5.0	1	1.4	COTTON OK		-72			26
6	5.2	3	2.5	1	.7	CRAIG OK	-52	-71	-86	01	18
10	5.4	6	2.4	10	3.0	CREEK OK	-55	23	-45	41	59
6	6.2	2	1.4	2	.9	CUSTER OK	-78	-38	-86	01	19
3	4.4	2	2.2	4	3.1	DELAWARE OK	-49	40	-28	61	63
2	3.8	1	1.5	2	2.2	DEWEY OK	-61	49	-43	44	42
		4	8.8	1	3.0	ELLIS OK		-66			61
12	4.3	14	3.9	14	2.9	GARFIELD OK	-9	-26	-33	57	57
6	4.4	6	3.1	3	1.6	GARVIN OK	-29	-49	-64	18	29
2	1.1	4	2.1	5	1.7	GRADY OK	92	-16	60	90	32
2	2.8	3	3.4	3	2.5	GRANT OK	21	-26	-11	72	49
3	4.0			3	3.6	GREER OK			-10	73	72
1	2.8	2	4.2			HARMON OK	46				
1	1.6	3	5.9	2	2.5	HASKELL OK	273	-57	60	90	49
				2	1.3	HUGHES OK					23
3	2.4	5	3.7	4	2.3	JACKSON OK	50	-38	-7	74	44
4	5.9	1	1.6	2	2.1	JEFFERSON OK	-72	25	-65	17	38
4	7.8					JOHNSTON OK					
11	4.4	8	2.3	14	3.5	KAY OK	-48	53	-21	66	69
				3	2.7	KINGFISHER OK					52
5	4.7	5	3.8	3	2.1	KIOWA OK	-20	-44	-55	27	39
		3	5.8	2	3.9	LATIMER OK		-33			77
3	1.6	6	2.9	6	2.3	LE FLORE OK	77	-22	39	87	44
2	1.6	4	3.5	5	2.4	LINCOLN OK	116	-32	48	89	46
7	5.4	4	2.2	1	.7	LOGAN OK	-59	-71	-88	01	18
				2	4.1	LOVE OK					79
2	3.1	1	.9	3	2.7	MC CLAIN OK	-70	183	-14	70	52
5	5.1	4	2.9	4	2.1	MC CURTAIN OK	-42	-30	-60	22	39
1	1.6	1	1.3	1	1.1	MC INTOSH OK	-20	-18	-34	55	21
2	3.6	1	1.9	1	2.6	MAJOR OK	-49	42	-28	62	51
		2	3.6	1	.9	MARSHALL OK		-75			19
3	3.2	6	4.1	3	2.0	MAYES OK	28	-51	-37	52	38
1	1.9	2	2.1	2	3.0	MURRAY OK	11	44	60	91	61
15	5.7	9	2.6	16	4.3	MUSKOGEE OK	-55	66	-26	63	82
3	4.7	1	1.9			NOBLE OK	-60				
2	3.4	3	4.6			NOWATA OK	37				
3	4.7					OKFUSKEE OK					
65	4.3	85	3.9	64	2.2	OKLAHOMA OK	-9	-44	-49	35	42
6	3.3	7	3.3	11	4.2	OKMULGEE OK	2	28	30	85	82
3	1.9	6	3.4	7	3.4	OSAGE OK	78	1	80	92	69
7	5.0	9	5.0	3	1.6	OTTAWA OK	0	-68	-68	14	29
6	7.8	5	4.9	3	3.1	PAWNEE OK	-37	-37	-61	22	61
6	2.9	6	2.5	7	2.2	PAYNE OK	-14	-12	-24	64	42
4	2.1	4	1.6	3	.8	PITTSBURG OK	-22	-52	-63	20	18
7	4.3	5	3.1	6	2.2	PONTOTOC OK	-29	-27	-48	36	43
7	2.8	11	4.0	10	2.7	POTTAWATOMIE OK	42	-31	-1	76	54
2	3.7	2	3.2	1	1.0	PUSHMATAHA OK	-13	-70	-73	09	20
3	9.6					ROGER MILLS OK					
6	6.2	3	2.0	5	3.1	ROGERS OK	-67	56	-49	35	63
3	2.2	4	2.5	5	2.4	SEMINOLE OK	13	-4	8	80	46
2	2.8	2	1.9	2	1.1	SEQUOYAH OK	-32	-43	-61	21	21
3	1.7	8	3.4	2	.6	STEPHENS OK	105	-82	-63	20	17
1	1.9	4	5.7	1	1.0	TEXAS OK	200	-83	-50	33	20
1	1.3	2	1.9	3	2.3	TILLMAN OK	44	22	76	92	44
40	3.3	52	3.2	64	2.9	TULSA OK	-4	-10	-13	71	57
2	3.7	1	1.3	3	2.3	WAGONER OK	-65	75	-38	50	44
9	5.9	10	4.6	9	2.8	WASHINGTON OK	-22	-40	-53	30	55
6	7.4	3	2.9	3	2.4	WASHITA OK	-61	-16	-67	15	48
		1	1.2	2	1.6	WOODS OK		39			30
4	4.6	1	.8	2	1.2	WOODWARD OK	-83	52	-75	08	22
400	5.1	402	3.8	422	2.9	OREGON OR	-26	-22	-43		
5	5.7	3	3.4	2	1.6	BAKER OR	-40	-54	-72	10	29
4	2.8	7	3.8	12	5.2	BENTON OR	33	37	82	93	90
25	5.4	29	4.3	42	4.0	CLACKAMAS OR	-20	-6	-25	63	79
6	3.6	8	3.9	6	2.3	CLATSOP OR	9	-40	-35	54	45
4	3.6	4	3.3	5	2.9	COLUMBIA OR	-8	-14	-21	66	57
4	2.0	9	4.0	22*	6.4	COOS OR	105	59	226	98	96
3	13.4	2	3.5	2	2.9	CROOK OR	-74	-19	-79	05	57
1	4.2	1	3.1	1	1.1	CURRY OR	-25	-66	-74	08	21
4	5.3	5	3.6	8	3.9	DESCHUTES OR	-33	8	-27	62	77
9	4.6	11	3.8	9	2.1	DOUGLAS OR	-19	-45	-55	27	39
		1	7.4			GILLIAM OR					
1	3.6	1	3.0	1	2.2	GRANT OR	-16	-26	-38	50	42
1	3.6	2	7.5			HARNEY OR	107				
3	4.4	2	2.8	1	1.2	HOOD RIVER OR	-36	-57	-73	10	23
18	5.6	19	3.9	21	2.9	JACKSON OR	-31	-24	-47	38	59
		1	3.9			JEFFERSON OR					
10	7.1	7	3.6	15	4.4	JOSEPHINE OR	-49	23	-38	50	84
11	7.1	9	4.6	11	4.1	KLAMATH OR	-36	-10	-42	45	80

* SIGNIFICANTLY MORE DEATHS THAN EXPECTED

WHITE FEMALE

CANCER OF THE RECTUM

ICD 154 EXCEPT 154.3

COUNTY AND STATE	1950-1959 DEATHS	1950-1959 RATE	1960-1969 DEATHS	1960-1969 RATE	1970-1979 DEATHS	1970-1979 RATE	PERCENT CHANGE 60S/50S	PERCENT CHANGE 70S/60S	PERCENT CHANGE 70S/50S	PERCENTILE %CHANGE 70S/50S	PERCENTILE RATES 1970S
LAKE OR	2	10.0			1	2.9			-71	11	57
LANE OR	25	4.5	25	2.9	28	2.1	-35	-28	-53	29	40
LINCOLN OR	5	4.8	9	5.3	8	3.1	12	-41	-34	55	63
LINN OR	14	6.2	10	3.4	11	2.4	-44	-30	-61	21	47
MALHEUR OR	1	2.4	2	2.2	4	3.4	-12	60	41	88	68
MARION OR	37	6.2	36	4.2	23	2.0	-32	-54	-68	13	36
MORROW OR			2	8.8	2	5.2		-41			90
MULTNOMAH OR	153	5.3	154	4.3	121	2.9	-19	-33	-45	41	58
POLK OR	4	3.0	2	1.1	7	3.0	-61	161	1	77	60
SHERMAN OR	1	10.1	1	10.0	1	9.0	-1	-10	-11	72	98
TILLAMOOK OR	4	5.1	3	2.9	3	1.9	-44	-35	-64	19	35
UMATILLA OR	9	4.4	8	3.3	16	5.2	-26	58	16	82	90
UNION OR	8	8.2	2	1.6	2	1.5	-81	-4	-82	03	28
WALLOWA OR			1	2.6	1	2.4		-7			47
WASCO OR	4	4.4	2	1.7	8	5.2	-62	215	18	82	90
WASHINGTON OR	11	3.6	9	1.5	18	2.0	-58	31	-45	42	36
YAMHILL OR	13	7.4	15	6.3	10	3.2	-16	-49	-57	25	64
PENNSYLVANIA PA	3776*	7.2	3635*	5.7	2986*	3.9	-20	-32	-46		
ADAMS PA	23*	9.3	15	5.1	13	3.2	-45	-37	-65	17	64
ALLEGHENY PA	489*	7.0	516*	5.9	418*	4.1	-15	-31	-42	45	79
ARMSTRONG PA	19	5.1	20	4.5	22	4.1	-11	-9	-19	67	80
BEAVER PA	45	5.9	45	4.8	45	3.7	-19	-23	-38	51	73
BEDFORD PA	6	3.2	21*	9.6	9	3.4	195	-64	5	79	68
BERKS PA	106	7.2	112*	6.2	110*	5.0	-14	-19	-30	59	89
BLAIR PA	54	6.9	48	5.0	35	3.2	-27	-36	-53	29	65
BRADFORD PA	24	8.0	22	6.8	7	1.9	-15	-72	-76	07	36
BUCKS PA	47	5.3	55	4.1	70	3.5	-22	-14	-33	56	70
BUTLER PA	38	7.5	35	5.7	26	3.3	-24	-43	-57	26	66
CAMBRIA PA	63	7.0	54	5.0	34	2.6	-28	-48	-62	20	51
CAMERON PA	1	3.0	1	5.0			64				
CARBON PA	23	7.7	24*	7.1	22*	5.7	-8	-19	-26	63	92
CENTRE PA	21	7.9	16	4.8	13	3.1	-39	-37	-61	21	62
CHESTER PA	59	7.3	48	4.5	56	3.8	-39	-15	-48	37	75
CLARION PA	11	5.8	7	3.2	10	4.1	-44	26	-30	60	79
CLEARFIELD PA	26	6.3	39*	8.6	21	4.0	38	-54	-36	52	78
CLINTON PA	8	4.5	17*	8.3	7	2.9	87	-65	-35	54	57
COLUMBIA PA	25	8.5	18	5.2	10	2.4	-38	-54	-72	10	47
CRAWFORD PA	35	8.1	21	4.4	19	3.2	-46	-26	-60	23	65
CUMBERLAND PA	32	6.1	45	6.3	37	3.9	4	-39	-36	52	77
DAUPHIN PA	73	6.6	59	4.6	49	3.3	-31	-27	-50	34	67
DELAWARE PA	157*	7.6	177*	6.1	117	3.2	-21	-48	-59	24	63
ELK PA	9	5.4	16*	8.4	15*	6.9	55	-18	27	85	96
ERIE PA	85*	7.6	83*	6.3	71*	4.4	-17	-29	-41	46	84
FAYETTE PA	56	6.8	62*	6.7	42	3.9	-2	-42	-43	44	77
FOREST PA			1	3.5	1	3.3		-7			66
FRANKLIN PA	27	6.8	27	5.1	29	4.1	-25	-20	-40	49	80
FULTON PA	2	5.2	2	3.7	3	4.4	-29	20	-15	69	84
GREENE PA	10	4.8	15	6.2	12	4.5	29	-27	-6	74	85
HUNTINGDON PA	16	8.5	13	6.5	10	3.9	-23	-40	-54	28	77
INDIANA PA	21	5.8	26	6.2	30*	5.8	7	-7	0	77	93
JEFFERSON PA	15	5.7	17	5.7	22*	5.8	0	2	2	78	93
JUNIATA PA	6	7.6	12*	13.1	7	5.7	73	-56	-24	64	93
LACKAWANNA PA	124*	8.7	129*	7.9	94*	4.9	-9	-38	-43	43	88
LANCASTER PA	74	5.4	86	4.9	73	3.3	-8	-34	-39	49	66
LAWRENCE PA	43	8.4	27	4.4	15	2.2	-47	-51	-74	08	41
LEBANON PA	34	7.9	40*	7.4	31	4.7	-6	-36	-40	47	87
LEHIGH PA	66	5.7	93*	6.3	58	3.1	9	-51	-47	39	62
LUZERNE PA	172*	8.5	185*	7.8	152*	5.3	-8	-32	-38	51	91
LYCOMING PA	39	6.4	38	5.5	32	4.0	-14	-26	-37	51	79
MC KEAN PA	16	5.4	18	5.3	21*	5.6	-2	5	3	78	92
MERCER PA	38	6.6	34	4.9	33	3.9	-26	-20	-41	46	77
MIFFLIN PA	25*	13.2	15	6.3	14	4.3	-52	-31	-67	14	83
MONROE PA	20*	10.3	16	6.2	21*	5.6	-40	-10	-46	40	92
MONTGOMERY PA	132	6.1	160	5.2	158	3.7	-13	-29	-39	50	74
MONTOUR PA	4	3.1	4	3.5	3	2.1	13	-40	-32	57	39
NORTHAMPTON PA	71	7.3	55	4.6	59	3.8	-37	-16	-48	37	76
NORTHUMBERLAND PA	40	6.5	40	5.8	36	4.4	-11	-24	-32	57	84
PERRY PA	12	9.4	10	6.8	1	.7	-27	-90	-93	00	18
PHILADELPHIA PA	815*	8.6	625*	6.2	405*	3.9	-28	-37	-55	27	77
PIKE PA	6	9.4			11*	8.5			-9	73	98
POTTER PA	1	.9	6	5.3	3	2.5	496	-52	185	97	49
SCHUYLKILL PA	70	6.9	76*	6.8	54	3.9	-1	-42	-43	44	77
SNYDER PA	8	6.5	8	5.3	6	3.6	-19	-33	-45	41	71
SOMERSET PA	18	4.8	21	4.5	22	3.9	-6	-12	-17	68	77
SULLIVAN PA	4	9.9	4	9.6			-3				
SUSQUEHANNA PA	14	7.4	5	2.4	9	3.9	-68	63	-47	38	77
TIOGA PA	10	5.2	10	4.7	8	3.0	-10	-36	-42	45	61
UNION PA	10	7.6	9	6.0	13*	6.2	-20	3	-18	68	95
VENANGO PA	25	7.2	19	5.0	16	3.6	-30	-27	-49	34	72
WARREN PA	16	5.6	12	3.9	10	2.8	-30	-29	-51	32	55
WASHINGTON PA	50	5.2	50	4.3	63*	4.6	-18	6	-13	71	85
WAYNE PA	9	4.9	9	3.9	7	2.9	-20	-25	-40	48	59
WESTMORELAND PA	88	6.0	71	3.8	82	3.4	-37	-10	-43	43	68
WYOMING PA	15*	15.0	7	6.0	5	3.9	-60	-34	-74	08	78
YORK PA	75	6.8	64	4.6	79*	4.2	-32	-8	-37	51	81
RHODE ISLAND RI	387*	8.8	328*	6.1	323*	4.9	-31	-20	-45		
BRISTOL RI	12	7.4	15	6.8	7	2.3	-8	-67	-69	12	44
KENT RI	46*	11.1	42*	6.7	40	4.5	-39	-34	-60	22	85
NEWPORT RI	22	7.6	19	5.3	23	4.8	-31	-9	-37	52	88
PROVIDENCE RI	286*	8.7	233*	6.1	234*	5.2	-30	-14	-40	48	90
WASHINGTON RI	21	8.8	19	5.9	19	4.1	-33	-30	-53	29	80
SOUTH CAROLINA SC	179	3.4	208	2.8	249	2.4	-17	-14	-29		
ABBEVILLE SC	2	2.7	1	1.2	1	1.3	-54	4	-53	30	24
AIKEN SC	6	4.8	6	2.5	6	1.5	-48	-41	-69	12	27
ALLENDALE SC					1	3.4					67

* SIGNIFICANTLY MORE DEATHS THAN EXPECTED

WHITE FEMALE

CANCER OF THE RECTUM

ICD 154 EXCEPT 154.3

1950-1959 DEATHS	1950-1959 RATE	1960-1969 DEATHS	1960-1969 RATE	1970-1979 DEATHS	1970-1979 RATE	COUNTY AND STATE	PERCENT CHANGE 60S/50S	PERCENT CHANGE 70S/60S	PERCENT CHANGE 70S/50S	PERCENTILE %CHANGE 70S/50S	PERCENTILE RATES 1970S
10	3.8	8	1.9	15	2.6	ANDERSON SC	-49	36	-31	59	52
2	6.2	1	2.4	1	2.1	BAMBERG SC	-62	-14	-67	15	39
				1	1.8	BARNWELL SC					34
				5	3.9	BERKELEY SC					76
1	4.3	1	3.0			CALHOUN SC	-31				
19	5.3	18	3.4	20	2.8	CHARLESTON SC	-36	-17	-47	39	56
7	5.7	5	3.5	2	1.2	CHEROKEE SC	-38	-66	-79	04	23
3	3.8	5	5.3	2	1.6	CHESTER SC	37	-70	-59	23	28
3	3.0	6	5.5	2	1.6	CHESTERFIELD SC	84	-70	-45	41	30
1	2.6	1	2.2			CLARENDON SC	-16				
2	3.0			3	2.9	COLLETON SC			-6	74	56
				4	2.0	DARLINGTON SC					37
2	2.9	3	4.4	2	1.9	DILLON SC	52	-57	-34	55	35
2	3.9	1	1.5	4	3.1	DORCHESTER SC	-61	109	-19	67	62
1	2.8					EDGEFIELD SC					
1	2.5	2	4.4	1	1.5	FAIRFIELD SC	73	-66	-42	46	27
9	5.8	11	4.6	9	2.8	FLORENCE SC	-22	-39	-52	31	55
1	3.1	3	4.8	5	4.8	GEORGETOWN SC	55	0	55	90	88
13	2.7	27	3.1	38	3.2	GREENVILLE SC	15	5	20	83	65
10	9.4	5	3.2	4	1.8	GREENWOOD SC	-66	-43	-81	04	33
		1	2.2	2	3.4	HAMPTON SC		57			68
3	1.9	9	4.8	8	2.2	HORRY SC	151	-53	17	82	43
		1	4.0			JASPER SC					
2	2.6			2	1.8	KERSHAW SC			-29	60	34
4	3.8	4	3.1	2	1.0	LANCASTER SC	-20	-67	-74	08	21
10	8.0	4	2.1	3	1.2	LAURENS SC	-74	-41	-85	02	23
1	2.7	2	5.1	1	1.7	LEE SC	90	-67	-37	52	31
2	1.4	5	2.2	10	2.1	LEXINGTON SC	62	-2	58	90	40
		2	11.4			MC CORMICK SC					
3	5.8	2	2.4	2	2.0	MARION SC	-58	-15	-65	18	38
2	2.8	2	2.5	3	3.0	MARLBORO SC	-11	20	7	79	60
3	3.1	7	5.3	7	4.3	NEWBERRY SC	72	-19	39	87	82
1	.7	5	3.1	5	2.0	OCONEE SC	320	-34	178	97	38
5	3.9	5	3.1	3	1.3	ORANGEBURG SC	-21	-57	-66	16	24
5	3.2	1	.5	5	1.6	PICKENS SC	-84	207	-51	32	28
12	2.7	15	2.4	22	2.8	RICHLAND SC	-10	16	4	78	55
				3	4.0	SALUDA SC					78
16	3.2	21	3.4	21	2.3	SPARTANBURG SC	6	-31	-27	62	45
4	3.5	7	5.5	6	3.4	SUMTER SC	58	-39	-3	75	67
6	6.6	3	3.3	4	2.7	UNION SC	-50	-17	-58	24	54
				5	7.1	WILLIAMSBURG SC					96
5	2.6	8	3.1	9	2.4	YORK SC	20	-22	-6	74	47
154	5.2	130	3.5	113	2.4	SOUTH DAKOTA SD	-34	-30	-54		
2	10.6	2	6.1			AURORA SD	-43				
6	6.6	3	2.2	6	4.1	BEADLE SD	-67	91	-37	51	80
				2*	19.3	BENNETT SD					99
6	11.5			1	2.1	BON HOMME SD			-81	03	40
4	4.6	2	2.0	2	1.8	BROOKINGS SD	-56	-13	-62	20	32
8	5.1	9	4.4	7	2.9	BROWN SD	-14	-35	-44	42	57
1	3.3	1	2.0	1	1.4	BRULE SD	-40	-27	-56	26	26
				1*	212.0	BUFFALO SD					99
1	2.4	3	5.1			BUTTE SD	116				
1	1.8	1	1.8	1	1.8	CHARLES MIX SD	-2	2	0	77	34
2	4.6	3	7.8	2	4.0	CLARK SD	68	-49	-14	70	78
3	6.3	1	1.8	1	2.0	CLAY SD	-71	9	-69	13	37
8	8.4	7	6.1	4	2.7	CODINGTON SD	-28	-56	-68	13	53
		1	4.0			CUSTER SD					
7	7.4	3	2.4	4	2.2	DAVISON SD	-67	-10	-71	11	41
6	11.0	1	1.0	6*	7.0	DAY SD	-91	602	-37	52	96
3	8.5	1	2.1	3	6.8	DEUEL SD	-75	229	-19	67	96
1	6.7					DEWEY SD					
2	9.3	1	2.5	1	2.8	DOUGLAS SD	-73	11	-70	11	55
1	4.5	1	2.7			EDMUNDS SD	-40				
3	7.0	2	2.9	3	4.1	FALL RIVER SD	-58	39	-42	45	80
1	4.0			1	2.2	FAULK SD			-45	41	40
2	4.4					GRANT SD					
		1	2.2			GREGORY SD					
		1	7.0	1	9.6	HAAKON SD		37			98
2	4.7	3	6.1	1	2.2	HAMLIN SD	30	-65	-55	28	40
1	4.5					HAND SD					
5	13.1	4	7.9	2	2.8	HUGHES SD	-40	-65	-79	05	55
2	2.9	2	2.4			HUTCHINSON SD	-18				
		3*	26.1			HYDE SD					
2	7.8	4*	13.5	1	2.4	JERAULD SD	73	-82	-69	12	46
		1	9.4			JONES SD					
2	4.1	1	1.8			KINGSBURY SD	-57				
5	8.4	1	1.6	2	1.8	LAKE SD	-80	10	-79	05	33
4	4.8	3	3.1			LAWRENCE SD	-36				
5	8.0	6	6.4	3	2.4	LINCOLN SD	-20	-63	-70	11	46
4	9.5	4	7.1			MC COOK SD	-25				
				1	1.9	MC PHERSON SD					34
2	5.6	1	2.8	2	3.2	MARSHALL SD	-50	15	-43	44	64
		2	3.4	4	5.4	MEADE SD		62			91
1	10.3					MELLETTE SD					
4	12.0	1	2.4	2	27.1	MINER SD	-80	1035	126	95	99
12	3.6	14	3.1	12	1.7	MINNEHAHA SD	-12	-47	-53	29	30
2	4.2	1	1.9			MOODY SD	-55				
11	7.2	7	3.1	9	2.9	PENNINGTON SD	-56	-7	-59	23	58
1	4.2	2	6.1	1	2.3	PERKINS SD	45	-62	-44	43	45
1	4.1	1	2.7			POTTER SD	-35				
4	6.5	4	5.3	3	3.6	ROBERTS SD	-18	-32	-45	42	71
		1	3.4	1	3.7	SANBORN SD		8			73
1*	74.3					SHANNON SD					
1	1.6	6	8.2	3	2.6	SPINK SD	414	-68	63	91	51
		1	8.6	2*	18.8	STANLEY SD		117			99
		3*	37.4			SULLY SD					
				1	12.9	TODD SD					99
4	9.9	2	4.5	1	3.0	TRIPP SD	-55	-33	-70	12	60
2	3.5	2	3.0	5	5.6	TURNER SD	-14	85	59	90	92

* SIGNIFICANTLY MORE DEATHS THAN EXPECTED

WHITE FEMALE — CANCER OF THE RECTUM — ICD 154 EXCEPT 154.3

1950-1959 DEATHS	1950-1959 RATE	1960-1969 DEATHS	1960-1969 RATE	1970-1979 DEATHS	1970-1979 RATE	COUNTY AND STATE	PERCENT CHANGE 60S/50S	PERCENT CHANGE 70S/60S	PERCENT CHANGE 70S/50S	PERCENTILE %CHANGE 70S/50S	PERCENTILE RATES 1970S
3	4.7	1	1.2	4	3.7	UNION SD	-75	220	-21	66	74
1	2.8	2	4.7	2	3.0	WALWORTH SD	69	-37	7	79	60
3	3.1	3	2.8	3	2.1	YANKTON SD	-11	-25	-33	56	39
1	74.3			1*	18.9	ZIEBACH SD			-75	08	99
436	3.5	507	3.1	510	2.3	TENNESSEE TN	-12	-24	-34		
4	2.3	9	3.8	11	3.1	ANDERSON TN	69	-19	38	87	63
3	2.6	5	3.4	3	1.6	BEDFORD TN	30	-53	-39	49	30
				1	1.8	BLEDSOE TN					33
3	1.6	5	1.8	8	1.9	BLOUNT TN	17	3	21	83	35
5	3.6	6	3.4	7	2.8	BRADLEY TN	-5	-17	-21	66	56
4	3.0	1	.6	7	3.1	CAMPBELL TN	-79	383	3	78	62
		1	3.1	1	1.3	CANNON TN		-57			24
3	2.6	7	4.7	5	2.5	CARROLL TN	78	-47	-6	74	48
2	1.4	7	3.7	9	3.2	CARTER TN	159	-12	128	95	65
2	6.5	1	1.9			CHEATHAM TN	-71				
1	2.0	1	1.6	3	4.7	CHESTER TN	-22	203	135	95	87
5	6.2	3	3.3	3	2.6	CLAIBORNE TN	-46	-22	-58	24	51
				2	3.6	CLAY TN					71
7	9.7	1	1.2	3	2.2	COCKE TN	-88	81	-78	05	41
7	6.5	8	5.9	2	1.0	COFFEE TN	-9	-83	-85	02	20
1	1.2	1	.9			CROCKETT TN	-22				
5	7.0	3	3.2	3	1.9	CUMBERLAND TN	-55	-42	-74	08	34
54	4.1	61	3.5	41	1.8	DAVIDSON TN	-15	-49	-57	26	33
				1	2.0	DECATUR TN					38
1	1.5	3	4.5	3	3.1	DE KALB TN	203	-32	107	94	61
4	4.2	8	7.0			DICKSON TN	66				
2	1.8	2	1.1	6	2.9	DYER TN	-35	157	66	91	58
		1	2.0			FAYETTE TN					
		1	1.6	2	2.6	FENTRESS TN		61			50
3	2.8	5	3.7			FRANKLIN TN	30				
7	3.3	7	2.7	8	2.5	GIBSON TN	-17	-10	-25	63	48
3	2.6	4	2.7	5	3.2	GILES TN	5	19	24	84	65
1	1.9			2	2.4	GRAINGER TN			23	83	46
5	3.7	7	3.5	7	2.3	GREENE TN	-6	-34	-38	51	45
2	4.0	2	3.5	1	1.1	GRUNDY TN	-11	-70	-73	09	21
6	5.8	7	4.5	2	1.1	HAMBLEN TN	-22	-77	-82	03	21
30	3.9	29	2.8	40	2.9	HAMILTON TN	-29	3	-27	62	57
1	2.9					HANCOCK TN					
3	3.2	1	.8	1	.6	HARDEMAN TN	-76	-26	-82	03	17
2	2.5	2	2.1	1	.7	HARDIN TN	-17	-66	-72	10	18
1	.9	7	4.7	3	1.4	HAWKINS TN	451	-71	61	91	25
2	4.0	1	2.1	8*	10.3	HAYWOOD TN	-48	398	160	96	99
		3	3.2	2	1.4	HENDERSON TN		-56			26
5	4.0	2	1.4	8	4.3	HENRY TN	-64	201	8	80	82
2	4.3	1	1.3	4	4.9	HICKMAN TN	-69	266	14	81	88
		1	2.9	3	9.0	HOUSTON TN		207			98
1	2.0	1	1.8	1	1.0	HUMPHREYS TN	-10	-46	-51	32	20
3	5.9	1	1.5			JACKSON TN	-74				
1	1.2	5	4.5	1	.7	JEFFERSON TN	268	-85	-46	41	18
1	1.9	1	1.8	1	1.1	JOHNSON TN	-6	-42	-45	41	21
36	4.0	38	3.0	37	2.2	KNOX TN	-26	-26	-45	42	42
1	2.6	1	3.1	2	4.8	LAKE TN	18	54	81	92	87
2	2.7	4	4.6	1	.9	LAUDERDALE TN	71	-81	-67	15	19
5	4.3	4	2.7	3	1.3	LAWRENCE TN	-37	-51	-69	12	24
3	10.1	1	2.8	1	1.9	LEWIS TN	-72	-33	-81	03	35
3	2.8	3	2.3	6	3.6	LINCOLN TN	-17	53	28	85	71
4	5.2	2	1.5	3	1.4	LOUDON TN	-72	0	-72	10	27
6	4.9	5	3.0	4	1.7	MC MINN TN	-39	-42	-65	17	32
1	1.2	2	2.0	2	1.5	MC NAIRY TN	75	-24	33	86	28
1	1.6	1	1.4			MACON TN	-12				
12	5.3	8	2.5	4	1.2	MADISON TN	-53	-52	-77	06	23
		5	5.8	1	.8	MARION TN		-86			18
3	3.2	2	1.4	1	.6	MARSHALL TN	-55	-60	-82	03	17
7	4.7	2	1.4	10	3.9	MAURY TN	-71	188	-16	69	77
		3*	13.2	1	5.0	MEIGS TN		-62			89
3	3.9	5	4.5	4	2.9	MONROE TN	16	-34	-24	64	59
5	4.2	5	2.7	13	5.4	MONTGOMERY TN	-35	101	30	85	91
1	5.3	2	9.4			MOORE TN	78				
		3	4.6	4	4.6	MORGAN TN		-1			86
3	1.9	7	3.6	5	2.0	OBION TN	92	-44	8	80	38
2	3.0	3	3.6	5	4.8	OVERTON TN	19	34	59	90	87
		1	5.0			PERRY TN					
2	5.5	3	4.6	3	4.4	POLK TN	-17	-4	-20	66	84
2	1.5	4	2.0	5	1.7	PUTNAM TN	37	-18	12	81	30
2	2.9	2	2.5	2	1.8	RHEA TN	-13	-29	-38	50	32
4	3.0	9	5.4	6	2.5	ROANE TN	80	-53	-15	69	49
1	.8	5	3.6	3	1.5	ROBERTSON TN	371	-58	98	93	27
4	2.5	6	2.7	4	1.2	RUTHERFORD TN	11	-55	-50	33	23
1	2.0	1	1.6	2	2.0	SCOTT TN	-20	22	-3	75	36
2	8.3			2	4.2	SEQUATCHIE TN			-49	36	82
3	4.2	4	3.0	8	4.2	SEVIER TN	-28	42	1	77	81
58	3.9	84	4.0	80	3.0	SHELBY TN	3	-25	-23	65	60
4	5.5	2	2.8	1	.8	SMITH TN	-49	-71	-85	02	19
1	2.2	2	4.2			STEWART TN	91				
11	3.5	15	2.9	18	2.4	SULLIVAN TN	-18	-17	-32	57	46
8	5.6	6	2.8	7	2.4	SUMNER TN	-50	-14	-57	25	47
2	2.6	2	2.2	4	3.6	TIPTON TN	-16	66	40	87	72
2	2.8	3	4.2	1	.7	UNICOI TN	52	-83	-74	08	18
1	2.9			1	2.0	UNION TN			-31	58	37
4	4.1	2	1.3	1	.6	WARREN TN	-67	-53	-85	02	17
12	4.9	10	3.1	14	3.3	WASHINGTON TN	-37	5	-34	55	66
2	4.4	3	4.9	2	3.0	WAYNE TN	11	-39	-32	57	59
5	3.1	4	2.1	7	2.8	WEAKLEY TN	-32	33	-9	73	56
2	2.7	2	2.3	3	2.3	WHITE TN	-14	-3	-17	69	43
6	6.3	3	2.3	4	1.9	WILLIAMSON TN	-64	-16	-70	12	35
7	5.5	6	3.4	6	2.7	WILSON TN	-38	-22	-51	32	53
1076	3.5	1197	2.8	1307	2.2	TEXAS TX	-22	-20	-37		

* SIGNIFICANTLY MORE DEATHS THAN EXPECTED

WHITE FEMALE

CANCER OF THE RECTUM

ICD 154 EXCEPT 154.3

1950-1959 DEATHS	1950-1959 RATE	1960-1969 DEATHS	1960-1969 RATE	1970-1979 DEATHS	1970-1979 RATE	COUNTY AND STATE	PERCENT CHANGE 60S/50S	PERCENT CHANGE 70S/60S	PERCENT CHANGE 70S/50S	PERCENTILE %CHANGE 70S/50S	PERCENTILE RATES 1970S
5	4.0	8	4.4	6	3.2	ANDERSON TX	10	-28	-21	66	63
				1	2.0	ANDREWS TX					38
		7	3.3	8	2.9	ANGELINA TX		-12			57
		1	2.1	3	3.9	ARANSAS TX		88			77
2	6.5	1	2.6	1	1.8	ARCHER TX	-60	-31	-72	10	34
2	18.8	1	6.2			ARMSTRONG TX	-67				
1	1.4			2	1.5	ATASCOSA TX			10	80	28
2	2.2	3	3.1	4	2.8	AUSTIN TX	43	-11	28	85	56
1	2.7					BAILEY TX					
1	3.4	1	2.0			BANDERA TX	-39				
6	7.7	3	3.0	1	.6	BASTROP TX	-61	-80	-92	00	17
2	2.7	2	2.0	3	2.5	BEE TX	-23	24	-5	74	49
3	1.1	7	1.8	8	1.5	BELL TX	63	-13	41	88	28
74	3.5	97	3.3	83	2.1	BEXAR TX	-6	-37	-40	47	39
1	3.9					BLANCO TX					
		1*	53.6			BORDEN TX					
1	.9	5	3.9	5	3.3	BOSQUE TX	327	-16	257	98	66
9	4.3	8	2.7	7	1.7	BOWIE TX	-37	-37	-61	21	31
4	2.7	5	1.9	10	2.8	BRAZORIA TX	-29	45	3	78	55
3	3.1			7	3.0	BRAZOS TX			-4	75	59
2	8.1	1	3.1	1	2.5	BREWSTER TX	-62	-20	-70	12	48
1	4.1			2	5.4	BROOKS TX			32	86	91
5	2.7	5	2.4	1	.5	BROWN TX	-13	-80	-82	03	17
3	5.5	2	3.0			BURLESON TX	-46				
1	2.0	2	2.1	1	.6	BURNET TX	5	-69	-68	14	18
2	2.1	1	.8	2	1.6	CALDWELL TX	-62	99	-24	64	30
				3	4.6	CALHOUN TX					86
2	2.9	2	2.4	1	.9	CALLAHAN TX	-18	-61	-68	13	20
16	4.3	10	1.9	7	.9	CAMERON TX	-57	-50	-78	05	20
2	6.2	2	5.5	1	1.7	CAMP TX	-11	-69	-73	10	31
1	3.7	1	3.2			CARSON TX	-12				
4	4.1	2	1.7	1	.7	CASS TX	-60	-57	-83	02	18
1	6.4					CASTRO TX					
		1	3.4			CHAMBERS TX					
5	3.1	9	4.3	3	1.3	CHEROKEE TX	38	-69	-58	24	24
1	2.1			2	2.6	CHILDRESS TX			22	83	51
2	3.4	2	3.1	1	1.1	CLAY TX	-11	-65	-69	13	21
				2	6.5	COKE TX					96
		1	.9	4	2.5	COLEMAN TX		182			48
6	2.7	15	4.8	10	2.4	COLLIN TX	79	-50	-11	72	46
2	5.0	1	2.0			COLLINGSWORTH TX	-60				
3	3.9	2	2.3			COLORADO TX	-39				
5	5.4	3	2.1	7	3.7	COMAL TX	-61	80	-30	59	74
		3	2.0	4	2.9	COMANCHE TX		44			58
2	2.1	1	.6	4	2.3	COOKE TX	-71	272	9	80	44
1	1.0	4	3.7	4	2.4	CORYELL TX	276	-33	152	96	48
1	10.6					CRANE TX					
3	7.3			2	1.9	CROSBY TX			-73	09	36
2*	34.0			1	9.3	CULBERSON TX			-73	09	98
				1	2.3	DALLAM TX					43
108	4.0	117	2.8	112	2.0	DALLAS TX	-29	-31	-51	32	36
2	2.8	2	2.6	1	.9	DAWSON TX	-6	-65	-68	14	20
		2	3.8	2	2.4	DEAF SMITH TX		-36			47
3	6.3	2	3.4	1	1.5	DELTA TX	-47	-56	-77	06	27
9	4.3	7	2.3	13	3.0	DENTON TX	-46	30	-30	60	61
5	4.0	2	1.7	2	1.0	DE WITT TX	-56	-42	-75	08	21
1	3.5					DICKENS TX					
		1	5.2	1	2.3	DONLEY TX		-55			45
1	2.3					DUVAL TX					
5	3.4	6	4.2	7	3.4	EASTLAND TX	24	-18	2	78	69
3	2.2	7	2.6	8	2.3	ECTOR TX	19	-12	5	79	43
				1	5.8	EDWARDS TX					93
4	1.9	9	3.0	3	.7	ELLIS TX	59	-76	-62	20	18
25	3.7	25	2.7	47	3.6	EL PASO TX	-25	34	0	77	73
6	4.7	3	1.6	6	3.6	ERATH TX	-65	124	-22	65	73
4	3.7	3	2.6	1	1.1	FALLS TX	-30	-59	-71	10	21
6	3.0	5	2.8	4	1.4	FANNIN TX	-7	-50	-53	30	26
3	2.3	7	4.9	5	2.7	FAYETTE TX	115	-46	17	82	52
4	8.7	4	7.6	3	4.2	FISHER TX	-13	-45	-52	31	81
		1	1.8			FLOYD TX					
		2	10.3			FOARD TX					
5	6.1	3	2.2	9	3.4	FORT BEND TX	-64	52	-45	42	68
1	2.2	1	3.1			FRANKLIN TX	43				
4	5.7	2	1.9	5	6.1	FREESTONE TX	-66	212	7	79	94
		1	1.9			FRIO TX					
		1	2.6	1	2.1	GAINES TX		-20			39
19	5.5	18	3.6	28	4.0	GALVESTON TX	-35	12	-28	61	78
2	9.6			1	2.9	GARZA TX			-70	12	58
2	2.7	2	1.7	3	2.5	GILLESPIE TX	-35	47	-4	75	50
1*	37.5	2*	60.1			GLASSCOCK TX	60				
3	11.0			1	3.5	GOLIAD TX			-68	14	70
5	5.4	2	2.0	4	4.0	GONZALES TX	-62	97	-26	63	79
4	4.8	3	2.6	3	2.0	GRAY TX	-46	-24	-59	23	37
13	3.5	17	3.4	18	2.6	GRAYSON TX	-4	-21	-24	64	52
7	3.5	12	4.2	12	2.8	GREGG TX	20	-33	-20	67	55
5	8.6	2	2.7	2	3.5	GRIMES TX	-69	31	-59	23	70
2	2.0	3	2.2	5	2.0	GUADALUPE TX	10	-9	0	77	37
3	2.7	5	2.9	5	2.7	HALE TX	9	-8	1	77	53
1	2.2					HALL TX					
3	4.7	3	3.1	3	1.9	HAMILTON TX	-33	-39	-59	23	35
2	11.9					HANSFORD TX					
		2	3.0	2	2.5	HARDEMAN TX		-18			49
2	4.0	2	1.9	1	.6	HARDIN TX	-51	-72	-86	01	17
130	4.9	142	3.2	168	2.6	HARRIS TX	-35	-19	-47	38	51
2	1.7	3	2.0	5	2.2	HARRISON TX	12	15	28	85	43
2	3.3	1	1.1	1	1.2	HASKELL TX	-66	8	-64	19	23
4	5.4	1	.9	5	4.1	HAYS TX	-83	338	-24	64	79
		1	5.6			HEMPHILL TX					
		2	1.3	8	3.3	HENDERSON TX		155			66
6	1.3	14	2.1	11	1.1	HIDALGO TX	60	-48	-17	68	22
5	2.7	2	1.1	7	2.6	HILL TX	-60	137	-4	75	51

* SIGNIFICANTLY MORE DEATHS THAN EXPECTED

WHITE FEMALE — CANCER OF THE RECTUM — ICD 154 EXCEPT 154.3

1950-1959 DEATHS	1950-1959 RATE	1960-1969 DEATHS	1960-1969 RATE	1970-1979 DEATHS	1970-1979 RATE	COUNTY AND STATE	PERCENT CHANGE 60S/50S	PERCENT CHANGE 70S/60S	PERCENT CHANGE 70S/50S	PERCENTILE %CHANGE 70S/50S	PERCENTILE RATES 1970S
2	2.6	1	1.2			HOCKLEY TX	-52				
				3	3.4	HOOD TX					68
9	6.7	1	.6	3	2.5	HOPKINS TX	-91	318	-63	19	48
1	1.5			2	2.5	HOUSTON TX			69	92	50
1	1.0	5	3.6	7	3.9	HOWARD TX	278	6	302	99	76
				1	7.3	HUDSPETH TX					97
9	4.2	4	1.4	9	2.6	HUNT TX	-66	83	-38	50	51
		4	5.2	4	3.0	HUTCHINSON TX		-42			59
1	13.2					IRION TX					
		2	4.3	2	4.3	JACK TX		0			83
1	1.7	2	4.4			JACKSON TX	154				
3	4.1	3	2.6			JASPER TX	-37				
22	3.4	27	3.2	30	2.7	JEFFERSON TX	-7	-15	-21	66	54
				1	4.1	JIM HOGG TX					80
2	2.9	1	.6	6	3.8	JIM WELLS TX	-78	486	30	85	75
10	5.0	6	2.5	8	2.4	JOHNSON TX	-50	-4	-52	31	46
1	1.0	6	4.3	3	2.3	JONES TX	320	-46	128	95	45
4	7.1	2	2.3	2	1.9	KARNES TX	-67	-18	-73	09	35
5	3.2	5	2.9	6	2.8	KAUFMAN TX	-9	-6	-15	70	54
3	8.3	5	11.4	1	1.4	KENDALL TX	36	-88	-84	02	25
3	2.7	1	.7	10	4.1	KERR TX	-76	520	48	89	79
		1	3.0	1	4.2	KIMBLE TX		39			81
1*	112.4					KING TX					
4	5.8	3	3.7	4	3.7	KLEBERG TX	-37	1	-36	53	74
5	2.4	8	3.0	6	2.2	LAMAR TX	22	-27	-11	72	41
2	4.1	2	2.1			LAMB TX	-49				
2	3.7	3	4.2	1	1.1	LAMPASAS TX	14	-73	-69	12	22
		1	4.0			LA SALLE TX					
7	5.7	2	1.1	7	3.2	LAVACA TX	-81	199	-43	43	65
1	1.9	2	4.2	1	1.3	LEE TX	119	-68	-31	59	24
3	6.0	2	3.5	1	.9	LEON TX	-42	-74	-85	02	19
		2	1.7	5	2.4	LIBERTY TX		41			47
3	2.2	6	4.1	5	2.7	LIMESTONE TX	85	-35	21	83	53
1	4.9	1	4.6			LIPSCOMB TX	-5				
		1	1.8	1	.8	LLANO TX		-58			18
15	4.0	16	2.9	14	1.8	LUBBOCK TX	-27	-37	-54	28	34
1	2.4	2	4.5			LYNN TX	89				
1	1.7	2	2.3	1	.9	MC CULLOCH TX	33	-61	-49	36	19
26	4.7	18	2.3	21	2.2	MC LENNAN TX	-51	-4	-53	30	42
				2	2.8	MADISON TX					55
		1	3.1			MARION TX					
		1	2.5	1	3.5	MASON TX		38			69
2	3.3	2	1.8	3	2.3	MATAGORDA TX	-45	30	-29	61	45
1	2.3			1	.7	MAVERICK TX			-69	13	18
2	2.7	3	3.0	2	1.5	MEDINA TX	12	-50	-44	43	27
				1	3.7	MENARD TX					73
3	4.5	7	3.5	8	3.2	MIDLAND TX	-23	-7	-28	61	65
5	4.0	5	3.9	4	1.9	MILAM TX	-2	-51	-52	31	35
1	2.3	4	8.6	3	4.4	MILLS TX	270	-48	91	93	84
1	1.6	3	5.7	1	1.1	MITCHELL TX	252	-81	-33	57	22
5	4.8	4	2.7	1	.5	MONTAGUE TX	-43	-83	-90	00	17
2	2.6	1	.9	4	1.3	MONTGOMERY TX	-67	51	-50	34	24
1	2.2	2	4.6			MOORE TX	109				
1	2.9	1	1.8	6*	8.0	MORRIS TX	-40	354	173	97	98
		1	4.9			MOTLEY TX					
3	2.4	7	4.6	3	1.5	NACOGDOCHES TX	90	-67	-38	50	28
8	4.1	6	2.6	9	3.1	NAVARRO TX	-36	17	-25	63	62
		3	2.9	4	3.3	NOLAN TX		12			66
22	4.5	19	2.4	29	2.9	NUECES TX	-46	20	-35	53	58
1	3.7					OCHILTREE TX					
		1	12.2			OLDHAM TX					
6	5.2	6	3.3	1	.3	ORANGE TX	-36	-90	-94	00	16
5	4.4	9	7.2	5	2.9	PALO PINTO TX	62	-60	-35	54	58
2	3.2	2	2.3			PANOLA TX	-29				
5	4.3	5	2.7	7	2.6	PARKER TX	-37	-1	-38	51	52
2	3.5	3	4.5	1	1.1	POLK TX	26	-76	-70	12	21
12	3.5	13	3.1	8	1.4	POTTER TX	-10	-56	-61	22	25
1	5.2	1	6.3	1	2.2	RAINS TX	21	-65	-57	25	42
		3	1.8	7	2.7	RANDALL TX		49			52
				1	8.6	REAGAN TX					98
4	4.1	2	1.9	4	2.9	RED RIVER TX	-53	51	-28	61	59
		1	1.9			REEVES TX					
2	6.3			1	1.9	REFUGIO TX			-70	11	35
5	8.0	2	2.8	2	1.4	ROBERTSON TX	-65	-50	-82	03	26
		3	8.5	1	1.2	ROCKWALL TX		-86			23
6	6.4	1	.7			RUNNELS TX	-89				
3	2.3	4	2.0	4	1.6	RUSK TX	-14	-21	-31	58	28
		1	2.5			SAN AUGUSTINE TX					
				1	3.0	SAN JACINTO TX					59
1	.6	4	2.7	3	1.4	SAN PATRICIO TX	388	-47	157	96	27
1	1.7	1	2.3	2	1.9	SAN SABA TX	34	-19	9	80	35
1	3.1	2	2.1	2	2.0	SCURRY TX	-31	-7	-36	53	37
		3	8.9			SHACKELFORD TX					
3	3.0	3	2.5	4	2.5	SHELBY TX	-16	-4	-19	67	48
6	2.0	10	2.5	13	2.2	SMITH TX	26	-11	12	81	42
2	10.3	3	9.9	1	2.4	SOMERVELL TX	-5	-76	-77	06	46
2	4.9	3	4.5			STARR TX	-8				
2	2.7	4	5.2			STEPHENS TX	93				
1	74.3			2*	19.0	STERLING TX			-74	08	99
1	6.4	1	5.7			STONEWALL TX	-11				
		2	11.7			SUTTON TX					
1	3.3	1	2.2	1	1.3	SWISHER TX	-31	-40	-59	24	25
65	4.2	91	3.7	90	2.6	TARRANT TX	-12	-29	-37	51	51
16	5.1	4	.9	13	2.2	TAYLOR TX	-82	134	-58	25	41
1	17.2					TERRELL TX					
1	2.4			3	4.5	TERRY TX			84	93	85
2	2.7	5	4.5			TITUS TX	66				
10	3.7	9	2.7	11	2.4	TOM GREEN TX	-26	-13	-36	53	46
21	3.4	22	2.3	23	1.7	TRAVIS TX	-33	-26	-50	33	31
2	5.2	1	1.9	1	1.2	TRINITY TX	-64	-34	-76	07	23
		1	2.3	1	.9	TYLER TX		-62			19

* SIGNIFICANTLY MORE DEATHS THAN EXPECTED

WHITE FEMALE

CANCER OF THE RECTUM

ICD 154 EXCEPT 154.3

1950-1959 DEATHS	1950-1959 RATE	1960-1969 DEATHS	1960-1969 RATE	1970-1979 DEATHS	1970-1979 RATE	COUNTY AND STATE	PERCENT CHANGE 60S/50S	PERCENT CHANGE 70S/60S	PERCENT CHANGE 70S/50S	PERCENTILE %CHANGE 70S/50S	PERCENTILE RATES 1970S
5	5.7	3	3.9	6	3.9	UPSHUR TX	-32	2	-31	59	78
2	2.6	2	2.1	1	.8	UVALDE TX	-20	-63	-70	12	18
1	1.4	2	2.5	1	.8	VAL VERDE TX	80	-67	-41	46	19
3	2.6	6	4.3	5	2.1	VAN ZANDT TX	68	-50	-17	69	40
2	1.5	2	1.0	13	5.3	VICTORIA TX	-31	418	257	98	91
2	5.0	1	1.5	3	2.5	WALKER TX	-71	75	-49	35	50
				2	2.9	WALLER TX					58
		1	1.2	1	1.7	WARD TX		39			31
6	6.8	3	2.1	9	4.9	WASHINGTON TX	-70	138	-28	61	88
6	2.7	5	1.7	8	2.0	WEBB TX	-38	18	-27	62	37
4	3.3	4	2.9	9	4.7	WHARTON TX	-13	64	43	88	87
1	1.9	3	4.7	3	4.4	WHEELER TX	141	-6	127	95	84
8	2.0	16	2.9	18	2.6	WICHITA TX	47	-9	34	86	52
4	3.9	1	.7	3	1.7	WILBARGER TX	-82	137	-57	25	31
4	9.7	1	1.7			WILLACY TX	-82				
4	2.1	6	2.7	8	2.1	WILLIAMSON TX	30	-24	-2	76	39
3	5.3	3	4.2	2	2.7	WILSON TX	-21	-34	-48	36	54
3	24.0	1	3.3	4*	9.8	WINKLER TX	-86	194	-59	23	98
4	3.8	11*	9.5	4	2.7	WISE TX	153	-72	-28	61	53
3	3.8	3	2.6	4	1.9	WOOD TX	-32	-28	-51	32	34
		2	5.9			YOAKUM TX					
		5	4.1	2	1.1	YOUNG TX		-72			22
112	3.9	98	2.6	105	2.0	UTAH UT	-35	-21	-49		
1	5.7					BEAVER UT					
4	5.8	5	5.2	2	1.4	BOX ELDER UT	-10	-72	-75	07	26
2	1.6	7	3.9	4	1.5	CACHE UT	141	-63	-10	72	27
2	2.2	3	3.5	1	1.4	CARBON UT	63	-61	-36	53	26
1*	112.4					DAGGETT UT					
4	4.0	2	1.1	3	.8	DAVIS UT	-72	-29	-80	04	18
1	3.3	5*	21.8	1	3.0	DUCHESNE UT	559	-86	-10	72	59
		1	3.4			EMERY UT					
1	5.5					GRAND UT					
				2	3.4	IRON UT					68
2	9.2			1	3.6	JUAB UT			-61	21	71
1	11.7					KANE UT					
1	3.2					MILLARD UT					
2	22.3			2*	11.8	MORGAN UT			-47	38	99
1	16.4					PIUTE UT					
59	4.6	51	2.8	60	2.6	SALT LAKE UT	-38	-10	-44	43	50
1	3.8					SAN JUAN UT					
2	3.1	2	2.5	3	3.9	SANPETE UT	-19	55	25	84	77
3	5.7	1	1.9			SEVIER UT	-68				
2	6.5					SUMMIT UT					
4	8.3			3	3.6	TOOELE UT			-57	26	72
10	3.4	10	2.5	9	1.5	UTAH UT	-28	-40	-57	26	27
1	4.7					WASATCH UT					
				1	.7	WASHINGTON UT					18
7	2.0	11	2.3	13	2.0	WEBER UT	13	-14	-2	76	37
151	7.1	187*	7.4	119	3.9	VERMONT VT	4	-47	-45		
6	5.9	10*	9.0	4	2.2	ADDISON VT	53	-75	-62	20	42
15*	10.1	15*	8.5	12	6.0	BENNINGTON VT	-17	-29	-41	47	94
7	4.9	19*	11.7	7	3.9	CALEDONIA VT	138	-67	-21	66	76
29*	9.4	28*	6.9	18	3.8	CHITTENDEN VT	-27	-45	-59	23	76
3	9.7	2	6.1	2	5.3	ESSEX VT	-37	-14	-46	40	90
13	8.3	9	5.0	3	1.1	FRANKLIN VT	-40	-78	-87	01	22
1	5.5	2	13.2	1	4.4	GRAND ISLE VT	142	-67	-20	67	83
2	3.2	2	3.5	4	4.3	LAMOILLE VT	8	25	35	87	83
4	3.8	13*	12.0	4	2.9	ORANGE VT	221	-76	-22	65	58
8	7.2	8	6.3	10*	5.9	OPLEANS VT	-12	-7	-19	68	93
15	5.0	22	6.4	18	4.6	RUTLAND VT	28	-28	-8	74	86
17	7.0	20	7.1	15	4.3	WASHINGTON VT	2	-40	-38	50	82
18	8.3	18*	7.9	13	4.9	WINDHAM VT	-5	-37	-40	47	89
13	5.8	19	6.3	8	2.2	WINDSOR VT	8	-65	-62	20	43
487	4.2	585	3.7	666	3.1	VIRGINIA VA	-12	-17	-27		
4	3.1	7	3.6	8	3.1	ACCOMACK VA	15	-13	0	77	63
11	4.9	9	3.2	7	1.8	ALBEMARLE#VA	-36	-44	-64	18	33
5	4.8	6	4.2	5	2.7	ALLEGHANY#VA	-12	-37	-44	42	52
1	5.9	1	3.2	1	3.1	AMELIA VA	-45	-4	-47	38	62
4	5.5			4	3.0	AMHERST VA			-44	42	61
1	3.1			3	5.1	APPOMATTOX VA			63	91	89
40	7.1	31	3.7	28	2.9	ARLINGTON#VA	-48	-22	-59	23	57
19	5.7	9	2.1	30*	5.9	AUGUSTA#VA	-63	176	3	78	93
				1	2.2	BATH VA					40
17	3.7	15	2.6	30	3.9	BEDFORD#VA	-30	51	5	79	77
1	4.2			1	3.1	BLAND VA			-26	63	62
1	1.3	4	4.7	5	4.8	BOTETOURT VA	253	2	261	98	87
2	4.8			4	7.6	BRUNSWICK VA			57	90	97
3	2.2	1	.8	4	4.4	BUCHANAN VA	-62	443	105	94	84
1	2.7	2	4.9			BUCKINGHAM VA	85				
3	8.5	3	8.8	2	4.3	CAROLINE VA	3	-50	-49	35	83
8	3.8	14	5.9	5	1.7	CARROLL#VA	54	-72	-56	26	30
		2	4.0			CHARLOTTE VA					
47	4.1	40	2.8	44	2.6	CHESTERFIELD#VA	-32	-6	-36	53	51
1	3.0	2	6.3	1	2.0	CLARKE VA	108	-69	-35	54	36
2	9.6	2	7.8			CRAIG VA	-19				
2	3.1	3	3.6	3	2.6	CULPEPER VA	15	-28	-18	68	50
1	2.4	1	1.5	1	1.1	DICKENSON VA	-36	-26	-52	31	22
11	5.1	10	3.6	4	1.0	DINWIDDIE#VA	-30	-73	-81	04	20
		1	3.8	3	7.8	ESSEX VA		103			97
32	5.7	71	6.2	60	3.0	FAIPFAX#VA	9	-52	-48	36	59
2	2.4	4	3.8	1	.9	FAUQUIER VA	59	-77	-63	19	19
1	1.7	1	1.5	4	4.3	FLOYD VA	-15	197	151	96	83
2	7.6	1	3.1	2	5.1	FLUVANNA VA	-59	63	-33	57	89
2	2.0	4	3.6	2	1.2	FRANKLIN VA	82	-67	-40	48	23
6	3.7	14	7.0	13	4.3	FREDERICK#VA	86	-39	14	81	82

* SIGNIFICANTLY MORE DEATHS THAN EXPECTED

COUNTIES COMBINED; SEE TECHNICAL NOTES

WHITE FEMALE — CANCER OF THE RECTUM — ICD 154 EXCEPT 154.3

1950-1959 DEATHS	1950-1959 RATE	1960-1969 DEATHS	1960-1969 RATE	1970-1979 DEATHS	1970-1979 RATE	COUNTY AND STATE	PERCENT CHANGE 60S/50S	PERCENT CHANGE 70S/60S	PERCENT CHANGE 70S/50S	PERCENTILE %CHANGE 70S/50S	PERCENTILE RATES 1970S
2	3.9	4	6.0	2	1.8	GILES VA	55	-69	-52	30	34
2	5.2			1	1.3	GLOUCESTER VA			-75	07	24
1	3.7			1	2.5	GOOCHLAND VA			-34	56	48
		1	4.9			GREENE VA					
2	8.2	1	2.8			GREENSVILLE#VA	-66				
5	5.3	4	3.1	8	5.0	HALIFAX#VA	-41	60	-5	75	89
3	3.9	3	2.2	6	3.2	HANOVER VA	-44	47	-18	68	64
5	1.5	9	2.0	25	3.0	HENRICO VA	38	45	100	94	59
1	.9	9	4.6	9	3.6	HENRY#VA	426	-23	307	99	71
2	9.0			2	8.8	HIGHLAND VA			-3	76	98
		2	4.0	5	7.5	ISLE OF WIGHT VA		89			97
8	7.1	4	2.2	6	2.4	JAMES CITY#VA	-69	11	-66	16	48
		1	3.8			KING GEORGE VA					
		1	4.3	2	6.2	KING WILLIAM VA		45			95
3	10.8	1	1.6			LANCASTER VA	-85				
7	6.2	4	3.2	5	2.8	LEE VA	-48	-12	-54	28	56
6	6.1	2	1.5	3	1.8	LOUDOUN VA	-75	20	-70	11	33
2	5.4	1	1.7	4	5.0	LOUISA VA	-68	188	-7	74	89
1	2.0	1	1.9			LUNENBURG VA	-6				
2	6.1	3	7.8			MADISON VA	27				
		3	6.7	5	7.3	MATHEWS VA		9			97
		4	3.9	7	5.7	MECKLENBURG VA		44			92
		1	3.6	1	2.0	MIDDLESEX VA		-44			37
7	5.1	6	3.2	9	3.5	MONTGOMERY#VA	-37	10	-31	59	70
1	2.1	4	7.4			NELSON VA	245				
		2	3.2	1	1.2	NORTHAMPTON VA		-63			23
1	2.3	2	3.9	1	1.2	NORTHUMBERLAND VA	71	-69	-46	39	23
1	2.1	3	3.2	2	2.5	NOTTOWAY VA	54	-23	20	83	48
				3	2.8	ORANGE VA					55
2	2.9	5	5.4	4	3.3	PAGE VA	84	-39	12	81	65
1	1.6	1	1.6	5	4.2	PATRICK VA	1	168	170	97	81
10	3.0	18	4.4	15	2.6	PITTSYLVANIA#VA	50	-40	-10	72	52
2	6.2			4	6.7	PRINCE EDWARD VA			8	80	96
6	7.2	4	2.6	9	3.7	PRINCE WILLIAM#VA	-63	41	-48	36	74
3	3.3	4	3.4	4	2.3	PULASKI VA	3	-33	-31	59	44
3	14.7	1	3.4	2	6.1	RAPPAHANNOCK VA	-77	82	-58	24	95
1	6.4	1	3.7	1	1.9	RICHMOND VA	-42	-49	-70	11	35
24	3.8	40	4.6	40	3.2	ROANOKE#VA	23	-29	-14	71	65
6	5.2	5	3.6	11*	6.8	ROCKBRIDGE#VA	-30	88	31	86	96
14	5.9	16	5.3	12	2.9	ROCKINGHAM#VA	-9	-45	-50	34	59
		5	4.9	2	1.3	RUSSELL VA		-74			24
3	2.9	5	3.6	5	3.2	SCOTT VA	25	-12	10	80	63
9	7.4	6	4.0	4	2.1	SHENANDOAH VA	-46	-47	-71	10	40
3	3.3	2	1.1	4	1.8	SMYTH VA	-66	58	-46	40	32
6	10.8	4	6.2	3	4.0	SOUTHAMPTON#VA	-43	-36	-63	19	78
5	5.6	7	5.5	4	2.1	SPOTSYLVANIA#VA	-2	-62	-63	19	39
2	4.3	2	2.5	3	3.4	STAFFORD VA	-43	35	-22	65	67
1	7.6	1	5.2			SURRY VA	-31				
				4*	12.6	SUSSEX VA					99
6	4.5	7	4.1	16*	6.8	TAZEWELL VA	-8	66	52	89	96
6	11.5	3	4.1	5	4.4	WARREN VA	-64	6	-62	20	83
10	5.0	15	5.5	12	3.0	WASHINGTON#VA	11	-46	-40	48	60
1	2.7	2	3.8	2	2.5	WESTMORELAND VA	41	-34	-7	74	48
9	5.8	7	3.3	9	3.5	WISE#VA	-44	8	-40	48	70
4	3.7	3	2.4	5	3.3	WYTHE VA	-35	35	-13	71	65
47	4.1	53	3.3	69	3.1	CHESAPEAKE CITY#VA	-20	-6	-25	63	62
4	2.4	12	4.5	8	2.4	HAMPTON CITY VA	85	-46	0	77	47
4	1.8	15	5.0	18	4.2	NEWPORT NEWS CITY VA	174	-17	128	95	81
3	3.4	7	5.7	2	1.5	SUFFOLK CITY VA	69	-73	-55	27	28
677	5.7	590	3.7	667	3.1	WASHINGTON WA	-35	-15	-44		
1	4.3	1	2.8			ADAMS WA	-36				
3	4.2	3	3.4	7	6.5	ASOTIN WA	-19	93	56	90	96
4	2.4	6	2.4	10	2.7	BENTON WA	-3	16	13	81	54
7	3.5	6	2.6	8	2.8	CHELAN WA	-26	6	-22	65	54
1	.8	8	5.2	8	2.6	CLALLAM WA	575	-50	237	98	51
23	5.9	17	3.0	25	3.1	CLARK WA	-49	2	-48	36	62
5*	18.6	1	2.2	3	7.9	COLUMBIA WA	-88	259	-58	24	97
7	3.0	19	6.0	13	3.4	COWLITZ WA	97	-44	11	80	67
				1	.9	DOUGLAS WA					19
3	6.5	3	3.9	3	2.5	FRANKLIN WA	-41	-37	-62	20	48
1	4.7	1	7.1	1	2.3	GARFIELD WA	50	-67	-51	32	45
2	1.2	1	.9	2	1.0	GRANT WA	-22	11	-13	71	21
14	5.8	20	6.4	19	5.0	GRAYS HARBOR WA	10	-23	-15	70	88
3	4.3	4	4.3	6	3.4	ISLAND WA	1	-21	-20	67	68
				2	2.1	JEFFERSON WA					40
261	6.7	209	3.9	200	2.9	KING WA	-41	-27	-57	25	57
26	7.7	12	2.5	26	3.9	KITSAP WA	-67	57	-49	36	78
8	8.5	4	3.8	4	3.3	KITTITAS WA	-56	-13	-62	21	66
4	7.1	2	3.1	3	3.9	KLICKITAT WA	-57	28	-44	42	78
12	4.8	9	2.8	7	1.9	LEWIS WA	-42	-32	-61	22	35
1	1.9	3	5.2			LINCOLN WA	168				
2	3.4	2	2.4	4	2.4	MASON WA	-28	-2	-29	60	47
4	3.8	5	3.7	7	4.3	OKANOGAN WA	-1	15	14	81	82
5	6.0	3	2.8	4	3.1	PACIFIC WA	-54	12	-48	36	63
1	2.7			2	4.3	PEND OREILLE WA			59	90	83
72	5.3	53	3.1	78	3.5	PIERCE WA	-42	14	-34	56	70
3	14.5	2	6.4	3	6.3	SAN JUAN WA	-56	-1	-57	25	95
13	5.3	13	4.3	12	2.7	SKAGIT WA	-18	-36	-48	37	54
1	4.6	1	4.3			SKAMANIA WA	-7				
34	5.2	36	4.0	39	2.8	SNOHOMISH WA	-24	-31	-47	38	55
74	5.8	69	4.2	77	3.8	SPOKANE WA	-27	-11	-34	55	75
2	2.5	2	2.1	2	1.8	STEVENS WA	-17	-15	-29	60	32
16	6.5	18	5.6	19	3.3	THURSTON WA	-14	-42	-50	33	66
11	5.3	7	2.6	18*	4.8	WALLA WALLA WA	-51	84	-10	72	87
19	4.7	16	3.1	20	3.4	WHATCOM WA	-34	9	-29	61	67
7	5.1	3	1.7	6	3.3	WHITMAN WA	-66	92	-34	55	67
27	4.6	31	4.1	28	2.9	YAKIMA WA	-10	-29	-37	52	58

* SIGNIFICANTLY MORE DEATHS THAN EXPECTED

COUNTIES COMBINED; SEE TECHNICAL NOTES

WHITE FEMALE — CANCER OF THE RECTUM — ICD 154 EXCEPT 154.3

1950-1959 DEATHS	1950-1959 RATE	1960-1969 DEATHS	1960-1969 RATE	1970-1979 DEATHS	1970-1979 RATE	COUNTY AND STATE	PERCENT CHANGE 60S/50S	PERCENT CHANGE 70S/60S	PERCENT CHANGE 70S/50S	PERCENTILE %CHANGE 70S/50S	PERCENTILE RATES 1970S
338	4.4	387	4.2	379	3.4	WEST VIRGINIA WV	-6	-19	-24		
2	2.5	5	5.0	5	4.2	BARBOUR WV	100	-18	65	91	81
9	5.8	13	7.2	18*	8.0	BERKELEY WV	23	11	37	87	98
2	2.2	2	1.9	4	2.9	BOONE WV	-14	56	34	87	58
4	5.0	6	7.4	1	1.5	BRAXTON WV	48	-80	-70	12	28
4	4.0	8	5.9	2	1.1	BROOKE WV	49	-81	-72	10	22
22	4.3	20	3.1	29	3.9	CABELL WV	-28	26	-10	73	77
1	2.4			2	3.5	CALHOUN WV			49	89	70
2	3.5			2	3.7	CLAY WV			4	78	73
3	7.1	1	1.9	4	6.6	DODDRIDGE WV	-73	250	-7	74	96
6	2.9	12	4.7	11	3.2	FAYETTE WV	65	-32	13	81	65
2	4.8	2	4.7			GILMER WV	-3				
		1	2.1	2	3.7	GRANT WV		78			73
1	.8	6	3.3	7	3.0	GREENBRIER WV	329	-7	300	99	61
4	6.9	1	1.9	1	1.2	HAMPSHIRE WV	-73	-33	-82	03	23
10	7.7	13*	7.7	4	2.1	HANCOCK WV	0	-73	-73	09	38
4	10.1	2	4.4	4	6.6	HARDY WV	-56	50	-34	55	96
23	5.6	26	5.7	32*	5.3	HARRISON WV	2	-6	-4	75	91
		4	4.1	4	2.9	JACKSON WV		-30			57
2	2.6	1	1.1	3	2.1	JEFFERSON WV	-57	88	-20	67	39
41	5.2	39	3.6	36	2.7	KANAWHA WV	-30	-26	-48	36	52
9	6.1	7	4.7	4	2.8	LEWIS WV	-22	-40	-53	29	56
3	3.6	2	2.3			LINCOLN WV	-35				
3	2.3	6	3.6	4	1.9	LOGAN WV	55	-46	-16	69	36
8	6.6	6	3.9	9	4.3	MC DOWELL WV	-41	9	-35	54	82
17	5.3	28*	7.2	15	3.4	MARION WV	36	-53	-36	53	68
5	3.0	8	3.9	5	2.2	MARSHALL WV	32	-44	-26	62	41
1	.9	7	5.9	3	2.1	MASON WV	528	-64	124	95	39
10	3.8	3	.9	13	3.2	MERCER WV	-76	257	-16	69	63
3	3.6	6	5.0	12*	7.9	MINERAL WV	41	57	121	95	98
7	5.5	8	5.7	4	2.6	MINGO WV	4	-55	-53	29	50
14	5.8	10	3.5	7	1.9	MONONGALIA WV	-40	-44	-67	15	36
5	8.7	2	2.5	4	3.7	MONROE WV	-72	48	-58	24	73
3	7.4	2	3.6	6*	10.2	MORGAN WV	-51	179	37	87	99
4	4.2	4	3.5	4	3.1	NICHOLAS WV	-16	-11	-26	63	62
22	5.5	27	5.8	26*	5.4	OHIO WV	6	-7	-1	76	91
1	2.3	1	2.0	4	6.4	PENDLETON WV	-12	226	185	97	95
1	2.8	5*	12.2			PLEASANTS WV	340				
4	7.8	3	4.8	1	1.1	POCAHONTAS WV	-39	-76	-85	02	22
5	3.8	8	5.1	7	3.6	PRESTON WV	35	-29	-4	75	72
1	1.2	5	4.6	5	3.2	PUTNAM WV	294	-30	177	97	64
8	3.9	7	2.2	18	4.1	RALEIGH WV	-45	88	4	78	80
3	2.5	9	6.4	4	2.1	RANDOLPH WV	157	-68	-17	68	39
		7	8.6	2	3.3	RITCHIE WV		-61			67
2	2.7	2	2.2	6	5.0	ROANE WV	-19	130	86	93	89
7	9.2	2	2.0	3	3.2	SUMMERS WV	-78	54	-66	16	63
3	3.1	8	7.7	4	3.1	TAYLOR WV	146	-60	-1	77	62
4	8.7			4	5.3	TUCKER WV			-39	49	91
6	9.3	3	3.5	4	5.4	TYLER WV	-62	55	-41	46	92
4	3.7	3	2.7	2	1.3	UPSHUR WV	-28	-52	-65	17	24
4	2.4	2	1.1	4	1.8	WAYNE WV	-56	70	-25	63	34
6	10.9	3	4.9	1	1.5	WEBSTER WV	-55	-69	-86	01	27
6	6.5	6	5.1	3	2.7	WETZEL WV	-22	-48	-59	23	52
1	3.9	1	4.1	1	3.1	WIRT WV	5	-24	-21	66	62
14	3.6	23	5.2	13	2.4	WOOD WV	43	-54	-35	54	45
2	1.7	1	1.1	6	4.4	WYOMING WV	-36	308	162	96	84
1239*	6.8	1132*	5.0	1008	3.6	WISCONSIN WI	-27	-28	-47		
1	1.8	1	1.5	2	1.6	ADAMS WI	-18	8	-12	72	29
6	6.1	4	3.1	9*	7.9	ASHLAND WI	-50	156	28	85	97
9	5.2	12	5.2	8	2.9	BARRON WI	0	-44	-44	43	59
6	8.5	2	2.6	5	5.3	BAYFIELD WI	-70	106	-37	51	91
29	6.0	50*	7.8	28	3.4	BROWN WI	31	-57	-44	43	68
8	10.3	5	7.2	3	3.0	BUFFALO WI	-30	-58	-71	11	60
4	7.8	5	7.2	3	2.6	BURNETT WI	-8	-64	-67	15	51
12*	13.2	6	5.5	14*	9.6	CALUMET WI	-58	75	-27	62	98
9	4.4	16	6.4	11	3.6	CHIPPEWA WI	44	-43	-18	68	72
9	5.5	10	5.2	10	4.4	CLARK WI	-5	-15	-20	67	84
17	7.5	14	5.3	10	3.2	COLUMBIA WI	-30	-40	-58	24	64
5	6.3	1	1.4			CRAWFORD WI	-79				
41	4.8	55	4.9	39	2.5	DANE WI	1	-48	-47	38	50
23	7.6	17	4.5	16	3.1	DODGE WI	-40	-30	-58	24	63
10	9.2	4	3.2	3	1.6	DOOR WI	-65	-49	-83	03	29
16	6.9	16	5.9	15	4.8	DOUGLAS WI	-14	-18	-30	59	87
6	4.3	3	1.7	4	1.6	DUNN WI	-61	-7	-64	19	29
17	5.9	21	6.1	16	3.8	EAU CLAIRE WI	4	-38	-35	54	75
1	7.3	1	5.3			FLORENCE WI	-28				
29	6.7	32	6.8	31*	5.6	FOND DU LAC WI	0	-16	-16	69	92
1	3.3	2	4.6	3	5.4	FOREST WI	39	17	62	91	91
12	5.3	8	2.8	5	1.6	GRANT WI	-48	-42	-70	12	29
9	5.9	3	1.8	8	3.5	GREEN WI	-69	96	-40	48	70
7	7.6	8	6.7	6	4.1	GREEN LAKE WI	-12	-39	-47	39	80
7	6.2	8	5.9	4	2.8	IOWA WI	-5	-52	-54	28	56
		4	7.8	2	2.8	IRON WI		-65			55
5	5.8	4	4.3	2	2.9	JACKSON WI	-26	-34	-51	33	57
19	6.6	11	3.1	6	1.3	JEFFERSON WI	-53	-57	-80	04	24
4	4.0	6	4.7	6	3.8	JUNEAU WI	18	-21	-6	74	74
32	8.4	25	5.0	29	4.2	KENOSHA WI	-41	-15	-50	33	81
4	4.5	5	4.2	6	4.1	KEWAUNEE WI	-6	-3	-9	73	80
24	6.6	18	3.8	28	4.9	LA CROSSE WI	-42	29	-25	63	88
11*	10.7	6	5.4	1	1.0	LAFAYETTE WI	-50	-82	-91	00	21
8	7.5	8	6.5	7	4.3	LANGLADE WI	-14	-33	-43	45	83
14*	11.9	9	6.2	8	4.2	LINCOLN WI	-48	-32	-64	18	81
20	5.6	16	3.7	23	4.5	MANITOWOC WI	-35	22	-20	67	84
20	5.5	18	3.7	30*	5.1	MARATHON WI	-32	36	-7	74	89
16	8.7	13	5.8	12	3.9	MARINETTE WI	-34	-32	-55	27	77
3	4.3	4	5.3	2	3.2	MARQUETTE WI	25	-41	-26	63	64
19	6.1	12	3.7	11	2.5	MENOMINEE#WI	-39	-34	-60	22	48

* SIGNIFICANTLY MORE DEATHS THAN EXPECTED

COUNTIES COMBINED; SEE TECHNICAL NOTES

WHITE FEMALE CANCER OF THE RECTUM ICD 154 EXCEPT 154.3

1950-1959 DEATHS	1950-1959 RATE	1960-1969 DEATHS	1960-1969 RATE	1970-1979 DEATHS	1970-1979 RATE	COUNTY AND STATE	PERCENT CHANGE 60S/50S	PERCENT CHANGE 70S/60S	PERCENT CHANGE 70S/50S	PERCENTILE %CHANGE 70S/50S	PERCENTILE RATES 1970S
382*	8.4	317*	5.7	235	3.6	MILWAUKEE WI	-32	-36	-57	26	72
7	4.1	9	4.2	5	2.2	MONROE WI	4	-48	-46	40	42
5	5.0	4	3.1	12*	5.7	ONEIDA WI	-39	87	14	82	93
25	6.1	19	3.9	23	3.6	OUTAGAMIE WI	-37	-6	-41	46	72
15*	10.9	6	3.4	7	2.4	OZAUKEE WI	-69	-29	-78	05	47
1	1.9	1	3.2			PEPIN WI	71				
8	6.4	2	1.2	3	2.2	PIERCE WI	-81	80	-65	17	42
7	5.5	3	1.8	5	1.7	POLK WI	-68	-5	-69	12	31
10	5.6	14	6.9	11	4.3	PORTAGE WI	22	-38	-24	64	82
4	4.9	3	3.5	4	3.4	PRICE WI	-30	-1	-31	59	68
53*	9.2	41	5.4	32	3.4	RACINE WI	-41	-38	-63	19	68
4	4.2	4	3.7	4	2.4	RICHLAND WI	-11	-36	-43	44	46
37	7.0	36	5.5	29	3.7	ROCK WI	-21	-33	-47	38	73
3	4.6	2	2.5	3	2.7	RUSK WI	-45	4	-42	45	52
5	3.5	7	4.0	4	1.7	ST CROIX WI	14	-57	-51	32	32
7	3.4	12	4.7	10	3.0	SAUK WI	38	-36	-12	72	60
1	1.9	3	5.3	3	3.2	SAWYER WI	185	-40	71	92	64
33	7.2	32	5.5	41*	5.7	SHEBOYGAN WI	-24	4	-21	66	92
1	1.3	4	4.0	5	4.7	TAYLOR WI	209	18	265	98	86
7	4.7	6	4.5	10	5.3	TREMPEALEAU WI	-4	18	13	81	90
8	5.4	11	6.5	6	2.2	VERNON WI	20	-66	-59	23	42
4	7.6	2	3.1	8*	6.1	VILAS WI	-59	96	-19	67	95
24	8.8	18	5.3	13	2.7	WALWORTH WI	-40	-48	-69	12	53
1	2.2	3	5.2	3	3.3	WASHBURN WI	138	-36	52	89	66
9	4.8	9	3.6	8	2.3	WASHINGTON WI	-25	-37	-53	30	44
21	4.3	41	5.4	27	2.4	WAUKESHA WI	26	-55	-43	43	47
8	3.8	10	3.4	11	3.0	WAUPACA WI	-12	-11	-22	65	60
2	2.0	5	5.6	4	3.0	WAUSHARA WI	180	-46	52	89	61
42	8.0	27	4.0	32	4.0	WINNEBAGO WI	-51	0	-51	32	78
12	5.2	18	5.7	24*	6.0	WOOD WI	8	6	15	82	94
39	3.7	36	2.5	51	2.7	WYOMING WY	-33	9	-27		
4	6.4	2	2.2	2	2.1	ALBANY WY	-65	-7	-67	15	39
1	2.7			1	1.8	BIG HORN WY			-33	56	33
1	4.1	2	6.5	1	2.1	CAMPBELL WY	59	-68	-49	35	39
1	1.1			3	3.9	CARBON WY			260	98	77
				1	2.8	CONVERSE WY					56
		2	9.9			CROOK WY					
1	1.7	2	2.4	4	3.4	FREMONT WY	40	39	95	93	67
				3	3.9	GOSHEN WY					77
		4	11.4			HOT SPRINGS WY					
		1	2.1	2	4.7	JOHNSON WY		121			86
8	4.9	7	3.0	5	1.6	LARAMIE WY	-39	-46	-67	15	29
1	3.1					LINCOLN WY					
8	5.9	3	1.4	9	3.8	NATRONA WY	-76	166	-35	54	75
2	27.3					NIOBRARA WY					
4	6.8			6	4.5	PARK-TETON#WY			-34	56	85
1	2.5	1	2.5			PLATTE WY	-2				
4	4.1	3	2.8	7	4.0	SHERIDAN WY	-31	43	-1	76	79
2	2.9	4	5.1	4	3.7	SWEETWATER WY	76	-28	27	85	73
		2	5.5	2	5.2	UINTA WY		-6			90
		2	7.5	1	2.7	WASHAKIE WY		-64			53

NON-WHITE MALE

CANCER OF THE RECTUM

ICD 154 EXCEPT 154.3

1950-1959 DEATHS	1950-1959 RATE	1960-1969 DEATHS	1960-1969 RATE	1970-1979 DEATHS	1970-1979 RATE	COUNTY AND STATE	PERCENT CHANGE 60S/50S	PERCENT CHANGE 70S/60S	PERCENT CHANGE 70S/50S	PERCENTILE %CHANGE 70S/50S	PERCENTILE RATES 1970S
3637	6.1	4320	5.8	4742	5.0	UNITED STATES	-5	-15	-19		
143	4.3	120	3.4	127	3.4	ALABAMA AL	-21	0	-21		
3	10.5			2	6.6	AUTAUGA AL			-37	34	86
4	19.8	1	2.7	2	4.7	BALDWIN AL	-87	77	-76	05	81
2	4.4			2	4.4	BARBOUR AL			1	58	80
1	6.5			1	7.4	BIBB AL			15	65	88
		1	2.5	1	2.5	BULLOCK AL		1			74
1	2.8	2	5.3	1	2.8	BUTLER AL	88	-46	2	59	75
				3	5.9	CALHOUN AL					85
		5	9.7	3	6.6	CHAMBERS AL		-32			86
				1	7.0	CHILTON AL					87
2	6.0					CHOCTAW AL					
1	2.0	1	2.0	3	6.4	CLARKE AL	-3	227	216	94	86
		1	4.8			COFFEE AL					
2	6.7			5*	14.4	COLBERT AL			114	86	95
1	3.0			4	11.5	CONECUH AL			284	97	93
1	9.6					COOSA AL					
				1	6.0	COVINGTON AL					85
		1	6.6	1	5.6	CRENSHAW AL		-16			84
		1	6.3			DALE AL					
2	1.8	2	1.7	4	3.9	DALLAS AL	-7	134	118	87	78
1*	30.3	1*	29.8	1*	22.9	DE KALB AL	-2	-23	-24	45	97
2	4.8	3	8.9			ELMORE AL	85				
1	2.6	1	2.1	1	2.5	ESCAMBIA AL	-19	20	-3	56	74
2	3.5	2	3.3			ETOWAH AL	-7				
1	11.6			1	8.2	FAYETTE AL			-30	42	90
		1	18.7			FRANKLIN AL					
				1	5.9	GENEVA AL					85
2	4.1	1	2.9			GREENE AL	-28				
1	1.5	1	1.9	1	2.0	HALE AL	31	3	35	72	73
1	5.2			1	3.0	HENRY AL			-43	29	75
		3	6.5	1	1.6	HOUSTON AL		-76			72
				1	7.4	JACKSON AL					88
50	6.5	32	3.8	40	4.5	JEFFERSON AL	-42	18	-32	39	80
		1	10.7			LAMAR AL					
3	9.8	2	6.2	2	7.6	LAUDERDALE AL	-36	22	-22	46	89
2	3.8	2	3.4	3	4.7	LEE AL	-9	36	24	69	81
		1	3.4			LIMESTONE AL					
3	3.5	3	2.9			MACON AL	-18				
2	2.9	3	3.9	1	1.1	MADISON AL	32	-73	-64	13	71
5	6.7	7	10.6	2	2.4	MARENGO AL	58	-77	-64	13	74
14	5.4	18	5.1	13	3.3	MOBILE AL	-6	-35	-38	33	77
1	2.7					MONROE AL					
8	3.3	6	2.7	11	4.6	MONTGOMERY AL	-17	70	41	74	81
3	9.8			1	3.5	MORGAN AL			-64	14	77
		2	5.7			PICKENS AL					
2	4.3					PIKE AL					
3	7.1	3	4.2	5	6.1	RUSSELL AL	-40	44	-14	52	85
1	4.5			1	3.2	SHELBY AL			-30	41	76
3	4.5	3	4.1			SUMTER AL	-9				
2	3.1	2	3.6	1	1.6	TALLADEGA AL	17	-56	-48	26	72
1	2.9	2	6.7			TALLAPOOSA AL	128				
3	4.3	2	2.4	4	3.1	TUSCALOOSA AL	-45	31	-28	43	76
3	10.5	2	7.7	1	3.4	WALKER AL	-26	-55	-67	10	77
2	13.1	1	6.4			WASHINGTON AL	-51				
1	1.4					WILCOX AL					
		7	6.0	14	7.5	ALASKA AK		24			
13	3.6	13	3.1	24	3.0	ARIZONA AZ	-12	-4	-15		
1	1.9	1	1.7	1	1.4	APACHE AZ	-8	-19	-25	44	71
		1	23.6			COCHISE AZ					
		1	2.2	1	2.1	COCONINO AZ		-4			73
7	7.2	5	3.4	10	3.9	MARICOPA AZ	-52	13	-46	27	78
		1	1.8	1	1.1	NAVAJO AZ		-37			71
2	3.7	1	1.5	11	7.2	PIMA AZ	-59	368	92	83	88
1	3.7	3	9.5			PINAL AZ	157				
2	13.8					YUMA AZ					
58	3.3	69	3.8	66	3.7	ARKANSAS AR	15	-2	13		
		3	10.7			ARKANSAS AR					
1	2.7	1	2.6	1	2.1	ASHLEY AR	-7	-17	-22	46	73
		1	3.8	1	3.4	BRADLEY AR		-11			77
2	15.6					CALHOUN AR					
4	5.7	4	5.1	2	2.6	CHICOT AR	-11	-50	-56	20	74
				1	3.7	CLARK AR					78
				1	9.1	CLEVELAND AR					91
2	4.8	2	4.0	1	3.2	COLUMBIA AR	-17	-19	-33	39	76
		1	5.7			CONWAY AR					
1	18.4					CRAIGHEAD AR					
5	4.0	2	2.0	4	4.3	CRITTENDEN AR	-51	121	9	62	80
				1	2.7	CROSS AR					75
2	11.3					DALLAS AR					
2	4.3	2	5.6	3	5.4	DESHA AR	31	-5	25	69	83
1	5.5			2	6.9	DREW AR			26	70	87
		2	19.4			FAULKNER AR					
2	7.3	1	3.7	3	10.1	GARLAND AR	-50	175	38	73	92
2	4.1			2	3.3	HEMPSTEAD AR			-21	47	77
				1	7.0	HOWARD AR					87
		1*	23.6			INDEPENDENCE AR					
1	12.4	1	5.9	1	4.6	JACKSON AR	-52	-21	-62	15	81
4	2.7	10	6.3	11	6.4	JEFFERSON AR	135	2	140	89	86
		2	3.5	4	12.0	LEE AR		237			93
2	6.2	3	9.1			LINCOLN AR	46				

* SIGNIFICANTLY MORE DEATHS THAN EXPECTED

NON-WHITE MALE — CANCER OF THE RECTUM — ICD 154 EXCEPT 154.3

COUNTY AND STATE	1950-1959 DEATHS	1950-1959 RATE	1960-1969 DEATHS	1960-1969 RATE	1970-1979 DEATHS	1970-1979 RATE	PERCENT CHANGE 60S/50S	PERCENT CHANGE 70S/60S	PERCENT CHANGE 70S/50S	PERCENTILE %CHANGE 70S/50S	PERCENTILE RATES 1970S
LITTLE RIVER AR			1	4.8							
LONOKE AR	1	3.1	1	3.3			8				
MILLER AR			1	2.8	3	7.9		183			89
MISSISSIPPI AR	1	1.1	3	3.9			255				
MONROE AR	1	1.9			1	2.0			5	60	73
NEVADA AR	1	4.2	1	3.5			-16				
OUACHITA AR	1	2.6	2	3.5	3	5.3	35	52	106	85	83
PHILLIPS AR	3	2.3	8	6.0	3	2.9	162	-52	25	69	75
POINSETT AR	1	7.9			1	8.2			4	59	90
PULASKI AR	12	5.9	12	5.2	6	2.6	-13	-50	-57	19	74
ST FRANCIS AR	2	2.2	2	2.3	4	5.4	5	137	149	90	83
SALINE AR			1	6.1							
SEVIER AR	1	14.6									
UNION AR	3	6.1	1	1.7	5	6.9	-73	315	13	64	87
WOODRUFF AR					1	3.3					77
CALIFORNIA CA	218	7.5	319	6.5	415	4.5	-13	-31	-40		
ALAMEDA CA	27	9.3	28	5.5	45	5.3	-41	-5	-43	29	83
ALPINE CA					1*	5.0					82
BUTTE CA	2*	31.7	2	15.6	1	4.2	-51	-73	-87	01	79
CONTRA COSTA CA	4	4.3	3	3.7	7	4.3	-13	16	1	57	80
FRESNO CA	9	8.2	5	3.8	8	3.7	-54	-1	-55	21	78
GLENN CA			1*	67.0							
HUMBOLDT CA	2*	21.3									
IMPERIAL CA	2	5.4	1	4.4	2	3.1	-18	-30	-43	30	76
INYO CA					1	37.2					98
KERN CA	5	5.9	5	4.7	9	5.3	-20	11	-11	53	83
KINGS CA	1	7.3			1	1.5			-79	03	72
LOS ANGELES CA	96	7.9	156*	7.5	194	5.1	-4	-33	-36	36	82
MARIN CA			1	13.2	2	12.7		-4			94
MERCED CA	1	10.2			1	2.8			-72	08	75
MONTEREY CA	3	3.6	6	11.5	8	5.5	224	-52	56	76	84
ORANGE CA	1	5.3	1	.8	5	2.3	-86	201	-57	18	73
PLACER CA			1	6.9	2	10.3		50			92
RIVERSIDE CA	1	2.8	1	1.3	5	2.1	-55	65	-25	44	73
SACRAMENTO CA	4	4.0	10	7.7	16	5.3	92	-31	33	72	83
SAN BENITO CA			1*	56.2							
SAN BERNARDINO CA	2	2.6	4	5.4	6	2.7	107	-51	1	58	75
SAN DIEGO CA	6	9.3	7	5.4	9	2.6	-41	-52	-72	08	74
SAN FRANCISCO CA	30	7.7	51*	8.3	52	5.5	9	-34	-29	42	83
SAN JOAQUIN CA	4	3.1	13	7.5	13	5.6	141	-25	80	81	84
SAN MATEO CA			3	5.4	7	4.3		-19			80
SANTA BARBARA CA			4	14.2	1	1.5		-90			72
SANTA CLARA CA	6	12.0	4	4.9	8	2.5	-59	-49	-79	04	74
SANTA CRUZ CA			3	14.8							
SHASTA CA	1	26.0									
SOLANO CA	1	1.6	2	3.3	3	3.0	112	-11	89	83	75
SONOMA CA	3*	42.5	2	14.0	1	4.1	-67	-71	-90	00	79
STANISLAUS CA	1	17.9			1	1.9			-89	01	72
SUTTER CA	1	18.6	1	12.4			-33				
TULARE CA	5	17.4	2	5.9	3	3.1	-66	-47	-82	02	76
YOLO CA					1	3.0					75
YUBA CA			1	20.2	2	9.8		-51			92
COLORADO CO	13	8.7	10	5.5	7	2.0	-36	-64	-77		
ADAMS CO			1	23.6							
DENVER-ARAPAHOE#CO	12*	12.6	7	5.9	7	3.3	-53	-44	-73	07	77
PUEBLO CO	1	6.6	1	6.2			-6				
WELD CO			1	14.7							
CONNECTICUT CT	23*	13.1	19	6.4	28	5.5	-51	-15	-58		
FAIRFIELD CT	4	6.6	5	6.4	6	4.1	-2	-36	-37	34	79
HARTFORD CT	6	12.9	6	6.5	6	4.6	-50	-28	-64	13	81
MIDDLESEX CT					1	11.6					93
NEW HAVEN CT	10*	18.3	6	6.1	12	7.5	-67	24	-59	16	89
NEW LONDON CT	1	9.2	2	16.1	2	5.9	74	-63	-36	36	85
TOLLAND CT	1*	112.4			1*	28.7			-74	07	97
WINDHAM CT	1*	86.0									
DELAWARE DE	14	6.4	20	8.7	16	5.2	35	-40	-19		
KENT DE			1	2.6	4	8.8		235			90
NEW CASTLE DE	12*	9.2	11	6.9	7	3.9	-24	-44	-57	18	78
SUSSEX DE	2	4.2	8*	17.7	5	6.6	327	-63	59	78	86
DIST. OF COLUMBIA DC	98*	10.6	97	7.8	134*	8.4	-26	8	-20		
WASHINGTON DC	98*	10.6	97	7.8	134*	8.4	-26	8	-20	48	90
FLORIDA FL	105	4.3	149	4.7	174	4.3	11	-10	0		
ALACHUA FL	1	1.5	3	4.1	3	3.6	175	-12	143	90	77
BAKER FL					1	12.8					94
BAY FL			1	2.8	1	2.5		-11			74
BRADFORD FL	1	15.7	1	14.1	1	7.7	-11	-46	-51	24	89
BREVARD FL	2	5.2			3	4.8			-8	54	81
BROWARD FL	4	4.0	7	3.6	6	1.8	-9	-49	-54	22	72
CALHOUN FL	1*	28.9									
CHARLOTTE FL			2*	48.1	1	47.2		-2			99
CITRUS FL			1	12.3	2*	26.9		119			97
CLAY FL	1	11.1			1	7.8			-30	42	89
COLLIER FL	1	10.1			1	8.1			-20	48	89
DADE FL	10	4.1	23	5.3	24	3.4	29	-36	-17	50	77
DUVAL FL	14	4.8	32*	8.9	33*	7.4	86	-17	56	77	88
ESCAMBIA FL	6	6.6	3	2.8	7	5.3	-57	89	-19	49	83
GADSDEN FL	3	3.5	2	2.0	2	2.7	-43	35	-23	45	75
GULF FL					2*	23.5					97
HAMILTON FL					1	6.7					87

* SIGNIFICANTLY MORE DEATHS THAN EXPECTED

\# COUNTIES COMBINED; SEE TECHNICAL NOTES

NON-WHITE MALE — CANCER OF THE RECTUM — ICD 154 EXCEPT 154.3

1950-1959 DEATHS	1950-1959 RATE	1960-1969 DEATHS	1960-1969 RATE	1970-1979 DEATHS	1970-1979 RATE	COUNTY AND STATE	PERCENT CHANGE 60S/50S	PERCENT CHANGE 70S/60S	PERCENT CHANGE 70S/50S	PERCENTILE %CHANGE 70S/50S	PERCENTILE RATES 1970S
				1	7.0	HENDRY FL					87
1	9.5					HERNANDO FL					
1	4.2					HIGHLANDS FL					
15	7.6	9	3.9	4	1.5	HILLSBOROUGH FL	-49	-60	-80	03	72
2	13.1	1	4.0	1	2.9	INDIAN RIVER FL	-69	-27	-78	04	75
		2	5.4	2	3.8	JACKSON FL		-28			78
		1	4.3	1	7.5	JEFFERSON FL		74			89
		1	1.6	4	6.8	LAKE FL		334			87
1	5.0	2	6.5	2	5.2	LEE FL	30	-19	5	60	82
4	6.0	6	8.0	3	3.3	LEON FL	33	-59	-45	28	77
				1	6.5	LEVY FL					86
		1	2.8	2	4.5	MANATEE FL		59			80
2	3.1	3	3.9	4	4.7	MARION FL	27	21	53	76	81
		2	12.8	1	5.8	MARTIN FL		-55			84
				1	7.0	OKALOOSA FL					87
1*	19.1					OKEECHOBEE FL					
3	4.8	5	3.9	11	5.3	ORANGE FL	-19	34	9	62	83
3	2.4	14	7.9	12	4.8	PALM BEACH FL	228	-39	101	84	82
		1	5.1			PASCO FL					
4	6.3	8	6.5	6	3.3	PINELLAS FL	3	-49	-47	26	77
9	7.9	8	5.5	10	5.4	POLK FL	-30	-2	-32	40	83
1	2.6					PUTNAM FL					
2	6.6	1	3.9			ST JOHNS FL	-42				
1	18.6			2	3.1	ST LUCIE FL			-83	02	76
		2*	30.5	1	11.7	SANTA ROSA FL		-62			93
2	10.0	2	6.0	2	6.9	SARASOTA FL	-40	16	-31	41	87
2	4.3	2	3.1	4	5.9	SEMINOLE FL	-27	88	37	72	84
2	14.1			2	12.1	SUMTER FL			-14	51	93
1	5.6			1	4.5	SUWANNEE FL			-20	48	80
		1	5.6	1	5.7	TAYLOR FL		2			84
2	7.3	2	2.4	6	5.7	VOLUSIA FL	-68	142	-22	46	84
1	14.5					WAKULLA FL					
1	18.2					WALTON FL					
150	4.3	139	3.7	167	4.0	GEORGIA GA	-14	6	-8		
		1	8.8			APPLING GA					
1	1.8	1	1.7	2	4.0	BALDWIN GA	-6	137	122	87	79
1	4.6	1	10.2			BARTOW GA	121				
1	5.4					BEN HILL GA					
1	28.7					BERRIEN GA					
7	4.8	1	.9	5	2.6	BIBB GA	-82	200	-45	28	74
1	4.0	1	2.9	2	7.4	BROOKS GA	-28	155	85	82	88
1	2.6	1	2.5			BULLOCH GA	-2				
1	1.8	1	2.5			BURKE GA	41				
1	6.5	4*	22.8	1	4.7	CARROLL GA	251	-79	-28	43	81
				1	16.7	CHARLTON GA					96
7	3.4	15	7.5	11	4.3	CHATHAM GA	118	-42	26	70	80
1	18.9			1	10.4	CHATTOOGA GA			-45	28	92
				1*	13.4	CHEROKEE GA					94
2	6.6	2	5.6	5*	9.7	CLARKE GA	-15	74	49	75	91
1	12.5			1	5.4	CLAYTON GA			-57	18	83
		3	9.9	2	6.6	COBB GA		-33			86
1	4.4			1	4.7	COFFEE GA			6	61	81
1	3.4	2	9.9			COLQUITT GA	195				
				1	8.8	COOK GA					91
3	9.9	1	2.8	2	5.9	COWETA GA	-72	110	-41	32	84
				2*	18.5	CRAWFORD GA					96
1	2.7			1	3.7	CRISP GA			38	73	78
2	6.5					DECATUR GA					
2	3.8	4	4.3	5	2.3	DE KALB GA	13	-46	-39	33	73
		1	6.0			DODGE GA					
4	7.4	3	3.8	8	8.0	DOUGHERTY GA	-49	111	7	61	89
		1	3.2			EARLY GA					
		1	6.9			EFFINGHAM GA					
1	5.9	1	3.8	1	7.3	ELBERT GA	-35	90	23	68	88
		1	4.0	1	3.7	EMANUEL GA		-7			78
2*	27.2			1	12.5	EVANS GA			-54	22	94
1	14.8					FAYETTE GA					
2	4.3			3	8.0	FLOYD GA			86	82	89
				1*	21.0	FRANKLIN GA					96
33	5.8	31	4.9	40	5.6	FULTON GA	-15	14	-4	56	84
2	5.2	5	9.4			GLYNN GA	80				
1*	26.8					GORDON GA					
1	4.1	1	4.4	1	4.1	GRADY GA	8	-6	1	58	79
				1	3.4	GREENE GA					77
1	5.0	1	2.3	2	7.3	HANCOCK GA	-55	223	45	74	88
1	26.8					HARALSON GA					
1	3.8			3*	12.3	HARRIS GA			225	94	93
		1	11.2			HART GA					
1	4.5	1	3.8	1	2.8	HOUSTON GA	-15	-28	-38	33	75
3*	23.5	2*	17.4			JASPER GA	-26				
1	2.8	1	2.9	1	2.4	JEFFERSON GA	2	-17	-15	50	74
1	5.8			1	10.2	JENKINS GA			78	80	92
3*	26.7	1	37.5			JOHNSON GA	41				
		1	8.6	1	8.6	JONES GA		0			90
				1	9.4	LAMAR GA					91
1	2.9	1	2.2			LAURENS GA	-25				
1	8.2	1	7.7	1	4.7	LEE GA	-7	-38	-43	30	81
		1	5.5			LIBERTY GA					
1	37.2	1	12.8			LINCOLN GA	-65				
1	2.3	2	3.8	1	1.5	LOWNDES GA	67	-60	-34	37	72
		1	5.2			MC DUFFIE GA					
				1	8.0	MC INTOSH GA					89
				1	37.2	MADISON GA					98
1	7.9					MARION GA					
		1	3.5	2	5.4	MERIWETHER GA		52			83
				1	13.2	MILLER GA					94
		1	3.5			MITCHELL GA					
				3*	20.4	MONROE GA					96
		1	13.2			MONTGOMERY GA					

* SIGNIFICANTLY MORE DEATHS THAN EXPECTED

NON-WHITE MALE — CANCER OF THE RECTUM — ICD 154 EXCEPT 154.3

1950-1959 DEATHS	1950-1959 RATE	1960-1969 DEATHS	1960-1969 RATE	1970-1979 DEATHS	1970-1979 RATE	COUNTY AND STATE	PERCENT CHANGE 60S/50S	PERCENT CHANGE 70S/60S	PERCENT CHANGE 70S/50S	PERCENTILE %CHANGE 70S/50S	PERCENTILE RATES 1970S
2	2.7	4	3.5	10	6.1	MUSCOGEE#GA	29	72	123	88	85
1	4.8					NEWTON GA					
		2*	44.8			OCONEE GA					
		1	9.6			OGLETHORPE GA					
				1*	21.0	PAULDING GA					96
		1	16.1			PIKE GA					
		1	7.0			POLK GA					
1	6.1	1	14.9			PULASKI GA	142				
2	7.4					RANDOLPH GA					
15*	11.9	4	2.7	7	4.0	RICHMOND GA	-78	49	-66	10	79
1	9.8					ROCKDALE GA					
				1	37.2	SCHLEY GA					98
		1	4.2	1	4.4	SCREVEN GA		4			80
1	3.4			2	5.0	SPALDING GA			50	75	82
		1	16.7	1	10.2	STEPHENS GA		-39			92
2	11.1			1	9.0	STEWART GA			-19	49	91
2	4.3	1	2.4	4	10.9	SUMTER GA	-44	349	152	91	93
1	7.0					TALBOT GA					
1	4.8	1	5.2			TATTNALL GA	9				
2	15.1			1	8.8	TAYLOR GA			-42	30	91
1	8.9					TELFAIR GA					
3	13.1					TERRELL GA					
		5	9.4	7*	12.5	THOMAS GA		33			94
		2	8.9	1	3.2	TIFT GA		-64			76
2	15.7	1	6.4	1	7.2	TOOMBS GA	-59	13	-54	22	88
4	8.3	3	5.7			TROUP GA	-31				
1	14.1					TURNER GA					
		1	6.9	1	5.5	TWIGGS GA		-20			83
1	8.8	1	9.8			WALKER GA	11				
		1	6.0			WALTON GA					
2	5.9	1	2.5	1	2.6	WARE GA	-57	3	-56	20	74
1	6.5	2	11.8	2	13.3	WARREN GA	82	13	105	85	94
		1	2.3			WASHINGTON GA					
1	7.9			1	5.4	WAYNE GA			-32	39	83
1	27.3	1	13.7			WHITFIELD GA	-50				
				2*	19.0	WILCOX GA					96
1	3.4			1	4.4	WILKES GA			30	71	80
1	3.2	1	4.7			WORTH GA	45				
		125*	7.6	152*	6.5	HAWAII HI		-14			
		20	7.2	26*	8.6	HAWAII HI		20			90
		85	8.1	100	6.1	HONOLULU HI		-25			85
		7	5.0	12	7.1	KAUAI HI		42			88
		13	7.0	14	5.8	MAUI#HI		-18			84
4	10.4	2	5.8	2	3.2	IDAHO ID	-45	-45	-69		
				1	14.6	ADA ID					95
1	13.4					BANNOCK ID					
				1*	38.5	BONNEVILLE ID					98
1*	112.4	1*	141.3			LEWIS ID	26				
1*	546.4					LINCOLN ID					
1*	37.2					POWER ID					
		1*	56.2			WASHINGTON ID					
250*	9.8	281*	7.8	300*	6.2	ILLINOIS IL	-21	-20	-37		
1	13.4	1	24.8	1	13.7	ADAMS IL	85	-45	2	59	94
1	2.8					ALEXANDER IL					
3*	22.3	3	14.3	2	5.0	CHAMPAIGN IL	-36	-65	-78	04	82
				1*	37.2	CLINTON IL					98
207*	10.7	240*	8.3	254*	6.5	COOK IL	-22	-22	-39	33	86
1*	68.8					EDGAR IL					
				1*	74.3	HARDIN IL					99
		1*	34.4			HENRY IL					
		2	15.4	4*	30.6	JACKSON IL		99			98
		1	30.4			JEFFERSON IL					
2	18.2			2	8.0	KANE IL			-56	20	89
		1	1.6	3	5.3	KANKAKEE IL		235			83
1	10.4			1	10.6	KNOX IL			2	58	92
		1	3.0			LAKE IL					
2*	56.6					MC LEAN IL					
1	5.6			5*	19.1	MACON IL			239	95	96
5	11.5	2	4.0	3	5.2	MADISON IL	-65	30	-54	21	82
1	17.2					MARION IL					
1	9.6					MASSAC IL					
1*	133.9					MONTGOMERY IL					
				2*	30.9	MORGAN IL					98
5*	16.7	4	9.9	2	4.5	PEORIA IL	-41	-55	-73	07	80
3	11.4	3	13.6			PULASKI IL	20				
		1	7.2	1	2.0	ROCK ISLAND IL		-72			73
10	7.5	9	4.2	9	4.2	ST CLAIR IL	-44	1	-44	29	79
		1	8.3			SALINE IL					
3	11.8	7*	26.8	1	4.1	SANGAMON IL	127	-85	-65	12	79
1	4.5			2	7.8	VERMILION IL			74	80	89
1	2.0	1	3.3	3	3.1	WILL IL	65	-7	53	76	76
		1	14.1			WILLIAMSON IL					
		2	18.8	3	7.5	WINNEBAGO IL		-60			89
71*	9.5	88*	8.5	78	6.3	INDIANA IN	-10	-27	-34		
		2	5.9	4	9.1	ALLEN IN		54			91
2	19.7	3*	23.2	3*	24.0	CLARK IN	18	3	22	68	97
1	3.4					DELAWARE IN					
				2	17.0	ELKHART IN					96
				1*	42.4	FAYETTE IN					99
				1*	172.0	FULTON IN					99
		1*	18.9			GIBSON IN					
				1	10.3	HOWARD IN					92
		1*	44.6			JOHNSON IN					

* SIGNIFICANTLY MORE DEATHS THAN EXPECTED

\# COUNTIES COMBINED; SEE TEC NICAL NOTES

NON-WHITE MALE — CANCER OF THE RECTUM — ICD 154 EXCEPT 154.3

1950-1959 DEATHS	1950-1959 RATE	1960-1969 DEATHS	1960-1969 RATE	1970-1979 DEATHS	1970-1979 RATE	COUNTY AND STATE	PERCENT CHANGE 60S/50S	PERCENT CHANGE 70S/60S	PERCENT CHANGE 70S/50S	PERCENTILE %CHANGE 70S/50S	PERCENTILE RATES 1970S
2*	75.3	1*	188.7			KNOX IN	151				
21*	11.1	30*	10.6	27	7.3	LAKE IN	-5	-31	-34	37	88
1	10.6			1	7.0	LA PORTE IN			-34	37	87
		1	7.9	1	4.3	MADISON IN		-46			79
35*	12.4	35*	8.4	26	5.5	MARION IN	-32	-34	-56	20	84
		2*	60.8	1*	19.2	MONROE IN		-68			96
		2*	188.7			OHIO IN					
				1*	74.3	PUTNAM IN					99
1	2.3	3	5.1	4	5.9	ST JOSEPH IN	126	14	158	91	84
1*	424.0					STEUBEN IN					
5	11.8			3	6.0	VANDERBURGH IN			-49	25	85
2	7.9	1	3.7	1	3.7	VIGO IN	-53	-1	-53	23	78
		6*	42.8	1	5.9	WAYNE IN		-86			85
9	8.3	11	8.8	9	6.1	IOWA IA	6	-30	-26		
				1*	344.1	ALLAMAKEE IA					99
		2	18.5			BLACK HAWK IA					
		1*	141.3			DALLAS IA					
		2*	69.7			DES MOINES IA					
		2*	35.9			LEE IA					
2*	55.4					LINN IA					
4	9.0	2	4.2	7*	14.3	POLK IA	-54	244	59	78	95
1*	34.2	1*	26.5			POTTAWATTAMIE IA	-23				
2*	24.6					SCOTT IA					
				1*	47.2	TAMA IA					99
		1	11.8			WOODBURY IA					
23	5.9	31	7.3	16	3.3	KANSAS KS	23	-55	-45		
1	9.4					ATCHISON KS					
		1*	37.2			BARTON KS					
1*	28.7					BROWN KS					
		1*	188.7			CHAUTAUQUA KS					
1	14.9					COWLEY KS					
1	9.1	1	4.7			DOUGLAS KS	-48				
		1*	37.2			FORD KS					
		2*	31.8			GEARY KS					
				1*	490.8	GREENWOOD KS					99
2*	30.3	1	9.4			LABETTE KS	-69				
1	2.8			1	4.2	LEAVENWORTH KS			48	75	79
1*	172.0					LINN KS					
1*	24.6	1*	26.8			LYON KS	9				
		1*	24.9			MIAMI KS					
		1	3.9			MONTGOMERY KS					
		1*	31.5			NEOSHO KS					
1*	424.0					NORTON KS					
				1	20.6	SALINE KS					96
3	6.6	3	5.7			SEDGWICK KS	-14				
4	9.8	5	11.5	1	1.8	SHAWNEE KS	17	-84	-82	02	72
				1*	37.5	STEVENS KS					98
		1*	74.3			WABAUNSEE KS					
6	4.5	11	7.2	11	6.6	WYANDOTTE KS	61	-9	46	74	86
65	6.6	72	6.9	74*	7.1	KENTUCKY KY	5	2	7		
1*	37.2					ADAIR KY					
				1*	37.5	ALLEN KY					98
1*	18.6					ANDERSON KY					
		1*	37.8			BALLARD KY					
3*	29.9	1	10.1	1	10.1	BARREN KY	-66	0	-66	10	92
1*	122.7			1*	106.0	BOONE KY			-14	52	99
2	9.6	4*	28.6	2	20.1	BOURBON KY	197	-30	108	85	96
1	27.3	1	12.4			BOYD KY	-55				
1	7.3					BOYLE KY					
				1*	133.9	BRACKEN KY					99
				1*	188.7	BREATHITT KY					99
1*	22.9					BRECKINRIDGE KY					
		1*	17.9	1*	14.9	CALLOWAY KY		-17			95
2	3.3	4	8.1	1	2.2	CHRISTIAN KY	143	-73	-34	38	73
3*	26.9			1	7.5	CLARK KY			-72	08	89
		1*	38.3			CUMBERLAND KY					
		1	5.0	1	7.2	DAVIESS KY		44			88
4	4.1	6	6.2	5	4.7	FAYETTE KY	52	-24	15	65	81
				1*	37.7	FLOYD KY					98
				1	15.7	HARDIN KY					95
				1*	29.8	HARRISON KY					98
3	18.7	2	13.8	6*	41.3	HENDERSON KY	-26	199	121	87	98
		1*	31.5			HENRY KY					
1*	24.6			1*	26.5	HICKMAN KY			8	62	97
2	10.5	1	3.7	2	7.0	HOPKINS KY	-65	90	-33	38	88
24	8.2	36*	10.6	30*	7.5	JEFFERSON KY	29	-29	-8	54	89
1	13.8					JESSAMINE KY					
2	8.9	1	6.1	2	14.9	KENTON KY	-31	142	67	79	95
1*	74.3					LAUREL KY					
				1	9.9	LINCOLN KY					92
1	6.4			1	8.2	LOGAN KY			28	71	90
3	9.7	2	5.5	2	6.4	MC CRACKEN KY	-43	16	-34	37	86
				1	24.8	MARION KY					97
1	9.1			1	13.3	MASON KY			46	74	94
1	8.2	1	12.3	1	12.8	MUHLENBERG KY	50	4	56	77	94
				1*	26.5	OLDHAM KY					97
1*	26.8					OWEN KY					
				1*	23.6	PERRY KY					97
1	9.6					SCOTT KY					
1	10.2					SHELBY KY					
1	10.7	1	13.3			SIMPSON KY	24				
		1	17.7			TAYLOR KY					
		1	16.7	1	17.9	UNION KY		7			96
1	5.0	3	14.5	3	13.5	WARREN KY	193	-7	173	92	94
		1*	30.3	1*	19.1	WASHINGTON KY		-37			96

* SIGNIFICANTLY MORE DEATHS THAN EXPECTED

CANCER OF THE RECTUM — ICD 154 EXCEPT 154.3

NON-WHITE MALE

1950-1959 DEATHS	1950-1959 RATE	1960-1969 DEATHS	1960-1969 RATE	1970-1979 DEATHS	1970-1979 RATE	COUNTY AND STATE	PERCENT CHANGE 60S/50S	PERCENT CHANGE 70S/60S	PERCENT CHANGE 70S/50S	PERCENTILE %CHANGE 70S/50S	PERCENTILE RATES 1970S
		1	10.7			WOODFORD KY					
180	5.4	212	5.6	189	4.5	LOUISIANA LA	4	-20	-17		
3	11.3	3	8.6	1	2.5	ACADIA LA	-24	-71	-78	04	74
1	8.6					ALLEN LA					
1	2.0	1	2.6	3	8.3	ASCENSION LA	31	215	311	98	90
2	9.1			1	4.6	ASSUMPTION LA			-50	25	80
1	2.5	2	5.2			AVOYELLES LA	107				
1	4.2			1	7.0	BEAUREGARD LA			65	78	87
		3	7.3	2	4.8	BIENVILLE LA		-34			81
1	1.8	3	5.8	2	3.1	BOSSIER LA	216	-47	68	79	76
11	4.5	15	4.7	11	3.3	CADDO LA	5	-31	-27	43	77
5	6.0	5	5.6	6	4.7	CALCASIEU LA	-7	-15	-22	46	81
				3*	21.1	CALDWELL LA					97
				2	19.3	CATAHOULA LA					96
1	2.1	4	10.0			CLAIBORNE LA	381				
1	3.1	2	6.7	1	1.9	CONCORDIA LA	121	-71	-37	35	73
		6	8.3	3	5.0	DE SOTO LA		-39			82
11	5.5	10	4.2	7	2.2	EAST BATON ROUGE LA	-23	-49	-61	16	73
2	5.5			1	2.4	EAST CARROLL LA			-57	18	74
1	2.2	3	6.7	2	4.9	EAST FELICIANA LA	208	-27	124	88	82
1	4.3					EVANGELINE LA					
		1	2.3	1	2.5	FRANKLIN LA		7			74
1	5.4	2	13.6	1	4.6	GRANT LA	154	-66	-14	52	81
2	4.8	2	3.6	1	1.6	IBERIA LA	-25	-56	-67	10	72
2	3.7	1	1.6	2	3.6	IBERVILLE LA	-56	125	-2	57	77
				1	3.8	JACKSON LA					78
4	6.5	9	9.1	3	2.1	JEFFERSON LA	40	-77	-67	10	73
1	2.9			2	10.1	JEFFERSON DAVIS LA			254	96	92
1	3.7	5	6.6	1	1.5	LAFAYETTE LA	77	-76	-58	17	72
2	10.1					LAFOURCHE LA					
1	2.4	2	5.3	2	4.3	LINCOLN LA	122	-20	78	81	79
1	14.5					LIVINGSTON LA					
3	4.1	5	11.5	3	12.9	MADISON LA	177	12	211	93	94
1	1.2	4	6.0	4	4.9	MOREHOUSE LA	384	-18	298	98	82
5	7.5	2	2.8	2	2.8	NATCHITOCHES LA	-63	-2	-63	14	75
56*	9.1	66*	7.9	67*	7.2	ORLEANS LA	-14	-8	-21	47	88
6	5.2	4	3.7	8	6.5	OUACHITA LA	-28	75	26	69	86
		4*	24.6	3*	14.3	PLAQUEMINES LA		-42			95
		2	4.4			POINTE COUPEE LA					
15*	11.3	8	5.7	4	2.7	RAPIDES LA	-49	-52	-76	06	75
1	5.3	1	5.5			RED RIVER LA	5				
2	5.7	3	8.2	1	2.0	RICHLAND LA	44	-75	-64	13	73
1	6.2	2	9.3	1	7.2	SABINE LA	50	-23	16	66	88
1	14.0	1	14.4	1	18.1	ST BERNARD LA	3	25	29	71	96
		1	4.6	1	3.6	ST CHARLES LA		-21			77
				1	4.0	ST HELENA LA					79
		4	13.0	2	5.1	ST JAMES LA		-60			82
1	3.0	1	3.0	1	2.7	ST JOHN THE BAPT LA	1	-10	-9	53	75
6	6.1	4	3.6	6	5.2	ST LANDRY LA	-41	44	-14	51	82
2	6.7	1	3.0			ST MARTIN LA	-56				
1	1.8	4	6.6	3	5.1	ST MARY LA	263	-22	181	93	82
2	7.0	1	2.8	1	2.3	ST TAMMANY LA	-60	-17	-66	10	74
3	4.1	2	2.6	1	1.0	TANGIPAHOA LA	-38	-61	-76	06	71
2	4.6			1	2.9	TENSAS LA			-36	35	75
1	2.2	1	3.6	3	7.6	TERREBONNE LA	67	110	250	95	89
2	9.5			1	3.3	UNION LA			-65	12	77
1	5.2	1	5.5			VERMILION LA	6				
		1	10.3			VERNON LA					
3	5.4	5	10.1	5	9.4	WASHINGTON LA	86	-8	72	80	91
1	1.6	4	7.3	4	5.2	WEBSTER LA	367	-28	237	95	83
2	8.1			2	7.8	WEST BATON ROUGE LA			-4	56	89
1	8.5	1	8.9			WEST CARROLL LA	5				
2	6.9					WEST FELICIANA LA					
1	4.3			3	11.8	WINN LA			174	92	93
3*	28.6	2	14.4			MAINE ME	-50				
1*	38.2	1*	32.6			CUMBERLAND ME	-15				
1*	67.0	1*	267.9			WASHINGTON ME	300				
1*	94.4					YORK ME					
103	7.3	134*	6.7	150*	6.0	MARYLAND MD	-8	-11	-18		
1	16.7					ALLEGANY MD					
5	6.1	6	4.9	4	3.2	ANNE ARUNDEL MD	-20	-34	-47	27	76
3	6.6	7	10.5	5	4.8	BALTIMORE MD	59	-54	-26	43	82
2	11.4			2	7.6	CALVERT MD			-34	38	89
2	13.2	2	11.9	3*	14.0	CAROLINE MD	-10	18	6	60	94
1	6.7					CARROLL MD					
				1	6.3	CECIL MD					86
		3	10.0	1	4.3	CHARLES MD		-57			80
1	2.6	4	9.2	1	2.2	DORCHESTER MD	252	-76	-16	50	73
		4*	18.0	2	7.7	FREDERICK MD		-57			89
				2	5.2	HARFORD MD					82
				1	3.4	HOWARD MD					77
1	5.2	2	10.6	2	9.5	KENT MD	105	-10	84	81	91
3	8.0	3	5.0	3	3.1	MONTGOMERY MD	-37	-38	-61	16	76
1	1.3	3	2.0	9	4.1	PRINCE GEORGE'S MD	57	107	226	95	79
3	15.9	1	4.3			QUEEN ANNE'S MD	-73				
1	5.4	4*	15.7			ST MARY'S MD	192				
3	9.2			2	6.6	SOMERSET MD			-29	42	86
		1	3.3	1	3.1	TALBOT MD		-4			76
		1	9.8	1	9.4	WASHINGTON MD		-4			91
3	9.1	2	4.2	3	6.1	WICOMICO MD	-54	44	-33	38	85
2	6.3	2	6.1	2	6.0	WORCESTER MD	-3	-2	-5	56	85
71*	8.9	89*	7.9	105*	7.4	BALTIMORE CITY MD	-12	-6	-17	50	88
38*	10.3	33	6.9	46	6.5	MASSACHUSETTS MA	-33	-6	-37		

* SIGNIFICANTLY MORE DEATHS THAN EXPECTED

NON-WHITE MALE

CANCER OF THE RECTUM

ICD 154 EXCEPT 154.3

1950-1959 DEATHS	1950-1959 RATE	1960-1969 DEATHS	1960-1969 RATE	1970-1979 DEATHS	1970-1979 RATE	COUNTY AND STATE		PERCENT CHANGE 60S/50S	PERCENT CHANGE 70S/60S	PERCENT CHANGE 70S/50S	PERCENTILE %CHANGE 70S/50S	PERCENTILE RATES 1970S
2*	19.1	2*	21.4	3*	14.6	BARNSTABLE	MA	12	-32	-24	45	95
1	19.1	1	11.4	1	10.9	BERKSHIRE	MA	-40	-5	-43	29	93
5*	20.1			2	4.0	BRISTOL	MA			-80	02	78
				1*	38.2	DUKES	MA					98
2	23.7	1	6.6			ESSEX	MA	-72				
3	13.4	2	4.6	5	6.4	HAMPDEN	MA	-66	40	-52	24	86
2	3.5	3	4.0	6	6.6	MIDDLESEX	MA	14	65	88	82	86
1*	141.3					NANTUCKET	MA					
		2	13.2			NORFOLK	MA					
		2	8.6	3	10.5	PLYMOUTH	MA		21			92
21*	10.9	20	8.4	24	7.0	SUFFOLK	MA	-23	-17	-36	35	87
1	7.9			1	3.1	WORCESTER	MA			-61	16	76
125	7.6	171	6.7	177	5.0	MICHIGAN	MI	-12	-26	-34		
		1	17.0			ALLEGAN	MI					
1	1.4	1	1.7	5	7.9	BERRIEN	MI	22	369	470	99	89
2	11.3	3	17.6	1	2.9	CALHOUN	MI	55	-84	-75	06	75
1	7.4	1	4.9			CASS	MI	-33				
6	9.6	3	2.3	4	2.4	GENESEE	MI	-76	3	-75	06	74
2	11.5	2	8.6	3	9.2	INGHAM	MI	-26	7	-21	47	91
		1*	10.1			ISABELLA	MI					
				2	7.0	JACKSON	MI					88
2	16.7	3	19.1	1	7.0	KALAMAZOO	MI	15	-63	-58	17	87
3	10.6	3	6.8	1	1.1	KENT	MI	-36	-84	-89	01	71
		1	14.9			LAKE	MI					
1*	163.6					LAPEER	MI					
1	7.9			1	2.8	MACOMB	MI			-65	12	75
1	1.3			4	7.1	MUSKEGON	MI			431	99	88
3*	162.9	1	42.0	1	12.2	NEWAYGO	MI	-74	-71	-93	00	93
4	16.0	4	5.1	4	3.7	OAKLAND	MI	-68	-28	-77	05	78
		1	1.5	4	4.7	SAGINAW	MI		219			81
		2	15.3	2	14.2	ST CLAIR	MI		-7			94
		1	14.6			ST JOSEPH	MI					
		4*	13.9			VAN BUREN	MI					
1	2.6	4	8.2	2	4.9	WASHTENAW	MI	219	-41	89	83	82
97*	8.1	135	7.2	142	5.5	WAYNE	MI	-11	-23	-32	40	84
14*	12.7	7	4.9	6	2.5	MINNESOTA	MN	-61	-50	-81		
1	37.5					BECKER	MN					
1	11.8					BELTRAMI	MN					
1*	44.6					CARLTON	MN					
1	21.0					CASS	MN					
				1*	73.7	CLEARWATER	MN					99
7*	19.3	1	2.1	2	1.5	HENNEPIN	MN	-89	-30	-92	00	72
		1*	119.2			KOOCHICHING	MN					
		2*	61.0			MAHNOMEN	MN					
1	3.9	3	8.5	3	6.0	RAMSEY	MN	118	-30	53	76	85
1*	86.0					REDWOOD	MN					
1	10.6					ST LOUIS	MN					
115	3.4	138	4.2	126	3.6	MISSISSIPPI	MS	21	-13	6		
2	3.4	5	7.6	9*	11.3	ADAMS	MS	127	48	235	95	93
				1	5.4	ALCORN	MS					83
1	2.9			2	7.3	AMITE	MS			147	90	88
2	6.4					ATTALA	MS					
5	3.5	6	4.5	8	5.9	BOLIVAR	MS	30	32	72	80	85
1	9.6	1	7.4			CALHOUN	MS	-22				
1	3.3					CARROLL	MS					
1	4.0	1	3.9	1	3.7	CHICKASAW	MS	-4	-4	-8	54	78
4	15.7					CLAIBORNE	MS					
1	3.5	1	2.9	1	4.6	CLARKE	MS	-16	59	34	72	81
1	3.1	2	5.3			CLAY	MS	72				
3	2.2	7	5.9	9	8.6	COAHOMA	MS	164	44	281	97	90
2	4.5	1	2.0	1	2.0	COPIAH	MS	-57	1	-56	19	73
				1	5.3	COVINGTON	MS					83
2	3.3	4	7.9	3	6.9	DE SOTO	MS	140	-12	112	86	87
2	3.7	1	2.7	4	8.3	FORREST	MS	-27	212	127	88	90
		1	5.6			FRANKLIN	MS					
		1	22.5			GREENE	MS					
		2	5.9	2	5.6	GRENADA	MS		-4			84
1	10.2					HANCOCK	MS					
2	4.6	7	10.1	3	3.5	HARRISON	MS	120	-65	-24	45	77
9	4.9	18	7.3	9	2.7	HINDS	MS	50	-63	-45	28	75
1	1.9	1	1.1	2	2.2	HOLMES	MS	-42	107	20	67	73
2	3.3	2	5.0			HUMPHREYS	MS	50				
				1	19.3	ISSAQUENA	MS					96
				1	37.2	ITAWAMBA	MS					98
2	10.8	1	2.3	1	2.4	JACKSON	MS	-78	3	-78	05	74
		1	3.9	3	8.1	JASPER	MS		111			90
1	2.8	1	3.2	2	5.7	JEFFERSON	MS	16	76	105	84	84
4*	18.4	3	10.6	1	6.4	JEFFERSON DAVIS	MS	-43	-39	-65	12	86
5	10.0	2	3.6	3	4.9	JONES	MS	-64	35	-51	24	82
3	12.1	1	4.1	1	3.7	KEMPER	MS	-66	-9	-69	09	78
				1	3.0	LAFAYETTE	MS					75
		1	17.6			LAMAR	MS					
1	1.1	3	3.7	2	2.2	LAUDERDALE	MS	246	-42	102	84	73
				1	5.6	LAWRENCE	MS					84
2	6.3	1	3.4	1	2.9	LEAKE	MS	-46	-16	-54	22	75
3	8.6	2	6.2	3	7.8	LEE	MS	-28	27	-9	53	89
5	4.5	5	4.8	3	3.0	LEFLORE	MS	7	-38	-33	38	76
1	3.2					LINCOLN	MS					
1	2.0	3	5.5	3	5.3	LOWNDES	MS	170	-3	161	91	83
6	6.9	1	1.3	2	2.2	MADISON	MS	-81	71	-68	09	73
1	3.8			1	2.7	MARION	MS			-29	42	75
1	1.9	4	7.0	3	4.6	MARSHALL	MS	272	-34	146	90	81
				1	1.8	MONROE	MS					72
1	4.1					MONTGOMERY	MS					
1	3.5	1	4.8	1	3.5	NEWTON	MS	35	-26	0	57	77

* SIGNIFICANTLY MORE DEATHS THAN EXPECTED

NON-WHITE MALE — CANCER OF THE RECTUM — ICD 154 EXCEPT 154.3

1950-1959 DEATHS	1950-1959 RATE	1960-1969 DEATHS	1960-1969 RATE	1970-1979 DEATHS	1970-1979 RATE	COUNTY AND STATE	PERCENT CHANGE 60S/50S	PERCENT CHANGE 70S/60S	PERCENT CHANGE 70S/50S	PERCENTILE %CHANGE 70S/50S	PERCENTILE RATES 1970S
2	4.3	1	2.4			NOXUBEE MS	-45				
1	2.2	1	2.7	4	11.3	OKTIBBEHA MS	23	318	414	99	93
1	1.8					PANOLA MS					
		1	9.4			PEARL RIVER MS					
				1	14.8	PERRY MS					95
5	9.5	3	5.0	3	4.3	PIKE MS	-47	-13	-54	22	80
1	4.8	1	7.6	1	5.9	PONTOTOC MS	58	-23	22	68	85
2*	32.9					PRENTISS MS					
1	1.8					QUITMAN MS					
		1	2.2			RANKIN MS					
1	3.1	1	2.6			SCOTT MS	-15				
2	7.8	2	9.3			SHARKEY MS	19				
		1	3.7	2	8.2	SIMPSON MS		118			90
		1	13.2			STONE MS					
2	2.3	9	8.8	1	1.0	SUNFLOWER MS	278	-89	-58	17	71
1	1.6	1	1.8	1	1.8	TALLAHATCHIE MS	13	4	17	66	72
				1	6.4	TIPPAH MS					86
5	8.1	3	5.9	1	3.7	TUNICA MS	-27	-38	-55	21	78
1	7.5					UNION MS					
				1	5.0	WALTHALL MS					82
3	3.5	8	9.0	7	7.2	WARREN MS	159	-20	106	85	88
3	1.6	9	5.8	3	1.8	WASHINGTON MS	269	-69	13	64	72
				1	3.1	WAYNE MS					76
				1	8.2	WEBSTER MS					90
1	2.9	2	4.9	2	4.6	WILKINSON MS	67	-6	58	77	81
1	3.3	1	3.4	2	5.8	WINSTON MS	1	72	74	80	84
1	1.5	1	1.2	3	4.7	YAZOO MS	-15	281	225	94	81
110*	7.7	112	6.6	93	5.0	MISSOURI MO	-14	-25	-35		
1	9.6	1	14.3	1	9.9	AUDRAIN MO	49	-31	3	59	92
		2	11.4			BOONE MO					
2	12.3	1	7.4	4*	21.1	BUCHANAN MO	-40	183	71	79	97
1	5.6			1	5.2	BUTLER MO			-7	54	83
1	5.6					CALLAWAY MO					
1	18.9	1	17.2			CHARITON MO	-9				
1*	37.7					CLINTON MO					
1	10.4	2*	47.8	1	14.9	COLE MO	358	-69	43	74	95
2*	20.0					COOPER MO					
1*	490.8					DADE MO					
		1*	18.6	1*	17.9	FRANKLIN MO		-4			96
1	8.4			1	6.6	GREENE MO			-21	47	87
		1	10.3	1	15.0	HOWARD MO		45			95
22	6.8	24	6.8	17	3.9	JACKSON MO	0	-42	-42	31	78
		1	13.8	1	13.8	JASPER MO		0			94
1*	23.8					JEFFERSON MO					
1	11.2			1	14.3	LAFAYETTE MO			28	70	95
1	14.5			1*	26.5	LINCOLN MO			82	81	97
		1*	94.4			LINN MO					
1*	27.0	1*	84.8			MACON MO	215				
1	3.9			1	6.0	MISSISSIPPI MO			56	77	85
1	3.2	1	4.1			NEW MADRID MO	26				
				1*	42.4	NEWTON MO					99
2	3.8	1	2.5	1	1.9	PEMISCOT MO	-35	-25	-51	24	72
1	9.0	3*	26.5			PETTIS MO	195				
		1	6.8	1	9.4	PIKE MO		40			91
		1*	49.7			PLATTE MO					
		1*	43.0			RALLS MO					
				2*	31.8	RANDOLPH MO					98
				1	18.6	ST CHARLES MO					96
7	9.9	12*	11.0	6	2.7	ST LOUIS COUNTY MO	11	-76	-73	07	75
1	7.2	1	5.0	1	15.6	SALINE MO	-31	215	118	86	95
		1*	163.6			SHELBY MO					
59*	8.8	54	6.5	49	5.5	ST LOUIS CITY MO	-27	-15	-38	33	83
4	5.8			2	2.2	MONTANA MT			-62		
1	10.7					BIG HORN MT					
				1	24.9	BLAINE MT					97
1*	344.1					GRANITE MT					
1*	61.3					LEWIS AND CLARK MT					
				1	9.2	ROOSEVELT MT					91
1*	546.4					TOOLE MT					
5	4.1	8	5.7	15*	9.1	NEBRASKA NB	39	59	121		
				1*	245.4	ADAMS NB					99
				1*	112.4	BURT NB					99
3	3.2	7	6.6	10	8.5	DOUGLAS NB	105	29	164	92	90
		1	16.3			LANCASTER NB					
1*	182.1					LINCOLN NB					
				1*	89.3	RICHARDSON NB					99
				2*	33.4	SCOTTS BLUFF NB					98
1	15.7					THURSTON NB					
2	3.9	2	3.3	7	4.4	NEVADA NV	-16	35	14		
2	8.7	1	9.3	3	3.3	CLARK NV	6	-64	-62	15	77
				1*	172.0	DOUGLAS NV					99
				1	15.8	ELKO NV					95
		1	4.2	2	7.5	WASHOE NV		77			89
1	15.2					NEW HAMPSHIRE NH					
1*	54.6					ROCKINGHAM NH					
118*	9.9	157*	8.7	162*	6.6	NEW JERSEY NJ	-12	-25	-34		
13*	11.2	5	3.6	10	6.5	ATLANTIC NJ	-68	82	-42	31	86
3	8.6	5	5.7	7	5.0	BERGEN NJ	-34	-12	-41	31	82
3	8.1	3	7.4	3	5.2	BURLINGTON NJ	-9	-30	-36	36	82

* SIGNIFICANTLY MORE DEATHS THAN EXPECTED

NON-WHITE MALE — CANCER OF THE RECTUM — ICD 154 EXCEPT 154.3

1950-1959 DEATHS	1950-1959 RATE	1960-1969 DEATHS	1960-1969 RATE	1970-1979 DEATHS	1970-1979 RATE	COUNTY AND STATE		PERCENT CHANGE 60S/50S	PERCENT CHANGE 70S/60S	PERCENT CHANGE 70S/50S	PERCENTILE %CHANGE 70S/50S	PERCENTILE RATES 1970S
12*	12.0	22*	15.8	11	6.6	CAMDEN	NJ	31	-58	-45	28	86
1	6.3					CAPE MAY	NJ					
2	6.4	4	8.5	4	6.0	CUMBERLAND	NJ	32	-29	-6	55	85
43*	12.3	56*	10.1	57*	7.5	ESSEX	NJ	-18	-25	-39	33	89
2	6.1	4	6.7	5	6.8	GLOUCESTER	NJ	11	2	12	63	87
7	11.1	13*	9.0	10	4.1	HUDSON	NJ	-20	-55	-63	14	79
1*	24.6					HUNTERDON	NJ					
3	8.2	5	4.1	12	8.6	MERCER	NJ	-51	111	4	59	90
1	3.2			2	2.5	MIDDLESEX	NJ			-22	46	74
7	8.0	13*	11.9	6	4.6	MONMOUTH	NJ	50	-62	-42	30	81
1	4.6	3	10.3	3	10.8	MORRIS	NJ	125	4	135	89	93
1	8.4	2	10.1	1	3.2	OCEAN	NJ	21	-69	-62	15	76
5	11.7	4	8.6	8	6.3	PASSAIC	NJ	-26	-27	-46	27	86
4	11.8	3	9.4	3	7.2	SALEM	NJ	-20	-24	-39	32	88
				2	7.3	SOMERSET	NJ					88
9	9.9	15*	13.1	18*	8.5	UNION	NJ	32	-35	-14	51	90
7	4.2	2	.9	8	1.7	NEW MEXICO	NM	-80	94	-61		
1	7.6			3	2.4	BERNALILLO	NM			-69	09	74
				1	6.7	LEA	NM					87
1*	54.6					LUNA	NM					
2	4.2			1	1.3	MC KINLEY	NM			-68	09	71
				1	3.9	RIO ARRIBA	NM					78
2	10.0	1	5.3			SANDOVAL	NM	-47				
		1	2.3	1	1.2	SAN JUAN	NM		-47			71
1	26.5					TAOS	NM					
				1	2.0	VALENCIA	NM					73
374*	12.0	401*	8.3	488*	6.6	NEW YORK	NY	-31	-21	-45		
1	2.2	3	8.6	3	4.8	ALBANY	NY	288	-43	119	87	82
47*	14.6	38	6.3	59	5.3	BRONX	NY	-57	-17	-64	12	83
1	11.8					BROOME	NY					
				1	12.5	CATTARAUGUS	NY					94
		1	47.2			CAYUGA	NY					
1*	18.9			1	37.5	CHAUTAUQUA	NY			99	84	98
		2*	28.2			CHEMUNG	NY					
				1*	74.3	CHENANGO	NY					99
1	16.6			2*	25.4	COLUMBIA	NY			54	76	97
1	3.9	2	3.8	3	8.4	DUTCHESS	NY	-4	122	114	86	90
15	11.0	29*	13.6	16	4.9	ERIE	NY	24	-64	-56	20	82
				1*	39.0	HERKIMER	NY					98
66*	12.1	70	6.2	111*	7.1	KINGS(BROOKLYN)	NY	-49	14	-41	31	88
		1*	38.2			LIVINGSTON	NY					
5	16.8	3	4.8	7	6.1	MONROE	NY	-71	26	-64	14	85
7	10.0	9	6.2	10	4.3	NASSAU	NY	-38	-29	-57	19	80
165*	12.8	174*	11.0	162*	8.0	NEW YORK(MANHATTAN)	NY	-14	-27	-37	34	89
2	18.9	1	6.0	2	8.9	NIAGARA	NY	-69	49	-53	23	91
		1	7.4			ONEIDA	NY					
				1	1.6	ONONDAGA	NY					72
2	11.8			2	3.4	ORANGE	NY			-71	08	77
				1	9.3	ORLEANS	NY					91
				1*	133.9	PUTNAM	NY					99
27*	13.1	34	7.2	59	5.9	QUEENS	NY	-45	-18	-55	21	85
				1	8.0	RENSSELAER	NY					89
2	9.9	2	8.7	3	5.6	RICHMOND	NY	-12	-35	-43	29	84
4*	24.4			1	2.4	ROCKLAND	NY			-90	00	74
2*	71.1	1	12.8	1	10.1	SARATOGA	NY	-82	-21	-86	01	92
1	38.3					SCHENECTADY	NY					
5	7.7	11	6.8	15	8.1	SUFFOLK	NY	-11	19	6	60	90
		1	13.3			TOMPKINS	NY					
1	6.7					ULSTER	NY					
18*	13.5	18	9.6	24*	7.4	WESTCHESTER	NY	-29	-23	-45	28	88
114	3.7	154	4.3	182	4.3	NORTH CAROLINA	NC	16	1	18		
		1	3.0	2	3.0	ALAMANCE	NC		1			76
		3	7.3			ANSON	NC					
		2	3.7	1	1.6	BEAUFORT	NC		-58			72
		1	2.4			BERTIE	NC					
3	8.8	1	2.7	4	9.5	BLADEN	NC	-70	257	7	62	91
				2	6.8	BRUNSWICK	NC					87
4	5.5	5	7.7	3	4.3	BUNCOMBE	NC	39	-43	-21	47	80
		1	6.4	1	5.9	BURKE	NC		-8			85
1	2.5	5*	16.6	4	11.0	CABARRUS	NC	565	-33	342	98	93
				1	8.8	CALDWELL	NC					90
1	16.1					CAMDEN	NC					
1	11.6			1	5.8	CARTERET	NC			-50	25	84
1	3.0	1	2.7	1	3.9	CASWELL	NC	-10	43	29	71	78
1	8.0					CATAWBA	NC					
1	2.1					CHATHAM	NC					
1	6.8					CHOWAN	NC					
		2	4.3	2	5.2	CLEVELAND	NC		22			82
				2	2.9	COLUMBUS	NC					75
		1	1.9	1	1.9	CRAVEN	NC		3			73
4	7.1	4	4.0	8	6.2	CUMBERLAND	NC	-44	57	-13	53	85
		1	10.7	1	24.8	CURRITUCK	NC		131			97
1	4.6	3	10.2	1	2.5	DAVIDSON	NC	120	-75	-45	27	74
1	11.2					DAVIE	NC					
				2	3.5	DUPLIN	NC					77
7	7.0	3	1.9	13	8.4	DURHAM	NC	-72	336	21	67	90
1	1.3	3	4.0	6	6.3	EDGECOMBE	NC	205	58	380	99	86
12	9.9	14	8.7	7	3.7	FORSYTH	NC	-12	-58	-63	14	78
		1	2.2	2	4.4	FRANKLIN	NC		100			80
3	6.8	1	3.1	1	1.3	GASTON	NC	-54	-57	-80	03	71
1	8.6			1	4.7	GATES	NC			-46	27	81
1	2.4			3	5.1	GRANVILLE	NC			114	86	82
				2	11.5	GREENE	NC					93
5	3.9	6	3.4	7	3.5	GUILFORD	NC	-11	3	-8	54	77
2	2.7	2	2.1	2	2.1	HALIFAX	NC	-23	-2	-25	44	73

* SIGNIFICANTLY MORE DEATHS THAN EXPECTED

NON-WHITE MALE — CANCER OF THE RECTUM — ICD 154 EXCEPT 154.3

1950-1959 DEATHS	1950-1959 RATE	1960-1969 DEATHS	1960-1969 RATE	1970-1979 DEATHS	1970-1979 RATE	COUNTY AND STATE	PERCENT CHANGE 60S/50S	PERCENT CHANGE 70S/60S	PERCENT CHANGE 70S/50S	PERCENTILE %CHANGE 70S/50S	PERCENTILE RATES 1970S
1	2.2	2	5.5	3	8.2	HARNETT NC	153	50	280	96	90
		2*	31.6	1	8.2	HENDERSON NC		-74			90
1	1.6	3	6.5	5	10.4	HERTFORD NC	305	61	553	99	92
				3	10.2	HOKE NC					92
1	2.9	2	5.0	1	2.1	IREDELL NC	73	-58	-28	43	73
2	5.2	6*	14.7	3	5.9	JOHNSTON NC	181	-60	13	64	85
		1	4.3			LEE NC					
1	2.0	2	2.8	4	4.4	LENOIR NC	40	58	122	87	80
1	12.5					LINCOLN NC					
				1	17.0	MC DOWELL NC					96
1*	68.8					MACON NC					
1	2.3					MARTIN NC					
6	3.3	12	6.1	17	6.6	MECKLENBURG NC	87	8	102	84	87
1	9.2			1	4.0	MONTGOMERY NC			-56	19	79
1	3.1			4	11.1	MOORE NC			258	96	93
6	9.4	3	4.5	1	1.3	NASH NC	-52	-71	-86	01	71
4	5.8	6	10.9	2	2.4	NEW HANOVER NC	88	-78	-58	17	74
		1	2.2	1	1.8	NORTHAMPTON NC		-18			72
		1	1.7	1	2.4	ONSLOW NC		37			74
1	3.9	3	8.2	2	4.7	ORANGE NC	110	-43	20	67	81
1	3.6	2	5.8	2	5.8	PASQUOTANK NC	60	0	60	78	84
1	4.7			2	5.6	PENDER NC			18	66	84
				2	12.4	PERQUIMANS NC					94
1	4.8	2	9.0	2	5.9	PERSON NC	88	-35	23	69	84
2	3.1	5	5.8	3	3.1	PITT NC	87	-48	-2	56	76
1	20.4					POLK NC					
1	8.8	1	5.9	1	4.1	RANDOLPH NC	-33	-31	-53	23	79
1	1.9	2	6.1	2	4.3	RICHMOND NC	220	-30	123	88	79
2	1.4	5	4.0	4	2.6	ROBESON NC	192	-35	88	83	74
2	5.0			1	1.8	ROCKINGHAM NC			-64	13	72
2	4.2	4	7.8	1	1.7	ROWAN NC	85	-78	-60	16	72
1	2.0	2	10.1			RUTHERFORD NC	397				
3	7.4	3	5.9			SAMPSON NC	-21				
1	1.8	1	2.7	2	7.1	SCOTLAND NC	49	168	298	98	88
				1	10.1	SURRY NC					92
1	28.1					SWAIN NC					
		1*	32.7			TRANSYLVANIA NC					
2	8.6	1	4.6	1	3.0	UNION NC	-47	-35	-65	11	76
		8*	16.3	2	3.8	VANCE NC		-77			78
5	3.7	8	5.5	7	3.4	WAKE NC	49	-38	-8	54	77
2	3.8	1	2.4	7*	14.8	WARREN NC	-35	507	293	97	95
1	3.3					WASHINGTON NC					
2	2.2	2	1.7	5	4.8	WAYNE NC	-22	186	123	87	81
				1	10.1	WILKES NC					92
2	5.8	1	1.8	5	6.6	WILSON NC	-68	256	13	64	86
3	12.8	2	7.8	3	5.9	NORTH DAKOTA ND	-39	-25	-54		
				1*	17.9	CASS ND					96
1*	188.7					OLIVER ND					
2	19.5	2	17.2	2	12.8	ROLETTE ND	-12	-25	-34	37	94
190*	9.5	229*	7.8	277*	7.9	OHIO OH	-18	1	-17		
2	12.2	2	7.6	1	2.6	ALLEN OH	-37	-66	-79	04	74
1	15.3					ASHTABULA OH					
		1	14.1			ATHENS OH					
				1	9.4	BELMONT OH					91
1	18.7			1*	44.6	BROWN OH			138	89	99
1	3.0	1	2.6	4	11.0	BUTLER OH	-14	322	261	96	93
2*	29.1					CHAMPAIGN OH					
4	8.5	3	5.1	6	10.4	CLARK OH	-40	104	22	68	92
		1	20.2			CLERMONT OH					
1*	28.7					CLINTON OH					
1	27.0			1	10.8	COLUMBIANA OH			-60	16	93
1*	122.7					CRAWFORD OH					
55*	9.8	66	7.3	88*	8.0	CUYAHOGA OH	-26	10	-18	49	89
		1*	91.1			DARKE OH					
		3*	25.8	2	13.1	ERIE OH		-49			94
				1*	26.5	FAYETTE OH					97
26*	12.5	20	6.4	21	5.8	FRANKLIN OH	-49	-9	-54	22	84
1*	424.0					FULTON OH					
				2*	30.4	GALLIA OH					98
4*	17.6	2	7.5	6*	23.2	GREENE OH	-57	209	32	72	97
		1	27.3	1*	21.5	GUERNSEY OH		-21			97
34*	10.2	51*	10.2	52*	9.5	HAMILTON OH	0	-7	-7	55	91
1	12.6			1*	18.7	HIGHLAND OH			49	75	96
				1*	114.7	HOCKING OH					99
2	9.6	2	7.0	3	14.6	JEFFERSON OH	-27	107	51	75	95
				1	5.3	LAKE OH					83
2*	22.5					LAWRENCE OH					
				1	12.2	LICKING OH					93
1*	14.9	1*	22.3	1*	19.1	LOGAN OH	50	-14	29	71	96
1	3.4	4	10.9			LORAIN OH	222				
8	6.0	15	10.3	12	6.3	LUCAS OH	72	-38	6	61	86
11	12.8	13	8.8	11	6.9	MAHONING OH	-32	-21	-46	27	87
1*	28.1					MEIGS OH					
2*	28.5			1	14.2	MIAMI OH			-50	25	94
9	5.3	15	7.3	25*	8.4	MONTGOMERY OH	39	14	58	77	90
		1	8.6	3*	22.0	MUSKINGUM OH		156			97
1*	62.9	2*	116.5			PAULDING OH	85				
		1*	112.4			PERRY OH					
1*	62.9					PICKAWAY OH					
		2*	30.6			PORTAGE OH					
1	8.9			3	9.9	RICHLAND OH			11	63	92
2	13.8	1	6.7	3*	16.7	ROSS OH	-51	149	21	67	96
				3*	30.9	SCIOTO OH					98
2	2.9	7	9.7	5	6.1	STARK OH	236	-37	111	86	85
6	6.3	9	6.6	12	6.5	SUMMIT OH	6	-1	4	60	86
4	13.1	3	6.7	3	6.2	TRUMBULL OH	-49	-8	-53	23	85
1*	109.3					VAN WERT OH					

* SIGNIFICANTLY MORE DEATHS THAN EXPECTED

NON-WHITE MALE — CANCER OF THE RECTUM — ICD 154 EXCEPT 154.3

1950-1959 DEATHS	1950-1959 RATE	1960-1969 DEATHS	1960-1969 RATE	1970-1979 DEATHS	1970-1979 RATE	COUNTY AND STATE	PERCENT CHANGE 60S/50S	PERCENT CHANGE 70S/60S	PERCENT CHANGE 70S/50S	PERCENTILE %CHANGE 70S/50S	PERCENTILE RATES 1970S
		1*	38.3			WARREN OH					
				1*	14.8	WAYNE OH					95
44	5.2	35	3.4	57	4.4	OKLAHOMA OK	-34	28	-16		
		3*	23.4	1	4.7	ADAIR OK		-80			81
1	17.9	1	12.8			ATOKA OK	-29				
				2	7.3	CARTER OK					88
		1	6.7			CHEROKEE OK					
				2	8.7	CHOCTAW OK					90
				1	16.7	CLEVELAND OK					96
1	7.4	1	5.9	1	4.7	COMANCHE OK	-20	-21	-37	34	81
		1	5.5	1	3.9	CREEK OK		-29			78
				1	14.9	CUSTER OK					95
				3*	14.3	DELAWARE OK					95
1	19.1	1	13.3	1	9.9	GARFIELD OK	-31	-26	-48	26	92
				1	11.1	GARVIN OK					93
2*	22.1					GRADY OK					
1	12.5					HUGHES OK					
		1	8.0	3*	22.8	KAY OK		184			97
1	14.9					KINGFISHER OK					
2*	35.3					KIOWA OK					
		1	8.0			LE FLORE OK					
1	9.9					LINCOLN OK					
2	7.0	2	9.5			LOGAN OK	35				
1	2.8			4	13.6	MC CURTAIN OK			376	98	94
1	5.2	2	11.6			MC INTOSH OK	122				
				1	7.1	MAYES OK					88
3	3.2	3	3.6	4	5.4	MUSKOGEE OK	11	49	65	78	83
1	9.0					NOWATA OK					
		3	16.8	1	3.2	OKFUSKEE OK		-81			76
8	7.5	4	2.5	9	3.9	OKLAHOMA OK	-67	57	-49	26	78
5	13.0			2	4.0	OKMULGEE OK			-70	08	78
				1	6.7	OTTAWA OK					87
				2*	24.7	PAYNE OK					97
				1	4.0	PITTSBURG OK					79
1	12.2			1	7.0	PONTOTOC OK			-42	30	87
1	9.4			2	12.9	POTTAWATOMIE OK			37	73	94
1	21.3					PUSHMATAHA OK					
2	8.5					SEMINOLE OK					
				2	7.9	SEQUOYAH OK					89
1	14.9					TILLMAN OK					
7	7.1	10	6.8	9	4.6	TULSA OK	-5	-32	-35	36	81
		1	6.7	1	9.4	WASHINGTON OK		39			91
9	8.1	12	8.6	13	5.0	OREGON OR	6	-42	-38		
1*	70.1					CLACKAMAS OR					
1*	54.6					COOS OR					
1	13.3					KLAMATH OR					
		1*	89.3			LINCOLN OR					
		2*	19.4			MALHEUR OR					
				1	4.4	MARION OR					80
5	8.2	9	10.4	12	8.3	MULTNOMAH OR	27	-20	1	57	90
1*	133.9					UNION OR					
233*	8.9	286*	8.8	253*	6.5	PENNSYLVANIA PA	0	-27	-27		
43	8.8	53*	8.8	48*	7.0	ALLEGHENY PA	-1	-20	-20	48	88
1	14.1					ARMSTRONG PA					
1	1.6	4	9.8	2	4.3	BEAVER PA	509	-57	165	92	79
2	16.8	2	10.3	5*	18.3	BERKS PA	-38	77	9	63	96
		1	18.9			BLAIR PA					
3*	26.9	2	12.8	2	7.0	BUCKS PA	-52	-45	-74	07	87
				1*	26.5	BUTLER PA					97
1	28.1	2	11.5	2	14.4	CAMBRIA PA	-59	26	-49	26	95
		1*	94.4			CENTRE PA					
1	1.4	6	9.7	1	1.0	CHESTER PA	611	-90	-28	42	71
1*	49.2					CLEARFIELD PA					
1	19.3					CRAWFORD PA					
3*	57.3	1	24.8			CUMBERLAND PA	-57				
5	7.1	6	7.1	3	3.0	DAUPHIN PA	0	-57	-57	18	76
17*	14.3	14	9.2	15*	9.8	DELAWARE PA	-36	6	-32	39	91
		1	5.6	1	2.5	ERIE PA		-56			74
7*	14.2	4	8.6	4	12.1	FAYETTE PA	-39	40	-15	51	93
		1	12.1	1	11.8	FRANKLIN PA		-3			93
		1*	78.1			GREENE PA					
1*	31.3					INDIANA PA					
				1	18.2	LACKAWANNA PA					96
		1	7.3	2*	16.5	LAWRENCE PA		126			96
1	22.8					LUZERNE PA					
		1	15.0			LYCOMING PA					
		2	10.3	2	9.1	MERCER PA		-11			91
1*	133.9					MIFFLIN PA					
2*	61.1			1*	74.3	MONROE PA			22	68	99
4	6.7	4	4.9	5	4.9	MONTGOMERY PA	-26	-2	-27	43	82
1	10.9	1	8.6			NORTHAMPTON PA	-21				
128*	8.8	167*	8.9	144*	6.1	PHILADELPHIA PA	2	-32	-31	40	85
				1*	37.5	SCHUYLKILL PA					98
1*	122.7					SOMERSET PA					
5	10.1	6*	13.4	6*	12.0	WASHINGTON PA	33	-11	19	66	93
2	6.7	3	8.3	2	6.5	WESTMORELAND PA	23	-22	-3	56	86
1	5.5	2	7.4	4*	15.5	YORK PA	34	111	182	93	95
6	10.3	3	4.1	8	8.7	RHODE ISLAND RI	-60	111	-15		
		1*	53.6			KENT RI					
				1	10.7	NEWPORT RI					93
4	9.7	2	3.6	6	8.5	PROVIDENCE RI	-62	133	-13	53	90
2*	50.3			1	12.5	WASHINGTON RI			-75	06	94

* SIGNIFICANTLY MORE DEATHS THAN EXPECTED

CANCER OF THE RECTUM — ICD 154 EXCEPT 154.3

NON-WHITE MALE

1950-1959 DEATHS	1950-1959 RATE	1960-1969 DEATHS	1960-1969 RATE	1970-1979 DEATHS	1970-1979 RATE	COUNTY AND STATE		PERCENT CHANGE 60S/50S	PERCENT CHANGE 70S/60S	PERCENT CHANGE 70S/50S	PERCENTILE %CHANGE 70S/50S	PERCENTILE RATES 1970S
71	3.2	87	3.6	88	3.2	SOUTH CAROLINA	SC	13	-10	1		
1	1.7	2	3.2	3	4.4	AIKEN	SC	91	38	165	92	80
2	8.9	1	4.8			ALLENDALE	SC	-46				
4	5.7	3	4.6	7	10.2	ANDERSON	SC	-20	123	79	81	92
1	3.8	1	3.8			BAMBERG	SC	1				
		1	3.4	2	6.0	BARNWELL	SC		77			85
1	1.4			1	1.6	BEAUFORT	SC			13	64	72
1	1.6	1	1.8	3	3.8	BERKELEY	SC	9	114	133	89	78
				2	9.1	CALHOUN	SC					91
8	5.3	8	3.7	5	1.8	CHARLESTON	SC	-31	-50	-66	11	72
4*	19.0			2	7.4	CHEROKEE	SC			-61	15	88
1	2.0			1	2.5	CHESTER	SC			28	70	74
1	2.7	2	6.8	1	5.4	CHESTERFIELD	SC	149	-21	96	84	83
2	4.1			1	1.7	CLARENDON	SC			-57	18	72
		2	4.3			COLLETON	SC					
3	5.5	2	3.2			DARLINGTON	SC	-42				
2	4.4	1	3.4	1	2.6	DILLON	SC	-22	-24	-41	32	74
				3	5.6	DORCHESTER	SC					84
1	3.7			3	11.2	EDGEFIELD	SC			202	93	93
		1	2.9	1	3.2	FAIRFIELD	SC		10			76
1	1.3	8	8.2	4	4.0	FLORENCE	SC	555	-52	216	94	78
1	2.4	2	5.9	1	1.6	GEORGETOWN	SC	145	-73	-35	36	72
2	2.5	9	7.6	6	4.7	GREENVILLE	SC	208	-38	92	83	81
1	1.7	2	5.1	1	2.0	GREENWOOD	SC	204	-61	18	66	73
		3	10.1	1	2.8	HAMPTON	SC		-73			75
		4	10.8			HORRY	SC					
		1	3.5			JASPER	SC					
1	2.1			1	2.4	KERSHAW	SC			16	66	74
1	3.3			2	5.2	LANCASTER	SC			56	77	82
3	8.3	3	6.9	2	3.9	LAURENS	SC	-17	-44	-53	23	78
1	1.5	1	3.0	2	4.8	LEE	SC	93	61	211	93	81
				4	10.9	LEXINGTON	SC					93
1	7.0					MC CORMICK	SC					
3	5.5	3	6.3	2	3.4	MARION	SC	14	-46	-39	32	77
				2	4.0	MARLBORO	SC					79
1	3.1	1	2.8	1	2.9	NEWBERRY	SC	-9	2	-7	55	75
5	4.8	2	1.5	4	3.2	ORANGEBURG	SC	-69	117	-34	37	76
		2	9.6	1	4.3	PICKENS	SC		-55			80
6	3.3	8	4.1	11	4.5	RICHLAND	SC	25	11	38	73	80
2	2.1	3	2.2	3	2.3	SPARTANBURG	SC	3	7	11	63	73
3	3.7	4	4.4	2	1.7	SUMTER	SC	20	-62	-55	21	72
3	10.7					UNION	SC					
3	4.3	3	6.0	1	2.0	WILLIAMSBURG	SC	38	-66	-53	24	73
1	1.6	3	5.1	1	1.3	YORK	SC	219	-75	-20	48	71
2	2.5	4	4.0	6	6.0	SOUTH DAKOTA	SD	58	51	138		
1*	14.9					CHARLES MIX	SD					
		1	14.9			CORSON	SD					
				1*	122.7	DAY	SD					99
		1*	42.4			JACKSON	SD					
				1*	106.0	MEADE	SD					99
				1*	47.2	MELLETTE	SD					99
1*	136.6					MOODY	SD					
		2	7.8	1	3.8	SHANNON	SD		-51			78
				2*	16.5	TODD	SD					95
101	4.5	120	4.9	116	4.5	TENNESSEE	TN	10	-9	0		
1	7.8	1	9.0			BEDFORD	TN	16				
		2	16.5	2	15.6	BLOUNT	TN		-5			95
1	15.8					BRADLEY	TN					
1	7.5	1	5.7	1	6.5	CARROLL	TN	-25	15	-14	52	86
		1	12.8			CHESTER	TN					
				1*	27.0	COCKE	TN					97
1	7.7					CROCKETT	TN					
16	5.4	15	4.7	15	4.3	DAVIDSON	TN	-13	-8	-21	48	79
2	10.0	1	4.9			DYER	TN	-51				
		1	1.5			FAYETTE	TN					
1	10.2	1	10.2			FRANKLIN	TN	0				
1	2.2	4	9.9			GIBSON	TN	340				
1	5.6	1	4.3	3*	19.0	GILES	TN	-24	342	237	95	96
1	14.3			1	13.3	HAMBLEN	TN			-7	55	94
6	4.0	17*	10.5	7	3.4	HAMILTON	TN	165	-67	-13	52	77
		1*	188.7			HANCOCK	TN					
1	3.4	1	3.3	1	2.8	HARDEMAN	TN	-5	-13	-17	49	75
1	11.9			1	29.8	HARDIN	TN			151	90	98
				1*	21.5	HAWKINS	TN					97
2	4.3	1	1.8	2	4.1	HAYWOOD	TN	-58	121	-7	55	79
1	20.2					HENDERSON	TN					
1	6.0			2	9.9	HENRY	TN			66	79	92
		1*	86.0	1*	38.2	HOUSTON	TN		-56			98
				1*	30.3	JEFFERSON	TN					98
3	3.0	2	1.8	3	3.1	KNOX	TN	-41	77	5	60	76
2	5.5	2	6.0			LAUDERDALE	TN	9				
1	7.2	1	8.6	1	7.7	LINCOLN	TN	19	-11	6	61	89
				1*	37.2	LOUDON	TN					98
				2*	21.9	MC MINN	TN					97
5	6.7	2	2.7	4	4.0	MADISON	TN	-60	51	-40	32	79
1	20.2					MARION	TN					
				1	8.7	MARSHALL	TN					90
3	6.5	2	5.1	1	2.2	MAURY	TN	-21	-57	-66	11	73
		1*	23.4			MONROE	TN					
2	5.1	1	2.3	4	9.4	MONTGOMERY	TN	-56	311	82	81	91
		1	6.2	1	5.7	OBION	TN		-9			84
		1*	78.1			PUTNAM	TN					
		1*	62.9	1*	54.6	RHEA	TN		-13			99
2	7.8					RUTHERFORD	TN					
36	4.7	51	5.4	48	4.7	SHELBY	TN	15	-13	0	57	81

* SIGNIFICANTLY MORE DEATHS THAN EXPECTED

NON-WHITE MALE

CANCER OF THE RECTUM

ICD 154 EXCEPT 154.3

1950-1959 DEATHS	1950-1959 RATE	1960-1969 DEATHS	1960-1969 RATE	1970-1979 DEATHS	1970-1979 RATE	COUNTY AND STATE	PERCENT CHANGE 60S/50S	PERCENT CHANGE 70S/60S	PERCENT CHANGE 70S/50S	PERCENTILE %CHANGE 70S/50S	PERCENTILE RATES 1970S
1	8.6	2*	16.4	2*	27.5	SULLIVAN TN	90	68	220	94	97
1	4.6	1	4.7			SUMNER TN	1				
4	9.2	1	2.7	3	6.3	TIPTON TN	-71	137	-32	39	86
				1*	37.5	TROUSDALE TN					98
				1	14.5	WARREN TN					95
1	3.4	1	4.5			WASHINGTON TN	31				
		1	5.2	3	14.2	WILLIAMSON TN		171			94
1	5.0					WILSON TN					
176	4.3	201	4.1	218	3.2	TEXAS TX	-6	-21	-26		
2	4.0	1	2.0	1	3.1	ANDERSON TX	-49	51	-23	45	76
2	14.6	1	5.4			AUSTIN TX	-63				
1	3.3	2	6.4	2	6.8	BASTROP TX	96	7	109	85	87
7	5.4	16	8.4	9	2.8	BEXAR TX	55	-67	-49	25	75
1	1.6	2	2.9	4	6.4	BOWIE TX	81	117	293	97	86
		1	2.6	1	2.1	BRAZORIA TX		-19			73
3	6.9	1	1.9	1	1.6	BRAZOS TX	-72	-17	-77	05	72
1*	38.2					BROWN TX					
		1	6.3			BURLESON TX					
1	6.5	1	6.2	2	15.6	CAMP TX	-4	151	140	90	95
		1	2.5	1	4.1	CASS TX		61			79
				1	7.8	CHAMBERS TX					89
		3	6.8	1	2.1	CHEROKEE TX		-68			73
1*	22.9					CHILDRESS TX					
1	5.8	1	4.5	2	8.8	COLLIN TX	-23	97	51	75	90
		2	11.0			COLORADO TX					
		1*	80.4			COTTLE TX					
				2*	35.9	CROSBY TX					98
9	2.4	22	4.2	24	3.3	DALLAS TX	75	-21	37	73	77
				1	12.3	DAWSON TX					93
1*	37.2					DELTA TX					
1	7.7					DENTON TX					
		2	8.4	3*	17.6	DE WITT TX		108			96
		1	12.2	1	9.4	ECTOR TX		-23			91
2	4.4	1	2.1	3	5.9	ELLIS TX	-52	179	34	72	84
2	11.4			3	1.4	EL PASO TX			-88	01	71
				1*	112.4	ERATH TX					99
2	6.1			2	7.0	FALLS TX			15	65	87
3*	21.0	1	6.7			FANNIN TX	-68				
2	7.0	3	15.1	1	4.4	FAYETTE TX	114	-71	-38	34	80
		1*	28.3			FLOYD TX					
2	7.9			1	1.9	FORT BEND TX			-76	06	72
		3	8.8	3	8.2	FREESTONE TX		-7			90
6	11.5	9	7.1	8	6.8	GALVESTON TX	-38	-5	-41	31	87
				1*	30.3	GARZA TX					98
		2	10.6			GONZALES TX					
1*	34.4					GRAY TX					
2	5.5			3	10.6	GRAYSON TX			92	83	92
2	4.4	3	4.4	4	5.4	GREGG TX	1	21	23	68	83
1	3.5	2	5.2	1	9.8	GRIMES TX	50	86	179	92	91
1	6.3	1	6.0			GUADALUPE TX	-5				
		1	15.6			HALE TX					
		2*	44.8			HARDEMAN TX					
1	9.4			1	4.0	HARDIN TX			-57	19	79
40	6.5	36	3.9	59	4.5	HARRIS TX	-40	14	-32	40	80
		5	5.5	3	3.3	HARRISON TX		-41			77
		2	10.2			HENDERSON TX					
1	5.7	1	5.4			HILL TX	-6				
		1	11.2			HOPKINS TX					
2	6.5	1	2.6	1	2.1	HOUSTON TX	-60	-19	-68	09	73
3	11.5	1	4.1	2	8.0	HUNT TX	-65	97	-30	41	89
3*	36.6	1	11.2			JACKSON TX	-70				
1	4.1					JASPER TX					
7	4.4	7	3.1	8	3.0	JEFFERSON TX	-28	-3	-31	41	76
		1	14.0			JOHNSON TX					
1	2.5					KAUFMAN TX					
		1	37.7	1	16.7	KERR TX		-56			96
				1*	133.9	KINNEY TX					99
				1	12.8	KLEBERG TX					94
		5*	15.9	1	5.5	LAMAR TX		-66			83
		1	10.8			LEE TX					
2	6.7					LIBERTY TX					
1	2.4	1	3.9			LIMESTONE TX	59				
1	3.2					LUBBOCK TX					
6	5.8	2	1.8	7	5.9	MC LENNAN TX	-69	230	2	58	84
1	8.5	2	15.3	1	5.9	MADISON TX	80	-62	-31	40	85
		1	3.2			MARION TX					
		1	3.0	1	3.5	MATAGORDA TX		14			77
1	9.4	2	8.6	1	4.0	MIDLAND TX	-8	-54	-58	17	79
2	8.6	2	11.8			MILAM TX	37				
2	27.5					MORRIS TX					
3	10.5	1	3.6			NACOGDOCHES TX	-65				
3	6.3	1	2.6	1	1.4	NAVARRO TX	-60	-44	-77	05	71
		1	1.7			NUECES TX					
1	5.4			1	9.4	ORANGE TX			74	80	91
				1	16.1	PALO PINTO TX					95
1	5.3			1	4.4	PANOLA TX			-17	50	80
2	10.1			1	3.8	POLK TX			-62	15	78
1	9.4					POTTER TX					
3	14.9	1	5.5			RED RIVER TX	-63				
		1	4.4	2	5.8	ROBERTSON TX		32			84
1	15.6					ROCKWALL TX					
1	2.4	1	1.6	2	3.0	RUSK TX	-36	90	21	67	75
				1	11.2	SABINE TX					93
				3*	23.5	SAN AUGUSTINE TX					97
		1	5.1			SAN JACINTO TX					
				2	10.0	SHELBY TX					92
3	3.5	1	1.1	1	.7	SMITH TX	-69	-35	-80	03	71
1*	119.2					SWISHER TX					
2	1.0	11	5.0	12	4.1	TARRANT TX	398	-18	309	98	79

* SIGNIFICANTLY MORE DEATHS THAN EXPECTED

NON-WHITE MALE — CANCER OF THE RECTUM — ICD 154 EXCEPT 154.3

1950-1959 DEATHS	1950-1959 RATE	1960-1969 DEATHS	1960-1969 RATE	1970-1979 DEATHS	1970-1979 RATE	COUNTY AND STATE	PERCENT CHANGE 60S/50S	PERCENT CHANGE 70S/60S	PERCENT CHANGE 70S/50S	PERCENTILE %CHANGE 70S/50S	PERCENTILE RATES 1970S
1	7.6					TAYLOR TX					
				1*	26.0	TERRY TX					97
				1	8.8	TITUS TX					91
1	12.8	2	13.0			TOM GREEN TX	1				
4	4.2	4	3.7	5	3.3	TRAVIS TX	-11	-13	-23	45	77
1	8.6	1	7.9			TRINITY TX	-9				
2	7.8	1	4.2	1	3.7	UPSHUR TX	-46	-11	-52	24	78
1	6.1	2	11.0	2	6.8	VICTORIA TX	81	-38	12	63	87
3	9.5					WALKER TX					
1	4.3			2	7.0	WALLER TX			65	78	87
4	9.1	3	9.9	1	3.1	WASHINGTON TX	9	-68	-66	11	76
1	2.9	3	7.3			WHARTON TX	149				
		1*	49.2			WHEELER TX					
2	9.0	2	10.1			WICHITA TX	12				
1	8.0	1	3.8	1	3.6	WILLIAMSON TX	-53	-4	-55	21	78
				1	7.3	WOOD TX					88
6	9.7	3	3.5	2	1.7	UTAH UT	-64	-51	-83		
1*	26.8					BOX ELDER UT					
		1*	34.2			DAVIS UT					
2	7.8	1	2.3	1	2.0	SALT LAKE UT	-70	-14	-74	07	73
3*	24.1	1	4.9	1	4.1	WEBER UT	-80	-15	-83	02	79
164	6.0	182	5.9	183	5.3	VIRGINIA VA	-1	-11	-12		
2	3.8	4	7.3	3	6.3	ACCOMACK VA	94	-15	66	79	85
3	7.9	5	10.5	3	5.4	ALBEMARLE#VA	34	-49	-32	40	83
1	7.2			1	5.9	ALLEGHANY#VA			-18	49	85
1	6.2	1	5.3	1	6.7	AMELIA VA	-15	26	7	61	87
1	3.9			1	4.2	AMHERST VA			8	62	79
1	12.2			1	8.5	APPOMATTOX VA			-30	41	90
3	13.4	4	11.1	1	2.8	ARLINGTON#VA	-17	-75	-79	03	75
1	5.1	2	8.5	2	6.6	AUGUSTA#VA	65	-22	28	71	86
6	7.1	15*	15.0	8	7.2	BEDFORD#VA	112	-52	1	57	88
				1	13.2	BOTETOURT VA					94
				4	10.7	BRUNSWICK VA					92
				1	4.7	BUCKINGHAM VA					81
1	4.2	2	7.1	1	3.1	CAROLINE VA	70	-56	-25	44	76
1	7.5			2	13.9	CHARLES CITY VA			85	82	94
		2	10.3	4*	16.4	CHARLOTTE VA		59			95
26	8.0	29	7.4	22	5.1	CHESTERFIELD#VA	-7	-31	-36	35	82
		1	13.2			CLARKE VA					
2	16.7			1	5.2	CULPEPER VA			-69	09	83
1	6.4	1	4.0	2	11.9	CUMBERLAND VA	-37	194	85	82	93
4	2.7	7	4.5	10	6.8	DINWIDDIE#VA	66	52	153	91	87
2	20.7	1	6.6	2	16.1	ESSEX VA	-68	146	-22	46	95
3	7.1	7	6.8	6	6.3	FAIRFAX#VA	-5	-7	-11	53	86
1	4.5	4*	14.5	3	10.4	FAUQUIER VA	224	-28	132	88	92
		1	7.6	1	5.8	FRANKLIN VA		-23			84
1	10.5					FREDERICK#VA					
		1*	57.3			GILES VA					
1	7.4			1	4.4	GLOUCESTER VA			-41	31	80
1	4.3			1	4.8	GOOCHLAND VA			10	63	81
				1*	24.7	GREENE VA					97
1	2.8			1	2.9	GREENSVILLE#VA			4	59	75
1	1.8	4	6.2	3	5.1	HALIFAX#VA	247	-18	185	93	82
1	4.1			1	2.8	HANOVER VA			-32	39	75
2	3.7	2	3.3			HENRY#VA	-10				
1*	172.0					HIGHLAND VA					
2	7.2	3	9.3	1	5.4	ISLE OF WIGHT VA	30	-42	-25	44	83
7*	18.2	2	4.1	1	1.8	JAMES CITY#VA	-77	-57	-90	00	72
1	5.2	1	5.1	2	13.2	KING AND QUEEN VA	-3	160	153	91	94
1	11.6	1	13.6			KING GEORGE VA	17				
1	7.9			1	6.7	KING WILLIAM VA			-14	51	87
				2	11.7	LANCASTER VA					93
				1*	122.7	LEE VA					99
		1	6.5	3*	15.8	LOUDOUN VA		143			95
3	10.3	1	4.1	1	3.5	LOUISA VA	-60	-15	-66	11	77
1	3.9					LUNENBURG VA					
				1	10.6	MADISON VA					92
4	9.3			1	1.9	MECKLENBURG VA			-80	03	72
4*	27.1	1	6.3			MIDDLESEX VA	-77				
1	14.5					MONTGOMERY#VA					
				1	9.4	NELSON VA					91
6*	15.4	1	2.5			NORTHAMPTON VA	-84				
				1	4.5	NORTHUMBERLAND VA					80
2	8.3	1	3.8	3	10.4	NOTTOWAY VA	-54	171	26	70	92
				4*	31.0	ORANGE VA					98
3	2.5	10*	10.6	6	5.9	PITTSYLVANIA#VA	316	-44	133	89	85
2	15.8					POWHATAN VA					
		1	4.0			PRINCE EDWARD VA					
		1	8.9			PRINCE WILLIAM#VA					
1	11.2					PULASKI VA					
2*	29.9					RICHMOND VA					
8	9.7	7	7.3	4	3.6	ROANOKE#VA	-25	-50	-63	15	78
1	7.2			1	8.2	ROCKBRIDGE#VA			14	65	90
				1*	49.2	SHENANDOAH VA					99
2	4.5	3	7.6	4	6.6	SOUTHAMPTON#VA	67	-14	45	74	86
2	10.9	3	14.4			SPOTSYLVANIA#VA	32				
		2*	38.7			STAFFORD VA					
				2	12.8	SURRY VA					94
		2	7.0	1	5.2	SUSSEX VA		-25			83
1	8.6	1	14.9	1	9.9	TAZEWELL VA	72	-33	15	65	92
1*	18.1					WARREN VA					
1	5.1			1	4.3	WESTMORELAND VA			-16	50	79
				2*	32.8	WISE#VA					98
		1	35.1			WYTHE VA					
27	6.5	27	5.3	33	5.6	CHESAPEAKE CITY#VA	-18	5	-14	51	84
5	6.9	8	7.7	6	4.3	HAMPTON CITY VA	11	-43	-37	34	80
2	1.3	9	8.3	8	5.0	NEWPORT NEWS CITY VA	528	-40	278	96	82

* SIGNIFICANTLY MORE DEATHS THAN EXPECTED

COUNTIES COMBINED; SEE TECHNICAL NOTES

NON-WHITE MALE — CANCER OF THE RECTUM — ICD 154 EXCEPT 154.3

1950-1959 DEATHS	1950-1959 RATE	1960-1969 DEATHS	1960-1969 RATE	1970-1979 DEATHS	1970-1979 RATE	COUNTY AND STATE	PERCENT CHANGE 60S/50S	PERCENT CHANGE 70S/60S	PERCENT CHANGE 70S/50S	PERCENTILE %CHANGE 70S/50S	PERCENTILE RATES 1970S
4	4.2	2	2.5	1	1.5	SUFFOLK CITY VA	-40	-40	-64	13	72
23	8.4	23	6.2	25	4.5	WASHINGTON WA	-26	-27	-46		
1*	53.6					CLARK WA					
1*	596.2					COLUMBIA WA					
17*	12.0	18	8.2	19	5.9	KING WA	-31	-28	-51	25	84
1	37.2					KITSAP WA					
1	1.2	4	14.3	2	4.5	PIERCE WA	1123	-68	288	97	80
2	10.1			3	10.9	SPOKANE WA			7	62	93
		1	20.2			WHATCOM WA					
				1	2.5	YAKIMA WA					74
27	5.7	24	5.3	22	5.7	WEST VIRGINIA WV	-7	7	0		
		2*	18.4			BERKELEY WV					
1*	49.7					BROOKE WV					
2	7.9			2	9.9	CABELL WV			26	69	92
4	8.5			1	3.0	FAYETTE WV			-64	12	76
1	11.4					GREENBRIER WV					
		1	33.5			HANCOCK WV					
		1	7.2	1	16.1	HARRISON WV		124			95
2	16.8	1	7.0			JEFFERSON WV	-58				
3	4.1	6	9.3	6	9.6	KANAWHA WV	130	3	137	89	91
1	4.4			1	5.1	LOGAN WV			15	65	82
2	2.2	3	4.5	5	8.0	MC DOWELL WV	102	80	263	96	89
1	7.0	1	5.6	1	4.5	MARION WV	-20	-20	-36	35	80
2	4.4	2	5.3	1	2.6	MERCER WV	20	-51	-42	30	74
3*	34.1					MINGO WV					
		2	16.7	1	14.4	OHIO WV		-14			95
1*	38.3					POCAHONTAS WV					
3	4.9	4	8.1	2	3.2	RALEIGH WV	65	-61	-36	36	76
1*	24.8					UPSHUR WV					
				1*	24.6	WOOD WV					97
		1	15.3			WYOMING WV					
12	8.0	3	1.1	16	3.8	WISCONSIN WI	-87	254	-53		
1*	74.3					LANGLADE WI					
9	10.2	3	1.7	13	4.5	MILWAUKEE WI	-84	165	-56	19	80
				1	9.3	RACINE WI					91
1	27.0			1	6.0	ROCK WI			-78	04	85
				1*	31.5	VILAS WI					98
1*	30.4					WAUKESHA WI					
		1	4.4	1	1.7	WYOMING WY		-61			
		1*	106.0			ALBANY WY					
				1*	20.4	SWEETWATER WY					96

* SIGNIFICANTLY MORE DEATHS THAN EXPECTED

CANCER OF THE RECTUM ICD 154 EXCEPT 154.3

NON-WHITE FEMALE 1950-1959 DEATHS	RATE	1960-1969 DEATHS	RATE	1970-1979 DEATHS	RATE	COUNTY AND STATE	PERCENT CHANGE 60S/50S	70S/60S	70S/50S	PERCENTILE %CHANGE 70S/50S	RATES 1970S
3334	5.1	3733	4.3	3892	3.2	UNITED STATES	-16	-27	-38		
168	4.3	149	3.4	155	3.0	ALABAMA AL	-22	-11	-31		
1	3.1					AUTAUGA AL					
1	3.9	1	2.4	1	2.3	BALDWIN AL	-39	-3	-41	41	79
				3	5.0	BARBOUR AL					90
1	5.1	1	5.7	1	5.4	BIBB AL	12	-6	5	73	91
				1*	20.2	BLOUNT AL					98
4	8.2	1	2.6			BULLOCK AL	-68				
2	4.2	2	5.4	2	2.5	BUTLER AL	30	-53	-39	43	80
1	2.0	5	6.7	2	2.2	CALHOUN AL	233	-67	9	74	79
3	6.3	1	1.6	2	3.4	CHAMBERS AL	-74	104	-47	36	84
1	8.6			1	4.1	CHILTON AL			-52	30	87
1	3.2	1	2.5			CHOCTAW AL	-23				
1	2.4	2	3.9	2	4.1	CLARKE AL	64	4	70	88	87
		3*	47.0			CLAY AL					
		1	4.3			COFFEE AL					
2	5.7			3	6.6	COLBERT AL			14	78	93
		2	5.1	1	1.9	CONECUH AL		-63			78
		1	2.8			COVINGTON AL					
1	4.8			1	3.6	CRENSHAW AL			-24	57	85
				1*	49.1	CULLMAN AL					99
1	5.5			1	3.6	DALE AL			-34	48	85
8	5.0	5	3.0	8	4.8	DALLAS AL	-40	63	-3	69	89
2	5.5			1	3.1	ELMORE AL			-44	38	82
2	4.1			3	4.0	ESCAMBIA AL			-4	68	86
2	3.9	2	3.3	4	5.0	ETOWAH AL	-17	51	26	82	90
1	10.6					FAYETTE AL					
1	13.8					FRANKLIN AL					
4	7.2	1	2.3	1	1.8	GREENE AL	-68	-21	-75	09	77
1	1.4	4	6.0	2	2.7	HALE AL	318	-55	89	90	81
2	2.8	2	3.5	2	2.7	HOUSTON AL	25	-23	-3	68	81
		1	8.6			JACKSON AL					
57	6.4	46	4.3	48	3.7	JEFFERSON AL	-32	-15	-42	40	85
		1	11.8			LAMAR AL					
1	3.3	1	2.8	2	4.7	LAUDERDALE AL	-15	70	45	85	89
		1	4.3	1	1.7	LAWRENCE AL		-61			77
3	4.8	2	2.2	1	1.3	LEE AL	-54	-42	-74	10	76
2	4.3	3	6.0			LOWNDES AL	40				
4	4.9	2	2.0	4	3.1	MACON AL	-58	51	-37	46	82
1	1.4	4	4.3	4	3.5	MADISON AL	211	-19	152	95	84
3	3.3	1	1.2			MARENGO AL	-64				
17	5.5	17	4.3	17	3.3	MOBILE AL	-23	-22	-40	43	83
1	1.7			1	1.7	MONROE AL			-5	67	77
9	3.1	13	4.2	9	2.5	MONTGOMERY AL	36	-41	-19	58	80
3	9.0	2	4.4	2	4.6	MORGAN AL	-51	4	-50	33	88
		2	4.2	4	5.7	PERRY AL		38			91
1	1.9			1	1.1	PICKENS AL			-42	40	75
3	5.2	4	8.0	2	4.3	PIKE AL	53	-46	-17	60	88
				1	6.3	RANDOLPH AL					92
2	2.6	3	3.2	2	2.1	RUSSELL AL	21	-33	-18	60	79
		1	5.5			ST CLAIR AL					
1	4.3	1	3.5	2	4.8	SHELBY AL	-17	38	14	77	89
1	1.4	2	2.7			SUMTER AL	94				
6	8.2	3	4.0	1	1.0	TALLADEGA AL	-51	-74	-87	02	75
		1	2.4			TALLAPOOSA AL					
7	6.8	1	.9	7	4.3	TUSCALOOSA AL	-87	394	-36	46	87
1	3.0	1	3.3	1	2.6	WALKER AL	9	-20	-13	63	81
1	5.4			1	4.5	WASHINGTON AL			-17	61	88
1	1.6	1	1.5	1	1.6	WILCOX AL	-5	6	1	71	77
		1	1.1	7	5.3	ALASKA AK		377			
6	1.9	7	1.8	6	.6	ARIZONA AZ	-7	-68	-70		
				1	1.0	APACHE AZ					75
1	4.4					COCONINO AZ					
3	2.9	5	3.8	4	1.0	MARICOPA AZ	30	-74	-66	18	75
		2	2.6	1	.4	PIMA AZ		-83			75
2	8.3					PINAL AZ					
51	3.0	69	3.6	55	2.5	ARKANSAS AR	20	-31	-18		
		4*	14.8	1	3.5	ARKANSAS AR		-76			84
		1	2.4	1	2.1	ASHLEY AR		-13			78
1	5.1					BRADLEY AR					
				1	10.5	CALHOUN AR					96
2	3.3	5	7.2	2	2.0	CHICOT AR	117	-72	-40	43	78
1	5.4					CLARK AR					
1	9.3					CLEVELAND AR					
2	5.4	1	2.5			COLUMBIA AR	-53				
1	6.1			1	5.2	CONWAY AR			-15	62	90
1*	21.3					CRAWFORD AR					
4	4.4	3	2.4			CRITTENDEN AR	-45				
1	4.2					CROSS AR					
1	3.9					DALLAS AR					
1	1.5	3	5.8	3	4.7	DESHA AR	293	-18	221	99	89
2	7.3	1	3.3			DREW AR	-55				
1	2.5	4*	13.3			GARLAND AR	442				
		3	7.8			HEMPSTEAD AR					
1	10.4					HOT SPRING AR					
		1	6.2			HOWARD AR					
		3*	19.9			JACKSON AR					
3	1.6	3	1.8	12*	5.6	JEFFERSON AR	10	209	240	99	91
1	5.6	2	9.5	1	2.3	LAFAYETTE AR	71	-76	-59	25	79
		1	1.4			LEE AR					
				1	4.8	LINCOLN AR					89

* SIGNIFICANTLY MORE DEATHS THAN EXPECTED

NON-WHITE FEMALE — CANCER OF THE RECTUM — ICD 154 EXCEPT 154.3

1950-1959 DEATHS	1950-1959 RATE	1960-1969 DEATHS	1960-1969 RATE	1970-1979 DEATHS	1970-1979 RATE	COUNTY AND STATE	PERCENT CHANGE 60S/50S	PERCENT CHANGE 70S/60S	PERCENT CHANGE 70S/50S	PERCENTILE %CHANGE 70S/50S	PERCENTILE RATES 1970S
				1	3.3	LONOKE AR					83
2	5.0	1	2.3	2	3.3	MILLER AR	-54	43	-34	49	83
2	2.6	1	.9	2	2.0	MISSISSIPPI AR	-64	110	-25	55	78
1	1.9	5*	12.1	3	5.5	MONROE AR	548	-54	198	98	91
2	9.5					NEVADA AR					
4	9.5	2	2.7	1	1.5	OUACHITA AR	-71	-46	-85	03	76
		6	4.9	4	3.5	PHILLIPS AR		-28			84
1	10.2					POINSETT AR					
10	4.7	9	3.3	11	3.3	PULASKI AR	-29	-1	-30	52	83
2	2.8	4	4.9	2	2.0	ST FRANCIS AR	74	-60	-30	51	78
				2	5.4	SEBASTIAN AR					91
1	18.6					SEVIER AR					
1	1.9	2	3.3	2	2.4	UNION AR	72	-26	27	82	80
1	22.5	1	17.0			WHITE AR	-25				
		3	9.9	2	8.5	WOODRUFF AR		-14			95
140	5.7	191	3.8	269	2.4	CALIFORNIA CA	-33	-37	-57		
15	5.8	27	4.7	22	2.1	ALAMEDA CA	-18	-55	-63	20	78
				1	4.6	BUTTE CA					89
1	6.2	1	.6	5	2.0	CONTRA COSTA CA	-91	249	-67	16	78
5	6.2	3	3.6	6	2.3	FRESNO CA	-42	-34	-62	21	80
1	8.2	1	6.2	2	9.3	IMPERIAL CA	-24	49	14	77	95
3	6.1	4	5.7	2	1.1	KERN CA	-5	-81	-82	04	75
80	6.6	101	4.0	145	2.8	LOS ANGELES CA	-40	-28	-57	25	81
1	4.4	3	7.9			MONTEREY CA	80				
		1*	22.3			NAPA CA					
		3	8.3	3	.9	ORANGE CA		-89			75
				1	4.2	PLACER CA					87
		1	1.4	2	.7	RIVERSIDE CA		-48			75
4	4.7	3	2.2	8	2.0	SACRAMENTO CA	-53	-9	-57	25	78
2	5.8	3	4.1	6	2.6	SAN BERNARDINO CA	-29	-37	-55	28	80
2	2.5	8	4.7	11	2.2	SAN DIEGO CA	91	-52	-9	65	79
13	4.6	20	3.6	32	3.1	SAN FRANCISCO CA	-21	-14	-32	49	83
4	7.0	3	4.2	3	1.8	SAN JOAQUIN CA	-40	-56	-74	09	77
1	2.7	1	1.0	6	3.0	SAN MATEO CA	-64	210	11	76	82
		2	6.8	2	4.5	SANTA BARBARA CA		-34			88
2	6.0	1	1.8	5	1.0	SANTA CLARA CA	-70	-41	-83	04	75
2*	29.2	1	5.4	1	1.7	SANTA CRUZ CA	-81	-68	-94	00	77
		1*	14.9			SISKIYOU CA					
		2	3.4	2	2.1	SOLANO CA		-38			79
1	16.4			1	1.9	SONOMA CA			-89	01	78
1	14.9					STANISLAUS CA					
				2	4.2	TULARE CA					87
		1	18.7	1	.6	VENTURA CA		-97			75
2*	44.7					YOLO CA					
8	5.6	14	6.0	7	1.6	COLORADO CO	8	-73	-71		
6	6.4	12*	7.4	2	.8	DENVER-ARAPAHOE#CO	15	-90	-88	01	75
		1	6.0	2	5.7	EL PASO CO		-6			91
				1*	7.7	LARIMER CO					94
1*	267.9					LAS ANIMAS CO					
1	9.8			1	6.2	PUEBLO CO			-37	46	92
		1*	45.9	1	10.0	WELD CO		-78			96
18	8.4	21	5.3	20	3.2	CONNECTICUT CT	-37	-39	-62		
5	6.5	8	6.8	8	4.3	FAIRFIELD CT	5	-37	-35	48	87
7	13.8	5	4.6	4	1.5	HARTFORD CT	-67	-67	-89	01	76
1*	28.3	1*	38.5	1	17.2	LITCHFIELD CT	36	-55	-39	44	98
		1	7.1			MIDDLESEX CT					
4	5.4	5	4.2	5	3.1	NEW HAVEN CT	-22	-27	-43	39	82
1	6.9			2	7.7	NEW LONDON CT			12	76	94
		1*	29.5			TOLLAND CT					
10	4.9	15	5.8	12	3.4	DELAWARE DE	17	-40	-30		
3	8.7	2	4.8	1	1.5	KENT DE	-44	-68	-82	04	77
4	2.9	11	6.7	6	2.9	NEW CASTLE DE	129	-57	-2	69	81
3	6.8	2	3.5	5	6.9	SUSSEX DE	-49	97	0	70	93
94*	8.1	109*	6.5	121*	5.2	DIST. OF COLUMBIA DC	-20	-20	-36		
94*	8.1	109*	6.5	121*	5.2	WASHINGTON DC	-20	-20	-36	46	90
103	3.7	146	4.2	154	3.0	FLORIDA FL	14	-29	-20		
2	2.9	6	6.4	2	1.9	ALACHUA FL	119	-70	-34	48	78
1	4.3					BAY FL					
				1	7.3	BRADFORD FL					94
4*	11.6	2	6.6			BREVARD FL	-43				
6	4.6	3	1.8	9	3.0	BROWARD FL	-60	65	-34	48	82
1	14.9					CITRUS FL					
1	6.9					COLLIER FL					
2	6.5	1	3.8			COLUMBIA FL	-41				
11	3.3	25	5.9	29	3.0	DADE FL	77	-49	-10	64	82
		1	10.1			DE SOTO FL					
16	4.1	29	6.0	23	3.7	DUVAL FL	46	-39	-11	64	85
5	5.3	3	2.1	11	5.8	ESCAMBIA FL	-59	169	9	75	91
1	10.4					FRANKLIN FL					
1	1.0	4	3.6	1	.8	GADSDEN FL	275	-79	-22	58	75
1*	60.7	1*	67.0			GILCHRIST FL	10				
1*	8.5					HENDRY FL					
				1	9.6	HERNANDO FL					96
1	6.7					HIGHLANDS FL					
15*	7.2	6	2.5	13	3.7	HILLSBOROUGH FL	-65	45	-49	34	85
1	1.9	4	7.2	1	1.6	JACKSON FL	283	-77	-14	62	77
1	4.9			1	4.0	JEFFERSON FL			-19	59	86
		2	4.8	3	5.3	LAKE FL		11			90
		1	4.6	1	1.7	LEE FL		-63			77

* SIGNIFICANTLY MORE DEATHS THAN EXPECTED

\# COUNTIES COMBINED; SEE TECHNICAL NOTES

NON-WHITE FEMALE — CANCER OF THE RECTUM — ICD 154 EXCEPT 154.3

1950-1959 DEATHS	1950-1959 RATE	1960-1969 DEATHS	1960-1969 RATE	1970-1979 DEATHS	1970-1979 RATE	COUNTY AND STATE	PERCENT CHANGE 60S/50S	PERCENT CHANGE 70S/60S	PERCENT CHANGE 70S/50S	PERCENTILE %CHANGE 70S/50S	PERCENTILE RATES 1970S
		3	3.6	2	1.1	LEON FL		-70			75
		1	7.6			LEVY FL					
1	3.6	4*	13.7			MADISON FL	281				
		2	4.4	3	6.0	MANATEE FL		37			92
1	1.2	4	4.9	6	5.1	MARION FL	318	2	329	99	90
		1	7.0	1	6.5	MONROE FL		-7			93
1	3.5					NASSAU FL					
7	7.4	7	4.2	7	3.0	ORANGE FL	-43	-28	-59	24	82
5	2.8	10	4.9	14	5.2	PALM BEACH FL	72	7	85	90	90
		1	8.2			PASCO FL					
4	4.2	5	3.4	6	2.6	PINELLAS FL	-20	-24	-39	44	80
1	.9	6	4.5	4	2.1	POLK FL	382	-53	125	93	78
3	9.5	3	6.2	1	2.0	PUTNAM FL	-35	-67	-79	06	78
3	9.5	1	2.6	1	2.6	ST JOHNS FL	-72	1	-72	11	81
				3	4.1	ST LUCIE FL					87
		1	16.3			SANTA ROSA FL					
1	3.0	1	3.0	2	4.2	SARASOTA FL	-1	41	40	84	87
1	1.4	3	5.4	1	1.1	SEMINOLE FL	278	-80	-24	57	75
				1	6.7	SUMTER FL					93
				1	4.5	SUWANNEE FL					88
				1	5.7	TAYLOR FL					91
3	3.0	4	3.8	3	2.1	VOLUSIA FL	27	-44	-29	52	79
1	18.6			1	10.5	WAKULLA FL			-44	38	96
		1	10.6			WASHINGTON FL					
183	4.4	153	3.0	141	2.3	GEORGIA GA	-30	-24	-47		
				1*	13.8	ATKINSON GA					97
1*	34.2					BACON GA					
1	1.2	1	1.2	1	1.3	BALDWIN GA	-3	14	10	75	76
1	5.1					BARTOW GA					
1	4.7	1	2.0			BEN HILL GA	-58				
1	15.7					BERRIEN GA					
4	2.0	5	2.2	6	2.1	BIBB GA	12	-5	7	74	79
		1	3.3	2	5.2	BROOKS GA		58			90
1	3.0	1	1.8			BULLOCH GA	-40				
4	5.2	1	1.1			BURKE GA	-78				
1	7.0			1	4.3	BUTTS GA			-38	45	88
1	4.3					CALHOUN GA					
		1	7.2			CAMDEN GA					
				2	6.8	CARROLL GA					93
		1*	12.5			CHARLTON GA					
12	5.3	8	2.5	5	1.4	CHATHAM GA	-53	-44	-74	10	76
1	14.5					CHATTOOGA GA					
				4	6.4	CLARKE GA					92
		1	6.1			CLAY GA					
1	8.6			1	4.8	CLAYTON GA			-44	37	89
1	2.2	2	5.6	1	2.2	COBB GA	153	-60	1	71	79
2*	16.2					COOK GA					
2	4.4	3	6.2	2	4.1	COWETA GA	43	-33	-5	68	87
3	9.7			1	1.8	CRISP GA			-81	04	77
2	5.2			1	1.5	DECATUR GA			-72	12	76
4	5.0	7	4.8	7	2.9	DE KALB GA	-4	-39	-42	40	82
				1	4.1	DOOLY GA					87
2	1.6	3	2.6	6	3.9	DOUGHERTY GA	65	49	147	95	86
1	4.3	2	5.6			EARLY GA	32				
1	9.3					EVANS GA					
				1*	12.8	FAYETTE GA					97
				3	5.2	FLOYD GA					90
50*	7.4	44	4.7	36	3.1	FULTON GA	-37	-35	-59	24	82
1	1.3	3	5.4	2	3.0	GLYNN GA	317	-45	130	94	82
1	4.8					GRADY GA					
1	4.5					GREENE GA					
2*	60.6					HABERSHAM GA					
1	4.4	1	4.6	1	3.7	HALL GA	3	-20	-18	60	85
		1	3.1			HANCOCK GA					
		1*	16.4			HARALSON GA					
2	7.4	1	3.5	1	4.0	HARRIS GA	-52	13	-46	36	86
1	11.8			3*	17.0	HART GA			44	84	98
1*	15.0					HEARD GA					
1	5.7			1	3.2	HENRY GA			-43	39	83
1	3.0	2	9.2	1	2.4	HOUSTON GA	212	-74	-19	59	80
1	9.1					IRWIN GA					
		2*	18.0			JACKSON GA					
1	17.7					JEFF DAVIS GA					
1	2.3	1	2.9			JEFFERSON GA	31				
2	7.5			1	4.5	JENKINS GA			-40	43	88
		1	12.4			JOHNSON GA					
1	5.5			1	4.8	JONES GA			-13	62	89
2	16.2			1	5.2	LAMAR GA			-68	15	90
2	3.4					LAURENS GA					
2	3.1	1	1.4	1	1.2	LOWNDES GA	-54	-15	-61	22	76
2	10.9			2	6.6	MC DUFFIE GA			-39	44	93
				1	6.2	MC INTOSH GA					92
1	2.1	2	6.0			MACON GA	179				
3	8.6	1	2.4			MERIWETHER GA	-72				
				1	13.6	MILLER GA					97
		1	2.7	1	1.9	MITCHELL GA		-29			78
		1	4.1			MONROE GA					
		1	9.3			MONTGOMERY GA					
3	2.6	7	3.9	6	2.3	MUSCOGEE#GA	47	-39	-11	64	80
1	4.0			1	2.0	NEWTON GA			-49	34	78
1	7.9					OGLETHORPE GA					
4*	15.8	1	3.0			PEACH GA	-81				
				1	5.7	PULASKI GA					91
1	2.7					RANDOLPH GA					
11	8.2	18*	8.3	13	5.3	RICHMOND GA	2	-37	-36	47	90
		1	10.1			ROCKDALE GA					
2	5.9	2	5.5			SCREVEN GA	-8				
3	9.1	2	4.5			SPALDING GA	-51				
2	9.2					STEWART GA					

* SIGNIFICANTLY MORE DEATHS THAN EXPECTED

\# COUNTIES COMBINED; SEE TECHNICAL NOTES

NON-WHITE FEMALE — CANCER OF THE RECTUM — ICD 154 EXCEPT 154.3

1950-1959 DEATHS	1950-1959 RATE	1960-1969 DEATHS	1960-1969 RATE	1970-1979 DEATHS	1970-1979 RATE	COUNTY AND STATE	PERCENT CHANGE 60S/50S	PERCENT CHANGE 70S/60S	PERCENT CHANGE 70S/50S	PERCENTILE %CHANGE 70S/50S	PERCENTILE RATES 1970S
3	6.3	2	2.6	1	1.5	SUMTER GA	-59	-42	-76	08	76
				1	3.8	TALBOT GA					85
1	10.3					TALIAFERRO GA					
				2*	14.6	TATTNALL GA					97
				1	4.5	TELFAIR GA					88
2	6.0			1	2.9	TERRELL GA			-51	32	82
3	4.6	1	1.7	2	2.7	THOMAS GA	-63	61	-41	42	81
2	8.2	4*	19.8			TOOMBS GA	143				
1	11.6					TREUTLEN GA					
3	5.2	4	5.8	1	1.3	TROUP GA	10	-78	-75	08	76
2	14.0					TWIGGS GA					
2	10.7	1	3.9	1	3.9	UPSON GA	-64	0	-64	20	86
		1	7.8			WALKER GA					
1	4.1			2	8.3	WALTON GA			105	92	95
1	2.0	2	4.3	2	3.6	WARE GA	117	-16	82	89	85
1	8.7			1	3.8	WARREN GA			-56	26	85
		2	4.4	3	5.4	WASHINGTON GA		21			91
1	8.3					WAYNE GA					
				1	8.8	WHITFIELD GA					95
1	3.2			1	4.4	WILKES GA			37	83	88
		1	6.0			WILKINSON GA					
3	15.4	2	8.6			WORTH GA	-44				
		56	4.1	56	2.5	HAWAII HI		-39			
		7	3.9	5	1.9	HAWAII HI		-50			78
		43	4.5	37	2.2	HONOLULU HI		-51			79
		1	1.3	7	6.9	KAUAI HI		444			93
		5	4.1	7	3.3	MAUI#HI		-19			83
239*	7.9	256*	6.0	251*	4.1	ILLINOIS IL	-23	-33	-48		
		1	12.7	1	12.7	ADAMS IL		0			97
1	2.2	1	2.6	1	2.7	ALEXANDER IL	15	5	21	80	81
		1*	53.6	1*	59.6	BOND IL		11			99
		2	8.8	2	7.1	CHAMPAIGN IL		-19			94
211*	8.7	220*	6.3	214*	4.3	COOK IL	-27	-33	-51	32	87
		1*	172.0			GREENE IL					
2	12.4	2	11.9	1	3.9	JACKSON IL	-4	-67	-69	14	86
1	7.7	1	3.1			KANE IL	-60				
1	4.6			4	5.1	LAKE IL			11	75	90
1	22.9			1	9.2	MC LEAN IL			-60	23	95
		1	5.5			MACON IL					
1	2.3	3	5.0	3	3.8	MADISON IL	122	-24	69	88	86
				2*	13.6	MARION IL					97
				1	7.5	MASSAC IL					94
2	11.3	2	5.1	4	7.0	PEORIA IL	-54	37	-38	45	93
1*	26.8	1*	28.9			PERRY IL	8				
2	6.5	1	3.7			PULASKI IL	-42				
1*	27.0					RANDOLPH IL					
1	10.8	1	7.0			ROCK ISLAND IL	-34				
5	2.8	8	4.0	9	3.0	ST CLAIR IL	44	-24	8	74	82
		1	14.1			SALINE IL					
3	10.9	4*	11.2	2	3.3	SANGAMON IL	3	-70	-70	14	83
2	10.8	1	4.3	1	3.4	VERMILION IL	-61	-20	-69	14	84
3*	16.9	2	10.2	2	2.5	WILL IL	-40	-75	-85	02	80
		1	13.8			WILLIAMSON IL					
1	5.6	1	2.7	2	4.0	WINNEBAGO IL	-52	47	-30	52	86
55	6.5	47	4.0	60	3.7	INDIANA IN	-39	-6	-43		
1	1.9			2	4.0	ALLEN IN			111	93	86
1	12.4			2*	10.4	CLARK IN			-16	61	96
		1	3.1			DELAWARE IN					
				1	7.4	ELKHART IN					94
1*	24.4					FAYETTE IN					
1	9.3	1	5.3	1	5.1	GRANT IN	-43	-3	-45	37	90
1	11.5					HOWARD IN					
				1*	37.2	JACKSON IN					99
1*	54.5					JEFFERSON IN					
		1*	56.2			KOSCIUSKO IN					
13	6.5	19	6.4	20	4.4	LAKE IN	-1	-32	-33	49	88
1	12.8					LA PORTE IN					
2*	21.3			1	5.4	MADISON IN			-75	09	91
23	6.3	18	3.5	18	2.8	MARION IN	-45	-21	-56	26	81
		1*	32.6			MONROE IN					
		1*	81.8			RUSH IN					
5*	15.5	3	5.0	6*	6.9	ST JOSEPH IN	-67	37	-55	28	93
1*	34.4					SHELBY IN					
1	2.2	1	2.2	5*	7.5	VANDERBURGH IN	4	232	247	99	94
2	8.2	1	3.4	2	6.3	VIGO IN	-59	86	-23	57	92
1	4.5			1	5.1	WAYNE IN			14	78	90
8	8.5	4	3.3	8	4.5	IOWA IA	-62	39	-47		
1	37.7			1	4.1	BLACK HAWK IA			-89	01	87
1*	74.3					CLINTON IA					
1*	188.7					DUBUQUE IA					
1	14.1					LINN IA					
		1*	94.4			MONROE IA					
2	5.6	3	6.5	4	6.9	POLK IA	15	7	22	81	93
				1*	74.3	STORY IA					99
				1*	35.1	WAPELLO IA					99
2*	36.0			1	16.1	WOODBURY IA			-55	27	98
18	4.4	30*	6.3	12	2.0	KANSAS KS	42	-68	-54		
1*	49.2					ALLEN KS					
1*	17.2					BOURBON KS					
				1*	49.7	CHEROKEE KS					99
1	16.8					COWLEY KS					

* SIGNIFICANTLY MORE DEATHS THAN EXPECTED

\# COUNTIES COMBINED; SEE TECHNICAL NOTES

NON-WHITE FEMALE — CANCER OF THE RECTUM — ICD 154 EXCEPT 154.3

1950-1959 DEATHS	1950-1959 RATE	1960-1969 DEATHS	1960-1969 RATE	1970-1979 DEATHS	1970-1979 RATE	COUNTY AND STATE	PERCENT CHANGE 60S/50S	PERCENT CHANGE 70S/60S	PERCENT CHANGE 70S/50S	PERCENTILE %CHANGE 70S/50S	PERCENTILE RATES 1970S
1	6.4			2*	12.9	DOUGLAS KS			102	91	97
		1*	188.7			FINNEY KS					
		1*	22.5			FRANKLIN KS					
		1	10.5			GEARY KS					
		1*	27.0			JACKSON KS					
1	23.6					JOHNSON KS					
		1	7.0			LABETTE KS					
2	11.0	4*	23.5	2*	9.7	LEAVENWORTH KS	113	-59	-12	63	96
		4*	19.0	1	4.6	MONTGOMERY KS		-76			88
2	3.3	2	2.6			SEDGWICK KS	-22				
1	2.1	4	8.3	3	4.9	SHAWNEE KS	288	-41	130	94	89
8	4.9	11	6.4	3	1.5	WYANDOTTE KS	30	-77	-70	13	76
56	5.3	65	5.5	64*	4.7	KENTUCKY KY	4	-14	-10		
				1*	19.1	ADAIR KY					98
		1*	106.0			ALLEN KY					
1	7.8			2*	19.0	BARREN KY			143	95	98
		1	10.6	1*	9.6	BELL KY		-10			96
		1	15.6			BOYD KY					
		1*	44.6			BRECKINRIDGE KY					
1*	39.0			1*	15.6	CALLOWAY KY			-60	23	98
		1*	40.9			CARROLL KY					
3	5.6	1	2.3	3	5.8	CHRISTIAN KY	-59	154	5	72	91
1	6.2			1	5.8	DAVIESS KY			-7	67	91
5	5.2	8	7.4	5	3.7	FAYETTE KY	43	-50	-29	52	85
1	8.4	1	8.0			FRANKLIN KY	-5				
1	8.4					FULTON KY					
				1*	56.2	GRANT KY					99
		1*	56.2			HANCOCK KY					
1	17.9					HARDIN KY					
				2*	12.5	HARLAN KY					97
1*	37.2					HARRISON KY					
		2	9.8	2	7.4	HENDERSON KY		-24			94
1	3.4	4*	13.8			HOPKINS KY	305				
20	5.4	29*	6.9	27*	5.1	JEFFERSON KY	26	-26	-7	67	90
				1	16.4	JESSAMINE KY					98
3	13.8	1	5.6	2	7.9	KENTON KY	-59	42	-42	40	94
1	12.2					LINCOLN KY					
				1	5.9	LOGAN KY					92
1*	24.8					LYON KY					
3	8.6			4*	8.9	MC CRACKEN KY			3	71	95
		2	12.5	2*	8.9	MADISON KY		-29			95
		1	9.4	1	14.8	MASON KY		57			98
		2*	42.3	2*	29.4	MERCER KY		-31			99
3*	37.9			1	14.5	MONTGOMERY KY			-62	22	97
		1	7.7			MUHLENBERG KY					
1*	490.8					PENDLETON KY					
		2*	15.7			SCOTT KY					
1	8.9					SIMPSON KY					
		1	6.5			TODD KY					
		1	22.5			TRIGG KY					
		1	9.9			UNION KY					
2	8.3			2	7.4	WARREN KY			-11	64	94
		1	12.6	1	4.4	WEBSTER KY		-65			88
1*	98.3					WHITLEY KY					
4*	37.1	1	7.0	1	11.8	WOODFORD KY	-81	68	-68	15	97
186	5.0	188	4.1	199*	3.7	LOUISIANA LA	-17	-10	-25		
2	5.8	1	3.4	2	3.9	ACADIA LA	-42	17	-32	49	86
		2	9.3	3*	10.9	ALLEN LA		17			96
		1	2.4	4*	8.8	ASCENSION LA		264			95
1	5.0			3*	9.0	ASSUMPTION LA			82	89	95
1	2.0			1	1.3	AVOYELLES LA			-36	46	76
		1	3.2	1	1.4	BIENVILLE LA		-55			76
1	2.0	3	4.2	2	1.9	BOSSIER LA	113	-56	-6	67	77
8	2.7	10	2.5	19	4.2	CADDO LA	-8	68	54	87	87
2	2.7	7	7.1	5	3.5	CALCASIEU LA	161	-51	29	82	84
				1	5.3	CALDWELL LA					90
				1	4.6	CATAHOULA LA					88
		4	7.7			CLAIBORNE LA					
1	1.6	2	3.6	1	1.7	CONCORDIA LA	120	-52	6	73	77
		2	2.7			DE SOTO LA					
8	2.9	6	2.0	7	1.7	EAST BATON ROUGE LA	-30	-16	-41	41	77
4	8.0	2	3.6	1	1.6	EAST CARROLL LA	-54	-55	-79	06	77
		1	1.5			EAST FELICIANA LA					
2	11.4	1	3.2			EVANGELINE LA	-72				
1	2.5			2	4.2	FRANKLIN LA			65	88	87
				1	7.1	GRANT LA					94
5	10.2	3	3.9	1	1.5	IBERIA LA	-62	-61	-85	02	77
2	3.2	4	5.8	5	6.5	IBERVILLE LA	84	11	105	91	92
9*	12.9	4	2.8	1	.6	JEFFERSON LA	-79	-80	-96	00	75
1	4.8	3*	14.4	1	4.0	JEFFERSON DAVIS LA	201	-72	-16	61	86
2	5.5	3	4.5	2	1.9	LAFAYETTE LA	-17	-57	-65	19	78
2	11.3	4*	16.7	1	3.7	LAFOURCHE LA	47	-78	-68	16	85
2*	25.8			1	8.6	LA SALLE LA			-66	17	95
1	2.3	2	4.0	2	3.6	LINCOLN LA	73	-9	56	88	85
				1	7.4	LIVINGSTON LA					94
1	2.0	1	1.4	2	2.3	MADISON LA	-31	61	11	76	79
		3	4.2			MOREHOUSE LA					
4	4.7			2	2.1	NATCHITOCHES LA			-56	27	78
64*	7.4	69*	6.6	70*	5.5	ORLEANS LA	-11	-16	-25	54	91
9	7.7	3	1.7	3	1.5	OUACHITA LA	-78	-13	-81	05	76
1	5.8	1	5.0			PLAQUEMINES LA	-15				
2	5.0	1	1.7			POINTE COUPEE LA	-65				
3	2.3	3	1.7	8	4.5	RAPIDES LA	-29	167	90	91	88
4	11.2			4	6.2	RICHLAND LA			-45	37	92
2	14.4	1	4.3			SABINE LA	-70				
1	12.3			1	5.9	ST BERNARD LA			-52	30	92
1	4.1			1	3.7	ST CHARLES LA			-9	65	85

* SIGNIFICANTLY MORE DEATHS THAN EXPECTED

NON-WHITE FEMALE | CANCER OF THE RECTUM | ICD 154 EXCEPT 154.3

1950-1959 DEATHS	1950-1959 RATE	1960-1969 DEATHS	1960-1969 RATE	1970-1979 DEATHS	1970-1979 RATE	COUNTY AND STATE	PERCENT CHANGE 60S/50S	PERCENT CHANGE 70S/60S	PERCENT CHANGE 70S/50S	PERCENTILE %CHANGE 70S/50S	PERCENTILE RATES 1970S
				1	3.6	ST HELENA LA					85
		2	4.5	1	1.9	ST JAMES LA		-58			77
1	3.9	2	5.8	1	2.3	ST JOHN THE BAPT LA	48	-60	-40	42	80
5	5.5	2	1.6	4	3.1	ST LANDRY LA	-71	93	-44	38	82
2	7.4			4*	9.2	ST MARTIN LA			26	81	95
5	9.0	9*	13.9	8*	10.9	ST MARY LA	55	-22	21	80	96
1	2.7	2	4.5	2	4.0	ST TAMMANY LA	63	-11	46	85	86
5	6.0	3	3.1	1	1.2	TANGIPAHOA LA	-48	-63	-80	05	76
2	4.4	4	12.4	1	2.1	TENSAS LA	183	-83	-53	29	78
3	8.3	4	8.1	1	2.6	TERREBONNE LA	-2	-68	-69	14	80
2	8.9	2	6.5	3	6.7	UNION LA	-26	3	-24	56	93
2	13.0	1	3.7	1	3.3	VERMILION LA	-72	-12	-75	09	83
1	13.5					VERNON LA					
2	3.8	2	3.0	3	4.6	WASHINGTON LA	-20	50	19	79	88
3	6.0	4	5.9	3	3.7	WEBSTER LA	-2	-37	-38	45	85
2	9.5	1	3.5			WEST BATON ROUGE LA	-63				
1	5.2	1	6.8	1	5.0	WEST CARROLL LA	30	-27	-5	68	89
1	5.0	1	4.5	3*	14.2	WEST FELICIANA LA	-10	214	182	97	97
1	5.7			1	4.4	WINN LA			-22	57	88
				1	3.3	MAINE ME					
				1*	49.2	WASHINGTON ME					99
87	5.8	116*	5.5	137*	4.6	MARYLAND MD	-6	-17	-21		
				1*	12.8	ALLEGANY MD					97
4	4.1	4	3.5	6	4.1	ANNE ARUNDEL MD	-14	16	-1	70	87
4	7.7	3	3.7	4	2.9	BALTIMORE MD	-51	-23	-63	21	81
		2	10.1	2	8.2	CALVERT MD		-19			94
1	5.5	2	14.1			CAROLINE MD	155				
1	13.8	1	10.6			CARROLL MD	-23				
1	8.8	1	22.5			CECIL MD	157				
2	9.1	2	8.4	2	4.6	CHARLES MD	-8	-45	-49	34	88
3	8.8	3	6.9	3	6.7	DORCHESTER MD	-21	-4	-24	56	93
3*	15.0	2	9.4	2	9.0	FREDERICK MD	-37	-4	-40	43	95
				1	3.3	HARFORD MD					83
2	16.7			2	5.8	HOWARD MD			-65	18	91
		1	8.3			KENT MD					
6*	15.4	1	1.2	8	6.4	MONTGOMERY MD	-92	424	-58	25	92
7	9.1	3	2.6	5	1.5	PRINCE GEORGE'S MD	-71	-42	-83	03	76
2	11.7			1	4.0	QUEEN ANNE'S MD			-66	18	86
		1	5.7	3*	9.9	ST MARY'S MD		74			96
1	3.3	3	9.8	2	5.6	SOMERSET MD	194	-43	67	88	91
				1	2.7	TALBOT MD					81
1	9.2	1	8.2			WASHINGTON MD	-11				
4	11.4	3	6.2	1	1.7	WICOMICO MD	-45	-73	-85	02	77
		1	3.4	4*	11.2	WORCESTER MD		229			97
45	5.2	82*	6.0	89*	5.1	BALTIMORE CITY MD	15	-14	-1	70	90
31*	8.4	32	5.5	23	2.5	MASSACHUSETTS MA	-35	-55	-71		
2*	46.7	1	10.3	1	4.5	BARNSTABLE MA	-78	-57	-90	00	88
2*	27.8			1	13.5	BERKSHIRE MA			-52	31	97
1	6.0	1	3.7	3	6.9	BRISTOL MA	-38	84	14	77	93
3	8.1	3	5.1	3	3.3	HAMPDEN MA	-38	-35	-59	24	83
5	9.7	2	2.1	3	1.9	MIDDLESEX MA	-78	-12	-81	05	77
1	6.9	2	10.6	1	3.3	PLYMOUTH MA	54	-69	-53	29	83
17	8.6	22*	6.8	10	2.2	SUFFOLK MA	-21	-68	-75	09	79
		1	6.4	1	3.4	WORCESTER MA		-47			84
101	5.4	142	4.9	147	3.5	MICHIGAN MI	-10	-28	-36		
		2*	32.7			ALLEGAN MI					
1*	86.0					BARAGA MI					
2	5.6	4	8.2	5	6.9	BERRIEN MI	45	-16	22	80	93
		2	6.6	3	6.6	CALHOUN MI		-1			93
				2	7.0	CASS MI					93
5	18.7	5	4.1	6	4.0	GENESEE MI	-78	-4	-79	06	86
1	3.4	1	2.4	1	1.8	INGHAM MI	-30	-24	-47	36	77
				2*	49.5	ISABELLA MI					99
		1	5.7			KALAMAZOO MI					
1	3.7	2	3.5	4	3.7	KENT MI	-5	4	-1	69	85
				1	3.7	LAKE MI					85
1	7.4			3	6.8	MACOMB MI			-9	65	93
1	3.6	2	6.5	1	2.5	MUSKEGON MI	80	-62	-31	51	80
				1*	12.4	NEWAYGO MI					97
2	4.3	2	1.8	3	2.1	OAKLAND MI	-58	21	-50	33	79
1	2.5			2	2.3	SAGINAW MI			-8	66	79
1*	21.0					ST JOSEPH MI					
		1	2.8			VAN BUREN MI					
1	3.9	3	5.5	2	1.8	WASHTENAW MI	41	-67	-54	28	77
84	5.8	117*	5.3	111	3.5	WAYNE MI	-8	-33	-39	44	84
7	6.6	4	2.7	5	1.9	MINNESOTA MN	-59	-30	-71		
2*	27.4					CASS MN					
2	5.7	2	3.4	3	2.9	HENNEPIN MN	-40	-14	-49	35	82
		1*	37.3			MAHNOMEN MN					
1	4.9	1	2.3	2	2.5	RAMSEY MN	-54	12	-48	35	80
2*	38.1					ST LOUIS MN					
134	3.7	117	3.0	128	3.0	MISSISSIPPI MS	-19	-1	-20		
5	6.9	5	5.3	4	3.5	ADAMS MS	-23	-35	-50	33	84
2	6.9					AMITE MS					
2	5.5	3	7.7	1	1.8	ATTALA MS	40	-77	-68	16	77
1	8.6					BENTON MS					
9	5.7	4	2.6	5	3.3	BOLIVAR MS	-55	27	-42	39	83
2	6.9	1	4.0	1	4.1	CHICKASAW MS	-43	4	-41	42	87
				1	8.2	CHOCTAW MS					94

* SIGNIFICANTLY MORE DEATHS THAN EXPECTED

NON-WHITE FEMALE — CANCER OF THE RECTUM — ICD 154 EXCEPT 154.3

1950-1959 DEATHS	1950-1959 RATE	1960-1969 DEATHS	1960-1969 RATE	1970-1979 DEATHS	1970-1979 RATE	COUNTY AND STATE	PERCENT CHANGE 60S/50S	PERCENT CHANGE 70S/60S	PERCENT CHANGE 70S/50S	PERCENTILE %CHANGE 70S/50S	PERCENTILE RATES 1970S
1	3.2	1	2.6			CLAIBORNE MS	-19				
		1	4.0	1	4.9	CLARKE MS		22			89
1	3.3	2	3.9	1	2.4	CLAY MS	17	-39	-28	53	80
4	3.2	5	3.3	5	3.1	COAHOMA MS	3	-4	-2	69	83
2	3.4	2	3.1	3	3.5	COPIAH MS	-9	15	5	72	84
1	6.7	2	10.1			COVINGTON MS	50				
1	1.4	1	1.7			DE SOTO MS	19				
1	1.7	2	3.0	2	2.4	FORREST MS	74	-20	39	84	80
1	6.4			2*	8.1	FRANKLIN MS			27	82	94
1*	23.8	1*	21.2			GEORGE MS	-11				
1	2.3					GRENADA MS					
6	9.4	2	2.3	7*	7.0	HARRISON MS	-76	210	-25	54	93
10	3.9	7	2.2	14	3.4	HINDS MS	-43	53	-13	62	84
4	4.7	1	1.2	3	3.1	HOLMES MS	-74	161	-33	49	83
		2	5.0			HUMPHREYS MS					
		1	8.6			ISSAQUENA MS					
				1*	9.2	ITAWAMBA MS					95
		1	3.9	2	3.4	JACKSON MS		-13			84
1	3.2	2	6.0	4*	9.0	JEFFERSON MS	87	51	181	97	95
3	9.1					JEFFERSON DAVIS MS					
1	2.3			2	2.6	JONES MS			17	79	81
2	6.1			1	3.6	KEMPER MS			-41	42	85
1	3.7			1	2.5	LAFAYETTE MS			-32	50	80
2	2.1	2	1.4	6	4.8	LAUDERDALE MS	-32	243	133	94	89
2	11.1					LAWRENCE MS					
		1	2.4	1	3.0	LEAKE MS		25			82
2	5.3	3	6.6			LEE MS	25				
5	4.0	5	3.9	2	1.5	LEFLORE MS	-3	-62	-63	20	76
1	2.6					LINCOLN MS					
2	2.3	1	1.3	3	3.0	LOWNDES MS	-44	133	30	83	82
5	5.3	4	4.0	5	4.9	MADISON MS	-25	22	-8	66	89
2	9.2	2	6.0	1	2.6	MARION MS	-35	-56	-71	12	81
1	1.4	1	1.8			MARSHALL MS	29				
1	1.9	2	3.9			MONROE MS	108				
1	4.3					MONTGOMERY MS					
1	4.7					NESHOBA MS					
2	8.1	3	9.2	1	2.1	NEWTON MS	14	-77	-74	10	79
		1	2.3			NOXUBEE MS					
1	2.6	1	2.6	1	2.1	OKTIBBEHA MS	-1	-17	-18	60	79
4	6.1	2	2.8	5	6.5	PANOLA MS	-54	131	7	74	92
2	10.0					PEARL RIVER MS					
1	1.4	2	3.8			PIKE MS	168				
		1	7.0			PONTOTOC MS					
2	3.2					QUITMAN MS					
		2	5.2	2	5.6	SCOTT MS		9			91
1	2.4	1	4.2	1	3.6	SHARKEY MS	75	-14	50	86	85
2	7.1	1	7.0	1	3.2	SIMPSON MS	-1	-54	-54	28	83
		1	12.0			SMITH MS					
2*	26.0			1	11.2	STONE MS			-57	26	97
1	1.0	6	5.8	3	3.1	SUNFLOWER MS	484	-47	211	98	82
2	3.8	1	1.8	4	6.9	TALLAHATCHIE MS	-52	287	84	90	93
1	3.4	1	3.2	2	4.1	TATE MS	-6	27	20	80	87
1	2.1	2	4.4	1	1.9	TUNICA MS	112	-57	-10	64	78
3*	18.0	2	11.8	1	5.0	UNION MS	-34	-58	-72	11	90
1	5.8	1	4.7			WALTHALL MS	-20				
4	3.5	4	4.0	7	6.2	WARREN MS	13	55	75	89	92
8	4.0	13	6.6	17*	8.4	WASHINGTON MS	65	28	110	92	95
1	4.3	1	3.6			WAYNE MS	-16				
		1	7.3			WEBSTER MS					
1	2.4			1	2.2	WILKINSON MS			-8	66	79
2	10.0	1	3.0			WINSTON MS	-70				
2	8.4					YALOBUSHA MS					
2	2.8	2	2.3	1	1.6	YAZOO MS	-20	-27	-41	41	77
105*	6.9	88	4.4	87	3.6	MISSOURI MO	-35	-20	-48		
		1*	196.5			BENTON MO					
3*	18.0	1	4.3			BOONE MO	-76				
1	4.8			1	6.5	BUCHANAN MO			35	83	93
2	16.7			3*	13.5	BUTLER MO			-19	59	97
		1	8.2			CALLAWAY MO					
				1	18.2	CAPE GIRARDEAU MO					98
				1*	21.2	CHARITON MO					98
		1*	30.7			CLAY MO					
1	9.8					COOPER MO					
1	8.5					GREENE MO					
1	10.5			1	6.2	HOWARD MO			-41	41	92
28*	8.6	18	4.5	19	3.5	JACKSON MO	-48	-22	-59	23	84
		1	8.6			JASPER MO					
1*	61.3					LINN MO					
1*	37.5					LIVINGSTON MO					
1	8.9			1	10.4	MARION MO			17	78	96
1*	20.2					MONTGOMERY MO					
2	4.4	1	2.8	4*	9.7	PEMISCOT MO	-36	250	122	93	96
1	7.5	1	7.6	1	10.6	PETTIS MO	1	39	40	84	96
				1	9.3	RANDOLPH MO					95
3	3.3	8	6.3	6	1.9	ST LOUIS COUNTY MO	92	-71	-44	38	77
		1	7.7	1	6.9	SCOTT MO		-11			93
58*	7.7	54	5.1	47	3.8	ST LOUIS CITY MO	-34	-26	-51	32	86
		2	3.2	3	3.5	MONTANA MT		9			
				1*	16.4	BLAINE MT					98
		1	7.6			GLACIER MT					
				1*	42.4	LEWIS AND CLARK MT					99
		1*	44.6			MISSOULA MT					
				1	9.8	ROOSEVELT MT					96
10	9.3	4	3.0	6	2.9	NEBRASKA NB	-68	-4	-69		
1*	136.6					DAWES NB					

* SIGNIFICANTLY MORE DEATHS THAN EXPECTED

NON-WHITE FEMALE — CANCER OF THE RECTUM — ICD 154 EXCEPT 154.3

1950-1959 DEATHS	1950-1959 RATE	1960-1969 DEATHS	1960-1969 RATE	1970-1979 DEATHS	1970-1979 RATE	COUNTY AND STATE		PERCENT CHANGE 60S/50S	PERCENT CHANGE 70S/60S	PERCENT CHANGE 70S/50S	PERCENTILE %CHANGE 70S/50S	PERCENTILE RATES 1970S
8	9.8	4	3.9	5	3.3	DOUGLAS	NB	-60	-16	-67	16	83
				1	5.7	LANCASTER	NB					91
1*	62.9					SCOTTS BLUFF	NB					
1	2.4	2	3.6	1	.6	NEVADA	NV	51	-82	-73		
1	6.4			1	1.0	CLARK	NV			-84	03	75
		1*	18.9			ELKO	NV					
		1	13.4			WASHOE	NV					
127*	9.1	114	5.1	123	3.5	NEW JERSEY	NJ	-44	-32	-62		
13*	9.6	7	3.9	10	4.8	ATLANTIC	NJ	-59	22	-50	33	89
7*	12.5	5	6.1	10	6.1	BERGEN	NJ	-51	-1	-51	31	92
4	11.6	2	4.0	2	2.3	BURLINGTON	NJ	-65	-43	-80	05	79
11*	10.3	13*	8.2	7	2.8	CAMDEN	NJ	-20	-66	-73	11	81
				3*	9.1	CAPE MAY	NJ					95
3	9.7	3	5.4	2	2.2	CUMBERLAND	NJ	-44	-58	-77	07	79
38*	9.2	49*	6.7	39	3.6	ESSEX	NJ	-27	-47	-61	23	84
2	5.3	5	9.5	2	2.4	GLOUCESTER	NJ	79	-75	-55	28	80
10*	9.3	9	5.0	9	2.9	HUDSON	NJ	-46	-41	-68	15	82
6	5.6	3	1.8	9	4.2	MERCER	NJ	-67	127	-26	54	87
5*	14.2	1	1.7	4	3.3	MIDDLESEX	NJ	-88	91	-77	07	83
8	8.2	2	1.4	5	2.4	MONMOUTH	NJ	-83	74	-71	12	80
2	14.5	1	3.0	2	3.0	MORRIS	NJ	-79	0	-79	06	82
1	7.4			2	3.8	OCEAN	NJ			-48	35	86
6*	10.9	6	5.6	2	.8	PASSAIC	NJ	-49	-86	-93	00	75
1	4.3			1	2.8	SALEM	NJ			-34	48	81
1	9.1			2	5.8	SOMERSET	NJ			-36	47	91
9	8.5	8	5.5	12	4.0	UNION	NJ	-35	-27	-52	29	86
4	3.1	5	2.4	5	.9	NEW MEXICO	NM	-23	-63	-71		
1	7.4	1	5.8			BERNALILLO	NM	-22				
		1*	16.1			CURRY	NM					
				2*	12.9	EDDY	NM					97
		1	6.9			LEA	NM					
		1	2.9	1	1.1	MC KINLEY	NM		-61			75
2*	64.2					OTERO	NM					
				1	2.5	RIO ARRIBA	NM					80
1	8.2			1	2.7	SANDOVAL	NM			-67	16	81
		1	1.3			SAN JUAN	NM					
292*	7.1	380*	5.7	407*	3.8	NEW YORK	NY	-20	-33	-47		
2	4.2			4	6.2	ALBANY	NY			47	86	92
35*	9.2	44	5.2	57	3.6	BRONX	NY	-44	-30	-61	23	85
1	37.5	1	11.2	1	8.7	BROOME	NY	-70	-22	-77	07	95
		1	13.4	1	9.5	CATTARAUGUS	NY		-29			96
				1*	37.5	CAYUGA	NY					99
		1	9.6	2*	16.9	CHEMUNG	NY		75			98
1*	490.8					DELAWARE	NY					
1	1.9	3	7.2	3	5.0	DUTCHESS	NY	283	-30	167	96	90
8	5.3	14	4.7	16	4.2	ERIE	NY	-11	-12	-21	58	87
				1*	267.9	ESSEX	NY					99
		1*	53.0			FULTON	NY					
		1*	32.6			GREENE	NY					
71*	8.8	68	3.8	102	4.0	KINGS(BROOKLYN)	NY	-57	8	-54	29	87
1	2.2	4	4.7	2	2.0	MONROE	NY	111	-57	-9	65	78
		1*	84.8			MONTGOMERY	NY					
4	4.2	4	2.1	11	4.0	NASSAU	NY	-51	96	-4	68	86
120*	7.1	164*	8.0	117*	4.1	NEW YORK(MANHATTAN)	NY	12	-48	-42	40	87
		1	5.1	3	7.3	NIAGARA	NY		41			94
				1	4.7	ONEIDA	NY					89
2	14.7			2	2.9	ONONDAGA	NY			-80	05	82
1	5.5	4*	11.0			ORANGE	NY	101				
		1*	106.0			OSWEGO	NY					
28*	6.7	25	2.7	44	3.0	QUEENS	NY	-60	13	-55	27	82
1	20.6	1	14.9	1	9.6	RENSSELAER	NY	-28	-35	-53	29	96
2	7.3	2	7.0			RICHMOND	NY	-4				
1	6.3	2	4.8	4	6.5	ROCKLAND	NY	-23	33	3	71	92
				1	8.6	SARATOGA	NY					95
1*	38.3					STEUBEN	NY					
2	1.7	4	2.2	12	4.8	SUFFOLK	NY	29	119	183	97	89
1*	24.6			1	7.0	SULLIVAN	NY			-71	12	93
1	23.6					TOMPKINS	NY					
				1*	188.7	WARREN	NY					99
		1*	67.0			WASHINGTON	NY					
8	4.1	32*	10.0	19	3.6	WESTCHESTER	NY	144	-64	-12	63	85
148	4.3	135	3.1	160	2.8	NORTH CAROLINA	NC	-27	-11	-34		
2	4.2	3	5.9	3	4.6	ALAMANCE	NC	42	-22	10	75	88
1*	49.2					ALLEGHANY	NC					
				2	2.4	ANSON	NC					80
1	3.7	2	3.6	1	1.4	BEAUFORT	NC	-4	-62	-64	20	76
1	1.9	1	2.2	3	4.8	BERTIE	NC	17	122	159	96	89
1	2.9			1	2.1	BLADEN	NC			-29	52	78
1	5.1					BRUNSWICK	NC					
2	2.3	2	2.9	2	1.9	BUNCOMBE	NC	24	-34	-18	59	78
		1	2.3	2	3.4	CABARRUS	NC		46			84
		1	6.5			CALDWELL	NC					
		1	7.7	1	5.4	CARTERET	NC		-31			91
1	4.3	1	2.7			CASWELL	NC	-37				
2	8.3	1	4.2	1	2.6	CATAWBA	NC	-49	-38	-68	15	81
2	7.2					CHATHAM	NC					
		1*	60.7			CHEROKEE	NC					
3*	17.7	1	5.6			CHOWAN	NC	-69				
1	1.7			4	5.0	CLEVELAND	NC			197	98	90
1	1.7			4	4.8	COLUMBUS	NC			178	97	89
6	9.4	4	5.5	2	2.3	CRAVEN	NC	-41	-59	-76	08	79

* SIGNIFICANTLY MORE DEATHS THAN EXPECTED

NON-WHITE FEMALE

CANCER OF THE RECTUM

ICD 154 EXCEPT 154.3

1950-1959 DEATHS	1950-1959 RATE	1960-1969 DEATHS	1960-1969 RATE	1970-1979 DEATHS	1970-1979 RATE	COUNTY AND STATE	PERCENT CHANGE 60S/50S	PERCENT CHANGE 70S/60S	PERCENT CHANGE 70S/50S	PERCENTILE %CHANGE 70S/50S	PERCENTILE RATES 1970S
2	2.1	4	3.2	4	2.2	CUMBERLAND NC	52	-31	4	72	79
1	4.0	1	4.3	1	3.3	DAVIDSON NC	7	-25	-20	58	83
		1	1.7	3	4.8	DUPLIN NC		187			89
6	4.9	6	3.4	14*	6.0	DURHAM NC	-31	77	23	81	92
4	4.2	1	1.4	3	2.0	EDGECOMBE NC	-68	50	-52	31	78
8	4.9	15	7.2	5	1.8	FORSYTH NC	48	-75	-62	21	77
3	8.6	1	2.1	2	3.5	FRANKLIN NC	-75	66	-59	24	84
1	1.6	1	1.2	4	4.1	GASTON NC	-24	246	163	96	87
1	6.5					GATES NC					
		1	3.4	1	1.3	GRANVILLE NC		-60			76
		1	4.1	2	6.6	GREENE NC		60			93
10	6.6	10	5.5	6	2.2	GUILFORD NC	-16	-60	-67	17	79
2	2.4	5	5.0	1	.7	HALIFAX NC	108	-87	-72	11	75
1	2.3	1	2.3	1	1.2	HAPNETT NC	0	-48	-48	35	76
1*	30.4					HAYWOOD NC					
		1	8.4	1	6.1	HENDERSON NC		-27			92
1	5.7	3	6.9	2	3.2	HERTFORD NC	20	-54	-45	37	83
		1	3.7	3	9.2	HOKE NC		148			95
1	10.3	1	9.6			HYDE NC	-7				
4	14.1	1	2.0	2	3.5	IREDELL NC	-86	76	-75	08	84
		1	16.3			JACKSON NC					
4	9.9			1	1.2	JOHNSTON NC			-88	01	76
2	15.9					JONES NC					
3*	15.9			2	5.5	LEE NC			-66	18	91
1	1.9	3	3.5	2	2.0	LENOIR NC	87	-44	6	73	78
				1	5.3	LINCOLN NC					90
16*	7.5	8	3.0	13	3.5	MECKLENBURG NC	-60	16	-54	29	84
				2	9.5	MONTGOMERY NC					96
1	3.2			2	3.2	MOORE NC			-1	70	83
4	5.3	1	1.5	2	1.9	NASH NC	-72	27	-65	19	77
6	6.9	4	3.9	1	.7	NEW HANOVER NC	-44	-81	-89	00	75
2	3.9	1	1.9			NORTHAMPTON NC	-51				
2	15.4	1	3.0	1	2.3	ONSLOW NC	-80	-24	-85	02	80
1	2.9	2	5.5	3	6.0	ORANGE NC	91	9	109	92	92
				1	7.1	PAMLICO NC					94
3	9.5	2	5.6			PASQUOTANK NC	-40				
		2	5.8	1	2.0	PENDER NC		-66			78
				1	5.3	PERQUIMANS NC					90
		1	2.6			PERSON NC					
1	1.1	2	1.7	4	3.1	PITT NC	63	77	188	98	82
2	17.7	1	4.8			RANDOLPH NC	-73				
1	2.6	1	1.7	2	2.6	RICHMOND NC	-33	50	0	70	80
3	2.3	1	.6	10	4.9	RCBESON NC	-74	715	110	93	89
4	6.6			1	1.2	ROCKINGHAM NC			-82	04	76
2	4.7	2	3.2	1	1.4	ROWAN NC	-31	-57	-70	13	76
		1	4.7	1	3.6	RUTHERFORD NC		-22			85
1	1.5	2	2.8	3	3.8	SAMPSON NC	87	34	150	95	85
2	5.0			2	3.8	SCOTLAND NC			-24	56	86
				1	3.4	STANLY NC					84
1	21.0					SURRY NC					
1*	9.6	1	25.8	1	11.5	SWAIN NC	169	-56	20	80	97
		1*	37.7			TRANSYLVANIA NC					
3	9.4					UNION NC					
2	4.2	2	3.2			VANCE NC	-22				
7	5.5	8	4.3	10	4.1	WAKE NC	-20	-5	-25	56	87
1	2.3	1	1.7	3	4.7	WARREN NC	-29	184	103	91	89
		2	9.6			WASHINGTON NC					
1	.8	9	7.7	3	2.3	WAYNE NC	851	-71	177	97	79
		1	1.5	4	3.9	WILSON NC		169			86
1	3.7	2	6.3	5*	9.7	NORTH DAKOTA ND	73	53	165		
1*	27.3	1*	12.4			BENSON ND	-55				
				1*	85.2	CASS ND					99
				1*	34.4	MOUNTRAIL ND					99
		1	9.0	2*	12.6	ROLETTE ND		40			97
				1*	17.5	SIOUX ND					98
151*	6.9	183*	5.7	178*	3.9	OHIO OH	-18	-31	-44		
		1	3.0	1	3.5	ALLEN OH		17			84
		1	9.6	1	7.3	ASHTABULA OH		-25			94
1	10.7	4*	35.2			BELMONT OH	228				
1*	27.0					BROWN OH					
2	6.0	4	8.6	1	1.7	BUTLER OH	45	-80	-71	13	77
1	2.7	2	2.8	3	3.9	CLARK OH	6	37	45	85	86
1*	26.5					CLERMONT OH					
		2*	58.1			CLINTON OH					
		1	11.2			COLUMBIANA OH					
56*	8.7	66*	6.8	59	3.9	CUYAHOGA OH	-22	-43	-55	27	86
1*	67.0	1*	106.0			DARKE OH	58				
2*	23.7			2*	12.0	ERIE OH			-49	34	97
		1*	28.8			FAYETTE OH					
19	8.4	16	4.7	13	2.9	FRANKLIN OH	-44	-40	-66	17	81
		1*	112.4			GEAUGA OH					
2	9.0			4*	13.9	GREENE OH			54	87	97
29	6.3	30	5.2	35*	4.6	HAMILTON OH	-18	-12	-28	53	88
				1*	24.8	HANCOCK OH					98
				2*	62.4	HURON OH					99
1*	62.9					JACKSON OH					
2	13.9	1	4.2			JEFFERSON OH	-70				
1	21.0			1	6.5	LAKE OH			-69	14	93
				1	7.9	LICKING OH					94
		1*	35.3			LOGAN OH					
1	3.1	2	5.0	3	4.6	LORAIN OH	62	-7	50	87	89
8	6.4	9	5.1	12	4.4	LUCAS OH	-21	-14	-32	50	88
9	9.8	8	5.4	6	3.3	MAHONING OH	-45	-40	-67	17	83
				1*	22.9	MEDINA OH					98
		1	10.5			MIAMI OH					
5	2.9	12	4.9	18	5.3	MONTGOMERY OH	67	9	81	89	90
		1*	44.6			MORGAN OH					

* SIGNIFICANTLY MORE DEATHS THAN EXPECTED

NON-WHITE FEMALE

CANCER OF THE RECTUM

ICD 154 EXCEPT 154.3

1950-1959 DEATHS	1950-1959 RATE	1960-1969 DEATHS	1960-1969 RATE	1970-1979 DEATHS	1970-1979 RATE	COUNTY AND STATE	PERCENT CHANGE 60S/50S	PERCENT CHANGE 70S/60S	PERCENT CHANGE 70S/50S	PERCENTILE %CHANGE 70S/50S	PERCENTILE RATES 1970S
2*	18.7					MUSKINGUM OH					
		1	13.8	1	8.2	PORTAGE OH		-41			94
		1	12.4	1	8.0	ROSS OH		-35			94
				1	10.2	SCIOTO OH					96
2	8.8	7*	10.5	3	2.5	STARK OH	19	-76	-72	12	80
3	2.5	6	3.6	7	3.1	SUMMIT OH	46	-13	27	82	83
2	11.7	1	2.1			TRUMBULL OH	-82				
		1*	22.3			WARREN OH					
				1*	12.4	WASHINGTON OH					97
		1*	38.3			WOOD OH					
46	4.9	34	2.8	36	2.2	OKLAHOMA OK	-42	-22	-55		
2*	22.8					ADAIR OK					
				1*	38.5	BECKHAM OK					99
1	9.3					BRYAN OK					
				1	3.5	CADDO OK					84
		1	11.8	1	9.1	CANADIAN OK		-23			95
2	9.2					CARTER OK					
1	8.0	1	4.9			CHEROKEE OK	-39				
		1	3.9			CHOCTAW OK					
1	3.1	1	3.0	1	1.5	COMANCHE OK	-5	-49	-51	31	77
1	15.7	1	7.4			CRAIG OK	-53				
1	4.5	1	4.5			CREEK OK	-1				
				1*	14.9	CUSTER OK					98
		1	8.8			DELAWARE OK					
2*	24.7					GARFIELD OK					
				1	7.0	GARVIN OK					93
				2*	21.2	JACKSON OK					98
		1*	37.2			JOHNSTON OK					
1	5.6	1	4.1			LOGAN OK	-27				
3	7.9	1	2.0			MC CURTAIN OK	-75				
1	5.7	1	6.7			MC INTOSH OK	18				
				2*	8.5	MAYES OK					95
1*	35.3					MURRAY OK					
2	2.0	4	4.0	3	2.3	MUSKOGEE OK	99	-43	14	77	79
1*	28.1	1*	19.1			NOBLE OK	-32				
1	21.0					NOWATA OK					
2	8.0					OKFUSKEE OK					
8	6.0	6	2.7	7	2.3	OKLAHOMA OK	-54	-17	-62	21	79
2	5.1	1	2.1	3	5.3	OKMULGEE OK	-59	154	3	72	90
1	8.0	1	6.7			OSAGE OK	-16				
				1*	11.1	PAWNEE OK					96
1	20.6	2*	91.3			PAYNE OK	343				
		1	9.4	1	7.4	PONTOTOC OK		-21			94
2*	17.2	1	5.4	1	5.8	POTTAWATOMIE OK	-69	8	-66	17	91
				1*	11.2	PUSHMATAHA OK					97
1	9.1			1	4.6	ROGERS OK			-49	34	88
2	6.5	2	7.0			SEMINOLE OK	8				
2	2.2	4	2.3	8	3.0	TULSA OK	6	28	36	83	82
4*	18.8					WAGONER OK					
6	7.7	5	4.2	2	1.0	OREGON OR	-45	-77	-88		
				1	15.7	CLACKAMAS OR					98
		1*	188.7			DOUGLAS OR					
6*	12.5	4	4.0			MULTNOMAH OR	-68				
				1*	94.4	TILLAMOOK OR					99
202*	7.0	225*	5.8	231*	4.4	PENNSYLVANIA PA	-18	-23	-37		
34	6.4	50*	8.1	42*	5.1	ALLEGHENY PA	26	-36	-19	59	90
				1*	18.9	ARMSTRONG PA					98
1	3.4			5*	8.5	BEAVER PA			150	95	95
		2	9.1	1	2.8	BERKS PA		-69			81
		1	12.2			BLAIR PA					
		1	4.3	1	2.8	BUCKS PA		-36			81
2*	24.9					CAMBRIA PA					
2	4.3	5	7.0	4	4.2	CHESTER PA	62	-40	-3	69	87
1*	18.2			1*	10.6	CRAWFORD PA			-42	41	96
		2*	22.2			CUMBERLAND PA					
5	8.9	3	4.0	5	3.4	DAUPHIN PA	-54	-15	-61	23	84
8	5.9	9	4.9	7	2.8	DELAWARE PA	-17	-42	-52	30	81
1	6.6	2	7.1			ERIE PA	7				
2	7.6	5*	13.2			FAYETTE PA	74				
1	15.7			2*	28.8	FRANKLIN PA			84	90	99
1*	40.9					HUNTINGDON PA					
		1	15.8			LACKAWANNA PA					
1	7.5	2	27.3	2	6.3	LANCASTER PA	266	-77	-15	61	92
				1	5.4	LAWRENCE PA					91
		1*	20.6			LEHIGH PA					
1	24.8			1	9.6	LUZERNE PA			-61	22	96
1	16.7	2*	22.0			LYCOMING PA	32				
				1*	74.3	MC KEAN PA					99
1	8.4	1	3.9			MERCER PA	-53				
2	3.0	5	4.5	5	3.6	MONTGOMERY PA	48	-21	17	79	84
		1	11.2			NORTHAMPTON PA					
				1*	59.6	NORTHUMBERLAND PA					99
134*	7.7	127	5.1	141*	4.4	PHILADELPHIA PA	-34	-14	-43	39	88
				1*	20.6	SCHUYLKILL PA					98
				1*	57.3	VENANGO PA					99
2	4.6	1	2.0	4*	8.9	WASHINGTON PA	-55	334	93	91	95
2	7.2	3	8.9	2	4.6	WESTMORELAND PA	23	-48	-36	47	89
		1	4.3	2	8.2	YORK PA		91			94
4	7.1	5	5.4	4	2.6	RHODE ISLAND RI	-24	-53	-64		
1*	60.6			1*	21.9	KENT RI			-64	20	98
1	12.6	1	7.5			NEWPORT RI	-40				
1	2.3	4	6.0	3	2.3	PROVIDENCE RI	160	-61	2	71	80
1*	19.0					WASHINGTON RI					

* SIGNIFICANTLY MORE DEATHS THAN EXPECTED

NON-WHITE FEMALE — CANCER OF THE RECTUM — ICD 154 EXCEPT 154.3

1950-1959 DEATHS	1950-1959 RATE	1960-1969 DEATHS	1960-1969 RATE	1970-1979 DEATHS	1970-1979 RATE	COUNTY AND STATE	PERCENT CHANGE 60S/50S	PERCENT CHANGE 70S/60S	PERCENT CHANGE 70S/50S	PERCENTILE %CHANGE 70S/50S	PERCENTILE RATES 1970S
86	3.3	93	3.1	91	2.4	SOUTH CAROLINA SC	-6	-22	-27		
1	4.2	3	10.3	4*	9.8	ABBEVILLE SC	145	-4	134	94	96
2	2.5			3	2.7	AIKEN SC			12	76	81
		2	6.9	1	2.2	ALLENDALE SC		-68			79
4	7.8	3	3.5	3	2.8	ANDERSON SC	-55	-20	-64	19	81
2	6.1	2	6.6			BAMBERG SC	8				
		1	4.1	1	2.9	BARNWELL SC		-30			81
3	5.2	3	4.8			BEAUFORT SC	-8				
		1	1.5	1	1.0	BERKELEY SC		-36			75
3	8.8					CALHOUN SC					
9	3.4	13	4.5	11	2.8	CHARLESTON SC	30	-36	-17	60	81
		2	7.7			CHEPOKEE SC					
1	1.9	2	3.6	2	3.9	CHESTER SC	95	6	107	92	86
3	8.9	2	4.5			CHESTERFIELD SC	-50				
		2	4.0	1	1.8	CLARENDON SC		-54			77
2	4.3			1	1.5	COLLETON SC			-65	19	76
3	2.8	3	3.9	2	1.5	DARLINGTON SC	39	-63	-48	35	76
3	10.2	1	2.1	1	2.0	DORCHESTER SC	-80	-2	-80	05	78
1	4.1			1	3.7	EDGEFIELD SC			-9	65	85
2	4.3			2	3.8	FAIRFIELD SC			-10	64	86
1	.9	7	5.5	4	2.8	FLORENCE SC	496	-50	200	98	81
1	1.6			1	1.2	GEORGETOWN SC			-28	53	75
5	4.8	9	5.4	5	2.7	GREENVILLE SC	13	-50	-43	39	81
1	2.2	2	3.6			GREENWOOD SC	63				
2	6.1	1	3.2	4*	9.9	HAMPTON SC	-47	205	62	88	96
				2	2.9	HORRY SC					82
1	4.9			1	2.3	JASPER SC			-52	30	80
1	2.1					KERSHAW SC					
1	2.2					LANCASTER SC					
3	5.4			1	1.5	LAURENS SC			-73	11	76
		1	2.4			LEE SC					
1	3.1	5*	12.1	1	1.7	LEXINGTON SC	292	-86	-44	37	77
				1	6.0	MC CORMICK SC					92
2	3.6	2	3.9	3	3.2	MARION SC	9	-19	-11	63	83
1	1.7	1	2.1	1	1.2	MARLBORO SC	25	-43	-28	53	76
1	2.2	1	2.5	1	1.6	NEWBERRY SC	14	-36	-27	54	77
4	3.5	5	3.4	7	4.0	ORANGEBURG SC	-2	17	15	78	86
1	5.7			2	7.3	PICKENS SC			28	82	94
9	4.3	9	3.9	3	1.0	RICHLAND SC	-10	-75	-78	07	75
		1	6.6	2	9.8	SALUDA SC		48			96
3	2.6	5	3.3	6	3.9	SPARTANBURG SC	26	19	49	86	86
2	1.9	2	1.5	5	2.9	SUMTER SC	-21	92	52	87	82
2	4.4	1	2.0			UNION SC	-53				
1	1.3	1	1.2	1	.8	WILLIAMSBURG SC	-13	-30	-39	44	75
4	6.5			6	5.7	YORK SC			-11	63	91
4	6.7	5	6.3	2	1.8	SOUTH DAKOTA SD	-7	-71	-73		
1	18.2					DEWEY SD					
		1*	91.1			MOODY SD					
				1	13.5	PENNINGTON SD					97
2*	53.8					ROBERTS SD					
		2	12.2			SHANNON SD					
1	12.8			1	6.7	TODD SD			-47	36	93
		1*	267.9			TURNER SD					
		1*	119.2			WALWORTH SD					
112	4.7	115	4.1	123	3.6	TENNESSEE TN	-14	-12	-25		
		1	18.1			ANDERSON TN					
2	14.9			1	5.0	BEDFORD TN			-67	17	90
		1	11.1	1	5.3	BLOUNT TN		-52			90
		1	11.5			BRADLEY TN					
2	12.3	1	4.9			CARROLL TN	-60				
		2*	16.1			CROCKETT TN					
24	8.0	16	3.9	30*	6.0	DAVIDSON TN	-51	53	-25	55	92
		2*	28.8			DICKSON TN					
		1	3.4			DYER TN					
		2	3.3			FAYETTE TN					
1	9.3	1	10.7			FRANKLIN TN	15				
1	19.3			1	12.4	HAMBLEN TN			-36	47	97
12	6.2	9	3.5	19*	6.9	HAMILTON TN	-44	97	11	75	93
		1	1.6			HAYWOOD TN					
1	5.4	1	5.1	1	4.0	HENRY TN	-6	-21	-25	55	86
5	4.6	5	3.9	3	1.9	KNOX TN	-14	-52	-59	24	77
				1	11.6	LAKE TN					97
		1	3.7			LAUDERDALE TN					
1	7.9	2*	12.9	1	3.9	LINCOLN TN	65	-70	-50	32	86
1*	38.2					LOUDON TN					
		1	10.4	1	8.4	MC MINN TN		-19			95
3	3.1	7	7.2	4	3.5	MADISON TN	131	-51	13	77	84
1	6.6					MARSHALL TN					
3	7.6	3	6.7	5*	9.0	MAURY TN	-12	34	19	79	95
				1*	16.7	MONROE TN					98
2	5.2	1	2.1	2	3.3	MONTGOMERY TN	-60	57	-37	45	83
1	5.7			2*	10.5	OBION TN			85	90	96
1*	74.3					OVERTON TN					
1	5.3	1	4.5			ROBERTSON TN	-16				
2	6.0			2	4.7	RUTHERFORD TN			-23	57	89
42	5.0	47	4.4	47	3.4	SHELBY TN	-11	-22	-31	51	84
		1*	26.8			SMITH TN					
		1	8.7			SULLIVAN TN					
1	5.6					SUMNER TN					
1	3.3	3	7.8			TIPTON TN	140				
1	22.3					WARREN TN					
1	11.8					WASHINGTON TN					
1	6.1	1	4.5	1	3.6	WILLIAMSON TN	-26	-19	-40	42	85
1	5.4	2	10.0			WILSON TN	86				

* SIGNIFICANTLY MORE DEATHS THAN EXPECTED

NON-WHITE FEMALE — CANCER OF THE RECTUM — ICD 154 EXCEPT 154.3

1950-1959 DEATHS	1950-1959 RATE	1960-1969 DEATHS	1960-1969 RATE	1970-1979 DEATHS	1970-1979 RATE	COUNTY AND STATE	PERCENT CHANGE 60S/50S	PERCENT CHANGE 70S/60S	PERCENT CHANGE 70S/50S	PERCENTILE %CHANGE 70S/50S	PERCENTILE RATES 1970S
156	3.6	192	3.5	167	2.0	TEXAS TX	-4	-42	-44		
1	2.3	1	1.5	2	2.9	ANDERSON TX	-33	87	25	81	82
				1	5.4	AUSTIN TX					91
				1	2.0	BELL TX					78
2	1.6	15	6.1	6	1.2	BEXAR TX	285	-80	-22	58	76
7	11.1	2	2.3	3	3.1	BOWIE TX	-79	31	-73	11	82
1	3.6	3	8.0	2	3.6	BRAZORIA TX	121	-55	-1	70	84
				1	1.0	BRAZOS TX					75
2	8.8	2	9.2	2	8.2	BURLESON TX	4	-11	-7	66	94
1	7.0					CALDWELL TX					
1	6.0					CAMP TX					
3	9.3			2	6.6	CASS TX			-29	52	93
		2	4.7	2	3.4	CHEROKEE TX		-28			84
				1*	37.5	CHILDRESS TX					99
				1*	14.9	COLEMAN TX					98
1	5.9	4*	22.6	3*	12.7	COLLIN TX	281	-44	113	93	97
2	8.2	1	2.6	1	5.6	COLORADO TX	-68	115	-32	50	91
		1*	114.7			CRANE TX					
10	2.5	21	3.5	21	2.3	DALLAS TX	38	-33	-8	66	79
				1*	19.6	DAWSON TX					98
		1*	20.6			DELTA TX					
1	9.1					DENTON TX					
		1	3.7			DE WITT TX					
4	8.7	3	7.1	3	6.3	ELLIS TX	-18	-11	-28	53	92
1	9.3	1	4.1	1	2.5	EL PASO TX	-56	-39	-73	10	80
2	5.0	1	2.1	2	3.2	FALLS TX	-58	49	-37	45	83
2	11.2	1	5.0			FANNIN TX	-56				
3	8.3					FORT BEND TX					
		1*	57.3			FRANKLIN TX					
1	4.0	2	7.1	1	3.1	FREESTONE TX	78	-56	-22	58	82
5	4.6	4	3.1	6	3.2	GALVESTON TX	-33	2	-31	50	83
		2	4.9			GRAYSON TX					
2	3.0	7*	8.9	2	1.9	GREGG TX	196	-78	-35	47	78
		2	5.9	1	2.3	GRIMES TX		-61			79
1	5.8					GUADALUPE TX					
				1	4.8	HALE TX					89
1	6.3	1	6.6			HARDIN TX	5				
38	6.2	36	3.6	31	1.9	HARRIS TX	-42	-48	-70	13	78
4	4.3	2	1.5	1	.7	HARRISON TX	-64	-54	-84	03	75
		3	11.4	2	5.7	HENDERSON TX		-50			91
2	12.5	1	5.1	1	4.0	HILL TX	-60	-21	-68	15	86
1	2.2	2	4.2			HOUSTON TX	94				
				1	8.9	HOWARD TX					95
		1	3.3	1	2.8	HUNT TX		-17			81
		2*	25.6			JACKSON TX					
1	7.5					JASPER TX					
6	3.8	4	1.5	10	3.2	JEFFERSON TX	-62	120	-16	61	83
1*	17.0					JONES TX					
		1	3.1			KAUFMAN TX					
		2	5.6	2	3.6	LAMAR TX		-36			85
1	7.9			1	3.9	LAVACA TX			-50	32	86
		1	3.0			LEON TX					
		1	2.8	1	2.3	LIBERTY TX		-17			80
		2	5.1	1	2.1	LIMESTONE TX		-60			78
				1*	9.4	MC CULLOCH TX					96
2	2.1	6	5.3	7	4.3	MC LENNAN TX	154	-18	108	92	88
1	11.6					MADISON TX					
2	6.2	2	7.3			MARION TX	18				
3	13.2	3	11.0			MATAGORDA TX	-16				
1	4.0					MILAM TX					
1*	35.1					MITCHELL TX					
				2	4.8	MONTGOMERY TX					89
		1	6.6			MORRIS TX					
4*	13.2	3	8.2			NACOGDOCHES TX	-38				
				3	4.6	NAVARRO TX					88
1	5.7	1	5.0	1	4.3	NEWTON TX	-13	-14	-25	55	87
1*	74.3			1*	14.8	NOLAN TX			-80	06	98
		1	3.7			NUECES TX					
		2	10.1	1	3.7	ORANGE TX		-63			85
1*	47.2					PARKER TX					
2	10.7			1	3.2	POLK TX			-71	13	83
		1	3.2			POTTER TX					
1	4.1					RED RIVER TX					
1*	19.1					REFUGIO TX					
2	4.9			1	2.9	ROBERTSON TX			-41	41	82
1*	24.6					ROCKWALL TX					
1	2.0	2	3.4	2	2.7	RUSK TX	68	-19	36	83	81
		1	6.3	1	5.2	SAN AUGUSTINE TX		-16			90
1	4.1					SAN JACINTO TX					
				1*	53.0	SAN SABA TX					99
				1*	22.5	SCURRY TX					98
		1	2.8	1	4.1	SHELBY TX		49			87
1	1.1	4	3.7	4	2.8	SMITH TX	251	-23	170	96	81
				1*	74.3	STONEWALL TX					99
12	6.2	6	1.9	9	2.4	TARRANT TX	-68	22	-62	22	80
		2	11.4	1	5.0	TAYLOR TX		-57			89
		1	4.0			TOM GREEN TX					
1	.8	7	5.0	2	1.1	TRAVIS TX	551	-77	48	86	75
1	3.6					UPSHUR TX					
		2	9.1	1	4.5	VICTORIA TX		-50			88
		2	5.2			WALKER TX					
				2	6.0	WALLER TX					92
2	4.3	3	7.7	1	2.1	WASHINGTON TX	80	-73	-52	30	78
2	4.7	1	1.7	1	2.3	WHARTON TX	-63	34	-51	31	80
2	4.9	1	2.4	2	3.7	WICHITA TX	-52	58	-24	56	85
1	3.1	1	2.7			WILLIAMSON TX	-13				
		1	9.0			WOOD TX					

* SIGNIFICANTLY MORE DEATHS THAN EXPECTED

NON-WHITE FEMALE CANCER OF THE RECTUM ICD 154 EXCEPT 154.3

1950-1959 DEATHS	1950-1959 RATE	1960-1969 DEATHS	1960-1969 RATE	1970-1979 DEATHS	1970-1979 RATE	COUNTY AND STATE	PERCENT CHANGE 60S/50S	PERCENT CHANGE 70S/60S	PERCENT CHANGE 70S/50S	PERCENTILE %CHANGE 70S/50S	PERCENTILE RATES 1970S
1	2.8	3	6.8	3	2.6	UTAH UT	143	-61	-6		
				1*	21.0	DAVIS UT					98
1	6.8	2	10.7	2	3.0	SALT LAKE UT	56	-72	-57	26	82
		1	11.8			WEBER UT					
				1	11.5	VERMONT VT					
				1*	68.8	CHITTENDEN VT					99
138	4.8	159	4.7	163	3.7	VIRGINIA VA	-3	-20	-22		
1	1.8	3	6.8	2	4.2	ACCOMACK VA	271	-38	128	94	87
		4	7.1	1	1.4	ALBEMARLE#VA		-81			76
1	7.0	2	12.8			ALLEGHANY#VA	82				
2	15.2			1	6.5	AMELIA VA			-57	26	93
				2	6.1	AMHERST VA					92
2	9.7	2	6.3	2	3.4	ARLINGTON#VA	-35	-46	-65	18	84
2	9.3	1	2.8	2	4.9	AUGUSTA#VA	-69	72	-47	36	89
1*	47.2					BATH VA					
3	3.0	4	3.6	6	3.8	BEDFORD#VA	18	6	25	81	85
		1	10.4			BOTETOURT VA					
1	2.8	1	2.6	3	6.7	BRUNSWICK VA	-7	159	141	94	93
				1	4.5	BUCKINGHAM VA					88
2	7.7			3*	8.4	CAROLINE VA			8	74	95
		1*	31.5			CARROLL#VA					
1	5.2					CHARLOTTE VA					
25	7.0	27	6.1	28*	4.8	CHESTERFIELD#VA	-13	-21	-31	51	89
1	22.3					CLARKE VA					
1	7.3					CULPEPER VA					
2	12.9					CUMBERLAND VA					
8	5.2	5	3.1	6	3.4	DINWIDDIE#VA	-41	10	-35	47	84
2	4.7	4	4.8	7	4.4	FAIRFAX#VA	2	-7	-5	67	88
1	3.8	1	4.1			FAUQUIER VA	10				
1	8.7			1	4.4	FLUVANNA VA			-50	33	88
		2*	12.3			FRANKLIN VA					
2*	25.8			1	9.0	FREDERICK#VA			-65	19	9!
1	7.6			1	5.2	GLOUCESTER VA			-31	51	9(
		2*	43.0			GREENE VA					
				1	2.3	GREENSVILLE#VA					79
1	1.7	3	5.0	3	4.2	HALIFAX#VA	200	-16	153	96	87
1	4.8			2	7.0	HANOVER VA			45	85	93
1	4.3	1	3.0			HENRICO VA	-31				
1	2.8	4	7.5	7*	9.6	HENRY#VA	165	29	242	99	96
				1	3.2	ISLE OF WIGHT VA					83
3	8.1	3	5.6	2	3.1	JAMES CITY#VA	-30	-44	-61	22	83
1	6.5					KING AND QUEEN VA					
		1	11.2			KING GEORGE VA					
2	13.7					LANCASTER VA					
1	5.9	1	14.9	2	8.6	LOUDOUN VA	152	-42	46	85	95
1	5.3			1	4.5	LOUISA VA			-15	62	88
				1	4.2	LUNENBURG VA					87
		1	15.7	1	9.4	MADISON VA		-40			96
2	5.1	4	7.1	4	5.8	MECKLENBURG VA	37	-18	12	76	91
1	8.9					MONTGOMERY#VA					
1	7.1			1	7.5	NELSON VA			6	73	94
		2	5.4			NORTHAMPTON VA					
1	6.4	2	11.4			NORTHUMBERLAND VA	79				
1	4.3	3	9.1	2	6.0	NOTTOWAY VA	109	-34	39	84	92
1	5.3	1	7.0	1	5.5	ORANGE VA	32	-21	5	72	91
7	5.9	5	4.5	2	1.3	PITTSYLVANIA#VA	-24	-70	-77	07	76
2	7.8	2	7.8	1	3.4	PRINCE EDWARD VA	0	-56	-56	27	84
		1	8.0	2	7.6	PRINCE WILLIAM#VA		-6			94
1	10.3	1	9.6			PULASKI VA	-7				
				1*	22.9	RAPPAHANNOCK VA					98
		7	6.0	3	2.5	ROANOKE#VA		-58			80
1	7.0	1	5.6	2*	10.9	ROCKBRIDGE#VA	-20	95	56	87	96
1	16.4	1	19.9	2*	28.2	ROCKINGHAM#VA	21	42	72	89	98
1*	30.4					SMYTH VA					
4	8.2	4	6.8	4	6.1	SOUTHAMPTON#VA	-17	-9	-25	55	92
		1	3.7	2	6.3	SPOTSYLVANIA#VA		71			92
		1	7.3	1	6.2	SURRY VA		-15			92
2	9.5	2	7.5			SUSSEX VA	-21				
		1*	23.6			WARREN VA					
		1	7.5			WASHINGTON#VA					
		1	5.0			WESTMORELAND VA					
1	18.7					WYTHE VA					
26	5.1	29	4.8	29	3.5	CHESAPEAKE CITY#VA	-6	-27	-32	50	84
2	3.8	2	2.5	7	5.7	HAMPTON CITY VA	-35	130	50	86	91
8	6.0	8	5.5	7	4.2	NEWPORT NEWS CITY VA	-31	-24	-48	35	87
3	3.1	5	4.3	4	3.7	SUFFOLK CITY VA	36	-14	18	79	85
10	4.4	15	5.0	20	2.9	WASHINGTON WA	15	-42	-34		
				1*	11.4	BENTON WA					97
		1*	38.2			CLALLAM WA					
7	5.8	10	6.5	9	2.5	KING WA	11	-62	-58	25	80
1	6.7					KITSAP WA					
2	9.7	1	.6	7*	7.3	PIERCE WA	-94	1128	-25	54	94
		3*	16.0			SPOKANE WA					
				1*	10.2	WHATCOM WA					96
				2	4.8	YAKIMA WA					89
20	5.2	27	5.8	13	2.7	WEST VIRGINIA WV	10	-53	-48		
		1	17.6			BERKELEY WV					
1	4.4			2	5.1	CABELL WV			16	78	90
2	6.0	2	5.6	1	2.3	FAYETTE WV	-6	-59	-62	21	79
2*	25.5	1	7.3			GREENBRIER WV	-72				
1	10.0			1	10.6	JEFFERSON WV			6	73	96

* SIGNIFICANTLY MORE DEATHS THAN EXPECTED   # COUNTIES COMBINED; SEE TECHNICAL NOTES

NON-WHITE FEMALE

CANCER OF THE RECTUM

ICD 154 EXCEPT 154.3

1950-1959 DEATHS	1950-1959 RATE	1960-1969 DEATHS	1960-1969 RATE	1970-1979 DEATHS	1970-1979 RATE	COUNTY AND STATE	PERCENT CHANGE 60S/50S	PERCENT CHANGE 70S/60S	PERCENT CHANGE 70S/50S	PERCENTILE %CHANGE 70S/50S	PERCENTILE RATES 1970S
4	5.2	2	2.4	1	1.4	KANAWHA WV	-53	-44	-74	10	76
3*	25.3	2	7.1	1	6.2	LOGAN WV	-72	-13	-76	08	92
		6*	11.2	4	8.3	MC DOWELL WV		-26			94
1	5.8	1	5.7			MARION WV	-2				
3	7.4	2	4.0	2	4.1	MERCER WV	-46	3	-44	38	87
1	13.8	1	15.7			MONONGALIA WV	14				
		1	8.8			OHIO WV					
		7*	11.9			RALEIGH WV					
				1*	47.2	RANDOLPH WV					99
		1*	37.2			SUMMERS WV					
1*	188.7					TUCKER WV					
1	22.9					WYOMING WV					
7	3.8	7	2.5	22	4.4	WISCONSIN WI	-34	75	15		
				1*	28.1	ASHLAND WI					98
				1	8.6	BROWN WI					95
				1*	47.2	DODGE WI					99
1*	84.8			1	13.2	KENOSHA WI			-84	03	97
5	3.6	6	2.9	13	4.0	MILWAUKEE WI	-19	38	11	76	86
1*	424.0					PORTAGE WI					
				1	5.0	RACINE WI					89
				1	5.1	ROCK WI					90
		1*	94.4			SAUK WI					
				1*	2.7	WAUKESHA WI					81
				2*	106.5	WOOD WI					99
		1	4.2	1	2.5	WYOMING WY		-42			
		1	10.5	1	7.6	FREMONT WY		-27			94

* SIGNIFICANTLY MORE DEATHS THAN EXPECTED

1950-1959 DEATHS	1950-1959 RATE	1960-1969 DEATHS	1960-1969 RATE	1970-1979 DEATHS	1970-1979 RATE	COUNTY AND STATE	PERCENT CHANGE 60S/50S	PERCENT CHANGE 70S/60S	PERCENT CHANGE 70S/50S	PERCENTILE %CHANGE 70S/50S	PERCENTILE RATES 1970S
19404	3.0	27997	3.7	39785	4.6	UNITED STATES	24	26	55		
362*	4.5	354	3.6	576	4.7	ALABAMA AL	-21	32	4		
3	6.8	4	7.7	4	5.1	AUTAUGA AL	13	-33	-24	16	69
5	3.2	5	2.1	15	4.7	BALDWIN AL	-35	126	47	47	61
		6*	11.0	1	1.3	BARBOUR AL		-88			12
3	6.2	3	4.9	3	4.8	BIBB AL	-20	-3	-22	17	63
3	2.7	2	1.5	9	5.8	BLOUNT AL	-45	293	115	69	78
		1	4.1			BULLOCK AL					
4	5.6	4	5.5	5	6.0	BUTLER AL	-1	8	7	30	79
6	2.7	6	2.3	25*	7.1	CALHOUN AL	-13	211	169	78	88
7*	6.9	2	1.8	3	1.8	CHAMBERS AL	-74	0	-74	01	16
1	1.4			1	1.1	CHEROKEE AL			-20	19	11
1	1.3	1	.8	4	3.1	CHILTON AL	-37	286	143	75	32
		1	2.7	1	2.2	CHOCTAW AL		-18			20
7*	11.4			5	6.1	CLARKE AL			-46	08	81
5*	7.7			4	6.1	CLAY AL			-21	18	80
1	2.2	2	4.1	4	6.7	CLEBURNE AL	84	65	203	83	86
4	3.8	1	1.0	7	5.0	COFFEE AL	-73	397	32	41	67
5	4.9	6	4.4	9	4.9	COLBERT AL	-11	12	-1	27	65
1	2.0	2	3.2	3	5.6	CONECUH AL	62	74	182	80	75
1	2.7	1	2.3	2	4.3	COOSA AL	-14	82	57	50	53
11*	8.3	5	3.1	6	3.4	COVINGTON AL	-63	11	-59	04	37
1	2.2	2	5.0	1	1.2	CRENSHAW AL	129	-76	-44	09	12
8	4.8	4	1.8	14	5.1	CULLMAN AL	-62	178	6	29	68
3	4.6	1	.6	9	7.1	DALE AL	-88	1184	54	49	88
4	5.4	4	4.6	2	1.7	DALLAS AL	-15	-62	-68	02	16
6	3.6	2	1.0	6	2.3	DE KALB AL	-73	134	-37	10	21
4	4.3	3	2.5	5	3.2	ELMORE AL	-42	29	-26	16	34
3	3.4	6	6.2	3	2.5	ESCAMBIA AL	81	-59	-26	15	24
12	4.8	9	2.7	24	5.8	ETOWAH AL	-45	117	20	35	77
1	1.6	4	5.1	4	4.5	FAYETTE AL	226	-12	188	81	58
2	1.9	2	1.7	7	5.4	FRANKLIN AL	-6	209	189	81	72
3	3.6	4	4.2	6	5.6	GENEVA AL	16	33	54	49	75
1	8.6	1	9.4			GREENE AL	9				
1	3.3	2	7.0			HALE AL	111				
2	5.0			2	4.9	HENRY AL			-2	26	65
2	1.8	6	3.8	14	5.9	HOUSTON AL	117	54	233	86	78
6	4.0	6	3.8	17*	8.8	JACKSON AL	-5	128	116	69	94
78*	5.8	66	3.9	91	4.6	JEFFERSON AL	-32	17	-21	18	59
2	3.6	6*	9.1	3	3.9	LAMAR AL	150	-57	7	30	45
8	6.3	8	4.2	16	5.5	LAUDERDALE AL	-33	29	-13	22	73
1	1.3			8	7.0	LAWRENCE AL			417	95	87
		6	4.7	11	7.7	LEE AL		62			91
5	3.8	2	1.4	4	2.9	LIMESTONE AL	-62	107	-22	17	29
		1	6.4	2	15.0	LOWNDES AL		135			99
1	4.0	1	3.0	1	2.9	MACON AL	-26	-3	-28	14	29
8	3.5	14	4.2	23	5.7	MADISON AL	18	37	62	52	77
5*	12.3	3	6.4	4	7.5	MARENGO AL	-48	18	-39	10	90
4	3.6	2	1.9	3	1.9	MARION AL	-48	2	-47	08	17
8	5.2	6	2.9	10	3.8	MARSHALL AL	-43	31	-26	15	44
34*	6.2	42*	6.0	54	5.7	MOBILE AL	-3	-4	-7	24	77
2	3.6			2	3.2	MONROE AL			-10	23	34
20*	7.3	14	4.2	17	3.9	MONTGOMERY AL	-43	-6	-46	08	45
7	4.0	7	3.1	13	5.2	MORGAN AL	-22	65	29	39	69
1	4.5	1	3.3			PERRY AL	-27				
3	5.1			2	2.9	PICKENS AL			-44	09	28
1	1.3	4	5.6	5	5.8	PIKE AL	321	2	331	92	77
1	1.1	3	3.5	3	3.2	RANDOLPH AL	201	-7	180	80	34
3	5.0	6	7.0	8	6.9	RUSSELL AL	41	-2	39	43	87
4	4.0	2	2.0	9	6.4	ST CLAIR AL	-50	219	59	51	84
8*	8.4	7	6.1	6	4.6	SHELBY AL	-27	-24	-45	08	59
		1	3.5	4*	12.7	SUMTER AL		263			98
4	2.8	13*	7.3	14	6.4	TALLADEGA AL	166	-12	133	73	84
4	3.5	6	5.2	6	4.2	TALLAPOOSA AL	51	-20	21	36	51
4	1.5	10	2.8	8	2.0	TUSCALOOSA AL	84	-28	33	41	18
11*	5.3	9	3.7	17	6.0	WALKER AL	-29	60	14	33	79
		2	4.3			WASHINGTON AL					
3*	10.5	1	4.1	1	3.9	WILCOX AL	-61	-4	-63	03	46
5*	7.1	3	4.1	6	6.8	WINSTON AL	-43	66	-5	25	86
		9	3.3	26	4.3	ALASKA AK		32			
93	2.9	196	3.5	421	4.5	ARIZONA AZ	20	28	55		
				1	4.4	APACHE AZ					55
3	2.8	5	2.8	16	4.9	COCHISE AZ	0	74	74	57	64
2	4.6	3	4.2	7	6.5	COCONINO AZ	-8	55	43	45	85
2	2.0	6	6.2	8	4.2	GILA AZ	206	-33	104	66	51
1	1.9	1	2.0	4	5.3	GRAHAM AZ	7	163	182	80	71
				2	3.9	GREENLEE AZ					45
48	2.9	112	3.5	233	4.4	MARICOPA AZ	19	26	51	48	55
1	2.2	3	4.1	12	5.3	MOHAVE AZ	91	30	148	76	71
1	1.5	1	1.8	3	2.5	NAVAJO AZ	18	44	70	56	24
24	3.9	46	4.0	88	4.7	PIMA AZ	2	18	20	35	61
2	1.1	3	1.4	17	5.1	PINAL AZ	29	263	370	93	67
2	4.6			4	7.4	SANTA CRUZ AZ			60	52	90
2	1.1	10	5.2	17	4.3	YAVAPAI AZ	370	-17	291	90	53
5	4.7	6	2.8	9	3.0	YUMA AZ	-40	7	-36	11	30
198	2.9	233	2.9	422	4.4	ARKANSAS AR	1	52	54		
6*	7.0	4	4.1	6	5.9	ARKANSAS AR	-42	46	-16	20	79
1	1.7	3	3.9	5	5.3	ASHLEY AR	126	35	206	83	71
1	1.4	3	1.8	9	6.2	BAXTER AR	36	237	357	93	82
5	2.3	5	1.8	21	5.0	BENTON AR	-23	185	120	70	67

* SIGNIFICANTLY MORE DEATHS THAN EXPECTED

WHITE MALE

CANCER OF THE LIVER AND GALLBLADDER INCLUDING BILE DUCTS

ICD 155,156

1950-1959 DEATHS	1950-1959 RATE	1960-1969 DEATHS	1960-1969 RATE	1970-1979 DEATHS	1970-1979 RATE	COUNTY AND STATE	PERCENT CHANGE 60S/50S	PERCENT CHANGE 70S/60S	PERCENT CHANGE 70S/50S	PERCENTILE %CHANGE 70S/50S	PERCENTILE RATES 1970S
1	.9	5	3.9	4	2.6	BOONE AR	315	-33	176	80	24
1	1.9	2	3.3	5	7.7	BRADLEY AR	73	134	304	90	91
1	4.5	1	4.2	2	6.7	CALHOUN AR	-7	59	48	47	86
2	2.5	3	3.5			CARROLL AR	41				
		2	4.7	6*	13.5	CHICOT AR		184			98
2	2.3	1	1.2	4	4.1	CLARK AR	-46	228	77	58	49
3	2.4	4	3.2	10	7.3	CLAY AR	36	126	207	84	89
2	3.6	1	1.4	4	4.5	CLEBURNE AR	-60	214	24	37	56
				2	4.2	CLEVELAND AR					52
1	1.2	1	1.1	3	2.9	COLUMBIA AR	-12	168	136	73	29
		2	3.7	1	1.0	CONWAY AR		-72			11
5	2.7	7	3.4	15	6.3	CRAIGHEAD AR	24	88	132	73	83
3	2.4	3	2.2	4	2.6	CRAWFORD AR	-9	20	8	30	25
2	3.4	2	2.3	5	5.2	CRITTENDEN AR	-33	129	54	49	70
2	3.3	1	1.5	6	8.0	CROSS AR	-55	436	141	74	92
2	4.6	1	2.2	1	2.0	DALLAS AR	-53	-9	-57	04	18
1	1.9	3	6.4	3	5.3	DESHA AR	241	-16	186	81	71
1	2.0	1	1.9	4	7.8	DREW AR	-7	310	282	89	91
5	4.9	3	2.4	5	3.2	FAULKNER AR	-52	35	-35	11	34
2	3.1	4	5.1	2	2.3	FRANKLIN AR	67	-55	-25	16	21
1	1.8	1	1.6	3	3.9	FULTON AR	-9	145	123	70	45
10	3.8	12	4.1	37*	8.1	GARLAND AR	5	99	110	67	92
2	3.8	1	1.7	1	1.7	GRANT AR	-54	-2	-55	05	15
7	5.1	2	1.4	6	3.8	GREENE AR	-74	179	-26	15	44
		8*	9.4	5	4.6	HEMPSTEAD AR		-51			59
		2	1.9	5	4.2	HOT SPRING AR		125			51
2	3.8	2	2.6	4	6.0	HOWARD AR	-31	128	57	50	80
6	5.8	2	1.5	4	2.7	INDEPENDENCE AR	-75	83	-53	06	26
1	2.4	4	6.1	3	3.6	IZARD AR	153	-40	51	48	41
5	5.7	2	2.2	3	2.5	JACKSON AR	-62	16	-56	05	24
6	3.5	11	5.3	12	4.9	JEFFERSON AR	51	-7	41	44	65
2	3.1			2	1.5	JOHNSON AR			-50	07	14
2	5.4	2	5.3			LAFAYETTE AR	-1				
3	3.2	4	3.8	4	4.1	LAWRENCE AR	16	8	25	38	48
2	5.1	2	5.2	4	10.8	LEE AR	2	107	110	67	96
2	8.3	1	2.6	1	3.0	LINCOLN AR	-69	17	-63	03	31
		1	2.2	4	8.3	LITTLE RIVER AR		271			93
6	6.4	4	3.2	8	7.0	LOGAN AR	-50	119	10	31	88
2	2.1	2	1.8	6	4.8	LONOKE AR	-12	162	130	72	63
		1	1.3	6	8.8	MADISON AR		574			94
1	2.1			2	1.9	MARION AR			-7	24	17
6	5.7	6	4.8	8	5.2	MILLER AR	-17	9	-9	24	70
8	5.1	4	2.2	8	4.0	MISSISSIPPI AR	-57	80	-23	17	46
1	2.5	2	4.1	4	8.5	MONROE AR	64	107	240	86	93
		3	7.7	2	3.0	MONTGOMERY AR		-61			31
				4	7.0	NEVADA AR					87
		1	3.0			NEWTON AR					
1	1.3	1	1.2	6	5.9	OUACHITA AR	-9	410	365	93	79
2	7.8	1	2.7	2	4.9	PERRY AR	-66	85	-37	11	65
1	1.0	2	2.3	1	1.3	PHILLIPS AR	136	-44	33	41	12
2	3.7	2	3.7	2	3.2	PIKE AR	1	-15	-14	21	33
4	3.3	2	1.7	7	5.7	POINSETT AR	-46	229	76	58	77
1	1.1	7	7.4	11*	9.2	POLK AR	601	24	770	99	95
2	1.6	2	1.5	8	5.0	POPE AR	-6	221	201	83	66
2	4.1	3	7.1	1	1.7	PRAIRIE AR	72	-76	-59	04	15
21	3.0	18	2.2	32	3.2	PULASKI AR	-27	48	8	30	34
4	5.9	3	3.5	4	4.0	RANDOLPH AR	-41	14	-32	13	46
2	3.4	2	2.8	5	7.9	ST FRANCIS AR	-20	187	131	72	92
5	4.4	2	1.4	5	3.0	SALINE AR	-69	116	-33	13	30
2	3.9	2	4.0	2	2.6	SCOTT AR	3	-36	-33	12	25
2	4.2	1	1.7	1	1.5	SEARCY AR	-60	-13	-65	02	13
4	1.4	10	3.0	9	2.4	SEBASTIAN AR	118	-20	75	57	23
1	1.8	3	4.4	4	5.3	SEVIER AR	149	19	196	82	70
2	4.4	2	3.4	4	3.0	SHARP AR	-22	-13	-32	13	30
1	2.8					STONE AR					
1	.7	4	2.1	8	4.1	UNION AR	204	94	492	96	49
3	6.2			4	4.4	VAN BUREN AR			-29	14	55
2	.8	17	5.4	17	4.5	WASHINGTON AR	580	-18	459	96	56
7	3.8	5	2.6	10	4.3	WHITE AR	-33	67	13	32	53
1	2.0			2	4.9	WOODRUFF AR			141	74	65
1	1.6	2	2.3	4	4.5	YELL AR	48	95	188	81	57
1713*	3.3	2595	3.9	3658	4.6	CALIFORNIA CA	16	18	37		
103	3.3	152	4.3	167	4.7	ALAMEDA CA	30	9	42	45	60
		3	4.5	2	1.6	AMADOR CA		-64			14
8	1.9	20	3.6	24	3.1	BUTTE CA	93	-13	69	55	33
4	6.3	3	3.1	3	2.8	CALAVERAS CA	-51	-10	-56	05	27
2	2.8	2	2.9	4	5.5	COLUSA CA	3	90	97	63	74
35	3.7	59	4.0	100	4.8	CONTRA COSTA CA	8	21	30	40	63
1	1.5	1	1.1	3	3.0	DEL NORTE CA	-26	164	97	63	30
2	1.6	4	2.8	7	2.6	EL DORADO CA	74	-7	63	53	25
41	3.4	58	3.9	76	4.5	FRESNO CA	12	15	29	40	56
4	4.6	4	4.3	9	8.8	GLENN CA	-6	105	92	62	94
6	1.5	12	3.1	18	3.9	HUMBOLDT CA	107	27	162	77	45
7	2.6	8	2.9	15	5.4	IMPERIAL CA	12	84	105	66	72
1	1.5	1	1.6	5	4.7	INYO CA	4	197	210	84	60
37*	4.5	38	3.6	76	5.9	KERN CA	-18	61	31	41	78
5	2.7	8	3.9	8	3.7	KINGS CA	44	-5	36	42	42
2	1.5	6	3.6	11	4.4	LAKE CA	142	23	196	82	55
1	1.3	3	4.6	2	2.2	LASSEN CA	267	-52	74	57	20
723*	3.7	932	3.9	1187	4.8	LOS ANGELES CA	6	23	31	40	62
4	3.8	13*	6.6	15	6.7	MADERA CA	76	2	79	58	86
7	1.8	22	3.7	40	4.8	MARIN CA	111	30	173	79	62
1	2.4					MARIPOSA CA					
4	1.7	11	4.3	13	4.4	MENDOCINO CA	161	1	163	78	54
5	1.7	8	2.3	20	5.0	MERCED CA	34	122	197	82	67
2	3.9	1	2.4			MODOC CA	-38				
				1	4.9	MONO CA					64
19	3.8	20	2.9	37	4.1	MONTEREY CA	-24	44	9	31	50

* SIGNIFICANTLY MORE DEATHS THAN EXPECTED

WHITE MALE | CANCER OF THE LIVER AND GALLBLADDER INCLUDING BILE DUCTS | ICD 155,156

1950-1959 DEATHS	1950-1959 RATE	1960-1969 DEATHS	1960-1969 RATE	1970-1979 DEATHS	1970-1979 RATE	COUNTY AND STATE	PERCENT CHANGE 60S/50S	PERCENT CHANGE 70S/60S	PERCENT CHANGE 70S/50S	PERCENTILE %CHANGE 70S/50S	PERCENTILE RATES 1970S
7	1.8	18	3.9	23	4.4	NAPA CA	110	14	139	74	54
5	3.7			5	2.1	NEVADA CA			-44	08	18
39	2.6	105	3.3	226	4.2	ORANGE CA	27	27	61	52	51
5	1.9	10	2.9	20	4.6	PLACER CA	56	57	145	75	59
3	4.3	1	1.2	1	1.1	PLUMAS CA	-71	-13	-75	01	11
35	3.4	57	2.9	131	4.2	RIVERSIDE CA	-16	45	23	37	51
38	2.8	87	4.7	111	4.9	SACRAMENTO CA	68	5	77	58	64
4	5.3	4	4.9	6	7.0	SAN BENITO CA	-8	44	32	41	87
53	3.3	74	3.0	147	4.8	SAN BERNARDINO CA	-7	58	46	46	62
93	3.4	154	3.7	233	3.9	SAN DIEGO CA	8	5	12	32	45
153*	4.1	204*	5.8	185*	6.0	SAN FRANCISCO CA	41	4	47	47	80
31	3.1	58	4.8	81*	6.2	SAN JOAQUIN CA	54	29	98	64	81
10	2.8	16	3.2	30	4.6	SAN LUIS OBISPO CA	12	45	62	52	59
41	3.5	72	4.3	91	4.2	SAN MATEO CA	23	-3	20	35	50
18	3.3	30	3.7	44	4.0	SANTA BARBARA CA	11	7	19	35	47
42	2.8	110	4.4	134	4.2	SANTA CLARA CA	60	-4	53	49	52
22	4.1	23	3.7	31	3.7	SANTA CRUZ CA	-12	2	-10	23	42
3	1.4	9	3.0	24	5.4	SHASTA CA	109	78	273	88	72
		2*	18.9			SIERRA CA					
3	1.6	3	1.5	6	3.3	SISKIYOU CA	-7	116	101	65	35
9	2.0	16	3.7	28	4.8	SOLANO CA	84	30	139	74	63
23	3.4	30	3.1	56	4.4	SONOMA CA	-10	41	28	39	54
9	1.3	27	3.3	53	5.3	STANISLAUS CA	152	63	309	91	71
4	3.2	9	5.7	5	3.0	SUTTER CA	80	-48	-7	25	30
3	2.7	4	2.6	6	2.8	TEHAMA CA	-3	8	5	29	28
3	11.6			2	3.5	TRINITY CA			-70	01	39
11	1.7	33	4.3	55	6.2	TULARE CA	153	42	261	88	81
1	1.0	2	2.2	2	1.2	TUOLUMNE CA	122	-45	23	37	12
16	2.8	30	2.9	57	4.0	VENTURA CA	5	40	47	46	48
4	1.8	13	4.4	20	5.7	YOLO CA	147	30	221	85	76
1	.8	5	4.2	2	1.2	YUBA CA	457	-71	61	52	12
251*	4.0	259	3.4	395	4.3	COLORADO CO	-15	28	9		
5	3.6	11	3.4	24	4.9	ADAMS CO	-6	46	37	43	65
2	4.4	2	4.9	2	4.4	ALAMOSA CO	13	-11	0	27	55
1	7.3	2*	18.8	1	7.5	ARCHULETA CO	157	-60	2	28	90
1	3.0	1	2.5	2	6.2	BACA CO	-18	151	104	66	82
		2	4.4	7*	17.2	BENT CO		293			99
13*	5.4	6	1.8	7	1.6	BOULDER CO	-67	-12	-71	01	14
1	1.9	1	2.5	5	9.6	CHAFFEE CO	31	288	407	94	95
1	6.3					CLEAR CREEK CO					
		1	3.4	3	11.3	CONEJOS CO		233			97
2*	12.5	2	11.7	1	5.8	COSTILLA CO	-6	-50	-53	06	78
1	4.9	1	4.1			CROWLEY CO	-16				
				1	14.5	CUSTER CO					98
8*	8.1	3	2.4	2	2.0	DELTA CO	-70	-18	-75	00	18
87*	4.0	93	3.7	126	4.7	DENVER-ARAPAHOE#CO	-6	26	18	34	61
1*	20.6					DOLORES CO					
				4	10.2	DOUGLAS CO					96
				2	6.4	EAGLE CO					84
1	5.9	1	3.8	2	6.6	ELBERT CO	-35	73	12	32	85
13	3.4	16	3.0	45	6.0	EL PASO CO	-11	100	78	58	60
5	3.4	3	2.2	8	4.5	FREMONT CO	-38	108	30	40	57
2	3.0	3	3.8	2	2.6	GARFIELD CO	26	-31	-13	22	25
		1*	19.9	1	31.5	GILPIN CO		58			99
1	14.9					GRAND CO					
				1	4.8	GUNNISON CO					62
4	7.2	1	1.5	1	1.9	HUERFANO CO	-80	31	-74	01	17
12	4.4	18	3.5	40	4.5	JEFFERSON CO	-22	30	2	28	58
				3	6.3	KIT CARSON CO					82
1	5.0					LAKE CO					
5*	6.9			5	5.4	LA PLATA CO			-22	17	72
10	4.1	13	4.0	12	2.6	LARIMER CO	-3	-34	-36	11	25
9*	7.5	7	5.4	2	1.7	LAS ANIMAS CO	-28	-68	-77	00	15
		2	6.6	1	3.5	LINCOLN CO		-47			38
4	5.3	3	3.3	6	6.2	LOGAN CO	-37	87	18	34	82
11	5.4	9	3.4	7	2.1	MESA CO	-37	-38	-61	03	19
2	8.4	1	4.2			MOFFAT CO	-50				
2	6.6	4	8.0	4	7.7	MONTEZUMA CO	22	-3	18	34	91
6*	8.1	2	2.3	5	4.6	MONTROSE CO	-72	100	-44	09	58
4	4.2	2	1.9	6	5.0	MORGAN CO	-54	164	21	36	67
1	.9	2	1.7	5	4.2	OTERO CO	88	152	373	94	51
1	10.7			1	9.3	OURAY CO			-14	21	95
1	8.3	1	14.9	1	5.1	PARK CO	79	-66	-39	10	67
2	7.2	1	2.7			PHILLIPS CO	-62				
3	4.7			3	4.5	PROWERS CO			-5	26	57
14	3.1	15	2.9	29	5.5	PUEBLO CO	-4	88	80	59	74
1	8.6			1	5.5	RIO BLANCO CO			-36	11	74
1	2.0	2	3.5	1	1.9	RIO GRANDE CO	76	-46	-5	25	17
2	5.7	2	5.6	1	4.1	ROUTT CO	-2	-27	-28	14	49
		1	5.6	3*	17.1	SAGUACHE CO		203			99
				1*	44.6	SAN JUAN CO					99
1	4.2			1	9.3	SEDGWICK CO			122	70	95
		1	6.1	3*	10.3	TELLER CO		67			96
		3	7.3			WASHINGTON CO					
8	2.5	17	5.0	5	1.3	WELD CO	98	-74	-49	07	12
1	1.6	3	4.7	2	3.1	YUMA CO	191	-35	90	62	31
280	2.8	462	4.0	630	4.8	CONNECTICUT CT	41	19	69		
71	2.8	115	4.0	150	4.5	FAIRFIELD CT	40	14	60	52	58
73	2.8	133	4.4	161	4.6	HARTFORD CT	54	5	61	52	59
13	2.3	22	3.5	33	4.5	LITCHFIELD CT	50	29	93	62	57
7	1.7	12	2.6	31	5.7	MIDDLESEX CT	53	124	243	87	77
78	3.0	128	4.2	175	5.2	NEW HAVEN CT	38	23	70	56	69
24	3.4	24	3.0	41	4.4	NEW LONDON CT	-12	48	30	40	56
8	3.8	8	2.9	17	5.0	TOLLAND CT	-23	72	32	41	67
6	2.0	20	5.7	22	5.5	WINDHAM CT	186	-4	174	80	73

* SIGNIFICANTLY MORE DEATHS THAN EXPECTED

\# COUNTIES COMBINED; SEE TECHNICAL NOTES

WHITE MALE

CANCER OF THE LIVER AND GALLBLADDER INCLUDING BILE DUCTS

ICD 155,156

1950-1959 DEATHS	1950-1959 RATE	1960-1969 DEATHS	1960-1969 RATE	1970-1979 DEATHS	1970-1979 RATE	COUNTY AND STATE	PERCENT CHANGE 60S/50S	PERCENT CHANGE 70S/60S	PERCENT CHANGE 70S/50S	PERCENTILE %CHANGE 70S/50S	PERCENTILE RATES 1970S
44	3.3	63	3.7	81	4.2	DELAWARE DE	13	15	29		
6	3.1	5	2.1	9	3.3	KENT DE	-30	55	8	30	36
32	3.6	46	3.9	56	4.5	NEW CASTLE DE	9	15	25	38	57
6	2.2	12	3.8	16	4.0	SUSSEX DE	69	8	82	59	48
81*	4.0	76	4.7	49	4.3	DIST. OF COLUMBIA DC	17	-8	8		
81*	4.0	76	4.7	49	4.3	WASHINGTON DC	17	-8	8	30	54
366	2.2	1044	3.5	1967	4.1	FLORIDA FL	60	19	91		
3	2.0	4	2.2	13	4.6	ALACHUA FL	12	110	134	73	59
				2	4.2	BAKER FL					52
2	1.9	9	4.0	16	5.7	BAY FL	114	43	207	83	77
		2	4.1	5	7.5	BRADFORD FL		81			90
5	1.7	15	3.4	40	3.7	BREVARD FL	97	9	115	69	42
20	1.9	99	3.0	263	4.5	BROWARD FL	56	48	131	73	56
1	2.9			2	5.9	CALHOUN FL			102	65	79
1	1.0	8	1.8	19	3.9	CHARLOTTE FL	77	118	287	89	45
1	1.4			13	4.1	CITRUS FL			203	83	49
1	1.5	1	1.9	10	5.8	CLAY FL	21	212	277	89	77
1	2.6	4	2.3	20	4.3	COLLIER FL	-10	87	68	55	54
		1	1.4	7	6.9	COLUMBIA FL		383			87
82	2.5	250*	4.6	393	5.1	DADE FL	83	12	106	66	69
		3	5.5	1	1.4	DE SOTO FL		-74			13
21	2.1	40	3.0	68	4.4	DUVAL FL	44	43	106	67	54
6	1.7	17	3.7	37	5.7	ESCAMBIA FL	119	53	235	86	76
				3	3.5	FLAGLER FL					39
1	4.7	1	3.7	2	5.1	FRANKLIN FL	-21	38	9	31	68
		2	2.1	5	5.2	GADSDEN FL		146			70
		1	4.9	2	8.1	GILCHRIST FL		64			92
				3	12.7	GLADES FL					98
		2	9.6			GULF FL					
1	4.6	1	4.5	1	2.8	HAMILTON FL	-2	-37	-38	10	28
4*	9.2	2	3.4	4	5.1	HARDEE FL	-64	51	-45	08	68
1	3.6	1	2.3	2	4.0	HENDRY FL	-37	77	12	32	47
		8*	8.7	16	5.9	HERNANDO FL		-32			78
3	3.9	8	3.9	12	3.4	HIGHLANDS FL	0	-13	-13	22	37
34	2.5	75	4.0	104	4.5	HILLSBOROUGH FL	62	10	78	58	56
		3	4.2	3	3.7	HOLMES FL		-12			42
2	1.9	3	1.9	9	2.6	INDIAN RIVER FL	1	35	36	42	25
4	4.2	3	2.6	9	6.9	JACKSON FL	-39	167	64	53	87
		1	5.1	3	11.2	JEFFERSON FL		119			96
				1	3.9	LAFAYETTE FL					45
4	1.6	13	3.2	35	5.4	LAKE FL	103	65	236	86	72
5	3.0	7	1.2	33	3.0	LEE FL	-61	153	-2	26	30
7*	6.6	11*	7.4	11	3.5	LEON FL	12	-52	-46	08	40
4*	9.7	3	5.6	4	3.8	LEVY FL	-42	-31	-60	03	44
				1	7.0	LIBERTY FL					88
1	2.6			3	5.5	MADISON FL			112	68	74
5	1.0	29	4.1	28	2.8	MANATEE FL	319	-31	188	81	28
2	1.2	3	1.3	12	2.6	MARION FL	9	104	122	70	25
3	3.9	1	.3	14	3.1	MARTIN FL	-91	831	-19	19	32
2	1.5	4	2.3	7	2.4	MONROE FL	52	5	59	51	22
1	3.1	1	1.4	3	3.1	NASSAU FL	-54	115	0	27	31
1	1.2	5	3.0	7	2.8	OKALOOSA FL	162	-7	144	75	28
				5	7.6	OKEECHOBEE FL					91
19	2.8	32	2.9	63	4.1	ORANGE FL	2	42	45	46	49
2	1.4	2	1.3	11	4.4	OSCEOLA FL	-8	244	217	85	55
15	1.6	55	3.2	101	3.3	PALM BEACH FL	93	3	99	64	35
3	1.0	17	2.8	49	2.9	PASCO FL	195	2	202	83	29
58	2.4	152	3.2	242	3.9	PINELLAS FL	33	20	60	52	45
12	1.7	39	4.1	55	4.2	POLK FL	142	2	148	75	50
		1	.7	6	2.7	PUTNAM FL		280			26
1	.9	2	1.6	8	4.1	ST JOHNS FL	79	164	371	93	49
2	1.4	14*	5.7	18	5.0	ST LUCIE FL	300	-13	247	87	66
2	3.7	3	2.5	2	1.5	SANTA ROSA FL	-33	-39	-59	04	14
10	2.2	32	3.6	68	4.5	SARASOTA FL	62	23	99	64	56
		10	4.0	16	3.4	SEMINOLE FL		-15			37
1	1.9	2	2.7	6	4.6	SUMTER FL	41	71	141	74	59
		2	3.7	4	5.9	SUWANNEE FL		59			78
				3	5.6	TAYLOR FL					75
9	1.3	37	3.0	51	3.1	VOLUSIA FL	136	2	140	74	32
		2	7.7	4*	9.5	WAKULLA FL		23			95
3	4.2	1	1.1	3	2.7	WALTON FL	-74	151	-35	11	27
				6	9.0	WASHINGTON FL					95
313*	3.6	309	2.7	614	4.3	GEORGIA GA	-25	61	21		
1	2.3	1	2.1	4	10.3	APPLING GA	-9	400	357	93	96
2*	8.2			3*	13.9	ATKINSON GA			68	55	98
1	4.0	1	3.0	2	6.3	BACON GA	-25	111	59	51	83
				2*	18.7	BAKER GA					99
1	1.0	2	1.7	6	5.3	BALDWIN GA	76	205	436	95	71
		1	4.0	4	9.7	BANKS GA		140			96
		1	1.4	1	1.7	BARROW GA		14			15
1	1.2	4	4.0	9	7.4	BARTOW GA	228	83	500	96	90
2	5.2	2	4.1	4	8.5	BEN HILL GA	-21	107	63	53	94
1	2.0	1	2.2	3	5.4	BERRIEN GA	11	141	167	78	73
13	5.3	13	3.6	22	5.6	BIBB GA	-32	55	6	29	75
1	12.4			4*	10.6	BLECKLEY GA			-15	21	96
1	4.5			1	4.4	BRANTLEY GA			-2	26	55
				1	1.9	BROOKS GA					17
		1	6.1	1	8.6	BRYAN GA		42			94
1	1.4	2	2.2	2	1.9	BULLOCH GA	58	-13	37	43	16
		1	2.7	4	9.1	BURKE GA		240			95
1	3.2	2	6.5	2	6.2	BUTTS GA	103	-5	92	62	81
1	6.5			2	14.7	CALHOUN GA			126	71	99

* SIGNIFICANTLY MORE DEATHS THAN EXPECTED

1950-1959 DEATHS	1950-1959 RATE	1960-1969 DEATHS	1960-1969 RATE	1970-1979 DEATHS	1970-1979 RATE	COUNTY AND STATE	PERCENT CHANGE 60S/50S	PERCENT CHANGE 70S/60S	PERCENT CHANGE 70S/50S	PERCENTILE %CHANGE 70S/50S	PERCENTILE RATES 1970S
1	3.7	2	10.4	2	4.7	CAMDEN GA	182	-55	27	38	61
1	4.4	1	3.9	1	3.0	CANDLER GA	-12	-23	-32	13	30
9*	6.9	7	4.6	6	3.5	CARROLL GA	-34	-23	-49	07	39
2	4.7			2	1.2	CATOOSA GA			-75	00	11
		2*	11.8	1	4.2	CHARLTON GA		-65			50
18*	6.5	9	2.2	19	3.8	CHATHAM GA	-67	78	-41	10	44
		1	1.1	3	3.3	CHATTOOGA GA		193			35
4	5.3	2	1.8	4	2.6	CHEROKEE GA	-66	46	-50	07	25
5	6.3	4	3.3	10	7.1	CLARKE GA	-48	117	12	32	88
4	5.0	5	3.2	13	7.1	CLAYTON GA	-37	123	41	44	88
				1	5.4	CLINCH GA					72
7	3.6	9	3.3	20	3.1	COBB GA	-9	-8	-16	20	31
2	3.4	2	2.3	3	3.0	COFFEE GA	-31	26	-13	22	30
5	5.3	3	2.7	6	4.6	COLQUITT GA	-50	73	-13	22	59
1	6.5	1	3.6	3	2.8	COLUMBIA GA	-45	-22	-57	04	27
4	5.1	4	4.3	8	7.2	COWETA GA	-16	68	41	45	88
		2	3.5	4	8.2	CRISP GA		135			93
1	3.6			2	5.0	DADE GA			42	45	67
1	6.7					DAWSON GA					
2	3.7			2	2.6	DECATUR GA			-31	14	24
12	2.8	26	3.8	47	4.1	DE KALB GA	37	7	47	47	49
1	3.4	2	3.4	2	3.1	DODGE GA	1	-11	-10	23	31
3*	9.8	1	3.1	1	3.0	DOOLY GA	-68	-4	-70	01	30
6*	8.5	4	4.0	3	1.7	DOUGHERTY GA	-53	-56	-79	00	15
1	2.3	2	2.4	7	5.5	DOUGLAS GA	5	133	144	75	74
		2	5.6	4	12.9	EARLY GA		130			98
1*	37.2					ECHOLS GA					
1	3.8	1	2.7	2	4.9	EFFINGHAM GA	-29	83	29	39	64
1	1.9	1	1.6	3	4.2	ELBERT GA	-15	160	121	70	51
		1	1.7	2	2.5	EMANUEL GA		43			23
1	3.5			1	4.3	EVANS GA			23	37	53
1	1.7	1	1.5	1	1.5	FANNIN GA	-15	0	-15	21	13
1	4.3			2	3.2	FAYETTE GA			-27	15	34
3	1.2	7	3.0	11	3.4	FLOYD GA	150	11	179	80	36
4*	8.5	3	5.0	2	2.7	FORSYTH GA	-41	-46	-68	01	26
4*	7.4			4	7.1	FRANKLIN GA			-5	25	88
49	4.1	47	3.2	73	5.1	FULTON GA	-23	62	25	38	69
2	4.8	2	4.7	2	3.8	GILMER GA	-2	-20	-21	18	43
				1	7.2	GLASCOCK GA					89
2	2.4	3	2.8	8	5.0	GLYNN GA	15	80	106	67	66
2	2.4	2	2.0	6	6.3	GORDON GA	-17	217	164	78	82
1	1.9	3	4.8			GRADY GA	159				
1	3.7	1	3.2	1	3.8	GREENE GA	-12	17	3	28	43
4	2.9	1	.5	12	4.4	GWINNETT GA	-83	801	51	48	54
4	7.6	2	3.0	4	3.8	HABERSHAM GA	-61	26	-50	07	43
1	.7	4	2.1	8	3.3	HALL GA	188	56	349	92	35
		1	5.8	2	12.8	HANCOCK GA		120			98
2	3.2	1	1.5	1	1.2	HARALSON GA	-52	-19	-61	03	12
				3	6.8	HARRIS GA					86
3	6.9	1	1.8	7*	9.5	HART GA	-74	428	38	43	95
1	3.8					HEARD GA					
		4	7.1	2	2.5	HENRY GA		-64			24
		1	1.3	2	2.1	HOUSTON GA		65			19
1	3.3	1	3.8			IRWIN GA	14				
1	1.6	2	2.8	3	3.8	JACKSON GA	75	37	139	74	43
				1	3.2	JASPER GA					33
				3	8.3	JEFF DAVIS GA					93
2	6.0	2	5.3	1	2.1	JEFFERSON GA	-11	-60	-65	02	19
		1	3.9	1	3.5	JENKINS GA		-10			39
				1	4.3	JOHNSON GA					53
				1	2.3	JONES GA					21
		2	5.8	2	5.0	LAMAR GA		-14			66
		1	7.2	1	5.8	LANIER GA		-20			77
2	3.7	3	2.6	9	6.7	LAURENS GA	-31	162	82	60	86
1	11.2			2	8.4	LIBERTY GA			-25	16	93
		1	5.7			LINCOLN GA					
1	10.7	1	8.5			LONG GA	-21				
1	1.1	2	2.0	3	2.1	LOWNDES GA	82	5	91	62	19
1	4.0	1	3.2	1	2.9	LUMPKIN GA	-20	-9	-28	15	29
1	4.3			1	1.6	MC DUFFIE GA			-63	03	14
1	4.0	2	6.7	1	2.7	MACON GA	69	-60	-33	13	26
1	2.5			2	3.9	MADISON GA			58	51	46
2*	15.4					MARION GA					
		1	2.4	3	4.6	MERIWETHER GA		91			59
				3	11.7	MILLER GA					97
1	2.3			2	3.9	MITCHELL GA			72	56	45
2	7.8			2	6.2	MONROE GA			-21	18	81
1	5.1	1	4.5	1	4.1	MONTGOMERY GA	-10	-10	-19	19	48
1	3.9			1	3.1	MORGAN GA			-19	19	32
1	2.5			3	5.2	MURRAY GA			106	67	69
10	5.0	9	2.4	23	6.2	MUSCOGEE#GA	-51	156	24	37	81
3	5.2	2	2.9	3	3.1	NEWTON GA	-44	6	-40	10	32
		1	3.9			OCONEE GA					
1	3.2			1	3.6	OGLETHORPE GA			14	33	40
3	6.7	2	3.2	3	4.1	PAULDING GA	-52	29	-39	10	50
		1	3.5	3	7.1	PEACH GA		101			88
1	3.2	3	7.5	2	4.7	PICKENS GA	133	-37	46	46	62
3*	10.6	2	7.2	1	4.1	PIERCE GA	-32	-43	-61	03	50
2	8.0			2	7.1	PIKE GA			-12	23	88
		1	.9	6	4.7	POLK GA		414			60
1	3.6			2	6.7	PULASKI GA			87	61	86
2*	14.4	1	6.2			PUTNAM GA	-57				
2*	44.8					QUITMAN GA					
		1	2.6	1	1.3	RABUN GA		-47			13
2	8.3	2	12.7			RANDOLPH GA	52				
4	1.7	5	2.6	17	4.5	RICHMOND GA	54	74	167	78	57
1	3.0			2	2.2	ROCKDALE GA			-26	16	20
1	2.8			1	2.2	SCREVEN GA			-22	17	20
				2	6.3	SEMINOLE GA					83
4	4.4	3	2.6	6	4.3	SPALDING GA	-41	65	-2	26	53

* SIGNIFICANTLY MORE DEATHS THAN EXPECTED

\# COUNTIES COMBINED; SEE TECHNICAL NOTES

WHITE MALE

CANCER OF THE LIVER AND GALLBLADDER INCLUDING BILE DUCTS

ICD 155,156

1950-1959 DEATHS	1950-1959 RATE	1960-1969 DEATHS	1960-1969 RATE	1970-1979 DEATHS	1970-1979 RATE	COUNTY AND STATE	PERCENT CHANGE 60S/50S	PERCENT CHANGE 70S/60S	PERCENT CHANGE 70S/50S	PERCENTILE %CHANGE 70S/50S	PERCENTILE RATES 1970S
2	4.6	1	1.7			STEPHENS GA	-62				
1	6.7					STEWART GA					
1	2.1	4	7.1	3	4.2	SUMTER GA	237	-41	99	64	51
1	7.0	1	7.9	1	8.3	TALBOT GA	12	5	18	35	93
				1	1.5	TATTNALL GA					14
				2	5.0	TELFAIR GA					67
1	3.9			2	6.8	TERRELL GA			73	57	87
1	1.1	4	3.9	4	3.8	THOMAS GA	260	-4	246	87	43
1	1.6	1	1.8	7	8.1	TIFT GA	13	352	410	94	93
2	3.8			3	4.2	TOOMBS GA			10	31	51
				1	2.9	TOWNS GA					30
2*	12.8			2	14.0	TREUTLEN GA			9	31	98
4	4.4	6	4.0	5	3.7	TROUP GA	-8	-8	-16	20	41
		1	3.6	1	3.0	TURNER GA		-15			31
		1	10.6	1	4.1	TWIGGS GA		-62			48
				2	6.5	UNION GA					65
1	1.8	4	5.0	5	5.8	UPSON GA	172	16	216	84	77
9*	6.0	4	2.4	9	4.1	WALKER GA	-60	70	-33	13	48
5*	10.0	3	3.8	2	2.1	WALTON GA	-62	-44	-79	00	19
2	2.7	2	1.9	5	3.9	WARE GA	-28	101	44	45	45
				1	5.7	WARREN GA					77
1	3.4	1	2.0	2	4.1	WASHINGTON GA	-40	100	20	35	49
2	5.4	1	4.6	1	1.9	WAYNE GA	-14	-59	-65	02	17
				2	9.4	WHEELER GA					95
1	3.9			1	2.3	WHITE GA			-43	09	21
5	6.2	9	5.6	10	4.9	WHITFIELD GA	-9	-12	-20	19	65
2	7.4	1	3.7	2	6.7	WILCOX GA	-50	83	-9	24	86
2	7.8			1	3.5	WILKES GA			-55	05	38
		1	3.8	1	3.4	WILKINSON GA		-11			37
				3	5.2	WORTH GA					69
		30	5.8	44	5.5	HAWAII HI		-4			
		6*	12.5	2	1.9	HAWAII HI		-85			16
		22	5.4	36	6.0	HONOLULU HI		11			79
				2	5.2	KAUAI HI					69
		2	5.5	4	6.2	MAUI#HI		12			82
60	2.2	66	2.1	97	2.6	IDAHO ID	-5	24	17		
12	3.6	10	2.2	18	3.3	ADA ID	-39	48	-10	23	35
		1	5.2			ADAMS ID					
3	2.1	4	2.1	3	1.5	BANNOCK ID	2	-29	-28	14	14
		1	3.1			BEAR LAKE ID					
1	1.1	2	1.9	4	3.8	BINGHAM ID	71	103	247	87	43
		1	3.7	1	3.2	BLAINE ID		-14			34
1	1.2			1	1.2	BONNER ID			6	29	12
3	2.4	2	1.4	6	2.9	BONNEVILLE ID	-41	105	20	36	29
		2	6.0			BOUNDARY ID					
6	2.3	7	2.4	17	4.6	CANYON ID	4	94	101	65	59
		1	3.4			CARIBOU ID					
2	3.3					CASSIA ID					
3	5.6					CLEARWATER ID					
2	7.1	3	8.9	2	4.4	ELMORE ID	25	-51	-39	10	54
		1	2.1	1	1.9	FRANKLIN ID		-10			17
				1	2.2	FREMONT ID					20
3	5.9			4	6.0	GEM ID			3	28	80
1	1.9	1	1.9			GOODING ID	1				
2	3.6	1	1.5	2	2.7	IDAHO ID	-60	84	-26	15	26
		2	4.0			JEFFERSON ID					
1	2.1	1	2.1	1	1.4	JEROME ID	-1	-33	-33	12	13
2	1.4	4	2.2	8	3.4	KOOTENAI ID	52	59	141	74	37
2	2.5	4	4.6	1	1.0	LATAH ID	85	-78	-59	04	11
				1	2.1	LEMHI ID					19
				2	7.8	LEWIS ID					91
1	2.9			1	2.4	MADISON ID			-17	20	22
		1	1.2			MINIDOKA ID					
4	3.2	3	2.2	2	1.4	NEZ PERCE ID	-31	-37	-56	05	13
2	7.5	1	3.3	1	2.7	OWYHEE ID	-55	-20	-64	02	25
1	1.6	3	3.3	6	6.4	PAYETTE ID	106	93	297	90	84
		3*	14.1	1	3.8	POWER ID		-73			43
4	4.9	2	2.5	2	2.1	SHOSHONE ID	-49	-17	-57	04	19
3	1.7	5	2.5	11	4.4	TWIN FALLS ID	48	79	166	78	55
1	6.2					VALLEY ID					
1432*	3.5	1780*	4.1	2411*	5.4	ILLINOIS IL	16	33	53		
14	3.5	14	3.6	15	3.6	ADAMS IL	2	1	3	28	41
2	3.0	3	4.8	4	6.7	ALEXANDER IL	59	38	120	70	85
4	4.6	2	2.4	4	4.2	BOND IL	-48	72	-10	24	51
5	5.0	2	2.0	7	6.0	BOONE IL	-61	203	19	35	80
		1	1.7	2	5.1	BROWN IL		207			69
10	4.6	4	1.8	15	6.4	BUPEAU IL	-61	255	40	44	83
				6*	13.0	CALHOUN IL					98
2	1.6	2	1.8	10*	8.7	CARROLL IL	7	395	427	95	94
3	4.4	2	1.9	3	3.3	CASS IL	-58	78	-25	16	35
12	3.4	10	2.7	11	2.7	CHAMPAIGN IL	-21	-1	-21	18	25
5	2.2	8	3.6	8	3.8	CHRISTIAN IL	65	5	72	57	43
		3	2.6	3	2.8	CLARK IL		9			28
6	6.0	2	1.9	3	3.4	CLAY IL	-69	80	-44	09	37
6	4.8			3	2.1	CLINTON IL			-57	04	19
3	1.4	1	.4	12	5.5	COLES IL	-70	1174	279	89	73
820*	4.4	1001*	5.1	1265*	6.7	COOK IL	17	30	53	49	86
3	2.9	2	1.6	7	5.6	CRAWFORD IL	-46	251	90	62	75
		1	1.4	3	4.7	CUMBERLAND IL		227			61
2	.9	4	1.7	5	2.0	DE KALB IL	82	18	115	69	18
2	2.3	2	1.8	4	3.9	DE WITT IL	-22	121	71	56	46
3	3.1			6	6.0	DOUGLAS IL			91	62	80
27	3.4	55	4.7	89	5.5	DU PAGE IL	39	18	65	54	74
3	2.1	3	2.3	3	2.6	EDGAR IL	9	11	22	36	25

* SIGNIFICANTLY MORE DEATHS THAN EXPECTED

COUNTIES COMBINED; SEE TECHNICAL NOTES

WHITE MALE

CANCER OF THE LIVER AND GALLBLADDER INCLUDING BILE DUCTS

ICD 155,156

1950-1959 DEATHS	1950-1959 RATE	1960-1969 DEATHS	1960-1969 RATE	1970-1979 DEATHS	1970-1979 RATE	COUNTY AND STATE	PERCENT CHANGE 60S/50S	PERCENT CHANGE 70S/60S	PERCENT CHANGE 70S/50S	PERCENTILE %CHANGE 70S/50S	PERCENTILE RATES 1970S
				1	1.6	EDWARDS IL					15
1	1.7	7	5.7	8	5.4	EFFINGHAM IL	236	-5	219	85	72
3	2.2	2	1.3	6	4.1	FAYETTE IL	-39	205	87	61	50
4	4.5	3	3.7	3	3.1	FORD IL	-18	-15	-30	14	32
15*	4.9	19*	6.2	14	4.7	FRANKLIN IL	25	-24	-4	26	61
3	1.3	7	2.9	15	5.7	FULTON IL	128	97	350	92	77
1	2.5	1	1.9	1	2.3	GALLATIN IL	-24	21	-8	24	21
6	4.8	2	1.5	3	2.7	GREENE IL	-70	85	-44	09	26
2	2.4	3	2.7	5	3.6	GRUNDY IL	13	38	55	50	41
		5	8.9	6	8.1	HAMILTON IL		-9			92
4	2.7	2	1.3	5	3.6	HANCOCK IL	-51	166	31	41	40
		1	2.6	2	5.0	HARDIN IL		92			67
1	3.2			2	3.6	HENDERSON IL			11	32	40
9	3.7	6	2.1	12	4.2	HENRY IL	-43	99	13	33	51
3	1.6	3	1.7	9	4.9	IROQUOIS IL	1	195	198	82	64
5	2.6	12	6.2	16*	8.0	JACKSON IL	136	29	204	83	92
2	2.9	3	3.6	3	3.5	JASPER IL	27	-4	21	36	38
8	4.5	7	3.6	10	5.4	JEFFERSON IL	-21	51	20	35	72
2	2.4	4	4.7	4	4.4	JERSEY IL	93	-7	79	58	54
1	1.3	7	5.3	8	6.4	JO DAVIESS IL	316	20	400	94	84
1	1.6	3	5.7	2	4.5	JOHNSON IL	248	-22	172	79	56
20	2.4	25	2.9	49	5.2	KANE IL	21	80	118	69	69
13	3.2	14	3.2	24	6.0	KANKAKEE IL	1	87	90	61	79
3	4.3	2	2.6	3	3.1	KENDALL IL	-40	19	-29	14	32
8	2.6	9	2.9	6	1.8	KNOX IL	8	-36	-31	13	16
26	3.4	35	3.2	67	5.5	LAKE IL	-5	73	63	53	73
19	3.5	21	3.7	25	4.3	LA SALLE IL	4	18	22	37	54
5	4.7	1	.9	9	8.1	LAWRENCE IL	-82	847	72	57	93
4	2.1	7	3.6	3	1.7	LEE IL	72	-54	-21	18	15
2	.8	2	.9	8	3.9	LIVINGSTON IL	4	345	361	93	45
3	1.8	4	2.4	2	1.2	LOGAN IL	35	-51	-34	12	11
3	1.9	3	2.0	4	2.5	MC DONOUGH IL	5	27	33	41	24
12	4.1	10	2.2	26	5.0	MC HENRY IL	-45	125	23	37	67
8	2.0	11	2.7	21	5.0	MC LEAN IL	33	86	148	76	66
18	3.8	17	3.3	18	3.3	MACON IL	-13	-2	-15	21	35
6	1.9	8	2.5	11	3.9	MACOUPIN IL	33	52	102	65	44
20	2.4	28	3.0	47	4.5	MADISON IL	26	47	85	61	57
6	2.6	12	5.3	13	5.7	MARION IL	101	8	118	69	77
2	2.5	1	1.2	3	3.9	MARSHALL IL	-51	220	58	51	46
2	2.4	3	3.2	3	3.0	MASON IL	32	-7	23	37	30
		1	1.3	6	7.0	MASSAC IL		463			88
1	1.7	1	1.6	1	2.0	MENARD IL	-11	28	14	33	18
4	3.7	4	4.1	3	2.8	MERCER IL	12	-32	-24	16	27
3	4.7	3	3.1	2	2.2	MONROE IL	-34	-29	-53	06	20
6	2.8	8	3.5	7	3.7	MONTGOMERY IL	25	4	30	40	41
2	1.0	7	3.4	11	6.0	MORGAN IL	257	77	530	97	80
3	4.1	2	1.9	3	4.3	MOULTRIE IL	-55	129	4	29	53
5	2.8	7	3.8	7	3.3	OGLE IL	37	-15	17	34	35
22	2.7	31	3.6	32	3.7	PEORIA IL	37	2	40	44	42
3	2.5	4	3.1	8	6.8	PERRY IL	22	121	170	79	86
3	3.8	2	2.5	2	2.4	PIATT IL	-35	-4	-37	11	22
4	2.9	2	1.5	3	2.1	PIKE IL	-48	41	-27	15	19
1	3.5	1	3.4	1	2.7	POPE IL	-2	-22	-24	16	26
2	3.7	3	6.3	7*	15.9	PULASKI IL	72	153	335	92	99
2	7.5	3	9.2	3	11.1	PUTNAM IL	22	21	48	47	96
3	1.7	6	3.3	15*	8.1	RANDOLPH IL	97	144	379	94	92
2	2.1	3	3.5	1	.9	RICHLAND IL	66	-75	-59	04	10
14	2.1	27	3.7	33	4.6	ROCK ISLAND IL	73	25	117	69	59
20	2.6	30	3.2	45	4.7	ST CLAIR IL	25	44	80	59	60
4	2.3	6	3.3	12	6.8	SALINE IL	41	106	191	81	86
27	4.2	27	3.9	41	5.5	SANGAMON IL	-9	43	30	40	74
3	4.6	2	4.0	1	1.6	SCHUYLER IL	-13	-60	-65	02	15
2	5.3			3	7.4	SCOTT IL			38	43	90
5	3.2	5	3.3	2	1.3	SHELBY IL	2	-60	-59	04	12
		1	2.2	1	2.2	STARK IL		1			20
3	1.4	7	2.9	10	4.2	STEPHENSON IL	115	42	206	83	50
8	2.6	11	2.7	10	2.0	TAZEWELL IL	1	-27	-26	16	18
1	.8	6	4.8	8	7.6	UNION IL	503	59	859	99	91
14	2.9	22	4.5	30	6.2	VERMILION IL	56	37	115	69	82
7*	9.2	3	4.0	2	2.5	WABASH IL	-56	-38	-73	01	24
6	5.0	5	4.4	4	3.3	WARREN IL	-12	-25	-34	12	35
4	4.3	4	3.3	7	5.5	WASHINGTON IL	-23	65	27	38	74
2	1.7	5	4.6	7	5.9	WAYNE IL	179	27	256	87	79
4	3.1	6	5.0	7	5.7	WHITE IL	62	13	83	60	76
3	1.4	12	4.6	11	3.7	WHITESIDE IL	227	-19	166	78	42
19	2.7	31	3.9	37	4.2	WILL IL	41	9	53	49	51
5	1.7	3	1.1	17	5.8	WILLIAMSON IL	-35	443	251	87	78
28	3.9	43	4.9	42	4.4	WINNEBAGO IL	26	-10	13	33	55
4	3.6	4	3.2	4	2.7	WOODFORD IL	-11	-15	-25	16	26
490	2.6	672	3.3	988	4.6	INDIANA IN	26	39	76		
		3	2.4	5	4.1	ADAMS IN		74			50
27	3.8	31	3.3	39	3.8	ALLEN IN	-12	13	-1	27	43
7	4.5	8	4.2	12	5.8	BARTHOLOMEW IN	-7	37	28	39	77
2	3.1	3	5.0	4	6.6	BENTON IN	64	31	115	69	85
1	1.2	5	7.3	2	2.5	BLACKFORD IN	523	-66	113	68	24
5	3.4	8	5.6	3	1.9	BOONE IN	63	-66	-45	08	17
				2	3.5	BROWN IN					39
				4	4.2	CARROLL IN					51
5	2.1	10	4.8	6	2.9	CASS IN	130	-40	39	43	29
6	2.8	8	3.3	14	5.3	CLARK IN	19	64	94	63	71
3	2.2	5	3.4	9	6.0	CLAY IN	51	79	171	79	80
1	.7	4	2.3	10	6.3	CLINTON IN	236	175	822	99	83
1	1.4	2	3.3	4	7.9	CRAWFORD IN	131	139	452	95	92
3	1.9	6	4.0	13*	8.9	DAVIESS IN	111	120	364	93	94
5	3.8	2	1.4	4	2.8	DEARBORN IN	-63	98	-27	15	27
2	1.6	6	5.8	4	3.5	DECATUR IN	266	-39	121	70	39
3	2.1	1	.7	5	3.4	DE KALB IN	-67	389	63	53	38
10	2.5	23	5.5	14	3.2	DELAWARE IN	117	-43	24	37	33

* SIGNIFICANTLY MORE DEATHS THAN EXPECTED

WHITE MALE

CANCER OF THE LIVER AND GALLBLADDER
INCLUDING BILE DUCTS

ICD 155,156

1950-1959 DEATHS	1950-1959 RATE	1960-1969 DEATHS	1960-1969 RATE	1970-1979 DEATHS	1970-1979 RATE	COUNTY AND STATE		PERCENT CHANGE 60S/50S	PERCENT CHANGE 70S/60S	PERCENT CHANGE 70S/50S	PERCENTILE %CHANGE 70S/50S	PERCENTILE RATES 1970S
5	4.2	6	4.6	4	2.8	DUBOIS	IN	9	-39	-33	12	28
6	1.5	17	3.8	19	3.8	ELKHART	IN	160	1	163	78	44
		5	4.6	5	4.2	FAYETTE	IN		-7			52
9	4.8	10	4.2	15	6.1	FLOYD	IN	-13	45	26	38	80
1	1.1	3	3.0	6	6.0	FOUNTAIN	IN	174	99	446	95	79
		1	1.1	4	4.7	FRANKLIN	IN		314			61
2	2.2	5	5.2	5	5.0	FULTON	IN	137	-3	131	72	67
2	1.0	8	4.5	9	5.3	GIBSON	IN	359	19	446	95	71
10	3.0	14	4.1	17	4.9	GRANT	IN	39	19	66	54	65
8	4.5	5	3.0	11	6.3	GREENE	IN	-34	109	38	43	82
7	4.3	4	2.1	10	4.2	HAMILTON	IN	-50	98	-2	26	52
3	2.6	1	.8	4	2.8	HANCOCK	IN	-69	243	8	30	27
		4	3.5	7	6.3	HARRISON	IN		77			82
5	3.6	7	4.0	6	3.2	HENDRICKS	IN	11	-21	-12	22	33
3	1.3	9	3.9	12	5.0	HENRY	IN	211	26	293	90	66
4	1.6	15	5.5	12	4.7	HOWARD	IN	248	-15	195	82	60
6	3.5	4	2.3	5	2.9	HUNTINGTON	IN	-35	29	-16	20	29
3	2.0	5	3.4	6	3.7	JACKSON	IN	68	6	78	58	41
1	1.2	3	3.1	5	5.2	JASPER	IN	157	64	321	91	69
				5	4.5	JAY	IN					57
3	2.3			6	4.6	JEFFERSON	IN			99	64	58
1	1.2	2	2.4	3	3.2	JENNINGS	IN	107	33	176	80	34
3	2.1	4	2.2	7	3.5	JOHNSON	IN	3	59	64	53	39
6	2.6	10	4.5	8	3.4	KNOX	IN	76	-25	31	41	36
		3	1.4	10	4.4	KOSCIUSKO	IN		213			54
1	1.1	6	7.4	3	3.0	LAGRANGE	IN	595	-59	186	81	31
47	3.3	63	3.9	100*	6.2	LAKE	IN	18	59	89	61	81
10	2.4	12	2.7	20	4.4	LA PORTE	IN	9	66	81	59	55
10*	5.9	4	2.3	14	7.1	LAWRENCE	IN	-62	211	19	35	88
15	3.1	14	2.7	25	4.6	MADISON	IN	-13	68	47	47	60
69	3.5	91	3.8	115	4.6	MARION	IN	9	21	32	41	60
1	.7	4	2.5	7	3.9	MARSHALL	IN	273	59	495	96	46
1	2.3	1	2.2	3	5.4	MARTIN	IN	-6	152	137	73	73
3	1.7	3	1.7	12	7.8	MIAMI	IN	0	355	353	92	92
2	.6	2	1.0	10	3.9	MONROE	IN	49	307	505	97	45
2	1.2	5	3.1	2	1.1	MONTGOMERY	IN	164	-64	-5	25	11
		9	6.1	8	4.1	MORGAN	IN		-32			49
		2	3.1	6	8.9	NEWTON	IN		188			94
1	.6	6	4.2	6	4.1	NOBLE	IN	558	-3	535	97	48
2	8.2	1	3.9	2	7.6	OHIO	IN	-52	96	-7	24	90
		2	1.9			ORANGE	IN					
1	1.9	3	3.9	6	8.2	OWEN	IN	104	113	334	92	93
1	.9	2	2.3	8	8.3	PARKE	IN	148	258	788	99	93
3	3.2	4	4.1	2	2.1	PERRY	IN	26	-48	-35	11	19
1	1.1	1	1.1	3	4.0	PIKE	IN	0	261	262	88	47
1	.5	12	4.6	13	4.0	PORTER	IN	810	-13	688	98	47
4	3.8	4	4.1	3	2.7	POSEY	IN	7	-33	-29	14	27
3	4.0	2	2.9	4	5.8	PULASKI	IN	-26	99	47	46	77
2	1.6	4	3.1	2	1.5	PUTNAM	IN	96	-52	-5	25	14
6	3.7	1	.7	7	4.9	RANDOLPH	IN	-82	638	34	42	65
4	3.5	2	1.7	4	3.2	RIPLEY	IN	-52	92	-8	24	34
1	.8	2	2.1	4	3.9	RUSH	IN	167	91	410	95	46
19	2.2	30	3.0	54	5.3	ST JOSEPH	IN	35	77	140	74	71
2	3.0	2	2.9	8*	9.1	SCOTT	IN	-4	218	204	83	95
5	3.3	3	1.9	4	2.3	SHELBY	IN	-44	24	-30	14	21
7*	7.2	3	3.4	3	3.1	SPENCER	IN	-53	-9	-57	04	31
5	5.8	4	3.7	4	3.4	STARKE	IN	-36	-7	-40	10	38
2	1.9	1	1.0	4	3.8	STEUBEN	IN	-48	295	105	66	44
10*	6.4	1	.6	7	5.3	SULLIVAN	IN	-91	779	-17	20	71
				3	7.6	SWITZERLAND	IN					91
11	3.5	10	3.0	17	4.7	TIPPECANOE	IN	-16	57	33	41	60
		1	1.1	5	6.4	TIPTON	IN		491			84
25	3.9	22	3.0	44	5.7	VANDERBURGH	IN	-22	89	46	46	76
2	1.6	5	5.3	4	4.1	VERMILLION	IN	235	-22	161	77	49
17	3.0	27	5.1	34*	6.7	VIGO	IN	68	33	123	70	86
2	1.4	7	4.5	6	3.6	WABASH	IN	210	-20	149	76	40
2	3.8			1	1.7	WARREN	IN			-54	05	15
4	3.7	3	2.4	10	7.7	WARRICK	IN	-35	218	107	67	91
		1	.9	7	6.2	WASHINGTON	IN		599			82
11	3.4	14	4.1	15	4.4	WAYNE	IN	20	8	29	39	55
2	1.8	4	3.7	4	3.4	WELLS	IN	102	-9	83	60	37
3	2.8			5	4.3	WHITE	IN			55	50	53
1	.9	2	2.2	5	4.3	WHITLEY	IN	138	94	362	93	52
331	2.4	428	3.0	566	3.9	IOWA	IA	24	30	60		
2	2.7			4	5.6	ADAIR	IA			103	66	75
1	1.8			3	6.4	ADAMS	IA			246	87	83
4	4.4	5	5.3	3	2.9	ALLAMAKEE	IA	20	-45	-34	12	29
4	2.9	9*	7.0	4	3.4	APPANOOSE	IA	142	-51	17	34	37
7*	11.7	2	4.6	5	9.3	AUDUBON	IA	-60	103	-20	19	95
2	1.5	5	3.8	4	3.0	BENTON	IA	159	-22	104	66	31
6	1.3	22	4.3	23	4.4	BLACK HAWK	IA	240	2	247	87	55
5	3.1	4	2.4	6	3.7	BOONE	IA	-23	57	21	36	42
4	4.1	1	.9	10	8.1	BREMER	IA	-78	799	99	64	92
5	4.4	1	.7	1	1.1	BUCHANAN	IA	-83	50	-75	01	11
4	3.5	6	5.0	3	2.2	BUENA VISTA	IA	44	-56	-37	11	20
1	1.0	2	1.9	2	2.0	BUTLER	IA	89	9	106	66	18
3	3.3	3	3.6			CALHOUN	IA	11				
3	2.6	6	5.1	7	5.3	CARROLL	IA	92	4	100	65	70
4	3.8	5	4.3	3	2.7	CASS	IA	16	-37	-27	15	27
2	2.0	2	2.1	3	2.6	CEDAR	IA	4	24	30	40	25
5	2.1	6	2.5	16	6.6	CERRO GORDO	IA	15	167	206	83	85
1	.9			5	5.4	CHEROKEE	IA			500	96	72
3	3.5	3	3.4	4	4.8	CHICKASAW	IA	-3	40	36	42	63
				1	1.6	CLARKE	IA					14
3	3.1			6	6.1	CLAY	IA			99	64	81
1	.8	1	.7	5	3.4	CLAYTON	IA	-21	422	312	91	37
5	1.8	12	4.2	6	2.1	CLINTON	IA	125	-51	11	32	18
6	6.1	7	6.6	4	3.7	CRAWFORD	IA	9	-44	-39	10	42

* SIGNIFICANTLY MORE DEATHS THAN EXPECTED

WHITE MALE

CANCER OF THE LIVER AND GALLBLADDER INCLUDING BILE DUCTS

ICD 155,156

1950-1959 DEATHS	1950-1959 RATE	1960-1969 DEATHS	1960-1969 RATE	1970-1979 DEATHS	1970-1979 RATE	COUNTY AND STATE	PERCENT CHANGE 60S/50S	PERCENT CHANGE 70S/60S	PERCENT CHANGE 70S/50S	PERCENTILE %CHANGE 70S/50S	PERCENTILE RATES 1970S
3	2.2	1	.8	4	2.4	DALLAS IA	-64	210	12	32	22
		2	3.1	4	6.4	DAVIS IA		106			84
1	1.3	1	2.2	1	1.4	DECATUR IA	72	-36	9	31	13
4	4.2	3	3.1	2	1.9	DELAWARE IA	-27	-38	-55	05	17
8	3.7	13	5.6	11	4.6	DES MOINES IA	53	-19	24	37	59
		3	3.6	4	5.4	DICKINSON IA		49			72
7	2.3	10	3.1	19	5.0	DUBUQUE IA	35	63	120	70	67
		1	1.6	2	2.4	EMMET IA		54			22
5	3.2	5	3.3	6	3.9	FAYETTE IA	6	16	22	37	44
5	4.2			4	3.4	FLOYD IA			-20	19	37
1	1.3	2	2.1	4	4.5	FRANKLIN IA	59	116	244	87	58
		4	5.9	1	1.4	FREMONT IA		-76			13
2	2.2	5	5.7	5	5.5	GREENE IA	159	-4	148	75	73
		3	3.9	1	1.3	GRUNDY IA		-67			12
1	1.1	2	2.7	2	2.5	GUTHRIE IA	139	-9	117	69	23
		1	1.0	2	1.9	HAMILTON IA		89			17
4	5.3	2	2.6	1	1.2	HANCOCK IA	-50	-55	-78	00	11
5	4.1	2	1.5	5	4.1	HARDIN IA	-63	171	1	28	50
1	1.0	6	5.7	6	5.8	HARRISON IA	450	2	461	96	78
		4	3.7	2	2.0	HENRY IA		-47			18
1	1.3	1	1.2	5	5.6	HOWARD IA	-13	380	316	91	75
2	3.1	1	1.4	3	4.1	HUMBOLDT IA	-54	181	30	40	48
		1	1.6	3	6.4	IDA IA		312			84
4	4.5	2	2.4	1	1.1	IOWA IA	-47	-55	-76	00	11
2	1.6	4	3.5	5	4.1	JACKSON IA	116	19	158	77	50
3	1.7	3	1.7	5	2.7	JASPER IA	-2	61	57	50	26
				2	2.0	JEFFERSON IA					18
4	2.2	9	4.7	9	4.2	JOHNSON IA	113	-11	88	61	51
		4	3.9	5	4.5	JONES IA		15			57
5	4.8	4	3.8	3	2.9	KEOKUK IA	-20	-26	-40	10	28
				6	4.6	KOSSUTH IA					59
10	4.9	11	5.2	7	3.5	LEE IA	6	-33	-29	14	38
19	3.6	22	3.7	31	4.8	LINN IA	4	28	34	42	62
1	1.5	1	1.9	3	4.3	LOUISA IA	28	127	191	82	52
1	1.1	2	3.1	3	4.9	LUCAS IA	192	55	352	92	64
1	1.4	1	1.1	5	6.2	LYON IA	-23	466	335	92	81
2	2.3	1	1.1	4	4.5	MADISON IA	-50	305	101	65	58
5	3.7	4	2.5	4	2.4	MAHASKA IA	-34	-2	-35	11	22
3	1.7	2	1.2	4	2.3	MARION IA	-29	84	31	40	21
7	3.6	3	1.3	9	4.1	MARSHALL IA	-63	203	13	32	48
				3	4.1	MILLS IA					49
2	2.4	2	2.4	4	4.3	MITCHELL IA	0	81	82	59	53
1	1.1			4	5.7	MONONA IA			440	95	76
		2	2.2	3	4.1	MONROE IA		83			48
2	2.5	1	.9	3	3.1	MONTGOMERY IA	-62	229	24	37	32
4	2.3	7	3.5	3	1.6	MUSCATINE IA	50	-54	-31	13	15
2	2.1	3	2.4	2	1.7	O'BRIEN IA	14	-28	-18	19	15
		2	4.2			OSCEOLA IA					
2	1.4	2	1.4	4	2.8	PAGE IA	-5	108	98	63	28
		1	1.2	3	3.2	PALO ALTO IA		163			34
5	4.4	2	1.4	7	4.8	PLYMOUTH IA	-67	237	10	31	63
2	2.9	2	2.3	4	5.7	POCAHONTAS IA	-21	150	99	64	77
33	3.3	36	3.3	53	4.7	POLK IA	1	41	42	45	61
4	1.2	14	3.8	20	5.1	POTTAWATTAMIE IA	226	35	340	92	68
1	.9	5	4.9	4	3.6	POWESHIEK IA	414	-26	281	89	40
1	2.0					RINGGOLD IA					
1	1.1	2	2.0	3	2.9	SAC IA	84	43	162	77	28
15	2.9	20	3.8	21	3.8	SCOTT IA	29	-1	27	39	43
		3	3.1	4	4.0	SHELBY IA		28			47
4	3.5	4	3.1	2	1.4	SIOUX IA	-11	-55	-61	03	13
5	2.8	5	2.4	9	4.0	STORY IA	-14	70	46	46	48
3	2.3	2	1.4	5	3.5	TAMA IA	-38	154	56	50	39
2	3.1			1	2.1	TAYLOR IA			-32	13	19
3	2.7	4	4.0	5	5.4	UNION IA	47	33	96	63	72
1	1.7	3	4.1	3	3.7	VAN BUREN IA	147	-8	127	71	43
4	1.8	10	4.1	12	4.8	WAPELLO IA	131	17	169	79	62
4	3.8	5	4.6	4	3.7	WARREN IA	20	-19	-4	26	42
1	.8	1	.8	3	2.9	WASHINGTON IA	1	269	272	88	29
1	1.0	3	3.6			WAYNE IA	245				
5	2.3	4	1.7	7	3.1	WEBSTER IA	-26	85	36	42	32
3	3.8	7*	9.0	4	4.8	WINNEBAGO IA	136	-47	25	38	62
3	2.7	2	1.8	4	3.3	WINNESHIEK IA	-35	85	21	36	35
14	2.5	21	4.0	26	5.1	WOODBURY IA	61	26	103	66	68
2	3.8	1	1.8	1	1.5	WORTH IA	-53	-16	-60	03	14
3	2.7	4	3.9	4	3.6	WRIGHT IA	46	-8	33	41	40
221	2.3	303	2.9	410	3.7	KANSAS KS	27	29	64		
1	.8	2	2.1	3	2.4	ALLEN KS	155	16	195	82	23
2	2.6	1	1.3			ANDERSON KS	-49				
1	1.0	3	2.8	5	5.0	ATCHISON KS	199	76	427	95	67
1	2.1	2	3.6	2	3.5	BARBER KS	75	-4	69	55	38
5	4.1	5	3.5	3	2.0	BARTON KS	-16	-43	-53	06	17
		3	2.1	3	2.4	BOURBON KS		16			23
		1	.8			BROWN KS					
1	.5	7	3.7	7	3.4	BUTLER KS	582	-8	528	97	37
		2	7.6	2	6.2	CHASE KS		-18			82
1	2.0	3	7.0	4	10.9	CHAUTAUQUA KS	244	57	439	95	96
6	4.4			4	3.1	CHEROKEE KS			-31	14	31
1	4.8					CLARK KS					
3	4.0	4	4.7	1	2.4	CLAY KS	19	-48	-39	10	23
1	.8	1	.9	2	1.7	CLOUD KS	7	86	98	64	15
4	4.9			3	4.3	COFFEY KS			-14	21	52
				2	12.2	COMANCHE KS					97
8	3.8	7	3.4	10	4.5	COWLEY KS	-11	31	17	34	56
6	2.7	15	6.1	5	1.7	CRAWFORD KS	124	-72	-36	11	16
5	3.5	5	3.6	3	2.3	DICKINSON KS	2	-37	-36	11	21
1	1.3			4	7.6	DONIPHAN KS			480	96	90
6	4.1	7	4.3	5	2.9	DOUGLAS KS	5	-33	-30	14	28
1	3.5			1	3.5	EDWARDS KS			-1	27	39

* SIGNIFICANTLY MORE DEATHS THAN EXPECTED

WHITE MALE

CANCER OF THE LIVER AND GALLBLADDER INCLUDING BILE DUCTS

ICD 155,156

1950-1959 DEATHS	1950-1959 RATE	1960-1969 DEATHS	1960-1969 RATE	1970-1979 DEATHS	1970-1979 RATE	COUNTY AND STATE	PERCENT CHANGE 60S/50S	PERCENT CHANGE 70S/60S	PERCENT CHANGE 70S/50S	PERCENTILE %CHANGE 70S/50S	PERCENTILE RATES 1970S
3	6.0	2	5.5	1	2.0	ELK KS	-8	-64	-67	02	18
3	4.3	5	6.1	3	3.3	ELLIS KS	42	-47	-25	16	35
2	3.8	2	3.6	3	5.6	ELLSWORTH KS	-4	56	49	47	76
2	3.5			2	3.1	FINNEY KS			-12	22	31
2	2.0	2	2.1	1	.9	FORD KS	8	-56	-53	06	10
1	.7	1	.8	6	4.6	FRANKLIN KS	15	483	569	97	60
2	3.0	5	6.6	3	4.7	GEARY KS	120	-29	56	50	60
				1	4.2	GOVE KS					50
		2	7.8	2	8.1	GRAHAM KS		4			92
		2	6.8	2	6.7	GRANT KS		-1			86
1*	14.1			1	8.9	GREELEY KS			-37	11	94
2	2.9	5	5.8	4	5.0	GREENWOOD KS	97	-13	71	56	66
				1	7.5	HAMILTON KS					90
2	3.5	1	1.7	2	3.0	HARPER KS	-51	76	-14	21	30
7	5.9	4	2.8	1	.8	HARVEY KS	-52	-70	-86	00	10
				1	7.2	HASKELL KS					88
		1	5.5	1	6.6	HODGEMAN KS		20			85
		3	3.2	3	4.0	JACKSON KS		26			48
4	5.1	1	.9	2	2.0	JEFFERSON KS	-82	120	-60	04	18
1	2.4			1	2.4	JEWELL KS			0	27	22
10	3.7	14	2.9	27	3.5	JOHNSON KS	-23	22	-5	25	39
				1	6.3	KEARNY KS					83
2	3.6	1	1.8	2	3.2	KINGMAN KS	-52	81	-13	22	33
3	1.9	4	2.8	7	4.3	LABETTE KS	46	57	129	72	54
6	2.6	8	3.5	6	2.6	LEAVENWORTH KS	34	-26	-1	27	25
2	4.9			3	6.3	LINCOLN KS			29	39	82
1	1.1	1	1.1	1	1.4	LINN KS	-1	29	27	38	13
				1	5.7	LOGAN KS					76
4	2.6	2	1.1	5	3.1	LYON KS	-58	178	17	34	32
				2	1.3	MC PHERSON KS					12
2	2.0	5	5.3	3	3.5	MARION KS	160	-34	71	56	38
3	2.5	6	5.6	1	.8	MARSHALL KS	126	-85	-67	02	10
		1	7.6	1	3.2	MEADE KS		-58			33
		2	1.5	1	.9	MIAMI KS		-38			10
2	3.5	5	6.9	3	5.4	MITCHELL KS	99	-22	55	50	72
9	3.7	8	3.2	13	5.2	MONTGOMERY KS	-12	60	40	44	69
		2	4.2	8*	12.0	MORRIS KS		188			97
1	9.9					MORTON KS					
1	1.3	1	1.1	1	1.1	NEMAHA KS	-19	1	-17	19	11
1	.7	2	1.5	7	6.2	NEOSHO KS	102	323	756	99	81
		1	2.7	2	5.3	NESS KS		96			71
1	1.8			2	3.6	NORTON KS			100	65	41
2	2.4	1	.9	1	1.6	OSAGE KS	-62	77	-33	13	15
		3	4.5	2	3.1	OSBORNE KS		-30			33
2	4.3	1	2.2	1	2.1	OTTAWA KS	-50	-1	-51	07	19
1	1.4	1	1.7	1	1.7	PAWNEE KS	17	1	18	34	15
1	1.8	5	8.1	2	2.5	PHILLIPS KS	347	-69	38	43	24
2	2.4	1	1.2	4	4.5	POTTAWATOMIE KS	-52	282	82	59	56
				2	2.9	PRATT KS					29
				2	5.3	RAWLINS KS					71
4	1.5	5	1.7	20	6.3	RENO KS	17	263	325	91	82
3	3.9	1	1.2	4	5.1	REPUBLIC KS	-69	325	31	40	68
		3	3.6			RICE KS					
1	.8	6	5.0	4	3.1	RILEY KS	519	-38	286	89	33
1	2.3	2	3.7	3	7.2	ROOKS KS	60	92	208	84	89
2	4.9			2	4.3	RUSH KS			-12	22	54
		1	1.4	6	9.1	RUSSELL KS		537			95
5	3.1	8	4.6	11	5.3	SALINE KS	48	17	73	57	71
1	5.7			1	3.9	SCOTT KS			-32	13	45
17	1.8	29	2.6	58	4.6	SEDGWICK KS	44	80	160	77	60
2	4.6			6*	11.5	SEWARD KS			149	76	97
10	2.0	24	4.2	18	2.9	SHAWNEE KS	111	-32	43	45	28
		1	2.4			SHERMAN KS					
3	5.6	1	1.2	2	4.3	SMITH KS	-78	251	-24	16	52
1*	27.0					STANTON KS					
				1	3.9	STEVENS KS					46
3	1.8	5	3.6	9	5.5	SUMNER KS	96	53	200	83	74
2	6.7	1	2.4	2	4.9	THOMAS KS	-64	103	-27	15	64
		2	12.0			TREGO KS					
4*	7.2			3	7.7	WABAUNSEE KS			7	30	91
		1	10.1			WALLACE KS					
2	2.5	6	7.9	2	2.3	WASHINGTON KS	214	-71	-7	24	21
1	.9	1	1.0	1	1.4	WILSON KS	11	43	59	51	13
		2	5.0	1	1.6	WOODSON KS		-69			14
17	2.7	25	4.1	34	5.5	WYANDOTTE KS	49	36	103	66	74
246	2.1	459	3.5	717	4.9	KENTUCKY KY	68	42	138		
1	1.3	2	2.4	8*	10.2	ADAIR KY	82	319	661	98	96
		1	1.2	4	4.8	ALLEN KY		285			63
1	2.0	2	4.3	4	7.2	ANDERSON KY	118	65	261	88	88
5*	9.4			1	1.6	BALLARD KY			-83	00	15
1	.8	3	2.1	3	1.9	BARREN KY	154	-9	132	73	17
2	4.0			2	3.4	BATH KY			-14	21	38
1	.7	5	3.4	9	5.7	BELL KY	412	66	750	99	76
5	6.6	2	2.4	1	.7	BOONE KY	-63	-70	-89	00	10
1	1.4	4	5.1	4	5.0	BOURBON KY	263	-3	253	87	66
2	1.4	11	5.0	14	5.7	BOYD KY	253	13	300	90	76
1	1.1	3	3.1	7	7.5	BOYLE KY	171	140	549	97	90
1	1.9			2	5.2	BRACKEN KY			173	79	70
2	3.0	4	6.2	4	5.8	BREATHITT KY	105	-6	92	62	78
1	1.1	5	5.8	5	4.8	BRECKINRIDGE KY	415	-16	331	92	63
4	6.9	3	5.1	1	1.3	BULLITT KY	-26	-75	-81	00	12
1	1.5	1	1.7	2	2.9	BUTLER KY	11	74	94	63	29
		2	2.8	6	8.4	CALDWELL KY		197			93
1	.8	5	3.7	8	5.8	CALLOWAY KY	384	59	668	98	78
11	3.2	17	4.2	18	4.5	CAMPBELL KY	33	6	42	45	58
		1	3.7	1	2.3	CARLISLE KY		-37			21
3	6.5	1	2.6	1	2.5	CARROLL KY	-61	-2	-62	03	24
4	4.6	2	2.0	3	2.3	CARTER KY	-56	14	-50	07	21

* SIGNIFICANTLY MORE DEATHS THAN EXPECTED

WHITE MALE

CANCER OF THE LIVER AND GALLBLADDER INCLUDING BILE DUCTS

ICD 155,156

1950-1959 DEATHS	1950-1959 RATE	1960-1969 DEATHS	1960-1969 RATE	1970-1979 DEATHS	1970-1979 RATE	COUNTY AND STATE	PERCENT CHANGE 60S/50S	PERCENT CHANGE 70S/60S	PERCENT CHANGE 70S/50S	PERCENTILE %CHANGE 70S/50S	PERCENTILE RATES 1970S
		3	6.8	5	6.3	CASEY KY		-8			82
6	3.9	2	1.2	9	5.1	CHRISTIAN KY	-69	322	29	40	68
		4	4.2	6	5.7	CLARK KY		36			76
		1	1.3			CLAY KY					
1	2.4			4	7.7	CLINTON KY			224	85	91
1	1.9	1	2.1	2	3.1	CRITTENDEN KY	9	51	64	54	33
1	2.7	1	2.8	2	4.4	CUMBERLAND KY	4	58	63	53	55
5	2.1	8	2.9	18	5.6	DAVIESS KY	43	90	172	79	75
		1	2.3	2	4.9	EDMONSON KY		112			64
2	7.0	1	2.9	2	20.6	ELLIOTT KY	-59	620	196	82	99
2	3.1			2	2.6	ESTILL KY			-15	21	25
14	3.8	18	3.9	25	4.4	FAYETTE KY	2	13	16	34	56
				1	1.4	FLEMING KY					13
1	.7	5	3.2	12	6.4	FLOYD KY	342	104	801	99	84
2	1.6	3	2.4	4	2.5	FRANKLIN KY	45	6	53	49	24
1	1.7	1	1.7	2	3.5	FULTON KY	2	107	112	68	39
1	4.2	1	4.6	1	4.4	GALLATIN KY	8	-5	3	28	55
		3	6.2	1	1.9	GARRARD KY		-70			16
1	1.6	3	4.8	3	4.9	GRANT KY	203	3	212	84	66
4	2.5	5	2.7	10	5.0	GRAVES KY	6	88	99	64	67
2	1.8	3	2.7	6	5.1	GRAYSON KY	52	88	186	81	68
2	3.5	1	3.9	4	6.2	GREEN KY	12	60	79	59	82
1	.8	4	3.3	9	6.3	GREENUP KY	303	91	670	98	82
2	7.3	1	3.2	1	3.6	HANCOCK KY	-56	12	-50	07	41
3	2.1	10	5.5	10	4.8	HARDIN KY	161	-11	131	73	63
2	.9	9	5.2	15*	8.4	HARLAN KY	455	62	799	99	93
		4	5.1	12*	13.2	HARRISON KY		158			98
3	3.8	1	1.1	4	5.5	HART KY	-71	398	45	46	74
2	1.6	3	2.2	4	2.3	HENDERSON KY	36	3	40	44	21
3	4.7	2	3.6	1	1.3	HENRY KY	-24	-64	-73	01	12
				4	10.4	HICKMAN KY					96
5	2.7	11	5.7	4	2.0	HOPKINS KY	110	-65	-27	15	18
		2	3.7	2	3.7	JACKSON KY		0			41
41	2.2	90	4.1	146*	6.1	JEFFERSON KY	89	48	180	80	81
		2	3.3	5	6.7	JESSAMINE KY		106			86
1	1.3	2	1.9	2	1.7	JOHNSON KY	49	-14	28	39	15
19	3.9	25	4.7	27	5.1	KENTON KY	20	6	28	39	67
				2	2.6	KNOTT KY					25
2	2.0	2	1.7	5	4.3	KNOX KY	-12	147	117	69	53
2	3.7	2	3.3	8*	12.6	LARUE KY	-11	281	241	87	98
1	.7	4	2.9	6	3.6	LAUREL KY	297	22	386	94	40
1	1.4			7*	11.3	LAWRENCE KY			712	99	97
		1	3.0	3	7.2	LEE KY		139			89
				1	2.1	LESLIE KY					19
1	1.6	3	3.0	13*	11.4	LETCHER KY	89	287	632	98	97
1	1.8	3	4.6	1	1.4	LEWIS KY	160	-69	-21	18	13
2	2.4	3	3.4	6	5.8	LINCOLN KY	42	72	143	75	77
		2	4.3	3	6.1	LIVINGSTON KY		41			81
1	1.0	3	2.5	4	3.0	LOGAN KY	157	17	202	83	30
		1	3.7	3	7.8	LYON KY		109			91
4	1.7	12	4.9	12	4.5	MC CRACKEN KY	185	-9	161	77	57
2	3.7	3	4.8	4	6.9	MC CREARY KY	31	44	89	61	87
		2	4.4	3	5.6	MC LEAN KY		26			75
2	1.6	4	3.0	8	4.8	MADISON KY	87	62	202	83	63
		4	8.5	2	3.9	MAGOFFIN KY		-55			44
		3	4.7	3	4.2	MARION KY		-11			51
1	1.3	4	3.4	5	4.0	MARSHALL KY	170	16	213	84	47
		1	2.5	6*	13.2	MARTIN KY		419			98
		5	5.8	4	4.5	MASON KY		-23			57
1	2.3	2	4.4	2	3.3	MEADE KY	89	-25	43	45	35
				1	3.5	MENIFEE KY					38
2	3.0	1	1.3	4	5.0	MERCER KY	-55	276	68	55	66
				2	3.7	METCALFE KY					41
1	1.6	4	5.9			MONROE KY	263				
2	3.1	3	4.8	5	7.1	MONTGOMERY KY	55	46	127	71	88
6*	11.4	6*	9.5	3	5.6	MORGAN KY	-17	-41	-51	07	75
4	2.7	9	5.8	5	3.3	MUHLENBERG KY	114	-43	22	36	35
3	3.6			4	3.9	NELSON KY			9	30	45
1	2.2	3	7.5	3	7.6	NICHOLAS KY	240	1	243	87	90
1	.8	3	2.6	5	3.8	OHIO KY	211	47	355	92	44
1	2.2	1	1.5	3	3.9	OLDHAM KY	-32	166	81	59	46
		1	1.9	2	3.8	OWEN KY		94			43
		2	3.0	2	3.6	PENDLETON KY		18			40
1	.7	7	5.8	8	6.1	PERRY KY	689	6	737	99	81
3	1.6	5	1.9	11	3.8	PIKE KY	17	104	138	74	44
		1	2.8	2	5.4	POWELL KY		93			72
2	1.2	5	2.8	10	4.6	PULASKI KY	141	63	293	90	58
				1	4.5	ROBERTSON KY					58
		1	1.5	7*	9.5	ROCKCASTLE KY		525			95
1	2.2	2	3.6	4	6.7	ROWAN KY	60	87	199	82	86
		2	3.3	5	7.3	RUSSELL KY		120			89
1	1.3	2	3.0	4	5.0	SCOTT KY	134	68	293	90	66
4	4.5	1	1.1	3	3.1	SHELBY KY	-75	176	-31	13	32
2	3.9			6	9.5	SIMPSON KY			146	75	95
				1	3.7	SPENCER KY					42
2	2.8	1	1.5	4	4.5	TAYLOR KY	-49	208	58	51	57
		4	6.8	1	2.4	TODD KY		-65			22
1	2.5	1	2.8	5	9.1	TRIGG KY	12	226	264	88	95
				2	2.4	UNION KY					22
4	2.7	7	3.5	5	2.2	WARREN KY	29	-38	-19	19	19
1	1.9	2	3.9	2	4.0	WASHINGTON KY	107	3	114	68	47
		4	5.0	3	3.9	WAYNE KY		-21			46
1	1.1	1	1.0			WEBSTER KY	-9				
1	.7	6	4.2	4	2.8	WHITLEY KY	480	-32	292	90	28
		1	3.1	1	2.8	WOLFE KY		-10			27
4*	8.7	2	3.5	3	4.6	WOODFORD KY	-60	32	-47	08	59
172	2.4	311	3.4	569*	5.3	LOUISIANA LA	44	55	124		
7	4.6	8	4.2	9	5.0	ACADIA LA	-9	18	7	30	66

* SIGNIFICANTLY MORE DEATHS THAN EXPECTED

WHITE MALE

CANCER OF THE LIVER AND GALLBLADDER INCLUDING BILE DUCTS

ICD 155,156

1950-1959 DEATHS	1950-1959 RATE	1960-1969 DEATHS	1960-1969 RATE	1970-1979 DEATHS	1970-1979 RATE	COUNTY AND STATE	PERCENT CHANGE 60S/50S	PERCENT CHANGE 70S/60S	PERCENT CHANGE 70S/50S	PERCENTILE %CHANGE 70S/50S	PERCENTILE RATES 1970S
1	1.7	2	2.6	4	4.6	ALLEN LA	58	76	178	80	60
1	1.0	4	5.2	6	6.1	ASCENSION LA	423	16	508	97	80
1	2.2	1	2.1	3	5.4	ASSUMPTION LA	-4	158	148	76	72
5	3.9	4	2.6	4	2.6	AVOYELLES LA	-34	1	-34	12	25
1	1.2	5	6.4	4	4.5	BEAUREGARD LA	420	-30	266	88	57
		1	1.8	2	2.7	BIENVILLE LA		49			27
2	1.5	3	2.7	7	4.2	BOSSIER LA	74	55	170	79	50
7	1.5	22	3.8	33	4.9	CADDO LA	152	27	219	85	64
6	2.9	15	4.2	15	3.2	CALCASIEU LA	43	-24	9	31	34
2	6.8	1	2.1	3	6.0	CALDWELL LA	-68	180	-11	23	79
1	5.8	3*	8.9	2	4.7	CAMERON LA	53	-47	-19	19	61
1	2.4	1	2.2	2	4.0	CATAHOULA LA	-8	85	70	56	47
2	3.4	1	1.9	2	3.0	CLAIBORNE LA	-44	58	-12	22	30
		2	4.2	2	4.1	CONCORDIA LA		-4			48
2	3.1	4	6.6	4	5.1	DE SOTO LA	115	-22	68	55	69
11	3.7	14	3.0	32	5.1	EAST BATON ROUGE LA	-19	71	39	43	69
2*	9.3					EAST CARROLL LA					
1	1.4	1	1.5	4	7.3	EAST FELICIANA LA	8	383	422	95	89
2	1.8	2	1.6	11*	11.2	EVANGELINE LA	-11	588	511	97	96
1	1.4	1	1.2	5	5.5	FRANKLIN LA	-19	378	289	90	74
1	2.0	2	5.1	5	7.0	GRANT LA	151	39	249	87	88
		5	4.3	10	6.1	IBERIA LA		43			81
1	1.6	2	2.5	2	2.7	IBERVILLE LA	61	6	71	56	26
		1	1.7	2	2.9	JACKSON LA		78			29
8	2.7	22	3.6	43	4.8	JEFFERSON LA	34	34	79	59	63
3	4.0	2	2.9	9	7.9	JEFFERSON DAVIS LA	-28	174	98	64	92
4	2.6	12	5.4	15	5.2	LAFAYETTE LA	111	-4	103	66	70
3	2.5	3	1.7	18*	9.2	LAFOURCHE LA	-35	455	261	88	95
2	3.8	1	1.6			LA SALLE LA	-59				
2	3.2	3	3.8	2	2.2	LINCOLN LA	19	-42	-31	13	20
2	2.9	3	2.5	4	2.9	LIVINGSTON LA	-15	16	-1	27	28
				5*	15.5	MADISON LA					99
		4	5.0	4	4.4	MOREHOUSE LA		-12			55
1	1.1	2	1.9	4	3.3	NATCHITOCHES LA	72	68	189	81	35
46	3.0	78	4.7	111*	7.2	ORLEANS LA	60	52	142	75	89
7	2.8	8	2.6	17	5.5	OUACHITA LA	-6	107	94	63	73
1	3.0	2	3.4	6*	9.6	PLAQUEMINES LA	13	185	221	85	95
1	2.3	3	5.6	8*	12.1	POINTE COUPEE LA	142	115	419	95	97
6	2.5	7	2.2	11	2.8	RAPIDES LA	-12	29	14	33	28
		1	2.8	3	8.5	RED RIVER LA		205			93
				6	7.8	RICHLAND LA					91
2	2.6	1	1.0	5	4.8	SABINE LA	-62	384	84	60	62
1	2.5	2	2.1	8	5.2	ST BERNARD LA	-16	146	106	66	69
		2	2.5	1	1.2	ST CHARLES LA		-54			11
		1	6.2			ST HELENA LA					
1	3.7			6*	14.3	ST JAMES LA			288	90	98
		1	2.7	2	4.1	ST JOHN THE BAPT LA		52			50
3	3.6	6	3.5	11	5.0	ST LANDRY LA	-5	44	37	43	66
1	1.2	5	7.7	7	7.3	ST MARTIN LA	531	-5	501	96	89
1	.8	4	3.7	4	2.6	ST MARY LA	360	-28	230	86	25
		2	1.1	11	4.6	ST TAMMANY LA		329			59
4	2.6	3	1.6	15	6.4	TANGIPAHOA LA	-37	298	150	76	84
5	3.5	4	2.2	10	5.5	TERREBONNE LA	-36	147	57	50	74
2	3.4	3	3.8	7	7.9	UNION LA	12	108	134	73	92
3	2.0	7	4.3	14	7.2	VERMILION LA	112	67	254	87	89
2	2.4	2	2.3	7	5.9	VERNON LA	-7	162	143	75	79
2	2.3	5	4.0	8	5.7	WASHINGTON LA	78	42	152	76	77
		1	.6	6	3.8	WEBSTER LA		487			43
		4*	12.4	1	2.7	WEST BATON ROUGE LA		-79			26
1	1.6	1	1.9	4	6.3	WEST CARROLL LA	14	234	281	89	82
		1	1.4	3	4.0	WINN LA		178			47
101	2.2	147	3.1	215	4.2	MAINE ME	38	38	90		
14	3.4	14	3.2	26	5.8	ANDROSCOGGIN ME	-6	80	70	56	78
5	1.5	7	1.7	14	3.7	AROOSTOOK ME	12	116	141	74	41
15	1.9	42	4.8	33	3.5	CUMBERLAND ME	148	-27	82	59	38
2	1.7	4	3.6	6	5.3	FRANKLIN ME	114	46	213	84	70
3	1.6	5	2.4	9	4.0	HANCOCK ME	53	67	155	77	48
9	2.0	12	2.7	16	3.2	KENNEBEC ME	32	20	59	51	34
7	4.0	3	1.7	9	4.8	KNOX ME	-58	185	20	35	62
1	.8	6	4.9	4	3.2	LINCOLN ME	545	-35	321	91	33
6	2.4	8	3.4	13	5.1	OXFORD ME	40	48	107	67	68
11	2.1	16	3.0	22	4.1	PENOBSCOT ME	40	36	90	61	48
4	3.9	5	4.9	8	8.5	PISCATAQUIS ME	27	74	121	70	94
2	1.7	3	2.5			SAGADAHOC ME	42				
3	1.4	7	3.4	11	5.1	SOMERSET ME	148	51	275	89	68
1	.7	1	.9	5	3.7	WALDO ME	18	327	402	94	42
4	1.9	3	1.6	6	2.9	WASHINGTON ME	-18	86	53	49	29
14	3.0	11	2.2	33	5.7	YORK ME	-27	161	91	62	77
222	2.6	383	3.5	648*	5.1	MARYLAND MD	34	44	92		
5	1.1	11	2.7	28	6.3	ALLEGANY MD	137	131	447	95	82
11	2.8	21	3.2	54	5.7	ANNE ARUNDEL MD	13	78	101	65	76
35	3.3	67	3.8	118	4.7	BALTIMORE MD	15	23	41	45	61
		2	4.1	8*	9.1	CALVERT MD		121			95
		4	4.0	3	3.4	CAROLINE MD		-15			37
5	2.2	6	2.1	14	4.0	CARROLL MD	-4	88	81	59	46
6	3.9	4	2.2	4	1.9	CECIL MD	-43	-14	-51	06	17
1	1.5	7*	7.9	5	5.0	CHARLES MD	432	-37	235	86	66
1	.7	3	2.2	4	2.8	DORCHESTER MD	203	25	278	89	27
6	2.2	7	2.3	19	5.0	FREDERICK MD	4	118	127	71	66
1	1.0	2	1.9	3	2.5	GARRETT MD	85	30	141	74	23
4	2.5	7	2.6	21	6.2	HARFORD MD	2	140	145	75	82
		3	2.1	9	4.9	HOWARD MD		136			64
1	1.7			1	1.3	KENT MD			-22	17	12
13	2.1	40	3.2	95	5.6	MONTGOMERY MD	55	75	172	79	75
17	3.8	41	3.6	70	5.6	PRINCE GEORGE'S MD	-7	57	45	46	75
2	3.6	1	1.2	3	3.1	QUEEN ANNE'S MD	-67	169	-12	22	33

* SIGNIFICANTLY MORE DEATHS THAN EXPECTED

1950-1959 DEATHS	1950-1959 RATE	1960-1969 DEATHS	1960-1969 RATE	1970-1979 DEATHS	1970-1979 RATE	COUNTY AND STATE	PERCENT CHANGE 60S/50S	PERCENT CHANGE 70S/60S	PERCENT CHANGE 70S/50S	PERCENTILE %CHANGE 70S/50S	PERCENTILE RATES 1970S
4	6.5	4	4.8	4	3.2	ST MARY'S MD	-26	-33	-50	07	35
3	3.8	2	2.6	1	1.3	SOMERSET MD	-31	-51	-66	02	12
3	3.5	1	1.1	4	3.0	TALBOT MD	-69	177	-13	22	31
7	2.1	23*	5.7	15	3.1	WASHINGTON MD	176	-45	52	49	33
6	3.6	3	1.7	8	3.5	WICOMICO MD	-51	102	-1	27	39
		3	3.3	2	1.3	WORCESTER MD		-59			12
91	3.0	121	4.2	155*	6.1	BALTIMORE CITY MD	41	46	106	67	80
671	3.0	924	3.8	1266*	4.9	MASSACHUSETTS MA	27	30	66		
7	2.2	23	4.9	37	4.4	BARNSTABLE MA	116	-10	95	63	54
6	.8	16	2.3	48	6.3	BERKSHIRE MA	168	176	641	98	83
54	3.1	66	3.2	116	5.3	BRISTOL MA	3	67	72	56	71
		3	8.7	4	8.7	DUKES MA		-1			94
88	3.2	119	4.1	146	4.8	ESSEX MA	27	18	50	48	63
6	2.1	7	2.2	11	3.3	FRANKLIN MA	6	51	59	51	36
56	3.1	80	4.1	91	4.5	HAMPDEN MA	32	8	42	45	57
13	3.1	13	2.7	27	4.9	HAMPSHIRE MA	-13	84	60	52	64
145	2.9	222	4.2	298	5.2	MIDDLESEX MA	41	26	78	58	70
				4*	16.0	NANTUCKET MA					99
52	2.7	65	2.8	117	4.4	NORFOLK MA	6	57	67	54	56
20	1.9	40	3.2	61	4.2	PLYMOUTH MA	69	32	123	70	51
146*	3.8	183*	5.3	179*	6.1	SUFFOLK MA	39	16	62	53	81
78	2.9	87	2.9	127	4.1	WORCESTER MA	1	41	42	45	50
804	2.9	1254*	4.1	1582	4.7	MICHIGAN MI	40	16	64		
2	5.9			8*	10.4	ALCONA MI			77	58	96
1	1.4	1	1.7	6*	12.5	ALGER MI	21	617	771	99	98
3	1.2	12	4.2	19	5.7	ALLEGAN MI	250	34	370	93	77
7*	6.0	6	5.0	9	6.8	ALPENA MI	-17	36	13	32	87
1	1.2	2	2.5	2	2.1	ANTRIM MI	101	-14	72	57	19
		4	6.3	5	5.4	ARENAC MI		-15			72
1	2.4			4	6.6	BARAGA MI			179	80	85
5	3.5	5	3.1	6	3.1	BARRY MI	-11	0	-11	23	32
10	2.5	12	2.7	15	3.2	BAY MI	11	18	31	40	34
1	2.0	3	7.2	2	2.7	BENZIE MI	271	-62	40	44	27
25*	4.8	22	3.7	31	4.8	BERRIEN MI	-23	31	1	27	62
2	1.3	6	3.6	10	5.4	BPANCH MI	184	50	324	91	73
8	1.6	27	4.8	35	5.9	CALHOUN MI	195	23	264	88	78
4	2.6	4	2.4	9	4.6	CASS MI	-6	92	81	59	60
		2	2.3	6	6.3	CHARLEVOIX MI		168			83
		4	4.4	6	5.6	CHEBOYGAN MI		27			75
2	1.5	6	4.5	8	6.0	CHIPPEWA MI	194	35	296	90	80
1	1.7	2	2.2	9	6.4	CLARE MI	29	197	283	89	84
5	3.6	6	3.8	6	3.5	CLINTON MI	6	-6	0	27	39
4*	15.9	2	6.8	1	2.7	CRAWFORD MI	-58	-60	-83	00	27
2	1.1	7	3.6	4	2.0	DELTA MI	235	-45	83	60	18
1	.7	7	4.5	7	4.3	DICKINSON MI	523	-4	495	96	52
5	2.4	8	3.7	11	3.8	EATON MI	50	3	55	49	43
2	2.6	6	7.6	2	1.9	EMMET MI	195	-75	-26	15	17
16	1.7	43	3.4	60	4.5	GENESEE MI	102	32	167	78	57
1	1.7	1	1.1	3	2.6	GLADWIN MI	-37	140	51	48	24
3	2.2	4	2.5	11	7.0	GOGEBIC MI	14	183	221	85	88
4	2.4	3	1.7	8	4.0	GRAND TRAVERSE MI	-29	138	70	56	47
4	2.2	8	4.7	7	3.9	GRATIOT MI	116	-16	81	59	45
2	1.2	8	4.3	3	1.6	HILLSDALE MI	274	-64	36	42	14
9	3.8	8	3.4	13	5.7	HOUGHTON MI	-11	70	51	48	77
		6	3.1	7	3.4	HURON MI		8			37
10	1.5	28	3.9	37	4.5	INGHAM MI	151	17	195	82	58
2	.9	5	2.5	6	3.0	IONIA MI	179	21	237	86	31
2	2.6	4	4.0	7	5.1	IOSCO MI	59	27	101	65	69
4	4.3	8*	7.9	5	4.6	IRON MI	84	-41	7	30	60
4	3.1	4	3.0	4	2.0	ISABELLA MI	-2	-33	-35	12	18
15	3.0	19	3.5	24	4.3	JACKSON MI	16	24	43	45	53
18	3.0	19	2.8	28	4.1	KALAMAZOO MI	-6	43	35	42	48
		2	7.3			KALKASKA MI					
59*	4.3	48	3.2	78	4.8	KENT MI	-26	49	11	31	62
1	5.9			2	7.3	KEWEENAW MI			25	37	89
1	2.9			1	2.3	LAKE MI			-22	18	21
		9	4.7	10	4.7	LAPEER MI		1			62
1	1.6	3	4.6	3	4.0	LEELANAU MI	181	-12	146	75	48
7	2.2	16	4.9	15	4.3	LENAWEE MI	125	-11	101	65	53
10*	6.8	6	3.3	19	7.7	LIVINGSTON MI	-52	136	12	32	91
		1	2.1	5*	13.1	LUCE MI		525			98
2	4.6	1	2.1	3	5.0	MACKINAC MI	-54	136	9	30	67
21	2.3	60	4.3	95	4.9	MACOMB MI	90	14	116	69	65
5	4.8	1	.8	4	3.4	MANISTEE MI	-84	336	-28	14	37
4	1.4	12	4.6	7	2.5	MARQUETTE MI	218	-45	74	57	24
7	5.4	2	1.7	3	1.9	MASON MI	-69	13	-65	02	17
3	2.8	2	1.7	3	2.8	MECOSTA MI	-41	70	0	27	28
1	.8	4	3.0	3	2.0	MENOMINEE MI	293	-34	158	77	17
2	1.1	5	2.8	12	5.3	MIDLAND MI	156	88	382	94	71
1	2.5	1	2.3	1	2.0	MISSAUKEE MI	-11	-11	-21	18	18
7	1.8	22	5.2	28	6.1	MONROE MI	195	19	250	87	81
4	2.2	6	2.9	9	4.0	MONTCALM MI	34	37	84	60	47
		2	5.7	2	4.1	MONTMORENCY MI		-29			48
21	4.4	20	3.6	32	5.4	MUSKEGON MI	-19	51	23	37	72
3	2.2	5	3.9	15*	8.9	NEWAYGO MI	78	131	312	91	94
42	2.7	84	3.3	139	4.0	OAKLAND MI	23	20	48	47	46
2	2.4	8*	8.0	7	6.9	OCEANA MI	235	-14	189	81	87
1	2.3	1	2.1	1	1.4	OGEMAW MI	-12	-34	-41	09	13
		3	5.1	7*	11.6	ONTONAGON MI		125			97
1	1.7	3	3.3	6	6.2	OSCEOLA MI	100	86	272	88	82
3*	13.4	1	5.2	2	4.0	OSCODA MI	-61	-23	-70	01	47
2	5.6			1	2.3	OTSEGO MI			-59	04	21
8	2.2	12	2.9	12	2.4	OTTAWA MI	32	-19	6	29	22
2	3.2	4	5.6	1	1.1	PRESQUE ISLE MI	77	-80	-65	02	11
		2	2.5	8	9.1	ROSCOMMON MI		259			95
13	1.9	28	4.1	32	4.2	SAGINAW MI	112	3	120	70	51

* SIGNIFICANTLY MORE DEATHS THAN EXPECTED

WHITE MALE

CANCER OF THE LIVER AND GALLBLADDER INCLUDING BILE DUCTS

ICD 155,156

1950-1959 DEATHS	1950-1959 RATE	1960-1969 DEATHS	1960-1969 RATE	1970-1979 DEATHS	1970-1979 RATE	COUNTY AND STATE	PERCENT CHANGE 60S/50S	PERCENT CHANGE 70S/60S	PERCENT CHANGE 70S/50S	PERCENTILE %CHANGE 70S/50S	PERCENTILE RATES 1970S
10	2.2	21	4.1	31	5.5	ST CLAIR MI	90	33	153	76	74
7	3.5	13	6.1	7	2.8	ST JOSEPH MI	74	-53	-19	19	28
4	2.3	4	2.3	10	4.7	SANILAC MI	0	102	102	65	62
				2	3.4	SCHOOLCRAFT MI					37
8	3.3	8	3.3	11	4.6	SHIAWASSEE MI	0	38	38	43	58
2	1.0	5	2.3	10	4.6	TUSCOLA MI	132	103	371	93	60
5	2.0	7	2.7	10	3.5	VAN BUREN MI	34	31	75	57	39
5	1.2	19	3.5	29	4.6	WASHTENAW MI	191	32	285	89	59
342*	3.7	481*	5.2	456*	5.4	WAYNE MI	39	4	44	46	73
1	.9			6	5.1	WEXFORD MI			481	96	68
469	3.1	608	3.6	780	4.2	MINNESOTA MN	16	18	37		
1	1.3	2	1.7	3	2.8	AITKIN MN	29	64	112	68	27
4	2.6	7	1.9	13	3.9	ANOKA MN	-30	111	48	47	46
4	2.8	2	1.3	6	3.7	BECKER MN	-54	188	32	41	42
5	4.0	4	3.2	4	3.0	BELTRAMI MN	-22	-5	-26	16	30
2	2.4	3	2.8	5	4.8	BENTON MN	16	72	100	65	63
2	3.4	3	4.7	4	8.5	BIG STONE MN	36	82	146	75	93
4	1.8	2	.9	9	4.2	BLUE EARTH MN	-50	356	130	72	52
4	3.2	6	3.9	6	3.5	BROWN MN	21	-8	12	32	39
5	3.7	8	5.4	8	5.5	CARLTON MN	47	2	51	48	74
2	2.0	6	5.5	8	5.8	CARVER MN	174	7	193	82	78
2	1.8	6	4.3	7	4.1	CASS MN	144	-5	131	72	49
3	3.3	3	3.0	7	7.6	CHIPPEWA MN	-9	152	128	72	90
2	2.0	4	3.7	6	4.9	CHISAGO MN	90	31	150	76	64
		3	1.7	9	4.6	CLAY MN		171			59
2	3.3	4	6.3	4	5.4	CLEARWATER MN	88	-14	62	53	73
				2	8.0	COOK MN					92
3	3.3	2	2.2	3	2.8	COTTONWOOD MN	-32	23	-16	20	27
10*	5.5	7	3.3	8	3.2	CROW WING MN	-41	-3	-42	09	33
4	1.9	9	3.3	13	3.3	DAKOTA MN	71	1	72	56	35
2	2.8			3	3.9	DODGE MN			42	45	46
2	1.5	1	.8	7	4.1	DOUGLAS MN	-49	425	169	78	49
1	.7	4	3.2	3	2.3	FARIBAULT MN	339	-28	215	84	21
1	.9	3	2.0	7	4.5	FILLMORE MN	123	129	410	94	56
4	2.3	8	4.1	2	1.0	FREEBORN MN	80	-75	-55	05	11
4	2.1	4	1.9	9	3.8	GOODHUE MN	-8	95	79	58	43
1	1.9	3	5.7	4	5.3	GRANT MN	197	-6	178	80	71
113	3.7	149	4.2	167	4.4	HENNEPIN MN	13	7	21	36	56
4	5.0	2	2.1	5	5.1	HOUSTON MN	-58	144	3	28	68
2	2.5	3	3.3	5	5.6	HUBBARD MN	33	70	126	71	75
1	1.1	5	5.8	3	2.8	ISANTI MN	432	-52	154	76	27
9	6.6	12	6.3	9	4.1	ITASCA MN	-4	-36	-39	10	48
2	2.3	5	5.3	3	3.7	JACKSON MN	132	-31	60	52	41
		1	2.0			KANABEC MN					
		8	4.6	7	3.5	KANDIYOHI MN		-24			38
3	5.4	1	1.8	2	3.8	KITTSON MN	-67	115	-29	14	44
2	2.4	5	5.4	3	3.4	KOOCHICHING MN	129	-37	44	45	36
1	1.3	3	3.3	3	3.2	LAC QUI PARLE MN	150	-5	139	74	33
1	4.4	3	5.6			LAKE MN	28				
1	5.3			2	6.9	LAKE OF THE WOODS MN			29	40	87
3	2.5	3	2.3	6	4.0	LE SUEUR MN	-9	77	62	53	47
2	3.6	3	4.7	1	1.5	LINCOLN MN	31	-67	-57	05	14
4	3.0	1	.9	3	2.4	LYON MN	-72	176	-22	17	22
3	2.4	3	2.0	6	3.8	MC LEOD MN	-17	90	57	50	43
		3	10.1			MAHNOMEN MN					
2	2.4	2	2.5	6	6.8	MARSHALL MN	1	177	180	80	87
2	1.4	2	1.3	3	2.0	MARTIN MN	-8	51	39	44	18
4	3.7	5	4.4	8	6.0	MEEKER MN	22	35	64	53	79
2	2.1	3	3.3	6	5.8	MILLE LACS MN	60	76	182	80	78
10*	7.5	4	2.3	9	4.9	MORRISON MN	-69	110	-35	11	64
6	3.5	3	1.5	3	1.1	MOWER MN	-59	-21	-67	02	11
2	2.9	1	1.5			MURRAY MN	-47				
3	2.5	5	4.4	10*	9.2	NICOLLET MN	75	109	267	88	95
2	1.8	2	1.7	3	2.3	NOBLES MN	-8	40	29	39	21
3	3.6	4	4.5	4	4.4	NORMAN MN	24	-1	23	37	56
3	1.4	7	2.6	18	5.8	OLMSTED MN	83	126	313	91	78
8	2.5	10	3.0	13	3.3	OTTER TAIL MN	19	9	29	40	35
1	1.3	1	1.3	4	4.4	PENNINGTON MN	4	227	241	86	54
5	4.3	5	4.0	10	7.3	PINE MN	-8	83	68	55	89
		1	1.0	2	2.6	PIPESTONE MN		153			25
9	4.9	6	2.9	8	3.6	POLK MN	-41	22	-28	15	40
1	1.1	5	5.7	6	7.5	POPE MN	405	31	564	97	90
76*	4.9	86*	4.9	80	4.3	RAMSEY MN	-1	-12	-13	22	52
2	1.6	5	4.0	3	2.5	REDWOOD MN	153	-36	62	52	24
6	4.5	1	.6	2	1.2	RENVILLE MN	-86	89	-74	01	11
4	2.5	8	4.3	12	6.0	RICE MN	74	40	143	75	79
2	2.9	3	4.5	2	3.5	ROCK MN	58	-23	22	36	38
3	4.7	1	1.1	6	6.8	ROSEAU MN	-77	528	45	46	87
31	2.5	54	4.5	53	4.5	ST LOUIS MN	78	0	77	58	56
4	4.6	2	1.8	10	7.6	SCOTT MN	-60	320	66	54	90
		4	5.0	6	6.4	SHERBURNE MN		29			84
3	3.2	4	4.2	1	1.5	SIBLEY MN	30	-64	-53	06	14
8	2.4	14	3.6	19	4.3	STEARNS MN	50	19	79	58	52
4	3.3	7	5.5	6	4.2	STEELE MN	68	-24	28	39	51
2	3.6	1	1.6	6	9.4	STEVENS MN	-55	471	160	77	95
4	4.8	2	2.2	5	4.9	SWIFT MN	-53	118	2	28	65
3	2.3	5	3.0	8	5.3	TODD MN	30	77	129	72	71
1	2.2	5*	12.0	2	3.6	TRAVERSE MN	442	-70	64	54	41
1	1.1	1	.9			WABASHA MN	-15				
2	3.1			6	6.7	WADENA MN			117	69	86
1	1.0	1	1.2	5	5.0	WASECA MN	15	325	388	94	66
6	3.3	6	2.9	18	6.3	WASHINGTON MN	-14	121	91	62	83
1	1.1	1	1.1	5	6.4	WATONWAN MN	-3	470	455	95	83
5*	11.4	2	3.9	4	7.3	WILKIN MN	-66	90	-36	11	89
6	2.8	7	3.2	9	3.6	WINONA MN	12	13	28	39	40
3	1.7	7	3.4	8	3.7	WRIGHT MN	100	11	123	70	42
3	3.3	1	1.2	6	5.5	YELLOW MEDICINE MN	-64	369	67	55	73

* SIGNIFICANTLY MORE DEATHS THAN EXPECTED

1950-1959 DEATHS	1950-1959 RATE	1960-1969 DEATHS	1960-1969 RATE	1970-1979 DEATHS	1970-1979 RATE	COUNTY AND STATE	PERCENT CHANGE 60S/50S	PERCENT CHANGE 70S/60S	PERCENT CHANGE 70S/50S	PERCENTILE %CHANGE 70S/50S	PERCENTILE RATES 1970S
96	1.8	200	3.4	299	4.3	MISSISSIPPI MS	82	28	132		
		1	1.7	6	7.8	ADAMS MS		363			91
2	2.0	4	3.4	7	5.3	ALCORN MS	69	53	159	77	71
		7*	9.2	1	1.1	ATTALA MS		-88			11
3	3.3	3	4.1	1	1.3	BOLIVAR MS	23	-68	-61	03	12
3	5.3	1	2.1	2	2.4	CALHOUN MS	-60	16	-54	05	23
1	3.0	1	4.4	1	3.4	CARROLL MS	46	-23	13	33	37
1	2.2	5*	9.2	4	6.8	CHICKASAW MS	326	-26	215	84	87
				2	4.7	CHOCTAW MS					60
2	3.4	1	1.6	1	1.6	CLARKE MS	-52	-2	-53	06	14
1	3.0	3	6.5	1	2.4	CLAY MS	119	-64	-20	18	22
1	2.2	3	4.6	3	4.9	COAHOMA MS	109	8	127	71	65
		1	1.7	3	4.4	COPIAH MS		163			56
1	2.2	2	5.0	2	3.0	COVINGTON MS	133	-40	40	44	31
		4	6.5	1	1.5	DE SOTO MS		-77			14
3	2.3	3	1.8	5	2.7	FORREST MS	-23	52	16	34	25
		2	7.0	2	7.4	FRANKLIN MS		5			90
		5*	9.9	4	7.7	GEORGE MS		-23			91
		1	5.3	2	5.7	GREENE MS		7			76
1	2.6	1	1.9	4	6.7	GRENADA MS	-24	242	159	77	86
				3	3.4	HANCOCK MS					36
8	2.5	15	4.4	25	5.6	HARRISON MS	79	27	127	71	75
6	1.7	15	3.4	25	5.0	HINDS MS	97	46	189	81	67
1	2.0	2	4.4	2	4.7	HOLMES MS	115	7	131	72	61
				1	4.6	HUMPHREYS MS					59
1*	18.9					ISSAQUENA MS					
		3	3.6	4	4.1	ITAWAMBA MS		16			49
4	3.9	10	5.7	9	3.3	JACKSON MS	45	-42	-15	20	36
1	2.5			1	1.8	JASPER MS			-29	14	16
		3*	17.6	1	6.5	JEFFERSON MS		-63			84
2	7.1	1	2.8			JEFFERSON DAVIS MS	-60				
3	1.7	5	2.5	12	5.6	JONES MS	47	121	224	85	74
1	3.4	1	2.9	2	5.6	KEMPER MS	-16	94	63	53	75
2	3.2	2	3.1	2	2.6	LAFAYETTE MS	-3	-17	-19	19	24
		5*	8.4	4	5.7	LAMAR MS		-33			76
3	1.6	4	2.0	7	3.2	LAUDERDALE MS	23	60	97	63	33
1	3.6			2	4.4	LAWRENCE MS			24	37	55
1	1.4	1	1.3	5	6.5	LEAKE MS	-3	381	368	93	84
2	1.7	4	2.7	5	2.9	LEE MS	55	9	69	55	30
1	2.1	3	3.3	4	4.6	LEFLORE MS	58	40	122	70	58
		2	2.2	5	5.1	LINCOLN MS		130			69
		7	6.4	6	5.4	LOWNDES MS		-15			73
1	2.0			2	3.1	MADISON MS			55	49	32
		1	1.3			MARION MS					
		3	7.5			MARSHALL MS					
3	2.8	3	3.4	12*	8.9	MONROE MS	19	166	217	85	94
2	4.6	1	2.0	3	5.3	MONTGOMERY MS	-56	157	14	33	70
1	1.6	2	2.5	2	2.3	NESHOBA MS	54	-8	42	45	21
1	1.3	2	2.4	5	5.8	NEWTON MS	80	140	331	92	78
				1	4.4	NOXUBEE MS					56
				1	1.5	OKTIBBEHA MS					14
1	1.5	3	4.0	5	5.7	PANOLA MS	165	41	274	88	76
1	1.3	2	2.1	4	3.6	PEARL RIVER MS	59	68	168	78	40
3	3.4	3	3.0	5	4.7	PIKE MS	-13	57	37	43	61
2	2.5	4	4.9	1	1.2	PONTOTOC MS	94	-75	-52	06	12
3	4.1	2	2.0	6	6.1	PRENTISS MS	-51	204	50	48	81
3*	9.2			1	3.6	QUITMAN MS			-61	03	41
1	1.3	2	2.1	8	5.5	RANKIN MS	63	165	334	92	74
		2	2.9	4	5.4	SCOTT MS		89			73
		2	2.6	1	1.1	SIMPSON MS		-56			11
1	1.7	1	1.6	2	2.8	SMITH MS	-7	72	60	52	27
1	4.0			1	3.4	STONE MS			-15	21	37
		3	4.3	7*	9.2	SUNFLOWER MS		114			95
1	2.5	4	8.3			TALLAHATCHIE MS	233				
1	2.2	3	7.9	2	4.0	TATE MS	254	-50	77	58	46
		1	1.4	5	5.7	TIPPAH MS		303			76
2	2.8	1	1.1	7	8.0	TISHOMINGO MS	-60	604	184	81	92
		1	14.1	3*	19.3	TUNICA MS		38			99
				3	3.1	UNION MS					32
		2	5.0	3	6.3	WALTHALL MS		27			83
1	1.2	6	6.4	6	5.3	WARREN MS	435	-18	340	92	70
3	3.4	10*	10.3	5	3.9	WASHINGTON MS	204	-62	15	33	45
3	8.0	1	1.8	1	1.7	WAYNE MS	-77	-6	-79	00	15
				6*	11.6	WEBSTER MS					97
				1	5.5	WILKINSON MS					73
				4	5.9	WINSTON MS					78
1	2.0	3	5.7	4	7.8	YALOBUSHA MS	182	36	283	89	91
4	6.3	1	1.7	1	1.5	YAZOO MS	-73	-13	-77	00	13
635	3.3	782	3.7	1067	4.8	MISSOURI MO	13	30	47		
1	1.0	4	3.4	5	3.8	ADAIR MO	233	12	272	88	43
1	1.2	2	2.2	1	1.2	ANDREW MO	85	-46	0	27	12
2	3.0	4	6.7	1	1.1	ATCHISON MO	125	-84	-63	03	11
5	3.9	7	5.1	4	2.9	AUDRAIN MO	32	-44	-26	15	28
4	3.1	3	2.0	7	4.0	BARRY MO	-38	104	28	39	47
		2	2.0	5	6.4	BARTON MO		212			83
3	2.0	4	3.4	5	3.8	BATES MO	66	13	88	61	44
1	1.2	2	2.3	3	2.8	BENTON MO	93	22	135	73	27
2	3.8			1	1.4	BOLLINGER MO			-63	03	13
4	1.9	7	3.3	8	3.3	BOONE MO	71	-2	67	55	35
13	2.7	18	3.5	26	5.6	BUCHANAN MO	30	59	106	67	74
5	3.2	7	3.4	13	6.3	BUTLER MO	7	87	100	64	83
		5	7.6	5	7.0	CALDWELL MO		-8			88
5	3.5	5	3.7	5	3.7	CALLAWAY MO	4	0	4	29	41
2	3.2	2	2.5	8	6.6	CAMDEN MO	-23	167	107	67	85
4	2.1	7	3.2	17	6.7	CAPE GIRARDEAU MO	54	107	220	85	86
4	3.8	3	2.7	4	4.8	CARROLL MO	-28	79	29	39	63

* SIGNIFICANTLY MORE DEATHS THAN EXPECTED

WHITE MALE

CANCER OF THE LIVER AND GALLBLADDER INCLUDING BILE DUCTS

ICD 155,156

1950-1959 DEATHS	1950-1959 RATE	1960-1969 DEATHS	1960-1969 RATE	1970-1979 DEATHS	1970-1979 RATE	COUNTY AND STATE	PERCENT CHANGE 60S/50S	PERCENT CHANGE 70S/60S	PERCENT CHANGE 70S/50S	PERCENTILE %CHANGE 70S/50S	PERCENTILE RATES 1970S
		2	6.9	3	11.5	CARTER MO		66			97
3	2.1	5	3.3	6	3.3	CASS MO	60	0	61	52	36
1	1.2	1	1.1	5	4.8	CEDAR MO	-7	338	309	91	62
2	2.2	5	5.1	2	2.3	CHARITON MO	129	-55	2	28	21
2	2.5	3	3.6	5	5.6	CHRISTIAN MO	46	54	125	71	75
		3	4.5			CLARK MO					
7	2.7	10	3.0	17	4.1	CLAY MO	8	38	49	47	49
2	2.0	3	4.4	2	2.4	CLINTON MO	123	-46	20	36	22
6	3.4	4	2.2	6	2.8	COLE MO	-35	28	-17	19	28
4	4.0	1	1.2	4	4.4	COOPER MO	-71	280	10	31	55
1	1.1	4	4.8	6	5.1	CRAWFORD MO	333	7	363	93	69
3	4.2	3	4.6	1	2.0	DADE MO	10	-57	-53	06	17
3	4.2	3	3.6	5	6.2	DALLAS MO	-14	72	48	47	82
		2	2.3	4	6.2	DAVIESS MO		170			81
2	4.2	2	4.0	3	6.8	DE KALB MO	-4	71	65	54	87
3	4.3	1	1.0	4	4.1	DENT MO	-77	300	-6	25	48
2	3.3	3	4.4	2	3.4	DOUGLAS MO	32	-22	3	28	38
2	1.2	4	2.1	12	6.3	DUNKLIN MO	83	198	445	95	82
9	4.0	15	5.8	13	4.3	FRANKLIN MO	48	-27	8	30	53
2	3.6	6	6.7	7	7.0	GASCONADE MO	84	5	93	62	88
1	1.3	2	2.4	1	1.1	GENTRY MO	90	-56	-16	20	11
15	2.7	16	2.4	37	5.1	GREENE MO	-11	108	85	60	68
3	2.6	1	1.3	2	2.5	GRUNDY MO	-49	86	-4	26	23
3	3.4	4	3.7	2	2.1	HARRISON MO	10	-42	-37	11	19
3	2.1	5	3.4	7	4.5	HENRY MO	61	31	112	68	56
2	4.6	2	4.1	3	7.4	HICKORY MO	-13	83	60	52	90
3	4.9	2	2.9	2	5.2	HOLT MO			5	29	69
7	5.6	5	3.5	8	4.9	HOWARD MO	-37	38	-13	21	64
		1	2.3	3	4.3	HOWELL MO		92			54
						IRON MO					
103*	4.6	108	4.4	101	4.1	JACKSON MO	-6	-5	-11	23	50
11	2.6	24*	5.6	36*	8.1	JASPER MO	115	46	214	84	92
7	3.6	7	2.5	29*	6.6	JEFFERSON MO	-30	165	85	60	85
7	5.4	3	2.1	1	.6	JOHNSON MO	-60	-74	-90	00	10
		1	2.7	3	8.7	KNOX MO		226			94
1	.8	4	3.5	3	2.3	LACLEDE MO	311	-34	172	79	21
8	5.4	3	1.9	6	3.4	LAFAYETTE MO	-65	84	-37	11	37
9*	6.1	5	3.2	14	7.8	LAWRENCE MO	-48	143	27	39	91
1	1.2			4	6.7	LEWIS MO			457	95	86
2	2.1	2	1.6	4	3.6	LINCOLN MO	-26	126	67	55	40
6	4.2	2	1.5	3	4.1	LINN MO	-64	173	-1	27	50
2	2.2			7	7.7	LIVINGSTON MO			255	87	91
1	1.1	4	4.5	3	2.9	MC DONALD MO	318	-36	166	78	29
4	3.0	2	1.5	1	.8	MACON MO	-50	-50	-75	01	10
		2	3.1	2	2.5	MADISON MO		-20			23
1	2.6	2	4.5	2	3.8	MARIES MO	74	-15	49	47	43
1	.6	7	4.1	9	5.9	MARION MO	546	44	827	99	78
1	1.1	2	1.9	5	5.0	MILLER MO	80	158	366	93	66
2	2.3	1	1.4	2	2.7	MISSISSIPPI MO	-40	96	17	34	26
2	2.6	3	4.2	2	2.8	MONITEAU MO	63	-33	9	30	28
2	2.3	2	2.6	6	8.5	MONROE MO	12	231	271	88	94
4	4.4	2	2.1	4	4.2	MONTGOMERY MO	-54	103	-6	25	51
		3	3.8	7	6.3	MORGAN MO		65			82
2	1.9	4	3.8	7	6.3	NEW MADRID MO	105	66	240	86	83
6	3.9	5	3.2	14	6.9	NEWTON MO	-18	117	78	58	87
		1	.6	3	2.8	NODAWAY MO		355			28
		2	3.4	6	7.5	OREGON MO		124			90
2	3.0	1	1.9	2	2.9	OSAGE MO	-36	52	-3	26	29
		1	1.8	2	4.5	OZARK MO		145			57
2	1.7	1	.8	10*	9.5	PEMISCOT MO	-57	1158	447	95	95
6*	7.5			8	8.0	PERRY MO			7	29	92
4	2.3	8	4.1	14	7.2	PETTIS MO	82	75	217	85	88
3	2.6	2	1.5	12	8.4	PHELPS MO	-42	447	217	85	93
2	2.1	5	4.8	1	.9	PIKE MO	126	-81	-57	04	10
3	3.4			4	2.7	PLATTE MO			-20	18	26
3	2.5	5	4.8	5	4.1	POLK MO	95	-15	66	54	49
4	8.0	6	7.1	9*	10.2	PULASKI MO	-11	42	27	39	96
				7*	11.9	PUTNAM MO					97
1	1.8	3	4.8			RALLS MO	172				
2	1.1	2	1.1	11	7.1	RANDOLPH MO	7	527	570	98	88
1	.8			2	1.8	RAY MO			128	72	16
1	2.4	2	5.5	1	2.3	REYNOLDS MO	131	-59	-6	25	20
		3	3.6	6	6.9	RIPLEY MO		90			87
3	1.9	16*	7.7	9	3.0	ST CHARLES MO	309	-61	59	51	30
2	2.7	2	2.4	3	3.7	ST CLAIR MO	-13	57	36	42	41
1	1.9	7*	10.5	6	8.3	STE GENEVIEVE MO	449	-21	335	92	93
2	1.0	7	3.3	11	4.7	ST FRANCOIS MO	230	43	372	94	61
63	3.7	105	3.9	161	4.7	ST LOUIS COUNTY MO	4	21	26	38	61
3	2.0	4	2.7	8	5.3	SALINE MO	33	97	162	77	71
				2	4.7	SCHUYLER MO					61
1	1.8	1	1.7			SCOTLAND MO	-10				
1	.8	6	4.0	6	3.8	SCOTT MO	423	-7	386	94	43
		1	2.6			SHANNON MO					
3	3.6			1	1.4	SHELBY MO			-62	03	13
3	2.1	7	4.6	8	4.8	STODDARD MO	120	3	127	71	62
4*	7.5	3	4.5	6	5.2	STONE MO	-40	15	-31	13	69
4	5.9	3	4.3	3	6.5	SULLIVAN MO	-27	52	11	31	84
				8	6.1	TANEY MO					80
3	2.6	2	1.8	4	3.1	TEXAS MO	-32	74	19	35	32
2	1.3	5	3.1	4	3.3	VERNON MO	137	6	150	76	35
3	4.6	2	2.9	4	5.0	WARREN MO	-36	73	10	31	67
2	2.8	4	5.1	3	3.5	WASHINGTON MO	78	-30	25	38	39
1	2.0	2	3.1	3	3.2	WAYNE MO	56	1	58	51	33
3	3.4	4	3.8	4	4.1	WEBSTER MO	13	7	20	35	48
1	4.2	1	2.4	1	6.9	WORTH MO	-43	188	65	54	87
3	2.7	8*	7.6	7	6.0	WRIGHT MO	178	-21	119	70	79
163*	5.1	152*	5.6	137*	6.6	ST LOUIS CITY MO	9	18	28	39	85
75	2.4	109	3.4	136	3.9	MONTANA MT	41	16	63		

* SIGNIFICANTLY MORE DEATHS THAN EXPECTED

COUNTY AND STATE	1950-1959 DEATHS	1950-1959 RATE	1960-1969 DEATHS	1960-1969 RATE	1970-1979 DEATHS	1970-1979 RATE	PERCENT CHANGE 60S/50S	PERCENT CHANGE 70S/60S	PERCENT CHANGE 70S/50S	PERCENTILE %CHANGE 70S/50S	PERCENTILE RATES 1970S
BEAVERHEAD MT			2	4.2	3	7.3		76			89
BIG HORN MT	1	3.1	2	7.2	1	3.4	133	-52	11	31	37
BLAINE MT	1	3.0	2	5.0	5*	16.1	67	222	438	95	99
BROADWATER MT			1	6.4	2	12.0		87			97
CARBON MT	2	3.5	1	1.6	3	5.6	-55	247	57	50	75
CARTER MT					1	6.5					84
CASCADE MT	9	3.8	8	2.8	12	3.7	-24	30	-2	26	42
CHOUTEAU MT	3	6.2	1	3.0	2	5.0	-52	67	-20	18	66
CUSTER MT	1	1.2	2	2.8	1	1.3	129	-55	3	28	12
DANIELS MT					5*	18.4					99
DAWSON MT	1	1.9			3	6.0			214	84	79
DEER LODGE MT	3	2.9	3	3.3	2	1.8	16	-45	-36	11	16
FERGUS MT	3	4.4	4	4.4	4	4.8	0	8	8	30	62
FLATHEAD MT	6	3.8	3	1.7	10	4.5	-56	168	18	34	57
GALLATIN MT	3	2.9	2	2.0	2	1.7	-33	-14	-42	09	15
GLACIER MT	1	3.3	1	3.3			2				
GOLDEN VALLEY MT			1	13.8							
GRANITE MT			1	4.9							
HILL MT	3	4.3	5	7.5	1	1.4	76	-81	-66	02	13
JEFFERSON MT			1	4.1	1	3.2		-22			34
LAKE MT					2	2.1					19
LEWIS AND CLARK MT	4	3.1	10*	7.1	8	5.2	130	-26	69	55	70
LIBERTY MT			1	11.1							
LINCOLN MT					3	4.6					58
MC CONE MT			1	6.0							
MADISON MT			4*	10.5	2	6.2		-41			81
MEAGHER MT	1	4.2									
MISSOULA MT	2	1.1	7	3.5	7	2.9	220	-18	161	77	29
MUSSELSHELL MT	3*	7.5	2	6.1			-18				
PARK#MT	1	1.6	4	6.7	3	3.9	323	-41	150	76	46
PETROLEUM MT			1*	33.5							
PHILLIPS MT			1	2.4	1	2.0		-15			18
PONDERA MT			1	3.2	1	2.7		-15			26
POWDER RIVER MT	1	7.2									
POWELL MT	2	5.7	3	8.2			45				
PRAIRIE MT					1	6.7					86
RAVALLI MT			1	1.0	2	1.9		90			17
RICHLAND MT	1	1.8	1	2.2	1	2.0	27	-8	17	34	18
ROOSEVELT MT	3	7.3			4	10.1			37	43	96
ROSEBUD MT			1	4.4							
SANDERS MT			2	4.0	2	3.7		-7			42
SHERIDAN MT	2	5.8			2	5.1			-12	22	69
SILVER BOW MT	9	3.3	12	5.8	12	5.4	75	-8	62	52	72
STILLWATER MT	1	3.4									
SWEET GRASS MT	1	3.6	1	3.6	1	4.2	-1	19	17	34	52
TETON MT					3	7.3					89
TOOLE MT	2	6.0									
TREASURE MT					1	28.1					99
VALLEY MT	1	1.6			3	6.5			313	91	84
WIBAUX MT					2*	21.2					99
YELLOWSTONE MT	4	1.6	16	5.1	17	4.3	223	-16	171	79	52
NEBRASKA NB	216	3.2	264	3.5	266	3.4	12	-3	9		
ADAMS NB	5	3.0	5	2.9	3	1.9	-4	-33	-36	11	17
ANTELOPE NB	2	3.5	2	2.7	4	5.6	-24	107	57	50	74
BOONE NB	2	3.2	2	4.1	2	3.4	30	-18	7	30	37
BOX BUTTE NB	2	3.3	2	3.0	2	2.7	-7	-10	-16	20	27
BOYD NB			2	6.1	2	5.2		-15			70
BROWN NB			1	3.0							
BUFFALO NB	1	.7	6	3.9	7	4.3	466	11	531	97	53
BURT NB	4	5.2			1	1.6			-69	01	14
BUTLER NB	2	2.9	2	2.8	5	6.6	-1	133	132	73	85
CASS NB	2	2.0	3	3.0	3	3.1	47	5	54	49	32
CEDAR NB	3	4.4	2	3.0	5	5.9	-31	94	34	42	79
CHASE NB			1	3.6	1	4.3		20			54
CHERRY NB	1	2.6	1	2.4			-8				
CHEYENNE NB	2	4.1	1	1.5	1	1.5	-63	-3	-64	02	14
CLAY NB	1	1.2			4	5.6			371	93	75
COLFAX NB	1	1.6	5	6.8	2	2.2	325	-67	39	44	20
CUMING NB			3	4.0	4	5.7		43			76
CUSTER NB	2	1.6	2	1.7	5	4.0	7	132	148	76	47
DAKOTA NB			1	1.5	2	3.2		110			34
DAWES NB	2	4.5	2	3.3			-26				
DAWSON NB	3	2.9	3	2.8	1	1.0	-1	-63	-63	02	11
DEUEL NB			3*	13.2							
DIXON NB	1	2.2			1	2.4			10	31	22
DODGE NB	1	.8	7	3.9	2	1.0	413	-74	35	42	11
DOUGLAS NB	59*	4.9	52	3.8	62	4.3	-22	12	-13	22	52
DUNDY NB	1	4.5									
FILLMORE NB	4	5.9	4	5.7	2	3.9	-3	-31	-33	12	46
FRANKLIN NB			2	4.6	2	4.6		1			60
FRONTIER NB					1	3.9					44
FURNAS NB	1	1.7	2	3.4	2	3.6	99	8	115	69	40
GAGE NB	3	1.9	9	5.4	3	1.8	178	-66	-6	25	16
GARDEN NB					1	12.5					98
GARFIELD NB	1	5.9	1	5.6			-7				
GOSPER NB			1	5.9							
GRANT NB					1*	19.1					99
GREELEY NB	1	3.0			1	5.4			80	59	72
HALL NB	6	3.3	5	2.6	5	2.2	-22	-13	-32	13	20
HAMILTON NB	1	2.3	2	3.8	1	1.8	62	-53	-24	16	16
HARLAN NB			1	3.9							
HAYES NB	1	8.3									
HITCHCOCK NB	1	3.4	2	8.7	2	5.8	152	-33	68	55	77
HOLT NB	2	2.6	4	4.9	5	6.2	91	27	143	75	82
HOOKER NB			1	16.5							
HOWARD NB			1	1.9	1	1.8		-5			16
JEFFERSON NB	6*	7.7	5	5.9	2	2.3	-23	-61	-70	01	21
JOHNSON NB	1	2.6			2	4.2			66	54	52

* SIGNIFICANTLY MORE DEATHS THAN EXPECTED

COUNTIES COMBINED; SEE TECHNICAL NOTES

WHITE MALE

CANCER OF THE LIVER AND GALLBLADDER INCLUDING BILE DUCTS

ICD 155,156

1950-1959 DEATHS	1950-1959 RATE	1960-1969 DEATHS	1960-1969 RATE	1970-1979 DEATHS	1970-1979 RATE	COUNTY AND STATE		PERCENT CHANGE 60S/50S	PERCENT CHANGE 70S/60S	PERCENT CHANGE 70S/50S	PERCENTILE %CHANGE 70S/50S	PERCENTILE RATES 1970S
1	2.8	2	4.1	1	2.6	KEARNEY	NB	45	-35	-6	25	25
		4*	10.1	1	2.0	KEITH	NB		-80			18
2*	18.1			1*	23.8	KEYA PAHA	NB			32	41	99
1	3.6	1	5.0	1	5.1	KIMBALL	NB	36	3	41	44	68
2	2.3			1	.9	KNOX	NB			-58	04	10
19	3.2	27	4.3	32	4.7	LANCASTER	NB	31	10	44	46	61
2	1.8	3	1.9	5	2.9	LINCOLN	NB	6	50	59	51	29
				1*	14.9	LOUP	NB					99
		1*	24.6			MC PHERSON	NB					
2	1.2	7	4.4	5	2.8	MADISON	NB	256	-36	129	72	27
1	2.4	3	5.5			MERRICK	NB	127				
2	4.9	2	4.7			MOPRILL	NB	-4				
		3	8.2	2	6.2	NANCE	NB		-24			82
1	1.5	2	3.5	1	1.9	NEMAHA	NB	132	-48	21	36	16
2	3.8	2	3.9	2	4.2	NUCKOLLS	NB	2	8	10	31	51
1	1.0	4	3.7	9	7.9	OTOE	NB	283	112	711	98	92
2	4.4			2	5.0	PAWNEE	NB			13	33	66
				2	8.6	PERKINS	NB					94
5*	8.9	2	2.7	3	4.3	PHELPS	NB	-69	58	-51	07	54
2	3.9	1	2.4	4	7.3	PIERCE	NB	-38	206	89	61	89
3	3.0	5	4.1	3	2.2	PLATTE	NB	36	-46	-27	15	20
		2	4.0	2	3.5	POLK	NB		-12			38
5*	8.0	1	1.3	1	1.3	RED WILLOW	NB	-83	-4	-84	00	12
2	2.1	3	3.1			RICHARDSON	NB	49				
1	7.2					ROCK	NB					
4	4.2	4	3.4	4	3.4	SALINE	NB	-19	-2	-20	18	36
1	2.0	4	4.9	5	4.2	SARPY	NB	149	-14	114	68	51
4	3.7	5	4.1	3	2.2	SAUNDERS	NB	10	-46	-40	10	20
7	5.2	6	4.1	5	3.0	SCOTTS BLUFF	NB	-21	-25	-41	09	31
3	4.1	5	5.8	3	3.0	SEWARD	NB	39	-48	-28	15	30
3	6.5	3	6.0			SHERIDAN	NB	-7				
		1	2.6	3	7.9	SHERMAN	NB		211			92
		1	8.0			SIOUX	NB					
1	4.0					STANTON	NB					
		4	5.9	5	7.1	THAYER	NB		21			88
		3	10.1	3	9.8	THURSTON	NB		-3			96
1	1.8					VALLEY	NB					
2	2.8	2	2.7	2	2.4	WASHINGTON	NB	-6	-9	-14	21	23
		1	1.8	2	3.3	WAYNE	NB		86			35
3	5.4	1	1.7	1	2.7	WEBSTER	NB	-67	56	-49	07	26
6*	6.7	1	1.0	1	.9	YORK	NB	-85	-8	-86	00	10
21	2.5	47	3.4	99	4.2	NEVADA	NV	38	26	73		
2	5.6	1	2.3	3	4.3	CHURCHILL	NV	-59	89	-22	17	54
8	3.4	24	3.2	54	4.7	CLARK	NV	-6	46	38	43	61
2	4.3	1	1.3	1	1.3	ELKO	NV	-69	-4	-70	01	12
1*	18.9					ESMERALDA	NV					
				1	26.8	EUREKA	NV					99
1	3.1			2	5.7	HUMBOLDT	NV			85	60	76
				3*	17.0	LINCOLN	NV					99
		1	3.9	1	1.4	LYON	NV		-65			13
1	5.2					MINERAL	NV					
1	4.1	1	6.5	2	8.3	NYE	NV	59	27	102	65	93
				1	7.6	PERSHING	NV					90
4	1.3	16	4.3	26	4.2	WASHOE	NV	229	-2	223	85	51
		3	6.6			WHITE PINE	NV					
1	4.2			5	4.1	ORMSBY-CARSON CITY	NV			-3	26	48
72	2.6	112	3.6	151	4.2	NEW HAMPSHIRE	NH	39	17	63		
3	2.0	5	2.8	7	3.3	BELKNAP	NH	41	17	64	53	35
3	2.8	8	6.9	6	4.4	CARROLL	NH	145	-36	56	50	55
1	.5	6	2.8	8	3.3	CHESHIRE	NH	456	18	556	97	35
4	2.3	10	5.1	4	2.3	COOS	NH	120	-55	-2	26	21
6	2.4	6	2.3	14	5.1	GRAFTON	NH	-4	119	110	68	68
31	3.9	27	3.1	41	4.0	HILLSBOROUGH	NH	-21	29	3	28	47
8	2.3	14	3.7	15	3.5	MERRIMACK	NH	62	-6	53	49	38
8	2.0	13	2.6	29	4.6	ROCKINGHAM	NH	33	78	137	73	60
5	2.2	15	5.4	21	6.9	STRAFFORD	NH	146	27	213	84	87
3	2.0	8	5.0	6	3.4	SULLIVAN	NH	146	-32	67	55	37
764*	3.4	1097*	4.1	1591*	5.3	NEW JERSEY	NJ	20	29	55		
26	3.8	46*	5.3	58	6.0	ATLANTIC	NJ	38	15	60	52	80
86	3.3	139	4.1	219*	5.8	BERGEN	NJ	24	40	73	57	77
15	2.5	30	4.2	59	5.8	BURLINGTON	NJ	67	38	132	73	77
35	2.5	60	4.0	96	5.5	CAMDEN	NJ	59	38	120	70	73
6	1.9	17	4.4	21	4.1	CAPE MAY	NJ	131	-6	118	69	50
10	2.4	15	3.4	24	4.7	CUMBERLAND	NJ	45	36	98	63	60
147*	4.0	180*	4.9	162	5.0	ESSEX	NJ	23	2	26	38	66
13	3.1	14	2.6	36	5.7	GLOUCESTER	NJ	-14	118	88	61	77
111*	3.9	110	3.8	162*	6.1	HUDSON	NJ	-1	60	57	50	80
7	3.0	12	4.0	14	4.1	HUNTERDON	NJ	31	4	36	42	50
24	2.3	49	4.4	72	5.9	MERCER	NJ	93	33	157	77	79
50*	4.2	66	4.2	106	5.4	MIDDLESEX	NJ	0	29	30	40	73
39	3.3	53	3.5	95	4.9	MONMOUTH	NJ	6	39	47	47	64
30	3.4	43	3.7	64	4.9	MORRIS	NJ	9	32	43	45	65
13	3.1	37	3.6	98	5.3	OCEAN	NJ	17	46	71	56	71
68*	4.1	71	3.7	98	4.9	PASSAIC	NJ	-10	33	21	36	64
4	1.6	6	2.3	8	3.2	SALEM	NJ	40	38	93	62	33
12	2.4	20	3.2	36	4.8	SOMERSET	NJ	34	52	103	66	63
6	3.0	12	3.9	18	4.8	SUSSEX	NJ	31	23	61	52	63
52	3.0	107*	4.9	120	5.0	UNION	NJ	64	3	68	55	67
10	3.2	10	2.8	25	6.5	WARREN	NJ	-11	133	106	67	84
60	2.7	105	3.6	200	5.4	NEW MEXICO	NM	34	50	101		
8	1.6	34	4.4	54	5.3	BERNALILLO	NM	170	21	226	86	70
				1	6.7	CATRON	NM					86

* SIGNIFICANTLY MORE DEATHS THAN EXPECTED

WHITE MALE

CANCER OF THE LIVER AND GALLBLADDER INCLUDING BILE DUCTS

COUNTY AND STATE

ICD 155,156

COUNTY AND STATE	1950-1959 DEATHS	1950-1959 RATE	1960-1969 DEATHS	1960-1969 RATE	1970-1979 DEATHS	1970-1979 RATE	PERCENT CHANGE 60S/50S	PERCENT CHANGE 70S/60S	PERCENT CHANGE 70S/50S	PERCENTILE %CHANGE 70S/50S	PERCENTILE RATES 1970S
CHAVES NM	2	1.5	5	3.3	13	6.1	123	87	316	91	81
COLFAX NM	3	4.3	1	1.1	5	7.3	-73	546	72	56	89
CURRY NM	3	3.5	2	2.0	7	5.4	-42	170	55	50	73
DONA ANA NM	6	5.5	10	6.5	15	6.7	19	2	21	36	86
EDDY NM	3	1.9	8	5.3	8	4.3	177	-18	127	71	54
GRANT NM	2	2.9			6	6.7			127	71	85
GUADALUPE NM	2	9.0			1	3.7			-58	04	42
LEA NM	2	1.5	8	7.3	10	6.7	392	-9	349	92	86
LINCOLN NM	1	2.4	1	2.1			-11				
LOS ALAMOS NM			2	11.6							
LUNA NM					2	2.9					30
MC KINLEY NM	1	5.1	2	4.6	1	2.9	-11	-37	-44	09	29
MORA NM			3	9.9	1	3.1		-69			32
OTERO NM			1	.4	4	4.8		1125			62
QUAY NM			2	3.7	9*	14.2		281			98
RIO ARRIBA NM	3	4.6	5	6.6	7*	9.7	44	46	110	67	96
ROOSEVELT NM			1	1.4							
SANDOVAL NM	1	3.2			3	4.9			52	49	64
SAN JUAN NM	1	1.7	1	.8	2	1.2	-53	50	-29	14	12
SAN MIGUEL NM	3	3.4	1	1.3	10*	11.2	-63	789	232	86	96
SANTA FE NM	6	4.0	6	3.4	9	4.3	-15	25	6	29	52
SIERRA NM	4	6.8	3	4.9	5	8.1	-28	65	19	35	92
SOCORRO NM	1	3.4	2	5.1	3	7.3	50	42	113	68	89
TAOS NM	1	1.5	4	5.2	7*	11.6	249	125	685	98	97
TORRANCE NM	4*	15.1	1	2.8	4*	14.4	-82	419	-5	25	98
UNION NM	1	2.7			3	10.1			275	89	96
VALENCIA NM	2	3.3	2	2.0	10	7.4	-38	267	127	71	90
NEW YORK NY	2418*	3.5	3213*	4.2	3935*	5.1	21	22	48		
ALBANY NY	41	3.7	49	3.8	55	4.0	5	4	9	31	47
ALLEGANY NY	5	2.2	4	2.0	8	3.4	-9	69	54	49	37
BRONX NY	209	3.2	246	3.9	287*	5.6	22	45	78	58	75
BROOME NY	21	2.2	45	4.5	47	4.6	106	2	111	68	60
CATTARAUGUS NY	4	1.0	13	3.2	13	3.2	226	-1	222	85	34
CAYUGA NY	11	2.8	12	3.2	14	3.4	14	5	20	35	37
CHAUTAUQUA NY	21	2.9	24	3.0	20	2.5	3	-16	-14	21	24
CHEMUNG NY	8	1.9	10	2.2	18	4.0	16	78	106	67	47
CHENANGO NY	6	3.0	9	4.0	8	3.4	33	-15	13	32	36
CLINTON NY	3	1.2	4	1.9	6	2.3	57	19	87	61	20
COLUMBIA NY	6	2.1	11	3.7	16	5.0	78	34	139	74	67
CORTLAND NY	3	1.4	6	3.0	5	2.4	112	-20	71	56	23
DELAWARE NY	3	1.3	4	1.8	11	4.5	45	143	252	87	56
DUTCHESS NY	15	2.0	25	2.8	47	4.8	40	70	138	73	62
ERIE NY	197*	4.9	239*	5.2	276*	6.1	7	17	25	37	80
ESSEX NY	4	2.4	12*	6.5	12	6.4	177	-2	171	79	83
FRANKLIN NY	13*	6.3	7	3.4	15	7.2	-46	110	14	33	89
FULTON NY	8	2.7	12	4.0	24*	7.7	45	94	182	80	91
GENESEE NY	13*	5.1	14	5.4	14	5.1	7	-5	1	28	69
GREENE NY	4	2.1	8	3.9	6	2.6	85	-34	21	36	25
HAMILTON NY	1	4.0			1	3.2			-19	19	34
HERKIMER NY	15	4.3	11	2.9	16	4.4	-33	52	2	28	54
JEFFERSON NY	14	2.9	17	3.7	24	5.6	27	50	90	61	74
KINGS(BROOKLYN) NY	358	3.2	391	3.5	552*	5.6	9	61	75	57	75
LEWIS NY	1	.8	3	2.9	5	4.3	273	49	456	95	53
LIVINGSTON NY	2	1.0	4	1.9	6	2.3	87	22	128	72	21
MADISON NY	9	3.8	7	2.7	8	2.9	-27	6	-23	17	29
MONROE NY	116*	4.8	133*	4.8	118	4.1	0	-13	-14	21	49
MONTGOMERY NY	10	2.8	6	1.6	16	4.6	-41	183	68	55	60
NASSAU NY	118*	4.0	183	4.2	269	5.1	6	19	27	38	68
NEW YORK(MANHATTAN) NY	342*	4.2	583*	7.8	398*	6.7	84	-13	59	51	86
NIAGARA NY	34	3.9	49	5.0	50	4.8	28	-3	23	37	63
ONEIDA NY	42	3.4	62	4.9	57	4.5	44	-8	32	41	56
ONONDAGA NY	71*	4.2	78	4.4	90	4.7	4	7	12	32	61
ONTARIO NY	11	3.0	12	3.3	16	3.9	8	19	28	39	45
ORANGE NY	30	3.6	45	4.8	63	6.0	32	26	67	54	80
ORLEANS NY	9	4.9	2	1.2	8	4.4	-76	279	-9	24	55
OSWEGO NY	13	3.2	21	4.9	18	3.8	53	-22	19	35	44
OTSEGO NY	6	2.2	9	3.0	15	5.0	37	69	131	72	67
PUTNAM NY	4	2.6	10	5.2	14	5.2	96	1	98	64	70
QUEENS NY	238	3.3	309	3.5	560*	5.9	6	69	80	59	79
RENSSELAER NY	21	3.0	24	3.4	43	6.1	15	78	104	66	80
RICHMOND NY	38	4.2	45	4.1	66	5.3	-2	29	27	38	71
ROCKLAND NY	12	2.7	28	4.5	34	4.3	64	-4	57	50	52
ST LAWRENCE NY	6	1.2	11	2.3	13	2.8	91	21	131	72	27
SARATOGA NY	5	1.3	16	3.7	27	5.0	188	36	292	90	67
SCHENECTADY NY	28	3.8	33	4.3	33	4.2	14	-4	10	31	51
SCHOHARIE NY	4	2.9	4	2.9	3	1.9	0	-35	-35	11	16
SCHUYLER NY	2	2.4	2	2.6	3	3.3	8	30	41	44	36
SENECA NY	5	2.7	1	.6	9	5.2	-79	826	93	63	69
STEUBEN NY	21	4.3	20	3.8	23	4.6	-11	19	6	29	58
SUFFOLK NY	52	2.6	103	3.2	187	4.5	25	39	75	57	57
SULLIVAN NY	8	3.1	10	3.6	22	6.5	16	83	112	68	84
TIOGA NY	3	1.8	9	5.7	9	5.1	214	-10	184	81	68
TOMPKINS NY	5	2.1	7	2.9	11	4.2	39	46	104	66	51
ULSTER NY	20	3.6	17	2.6	23	3.1	-28	18	-15	21	32
WARREN NY	7	3.1	8	3.7	12	4.8	21	30	57	50	63
WASHINGTON NY	5	1.9	6	2.4	10	3.9	22	65	102	65	45
WAYNE NY	4	1.1	7	2.0	15	4.1	87	101	275	89	48
WESTCHESTER NY	124*	4.2	167*	4.6	171	4.3	10	-5	4	29	54
WYOMING NY	7	3.9	5	2.8	12	6.6	-28	134	69	55	85
YATES NY	2	1.8	1	.8	3	2.6	-53	211	46	46	24
NORTH CAROLINA NC	352	3.4	379	2.8	665	3.8	-17	38	14		
ALAMANCE NC	6	3.4	2	.9	8	2.2	-72	136	-33	12	20
ALEXANDER NC	1	1.8	1	1.4	3	3.9	-23	181	117	69	45
ALLEGHANY NC	1	2.5	1	2.4	3	5.6	-3	135	129	72	75
ANSON NC	1	1.8	2	3.3			85				
ASHE NC			3	3.2	6	5.0		59			67

* SIGNIFICANTLY MORE DEATHS THAN EXPECTED

WHITE MALE

CANCER OF THE LIVER AND GALLBLADDER INCLUDING BILE DUCTS

ICD 155,156

1950-1959 DEATHS	1950-1959 RATE	1960-1969 DEATHS	1960-1969 RATE	1970-1979 DEATHS	1970-1979 RATE	COUNTY AND STATE		PERCENT CHANGE 60S/50S	PERCENT CHANGE 70S/60S	PERCENT CHANGE 70S/50S	PERCENTILE %CHANGE 70S/50S	PERCENTILE RATES 1970S
		2	3.5	4	6.3	AVERY	NC		82			82
1	.6	2	1.6	8	5.6	BEAUFORT	NC	155	253	801	99	75
2	4.6	1	1.9			BERTIE	NC	-58				
				4	4.5	BLADEN	NC					56
		4	6.0			BRUNSWICK	NC					
18	4.1	25	4.2	35	4.9	BUNCOMBE	NC	2	16	19	35	65
3	1.3	2	1.2	5	2.0	BURKE	NC	-13	72	50	48	18
4	1.9	11	4.4	10	3.4	CABARRUS	NC	135	-24	79	59	36
3	2.4	7	5.0	7	3.3	CALDWELL	NC	105	-34	35	42	35
2*	14.1	1	4.8	1	4.6	CAMDEN	NC	-66	-5	-67	02	59
6*	6.7	3	2.4	3	2.1	CARTERET	NC	-64	-13	-69	01	18
1	2.1	1	2.0	1	1.8	CASWELL	NC	-7	-6	-12	22	16
7	3.3	2	.7	9	2.3	CATAWBA	NC	-80	251	-30	14	21
5	7.0			4	3.4	CHATHAM	NC			-51	07	37
3	4.0	2	1.9	6	6.3	CHEROKEE	NC	-51	225	59	51	83
		1	2.9	1	2.1	CHOWAN	NC		-28			19
		1	2.6	2	4.7	CLAY	NC		80			61
8	4.9	3	1.6	14	5.7	CLEVELAND	NC	-68	263	15	33	76
1	.9	5	4.0	11	7.4	COLUMBUS	NC	320	86	681	98	90
4	4.1	3	3.1	2	2.1	CRAVEN	NC	-24	-33	-49	07	19
2	1.4	6	3.3	12	3.2	CUMBERLAND	NC	140	-2	135	73	34
		1	3.3	2	4.5	CURRITUCK	NC		36			58
		1	3.2	3	4.7	DARE	NC		49			62
8	4.6	4	1.3	5	1.3	DAVIDSON	NC	-72	-4	-73	01	12
2	4.0	1	1.4	2	1.9	DAVIE	NC	-64	33	-53	06	17
3	3.1	2	2.4	7	5.4	DUPLIN	NC	-24	127	71	56	72
10	4.8	9	3.1	9	2.4	DURHAM	NC	-36	-22	-50	07	22
1	1.1	3	2.9	4	3.2	EDGECOMBE	NC	161	8	181	80	33
10	2.6	17	3.6	33	4.9	FORSYTH	NC	39	37	91	62	65
2	2.7	5	6.9	1	1.2	FRANKLIN	NC	153	-82	-55	05	12
9	2.8	17	4.4	29	5.9	GASTON	NC	56	35	109	67	79
				1	5.7	GATES	NC					77
		2	6.0	4	10.7	GRAHAM	NC		79			96
1	2.9	1	1.5	3	3.4	GRANVILLE	NC	-47	122	18	35	36
1	2.9	1	2.3	1	2.4	GREENE	NC	-21	7	-15	20	22
24*	6.0	28	4.0	51	5.4	GUILFORD	NC	-33	36	-9	24	73
3	2.7	3	3.1	6	4.3	HALIFAX	NC	15	39	60	52	54
4	3.1	3	2.2	3	1.8	HARNETT	NC	-28	-18	-41	09	16
5	3.3	6	3.1	12	5.0	HAYWOOD	NC	-9	63	49	47	66
6	4.0	12	5.8	20	7.1	HENDERSON	NC	45	21	76	57	88
1	3.8			1	1.6	HERTFORD	NC			-57	04	15
		1	3.1			HOKE	NC					
2	9.1	2	10.4	4*	21.6	HYDE	NC	14	108	138	74	99
8	4.2	5	2.3	11	3.7	IREDELL	NC	-45	59	-12	22	42
4	5.0	1	1.4	1	.9	JACKSON	NC	-71	-34	-81	00	10
11*	7.4	3	1.2	7	3.3	JOHNSTON	NC	-83	166	-55	05	35
2	8.6	1	4.1	1	4.1	JONES	NC	-52	0	-52	06	49
3	4.0	3	3.6	3	2.5	LEE	NC	-9	-31	-37	10	23
5	6.5	2	1.7	6	3.8	LENOIR	NC	-74	128	-42	09	44
2	2.6			7	4.9	LINCOLN	NC			87	61	65
1	1.2	2	1.4	5	3.4	MC DOWELL	NC	23	140	194	82	37
2	2.6	1	1.0	7	5.0	MACON	NC	-62	392	89	61	66
2	2.6	2	2.8	1	1.0	MADISON	NC	9	-64	-60	03	11
		2	2.8	2	2.7	MARTIN	NC		-1			26
22	4.6	29	4.0	38	4.2	MECKLENBURG	NC	-12	4	-9	24	50
3	5.5	2	3.1	4	4.9	MITCHELL	NC	-44	59	-11	23	64
1	1.7			1	1.3	MONTGOMERY	NC			-24	16	12
5	6.0	3	2.9	6	2.9	MOORE	NC	-52	3	-51	07	30
4	3.3	4	2.8	6	4.2	NASH	NC	-15	49	26	38	51
3	1.9	8	3.2	17	6.1	NEW HANOVER	NC	64	93	216	84	81
				1	1.6	NORTHAMPTON	NC					14
4	6.3	4	4.4	7	4.7	ONSLOW	NC	-30	8	-25	16	62
3	3.1			4	2.9	ORANGE	NC			-6	25	29
2	6.5	1	2.7	2	5.0	PAMLICO	NC	-58	83	-22	17	67
2	3.8	1	1.3	3	3.5	PASQUOTANK	NC	-67	178	-8	24	39
1	2.5	2	4.4	1	1.4	PENDER	NC	78	-68	-43	09	13
		1	3.2	1	2.7	PERQUIMANS	NC		-17			26
				3	3.7	PERSON	NC					41
5	5.9	1	.6	7	3.2	PITT	NC	-90	435	-46	08	34
		2	3.4	2	3.6	POLK	NC		5			40
9	4.6	7	2.6	15	4.6	RANDOLPH	NC	-43	78	1	27	59
3	3.9	5	4.1	5	4.0	RICHMOND	NC	6	-4	2	28	46
4	2.8	6	4.2	7	3.9	ROBESON	NC	48	-9	36	42	45
8	5.1	7	3.1	13	4.3	ROCKINGHAM	NC	-40	40	-15	20	54
7	2.8	12	3.8	9	2.4	ROWAN	NC	33	-37	-17	20	22
5	3.2	6	3.3	3	1.5	RUTHERFORD	NC	3	-53	-52	06	14
		2	2.0	7	4.4	SAMPSON	NC		114			54
				4	7.3	SCOTLAND	NC					89
7	5.5	6	4.0	8	4.6	STANLY	NC	-27	16	-15	20	60
2	2.4	2	2.1	1	.9	STOKES	NC	-13	-58	-63	02	10
4	2.3	3	1.4	6	2.6	SURRY	NC	-40	88	13	32	25
				1	2.2	SWAIN	NC					20
3	6.0	2	2.5	3	3.0	TRANSYLVANIA	NC	-59	22	-50	07	30
				1	6.3	TYRRELL	NC					83
3	2.5	6	3.6	8	3.5	UNION	NC	42	-3	38	43	38
1	1.2	1	1.5	1	1.3	VANCE	NC	21	-10	9	31	13
16	4.3	5	1.4	23	3.7	WAKE	NC	-68	167	-15	21	42
1	3.4					WARREN	NC					
				4	10.6	WASHINGTON	NC					96
1	1.8	2	2.3	4	3.5	WATAUGA	NC	29	52	96	63	38
3	2.3	4	2.6	9	4.1	WAYNE	NC	13	56	76	57	49
4	2.8	1	.6	6	2.8	WILKES	NC	-80	405	0	27	28
3	3.0	8	5.4	4	2.5	WILSON	NC	80	-55	-18	19	23
2	2.7	4	4.1	6	4.5	YADKIN	NC	52	9	66	54	57
				3	4.2	YANCEY	NC					51
91	3.2	111	3.7	101	3.1	NORTH DAKOTA	ND	16	-16	-2		
1	1.2	5	5.6	4	3.8	BARNES	ND	348	-32	204	83	43
2	4.0	1	2.3	1	2.1	BENSON	ND	-43	-9	-47	08	19

* SIGNIFICANTLY MORE DEATHS THAN EXPECTED

WHITE MALE

CANCER OF THE LIVER AND GALLBLADDER INCLUDING BILE DUCTS

ICD 155,156

1950-1959 DEATHS	1950-1959 RATE	1960-1969 DEATHS	1960-1969 RATE	1970-1979 DEATHS	1970-1979 RATE	COUNTY AND STATE	PERCENT CHANGE 60S/50S	PERCENT CHANGE 70S/60S	PERCENT CHANGE 70S/50S	PERCENTILE %CHANGE 70S/50S	PERCENTILE RATES 1970S
				2	6.5	BOWMAN ND					84
				2	6.9	BURKE ND					87
8*	7.2	4	2.7	3	2.1	BURLEIGH ND	-62	-24	-72	01	18
14*	4.9	13	4.3	9	2.7	CASS ND	-12	-37	-45	08	26
1	2.2			3	5.2	CAVALIER ND			134	73	69
		2	4.2	2	4.8	DICKEY ND		15			63
		1	2.8	2	7.0	DIVIDE ND		153			87
2	8.4	2	9.7	1	4.0	DUNN ND	16	-58	-52	06	48
1	3.5	1	3.4			EDDY ND	-4				
3*	10.6	1	2.6			EMMONS ND	-76				
1	4.3	1	3.2	1	3.7	FOSTER ND	-26	16	-15	21	42
				1	8.1	GOLDEN VALLEY ND					92
5	3.0	5	2.6	12	5.9	GRAND FORKS ND	-15	129	95	63	78
1	2.4	3*	12.2	2	8.4	GRANT ND	402	-32	243	87	93
2	5.9	1	2.6			GRIGGS ND	-57				
2	7.4					HETTINGER ND					
		1	3.7	2	9.0	KIDDER ND		143			94
1	2.0	3	7.3	1	1.9	LA MOURE ND	269	-75	-6	25	16
		7*	12.3	2	3.5	MC HENRY ND		-72			38
2	7.1			2	5.8	MC INTOSH ND			-18	19	77
1	2.2	2	7.7	2	5.8	MC KENZIE ND	253	-24	168	78	78
1	1.3	4	5.4	4	4.7	MC LEAN ND	308	-12	259	88	62
1	4.3	3	9.1	1	3.4	MERCER ND	110	-62	-21	18	37
3	3.4	1	1.2	1	.8	MORTON ND	-63	-32	-75	01	10
2	4.0	5*	10.0	4	6.6	MOUNTRAIL ND	153	-34	68	55	85
2	3.8	3	5.3	1	2.7	NELSON ND	38	-49	-29	14	26
		2	3.1	3	3.6	PEMBINA ND		16			40
1	3.3	2	7.1			PIERCE ND	116				
		7*	8.8	1	1.2	RAMSEY ND		-86			12
2	3.7	3	5.9	2	3.5	RANSOM ND	62	-41	-5	25	38
1	2.4	2	9.5	1	3.9	RENVILLE ND	299	-59	63	53	45
3	2.9	4	3.3	2	1.7	RICHLAND ND	14	-47	-40	10	15
2	6.2	1	3.2	1	4.6	ROLETTE ND	-49	46	-25	16	60
				2	5.0	SARGENT ND					66
1	5.0	3*	14.9	1	4.0	SHERIDAN ND	194	-73	-21	18	47
1*	37.5	1*	13.7			SIOUX ND	-63				
1	9.3					SLOPE ND					
2	3.1			2	2.4	STARK ND			-22	17	22
				1	4.2	STEELE ND					51
2	1.5	1	.8	2	1.5	STUTSMAN ND	-50	94	-2	26	13
		1	3.0	1	2.9	TOWNER ND		-3			29
3	4.6	1	1.3			TRAILL ND	-71				
3	3.0			3	2.6	WALSH ND			-14	21	24
7	4.5	8	4.2	10	4.7	WARD ND	-5	12	7	30	62
3	6.6	3	6.0	2	3.4	WELLS ND	-9	-43	-49	08	36
3	2.8	3	2.9	2	1.8	WILLIAMS ND	4	-37	-34	12	16
1093	2.9	1606*	4.0	2080*	4.9	OHIO OH	35	22	66		
3	2.3	6	4.9	5	3.6	ADAMS OH	115	-26	58	51	40
9	2.3	20	4.5	21	4.5	ALLEN OH	99	1	101	65	58
2	1.2	3	1.5	5	2.5	ASHLAND OH	28	62	108	67	23
12	3.0	18	4.1	20	4.3	ASHTABULA OH	40	5	46	46	53
4	1.9	8	3.8	23*	10.8	ATHENS OH	102	186	476	96	96
2	1.2	4	2.3	5	2.7	AUGLAIZE OH	91	17	124	71	26
15	3.2	13	2.9	20	4.3	BELMONT OH	-11	49	33	41	53
2	1.5	3	2.2	3	1.9	BROWN OH	40	-12	24	37	17
20	3.3	30	4.2	49	5.5	BUTLER OH	28	32	69	56	74
6	5.9	1	1.0	5	4.5	CARROLL OH	-84	367	-23	17	58
5	3.3	2	1.4	8	5.5	CHAMPAIGN OH	-58	299	68	55	74
15	3.0	13	2.4	24	3.9	CLARK OH	-22	66	29	40	46
6	2.4	9	3.1	16	4.8	CLERMONT OH	25	58	98	63	63
5	3.1	7	4.8	9	6.1	CLINTON OH	55	27	96	63	81
18	3.6	20	4.0	27	4.9	COLUMBIANA OH	10	24	36	42	65
1	.5	5	2.9	7	4.1	COSHOCTON OH	424	41	641	98	48
3	1.4	7	3.1	10	4.6	CRAWFORD OH	123	46	225	86	59
223*	3.8	278*	4.3	358*	5.6	CUYAHOGA OH	14	30	48	47	75
8	3.5	11	4.5	9	3.5	DARKE OH	29	-23	-1	27	39
1	.7	1	.8	5	3.5	DEFIANCE OH	14	333	393	94	39
5	3.0	1	.6	13	6.8	DELAWARE OH	-80	1019	129	72	86
8	2.5	13	3.9	17	5.2	ERIE OH	56	32	106	66	69
5	1.7	12	3.8	10	2.8	FAIRFIELD OH	118	-27	60	52	27
2	1.7	6	4.8	4	3.1	FAYETTE OH	162	-36	81	59	31
72	3.5	107*	4.5	125	4.9	FRANKLIN OH	29	10	42	45	66
		2	1.4	6	3.9	FULTON OH		189			46
6	5.2	5	3.7	6	4.4	GALLIA OH	-29	19	-15	20	55
2	1.8	6	3.6	9	3.9	GEAUGA OH	99	7	114	68	44
3	1.2	8	3.0	14	4.2	GREENE OH	140	41	237	86	50
5	2.2	8	3.3	10	4.3	GUERNSEY OH	51	31	98	63	53
124*	4.3	175*	5.4	179	5.4	HAMILTON OH	26	0	26	38	72
1	.4	4	1.6	10	3.6	HANCOCK OH	254	135	733	99	41
4	2.6	2	1.4	5	3.2	HARDIN OH	-47	133	23	37	33
5	4.6	3	3.0	6	6.4	HARRISON OH	-35	114	38	43	83
3	2.5	4	3.0	9	6.9	HENRY OH	19	133	179	80	87
3	1.7	5	2.9	9	5.3	HIGHLAND OH	71	82	212	84	70
2	1.8	5	5.2	9	7.9	HOCKING OH	189	53	344	92	92
2	2.3	8*	7.8	2	1.9	HOLMES OH	241	-75	-15	20	17
1	.4	8	3.7	19*	8.7	HURON OH	798	133	1993	99	94
2	1.2	5	3.4	6	3.8	JACKSON OH	183	14	222	85	44
11	2.5	22	4.8	33*	7.2	JEFFERSON OH	96	49	192	82	89
6	3.1	2	1.0	3	1.5	KNOX OH	-67	42	-53	06	13
6	1.6	14	2.8	27	4.0	LAKE OH	74	43	149	76	47
8	3.5	15*	6.2	13	5.0	LAWRENCE OH	75	-19	41	45	66
8	2.0	20	4.6	24	5.0	LICKING OH	133	8	152	76	66
3	1.8	2	1.1	10	5.1	LOGAN OH	-40	369	179	80	68
24	3.2	28	3.7	45	5.0	LORAIN OH	15	35	55	50	66
58	3.3	72	3.7	93	4.8	LUCAS OH	13	29	46	46	63
		6	5.6	11*	9.1	MADISON OH		63			95
31	2.4	74*	5.8	83*	6.3	MAHONING OH	138	10	162	77	83
4	1.6	7	2.8	10	3.9	MARION OH	73	39	139	74	45

* SIGNIFICANTLY MORE DEATHS THAN EXPECTED

WHITE MALE

CANCER OF THE LIVER AND GALLBLADDER INCLUDING BILE DUCTS

ICD 155,156

1950-1959 DEATHS	1950-1959 RATE	1960-1969 DEATHS	1960-1969 RATE	1970-1979 DEATHS	1970-1979 RATE	COUNTY AND STATE	PERCENT CHANGE 60S/50S	PERCENT CHANGE 70S/60S	PERCENT CHANGE 70S/50S	PERCENTILE %CHANGE 70S/50S	PERCENTILE RATES 1970S
2	.8	13	4.9	10	3.3	MEDINA OH	481	-32	294	90	35
6	4.9	1	.7	3	2.5	MEIGS OH	-86	252	-50	07	23
3	1.9	4	2.4	4	2.4	MERCER OH	30	-2	27	38	22
8	2.8	10	3.0	17	4.7	MIAMI OH	9	54	69	55	60
1	1.1	3	3.3	3	2.9	MONROE OH	199	-10	169	78	29
52	3.3	81	4.5	92	4.6	MONTGOMERY OH	37	2	41	44	59
2	2.8	4	5.5	1	1.2	MORGAN OH	95	-77	-56	05	12
2	1.9			6	5.5	MORROW OH			186	81	74
9	2.4	11	3.0	14	3.7	MUSKINGUM OH	24	25	55	50	42
3	4.4	3	4.2	2	3.0	NOBLE OH	-6	-29	-33	12	30
4	2.5	5	3.0	7	3.7	OTTAWA OH	22	23	50	48	42
1	1.6	5	6.6	5	6.1	PAULDING OH	322	-8	289	90	80
3	1.6	5	3.3	8	5.3	PERRY OH	101	63	227	86	71
13*	8.7	3	2.0	8	4.9	PICKAWAY OH	-77	142	-43	09	65
2	2.7			3	3.1	PIKE OH			13	32	32
9	3.0	7	1.9	18	4.4	PORTAGE OH	-36	132	49	48	55
1	.8	2	1.4	4	2.5	PREBLE OH	74	74	204	83	23
3	2.4	5	3.7	10	7.3	PUTNAM OH	51	101	204	83	89
15	3.6	17	3.5	20	3.8	RICHLAND OH	-3	9	6	29	44
5	2.0	13	4.5	14	4.5	ROSS OH	128	0	128	72	58
6	2.5	8	3.1	14	5.1	SANDUSKY OH	26	64	107	67	68
10	2.9	22	5.5	17	4.0	SCIOTO OH	94	-28	41	44	47
8	3.2	7	2.6	11	4.2	SENECA OH	-18	60	31	41	50
1	.7	6	4.0	8	4.9	SHELBY OH	479	21	600	98	64
42	3.1	53	3.4	72	4.5	STARK OH	11	30	44	46	56
38	2.1	80	4.1	91	4.2	SUMMIT OH	96	3	102	65	51
27	3.7	58*	7.0	56	5.8	TRUMBULL OH	88	-17	56	50	78
8	2.1	15	3.9	21	5.2	TUSCARAWAS OH	87	34	150	76	69
5	4.1	5	4.5	4	3.7	UNION OH	11	-17	-9	24	42
2	1.2	4	2.6	5	3.2	VAN WERT OH	107	26	159	77	34
1	2.7	2	3.2	3	5.4	VINTON OH	20	70	104	66	73
6	3.4	6	2.6	13	4.5	WARREN OH	-23	73	33	41	56
4	1.5	11	4.2	20*	7.2	WASHINGTON OH	177	71	372	93	88
6	1.9	9	2.7	15	4.1	WAYNE OH	41	51	112	68	49
2	1.3	4	2.5	6	3.8	WILLIAMS OH	96	53	200	83	43
7	2.5	8	2.5	16	4.6	WOOD OH	4	80	86	61	58
2	2.0	8*	7.2	10*	9.0	WYANDOT OH	270	24	358	93	94
206	2.1	338	3.1	519	4.2	OKLAHOMA OK	47	37	102		
1	1.6	2	3.3	5	6.4	ADAIR OK	109	97	311	91	84
6*	9.0	2	3.3	1	1.8	ALFALFA OK	-63	-47	-80	00	16
2	2.6	2	3.5	1	1.4	ATOKA OK	36	-62	-48	08	13
		1	.8	3	2.3	BECKHAM OK		190			21
1	1.5	2	2.6	1	1.2	BLAINE OK	68	-54	-23	17	11
2	1.7	8	5.3	7	4.1	BRYAN OK	223	-24	146	75	48
1	.6	3	1.8	16*	9.9	CADDO OK	194	446	1505	99	96
2	1.5	5	3.8	6	3.7	CANADIAN OK	148	-4	138	73	41
3	1.9	10	5.0	6	3.0	CARTER OK	165	-40	58	51	30
		4	4.0	6	5.4	CHEROKEE OK		33			72
		2	2.0	4	4.2	CHOCTAW OK		117			52
1	4.6					CIMARRON OK					
2	1.3	6	2.5	6	2.1	CLEVELAND OK	101	-17	66	54	19
2	5.5	1	2.5	2	4.2	COAL OK	-55	73	-23	17	52
1	.6	11	6.0	11	4.4	COMANCHE OK	971	-27	684	98	55
2	3.9	3	5.7	1	1.9	COTTON OK	44	-67	-52	06	17
		1	1.0	4	4.2	CRAIG OK		330			51
5	2.3	9	4.6	11	4.8	CREEK OK	98	4	106	67	63
1	1.0	2	1.7			CUSTER OK	71				
1	1.3	3	2.6	6	5.2	DELAWARE OK	103	101	308	91	69
1	2.3			1	1.9	DEWEY OK			-19	19	16
2	5.5			1	2.2	ELLIS OK			-59	04	20
1	.4	3	1.2	11	3.9	GARFIELD OK	167	235	794	99	45
3	2.1	1	.7	9	5.5	GARVIN OK	-67	679	157	77	73
3	1.8	4	2.3	5	2.7	GRADY OK	29	17	51	48	26
2	3.0	3	5.0	2	3.3	GRANT OK	69	-34	11	31	35
2	3.1	2	4.0	11*	14.3	GREER OK	27	259	355	92	98
1	2.4	1	2.6	1	2.5	HARMON OK	8	-3	4	29	24
				1	3.0	HARPER OK					31
		1	1.7	6	9.3	HASKELL OK		456			95
2	1.9	2	2.9	6	6.2	HUGHES OK	47	117	220	85	82
		2	1.8	7	6.0	JACKSON OK		232			79
1	1.6	3	4.4			JEFFERSON OK	174				
		3	4.9	2	2.7	JOHNSTON OK		-44			26
4	1.7	8	3.0	13	4.6	KAY OK	76	55	173	79	59
1	1.4	1	1.6	2	2.6	KINGFISHER OK	13	61	82	59	25
2	2.3			2	2.1	KIOWA OK			-8	24	19
		3	6.8	3	5.5	LATIMER OK		-19			74
6	3.6	6	3.7	10	5.1	LE FLORE OK	2	38	42	45	68
2	1.9	8	6.5	7	4.5	LINCOLN OK	251	-30	145	75	58
3	3.2	1	.9	3	2.6	LOGAN OK	-72	191	-19	19	25
		1	2.9	3	6.6	LOVE OK		131			85
1	2.1					MC CLAIN OK					
2	1.7	3	2.3	7	4.7	MC CURTAIN OK	37	99	173	79	60
		1	1.5	6	7.5	MC INTOSH OK		394			90
		1	2.1	3	4.8	MAJOR OK		131			63
		2	2.9	4	7.6	MARSHALL OK		159			90
2	1.8	5	3.7	2	1.4	MAYES OK	101	-61	-22	17	13
1	1.2	1	1.0	4	4.8	MURRAY OK	-17	365	288	90	63
7	3.0	7	2.7	16	5.8	MUSKOGEE OK	-12	116	91	62	77
1	1.7	1	1.6	8*	12.3	NOBLE OK	-2	658	646	98	97
				2	2.8	NOWATA OK					27
		3	5.1	2	2.7	OKFUSKEE OK		-47			26
46	3.5	62	3.9	69	3.5	OKLAHOMA OK	12	-11	-1	27	38
5	2.7	5	2.8	10	4.9	OKMULGEE OK	6	72	83	60	64
4	2.6	3	1.9	5	3.0	OSAGE OK	-28	61	15	33	30
4	2.8	4	2.9	6	3.7	OTTAWA OK	3	28	33	41	41
2	2.7	2	2.3	3	3.5	PAWNEE OK	-16	57	32	41	39
4	2.4	3	1.5	6	2.5	PAYNE OK	-38	66	2	28	23
8	4.2	13*	6.5	21*	10.1	PITTSBURG OK	55	56	142	74	96

* SIGNIFICANTLY MORE DEATHS THAN EXPECTED

1950-1959 DEATHS	1950-1959 RATE	1960-1969 DEATHS	1960-1969 RATE	1970-1979 DEATHS	1970-1979 RATE	COUNTY AND STATE		PERCENT CHANGE 60S/50S	PERCENT CHANGE 70S/60S	PERCENT CHANGE 70S/50S	PERCENTILE %CHANGE 70S/50S	PERCENTILE RATES 1970S
4	2.6	7	4.7	9	5.7	PONTOTOC	OK	79	20	115	69	76
5	2.2	3	1.2	10	3.8	POTTAWATOMIE	OK	-45	209	70	56	44
		6*	9.4	4	5.2	PUSHMATAHA	OK		-44			70
				1	2.4	ROGER MILLS	OK					23
1	1.0	3	2.5	8	5.7	ROGERS	OK	149	125	461	96	76
2	1.3	6	4.5	7	4.6	SEMINOLE	OK	260	2	266	88	59
2	2.1	2	2.1	7	6.1	SEQUOYAH	OK	-2	196	189	81	80
3	1.8	2	1.0	9	4.0	STEPHENS	OK	-45	297	119	70	47
1	1.8	2	2.6	1	1.5	TEXAS	OK	45	-42	-15	20	14
1	1.1			1	1.2	TILLMAN	OK			9	30	12
28	2.8	48	4.0	61	4.2	TULSA	OK	44	4	49	47	50
		1	1.1	4	3.0	WAGONER	OK		172			31
1	.7	8	5.3	11	4.9	WASHINGTON	OK	655	-7	606	98	65
		2	3.0	3	3.1	WASHITA	OK		4			33
1	1.4	1	1.3	2	3.1	WOODS	OK	-9	141	119	69	31
1	1.1	1	1.4	4	4.4	WOODWARD	OK	30	222	319	91	55
199	2.6	318	3.4	390	3.5	OREGON	OR	33	2	36		
1	.9	6	6.1	6	6.1	BAKER	OR	548	0	545	97	81
3	2.3	5	3.6	2	1.1	BENTON	OR	52	-70	-54	05	11
9	1.8	16	2.6	14	1.8	CLACKAMAS	OR	42	-33	-4	26	16
5	2.7	4	2.2	8	4.1	CLATSOP	OR	-17	82	52	48	48
4	2.8	9	6.3	3	1.9	COLUMBIA	OR	122	-69	-32	13	17
6	2.4	7	2.7	20	6.6	COOS	OR	13	142	174	79	85
				2	2.9	CROOK	OR					29
		1	4.6	1	.8	CURRY	OR		-83			10
2	1.7	6	4.4	7	3.3	DESCHUTES	OR	159	-23	99	64	36
5	2.4	13	4.4	18	4.5	DOUGLAS	OR	79	3	84	60	57
1	7.4	1	10.6			GILLIAM	OR	43				
		2	4.7	2	4.5	GRANT	OR		-4			58
		1	2.4			HARNEY	OR					
2	2.8			3	3.5	HOOD RIVER	OR			26	38	38
9	2.6	15	3.3	23	4.0	JACKSON	OR	24	22	52	48	47
		2	7.3			JEFFERSON	OR					
1	.6	8	4.2	9	3.3	JOSEPHINE	OR	617	-22	463	96	35
3	1.6	6	2.7	10	3.9	KLAMATH	OR	64	47	140	74	46
		2	6.9	1	2.7	LAKE	OR		-61			26
16	2.5	30	3.7	31	3.1	LANE	OR	49	-17	24	37	31
1	.6	2	1.1	8	3.9	LINCOLN	OR	84	255	553	97	45
5	2.1	7	2.3	14	3.7	LINN	OR	9	61	76	57	42
3	3.2	2	2.0	3	2.3	MALHEUR	OR	-38	16	-28	15	21
8	1.3	28	4.1	19	2.2	MARION	OR	204	-46	65	54	20
		1	3.4	1	4.9	MORROW	OR		42			64
85	3.5	106	3.9	108	4.0	MULTNOMAH	OR	10	3	14	33	47
1	.7	4	2.6	9	4.4	POLK	OR	266	69	519	97	56
1	9.0					SHERMAN	OR					
3	3.5	4	4.9	6	4.9	TILLAMOOK	OR	40	-1	39	43	64
3	2.4	6	2.6	12	4.5	UMATILLA	OR	12	72	92	62	58
3	2.8	2	2.0	3	2.5	UNION	OR	-30	25	-12	22	23
				2	5.0	WALLOWA	OR					66
5	6.4	5	5.7	4	3.5	WASCO	OR	-11	-39	-46	08	38
7	2.1	10	2.0	30	4.2	WASHINGTON	OR	-5	110	99	64	51
3*	19.5			1	9.0	WHEELER	OR			-54	05	94
4	2.0	7	3.3	10	3.9	YAMHILL	OR	67	19	99	64	46
1322	2.8	1822	3.6	2574	4.8	PENNSYLVANIA	PA	27	33	69		
2	1.0	8	3.3	13	4.5	ADAMS	PA	236	35	354	92	56
210	3.3	250	3.5	419*	5.9	ALLEGHENY	PA	7	65	77	58	78
10	2.7	10	2.6	25	5.9	ARMSTRONG	PA	-2	124	120	70	79
27	3.5	42	4.8	54	5.5	BEAVER	PA	39	14	58	51	74
1	.5	4	1.9	7	3.1	BEDFORD	PA	277	64	517	97	33
41	3.1	52	3.5	76	4.7	BERKS	PA	15	32	51	48	61
23	3.2	18	2.5	22	3.0	BLAIR	PA	-23	22	-6	25	31
4	1.3	7	2.6	15	5.0	BRADFORD	PA	94	95	278	89	67
18	2.4	41	3.6	49	3.3	BUCKS	PA	48	-8	36	42	35
9	1.9	18	3.3	29	4.7	BUTLER	PA	72	42	145	75	61
24	2.6	31	3.3	51	5.2	CAMBRIA	PA	27	59	101	65	70
9	2.9	13	4.9	18	5.6	CARBON	PA	68	13	91	62	75
4	1.5	10	3.7	15	4.8	CENTRE	PA	144	32	224	85	63
24	3.2	26	3.0	33	3.0	CHESTER	PA	-4	-1	-4	26	31
3	1.6	4	2.1	4	2.0	CLARION	PA	31	-1	29	39	18
12	3.0	12	2.9	27	6.4	CLEARFIELD	PA	-2	122	116	69	84
5	2.9	5	2.7	8	4.3	CLINTON	PA	-10	62	47	47	53
9	3.5	10	3.6	18	5.9	COLUMBIA	PA	2	66	69	56	79
7	1.8	13	3.3	18	4.2	CRAWFORD	PA	79	28	129	72	52
8	1.6	16	3.2	25	3.6	CUMBERLAND	PA	97	13	121	70	41
20	2.3	26	2.7	36	3.6	DAUPHIN	PA	21	31	58	51	40
47	2.7	86	3.9	102	4.1	DELAWARE	PA	44	6	53	49	50
3	1.8	10	5.8	8	4.2	ELK	PA	229	-29	135	73	50
30	3.0	42	4.0	70	6.1	ERIE	PA	35	53	106	67	80
34	4.2	45*	5.3	47	5.6	FAYETTE	PA	27	5	33	41	74
				1	2.4	FOREST	PA					22
13	3.6	15	3.8	40*	8.0	FRANKLIN	PA	6	112	125	71	92
1	2.2			3	5.3	FULTON	PA			142	75	71
4	1.7	10	4.6	13	5.9	GREENE	PA	167	28	240	86	78
5	2.8	9	5.1	10	5.0	HUNTINGDON	PA	79	0	78	58	67
4	1.1	13	3.5	13	3.2	INDIANA	PA	210	-7	189	81	34
4	1.5	8	3.5	22*	8.4	JEFFERSON	PA	130	143	460	96	93
2	2.7	1	1.3	7	8.0	JUNIATA	PA	-53	534	196	82	92
15	1.2	37	3.1	56	4.3	LACKAWANNA	PA	165	39	270	88	52
25	2.2	36	2.8	47	3.1	LANCASTER	PA	23	11	37	43	32
20	3.8	17	3.2	25	4.5	LAWRENCE	PA	-15	41	20	35	58
9	2.3	12	2.8	18	3.5	LEBANON	PA	22	25	52	48	39
26	2.5	35	3.0	39	3.1	LEHIGH	PA	16	3	20	35	31
46	2.6	56	3.0	98	5.1	LUZERNE	PA	17	70	99	64	68
12	2.4	19	3.5	31	5.5	LYCOMING	PA	48	56	131	72	73
8	2.8	7	2.6	12	4.2	MC KEAN	PA	-8	62	49	47	52
12	2.3	20	3.5	21	3.3	MERCER	PA	54	-6	45	46	36

* SIGNIFICANTLY MORE DEATHS THAN EXPECTED

CANCER OF THE LIVER AND GALLBLADDER INCLUDING BILE DUCTS

1950-1959 DEATHS	1950-1959 RATE	1960-1969 DEATHS	1960-1969 RATE	1970-1979 DEATHS	1970-1979 RATE	COUNTY AND STATE	PERCENT CHANGE 60S/50S	PERCENT CHANGE 70S/60S	PERCENT CHANGE 70S/50S	PERCENTILE %CHANGE 70S/50S	PERCENTILE RATES 1970S
4	2.4	7	3.8	13	6.4	MIFFLIN PA	56	70	165	78	84
2	1.1	8	3.6	12	4.0	MONROE PA	232	11	270	88	47
37	2.1	70	3.0	131	4.5	MONTGOMERY PA	46	51	120	70	58
1	1.2	1	1.0	1	1.0	MONTOUR PA	-12	1	-11	23	11
30	3.2	48	4.9	47	4.5	NORTHAMPTON PA	50	-7	39	44	57
21	3.9	26	4.8	31	5.3	NORTHUMBERLAND PA	24	11	38	43	71
2	1.7			10	6.8	PERRY PA			307	90	86
258	3.4	338*	4.6	318	4.8	PHILADELPHIA PA	34	5	40	44	62
1	1.7	4	4.2	3	2.6	PIKE PA	142	-38	51	48	25
		2	2.3	8	8.8	POTTER PA		282			94
36	3.9	39	4.1	45	4.6	SCHUYLKILL PA	6	12	19	35	60
3	2.7	2	1.5	8	5.9	SNYDER PA	-43	281	116	69	78
10	2.3	14	3.3	19	4.5	SOMERSET PA	43	36	94	63	58
4*	12.2			4	9.6	SULLIVAN PA			-22	17	95
2	1.1	5	3.0	6	3.1	SUSQUEHANNA PA	173	4	184	81	32
2	1.0	5	2.8	11	5.7	TIOGA PA	177	106	471	96	76
		3	2.6	6	4.9	UNION PA		91			65
12	3.7	15	5.0	21	6.8	VENANGO PA	35	36	85	60	87
1	.5	5	1.9	9	3.6	WARREN PA	309	86	660	98	40
36	3.6	38	3.6	71*	6.5	WASHINGTON PA	0	79	80	59	85
3	1.6	4	2.2	5	2.7	WAYNE PA	33	24	65	54	26
42	3.0	60	3.5	97	5.3	WESTMORELAND PA	19	51	79	59	71
		3	3.3	7	6.3	WYOMING PA		90			83
25	2.7	31	2.8	46	3.4	YORK PA	2	22	25	38	36
83	2.3	166	4.1	231	5.2	RHODE ISLAND RI	80	27	129		
3	2.2	7	3.9	12	5.6	BRISTOL RI	78	42	153	76	75
8	1.9	19	4.0	32	5.1	KENT RI	108	29	168	78	68
6	2.4	8	3.1	12	3.9	NEWPORT RI	31	23	62	52	45
66	2.5	122	4.3	155	5.3	PROVIDENCE RI	70	23	109	67	70
		10	4.2	20	6.1	WASHINGTON RI		45			80
89	2.0	156	2.7	299	4.1	SOUTH CAROLINA SC	33	52	103		
1	1.6	5	7.3	1	1.1	ABBEVILLE SC	348	-84	-29	14	11
2	1.1	10	4.8	16	5.3	AIKEN SC	339	11	386	94	71
5	1.8	6	1.5	14	3.1	ANDERSON SC	-14	104	75	57	33
1	3.2	3	8.1	1	3.2	BAMBEPG SC	155	-60	2	28	34
		1	3.0			BARNWELL SC					
1	3.5	3	7.8	3	2.7	BEAUFORT SC	124	-65	-22	18	26
		2	1.3	4	3.9	BERKELEY SC		200			45
				4*	19.4	CALHOUN SC					99
13	5.1	14	3.4	21	4.4	CHARLESTON SC	-34	31	-13	22	56
		1	.7	3	2.5	CHEROKEE SC		251			24
2	2.7	3	3.7	5	5.5	CHESTER SC	34	49	100	65	74
1	1.4	4	4.9	4	3.7	CHESTERFIELD SC	254	-25	167	78	42
1	3.2	3	7.3	1	1.8	CLARENDON SC	126	-76	-46	08	16
2	3.7	2	3.0	2	3.3	COLLETON SC	-19	11	-10	23	36
		1	1.6	3	2.4	DARLINGTON SC		50			23
		1	1.3	4	5.3	DILLON SC		292			71
				5	7.5	DORCHESTER SC					90
				1	2.3	EDGEFIELD SC					21
				2	4.6	FAIRFIELD SC					60
3	1.7	5	3.8	10	4.1	FLORENCE SC	121	9	141	74	49
2	4.2			6	7.1	GEORGETOWN SC			70	56	88
12	3.2	21	3.3	42	5.1	GREENVILLE SC	3	54	59	51	69
3	2.6	1	.7	9	5.5	GREENWOOD SC	-73	700	112	68	73
				2	3.9	HAMPTON SC					45
1	.8	3	1.5	7	2.6	HORRY SC	93	69	226	86	24
2*	12.6			2	8.2	JASPER SC			-35	12	93
1	1.6	2	2.1	3	3.1	KERSHAW SC	33	47	95	63	32
3	3.2	3	2.2	2	1.2	LANCASTER SC	-30	-45	-61	03	12
1	.6	1	.9	7	4.3	LAURENS SC	39	399	592	98	53
1	3.2	2	5.9	2	4.7	LEE SC	86	-21	47	46	61
2	1.5	7	2.9	15	4.9	LEXINGTON SC	98	71	239	86	65
				1	4.9	MC CORMICK SC					65
				5	9.0	MARION SC					94
3	5.4	2	7.6	2	3.1	MARLBORO SC	41	-59	-42	09	33
				2	1.7	NEWBERRY SC					15
		4	2.7	4	2.7	OCONEE SC		1			27
5	4.7	5	4.6	5	2.7	ORANGEBURG SC	-3	-42	-43	09	26
3	2.5	7	3.4	10	4.9	PICKENS SC	36	41	92	62	64
3	1.2	11	2.7	16	2.9	RICHLAND SC	132	7	148	76	29
				2	3.8	SALUDA SC					44
10	2.4	12	2.5	26	4.1	SPARTANBURG SC	2	64	67	54	48
2	2.2	3	2.4	5	3.5	SUMTER SC	11	46	62	53	39
				4	4.3	UNION SC					53
1	2.8			1	2.6	WILLIAMSBURG SC			-5	25	25
2	1.4	8	4.4	15	5.7	YORK SC	221	30	318	91	76
67	2.1	118	3.4	122	3.5	SOUTH DAKOTA SD	62	2	65		
		1	4.2	1	4.0	AURORA SD		-5			47
		1	.8	7	6.4	BEADLE SD		659			83
1	2.0	2	3.2	1	1.5	BON HOMME SD	60	-54	-25	16	14
		2	2.1	7	7.3	BROOKINGS SD		253			89
4	2.4	2	1.7	3	1.7	BROWN SD	-30	-1	-31	13	15
1	3.1			1	3.1	BRULE SD			0	27	31
1	3.1	1	2.2	2	3.8	BUTTE SD	-28	68	21	36	43
3	4.4			1	1.7	CHARLES MIX SD			-61	03	15
1	3.5	2	4.5	2	4.5	CLARK SD	28	0	28	39	58
		3	6.2	2	4.3	CLAY SD		-31			53
		1	1.2	2	1.8	CODINGTON SD		48			16
				2	11.6	CORSON SD					97
		1	3.6			CUSTER SD					
3	3.4	2	2.2	2	2.5	DAVISON SD	-34	13	-26	16	24
		2	2.6	3	3.7	DAY SD		43			41
		1	2.5	2	3.8	DEUEL SD		54			44
1	5.2	2*	26.1	1	7.5	DEWEY SD	400	-71	45	46	90

* SIGNIFICANTLY MORE DEATHS THAN EXPECTED

1950-1959		1960-1969		1970-1979		COUNTY AND STATE		PERCENT CHANGE			PERCENTILE	
											%CHANGE	RATES
DEATHS	RATE	DEATHS	RATE	DEATHS	RATE			60S/50S	70S/60S	70S/50S	70S/50S	1970S
		1	3.7	2	7.3	DOUGLAS	SD		95			89
		2	5.5	1	3.0	EDMUNDS	SD		-46			30
1	.9	6	5.9	5	6.3	FALL RIVER	SD	548	7	595	98	83
1	2.4	2	5.2	1	3.5	GREGORY	SD	121	-33	49	47	39
3	6.7	1	2.0	1	2.4	HAMLIN	SD	-70	19	-64	02	23
1	4.3	1	2.9	1	2.9	HAND	SD	-33	-2	-34	12	28
2	8.3					HANSON	SD					
1	2.1	1	2.1	2	3.6	HUGHES	SD	-1	71	71	56	40
1	1.7	3	5.2			HUTCHINSON	SD	208				
		2	7.2	1	3.4	JERAULD	SD		-53			36
1	11.2	1	9.6	1	12.4	JONES	SD	-14	30	11	32	98
1	1.4			1	1.4	KINGSBURY	SD			-4	26	13
3	5.9	6*	8.2	3	4.1	LAKE	SD	38	-50	-31	13	49
2	2.5	4	5.3	5	5.5	LAWRENCE	SD	114	4	122	70	73
		1	1.2	2	2.1	LINCOLN	SD		69			19
		1	4.9	1	4.8	LYMAN	SD		-3			62
				3	7.1	MC COOK	SD					88
1	2.7	3	8.5	2	5.4	MC PHERSON	SD	216	-36	101	65	72
1	2.2	1	1.9	1	1.7	MARSHALL	SD	-16	-8	-23	17	15
1	2.8	3	5.6			MEADE	SD	102				
		1	12.5	1	12.0	MELLETTE	SD		-4			97
		1	2.6	1	3.2	MINER	SD		22			34
7	2.1	22*	6.1	18	4.1	MINNEHAHA	SD	194	-32	100	65	50
2	4.3	2	4.5	1	2.2	MOODY	SD	5	-50	-48	08	20
7	4.5	2	.7	10	4.3	PENNINGTON	SD	-84	473	-6	25	52
1	3.8			3	9.3	PERKINS	SD			146	75	95
1	6.2	1	3.5			POTTER	SD	-44				
1	1.6	1	1.3	3	5.3	ROBERTS	SD	-18	314	241	86	71
		2	7.2			SANBORN	SD					
1*	56.2					SHANNON	SD					
		6*	9.1	2	3.5	SPINK	SD		-62			38
		2*	17.0			STANLEY	SD					
				1	8.1	SULLY	SD					92
2	5.0	2	4.1	3	7.6	TRIPP	SD	-18	85	52	48	91
3	3.9	1	1.0	2	3.0	TURNER	SD	-74	190	-24	17	30
2	3.2	4	6.1	2	3.2	UNION	SD	93	-48	1	27	34
1	2.8	3	6.3			WALWORTH	SD	122				
4	4.2	6	5.5	3	3.0	YANKTON	SD	31	-45	-27	15	31
269	2.4	430	3.2	720	4.5	TENNESSEE	TN	30	44	87		
3	2.3	2	1.2	8	2.9	ANDERSON	TN	-49	146	25	38	29
		4	3.6	4	3.1	BEDFORD	TN		-13			32
		1	1.4	4	5.2	BENTON	TN		286			70
2	6.5	2	5.0	2	5.4	BLEDSOE	TN	-22	6	-17	20	72
7	4.3	9	3.7	13	3.9	BLOUNT	TN	-14	7	-9	24	46
		2	1.2	14	6.4	BRADLEY	TN		435			84
1	.9	6	4.8	9	6.0	CAMPBELL	TN	451	26	594	98	79
1	2.6	3	6.9			CANNON	TN	165				
2	2.1	1	.7	8	6.5	CARROLL	TN	-67	814	203	83	84
2	1.5	9	4.8	8	3.4	CARTER	TN	228	-30	130	72	36
				3	3.8	CHEATHAM	TN					44
		2	4.2	3	5.7	CHESTER	TN		36			76
1	1.0	4	4.0	9	8.4	CLAIBORNE	TN	288	111	718	99	93
				3	8.1	CLAY	TN					92
2	1.9	5	4.7	2	2.3	COCKE	TN	143	-52	17	34	21
4	4.0	2	1.4	3	2.0	COFFEE	TN	-66	48	-49	07	18
2	2.8	4	6.4	2	2.3	CROCKETT	TN	133	-65	-18	19	21
2	2.7	3	3.4	4	3.1	CUMBERLAND	TN	28	-9	16	34	32
31	3.2	46	3.5	68	4.5	DAVIDSON	TN	11	27	41	44	57
		2	4.5	1	1.8	DECATUR	TN		-60			16
1	2.0	6*	9.6	2	2.8	DE KALB	TN	377	-71	39	44	27
2	2.0	1	.9	7	5.2	DICKSON	TN	-56	492	160	77	70
2	2.3	7	5.0	8	5.5	DYER	TN	120	10	140	74	74
				2	4.1	FAYETTE	TN					49
2	3.9	4	6.6	5	7.3	FENTRESS	TN	71	10	89	61	89
5	6.0			8	6.0	FRANKLIN	TN			0	27	80
4	1.8	4	2.0	7	3.0	GIBSON	TN	11	47	63	53	30
2	1.9	3	2.9	5	4.5	GILES	TN	52	57	138	74	57
		1	1.7	4	5.5	GRAINGER	TN		217			73
5	3.2	8	4.3	8	3.5	GREENE	TN	35	-20	8	30	38
				5	8.5	GRUNDY	TN					93
				9	5.3	HAMBLEN	TN					70
11	2.0	30	3.8	52	5.8	HAMILTON	TN	91	53	193	82	78
		2	5.9			HANCOCK	TN					
2	2.5	2	2.8	1	1.1	HARDEMAN	TN	13	-61	-56	05	11
		2	2.3	3	3.0	HARDIN	TN		27			30
2	1.6	1	.6	3	1.6	HAWKINS	TN	-63	158	-4	26	14
1	2.0	2	4.3			HAYWOOD	TN	111				
1	1.0	2	2.3	1	1.1	HENDERSON	TN	135	-51	15	33	11
2	1.6	3	2.7	7	4.4	HENRY	TN	68	62	171	79	54
1	1.6	1	1.6	4	6.1	HICKMAN	TN	2	275	283	89	81
		2	2.9	2	2.9	HUMPHREYS	TN		0			29
1	2.0	2	4.1			JACKSON	TN	108				
1	.8	4	4.1	6	5.2	JEFFERSON	TN	394	27	529	97	69
2	4.5	2	3.3	3	4.1	JOHNSON	TN	-27	27	-8	24	50
16	2.1	25	2.6	54	4.8	KNOX	TN	22	87	127	71	63
		1	3.4			LAKE	TN					
3	5.0	3	4.3	3	3.9	LAUDERDALE	TN	-14	-9	-22	17	46
3	2.8	2	1.4	5	3.4	LAWRENCE	TN	-49	139	22	36	38
		1	3.4	2	4.9	LEWIS	TN		44			65
2	2.4	1	1.0	4	3.5	LINCOLN	TN	-56	236	48	47	39
4	4.6	7	6.9	5	4.4	LOUDON	TN	51	-37	-5	25	54
5	3.7	3	2.0	4	2.1	MC MINN	TN	-47	6	-44	09	19
		3	3.7	7	6.5	MC NAIRY	TN		76			84
3	4.6	4	5.9	2	2.4	MACON	TN	27	-59	-48	08	22
7	3.6	1	.5	6	2.5	MADISON	TN	-86	373	-32	13	23
2	2.4	2	2.6	6	7.1	MARION	TN	8	176	197	82	88
1	1.2	3	4.5	3	3.4	MARSHALL	TN	268	-26	174	79	36
2	1.6	3	1.9	9	4.8	MAURY	TN	19	153	200	83	63

* SIGNIFICANTLY MORE DEATHS THAN EXPECTED

WHITE MALE

CANCER OF THE LIVER AND GALLBLADDER
INCLUDING BILE DUCTS

ICD 155,156

1950-1959 DEATHS	1950-1959 RATE	1960-1969 DEATHS	1960-1969 RATE	1970-1979 DEATHS	1970-1979 RATE	COUNTY AND STATE		PERCENT CHANGE 60S/50S	PERCENT CHANGE 70S/60S	PERCENT CHANGE 70S/50S	PERCENTILE %CHANGE 70S/50S	PERCENTILE RATES 1970S
1	5.1	2	8.8	1	3.4	MEIGS	TN	72	-62	-34	12	36
3	3.2	5	4.6	3	2.5	MONROE	TN	45	-45	-21	18	23
2	1.5	3	2.4	4	2.1	MONTGOMERY	TN	62	-13	41	44	18
		2	9.9	1	4.8	MOORE	TN		-51			63
		3	4.6	5	7.8	MORGAN	TN		69			91
1	.7	2	1.3	11	6.5	OBION	TN	88	402	844	99	84
1	1.3	6	7.8	7	7.7	OVERTON	TN	516	-1	512	97	91
		2	6.1	2	6.0	PERRY	TN		-1			80
		1	8.0	3	11.7	PICKETT	TN		46			97
		4	7.8	6*	12.0	POLK	TN		53			97
2	1.6	4	2.7	8	4.5	PUTNAM	TN	69	67	182	80	57
1	2.1	6*	8.3	2	2.4	RHEA	TN	300	-71	17	34	22
2	1.7	4	2.8	8	5.4	ROANE	TN	68	90	218	85	72
4	3.5	1	.8	6	4.0	ROBERTSON	TN	-78	417	13	32	46
7	5.0	9	5.0	10	4.0	RUTHERFORD	TN	0	-20	-20	18	46
2	3.6	5	7.4	10*	13.4	SCOTT	TN	104	82	271	88	98
		1	3.7			SEQUATCHIE	TN					
2	2.4	3	2.3	6	3.5	SEVIER	TN	-2	55	51	48	40
50*	4.4	65	4.4	88	5.5	SHELBY	TN	1	25	26	38	74
1	1.4			5	6.9	SMITH	TN			387	94	87
		1	2.3			STEWART	TN					
7	2.0	13	3.2	15	2.7	SULLIVAN	TN	57	-14	35	42	26
2	1.4	6	3.2	10	4.0	SUMNER	TN	131	23	185	81	46
2	3.0			3	4.0	TIPTON	TN			35	42	47
				1	3.3	TROUSDALE	TN					36
				4	6.6	UNICOI	TN					85
		2	4.9	3	5.2	UNION	TN		7			69
		1	5.6	1	5.1	VAN BUREN	TN		-10			68
2	2.0	5	4.1	6	4.1	WARREN	TN	109	0	108	67	49
9	3.1	7	2.0	17	4.7	WASHINGTON	TN	-36	136	51	48	61
1	2.0	1	1.5	5	8.1	WAYNE	TN	-25	431	299	90	92
1	.7	4	2.2	5	2.8	WEAKLEY	TN	222	28	311	91	28
1	1.6	1	1.2	4	4.2	WHITE	TN	-24	256	170	79	51
3	3.6	4	3.6	11	6.5	WILLIAMSON	TN	0	83	82	60	85
7*	5.9	2	1.4	19*	11.7	WILSON	TN	-77	755	100	65	97
946*	3.3	1576*	4.5	2228*	5.1	TEXAS	TX	34	15	55		
3	2.7	4	2.8	7	4.4	ANDERSON	TX	6	55	65	54	55
				2	5.7	ANDREWS	TX					77
6	4.1	9	5.1	13	5.8	ANGELINA	TX	26	12	41	44	77
1	3.1	2	3.9	4	4.7	ARANSAS	TX	28	19	53	49	61
		1	2.5	1	2.2	ARCHER	TX		-15			20
1	10.2					ARMSTRONG	TX					
2	3.2	2	2.0	7	6.1	ATASCOSA	TX	-38	203	87	61	80
3	6.0	3	3.1	3	2.6	AUSTIN	TX	-49	-17	-58	04	24
1	5.1	1	4.5			BAILEY	TX	-12				
2	6.4			2	3.8	BANDERA	TX			-40	10	44
1	1.1	2	2.0	4	3.2	BASTROP	TX	71	62	177	80	33
		2	5.3	2	4.1	BAYLOR	TX		-23			49
2	3.2	3	3.8	5	5.9	BEE	TX	16	58	83	60	79
8	2.8	22*	6.1	23	5.4	BELL	TX	118	-13	91	62	72
71*	4.1	141*	6.3	212*	7.7	BEXAR	TX	55	23	90	62	91
1	5.7	1	3.5	2	4.6	BLANCO	TX	-38	29	-19	19	59
3	3.6			1	.6	BOSQUE	TX			-82	00	10
8	4.1	13	5.0	14	4.9	BOWIE	TX	24	-3	20	35	64
7	4.6	10	5.5	12	3.1	BRAZORIA	TX	19	-44	-34	12	31
1	.9	4	3.4	8	4.5	BRAZOS	TX	269	32	388	94	57
		2	7.3	1	3.7	BREWSTER	TX		-49			43
2*	11.7					BRISCOE	TX					
3*	7.9	2	7.9	8*	25.0	BROOKS	TX	0	218	218	85	99
3	2.2	5	3.3	9	4.7	BROWN	TX	51	43	116	69	61
		2	3.3	1	1.4	BURLESON	TX		-56			13
		3	3.6	6	5.1	BURNET	TX		43			68
2	2.8	10*	11.0	3	2.9	CALDWELL	TX	298	-74	4	29	29
		3	6.8	5	10.3	CALHOUN	TX		51			96
2	3.3	1	1.4	5	6.8	CALLAHAN	TX	-59	393	105	66	87
12	3.1	31*	6.3	32	5.1	CAMERON	TX	103	-20	63	53	68
1	5.7			6*	15.2	CAMP	TX			166	78	99
				2	5.6	CARSON	TX					75
2	2.0	2	1.7	9	7.0	CASS	TX	-15	312	251	87	88
				4*	14.5	CASTRO	TX					98
				2	3.7	CHAMBERS	TX					42
2	1.4	6	3.5	10	4.9	CHEROKEE	TX	150	42	254	87	65
2	3.9					CHILDRESS	TX					
2	3.3	4	6.3			CLAY	TX	94				
		1	5.8	1	6.6	COCHRAN	TX		14			85
1	5.9			1	3.9	COKE	TX			-33	12	46
3	3.5	5	6.1	8	9.8	COLEMAN	TX	73	62	181	80	96
6	3.1	9	4.0	15	5.1	COLLIN	TX	28	27	63	53	68
2	5.1	1	2.9	1	2.4	COLLINGSWORTH	TX	-44	-16	-53	06	22
1	1.3	4	5.2	3	3.6	COLORADO	TX	315	-32	184	81	40
4	4.8	2	1.8	15*	9.0	COMAL	TX	-63	395	85	60	94
1	.9	6	7.1	5	5.6	COMANCHE	TX	720	-21	544	97	75
				1	4.3	CONCHO	TX					53
3	3.1	7	5.7	2	1.5	COOKE	TX	84	-73	-51	07	14
3	3.7	1	1.1	8	7.1	CORYELL	TX	-70	558	95	63	88
				1	4.5	COTTLE	TX					56
				1	6.5	CROCKETT	TX					85
1	2.1	3	7.0	1	2.6	CROSBY	TX	237	-63	23	37	24
				1	3.2	DALLAM	TX					34
73	3.4	118	3.7	192	4.9	DALLAS	TX	9	31	43	45	65
		2	2.0	1	1.4	DAWSON	TX		-31			13
3	5.8	2	2.8	3	5.5	DEAF SMITH	TX	-53	98	-7	24	73
2	4.5	1	2.3	1	2.7	DELTA	TX	-48	13	-42	09	25
9	4.9	8	3.7	8	2.7	DENTON	TX	-23	-27	-44	09	27
1	1.0	7	6.6	7	6.0	DE WITT	TX	572	-10	503	97	79
		2	7.1	1	3.8	DICKENS	TX		-47			44
2	5.5	1	2.6	3	6.9	DIMMIT	TX	-53	166	25	38	87
1	2.8					DONLEY	TX					

* SIGNIFICANTLY MORE DEATHS THAN EXPECTED

1950-1959 DEATHS	1950-1959 RATE	1960-1969 DEATHS	1960-1969 RATE	1970-1979 DEATHS	1970-1979 RATE	COUNTY AND STATE	PERCENT CHANGE 60S/50S	PERCENT CHANGE 70S/60S	PERCENT CHANGE 70S/50S	PERCENTILE %CHANGE 70S/50S	PERCENTILE RATES 1970S
3	4.9	4	7.2	8*	12.0	DUVAL TX	48	66	146	75	97
3	2.9	10	6.7	8	4.4	EASTLAND TX	137	-35	53	49	54
4	4.7	8	3.8	16	5.3	ECTOR TX	-18	39	14	33	71
3	1.6	5	2.3	13	5.9	ELLIS TX	47	154	273	88	78
29*	4.8	46*	5.8	72*	7.6	EL PASO TX	20	31	57	50	91
1	.9	1	.7	6	4.3	ERATH TX	-18	506	400	94	53
2	1.6	3	2.6			FALLS TX	63				
4	2.5	14*	8.4	10	5.6	FANNIN TX	238	-33	127	71	75
2	1.7	6	4.3	6	3.6	FAYETTE TX	151	-16	112	68	40
1	2.1	1	2.1			FISHER TX	-1				
		1	2.2	1	1.7	FLOYD TX		-21			16
1	6.1	1	4.6			FOARD TX	-24				
1	.9	4	2.7	10	5.2	FORT BEND TX	192	93	466	96	70
				3	5.7	FRANKLIN TX					76
		1	1.4	2	2.5	FREESTONE TX		85			24
		2	4.4	3	5.8	FRIO TX		32			78
2	8.8			1	1.8	GAINES TX			-80	00	16
21*	7.1	17	3.3	34	6.3	GALVESTON TX	-54	91	-12	22	82
1	4.0	1	3.6	1	4.5	GARZA TX	-10	24	12	32	57
1	2.5	2	3.1	2	1.8	GILLESPIE TX	27	-43	-28	14	16
				1*	23.6	GLASSCOCK TX					99
1	4.0	3*	10.5	3	11.9	GOLIAD TX	163	13	198	82	97
1	1.2	6	6.9	5	4.8	GONZALES TX	471	-31	296	90	62
6*	6.8	5	4.0	4	2.7	GRAY TX	-41	-32	-60	04	26
12	3.5	23*	6.4	14	3.2	GRAYSON TX	82	-50	-10	23	33
4	1.6	6	2.3	18	6.4	GREGG TX	43	172	288	90	83
		2	3.2	4	6.0	GRIMES TX		89			79
4	4.6	10*	7.9	7	3.8	GUADALUPE TX	73	-52	-17	20	43
4	3.8	6	4.4	5	3.2	HALE TX	17	-27	-15	21	34
		3	7.0	1	4.6	HALL TX		-34			59
1	1.6	4	10.0	4	4.7	HAMILTON TX	538	-54	196	82	60
1	7.5	1	3.1			HANSFORD TX	-59				
1	2.1	1	1.9	3	6.0	HARDEMAN TX	-8	209	183	81	79
2	2.5	6	5.4	5	3.4	HARDIN TX	119	-36	40	44	37
111*	4.7	180*	4.9	247*	5.3	HARRIS TX	3	8	11	32	70
5	4.2	13*	10.1	5	3.0	HARRISON TX	141	-70	-28	14	30
				1	5.4	HARTLEY TX					72
2	3.4	1	1.6	3	5.3	HASKELL TX	-52	225	54	49	70
4	5.5	6	7.5	8	7.4	HAYS TX	36	-1	34	42	90
1	5.6	1	6.6	4*	21.6	HEMPHILL TX	19	227	288	90	99
5	4.4	4	2.9	12	5.7	HENDERSON TX	-33	95	31	40	77
10	2.0	28	4.7	44	5.4	HIDALGO TX	128	16	163	78	72
5	3.6	4	2.4	7	3.9	HILL TX	-34	64	9	30	45
1	1.5			3	3.6	HOCKLEY TX			131	73	40
1	2.5	5*	12.2	1	1.1	HOOD TX	388	-91	-54	05	11
4	3.7	4	3.5	6	3.8	HOPKINS TX	-5	11	5	29	44
4	5.6	5	6.3	3	3.7	HOUSTON TX	13	-41	-33	12	42
4	4.0	5	4.3	3	2.2	HOWARD TX	7	-49	-45	08	20
4	2.0	8	3.6	8	3.6	HUNT TX	85	-2	82	60	40
2	1.8	1	.7	4	4.7	HUTCHINSON TX	-63	609	160	77	61
1*	14.9	1	10.8			IRION TX	-28				
		1	2.6	4	7.6	JACK TX		189			90
1	2.2	2	5.3	5	9.3	JACKSON TX	143	77	330	91	95
4	4.7	6	6.8	12*	9.7	JASPER TX	43	43	104	66	96
21	4.3	39*	5.4	34	4.4	JEFFERSON TX	25	-19	1	27	54
		2	9.0	3*	11.3	JIM HOGG TX		26			97
2	2.1	12*	10.3	3	2.3	JIM WELLS TX	398	-78	9	31	21
5	3.0	12	5.9	9	3.3	JOHNSON TX	100	-43	13	33	36
4	4.6	2	1.6	4	3.6	JONES TX	-66	128	-22	18	40
3	7.0	4	5.9	3	4.1	KARNES TX	-16	-29	-41	10	50
2	1.7	2	1.3	1	.6	KAUFMAN TX	-24	-56	-66	02	10
1	2.2	4*	9.0	1	1.7	KENDALL TX	320	-81	-20	19	15
1	10.1	1	10.1			KENT TX	0				
3	3.1	6	5.1	4	1.7	KERR TX	62	-67	-46	08	15
		3*	9.8	5*	15.1	KIMBLE TX		54			99
1*	23.8					KING TX					
1	8.4	4*	44.4	2*	17.1	KINNEY TX	429	-62	103	66	99
1	2.1	6	11.5	9*	12.0	KLEBERG TX	437	4	458	96	97
1	3.0	3	6.2	3	6.5	KNOX TX	109	5	120	70	85
5	3.3	7	3.9	8	3.7	LAMAR TX	19	-5	13	33	41
2	2.9	2	2.5	5	5.3	LAMB TX	-13	111	84	60	71
1	1.4	5*	8.3	3	4.4	LAMPASAS TX	479	-47	209	84	55
2	7.1			3	12.6	LA SALLE TX			78	58	98
1	.9	1	.6	6	4.9	LAVACA TX	-25	659	468	96	65
4*	9.1	5*	11.1	2	3.5	LEE TX	23	-68	-61	03	39
		4	9.2	2	3.7	LEON TX		-60			41
1	1.1	10*	8.1	8	5.2	LIBERTY TX	630	-36	369	93	70
3	2.8	4	4.3	8	7.5	LIMESTONE TX	54	74	168	78	90
2*	11.0					LIPSCOMB TX					
		5*	14.8	2	3.8	LIVE OAK TX		-74			43
1	2.7	2	3.5	2	2.4	LLANO TX	29	-31	-10	23	22
7	1.9	7	1.8	20	3.7	LUBBOCK TX	-5	100	90	61	41
				1	2.2	LYNN TX					19
		1	1.3	3	3.6	MC CULLOCH TX		171			40
16	2.9	32*	5.2	30	4.3	MC LENNAN TX	77	-17	47	46	53
1	2.3	1	2.3	2	4.3	MADISON TX	-1	87	85	60	53
		2	7.0	3	6.3	MARION TX		-11			82
2*	8.2					MARTIN TX					
1	4.2	1	4.3	1	4.4	MASON TX	3	2	5	29	54
3	4.4	3	3.3	5	4.4	MATAGORDA TX	-23	32	1	28	55
2	4.7	7*	15.2	11*	14.6	MAVERICK TX	224	-4	210	84	98
2	2.2	3	3.4	10*	8.5	MEDINA TX	55	150	287	89	94
3*	15.3	1	6.2	2	12.9	MENARD TX	-59	107	-16	20	98
3	5.1	5	5.0	6	2.5	MIDLAND TX	-2	-51	-52	06	23
2	2.2	4	3.1	9	7.7	MILAM TX	42	148	254	87	91
		3	7.9	3	5.4	MILLS TX		-32			72
1	1.9	1	1.7			MITCHELL TX	-12				
2	1.9	4	4.0	7	7.1	MONTAGUE TX	107	78	268	88	88
5	4.8	11*	7.8	8	3.7	MONTGOMERY TX	61	-52	-22	17	43
		2	4.7	2	6.0	MOORE TX		26			79

* SIGNIFICANTLY MORE DEATHS THAN EXPECTED

WHITE MALE

CANCER OF THE LIVER AND GALLBLADDER INCLUDING BILE DUCTS

ICD 155,156

1950-1959 DEATHS	1950-1959 RATE	1960-1969 DEATHS	1960-1969 RATE	1970-1979 DEATHS	1970-1979 RATE	COUNTY AND STATE	PERCENT CHANGE 60S/50S	PERCENT CHANGE 70S/60S	PERCENT CHANGE 70S/50S	PERCENTILE %CHANGE 70S/50S	PERCENTILE RATES 1970S
1	2.2	1	2.3	1	1.7	MORRIS TX	3	-26	-24	16	15
		2	10.6			MOTLEY TX					
4	3.6	8	5.8	4	2.7	NACOGDOCHES TX	61	-53	-24	16	27
6	3.5	5	3.1	12	6.2	NAVARRO TX	-13	101	76	57	81
2	5.4			1	2.5	NEWTON TX			-54	05	24
3	3.8	3	3.4	5	5.9	NOLAN TX	-10	71	53	49	78
19	4.2	45*	8.2	64*	8.9	NUECES TX	96	7	111	68	94
		2	6.4	1	2.2	OCHILTREE TX		-65			20
10*	10.2	8	4.3	10	3.9	ORANGE TX	-57	-10	-62	03	45
4	3.5	4	3.6	9	7.2	PALO PINTO TX	1	103	104	66	89
2	3.0	2	2.7	1	1.4	PANOLA TX	-8	-48	-52	06	13
5	4.0	8	5.5	8	4.1	PARKER TX	39	-25	4	29	49
1	4.1			3	8.3	PARMER TX			102	65	93
1	2.2	1	3.3	1	3.4	PECOS TX	53	3	58	51	37
2	3.8	4	5.9	6	5.5	POLK TX	52	-6	44	45	74
2	.6	7	1.7	18	4.7	POTTER TX	203	177	742	99	60
		1	3.4	1	3.9	PRESIDIO TX		16			46
1	3.1	2	7.5	3	8.4	RAINS TX	140	13	170	79	93
3	5.6	4	2.9	8	3.9	RANDALL TX	-48	32	-31	13	44
				2*	13.4	REAGAN TX					98
		1	7.4	1	7.8	REAL TX		5			91
4	4.2	4	4.0	4	4.8	RED RIVER TX	-6	22	15	33	63
1	2.5	3	6.5	4	9.5	REEVES TX	162	47	286	89	95
		1	2.2			REFUGIO TX					
2	2.8	4	5.4	3	6.3	ROBERTSON TX	94	17	127	71	83
1	3.5	2	6.8	3	6.6	ROCKWALL TX	93	-4	86	61	85
2	2.3	2	2.2			RUNNELS TX	-5				
3	1.7	5	2.7	15	7.9	RUSK TX	58	192	363	93	92
2	5.1			1	2.5	SABINE TX			-51	06	23
1	3.5	2	5.1			SAN AUGUSTINE TX	43				
				1	4.9	SAN JACINTO TX					64
2	1.7	7	4.7	8	4.3	SAN PATRICIO TX	176	-8	155	76	53
1	1.7					SAN SABA TX					
1	7.7					SCHLEICHER TX					
1	2.1	1	1.4	6	8.2	SCURRY TX	-34	486	285	89	93
1	4.5			1	2.8	SHACKELFORD TX			-37	11	28
4	3.9	1	.8	8	7.2	SHELBY TX	-80	796	83	60	89
10	3.8	11	3.5	15	3.5	SMITH TX	-7	-3	-9	24	38
1	4.8					SOMERVELL TX					
4*	7.9	5	7.7	4	4.8	STARR TX	-3	-38	-39	10	62
3	5.3	3	4.9	2	3.2	STEPHENS TX	-9	-34	-40	10	34
1	6.3			2	11.2	STONEWALL TX			79	59	96
		2*	11.7	1	5.9	SUTTON TX		-50			78
		1	1.9	5*	10.3	SWISHER TX		452			96
45	3.0	60	3.1	115	4.9	TARRANT TX	4	57	62	53	65
6	2.2	15	4.6	13	3.4	TAYLOR TX	109	-27	53	49	37
3	6.2	1	1.7	3	5.2	TERRY TX	-72	208	-15	20	70
2*	9.8			1	6.5	THROCKMORTON TX			-33	12	85
3	4.2	3	4.7	3	2.7	TITUS TX	14	-42	-34	12	27
11	4.5	9	3.1	15	4.4	TOM GREEN TX	-32	42	-3	26	55
22	4.1	35	5.0	36	3.7	TRAVIS TX	20	-25	-10	23	42
2	4.7	3	8.9	5	8.2	TRINITY TX	91	-8	76	57	93
2	3.1	3	4.6	3	3.5	TYLER TX	49	-23	15	33	40
1	1.2	4	4.0	8	6.4	UPSHUR TX	220	59	409	94	83
		1	3.9	2	12.4	UPTON TX		213			97
2	3.2	3	3.5	12*	13.0	UVALDE TX	8	275	305	90	98
2	2.9	6*	9.7	13*	13.7	VAL VERDE TX	230	42	368	93	98
3	2.6	3	2.2	8	4.5	VAN ZANDT TX	-12	102	77	58	58
5	5.2	5	2.9	15*	8.8	VICTORIA TX	-44	203	68	55	94
		2	2.4	3	3.7	WALKER TX		55			42
		2	5.6	2	3.5	WALLER TX		-37			39
		1	1.9	1	1.6	WARD TX		-17			14
3	3.3	4	4.1	5	4.3	WASHINGTON TX	22	6	30	40	53
12*	6.4	19*	8.6	30*	11.5	WEBB TX	35	34	80	59	97
2	1.7	6	4.7	8	4.9	WHARTON TX	178	5	192	82	64
2	4.8	1	1.8			WHEELER TX	-63				
8	2.2	16	3.9	21	4.6	WICHITA TX	77	16	106	67	59
8*	10.3	2	2.1	3	3.7	WILBARGER TX	-80	75	-64	02	41
1	1.9	4	6.5	5	8.3	WILLACY TX	243	27	335	92	93
5	2.8	3	1.5	15	6.3	WILLIAMSON TX	-48	333	124	71	83
2	3.0	1	1.7	9*	11.8	WILSON TX	-44	603	297	90	97
1	4.0					WINKLER TX					
2	1.8	6	5.1	6	4.5	WISE TX	183	-10	154	76	58
2	1.9	8	7.3	5	3.6	WOOD TX	294	-51	94	63	40
1	5.2			2	6.9	YOAKUM TX			34	42	87
2	2.0	3	3.0	5	3.8	YOUNG TX	44	29	86	61	44
		2	8.4	2	6.4	ZAPATA TX		-25			83
2	6.9	1	3.0	4	9.5	ZAVALA TX	-56	217	39	43	95
66	2.6	106	3.2	113	2.7	UTAH UT	22	-15	4		
				1	4.1	BEAVER UT					49
2	3.0	6	6.4	2	1.7	BOX ELDER UT	115	-73	-42	09	15
		4	2.4	4	2.2	CACHE UT		-10			20
3	3.0	2	2.2	4	4.1	CARBON UT	-25	81	36	42	48
2	3.3	9	6.1	4	1.7	DAVIS UT	86	-72	-49	07	15
		2	7.4			DUCHESNE UT					
1	3.9	1	3.7	1	2.8	EMERY UT	-4	-24	-27	15	28
		1	3.4	1	3.4	GRAND UT		2			37
2	6.0	3	6.8	1	2.2	IRON UT	14	-68	-64	02	20
		1	4.2			JUAB UT					
1	2.6					MILLARD UT					
33	3.2	51	3.6	50	2.9	SALT LAKE UT	15	-21	-9	24	28
		1	2.9	4	7.4	SANPETE UT		158			90
2	4.0	1	1.9			SEVIER UT	-52				
1	3.1			2	6.2	SUMMIT UT			99	64	82
1	2.0			2	3.3	TOOELE UT			66	54	36
1	2.8	2	6.0	1	2.2	UINTAH UT	116	-64	-21	18	20
7	2.5	9	2.4	20	4.1	UTAH UT	-6	76	66	54	50
1	5.0	1	3.8			WASATCH UT	-24				

* SIGNIFICANTLY MORE DEATHS THAN EXPECTED

WHITE MALE

CANCER OF THE LIVER AND GALLBLADDER INCLUDING BILE DUCTS

ICD 155,156

1950-1959 DEATHS	1950-1959 RATE	1960-1969 DEATHS	1960-1969 RATE	1970-1979 DEATHS	1970-1979 RATE	COUNTY AND STATE	PERCENT CHANGE 60S/50S	PERCENT CHANGE 70S/60S	PERCENT CHANGE 70S/50S	PERCENTILE %CHANGE 70S/50S	PERCENTILE RATES 1970S
1	1.9	4	6.9	3	2.0	WASHINGTON UT	256	-71	4	29	18
		1	10.2	1	7.9	WAYNE UT		-23			92
8	2.7	7	1.7	12	2.4	WEBER UT	-37	41	-11	23	22
44	2.3	40	2.1	74	3.5	VERMONT VT	-10	67	50		
3	3.1	1	1.2	6	5.7	ADDISON VT	-61	374	83	60	76
6	4.5	3	2.4	5	3.0	BENNINGTON VT	-48	26	-34	12	30
1	.7	2	1.6	4	3.1	CALEDONIA VT	111	95	312	91	31
9	3.7	5	2.0	8	2.5	CHITTENDEN VT	-47	28	-33	13	24
1	2.6					ESSEX VT					
5	3.5	4	2.6	3	2.2	FRANKLIN VT	-27	-15	-38	10	20
		1	4.7			GRAND ISLE VT					
1	1.8	1	1.6	2	3.0	LAMOILLE VT	-12	92	68	55	31
		1	1.0	2	1.9	ORANGE VT		80			17
4	3.6	2	1.9	8	7.3	ORLEANS VT	-46	282	106	66	89
3	1.2	9	3.8	14	5.1	RUTLAND VT	206	35	314	91	68
2	.9	5	2.5	6	2.7	WASHINGTON VT	177	12	209	84	27
3	1.8	2	1.2	9	4.9	WINDHAM VT	-35	322	173	79	65
6	2.6	4	1.8	7	2.8	WINDSOR VT	-33	59	7	30	28
202	2.0	365	2.9	670	4.4	VIRGINIA VA	45	52	121		
1	.8	1	.7	3	2.1	ACCOMACK VA	-13	202	163	78	19
2	1.2	5	2.4	16	6.8	ALBEMARLE#VA	106	184	486	96	86
3	2.9	4	3.0	8	6.1	ALLEGHANY#VA	6	101	114	68	81
		1	4.7	3	12.2	AMELIA VA		159			97
1	1.5	1	1.2	3	3.2	AMHERST VA	-25	180	111	68	34
		1	2.2			APPOMATTOX VA					
12	3.3	18	2.9	29	4.8	ARLINGTON#VA	-12	68	47	47	64
1	.3	5	1.7	7	1.9	AUGUSTA#VA	385	10	435	95	17
		3*	12.6	2	6.3	BATH VA		-50			83
7	1.8	13	2.8	23	4.2	BEDFORD#VA	52	51	130	72	52
		2	7.3			BLAND VA					
		7*	8.6			BOTETOURT VA					
				1	3.1	BRUNSWICK VA					33
3	3.2	5	4.1	1	1.2	BUCHANAN VA	29	-71	-62	03	11
		1	2.8	1	2.3	BUCKINGHAM VA		-17			21
1	3.4	1	2.9	2	5.1	CAROLINE VA	-14	74	49	48	67
				8	2.9	CARROLL#VA					29
		1	2.0	2	4.6	CHARLOTTE VA		126			59
27	3.5	36	3.8	30	3.1	CHESTERFIELD#VA	10	-19	-11	23	32
1	4.0	1	2.8	1	1.9	CLARKE VA	-31	-30	-52	06	17
		1	5.2	1	5.1	CRAIG VA		-2			68
1	1.7	2	3.4	2	2.1	CULPEPER VA	102	-38	26	38	19
				2	8.1	CUMBERLAND VA					92
2	2.7	1	1.0	4	5.7	DICKENSON VA	-61	454	114	68	76
2	1.0	6	2.8	14	4.7	DINWIDDIE#VA	179	69	372	93	62
1	6.2			2	6.4	ESSEX VA			3	28	84
10	1.9	30	3.1	59	3.8	FAIRFAX#VA	61	23	98	64	43
		5	5.8	4	3.7	FAUQUIER VA		-37			41
		2	3.3	3	4.6	FLOYD VA		38			59
		2	6.5	1	3.3	FLUVANNA VA		-49			35
2	2.0	1	.9	5	3.9	FRANKLIN VA	-54	313	91	62	45
13*	10.3	4	2.3	8	3.8	FREDERICK#VA	-78	64	-63	03	43
1	2.0	1	2.4			GILES VA	21				
				5	7.3	GLOUCESTER VA					89
				2	6.1	GOOCHLAND VA					80
1	6.1	1	4.1	1	5.4	GREENE VA	-32	30	-12	23	72
		1	2.5	2	5.9	GREENSVILLE#VA		133			78
				11*	9.0	HALIFAX#VA					94
		1	1.7	4	2.4	HANOVER VA		44			22
9	3.1	10	2.7	23	4.8	HENRICO VA	-11	76	58	51	63
3	2.6	5	3.9	13	6.1	HENRY#VA	50	56	134	73	81
				1	4.9	HIGHLAND VA					64
1	2.6			1	2.2	ISLE OF WIGHT VA			-15	20	20
1	1.1	1	.8	4	2.5	JAMES CITY#VA	-29	211	122	70	23
				2	12.7	KING AND QUEEN VA					98
				2	7.2	KING GEORGE VA					89
		1	4.3			KING WILLIAM VA					
1	2.7	1	1.8	2	3.2	LANCASTER VA	-33	77	18	35	34
2	2.6	1	.8	3	2.3	LEE VA	-69	193	-9	24	21
1	1.3	2	2.0	8	6.4	LOUDOUN VA	54	223	398	94	83
2	4.5	2	3.9	2	3.5	LOUISA VA	-14	-10	-22	17	39
		2	4.7	3	6.5	LUNENBURG VA		37			84
		1	2.9			MADISON VA					
		2	7.5	1	2.1	MATHEWS VA		-72			19
		3	3.6	3	2.8	MECKLENBURG VA		-22			27
		1	4.6	2	5.0	MIDDLESEX VA		10			67
4	3.4	1	.6	8	4.1	MONTGOMERY#VA	-82	580	21	36	49
3	6.4			1	1.6	NELSON VA			-75	01	15
3	6.0			2	3.7	NORTHAMPTON VA			-37	11	43
				1	1.4	NORTHUMBERLAND VA					13
		2	4.0	1	1.8	NOTTOWAY VA		-55			16
1	2.3	2	3.7	2	2.8	ORANGE VA	60	-24	22	36	28
1	1.8			2	2.1	PAGE VA			16	34	19
1	1.5	1	1.5	8*	9.9	PATRICK VA	1	552	558	97	96
1	.4	12	3.5	26*	6.9	PITTSYLVANIA#VA	808	98	1694	99	87
2	5.2			1	2.4	PRINCE EDWARD VA			-53	05	22
		2	1.0	4	2.2	PRINCE WILLIAM#VA		113			20
1	1.5	3	2.8	4	3.7	PULASKI VA	88	34	152	76	42
		1	4.1	4*	13.4	RAPPAHANNOCK VA		229			98
				2	6.4	RICHMOND VA					84
19	3.5	23	3.8	33	4.2	ROANOKE#VA	10	11	22	36	52
1	1.0	3	3.2	3	2.4	ROCKBRIDGE#VA	215	-24	140	74	23
2	1.0	5	2.4	12	4.3	ROCKINGHAM#VA	140	79	331	92	53
		7	6.4	7	5.4	RUSSELL VA		-15			73
1	.9	3	2.5	6	4.6	SCOTT VA	168	83	388	94	59
3	2.9	1	1.0	1	.7	SHENANDOAH VA	-66	-27	-75	01	10
1	1.3	5	3.8	8	5.2	SMYTH VA	189	38	297	90	69

* SIGNIFICANTLY MORE DEATHS THAN EXPECTED

COUNTIES COMBINED; SEE TECHNICAL NOTES

1950-1959 DEATHS	1950-1959 RATE	1960-1969 DEATHS	1960-1969 RATE	1970-1979 DEATHS	1970-1979 RATE	COUNTY AND STATE	PERCENT CHANGE 60S/50S	PERCENT CHANGE 70S/60S	PERCENT CHANGE 70S/50S	PERCENTILE %CHANGE 70S/50S	PERCENTILE RATES 1970S
2	4.0	2	4.6	2	3.5	SOUTHAMPTON#VA	15	-23	-12	22	39
1	1.4	1	.9	2	1.6	SPOTSYLVANIA#VA	-38	83	13	32	14
1	2.4	1	1.8	2	1.6	STAFFORD VA	-26	-11	-34	12	14
				2	15.2	SURRY VA					99
1	4.8			1	4.2	SUSSEX VA			-12	22	52
4	2.8	9	5.7	15*	8.0	TAZEWELL VA	104	41	186	81	92
2	4.5	2	3.1	3	4.3	WARREN VA	-30	37	-5	25	52
3	1.5	10	4.1	12	4.3	WASHINGTON#VA	183	4	195	82	53
		1	2.1	2	3.3	WESTMORELAND VA		61			36
2	1.2	5	3.3	13	6.8	WISE#VA	180	102	465	96	86
4	4.5	2	2.2	3	2.7	WYTHE VA	-51	22	-41	10	25
19	1.7	48	3.9	93*	6.5	CHESAPEAKE CITY#VA	123	66	270	88	84
3	1.2	6	2.9	15	6.8	HAMPTON CITY VA	150	136	489	96	87
3	1.8	8	3.3	21*	6.8	NEWPORT NEWS CITY VA	82	107	277	89	87
1	1.3	3	3.2	8	6.7	SUFFOLK CITY VA	154	109	430	95	86
331	2.7	476	3.4	611	3.9	WASHINGTON WA	25	13	41		
2	3.7	2	4.5	5	12.4	ADAMS WA	21	178	235	86	97
3	3.8	1	1.1	4	5.0	ASOTIN WA	-70	339	31	40	67
7	4.3	9	3.3	8	2.0	BENTON WA	-23	-40	-54	05	18
1	.5	8	3.6	9	3.5	CHELAN WA	629	-2	618	98	39
4	2.9	9	5.1	7	3.1	CLALLAM WA	74	-38	7	30	32
11	2.6	12	2.5	24	3.8	CLARK WA	-5	53	45	46	43
1	3.2	1	3.8	1	3.4	COLUMBIA WA	22	-12	7	30	36
2	1.1	6	2.0	16	5.0	COWLITZ WA	86	150	365	93	66
1	1.7	2	2.5	4	4.7	DOUGLAS WA	42	92	173	79	61
		1	6.7	1	3.8	FERRY WA		-43			44
2	3.1	5	5.4	6	6.0	FRANKLIN WA	75	12	97	63	80
1	6.9			1	4.7	GARFIELD WA			-32	13	61
6	5.9	6	3.4	5	2.6	GRANT WA	-43	-23	-56	05	25
10	3.0	9	3.0	16	4.7	GRAYS HARBOR WA	-1	58	56	50	60
2	3.3	3	2.8	9	5.6	ISLAND WA	-15	103	72	57	75
3	7.3	2	2.9	2	1.6	JEFFERSON WA	-60	-44	-78	00	15
120	3.2	175	4.2	206	4.5	KING WA	30	7	39	44	57
5	1.2	9	2.0	21	4.1	KITSAP WA	67	98	232	86	48
3	2.4	2	1.9	5	4.5	KITTITAS WA	-22	137	85	60	57
2	2.9	3	4.2	1	1.5	KLICKITAT WA	46	-65	-49	07	13
6	2.3	3	1.0	7	2.4	LEWIS WA	-54	128	5	29	22
1	1.8	5	8.1	2	3.1	LINCOLN WA	353	-62	73	57	32
1	1.4	4	4.2	1	.8	MASON WA	211	-82	-43	09	10
4	3.3	5	3.8			OKANOGAN WA	13				
7*	7.4	1	.9	9	7.7	PACIFIC WA	-88	744	5	29	91
		1	2.0	1	3.1	PEND OREILLE WA		54			32
27	2.1	44	3.0	56	3.4	PIERCE WA	46	11	63	53	37
6	2.4	12	4.3	17	5.6	SKAGIT WA	80	31	136	73	75
1	5.7	2	7.6	1	2.7	SKAMANIA WA	32	-64	-52	06	26
16	2.3	22	2.6	40	3.6	SNOHOMISH WA	13	43	62	52	41
36	3.0	44	3.2	44	3.0	SPOKANE WA	7	-5	1	28	31
3	2.6	3	2.9	4	3.4	STEVENS WA	9	20	31	40	37
3	1.3	6	1.9	9	2.1	THURSTON WA	46	11	62	53	19
		2	8.0	2	8.4	WAHKIAKUM WA		6			93
3	1.3	9	3.9	10	4.3	WALLA WALLA WA	201	10	231	86	53
13	3.5	19	4.7	17	3.9	WHATCOM WA	35	-16	13	33	46
4	2.8	3	2.4	5	3.6	WHITMAN WA	-15	51	28	39	41
14	2.1	26	3.6	35	4.5	YAKIMA WA	71	25	113	68	57
198	2.6	328	4.0	491*	5.6	WEST VIRGINIA WV	54	41	117		
1	1.2	6	6.6	6	7.2	BARBOUR WV	437	9	484	96	89
5	3.1	7	3.7	8	4.1	BERKELEY WV	18	12	32	41	50
2	1.7	3	2.1	6	5.1	BOONE WV	23	140	195	82	68
2	2.3	1	1.2	6	7.3	BRAXTON WV	-46	487	219	85	89
5	4.2	7	5.8	7	4.9	BROOKE WV	39	-15	18	34	65
12	2.7	23	4.7	31	6.1	CABELL WV	75	30	127	71	81
		1	2.8	6*	12.0	CALHOUN WV		335			97
1	1.9	1	1.8	4	7.4	CLAY WV	-8	321	287	89	90
1	1.7	1	1.8	5	10.5	DODDRIDGE WV	8	470	513	97	96
11	5.1	15*	6.0	16	6.0	FAYETTE WV	18	-1	16	34	79
		1	2.6	1	2.7	GILMER WV		3			25
4*	10.6	1	2.3	2	4.4	GRANT WV	-78	89	-59	04	55
4	2.5	3	1.9	6	3.2	GREENBRIER WV	-24	68	27	39	34
2	3.4	4	5.9	4	6.1	HAMPSHIRE WV	73	2	77	58	80
6	3.9	9	5.8	11	6.1	HANCOCK WV	51	5	58	51	81
1	2.0	1	2.0	1	1.8	HARDY WV	-1	-9	-10	23	16
9	2.3	15	3.8	16	3.7	HARRISON WV	66	-1	65	54	42
1	1.3	4	5.0	7	6.8	JACKSON WV	299	37	444	95	87
2	3.0	1	1.2	8	8.2	JEFFERSON WV	-61	608	174	80	93
16	2.1	45	5.1	63*	6.5	KANAWHA WV	139	29	208	84	85
2	1.7	3	2.5	10*	8.2	LEWIS WV	43	233	377	94	93
1	1.2	1	1.0	9	8.3	LINCOLN WV	-19	727	569	98	93
3	1.2	8	4.2	15*	7.7	LOGAN WV	237	83	518	97	91
5	3.1	6	3.6	8	4.4	MC DOWELL WV	19	22	45	46	56
11	3.4	16	4.9	19	5.6	MARION WV	43	16	65	54	76
3	1.5	9	5.1	7	3.5	MARSHALL WV	234	-31	130	72	39
1	1.0	4	3.5	4	3.2	MASON WV	243	-9	212	84	33
4	1.4	9	3.2	20	6.6	MERCER WV	127	103	360	93	65
		3	2.8	4	3.1	MINERAL WV		11			33
3	2.0	7	4.6	9	6.2	MINGO WV	130	34	208	84	81
14*	5.9	3	1.2	15	5.7	MONONGALIA WV	-80	394	-3	26	76
2	3.5	2	2.5	1	1.5	MONROE WV	-29	-39	-57	05	14
1	2.3	3	8.4	3	6.4	MORGAN WV	268	-24	181	80	83
1	1.0	4	3.5	5	4.1	NICHOLAS WV	265	15	320	91	49
16	4.5	24*	6.8	20	6.2	OHIO WV	52	-10	38	43	81
				1	1.9	PENDLETON WV					17
1	3.3	1	3.0	1	2.6	PLEASANTS WV	-7	-15	-22	18	24
2	3.6	3	4.8	2	3.2	POCAHONTAS WV	35	-34	-11	23	34
5	3.6	7	5.3	7	5.0	PRESTON WV	44	-6	36	43	66
		3	2.8	8	6.6	PUTNAM WV		139			85
9	3.4	16	5.3	28*	8.5	RALEIGH WV	54	61	147	75	94

* SIGNIFICANTLY MORE DEATHS THAN EXPECTED

\# COUNTIES COMBINED; SEE TECHNICAL NOTES

1950-1959 DEATHS	1950-1959 RATE	1960-1969 DEATHS	1960-1969 RATE	1970-1979 DEATHS	1970-1979 RATE	COUNTY AND STATE		PERCENT CHANGE 60S/50S	PERCENT CHANGE 70S/60S	PERCENT CHANGE 70S/50S	PERCENTILE %CHANGE 70S/50S	PERCENTILE RATES 1970S
1	.7	5	4.2	5	3.5	RANDOLPH	WV	500	-18	390	94	38
2	3.2	1	1.3	2	2.4	RITCHIE	WV	-59	79	-26	16	22
1	1.1	4	4.3	2	2.1	ROANE	WV	304	-51	98	64	19
		1	1.4	7	8.8	SUMMERS	WV		538			94
2	2.1	3	3.4	1	1.5	TAYLOR	WV	61	-55	-27	15	14
		1	1.5			TUCKER	WV					
1	2.5			3	4.8	TYLER	WV			93	62	62
2	2.1	3	3.2	5	4.9	UPSHUR	WV	53	56	137	73	65
5	3.2	6	3.4	9	4.7	WAYNE	WV	7	38	48	47	60
1	1.0	2	3.3	5	7.4	WEBSTER	WV	214	124	604	98	90
		2	1.9	7	6.6	WETZEL	WV		256			85
10	3.2	15	4.2	21	5.3	WOOD	WV	29	28	65	54	72
4	4.5	4	3.0	14*	14.5	WYOMING	WV	-33	384	225	85	98
532	3.0	733	3.8	969	4.5	WISCONSIN	WI	24	21	49		
1	1.3	2	4.0	3	3.3	ADAMS	WI	202	-17	151	76	36
2	2.0	4	3.4	6	5.5	ASHLAND	WI	69	59	168	78	73
1	.5	11	4.9	12	4.8	BARRON	WI	853	-3	829	99	62
3	3.0	1	1.2	6	6.0	BAYFIELD	WI	-61	407	99	64	79
17	4.0	35*	6.4	32	5.2	BROWN	WI	58	-18	30	40	70
1	1.2	8*	9.7	3	3.6	BUFFALO	WI	739	-63	210	84	40
2	2.5	1	1.5	2	2.4	BURNETT	WI	-38	59	-2	26	23
1	1.3	4	3.6	6	4.9	CALUMET	WI	183	36	285	89	64
12*	5.6	7	2.9	14	5.3	CHIPPEWA	WI	-48	81	-5	25	70
5	2.8	6	3.1	9	4.5	CLARK	WI	13	43	61	52	56
4	1.9	13	5.8	8	3.2	COLUMBIA	WI	209	-45	70	56	33
1	1.1	5	5.1	2	2.5	CRAWFORD	WI	349	-52	114	69	23
17	2.3	34	4.0	46	4.5	DANE	WI	72	14	96	63	58
7	2.3	6	1.7	15	4.0	DODGE	WI	-26	138	76	57	47
2	1.7	3	2.1	7	5.2	DOOR	WI	27	146	212	84	70
6	2.2	10	3.9	10	4.2	DOUGLAS	WI	73	7	85	60	50
4	2.6	2	1.4	9	5.1	DUNN	WI	-46	257	93	62	68
11	4.1	10	3.6	16	5.0	EAU CLAIRE	WI	-13	38	21	36	66
		1	4.2			FLORENCE	WI					
12	3.2	16	4.2	10	2.4	FOND DU LAC	WI	29	-43	-27	15	22
1	1.6	3	5.9	2	3.1	FOREST	WI	259	-47	89	61	32
2	.8	3	1.4	8	3.3	GRANT	WI	77	136	319	91	36
1	.7	6	3.8	5	2.9	GREEN	WI	465	-23	335	92	29
1	1.1	2	1.7	3	2.2	GREEN LAKE	WI	57	26	98	64	20
2	1.7	3	2.6	2	2.0	IOWA	WI	52	-22	18	34	18
2	3.7	2	4.3	2	3.6	IRON	WI	18	-17	-2	26	40
2	2.1	4	4.3	2	1.9	JACKSON	WI	103	-56	-11	23	16
6	2.2	6	2.0	8	2.5	JEFFERSON	WI	-12	28	13	32	24
3	2.4	8	7.6	6	5.4	JUNEAU	WI	216	-29	126	71	73
11	2.5	13	2.6	29	5.6	KENOSHA	WI	6	114	127	71	76
1	1.1	3	3.2	3	2.8	KEWAUNEE	WI	190	-12	156	77	27
6	1.9	10	2.8	21	5.3	LA CROSSE	WI	48	90	181	80	71
		5	5.3	3	3.2	LAFAYETTE	WI		-41			33
2	1.7	6	5.0	4	2.8	LANGLADE	WI	188	-43	64	53	28
4	3.1	2	1.4	10	6.7	LINCOLN	WI	-54	362	114	68	86
13	4.3	14	3.6	14	3.1	MANITOWOC	WI	-15	-15	-28	15	32
9	2.2	15	3.4	29	6.1	MARATHON	WI	49	82	171	79	80
6	3.2	8	3.7	8	3.5	MARINETTE	WI	16	-5	11	31	39
1	1.3	1	1.2	3	4.1	MARQUETTE	WI	-8	232	206	83	49
11	3.3	14	3.8	30*	7.4	MENOMINEE#	WI	15	96	124	71	90
165*	4.1	209*	4.9	238*	5.4	MILWAUKEE	WI	17	11	30	40	72
2	1.1	5	2.6	7	3.4	MONROE	WI	127	32	201	83	36
6	6.1	3	2.3	6	3.3	ONEIDA	WI	-61	43	-45	08	36
9	2.6	17	4.1	23	4.7	OUTAGAMIE	WI	60	15	83	60	61
2	1.6	3	1.6	13	5.5	OZAUKEE	WI	6	235	254	87	74
				3	6.1	PEPIN	WI					80
3	2.4	3	2.3	3	2.0	PIERCE	WI	-6	-10	-15	21	18
5	3.2	3	1.6	10	4.9	POLK	WI	-50	205	53	49	64
7	4.4	8	4.2	5	2.4	PORTAGE	WI	-5	-42	-45	08	22
1	.9	2	1.8	1	.7	PRICE	WI	94	-60	-22	17	10
23	4.8	24	3.6	29	4.2	RACINE	WI	-24	17	-11	23	52
4	3.8	2	1.9	3	2.6	RICHLAND	WI	-49	36	-30	14	25
19	4.2	20	4.0	20	3.7	ROCK	WI	-5	-9	-14	21	41
4	3.8	4	4.9	4	4.4	RUSK	WI	27	-10	14	33	55
2	1.6	6	4.0	8	4.4	ST CROIX	WI	157	11	185	81	55
6	3.0	5	2.5	10	4.0	SAUK	WI	-16	61	35	42	47
4	5.7	3	4.0	5	5.8	SAWYER	WI	-30	47	3	28	78
13	3.2	14	2.7	24	4.8	SHEBOYGAN	WI	-15	76	50	48	62
5	4.6	9*	8.6	6	5.1	TAYLOR	WI	86	-41	10	31	68
2	1.2	7	4.1	6	3.7	TREMPEALEAU	WI	234	-11	199	82	42
5	3.1	1	.5	9	4.8	VERNON	WI	-83	814	57	50	63
1	3.0	1	.9	6	7.6	VILAS	WI	-68	702	157	77	91
5	2.1	7	2.2	15	4.5	WALWORTH	WI	6	100	112	68	57
1	1.2	1	1.4			WASHBURN	WI	18				
3	1.5	7	3.1	13	4.9	WASHINGTON	WI	102	60	223	85	65
18	3.6	26	3.9	34	3.8	WAUKESHA	WI	7	-2	5	29	43
9	3.9	5	1.8	11	4.5	WAUPACA	WI	-53	147	16	34	56
		2	2.0	3	3.3	WAUSHARA	WI		62			35
10	2.1	19	3.8	31	5.4	WINNEBAGO	WI	79	43	156	77	73
5	2.1	10	3.4	15	4.5	WOOD	WI	61	30	109	67	57
37	2.9	35	2.5	52	3.3	WYOMING	WY	-15	34	13		
		4	5.3	3	3.9	ALBANY	WY		-28			44
		1	1.9	4	6.7	BIG HORN	WY		256			86
1	3.7			3	6.6	CAMPBELL	WY			79	58	85
2	3.3	2	3.0	3	4.0	CARBON	WY	-8	32	22	36	47
1	3.9	1	3.0	1	3.1	CONVERSE	WY	-23	1	-22	18	31
1	3.9	1	5.6			CROOK	WY	45				
				3	2.6	FREMONT	WY					24
2	3.1	3	4.6	2	2.8	GOSHEN	WY	50	-38	-7	24	28
1	2.6			3	8.5	HOT SPRINGS	WY			225	86	94
				1	1.9	JOHNSON	WY					17
5	3.3	7	3.7	7	3.0	LARAMIE	WY	11	-19	-10	23	30

* SIGNIFICANTLY MORE DEATHS THAN EXPECTED

COUNTIES COMBINED; SEE TECHNICAL NOTES

WHITE MALE

CANCER OF THE LIVER AND GALLBLADDER INCLUDING BILE DUCTS

ICD 155,156

1950-1959 DEATHS	1950-1959 RATE	1960-1969 DEATHS	1960-1969 RATE	1970-1979 DEATHS	1970-1979 RATE	COUNTY AND STATE	PERCENT CHANGE 60S/50S	PERCENT CHANGE 70S/60S	PERCENT CHANGE 70S/50S	PERCENTILE %CHANGE 70S/50S	PERCENTILE RATES 1970S
2	5.3	1	2.2	2	4.4	LINCOLN WY	-59	104	-17	20	55
9*	5.6	4	2.1	6	3.4	NATRONA WY	-63	63	-39	10	38
2	8.4					NIOBRARA WY					
3	4.6	3	3.3	2	1.9	PARK-TETON#WY	-28	-42	-58	04	17
1	2.0	1	2.9			PLATTE WY	45				
3	2.4	1	.9	1	.8	SHERIDAN WY	-63	-6	-65	02	10
		-		1	5.7	SUBLETTE WY					76
3	4.7	2	2.2	4	4.1	SWEETWATER WY	-54	88	-14	21	48
		1	2.5	2	4.7	UINTA WY		85			61
1	3.2	1	4.6	3	9.6	WASHAKIE WY	46	108	203	83	95

* SIGNIFICANTLY MORE DEATHS THAN EXPECTED

COUNTIES COMBINED; SEE TECHNICAL NOTES

WHITE FEMALE — CANCER OF THE LIVER AND GALLBLADDER INCLUDING BILE DUCTS — ICD 155,156

1950-1959 DEATHS	1950-1959 RATE	1960-1969 DEATHS	1960-1969 RATE	1970-1979 DEATHS	1970-1979 RATE	COUNTY AND STATE	PERCENT CHANGE 60S/50S	PERCENT CHANGE 70S/60S	PERCENT CHANGE 70S/50S	PERCENTILE %CHANGE 70S/50S	PERCENTILE RATES 1970S
30443	4.2	36360	3.9	44579	3.8	UNITED STATES	-8	-2	-10		
359	4.1	361	3.0	558	3.4	ALABAMA AL	-27	15	-16		
1	2.3			2	2.6	AUTAUGA AL			13	62	29
8	5.0	5	2.1	13	3.5	BALDWIN AL	-58	66	-30	31	50
3	4.5	1	1.0	2	2.2	BARBOUR AL	-77	113	-51	15	23
2	3.6	1	1.5	4	5.2	BIBB AL	-60	256	43	75	80
4	6.3	3	2.0	6	3.2	BLOUNT AL	-68	58	-49	16	43
1	4.3	2	8.8			BULLOCK AL	104				
6	7.7	4	4.6	2	1.5	BUTLER AL	-40	-68	-81	01	15
15	6.2	4	1.2	16	3.3	CALHOUN AL	-80	165	-47	16	45
3	2.3	1	.6	15*	7.2	CHAMBERS AL	-74	1098	216	95	92
2	3.2	2	2.8	4	4.5	CHEROKEE AL	-13	61	40	74	69
2	1.7	4	2.7	5	2.9	CHILTON AL	56	9	70	82	36
1	2.6	2	4.1	4	7.4	CHOCTAW AL	55	61	180	93	93
3	4.6	3	3.5	5	4.2	CLARKE AL	-23	19	-8	49	65
2	3.3	3	3.8	3	2.8	CLAY AL	16	-26	-14	44	34
1	2.2	2	3.8	2	3.0	CLEBURNE AL	72	-20	38	74	38
2	1.7	1	.6	5	2.7	COFFEE AL	-65	350	59	80	32
1	.7	9	5.3	6	2.4	COLBERT AL	706	-55	266	96	27
1	1.9	1	1.3			CONECUH AL	-31				
1	2.7	1	2.3	3	5.1	COOSA AL	-15	128	94	86	79
5	3.0	7	3.6	3	1.4	COVINGTON AL	20	-60	-52	14	14
1	1.7	3	3.9	1	1.1	CRENSHAW AL	133	-73	-37	25	11
6	3.3	11	4.3	10	3.0	CULLMAN AL	31	-31	-9	48	37
2	2.7	5	4.4	3	2.1	DALE AL	65	-53	-22	38	21
3	3.1	1	.6	7	4.4	DALLAS AL	-60	614	40	74	68
4	2.1	5	2.4	9	3.0	DE KALB AL	12	26	41	75	37
3	2.9	4	3.4	3	1.8	ELMORE AL	18	-48	-39	23	18
3	3.1	1	1.0	4	2.6	ESCAMBIA AL	-69	169	-16	42	29
21*	7.4	10	2.3	18	3.0	ETOWAH AL	-68	27	-60	09	36
4	5.5	3	3.1	2	2.0	FAYETTE AL	-43	-36	-64	08	20
5	6.3	14*	10.9	9	4.6	FRANKLIN AL	73	-58	-27	34	72
2	2.4	3	2.5	5	3.2	GENEVA AL	4	26	32	71	42
1	2.2	1	2.0	2	3.7	HALE AL	-11	86	66	81	55
2	4.5	3	5.2	1	1.8	HENRY AL	16	-65	-59	10	18
2	1.3	3	1.7	11	3.7	HOUSTON AL	29	125	190	94	55
5	3.7	7	3.8	12	5.1	JACKSON AL	5	32	38	74	78
70	4.4	67	3.0	101	3.4	JEFFERSON AL	-33	14	-23	37	47
6*	8.5	1	1.0	2	2.1	LAMAR AL	-88	102	-75	03	22
7	3.4	12	4.5	6	1.5	LAUDERDALE AL	33	-67	-56	11	15
2	2.6	4	4.2	7	6.1	LAWRENCE AL	62	42	131	90	87
1	1.1	3	2.2	6	3.1	LEE AL	106	39	188	94	40
8	8.3	5	3.5	10	4.9	LIMESTONE AL	-57	41	-40	22	77
		1	4.6	2	6.6	LOWNDES AL		43			90
				1	3.1	MACON AL					38
12	5.0	12	2.8	22	3.6	MADISON AL	-45	32	-28	33	53
2	3.7	2	3.4	2	3.1	MARENGO AL	-8	-10	-18	41	40
1	1.0	4	3.2	8	4.6	MARION AL	227	42	365	98	71
6	4.2	11	4.2	16	4.2	MARSHALL AL	0	1	1	55	65
23	3.8	24	2.6	37	3.0	MOBILE AL	-30	15	-20	39	38
3	6.1	5	7.1	6	7.3	MONROE AL	15	3	19	65	93
17	4.5	11	1.9	19	2.9	MONTGOMERY AL	-58	50	-37	25	34
14*	6.8	8	3.3	8	2.1	MORGAN AL	-51	-35	-68	05	22
1	2.7			1	1.8	PERRY AL			-35	27	17
4	6.1	1	1.2	2	2.2	PICKENS AL	-81	94	-64	08	23
4	5.7	5	5.2	5	4.1	PIKE AL	-9	-22	-29	32	62
5	6.7	2	1.9	2	1.4	RANDOLPH AL	-72	-27	-79	02	14
6	7.5	9*	8.6	9	5.7	RUSSELL AL	15	-34	-24	36	84
2	1.9	1	.9	3	1.9	ST CLAIR AL	-55	119	-2	53	19
4	3.7	6	4.3	6	2.6	SHELBY AL	17	-39	-29	32	30
3	9.9			2	5.9	SUMTER AL			-41	21	86
3	2.0	7	3.1	26*	8.9	TALLADEGA AL	58	188	354	98	96
6	5.5	5	3.2	9	4.4	TALLAPOOSA AL	-42	38	-21	39	68
5	1.8	5	1.1	21	4.3	TUSCALOOSA AL	-36	274	137	90	66
9	4.1	12	4.9	13	3.4	WALKER AL	19	-30	-17	42	48
1	2.3	4	10.6	2	3.5	WASHINGTON AL	349	-67	50	78	50
				3	8.5	WILCOX AL					96
6*	10.8	4	5.4	4	3.4	WINSTON AL	-50	-38	-69	05	46
		3	.6	15	3.5	ALASKA AK		486			
113	3.5	259	4.2	397	3.5	ARIZONA AZ	20	-16	1		
				1	3.4	APACHE AZ					47
11	7.1	14	6.1	24*	6.6	COCHISE AZ	-13	8	-7	50	90
4	7.3	6	7.9	1	1.0	COCONINO AZ	9	-88	-87	00	11
2	2.3	6	5.2	8	4.9	GILA AZ	123	-5	112	88	76
2	4.1	3	4.5	3	4.0	GRAHAM AZ	10	-12	-4	52	61
2	4.5	5*	14.8	3	9.7	GREENLEE AZ	226	-34	114	88	97
50	3.0	128	3.8	210	3.2	MARICOPA AZ	29	-17	7	59	42
		2	4.1	1	.6	MOHAVE AZ		-86			10
5*	12.0	2	4.6	1	.9	NAVAJO AZ	-62	-81	-93	00	10
25	3.2	60	4.1	93	3.9	PIMA AZ	27	-6	19	65	57
4	5.2	11	4.7	14	4.7	PINAL AZ	-9	-1	-9	48	73
2	4.2	3	5.9	11*	14.3	SANTA CRUZ AZ	39	144	240	95	99
3	2.3	12	5.3	15	3.7	YAVAPAI AZ	131	-31	60	80	54
3	3.8	7	4.5	12	4.0	YUMA AZ	18	-11	5	58	61
258	3.8	249	2.7	399	3.3	ARKANSAS AR	-29	24	-12		
2	2.4	4	3.7	5	4.0	ARKANSAS AR	53	7	64	81	61
4	6.7	4	5.1	3	3.1	ASHLEY AR	-24	-40	-55	12	39
1	1.0	1	.5	4	1.9	BAXTER AR	-48	256	86	85	19
10	3.9	8	2.5	11	2.6	BENTON AR	-36	2	-35	27	29
6	7.2	6	4.8	7	3.7	BOONE AR	-34	-23	-49	16	54

* SIGNIFICANTLY MORE DEATHS THAN EXPECTED

WHITE FEMALE — CANCER OF THE LIVER AND GALLBLADDER INCLUDING BILE DUCTS — ICD 155,156

1950-1959 DEATHS	1950-1959 RATE	1960-1969 DEATHS	1960-1969 RATE	1970-1979 DEATHS	1970-1979 RATE	COUNTY AND STATE		PERCENT CHANGE 60S/50S	PERCENT CHANGE 70S/60S	PERCENT CHANGE 70S/50S	PERCENTILE %CHANGE 70S/50S	PERCENTILE RATES 1970S
1	2.1	3	3.3	2	2.0	BRADLEY	AR	59	-41	-6	51	20
1	3.9	1	2.9	1	2.8	CALHOUN	AR	-27	-4	-30	31	33
4	4.5			4	2.9	CARROLL	AR			-36	25	35
		1	2.1	3	5.1	CHICOT	AR		141			79
4	4.9	1	1.0	5	3.9	CLARK	AR	-79	280	-20	39	58
5	4.3	2	1.4	13*	7.3	CLAY	AR	-68	432	70	82	93
3	5.9	1	1.5			CLEBURNE	AR	-74				
1	3.3	1	2.5	3	7.5	CLEVELAND	AR	-25	202	128	89	93
3	3.6	3	2.8	3	1.9	COLUMBIA	AR	-23	-31	-47	17	19
3	4.3	2	2.4	2	2.1	CONWAY	AR	-44	-14	-52	14	21
8	4.2	8	3.2	16	5.2	CRAIGHEAD	AR	-24	60	22	67	79
3	2.9	6	4.2	8	4.2	CRAWFORD	AR	44	0	44	76	65
2	4.2	2	2.3	1	.8	CRITTENDEN	AR	-45	-66	-81	01	10
3	4.4	2	2.4	6	5.7	CROSS	AR	-45	135	30	70	84
3	7.3	1	1.8	1	1.7	DALLAS	AR	-76	-4	-77	02	16
2	5.2	3	5.0	2	3.3	DESHA	AR	-4	-33	-36	26	46
3	5.3	3	5.0	2	2.6	DREW	AR	-6	-49	-52	14	29
1	1.0	3	1.9	3	1.6	FAULKNER	AR	94	-16	63	81	16
3	4.8			3	3.5	FRANKLIN	AR			-28	33	49
2	4.1	2	2.8			FULTON	AR	-31				
14	5.8	11	2.8	25	4.6	GARLAND	AR	-51	63	-21	38	71
1	2.2	2	3.6	5	5.6	GRANT	AR	68	57	163	92	84
7	5.7	6	3.4	9	4.2	GREENE	AR	-41	25	-26	34	65
4	5.6	5	4.9	5	4.0	HEMPSTEAD	AR	-12	-20	-29	31	60
1	1.0	1	.7	8	5.1	HOT SPRING	AR	-26	602	418	98	78
2	3.4	3	3.9	4	6.3	HOWARD	AR	15	60	84	84	88
6	5.9	4	2.4	7	4.2	INDEPENDENCE	AR	-59	70	-30	31	64
1	2.0	1	1.6	3	5.6	IZARD	AR	-20	255	183	93	83
4	4.5	1	1.0	8	6.0	JACKSON	AR	-79	522	33	72	87
6	3.3	7	2.8	10	3.1	JEFFERSON	AR	-13	9	-5	51	39
2	2.5			4	3.8	JOHNSON	AR			52	78	57
1	2.7	1	2.4	1	1.5	LAFAYETTE	AR	-8	-37	-42	20	15
4	4.0	4	3.9	5	3.6	LAWRENCE	AR	-4	-8	-12	46	51
1	2.6	1	2.3	1	2.0	LEE	AR	-9	-16	-24	36	20
2	6.2			1	1.7	LINCOLN	AR			-73	04	16
				2	2.6	LITTLE RIVER	AR					30
		7	5.5	5	3.0	LOGAN	AR		-45			38
3	2.8	4	3.3	7	4.4	LONOKE	AR	19	32	57	79	68
2	3.4	5	7.8	3	4.3	MADISON	AR	126	-45	25	68	66
3	8.8	4	7.3			MARION	AR	-17				
5	4.6	4	2.7	7	3.6	MILLER	AR	-42	33	-23	37	51
6	3.4	5	2.2	7	2.8	MISSISSIPPI	AR	-34	26	-16	42	34
		3	5.3	3	4.4	MONROE	AR		-17			69
2	5.8	5*	13.2	1	3.3	MONTGOMERY	AR	127	-75	-44	18	44
3	5.8	1	2.7	8*	10.6	NEVADA	AR	-52	288	85	85	98
2	6.4	2	5.8	2	4.3	NEWTON	AR	-10	-26	-33	28	66
3	5.3	3	2.6	8	5.1	OUACHITA	AR	-51	93	-5	51	78
1	3.5	2	5.2			PERRY	AR	48				
3	4.0	1	1.3	3	2.9	PHILLIPS	AR	-69	133	-27	33	36
1	2.0			5	6.7	PIKE	AR			235	95	90
3	2.6	3	2.4	10	6.8	POINSETT	AR	-6	179	164	92	91
8*	9.7	3	3.7	6	6.7	POLK	AR	-62	81	-31	30	90
3	2.8	3	1.9	4	1.8	POPE	AR	-33	-6	-37	25	18
1	2.2	1	2.6	2	3.6	PRAIRIE	AR	20	40	68	82	53
20	2.5	31	2.8	37	2.5	PULASKI	AR	10	-10	-1	54	27
4	5.7	3	3.5	5	4.7	RANDOLPH	AR	-39	35	-17	41	73
		2	2.6			ST FRANCIS	AR					
7	5.6	5	3.0	8	3.5	SALINE	AR	-47	17	-38	24	49
2	4.3	5*	9.5	4	6.3	SCOTT	AR	122	-34	46	76	88
		1	1.6	4	6.6	SEARCY	AR		321			90
14	4.3	14	3.3	13	2.4	SEBASTIAN	AR	-23	-28	-44	18	26
3	4.8			2	2.1	SEVIER	AR			-56	12	22
1	2.5			3	2.8	SHARP	AR			11	61	34
2	6.1			1	1.8	STONE	AR			-70	05	18
2	1.3	1	.5	8	3.1	UNION	AR	-63	537	136	90	40
4	9.9	1	1.4			VAN BUREN	AR	-86				
9	3.8	9	2.5	12	2.3	WASHINGTON	AR	-34	-10	-41	21	24
5	3.1	3	1.2	13	4.2	WHITE	AR	-61	249	35	73	65
2	5.8	1	1.8	1	1.8	WOODRUFF	AR	-68	-2	-69	05	18
		2	1.9	1	.8	YELL	AR		-58			10
2526	4.2	3080	3.6	3958	3.6	CALIFORNIA	CA	-15	-1	-16		
158	4.0	160	3.3	181	3.3	ALAMEDA	CA	-18	2	-16	42	46
1	1.8	3	5.0	3	2.4	AMADOR	CA	171	-51	32	71	27
10	2.6	13	1.9	30	3.3	BUTTE	CA	-25	71	27	69	45
2	3.5	1	.9	5	3.8	CALAVERAS	CA	-74	309	8	60	56
				4	5.4	COLUSA	CA					81
36	3.5	55	3.2	77	2.8	CONTRA COSTA	CA	-10	-12	-21	38	33
		2	2.9	4	5.2	DEL NORTE	CA		78			80
2	2.1	5	2.9	9	3.1	EL DORADO	CA	36	7	46	76	40
67*	5.7	69	4.1	107*	4.8	FRESNO	CA	-28	18	-15	43	74
		4	3.9	7	5.5	GLENN	CA		41			83
8	2.3	9	2.0	15	2.8	HUMBOLDT	CA	-12	38	21	66	33
11	6.3	8	3.3	16	5.7	IMPERIAL	CA	-47	71	-10	48	84
1	3.2	1	1.5	4	3.8	INYO	CA	-53	151	17	64	56
36	5.1	47	4.2	55	3.4	KERN	CA	-17	-19	-33	28	48
3	1.7	7	3.3	15	5.4	KINGS	CA	89	64	211	94	81
2	2.1	3	3.3	7	2.5	LAKE	CA	58	-25	19	65	27
1	1.7	2	2.9	1	1.2	LASSEN	CA	76	-58	-26	35	12
1157	4.5	1286	3.8	1357	3.6	LOS ANGELES	CA	-15	-5	-19	40	53
7	5.1	6	3.1	15	5.9	MADERA	CA	-39	90	16	63	86
16	3.3	18	2.3	42	3.9	MARIN	CA	-31	68	16	63	58
				3	5.8	MARIPOSA	CA					85
6	2.9	7	2.6	14	4.2	MENDOCINO	CA	-11	65	47	77	65
3	1.2	10	2.9	36*	7.8	MERCED	CA	136	169	537	99	94
4*	12.8			1	2.1	MODOC	CA			-83	01	22
				1	4.2	MONO	CA					64
12	2.3	19	2.2	38	3.3	MONTEREY	CA	-5	52	45	76	45
11	3.3	16	3.4	15	2.6	NAPA	CA	2	-25	-23	37	29

* SIGNIFICANTLY MORE DEATHS THAN EXPECTED

WHITE FEMALE

CANCER OF THE LIVER AND GALLBLADDER INCLUDING BILE DUCTS

ICD 155,156

1950-1959 DEATHS	1950-1959 RATE	1960-1969 DEATHS	1960-1969 RATE	1970-1979 DEATHS	1970-1979 RATE	COUNTY AND STATE	PERCENT CHANGE 60S/50S	PERCENT CHANGE 70S/60S	PERCENT CHANGE 70S/50S	PERCENTILE %CHANGE 70S/50S	PERCENTILE RATES 1970S
4	3.1	4	2.4	9	3.5	NEVADA CA	-22	44	13	62	49
67	3.6	149	3.3	257	3.2	ORANGE CA	-9	-1	-10	48	44
6	2.4	14	3.7	17	3.0	PLACER CA	54	-21	22	67	36
1	1.7	1	1.5	1	1.4	PLUMAS CA	-9	-8	-17	42	14
54	4.4	77	3.7	95	2.6	RIVERSIDE CA	-17	-30	-42	20	29
51	3.7	77	3.7	127	3.9	SACRAMENTO CA	-1	7	6	59	59
1	1.5	4	4.7	5	4.9	SAN BENITO CA	203	4	216	94	76
81	4.8	97	3.3	135	3.5	SAN BERNARDINO CA	-31	4	-28	33	49
111	3.4	181	3.4	279	3.4	SAN DIEGO CA	-1	0	-1	54	48
205	4.7	210	4.3	177	4.1	SAN FRANCISCO CA	-8	-5	-13	45	62
46	5.1	49	3.8	62	3.9	SAN JOAQUIN CA	-25	3	-23	37	60
12	3.6	16	3.2	27	3.3	SAN LUIS OBISPO CA	-13	6	-8	49	46
41	3.1	71	3.3	97	3.2	SAN MATEO CA	7	-4	3	57	42
25	3.7	38	3.5	55	3.4	SANTA BARBARA CA	-5	-4	-8	49	47
87	4.6	125	3.9	196	4.1	SANTA CLARA CA	-14	5	-9	48	63
20	3.3	25	2.9	40	3.4	SANTA CRUZ CA	-14	18	2	56	47
5	3.7	12	4.0	15	3.0	SHASTA CA	8	-27	-21	38	37
3	2.2	5	3.2	6	2.8	SISKIYOU CA	43	-12	25	68	33
13	3.8	13	2.8	24	3.2	SOLANO CA	-27	16	-16	42	43
31	4.1	30	2.7	62	3.6	SONOMA CA	-35	34	-13	45	52
22	3.4	34	3.5	52	3.9	STANISLAUS CA	3	9	13	62	58
2	1.9	9	5.4	7	3.1	SUTTER CA	183	-43	62	80	40
2	2.0	3	2.2	10	4.3	TEHAMA CA	9	98	115	88	67
1	4.3	2	7.1	3	6.2	TRINITY CA	64	-12	45	76	88
27	4.5	30	3.7	51	4.6	TULARE CA	-17	25	3	57	71
		5	4.4	4	2.1	TUOLUMNE CA		-52			22
42*	7.2	35	3.3	64	3.5	VENTURA CA	-54	6	-51	14	50
10	5.5	12	4.5	15	3.5	YOLO CA	-17	-23	-36	25	49
2	1.9	1	.8	4	2.3	YUBA CA	-59	191	20	66	24
350*	5.0	354	3.7	425	3.5	COLORADO CO	-25	-7	-30		
9	6.8	16	4.8	29	5.2	ADAMS CO	-30	7	-25	36	79
3	10.2	4	9.1	2	4.0	ALAMOSA CO	-11	-56	-61	09	60
1	8.5					ARCHULETA CO					
2	8.1	1	2.7	1	2.2	BACA CO	-67	-16	-72	04	24
1	3.0	2	6.0	2	5.7	BENT CO	96	-4	88	85	85
9	2.8	6	1.1	18	2.9	BOULDER CO	-61	166	4	57	35
1	2.4			2	3.0	CHAFFEE CO			24	68	37
1	8.6	1	5.7	1	5.9	CHEYENNE CO	-34	3	-32	29	86
1	6.3			1	5.5	CLEAR CREEK CO			-13	45	82
1	2.6	4*	11.4	1	3.5	CONEJOS CO	342	-69	37	73	50
1	5.8	3*	15.4	2	17.3	COSTILLA CO	165	13	199	94	99
		2	8.0	1	3.8	CROWLEY CO		-52			57
2*	24.9			1	8.6	CUSTER CO			-65	07	96
11*	11.2	7	5.8	4	2.8	DELTA CO	-48	-51	-75	03	34
116	4.3	117	3.3	139	3.6	DENVER-ARAPAHOE#CO	-22	7	-17	42	51
1	5.0	1	4.2	2	3.7	DOUGLAS CO	-16	-11	-25	35	54
3*	21.6			1	3.7	EAGLE CO			-83	01	55
		2	8.1			ELBERT CO					
28	5.2	31	4.2	42	3.9	EL PASO CO	-18	-7	-24	36	59
14*	10.1	7	3.7	9	3.0	FREMONT CO	-64	-19	-71	05	37
1	2.0	2	2.6	3	2.7	GARFIELD CO	29	6	37	74	31
1	11.2					GRAND CO					
4	8.9	2	4.1	5*	10.2	HUERFANO CO	-55	151	14	62	98
11	3.7	16	2.9	26	2.4	JEFFERSON CO	-22	-19	-37	25	25
		2*	12.0	1	10.3	KIOWA CO		-14			98
1	2.8	2	4.9	3	5.2	KIT CARSON CO	78	5	87	85	79
2	10.6	2	20.4			LAKE CO	92				
4	6.1	7	7.9	6	5.9	LA PLATA CO	31	-25	-3	53	86
17	6.6	11	2.5	16	2.9	LARIMER CO	-62	15	-56	12	35
8	8.0	8	7.2	14*	10.7	LAS ANIMAS CO	-11	49	33	72	98
		1	3.1	2	5.3	LINCOLN CO		71			81
6	8.1	3	2.8	4	3.4	LOGAN CO	-65	21	-58	11	48
12	5.5	14	4.7	10	2.3	MESA CO	-14	-52	-59	10	24
		1*	37.7			MINERAL CO					
1	3.5	2	7.4			MOFFAT CO	113				
1	2.0	4	7.3	2	3.2	MONTEZUMA CO	261	-56	59	80	43
5	7.5	3	3.4	6	4.9	MONTROSE CO	-55	44	-35	27	75
5	6.9	3	2.6	6	5.4	MORGAN CO	-63	110	-22	37	82
7	6.1	4	2.4	11	7.5	OTERO CO	-60	211	23	67	93
		2*	20.9			OURAY CO					
				2*	15.1	PARK CO					99
1	3.8	2	6.7	2	3.5	PHILLIPS CO	77	-48	-8	49	49
7*	10.9	3	4.1	3	3.8	PROWERS CO	-63	-7	-65	07	56
24	4.9	34*	5.8	18	2.6	PUEBLO CO	18	-54	-46	17	30
		1	5.3			RIO BLANCO CO					
3	6.9	2	3.5	1	1.5	RIO GRANDE CO	-50	-56	-78	02	15
2	6.7	2	6.3	1	3.1	ROUTT CO	-5	-51	-53	13	41
3*	19.7			2	11.2	SAGUACHE CO			-43	19	98
				1	8.8	SAN MIGUEL CO					96
				1	3.8	SEDGWICK CO					57
				1	8.5	SUMMIT CO					96
		1	5.6	2	8.9	TELLER CO		59			97
1	3.5	1	2.9	1	1.7	WASHINGTON CO	-17	-43	-53	13	16
18	6.0	14	3.5	16	2.9	WELD CO	-42	-18	-52	14	35
		1	1.3	1	.9	YUMA CO		-31			10
543*	4.8	542	3.6	706	3.8	CONNECTICUT CT	-24	4	-21		
147*	5.1	164	4.2	188	3.9	FAIRFIELD CT	-17	-7	-23	37	59
141	4.8	148	3.8	186	3.8	HARTFORD CT	-21	0	-21	38	56
27	4.3	20	2.5	37	3.6	LITCHFIELD CT	-42	42	-18	41	51
16	3.6	14	2.4	30	4.1	MIDDLESEX CT	-33	70	15	62	63
152	5.0	132	3.4	192	3.9	NEW HAVEN CT	-32	15	-22	38	58
31	3.7	35	3.4	37	2.8	NEW LONDON CT	-10	-15	-24	36	34
14	5.9	14	4.5	15	3.6	TOLLAND CT	-24	-20	-39	23	52
15	3.9	15	3.2	21	3.6	WINDHAM CT	-19	13	-8	49	52

* SIGNIFICANTLY MORE DEATHS THAN EXPECTED

COUNTIES COMBINED; SEE TECHNICAL NOTES

WHITE FEMALE

CANCER OF THE LIVER AND GALLBLADDER INCLUDING BILE DUCTS

ICD 155,156

1950-1959 DEATHS	1950-1959 RATE	1960-1969 DEATHS	1960-1969 RATE	1970-1979 DEATHS	1970-1979 RATE	COUNTY AND STATE	PERCENT CHANGE 60S/50S	PERCENT CHANGE 70S/60S	PERCENT CHANGE 70S/50S	PERCENTILE %CHANGE 70S/50S	PERCENTILE RATES 1970S
61	4.0	92	4.5	101	3.6	DELAWARE DE	10	-19	-10		
6	3.0	10	4.3	11	3.3	KENT DE	42	-24	9	60	44
40	3.9	68	4.8	76	4.1	NEW CASTLE DE	24	-15	6	59	63
15	5.4	14	3.4	14	2.4	SUSSEX DE	-37	-31	-57	11	26
83	2.9	81	3.0	72	3.3	DIST. OF COLUMBIA DC	4	10	14		
83	2.9	81	3.0	72	3.3	WASHINGTON DC	4	10	14	62	45
388	2.3	969	2.8	1835	3.0	FLORIDA FL	24	9	35		
3	2.4	8	3.4	6	1.4	ALACHUA FL	45	-58	-40	22	14
				3	6.4	BAKER FL					89
1	1.0	9	4.1	10	2.4	BAY FL	311	-41	141	90	26
1	2.2	1	1.8	2	3.0	BRADFORD FL	-16	64	37	74	38
10	4.6	14	2.4	37	3.2	BREVARD FL	-48	32	-31	30	42
23	2.2	88	2.7	197	2.7	BROWARD FL	26	0	26	68	32
				2	2.9	CALHOUN FL					35
		2	1.3	11	1.8	CHARLOTTE FL		42			18
		3	2.3	12	2.9	CITRUS FL		25			36
1	2.7	1	.7	6	3.3	CLAY FL	-76	400	21	66	44
2	7.3	2	1.3	6	1.3	COLLIER FL	-82	2	-82	01	13
		1	1.1	1	.6	COLUMBIA FL		-43			10
89	2.7	208	3.2	434*	4.3	DADE FL	17	33	56	79	66
2	3.5			4	4.4	DE SOTO FL			24	67	67
		1	7.3	1	3.9	DIXIE FL		-46			59
33	3.1	48	2.8	72	3.1	DUVAL FL	-10	10	0	54	41
10	3.1	14	2.4	25	2.9	ESCAMBIA FL	-23	23	-5	52	36
		1	4.9	2	3.2	FLAGLER FL		-34			43
		3	8.7	3	6.0	FRANKLIN FL		-32			87
1	.8	2	1.2	2	1.2	GADSDEN FL	39	6	48	77	13
		1	4.4			GILCHRIST FL					
1	6.1					GULF FL					
1	3.7	1	4.1	1	2.3	HAMILTON FL	12	-44	-37	24	25
		2	2.7	2	2.5	HARDEE FL		-7			27
1	2.6	1	2.0	5	2.5	HERNANDO FL	-22	26	-2	53	27
2	2.7	7	3.0	8	1.8	HIGHLANDS FL	10	-40	-34	27	18
31	2.3	67	3.1	110	3.4	HILLSBOROUGH FL	33	12	49	77	48
1	1.6	5	7.0	5	4.9	HOLMES FL	329	-30	199	94	76
1	.9	7	3.4	12	3.7	INDIAN RIVER FL	286	9	319	97	55
2	2.0	2	1.4	4	2.2	JACKSON FL	-29	56	11	60	23
		2	7.1			JEFFERSON FL					
				2	11.1	LAFAYETTE FL					98
7	2.5	16	3.0	16	2.0	LAKE FL	20	-33	-20	39	20
5	2.3	16	3.1	29	2.3	LEE FL	34	-27	-2	53	24
2	1.8	8	3.8	9	2.5	LEON FL	116	-34	41	75	28
		2	3.3			LEVY FL					
1	8.5			1	5.9	LIBERTY FL			-30	31	86
1	2.8			1	1.8	MADISON FL			-37	24	17
5	1.3	18	1.8	40	2.3	MANATEE FL	35	28	73	82	25
1	.6	5	1.6	16	2.9	MARION FL	166	75	367	98	34
		5	2.6	12	2.7	MARTIN FL		5			32
2	1.5	9	6.1	12	3.9	MONROE FL	310	-37	161	92	58
1	3.2	2	3.7			NASSAU FL	16				
		3	1.7	9	2.8	OKALOOSA FL		66			33
				5	6.4	OKEECHOBEE FL					89
19	2.1	45	3.0	84	3.9	ORANGE FL	39	30	81	84	58
2	1.1	9	4.6	6	1.7	OSCEOLA FL	311	-64	49	77	16
19	2.1	54	2.9	100	2.8	PALM BEACH FL	38	-6	30	70	32
4	1.8	15	2.2	53	3.5	PASCO FL	23	61	97	87	49
52	2.0	138	2.6	224	2.7	PINELLAS FL	30	6	38	74	32
17	2.5	33	3.1	51	3.0	POLK FL	26	-4	21	66	37
2	2.2	8	6.2	8	3.2	PUTNAM FL	183	-49	45	76	42
3	2.4	1	.5	8	3.1	ST JOHNS FL	-78	509	31	71	40
4	4.6	12	5.3	18	3.4	ST LUCIE FL	17	-36	-25	36	48
2	3.4	5	6.0	7	3.8	SANTA ROSA FL	78	-37	12	61	56
6	2.2	20	2.1	58	2.6	SARASOTA FL	-5	25	20	65	30
3	2.1	6	2.7	19	3.3	SEMINOLE FL	25	23	53	78	45
2	4.1	2	3.5	4	3.6	SUMTER FL	-15	5	-11	47	53
1	1.8	1	1.4	2	2.0	SUWANNEE FL	-20	40	12	61	20
		2	4.7	2	3.1	TAYLOR FL		-33			41
		1	5.0	1	4.4	UNION FL		-12			68
10	1.5	28	2.1	48	2.1	VOLUSIA FL	42	0	41	75	22
				2	7.8	WAKULLA FL					94
		2	2.4	3	2.5	WALTON FL		5			28
1	2.2	2	5.0	2	2.3	WASHINGTON FL	132	-55	4	58	24
331	3.2	358	2.4	602	3.0	GEORGIA GA	-24	24	-5		
1	2.4			1	2.1	APPLING GA			-14	44	21
1	6.7			2	6.7	ATKINSON GA			0	54	91
1	4.8					BACON GA					
		1	.5	1	.6	BALDWIN GA		18			10
1	3.6	1	2.4	4	10.4	BANKS GA	-33	326	187	93	98
1	1.5	6	7.2	4	3.4	BARROW GA	385	-53	127	89	47
6	6.6	3	2.2	2	1.2	BARTOW GA	-66	-45	-81	01	13
		1	1.6			BEN HILL GA					
1	2.9			1	1.7	BERRIEN GA			-40	22	17
10	3.0	10	2.0	17	2.9	BIBB GA	-33	43	-3	52	36
				1	1.4	BLECKLEY GA					14
3*	19.9			1	3.7	BRANTLEY GA			-81	01	55
1	2.2	1	1.9	1	1.2	BROOKS GA	-16	-37	-47	17	12
				2	8.5	BRYAN GA					96
2	2.8			1	.7	BULLOCH GA			-75	03	10
		2	5.2	2	4.4	BURKE GA		-15			68
2	7.6			3	4.8	BUTTS GA			-37	25	74
1	8.0	1	5.9			CALHOUN GA	-26				
1	13.5					CAMDEN GA					

* SIGNIFICANTLY MORE DEATHS THAN EXPECTED

WHITE FEMALE

CANCER OF THE LIVER AND GALLBLADDER INCLUDING BILE DUCTS

ICD 155,156

1950-1959 DEATHS	1950-1959 RATE	1960-1969 DEATHS	1960-1969 RATE	1970-1979 DEATHS	1970-1979 RATE	COUNTY AND STATE	PERCENT CHANGE 60S/50S	PERCENT CHANGE 70S/60S	PERCENT CHANGE 70S/50S	PERCENTILE %CHANGE 70S/50S	PERCENTILE RATES 1970S
		1	4.1	1	2.2	CANDLER GA		-46			23
5	4.3	5	2.6	8	3.3	CARROLL GA	-38	24	-23	37	45
2	4.0	4	4.7	7	4.4	CATOOSA GA	20	-6	12	61	69
				2	9.1	CHARLTON GA					97
7	1.9	17	2.8	17	2.4	CHATHAM GA	45	-14	25	68	26
2	2.8	2	2.0	8	7.0	CHATTOOGA GA	-29	255	152	91	92
4	5.1	6	4.8	4	2.1	CHEROKEE GA	-5	-56	-58	11	22
5	4.1	1	.6	6	2.4	CLARKE GA	-86	305	-42	20	26
		1	8.4			CLAY GA					
1	1.1	3	1.8	8	2.8	CLAYTON GA	55	57	143	90	33
1	10.2			1	5.6	CLINCH GA			-45	18	83
12	5.7	11	2.3	25	3.1	COBB GA	-59	33	-46	17	39
1	2.0	3	3.9	5	5.3	COFFEE GA	97	36	169	92	80
1	1.0	4	3.0	6	3.6	COLQUITT GA	198	22	263	96	53
		2	6.2	2	2.8	COLUMBIA GA		-55			33
1	2.5	1	2.3	1	1.7	COOK GA	-8	-26	-32	29	16
4	4.7	3	2.6	7	4.4	COWETA GA	-44	68	-6	51	68
1	5.8					CRAWFORD GA					
				3	3.7	CRISP GA					55
1	3.3			1	2.0	DADE GA			-40	22	20
1	7.1	1	4.8	4*	18.3	DAWSON GA	-31	278	159	92	99
2	3.6			1	1.2	DECATUR GA			-67	06	12
25	3.8	21	1.8	39	2.4	DE KALB GA	-53	36	-36	25	27
1	1.6	2	3.2	3	3.3	DODGE GA	99	4	107	87	46
2	6.9	2	5.3	1	1.9	DOOLY GA	-23	-64	-73	04	19
4	2.6	2	1.2	9	3.7	DOUGHERTY GA	-54	215	46	76	55
1	2.3	3	3.6	5	3.7	DOUGLAS GA	56	1	58	79	54
1	3.3	2	4.7			EARLY GA	41				
1	3.5	1	3.2	2	3.5	EFFINGHAM GA	-8	10	1	55	50
1	1.3			5	5.4	ELBERT GA			310	97	82
4	9.6	1	2.1	4	4.4	EMANUEL GA	-78	109	-54	13	69
		2	9.2	1	3.1	EVANS GA		-66			40
2	4.3	2	2.6	3	3.1	FANNIN GA	-39	16	-29	32	39
		3	7.1			FAYETTE GA					
11	4.8	8	2.5	8	1.8	FLOYD GA	-47	-27	-61	09	18
3	6.7	2	3.3	5	5.1	FORSYTH GA	-51	54	-24	36	79
4	6.6	4	5.9	4	4.3	FRANKLIN GA	-10	-27	-35	27	67
52	3.2	54	2.6	56	2.3	FULTON GA	-20	-11	-29	32	25
2	4.9	1	2.3	7*	10.4	GILMER GA	-54	359	110	88	98
2	1.6			7	3.0	GLYNN GA			92	86	37
3	3.8	1	.9	9	6.8	GORDON GA	-75	623	80	84	91
				2	2.3	GRADY GA					25
1	2.9	1	2.3			GREENE GA	-22				
1	.7	8	3.9	12	3.2	GWINNETT GA	495	-19	383	98	42
2	3.0	3	2.8	1	.8	HABERSHAM GA	-5	-73	-74	03	10
7	4.4	6	2.8	15	4.5	HALL GA	-36	59	2	56	70
4	6.3	1	1.1	5	5.1	HARALSON GA	-82	344	-20	39	78
1	2.1	3	4.8	3	3.2	HART GA	131	-33	56	79	44
1	5.9	1	4.3	2	5.3	HEARD GA	-27	23	-10	47	81
4	7.6	1	1.3	4	3.1	HENRY GA	-83	145	-59	10	39
4	10.4	4	3.9	8	4.6	HOUSTON GA	-62	18	-55	12	72
1	1.8	2	2.1	10*	7.4	JACKSON GA	17	258	319	97	93
				1	2.6	JASPER GA					29
2	7.1	1	5.0	5*	11.3	JEFF DAVIS GA	-30	128	60	80	99
1	2.0			1	1.3	JEFFERSON GA			-34	27	13
		1	4.2	1	3.5	JENKINS GA		-17			49
		1	2.5			JOHNSON GA					
1	6.7			1	2.2	JONES GA			-67	06	24
2	7.6	1	2.2	2	3.1	LAMAR GA	-71	39	-59	10	40
		1	6.2	1	4.7	LANIER GA		-25			73
3	3.5	3	2.7	7	4.4	LAURENS GA	-24	66	27	69	68
2*	23.3			1	2.8	LEE GA			-88	00	33
				2	3.4	LIBERTY GA					47
				2	9.1	LINCOLN GA					97
		1	8.5			LONG GA					
		3	2.2	4	1.7	LOWNDES GA		-22			17
1	3.0	1	3.6	2	4.7	LUMPKIN GA	19	29	53	78	72
		1	2.1	5	8.8	MC DUFFIE GA		319			96
				1	1.6	MACON GA					16
		1	1.9	4	5.8	MADISON GA		197			85
1	1.4	2	2.3	7*	9.1	MERIWETHER GA	61	294	533	99	97
1	4.3			1	2.7	MILLER GA			-36	26	32
		2	3.0	1	1.2	MITCHELL GA		-61			12
1	4.4	1	3.1	1	1.6	MONROE GA	-29	-49	-64	08	15
		1	3.8			MONTGOMERY GA					
1	3.0	2	4.7	1	2.2	MORGAN GA	55	-53	-27	34	23
1	2.4	1	2.1	3	4.1	MURRAY GA	-12	92	69	82	62
5	1.4	15	3.2	16	2.6	MUSCOGEE#GA	126	-19	84	84	30
1	1.8	1	1.0	1	.8	NEWTON GA	-41	-26	-56	11	10
2	7.2	1	3.7	3	7.1	OCONEE GA	-49	94	-1	54	92
		1	3.5	1	4.0	OGLETHORPE GA		15			60
2	4.5	2	3.3	1	.8	PAULDING GA	-26	-76	-83	01	10
		2	5.7	1	2.1	PEACH GA		-63			22
		1	2.2	2	3.2	PICKENS GA		46			42
1	3.7					PIERCE GA					
1	3.3	1	3.0			PIKE GA	-9				
5	5.4	2	1.5	8	4.4	POLK GA	-73	196	-19	40	68
1	5.4			2	6.5	PULASKI GA			20	65	89
2	11.5					PUTNAM GA					
3	9.2			2	2.7	RABUN GA			-70	05	32
2	7.1	1	2.8	1	2.3	RANDOLPH GA	-61	-17	-68	06	25
10	3.4	9	2.4	23	4.2	RICHMOND GA	-30	77	24	67	64
1	4.5	1	1.7	1	1.0	ROCKDALE GA	-62	-41	-77	02	11
				3*	23.0	SCHLEY GA					99
				1	1.8	SCREVEN GA					18
1	5.2	2	7.7			SEMINOLE GA	48				
4	3.5	3	2.0	4	2.0	SPALDING GA	-43	-1	-44	19	20
1	1.6	3	3.5			STEPHENS GA	115				
1	5.9			2	9.4	STEWART GA			60	80	97
2	3.0			4	3.7	SUMTER GA			24	68	55

* SIGNIFICANTLY MORE DEATHS THAN EXPECTED

COUNTIES COMBINED; SEE TECHNICAL NOTES

WHITE FEMALE

CANCER OF THE LIVER AND GALLBLADDER INCLUDING BILE DUCTS

ICD 155,156

1950-1959 DEATHS	1950-1959 RATE	1960-1969 DEATHS	1960-1969 RATE	1970-1979 DEATHS	1970-1979 RATE	COUNTY AND STATE	PERCENT CHANGE 60S/50S	PERCENT CHANGE 70S/60S	PERCENT CHANGE 70S/50S	PERCENTILE %CHANGE 70S/50S	PERCENTILE RATES 1970S
1	7.7			3*	14.6	TALBOT GA			90	86	99
4	8.8	1	1.8	3	3.8	TATTNALL GA	-80	109	-58	11	56
		1	4.9	1	2.7	TAYLOR GA		-46			30
		1	1.9	2	3.1	TELFAIR GA		60			39
		2	6.7	3	5.1	TERRELL GA		-24			78
3	3.7	1	.7	4	3.1	THOMAS GA	-82	362	-17	41	40
				4	3.4	TIFT GA					47
1	2.2	1	1.1	2	2.1	TOOMBS GA	-52	101	-3	52	22
1	5.6	2	7.9	2	5.1	TOWNS GA	42	-36	-9	48	78
1	5.3	1	5.1	2	7.4	TREUTLEN GA	-4	45	39	74	93
1	.6	4	2.2	9	4.4	TROUP GA	279	97	648	99	69
		2	5.6	1	1.9	TURNER GA		-66			19
		1	5.2			TWIGGS GA					
		3	8.0	3	3.8	UNION GA		-53			56
5	6.8	2	2.0	4	3.2	UPSON GA	-71	58	-53	13	41
4	2.5	7	3.2	8	2.8	WALKER GA	28	-12	12	61	33
2	3.4	2	1.9	3	2.1	WALTON GA	-45	11	-39	23	21
		6	4.7	2	1.1	WARE GA		-78			11
		1	3.6			WARREN GA					
2	3.8	1	1.5	2	2.8	WASHINGTON GA	-59	80	-26	34	33
1	2.8	3	4.7	4	4.4	WAYNE GA	69	-6	59	80	69
		2	8.1	2	8.3	WHEELER GA		2			95
1	3.3	1	2.9	2	4.4	WHITE GA	-11	50	33	72	67
12*	8.8	8	4.5	11	4.1	WHITFIELD GA	-49	-7	-53	13	63
1	3.4			1	2.4	WILKES GA			-31	30	25
		1	2.9	2	5.8	WILKINSON GA		100			86
1	4.4	3	6.2	2	2.8	WORTH GA	41	-55	-36	25	33
		22	3.8	37	3.9	HAWAII HI		3			
		3	7.0	6	5.8	HAWAII HI		-17			85
		15	3.1	25	3.4	HONOLULU HI		9			46
		1	5.2	1	4.1	KAUAI HI		-20			63
		3	7.9	5	6.8	MAUI#HI		-14			91
109	4.5	114	3.5	131	3.0	IDAHO ID	-21	-15	-33		
20	5.6	15	2.7	19	2.6	ADA ID	-52	-2	-53	13	30
		2*	28.6	1	6.3	ADAMS ID		-78			88
4	2.7	6	3.4	8	3.0	BANNOCK ID	26	-12	11	61	37
4*	13.9	1	3.0	1	3.4	BEAR LAKE ID	-78	12	-76	03	47
		1	3.9	1	1.9	BENEWAH ID		-51			19
5	6.2	2	1.8	5	4.0	BINGHAM ID	-71	123	-35	26	61
				1	3.1	BLAINE ID					41
				1	9.9	BOISE ID					98
1	2.1	5	6.1	3	2.7	BONNER ID	194	-56	29	70	30
8	7.4	9	5.8	6	2.6	BONNEVILLE ID	-22	-54	-64	07	30
1	4.6			2	6.3	BOUNDARY ID			36	73	89
17	6.2	12	3.2	12	2.5	CANYON ID	-49	-22	-60	09	27
		1	4.8	2	6.5	CARIBOU ID		34			89
2	3.6	2	2.8	1	1.2	CASSIA ID	-22	-56	-66	07	13
1	3.8					CLEARWATER ID					
1	10.5			1	6.2	CUSTER ID			-40	21	88
2	9.3					ELMORE ID					
		4	8.7			FRANKLIN ID					
		1	2.5	1	2.6	FREMONT ID		1			29
2	4.4	1	1.6	6*	7.2	GEM ID	-63	343	64	81	92
1	2.2	3	6.9	3	4.2	GOODING ID	212	-39	89	85	64
3	6.9	1	2.0	2	2.3	IDAHO ID	-72	16	-67	06	24
		3	7.5	3	5.8	JEFFERSON ID		-23			85
		2	3.4	4	5.6	JEROME ID		63			83
7	4.6	5	2.9	5	1.9	KOOTENAI ID	-38	-33	-59	10	19
6	7.6	3	2.9	5	4.0	LATAH ID	-62	37	-47	17	61
				3	8.2	LEMHI ID					95
		1	5.2			LEWIS ID					
		1	2.8	2	3.0	MADISON ID		10			38
2	4.8			2	2.9	MINIDOKA ID			-39	22	35
9	7.8	3	1.9	2	1.0	NEZ PERCE ID	-75	-46	-87	00	11
1	4.8	1	3.4	2	5.6	OWYHEE ID	-28	63	16	63	83
1	2.4	4	4.8	6	5.4	PAYETTE ID	101	12	124	89	82
1	6.3	3*	16.7	1	3.5	POWER ID	166	-79	-45	18	49
3	3.9	1	1.7	2	2.1	SHOSHONE ID	-57	23	-47	17	21
5	3.0	18*	7.5	14	4.2	TWIN FALLS ID	154	-45	40	74	64
		1	6.7	2	10.8	VALLEY ID		60			98
2	3.7	2	2.6	2	2.1	WASHINGTON ID	-30	-18	-43	19	22
2258*	5.0	2578*	4.7	2765*	4.4	ILLINOIS IL	-5	-7	-12		
17	3.9	18	3.8	18	3.4	ADAMS IL	-3	-9	-11	46	48
5	6.3	2	2.5	5	5.3	ALEXANDER IL	-60	108	-16	42	81
5	5.0	8	8.5	2	1.4	BOND IL	70	-84	-73	04	13
2	1.9	4	2.8	8	5.2	BOONE IL	45	84	166	92	80
4	7.0	1	1.6	3	3.5	BROWN IL	-78	126	-49	15	51
13	5.6	10	3.8	11	4.2	BUREAU IL	-32	12	-24	36	66
1	2.5	3	6.2	1	1.7	CALHOUN IL	146	-73	-34	28	16
5	4.2	5	3.5	8	5.8	CARROLL IL	-15	65	40	74	85
4	4.4	2	1.7	3	2.5	CASS IL	-61	43	-44	18	27
24	5.7	17	3.1	22	3.4	CHAMPAIGN IL	-45	8	-41	21	46
2	.8	6	2.3	13	4.0	CHRISTIAN IL	188	77	409	98	62
4	3.3	5	2.8	4	2.3	CLARK IL	-16	-16	-29	31	25
8	7.6	7	5.7	7	5.6	CLAY IL	-25	-2	-26	34	83
8	6.4	11*	7.2	4	2.1	CLINTON IL	12	-71	-68	06	21
9	3.6	12	3.9	9	3.1	COLES IL	10	-20	-13	45	40
1178*	5.5	1330*	5.3	1308*	4.8	COOK IL	-4	-10	-13	44	74
3	2.4	3	1.8	12*	6.8	CRAWFORD IL	-27	280	178	93	91
4	6.8	6	8.8	4	4.3	CUMBERLAND IL	30	-52	-37	25	66
13	5.5	14	4.6	14	3.6	DE KALB IL	-16	-22	-34	27	53
2	1.7	3	2.2	2	1.7	DE WITT IL	25	-22	-3	53	16
6	5.6	2	1.8	7	5.5	DOUGLAS IL	-69	211	-3	53	82
33	3.7	62	4.4	89	3.7	DU PAGE IL	17	-15	0	55	55

* SIGNIFICANTLY MORE DEATHS THAN EXPECTED

COUNTIES COMBINED; SEE TECHNICAL NOTES

WHITE FEMALE — CANCER OF THE LIVER AND GALLBLADDER INCLUDING BILE DUCTS — ICD 155,156

1950-1959 DEATHS	1950-1959 RATE	1960-1969 DEATHS	1960-1969 RATE	1970-1979 DEATHS	1970-1979 RATE	COUNTY AND STATE	PERCENT CHANGE 60S/50S	PERCENT CHANGE 70S/60S	PERCENT CHANGE 70S/50S	PERCENTILE %CHANGE 70S/50S	PERCENTILE RATES 1970S
4	2.3	7	4.0	6	3.1	EDGAR IL	70	-22	33	72	40
		5	7.6	2	2.8	EDWARDS IL		-63			34
4	3.5	5	3.1	10	4.7	EFFINGHAM IL	-13	53	34	72	73
6	4.9	9	5.5	2	1.0	FAYETTE IL	11	-82	-80	02	11
5	4.8			6	4.8	FORD IL			0	55	75
20*	7.2	13	3.5	17	4.1	FRANKLIN IL	-52	19	-43	19	64
6	2.4	16	5.6	20	6.6	FULTON IL	137	18	179	93	90
2	4.6	4	8.6			GALLATIN IL	86				
9	7.8	6	4.0	10	7.4	GREENE IL	-49	86	-5	51	93
3	2.7	2	1.5	5	3.2	GPUNDY IL	-43	109	19	65	42
1	1.3	3	3.5	2	3.0	HAMILTON IL	166	-15	126	89	37
13	6.9	9	5.2	9	4.2	HANCOCK IL	-24	-19	-38	23	65
1	2.9	4*	10.3	2	5.0	HARDIN IL	249	-51	70	82	77
3	6.5			2	3.1	HENDERSON IL			-52	14	41
10	3.4	16	4.7	18	4.5	HENRY IL	35	-3	31	70	70
7	4.0	9	4.1	8	3.4	IROQUOIS IL	2	-16	-15	44	48
8	4.1	16*	7.0	21*	7.7	JACKSON IL	71	9	87	85	94
2	2.6	4	4.1	2	2.5	JASPER IL	60	-39	-2	53	28
8	4.1	5	2.2	6	2.1	JEFFERSON IL	-47	-3	-48	16	22
3	3.3	9*	9.1	5	3.7	JERSEY IL	174	-59	11	61	54
7	5.2	5	3.1	11*	7.6	JO DAVIESS IL	-41	148	46	76	94
1	2.2	1	1.8	4	7.6	JOHNSON IL	-18	321	245	95	94
36	3.7	51	4.3	65	4.7	KANE IL	16	9	27	69	72
18	3.9	19	3.6	20	3.7	KANKAKEE IL	-9	2	-7	50	53
2	2.9	11*	11.3	3	2.0	KENDALL IL	286	-82	-30	31	21
21	6.1	18	4.2	24	5.0	KNOX IL	-31	19	-18	40	77
40	4.8	55	4.4	67	4.1	LAKE IL	-7	-6	-13	45	63
26	4.3	34	4.7	35	4.0	LA SALLE IL	9	-15	-8	49	61
3	2.3	3	1.9	8	4.1	LAWRENCE IL	-18	122	82	84	63
9	4.6	8	3.4	4	1.4	LEE IL	-26	-60	-70	05	14
5	2.1	9	3.6	13	4.1	LIVINGSTON IL	76	12	97	87	62
5	2.8	11	4.9	13	4.9	LOGAN IL	73	2	76	83	76
10	5.4	6	2.8	4	1.3	MC DONOUGH IL	-47	-53	-75	03	13
16	4.8	35*	7.2	33	5.0	MC HENRY IL	51	-30	6	59	77
22	4.7	26	4.6	27	4.0	MC LEAN IL	-3	-12	-15	43	61
32	5.6	23	3.3	32	4.0	MACON IL	-41	21	-29	32	60
11	3.4	10	2.5	19	4.5	MACOUPIN IL	-26	77	31	71	70
52*	5.8	58	5.1	67	4.5	MADISON IL	-13	-12	-23	37	70
12	4.9	10	3.5	11	3.1	MARION IL	-30	-9	-36	25	41
2	2.2	2	2.1	3	2.8	MARSHALL IL	-5	35	28	69	34
5	4.9	4	3.7	6	4.9	MASON IL	-24	34	1	55	76
2	2.3	2	2.5	3	2.8	MASSAC IL	7	12	20	65	33
4	6.8			3	3.1	MENARD IL			-54	13	41
2	2.1			7	5.0	MERCER IL			136	90	77
7*	8.7	2	2.0	7	5.1	MONROE IL	-77	161	-41	21	78
9	3.7	11	4.6	14	4.8	MONTGOMERY IL	24	4	29	70	74
7	2.7	8	3.2	11	3.5	MORGAN IL	19	11	32	71	49
2	2.0	3	4.1	2	1.5	MOULTRIE IL	104	-63	-24	36	15
6	3.0	15*	6.4	12	4.7	OGLE IL	112	-27	54	78	72
35	3.7	44	3.9	49	3.9	PEORIA IL	7	0	7	59	59
2	1.7	7	4.5	7	4.5	PERRY IL	173	-1	171	92	70
5	5.6	5	5.2	4	3.7	PIATT IL	-7	-29	-34	27	54
6	4.5	7	4.4	8	3.9	PIKE IL	-4	-10	-13	44	59
1	2.8			1	3.1	POPE IL			9	60	39
4	10.0	2	3.7	4	8.5	PULASKI IL	-63	134	-15	44	96
2	7.2	4*	17.5	2	5.4	PUTNAM IL	143	-69	-24	36	82
7	4.1	7	3.5	15	6.4	RANDOLPH IL	-16	85	56	79	89
5	4.8	6	4.3	3	2.3	RICHLAND IL	-9	-47	-52	14	24
34	4.7	36	4.2	52	5.2	ROCK ISLAND IL	-11	25	12	61	80
50	5.5	41	3.5	50	3.7	ST CLAIR IL	-36	5	-33	28	54
11	6.3	21*	9.2	17*	5.5	SALINE IL	46	-40	-12	45	82
42	5.5	37	3.6	41	3.3	SANGAMON IL	-34	-10	-41	21	45
4	6.9	4	4.8	1	1.1	SCHUYLER IL	-31	-77	-84	01	11
2	3.9	3	5.3	4	6.8	SCOTT IL	35	29	75	83	91
15*	9.0	9	4.7	11	6.3	SHELBY IL	-48	35	-30	31	88
2	3.4	3	5.5	6*	8.7	STARK IL	61	59	155	91	96
8	2.8	15	4.4	16	5.0	STEPHENSON IL	56	14	77	83	77
18	5.3	19	4.2	22	3.6	TAZEWELL IL	-22	-14	-33	28	51
3	2.0	9	5.6	7	4.6	UNION IL	187	-18	136	90	72
24	4.9	21	3.8	34	5.1	VERMILION IL	-23	35	4	57	78
		7	7.3	2	2.0	WABASH IL		-72			20
5	3.3	7	4.9	6	3.1	WARREN IL	49	-36	-5	51	40
7	7.0	8	5.5	8	5.3	WASHINGTON IL	-21	-4	-25	35	81
5	4.7	6	5.0	6	4.0	WAYNE IL	6	-19	-14	44	62
10*	7.5	12*	7.5	10	5.4	WHITE IL	-1	-28	-29	32	81
29*	10.8	15	4.8	17	4.1	WHITESIDE IL	-56	-14	-62	08	62
30	4.2	41	4.6	62*	5.1	WILL IL	8	12	21	66	79
12	4.1	18	5.3	26*	6.0	WILLIAMSON IL	28	14	46	76	87
34	4.2	43	4.0	35	2.6	WINNEBAGO IL	-6	-34	-38	23	29
7	5.9	8	4.8	4	2.2	WOODFORD IL	-18	-55	-63	08	23
955*	4.6	1056*	4.2	1215*	4.0	INDIANA IN	-9	-4	-12		
9	7.3	9	6.5	4	2.4	ADAMS IN	-11	-64	-68	06	26
40	4.1	55	4.5	55	3.6	ALLEN IN	9	-19	-12	46	53
15*	7.9	9	3.8	8	2.7	BARTHOLOMEW IN	-51	-30	-66	07	31
1	1.7	2	2.8	5	6.3	BENTON IN	66	128	278	96	89
3	3.6	7	8.1	5	4.3	BLACKFORD IN	122	-46	19	65	67
7	3.9	4	2.2	9	3.9	BOONE IN	-44	78	-1	54	59
3	9.9	2	4.1	2	3.7	BROWN IN	-58	-11	-63	08	53
5	4.7	2	1.6	5	2.9	CARROLL IN	-66	84	-38	24	35
8	3.0	10	3.4	13	4.5	CASS IN	14	30	48	77	69
9	4.3	9	3.0	15	3.8	CLARK IN	-31	27	-12	45	57
11	6.6	19*	10.5	11	4.3	CLAY IN	59	-59	-35	27	67
8	4.4	5	2.5	10	4.1	CLINTON IN	-43	61	-8	49	62
3	5.3	2	3.2	2	3.0	CRAWFORD IN	-40	-7	-44	18	37
9	6.2	8	4.6	13*	6.1	DAVIESS IN	-26	32	-2	53	87
7	5.0	8	5.3	12	5.9	DEARBORN IN	6	11	17	64	86
2	1.9	6	3.7	4	2.9	DECATUR IN	100	-22	55	79	36
1	.5	4	2.3	9	4.3	DE KALB IN	335	87	714	99	67

* SIGNIFICANTLY MORE DEATHS THAN EXPECTED

WHITE FEMALE

CANCER OF THE LIVER AND GALLBLADDER
INCLUDING BILE DUCTS

ICD 155,156

1950-1959 DEATHS	1950-1959 RATE	1960-1969 DEATHS	1960-1969 RATE	1970-1979 DEATHS	1970-1979 RATE	COUNTY AND STATE	PERCENT CHANGE 60S/50S	PERCENT CHANGE 70S/60S	PERCENT CHANGE 70S/50S	PERCENTILE %CHANGE 70S/50S	PERCENTILE RATES 1970S
28*	6.6	23	4.1	19	2.8	DELAWARE IN	-38	-32	-58	11	33
10*	8.2	5	3.3	10	5.2	DUBOIS IN	-60	56	-37	24	79
22	4.6	21	3.6	30	4.1	ELKHART IN	-23	15	-11	46	62
5	4.4	8	6.0	6	3.5	FAYETTE IN	36	-42	-22	38	49
12	5.1	12	4.0	20	5.6	FLOYD IN	-22	41	11	61	83
7	6.1	7	5.7	4	2.3	FOUNTAIN IN	-6	-60	-63	08	24
3	3.1	6	6.1	8	7.7	FRANKLIN IN	101	27	154	91	94
5	4.4	6	5.5	5	3.9	FULTON IN	26	-29	-11	46	59
9	5.2	9	4.0	16*	6.0	GIBSON IN	-23	52	17	64	87
18	5.4	22	5.7	23	5.0	GRANT IN	7	-13	-7	50	77
12	6.9	5	2.8	9	3.4	GREENE IN	-59	21	-51	15	47
3	1.7	12	4.9	12	3.4	HAMILTON IN	184	-32	93	86	46
6	4.6	6	4.3	12	6.2	HANCOCK IN	-7	46	36	73	88
8*	8.7	2	1.9	7	5.2	HARRISON IN	-79	181	-40	22	80
3	2.1	8	4.2	5	2.0	HENDRICKS IN	99	-52	-4	52	20
10	4.0	12	4.3	9	2.4	HENRY IN	6	-43	-40	22	27
14	4.8	13	3.8	18	4.4	HOWARD IN	-22	16	-9	48	68
14	6.6	12	4.1	12	4.5	HUNTINGTON IN	-38	9	-32	29	69
8	5.2	12	6.8	6	2.2	JACKSON IN	31	-67	-57	11	23
3	3.6	2	1.8	4	3.3	JASPER IN	-49	78	-9	48	45
4	2.9	3	1.8	5	3.2	JAY IN	-37	74	9	60	41
5	3.8	2	1.5	5	3.2	JEFFERSON IN	-59	108	-15	43	42
5	7.8	6	6.8	2	2.1	JENNINGS IN	-13	-68	-72	04	23
7	3.8	8	3.0	15	3.6	JOHNSON IN	-21	17	-7	50	51
13	5.4	15	5.1	16	4.6	KNOX IN	-5	-10	-15	43	71
10	4.7	13	5.2	12	4.1	KOSCIUSKO IN	10	-22	-14	44	62
3	4.0	2	2.0	4	3.2	LAGRANGE IN	-51	64	-20	39	43
71*	5.5	83*	5.2	89	4.3	LAKE IN	-6	-16	-21	38	67
19	4.9	17	3.4	14	2.3	LA PORTE IN	-31	-32	-53	13	25
14*	7.4	7	3.1	19*	6.9	LAWRENCE IN	-58	124	-7	50	92
32*	6.2	20	3.1	34	4.2	MADISON IN	-49	35	-31	30	65
120	4.4	147	4.4	123	3.2	MARION IN	0	-27	-27	33	43
4	2.4	5	2.5	12	4.3	MARSHALL IN	4	70	78	83	66
		2	3.8	5	8.5	MARTIN IN		124			96
4	2.1	5	2.4	8	3.3	MIAMI IN	19	34	60	80	45
11	5.6	9	3.3	13	4.0	MONROE IN	-41	21	-29	32	60
6	3.3	8	3.3	9	3.4	MONTGOMERY IN	0	2	2	56	47
7	4.8	4	2.1	11	4.6	MORGAN IN	-55	113	-4	52	71
6*	10.5	3	3.8	7*	7.2	NEWTON IN	-64	91	-31	29	93
1	.7	4	2.2	6	2.8	NOBLE IN	203	30	293	96	34
2	7.9	1	3.5	2	9.1	OHIO IN	-55	158	16	63	97
7	6.8	2	1.8	9*	7.3	ORANGE IN	-73	299	7	59	93
1	1.4	4	3.9	6	5.6	OWEN IN	185	44	310	97	83
2	2.2	4	4.2	9*	8.4	PARKE IN	91	101	285	96	96
2	2.1	1	.9	7	5.2	PERRY IN	-55	463	151	91	79
4	4.7	12*	12.6	9*	8.7	PIKE IN	166	-31	84	65	96
9	4.3	11	4.1	14	3.6	PORTER IN	-3	-13	-16	42	51
6	5.2	4	2.9	7	5.0	POSEY IN	-43	71	-3	52	77
1	1.2	3	3.5	9*	9.5	PULASKI IN	191	168	682	99	97
5	4.0	9	5.3	8	5.1	PUTNAM IN	32	-3	29	70	79
10	5.8	5	2.7	7	3.6	RANDOLPH IN	-52	30	-38	24	51
6	4.8	12*	8.7	6	4.2	RIPLEY IN	83	-52	-12	46	64
5	4.2	6	4.7	6	4.1	RUSH IN	12	-12	-2	53	63
40	4.4	47	3.9	59	4.0	ST JOSEPH IN	-13	3	-10	47	60
1	1.9	4	5.4	10*	10.3	SCOTT IN	187	89	443	99	98
7	3.9	11	5.2	12	4.7	SHELBY IN	35	-10	22	67	73
4	4.4	7	6.9	3	2.5	SPENCER IN	57	-64	-44	18	27
5	6.4	7	6.3	7	4.8	STARKE IN	-1	-24	-25	35	74
5	5.1	3	3.1	6	4.1	STEUBEN IN	-39	32	-20	39	62
10	6.3	14*	8.0	9	4.5	SULLIVAN IN	27	-45	-30	31	69
		1	2.6			SWITZERLAND IN					
17	4.6	13	3.1	13	2.7	TIPPECANOE IN	-33	-12	-41	21	31
7	8.0	5	4.8	5	4.7	TIPTON IN	-39	-3	-41	21	73
		1	2.2	3	8.6	UNION IN		284			96
38	4.6	40	4.2	58	4.8	VANDERBURGH IN	-10	16	5	58	75
6	5.5	4	2.8	5	3.2	VERMILLION IN	-50	15	-42	20	42
25	3.9	40	5.6	37	4.4	VIGO IN	43	-21	14	62	68
5	2.6	8	3.2	10	4.3	WABASH IN	22	35	64	81	66
1	2.2	1	1.8	2	2.9	WARREN IN	-19	64	33	72	36
4	3.1	4	2.9	8	4.7	WARRICK IN	-6	63	53	78	73
1	1.0	3	2.7	3	2.5	WASHINGTON IN	168	-6	151	91	28
26*	6.7	23	5.3	20	3.9	WAYNE IN	-20	-27	-42	20	58
3	2.5	5	3.9	7	4.4	WELLS IN	52	15	74	83	68
3	2.8	5	3.7	4	2.4	WHITE IN	33	-34	-12	46	27
4	3.4	9	7.1	9	5.7	WHITLEY IN	109	-19	69	82	84
675	4.4	710	3.9	783	3.8	IOWA IA	-10	-4	-14		
2	2.7	3	3.1	6	6.1	ADAIR IA	18	96	131	90	87
2	4.0			5*	8.4	ADAMS IA			112	88	95
8	8.4	2	1.8	4	3.5	ALLAMAKEE IA	-79	96	-58	11	50
5	3.4	6	3.5	5	4.4	APPANOOSE IA	4	26	31	71	69
6*	9.3	3	3.8	1	1.9	AUDUBON IA	-59	-51	-80	02	19
5	3.7	9	5.1	6	3.5	BENTON IA	35	-30	-5	51	50
17	3.2	28	4.5	25	3.3	BLACK HAWK IA	39	-27	2	56	45
5	2.7	6	2.6	8	2.9	BOONE IA	-4	10	5	58	35
11*	10.3	5	3.7	10	6.7	BREMER IA	-64	80	-35	26	90
6	4.6	8	6.3	4	3.0	BUCHANAN IA	37	-52	-34	27	38
7	5.4	7	5.1	5	3.0	BUENA VISTA IA	-6	-40	-44	19	38
9*	8.9	4	3.2	1	.5	BUTLER IA	-64	-84	-94	00	09
2	1.8	11*	9.0	7	6.1	CALHOUN IA	396	-32	237	95	87
6	4.8	10	6.3	7	3.7	CARROLL IA	32	-42	-23	37	54
5	3.7	6	4.7	4	1.9	CASS IA	29	-59	-47	16	20
7	6.8	6	4.8	9	8.0	CEDAR IA	-30	66	17	64	95
13	5.0	13	4.0	15	3.9	CERRO GORDO IA	-20	-3	-23	37	58
3	2.4	3	2.3	5	3.8	CHEROKEE IA	-5	69	60	80	57
5	5.9	7	6.7	4	3.0	CHICKASAW IA	13	-55	-50	15	38
4	5.7	3	4.2	5	5.7	CLARKE IA	-27	35	-1	53	84
1	1.1	3	2.4	6	4.5	CLAY IA	123	86	314	97	69
3	2.3	14*	9.2	6	3.6	CLAYTON IA	311	-61	61	80	52

* SIGNIFICANTLY MORE DEATHS THAN EXPECTED

WHITE FEMALE — CANCER OF THE LIVER AND GALLBLADDER INCLUDING BILE DUCTS — ICD 155,156

1950-1959 DEATHS	1950-1959 RATE	1960-1969 DEATHS	1960-1969 RATE	1970-1979 DEATHS	1970-1979 RATE	COUNTY AND STATE	PERCENT CHANGE 60S/50S	PERCENT CHANGE 70S/60S	PERCENT CHANGE 70S/50S	PERCENTILE %CHANGE 70S/50S	PERCENTILE RATES 1970S
15	5.1	19	5.6	12	3.1	CLINTON IA	9	-44	-39	23	41
4	3.8	7	5.8	8	6.0	CRAWFORD IA	51	3	57	79	87
4	2.8	10	5.2	4	1.9	DALLAS IA	87	-64	-32	29	19
2	3.4	6*	6.6	1	1.8	DAVIS IA	157	-79	-46	17	18
5	5.5	5	5.7	4	4.2	DECATUR IA	3	-27	-24	36	64
1	1.0	2	1.6	6	5.9	DELAWARE IA	50	276	464	99	86
13	4.9	12	3.8	18	4.9	DES MOINES IA	-22	28	0	55	75
2	3.0	3	3.3	5	3.6	DICKINSON IA	11	7	19	65	51
21	4.9	23	4.8	19	3.1	DUBUQUE IA	-2	-35	-36	26	41
4	4.8	3	2.8	4	3.3	EMMET IA	-42	20	-31	30	45
12	7.0	6	3.4	13	6.5	FAYETTE IA	-51	92	-6	51	90
4	3.2	6	3.7	5	4.0	FLOYD IA	16	8	26	68	61
9*	10.3	5	4.5	3	2.3	FRANKLIN IA	-57	-49	-78	02	24
1	1.3	3	4.2			FREMONT IA	221				
5	5.5	6	4.7	4	4.2	GREENE IA	-14	-11	-23	36	64
4	4.9	4	3.6	8	7.6	GRUNDY IA	-26	107	53	78	93
5	5.0	2	1.6	3	3.0	GUTHRIE IA	-68	81	-41	21	37
5	4.1	4	2.9	6	3.8	HAMILTON IA	-29	32	-6	50	57
2	2.4	4	5.0	6	4.0	HANCOCK IA	109	-20	67	81	61
9	6.2	8	4.7	9	4.6	HARDIN IA	-23	-3	-25	35	72
6	5.2	4	2.7	6	3.7	HARRISON IA	-47	34	-29	31	53
2	1.6	6	4.4	3	2.0	HENRY IA	176	-55	24	67	20
5	6.2	1	1.2	4	3.7	HOWARD IA	-80	199	-40	22	54
8*	11.3	4	5.2	2	1.9	HUMBOLDT IA	-54	-64	-83	01	19
5	7.9	1	1.3	5	4.8	IDA IA	-84	282	-39	23	74
5	5.0	4	3.3	4	3.1	IOWA IA	-34	-7	-39	23	39
3	2.7	5	3.9	7	5.1	JACKSON IA	45	30	89	85	79
7	3.9	3	1.4	10	3.9	JASPER IA	-65	185	0	55	60
6	6.0	5	3.8	6	3.6	JEFFERSON IA	-36	-4	-39	23	53
6	2.8	6	2.4	5	1.4	JOHNSON IA	-13	-41	-49	15	14
7	6.2	7	5.4	4	2.8	JONES IA	-14	-48	-55	12	33
2	1.4	6	5.2	5	2.7	KEOKUK IA	273	-48	94	86	32
7	5.9	10	7.1	6	3.7	KOSSUTH IA	20	-48	-37	24	55
8	3.0	12	4.1	13	4.2	LEE IA	36	2	39	74	64
31	4.7	29	3.6	32	3.2	LINN IA	-24	-11	-32	29	42
2	3.1	3	5.1	4	4.7	LOUISA IA	65	-7	54	78	73
4	3.9	6	5.6	1	1.1	LUCAS IA	43	-80	-72	04	11
1	1.4	3	3.6	4	4.0	LYON IA	152	12	182	93	61
5	5.6	7	6.8	9*	6.3	MADISON IA	21	-8	12	61	88
4	2.4	7	3.6	6	3.0	MAHASKA IA	49	-15	27	69	38
6	4.0	7	4.0	11	5.4	MARION IA	2	34	36	73	82
12	5.5	7	2.9	10	3.3	MARSHALL IA	-48	15	-40	22	46
4	5.4	5	6.6	2	2.1	MILLS IA	21	-69	-62	08	21
2	1.8	2	2.1	5	4.2	MITCHELL IA	15	102	132	90	65
4	4.2	4	4.7	6	5.9	MONONA IA	11	24	38	74	86
1	1.3	9*	9.6	3	3.8	MONROE IA	656	-60	203	94	57
1	.9	4	2.9	7	4.6	MONTGOMERY IA	238	55	423	98	71
10	5.1	9	4.0	10	3.5	MUSCATINE IA	-20	-13	-31	30	50
6	5.5	7	5.1	7	3.4	O'BRIEN IA	-7	-33	-37	24	48
1	2.4	1	1.7			OSCEOLA IA	-30				
9	4.9	3	1.6	13*	6.0	PAGE IA	-66	266	24	67	87
1	1.6	3	3.3	5	5.5	PALO ALTO IA	109	67	248	96	83
3	2.7	5	3.4	8	4.1	PLYMOUTH IA	24	22	51	78	63
3	3.8	1	1.0	1	1.4	POCAHONTAS IA	-73	36	-63	08	14
63	5.0	48	3.1	58	3.2	POLK IA	-37	3	-35	26	43
23	5.9	16	3.4	22	4.1	POTTAWATTAMIE IA	-43	19	-32	29	62
1	1.0	6	4.7	5	2.9	POWESHIEK IA	387	-39	198	94	35
1	1.6	2	2.7	1	2.7	RINGGOLD IA	65	1	68	81	32
5	5.2	6	4.3	6	3.7	SAC IA	-18	-14	-30	31	53
26	4.7	21	2.9	32	4.1	SCOTT IA	-38	40	-14	44	62
3	3.3	3	3.8	1	.4	SHELBY IA	14	-91	-89	00	09
8	6.6	5	3.0	12	5.7	SIOUX IA	-55	92	-13	45	85
13	5.5	8	2.9	15	4.1	STORY IA	-47	38	-26	34	62
4	3.0	9	5.6	5	2.7	TAMA IA	89	-52	-9	48	32
9*	10.5	2	1.7	5	7.4	TAYLOR IA	-84	337	-29	32	93
4	3.2	4	3.5	3	2.3	UNION IA	9	-35	-29	31	24
1	1.2	8*	8.8	4	6.0	VAN BUREN IA	621	-32	389	98	86
6	2.1	10	3.3	10	3.0	WAPELLO IA	56	-11	40	74	37
8	6.6	5	4.0	9	5.6	WARREN IA	-39	40	-15	43	83
7	4.8	7	4.5	8	4.6	WASHINGTON IA	-6	2	-4	52	71
2	2.2	5	4.6	3	3.3	WAYNE IA	112	-28	52	78	45
10	3.9	11	3.2	15	3.3	WEBSTER IA	-17	4	-14	44	46
4	5.2	1	1.1	7	4.5	WINNEBAGO IA	-79	318	-14	44	70
6	4.7	6	3.7	4	2.4	WINNESHIEK IA	-21	-36	-50	15	26
12	2.1	27	4.0	41*	5.8	WOODBURY IA	89	45	174	93	86
1	1.7	3	4.4			WORTH IA	154				
12*	10.4	6	4.4	7	4.3	WRIGHT IA	-57	-3	-59	10	67
353	3.2	460	3.4	524	3.4	KANSAS KS	5	-1	4		
3	2.1	6	4.1	3	1.5	ALLEN KS	100	-63	-25	35	15
3	4.1	3	4.7	3	2.9	ANDERSON KS	14	-39	-31	31	34
2	1.4	5	3.7	4	2.8	ATCHISON KS	160	-25	96	86	33
		4	5.9	2	2.6	BARBER KS		-56			29
2	1.9	4	2.5	6	2.4	BARTON KS	32	-2	30	70	26
4	3.0	8	4.2	2	1.7	BOURBON KS	43	-60	-43	19	16
5	4.4	3	1.9	4	2.5	BROWN KS	-56	31	-43	19	28
2	1.1	8	3.6	9	3.1	BUTLER KS	227	-14	163	93	40
1	3.2	3	6.7	1	2.8	CHASE KS	111	-59	-13	44	32
1	2.3	3	9.4	2	3.1	CHAUTAUQUA KS	310	-67	36	73	40
3	1.8	4	2.0	10	4.9	CHEROKEE KS	9	150	172	92	76
				1	2.4	CHEYENNE KS					26
1	3.5			2	6.9	CLARK KS			97	87	91
3	3.2	5	4.9	2	1.9	CLAY KS	53	-62	-42	20	19
2	1.8	6	3.9	5	3.2	CLOUD KS	119	-18	81	84	43
8*	11.1	3	4.4			COFFEY KS	-60				
		1	3.5			COMANCHE KS					
16	6.4	12	4.0	7	1.8	COWLEY KS	-37	-55	-72	04	18
6	2.0	8	3.0	12	3.6	CRAWFORD KS	49	21	79	83	52
				3	4.7	DECATUR KS					72

* SIGNIFICANTLY MORE DEATHS THAN EXPECTED

1950-1959 DEATHS	1950-1959 RATE	1960-1969 DEATHS	1960-1969 RATE	1970-1979 DEATHS	1970-1979 RATE	COUNTY AND STATE	PERCENT CHANGE 60S/50S	PERCENT CHANGE 70S/60S	PERCENT CHANGE 70S/50S	PERCENTILE %CHANGE 70S/50S	PERCENTILE RATES 1970S
		9	4.4	6	2.7	DICKINSON KS		-38			31
1	1.8	2	2.8	6*	7.8	DONIPHAN KS	55	179	333	97	94
6	3.3	4	2.1	13	4.8	DOUGLAS KS	-36	131	47	77	74
		1	2.1	2	4.3	EDWARDS KS		104			66
1	1.8	2	3.3	2	2.0	ELK KS	81	-39	11	61	20
1	1.6	2	2.1	7	5.2	ELLIS KS	34	145	230	95	80
3	5.5	4	6.4	1	1.5	ELLSWORTH KS	17	-76	-72	04	15
		6	8.3	3	3.2	FINNEY KS		-61			43
5	5.2	2	1.7	7	4.3	FORD KS	-67	152	-18	41	66
4	2.7	5	3.0	7	3.8	FRANKLIN KS	11	25	39	74	56
3	4.0	5	5.8	5	5.2	GEARY KS	42	-10	28	69	80
1	6.1					GOVE KS					
2	7.5					GRAHAM KS					
		1	4.4	3*	9.3	GRANT KS		111			97
1	4.2			1	2.2	GRAY KS			-49	16	23
1	16.4	1	13.5			GREELEY KS	-18				
4	4.4	6	5.6	2	1.6	GREENWOOD KS	27	-72	-65	07	15
				1	5.3	HAMILTON KS					80
1	1.4	5	5.5	2	1.7	HARPER KS	287	-69	18	65	16
4	2.5	9	5.7	6	2.0	HARVEY KS	130	-65	-20	39	20
		2*	18.3	1	6.4	HASKELL KS		-65			89
2	2.3	5	6.5	3	2.9	JACKSON KS	182	-56	25	68	35
1	1.0	4	4.6	4	2.8	JEFFERSON KS	350	-40	171	92	33
4	6.0	2	2.2	3	4.5	JEWELL KS	-63	102	-26	35	70
9	2.4	14	2.2	33	3.2	JOHNSON KS	-7	45	35	73	41
1	6.8					KEARNY KS					
3	5.1	1	1.1	3	3.3	KINGMAN KS	-78	191	-36	26	44
		2	5.7	1	3.7	KIOWA KS		-34			55
5	2.7	5	2.3	7	2.8	LABETTE KS	-14	22	5	58	34
1	12.6	1	5.0			LANE KS	-61				
6	3.1	14	5.9	13	5.5	LEAVENWORTH KS	92	-7	78	83	82
1	2.5	1	1.8	6*	10.1	LINCOLN KS	-27	456	309	97	98
4	4.8	3	3.1	3	2.8	LINN KS	-37	-9	-43	19	33
1	5.0	3*	14.9			LOGAN KS	199				
		6	3.7			LYON KS					
5	3.5	6	2.9	6	3.0	MC PHERSON KS	-18	3	-16	42	37
5	4.7	4	2.9	4	3.5	MARION KS	-38	19	-26	34	49
4	3.6	4	3.0	6	4.2	MARSHALL KS	-16	39	17	64	65
2	6.3	1	3.2	1	1.9	MEADE KS	-49	-40	-70	05	19
2	1.4	5	3.2	5	3.8	MIAMI KS	128	19	171	92	56
2	3.3	1	1.1	1	1.0	MITCHELL KS	-66	-14	-70	05	11
11	3.7	15	4.6	12	3.7	MONTGOMERY KS	25	-19	1	55	54
3	6.5	1	1.1	4	3.8	MORRIS KS	-83	244	-41	21	56
1	9.3	3*	23.0			MORTON KS	148				
2	2.2	1	.9	6	5.0	NEMAHA KS	-61	474	123	89	77
2	1.6	5	3.1	3	1.4	NEOSHO KS	95	-55	-13	45	14
		2	4.9	1	2.8	NESS KS		-43			34
3	5.3	4	6.4	4	5.4	NORTON KS	22	-16	2	56	82
4	5.0	2	2.6	5	2.8	OSAGE KS	-47	8	-43	19	34
3	5.3	4	5.4	3	4.9	OSBORNE KS	3	-10	-8	50	75
		2	3.1			OTTAWA KS					
4	5.1	1	1.3	1	1.3	PAWNEE KS	-75	-2	-75	03	13
1	1.6	4	6.4	4	3.9	PHILLIPS KS	313	-40	148	91	57
2	2.6	2	1.6	4	2.9	POTTAWATOMIE KS	-37	76	11	61	35
4	5.9	1	1.0	4	4.0	PPATT KS	-83	305	-31	30	61
1	2.8	3	8.4			RAWLINS KS	194				
10	3.1	10	2.4	19	4.2	RENO KS	-23	77	35	73	64
5	5.6	2	3.3	3	3.7	REPUBLIC KS	-42	14	-34	28	54
1	1.1	4	3.7	5	3.9	RICE KS	246	6	268	96	60
4	3.0	3	2.2	8	3.9	RILEY KS	-28	80	29	70	59
3	6.5	1	1.0	4	6.4	ROOKS KS	-84	525	-2	53	89
		2	4.4	3	4.2	RUSH KS		-3			65
				4	4.6	RUSSELL KS					71
9	4.4	5	2.0	4	1.3	SALINE KS	-56	-36	-71	04	13
				1	2.7	SCOTT KS					30
30	3.0	53	3.5	71	4.0	SEDGWICK KS	17	13	33	72	60
2	5.6	6*	11.2	2	2.6	SEWARD KS	101	-77	-53	13	30
26	4.2	23	2.9	29	3.2	SHAWNEE KS	-31	9	-25	35	41
3*	18.8	4*	17.3			SHERIDAN KS	-8				
4*	10.9	1	2.6	1	1.4	SHERMAN KS	-76	-44	-87	00	14
2	3.3	2	2.4	2	1.4	SMITH KS	-29	-42	-59	10	14
1	1.6	5	8.1	2	2.4	STAFFORD KS	397	-70	50	78	27
		1	4.5	1	4.9	STEVENS KS		9			76
8	4.8	8	3.6	14*	5.8	SUMNER KS	-24	61	22	67	85
1	3.3	1	2.7	3	5.7	THOMAS KS	-20	114	71	82	84
1	4.4	1	3.5	3	8.1	TREGO KS	-20	130	85	85	95
2	3.5	4	6.3	2	3.7	WABAUNSEE KS	78	-41	5	58	55
				1	8.0	WALLACE KS					95
2	2.2	3	2.5			WASHINGTON KS	16				
		3*	22.3			WICHITA KS					
5	5.1	5	4.6	4	4.1	WILSON KS	-11	-11	-20	39	62
		3	4.6	1	1.6	WOODSON KS		-65			16
35	5.1	26	3.2	34	3.8	WYANDOTTE KS	-39	20	-26	34	56
473	3.7	642	4.1	872*	4.6	KENTUCKY KY	9	13	22		
2	2.6	4	5.2	5	5.2	ADAIR KY	96	0	96	87	79
6	7.9	2	1.9	3	3.4	ALLEN KY	-75	74	-57	11	47
1	2.3	1	2.3	5	7.3	ANDERSON KY	1	215	218	95	93
2	5.1	2	4.0			BALLARD KY	-21				
1	1.0	8	5.2	11	5.5	BARREN KY	444	5	468	99	82
3	6.4	4	8.4	4	6.1	BATH KY	32	-27	-3	52	88
9	6.6	9	5.4	12	6.0	BELL KY	-17	10	-9	49	87
3	4.0	8*	8.2	4	2.5	BOONE KY	104	-69	-37	25	28
1	1.4	4	4.4	6	4.9	BOURBON KY	214	13	256	96	76
13	6.1	16	5.9	27*	8.0	BOYD KY	-3	35	31	71	95
		7	6.0	4	2.7	BOYLE KY		-55			31
1	2.5	3	6.4	2	3.6	BRACKEN KY	154	-44	42	75	52
1	1.6	3	4.3	5	6.4	BREATHITT KY	164	50	295	96	89
3	4.1	1	1.3	2	1.7	BRECKINRIDGE KY	-69	32	-59	10	17

* SIGNIFICANTLY MORE DEATHS THAN EXPECTED

WHITE FEMALE

CANCER OF THE LIVER AND GALLBLADDER INCLUDING BILE DUCTS

ICD 155,156

1950-1959 DEATHS	1950-1959 RATE	1960-1969 DEATHS	1960-1969 RATE	1970-1979 DEATHS	1970-1979 RATE	COUNTY AND STATE	PERCENT CHANGE 60S/50S	PERCENT CHANGE 70S/60S	PERCENT CHANGE 70S/50S	PERCENTILE %CHANGE 70S/50S	PERCENTILE RATES 1970S
3	5.9	4	4.9	9*	7.8	BULLITT KY	-16	59	33	72	94
1	1.8	1	1.4	6*	8.6	BUTLER KY	-23	509	371	98	96
5	6.5	2	1.8	3	2.5	CALDWELL KY	-72	42	-61	09	28
6	4.7	6	3.4	4	2.1	CALLOWAY KY	-27	-40	-56	11	21
18	4.0	20	3.6	19	3.4	CAMPBELL KY	-9	-5	-14	44	48
1	3.3			2	3.2	CARLISLE KY			-1	54	44
3	7.3			6*	10.6	CARROLL KY			46	76	98
8*	8.9	3	2.9	10*	8.2	CARTER KY	-68	184	-8	49	95
2	3.0	3	3.7			CASEY KY	22				
7	4.1	11	5.2	13	5.4	CHRISTIAN KY	27	4	32	71	81
4	4.7	5	4.6	6	3.8	CLARK KY	-3	-17	-19	40	57
2	3.2			13*	14.5	CLAY KY			352	98	99
2	4.7	4	8.1	3	4.7	CLINTON KY	71	-43	-2	53	72
1	1.7	3	5.5	2	1.7	CRITTENDEN KY	219	-69	-1	54	17
3	8.0	1	2.4	3	6.3	CUMBERLAND KY	-69	158	-21	38	88
14	5.0	12	3.2	23	5.2	DAVIESS KY	-35	61	5	58	80
2	5.0			2	3.4	EDMONSON KY			-32	29	47
		2	7.3			ELLIOTT KY					
5	7.9	3	3.9	8*	8.9	ESTILL KY	-50	126	13	62	96
25	5.2	15	2.3	20	2.4	FAYETTE KY	-56	6	-54	13	27
2	1.9	7	4.1	12	5.8	FLOYD KY	110	42	198	94	85
2	1.6	4	2.6	2	1.0	FRANKLIN KY	61	-63	-40	22	11
1	1.5	4	5.3	4	5.8	FULTON KY	268	9	299	97	85
3*	14.5	1	4.5	1	4.8	GALLATIN KY	-69	5	-67	06	74
1	1.9	2	4.0	1	1.3	GARRARD KY	114	-67	-29	32	13
2	3.6	4	6.5	4	6.4	GRANT KY	80	-2	76	83	89
5	2.8	7	3.1	12	4.5	GRAVES KY	11	46	62	80	69
3	3.1	3	2.9	8	6.7	GRAYSON KY	-7	131	115	88	90
3	6.4	4	6.2	4	6.4	GREEN KY	-4	3	0	54	89
5	5.2	15*	11.9	12*	7.0	GREENUP KY	127	-41	33	72	92
		1	4.1	1	2.6	HANCOCK KY		-36			29
4	2.8	8	5.0	21*	8.7	HARDIN KY	74	77	207	94	96
1	.6	15*	7.9	16*	7.1	HARLAN KY	1213	-10	1081	99	92
2	1.9	4	3.5	4	3.7	HARRISON KY	85	3	91	86	53
1	1.6	5	6.4	4	4.5	HART KY	293	-30	177	93	69
5	3.3	5	2.6	6	2.6	HENDERSON KY	-22	0	-22	37	29
2	3.1	3	4.7	2	2.7	HENRY KY	52	-43	-13	45	31
		2	4.0	3	5.5	HICKMAN KY		37			62
3	1.6	6	2.6	11	3.9	HOPKINS KY	62	51	144	91	59
2	4.2	3	5.7	3	4.9	JACKSON KY	38	-15	17	64	76
95	4.2	110	3.7	138	3.8	JEFFERSON KY	-12	2	-10	47	56
1	2.3	2	2.9	2	2.3	JESSAMINE KY	28	-21	1	55	24
1	1.2	7	6.2	12*	9.9	JOHNSON KY	405	61	714	99	98
36*	6.6	31	4.4	37	4.5	KENTON KY	-34	4	-32	29	71
				9*	11.0	KNOTT KY					98
2	1.9	5	4.0	9	4.8	KNOX KY	111	22	158	91	75
2	4.3	2	3.2	4	5.2	LARUE KY	-25	60	21	66	80
4	4.1	3	2.2	10	5.4	LAUREL KY	-47	145	30	70	81
5	7.6			6	6.7	LAWRENCE KY			-11	46	90
2	6.4	2	4.8	3	6.0	LEE KY	-24	24	-6	51	87
		2	5.0			LESLIE KY					
5	5.0	6	5.2	2	1.5	LETCHER KY	4	-71	-70	05	15
		4	6.6	2	2.9	LEWIS KY		-56			34
6	7.6	2	2.1	3	3.1	LINCOLN KY	-72	47	-59	10	41
2	4.5	2	3.8	6*	9.4	LIVINGSTON KY	-17	151	108	88	97
2	1.8	4	2.9	10	8.1	LOGAN KY	64	177	354	98	95
1	3.5					LYON KY					
15	6.1	13	4.1	23	5.7	MC CRACKEN KY	-33	41	-6	51	85
4	13.7	1	1.4	2	3.3	MC CREARY KY	-90	140	-76	03	44
3	5.4	7*	11.7	4	4.8	MC LEAN KY	115	-59	-12	45	74
5	3.8	4	2.3	12	5.3	MADISON KY	-39	130	40	75	81
		2	5.1	3	5.2	MAGOFFIN KY		1			80
		7*	8.2	5	5.0	MARION KY		-39			77
1	1.1	3	2.7	9	5.9	MARSHALL KY	153	115	444	99	86
		3	7.3	2	3.6	MARTIN KY		-51			52
4	4.6	3	2.9	5	4.1	MASON KY	-37	43	-10	48	63
		2	4.5	3	5.5	MEADE KY		23			83
		1	4.9	3*	12.4	MENIFEE KY		155			99
		4	4.1	5	4.4	MERCER KY		7			68
1	2.2			1	1.4	METCALFE KY			-37	25	14
1	1.6	2	2.8	1	1.5	MONROE KY	78	-47	-6	50	15
6*	8.9	7*	8.5	7	7.4	MONTGOMERY KY	-4	-14	-17	41	93
		7*	12.9	7*	9.8	MORGAN KY		-24			97
4	2.6	7	3.8	9	4.1	MUHLENBERG KY	47	8	58	79	62
4	4.8	1	1.0	5	4.2	NELSON KY	-79	331	-12	46	65
		4	8.7	3	6.9	NICHOLAS KY		-21			91
		5	3.8	10	7.1	OHIO KY		85			92
		2	2.9	2	2.4	OLDHAM KY		-18			26
1	1.8			1	1.2	OWEN KY			-33	28	12
		1	3.7	1	2.3	OWSLEY KY		-37			24
		3	3.9	6*	10.0	PENDLETON KY		158			98
3	2.7	12*	9.4	6	4.2	PERRY KY	254	-55	59	80	65
6	2.6	15	6.0	14	4.3	PIKE KY	129	-29	62	80	66
1	3.3	1	3.2	5*	10.7	POWELL KY	-3	229	219	95	98
5	3.1	11	5.5	11	4.2	PULASKI KY	75	-23	34	72	64
3	5.3	3	4.7	5	5.7	ROCKCASTLE KY	-11	21	8	59	84
1	2.7	1	1.2			ROWAN KY	-54				
1	2.4	2	3.4	4	4.6	RUSSELL KY	39	35	87	85	71
4	5.6	1	1.4	6	5.7	SCOTT KY	-76	318	2	56	85
1	1.0	3	3.0	5	3.9	SHELBY KY	214	29	304	97	58
2	3.4	1	1.4	8*	10.4	SIMPSON KY	-59	660	208	94	98
2	8.3	1	3.4	2	7.6	SPENCER KY	-59	122	-8	49	94
1	1.4	5	5.3	4	3.9	TAYLOR KY	293	-27	187	93	58
1	1.6	2	3.4	1	1.3	TODD KY	111	-61	-18	40	13
		2	3.7	5	7.0	TRIGG KY		92			92
		5*	19.6			TRIMBLE KY					
7*	9.2	3	3.9	2	2.0	UNION KY	-58	-47	-78	02	20
8	4.0	5	1.9	11	3.4	WARREN KY	-53	81	-15	43	47
		6*	9.7	2	2.9	WASHINGTON KY		-70			35
4	6.6	4	5.3	11*	11.0	WAYNE KY	-19	107	67	81	98

* SIGNIFICANTLY MORE DEATHS THAN EXPECTED

WHITE FEMALE

CANCER OF THE LIVER AND GALLBLADDER INCLUDING BILE DUCTS

ICD 155,156

1950-1959 DEATHS	1950-1959 RATE	1960-1969 DEATHS	1960-1969 RATE	1970-1979 DEATHS	1970-1979 RATE	COUNTY AND STATE	PERCENT CHANGE 60S/50S	PERCENT CHANGE 70S/60S	PERCENT CHANGE 70S/50S	PERCENTILE %CHANGE 70S/50S	PERCENTILE RATES 1970S
2	2.1	6	5.8	2	2.0	WEBSTER KY	174	-65	-5	51	20
1	.7	6	3.8	6	2.7	WHITLEY KY	460	-29	298	96	32
1	4.3	2	5.4	1	2.7	WOLFE KY	25	-50	-37	25	32
1	2.4	4	6.4	6	7.8	WOODFORD KY	169	23	231	95	94
188	2.3	307	2.7	481	3.2	LOUISIANA LA	20	18	42		
2	1.4	7	3.7	11	4.2	ACADIA LA	163	14	201	94	66
1	1.6	1	1.3	6	6.8	ALLEN LA	-23	440	314	97	91
1	1.5	3	3.4	1	1.0	ASCENSION LA	122	-72	-38	24	11
		4	7.0	3	4.5	ASSUMPTION LA		-36			70
2	1.7	3	1.8	6	2.8	AVOYELLES LA	5	57	65	81	33
3	4.3			2	1.6	BEAUREGARD LA			-63	08	16
		2	3.0	1	1.0	BIENVILLE LA		-67			11
1	1.2	3	2.8	3	1.5	BOSSIER LA	128	-45	25	68	15
12	2.3	24	3.0	34	3.2	CADDO LA	28	7	37	74	43
11	4.1	11	3.0	18	3.1	CALCASIEU LA	-27	4	-25	35	39
2	5.7			4	6.8	CALDWELL LA			19	65	91
		2	6.0			CAMERON LA					
				2	4.2	CATAHOULA LA					65
2	3.0	1	3.2	3	3.1	CLAIBORNE LA	9	-5	4	57	39
		1	3.9	2	4.0	CONCORDIA LA		1			61
1	1.3	2	2.3	2	1.9	DE SOTO LA	82	-20	47	77	19
13	3.5	19	3.0	20	2.1	EAST BATON ROUGE LA	-15	-30	-41	21	21
		1	3.8	1	3.5	EAST CARROLL LA		-9			49
				2	2.5	EAST FELICIANA LA					27
2	1.8	7	6.1	8	5.0	EVANGELINE LA	247	-18	184	93	77
2	3.0	2	2.1	2	2.1	FRANKLIN LA	-32	2	-30	31	22
3	6.1	1	1.3	2	2.1	GRANT LA	-79	66	-65	07	22
2	2.1	1	.6	8	3.7	IBERIA LA	-71	502	76	83	55
		1	1.2	5	5.3	IBERVILLE LA		330			81
1	2.4	1	1.4	1	1.1	JACKSON LA	-42	-24	-56	12	11
10	3.2	24	3.7	43	3.3	JEFFERSON LA	14	-10	2	56	46
1	1.0			9	5.8	JEFFERSON DAVIS LA			500	99	85
2	1.5	6	2.2	18	4.5	LAFAYETTE LA	43	105	193	94	70
4	2.8	12	6.4	9	3.4	LAFOURCHE LA	127	-47	21	66	48
1	1.8	1	1.6	7	6.9	LA SALLE LA	-7	322	290	96	91
1	1.3	2	1.5	2	1.7	LINCOLN LA	15	16	33	72	17
3	3.9	4	4.3	3	1.8	LIVINGSTON LA	9	-57	-53	13	18
				2	4.1	MADISON LA					63
3	4.4	3	3.5	3	2.5	MOREHOUSE LA	-20	-30	-44	19	27
2	2.2	3	2.6	4	2.7	NATCHITOCHES LA	17	3	21	66	30
49	2.2	76	3.1	83	3.3	ORLEANS LA	40	7	50	77	45
3	1.2	6	1.6	10	2.3	OUACHITA LA	32	41	85	85	24
		1	2.1	1	2.2	PLAQUEMINES LA		4			23
1	2.4	6*	10.1	6*	8.7	POINTE COUPEE LA	321	-14	263	96	96
6	2.2	9	2.3	14	2.8	RAPIDES LA	1	24	25	68	33
				2	3.4	RED RIVER LA					48
		1	1.3	3	3.2	RICHLAND LA		150			43
		4	3.6	5	3.7	SABINE LA		4			55
		3	2.8	7	3.3	ST BERNARD LA		15			45
		2	4.1	2	2.0	ST CHARLES LA		-50			21
				2	8.4	ST HELENA LA					96
5*	12.4	1	2.0	1	1.6	ST JAMES LA	-84	-21	-87	00	15
		2	4.4	4	5.2	ST JOHN THE BAPT LA		18			80
4	1.9	3	1.3	11	3.8	ST LANDRY LA	-34	198	97	87	57
2	3.0	3	3.2	7	5.6	ST MARTIN LA	6	76	87	85	83
2	2.2	4	3.3	2	1.2	ST MARY LA	48	-64	-47	17	12
1	.8	4	2.2	9	3.1	ST TAMMANY LA	168	40	276	96	38
3	2.1	3	1.3	14	4.7	TANGIPAHOA LA	-40	269	123	89	72
		1	5.9	3*	11.3	TENSAS LA		91			99
6	6.0	3	1.7	11	4.4	TERREBONNE LA	-72	165	-27	34	68
3	4.5	1	1.0	6	6.1	UNION LA	-79	541	35	73	87
4	2.9	8	4.2	11	4.6	VERMILION LA	46	10	62	80	71
2	2.8	1	1.1	4	2.9	VERNON LA	-61	163	4	57	36
1	.8	2	1.4	5	2.6	WASHINGTON LA	82	81	230	95	29
1	.6	4	2.6	6	2.9	WEBSTER LA	350	10	397	98	35
2	9.9	2	5.8	2	5.0	WEST BATON ROUGE LA	-42	-13	-49	15	78
1	1.8	2	3.1	1	1.0	WEST CARROLL LA	68	-67	-45	18	11
1	7.0					WEST FELICIANA LA					
3	5.1	3	4.2	2	2.1	WINN LA	-18	-49	-58	11	22
168	3.3	200	3.4	258	3.6	MAINE ME	2	7	10		
17	3.6	26	4.8	30	4.6	ANDROSCOGGIN ME	34	-2	30	70	72
7	2.3	17	4.4	18	3.8	AROOSTOOK ME	90	-13	66	81	57
36	3.5	34	2.8	47	3.3	CUMBERLAND ME	-21	19	-5	51	45
5	4.6	4	3.3	2	1.3	FRANKLIN ME	-29	-59	-70	05	13
5	2.4	8	3.0	12	3.4	HANCOCK ME	27	12	43	75	47
20	4.2	17	2.9	19	2.7	KENNEBEC ME	-31	-7	-36	26	30
5	2.6	11	4.8	7	2.4	KNOX ME	86	-49	-6	51	27
3	2.1	5	3.4	6	3.1	LINCOLN ME	65	-11	47	77	39
6	2.7	9	3.3	17	4.9	OXFORD ME	23	52	87	85	76
26	4.5	27	4.1	29	3.8	PENOBSCOT ME	-8	-7	-15	43	57
5	4.5	3	2.6	5	3.9	PISCATAQUIS ME	-43	52	-13	45	59
4	3.1	5	3.3	5	3.0	SAGADAHOC ME	7	-10	-3	52	38
8	3.9	8	3.0	13	4.7	SOMERSET ME	-22	54	19	65	72
6	4.6	8	6.3	5	3.6	WALDO ME	36	-42	-22	38	52
5	2.6	3	1.4	7	2.7	WASHINGTON ME	-47	100	5	58	32
10	1.8	15	2.2	36	4.2	YORK ME	23	88	131	90	64
398	4.0	488	3.5	611	3.4	MARYLAND MD	-11	-3	-14		
16	3.7	25	4.6	25	4.3	ALLEGANY MD	25	-8	15	63	66
10	2.6	30	4.4	51	4.5	ANNE ARUNDEL MD	73	1	75	83	69
56	4.2	65	2.9	109	3.3	BALTIMORE MD	-29	12	-21	38	45
		1	1.4	3	3.5	CALVERT MD		157			50
8*	8.4	12*	10.3	11*	8.0	CAROLINE MD	23	-22	-4	52	95
8	2.8	9	2.6	11	2.5	CARROLL MD	-7	-6	-13	45	27
4	3.0	8	4.3	10	4.0	CECIL MD	44	-7	33	72	61

* SIGNIFICANTLY MORE DEATHS THAN EXPECTED

1950-1959 DEATHS	1950-1959 RATE	1960-1969 DEATHS	1960-1969 RATE	1970-1979 DEATHS	1970-1979 RATE	COUNTY AND STATE	PERCENT CHANGE 60S/50S	PERCENT CHANGE 70S/60S	PERCENT CHANGE 70S/50S	PERCENTILE %CHANGE 70S/50S	PERCENTILE RATES 1970S
5	12.7	2	2.1	2	1.4	CHARLES MD	-83	-33	-89	00	14
6	4.4	4	2.8	5	2.4	DORCHESTER MD	-35	-15	-45	18	26
9	2.8	14	3.6	16	3.2	FREDERICK MD	31	-12	15	63	42
2	2.7	4	3.7	9	5.7	GARRETT MD	36	56	112	88	84
8	4.5	7	2.5	12	2.9	HARFORD MD	-44	15	-36	26	35
3	2.9	5	3.6	11	4.1	HOWARD MD	25	12	40	75	62
1	1.5			4	4.4	KENT MD			186	93	68
36	4.2	53	3.3	82	3.3	MONTGOMERY MD	-23	0	-22	37	45
14	2.2	35	2.8	58	3.0	PRINCE GEORGE'S MD	25	10	38	74	38
1	1.5			6	5.2	QUEEN ANNE'S MD			248	95	80
4	5.6	4	4.3	4	2.9	ST MARY'S MD	-24	-34	-49	15	34
6	7.3	3	3.0	4	3.7	SOMERSET MD	-59	24	-49	15	54
3	2.8	4	3.2	7	4.2	TALBOT MD	15	31	50	78	64
24	6.0	29	5.6	26	3.8	WASHINGTON MD	-8	-32	-37	24	56
13	6.5	4	1.5	4	1.2	WICOMICO MD	-76	-20	-81	01	12
4	4.3	4	3.5	3	1.6	WORCESTER MD	-19	-55	-64	08	15
157	4.0	166	4.1	138	3.6	BALTIMORE CITY MD	4	-13	-10	48	52
1066	3.7	1172	3.4	1445	3.6	MASSACHUSETTS MA	-7	5	-2		
14	3.9	18	2.7	39	3.2	BARNSTABLE MA	-30	19	-16	42	43
25	2.9	49	5.1	50	4.2	BERKSHIRE MA	72	-16	44	76	65
89	3.7	96	3.4	143	4.3	BRISTOL MA	-8	24	14	62	66
		1	2.2	2	2.1	DUKES MA		-3			21
129	3.7	126	3.0	172	3.5	ESSEX MA	-18	18	-3	52	50
12	3.5	18	4.0	19	4.0	FRANKLIN MA	15	1	16	63	61
80	3.5	99	3.8	96	3.0	HAMPDEN MA	6	-20	-16	43	37
22	4.2	17	2.5	34	4.1	HAMPSHIRE MA	-42	67	-3	53	63
237	3.6	275	3.5	356	3.9	MIDDLESEX MA	-3	11	8	59	58
1	5.1	2	7.5	1	2.5	NANTUCKET MA	48	-66	-50	15	28
89	3.4	117	3.4	134	3.1	NORFOLK MA	0	-9	-9	49	41
47	3.6	66	4.0	63	2.8	PLYMOUTH MA	11	-29	-22	38	33
208	4.2	159	3.2	188	3.9	SUFFOLK MA	-24	22	-8	49	58
113	3.3	129	3.2	148	3.1	WORCESTER MA	-3	-3	-6	51	41
1337*	4.7	1598*	4.4	1771*	4.0	MICHIGAN MI	-5	-9	-13		
1	2.7	2	5.4	2	2.4	ALCONA MI	105	-56	-10	47	26
2	4.8	1	2.1			ALGER MI	-57				
19*	7.4	23*	7.2	19	5.1	ALLEGAN MI	-2	-30	-32	29	78
4	4.1	8	6.1	12*	7.6	ALPENA MI	49	25	86	85	93
4	6.0	1	1.5	8*	7.6	ANTRIM MI	-75	412	27	69	94
3	7.6	4	6.8	3	3.3	ARENAC MI	-11	-51	-56	11	46
1	4.0	1	2.2	3	5.3	BARAGA MI	-45	141	32	71	81
8	4.9	8	4.3	9	3.7	BARRY MI	-12	-12	-23	37	55
17	3.9	24	4.6	26	4.1	BAY MI	19	-10	7	59	63
3	6.9	3	6.0	2	3.8	BENZIE MI	-12	-37	-45	18	56
30	5.1	31	4.2	26	2.9	BERRIEN MI	-17	-32	-44	19	35
10	5.5	7	3.3	8	3.2	BRANCH MI	-41	-3	-42	20	42
23	3.9	18	2.6	31	3.9	CALHOUN MI	-33	48	0	54	58
7	4.4	4	2.1	3	1.4	CASS MI	-52	-34	-69	05	14
3	4.0	4	5.2	7	6.2	CHARLEVOIX MI	32	19	57	79	88
4	5.6	5	5.5	8	6.2	CHEBOYGAN MI	-1	12	10	60	88
6	5.1	4	3.1	6	3.9	CHIPPEWA MI	-39	24	-24	36	58
2	3.7	2	3.1	4	2.8	CLARE MI	-15	-11	-24	36	33
10	6.2	8	4.4	2	.8	CLINTON MI	-28	-82	-87	00	10
		1	3.1	1	1.8	CRAWFORD MI		-42			17
8	4.8	7	3.6	19*	8.6	DELTA MI	-25	141	80	84	96
6	4.1	6	3.3	9	3.6	DICKINSON MI	-21	11	-12	46	52
12	5.4	14	5.1	13	3.7	EATON MI	-6	-27	-32	29	54
3	3.4	5	4.1	5	3.1	EMMET MI	20	-25	-10	48	40
44	4.0	57	4.0	60	3.4	GENESEE MI	1	-15	-15	43	46
3	5.7	3	4.2	7	7.0	GLADWIN MI	-27	66	21	66	92
13*	9.8	7	4.5	11	5.1	GOGEBIC MI	-55	16	-47	16	79
11	5.4	8	3.7	10	4.0	GRAND TRAVERSE MI	-32	8	-27	34	60
14*	7.2	11	4.5	9	3.5	GRATIOT MI	-38	-21	-51	15	50
8	4.1	6	2.8	4	1.2	HILLSDALE MI	-33	-57	-71	04	12
8	3.5	8	3.2	19*	7.1	HOUGHTON MI	-9	121	101	87	92
5	3.0	10	4.7	7	3.0	HURON MI	59	-37	0	55	37
36	4.6	41	4.2	41	3.6	INGHAM MI	-9	-14	-22	38	53
10	4.8	7	2.8	18*	7.1	IONIA MI	-41	150	48	77	92
2	3.1			4	2.7	IOSCO MI			-13	45	31
4	4.3	1	.9	5	3.8	IRON MI	-78	303	-12	46	56
2	1.6	9	5.9	7	3.8	ISABELLA MI	272	-35	143	90	57
19	3.5	24	3.7	34	4.3	JACKSON MI	5	16	22	66	66
25	3.8	32	3.8	26	2.4	KALAMAZOO MI	0	-38	-37	24	26
		6*	18.0			KALKASKA MI					
96*	6.0	96	4.8	102	4.2	KENT MI	-20	-13	-30	31	64
		1	6.3	1	4.1	KEWEENAW MI		-34			63
1	4.2	1	2.8	1	2.6	LAKE MI	-31	-10	-39	23	28
4	2.3	6	2.6	16*	6.6	LAPEER MI	12	157	187	94	90
7*	16.7	3	4.5	5	6.1	LEELANAU MI	-73	34	-64	08	87
17	4.9	31*	7.6	23	4.8	LENAWEE MI	56	-37	-2	53	74
5	3.3	4	1.8	12	3.7	LIVINGSTON MI	-47	106	10	60	54
		3	6.5			LUCE MI					
1	3.3	2	3.6	1	1.7	MACKINAC MI	9	-54	-49	15	16
30	3.9	62	4.0	124*	5.2	MACOMB MI	3	29	33	72	80
6	5.6	10*	7.7	11	7.3	MANISTEE MI	37	-5	30	70	93
10	4.3	18*	6.4	18	5.7	MARQUETTE MI	48	-12	31	71	84
6	5.2	6	4.3	4	2.5	MASON MI	-17	-42	-52	14	27
2	2.0	6	5.3	6	3.7	MECOSTA MI	167	-31	83	84	53
6	4.7	7	4.7	10	5.6	MENOMINEE MI	1	20	20	66	83
3	2.2	7	4.3	7	2.8	MIDLAND MI	100	-36	28	69	32
3	9.0	1	2.4	3	5.8	MISSAUKEE MI	-74	146	-35	26	86
13	4.0	15	3.4	25	4.5	MONROE MI	-15	34	15	63	71
12	6.5	7	3.1	12	4.5	MONTCALM MI	-53	47	-31	30	70
1	9.4	1	3.0			MONTMORENCY MI	-68				
25	4.8	37	5.6	47*	5.9	MUSKEGON MI	16	5	23	67	86
5	4.2	11*	8.1	6	3.0	NEWAYGO MI	92	-63	-29	32	38
76	4.4	108	3.8	149	3.3	OAKLAND MI	-15	-11	-25	35	46

* SIGNIFICANTLY MORE DEATHS THAN EXPECTED

WHITE FEMALE

CANCER OF THE LIVER AND GALLBLADDER INCLUDING BILE DUCTS

ICD 155,156

1950-1959 DEATHS	1950-1959 RATE	1960-1969 DEATHS	1960-1969 RATE	1970-1979 DEATHS	1970-1979 RATE	COUNTY AND STATE	PERCENT CHANGE 60S/50S	PERCENT CHANGE 70S/60S	PERCENT CHANGE 70S/50S	PERCENTILE %CHANGE 70S/50S	PERCENTILE RATES 1970S
5	6.0	9*	8.5	3	2.5	OCEANA MI	40	-70	-58	11	28
4	7.9	4	5.7	1	.8	OGEMAW MI	-28	-86	-90	00	10
		3	5.8	4	6.6	ONTONAGON MI		14			90
4	5.6	9*	10.9	5	5.1	OSCEOLA MI	95	-53	-8	49	79
5*	32.2			1	3.4	OSCODA MI			-90	00	46
2	6.2	2	4.6	2	3.1	OTSEGO MI	-25	-33	-50	15	39
21	6.0	27	5.7	30	4.5	OTTAWA MI	-5	-21	-25	35	69
1	2.2	5	7.2	5	5.3	PRESQUE ISLE MI	231	-26	146	91	81
5*	14.9	3	6.0	3	2.4	ROSCOMMON MI	-59	-61	-84	01	26
35	5.1	46	5.2	44	4.5	SAGINAW MI	2	-13	-12	46	70
20	4.2	18	3.0	23	3.2	ST CLAIR MI	-30	7	-25	35	41
9	4.0	16	6.2	5	1.4	ST JOSEPH MI	54	-77	-64	07	14
11	6.0	13	6.0	9	3.8	SANILAC MI	1	-37	-36	26	57
2	4.7	4	8.5	3	6.4	SCHOOLCRAFT MI	82	-25	36	73	89
8	3.2	13	4.5	12	3.6	SHIAWASSEE MI	38	-18	12	61	53
6	2.8	4	1.5	9	3.2	TUSCOLA MI	-48	114	11	61	42
7	2.9	16	5.3	13	3.6	VAN BUREN MI	86	-33	24	68	51
19	3.2	20	2.7	25	2.9	WASHTENAW MI	-14	7	-9	48	35
447*	4.7	519*	4.8	496*	4.4	WAYNE MI	1	-8	-8	50	68
7	7.6	3	2.7	7	4.9	WEXFORD MI	-65	86	-35	26	76
845*	5.4	957*	4.8	1029*	4.1	MINNESOTA MN	-11	-13	-23		
3	3.9	3	3.7	10*	9.9	AITKIN MN	-4	166	157	91	97
8	4.8	11	4.6	18	3.9	ANOKA MN	-4	-15	-18	40	59
5	5.1	6	4.2	3	1.8	BECKER MN	-17	-59	-65	07	17
6	6.1	3	2.4	10	6.7	BELTRAMI MN	-61	184	11	61	91
6*	9.0	7*	8.4	5	4.1	BENTON MN	-7	-51	-54	12	63
2	4.2	7*	11.6	3	4.8	BIG STONE MN	174	-58	14	62	74
10	4.3	11	3.7	11	3.5	BLUE EARTH MN	-14	-5	-19	40	49
10	6.7	8	4.8	6	3.2	BROWN MN	-29	-34	-53	13	42
10*	8.5	4	2.6	15*	8.9	CARLTON MN	-69	237	4	58	96
9*	9.1	13*	10.1	10	6.6	CARVER MN	11	-35	-28	33	90
5	5.3	3	3.2	4	2.6	CASS MN	-39	-19	-50	15	30
4	4.7	4	3.9	4	2.7	CHIPPEWA MN	-18	-30	-42	20	32
5	5.9	6	5.4	1	.8	CHISAGO MN	-8	-85	-87	00	10
10	8.0	9	4.8	9	3.5	CLAY MN	-40	-27	-56	11	50
1	2.3	3	6.6	1	.8	CLEARWATER MN	192	-88	-65	07	10
1	9.6					COOK MN					
8*	9.5	3	3.6	8	5.7	COTTONWOOD MN	-61	56	-40	22	84
8	6.1	15*	7.2	13	4.5	CROW WING MN	17	-37	-27	34	70
9	3.9	19	5.3	19	3.3	DAKOTA MN	36	-38	-15	43	45
2	2.7	4	4.3	6	6.1	DODGE MN	58	41	123	89	88
6	5.1	9	5.4	9	4.8	DOUGLAS MN	6	-11	-6	51	75
4	3.0	9	5.8	9	4.6	FARIBAULT MN	91	-20	52	78	71
5	3.2	9	5.7	7	3.3	FILLMORE MN	78	-42	2	56	45
14*	8.0	12	5.3	9	3.0	FREEBORN MN	-34	-43	-62	08	38
7	3.5	15	6.3	13	4.3	GOODHUE MN	80	-32	23	67	66
2	3.8	6*	9.6	5	7.3	GRANT MN	154	-24	93	86	93
203*	5.3	207	4.1	208	3.5	HENNEPIN MN	-22	-14	-33	28	50
4	4.7	3	3.6	6	5.1	HOUSTON MN	-25	43	7	59	78
2	4.1	4	5.2	7	6.8	HUBBARD MN	27	33	68	82	91
		8*	9.1	2	1.3	ISANTI MN		-85			13
6	5.3	7	4.5	9	4.1	ITASCA MN	-15	-10	-23	37	63
5	6.3	2	1.9	5	5.2	JACKSON MN	-70	177	-17	41	80
1	1.8	1	3.5	3	3.3	KANABEC MN	89	-5	80	84	46
3	2.0	6	3.0	13	5.7	KANDIYOHI MN	50	86	179	93	84
1	2.5	3	6.8	4	7.2	KITTSON MN	172	5	186	93	92
6*	12.2			3	2.9	KOOCHICHING MN			-76	03	36
4	5.1	3	3.3	5	5.4	LAC QUI PARLE MN	-36	66	6	58	82
1	2.1	4	7.6	6*	8.3	LAKE MN	259	9	293	96	95
		1	3.4			LAKE OF THE WOODS MN					
2	1.7	14*	10.6	5	3.1	LE SUEUR MN	525	-71	84	84	40
		2	3.6	6	7.9	LINCOLN MN		124			95
7	6.5	5	3.4	5	2.5	LYON MN	-48	-26	-61	09	28
10	7.8	5	3.0	5	2.6	MC LEOD MN	-61	-16	-67	06	28
1	3.3	5*	19.2	4*	13.3	MAHNOMEN MN	480	-31	302	97	99
2	3.6	3	4.1	5	4.9	MARSHALL MN	14	20	36	73	76
6	5.5	6	3.5	8	3.7	MARTIN MN	-37	7	-33	28	55
6	6.2	8	7.4	6	3.8	MEEKER MN	19	-49	-39	23	56
1	1.3	6	6.6	11*	8.6	MILLE LACS MN	395	30	543	99	96
10*	8.7	9	5.7	7	3.8	MORRISON MN	-34	-34	-57	11	55
16*	8.2	18*	7.4	14	4.2	MOWER MN	-10	-43	-49	16	65
2	3.5	5	7.0	2	2.4	MURRAY MN	100	-66	-31	30	27
3	2.7	9	6.5	7	4.5	NICOLLET MN	137	-31	63	81	69
4	4.0	6	4.4	8	4.7	NOBLES MN	10	7	18	64	73
2	3.0	6	6.8	4	4.1	NORMAN MN	129	-39	39	74	63
11	3.9	12	3.1	15	3.1	OLMSTED MN	-20	0	-19	40	41
8	2.8	13	3.7	24*	5.0	OTTER TAIL MN	29	36	76	83	77
6*	9.4	2	1.9	4	3.8	PENNINGTON MN	-80	104	-59	10	57
4	4.1	10*	8.7	7	4.5	PINE MN	112	-49	8	60	69
5	7.2	7*	7.4	7	7.6	PIPESTONE MN	3	2	5	58	94
11	6.4	10	4.5	18*	6.6	POLK MN	-30	47	3	57	90
4	6.0	2	2.6	5	4.9	POPE MN	-57	89	-19	39	75
121*	6.2	133*	5.1	112	3.8	RAMSEY MN	-17	-26	-38	23	56
1	4.6			1	2.8	RED LAKE MN			-40	22	32
6	5.0	6	4.3	4	1.7	REDWOOD MN	-14	-62	-67	06	16
10*	9.1	8	6.0	10	5.7	RENVILLE MN	-34	-5	-37	24	84
10	5.1	9	3.6	12	4.7	RICE MN	-29	30	-8	49	73
2	3.7	5	6.8	3	4.3	ROCK MN	85	-37	16	63	67
5	9.5	3	5.0	5	6.0	ROSEAU MN	-47	18	-38	24	87
70*	6.3	69*	5.1	86*	5.3	ST LOUIS MN	-20	5	-16	42	81
6	7.3	6	5.5	8	5.2	SCOTT MN	-24	-5	-28	33	80
3	5.1	5	6.9	5	4.2	SHERBURNE MN	36	-39	-17	41	65
1	1.3	4	4.1	4	3.0	SIBLEY MN	207	-26	128	89	38
17	5.5	20	4.9	32*	6.2	STEARNS MN	-11	27	13	62	88
4	3.3	6	4.2	9	4.2	STEELE MN	29	0	29	70	65
2	4.2	2	3.3	4	4.9	STEVENS MN	-21	49	18	65	76
5	6.6	4	4.5	2	1.4	SWIFT MN	-32	-69	-79	02	14
7	5.9	7	4.8	5	2.6	TODD MN	-19	-45	-56	12	29

* SIGNIFICANTLY MORE DEATHS THAN EXPECTED

WHITE FEMALE

CANCER OF THE LIVER AND GALLBLADDER INCLUDING BILE DUCTS

ICD 155,156

1950-1959 DEATHS	1950-1959 RATE	1960-1969 DEATHS	1960-1969 RATE	1970-1979 DEATHS	1970-1979 RATE	COUNTY AND STATE		PERCENT CHANGE 60S/50S	PERCENT CHANGE 70S/60S	PERCENT CHANGE 70S/50S	PERCENTILE %CHANGE 70S/50S	PERCENTILE RATES 1970S
1	3.1	1	1.8	7*	11.0	TRAVERSE	MN	-42	512	253	96	98
7	6.2	4	3.1	8	5.7	WABASHA	MN	-51	86	-8	49	85
3	5.1	5	6.7	7	8.1	WADENA	MN	30	22	59	80	95
2	2.2	4	3.7	6	4.5	WASECA	MN	71	21	108	88	70
8	4.3	17*	7.4	12	3.5	WASHINGTON	MN	72	-53	-20	39	48
4	5.1	2	2.0	3	2.5	WATONWAN	MN	-61	28	-51	15	28
2	4.9	4	7.5	3	5.3	WILKIN	MN	51	-30	6	59	80
18*	8.2	14	4.9	16	4.7	WINONA	MN	-40	-4	-42	20	73
12*	7.9	7	3.8	8	2.6	WRIGHT	MN	-52	-33	-68	06	28
1	1.3	1	1.0	1	.7	YELLOW MEDICINE	MN	-22	-27	-43	19	10
99	1.8	176	2.4	287	3.1	MISSISSIPPI	MS	38	28	77		
		1	1.2	2	1.6	ADAMS	MS		27			15
2	1.8	4	2.9	9	5.3	ALCORN	MS	58	84	191	94	81
2	4.5	1	1.9	1	2.2	AMITE	MS	-58	15	-52	14	23
1	1.3	2	2.0	1	.8	ATTALA	MS	53	-62	-41	21	10
				1	2.5	BENTON	MS					28
3	3.8	2	2.0	3	2.7	BOLIVAR	MS	-47	36	-29	32	31
		1	1.3	6	5.8	CALHOUN	MS		351			85
				2	5.0	CARROLL	MS					77
		1	2.0	5	7.3	CHICKASAW	MS		260			93
2	5.1	3	5.6	2	4.6	CHOCTAW	MS	9	-18	-10	47	71
2	11.6			1	3.2	CLAIBORNE	MS			-72	04	43
1	1.7	1	1.6	5	6.9	CLARKE	MS	-2	327	317	97	92
		3	4.9	4	5.7	CLAY	MS		16			84
1	1.4	3	3.4	2	2.0	COAHOMA	MS	142	-40	46	76	21
2	2.3	4	3.8	2	1.5	COPIAH	MS	68	-59	-32	29	15
		1	1.5	6	5.1	DE SOTO	MS		239			79
3	2.4	6	2.9	9	3.2	FORREST	MS	17	13	32	71	43
		2	5.3	2	5.9	FRANKLIN	MS		10			86
		3	7.4	2	2.7	GEORGE	MS		-63			32
1	3.7	2	7.4	1	2.4	GREENE	MS	98	-68	-36	26	26
3	6.9	3	4.7			GRENADA	MS	-32				
1	2.3	1	1.5	2	1.9	HANCOCK	MS	-37	28	-19	40	19
2	.9	12	3.0	21	4.0	HARRISON	MS	232	32	339	97	60
8	2.0	14	2.4	21	2.6	HINDS	MS	22	8	31	71	29
				1	1.9	HOLMES	MS					18
		2	7.8	2	6.1	HUMPHREYS	MS		-23			87
1	1.3	2	2.4	7	6.7	ITAWAMBA	MS	83	175	402	98	90
2	1.8	4	2.2	10	3.6	JACKSON	MS	22	64	100	87	51
1	1.9			2	2.4	JASPER	MS			25	68	26
		1	3.6	1	8.8	JEFFERSON	MS		143			96
2	6.0	1	2.6	1	1.7	JEFFERSON DAVIS	MS	-56	-34	-71	04	17
3	1.6	4	1.8	7	2.3	JONES	MS	8	31	42	75	25
		1	2.0			KEMPER	MS					
2	2.6	3	3.7	3	3.4	LAFAYETTE	MS	42	-9	30	70	47
		1	1.5	5	5.7	LAMAR	MS		286			85
3	1.3	12	3.9	9	2.6	LAUDERDALE	MS	203	-35	98	87	29
2	6.3	1	2.3	5*	9.3	LAWRENCE	MS	-63	298	48	77	97
2	3.2	1	1.1	1	1.0	LEAKE	MS	-65	-6	-67	06	11
4	2.9	5	2.6	7	2.7	LEE	MS	-8	2	-6	50	31
1	.9	3	2.9	4	3.3	LEFLORE	MS	231	12	270	96	45
2	2.1	2	1.8	2	1.2	LINCOLN	MS	-15	-31	-42	20	12
2	1.9	4	3.1	8	4.7	LOWNDES	MS	59	53	143	91	73
		2	3.4	3	3.9	MADISON	MS		18			60
1	1.5	4	4.5			MARION	MS	197				
4	10.7	2	3.4			MARSHALL	MS	-68				
1	1.2	2	1.5	5	3.1	MONROE	MS	24	109	159	92	40
1	1.9			1	1.2	MONTGOMERY	MS			-38	24	12
		1	1.1	5	3.9	NESHOBA	MS		253			59
2	2.4	3	3.0	4	3.0	NEWTON	MS	25	-3	22	66	37
1	3.4					NOXUBEE	MS					
2	2.8	1	1.6	1	1.3	OKTIBBEHA	MS	-44	-15	-52	13	13
1	1.3	1	1.3	4	3.4	PANOLA	MS	4	161	171	92	48
1	1.6	5	5.2	4	2.9	PEARL RIVER	MS	225	-45	80	84	35
2	6.8			2	4.6	PERRY	MS			-32	28	71
4	3.5	1	.6	3	1.9	PIKE	MS	-82	201	-46	17	19
		3	2.9	4	3.3	PONTOTOC	MS		14			45
2	2.9	5	4.9	5	3.9	PRENTISS	MS	68	-20	35	72	59
1	2.5			2	4.8	QUITMAN	MS			87	85	74
		1	.7	6	3.2	RANKIN	MS		362			43
				5	5.5	SCOTT	MS					82
		1	5.8			SHARKEY	MS					
1	1.4	1	1.3	6	5.8	SIMPSON	MS	-7	358	324	97	85
		2	3.6	1	1.2	SMITH	MS		-68			12
3	5.4	4	5.4	5	5.4	SUNFLOWER	MS	1	-1	0	55	82
		1	1.7	1	1.1	TATE	MS		-35			11
		3	3.4	3	3.5	TIPPAH	MS		4			50
2	2.7	1	1.0	2	2.0	TISHOMINGO	MS	-62	97	-25	35	20
				1	5.6	TUNICA	MS					83
2	2.4	1	1.0	5	4.0	UNION	MS	-57	293	69	82	61
2	1.8	3	2.1	5	3.3	WARREN	MS	17	53	79	83	45
		2	1.6	6	3.2	WASHINGTON	MS		102			43
		3	5.0	1	1.3	WAYNE	MS		-74			13
		1	1.7	3	3.4	WEBSTER	MS		100			46
				1	4.7	WILKINSON	MS					72
		2	2.7	4	5.3	WINSTON	MS		98			80
1	1.9	1	2.5	2	2.1	YALOBUSHA	MS	27	-13	10	60	23
4	6.3	2	2.5	2	2.5	YAZOO	MS	-61	0	-61	09	27
1000	4.4	1005	3.7	1259	4.0	MISSOURI	MO	-17	7	-10		
7	5.3	6	4.4	8	4.7	ADAIR	MO	-18	7	-12	46	73
		2	1.9			ANDREW	MO					
6*	8.9	1	1.2	4	4.5	ATCHISON	MO	-87	290	-50	15	70
9	6.6	5	2.7	14*	6.2	AUDRAIN	MO	-59	129	-7	50	88
7	5.1	10	5.9	13	7.0	BARRY	MO	16	17	36	73	92
10*	11.0	2	1.7	5	3.6	BARTON	MO	-85	116	-68	06	51
3	2.1	7	4.1	3	1.6	BATES	MO	101	-61	-22	38	16

* SIGNIFICANTLY MORE DEATHS THAN EXPECTED

WHITE FEMALE

CANCER OF THE LIVER AND GALLBLADDER INCLUDING BILE DUCTS

ICD 155,156

1950-1959 DEATHS	1950-1959 RATE	1960-1969 DEATHS	1960-1969 RATE	1970-1979 DEATHS	1970-1979 RATE	COUNTY AND STATE	PERCENT CHANGE 60S/50S	PERCENT CHANGE 70S/60S	PERCENT CHANGE 70S/50S	PERCENTILE %CHANGE 70S/50S	PERCENTILE RATES 1970S
3	4.1	4	4.7	5	4.2	BENTON MO	14	-10	2	56	65
2	3.7	4	7.9	6*	9.1	BOLLINGER MO	113	15	145	91	97
11	4.9	11	3.6	15	3.9	BOONE MO	-27	8	-21	38	57
36	5.6	17	2.3	26	3.3	BUCHANAN MO	-58	40	-42	20	44
4	2.4	7	3.7	11	4.4	BUTLER MO	52	17	78	83	67
4	4.4	5	7.0	6	6.9	CALDWELL MO	58	-1	57	79	92
9	6.1	9	6.0	7	3.4	CALLAWAY MO	-1	-43	-43	19	48
4	7.1	2	2.6	4	3.5	CAMDEN MO	-63	32	-51	15	49
10	4.7	16	5.9	11	3.2	CAPE GIRARDEAU MO	26	-46	-33	28	42
2	1.6	4	3.2	4	2.6	CARROLL MO	96	-18	61	80	30
		2	8.8	3	8.0	CARTER MO		-9			95
5	3.6	7	3.9	8	3.6	CASS MO	9	-8	0	55	52
1	1.3	2	1.8	4	4.9	CEDAR MO	45	168	289	96	76
3	2.7	4	3.9	5	4.8	CHARITON MO	46	23	79	83	74
3	3.8	3	3.3			CHRISTIAN MO	-15				
2	3.2	1	1.3	5	8.1	CLARK MO	-59	520	157	91	95
9	3.3	20	4.9	21	3.6	CLAY MO	51	-26	12	61	52
3	3.7	3	3.4	6	6.8	CLINTON MO	-7	97	83	84	91
7	3.7	8	3.6	11	3.8	COLE MO	-2	5	3	57	56
1	1.1	6	5.6	4	2.6	COOPER MO	410	-53	139	90	29
10*	14.6			6	4.1	CRAWFORD MO			-72	04	62
3	4.7	3	3.9	1	1.1	DADE MO	-16	-71	-76	03	12
4	6.0	6	7.2	6	5.4	DALLAS MO	19	-25	-11	47	81
1	1.2	2	2.2	6	4.6	DAVIESS MO	83	107	278	96	72
		3	4.4	2	1.7	DE KALB MO		-62			16
4	5.4	4	4.6	4	3.2	DENT MO	-15	-29	-40	22	44
2	3.5	2	3.2	1	2.1	DOUGLAS MO	-9	-35	-41	21	21
7	4.2	7	3.1	10	3.5	DUNKLIN MO	-26	14	-15	43	50
6	2.8	9	3.1	13	3.6	FRANKLIN MO	9	16	27	69	52
3	3.3	2	1.8	4	3.9	GASCONADE MO	-45	118	20	65	59
6	6.7	2	2.7	4	3.1	GENTRY MO	-60	13	-54	12	38
27	4.2	27	3.2	36	3.3	GREENE MO	-24	4	-21	38	45
5	4.7			5	4.0	GRUNDY MO			-15	44	60
2	2.1	1	.9	5	5.7	HARRISON MO	-59	561	169	92	84
8	4.9	4	2.3	8	4.8	HENRY MO	-53	113	-1	54	75
1	2.8	1	2.0	2	3.7	HICKORY MO	-31	88	30	70	54
		2	6.5	2	1.9	HOLT MO		-70			20
7*	9.0	3	2.7	3	3.1	HOWARD MO	-70	16	-66	07	39
4	3.3	5	3.3	14*	7.1	HOWELL MO	2	114	119	89	92
2	4.0	1	.9	8*	13.3	IRON MO	-76	1303	232	95	99
141	4.8	116	3.3	135	3.4	JACKSON MO	-31	5	-28	33	48
9	1.6	21	3.5	27	4.0	JASPER MO	123	16	159	92	61
13	6.0	20*	6.3	12	2.4	JEFFERSON MO	4	-62	-61	09	26
13*	9.0	3	1.4	14*	7.2	JOHNSON MO	-85	432	-19	40	93
2	3.3	2	3.7	5	6.8	KNOX MO	14	84	110	88	91
		5	3.8	10	5.8	LACLEDE MO		54			85
5	3.1	10	4.9	5	2.6	LAFAYETTE MO	60	-46	-13	45	30
7	4.5	5	2.4	13	5.7	LAWRENCE MO	-46	136	28	69	84
6	7.7	2	1.9	1	1.2	LEWIS MO	-75	-35	-84	01	12
6	5.7	4	3.1	5	3.1	LINCOLN MO	-45	-1	-46	17	39
3	2.4	6	3.2	5	2.5	LINN MO	32	-20	5	58	28
2	1.5	5	3.4	4	2.3	LIVINGSTON MO	129	-32	56	79	25
3	4.0	3	3.2	4	3.7	MC DONALD MO	-21	16	-9	48	54
4	2.7	4	2.3	12*	6.9	MACON MO	-11	194	160	92	91
2	3.4	3	3.5	7	7.6	MADISON MO	3	114	121	89	93
2	4.9	5*	11.1	2	3.4	MARIES MO	125	-69	-31	30	47
5	2.7	8	3.1	8	3.4	MARION MO	15	8	24	68	47
1	1.8	1	2.2			MERCER MO	17				
2	2.5	3	2.4	8	6.1	MILLER MO	-1	152	149	91	87
4	5.5	6	7.3	4	3.2	MISSISSIPPI MO	32	-55	-41	21	44
2	2.7	1	.8	4	3.2	MONITEAU MO	-69	276	17	64	42
4	4.1	3	4.3	4	2.8	MONROE MO	6	-35	-31	30	33
6	6.8	3	2.9	6	4.9	MONTGOMERY MO	-57	65	-29	32	75
4	5.7	3	3.8	8	7.7	MORGAN MO	-33	103	36	73	94
5	5.0	2	1.6	7	5.0	NEW MADRID MO	-68	213	1	56	77
5	2.9	4	2.3	16*	6.2	NEWTON MO	-20	169	114	88	88
12*	8.0	5	2.2	10	5.2	NODAWAY MO	-72	139	-34	27	80
1	1.6	1	1.2	8*	8.0	OREGON MO	-25	559	395	98	95
4	6.2	6*	8.2	3	4.3	OSAGE MO	33	-48	-30	31	67
		2	4.0	2	3.2	OZARK MO		-20			43
6	5.2	3	2.1	6	5.2	PEMISCOT MO	-59	145	0	54	80
2	2.4	3	3.6	5	4.3	PERRY MO	53	19	82	84	67
5	2.3	9	3.6	16	4.8	PETTIS MO	56	35	111	88	75
5	4.1	5	3.6	9	4.7	PHELPS MO	-12	31	15	63	72
3	3.0	2	2.0	4	3.8	PIKE MO	-33	86	25	68	56
1	1.0	5	4.7	3	1.9	PLATTE MO	359	-59	87	85	19
2	2.0	2	1.7	4	2.7	POLK MO	-14	59	36	73	31
3	4.9	4	4.4	3	2.9	PULASKI MO	-11	-33	-40	21	36
		3	4.9	4	5.3	PUTNAM MO		9			81
2	3.6	2	2.5	4	7.3	RALLS MO	-31	191	101	87	93
8	4.6	4	1.7	14*	6.2	RANDOLPH MO	-64	275	34	72	88
6	5.2	7	4.5	6	3.6	RAY MO	-14	-20	-31	29	52
		1	3.8	3	7.1	REYNOLDS MO		85			92
2	3.6	3	4.2	2	1.5	RIPLEY MO	17	-65	-58	10	15
15*	8.1	7	2.5	10	2.2	ST CHARLES MO	-69	-14	-73	03	23
5	6.2	2	2.3	2	2.1	ST CLAIR MO	-63	-7	-65	07	23
4	7.1	6*	8.3	5	5.8	STE GENEVIEVE MO	16	-29	-18	40	86
8	3.6	12	4.3	13	3.5	ST FRANCOIS MO	18	-20	-5	52	48
102	4.5	149	4.1	214	4.2	ST LOUIS COUNTY MO	-9	2	-7	50	64
5	3.0	12	6.1	8	3.2	SALINE MO	105	-47	8	60	43
		1	2.2	1	1.2	SCHUYLER MO		-47			12
		1	1.5	2	2.3	SCOTLAND MO		50			24
5	3.9	10	6.2	8	3.6	SCOTT MO	59	-41	-7	50	53
1	3.4	2	4.7	2	3.7	SHANNON MO	39	-21	10	60	54
7	6.8	4	3.4	3	3.8	SHELBY MO	-51	13	-44	18	57
5	4.0	9	5.1	16*	7.8	STODDARD MO	27	52	93	86	94
2	4.3			3	3.2	STONE MO			-24	36	44
3	4.2	8*	10.6	6	7.6	SULLIVAN MO	155	-29	82	84	94
1	2.1	2	2.3	7	4.3	TANEY MO	10	91	111	88	67
1	.9	6	4.2	7	4.5	TEXAS MO	361	6	390	98	69

* SIGNIFICANTLY MORE DEATHS THAN EXPECTED

1950-1959 DEATHS	1950-1959 RATE	1960-1969 DEATHS	1960-1969 RATE	1970-1979 DEATHS	1970-1979 RATE	COUNTY AND STATE	PERCENT CHANGE 60S/50S	PERCENT CHANGE 70S/60S	PERCENT CHANGE 70S/50S	PERCENTILE %CHANGE 70S/50S	PERCENTILE RATES 1970S
1	.5	5	2.4	7	4.0	VERNON MO	438	63	778	99	60
1	1.5	3	5.0	5	6.3	WARREN MO	231	27	320	97	89
6*	8.6	5	7.0	6	5.9	WASHINGTON MO	-18	-15	-31	30	86
2	3.1	2	3.0	4	5.2	WAYNE MO	-4	72	66	81	80
5	4.7	4	4.0	1	.6	WEBSTER MO	-16	-84	-87	00	10
1	2.3					WORTH MO					
6	6.6	1	1.2	4	3.6	WRIGHT MO	-82	196	-46	17	51
233*	5.4	187	4.4	144	3.7	ST LOUIS CITY MO	-20	-15	-32	29	54
108	4.1	115	3.4	130	3.2	MONTANA MT	-15	-7	-21		
2	5.9	2	5.2	2	4.3	BEAVERHEAD MT	-13	-16	-27	33	67
1	3.0	2	7.1	1	3.5	BIG HORN MT	136	-51	16	63	49
1	3.3	1	2.3			BLAINE MT	-30				
2	3.7	3	5.4	1	1.6	CARBON MT	46	-71	-58	11	15
1	12.4					CARTER MT					
12	5.3	16	5.1	16	4.0	CASCADE MT	-3	-22	-25	36	61
2	5.6	2	6.3			CHOUTEAU MT	12				
6*	9.7	4	5.9	3	3.6	CUSTER MT	-40	-39	-63	08	51
3	7.3	4	7.5	1	1.5	DAWSON MT	4	-80	-79	02	15
3	3.6	4	4.2	6	6.1	DEER LODGE MT	17	45	71	82	87
		2	9.2	1	4.0	FALLON MT		-56			61
3	4.2	2	2.0			FERGUS MT	-54				
5	3.4	9	4.6	6	2.3	FLATHEAD MT	37	-50	-31	30	25
9*	9.0	1	.8	6	3.7	GALLATIN MT	-91	354	-59	10	54
1	4.5			1	2.9	GLACIER MT			-36	26	35
		1	14.1			GOLDEN VALLEY MT					
1	1.7			3	3.9	HILL MT			130	90	58
1	4.1					JEFFERSON MT					
		1	7.0			JUDITH BASIN MT					
2	3.3	1	1.2	6	5.8	LAKE MT	-65	396	76	83	85
5	4.0	6	3.6	10	4.7	LEWIS AND CLARK MT	-10	31	18	64	73
				1	5.9	LIBERTY MT					86
				1	1.6	LINCOLN MT					16
1	8.0	2*	14.2	1	7.2	MC CONE MT	77	-50	-11	47	92
1	3.4			1	2.1	MADISON MT			-39	23	21
				1	5.5	MINERAL MT					82
6	4.0	8	3.4	5	1.8	MISSOULA MT	-15	-46	-54	13	18
1	3.7					MUSSELSHELL MT					
5	7.3			3	2.8	PARK#MT			-61	09	34
		1*	18.7			PETROLEUM MT					
3	8.6	1	4.8			PHILLIPS MT	-44				
		2	6.2	4*	10.0	PONDERA MT		60			98
		1	9.2			POWDER RIVER MT					
		1	6.5	1	4.6	PRAIRIE MT		-29			72
2	2.6	1	1.0	3	1.9	RAVALLI MT	-60	81	-28	33	19
2	4.5	2	3.7	2	3.6	RICHLAND MT	-18	-2	-20	39	52
		1	2.6	2	4.4	ROOSEVELT MT		66			68
1	4.4	2	8.8			ROSEBUD MT	101				
3	8.4	2	5.8	2	4.0	SANDERS MT	-31	-30	-52	14	61
		1	2.5	1	2.7	SHERIDAN MT		9			31
12	4.9	8	2.9	18*	5.7	SILVER BOW MT	-42	101	16	63	85
		1	3.5	1	2.4	STILLWATER MT		-33			26
1	4.5					SWEET GRASS MT					
2	7.2			1	2.5	TETON MT			-65	07	28
		1*	20.4			TREASURE MT					
2	5.7	1	2.1	4	6.5	VALLEY MT	-62	205	14	62	89
1	5.9	2	12.6	1	3.6	WHEATLAND MT	112	-71	-39	23	52
5	2.7	16	4.2	14	3.0	YELLOWSTONE MT	60	-29	13	62	37
388*	5.2	402*	4.5	332	3.1	NEBRASKA NB	-14	-30	-40		
12	6.4	11	4.5	4	1.5	ADAMS NB	-30	-67	-77	02	14
3	4.5	6	7.7	3	3.2	ANTELOPE NB	70	-59	-30	31	42
		1*	13.8			BLAINE NB					
3	5.0	3	5.8	4	4.5	BOONE NB	15	-22	-10	48	70
5	8.6			3	3.9	BOX BUTTE NB			-54	12	59
2	9.2	1	3.0	5*	16.7	BOYD NB	-67	455	82	84	99
		2	8.1	1	1.4	BROWN NB		-83			13
6	3.7	11	5.3	8	4.2	BUFFALO NB	44	-21	13	62	65
4	5.8	9*	10.2	5	5.2	BURT NB	77	-49	-10	47	79
1	1.0	7*	9.3			BUTLER NB	845				
3	2.7	3	3.2	5	4.0	CASS NB	17	27	48	77	62
9*	13.4	4	5.1	3	3.2	CEDAR NB	-62	-38	-76	03	42
1	5.0			2	4.3	CHASE NB			-13	45	67
1	2.8	1	2.1			CHERRY NB	-24				
2	3.8	3	4.4	1	1.2	CHEYENNE NB	16	-72	-68	06	12
4	6.1	2	3.5	2	2.2	CLAY NB	-43	-38	-64	07	23
5	7.1	4	4.3	7	6.7	COLFAX NB	-39	55	-6	51	90
2	3.2	5	5.8	3	3.1	CUMING NB	83	-46	-1	54	41
7	6.4	2	1.6	1	.5	CUSTER NB	-75	-67	-92	00	09
3	6.7	3	4.9	2	2.4	DAKOTA NB	-27	-51	-64	07	26
2	4.1	3	3.8	1	1.2	DAWES NB	-8	-70	-72	04	12
8	8.3	3	2.5	3	2.0	DAWSON NB	-70	-20	-76	03	20
1	4.5	1	4.0	2	8.9	DEUEL NB	-11	124	98	87	96
2	4.0	3	4.8	2	3.7	DIXON NB	19	-24	-9	48	53
9	5.2	14	6.6	8	3.1	DODGE NB	27	-53	-41	21	39
79*	5.3	80	4.4	56	2.5	DOUGLAS NB	-17	-43	-53	13	28
1	4.1	1	3.9	1	2.1	DUNDY NB	-5	-47	-49	15	21
2	2.3	7*	9.1	3	3.7	FILLMORE NB	288	-59	59	80	55
1	2.8	3	6.1	2	3.6	FRANKLIN NB	117	-41	28	70	52
1	3.8			1	1.9	FRONTIER NB			-51	14	18
2	3.1	3	3.5	3	3.3	FURNAS NB	13	-7	6	59	44
5	2.6	11	5.8	4	1.5	GAGE NB	125	-74	-43	20	14
2	9.3			1	3.6	GARDEN NB			-62	08	51
		2	8.6	1	5.9	GARFIELD NB		-31			86
2	7.4			1	2.2	GREELEY NB			-70	05	23
9	4.6	8	3.3	18	6.1	HALL NB	-29	85	32	71	87
3	4.8	1	1.3	5	9.0	HAMILTON NB	-73	598	89	86	97
1	2.7	1	2.0	2	3.5	HARLAN NB	-26	74	28	69	49

* SIGNIFICANTLY MORE DEATHS THAN EXPECTED

\# COUNTIES COMBINED; SEE TECHNICAL NOTES

WHITE FEMALE — CANCER OF THE LIVER AND GALLBLADDER INCLUDING BILE DUCTS — ICD 155,156

1950-1959 DEATHS	1950-1959 RATE	1960-1969 DEATHS	1960-1969 RATE	1970-1979 DEATHS	1970-1979 RATE	COUNTY AND STATE	PERCENT CHANGE 60S/50S	PERCENT CHANGE 70S/60S	PERCENT CHANGE 70S/50S	PERCENTILE %CHANGE 70S/50S	PERCENTILE RATES 1970S
		1	16.7	1*	54.5	HAYES NB		226			99
1	3.5	2	6.0	1	2.6	HITCHCOCK NB	71	-57	-26	35	30
4	5.4	4	4.7	2	1.7	HOLT NB	-11	-64	-68	06	17
4	9.5	1	1.9	1	1.7	HOWARD NB	-80	-8	-82	01	17
		4	4.6	5	3.8	JEFFERSON NB		-19			56
1	2.2	5*	9.3			JOHNSON NB	316				
2	4.0	1	2.1	5	4.9	KEARNEY NB	-48	136	23	67	75
		2	3.9	1	1.3	KEITH NB		-67			13
		2	7.2	1	2.9	KIMBALL NB		-59			36
4	5.1	4	4.6	7	6.8	KNOX NB	-9	46	32	71	91
47*	6.1	36	3.9	29	2.6	LANCASTER NB	-36	-35	-58	11	28
6	4.7	5	3.3	9	4.2	LINCOLN NB	-30	28	-11	47	64
1*	20.6					LOGAN NB					
12*	7.0	9	4.4	10	4.6	MADISON NB	-37	4	-35	27	71
1	1.9	1	1.5	1	1.7	MERRICK NB	-23	12	-14	44	16
2	5.6	2	5.2	1	1.4	MORRILL NB	-8	-73	-75	03	14
4*	11.5					NANCE NB					
4	5.9			4	5.0	NEMAHA NB			-15	43	77
4	6.9	4	5.0			NUCKOLLS NB	-28				
3	3.0	2	1.4	3	1.7	OTCE NB	-54	21	-44	19	16
4	7.9	3	5.9	2	6.9	PAWNEE NB	-25	16	-14	44	91
1	6.1	3*	12.0	2	7.1	PERKINS NB	98	-41	16	63	92
4	7.5	7*	10.0	3	4.0	PHELPS NB	34	-60	-47	17	61
5*	10.1	5	8.3	2	2.4	PIERCE NB	-17	-72	-76	03	26
3	3.3	5	3.7	8	4.8	PLATTE NB	9	30	42	75	74
2	3.3	5	6.9	2	2.9	POLK NB	107	-57	-11	47	36
3	4.0	4	4.2	4	4.7	RED WILLOW NB	4	13	18	64	73
2	1.6	5	4.8	3	1.6	RICHARDSON NB	198	-67	-1	54	16
3*	20.0	2	10.0			ROCK NB	-50				
8*	8.0	9*	7.8	8	4.8	SALINE NB	-3	-38	-40	21	74
		3	3.0	5	2.6	SARPY NB		-15			29
4	4.1	6	5.3	5	2.8	SAUNDERS NB	30	-48	-32	29	33
7	5.2	13*	7.1	5	2.3	SCOTTS BLUFF NB	37	-67	-55	12	25
7*	8.2			3	2.9	SEWARD NB			-64	07	36
2	4.1			3	5.8	SHERIDAN NB			41	75	85
1	3.1	2	5.2	1	2.5	SHERMAN NB	69	-51	-18	40	28
2	8.7	1	2.7	2	4.9	STANTON NB	-69	83	-44	18	75
3	4.1	4	5.5			THAYER NB	34				
1	2.6	2	6.0	1	1.8	THURSTON NB	127	-70	-31	30	18
7*	15.7	4	8.2	1	2.0	VALLEY NB	-48	-75	-87	00	21
3	4.1	9*	9.2	5	6.1	WASHINGTON NB	125	-33	51	78	88
1	2.0	1	1.6	3	4.8	WAYNE NB	-20	209	145	91	75
4	7.4	1	1.3	3	5.1	WEBSTER NB	-83	302	-32	29	78
		1*	17.2			WHEELER NB					
8*	8.5	3	2.7	7	6.1	YORK NB	-68	121	-28	33	87
16	2.5	41	3.3	75	3.1	NEVADA NV	33	-6	24		
		2	4.8	3	5.1	CHURCHILL NV		7			79
3	2.1	16	2.9	31	2.6	CLARK NV	43	-11	27	69	29
				4	6.6	DOUGLAS NV					90
1	1.9	2	4.7	3	5.0	ELKO NV	146	6	162	92	77
		1*	37.7			EUREKA NV					
1	4.9	2	10.2			HUMBOLDT NV	108				
				1	8.8	LANDER NV					96
1	4.9	1	3.6	2	3.3	LYON NV	-26	-8	-32	29	46
				3*	11.8	MINERAL NV					99
1	6.7					NYE NV					
		2*	16.7	1	7.7	PERSHING NV		-54			94
		1*	14.6			STOREY NV					
9	4.0	11	2.7	17	2.6	WASHOE NV	-33	-3	-35	27	29
		1	2.9	2	4.7	WHITE PINE NV		61			74
		2	5.1	8	6.6	ORMSBY-CARSON CITY NV		28			90
129	3.8	145	3.5	179	3.4	NEW HAMPSHIRE NH	-8	-4	-12		
9	5.3	12	5.4	8	2.7	BELKNAP NH	1	-50	-49	15	31
7	6.5	7	5.0	9	5.1	CARROLL NH	-23	2	-22	38	78
11	4.6	14	4.9	6	1.5	CHESHIRE NH	6	-70	-68	06	15
9	4.9	4	1.9	12	4.6	COOS NH	-62	145	-7	50	71
7	2.4	16	4.7	15	3.4	GRAFTON NH	98	-27	44	76	48
40	4.0	46	3.8	62	4.1	HILLSBOROUGH NH	-6	8	2	56	62
11	2.2	11	1.8	22	3.1	MERRIMACK NH	-17	67	39	74	39
21	4.6	21	3.5	21	2.4	ROCKINGHAM NH	-24	-32	-49	16	26
9	3.1	9	2.7	15	3.4	STRAFFORD NH	-14	29	11	60	48
5	3.2	5	2.9	9	4.5	SULLIVAN NH	-9	55	40	75	70
1106	4.2	1436*	4.2	1798*	4.3	NEW JERSEY NJ	-1	1	1		
35	4.0	45	3.9	66	4.9	ATLANTIC NJ	-3	27	22	67	76
129	4.2	182	4.1	237	4.2	BERGEN NJ	-3	3	-1	54	65
26	4.1	43	4.7	58	4.2	BURLINGTON NJ	16	-11	4	57	65
88*	5.9	81	3.9	94	3.8	CAMDEN NJ	-34	-1	-35	26	57
10	3.2	16	3.3	25	3.7	CAPE MAY NJ	4	13	18	64	55
21	4.5	23	4.1	35	4.9	CUMBERLAND NJ	-9	19	8	59	75
166	3.6	207	4.2	218*	4.5	ESSEX NJ	17	9	27	69	71
18	3.6	30	4.7	32	3.6	GLOUCESTER NJ	30	-24	0	54	52
154	4.7	161	4.4	222*	5.8	HUDSON NJ	-6	32	23	67	85
14	5.2	18	5.2	16	3.5	HUNTERDON NJ	0	-31	-31	30	51
47	4.0	57	3.9	61	3.5	MERCER NJ	-2	-10	-12	46	50
71*	5.2	91	4.7	128*	4.9	MIDDLESEX NJ	-11	4	-7	50	75
54	3.8	79	3.7	113	3.9	MONMOUTH NJ	-2	6	4	57	60
36	3.5	63	4.2	77	3.9	MORRIS NJ	23	-9	12	61	58
17	3.2	39	4.0	81	3.4	OCEAN NJ	25	-17	4	57	46
93	5.0	95	3.9	129*	4.7	PASSAIC NJ	-21	21	-5	51	73
12	5.7	12	4.7	9	2.9	SALEM NJ	-17	-38	-49	16	36
12	2.3	34	4.7	42	4.3	SOMERSET NJ	105	-8	88	85	66
6	2.8	15	4.7	20	4.1	SUSSEX NJ	69	-12	49	77	63
84	4.2	127	4.6	120	3.6	UNION NJ	10	-22	-13	44	52
13	3.6	18	4.0	15	2.9	WARREN NJ	12	-29	-20	39	34

* SIGNIFICANTLY MORE DEATHS THAN EXPECTED

WHITE FEMALE

CANCER OF THE LIVER AND GALLBLADDER INCLUDING BILE DUCTS

ICD 155,156

1950-1959 DEATHS	1950-1959 RATE	1960-1969 DEATHS	1960-1969 RATE	1970-1979 DEATHS	1970-1979 RATE	COUNTY AND STATE	PERCENT CHANGE 60S/50S	PERCENT CHANGE 70S/60S	PERCENT CHANGE 70S/50S	PERCENTILE %CHANGE 70S/50S	PERCENTILE RATES 1970S
98	4.4	165*	5.2	252*	5.5	NEW MEXICO NM	19	4	23		
19	3.4	40	4.2	73*	5.1	BERNALILLO NM	23	20	48	77	78
1	8.2			1	8.6	CATRON NM			5	58	96
2	1.4	12	6.5	11	3.8	CHAVES NM	350	-42	162	92	56
5	8.0	7*	11.1	4	4.5	COLFAX NM	39	-59	-44	19	70
6	6.0	3	2.5	1	.6	CURRY NM	-57	-75	-89	00	10
		1	6.0			DE BACA NM					
11*	8.6	9	5.1	28*	10.1	DONA ANA NM	-40	98	18	65	98
3	2.2	8	4.1	12	4.9	EDDY NM	88	21	128	89	76
4	5.8	5	6.1	4	3.4	GRANT NM	4	-45	-42	20	46
1	9.3	4*	19.3	3*	14.5	GUADALUPE NM	108	-25	56	79	99
1	11.2					HARDING NM					
				2	6.3	HIDALGO NM					88
2	2.8	5	4.2	7	3.5	LEA NM	49	-17	23	67	49
2	5.3	1	2.4	3	5.6	LINCOLN NM	-55	133	6	58	83
		4*	19.9			LOS ALAMOS NM					
		2	5.2	6	6.6	LUNA NM		27			90
1	2.0	5*	12.4	4	6.4	MC KINLEY NM	512	-48	216	95	89
1	3.7	2	7.5	1	5.7	MORA NM	101	-24	53	78	84
2	5.1	6	6.8	9*	6.9	OTERO NM	32	2	35	73	91
2	3.7	1	1.6	1	2.1	QUAY NM	-57	33	-43	19	22
3	4.6	10*	12.7	13*	17.6	RIO ARRIBA NM	174	38	280	96	99
		3	4.0	5	5.3	ROOSEVELT NM		32			81
		1	4.1	3	3.9	SANDOVAL NM		-7			57
2	5.9	2	1.5	7	4.3	SAN JUAN NM	-74	177	-27	33	66
5	5.6	5	5.0	11*	10.2	SAN MIGUEL NM	-10	104	83	84	98
12*	8.9	11	5.9	12	4.1	SANTA FE NM	-34	-30	-54	13	63
				2	2.9	SIERRA NM					36
1	2.9	2	6.0	11*	23.8	SOCORRO NM	102	299	707	99	99
2	3.7	6*	9.1	4	5.8	TAOS NM	146	-36	58	79	86
3*	13.0	2	6.9	2	5.2	TORRANCE NM	-47	-25	-60	10	80
2	6.8	1	2.4	2	5.7	UNION NM	-64	134	-17	42	84
5	8.3	7*	7.6	10*	7.9	VALENCIA NM	-9	4	-5	52	94
3759*	4.7	4165*	4.3	4476*	4.1	NEW YORK NY	-8	-4	-12		
56	3.8	68	3.8	82	4.1	ALBANY NY	1	8	9	60	63
16	6.6	10	3.6	14	4.3	ALLEGANY NY	-46	19	-35	26	66
370*	5.1	331	4.0	389*	5.1	BRONX NY	-21	26	0	55	78
35	3.3	46	3.6	54	3.7	BROOME NY	9	0	10	60	53
17	3.8	14	2.7	16	3.0	CATTARAUGUS NY	-28	11	-21	39	38
13	2.9	14	2.9	24	4.3	CAYUGA NY	-1	49	47	77	66
26	3.1	41	4.3	42	3.9	CHAUTAUQUA NY	40	-10	26	68	57
15	2.9	30	5.0	27	3.9	CHEMUNG NY	76	-22	36	73	58
15	6.2	11	4.2	12	3.8	CHENANGO NY	-33	-10	-40	22	56
9	3.7	7	2.4	7	1.9	CLINTON NY	-34	-20	-47	17	20
6	2.0	9	2.3	18	4.1	COLUMBIA NY	16	76	105	87	63
10	4.3	9	3.5	5	1.7	CORTLAND NY	-19	-50	-60	09	17
9	3.4	17	6.0	11	3.1	DELAWARE NY	74	-48	-10	48	40
26	2.8	48	4.2	55	4.1	DUTCHESS NY	49	-3	44	76	62
269*	5.7	276*	4.8	287*	4.4	ERIE NY	-17	-9	-24	36	67
4	2.2	6	2.9	13	4.7	ESSEX NY	31	60	109	88	72
11	4.6	6	2.3	16	6.1	FRANKLIN NY	-51	170	33	72	87
18	5.0	15	3.8	14	3.3	FULTON NY	-23	-15	-35	27	44
19*	6.7	21*	6.5	14	4.2	GENESEE NY	-3	-37	-38	24	64
4	1.9	12	5.2	11	3.6	GREENE NY	175	-30	91	86	52
		2	5.8	1	2.4	HAMILTON NY		-58			27
25*	6.2	18	3.9	20	4.4	HERKIMER NY	-37	12	-29	32	68
28	5.2	33*	5.2	35*	5.6	JEFFERSON NY	0	8	9	60	83
680*	5.3	564	4.0	605*	4.5	KINGS(BROOKLYN) NY	-24	12	-15	43	70
3	2.2	8	6.0	6	3.9	LEWIS NY	176	-35	78	83	58
10	4.2	15	5.7	13	4.1	LIVINGSTON NY	36	-28	-2	53	62
13	5.0	6	2.2	13	3.7	MADISON NY	-56	71	-25	35	55
156*	5.2	150	4.1	153	3.5	MONROE NY	-21	-14	-32	29	50
15	3.9	13	3.1	19	3.5	MONTGOMERY NY	-20	14	-9	49	50
151	4.1	245	4.2	296	3.9	NASSAU NY	4	-9	-5	52	58
465*	4.9	656*	6.7	391*	4.2	NEW YORK(MANHATTAN) NY	35	-38	-16	42	64
49	5.2	61*	5.3	59	4.2	NIAGARA NY	3	-21	-18	40	65
45	3.4	71	4.5	70	3.8	ONEIDA NY	32	-15	13	62	57
106*	5.5	111	4.7	88	3.2	ONONDAGA NY	-15	-31	-41	21	43
18	4.8	12	2.7	16	3.0	ONTARIO NY	-44	13	-37	24	38
42	4.2	40	3.4	51	3.7	ORANGE NY	-20	10	-12	45	54
10	4.9	11	4.9	6	2.4	ORLEANS NY	0	-52	-52	14	25
19	4.3	26	5.3	23	4.0	OSWEGO NY	25	-25	-6	51	61
16	4.5	11	2.9	12	3.0	OTSEGO NY	-36	5	-33	28	38
5	3.3	10	4.6	8	2.5	PUTNAM NY	40	-46	-24	36	28
368	4.4	426	3.6	617*	4.5	QUEENS NY	-18	27	4	57	70
38	4.6	40	4.2	39	3.6	RENSSELAER NY	-8	-14	-21	39	53
33	3.5	47	3.5	86*	5.0	RICHMOND NY	0	41	41	75	77
20	3.7	24	3.0	40	3.7	ROCKLAND NY	-18	23	1	55	55
18	3.4	17	2.8	25	3.7	ST LAWRENCE NY	-16	32	10	60	55
19	4.5	20	3.6	19	2.7	SARATOGA NY	-18	-26	-40	22	31
41	4.7	42	4.2	36	3.2	SCHENECTADY NY	-12	-24	-33	28	42
8	5.7	1	.4	11	5.9	SCHOHARIE NY	-92	1226	4	57	86
5	5.5	6	6.5	2	2.0	SCHUYLER NY	17	-69	-63	08	20
5	2.7	2	1.0	4	1.8	SENECA NY	-65	83	-35	26	17
16	2.9	27	4.7	19	2.9	STEUBEN NY	61	-37	1	56	36
97	4.5	126	3.3	218	3.9	SUFFOLK NY	-27	19	-13	45	58
11	4.6	13	4.4	12	3.4	SULLIVAN NY	-5	-22	-25	35	48
9	4.9	10	4.7	4	1.6	TIOGA NY	-5	-66	-68	06	15
10	3.5	12	4.0	11	3.3	TOMPKINS NY	13	-18	-7	50	44
38	5.7	39	4.9	48	5.1	ULSTER NY	-15	4	-12	46	78
12	4.4	17	5.2	11	3.1	WARREN NY	18	-40	-29	32	41
8	2.9	10	3.1	15	4.0	WASHINGTON NY	6	28	35	73	60
16	4.2	18	4.1	23	4.9	WAYNE NY	-2	20	17	64	75
169	4.6	204	4.3	225	3.9	WESTCHESTER NY	-7	-9	-15	43	58
6	3.0	8	3.6	10	4.2	WYOMING NY	22	16	42	75	65
8	6.1	2	1.3	5	2.7	YATES NY	-78	104	-55	12	32

* SIGNIFICANTLY MORE DEATHS THAN EXPECTED

WHITE FEMALE

CANCER OF THE LIVER AND GALLBLADDER INCLUDING BILE DUCTS

ICD 155,156

1950-1959 DEATHS	1950-1959 RATE	1960-1969 DEATHS	1960-1969 RATE	1970-1979 DEATHS	1970-1979 RATE	COUNTY AND STATE	PERCENT CHANGE 60S/50S	PERCENT CHANGE 70S/60S	PERCENT CHANGE 70S/50S	PERCENTILE %CHANGE 70S/50S	PERCENTILE RATES 1970S
491	4.1	444	2.6	671	2.8	NORTH CAROLINA NC	-36	8	-30		
7	3.0	13	3.9	9	1.9	ALAMANCE NC	31	-53	-38	24	18
1	1.8			2	2.1	ALEXANDER NC			14	62	21
4	10.6	1	2.5	2	3.3	ALLEGHANY NC	-76	28	-69	05	44
4	6.5	1	1.6	1	1.2	ANSON NC	-76	-25	-82	01	12
6	6.4	3	2.7	3	2.3	ASHE NC	-58	-16	-64	07	24
2	4.0	2	3.9	5	6.2	AVERY NC	-2	59	56	79	88
3	2.8	2	1.5	3	1.5	BEAUFORT NC	-45	0	-46	18	15
1	1.6			5	6.5	BERTIE NC			314	97	89
3	5.1	1	1.2	1	1.2	BLADEN NC	-77	2	-77	03	12
1	1.9	3	4.9	7	5.1	BRUNSWICK NC	160	4	171	92	78
19	3.4	24	3.2	23	2.4	BUNCOMBE NC	-7	-24	-29	32	26
5	3.1	5	2.1	9	2.9	BURKE NC	-30	35	-6	51	35
10	5.0	6	2.1	16	4.1	CABARRUS NC	-58	96	-17	41	63
8	5.8	4	2.1	14	4.9	CALDWELL NC	-65	139	-16	43	76
1	5.0	1	5.1	1	3.5	CAMDEN NC	3	-31	-29	32	50
4	4.6	2	1.6	7	3.8	CARTERET NC	-65	138	-17	42	56
1	2.2	1	1.8	2	3.0	CASWELL NC	-15	66	41	75	38
9	4.3	5	1.5	8	1.7	CATAWBA NC	-66	16	-61	09	17
7	8.3	1	1.0	5	3.4	CHATHAM NC	-88	225	-60	10	46
6	7.8	2	2.1	4	3.2	CHEROKEE NC	-72	50	-59	10	43
1	3.6			3	5.3	CHOWAN NC			46	76	81
1	4.0	4*	11.7	2	5.8	CLAY NC	193	-50	46	76	85
8	4.5	6	2.1	11	2.8	CLEVELAND NC	-53	33	-37	25	34
4	3.4	4	3.5	3	1.7	COLUMBUS NC	1	-52	-51	15	16
5	5.3			5	2.4	CRAVEN NC			-56	12	25
6	3.4	7	2.5	10	2.3	CUMBERLAND NC	-26	-8	-32	29	25
1	4.4	2	6.8			CURRITUCK NC	56				
3	10.4			2	2.1	DARE NC			-80	02	22
10	5.3	6	2.0	12	2.6	DAVIDSON NC	-62	29	-51	15	29
4	6.7	1	1.3	3	3.0	DAVIE NC	-80	120	-56	12	37
4	4.4	2	1.7	1	.8	DUPLIN NC	-61	-54	-82	01	10
11	3.6	9	2.3	11	2.0	DURHAM NC	-36	-12	-44	18	20
4	5.3	4	3.4	3	1.8	EDGECOMBE NC	-36	-47	-66	07	18
22	5.1	29	4.2	27	2.6	FORSYTH NC	-18	-38	-49	16	30
2	3.7	3	3.1	1	.8	FRANKLIN NC	-16	-74	-78	02	10
16	5.2	9	1.9	20	2.8	GASTON NC	-64	47	-47	17	33
		1	3.4	3	7.1	GATES NC		109			92
1	4.5			1	2.3	GRAHAM NC			-48	16	25
2	2.1	1	.8	1	.9	GRANVILLE NC	-59	9	-55	12	10
		4*	11.6	1	1.6	GREENE NC		-87			15
35	5.3	31	3.3	43	3.2	GUILFORD NC	-38	-3	-40	22	42
		3	1.9	9	4.5	HALIFAX NC		131			70
7	6.4	1	.6	5	2.1	HARNETT NC	-90	243	-67	06	22
12*	8.1	7	3.3	4	1.5	HAYWOOD NC	-59	-55	-82	01	15
6	3.2	7	2.7	10	2.6	HENDERSON NC	-16	-4	-19	40	29
		3	4.8	3	4.1	HERTFORD NC		-15			62
2	7.7					HOKE NC					
1	4.0	1	3.7			HYDE NC	-8				
12	5.7	6	2.0	10	2.4	IREDELL NC	-65	18	-58	10	26
6	8.1	3	3.5	3	2.6	JACKSON NC	-56	-25	-67	06	30
8	5.8	4	1.5	13	4.5	JOHNSTON NC	-74	198	-22	37	70
1	4.3	1	4.0	2	6.5	JONES NC	-7	63	52	78	90
5	6.2	2	1.8	4	2.5	LEE NC	-71	37	-60	10	28
5	5.8	3	2.1	3	1.3	LENOIR NC	-64	-40	-78	02	13
5	4.8	6	4.8	2	1.0	LINCOLN NC	1	-79	-79	02	11
4	5.4			8	4.4	MC DOWELL NC			-19	40	68
3	4.2	5	5.0	2	1.4	MACON NC	21	-72	-66	06	14
		2	2.3	3	2.1	MADISON NC		-7			22
2	3.0	1	2.2	4	4.4	MARTIN NC	-24	94	47	76	67
15	2.4	22	2.1	36	2.4	MECKLENBURG NC	-12	16	2	56	26
1	1.3	2	2.8	2	1.8	MITCHELL NC	107	-35	33	72	17
2	3.7	4	5.1	2	1.6	MONTGOMERY NC	39	-70	-58	11	15
4	3.2	3	1.8	3	1.0	MOORE NC	-43	-44	-68	05	11
1	.6	4	2.0	3	1.2	NASH NC	221	-39	96	86	13
7	3.3	12	4.0	20	5.2	NEW HANOVER NC	21	30	58	79	80
3	5.3	1	1.2	3	3.1	NORTHAMPTON NC	-77	147	-42	20	40
3	3.8	3	2.7	4	2.2	ONSLOW NC	-29	-19	-43	20	23
9*	10.4	4	2.5	7	3.7	ORANGE NC	-76	46	-65	07	54
3	9.4			1	1.9	PAMLICO NC			-80	02	19
2	2.9	2	2.2	4	3.1	PASQUOTANK NC	-23	41	8	59	40
1	2.2	1	1.6	3	3.9	PENDER NC	-29	143	72	82	58
1	3.6	1	2.7			PERQUIMANS NC	-23				
1	2.3	4	4.5	2	1.7	PERSON NC	92	-61	-26	34	17
3	2.3	5	2.8	9	3.6	PITT NC	23	27	56	79	51
		4	4.4	3	2.7	POLK NC		-39			31
15*	8.7	6	2.0	19	4.5	RANDOLPH NC	-77	130	-48	16	70
1	1.1	7	5.3	4	2.0	RICHMOND NC	398	-61	93	86	21
2	1.2	8	4.3	8	3.4	ROBESON NC	246	-20	175	93	47
9	3.9	11	4.1	18	4.6	ROCKINGHAM NC	6	11	17	64	71
9	3.7	7	1.9	15	2.8	ROWAN NC	-50	52	-24	36	34
8	5.1	3	1.4	7	2.0	RUTHERFORD NC	-73	44	-61	09	20
6	5.0	5	3.4	6	2.9	SAMPSON NC	-32	-14	-41	21	35
				3	3.3	SCOTLAND NC					44
5	3.7	4	2.3	16*	6.1	STANLY NC	-38	169	66	81	87
4	5.6	5	4.8	3	2.1	STOKES NC	-14	-56	-62	08	22
10	6.5	9	3.9	8	2.5	SURRY NC	-40	-35	-61	09	28
1	3.1	1	2.5	3	6.4	SWAIN NC	-19	156	108	88	89
1	2.0	3	4.4	4	3.9	TRANSYLVANIA NC	125	-12	99	87	58
		1	5.6	1	4.8	TYRRELL NC		-14			75
7	4.8	3	1.4	10	4.1	UNION NC	-71	199	-14	44	63
5	5.8	6	5.0	1	1.0	VANCE NC	-13	-81	-83	01	11
14	3.3	10	1.7	21	2.3	WAKE NC	-49	34	-32	29	24
2	4.7			2	2.8	WARREN NC			-40	22	34
1	3.2			1	1.7	WASHINGTON NC			-48	16	16
		2	2.5	4	3.1	WATAUGA NC		27			40
4	2.4	5	2.3	3	1.1	WAYNE NC	-3	-54	-55	12	11
4	2.5	6	2.7	21*	7.7	WILKES NC	10	180	208	94	94

* SIGNIFICANTLY MORE DEATHS THAN EXPECTED

WHITE FEMALE

CANCER OF THE LIVER AND GALLBLADDER INCLUDING BILE DUCTS

ICD 155,156

1950-1959 DEATHS	1950-1959 RATE	1960-1969 DEATHS	1960-1969 RATE	1970-1979 DEATHS	1970-1979 RATE	COUNTY AND STATE	PERCENT CHANGE 60S/50S	PERCENT CHANGE 70S/60S	PERCENT CHANGE 70S/50S	PERCENTILE %CHANGE 70S/50S	PERCENTILE RATES 1970S
1	.6	6	3.5	10	4.4	WILSON NC	453	25	592	99	69
5	5.5	7	6.3	5	3.1	YADKIN NC	14	-50	-43	19	40
		2	2.9	1	1.3	YANCEY NC		-56			13
119	4.7	144	4.6	170*	4.3	NORTH DAKOTA ND	-2	-6	-8		
		1	4.8	1	3.4	ADAMS ND		-30			47
3	3.6	3	3.2	8	7.0	BARNES ND	-12	116	91	86	92
2	5.1			3	6.4	BENSON ND			24	68	89
1	2.1	3	5.2	5	6.1	BOTTINEAU ND	154	17	197	94	87
2	8.9			1	5.0	BOWMAN ND			-44	18	77
3	11.2	4*	12.3	3	9.3	BURKE ND	10	-24	-17	41	97
6	5.4	9	5.6	7	3.1	BURLEIGH ND	5	-44	-42	20	40
15	5.7	16	4.4	24	5.1	CASS ND	-23	16	-10	47	79
4	8.9	3	6.5	8*	15.0	CAVALIER ND	-27	129	68	82	99
4*	11.0	2	3.7	3	4.7	DICKEY ND	-66	27	-57	11	73
1	2.8	2	5.1	2	6.9	DIVIDE ND	79	36	143	91	91
1	8.2			1	4.2	DUNN ND			-49	15	64
1	4.7	1	3.4	2	4.6	EDDY ND	-26	33	-2	53	71
3	11.3	2	6.0	2	5.1	EMMONS ND	-47	-14	-55	12	78
1	4.7	1	4.2	2	5.7	FOSTER ND	-11	35	21	66	84
		1	6.2			GOLDEN VALLEY ND					
7	3.8	13	5.6	9	3.0	GRAND FORKS ND	45	-47	-23	37	37
				3	10.4	GRIGGS ND					98
4*	19.4			2	6.5	HETTINGER ND			-66	06	90
1	5.4	2	9.6			KIDDER ND	80				
3	7.1	1	1.8	3	3.9	LA MOURE ND	-74	113	-45	18	58
2	12.6			1	5.6	LOGAN ND			-56	12	83
2	3.6			2	3.3	MC HENRY ND			-8	49	45
4*	14.1	3	8.1	1	2.2	MC INTOSH ND	-42	-73	-85	00	23
1	3.9			3	8.8	MC KENZIE ND			129	89	96
1	1.6	3	4.7	2	2.7	MC LEAN ND	198	-42	73	83	32
1	3.3			1	2.2	MERCER ND			-31	30	24
2	2.4	6	6.5	7	5.8	MORTON ND	171	-10	146	91	85
2	5.0	4	8.5	2	3.2	MOUNTRAIL ND	72	-62	-35	27	44
2	5.1	4	7.6			NELSON ND	49				
2*	22.9					OLIVER ND					
1	1.6	4	5.5	3	4.6	PEMBINA ND	252	-17	191	94	71
1	3.2	4*	11.5	3	7.0	PIERCE ND	260	-39	120	89	92
		5	5.6	3	2.6	RAMSEY ND		-53			30
2	3.9	4	7.5	3	4.8	RANSOM ND	92	-36	22	66	74
1	3.8	1	4.9	2	8.3	RENVILLE ND	30	68	119	89	95
3	2.8	4	3.9	9	6.1	RICHLAND ND	38	57	117	88	88
1	5.3	1	3.3	4*	9.9	ROLETTE ND	-37	199	88	85	98
2	6.3			1	2.1	SARGENT ND			-66	06	22
1	5.0	1	7.2			SHERIDAN ND	44				
1	11.5	1	12.6			SLOPE ND	10				
4	6.3	3	3.7	4	4.0	STARK ND	-42	10	-36	26	61
		1	4.0			STEELE ND					
3	2.8	5	4.2	8	4.6	STUTSMAN ND	49	9	62	80	71
		2	7.4	1	2.1	TOWNER ND		-71			22
		2	3.1	3	2.3	TRAILL ND		-24			25
5	5.6	3	2.6	4	2.9	WALSH ND	-53	11	-48	16	35
8	5.5	10	4.9	8	3.2	WARD ND	-11	-35	-42	20	43
4	9.5	4	8.5	2	2.4	WELLS ND	-10	-72	-75	03	26
1	1.0	5	5.0	4	3.2	WILLIAMS ND	389	-36	215	94	43
1964*	4.8	2266*	4.5	2541*	4.3	OHIO OH	-5	-5	-9		
3	2.3	14*	11.0	8	5.2	ADAMS OH	381	-53	126	89	79
22	4.6	17	3.0	32	4.8	ALLEN OH	-35	61	4	58	75
7	3.3	10	4.5	9	3.2	ASHLAND OH	35	-27	-1	53	44
26	5.7	21	4.2	18	3.0	ASHTABULA OH	-26	-29	-48	16	37
14	5.3	12	4.6	18*	6.6	ATHENS OH	-13	42	24	68	90
9	4.8	9	4.1	6	2.5	AUGLAIZE OH	-13	-39	-47	17	28
16	3.3	19	3.4	23	3.8	BELMONT OH	3	12	15	63	57
6	4.2	3	1.8	8	4.2	BROWN OH	-57	137	1	55	66
31	4.5	41	4.5	53	4.5	BUTLER OH	2	-1	1	55	70
8	7.6	4	3.8	7	5.2	CARROLL OH	-50	36	-32	29	80
8	5.6	7	4.3	9	4.2	CHAMPAIGN OH	-23	-3	-25	35	64
38*	6.6	33	4.5	44	5.3	CLARK OH	-31	16	-19	40	81
7	2.9	9	2.7	21	4.7	CLERMONT OH	-7	74	62	81	73
6	3.8	9	5.0	15*	6.7	CLINTON OH	32	34	77	83	90
30	5.5	32	5.0	38	5.3	COLUMBIANA OH	-8	5	-3	53	81
5	3.0	11	5.1	10	4.3	COSHOCTON OH	73	-15	47	76	67
11	4.6	11	3.9	10	3.2	CRAWFORD OH	-16	-17	-31	30	42
341*	5.1	377*	4.6	418*	4.5	CUYAHOGA OH	-9	-3	-11	46	70
16	6.5	1	.4	15	4.2	DARKE OH	-93	830	-36	26	64
5	3.6	10	5.6	12	5.7	DEFIANCE OH	57	1	58	79	84
6	3.7	10	4.9	6	2.4	DELAWARE OH	34	-51	-34	27	27
19	6.4	18	5.3	19	4.7	ERIE OH	-17	-12	-27	34	72
10	3.2	17	4.4	15	3.0	FAIRFIELD OH	39	-33	-6	50	37
3	2.0	7	4.5	20*	11.8	FAYETTE OH	127	165	502	99	99
125	4.9	135	4.2	136	3.5	FRANKLIN OH	-13	-18	-29	32	49
2	1.2	2	1.1	12	6.0	FULTON OH	-9	445	397	98	87
7	5.5	5	3.5	6	3.4	GALLIA OH	-36	-3	-38	23	47
7	5.2	12	6.1	11	3.8	GEAUGA OH	18	-37	-26	34	57
7	3.1	12	3.7	28*	6.3	GREENE OH	18	71	102	87	89
6	2.6	19*	6.3	10	3.0	GUERNSEY OH	144	-53	15	63	37
198*	5.1	252*	5.4	208	4.0	HAMILTON OH	7	-27	-22	38	60
9	3.4	16	4.7	9	2.3	HANCOCK OH	40	-52	-33	28	24
12	7.6	14*	8.0	7	3.0	HARDIN OH	6	-62	-60	09	38
6	5.6	5	4.9	2	1.5	HARRISON OH	-14	-69	-73	04	15
12*	9.8	7	4.4	12	6.8	HENRY OH	-55	54	-31	30	91
5	2.9	10	5.0	14	6.7	HIGHLAND OH	72	33	128	89	90
5	5.1	5	4.5	5	3.6	HOCKING OH	-11	-20	-29	32	51
3	3.4	6	5.1	3	2.0	HOLMES OH	48	-61	-42	20	20
7	3.0	14	5.0	14	4.6	HURON OH	70	-8	56	79	72
14*	9.0	13*	7.4	9	4.8	JACKSON OH	-19	-34	-46	17	75
18	4.3	21	4.3	48*	8.2	JEFFERSON OH	-1	93	91	86	95

* SIGNIFICANTLY MORE DEATHS THAN EXPECTED

WHITE FEMALE

CANCER OF THE LIVER AND GALLBLADDER INCLUDING BILE DUCTS

ICD 155,156

1950-1959 DEATHS	1950-1959 RATE	1960-1969 DEATHS	1960-1969 RATE	1970-1979 DEATHS	1970-1979 RATE	COUNTY AND STATE		PERCENT CHANGE 60S/50S	PERCENT CHANGE 70S/60S	PERCENT CHANGE 70S/50S	PERCENTILE %CHANGE 70S/50S	PERCENTILE RATES 1970S
6	2.9	6	2.3	15	4.8	KNOX	OH	-20	111	68	82	74
13	3.1	15	2.6	40	4.6	LAKE	OH	-14	76	52	78	72
8	3.8	10	3.8	13	3.9	LAWRENCE	OH	-2	3	1	56	58
16	3.7	19	3.6	28	4.4	LICKING	OH	-3	20	17	64	68
6	3.3	6	2.6	10	3.9	LOGAN	OH	-19	49	20	66	59
32	4.7	43	4.6	57	4.8	LORAIN	OH	-3	4	1	56	74
115*	5.6	109	4.3	110	3.9	LUCAS	OH	-23	-9	-30	31	59
2	2.0	2	1.5	2	1.2	MADISON	OH	-25	-20	-40	22	12
59	4.8	96*	6.4	83	4.7	MAHONING	OH	33	-26	-2	53	74
12	4.1	12	3.5	8	2.3	MARION	OH	-16	-34	-45	18	24
12	5.0	13	4.2	12	3.3	MEDINA	OH	-15	-23	-35	27	44
7	5.6	6	4.6	8	4.7	MEIGS	OH	-18	2	-16	42	73
9	5.9	7	3.7	9	4.3	MERCER	OH	-37	16	-27	33	66
26*	7.4	18	4.2	23	4.7	MIAMI	OH	-43	11	-37	25	72
3	3.3	6	5.8	7	6.7	MONROE	OH	76	16	105	87	90
79	4.3	103	4.4	136*	4.8	MONTGOMERY	OH	1	11	11	61	75
1	1.8	3	3.6	2	1.2	MORGAN	OH	106	-66	-30	31	12
2	2.0	2	1.6	4	2.9	MORROW	OH	-21	79	41	75	34
21	5.0	14	2.8	24	4.4	MUSKINGUM	OH	-44	58	-11	47	68
7*	10.3	3	3.6	10*	12.7	NOBLE	OH	-65	249	24	67	99
9	5.4	8	4.2	15*	6.3	OTTAWA	OH	-23	50	15	63	88
6	7.3	7	8.1	5	4.5	PAULDING	OH	12	-44	-37	24	71
7	4.4	10	5.5	9	5.6	PERRY	OH	25	2	27	69	83
8	4.9	9	5.3	12	5.9	PICKAWAY	OH	8	11	20	66	86
5	7.7	6	6.6	7	6.1	PIKE	OH	-15	-8	-21	38	87
18	5.9	17	4.4	21	4.3	PORTAGE	OH	-25	-4	-28	33	66
8	5.0	3	1.7	8	3.8	PREBLE	OH	-66	124	-23	37	57
8	6.1	9	5.9	8	5.2	PUTNAM	OH	-3	-12	-15	43	79
24	5.2	21	3.8	31	4.7	RICHLAND	OH	-27	25	-9	48	73
9	3.4	9	3.1	16	4.7	ROSS	OH	-10	53	38	74	73
14	5.0	15	4.5	12	3.4	SANDUSKY	OH	-11	-24	-32	29	48
31*	7.8	30*	6.2	32*	6.0	SCIOTO	OH	-20	-3	-22	37	87
14	4.6	14	4.3	13	3.7	SENECA	OH	-8	-14	-21	38	53
7	4.4	23*	12.6	10	4.9	SHELBY	OH	189	-61	12	61	75
75	5.1	77	4.2	85	4.0	STARK	OH	-17	-6	-22	38	61
84	4.4	113	4.5	111	3.6	SUMMIT	OH	3	-19	-16	42	53
36	4.8	45	4.8	48	3.9	TRUMBULL	OH	0	-19	-18	40	59
14	3.6	24	5.4	28	5.1	TUSCARAWAS	OH	49	-6	40	75	79
6	4.7	9	5.9	5	2.9	UNION	OH	26	-50	-37	25	36
7	3.8	13*	6.3	9	4.1	VAN WERT	OH	63	-35	7	59	62
2	4.4	1	1.9	1	1.1	VINTON	OH	-58	-40	-75	03	11
9	4.4	5	1.9	6	1.6	WARREN	OH	-58	-17	-65	07	15
10	3.4	8	2.6	19	5.5	WASHINGTON	OH	-24	113	63	81	83
13	4.0	17	4.0	19	4.0	WAYNE	OH	1	-1	0	55	60
5	2.8	6	3.2	13	6.1	WILLIAMS	OH	14	94	122	89	88
8	2.6	19	5.1	15	3.1	WOOD	OH	96	-40	18	64	39
1	.6	5	3.9	14*	8.2	WYANDOT	OH	524	109	1208	99	95
285	2.7	395	2.9	563	3.3	OKLAHOMA	OK	7	13	22		
		3	4.4	2	2.1	ADAIR	OK		-53			21
3	4.2	4	4.4	7*	5.9	ALFALFA	OK	5	34	41	75	86
1	1.6	3	4.8	3	4.6	ATOKA	OK	195	-4	182	93	72
1	3.0	1	2.8			BEAVER	OK	-5				
1	.9	1	.4	6	3.0	BECKHAM	OK	-57	654	222	95	36
		2	2.0	5	4.4	BLAINE	OK		117			69
2	1.0	5	2.4	5	3.3	BRYAN	OK	132	36	216	94	44
8	5.0	2	1.2	4	1.5	CADDO	OK	-76	21	-71	04	14
4	3.5	6	3.6	9	3.7	CANADIAN	OK	2	3	4	57	54
3	1.5	3	1.0	10	3.5	CARTER	OK	-32	252	140	90	50
		3	3.1	5	4.1	CHEROKEE	OK		34			63
2	2.3	6	7.3	7	5.5	CHOCTAW	OK	214	-24	137	90	82
2	.9	4	1.5	9	2.3	CLEVELAND	OK	60	56	148	91	25
1	2.5	2	4.8	1	2.3	COAL	OK	92	-52	-8	49	24
8	4.3	11	4.6	7	2.1	COMANCHE	OK	6	-55	-52	14	21
				2	3.7	COTTON	OK					53
3	2.6	5	4.2	1	.8	CRAIG	OK	60	-82	-70	05	10
3	1.8	7	2.9	8	2.7	CREEK	OK	63	-5	55	79	32
3	2.9	3	2.6	5	3.0	CUSTER	OK	-11	15	2	56	37
1	1.4	2	1.8	4	2.8	DELAWARE	OK	32	54	102	87	32
				4	7.2	DEWEY	OK					93
		1	2.8			ELLIS	OK					
8	2.5	15	4.3	20	4.5	GARFIELD	OK	68	5	76	83	69
2	1.5	5	2.9	4	1.7	GARVIN	OK	91	-41	12	61	17
6	3.4	3	1.0	2	1.0	GRADY	OK	-69	-6	-71	04	11
2	2.9			1	2.3	GRANT	OK			-21	39	25
3	4.5	4	4.6	2	1.7	GREER	OK	2	-63	-63	08	17
1	2.4					HARMON	OK					
		2	2.7	3	3.4	HASKELL	OK		26			48
3	2.9	5	4.0	3	2.9	HUGHES	OK	38	-27	1	55	36
1	.8	6	4.3	3	1.8	JACKSON	OK	424	-57	124	89	18
4	6.4	3	5.1	5	5.4	JEFFERSON	OK	-20	6	-15	43	82
2	3.7	5	9.0	2	2.3	JOHNSTON	OK	142	-75	-38	24	25
7	2.6	8	2.5	17	4.0	KAY	OK	-6	59	50	77	60
2	3.3	3	3.1	3	2.7	KINGFISHER	OK	-5	-14	-18	40	31
2	1.9	2	2.2	7	5.5	KIOWA	OK	18	149	194	94	82
1	3.0	3	6.5	5	6.9	LATIMER	OK	116	7	132	90	92
3	1.7	11	4.9	9	3.5	LE FLORE	OK	192	-29	107	87	50
1	.9	4	3.2	8	4.7	LINCOLN	OK	246	46	405	98	73
6	4.5	4	2.7	6	4.0	LOGAN	OK	-39	44	-12	45	60
1	3.6	1	2.0			LOVE	OK	-46				
4	5.3			5	4.9	MC CLAIN	OK			-7	50	76
3	2.8	5	4.7	9	5.3	MC CURTAIN	OK	70	13	92	86	81
2	3.4	5	6.4	4	3.4	MC INTOSH	OK	89	-47	0	54	47
2	4.0	2	2.9	2	3.2	MAJOR	OK	-28	11	-20	39	42
3	6.7	2	3.4	2	2.2	MARSHALL	OK	-49	-36	-67	06	23
2	2.0	3	2.3	2	1.0	MAYES	OK	14	-57	-51	14	11
1	1.1	4	4.3	4	5.0	MURRAY	OK	275	18	341	98	77
7	2.8	14	4.1	11	2.6	MUSKOGEE	OK	49	-36	-5	52	30
1	1.5	6	8.1	2	1.7	NOBLE	OK	422	-78	13	62	17

* SIGNIFICANTLY MORE DEATHS THAN EXPECTED

WHITE FEMALE

CANCER OF THE LIVER AND GALLBLADDER INCLUDING BILE DUCTS

COUNTY AND STATE

ICD 155,156

1950-1959 DEATHS	1950-1959 RATE	1960-1969 DEATHS	1960-1969 RATE	1970-1979 DEATHS	1970-1979 RATE	COUNTY AND STATE		PERCENT CHANGE 60S/50S	PERCENT CHANGE 70S/60S	PERCENT CHANGE 70S/50S	PERCENTILE %CHANGE 70S/50S	PERCENTILE RATES 1970S
1	1.6	1	1.1	3	3.8	NOWATA	OK	-32	246	136	90	56
2	2.9	1	1.4	5	5.1	OKFUSKEE	OK	-51	258	76	83	78
41	2.8	61	2.7	93	3.3	OKLAHOMA	OK	-2	22	19	65	46
3	1.5	9	3.9	9	3.3	OKMULGEE	OK	170	-16	128	69	46
5	3.4	3	1.5	8	3.9	OSAGE	OK	-54	151	15	63	58
2	1.4	8	4.2	14*	6.0	OTTAWA	OK	198	41	319	97	86
2	2.3	1	.9	2	1.4	PAWNEE	OK	-60	56	-38	24	14
9	4.5	5	1.7	9	2.9	PAYNE	OK	-62	70	-35	26	36
8	4.0	4	1.5	6	1.9	PITTSBURG	OK	-62	22	-54	13	18
9	5.6	8	3.8	9	4.1	PONTOTOC	OK	-31	8	-25	35	64
6	2.6	9	3.1	14	3.6	POTTAWATOMIE	OK	19	16	38	74	51
1	1.6	5	7.6	3	3.8	PUSHMATAHA	OK	377	-50	141	90	57
1	3.2	1	3.0	2	4.9	ROGER MILLS	OK	-6	62	53	78	75
4	3.8	6	4.8	1	.6	ROGERS	OK	28	-88	-85	00	10
3	2.0	3	1.7	6	2.6	SEMINOLE	OK	-15	55	32	71	29
		5	4.6	5	3.6	SEQUOYAH	OK		-23			51
5	3.0	5	1.9	12	3.8	STEPHENS	OK	-37	96	24	68	55
5	8.3	6	7.4	5	5.7	TEXAS	OK	-11	-23	-31	30	84
2	2.1	3	3.4	3	2.1	TILLMAN	OK	62	-39	-2	53	21
42	3.8	36	2.1	89	3.8	TULSA	OK	-44	78	1	55	57
1	1.3	2	2.3	3	2.4	WAGONER	OK	80	2	84	85	26
4	2.4	8	3.7	9	3.1	WASHINGTON	OK	55	-17	29	70	38
		1	1.1			WASHITA	OK					
3	3.8	4	4.2	3	2.2	WOODS	OK	11	-47	-41	21	23
		1	1.1	5	4.3	WOODWARD	OK		288			67
261	3.3	381	3.5	402	2.9	OREGON	OR	8	-19	-12		
3	3.9	4	3.8	3	3.4	BAKER	OR	-3	-9	-12	46	48
4	3.3	11	5.7	7	2.5	BENTON	OR	73	-55	-23	37	28
13	2.6	25	3.6	26	2.5	CLACKAMAS	OR	37	-31	-6	51	27
2	1.5	2	1.0	4	2.0	CLATSOP	OR	-37	112	33	72	20
2	2.3	2	1.4	4	2.2	COLUMBIA	OR	-39	63	0	54	23
2	1.0	5	2.2	14	4.4	COOS	OR	115	102	336	97	69
2	8.4					CROOK	OR					
1	2.4	2	5.0	3	2.8	CURRY	OR	108	-44	17	64	34
3	3.4	7	4.8	4	1.8	DESCHUTES	OR	41	-63	-48	16	17
9	4.2	4	1.6	13	3.1	DOUGLAS	OR	-61	91	-26	34	40
				1	9.2	GILLIAM	OR					97
1	3.1	1	2.9			GRANT	OR	-6				
1	4.6			3	8.3	HARNEY	OR			80	84	95
		3	4.2	3	3.4	HOOD RIVER	OR		-18			48
14	4.2	20	4.0	18	2.8	JACKSON	OR	-5	-30	-34	28	34
				1	2.7	JEFFERSON	OR					30
3	2.0	7	3.3	13	3.6	JOSEPHINE	OR	65	9	80	84	52
8	6.1	8	4.5	11	3.9	KLAMATH	OR	-26	-13	-36	26	58
		3*	12.5	3	9.4	LAKE	OR		-25			97
17	2.9	33	3.9	38	3.2	LANE	OR	34	-17	12	61	43
4	4.0	5	2.9	7	3.0	LINCOLN	OR	-29	6	-25	35	38
9	3.7	10	3.1	25	5.6	LINN	OR	-17	83	53	78	84
2	3.2	5	5.2	8	5.3	MALHEUR	OR	61	2	64	81	81
15	2.5	31	3.7	29	2.7	MARION	OR	50	-27	10	60	31
1	4.0	3*	11.0			MORROW	OR	178				
100	3.3	133	3.6	105	2.7	MULTNOMAH	OR	8	-25	-20	39	31
5	4.9	3	1.6	9	3.0	POLK	OR	-67	90	-37	24	38
1	10.1					SHERMAN	OR					
2	2.4	4	4.1	1	.5	TILLAMOOK	OR	69	-87	-78	02	10
14*	7.5	10	4.2	11	3.6	UMATILLA	OR	-44	-13	-52	14	53
6	6.5	4	3.4	2	1.7	UNION	OR	-47	-51	-74	03	17
4*	13.9	3	7.7	1	1.9	WALLOWA	OR	-45	-75	-86	00	19
1	1.3	1	.9	2	1.4	WASCO	OR	-29	50	6	59	14
4	1.1	22	3.8	22	2.4	WASHINGTON	OR	237	-38	108	88	25
1	9.6					WHEELER	OR					
7	3.9	10	4.5	11	3.5	YAMHILL	OR	16	-21	-9	48	50
2494*	4.7	2764*	4.3	3251*	4.3	PENNSYLVANIA	PA	-8	0	-8		
16	6.6	6	2.1	15	4.3	ADAMS	PA	-69	111	-34	27	67
356*	5.0	413*	4.7	427*	4.2	ALLEGHENY	PA	-6	-10	-16	42	64
17	4.4	14	3.2	35*	6.5	ARMSTRONG	PA	-28	105	48	77	89
30	4.1	49	5.1	63*	5.3	BEAVER	PA	23	4	28	69	81
11	5.6	10	4.4	11	4.1	BEDFORD	PA	-20	-7	-26	34	63
91*	6.0	67	3.7	80	3.7	BERKS	PA	-39	1	-38	24	55
39	4.8	39	4.1	51	5.0	BLAIR	PA	-14	22	5	58	77
14	4.6	22*	6.2	20	5.4	BRADFORD	PA	35	-14	17	63	81
44	5.1	51	3.8	82	4.2	BUCKS	PA	-25	10	-17	41	65
16	3.1	27	4.2	38	4.6	BUTLER	PA	35	10	49	77	72
53*	5.7	71*	6.6	73*	5.7	CAMBRIA	PA	16	-13	0	55	84
2	7.6	1	2.6			CAMERON	PA	-66				
25*	8.4	16	4.4	21	5.1	CARBON	PA	-48	18	-39	23	79
9	3.4	9	2.6	21	4.9	CENTRE	PA	-24	91	45	76	76
35	4.3	35	3.3	64	4.6	CHESTER	PA	-24	39	6	58	71
13	6.7	8	3.6	13	5.4	CLARION	PA	-46	49	-20	39	82
28*	6.6	31*	6.8	27	4.9	CLEARFIELD	PA	2	-27	-26	34	76
12	6.6	6	3.0	9	3.9	CLINTON	PA	-54	30	-41	21	59
14	4.7	16	4.8	20	4.9	COLUMBIA	PA	2	3	5	58	76
14	3.3	13	2.6	32*	5.6	CRAWFORD	PA	-20	111	69	82	83
14	2.8	32	4.6	34	3.5	CUMBERLAND	PA	64	-24	24	68	49
45	4.1	55	4.2	51	3.5	DAUPHIN	PA	3	-17	-15	43	49
86	4.2	102	3.4	132	3.6	DELAWARE	PA	-18	4	-14	44	51
8	4.8	6	3.2	12	5.1	ELK	PA	-34	60	5	58	78
51	4.5	66	4.9	74	4.6	ERIE	PA	11	-7	3	57	71
55*	6.6	58*	6.2	63*	5.8	FAYETTE	PA	-7	-6	-12	46	85
1	3.4	1	3.4			FOREST	PA	0				
21	5.1	17	3.4	35	5.0	FRANKLIN	PA	-34	50	-1	54	78
1	2.4	5*	9.1	4	6.5	FULTON	PA	271	-29	165	92	89
13	6.2	20*	8.4	20*	7.7	GREENE	PA	35	-8	24	67	94
7	3.6	10	4.8	12	4.7	HUNTINGDON	PA	33	-2	30	70	72
18	5.0	17	4.2	26	5.4	INDIANA	PA	-16	30	10	60	82
18*	6.7	12	4.1	9	2.8	JEFFERSON	PA	-38	-32	-58	11	33

* SIGNIFICANTLY MORE DEATHS THAN EXPECTED

WHITE FEMALE

CANCER OF THE LIVER AND GALLBLADDER INCLUDING BILE DUCTS

ICD 155,156

1950-1959 DEATHS	1950-1959 RATE	1960-1969 DEATHS	1960-1969 RATE	1970-1979 DEATHS	1970-1979 RATE	COUNTY AND STATE	PERCENT CHANGE 60S/50S	PERCENT CHANGE 70S/60S	PERCENT CHANGE 70S/50S	PERCENTILE %CHANGE 70S/50S	PERCENTILE RATES 1970S
1	1.2	5	5.5	8	6.4	JUNIATA PA	360	16	433	99	89
48	3.3	72	4.2	82	4.0	LACKAWANNA PA	26	-5	20	66	61
49	3.6	64	3.7	76	3.6	LANCASTER PA	3	-2	1	55	52
34*	6.4	36*	5.9	31	4.2	LAWRENCE PA	-8	-29	-35	27	64
19	4.5	21	4.0	20	2.9	LEBANON PA	-10	-27	-35	27	36
44	3.9	51	3.5	79	4.3	LEHIGH PA	-10	21	9	60	66
105*	5.0	107	4.3	127	4.4	LUZERNE PA	-13	0	-13	45	68
31	4.9	26	3.7	23	2.7	LYCOMING PA	-24	-28	-45	18	31
9	2.9	16	4.6	17	4.2	MC KEAN PA	56	-9	42	75	64
24	4.1	33	4.8	30	3.6	MERCER PA	18	-25	-11	46	53
14*	7.0	10	4.0	14	4.6	MIFFLIN PA	-43	15	-34	27	72
8	3.9	13	4.9	11	3.4	MONROE PA	26	-31	-13	45	48
94	4.4	121	3.9	135	3.3	MONTGOMERY PA	-11	-16	-25	35	44
6	5.5	3	2.3	3	2.0	MONTOUR PA	-58	-14	-64	07	20
48	4.9	58	4.7	67	4.4	NORTHAMPTON PA	-4	-6	-10	47	68
34	5.3	31	4.4	51*	6.5	NORTHUMBERLAND PA	-18	48	21	66	89
4	2.8	6	4.0	8	4.4	PERRY PA	40	12	57	79	69
415	4.3	431	4.2	450*	4.4	PHILADELPHIA PA	-2	5	2	56	68
3	5.0	2	1.9	4	2.8	PIKE PA	-62	49	-43	19	34
5	4.9	2	2.1	4	3.0	POTTER PA	-58	45	-39	23	37
70*	6.9	69*	6.0	76*	5.5	SCHUYLKILL PA	-13	-8	-20	39	83
6	5.5	4	2.6	10	5.5	SNYDER PA	-52	111	1	55	83
17	4.5	19	4.0	19	3.5	SOMERSET PA	-12	-11	-22	38	50
1	2.6					SULLIVAN PA					
6	3.4	6	3.4	12	5.5	SUSQUEHANNA PA	-1	63	62	80	83
6	3.1	10	4.8	7	2.7	TIOGA PA	57	-45	-13	45	31
3	2.2	5	3.4	5	2.1	UNION PA	54	-37	-3	53	22
16	4.6	23*	6.2	25*	5.7	VENANGO PA	36	-8	25	68	84
13	4.7	8	2.5	9	2.7	WARREN PA	-46	7	-43	19	31
55*	5.5	70*	6.1	68	4.9	WASHINGTON PA	11	-19	-11	47	76
12	6.6	8	3.8	8	2.8	WAYNE PA	-43	-26	-58	11	33
59	4.0	73	3.8	123*	5.1	WESTMORELAND PA	-5	34	27	69	79
3	3.1	1	1.0	7	4.9	WYOMING PA	-67	385	59	80	76
55	4.9	55	3.9	68	3.8	YORK PA	-22	-2	-23	37	56
166	3.7	217	4.0	287*	4.3	RHODE ISLAND RI	8	9	18		
5	3.2	9	4.4	7	2.3	BRISTOL RI	36	-47	-28	33	25
15	3.4	24	3.9	26	3.0	KENT RI	16	-24	-12	46	37
14	4.9	11	3.1	19	4.0	NEWPORT RI	-38	32	-18	40	61
128	3.8	164	4.2	221*	4.8	PROVIDENCE RI	11	16	28	70	75
4	1.9	9	2.8	14	3.2	WASHINGTON RI	49	13	68	82	41
120	2.3	176	2.4	324	3.1	SOUTH CAROLINA SC	1	33	34		
1	1.5	1	1.4	1	1.0	ABBEVILLE SC	-11	-25	-34	28	11
4	1.6	3	1.3	19	5.1	AIKEN SC	-19	305	228	95	78
1	4.3					ALLENDALE SC					
5	1.6	13	3.1	26	4.4	ANDERSON SC	93	43	177	93	68
		1	2.3	2	3.5	BAMBERG SC		51			50
1	2.9			2	3.1	BARNWELL SC			6	58	39
1	3.3	1	1.2	2	1.2	BEAUFORT SC	-63	-2	-64	08	12
1	1.5	1	1.2	5	3.9	BERKELEY SC	-23	233	156	91	60
1	4.4			2	4.5	CALHOUN SC			3	57	71
18	5.0	16	3.0	17	2.4	CHARLESTON SC	-40	-21	-53	13	25
1	.9	6	4.3	7	4.1	CHEROKEE SC	394	-3	378	98	63
2	2.3	2	2.1	2	1.8	CHESTER SC	-10	-14	-23	37	17
		2	1.9	5	4.1	CHESTERFIELD SC		114			62
1	2.4	1	2.1			CLARENDON SC	-10				
1	1.3	5	6.5	2	2.3	COLLETON SC	404	-64	79	84	25
		3	1.8	10	4.7	DARLINGTON SC		164			73
2	2.5	1	1.2	4	4.3	DILLON SC	-53	259	69	82	66
				3	2.5	DORCHESTER SC					27
1	2.3					EDGEFIELD SC					
		1	2.2	4	6.7	FAIRFIELD SC		200			90
4	2.0	13	5.6	11	3.5	FLORENCE SC	181	-37	77	83	50
		3	4.2	4	3.9	GEORGETOWN SC		-8			59
19	3.8	15	1.6	32	2.7	GREENVILLE SC	-57	63	-29	31	31
5	4.0	5	2.9	8	3.3	GREENWOOD SC	-28	17	-16	42	46
				2	3.7	HAMPTON SC					55
2	1.4	4	2.0	8	2.1	HORRY SC	44	5	50	78	22
		1	4.5			JASPER SC					
2	2.8	1	1.1	3	2.4	KERSHAW SC	-62	126	-14	44	27
4	5.1	3	2.3	1	.4	LANCASTER SC	-55	-82	-92	00	09
1	.7	4	2.0	7	3.0	LAURENS SC	198	50	348	98	37
1	2.5	1	2.2	2	3.4	LEE SC	-9	52	38	74	47
4	2.5	7	2.5	9	2.1	LEXINGTON SC	0	-15	-15	43	22
1	5.9					MC CORMICK SC					
				1	1.1	MARION SC					11
1	1.4	3	3.8	5	5.1	MARLBORO SC	162	37	258	96	79
2	1.9	3	2.3	2	1.3	NEWBERRY SC	20	-44	-33	28	13
3	2.3	3	1.9	11	4.7	OCONEE SC	-20	151	100	87	73
6	5.3	6	3.3	13	5.5	ORANGEBURG SC	-37	65	4	57	82
3	3.0	6	3.0	11	3.7	PICKENS SC	1	25	26	68	55
5	1.2	8	1.4	23	2.9	RICHLAND SC	14	108	137	90	36
		1	1.6	2	2.7	SALUDA SC		71			32
3	.8	18	2.7	33	3.7	SPARTANBURG SC	234	36	354	98	54
6	6.4	4	3.3	9	4.7	SUMTER SC	-49	44	-26	34	74
1	.9	4	3.6	3	2.1	UNION SC	295	-43	125	89	21
3	5.6	2	3.8	3	3.4	WILLIAMSBURG SC	-33	-10	-39	22	47
3	1.4	4	1.6	8	2.1	YORK SC	12	28	44	76	21
131	4.4	180*	4.8	143	3.1	SOUTH DAKOTA SD	10	-35	-29		
1	3.9	4*	12.8	1	2.3	AURORA SD	232	-82	-39	22	25
4	3.8	5	4.1			BEADLE SD	8				
2	3.6	4	6.9	1	1.2	BON HOMME SD	89	-83	-68	05	12
2	2.4	5	4.7	6	3.9	BROOKINGS SD	96	-17	62	80	59
6	4.0	17*	8.9	7	3.1	BROWN SD	126	-66	-23	37	39
1	4.2	1	3.0			BRULE SD	-29				

* SIGNIFICANTLY MORE DEATHS THAN EXPECTED

WHITE FEMALE

CANCER OF THE LIVER AND GALLBLADDER INCLUDING BILE DUCTS

ICD 155,156

1950-1959 DEATHS	1950-1959 RATE	1960-1969 DEATHS	1960-1969 RATE	1970-1979 DEATHS	1970-1979 RATE	COUNTY AND STATE	PERCENT CHANGE 60S/50S	PERCENT CHANGE 70S/60S	PERCENT CHANGE 70S/50S	PERCENTILE %CHANGE 70S/50S	PERCENTILE RATES 1970S
		1*	31.5			BUFFALO SD					
				1	2.1	BUTTE SD					21
1	1.9	9*	16.9	1	1.2	CHARLES MIX SD	810	-93	-34	27	12
1	5.0			3	6.5	CLARK SD			31	71	89
7*	13.3	3	4.7	3	3.2	CLAY SD	-64	-32	-76	03	42
5	5.0	6	5.0	1	.9	CODINGTON SD	1	-83	-83	01	10
		3*	17.4	1	7.4	CORSON SD		-57			93
2	11.0	1	3.2			CUSTER SD	-71				
1	1.0	7	5.8	5	2.7	DAVISON SD	492	-53	177	93	31
5	9.2	1	2.0	1	1.0	DAY SD	-78	-50	-89	00	11
2	5.4	3	7.6	1	2.5	DEUEL SD	42	-67	-53	13	28
1	6.7	1	10.2			DEWEY SD	53				
1	4.3	2	6.5	1	2.2	DOUGLAS SD	50	-65	-48	16	24
1	3.2	2	7.6	2	4.6	EDMUNDS SD	135	-39	43	75	71
5*	10.5	5	8.7	1	2.3	FALL RIVER SD	-17	-74	-78	02	24
		1	3.8	2	9.0	FAULK SD		136			97
4	7.3	3	4.7	3	3.5	GRANT SD	-36	-26	-52	13	49
3	9.3	3	5.7	3	3.7	GREGORY SD	-38	-35	-60	10	55
		2*	15.3			HAAKON SD					
				3	6.6	HAMLIN SD					90
		1	2.0	2	5.4	HAND SD		167			81
				2	5.6	HANSON SD					83
		2*	36.4			HARDING SD					
4	11.3	3	5.3	2	3.1	HUGHES SD	-53	-42	-73	04	39
4	8.4	1	1.7	8*	9.0	HUTCHINSON SD	-80	433	7	59	97
		1	5.9	1	4.6	HYDE SD		-22			71
2	7.7			1	2.7	JERAULD SD			-65	07	30
1	12.8					JONES SD					
				7*	8.6	KINGSBURY SD					96
2	2.9	1	1.2	4	4.1	LAKE SD	-59	246	41	75	63
1	1.2	5	4.8	2	1.3	LAWRENCE SD	309	-74	8	60	13
4	5.7	6	5.9	6	7.3	LINCOLN SD	4	24	29	70	93
1	7.5					LYMAN SD					
4	9.3	1	1.3			MC COOK SD	-85				
4*	14.8	2	5.4	2	3.8	MC PHERSON SD	-63	-31	-75	03	56
2	5.8	2	5.4	3	6.7	MARSHALL SD	-8	25	15	62	90
3	6.5	3	5.0	3	4.5	MEADE SD	-23	-10	-31	31	70
2	11.2			2	5.7	MINER SD			-49	15	84
15	4.1	20	4.1	14	2.3	MINNEHAHA SD	-2	-43	-44	18	24
3	6.2	4	8.0	6*	10.3	MOODY SD	30	29	68	82	98
6	4.3	9	4.0	11	3.5	PENNINGTON SD	-6	-11	-17	42	51
2	6.7	1	3.5	3	8.3	PERKINS SD	-47	137	25	68	95
1	5.3			1	3.3	POTTER SD			-39	23	44
2	3.5	3	4.4	5	6.0	ROBERTS SD	26	36	72	82	87
1	3.9	3*	14.1	2	7.7	SANBORN SD	263	-45	98	87	94
2	3.3	4	4.6			SPINK SD	37				
1	14.9			1	16.1	STANLEY SD			8	59	99
		1	9.6			TODD SD					
3	7.4	1	1.6	1	2.2	TRIPP SD	-79	38	-71	04	23
2	2.7	3	3.7	1	1.5	TURNER SD	37	-58	-43	19	15
1	1.6	7*	9.5	1	1.3	UNION SD	480	-86	-18	40	13
		3	6.7	2	3.8	WALWORTH SD		-43			57
3	2.5	4	2.6	3	2.0	YANKTON SD	4	-25	-22	37	20
410	3.2	577	3.5	792	3.6	TENNESSEE TN	7	4	12		
3	2.2	6	3.2	10	3.2	ANDERSON TN	44	-1	43	75	41
4	3.5	5	3.8	12*	7.5	BEDFORD TN	9	97	114	88	93
1	1.5	3	3.4	5	5.2	BENTON TN	131	50	246	95	79
				4	8.7	BLEDSOE TN					96
7	3.3	7	2.5	17	4.2	BLOUNT TN	-24	67	27	69	64
6	4.1	4	2.1	8	3.2	BRADLEY TN	-48	50	-22	38	43
4	3.6	5	3.0	13*	6.9	CAMPBELL TN	-18	130	88	85	91
3	7.0	1	1.9	1	1.5	CANNON TN	-73	-23	-79	02	14
2	1.4	7	4.4	7	3.6	CARROLL TN	209	-19	152	91	51
3	1.9	11	5.5	11	3.6	CARTER TN	195	-36	89	86	51
2	4.5			2	2.6	CHEATHAM TN			-43	19	29
3	6.4	8*	13.9	3	5.2	CHESTER TN	116	-63	-20	39	80
4	4.9	4	4.0	4	3.2	CLAIBORNE TN	-19	-18	-33	28	44
		2	5.4	1	2.1	CLAY TN		-61			21
		1	.8	5	3.5	COCKE TN		320			49
6	5.3	5	3.6	6	3.2	COFFEE TN	-32	-10	-39	23	44
2	2.9	2	2.6	3	2.8	CROCKETT TN	-10	10	-1	53	34
5	6.9	2	2.1	9	5.7	CUMBERLAND TN	-69	166	-18	41	84
52	3.8	53	3.0	77	3.4	DAVIDSON TN	-23	15	-11	47	48
		2	3.6	3	3.9	DECATUR TN		9			59
		1	1.3	3	3.1	DE KALB TN		143			40
2	2.3	5	4.2	7	3.8	DICKSON TN	82	-11	62	80	56
7	5.3	4	2.4	7	3.4	DYER TN	-55	42	-36	26	47
1	2.8	1	2.0			FAYETTE TN	-30				
2	5.6	1	1.5	4	4.7	FENTRESS TN	-73	212	-15	43	74
3	3.1	6	4.5	7	4.1	FRANKLIN TN	48	-11	32	71	62
12	5.4	7	2.6	16	4.7	GIBSON TN	-52	83	-13	45	73
2	1.6	11*	8.1	2	1.2	GILES TN	393	-85	-25	35	12
		1	1.7	3	3.6	GRAINGER TN		110			51
5	2.9	5	2.4	18*	6.2	GREENE TN	-18	162	115	88	88
2	4.2	6*	9.7	2	2.7	GRUNDY TN	132	-72	-35	26	31
3	3.1	1	.8	8	3.9	HAMBLEN TN	-75	411	28	69	59
32	4.1	30	2.9	58	4.3	HAMILTON TN	-30	49	4	57	67
3	9.3	1	3.9	5*	11.5	HANCOCK TN	-58	196	23	67	99
2	2.7	3	2.5	3	2.7	HARDEMAN TN	-10	7	-3	52	30
1	1.3	2	2.2	3	2.3	HARDIN TN	76	4	83	84	25
2	1.7	6	4.1	7	3.8	HAWKINS TN	136	-7	120	89	57
				5	7.7	HAYWOOD TN					94
2	3.0	7	7.2	4	3.1	HENDERSON TN	142	-58	2	56	39
1	.9	4	2.6	8	3.9	HENRY TN	194	51	343	98	59
2	4.4	2	3.6	4	5.0	HICKMAN TN	-18	41	16	63	78
				2	4.5	HOUSTON TN					69
3	5.9			4	4.4	HUMPHREYS TN			-25	35	69
1	2.2					JACKSON TN					

* SIGNIFICANTLY MORE DEATHS THAN EXPECTED

WHITE FEMALE

CANCER OF THE LIVER AND GALLBLADDER INCLUDING BILE DUCTS

ICD 155,156

1950-1959 DEATHS	1950-1959 RATE	1960-1969 DEATHS	1960-1969 RATE	1970-1979 DEATHS	1970-1979 RATE	COUNTY AND STATE		PERCENT CHANGE 60S/50S	PERCENT CHANGE 70S/60S	PERCENT CHANGE 70S/50S	PERCENTILE %CHANGE 70S/50S	PERCENTILE RATES 1970S
2	2.1	2	1.9	3	1.7	JEFFERSON	TN	-8	-9	-16	42	17
2	4.0	4	6.4	2	2.7	JOHNSON	TN	59	-58	-33	28	31
21	2.3	36	2.7	46	2.7	KNOX	TN	22	-1	20	66	31
1	3.1			2	5.0	LAKE	TN			63	81	77
1	1.6	2	2.0	5	4.9	LAUDERDALE	TN	28	138	204	94	75
1	.7	6	3.9	6	3.1	LAWRENCE	TN	430	-20	323	97	39
1	3.7	3	7.8			LEWIS	TN	114				
		4	3.4	8	5.0	LINCOLN	TN		50			77
4	3.9	9	6.9	6	3.5	LOUDON	TN	76	-50	-12	46	49
1	.9	4	2.1	9	3.6	MC MINN	TN	127	68	281	96	52
3	4.3	4	4.2	6	4.4	MC NAIRY	TN	-1	3	2	56	68
		1	1.0	6	5.8	MACON	TN		455			85
7	3.1	11	3.6	6	1.6	MADISON	TN	14	-54	-47	16	16
2	3.2	1	1.0	8	6.6	MARION	TN	-68	555	107	87	90
3	3.4	6	5.6	5	3.9	MARSHALL	TN	63	-30	15	63	60
3	1.9	5	2.7	5	1.9	MAURY	TN	39	-30	-3	52	19
1	4.6	1	3.8	1	3.3	MEIGS	TN	-18	-12	-28	33	45
7	7.9	5	4.2	8	5.0	MONROE	TN	-47	19	-37	25	77
1	.7	4	2.1	11	4.7	MONTGOMERY	TN	191	124	553	99	73
				1	5.0	MOORE	TN					77
				3	3.2	MORGAN	TN					41
4	2.6	11	5.5	8	4.0	OBION	TN	107	-27	51	78	61
4	5.3	6	7.2	10*	9.5	OVERTON	TN	36	32	80	84	97
1	3.7	3	8.2	2	4.1	PERRY	TN	123	-50	12	61	63
1	5.8	3*	14.5	2	7.4	PICKETT	TN	149	-49	28	69	93
		4	6.2	1	1.3	POLK	TN		-80			13
4	2.7	7	4.1	7	2.7	PUTNAM	TN	49	-33	0	55	32
2	2.6	6	6.8	2	1.9	RHEA	TN	162	-72	-26	35	19
8	6.4	6	3.6	9	3.6	ROANE	TN	-43	0	-44	19	52
6	6.0	6	4.6	5	2.8	ROBERTSON	TN	-23	-39	-53	13	34
2	1.2	8	3.9	7	2.3	RUTHERFORD	TN	230	-40	97	87	25
6*	10.0	1	2.0	6	6.3	SCOTT	TN	-80	219	-37	25	89
2	8.3	1	4.3	1	2.3	SEQUATCHIE	TN	-48	-48	-73	04	24
2	1.8	4	2.8	6	2.9	SEVIER	TN	53	3	57	79	35
54	3.6	70	3.4	96	3.7	SHELBY	TN	-7	9	1	56	54
3	4.5	2	2.7	4	4.4	SMITH	TN	-39	64	-1	54	69
2	4.1	2	4.2	3	4.8	STEWART	TN	2	15	17	64	74
16	5.0	30*	5.8	26	3.6	SULLIVAN	TN	15	-39	-29	32	51
3	2.0	10	5.0	13	4.3	SUMNER	TN	155	-14	119	89	66
3	4.0	5	5.8	6	4.8	TIPTON	TN	46	-17	22	67	75
		1	3.2			TROUSDALE	TN					
4	5.9	2	2.7	4	3.6	UNICOI	TN	-55	33	-40	22	51
		2	4.4	1	1.8	UNION	TN		-58			18
2*	13.2	1	6.0	2	9.4	VAN BUREN	TN	-55	58	-29	32	97
3	3.0	3	2.3	4	2.3	WARREN	TN	-23	-1	-24	36	24
7	2.9	25*	7.9	10	2.2	WASHINGTON	TN	174	-72	-24	36	23
1	1.2	2	3.4			WAYNE	TN	177				
3	2.2	5	2.8	11	4.7	WEAKLEY	TN	28	70	118	88	73
2	2.8	2	2.1	4	3.3	WHITE	TN	-27	61	18	64	46
3	3.6	3	2.7	7	3.3	WILLIAMSON	TN	-26	24	-8	49	45
6	4.7	6	3.9	7	3.0	WILSON	TN	-19	-21	-36	25	38
1070	3.4	1674	3.8	2175	3.7	TEXAS	TX	14	-3	10		
6	4.2	10	5.5	10	4.9	ANDERSON	TX	31	-11	17	64	75
		1	3.7	1	2.0	ANDREWS	TX		-45			21
8	5.4	6	2.7	11	3.6	ANGELINA	TX	-51	36	-33	28	52
1	3.1	3	7.3	3	4.3	ARANSAS	TX	136	-41	40	74	67
1	4.4	1	2.6	2	4.2	ARCHER	TX	-41	60	-5	52	64
3	3.9	4	3.9	6	5.1	ATASCOSA	TX	-1	32	31	71	78
4	4.3	3	3.2	6	5.0	AUSTIN	TX	-25	57	18	64	77
		2	4.7	1	3.1	BAILEY	TX		-36			39
1	3.9	1	2.3			BANDERA	TX	-39				
2	2.6	5	4.5	4	2.6	BASTROP	TX	72	-42	-1	54	29
7*	19.0	4	8.6	2	7.1	BAYLOR	TX	-55	-17	-63	08	92
5	6.8	3	3.3	11*	9.0	BEE	TX	-52	177	33	72	97
9	3.0	13	3.4	17	3.2	BELL	TX	11	-5	5	58	42
94	4.2	185*	6.0	218*	5.5	BEXAR	TX	43	-9	30	70	82
1	3.4			1	2.3	BLANCO	TX			-32	29	25
		1	.9	2	1.0	BOSQUE	TX		16			11
8	4.1	9	3.0	15	3.9	BOWIE	TX	-27	28	-6	50	58
4	3.2	11	4.3	13	3.1	BRAZORIA	TX	33	-28	-4	52	39
1	.9	1	.6	6	2.9	BRAZOS	TX	-26	357	238	95	36
2	6.9	3	9.1	3	7.8	BREWSTER	TX	32	-14	13	62	94
1	4.0	1	3.4	2	5.3	BROOKS	TX	-14	57	35	73	81
2	1.1	9	3.8	7	2.9	BROWN	TX	231	-24	152	91	35
2	3.2	2	2.9	1	1.3	BURLESON	TX	-10	-56	-61	09	13
1	1.9	2	1.9	2	1.4	BURNET	TX	2	-28	-27	33	14
4	4.7	8	7.1	8	5.1	CALDWELL	TX	53	-28	10	60	79
2	7.7	4	9.1	4	5.8	CALHOUN	TX	18	-37	-25	35	85
2	2.8	1	1.7	1	1.6	CALLAHAN	TX	-41	-4	-44	18	16
9	2.3	35*	6.4	48*	6.4	CAMERON	TX	174	0	174	93	89
2	5.5			1	2.1	CAMP	TX			-61	09	22
		3	8.4			CARSON	TX					
3	3.1	4	3.0	8	4.5	CASS	TX	-4	53	48	77	70
				2	2.8	CHAMBERS	TX					33
4	2.3	7	3.5	9	3.5	CHEROKEE	TX	52	2	54	78	50
1	1.9					CHILDRESS	TX					
3	6.1			2	2.3	CLAY	TX			-62	08	25
				1	3.9	COCHRAN	TX					59
		2	6.9	2	5.4	COKE	TX		-23			81
3	3.0	1	.7	2	1.3	COLEMAN	TX	-78	90	-57	11	13
6	2.5	8	2.7	11	2.6	COLLIN	TX	10	-4	6	58	30
		2	4.3	3	5.7	COLLINGSWORTH	TX		31			84
3	4.0	2	1.9	1	.7	COLORADO	TX	-53	-63	-83	01	10
1	1.0	6	4.4	8	3.2	COMAL	TX	349	-26	231	95	43
5	4.2	5	6.6	2	1.6	COMANCHE	TX	57	-75	-61	09	16
2	8.5	2	6.9	1	3.2	CONCHO	TX	-19	-54	-63	08	42
6	4.7	11*	7.2	4	2.3	COOKE	TX	53	-68	-52	14	24
5	5.4	2	1.6	6	3.9	CORYELL	TX	-71	147	-28	33	59

* SIGNIFICANTLY MORE DEATHS THAN EXPECTED

WHITE FEMALE

CANCER OF THE LIVER AND GALLBLADDER INCLUDING BILE DUCTS

ICD 155,156

1950-1959 DEATHS	1950-1959 RATE	1960-1969 DEATHS	1960-1969 RATE	1970-1979 DEATHS	1970-1979 RATE	COUNTY AND STATE	PERCENT CHANGE 60S/50S	PERCENT CHANGE 70S/60S	PERCENT CHANGE 70S/50S	PERCENTILE %CHANGE RATES 70S/50S	PERCENTILE RATES 1970S
				1	2.8	COTTLE TX					33
1	19.1			2	9.3	CRANE TX			-52	14	97
1	9.4	1	4.7	2	10.6	CROCKETT TX	-50	125	13	62	98
1	2.7	2	4.6	3	5.1	CROSBY TX	73	11	92	86	78
1*	47.2			1	9.4	CULBERSON TX			-80	02	97
1	3.2	3	9.1			DALLAM TX	188				
96	3.4	112	2.7	167	2.9	DALLAS TX	-20	6	-15	43	36
1	2.1	2	2.6	2	1.8	DAWSON TX	21	-30	-15	43	18
3	7.9			4	4.2	DEAF SMITH TX			-47	17	65
1	2.3	1	1.6	3	4.3	DELTA TX	-30	168	88	85	67
8	3.9	8	2.9	7	1.8	DENTON TX	-26	-40	-55	12	17
2	1.8	3	1.9	3	2.3	DE WITT TX	5	18	25	68	24
1	3.3	2	5.7			DICKENS TX	71				
1	3.1	3	8.5	3	6.1	DIMMIT TX	171	-28	96	86	87
1	2.5			2	4.7	DONLEY TX			91	86	73
2	4.1	6*	10.1	3	4.1	DUVAL TX	145	-59	-1	54	62
7	3.9	12	6.8	8	3.7	EASTLAND TX	73	-46	-6	51	54
1	.7	7	2.8	9	2.9	ECTOR TX	281	1	284	96	35
1	10.1	1	9.6	1	6.8	EDWARDS TX	-5	-29	-33	28	91
6	2.8	9	3.8	12	3.9	ELLIS TX	35	3	39	74	58
59*	7.8	89*	8.7	117*	9.4	EL PASO TX	11	8	19	65	97
3	2.0	2	1.1	8	3.6	ERATH TX	-45	227	79	84	52
5	4.4	7	6.6	3	2.4	FALLS TX	51	-64	-46	17	26
7	3.7	9	3.8	6	2.8	FANNIN TX	2	-26	-25	36	33
2	1.7	4	2.4	10	4.5	FAYETTE TX	44	87	169	92	71
		2	4.2	2	3.1	FISHER TX		-27			40
3	5.7	1	1.7	1	1.2	FLOYD TX	-71	-27	-79	02	12
		1	4.3			FOARD TX					
4	4.3	6	3.8	9	4.0	FORT BEND TX	-12	5	-8	49	60
3	8.4	2	4.4	1	1.8	FRANKLIN TX	-47	-59	-78	02	18
2	2.4			3	3.2	FREESTONE TX			32	71	42
4	9.9	5*	9.7	2	2.3	FRIO TX	-2	-77	-77	02	24
1	7.4	3	7.8	4	7.9	GAINES TX	6	1	7	59	94
11	3.0	23	4.6	33	4.5	GALVESTON TX	52	-3	48	77	69
2	8.0	1	3.2	1	2.4	GARZA TX	-60	-25	-70	05	26
1	1.3	3	2.8	3	2.1	GILLESPIE TX	122	-23	71	82	22
		1*	17.2			GLASSCOCK TX					
1	3.3			1	2.4	GOLIAD TX			-27	34	27
2	2.0	1	1.3	7	4.8	GONZALES TX	-36	270	137	90	74
4	4.7	11*	9.2	4	2.5	GRAY TX	93	-73	-47	17	28
10	2.7	20	3.8	16	2.5	GRAYSON TX	42	-34	-7	50	28
4	2.0	16	5.3	18	3.9	GREGG TX	166	-26	97	87	59
1	3.3	5	8.3	3	3.0	GRIMES TX	150	-64	-10	47	37
6	5.1	7	4.5	11	4.8	GUADALUPE TX	-12	7	-6	50	74
8	6.1	5	3.2	5	2.7	HALE TX	-48	-15	-56	12	31
		3	5.5	2	2.9	HALL TX		-48			36
		7	7.9	2	1.3	HAMILTON TX		-83			13
		2	9.4			HANSFORD TX					
		5	7.5	5	8.0	HARDEMAN TX		7			95
1	1.5	3	2.5	5	3.3	HARDIN TX	71	28	119	89	44
104	4.0	163	3.8	216	3.4	HARRIS TX	-5	-10	-15	43	47
1	.7	6	3.6			HARRISON TX	440				
3	4.6	1	1.4	2	3.4	HASKELL TX	-69	134	-28	33	46
4	5.1	5	4.7	6	4.3	HAYS TX	-8	-9	-16	42	67
5	4.6			4	1.9	HENDERSON TX			-60	09	19
12	2.6	41*	6.1	59*	6.1	HIDALGO TX	135	0	136	90	87
7	4.2	6	3.2	8	3.5	HILL TX	-24	9	-17	41	49
2	4.8	4	5.5	4	3.7	HOCKLEY TX	15	-33	-23	36	54
2	5.3	5*	8.2	1	1.7	HOOD TX	55	-79	-68	05	16
4	2.8	5	2.9	3	1.4	HOPKINS TX	6	-51	-48	16	14
1	1.1	2	2.5	3	2.2	HOUSTON TX	128	-10	107	87	23
3	3.0	5	3.4	3	1.6	HOWARD TX	14	-54	-47	17	16
		2*	23.9	1	11.6	HUDSPETH TX		-51			99
3	1.2	7	2.7	13	4.0	HUNT TX	121	48	227	95	61
3	3.7	4	3.8	6	4.0	HUTCHINSON TX	2	6	8	59	62
		1	9.2	2*	18.1	IRION TX		96			99
2	4.0			4	6.5	JACK TX			62	80	89
		5	8.5	2	2.7	JACKSON TX		-69			31
1	1.3	6	5.3	9	6.0	JASPER TX	313	14	371	98	87
11	1.8	31	3.7	40	3.5	JEFFERSON TX	107	-6	95	86	50
2	11.6			4*	13.4	JIM HOGG TX			16	63	99
2	3.3	7	4.9	9	5.9	JIM WELLS TX	51	20	82	84	86
8	3.8	12	4.6	10	2.5	JOHNSON TX	21	-45	-33	28	28
3	2.7	2	1.5	6	3.1	JONES TX	-44	103	15	62	41
3	5.2	2	2.6	6	6.8	KARNES TX	-50	161	29	70	91
4	2.4	4	1.9	5	2.1	KAUFMAN TX	-21	13	-11	47	22
1	3.0			2	2.6	KENDALL TX			-12	46	30
		1	9.6			KENT TX					
2	1.3	6	2.5	6	1.9	KERR TX	99	-25	49	77	19
		1	3.2			KIMBLE TX					
1	9.6					KINNEY TX					
		9*	9.6	6	5.6	KLEBERG TX		-42			83
		1	2.4	1	1.7	KNOX TX		-28			17
8	4.4	8	3.9	9	3.1	LAMAR TX	-10	-20	-28	33	41
1	1.2	2	2.2	5	4.0	LAMB TX	75	82	218	95	60
		3	6.5	1	1.9	LAMPASAS TX		-71			19
1	4.1	2	8.6	3*	9.4	LA SALLE TX	109	10	131	90	97
3	2.1	4	2.5	4	1.9	LAVACA TX	19	-25	-11	47	19
		1	1.5	2	3.2	LEE TX		117			44
1	1.7	2	3.1			LEON TX	80				
5	5.2	3	2.2	8	4.1	LIBERTY TX	-58	92	-20	39	64
2	1.4	6	4.2	3	1.8	LIMESTONE TX	208	-58	30	70	18
		1	4.5			LIPSCOMB TX					
		1	2.6	3	5.0	LIVE OAK TX		95			77
		4	5.7	3	2.3	LLANO TX		-60			24
11	2.5	9	1.6	21	2.7	LUBBOCK TX	-35	69	10	60	32
				2	3.7	LYNN TX					54
2	3.0	4	4.7	2	1.5	MC CULLOCH TX	58	-67	-48	16	15
13	2.1	27	3.5	29	3.2	MC LENNAN TX	65	-7	53	78	43
		1	15.1			MC MULLEN TX					

* SIGNIFICANTLY MORE DEATHS THAN EXPECTED

WHITE FEMALE

CANCER OF THE LIVER AND GALLBLADDER INCLUDING BILE DUCTS

ICD 155,156

1950-1959 DEATHS	1950-1959 RATE	1960-1969 DEATHS	1960-1969 RATE	1970-1979 DEATHS	1970-1979 RATE	COUNTY AND STATE	PERCENT CHANGE 60S/50S	PERCENT CHANGE 70S/60S	PERCENT CHANGE 70S/50S	PERCENTILE %CHANGE 70S/50S	PERCENTILE RATES 1970S
		1	1.5	2	3.1	MADISON TX		106			40
1	5.4					MARION TX					
				1	5.7	MARTIN TX					84
		3	6.3	2	8.4	MASON TX		33			95
4	5.0	5	5.3	7	5.1	MATAGORDA TX	7	-3	3	57	79
2	4.9	5*	8.3	11*	13.5	MAVERICK TX	68	63	174	93	99
6	7.5	4	4.5	8	6.2	MEDINA TX	-40	37	-17	41	88
1	4.1			1	2.2	MENARD TX			-47	17	23
2	2.1	5	3.3	11	3.7	MIDLAND TX	56	13	76	83	54
6	4.9	5	3.9	2	.9	MILAM TX	-21	-78	-82	01	10
1	2.5			3	3.2	MILLS TX			31	70	44
1	1.8			2	2.6	MITCHELL TX			49	77	29
1	.8	5	3.3	6	3.1	MONTAGUE TX	325	-7	296	96	39
2	2.2			12	3.6	MONTGOMERY TX			65	81	52
1	7.5	1	2.9	1	2.4	MOORE TX	-61	-18	-68	05	26
2	4.1			3	3.4	MORRIS TX			-18	41	47
2	14.4	1	5.6			MOTLEY TX	-61				
2	1.4	4	2.6	8	3.8	NACOGDOCHES TX	83	43	162	92	56
4	2.4	5	2.1	12	4.1	NAVARRO TX	-13	91	67	81	62
1	3.2	3	6.6	2	3.4	NEWTON TX	106	-48	6	59	47
2	2.4	1	1.0	5	3.1	NOLAN TX	-58	211	32	71	40
24	4.9	43*	5.9	59*	6.0	NUECES TX	18	2	21	66	87
1	3.7			3	6.2	OCHILTREE TX			68	82	88
4	3.7	3	1.2	14	5.0	ORANGE TX	-67	305	36	73	77
5	4.3	5	3.5	8	4.3	PALO PINTO TX	-19	23	0	54	67
5	7.3	1	1.1			PANOLA TX	-85				
6	4.4	5	2.7	7	3.1	PARKER TX	-39	16	-29	32	40
		2	5.1	2	5.1	PARMER TX		2			79
		2	5.5	5*	9.4	PECOS TX		71			97
4	7.3	3	3.8	5	3.5	POLK TX	-47	-10	-52	14	49
11	3.8	12	2.7	10	1.8	POTTER TX	-30	-33	-54	13	18
		3	11.6	1	3.3	PRESIDIO TX		-71			45
		1	4.2	1	3.1	RAINS TX		-27			39
1	1.7	1	.4	3	1.1	RANDALL TX	-77	196	-32	29	12
		1	6.6	1	8.3	REAGAN TX		25			95
3	3.0	5	3.7	3	1.8	RED RIVER TX	25	-52	-40	22	17
1	3.0	1	2.2	2	3.6	REEVES TX	-26	65	23	67	53
2	4.5	1	2.8	2	4.0	REFUGIO TX	-37	41	-11	47	61
3	5.4	3	4.3	1	1.0	ROBERTSON TX	-21	-77	-82	01	11
1	2.4	1	3.3	3	7.5	ROCKWALL TX	39	129	217	95	93
2	2.2	1	1.0	1	.9	RUNNELS TX	-57	-4	-58	10	10
3	2.1	6	3.0	7	3.1	RUSK TX	44	2	47	77	39
1	2.8	4*	10.5	5*	7.6	SABINE TX	276	-28	170	92	93
				1	1.9	SAN AUGUSTINE TX					19
		2	7.5	4*	7.9	SAN JACINTO TX		6			94
11*	10.0	4	2.4	12	5.7	SAN PATRICIO TX	-76	138	-43	19	84
1	1.9	1	5.3	4	5.3	SAN SABA TX	184	-1	179	93	81
				1	4.8	SCHLEICHER TX					74
1	1.3	3	4.4	3	2.9	SCURRY TX	231	-35	116	88	35
		1	2.9	1	1.4	SHACKELFORD TX		-51			14
2	1.8	4	3.7	6	3.3	SHELBY TX	109	-11	87	85	46
10	3.7	12	3.2	15	2.7	SMITH TX	-13	-17	-28	33	31
		1	3.3			SOMERVELL TX					
1	2.1	2	3.2	5	5.0	STARR TX	55	54	139	90	77
3	5.7	4	4.3	2	2.1	STEPHENS TX	-25	-51	-63	08	22
				1	18.2	STERLING TX					99
		1	4.9	1	3.6	STONEWALL TX		-27			52
1	7.6			3*	17.1	SUTTON TX			124	89	99
2	4.7	2	4.2	3	5.6	SWISHER TX	-12	34	19	65	83
42	2.7	73	3.0	75	2.2	TARRANT TX	11	-26	-18	40	23
5	1.7	9	2.1	12	2.2	TAYLOR TX	26	4	32	71	23
1	2.5			3	4.1	TERRY TX			67	81	63
				1	3.4	THROCKMORTON TX					48
6	7.6			10*	8.4	TITUS TX			11	61	95
8	3.4	10	2.8	18	3.9	TOM GREEN TX	-16	39	16	63	59
25	3.7	34	3.5	45	3.3	TRAVIS TX	-4	-7	-11	47	44
1	3.0	2	3.4	4	5.8	TRINITY TX	13	70	92	86	85
1	1.7	1	1.1	2	1.7	TYLER TX	-37	51	-4	52	16
5	5.6	4	3.2	2	1.2	UPSHUR TX	-43	-62	-79	02	12
		2	11.6	1	4.5	UPTON TX		-61			71
3	3.5	6	6.6	2	1.7	UVALDE TX	87	-74	-52	14	17
11*	16.8	5	6.1	11*	10.1	VAL VERDE TX	-64	65	-40	22	98
5	4.4	3	2.0	5	1.8	VAN ZANDT TX	-56	-7	-59	10	18
2	1.4	10	5.4	14	5.5	VICTORIA TX	274	2	283	96	82
3	6.0	1	1.8	4	4.6	WALKER TX	-70	161	-22	37	72
1	2.8	1	1.8	2	3.5	WALLER TX	-34	92	26	68	49
2	6.5	1	1.3	3	4.7	WARD TX	-80	274	-27	34	74
2	2.0	10*	7.9	6	3.7	WASHINGTON TX	286	-54	79	83	53
16*	7.3	30*	10.9	38*	9.9	WEBB TX	49	-9	36	73	98
1	.8	9	5.7	12	6.2	WHARTON TX	653	10	732	99	88
		2	2.9	2	3.1	WHEELER TX		6			39
10	2.4	15	2.6	26	3.9	WICHITA TX	9	50	64	81	59
6	6.2	3	2.4	4	2.8	WILBARGER TX	-61	14	-55	12	33
2	5.2	2	3.1	6*	8.4	WILLACY TX	-40	174	63	81	96
5	3.0	9	4.3	14	3.9	WILLIAMSON TX	46	-9	33	72	60
		4	5.4	6	8.0	WILSON TX		49			95
		2	12.9	2	4.9	WINKLER TX		-62			76
		2	1.3	6	3.9	WISE TX		204			60
8	7.6	8	5.9	8	3.9	WOOD TX	-23	-34	-49	16	58
				1	4.2	YOAKUM TX					64
2	2.0	2	1.2	3	2.0	YOUNG TX	-38	66	3	57	21
1	5.7	2	8.8	1	2.0	ZAPATA TX	55	-77	-65	07	20
1	2.9	2	5.2	3	7.1	ZAVALA TX	81	36	146	91	92
134	4.8	136	3.6	155	3.0	UTAH UT	-24	-18	-38		
1	4.9	1	5.1	1	3.3	BEAVER UT	5	-36	-33	28	44
2	2.6	5	4.8	4	3.0	BOX ELDER UT	81	-37	14	62	38
7	4.9	11	6.2	1	.3	CACHE UT	28	-94	-93	00	09
3	8.1	1	1.0	4	4.4	CARBON UT	-88	353	-46	17	68

* SIGNIFICANTLY MORE DEATHS THAN EXPECTED

1950-1959 DEATHS	1950-1959 RATE	1960-1969 DEATHS	1960-1969 RATE	1970-1979 DEATHS	1970-1979 RATE	COUNTY AND STATE	PERCENT CHANGE 60S/50S	PERCENT CHANGE 70S/60S	PERCENT CHANGE 70S/50S	PERCENTILE %CHANGE 70S/50S	PERCENTILE RATES 1970S
9*	9.0	6	3.2	9	2.6	DAVIS UT	-64	-20	-71	04	29
2	10.3	1	5.1	1	2.5	DUCHESNE UT	-50	-51	-76	03	27
4*	17.0	3*	11.0	2	5.6	EMERY UT	-35	-49	-67	06	83
1	7.5					GARFIELD UT					
1	10.1	1	6.6	3*	12.3	GRAND UT	-34	87	22	67	99
		2	5.3	3	4.9	IRON UT		-8			75
		1	4.8	2	6.6	JUAB UT		37			90
1	24.8					KANE UT					
1	3.1	1	2.7	1	2.0	MILLARD UT	-15	-26	-37	24	20
66	5.1	68	3.9	70	2.9	SALT LAKE UT	-24	-25	-42	20	36
2*	22.9					SAN JUAN UT					
1	1.2	5	6.6	2	2.1	SANPETE UT	432	-68	69	82	22
5*	10.1	2	3.4	3	3.9	SEVIER UT	-67	15	-62	09	59
3*	11.6	2	7.2	1	3.5	SUMMIT UT	-37	-52	-70	05	49
1	2.2	1	1.6	2	2.7	TOOELE UT	-27	68	23	67	30
		1	3.1			UINTAH UT					
11	3.9	11	2.8	21	3.7	UTAH UT	-28	32	-5	52	54
1	2.3	4	6.5	5	4.3	WASHINGTON UT	177	-33	85	85	67
12	3.4	9	2.0	20	3.2	WEBER UT	-43	64	-7	50	43
77	3.5	94	3.7	99	3.1	VERMONT VT	8	-16	-9		
4	4.3	4	3.9	2	1.4	ADDISON VT	-9	-65	-68	06	14
6	3.8	7	3.6	9	3.3	BENNINGTON VT	-5	-8	-13	45	45
5	3.4	9	6.2	5	2.8	CALEDONIA VT	82	-55	-17	41	34
9	2.7	13	3.3	12	2.4	CHITTENDEN VT	21	-29	-13	44	26
		3	9.9	1	2.5	ESSEX VT		-75			27
8	4.8	4	2.1	4	1.9	FRANKLIN VT	-56	-10	-60	09	19
1	7.0					GRAND ISLE VT					
3	4.7	1	1.3	4	4.3	LAMOILLE VT	-73	234	-10	47	66
3	3.4	5	3.6	6	4.3	ORANGE VT	8	17	27	69	66
5	4.4	6	5.2	7	4.7	ORLEANS VT	18	-9	7	59	73
8	2.6	13	3.6	16	3.6	RUTLAND VT	37	0	37	73	52
9	3.5	15	5.0	11	3.2	WASHINGTON VT	43	-36	-9	49	41
7	3.4	8	3.7	11	4.0	WINDHAM VT	8	8	17	63	61
9	3.8	6	2.2	11	3.4	WINDSOR VT	-41	54	-10	48	48
370	3.2	528	3.3	679	3.2	VIRGINIA VA	3	-4	0		
		6	3.3	6	3.1	ACCOMACK VA		-7			39
11	4.9	8	2.6	11	2.9	ALBEMARLE#VA	-46	9	-41	21	35
5	3.7	1	.7	7	3.9	ALLEGHANY#VA	-82	468	4	57	57
		1	3.9	1	2.6	AMELIA VA		-32			29
1	1.3	2	2.0	2	1.8	AMHERST VA	52	-11	35	73	18
		2	4.7			APPOMATTOX VA					
17	3.5	34	4.1	31	3.1	ARLINGTON#VA	17	-24	-11	46	41
13	3.9	21	4.8	13	2.6	AUGUSTA#VA	21	-46	-35	27	29
3	11.3	1	3.3	2	5.1	BATH VA	-71	56	-55	12	79
11	2.4	15	2.6	21	2.7	BEDFORD#VA	8	7	16	63	32
		1	2.9	1	2.4	BLAND VA		-19			25
2	2.8	5	6.0	3	2.5	BOTETOURT VA	114	-58	-9	48	28
1	2.5	2	4.0	1	3.1	BRUNSWICK VA	60	-24	21	66	39
3	3.5	6	7.2	12*	10.1	BUCHANAN VA	105	39	184	93	98
		1	2.7	2	4.2	BUCKINGHAM VA		60			65
3	9.1	1	3.0	2	4.4	CAROLINE VA	-67	44	-52	14	68
5	2.3	4	1.6	11	3.5	CARROLL#VA	-32	123	51	78	50
3	7.8	1	2.3	2	3.3	CHARLOTTE VA	-71	45	-58	11	45
25	2.1	31	2.2	40	2.5	CHESTERFIELD#VA	2	15	18	65	27
1	2.8	1	2.5			CLARKE VA	-11				
				3*	11.1	CRAIG VA					98
3	5.3			3	2.1	CULPEPER VA			-60	09	22
				1	3.1	CUMBERLAND VA					40
2	3.5	2	3.0	3	3.4	DICKENSON VA	-15	14	-3	53	48
4	1.6	4	1.4	12	3.3	DINWIDDIE#VA	-14	144	111	88	46
17	3.2	29	1.9	54	2.7	FAIRFAX#VA	-40	39	-17	41	30
3	3.5	4	3.7	7	5.5	FAUQUIER VA	5	49	56	79	82
		3	4.1	2	2.0	FLOYD VA		-50			21
2	8.7	2	6.5	2	4.8	FLUVANNA VA	-25	-27	-45	18	74
1	1.1	2	1.7	6	3.7	FRANKLIN VA	59	118	248	95	55
7	4.4	7	3.0	5	1.8	FREDERICK#VA	-33	-38	-58	10	18
3	4.5	6	7.9	5	5.2	GILES VA	78	-34	18	64	80
1	2.6	1	2.1	3	3.5	GLOUCESTER VA	-22	70	33	72	50
1	4.4			1	3.0	GOOCHLAND VA			-31	30	38
2	11.2	2	8.3			GREENE VA	-26				
				3	4.8	GREENSVILLE#VA					74
7	6.7	4	3.0	6	3.6	HALIFAX#VA	-55	21	-46	17	53
2	2.7	3	2.3	6	3.2	HANOVER VA	-14	36	18	64	41
10	3.7	15	2.9	28	3.4	HENRICO VA	-21	15	-9	48	47
5	5.0	12	7.1	5	1.8	HENRY#VA	42	-74	-63	08	18
		1	3.7			HIGHLAND VA					
1	2.5	1	2.0	1	1.3	ISLE OF WIGHT VA	-20	-35	-48	16	13
		2	1.2	4	1.9	JAMES CITY#VA		62			19
				1	4.5	KING AND QUEEN VA					70
2	9.1			1	2.9	KING GEORGE VA			-68	05	34
				1	2.9	KING WILLIAM VA					35
		2	3.7	1	1.2	LANCASTER VA		-69			12
7	6.5	3	2.1	11	6.3	LEE VA	-67	196	-2	53	88
3	3.1	4	3.1	9	5.4	LOUDOUN VA	-1	78	76	83	82
1	2.1	1	2.2	6*	8.7	LOUISA VA	4	295	311	97	96
		2	3.7	1	1.8	LUNENBURG VA		-49			18
		5*	12.4	1	2.2	MADISON VA		-83			23
2	4.0	2	7.4	1	1.1	MATHEWS VA	83	-85	-73	03	11
7*	9.9	3	3.0	5	4.5	MECKLENBURG VA	-70	51	-54	13	70
6	4.2	10	4.9	9	3.2	MONTGOMERY#VA	17	-35	-24	36	42
2	3.6	5	8.8	2	3.0	NELSON VA	144	-66	-17	41	37
2	3.4	2	2.6	2	2.0	NORTHAMPTON VA	-25	-21	-41	21	21
1	2.9	1	1.8			NORTHUMBERLAND VA	-37				
				2	1.8	NOTTOWAY VA					18
2	3.7	3	4.6	6	7.6	ORANGE VA	25	66	107	87	94
1	1.2	4	4.6	6	5.0	PAGE VA	284	9	318	97	77

* SIGNIFICANTLY MORE DEATHS THAN EXPECTED

COUNTIES COMBINED; SEE TECHNICAL NOTES

1950-1959 DEATHS	1950-1959 RATE	1960-1969 DEATHS	1960-1969 RATE	1970-1979 DEATHS	1970-1979 RATE	COUNTY AND STATE	PERCENT CHANGE 60S/50S	PERCENT CHANGE 70S/60S	PERCENT CHANGE 70S/50S	PERCENTILE %CHANGE 70S/50S	PERCENTILE RATES 1970S
		2	2.6	6	6.4	PATRICK VA		144			89
7	2.2	14	3.3	16	3.0	PITTSYLVANIA#VA	50	-10	35	73	37
				1	3.4	POWHATAN VA					47
1	2.1	1	2.1	3	5.7	PRINCE EDWARD VA	-2	175	169	92	84
		4	2.6	8	3.3	PRINCE WILLIAM#VA		29			46
9*	9.3	6	4.7	10	5.2	PULASKI VA	-49	10	-44	18	80
				2	6.0	RAPPAHANNOCK VA					87
1	4.4			3	7.1	RICHMOND VA			61	80	92
28	4.6	27	3.1	35	2.9	ROANOKE#VA	-33	-5	-36	25	36
5	4.4	10*	7.2	3	1.7	ROCKBRIDGE#VA	66	-76	-61	09	17
10	4.1	17	5.9	11	2.8	ROCKINGHAM#VA	43	-53	-33	28	32
4	4.0	6	5.3	9	6.7	RUSSELL VA	33	28	70	82	91
1	.9	10*	7.6	5	3.0	SCOTT VA	737	-61	229	95	38
4	3.4	4	2.7	5	2.5	SHENANDOAH VA	-18	-9	-26	34	27
6	4.8	8	5.0	4	1.9	SMYTH VA	3	-62	-61	09	19
2	5.1	3	4.1	5	5.6	SOUTHAMPTON#VA	-20	35	9	60	83
3	3.0	6	4.7	3	1.5	SPOTSYLVANIA#VA	57	-68	-49	15	15
2	4.1	2	3.3	1	1.2	STAFFORD VA	-20	-63	-71	05	12
				1	3.2	SURRY VA					43
		1	4.9			SUSSEX VA					
10	7.0	8	4.3	18*	7.6	TAZEWELL VA	-38	76	9	60	94
1	2.2	3	4.3	3	2.7	WARREN VA	94	-38	21	66	30
7	3.2	14	4.8	10	2.6	WASHINGTON#VA	51	-45	-17	41	30
2	6.7	2	4.0	4	7.0	WESTMORELAND VA	-40	74	4	57	92
4	2.5	6	3.1	20*	7.9	WISE#VA	23	152	209	94	94
2	2.0	1	.8	4	2.9	WYTHE VA	-58	245	45	76	35
45	3.8	58	3.5	70	3.1	CHESAPEAKE CITY#VA	-9	-10	-18	41	41
3	2.0	10	4.8	6	1.9	HAMPTON CITY VA	144	-60	-3	52	19
1	.6	10	3.0	12	2.6	NEWPORT NEWS CITY VA	399	-15	324	97	28
1	2.0	4	3.2	8	4.9	SUFFOLK CITY VA	66	52	151	91	76
499	4.2	556	3.5	717	3.4	WASHINGTON WA	-17	-1	-17		
1	3.3	2	4.6	2	3.0	ADAMS WA	39	-35	-10	48	37
4	4.1	3	3.6	7	5.0	ASOTIN WA	-12	40	24	67	77
6	5.5	7	3.1	12	3.5	BENTON WA	-44	16	-36	26	50
8	4.1	9	3.5	11	3.3	CHELAN WA	-14	-7	-20	39	45
4	3.4	7	3.8	13	4.3	CLALLAM WA	13	13	27	69	66
13	3.7	10	1.9	27	3.3	CLARK WA	-49	77	-9	48	46
1	3.6	4*	15.5	1	3.7	COLUMBIA WA	331	-76	3	57	55
9	4.3	12	3.9	10	2.7	COWLITZ WA	-9	-30	-37	25	32
1	2.5	3	4.2	1	1.0	DOUGLAS WA	69	-76	-59	10	11
		1	7.4			FERRY WA					
2	4.9	4	4.5	4	3.2	FRANKLIN WA	-7	-29	-34	27	43
2	11.1	2	11.4			GARFIELD WA	2				
2	1.9	6	5.1	4	2.2	GRANT WA	172	-57	17	64	23
7	2.7	16	4.9	16	4.2	GRAYS HARBOR WA	82	-13	58	79	65
2	5.0	4	3.6	6	3.2	ISLAND WA	-28	-12	-36	25	42
				3	3.7	JEFFERSON WA					53
166	4.2	179	3.4	230	3.5	KING WA	-20	2	-18	41	49
10	2.8	14	3.0	22	3.2	KITSAP WA	8	5	13	62	42
3	3.2	4	3.3	5	3.8	KITTITAS WA	4	14	18	65	56
4	7.8	4	6.1	1	1.3	KLICKITAT WA	-22	-79	-83	01	13
13	4.9	15	5.0	15	4.2	LEWIS WA	3	-16	-14	44	65
2	3.8	4	5.4	4	5.3	LINCOLN WA	41	-2	38	74	81
2	2.7	2	2.1	5	3.0	MASON WA	-23	45	11	61	38
5	5.3	6	4.6	6	3.9	OKANOGAN WA	-13	-15	-26	34	60
7	8.2	4	3.7	2	1.4	PACIFIC WA	-55	-62	-82	01	14
1	3.4	1	2.9	1	2.0	PEND OREILLE WA	-16	-32	-43	20	20
63	4.7	67	3.9	74	3.4	PIERCE WA	-18	-11	-27	34	48
		1	3.2	3	6.8	SAN JUAN WA		110			91
10	3.8	11	3.3	17	4.4	SKAGIT WA	-13	35	18	65	69
		1	5.9			SKAMANIA WA					
38	5.9	26	2.8	39	2.8	SNOHOMISH WA	-52	0	-52	14	34
35	2.7	53	3.1	72	3.5	SPOKANE WA	17	12	31	71	50
3	4.8	1	1.0	5	4.4	STEVENS WA	-79	352	-7	50	68
8	3.0	11	3.2	25	4.4	THURSTON WA	7	39	49	77	69
1	5.4					WAHKIAKUM WA					
11	5.1	8	3.0	12	3.7	WALLA WALLA WA	-42	23	-29	32	53
22	5.3	18	3.6	21	3.3	WHATCOM WA	-32	-10	-38	24	45
5	3.7	7	4.4	8	4.5	WHITMAN WA	18	2	21	66	70
28	4.8	29	3.8	33	3.5	YAKIMA WA	-21	-9	-28	33	48
363*	4.7	381	4.1	529*	4.6	WEST VIRGINIA WV	-14	14	-1		
6	7.2	2	2.1	4	3.5	BARBOUR WV	-70	61	-52	14	49
6	4.5	5	2.7	11	4.5	BERKELEY WV	-41	69	-1	54	70
8*	7.0	3	2.7	7	4.9	BOONE WV	-61	80	-30	31	75
3	3.6	1	1.2	6	5.9	BRAXTON WV	-68	414	66	81	86
4	3.3	4	3.0	8	4.7	BROOKE WV	-10	59	43	75	73
25	4.8	22	3.4	21	2.9	CABELL WV	-29	-14	-39	23	35
		3	6.8	3	5.1	CALHOUN WV		-24			78
2	4.7			1	1.6	CLAY WV			-66	07	16
		2	4.1	4	5.5	DODDRIDGE WV		35			83
19*	8.1	18*	7.0	23*	6.9	FAYETTE WV	-15	0	-15	43	91
1	2.3	2	4.2	4	6.5	GILMER WV	81	55	180	93	89
2	5.7	2	4.7	3	5.1	GRANT WV	-18	9	-11	47	78
13*	9.0	10	5.3	11	4.3	GREENBRIER WV	-41	-19	-52	14	66
		3	4.7	1	1.3	HAMPSHIRE WV		-73			13
4	2.7	5	3.0	15*	6.9	HANCOCK WV	13	128	158	91	92
3	6.7	3	6.2	2	3.2	HARDY WV	-7	-48	-52	14	42
9	2.3	17	3.6	39*	6.8	HARRISON WV	60	86	198	94	91
2	2.4	5	5.6	2	1.6	JACKSON WV	136	-71	-30	31	16
1	1.3	6	6.4	5	3.8	JEFFERSON WV	388	-41	187	93	56
47	5.8	42	3.9	55	4.1	KANAWHA WV	-34	7	-29	32	63
6	4.1	5	3.8	11*	6.9	LEWIS WV	-7	82	69	82	91
4	4.7	4	4.1	9*	7.8	LINCOLN WV	-12	90	67	81	94
7	3.9	12	6.3	19*	8.4	LOGAN WV	64	33	118	88	95
4	2.1	10	5.4	6	2.8	MC DOWELL WV	157	-49	32	71	33
19	6.0	16	4.1	20	4.2	MARION WV	-31	3	-30	31	65

* SIGNIFICANTLY MORE DEATHS THAN EXPECTED

COUNTIES COMBINED; SEE TECHNICAL NOTES

1950-1959 DEATHS	1950-1959 RATE	1960-1969 DEATHS	1960-1969 RATE	1970-1979 DEATHS	1970-1979 RATE	COUNTY AND STATE	PERCENT CHANGE 60S/50S	PERCENT CHANGE 70S/60S	PERCENT CHANGE 70S/50S	PERCENTILE %CHANGE 70S/50S	PERCENTILE RATES 1970S
16*	9.1	12	5.7	19*	7.5	MARSHALL WV	-38	33	-18	41	93
5	5.0	3	2.4	6	4.1	MASON WV	-52	69	-18	40	62
16	6.1	4	1.1	22	5.0	MERCER WV	-82	358	-19	40	77
6	5.8	4	3.4	8	5.3	MINERAL WV	-42	58	-8	49	81
5	3.9	16*	10.9	6	3.6	MINGO WV	178	-67	-9	49	51
14	5.7	12	4.3	19	5.4	MONONGALIA WV	-25	25	-6	51	82
2	2.6	2	2.6	7*	8.4	MONROE WV	0	220	221	95	96
		1	1.9	2	2.8	MORGAN WV		48			32
4	4.2	1	1.0	11*	7.6	NICHOLAS WV	-77	685	83	84	94
17	4.1	17	3.9	20	4.1	OHIO WV	-5	7	2	56	63
1	2.3	1	2.0	1	1.3	PENDLETON WV	-13	-37	-45	18	13
		1	2.8	3	7.0	PLEASANTS WV		148			92
1	2.1	1	1.6	2	2.8	POCAHONTAS WV	-24	74	32	71	34
4	3.1	8	5.4	7	3.7	PRESTON WV	73	-32	18	65	54
6	7.4	2	1.9	8	5.1	PUTNAM WV	-74	170	-31	30	78
10	5.1	15	4.6	14	3.1	RALEIGH WV	-9	-32	-38	23	41
10*	7.6	7	5.0	7	4.3	RANDOLPH WV	-34	-13	-43	19	67
3	3.5	7*	7.2	2	1.9	RITCHIE WV	107	-73	-44	18	20
4	4.3	6	5.7	3	2.7	ROANE WV	33	-54	-38	23	30
		3	3.4	4	4.4	SUMMERS WV		29			69
4	4.1	3	3.0	5	3.8	TAYLOR WV	-26	26	-7	50	56
3	6.0	7*	12.0	5	9.3	TUCKER WV	101	-22	57	79	97
1	1.4	1	1.3			TYLER WV	-8				
7	7.1	8	6.4	3	2.9	UPSHUR WV	-10	-54	-59	10	36
7	5.1	5	2.8	9	3.7	WAYNE WV	-44	32	-26	34	55
1	1.7	2	3.5	2	3.4	WEBSTER WV	101	-4	93	86	47
3	3.0	5	4.4	9	6.9	WETZEL WV	48	58	133	90	92
1	3.1			2	5.9	WIRT WV			91	86	66
16	4.6	19	4.0	26	4.6	WOOD WV	-12	16	1	56	72
1	1.3	6	5.4	7	5.8	WYOMING WV	315	7	344	98	85
905*	4.9	1090*	4.7	1253*	4.4	WISCONSIN WI	-4	-6	-10		
1	3.3	1	1.2	4	3.9	ADAMS WI	-65	237	19	65	60
6	6.2	4	2.8	7	5.7	ASHLAND WI	-55	106	-8	49	84
9	4.9	7	3.4	10	3.6	BARRON WI	-31	4	-28	33	51
		7*	9.8	3	2.9	BAYFIELD WI		-71			35
28	6.0	21	3.2	37	4.4	BROWN WI	-47	37	-28	33	68
5	7.4	7	8.0	7	7.9	BUFFALO WI	8	-1	8	59	94
5	9.3	3	4.7	4	4.3	BURNETT WI	-49	-9	-54	13	67
3	3.1	6	5.1	1	.8	CALUMET WI	65	-85	-75	03	10
13	6.3	8	3.2	20*	6.3	CHIPPEWA WI	-48	95	0	55	88
10	5.9	13	6.3	11	4.9	CLARK WI	7	-22	-16	42	76
11	4.7	13	4.6	16	4.4	COLUMBIA WI	-3	-3	-6	51	69
8*	9.8	8*	8.7	6	4.4	CRAWFORD WI	-11	-50	-55	12	68
39	4.5	49	4.1	45	3.1	DANE WI	-7	-26	-31	30	39
18	5.4	20	5.1	15	3.1	DODGE WI	-5	-40	-43	20	40
4	3.4	7	5.0	8	4.9	DOOR WI	47	-1	46	76	76
10	4.1	20*	7.3	15	4.6	DOUGLAS WI	78	-36	13	62	72
5	3.8	12*	7.4	9	5.1	DUNN WI	95	-32	33	72	78
13	4.6	18	4.7	17	3.7	EAU CLAIRE WI	3	-22	-20	39	54
2	10.7					FLORENCE WI					
19	4.2	27	5.3	25	4.0	FOND DU LAC WI	26	-25	-5	51	61
		1	2.5	3	5.3	FOREST WI		107			80
15	5.8	14	5.0	16	4.6	GRANT WI	-13	-9	-21	38	71
8	5.2	6	3.3	10	5.2	GREEN WI	-36	60	1	56	80
5	5.1	5	3.8	10	6.7	GREEN LAKE WI	-25	76	32	72	91
5	4.3	4	3.6	8	5.5	IOWA WI	-15	52	30	70	83
3	7.0	5*	9.6	3	4.9	IRON WI	38	-49	-29	32	76
1	1.1	4	4.0	7	5.9	JACKSON WI	280	46	454	99	86
12	4.1	12	3.2	20	4.8	JEFFERSON WI	-21	48	17	64	74
1	1.0	7	5.8	8	5.9	JUNEAU WI	474	2	488	99	86
22	5.3	27	5.0	36	5.1	KENOSHA WI	-6	3	-4	52	78
11*	12.3	5	4.9	8	6.6	KEWAUNEE WI	-61	35	-47	17	90
18	4.7	21	4.6	24	4.2	LA CROSSE WI	-3	-8	-11	47	65
4	3.9	11*	10.9	4	3.2	LAFAYETTE WI	181	-71	-17	41	43
5	4.5	10*	8.0	2	1.2	LANGLADE WI	77	-85	-73	04	12
1	.8	10	7.0	14*	8.2	LINCOLN WI	753	17	899	99	95
17	4.8	9	2.1	27	4.9	MANITOWOC WI	-57	137	3	57	76
24*	7.4	22	4.6	27	4.7	MARATHON WI	-38	2	-37	25	72
7	3.8	13	5.4	15	4.9	MARINETTE WI	42	-10	28	69	75
4	5.9	6*	8.4	7*	7.5	MARQUETTE WI	43	-10	28	69	93
15	5.0	19	5.2	20	4.3	MENOMINEE#WI	5	-18	-13	44	67
225*	5.0	289*	5.1	268	4.1	MILWAUKEE WI	3	-20	-18	41	62
6	3.5	9	4.7	11	4.8	MONROE WI	34	1	35	73	74
5	5.4	11*	8.1	13*	6.5	ONEIDA WI	50	-19	21	66	89
24	6.4	22	4.1	34	5.5	OUTAGAMIE WI	-36	34	-15	44	82
7	5.2	1	.5	10	3.6	OZAUKEE WI	-91	653	-31	30	52
				3	3.9	PEPIN WI					60
6	5.2	5	3.6	4	2.7	PIERCE WI	-30	-25	-47	16	31
3	2.3	8	4.6	8	3.4	POLK WI	99	-28	44	76	46
10	5.2	9	4.2	9	3.2	PORTAGE WI	-19	-24	-39	23	43
2	2.4	6	5.6	6	4.4	PRICE WI	135	-21	86	85	68
38*	6.3	25	3.3	33	3.6	RACINE WI	-49	12	-43	20	53
4	3.8	6	4.9	2	1.2	RICHLAND WI	29	-75	-68	05	12
22	4.1	34	5.2	40	4.8	ROCK WI	28	-7	19	65	75
4	5.5	8*	8.4	7	5.9	RUSK WI	51	-29	7	59	86
8	5.3	6	3.5	6	2.1	ST CROIX WI	-34	-40	-61	09	21
19*	8.6	11	4.3	17	5.2	SAUK WI	-51	23	-39	22	80
1	2.2	3	4.9	1	1.5	SAWYER WI	126	-69	-31	30	15
21	4.5	24	4.1	37	5.5	SHEBOYGAN WI	-8	33	22	67	82
3	3.9	5	5.2	6	5.5	TAYLOR WI	34	6	42	75	83
5	3.5	10	5.3	6	2.9	TREMPEALEAU WI	50	-44	-17	42	36
2	1.3	2	.9	11	5.8	VERNON WI	-29	522	339	97	85
5*	10.2	6*	9.2	7	6.2	VILAS WI	-10	-32	-39	23	88
9	3.3	15	4.5	20	4.3	WALWORTH WI	35	-6	27	69	66
2	3.1	1	1.3	8*	8.3	WASHBURN WI	-58	535	170	92	95
13	6.9	12	4.6	14	3.9	WASHINGTON WI	-33	-15	-42	20	60
24	4.5	42*	5.7	49	4.6	WAUKESHA WI	25	-18	2	56	72
12	5.2	13	4.8	18	5.5	WAUPACA WI	-8	15	5	58	82

* SIGNIFICANTLY MORE DEATHS THAN EXPECTED

\# COUNTIES COMBINED; SEE TECHNICAL NOTES

WHITE FEMALE

CANCER OF THE LIVER AND GALLBLADDER
INCLUDING BILE DUCTS

ICD 155,156

1950-1959 DEATHS	1950-1959 RATE	1960-1969 DEATHS	1960-1969 RATE	1970-1979 DEATHS	1970-1979 RATE	COUNTY AND STATE	PERCENT CHANGE 60S/50S	PERCENT CHANGE 70S/60S	PERCENT CHANGE 70S/50S	PERCENTILE %CHANGE 70S/50S	PERCENTILE RATES 1970S
4	4.4	3	3.1	13*	9.4	WAUSHARA WI	-29	199	112	88	97
25	4.7	27	4.2	41	4.9	WINNEBAGO WI	-11	18	5	58	76
9	3.8	15	4.8	22	5.1	WOOD WI	25	8	34	72	79
44	4.1	55	3.9	50	2.7	WYOMING WY	-7	-29	-34		
		4	4.8	3	2.6	ALBANY WY		-47			29
		2	3.6	3	3.5	BIG HORN WY		-3			50
1	4.5			1	2.7	CAMPBELL WY			-41	21	31
1	1.7			5	7.3	CARBON WY			331	97	93
2	7.6	2	6.3	1	2.1	CONVERSE WY	-17	-67	-72	04	22
1	5.6			2	8.3	CROOK WY			46	76	95
		5	5.4	3	2.7	FREMONT WY		-50			32
1	1.8	3	4.5	1	1.4	GOSHEN WY	152	-68	-19	40	14
		3	7.2	1	3.0	HOT SPRINGS WY		-58			37
1	3.4					JOHNSON WY					
9	5.9	12	5.5	5	1.7	LARAMIE WY	-7	-69	-71	04	17
		2	6.0	1	2.2	LINCOLN WY		-63			23
8	6.7	5	2.5	7	2.7	NATRONA WY	-63	9	-60	10	31
		2	9.2	1	3.8	NIOBRARA WY		-58			57
3	5.6	2	2.6	1	.6	PARK-TETON#WY	-54	-78	-90	00	10
3	8.0	4*	9.3	2	4.6	PLATTE WY	16	-50	-42	20	72
4	4.2	2	1.4	3	1.9	SHERIDAN WY	-66	33	-55	12	19
2*	17.1					SUBLETTE WY					
5	9.1	1	1.1	8*	8.0	SWEETWATER WY	-88	633	-12	46	95
1	2.8					UINTA WY					
		4*	13.7	2	5.0	WASHAKIE WY		-63			78

* SIGNIFICANTLY MORE DEATHS THAN EXPECTED

COUNTIES COMBINED; SEE TECHNICAL NOTES

NON-WHITE MALE

CANCER OF THE LIVER AND GALLBLADDER INCLUDING BILE DUCTS

ICD 155,156

1950-1959 DEATHS	1950-1959 RATE	1960-1969 DEATHS	1960-1969 RATE	1970-1979 DEATHS	1970-1979 RATE	COUNTY AND STATE	PERCENT CHANGE 60S/50S	PERCENT CHANGE 70S/60S	PERCENT CHANGE 70S/50S	PERCENTILE %CHANGE 70S/50S	PERCENTILE RATES 1970S
1940	3.0	4026	5.1	6962	6.9	UNITED STATES	69	35	129		
165*	4.8	167	4.7	276	7.4	ALABAMA AL	-1	57	56		
1	3.8			1	4.4	AUTAUGA AL			18	37	74
4*	12.7	1	2.3	7*	16.9	BALDWIN AL	-82	627	33	43	95
				3	8.2	BULLOCK AL					85
4*	9.8	2	4.8	2	5.1	BUTLER AL	-51	6	-48	10	76
1	2.3	3	4.4	3	4.2	CALHOUN AL	90	-5	81	58	73
2	3.8			2	3.5	CHAMBERS AL			-7	25	71
1*	19.1			1	16.1	CHEROKEE AL			-16	22	95
2*	12.1	1	6.2			CHILTON AL	-49				
3*	7.6	1	3.4	2	8.4	CHOCTAW AL	-56	151	12	35	85
4*	8.6	2	4.6	4	8.4	CLARKE AL	-46	81	-3	28	85
				1*	45.4	CLEBURNE AL					99
1	4.8			1	5.4	COFFEE AL			12	35	77
2	5.5	1	2.9	3	8.9	COLBERT AL	-47	205	61	50	86
				1	4.3	CONECUH AL					74
				1	5.5	COOSA AL					77
		3*	12.5	3	15.6	COVINGTON AL		25			94
3*	18.5	2	9.2			CRENSHAW AL	-50				
2*	9.5			1	3.9	DALE AL			-58	07	72
1	.9	6	6.1	1	1.0	DALLAS AL	576	-84	9	34	67
1	3.0			1	2.6	ELMORE AL			-14	23	69
2	4.9	3	6.5	2	5.2	ESCAMBIA AL	31	-20	6	32	76
3	7.0	1	1.6	6	9.5	ETOWAH AL	-77	482	35	44	88
1*	18.1	1	26.5			FRANKLIN AL	46				
2	3.7	1	1.9			GREENE AL	-49				
1	2.2	1	1.7	3	5.3	HALE AL	-21	210	144	69	76
		1	3.8	2	11.1	HENRY AL		194			90
1	2.1			1	1.9	HOUSTON AL			-13	24	68
42*	5.5	50	5.6	72	7.9	JEFFERSON AL	2	40	43	46	84
				1	11.1	LAMAR AL					90
1	3.5			1	3.5	LAUDERDALE AL			1	30	71
		1	4.5	1	5.1	LAWRENCE AL		13			76
4	6.5	1	2.1	5	7.4	LEE AL	-67	250	14	36	83
				1	3.5	LIMESTONE AL					71
2	3.8			5	14.3	LOWNDES AL			281	84	94
7*	5.9	3	2.7	5	4.6	MACON AL	-54	67	-23	19	74
		4	5.9	6	6.9	MADISON AL		18			81
9*	11.0	4	6.8	7	11.6	MARENGO AL	-38	71	6	32	91
				1*	28.7	MARION AL					98
				1	22.9	MARSHALL AL					97
21*	7.0	31*	9.0	50*	13.4	MOBILE AL	28	50	92	60	93
		2	3.9			MONROE AL					
8	3.3	12	5.5	37*	15.6	MONTGOMERY AL	65	184	370	90	94
2	6.4	2	5.4	2	7.0	MORGAN AL	-15	30	10	34	82
1	2.5	1	1.8	3	6.4	PERRY AL	-28	253	153	70	80
		1	3.7			PICKENS AL					
		1	4.6	1	4.0	PIKE AL		-14			73
3*	14.2			1	5.1	RANDOLPH AL			-64	05	76
4	6.3	3	4.2	3	4.3	RUSSELL AL	-33	1	-32	15	74
		4*	25.6			ST CLAIR AL					
4*	17.0	2	6.6	1	4.0	SHELBY AL	-61	-40	-76	02	73
3	4.9	2	4.4	3	7.0	SUMTER AL	-10	60	44	47	82
1	2.4	3	4.8	4	5.8	TALLADEGA AL	101	19	139	68	78
1	2.5	2	5.5	5	13.7	TALLAPOOSA AL	119	149	447	93	93
5	5.6	5	4.7	2	1.6	TUSCALOOSA AL	-17	-66	-72	03	67
1	3.0	2	6.3	1	3.7	WALKER AL	106	-41	22	39	72
1	5.6			3	17.1	WASHINGTON AL			208	77	95
3	5.4	1	3.0	2	7.0	WILCOX AL	-44	131	29	42	82
		4	2.8	22*	11.4	ALASKA AK		306			
18*	5.3	46*	11.3	52	7.1	ARIZONA AZ	113	-37	33		
6*	11.3	7*	12.7	10	11.5	APACHE AZ	12	-9	2	30	91
		2	5.8	1	2.5	COCONINO AZ		-58			69
		3*	26.0	2	16.1	GILA AZ		-38			95
1*	28.1	1	18.7			GRAHAM AZ	-33				
6	6.0	10	6.6	19	6.5	MARICOPA AZ	11	-1	9	33	80
		1*	18.6			MOHAVE AZ					
1	2.7	1	1.8	5	6.9	NAVAJO AZ	-31	273	157	71	81
2	4.0	9*	15.6	9	7.0	PIMA AZ	288	-55	74	55	81
2	4.9	8*	26.3	3	8.9	PINAL AZ	434	-66	82	58	86
				1	11.1	YAVAPAI AZ					90
		4*	29.5	2	7.8	YUMA AZ		-74			84
44	2.4	59	3.5	85	4.8	ARKANSAS AR	46	36	98		
				2	6.2	ARKANSAS AR					80
1	3.2	3	9.4	1	2.4	ASHLEY AR	188	-74	-26	17	69
				2	8.1	BRADLEY AR					84
2	3.2	1	1.5	1	1.2	CHICOT AR	-52	-21	-62	06	67
1	2.3	2	3.5	3	7.8	COLUMBIA AR	52	123	240	79	84
1	7.0	1	5.8			CONWAY AR	-17				
1*	19.1					CRAIGHEAD AR					
				1	23.8	CRAWFORD AR					97
1	.6	2	2.1	7	8.6	CRITTENDEN AR	254	313	1363	99	86
1	3.7					CROSS AR					
				1	5.6	DALLAS AR					77
1	2.1	2	4.3	2	3.0	DESHA AR	108	-31	44	46	70
1	3.5	1	3.6	4	15.5	DREW AR	3	328	342	88	94
				1	6.5	FAULKNER AR					80
2	7.3	1	3.8	3	12.6	GARLAND AR	-47	228	73	54	92
1*	38.2					GRANT AR					
				3	7.7	HEMPSTEAD AR					83

* SIGNIFICANTLY MORE DEATHS THAN EXPECTED

NON-WHITE MALE

CANCER OF THE LIVER AND GALLBLADDER INCLUDING BILE DUCTS

ICD 155,156

1950-1959 DEATHS	1950-1959 RATE	1960-1969 DEATHS	1960-1969 RATE	1970-1979 DEATHS	1970-1979 RATE	COUNTY AND STATE	PERCENT CHANGE 60S/50S	PERCENT CHANGE 70S/60S	PERCENT CHANGE 70S/50S	PERCENTILE %CHANGE 70S/50S	PERCENTILE RATES 1970S
		1	8.4			HOT SPRING AR					
		1	7.0	1	1.8	HOWARD AR		-75			68
		1	3.9	2	9.0	JACKSON AR		133			86
2	1.1	7	4.7	9	5.6	JEFFERSON AR	306	20	388	90	77
				1*	119.2	JOHNSON AR					99
		3	17.6			LAFAYETTE AR					
2	3.1			2	2.9	LEE AR			-7	25	70
1	2.8					LINCOLN AR					
		1	7.2	1	5.0	LITTLE RIVER AR		-31			76
		1	3.3	1	3.7	LONOKE AR		12			72
1	2.6			1	2.3	MILLER AR			-13	23	68
1	1.2	2	2.1	2	2.3	MISSISSIPPI AR	68	8	82	58	68
2	4.8	2	7.1	2	4.9	MONROE AR	48	-32	1	30	75
				4*	16.8	NEVADA AR					95
		3	5.6	3	4.7	OUACHITA AR		-17			75
5	4.3	1	.7	3	2.6	PHILLIPS AR	-84	263	-41	14	69
		1	7.4			POINSETT AR					
				1	9.9	PRAIRIE AR					88
7	3.1	13	5.8	16	6.3	PULASKI AR	88	8	102	62	80
		1*	53.6			RANDOLPH AR					
4	4.8	2	3.1	1	1.2	ST FRANCIS AR	-36	-61	-75	03	67
2*	8.4	1	4.3			SEBASTIAN AR	-49				
				1	14.9	SEVIER AR					94
3	4.5	3	6.0	1	1.3	UNION AR	32	-78	-71	03	67
1*	26.5	1	22.7			WHITE AR	-14				
				1	4.0	WOODRUFF AR					73
		1*	37.2	1*	56.2	YELL AR		51			99
175*	5.7	418*	7.9	697	6.7	CALIFORNIA CA	38	-16	16		
9	3.4	33	6.6	73	8.0	ALAMEDA CA	94	22	136	68	84
		2*	17.0	4*	20.0	BUTTE CA		18			96
3	2.9	6	7.2	15	6.1	CONTRA COSTA CA	145	-15	108	64	79
6	6.4	11	8.2	10	3.5	FRESNO CA	28	-57	-45	12	71
				1	4.1	HUMBOLDT CA					73
		3	10.2	2	2.2	IMPERIAL CA		-79			68
		1*	21.2			INYO CA					
4	6.7	7	6.9	11	6.3	KERN CA	3	-9	-7	26	80
		1	5.3	1	1.9	KINGS CA		-64			68
74*	5.4	145	6.4	264	6.0	LOS ANGELES CA	17	-6	11	34	79
1	5.7	1	4.9			MADERA CA	-14				
		2	21.5	1	4.9	MARIN CA		-77			75
1	16.7	1	10.6			MENDOCINO CA	-37				
2	9.8	3	10.9	2	3.7	MERCED CA	11	-66	-62	05	72
2	7.3	5	6.3	6	4.7	MONTEREY CA	-14	-26	-36	15	75
		1	10.9			NAPA CA					
				1*	25.2	NEVADA CA					97
1	4.4			11	3.7	ORANGE CA			-16	22	72
1	1.6	5	5.8	12	6.2	RIVERSIDE CA	256	6	277	84	79
10*	12.0	19*	11.2	25	8.2	SACRAMENTO CA	-7	-27	-32	16	85
1	2.9	3	3.6	12	5.3	SAN BERNARDINO CA	25	44	80	57	76
4	4.8	6	4.0	18	4.5	SAN DIEGO CA	-16	12	-6	26	74
32*	8.0	119*	18.7	143*	13.8	SAN FRANCISCO CA	135	-26	73	54	93
4	6.0	14	7.1	28*	9.8	SAN JOAQUIN CA	17	38	63	51	88
				1	2.9	SAN LUIS OBISPO CA					70
3	9.5	4	5.1	11	5.3	SAN MATEO CA	-47	5	-44	13	77
3*	17.8	2	7.3	3	2.5	SANTA BARBARA CA	-59	-66	-86	01	69
6*	11.0	13*	15.0	19	6.1	SANTA CLARA CA	36	-60	-45	12	79
		3	11.5	2	2.8	SANTA CRUZ CA		-75			70
				2	14.4	SHASTA CA					94
1*	11.9					SISKIYOU CA					
1	4.6	2	6.1	8	7.9	SOLANO CA	33	31	73	54	84
		1	6.1	3	10.9	SONOMA CA		78			90
				1	2.3	STANISLAUS CA					69
		1*	68.3			TEHAMA CA					
3	8.0	1	2.1	3	2.7	TULARE CA	-73	26	-66	04	69
1	6.8	2	4.6	2	1.4	VENTURA CA	-33	-69	-79	02	67
1	5.2			1	4.1	YOLO CA			-20	20	73
1*	13.3	1	6.4	1	5.7	YUBA CA	-52	-11	-57	08	78
10*	7.1	9	3.3	19	3.9	COLORADO CO	-53	18	-45		
1*	20.6	1	19.1			ADAMS CO	-7				
7*	8.6	6	3.3	15	5.7	DENVER-ARAPAHOE#CO	-61	71	-34	15	78
2*	15.8	2	3.4			EL PASO CO	-78				
				1	10.3	LA PLATA CO					89
				1	9.5	LAPIMER CO					88
				2	13.4	PUEBLO CO					93
7	3.5	20	5.0	26	4.6	CONNECTICUT CT	44	-9	32		
1	3.8	4	3.0	6	2.9	FAIRFIELD CT	-21	-4	-24	18	70
		4	3.2	8	4.5	HARTFORD CT		42			74
1*	18.2	1	12.1			MIDDLESEX CT	-33				
4	4.9	9	8.1	10	6.4	NEW HAVEN CT	64	-21	30	42	80
		2	10.0	1	5.1	NEW LONDON CT		-49			76
1*	53.6			1	30.3	WINDHAM CT			-43	13	98
5	2.3	16	6.3	20	6.6	DELAWARE DE	178	4	191		
1	2.9	4	10.2	2	4.1	KENT DE	257	-60	43	45	73
4	2.9	12	7.7	15	8.0	NEW CASTLE DE	162	4	171	74	84
				3	4.4	SUSSEX DE					74
40	3.6	115*	7.9	213*	12.5	DIST. OF COLUMBIA DC	121	59	251		
40	3.6	115*	7.9	213*	12.5	WASHINGTON DC	121	59	251	81	92
58	2.0	135	3.9	262	6.2	FLORIDA FL	91	60	206		

* SIGNIFICANTLY MORE DEATHS THAN EXPECTED

COUNTIES COMBINED; SEE TECHNICAL NOTES

NON-WHITE MALE

CANCER OF THE LIVER AND GALLBLADDER INCLUDING BILE DUCTS

ICD 155,156

1950-1959 DEATHS	1950-1959 RATE	1960-1969 DEATHS	1960-1969 RATE	1970-1979 DEATHS	1970-1979 RATE	COUNTY AND STATE	PERCENT CHANGE 60S/50S	PERCENT CHANGE 70S/60S	PERCENT CHANGE 70S/50S	PERCENTILE %CHANGE 70S/50S	PERCENTILE RATES 1970S
1	1.4	3	3.7	4	5.2	ALACHUA FL	162	40	266	83	76
		1	13.2			BAKER FL					
2	6.2	1	4.2	1	2.6	BAY FL	-33	-38	-59	07	69
		1	8.3	3*	17.3	BRADFORD FL		110			95
2	6.7	1	2.0	3	3.9	BREVARD FL	-70	98	-41	14	72
1	.6	9	4.6	18	6.8	BROWARD FL	664	48	1034	99	81
				1	24.8	CHARLOTTE FL					97
		1	14.8			CITRUS FL					
				2	15.5	COLLIER FL					94
11	2.6	34*	8.2	46	6.0	DADE FL	219	-27	133	67	79
14	3.9	15	3.4	34	7.0	DUVAL FL	-13	107	79	56	81
		2	1.7	9	5.6	ESCAMBIA FL		225			77
1*	17.5			1	10.6	FRANKLIN FL			-39	14	90
		2	2.4	6	6.1	GADSDEN FL		156			79
		1	7.3	1	9.6	GULF FL		31			88
		1	8.7			HAMILTON FL					
		1	15.0			HARDEE FL					
				1	6.0	HENDRY FL					79
				2	20.4	HERNANDO FL					96
1	6.3	1	3.5	5*	16.9	HIGHLANDS FL	-44	386	170	73	95
4	1.6	8	3.0	9	2.9	HILLSBOROUGH FL	80	-1	77	56	70
				1*	49.1	HOLMES FL					99
		1	2.3	7*	14.2	JACKSON FL		522			93
		1	5.6			JEFFERSON FL					
		1*	70.3			LAFAYETTE FL					
2	5.9	2	3.5	3	5.9	LAKE FL	-40	66	-1	29	78
				2	4.1	LEE FL					73
1	.9	4	4.8	6	7.4	LEON FL	461	53	761	98	82
				2	13.5	LEVY FL					93
		1	4.5	1	4.9	MADISON FL		7			75
		1	2.0	6	12.3	MANATEE FL		514			92
1	1.3	1	1.5	8	9.7	MARION FL	14	550	642	97	88
				2	11.3	MARTIN FL					91
		1	8.4	4*	32.3	MONROE FL		286			98
1	5.8					NASSAU FL					
				2	11.3	OKALOOSA FL					91
				2*	21.4	OKEECHOBEE FL					96
2	2.1	4	2.1	11	5.8	ORANGE FL	0	181	180	74	78
		1	6.6			OSCEOLA FL					
2	2.4	7	3.3	6	2.3	PALM BEACH FL	35	-30	-6	26	69
				2	9.0	PASCO FL					86
5	5.4	5	3.8	14	8.6	PINELLAS FL	-31	129	58	49	86
1	.6	8	5.0	9	5.6	POLK FL	732	11	820	98	77
		1	2.2	3	6.7	PUTNAM FL		207			81
1	2.7					ST JOHNS FL					
		1	3.4	3	6.2	ST LUCIE FL		85			80
				2*	26.5	SANTA ROSA FL					97
1	3.4	3	10.4	4	14.1	SARASOTA FL	205	36	315	87	93
		3	5.6	1	1.3	SEMINOLE FL		-78			67
				2	12.2	SUMTER FL					92
1	3.7			2	12.9	SUWANNEE FL			250	81	92
		2	12.1	1	5.7	TAYLOR FL		-53			78
3	4.1	4	4.2	9	9.2	VOLUSIA FL	3	122	128	67	87
		1	20.1			WAKULLA FL					
				1	11.5	WASHINGTON FL					91
123	3.4	115	2.9	192	4.4	GEORGIA GA	-16	51	28		
				1	8.2	APPLING GA					85
				2*	35.8	ATKINSON GA					98
1*	24.6					BACON GA					
				1	15.1	BAKER GA					94
3	6.3					BALDWIN GA					
				1	10.2	BARTOW GA					89
5	3.6	7	4.0	12	6.5	BIBB GA	13	61	82	58	80
1	3.4			2	7.4	BROOKS GA			115	65	82
		1	14.0			BRYAN GA					
		1	4.0	1	3.3	BULLOCH GA		-19			71
				2	4.5	BURKE GA					74
1	6.8			1	5.0	CAMDEN GA			-27	16	76
		2	12.3	1	2.4	CARROLL GA		-80			69
				1*	106.0	CATOOSA GA					99
		1	13.8	1	13.3	CHARLTON GA		-4			93
7	2.9	7	2.7	22	8.5	CHATHAM GA	-6	210	191	76	85
2	5.3			1	2.0	CLARKE GA			-63	05	68
2*	17.7			1	9.6	CLAY GA			-46	12	88
				1	3.0	CLAYTON GA					70
		2	7.8	2	5.6	COBB GA		-28			78
				1	8.3	COFFEE GA					85
1	4.8			2	8.4	COLQUITT GA			73	54	85
1	6.7					COLUMBIA GA					
1	7.7					COOK GA					
1	2.8					COWETA GA					
				2	21.2	CRAWFORD GA					96
				1	4.1	CRISP GA					73
1	2.3	3	8.8	3	7.9	DECATUR GA	287	-10	247	80	84
		2	1.2	19	7.4	DE KALB GA		518			83
2*	11.5					DODGE GA					
2*	11.7			3	14.3	DOOLY GA			23	39	94
1	1.2	3	4.0	8	8.2	DOUGHERTY GA	233	103	576	96	85
		2	10.3	2	8.6	EARLY GA		-16			86
1	2.6	1	4.0	1	3.4	ELBERT GA	54	-13	34	43	71
2*	23.7	2*	22.8	1	9.9	EVANS GA	-4	-57	-58	07	88
		1	2.7	1	3.1	FLOYD GA		16			70
33*	6.6	34	4.9	27	3.4	FULTON GA	-25	-32	-49	10	71
2	3.3	1	1.9			GLYNN GA	-42				
		1*	19.2			GORDON GA					
2*	6.0					GRADY GA					
1	4.3					GREENE GA					
1	9.4	2*	17.6	1	7.9	GWINNETT GA	87	-55	-16	21	84

* SIGNIFICANTLY MORE DEATHS THAN EXPECTED

NON-WHITE MALE

CANCER OF THE LIVER AND GALLBLADDER INCLUDING BILE DUCTS

ICD 155,156

1950-1959 DEATHS	1950-1959 RATE	1960-1969 DEATHS	1960-1969 RATE	1970-1979 DEATHS	1970-1979 RATE	COUNTY AND STATE	PERCENT CHANGE 60S/50S	PERCENT CHANGE 70S/60S	PERCENT CHANGE 70S/50S	PERCENTILE %CHANGE 70S/50S	PERCENTILE RATES 1970S
				1	3.4	HANCOCK GA					71
1	4.6	1	5.4			HARRIS GA	16				
				1	4.4	HENRY GA					74
		1	3.4	1	1.7	HOUSTON GA		-50			67
		1	9.6			IRWIN GA					
2*	21.7			1	13.4	JACKSON GA			-38	14	93
1	8.5					JASPER GA					
				3	8.3	JEFFERSON GA					85
2*	8.7	1	6.3	2	8.5	JENKINS GA	-28	36	-2	28	85
1	14.9					JOHNSON GA					
1	10.1					JONES GA					
				2	12.7	LAMAR GA					92
2	4.3	1	2.8	2	4.6	LAURENS GA	-33	63	9	33	75
1	4.0	1	4.8	2	10.7	LIBERTY GA	21	123	169	73	90
1	15.6					LINCOLN GA					
3	5.6	4	7.0	3	5.2	LOWNDES GA	25	-25	-7	26	76
1*	490.8					LUMPKIN GA					
				1	2.8	MC DUFFIE GA					70
		1	3.1	1	4.6	MACON GA		46			74
				1	14.3	MADISON GA					94
1	7.9					MARION GA					
1	3.3	1	1.4	2	5.7	MERIWETHER GA	-57	303	73	54	78
		1	3.2	2	5.7	MITCHELL GA		76			78
		1	6.8	2	11.5	MORGAN GA		70			91
4	4.5	6	4.4	3	1.9	MUSCOGEE#GA	-2	-58	-59	06	68
1	6.3	3	11.1			NEWTON GA	75				
				1	9.4	OGLETHORPE GA					87
				1	3.5	PEACH GA					71
1	8.5	1	10.7			PIKE GA	26				
2*	15.0			1	9.4	POLK GA			-38	15	87
		1	7.3	1	6.6	PULASKI GA		-10			81
				1	5.8	PUTNAM GA					78
4	3.2	8	5.4	10	5.6	RICHMOND GA	65	4	72	53	77
1*	14.1					SCHLEY GA					
1	3.0	1	5.4	1	4.5	SCREVEN GA	79	-17	48	47	74
2	6.7			2	5.3	SPALDING GA			-21	20	77
1	2.2	2	5.1			SUMTER GA	130				
1	7.0					TALBOT GA					
1	6.1			1	7.0	TATTNALL GA			15	37	81
1	10.1	1	7.2	2	13.2	TELFAIR GA	-28	82	31	42	93
				2	7.1	TERRELL GA					82
		1	1.7	6	11.6	THOMAS GA		565			91
		1	4.0			TIFT GA					
1*	9.2			1	22.9	TREUTLEN GA			149	69	97
1	1.6	1	2.0	1	1.8	TROUP GA	30	-14	12	35	68
		1	10.3			TURNER GA					
1	10.9					WALKER GA					
2*	10.5					WALTON GA					
				1	3.4	WARE GA					71
1	6.1			1	6.0	WARREN GA			-1	29	79
1*	19.1			1	6.6	WHITFIELD GA			-66	05	81
2*	17.9			1	9.6	WILCOX GA			-46	11	88
		1	5.0	2	8.6	WILKES GA		71			86
2	5.5					WORTH GA					
		189*	11.1	232*	9.6	HAWAII HI		-13			
		27*	10.4	22	7.3	HAWAII HI		-30			82
		126*	11.0	171*	9.9	HONOLULU HI		-10			88
		16*	12.8	15	9.1	KAUAI HI		-28			87
		20*	11.0	24	10.9	MAUI#HI		-1			90
		4	10.5	1	.4	IDAHO ID		-96			
		1	16.6			BANNOCK ID					
		1	16.3			BINGHAM ID					
		1*	28.1	1*	11.3	BONNEVILLE ID		-60			91
		1*	89.3			TWIN FALLS ID					
107	3.6	266*	6.6	515*	9.9	ILLINOIS IL	84	49	174		
1*	13.8			3*	39.3	ADAMS IL			185	75	98
		1	3.0			ALEXANDER IL					
		2	9.7	3	13.0	CHAMPAIGN IL		34			92
94*	4.1	236*	7.3	442*	10.4	COOK IL	79	42	154	70	89
				1*	267.9	CUMBERLAND IL					99
				3	7.3	DU PAGE IL					82
		1	5.1	2	10.0	JACKSON IL		97			89
				1	21.2	JEFFERSON IL					96
		1	3.5	1	6.1	KANE IL		72			79
		2	4.1	3	5.4	KANKAKEE IL		32			77
1	5.3	1	3.9	7	7.5	LAKE IL	-26	93	43	46	83
		1*	26.5			LA SALLE IL					
		1*	68.8			LAWRENCE IL					
		1	56.2	1	17.5	MC LEAN IL		-69			95
		2	8.4	2	8.3	MACON IL		-1			85
		4	7.6	3	5.2	MADISON IL		-31			76
				2*	41.2	MORGAN IL					98
3*	10.8	1	3.3	2	4.6	PEORIA IL	-70	40	-58	08	74
1	3.2	1	4.1	3	11.5	PULASKI IL	31	180	266	82	91
2*	17.8			2*	37.0	RANDOLPH IL			108	64	98
1	5.5	1	6.7	3	6.3	ROCK ISLAND IL	22	-6	15	36	80
1	1.0	2	1.1	18	7.9	ST CLAIR IL	9	611	673	98	84
				1	9.3	SALINE IL					87
1	3.8	1	3.1	2	4.9	SANGAMON IL	-19	59	28	41	75
				1	12.1	STEPHENSON IL					91
		1*	45.5			UNION IL					
		1	3.8	2	8.2	VERMILION IL		115			85
				1*	56.2	WARREN IL					99
		1*	78.1			WHITESIDE IL					

* SIGNIFICANTLY MORE DEATHS THAN EXPECTED

COUNTIES COMBINED; SEE TECHNICAL NOTES

NON-WHITE MALE

CANCER OF THE LIVER AND GALLBLADDER INCLUDING BILE DUCTS

ICD 155.156

1950-1959 DEATHS	1950-1959 RATE	1960-1969 DEATHS	1960-1969 RATE	1970-1979 DEATHS	1970-1979 RATE	COUNTY AND STATE	PERCENT CHANGE 60S/50S	PERCENT CHANGE 70S/60S	PERCENT CHANGE 70S/50S	PERCENTILE %CHANGE 70S/50S	PERCENTILE RATES 1970S
		1	2.9	1	1.5	WILL IL		-49			67
2*	9.9	3	7.7	5	9.8	WINNEBAGO IL	-22	27	-1	29	88
21	2.2	60	5.2	103	7.6	INDIANA IN	139	46	250		
2	7.3	3	6.3	7	9.3	ALLEN IN	-14	48	28	41	87
1	8.4			1	6.5	CLARK IN			-22	19	80
		2	7.0	3	10.2	DELAWARE IN		46			89
1*	12.4			2	9.1	ELKHART IN			-27	17	87
		2*	20.6	1	13.0	FLOYD IN		-37			92
1*	47.1					HENRY IN					
				1	9.6	HOWARD IN					88
		1*	54.7			JOHNSON IN					
4	1.6	12	3.7	26	6.5	LAKE IN	138	75	317	87	80
				2	8.7	LA PORTE IN					86
		2	12.7	4*	19.2	MADISON IN		52			96
10	2.5	30	6.7	43	8.5	MARION IN	163	28	235	78	86
		1*	37.7			MONROE IN					
1	3.1	5	8.5	8	11.2	ST JOSEPH IN	170	32	257	81	90
		1*	29.8			TIPPECANOE IN					
				4	8.3	VANDERBURGH IN					85
1	2.9	1	3.6	1	4.4	VIGO IN	25	20	51	48	74
		3	2.4	13	7.8	IOWA IA		221			
				1*	117.9	APPANOOSE IA					99
				3	12.6	BLACK HAWK IA					92
				1*	13.9	DUBUQUE IA					93
		1*	45.4			JOHNSON IA					
				1	15.0	LEE IA					94
		2	4.5	4	7.6	POLK IA		67			83
				3*	17.7	SCOTT IA					95
9	2.5	9	2.1	37	7.6	KANSAS KS	-17	270	206		
				1*	37.7	ALLEN KS					98
		1	12.0	2	17.6	DOUGLAS KS		47			95
1*	22.5					GEARY KS					
				2	23.1	JOHNSON KS					97
				2*	31.3	LABETTE KS					98
3*	12.6	1	3.6	5*	22.7	LEAVENWORTH KS	-71	528	81	57	97
1*	141.3					LINN KS					
		1*	31.5			LYON KS					
				1*	18.7	MIAMI KS					96
1	5.1	1	4.5	3	12.8	MONTGOMERY KS	-13	187	150	69	92
1*	22.3					RENO KS					
		1	1.2	5	4.8	SEDGWICK KS		298			75
1	2.5			2	3.4	SHAWNEE KS			37	44	71
1	1.1	4	2.8	14	8.7	WYANDOTTE KS	144	214	665	97	86
20	2.0	37	3.5	66	6.3	KENTUCKY KY	78	81	221		
		1	22.9			BELL KY					
				2	13.0	BOYLE KY					92
		1	14.5	1	26.5	CALDWELL KY		82			97
1	1.1			1	2.6	CHRISTIAN KY			138	68	69
				1	8.0	CLARK KY					84
		1*	67.0			CRITTENDEN KY					
1*	112.4					CUMBERLAND KY					
1	4.7	1	6.6			DAVIESS KY	41				
2	2.0	6	6.3	11	10.8	FAYETTE KY	212	70	430	92	90
				1	7.6	GRAVES KY					83
		2*	21.2	1	.4	HARDIN KY		-98			67
				1	19.7	HARLAN KY					96
		2	9.4	1	9.8	HENDERSON KY		4			88
		2	9.8	3	20.3	HOPKINS KY		107			96
11	3.3	16	4.3	27	6.8	JEFFERSON KY	31	58	107	63	81
				1	24.5	JESSAMINE KY					97
1	4.4			2	11.4	KENTON KY			162	72	91
				1	17.2	LINCOLN KY					95
				1	3.0	MC CRACKEN KY					70
				2	12.4	MADISON KY					92
		1	24.8	1	21.0	MARION KY		-16			96
		1*	47.2			OHIO KY					
1*	70.3					OWEN KY					
1*	14.9	1*	19.1			PERRY KY	29				
1	10.0	1	9.6			SCOTT KY	-4				
		1	9.6			SHELBY KY					
				2	25.7	TODD KY					97
				1	12.4	UNION KY					92
				4*	17.7	WARREN KY					95
				1	15.0	WOODFORD KY					94
80	2.3	144	3.7	267	6.4	LOUISIANA LA	63	73	183		
1	4.2	2	5.3	2	6.7	ACADIA LA	28	25	60	50	81
1	5.1			1	5.1	ALLEN LA			1	30	76
		2	5.5	1	2.6	ASCENSION LA		-52			69
				1	3.9	ASSUMPTION LA					72
		2	5.1	2	5.5	AVOYELLES LA		8			77
				1	6.0	BEAUREGARD LA					79
				3	9.2	BIENVILLE LA					87
3	5.8	3	5.0	1	1.4	BOSSIER LA	-14	-72	-76	03	67
15*	5.7	16	5.1	13	3.9	CADDO LA	-11	-22	-31	16	72
2	2.4	1	.8	6	5.4	CALCASIEU LA	-68	610	126	66	77
		1	9.1			CALDWELL LA					
				1	6.3	CATAHOULA LA					80
1	1.7	3	5.6	3	7.2	CLAIBORNE LA	241	28	338	88	82
1	3.0					CONCOPDIA LA					
1	1.2			2	2.1	DE SOTO LA			73	55	68

* SIGNIFICANTLY MORE DEATHS THAN EXPECTED

NON-WHITE MALE

CANCER OF THE LIVER AND GALLBLADDER INCLUDING BILE DUCTS

ICD 155,156

1950-1959 DEATHS	1950-1959 RATE	1960-1969 DEATHS	1960-1969 RATE	1970-1979 DEATHS	1970-1979 RATE	COUNTY AND STATE	PERCENT CHANGE 60S/50S	PERCENT CHANGE 70S/60S	PERCENT CHANGE 70S/50S	PERCENTILE %CHANGE 70S/50S	PERCENTILE RATES 1970S
2	1.0	7	3.2	16	5.3	EAST BATON ROUGE LA	225	69	447	93	77
		1	2.2	1	2.0	EAST CARROLL LA		-10			68
				1	2.6	EAST FELICIANA LA					69
		2	10.2	2	8.1	EVANGELINE LA		-21			84
1	6.6			1	10.0	GRANT LA			51	48	69
		2	3.3	7	12.1	IBERIA LA		262			91
2	2.8	3	5.8	1	2.1	IBERVILLE LA	103	-63	-25	17	68
		3	16.0	1	4.0	JACKSON LA		-75			73
		6	4.7	9	5.5	JEFFERSON LA		19			77
		2	10.9			JEFFERSON DAVIS LA					
1	1.9	1	2.1	3	4.0	LAFAYETTE LA	8	92	107	64	73
		3*	15.4	3	13.6	LAFOURCHE LA		-12			93
				1	8.9	LA SALLE LA					86
				4	8.0	LINCOLN LA					84
1	2.6	1	1.5	1	4.4	MADISON LA	-41	194	74	55	74
		1	1.7	2	3.7	MOREHOUSE LA		113			72
3	5.3	2	4.2	3	5.1	NATCHITOCHES LA	-20	21	-3	27	76
24	3.6	45	5.4	73	7.6	ORLEANS LA	53	39	112	64	83
1	1.0	4	3.3	5	4.2	OUACHITA LA	236	25	320	87	73
				4*	20.5	PLAQUEMINES LA					96
		1	3.1	4	9.4	POINTE COUPEE LA		204			87
2	1.4	3	2.3	17*	12.1	RAPIDES LA	57	437	744	98	92
				1	6.1	RED RIVER LA					79
		1	4.2	1	4.7	RICHLAND LA		12			75
1	5.3			2	9.7	SABINE LA			83	58	88
		1	9.6	3*	30.5	ST BERNARD LA		216			98
1	4.8			1	4.1	ST CHARLES LA			-15	22	73
				1	6.3	ST HELENA LA					80
		1	3.7	3	10.2	ST JAMES LA		175			89
1	2.8			7*	17.8	ST JOHN THE BAPT LA			540	96	95
3	2.6	6	6.2	9	7.5	ST LANDRY LA	134	22	185	75	83
		1	2.6	3	7.6	ST MARTIN LA		198			83
		1	2.0	6	9.3	ST MARY LA		378			87
2	5.5	4	11.1	4	8.6	ST TAMMANY LA	103	-22	58	49	86
1	1.4	3	3.9	6	7.7	TANGIPAHOA LA	168	99	432	92	83
		3	10.6	2	5.6	TENSAS LA		-47			77
1	2.8			4	11.1	TERREBONNE LA			295	86	90
		1	4.0	1	5.1	UNION LA		26			76
				1	4.7	VERMILION LA					75
				1	8.1	VERNON LA					84
2	3.9	2	4.2	1	1.9	WASHINGTON LA	8	-56	-52	09	68
3	5.4			9*	14.0	WEBSTER LA			157	71	93
2	6.8	2	6.3			WEST BATON ROUGE LA	-8				
1	7.0			3*	25.7	WEST CARROLL LA			264	82	97
		1	3.5			WEST FELICIANA LA					
				2	10.0	WINN LA					89
		2	13.4	2	6.3	MAINE ME		-53			
				1*	78.1	ANDROSCOGGIN ME					99
		1*	61.3			AROOSTOOK ME					
		1*	62.9			WASHINGTON ME					
				1*	42.0	YORK ME					99
33	2.0	112	5.5	240*	9.0	MARYLAND MD	181	63	358		
4	4.0	3	2.5	10	7.6	ANNE ARUNDEL MD	-38	210	92	60	83
2	3.5	1	1.4	8	7.9	BALTIMORE MD	-60	463	125	66	84
		1	5.3	2	5.8	CALVERT MD		8			78
				1	4.5	CAROLINE MD					74
				2	14.9	CARROLL MD					94
		2	13.6	1	2.7	CECIL MD		-81			69
				3	7.2	CHARLES MD					82
		2	4.9	1	2.0	DORCHESTER MD		-59			68
1	4.1	2	12.8	1	4.0	FREDERICK MD	213	-69	-3	27	72
				1	2.3	HARFORD MD					69
				1	1.4	HOWARD MD					67
				1	5.6	KENT MD					77
1	2.5	2	2.3	5	4.4	MONTGOMERY MD	-9	95	77	56	74
2	1.9	8	6.0	24	9.9	PRINCE GEORGE'S MD	216	65	420	92	88
		2	10.2	1	3.5	QUEEN ANNE'S MD		-65			71
		1	3.9	3	9.6	ST MARY'S MD		147			88
1	2.9			4	16.2	SOMERSET MD			457	94	95
1	6.6			1	2.9	TALBOT MD			-56	09	70
		1	10.7			WASHINGTON MD					
3	7.0	2	4.2	3	5.7	WICOMICO MD	-39	33	-19	20	78
		2	6.5	4	11.5	WORCESTER MD		76			91
18	1.8	83*	6.9	163*	11.0	BALTIMORE CITY MD	284	58	507	95	90
23*	5.6	38	7.7	65	8.4	MASSACHUSETTS MA	37	9	49		
				2	21.4	BERKSHIRE MA					96
1	3.9	2	7.6	3	6.1	BRISTOL MA	98	-20	58	49	79
		1*	26.8			DUKES MA					
1	9.3	3*	24.3	1	3.9	ESSEX MA	161	-84	-58	07	72
1	4.4	1	2.9	4	3.9	HAMPDEN MA	-34	34	-12	24	72
3	5.9	3	4.2	5	4.4	MIDDLESEX MA	-28	3	-26	17	74
		1	11.2	2	1.5	NORFOLK MA		-87			67
				3	10.4	PLYMOUTH MA					89
17*	7.4	27*	10.2	45*	12.1	SUFFOLK MA	39	18	65	52	91
67	3.4	135	4.6	265	7.1	MICHIGAN MI	36	55	111		
		1	19.7			ALLEGAN MI					
2	8.7			7	10.3	BERRIEN MI			18	38	89
1	3.7	7*	24.2	5	14.5	CALHOUN MI	560	-40	295	86	94
				4	13.6	CASS MI					93
4	5.3	3	2.1	9	5.7	GENESEE MI	-61	178	8	33	78
				1*	74.3	GRAND TRAVERSE MI					99
1	9.9			4	9.6	INGHAM MI			-3	27	88

* SIGNIFICANTLY MORE DEATHS THAN EXPECTED

NON-WHITE MALE

CANCER OF THE LIVER AND GALLBLADDER INCLUDING BILE DUCTS

ICD 155,156

1950-1959 DEATHS	1950-1959 RATE	1960-1969 DEATHS	1960-1969 RATE	1970-1979 DEATHS	1970-1979 RATE	COUNTY AND STATE	PERCENT CHANGE 60S/50S	PERCENT CHANGE 70S/60S	PERCENT CHANGE 70S/50S	PERCENTILE %CHANGE 70S/50S	PERCENTILE RATES 1970S
		2	8.5	1	2.2	JACKSON MI		-75			68
				3	12.1	KALAMAZOO MI					91
3*	5.8	4	7.0	4	7.9	KENT MI	21	12	36	44	84
				3*	22.1	LAKE MI					96
2*	11.8	2	9.0	1	2.1	MACOMB MI	-24	-77	-82	01	68
				1*	70.7	MECOSTA MI					99
				2	19.8	MONROE MI					96
1	5.2	3	5.8	6	8.5	MUSKEGON MI	11	46	62	50	85
		3	3.0	7	6.1	OAKLAND MI		103			79
				1*	74.3	OCEANA MI					99
1	2.1	6	10.4	6	6.5	SAGINAW MI	388	-37	206	77	80
				1*	112.4	SHIAWASSEE MI					99
				1	4.5	VAN BUREN MI					74
1	1.0	4	9.0	5	7.3	WASHTENAW MI	760	-18	602	97	82
51	3.5	99	4.5	193	6.9	WAYNE MI	30	53	99	61	81
		1*	267.9			WEXFORD MI					
6	4.8	8	4.9	17	6.9	MINNESOTA MN	2	40	42		
1*	490.8					AITKIN MN					
1*	188.7			1	15.6	ANOKA MN			-92	00	94
				1	17.2	CASS MN					95
				1*	33.5	CLEARWATER MN					98
		1*	84.8	2*	9.9	DAKOTA MN		-88			89
		1*	212.0			GOODHUE MN					
3	5.4	4	6.3	6	5.6	HENNEPIN MN	16	-11	3	31	77
				1*	267.9	NOBLES MN					99
1	2.6			5	9.6	RAMSEY MN			275	83	88
		2*	43.2			ST LOUIS MN					
35	1.0	85	2.7	200	5.9	MISSISSIPPI MS	162	120	476		
		3	4.6	3	3.6	ADAMS MS		-22			71
		1	5.1			ALCORN MS					
				2	6.1	AMITE MS					79
				1	12.3	BENTON MS					92
2	1.4	2	1.5	7	5.8	BOLIVAR MS	2	295	303	86	78
		2	7.0			CARROLL MS					
				1	6.8	CHOCTAW MS					81
				3	9.2	CLARKE MS					87
		2	5.0	3	7.7	CLAY MS		55			83
2	1.6	5	4.2	4	4.3	COAHOMA MS	167	1	169	73	73
				7	13.2	COPIAH MS					93
		1	3.5	1	4.7	COVINGTON MS		33			75
				1	1.9	DE SOTO MS					68
		5	8.0	3	5.7	FORREST MS		-29			78
		1	6.5	2	12.2	FRANKLIN MS		87			92
				1	11.2	GREENE MS					90
		1	2.8	1	2.3	GRENADA MS		-17			69
		1	10.7	1	7.7	HANCOCK MS		-29			83
1	1.9	4	6.4	4	4.6	HARRISON MS	229	-28	137	68	75
6	3.0	10	3.8	30	9.1	HINDS MS	27	136	201	77	87
1	1.1			6	8.5	HOLMES MS			644	97	85
		2	6.0	3	9.4	HUMPHREYS MS		55			87
		1	12.6			ISSAQUENA MS					
				1	14.1	ITAWAMBA MS					93
1	3.3			3	5.3	JACKSON MS			58	49	76
		1	4.8	3	8.7	JEFFERSON DAVIS MS		80			86
				4	7.2	JONES MS					82
				1	3.1	KEMPER MS					70
				1	2.7	LAFAYETTE MS					69
1	12.3			1	10.4	LAMAR MS			-15	22	89
		3	3.9	3	2.9	LAUDERDALE MS		-26			70
				1	2.7	LEAKE MS					69
				3	7.5	LEE MS					83
2	1.8	4	4.9	9	10.1	LEFLORE MS	176	105	466	94	89
1	1.5	1	2.1	5	7.9	LOWNDES MS	40	271	418	91	84
		2	2.8	4	4.3	MADISON MS		51			73
		3	6.0	4	6.1	MARSHALL MS		3			79
				5	12.5	MONROE MS					92
1	4.8	2	7.5			MONTGOMERY MS	57				
		1	7.4	3	14.7	NESHOBA MS		98			94
		2	8.5	1	3.2	NEWTON MS		-62			70
2	4.0			2	6.5	NOXUBEE MS			64	51	80
				1	3.5	OKTIBBEHA MS					71
1	1.6	2	4.9	3	4.6	PANOLA MS	217	-7	196	76	75
				1	5.7	PEARL RIVER MS					78
				2	3.1	PIKE MS					70
1	6.3					PONTOTOC MS					
1	1.7	1	1.6	4	6.3	RANKIN MS	-3	284	271	83	80
				3	8.9	SCOTT MS					86
		1	6.0	1	4.2	SHARKEY MS		-30			73
1	3.6	1	3.2	1	5.2	SIMPSON MS	-10	62	46	47	76
				1	11.4	SMITH MS					91
				1	12.2	STONE MS					92
4	3.3	4	4.1	8	10.7	SUNFLOWER MS	22	161	219	78	90
				2	5.7	TALLAHATCHIE MS					78
				2	8.7	TATE MS					86
				1	35.3	TISHOMINGO MS					98
		1	2.2	6*	21.3	TUNICA MS		864			96
				1	4.5	WALTHALL MS					74
4	4.1	6	6.3	2	2.0	WARREN MS	53	-69	-52	09	68
3	1.9	4	2.7	15	9.3	WASHINGTON MS	43	244	391	90	87
		1	4.8	2	6.7	WAYNE MS		41			81
				3	9.2	WILKINSON MS					87
		1	3.5	1	3.8	WINSTON MS		8			72
		3	4.7	1	1.8	YAZOO MS		-62			68
54	3.6	105	6.0	186*	9.7	MISSOURI MO	66	62	169		

* SIGNIFICANTLY MORE DEATHS THAN EXPECTED

NON-WHITE MALE

CANCER OF THE LIVER AND GALLBLADDER INCLUDING BILE DUCTS

ICD 155,156

1950-1959 DEATHS	1950-1959 RATE	1960-1969 DEATHS	1960-1969 RATE	1970-1979 DEATHS	1970-1979 RATE	COUNTY AND STATE	PERCENT CHANGE 60S/50S	PERCENT CHANGE 70S/60S	PERCENT CHANGE 70S/50S	PERCENTILE %CHANGE 70S/50S	PERCENTILE RATES 1970S
1	8.9					AUDRAIN MO					
		1	6.4	4*	23.4	BOONE MO		265			97
1	4.6			1	5.9	BUCHANAN MO			27	40	78
				1	7.0	BUTLER MO					82
				1	12.5	CAPE GIRARDEAU MO					92
		1*	122.7			CASS MO					
		1	16.4			CHARITON MO					
		1*	546.4			CHRISTIAN MO					
				1	10.6	COLE MO					90
		1	12.7	1	11.5	DUNKLIN MO		-10			91
1	6.9					GREENE MO					
		1*	112.4			HOLT MO					
1*	21.3	1	10.3			HOWARD MO	-52				
15	4.7	21	5.4	46*	10.5	JACKSON MO	15	95	123	66	89
				1	15.6	JASPER MO					94
1*	26.5					JEFFERSON MO					
				1	12.7	MARION MO					92
2	9.4			1	12.0	MISSISSIPPI MO			27	41	91
		1*	212.0			MORGAN MO					
				5*	24.8	NEW MADRID MO					97
				2*	110.0	NEWTON MO					99
1	2.9	1	2.0	2	4.4	PEMISCOT MO	-33	123	50	47	74
1	8.4					PETTIS MO					
				1	10.6	PIKE MO					90
				2*	27.9	RANDOLPH MO					98
				1	12.6	ST CHARLES MO					92
4	5.7	5	5.6	10	5.6	ST LOUIS COUNTY MO	-2	1	-1	29	78
1	7.5					SALINE MO					
				1	6.9	SCOTT MO					81
25	3.4	70*	7.7	104*	11.3	ST LOUIS CITY MO	130	46	237	79	91
6*	8.8	2	2.8	4	5.0	MONTANA MT	-68	75	-44		
				1	11.2	BIG HORN MT					90
2*	18.1	1	12.6	1	15.7	GLACIER MT	-31	25	-13	23	94
1*	56.2					PONDERA MT					
1*	11.8			1	10.0	ROOSEVELT MT			-15	22	89
1*	14.9			1	53.6	ROSEBUD MT			260	82	99
1*	86.0	1*	53.0			SILVER BOW MT	-38				
12*	10.6	8	5.5	7	3.6	NEBRASKA NB	-48	-35	-66		
				1*	62.9	ADAMS NB					99
1*	188.7					BOX BUTTE NB					
7*	7.7	5	4.4	4	3.2	DOUGLAS NB	-42	-27	-58	07	71
1*	112.4					KNOX NB					
2*	19.3	1	11.5	1	1.6	LANCASTER NB	-41	-86	-92	00	67
		1*	67.0			MADISON NB					
				1*	10.3	SARPY NB					89
1*	16.3					THURSTON NB					
		1*	112.4			WASHINGTON NB					
2	3.8	5	7.8	13	7.9	NEVADA NV	104	1	106		
1	2.8	3	7.8	11	11.4	CLARK NV	177	45	302	86	91
		1	18.9			ELKO NV					
				1	28.1	MINERAL NV					98
1	13.7	1	5.7	1	1.9	WASHOE NV	-58	-66	-86	01	68
39	2.7	93	4.4	174	6.1	NEW JERSEY NJ	62	37	123		
4	3.0	4	2.9	10	7.1	ATLANTIC NJ	-2	146	140	69	82
4	7.1	4	4.3	9	6.8	BERGEN NJ	-39	56	-4	27	81
		2	4.0	2	1.1	BURLINGTON NJ		-72			67
5	4.3	7	4.8	14	7.4	CAMDEN NJ	12	52	70	53	82
1	6.7					CAPE MAY NJ					
1	1.7	1	2.7	8	10.4	CUMBERLAND NJ	61	281	511	95	89
9	2.5	31	4.7	52	5.4	ESSEX NJ	87	16	116	65	77
1	2.7	3	5.1	5	7.6	GLOUCESTER NJ	87	50	180	74	83
2	2.0	7	4.4	9	4.0	HUDSON NJ	116	-9	96	61	73
2	2.5	5	2.8	12	7.5	MERCER NJ	11	171	201	77	83
		1	1.1	6	6.4	MIDDLESEX NJ		512			80
6	5.3	5	3.4	15	9.4	MONMOUTH NJ	-37	179	75	56	87
		2	12.5	3	8.3	MORRIS NJ		-33			85
		1	5.3	3	11.3	OCEAN NJ		112			90
2	4.8	10*	9.8	10	7.6	PASSAIC NJ	104	-22	59	49	83
		2	6.4	2	5.0	SALEM NJ		-23			76
		1	5.9	2	10.1	SOMERSET NJ		71			89
2	1.2	7	3.6	11	4.7	UNION NJ	206	31	301	86	75
				1	14.5	WARREN NJ					94
7	3.9	18*	9.0	27	5.7	NEW MEXICO NM	133	-37	47		
1	6.5			3	5.9	BERNALILLO NM			-9	25	78
				2	11.4	EDDY NM					91
1	1.1	13*	26.7	9	11.3	MC KINLEY NM	2300	-58	911	99	90
1*	47.2					OTERO NM					
				1	2.4	RIO ARRIBA NM					69
1	5.7	1	4.2	5	17.7	SANDOVAL NM	-26	317	209	78	95
1	2.5	3	9.0	4	8.8	SAN JUAN NM	258	-2	251	81	86
1*	34.4					SANTA FE NM					
				1	11.8	TAOS NM					91
1	5.8	1	5.0	2	6.2	VALENCIA NM	-15	26	7	33	80
145	3.6	373*	6.4	690*	8.6	NEW YORK NY	80	34	140		
1	3.8	4	9.4	4	6.6	ALBANY NY	150	-30	74	55	81
12	5.2	47	6.3	85	8.2	BRONX NY	22	31	59	50	85
1*	13.6					BROOME NY					
1*	17.9					CATTARAUGUS NY					

* SIGNIFICANTLY MORE DEATHS THAN EXPECTED

NON-WHITE MALE | CANCER OF THE LIVER AND GALLBLADDER INCLUDING BILE DUCTS | ICD 155,156

1950-1959 DEATHS	1950-1959 RATE	1960-1969 DEATHS	1960-1969 RATE	1970-1979 DEATHS	1970-1979 RATE	COUNTY AND STATE		PERCENT CHANGE 60S/50S	PERCENT CHANGE 70S/60S	PERCENT CHANGE 70S/50S	PERCENTILE %CHANGE 70S/50S	PERCENTILE RATES 1970S
1*	19.6					CHAUTAUQUA	NY					
1*	16.1			2	21.5	CHEMUNG	NY			34	43	96
				1*	37.8	CHENANGO	NY					98
				1	1.7	DUTCHESS	NY					67
9	4.2	27*	10.9	40*	10.6	ERIE	NY	162	-3	155	70	90
		1	11.2			FRANKLIN	NY					
17	1.8	48	4.0	148	7.7	KINGS(BROOKLYN)	NY	123	95	336	88	84
				1	21.5	LIVINGSTON	NY					96
2	6.4	6	5.3	11	9.2	MONROE	NY	-18	75	43	46	87
5	4.3	8	5.2	14	5.4	NASSAU	NY	20	4	25	40	77
71*	4.3	173*	9.2	246*	11.8	NEW YORK(MANHATTAN)	NY	112	28	171	74	91
2*	8.5	2	7.1	2	4.3	NIAGARA	NY	-17	-39	-49	10	74
1	7.9	2	10.8	2	8.9	ONEIDA	NY	37	-17	13	35	86
		1	.8	2	2.6	ONONDAGA	NY		242			69
1	5.6	2	6.9	4	8.9	ORANGE	NY	23	30	59	50	86
		1*	15.8	4*	95.3	ORLEANS	NY		502			99
12	3.0	26	3.6	78	7.1	QUEENS	NY	20	96	134	67	82
		2*	26.7	1	11.2	RENSSELAER	NY		-58			90
1	2.0			6	9.9	RICHMOND	NY			389	90	88
				6	8.8	ROCKLAND	NY					86
				1	11.1	SARATOGA	NY					90
				2	12.1	SCHENECTADY	NY					92
2	1.2	6	3.1	10	4.6	SUFFOLK	NY	160	49	287	85	75
		1	20.2	4*	42.5	SULLIVAN	NY		110			99
		2	14.3	1	4.5	ULSTER	NY		-68			74
1*	30.7					WAYNE	NY					
4	2.4	14	6.2	14	4.0	WESTCHESTER	NY	152	-35	65	52	73
96	3.0	82	2.1	201	4.6	NORTH CAROLINA	NC	-29	115	52		
3	11.2	1	1.5	1	1.5	ALAMANCE	NC	-87	5	-86	00	67
1*	67.0					ALLEGHANY	NC					
1	2.5			1	2.5	ANSON	NC			-2	28	69
3	6.8	1	2.1			BEAUFORT	NC	-69				
1	2.2	2	4.7	3	6.4	BERTIE	NC	108	38	187	76	80
1	2.1			4	9.0	BLADEN	NC			331	88	86
		1	3.6			BRUNSWICK	NC					
3	4.3	1	1.3	2	3.1	BUNCOMBE	NC	-69	131	-28	16	70
1	3.6			1	1.9	CABARRUS	NC			-46	11	68
				1	7.4	CALDWELL	NC					83
				1	3.6	CASWELL	NC					71
				1	3.8	CATAWBA	NC					72
		1	3.7	1	2.8	CHATHAM	NC		-25			70
				1	3.9	CHOWAN	NC					72
1	2.4	1	1.6	1	2.9	COLUMBUS	NC	-32	75	19	38	70
1	1.7	1	1.4	4	6.2	CRAVEN	NC	-14	331	271	83	79
1	1.3	4	4.3	8	4.9	CUMBERLAND	NC	244	15	295	85	75
1	10.3	1	11.8	1	9.2	CURRITUCK	NC	15	-22	-10	24	87
				1*	45.5	DARE	NC					99
2	15.3	1	4.6	1	5.1	DAVIDSON	NC	-70	11	-67	04	76
1*	15.0			1	14.1	DAVIE	NC			-6	26	93
1	2.0			3	5.5	DUPLIN	NC			171	73	77
2	1.5	4	2.8	17	9.2	DURHAM	NC	80	228	491	95	87
1	2.0			1	1.1	EDGECOMBE	NC			-47	11	67
8	5.1	7	3.7	11	5.6	FORSYTH	NC	-29	54	10	34	78
2	5.2	2	4.9			FRANKLIN	NC	-6				
2	4.3	1	1.4	6	9.6	GASTON	NC	-67	563	121	65	88
				1	3.9	GATES	NC					72
1	2.4	2	3.7	2	3.5	GRANVILLE	NC	52	-5	44	47	71
		1	4.5	2	7.2	GREENE	NC		61			82
8*	6.2	2	1.1	14	6.6	GUILFORD	NC	-82	489	5	31	80
2	2.4	1	1.2	6	6.1	HALIFAX	NC	-49	406	156	71	79
				2	4.3	HARNETT	NC					74
		3*	49.6	2	25.3	HENDERSON	NC		-49			97
4*	10.4	2	4.4	1	1.8	HERTFORD	NC	-57	-60	-83	01	68
				1	7.5	HYDE	NC					83
		1	2.1	1	2.0	IREDELL	NC		-7			68
				1	14.1	JACKSON	NC					93
		2	4.4			JOHNSTON	NC					
1	2.6	1	5.6			LEE	NC	114				
		1	1.3	3	3.6	LENOIR	NC		178			71
				1	9.2	LINCOLN	NC					87
1	1.9			2	5.6	MARTIN	NC			195	76	78
5	2.5	12	6.0	11	4.5	MECKLENBURG	NC	135	-25	76	56	74
				2	11.7	MONTGOMERY	NC					91
		1	2.7	2	5.8	MOORE	NC		117			78
4	5.5			2	2.3	NASH	NC			-57	08	69
3	5.2	4	5.1	8	10.6	NEW HANOVER	NC	-1	108	106	63	90
				4	7.6	NORTHAMPTON	NC					83
				1	2.7	ONSLOW	NC					69
5*	18.8	1	2.0	1	1.6	ORANGE	NC	-89	-22	-92	00	67
				1	7.0	PAMLICO	NC					81
				2	5.0	PASQUOTANK	NC					76
		1	3.0	3	10.5	PENDER	NC		252			89
		1	2.9	3	9.6	PERSON	NC		234			88
1	1.1	1	1.2	6	6.0	PITT	NC	8	386	426	92	79
				2	10.6	RANDOLPH	NC					90
				1	1.7	RICHMOND	NC					68
2	1.7	2	1.2	4	2.7	ROBESON	NC	-28	118	57	48	69
1	2.8	2	2.7	2	3.2	ROCKINGHAM	NC	-3	17	13	35	70
2	4.6					ROWAN	NC					
1	5.7					RUTHERFORD	NC					
				7	9.6	SAMPSON	NC					88
		3	8.2	2	5.3	SCOTLAND	NC		-36			77
1	4.5			1	4.1	STANLY	NC			-10	25	73
2*	44.8			1	13.4	STOKES	NC			-70	03	93
1	9.1					SURRY	NC					
1*	19.3	1	14.9	1	16.1	SWAIN	NC	-23	8	-16	21	95
		1	1.8	4	7.1	VANCE	NC		288			82
5	4.2	2	1.1	10	5.2	WAKE	NC	-73	362	23	39	76

* SIGNIFICANTLY MORE DEATHS THAN EXPECTED

NON-WHITE MALE

CANCER OF THE LIVER AND GALLBLADDER INCLUDING BILE DUCTS

ICD 155.156

1950-1959 DEATHS	1950-1959 RATE	1960-1969 DEATHS	1960-1969 RATE	1970-1979 DEATHS	1970-1979 RATE	COUNTY AND STATE		PERCENT CHANGE 60S/50S	PERCENT CHANGE 70S/60S	PERCENT CHANGE 70S/50S	PERCENTILE %CHANGE 70S/50S	PERCENTILE RATES 1970S
2	4.7			1	2.1	WARREN	NC			-56	09	68
				1	5.6	WASHINGTON	NC					77
1	1.1	2	1.8	3	2.9	WAYNE	NC	55	62	150	69	70
2*	21.3			1	7.0	WILKES	NC			-67	04	82
3	4.7	3	4.3	3	3.9	WILSON	NC	-9	-8	-17	21	72
1	3.8	3	10.0	4	9.1	NORTH DAKOTA	ND	165	-9	142		
		1*	22.3			BENSON	ND					
				1*	133.9	MC LEAN	ND					99
1*	188.7					MOUNTRAIL	ND					
		1	7.6	2	10.8	ROLETTE	ND		43			90
		1	20.6	1	15.6	SIOUX	ND		-24			94
92	4.0	195*	6.1	290	7.6	OHIO	OH	53	24	90		
		2	6.2	3	8.3	ALLEN	OH		35			85
		1	3.0	3	7.2	BUTLER	OH		140			82
1*	17.0					CHAMPAIGN	OH					
2	4.5	3	5.5	4	7.6	CLARK	OH	22	37	66	52	83
		1*	23.4			CLERMONT	OH					
				1*	26.8	CLINTON	OH					97
				1	9.9	COLUMBIANA	OH					88
21	3.0	54	5.7	104	8.2	CUYAHOGA	OH	89	42	168	72	85
				1	21.5	DELAWARE	OH					96
1	8.0	1	5.1	1	2.9	ERIE	OH	-37	-43	-64	05	70
1*	27.3	1*	27.3			FAYETTE	OH	0				
13	5.3	32*	9.8	39	10.1	FRANKLIN	OH	87	3	93	60	89
				1	20.2	GEAUGA	OH					96
1	4.3					GREENE	OH					
				1	26.5	GUERNSEY	OH					97
17	4.1	39	7.5	43	7.4	HAMILTON	OH	81	-1	79	57	83
1*	18.1					HIGHLAND	OH					
1*	47.1					HURON	OH					
		1	3.7	2	5.6	JEFFERSON	OH		52			77
1*	26.8					KNOX	OH					
1	13.3					LAWRENCE	OH					
				1	9.1	LICKING	OH					87
		3	5.7	3	5.7	LORAIN	OH		-1			78
5	4.7	7	3.4	12	5.6	LUCAS	OH	-27	63	19	38	77
		1	8.3	1	21.3	MADISON	OH		158			96
3	2.1	11	6.7	13	7.6	MAHONING	OH	222	13	264	82	83
		3*	110.2			MARION	OH					
				1*	30.3	MEDINA	OH					98
1*	34.4					MEIGS	OH					
1*	18.4	1	15.0			MIAMI	OH	-19				
8	4.7	20*	7.9	31*	9.3	MONTGOMERY	OH	67	17	96	61	87
		2	15.2	1	8.2	MUSKINGUM	OH		-46			85
		1	12.3	1	3.5	PORTAGE	OH		-72			71
1	7.5	1	4.2	1	2.9	RICHLAND	OH	-44	-30	-61	06	70
		1	5.3	1	6.0	ROSS	OH		13			79
1*	28.3					SANDUSKY	OH					
1	15.1			1	12.6	SCIOTO	OH			-17	20	92
4	8.8	3	4.0	5	5.1	STARK	OH	-55	30	-42	13	76
3	3.4	5	2.9	7	3.6	SUMMIT	OH	-13	24	8	33	72
1	2.4	1	2.1	5	10.8	TRUMBULL	OH	-10	405	357	89	90
1*	20.2					WARREN	OH					
1*	42.6			1	37.2	WAYNE	OH			-13	24	98
				1*	29.8	WOOD	OH					98
21	2.4	53	5.1	77	6.1	OKLAHOMA	OK	109	21	152		
1	7.6	3*	17.6			ADAIR	OK	132				
		1	18.7			ATOKA	OK					
		1	11.8			BLAINE	OK					
				1	5.6	CADDO	OK					77
				1	11.6	CANADIAN	OK					91
1	7.8			1	4.5	CHEROKEE	OK			-42	13	74
1	4.5			3	12.2	CHOCTAW	OK			171	73	92
		1*	424.0			CIMARRON	OK					
				1*	35.3	COAL	OK					98
2*	11.5	1	6.2	2	5.6	COMANCHE	OK	-46	-11	-51	09	77
		2	7.8	2	7.3	CREEK	OK		-6			82
		1	7.4	3	14.4	DELAWARE	OK		96			94
		1*	267.9			DEWEY	OK					
				1	12.6	GARFIELD	OK					92
		1	14.9			GARVIN	OK					
1	11.9	2*	23.2	2*	21.9	GRADY	OK	95	-5	85	59	96
				1	13.3	JACKSON	OK					93
		1	11.8	3*	24.2	KAY	OK		105			97
				1	16.1	KINGFISHER	OK					95
				2*	28.7	KIOWA	OK					98
				1	6.7	LE FLORE	OK					81
1*	9.9					LINCOLN	OK					
		1	4.2	2	7.6	LOGAN	OK		80			83
1	1.9	4	11.5	3	11.6	MC CURTAIN	OK	510	1	519	96	91
				1	6.1	MAYES	OK					79
		1*	26.5			MURRAY	OK					
2	2.0	2	3.8	4	4.0	MUSKOGEE	OK	87	7	100	62	73
		1	4.5			OKFUSKEE	OK					
2	1.4	11	5.9	18	7.4	OKLAHOMA	OK	311	26	417	91	82
1	2.0	1	3.0	3	9.4	OKMULGEE	OK	51	212	370	89	87
		1	7.4	1	3.9	OSAGE	OK		-48			72
		1	11.4	1	10.6	OTTAWA	OK		-7			90
1*	14.0	1	14.5			PAYNE	OK	3				
		3*	13.4			PITTSBURG	OK					
		1	13.7			PONTOTOC	OK					
1	6.9					POTTAWATOMIE	OK					
1*	28.3					PUSHMATAHA	OK					
				1*	89.3	ROGER MILLS	OK					99

* SIGNIFICANTLY MORE DEATHS THAN EXPECTED

NON-WHITE MALE

CANCER OF THE LIVER AND GALLBLADDER INCLUDING BILE DUCTS

ICD 155,156

1950-1959 DEATHS	1950-1959 RATE	1960-1969 DEATHS	1960-1969 RATE	1970-1979 DEATHS	1970-1979 RATE	COUNTY AND STATE	PERCENT CHANGE 60S/50S	PERCENT CHANGE 70S/60S	PERCENT CHANGE 70S/50S	PERCENTILE %CHANGE 70S/50S	PERCENTILE RATES 1970S
1	4.5					SEMINOLE OK					
		1	10.6	2	12.0	SEQUOYAH OK		13			91
				1	19.9	STEPHENS OK					96
1	10.2			1	14.4	TILLMAN OK			41	45	94
3	2.4	9	5.9	12	6.2	TULSA OK	144	6	157	71	79
				2	10.4	WAGONER OK					89
7	6.2	12	7.2	14	5.1	OREGON OR	16	-29	-17		
1*	15.8					HOOD RIVER OR					
		1*	89.3			JEFFERSON OR					
				1*	56.2	JOSEPHINE OR					99
				2	15.0	LANE OR					94
				1	7.5	MALHEUR OR					83
				1	4.5	MARION OR					74
5	6.5	11*	10.0	9	5.4	MULTNOMAH OR	53	-46	-17	20	77
1*	28.1					UMATILLA OR					
91	3.2	187	5.4	295	7.1	PENNSYLVANIA PA	67	32	120		
24	4.6	34	5.3	65*	9.5	ALLEGHENY PA	16	77	106	63	88
				1	22.5	ARMSTRONG PA					97
2	4.9	2	4.7	2	3.8	BEAVER PA	-5	-18	-22	19	72
1	4.5	1	4.5	1	4.4	BERKS PA	1	-3	-2	28	74
1	3.8	1	2.2			BUCKS PA	-42				
1	6.0	1	6.5			CAMBRIA PA	9				
		2	2.8	5	6.1	CHESTER PA		116			79
				1	18.7	CRAWFORD PA					96
				2*	31.6	CUMBERLAND PA					98
5*	6.0	3	3.4	8	7.5	DAUPHIN PA	-43	119	25	40	83
1	.6	10	7.5	12	5.9	DELAWARE PA	1161	-21	900	99	78
				2	8.1	ERIE PA					84
1	1.8	1	2.2	4	9.8	FAYETTE PA	18	352	434	93	88
				2	20.6	FRANKLIN PA					96
		1	19.6			LACKAWANNA PA					
1	5.8			1	3.6	LANCASTER PA			-37	15	71
				2	17.4	LAWRENCE PA					95
				1	18.8	LEBANON PA					96
		1	19.0			LUZERNE PA					
		1	17.6			LYCOMING PA					
		1	4.9	3	13.3	MERCER PA		170			93
				1	21.8	MONROE PA					96
1	1.8	3	3.4	8	7.9	MONTGOMERY PA	91	134	345	88	84
		1	10.1	1	6.1	NORTHAMPTON PA		-39			79
50	3.3	116	5.6	166	6.6	PHILADELPHIA PA	70	17	100	62	81
				1*	112.4	SCHUYLKILL PA					99
2	3.8	3	6.8	4	10.2	WASHINGTON PA	79	50	168	72	89
		1*	9.9			WAYNE PA					
1	3.2	1	2.7	1	3.2	WESTMORELAND PA	-16	19	0	30	70
		3*	28.4	1	2.8	YORK PA		-90			70
2	3.4	1	1.3	8	7.7	RHODE ISLAND RI	-61	472	126		
1	9.9					NEWPORT RI					
1	1.9	1	1.7	8	10.0	PROVIDENCE RI	-14	497	412	91	89
25	1.1	60	2.4	122	4.2	SOUTH CAROLINA SC	111	78	276		
		1	1.2			AIKEN SC					
		1	4.5			ALLENDALE SC					
		1	1.7	5	6.3	ANDERSON SC		266			80
				1	3.3	BAMBERG SC					71
		1	2.9	1	3.8	BARNWELL SC		34			72
				2	3.8	BEAUFORT SC					72
		3	5.3			BERKELEY SC					
3	1.0	7	3.2	10	4.1	CHARLESTON SC	209	27	291	85	73
1	8.3			1	3.5	CHEROKEE SC			-57	08	71
3	7.3					CHESTER SC					
				1	2.5	CHESTERFIELD SC					69
		1	2.0	1	1.7	CLARENDON SC		-16			67
1	2.5	2	3.9	3	4.2	COLLETON SC	55	8	66	52	73
		1	1.5	5	7.1	DARLINGTON SC		388			82
		1	2.7	3	7.7	DILLON SC		183			84
1	2.5	1	2.5	2	3.3	DORCHESTER SC	0	32	32	43	71
				3	10.1	EDGEFIELD SC					89
		1	2.2	3	6.9	FAIRFIELD SC		208			81
2	2.1	1	1.0	9	7.1	FLORENCE SC	-54	634	241	80	82
1	2.6	1	2.0	2	3.2	GEORGETOWN SC	-24	60	23	39	71
2	2.4	6	5.5	6	4.5	GREENVILLE SC	131	-18	90	60	74
		2	4.3	4	7.4	HORRY SC		71			83
				3	8.7	JASPER SC					86
1	2.5	2	4.8	5	12.3	KERSHAW SC	95	154	395	90	92
2	5.0	2	5.7	4	10.2	LANCASTER SC	13	79	102	62	89
		3	6.0	1	1.1	LAURENS SC		-81			67
		1	3.0	1	4.2	LEE SC		40			73
		2	6.6	2	5.3	LEXINGTON SC		-19			77
		1	5.1			MC CORMICK SC					
		2	3.6	2	3.3	MARION SC		-9			71
		1	2.4	1	1.8	MARLBORO SC		-26			68
				4	9.6	NEWBERRY SC					88
				2	12.0	OCONEE SC					91
1	1.1	3	2.4	4	3.7	ORANGEBURG SC	128	53	249	80	72
1	8.6			2	8.4	PICKENS SC			-3	28	85
2	1.1	4	1.8	10	4.2	RICHLAND SC	59	131	266	83	73
				2	12.3	SALUDA SC					92
		3	2.7	4	3.2	SPARTANBURG SC		17			70
3	3.3	2	2.0	6	6.0	SUMTER SC	-38	196	85	59	79
		2	2.6	2	2.8	WILLIAMSBURG SC		5			69
1	1.6	1	1.5	5	7.1	YORK SC	-5	374	351	89	82

* SIGNIFICANTLY MORE DEATHS THAN EXPECTED

1950-1959 DEATHS	1950-1959 RATE	1960-1969 DEATHS	1960-1969 RATE	1970-1979 DEATHS	1970-1979 RATE	COUNTY AND STATE	PERCENT CHANGE 60S/50S	PERCENT CHANGE 70S/60S	PERCENT CHANGE 70S/50S	PERCENTILE %CHANGE 70S/50S	PERCENTILE RATES 1970S
6*	7.9	5	5.7	4	4.2	SOUTH DAKOTA SD	-28	-27	-47		
				1*	27.0	BENNETT SD					97
		1	11.8			CORSON SD					
				1	15.7	DEWEY SD					94
1*	112.4					HUGHES SD					
		1*	19.1			MELLETTE SD					
1*	56.2					PENNINGTON SD					
1*	14.1					ROBERTS SD					
2*	10.4	1	4.7	2	7.9	SHANNON SD	-55	69	-24	18	84
1	10.2	1	8.6			TODD SD	-16				
		1*	22.5			ZIEBACH SD					
47	2.0	108	4.3	187	7.0	TENNESSEE TN	112	61	242		
				2*	36.2	ANDERSON TN					98
				2	14.2	BEDFORD TN					94
1*	89.3					BLEDSOE TN					
				1	8.1	BLOUNT TN					84
				1	6.6	BRADLEY TN					81
				1	8.2	CARROLL TN					85
				1	16.3	CHESTER TN					95
				1	7.2	CROCKETT TN					82
4	1.5	9	2.8	17	4.7	DAVIDSON TN	86	68	212	78	75
		1	19.6			DICKSON TN					
		1	3.6	2	8.4	DYER TN		132			85
1	1.3	2	3.2	1	1.7	FAYETTE TN	152	-48	31	43	67
				1*	344.1	FENTRESS TN					99
		1	10.7			FRANKLIN TN					
		1	2.2	2	5.3	GIBSON TN		142			76
				1	5.0	GILES TN					76
5	2.6	2	1.2	14	6.8	HAMILTON TN	-54	464	161	72	81
		2	5.9	2	6.5	HARDEMAN TN		9			80
				1	15.7	HARDIN TN					94
				1	19.1	HAWKINS TN					96
1	1.8			4	9.3	HAYWOOD TN			419	92	87
				1	4.5	HENRY TN					74
2	2.2	6	6.3	12	10.9	KNOX TN	191	74	406	91	90
1	2.8	2	4.6	3	9.4	LAUDERDALE TN	64	104	236	79	87
				1	13.3	MC MINN TN					93
				1	18.6	MC NAIRY TN					96
1	1.5	3	3.2	6	5.8	MADISON TN	112	80	282	84	78
		1	15.6			MARION TN					
				1	8.1	MARSHALL TN					84
				1	2.6	MAURY TN					69
1	2.8	1	2.8	2	4.7	MONTGOMERY TN	-1	69	68	53	75
		1*	344.1			POLK TN					
1	4.0					ROBERTSON TN					
1	3.5					RUTHERFORD TN					
22	2.8	73*	7.6	88	8.4	SHELBY TN	177	9	203	77	85
				1*	24.8	SMITH TN					97
1	10.7	2*	17.9	2	21.8	SULLIVAN TN	68	22	104	63	96
2*	11.2			2	9.3	SUMNER TN			-17	21	87
1	2.6			3	5.9	TIPTON TN			128	67	78
				1	17.2	WARREN TN					95
1	5.5			3*	20.8	WASHINGTON TN			281	84	96
				1	13.2	WEAKLEY TN					93
1	3.3					WILLIAMSON TN					
				3	16.1	WILSON TN					95
159*	3.6	324*	6.5	439	6.3	TEXAS TX	83	-3	77		
		4	13.4	1	1.8	ANDERSON TX		-87			68
1	3.5	2	6.1	2	4.9	ANGELINA TX	77	-21	40	45	75
		1	3.2	1	2.7	BASTROP TX		-17			69
2	6.5	1	1.8			BELL TX	-72				
8	6.2	20*	11.6	20	4.7	BEXAR TX	89	-60	-23	18	75
1	1.6	4	6.0	7	9.4	BOWIE TX	268	56	475	94	87
1	1.2	3	8.9	2	4.2	BRAZORIA TX	620	-53	235	79	73
		1	2.5	2	5.7	BRAZOS TX		134			78
				1	3.6	BURLESON TX					72
2*	13.5			3	17.6	CALDWELL TX			30	42	95
		1	26.5	2	6.0	CAMERON TX		-77			79
1	5.0			1	6.0	CAMP TX			21	39	79
		1	2.7	2	6.1	CASS TX		129			79
2*	46.2			1	9.2	CHAMBERS TX			-80	02	87
1	2.1	3	6.6	2	4.1	CHEROKEE TX	221	-38	100	62	73
		1	5.2	2	9.2	COLLIN TX		77			87
2	10.4	1	4.9	4*	17.0	COLORADO TX	-53	250	63	51	95
1*	80.4					COMAL TX					
16	4.5	37	6.8	65	8.4	DALLAS TX	50	25	87	59	85
1*	16.1					DAWSON TX					
		1*	24.4	1*	28.3	DELTA TX		16			98
				1	5.3	DENTON TX					76
1	4.7	1	4.3			DE WITT TX	-8				
1*	68.3					DICKENS TX					
		1	6.0	1	9.6	ECTOR TX		60			88
2	3.6			4	10.2	ELLIS TX			186	75	89
2	9.2	1	1.2	5	1.3	EL PASO TX	-87	13	-85	01	67
1	3.3	3	12.1	2	6.4	FALLS TX	265	-47	94	61	80
		2	13.0	2	13.4	FANNIN TX		3			93
1	3.7			1	5.1	FAYETTE TX			37	44	76
1	2.9	3	8.1	2	3.0	FORT BEND TX	183	-63	3	31	70
		1*	61.3			FRANKLIN TX					
		1	3.3			FREESTONE TX					
5	5.3	8	6.8	8	4.8	GALVESTON TX	27	-30	-11	24	75
		2	12.5	1	4.7	GONZALES TX		-62			75
3*	9.9			1	3.2	GRAYSON TX			-68	04	70
1	1.5	4	6.0	4	5.7	GREGG TX	306	-5	284	84	78

* SIGNIFICANTLY MORE DEATHS THAN EXPECTED

1950-1959 DEATHS	1950-1959 RATE	1960-1969 DEATHS	1960-1969 RATE	1970-1979 DEATHS	1970-1979 RATE	COUNTY AND STATE	PERCENT CHANGE 60S/50S	PERCENT CHANGE 70S/60S	PERCENT CHANGE 70S/50S	PERCENTILE %CHANGE 70S/50S	PERCENTILE RATES 1970S
1	3.1	2	6.0	1	2.4	GRIMES TX	94	-60	-22	19	69
1	4.4	2	9.2	1	4.6	GUADALUPE TX	112	-50	6	32	75
1	5.5	2	9.4	2	13.9	HARDIN TX	72	48	155	70	93
39*	5.4	73*	8.2	104	7.5	HARRIS TX	50	-8	38	45	83
1	1.2	2	3.1	3	4.3	HARRISON TX	155	36	247	80	73
		1*	94.4			HASKELL TX					
				1	17.0	HAYS TX					95
1	4.0	1	5.7	2	9.1	HENDERSON TX	43	59	128	66	87
1	4.2			3	23.7	HILL TX			457	94	97
		1*	23.8			HOCKLEY TX					
				1	9.8	HOPKINS TX					88
				3	6.4	HOUSTON TX					80
1*	44.6					HOWARD TX					
2	8.0			1	3.3	HUNT TX			-59	06	71
		1	12.9	1	10.6	JACKSON TX		-18			90
11*	5.4	13	6.1	18	6.8	JEFFERSON TX	14	11	27	41	81
1*	49.7					JIM WELLS TX					
1*	10.8	1	16.1			JOHNSON TX	50				
		2*	34.8			JONES TX					
1	2.6	2	4.3	2	4.7	KAUFMAN TX	64	11	81	57	75
				1	24.8	KERR TX					97
1*	32.8					KNOX TX					
		1	2.3	4	12.1	LAMAR TX		414			91
				1	12.7	LAMB TX					92
1	9.0	3*	25.8	1	6.9	LAVACA TX	188	-73	-23	18	81
				1	5.7	LEE TX					78
1	4.9	1	4.0	2	20.0	LEON TX	-18	399	307	87	96
3*	10.3	2	6.5	3	7.9	LIBERTY TX	-37	22	-23	18	84
2	6.5	1	2.8			LIMESTONE TX	-57				
		1	5.3	3	5.1	LUBBOCK TX		-5			76
		1*	32.8	1	27.0	LYNN TX		-18			97
2	2.0	8	8.1	16*	13.3	MC LENNAN TX	313	64	577	96	93
				1	12.9	MADISON TX					92
1	4.9	2	7.6			MARION TX	53				
				3	8.9	MATAGORDA TX					86
				2	6.6	MIDLAND TX					81
1	3.6	2	9.3	1	3.8	MILAM TX	162	-60	6	32	72
		2*	81.8			MITCHELL TX					
		2	6.1	2	6.6	MONTGOMERY TX		8			80
		1	5.6	1	6.0	MORRIS TX		6			79
		2	7.0	2	4.8	NACOGDOCHES TX		-31			75
3	6.1	5*	13.2	4	10.2	NAVARRO TX	116	-23	67	53	89
		2	14.9	1	8.8	NEWTON TX		-41			86
		1*	28.1	1	17.2	NOLAN TX		-39			95
1	3.3	5*	12.8	3	3.8	NUECES TX	289	-70	15	36	72
2*	23.7			3	13.6	ORANGE TX			-42	13	93
		2*	73.4	1	16.1	PALO PINTO TX		-78			95
1	4.6					PANOLA TX					
				1*	18.2	PARKER TX					95
				1	3.6	POLK TX					72
		2	8.9	2	7.8	POTTER TX		-12			84
		1	5.6	1	3.9	RED RIVER TX		-31			72
1*	49.1			1	25.8	REEVES TX			-47	11	97
				1	17.7	REFUGIO TX					95
1	2.3	2	6.3	1	2.7	ROBERTSON TX	174	-57	18	38	69
				2*	42.6	ROCKWALL TX					99
		1*	57.3			RUNNELS TX					
		2	4.0	3	5.8	RUSK TX		43			78
		1	14.0			SABINE TX					
		1	4.1	2	7.9	SAN JACINTO TX		96			84
		1*	32.1			SAN PATRICIO TX					
		1	5.4	1	4.3	SHELBY TX		-21			73
3	3.0	8	8.3	8	6.4	SMITH TX	181	-23	116	65	80
8	4.0	10	3.9	26	8.3	TARRANT TX	-3	116	108	64	85
		3*	20.6	2	6.6	TAYLOR TX		-68			81
				1	24.9	TERRY TX					97
				2	13.1	TITUS TX					92
1	6.5	1	6.9	2	5.6	TOM GREEN TX	5	-19	-14	23	77
2	2.0	4	3.9	8	4.4	TRAVIS TX	101	13	126	66	74
		2	17.6	1	14.8	TRINITY TX		-16			94
1	9.6			1	12.0	TYLER TX			25	40	91
		1	3.9			UPSHUR TX					
		1*	78.1			UPTON TX					
1	5.4	3*	16.2	1	2.9	VICTORIA TX	199	-82	-46	11	70
		6*	14.1	3	6.3	WALKER TX		-55			80
		1	4.4	1	2.4	WALLER TX		-45			69
				1	18.2	WARD TX					95
1	3.4	8*	21.9	4	13.1	WASHINGTON TX	542	-40	285	85	92
		2	4.2	2	4.7	WHARTON TX		10			75
1	6.5	1	2.7	3	7.6	WICHITA TX	-58	178	16	37	83
		1	15.8	1	13.4	WILBARGER TX		-15			93
				1	32.7	WILLACY TX					98
1	4.0	2	6.9	3	10.4	WILLIAMSON TX	72	49	157	71	89
		1*	267.9			WINKLER TX					
		1	5.3			WOOD TX					
		1*	54.5			YOUNG TX					
1	1.4	4	6.6	6	6.3	UTAH UT	359	-4	339		
				1*	22.5	CARBON UT					97
				1	27.0	DAVIS UT					97
1	3.3	2	7.5	3	6.3	SALT LAKE UT	130	-16	93	60	80
		1	9.3			SAN JUAN UT					
		1	15.7	1	3.6	WEBER UT		-77			72
41	1.4	129	4.1	232	6.4	VIRGINIA VA	192	56	357		
		1	2.1	1	1.8	ACCOMACK VA		-16			68
		3	6.5			ALBEMARLE#VA					
1	6.8			1	7.9	ALLEGHANY#VA			16	37	84

* SIGNIFICANTLY MORE DEATHS THAN EXPECTED

COUNTIES COMBINED; SEE TECHNICAL NOTES

NON-WHITE MALE

CANCER OF THE LIVER AND GALLBLADDER INCLUDING BILE DUCTS

ICD 155,156

1950-1959 DEATHS	1950-1959 RATE	1960-1969 DEATHS	1960-1969 RATE	1970-1979 DEATHS	1970-1979 RATE	COUNTY AND STATE	PERCENT CHANGE 60S/50S	PERCENT CHANGE 70S/60S	PERCENT CHANGE 70S/50S	PERCENTILE %CHANGE 70S/50S	PERCENTILE RATES 1970S
				1	4.9	AMELIA VA					75
				1	3.5	AMHERST VA					71
1	8.9					APPOMATTOX VA					
				5	8.8	ARLINGTON#VA					86
				1	2.9	AUGUSTA#VA					70
2	2.1	10*	10.1	8	7.4	BEDFORD#VA	384	-27	253	81	82
		1	12.5			BOTETOURT VA					
2	5.6			3	7.1	BRUNSWICK VA			27	41	82
		2	10.9			BUCKINGHAM VA					
				1	26.8	CARROLL#VA					97
				2	9.5	CHARLES CITY VA					88
				1	4.6	CHARLOTTE VA					75
3	1.0	21	4.9	27	6.0	CHESTERFIELD#VA	398	22	506	95	79
		1	22.5			CLARKE VA					
				1	7.4	CUMBERLAND VA					83
5	3.1	3	1.7	12	6.9	DINWIDDIE#VA	-45	317	128	67	61
1	7.6					ESSEX VA					
		4	4.5	13	9.0	FAIRFAX#VA		99			86
		2	10.4	2	7.5	FAUQUIER VA		-28			83
		3*	26.2	1	4.8	FRANKLIN VA		-81			75
				2	10.2	GLOUCESTER VA					89
				1*	28.3	GREENE VA					98
		1	3.4	3	9.5	GREENSVILLE#VA		179			88
3	5.2	4	6.8	2	3.6	HALIFAX#VA	32	-47	-30	16	72
		1	5.1	1	3.4	HANOVER VA		-33			71
		4*	13.4	1	1.8	HENRICO VA		-87			68
				3	6.3	HENRY#VA					80
		1	2.7	2	4.9	ISLE OF WIGHT VA		82			76
1	3.0			6	8.5	JAMES CITY#VA			186	75	85
				1	6.3	KING AND QUEEN VA					80
				1	8.3	KING GEORGE VA					85
				1	5.3	LANCASTER VA					77
				2	13.7	LOUDOUN VA					93
				1	3.5	LOUISA VA					71
1	4.0			1	4.2	LUNENBURG VA			6	31	73
				1	10.2	MADISON VA					89
1	3.0	1	2.0	2	3.4	MECKLENBURG VA	-32	71	16	37	71
				1	7.9	MONTGOMERY#VA					84
				1	3.0	NORTHAMPTON VA					70
		1	5.2	1	5.4	NORTHUMBERLAND VA		3			77
				1	6.8	ORANGE VA					81
2	3.3	4	4.4	6	5.2	PITTSYLVANIA#VA	33	17	56	48	76
				1	8.3	POWHATAN VA					85
				2	7.2	PRINCE EDWARD VA					82
		1	8.4	1	6.0	PRINCE WILLIAM#VA		-29			79
		1	12.8	1	13.8	PULASKI VA		8			93
1	.9	5	7.0	5	4.9	ROANOKE#VA	684	-30	452	93	75
				1	10.2	ROCKBRIDGE#VA					89
				1	14.9	ROCKINGHAM#VA					94
				1*	34.4	SMYTH VA					98
		1	1.8	5	7.8	SOUTHAMPTON#VA		345			84
		1	4.8	2	6.9	SPOTSYLVANIA#VA		42			81
				2	14.5	SURRY VA					94
1	4.2			2	6.9	SUSSEX VA			64	52	81
				2	26.0	WASHINGTON#VA					97
				1	4.5	WESTMORELAND VA					74
8	1.4	35	6.1	53	8.0	CHESAPEAKE CITY#VA	345	32	488	94	84
2	2.0	6	6.8	8	8.8	HAMPTON CITY VA	245	30	347	89	86
5	5.4	5	3.1	12	8.8	NEWPORT NEWS CITY VA	-42	181	63	51	86
1	1.3	6	6.2	10	10.7	SUFFOLK CITY VA	379	73	728	98	90
15	6.4	33*	7.7	45	6.7	WASHINGTON WA	22	-14	5		
1*	27.0					CLALLAM WA					
				1	16.4	CLARK WA					95
				1	11.2	FRANKLIN WA					90
		3*	55.3			GRANT WA					
				2*	28.5	GRAYS HARBOR WA					98
11*	8.0	18	6.7	34	9.0	KING WA	-16	34	13	36	86
		1	7.6	2	8.9	KITSAP WA		18			86
		1	19.1			OKANOGAN WA					
1	7.0	2	6.0	3	3.6	PIERCE WA	-14	-41	-49	10	71
		2*	42.1	1	8.6	SNOHOMISH WA		-80			86
2*	9.1	4*	16.6			SPOKANE WA	83				
		2	7.1	1	2.0	YAKIMA WA		-73			68
19	3.7	13	3.5	20	4.7	WEST VIRGINIA WV	-5	33	26		
		1	17.0			BERKELEY WV					
				1*	44.6	BOONE WV					99
		1*	273.2			BRAXTON WV					
		1*	74.3			BROOKE WV					
2	7.1			1	3.9	CABELL WV			-45	12	72
1	1.8	1	2.9	2	5.1	FAYETTE WV	60	77	183	75	76
		1*	112.4			GRANT WV					
		1	14.9			HANCOCK WV					
1	13.0					HARRISON WV					
2*	15.3					JEFFERSON WV					
2	2.7			7	9.1	KANAWHA WV			237	79	87
1	2.2			1	4.1	LOGAN WV			89	59	73
3	4.5	2	4.8	2	4.8	MC DOWELL WV	8	-1	7	32	75
3	7.5	2	4.7	3	10.6	MERCER WV	-37	124	41	45	90
		1*	32.6			MINERAL WV					
				1	30.3	MONONGALIA WV					98
3	8.7	2	3.0	1	1.9	RALEIGH WV	-65	-38	-78	02	68
				1*	112.4	TYLER WV					99
1*	10.8					WYOMING WV					
5	2.1	17	5.6	29	6.5	WISCONSIN WI	169	15	208		

* SIGNIFICANTLY MORE DEATHS THAN EXPECTED

\# COUNTIES COMBINED; SEE TECHNICAL NOTES

1950-1959 DEATHS	1950-1959 RATE	1960-1969 DEATHS	1960-1969 RATE	1970-1979 DEATHS	1970-1979 RATE	COUNTY AND STATE	PERCENT CHANGE 60S/50S	PERCENT CHANGE 70S/60S	PERCENT CHANGE 70S/50S	PERCENTILE %CHANGE 70S/50S	PERCENTILE RATES 1970S
				1*	37.2	ASHLAND WI					98
		1	18.9	2	27.5	DANE WI		46			98
		1*	106.0			JUNEAU WI					
		1	6.7			MENOMINEE#WI					
2	1.2	10	3.9	23	7.2	MILWAUKEE WI	239	83	521	96	82
1*	28.4	2*	94.6			OUTAGAMIE WI	233				
1	3.3	1	14.9			RACINE WI	357				
				1	13.5	ROCK WI					93
		1*	141.3			SHEBOYGAN WI					
				1*	28.7	VILAS WI					98
				1	15.9	WAUKESHA WI					94
1*	86.0					WOOD WI					
1	5.1			1	3.3	WYOMING WY			-34		
				1*	44.6	ALBANY WY					99
1*	38.3					SWEETWATER WY					

* SIGNIFICANTLY MORE DEATHS THAN EXPECTED

\# COUNTIES COMBINED; SEE TECHNICAL NOTES

NON-WHITE FEMALE

CANCER OF THE LIVER AND GALLBLADDER
INCLUDING BILE DUCTS

ICD 155,156

1950-1959 DEATHS	1950-1959 RATE	1960-1969 DEATHS	1960-1969 RATE	1970-1979 DEATHS	1970-1979 RATE	COUNTY AND STATE	PERCENT CHANGE 60S/50S	PERCENT CHANGE 70S/60S	PERCENT CHANGE 70S/50S	PERCENTILE %CHANGE 70S/50S	PERCENTILE RATES 1970S
1569	2.4	2699	3.1	4581	3.7	UNITED STATES	27	20	52		
121*	3.2	99	2.2	156	3.0	ALABAMA AL	-30	34	-5		
				2	5.9	AUTAUGA AL					88
3*	9.7			3	5.4	BALDWIN AL			-44	23	87
2	4.3	1	1.6	3	4.0	BARBOUR AL	-62	146	-7	42	82
1	1.5	1	2.1			BULLOCK AL	36				
		2	4.3	3	6.8	BUTLER AL		57			90
8*	13.2			3	3.0	CALHOUN AL			-77	04	77
1	2.4	2	3.0	1	1.4	CHAMBERS AL	23	-53	-42	26	71
				1	6.9	CHILTON AL					90
		1	2.6	1	1.7	CHOCTAW AL		-33			72
6*	11.5	1	1.7	1	1.4	CLARKE AL	-86	-15	-88	00	71
		1*	18.9			CLAY AL					
		1	3.7	3	7.7	COFFEE AL		111			91
1	2.5			2	3.9	COLBERT AL			58	68	81
1	3.0			2	4.0	CONECUH AL			33	58	82
		2*	14.9			COOSA AL					
1	3.0			1	3.8	CRENSHAW AL			26	56	81
1	7.0	1	5.1	2	7.8	DALE AL	-27	53	12	52	91
5	3.3	3	1.6	5	2.3	DALLAS AL	-51	45	-30	30	74
3*	8.9	3	6.3	4	7.8	ELMORE AL	-29	24	-12	38	91
3*	8.6	1	1.7			ESCAMBIA AL	-80				
3	4.6			1	1.1	ETOWAH AL			-77	05	70
				1	6.9	FAYETTE AL					90
				2	15.3	GENEVA AL					96
2	3.5					GREENE AL					
		2	3.8	2	2.7	HALE AL		-28			76
1	2.0	2	3.7			HOUSTON AL	84				
1*	8.9	1	11.9			JACKSON AL	34				
26	2.6	32	2.9	33	2.6	JEFFERSON AL	13	-12	0	45	75
1*	28.1	2*	19.1			LAMAR AL	-32				
1	4.3	1	1.9	2	4.3	LAUDERDALE AL	-56	128	0	46	83
				2	7.0	LAWRENCE AL					90
2	2.5	2	2.5	8*	9.3	LEE AL	-1	276	272	95	93
1	2.2	1	2.2	1	1.8	LOWNDES AL	-1	-19	-19	34	73
1	1.6	2	2.1	2	1.4	MACON AL	32	-36	-16	36	71
2	2.3	1	.8	4	3.8	MADISON AL	-64	342	60	69	80
9*	10.4	2	2.7	1	1.3	MARENGO AL	-74	-52	-88	00	71
				1*	14.5	MARSHALL AL					95
9	3.7	10	2.5	22	4.2	MOBILE AL	-34	72	13	52	82
1	2.5			1	1.8	MONROE AL			-28	31	72
7	2.7	3	.9	18	5.0	MONTGOMERY AL	-66	438	84	76	85
1	2.5	3	6.6	3	7.7	MORGAN AL	169	16	213	91	91
1	1.3	2	3.7			PERRY AL	180				
1	1.9	1	2.3			PICKENS AL	23				
3	5.5			1	1.5	PIKE AL			-73	08	72
		1	4.4			RANDOLPH AL					
		2	2.1	4	3.8	RUSSELL AL		84			81
1	6.6					ST CLAIR AL					
1	1.4	2	2.9	1	.9	SUMTER AL	99	-68	-36	28	70
2	3.0	3	3.9	2	2.0	TALLADEGA AL	29	-48	-33	30	73
				1	2.4	TALLAPOOSA AL					75
5	4.5	4	3.0	2	1.4	TUSCALOOSA AL	-34	-53	-69	09	71
1	3.4			3	8.8	WALKER AL			159	86	92
2	2.6			1	1.3	WILCOX AL			-52	18	71
		7*	6.9	13*	8.5	ALASKA AK		25			
33*	11.9	52*	14.2	73*	8.8	ARIZONA AZ	20	-38	-26		
7*	15.6	17*	31.1	13*	18.5	APACHE AZ	99	-40	18	54	96
2*	11.1	2	6.2	2	5.1	COCONINO AZ	-44	-19	-54	15	85
2*	30.5	1	9.4	4*	24.9	GILA AZ	-69	164	-18	35	97
1*	94.4	1*	20.2			GRAHAM AZ	-79				
4	3.7	6	5.6	15	3.7	MARICOPA AZ	52	-33	1	47	80
				1*	9.8	MOHAVE AZ					93
4*	9.0	10*	22.0	11*	15.9	NAVAJO AZ	146	-28	78	75	96
4*	7.7	10*	15.6	12*	8.2	PIMA AZ	104	-48	7	50	92
7*	41.3	4*	18.6	14*	26.8	PINAL AZ	-55	44	-35	29	97
2*	29.9	1	11.5	1	3.9	YUMA AZ	-62	-66	-87	01	81
42	2.4	39	2.0	76	3.3	ARKANSAS AR	-17	66	39		
				2	8.9	ARKANSAS AR					92
		3	7.3	1	1.8	ASHLEY AR		-75			73
1	4.3	1	4.5	2	6.7	BRADLEY AR	5	48	55	66	89
1	1.7			3	3.0	CHICOT AR			74	74	78
1	3.7	2	9.2			CLARK AR	151				
1*	7.3			1	10.6	CLEVELAND AR			45	62	94
		2	4.0	1	1.5	COLUMBIA AR		-62			72
				1	6.1	CONWAY AR					88
		1*	10.2			CRAWFORD AR					
3	2.7	1	.8	2	1.3	CRITTENDEN AR	-70	62	-51	18	71
		1	4.0	2	5.7	CROSS AR		40			87
		2	6.5	1	3.7	DALLAS AR		-42			80
				2	2.5	DESHA AR					75
1	7.4					FAULKNER AR					
				2	5.2	GARLAND AR					86
		3	7.7	2	4.8	HEMPSTEAD AR		-37			85
				3*	16.8	HOT SPRING AR					96
				1	10.7	HOWARD AR					94
1*	20.2					INDEPENDENCE AR					
2	1.1	4	2.3	5	1.9	JEFFERSON AR	117	-19	76	74	73
				2	4.7	LAFAYETTE AR					84
		1	1.3	2	3.0	LEE AR		138			78

* SIGNIFICANTLY MORE DEATHS THAN EXPECTED

1950-1959 DEATHS	1950-1959 RATE	1960-1969 DEATHS	1960-1969 RATE	1970-1979 DEATHS	1970-1979 RATE	COUNTY AND STATE	PERCENT CHANGE 60S/50S	PERCENT CHANGE 70S/60S	PERCENT CHANGE 70S/50S	PERCENTILE %CHANGE 70S/50S	PERCENTILE RATES 1970S
		1	2.8			LINCOLN AR					
				2	14.6	LITTLE RIVER AR					95
1*	38.3					LOGAN AR					
1	4.4			1	5.4	LONOKE AR			22	55	87
2	5.2	1	2.6	1	1.4	MILLER AR	-51	-45	-73	07	71
2	2.6	2	2.4	4	5.2	MISSISSIPPI AR	-5	111	100	78	86
		1	2.0	3	7.3	MONROE AR		269			90
3*	19.7			1	4.6	NEVADA AR			-77	04	84
1	1.4	2	2.9	6*	7.8	OUACHITA AR	109	165	454	97	91
8*	8.0			5	3.7	PHILLIPS AR			-54	16	80
				1	8.1	PRAIRIE AR					92
7	2.9	3	1.0	11	3.4	PULASKI AR	-64	224	17	54	79
1	1.3	4	5.0	3	3.5	ST FRANCIS AR	271	-30	158	86	79
1*	424.0					SCOTT AR					
2*	9.9	1	3.6	1	2.6	SEBASTIAN AR	-63	-28	-73	06	76
1	1.5	3	4.5	2	2.6	UNION AR	208	-43	76	74	75
1	3.7					WOODRUFF AR					
76	2.9	214*	4.3	353	3.1	CALIFORNIA CA	49	-27	8		
6	2.0	26	4.9	32	3.0	ALAMEDA CA	145	-38	53	66	78
				2	7.5	BUTTE CA					91
3	9.2	2	1.2	8	3.3	CONTRA COSTA CA	-87	173	-65	11	78
4	5.0	4	4.7	9	2.9	FRESNO CA	-6	-39	-43	23	77
1*	7.5	3*	22.3			HUMBOLDT CA	197				
		3*	22.3			IMPERIAL CA					
				1*	12.4	INYO CA					94
4*	6.2	5	5.1	2	1.7	KERN CA	-18	-66	-72	08	72
				1	2.8	KINGS CA					77
34	2.7	109*	4.3	154	3.0	LOS ANGELES CA	56	-30	9	50	77
1*	20.2	1	7.2	1	5.0	MARIN CA	-64	-31	-75	05	85
		1	5.5	4	4.9	MONTEREY CA		-10			85
		1	.8	9	3.4	ORANGE CA		309			79
		1	12.2	1	5.7	PLACER CA		-53			87
		1*	43.3			PLUMAS CA					
3*	7.2	2	1.9	8	4.4	RIVERSIDE CA	-73	124	-40	26	83
3	3.9	7	5.5	13	4.7	SACRAMENTO CA	39	-15	18	54	84
				1	12.4	SAN BENITO CA					94
2	5.0	1	1.4	3	.9	SAN BERNARDINO CA	-71	-37	-82	03	70
2	3.1	9	5.1	14	3.0	SAN DIEGO CA	67	-42	-4	44	77
8	2.9	21	4.3	51	4.7	SAN FRANCISCO CA	48	10	63	70	84
1	1.3	3	3.3	4	1.9	SAN JOAQUIN CA	147	-44	39	60	73
				1	7.5	SAN LUIS OBISPO CA					91
		1	1.3	4	1.8	SAN MATEO CA		35			73
1	7.7			4	6.1	SANTA BARBARA CA			-20	34	68
1	1.6	6	6.4	13	4.2	SANTA CLARA CA	306	-33	170	88	82
1*	20.2			2*	38.0	SISKIYOU CA			88	77	98
		2	4.2	1	.7	SOLANO CA		-83			70
1*	13.2			2	5.6	SONOMA CA			-58	14	87
		4*	19.2	4	4.6	TULARE CA		-76			84
		1	5.9	1	.8	VENTURA CA		-86			70
				2	4.3	YOLO CA					83
				1	10.1	YUBA CA					93
1	.9	8	3.4	10	2.2	COLORADO CO	277	-36	140		
1	1.3	5	2.8	6	2.1	DENVER-ARAPAHOE#CO	111	-24	60	69	74
				2	6.4	EL PASO CO					89
		1*	74.3			LA PLATA CO					
		2*	259.4	2*	56.3	MONTEZUMA CO		-78			98
6	2.2	18	4.3	24	3.0	CONNECTICUT CT	100	-31	38		
		7	5.7	2	.7	FAIRFIELD CT		-88			70
4	5.9	2	1.8	7	2.8	HARTFORD CT	-69	54	-52	17	77
1*	13.3					MIDDLESEX CT					
1	.5	7	5.1	14*	6.3	NEW HAVEN CT	862	23	1081	99	88
		2*	10.5	1	1.8	NEW LONDON CT		-83			72
9*	4.4	9	3.4	12	3.3	DELAWARE DE	-23	-3	-25		
1	3.2			1	1.7	KENT DE			-46	21	72
6*	4.8	5	2.8	5	2.2	NEW CASTLE DE	-42	-21	-54	16	74
2	5.0	4	7.8	6*	8.4	SUSSEX DE	56	7	67	72	92
29	2.6	61	3.5	125*	5.4	DIST. OF COLUMBIA DC	33	54	105		
29	2.6	61	3.5	125*	5.4	WASHINGTON DC	33	54	105	79	87
38	1.5	85	2.3	162	3.1	FLORIDA FL	53	34	105		
2	2.3	2	2.1	8	7.6	ALACHUA FL	-8	256	230	92	91
1	1.8	1	2.0	1	1.9	BAY FL	10	-5	4	49	73
1*	16.1	1	7.3			BRADFORD FL	-54				
3	4.7	4	1.8	9	2.3	BROWARD FL	-61	23	-52	17	74
		1*	22.3			COLLIER FL					
5	2.1	10	1.9	22	2.1	DADE FL	-10	12	1	46	74
				1	6.7	DE SOTO FL					90
8	2.0	13	2.5	20	3.2	DUVAL FL	24	27	57	68	78
2	2.0	3	2.1	11	5.3	ESCAMBIA FL	4	151	161	87	86
		1	1.0	5	4.5	GADSDEN FL		370			83
				1	4.0	HAMILTON FL					82
				1	38.3	HARDEE FL					98
		2*	9.4	2	6.9	HIGHLANDS FL		-27			90
2	1.1	3	1.1	11	3.2	HILLSBOROUGH FL	-1	201	198	90	78
		1*	24.6			HOLMES FL					
1	6.8			2	7.5	INDIAN RIVER FL			10	51	91
		1	2.0	1	1.5	JACKSON FL		-24			72
1	2.8	1	2.3	1	3.5	JEFFERSON FL	-17	51	25	56	79
				2	2.8	LAKE FL					77

* SIGNIFICANTLY MORE DEATHS THAN EXPECTED

\# COUNTIES COMBINED; SEE TECHNICAL NOTES

NON-WHITE FEMALE

CANCER OF THE LIVER AND GALLBLADDER INCLUDING BILE DUCTS

ICD 155,156

1950-1959 DEATHS	1950-1959 RATE	1960-1969 DEATHS	1960-1969 RATE	1970-1979 DEATHS	1970-1979 RATE	COUNTY AND STATE	PERCENT CHANGE 60S/50S	PERCENT CHANGE 70S/60S	PERCENT CHANGE 70S/50S	PERCENTILE %CHANGE 70S/50S	PERCENTILE RATES 1970S
				2	6.3	LEE FL					89
1	1.4	4	4.0	7	5.7	LEON FL	192	42	313	96	87
				2	14.1	LEVY FL					95
1	4.9	1	3.3			MADISON FL	-33				
		1	2.8	1	1.9	MANATEE FL		-32			73
		4	4.9	6	5.7	MARION FL		16			87
				1	4.2	MONROE FL					82
		1	6.1			NASSAU FL					
		1	6.7	1	5.5	OKALOOSA FL		-18			87
1	1.0	8	6.3	7	3.0	ORANGE FL	505	-52	191	90	78
4	5.0	7	4.5	8	2.9	PALM BEACH FL	-12	-35	-43	24	77
1*	11.2			1	6.5	PASCO FL			-42	25	89
2	1.4	4	2.3	9	4.0	PINELLAS FL	62	72	179	89	82
1	1.2	7	5.3	4	1.9	POLK FL	336	-63	61	69	73
				2	3.0	PUTNAM FL					78
				1	2.4	ST LUCIE FL					75
				1	11.1	SANTA ROSA FL					94
		1	3.3	2	4.2	SARASOTA FL		28			82
		2	2.4	4	4.1	SEMINOLE FL		68			82
				1	4.0	SUWANNEE FL					81
1	1.1			4	2.6	VOLUSIA FL			129	82	75
106	2.6	78	1.5	164	2.7	GEORGIA GA	-41	77	4		
				1	9.6	BAKER GA					93
				1	8.4	BARROW GA					92
				1	3.6	BARTOW GA					80
		1	4.3			BEN HILL GA					
3	1.7	3	1.5	8	2.8	BIBB GA	-10	83	64	70	77
		1	6.2			BLECKLEY GA					
1	5.9			3	8.6	BROOKS GA			46	63	92
				3	6.7	BULLOCH GA					89
3	5.0	2	3.9	2	3.7	BURKE GA	-22	-5	-26	33	80
3*	12.7	1	4.5	3*	12.5	CALHOUN GA	-64	174	-2	45	94
		2*	18.9			CANDLER GA					
8	3.1	5	1.5	10	2.7	CHATHAM GA	-50	79	-11	39	77
1	1.6			2	2.8	CLARKE GA			69	72	77
		1	5.5	3*	14.2	CLAYTON GA		155			95
				1*	13.4	CLINCH GA					95
1	2.6	1	2.8	1	2.1	COBB GA	8	-24	-18	35	74
				2	10.7	COFFEE GA					94
		1	3.8	1	2.5	COLQUITT GA		-33			75
				1	2.7	COLUMBIA GA					76
2	4.0	1	2.5	1	2.2	COWETA GA	-38	-13	-46	22	74
				2	5.1	CRISP GA					86
2	3.6			2	4.4	DECATUR GA			21	55	83
3	5.0	6	3.7	12	4.5	DE KALB GA	-25	20	-10	41	84
		1	4.1			DODGE GA					
1	4.4					DOOLY GA					
2	2.7			7	4.5	DOUGHERTY GA			70	72	84
1	4.3			1	3.8	EARLY GA			-11	39	81
1	4.3	1	4.5	2	6.3	EMANUEL GA	3	41	46	63	89
1	3.3			2	3.8	FLOYD GA			15	53	80
1*	28.1	1*	14.9			FRANKLIN GA	-47				
18	2.5	16	1.6	34	2.9	FULTON GA	-35	80	16	53	77
1	2.4	4	6.2	2	2.5	GLYNN GA	155	-60	2	48	75
				2*	29.6	GORDON GA					98
		1	4.0	1	3.1	GREENE GA		-24			78
		1	6.3	1	6.1	GWINNETT GA		-2			88
1	3.3	2	9.7			HARRIS GA	193				
3*	19.2			1	3.3	HENRY GA			-83	02	79
		1	3.0			HOUSTON GA					
				1	7.2	JACKSON GA					90
				2*	13.8	JASPER GA					95
		2	5.5	1	2.2	JEFFERSON GA		-60			74
1	5.8					JENKINS GA					
1	6.2					JONES GA					
				1	4.9	LAMAR GA					85
2	4.9	1	2.2	1	1.3	LAURENS GA	-55	-43	-74	06	71
1	5.6					LIBERTY GA					
				1*	21.3	LONG GA					97
1	1.4	1	1.4	1	1.4	LOWNDES GA	-2	0	-2	45	71
1	2.5			1	2.1	MACON GA			-15	37	74
1*	12.4					MADISON GA					
1	10.7					MARION GA					
				1	2.4	MERIWETHER GA					75
1	2.5			3	6.2	MITCHELL GA			145	84	88
1*	10.3					MONTGOMERY GA					
3*	13.4	1	5.8	1	4.6	MORGAN GA	-57	-21	-66	10	84
3	2.0	2	1.0	7	3.1	MUSCOGEE#GA	-49	197	52	65	78
1	3.3	2	5.8			NEWTON GA	76				
2*	12.5					OGLETHORPE GA					
1	4.4			2	3.9	PEACH GA			-13	37	81
1	6.8					POLK GA					
1	9.0					PUTNAM GA					
1	2.8					RANDOLPH GA					
3	2.1	4	2.0	5	2.0	RICHMOND GA	-4	2	-1	45	73
1*	10.7					ROCKDALE GA					
1*	15.7			1	8.9	SCHLEY GA			-43	23	92
1	3.3	1	3.5			SCREVEN GA	3				
3*	7.7			2	3.5	SPALDING GA			-54	16	79
1*	18.9	1	9.8	1	10.5	STEPHENS GA	-48	7	-44	23	93
		2	3.2	2	3.7	SUMTER GA		15			80
		1	8.0	1	8.0	TALIAFERRO GA		0			91
				2	11.4	TAYLOR GA					94
				1	5.0	TELFAIR GA					85
1	2.7					TERRELL GA					
2	3.3			1	1.3	THOMAS GA			-62	12	71
				1	3.8	TOOMBS GA					80
1*	10.5					TREUTLEN GA					

* SIGNIFICANTLY MORE DEATHS THAN EXPECTED

COUNTIES COMBINED; SEE TECHNICAL NOTES

NON-WHITE FEMALE

CANCER OF THE LIVER AND GALLBLADDER INCLUDING BILE DUCTS

ICD 155,156

1950-1959 DEATHS	1950-1959 RATE	1960-1969 DEATHS	1960-1969 RATE	1970-1979 DEATHS	1970-1979 RATE	COUNTY AND STATE	PERCENT CHANGE 60S/50S	PERCENT CHANGE 70S/60S	PERCENT CHANGE 70S/50S	PERCENTILE %CHANGE 70S/50S	PERCENTILE RATES 1970S
2	2.5	1	1.5	1	1.3	TROUP GA	-40	-15	-48	19	71
		2	6.4	1	3.1	UPSON GA		-51			78
1	4.8					WALTON GA					
1	3.1					WARE GA					
1	5.8			1	5.9	WARREN GA			1	47	88
1	1.9	2	3.7	1	2.4	WASHINGTON GA	98	-35	29	57	75
				1	5.4	WAYNE GA					86
1*	68.8					WHITE GA					
1*	15.1	1	11.2	1	8.0	WHITFIELD GA	-26	-28	-47	21	91
				1	7.4	WILCOX GA					91
2	7.1			3	8.2	WILKES GA			16	53	92
1	5.7					WILKINSON GA					
		1	4.8	1	3.9	WORTH GA		-18			81
		82*	6.1	144*	6.7	HAWAII HI		10			
		10	5.2	18*	7.4	HAWAII HI		42			90
		63*	6.7	110*	6.8	HONOLULU HI		2			90
		3	3.6	8	6.7	KAUAI HI		85			89
		6	4.7	8	4.4	MAUI#HI		-5			83
1	3.3	3*	10.8	3	5.0	IDAHO ID	225	-54	50		
		2*	40.3	1	9.9	BANNOCK ID		-75			93
1*	24.6			1	9.9	BINGHAM ID			-60	13	93
				1*	35.3	BONNEVILLE ID					98
		1*	267.9			OWYHEE ID					
82	2.9	158	3.5	296*	4.6	ILLINOIS IL	22	31	59		
		2	6.1			ALEXANDER IL					
		1	4.7	2	4.8	CHAMPAIGN IL		1			85
1*	149.1					COLES IL					
69	3.0	127	3.4	245*	4.7	COOK IL	14	37	57	68	84
				3	17.0	DU PAGE IL					96
2*	16.2					JACKSON IL					
		1*	14.5			JEFFERSON IL					
1	7.2	1	3.0	1	2.0	KANE IL	-58	-36	-73	07	73
		1	1.4	3	4.6	KANKAKEE IL		216			84
				1	6.6	KNOX IL					89
		1	2.1	1	.4	LAKE IL		-79			70
		1*	18.7	2*	22.2	MC LEAN IL		18			97
1	5.8			6*	14.0	MACON IL			140	83	95
1	2.0	3	5.9			MADISON IL	200				
		2*	20.5			MARION IL					
2*	10.1	1	2.8	1	2.2	PEORIA IL	-72	-20	-78	04	74
		1*	32.6			PERRY IL					
				1*	546.6	PIKE IL					99
				1*	112.4	POPE IL					99
1	5.0	1	4.9	8*	38.1	PULASKI IL	-1	679	668	98	98
1*	27.3					RANDOLPH IL					
		1	10.5			ROCK ISLAND IL					
1	.8	9	3.9	11	3.9	ST CLAIR IL	399	-1	394	96	81
		1	2.5	6*	13.6	SANGAMON IL		445			95
1	6.3					VERMILION IL					98
		1*	33.5	1*	53.0	WARREN IL		58			
		1*	94.4			WHITE IL					
		1	1.9	1	1.2	WILL IL		-33			71
1	11.9	1	5.2	2	3.2	WINNEBAGO IL	-56	-39	-73	06	78
19	2.4	45	3.7	68	4.1	INDIANA IN	56	10	72		
1	3.3	2	3.3	2	2.4	ALLEN IN	0	-28	-27	32	75
				1*	37.7	CASS IN					98
				1	6.4	CLARK IN					89
2*	9.8	2	7.2	3	7.9	DELAWARE IN	-26	10	-19	34	91
1*	27.0					FAYETTE IN					
				1	9.7	FLOYD IN					93
		1*	27.3			GIBSON IN					
		1	7.2	1	5.6	GRANT IN		-22			87
				1	5.6	HOWARD IN					87
		1*	89.3			JACKSON IN					
3	.9	19*	5:8	22	4.9	LAKE IN	520	-16	418	96	85
		1	5.3			LA PORTE IN					
		1	4.0	4*	13.2	MADISON IN		228			95
11	3.5	10	2.0	22	3.3	MARION IN	-42	63	-5	44	79
		1*	589.6			MARSHALL IN					
				1*	53.0	POSEY IN					98
1	3.9	2	4.9	3	3.8	ST JOSEPH IN	28	-23	-2	45	81
				1*	20.2	TIPPECANOE IN					97
		1	1.9	3	4.5	VANDERBURGH IN		144			84
		1	3.2	1	4.4	VIGO IN		36			83
		2*	12.4	1	4.7	WAYNE IN		-62			84
3	3.3	6	5.0	3	1.6	IOWA IA	51	-68	-51		
1*	62.9					APPANOOSE IA					
		1	5.6			BLACK HAWK IA					
		1*	60.7			CLINTON IA					
		1*	11.5			LEE IA					
		1*	21.0			LINN IA					
2	5.7	2	4.3	2	2.6	POLK IA	-25	-39	-54	16	76
				1*	33.5	WAPELLO IA					98
7	1.9	14	2.9	30	4.6	KANSAS KS	50	61	141		
				1*	30.3	BOURBON KS					98
		1*	28.1			BROWN KS					
1*	22.3					COWLEY KS					
				2*	16.4	DOUGLAS KS					96

* SIGNIFICANTLY MORE DEATHS THAN EXPECTED

\# COUNTIES COMBINED; SEE TECHNICAL NOTES

NON-WHITE FEMALE

CANCER OF THE LIVER AND GALLBLADDER
INCLUDING BILE DUCTS

ICD 155,156

1950-1959 DEATHS	1950-1959 RATE	1960-1969 DEATHS	1960-1969 RATE	1970-1979 DEATHS	1970-1979 RATE	COUNTY AND STATE	PERCENT CHANGE 60S/50S	PERCENT CHANGE 70S/60S	PERCENT CHANGE 70S/50S	PERCENTILE %CHANGE 70S/50S	PERCENTILE RATES 1970S
				1*	74.3	EDWARDS KS					99
		1*	31.5			FRANKLIN KS					
		1	18.6			GEARY KS					
				1*	24.8	JACKSON KS					97
1*	24.9	1*	21.3	1	6.7	JOHNSON KS	-14	-68	-73	07	90
				1	6.7	LABETTE KS					90
1	5.9			2	9.8	LEAVENWORTH KS			67	72	93
				1	4.7	MONTGOMERY KS					85
				1	5.7	RENO KS					87
				1*	10.6	RILEY KS					94
		1	2.3	2	1.3	SEDGWICK KS		-42			71
1	3.0	2	3.7	5	7.4	SHAWNEE KS	23	100	147	85	90
3	2.4	7	3.8	11	5.3	WYANDOTTE KS	61	38	122	81	86
24	2.2	33	2.7	57	4.3	KENTUCKY KY	26	57	98		
1*	22.9					ANDERSON KY					
				1	7.5	BARREN KY					91
				1	9.2	BELL KY					93
				2	10.4	BOYLE KY					93
1*	22.1	1*	21.1	2*	18.4	CALDWELL KY	-5	-13	-17	36	96
1	1.9			1	1.9	CHRISTIAN KY			4	49	73
1	6.4	1	5.7	1	5.3	CLARK KY	-10	-7	-17	36	86
				2	10.8	DAVIESS KY					94
4	3.8	3	2.7	5	3.5	FAYETTE KY	-27	29	-6	43	80
1*	9.9					FULTON KY					
1	9.6			2*	26.9	GRAVES KY			180	89	97
				1	10.5	HARDIN KY					93
1	7.2			2*	16.8	HARLAN KY			135	83	96
1*	17.2					HART KY					
1	4.9	1	3.8			HENDERSON KY	-22				
				1	3.6	HOPKINS KY					80
7	2.1	19	4.3	24	4.6	JEFFERSON KY	102	5	113	80	84
		1*	34.4			LARUE KY					
		1	7.0	1	8.3	LOGAN KY		20			92
				2	4.9	MC CRACKEN KY					85
2*	12.9					MADISON KY					
				2*	19.3	MASON KY					96
		1	10.4			MONTGOMERY KY					
		1	11.1	1	14.9	NELSON KY		34			96
				1*	60.7	OHIO KY					98
1	9.9			1	8.4	SCOTT KY			-16	36	92
		1	9.6	1	8.8	SHELBY KY		-8			92
				1	15.7	TAYLOR KY					96
				1	9.6	TODD KY					93
1	3.3	1	3.8	1	5.1	WARREN KY	15	32	51	65	85
		2*	10.8			WEBSTER KY					
45	1.2	83	1.8	188	3.5	LOUISIANA LA	54	93	197		
1	2.6			2	4.0	ACADIA LA			55	67	81
		2	9.0	1	4.2	ALLEN LA		-53			82
		3	6.1	5*	10.7	ASCENSION LA		74			94
				2	5.2	ASSUMPTION LA					86
				1	2.6	AVOYELLES LA					75
1	4.0	1	2.6	2	3.7	BIENVILLE LA	-35	40	-8	41	80
		1	1.7	1	1.1	BOSSIER LA		-36			70
3	1.1	4	1.0	13	2.7	CADDO LA	-6	169	153	85	76
1	1.3			6	4.3	CALCASIEU LA			238	92	83
		1	10.3			CALDWELL LA					
		1	1.4	1	1.5	CLAIBORNE LA		5			72
		2	5.0			CONCORDIA LA					
				1	1.7	DE SOTO LA					72
4	1.4	5	1.8	13	3.2	EAST BATON ROUGE LA	25	81	127	82	78
		1	2.0	3	7.0	EAST CARROLL LA		242			90
		1	1.8	1	1.5	EAST FELICIANA LA		-14			72
		1	3.9			EVANGELINE LA					
				2	3.6	FRANKLIN LA					80
1	2.3	4	6.5	4	5.4	IBERIA LA	180	-18	129	82	86
		1	1.8	4	5.5	IBERVILLE LA		210			87
1	8.6					JACKSON LA					
1	1.0	2	1.8	3	1.8	JEFFERSON LA	79	-1	77	75	73
				2	7.7	JEFFERSON DAVIS LA					91
		2	3.0	5	5.7	LAFAYETTE LA		88			88
				2	7.4	LAFOURCHE LA					90
				1	7.2	LA SALLE LA					90
1	2.2			1	1.4	LINCOLN LA			-37	28	71
		1	7.3	1	5.9	LIVINGSTON LA		-18			88
1	1.8	1	1.4	4	6.3	MADISON LA	-19	339	256	94	89
		1	1.2	1	.9	MOREHOUSE LA		-21			70
1	1.1	2	2.9	4	4.0	NATCHITOCHES LA	166	38	267	94	82
19	2.1	19	1.7	55	4.2	ORLEANS LA	-19	145	99	78	82
		5	3.2	2	1.2	OUACHITA LA		-62			71
1	5.1	1	5.1			PLAQUEMINES LA	-2				
				3	5.3	POINTE COUPEE LA					86
1	.6	2	1.2	3	1.8	RAPIDES LA	103	47	197	90	73
				1	1.6	RICHLAND LA					72
				1	4.0	SABINE LA					82
		1	4.8	3*	10.8	ST CHARLES LA		122			94
		3	8.6	2	5.3	ST JAMES LA		-38			86
1	2.1	1	2.7	1	2.7	ST JOHN THE BAPT LA	27	1	28	57	76
3	3.0	3	1.9	5	3.2	ST LANDRY LA	-36	66	6	50	78
1	3.8			2	4.5	ST MARTIN LA			19	55	84
		1	1.4	5	7.3	ST MARY LA		402			90
				2	4.3	ST TAMMANY LA					83
		1	1.2			TANGIPAHOA LA					
		1	1.8	1	1.8	TENSAS LA		-4			72
1	2.0	1	2.1			TERREBONNE LA	4				
		1	2.7	1	3.0	UNION LA		11			77
				2	7.1	VERMILION LA					90

* SIGNIFICANTLY MORE DEATHS THAN EXPECTED

NON-WHITE FEMALE

CANCER OF THE LIVER AND GALLBLADDER INCLUDING BILE DUCTS

ICD 155,156

1950-1959 DEATHS	1950-1959 RATE	1960-1969 DEATHS	1960-1969 RATE	1970-1979 DEATHS	1970-1979 RATE	COUNTY AND STATE	PERCENT CHANGE 60S/50S	PERCENT CHANGE 70S/60S	PERCENT CHANGE 70S/50S	PERCENTILE %CHANGE 70S/50S	PERCENTILE RATES 1970S
		1	5.7			VERNON LA					
1	1.5	1	1.5	1	1.5	WASHINGTON LA	4	-3	1	47	72
1	1.9	1	1.1	2	2.4	WEBSTER LA	-41	111	25	56	75
				2	6.6	WEST BATON ROUGE LA					89
		2*	15.6	2*	15.7	WEST CARROLL LA		1			96
		1	5.1	1	6.2	WEST FELICIANA LA		22			88
				5*	18.7	WINN LA					96
2*	17.9			1	3.4	MAINE ME			-81		
				1*	22.3	CUMBERLAND ME					97
2*	135.7					WASHINGTON ME					
33	2.1	74	3.4	130	4.2	MARYLAND MD	58	25	98		
		1	.8	9	6.6	ANNE ARUNDEL MD		735			89
1	2.2	2	2.4	7	5.7	BALTIMORE MD	12	135	162	87	87
		2	10.4			CALVERT MD					
1	7.0					CAROLINE MD					
				1	8.2	CECIL MD					92
				1	2.7	CHARLES MD					76
		1	2.1	3	6.4	DORCHESTER MD		208			89
				2	6.5	FREDERICK MD					89
1	5.4			1	2.0	HARFORD MD			-64	11	73
1	6.1	1	5.2	2	8.0	KENT MD	-15	54	30	57	91
		1	1.3	5	3.5	MONTGOMERY MD		163			79
		3	2.9	10	2.7	PRINCE GEORGE'S MD		-8			76
1	5.0			1	4.2	QUEEN ANNE'S MD			-15	37	82
		1	3.7	2	5.9	ST MARY'S MD		57			88
				2	4.6	SOMERSET MD					84
1	4.4			2	5.2	TALBOT MD			19	55	86
				2*	14.5	WASHINGTON MD					95
2	7.7	2	4.1	2	3.0	WICOMICO MD	-47	-27	-61	12	78
		1	3.3	2	7.1	WORCESTER MD		113			90
25	2.6	59*	4.4	76	4.3	BALTIMORE CITY MD	67	-2	63	70	82
22*	5.8	29*	5.3	30	3.2	MASSACHUSETTS MA	-8	-40	-45		
				1	9.5	BERKSHIRE MA					93
				1	2.1	BRISTOL MA					74
3*	10.1	2	3.9	5	5.4	HAMPDEN MA	-61	37	-47	21	87
6*	9.7	5	6.4	2	1.4	MIDDLESEX MA	-34	-78	-86	01	71
		1	5.9	1	2.2	NORFOLK MA		-64			74
		3*	19.8	2	4.0	PLYMOUTH MA		-80			81
13*	6.6	18*	6.0	18	3.8	SUFFOLK MA	-9	-37	-42	25	81
46	2.8	120*	3.9	142	3.2	MICHIGAN MI	41	-19	15		
				1	10.5	ALLEGAN MI					93
1	1.8	3	6.2	2	2.6	BERRIEN MI	239	-58	42	62	76
1	5.2			2	3.8	CALHOUN MI			-27	33	81
1	4.3	1	5.5	4*	13.0	CASS MI	26	138	201	91	95
1*	23.6					EMMET MI					
		3	1.6	6	3.8	GENESEE MI		136			81
				2	3.7	INGHAM MI					80
1	6.4					JACKSON MI					
				1	3.4	KALAMAZOO MI					79
1	5.7	3	4.2	3	3.7	KENT MI	-27	-10	-35	29	80
1	4.6					LAKE MI					
				1*	44.6	MACKINAC MI					98
1	6.2	2	5.8	1	1.1	MACOMB MI	-7	-81	-83	02	70
1*	62.9					MECOSTA MI					
1	2.8	3	7.8	3	4.0	MUSKEGON MI	176	-49	40	60	81
		3	3.9	4	2.6	OAKLAND MI		-34			76
1	1.0	5*	7.7	3	3.4	SAGINAW MI	660	-55	241	93	79
1*	7.8					ST CLAIR MI					
1*	38.3	1*	12.1			ST JOSEPH MI	-68				
				1	2.2	VAN BUREN MI					74
1	3.5	3	6.9	1	1.1	WASHTENAW MI	98	-84	-69	10	70
32	2.5	93*	4.1	107	3.3	WAYNE MI	61	-20	28	57	79
4	3.4	9	5.8	16*	6.1	MINNESOTA MN	72	4	79		
		1*	12.4	2*	20.2	BELTRAMI MN		63			97
				1*	267.9	BLUE EARTH MN					99
				2*	84.6	CARLTON MN					99
2*	45.8	1*	7.4	1*	17.1	CASS MN	-84	130	-63	11	96
		4	6.4	2	1.2	HENNEPIN MN		-82			71
				1*	74.3	ITASCA MN					99
1*	172.0			1*	188.7	KOOCHICHING MN			10	51	99
		1*	35.1			MAHNOMEN MN					
1*	172.0	1*	267.9			PINE MN	56				
				6*	7.9	RAMSEY MN					91
		1*	94.4			YELLOW MEDICINE MN					
33	.9	69	1.8	138	3.1	MISSISSIPPI MS	89	74	229		
1	1.5	2	2.3	5	3.8	ADAMS MS	54	68	160	87	81
1	5.9					ALCORN MS					
		2	4.8	4	8.6	ATTALA MS		77			92
1	.6	2	1.1	4	2.3	BOLIVAR MS	77	105	263	94	75
1	3.2	1	3.7	1	4.1	CARROLL MS	18	11	30	58	82
				2	5.6	CHICKASAW MS					87
				1	6.8	CHOCTAW MS					90
1	2.9	1	2.8	3	7.5	CLAIBORNE MS	-3	166	159	86	91
				1	2.7	CLARKE MS					76
		1	1.7	2	4.9	CLAY MS		184			85
2	1.2	1	.7	5	3.3	COAHOMA MS	-42	363	169	87	79
2	4.0			3	3.5	COPIAH MS			-12	38	79
				1	3.8	COVINGTON MS					80

* SIGNIFICANTLY MORE DEATHS THAN EXPECTED

NON-WHITE FEMALE

CANCER OF THE LIVER AND GALLBLADDER INCLUDING BILE DUCTS

ICD 155,156

COUNTY AND STATE	1950-1959 DEATHS	1950-1959 RATE	1960-1969 DEATHS	1960-1969 RATE	1970-1979 DEATHS	1970-1979 RATE	PERCENT CHANGE 60S/50S	PERCENT CHANGE 70S/60S	PERCENT CHANGE 70S/50S	PERCENTILE %CHANGE 70S/50S	PERCENTILE RATES 1970S
DE SOTO MS					1	1.4					71
FORREST MS			1	1.2	1	1.3		7			71
FRANKLIN MS					3*	14.6					96
GRENADA MS			1	2.4	2	3.7		55			80
HANCOCK MS					1	8.9					92
HARRISON MS			2	2.5	5	4.3		69			83
HINDS MS	1	.4	7	2.6	19	4.5	589	74	1100	99	83
HOLMES MS	2	1.7	2	2.0	3	3.8	20	92	130	82	81
HUMPHREYS MS	1	2.1									
JACKSON MS	1	5.1	1	2.9			-44				
JASPER MS			1	2.8							
JEFFERSON MS			1	2.0							
JEFFERSON DAVIS MS			1	4.0							
JONES MS					2	2.5					75
KEMPER MS			1	2.4							
LAFAYETTE MS			1	3.2	2	5.5		72			87
LAMAR MS			1	9.0	1	7.6		-15			91
LAUDERDALE MS			5	4.3	2	1.7		-60			72
LAWRENCE MS			1	4.9							
LEAKE MS			2	7.0							
LEE MS	1	2.2	1	2.1	2	3.6	-1	67	65	71	80
LEFLORE MS	1	1.6	1	.8	4	2.7	-46	222	73	73	76
LINCOLN MS					2	5.0					85
LOWNDES MS	2	2.9	1	1.5	3	2.7	-49	83	-6	43	76
MADISON MS	1	1.0	1	.7	3	2.4	-26	224	140	83	75
MARION MS	1	6.2	1	3.2			-49				
MARSHALL MS			1	1.5	2	3.0		103			77
MONROE MS	2	3.3			4	5.8			76	74	88
MONTGOMERY MS	1	3.4									
NESHOBA MS					2	8.1					91
NEWTON MS	1	4.4			1	2.3			-47	20	74
NOXUBEE MS	1	2.2									
OKTIBBEHA MS	2	5.2	1	2.3			-56				
PANOLA MS	1	1.5	2	3.2	4	5.9	106	87	286	95	88
PIKE MS			2	3.0	4	4.9		63			85
PRENTISS MS			1	11.2	1	8.3		-26			92
QUITMAN MS	1	2.3									
RANKIN MS	1	1.9									
SCOTT MS					2	5.2					86
SHARKEY MS					1	4.4					83
SIMPSON MS			2	8.0	1	2.6		-68			75
STONE MS			1	14.0	1	12.4		-12			94
SUNFLOWER MS	1	.7			2	2.4			266	94	75
TALLAHATCHIE MS					3	4.4					83
TATE MS					2	6.4					89
TIPPAH MS					1	9.1					93
TUNICA MS			4*	8.0	4*	8.2		3			92
WARREN MS	1	.7	3	2.7	2	1.3	300	-50	98	78	71
WASHINGTON MS			5	2.4	7	3.1		32			78
WILKINSON MS			1	2.9	2	5.2		80			86
YALOBUSHA MS	1	5.0									
YAZOO MS			3	4.2	4	4.4		4			83
MISSOURI MO	50	3.3	80*	4.0	111	4.5	22	12	36		
AUDRAIN MO	1*	10.6			2*	15.0			41	61	96
BOONE MO			3*	14.7	1	4.7		-68			85
BUCHANAN MO	1	5.4	1	5.4	1	4.8	0	-10	-10	41	85
BUTLER MO			1	11.8	1	14.9		26			96
CALLAWAY MO			2*	16.0	1	7.7		-52			91
CAPE GIRARDEAU MO			1*	20.6							
CHARITON MO			1*	26.5							
CLINTON MO	1*	49.2									
COOPER MO	2*	25.3									
DUNKLIN MO					1	9.6					93
FRANKLIN MO			1*	28.3							
HENRY MO	1*	37.7									
JACKSON MO	11	3.5	23*	5.3	33*	5.9	54	11	70	73	88
JASPER MO	1*	15.5			1	8.9			-42	25	92
JEFFERSON MO			1*	37.7							
MARION MO	1	7.8									
MONTGOMERY MO	1*	23.6									
MORGAN MO					1*	62.9					98
NEW MADRID MO			1	5.0	1	6.5		29			89
PETTIS MO			1	5.7							
PHELPS MO			1*	62.9							
RANDOLPH MO	2*	21.2									
RAY MO	1*	23.6			1*	26.5			12	52	97
ST LOUIS COUNTY MO	3	3.8	1	.6	7	3.5	-85	537	-6	43	80
WARREN MO	1*	81.8									
ST LOUIS CITY MO	23	2.9	42	4.0	60	4.8	39	21	68	72	85
MONTANA MT	10*	18.7	8*	12.1	13*	14.7	-35	21	-22		
BEAVERHEAD MT					1*	424.0					99
BIG HOPN MT	2*	22.1	2*	22.8			3				
BLAINE MT			2*	44.8	2*	39.5		-12			98
CASCADE MT	1*	36.8									
CUSTER MT	1*	490.8									
GLACIER MT	1	10.1	2*	17.7	2*	13.5	76	-24	34	59	95
HILL MT					1*	31.3					98
LAKE MT	1*	17.2			1	12.6			-27	32	95
LEWIS AND CLARK MT	1*	68.8									
PHILLIPS MT					1*	49.2					98
ROOSEVELT MT	1*	18.2	1	15.8			-13				
ROSEBUD MT					2*	31.5					98
SANDERS MT			1*	74.3							
TOOLE MT					2*	565.1					99
VALLEY MT	1*	133.9			1*	74.3			-44	22	99
YELLOWSTONE MT	1*	56.2									

* SIGNIFICANTLY MORE DEATHS THAN EXPECTED

NON-WHITE FEMALE

CANCER OF THE LIVER AND GALLBLADDER INCLUDING BILE DUCTS

ICD 155,156

1950-1959 DEATHS	1950-1959 RATE	1960-1969 DEATHS	1960-1969 RATE	1970-1979 DEATHS	1970-1979 RATE	COUNTY AND STATE	PERCENT CHANGE 60S/50S	PERCENT CHANGE 70S/60S	PERCENT CHANGE 70S/50S	PERCENTILE %CHANGE 70S/50S	PERCENTILE RATES 1970S
8*	7.1	10*	6.9	12	6.1	NEBRASKA NB	-2	-12	-14		
		1*	596.2			CASS NB					
3	3.3	4	3.8	7	4.9	DOUGLAS NB	15	27	46	63	85
3*	184.6	2*	175.3			KNOX NB	-5				
1*	267.9					POLK NB					
1*	344.1					RICHARDSON NB					
		3*	49.8	5*	65.0	THURSTON NB		31			98
3*	8.2	7*	14.2	5	3.1	NEVADA NV	74	-78	-62		
1*	43.0					CHURCHILL NV					
		1	1.9	4	4.4	CLARK NV		128			83
1*	18.6	1*	21.0			ELKO NV	13				
1*	212.0	1*	74.3			LANDER NV	-65				
		2*	20.5	1	2.1	WASHOE NV		-90			74
		1*	112.4			WHITE PINE NV					
		1*	53.6			ORMSBY-CARSON CITY NV					
41	2.7	70	3.1	116	3.3	NEW JERSEY NJ	13	7	21		
3	2.5	6	3.4	6	2.7	ATLANTIC NJ	37	-20	10	51	77
2	3.7	4	3.7	5	2.3	BERGEN NJ	1	-38	-38	27	74
1	3.6			3	3.6	BURLINGTON NJ			2	48	80
3	2.4	5	3.1	10	4.0	CAMDEN NJ	30	27	66	71	81
1	9.4	1	4.8	1	3.3	CAPE MAY NJ	-48	-32	-65	11	79
1	1.6	2	5.7	4	4.5	CUMBERLAND NJ	246	-20	178	88	84
11	2.3	25	3.3	46	4.2	ESSEX NJ	44	28	85	76	82
2	4.0	1	1.6	2	2.5	GLOUCESTER NJ	-59	54	-37	27	75
6*	6.3	5	3.6	5	1.6	HUDSON NJ	-43	-56	-75	06	72
		7	5.2	9	4.7	MERCER NJ		-9			85
2	7.8	2	3.0	3	3.5	MIDDLESEX NJ	-62	17	-55	15	79
3	2.5	2	1.4	4	1.9	MONMOUTH NJ	-45	33	-27	32	73
				4	8.2	MORRIS NJ					92
1*	8.8					OCEAN NJ					
1	.9	3	4.5	4	1.8	PASSAIC NJ	379	-60	90	77	73
				2	5.1	SALEM NJ					86
		1	5.3	1	3.7	SOMERSET NJ		-31			80
4	2.6	6	3.7	7	2.3	UNION NJ	43	-38	-11	39	74
14*	11.2	27*	14.5	43*	7.8	NEW MEXICO NM	29	-46	-30		
1	7.3	4*	12.6	3	4.6	BERNALILLO NM	72	-64	-37	27	84
				1	6.6	EDDY NM					89
				1	5.9	LEA NM					88
8*	23.8	9*	18.8	15*	18.7	MC KINLEY NM	-21	-1	-21	34	96
		2*	73.4	1	7.0	OTERO NM		-90			90
				1*	47.1	QUAY NM					98
1*	16.4	1*	14.1	1	1.5	RIO ARRIBA NM	-14	-89	-91	00	72
				4*	12.1	SANDOVAL NM					94
		8*	27.0	8*	13.7	SAN JUAN NM		-49			95
				1*	9.5	SOCORRO NM					93
2*	48.8			4*	14.5	TAOS NM			-70	09	95
2*	14.2	3*	18.6	3	8.1	VALENCIA NM	31	-56	-43	24	91
114	2.9	217	3.2	427	3.8	NEW YORK NY	10	19	31		
2	6.9	1	2.4	2	2.9	ALBANY NY	-65	21	-58	14	77
4	.7	17	1.7	56	3.1	BRONX NY	144	85	351	96	78
		2*	25.2			CATTARAUGUS NY					
				1*	109.3	CORTLAND NY					99
		1	2.1			DUTCHESS NY					
8	4.7	6	1.9	19	4.8	ERIE NY	-59	146	1	46	85
		1*	25.8			GENESEE NY					
17	2.4	46	2.8	116	4.1	KINGS(BROOKLYN) NY	16	48	72	73	82
		1*	14.7	1*	13.7	LIVINGSTON NY		-7			95
		2	2.9	4	2.4	MONROE NY		-15			75
5	4.6	7	4.5	11	3.4	NASSAU NY	-4	-24	-27	32	79
47	2.8	93*	4.3	120	4.3	NEW YORK(MANHATTAN) NY	54	-2	51	64	83
				2	4.5	NIAGARA NY					84
1*	23.6			2	7.0	ONEIDA NY			-70	09	90
2*	6.9	1	2.4	4	4.6	ONONDAGA NY	-66	96	-33	29	84
		1	1.7	2	4.4	ORANGE NY		165			83
15*	5.5	17	1.8	55	3.4	QUEENS NY	-67	84	-38	27	79
1*	15.3					RENSSELAER NY					
1	8.2	3	10.2	2	2.5	RICHMOND NY	24	-76	-70	09	75
		1	4.8	2	3.4	ROCKLAND NY		-29			79
1*	16.3					SARATOGA NY					
		1	7.9	1	4.3	SCHENECTADY NY		-46			83
		5	3.4	8	2.8	SUFFOLK NY		-18			77
				1	9.6	SULLIVAN NY					93
				1	9.4	TOMPKINS NY					93
		1	6.2	3*	14.0	ULSTER NY		125			95
10*	5.2	10	3.0	14	2.7	WESTCHESTER NY	-41	-11	-48	19	76
88	2.4	92	2.1	176	3.1	NORTH CAROLINA NC	-13	47	29		
2	3.4	4	6.9	4	4.4	ALAMANCE NC	103	-35	31	58	83
				1	14.0	ALEXANDER NC					95
3*	5.9	2	4.0	1	2.0	ANSON NC	-33	-50	-67	10	73
				2	2.9	BEAUFORT NC					77
1	1.9			3	4.8	BERTIE NC			156	86	85
		2	5.1	3	4.6	BLADEN NC		-9			84
		1	3.5	1	2.8	BRUNSWICK NC		-21			77
1	1.8	3	3.4	3	2.7	BUNCOMBE NC	87	-19	51	64	77
1	8.6			1	3.6	BURKE NC			-58	14	80
1	2.6			1	2.3	CABARRUS NC			-11	39	75
		1	6.9	1	7.9	CALDWELL NC		15			91
3*	21.9	1	24.8	1	5.2	CARTERET NC	13	-79	-76	05	86

* SIGNIFICANTLY MORE DEATHS THAN EXPECTED

NON-WHITE FEMALE

CANCER OF THE LIVER AND GALLBLADDER INCLUDING BILE DUCTS

ICD 155,156

1950-1959 DEATHS	1950-1959 RATE	1960-1969 DEATHS	1960-1969 RATE	1970-1979 DEATHS	1970-1979 RATE	COUNTY AND STATE	PERCENT CHANGE 60S/50S	PERCENT CHANGE 70S/60S	PERCENT CHANGE 70S/50S	PERCENTILE %CHANGE 70S/50S	PERCENTILE RATES 1970S
		1	3.5	2	5.3	CASWELL NC		48			86
		1	3.2	1	3.3	CATAWBA NC		1			78
1	2.3	1	3.4	1	3.9	CHATHAM NC	44	15	65	71	81
		1	5.0	1	5.1	CHOWAN NC		2			85
1	2.2	1	2.0	4	6.4	CLEVELAND NC	-9	217	190	90	89
5*	8.9			5	6.1	COLUMBUS NC			-32	30	88
1	1.9	1	1.6	4	5.4	CRAVEN NC	-17	236	179	89	86
		2	1.2	6	3.6	CUMBERLAND NC		206			80
				2	4.3	DAVIDSON NC					83
		1	2.1	1	1.6	DUPLIN NC		-23			72
		2	1.4	10	4.3	DURHAM NC		199			83
2	2.2			4	3.2	EDGECOMBE NC			48	63	78
5	3.0	11	5.1	7	2.6	FORSYTH NC	71	-48	-12	38	76
2	5.1			2	2.2	FRANKLIN NC			-57	14	74
2	4.1			3	3.6	GASTON NC			-12	38	80
		1	1.5	4	5.6	GRANVILLE NC		265			87
1	6.7					GREENE NC					
4	2.4	6	2.6	10	3.4	GUILFORD NC	8	31	41	61	79
3	3.6	1	1.0	2	1.6	HALIFAX NC	-72	59	-55	15	72
1	3.3	3	5.9			HARNETT NC	77				
		1	9.6			HENDERSON NC					
		1	2.1	2	3.1	HERTFORD NC		48			78
1	3.8					HOKE NC					
		3	7.0	2	3.2	IREDELL NC		-55			78
		1*	18.7	1	11.8	JACKSON NC		-37			94
				1	1.5	JOHNSTON NC					72
2	2.7			1	.8	LENOIR NC			-71	08	70
				1	5.5	LINCOLN NC					87
3*	42.8					MC DOWELL NC					
1	2.9			5*	10.0	MARTIN NC			239	92	93
15*	8.3	5	1.8	10	2.6	MECKLENBURG NC	-79	48	-68	10	76
2	5.2					MOORE NC					
2	3.2	2	2.5	1	1.0	NASH NC	-22	-60	-69	09	70
1	1.4	3	3.5	7	5.7	NEW HANOVER NC	148	63	304	95	87
		1	2.1	4	5.8	NORTHAMPTON NC		172			88
1	3.5			1	1.9	ORANGE NC			-45	22	73
		3	8.0			PASQUOTANK NC					
1	2.9					PENDER NC					
1	6.5			1	3.4	PERQUIMANS NC			-47	20	79
				4*	9.1	PERSON NC					93
5	6.0	1	1.0			PITT NC	-84				
				1	9.2	POLK NC					93
1	7.9			4*	13.0	RANDOLPH NC			65	71	95
1	1.8	1	1.8	2	2.9	RICHMOND NC	1	62	64	70	77
1	.7	2	1.3	5	2.5	ROBESON NC	77	95	246	93	75
1	1.1	3	5.6	5	7.5	ROCKINGHAM NC	417	33	589	98	91
1	2.3	2	3.4	3	3.8	ROWAN NC	44	11	60	68	80
1	5.2	1	4.1	1	2.7	RUTHERFORD NC	-20	-34	-48	19	76
				1	1.3	SAMPSON NC					71
1	2.7	2	4.3	2	4.3	SCOTLAND NC	59	1	62	69	83
				1	2.6	STANLY NC					76
		1*	15.7			STOKES NC					
		1*	25.8			SWAIN NC					
				1	8.6	TYRRELL NC					92
				1	2.1	UNION NC					74
1	2.2	2	3.3	2	2.6	VANCE NC	48	-20	19	54	76
		1	.6	7	2.7	WAKE NC		388			76
2	4.2			1	1.8	WARREN NC			-57	15	73
				3*	12.1	WASHINGTON NC					94
2	1.4	3	2.3	2	1.5	WAYNE NC	71	-36	10	51	72
		2*	18.4	1	6.8	WILKES NC		-63			90
1	1.1	3	3.3	1	1.0	WILSON NC	193	-69	-11	40	70
2	7.6	7*	21.2	8*	16.0	NORTH DAKOTA ND	180	-24	111		
1*	81.8					DUNN ND					
		1*	112.4	2*	89.3	MC LEAN ND		-21			99
				2*	76.0	MOUNTRAIL ND					99
		1*	117.9			RICHLAND ND					
1	10.3			3*	14.3	ROLETTE ND			39	60	95
		4*	88.6	1*	17.9	SIOUX ND		-80			96
		1*	344.1			WARD ND					
90*	4.1	150*	4.5	194	4.2	OHIO OH	11	-6	5		
		1	5.4			ALLEN OH					
				1*	68.8	ASHLAND OH					99
1	8.4			2*	14.2	BELMONT OH			70	73	95
2	9.0	1	2.0	2	3.5	BUTLER OH	-77	70	-62	12	79
1	1.9	1	1.6	5	6.7	CLARK OH	-17	318	247	93	89
		1*	23.6			CLINTON OH					
		1*	26.0			COSHOCTON OH					
		1*	33.5			CRAWFORD OH					
29*	3.9	40	3.6	68	4.7	CUYAHOGA OH	-10	31	18	54	84
				1*	18.1	DELAWARE OH					96
		1*	38.3	1*	17.2	FAYETTE OH		-55			96
9	3.9	16	4.4	17	3.6	FRANKLIN OH	12	-17	-8	42	80
1*	11.2					GALLIA OH					
3*	12.2	3*	12.0	2	5.7	GREENE OH	-1	-53	-53	17	87
19*	4.6	26	4.6	35	4.6	HAMILTON OH	0	2	2	48	84
1*	56.2					HANCOCK OH					
1*	28.1					HARDIN OH					
1*	18.9	1*	28.1			HIGHLAND OH	49				
				3*	10.6	JEFFERSON OH					94
				1	8.6	LAKE OH					92
1*	12.6			1	10.5	LAWRENCE OH			-17	35	93
		1*	16.7			LICKING OH					
1	5.1	3	6.9			LORAIN OH	35				
3	2.6	11*	6.0	13	4.9	LUCAS OH	129	-19	86	76	85
		3*	72.1			MADISON OH					

* SIGNIFICANTLY MORE DEATHS THAN EXPECTED

1950-1959 DEATHS	1950-1959 RATE	1960-1969 DEATHS	1960-1969 RATE	1970-1979 DEATHS	1970-1979 RATE	COUNTY AND STATE	PERCENT CHANGE 60S/50S	PERCENT CHANGE 70S/60S	PERCENT CHANGE 70S/50S	PERCENTILE %CHANGE 70S/50S	PERCENTILE RATES 1970S
2	1.7	7	5.4	8	4.1	MAHONING OH	225	-24	147	85	82
				1	9.4	MIAMI OH					93
5	2.6	14*	6.6	15	3.8	MONTGOMERY OH	148	-42	44	62	81
1	9.4	1	6.5			MUSKINGUM OH	-32				
		1*	60.7			PAULDING OH					
1*	15.6					PORTAGE OH					
1	9.3			1	1.7	RICHLAND OH			-81	03	72
				1	5.9	ROSS OH					88
1*	9.4					SCIOTO OH					
3	5.4	4	6.7	1	.9	STARK OH	24	-87	-84	01	70
2	2.2	7	5.0	10	4.8	SUMMIT OH	127	-4	118	81	85
1	1.9	2	3.7	3	3.5	TRUMBULL OH	95	-5	86	77	79
				1*	23.6	TUSCARAWAS OH					97
		1*	26.8			WARREN OH					
				1*	14.1	WASHINGTON OH					95
		2*	35.8			WAYNE OH					
26	2.8	54*	4.6	78	4.8	OKLAHOMA OK	62	5	71		
		2*	12.6	1	3.9	ADAIR OK		-69			81
		1*	23.8			ATOKA OK					
		1*.	27.0			BECKHAM OK					
		2*	25.4			BLAINE OK					
		1	7.4			BRYAN OK					
1	7.7	2*	10.5	7*	29.0	CADDO OK	37	176	278	95	98
				1	7.5	CANADIAN OK					91
		2	8.0			CARTER OK					
1	6.1	2*	12.2			CHEROKEE OK	100				
1	4.9	1	3.3	1	4.0	CHOCTAW OK	-33	22	-18	35	81
2*	10.2	1	3.4	2	5.4	COMANCHE OK	-67	58	-47	20	86
		1	5.6			CRAIG OK					
1	4.5	1	3.4	1	4.2	CREEK OK	-24	21	-8	42	82
1*	10.6					CUSTER OK					
		1	9.0			DELAWARE OK					
				1*	78.1	DEWEY OK					99
		1	28.1	2*	15.6	GARFIELD OK		-44			96
		1*	8.4			GARVIN OK					
1	12.5			1	6.1	GRADY OK			-51	18	88
1	10.3	1	7.7	1	5.9	HUGHES OK	-26	-23	-43	24	88
		2*	16.6	2*	11.0	KAY OK		-34			94
1*	20.4			1	11.8	KIOWA OK			-42	25	94
				1*	11.2	LATIMER OK					94
		1	9.0	1	3.4	LE FLORE OK		-62			79
2	5.4	2	6.2	5*	15.0	MC CURTAIN OK	15	142	177	88	96
1	6.4	1	5.2	2	5.4	MC INTOSH OK	-17	3	-15	36	86
				1*	21.2	MARSHALL OK					97
		3*	28.1	1	6.4	MAYES OK		-77			89
		3	2.4	8	6.9	MUSKOGEE OK		191			90
				1*	23.6	NOBLE OK					97
				1	5.6	NOWATA OK					87
		1	4.0	3*	13.0	OKFUSKEE OK		225			95
5	3.4	7	3.0	4	1.4	OKLAHOMA OK	-12	-53	-59	13	71
		2	4.4	2	4.3	OKMULGEE OK		-1			83
1	8.5			1	3.5	OSAGE OK			-59	13	80
1*	14.8			1	5.5	OTTAWA OK			-63	12	87
1*	22.7					PAWNEE OK					
				1	10.8	PAYNE OK					94
		3*	14.1	1	2.1	PITTSBURG OK		-85			74
		1	8.4			PONTOTOC OK					
		1	6.6	2	8.6	POTTAWATOMIE OK		31			92
2*	9.4	1	3.7	5*	19.7	SEMINOLE OK	-60	429	110	79	96
				1	6.5	SEQUOYAH OK					89
				2*	24.6	STEPHENS OK					97
				1	10.6	TILLMAN OK					94
3	3.0	3	1.7	11	4.3	TULSA OK	-45	159	42	61	83
		2*	10.7	1	4.2	WASHINGTON OK		-61			82
2	2.7	6	4.9	10	4.0	OREGON OR	79	-18	47		
				1	11.1	CLACKAMAS OR					94
		1*	42.4			COOS OR					
1*	89.3					JEFFERSON OR					
		2*	32.5			KLAMATH OR					
		1*	89.3			LINCOLN OR					
				1	16.1	MARION OR					96
1	2.2	2	2.7	5	3.4	MULTNOMAH OR	18	27	50	64	79
				3*	42.1	UMATILLA OR					98
86	2.8	132	3.3	208	3.9	PENNSYLVANIA PA	16	20	39		
19*	3.3	19	3.1	39	4.6	ALLEGHENY PA	-8	51	39	60	84
1	2.5			2	3.9	BEAVER PA			55	67	81
		1	6.0	1	2.4	BERKS PA		-59			75
1*	17.5			1	11.5	BLAIR PA			-35	29	94
				1*	133.9	BRADFORD PA					99
1	9.9			1	1.1	BUCKS PA			-89	00	70
				1*	15.2	BUTLER PA					96
		1	7.3	1	5.4	CAMBRIA PA		-26			87
2	3.6	4	5.2	5	5.6	CHESTER PA	47	6	56	67	87
		1*	86.0			CLEARFIELD PA					
				1	5.7	CUMBERLAND PA					87
2	1.5	5	5.0	2	1.4	DAUPHIN PA	223	-72	-10	41	71
4	3.6	11*	6.5	9	3.8	DELAWARE PA	80	-41	6	49	81
1	3.0	1	5.6	1	4.5	ERIE PA	89	-20	51	65	83
		2	5.3	1	2.1	FAYETTE PA		-60			74
				1*	596.2	FOREST PA					99
				1	9.0	FRANKLIN PA					92
1	7.8	2*	10.9	1	3.7	LANCASTER PA	39	-66	-52	17	80
1	5.8					LAWRENCE PA					
				1	12.6	LEHIGH PA					95

* SIGNIFICANTLY MORE DEATHS THAN EXPECTED

NON-WHITE FEMALE

CANCER OF THE LIVER AND GALLBLADDER INCLUDING BILE DUCTS

ICD 155,156

1950-1959 DEATHS	1950-1959 RATE	1960-1969 DEATHS	1960-1969 RATE	1970-1979 DEATHS	1970-1979 RATE	COUNTY AND STATE	PERCENT CHANGE 60S/50S	PERCENT CHANGE 70S/60S	PERCENT CHANGE 70S/50S	PERCENTILE %CHANGE 70S/50S	PERCENTILE RATES 1970S
				1*	94.4	MC KEAN PA					99
		2*	10.3	4*	19.0	MERCER PA		85			96
		3	3.4	8	5.3	MONTGOMERY PA		58			86
		1	9.6	1	5.3	NORTHAMPTON PA		-44			86
				2*	108.1	NORTHUMBERLAND PA					99
51	2.9	77	3.0	117	3.5	PHILADELPHIA PA	5	18	24	56	79
				1*	22.5	VENANGO PA					97
1	2.8	1	1.9	3	6.0	WASHINGTON PA	-34	221	112	80	88
1	3.1					WESTMORELAND PA					
		1	8.3	1	3.1	YORK PA		-62			78
2	3.3	2	2.9	6	4.4	RHODE ISLAND RI	-13	51	31		
				1*	32.7	KENT RI					98
				1	6.7	NEWPORT RI					89
1	2.0	2	3.9	4	3.6	PROVIDENCE RI	92	-7	78	75	80
1*	14.9					WASHINGTON RI					
31	1.1	56	1.8	134	3.5	SOUTH CAROLINA SC	58	97	212		
2	2.8	1	1.2	3	2.7	AIKEN SC	-57	118	-6	44	76
		2	7.0	1	2.0	ALLENDALE SC		-71.			73
1	1.5			3	3.3	ANDERSON SC			117	80	79
		1	2.4	2	5.2	BAMBERG SC		114			86
1	3.2	1	2.5			BARNWELL SC	-21				
		1	1.5	2	2.7	BEAUFORT SC		81			76
		1	1.8	2	2.0	BERKELEY SC		13			73
1	2.4			2	5.6	CALHOUN SC			131	83	87
1	.3	6	2.3	9	2.3	CHARLESTON SC	656	2	673	98	74
2	6.6			3	10.1	CHEROKEE SC			52	66	93
		1	1.7	1	1.6	CHESTER SC		-4			72
1	2.6	1	2.4	3	5.2	CHESTERFIELD SC	-6	116	102	79	86
		1	1.7	4	5.2	CLARENDON SC		205			86
		1	1.8	4	6.2	COLLETON SC		248			88
		2	2.4	3	2.8	DARLINGTON SC		18			77
				1	2.0	DILLON SC					73
1	2.9			2	3.3	DORCHESTER SC			13	52	78
				2	5.8	EDGEFIELD SC					88
		1	1.9	4	7.1	FAIRFIELD SC		273			90
		3	2.1	9	5.3	FLORENCE SC		156			86
1	1.5			4	5.2	GEORGETOWN SC			240	92	86
1	.7	3	1.9	9	4.5	GREENVILLE SC	169	139	543	97	84
1	2.5			1	1.2	GREENWOOD SC			-52	18	71
				3	6.7	HAMPTON SC					89
				1	1.4	HORRY SC					71
				2	5.1	JASPER SC					86
2	4.2	4	6.8	4	6.5	KERSHAW SC	60	-4	53	66	89
		1	2.1	2	3.4	LANCASTER SC		63			79
1	1.9			2	2.7	LAURENS SC			42	62	76
1	2.5	1	2.5			LEE SC	1				
		1	1.4	2	2.2	MARION SC		59			74
		1	2.0	2	3.0	MARLBORO SC		50			78
1	2.3	2	4.4	1	2.1	NEWBERRY SC	91	-53	-10	40	74
		2*	10.3			OCONEE SC					
2	1.6			4	2.3	ORANGEBURG SC			49	64	74
4	1.9	5	2.2	19*	5.4	RICHLAND SC	21	143	193	90	87
				1	4.3	SALUDA SC					83
2	1.4	5	3.7	7	4.0	SPARTANBURG SC	161	7	178	88	81
3	2.3	3	2.4	2	1.2	SUMTER SC	1	-47	-47	21	71
		2	5.1	2	5.0	UNION SC		-1			85
				2	1.8	WILLIAMSBURG SC					73
2	3.0	3	2.7	4	4.2	YORK SC	-10	53	38	59	82
6*	8.6	20*	24.7	16*	14.9	SOUTH DAKOTA SD	186	-40	72		
1*	60.6	1*	32.8	1*	24.4	BENNETT SD	-46	-26	-60	13	97
		1*	26.8	1*	28.3	BUFFALO SD		6			97
		2*	51.7	1*	20.6	CHARLES MIX SD		-60			97
		2*	31.3			CORSON SD					
		1*	106.0			DAY SD					
2*	38.5	1*	18.6	2*	27.0	DEWEY SD	-52	45	-30	30	97
				1*	44.6	FALL RIVER SD					98
		1*	141.3	1*	78.1	GREGORY SD		-45			99
				1*	112.4	HUGHES SD					99
1*	49.2					JACKSON SD					
		2*	89.4			MELLETTE SD					
		1*	112.4			MINNEHAHA SD					
				1*	98.2	MOODY SD					99
1*	14.9	1*	20.2			PENNINGTON SD	36				
				2*	35.4	ROBERTS SD					98
1	6.2	4*	20.6	1	4.5	SHANNON SD	232	-78	-28	31	84
		1	10.8	3*	22.0	TODD SD		105			97
		1*	106.0			TRIPP SD					
		1*	26.8	1*	74.3	ZIEBACH SD		178			99
38	1.5	76	2.6	133	3.8	TENNESSEE TN	71	48	152		
		1	7.9	1	4.5	BEDFORD TN		-43			83
				2	12.3	BLOUNT TN					94
1*	9.9					BRADLEY TN					
		1*	74.3			CAMPBELL TN					
		1	5.3	2	8.0	CARROLL TN		51			91
				1*	28.7	CHEATHAM TN					98
8	2.3	13	3.3	15	3.1	DAVIDSON TN	47	-8	35	59	78
		1	3.9	1	2.7	DYER TN		-32			76
		1	1.6	1	1.7	FAYETTE TN		0			72
		1	1.8	2	4.0	GIBSON TN		117			81
				1	2.1	GILES TN					74
				1	13.2	GREENE TN					95
				1	7.2	HAMBLEN TN					90

* SIGNIFICANTLY MORE DEATHS THAN EXPECTED

NON-WHITE FEMALE — CANCER OF THE LIVER AND GALLBLADDER INCLUDING BILE DUCTS — ICD 155,156

1950-1959 DEATHS	1950-1959 RATE	1960-1969 DEATHS	1960-1969 RATE	1970-1979 DEATHS	1970-1979 RATE	COUNTY AND STATE	PERCENT CHANGE 60S/50S	PERCENT CHANGE 70S/60S	PERCENT CHANGE 70S/50S	PERCENTILE %CHANGE 70S/50S	PERCENTILE RATES 1970S
1	.4	10	4.2	13	4.3	HAMILTON TN	1070	1	1083	99	83
2	5.0	1	1.4	1	2.6	HARDEMAN TN	-72	85	-48	19	76
		1	2.1	1	1.5	HAYWOOD TN		-29			71
		1	7.4			HENDERSON TN					
				1*	188.7	HOUSTON TN					99
		1*	23.6			JEFFERSON TN					
1	.8	2	1.4	10	6.5	KNOX TN	66	370	682	98	89
				2	4.0	LAUDERDALE TN					82
				1	8.4	LINCOLN TN					92
1	1.1	3	2.9	3	2.8	MADISON TN	164	-7	147	85	77
				2*	21.6	MARSHALL TN					97
		2	4.4	4	7.8	MAURY TN		80			91
				1*	12.5	MONROE TN					94
				1	1.9	MONTGOMERY TN					73
		1	5.1			OBION TN					
1	4.8	1	4.3			ROBERTSON TN	-10				
1	2.9			1	2.1	RUTHERFORD TN			-29	31	74
16	1.9	33	3.0	58	4.3	SHELBY TN	52	45	121	81	83
2*	26.6			1	6.4	SULLIVAN TN			-76	05	89
				1	4.5	SUMNER TN					83
1	2.2	1	2.6	2	4.7	TIPTON TN	17	85	117	81	84
1*	24.5					TROUSDALE TN					
2*	9.5			2	8.3	WILLIAMSON TN			-13	37	92
93	2.1	163	2.9	243	2.9	TEXAS TX	43	-2	39		
1	3.0	2	4.0	3	4.5	ANDERSON TX	36	12	52	65	84
		2	5.5	2	4.5	ANGELINA TX		-18			84
		3*	13.2			AUSTIN TX					
1	3.3	3*	9.9	3	6.5	BASTROP TX	201	-35	97	78	89
1	3.3	1	2.5			BELL TX	-25				
3	1.4	11	4.4	22	4.3	BEXAR TX	210	-1	208	91	83
		9*	11.6	3	2.9	BOWIE TX		-75			77
1	2.4	1	2.3	1	1.9	BRAZORIA TX	-6	-18	-23	33	73
2	4.1	1	1.5			BRAZOS TX	-64				
1*	18.9					BROWN TX					
1	3.2			1	1.7	BURLESON TX			-47	20	72
		2*	9.8			CAMP TX					
1	4.2					CASS TX					
				1	10.2	CHAMBERS TX					93
		1	2.2	1	1.6	CHEROKEE TX		-27			72
1	4.1			1	5.5	COLORADO TX			35	59	87
1*	12.5	1*	20.6	1*	12.5	COOKE TX	65	-39	0	46	94
				1*	18.6	CORYELL TX					96
				1*	22.9	CROSBY TX					97
7	1.7	13	2.0	25	2.7	DALLAS TX	15	35	56	67	76
		1	6.9	1	5.3	DENTON TX		-24			86
1*	7.7	1	6.6			ECTOR TX	-14				
		1	2.3	1	1.6	ELLIS TX		-27			72
		2	6.9	2	.4	EL PASO TX		-94			70
		2	4.1	1	1.9	FALLS TX		-53			73
		1	7.3	1	10.2	FANNIN TX		40			93
		1	3.0	2	3.2	FORT BEND TX		6			78
1	2.2	4	4.4	4	2.0	GALVESTON TX	103	-53	-6	43	73
1*	39.3					GOLIAD TX					
1	3.7					GONZALES TX					
1	3.2			1	1.9	GRAYSON TX			-40	26	73
2	2.8	3	3.8	4	3.9	GREGG TX	38	2	41	61	81
1	2.3	2	5.8	3	7.4	GRIMES TX	156	27	225	91	90
				2	9.9	HARDIN TX					93
25*	3.1	30	3.1	48	2.8	HARRIS TX	0	-10	-10	40	77
5	5.1	5	4.5	3	2.8	HARRISON TX	-12	-38	-46	22	77
		1	4.7	4*	11.4	HENDERSON TX		144			94
		1	4.0	1	4.0	HILL TX		2			82
1	2.6			1	2.4	HOUSTON TX			-8	42	75
1	5.9			2	6.3	HUNT TX			7	50	89
		1	3.0			JASPER TX					
1	.5	6	2.5	11	4.2	JEFFERSON TX	390	67	716	99	82
		1	2.1	4	8.1	KAUFMAN TX		281			92
				1	14.5	KERR TX					95
				1*	67.0	KINNEY TX					99
2	3.9	1	2.1	1	2.2	LAMAR TX	-45	4	-43	24	74
				1*	21.1	LAMPASAS TX					97
1	8.5					LAVACA TX					
				1	5.5	LEE TX					87
		2	5.8	4*	11.3	LIBERTY TX		94			94
1	2.9			3	5.4	LIMESTONE TX			89	77	87
1	3.2			2	3.2	LUBBOCK TX			1	47	78
2	2.2	3	2.3	10	6.1	MC LENNAN TX	6	165	181	89	88
		2*	13.6			MADISON TX					
1	4.6					MARION TX					
1*	6.7					MIDLAND TX					
		2	10.7	1	2.7	MILAM TX		-75			76
				1*	24.4	MITCHELL TX					97
		2	7.5			MONTGOMERY TX					
		1	6.1			MORRIS TX					
				1	2.2	NACOGDOCHES TX					74
1	1.5	3	5.1	1	1.4	NAVARRO TX	235	-73	-10	40	71
				1	4.2	NEWTON TX					82
		1*	21.9			NOLAN TX					
		3	7.2	3	3.2	NUECES TX		-56			78
3*	15.2			1	2.6	PANOLA TX			-83	02	76
				1*	49.1	PARMER TX					98
3*	15.2	1	5.3			POLK TX	-65				
		1	3.3	3	7.4	POTTER TX		124			90
		2	8.0			RED RIVER TX					
1	2.6	1	2.6			ROBERTSON TX	1				
1	1.9	3	5.5			RUSK TX	187				
		1	10.1	1	8.3	SABINE TX		-18			92
1	5.6					SAN JACINTO TX					

* SIGNIFICANTLY MORE DEATHS THAN EXPECTED

1950-1959 DEATHS	1950-1959 RATE	1960-1969 DEATHS	1960-1969 RATE	1970-1979 DEATHS	1970-1979 RATE	COUNTY AND STATE	PERCENT CHANGE 60S/50S	PERCENT CHANGE 70S/60S	PERCENT CHANGE 70S/50S	PERCENTILE %CHANGE 70S/50S	PERCENTILE RATES 1970S
1	1.1			9	6.4	SMITH TX			478	97	89
3	1.8	8	2.9	16	3.7	TARRANT TX	67	28	113	80	80
				1	2.9	TAYLOR TX					77
				1*	38.2	TERRY TX					98
1	7.9			1	5.9	TITUS TX			-25	33	88
		1	4.3			TOM GREEN TX					
1	.9	6	4.1	4	2.1	TRAVIS TX	371	-49	140	84	74
				1	14.1	TYLER TX					95
1	3.4	1	4.1			UPSHUR TX	20				
				1	3.9	VICTORIA TX					81
1	2.6	2	4.8			WALKER TX	82				
1	8.3	1	3.0	1	1.8	WALLER TX	-64	-41	-79	03	72
1	1.6					WASHINGTON TX					
		1	2.1	1	1.5	WHARTON TX		-27			72
		1	2.5	2	3.8	WICHITA TX		54			81
2*	26.9					WILBARGER TX					
				2	5.8	WILLIAMSON TX					88
				1*	100.7	WISE TX					99
				1	4.8	WOOD TX					85
				1*	37.3	YOUNG TX					98
1	2.1	3	8.7	6	5.7	UTAH UT	316	-34	176		
		1	5.1	4	8.3	SALT LAKE UT		62			92
		1*	28.1	1	11.8	SAN JUAN UT		-58			94
1*	26.8	1*	14.9	1*	27.3	UINTAH UT	-45	83	2	48	97
38	1.4	82	2.4	152	3.4	VIRGINIA VA	72	43	145		
		1	2.0	1	1.5	ACCOMACK VA		-26			71
2	4.8	4	7.3	2	2.9	ALBEMARLE#VA	52	-60	-39	26	77
				2	10.5	ALLEGHANY#VA					94
1	4.9	1	5.0			AMHERST VA	2				
3*	14.2	2	6.3	2	2.4	ARLINGTON#VA	-55	-62	-83	02	75
1	3.3					AUGUSTA#VA					
2	1.9	7	5.9	4	2.5	BEDFORD#VA	216	-58	33	58	75
		2*	20.4			BOTETOURT VA					
2	5.9	1	3.0			BRUNSWICK VA	-50				
1	5.7			2	8.3	BUCKINGHAM VA			45	63	92
1	3.3			1	2.4	CAROLINE VA			-28	31	75
				3*	55.3	CARROLL#VA					98
				1	4.0	CHARLOTTE VA					81
4	1.2	9	1.9	18	2.9	CHESTERFIELD#VA	62	51	145	84	77
		1	5.5	1	4.6	CULPEPER VA		-16			84
		1	6.5			CUMBERLAND VA					
		2	1.3	11	5.7	DINWIDDIE#VA		354			88
3	6.0	2	2.1	8	6.1	FAIRFAX#VA	-65	198	3	49	88
				1	7.4	FLUVANNA VA					90
		1	14.1	1	12.8	FREDERICK#VA		-9			95
				1	4.2	GLOUCESTER VA					82
1	6.9			1	3.5	GOOCHLAND VA			-49	18	80
2	3.7			7*	8.9	HALIFAX#VA			142	84	92
		1	3.0	1	3.1	HANOVER VA		5			78
		2	6.6	1	1.3	HENRICO VA		-81			71
		1	1.4	6*	7.4	HENRY#VA		426			90
				1	2.3	ISLE OF WIGHT VA					75
				3	4.2	JAMES CITY#VA					82
				3*	18.9	KING AND QUEEN VA					96
		1	6.5	1	5.2	KING WILLIAM VA		-21			86
		1	4.3	2	10.2	LOUISA VA		137			93
		1	9.8	1	6.7	MATHEWS VA		-32			89
1	1.8					MECKLENBURG VA					
				1	2.7	MIDDLESEX VA					76
		1	2.4	1	2.3	NORTHAMPTON VA		-5			74
				1	7.7	ORANGE VA					91
2	1.7	2	1.8	5	3.4	PITTSYLVANIA#VA	6	91	101	79	79
				1	7.3	POWHATAN VA					90
		1	3.8	2	7.8	PRINCE EDWARD VA		104			91
6*	8.1	2	1.7	7	5.2	ROANOKE#VA	-79	206	-36	28	86
1*	10.0					ROCKINGHAM#VA					
				1*	60.7	SCOTT VA					98
				1	1.6	SOUTHAMPTON#VA					72
1	9.4			1	2.7	SPOTSYLVANIA#VA			-71	08	76
		2	6.1			SUSSEX VA					
				1	6.5	WASHINGTON#VA					89
		2*	9.7	1	4.7	WESTMORELAND VA		-52			84
				1	24.6	WISE#VA					97
3	.6	18	2.7	28	3.4	CHESAPEAKE CITY#VA	329	24	432	97	79
		4	4.6	3	2.4	HAMPTON CITY VA		-48			75
		4	3.1	6	3.1	NEWPORT NEWS CITY VA		-1			78
1	1.2	5	5.1	5	4.2	SUFFOLK CITY VA	330	-19	249	93	82
11*	5.7	13	4.4	26	4.1	WASHINGTON WA	-22	-8	-29		
		1*	21.0			FERRY WA					
3	2.7	6	3.2	17	4.9	KING WA	21	52	84	76	85
1*	38.2	2*	30.4			KITSAP WA	-20				
				1*	172.0	KLICKITAT WA					99
2*	48.1					OKANOGAN WA					
				1*	188.7	PEND OREILLE WA					99
		2	9.6	1	1.8	PIERCE WA		-81			73
				1	2.5	SNOHOMISH WA					75
		1	4.4	2	4.5	SPOKANE WA		4			84
1*	60.6					STEVENS WA					
		1*	114.7	1	4.8	THURSTON WA		-96			85
4*	23.9			2	3.0	YAKIMA WA			-87	01	77
24*	6.0	14	2.9	25	4.7	WEST VIRGINIA WV	-51	61	-21		
				1*	18.6	BOONE WV					96

* SIGNIFICANTLY MORE DEATHS THAN EXPECTED

\# COUNTIES COMBINED; SEE TECHNICAL NOTES

1950-1959 DEATHS	1950-1959 RATE	1960-1969 DEATHS	1960-1969 RATE	1970-1979 DEATHS	1970-1979 RATE	COUNTY AND STATE	PERCENT CHANGE 60S/50S	PERCENT CHANGE 70S/60S	PERCENT CHANGE 70S/50S	PERCENTILE %CHANGE 70S/50S	PERCENTILE RATES 1970S
		1*	33.5			BROOKE WV					
1	2.8	1	2.8			CABELL WV	-2				
3*	9.6	1	2.6	2	5.9	FAYETTE WV	-73	129	-38	27	88
		1	8.9	3*	26.9	GREENBRIER WV		202			97
				1*	14.0	HANCOCK WV					95
1*	10.3					HARRISON WV					
1	7.2	2*	14.8	2*	13.1	JEFFERSON WV	107	-11	84	75	95
2	2.5	1	1.3	3	2.8	KANAWHA WV	-48	120	14	53	77
1	13.5					LOGAN WV					
6*	11.4	1	1.5	2	3.1	MC DOWELL WV	-87	102	-73	07	78
				2	6.2	MARION WV					88
3*	8.8	1	2.1	1	1.7	MERCER WV	-76	-18	-80	03	72
1*	18.9					MONONGALIA WV					
				2*	15.7	OHIO WV					96
				1*	188.7	PENDLETON WV					99
2	3.4	3	4.6	5	7.8	RALEIGH WV	34	70	127	81	91
1*	89.3	1*	49.2			RANDOLPH WV	-45				
1*	344.1					TAYLOR WV					
1*	22.3	1*	14.1			WOOD WV	-37				
8*	7.5	13	4.6	20	4.1	WISCONSIN WI	-38	-11	-46		
1*	31.5					ASHLAND WI					
				1	9.3	BROWN WI					93
1*	74.3					CHIPPEWA WI					
		1*	109.3			FOREST WI					
1*	267.9					JACKSON WI					
1*	245.4					JUNEAU WI					
		1	11.8			MENOMINEE#WI					
3	4.4	8	3.7	14	4.2	MILWAUKEE WI	-16	14	-4	44	82
				1*	74.3	MONROE WI					99
		1	7.3	1	3.3	RACINE WI		-54			79
				1	5.0	ROCK WI					85
1*	32.7			1*	21.0	SAWYER WI			-36	28	97
		2*	81.8	1*	22.5	VILAS WI		-72			97
1	4.7	2	10.5	5*	10.0	WYOMING WY	122	-4	112		
		2*	28.6	4*	28.5	FREMONT WY		0			97
1*	37.8			1	8.2	LARAMIE WY			-78	04	92

* SIGNIFICANTLY MORE DEATHS THAN EXPECTED

COUNTIES COMBINED; SEE TECHNICAL NOTES

www.ingramcontent.com/pod-product-compliance
Lightning Source LLC
LaVergne TN
LVHW081959200926
848481LV00010B/2772